Springer Collected Works in Mathematics

More information about this series at http://www.springer.com/series/11104

Moscow, April 30, 1982

Israel M. Gelfand

Collected Papers III

Editors
Semen G. Gindikin
Victor W. Guillemin
Alexandr A. Kirillov
Bertram Kostant
Shlomo Sternberg

Reprint of the 1989 Edition

Springer

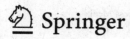

Author
Israel M. Gelfand (1913 – 2009)
Department of Mathematics
Rutgers State University of New Jersey
New Brunswick, NJ
USA

Editors
Semen G. Gindikin
Department of Mathematics
Rutgers State University of New Jersey
New Brunswick, NJ
USA

Victor W. Guillemin
Massachusetts Institute of Technology
Cambridge, MA
USA

Alexandr A. Kirillov
University of Pennsylvania
Philadelphia, PA
USA

Bertram Kostant
Massachusetts Institute of Technology
Cambridge, MA
USA

Shlomo Sternberg
Harvard University
Cambridge, MA
USA

ISSN 2194-9875
Springer Collected Works in Mathematics
ISBN 978-3-642-30813-0 (Softcover)
978-3-540-19399-9 (Hardcover)

Library of Congress Control Number: 2012954381

Mathematics Subject Classification (2010): 22 E 46, 22 E 65, 28 C 20, 33 A 35, 46 F 25, 53 C 65, 55 N, 57 Q, 57 R, 60 G 20, 90 D 40, 92 A 09, 94 A 17

Springer Heidelberg New York Dordrecht London

Printed on acid-free paper

Springer-Verlag GmbH Berlin Heidelberg is part of Springer Science+Business Media
(www.springer.com)

IZRAIL M. GELFAND

COLLECTED PAPERS

VOLUME III

Edited by

S.G. Gindikin V.W. Guillemin A.A. Kirillov
B. Kostant S. Sternberg

SPRINGER-VERLAG
BERLIN HEIDELBERG NEW YORK
LONDON PARIS TOKYO

Professor Izrail M. Gelfand
Member of the Academy of Sciences of the USSR
A. N. Belozersky Laboratory of Molecular Biology and
Bioorganic Chemistry, Building "A", Moscow State University
Moscow, GSP-234, 119899, USSR

Editors:

Professor Semen G. Gindikin
A. N. Belozersky Laboratory of Molecular Biology and
Bioorganic Chemistry, Building "A", Moscow State University
Moscow, GSP-234, 119899, USSR

Professor Victor W. Guillemin
Department of Mathematics, Massachusetts Institute of Technology
Cambridge, MA 02139, USA

Professor Aleksandr A. Kirillov
Moskovskii Universitet, Mehmat, Moscow GSP-234, 119899, USSR

Professor Bertram Kostant
Department of Mathematics, Massachusetts Institute of Technology
Cambridge, MA 02139, USA

Professor Shlomo Sternberg
Department of Mathematics, Harvard University, Science Center
One Oxford Street, Cambridge, MA 02138, USA

1980 Mathematics Subject Classification (1985 Revision)
22 E 46, 22 E 65, 28 C 20, 33 A 35, 46 F 25, 53 C 65, 55 N, 57 Q, 57 R, 60 G 20,
90 D 40, 92 A 09, 94 A 17

ISBN 978-3-540-19399-9 Springer-Verlag Berlin Heidelberg New York

Library of Congress Cataloging-in-Publication Data
(Revised for vols. 2–3)
Gel'fand, I. M. (Izrail' Moiseevich)
Collected papers.
In English, French, and German. Includes bibliographies;
1. Mathematics. 2. Gel'fand, I. M. (Izrail' Moiseevich). 3. Mathematicians–Soviet Union–Biography.
I. Gindikin, S. G. (Semen Grigor'evich)
II. Title. QA3.G38 1987 510 87-32254
ISBN 978-3-540-19399-9

2141/3140-543210 Printed on acid-free paper

Preface

I am very grateful to Springer-Verlag for publishing my three volumes of Collected Papers. I was asked to write a survey of the papers included: I do not think that the author has this right. It seems to me that whatever a man achieves in science comes from above. For this reason, he cannot be the judge of his own work. But I hope that at the end of the third volume I will be able to write what I have tried to do in these papers and did not succeed or did not fully succeed to do. It may be that some of these problems remain interesting today. From the bibliography you can see that a lot of papers are written together with some of my colleagues and friends. I want to express my deep gratitude to them. I very much enjoyed working with them and I learned a lot from my contact with them.

I am very grateful to all my friends who did a lot of work, especially in translating, correcting and generally editing these three volumes. Without them it would not have been possible to get this edition ready.

Thanks also very much to Professors S. G. Gindikin, V. W. Guillemin, A. A. Kirillov, B. Konstant and S. Sternberg.

Moscow, September 1987 I. M. Gelfand

Foreword

A considerable part of the last volume of Gelfand's Selecta is devoted to two areas of mathematics he has been systematically working in over the last three decades. At the end of the 50s he started developing integral geometry, and a decade later he published the first in a series of papers on the cohomology of infinite-dimensional Lie algebras.

Integral geometry in Gelfand's sense developed as a part of representation theory. It turned out that the problem of finding a Plancherel formula for complex semisimple Lie groups (as well as for many other problems of harmonic analysis) can be reduced to the problem of reconstructing a function on the group (or a homogeneous space) from its integrals over orispheres. A number of papers on this subject appear in the second volume of these Selecta. It turned out, however, that similar formulas are true for a very large class of general integral transformations of geometric type, and that they need not be associated to groups or homogeneous spaces. A problem which arose naturally was that of asking what are the geometric structures (more general than the group structures) responsible for the existence of such formulas, and what are the possibilities of formulating some kind of multidimensional harmonic analysis outside the traditional framework of analysis on homogeneous spaces. Gelfand's approach to integral geometry always emphasised the search for explicit formulas and in that respect it is very close to the modern system of values in the multidimensional analysis.

The series of papers devoted to the cohomology of infinite-dimensional Lie algebras started with an unexpected observation that it can be both finite-dimensional and non-trivial. The cohomology of the Lie algebra of smooth vector fields on any smooth manifold turned out to be finite-dimensional (with coefficients either trivial or in many non-trivial modules). The fact that there is a non-trivial cohomology theory for infinite-dimensional Lie algebras has led to important analytical and topological applications. One series of results is that on characteristic classes for foliations. The same section of this volume includes papers on "formal" geometry: localisation of topological invariants, combinatorial formulas for characteristic classes.

The volume is concluded by Gelfand's most recent works on the general theory of hypergeometric functions (which appeared after the main body of the work on preparing these Selecta had been completed). This series of papers is also very close to integral geometry. The starting-point for this new extensive program of Gelfand's was an observation that the Gauss hypergeometric function can be naturally interpreted in the language of the John transformation which associates to each function in a 3-space its integrals over lines. The hypergeometric equation is a consequence of the ultrahyperbolic equation describing the image of that transformation. This makes it possible to consider multidimensional versions of such constructions and to develop a unified approach to various seemingly unrelated specific examples of hypergeometric functions. A new elegant and very promising theory has developed with outlets to other areas of mathematics – a situation typical for Gelfand's way of working. Combinatorics and geometry of Grassmannians are among these other areas.

A small section of the volume includes papers on information theory and the theory of functional integration. Gelfand's work in these areas did not last long, but the results of his activity were of intrinsic importance for their initial development. Finally, we deemed it necessary to include at least some examples of his works on computational mathematics, automata theory, and biology. With respect to biology it is absolutely necessary to emphasize that Gelfand has always had very little interest in any attempts at application of mathematics or in developing mathematical models in biology (only his very first works could be interpreted in such a way). Instead, for 30 years he has been working as a professional biologist in neurophysiology and cell biology doing experimental work with his colleagues in this field.

The last volume of Gelfand's Selecta goes to press in the year of his 75th birthday. The Editors extend to him their warmest greetings and congratulations. We are very happy that he has met this date in his customary state of full creative activity: His latest works on hypergeometric functions bear the best witness to this. We have no doubt that a large number of remarkable new works of his have yet to appear making it necessary to continue the publication of these Selecta.

S. G. Gindikin

Table of contents

Part I. Integral geometry

1. Integral transformations connected with straight line complexes in a complex affine space (with M.I. Graev) 3
2. Integral geometry on k-dimensional planes (with M.I. Graev and Z.Ya. Shapiro) 7
3. Complexes of k-dimensional planes in the space $\mathbf{C}^n$ and Plancherel's formula for the group $GL(n, C)$ (with M.I. Graev) 21
4. Complexes of straight lines in the space $\mathbf{C}^n$ (with M.I. Graev) 26
5. Differential forms and integral geometry (with M.I. Graev and Z.Ya. Shapiro) 37
6. Integral geometry in projective space (with M.I. Graev and Z.Ya. Shapiro) 51
7. A problem of integral geometry connected with a pair of Grassmann manifolds (with M.I. Graev and Z.Ya. Shapiro) 68
8. Nonlocal inversion formulas in real integral geometry (with S.G. Gindikin) 73
9. A local problem of integral geometry in a space of curves (with S.G. Gindikin and Z.Ya. Shapiro) 80
10. A problem of integral geometry in RP^n connected with the integration of differential forms (with S.G. Gindikin and M.I. Graev) 96
11. Integral geometry in affine and projective spaces (with S.G. Gindikin and M.I. Graev) 99
12. Geometric structures of double bundles and their relation to certain problems in integral geometry (with G.S. Shmelev) 228
13. The problem of integral geometry and intertwining operators for a pair of real Grassmannian manifolds (with M.I. Graev and R. Rosu) 241
14. On a characterization of Grassmann manifolds (with A.B. Goncharov) 266

Part II. Cohomology and characteristic classes

1. On classifying spaces for principal fiberings with Hausdorff bases (with D.B. Fuks) 273
2. Topology of noncompact Lie groups (with D.B. Fuks) 277
3. The cohomologies of the Lie algebra of the vector fields in a circle (with D.B. Fuks) 288

4. Cohomologies of the Lie algebra of tangential vector fields of a smooth manifold (with D. B. Fuks) 290
5. Cohomology of the Lie algebra of formal vector fields (with D. B. Fuks) 307
6. Cohomologies of Lie algebra of tangential vector fields II (with D. B. Fuks) 323
7. Cohomologies of Lie algebra of vector fields with nontrivial coefficients (with D. B. Fuks) 330
8. Quadratic forms over commutative group rings and the *K*-theory (with A. S. Mishchenko) 342
9. Upper bounds for cohomology of infinite-dimensional Lie algebras (with D. B. Fuks) 347
10. The actions of infinite-dimensional Lie algebras (with D. B. Fuks and D. A. Kazhdan) 349
11. Cohomology of the Lie algebra of Hamiltonian formal vector fields (with D. B. Fuks and D. I. Kalinin) 354
12. PL Foliations (with D. B. Fuks) 358
13. Cohomologies of the Lie algebra of formal vector fields with coefficients in its adjoint space and variations of characteristic classes of foliations (with B. L. Feigin and D. B. Fuks) 365
14. The Gauss-Bonnet theorem and the Atiyah-Patodi-Singer functionals for the characteristic classes of foliations (with D. B. Fuks and A. M. Gabrielov) 379
15. PL Foliations II (with D. B. Fuks) 403
16. Combinatorial computation of characteristic classes (with A. M. Gabrielov and M. V. Losik) 407
17. Combinatorial computation of characteristic classes (with A. M. Gabrielov and M. V. Losik) 420
18. A local combinatorial formula for the first class of Pontryagin (with A. M. Gabrielov and M. V. Losik) 437
19. Atiyah-Patodi-Singer functionals for characteristic functionals for tangent bundles (with A. M. Gabrielov and M. V. Losik) 441
20. Computing of characteristic classes of combinatorial vector bundles (with M. V. Losik) 454
21. Cohomology of infinite-dimensional Lie algebras and Laplace operators (with B. L. Feigin and D. B. Fuks) 487
22. Geometry in Grassmannians and a generalization of the dilogarithm (with R. D. MacPherson) 492

Part III. Functional integration; probability; information theory

1. Generalized random processes 529
2. Integration in functional spaces and its applications in quantum physics (with A. M. Yaglom) 534

3. To the general definition of the amount of information (with A. N. Kolmogorov and A. M. Yaglom) 556
4. Calculation of the amount of information about a random function contained in another such function (with A. M. Yaglom) 560

Part IV. Mathematics of computation; cybernetics; biology

1. On the application of the method of random tests (the Monte Carlo method) for the solution of the kinetic equation (with N. N. Chentsov, S. M. Fejnberg, and A. S. Frolov) 611
2. On difference schemes for the solution of the equation of thermal conductivity. The double sweep method for solution of difference equations (with O. V. Lokytsievskij) 617
3. Some methods of control for complex systems (with M. L. Tsetlin) . . 648
4. Determination of crystal structure by the method of nonlocal search (with Yu. G. Fedorov and I. I. Piatetski-Shapiro) 671
5. Determination of crystal structures by the method of the *R*-factor minimization (with Yu. G. Fedorov, R. A. Kayushina, and B. K. Vainstein) 675
6. On certain classes of games and automata games (with I. I. Piatetski-Shapiro and M. L. Tsetlin) 680
7. Some problems in the analysis of movements (with V. S. Gurfinkel, M. L. Shik, and M. L. Tsetlin) 686
8. Recordings of neurones of the dorsal spinocerebellar tract during evoked locomotion (with Yu. I. Arshavskij, M. B. Berkenblit, O. I. Fukson, and G. N. Orlovskij) 703
9. Origin of modulation in neurones of the ventral spinocerebellar tract during locomotion (with Yu. I. Arshavskij, M. B. Berkenblit, O. I. Fukson, and G. N. Orlovskij) 707
10. Generation of scratching I. Activity of spinal interneurons during scratching (with M. B. Berkinblit, T. G. Delyagina, A. G. Fel'dman, and G. N. Orlovskij) 711
11. The cerebellum and control of rhythmical movements (with Yu. I. Arshavskij and G. N. Orlovskij) 729
12. Interrelationships of contacting cells in the cell complexes of mouse ascites hepatoma (with V. I. Guelstein, A. G. Malenkov, and Yu. M. Vasil'ev) 735
13. Initiation of DNA synthesis in cell cultures by colcemid (with V. I. Guelstein and Yu. M. Vasil'ev) 747
14. Mechanisms of morphogenesis in cell cultures (with Yu. M. Vasil'ev) 750
15. Possible common mechanism of morphological and growth-related alterations accompanying neoplastic transformation (with Yu. M. Vasil'ev) 866
16. Special type of morphological reorganization induced by phorbol ester: Reversible partition of cell into mobile and stable domains (with V. B. Dugina, T. M. Svitkina, and Yu. M. Vasil'ev) 870

Part V. General theory of hypergeometric functions

1. General theory of hypergeometric functions 877
2. Generalized hypergeometric equations (with S.I. Gelfand) 882
3. A duality theorem for general hypergeometric functions (with M.I. Graev) 886
4. Algebraic and combinatorial aspects of the general theory of hypergeometric functions (with A.V. Zelevinskij) 891
5. Combinatorial geometries, convex polyhedra, and Schubert cells (with M. Goresky, R.D. MacPherson, and V.V. Serganova) 906
6. Strata of a maximal torus in a compact homogeneous space (with V.V. Serganova) 922
7. Combinatorial geometries and torus strata on homogeneous compact manifolds (with V.V. Serganova) 926
8. General hypergeometric functions on complex Grassmannians (with V.A. Vasiliev and A.V. Zelevinskij) 959
9. Holonomic systems of equations and series of hypergeometric type (with M.I. Graev and A.V. Zelevinskij) 977
10. On Heaviside functions of configuration of hyperplanes (with A.N. Varchenko) 983
11. Arrangement of real hyperplanes and the associated partition function (with T.V. Alekseevskaya and A.V. Zelevinskij) 1006
12. Generalized Airey functions, Schubert cells and Jordan groups (with V.S. Retakh, V.V. Serganova) 1011

Appendix

Cardiac arrhythmias and circle mappings (by V.I. Arnold) 1019

An Editorial Perspective (by B. Kostant) 1025

Table of contents for volumes I and II 1027

Bibliography 1033

Acknowledgements 1075

Part I

Integral geometry

1.

(with M. I. Graev)

Integral transformations connected with straight line complexes in a complex affine space

Dokl. Akad. Nauk SSSR **138** (1961) 1266–1269 [Sov. Math., Dokl. **2** (1961) 809–812].
Zbl. **109**:151

We consider a three-dimensional complex affine space of points $z = (z^1, z^2, z^3)$ and in it all possible complex straight lines. We shall indicate these lines by the Plücker coordinates $\alpha = (\alpha^1, \alpha^2, \alpha^3)$ and $p = (p_1, p_2, p_3)$. In this definition, α is the directed vector of the line, and p is the vector product $p = [\alpha, z]$ of the vector α and the radius-vector of an arbitrary point z of the line.*

The Plücker coordinates are connected with each other by the relation $\alpha^1 p_1 + \alpha^2 p_2 + \alpha^3 p_3 = 0$. Any homogeneous equation $F(\alpha, p) = 0$ between the coordinates α, p which is not a consequence of the equation $\alpha^1 p_1 + \alpha^2 p_2 + \alpha^3 p_3 = 0$ determines some three-parameter family of lines, called a *complex* [1]. In the following we suppose that F means a homogeneous polynomial in α and p.

We shall formulate here a problem in the integral geometry for line complexes. Let the function $f(z) = f(z^1, z^2, z^3)$ be given. We shall suppose that it is a sufficient number of times differentiable (in z and $\bar{z}$) and decreases sufficiently fast toward infinity. We define the integral of the function $f(z)$ along the line α, p of the complex in the following way:

$$\phi(\alpha, p) = \frac{i}{2} \int f(\alpha t + \beta)\, dt\, d\bar{t}, \tag{1}$$

* We can therefore consider $p = [\alpha, z]$ as the vector equation of the line given by the Plücker coordinates α and p.

where β is an arbitrary point of the line α, p, i.e., $p = [\alpha, \beta]$.* Our problem consists in restoring for $\phi(\alpha, p)$ the original function $f(z)$, i.e., in obtaining the inversion of formula (1).

In this paper we shall solve this problem for complexes satisfying the following additional condition.

Condition of tangency. Let α, p be an arbitrary line of the complex; z', z'' be arbitrary points on the line α, p. We consider *all* lines of the complex which pass through z' or z''. These lines form two conical surfaces with α, p as their common generator.** It is now required that these conical surfaces always be tangent to each other along the common generator α, p.

The following complexes satisfy the condition of tangency: the complex of all lines which are tangent to some (algebraic) surface; the complex of all lines which intersect some (algebraic) surface; the complex of all lines which intersect some (algebraic) curve; the complex of all lines which are parallel to the generators of some (algebraic) conical surface.

Necessary and sufficient condition that the complex, given by the equation $F(\alpha, p) = 0$, satisfies the condition of tangency, is that the Plücker coordinates of the lines of the complex satisfy the relation

$$F'_{\alpha^1} F'_{p_1} + F'_{\alpha^2} F'_{p_2} + F'_{\alpha^3} F'_{p_3} = 0.$$

For complexes which satisfy the condition of tangency the problem of integral geometry has a unique solution. This solution is a local one. Indeed, in order to restore the value of the function f at some point z_0, it is sufficient to know only the integral along the lines of the complex passing through the point z_0 and along the lines of the complex infinitely near to it (besides, we can consider only the infinitely near *parallel* lines). It appears that the considered complexes are in a sense the most general complexes, for which the solution of our problem of integral geometry is unique and local (see the note below). We shall now give the solution of our problem.

Let the complex be given by $F(\alpha, p) = 0$. We intend to seek the value of the function $f(z)$ at some point z_0. For this we lay at first all possible lines of the complex through the point z_0. The directed vector α of these lines satisfies the equation

$$G(\alpha; z_0) \equiv F(\alpha, [\alpha, z_0]) = 0. \tag{2}$$

We introduce the notation $a_i(\alpha; z_0) = G'_{\alpha^i}(\alpha; z_0)$, $i = 1, 2, 3$. We then displace in the complex every one of the lines α, p which pass through the point z_0, parallel to itself along an infinitely small distance. The Plücker coordinates α do not change under this displacement, but the coordinates p take an increase $dp = (dp_1, dp_2, dp_3)$. It can be shown that in a complex which satisfies the condition of tangency, the vector (dp_1, dp_2, dp_3) is proportional to the vector (a_1, a_2, a_3).

We introduce the operators "of the infinitely small parallel displacement of the lines of the complex passing through the point z_0":

$$L_{z_0} \varphi = \sum a_i(\alpha; z_0) \varphi'_{p_i}(\alpha, [\alpha, z_0]),$$
$$\bar{L}_{z_0} \varphi = \sum \overline{a_i(\alpha; z_0)} \varphi'_{\bar{p}_i}(\alpha, [\alpha, z_0]),$$

where α satisfies the equation (2).

Theorem. *We suppose that the complex $F(\alpha, p) = 0$ satisfies the condition of tangency, and that on the cone formed by the lines of the complex passing through a point z_0, the point z_0 is a unique singular point. Then the value of the function $f(z)$ at the point z_0 is expressed through the integral*

---

* By definition, $\frac{i}{2} dt\, d\bar{t} = d\sigma\, d\tau$, where $t = \sigma + i\tau$.

** Not counting the singular case, when all lines passing through one of the points z', z'', belong to the complex. In this case we shall suppose that the condition of tangency is fulfilled.

$\phi(\alpha, p)$ of the function $f(z)$ along a line of the complex by the following formula of inversion.

$$f(z_0) = C_{z_0}\int_\Gamma L_{z_0}\bar{L}_{z_0}\varphi(\alpha, [\alpha, z_0])\,\omega_{z_0}(\alpha)\,\overline{\omega_{z_0}(\alpha)}. \tag{3}$$

The integration is conducted along an arbitrary contour Γ on the cone (2) which intersects every generator of the cone only once; $\omega_{z_0}(\alpha)$ is a differential form on the cone (2) of the form*

$$\omega_{z_0}(\alpha) = a_3^{-1}(\alpha^1 d\alpha^2 - \alpha^2 d\alpha^1) = a_1^{-1}(\alpha^2 d\alpha^3 - \alpha^3 d\alpha^2) = a_2^{-1}(\alpha^3 d\alpha^1 - \alpha^1 d\alpha^3).$$

The constant C_{z_0} in the formula (3) is expressed by means of the formula

$$C_{z_0}^{-1} = \pi\Delta\int_\Gamma B(a_1, a_2, a_3)\,A^{-2}(\alpha^1, \alpha^2, \alpha^3)\,\omega_{z_0}(\alpha)\,\overline{\omega_{z_0}(\alpha)},$$

where A, B are an arbitrary pair of conjugate positive definite Hermitian forms, and Δ is the discriminant of the form A.

The conversion formula was previously obtained by another method for special cases, namely in [2] for the complex of lines which intersect a circle, and in [3] for a complex of lines which intersect an arbitrary algebraic curve.

Remark. We can also obtain the conversion formula in a more general form.

$$f(z_0) = C_{z_0}\int_\Gamma (\mathcal{L}_{z_0} + w)(\bar{\mathcal{L}}_{z_0} + \bar{w})\,\varphi(\alpha, [\alpha, z_0])\,\omega_{z_0}(\alpha)\,\overline{\omega_{z_0}(\alpha)}. \tag{4}$$

where $\mathcal{L}_{z_0}$, $\bar{\mathcal{L}}_{z_0}$ are operators of the infinitely small (not necessarily parallel) displacement of the lines of the complex passing through the point z_0:

$$\mathcal{L}_{z_0}\varphi = \sum (u^i\varphi'_{\alpha^i} + v_i\varphi'_{p_i}),$$

$$\bar{\mathcal{L}}_{z_0}\varphi = \sum (\bar{u}^i\varphi'_{\bar{\alpha}^i} + \bar{v}_i\varphi'_{\bar{p}_i}),$$

where $w = w(\alpha)$ is some function. It is not difficult to find necessary and sufficient conditions on the functions u^i, v_i, w for which the inversion formula (4) holds. We shall not give them here.

The inversion formula (4) is only valid for those complexes, which satisfy the condition of tangency.

There exist line complexes, for which our problem of integral geometry has a nonunique solution. By way of example we take the complex of the first degree $k_1\alpha^1 + k_2\alpha^2 + k_3\alpha^3 + l^1p_1 + l^2p_2 + l^3p_3 = 0$, where $k_1l^1 + k_2l^2 + k_3l^3 \neq 0$.**

For this complex of the first degree we ask another problem of integral geometry equivalent to the preceding: knowing the integrals of the function $f(z)$ along the lines of the complex, we ask to compute the integral of the function $f(z)$ along an arbitrary line.

With the complexes of the first order is connected an involutionary transformation in the space of all lines, preserving the lines of the complex. Indeed, if we draw the lines of the complex which intersect a given line π then these lines intersect also some other line π'. This π' is called the *conjugate* line of π with respect to the complex. It can be shown *that if the integrals of the function $f(z)$ along the lines of the complex $k_1\alpha^1 + k_2\alpha^2 + k_3\alpha^3 + l^1p_1 + l^2p_2 + l^3p_3 = 0$, where $k_1l^1 + k_2l^2 + k_3l^3 \neq 0$, are known, then only the sum of the integrals of the function $f(z)$ along an arbitrary pair of conjugate*

---

* Such a contour Γ, which intersects every generator of the cone only once, does in fact not exist, and the expression mentioned must be understood as follows. The space of the generators of the cone can be decomposed into sufficiently small regions, and for every region is taken a contour Γ_i, which intersects the generators of this region only once. The integral (3) is defined as the sum of the integrals along the Γ_i. It is easy to show that the integrals along the Γ_i do not depend on the nature of the contours Γ_i, and that their sum does not depend on the method of subdivision into parts of the space of generators.

** The geometry of complexes of the first degree is considered in detail in [1].

lines is uniquely determined. The problem of integral geometry has by the same token a unique solution in the class of such functions $f(z)$ for which their integrals along conjugate lines coincide.

We consider for simplicity's sake the complex $\alpha^1 = \lambda p_1$, where $\lambda \neq 0$ (to the form $\alpha^1 = \lambda p_1$ can the equation of any complex of the first degree be reduced by a suitable affine transformation of space). The line conjugate to the line $(\alpha^1, \alpha^2, \alpha^3; p_1, p_2, p_3)$ with respect to the complex $\alpha^1 = \lambda p_1$ has the Plücker coordinates $(\lambda p_1, \alpha^2, \alpha^3; \lambda^{-1}\alpha^1, p_2, p_3)$. *The sum of the integrals of the function $f(z)$ along a pair of conjugate lines with respect to the line complex $\alpha^1 = \lambda p_1$ is expressed by the integrals of the function $f(z)$ along the lines of the complex in the following form*

$$\varphi(\alpha_0^1, \alpha_0^2, \alpha_0^3; p_1^0, p_2^0, p_3^0) + \varphi(\lambda p_1^0, \alpha_0^2, \alpha_0^3; \lambda^{-1}\alpha_0^1, p_2^0, p_3^0) =$$
$$= C|\alpha_0^1 - \lambda p_1^0|^2 \int L\overline{L}\varphi(\lambda, \alpha^2, \alpha^3; 1, p_2, p_3)\,|\alpha_0^2\alpha^3 - \alpha_0^3\alpha^2|^{-4}\, d\alpha^2\, d\alpha^3\, d\overline{\alpha}^2\, d\overline{\alpha}^3. \tag{5}$$

Here $L\phi = \alpha^2\phi'_{p_3} - \alpha^3\phi'_{p_2}$, $\overline{L}\phi = \overline{\alpha}^2\phi'_{\overline{p}_3} - \overline{\alpha}^3\phi'_{\overline{p}_2}$ are operators of the infinitely small parallel displacement in the complex; the integral is taken along the manifold of all those lines of the complex which intersect the given pair of conjugate lines $(\alpha_0^1, \alpha_0^2, \alpha_0^3; p_1^0, p_2^0, p_3^0)$ and $(\lambda p_1^0, \alpha_0^2, \alpha_0^3; \lambda^{-1}\alpha_0^1, p_2^0, p_3^0)$.*

Received 8/MAR/61

BIBLIOGRAPHY

[1] F. Klein, *Vorlesungen über höhere geometrie*, J. Springer, Berlin, 1926.

[2] I. M. Gel'fand, Uspehi Mat. Nauk 15 (1960), no. 2 (92), 155 = Russian Math. Surveys 15 (1960), no. 2, 143.

[3] A. A. Kirillov, Dokl. Akad. Nauk SSSR 137 (1961), 276 = Soviet Math. Dokl. 2 (1961), 268.

Translated by:
D. J. Struik

2.

(with M. I. Graev and Z. Ya. Shapiro)

Integral geometry on k-dimensional planes

Funkts. Anal. Prilozh. 1 (1) (1967) 15-31 [Funct. Anal. Appl. 1 (1967) 14-27].
MR 35:3620. Zbl. 164:231

1. Statement of the Problem

Let C^n be n-dimensional complex space and let $H_{n,k}$ be the set of k-dimensional complex planes in C^n. Let $\alpha_1, \ldots, \alpha_k, \beta$ be vectors in C^n where the first k are linearly independent. Then a k-dimensional plane h is given by the equation

$$x = \alpha t + \beta, \tag{1}$$

where α is a matrix with columns $\alpha_1, \ldots, \alpha_k$, and t is in C^k. Two pairs (α, β) and (α', β') define the same plane h if and only if

$$\alpha' = \alpha\mu, \quad \beta' = \beta + \alpha t_0, \tag{2}$$

where μ is a nonsingular linear transformation into C^k, and t_0 is in C^k.

Let $g(\alpha)$ be a fixed infinitely differentiable function which does not vanish and which satisfies the condition

$$g(\alpha\mu) = g(\alpha)\,|\det \mu|^2. \tag{3}$$

With every function $f(x)$ belonging to S (S is the space of infinitely differentiable and rapidly decreasing functions in C^n) we associate a function $\varphi(h)$, h belonging to $H_{n,k}$, according to the formula

$$\varphi(h) = g(\alpha) \int f(\alpha t + \beta)\, dt\, d\bar{t}, \tag{4}$$

where $dt\, d\bar{t} = \left(\frac{i}{2}\right)^k dt_1 \ldots dt_k\, d\bar{t}_1 \ldots d\bar{t}_k$.

In view of condition (3) this integral depends only on the plane h and not on the way it is represented.

In this paper we are considering the following problems.

Problem 1. A function $g(\alpha)$, is given satisfying condition (3). What are necessary and sufficient conditions in order that the function $\varphi(h)$ be representable in form (4) where $f(x)$ belongs to S?

Problem 2. To give a formula for inverting the integral transform (4) that is a formula reconstructing $f(x)$ from the functions $\varphi(h)$ and $g(\alpha)$.

We make the following remark. Of course the value of $\varphi(h)$ on the set of all k-dimensional planes in C^n contains redundant information about the function $f(x)$. We shall be interested in formulas using the values of $\varphi(h)$ on the narrowest possible class of such planes.

2. The Manifold $H_{n,k}$

We denote by $G_{n,k}$ the complex Grassman manifold of k-dimensional subspaces of C^n, that is of k-dimensional planes passing through the origin. Let us give the mapping

$$\pi\colon H_{n,k} \to G_{n,k}, \tag{5}$$

associating with each plane h belonging to $H_{n,k}$ the subspace a in $G_{n,k}$ parallel to it. Under this mapping $H_{n,k}$ becomes a fiber space with base $G_{n,k}$. The fiber H_a over the point a in $G_{n,k}$ is the set of all k-dimensional planes parallel to the subspace a. Obviously there exists a natural isomorphism: $H_a \cong C^n/a$.

Moscow State University; Institute of Applied Mechanics, Academy of Sciences of the USSR. Translated from Funktsional'nyi Analiz i Ego Prilozheniya, Vol. 1, No. 1, pp. 15-31, January-March. Original article submitted November 4, 1966.

The space $H_{n,k}$ is given in a natural way the structure of the C^∞-manifold. In $H_{n,k}$ as in any fiber space with compact base and with C^∞-structure one can introduce the class Φ of infinitely differentiable and rapidly decreasing functions. For the direct product $U \times C^{n-k}$ functions infinitely differentiable with respect to arguments from U and C^{n-k} and for which all partial derivatives are rapidly decreasing belong to the class Φ. We note that if two products $U \times C^{n-k}$ are C^∞-isomorphic the classes of functions Φ on them coincide. A function on $H_{n,k}$ is said to belong to class Φ if its restriction to any trivial sub-bundle belongs to class Φ.

Let us introduce in C^n an arbitrary Hermitian metric and set $||h||$ equal to the distance of the plane h from the origin. Then the fact that a function rapidly decreases means that $|\varphi(h)| < C(N)||h||^{-N}$. Since in $H_{n,k}$ the group of affine transformations in C^n acts transitively, differential operators on $H_{n,k}$ can be associated with Lie operators on this group. The rapid decreasing of the derivatives which was defined above means that $|D\varphi| < C_D(N)||h||^{-N}$, where D is any polynomial in these differential operators.

We assigned an element h in $H_{n,k}$ with the help of the pair (α, β) where α belongs to the space E of $n \times k$ matrices of rank k and β is in C^n. Hence we have the mapping σ: $E \times C^n \to H_{n,k}$. By (2) the preimage of the point h in $H_{n,k}$ is the set of pairs $(\alpha\mu, \beta + \alpha t_1)$, where α and β are given, μ runs through the set of nonsingular linear transformations in C^k and t_1 runs through C^k. The mapping σ gives the mapping σ^*: $\varphi(h) \to \varphi(\alpha, \beta) \equiv \varphi[\sigma(\alpha, \beta)]$. The image of σ^* is the set of functions on $E \times C^n$, which are invariant with respect to transformations (2).

3. Construction of $\mathrm{Det}_B(A, A')$

We need an algebraic construction which will generalize the concept of the determinant of a numerical matrix to the case where the elements of the matrix are vectors in C^n.

We will call a multilinear form of two groups of vectors $(e_1, \dots, e_k)$ and $(e'_1, \dots, e'_k)$ a symmetric form of type (k, k) if it is symmetric with respect to the vectors of each group, is linear with respect to the first group and antilinear with respect to the second. If we denote by η^i_j and η'^i_j the coordinates of the vectors e_j and e_j' we can write a form of type (k, k) in the following way

$$B(e_1, \dots, e_k; e'_1, \dots, e'_k) = \sum_{p,q} b_{p,q}\, \eta_1^{p_1} \dots \eta_k^{p_k}\, \overline{\eta'^{q_1}_1} \dots \overline{\eta'^{q_k}_k}, \tag{6}$$

where $p = (p_1, \dots, p_k)$, $q = (q_1, \dots, q_k)$, p_j, $q_j = 1, \dots, n$, and the coefficients $b_{p,q}$ have the corresponding properties of symmetry.

Let us assign two quadratic matrices $A = (e_{ij})$, $A' = (e'_{ij})$ of order k whose elements are vectors in C^n. We set

$$\mathrm{Det}_B(A, A') = \sum_{\sigma,\sigma'} (-1)^{\mu+\nu} B(e_{1,s_1}, \dots, e_{k,s_k}; e'_{1,t_1}, \dots, e'_{k,t_k}). \tag{7}$$

The summation in (7) is taken over all permutations $\sigma = (s_1, \dots, s_k)$, $\sigma' = (t_1, \dots, t_k)$ of the numbers $1, \dots, k$; μ and ν are respectively the number of inversions in these permutations.

It is obvious that $\mathrm{Det}_B(A, A')$ is itself a multilinear form with respect to the rows and columns of A and A'. Let us write out a formula expressing $\mathrm{Det}_B(A, A')$ in terms of numerical determinants. To this end let us fix multiindices $p = (p, \dots, p_k)$ and $q = (q_1, \dots, q_k)$ and construct matrices with elements $\eta_{ij}(p) = \eta^{p_i}_{ij}$ and analogously $\eta'_{ij}(q) = \eta_{ij}'^{q_i}$. With the help of these peculiarly constructed matrices $\mathrm{Det}_B(A, A')$ is given by the formula

$$\mathrm{Det}_B(A, A') = \sum_{p,q} b_{p,q} \det(\eta_{ij}(p)) \det(\overline{\eta'_{ij}(q)}). \tag{8}$$

From the definition of the form $\mathrm{Det}_B(A, A')$ it is clear that it changes sign under any permutation of the rows or any permutation of the columns of either of the matrices A or A'.

4. Definition of the Operator $\varkappa_\beta$

We define now the operator which transforms functions $\varphi(h)$ in Φ into differential forms on the manifold $G_{n,k}$. In order to define this operator we take an arbitrary point β in C^n and construct the map

$$s_\beta : G_{n,k} \to H_{n,k},$$

which corresponds to every subspace a in $G_{n,k}$ the plane parallel to it passing through the point β. This gives the mapping

$$\overset{*}{s}_\beta:\ \varphi(h) \to \varphi(a;\ \beta) \equiv \varphi(s_\beta(a)).$$

If $\varphi(h)$ is in Φ, φ $(a;\ \beta)$ is an infinitely differentiable function. In addition if $\beta_1 - \beta_2$ is in a, to $S_{\beta_1}(a) = S_{\beta_2}(a)$ and so $\varphi(a;\ \beta_1) = \varphi(a;\ \beta_2)$.

We construct now from the function $\varphi(a;\ \beta)$ a multilinear form of type (k, k) from the 2k vectors $(e_1, \ldots, e_k)$ and $(e'_1, \ldots, e'_k)$ in C^n:

$$B(e_1, \ldots, e_k;\ e'_1, \ldots, e'_k) = \frac{1}{g(\alpha)} \sum_{p,q} \frac{\partial^{2k} \varphi(a;\ \beta)}{\partial \beta^p\, \partial \overline{\beta}^q}\, \eta_1^{p_1} \ldots \eta_k^{p_k}\, \overline{\eta_1'^{q_1} \ldots \eta_k'^{q_k}}, \tag{9}$$

where as above η_j^i and $\eta_j'^i$ denote the coordinates of the vectors e_j and e'_j and

$$\frac{\partial^{2k}}{\partial \beta^p\, \partial \overline{\beta}^q} = \frac{\partial^{2k}}{\partial \beta^{p_1} \ldots \partial \beta^{p_k}\, \partial \overline{\beta}^{q_1} \ldots \partial \overline{\beta}^{q_k}}.$$

It is convenient to write (9) also as

$$B(e_1, \ldots, e_k;\ e'_1, \ldots, e'_k) = \frac{1}{g(\alpha)} \left.\frac{\partial^{2k} \varphi(a;\ \beta + t_1 e_1 + \ldots + t_k e_k + s_1 e'_1 + \ldots + s_k e'_k)}{\partial t_1 \ldots \partial t_k\, \partial \overline{s}_1 \ldots \partial \overline{s}_k}\right|_{t=s=0} \tag{10}$$

With the help of form (9) let us construct the differential form of type (k, k) on the Grassman manifold which plays a fundamental role in this paper. Let us agree to assign a subspace a in $G_{n,k}$ in the following way. As in section 2 we denote by E the set of all k-dimensional hedra in C^n. Every such hedron is given by an $n \times k$ matrix $\alpha = (\alpha_j^i)$ (in the j-th column are the coordinates of the j-th vector). We associate with every α in E the subspace a in $G_{n,k}$ consisting of those planes in $G_{n,k}$ which are parallel to faces of α. We obtain the mapping

$$\tau:\ E \to G_{n,k}.$$

From (9) we construct first a differential form in the space E and then we show that this form is the image of a differential form on $G_{n,k}$. To this end let us fix a point α in E and assign at this point 2k vectors $l_1, \ldots, l_k;\ l'_1, \ldots, l'_k$ of the tangent space T_α. We write the vector l_j in T_α as a column of vectors e_{ij} in C^n. We obtain two matrices $A = (e_{ij})$ and $A' = (e'_{ij})$. From these and from the form B given by (9) we construct the form

$$\mathrm{Det}_B(A, A') = \frac{1}{g(\alpha)} \sum_{p,q} \frac{\partial^{2k} \varphi(a;\ \beta)}{\partial \beta^p\, \partial \overline{\beta}^q} \det(\eta_{ij}(p)) \det(\overline{\eta'_{ij}(q)}), \tag{11}$$

defined in section 3. Here $\eta_{ij}(p)$ denotes the p_i-th coordinate of the vector e_{ij}. It follows from the definition of the form $\mathrm{Det}_B(A, A')$ that at every point α in E form (11) is a skew-symmetric multilinear form of the vectors $l_1, \ldots, l_k,\ l'_1, \ldots, l'_k$ in T_α, where T_α is the tangent space to E at the point α.

Our goal is to show that form (11) is defined essentially in $G_{n,k}$, that is, is the image of a form in $G_{n,k}$ under the mapping τ^* induced by the mapping τ: $E \to G_{n,k}$. This assertion is proved by several simple lemmas.

Let T_a be the tangent space to $G_{n,k}$ at the point a. The mapping τ induces a mapping of the corresponding tangent spaces τ: $T_\alpha \to T_{\tau(\alpha)}$

LEMMA 1. The kernel of the mapping τ: $T_\alpha \to T_{\tau(\alpha)}$ consists of all vectors $l = (e_1, \ldots, e_k)$ such that e_j is in $a = \tau(\alpha)$, $j = 1, \ldots, k$.

Proof. Let us take an arbitrary vector l in T_α and consider the curve in E which is described by the point $\alpha + sl$ as s varies. The vector l is mapped to zero by τ if and only if $\frac{d}{ds}\tau(\alpha + sl)|_{s=0} = 0$. This condition means that to within o(s) the matrices $\alpha + sl$ and α define the same point a in $G_{n,k}$; that is, the equation $\alpha + sl = \alpha\mu(s) + o(s)$, holds where $\mu(s)$ is a nonsingular $k \times k$ matrix. From this equality we have: $l = [\alpha\mu(s) - \alpha]s^{-1} + o(1)$. As $s \to 0$, $o(1) \to 0$ and consequently $l = \lim_{|s \to 0} (\alpha\mu(s) - \alpha)s^{-1}$. Since $\alpha = (\alpha_1, \ldots, \alpha_k)$ is a set of k vectors of C^n each of which belongs to a, it is apparent that all k of the vectors which comprise $l = (e_1, \ldots, e_k)$ also belong to a.

LEMMA 2. If in the form $B(e_1, \ldots, e_k; e'_1, \ldots, e'_k)$ given by formula (9), where e_j, e'_j are in C^n, at least one of the vectors belongs to a, $B(e_1, \ldots, e_k; e'_1, \ldots, e'_k) = 0$.

Proof. Let us use expression (10) for the form $B(e_1, \ldots, e_k; e'_1, \ldots, e'_k)$. If one of the vectors e_j, e_j' belongs to a, say e_{j_1}, the function $\varphi(a; \beta + t_1e_1 + \ldots + s_ke_k)$ is in fact independent of the variable t_{j_1}; consequently by (10) we have $B(e_1, \ldots, e_k; e'_1, \ldots, e'_k) = 0$.

LEMMA 3. If form $B(e_1, \ldots, e'_k)$ is defined by (9) and if in the matrix $A = (e_{ij})$ or $A' = (e'_{ij})$ all the vectors of at least one column belong to a, $\mathrm{Det}_B(A, A') = 0$.

Proof. From the definition of $\mathrm{Det}_B(A, A')$ [cf. formula (7)] it follows that it is the sum of the values of the form B on the sets of 2k vectors each of which contains one vector from each column of the matrices A, A'. Hence by Lemma 2 each summand in this sum is equal to zero.

It follows from Lemmas 1-3 that form (11) is equal to zero on the kernel of the mapping $\tau: T_\alpha \to T_{\tau(\alpha)}$ and consequently is a skew-symmetric multilinear form in $T_{\tau(\alpha)}$.

LEMMA 4. If $\tilde{\alpha} = \alpha\mu$, $\tilde{A} = \mu$ and $\tilde{A}' = A'\mu$, $\mathrm{Det}_B(A, A') = \mathrm{Det}_{\tilde{B}}(\tilde{A}, \tilde{A}')$.

Proof. It is obvious from formula (11) that when A and A' are replaced respectively by $\tilde{A} = A\mu$ and $\tilde{A}' = A'\mu$ all the summands in the sum on the right are multiplied by $|\det\mu|^2$. On the other hand the function $g(\alpha)$ satisfies condition (3): $g(\alpha\mu) = g(\alpha)|\det\mu|^2$. The lemma is proved.

To conclude the proof of the fundamental assertion it remains to show that if $\tau(\tilde{\alpha}) = \tau(\alpha) = a$, the forms induced in T_a from T_α and $T_{\tilde{\alpha}}$ are the same. We choose 2k vectors in T_a and let $l_1, \ldots, l_k, l_1', \ldots, l_k'$ be any of their preimages in T_α. If $\tau(\tilde{\alpha}) = \tau(\alpha) = a$, then $\tilde{\alpha} = \alpha\mu$ for some μ and at the point $\tilde{\alpha}$ we can take, as the preimages of the vectors from T_a, $l_{j\mu}$ and $l'_{j\mu}$. According to Lemma 4 the value of the form in T_a is not changed.

We denote by $\varkappa_\beta\varphi(h)$ the multilinear skew-symmetric differential form in $G_{n,k}$ which is obtained in this way. The form $\varkappa_\beta\varphi(h)$ will play a fundamental role in our investigation.

It follows immediately from (11) that in E the form $\varkappa_\beta\varphi(h)$ can be written

$$\varkappa_\beta \varphi(h) = \frac{1}{g(\alpha)} \sum_{p,q} \frac{\partial^{2k}\varphi(a;\beta)}{\partial\beta^p\, \partial\bar{\beta}^q} d\alpha_1^{p_1} \wedge \ldots \wedge d\alpha_k^{p_k} \wedge d\bar{\alpha}_1^{q_1} \wedge \ldots \wedge d\bar{\alpha}_k^{q_k}, \tag{12}$$

where

$$\frac{\partial^{2k}}{\partial\beta^p\, \partial\bar{\beta}^q} = \frac{\partial^{2k}}{\partial\beta^{p_1} \ldots \partial\beta^{p_k}\, \partial\bar{\beta}^{q_1} \ldots \partial\bar{\beta}^{q_k}}.$$

Although the individual factors of the inner product (12) are forms only in E and cannot be carried over into $G_{n,k}$, the form in its entirety as we proved is a differential form in $G_{n,k}$.

Let us clarify the connection between the operators $\varkappa_\beta\varphi$ in spaces of different dimensionality. Let $m < n$ and let the embedding $C^m \to C^n$ be given. It induces embeddings $i: G_{m,k} \to G_{n,k}$; $H_{m,k} \to H_{n,k}$. We have the commutative diagram

$$\begin{array}{ccc} H_{m,k} & \xrightarrow{i} & H_{n,k} \\ \downarrow\pi & & \downarrow\pi \\ G_{m,k} & \xrightarrow{i} & G_{n,k}. \end{array}$$

The mapping i defines a dual mapping i^* of functions and differential forms. The following sample lemma which we shall require later is true.

LEMMA. Let $\varphi(h)$ be a function from $H_{n,k}$ and let β be in C^m. Then

$$\varkappa_\beta i^*\varphi = i^*\varkappa_\beta\varphi.$$

Proof. Let us choose a basis in C^n so that the image of C^m in C^n is given by the equations $x_{m+1} = \ldots = x_n = 0$. Let h be in $H_{n,k}$ and β be in C^m. The form $\varkappa_\beta \varphi$ is given by formula (12). To obtain the expression for $i^* \varkappa_\beta \varphi$ from this formula we must confine ourselves to points a in $G_{m,k}$ and to the tangent vectors to the points of this manifold. For this we must suppose that for $\nu > m$ $\alpha_j^\nu = 0$ and $d\alpha_j^\nu = 0$. The only summands which remain in formula (12) are those for which all $p_i, q_i \le m$. Hence this form coincides with the form $\varkappa_\beta i^* \varphi$, defined on $G_{m,k}$.

It is apparent that if β is in C^m the form $\varkappa_\beta \varphi$ is independent of the dimension of the space C^n. It is most natural therefore to consider it in $G_{\infty, k}$.

5. Local Coordinates in $H_{n,k}$; a Formula for $\varkappa_\beta \varphi$ in Local Coordinates

We shall divide $G_{n,k}$ into a finite number of neighborhoods in each of which we shall construct a local system of coordinates.

Let $\nu = (i_1, \ldots, i_k)$ be a set of k distinct indices $(1 \le i_l \le n)$. We shall say that a is in $U_\nu \subset G_{n,k}$ if $\det \alpha_\nu \ne 0$, where α_ν is the $k \times k$ matrix composed of the rows $(i_1, \ldots, i_k)$ of the matrix $\alpha = \tau^{-1}(a)$. Since $\tau(\alpha) = \tau(\alpha\mu)$, setting $\mu = \alpha_\nu^{-1}$, we can assume that an element a in U_ν is given by the matrix $\hat{\alpha}$ for which $\hat{\alpha}_\nu$ is a unit matrix. The correspondence between a in U_ν and $\hat{\alpha}$ is one to one. Since all sets of indices are equivalent we shall for the sake of definiteness carry out the discussion with $\nu = (n-k+1, \ldots, n)$. In this case $\hat{\alpha}$ can be written

$$\hat{\alpha} = \begin{pmatrix} \alpha_1^1 & \ldots & \alpha_k^1 \\ \cdot & \cdot\cdot\cdot & \cdot \\ \alpha_1^{n-k} & \ldots & \alpha_k^{n-k} \\ 1 & \ldots & 0 \\ \cdot & \cdot\cdot\cdot & \cdot \\ 0 & \ldots & 1 \end{pmatrix}. \tag{13}$$

Let us take $k(n-k)$ elements α_j^i of the matrix $\hat{\alpha}$ as coordinates in U_ν. We introduce now coordinates in $\pi^{-1}(U_\nu) \subset H_{n,k}$. An element h in $H_{n,k}$ is a plane with equation $x = \alpha t + \beta$. Since $\alpha_\nu = I$ we can without changing the plane choose a vector $\hat{\beta}$ for which $\beta^{i_1} = \ldots = \beta^{i_k} = 0$. The α_j^i chosen previously and the nonzero coordinates of the vector $\hat{\beta}$ define a system of coordinates in $\pi^{-1}(U_\nu)$.

From the condition that $\varphi(h)$ belongs to Φ it follows that for any fixed α the function $\varphi(h) = \varphi(\hat{\alpha}; \hat{\beta})$ is an infinitely differentiable and rapidly decreasing function of $\beta^1, \ldots, \beta^{n-k}$.

We shall now write an expression for $\varkappa_\beta \varphi$ in these coordinates. Let us note that matrices of type (13) form a submanifold V_ν in the space E of all $n \times k$ matrices and this set V_ν as part of E is diffeomorphic to U_ν. Since the form $\varkappa_\beta \varphi$ which we first constructed on E is actually defined on $G_{n,k}$, the restriction of the form $\varkappa_\beta \varphi$ in E to this subset (the collection of matrices $\hat{\alpha}$ of type (13) and their differentials on the tangent manifold V_ν) gives us the form $\varkappa_\beta \varphi$ in $U_\nu \subset G_{n,k}$. This leads us to the formula:

$$\varkappa_\beta \varphi(h) = \frac{1}{g(\hat{\alpha})} \sum_{p,q}{}' \frac{\partial^{2k} \varphi(\alpha; \beta)}{\partial \beta^p \partial \bar{\beta}^q} d\alpha_1^{p_1} \wedge \ldots \wedge d\alpha_k^{p_k} \wedge d\bar{\alpha}_1^{q_1} \wedge \ldots \wedge d\bar{\alpha}_k^{q_k}, \tag{14}$$

where $p = (p_1, \ldots, p_k)$, $q = (q_1, \ldots, q_k)$, the indices p_j, q_j taking on only values from 1 to n−k. We have written the form $\varkappa_\beta \varphi$ in a "simplified" way, having chosen a system of local coordinates in $G_{n,k}$. But in a neighborhood of the same point a in $G_{n,k}$ it is possible to choose different systems of coordinates and to obtain forms in different "simplified" representations. From our discussion it follows that the forms are equal although this is not immediately obvious.

6. Fourier Transform of the Function $\varphi(h)$

We shall introduce a fiber space $H'_{n,k}$ which is the dual of $H_{n,k}$. We recall that the manifold $G_{n,k}$ of k-dimensional subspaces of C^n is the base of a fiber space; the factor space $H_a = C^n/a$ is the fiber over the point a.

The fiber space $H'_{n,k}$, the dual of $H_{n,k}$, is defined as a space with the same base $G_{n,k}$. The fiber in $H'_{n,k}$ over the point a in $G_{n,k}$ is the space $H'_a = (C^n/a)'$, that is, the space of all ξ in $(C^n)'$ which are equal to zero on a. Hence the elements of the space $H'_{n,k}$ are all possible pairs (a, ξ), where a is in

$G_{n,k}$, ξ is in $(C^n/a)'$, such that ξ reduces to zero on a. Let us note that the natural isomorphism $H'_{n,k} \to G_{n,k} \times (C^n)'$ holds.* The topology in $H'_{n,k}$ is that induced by the topology on $G_{n,k} \times (C^n)'$.

We shall define the Fourier transform $\widetilde{\varphi}(h')$ of the function $\varphi(h)$ in Φ. Let $\widetilde{\beta}$ belong to $H_a = C^n/a$. We shall denote by $\varphi(a; \widetilde{\beta})$ the value of the function $\varphi(h)$ at the point $\widetilde{\beta}$. By hypothesis $\varphi(a, \widetilde{\beta})$ is a rapidly decreasing function of $\widetilde{\beta}$ in $H_a = C^n/a$.

Let us introduce a measure in C^n/a. The following measure is given on the subspace a: $x = \alpha t$: $d\mu_a(x) = g(\alpha) dt\, d\bar{t}$. Let us assign on all of C^n an element of volume $dv = (i/2)^n dx_1 \dots dx_n d\bar{x}_1 \dots d\bar{x}_n$ and define a measure $d\mu_{H_a}(\widetilde{\beta})$ on H_a by the relationship

$$dv = d\mu_a(x) \cdot d\mu_{H_a}(\widetilde{\beta}).$$

We set

$$\widetilde{\varphi}(a; \xi) = (2\pi)^{-(n-k)} \int_{H_a} \varphi(a; \widetilde{\beta}) e^{i((\xi,\beta)+(\bar{\xi},\bar{\beta}))/2} d\mu_{H_a}(\widetilde{\beta}), \tag{15}$$

where ξ is in H'_a. We shall call the function $\widetilde{\varphi}(h') = \widetilde{\varphi}(a; \xi)$ defined on $H'_{n,k}$ the Fourier transform of the original function $\varphi(h)$.

Let us formulate the basic properties of the function $\widetilde{\varphi}(h')$.

1) If $\varphi(h)$ belongs to Φ_H, $\widetilde{\varphi}(h')$ belongs to $\Phi_{H'}$; that is, $\widetilde{\varphi}(h')$ is an infinitely differentiable rapidly decreasing function on $H'_{n,k}$.

2) The function $\varphi(h)$ is expressed in terms of its Fourier transform $\widetilde{\varphi}(h')$ by means of the following inversion formula:

$$\varphi(a; \widetilde{\beta}) = (2\pi)^{-(n-k)} \int_{H'_a} \widetilde{\varphi}(a; \xi) e^{-i((\xi,\beta)+(\bar{\xi},\bar{\beta}))/2} d\mu_{H'_a}(\xi), \tag{16}$$

where $d\mu_{H'_a}(\xi)$ is the measure on H'_a, which is the dual of the measure $d\mu_{H_a}(\widetilde{\beta})$ on H_a.

Propositions 1 and 2 follow immediately from properties of the usual Fourier transform in C^{n-k}.

Let us compute the Fourier transform $\widetilde{\varkappa_\beta \varphi}$ of the form $\varkappa_\beta \varphi$.

The following holds:

$$\widetilde{\varkappa_\beta \varphi} = \frac{\widetilde{\varphi}(a; \xi)}{g(\alpha)} \sum_{p,q} \xi^{(p)} \bar{\xi}^{(q)} d\alpha_1^{p_1} \wedge \dots \wedge d\alpha_k^{p_k} \wedge d\bar{\alpha}_1^{q_1} \wedge \dots \wedge d\bar{\alpha}_k^{q_k}, \tag{17}$$

where $p = (p_1, \dots, p_k)$, $q = (q_1, \dots, q_k)$; $1 \le p_j, q_j \le n$; $\xi^{(p)} = \xi_{p_1} \dots \xi_{p_k}$, $\bar{\xi}^{(q)} = \bar{\xi}_{q_1} \dots \bar{\xi}_{q_k}$. Here the ξ_i are coordinates of the vector ξ. Let us recall that ξ reduces to zero on the subspace a.

Formula (17) is immediately obtained if one uses formulas (15) and (16) for the Fourier transform and the definition of $\varkappa_\beta \varphi$ in formula (12).

We note that (17) may also be written in the following convenient way:

$$\widetilde{\varkappa_\beta \varphi} = \frac{\widetilde{\varphi}(a; \xi)}{g(\alpha)} \bigwedge_{j=1}^{k} (\xi_1 d\alpha_j^1 + \dots + \xi_n d\alpha_j^n) \wedge \bigwedge_{j=1}^{k} (\bar{\xi}_1 d\bar{\alpha}_j^1 + \dots + \bar{\xi}_n d\bar{\alpha}_j^n). \tag{17'}$$

In conclusion let us indicate the appearance of the expressions in local coordinates for $\widetilde{\varphi}(a; \xi)$ and $\widetilde{\varkappa_\beta \varphi}$. For definiteness let us consider the system of local coordinates $(\hat{\alpha}, \hat{\beta})$, introduced on page 18. We note that the measure $d\mu_{H_a}(\widetilde{\beta})$ can be written in the coordinates of $\hat{\beta}$ as follows:

$$d\mu_{H_a}(\widetilde{\beta}) = \frac{1}{g(\hat{\alpha})} d\hat{\beta}\, d\bar{\hat{\beta}} = \left(\frac{i}{2}\right)^{n-k} \frac{1}{g(\hat{\alpha})} d\hat{\beta}^1 \dots d\hat{\beta}^{n-k} d\bar{\hat{\beta}}^1 \dots d\bar{\hat{\beta}}^{n-k}.$$

* We note that there is defined also a natural mapping $H'_{n,k} \to (C^n)'$ associating (a, ξ) in $H'_{n,k}$ with the point ξ in $(C^n)'$

Hence we have

$$\widetilde{\varphi}(a;\xi)=\frac{(2\pi)^{-(n-k)}}{g(\hat{\alpha})}\int\varphi(\hat{\alpha};\hat{\beta})\,e^{i((\xi,\hat{\beta})+(\overline{\xi},\overline{\hat{\beta}}))/2}\,d\hat{\beta}\,d\overline{\hat{\beta}}, \tag{18}$$

where $(\xi, \hat{\beta}) = \xi_1\beta^1 + \ldots + \xi_{n-k}\beta^{n-k}$.

The formula for the inverse Fourier transformation is:

$$\varphi(\hat{\alpha};\hat{\beta})=(2\pi)^{-(n-k)}\,g(\hat{\alpha})\int\widetilde{\varphi}(a;\xi)\,e^{-i((\xi,\hat{\beta})+(\overline{\xi},\overline{\hat{\beta}}))/2}\,d\xi'\,d\overline{\xi}', \tag{19}$$

where $d\xi'\,d\overline{\xi}' = \left(\frac{i}{2}\right)^{n-k} d\xi_1 \ldots d\xi_{n-k}\, d\overline{\xi}_1 \ldots d\overline{\xi}_{n-k}$. Further we have

$$\widetilde{\varkappa_\beta\varphi}=\frac{\widetilde{\varphi}(a;\xi)}{g(\hat{\alpha})}\sum_{p,q}\xi^{(p)}\overline{\xi}^{(q)}\,d\alpha_1^{p_1}\wedge\ldots\wedge d\alpha_k^{p_k}\wedge d\overline{\alpha}_1^{q_1}\wedge\ldots\wedge d\overline{\alpha}_k^{q_k}, \tag{20}$$

where $p = (p_1, \ldots, p_k)$, $q = (q_1, \ldots, q_k)$, p_i, q_i taking on only the values from 1 to n-k as opposed to (17). We note that (20) can also be written in the following convenient way:

$$\widetilde{\varkappa_\beta\varphi}=\frac{\varphi(a;\xi)}{g(\hat{\alpha})}\bigwedge_{j=1}^{k}(\xi_1\,d\alpha_j^1+\ldots+\xi_{n-k}\,d\alpha_j^{n-k})\wedge\bigwedge_{j=1}^{k}(\overline{\xi}_1\,d\overline{\alpha}_j^1+\ldots+\overline{\xi}_{n-k}\,d\overline{\alpha}_j^{n-k}). \tag{20'}$$

7. Formulation and Proof of Theorem 1

Let $g(\alpha)$ be an infinitely differentiable nonvanishing function on E, satisfying the condition $g(\alpha\mu) = g(\alpha)|\det\mu|^2$ for every nonsingular $k \times k$ matrix μ.

THEOREM 1. In order that a function $\varphi(h)$ on $H_{n,k}$, where $k < n-1$,* be representable as

$$\varphi(h)=g(\alpha)\int f(\alpha t+\beta)\,dt\,d\overline{t}, \tag{21}$$

where $f(x)$ is in S, it is necessary and sufficient that the following conditions be satisfied:

1. $\varphi(h)$ belongs to Φ.
2. The form $\varkappa_\beta\varphi$ defined by formula (12) be closed on the manifold $G_{n,k}$ for any fixed β.†

Proof of the Necessity of Condition 1. Let us choose a metric in $H_{n,k}$ for which $\|h\|$ is the distance of the plane h from the origin. It is clear that the rapid decreasing of the function $\varphi(h)$ and of its derivatives is an immediate consequence of the rapid decreasing of the function $f(x)$.

Proof of the Necessity of Condition 2. Let us assume that the function $\varphi(h)$ is representable as in (21). We prove that $d(\varkappa_\beta\varphi) = 0$.

We note beforehand that the function

$$u(\alpha;\beta)\equiv\frac{\varphi(h)}{g(\alpha)}=\int f(\alpha t+\beta)\,dt\,d\overline{t} \tag{22}$$

satisfies the system of equations

$$\frac{\partial^2 u}{\partial\alpha_j^i\,\partial\beta^{i'}}-\frac{\partial^2 u}{\partial\alpha_j^{i'}\,\partial\beta^i}=0,\quad \frac{\partial^2 u}{\partial\overline{\alpha}_j^i\,\partial\overline{\beta}^{i'}}-\frac{\partial^2 u}{\partial\overline{\alpha}_j^{i'}\,\partial\overline{\beta}^i}=0 \tag{23}$$

$$(j=1,\ldots,k;\ i,i'=1,\ldots,n).$$

These equations follow directly from (22) [since $f(x)$ is in S, the differentiation on the right hand side of equation (22) may be carried out under the integral sign].

*The case $k = n-1$ is distinguished from the case $k < n-1$ and is considered in [1].

†Theorem 1 is valid for the real space R^n also ($\varkappa_\beta\varphi$ in R^n is defined by the formula

$$\varkappa_\beta\varphi=\frac{i}{g^{(\alpha)}}\sum_p\frac{\partial_\varphi^k(\alpha;\beta)}{\partial\beta^{p_1}\ldots\partial\beta^{p_k}}\,d\alpha_1^{p_1}\wedge\ldots\wedge d\alpha_k^{p_k});$$

the proof is the same as that for C^n.

Let us introduce the notation

$$d_j(\varkappa_\beta\varphi)=\sum_{i=1}^{n}\frac{\partial(\varkappa_\beta\varphi)}{\partial\alpha_j^i}d\alpha_j^i,\qquad \bar d_j(\varkappa_\beta\varphi)=\sum_{i=1}^{n}\frac{\partial(\varkappa_\beta\varphi)}{\partial\bar\alpha_j^i}d\bar\alpha_j^i.$$

Then $d(\varkappa_\beta\varphi)=\sum_{j=1}^{k}(d_j(\varkappa_\beta\varphi)+\bar d_j(\varkappa_\beta\varphi))$ and for the proof of the closedness of the form $\varkappa_\beta\varphi$ it suffices to prove the following equalities: $d_j(\varkappa_\beta\varphi)=0$, $\bar d_j(\varkappa_\beta\varphi)=0$. For definiteness let us assume for example that $j=1$. Replacing $\varkappa_\beta\varphi$ with the expression for it in (12) we obtain

$$d_1(\varkappa_\beta\varphi)=\sum_{i,p,q}\frac{\partial^{2k+1}u}{\partial\alpha_1^i\,\partial\beta^p\,\partial\bar\beta^q}d\alpha_1^i\wedge d\alpha_1^{p_1}\wedge\ldots\wedge d\alpha_k^{p_k}\wedge d\bar\alpha_1^{q_1}\wedge\ldots\wedge d\bar\alpha_k^{q_k}. \tag{24}$$

We denote by p' the multiindex $(p_2,\ldots,p_k)$. Regrouping the summands we can rewrite expression (24) for $d_1(\varkappa_\beta\varphi)$ in the following way:

$$d_1(\varkappa_\beta\varphi)=\sum_{p',q}\frac{\partial^{2k-1}}{\partial\beta^{p'}\,\partial\bar\beta^q}\left\{\sum_{i,p_1=1}^{n}\frac{\partial^2u}{\partial\alpha_1^i\,\partial\beta^{p_1}}d\alpha_1^i\wedge d\alpha_1^{p_1}\right\}\wedge d\alpha_2^{p_2}\wedge\ldots\wedge d\bar\alpha_k^{q_k}.$$

From equation (23) the expression in brackets is equal to zero. Consequently, $d_1(\varkappa_\beta\varphi)=0$. Analogously $d_j(\varkappa_\beta\varphi)=0$, $\bar d_j(\varkappa_\beta\varphi)=0$ for the remaining value of j.

Proof of the Sufficiency of the Conditions of Theorem 1. Let $\varphi(h)$ satisfy the conditions of the theorem; that is, let $\varphi(h)$ belong to Φ and let the form $\varkappa_\beta\varphi$ be closed. We proceed from the function $\varphi(h)$ to its Fourier transform $\tilde\varphi(h')=\tilde\varphi(a;\xi)$. Let us recall that the space $H'_{n,k}$ on which the function $\varphi(a;\xi)$ is defined consists of the pair (a,ξ) where a is in $G_{n,k}$ and ξ is in $(C^n/a)'$, in other words, $(e,\xi)=0$ for any vector e lying in the space a. The following important lemma is a fundamental step in the proof.

LEMMA 1. For any $\xi\neq 0$, $\tilde\varphi(a;\xi)=F(\xi)$. In other words, for any fixed vector ξ in $(C^n)'$ $(\xi\neq 0)$ the function $\varphi(a;\xi)$ is constant on the set of subspaces a in $G_{n,k}$ such that $(a,\xi)=0$.*

Proof. The lemma will be proved if we prove it in every neighborhood U_ν belonging to a cover of $G_{n,k}$ (cf. section 5).

For definiteness let us choose the neighborhood defined by the coordinates

$$\hat\alpha=\begin{Vmatrix}\alpha_1^1 & \ldots & \alpha_k^1\\ \cdots & \cdots & \cdots\\ \alpha_1^{n-k} & \ldots & \alpha_k^{n-k}\\ 1 & \ldots & 0\\ \cdots & \cdots & \cdots\\ 0 & \ldots & 1\end{Vmatrix}.$$

The neighborhood $\pi^{-1}(U_\nu)\subset H_{n,k}$ is then given by the coordinates of $\hat\alpha$ and by the nonzero coordinates of the vector $\hat\beta=(\hat\beta^1,\ldots,\hat\beta^{n-k},0,\ldots,0)$.

To prove the theorem we proceed in the condition on the closedness of the form $\varkappa_\beta\varphi$ to its Fourier transform. However, first let us consider in more detail the Fourier transform itself. It is given by the formula

$$\tilde\varphi(a;\xi)=\frac{(2\pi)^{-(n-k)}}{g(\hat\alpha)}\int\varphi(\hat\alpha;\hat\beta)\,e^{i((\xi,\hat\beta)+(\bar\xi,\bar{\hat\beta}))/2}\,d\hat\beta\,d\bar{\hat\beta}, \tag{25}$$

where $(a,\xi)=0$. Since the columns of the matrix $\hat\alpha$ give the coordinates of the basis vectors in a, the condition that $(a,\xi)=0$ is given by the system of equations

$$\xi_1\alpha_j^1+\ldots+\xi_{n-k}\alpha_j^{n-k}+\xi_{n-k+j}=0,\qquad j=1,\ldots,k. \tag{26}$$

---

*We say that $(a,\xi)=0$ if for every x in $a\subset C^n$, $(x,\xi)=0$.

Hence we can choose $\xi_1, \ldots, \xi_{n,k}$ for a in U_v as the independent parameters in $(C^n/a)'$. The remaining ξ_i are defined by formula (26). We note that since $(\xi, \widehat{\beta}) = \xi_1\widehat{\beta}^1 + \ldots + \xi_{n-k}\widehat{\beta}^{n-k}$, only the components $\xi' = (\xi_1, \ldots, \xi_{n-k})$ enter into formula (25) for the Fourier transform. We shall keep in mind that $\widetilde{\varphi}$ depends on the variables $\widehat{\alpha}$ and ξ'.

We proceed now in the condition of the closedness of the form $\varkappa_\beta\varphi$ to its Fourier transform. Since

$$g(\widehat{\alpha})\,\widetilde{\varkappa_\beta\varphi} = (2\pi)^{-(n-k)}\int \varkappa_\beta\varphi \cdot e^{i((\xi,\widehat{\beta})+(\overline{\xi},\overline{\widehat{\beta}}))/2}\,d\widehat{\beta}\,d\overline{\widehat{\beta}},$$

we have

$$d_{\widehat{\alpha}}\,(g(\widehat{\alpha})\,\widetilde{\varkappa_\beta\varphi}) = (2\pi)^{-(n-k)}\int d_{\widehat{\alpha}}\,(\varkappa_\beta\varphi)\,e^{i((\xi,\widehat{\beta})+(\overline{\xi},\overline{\widehat{\beta}}))/2}\,d\widehat{\beta}\,d\overline{\widehat{\beta}},$$

and so for any fixed ξ', $d_{\widehat{\alpha}}(g(\widehat{\alpha})\widetilde{\varkappa_\beta\varphi}) = 0$

Let us write $d_{\widehat{\alpha}}$ as

$$d_{\widehat{\alpha}} = \sum_{l=1}^{k}(d_l + \overline{d}_l),$$

where $d_j = \sum_{l=1}^{n-k}\frac{\partial}{\partial\alpha_j^l}\,d\alpha_j^l$, $j = 1, \ldots, k$.

We show that

$$d_l\,(g\,(\widehat{\alpha})\,\widetilde{\varkappa_\beta\varphi}) = 0, \qquad \overline{d}_l\,(g\,(\widehat{\alpha})\,\widetilde{\varkappa_\beta\varphi}) = 0. \tag{27}$$

Let us show for example that $d_1(g(\widehat{\alpha})\widetilde{\varkappa_\beta\varphi}) = 0$. We recall that

$$g(\widehat{\alpha})\widetilde{\varkappa_\beta\varphi} = \widetilde{\varphi}\,(a;\,\xi)\bigwedge_{l=1}^{k}(\xi_1\,d\alpha_l^1 + \ldots + \xi_{n-k}\,d\alpha_l^{n-k})\wedge\bigwedge_{l=1}^{k}(\overline{\xi}_1\,d\overline{\alpha}_l^1 + \ldots + \overline{\xi}_{n-k}\,d\overline{\alpha}_l^{n-k}).$$

We note that the form $d_1(g(\alpha)\widetilde{\varkappa_\beta\varphi})$ is of degree 2 with respect to $d\alpha_1^1, \ldots, d\alpha_1^{n-k}$, while the forms $d_j(g(\widehat{\alpha})\widetilde{\varkappa_\beta\varphi})$ for $j \neq 1$ and all the $\overline{d}_l(g(\widehat{\alpha})\widetilde{\varkappa_\beta\varphi})$ are of the degree 1. So from the equation $(\Sigma d_j + \Sigma\overline{d}_j)\times(g(\widehat{\alpha})\widetilde{\varkappa_\beta\varphi}) = 0$ it follows that $d_1(g(\widehat{\alpha})\widetilde{\varkappa_\beta\varphi}) = 0$. Analogously $d_j(g(\widehat{\alpha})\widetilde{\varkappa_\beta\varphi}) = 0$, $\overline{d}_j(g(\widehat{\alpha})\widetilde{\varkappa_\beta\varphi}) = 0$ for the remaining values of j.

Let us write Eq. (27) in more detail. First of all let us get $\widetilde{\varphi}(a;\,\xi) = \psi\,(\widehat{\alpha};\,\xi')$. Putting the expression for $g(\widehat{\alpha})\widetilde{\varkappa_\beta\varphi}$, in the first of Eq. (27) we obtain

$$\sum_{\mu<\nu}\left(\xi_\mu\frac{\partial\psi}{\partial\alpha_l^\nu} - \xi_\nu\frac{\partial\psi}{\partial\alpha_l^\mu}\right)d\alpha_l^\nu\wedge d\alpha_l^\mu$$

$$\wedge\bigwedge_{m\neq l}(\xi_1\,d\alpha_m^1 + \ldots + \xi_{n-k}\,d\alpha_m^{n-k})\wedge\bigwedge_{m=1}^{k}(\overline{\xi}_1\,d\overline{\alpha}_m^1 + \ldots + \overline{\xi}_{n-k}\,d\overline{\alpha}_m^{n-k}) = 0,$$

and so

$$\xi_\mu\frac{\partial\psi}{\partial\alpha_l^\nu} - \xi_\nu\frac{\partial\psi}{\partial\alpha_l^\mu} = 0 \qquad (l = 1, \ldots, k;\ \mu, \nu = 1, \ldots, n-k). \tag{28}$$

In an analogous way one can verify the equations

$$\overline{\xi}_\mu\frac{\partial\psi}{\partial\overline{\alpha}_l^\nu} - \overline{\xi}_\nu\frac{\partial\psi}{\partial\overline{\alpha}_l^\mu} = 0 \qquad (l = 1, \ldots, k;\ \mu, \nu = 1, \ldots, n-k). \tag{28'}$$

We obtained a system of equations for $\widetilde{\varphi} = \psi\,(\widehat{\alpha};\,\xi')$ with constant coefficients $\xi_1, \ldots, \xi_{n-k}$. Its general solution has the following format:

$$\psi(\widehat{\alpha};\,\xi') = F\,(\xi_1, \ldots, \xi_{n-k}, -(\xi_1\alpha_1^1 + \ldots + \xi_{n-k}\,\alpha_1^{n-k}), -(\xi_1\alpha_k^1 + \ldots + \xi_{n-k}\,\alpha_k^{n-k})). \tag{29}$$

On the basis of (26) we conclude that $\psi(\widehat{\alpha};\,\xi') = F(\xi_1, \ldots, \xi_{n-k}, \xi_{n-k+1}, \ldots, \xi_n)$. The lemma is proved.

LEMMA 2. Let a function $F(\xi)$, ξ in $(C^n)'$ be given and let there exist a function $\widetilde{\varphi}(a;\,\xi)$ in Φ_H'* such that for every point $\xi \neq 0$ the equation

* That is $\widetilde{\varphi}$ is an infinitely differentiable and rapidly decreasing function on the fiber space $H'_{h,k}$.

$$F(\xi) = \widetilde{\varphi}(a; \xi) \tag{30}$$

holds for all a in $G_{n,k}$ such that $(a, \xi) = 0$. Then $F(\xi)$ is in S.

Proof. I. First we prove that $F(\xi)$ is infinitely differentiable at the point $\xi \neq 0$. If $\xi = (\xi_1, \ldots, \xi_n) \neq 0$, $\xi_i \neq 0$ for at least one i; for the sake of definiteness we assume that $\xi_1 \neq 0$. For every point ξ with $\xi_1 \neq 0$ we consider the subspace a in $G_{n,k}$ given by the matrix $\hat{\alpha}$ of type (13) for which $\alpha_j^i = 0$ for $i = 2, \ldots, n-k$; $j = 1, \ldots, k$ and $\alpha_j^1 = -\xi_1^{-1}\xi_{n-k+j}$, $j = 1, \ldots, k$. The condition $(a, \xi) = 0$ is then satisfied. Since the right hand side of Eq. (30) is now an infinitely differentiable function of ξ for $\xi_1 \neq 0$, the infinite differentiability of $F(\xi)$ for $\xi \neq 0$ is proved.

II. We prove now the somewhat more difficult assertion:

It is possible to define $F(\xi)$ at the point $\xi = 0$ in such a way that it is infinitely differentiable there.

It is convenient to preface the proof of this assertion with two auxiliary propositions.

PROPOSITION 1. Let the function $F(\xi)$, ξ in $(C^n)'$ be given. We shall assume that there exists an infinitely differentiable function $\widetilde{\varphi}(a, \xi)$ on all of $H'_{n,k}$ such that for $\xi \neq 0$ the following equality holds:

$$F(\xi) = \widetilde{\varphi}(a; \xi), \quad (a, \xi) = 0.$$

Then for $\xi \neq 0$ the partial derivatives $F'_{\xi_i}(\xi)$ of the function $F(\xi)$ can be represented as

$$F'_{\xi_i}(\xi) = \varphi_i(a; \xi), \tag{31}$$

where $\varphi_i(a; \xi)$ is an infinitely differentiable function on all of $H'_{n,k}$. The analogous proposition for $F'_{\bar{\xi}_i}$ holds.

Proof. We cover $G_{n,k}$ with the neighborhoods U_ν introduced in section 5 and introduce in each of these neighborhoods the local coordinates given by the elements of the matrix $\hat{\alpha}$. The condition $(a, \xi) = 0$ means that*

$$\xi_{n-k+j} = -(\xi_1\alpha_j^1 + \ldots + \xi_{n-k}\alpha_j^{n-k}), \quad j = 1, \ldots, k. \tag{32}$$

Hence $\widetilde{\varphi}(a; \xi)$ is an infinitely differentiable function of $\hat{\alpha}$ and $\xi' = (\xi_1, \ldots, \xi_{n-k})$.

We first prove the assertion for $F'_{\xi_{n-k+j}}(\xi)$. Since the function $F(\xi)$ is differentiable for $\xi \neq 0$ we have by (31) and (32)

$$F'_{\xi_{n-k+j}}(\xi) = -\frac{1}{\xi_l}\frac{\partial\widetilde{\varphi}(\hat{\alpha}; \xi')}{\partial\alpha_j^l}, \quad l = 1, \ldots, n-k. \tag{33}$$

We note that $n-k \geq 2$. The function $F'_{\xi_{n-k+j}}(\xi)$ is defined for all $\xi \neq 0$ by this equation since in our neighborhood it follows from the fact that $\xi \neq 0$ that all of ξ_l, $l = 1, \ldots, n-k$ are nonzero.

The assertion that the right hand side of (33) can be extended to an infinitely differentiable function of $\hat{\alpha}$, ξ' everywhere including $\xi' = 0$ is a consequence of the following observation. If an infinitely differentiable function $u_l(\hat{\alpha}; \xi')$ $(l = 1, \ldots, n-k)$ is given and if the equations $\xi_{l'}u_l(\hat{\alpha}; \xi') = \xi_l u_{l'}(\hat{\alpha}; \xi')$, $l, l' = 1, \ldots, n-k$ are satisfied, there exists an infinitely differentiable function $u(\hat{\alpha}; \xi')$, defined for all ξ', such that $u_l(\hat{\alpha}; \xi') = \xi_l u(\hat{\alpha}; \xi')$.†

Since for all $l \leq n-k$, F'_{ξ_l}

$$F'_{\xi_l}(\xi) = \frac{\partial\widetilde{\varphi}(\hat{\alpha}; \xi')}{\partial\xi_l} + \sum_{j=1}^{k}\alpha_j^l F'_{\xi_{n-k+j}}(\xi), \tag{34}$$

the assertion is proved for all partial derivatives F'_{ξ_i}. Since the neighborhoods U_ν form a cover of $G_{n,k}$, Proposition 1 is proved.

---

* For definiteness we chose $\nu = (n-k+1, \ldots, n)$.

† In fact, in view of the equation $\xi_{l'}u_l = \xi_l u_{l'}$, we have $u_l|_{\xi_l=0} = 0$. Consequently, $u_l(\hat{\alpha}; \xi') = \xi_l \hat{u}_l\cdot(\hat{\alpha}; \xi')$, where $\hat{u}_l(\hat{\alpha}; \xi')$ is an infinitely differentiable function including $\xi' = 0$. From the equation $\xi_{l'}u_l = \xi_l u_{l'}$, it follows that $\hat{u}_l = \hat{u}_{l'}$.

PROPOSITION 2. If the function $f(\xi)$, ξ in $(C^n)'$ satisfies the hypothesis of Proposition 1, that is, is representable for $\xi \neq 0$ as $f(\xi) = \psi(a; \xi)$ [$\widetilde{\psi}(a; \xi)$ is an infinitely differentiable function on all of $H'_{n,k}$], then the limit $\lim_{\xi\to 0} f(\xi)$ exists.

Proof. We choose $U_\nu \subset G_{n,k}$. We take in U_ν the set $U_{\nu, N}$, consisting of those elements $\widehat{\alpha}$ in U_ν which satisfy the additional condition that $|\alpha_j^i| < N$. One can show that for sufficiently large N the neighborhoods $U_{\nu, N}$ give a cover of $G_{n,k}$. In the neighborhood $U_{\nu,N}$ the function $\psi(a; \xi)$ can be written $\psi(a; \xi) = \psi(\widehat{\alpha}; \xi')$, $\xi' = (\xi_1, \ldots, \xi_{n-k})$. According to Taylor's formula $\psi(\widehat{\alpha}; \xi') = \psi(\widehat{\alpha}; 0) + o(||\xi'||)$, where in view of the boundedness of the α_j^i the estimates are uniform with respect to $\widehat{\alpha}$. We show that $\psi(\widehat{\alpha}; 0)$ does not depend on $\widehat{\alpha}$. In fact it follows from the equation $f(\xi_1, \ldots, \xi_{n-k}, -(\xi_1\alpha_1^1 + \ldots + \xi_{n-k} \cdot \alpha_1^{n-k}), \ldots) = \psi(\widehat{\alpha}; \xi')$ that for $\xi_1 = 0$ the function $\psi(\widehat{\alpha}; \xi')$ does not depend on $\alpha_1^1, \ldots \alpha_k^1$. Hence $\psi(\widehat{\alpha}; 0)$ does not depend on $\alpha_1^1, \ldots, \alpha_k^1$. One can show analogously that $\psi(\widehat{\alpha}; 0)$ does not depend on the remaining α_j^i. Hence it is proved that $f(\xi) = c + o(||\xi'||)$; our assertion follows immediately.

Now we are able to prove that $F(\xi)$ is infinitely differentiable at the point $\xi = 0$. By definition, $F(\xi)$ satisfies the condition of Proposition 1. Using induction we can establish on the basis of Proposition 1 that all the partial derivatives $\partial^{(p)}\bar{\partial}^{(q)}F$ of the function $F(\xi)$ satisfy this condition. Consequently, by Proposition 2 $\lim_{\xi\to 0} \partial^{(p)}\bar{\partial}^{(q)}F(\xi)$ exists.

Let us note that in our discussion we relied heavily on the fact that $k < n-1$. For $k = n-1$ the requirement that $F(\xi)$ be infinitely differentiable at $\xi = 0$ leads to an additional condition on $\varphi(h)$ (cf. [1]).

III. Now we prove that $F(\xi)$ together with all its partial derivatives rapidly decreases as $||\xi|| \to \infty$. We take the neighborhood $U_{\nu,N} \subset G_{n,k}$ introduced above and its preimage $V_{\nu,N} \subset H'_{n,k}$ under the mapping $H'_{n,k} \to G_{n,k}$. If $H'_{n,k} \to (C^n)'$ is the natural mapping introduced in section 6 we denote $K_{\nu,N}$ the image of $V_{\nu,N}$ under this mapping. Let us choose for example $\nu = (n-k+1, \ldots, n)$. It can be shown that the point ξ in $K_{\nu,N}$ satisfies the inequality $||\xi'|| > C\,||\xi||$, where $\xi' = (\xi_1, \ldots, \xi_{n-k})$; furthermore there exists $C_1 > 0$ such that every point ξ satisfying the inequality $||\xi'|| > C_1 ||\xi||$ belongs to $K_{\nu,N}$.

By the hypothesis of Lemma 2 and the definition of a rapidly decreasing function on the fiber space $H'_{n,k}$, the function $\widetilde{\varphi}(a; \xi)$; $= \psi(\widehat{\alpha}; \xi')$ is rapidly decreasing with respect to ξ' and uniformly decreasing with respect to $\widehat{\alpha}$ in $U_{\nu,N}$. Hence $F(\xi) = \widetilde{\varphi}(a; \xi)$ is a rapidly decreasing function of ξ in the domain $K_{\nu,N} \subset (C^n)'$.

From the formulas for the partial derivatives $F'_{\xi_i}(\xi)$ [cf. (33) and (34)] it is easily verified that for any $p = (p_1, \ldots, p_n)$, $q = (q_1, \ldots, q_n)$ the function $\psi_{p,q}(\widehat{\alpha}, \xi') = \partial^{(p)}\partial^{(q)}F(\xi)$ is rapidly decreasing with respect to ξ' and is uniformly decreasing with respect to $\widehat{\alpha}$ in $U_{\nu,N}$. It follows that the partial derivatives $\partial^{(p)}\partial^{(q)}F(\xi)$ decrease rapidly with respect to ξ in the domain $K_{\nu,N}$. Since the $K_{\nu,N}$ form a finite cover of $(C^n)'$, we have proved that $F(\xi)$ together with all its partial derivativesdecreases rapidlyon all of $(C^n)'$. Lemma 2 is proved.

Let us proceed to the proof of Theorem 1. Let $F(\xi)$ be the function in $(C^n)'$ defined by Lemma 1. By Lemma 2 $F(\xi)$ belongs toS. Consequently, its Fourier transform

$$f(x) = (2\pi)^{-n} \int F(\xi)\, e^{-i((\xi,x)+(\bar{\xi},\bar{x}))/2}\, d\xi\, d\bar{\xi}$$

belongs to S.

The theorem will be proved if we show that

$$\varphi_1(a; \beta) = g(\alpha) \int f(\alpha t + \beta)\, dt\, d\bar{t}$$

coincides with $\varphi(a; \beta)$. To do so we proceed from the function $\varphi_1(a; \beta)$ to its Fourier transform:

$$\widetilde{\varphi}_1(a; \xi) = (2\pi)^{-n} g(\alpha) \iint f(\alpha t + \beta)\, e^{i((\xi,\beta)+(\bar{\xi},\bar{\beta}))/2}\, dt\, d\bar{t} \cdot d\mu_{C^n/a}(\beta).$$

Keeping in mind that $(a, \xi) = 0$ we transform this integral in the following way:

$$\widetilde{\varphi}_1(a; \xi) = (2\pi)^{-n} \int\limits_{C^n/a} \int\limits_a f(x+\beta)\, e^{i((\xi,x+\beta)+(\bar{\xi},\overline{x+\beta}))/2}\, d\mu_a(x)\, d\mu_{C^n/a}(\beta) = (2\pi)^{-n} \int\limits_{C^n} f(x)\, e^{i((\xi,x)+(\bar{\xi},\bar{x}))/2}\, dx\, d\bar{x} = F(\xi) = \widetilde{\varphi}(a; \xi).$$

It follows that $\varphi_1 = \varphi$. The theorem is proved.

8. An Inversion Formula

Let the function $\varphi(h)$, h in $H_{n,k}$ satisfy the conditions of Theorem 1. Then by Theorem 1 there exists a function $f(x)$ in S such that

$$\varphi(h) = g(\alpha) \int f(\alpha t + \beta)\, dt\, d\bar{t}.$$

We shall indicate a formula which expresses $f(x)$ in terms of $\varphi(h)$.

THEOREM 2. Let γ be an arbitrary cycle in $G_{n,k}$ of real dimension 2k. Then

$$\int_\gamma \varkappa_\beta \varphi = c_\gamma f(\beta). \tag{35}$$

In addition there exist cycles γ for which $c_\gamma \neq 0$.

Proof. Since $\varkappa_\beta \varphi$ is a closed form it suffices to compute integral (35) along cycles forming the base of a 2k-dimensional homology of the manifold $G_{n,k}$. Let us determine the elements of this basis [3]. We denote by $\nu = (a_1, \ldots, a_k)$ the set of k integers such that $0 \le a_1 \le \ldots \le a_k \le n-k$ and $a_1 + \ldots + a_k = k$, and consider the chain of k subspaces of $\mathbf{C}^n$:

$$E^{a_1+1} \subset E^{a_2+2} \subset \ldots \subset E^{a_k+k}$$

of respective dimensions $a_1 + 1, \ldots, a_k + k$. The cycle γ_ν corresponding to the set ν, consists of all a belonging to $G_{n,k}$ for which

$$\dim (a \cap E^{a_i+i}) \geq i, \quad i = 1, \ldots, k$$

(complex dimension).

All the cycles γ_ν form the basis of a 2k-dimensional homology in $G_{n,k}$. In particular the cycle γ_{ν_0} corresponding to the set $\nu_0 = (1, 1, \ldots, 1)$, consists of all k-dimensional subspaces belonging to a subspace E^{k+1}.

Let us compute first the integral along the chain γ_ν for $\nu \neq \nu_0$. For such a cycle $a_1 = 0$ (since $a_1 + \ldots + a_k = k$ and $a_i > 1$ for at least one i). That is, every a in γ_ν contains fixed complex line E^1. Let us give every subspace a in γ_ν by means of a basis of vectors $\alpha_1, \ldots, \alpha_k$, where $\alpha_1 \subset E^1$ is fixed. Then on γ_ν we have $d\alpha_1^1 = d\alpha_1^2 = \ldots = d\alpha_1^n = 0$. Consequently on γ_ν the form $\varkappa_\beta \varphi = 0$ and hence $\int_{\gamma_\nu} \varkappa_\beta \varphi = 0$.

We compute now integral (35) along the cycle γ_{ν_0}. We take an arbitrary point β in $\mathbf{C}^n$ and fix some subspace $E^{k+1} \subset \mathbf{C}^n$ which contains the point β. We take the cycle γ_{ν_0}, consisting of all k-dimensional subspaces of E^{k+1}, that is, $\gamma_{\nu_0} = G_{k+1,k}$. Now we put $\mathbf{C}^n$ back on E^{k+1}, that is, we consider the embedding $i: E^{k+1} \to \mathbf{C}^n$; $G_{k+1,k} \to G_{n,k}$. Since β is in E^{k+1} it follows from the lemma proved at the end of section 4 that

$$\int_{\gamma_{\nu_0}} \varkappa_\beta \varphi = \int_{G_{k+1,k}} \varkappa_\beta i^* \varphi. \tag{36}$$

The expression on the right hand side of (36) is defined in E^{k+1}. In the sequel we shall assume that the entire discussion is being carried out in E^{k+1} and instead of the integral $\int_{G_{k+1,k}} \varkappa_\beta i^* \varphi$ we shall write simply $\int_{G_{k+1,k}} \varkappa_\beta \varphi$. However, for the case of $G_{k+1,k}$ Theorem 2 has already been essentially examined in [1]. Let us repeat briefly the results obtained there.

We assign a plane h in $H_{k+1,k}$ by the equation $(\zeta, x) \equiv \zeta_1 x^1 + \ldots + \zeta_{k+1} x^{k+1} = s$. In this way plane h is given by the pair ζ, s, where $\zeta \neq 0$. In addition the pairs ζ, s and $\lambda\zeta$, $\lambda s (\lambda \neq 0)$ define the same plane. The integral of the function $f(x)$ is given by the formula $\check{f}(\zeta, s) = (\frac{i}{2})^k \int_h f(x)\, d_1 v \wedge d_1 \bar{v}$, where $d_1 v$ is a k-form defined by the condition $d(\zeta, x) \wedge d_1 v = dx^1 \wedge \ldots \wedge dx^{k+1}$. Below we shall indicate the connection between the function $\check{f}(\zeta, s)$ and the function $\varphi(h)$. The following formula holds [1]:

$$\int_{G_{k+1,k}} \frac{\partial^{2k}\check{f}(\zeta, s)}{\partial s^k \partial \bar{s}^k}\, \omega(\zeta)\wedge\overline{\omega(\zeta)} = (2i)^k \pi^{2k} f(x). \qquad (37)$$

In the right hand side of the formula we have to set $s = (\zeta, x)$. Here $\omega(\zeta) = \sum_{\nu=1}^{k+1} (-1)^\nu \zeta_\nu d\zeta_1\wedge\ldots\wedge d\zeta_{\nu-1}\wedge d\zeta_{\nu+1}\wedge\ldots\wedge d\zeta_{k+1}$. One can show that for fixed x the expression under the integral sign in (37) does not depend on ζ but only on the subspace a: $(\zeta, x) = 0$, and therefore the integration in (37) is carried out over all a in $G_{k+1, k}$.

Let us show that the expression under the integral sign (37) coincides with the form $\varkappa_x \varphi(h)$ where $\varphi(h)$ is defined in terms of the same function $f(x)$ as $\check{f}(\zeta, s)$ is. Since both forms depend only on the plane h and not on the manner in which it is represented, whether by α, β or by ζ, s, it suffices to verify that the forms coincide in local coordinates.* Let the point h in $H_{k+1, k}$ be such that in a neighborhood of h the local coordinates α, β can be introduced

$$\hat{\alpha} = \begin{pmatrix} \alpha_1^1 \ldots \alpha_k^1 \\ 1 \ \ldots 0 \\ \ldots\ldots \\ 0 \ \ldots 1 \end{pmatrix}, \quad \hat{\beta} = (\beta^1, 0, \ldots, 0).$$

Then

$$\zeta = \lambda(1, \alpha_1^1, \ldots, \alpha_k^1), \quad s = (\zeta, \hat{\beta}) = \lambda\beta^1;$$
$$\check{f}(\zeta, s) = \left(\frac{i}{2}\right)^k |\lambda|^{-2} \int f(x)\, dx^2\wedge\ldots\wedge dx^{k+1}\wedge d\bar{x}^2\wedge\ldots\wedge d\overline{x^{k+1}}, \qquad (38)$$

where $x = (\beta^1 - \alpha_1^1 x^2 - \ldots - \alpha_k^1 x^{k+1}, x^2, \ldots, x^{k+1})$. Comparing these with the expression for $\varphi(h)$ in these local coordinates we obtain

$$\check{f}(\zeta, s) = |\lambda|^{-2} g^{-1}(\alpha)\varphi(h). \qquad (39)$$

Since the form in the integral in (37) does not depend on ζ but only on h we can assume that $\lambda = 1$. Now the subintegral expression in (37) can be written as follows

$$\frac{\partial^{2k}\check{f}(\zeta, s)}{\partial s^k \partial \bar{s}^k} d\zeta_2\wedge\ldots\wedge d\zeta_{k+1}\wedge d\bar{\zeta}_2\wedge\ldots\wedge d\bar{\zeta}_{k+1}.$$

It follows from (38) and (39) that this expression coincides with the value of $\varkappa_x \varphi$ in local coordinates. Hence

$$\int_{G_{k+1,k}} \varkappa_x\varphi = (2i)^k \pi^{2k} f(x).$$

Theorem 2 has been proved completely. At the same time we have computed the constant c_γ.

COROLLARY. $c_{\gamma\nu} = 0$ for $\nu \neq \nu_0$; $c_{\gamma\nu_0} = (2i)^k \pi^{2k}$.

* The connection between $\check{f}(\zeta, s)$ and $\varphi(h)$ can be formulated in a more invariant way. To do so let us determine the connection between ζ, s and α, β. Let $\alpha = (\alpha_1, \ldots, \alpha_k)$, where α_j is in C^{k+1}. Then ζ is defined by

$$\alpha_1\wedge\ldots\wedge\alpha_k\wedge x = (\zeta, x)\, e_1\wedge\ldots\wedge e_{k+1},$$

where $e_1, \ldots, e_{k+1}$ is a fixed hedron in C^{k+1}. We abbreviate this equation

$$\zeta = (\alpha_1\wedge\ldots\wedge\alpha_k)'.$$

Further $s = (\zeta, \beta)$ and the connection between $\check{f}$ and φ is determined by the formula $\check{f}(\zeta, s) = g^{-1}(\alpha)\varphi(h)$.

The assertion we required on the equality of the forms now reduces to the following identity:

$$\sum_{\nu=1}^{k+1} (-1)^\nu \zeta_\nu d\zeta_1\wedge\ldots\wedge d\zeta_{\nu-1}\wedge d\zeta_{\nu+1}\wedge\ldots\wedge d\zeta_{k+1} = \sum_p \zeta_{p_1}\ldots\zeta_{p_k} d\alpha_1^{p_1}\wedge\ldots\wedge d\alpha_k^{p_k}.$$

LITERATURE CITED

1. I. M. Gel'fand, M. I. Graev, and N. Ya. Vilenkin, Integral geometry and related questions on the theory of representations ("Generalized Functions," section 5), Fizmatgiz, Moscow (1962).
2. I. M. Gel'fand, M. I. Graev, and Z. Ya. Shapiro, Integral geometry on the manifold of k-dimensional planes, DAN SSR, 168, No. 6, 1236-1238 (1966).
3. Shen-shen' Chzhen', Complex Manifolds [Russian translation], IL, Moscow (1961).
4. M. F. Atiyah, k-Theory, Cambridge, Mass. (1964).
5. F. John, The ultrahyperbolic differential equation with 4 independent variables, Duke Math. J., 4, 300-322 (1938).

All abbreviations of periodicals in the above bibliography are letter-by-letter transliterations of the abbreviations as given in the original Russian journal. *Some or all of this periodical literature may well be available in English translation.* A complete list of the cover-to-cover English translations appears at the back of the first issue of this year.

3.

(with M. I. Graev)

Complexes of k-dimensional planes in the space C^n and Plancherel's formula for the group $GL(n, C)$

Dokl. Akad. Nauk SSSR **179** (1968) 522–525 [Sov. Math., Dokl. **9** (1968) 394–398].
MR 37:4764. Zbl. **198**:271

1. Let $H = H_{n,k}$ be the set of all k-dimensional planes in C^n, provided in a natural way by the structure of a complex analytic manifold. By *complex* K of k-dimensional planes in C^n we shall mean an arbitrary closed irreducible algebraic submanifold in H whose (complex) dimension is equal to n, and such that through each point $x \in C^n$ (with the exception, perhaps, of a manifold of points having a lower dimension) passes at least one plane $h \in K$.*

If a complex K of k-dimensional planes in C^n is given, then there may be formulated for it the following problem of integral goemetry. Let us associate with each function $f(x) \in S$ (S is the space of infinitely differentiable and rapidly decreasing functions on C^n) its integrals $\phi(h)$ over the planes $h \in K$. It is required to obtain the inversion formula, i.e. to reconstruct the original function $f(x)$ from the function $\phi(h)$.

In this article we define a class of admissible complexes for which the inversion formula follows directly from the results of [1,2]. In §2, for the case $k = 1$ (i.e. for complexes of straight lines) we obtain a simple geometrical description of admissible complexes of the general situation.

In §3 we consider an example of a complex of $n(n-1)/2$-dimensional planes in C^{n^2}, which arises naturally in the theory of representations of the group $GL(n, C)$. We prove that this complex is admissible and obtain an inversion formula for it. This inversion formula was proved before by other methods in [3,4]. The method in the present article is, in the opinion of the authors, the most simple and natural one.

Let us present briefly the results of article [2]. We prescribe an arbitrary k-dimensional plane h in C^n by the equation $x = \alpha t + \beta$, where α is a matrix whose columns are k linearly independent vectors $\alpha_1, \cdots, \alpha_k \in C^n$, $\beta \in C^n$ and t runs through C^k. Let us further prescribe an infinitely differentiable function $g(\alpha)$ which does not vanish, and for an arbitrary nondegenerate linear transformation μ in C^k satisfies the following condition: $g(\alpha\mu) = g(\alpha)\,|\det \mu|^2$. With each function $f(x) \in S$ we associate a function $\phi(h)$ on $H_{n,k}$ by means of the formula

$$\varphi(h) = g(\alpha) \int f(\alpha t + \beta)\, dt\, d\bar{t} \tag{1}$$

(it is easy to verify that the integral is independent of the parametric form in which the plane h is prescribed).

In article [2] for each point $\beta \in C^n$ the linear map was defined

$$\varkappa_\beta\colon \Phi_H \to \Phi_{G_\beta}^{(k,k)}$$

---

* For simplicity we consider manifolds without singularities, even though all results remain valid also for manifolds with singularities.

of the space Φ_H of all infinitely differentiable functions on $H_{n,k}$ into the space $\Phi_{G_\beta}^{(k,k)}$ of differential forms of the type (k, k), defined on a submanifold G_β of k-dimensional planes passing through the point β. The following formula of inversion of the integral transformation (1) was established:

$$\int_\gamma \varkappa_\beta \varphi = c_\gamma f(\beta), \tag{2}$$

where the integral is taken over an arbitrary cycle $\gamma \subset G_\beta$ of real dimension $2k$, and the factor C_γ depends only on the class of homologies to which the cycle γ belongs.

Definition. Let K be a complex analytic submanifold in $H_{n,k}$ of dimension $\geq n$ (in particular, a complex) and let i^*: $\Phi_H \to \Phi_K$, $\Phi_{G_\beta}^{(k,k)} \to \Phi_{G_\beta \cap K}^{(k,k)}$ be the natural maps of functions and differential forms. We will call the submanifold K *admissible* if for almost each $\beta \in \mathbf{C}^n$ there exists a linear map κ'_β: $\Phi_K \to \Phi_{G_\beta \cap K}^{(k,k)}$ such that to commutative diagram

$$\begin{array}{ccc} \Phi_H & \xrightarrow{\varkappa_\beta} & \Phi_{G_\beta}^{(k,k)} \\ {\scriptstyle i^*}\downarrow & & \downarrow{\scriptstyle i^*} \\ \Phi_K & \xrightarrow{\varkappa'_\beta} & \Phi_{G_\beta \cap K}^{(k,k)} \end{array}$$

holds. (It is obvious, that in this case κ'_β is defined uniquely and is, like κ_β, a differential operator.) If K is an admissible submanifold, then, in view of (2), we have the following inversion of formula (1), in which h runs through K:

$$\int_\gamma \varkappa'_\beta \varphi = c_\gamma f(\beta),$$

where γ is an arbitrary cycle of real dimension $2k$ in $G_\beta \cap K$. In particular, if K is a complex, then for points $\beta \in \mathbf{C}^n$ of a general situation $\dim(G_\beta \cap K) = 2k$, and therefore it is possible to set $\gamma = G_\beta \cap K$. (It is clear, that this formula defines $f(\beta)$ only if $c_\gamma \neq 0$.)

2. Admissible complexes of straight lines in $\mathbf{C}^n$.

Theorem 1. *A complex of straight lines K is admissible if and only if for almost every straight line $h \in K$ the following condition is satisfied: if we delete from the straight line h a certain number of points and take the manifold of straight lines from K intersecting h in the remaining points, then the closure in K of this manifold will be nonsingular at the point $h \in K$.*

In the sequel we will assume that $\mathbf{C}^n$ is imbedded in the projective space $\mathbf{CP}^n$, and that thus each straight line h is completed by the point at infinity. Let K be the complex of straight lines $h \in K$ and let $x \in \mathbf{C}^n$ be a point on the line h. Let us consider in h the space τ_H, which is the tangent space to the manifold H of all straight lines, and let us take in it 2 subspaces, namely, the tangent subspace τ_K to the complex K and the tangent subspace τ_{G_x} to the manifold G_x of all straight lines passing through x. Since $\dim \tau_H = 2n-2$, $\dim \tau_K = n$ and $\dim \tau_{G_x} = n-1$, we have in the general case: $\dim(\tau_K \cap \tau_{G_x}) = 1$. We will call a point x of the straight line $h \in K$ a *critical point*, if $\dim(\tau_K \cap \tau_{G_x}) > 1$.

Let us find all critical points of a straight line $h \in K$. For definiteness, let h be nonparallel to the hyperplane $x^n = 0$. Then each straight line near h may be prescribed by the system of equations $x^i = \alpha^i x^n + \beta^i x^0$, $i = 1, \cdots, n-1$, where $(x^0, \cdots, x^n)$ are homogeneous coordinates in $\mathbf{CP}^n$. Let us take $2n-2$ complex numbers α^i, β^i as local coordinates on the manifold of all straight lines in the neighborhood of h, and let the complex K be prescribed in the neighborhood of h by the equations $u^1(\alpha, \beta) = 0, \cdots, u^{n-2}(\alpha, \beta) = 0$. Then, *in order for the point $x = (x^0, \cdots, x^n) \in h$ to be critical, it is necessary and sufficient that the rank of the matrix*

$$\|x^0 \partial u^i / \partial \alpha^j|_h - x^n \partial u^i / \partial \beta^j|_h\|_{i=1,\ldots,n-2;\ j=1,\ldots,n-1}$$

be less than $n-2$. Let $\Delta_i(x^0, x^n)$ be the minors of order $n-2$ of this matrix. It is possible to prove that in the case of an admissible complex they differ by factors which are independent of x. Thus, if $\Delta_i \not\equiv 0$ for at least one i, then (since Δ_i is a homogeneous polynomial of degree $n-2$ of x^0, x^n), on the projective straight line h there are, counting multiplicities, $n-2$ critical points. If, on the other hand, $\Delta_i \equiv 0$ for all i, then all points of the straight line h are critical (such straight lines form in K a manifold of lower dimension).

Theorem 2. *Critical points of straight lines of an admissible complex form in* CP^n *a manifold of dimension less than* n.

Let us call an admissible complex of straight lines K a *complex of a general situation* if on each straight line $h \in K$ (with the exception, perhaps, of a manifold of straight lines having a lower dimension) there are exactly $n-2$ pairwise different critical points. Let us describe the local structure of such complexes.

Theorem 3. *On a straight line* h_0 *of an admissible complex* K *let there exist exactly* $n-2$ *pairwise different critical points* $x_1, \cdots, x_{n-2}$. *Then critical points of straight lines* $h \in K$, *which are near to* h_0, *describe* $n-2$ *local algebraic surfaces* $M_1, \cdots, M_{n-2}$, *each of which has dimension* $n-1$ *or* $n-2$; *in addition, if* $\dim M_i = n-1$, *then the straight lines* h *are tangent to* M_i.

Let us further assume that the points $x_i \in M_i$ *are nonsingular, and let* P_i *be a hyperplane in* CP^n, *stretched over the plane tangent to* M_i *at the point* x_i *and the line* h_0 $(i = 1, \cdots, n-2)$. *Then, if the hyperplanes* P_i *appear in a general situation, the set of all straight lines in* C^n *which are tangent to each local surface* M_i *of dimension* $n-1$ *and intersecting each local surface* M_i *of dimension* $n-2$, *is contained in* K *and forms in* K *a neighborhood of the straight line* h_0. *Conversely, if almost every straight line* h_0 *of complex* K *possesses a neighborhood of the above-mentioned type, then* K *is an admissible complex of the general situation.*

Theorem 4. *For* $n = 3$ *an arbitrary admissible complex of straight lines is either a complex of straight lines tangent to some 2-dimensional algebraic submanifold or a complex of straight lines intersecting some algebraic curve.*

3. **Plancherel's theorem for the group** $GL(n, C)$. Here we consider a complex of planes which arises in the course of the derivation of Plancherel's formula on the group $GL(n, C)$. We will regard the matrices $x \in GL(n, C)$ as points of the space C^{n^2}. Let us consider in $GL(n, C)$ a family K of surfaces prescribed by the parametric equations $x = x_1^{-1} z x_2$, where x_1, x_2 are fixed matrices, and "the parameter" z runs through the lower triangular unipotent matrices. It is obvious that these surfaces are $n(n-1)/2$-dimensional planes in C^{n^2}. It is easy to verify that the manifold K of these surfaces is a complex. We will prove here that the complex K is admissible, and will simultaneously evaluate the differential form $\kappa'_\beta \phi$ relative to it. In view of homogeneity of the space $GL(n, C)$ it is sufficient to consider the case $\beta = e$, where e is the unit matrix.

Let us introduce the notations: x_+ is the matrix which is obtained from x if all elements x_{pq} appearing under the main diagonal are changed to zeros; $x_- = x - x_+$. An arbitrary $n(n-1)/2$-dimensional plane in C^{n^2} (with the exception of the manifold of planes having a lower dimension) may be prescribed by the following equation:

$$x_+ = \sum_{p>q} \alpha^{pq} x_{pq} + \beta, \tag{3}$$

where α^{pq}, β are upper triangular matrices. Let us take the elements of the matrices α^{pq}, β as the local coordinates on the manifold of all $n(n-1)/2$-dimensional planes in $\mathbf{C}^{n^2}$. In these coordinates the form $\kappa_e\phi$, defined in (2), has the form

$$\varkappa_e\varphi = \bigwedge_{p>q}\Big(\sum_{i\leqslant j} D_{ij}d\alpha_{ij}^{pq}\Big)\wedge\bigwedge_{p>q}\Big(\sum_{i\leqslant j}\overline{D}_{ij}d\overline{\alpha}_{ij}^{pq}\Big)\varphi\,|_{\beta=e} = \bigwedge_{p>q}\mathrm{tr}\,(D'd\alpha^{pq})\wedge\bigwedge_{p>q}\mathrm{tr}\,(\overline{D}'d\overline{\alpha}^{pq})\varphi\,|_{\beta=e},$$

where $D=\|D_{ij}\|$, $\overline{D}=\|\overline{D}_{ij}\|$; $D_{ij}=\partial/\partial\beta_{ij}$, $\overline{D}_{ij}=\partial/\partial\overline{\beta}_{ij}$ for $i\le j$; $D_{ij}=\overline{D}_{ij}=0$ for $i>j$; D' denotes a transposed matrix. This form is given on the manifold of planes passing through e.

Let us find the restriction $i^*\kappa_e\phi$ of the form $\kappa_e\phi$ to the submanifold of planes passing through e and belonging to the complex K. These planes (with the exception of a submanifold of planes having a lower dimension) are given by parametric equations $x=\zeta^{-1}z\zeta$, in which "the parameter" is the matrix z, and ζ is the upper triangular unipotent matrix which prescribes the plane. Let us reduce the equations of the planes to the form (3). Eliminating from the equations "the parameter" z, we obtain $x_+=\zeta^{-1}(\zeta x_-\zeta^{-1})_+\zeta+e$. Thus

$$\alpha^{pq}=\zeta^{-1}(\zeta e^{pq}\zeta^{-1})_+\zeta,$$

where e^{pq} is a matrix in which at the intersection of the pth row and qth column there appears 1, and all the remaining elements are zeros. Let us introduce the notations $\zeta'^{-1}D\zeta'=F$, $d\zeta\cdot\zeta^{-1}=u$. Then after elementary reductions we obtain: $\wedge_{p>q}\mathrm{tr}(D'd\alpha^{pq})=\wedge_{q<p}(\zeta^{-1}(F_+u'-u'F_+)'_-\zeta)_{qp}=\wedge_{q<p}(uF'_+-F'_+u)_{qp}$.* Since $(uF'_+-F'_+u)_{qp}=(F_{pp}-F_{qq})u_{qp}+\Sigma_{i>p}F_{pi}u_{qi}-\Sigma_{i<q}F_{iq}u_{ip}$, it is easy to see from this, that $\wedge_{q<p}(uF'_+-F'_+u)_{qp}=\Pi_{q<p}(F_{pp}-F_{qq})\cdot\wedge_{q<p}u_{qp}=\Pi_{q<p}(F_{pp}-F_{qq})\cdot\wedge_{q<p}d\zeta_{qp}$. This results in

$$i^*\varkappa_e\varphi=\prod_{q<p}(F_{pp}-F_{qq})(\overline{F}_{pp}-\overline{F}_{qq})\varphi\,|_{\beta=e}\cdot\bigwedge_{q<p}d\zeta_{qp}\wedge\bigwedge_{q<p}d\overline{\zeta}_{qp}. \tag{4}$$

It remains to show that the operators F_{pp} belong to the tangent space to the complex K and consequently the right-hand side of equation (4) depends only on $i^*\phi$. For this purpose we consider the manifold of planes of the complex K, defined by parametric equations $x=\zeta^{-1}z\delta\zeta$, where δ is a diagonal matrix with the diagonal elements $\delta_1,\cdots,\delta_n$. In the form (3) equations of these planes have the form $x_+=\zeta^{-1}(\zeta x_-\zeta^{-1})_+\zeta+\zeta^{-1}\delta\zeta$, which implies $\beta=\zeta^{-1}\delta\zeta$. Consequently, $\partial/\partial\delta_p=\Sigma_{i,j}(\zeta^{-1})_{ip}\zeta_{pj}\partial/\partial\beta_{ij}$, i.e. $\partial/\partial\delta_p=F_{pp}$. Thus, we finally obtain

$$i^*\varkappa_e\varphi=\prod_{q<p}\Big(\frac{\partial}{\partial\delta_p}-\frac{\partial}{\partial\delta_q}\Big)\Big(\frac{\partial}{\partial\overline{\delta}_p}-\frac{\partial}{\partial\overline{\delta}_q}\Big)(i^*\varphi)\,|_{\delta=e}\bigwedge_{q<p}d\zeta_{qp}\wedge\bigwedge_{q<p}d\overline{\zeta}_{qp}.$$

This form coincides with the differential form obtained in [3,4].

To find the coefficient c_γ in the inversion formula, it is necessary, by [2], to take the basis $\gamma_1,\cdots,\gamma_s$ in the group of homologies $H_{2k}(G_e)$, $k=n(n-1)/2$, consisting of Schubert manifolds, where γ_1 is the Euler cycle (i.e. a manifold of planes lying in a fixed $(k+1)$-dimensional plane). Let Let $\gamma\equiv G_e\cap K=\Sigma a_i\gamma_{ij}$ then $c_\gamma=(2i)^k\pi^{2k}a_1$. By means of simple arguments (for example, by computing the intersection index) it is possible to show that $a_1=n!$.

The authors express their gratitude to I. N. Bernšteĭn, who read the manuscript and made a number of useful remarks.

Received 26/DEC/67

* By x'_- (x'_+) we denote the transpose of x_- (respectively x_+).

BIBLIOGRAPHY

[1] I. M. Gel'fand, M. I. Graev and Z. Ja. Šapiro, Dokl. Akad. Nauk SSSR 168 (1966), 1236 = Soviet Math. Dokl. 7 (1966), 801. MR 33 #6565.

[2] ———, Funkcional. Anal. i Priložen. 1 (1967), 15. MR 35 #3620.

[3] I. M. Gel'fand and M. A. Naĭmark, Trudy Mat. Inst. Steklov. 36 (1950), 13. MR 13, 722.

[4] I. M. Gel'fand and M. I. Graev, Trudy Moskov. Mat. Obšč. 4 (1955), 375; English transl., Amer. Math. Soc. Transl. (2) 9 (1958), 123. MR 17, 173; MR 19, 1181.

Translated by:
V. I. Filippenko

4.

(with M. I. Graev)

Complexes of straight lines in the space C^n

Funkts. Anal. Prilozh. 2(3) (1968) 39-52 [Funct. Anal. Appl. 2 (1968) 219-229].
MR 38:6522. Zbl. 179:509

1. Introduction

Let $H = H_{n,k}$ be the manifold of all k-planes in C^n. With each function $f(x) \in S$ (S is the space of infinitely differentiable functions on C) we associate the integral $\varphi(h)$ of $f(x)$ over the planes $h \in H$. The initial function can be expressed in terms of $\varphi(h)$; according to [1], to recover the value of f at the point $\beta \in C^n$ it is necessary to consider the manifold G_β of k-planes through β and to define in G_β an (in general, arbitrary) cycle γ of real dimension 2k. The inversion formula obtained in [1] expresses $f(\beta)$ in terms of the values of $\varphi(h)$ and of a finite number of its partial derivatives on γ.

Of course, to recover $f(x)$ it is not necessary to know $\varphi(h)$ on the whole of the manifold H of k-planes. In fact, it is sufficient to know $\varphi(h)$ on some n-dimensional submanifold of H. Thus, we come to the following problem in integral geometry; given an n-dimensional submanifold $K \subset H_{n,k}$ of k-planes in C^n, it is required to recover $f(x) \in S$, knowing its integral $\varphi(h)$ over the planes $h \in K$.

In paragraph 2 of the present article we introduce a class of n-dimensional submanifolds in $H_{n,k}$ which we call admissible complexes. The inversion formula for these manifolds follows immediately from the results of [1]. The main problem is to describe this class of manifolds.

In the present work we only consider the simplest case: the n-dimensional manifolds K of straight lines in C^n. In paragraph 4 we derive necessary and sufficient conditions for K to be an admissible complex. In paragraph 5 we obtain (though only for complexes in "general position") a simple geometrical description of admissible complexes. Finally, in paragraph 6 we give some examples of complexes of straight lines. A brief summary of the results of the present article is contained in [2].*

2. The Definition of an Admissible Complex

Let us first recall the results of [1]. We define a k-plane h in C^n by the equation h: $x = \alpha t + \beta$, where α is a matrix whose columns are k linearly independent vectors $\alpha_1, \ldots, \alpha_k \in C^n$, $\beta \in C^n$, and t runs over C^k. Let $g(\alpha)$ be a fixed infinitely differentiable function of α that does not vanish anywhere and satisfies the following condition for any non-degenerate linear transformation μ in C^k:

$$g(\alpha\mu) = g(\alpha)\,|\det \mu\,|^2. \tag{1}$$

With each function $f(x) \in S$ we associate the function $\varphi(h)$ on the manifold $H_{n,k}$ of k-planes in C^n defined by

$$\varphi(h) = g(\alpha) \int f(\alpha t + \beta)\, dt\, d\bar{t}. \tag{2}$$

[From (1) it is easy to see that the integral does not depend on the parametric representation of h.]

For each $\beta \in C^n$ an explicit formula was constructed in [1] for the continuous linear map

*For certain complexes of straight lines the problem of integral geometry was solved earlier: in [4] for the complexes of lines in C^3 that intersect a given hyperbola; in [5] for the complexes of lines in C^n that intersect a given arbitrary curve; in [4] for a certain class of complexes of lines in C^3. All these complexes are admissible in the sense considered here.

Institute of Applied Mathematics, Academy of Sciences of the USSR and Moscow State University.
Translated from Funktsional'nyi Analiz i Ego Prilozheniya, Vol. 2, No. 3, pp. 39-52, July-September, 1968.
Original article submitted March 6, 1968.

$$\varkappa_\beta\colon \Phi_H \to \Phi_{G_\beta}^{(k,k)}$$

of the space Φ_H of all infinitely differentiable functions on $H_{n,k}$ into the space $\Phi_{G_\beta}^{(k,k)}$ of differential forms of type (k, k) on G_β, where G_β is the manifold of k-planes that pass through $\beta \in C^n$. This map is a differential operator, that is, $\varkappa_\beta\varphi|_h$, $h \in G_\beta$, depends only on the values of φ and of a finite number of its partial derivatives at the point h.

We are going to give an expression for $\varkappa_\beta\varphi$ in the case of the manifold of straight lines $H_{n,1}$. First we define a local system of coordinates in $H_{n,1}$. Let $h \in G_\beta$. For the parameter t of the straight line $h \in G_\beta$ and of the straight lines close to h, we take one of the coordinates $x^1, \ldots, x^n$, for example x^n (this is possible if $x^n \neq$ const on h). Then the equations of the lines close to h become:

$$x^i = \alpha^i x^n + \beta^i, \quad i = 1, \ldots, n-1.$$

We take the coefficients α^i, β^i in these equations as the local system of coordinates on $H_{n,1}$ in the neighborhood of $h \in H_{n,1}$.

By [1], in these coordinates the form $\varkappa_\beta\varphi$ is as follows:

$$\varkappa_\beta\varphi\,|_h = \frac{1}{g(\alpha)} \sum_{i,j=1}^{n-1} \frac{\partial^2\varphi}{\partial\beta^i\,\partial\bar\beta^j}\Big|_h\, d\alpha^i \wedge d\bar\alpha^j, \quad h \in G_\beta. \tag{3}$$

If we do not take x^n for the parameter t of the straight lines but some other coordinate x^i, then we arrive at another local system of coordinates on $H_{n,1}$ in which $\varkappa_\beta\varphi$ has an analogous form.

The following formula for the inversion of the integral transformation (2) was established in [1]:

$$\int_\gamma \varkappa_\beta\varphi = c_\gamma f(\beta); \tag{4}$$

here the integral is taken over an arbitrary cycle $\gamma \subset G_\beta$ of real dimension 2k, and the multiplier c_γ depends only on the homology class to which γ belongs. (In [1], conditions for $c_\gamma \neq 0$ were also indicated.)

We now turn to the definition of an admissible complex.

Definition 1. Any closed irreducible algebraic submanifold K in $H_{n,k}$ whose (complex) dimension is n and that is such that, through each point $x \in C^n$ (with the exception, possibly, of a manifold of points of lower dimension), passes at least one plane $h \in K$, will be said to be a complex of k-planes in C^n.

Definition 2. A complex of k-planes K will be said to be admissible if, for almost any $\beta \in C^n$, there is a continuous linear mapping $\varkappa'_\beta\colon \Phi_K \to \Phi_{G_\beta \cap K}^{(k,k)}$ such that the following diagram is commutative:

$$\begin{array}{ccc} \Phi_H & \xrightarrow{\varkappa_\beta} & \Phi_{G_\beta}^{(k,k)} \\ \downarrow{\scriptstyle i^*} & & \downarrow{\scriptstyle i^*} \\ \Phi_K & \xrightarrow{\varkappa'_\beta} & \Phi_{G_\beta \cap K}^{(k,k)}. \end{array} \tag{5}$$

Here i* denote the mappings induced by the natural embeddings $i\colon K \to H$, $G_\beta \cap K \to G_\beta$.

It is evident that if $\varkappa'_\beta$ exists it is uniquely defined in terms of $\varkappa_\beta$; namely, for any $\varphi \in \Phi_K$ we have

$$\varkappa'_\beta\varphi = i^*\varkappa_\beta i^{*-1}\varphi, \tag{6}$$

where $i^{*-1}\varphi$ is any inverse image of φ in Φ_H; the right-hand side of the equation does not depend on the choice of this inverse image.

It follows from (6) that $\varkappa'_\beta$ is, as is $\varkappa_\beta$, a differential operator.

For any admissible complex K we have the following inversion formula that expresses $f(\beta)$ in terms of $\varphi(h)$, $h \in K$:

$$\int_\gamma \varkappa'_\beta \varphi = c_\gamma f(\beta)^{*}, \tag{7}$$

where γ is an arbitrary cycle in $G_\beta \cap K$ of real dimension 2k.

In fact, the formula follows immediately from (4) and from the commutativity of the diagram (5).

Let us note that, for almost all $\beta \in C^n$, $G_\beta \cap K$ itself has real dimension 2k and hence, in (7), we may take $\gamma = G_\beta \cap K$.

We are going to give an explicit expression for $\varkappa'_\beta \varphi$ in the case of a complex K of straight lines. Suppose, for example, that on the manifold $H_{n,1}$ a system of coordinates $\alpha^1, \ldots, \alpha^{n-1}, \beta^1, \ldots, \beta^{n-1}$ (see p. 220) is introduced into a neighborhood of $h \in G_\beta \cap K$, and suppose that $\alpha^1, \ldots, \alpha^{n-1}, \beta^1$ can be taken as local coordinates on K in the neighborhood of h. Then, in the coordinates $\alpha^1, \ldots, \alpha^{n-1}, \beta^1$, the form $\varkappa'_\beta \varphi$ has the following representation:

$$\varkappa'_\beta \varphi \,|_h = \frac{1}{g(\alpha)} \left. \frac{\partial^2 \varphi(\alpha^1, \ldots, \alpha^{n-1}, \beta^1)}{\partial \beta^1 \, \partial \overline{\beta^1}} \right|_h d\alpha^1 \wedge d\overline{\alpha^1}. \tag{8}$$

This form has an analogous representation in other local coordinates on K.

To prove (8) we augment the system of coordinates $\alpha^1, \ldots, \alpha^{n-1}, \beta^1$ on K to a local system of coordinates $\alpha^1, \ldots, \alpha^{n-1}, \beta^1, u^1, \ldots, u^{n-2}$ on H such that $u^1 = \ldots = u^{n-2} = 0$ on K.

Let $\varphi \in \Phi_K$ and let φ_1 be any inverse image of φ in Φ_H. We express in (4) for $\varkappa_\beta \varphi_1$ the derivatives with respect to β^i in terms of the derivatives with respect to the new variables $\beta^1, u^1, \ldots, u^{n-2}$. We obtain

$$\varkappa_\beta \varphi_1 \,|_h = \frac{1}{g(\alpha)} \left. \frac{\partial^2 \varphi_1}{\partial \beta^1 \, \partial \overline{\beta^1}} \right|_h d\alpha^1 \wedge d\overline{\alpha^1} + \tilde{\varkappa}_\beta \varphi_1 \,|_h,$$

where all the coefficients of the form $\tilde{\varkappa}_\beta \varphi_1$ contain differentiations with respect to $u^i, \overline{u^j}$. It follows from here that

$$\varkappa'_\beta \varphi \,|_h \equiv i^* \varkappa_\beta \varphi_1 \,|_h = \frac{1}{g(\alpha)} \left. \frac{\partial^2 \varphi(\alpha^1, \ldots, \alpha^{n-1}, \beta^1)}{\partial \beta^1 \, \partial \overline{\beta^1}} \right|_h d\alpha^1 \wedge d\overline{\alpha^1} + i^* \tilde{\varkappa}_\beta \varphi_1 \,|_h. \tag{9}$$

Since the complex K is admissible, the right-hand side of (9) cannot contain the partial derivatives of φ_1 with respect to $u^i, \overline{u}^j$ (otherwise it would depend on the choice of the inverse image φ_1 of φ). Consequently $i^* \tilde{\varkappa}_\beta \varphi_1 |_h = 0$, which proves (8).

3. Critical Points

From now on we will exclusively consider complexes of straight lines. First we introduce the concept of a critical point on a straight line of the complex K which, as will be seen in what follows (see paragraphs 4 and 5), plays an essential role in the description of admissible complexes of straight lines.

We will always assume that the space $\mathbf{C}^n$ is embedded in the projective space $\mathbf{CP}^n$ and so each straight line in $\mathbf{C}^n$ is augmented by a point at infinity. The points of $\mathbf{CP}^n$ will be given by the homogeneous coordinates $x = (x^0, x^1, \ldots, x^n)$.

Let K be an arbitrary complex of straight lines in $\mathbf{C}^n$, $h \in K$ and let x be a point of the line h. We denote the manifold of all the straight lines in $\mathbf{CP}^n$ that pass through x by G_x.

We consider the tangent space $T_h(H)$ at the point $h \in H$ of the (2n-2)-dimensional manifold H of all straight lines. In it we distinguish two subspaces: the tangent subspace $T_h(K)$ to K, and the tangent subspace $T_h(G_k)$ to the manifold G_x. Since $\dim T_h(H) = 2n-2$, $\dim T_h(K) = n$ and $\dim T_h(G_x) = n-1$ then, in general, we have $\dim (T_h(K) \cap T_h(G_x)) = 1$.

__Definition.__ A point x of the straight line $h \in K$ is said to be critical if $\dim (T_h(K) \cap T_h(G_x)) > 1$.

---

*It is clear that $f(\beta)$ is determined from (7) only when $c_\gamma \neq 0$.

It is evident that if x is not a critical point of $h \in K$, then the manifold $G_x \cap K$ is one-dimensional in a neighborhood of h. If it is a critical point then, in a neighborhood of h, the dimension of $G_x \cap K$ can, in general, be arbitrary.

We are going to find necessary and sufficient conditions for the point x of the straight line $h \in K$ to be a critical point.

As a preliminary we introduce a local system of coordinates on H in the neighborhood of h. It is obvious that at least one of the coordinates x^i, $i = 1, \ldots, n$ is non-constant on h; suppose for definiteness that $x^n \neq$ constant on h. Then any straight line sufficiently close to h can be defined by the system of equations

$$x^i = \alpha^i x^n + \beta^i x^0, \quad i = 1, \ldots, n-1. \tag{1}$$

We take the coefficients $(\alpha, \beta) = (\alpha^1, \ldots, \alpha^{n-1}; \beta^1, \ldots, \beta^{n-1})$ of this system as local coordinates on H.

LEMMA 1. Let $u^1, \ldots, u^{2n-2}$ be a local system of coordinates on H in a neighborhood of $h \in K$ such that the complex K is defined in this system by the equations

$$u^1 = 0, \ \ldots, \ u^{n-2} = 0. \tag{2}$$

Then, for the point $x = (x^0, x^1, \ldots, x^n)$ of $h \in K$ to be a critical point it is necessary and sufficient that the rank of the matrix

$$\left\| x^0 \frac{\partial u^i}{\partial \alpha^j}\bigg|_h - x^n \frac{\partial u^i}{\partial \beta^j}\bigg|_h \right\|_{\substack{i=1,\ldots,n-2 \\ j=1,\ldots,n-1}} \tag{3}$$

be less than $n-2$.

Proof. Let $T_h(H)$ be the tangent space to K at h; we take $\frac{\partial}{\partial \alpha^1}, \ldots, \frac{\partial}{\partial \alpha^{n-1}}, \frac{\partial}{\partial \beta^1}, \ldots, \frac{\partial}{\partial \beta^{n-1}}$ to be a basis in $T_h(H)$ and will define vectors $\tau \in T_h(H)$ by their coordinates in this basis: $\tau = (\xi, \eta) = (\xi^1, \ldots, \xi^{n-1}; \eta^1, \ldots, \eta^{n-1})$. Obviously, $\tau \in T_h(K)$ if and only if the coordinates ξ, η satisfy the following system of equations

$$\sum_{j=1}^{n-1} \frac{\partial u^i}{\partial \alpha^j}\bigg|_h \xi^j + \sum_{j=1}^{n-1} \frac{\partial u^i}{\partial \beta^j}\bigg|_h \eta^j = 0, \quad i = 1, \ldots, n-2. \tag{4}$$

On the other hand, it follows from the conditions $x^i = \alpha^i x^n + \beta^i x^0$ that the subspace $T_h(G_x)$, tangent to G_x, consists of vectors of the form $\tau = (x^0\xi, -x^n\xi)$. It follows from here that $T_h(K) \cap T_h(G_x)$ consists of all of the vectors of the form $\tau = (x^0\xi, -x^n\xi)$, where $\xi = (\xi^1, \ldots, \xi^{n-1})$ runs over the solutions of the system of equations

$$\sum_{j=1}^{n-1} \left(x^0 \frac{\partial u^i}{\partial \alpha^j}\bigg|_h - x^n \frac{\partial u^i}{\partial \beta^j}\bigg|_h \right) \xi^j = 0, \quad i = 1, \ldots, n-2. \tag{5}$$

The assertion in the Lemma now follows immediately from the definition of a critical point.

LEMMA 2. Let $t^1, \ldots t^n$ be a local system of coordinates on K in a neighborhood of $h \in K$. Then, for the point $x = (x^0, \ldots, x^n)$ of the straight line $h \in K$ to be a critical point it is necessary and sufficient that the rank of the matrix

$$\left\| \frac{\partial \alpha^i}{\partial t^j}\bigg|_h x^n + \frac{\partial \beta^i}{\partial t^j}\bigg|_h x^0 \right\|_{\substack{i=1,\ldots,n-1 \\ j=1,\ldots,n}} \tag{6}$$

be less than $n-1$.

Proof. We augment the local system of coordinates $t^1, \ldots, t^n$ on K to the local system of coordinates $t^1, \ldots, t^n, u^1, \ldots, u^{n-2}$ on the manifold H of straight lines so that, in this coordinate system, K is defined by the equations $u^1 = 0, \ldots, u^{n-2} = 0$.

We take $\frac{\partial}{\partial t^1}, \ldots, \frac{\partial}{\partial t^n}, \frac{\partial}{\partial u^1}, \ldots, \frac{\partial}{\partial u^{n-2}}$ as a basis in $T_h(H)$ and will define vectors $\tau \in T_h(H)$ by their coordinates in this basis: $\tau = (\lambda, \mu) = (\lambda^1, \ldots, \lambda^n; \mu^1, \ldots, \mu^{n-2})$. It is obvious that the subspace $T_h(K) \subset T_h(H)$, tangential to K at h, consists of vectors of the form $\tau = (\lambda, 0)$. On the other hand, in view of the relations $x^i = \alpha^i x^n + \beta^i x^0$, the subspace $T_h(G_x) \subset T_h(H)$, tangential to G_x, consists of vectors τ whose coordinates satisfy the following system of equations:

$$\sum_{j=1}^{n} \left(\frac{\partial \alpha^i}{\partial t^j} \bigg|_h x^n + \frac{\partial \beta^i}{\partial t^j} \bigg|_h x^0 \right) \lambda^j + \sum_{j=1}^{n-2} \left(\frac{\partial \alpha^i}{\partial u^j} \bigg|_h x^n + \frac{\partial \beta^i}{\partial u^j} \bigg|_h x^0 \right) \mu^j = 0, \tag{7}$$

$$i = 1, \ldots, n-1.$$

It follows from here that $T_h(K) \cap T_h(H)$ consists of all vectors of the form $\tau = (\lambda, 0)$, where $\lambda = (\lambda^1, \ldots, \lambda^n)$ runs over the solutions of the system of equations:

$$\sum_{j=1}^{n} \left(\frac{\partial \alpha^i}{\partial t^j} \bigg|_h x^n + \frac{\partial \beta^i}{\partial t^j} \bigg|_h x^0 \right) \lambda^j = 0, \quad i = 1, \ldots, n-1. \tag{8}$$

The assertion in the Lemma now follows immediately from the definition of a critical point.

In passing we have proved the following:

LEMMA 3. The rank of the $(n-1) \times n$ matrix (6) is greater than the rank of the $(n-2) \times (n-1)$ matrix (3) by unity.

In fact, the spaces of the solutions of a homogeneous system of equations with the matrix (6) and of the solutions of a homogeneous system of equations with the matrix (3) have the same dimension, dim $(T_h(K) \cap T_h(G_x))$.

THEOREM 1. Let K be an arbitrary complex of straight lines in C^n. On each straight line $h \in K$ either there are not more than $n-2$ critical points or all the points of h are critical.

Proof. Suppose that the same local system of coordinates as we used in Lemma 1 is introduced in a neighborhood of h. We denote by $\Delta_i(x^0, x^n)$ the minors of order $n-2$ of the matrix (3). If $\Delta_i \equiv 0$ for all i then, in view of Lemma 1, all the points of h are critical. Next suppose that $\Delta_i \not\equiv 0$ for at least one i. By Lemma 3 the coordinates x^0, x^n of critical points must satisfy the system of equations $\Delta_i(x^0, x^n) = 0$, $i = 1, \ldots, n-1$. Since Δ_i is a homogeneous polynomial of degree $n-2$ in x^0, x^n then, consequently, there are not more than $n-2$ critical points on the line.

Definition. A straight line $h \in K$ is said to be a critical line of K if all of its points are critical points.

LEMMA 4. The dimension of the manifold of critical lines of the complex K is less than n.

Proof. Let us assume that the dimension of the manifold of critical lines is n. It is obvious that this manifold is algebraic. Hence, since K is an irreducible manifold, this manifold coincides with K. Thus, under our assumption, any straight line of K is critical. We are going to show that this is impossible.

Let h be a fixed straight line from K; we introduce in H, in a neighborhood of h, the local system of coordinates (α, β) that we used in Lemma 1. Suppose that $t^1, \ldots, t^n$ is a local system of coordinates on K in the neighborhood of h. Then the coordinates α, β of the straight lines of the complex K that are sufficiently close to h, are given functions of t:

$$\alpha^i = \alpha^i(t), \quad \beta^i = \beta^i(t); \quad t = (t^1, \ldots, t^n); \quad i = 1, \ldots, n-1.$$

Consider the manifold $M \subset C^n$ of points that lie on the straight lines of the complex sufficiently close to h. In view of our definition of a complex, dim $M = n$. Hence, we will obtain a contradiction if we can prove that dim $M < n$.

M consists of points x with the coordinates

$$x^0 = 1, \quad x^1 = \alpha^1(t)\, x^n + \beta^1(t), \; \ldots, \; x^{n-1} = \alpha^{n-1}(t)\, x^n + \beta^{n-1}(t), \quad x^n,$$

where t runs over some domain in C^n and x^n runs over C. In view of Lemma 2 the rank of the matrix

$$\left\| \frac{\partial x^i}{\partial t^j} \right\| = \left\| \frac{\partial \alpha^i}{\partial t^j} x^n + \frac{\partial \beta^i}{\partial t^j} \right\|_{\substack{i=1,\ldots,n-1 \\ j=1,\ldots,n}}$$

is less than n−1. It follows directly that dim M < n, and the Lemma is proved.

LEMMA 5. The dimension of the manifold of the points of C^n that belong to at at least one critical straight line of the complex K is less than n.

Proof. In view of Lemma 4 the manifold L of critical lines of K has dimension less than n. If dim L < n−1, then the assertion of the Lemma is obviously true; hence it is sufficient to consider the case when dim L = n−1.

Let h be a non-singular point of L. We define a local system of coordinates $t^1, \ldots, t^n$ in K in a neighborhood of h such that, in this neighborhood, L is given by the equation $t^n = 0$. The coordinates α, β of the lines of K sufficiently close to h are given functions of $t = (t^1, \ldots, t^n)$.

We consider the manifold $M \subset C^n$ of points lying on the straight lines $h' \in L$ sufficiently close to h. The points of M are given by the coordinates

$$x^0 = 1,\ x^1 = \alpha^1(t')x^n + \beta^1(t'),\ \ldots,\ x^{n-1} = \alpha^{n-1}(t')x^n + \beta^{n-1}(t'),\ x^n,$$

where $t' = (t^1, \ldots, t^{n-1}, 0)$; here $(t^1, \ldots, t^{n-1})$ runs over some domain in C^{n-1} and x^n runs over C.

We must prove that dim M < n. To do this we consider the Jacobian $D(x^1, \ldots, x^{n-1})/D(t^1, \ldots, t^{n-1})$. This Jacobian is one of the minors of order (n−1) of the matrix

$$\left\| \left. \frac{\partial \alpha^i}{\partial t^j} \right|_{t^n=0} x^n + \left. \frac{\partial \beta^i}{\partial t^j} \right|_{t^n=0} \right\|_{\substack{i=1,\ldots,n-1 \\ j=1,\ldots,n}}.$$

According to Lemma 2 the rank of this matrix is less then n−1 and hence $D(x^1, \ldots, x^{n-1})/D(t^1, \ldots, t^{n-1}) = 0$. The Lemma is proved.

THEOREM 2. The dimension of the manifold of points of C^n that are critical points for at least one straight line of the complex K is less than n.

Proof. Let us note that, by Lemma 5, the dimension of the manifold of points that lie on the critical lines of K is less than n. Hence, instead of K, it is sufficient to consider the manifold of straight lines K' K obtained by removing all the critical lines from K.

Let $h_0 \in K'$. As before we may assume that the straight lines sufficiently close to h_0 are given by the equations $x^i = \alpha^i x^n + \beta^i$, i = 1, ..., n−1. We introduce a local system of coordinates $t^1, \ldots, t^n$ in K', in a neighborhood of h_0; we are going to consider the submanifold $M \subset C^n$ of critical points of the lines h = h(t) ∈ K', sufficiently close to h_0. We need to prove that dim M < n.

The manifold M consists of points $x \in C^n$ whose coordinates satisfy the following relations

$$x^0 = 1,\ x^i = \alpha^i(t)x^n + \beta^i(t),\ \ i = 1, \ldots, n-1, \tag{9}$$

$$\Delta_j(t, x^n) = 0,\ \ j = 1, \ldots, n. \tag{10}$$

Here α^i, β^i are given functions of t, Δ_j is a minor of order (n−1) of the matrix $\left\| \frac{\partial \alpha^i}{\partial t^j} x^n + \frac{\partial \beta^i}{\partial t^j} \right\|$; t = $(t^1, \ldots, t^n)$ runs over some small domain $U \subset C^n$.

Since h_0 is not a critical line, $\Delta_j(t, x^n) \not\equiv 0$ for at least one $j = j_0$. We replace the system (10) by the single relation

$$\Delta_{j_0}(t, x^n) = 0; \tag{11}$$

such a substitution can only enlarge the manifold M.

Equation (11) defines x^n as an algebraic function of t. Suppose that $x^n = f(t)$ is one of the branches of this function. By substituting this expression for x^n into (9) we will define $x^1, \ldots, x^n$ as functions of $t^1, \ldots, t^n$.

To prove the Theorem it is sufficient to make certain that the Jacobian D(x)/D(t) is zero. We have

$$\frac{D(x)}{D(t)} = \begin{vmatrix} \frac{\partial\alpha^1}{\partial t^1}x^n + \alpha^1\frac{\partial x^n}{\partial t^1} + \frac{\partial\beta^1}{\partial t^1} & \ldots & \frac{\partial\alpha^{n-1}}{\partial t^1}x^n + \alpha^{n-1}\frac{\partial x^n}{\partial t^1} + \frac{\partial\beta^{n-1}}{\partial t^1} & \frac{\partial x^n}{\partial t^1} \\ \cdots & \cdots & \cdots & \cdots \\ \frac{\partial\alpha^1}{\partial t^n}x^n + \alpha^1\frac{\partial x^n}{\partial t^n} + \frac{\partial\beta^1}{\partial t^n} & \ldots & \frac{\partial\alpha^{n-1}}{\partial t^n}x^n + \alpha^{n-1}\frac{\partial x^n}{\partial t^n} + \frac{\partial\beta^{n-1}}{\partial t^n} & \frac{\partial x^n}{\partial t^n} \end{vmatrix} = \begin{vmatrix} \frac{\partial\alpha^1}{\partial t^1}x^n + \frac{\partial\beta^1}{\partial t^1} & \ldots & \frac{\partial\alpha^{n-1}}{\partial t^1}x^n + \frac{\partial\beta^{n-1}}{\partial t^1} & \frac{\partial x^n}{\partial t^1} \\ \cdots & \cdots & \cdots & \cdots \\ \frac{\partial\alpha^1}{\partial t^n}x^n + \frac{\partial\beta^1}{\partial t^n} & \ldots & \frac{\partial\alpha^{n-1}}{\partial t^n}x^n + \frac{\partial\beta^{n-1}}{\partial t^n} & \frac{\partial x^n}{\partial t^n} \end{vmatrix}.$$

By Lemma 2 the first $n-1$ columns of this determinant are linearly dependent and so $D(x)/D(t) = 0$ as required.

4. Necessary and Sufficient Conditions for a Complex of Straight Lines to Be Admissible

THEOREM 3. For the complex of straight lines K to be admissible it is necessary and sufficient that, for any non-critical line $h_0 \in K$, the following holds: the closure of the manifold of straight lines $h \in K$ that intersect h_0 at its non-critical points does not have singularities in h_0.

Proof. We will assume that the straight lines close to h_0 are given by the equations $x^i = \alpha^i x^n + \beta^i x^0$, $i = 1, \ldots, n-1$; we take the coefficients α^i, β^i in these equations as the local coordinates on the manifold of all straight lines.

Let $x \in h_0$ be a non-critical point; then $h_0 \in G_x$. We consider the form $\varkappa_x\varphi$ at the point $h_0 \in G_x$ (see paragraph 2)

$$\varkappa_x\varphi\,|_{h_0} = \frac{1}{g(\alpha)}\sum_{i,j=1}^{n-1}\frac{\partial^2\varphi}{\partial\beta^i\,\partial\bar\beta^j}\bigg|_{h_0} d\alpha^i \wedge d\bar\alpha^j. \tag{1}$$

We introduce another local system of coordinates $u^1, \ldots, u^{2n-2}$ into a neighborhood of h_0 such that, in these coordinates, K is given by the equations $u^1 = 0, \ldots, u^{n-2} = 0$. In these coordinates the form (1) becomes

$$\varkappa_x\varphi\,|_{h_0} = \frac{1}{g(\alpha)}\sum_{k,l=1}^{2n-2}\left(\frac{\partial^2\varphi}{\partial u^k\,\partial\bar u^l}\bigg|_{h_0}\left(\sum_{i=1}^{n-1}\frac{\partial u^k}{\partial\beta^i}\bigg|_{h_0} d\alpha^i\right)\wedge\left(\overline{\sum_{j=1}^{n-1}\frac{\partial u^l}{\partial\beta^j}\bigg|_{h_0} d\alpha^j}\right)\right). \tag{2}$$

We consider the restriction $i^*\varkappa_x\varphi|_{h_0}$ of this form to the manifold $G_x \cap K$. If the complex K is admissible then this restriction does not contain the partial derivatives with respect to u^i, $\bar u^j$, $i, j = 1, \ldots, n-2$. It follows from this that

$$i^*\left(\sum_{i=1}^{n-1}\frac{\partial u^k}{\partial\beta^i}\bigg|_{h_0} d\alpha^i\right) = 0, \quad k = 1, \ldots, n-2. \tag{3}$$

Conversely, if (3) holds for any non-critical straight line $h_0 \in K$ and for any non-critical point $x \in h_0$, then it is obvious that K is admissible. Thus, (3) is a necessary and sufficient condition for a complex to be admissible. Let us clarify its geometrical significance.

We consider the tangent space $T_h(H)$ to the manifold of all straight lines at $h_0 \in H$; we take $\frac{\partial}{\partial\alpha^1}, \ldots, \frac{\partial}{\partial\alpha^{n-1}}, \frac{\partial}{\partial\beta^1}, \ldots, \frac{\partial}{\partial\beta^{n-1}}$ as a basis in $T_h(H)$ and will define vectors $\tau \in T_h(H)$ by their coordinates in this basis: $\tau = (\xi, \eta) = (\xi^1, \ldots, \xi^{n-1}; \eta^1, \ldots, \eta^{n-1})$.

Let $x = (x^0, \ldots, x^n)$ be a non-critical point on h_0. Then $G_x \cap K$ is one-dimensional. According to paragraph 3, a tangent vector to $G_x \cap K$ is of the form $(x^0\xi_x, -x^n\xi_x)$, where the $(n-1)$-dimensional vector ξ_x is determined uniquely, with accuracy up to a multiplier, from the system of equations

$$\sum_{j=1}^{n-1}\left(x^0\frac{\partial u^i}{\partial\alpha^j}\Big|_{h_0}-x^n\frac{\partial u^i}{\partial\beta^j}\Big|_{h_0}\right)\xi_x^j=0,\quad i=1,\ldots,n-2. \tag{4}$$

If the complex K is admissible then, by (3), ξ_x also satisfies the system of equations

$$\sum_{j=1}^{n-1}\frac{\partial u^i}{\partial\beta^j}\Big|_{h_0}\xi_x^j=0,\quad i=1,\ldots,n-2. \tag{5}$$

It follows from (4) and (5) that ξ_x satisfies the system

$$\sum_{j=1}^{n-1}\left(a\frac{\partial u^i}{\partial\alpha^j}\Big|_{h_0}-\frac{\partial u^i}{\partial\beta^j}\Big|_{h_0}\right)\xi_x^j=0,\quad i=1,\ldots,n-2, \tag{6}$$

for any fixed $a\neq 0$; consequently it does not depend on $x\in h_0$. This proves that, when K is admissible, the vectors tangential to $G_x\cap K$, where $x\in C^n$ runs over the non-critical points of the straight line h_0, lie in a two-dimensional tangent plane, the plane formed by the vectors $(\lambda\xi,\mu\xi)$, where ξ is a fixed (n−1)-dimensional vector. The necessity of the condition in the Theorem follows immediately from this last result.

Conversely, suppose the condition in the Theorem holds. Then the vectors $(x^0\xi_x,-x^n\xi_x)$, tangential to $G_x\cap K$, where $x\in C^n$ runs over all the non-critical points of the straight line h_0, lie in a fixed two-dimensional plane. It is obvious that this is possible only when the vectors ξ_x are proportional to a fixed vector ξ. This proves that, for different $x\in h_0$, the systems (4) are equivalent. But then ξ_x satisfies (5), that is, condition (3) holds for K; thus the complex K is admissible.

We are going to state Theorem 3 in another, equivalent form.

THEOREM 3'. For the complex K of straight lines in C^n to be admissible it is necessary and sufficient that, for any non-critical line $h_0\in K$, the following holds:

Let $\mathcal{K}_x\subset C^n$ be the two-dimensional cone formed by the points of the straight lines $h\in K$ sufficiently close to h_0 and intersecting h_0 in a non-critical point x of this line. Then the tangent planes to all the cones $\mathcal{K}_x$ along their common generator h_0 coincide.

To prove this it is sufficient to note that the plane tangential to $\mathcal{K}_x$ along h_0 is parallel to the vector $(\xi^1,\ldots,\xi^{n-1},0)$ where $\xi=(\xi^1,\ldots,\xi^{n-1})$ is determined from the system (4).

We next formulate the condition for a complex to be admissible in analytical form by taking, as before, $\alpha^1,\ldots,\alpha^{n-1},\beta^1,\ldots,\beta^{n-1}$ as a local system of coordinates on the manifold of all straight lines.

THEOREM 4. For the complex K to be admissible it is necessary and sufficient that the following condition holds for any non-critical line $h_0\in K$:

Let $u^1=0,\ldots,u^{n-2}=0$ be the equations of K in a neighborhood of h_0. Then, for different x^0, x^n, the systems of equations

$$\sum_{j=1}^{n-1}\left(x^0\frac{\partial u^i}{\partial\alpha^j}\Big|_{h_0}-x^n\frac{\partial u^i}{\partial\beta^j}\Big|_{h_0}\right)\xi^j=0,\quad i=1,\ldots,n-1, \tag{7}$$

are equivalent to one another when $x=(x^0,\ldots,x^n)$ runs over the non-critical points of h_0.

Proof. As we already know, $(x^0\xi,-x^n\xi)$ [ξ is the (n−1)-dimensional vector determined from the system (7)] is a tangent vector to $G_x\cap K$. Under the conditions of the Theorem ξ is independent of x; but then K satisfies the conditions of Theorem 3 and, consequently, is admissible. Conversely, if K is admissible, then the tangent vectors $(x^0\xi,-x^n\xi)$ to $G_x\cap K$ lie in a fixed two-dimensional plane. It follows from this result that ξ is independent of x and hence the equations at (7) are equivalent.

We are going to give an equivalent formulation of Theorem 4. Let $\Delta_i(x^0,x^n)$ denote the minors of order n−2 of the matrix

$$\left\|x^0\frac{\partial u^i}{\partial\alpha^j}\Big|_{h_0}-x^n\frac{\partial u^i}{\partial\beta^j}\Big|_{h_0}\right\|_{\substack{i=1,\ldots,n-2\\ j=1,\ldots,n-1}} \tag{8}$$

Note that these minors are homogeneous polynomials in x^0, x^n of degree $n-2$; also, since h_0 is a non-critical line, they cannot all be identically zero.

THEOREM 4'. For the complex K to be admissible it is necessary and sufficient that, for any non-critical straight line $h_0 \in K$, the minors $\Delta_i(x^0, x^n)$ only differ from one another by constant multipliers.

In fact, it follows from system (7) that

$$\frac{\xi^1}{\Delta_1(x^0, x^n)} = \frac{\xi^2}{\Delta_2(x^0, x^n)} = \ldots = \frac{\xi^n}{\Delta_n(x^0, x^n)}. \tag{9}$$

Hence our assertion follows immediately from Theorem 4.

THEOREM 5. On each non-critical straight line of an admissible complex K there is, counting multiplicities, $n-2$ critical points.*

Proof. Let $\Delta_{i_0}(x^0, x^n)$ be a minor of order $n-2$ of the matrix (8) that is not identically zero. By Theorem 4' the remaining minors $\Delta_i(x^0, x^n)$ are proportional to Δ_{i_0}. Consequently the point $x = (x^0, \ldots, x^n) \in h_0$ is a critical point if and only if its coordinates satisfy

$$\Delta_{i_0}(x^0, x^n) = 0. \tag{10}$$

The required result follows immediately from here.

5. A Description of Admissible Complexes

We are going to describe the local structure of the complexes for which there are exactly $n-2$ pairwise distinct critical points on almost every straight line h.

THEOREM 6. Suppose that there are exactly $(n-2)$ pairwise distinct critical points $x_1, \ldots, x_{n-2}$ on the straight line h_0 of the admissible complex K. Then the critical points of the lines $h \in K$ close to h_0 describe $n-2$ local algebraic surfaces $M_1, \ldots M_{n-2}$, each of which is $(n-1)$- or $(n-2)$-dimensional; if $\dim M_i = n-1$, then the lines h touch M_i.

We next assume that the points $x_i \in M_i$ are non-singular. Let P_i be a hyperplane in CP^n spanned by the tangent plane to M_i at x_i and by the straight line h_0 $(i = 1, \ldots, n-2)$; we assume that the P_i are in general position [that is , $\dim \left(\bigcap_{i=1}^{n-2} P_i\right) = 2$]. Then the set of all straight lines in C^n that touch each $(n-1)$-dimensional local surface M_i and that intersect each $(n-2)$-dimensional local surface M_i, is contained in K and, in K, forms a neighborhood of h_0. Conversely, if almost every straight line of the complex K has a neighborhood of this form, then K is admissible.

Proof. By Theorem 2, $\dim M_i < n$. We will show that either $\dim M_i = n-1$ or that $\dim M_i = n-2$. It is sufficient to confine ourselves to the case when M_i does not contain points at infinity; in this case we will define points $x \in M_i$ by non-homogeneous coordinates $x^1, \ldots, x^n$, putting $x^0 = 1$.

Let $t^1, \ldots, t^n$ be a local system of coordinates on K in a neighborhood of h_0. Then the coordinates α^i, β^i of the straight lines $h \in K$ are functions of $t^1, \ldots, t^n$. The coordinates of points $x \in M_i$ are also functions of $t^1, \ldots, t^n$: x^n is determined from (10) of paragraph 4, $x^0 = 1$ and the remaining coordinates are given by $x^i = \alpha^i x^n + \beta^i$. The dimension of M_i is equal to the rank of the matrix $\| \partial x^i / \partial t^j \|$ which we now calculate.

It is easy to see that the matrix $\| \partial x^i / \partial t^j \|$ is equivalent to the following matrix (cf. the proof of Lemma 4):

$$\left\| \begin{matrix} \frac{\partial \alpha^1}{\partial t^1} x^n + \frac{\partial \beta^1}{\partial t^1} & \cdots & \frac{\partial \alpha^1}{\partial t^n} x^n + \frac{\partial \beta^1}{\partial t^n} \\ \cdots\cdots & \cdots & \cdots\cdots \\ \frac{\partial \alpha^{n-1}}{\partial t^1} x^n + \frac{\partial \beta^{n-1}}{\partial t^1} & \cdots & \frac{\partial \alpha^{n-1}}{\partial t^n} x^n + \frac{\partial \beta^{n-1}}{\partial t^n} \\ \frac{\partial x^n}{\partial t^1} & \cdots & \frac{\partial x^n}{\partial t^n} \end{matrix} \right\|.$$

*Let us note that if K is not admissible then, in general, there may be no critical points on its straight lines.

By Lemma 3 the rank of the matrix derived from that at (1) by removing the last row is greater than the rank of the matrix $\left\| \frac{\partial u^i}{\partial \alpha^j} - x^n \frac{\partial u^i}{\partial \beta^j} \right\|$ by unity; here $u^1 = 0, \ldots, u^{n-2} = 0$ are the equations of the complex. But, since x is a simple critical point, the rank of the last matrix is $n-3$. It follows immediately from this, that the rank of the matrix at (1) is $n-1$ or $n-2$. Thus, either $\dim M_i = n-1$ or $\dim M_i = n-2$.

We next show that if $\dim M_i = n-1$ then the straight line h_0: $x^i = \alpha^i x^n + \beta^i$ touches M_i, that is, the vector $(\alpha^1, \ldots, \alpha^{n-1}, 1)$ lies in the tangent plane to M_i at the point x_i. Note that the tangent plane to M_i is the plane spanned by the column vectors of matrix $\| \partial x^i / \partial t^j \|$; therefore it is sufficient to show that the rank of this matrix is unchanged under the addition to it of the column $(\alpha^1, \ldots, \alpha^{n-1}, 1)$.

It is easy to see that the matrix obtained by adding the column $(\alpha^1, \ldots, \alpha^{n-1}, 1)$ to $\| \partial x^i / \partial t^j \|$ is reduced by elementary transformations to the form

$$\left\| \begin{matrix} \frac{\partial \alpha^1}{\partial t^1} x^n + \frac{\partial \beta^1}{\partial t^1} & \ldots & \frac{\partial \alpha^1}{\partial t^n} x^n + \frac{\partial \beta^1}{\partial t^n} & 0 \\ \cdots & \cdots & \cdots & \cdots \\ \frac{\partial \alpha^{n-1}}{\partial t^1} x^n + \frac{\partial \beta^{n-1}}{\partial t^1} & \ldots & \frac{\partial \alpha^{n-1}}{\partial t^n} x^n + \frac{\partial \beta^{n-1}}{\partial t^n} & 0 \\ 0 & \ldots & 0 & 1 \end{matrix} \right\|.$$

Consequently its rank is $n-1$; that is, it is the same as the rank of the matrix $\| \partial x^i / \partial t^j \|$, as we required to prove.*

We are now going to show that the manifold $\mathfrak{M}$ of straight lines that touch the $(n-1)$-dimensional local surfaces M_i and that intersect the $(n-2)$-dimensional local surfaces M^i is contained in K and that it is a neighborhood of $h_0 \in K$. For this it is sufficient to make certain that $\dim \mathfrak{M} = n$.

Let us assume otherwise, that $\dim \mathfrak{M} > n$. Then, for almost every point x sufficiently close to h_0 we have $\dim(G_x \cap \mathfrak{M}) > 1$. Let x be one of these points. Without loss of generality we may assume that $x \in h_0$. Let $\mathcal{K}_x$ be the cone in C^n formed by the points of the straight lines $h \in \mathfrak{M}$ passing through x. By our assumption, $\dim \mathcal{K}_x \geq 3$. We consider the plane $\mathcal{H}_x$ tangential to $\mathcal{K}_x$, along h_0; its dimension satisfies $\dim \mathcal{H}_x \geq 3$. It follows from the definition of $\mathfrak{M}$ that $\mathcal{H}_x \subseteq P_i$, $i = 1, \ldots, n-2$, where the P_i are the hyperplanes defined in the statement of the Theorem. Consequently, $\mathcal{H}_x \subset \bigcap_i P_i$. But $\dim \left(\bigcap_i P_i \right) = 2$, and so $\dim \mathcal{H}_x \leq 2$, yielding a contradiction. Thus we have shown that $\dim \mathfrak{M} = n$.

Finally we are going to prove that, if a neighborhood of almost any straight line h_0 of the complex K has the described structure, then this complex is admissible. Let x be an arbitrary non-critical point on h_0. We consider the two-dimensional cone $\mathcal{K}_x \subset C^n$ formed by the points of the straight lines $h \in K$ close to h_0 and passing through x. As we noted above, the tangent plane $\mathcal{H}_x$ to this cone along h_0 is contained in $\bigcap_i P_i$. But $\dim \left(\bigcap_i P_i \right) = 2$; consequently, $\mathcal{H}_x = \bigcap_i P_i$ and therefore $\mathcal{H}_x$ does not depend on the choice of $x \in h_0$. Hence, in view of Theorem 3', the complex K is admissible.

COROLLARY. Any admissible complex of straight lines in C^3 is either a complex of lines intersecting a given curve or a complex of lines touching a given surface.

6. Examples of Complexes of Straight Lines

We have just given a description of the admissible complexes of straight lines in C^3. Let us note that the inversion formulas for these complexes were first given in [3].

We are going to present an example of a complex of lines in C^3 that is not admissible. We define a non-degenerate, symmetric bilinear form (x, y) in C^3, for example,

$$(x, y) = x^1 y^1 + x^2 y^2 + x^3 y^3; \tag{1}$$

*It also follows from this argument that when $\dim M_i = n-2$ the straight line h_0 does not touch the surface M_i.

we say that the pair of vectors x, y are orthogonal if $(x, y) = 0$.

Further, let $P(x, y)$ be another non-degenerate, symmetric bilinear form in C^3, for example,

$$P(x, y) = a_1 x^1 y^1 + a_2 x^2 y^2 + a_3 x^3 y^3. \tag{2}$$

We consider the second-order surface $P(x, x) = 1$. With each plane in C^3 we associate a pair of straight lines, the principal (that is, conjugate and mutually perpendicular) diameters of the curve of the second degree, that is, the section of the surface $P(x, x) = 1$ by this plane. The set of all straight lines obtained in this way forms a complex which we will call a Lie complex.

It is not difficult to find the equation of this complex. In terms of the Plücker coordinates α, p* it has the form

$$Q(\alpha, p) = 0, \tag{3}$$

where Q is the bilinear form conjugate to P. For example, if the bilinear forms (x, y) and P(x, y) are given by (1) and (2), then the equation of the Lie complex is $\frac{1}{a_1}\alpha^1 p^1 + \frac{1}{a_2}\alpha^2 p^2 + \frac{1}{a_3}\alpha^3 p^3 = 0$.

Next we give some examples of admissible complexes in C^n.

Suppose that, in C^n, we are given irreducible, (n−1)-dimensional algebraic surfaces $M_1, \ldots, M_s$ and (n−2)-dimensional surfaces $N_1, \ldots, N_t$, $s+t=n-2$.. We consider the manifold K of those straight lines in C^n that touch M_i and intersect N_j in pairwise distinct points. Then, if M_i and N_j are in general position, the closure of K is a complex; in addition, by Theorem 5, it is admissible.

Let us give yet another general construction for admissible complexes. Supose that, in C^n, we are given irreducible algebraic surfaces $N_1, \ldots, N_t$ of dimensions $k_1, \ldots, k_t$ respectively, where

$$\sum_{i=1}^{t} (n - k_i - 1) = n - 2. \tag{4}$$

We consider the manifold K of straight lines in C^n that intersect all the surfaces N_i in pairwise distinct points. Then, if N_i is in general position, the closure of K is a complex which, moreover, is admissible.

In fact, we will show that dim K = n. Let $\mathfrak{N}_i$ be the manifold of all lines intersecting N_i. Then $K = \bigcap_i \mathfrak{N}_i$... Hence, since N_i is in general position, codim $K = \sum_i \operatorname{codim} \mathfrak{N}_i = \sum_i (n - k_i - 1)$. By (4) this implies that dim K = n.

We establish that K is admissible by the same simple arguments as were used at the end of the proof of Theorem 5.

In view of this general construction, the following manifolds of straight lines are admissible complexes:

1) the manifold of all straight lines in C^n that intersect a given curve (the inversion formula for this complex was derived earlier in [5]);

2) the manifold of straight lines in C^n that intersect a pair of given surfaces of dimensions k and n−k, etc.

LITERATURE CITED

1. I. M. Gel'fand, M. I. Graev, and Z. Ya. Shapiro, "Integral geometry on k-planes," Funktsional. Analiz i Ego Prilozhen., 1, No. 1, 15–31 (1967).
2. I. M. Gel'fand and M. I. Graev, "Complexes of k-planes in C^n and the Plancherel formula for the group GL(n, C)," Dokl. Akad. Nauk SSSR, 179, No. 3 (1968).
3. I. M. Gel'fand, M. I. Graev, and N. Ya. Vilenkin, Integral Geometry and Representation Theory, Generalized Functions [in Russian], No. 5, Fizmatgiz, Moscow (1962).
4. I. M. Gel'fand and M. Ya. Naimark, "Unitary representations of the Lorentz group," Izv. Akad. Nauk SSSR, Ser. Matem., 11, 411–504 (1947).
5. A. A. Kirillov, "On a problem of M. I. Gel'fand, " Dokl. Akad. Nauk SSSR, 137, No. 2, 276–277 (1961).

*α is the direction vector of the line, $p = [\alpha, \beta]$, where β is any point of the line.

5.

(with M. I. Graev and Z. Ya. Shapiro)

Differential forms and integral geometry

Funkts. Anal. Prilozh. 3 (2) (1969) 24–40 [Funct. Anal. Appl. 3 (1969) 101–114].
MR **39**:6232. Zbl. **191**:528

§ 1. GENERAL STATEMENT OF PROBLEMS OF INTEGRAL GEOMETRY

I. Double Fiberings

1. Let two infinitely differentiable real manifolds B and C be given. We assume that a relation of incidence, satisfying the following axioms, is introduced between certain pairs of points $b \in B$ and $c \in C$. For every $c \in C$, we denote by $H_c \subset B$ the subset of points $b \in B$, incident to the point c; analogously, for every $b \in B$, we denote by $G_b \subset C$ the subset of points $c \in C$, incident to b.

Axiom 1. It follows from $H_{c_1} = H_{c_2}$ that $c_1 = c_2$; analogously, if $G_{b_1} = G_{b_2}$, then $b_1 = b_2$.

Let A be the subset of all pairs $(b, c) \in B \times C$, connected by the incidence relation.

Axiom 2. A is an infinitely differentiable submanifold in $B \times C$.

Axiom 3. The natural projections $\pi_1 : A \to B$ and $\pi_2 : A \to C$ are infinitely differentiable fiberings of the manifold A.

We note that the fiber $F_b^{(1)} = \pi_1^{-1}b$ consists of all pairs (b, c), where $c \in G_b$; thus, for constant b, the map $\pi_2 : F_b^{(1)} \equiv \pi_1^{-1}b \to G_b$ is diffeomorphism. Analogously, if the fiber $\pi_2^{-1}c$ is denoted by $F_c^{(2)}$, then $F_c^{(2)}$ is is obviously diffeomorphous with H_c.

The fibering of the submanifold $A \subset B \times C$, defined by Axiom 3, will be called a canonical double fibering connected with the given incidence relation.

Thus, the relation of incidence between elements of the two manifolds B and C is a representation of submanifold A in the direct product $B \times C$, such that Axioms 1 and 3 are satisfied.

This viewpoint, in which this submanifold in $B \times C$ is taken as a base, would be more convenient if we wished to set forth concepts of integral geometry schematically. However, since the main part of this article will be devoted to analysis of a concrete example (§ 2), this difference in approaches will not matter.

2. Now let us assume that we are given three manifolds $\tilde{A}$, B and C, and two fiberings $\tilde{\pi}_1 : \tilde{A} \to B$ and $\tilde{\pi}_2 : \tilde{A} \to C$. These two fiberings define the incidence relation between points of the manifolds B and C. Namely, the two points $b \in B$ and $c \in C$ will be called incident if an $\tilde{a} \in \tilde{A}$ exists such that $\tilde{\pi}_1\tilde{a} = b$ and $\tilde{\pi}_2\tilde{a} = c$.

If the incidence relation introduced in this way satisfies Axioms 1–3, the two fiberings $(\tilde{A}, \tilde{\pi}_1, B)$ and $(\tilde{A}, \tilde{\pi}_2, C)$ are called a double fibering of the manifold $\tilde{A}$.

Obviously, not every pair of fiberings forms a double fibering. Actually, by assuming, e.g., $\tilde{A} = B \times C$ and considering the natural projections of this product on B and C, we obtain an incidence relation which does not satisfy Axiom 1. We assert the following:

For every double fibering

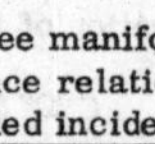

a map $\pi : \tilde{A} \to A$ exists, where A is a space of the corresponding canonical fibering, such that the following diagram is commutative:

Institute of Applied Mathematics, Academy of Sciences of the USSR; Moscow State University. Translated from Funktsional'nyi Analiz i Ego Prilozheniya, Vol. 3, No. 2, pp. 24–40, April–June, 1969. Original article submitted December 25, 1968.

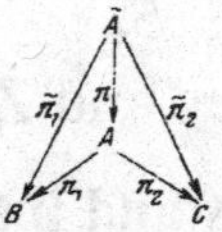

Actually, if $\tilde{a} \in \tilde{A}$, we obtain $\pi\tilde{a} = (\tilde{\pi}_1\tilde{a}, \tilde{\pi}_2\tilde{a})$. Since the elements $b = \tilde{\pi}_1\tilde{a}$ and $c = \tilde{\pi}_2\tilde{a}$ are incident, $(\tilde{\pi}_1\tilde{a}, \tilde{\pi}_2\tilde{a}) \in A$. Obviously, the map π meets the requirement that the diagram be commutative.

II. Differential Forms on Fiber Space

Relatively well-known facts regarding filtration, as well as integration of forms in fiber spaces, which will be required below, are set forth in this section. The newer concept of the integral of form in a fiber space, on which this article is based, is set forth in Section III.†

1. Let $\pi : A \to X$ denote a fibering of the manifold A. The projection π induces a map of tangent $\pi' : T_a(A) \to T_{\pi a}(X)$.

We shall consider the spaces of all differential forms with infinitely differentiable coefficients on each of the manifolds A and X, denoted by $\Omega(A)$ and $\Omega(X)$, respectively. The projection π induces the map π^* of functions and forms from $\Omega(X)$ to $\Omega(A)$:

$$\pi^* : \Omega(X) \to \Omega(A).$$

This map is called a lift of the form from $\Omega(X)$ to $\Omega(A)$.

Let $\omega \in \Omega(A)$ be a form of degree s. We denote by $\omega(a; \eta_1, \ldots, \eta_s)$ the value of this form on the vectors $\eta_1, \ldots, \eta_s \in T_a(A)$, where $T_a(A)$ is a space tangent to A at the point a.

The form ω will be called constant on the fibers of the fibering $\pi : A \to X$ if

$$\omega(a_1; \eta_1, \ldots, \eta_s) = \omega(a_2; \eta_1', \ldots, \eta_s')$$

whenever $\pi a_1 = \pi a_2$, $\pi' \eta_i = \pi' \eta_i'$, $i = 1, \ldots, s$.

It is readily seen that for the form $\omega \in \Omega(A)$ to be a lift of a certain form $\omega_1 \in \Omega(X)$, it is necessary and sufficient that it be constant on the fibers of the fibering $\pi : A \to X$.

2. We shall define the filtration of forms from $\Omega(A)$, which is connected with the fibering of the manifold A.

We denote by $\Omega^q(A) \subset \Omega(A)$ the subspace of forms $\omega(a; \eta_1, \ldots, \eta_s)$, which revert to zero when more than q vectors η_i lie in the space tangent to the fiber, i.e., $\omega(a; \eta_1, \ldots, \eta_s) = 0$ if $\eta_i \in T_a(F_{\pi a})$ at $i = 1, 2, \ldots, q+1$.

The spaces $\Omega^q(A)$ prescribe increasing filtration of $\Omega(A)$, i.e.,

$$0 \subset \Omega^0(A) \subset \ldots \subset \Omega^N(A) = \Omega(A), \tag{1}$$

where $N = \dim F_{\pi a}$.

We shall write the condition under which $\omega \in \Omega^q(A)$ in local coordinates on A. We introduce in the neighborhood of an arbitrary point $a \in A$ a local system of coordinates $\alpha^1, \ldots, \alpha^N, x^{N+1}, \ldots, x^{N+M}$ (N is the dimension of the fiber) such that the first N coordinates α^i are coordinates in the fiber, whereas the x^j are coordinates of the projection. The form ω in the neighborhood of the point a may be expressed as follows:

$$\omega = \sum_{\nu_1 < \cdots < \nu_s} \varphi_{\nu_1, \ldots, \nu_s}(\alpha^i, x^j)\, d\alpha^{\nu_1} \wedge \ldots \wedge d\alpha^{\nu_t} \wedge dx^{\nu_{t+1}} \wedge \ldots \wedge dx^{\nu_s}. \tag{2}$$

†The need for this concept arises, for instance, in connection with the fact that, although all groups of cohomologies are equal to zero in R^n, they must be treated as if they were different from zero in many problems of analysis, including integral geometry (for the definition of the corresponding cohomologies see, e.g., [2]).

Obviously, $\omega \in \Omega^q(A)$ when and only when the number of differentials $d\alpha^i$ (i.e., "differentials along the fiber") in each term of Eq. (2) does not exceed q.

It follows from formula (2) that if $\omega \in \Omega^q(A)$, then $d\omega \in \Omega^{q+1}(A)$.

Definition. We shall call the form $\omega \in \Omega^q(A)$ relatively closed in Ω^q (or closed in Ω^q with respect to the fibering $\pi: A \to X$) if $d\omega \in \Omega^q(A)$, i.e., taking the differential does not increase the filtration of the form.

Besides the filtration (1), a natural gradation in degrees of forms is prescribed in the space $\Omega(A)$: $\Omega(A) = \sum_{k=0}^{\infty} \Omega_k(A)$, where $\Omega_k(A)$ is a subspace of forms of degree k. We shall put $\Omega_k^q(A) = \Omega_k(A) \cap \Omega^q(A)$.

The filtration (1) is commensurable with the gradation, i.e., $\Omega^q(A) = \sum_{k=0}^{\infty} \Omega_k^q(A)$. We now introduce the spaces $E_0^{p,q} = \Omega_{p+q}^q(A)/\Omega_{p+q}^{q-1}(A)$.

We shall denote the elements of the space $E_0^{p,q}$ (i.e., the class of equivalent forms) by $\omega_0^{p,q}$. In the space $E_0^{p,q}$, the differential d_0 enters naturally. Namely, if $\omega_0^{p,q} \in E_0^{p,q}$, then $d_0\omega_0^{p,q}$ is an element of $E_0^{p,q+1}$, containing a differential of some form of the class $\omega_0^{p,q}$. (It is easily shown that this definition is proper, i.e., it does not depend on the choice of member of the class $\omega_0^{p,q}$.)

Then we have: $d_0^2 = 0$. Thereby, the inclusions $\operatorname{Im} d_0E_0^{p,q-1} \subset \operatorname{Ker} d_0E_0^{p,q}$ occur, and we can define the groups $E_1^{p,q} = \operatorname{Ker} d_0E_0^{p,q}/\operatorname{Im} d_0E_0^{p,q-1}$.

The following statement is valid:

The group $E_1^{p,q}$ is isomorphous with the group of forms of degree p on the base X with values in the q-dimensional group of cohomologies of the fiber: $E_1^{p,q} = \Omega_p(X, H^q(F_x))$.

Continuation of this construction results in the classical spectral sequence of Leray-Serre. Thus, problems of integral geometry are considered in the initial terms of this sequence, i.e., its "prehistory."

3. We shall define the operation of integration over q-dimensional manifolds in a fiber for the forms $\omega \in \Omega_{p+q}^q(A)$.

We specify p vectors $\xi_1, \ldots, \xi_p \in T_x(X)$. At the arbitrary point $a \in F_x$, we take p + q vectors from $T_a(A)$: $\eta_1, \ldots, \eta_q, \zeta_1, \ldots, \zeta_p$ such that $\eta_i \in T_a(F_x)$, $i = 1, \ldots, q$, and $\pi'\zeta_j = \xi_j$, $j = 1, \ldots, p$. Then the value of the form $\omega(a; \eta_1, \ldots, \eta_q; \zeta_1, \ldots, \zeta_p)$ for the given quantities $\xi_j = \pi'\zeta_j$ does not depend on the choice of $\zeta_1, \ldots, \zeta_p$. Actually, by replacing any vector ζ_j, e.g., ζ_1, with the vector $\zeta_1' = \zeta_1 + \eta'$, where $\eta' \in T_a(F_x)$, we add to $\omega(a; \eta_1, \ldots, \eta_q; \zeta_1, \ldots, \zeta_p)$ the quantity $\omega(a; \eta_1, \ldots, \eta_q; \eta', \zeta_2, \ldots, \zeta_p)$, which is equal to zero, since $\omega \in \Omega_{p+q}^q(A)$, and q + 1 vectors lie in $T_a(F_x)$.

Thus, if the $\xi_1, \ldots, \xi_p$ are fixed and a varies along the fiber $F_x(x = \pi a)$, then $\omega = \omega(a; \eta_1, \ldots, \eta_q; \zeta_1, \ldots, \zeta_p)$ is transformed to a form of degree q on F_x. We shall specify a subfibering of the fibering $A \to X$ with a fiber γ_x of dimension q. On integrating the resulting form over $\gamma_x \subset F_x$, we obtain a differential form $\omega_1(x; \xi_1, \ldots, \xi_p)$ of degree p on the manifold X. We shall retain the usual notation for this integral, i.e., we shall write: $\omega_1 = \int_{\gamma_x} \omega$. (If the fibers γ_x are not compact, further conditions must be imposed on the forms under consideration to ensure convergence of integrals.)

It is readily proved that Stokes' relative formula applies to the above concept of the integral:

If $\omega \in \Omega_{p+q-1}^{q-1}(A)$ and $\gamma_x \subset F_x$ is a q-dimensional submanifold with the boundary $\partial\gamma_x$, the following equation is valid:

$$\int_{\partial\gamma_x} \omega = \int_{\gamma_x} d\omega, \tag{3}$$

where both integrals belong to $\Omega_p(X)$.

III. General Statement of Problems of Integral Geometry

1. Let there be given the canonical double fibering

The symbols $\Omega(A)$, $\Omega(B)$, and $\Omega(C)$ denote spaces of differential forms on the corresponding manifolds.

The main role in integral geometry is played by the construction and study of operators which convert forms from $\Omega(B)$ to forms or functions from $\Omega(C)$.

These operators are constructed according to the following scheme. Let $\omega \in \Omega_p(B)$. We consider the form $\pi_1^*\omega$ on A and define the operation of integration of this form over fibers of the fibering $\pi_2 : A \to C$. Let k be the dimension of a fiber of this fibering.

If $p \geq k$, the form $\pi_1^*\omega$ is relatively closed in the fibering π_2 as a form of maximum dimension. Its integral over the fiber is defined in II, Paragraph 3, and is an element of $\Omega_{p-k}(C)$.

The map constructed thereby $\Omega_p(B) \to \Omega_{p-k}(C)$ will be the desired operator when $p \geq k$.

Study of the kernel, cokernel, and inversion formula for this operator (in the case $p \geq k$) is also a problem of integral geometry.

However, the example analyzed in this article refers to the case where $p \leq k$, which is much more interesting in many respects. In this case, the operator from $\Omega_p(B)$ to $\Omega(C)$ may be defined by two methods.

First Method. The form $\pi_1\omega$ is assumed to be closed in Ω_p with respect to the fibering $\pi_2 : A \to C$. We shall specify a subfibering of the fibering $A \to C$, the fibers of which are p-dimensional cycles in $F_c^{(2)}$. The desired form from $\Omega(C)$ is obtained by integrating $\pi_1^*\omega$ over these cycles.

We shall not use this method in defining the operator, but we note that the inversion operator in a specific case, considered in § 2, has just such a structure.

Second Method. This, the central method for defining the operator, is connected with the necessity of integrating over the fiber a form of degree less than the dimension of the latter. To define the operator in this way, one must introduce an additional structure in the fibering $\pi_2 : A \to C$. This structure will be defined in the next paragraph.

2. Let us assume that $\pi_2 : A \to C$ is a fibering with the structural group G. This means that a standard fiber F_0 is specified in which a group of automorphisms G is effective, and for each sufficiently small neighborhood $U \subset C$ of the element $c \in C$, the diffeomorphism $\pi_2^{-1}U$ into $U \times F_0$, transforming F_c into $c \times F_0$, is defined. In this case, the automorphism F_0, resulting from two different maps of the same fiber, belongs to the group G.

We shall specify on F_0 a finite-dimensional space M_0^q of differential forms of degree q, invariant with respect to the action of group G. In this space M_0^q, we construct a vector fibering over C.

Namely, each diffeomorphism $F_c \to F_0$ of the fiber over $c \in C$ into the standard fiber induces a map of M_0^q into some finite-dimensional subspace M_c^q of differential forms over F_c. In this case, owing to the invariance of M_0^q with respect to the group G, M_c^q does not depend on the choice of the diffeomorphism $F_c \to F_0$.

We introduce the space $M^q = \bigcup_c M_c^q$, provided in the standard way with a structure of vector fibering over C.

The action of the group G in the standard fiber F_0 of the fibering $A \to C$ naturally defines the representation of this group in the standard fiber M_0^q of the vector fibering M^q; the operators of this representation form a structural group of the fibering M^q.

The vector fibering $M^q \to C$ (as well as the fibering $L^q \to C$, doubled to it) will be called the L^q structure in the fibering $A \to C$.

The set of such fiberings for q = 1, 2, ... will be called briefly the L structure.

3. Below, we shall require the principal G fibering E, constructed on the fibering (A, π_2, C). We shall describe this fibering.

The points of space E are pairs (c, e_c), where $c \in C$ and e_c is a diffeomorphism of the standard fiber F_0 into F_c. The topology in E is naturally determined by the topology of the initial fibering (A, π_2, C). The effect of the group G on the point (c, e_c) is defined thus: $eg = (c, e_c g)$. In this case, the map $\pi : (c, e_e)| \to c$ prescribes in E the structure of the principal G fibering with the base C.

The above vector fibering L^q may be constructed directly on the principal fibering of E and the space $L_0^q = (M_0^q)^*$. For this, one must take the direct product $E \times L_0^q$. Since the action of the group G is defined in E and L_0^q, it is also defined in the direct product: The element $g \in G$ transforms $(e, l) \in E \times L_0^q$ into $(e, l)g = (eg, T_g l) = (c, e_c g, T_g l)$, where T_g is the operator of the representation of group G in the space L_0^q, corresponding to g. It is readily shown that our space L^q coincides with the space of G orbits of the direct product $E \times L_0^q$.

4. We shall define the operator of integration over a fiber of the fibering $\pi_2 : A \to C$ of a form having the degree $p \leq k$, where $k = \dim F_c$.

For this, we assume that the L^q structure is introduced in the fibering $A \to C$, where $q = k - p$. Thus, a finite-dimensional subspace M_c^q of differential forms of degree q is specified on every fiber F_c.

Let $\omega \in \Omega_p(B)$. Here $\pi_1^* \omega$ denotes the lift of the form ω to A, whereas $\pi_1^* \omega |_{F_c}$ represents the limitation of form $\pi_1^* \omega$ to the fiber F_c. For any differential form $m_c \in M_c^q$, then $\pi_1^* \omega |_{F_c} \wedge m_c$ is a form of degree k on the fiber F_c.

We put

$$(\varphi(c), m_c) = \int_{F_c} \pi_1^* \omega |_{F_c} \wedge m_c. \quad (4)$$

It is obvious from the definition that $\varphi(c)$ at any given value of c is a linear functional of M_c^q, i.e., $\varphi(c) \in L_c^q$. Thus, φ is a section of the vector fibering L^q.

The resulting element φ of the space of sections $\Gamma(L^q)$ will be called the integral of the initial form $\omega \in \Omega_p(B)$ over fibers of the fibering $\pi_2 : A \to C$ provided with the L^q structure, and denoted thus: $\varphi = \int \omega$.

In Paragraph 3, we constructed the L^q structure by using the action of the group G in the direct product $E \times L_0^q$, where E is a space of the principal G fibering over C. This construction enables one to lift the section $\varphi(c) \in \Gamma(L^q)$ to a section of the trivial fibering $E \times L_0^q \to E$, which is invariant with respect to the action of group G, i.e., to the function $\varphi(e)$ on E with values in the fixed space L_0^q: $\varphi(e) = e_c^{-1}\varphi(c)$ for any $e = (c, e_c)$. It follows directly from the definition that the function $\varphi(e)$ with values in L_0^q is a lift of some section $\varphi(c) \in \Gamma(L^q)$ when and only when $T_g \varphi(eg) = \varphi(e)$ for any $g \in G$.

§2. INTEGRATION OF DIFFERENTIAL FORMS IN R^n OVER k-DIMENSIONAL PLANES

In this section, the following example will be considered.

In an n-dimensional space R^n, a differential form ω of degree p ($p \leq k$) with coefficients from S (S is a space of infinitely differentiable, rapidly decreasing functions on R^n) is given. Let $H_{n,k}$ be a space of oriented k-dimensional planes (the definition of the integral will be given later in Paragraphs 2 and 3), we obtain the function $\varphi(h)$ on $H_{n,k}$. Our problem is to determine what sort of functions are obtained in this case and how to recover the form ω from them. The result may be formulated either in terms of $\varphi(h)$ or in more convenient terms, connected with the principal fibering. The final result is formulated in Theorems 1 and 2.

1. Thus, the above problem of integral geometry is considered for the following case.

$B = R^n$ is an n-dimensional real space; $C = H_{n,k}$ is the space of all oriented k-dimensional planes in R^n.

The point $\beta \in R^n$ and the plane $h \subset R^n$ are considered incident if $\beta \in h$. Thus, the space A in our example is a space of pairs (β, h), where $\beta \in h$, and $h \subset R^n$. It is readily shown that the introduced incidence relation satisfies Axioms 1-3 in §1, i.e., in set A of incident pairs (β, h), double fibering occurs:

$$B = R^n \xleftarrow{\pi_1} A \xrightarrow{\pi_2} C = H_{n,k}$$

The fiber $F^{(1)}_{\beta}$ of the first fibering is a manifold of pairs (β, h), where β is fixed, and h varies in the manifold of k-dimensional planes passing through the point β. The fiber $F^{(2)}_{h}$ of the second fibering is a manifold of pairs (β, h), where h is fixed, and β passes through points of h; thus $F^{(2)}_{h} \cong R^k$.

The fibering $A \to C$ (fibers of which are k-dimensional planes in R^n) is one whose structural group is the entire group G of k-dimensional affine transformations which retain their orientation (including parallel shifts).

If the space R^k is chosen as the standard fiber, it is convenient for elements of group G to be specified by the pair $g = (\mu, t_0)$, where μ belongs to the group $GL_0(k, R)$ of linear transformations in R^k, retaining their orientation, and $t_0 \in R^k$. The action of G in the standard fiber R^k is prescribed in this case by the formula $gt = \mu t + t_0$.

2. We shall introduce an L^q structure in the fibering $A \to C$, where $q = k - p$. According to the general definition given in §1, this requires that we specify a finite-dimensional space M^q_0 of differential forms of degree q on the standard fiber R^k, and then to "spread" this space over all fibers of the fibering $A \to C$. As a result, we obtain the vector fibering M^q over C. For M^q_0, we shall take a space of differential forms of degree q with constant coefficients on R^k; thus, $M^q_0 = \overset{q}{\wedge}(R^k)^*$.

Obviously, M^q_0 is invariant with respect to the action of group G. In this case, the operators of the representation of group G, corresponding to parallel shifts, act trivially in M^q_0. Thus, the representation of group G reduces to a representation of the subgroup $GL_0(k, R)$ and is the q-th external degree of the standard representation of group $GL_0(k, R)$ in the space $(R^k)^*$.

3. After prescribing the L^q structure, we can introduce the operator of integration of the form ω of degree $p \leq k$ on R^n over k-dimensional planes: $\varphi = \int \omega$, $\varphi \in \Gamma(L^q)$. This operator is defined in the following manner. In each fiber $h = F^{(2)}_c$ of the fibering $A \to C$, we are given a space M^q_c of differential forms of degree $q = k - p$ with constant coefficients. By definition, φ is a function on $C = H_{n,k}$ whose values are linear functions on M^q_c:

$$(\varphi(c), m_c) = \int_h \omega|_h \wedge m_c, \quad m_c \in M^q_c.$$

4. We shall describe the principal fibering $E \to C$, constructed on the fibering $A \to C$. The elements of the space E are the pairs $e = (\alpha, \beta)$, where $\alpha : R^k \to R^n$ is a nondegenerate linear transformation, and $\beta \in R^n$. The group G acts in E in the following manner: If $g = (\mu, t_0)$, then $(\alpha, \beta)g = (\alpha\mu, \beta + \alpha t_0)$.

We shall show that the base of the principal fibering of E is C. For this, we compare with each pair (α, β) a k-dimensional oriented plane $h: x = \alpha t + \beta$, where t passes through R^k. Obviously, the two pairs (α, β) and (α', β') correspond to the same plane h when and only when they belong to the same orbit of group G.

Let us recover the fibering $A \to C$ from the principal fibering $E \to C$. For this, we shall take the standard fiber R^k, in which group G acts as a group of affine transformations, including parallel shifts. Moreover, in accordance with the standard construction, we shall take the direct product $E \times R^k$ and identify in it points lying on one orbit of group G. We must show that A is obtained in this case. The elements of the product $E \times R^k$ are triples (α, β, t); the triples (α, β, t) and $(\alpha\mu, \beta + \alpha t_0, \mu t + t_0)$ lie on the same trajectory. On each trajectory, we shall choose triples in which $t = 0$, i.e., $(\alpha, \beta, 0)$. Two such triples $(\alpha, \beta, 0)$ and $(\alpha', \beta', 0)$ are equivalent when and only when $\alpha' = \alpha\mu$, $\beta' = \beta$.

Thus, every point of the resulting fibering may be specified by a pair (α, β), where α is defined with accuracy up to multiplication to the right by $\mu \in GL_0(k, R)$.

In order to show that this fibering coincides with A, we shall compare each pair (α, β) with the element $(h, \beta) \in A$, where h is a k-dimensional plane: $x = \alpha t + \beta$. In this case, it is obvious that the two pairs (α, β) and (α', β') correspond to the same element of A when and only when these pairs are equivalent.

Incidentally, we defined the map $E \to A$, which brings into correspondence to the pair $(\alpha, \beta) \in E$, the pair $(h, \beta) \in A$. Thus, we crossed the map $E \to C$ through $A: E \to A \to C$. This is possible in any fibering in which a group acts transitively on the standard fiber. To avoid misunderstanding, we emphasize that this construction depends on the choice of fixed point in the standard fiber. In our case, this was the point $O \in R^k$.

5. By lifting the L^q structure introduced above from the space C to E, we obtain the trivial structure: $E \times \overset{q}{\wedge} R^k$.

Let us lift $\varphi(c) \in \Gamma(L^q)$ to this structure. Since the latter is trivial, φ becomes a function $\varphi(\alpha, \beta)$ on E with values in the standard space $\overset{q}{\wedge} R^k$. In this case, for such a function to be a lift of some section $\varphi(\alpha, \beta)$, it is necessary and sufficient that it be constant on the orbits of group G (see §1, II, Paragraph 4), i.e., that the conditions

$$\varphi(\alpha, \beta + \alpha t_0) = \varphi(\alpha, \beta), \tag{5}$$

$$\varphi(\alpha\mu, \beta) = T^{(q)}_{\mu^{-1}} \varphi(\alpha, \beta) \tag{6}$$

be met for any values of $t_0 \in R^k$ and $\mu \in GL_0(k, R)$. Here $T^{(q)}_{\mu^{-1}}$ is the operator of the representation of group $GL_0(k, R)$ in the space $\overset{q}{\wedge} R^k$, corresponding to μ^{-1}.

Now we shall give a definition of the equation $\varphi(\alpha, \beta) = \int \omega$, which does not depend on the preceding.

We shall consider the form ω on R^n of degree p. By substituting $x = \alpha t + \beta$ and $dx = \alpha dt$ into it, we obtain a form on R^k, denoted by $\omega(\alpha t + \beta)$.

By definition, $\varphi(\alpha, \beta)$ at any fixed values of $(\alpha, \beta) \in E$ is a linear functional on the forms τ of degree $q = k - p$ in R^k with constant coeffficients. The value of this functional on the form τ is determined by the following formula:

$$(\varphi(\alpha, \beta), \tau) = \int_{R^k} \omega(\alpha t + \beta) \wedge \tau. \tag{7}$$

The problem of integral geometry consists in finding out what sort of functions $\varphi(\alpha, \beta)$ are obtained in this way, and how they can be used to recover the form ω.

6. The space of differential forms of degree r on E with infinitely differentiable coefficients will be denoted by $\Omega_r(E)$. By virtue of our definition, the function $\varphi(\alpha, \beta)$ is an element of the tensor product $\Omega_{0,q} = \Omega_0(E) \otimes \overset{q}{\wedge} R^k$.

To solve the problem of integral geometry, we also require the spaces $\Omega_{r,s} = \Omega_r(E) \otimes \overset{s}{\wedge} R^k$, $r = 0, 1, \ldots$; $s = 0, 1, \ldots, k$.

We shall construct the operators $d_1: \Omega_{r,s} \to \Omega_{r+1,s+1}$, $d_2: \Omega_{r,s} \to \Omega_{r+1,s-1}$, $d_0: \Omega_{r,s} \to \Omega_{r+1,s}$. For this, we introduce the basis $e_1, \ldots, e_k$ in R^k and denote the doubled basis in $(R^k)^*$ by $e^1, \ldots, e^k$. Analogously, vectors of the basis in R^n and the doubled basis in $(R^n)^*$ are denoted respectively by f_i and f^j, $i, j = 1, \ldots, n$.

Since $\alpha \in \mathrm{Hom}(R^k, R^n) = R^n \otimes (R^k)^*$, the coordinates $\beta^i = (\beta, f^i)$ of the vector $\beta \in R^n$ and the coordinates $\alpha^i_j = (\alpha e_j, f^i)$ of the map α are determined thereby.

We introduce operators acting from $\Omega_r(E)$ to $\Omega_{r+1}(E)$. Let $\sigma \in \Omega_r(E)$. Then we put

$$d^{\beta}_{\alpha_j}\sigma = \sum_{l=1}^{n} d\beta^l \wedge \frac{\partial\sigma}{\partial\alpha^l_j}, \quad d^{\alpha_i}_{\alpha_j}\sigma = \sum_{l=1}^{n} d\alpha^l_i \wedge \frac{\partial\sigma}{\partial\alpha^l_j}, \quad d^{\alpha_i}_{\beta}\sigma = \sum_{l=1}^{n} d\alpha^l_i \wedge \frac{\partial\sigma}{\partial\beta^l}, \quad d^{\beta}_{\beta}\sigma = \sum_{l=1}^{n} d\beta^l \wedge \frac{\partial\sigma}{\partial\beta^l},$$

$i, j = 1, \ldots, k$; differentiation of the form means differentiation of its coefficients.

Further, let $y \in R^k$ and $\eta \in (R^k)^*$.

The operator $a_y : \overset{s}{\wedge} R^k \to \overset{s+1}{\wedge} R^{k+1}$, acting in the formula $a_y(y_1 \wedge \ldots \wedge y_s) = y \wedge y_1 \wedge \ldots \wedge y_s$, is denoted a_y. The operator $a'_\eta : \overset{s}{\wedge} R^k \to \overset{s-1}{\wedge} R^k$, acting in the following formula, is denoted by a'_η:

$$a'_\eta (y_1 \wedge \ldots \wedge y_s) - \sum_{\nu=1}^{s} (-1)^{\nu-1} (\eta, y_\nu) y_1 \wedge \ldots \wedge \widehat{y_\nu} \wedge \ldots \wedge y_s.$$

We shall now define the operators d_1, d_2, and d_0 by the following formulas:

$$d_1 = \sum_{i=1}^{k} d^{\beta}_{\alpha_i} \otimes a_{e_i}, \tag{8}$$

$$d_2 = \sum_{i=1}^{k} d^{\alpha_i}_{\beta} \otimes a'_{e^i}, \tag{9}$$

$$d_0 = d^{\beta}_{\beta}. \tag{10}$$

It is readily shown that these operators do not depend on the choice of bases in R^k and R^n.

7. We shall prove that the operators d_1, d_2, and d_0 are commutative with the action of the group $GL_0(k, R)$, i.e., for any $\mu \in GL_0(k, R)$, the diagram

$$\begin{array}{ccc} \Omega_{r,s} & \xrightarrow{d_1} & \Omega_{r+1,s+1} \\ T_\mu \downarrow & & T_\mu \downarrow \\ \Omega_{r,s} & \xrightarrow{d_1} & \Omega_{r+1,s+1} \end{array}$$

is commutative, as are analogous diagrams for operators d_1 and d_0.

Let $\varphi(\alpha, \beta) \in \Omega_{r,s}$, i.e., $\varphi(\alpha, \beta) = \sum_{\nu} \varphi_{\nu_1,\ldots,\nu_s}(\alpha, \beta) e_{\nu_1} \wedge \ldots \wedge e_{\nu_s}$, where $\varphi_\nu \in \Omega_r(E)$. The action of the element $\mu \in GL_0(k, R)$ transforms $\varphi(\alpha, \beta)$ to $T_\mu \varphi(\alpha, \beta)$, where, by definition

$$T_\mu \varphi(\alpha, \beta) - \sum_{\nu} \varphi_{\nu_1,\ldots,\nu_s}(\alpha\mu, \beta) \cdot (T_\mu e_{\nu_1}) \wedge \ldots \wedge (T_\mu e_{\nu_s}), \text{ whereas } T_\mu e_i = \sum_{j=1}^{k} \mu^j_i e_j.$$

Obviously, $d^{\beta}_{\alpha_i} \varphi_\nu(\alpha\mu, \beta) = \sum_{j=1}^{k} \mu^i_j (d^{\beta}_{\alpha} \varphi_\nu)(\alpha\mu, \beta)$.

Since $d_1 = \sum_{i=1}^{k} d^{\beta}_{\alpha_i} \otimes a_{e_i}$ by definition, it follows that

$$d_1 T_\mu \varphi(\alpha, \beta) = \sum_{\nu} \sum_{i=1}^{k} d^{\beta}_{\alpha_i} \varphi_\nu(\alpha\mu, \beta) \cdot e_i \wedge (T_\mu e_{\nu_1}) \wedge \ldots \wedge (T_\mu e_{\nu_s})$$

$$= \sum_{\nu} \sum_{j=1}^{k} (d^{\beta}_{\alpha_j} \varphi_\nu)(\alpha\mu, \beta) \cdot (T_\mu e_j) \wedge (T_\mu e_{\nu_1}) \wedge \ldots \wedge (T_\mu e_{\nu_s}).$$

The equality obtained means that $d_1 T_\mu = T_\mu d_1$. Commutativity of operators d_2 and d_0 with the action of the group is proved analogously.

For any r, s values, we prescribe the subspace $\Omega^0_{r,s} \subset \Omega_{r,s}$ of functions $\varphi(\alpha, \beta) \in \Omega_{r,s}$, invariant with respect to the action of the group $GL_0(k, R)$, i.e., such that $T_\mu \varphi = \varphi$ for any $\mu \in GL_0(k, R)$.

It follows from the statement proved above that the operators d_1, d_2, and d_0 produce maps of these subspaces, namely:

$$d_1 : \Omega^0_{r,s} \to \Omega^0_{r+1,s+1}, \quad d_2 : \Omega^0_{r,s} \to \Omega^0_{r+1,s-1}, \quad d_0 : \Omega^0_{r,s} \to \Omega^0_{r+1,s}.$$

8. Using the introduced operators d_1, d_2, and d_0, we now construct the operator Δ, which plays the main role, after putting

$$\Delta = d_2^k d_0 d_1^p. \tag{11}$$

Since $d_1 : \Omega^0_{r,s} \to \Omega^0_{r+1,s+1}$, $d_2 : \Omega^0_{r,s} \to \Omega^0_{r+1,s-1}$, $d_0 : \Omega^0_{r,s} \to \Omega^0_{r+1,s}$, the operator Δ maps $\Omega^0_{0,q} \subset \Omega_{0,q}$ into $\Omega^0_{k+p+1,0} \subset \Omega_{k+p+1,0}$:

$$\Delta : \Omega^0_{0,q} \to \Omega^0_{k+p+1,0}.$$

LEMMA 1. If the function $\varphi(\alpha, \beta)$ is a lift of the section $\varphi \in \Gamma(L^q)$ to E, the form $\Delta\varphi = d_2^k d_0 d_1^p \varphi$ is constant on the fibers of the fibering $E \to A$, i.e., it is a lift of the form from A.

Proof. Since $\varphi(\alpha, \beta) \in \Omega^0_{0,q}$, it follows that $\Delta\varphi \in \Omega^0_{k+p+1,0}$. Further, it is readily shown that the form $(\Delta\varphi)$ $(\alpha, \beta; \eta_1, \ldots, \eta_{k+p+1})$, $\eta_i \in T_e(E)$ lies in a subspace tangent to a fiber of the fibering $E \to A$. The given lemma follows directly.

9. We shall find the necessary conditions for representing $\varphi \in \Gamma(L^q)$ as

$$\varphi = \int \omega, \tag{12}$$

where ω is a form of degree p on R^n with coefficients from S:

$$\omega = \sum_{\nu} f_{\nu}(x)\, dx^{\nu} = \sum_{1 \le \nu_1 < \ldots < \nu_p \le n} f_{\nu_1,\ldots,\nu_p}(x^1, \ldots, x^n)\, dx^{\nu_1} \wedge \ldots \wedge dx^{\nu_p}. \tag{13}$$

LEMMA 2. If $\varphi \in \Gamma(L^q)$ can be represented as in (12), the differential form $\Delta\varphi = d_2^k d_0 d_1^p \varphi$, regarded as a form on A, is closed with respect to the fibering $A \to B$.

Proof. To prove the lemma, we find an expression for $\Delta\varphi$ by coefficients of the form ω. Let $\tau \in M_0^q = \overset{q}{\wedge}(R^k)^*$ be an arbitrary differential form with constant coefficients:

$$\tau = \sum_r c_r\, dt^r = \sum_{1 \le r_{p+1} < \ldots < r_k \le k} c_{r_{p+1},\ldots,r_k}\, dt^{r_{p+1}} \wedge \ldots \wedge dt^{r_k}.$$

Then $(\varphi(\alpha, \beta), \tau)$ is defined in the following manner (see formula (7)). The quantities $x = \alpha t + \beta$ and $dx = \alpha dt$ are substituted into the expression for the form ω, this form, now a function of t, is multiplied externally by τ, and the resulting form is integrated over the standard fiber R^k. As a result, we obtain:†

$$(\varphi(\alpha, \beta), \tau) = \sum_{\nu, r} (-1)^{\sigma} A^{\nu_1,\ldots,\nu_p}_{r_1,\ldots,r_p}(\alpha)\, c_{r_{p+1},\ldots,r_k} \psi_{\nu_1,\ldots,\nu_p}(\alpha, \beta), \tag{14}$$

where

$$\psi_{\nu_1,\ldots,\nu_p}(\alpha, \beta) = \int f_{\nu_1,\ldots,\nu_p}(\alpha t + \beta)\, dt^1 \wedge \ldots \wedge dt^k, \tag{15}$$

$$A^{\nu_1,\ldots,\nu_p}_{r_1,\ldots,r_p}(\alpha) = \begin{vmatrix} \alpha^{\nu_1}_{r_1} & \ldots & \alpha^{\nu_1}_{r_p} \\ \cdot & \cdot\;\cdot & \cdot \\ \alpha^{\nu_p}_{r_1} & \ldots & \alpha^{\nu_p}_{r_p} \end{vmatrix}; \tag{16}$$

summation is carried out over all $\nu = (\nu_1, \ldots, \nu_p)$, $1 \le \nu_1 < \ldots < \nu_p \le n$ and over all permutations $r = (r_1, \ldots, r_k)$ of indices $1, \ldots, k$, where $r_1 < \ldots < r_p$ and $r_{p+1} < \ldots < r_k$; σ is the evenness of the permutation r.

It follows from Eq. (14) that

$$\varphi(\alpha, \beta) = \sum_{\nu, r} (-1)^{\sigma} A^{\nu_1,\ldots,\nu_p}_{r_1,\ldots,r_p}(\alpha)\, \psi_{\nu_1,\ldots,\nu_p}(\alpha, \beta) \cdot e_{r_{p+1}} \wedge \ldots \wedge e_{r_k}. \tag{17}$$

---

†Using the following simple relation:

$$dx^{\nu_1} \wedge \ldots \wedge dx^{\nu_p} = \sum_{1 \le r_1 < \ldots < r_p \le k} A^{\nu_1,\ldots,\nu_p}_{r_1,\ldots,r_p}(\alpha)\, dt^{r_1} \wedge \ldots \wedge dt^{r_p}.$$

We apply to $\varphi(\alpha, \beta)$ the operator $d_0 d_1^p$. In this case we use the following, easily proved equations:

1) $d_0 d_1 \psi_{v_1,\ldots,v_p}(\alpha, \beta) = 0;$

2) $d_1^p (A_{r_1,\ldots,r_p}^{v_1,\ldots,v_p}(\alpha) e_{r_{p+1}} \wedge \ldots \wedge e_{r_k}) = (-1)^\sigma (p!)^2 (d\beta^{v_1} \wedge \ldots \wedge d\beta^{v_p}) e_1 \wedge \ldots \wedge e_k.$

As a result, we obtain

$$d_0 d_1^p \varphi(\alpha, \beta) = (p!)^2 d_\beta \left(\sum_v \psi_v(\alpha, \beta) d\beta^v \right) e_1 \wedge \ldots \wedge e_k, \tag{18}$$

where d_β is the operator of outer differentiation with respect to β.

This equation may be written in the following manner:

$$d_0 d_1^p \varphi(\alpha, \beta) = (p!)^2 \left(\sum_v \Psi_v(\alpha, \beta) d\beta^v \right) e_1 \wedge \ldots \wedge e_k, \tag{19}$$

where

$$v = (v_0, \ldots, v_p), \ 1 \leqslant v_0 < \ldots < v_p \leqslant n, \ d\beta^v = d\beta^{v_0} \wedge \ldots \wedge d\beta^{v_p},$$
$$\Psi_v(\alpha, \beta) = \int F_v(\alpha t + \beta) dt^1 \wedge \ldots \wedge dt^k, \tag{20}$$

and the $F_\nu(x)$ are coefficients of the form $d\omega$. Finally, we apply the operator d_2^k to this form. We note that $d_2^k = k!\, \varkappa\, e^1 \wedge \ldots \wedge e^k$, where $\varkappa = d_\beta^{\alpha_1} \ldots d_\beta^{\alpha_k}$.

As a result, we obtain

$$d_2^k d_0 d_1^p \varphi(\alpha, \beta) = (k!)(p!)^2 \sum_v \varkappa \Psi_v(\alpha, \beta) d\beta^v. \tag{21}$$

To prove that this form is closed with respect to the fibering $A \to B$, it is sufficient to show that each of the forms $\varkappa\Psi_\nu(\alpha, \beta)$ is relatively closed, i.e., that $(d_{\alpha_1}^{\alpha_1} + \ldots + d_{\alpha_k}^{\alpha_k}) d_\beta^{\alpha_1} \ldots d_\beta^{\alpha_k} \Psi_\nu = 0$.

However, this equality follows from the relations $d_{\alpha_i}^{\alpha_i} d_\beta^{\alpha_i} \Psi_\nu = 0$, $i = 1, \ldots, k$, which are easily proved by using formula (20).

10. The concept of the rapidly decreasing function on $C = H_{n,k}$ was introduced in [1]. We shall extend this concept to the sections $\varphi(c) \in \Gamma(L^q)$.

Let $\varphi(e)$ be a lift of $\varphi(c)$ to a vector fibering over E. We shall consider the fibering $\rho: H_{n,k} \to G_{n,k}$, where $G_{n,k}$ is a Grassman manifold of k-dimensional subspaces in R^n. We take the finite covering of $G_{n,k}$ by neighborhoods U satisfying the following two requirements: 1) the fibering $\rho^{-1}U \to U$ is trivial; 2) in the fibering $E \to H_{n,k}$, there exists a section over $\rho^{-1}U$: $s_U: \rho^{-1}U \to E$. We shall compare the function on $\rho^{-1}U \subset H_{n,k}: (s_U^*\varphi)(h) = \varphi(s_U h)$ with each U. We call $\varphi \in \Gamma(L^q)$ rapidly decreasing if each of the functions $s_U^*\varphi$ decreases rapidly together with all partial derivatives on $\rho^{-1}U \cong U \times R^k$.

It is readily shown that this definition does not depend on either the choice of covering $\{U\}$ or the choice of sections s_U.

<u>LEMMA 3.</u> If $\varphi = \int \omega$, where ω is a differential form of degree p on R^n with coefficients from S, $\varphi \in \Gamma(L^q)$ is an infinitely differentiable, rapidly decreasing section.

This statement follows directly from the explicit expression (17) for φ by coefficients of the form ω.

11. We shall formulate the principal theorem of this section.

<u>Theorem 1.</u> For $\varphi \in \Gamma(L^q)$ to be representable as an integral over k-dimensional planes, $k < n-1$,

$$\varphi = \int \omega,$$

where ω is a differential form of degree p on R^n with coefficients from S, it is necessary and sufficient to fulfill the following conditions:

1) the quantity φ is infinitely differentiable and decreases rapidly;

2) the form $\Delta\varphi \in \Omega_{k+p+1}(A)$ is closed with respect to the fibering $A \to B$.

We recall that, according to Lemma 1, if the function $\varphi(\alpha, \beta)$ on E with values in $L_0^q = \overset{q}{\wedge} R^k$ is a lift of some section $\varphi(c) \in \Gamma(L^q)$, the form $\Delta\varphi = d_2^k d_0 d_1^p \varphi$, defined in Paragraph 8, is constant on fibers of the fibering $E \to A$, i.e., it may be regarded as an element of $\Omega_{k+p+1}(A)$.

We proved the necessity of conditions 1) and 2) in Paragraphs 8 and 10 (Lemmas 2 and 3). The proof of their sufficiency will be based on another lemma, which we shall now formulate and prove.

12. Let the function

$$\varphi(\alpha, \beta) = \sum_{1 \leqslant r_{p+1} < \ldots < r_k \leqslant k} \varphi_{r_{p+1},\ldots,r_k}(\alpha, \beta)\, e_{r_{p+1}} \wedge \ldots \wedge e_{r_k} \tag{22}$$

(here $\varphi_{r_{p+1},\ldots,r_k}(\alpha, \beta) = (\varphi(\alpha, \beta), e^{r_{p+1}} \wedge \ldots \wedge e^{r_k})$) satisfy the conditions of the theorem. We shall consider $d_0 d_1^p \varphi$.

It follows from the definition of the operators d_0 and d_1 that this is a form of degree p + 1 on E, whose value lies in the one-dimensional space $\overset{k}{\wedge} R^k$, i.e., we may write $d_0 d_1^p \varphi$ as

$$(d_0 d_1^p \varphi)(\alpha, \beta;\ \eta_0, \ldots, \eta_p) = u(\alpha, \beta;\ \eta_0, \ldots, \eta_p)\, e_1 \wedge \ldots \wedge e_k,$$

where $\eta_i \in T_e(E)$, $e = (\alpha, \beta)$, and $u \in \Omega_{p+1}(E)$. Let σ be the fibering $E \to B = R^n$ (it is defined, since the fiberings $E \to A$ and $A \to B$), σ' the induced map of tangent spaces: $\sigma' : T_e(E) \to T_{\sigma e}(R^n) = R^n$.

If $\sigma'\eta_i = \sigma'\eta_i'$, i = 0, 1, ..., p, then

$$(d_0 d_1^p \varphi)(\alpha, \beta;\ \eta_0, \ldots, \eta_p) = (d_0 d_1^p \varphi)(\alpha, \beta;\ \eta_0', \ldots, \eta_p')$$

(see formulas (8) and (10), which specify d_1 and d_0). Hence, one may presume that $\eta_i \in R^n$.

LEMMA 4. If the function $\varphi(\alpha, \beta)$ is a lift to $E \times L_0^q$ in some section $\varphi(c) \in \Gamma(L^q)$, the form $d_0 d_1^p \varphi$ satisfies the equation:

$$(d_0 d_1^p \varphi)(\alpha, \beta;\ d\beta, \alpha e_{r_1}, \ldots, \alpha e_{r_p}) = c_r d_\beta \varphi_{r_{p+1},\ldots,r_k}(\alpha, \beta) \cdot e_1 \wedge \ldots \wedge e_k, \tag{23}$$

where $r = (r_1, \ldots, r_k)$ is an arbitrary permutation of indices 1, ..., k; c_r is a constant. In this case $c_r \neq 0$.

Thus, if $d_0 d_1^p \varphi = 0$, then $d_\beta \varphi = 0$.

Proof. We shall introduce the operators $X_i^j = \sum_{s=1}^n \alpha_i^s (\partial/\partial\alpha_j^s)$ and $X_0^j = \sum_{s=1}^n d\beta^s (\partial/\partial\alpha_j^s)$. We put $D_i = \sum_{j=1}^k X_{r_i}^j \otimes a_{e_j}$, i = 1, ..., p; $D_0 = \sum_{j=1}^k X_0^j \otimes a_{e_j}$. Here the a_{e_j} are the operators from $\overset{s}{\wedge} R^k$ to $\overset{s+1}{\wedge} R^k$: $a_{e_j}(y_1 \wedge \ldots \wedge y_s) = e_j \wedge y_1 \wedge \ldots \wedge y_s$, defined in Paragraph 6. Further, we put

$$\hat{D}_i = \sum_{s=1}^n \alpha_{r_i}^s \frac{\partial}{\partial\beta^s}, \quad i = 1, \ldots, p; \qquad \hat{D}_0 = \sum_{s=1}^n d\beta^s \frac{\partial}{\partial\beta^s}.$$

The form $(d_0 d_1^p \varphi)(\alpha, \beta;\ \alpha e_{r_1}, \ldots, \alpha e_{r_p})$ is expressed in this notation in the following manner:

$$(d_0 d_1^p \varphi)(\alpha, \beta;\ d\beta, \alpha e_{r_1}, \ldots, \alpha e_{r_p}) = \sum (-1)^\sigma \hat{D}_{i_0} \circ D_{i_1} \circ \ldots \circ D_{i_p} \varphi, \tag{24}$$

where summation is carried out over all permutations of the indices $(0, r, \ldots, r_p)$, and $\circ$ means that the operators are multiplied out as operators with constant coefficients.

It is readily shown that $D_i \circ D_j = -D_j \circ D_i$; hence, terms in (24), differing by the permutation of the indices $i_1, \ldots, i_p$, are equal. Therefore,

$$(d_0 d_1^p \varphi)(\alpha, \beta; d\beta, ae_{r_1}, \ldots, ae_{r_p})$$
$$= p! \sum_{s=0}^{p} (-1)^s D_0 \circ \ldots \circ D_{s-1} \circ D_{s+1} \circ \ldots \circ D_p \circ \hat{D}_s \varphi .$$

We shall express the symbolic product of the operators in the usual way.

PROPOSITION 1. Each of the operators $D_0 \circ \ldots \circ D_{s-1} \circ D_{s+1} \circ \ldots \circ D_p \circ \hat{D}_s$ is a linear combination, with the coefficients +1 or −1, of operators, having the form

$$a_{r_{i_1}} \ldots a_{r_{i_t}} D_{i_{t+1}} \ldots D_{i_p} \hat{D}_{i_{p+1}}, \tag{25}$$

where $(i_1, \ldots, i_{p+1})$ is a permutation of numbers $0, 1, \ldots, p$, $i_1, \ldots, i_t$ being different from zero (to simplify the notation, we write a_i instead of a_{e_i}).

The proof follows directly from the relations:

$$D_i \circ D_j = D_i D_j - a_{r_j} D_i \quad \text{at} \quad j \neq 0;$$

$$D_i \circ (D_{i_1} \circ \ldots \circ D_{i_s}) = D_i (D_{i_1} \circ \ldots \circ D_{i_s}) - \sum_{t=1}^{s} a_{r_{i_t}} D_i \circ D_{i_1} \circ \ldots \circ D_{i_{t-1}} \circ D_{i_{t+1}} \circ \ldots \circ D_{i_s},$$

where $i_1, \ldots, i_s$ are different from 0. Analogous relations apply when D_{i_s} is replaced by $\hat{D}_{i_s}$.

PROPOSITION 2. When $i \neq 0$, the following equations apply:

$$\hat{D}_i \varphi = 0; \quad D_i \varphi = q a_{r_i} \varphi. \tag{26}$$

The first equation follows directly from the relation $\varphi(\alpha, \beta + \alpha t_0) = \varphi(\alpha, \beta)$. We shall prove the second equation.

We recall that $D_i = \sum_{j=1}^{k} X^j_{r_i} \otimes a_j$, where X^j_i is a Lie operator on E, corresponding to an infinitesimal matrix which has a figure 1 at the intersection of the j-th row and i-th column and zeros everywhere else. Here T^j_i is the Lie operator in $\mathbf{R}^k$, corresponding to the same infinitesimal matrix; this operator transforms e_j to e_i and the rest of the basis vectors to zero.

Since the function φ satisfies the equation $\varphi(\alpha\mu, \beta) = T^{(q)}_{\mu^{-1}} \varphi(\alpha, \beta)$, it follows that

$$X^j_i \varphi(\alpha, \beta) = -(T^j_i)^{(q)} \varphi(\alpha, \beta),$$

where $(T^j_i)^{(q)}$ is the operator in $\overset{q}{\wedge} \mathbf{R}^k$, corresponding to the operator T^j_i in $\mathbf{R}^k$, namely:

$$(T^j_i)^{(q)} (y_1 \wedge \ldots \wedge y_q) = \sum_{s=1}^{q} y_1 \wedge \ldots \wedge (T^j_i y_s) \wedge \ldots \wedge y_q.$$

Hence, $D_i \varphi = -\sum_{j=1}^{k} a_j (T^j_{r_i})^{(q)} \varphi$. But it is readily shown that $\sum_{j=1}^{k} a_j (T^j_i)^{(q)} = -q a_i$. Hence, $D_i \varphi = q a_{r_i} \varphi$.

Let us turn to proof of the lemma. By virtue of Proposition 1, it is sufficient to find the result of applying each operator $a_{r_{i_1}} \ldots a_{r_{i_t}} D_{i_{t+1}} \ldots D_{i_p} \tilde{D}_{i_{p+1}}$ to φ. If $i_{p+1} \neq 0$, then, by virtue of Proposition 2, applying this operator to φ gives 0. If $i_{p+1} = 0$, then, again by virtue of Proposition 2, applying this operator to φ gives, with accuracy up to the constant factor,

$$a_{r_1} \ldots a_{r_p} \hat{D}_0 \varphi = (-1)^{\sigma} d_\beta \varphi_{r_{p+1}, \ldots, r_k} (\alpha, \beta) e_1 \wedge \ldots \wedge e_k.$$

Eq. (23) is proved thereby.

It remains to be shown that the coefficient c_r in Eq. (23) is different from zero. For this, we apply the operator $d_0 d_1^p$ to a function of the form

$$\varphi(\alpha, \beta) = \sum_r (-1)^\sigma A^{1,\dots,p}_{r_1,\dots,r_p}(\alpha)\,\psi(\alpha, \beta)\, e_{r_{p+1}} \wedge \dots \wedge e_{r_k}, \tag{27}$$

where $\psi(\alpha, \beta) = \int f(\alpha t + \beta)\, dt^1 \wedge \dots \wedge dt^k$.

We obtain (see the proof of Lemma 2)

$$d_0 d_1^p \varphi = (p!)^2 d_\beta (\psi(\alpha, \beta)\, d\beta^1 \wedge \dots \wedge d\beta^p)\, e_1 \wedge \dots \wedge e_k.$$

Hence

$$(d_0 d_1^p \varphi)(\alpha, \beta;\ d\beta,\ \alpha e_{r_1}, \dots, \alpha e_{r_p}) = (p!)^2 d_\beta (A^{1,\dots,p}_{r_1,\dots,r_p}(\alpha)\,\psi(\alpha, \beta))\, e_1 \wedge \dots e_k.$$

After comparing this with (27), we conclude that $c_r = (-1)^\sigma (p!)^2$. The lemma is proved.

13. Proof of Sufficiency of Conditions of the Theorem. Let the function φ satisfy the conditions of the theorem. We put

$$d_0 d_1^p \varphi(\alpha, \beta) = \Big(\sum_\nu \Psi_\nu(\alpha, \beta)\, d\beta^\nu\Big) e_1 \wedge \dots \wedge e_k, \tag{28}$$

where $\nu = (\nu_0, \dots, \nu_p)$, $d\beta^\nu = d\beta^{\nu_0} \wedge \dots \wedge d\beta^{\nu_p}$; $1 \le \nu_0 < \dots < \nu_p \le n$. Then

$$d_2^k d_0 d_1^p \varphi(\alpha, \beta) = k! \sum_\nu \varkappa \Psi_\nu(\alpha, \beta)\, d\beta^\nu,$$

where $\varkappa = d_\beta^{\alpha_1} \dots d_\beta^{\alpha_k}$. In this case, it follows from the conditions of the theorem that the functions $\Psi_\nu(\alpha, \beta)$ satisfy the following conditions:

1) $\Psi_\nu(\alpha\mu, \beta + \alpha t_0) = (\det \mu)^{-1} \Psi_\nu(\alpha, \beta)$;

2) $\Psi_\nu(\alpha, \beta)$ is an infinitely differentiable, rapidly decreasing function;

3) the form $\varkappa \Psi_\nu(\alpha, \beta)$ is a form on A closed with respect to the fibering $A \to B$.

It was shown in [1] that the function $\Psi_\nu(\alpha, \beta)$, satisfying these conditions, may be expressed as

$$\Psi_\nu(\alpha, \beta) = \int F_\nu(\alpha t + \beta)\, dt^1 \wedge \dots dt^k, \tag{29}$$

where $F_\nu(x) \in S$.† We shall consider a differential form of degree $p+1$ on $\mathbf{R}^n$: $\omega_1 = \sum_\nu F_\nu(x) dx^\nu$. We shall prove that it is closed, i.e., $d\omega_1 = 0$. In fact, it follows from Eq. (28) that

$$d_\beta \Big(\sum_\nu \Psi_\nu(\alpha, \beta)\, d\beta^\nu\Big) = 0. \tag{30}$$

Let $F'_{\nu'}(x)$, $\nu' = (\nu_0, \dots, \nu_{p+1})$ be coefficients of the form $d\omega_1$. It follows from Eqs. (29) and (30) that $\int F'_{\nu'}(\alpha t + \beta)\, dt^1 \wedge \dots \wedge dt^k \equiv 0$; hence, $F'_{\nu'}(x) \equiv 0$ for any ν', and therefore $d\omega_1 = 0$.

We shall use the following hypothesis: If ω_1 is a closed form of degree $p + 1 < n$, $p \ge 0$ in $\mathbf{R}^n$ with coefficients from S, there exists a form ω of degree p with coefficients from S such that $d\omega = \omega_1$.

By virtue of this hypothesis, there exists a form ω of degree p with coefficients from S:

$$\omega = \sum_\nu f_\nu(x)\, dx^\nu, \quad \nu = (\nu_1, \dots, \nu_p),$$

such that $d\omega = \sum F_\nu(x) dx^\nu$, $\nu = (\nu_0, \dots, \nu_p)$. Let $\varphi_1 = \int \omega$.

† The proof is carried out in detail in [1] for the case of a complex space. However, as was pointed out in [1] in a footnote on p. 23, this statement also applies to the real case. When k is even in the real case, an inversion formula is also obtained.

The theorem will be proved if we show that φ_1 coincides with the initial function φ with accuracy up to the factor.

Expressing φ_1 by coefficients of the form ω, we obtain

$$d_0 d_1^p \varphi_1 = (p!)^2 \Big(\sum_\nu \Psi_\nu(\alpha, \beta)\, d\beta^\nu\Big) e_1 \wedge \ldots \wedge e_k = (p!)^2 d_0 d_1^p \varphi.$$

According to Lemma 4, it follows from this that $d_\beta \varphi_1 = (p!)^2 d_\beta \varphi$. But then, since φ_1 and φ are rapidly decreasing functions, $\varphi_1 = (p!)^2 \varphi$. The theorem is proved.

14. Inversion Formula. Let the function φ satisfy the conditions of Theorem 1. Then $\varphi = \int \omega$, where ω is a form of degree $p < n-1$ in the space R^n with coefficients from S. Our problem is to recover ω in terms of φ. The solution of this problem is ambiguous. Namely, we shall show that if the form ω is closed, i.e., $d\omega = 0$, then $\int \omega = 0$.

As we noted above, it follows from the closed character of the form ω that $\omega = d\omega_1$, where ω_1 is a form of degree $p-1$, also with coefficients from S. Hence, $\int \omega = \int d\omega_1$. But from the definition of integrals of differential forms over k-dimensional planes, it follows directly that $\int d\omega_1 = 0$.

We shall see that $d\omega$ is already expressed unambiguously in terms of φ. Hence, it will follow, in particular, that the kernel of the map $\omega \mapsto \varphi$ coincides exactly with a subspace of closed forms.

THEOREM 2. Let ω be a differential form of degree p on R^n with coefficients from S, and let $\varphi = \int \omega$ be a function on $H_{n,k}$, obtained by integrating ω over k-dimensional planes, $p \leq k$. Then, if k is even, the following inversion formula applies:

$$\int_{\gamma_x} \Delta\varphi = c_{\gamma_x} d\omega, \tag{31}$$

where the integral is taken over an arbitrary k-dimensional cycle in a fiber of the fiber space over the point $x \in B = R^n$. The factor c_{γ_x} depends only on the class of homologies, to which the cycle γ belongs, it not being identically equal to zero.

Proof. Earlier (see p. 111), we derived the following expression for the form $\Delta\varphi = d_2^k d_0 d_1^p \varphi$:

$$\Delta\varphi = k!\,(p!)^2 \sum \varkappa \Psi_\nu(\alpha, x)\, dx^\nu, \quad \nu = (\nu_0, \ldots, \nu_p), \tag{32}$$

where $\Psi_\nu(\alpha, x) = \int F_\nu(\alpha t + x) dt^1 \wedge \ldots \wedge dt^k$, and the $F_\nu(x)$ are coefficients of the form

$$d\omega = \sum_\nu F_\nu(x)\, dx^\nu.$$

We shall use the inversion formula for integrals of functions which was derived in [1].† According to this formula,

$$\int_{\gamma_x} \varkappa \Psi_\nu(\alpha, x) = c_{\gamma_x} F_\nu(x).$$

Hence, integration of both parts of Eq. (32) over γ_x gives (31).

The authors sincerely thank I. N. Bernshtein for useful advice and a discussion of this study.

LITERATURE CITED

1. I. M. Gel'fand, M. I. Graev, and Z. Ya. Shapiro, "Integral geometry in k-dimensional planes," Funktsional'. Analiz i Ego Prilozhen., 1, No. 1, 15-31 (1967).
2. I. M. Gel'fand and D. B. Fuks, "Topology of noncompact Lie groups," Funktsional'. Analiz i Ego Prilozhen., 1, No. 4, 33-45 (1967).

† See footnote on p. 33.

6.

(with M. I. Graev and Z. Ya. Shapiro)

Integral geometry in projective space

Funkts. Anal. Prilozh. 4 (1) (1970) 14–32 [Funct. Anal. Appl. 4 (1970) 12–28].
MR 43:6856. Zbl. 199:255

§1. INTRODUCTION

The purpose of this work is the posing and the solving of the problem of integral geometry for the case of projective space. Before proceeding to pose the problem, however, we make at least the following remark. The apparatus which comes into play in this is related to a peculiar "differential geometry" on Grassman manifolds. Although the authors here do not extend application of this apparatus to questions other than integral geometry, the elegance and naturalness of this calculus leads one to think that it might be useful in other connections as well.

We proceed to describe the apparatus involved. Consider the manifold $G_{n+1,k+1}$ of orientable (k+1)-dimensional subspaces of the space R^{n+1}. Introduce the standard vector fibering in $G_{n+1,k+1}$:

$$L \to G_{n+1,k+1}.$$

(L, the space of pairs (h, x) where $h \in G_{n+1,k+1}$ and x is a point of h.) Let L' be the fibering dual to L, i.e., the space of pairs (h, β) where $h \in G_{n+1,k+1}$ is the (k+1)-dimensional oriented subspace of R^{n+1}, and $\beta \in h'$. We shall need also the direct sum of these fiberings: $L^{(m)} = \overset{m}{\oplus} L$, $L'^{(m)} = \overset{m}{\oplus} L'$.

We further introduce the space E of (k+1)-frames $u = (u_1, \ldots, u_{k+1})$, $u_i \in R^{n+1}$. We remark that it is the principal fiber space of the group $GL_0(k+1, R)$ (i.e., the group of nondegenerate linear mappings to R^{n+1} preserving orientation): the element $\mu \in GL_0(k+1, R)$ forms the mapping in E carrying the frame u to the frame $u\mu = (u_1', \ldots, u_{k+1}')$, where $u_k' = \sum_{i=1}^{k+1} u_j \mu_i^j$. It is obvious that two frames u and u' belong to the same fiber if and only if the natural orientable subspaces on it coincide. Accordingly, the basis of the fibering E is the manifold $G_{n+1,k+1}$.

Consider the function $\varphi(u)$ on the principal fibering E satisfying the condition $\varphi(u\mu) = (\det \mu)^{-1}\varphi(u)$. This function $\varphi(u)$ can be treated also as a section of the one-dimensional fibering on $G_{n+1,k+1}$.*

In the first part of the article we construct some peculiar collections of differential forms $\omega_0, \ldots, \omega_k$.

The form ω_k is a form of degree k which is constructed from the derivatives of the function φ up to the order k inclusive. It is a form in L', i.e., it depends on $h \in G_{n+1,k+1}$, a subspace in R^{n+1} and $\beta \in h'$. This form we shall also denote by $\omega_k(\beta)$.

***This fibering arises if in the space of pairs (u, t), $u \in E$, $t \in R$ there are identified respectively the pairs (u, t) and $(u\mu, (\det \mu)^{-1}t)$, $\mu \in GL_0(k+1, R)$.**

Institute of Applied Mathematics, Academy of Sciences of the USSR. Moscow State University. Translated from Funktsional'nyi Analiz i Ego Prilozheniya, Vol. 4, No. 1, pp. 14–32, January–March, 1970. Original article submitted September 30, 1969.

The form ω_{k-1} is defined on $L'^{(2)}$, i.e., it depends on $h \in G_{n+1,k+1}$ and the pair of vectors $\beta^{(1)}, \beta^{(2)} \in h'$. We shall denote it by $\omega_{k-1}(\beta^{(1)}, \beta^{(2)})$. This form is of degree $k-1$ and is constructed from the derivatives of $\varphi(u)$ up to order $k-1$ inclusive.

In general, ω_{k+1-m} ($m = 1, \ldots, k+1$) is a form of degree $k+1-m$ in $L'^{(m)}$. It is constructed from the derivatives of $\varphi(u)$ up to order $k+1-m$. For expressions for ω_{k+1-m} see § 3, (5).

Besides the forms ω_i there will be constructed from the function φ two additional sets of forms: $K_0, K_1, \ldots, K_k$ and $T_0, T_1, \ldots, T_k$. The central feature is the link between $d\omega_i$, the differential of the form ω_i, and the form ω_{i+1}. This link is expressed (Theorem 2) as follows:

$$d\omega_i = \nabla\omega_{i+1} + K_i + T_i$$

(for the definition of the operator ∇ see § 4). The differential form K_i is itself an important characteristic of the function $\varphi(u)$ (see Theorems 4 and 5). As to T_i, we remark that it is equal to zero on the manifolds of the subspaces passing through an arbitrary fixed point x (Theorem 3).

All fundamental definitions and results of this work are formulated in § 4 immediately following the definition of the form ω_i.

Though this article is closely related to [1] and [2], we regard it as independent.

§ 2. THE PROBLEM OF INTEGRAL GEOMETRY POSED

Consider, in $R^{n+1}\setminus 0$ an even functions $f(x)$ infinitely differentiable and homogeneous of degree $-(k+1)$:

$$f(\lambda x) = |\lambda|^{-k-1} f(x) \quad \text{for any} \quad \lambda \neq 0. \tag{1}$$

The functions f can also be treated as sections of a one-dimensional fibering in P^n. (This fibering arises if in the space of pairs (x, t) ($x \in R^{n+1}\setminus 0$, $t \in R$) there are identified respectively the pairs (x, t) and $(\lambda x, |\lambda|^{-k-1}t)$.)

We define the integrals for the functions $f(x)$ with respect to the subspaces $h \in G_{n+1,k+1}$. To do this we introduce for h a basis $u_1, \ldots, u_{k+1}$ and we shall give points $x \in h$ their coordinates $t^1, \ldots, t^{k+1}$ in this basis: $x = u_1t^1 + \ldots + u_{k+1}t^{k+1}$.

In standard oriented space R^{k+1} we shall assign the following differential form:

$$\sigma(t) = \sum_{i=1}^{k+1} (-1)^{i-1} t^i dt^1 \wedge \ldots \wedge \widehat{dt^i} \wedge \ldots \wedge dt^{k+1}. \tag{2}$$

Consider the differential form in $R^{k+1}\setminus 0$:

$$f(ut)\,\sigma(t) \equiv f(u_1t^1 + \ldots + u_{k+1}t^{k+1})\,\sigma(t). \tag{3}$$

It follows from Condition (1) on the functions $f(x)$ that this form can be dropped from $R^{k+1}\setminus 0$ to the sphere S^k under the natural fibering $R^{k+1}\setminus 0 \to S^k$.

We define

$$\varphi(u) = \frac{1}{2}\int_{S^k} f(ut)\,\sigma(t)\,^{*}). \tag{4}$$

*Since f is an even function, under the mapping $t \to -t$ the form $f(ut)\sigma(t)$ is multiplied by $(-1)^{k+1}$. Hence (4) can be considered as the integral with respect to the projective space P^k. For odd k (i.e., where P^k is an oriented space) the integral with respect to P^k follows the usually understood meaning; for even k (i.e., where P^k is a non-oriented space) it follows the understanding for integrals of odd forms (see J. de Rham, Differentiable Manifolds [Russian translation], Moscow, 1956, p. 44).

Inasmuch as the integral in (4) depends not only on $h \in G_{n+1,k+1}$ but also on the choice of basis $u = (u_1, \ldots, u_{k+1})$ in h, the structure of the function $\varphi(u)$ is determined not in $G_{n+1,k+1}$ itself but in the space of frames E.

From Formula (4) it follows that the function $\varphi(u)$ under the operations of the group $GL_0(k+1, R)$ transforms in the following way: $\varphi(u\mu) = (\det \mu)^{-1}\varphi(u)$. Accordingly, it may be treated as the section of a one-dimensional fibering in $G_{n+1,k+1}$.

In this work will be obtained the formula for the conversion expressing the original function $f(x)$ by $\varphi(u)$ (for the case of even k). The final result is formulated in § 4 (Theorem 6) and proved in § 8.

§ 3. CONSTRUCTION OF THE FORMS $\omega_0, \omega_1, \ldots, \omega_k$

Let $\varphi(u)$ be an arbitrary infinitely differentiable function on the principal fiber space $E \to G_{n+1,k+1}$, satisfying the condition

$$\varphi(u\mu) = (\det \mu)^{-1}\varphi(u) \qquad \text{for any} \qquad \mu \in GL_0(k+1, R). \tag{1}$$

With respect to this function we construct the chain of differential forms $\omega_0, \ldots, \omega_k$.

The form ω_{k+1-m} $(m = 1, \ldots, k+1)$ is its differential form of degree $k+1-m$ in the vector fibering $L'^{(m)} = \overset{m}{\oplus} L'$ where we consider L as the standard fibering in $G_{n+1,k+1}$ and L' as the dual fibering.

For describing the form ω_{k+1-m} we raise the fibering $L'^{(m)}$ with base $G_{n+1,k+1}$ in E. The fiber space thereby obtained we denote by $\tilde{L}'^{(m)}$. The form ω_{k+1-m} will first be defined as a differential form on $\tilde{L}'^{(m)}$; then it will be shown that it can be dropped from $\tilde{L}'^{(m)}$ to our space $L'^{(m)}$.

To begin with, we give a description of the fibering of the space $\tilde{L}'^{(m)}$. By definition, the basis of the space $\tilde{L}'^{(m)}$ serves E, i.e., the space of frames $u = (u_1, \ldots, u_{k+1})$, $u_i \in R^{n+1}$. Suppose that h is the $(k+1)$-dimensional subspace spanned on the frame u, and that h' is its conjugate subspace. Then, as points of the fiber in $\tilde{L}'^{(m)}$ on u we have sequences $(\beta^{(1)}, \ldots, \beta^{(m)})$ of vectors from h'. Accordingly, as points of the space $\tilde{L}'^{(m)}$ we have sequences $(u; \beta^{(1)}, \ldots, \beta^{(m)})$ where $u \in E$, $\beta^{(i)} \in h'$.

We remark that the fibering $\tilde{L}'^{(m)} \to E$ is trivial: $\tilde{L}'^{(m)} \cong E \times (R'^{k+1})^m$; therefore $\beta^{(1)}, \ldots, \beta^{(m)}$ can considered as vectors of the standard space $(R')^{k+1}$.

In fact, relating to each $\beta^{(i)} \in h'$ the vector $(\beta_1^{(i)}, \ldots, \beta_{k+1}^{(i)}) \in (R')^{k+1}$, where $\beta_j^{(i)} = (\beta^{(i)}, u_j)$, we get the image of h' in the standard space $(R')^{k+1}$.

In order to express the form ω_{k+1-m} in the space $L'^{(m)}$, we employ the following useful notation:

Let the letters a_j^i denote functions on some manifold and σ_j^i differential forms whose coefficients are linear differential operators such as the forms $\sum_{s,t} a_{st}(x)(\partial/\partial x_s)dx_t$, for example.

We agree to use the symbol

$$\begin{vmatrix} a_1^1 & \ldots & a_1^p & \sigma_1^1 & \ldots & \sigma_1^q \\ \cdot & \cdot & \cdot & \cdot & \cdot & \cdot \\ a_{p+q}^1 & \ldots & a_{p+q}^p & \sigma_{p+q}^1 & \ldots & \sigma_{p+q}^q \end{vmatrix}^{\wedge} \tag{2}$$

to denote the differential form of the following type:

$$\sum_{(i_1, \ldots, i_{p+q})} (-1)^s a_{i_1}^1 \ldots a_{i_p}^p \sigma_{i_{p+1}}^1 \wedge \ldots \wedge \sigma_{i_{p+q}}^q,$$

where the sum extends over all permutations $(i_1, \ldots, i_{p+q})$ of the indices $1, \ldots, p+q$ and s indicates the parity of corresponding permutation.

In that way, Expression (2) turns out to be the same as in the usual determinant. Here the only requirement is that the cofactors in each term be laid out in the increasing order of the column numbers.

It is obvious that the form (2) is antisymmetric with respect to the rows and with respect to the first p columns. Usually we shall consider Expression (2) as having the σ_j^i as homogeneous differential forms of degree 1. In this case the form (2) will not change upon permuting any two of the columns made up of differential forms.

If the σ_j^i are forms with operator coefficients, then in each term of the "determinant" (2) there enters a product of operators. In such a case we shall always assume that the operators do not act on the coefficients of the following factors, i.e., for example,

$$\left(a(x)\frac{\partial}{\partial x_1}dx_1\right)\wedge\left(b(x)\frac{\partial}{\partial x_2}dx_2\right)=a(x)\,b(x)\frac{\partial^2}{\partial x_1\partial x_2}dx_1\wedge dx_2.$$

Definition. Introduce the notation:

$$D_j^{(i)}=\sum_{l=1}^{k+1}\sum_{s=1}^{n+1}\beta_l^{(i)}\frac{\partial}{\partial u_l^s}du_j^s,\qquad \beta_l^{(i)}=(\beta^{(i)},u_l),\tag{3}$$

$$\tau_{p,q}(\beta^{(1)},\ldots,\beta^{(m)})=\begin{vmatrix}\beta_1^{(1)}\ldots\beta_1^{(m)} & d\beta_1^{(1)}\ldots d\beta_1^{(1)}\ldots d\beta_1^{(m)}\ldots d\beta_1^{(m)} & D_1^{(1)}\ldots D_1^{(1)}\ldots D_1^{(m)}\ldots D_1^{(m)}\\ \cdots & \cdots & \cdots\\ \beta_{k+1}^{(1)}\ldots\beta_{k+1}^{(m)} & \underbrace{d\beta_{k+1}^{(1)}\ldots d\beta_{k+1}^{(1)}}_{p_1\text{ times}}\ldots\underbrace{d\beta_{k+1}^{(m)}\ldots d\beta_{k+1}^{(m)}}_{p_m\text{ times}} & \underbrace{D_{k+1}^{(1)}\ldots D_{k+1}^{(1)}}_{q_1\text{ times}}\ldots\underbrace{D_{k+1}^{(m)}\ldots D_{k+1}^{(m)}}_{q_m\text{ times}}\end{vmatrix}^{\wedge},\tag{4}$$

where $p = (p_1, \ldots, p_m)$, $q = (q_1, \ldots, q_m)$.

The differential form $\omega_{k+1-m}(\beta^{(1)}, \ldots, \beta^{(m)})$ we define by the following formula:

$$\omega_{k+1-m}(\beta^{(1)},\ldots,\beta^{(m)})=(-1)^{k+1-m}\sum_{|p+q|=k+1-m}c_{p,q}\tau_{p,q}(\beta^{(1)},\ldots,\beta^{(m)})\,\varphi,\tag{5}$$

where

$$c_{p,q}=\frac{1}{|q|!}\cdot\frac{(p+q)!}{p!\,q!}.\tag{6}$$

Here we denote $|p| = p_1+\ldots+p_m$, $p! = p_1!\ldots p_m!$. The sum is extended over all p, q for which $|p+q| = k+1-m$.

§ 4. THE MAIN RESULTS FORMULATED

THEOREM 1. The differential form

$$\omega_{k+1-m}=(-1)^{k+1-m}\sum_{|p+q|=k+1-m}c_{p,q}\tau_{p,q}\varphi\tag{1}$$

can be dropped from $\widetilde{L}'^{(m)}$ to $L'^{(m)}$, $m = 1, \ldots, k+1$.

In order to formulate the following results we introduce new definitions.

First we introduce the operator ∇ which relates to the form $\omega_{k+2-m}(\beta^{(1)}, \ldots, \beta^{(m-1)})$ in $L'^{(m-1)}$ the form $\nabla\omega_{k+2-m}$ of the same degree in $L'^{(m)}$. With this we give the mapping $\pi_i : L'^{(m)} \to L'^{(m-1)}$ relating to each $(h; \beta^{(1)}, \ldots, \beta^{(m)}) \in L'^{(m)}$ the element $(h; \beta^{(1)}, \ldots, \hat{\beta}^{(i)}, \ldots, \beta^{(m)}) \in L'^{(m-1)}$ $(i = 1, \ldots, m)$. It induces the inverse mapping $\pi_i{}^*$ of the differential forms:

$$(\pi_i^*\omega_{k+2-m})(\beta^{(1)},\ldots,\beta^{(m)})=\omega_{k+2-m}(\beta^{(1)},\ldots,\widehat{\beta^{(i)}},\ldots,\beta^{(m)})\tag{2}$$

$(i = 1, \ldots, m)$. The operator ∇ we define so that:

$$\nabla\omega_{k+2-m} = \sum_{i=1}^{m}(-1)^{i-1}\pi_i^*\omega_{k+2-m}. \tag{3}$$

Now we construct the differential forms K_{k+1-m} and T_{k+1-m} in $L'^{(m)}$.

The differential form T_{k+1-m} we define by the formula

$$T_{k+1-m} = (-1)^k \sum_{|p+q|=k+2-m} \sum_{l=1}^{m} (-1)^{l-1} c_{p,q} \overset{l}{\tau}_{p,q}\varphi, \tag{4}$$

where $\overset{l}{\tau}_{p,q}$ is the differential form of the type (4) § 3 not containing the column $\beta^{(l)}$ ($l = 1, \ldots, m$).

We proceed now to define the form K_{k+1-m}. Put

$$B_s^{(i)} = \sum_{\nu=1}^{k+1} \beta_\nu^{(i)} \frac{\partial}{\partial u_\nu^s}, \quad d_s^{s'} = \sum_{\nu=1}^{k+1} \frac{\partial}{\partial u_\nu^s} du_\nu^{s'}, \quad d_u = \sum_{s=1}^{n+1} d_s^s$$

($i = 1, \ldots, m$; $s, s' = 1, \ldots, n+1$). Construct with $B_s^{(i)}$ and $d_s^{s'}$ the following differential forms of second degree:

$$E_j^{(i)} = d_u \wedge D_j^{(i)} = \sum_{s_1, s_2=1}^{n+1} B_{s_1}^{(i)} d_{s_2}^{s_2} \wedge du_j^{s_1}, \tag{5}$$

$$F_j^{(i)} = \sum_{s_1, s_2=1}^{n+1} B_{s_2}^{(i)} d_{s_2}^{s_1} \wedge du_j^{s_2}. \tag{6}$$

Now we introduce differential forms $\rho_{p,q}$. Let $D^{(\alpha_1)}, \ldots, D^{(\alpha_{|q|})}$ be the last $|q|$ columns in the determinant (4) § 3 specifying the form $\tau_{p,q}$. Denote by $\overset{i}{\rho}_{p,q}$ the differential form which arises out of $\tau_{p,q}$ by replacing the column $D^{(\alpha_i)}$ by the column $E^{(\alpha_i)} - F^{(\alpha_i)}$. Further, put

$$\rho_{p,q} = \frac{1}{|q|} \sum_{i=1}^{|q|} (-1)^{i-1} \overset{i}{\rho}_{p,q}. \tag{7}$$

We define the form K_{k+1-m} by the following formula

$$K_{k+1-m} = (-1)^{k+1-m} \sum_{|p+q|=k+1-m} c_{p,q}\, \rho_{p,q}\varphi. \tag{8}$$

In § 6 it will be shown that the differential forms K_{k+1-m} and T_{k+1-m} which we defined in $\widetilde{L}'^{(m)}$ can be dropped from $\widetilde{L}'^{(m)}$ to $L'^{(m)}$.

<u>THEOREM 2.</u> The differential form ω_{k+1-m} is expressible in the following way:

$$d\omega_{k+1-m} = \nabla\omega_{k+2-m} + K_{k+1-m} + T_{k+1-m} \tag{9}$$

($m = 1, \ldots, k+1$). When $m = 1$ we take it, in this equality, that $\nabla\omega_{k+2-m} = 0$.

For each fixed point $x \in R^{n+1}\setminus 0$ we denote by G_x the submanifold of all $(k+1)$-dimensional subspaces $h \in G_{n+1,k+1}$ passing through x. Suppose that $M_x^{(m)}$ is the submanifold in $L'^{(m)}$ consisting of all $(h; \beta^{(1)}, \ldots, \beta^{(m)}) \in L'^{(m)}$ such that $h \in G_x$, $(\beta^{(i)}, x) = 1$ ($i = 1, \ldots, m$).

<u>THEOREM 3.</u> $T_{k+1-m} = 0$ on each submanifold $M_m^{(m)}$ ($m = 1, \ldots, k+1$).

<u>THEOREM 4.</u> If the function $\varphi(u)$ is representable in the form of an integral

$$\varphi(u) = \frac{1}{2} \int_{S^k} f(ut)\,\sigma(t), \tag{10}$$

then $K_{k+1-m} = 0$ ($m = 1, \ldots, k+1$).

COROLLARY. If the function $\varphi(u)$ is representable in the form of the integral (10), then in each $M_x^{(m)} \subset L'^{(m)}$ there holds the equality

$$d\omega_{k+1-m} = \nabla\omega_{k+2-m} \ (m = 1, \ldots, k+1), \tag{11}$$

and in particular

$$d\omega_k = 0. \tag{12}$$

THEOREM 5. Let $\varphi(u)$ be an infinitely differentiable function in space E of $(k+1)$-frames satisfying the condition

$$\varphi(u\mu) = |\det\mu|^{-1}\varphi(u)$$

for an arbitrary $\mu \in GL(k+1, R)$, where $k < n-1$. In order that it be representable in the form (10) where $f(x)$ is an infinitely differentiable function in $R^{n+1}\setminus 0$ satisfying the condition $f(\lambda x) = |\lambda|^{-k-1}f(x)$ $(\lambda \neq 0)$, it is necessary and sufficient that $K_k = 0$.

Accordingly, on the strength of Theorems 4 and 5, the vanishing of the form K_k implies the vanishing of all the remaining forms K_i $(i = 0, 1, \ldots, k-1)$.†

We construct, for the form $\omega_k(\beta)$ the differential form on G_x. We give, for this, the section $s : G_x \to M_x^{(1)}$. It induces the inverse mapping s^* of differential forms. That is, we get a differential form $s^*\omega_k$ on G_x.

THEOREM 6. Let k be an even number. Then if a function is representable by the integral (10), then the following inverse formula holds

$$\int_\gamma s^*\omega_k = c_\gamma f(x) \tag{13}$$

where $\gamma \subset G_x$ is an arbitrary k-dimensional cycle. Here $c_\gamma \neq 0$ if γ is an Euler cycle‡ and $c_\gamma = 0$ for all other cycles from the natural basis of homologies on G_x.

Remark. Of course, even the form $s^*\omega_k$ on G_x depends on the choice of the section s. However, from the theorems formulated above it readily follows that two forms $s_1^*\omega_k$ and $s_2^*\omega_k$, constructed on different sections, differ in precise form. Accordingly, Integral (13) does not depend on choice of the section s.

In conclusion, we introduce explicit expressions for ω_i, K_i, T_i in the case $k = 1$:

$$\omega_0 = (\beta_1^{(1)}\beta_2^{(2)} - \beta_2^{(1)}\beta_1^{(2)})\varphi;$$

$$\omega_1 = -\sum_{s=1}^{n+1}\left(\beta_1\frac{\partial\varphi}{\partial u_1^s} + \beta_2\frac{\partial\varphi}{\partial u_2^s}\right)(\beta_1 du_2^s - \beta_2 du_1^s) - (\beta_1 d\beta_2 - \beta_2 d\beta_1)\varphi;$$

$$K_0 = 0;\ K_1 = -\sum_{s,t=1}^{n+1}\left(\frac{\partial^2\varphi}{\partial u_2^s\partial u_1^t} - \frac{\partial^2\varphi}{\partial u_2^t\partial u_1^s}\right)(\beta_1 du_2^s - \beta_2 du_1^s)\wedge(\beta_1 du_2^t - \beta_2 du_1^t);$$

$$T_0 = -\sum_{s=1}^{n+1}\left((\beta_1^{(1)} + \beta_1^{(2)})\frac{\partial\varphi}{\partial u_1^s} + (\beta_2^{(1)} + \beta_2^{(2)})\frac{\partial\varphi}{\partial u_2^s}\right)((\beta_1^{(2)} - \beta_1^{(1)})du_2^s - (\beta_2^{(2)} - \beta_2^{(1)})du_1^s)$$
$$- [(\beta_1^{(2)} - \beta_1^{(1)})\cdot d(\beta_2^{(1)} + \beta_2^{(2)}) - (\beta_2^{(2)} - \beta_2^{(1)})\cdot d(\beta_1^{(1)} + \beta_1^{(2)})]\varphi;$$

$$T_1 = -\sum_{s,t=1}^{n+1}\sum_{i,j=1}^{2}\beta_i\beta_j\frac{\partial^2\varphi}{\partial u_i^s\partial u_j^t}du_1^s\wedge du_2^t - 2\sum_{s=1}^{n+1}\left(\beta_1\frac{\partial\varphi}{\partial u_1^s} + \beta_2\frac{\partial\varphi}{\partial u_2^s}\right)(d\beta_1\wedge du_2^s$$
$$- d\beta_2\wedge du_1^s) - 2\varphi\cdot d\beta_1\wedge d\beta_2.$$

† The proof of Theorem 5 is not carried through in this article. The necessity of the hypothesis of Theorem 5 immediately follows from Theorem 4. The sufficiency of the hypothesis can be established by the consideration in the analogous theorem for the case of affine space proved in [1].

‡ That is, the manifold of all $(k+1)$-dimensional subspaces contained in a fixed $(k+2)$-dimensional subspace.

§5. PROOF OF THEOREM 1

The form ω_{k+1-m} is a sum of differential forms of the type

$$\omega^{r}_{k+1-m} = \sum_{p+q=r} c_{p,q}\tau_{p,q}\varphi, \qquad (1)$$

where $r = (r_1, \ldots, r_m)$ is a multiple index such that $|r| = k+1-m$. Hence it suffices that we show the theorem true for each term in (1).

Consider the fibering $\widetilde{L}'^{(m)} \to L'^{(m)}$. It is the principal fibering of the group $GL_0(k+1, R)$: two elements $(u, \beta^{(1)}, \ldots, \beta^{(m)})$ and $(u', \beta'^{(1)}, \ldots, \beta^{(m)})$ from $\widetilde{L}'^{(m)}$ lie in the same fiber if and only if they are transformed into one another by some element $\mu \in GL_0(k+1, R)$, i.e.

$$u' = u\mu, \quad \beta'^{(i)} = \beta^{(i)}\mu \quad (i = 1, \ldots, m). \qquad (2)$$

where $\beta^{(i)}\mu \in (R')^{k+1}$ is a vector with coordinates $\beta'^{(i)}_j = \sum_{l=1}^{k+1} \beta^{(i)}_l \mu^l_j$.

First we show that ω^r_{k+1-m} is an invariant form in $\widetilde{L}^{(m)}$. Indeed, under the Transformation (2) the operator $D^{(i)} = (D^{(i)}_1, \ldots, D^{(i)}_{k+1})$ transforms as a vector from $(R')^{k+1}$. Hence the determinant in the expression for $\tau_{p,q}$ is multiplied by $\det\mu$. Likewise the function $\varphi(u)$ under this transformation is multiplied by $(\det\mu)^{-1}$. Accordingly, the form ω^r_{k+1-m} is preserved under transformation (2).

To prove the theorem we use the following simple assertion:

<u>LEMMA 1.</u> If we are given a principal fibering $A \to B$ and suppose $\omega(a; \eta_1, \ldots, \eta_r)$ $(a \in A, \eta_i \in T_a)$ is an invariant differential form on A, then in order that this form be dropped from A to B it is necessary and sufficient that the following condition:

$$\omega(a; \eta_1 + \eta_1', \ldots, \eta_r + \eta_r') = \omega(a, \eta_1, \ldots, \eta_r)$$

be fulfilled for any $\eta_i' \in T_a$ lying in the subspace tangent to a fiber.

Apply this lemma to the form ω^r_{k+1-m}. Let $a = (u, \beta^{(1)}, \ldots, \beta^{(m)})$ be an arbitrary point from $\widetilde{L}'^{(m)}$. It is obvious that any vector tangent to a fiber at the point a has the form $(u\,d\mu, \beta^{(1)}d\mu, \ldots, \beta^{(m)}d\mu)$.

Hence, to demonstrate the theorem, it is sufficient to verify that the form ω^r_{k+1-m} is conserved upon replacing du by $du + u\,d\mu$ and $d\beta^{(i)}$ by $d\beta^{(i)} + \beta^{(i)}d\mu$ $(i = 1, \ldots, m)$. We show this assertion first. To begin with, we transform the expression for ω^r_{k+1-m}. We derive the following transform:

$$\tau^{s_1,\ldots,s_{|q|}}_{p_1,\ldots,p_m} = \begin{vmatrix} \beta^{(1)}_1 & \ldots & \beta^{(m)}_1 & d\beta^{(1)}_1 & \ldots & d\beta^{(1)}_1 & \ldots & d\beta^{(m)}_1 & \ldots & d\beta^{(m)}_1 & du^{s_1}_1 & \ldots & du^{s_{|q|}}_1 \\ \cdot & \cdot & \cdot & \cdot & \cdot & \cdot & \cdot & \cdot & \cdot & \cdot & \cdot & \cdot & \cdot \\ \beta^{(1)}_{k+1} & \ldots & \beta^{(m)}_{k+1} & \underbrace{d\beta^{(1)}_{k+1} & \ldots & d\beta^{(1)}_{k+1}}_{p_1} & \ldots & \underbrace{d\beta^{(m)}_{k+1} & \ldots & d\beta^{(m)}_{k+1}}_{p_m} & du^{s_1}_{k+1} & \ldots & du^{s_{|q|}}_{k+1} \end{vmatrix}^{\wedge}. \qquad (3)$$

We get, further

$$B^{(i)}_s = \sum_{\nu=1}^{k+1} \beta^{(i)}_\nu \frac{\partial}{\partial u^s_\nu}; \qquad (4)$$

and hence $D^{(i)}_j = \sum_{s=1}^{n+1} B^{(i)}_s du^s_j$.

Finally, setting $\xi_i = (\xi_i^{(1)}, \ldots, \xi_i^{(m)})$ we are led to the following polynomial:

$$P^q(\xi_1, \ldots, \xi_{|q|}) = \sum_{(i_1,\ldots,i_{|q|})} \xi_{i_1}^{(a_1)} \xi_{i_2}^{(a_2)} \ldots \xi_{i_{|q|}}^{(a_{|q|})}, \tag{5}$$

where the first q_1 indices a_ν are equal to 1, the following q_2 indices a_ν are equal to 2, etc., q_m indices are equal to m. The summation is extended over all permutations $(i_1, \ldots, i_{|q|})$ of the numbers $1, 2, \ldots, |q|$. (That is, each individual term enters the polynomial P^q with the coefficient $q! = q_1! \ldots q_m!$.)

In this notation the form ω^r_{k+1-m} is expressible, as may readily be checked, as follows:

$$\omega^r_{k+1-m} = \sum_{p+q=r} \sum_{s_i} \frac{1}{|q|!} c_{p,q} P^q(B_{s_1}, \ldots, B_{s_{|q|}}) \varphi \cdot \tau^{s_1,\ldots,s_{|q|}}_{p_1,\ldots,p_m}. \tag{6}$$

The summation is extended over the multiple indices p, q connected by the relation $p+q = r$ and over indices s_i, running through the values from 1 to $n+1$ $(i = 1, \ldots, |q|)$.

Now, in the form ω^r_{k+1-m} we replace du by $d\mu + u\,d\mu$ and $d\beta^{(i)}$ by $d\beta^{(i)} + \beta^{(i)} d\mu$ $(i = 1, \ldots, m)$, where $d\mu = \| d\mu^i_j \|$. As a result of these replacements, we are led to the following form:

$$\begin{aligned}\tilde{\omega}^r_{k+1-m} = \sum_{p+q=r} \sum_{p'+p''=p} \sum_{l=1}^{|q|} \sum_{s_l} c_{p,q} \frac{p!}{p'!\,p''!} \frac{1}{|p'|!\, l!\,(|q|-l)!} \\ \times P^{p'}(\beta_{t_{l+1}}, \ldots, \beta_{t_{l+|p'|}}) \cdot u^{s_1}_{t_1} \ldots u^{s_l}_{t_l} \cdot P^q(B_{s_1}, \ldots, B_{s_{|q|}}) \varphi \cdot \tau^{s_{l+1},\ldots,s_{|q|};t_1,\ldots,t_{l+|p'|}}_{p''_1,\ldots,p''_m}.\end{aligned} \tag{7}$$

We denote here

$$\tau^{s_{l+1},\ldots,s_{|q|};t_1,\ldots,t_{l+|p'|}}_{p''_1,\ldots,p''_m}$$

$$= \begin{vmatrix} \beta_1^{(1)} & \ldots \beta_1^{(m)} & d\beta_1^{(1)} \ldots d\beta_1^{(1)} & \ldots d\beta_1^{(m)} \ldots d\beta_1^{(m)} & du_1^{s_{l+1}} \ldots du_1^{s_{|q|}} d\mu_1^{t_1} \ldots d\mu_1^{t_{l+|p'|}} \\ \ldots & \ldots & \ldots & \ldots & \ldots \\ \beta_{k+1}^{(1)} & \ldots \beta_{k+1}^{(m)} & \underbrace{d\beta_{k+1}^{(1)} \ldots d\beta_{k+1}^{(1)}}_{p''_1} & \ldots \underbrace{d\beta_{k+1}^{(m)} \ldots d\beta_{k+1}^{(m)}}_{p''_m} & du_{k+1}^{s_{l+1}} \ldots du_{k+1}^{s_{|q|}} d\mu_{k+1}^{t_1} \ldots d\mu_{k+1}^{t_{l+|p'|}} \end{vmatrix}^{\wedge}.$$

Expression (7) can be simplified on the basis of the following two lemmas:

<u>LEMMA 2.</u> If the function $\varphi(u)$ on E satisfies the condition $\varphi(u\mu) = (\det \mu)^{-1} \varphi(u)$ for an arbitrary $\mu \in GL_0(k+1, R)$, then the following relation holds:

$$\sum_{s_1,\ldots,s_l} u^{s_1}_{t_1} \ldots u^{s_l}_{t_l} P^q(B_{s_1}, \ldots, B_{s_{|q|}}) \varphi = (-1)^l \frac{|q|!}{(|q|-l)!} P^q(\beta_{t_1}, \ldots, \beta_{t_l}, B_{s_{l+1}}, \ldots, B_{s_{|q|}}) \varphi. \tag{8}$$

<u>Proof of the Lemma.</u> We demonstrate the assertion first for the case $l = 1$, i.e., we show that

$$\sum_{s_1=1}^{n+1} u^{s_1}_{t_1} P^q(B_{s_1}, \ldots, B_{s_{|q|}}) \varphi = -|q| P^q(\beta_{t_1}, B_{s_2}, \ldots, B_{s_{|q|}}) \varphi. \tag{9}$$

From the condition that $\varphi(u\mu) = (\det \mu)^{-1} \varphi(u)$ there follows

$$\sum_{s_1=1}^{n+1} u^{s_1}_{t_1} \frac{\partial \varphi}{\partial u^{s_1}_{\nu}} = -\delta^{\nu}_{t_1} \varphi,$$

where $\delta^{\nu}_{t_1}$ is the Kronecker symbol. Hence, by induction on $|q|$ we readily obtain

$$\sum_{s_1=1}^{n+1} u^{s_1}_{t_1} \frac{\partial^{|q|} \varphi}{\partial u^{s_1}_{\nu_1} \ldots \partial u^{s_{|q|}}_{\nu_{|q|}}} = -\delta^{\nu_1}_{t_1} \frac{\partial^{|q|-1} \varphi}{\partial u^{s_2}_{\nu_2} \ldots \partial u^{s_{|q|}}_{\nu_{|q|}}} - \sum_{i=2}^{|q|} \delta^{\nu_i}_{t_1} \frac{\partial^{|q|-1} \varphi}{\partial u^{s_2}_{\nu_2} \ldots \partial u^{s_i}_{\nu_1} \ldots \partial u^{s_{|q|}}_{\nu_{|q|}}}. \tag{10}$$

Multiply both sides of Equality (10) by $\beta^{(l_1)}_{\nu_1}\beta^{(l_2)}_{\nu_2}\ldots\beta^{(l_{|q|})}_{\nu_{|q|}}$ and sum with respect to $\nu_1, \ldots, \nu_{|q|}$.

We get

$$\sum_{s_1=1}^{n+1} u^{s_1}_{t_1} B^{(l_1)}_{s_1} B^{(l_2)}_{s_2} \ldots B^{(l_{|q|})}_{s_{|q|}}\varphi = -\beta^{(l_1)}_{t_1} B^{(l_2)}_{s_2}\ldots B^{(l_{|q|})}_{s_{|q|}}\varphi - \sum_{i=2}^{|q|} \beta^{(l_i)}_{t_1} B^{(l_2)}_{s_2}\ldots B^{(l_1)}_{s_i}\ldots B^{(l_{|q|})}_{s_{|q|}}\varphi. \tag{11}$$

Let $(a_1, \ldots, a_{|q|})$ be a sequence of indices among which q_ν indices are equal to ν ($\nu = 1, \ldots, m$). Summing Equality (11) with respect to all $|q|!$ permutations $(l_1, \ldots, l_{|q|})$ of this sequence, we obtain the required equality (9).

In order to demonstrate the lemma for arbitrary l we remark that $P^q(\beta_{t_1}, B_{s_2}, \ldots, B_{s_{|q|}})$ is a linear combination of the polynomials $\beta^{(i)}_{t_1} P^{q_1, \ldots, q_i - 1, \ldots, q_m}(B_{s_2}, \ldots, B_{s_{|q|}})$. Therefore, repeatedly applying the relation (9) just demonstrated, we get the required equality (8).

<u>LEMMA 3.</u> For the polynomials $P^q(\xi_1, \ldots, \xi_{|q|})$ there holds the following relation:

$$P^q(\xi_1, \ldots, \xi_l, \eta_1, \ldots, \eta_{|q|-l}) = \sum_{|q'|=l} \frac{q!}{q'!\,(q-q')!} P^{q'}(\xi_1, \ldots, \xi_l) P^{q-q'}(\eta_1, \ldots, \eta_{|q|-l}) \tag{12}$$

(here $q'! = q'_1! \ldots q'_m!$).

For the proof it suffices, on left and right sides in the equality, to compare the coefficients with which the fixed monomial $\xi_1^{(a_1)} \ldots \xi^{(a_l)}\eta^{(b_1)} \ldots \eta^{(b_{|q|-l})}_{|q|-l}$ enters.

On the basis of the relations demonstrated in the lemmas, expression (7) for $\tilde{\omega}^r_{k+1-m}$ is brought to the form:

$$\begin{aligned}\tilde{\omega}^r_{k+1-m} = r! \sum_{p+q=r} \sum_{\substack{p'+p''=p\\ q'+q''=q}} \sum_{s_i} (p'!\,p''!\,q'!\,q''!\,(|q''|!)^2 |p'|!\,|q'|!)^{-1} \\ \times (-1)^{|q'|} P^{p'}(\beta_{t_{|q'|+1}}, \ldots, \beta_{t_{|q'|+|p'|}}) P^{q'}(\beta_{t_1}, \ldots \beta_{t_{|q'|}}) \\ \times P^{q''}(B_{s_1}, \ldots, B_{s_{|q''|}})\varphi \cdot \tau^{s_1, \ldots, s_{|q''|}; t_1, \ldots, t_{|q'|+|p'|}}_{p''_1, \ldots, p''_m}.\end{aligned} \tag{13}$$

Carrying out the summation here for $t_1, \ldots, t_{|q'|+|p'|}$ we get

$$\tilde{\omega}^r_{k+1-m} = r! \sum_{p'+p''+q'+q''=r} \sum_{s_i} (-1)^{|q'|} (p'!\,p''!\,q'!\,q''!\,(|q''|!)^2)^{-1} \times P^{q''}(B_{s_1}, \ldots, B_{s_{|q''|}})\varphi \cdot \tilde{\tau}^{s_1, \ldots, s_{|q''|}}_{p'', p'+q'}, \tag{14}$$

where we denote:

$$\tilde{\tau}^{s_1, \ldots, s_{|q''|}}_{p'', p'+q'} = \begin{vmatrix} \beta^{(1)}_1 \ldots \beta^{(m)}_1 & d\beta^{(1)}_1 \ldots d\beta^{(1)}_1 \ldots d\beta^{(m)}_1 \ldots d\beta^{(m)}_1 & \sigma^{(1)}_1 \ldots \sigma^{(1)}_1 \ldots \sigma^{(m)}_1 \ldots \sigma^{(m)}_1 \\ \cdots & \cdots & \cdots \\ \beta^{(1)}_{k+1} \ldots \beta^{(m)}_{k+1} & \underbrace{d\beta^{(1)}_{k+1} \ldots d\beta^{(1)}_{k+1}}_{p''_1} \ldots \underbrace{d\beta^{(m)}_{k+1} \ldots d\beta^{(m)}_{k+1}}_{p''_m} & \underbrace{\sigma^{(1)}_{k+1} \ldots \sigma^{(1)}_{k+1}}_{p'_1+q'_1} \ldots \underbrace{\sigma^{(m)}_{k+1} \ldots \sigma^{(m)}_{k+1}}_{p'_m+q'_m} \end{vmatrix}^{\wedge},$$

$$\sigma^{(i)}_j = \sum_{t=1}^{k+1} \beta^{(i)}_t d\mu^t_j.$$

For the proof it is sufficient to verify that in the expression for $\tilde{\omega}^r_{k+1-m}$ all terms depending on $d\mu$ cancel. For this, we calculate the coefficient for

$$\sum_{s_i} P^{q''}(B_{s_1}, \ldots, B_{s_{|q''|}})\varphi \cdot \tilde{\tau}^{s_1, \ldots, s_{|q''|}}_{p'', p'+q'}.$$

It is equal to $[r!/p''!q''!(|q''|!)^2]\Sigma(-1)^{|q'|}(1/p'!q'!)$ where the sum extends over all p', q' for which p' + q' is fixed.

Clearly, if $|p'+q'| \neq 0$ this sum is zero. The theorem is proved.

Remark 1. In the proof of Theorem 1, we nowhere used the fact that $|p+q| = k+1-m$. Consequently there has been proved a much more general assertion.

Denote by $\tau_{p,q}^{l_1,\dots,l_{m'}}$ $(0 \leq m' \leq m)$ the differential form defined by the same formulas (4) §3 as $\tau_{p,q}$ with the sole difference that the determinant (4) does not contain the columns $\beta^{(l_1)}, \dots, \beta^{(l_{m'})}$ but respectively the more general number $|p+q|$ of columns $d\beta^{(i)}$ and $D^{(j)}$ in Determinant (4) equal to $k+1-m+m'$. Then the differential form

$$\sum_{p+q=r} c_{p,q}\, \tau_{p,q}^{l_1,\dots,l_{m'}}\, \varphi,$$

where $r = (r_1, \dots, r_m)$ is any multiple index such that $|r| = k+1-m+m'$ can be dropped from $\widetilde{L}'^{(m)}$ to $L'^{(m)}$.

Remark 2. By reasoning analogous to that introduced in the proof of Theorem 1, we can also demonstrate the following uniqueness theorem:

Consider all possible differential forms in $\widetilde{L}'^{(m)}$ of the following type:

$$\sum_{p+q=r} a_{p,q}\tau_{p,q}\varphi, \tag{16}$$

where $\tau_{p,q}$ are forms defined by Formula (4) §3, $a_{p,q}$ are constant coefficients; the summation is extended over all p, q for which $p+q = r$ $(|r| = k+1-m)$.

THEOREM. For each multiple index $r = (r_1, \dots, r_m)$ the form ω^r_{k+1-m} is, up to a constant multiple, the unique form of type (16) which can be dropped from $\widetilde{L}'^{(m)}$ to $L'^{(m)}$.

§6. PROOF OF THEOREM 2

The form ω_{k+1-m} is given by the following formula (see (6) §5):

$$\omega_{k+1-m} = (-1)^{k+1-m} \sum_{s,|p+q|=k+1-m} \frac{1}{|q|!} c_{p,q} P^q(B_{s_1}, \dots, B_{s_{|q|}})\,\varphi \cdot \tau_{p_1,\dots,p_m}^{s_1,\dots,s_{|q|}}. \tag{1}$$

We recall the notation: $B_s = (B_s^{(1)}, \dots, B_s^{(m)})$, $B_s^{(i)} = \sum_{\nu=1}^{k+1} \beta_\nu^{(i)}(\partial/\partial u_\nu^s)$, the polynomial $P^q(\xi_1, \dots, \xi_{|q|})$ is defined by the formula (5) §5, and

$$\tau_{p_1,\dots,p_m}^{s_1,\dots,s_{|q|}} = \begin{vmatrix} \beta_1^{(1)} \dots \beta_1^{(m)} & \underbrace{d\beta_1^{(1)} \dots d\beta_1^{(1)}} & \dots & \underbrace{d\beta_1^{(m)} \dots d\beta_1^{(m)}} & du_1^{s_1} \dots du_1^{s_{|q|}} \\ \dots & \dots & \dots & \dots & \dots \\ \dots & \dots & \dots & \dots & \dots \\ \beta_{k+1}^{(1)} \dots \beta_{k+1}^{(m)} & \underbrace{d\beta_{k+1}^{(1)} \dots d\beta_{k+1}^{(1)}}_{p_1} & \dots & \underbrace{d\beta_{k+1}^{(m)} \dots d\beta_{k+1}^{(m)}}_{p_m} & du_{k+1}^{s_1} \dots du_{k+1}^{s_{|q|}} \end{vmatrix}$$

In order to write the differential form ω_{k+1-m} we use the following notation

$$B_s'^{(i)} = \sum_{\nu=1}^{k+1} \frac{\partial}{\partial u_\nu^s} d\beta_\nu^{(i)}, \quad d_{s'}^s = \sum_{\nu=1}^{k+1} \frac{\partial}{\partial u_\nu^{s'}} du_\nu^s, \quad d_u = \sum_{s=1}^{n+1} d_s^s.$$

Further, we agree to denote by $\tau^{l_{s_1,\ldots,s_{|q|}}}_{p_1,\ldots,p_l+1,\ldots,p_m}$ the differential form which arises from $\tau^{s_1,\ldots,s_{|q|}}_{p_1,\ldots,p_m}$ by removing the column $\beta^{(l)}$ and adding the column $d\beta^{(l)}$. In this notation, $d\omega_{k+1-m}$ will be expressed as follows:

$$
\begin{aligned}
d\omega_{k+1-m} = (-1)^{k+1-m} \sum_{s,|p+q|=k+1-m} \frac{1}{|q|!} c_{p,q} \Big\{ |q| P^{q}(B'_{s_1}, B_{s_2}, \ldots, B_{s_{|q|}}) \varphi \\
\times \tau^{s_1,\ldots,s_{|q|}}_{p_1,\ldots,p_m} + P^{q}(B_{s_1}, \ldots, B_{s_{|q|}}) \cdot d_u\varphi \wedge \tau^{s_1,\ldots,s_{|q|}}_{p_1,\ldots,p_m} \\
+ P^{q}(B_{s_1}, \ldots, B_{s_{|q|}}) \varphi \cdot \sum_{l=1}^{m} (-1)^{m-l} \tau^{l_{s_1,\ldots,s_{|q|}}}_{p_1,\ldots,p_l+1,\ldots,p_m} \Big\}.
\end{aligned}
\tag{2}
$$

We shall transform this expression. We have, by Lemma 3

$$
P^{q}(B'_{s_1}, B_{s_2}, \ldots, B_{s_{|q|}}) = \sum_{l=1}^{m} q_l P^{q_1,\ldots,q^l-1,\ldots,q_m}(B_{s_2}, \ldots, B_{s_{|q|}}) B'^{(l)}_{s_1}.
$$

Substitute this expression into Formula (2) and, in the resulting sum, make the changes of indices: q_l to q_l+1 and p_l to p_l-1. There results

$$
\begin{aligned}
d\omega_{k+1-m} = (-1)^{k+1-m} \sum_{s,p,q} \frac{1}{(|q|+1)!} c_{p,q} P^{q}(B_{s_2}, \ldots, B_{s_{|q|+1}}) \\
\times \Big(\sum_{l=1}^{m} p_l B'^{(l)}_{s_1} \varphi \wedge \tau^{s_1,\ldots,s_{|q|+1}}_{p_1,\ldots,p_l-1,\ldots,p_m} + (|q|+1) d^{s_1}_{s_1} \varphi \wedge \tau^{s_1,\ldots,s_{|q|+1}}_{p_1,\ldots,p_m} \Big) \\
+ (-1)^{k+1-m} \sum_{s,p,q} \Big\{ \frac{1}{|q|!} c_{p,q} P^{q}(B_{s_1}, \ldots, B_{s_{|q|}}) \varphi \cdot \sum_{l=1}^{m} (-1)^{m-l} \tau^{l_{s_1,\ldots,s_{|q|}}}_{p_1,\ldots,p_l+1,\ldots,p_m} \Big\}.
\end{aligned}
\tag{3}
$$

We now establish a lemma.

LEMMA 4. **The following relation holds:**

$$
\sum_{l=1}^{m} p_l B'^{(l)}_{s_1} \tau^{s_1,\ldots,s_{|q|+1}}_{p_1,\ldots,p_l-1,\ldots,p_m} = -\sum_{\nu=1}^{|q|+1} d^{s_\nu}_{s_1} \wedge \tau^{s_1,\ldots,\hat{s}_\nu,\ldots,s_{|q|+1}}_{p_1,\ldots,p_m} + \sum_{l=1}^{m} (-1)^{m-l} B^{(l)}_{s_1} \tau^{l_{s_1,\ldots,s_{|q|+1}}}_{p_1,\ldots,p_m}.
\tag{4}
$$

To demonstrate relation (4), consider the form

$$
\begin{vmatrix}
\beta^{(1)}_i & \ldots & \beta^{(m)}_i & d\beta^{(1)}_i & \ldots & d\beta^{(1)}_i & \ldots & d\beta^{(m)}_i & \ldots & d\beta^{(m)}_i & du^{s_1}_i & \ldots & du^{s_{|q|+1}}_i \\
\beta^{(1)}_1 & \ldots & \beta^{(m)}_1 & d\beta^{(1)}_1 & \ldots & d\beta^{(1)}_1 & \ldots & d\beta^{(m)}_1 & \ldots & d\beta^{(m)}_1 & du^{s_1}_1 & \ldots & du^{s_{|q|+1}}_1 \\
\beta^{(1)}_2 & \ldots & \beta^{(m)}_2 & d\beta^{(1)}_2 & \ldots & d\beta^{(1)}_2 & \ldots & d\beta^{(m)}_2 & \ldots & d\beta^{(m)}_2 & du^{s_1}_2 & \ldots & du^{s_{|q|+1}}_2 \\
\cdots & \cdots & \cdots & \cdots & \cdots & \cdots & \cdots & \cdots & \cdots & \cdots & \cdots & \cdots & \cdots \\
\beta^{(1)}_{k+1} & \ldots & \beta^{(m)}_{k+1} & \underbrace{d\beta^{(1)}_{k+1} & \ldots & d\beta^{(1)}_{k+1}}_{p_1} & \ldots & \underbrace{d\beta^{(m)}_{k+1} & \ldots & d\beta^{(m)}_{k+1}}_{p_m} & du^{s_1}_{k+1} & \ldots & du^{s_{|q|+1}}_{k+1}
\end{vmatrix}^{\wedge} = 0
$$

(The determinant is zero since its first and its (i+1)-st rows coincide.) Expanding the determinant by the elements of the first row we get

$$
\sum_{l=1}^{m} (-1)^{l-1} \beta^{(l)}_i \tau^{l_{s_1,\ldots,s_{|q|+1}}}_{p_1,\ldots,p_m} + (-1)^m \sum_{l=1}^{m} p_l d\beta^{(l)}_i \wedge \tau^{s_1,\ldots,s_{|q|+1}}_{p_1,\ldots,p_l-1,\ldots,p_m} + (-1)^m \sum_{\nu=1}^{|q|+1} du^{s_\nu}_i \wedge \tau^{s_1,\ldots,\hat{s}_\nu,\ldots,s_{|q|+1}}_{p_1,\ldots,p_m} = 0.
$$

Multiplying by $\partial/\partial u^{s_1}_i$ and summing with respect to i we get

$$
\sum_{l=1}^{m} (-1)^{l-1} B^{(l)}_{s_1} \tau^{l_{s_1,\ldots,s_{|q|+1}}}_{p_1,\ldots,p_m} + (-1)^m \sum_{l=1}^{m} p_l B'^{(l)}_{s_1} \wedge \tau^{s_1,\ldots,s_{|q|+1}}_{p_1,\ldots,p_l-1,\ldots,p_m} + (-1)^m \sum_{\nu=1}^{|q|+1} d^{s_\nu}_{s_1} \wedge \tau^{s_1,\ldots,\hat{s}_\nu,\ldots,s_{|q|+1}}_{p_1,\ldots,p_m} = 0.
$$

Hence relation (4) at once follows.

On the basis of relation (4) the expression for $d\omega_{k+1-m}$ assumes the following form:

$$d\omega_{k+1-m} = K_{k+1-m} + L_{k+1-m}, \tag{5}$$

where

$$K_{k+1-m} = (-1)^{k+1-m} \sum_{s,p,q} \frac{1}{|q|!} c_{p,q} P^q (B_{s_2}, \ldots, B_{s_{|q|+1}})\, d^{s_1}_{s_1}\varphi \wedge \tau^{s_2,\ldots,s_{|q|+1}}_{p_1,\ldots,p_m}$$

$$-(-1)^{k+1-m} \sum_{s,p,q} \sum_{\nu=1}^{|q|+1} \frac{1}{(|q|+1)!} c_{p,q} P^q (B_{s_2}, \ldots, B_{s_{|q|+1}})\, d^{s_\nu}_{s_1}\varphi \wedge \tau^{s_1,\ldots,\hat{s}_\nu,\ldots,s_{|q|+1}}_{p_1,\ldots,p_m}, \tag{6}$$

$$L_{k+1-m} = (-1)^k \sum_{l=1} (-1)^{l-1} \Big\{ \sum_{s,p,q} \frac{1}{(|q|+1)!} c_{p,q} P^q (B_{s_2}, \ldots, B_{s_{|q|+1}})\, B^{(l)}_{s_1}\varphi$$

$$\times \overset{l}{\tau}{}^{s_1,\ldots,s_{|q|+1}}_{p_1,\ldots,p_m} + \sum_{s,p,q} \frac{1}{|q|!} c_{p,q} P^q (B_{s_1}, \ldots, B_{s_{|q|}})\, \varphi \cdot \overset{l}{\tau}{}^{s_1,\ldots,s_{|q|}}_{p_1,\ldots,p_l+1,\ldots,p_m} \Big\}. \tag{7}$$

We transform the expressions for K_{k+1-m} and L_{k+1-m}. We begin with L_{k+1-m}. In the first sum interior we change the summation index q_l to $q_l - 1$ and, in the second, the index p_l to $p_l - 1$. There results

$$L_{k+1-m} = (-1)^k \sum_{l=1}^{m} \sum_{s,|p+q|=k+2-m} (-1)^{l-1} \frac{1}{|q|!} \Big\{ c'^{\,l}_{p,q} \times \sum_{\nu=1}^{|q|} P^{q_1,\ldots,q_l-1,\ldots,q_m} (B_{s_1}, \ldots, \hat{B}_{s_\nu}, \ldots, B_{s_{|q|}})\, B^{(l)}_{s_\nu}\varphi$$

$$+ c''^{\,l}_{p,q} P^q (B_{s_1}, \ldots, B_{s_{|q|}})\, \varphi \Big\} \overset{l}{\tau}{}^{s_1,\ldots,s_{|q|}}_{p_1,\ldots,p_m},$$

where

$$c'^{\,l}_{p,q} = \begin{cases} \dfrac{q_l}{p_l + q_l} c_{p,q} & \text{for } p_l + q_l \neq 0, \\ 0 & \text{for } p_l = q_l = 0; \end{cases} \qquad c''^{\,l}_{p,q} = \begin{cases} \dfrac{p_l}{p_l + q_l} c_{p,q} & \text{for } p_l + q_l \neq 0, \\ 0 & \text{for } p_l = q_l = 0. \end{cases}$$

We remark that $\sum_{\nu=1}^{|q|} P^{q_1,\ldots,q_l-1,\ldots,q_m}(B_{s_1}, \ldots, \hat{B}_{s_\nu}, \ldots, B_{s_{|q|}}) B^{(l)} = P^q(B_{s_1}, \ldots, B_{s_{|q|}})$. Hence the expression in the braces is equal to $c_{p,q}P^q(B_{s_1}, \ldots, B_{s_{|q|}})\varphi$ for $p_l + q_l \neq 0$ and 0 for $p_l = q_l = 0$. Thus

$$L_{k+1-m} = (-1)^k \sum_{l=1}^{m} \sum_{s,|p+q|=k+2-m} (-1)^{l-1} \frac{1}{|q|!} c_{p,q} P^q (B_{s_1}, \ldots, B_{s_{|q|}})\, \varphi\, \overset{l}{\tau}{}^{s_1,\ldots,s_{|q|}}_{p_1,\ldots,p_m}$$

$$+ (-1)^{k-m} \sum_{l=1}^{m} \sum_{\substack{s,|p+q|=k+2-m, \\ p_l = q_l = 0}} (-1)^{l-1} \frac{1}{|q|!} c_{p,q} P^q (B_{s_1}, \ldots, B_{s_{|q|}})\, \varphi \cdot \overset{l}{\tau}{}^{s_1,\ldots,s_{|q|}}_{p_1,\ldots,p_m}.$$

It is obvious that the second term in this expression is $\nabla\omega_{k+2-m}$. Summing with respect to s_i, we get

$$L_{k+1-m} = T_{k+1-m} + \nabla\omega_{k+2-m}, \tag{8}$$

where

$$T_{k+1-m} = (-1)^k \sum_{l=1}^{m} \sum_{|p+q|=k+2-m} (-1)^{l-1} c_{p,q} \overset{l}{\tau}_{p,q} \varphi \tag{9}$$

(the definition of the form $\overset{l}{\tau}_{p,q}$ is given following (4) § 4).

Now we simplify the form K_{k+1-m} in expression (6). After summing with s_i the first term in (6) assumes the following form: $(-1)^{k+1-m} d_u \wedge \tau_{p,q}\varphi$.

Introduce the new forms (of second degree):

$$E_j^{(i)} = d_u \wedge D_j^{(i)} = \sum_{s_1, s_2} B_{s_1}^{(i)} d_{s_2}^{s_1} \wedge du_j^{s_1}, \qquad F_j^{(i)} = \sum_{s_1, s_2} B_{s_1}^{(i)} d_{s_2}^{s_1} \wedge du_j^{s_1}.$$

Let $D^{(a_1)}, \ldots, D^{(a_{|q|})}$ be the last $|q|$ columns of the determinant $\tau_{p,q}$. Denote by $\overset{i}{\rho}{}'_{p,q}$ and $\overset{i}{\rho}{}''_{p,q}$ the differential forms which result from $\tau_{p,q}$ upon changing the column $D^{(a_i)}$ respectively to $E^{(a_i)}$ and to $F^{(a_i)}$. Further, put

$$\rho'_{p,q} = \frac{1}{|q|} \sum_{i=1}^{|q|} (-1)^{i-1} \overset{i}{\rho}{}'_{p,q}, \qquad \rho''_{p,q} = \frac{1}{|q|} \sum_{i=1}^{|q|} (-1)^{i-1} \overset{i}{\rho}{}''_{p,q}.$$

Then it is obvious that $d_u \wedge \tau_{p,q} = \rho'_{p,q}$. On the other hand, it is readily verified that the second term in (6) is equal to $-(-1)^{k+1-m} \rho''_{p,q}\varphi$. Hence

$$K_{k+1-m} = (-1)^{k+1-m} \sum_{|p+q|=k+1-m} c_{p,q} (\rho'_{p,q} - \rho''_{p,q}) \varphi. \tag{10}$$

Thus we have shown that

$$d\omega_{k+1-m} = \nabla \omega_{k+2-m} + K_{k+1-m} + T_{k+1-m}, \tag{11}$$

where the forms K_{k+1-m} and T_{k+1-m} are defined, respectively, by Formulas (10) and (9). Theorem 2 is proved.

We observe that the forms T_{k+1-m} and K_{k+1-m}, defined on $\widetilde{L}'^{(m)}$, can be dropped from $\widetilde{L}'^{(m)}$ to $L'^{(m)}$. In the case of the form T_{k+1-m} this follows from Remark 1, § 5, and in the case of the form K_{k+1-m}, from relation (11).

§ 7. PROOF OF THEOREMS 3 AND 4

Let $x \in R^{n+1} \setminus 0$. Consider the submanifold $M_x^{(m)} \subset L'^{(m)}$ consisting of the points $(h; \beta^{(1)}, \ldots, \beta^{(m)})$ where $h \in G_x$ – $(k+1)$-dimensional subspace passing through x, and $(\beta^{(i)}, x) = 1$ $(i = 1, \ldots, m)$. Theorem 3 asserts that the bounded differential form T_{k+1-m} in the submanifold $M_x^{(m)}$ is equal to zero. We recall that

$$T_{k+1-m} = (-1)^k \sum_{l=1}^{m} \sum_{s, |p+q|=k+2-m} (-1)^{l-1} \frac{1}{|q|!} c_{p,q} P^q (B_{s_1}, \ldots, B_{s_{|q|}}) \varphi \cdot \overset{l}{\tau}{}_{p_1, \ldots, p_m}^{s_1, \ldots, s_{|q|}}. \tag{1}$$

where

$$\overset{l}{\tau}{}_{p_1, \ldots, p_m}^{s_1, \ldots, s_{|q|}} = \begin{vmatrix} \beta_1^{(1)} & \ldots & \beta_1^{(m)} & d\beta_1^{(1)} & \ldots & d\beta_1^{(1)} & \ldots & d\beta_1^{(m)} & \ldots & d\beta_1^{(m)} & du_1^{s_1} & \ldots & du_1^{s_{|q|}} \\ \cdots & \cdots & \cdots & \cdots & \cdots & \cdots & \cdots & \cdots & \cdots & \cdots & \cdots & \cdots & \cdots \\ \beta_{k+1}^{(1)} & \ldots & \beta_{k+1}^{(m)} & \underbrace{d\beta_{k+1}^{(1)} & \ldots & d\beta_{k+1}^{(1)}}_{p_1} & \ldots & \underbrace{d\beta_{k+1}^{(m)} & \ldots & d\beta_{k+1}^{(m)}}_{p_m} & du_{k+1}^{s_1} & \ldots & du_{k+1}^{s_{|q|}} \end{vmatrix},$$

and where further, the column $\beta^{(l)}$ is absent from the determinant. (For the source of the rest of the notation, see § 5.)

To prove the Theorem we raise the fibering $M_x^{(m)}$ of basis G_x in the space E of frames $u = (u_1, \ldots, u_{k+1})$. The fibering obtained is denoted by $\widetilde{M}_x^{(m)}$. Accordingly, the points $\widetilde{M}_x^{(m)}$ are the $(u, \beta^{(1)}, \ldots, \beta^{(m)})$ such that the subspace spanned by u contains x and $(\beta^{(i)}, x) = 1$ $(i = 1, \ldots, m)$. We shall establish that the form T_{k+1-m} is equal to zero on $\widetilde{M}_x^{(m)}$.

By hypothesis, the point x lies in the subspace spanned by the frame u, i.e., $x = \lambda^1 u_1 + \ldots + \lambda^{k+1} u_{k+1}$. $(\beta^{(i)}, x) = 1$ signifies that

$$\lambda^1 \beta_1^{(i)} + \ldots + \lambda^{k+1} \beta_{k+1}^{(i)} = 1 \qquad (i = 1, \ldots, m), \tag{2}$$

where $\beta_j^{(i)} = (\beta^{(i)}, u_j)$.

From these equalities there follows:

$$\sum_j \lambda^j du_j^s = -\sum_j u_j^s d\lambda^j \qquad (s = 1, \ldots, n+1); \tag{3}$$

$$\sum_j \lambda^j d\beta_j^{(i)} = -\sum_j \beta_j^{(i)} d\lambda^j \qquad (i = 1, \ldots, k+1). \tag{4}$$

Take a small neighborhood of $\widetilde{M}_x^{(m)}$ in which one of the coordinates λ^i, say λ^{k+1}, differs from zero. We add to the last row of the determinant $\tau_{p_1, \ldots, p_m}^{l_{s_1, \ldots, s_{|q|}}}$ the previous rows multiplied respectively by λ^1/λ^{k+1}, ..., λ^k/λ^{k+1}. On the strength of Equalities (2), (3), (4) we get

$$\tau_{p_1, \ldots, p_m}^{l_{s_1, \ldots, s_{|q|}}} = \frac{1}{\lambda^{k+1}} \begin{vmatrix} \beta_1^{(1)} \ldots \beta_1^{(m)} & d\beta_1^{(1)} \ldots d\beta_1^{(1)} \ldots d\beta_1^{(m)} \ldots d\beta_1^{(m)} & du_1^{s_1} \ldots du_1^{s_{|q|}} \\ \ldots & \ldots & \ldots \\ \beta_k^{(1)} \ldots \beta_k^{(m)} & d\beta_k^{(1)} \ldots d\beta_k^{(1)} \ldots d\beta_k^{(m)} \ldots d\beta_k^{(m)} & du_k^{s_1} \ldots du_k^{s_{|q|}} \\ 1 \ldots 1 & \alpha^{(1)} \ldots \alpha^{(1)} \ldots \alpha^{(m)} \ldots \alpha^{(m)} & \sigma^{s_1} \ldots \sigma^{s_{|q|}} \end{vmatrix}$$

(the column $\beta^{(l)}$ is absent from the determinant). Here $\alpha^{(i)} = -\sum_j \beta_j^{(i)} d\lambda^j$, $\sigma^s = -\sum_j u_j^s d\lambda^j$.

Expand the determinant by the elements of the last row. For convenience, we introduce the following notation: Let $\tau_{p_1, \ldots, p_m}^{li_{s_1, \ldots, s_{|q|}}}$ be the cofactor of the element in the last row standing in the column of $\beta^{(i)}$, let $\tau_{p_1, \ldots, p_i - 1, \ldots, p_m}^{l_{s_1, \ldots, s_{|q|}}}$ be the cofactor of the element $\alpha^{(i)}$ $(i = 1, \ldots, m)$; and finally let $\tau_{p_1, \ldots, p_m}^{l_{s_1, \ldots, \hat{s}_i, \ldots, s_{|q|}}}$ be the cofactor of the element σ^{s_i}, $i = 1, \ldots, |q|$. Then we have

$$\tau_{p_1, \ldots, p_m}^{l_{s_1, \ldots, s_{|q|}}} = \sum_{i \neq l} = \tau_{p_1, \ldots, p_m}^{li_{s_1, \ldots, s_{|q|}}} + \sum_{i=1}^{m} p_i \alpha^{(l)} \tau_{p_1, \ldots, p_i - 1, \ldots, p_m}^{l_{s_1, \ldots, s_{|q|}}} + \sum_{i=1}^{k+1} \sigma^{s_i} \tau_{p_1, \ldots, p_m}^{l_{s_1, \ldots, \hat{s}_i, \ldots, s_{|q|}}}. \tag{5}$$

Substitute this expression into Formula (1) for T_{k+1-m}. We remark first that $\tau_{p_1, \ldots, p_m}^{li_{s_1, \ldots, s_{|q|}}} = (-1)^{i-l} \tau_{p_1, \ldots, p_m}^{l_{s_1, \ldots, s_{|q|}}}$. Hence $\sum_{l=1}^{m} \sum_{i \neq l} (-1)^{l-1} \tau_{p_1, \ldots, p_m}^{li_{s_1, \ldots, s_{|q|}}} = 0$. We note further, on the strength of Lemmas 2 and 3, that

$$\sum_{s_j=1}^{n+1} \sigma^{s_j} P^q (B_{s_1}, \ldots, B_{s_{|q|}}) \varphi = -|q| \sum_{i=1}^{m} q_i \alpha^{(i)} P^{q_1, \ldots, q_i - 1, \ldots, q_m} (B_{s_1}, \ldots \widehat{B_{s_j}}, \ldots, B_{s_{|q|}}) \varphi.$$

Hence, substituting expression (5) into the formula for T_{k+1-m} we get

$$T_{k+1-m} = \sum_{i,l=1}^{m} (-1)^{k+l-1} \alpha^{(l)} \wedge \frac{1}{\lambda^{k+1}} \Big\{ \sum_{s,p,q} \frac{1}{|q|!} c_{p,q} p_i P^q (B_{s_1}, \ldots, B_{s_{|q|}}) \varphi \cdot \tau_{p_1, \ldots, p_i - 1, \ldots, p_m}^{l_{s_1, \ldots, s_{|q|}}}$$
$$- \sum_{s,p,q} \frac{|q|}{(|q|-1)!} c_{p,q} q_i P^{q_1, \ldots, q_i - 1, \ldots, q_m} (B_1, \ldots, B_{s_{|q|-1}}) \varphi \cdot \tau_{p_1, \ldots, p_m}^{l_{s_1, \ldots, s_{|q|-1}}} \Big\}.$$

It is easy to confirm that the terms standing in the braces differ only in sign. For this it is enough in the first sum to change index p_i into p_i+1 and in the second the index q_i into q_i+1. Thus it is shown that $T_{k+1-m} = 0$ on $\tilde{M}_x^{(m)}$.

Proof of Theorem 4. Let the function $\varphi(u)$ be presented in integral form

$$\varphi(u) = \int_{S^k} f(ut)\,\sigma(t) = \int_{S^k} f(u_1 t^1 + \ldots + u_{k+1} t^{k+1})\,\sigma(t), \tag{6}$$

where $u_i = (u_i^1, \ldots, u_i^{n+1}) \in R^{n+1}$. We need to show that then $K_{k+1-m} = 0$ for all m.

It immediately follows from formula (6) that such a function $\varphi(u)$ satisfies the following system of equations:

$$\frac{\partial^2\varphi}{\partial u_{t_1}^{s_1}\,\partial u_{t_2}^{s_2}} = \frac{\partial^2\varphi}{\partial u_{t_1}^{s_2}\,\partial u_{t_2}^{s_1}}, \tag{7}$$

where $s_1, s_2 = 1, \ldots, n+1$; $t_1, t_2 = 1, \ldots, k+1$. Multiplying both sides of equality (7) by $\beta_{t_2}^{(i)} du_{t_1}^{s_1} \wedge du_j^{s_2}$ and summing with respect to t_1, t_2, s_1, s_2 we get

$$\sum_{s_1,s_2,t_1,t_2} \beta_{t_2}^{(i)} \frac{\partial^2\varphi}{\partial u_{t_1}^{s_1}\,\partial u_{t_2}^{s_2}}\, du_{t_1}^{s_1} \wedge du_j^{s_2} = \sum_{s_1,s_2,t_1,t_2} \beta_{t_2}^{(i)} \frac{\partial^2\varphi}{\partial u_{t_1}^{s_2}\,\partial u_{t_2}^{s_1}}\, du_{t_1}^{s_1} \wedge du_j^{s_2}.$$

In the notation introduced in § 4, this equality signifies that $(E_j^{(i)} - F_j^{(i)})\varphi = 0$ for any $i = 1, \ldots, m$; $j = 1, \ldots, k+1$. We now recall that the expression for K_{k+1-m} introduced in § 4 is a sum of terms into each of which enters as a factor one of the forms $E_j^{(i)} - F_j^{(i)}$. Accordingly, on the strength of what has been shown, $K_{k+1-m} = 0$ for any m.

§ 8. PROOF OF THEOREM 6

First we make the following remark. Suppose, as before, $M_x^{(1)}$ denotes a submanifold in L' consisting of elements (h, β) where $h \in G_x$ – (k+1)-dimensional subspace in R^{n+1} passing through a fixed point x, $\beta \in h'$ and where further $(\beta, x) = 1$.

Consider two sections: $s_1 : G_x \to M_x^{(1)}$ and $s_2 : G_x \to M_x^{(2)}$. These sections induce inverse mappings of the differential forms s_1^*, s_2^*.

We show that if the function $\varphi(u)$ is representable in integral form

$$\varphi(u) = \frac{1}{2}\int_{S^k} f(ut)\,\sigma(t), \tag{1}$$

then the difference $s_2^*\omega_k - s_1^*\omega_k$ is representable in integral form

Proof. The pair of sections s_1, s_2 defines the section $s = (s_1, s_2) : G_x \to M_x^{(2)}$. In fact, $(s_1, s_2)h = (s_1 h, s_2 h)$ for any $h \in G_x$. The mapping $s = (s_1, s_2)$ induces an inverse mapping s^* of the differential forms from $M_x^{(2)}$ to G_x, in particular, of the form $\nabla\omega_k = \pi_2^*\omega_k - \pi_1^*\omega_k$. Since, on the strength of the coroliary to Theorem 4, $\nabla\omega_k$ is an exact form on $M_x^{(2)}$, it follows that $s^*\nabla\omega_k = s^*\pi_2^*\omega_k - s^*\pi_1^*\omega_k$ is an exact form on G_x. But it is easily seen that $s^*\pi_2^*\omega_k = s_2^*\omega_k$ and $s^*\pi_1^*\omega_k = s_1^*\omega_k$. Thus it has been demonstrated that $s_2^*\omega - s_1^*\omega_k$ is an exact form on G_x.

We pass to the proof of Theorem 6. Let $f(x)$ be an infinitely differentiable function in $R^{n+1}\setminus 0$, satisfying the homogeneity condition

$$f(\lambda x) = |\lambda|^{-k-1} f(x). \tag{2}$$

We define the integral of the function with respect to the subspace $h \in G_{n+1,k+1}$ by Formula (1). (We recall that

$$\sigma(t) = \sum_{i=1}^{k+1} (-1)^{i-1} t^i dt^1 \wedge \ldots \wedge \widehat{dt^i} \wedge \ldots \wedge dt^{k+1}.)$$

Our problem is to obtain the inverse formula. We construct, in terms of the function $\varphi(u)$, the differential form ω_k on the vector fibering L' and take its restriction to the submanifold $M^{(1)}_{x_0} \subset L'$ where x_0 is an arbitrary fixed point in $R^{n+1} \setminus 0$.

Let $s: G_{x_0} \to M^{(1)}_{x_0}$ be an arbitrary section. Then $s^*\omega_k$ is a closed form on G_{x_0}. We must prove the following converse formula:

$$\int_\gamma s^*\omega_k = c_\gamma f(x_0), \tag{3}$$

where γ is an arbitrary k-dimensional cycle on G_{x_0}. This formula will be demonstrated here for the case of affine space.

We already showed that the forms $s^*\omega_k$ on G_{x_0} constructed with different sections $s: G_{x_0} \to M^{(1)}_{x_0}$ differ among themselves in the exact form. Accordingly, the integral (3) does not depend on the choice of section and hence formula (3) is in sufficiency shown for any one fixed section. We proceed to give such a section now.

We shall assign a vector $\beta \in (R^{n+1})'$ satisfying the condition $(\beta, x_0) = 1$. With this vector in natural form, there is determined the section $s: G_{n+1,k+1} \to L'$. (For let $h \in G_{n+1,k+1}$; then the associated space h' is the quotient space $(R^{n+1})'$ with annihilator h. The mapping s relates to each h the image of the vector β in this quotient space.)

Let R^n be the annihilator of β in R^{n+1}. Since the form ω_k does not depend on the choice of basis $u_1, \ldots, u_{k+1}$ in each subspace h, we can restrict ourselves to a basis of a special form. Namely, we can specify that $(\beta, u_i) = 0$, i.e., $u_i \in R^n$ for $i = 1, \ldots, k$ and $(\beta, u_{k+1}) = 1$.

Without loss of generality we can specify further that $x_0 = (0, \ldots, 0, 1)$ and that R^n is a vector space of the form $(x^1, \ldots, x^n, 0)$. Under these assumptions the expression for $s^*\omega_k$ assumes the following simple form:

$$s^*\omega_k = \sum_{i_1,\ldots,i_k=1}^{n} \left. \frac{\partial^k \varphi(u_1, \ldots, u_{k+1})}{\partial u^{i_1}_{k+1} \ldots \partial u^{i_k}_{k+1}} \right|_{u_{k+1}=x_0} du_1^{i_1} \wedge \ldots \wedge du_k^{i_k}. \tag{4}$$

That is, Formula (1) is equivalent with:

$$\int_\gamma \sum_{i_1,\ldots,i_k=1}^{n} \left. \frac{\partial^k \varphi(u_1, \ldots u_{k+1})}{\partial u^{i_1}_{k+1} \ldots \partial u^{i_k}_{k+1}} \right|_{u_{k+1}=x_0} du_1^{i_1} \wedge \ldots \wedge du_k^{i_k} = c_\gamma f(x_0), \tag{5}$$

where the integral is extended over an arbitrary k-dimensional cycle in G_{x_0}. We remark that the elements $h \in G_{x_0}$ uniquely determine the k-frames $(u_1, \ldots, u_k)$ in R^n. Thus G_{x_0} itself is here interpreted as the manifold $G_{n,k}$ of all k-dimensional subspaces of the space R^n.

We confirm that Formula (5) is equivalent to the formula for inversion in affine space proved in [1]. Indeed, we assign in R^n the function $f_1(x)$ by the formula

$$f_1(x^1, \ldots, x^n) = f(x^1, \ldots, x^n, 1)$$

and, for any vectors $u_1, \ldots, u_{k+1} \in R^n$, among which the first k are linearly independent, we put

$$\varphi_1(u_1, \ldots, u_{k+1}) = \int f_1(u_1 t^1 + \ldots + u_k t^k + u_{k+1})\, dt^1 \wedge \ldots \wedge dt^k. \tag{6}$$

Then it is obvious that

$$\varphi_1(u_1, \ldots, u_{k+1}) = \varphi(u_1, \ldots, u_k, u_{k+1} + x_0). \tag{7}$$

According to [1], there holds the following inversion formula:

$$\int_\gamma \sum_{i_1, \ldots, i_k = 1}^{n} \frac{\partial^k \varphi_1(u_1, \ldots, u_{k+1})}{\partial u_{k+1}^{i_1} \ldots \partial u_{k+1}^{i_k}} \Bigg|_{u_{k+1}=0} du_1^{i_1} \wedge \ldots \wedge du_k^{i_k} = c_\gamma f_1(0) = c_\gamma f(x_0), \tag{8}$$

where $\gamma \subset G_{n,k}$ is an arbitrary k-dimensional cycle. If, under the integral, we replace $\varphi_1(u_1, \ldots, u_{k+1})$ by $\varphi(u_1, \ldots, u_k, u_{k+1} + x_0)$, then we get precisely the inversion formula (5). The theorem is proved.

Remark. We used the inversion formula proved in [1] for rapidly decreasing functions. The formula and its proof, introduced in [1], is true also for functions f_1 in R^n more weakly decreasing in accord with the condition:

$$f_1(x) = O\left(((x^1)^2 + \ldots + (x^n)^2 + 1)^{-\frac{k+1}{2}}\right).$$

For the functions considered here $f_1(x^1, \ldots, x^n) = f(x^1, \ldots, x^n, 1)$, that condition is automatically satisfied (it easily follows from the condition of homogeneity of the function $f(x)$).

LITERATURE CITED

1. I. M. Gel'fand, M. I. Graev, and Z. Ya. Shapiro, "Integral geometry in k-dimensional spaces," Funktsional. Analiz., 1, No. 1, 15-31 (1967).
2. I. M. Gel'fand, M. I. Graev, and Z. Ya. Shapiro, "Differential forms and integral geometry," Funktsional. Analiz., 3, No. 2, 24-40 (1969).

7.

(with M. I. Graev and Z. Ya. Shapiro)

A problem of integral geometry connected with a pair of Grassmann manifolds

Dokl. Akad. Nauk SSSR **193** (1970) 259–262 [Sov. Math., Dokl. **11** (1970) 892–896].
MR **42**:3728. Zbl. **209**:267

The object of this article is to state and solve a problem of integral geometry for a pair of manifolds, the manifold $G_{n,k}$ of k-dimensional oriented subspaces of $\mathbf{R}^n$ and the manifold $G_{n,l}$ of l-dimensional oriented subspaces of $\mathbf{R}^n$. We shall suppose that $l < k$, $l + k \leq n$ and $k - l$ is an even number.* The results of the article can be carried over without any essential changes to the case of complex Grassmann manifolds (§4).

1. We consider the manifold $E_{n,l}$ of all l-tuples $v(v_1, \cdots, v_l)$ in $\mathbf{R}^n$; we provide it with the structure of the principal fiber space of the group $GL_0(l)$ with base $G_{n,l}$ ($GL_0(l)$ is the subgroup of linear transformations in $\mathbf{R}^l$ that preserve orientation**). Similarly, let $E_{n,k}$ be the manifold of all k-tuples $u = (u_1, \cdots, u_k)$ in $\mathbf{R}^n$, provided with the structure of the principal fiber space of the group $GL_0(k)$ with base $G_{n,k}$. We introduce the following spaces of functions: 1) the space $S_k(E_{n,l})$ of infinitely differentiable functions $f(v)$ on $E_{n,l}$ satisfying the relation

$$f(v\nu) = |\det \nu|^{-k} f(v) \qquad \text{for any} \quad \nu \in GL(l), \tag{1}$$

and 2) the space $S_l(E_{n,k})$ of infinitely differentiable functions $\phi(u)$ on $E_{n,k}$ satisfying the relation

$$\varphi(u\mu) = |\det \mu|^{-l} \varphi(u) \qquad \text{for any} \quad \mu \in GL(k).*** \tag{2}$$

We shall study a linear mapping $S_k(E_{n,l}) \to S_l(E_{n,k})$, which we now define.

Suppose that $u = (u_1, \cdots, u_k) \in E_{n,k}$ and h is the oriented k-dimensional subspace spanned by u. Consider the manifold $E_{k,l}$ of all l-tuples in $\mathbf{R}^n$ belonging to h. Since we have fixed a basis $u = (u_1, \cdots, u_k)$ in h, this determines a coordinate system in $E_{k,l}$: to each l-tuple $v = (v_1, \cdots, v_l)$ corresponds the matrix $\|t^i_j\|_{i=1,\cdots,k;j=1,\cdots,l}$ where $(t^1_j, \cdots, t^k_j)$ are the coordinates of v_j in the basis $u_1, \cdots, u_k$ $(1 \leq j \leq l)$. We define on $E_{k,l}$ the following differential form of degree $N = (k - l)l$:

$$\sigma(t) = \sigma_1(t) \wedge \ldots \wedge \sigma_l(t) \tag{3}$$

where

$$\sigma_i(t) = \sum_{(p_1, \ldots, p_k)} (-1)^s t_1^{p_1} \ldots t_l^{p_l} dt_i^{p_{l+1}} \wedge \ldots \wedge dt_i^{p_k} \tag{4}$$

* For the case $l = 1$ the problem of integral geometry was solved by somewhat different methods in [2]. The case $k + l = n$ (metric variant) was considered in [3].

** The projection $E_{n,l} \to G_{n,l}$ associates with each l-tuple v the l-dimensional oriented subspace spanned by it.

*** Functions of $S_k(E_{n,l})$ and $S_l(E_{n,k})$ can be interpreted as cross-sections of 1-dimensional vector fiberings over $G_{n,l}$ and $G_{n,k}$ respectively.

(the summation is carried out over all permutations $(p_1, \cdots, p_k)$ of the indices $1, \cdots, k$; s is the parity of the corresponding permutation).

Proposition 1. *Under transformations* $t \to \mu t \nu$, $\mu \in GL(k)$, $\nu \in GL(l)$, *the differential form* $\sigma(t)$ *is multiplied by* $(\det\mu)^l(\det\nu)^k$.

Proposition 2. *If* $f \in S_k(E_{n,l})$, *then the differential form* $f(ut)\sigma(t)$ *on* $E_{k,l}$ *can be lowered with* $E_{k,l}$ *onto the base* $G_{k,l}$ *of the natural fibering* $E_{k,l} \to G_{k,l}$.

We define the mapping $S_k(E_{n,l}) \to S_l(E_{n,k})$ *by the formula*

$$\varphi(u) = \int_{G_{k,l}} f(ut)\sigma(t). \tag{5}$$

The validity for $\phi(u)$ of relation (2) follows easily from Proposition 1.

Later we shall give a description of the image of the mapping $S_k(E_{n,l}) \to S_l(E_{n,k})$ (Theorem 2) and obtain the inversion of formula (5) (Theorem 4).

2. We introduce the space F of all possible pairs (h, β), where $h \in G_{n,k}$ and $\beta = (\beta^{(1)}, \cdots, \beta^{(l)})$ is an l-tuple in the space h' conjugate to h. In a natural way we provide the space F with the structure of a manifold and a fiber space over $G_{n,k}$ with fiber $E_{k,l}$.

We associate with a function $\phi(u) \in S_l(E_{n,k})$ a differential form on F of degree $N = (k-l)l$. For this we introduce the following differential form on $E_{n,k}$:

$$\omega_N = \varkappa_N \varphi = \Big(\bigwedge_{i=1}^{l} \bigwedge_{j=l+1}^{k} d_j^i\Big)\varphi(u) \tag{6}$$

where

$$d_j^i = \sum_{s=1}^{n} \frac{\partial}{\partial u_j^s}\, du_j^s. \tag{7}$$

We define a fibering $\pi\colon E_{n,k} \to F$, which relates $(u_1, \cdots, u_k) \in E_{n,k}$ to an element $(h; \beta^{(1)}, \cdots, \beta^{(l)}) \in F$, where h is the oriented k-dimensional subspace spanned by $u_1, \cdots, u_k$, and $\beta^{(i)}$ $(1 \le i \le l)$ are vectors of h' defined by the relations: $(\beta^{(i)}, u_j) = \delta_j^i$ for $j \le l$, $(\beta^{(i)}, u_j) = 0$ for $j > l$.

Theorem 1. *The differential form* ω_N *is the image of a differential form on* F *under the mapping* π^* *induced by the mapping* π; *in other words,* ω_N *can be lowered with* $E_{n,k}$ *into* F.*

We denote by $G_v \subset G_{n,k}$ the submanifold of subspaces $h \in G_{n,k}$ containing a given $v \in E_{n,l}$; suppose, furthermore, that $F_v \subset F$ is the totality of all $(h; \beta^{(1)}, \cdots, \beta^{(l)})$ such that $h \in G_v$ and $(\beta^{(i)}, v_j) = \delta_j^i$, $i, j = 1, \cdots, l$.

Theorem 2. *In order that a function* $\phi(u) \in S_l(E_{n,k})$ *should be the image of some function* $f(v) \in S_k(E_{n,l})$, *that is, that it should be representable in the form* (5), *it is necessary and sufficient that the differential form* $\omega_N = \kappa_N\phi$ *should be closed on each submanifold of* F_v.

3. In order to obtain the inversion formula, we construct, starting from the differential form ω_N,

---

* In [2], where the case $l = 1$ is considered, a differential form was defined not on F, but on the manifold $L' \ni F$ of all pairs (h, β), where $h \in G_{n,k}$, and β is an arbitrary (not necessarily nonzero) vector of h'. L' was prescribed as the base of the principal fibering $E_{n,k} \times (R^k)' \to L'$, and the form ω_N was defined by means of its lifting onto $E_{n,k} \times (R^k)'$. The method of defining the differential form ω_N given here is technically more convenient.

a differential form on G_v. For this we prescribe an arbitrary cross-section $s\colon G_v \to F_v$. It induces an inverse mapping s^* of differential forms. Thus $s^*\omega_N$ is a differential form on G_v.*

Theorem 3. *Let two arbitrary cross-sections $s_1\colon G_v \to F_v$ and $s_2\colon G_v \to F_v$ be given. Then, if the function $\phi(u)$ is representable in the form (5), the difference $s_1^*\omega_N - s_2^*\omega_N$ is an exact differential form on G_v.*

Theorem 4. *The inversion formula*

$$\int_\Gamma s^*\omega_N = c_\Gamma f(v) \tag{8}$$

is valid, where $s\colon G_v \to F_v$ is an arbitrary cross-section; the integral is taken over an arbitrary cycle $\Gamma \subset G_v$ of dimension $N = (k-l)l$. The coefficient c_Γ depends only on the homology class to which Γ belongs.

We outline the proof of formula (8). We show that the derivation of formula (8) is easily reduced to the inversion formula for the ordinary Radon transform.

Suppose that $v^0 = (v_1^0, \dots, v_l^0) \subset E_{n,l}$. We prescribe an l-tuple $b = (b^{(1)}, \dots, b^{(l)})$ in $(\mathbf{R}^n)'$ satisfying the following conditions: $(b^{(i)}, v_j^0) = \delta_j^i$, $i, j = 1, \dots, l$. Let $\mathbf{R}^{n-l} \subset \mathbf{R}^n$ be the annihilator of the set $\{b^{(1)}, \dots, b^{(l)}\}$. We define a function f_1 on $\mathbf{R}^{n-l}\underbrace{\times \cdots \times}_{l}\mathbf{R}^{n-l}$ by the formula

$$f_1(v_1', \dots, v_l') = f(v_1^0 + v_1', \dots, v_l^0 + v_l').$$

We consider the integrals of f_1 over manifolds $h_1 \times \cdots \times h_l$, where h_i is a $(k-l)$-dimensional plane in $\mathbf{R}^{n-l}$ $(1 \le i \le l)$:

$$\Phi(u_1, \alpha_{11}, \dots, \alpha_{1,k-l}; \dots; u_l', \alpha_{l,1}, \dots, \alpha_{l,k-l}) = \int f_1(u_1' + \alpha_{11}t_1^1 + \dots$$
$$\dots + \alpha_{1,k-l}t_1^{k-l}; \dots; u_l' + \alpha_{l,1}t_l^1 + \dots + \alpha_{l,k-l}t_l^{k-l}) \bigwedge_{j=1}^{l}\bigwedge_{i=1}^{k-l} dt_j^i,$$

$u_i', \alpha_{ij} \in \mathbf{R}^{n-l}$. By means of the function Φ we construct a differential form of degree N:

$$\tilde\omega = (\varkappa^{(1)} \wedge \dots \wedge \varkappa^{(l)})\,\Phi(0, \alpha_{11}, \dots, \alpha_{1,k-l}; \dots; 0, \alpha_{l,1}, \dots, \alpha_{l,k-l})$$

where

$$\varkappa^{(i)} = \bigwedge_{i=1}^{k-l}\Big(\sum_{s=1}^{n-l} \frac{\partial}{\partial u_i'^s}\, d\alpha_{ij}^s\Big), \quad 1 \leqslant i \leqslant l.$$

This differential form is defined on $G_{n-l,k-l}\underbrace{\times \cdots \times}_{l} G_{n-l,k-l}$ and is closed. On the basis of the inversion formula for the Radon transform in $\mathbf{R}^{n-l+1}$ [1] we obtain $\int_{\gamma\times\dots\times\gamma}\tilde\omega = c_{\gamma\times\dots\times\gamma} f_1(0)$,** where γ is the manifold of $(k-l)$-dimensional (oriented) subspaces contained in $\mathbf{R}^{k-l+1}$. On the other hand, if $\gamma_{i_1}, \dots, \gamma_{i_l}$ are arbitrary Schubert cells in $G_{n-l,k-l}$, such that $\dim\gamma_{i_1} + \dots + \dim\gamma_{i_l} = N$ and $\gamma_{i_1} \times \dots \times \gamma_{i_l} \neq \gamma \times \dots \times \gamma$, then the restriction of $\tilde\omega$ to $\gamma_{i_1} \times \dots \times \gamma_{i_l}$ is equal to zero (this follows immediately from the definition of $\tilde\omega$). Hence we obtain

---

* If $k + l > n$, then $\deg\omega_N > \dim G_v$, and so $s^*\omega_N \equiv 0$. Therefore the assertions of Theorems 3 and 4 are trivial in the case $k + l > n$.

** The coefficient $c_{\gamma\times\dots\times\gamma}$ is equal to $(c_\gamma)^l$, where c_γ is the coefficient in the inversion formula for the Radon transform in $\mathbf{R}^{k-l+1}$, see [1].

$$\int_{\Gamma}\tilde{\omega}=c_{\Gamma}f(v^0)$$

where $\Gamma\subset G_{n-l,k-l}\times\cdots\times G_{n-l,k-l}$ is an arbitrary cycle of dimension N. It is easy to verify that on the submanifold $\operatorname{diag}(\underbrace{G_{n-l,k-l}\times\cdots\times G_{n-l,k-l}}_{l})$ the differential form $\tilde{\omega}$ coincides with $s^*\omega_N$ for a suitable choice of the cross-section s. Therefore, if $\Gamma\subset\operatorname{diag}(G_{n-l,k-l}\times\cdots\times G_{n-l,k-l})$, then we obtain the required formula (8).

To calculate the coefficient c_Γ in the inversion formula (8) we construct a cellular decomposition of the manifold $G_v\cong G_{n-l,k-l}$. Let $\tilde{G}_{n-l,k-l}$ be the Grassmann manifold of the $(k-l)$-dimensional subspaces; we define a cellular decomposition of it into Schubert cells [4]. Our manifold $G_{n-l,k-l}$ of *oriented* subspaces is a double covering over $\tilde{G}_{n-l,k-l}$, and the inverse image of each Schubert sell splits into two connected components. The set of all these connected components is the required cellular decomposition of the manifold $G_v\cong G_{n-l,k-l}$. We take the Schubert cells of dimension $N=(k-l)l$. Among them is a cell Γ_0 whose closure is the manifold of all $(k-l)$-dimensional subspaces of $\mathbf{R}^k$; we denote the remaining cells of dimension N by $\Gamma_1,\dots,\Gamma_s$; furthermore, we denote by $\Gamma_i^{(1)},\Gamma_i^{(2)}$ the connected components of the inverse image of the cell Γ_i in $G_{n-l,k-l}$ $(0\le i\le s)$. It is easy to establish that: 1) the restrictions of $s^*\omega_N$ to $\Gamma_1^{(\epsilon)},\dots,\Gamma_s^{(\epsilon)}$ $(\epsilon=1,2)$ are equal to zero; 2) the integrals of $s^*\omega_N$ along $\Gamma_0^{(1)}$ and $\Gamma_0^{(2)}$ coincide.

Theorem 5. *Let $\Gamma\subset G_v$ be an arbitrary N-dimensional cycle. Then if Γ is homologous to the chain $\Sigma(n_i'\Gamma_i^{(1)}+n_i''\Gamma_i^{(2)})$, where $n_i', n_i''\in\mathbf{Z}$, we have*

$$c_\Gamma=c(n_0'+n_0'')$$

where $c\neq 0$ is a coefficient that does not depend on Γ.

4. The results given above can be carried over without essential changes to the case of complex Grassmann manifolds. In the definition of the spaces $S_k(E_{n,l})$ and $S_l(E_{n,k})$ the relations (1) and (2) must be replaced by the following:

$$f(v\nu)=|\det\nu|^{-2k}f(v)\qquad\text{for any}\qquad \nu\in GL(l,\mathbf{C});$$
$$\varphi(u\mu)=|\det\mu|^{-2l}\varphi(u)\qquad\text{for any}\qquad \mu\in GL(k,\mathbf{C}).$$

The mapping $S_k(E_{n,l})\to S_l(E_{n,k})$ is defined by the formula

$$\varphi(u)=\int_{G_{k,l}}f(ut)\,\sigma(t)\wedge\overline{\sigma(t)}.$$

Finally, the differential form $\omega_N=\kappa_N\phi$ must be replaced by $\omega_{N,N}=(\kappa_N\wedge\overline{\kappa_N})\phi$, where $\overline{\kappa_N}$ is obtained from κ_N by changing all the $\partial/\partial u_i^s$ and du_j^s to $\partial/\overline{\partial u_i^s}$ and $\overline{du_j^s}$ respectively.

Then all the theorems that we have stated for real manifolds remain valid in the complex case, and, moreover, without the further assumption that $k-l$ is even.

Institute of Applied Mathematics
Academy of Sciences of the USSR

Received 14/APR/70

BIBLIOGRAPHY

[1] I. M. Gel'fand, M. I. Graev and Z. Ja. Šapiro, *Integral geometry on k-dimensional planes*, Funkcional. Anal. i. Priložen. 1 (1967), 15–31. (Russian) MR 35 #3620.

[2] ———, *Integral geometry in a projective space*, Funkcional. Anal. i. Priložen. 4 (1970), 14–32. (Russian)

[3] E. E. Petrov, *The Radon transform in matrix spaces and in Grassmann manifolds*, Dokl. Akad. Nauk SSSR 177 (1967), 782–785 = Soviet Math. Dokl. 8 (1967), 1504–1507. MR 36 #7095.

[4] J. Schwartz, *Differential geometry and topology*, Gordon and Breach, New York, 1968; Russian transl., Moscow, 1970.

Translated by:
E. J. F. Primrose

8.
(with S. G. Gindikin)

Nonlocal inversion formulas in real integral geometry

Funkts. Anal. Prilozh. **11** (3) (1977) 12–19 [Funct. Anal. Appl. **11** (1977) 173–179].
MR **56**:16265. Zbl. **385**:53056

1. Introduction

Let R^n be n-dimensional real space; $G_{n,k}$ be the Grassmann manifold of k-dimensional nonoriented subspaces of R^n; $H_{n,k}$ be the manifold of k-dimensional nonoriented planes in $\mathbf{R}^n$; $\pi: H_{n,k} \to G_{n,k}$. Let, further, $U_{n,k}$ be the manifold of systems of vectors $\{\alpha_1, \ldots, \alpha_k; \beta\} = \{\alpha; \beta\}$, $\alpha_j \in \mathbf{R}^n$, $\beta \in \mathbf{R}^n$, where the vectors $\alpha_1, \ldots, \alpha_k$ are linearly independent. We set

$$\sigma: U_{n,k} \to H_{n,k}, \{\alpha_1, \ldots, \alpha_k; \beta\} \mapsto \{x: x = t_1\alpha_1 + \ldots + t_k\alpha_k + \beta, -\infty < t_j < \infty\}, \tag{1}$$

i.e., σ assigns to the collection $\{\alpha; \beta\}$ the plane $x = t\alpha + \beta$, $t = (t_1, \ldots, t_k)$.

Let $S(R^n)$ be the Schwartz space of rapidly decreasing, infinitely differentiable functions. For a function $f(x) \in S(\mathbf{R}^n)$ we consider its integrals over k-dimensional planes. More precisely, for $\{\alpha; \beta\} \in U_{n,k}$ we set

$$\hat{f}(\alpha_1, \ldots, \alpha_k; \beta) = \int_{-\infty}^{\infty} \ldots \int_{-\infty}^{\infty} f(t_1\alpha_1 + \ldots + t_k\alpha_k + \beta)\, dt_1 \ldots dt_k. \tag{2}$$

It is obvious that for $\alpha' = g\alpha$, $\beta' = \beta + s\alpha$, where g is a nondegenerate linear transformation of R^k, we have

$$\hat{f}(\alpha; \beta) = |\det g| \hat{f}(\alpha'; \beta'). \tag{3}$$

In $U_{n,k}$ we introduce an equivalence relation: $\{\alpha'; \beta'\} \sim \{\alpha; \beta\}$, if $\alpha' = g\alpha$, $\beta' = \beta + s\alpha$, $|\det g| = 1$. Let $V_{n,k}$ be obtained by taking the quotient with respect to this equivalence relation. Then $V_{n,k}$ admits a fibration $\tau: V_{n,k} \to U_{n,k}$ with homogeneous fibers, and the function $\hat{f}(\alpha; \beta)$ drops from $U_{n,k}$ to $V_{n,k}$.

We are interested in the question of reconstructing the function $f(x)$ from $\hat{f}(\alpha; \beta)$. In this problem the case $k = n - 1$ differs from the case $k < n - 1$. For $k = n - 1$ the map $f \to \hat{f}$ is called the Radon transform and the inversion formula is well known [1]. In this case $\hat{f}$ depends on n variables (such as the dimension of $H_{n,n-1}$), and to reconstruct f it is necessary to know $\hat{f}$ on a set of full dimension. For $k < n - 1$ we have: $\dim H_{n,k} > n$, and we arrive at an overdetermined problem. Here it is natural to speak of reconstructing f by restricting $\hat{f}$ to some n-dimensional submanifold of $H_{n,k}$. The inversion formula becomes nonunique.

In [2], for even k,* a scheme is given, allowing one to get a wide class of inversion formulas. For any $x \in \mathbf{R}^n$, we denote by G_x the collection of k-dimensional planes passing through x; G_x is a section of the bundle $\pi: H_{n,k} \to G_{n,k}$. For each smooth function $\varphi(\alpha; \beta)$ on $U_{n,k}$, satisfying (3), one constructs explicitly (see [2, Eq. (12)) a differential form $\varkappa_x\varphi$ of degree k on G_x. The coefficients of $\varkappa_x\varphi$ are expressed from φ with the aid of differential operators acting along the fibers of the bundle π. It turns out that for $k < n - 1$ the function $\varphi(\alpha; \beta)$ can be represented in the form (2) if and only if the form $\varkappa_x\varphi$ is closed for all $x \in \mathbf{R}^n$. Here for each k-dimensional cycle $\gamma \subset G_x$, we have

$$\int_\gamma \varkappa_x \hat{f} = c(\gamma) f(x). \tag{4}$$

*In [2] the case of a complex space is considered, but it is remarked that in the real case for even k the same scheme works.

Moscow State University. Translated from Funktsional'nyi Analiz i Ego Prilozheniya, Vol. 11, No. 3, pp. 12-19, July-September, 1977. Original article submitted March 25, 1977.

Both of these assertions are valid for any k (even or odd). However, while for even k there are cycles γ for which $c(\gamma) \neq 0$, for odd k for all cycles $c(\gamma) = 0$. As a result, (4) serves as an inversion formula only for even k.

What is to be done about inversion formulas for even k? First of all, it is natural to turn one's attention to the fact that the inversion formulas for the Radon transform are known for any k, but there are fundamental differences between the cases of even and odd k. For even k the inversion formula is local in the sense that to reconstruct the function f at the point x it suffices to know integrals only over hyperplanes close to x. At the same time, for odd k it is necessary to know integrals over all hyperplanes. For example, let $k = 1$. Then in (2) one can restrict oneself to $\alpha_1 = (1, p)$, $\beta = (0, q)$, i.e., $\hat{f}(\alpha_1; \beta) = \hat{f}(p,q) = \int_{-\infty}^{\infty} f(t, pt+q)\,dt$. We have:

$$f(x_1, x_2) = (2\pi)^{-2} \int_{-\infty}^{\infty} dp \int_{-\infty}^{\infty} \frac{\hat{f}'_q(p, q)\,dq}{x_2 - px_1 - q}, \tag{5}$$

where the inner integral is understood in the sense of principal values.

From what was said earlier about the form $\varkappa_x\varphi$, it follows that to calculate it, it is only necessary to know integrals over planes close to x, and hence the inversion formulas obtained in (4) are local. The situation with the Radon transform suggests that to consider the case of odd k one must have a scheme reducing to nonlocal formulas.

Subsequent analysis shows that also for even k there are nonlocal inversion formulas. We give the simplest example. Let the space $\mathbf{R}^4$ be split into the direct sum of subspaces $\mathbf{R}^2$, i.e., the points of $\mathbf{R}^4$ will be considered as pairs (x, y), $x \in \mathbf{R}^2$, $y \in \mathbf{R}^2$. We consider the family of two-dimensional planes $x = t_1\alpha_1 + \beta_1$, $y = t_2\alpha_2 + \beta_2$, where $\alpha_1, \alpha_2, \beta_1, \beta_2 \in \mathbf{R}^2$. This family depends on four parameters and one can assume $\alpha_1 = (1, p_1)$, $\alpha_2 = (1, p_2)$, $\beta_1 = (0, q_1)$, $\beta_2 = (0, q_2)$. We restrict $\hat{f}$ to the manifold introduced: $f(x, y) \mapsto \hat{f}(\alpha_1, \beta_1; \alpha_2, \beta_2) = \hat{f}(p_1, q_1; p_2, q_2)$. As is easy to note, this map is obtained by combining Radon transforms for the space $\mathbf{R}^2$: separately in x, $(\varphi(x) \mapsto \hat{\varphi}(p_1, q_1))$ and in y, $(\psi(y) \mapsto \hat{\psi}(p_2, q_2))$. Using (5), we get

$$f(x, y) = (2\pi)^{-4} \int_{-\infty}^{\infty}\int_{-\infty}^{\infty} dp_1 dp_2 \int_{-\infty}^{\infty}\int_{-\infty}^{\infty} \frac{\hat{f}''_{q_1q_2}(p_1, q_1; p_2, q_2)\,dq_1 dq_2}{(x_2 - p_1x_1 - q_1)(y_2 - p_2y_1 - q_2)}. \tag{6}$$

This inversion formula is nonlocal, and we have obtained an example of a four-dimensional manifold of two-dimensional planes in $\mathbf{R}^4$, by the restriction of $\hat{f}$ to which f is reconstructed by a nonlocal formula. We note that the integral over this manifold of $\varkappa_x\hat{f}$ is equal to zero, i.e., in (4), $c(\gamma) = 0$.

One can construct more complicated examples of manifolds having this property. A series of such examples is constructed in [3], where problems of integral geometry connected with Riemannian symmetric spaces of zero curvature are considered.*

In this paper a scheme is constructed which gives rise to a wide class of inversion formulas $\hat{f} \to f$, as a rule, nonlocal. From this scheme all nonlocal inversion formulas known to us arise. The results obtained are parallel to the results of [2] on local formulas. We have already remarked that in [2] different inversion formulas are obtained by integrating one differential form $\varkappa_x\hat{f}$ over different cycles. Now here from $\hat{f}$ one explicitly constructs a k-density $\chi_x\hat{f}$ (for the definition, see below), the integration of which over certain manifolds, which we call quasicycles, leads to different inversion formulas, as a rule nonlocal.

We give the definition of k-density. By a linear element of nonoriented volume in R^k is meant a function $\psi(u) = \psi(u_1, \ldots, u_k)$, $u_j \in \mathbf{R}^k$, such that

$$\psi(gu) = |\det g|\,\psi(u) \tag{7}$$

*We note that for symmetric spaces of nonnegative curvature the problem of integral geometry connected with orispheres has, in general, a nonlocal inversion formula [4].

for any linear transformation g of the space R^k; in particular, $\psi(u_1, \ldots, u_k) = 0$, if $u_1, \ldots, u_k$ are linearly dependent. Let M be a manifold, T_pM be the tangent space to M at the point p. We mean by a k-density (even) on M an infinitely differentiable map which assigns to each point $p \in M$ and to each k-dimensional subspace L in T_pM a linear volume element in L, i.e., a function $\psi(p; u_1, \ldots, u_k)$, $u_j \in L$, satisfying (7) in $u_1, \ldots, u_k$.

Remark. If in R^k one considers an oriented volume element (i.e., in (7) one should have det g instead of |det g|), then we arrive at the concept of an odd k-density, a special case of which is a differential form. The simplest examples of even k-densities are elements of arc-length (1-densities on the plane $\sqrt{dx_1^2 + dx_2^2}$,) or surface area.

The concept of integral of a k-density over a k-dimensional submanifold or a k-dimensional singular nonoriented chain is defined by the same scheme as the concept of integral of a differential form. Namely, if there is a k-density ψ in Euclidean space R^k, then it can be represented in the form $\psi(x; u_1, \ldots, u_k) = \varphi(x)\psi_0(u_1, \ldots, u_k)$, where ψ_0 is the canonical linear volume element ($\psi_0(u_1, \ldots, u_k) = 1$, if $\{u_j\}$ is an orthonormal basis). Then, if D is a convex bounded polyhedron, then $\int_D \psi(x; u) = \int_D \varphi(x)\, dx_1 \ldots dx_k$, where on the right is an ordinary Riemann integral.

By a singular nonoriented polyhedron of dimension k on a manifold M is meant the image of a polyhedron $D \subset \mathbf{R}^k$ under a differentiable map F: D → M. If on M there is a k-density $\psi(p; u)$, then its preimage $F^*\psi$ on D is defined. We set $\int_{FD} \psi = \int_D F^*\psi$. Integration extends by linearity from singular polyhedra to singular nonoriented chains.

We mention the structure of the rest of the paper. From a function $\varphi(\alpha; \beta)$ on $U_{n,k}$, satisfying (3), we construct for $x \in \mathbf{R}^n$, a k-density $\chi_x\varphi$ on G_x, the section of the bundle $\pi: H_{n,k} \to G_{n,k}$, corresponding to x. Further, we prove a basic theorem allowing one to obtain inversion formulas by integrating $\chi_x\hat{f}$, and we discuss some concrete inversion formulas arising in this way.

The authors thank M. I. Graev for useful discussions of the results of the paper.

2. Construction of the k-Density $\chi_x\varphi$

Although $\chi_x\varphi$ is defined invariantly, it is convenient to use coordinates in $H_{n,k}$ in its construction. Let R^n be split into the direct sum of two subspaces $\mathbf{R}^n = E \oplus F$, dim E = k, dim F = m = n − k. By H(F) we denote the collection of all k-dimensional planes $p \in H_{n,k}$, which intersect the subspace F in a point. It is clear that $\overline{H(F)} = H_{n,k}$.

In H(F) we introduce coordinates in the following way. Let $\{e_1, \ldots, e_k\}$ be a basis for E, $\{f_1, \ldots, f_m\}$ be a basis for F. To each rectangular matrix $A = \|a_{ij}\|$, $1 \leqslant i \leqslant m$, $1 \leqslant j \leqslant k$, and each collection of m numbers $(s_1, \ldots, s_m)$ we assign the collection of vectors

$$\alpha_i(A) = e_i + \sum_{j=1}^{m} a_{ji} f_j, \quad 1 \leqslant i \leqslant k; \qquad \beta(s) = \sum_{j=1}^{m} s_j f_j. \tag{8}$$

For fixed A and s, we denote by p (A, s) the k-dimensional plane consisting of all points of the form $\{t_1\alpha_1(A) + \ldots + t_k\alpha_k(A) + \beta(s)\}$. Here, as is easily seen, we get all planes from H(F) and only these. As a result, the elements of the matrix A and the components of s form a coordinate system in H(F); their number, N = m(k + 1), gives the dimension of $H_{n,k}$.

Simultaneously, for each plane $p \in H(F)$ we have fixed a corresponding collection $\{\alpha_1, \ldots, \alpha_k; \beta\} \in U_{n,k}$. Namely, $p(A, s) \mapsto \{\alpha_1(A), \ldots, \alpha_k(A); \beta(s)\}$. This allows one to carry $\varphi(\alpha; \beta)$ from $U_{n,k}$ to H(F). We set

$$\psi(A, s) = \varphi(\alpha(A), \beta(s)). \tag{9}$$

Remarks. 1. If, without changing F, $f_1, \ldots, f_m$, E is replaced by the subspace $p(A^\circ, 0)$, for some matrix A°, setting $e_i = \alpha_i(A^0)$, $1 \leqslant i \leqslant k$, then there arises a change of ordinates $A' = A - A^\circ$, s' = s. This allows one subsequently to carry over considerations in a neighborhood of the point $p(A^\circ, s)$ to H(F) in the neighborhood of the point p(0, s), where 0 is the matrix with all entries zero.

2. The choice of $\psi(A, s)$ is essentially determined by the subspace F. A change of basis $e \to ge$ in E leads to multiplication of ψ by $|\det g|^{-1}$ at all points (A, s). A change of E to p(A, 0) was considered in the previous remark. A change of basis in F does not change $\psi(A, s)$.

Let $x \in \mathbf{R}^n$, $x = (u, v)$, $u \in E$, $v \in F$. By s(A, x) we denote a collection $(s_1, \ldots, s_m)$ such that the plane p(A, s) passes through x. Then $s_j(A, x) = v_j - \sum_{i=1}^{k} a_{ji}u_i$, $1 \leqslant j \leqslant m$. Coordinates are taken with respect to the bases $\{e_1, \ldots, e_k\}$, $\{f_1, \ldots, f_m\}$. For short we write this as follows: $s(A, x) = v - Au$. Thus, $p(A, s(A, x)) \in G_x(F)$. The elements of matrix A will be coordinates in $G_x(F)$. We fix $x \in \mathbf{R}^n$ and for short we write A instead of p(A, s(A, x)). Tangent vectors to $G_x(F)$ at the point A are identified in a natural way with matrices $A^1, \ldots, A^k$, while linear independence of vectors corresponds to linear independence of matrices and conversely. Let $A^1, \ldots, A^k$ be linearly independent vectors from $T_A G_x(F)$. We set for $\eta = (\eta_1, \ldots, \eta_m)$

$$c_{ij}(A^1, \ldots, A^k; \eta) = \sum_{p=1}^{m} \eta_p a_{pi}^j, \quad 1 \leqslant i, j \leqslant k, \tag{10}$$
$$c(A^1, \ldots, A^k; \eta) = |\det \| c_{ij}(A^1, \ldots, A^k; \eta) \| |.$$

By $c(A^1, \ldots, A^k; D_s)$, $D_s = (D_{s_1}, \ldots, D_{s_m})$, $D_{s_j} = \frac{1}{i}\frac{\partial}{\partial s_j}$, we denote a pseudodifferential operator with symbol $c(A^1, \ldots, A^k; \eta)$; here, duality is established between $\mathbf{R}^k_\eta$, $\mathbf{R}^k_s$ with the aid of the form $\langle s, \eta \rangle = \sum_{j=1}^{m} s_j \eta_j$.

<u>Basic Definition.</u>

$$\chi_x \varphi(A; A^1, \ldots, A^k) = [c(A^1, \ldots, A^k; D_s)\psi(A, s)]_{s=v-Au}, \tag{11}$$

where φ and ψ are connected by (9), $x = (u, v)$.

On the right, there is a linear element of volume in the plane $L = L(A; A^1, \ldots, A^k)$ of the tangent space $T_A G_x$, spanned by the vectors corresponding to $A^1, \ldots, A^k$. That (7) holds follows immediately from (10). Thus, in (11) there is defined a k-density on $G_x(F)$. For completeness of exposition it should be proved that the k-density $\chi_x\varphi$ is independent of the basis in $\mathbf{R}^n$, $\{e_1, \ldots, e_k, f_1, \ldots, f_m\}$, which figures in its definition. One can see this directly by observing how the intermediate objects are changed by a change of basis. However, the invariance of $\chi_x\varphi$ will follow immediately from the basic theorem.

We note that the operator $c(A^1, \ldots, A^k; D_s)$ acts along the fibers of the bundle $\pi: H_{n,k} \to G_{n,k}$ (s is the coordinate in the fiber). In the following section we choose a plane depending on $A^1, \ldots, A^k$, along which this operator acts.

3. The Complexes M(A; L)

Following [1], we mean by complexes n-dimensional submanifolds of $H_{n,k}$. We connect with each k-dimensional subspace $L \subset T_A G_x$ a certain remarkable complex M(A; L). First of all, we consider the completely geodesic submanifold $M_x(A; L)$ in G_x tangent to L. The manifold $M_x(A; L)$ coincides with the closure of the set of points of the form $A + \lambda_1 A^1 + \ldots + \lambda_k A^k$. Its dimension is equal to k. We set $M(A; L) = \bigcup_{x \in \mathbf{R}^n} M_x(A; L)$, i.e., this is all the planes of $H_{n,k}$ obtained by parallel translation of the plane $p \in M_x(A; L)$. Here dim M(A; L) = n, since the fibers of the bundle π have dimension n − k. We note that the study of the complexes M(A; L) for arbitrary A reduces to A = 0 by virtue of Remark 1 on p. 15.

We denote by $Q_0(A; L)$ the linear span in $\mathbf{R}^n$ of the union of the k-dimensional subspaces $p \in M_0(A; L)$; we denote by q(L) the dimension of this subspace. By $Q_x(A; L)$ we denote the parallel translate of $Q_0(A; L)$ through the point $x \in \mathbf{R}^n$. The subspace $Q_0(A; L)$ is generated by vectors $\alpha_1(A), \ldots, \alpha_k(A)$, and also $f_{ip} = \sum_{j=1}^{m} a_{ji}^p f_j$, $1 \leqslant i \leqslant k$, $1 \leqslant p \leqslant k$. It follows from (8) that the subspaces p(A; 0) and F(L), spanned by these systems, do not intersect, and hence (q(L) − k) is equal to the number of linearly independent vectors among f_{ip}. This means

$$k+1 \leqslant q(L) \leqslant k(k+1). \tag{12}$$

For sufficiently large $n\ (n \geqslant k(k-1))$ the largest value of q(L) in (12) corresponds to the case of general position. From the form of the symbol $c(A^1, \ldots, A^k; \eta)$ it follows immediately that the operator $c(A^1, \ldots, A^k; D_s)$ acts along planes parallel to the subspace F(L) spanned by $\{f_{ip}\}$ ($\dim F(L) = q(L) - k$).

We consider some examples to find M(A; L), $c(A^1, \ldots, A^k; \eta)$. Here we assume A = 0, x = 0.

<u>Example 1.</u> We begin with the completely degenerate case when q(L) = q + 1. This case, after a change of basis in the initial space $\mathbf{R}^n$, leads to matrices for which $a^p_{1p} = 1$ for $1 \leqslant p \leqslant k$, $a^p_{ij} = 0$ in the remaining cases. Then $Q_0(0; L)$ is spanned by $\{e_1, \ldots, e_k; f_1\}$; $M_0(0; L)$ and consists of all k-dimensional subspaces of $Q_0(0; L)$, and M(0; L) consists of all k-dimensional planes parallel to $Q_0(0; L)$. We have: $\dim Q_0(0; L) = k+1$, $\dim M_0(0; L) = k$, $\dim M(0; L) = n$. Further, $c(A^1, \ldots, A^k; \eta) = |\eta_1|^k$. The operator $c(A^1, \ldots, A^k; D_s)$ will be differential for even k and nonlocal for k odd.

<u>Example 2.</u> We consider a generalization of Example 1. We choose r basis elements $f_1, \ldots, f_r$, and let $q_1, \ldots, q_r$ be natural numbers such that $q_1 + \ldots + q_r = k$. Let $a^p_{ip} = 1$ for $q_1 + \ldots + q_{i-1} < p \leqslant q_1 + \ldots + q_{i-1} + q_i$, $a^p_{ij} = 0$ in the other cases. Then $Q_0(0; L) = \mathbf{R}^{k+r}$ is spanned by $\{e_1, \ldots, e_k; f_1, \ldots, f_r\}$, and $M_0(0; L)$ contains all those k-dimensional subspaces p in $Q_0(0; L)$, such that p intersects each of the subspaces of $\mathbf{R}^{q_i+1}$ spanned by $\{e_{q_1+\ldots+q_{i-1}+1}, \ldots, e_{q_1+\ldots+q_{i-1}+q_i}, f_i\}$, in a subspace of dimension q_i. One verifies immediately that $\dim M_0(0; L) = k$ and $\dim M(0; L) = n$. Further, $q(L) = k + r \leqslant 2k$. We get the largest value for $q_1 = \ldots = q_r = 1$. We have: $c(A^1, \ldots, A^k; \eta) = |\eta_1|^{q_1} \ldots |\eta_r|^{q_r}$. The corresponding operator will be differential if all the q_j are even.

<u>Example 3.</u> In the case of general position, if n is sufficiently large, then all k^2 vectors $f_{ip} = \sum_{j=1}^{m} a^p_{ji} f_j$ are linearly independent, q(L) = k(k + 1). We change the basis $\{f_j\}$ so that the f_{ip} become its elements. One can assume that $a^p_{ij} = 1$, if $i = k(j-1) + p$, and the remaining $a^p_{ij} = 0$. In the subspace $Q_0(0; L) = \mathbf{R}^{k(k+1)}$, generated by $\{e_1, \ldots, e_k; f_1, \ldots, f_{k^2}\}$, we choose subspaces $Q_1, \ldots, Q_k$, spanned by $\{e_1; f_1, \ldots, f_k\}$, $\{e_2; f_{k+1}, \ldots, f_{2k}\}, \ldots, \{e_k; f_{k(k-1)+1}, \ldots, f_{k^2}\}$ respectively, and we establish isomorphisms among $Q_1, \ldots, Q_k$, induced by the mappings of bases $e_1 \leftrightarrow e_2 \leftrightarrow \ldots \leftrightarrow e_k$; $f_1 \leftrightarrow f_{k+1} \leftrightarrow \ldots \leftrightarrow f_{k(k-1)+1}; \ldots; f_k \leftrightarrow f_{2k} \leftrightarrow \ldots \leftrightarrow f_{k^2}$. Then $M_0(0; L)$ consists of those k-dimensional subspaces p of the space $Q_0(0; L)$, such that p intersects $Q_1, \ldots, Q_k$ along lines connected by the indicated isomorphisms. Then $\dim M_0(0; L) = k$, $\dim M(0; L) = n$. We have $c(A^1, \ldots, A^k; \eta) = |\det \|c_{ij}\||$, $c_{ij} = \eta_{k(i-1)+j}$. The corresponding operator will be nonlocal.

<u>Remark.</u> With the aid of the language introduced it is easy to describe the differential form $\varkappa_x\varphi$ from [2]. Let

$$d(A^1, \ldots, A^k; \eta) = \det \|c_{ij}(A^1, \ldots, A^k; \eta)\|$$

and $d(A^1, \ldots, A^k; D_s)$ be a differential operator with symbol $d(A^1, \ldots, A^k; D_s)$. Then

$$\varkappa_x\varphi(A; A^1, \ldots, A^k) = [d(A^1, \ldots, A^k; D_s)\psi(A, s)]_{s=v-Au}, \tag{13}$$

where for fixed A and subspace $L \subset T_A G_x$, $A^1, \ldots, A^k \in L$, on the right there is a linear element of oriented volume in L. Here $\varkappa_x\varphi$ turns out to be a differential form of degree k.

Since the differential form $\varkappa_x\varphi$, in distinction from the k-density $\chi_x\varphi$, is linear, for it there exists a more usual representation by means of the standard basis in the space of differential forms. In this form, the differential form $\varkappa_x\varphi$ is also constructed in [2]. Since there is no such possibility for k-densities, they must be written out differently in each k-dimensional subspace of the tangent space.

The complexes M(A; L) constructed in this section are invariant with respect to parallel translation. Hence from $\hat{f}|_{M(A;L)}$ one can compute $\varkappa_x\hat{f}$ on cycles of $M_x(A; L)$. However, $\int_{M_x(A;L)} \varkappa_x\hat{f}$, as a rule, is equal to zero [$c(M_x(A; L)) = 0$ in (4)]. The exception consists of only those degenerate collections $A^1, \ldots, A^k$ such that $\det \|c_{ij}(A^1, \ldots, A^k; \eta)\|$ is a nonnegative polynomial, i.e., $c(A^1, \ldots, A^k; \eta) = d(A^1, \ldots, A^k; \eta)$. This will happen in Example 2 when all q_j are even. Nevertheless, as will follow from the results of this paper, when it is local, it coincides with the formula containing $\varkappa_x\hat{f}$.

4. Basic Result

Let $\zeta \neq 0$, $\zeta \in (\mathbf{R}^n)'$ and $p \in H_{n,k}$. We say that p is incident to ξ, if $\langle \zeta, x \rangle =$ const for $x \in p$, i.e., p belongs to the hyperplane $\langle \zeta, x \rangle = c$ for some c. Let C be a singular nonoriented chain in G_x. By its Crofton symbol* we mean the function $Cr_C(\zeta)$ on $(\mathbf{R}^n)' \setminus 0$, equal to the number of points $p_1, \ldots, p_k \in C$, incident to ζ. This function is almost everywhere finite, piecewise constant, and assumes integral values. A chain C is called a quasicycle if its Crofton symbol is almost everywhere a constant Cr_C (the Crofton constant). By the Crofton operator $Cr_C(D_x)$ we mean the operator of convolution with $\check{C}r_C(x)$ (the Fourier transform of the Crofton symbol).

BASIC THEOREM. Let $f(x) \in S(\mathbf{R}^n)$ and C be a singular nonoriented chain of dimension k in G_x. Then

$$\int_C \chi_x \hat{f} = [\mathrm{Cr}_C(D_x)\hat{f}](x). \tag{14}$$

Proof. For the proof it suffices to consider the case when C is a small neighborhood of some point $p(A, s) \in G_x$ on some k-dimensional submanifold. The latter can always be replaced by the manifold $M_x(A; L)$, which at the point p(A, s) has the same tangent plane L. The general case reduces easily to A = 0, x = 0. Thus, let $C = C_\varepsilon = \{\lambda_1 A^1 + \ldots + \lambda_k A^k, |\lambda_j| < \varepsilon\}$. We have

$$\int_{C_\varepsilon} \varkappa_0 \hat{f} = \int_{|\lambda_j|<\varepsilon} [c(A^1, \ldots, A^k; D_s)\hat{f}(\alpha(\lambda_1 A^1 + \ldots + \lambda_k A^k), \beta(s))]_{s=0}\, d\lambda_1 \ldots d\lambda_k. \tag{15}$$

We take the Fourier transform of $\hat{f}(\alpha(A), \beta(s))$ with respect to s. We get

$$F_{s\to\eta}\hat{f}(\alpha(A), \beta(s)) = \tilde{f}(-\eta A, \eta), \text{ where } (\eta A)_j = \sum_{i=1}^m a_{ij}\eta_i, \langle(\xi, \eta), (u, v)\rangle = \sum \xi_i u_i + \sum \eta_j v_j.$$

This means that the right-hand side of (15) is equal to

$$\int_{|\lambda_j|<\varepsilon} \int_{\mathbf{R}^m} c(A^1, \ldots, A^k; \eta)\tilde{f}\Big(-\sum_{p=1}^k \lambda_p(\eta A^p)\Big)\, d\lambda\, d\eta = \int_{U_\varepsilon} \tilde{f}(\xi, \eta)\, d\xi\, d\eta, \tag{16}$$

where U_ε is the set

$$\{(\xi, \eta): \xi = -\sum_{p=1}^k \lambda_p(\eta A^p), |\lambda_j| < \varepsilon\}.$$

We have used the fact that $c(A^1, \ldots, A^k; \eta)$ is the Jacobian of the transition $\lambda \to \xi$. Taking account of the fact that the plane p(A, s) is incident with the points $(-\eta A, \eta)$, we get that the points of U_ε are incident with points of the neighborhood $\{|\lambda_j| < \varepsilon\}$ on $M_0(0; L)$. Thus, the right-hand side of (16) coincides with $[Cr_{C_\varepsilon}(D_x)f](x)$.

COROLLARY. For a quasicycle C we have $\int_C \chi_x \hat{f} = \mathrm{Cr}_C f(x)$.

As a result, quasicycles furnish inversion formulas for the integral transformation $f \to \hat{f}$.

Remark. In the case of oriented chains C, one can introduce an analog of the Crofton symbol $\delta_C(\zeta)$, which takes into account the number of incident points of a chain depending on orientations. Here $\int_C \varkappa_x \hat{f} = (\delta_C(D_x) f)(x)$. For a cycle γ we have $\delta_\gamma(\zeta)$ = const. However, it is possible that $\delta_\gamma(\zeta) \equiv 0$. Here it is not excluded that $Cr_\gamma(\zeta) = \text{const} \neq 0$. This is the situation in the examples given below.

5. Examples

1) The manifold $M_x(A; L)$ for any A, L is a quasicycle with Crofton number $Cr_C = 1$. Essentially, we have proved this already: The subspace $p(A + \lambda_1 A^1 + \ldots + \lambda_k A^k, 0) \in M_0(A, L)$ is incident with the points $(\xi, \eta) = (-\eta(A + \lambda_1 A^1 + \ldots + \lambda_k A^k), \eta)$, and the transition $\lambda \to \xi$

*An analogous function essentially figures in the known theorem of Crofton about the fact that the length of a curve coincides with the measure of the set of lines intersecting it.

is realized by a linear transformation (depending on η) with determinant $d(A^1, \ldots, A^k; \eta)$, which, being a polynomial in η, is almost everywhere different from zero.

2) Another method of constructing quasicycles is to consider some homogeneous compact submanifold in G_X. In [3] such examples are considered, which are connected with Riemannian symmetric spaces of zero curvature. We restrict ourselves to one type of example.

Let the space $\mathbf{R}^n$, $n = l(l+1)/2$, be realized as the space of symmetric matrices $x = \| x_{ij} \|$ of order l. Further, let $k = n - l$ and we shall construct in $G_{n,n-l}$ a submanifold C of dimension $k = n - l$.

We consider in R^n the k-dimensional subspace p(e) of matrices with diagonal elements equal to zero: $x_{11} = \ldots = x_{ll} = 0$, and also all subspaces p(u), which are obtained from p(e) by transformations $x \to uxu^{-1}$, $u \in SO(l)$. In $G_{n,n-l}$ we consider the submanifold C whose points are $p(u)$, $u \in SO(l)$. We show that C is a quasicycle. We identify $(R^n)'$ with the space of symmetric matrices with respect to the forms $\langle \zeta, x \rangle = \mathrm{sp}(\zeta x)$. Then the subspace p(e) is incident with the diagonal matrices h, and every matrix $\zeta = uhu^{-1}$, $u \in SO(l)$, h diagonal, is incident with the subspace p(u). As a result, $Cr_C = 1$.

LITERATURE CITED

1. I. M. Gel'fand, M. I. Graev, and N. Ya. Vilenkin, Integral Geometry and Representation Theory, Academic Press (1966).
2. I. M. Gel'fand, M. I. Graev, and Z. Ya. Shapiro, "Integral geometry on k-dimensional planes," Funkts. Anal. Prilozhen., 1, No. 1, 15-31 (1967).
3. S. G. Gindikin, "Unitary representations of automorphism groups of Riemannian symmetric spaces of zero curvature," Funkts. Anal. Prilozhen., 1, No. 1, 32-37 (1967).
4. S. G. Gindikin and F. I. Karpelevich, "On a problem of integral geometry," in: The Book in Memory of N. G. Chebotarev, Kazan Univ. (1964), pp. 30-43.

9.

(with S. G. Gindikin and Z. Ya. Shapiro)

A local problem of integral geometry in a space of curves

Funkts. Anal. Prilozh. **13** (2) (1979) 11-31 [Funct. Anal. Appl. **13** (1980) 87-102].
MR **80k**:53100. Zbl. **415**:53046

In integral geometry one studies the operator associating to analytical objects F on a manifold (functions, differential forms, densities) their integrals I_F over a certain set of submanifolds. Among the problems of integral geometry is the study of the kernel of this map and its image. In most of the cases studied the kernel is trivial, and then the question arises of constructing the inverse operator; it is desirable to describe as far as possible all possible constructions of this operator (all possible inversion formulas) and determine the sets of uniqueness for I_F. Local problems, in which F(x) is reconstructed in terms of the values of I_F for submanifolds close to x, occupy a special place among the problems of integral geometry.

The simplest such problem, the Radon problem [1], considers integrals of a function in a real or complex affine n-dimensional space over hyperplanes $h \in H_{n,n-1}$. This mapping has no kernal (e.g., for $F \in S$) and since dim $H_{n,n-1} = n$, the inversion formula is essentially unique and makes use of the values of I_F everywhere on $H_{n,n-1}$. To be sure, there exists no local inversion formula for R^n when n is even.

If we integrate functions over k-dimensional planes, $k < n-1$, then as before the map $F \mapsto I_F$ has no kernel, but now dim $H_{n,k} > n$ and functions on $H_{n,k}$ lying in the image of the map $F \mapsto I_F$ are given by a certain system of ultrahyperbolic differential equations which can be written out explicitly (cf. [2]). Moreover, it is possible here to describe in some sense all the possible local inversion formulas. Indeed, for each point x of the original space one can explicitly construct a differential operator which takes the functions I(h) on $H_{n,k}$ into differential forms $\varkappa_x I$ on the submanifold $G_{n,k}(x)$ of planes passing through x. The property of I to belong to the image (i.e., $I = I_F$ for some F(x)) turns out to be equivalent to the closedness of the form $\varkappa_x I$. The integrals $\varkappa_x I_F$ over the cycles $\gamma \subset G_{n,k}(x)$ are equal to $c(\gamma)F(x)$, where $c(\gamma)$ does not depend on F. However, for some cycles $c(\gamma)$ can be zero. Thus, in the real case $c(\gamma) = 0$ when k is odd for all cycles γ, and consequently local inversion formulas do not exist. In the complex case and in the real case for even k, the cycles γ for which $c(\gamma)$ does not equal 0 have been described and an inversion formula is thereby obtained.

The question of whether it is possible to reconstruct F in terms of the restriction of I_F to one or another submanifold M in $H_{n,k}$ reduces to the possibility of evaluating $\varkappa_x I$ in terms of $I(h)|_M$ and cycles with $c(\gamma) \neq 0$ for all x. Local conditions arise on M of the type of characteristicity conditions with respect to a certain system of differential equations. In concrete cases these conditions can usually be verified (cf. [3]). However, the problem of describing all such manifolds M (they are called admissible) is closely related to the definitive solution only when $k = 1$ (cf. [4, 5]). Nonlocal inversion formulas are studied in [6].

Inversion formulas in problems of integral geometry associated with integration over curved submanifolds are known only in a few special situations, mainly in problems on homogeneous manifolds (cf. [1, 7-10]). Although in each case these formulas are obtained from special arguments, they are remarkably similar to one another and are reminiscent of formulas for the flat case. Naturally, this agreement cannot be a coincidence. The reason for the similarity is that the situation described above for the flat case – the existence on a Grassmann manifold of a universal form $\varkappa_x I_F$ giving rise to all the local inversion formulas – generalizes to the case of arbitrary submanifolds. We restrict ourselves here to local considerations, and therefore the manifold may be assumed to be a domain D in $R^n(C^n)$, and the functions can be taken with compact support in D. Let $I_F(h)$ be the integrals of F(x) over arbitrary k-dimensional surfaces h in D. It turns out that the I_F can with the aid of a local operator be used to construct a closed variational 1-form on the infinite-dimensional space $E_{n,k}(x)$ of all k-dimensional surfaces passing through x which extends the form $\varkappa_x I_F$ to the Grassmann manifold $G_{n,k}(x)$. The integrals of $\varkappa_x I_F$ over cycles $\gamma \subset E_{n,k}(x)$ have, as before, the form $c(\gamma)F(x)$. More-

Institute of Applied Mathematics, Academy of Sciences of the USSR, Moscow State University. Translated from Funktsional'nyi Analiz i Ego Prilozheniya, Vol. 13, No. 2, pp. 11-31, April-June, 1979. Original article submitted December 25, 1978.

over, by considering cycles of surfaces which are homologous to cycles of planes in $G_{n,x}(x)$ giving inversion formulas, we obtain inversion formulas for surfaces.

This program clearly makes it possible to obtain in a unified way all the known inversion formulas. As in the flat case, there is the more difficult and very interesting problem of describing the admissible manifolds of surfaces.

In this paper this program is carried out for the case of curves. Section 1 is devoted to the construction of the form $\varkappa_x$ in the real case. In Sec. 2 the two-dimensional admissible manifolds of curves on the plane are described. This problem turns out to be closely related to the geometry of a second-order differential equation on the plane (cf. [11]). The families of curves $\Phi(x_1, x_2;\ \xi_1, \xi_2) = 0$ satisfying a local Desargues condition [11] turn out to be admissible, or, which is equivalent, the dual family of curves on the plane of parameters $\mathbf{R}^2_\xi$ is a family of geodesics for some affine or projective connection on the plane. This is related to a classical result of Cartan [12], according to which such families are characterized by the fact that they consist of the set of solutions of a second-order differential equation which is a polynomial of third degree with respect to the first derivatives.

The discussion is complicated by the fact that, as already noted, the Radon transform and in general the map $F \mapsto I_F(h)$, $h \in H_{n,1}(\mathbf{R})$, do not admit local inversion formulas. For this reason, complete results including inversion formulas are obtained only in the complex case. However, in the complex case the formulas are more cumbersome. The situation in integral geometry is usually such that even when in the real case local inversion formulas do not exist, it is nevertheless possible to write down real formulas which can then be rewritten automatically for the complex case. In our problem, this is reflected in the fact that the I_F can be used to construct a closed 1-form $\varkappa_x I_F$ (in the real situation) which, however, gives zero when integrated over all cycles. We carry out this construction in Sec. 1, and it seems to us that it is of independent interest and not just a method for preparing the way for the description of the corresponding complex formulas. In particular, the situation is that one can always describe the image locally, while local inversion formulas exist only for even k.

In Sec. 3 we write out formulas for the complex case without reduplicating the proofs, and only add some assertions concerning inversion formulas.

In conclusion, we note that there are a number of papers in which the problem of uniqueness sets in spaces of curves for I_F is studied (cf. [12-14]); for $k = n - 1$, a local variant of this problem is considered in [15].

1. General Problem of Integral Geometry for Families of Curves in $\mathbf{R}^n$

1. Fundamental Integral Transform. Let D be a bounded region in $\mathbf{R}^n$. Let Ξ denote the set of class C^2 curves in D. For simplicity we will first assume that the curves are defined by equations $x' = \varphi(x_1)$, $x' = (x_2, \ldots, x_n)$, $\varphi = (\varphi_2, \ldots, \varphi_n)$. We denote by $\Xi(x^0)$ the subset of curves passing through $x^0 \in D$. Let Π be the space of pairs (φ, ψ), where $\varphi \in \Xi$, and let $\psi(x_1)$ be a smooth density on this curve; correspondingly, let $\Pi(x^0)$ be the subspace of pairs corresponding to $\varphi \in \Xi(x^0)$.

To each class C^∞ function F(x) with support contained in D we associate the functional

$$I_F(\varphi, \psi) = \int F(x_1, \varphi(x_1))\, \psi(x_1)\, dx_1 \tag{1}$$

on Π.

2. First-Order Operators on Π. For a fixed $x^0 \in D$ these operators act from the space of functionals $\Phi(\Pi)$ on Π onto the space $\Omega_1(\Pi(x^0))$ of variational 1-forms on $\Pi(x^0)$. We denote by $T_{(\varphi,\psi)}\Pi$, $T_{(\varphi,\psi)}\Pi(x^0)$ the linear spaces of variations at the points $(\varphi, \psi) \in \Pi$, $(\varphi, \psi) \in \Pi(x^0)$, respectively. Let $M[x^0]$ be a linear operator from $T_{(\varphi,\psi)}\Pi(x^0)$ into $T_{(\varphi,\psi)}\Pi$; $c(\varphi, \psi;\ \delta\varphi, \delta\psi)$ a fixed 1-form. By first-order operators for $x^0 \in D$ we will mean operators of the form

$$L_{x^0} I(\varphi, \psi;\ \delta\varphi, \delta\psi) = \delta I(\varphi, \psi;\ M(\delta\varphi, \delta\psi)) + I(\varphi, \psi)\, c(\varphi, \psi;\ \delta\varphi, \delta\psi), \tag{2}$$

where $I(\varphi, \psi)$ is a functional on Π, $\delta I(\varphi, \psi;\ \delta\varphi, \delta\psi)$ is its variation. In other words, the first term on the right-hand side of Eq. (2) is obtained by substituting into δI the variations $(\delta_1\varphi, \delta_1\psi) = M[x^0](\delta\varphi, \delta\psi)$ (M depends on φ, ψ).

We note that the variation of functional (1) has the form

$$\delta I(\varphi, \psi; \delta\varphi, \delta\psi) = \int \langle F'_{x'}(x_1, \varphi(x_1)), \delta\varphi(x_1)\rangle \psi(x_1)\, dx_1 + I_F(\varphi, \delta\psi);$$
$$\langle F'_{x'}, \delta\varphi\rangle = F'_{x_2}\delta\varphi_2 + \ldots + F'_{x_n}\delta\varphi_n. \tag{3}$$

Let E be a submanifold in Π. We will assume without special mention that E is in general position and that dim $\pi E \geq n$; it follows that $\pi E \cap \Xi(x^0) \geqslant 1$. The operator L_{x^0} is said to be consistent with E if the corresponding operator $M[x^0]$ takes variations tangent to $E(x^0) = E \cap \Pi(x^0)$ into variations tangent to E. This means that the restriction of the form $L_{x^0}I|_{E(x^0)}$ can be evaluated in terms of the restriction of the functional I to E.

3. First-Order Admissible Operators. By a first-order operator on E, we will understand the restriction to E of an operator L_{x^0} consistent with E. A first-order operator on a manifold E contained in Π is said to be admissible if for every functional I_F of the form (1):

(i) The form $L_{x^0}I_F$ is closed on $E(x^0)$;

(ii) $\int_\gamma L_{x^0}I_F$ is determined solely by the values $F(x^0)$ for every 1-cycle γ on $E(x^0)$.

Remark. If $\pi E(x^0)$ is a projection of $E(x^0)$ onto Ξ which is sufficiently large (this will be spelled out later in more detail), then (ii) is a consequence of (i). Indeed, using expression (1), we can write $\int_\gamma L_{x^0}I_F$ as the value of some generalized function on F. Its support is contained in the union of curves corresponding to the points γ. Assume that γ can be replaced by a homologous cycle $\tilde{\gamma}$ in such a way that the supports of the generalized functions corresponding to γ and $\tilde{\gamma}$ intersect only in x^0. Then if (i) holds, the functional $\int_\gamma L_{x^0}I_F = \int_{\tilde{\gamma}} L_{x^0}I_F$ must be expressible in terms of $F(x^0)$. In particular, the argument goes through if $n > 2$ and $E = \Pi$. We note also that it can be shown that (i) and (ii) imply that $\int_\gamma L_{x^0}I_F = 0$ for all cycles γ. This is connected with the fact there are no local inversion formulas for curves on $\mathbf{R}^n$. At the same time, in the complex case the admissible operators admit inversion formulas, as we shall see.

Finally, we remark that if dim E = dim $\pi E = n$, then in general position dim $E(x^0) = 1$ and only condition (ii) remains.

An example of an admissible operator for every E is the operator which associates to a functional its variation $L_{x^0}I = \delta I(\varphi, \psi; \delta\varphi, \delta\psi)$ (in this case M is the identity operator and $c = 0$). This example can be made somewhat more elaborate. Let $p(\varphi, \psi)$ be a fixed functional on E. Consider the first-order operator

$$L_{x^0}I(\varphi, \psi; \delta\varphi, \delta\psi) = p(\varphi, \psi)\delta I(\varphi, \psi; \delta\varphi, \delta\psi) + I(\varphi, \psi)\delta p(\varphi, \psi; \delta\varphi, \delta\psi) \tag{4}$$

It corresponds to the operator M given by multiplication by the number $p(\varphi, \psi)$; $c = \delta p(\varphi, \psi; \delta\varphi, \delta\psi)$. The admissibility of (4) follows from the closedness of a form of type (4) on all of Π for every p and every functional I (and not only for a functional of the form I_F). Operators of form (4) will be called trivial admissible operators. We will be interested in nontrivial admissible operators, and such operators can exist on by no means any manifold E. It can be shown that a nontrivial admissible operator takes only functionals of the form I_F into closed forms.

4. Form $\varkappa_x I$. We start by explicitly giving a construction of an admissible operator on all of Π.

Proposition 1. Assume that each triple (x^0, φ, ψ), $x^0 \in D$, $(\varphi, \psi) \in \Pi$, has been placed in correspondence with a function $\mu(x_1) = \mu(x_1^0, \varphi, \psi; x_1)$ such that $(x_1 - x_1^0)\,\mu(x_1) \in C^2$. Then the first-order operator on Π corresponding to $M(\delta\varphi, \delta\psi) = (\mu\delta\varphi, \delta(\mu\psi))$, $c = 0$, i.e., the operator

$$\varkappa_{x^0}[\mu]\, I(\varphi, \psi; \delta\varphi, \delta\psi) = \delta I(\mu\delta\varphi, \delta(\mu\psi)), \quad \delta(\mu\psi) = \mu\delta\psi + \psi\delta\mu, \tag{5}$$

is admissible on Π.

The variation in (5) is taken for μ considered as an operator-valued function on Π. We also note that in spite of the singularity of μ at x_1^0, the operator obtained by multiplying $\delta\varphi$ by the function μ is everywhere defined on $T_\varphi\Xi(x^0)$.

For the proof it is necessary to verify conditions (i), (ii) for

$$\varkappa_{x^0}[\mu]I_F(\varphi, \psi; \delta\varphi, \delta\psi) = \int \langle F'_{x'}(x_1, \varphi(x_1)), \delta\varphi(x_1)\rangle \mu(x^0, \varphi, \psi; x_1)\psi(x_1)\, dx_1 + \int F(x_1, \varphi(x_1))\delta(\mu\psi)\, dx_1, \tag{5'}$$

and by the above remark it suffices to verify condition (i) for $n \geq 2$. In order to verify condition (i), it is convenient to regard (5), (5') as 1-forms on the bundle $\tilde{\Pi}(x^0)$ over $\Pi(x^0)$ consisting of triples (φ, ψ, μ), where the μ

are arbitrary functions satisfying the hypotheses of the proposition. After this, a direct calculation shows that the variation $\delta\varkappa_{x^0}I_F(\varphi, \psi, \mu; \delta\varphi, \delta\psi, \delta\mu; \delta_1\varphi, \delta_1\psi, \delta_1\mu)$ is equal to zero. We then obtain that $\varkappa_{x0}[\mu]I_F$ is closed from the fact that we are restricting a closed form to the section $\mu = \mu(\varphi, \psi)$. As has been noted, we only have to verify condition (ii) when n = 2. This verification can be given directly (cf. the argument in Sec. 2 on p. 93 or can be reduced to the case of a cycle of straight lines, for which it is well known. However, it is simpler to extend the functions F on $D \subset \mathbf{R}^2$ to functions on $\tilde{D} \subset \mathbf{R}^3$, and then condition (ii) for cycles γ of curves on D will be a special case of assertion (ii) for cycles of curves on $\tilde{D}$.

We have thus constructed a certain class of admissible operators L_{x0} on Π which are the restriction of the universal form $\varkappa_{x0}I$ on $\tilde{\Pi}$ to certain sections over Π. If this operator is consistent with $E \subset \Pi$ (i.e., $(\mu\delta\varphi, \delta(\mu\psi)) \in T_{(\varphi,\psi)}E$ for $(\delta\varphi, \delta\psi) \in T_{(\varphi,\psi)}E(x^0)$), then the restriction to E will be an admissible operator on E. It turns out that this construction exhausts all the possible admissible first-order operators on manifolds $E \subset \Pi$.

<u>THEOREM 1.</u> Every first-order admissible operator $L_{x0}I$ on $E \subset \Pi$ coincides with an operator of the form $\varkappa_{x0}[\mu]$ for some section $\mu(\varphi, \psi; x_1)$ of the bundle $\tilde{\Pi}$ over Π. Moreover,

$$M(\delta\varphi, \delta\psi) = (\mu\delta\varphi, \delta(\mu\psi) - c\psi). \tag{6}$$

<u>Proof.</u> Let L_{x0} be an arbitrary first-order operator on the manifold E. We study the condition for the form $L_{x0}I_F$ to be closed for I_F of the form (1). The variation $\delta(L_{x^0}I_F)(\varphi, \psi; \delta\varphi, \delta\psi; \delta_1\varphi, \delta_1\psi)$ is a linear function of F expressible as a sum of the form

$$\sum_{i,j}\int F''_{x_ix_j}(x_1, \varphi(x_1))\,\nu_{ij}(x_1)\,dx_1 + \sum_i \int F'_{x_i}(x_1, \varphi(x_1))\,\nu_i(x_1)\,dx_1 + \int F(x_1, \varphi(x_1))\,\nu(x_1)\,dx_1.$$

Saying that $\delta(L_{x0}I_F)$ is equal to zero for all F means that all the $\nu_{ij}, \nu_i, \nu, 2 \le i, j \le n$, are zero.

We study the equations $\nu_{ij}(x_1) \equiv 0,\ 2 \leqslant i, j \leqslant n$. Let $M^\varphi(\delta\varphi, \delta\psi)$, $M^\psi(\delta\varphi, \delta\psi)$ be components of the operator M taking values in the variations $\delta\varphi$, $\delta\psi$, respectively; $M^\varphi = (M_2^\varphi, \ldots, M_n^\varphi)$, where M_i^φ corresponds to the component $(\delta\varphi)_i$. Then the equality $\nu_{ij} \equiv 0$ can be written in the form

$$(\delta\varphi)_i M_j^\varphi(\delta_1\varphi, \delta_1\psi) + (\delta\varphi)_j M_i^\varphi(\delta_1\varphi, \delta_1\psi) = (\delta_1\varphi)_i M_j^\varphi(\delta\varphi, \delta\psi) + (\delta_1\varphi)_j M_i^\varphi(\delta\varphi, \delta\psi),$$

from which we obtain for a manifold in general position that $M^\varphi(\delta\varphi, \delta\psi) = \mu(\varphi, \psi)\delta\varphi$, where $\mu(\varphi, \psi)$ for fixed φ, ψ is a function of x_1 satisfying the conditions given above. Thus,

$$L_{x^0}I_F(\varphi, \psi; \delta\varphi, \delta\psi) = \int \langle F'_{x'}(x_1, \varphi(x_1)), \delta\varphi(x_1)\rangle\,\mu(x_1)\,\psi(x_1)\,dx_1 + \int F(x_1, \varphi(x_1))(M^\psi + c\psi)(x_1)\,dx_1. \tag{7}$$

From the fact that ν_i are zero, we obtain that for any variations $\delta\varphi, \delta_1\varphi$

$$\delta\varphi\,[\delta_1(\mu\psi) - M^\psi(\delta_1\varphi, \delta_1\psi) - c\psi] = \delta_1\varphi[\delta(\mu\psi) - M^\psi(\delta\varphi, \delta\psi) - c\psi], \tag{8}$$

where $\delta_1(\mu\psi)$ is the variation of $\mu\psi$ corresponding to $(\delta_1\varphi, \delta_1\psi)$. If the space of variations, as a module with respect to multiplication by functions, has a dimension greater than 1, we obtain (6). At the same time, we have found more precisely conditions under which (i) implies (ii).

We now determine in the general case when condition (ii) holds for (7). To do this we use μ appearing in (7) to construct $\varkappa_{x0}[\mu]$ and consider

$$L_{x^0}I_F - \varkappa_{x^0}[\mu]I_F = \int F(x_1, \varphi(x_1))(M^\psi + c\psi - \delta(\mu\psi))\,dx_1.$$

If condition (ii) holds for $L_{x0}I_F$, then (ii) will also hold for this difference. However, the regular functional on F, defined by $\int_\gamma I_F(\varphi, M^\psi + c\psi - \delta(\mu\psi))$ can be concentrated at a point only if it is equal to zero, i.e., (6) holds. This completes the proof of the necessity. The sufficiency follows from the fact already mentioned above that the restrictions of operator $\varkappa_{x0}[\mu]$ are admissible.

We note that, in general, a first-order operator cannot be represented uniquely in form (2), since the addition of $c\psi$ to M^ψ can leave the variation $M(\delta\varphi, \delta\psi)$ tangential to $E(x^0)$. On the other hand, the variation $\delta(\mu\psi)$ may fail to be tangential, and it becomes tangential only after $c\psi(x_1)$ has been subtracted, where $c(\varphi, \psi; \delta\varphi, \delta\psi)$ is a fixed 1-form on $E(x^0)$ (it does not depend on I).

<u>5. Examples.</u> 1. Let Π consist of all curves $\varphi \in \Xi$, and let $\psi \equiv 1$. Then $\delta\psi \equiv 0$ and by (6) $\delta\mu = c$ We take $\mu(x^0, \varphi; x_1)$ to be a function independent of φ. This can be done in various ways such that Eq. (6)

holds. Let, e.g., $\mu(x^0, x_1) = 1/(x_1 - x_1^0)$. Then multiplication by μ is everywhere defined on $T_\varphi \Xi(x^0)$, and we obtain on $\Xi(x^0)$ a closed 1-form

$$\delta I_F\left(\varphi, 1; \frac{\delta\varphi}{x_1 - x_1^0}, 0\right) = \int \langle F'_{x'}(x, \varphi(x_1)), \delta\varphi\rangle \frac{dx_1}{x_1 - x_1^0}.$$

2. We consider in Ξ the submanifold $\mathscr{L}$ of lines given by $\varphi(x_1) = \alpha x_1 + \beta$, $\alpha = (\alpha_2, \ldots, \alpha_n)$, $\beta = (\beta_2, \ldots, \beta_n)$. If $\psi \equiv 1$, the operator in the preceding example is consistent with $\mathscr{L}$, since multiplication by $1/(x_1 - x_1^0)$ does not take us outside $T_\varphi \mathscr{L}$ for $\delta\varphi \in T_\varphi \mathscr{L}(x^0)$. We can regard (α, β) as coordinates on $\mathscr{L}$. Moreover, the α are coordinates $\mathscr{L}(x^0)$ when $x^0 = 0 = (0, \ldots, 0)$. We have

$$L_0 I(\alpha, 0; x_1 d\alpha) = \sum \frac{\partial}{\partial \beta_i} I(\alpha, \beta)|_{\beta=0} d\alpha_i; \quad L_0 I_F = \sum \left(\int F'_{x_i}(x_1, \alpha x_1)\, dx_1\right) d\alpha_i.$$

We consider a somewhat more general solution. Let $\hat{\mathscr{L}}$ be the manifold of lines $\varphi(x_1) = \alpha x_1 + \beta$ and densities $\psi(x_1) = a x_1 + 1$. This is a $(2n+1)$-dimensional manifold, and (α, β, a) can be regarded as coordinates; $\hat{\mathscr{L}}(0)$ is defined by the condition $\beta = 0$. Let $x^0 = 0$; the 1-forms on $\hat{\mathscr{L}}(0)$ are combinations of $d\alpha_i$, da. We have $\delta\varphi = x_1 d\alpha$. $\delta\psi = x_1 da$. Assume as before that $\mu(0; x_1) = 1/x_1$. Then $\delta(\mu\psi) = da$. We put $M^\psi(\alpha, a; d\alpha, da) = -a\delta\psi = -a x_1 da$. and take da as the form c. Then $M^\psi + c\psi = -a x_1 da + (a x_1 + 1) da = da = \delta(\mu\psi)$. As a result we obtain a 1-form

$$\delta I(\alpha, 0, a; d\alpha, -a x_1 da) + I(\alpha, 0, a)\, da = \sum \frac{\partial}{\partial \beta_i} I(\alpha, 0, a)\, d\alpha_i + \left[I(\alpha, 0, a) - a \frac{\partial}{\partial a} I(\alpha, 0, a)\right] da.$$

Thus the form $\sum \left(\int F'_{x_i}(x_1, \alpha x_1)\, dx_1\right)(a x_1 + 1)\, d\alpha_i + \int F(x_1, \alpha x_1)\, dx_1\, da$ is closed on $\hat{\mathscr{L}}(0)$. (which, however, is easily verified directly). We note that the variation $\mu\delta\varphi = d\alpha$ does not depend on x_1.

6. Admissible First-Order Operators on Finite-Dimensional Manifolds of Curves. In this section we specialize the preceding formulas to the case when the manifold E is finite-dimensional. In this case the operators naturally become differential operators, and the variational forms become differential forms.

Assume given a family πE of curves $x' = \varphi(x_1; \xi_1, \ldots, \xi_k)$, depending on k parameters $\xi = (\xi_1, \ldots, \xi_k)$, $k > n$, and let the densities $\psi(x_1; \xi, \eta)$ depend in addition on m extra parameters $\eta_1, \ldots, \eta_m$. In the case when $m = 0$ there exists on each curve $x' = \varphi(x_1; \xi)$ a unique density $\psi(x_1; \xi)$. Thus, we will assume that $\pi E \subset \mathbf{R}^k$, $E \subset \mathbf{R}^{k+m}$. In this case, first-order operators are simply first-order differential operators taking the functions $I(\xi, \eta)$ into differential 1-forms on the submanifold $E(x^0) = \{(\varphi, \psi) : (x')^0 = \varphi(x_1^0; \xi)\}$. We note that

$$I_F(\xi, \eta) = \int F(x_1, \varphi(x_1, \xi))\psi(x_1; \xi, \eta)\, dx_1.$$

In addition, the variations have the form $\delta\varphi = \langle d\xi, \varphi'_\xi\rangle = d\xi_1 \frac{\partial\varphi(x_1, \xi)}{\partial\xi_1} + \ldots + d\xi_k \frac{\partial\varphi(x_1, \xi)}{\partial\xi_k}$, $\delta\psi = \langle d\xi, \psi'_\xi\rangle + \langle d\eta, \psi'_\eta\rangle$.

In the case of general position, $\dim \pi E(x^0) = k - n + 1$ and $d\xi$ such that $\delta\varphi(x_1^0) = \langle d\xi, \varphi'_\xi(x_1^0, \xi)\rangle = 0$ correspond to tangential variations to $\pi E(x^0)$. We construct a basis in the space of tangential variations. Consider the differential operators

$$\Delta_i\left(x_1^0, \xi; \frac{\partial}{\partial\xi}\right) = \det \begin{pmatrix} \frac{\partial\varphi_2}{\partial\xi_i}(x_1^0, \xi) \ldots \frac{\partial\varphi_n}{\partial\xi_i}(x_1^0, \xi) & \frac{\partial}{\partial\xi_i} \\ \frac{\partial\varphi_2}{\partial\xi_{i+1}}(x_1^0, \xi) \ldots \frac{\partial\varphi_n}{\partial\xi_{i+1}}(x_1^0, \xi) & \frac{\partial}{\partial\xi_{i+1}} \\ \cdots & \cdots \\ \frac{\partial\varphi_2}{\partial\xi_{i+n-1}}(x_1^0, \xi) \ldots \frac{\partial\varphi_n}{\partial\xi_{i+n-1}}(x_1^0, \xi) & \frac{\partial}{\partial\xi_{i+n-1}} \end{pmatrix}, \tag{9}$$

$$i = 1, \ldots, k - n + 1,$$

depending on x_1^0 as a parameter. Then it is obvious that

$$\delta^i(x_1^0, \xi; x_1) = \Delta_i\left(x_1^0, \xi; \frac{\partial}{\partial\xi}\right)\varphi(x_1, \xi), \quad i = 1, \ldots, k - n + 1, \tag{10}$$

is a basis in the space $T_\xi \pi E(x^0)$. Let $d\zeta = (d\zeta_1, \ldots, d\zeta_{k-n+1})$ be the corresponding coordinates. Then $(d\zeta, d\eta)$ are coordinates in $T_{(\xi,\eta)}E(x^0)$.

We now consider the operator $M(\delta\varphi, \delta\psi)$ in (2) corresponding to an admissible operator. Its components M^φ do not depend on $\delta\psi$ (Theorem 1). Let

$$M^\varphi(\delta^i(x_1)) = \langle a^i, \varphi'_\xi\rangle = \sum_{j=1}^{k} a^i_j(x^0_1, \xi, \eta)\frac{\partial\varphi(x_1,\xi)}{\partial\xi_j};$$

$M^\varphi(\delta^i)$ has $n-1$ components $M^\varphi(\delta^i)_2, \ldots, M^\varphi(\delta^i)_n$ (just like φ, δ^i). The requirement of admissibility also implies the independence of the ratios $M^\varphi(\delta^i)_j/\delta^i_j = \mu(x^0_1, \xi, \eta, x_1)$ for $1 \leqslant i \leqslant k-n+1$. $2 \leqslant j \leqslant n$. In addition, the requirement that the variations $M^\psi(\delta^i, d\eta)$ and $M^\varphi(\delta^i, d\eta) = M^\varphi(\delta^i)$ be consistent gives

$$M^\psi(\delta^i, d\eta) = \langle a^i(x^0_1, \xi, \eta), \psi'_\xi(x_1, \xi, \eta)\rangle + \langle b^i(x^0_1, \xi, \eta) + \sum d\eta_j d^j(x^0_1, \xi, \eta), \psi'_\eta(x_1, \xi, \eta)\rangle,$$

where a^i is a vector function of length k appearing in $M^\varphi(\delta^i)$, and b^i and d^j are vector functions of length m. Equation (6) becomes a system

$$\Delta_i(\mu\psi) = \langle a^i, \psi'_\xi\rangle + \langle b^i, \psi'_\eta\rangle + c^i, \quad \frac{\partial}{\partial\eta_j}(\mu\psi) = \langle d^j, \psi^j_\eta\rangle + e^j$$

for certain functions $c^i(\xi, \eta)$ and $e^j(\xi, \eta)$. We state the next corollary as a consequence of Theorem 1 for the finite-dimensional case.

COROLLARY. Assume given a finite-dimensional manifold E of curves $x_j = \varphi_j(x_1; \xi_1, \ldots \xi_k)$, $2 \leqslant j \leqslant n$, and densities $\psi(x_1; \xi_1, \ldots, \xi_k; \eta_1, \ldots, \eta_m)$ on which there is defined a first-order differential operator $I(\xi, \eta) \to L_{x^0}I$, where $L_{x^0}I$ is a differential 1-form on $E(x^0)$. The operator L_{x^0} is admissible if and only if it has the form

$$\begin{aligned}L_{x^0}I(\xi, \eta; d\xi, d\eta) &= dI\Big(\xi, \eta; \sum a^i(x^0_1, \xi, \eta)\, d\zeta_i, \sum b^i(x^0_1, \xi, \eta)\, d\zeta_i + \\ &+ \sum d^j(x^0_1, \xi, \eta)\, d\eta_j\Big) + I(\xi, \eta)\Big(\sum c^i(x^0_1, \xi, \eta)\, d\zeta_i + \sum e^j(x^0_1, \xi, \eta)\, d\eta_j\Big) = \\ &= \sum\left[\left\langle a^i, \frac{\partial}{\partial\xi}\right\rangle + \left\langle b^i, \frac{\partial}{\partial\eta}\right\rangle + c^i\right] I(\xi, \eta)\, d\zeta_i + \sum\left[\left\langle d^j, \frac{\partial}{\partial\eta}\right\rangle + e^j\right] I(\xi, \eta)\, d\eta_j,\end{aligned} \tag{11}$$

where $dI(\xi, \eta; d\xi, d\eta)$ is the differential of $I(\xi, \eta)$, $d\zeta_i$ are coordinates in the basis $\delta^i(x^0_1, \xi; x_1)$ (10) in $T_\xi\pi E(x^0)$, and the vector functions a^i, b^i, d^j, c^i, e^j satisfy the following system of differential equations

$$\mu(x^0_1, x_1, \xi, \eta) = \frac{\langle a^i(x^0_1, \xi, \eta), (\varphi_j)'_\xi(x_1, \xi)\rangle}{\Delta_i\left(x^0_1, \xi; \frac{\partial}{\partial\xi}\right)\varphi_j(x_1, \xi)} \tag{12}$$

which does not depend on $1 \le i \le k-n+1$, $2 \le j \le n$ (there are $(n-1)(k-n+1)-1$ equations); the operators Δ_i are defined in (9). In addition,

$$\begin{aligned}&\Delta_i\left(x^0_1, \xi; \frac{\partial}{\partial\xi}\right)[\mu(x^0_1, x_1, \xi, \eta)\psi(x_1, \xi, \eta)] = \\ &= \langle a^i(x^0_1, \xi, \eta), \psi'_\xi(x_1, \xi, \eta)\rangle + \langle b^i(x^0_1, \xi, \eta), \psi'_\eta(x_1, \xi, \eta)\rangle + c^i(x^0_1, \xi, \eta)\psi(x_1, \xi, \eta); \\ &\frac{\partial}{\partial\eta_j}[\mu(x^0_1, x_1, \xi, \eta)\psi(x_1, \xi_1, \eta)] = \langle d^j(x^0_1, \xi, \eta), \psi'_\eta(x_1, \xi, \eta)\rangle + e^j(x^0_1, \xi, \eta)\psi(x_1, \xi, \eta), \\ &\qquad 1 \leqslant i \leqslant k-n+1, \quad 1 \leqslant j \leqslant m.\end{aligned} \tag{13}$$

Thus, in the finite-dimensional case the definition of an admissible operator reduces to defining functions of x^0_1, ξ, η: a^i_l, b^i_p, d^j_p, c^i, e^j, $1 \leqslant i \leqslant k-n+1$, $1 \leqslant j \leqslant m$, $1 \leqslant l \leqslant k$, $1 \leqslant p \leqslant m$, satisfying Eqs. (12) and (13).

As has already been noted, not every manifold E is such that there exists at least one nontrivial admissible operator. In other words, system (12), (13) is solvable only under very stringent conditions on φ, ψ and even only on φ. A manifold $K \subset \Xi$ of curves $x' = \varphi(x_1)$ is called admissible if there exists over K a section $\Psi(\varphi)$ of the bundle Π (i.e., for each φ there exists a density $\psi(x_1)$ such that on the manifold of pairs $(\varphi, \Psi(\varphi))$, $\varphi \in K$, there exists at least one nontrivial admissible operator. In the following paragraph we give a description of two-dimensional admissible manifolds K of curves in $\mathbb{R}^2$. In addition, for this case we describe all possible densities $\Psi(\varphi)$ such that there exist nontrivial admissible operators on the $(\varphi, \Psi(\varphi))$, manifold $\varphi \in K$, and we describe all such operators.

2. Two-Dimensional Admissible Manifolds of Curves in R^2

1. Statement of the Problem. Let $D \subset R^2$ be a bounded domain, $x_2 = \varphi(x_1; \xi_1, \xi_2) = \varphi(x_1, \xi)$ curves in D, $\xi = (\xi_1, \xi_2) \in \Omega$. Let Ω also be a bounded domain in R^2. We will assume that this manifold of curves K is in general position. Then through each point $x^0 \in R^2$ there passes a one-parameter set $K(x^0)$ of curves $x_2^0 = \varphi(x_1^0, \xi)$. The tangential variations to $K(x^0)$ are proportional to

$$\delta^1(x_1^0, x_1, \xi_1, \xi_2) = \frac{\partial\varphi(x_1^0, \xi)}{\partial\xi_1}\frac{\partial\varphi(x_1, \xi)}{\partial\xi_2} - \frac{\partial\varphi(x_1^0, \xi)}{\partial\xi_2}\frac{\partial\varphi(x_1, \xi)}{\partial\xi_1}.$$

Assume we are given a map M such that

$$M^{\varphi}(\delta^1) = a_1(x_1^0, \xi)\frac{\partial\varphi(x_1, \xi)}{\partial\xi_1} + a_2(x_1^0, \xi)\frac{\partial\varphi(x_1\xi)}{\partial\xi_2}.$$

We put

$$\mu(x_1^0, x_1, \xi_1, \xi_2) = \frac{a_1(x_1^0, \xi)\varphi'_{\xi_1}(x_1, \xi) + a_2(x_1^0, \xi)\varphi'_{\xi_2}(x_1, \xi)}{\varphi'_{\xi_1}(x_1^0, \xi)\varphi'_{\xi_2}(x_1, \xi) - \varphi'_{\xi_2}(x_1^0, \xi)\varphi'_{\xi_1}(x_1, \xi)}. \tag{14}$$

Since k = n = 2, only the second group of conditions (13) remains, which reduces to a single equation. Thus, the operator

$$\begin{gathered} L_{x^0}I(\xi, -\varphi'_{\xi_2}(x_1^0, \xi)\,d\zeta, \varphi'_{\xi_1}(x_1^0, \xi)\,d\zeta)) = \\ = dI(\xi; a_1(x_1^0, \xi)\,d\zeta, a_2(x_2^0, \xi)\,d\zeta) + I(\xi)\,c(x_1^0, \xi)d\zeta = (\hat{L}_{x^0}I(\xi))d\zeta, \\ L_{x^0} = a_1(x_1^0, \xi)\frac{\partial}{\partial\xi_1} + a_2(x_1^0, \xi)\frac{\partial}{\partial\xi_2} + c(x_1^0, \xi), \qquad \xi \in K(x^0) \end{gathered} \tag{11'}$$

(where $d\zeta$ is the unique coordinate in $K(x^0)$ and $dI(\xi; d\xi_1, d\xi_2)$ is the differential of $I(\xi_1, \xi_2)$; $\varphi(x_1^0, \xi) = x_2^0$) is admissible if and only if

$$\begin{gathered} \varphi'_{\xi_1}(x_1^0, \xi)[\psi\cdot\mu]'_{\xi_2}(x_1^0, x_1, \xi) - \varphi'_{\xi_2}(x_1^0, \xi)[\psi\cdot\mu]'_{\xi_1}(x_1^0, x_1, \xi) = \\ = a_1(x_1^0, \xi)\psi'_{\xi_1}(x_1, \xi) + a_2(x_1^0, \xi)\psi'_{\xi_2}(x_1, \xi) + c(x_1^0, \xi)\psi(x_1, \xi), \end{gathered} \tag{13'}$$

where μ has form (14).

Our problem splits into two parts. We first need to find conditions on the family $\varphi(x_1, \xi_1, \xi_2)$ under which Eq. (13') has a nontrivial solution $\{\psi(x_1, \xi), a_1(x_1^0, \xi), a_2(x_1^0, \xi), c(x_1^0, \xi)\}$. Then under the assumption that these conditions are fulfilled, we have to describe the general solution $\{\psi, a_1, a_2, c\}$ of Eq. (13').

2. Direct Derivation of (13'). We recall that for n = 2 only condition (ii) remains for admissibility of L_{x^0}, i.e., for

$$I_F(\xi_1, \xi_2) = \int F(x_1, \varphi(x_1, \xi_1, \xi_2))\,\psi(x_1, \xi_1, \xi_2)\,dx_1 \tag{1'}$$

the integral

$$\int_{K(x^0)} L_{x^0}I_F(\xi_1, \xi_2; d\zeta) \tag{15}$$

depends only on $F(x^0)$ (in fact, it is equal to zero). This condition is equivalent to the integral in (15) being equal to zero if $F(x_1, x_2)$ is a function of compact support whose support does not contain x^0. The equivalence of this condition to condition (13') can be checked directly. We give here an equivalent, but not completely rigorous, calculation which shows that (13') is equivalent to the vanishing of the integral (15) when $F(x_1, x_2) = \delta(x_1 - a, x_2 - b)$, where $(a, b) \neq x^0$. Since x^0 is fixed, we may assume that $x^0 = 0 = (0, 0)$. In addition, we may assume by making diffeomorphic changes of variables in x and ξ that K(0) is defined by the condition $\xi_2 = 0$, i.e., $\varphi(0; \xi_1, 0) = 0$ and that $a_1 \equiv 0$, $a_2 \equiv 1$, i.e.,

$$L_0I(\xi_1, 0; d\xi_1) = dI(\xi_1, 0; 0, d\xi_1) + I(\xi_1, 0)\,c(0, \xi_1, 0)\,d\xi_1 = \left[\frac{\partial}{\partial\xi_2}I(\xi_1, 0) + c(0, \xi_1, 0)\,I(\xi_1, 0)\right]d\xi_1.$$

Then $\mu(0, x_1; \xi_1, 0) = \varphi'_{\xi_2}(x_1, \xi_1, 0)\,/\,\varphi'_{\xi_1}(x_1, \xi_1, 0)$ and Eq. (13') takes the form

$$\frac{\partial}{\partial\xi_1}\left[\psi(x_1; \xi_1, 0)\frac{\varphi'_{\xi_2}(x_1; \xi_1, 0)}{\varphi'_{\xi_1}(x_1; \xi_1, 0)}\right] = \psi'_{\xi_2}(x_1; \xi_1, 0) + c(\xi_1)\,\psi(x_1; \xi_1, 0). \tag{13''}$$

We show that Eq. (13") at the point $x_1 = a$ for a fixed ξ_1 is equivalent to (15) being equal to zero for $F(x_1, x_2) = \delta(x_1 - a, x_2 - b)$, where $b = \varphi(a; \xi_1, 0)$. Indeed, we write down (15) for $\delta(x_1 - a, x_2 - b)$

$$\int_{K'(0)} d\xi_1 \Big\{\int \delta'_{x_2}(x_1 - a, \varphi(x_1; \xi_1, 0) - b)\, \varphi'_{\xi_2}(x_1; \xi_1, 0)\, \psi_1(x_1; \xi_1, 0)\, dx_1 + \\ + \int \delta(x - a, \varphi(x_1; \xi_1, 0) - b)[\psi'_{\xi_2}(x_1; \xi_1, 0) + c(\xi_1)\psi(x_1; \xi_1, 0)]\, dx_1\Big\}.$$

We make the change of variable $\xi_1 \to x_2 = \varphi(x_1; \xi_1, 0)$, $\xi_1 = \tau(x_1, x_2)$:

$$(\delta'_{x_2}(x_1 - a, x_2 - b), \varphi'_{\xi_2}(x_1; \tau(x_1, x_2), 0)\, \psi(x_1; \tau(x_1, x_2), 0)/\varphi'_{\xi_1}(x_1; \tau(x_1, x_2), 0)) + \\ + (\delta(x_1 - a, x_2 - b), [\psi'_{\xi_2}(x_1; \tau(x_1, x_2), 0) + \\ + c(\tau(x_1, x_2))\psi(x_1; \tau(x_1, x_2)), 0)/\varphi'_{\xi_1}(x_1; \tau(x_1, x_2), 0)]).$$

The vanishing of this expression means that

$$-\frac{\partial}{\partial x_2}[\varphi'_{\xi_2}(a; \tau(a, b), 0)\, \psi(a; \tau(a, b), 0)/\varphi'_{\xi_1}(a; \tau(a, b), 0))] + \\ + [\psi'_{\xi_2}(a; \tau(a, b), 0) + c(\tau(a, b))\, \psi(a; \tau(a, b), 0)/\varphi'_{\xi_1}(a; \tau(a, b), 0)] = 0,$$

for $x_2 = \varphi(x_1; \xi_1, 0)$, and taking into account that $\tau(a, b) = \xi_1$, $\varphi'_{\xi_1}(x_1; \xi_1, 0)\, \partial/\partial x_2 = \partial/\partial \xi_1$, we obtain (13').

3. Solvability Condition for (13') (Necessary Conditions). **Thus, we are interested in conditions on the family $x_2 = \varphi(x_1, \xi)$ under which Eq. (13') is solvable for (a_1, a_2, c, ψ). Up to now we have assumed that the equations of the curves have been solved for x_2: $x_2 = \varphi(x_1, \xi)$. We now assume the differentiation with respect to ξ appears in Eq. (13'), and there are no differentiations with respect to x; here it is more convenient to turn to equations which are solved with respect to ξ_2. If $\Phi(x_1, x_2; \xi_1, \xi_2) = 0$ is another equation defining the family $x_2 = \varphi(x_1; \xi_1, \xi_2)$, then $\partial\varphi/\partial\xi_j = -\frac{\partial\Phi}{\partial\xi_j}\Big/\frac{\partial\Phi}{\partial x_2}$. We write the operator $\hat{L}_{x^0} = a_1\frac{\partial}{\partial\xi_1} + a_2\frac{\partial}{\partial\xi_2} + c$ in the form**

$$\hat{L}_{x_0} = -\frac{1}{\Phi'_{x_2}(x^0, \xi)}\left[A_1(x^0, \xi)\frac{\partial}{\partial\xi_1} + A_2(x^0, \xi)\frac{\partial}{\partial\xi_2} + C(x^0, \xi)\right], \\ A_i = -\Phi'_{x_2}a_i, \quad C = -\Phi'_{x_2}c. \tag{11"}$$

Then (13') is preserved for any equation defining the curves $\Phi(x, \xi) = 0$ if we replace φ by Φ, a_i by A_i, and c by C. Now assume that $\Phi(x, \xi) = \xi_2 + f(\xi_1; x_1, x_2)$, i.e., we have curves $\xi_2 + f(\xi_1; x_1, x_2) = 0$. Then*

$$\mu(x^0, x, \xi) = \frac{A_1(x^0, \xi)\, f'_{\xi_1}(x, \xi_1) + A_2(x^0, \xi)}{f'_{\xi_1}(x^0, \xi_1) - f'_{\xi_1}(x, \xi_1)},$$

and Eq. (13') takes the form

$$[A_1(x^0, \xi) f'(x, \xi_1) + A_2(x_0, \xi)][f'(x_0, \xi_1) - f'(x, \xi_1)][\psi'_{\xi_2}(x, \xi) f'(x^0, \xi_1) - \\ - \psi'_{\xi_1}(x, \xi)] + \psi(x, \xi)[f'(x^0, \xi_1) - f'(x, \xi_1)]\left[\left\{f'(x^0, \xi_1)\frac{\partial}{\partial\xi_2} - \frac{\partial}{\partial\xi_1}\right\}(A_1(x^0, \xi) f'(x, \xi_1) + A_2(x^0, \xi))\right] + \\ + \psi(x, \xi)[A_1(x^0, \xi) f'(x, \xi_1) + A_2(x_0)][f''(x^0, \xi_1) - f''(x, \xi_1)] = \\ = [A_1(x^0, \xi)\psi'_{\xi_1}(x, \xi) + A_2(x^0, \xi)\psi'_{\xi_2}(x, \xi) + c(x^0, \xi)\, \psi(x, \xi)]\, (f'(x^0, \xi_1) - f'(x, \xi_1))^2. \tag{16}$$

We expand the brackets, carry over all the terms into the right-hand part, and regroup the terms so that we obtain a system of expressions of the form $Q_j(x^0, \xi)R_j(x, \xi)$. We give the final result:

$$[\psi(x, \xi) f''(x, \xi_1) - \psi'_{\xi_1}(x, \xi) f'(x, \xi_1) + \\ + \psi'_{\xi_2}(x, \xi)(f'(x, \xi_1))^2]\{A_1(x^0, \xi) f'(x^0, \xi_1) + A_2(x^0, \xi)\} + \\ + [\psi'_{\xi_1}(x, \xi) - \psi'_{\xi_2}(x, \xi) f'(x, \xi_1)]\, f'(x^0, \xi_1)\{A_1(x^0, \xi) f'(x^0, \xi) + A_2(x^0, \xi)\} + \\ + \psi(x, \xi)\, (f'(x, \xi_1))^2\{(A_1)'_{\xi_2}(x^0, \xi) f'(x^0, \xi_1) - (A_1)'_{\xi_1}(x^0, \xi) + C(x^0, \xi)\} + \\ + \psi(x, \xi)\, f'(x, \xi_1)\{(A_1)'_{\xi_1}(x^0, \xi) f'(x^0, \xi_1) - A_1(x^0, \xi) f''(x^0, \xi) - \\ - (A_2)'_{\xi_1}(x^0, \xi) - (A_1)'_{\xi_2}(x^0, \xi)(f'(x^0, \xi_1))^2 + (A_2)'_{\xi_2}(x^0, \xi) f'(x^0, \xi_1) - \\ - 2C(x^0, \xi) f'(x^0, \xi)\} + \psi(x, \xi)\{C(x^0, \xi)(f'(x^0, \xi_1))^2 + (A_2)'_{\xi_1}(x^0, \xi) - \\ - A_2(x^0, \xi) f''(x^0, \xi) - (A_2)'_{\xi_2}(x^0, \xi)(f'(x^0, \xi))^2\} = 0.$$

*** It what follows we will often write f', f'' in place of f'_{ξ_1}, $f''_{\xi_1\xi_1}$, since we have no need of the notation f'_x, f''_{xx}.**

This identity must be satisfied for fixed $\xi = (\xi_1, \xi_2)$ for all $x^0 \in D$, $x \in D$. Taking ξ_1, ξ_2 as parameters, we separate the variables x^0, x.

Since we have an identity of the form

$$Q_1(x;\ \xi)\, R_1(x^0;\ \xi) + \ldots + Q_5\,(x;\ \xi) R_5\,(x^0;\ \xi) = 0,$$

if k among the Q_j are linearly independent, then at least 5 − k of the R_i are linearly independent, and the linear relations on the R_i are consequences of the relations on the Q_j. Of course, the coefficients in the linear relations are functions of the parameters ξ_1, ξ_2 (but not on x^0, x^1). We remark that the 10 functions Q_i, R_i are expressed in terms of the five functions f, ψ, A_1, A_2, C. Since the Q_i are expressed only in terms of the two functions f, ψ, it is natural to expect that there are at least two relations on the Q_i. We consider this possibility, i.e., we assume that there exist functions $\alpha_1(\xi), \ldots, \alpha_5(\xi), \beta_1(\xi), \ldots, \beta_5(\xi)$ such that

$$\alpha_1\,[\psi f'' - \psi'_{\xi_1} f' + \psi'_{\xi_1}\,(f')^2] + \alpha_2\,[\psi'_{\xi_1} - \psi'_{\xi_1} f'] + \alpha_3\psi\,(f')^2 + \alpha_4\psi f' + \alpha_5\psi = 0,$$
$$\beta_1[\psi f'' - \psi'_{\xi_1} f' + \psi'_{\xi_1}\,(f')^2] + \beta_2[\psi'_{\xi_1} - \psi'_{\xi_1} f'] + \beta_3\psi\,(f')^2 + \beta_4\psi f' + \beta_5\psi = 0$$

(the arguments x and ξ are omitted in f and ψ). We show that the assumption that the family of curves f is in general position allows us to assume that $\alpha_1 = \beta_2 \equiv 1,\ \alpha_2 = \beta_1 \equiv 0$, i.e.,

$$\begin{aligned}\psi f'' - \psi'_{\xi_1} f' + \psi'_{\xi_1}\,(f')^2 &= \alpha_3\psi\,(f')^2 + \alpha_4\psi f' + \alpha_5\psi = 0,\\ \psi'_{\xi_1} - \psi'_{\xi_1} f' &= \beta_3\psi\,(f')^2 + \beta_4\psi f' + \beta_5\psi.\end{aligned} \tag{17}$$

It suffices to show that we cannot have $\alpha_1\beta_2 - \alpha_2\beta_1 = 0$, or that we cannot have relations with $\alpha_1 = \alpha_2 \equiv 0$. If we had such a relation, then $\alpha_3(f')^2 + \alpha_4 f' + \alpha_5 = 0$ and our curves would satisfy a first-order equation with respect to ξ in which the coefficients are independent of x, and thus the curves would not depend essentially on the two parameters x_1 and x_2. Thus, our family is degenerate. We note that a finite number of curves of the family will pass through almost all the points. Thus, if there exists a pair of independent relations on Q, relations (17) hold.

We multiply the second of relations (17) by f', add the first, and divide both sides by ψ (we may assume because of "general position" that ψ does not vanish). As a result, we have eliminated ψ and obtain

$$\begin{gathered} f''_{\xi_1\xi_1}(x,\xi_1) = \gamma_1(\xi_1, \xi_2)(f'_{\xi_1}(x,\xi_1))^3 + \gamma_2(\xi_1, \xi_2)(f'_{\xi_1}(x,\xi_1))^2 + \gamma_3(\xi_1, \xi_2) f'_{\xi_1}(x,\xi_1) + \gamma_4(\xi_1, \xi_2),\\ \xi_2 + f(x, \xi_1) = 0,\quad \gamma_1 = \beta_3,\quad \gamma_2 = \beta_4 + \alpha_3,\quad \gamma_3 = \beta_5 + \alpha_4,\quad \gamma_4 = \alpha_5. \end{gathered} \tag{18}$$

Thus, if there are at least two independent relations among the Q_j, a necessary condition for (16) to be solvable is the possibility of defining a family $\xi_2 = -f(x, \xi_1)$ in the ξ plane by Eq. (18).

It remains to exclude the situation when the Q_j satisfy at most one relation. Then the $R_1, \ldots, R_5$ satisfy at least four independent linear relations, i.e., all the R_j are proportional (the coefficients of proportionality depend on ξ). It is convenient for us to transform the expressions for the R_j by introducing the functions

$$\begin{aligned} N\,(x^0, \xi_1, \xi_2) &= A_1\,(x^0, \xi)\, f'\,(x^0, \xi_1) + A_2\,(x^0, \xi),\\ P\,(x^0, \xi_1, \xi_2) &= C\,(x^0, \xi) - (A'_1)_{\xi_2}\,(x^0, \xi) - (A'_2)_{\xi_1}\,(x^0, \xi). \end{aligned} \tag{19}$$

We remark that the equalities

$$N \equiv 0,\quad P \equiv 0 \tag{20}$$

determine the trivial admissible operators (4)

$$\begin{gathered}\hat{L}_{x_1} = p\,(\xi_1, \xi_2)\left(\frac{\partial\varphi\,(x_1^0, \xi_1, \xi_2)}{\partial\xi_1}\,\frac{\partial}{\partial\xi_2} - \frac{\partial\varphi\,(x_1^0, \xi_1, \xi_2)}{\partial\xi_2}\,\frac{\partial}{\partial\xi_1}\right) + \frac{\partial\varphi\,(x_1^0, \xi_1, \xi_2)}{\partial\xi_1}\,\frac{\partial p\,(\xi_1, \xi_2)}{\partial\xi_2} - \frac{\partial\varphi\,(x_1^0, \xi_1, \xi_2)}{\partial\xi_2}\,\frac{\partial p\,(\xi_1, \xi_2)}{\partial\xi_1},\\ x_2^0 = \varphi\,(x_1^0, \xi),\end{gathered} \tag{4'}$$

where $p(\xi)$ is a fixed function. Functions R_j are expressed as follows in terms of N and P:

$$R_1 = N,\quad R_2 = Nf',\quad R_3 = N'_{\xi_2} + P,$$
$$R_4 = -N'_{\xi_2}\cdot f' - N'_{\xi_1} - 2P\cdot f',$$
$$R_5 = N'_{\xi_1}\cdot f - N\cdot f'' + P(f')^2. \tag{21}$$

In particular, the vanishing of all the R_j is equivalent to the triviality of the admissible operator. If, on the other hand, $M \neq 0$ and the Q_j are proportional, then $R_2/R_1 = f'$ cannot depend on x^0 and we have a degenerate case. If, on the other hand, $N \equiv 0$ but $P \not\equiv 0$ (otherwise the operator would be trivial), then the analogous conclusion is obtained by considering the ratios $R_4/R_3 = -2f'$.

Thus, we have proved that in order for Eq. (16) to have a nontrivial solution, it is necessary that the second-order equation defining $f(x, \xi_1)$ in the ξ plane have the form (18). We will prove in the following section that this condition is sufficient. Although to do this it would be sufficient to exhibit one nontrivial solution, we construct simultaneously the general solution $\{\psi, A_1, A_2, C\}$.

Remark. The $Q_1, \ldots, Q_5$ satisfy exactly two independent relations. If they satisfied three independent relations, then we would obtain a relation involving $(f')^2$, f', i.e., a second-order equation for f, from which it would follow that the family is degenerate.

4. Proof of the Sufficiency of (18). General Forms of the Solutions of (16) (of Admissable Operators). Let the family K of curves be defined by the equation $\xi_2 = -f(x, \xi_1)$, where $f(x, \xi_1)$ satisfies differential equation (18) for certain functions $\gamma_1(\xi), \ldots, \gamma_4(\xi)$. Assume that the same family can be defined by $x_2 = \varphi(x_1, \xi_1, \xi_2)$. There is an obvious relation between f and φ. In particular,

$$\varphi'_{\xi_1}(x_1, \xi_1, \xi_2) - f'_{\xi_1}(x_1, \varphi(x_1, \xi_1, \xi_2), \xi_1)\,\varphi'_{\xi_2}(x_1, \xi_1, \xi_2) = 0. \tag{22}$$

We are interested in all the possible ways of introducing densities ψ on curves in K such that admissible nontrivial operators exist. We note that the density $\psi(x_1, \xi_1, \xi_2)$ depends on the curve and the point on the curve. As arguments, we can take either (x_1, ξ_1, ξ_2) or (x_1, x_2, ξ_1) by considering $\psi(x_1, \xi_1, -f(x_1, x_2, \xi_1))$.

If Eq. (18) holds, then we can use any of the equivalent equations in (17) to define the density. It is more convenient to take the second relation,

$$\psi'_{\xi_1}(x_1, \xi_1, \xi_2) - f'_{\xi_1}(x_1, \varphi(x_1, \xi_1, \xi_2), \xi_1)\,\psi'_{\xi_2}(x_1, \xi_1, \xi_2) = \psi(x_1, \xi_1, \xi_2)\,H(\xi_1, \xi_2, x_1),$$
$$H = \gamma_1(\xi)\,[f'_{\xi_1}(x_1, \varphi(x_1, \xi), \xi_1)]^2 + \beta_4(\xi) f'_{\xi_1}(x_1, \varphi(x_1, \xi), \xi_1) + \beta_5(\xi). \tag{17'}$$

If desired, we can express $f'(x_1, \varphi(x_1, \xi), \xi_1)$ in terms of φ'_{ξ_1}, φ'_{ξ_2} in (22), and then (17') will contain only φ and ψ. We note that we make use of the fact that by Eq. (18), $\gamma_1(\xi) = \beta_3(\xi)$. In contrast, the two other functions $\beta_4(\xi)$, $\beta_5(\xi)$ can be defined arbitrarily. Instead of this pair of functions of ξ, we can choose a pair of arbitrary functions in another way. We remark that if we go from the solution ψ of Eq. (17') to the function $\psi(x_1, \xi)g(\xi)$, we obtain a solution of an analogous equation, except that β_4 is replaced by $\beta_4 - (g'_{\xi_2}/g)$ and β_5 is replaced by $\beta_5 + (g'_{\xi_1}/g)$ (γ_1 does not change). Making use of this fact, we can reduce the problem to the case $\beta_4(\xi) \equiv 0$, viz., seek ψ in the form $\psi(x_1, \xi) = d(\xi)\psi_0(x_1, \xi)$, where ψ_0 is a solution of the equation

$$\psi_{\xi_1} - f'\cdot\psi_{\xi_2} = H_{0,b}\cdot\psi,\quad H_{0,b} = \gamma_1(\xi)\,[f']^2 + b(\xi). \tag{17''}$$

We can now take $d(\xi)$ and $b(\xi)$ to be arbitrary functions. In addition, every function $e(x_1, x_2)$, considered as a function $e(x_1, \varphi(x_1, \xi))$ of x_1; ξ_1, ξ_2, is a solution of the homogeneous equation $e'_{\xi_1} - f'e'_{\xi_2} = 0$ (by (22)), and therefore solutions of (17') and (17'') remain solutions when multiplied by such a function. As a result,

$$\psi(x_1, \xi_1, \xi_2) = d(\xi_1, \xi_2)\,e_1(x_1, x_2)\exp(h(x_1, \xi_1, \xi_2)),$$
$$x_2 = \varphi(x_1, \xi_1, \xi_2), \tag{23}$$

where $d(\xi)$ and $e_1(x)$ are arbitrary functions, and $h(x_1, \xi)$ is a solution of the equation

$$h'_{\xi_1} - f'h'_{\xi_2} = H_{0,b}, \tag{24}$$

normalized, e.g., by the condition $h(x_1, 0, \xi_2) = 0$; $b(\xi_1, \xi_2)$ is a third arbitrary function.

We turn to the description of admissible operators, i.e., $A_1(x^0, \xi)$, $A_2(x^0, \xi)$, $C(x^0, \xi)$. These three functions are expressed essentially in terms of the two functions $N(x^0, \xi)$, $P(x^0, \xi)$ we find in (19).

If the $Q_1, \ldots, Q_5$ are related by Eqs. (17), then we have three linear relations involving five functions for the R_i: $-(\alpha_i R_1 + \beta_i R_2) = R_i$, $i = 3, 4, 5$, i.e.,

$$-N(\alpha_3+\beta_3 f') = N'_{\xi_1} + P, \quad -N(\alpha_4+\beta_4 f') = -N'_{\xi_2} f' - N'_{\xi_1} - 2Pf',$$
$$-N(\alpha_5+\beta_5 f') = N'_{\xi_2} f' - Nf'' + P(f')^2. \tag{25}$$

Three equations for the two functions N and P are obtained; however, these equations are not independent. If we multiply the second equation by f' and the first equation by $(f')^2$ and add them to the third equation, we obtain (18).

We now multiply the first equation by $2f'$ and add it to the second. We obtain an equation for $N(x^0, \xi)$ which is analogous to Eq. (17') for $\psi(x_1, \xi)$,

$$N'_{\xi_1} - f'N'_{\xi_2} = G\cdot N, \quad G = 2\beta_3 (f')^2 + (\beta_4 + 2\alpha_3) f' + \alpha_4. \tag{26}$$

Using (18), we express α_i in terms of γ_j and β_k,

$$G = 2\gamma_1 (f')^2 + (2\gamma_2 - \beta_4) f' + \gamma_3 - \beta_5. \tag{27}$$

Equation (26) is solved analogously to (17'). In particular, if we put $N(x^0, \xi) = d^{-1}(\xi) N_0(x^0, \xi)$, where $d(\xi)$ is the function which appeared in solution (17'), then N_0 is a solution of the equation

$$N'_{\xi_1} - f'N'_{\xi_2} = G_{0,b}\cdot N, \quad G_{0,b} = 2\gamma_1 (f')^2 + 2\gamma_2 f' + \gamma_3 - b, \tag{26'}$$

and we obtain that

$$N(x^0, \xi) = d^{-1}(\xi)\, e_2(x_1^0, x_2^0), \exp(g(x_1^0, \xi_1, \xi_2)), \quad x_2^0 = \varphi(x_1^0, \xi_1, \xi_2), \tag{28}$$

where $g(x_1^0, \xi)$ is a solution of the equation

$$g'_{\xi_1} - f' g'_{\xi_2} = G_{0,b}, \quad g(x_1^0, 0, \xi_2) = 0.$$

We next express P in terms of N using the first equation in (25); we obtain $P = -N'_{\xi_2} - N(\alpha_3 + \beta_3 f')$. It is necessary to use $\beta_3 = \gamma_1$, $\alpha_3 = \gamma_2 - \beta_4$, and the fact that β_4 is (by what was said earlier) expressable in terms of $d(\xi)$, viz., $\beta_4 = -d'_{\xi_2}/d$. As a result, we have

$$P = -N'_{\xi_2} - N(\gamma_1 f' + \gamma_2 + (d'_{\xi_2}/d)). \tag{29}$$

Next we can give $A_1(x^0, \xi)$ (or $A_2(x^0, \xi)$) arbitrarily, and use (19) to write

$$A_2(x^0, \xi) = N(x^0, \xi) - f'(x^0, \xi_1) A_1(x^0, \xi),$$
$$C(x^0, \xi) = P(x^0, \xi) + (A_1)'_{\xi_1}(x^0, \xi) + (A_2)'_{\xi_2}(x^0, \xi). \tag{30}$$

We summarize the results obtained.

THEOREM 2. The manifold K of curves $\xi_2 = -f(x_1, x_2, \xi_1)$ is admissible if and only if there exists $\gamma_1(\xi), \ldots, \gamma_4(\xi)$ such that $f(x_1, \xi)$ satisfies the equation

$$f'' = \gamma_1 (f')^3 + \gamma_2 (f')^2 + \gamma_3 f' + \gamma_4, \quad \xi_2 + f(x_1, \xi) = 0. \tag{18}$$

Under these conditions, the admissible densities are determined by giving two arbitrary functions $b(\xi)$, $d(\xi)$ of ξ and a single function of $x(e_1(x))$. Then

$$\psi(x_1, \xi) = d(\xi)\, e_1(x) \exp(h(x_1, \xi)),$$
$$h'_{\xi_1} - f'h'_{\xi_2} = \gamma_1 (f')^2 + b, \quad h(x_1, 0, \xi_2) = 0. \tag{23}$$

After the densities $\psi(x_1, \xi)$ have been fixed, in order to define the operator L_{x^0} it is necessary, in addition, to give an arbitrary function $e_2(x^0)$ and an arbitrary function $A_1(x_1^0, \xi)$. Then $A_2(x_1^0, \xi)$, $C(x_1^0, \xi)$ are expressed in terms of the set of given functions using (28)-(30).

We recall that for the coefficients of the operator L_{x^0} we have

$$a_i(x^0, \xi) = -\frac{1}{f_{x_2}} A_i(x^0; \xi), \quad i = 1, 2, \quad c(x^0, \xi) = -\frac{1}{f_{x_2}} C(x^0, \xi). \tag{11''}$$

Remarks. 1. We note that coefficient γ_4 in (18) does not appear in the calculation of ψ, A_1, A_2, C; γ_2 and γ_3 do not appear either in the calculation of ψ.

2. In (17') the main variables were taken to be x_1, ξ_1, ξ_2. If we make the change of variables $(x_1, \xi_1, \xi_2) \to (x_1, x_2, \xi_1)$, $x_2 = \varphi(x_1, \xi)$, then (17') takes the form

$$\widetilde{\psi}'_{\xi_1}(x_1, x_2, \xi_1) = \widetilde{\psi}(x_1, x_2, \xi_1)\cdot H(x_1, \xi_1, -f(x_1, x_2, \xi_1)),$$
$$\widetilde{\psi}(x_1, x_2, \xi_1) = \psi(x_1, \xi_1, -f(x_1, x_2, \xi_1)); \qquad (31)$$

in $\widetilde{\psi}$ is found by integrating $H(x_1, \xi)$ over the curves $\xi_2 = -f(x, \xi_1)$ in the ξ plane.

If $\gamma_4(\xi) \equiv 0$ in (18), then it is clear that $f(x_1, x_2, \xi_1)$ is a solution of (31) for $\beta_4 = \gamma_2, \beta_5 = \gamma_3$. Thus, when $\gamma_4 = 0$, we can take ψ to be

$$\psi(x_1, \xi_1, \xi_2) = f'_{\xi_1}(x_1, \varphi(x_1, \xi_1, \xi_2), \xi_1) = \frac{\varphi'_{\xi_1}(x_1, \xi)}{\varphi'_{\xi_2}(x_1, \xi)}. \qquad (32)$$

If we use the invariance of our constructions under diffeomorphisms, we obtain a certain class of admissible densities which do not require integration for their evaluation in terms of f. If, in addition, $\gamma_2 = \gamma_3$ is identically equal to zero, we can take $N(x_1^0, \xi)$ to be $[f'(x_1^0, \xi_1)]^2$, $x_2^0 = \varphi(x_1^0, \xi)$.

It can be shown that the coefficients $\gamma_2, \gamma_3, \gamma_4$ in (18) can be made zero by means of a diffeomorphic change of variables (11).

5. Admissible Operators on Manifolds of Curves and Circles. As an example of a two-dimensional manifold K, we consider the manifold of lines $x_2 = \xi_1 x_1 + \xi_2$, i.e., $\varphi(x_1, \xi) = \xi_1 x_1 + \xi_2$, $f(x, \xi_1) = \xi_1 x_1 - x_2$. In order to prove that this manifold is admissible it suffices to consider $\psi \equiv 1$, $\hat{L} = \partial/\partial\xi_2$. Then it is obvious that (13') is satisfied. However, we are also interested in other possibilities for ψ, $\hat{L}$.

We can fix a number a and put $\psi(x_1, \xi) = 1/(x_1 - a)$, $\hat{L} = (\partial/\partial\xi_1) - (a\partial/\partial\xi_2)$. Then $L\psi \equiv 0$, and, on the other hand, $\mu(x_1^0, x_1, \xi) = (x_1 - a)/(x_1 - x_1^0)$ does not depend on ξ and (13') holds. We emphasize that in this case, (13') is a consequence of the fact that both sides are equal to zero — $\mu\psi$ is constant for fixed x_1^0 and ψ does not depend on ξ. This example can be regarded as being obtained from the preceding example by passing to the line $x_1 = a$ "at infinity"; here $\hat{L} = \partial/\partial\xi_1 - a\partial/\partial\xi_2$ corresponds to an infinitesimal rotation about the point $(x_1 = a, x_2 = \xi_1 a + \xi_2)$ "at infinity."

A more general construction consists in fixing a curve $x_2 = \theta(x_1)$ and considering a small region around ξ in which the lines $x_2 = \xi_1 x_1 + \xi_2$ have a unique point of intersection $(a(\xi), \xi_1 a(\xi) + \xi_2)$ with the curve $x_2 = \theta(x_1)$ in D, i.e., the function $a(\xi)$ is defined by the equation $\theta(a(\xi_1, \xi_2)) = a(\xi_1, \xi_2)\xi_1 + \xi_2$. We put $\psi(x_1, \xi) = (x_1 - a(\xi))^{-1}$, $\hat{L} = \partial/\partial\xi_1 - (a(\xi)\partial/\partial\xi_2)$. In addition, L does not depend on x^0. Here $\mu(x_1^0, x_1, \xi) - (x_1 - a(\xi))/(x_1 - x_1^0)$ and $\mu\psi = 1/(x_1 - x_1^0)$ as before depend only on x_1^0. Furthermore, although $\psi(x_1, \xi)$ now depends on ξ, $L\psi \equiv 0$ since $\psi(x_1, \xi)$ does not change when we go to a line having the same point of intersection with the curve $x_2 = \theta(x_1)$. The geometric meaning of this construction is that the point "at infinity" of the line is its point of intersection with $x_2 = \theta(x_1)$, i.e., it is not necessary to fix a single affine structure on the plane. However, it is important that the form of $\psi(x_1, \xi)$ is determined by the point "at infinity," and $\hat{L}$ is an infinitesimal rotation about this point.

It is convenient to combine all these densities by considering over the manifold of lines $x_2 = \xi_1 x + \xi_2$ the bundle of densities $\psi(x_1, \xi_1, \xi_2, \eta_1, \eta_2) = \eta_1/(x_1 - \eta_2)$. We obtain a four-dimensional manifold E, and on the submanifolds $E(x^0)$: $x_2^0 = \xi_1 x_1^0 + \xi_2$ the form

$$L_{x^0} I_F = \frac{1}{\eta_1}\left(\frac{\partial I(\xi_1, \xi_2)}{\partial\xi_1} - \eta_2 \frac{\partial I(\xi_1, \xi_2)}{\partial\xi_2}\right) d\xi_1$$

is closed (the coefficients of $d\eta_1$, $d\eta_2$ in $L_{x^0} I_F$ are zero). We have

$$L_{x^0} I_F(\xi_1, \eta; d\xi_1, d\eta) = \left[\int F(x_1, \xi_1(x_1 - x_1^0) + x_2^0)\, dx_1\right] d\xi_1.$$

The set of measures $\psi(x_1, \xi, \eta)dx_1$ on the lines is remarkable in that it is invariant under projective transformations. A complex analogue of the form $L_{x^0} I$ was considered in [17].

The general form of admissible densities and operators is obtained from Theorem 2. The formulas are somewhat simpler than for arbitrary curves, since in the case of lines $f'' \equiv 0$, i.e., $\gamma_1 = \gamma_2 = \gamma_3 = \gamma_4 \equiv 0$. We give the formulas:

$$\psi(x_1, \xi) = d(\xi)\, e_1(x_1, \xi_1 x_1 + \xi_2) \exp(h_b(x_1, \xi)),$$
$$h_b(x_1, \xi) = \int_0^{\xi_1} b(t, \xi_1 x_1 + \xi_2 - \xi_1 t)\, dt,$$

$$N(x_1^0, \xi) = d^{-1}(\xi)\, e_2(x_1, \xi_1 x_1 + \xi_2) \exp(-h_b(x_1^0, \xi)),$$

$$P(x_1^0, \xi) = -N'_{\xi_2} + \beta_4 N = \frac{-N'_{\xi_1} - \beta_5 N}{x_1}, \quad \beta_5(\xi) = b(\xi) + (d'_{\xi_1}(\xi)/d(\xi)),$$

$$A_1(x_1^0, \xi) = a_1(x_1^0, \xi) = \frac{N(x_1^0, \xi) - A_2(x_1^0, \xi)}{x_1}, \quad A_2 = a_2,$$

$$C(x_1^0, \xi) = c(x_1^0, \xi) = P + (A_1)'_{\xi_1} + (A'_2)_{\xi_2} = \frac{-\beta_5 N - (A_2)'_{\xi_1} + x_1 (A_2)'_{\xi_2}}{x_1},$$

where $b(\xi)$, $d(\xi)$, $e_1(x)$, $e_2(x)$, $A_2(x_1^0, \xi)$ are arbitrary functions. We note that the admissible functions $h(x_1, \xi)$ form a linear space and $\psi(x_1, \xi)$ can be multiplied. We remark that if $b(\xi_1, \xi_2) = -\int a'_s(s, t)|_{s=\xi_1 t+\xi_2}\, dt$, then $h_b(x_1, \xi) = \int a(\xi_1 t + \xi_2, t) \frac{dt}{x_1 - t}$.*

We now investigate some two-dimensional manifolds of circles. Assume we are given a family $(x_1 - \xi_1)^2 + x_2^2 = \xi_2$ of circles with centers on the x_1 axis. Then $f'' = 2$ and Eq. (18) holds with $\gamma_1 = \gamma_2 = \gamma_3 \equiv 0$. For this reason, all the formulas will be analogous to the formulas for lines, except that in calculating $h_b(x_1, \xi)$ it is necessary to integrate not over lines but over translations of the parabola $\xi_2 = \xi_1^2$. In particular, we can take $\psi \equiv 1$, $\hat{L} = \partial/\partial\xi_2$. We note that by making the change of variables $\tilde{x}_2 = x_1^2 + x_2^2$, $\tilde{\xi}_2 = \xi_2 - \xi_1^2$, $\tilde{x}_1 = x_1$, $\tilde{\xi}_1 = \xi_1$, we transform our system of circles into a system of lines $\tilde{\xi}_2 = \tilde{x}_2 - 2\tilde{\xi}_1\tilde{x}_1$. We recall that these circles are geodesics in the Poincaré model of the Lobachevskii plane on the upper half plane.

We now consider the family of circles $(x_1 - \xi_1)^2 + x_2^2 - 2\xi_2 x_2 = 0$, which are horicycles in this model. Here

$$f(x, \xi_1) = -\frac{(x_1 - \xi_1)^2}{2x_2} - \frac{x_2}{2} \text{ and } f'' = -\frac{1}{2\xi_2}(f')^2 - \frac{1}{2\xi_2}.$$

Here we have again obtained a differential equation of the form (18); $\gamma_1 \equiv 0$, $\gamma_2(\xi) = \gamma_3(\xi) = 1/2\xi_2$, $\gamma_4 \equiv 0$. Since $\gamma_1 \equiv 0$, the formula for ψ will be similar to the preceding one, except that now we must integrate over the parabolas $\xi_2 = a(\xi_1 - b)^2 + a^{-1}$. In this case, as in the general case, for $\gamma_1 \equiv 0$, we can take $\psi \equiv 1$. Our family of circles will no longer be diffeomorphic to a family of lines, which can be seen either with the aid of a general criterion (cf. [11]) or by remarking that two circles in the family pass through any two points x^0.

Finally, if we consider a family of circles of a fixed radius, say $(x_1 - \xi_1)^2 + (x_2 - \xi_2)^2 = 1$, then we have for it $f'' = -[(f')^2 + 1]^{3/2}$ and since this equation does not have the form (18), this family is not admissible. We emphasize that the possibility of defining a family of curves by a differential equation of the form (18) is a differential invariant of the family. This is proved in [11], but is also a consequence of our arguments.

6. Geometric Meaning of the Admissibility Condition. The results obtained can be interpreted geometrically using well-known facts in the geometry of second-order differential equations in the plane (cf. [11, pp. 46-57]). Every two-parameter family K of curves $x_2 = \varphi(x_1, \xi_1, \xi_2)$ can be considered as a set of solutions of some second-order differentiation equation $x_2'' = A(x_1, x_2, x_2')$. To each two-parameter family of curves on the plane there is associated the dual family $x_2 - \varphi(x_1, \xi_1, \xi_2) = 0$ on the ξ plane, where x_1 and x_2 now play the role of parameters. We define this dual family $\hat{K}$ by means of equations which are solved for ξ_2, $\xi_2 = -f(x_1, x_2, \xi_1)$ and let $f(x, \xi_1)$ be a solution of the equation $\xi_2'' = \hat{A}(\xi_1, \xi_2, \xi_2')$. Theorem 2 proved above asserts that the family of curves is admissible if and only if $\hat{A}(\xi_1, \xi_2, \xi_2')$ is a polynomial of degree three in ξ_2' for the differential equation corresponding to the dual family. Because of the invariance of the notion of admissibility, this property of an equation is invariant under diffeomorphism (a direct proof is given in [11]).

It turns out that if $\hat{A}(\xi_1, \xi_2, \xi_2')$ is a polynomial of degree three in ξ_2', then by means of a diffeomorphism the equation $x_2'' = A(x_1, x_2, x_2')$ can in the neighborhood of any solution be brought to the form $x_2'' = O(|x_2|^3 + |x_2'|^3)$ in a neighborhood of $x_2 \equiv 0$. In the general case, we can by means of a diffeomorphism, bring the equation to the form $x_2'' = a(x_1)x_2^2 + O(|x_2|^3 + |x_2'|^3)$, where $a(x_1)$ measures the local departure from a Desargues plane (cf. [11]). In particular, if $a(x_1)$ vanishes, this means that the plane is locally Desargues. Thus, the condition that the family K be admissible coincides with the condition that it be locally Desargues. If, on the other hand, A is at the same time a polynomial of degree three in x_2' and $\hat{A}$ is a polynomial of degree three in ξ_2', the family of curves K is locally diffeomorphic to a family of lines.

Cartan [12] showed that second-order differential equations $\xi_2'' = A(\xi_1, \xi_2, \xi_2')$ in which the right-hand side is a polynomial of degree three in ξ_2' play an important role in geometry. Indeed, the solutions of these equa-

* The admissible operators on the manifold of lines in R^n were calculated jointly with M. I. Graev.

tions, and only they, are geodesics for the affine and projective connections on the plane. Thus, admissibility of the family K is equivalent to the condition that the dual family is the set of geodesics for an affine or projective connection. It turns out that these connections are closely related to the corresponding admissible operators, and the nonuniqueness of the connection for a fixed system of geodesics reflects the nonuniqueness of admissible operators L and densities ψ on an admissible manifold K. We clarify this relation.

Thus, to each ξ there corresponds a curve N_ξ in the x plane, and to each x there corresponds a curve M_x in the ξ plane. We have $x \in N_\xi \Leftrightarrow \xi \in M_x$. We fix ξ and consider the tangent space T_ξ as a linear space. To a point $x \in N_\xi$ we associate the line in T_ξ containing the origin and parallel to the tangent vector to M_x. This gives a local isomorphism between points N_ξ and lines in T_ξ passing through the origin, i.e., a local projective line structure is canonically introduced in N_ξ.

Assume that an affine connection is given on the ξ plane for which the M_x, and only they, are geodesics. This can be done if and only if $\hat{A}(\xi, \xi_2')$ is a polynomial of degree three in ξ_2'. Let ξ and η be two points in the ξ plane. To each curve joining ξ and η, the affine connection associates a linear transformation of tangent spaces. Under this transformation, lines passing through the origin go into lines passing through the origin, and we get a map from N_ξ and N_η compatible with their local projective line structures. If this curve is a geodesic M_x, then x is a point of intersection of N_ξ and N_η, and the point x is preserved under this map.

In order to pass to a problem in integral geometry, we define on each curve N_ξ the class of projective measures induced by the local projective line structure on N_ξ. These measures depend on two parameters, the first of which is related to the choice of the point at infinity, and the second of which is related to the choice of multiplicative constant (cf. p. 98). If on N_ξ we fix an arbitrary affine coordinate x_1 compatible with the projective structure, these measures have the form $(\eta_1/(x_1-\eta_2))dx_1$. As a result we obtain a fibering of the four-dimensional manifold of measures $\psi(x_1, \xi, \eta)dx_1$ over K. Thus, to each pair (ξ, η), where ξ parametrizes the curves N_ξ and the η are measures on them, there is associated the point $x_\infty(\xi, \eta)$ at infinity on N_ξ. We take the operator $\hat{L}_{x^0}$ corresponding to infinitesimal translations along $M_{x^0} \ni \xi$ at the point (ξ, η) to be the infinitesimal translation operator along $M_{x_\infty(\xi,\eta)}$. We recall that a translation along the geodesic M_x corresponds to a pointwise map of the curves N_ξ, $x \ni N_\xi$, preserving the point x and compatible with the projective structures on N_ξ. As a result, we obtain an admissible operator.

Indeed, let us verify (13') at some point (x^0, ξ, η). Let $\xi = 0 = (0, 0)$, and assume that an affine coordinate x_1 has been introduced on N_0. Let $(0, \eta)$ correspond to the measure $(\eta_1/(x_1-\eta_2))dx_1$. We introduce coordinates in a neighborhood of $\xi = 0$ in such a way that the parallel translation operator coincides with the identity operator up to terms of the third order of smallness. Let $\xi_2 = 0$ on M_x, where $x \in N_0$, $x_1 = 0$, $\xi_1 = 0$ on M_x when $x_1 = \infty$. By means of the connection we carry over the coordinate x_1 from N_0 to all the N_ξ, $\xi_1 = 0$. If $x \in N_{(0,\xi_2)}$, then we assign to it the coordinate x_1 thus obtained and $x_2 = \xi_2$. In these coordinates the N_ξ have the form $x_2 = \xi_1 x_1 + \xi_2 + O(|\xi|^3)$. Then

$$\psi(x, \xi, \eta) = \frac{\eta_1}{x_1-\eta_2} + O(|\xi|^2), \qquad \hat{L} = \frac{1}{\eta_1}\left(\frac{\partial}{\partial \xi_1} - \eta_2 \frac{\partial}{\partial \xi_2}\right),$$
$$\mu(x_1^0, x_1, \xi, \eta) = \frac{x_1-\eta_2}{\eta_1(x_1-x_1^0)} + O(|\xi|^2), \quad \hat{L}\psi = 1 + O(|\xi|),$$
$$\mu\psi = \frac{1}{x_1-x_1^0} + O(|\xi|)$$

and (13') holds with $\xi = 0$.

3. A Problem in Integral Geometry Associated with a Family of Curves in C^n

We give here some definitions and formulas for the complex case. The proofs are omitted since they do not differ from the proofs in the real case.

Let D be a bounded domain in C^n. Let Ξ denote the set of analytic curves in D: $z' = \varphi(z_1)$, $z' = (z_2, \ldots, z_n)$, $\varphi = (\varphi_2, \ldots, \varphi_n)$; $\Xi(z^0)$, the set of curves passing through $z^0 \in D$; $\psi(z_1)$, an analytic function, and let Π be the space of pairs (φ, ψ).

If F is a smooth function with compact support, we set

$$I_F(\varphi, \psi) = \int F(z_1, \varphi(z_1)) |\psi(z_1)|^2 dz_1 d\bar{z}_1. \tag{33}$$

We will consider on Π the holomorphic variations $(\delta\varphi, \delta\psi)$, the linear operators M on them, and the operators given in (1,1)-forms by

$$\begin{gathered} L_{z^0} I(\varphi, \psi; \delta\varphi, \delta\psi, \overline{\delta\varphi}, \overline{\delta\psi}) = \hat{L}_{z^0} \bar{\hat{L}}_{z^0} I, \\ \hat{L}_{z^0} I(\varphi, \psi; \delta\varphi, \delta\psi) = \delta I(\varphi, \psi; M(\delta\varphi, \delta\psi)) + I(\varphi, \psi)\, c(\varphi, \psi, \delta\varphi, \delta\psi), \\ \bar{\hat{L}}_{z^0} I(\varphi, \psi; \overline{\delta\varphi}, \overline{\delta\psi}) = \bar{\delta} I(\varphi, \psi; \overline{M(\delta\varphi, \delta\psi)}) + I(\varphi, \psi)\, \overline{c(\varphi, \psi, \delta\varphi, \delta\psi)} \end{gathered} \tag{34}$$

on analytic manifolds E in Π, where c is a fixed (1, 0)-form. For I_F we have

$$\begin{aligned} L_{z^0} I_F = & \int \{F''_{z'\bar{z}'}(z_1, \varphi(z_1)), M^{\varphi}(\delta\varphi, \delta\psi)\, \overline{M^{\varphi}(\delta\varphi, \delta\psi)}\}\, |\psi(z)|^2\, dz_1 d\bar{z}_1 + \\ & + \int \langle F'_{z'}(z_1, \varphi(z_1)), M^{\varphi}(\delta\varphi, \delta\psi)\rangle\, \psi(z_1)\, \overline{(M^{\psi}(\delta\varphi, \delta\psi) + c\psi(z_1))}\, dz_1 d\bar{z}_1 + \\ & + \int \langle F'_{\bar{z}'}(z_1, \varphi(z_1)), \overline{M^{\varphi}(\delta\varphi, \delta\psi)}\rangle\, \overline{\psi(z_1)}\, (M^{\psi}(\delta\varphi, \delta\varphi) + c\psi(z_1))\, dz_1 d\bar{z}_1 + \int F(z_1, \varphi(z_1))\, |M^{\psi} + c\psi|^2 dz_1 d\bar{z}_1 . \end{aligned}$$

Admissibility means that the (1,1)-form $L_{z^0} I_F$ on $E(z^0)$ is closed (condition (i)) and that for every analytic cycle which is one-dimensional (over C) $\int_\gamma L_{z^0} I_F$ is determined by the value $F(z^0)$ (condition (ii)).

We define the notion of a trivial operator analogously and define the form

$$\begin{aligned} \varkappa_{z^0} I_F = & \int \{F''_{z'\bar{z}'}(z_1, \varphi(z_1)),\ \delta\varphi(z_1)\, \overline{\delta\varphi(z_1)}\}\, |\mu(z_1^0, \varphi, \psi; z_1)|^2\, |\psi(z_1)|^2\, dz_1 d\bar{z}_1 + \\ & + \int \{F'_{z'}(z_1, \varphi(z_1)), \delta\varphi(z_1)\rangle\, \mu(z_1^0, \varphi, \psi; z_1)\, \psi(z_1)\, \overline{\delta(\mu\psi)}\, dz_1 d\bar{z}_1 + \\ & + \int \langle F'_{\bar{z}'}(z_1, \varphi(z_1)), \overline{\delta\varphi(z_1)}\rangle\, \overline{\mu(z_1^0, \varphi, \psi; z_1)}\, \overline{\psi(z_1)}\, \delta(\mu\psi)\, dz_1 d\bar{z}_1 + \int F(z_1, \varphi(z_1))\, |\delta(\mu\psi)|^2\, dz_1 d\bar{z}_1, \end{aligned}$$

where μ is a holomorphic function.

In the finite-dimensional version, the function $I(\xi, \eta)$ is used to construct the (1,1)-form (in the notation of (11)):

$$\begin{gathered} L_{z^0} I = \sum L_i \bar{L}_j I(\xi, \eta)\, d\zeta_i \overline{d\zeta_j} + \sum L_i \bar{M}_j I(\xi, \eta)\, d\zeta_i d\eta_j + \sum M_i \bar{L}_j I(\xi, \eta)\, d\eta_i d\zeta_j + \sum M_i \bar{M}_j I(\xi, \eta)\, d\eta_i \overline{d\eta_j}, \\ L_i = \left\langle a^i, \frac{\partial}{\partial\xi} \right\rangle + \left\langle b^i, \frac{\partial}{\partial\eta} \right\rangle + c^i, \quad \bar{L}_i = \left\langle \bar{a}^i, \frac{\partial}{\partial\bar{\xi}} \right\rangle + \left\langle \bar{b}^i, \frac{\partial}{\partial\bar{\eta}} \right\rangle + \bar{c}^i, \\ M_j = \left\langle d^j, \frac{\partial}{\partial\eta} \right\rangle + e^j, \quad \bar{M}_j = \left\langle \bar{d}^j, \frac{\partial}{\partial\bar{\eta}} \right\rangle + \bar{e}^j. \end{gathered} \tag{35}$$

The complex analog of Theorem 1 holds with the sole change that x must be replaced by z; ξ, η must be considered to be complex; and all the functions a^i, b^i, c^i, d^j, e^j are holomorphic.

The results of Sec. 2 carry over in the completely obvious way to the complex case. The analog of the operator (11') has the form $L_{z^0} I = \bar{\hat{L}}\hat{L} I(\xi)\, d\zeta d\bar{\zeta}$, where

$$\hat{L} = a_1(z_1^0, \xi) \frac{\partial}{\partial\xi_1} + a_2(z_1^0, \xi) \frac{\partial}{\partial\xi_2} + c(z_1^0, \xi),$$

a_1, a_2, and c are holomorphic. The analog of Eq. (13') has the same form, except that the functions which appear there are assumed to be holomorphic.

In order for a two-dimensional manifold of analytic curves $\xi_2 = -f(z_1, z_2, \xi_1)$, depending analytically on the parameters $(\xi_1, \xi_2) \in C^2$, to be admissible, it is necessary and sufficient that (18) hold with holomorphic $\gamma_j(\xi)$. The corresponding ψ, A_i, C are calculated as in the real case. The analysis of examples is similar.

The new fact here is that, as in the real case (γ is a cycle in $\Pi(z^0)$),

$$\int_\gamma \varkappa_{z^0} I_F = c(z^0, \gamma) F(z^0), \tag{36}$$

while at the same time in the complex case there exist cycles in $\Pi(z^0)$ for which $c(z^0, \gamma) \neq 0$ whereas in the real case $c(z^0, \gamma) \equiv 0$ for all cycles γ. An example is given by the cycles of lines $z' = \alpha z_1 + \beta$, which give inversion formulas. Here $\psi(z_1, \zeta) \equiv i/2$, $\mu = (i/2\pi) \cdot 1/(z_1 - z_1^0)$, and correspondingly $L = \frac{i}{2\pi} \sum \frac{\partial I(\alpha, \beta)}{\partial\beta_j} d\alpha_j$. In the set of lines, we can take $\pi\gamma \subset \Xi$ to be any one-dimensional cycle of lines $\pi\gamma$ such that for almost all complex hypersurfaces passing through z^0, the number of intersecting lines taking into account orientation is nonzero. In the case of a cycle, this number is the same for almost all hypersurfaces; this number gives

$c(z^0, \gamma)$. In this case when $n = 2$, it is necessary to take all lines, $c(z^0, \gamma) = 1$. In the case of curves it is simplest to evaluate $c(z^0, \gamma)$ by substituting a trial function. One can also pass to homologous cycles of lines.

On two-dimensional admissible manifolds of curves for ψ, L satisfying the analog of Theorem 2, we have

$$\int_{K(z^0)} L_z I_F = c(z^0) F(z^0). \tag{37}$$

If we assume that the cycle $K(z^0)$ is homologous to a cycle of lines passing through z^0, then it is easy to write down an expression for $c(z^0)$ in terms of ψ on the coefficients of L.

Indeed, let $\xi(z, z^0)$ be defined by the conditions $z_2 = \varphi(z_1, \xi(z, z^0))$, $z_2^0 = \varphi(z_1^0, \xi(z, z^0))$, i.e., the curve corresponding to $\xi(z, z^0)$ joins z and z^0. Put $k(z^0) = 2\pi \lim_{z \to z^0} (z_1 - z_1^0)\, \mu(z_1^0, \xi(z, z^0), z_1)\, \psi(z_1^0, \xi(z, z^0))$. Then $c(z^0) = |k(z^0)|^2$.

LITERATURE CITED

1. I. M. Gel'fand, M. I. Graev, and N. Ya. Vilenkin, Integral Geometry and Associated Problems in Theory of Representation [in Russian], Fizmatgiz, Moscow (1962).
2. I. M. Gel'fand, M. I. Graev, and Z. Ya. Shapiro, "Integral geometry on k-dimensional planes," Funkts. Anal. Prilozhen., 1, No. 1, 15-31 (1967).
3. I. M. Gel'fand and M. I. Graev, "Complexes of k-dimensional planes and the Plancherel formula for SL(n; C)," Dokl. Akad. Nauk SSSR, 179, No. 3, 522-525 (1968).
4. I. M. Gel'fand and M. I. Graev, "Admissible complexes of lines in C^n," Funkts. Anal. Prilozhen., 2, No. 3, 39-52 (1968).
5. K. Maius, "Admissible complexes with one critical point," Funkts. Anal. Prilozhen., 9, No. 2, 81-82 (1975).
6. I. M. Gel'fand and S. G. Gindikin, "Nonlocal inversion formulas in real integral geometry," Funkts. Anal. Prilozhen., 11, No. 3, 12-19 (1979).
7. S. Helgason, "Differential operators on homogeneous spaces," Acta Math., 102, 239-299 (1959).
8. S. Helgason, "Duality and Radon transform for symmetric spaces," Am. J. Math. 85, 667-692 (1963).
9. I. M. Gel'fand and M. I. Graev, "Geometry of homogeneous spaces, representations of groups in homogeneous spaces, and associated problems of integral geometry. I," Tr. Mosk. Mat. Ob-va, 8, 321-390 (1959).
10. S. G. Gindikin and F. I. Karpelevich, "On a problem in integral geometry," in: In Memory of N. G. Chebotarev [in Russian], Kazan. Univ. (1964), pp. 30-43.
11. V. I. Arnol'd, Supplementary Chapters to the Theory of Ordinary Differential Equations [in Russian], Nauka, Moscow (1978).
12. E. Cartan, "Sur les variétés à connection projective," Bull. Soc. Math. France, 52, 205-241 (1924); Oeuvres, III, 1, No. 70, Paris, 825-862 (1955).
13. R. G. Mukhometov, "On a problem of integral geometry," in: Mathematical Problems of Geophysics [in Russian], No. 6, Part 2, Computing Center of the Academy of Sciences of the USSR, Siberian Branch (1975), pp. 212-242.
14. V. G. Romanov, "Integral geometry on the geodesics of an isotropic Riemannian metric," Dokl. Akad. Nauk SSSR, 241, No. 2, 290-293 (1978).
15. I. N. Bernshtein and M. L. Gerver, "On a problem of integral geometry for a family of geodesics and the converse kinematic problem of seismics," Dokl. Akad. Nauk SSSR, 243, No. 2, 302-305 (1978).
16. V. Guillemin and S. Shternberg, Geometric Asymptotics, Am. Math. Soc., Providence, R. I. (1977).
17. I. M. Gel'fand, M. I. Graev, and Z. Ya. Shapiro, "Integral geometry in projective space," Funkts. Anal. Prilozhen., 4, No. 1, 14-32 (1970).

10.

(with S. G. Gindikin and M. I. Graev)

A problem of integral geometry in $\mathbf{RP}^n$, connected with the integration of differential forms

Funkts. Anal. Prilozh. **13** (4) (1979) 64–67 [Funct. Anal. Appl. **13** (1980) 288–290].
MR **83a**:43006. Zbl. **423**:58001

Let $\mathbf{RP}^n$ be real projective space, $S^n \to \mathbf{RP}^n$ be its two-sheeted covering by a sphere. We denote by $G_{p,n}$ the Grassman manifold of p-dimensional nonoriented planes in $\mathbf{RP}^n$, or, what is the same, of (p + 1)-dimensional subspaces of $\mathbf{R}^{n+1}$. For odd p, a p-dimensional plane in $\mathbf{RP}^n$ is orientable and one can consider the manifold $\tilde{G}_{p,n}$ of oriented p-dimensional planes in $\mathbf{RP}^n$; $\tilde{G}_{p,n} \to G_{p,n}$ is a two-sheeted covering.

We denote by $\Omega^r_+(\mathbf{RP}^n)$, $\Omega^r_+(G_{p,n})$ the spaces of C^∞-differential forms of degree r on the corresponding manifolds. By $\Omega^r_-(\mathbf{RP}^n)$, $\Omega^r_-(G_{p,n})$ we shall denote the spaces of odd C^∞-differential forms on the corresponding manifolds, i.e., differential forms on the covering manifolds S^n and $\tilde{G}_{n,p}$, respectively, odd with respect to the involution connected with the covering (cf. [1]).

1. Definition of the Integral Transformation I_p. We define for each p < n the operator I_p of integration of differential r-forms, $r \geq p$, with respect to p-dimensional planes in $\mathbf{RP}^n$. This definition depends on the oddness of p. For odd p we define the operator I_p: $\Omega^r_+(\mathbf{RP}^n) \to \Omega^{r-p}_-(G_{p,n})$.

Let $A = \{(x, h);\ x \in \mathbf{RP}^n,\ h \in \tilde{G}_{p,n},\ x \in h\}$; $A \to \mathbf{RP}^n$ and $A \to \tilde{G}_{p,n}$ be the natural projections. If one is given $h \in G_{p,n}$ and $b_1, \ldots, b_{r-p} \in T_hG_{p,n}$, then we choose at each point $x \in h$ vectors $\hat{b}_1, \ldots, \hat{b}_{r-p} \in T_x\mathbf{RP}^n$, such that b_j and $\hat{b}_j$ have common preimage in $T_{(x,h)}A$. Now if $\omega \in \Omega^r_+(\mathbf{RP}^n)$ and $\xi_1, \ldots, \xi_p \in T_xh \subset T_x\mathbf{RP}^n$, then one can show that

$$\omega_{b_1, \ldots, b_{r-p}}(x; \xi_1, \ldots, \xi_p) = \omega(x; \hat{b}_1, \ldots, \hat{b}_{r-p}, \xi_1, \ldots, \xi_p)$$

(for fixed $b_1, \ldots, b_{r-p}$) is independent of the choice of $\hat{b}_1, \ldots, \hat{b}_{r-p}$ and is a differential p-form on h. We set

$$I_p\omega(h; b_1, \ldots, b_{r-p}) = \int_h \omega_{b_1, \ldots, b_{r-p}}, \qquad h \in G_{p,n} \quad b_j \in T_h\tilde{G}_{p,n}. \tag{1}$$

One verifies directly that this expression is an (r − p)-form on $\tilde{G}_{p,n}$, odd with respect to the orientation. As a result, we have defined the operator I_p for odd p. For even p, if $\omega \in \Omega^r_-(\mathbf{RP}^n)$, then in an analogous way one defines an odd p-form $\omega_{b_1,\ldots,b_{r-p}}$ on the nonoriented p-dimensional plane h for $b_j \in T_hG_{p,n}$. Then its integral $I_p\omega(h; b_1, \ldots, b_{r-p})$ with respect to h makes sense and turns out to be an (r − p)-form on $G_{p,n}$. Thus, for even p one defines an operator I_p: $\Omega^r_-(\mathbf{RP}^n) \to \Omega^{r-p}_+(G_{p,n})$. Since for each p one chooses forms of completely definite parity, we shall usually omit the signs +, − and write I_p: $\Omega^r(\mathbf{RP}^n) \to \Omega^{r-p}(G_{p,n})$.

2. Description of the Image of the Operator I_p for r − p > 1. Injectivity of I_p. We shall call a form $\sigma \in \Omega^{r-p}(G_{p,n})$ admissible if for any (r − 1)-dimensional plane $L \subset \mathbf{RP}^n$ the restriction of σ to the submanifold of p-dimensional planes lying in L is equal to zero.

THEOREM 1. If r − p > 1, then the form $\sigma \in \Omega^{r-p}(G_{p,n})$ can be represented in the form $\sigma = I_p\omega$, $\omega \in \Omega^r(\mathbf{RP}^n)$ if and only if σ and $d\sigma$ are admissible.

For r − p = 1, any 1-form σ is admissible, and the admissibility of $d\sigma$ is a necessary but not sufficient condition for the representability in the form $\sigma = I_p\omega$. Necessary and sufficient conditions seem to be more complicated; they will be formulated below.

THEOREM 2. The operator I_p for r − p > 0 has no kernel.

3. Some Explicit Formulas. Our next goal is to construct the inverse operator for I_p for even p. This operator will be local in the sense that ω at a point $x \in \mathbf{RP}^n$ can be recovered by restricting $I_p\omega$ to an in-

Institute of Applied Mathematics, Academy of Sciences of the USSR. Moscow State University. Translated from Funktsionalinyi Analiz i Ego Prilozheniya, Vol. 13, No. 4, pp. 64–66, October-December, 1979. Original article submitted June 22, 1979.

finitesimal neighborhood of the submanifold of planes passing through x. For odd p the inverse operator is nonlocal and we shall not consider it here.

To construct the inverse operator, we need some auxiliary concepts which will be introduced here for all p and not just for even ones. First of all we introduce coordinates. We consider the bundle $R^{n+1} \setminus 0 \to RP^n$; coordinates in R^{n+1} are homogeneous coordinates in RP^n. On the other hand, we consider the Stiefel manifold of (p + 1)-frames $u = (u_1, \ldots, u_{p+1})$ in R^{n+1}; we denote it by $E_{p,n}$. There is a natural fibration $\pi: E_{p,n} \to G_{p,n}$, where π takes a frame into the subspace of R^{n+1} generated by it; for odd p there is also a natural fibration $E_{p,n} \to \tilde{G}_{p,n}$. This allows one to lift from $G_{p,n}$ to $E_{p,n}$ both even and odd differential forms. We denote by $\Omega_s(E_{p,n})$ the space of C^∞-differential s-forms on $E_{p,n}$, homogeneous of degree 0 in all u_i. We shall write them in the form

$$\sigma = \sum \sigma_{i_1, \ldots, i_s}^{j_1, \ldots, j_s}(u)\, du_{j_1}^{i_1} \wedge \ldots \wedge du_{j_s}^{i_s} = \sum \sigma_I^J(u)\, du_J^I, \tag{2}$$

where the coefficients are assumed to be antisymmetric in the pairs $(i_1, j_1), \ldots, (i_s, j_s)$; $I = (i_1, \ldots, i_s)$, $J = (j_1, \ldots, j_s)$. We lift to $E_{p,n}$ the forms $\sigma \in \Omega^s(G_{p,n})$.

<u>LEMMA 1.</u> If the form $\sigma \in \Omega^s(E_{p,n})$ drops to $G_{p,n}$, then it is admissible if and only if the coefficients $\sigma_{i_1, \ldots, i_s}^{j_1, \ldots, j_s}$ of this form are antisymmetric in $i_1, \ldots, i_s$ for any fixed $j_1, \ldots, j_s$.

We single out the last vector $u_{p+1} = x$ of the frame u; we set $u = (v, x)$, where $v = (v_1, \ldots, v_p)$ is a p-frame. We denote by Ω_0^s the subspace of forms in $\Omega^s(E_{p,n})$, containing differentials only with respect to $x^1, \ldots, x^{n+1}$. One defines a projection $\sigma \mapsto \partial_0$ onto Ω_0^s in the natural way.

We call a form $\tau = \sum_{i_1, \ldots, i_p} \tau_{i_1, \ldots, i_p}(v, x)\, dx^{i_1} \wedge \ldots \wedge dx^{i_p} \in \Omega_0^p$ p-flat if $p! \sum_{i_1, \ldots, i_p} \tau_{i_1, \ldots, i_p}(v, x)\, v_1^{i_1} \ldots v_p^{i_p} = 1$. We call a form $\nu = \sum_{i_1, \ldots, i_r} \nu_{i_1, \ldots, i_r}(v, x)\, dx^{i_1} \wedge \ldots \wedge dx^{i_r} \in \Omega_0^r\ (r > p)$ p-trivial if $\sum_{i_1, \ldots, i_p} \nu_{i_1, \ldots, i_r}(v, x)\, v_1^{i_1} \ldots v_p^{i_p} = 0$ for any $i_{p+1}, \ldots, i_r$ (by virtue of the stipulated assumptions, the coefficients of the form are antisymmetric). We introduce the operator $\varkappa: \Omega_0^r \to \Omega^{r+p}(E_{p,n})$:

$$\varkappa\theta = \sum_{i_1, \ldots, i_p} \frac{\partial^p}{\partial x^{i_1} \ldots \partial x^{i_p}}\, \theta \wedge dv_1^{i_1} \wedge \ldots \wedge dv_p^{i_p}. \tag{3}$$

We denote by Ξ_0^r the subspace of Ω_0^r of those forms θ such that the form $\varkappa\theta$ is closed with respect to $v = (v_1, \ldots, v_p)$.

If $\theta \in \Xi_0^r$, $\sigma \in \Omega^{r-p}(E_{p,n})$, then we shall write $\theta = \mathscr{K}\sigma$, if $\theta = \sigma_0 \wedge \tau + \nu$ for some p-trivial form ν and p-flat form τ. This notation is well defined since one has

<u>LEMMA 2.</u> For $\sigma \in \Omega^{r-p}(E_{p,n})$ there exists not more than one form $\theta \in \Xi_0^r$, for which $\theta = \mathscr{K}\sigma$.

We note that for a fixed pair (θ, σ), $\theta = \mathscr{K}\sigma$, $\theta \in \Xi_0^r$, $\sigma \in \Omega^{r-p}(E_{p,n})$, the pair τ, ν for which $\theta = \sigma_0 \wedge \tau + \nu$ is not unique. In addition, for any p-flat form τ one can choose a p-trivial form ν such that $\theta = \sigma_0 \wedge \tau + \nu$.

<u>THEOREM 3.</u> The operator $\mathscr{K}$ is defined on all $\sigma = I_p\omega$, $\omega \in \Omega^r(RP^n)$, while $\mathscr{K}|_{\operatorname{Im} I_p}$ is an isomorphism of $\operatorname{Im} I_p$ and Ξ_0^r.

Thus, the operator $\mathscr{K}$ gives another description of the image of the operator I_p for $r > p$, more precisely, of the projection of the image on Ω_0^{r-p}. In order to make this description more effective, we represent $\mathscr{K}$ as the restriction of some differential operator.

<u>THEOREM 4.</u> If $\sigma = I_p\omega$, $\omega \in \Omega^r(RP^n)$, $r > p$,

$$\sigma = \sum \sigma_{i_1, \ldots, i_{r-p}}^{j_1, \ldots, j_{r-p}}(u_1, \ldots, u_{p+1})\, du_{j_1}^{i_1} \wedge \ldots \wedge du_{j_p}^{i_p},$$

then

$$\mathscr{K}\sigma = \sum \delta_{i_1, \ldots, i_r}(v, x)\, dx^{i_1} \wedge \ldots \wedge dx^{i_r}, \tag{4}$$

where

$$\delta_{i_1, \ldots, i_r}(u) = \frac{(r-p)!}{r!p!} \sum_{k=1}^{p+1} (-1)^{k-1} \sigma_{i_{r-p+1}, \ldots, i_r}^{1, \ldots \hat{k} \ldots, p+1} \sigma_{i_1, \ldots, i_{r-p}}^{k, p+1, \ldots, p+1}(u),$$

$$\Delta_{i_1,\dots,i_p}^{j_1,\dots,j_p} = \det \| \partial/\partial u_{j_\beta}^{i_\alpha} \|_{\alpha,\,\beta=1,\dots,p}.$$

4. Description of the Image of the Operator I_p for $r - p = 1$. The operators introduced allow one to give a description of the forms $\sigma \in \Omega^1(G_{p,n})$ which can be represented in the form $\sigma = I_p\omega$, $\omega \in \Omega^{p+1}(RP^n)$.

THEOREM 5. A form $\sigma \in \Omega^1(G_{p,n})$ of the form $\sigma = \sum_{i,j} \sigma_i^j(u)du_j^i$ can be represented in the form $\sigma = I_p\omega$, $\omega \in \Omega^{p+1}(RP^n)$ if and only if $d\sigma$ is an admissible form and

$$\left(\frac{\partial^2}{\partial u_{l_1}^{k_1} \partial u_{l_2}^{k_2}} - \frac{\partial^2}{\partial u_{l_1}^{k_2} \partial u_{l_2}^{k_1}}\right) \sum_{k=1}^{p+1} (-1)^{k-1} \Delta_{i_1,\dots,i_p}^{1,\dots\hat{k}\dots,p+1} \sigma_q^k(u_1,\dots,u_{p+1}) = 0$$

for any collection of freely entering indices.

5. Vanishing Formula for Even p. We denote by $F_{p,n}$ the manifold of pairs (h, x), $h \in G_{p-1,n}$, $x \in RP^n$, $x \notin h$. We define a fibration of the Stiefel manifold $E_{p,n}$ over $F_{p,n}$, by associating with the frame $u = (v, x)$ the subspace h of dimension p spanned by the frame $v = (v_1, \dots, v_p)$ and the one-dimensional subspace generated by x; this pair of subspaces is canonically identified with the $(p-1)$-dimensional plane h in RP^n and the point $x \in RP^n$, $x \notin h$.

LEMMA 3. Let $\sigma = I_p\omega$, $\omega \in \Omega^r(RP^n)$, $r > p$. Then the form $\varkappa(\mathcal{K}\sigma)$ drops from $E_{p,n}$ to $F_{p,n}$. The dropped $(r + p)$-form has relative degree p in the bundle $F_{p,n} \to RP^n((h, x) \mapsto x)$ and is relatively closed in this bundle.

In connection with the general concepts of relative degree and relative closedness of forms in bundles, cf. [2]. In our situation, since $F_{p,n}$ is canonically imbedded in $G_{p-1,n} \times RP^n$, one is concerned simply with the degree and closedness with respect to dh, $h \in G_{p-1,n}$.

We consider the bundle $F_{p,n} \to RP^n$, $(h, x) \to x$. Its fiber F^x consists of $(p-1)$-dimensional planes not passing through x. In its own right, F^x fibers over the manifold $G^x \cong G_{p-1,n-1}$ of p-dimensional planes passing through x (the pair (h, x) is made to correspond with the plane generated by it). In the bundle $F^x \to G^x$ the fiber is contractible, and hence the homology of F^x is obtained by lifting the homology of G^x.

The bundle $F_{p,n} \to RP^n$ corresponds canonically to the bundle $(RP^n, H^p(F^x))$ over RP^n with fibers $H^p(F^x) = H^p(G^x)$. To each $(r + p)$-form on $F_{p,n}$ of relative degree p with respect to $h \in G_{p-1,n}$, relatively closed with respect to h, canonically corresponds an r-form on RP^n with coefficients in the bundle $(RP^n, H^p(F^x))$ (the integral over cycles in the fibers). In particular, this holds for forms of the form $\varkappa(\mathcal{K}\sigma)$, $\sigma = I_p\omega$. We shall study this map in more detail.

We consider the standard fiber $F \cong F^x$, $G \cong G^x$. We associate with each p-dimensional cycle $\gamma \subset F$ the collection of cycles $\gamma^x \subset F^x$, $x \in RP^n$, homologous with γ. Cycles in $G \cong G_{p-1,n-1}$ consisting of $(p-1)$-dimensional planes contained in a fixed p-dimensional plane will be called Eulerian, as also will any lift of such a cycle to F.

THEOREM 6. Let $\sigma = I_p\omega$, $\omega \in \Omega^r(RP^n)$, $r > p$, and γ be a p-dimensional cycle in F; $\mathcal{K}$, and $\varkappa$ are defined by formulas (3) and (4). Then

$$\int_{\gamma^x} \varkappa(\mathcal{K}\sigma) = c(\gamma)\,\omega, \tag{5}$$

where $c(\gamma)$ depends only on the homology class of γ. For odd p, $c(\gamma) = 0$ for all γ. For even p, $c(\gamma) = d(\gamma)c_e$, where $d(\gamma)$ is the index of intersection with an Eulerian cycle and $c_e \neq 0$.

LITERATURE CITED

1. G. de Rham, Differentiable Manifolds [Russian translation], IL, Moscow (1956).
2. I. M. Gel'fand, M. I. Graev, and Z. Ya. Shapiro, Funkts. Anal. Prilozhen., 3, No. 2, 24–40 (1969).
3. I. M. Gel'fand, S. G. Gindikin, and M. I. Graev, "Problems of integral geometry in projective space, connected with the integration of differential forms," Preprint IMP, No. 79, Moscow (1979).

11.

(with S. G. Gindikin and M. I. Graev)

Integral geometry in affine and projective spaces

Itogi Nauki Tekh., Ser. Sovrem. Probl. Mat. 16, 53-226, Moscow: VINITI (1980)
[J. Sov. Math. **18** (1980) 39-167]. MR **82m**:43017. Zbl. **465**:52005

A survey is given of results of integral geometry related to the integration of sections of one-dimensional bundles and differential forms over planes in affine and projective spaces.

INTRODUCTION

In the theory of differential equations, the problem consists in recovering functions on the basis of a given system of differential relations. However, there are problems in which the information regarding the unknown functions is prescribed in a nonlocal fashion. For example, suppose that for a function $f(x)$ the integral $\int_{-\infty}^{\infty} K(x, y)f(y)\,dy$ is known, where $K(x, y)$ is a given function. Problems thus arise of finding functions given their integral transforms. However, these problems are too general.

There is an important class of problems which we call problems of integral geometry which occupy an intermediate position between differential equations and the very general problems of arbitrary integral transforms. The first formulations of problems of integral geometry are due to Minkowski and Radon. In the work of Minkowski, the problem of recovering a function on a sphere from its integrals over larger disks was solved. Radon [64] studied the transform assigning to functions on Euclidean space their integrals over hyperplanes. This transformation, which we call the Radon transform, is closely related to the Fourier transform and to the method of plane waves; we discuss it in Chap. I.

The problem of integral geometry can generally be formulated as follows. Suppose there is a manifold X and a family of submanifolds ξ_α in it which depend on parameters α, e.g., a family of spheres of prescribed radius in Euclidean space, a family of lines in affine space which intersect a given curve, etc. We assume that this collection of submanifolds itself forms a manifold Y. Let f be a function on X. We denote by $\check{f}$ the integral

$$\check{f}(\alpha)=\int_{\xi_\alpha} f(x)\,d_\alpha x,$$

where $d_\alpha x$ is some measure on the manifold ξ_α. We thus obtain an integral transform $\mathcal{J}$, which assigns to functions on X functions on the collection Y of our submanifolds. The problem consists in describing the image of the mapping $\mathcal{J}$ and constructing the inverse operator, i.e., obtaining an inversion formula

---

Translated from Itogi Nauki i Tekhniki, Seriya Sovremennye Problemy Matematiki, Vol. 16, pp. 53-226, 1980.

For the Radon transform this problem was solved in [64]. It is interesting that in the case of odd-dimensional spaces the inversion formula for the Radon transform is local; this means that to find the original function at the point x it suffices to know its integrals over hyperplanes artitrarily close to x.

The works of Minkowski and Radon occasioned a number of extensions. In the works of John [29, 60] problems of recovering functions in terms of integrals over lines or integrals over spheres were solved. The general problem of recovering functions on Lobachevskii space and some other symmetric spaces in terms of their integrals over completely geodesic surfaces was first treated in the works of Helgason [50-53]. The problem of recovering functions on symmetric spaces in terms of their integrals over horospheres plays a major role in the development of the theory of representations of Lie groups; this problem was first considered in [12], and it was further developed in the works of Helgason [54-57]; see also [17, 18, 26].

All the problems enumerated above are homogeneous, i.e., the same group acts transitively both in the original manifolds and on the collection of submanifolds given in them, viz., in the case of Euclidean space this is the group of motions of Euclidean space, in the case of Lobachevskii space it is the group of motions of Lobachevskii space, etc.

An important step in the development of integral geometry was the discovery that explicit and local inversion formulas may also occur in inhomogeneous situations. A typical example of such a problem is the problem of recovering a function on $\mathbf{C}^3$ in terms of its integrals over complex lines which intersect a fixed curve [14, 17].* In [15, 33] all manifolds of lines in $\mathbf{C}^n$ were described for which there exists a local inversion formula (so-called admissible complexes of lines). For a generalization of the problem of integral geometry to the case of curves see [11].

We shall indicate a method of solving the problem of integral geometry for complexes of lines in $\mathbf{C}^n$. We replace this problem by another homogeneous problem which goes back to the works of John: recover a function on $\mathbf{C}^n$ if its integrals on all lines in $\mathbf{C}^n$ are known. Of course, this problem is overdetermined, since for $n > 2$ the lines in $\mathbf{C}^n$ form a manifold of dimension greater than n. The solvability condition for a given function φ on the manifold of lines can be written as a condition that a certain differential form $\varkappa\varphi$, which can be constructed explicitly from the function φ, be closed. The inversion formula can also be obtained by integrating this differential form. If the manifold of lines M is such that the form $\varkappa\varphi|_M$ on M can be uniquely recovered on the basis of the restriction $\varphi|_M$ of the function φ to M, then for M there exists a local inversion formula which can, moreover, be explicitly written down. Such a submanifold M is an admissible complex.†

This example indicates the usefulness of homogeneous, overdetermined problems. It may be said that essentially at the bottom of any problem of integral geometry there is a funda-

*The problem of lines intersecting a curve of second order is only inhomogeneous at first glance. Actually, it is a reformulation of a problem of harmonic analysis for the group SL(2, C).

†If the condition that the form $\varkappa\varphi$ be closed is represented in the form of a system of differential equations, then admissibility is essentially the property that the corresponding manifold M be characteristic.

mental, model, homogeneous problem. In those examples that we mentioned above these are problems connected with hyperplanes in $\mathbf{R}^n$ and lines in $\mathbf{C}^n$. The present paper is devoted to problems of integral geometry in which the space of points is n-dimensional affine or projective space while the submanfiolds are p-dimensional planes.

We note that the most general model problem that should be considered is the following. Suppose that in $\mathbf{RP}^n$ there is given a manifold of flags V_k, $k=(k_1, \ldots, k_r)$, $0\leqslant k_1<\ldots<k_r<n$, i.e., a manifold the points of which are collections of planes imbedded in one another: $v_{k_1}\subset\ldots\subset v_{k_r}$ ($\dim v_{k_i}=k_i$), and suppose there is given another manifold of flags V_l, $l=(l_1, \ldots, l_s)$. An incidence relation between V_k and V_l can be introduced in many different ways. The most natural definition of incidence is the following: $v_k\in V_k$ and $v_l\in V_l$ are incident if $v_k\times v_l$ belongs to a fixed orbit of the group $GL(n+1, \mathbf{R})$, acting in natural fashion on $V_k\times V_l$. (We remark that for complete flags these orbits are Schubert cells.) For example, we considered the space V_k, $k=(0, 1, 2)$ of complete flags in $\mathbf{RP}^3$; its elements are triples: a point, a line passing through it, and a plane passing through this line. If V_k, V_l are manifolds of complete flags in $\mathbf{RP}^3$, then between them there exist 6 distinct types of incidence.

The transformation of integral geometry connected with the pair V_k, V_l and the given incidence relation between V_k and V_l consists in integrating functions on V_k over submanifolds of flags incident with fixed flags of V_l; as a result functions on V_l are obtained. The present survey provides an introduction to this general problem, which we hope to clarify subsequently.

Until now we have spoken of the integration of functions for simplicity. Actually, it is possible to integrate any differential forms. Chapters III and IV are devoted to problems of integral geometry connected with the integration of differential forms over p-dimensional planes. We here restrict attention to forms of degree $r\geqslant p$; the case of forms of degree $r<p$ is treated in [21].

Chapters I and II are devoted to problems of integral geometry for functions on $\mathbf{R}^n$ and $\mathbf{RP}^n$ [9, 10]. In Chap. I we consider functions in the Schwartz space on $\mathbf{R}^n$ and integrate them over p-dimensional planes; as a result, functions are obtained on a manifold G_p of p-dimensional planes in $\mathbf{R}^n$. The integral transform thus defined is denoted by $\mathcal{I}_p$. The problem consists in describing the image of the operator $\mathcal{I}_p$ and on the basis of the function $\varphi=\mathcal{I}_pf$ recovering the original function f; the formula expressing f in terms of $\mathcal{I}_pf$ we call the inversion formula.

We first consider the solution of this problem for the case $p=n-1$, i.e., the Radon transform. The solution is well known, and we present it here, since the case $p=n-1$ is the simplest and also in order to make the exposition self-contained. The principal content of Chap. I is the treatment of the case $p<n-1$, which is considerably different from the case $p=n-1$. While for $p=n-1$ the dimension of the manifold of planes G_p is equal to n, i.e., to the dimension of the original space $\mathbf{R}^n$, for $p<n-1$ we have: $\dim G_p=(p+1)(n-p)>n$. This leads to the situation that functions φ on G_p, belonging to the image of the transform $\mathcal{I}_p$, satisfy additional relations (which "remove" the excess variables).

It is found that the added relations for functions in the image of the operator $\mathcal{I}_p$ can be obtained in the following manner. On the basis of a function φ on G_p we construct a certain differential p -form $\varkappa\varphi$. (We have already mentioned this form for the case of lines in $\mathbf{C}^n$.) It is proved that the function φ belongs to the image of the operator $\mathcal{I}_p$ if and only if the differential form $\varkappa\varphi$ is closed.

The differential form $\varkappa\varphi$ plays a fundamental role in the entire exposition. In the case of even p the transform inverse to $\mathcal{I}_p$ is expressed in terms of this form and as a result we obtain a local inversion formula (for odd p there is no local inversion formula). We remark that in the case of the complex space $\mathbf{C}^n$ a local inversion formula exists for any p.

Many of the results of Chap. I (in particular, the difference between the cases $p=n-1$ and $p<n-1$) become clearer on passage to projective-invariant language. This is done in Chap. II where we study the problem of integral geometry for the projective space $\mathbf{RP}^n$. Instead of functions on the space $\mathbf{RP}^n$ (which may be interpreted as the set of one-dimensional subspaces of $\mathbf{R}^{n+1}$) it is more natural to consider homogeneous functions on $\mathbf{R}^{n+1}$ of prescribed degree of homogeneity. On these functions the integral transform $\mathcal{I}_p$ is constructed. In the simplest case where the space consists of functions homogeneous of degree $-p-1$, the result of the integral transform $\mathcal{I}_p$ gives functions on the Grassmann manifold G_p of p -dimensional planes; for functions of other degrees of homogeneity we obtain tensor fields on G_p. In Chap. II a description is obtained of the image of the transformation $\mathcal{I}_p$ for all cases considered, and for even p an inversion formula is constructed.

Chapters III and IV are devoted to problems of integral geometry connected with the integration over p -dimensional planes in $\mathbf{R}^n$ and $\mathbf{RP}^n$ of differential forms of degree $r\geq p$ [3-8]. We have preferred to consider the simplest case separately in Chap. III — problems connected with the integration of differential forms over lines in three-dimensional space (i.e., the case $n=3$, $p=1$). Many definitions and formulas are here given in the simplest coordinate systems, although they actually have an invariant significance which appears later in Chap. IV. The content of Chap. III can be considered an extended introduction to the entire theory. At the same time, it is important to note that in a certain sense the three-dimensional case is contained as a most important component in the multidimensional case.

At the beginning of Chap. IV (Secs. 1, 2) the general problem of integral geometry connected with the integration of differential forms is expounded in the language of binary bundles [21, 5]. Roughly speaking, if there are given two manifolds B and C with an incidence relation between their elements, then we introduce the manifold A of incident pairs (b, c) together with the natural projections onto B and C. This is a binary bundle. If we are given differential r -forms on B, then we can go over from them to forms on C by means of standard operations of lifting and dropping forms; as a result, $(r-N)$ -forms on C are obtained, where N is the dimension of the fiber of the bundle $A\to C$. We establish various general properties of the operation thus introduced of integrating differential forms in binary bundles. The rest of the exposition in Chapter IV and the basic results are given as far as possible in the invariant terms of binary bundles (admissibility, partial closedness).

All the results of this paper presented for real spaces $\mathbf{R}^n$ and $\mathbf{RP}^n$, carry over automatically to the case of complex spaces $\mathbf{C}^n$ and $\mathbf{CP}^n$. The complex case for planes of arbitrary dimension p is analogous to the real case for planes of even dimension. In particular, in the complex case there is a local inversion formula for any p. At the end of each chapter a brief summary of the definitions and formulas for complex spaces is given.

CHAPTER I

INTEGRAL GEOMETRY FOR FUNCTIONS ON AFFINE SPACE

This chapter is devoted to the study of the most elementary transformation of integral geometry. We consider the real space $\mathbf{R}^n$ and the manifold $H_{p,n}$ of all p-dimensional planes in $\mathbf{R}^n$. The transformation $\mathcal{J}_p$ of integral geometry assigns to each smooth, sufficiently rapidly decreasing function $f(x)\in S(\mathbf{R}^n)$ its integrals over p-dimensional planes, i.e., the function

$$\hat{f}(\tilde{h})=\int_h f(x)\,d\mu_h(x),\quad \tilde{h}=(h,\mu_h),$$

where $h\in H_{p,n}$, and μ_h is a measure on h. How should the measure μ_h be chosen?

If a Euclidean structure is given on $\mathbf{R}^n$, then for μ_h it is natural to take the Euclidean measure on h. As a result we obtain a function $\hat{f}$ on $H_{p,n}$. In the case of affine space we prescribe μ_h as a measure invariant under parallel translations. Since this measure is unique only up to a multiplicative factor, $\hat{f}$ is a function not on $H_{p,n}$, but rather on the manifold $\tilde{H}_{p,n}$ of pairs (h,μ_h). Naturally, $\tilde{H}_{p,n}$ is a one-dimensional bundle over $H_{p,n}$.

In this chapter we first study the simplest case of $p=n-1$. The corresponding integral transform is usually called the Radon transform. It has been thoroughly studied [64, 17, 61] and is essentially equivalent to the classical method of plane waves in mathematical physics.

The method of the Radon transform is very close to the method of the Fourier transform, but in a number of cases it is more convenient. Essentially, the Fourier transform can be decomposed into a Radon transform and a subsequent Fourier transform in one-dimensional space. It is precisely this latter (one-dimensional Fourier transform) that is superfluous in a number of problems, and thus in many problems where the Fourier transform is used the Radon transform is essentially being used.

An advantage of transforms of Radon type to great extent becames apparent when, e.g., Euclidean space is replaced by Lobachevskii space. In this case an analogue of the Radon transform is naturally defined either by means of integration over non-Euclidean planes or (which turns out to be more productive) by integration over horospheres. The theory of these transforms is very closely related to the corresponding Euclidean theory (e.g., the inversion formulas are very similar). However, an analogue of the Euclidean Fourier transform — a representation of the group of motions of Lobachevskii space — is considerably more complicated than the Euclidean theory, and, probably, integral geometry is the simplest and most natural approach to constructing it.

The basic questions regarding the Radon transform are the description of the image of the space $S(\mathbf{R}^n)$ and the inversion formula. In contrast to the Fourier transform where the conditions on the image are only the conditions of smoothness and rapid decrease, here there are additional integral relations (the so-called Cavalieri conditions). The inversion formulas for the Radon transform are different in even- and odd-dimensional spaces. In the case of odd n to recover f at the point x it suffices to know the integrals of the function f on hyperplanes arbitrarily close to x (the inversion formula is local). For even n it is not local. This phenomenon is closely related to the fact that Huygens' principle holds only in odd-dimensional space.

We proceed to consider the case $p<n-1$. Since $\dim H_{p,n}=(p+1)(n-p)$, it follows that $\dim H_{p,n}>n$ for $p<n-1$. It is thus natural that functions in the image of the integral transform $\mathcal{J}_p$ should satisfy an appropriate number of conditions. The initial result is due to John [60]. He observed that the integrals of $f(x_1, x_2, x_3)$ in $\mathbf{R}^3$ over the lines $x_1=\alpha_1 x_3+\beta_1$, $x_2=\alpha_2 x_3+\beta_2$, i.e.,

$$\hat{f}(\alpha_1, \alpha_2, \beta_1, \beta_2)=\int_{-\infty}^{+\infty} f(\alpha_1 x_3+\beta_1, \alpha_2 x_3+\beta_2, x_3)\,dx_3, \tag{1}$$

satisfy the ultrahyperbolic equation $\frac{\partial^2 \hat{f}}{\partial\alpha_1\partial\beta_2}-\frac{\partial^2\hat{f}}{\partial\alpha_2\partial\beta_1}=0$ and that, conversely, any solution of this equation can be represented in the form (1). In Sec. 3, following [19], we present a system of differential equations describing the image of the integral transform $\mathcal{J}_p$ for $p<n-1$. It is found that this system of differential equations is a necessary and sufficient condition for the closedness of a certain differential p-form $\varkappa_x\hat{f}$, defined on the manifold G_x of planes passing through an arbitrary fixed point $x\in\mathbf{R}^n$. The differential form $\varkappa_x\hat{f}$ is constructed explicitly in Sec. 3 on the basis of the function $\hat{f}$ on $H_{p,n}$ and it plays an essential role in the entire exposition. It is important to note that in the description of the image of $\mathcal{J}_p$ for $p<n-1$ there are nonlocal conditions of the type of the Cavalieri conditions for $p=n-1$.

Following [19], in Sec. 3 an inversion formula is constructed for even p (for odd local inversion formulas do not exist; regarding nonlocal inversion formulas see [2]). It is proved that if $\hat{f}=\mathcal{J}_p f$, then, by integrating the closed differential form $\varkappa_x\hat{f}$ over any p-cycle in G_x, we obtain $f(x)$ up to a factor not depending on f. The cycles for which this factor is nonzero are described. Thus, by choosing various cycles we obtain various inversion formulas. We note that a certain boundary value problem for the system of differential equations describing the image is solved simultaneously.

Many of the questions considered in this chapter (e.g., the difference, in principle, between the cases $p=n-1$ and $p<n-1$) are clarified in the projective approach of Chap. II.

In the appendix we give the formulas pertaining to $\mathbf{C}^n$ without proof. The complex case resembles the real, even-dimensional case (in particular, the inversion formulas are always local).

1. Operator of Integration Over p-Dimensional Planes

1. Manifold H. We consider n-dimensional real affine space; we denote it by $\mathbf{R}^n$ (without assuming that a coordinate system has been fixed in $\mathbf{R}^n$). We denote by $H=H_{p,n}$ the set of all p-dimensional planes in $\mathbf{R}^n$ equipped with the natural structure of a manifold.

In $\mathbf{R}^n$ we prescribe the structure of a linear space, i.e., we fix an origin $0\in\mathbf{R}^n$, and denote by $G=G_{p,n}$ the Grassmann manifold of all p-dimensional linear subspaces of $\mathbf{R}^n$. For what follows it is important that H is a vector bundle over G. The projection $\pi: H\to G$ assigns to the plane $h\in H$ the subspace $a\in G$ parallel to it; thus, the fiber of the bundle over a is canonically isomorphic to the factor space $\mathbf{R}^n/a\cong\mathbf{R}^{n-p}$. The local trivialization of the bundle in a neighborhood of $a_0\in G$ can be realized as follows. We fix an $(n-p)$-dimensional subspace $b_0\subset\mathbf{R}^n$ transversal to a_0, and assign to each plane $h\in H$ transversal to b_0 its point of intersection with b_0. This mapping realizes a trivialization of the bundle $H\to G$ over any sufficiently small neighborhood of $a_0\in G$. We note that the Grassmann manifold G is compact (see, e.g., [45]).

We denote by $\tilde{H}=\tilde{H}_{p,n}$ the manifold of pairs (h, μ_h), where $h\in H$, and μ_h is an invariant measure on h. Since distinct measures μ_h on h differ by positive multiples, $\tilde{H}\to H$ is a principal bundle over H with structural group $(\mathbf{R}_+)^\times$ (the multiplicative group of positive, real numbers). The bundle $\tilde{H}\to H$ admits a global section. For example, if the structure of Euclidean space is introduced in $\mathbf{R}^n$, then for each plane $h\in H$ the Euclidean measure d_hx can be distinguished on h, and the section of Euclidean measures over H thus arises.

2. Integration over p-Dimensional Planes. To functions f on $\mathbf{R}^n$ we assign functions $\hat{f}$ on $\tilde{H}$ defined by the formula

$$\hat{f}(\tilde{h})=\int_h f(x)\,d\mu_h(x),\quad \tilde{h}=(h, \mu_h). \tag{1.1}$$

For $p=n-1$ the transformation (1.1) is called the Radon transform. It is closely related to the classical method of plane waves and has been studied in detail in a number of works (see, e.g., [17]). We observe that the function $\hat{f}$ on $\tilde{H}$ defined by formula (1.1) satisfies the following relation:

$$\hat{f}(h, \lambda\mu_h)=\lambda\hat{f}(h, \mu_h)\quad \text{for any}\quad \lambda>0. \tag{1.2}$$

If a Euclidean structure is given in $\mathbf{R}^n$, then in the definition of the transform (1.1) it is possible to restrict attention to Euclidean measures, i.e., to consider the transform

$$\hat{f}_e(h)=\int_h f(x)\,d_hx, \tag{1.1'}$$

where d_hx is the Euclidean measure on h. Because of formula (1.2), the affine transform $\hat{f}$ can be recovered from the Euclidean transform $\hat{f}_e$.

Integral transform (1.1) can be considered on various function spaces. In this chapter we restrict ourselves to the case of the Schwartz space $S(\mathbf{R}^n)$ of infinitely differentiable, rapidly decreasing functions on $\mathbf{R}^n$. For the Radon transform on other spaces see Chap. II and also [17] and [61].

It is obvious that for $f \in S(\mathbf{R}^n)$ the integral (1.1) converges. We shall be interested in the following two problems: 1) describe the image of the space $S(\mathbf{R}^n)$ under the transform (1.1) and 2) obtain an inversion formula for (1.1).

3. Space $S(H)$. It is not hard to see that functions $\hat{f}$ in the image of $S(\mathbf{R}^n)$ are infinitely differentiable on $\tilde{H}$ and satisfy a condition of rapid decrease. We shall make this concept more precise.

Let φ be a function on $\tilde{H}$ satisfying the condition (1.2). We fix a smooth section s of the bundle $\tilde{H} \to H$, e.g., the section $s=e$ of Euclidean measures. The restriction of the function φ to the section s can be considered a function φ_s on H, e.g., φ_e. Since H is a vector bundle over the compact manifold G, the concept of rapidly decreasing function on H can be defined in a natural way. We say that a function φ on $\tilde{H}$ satisfying condition (1.2) belongs to the space $S(H)$ if for a fixed section $s: H \to \tilde{H}$ the function φ_s is an infinitely differentiable, rapidly decreasing function on H. It is not hard to verify that this definition does not depend on the choice of the section s. Thus, formula (1.1) defines an operator $\mathcal{J}_p: S(\mathbf{R}^n) \to S(H)$.

2. Radon Transform of Rapidly Decreasing Functions

In this section we consider separately the case of the Radon transform (i.e., the transform (1.1) for the case $p=n-1$), due both to its independent interest and to the fact that the investigation of the operator $\mathcal{J}_p$ for arbitrary p (see Sec. 3) is based on the Radon transform.

1. An Expression for the Radon Transform in Coordinate Form. We denote by x points of the space $\mathbf{R}^n$ and by ξ points of the dual space $(\mathbf{R}^n)'$. If in $\mathbf{R}^n$ there is given a coordinate system $x=(x^1, \ldots, x^n)$, then we denote by $\xi=(\xi_1, \ldots, \xi_n)$ the dual coordinate system in $(\mathbf{R}^n)'$.

Let $F=((\mathbf{R}^n)' \setminus 0) \times \mathbf{R}^1$. To each pair $(\xi, s) \in F$ we assign the hyperplane h defined by the equation $\langle \xi, x \rangle = s$. On the basis of ξ, s we now construct a measure $d\mu_{\xi,s}$ on h. To this end we fix the volume element $dx=dx^1, \ldots, dx^n$ in $\mathbf{R}^n$ and define the measure $d\mu_{\xi,s}$ by the relation $dx=ds\,d\mu_{\xi,s}$. We have thus constructed a mapping $F \to \tilde{H}$ defined by the formula $(\xi, s) \mapsto (h, d\mu_{\xi,s})$, where $h \in H$, and $d\mu_{\xi,s}$ is the measure on h. We observe that this mapping makes F a two-sheeted covering over $\tilde{H}$, since to the pairs (ξ, s) and $(-\xi, -s)$ there corresponds the same element of $\tilde{H}$.

The Radon transform $\hat{f}$ is usually lifted from $\tilde{H}$ to F, i.e., we write

$$\hat{f}(\xi, s) = \int_{\langle \xi, x \rangle = s} f(x)\, d\mu_{\xi,s}(x). \tag{1.3}$$

Here $\hat{f}(\lambda\xi, \lambda s) = |\lambda|^{-1} \hat{f}(\xi, s)$ for any $\lambda \neq 0$; in particular, $\hat{f}(-\xi, -s) = \hat{f}(\xi, s)$.

If $\xi_i \neq 0$, then on the plane $\langle \xi, x \rangle = s$ it is possible to take as coordinates $x^1, \ldots, x^{i-1}, x^{i+1}, \ldots, x^n$; in these coordinates the measure $d\mu_{\xi,s}$ has the form

$$d\mu_{\xi,s}(x) = |\xi_i|^{-1}\, dx^1 \ldots dx^{i-1} dx^{i+1} \ldots dx^n. \tag{1.4}$$

Indeed, in $\mathbf{R}^n$ it suffices to go over from the coordinates $x^1, \ldots, x^n$ to the coordinates $s = \langle \xi, x \rangle, x^1, \ldots, x^{i-1}, x^{i+1}, \ldots, x^n$ and note that the Jacobian of the transformation is equal to ξ_i^{-1}.

Thus, in the coordinates $x^1, \dots, x^{i-1}, x^{i+1}, \dots, x^n$ the Radon transform of the function f is given by the formula

$$\hat{f}(\xi, s) = |\xi_i|^{-1} \int_{\langle \xi, x \rangle = s} f(x)\, dx^1 \dots dx^{i-1} dx^{i+1} \dots dx^n \tag{1.5}$$

(under the condition that $\xi_i \neq 0$). In other words

$$\hat{f}(\xi, s) = \int_{R^n} f(x)\delta(\langle \xi, x \rangle - s)\, dx. \tag{1.5'}$$

If a Euclidean structure is introduced in R^n then for $|\xi| = 1$ the corresponding measure $d\mu_{\xi,s}$ on the plane $\langle \xi, x \rangle = s$ coincides with Euclidean measure. As a result, to functions $\hat{f}$, there correspond functions $\hat{f}(\omega, s)$ on $S^{n-1} \times R^1$, where S^{n-1} is the unit sphere in $(R^n)' \simeq R^n$.

In § 1 we have introduced the space $S(H)$ of rapidly decreasing functions on $\tilde{H}$. It is easy to verify that in the coordinates (ω, s) this space is defined as follows: it consists of all C^∞ functions on $S^{n-1} \times R^1$ which satisfy the condition $\varphi(-\omega, -s) = \varphi(\omega, s)$ and are such that all their partial derivatives are rapidly decreasing in s uniformly with respect to ω.

<u>2. Connection of the Radon Transform with the Fourier Transform.</u> Let $\tilde{f}$ be the Fourier transform of the function $f \in S(R^n)$, i.e.,

$$\tilde{f}(\xi) = (2\pi)^{-n/2} \int_{R^n} f(x) e^{i\langle \xi, x \rangle}\, dx.$$

<u>Proposition 1.</u> The functions $\hat{f}(\xi, s)$ and $\tilde{f}(\xi)$ are connected by the following relations:

$$\tilde{f}(\xi) = (2\pi)^{-n/2} \int_{-\infty}^{+\infty} \hat{f}(\xi, s) e^{is} ds \text{ for } \xi \neq 0; \tag{1.6}$$

$$\hat{f}(\xi, s) = (2\pi)^{n/2-1} \int_{-\infty}^{+\infty} \tilde{f}(\lambda\xi) e^{-i\lambda s}\, d\lambda. \tag{1.7}$$

<u>Proof.</u> For fixed $\xi \neq 0$ we go over from the coordinates $(x^1, \dots, x^n)$ to coordinates on hyperplane $\langle \xi, x \rangle = s$ and s. Since $dx = ds\, d\mu_{\xi,s}$, it follows that

$$\tilde{f}(\xi) = (2\pi)^{-n/2} \int_{-\infty}^{+\infty} \int_{\langle \xi, x \rangle = s} f(x) e^{is} d\mu_{\xi,s}(x)\, ds = (2\pi)^{-n/2} \int_{-\infty}^{+\infty} \hat{f}(\xi, s) e^{is}\, ds.$$

Replacing (ξ, s) by $(\lambda\xi, \lambda s)$, we find

$$\tilde{f}(\lambda\xi) = (2\pi)^{-n/2} \int_{-\infty}^{+\infty} \hat{f}(\lambda\xi, \lambda s) e^{i\lambda s} |\lambda|\, ds = (2\pi)^{-n/2} \int_{-\infty}^{+\infty} \hat{f}(\xi, s) e^{i\lambda s}\, ds. \tag{1.8}$$

By the inversion formula for the one-dimensional Fourier transform, from this we obtain

$$\hat{f}(\xi, s) = (2\pi)^{n/2-1} \int_{-\infty}^{+\infty} \tilde{f}(\lambda\xi) e^{-i\lambda s} d\lambda.$$

3. Description of the Image of the Space $S(\mathbf{R}^n)$ Under the Radon Transform. It is well known that under the Fourier transform the space $S(\mathbf{R}^n)$ goes over into $S((\mathbf{R}^n)')$. In the case of the Radon transform the situation is more complicated: as already noted, the image of $S(\mathbf{R}^n)$ lies in $S(H)$; however, it does not coincide with $S(H)$. We reduce the problem of describing the image of $S(\mathbf{R}^n)$ under the Radon transform to an analogous problem for the Fourier transform.

THEOREM 1. In order that a function $\hat{f}(\xi, s) \in S(H)$ be the Radon transform of a function $f \in S(\mathbf{R}^n)$, it is necessary and sufficient that it satisfy the following Cavalieri conditions*:

$$I_k(\xi) = \int_{-\infty}^{+\infty} \hat{f}(\xi, s)\, s^k ds, \quad k = 0, 1, 2, \ldots$$

is a polynomial of degree   in ξ.

Proof. If $\hat{f}(\xi, s)$ is the Radon transform of a function $f \in S(\mathbf{R}^n)$, then we have

$$I_k(\xi) = \int_{-\infty}^{+\infty} \int_{\langle \xi, x \rangle = s} f(x)\, s^k d\mu_{\xi,s}(x)\, ds = \int_{\mathbf{R}^n} f(x) \langle \xi, x \rangle^k dx.$$

Hence $I_k(\xi)$ is a polynomial of degree $\leqslant k$ in ξ. Conversely, suppose that a function $\varphi(\xi, s) \in S(H)$ satisfies the Cavalieri conditions. We shall show that φ is then the Radon transform of a function $f \in S(\mathbf{R}^n)$.

We define a function $\tilde{f}$ on $(\mathbf{R}^n)'$ by the formula

$$\tilde{f}(\lambda\omega) = (2\pi)^{-n/2} \int_{-\infty}^{+\infty} \varphi(\omega, s)\, e^{i\lambda s} ds, \tag{1.9}$$

where $|\omega| = 1$. For $\xi = \lambda\omega \neq 0$ the correctness of the definition follows from the relation $\varphi(-\omega, -s) = \varphi(\omega, s)$. For $\xi = 0$ the correctness of the condition follows from the Cavalieri condition: the integral $I_0(\omega) = \int_{-\infty}^{+\infty} \varphi(\omega, s)\, ds$ does not depend on ω.

From (1.9) it follows immediatel y that the function $\tilde{f}(\xi)$ is infinitely differentiable everywhere with the possible exception of the point $\xi = 0$, and is rapidly decreasing together with all partial derivatives. We shall show that the function $\tilde{f}$ is infinitely differentiable at the point $\xi = 0$ also, and therefore $\tilde{f} \in S((\mathbf{R}^n)')$.

From Eq. (1.9) it follows that

$$\tilde{f}(\lambda\xi) = (2\pi)^{-n/2} \int_{-\infty}^{+\infty} \varphi(\xi, s)\, e^{i\lambda s} ds \tag{1.10}$$

for any $\xi \neq 0$. Differentiating (1.10) k times with respect to λ, we obtain

$$\sum_{k_1 + \ldots + k_n = k} \frac{k!}{k_1! \ldots k_n!} \xi_1^{k_1} \ldots \xi_n^{k_n} \left. \frac{\partial^k \tilde{f}(\xi)}{\partial \xi_1^{k_1} \ldots \partial \xi_n^{k_n}} \right|_{\xi \to \lambda\xi} = (2\pi)^{-n/2} i^k \int_{-\infty}^{+\infty} \varphi(\xi, s)\, s^k e^{i\lambda s} ds. \tag{1.11}$$

*We call these conditions Cavalieri conditions, because for $k=0$ the corresponding condition ($\int f(\xi, s) ds$ does not depend on ξ) is obviously related to Cavalieri's principle.

By the Cavalieri conditions the right side of Eq. (1.11) tends as $\lambda\to 0$ to a polynomial of degree $\leqslant k$ in ξ. This implies that each of the coefficients on the left side $\left.\frac{\partial^k \tilde{f}(\xi)}{\partial\xi_1^{k_1}\ldots\partial\xi_n^{k_n}}\right|_{\xi\to\lambda\xi}$ tends as $\lambda\to 0$ to a limit which does not depend on ξ. We have thus proved that $\tilde{f}$ is infinitely differentiable at $\xi=0$.

Let f be the inverse Fourier transform of a function $\tilde{f}$, and let $\hat{f}$ be the Radon transform of the function f. Since by what has been proved $\tilde{f}\in S((R^n)')$, it follows that $f\in S(R^n)$.

According to (1.7), we have

$$\hat{f}(\xi, s)=(2\pi)^{n/2-1}\int_{-\infty}^{+\infty}\tilde{f}(\lambda\xi)\,e^{-i\lambda s}d\lambda.$$

On the other hand, from (1.10) it follows that

$$\varphi(\xi, s)=(2\pi)^{n/2-1}\int_{-\infty}^{+\infty}\tilde{f}(\lambda\xi)\,e^{-i\lambda s}d\lambda.$$

Hence $\varphi=\hat{f}$, as required to prove.

Remark. It is evident from the theorem that the Cavalieri conditions are equivalent to the condition of infinite smoothness of the function $\tilde{f}$ at the point $\xi=0$.

4. Inversion Formula for the Radon Transform. The inversion formula for the Radon transform of a function f will here be obtained as a corollary of the inversion formula for the Fourier transform $\tilde{f}$:

$$f(x)=(2\pi)^{-n/2}\int_{(R^n)'}\tilde{f}(\xi)\,e^{-i\langle\xi,x\rangle}d\xi. \tag{1.12}$$

We write integral (1.12) in spherical coordinates $\xi=\lambda\omega$, $|\omega|=1$; here it is convenient to assume that $-\infty<\lambda<+\infty$, and hence ξ runs over $(R^n)'$ twice. Since $d\xi=|\lambda|^{n-1}d\lambda d\sigma$, where $d\sigma$ is the element of area of the unit sphere, we have

$$f(x)=\frac{1}{2}(2\pi)^{-n/2}\int_{|\omega|=1}\int_{-\infty}^{+\infty}\tilde{f}(\lambda\omega)\,e^{-i\lambda\langle\omega,x\rangle}|\lambda|^{n-1}d\lambda d\sigma. \tag{1.12'}$$

Substituting here the expression (1.8) for the function $\tilde{f}(\lambda\omega)$ in terms of the Radon transform $\hat{f}$, we obtain

$$f(x)=\frac{1}{2}(2\pi)^{-n}\int_{|\omega|=1}d\omega\left(\int_{-\infty}^{+\infty}|\lambda|^{n-1}\int_{-\infty}^{+\infty}\hat{f}(\omega, s)\,e^{i\lambda(\langle\omega,x\rangle-s)}ds\,d\lambda\right).$$

The expression in brackets can be simplified on the basis of the following relations (see [24]):

$$\int_{-\infty}^{+\infty}\int_{-\infty}^{+\infty}F(s)|\lambda|^{n-1}e^{-i\lambda(p-s)}ds\,d\lambda=\begin{cases}2\pi(-1)^{\frac{n-1}{2}}F^{(n-1)}(p) & \text{for odd } n;\\ 2(-1)^{n/2-1}\int_{-\infty}^{+\infty}F^{(n-1)}(s+p)\,s^{-1}ds & \text{for even } n,\end{cases}$$

where the integral is understood in the sense of principal value. We have thus proved the following result:

THEOREM 2. If $\hat{f}(\xi, s)$ is the Radon transform of a function $f \in S(\mathbf{R}^n)$, then there is the following inversion formula:

$$f(x)=\frac{(-1)^{\frac{n-1}{2}}}{2(2\pi)^{n-1}} \int\limits_{|\omega|=1} \left.\frac{\partial^{n-1}\hat{f}(\omega, s)}{\partial s^{n-1}}\right|_{s=\langle\omega, x\rangle} d\sigma, \tag{1.13}$$

where $d\sigma$ is the surface element of the sphere $|\omega|=1$; for even n

$$f(x)=\frac{(-1)^{n/2-1}}{2(2\pi)^n} \int\limits_{|\omega|=1} \int\limits_{-\infty}^{+\infty} \left.\frac{\partial^{n-1}\hat{f}(\omega, s)}{\partial s^{n-1}}\right|_{s=\langle\omega, x\rangle+p} p^{-1} dp\, d\sigma, \tag{1.14}$$

where the integral on p is understood in the sense of principal value.*

The basic difference between the inversion formulas for even and odd n is that for odd n the formula is local. This means to recover the function f at the point x it is necessary to know $\hat{f}$ only in an infinitesimal neighborhood of the manifold of planes passing through x. For even n the inversion formula is nonlocal.

5. Operator $\varkappa_x$. We wish to avoid Euclidean considerations and replace the spherical surface S^{n-1} by surfaces of more general form. To this end we note first that in coordinate notation the volume element on the sphere $|\omega|^2 \equiv \omega_1^2+\ldots+\omega_n^2=1$ has the form

$$d\sigma=\sum_{k=1}^{n}(-1)^{k-1}\omega_k d\omega_1 \ldots \widehat{d\omega_k} \ldots d\omega_n.$$

On $(\mathbf{R}^n)'$ we introduce the differential $(n-1)$-form

$$\sigma(\xi, d\xi)=\sum_{k=1}^{n}(-1)^{k-1}\xi_k d\xi_1\wedge\ldots\wedge\widehat{d\xi_k}\wedge\ldots\wedge d\xi_n \tag{1.15}$$

and in place of the integrand in (1.13) we consider the following differential $(n-1)$-form on $(\mathbf{R}^n)'\setminus 0$:

$$\varkappa_x\hat{f}=\left.\frac{\partial^{n-1}\hat{f}(\xi, s)}{\partial s^{n-1}}\right|_{s=\langle\xi, x\rangle}\sigma(\xi, d\xi). \tag{1.16}$$

Let $\tilde{S}^{n-1}$ be the space of rays in $(\mathbf{R}^n)'$ issuing from the point $\xi=0$, and let $\pi:(\mathbf{R}^n)'\to\tilde{S}^{n-1}$ be the canonical projection.

LEMMA. The differential form (1.16) drops from $(\mathbf{R}^n)'$ to the manifold of rays $\tilde{S}^{n-1}$.

Proof. It suffices to prove that $\varkappa_x\hat{f}$ is preserved by the change of variables

$$\xi_i\to\lambda\xi_i,\quad d\xi_i\to\lambda d\xi_i+\xi_i d\lambda\ (1\leqslant i\leqslant n), \text{where } \lambda>0. \tag{1.17}$$

From the relation $\hat{f}(\lambda\xi, \lambda s)=\lambda^{-1}\hat{f}(\xi, s)$ it follows that under the replacement $\xi_i\to\lambda\xi_i$ the expression $\left.\frac{\partial^{n-1}\hat{f}(\xi, s)}{\partial s^{n-1}}\right|_{s=\langle\xi, x\rangle}$ is multiplied by λ^{-n}. On the other hand, under the replacement $d\xi_i\to\lambda d\xi_i+\xi_i d\lambda\ (1\leqslant i\leqslant n)$ the form $\sigma(\xi, d\xi)$ is multiplied by λ^n. Hence, the form $\varkappa_x\hat{f}$ is preserved under the replacement (1.17).

COROLLARY. The form $\varkappa_x\hat{f}$ is closed on $(\mathbf{R}^n)'\setminus 0$.

Indeed, after dropping, it is a form of maximal degree on $\tilde{S}^{n-1}$.

*For another method of obtaining the inversion formula for the Radon transform not using the Fourier transform see, e.g., [17].

Remark. Since to each ray of $\tilde{S}^{n-1}$ there corresponds an oriented hyperplane in $\mathbf{R}^n$ passing through the point x, $\varkappa_x\hat{f}$ can be considered as a differential form on the manifold G_x of oriented hyperplanes passing through the point x. We emphasize that $\varkappa_x\hat{f}$ does not depend on the choice of measures on these hyperplanes.

The form $\varkappa_x\hat{f}$ is antisymmetric with respect to the involution $\xi\to-\xi$. The form $\varkappa_x\hat{f}$ therefore does not drop to the manifold $\tilde{G}_x$ of nonoriented hyperplanes passing through x, but it can be interpreted as an odd differential form on $\tilde{G}_x$ in the sense of de Rham [37].

THEOREM 3. For odd n there is the following inversion formula:

$$f(x)=\frac{(-1)^{\frac{n-1}{2}}}{2(2\pi)^{n-1}}\int_{\gamma}\varkappa_x\hat{f}, \tag{1.18}$$

where the integral is taken over any $(n-1)$-dimensional cycle $\gamma\subset(\mathbf{R}^n)'\setminus 0$ homologous to the sphere S^{n-1}. For even n and any $(n-1)$-dimensional cycle $\gamma\subset(\mathbf{R}^n)'\setminus 0$ we have

$$\int_{\gamma}\varkappa_x\hat{f}=0. \tag{1.19}$$

Proof. Since the form $\varkappa_x\hat{f}$ drops to the manifold of rays and is closed, the integral (1.18) does not depend on the choice of cycle γ homologous to the unit S^{n-1} sphere, and it may be assumed that $\gamma=S^{n-1}$. Now in the case $\gamma=S^{n-1}$ integrals (1.18) and (1.13) are equal; hence, for odd n equality (1.18) follows from (1.13).

As already noted, the form $\varkappa_x\hat{f}$ changes sign under the involution $\xi\to-\xi$. Since for even n the orientation of S^{n-1} under this involution does not change, it follows that $\int_{S^{n-1}}\varkappa_x\hat{f}=0$. Thus, since any $(n-1)$-dimensional cycle γ in $(\mathbf{R}^n)'\setminus 0$ is a multiple of S^{n-1}, it follows that $\int_{\gamma}\varkappa_x\hat{f}=0$. The proof of the theorem is complete.

The forms $\varkappa_x\hat{f}$ thus give an inversion formula only for odd n. However, it is convenient to consider them for any n.

Remark. Below it will also be convenient to write the inversion formula (1.18) in the form

$$f(x)=\frac{(-1)^{\frac{n-1}{2}}}{2(2\pi)^{n-1}}\int_{\gamma}\int_{-\infty}^{+\infty}\hat{f}(\xi,s)\,\delta^{(n-1)}(\langle\xi,x\rangle-s)\,ds\,\sigma(\xi,d\xi)$$

(for n an odd number). Similarly, for even n on the basis of formula (1.14) we obtain

$$f(x)=\frac{(-1)^{n/2}(n-1)!}{2(2\pi)^n}\int_{\gamma}\int_{-\infty}^{+\infty}\hat{f}(\xi,s)(\langle\xi,x\rangle-s)^{-n}\,ds\,\sigma(\xi,d\xi),$$

where $(\langle\xi,x\rangle-s)^{-n}$ is to be understood as a generalized function (see [24]).

6. An Expression for the Form $\varkappa_x\hat{f}$ in Other Coordinates. In the preceding paragraphs the hyperplanes were defined by the equations $\langle\xi,x\rangle=s$, i.e., essentially the coordinates (ξ,s) were prescribed, where $\xi\in(\mathbf{R}^n)'$, $s\in\mathbf{R}$. During the course of this paper, we shall also use other coordinates (α,β), which we now introduce.

Let $\mathbf{R}^{n-1}$ be the standard coordinate space with coordinates $t=(t^1,\dots,t^{n-1})$ and measure $dt=dt^1,\dots,dt^{n-1}$. We denote by E the manifold of pairs (α,β), where α is a nondegenerate linear mapping $\mathbf{R}^{n-1}\to\mathbf{R}^n$ and $\beta\in\mathbf{R}^n$. To each pair (α,β) we assign a hyperplane $h\in H$ defined by the equation

$$x=\alpha t+\beta.$$

The measure dt induces on h a measure $d\mu_{\alpha,\beta}=dt$. Thus, to each pair (α,β) there corresponds the pair $\tilde{h}=(h,d\mu_{\alpha,\beta})$, and a projection $E\to\tilde{H}$ has thus been constructed.

We shall write the Radon transform $\mathcal{J}_{n-1}$ in the coordinates α,β. By definition, if $\tilde{h}=(h,\mu_h)$, then $(\mathcal{J}_{n-1}f)(\tilde{h})=\hat{f}(\tilde{h})=\int_h f(x)\,d\mu_h(x)$. Hence, if $\tilde{h}$ is given by the pair (α,β), where α is a linear mapping $\mathbf{R}^{n-1}\to\mathbf{R}^n$ and $\beta\in\mathbf{R}^n$, then $\hat{f}(\tilde{h})=\int_h f(x)\,d\mu_h(x)=\int f(\alpha t+\beta)\,d\mu_{\alpha,\beta}=\int_{\mathbf{R}^{n-1}} f(\alpha t+\beta)\,dt$. Thus, the Radon transform in the coordinates (α,β) is given by the formula

$$\hat{f}(\alpha,\beta)=\int_{\mathbf{R}^{n-1}} f(\alpha t+\beta)\,dt. \tag{1.20}$$

We observe that the function $\hat{f}(\alpha,\beta)$ satisfies the following relation:

$$\hat{f}(\alpha g,\beta+\alpha t_0)=|\det g|^{-1}\hat{f}(\alpha,\beta) \tag{1.21}$$

for any $g\in GL(n-1,\mathbf{R})$ and $t_0\in\mathbf{R}^{n-1}$. This relation is a necessary and sufficient condition that a function defined on E drop from E to $\tilde{H}$ and satisfy on $\tilde{H}$ the condition (1.2).

Proposition 2. The pairs $(\xi,s)\in F$ and $(\alpha,\beta)\in E$ define the same element of $\tilde{H}$ if and only if $\xi_i=\varepsilon(-1)^i\Delta_i(\alpha)$ $(1\leqslant i\leqslant n)$ and $s=\langle\xi,\beta\rangle$, where $\Delta_i(\alpha)$ is the minor of the matrix α obtained by deleting the i-th row, and $\varepsilon=\pm1$.

Proof. It is obvious that the pairs (ξ,s) and (α,β) define the same hyperplane $h\in H$ if and only if they are connected by the relations

$$\langle\xi,\alpha_i\rangle=0\quad(1\leqslant i\leqslant n-1),\qquad\langle\xi,\beta\rangle=s. \tag{1.22}$$

For given (α,β) the pair (ξ,s) is determined uniquely from these relations up to a factor, viz.:

$$\xi_i=\lambda(-1)^i\Delta_i(\alpha)\quad(1\leqslant i\leqslant n),\quad s=\langle\xi,\beta\rangle.$$

It remains to observe that here the measures $d\mu_{\xi,s}$ and $d\mu_{\alpha,\beta}$ coincide if and only if $h\in H$. This follows from the explicit expressions for the measures $d\mu_{\xi,s}$ and $d\mu_{\alpha,\beta}$ in the coordinates $x^1,\dots,\hat{x}^i,\dots,x^n$ on h:

$$d\mu_{\xi,s}(x)=|\xi_i|^{-1}dx^1\dots\widehat{dx^i}\dots dx^n,$$
$$d\mu_{\alpha,\beta}(x)=|\Delta_i(\alpha)|^{-1}dx^1\dots\widehat{dx^i}\dots dx^n.$$

We now write the form $\varkappa_x\hat{f}$ in the coordinates (α,β). Let $x\in\mathbf{R}^n$, and let $\hat{f}(\alpha,\beta)$ be an arbitrary smooth function on E which satisfies the condition (1.21). To the function $\hat{f}$ we assign the following differential form $\tilde{\varkappa}_x\hat{f}$ on the manifold E_0 of all $n\times(n-1)$ matrices $\alpha=\|\alpha_j^i\|$ of rank $n-1$:

$$\tilde{\varkappa}_x\hat{f}=(-1)^{n-1}\sum_{i_1,\dots,i_{n-1}=1}^{n}\frac{\partial^{n-1}\hat{f}(\alpha,x)}{\partial x^{i_1}\dots\partial x^{i_{n-1}}}\,d\alpha_1^{i_1}\wedge\dots\wedge d\alpha_{n-1}^{i_{n-1}}. \tag{1.23}$$

We shall show that this form is the lift to E_0 of the form $\varkappa_x\hat{f}$ of part 5 defined on the manifold G_x of oriented hyperplanes passing through the point x.

To each matrix $\alpha=(\alpha_1,\ldots,\alpha_{n-1})\in E_0$ we assign the oriented hyperplane passing through x incident with the vectors α_i $(1<i<n-1)$ (the orientation is determined by the frame $\{\alpha_1,\ldots,\alpha_{n-1}\}$). Here to the matrices α and α' there corresponds the same oriented hyperplane if and only if $\alpha'=\alpha g$, where $g\in GL_+(n-1, \mathbf{R})$. We thus obtain the bundle

$$E_0\to G_x,$$

where G_x is the manifold of oriented hyperplanes passing through x ; it is a principal bundle with structural group $GL_+(n-1, \mathbf{R})$.

LEMMA. If a function $\hat{f}(\alpha, \beta)$ satisfies condition (1.21), then the form $\tilde{\varkappa}_x\hat{f}$ drops from E_0 to the manifold G_x and is antisymmetric with respect to a change of orientation.

Proof. It suffices to verify that 1) the form $\tilde{\varkappa}_x\hat{f}$ is invariant under the group $GL_+(n-1, \mathbf{R})$ and 2) the form $\tilde{\varkappa}_x\hat{f}$ is preserved by the replacement $d\alpha_j^i\to d\alpha_j^i+\sum_k \alpha_k^i dg_j^k$. Both properties follow easily from relation (1.21) for the function $\hat{f}(\alpha, \beta)$.

The antisymmetry of $\tilde{\varkappa}_x\hat{f}$ follows from the fact that this form changes sign under the replacement $\alpha_1\to-\alpha_1$.

Proposition 3. After dropping to G_x the form $\tilde{\varkappa}_x\hat{f}$ coincides with the form $\varkappa_x\hat{f}$.

Proof. We wish to prove the coincidence of the differential forms $\varkappa_x\hat{f}$ and $\tilde{\varkappa}_x\hat{f}$ on G_x obtained by dropping the forms defined, respectively, on $(\mathbf{R}^n)'\setminus 0$ and E_0. For this it suffices to prove that they coincide in a neighborhood of any $a_0\in G_x$. It may be assumed with no loss of generality that the hyperplane a_0 belongs to the set G_x^0 of hyperplanes passing through x and transversal to the line $x^1=\ldots=x^{n-1}=0$.

Any hyperplane in G_x^0 can be given by the equation

$$y^n+u_1y^1+\ldots+u_{n-1}y^{n-1}=s_x,$$

where $s_x=x^n+u_1x^1+\ldots+u_{n-1}x^{n-1}$. We choose $u_1,\ldots,u_{n-1}$ as coordinates on G_x^0. On the basis of these coordinates we construct local sections $\xi(u)$ and $\alpha(u)$ of the bundles $(\mathbf{R}^n)'\setminus 0\to G_x$ and $E_0\to G_x$. We set $\xi(u)=(u_1,\ldots,u_{n-1}, 1)$,

$$\alpha(u)=\begin{pmatrix}1 & \ldots & 0\\ 0 & \ldots & 1\\ -u_1 & \ldots & -u_{n-1}\end{pmatrix}.$$

The measures on the hyperplane $a(u)$ connected, respectively, with $\xi(u)$ and $\alpha(u)$ are both equal to $dy^1\ldots dy^{n-1}$, since $\xi_n(u)=1$ (see formula (1.4)) and $t^j=y^j\,(1\leqslant j\leqslant n-1)$. This implies that $\hat{f}(\alpha(u), x)=\hat{f}(\xi(u), \langle\xi(u), x\rangle)$. We denote this function by $\varphi(x)$.

We consider the restrictions of the forms $\varkappa_x\hat{f}$ and $\tilde{\varkappa}_x\hat{f}$ to the sections $\xi(u)$ and $\alpha(u)$, respectively. By formulas (1.16) and (1.23) we have

$$\varkappa_x\hat{f}=\tilde{\varkappa}_x\hat{f}=(-1)^{n-1}\frac{\partial^{n-1}\varphi(u, x)}{(\partial x^n)^{n-1}}du_1\wedge\ldots\wedge du_{n-1}.$$

The coincidence of $\varkappa_x\hat{f}$ and $\tilde{\varkappa}_x\hat{f}$ has thus been proved.

Below, in view of what has just been proved, we shall also denote formula (1.23) by $\varkappa_x \hat{f}$.

3. Integration Over p-Dimensional Planes in R^n, Where $p<n-1$

1. Two Methods of Defining the Operator $\mathcal{J}_p$. We recall that the operator $\mathcal{J}_p$ assigns to functions f on $\mathbf{R}^n$ functions $\hat{f}$ on the manifold $\tilde{H}=\tilde{H}_{p,n}$ of p-dimensional hyperplanes with measures. Since on $\tilde{H}$ there does not exist a global coordinate system, it is convenient to define the functions $\hat{f}$, on certain bundles over $\tilde{H}$ as in the case of Radon transform (see Sec. 2).

We denote by $x=(x^1,\ldots, x^n)$ coordinates in $\mathbf{R}^n$ and by $\xi=(\xi_1,\ldots,\xi_n)$ the dual coordinates in $(\mathbf{R}^n)'$. We introduce the manifold E of pairs (α, β), where $\alpha=(\alpha_1,\ldots, \alpha_p)$ is a p-frame in $\mathbf{R}^n$ and $\beta\in\mathbf{R}^n$. To each pair $(\alpha, \beta)\in E$ we assign a p-dimensional plane in $h\in H$:

$$x=\alpha t+\beta=\alpha_1 t^1+\ldots+\alpha_p t^p+\beta. \tag{1.24}$$

We further take $t=(t^1,\ldots, t^p)$ as coordinates on the plane h, and we prescribe on h the measure $d\mu_{\alpha,\beta}=dt=dt^1\ldots dt^p$. Thus, to each pair $(\alpha, \beta)\in E$ there corresponds a pair $(h, d\mu_{\alpha,\beta})\in\tilde{H}$, and a projection $E\to\tilde{H}$ has thus been constructed.

We note that the measure $d\mu_{\alpha,\beta}$ in terms of any set of coordinates $x^{i_1},\ldots, x^{i_p}$ can be written

$$d\mu_{\alpha,\beta}(x)=|p^{i_1,\ldots,i_p}(\alpha)|^{-1}dx^{i_1}\ldots dx^{i_p}, \tag{1.25}$$

where $p^{i_1,\ldots,i_p}(\alpha)$ is the minor of the matrix α composed of rows with indices $i_1,\ldots, i_p$ (under the condition that $p^{i_1,\ldots,i_p}(\alpha)\neq 0$).

The function $\hat{f}=\mathcal{J}_p f$, lifted to E, is given by the following formula:

$$\hat{f}(\alpha, \beta)=\int_h f(x)\,d\mu_{\alpha,\beta}(x)=\int_{\mathbf{R}^p} f(\alpha t+\beta)\,dt. \tag{1.26}$$

This function satisfies the following relation:

$$\hat{f}(\alpha g, \beta+\alpha t_0)=|\det g|^{-1}\hat{f}(\alpha, \beta)$$

for any $g\in GL(p, \mathbf{R})$ and $t_0\in\mathbf{R}^p$.

We now introduce another bundle over $\tilde{H}$. We denote by F the manifold of pairs (ξ, s) where $(\xi=\xi^1,\ldots,\xi^{n-p})$ is an $(n-p)$-frame in $(\mathbf{R}^n)'$ and $s=(s^1,\ldots, s^{n-p})\in\mathbf{R}^{n-p}$. To each pair $(\xi, s)\in F$ we assign the p-dimensional plane $h\in H$ given by the system of equations

$$\langle \xi^i, x\rangle=s^i,\ i=1,\ldots, n-p. \tag{1.27}$$

On the basis of (ξ, s) we now construct a measure $d\mu_{\xi,s}$ on h. For this we fix a volume element $dx=dx^1\ldots dx^n$ in $\mathbf{R}^n$ and define the measure $d\mu_{\xi,s}$ from the relation $dx=ds^1\ldots ds^{n-p}d\mu_{\xi,s}$. We have thus constructed a mapping $F\to\tilde{H}$ given by the formula $(\xi, s)\mapsto(h, d\mu_{\xi,s})$.

We shall write the expression for the measure $d\mu_{\xi,s}$ in terms of any set of coordinates $x^{i_1},\ldots,x^{i_p}$. We set $q^{i_1,\ldots,i_p}(\xi)=\operatorname{sgn}(i_1,\ldots, i_n)\,\Delta_{i_{p+1},\ldots,i_n}(\xi)$, where $(i_1,\ldots, i_n)$ is a permutation of the indices $1,\ldots, n$, and $\Delta_{i_{p+1},\ldots,i_n}(\xi)$ is the minor of the matrix ξ composed of rows with indices $i_{p+1},\ldots, i_n$. It is not hard to see that

$$d\mu_{\xi,s}(x)=|q^{i_1,\ldots,i_p}(\xi)|^{-1}dx^{i_1}\ldots dx^{i_p}. \tag{1.28}$$

The function $\hat{f}=\mathcal{J}_p f$, lifted from $\tilde{H}$ to F, is given by the following formula:

$$\hat{f}(\xi, s)=\int_{\langle\xi,x\rangle=s} f(x)\,d\mu_{\xi,s}(x). \tag{1.29}$$

In other words,

$$\hat{f}(\xi, s) = \int_{\mathbf{R}^n} f(x) \prod_{i=1}^{n-p} \delta(\langle \xi^i, x \rangle - s^i)\, dx. \tag{1.29'}$$

This function satisfies the following relation:

$$\hat{f}(\xi g, sg) = |\det g|^{-1} \hat{f}(\xi, s)$$

for any $g \in GL(n-p, \mathbf{R})$.

2. Plücker Coordinates and the Connection between the Bundles $E \to \tilde{H}$ and $F \to \tilde{H}$. We thus have two bundles over H, lifted to bundles over $\tilde{H}$:

There arises the question of in which cases do $(\alpha, \beta) \in E$ and $(\xi, s) \in F$ project onto the same point on H, i.e., determine the same plane. It is clear that for this the following conditions are necessary and sufficient:

$$\langle \xi^i, \alpha_j \rangle = 0 \quad (1 \leq i \leq n-p,\ 1 \leq j \leq p), \tag{1.30}$$

$$\langle \xi^i, \beta \rangle = s^i \quad (1 \leq i \leq n-p). \tag{1.31}$$

Condition (1.30) can be written very simply in Plücker coordinates. We recall that in part 1 we associated with each p-frame α in $\mathbf{R}^n$ a collection of numbers $p^{i_1,\ldots,i_p}(\alpha)$ (the minors of the matrix α) and with each $(n-p)$-frame ξ in $(\mathbf{R}^n)'$ a collection of numbers $q^{i_1,\ldots,i_p}(\xi)$; the numbers $p^{i_1,\ldots,i_p}(\alpha)$ are called the Plücker coordinates of the subspace in $\mathbf{R}^n$ spanned by the frame α. The following assertion is well known (see [44]).

LEMMA. 1) Two p-frames α and α' define the same subspace in $\mathbf{R}^n$ if and only if their corresponding Plücker coordinates are proportional.

2) The set of numbers $\{q^{i_1,\ldots,i_p}(\xi)\}$ coincides with the set of Plücker coordinates of a subspace in $\mathbf{R}^n$ orthogonal to the vectors $\xi^i \in (\mathbf{R}^n)'$, $i = 1, \ldots, n-p$.

Proposition 4. The pairs $(\xi, s) \in F$ and $(\alpha, \beta) \in E$ prescribe the same plane $h \in H$ if and only if $q^{i_1,\ldots,i_p}(\xi) = \lambda p^{i_1,\ldots,i_p}(\alpha)$ for any $i_1, \ldots, i_p$ and $s^i = \langle \xi^i, \beta \rangle$, $i = 1, \ldots, n-p$. Here $d\mu_{\xi,s} = d\mu_{\alpha,\beta}$ if and only if $\lambda = \pm 1$.

The assertion follows immediately from the lemma and a comparison of formulas (1.25) and (1.28) for $d\mu_{\alpha,\beta}$ and $d\mu_{\xi,s}$.

COROLLARY. To the pairs $(\xi, s) \in F$ and $(\alpha, \beta) \in E$ there corresponds the same element of $\tilde{H}$ if and only if the following relations are satisfied: $q^{i_1,\ldots,i_p}(\xi) = \varepsilon p^{i_1,\ldots,i_p}(\alpha)$ for any $i_1, \ldots, i_p$, where $\varepsilon = \pm 1$, and $s^i = \langle \xi^i, \beta \rangle$, $i = 1, \ldots, n-p$.

3. Operator $\varkappa_x$. Here for any $x \in \mathbf{R}^n$ we define an operator $\varkappa_x$ which assigns to functions $\varphi \in S(H)$ differential p-forms $\varkappa_x\varphi$ on the manifold G_x of p-dimensional, oriented planes in $\mathbf{R}^n$ passing through the point x. In the case $p = n-1$, this operator coincides with the operator $\varkappa_x$ of Sec. 2. For even p we obtain an inversion formula by means of the operator $\varkappa_x$. While for $p = n-1$, in the inversion formula the integrals of $\varkappa_x\varphi$ are taken over the entire manifold G_x, for $p < n-1$ the integrals are taken over certain p-dimensional cycles in the $p(n-p)$-dimensional manifold G_x.

We begin by presenting an explicit expression for the operator $\varkappa_x$ and investigating its properties; in part 4 we present the considerations which enable us to proceed to the required definition.

We denote by E_0 the manifold of $n\times p$ -matrices $\alpha=(\alpha_1, \ldots, \alpha_p)$ of rank p. Let $\varphi\in S(H)$, and let $\varphi(\alpha, \beta)$ be the lift of φ to the manifold E of pairs (α, β) (see part 1). To the function φ and an arbitrary point $x\in\mathbf{R}^n$ we assign the following differential p-form on E_0:

$$\varkappa_x\varphi=(-1)^p \sum_{i_1,\ldots,i_p=1}^{n} \frac{\partial^p\varphi(\alpha, x)}{\partial x^{i_1}\ldots\partial x^{i_p}} d\alpha_1^{i_1}\wedge\ldots\wedge d\alpha_p^{i_p}. \tag{1.32}$$

We observe that for $p=n-1$ this form coincides with the form of Sec. 2 (see formula (1.23)). We shall establish the properties of form (1.32).

We define the projection $E_0\to G_x$ by assigning to each matrix $\alpha=(\alpha_1,\ldots,\alpha_p)\in E_0$ the p-dimensional oriented plane passing through x and incident with the vectors $\alpha_1,\ldots,\alpha_p$ (the orientation is given by the frame α). It is obvious that two matrices α and α' define the same oriented plane if and only if $\alpha'=\alpha g$, where $g\in GL_+(p, \mathbf{R})$. Thus, $E_0\to G_x$ is a principal bundle with structural group $GL_+(p, \mathbf{R})$.

<u>LEMMA.</u> The differential form $\varkappa_x\varphi$ drops from E_0 to the manifold G_x and is antisymmetric with respect to an involution in G_x changing the orientation of the plane.

This assertion is proved by precisely the same arguments as in the case $p=n-1$ (see the lemma in part 6 of Sec. 2).

<u>LEMMA.</u> The differential form $\varkappa_x\varphi$ on G_x does not depend on the choice of coordinates in $\mathbf{R}^n$.

Indeed, on the one hand, the definition of the function φ does not depend on the choice of coordinates in $\mathbf{R}^n$. On the other hand, the operator $\varkappa_x$ can be represented as a product of operators of the form $\sum_{i=1}^{n}\frac{\partial}{\partial x^i} d\alpha_k^i$ $(1\leqslant k\leqslant p)$, each of which is preserved under affine transformations in $\mathbf{R}^n$.

We shall find the restrictions of the form $\varkappa_x\varphi$ to certain submanifolds in G_x.

<u>Proposition 5.</u> The restriction of the form $\varkappa_x\varphi$ to a submanifold $M\subset G_x$ of planes passing through an arbitrary fixed line is equal to zero.

Indeed, in the equations $x=\alpha t+\beta$ of the planes of M it may be assumed that α_1 is a fixed vector directed along the common line. Since each term of the form $\varkappa_x\varphi$ contains the factor $d\alpha_1^i$ $(1<i<n)$, it follows that its restriction to M is equal to zero.

Suppose now that $L\subset\mathbf{R}^n$ is an arbitrary $(p+1)$-dimensional plane passing through x, and let $G_x^L\subset G_x$ be the manifold of p-dimensional, oriented planes passing through x and belonging to L. With the plane L there is associated the operator $\varkappa_x$ defined in Sec. 2, which we now denote by $\varkappa_x^L$. This operator takes functions defined on the manifold $\tilde{H}^L$ of hyperplanes with measures in L into differential p-forms on G_x^L.

<u>Proposition 6.</u> For any $(p+1)$ -dimensional plane $L\subset\mathbf{R}^n$ passing through the point x, we have

$$\varkappa_x\varphi|_{G_x^L}=\varkappa_x^L\varphi^L, \text{ where } \varphi^L=\varphi|_{\tilde{H}^L}. \tag{1.33}$$

Proof. Since the form $\varkappa_x\varphi$ does not depend on the choice of coordinates in $\mathbf{R}^n$, it may be assumed with no loss of generality that $x=0$, and L is the coordinate hyperplane $x^{p+2}=\dots=x^n=0$. Then p-dimensional planes belonging to L are given by equations $x=\alpha' t'+\beta'$, where α' is a matrix whose rows beginning with the $(p+2)$-nd row are equal to zero and $\beta'=(\beta^1,\dots,\beta^{p+1},0,\dots,0)$. From the definition of the operator $\varkappa_x^L$ given in Sec. 2 it follows that $\varkappa_x^L\varphi^L=(-1)^p\sum\frac{\partial^p\varphi(\alpha,x)}{\partial x^{i_1}\dots\partial x^{i_p}}d\alpha_1^{i_1}\wedge\dots\wedge d\alpha_p^{i_p}$, where the sum is taken over $i_1,\dots,i_p=1,\dots,p+1$. It is obvious that this form coincides with the restriction of $\varkappa_x\varphi$ to the submanifold G_x^L.

We now introduce submanifolds in G_x more general than G_x^L. For this we consider an arbitrary decomposition $p=p_1+\dots+p_k$, $k\leqslant n-p$, of the number p into a sum of natural numbers, and we prescribe an arbitrary system of planes $L_1,\dots,L_k$ in $\mathbf{R}^n$ of the respective dimensions $p_1+1,\dots,p_k+1$ which pass through x and are in general position with respect to one another. We denote by $G_x^{L_1,\dots,L_k}$ the submanifold of all oriented planes $h\in G_x$ such that $\dim(h\cap L_i)=p_i$ $(1\leqslant i\leqslant k)$. It will be shown that the restriction of $\varkappa_x\varphi$ to $G_x^{L_1,\dots,L_k}$ reduces to the application of the operators $\varkappa_x^{L_i}$ connected with the planes L_i.

Let $\tilde H^{L_i}$ be the manifold of hyperplanes with measures in L_i. To each collection $h_i\in\tilde H^{L_i}$ $(1\leqslant i\leqslant k)$ we assign the p-dimensional plane $h\subset L$, where L is the plane generated by $L_1,\dots,L_k$ such that the projections of h onto L_i coincide with h_i; we equip h with the measure induced by the measures on h_i. We obtain a manifold of p-dimensional planes with measures which we denote by $\tilde H^{L_1,\dots,L_k}$.

According to Sec. 2, in the space of functions on $\tilde H^{L_i}$ there is defined an operator $\varkappa_x^{L_i}$ which takes these functions into differential forms on $G_x^{L_i}$. Thus, since $\tilde H^{L_1,\dots,L_k}=\tilde H^{L_1}\times\dots\tilde H^{L_k}$, in the space of functions on $\tilde H^{L_1,\dots,L_k}$ the operator $\varkappa_x^{L_1}\wedge\dots\wedge\varkappa_x^{L_k}$ is defined. It is easy to see that the image of this operator is differential p-forms on the p-dimensional manifold $G_x^{L_1,\dots,L_k}$.

Proposition 7. For any decomposition $p=p_1+\dots+p_k$ $(k\leqslant n-p)$ and any system of planes $L_1,\dots,L_k$ of the respective dimensions $p_1+1,\dots,p_k+1$ passing through x and in general position relative to one another there is the equality

$$\varkappa_x\varphi|_{G_x^{L_1,\dots,L_k}}=\varkappa_x^{L_1}\wedge\dots\wedge\varkappa_x^{L_k}\varphi^{L_1,\dots,L_k},\tag{1.34}$$

where $\varphi^{L_1,\dots,L_k}$ is the restriction of φ to $\tilde H^{L_1,\dots,L_k}$.

Proof. Since the form $\varkappa_x\varphi$ does not depend on the choice of coordinates in $\mathbf{R}^n$, it may be assumed with no loss of generality that $x=0$, and the L_i are the coordinate hyperplanes $L_i=\{x^{q_{i-1}+1},\dots,x^{q_i}\}$, where $q_i=p_1+\dots+p_i+i$ $(0\leqslant i\leqslant k)$. In this case the planes of the manifold $\tilde H^{L_1,\dots,L_k}$ are given by pairs $(\alpha',\beta')\in E$, where

$$\alpha'=\begin{pmatrix}\alpha^{(1)}\dots 0\\ \dots\dots\\ 0\dots\alpha^{(k)}\\ 0\dots 0\end{pmatrix},\quad \beta'=(\beta^1,\dots,\beta^{p+k},0,\dots,0),$$

and $\alpha^{(i)}$ is a matrix with p_i columns and (p_i+1) rows. It is easy to demonstrate the validity of Eq. (1.34) by directly writing out in the coordinates α' the expressions on the left and right sides.

4. Motivation for the Definition of the Operator $\varkappa_x$. We wish to assign to each function $\varphi\in S(H)$ a differential p-form $\varkappa_x\varphi$ on the manifold G_x of p-dimensional oriented planes

passing through the point $x \in \mathbb{R}^n$. In the case of even p we must obtain by means of this form an inversion of the integral transform $\mathcal{J}_p$; viz., if $\varphi = \mathcal{J}_p f$, then for any p-dimensional cycle $\gamma \subset G_x$ we should have the equality

$$\int_\gamma \varkappa_x \varphi = c_\gamma f(x). \tag{1.35}$$

In the case where all planes $h \in \gamma$ pass through a fixed line the cycle γ cannot "distinguish" the points of this line, and therefore Eq. (1.35) is possible only for $c_\gamma = 0$. In connection with this, we impose on $\varkappa_x$ the following condition:

1) the restriction of $\varkappa_x \varphi$ to a manifold of planes passing through an arbitrary fixed line is equal to zero.

We impose still another condition on the form $\varkappa_x \varphi$. Let $L_1, \ldots, L_k$ be any collection of planes of the respective dimensions $p_1+1, \ldots, p_k+1$, where $p_1 + \ldots + p_k = p$, $k \leqslant n-p$, which pass through x and are in general position relative to one another. According to part 3, these planes determine a manifold $\tilde{H}^{L_1, \ldots, L_k}$ of p-dimensional planes with measures, and $\tilde{H}^{L_1, \ldots, L_k} = \tilde{H}^{L_1} \times \ldots \times \tilde{H}^{L_k}$, where $\tilde{H}^{L_i}$ is the manifold of hyperplanes with measures in L_i. The restriction $\varphi^{L_1, \ldots, L_k}$ of the function $\varphi = \mathcal{J}_p f$ to $\tilde{H}^{L_1, \ldots, L_k}$ is given by the composition of Randon transforms connected with the hyperplanes $L_1, \ldots, L_k$. Therefore, if the dimensions of all the L_i are odd numbers, then from Sec. 2 we have the inversion formula

$$\int_{\gamma_1 \times \ldots \times \gamma_k} \varkappa_x^{L_1} \wedge \ldots \wedge \varkappa_x^{L_k} \varphi^{L_1, \ldots, L_k} = c f(x), \tag{1.36}$$

where $\varkappa_x^{L_i}$ is the operator contained in the inversion formula for the Radon transform connected with L_i, γ_i is the manifold of oriented hyperplanes in L_i passing through x, and $c \neq 0$. It is obvious that $\gamma_1 \times \ldots \times \gamma_k$ coincides with the submanifold $G_x^{L_1, \ldots, L_k}$ of oriented planes $h \in G_x$ such that $\dim(h \cap L_i) = p_i$, $1 \leqslant i \leqslant k$. It is natural to require that the restriction of $\varkappa_x \varphi$ to $G_x^{L_1, \ldots, L_k}$ coincide with the integrand in (1.36), i.e., that the following condition be satisfied:

2) $\varkappa_x \varphi|_{G_x^{L_1, \ldots, L_k}} = \varkappa_x^{L_1} \wedge \ldots \wedge \varkappa_x^{L_k} \varphi^{L_1, \ldots, L_k}$. We shall additionally require that condition 2 be satisfied for planes L_i of any dimensions and not only odd-dimensional planes.

According to part 3, the form $\varkappa_x \varphi$ constructed there satisfies conditions 1 and 2.

<u>Proposition 8.</u> A p-form ω on the manifold G_x is uniquely determined by conditions 1 and 2.

<u>Proof.</u> It suffices to show that ω is uniquely determined in a neighborhood of any $a_0 \in G_x$. It may be assumed with no loss of generality that $x = 0$ and a_0 belongs to the set G_x^0 of planes transversal to the $(n-p)$-dimensional coordinate plane $x^{n-p+1} = \ldots = x^n = 0$. Any plane of G_x^0 can be defined by the equations

$$y^i = \alpha_1^i y^{n-p+1} + \ldots + \alpha_p^i y^n \quad (1 \leqslant i \leqslant n-p),$$

with coefficients which we take as coordinates on G_x^0. In these coordinates the form ω has the form

$$\omega = \sum a_{i_1, \ldots, i_p}^{j_1, \ldots, j_p}(\alpha)\, d\alpha_{j_1}^{i_1} \wedge \ldots \wedge d\alpha_{j_p}^{i_p},$$

where the indices i_k run through the values $1,\ldots,n-p$, j_k runs through the values $1,\ldots,p$, and $j_1 \leqslant \ldots \leqslant j_p$.

We shall first show that if among the indices $j_1,\ldots,j_p$ there is at least one of the numbers $1,\ldots,p$, e.g., 1, then $a^{j_1,\ldots,j_p}_{i_1,\ldots,i_p}(\alpha)=0$. The form ω can thus contain only terms of the form $a^{1,\ldots,p}_{i_1,\ldots,i_p}(\alpha)\,d\alpha_1^{i_1}\wedge\ldots\wedge d\alpha_p^{i_p}$.

Indeed, we consider the submanifold M of planes for which the coordinates $\alpha_1^1,\ldots,\alpha_1^{n-p}$ are fixed numbers. It is obvious that all the planes of M pass through the line defined by the equations $y^i-\alpha_1^i y^{n-p+1}=0$ $(1\leqslant i\leqslant n-p)$, $y^{n-p+i}=0$ $(2\leqslant i\leqslant p)$; hence, by condition 1, $\omega|_M=0$. On the other hand, $\omega|_M$ is obtained from ω by dropping terms containing the factors $d\alpha_1^1,\ldots, d\alpha_1^{n-p}$. Since this term remains during this process, it is equal to zero.

We shall now show that each term $a_{i_1,\ldots,i_p}(\alpha)\,d\alpha_1^{i_1}\wedge\ldots\wedge d\alpha_p^{i_p}$ of the form ω is well defined by condition 2.

For any fixed collection $\{i_1,\ldots,i_p\}$ we denote by $M^{i_1,\ldots,i_p}$ the p-dimensional submanifold of planes for which all the coordinates α_j^i distinct from $\alpha_1^{i_1},\ldots,\alpha_p^{i_p}$ are fixed numbers. It is obvious that

$$\omega|_{M^{i_1,\ldots,i_p}}=a_{i_1,\ldots,i_p}(\alpha)\,d\alpha_1^{i_1}\wedge\ldots\wedge d\alpha_p^{i_p}.$$

On the other hand, we set $s^i=\{k;\ i_k\neq i\ (1\leqslant k\leqslant p)\}$, $i=1,\ldots,n-p$. In $\mathbf{R}^n$ we go over to the new coordinates $y'^i=y^i-\sum\limits_{k\in s^i}\alpha_k^i y^{n-p+k}$, $1\leqslant i\leqslant n-p$ (i.e., the sum is taken over those k for which α_k^i are distinct from $\alpha_1^{i_1},\ldots,\alpha_p^{i_p}$, and therefore fixed), $y'^{n-p+i}=y^{n-p+i}$, $1\leqslant i\leqslant p$. Let $r_1,\ldots,r_k$ be the set of those indices i for which $s^i\neq\varnothing$. We consider the coordinate planes $L_l=\{y'^{r_l}, y'^{n-p+j};\ j\in s^{r_l}\}$, $1\leqslant l\leqslant k$. It is not hard to verify that the planes L_l form a direct sum and that $G_x^{L_1,\ldots,L_k}=M^{i_1,\ldots,i_p}$. Thus, by condition 2 the form $a_{i_1,\ldots,i_p}(\alpha)\,d\alpha_1^{i_1}\wedge\ldots\wedge d\alpha_p^{i_p}=\omega|_{M^{i_1,\ldots,i_p}}$ is well defined, as was required to prove.

5. An Expression for the Operator $\varkappa_x$ in the Coordinates (ξ, s). Let F_0 be the manifold of $n\times(n-p)$-matrices $\xi=(\xi^1,\ldots,\xi^{n-p})$ of rank $n-p$. We introduce the following differential forms on F_0:

$$[\xi^1,\ldots,\xi^{n-p},d\xi^{i_1},\ldots,d\xi^{i_p}]=\sum\operatorname{sgn}(j_1,\ldots,j_n)\,\xi^1_{j_1}\ldots\xi^{n-p}_{j_{n-p}}d\xi^{i_1}_{j_{n-p+1}}\wedge\ldots\wedge d\xi^{i_p}_{j_n},$$

where the sum is taken over all permutations $(j_1,\ldots,j_n)$ of the indices $1,\ldots,n$. This expression can also be written in the form of a determinant

$$\begin{vmatrix}\xi_1^1\ldots\xi_1^{n-p} & d\xi_1^{i_1}\ldots d\xi_1^{i_p}\\ \cdot\ \cdot\ \cdot\ \cdot\ \cdot & \cdot\ \cdot\ \cdot\ \cdot\ \cdot\\ \xi_n^1\ldots\xi_n^{n-p} & d\xi_n^{i_1}\ldots d\xi_n^{i_p}\end{vmatrix},$$

in the expansion of which the multiplication of the differentials is by the exterior product.

To each $x\in\mathbf{R}^n$ and each function $\varphi\in S(H)$ we assign the following differential p-form on F_0:

$$\tilde{\varkappa}_x\varphi=\frac{1}{p!}\sum_{i_1,\ldots,i_p=1}^{n-p}\left.\frac{\partial^p\varphi(\xi,s)}{\partial s^{i_1}\ldots\partial s^{i_p}}\right|_{s=\langle\xi,x\rangle}[\xi^1,\ldots,\xi^{n-p},d\xi^{i_1},\ldots,d\xi^{i_p}],\qquad(1.37)$$

where $\varphi(\xi,s)$ is the lift of φ to the manifold F of pairs (ξ,s) (the notation $s=\langle\xi,x\rangle$ means $s^i=\langle\xi^i,x\rangle$, $i=1,\ldots,n-p$). We note that for $p=n-1$ this form coincides with that introduced in Sec. 2 (see formula (1.16)).

Remark. It is also convenient to write the expression for $\tilde{\varkappa}_x\varphi$ in the following symbolic form:

$$\tilde{\varkappa}_x\varphi = \frac{1}{p!}[\xi^1, \ldots, \xi^{n-p}, D, \ldots, D]\,\varphi(\xi, s)\Big|_{s=\langle \xi, x\rangle}, \qquad (1.37')$$

where $D = \sum_{i=1}^{n-p} \frac{\partial}{\partial s^i} d\xi^i$.

Let G_x be the manifold of oriented p-dimensional planes in $\mathbf{R}^n$ passing through x. We define the projection $F_0 \to G_x$ which assigns to each matrix $\xi = (\xi^1, \ldots, \xi^{n-p})$ the p-dimensional oriented plane passing through x and given by the equations $\langle \xi^i, y\rangle = s_x^i$, $i = 1, \ldots, n-p$, where $s_x^i = \langle \xi^i, x\rangle$ (the orientation of the plane is determined by the frame $\xi^1, \ldots, \xi^{n-p}$ in $(\mathbf{R}^n)'$). It is obvious that two matrices ξ and ξ' define the same oriented plane if and only if $\xi' = \xi g$, where $g \in GL_+(n-p, \mathbf{R})$. Thus, $F_0 \to G_x$ is a principal bundle with structural group $GL_+(n-p, \mathbf{R})$.

LEMMA. The differential form $\tilde{\varkappa}_x\varphi$ drops from F_0 to the manifold G_x and is antisymmetric with respect to change of orientation of a plane $h \in G_x$.

Proof. It suffices to verify that 1) the form $\tilde{\varkappa}_x\varphi$ is invariant under the group $GL_+(n-p, \mathbf{R})$ and 2) each of the differential forms $[\xi^1, \ldots, \xi^{n-p}, d\xi^{i_1}, \ldots, d\xi^{i_p}]$ is preserved under the replacement $d\xi^i \to d\xi^i + \sum_k \xi^k dg_k^i$.

The first property follows from formula (1.37') for $\tilde{\varkappa}_x\varphi$, if it is noted that the operator D is invariant under the transformation $(\xi, s) \to (\xi g, sg)$ and $\varphi(\xi g, sg) = (\det g)^{-1}\varphi(\xi, s)$. The second property is obvious. The antisymmetry of $\tilde{\varkappa}_x\varphi$ follows from the fact that this form changes sign under the substitution $\xi^1 \to -\xi^1$.

Proposition 9. After dropping to G_x the form $\tilde{\varkappa}_x\varphi$ coincides with the form $\varkappa_x\varphi$.

Proof. We wish to prove the coincidence of the differential forms $\varkappa_x\varphi$ and $\tilde{\varkappa}_x\varphi$ on G_x obtained by dropping the forms defined, respectively, on E_0 and F_0. It suffices to prove coincidence in a neighborhood of any $a_0 \in G_x$. It may be assumed with no loss of generality that the plane a_0 belongs to the set G_x^0 of planes passing through x and transversal to the $(n-p)$-dimensional plane $x^{n-p+1} = \ldots = x^n = 0$. Any plane in G_x^0 can be defined by equations

$$y^i = u_1^i y^{n-p+1} + \ldots + u_p^i y^n = s_x^i \quad (1 \leqslant i \leqslant n-p),$$

where $s_x^i = x^i + u_1^i x^{n-p+1} + \ldots + u_p^i x^n$; the coefficients u_j^i of these equations we take as coordinates on G_x^0. On the basis of these coordinates we construct local sections $\alpha(u)$ and $\xi(u)$ of the bundles $E_0 \to G_x$ and $F_0 \to G_x$ by setting

$$\alpha(u) = \begin{pmatrix} -u_1^1 & \ldots & -u_p^1 \\ \vdots & & \vdots \\ -u_1^{n-p} & \ldots & -u_p^{n-p} \\ 1 & \ldots & 0 \\ \vdots & & \vdots \\ 0 & \ldots & 1 \end{pmatrix}, \quad \xi(u) = \begin{pmatrix} 1 & \ldots & 0 \\ \vdots & & \vdots \\ 0 & \ldots & 1 \\ u_1^1 & \ldots & u_1^{n-p} \\ \vdots & & \vdots \\ u_p^1 & \ldots & u_p^{n-p} \end{pmatrix}.$$

The measures on the planes $a(u)$ connected, respectively, with $\alpha(u)$ and $\xi(u)$ are both equal to $dy^{n-p+1} \ldots dy^n$, since $p^{n-p+1,\ldots,n}(\alpha(u)) = q^{n-p+1,\ldots,n}(\xi(u)) = 1$ (see part 1). This implies that $\varphi(\alpha(u), x) = \varphi(\xi(u), \langle \xi(u), x\rangle)$. We denote this function by $\psi(u, x)$.

We note that

$$[\xi^1, \ldots, \xi^{n-p}, d\xi^{i_1}, \ldots, d\xi^{i_p}]\big|_{\xi(u)} = \sum \operatorname{sgn}(j_1, \ldots, j_p)\, du_{j_1}^{i_1} \wedge \ldots \wedge du_{j_p}^{i_p},$$

whence by formulas (1.32) and (1.37) we obtain

$$\tilde{\varkappa}_x\varphi=\varkappa_x\varphi=\sum_{i_1,\ldots,i_p=1}^{n-p}\frac{\partial^p\psi(u,x)}{\partial x^{i_1}\ldots\partial x^{i_p}}du_1^{i_1}\wedge\ldots\wedge du_p^{i_p}.$$

The coincidence of $\tilde{\varkappa}_x\varphi$ and $\varkappa_x\varphi$ on G_x has been proved. In view of what has been proved we shall henceforth also denote the form (1.37) by $\varkappa_x\varphi$.

6. Inversion Formula for the Case of Even p. Proposition 10. If the function can be represented in the form $\varphi=\mathcal{J}_pf$, $f\in S(\mathbf{R}^n)$, then the differential form $\varkappa_x\varphi$ on G_x is closed.

The proof reduces to two lemmas.

LEMMA 1. The function $\varphi=\mathcal{J}_pf$, i.e.,

$$\varphi(\alpha,\beta)=\int_{\mathbf{R}^p}f(\alpha t+\beta)\,dt, \tag{1.38}$$

satisfies the system of differential equations

$$\frac{\partial^2\varphi(\alpha,\beta)}{\partial\beta^i\partial\alpha_k^j}=\frac{\partial^2\varphi(\alpha,\beta)}{\partial\beta^j\partial\alpha_k^i}\ (1\leqslant i,\ j\leqslant n,\ 1\leqslant k\leqslant p). \tag{1.39}$$

The equations follow directly from (1.38).

LEMMA 2. If the function $\varphi(\alpha,\beta)\in S(H)$ satisfies the system of differential equations (1.39), then the form $\varkappa_x\varphi$ is closed.

Proof. From expression (1.32) for $\varkappa_x\varphi$ it follows that

$$d(\varkappa_x\varphi)=\sum_k\sum_{i_1,\ldots,i_p,j}\frac{\partial^{p+1}\varphi(\alpha,x)}{\partial x^{i_1}\ldots\partial x^{i_p}\partial\alpha_k^j}d\alpha_k^j\wedge d\alpha_1^{i_1}\wedge\ldots\wedge d\alpha_p^{i_p}. \tag{1.40}$$

Because of Eqs. (1.39), for any fixed k the terms of the sum (1.40) are antisymmetric with respect to i_k, j; hence, this sum is equal to zero.

Since the form $\varkappa_x\varphi$ is closed, the integral $\int_\gamma\varkappa_x\varphi$ over any p-dimensional cycle $\gamma\subset G_x$ depends only on the homology class of the cycle γ. The submanifold of p-dimensional oriented planes belonging to a fixed $(p+1)$-dimensional plane we call the Euler cycle in G_x.

THEOREM 4. If p is an even number, then for any p-dimensional cycle $\gamma\subset G_x$ there is the following inversion of the transformation $\varphi=\mathcal{J}_pf$, $f\in S(\mathbf{R}^n)$:

$$\int_\gamma\varkappa_x\varphi=(-1)^{p/2}2(2\pi)^p d(\gamma)f(x), \tag{1.41}$$

where $d(\gamma)$ is the index of intersection of γ with the Euler cycle. If p is an odd number, then for any p-dimensional cycle $\gamma\subset G_x$ we have

$$\int_\gamma\varkappa_x\varphi=0.$$

Proof. We consider a basis of the p-dimensional homology in G_x consisting of p-dimensional Schubert cells; it contains the Euler cycle γ_0. The proof of the theorem reduces to verifying the following two assertions:

$$1)\ \int_\gamma\varkappa_x\varphi=0$$

if γ is a Schubert cell distinct from the Euler cycle γ_0;

$$2)\ \int_{\gamma_0}\varkappa_x\varphi=\begin{cases}(-1)^{p/2}2(2\pi)^pf(x) & \text{for even } p; \\ 0 & \text{for odd } p.\end{cases} \tag{1.42, 1.42'}$$

If γ is a Schubert cell distinct from an Euler cycle, then it is known that all planes $h\in\gamma$ pass through a fixed line. Hence, by Proposition 5, $\varkappa_x\varphi|_\gamma=0$, and hence $\int_\gamma \varkappa_x\varphi=0$.

Suppose now that γ_0 is the Euler cycle consisting of p-dimensional oriented planes passing through x and belonging to a fixed $(p+1)$-dimensional plane L. By Proposition 6 we have $\varkappa_x\varphi|_{\gamma_0}=\varkappa_x^L\varphi^L$, where φ^L is the restriction of φ to the submanifold $\tilde{H}^L$ of all hyperplanes in L, and $\varkappa_x^L$ is the operator in the space of functions on $\tilde{H}^L$ defined in Sec. 2. The equalities (1.42) and (1.42') thus follow from Theorem 3.

7. Description of the Image of the Operator $\mathcal{I}_p$ for $p<n-1$. THEOREM 5. Let $\varphi\in S(H)$, $H=H_{p,n}$, where $p<n-1$. The following conditions are then equivalent:

(i) φ belongs to the image of the space $S(\mathbf{R}^n)$ under the transformation $\mathcal{I}_p$;

(ii) the lift $\varphi(\alpha,\beta)$ of the function φ to the manifold E satisfies the system of differential equations (1.39);

(iii) for any $x\in\mathbf{R}^n$ the differential form $\varkappa_x\varphi$ on G_x defined by formula (1.32) is closed.

The implications (i)$\to$(ii) and (ii)$\to$(iii) have already been established in Lemmas 1 and 2. Only the implication (iii)$\to$(i) must thus be proved. To this end it is convenient to introduce the Fourier transform of functions in $S(H)$, which is also of independent interest.

We recall that the manifold $H=H_{p,n}$ of p-dimensional planes in $\mathbf{R}^n$ is a vector bundle over the Grassmann manifold $G=G_{p,n}$ of p-dimensional subspaces where the fiber over $a\in G$ is the factor space $H_a=\mathbf{R}^n/a$. We consider the bundle $H'\to G$ dual to $H\to G$. We note that each fiber H_a' of the bundle $H'\to G$, i.e., the space dual to $\mathbf{R}^n/a$, is canonically imbedded in $(\mathbf{R}^n)'$ as a subspace of vectors ξ orthogonal to a. We denote by $S(H')$ the space of infinitely differentiable, rapidly decreasing functions on H'.

We shall define the Fourier transform of functions on $\tilde{H}$. Let $h=(a,\beta)$, where $a\in G$, $\beta\in H_a=\mathbf{R}^n/a$. On $\mathbf{R}^n$ we prescribe the volume element, and we associate with each invariant measure $dx=dx^1\ldots dx^n$ on the plane h the measure $d\mu_{H_a}$ on H_a defined from the relation $dx=d\mu_h d\mu_{H_a}$.

We consider an arbitrary function $\varphi(h,\mu_h)$ on $\tilde{H}$ satisfying the condition $\varphi(h,\lambda\mu_h)=\lambda\varphi(h,\mu_h)$. The Fourier transform of the function φ is the following function $\tilde{\varphi}$:

$$\tilde{\varphi}(a,\xi)=(2\pi)^{-\frac{n-p}{2}}\int_{H_a}\varphi(a,\beta;\mu_h)e^{i\langle\xi,\beta\rangle}\,d\mu_{H_a}(\beta), \tag{1.43}$$

where $\xi\in H'_a$, and $d\mu_{H_a}$ is the measure on H_a consistent with the measure $d\mu_h$ on h. It is obvious that the integral does not depend on the choice of the measure $d\mu_h$, and thus $\tilde{\varphi}$ is a function on H'. The next results follows immediately from the properties of the usual transform.

Proposition 11. The Fourier transform $\varphi\mapsto\tilde{\varphi}$ realizes an isomorphism of the spaces

$$S(H)\to S(H').$$

The inverse transform is given by the following formula:

$$\varphi(a,\beta;\mu_h)=(2\pi)^{-\frac{n-p}{2}}\int_{H'_a}\tilde{\varphi}(a,\xi)e^{-i\langle\xi,\beta\rangle}\,d\mu_{H'_a}(\xi), \tag{1.44}$$

where $d\mu_{H'_a}$ is the measure on H'_a dual to the measure $d\mu_{H_a}$ on H_a defined from the relation $dx = d\mu_h d\mu_{H_a}$.

Proposition 12. If a function $\varphi \in S(H)$ belongs to the image of the space $S(\mathbf{R}^n)$, i.e., $\varphi = \mathcal{J}_p f$, where $f \in S(\mathbf{R}^n)$, then

$$\tilde{\varphi}(a, \xi) = (2\pi)^{p/2} \tilde{f}(\xi),$$

where $\tilde{f}$ is the Fourier transform of the function f.

Proof. We have

$$\tilde{\varphi}(a, \xi) = (2\pi)^{-\frac{n-p}{2}} \int_{H_a} \varphi(a, \beta; \mu_h) e^{i\langle \xi, \beta\rangle} d\mu_{H_a}(\beta) =$$

$$= (2\pi)^{-\frac{n-p}{2}} \int_{H_a} \int_h f(x) e^{i\langle \xi, \beta\rangle} d\mu_h(x) d\mu_{H_a}(\beta) = (2\pi)^{-\frac{n-p}{2}} \int_{\mathbf{R}^n} f(x) e^{i\langle \xi, x\rangle} dx,$$

since $d\mu_h d\mu_{H_a} = dx$ and $\langle \xi, \beta\rangle = \langle \xi, x\rangle$, because of the orthogonality of a and ξ. Thus, $\tilde{\varphi}(a, \xi) = (2\pi)^{p/2} \tilde{f}(\xi)$.

Proposition 13. A function $\varphi \in S(H)$ can be represented in the form $\varphi = \mathcal{J}_p f$, where $f \in S(\mathbf{R}^n)$, if and only if the Fourier transform $\tilde{\varphi}$ of the function φ satisfies the following conditions:

1) for any fixed $\xi \in (\mathbf{R}^n)'$ the expression $\tilde{\varphi}(a, \xi)$ does not depend on the choice of subspace a orthogonal to ξ; thus, $\tilde{\varphi}(a, \xi) = F(\xi)$, where F is a function on $(\mathbf{R}^n)'$;

2) $F \in S((\mathbf{R}^n)')$.

Proof. The necessity of the conditions follows from Proposition 12. We shall prove that they are sufficient. Suppose that a function $\varphi \in S(H)$ satisfies the conditions of the proposition. We define a function $f \in S(\mathbf{R}^n)$ from the condition that its Fourier transform $\tilde{f}$ is given by the equality $\tilde{f}(\xi) = (2\pi)^{-p/2} F(\xi)$, where $F(\xi) = \tilde{\varphi}(a, \xi)$. Suppose, further, that $\varphi_1 = \mathcal{J}_p f$. By Proposition 12 we then have $\tilde{\varphi}_1(a, \xi) = (2\pi)^{p/2} \tilde{f}(\xi) = F(\xi) = \tilde{\varphi}(a, \xi)$, whence $\varphi = \varphi_1 = \mathcal{J}_p f$.

We return to the proof of Theorem 5. It suffices to prove that if the differential form $\varkappa_x \varphi$ on G_x is closed for any $x \in \mathbf{R}^n$, then the Fourier transform $\tilde{\varphi}$ of the function φ satisfies the conditions of Proposition 13. We give only a sketch of the proof (for a detailed proof according to this outline for the case of $\mathbf{C}^n$, which carries over without change to the case of $\mathbf{R}^n$, see [20]).

First, from the condition that $\varkappa_x \varphi$ is closed we deduce that for any $\xi \neq 0$ the function $\tilde{\varphi}(a, \xi)$ does not depend on a, i.e., $\tilde{\varphi}(a, \xi) = F(\xi)$ for $\xi \neq 0$. It is further established that $F(\xi)$ is infinitely differentiable for $\xi \neq 0$ and decreases rapidly together with all partial derivatives. Finally, we prove that $\tilde{\varphi}(a, 0)$ does not depend on a and if we set $F(0) = \tilde{\varphi}(a, 0)$, then the function F is infinitely differentiable at the point 0.

APPENDIX. A SUMMARY OF FORMULAS FOR THE COMPLEX SPACE $\mathbf{C}^n$

1. Definition of the operator $\mathcal{J}_p: f \mapsto \hat{f}$:

a) $\hat{f}(\alpha, \beta) = \left(\frac{i}{2}\right)^p \int_{\mathbf{C}^p} f(\alpha t + \beta) dt \wedge d\bar{t}$, where $t = (t^1, \ldots, t^p)$, $dt = dt^1 \wedge \ldots \wedge dt^p$, $d\bar{t} = d\bar{t}^1 \wedge \ldots \wedge d\bar{t}^p$.

b) $\hat{f}(\xi, s) = \left(\frac{i}{2}\right)^n \int_{\mathbf{C}^n} f(z) \prod_{k=1}^{n-p} \delta(\langle \xi^k, z\rangle - s^k) dz \wedge d\bar{z}$ (regarding the notation, see Sec. 3, part 1).

2. The inversion formula for the Radon transform (the case $p=n-1$):

$$f(z)=(-1)^{n-1}\pi^{-2n+2}\left(\frac{i}{2}\right)^{n-1}\int_{\Gamma}\hat{f}_s^{(n-1,n-1)}(\xi,\langle\xi,z\rangle)\sigma(\xi,d\xi)\wedge\sigma(\bar{\xi},d\bar{\xi}),$$

where $\sigma(\xi,d\xi)=\sum_{k=1}^{n}(-1)^{k-1}\xi_k d\xi_1\wedge\ldots\wedge\widehat{d\xi_k}\wedge\ldots\wedge d\xi_n$; the integral is taken over any surface in $\mathbf{C}^n$ which intersects once almost every complex line passing through the point $\xi=0$.

3. The inversion formula for arbitrary $p\leqslant n-1$:

$$\left(\frac{i}{2}\right)^p\int_{\Gamma}(\varkappa_z\wedge\bar{\varkappa}_z)\hat{f}=(-1)^p\pi^{2p}d(\gamma)f(z).$$

Here the operator $\varkappa_z$ is defined in the coordinates (α,β) and in the coordinates (ξ,s) by the respective formulas (1.32) and (1.37) where the partial derivatives with respect to real directions are replaced by the partial derivatives $\frac{\partial}{\partial z^i}$, $\frac{\partial}{\partial s^i}$ with respect to complex directions. Correspondingly, $\bar{\varkappa}_z$ contains the partial derivatives $\frac{\partial}{\partial \bar{z}^i}$, $\frac{\partial}{\partial \bar{s}^i}$. Thus, $(\varkappa_z\wedge\bar{\varkappa}_z)\varphi$ is a differential form of type (p, p). The integral is taken over any cycle γ of real dimension $2p$ in the manifold of complex p-dimensional planes passing through the point z; $d(\gamma)$ is the index of intersection of the cycle γ with the Euler cycle.

We emphasize that in the complex case the inversion formula is local for any p.

CHAPTER II

INTEGRAL GEOMETRY FOR ONE-DIMENSIONAL BUNDLES OVER PROJECTIVE SPACE

In this chapter we go over from the affine results of Chap. I to projective-invariant assertions. The first question that arises here is that of the function space on which the integral transform (integration over p planes) is to be constructed. The space $S(\mathbf{R}^n)$ is not suitable for this, since it is not projectively invariant in any sense. It is found that a suitable extension of it consists not of functions on the projective space $\mathbf{RP}^n$, but of even homogeneous functions of degree of homogeneity $(p+1)$ on the space $\mathbf{R}^{n+1}$, canonically covering $\mathbf{RP}^n$ (points of $\mathbf{RP}^n$ are one-dimensional subspaces in $\mathbf{R}^{n+1}$). We denote this space of functions by Φ_p.

The reason for introducing precisely the space Φ_p is related to the fact that on each $(p+1)$-dimensional subspace $\mathbf{R}^{p+1}$ there is a p-form which is unique up to a factor and is invariant under the group $SL(p+1,\mathbf{R})$ (the "projective analog" of the form $dx^1\wedge\ldots\wedge dx^p$); this is the form

$$\sigma_p(t,dt)=\sum_{k=1}^{p+1}(-1)^{k-1}t^k dt^1\wedge\ldots\wedge\widehat{dt^k}\wedge\ldots\wedge dt^{p+1},$$

where $t^1,\ldots,t^{p+1}$ are the coordinates in some basis. It is of fundamental importance that the form σ_p has the degree of homogeneity $p+1$; therefore, if $f\in\Phi_p$, then the form $f\sigma_p(t,dt)$ drops onto the corresponding p-dimensional plane in $\mathbf{RP}^n$, and we can integrate it over this plane.

We relate the transformation of functions $f\in\Phi_p$ thus defined to the transformation of functions in $S(\mathbf{R}^n)$ introduced in Chap. I. To this end we imbed $\mathbf{R}^n$ in $\mathbf{R}^{n+1}$ as a hyperplane not passing through 0. By restricting functions $f\in\Phi_p$ to this hyperplane, we realize Φ_p as a space of functions on $\mathbf{R}^n$ which decay at infinity that is somewhat broader than the space

$S(\mathbf{R}^n)$; the form σ_p under this restriction goes over into the form $c\,dx^1\wedge\cdots\wedge dx^p$. As a result, by integrating the form $f\sigma_p(t, dt)$, we obtain an integral transform which agrees on $S(\mathbf{R}^n)$ for the imbedding indicated with the integral transform of Chap. I. Thus, the integral transforms of Chap. II are essentially just the transforms of Chap. I on a somewhat broader space which are written in projectively invariant form.

The next important remark pertains to inversion formulas (for even p). In the inversion formulas of Chap. I the affine structure was essential, since they contained differential operators along families of parallel planes. The projective inversion formulas of this chapter possess considerably greater generality. In particular, in these formulas in each p plane passing through a fixed point x (at which f is to be recovered) it is possible to select "its own" affinization — the "infinitely distant" $(p-1)$ plane — and differentiate along families of p planes passing through this infinitely distant plane.

The difference, in principle, between the cases $p=n-1$ and $p<n-1$ in describing the image of the space $S(\mathbf{R}^n)$ becomes almost obvious in the framework of the projective transform. Namely, if $f\in S(\mathbf{R}^n)$, then for $p<n-1$ the projective transform of the function f vanishes on the "large" submanifold M_∞ ("infinitely distant" p planes). Since in the inversion formula for "infinitely distant" points it is possible to integrate only over this submanifold, it follows that, conversely, if the integral transform of a function f is equal to zero on M_∞ together with all partial derivatives, then $f\in S(\mathbf{R}^n)$. Moreover, for $p=n-1$ the projective transformation of a function $f\in S(\mathbf{R}^n)$ vanishes together with all derivatives at only one point (corresponding to the infinitely distance hyperplane), and this condition alone is insufficient in order that the function f belong to the space $S(\mathbf{R}^n)$ (i.e., vanish together with derivatives at infinitely distant points). The Cavalieri conditions are required.

An initial version of integral geometry in projective space was presented in [22]. The present exposition differs from [22] in many essential features.

The second part of the chapter (Sec. 3) is devoted to a generalization of the problem of integral geometry described above: integration of homogeneous functions on $\mathbf{R}^{n+1}$ of other degrees of homogeneity over p planes. Here the result of the integration is a tensor field on the bundle of a Stiefel manifold over the corresponding Grassmann manifold.

At the conclusion of the chapter, a summary of the formulas for complex projective space is given.

1. Integral Geometry for Certain One-Dimensional Bundles Over Projective Space

1. Space $\Phi_p(\mathbf{RP}^n)$. We denote by $\Phi_p(\mathbf{RP}^n)$ the space of C^∞ functions on $\mathbf{R}^{n+1}\setminus 0$ satisfying the homogeneity condition

$$f(\lambda x)=|\lambda|^{-p-1}f(x). \tag{2.1}$$

We note that for $p=-1$ $\Phi_p(\mathbf{RP}^n)$ consists of C^∞ functions on $\mathbf{RP}^n$; however, we shall be interested here in the case $p=1,\ldots,n-1$.

The space $\Phi_p(\mathbf{RP}^n)$ can be interpreted as the space of smooth sections of a certain one-dimensional vector bundle over $\mathbf{RP}^n$, which we now define.

We consider the manifold of all possible pairs (x, s), where $x\in\mathbf{R}^{n+1}\setminus 0$, $s\in\mathbf{R}^1$. On it we identify the pairs (x, s) and $(\lambda x, |\lambda|^{-p-1}s)$; we thus obtain a new manifold which we denote by B^{np}. We define the mapping

$$\pi_0: B^{np}\to \mathbf{RP}^n$$

by setting $\pi_0(x, s)=\pi x$, where $\pi: \mathbf{R}^{n+1}\setminus 0\to\mathbf{RP}^n$ is the standard mapping. π_0 is well defined, since under this mapping equivalent pairs (x, s) have the same image. It is not hard to see that $\pi_0: B^{np}\to\mathbf{RP}^n$ is a one-dimensional vector bundle over $\mathbf{RP}^n$.

We shall show that smooth sections of this bundle can be canonically identified with elements of the space $\Phi_p(\mathbf{RP}^n)$. Let $f\in\Phi_p(\mathbf{RP}^n)$. For each $\tilde{x}\in\mathbf{RP}^n$ we consider the set of pairs $(x, f(x))$ such that $\pi x=\tilde{x}$. By condition (2.1) it consists of equivalent pairs and thus determines a point of $B^{n,p}$ lying over $\tilde{x}$. Thus, to the function $f\in\Phi_p(\mathbf{RP}^n)$ there corresponds a section of the bundle $B^{np}\to\mathbf{RP}^n$. Conversely, suppose there is given a smooth section of the bundle π_0, i.e., to each point $\tilde{x}\in\mathbf{RP}^n$ there corresponds a class of equivalent pairs (x, s), where $\pi x=\tilde{x}$. Let $x\in\mathbf{R}^{n+1}\setminus 0$ and suppose that $\pi x=\tilde{x}$. We choose a class of equivalent pairs corresponding to $\tilde{x}$; for a given x it contains a single pair of form (x, s). Thus, to a section of the bundle π_0 there corresponds a function $x\mapsto s=f(x)$ on $\mathbf{R}^{n+1}\setminus 0$. Since here $(x, f(x))\sim(\lambda x, f(\lambda x))$, it follows that f satisfies the relation (2.1), and hence $f\in\Phi_p(\mathbf{RP}^n)$.

2. Definition of the Operator $\mathcal{I}_p$. We shall define the operator of integration over p-dimensional planes $\mathcal{I}_p$. To this end we assign to each $f\in\Phi_p(\mathbf{RP}^n)$ and each p-dimensional plane $h\subset\mathbf{RP}^n$ a differential p-form on h. We shall construct this form on the basis of a $(p+1)$-frame in $\mathbf{R}^{n+1}$ defining h.

We denote by G_p the manifold of $(p+1)$-dimensional, oriented subspaces in $\mathbf{R}^{n+1}$ and by E_p the manifold of all $(p+1)$-frames in $\mathbf{R}^{n+1}$, i.e., collections $u=(u_1,\dots,u_{p+1})$ of $p+1$ linearly independent vectors. Let

$$\pi: E_p\to G_p$$

be the canonical projection ($\pi u=h_u$ is the oriented subspace generated by the frame u). We note that $h_{u'}=h_u$ if and only if $u'=ug$, where $g\in GL_+(p+1, \mathbf{R})$; thus, $\pi: E_p\to G_p$ is a smooth bundle over G_p with structural group $GL_+(p+1, \mathbf{R})$.

In $\mathbf{R}^{p+1}$ we introduce the fixed p-form

$$\sigma_p(t, dt)=\sum_{k=1}^{p+1}(-1)^{k-1}t^k dt^1\wedge\dots\wedge\widehat{dt^k}\wedge\dots\wedge dt^{p+1}. \tag{2.2}$$

It is obvious that for any $\lambda>0$ we have

$$\sigma_p(\lambda t, \lambda dt+d\lambda t)=\lambda^{p+1}\sigma_p(t, dt) \tag{2.3}$$

Remark. Up to a factor, $\sigma_p(t, dt)$ is the only p-form on $\mathbf{R}^{p+1}$ which satisfies for any $g\in GL(p+1\ \mathbf{R})$ the relation $\sigma_p(gt, gdt)=\det g\cdot\sigma_p(t, dt)$. In particular, σ_p is invariant under $SL(p+1, \mathbf{R})$.

Suppose now that $f\in\Phi_p(\mathbf{RP}^n)$ and h_u is the $(p+1)$-dimensional subspace spanned by the frame $u\in E_p$. We denote by $t=(t^1,\dots,t^{p+1})$ the coordinates of points in h_u relative to the frame u, and we introduce the following differential p-form on h_u:

$$f(ut)\sigma_p(t, dt)\equiv f(u_1t^1+\dots+u_{p+1}t^{p+1})\sigma_p(t, dt). \tag{2.4}$$

This is the differential p-form corresponding to f on the given subspace h_u.

LEMMA. Form (2.4) drops onto the manifold of rays in h_u issuing from the point 0.

For the proof it suffices to show that form (2.4) is preserved under the transformations $t\to\lambda t$, $dt\to\lambda dt+d\lambda\cdot t$ $(\lambda>0)$. These properties follow easily from conditions (2.1) and (2.3).

Definition. We define the operator $\mathcal{J}_p$ on $\Phi_p(\mathbf{R}P^n)$ by the formula

$$(\mathcal{J}_p f)(u)=\frac{1}{2}\int_K f(ut)\,\sigma_p(t,dt), \tag{2.5}$$

where $K\subset h_u$ is an arbitrary surface in h_u intersecting once each ray issuing from the point 0.

Because of the lemma, the integral does not depend on the choice of the surface K (in particular, if a Euclidean structure is given in $\mathbf{R}^{n+1}$, then it may be assumed that $K=S^p$ is the p-dimensional unit sphere with center at the point 0).

Thus, to each function $f\in\Phi_p(\mathbf{R}P^n)$ there corresponds a function $\hat{f}=\mathcal{J}_p f$ on the manifold E_p of $(p+1)$-frames in $\mathbf{R}^{n+1}$.

Remark. Under the transformation $t\to -t$ the form $f(ut)\,\sigma_p(t,dt)$ is preserved in the case of odd p and changes sign in the case of even p. It can therefore be considered a form of the corresponding parity* on the projective plane $\mathbf{R}P^p$ corresponding to h_u, while the integral (2.5) can be written in the form

$$(\mathcal{J}_p f)(u)=\int_{\mathbf{R}P^p} f(ut)\,\sigma_p(t,dt). \tag{2.5'}$$

The next result follows immediately from the definition of $\hat{f}=\mathcal{J}_p f$.

LEMMA. The function $\hat{f}=\mathcal{J}_p f$ satisfies the following relation for any $g\in GL(p+1,\mathbf{R})$:

$$\hat{f}(ug)=|\det g|^{-1}\hat{f}(u). \tag{2.6}$$

In particular, it is invariant under the group $SL(p+1,\mathbf{R})$.

Since each equivalence class in E_p relative to the group $SL(p+1,\mathbf{R})$ can be prescribed by a multivector $u_1\wedge\ldots\wedge u_{p+1}$, $\hat{f}=\mathcal{J}_p f$ can thus be considered as a function on the manifold of multivectors $u_1\wedge\ldots\wedge u_{p+1}\neq 0$. We henceforth denote by $\Phi_0(E_p)$ the space of all C^∞ functions on E_p satisfying the condition (2.6). Thus, the operator $\mathcal{J}_p$ defines a mapping

$$\mathcal{J}_p:\Phi_p(\mathbf{R}P^n)\to\Phi_0(E_p).$$

3. Another Manner of Defining the Operator $\mathcal{J}_p$. In the preceding section we defined an operator $\mathcal{J}_p$ assigning to functions $f\in\Phi_p(\mathbf{R}P^n)$ functions on the manifold E_p of $(p+1)$-frames in $\mathbf{R}^{n+1}$. Here we consider a dual method of defining the operator $\mathcal{J}_p$.

We denote by F_p the manifold of all $(n-p)$-frames $\xi=(\xi^1,\ldots,\xi^{n-p})$ in the dual space $(\mathbf{R}^{n+1})'$. To each frame $\xi\in F_p$ we assign the $(p+1)$-dimensional subspace h_ξ in $\mathbf{R}^{n+1}$ orthogonal to the vectors of the frame, i.e., the subspace defined by the equations

$$\langle \xi^i, x\rangle=0,\quad i=1,\ldots,n-p.$$

In $\mathbf{R}^{n+1}$ we fix the differential form

*For even p the expression (2.5') can be understood as an integral of an odd form over a nonoriented manifold $\mathbf{R}P^p$ in the sense of de Rham [37].

$$\sigma_n(x, dx) = \sum_{k=1}^{n+1} (-1)^{k-1} x^k dx^1 \wedge \ldots \wedge \widehat{dx^k} \wedge \ldots \wedge dx^{n+1}.$$

As noted on p. 66, the form $\sigma_n(x, dx)$ is uniquely determined up to a factor by the condition of invariance under the group $SL(n+1, \mathbf{R})$. (In order to fix this factor, it suffices to fix an $(n+1)$-vector $e_1 \wedge \ldots \wedge e_{n+1}$.)

With the frame $\xi = (\xi^1, \ldots, \xi^{n-p})$ we associate the following differential p-form:

$$\langle \xi^1, dx \rangle \wedge \ldots \wedge \langle \xi^{n-p}, dx \rangle \,\lrcorner\, \sigma_n(x, dx).$$

By definition, this is any p-form $\omega_\xi(x, dx)$ on $\mathbf{R}^{n+1}$ satisfying the relation

$$\langle \xi^1, dx \rangle \wedge \ldots \wedge \langle \xi^{n-p}, dx \rangle \wedge \omega_\xi(x, dx) = \sigma_n(x, dx). \tag{2.7}$$

Although on all of $\mathbf{R}^{n+1}$ the condition (2.7) does not define a form ω_ξ on $\mathbf{R}^{n+1}$ uniquely, its restriction to the subspace h_ξ is uniquely defined.

To a function $f \in \Phi_p(\mathbf{RP}^n)$ we now assign the following differential p-form on $h = h_\xi$:

$$f(x)\, \omega_\xi(x, dx).$$

As in the preceding part, it is easy to verify that this differential form drops to the manifold of rays in h_ξ issuing from the point 0.

Definition. We set

$$(\mathcal{J}'_p f)(\xi) = \frac{1}{2} \int_K f(x)\, \omega_\xi(x, dx), \tag{2.8}$$

where K is any surface in h intersecting once each ray in h issuing from the point 0.

Using the delta function symbolism, the expression for $\mathcal{J}'_p f$ can be written

$$(\mathcal{J}'_p f)(\xi) = \frac{1}{2} \int_{K_1} f(x) \prod_{i=1}^{n-p} \delta(\langle \xi^i, x \rangle)\, \sigma_n(x, dx), \tag{2.8'}$$

where the integral is taken over any surface $K_1 \subset \mathbf{R}^{n+1}$ intersecting once each ray issuing from the point 0.

LEMMA. The function $\hat{f} = \mathcal{J}'_p f$ satisfies the following relation for any $g \in GL(n-p, \mathbf{R})$:

$$\hat{f}(\xi g) = |\det g|^{-1} \hat{f}(\xi). \tag{2.9}$$

The assertion follows immediately from the definition of the form $\omega_\xi(x, dx)$.

COROLLARY. The function $\hat{f}(\xi) = \mathcal{J}'_p f$ drops from the manifold F_p of all $(n-p)$-frames $\xi = (\xi^1, \ldots, \xi^{n-p})$ in $(\mathbf{R}^{n+1})'$ to the manifold $\tilde{F}_p$ of multivectors $\xi^1 \wedge \ldots \wedge \xi^{n-p} \neq 0$.

We denote by $\Phi_0(F_p)$ the space of all C^∞ functions on F_p satisfying the condition (2.9). By the lemma the image of the operator $\mathcal{J}'_p$ lies in this space, i.e.,

$$\mathcal{J}'_p : \Phi_p(\mathbf{RP}^n) \to \Phi_0(F_p).$$

We point out that in defining the operator $\mathcal{J}'_p$ we fixed the form $\sigma_n(x, dx)$ in $\mathbf{R}^{n+1}$; the definition of the operator $\mathcal{J}_p$, on the other hand, does not depend on the choice of the form σ_n.

4. Connection between the Operators $\mathcal{J}_p$ and $\mathcal{J}'_p$. We know that the function $(\mathcal{J}_p f)(u)$, where $u = (u_1, \ldots, u_{p+1})$, depends only on the $(p+1)$-vector $\hat{u} = u_1 \wedge \ldots \wedge u_{p+1}$ in $\mathbf{R}^{n+1}$. Similarly, if in $\mathbf{R}^{n+1}$ the form $\sigma_n(x, dx) = \sum (-1)^{k-1} x^k \cdot dx^1 \wedge \ldots \wedge \widehat{dx^k} \wedge \ldots \wedge dx^{n+1}$ is fixed, then the function $(\mathcal{J}'_p f)(\xi)$, where $\xi = (\xi^1, \ldots, \xi^{n-p})$, depends only on the $(n-p)$-vector $\hat{\xi} = \xi^1 \wedge \ldots \wedge \xi^{n-p}$ in $(\mathbf{R}^{n+1})'$. We can therefore write $(\mathcal{J}_p f)(\hat{u})$ and $(\mathcal{J}'_p f)(\hat{\xi})$.

We denote by $u^{i_1,\dots,i_{p+1}}$ and $\xi_{j_1,\dots,j_{n-p}}$ the coordinates of the polyvectors $\hat{u}=u_1\wedge\dots\wedge u_{p+1}$ and $\hat{\xi}=\xi^1\wedge\dots\wedge\xi^{n-p}$ relative to the dual bases in $\mathbf{R}^{n+1}$ and $(\mathbf{R}^{n+1})'$ consistent with a prescribed $(n+1)$-vector $e_1\wedge\dots\wedge e_{n+1}$. (If $u=\|u^i_j\|$ is the matrix composed of the coordinates of the vectors $u_1,\dots,u_{p+1}$, then $u^{i_1,\dots,i_{p+1}}$ is the minor of the matrix u formed from the rows with indices $i_1,\dots,i_{p+1}$.)*

Definition. Suppose that in $\mathbf{R}^{n+1}$ the $(n+1)$-vector $e_1\wedge\dots\wedge e_{n+1}$ is fixed. Suppose further that $\hat{u}=\{u^{i_1,\dots,i_{p+1}}\}$ is a $(p+1)$-vector in $\mathbf{R}^{n+1}$ and $\hat{\xi}=\{\xi_{j_1,\dots,j_{n-p}}\}$ is a $(n-p)$-vector in $(\mathbf{R}^{n+1})'$. We call the polyvectors $\hat{u}$ and $\hat{\xi}$ dual (and write $\hat{u}\sim\hat{\xi}$) if

$$u^{i_1,\dots,i_{p+1}}=\operatorname{sgn}(i_1,\dots,i_{n+1})\,\xi_{i_{p+2},\dots,i_{n+1}}$$

for any permutation $(i_1,\dots,i_{n+1})$ of the indices $1,\dots,n+1$.

Proposition 1. If $\hat{u}\sim\hat{\xi}$, then $(\mathcal{I}_pf)(\hat{u})=(\mathcal{I}'_pf)(\hat{\xi})$.

Proof. Let $\hat{u}=u_1\wedge\dots\wedge u_{p+1}$ and $\hat{\xi}=\xi^1\wedge\dots\wedge\xi^{n-p}$. We denote by h_u the subspace in $\mathbf{R}^{n+1}$ spanned by $u_1,\dots,u_{p+1}$ and by h_ξ the subspace in $\mathbf{R}^{n+1}$ orthogonal to the vectors $\xi^1,\dots,\xi^{n-p}$. It is known that if $\hat{u}\sim\hat{\xi}$, then $h_u=h_\xi$. Further, we choose $x^{i_1},\dots,x^{i_{p+1}}$ as coordinates on the subspace $h=h_u=h_\xi$ ($i_1,\dots,i_{p+1}$ are any pairwise distinct indices, $1\leqslant i_k\leqslant n+1$), and we write out the expressions for $(\mathcal{I}_pf)(u)$ and $(\mathcal{I}'_pf)(\xi)$ in these coordinates. From the relation $x=ut$ it follows that

$$\sigma_p(t,dt)=(u^{i_1,\dots,i_{p+1}})^{-1}\sigma_p^{i_1,\dots,i_{p+1}}(x,dx),$$

where $\sigma_p^{i_1,\dots,i_{p+1}}(x,dx)=\sum_k(-1)^{k-1}x^{i_k}dx^{i_1}\wedge\dots\wedge\widehat{dx^{i_k}}\wedge\dots\wedge dx^{i_{p+1}}$. On the other hand, from the definition of the differential form $\omega_\xi(x,dx)$ it follows easily that

$$\omega_\xi(x,dx)=(u^{i_1,\dots,i_{p+1}})^{-1}\sigma_p^{i_1,\dots,i_{p+1}}(x,dx),$$

where $\hat{u}\sim\hat{\xi}$. Thus, if $\hat{u}\sim\hat{\xi}$, then

$$(\mathcal{I}'_pf)(\xi)=(\mathcal{I}_pf)(u)=\frac{1}{2}(u^{i_1,\dots,i_{p+1}})^{-1}\int_K f(x)\,\sigma_p^{i_1,\dots,i_{p+1}}(x,dx),$$

provided that $u^{i_1,\dots,i_{p+1}}\neq0$; here K is a surface in h which intersects once each ray issuing from the point $x=0$. The proof of the assertion is complete.

5. Examples and Interpretations. Since functions $f\in\Phi_p(\mathbf{RP}^n)$ are uniquely determined by their values on any section K of the bundle $\mathbf{R}^{n+1}\setminus0\to\mathbf{RP}^n$, the operator $\mathcal{I}_p$ can be interpreted as the transform of functions defined on this section. We consider several examples.

1°. K is the hyperplane defined by the equation $x^{n+1}=1$. We set

$$F(x^1,\dots,x^n)=f(x^1,\dots,x^n,1). \tag{2.10}$$

In this case $\mathcal{I}_p$ can be interpreted as the integral transform which assigns to functions F on $K\cong\mathbf{R}^n$ their integrals over p-dimensional planes in $\mathbf{R}^n$. Each p-dimensional plane in K can be prescribed by means of a $(p+1)$-frame $u=(u_1,\dots,u_{p+1})$ in $\mathbf{R}^{n+1}$. By choosing these frames in various ways, we obtain various formulas for the integral transform of functions in the affine space K. We consider two examples.

a) We choose as the vectors $u_1,\dots,u_{p+1}$ the columns of the following matrix:

*The numbers $u^{i_1,\dots,i_{p+1}}$ are called the Plücker coordinates of the subspace in $\mathbf{R}^{n+1}$ spanned by $u_1,\dots,u_{p+1}$.

$$u=\begin{pmatrix} \alpha_1^1\ldots\alpha_p^1\ \beta^1 \\ \cdot\ \cdot\ \cdot\ \cdot\ \cdot\ \cdot \\ \alpha_1^n\ldots\alpha_p^n\ \beta^n \\ 0\ldots0\ 1 \end{pmatrix}.$$

Thus, the vectors $u_1, \ldots, u_p$ give the points of the "infinitely distant" plane on K, and u_{n+1} gives an arbitrary point in K. In the coordinates (α, β) the operator $\mathcal{I}_p$ is then given by the following formula:

$$(\mathcal{I}_pF)(\alpha, \beta)=\int F(\alpha_1t^1+\ldots+\alpha_pt^p+\beta)\,dt^1\wedge\ldots\wedge dt^p. \tag{2.11}$$

The operator obtained coincides with the operator $\mathcal{I}_p$ defined for affine space in Chap. I, formula (1.26).

b) We take as the vectors $u_1, \ldots, u_{p+1}$ the columns of the following matrix:

$$u=\begin{pmatrix} u_1^1\ldots u_{p+1}^1 \\ \cdot\ \cdot\ \cdot\ \cdot\ \cdot \\ u_1^n\ldots u_{p+1}^n \\ 1\ldots1 \end{pmatrix}.$$

Thus, these vectors give $p+1$ points on $K\cong\mathbf{R}^n$. The plane in $\mathbf{R}^n$ spanned by these points is given by the equation $x=u_1t^1+\ldots+u_{p+1}t^{p+1}\equiv ut$, where $t^1+\ldots+t^{p+1}=1$ (t^i are barycentric coordinates). The operator $\mathcal{I}_p$ is given in the coordinates u_j^i by the following formula:

$$(\mathcal{I}_pF)(u)=\int\limits_{t^1+\ldots+t^{p+1}=1} F(ut)\,dt^1\wedge\ldots\wedge dt^p. \tag{2.12}$$

We shall establish the connection between (2.11) and (2.12). We set $\beta=u_{p+1}$, $\alpha_i=u_i-u_{p+1}$ $(i=1,\ldots,p)$. It is then obvious that $(\mathcal{I}_pF)(u)=(\mathcal{I}_pF)(\alpha,\beta)$.

We shall determine functions F on which the operator $\mathcal{I}_p$. is defined. For this it is necessary in (2.10) to return to homogeneous coordinates. We obtain

$$f(x^1,\ldots,x^{n+1})=|x^{n+1}|^{-p-1}F(x^1/x^{n+1},\ldots,x^n/x^{n+1}).$$

The condition on F is that the corresponding function f defined for $x^{n+1}\neq0$ extends to a C^∞ function on $\mathbf{R}^{n+1}\setminus0$. We note that functions of the Schwartz space $S(\mathbf{R}^n$, (i.e., infinitely differentiable functions on $\mathbf{R}^n$ which decrease rapidly together with all partial derivatives) clearly satisfy this condition. Below the following simple assertion will be important.

For any $\eta\in(\mathbf{R}^{n+1})'\setminus0$ we denote by $\Phi_p^\eta(\mathbf{RP}^n)$ the subspace of functions $f\in\Phi_p(\mathbf{RP}^n)$ which together with all derivatives are equal to zero on the subspace $\langle\eta, x\rangle=0$.

LEMMA. The restriction of a function $f\in\Phi_p(\mathbf{RP}^n)$ to the hyperplane $\langle\eta, x\rangle=1$ belongs to the Schwartz space $S(\mathbf{R}^n)$ if and only if $f\in\Phi_p^\eta(\mathbf{RP}^n)$.

We thus obtain an imbedding of $S(\mathbf{R}^n)$ in $\Phi_p(\mathbf{RP}^n)$ as the subspace $\Phi_p^\eta(\mathbf{RP}^n)$.

2°. Suppose that a Euclidean metric is given in $\mathbf{R}^{n+1}$. Then $\mathcal{I}_p$ can be interpreted as an integral transform in the space of even C^∞ functions on the unit sphere S^n with center at the point 0. We denote by σ the p-dimensional sphere obtained as the intersection of S^n with a $(p+1)$-dimensional subspace in $\mathbf{R}^{n+1}$. The manifold of such spheres we denote by

The operator $\mathcal{I}_p$ takes even functions on S^n into functions on Σ_p according to the formula

$$(\mathcal{I}_pf)(\sigma)=\int\limits_\sigma f(x)\,d\sigma, \tag{2.13}$$

where $\sigma\in\Sigma_p$ and $d\sigma$ is the Euclidean measure on σ.

3°. Suppose that in $\mathbf{R}^{n+1}$ there is given the Minkowski metric $(x, x)=(x^1)^2+\ldots+(x^n)^2-(x^{n+1})^2$. As K we choose the pair of conjugate hyperboloids $(x, x)=\pm 1$, and we denote by Σ_p the sections of K by $(p+1)$-dimensional subspaces. The operator $\mathcal{J}_p$ is given by formula (2.13). It acts in the space of even C^∞ functions on K which satisfy the following added condition: the function $|(x,x)|^{-(p+1)/2}F(x^1|(x,x)|^{-1/2},\ldots,x^n|(x,x)|^{-1/2})$ defined for $(x,x)\neq 0$ extends to a C^∞ function on $\mathbf{R}^{n+1}\setminus 0$. We note that all rapidly decreasing functions on K satisfy this condition, and it would be possible to consider the operator $\mathcal{J}_p$ only on the subspace of such functions.

6. An Expression for the Operator $\mathcal{J}_p$ in Local Coordinates. The elements of the spaces $\Phi_0(E_p)$ and $\Phi_0(F_p)$ can be interpreted as functions on the Grassmann manifold G_p. For this it is necessary to give local sections of the bundles $E_p\to G_p$ and $F_p\to G_p$ and restrict functions of $\Phi_0(F_p)$ and $\Phi_0(F_p)$ to the corresponding sections.

We first introduce local coordinates on G_p. We decompose $\mathbf{R}^{n+1}$ into the direct sum of subspaces:

$$\mathbf{R}^{n+1}=L\oplus M, \text{ where } \dim L=n-p,\ \dim M=p+1,$$

and we fix bases $e_1,\ldots,e_{n-p}$ and $f_1,\ldots,f_{p+1}$ in L and M, respectively. We denote by G_L the set of all $(p+1)$-dimensional subspaces in $\mathbf{R}^{n+1}$ transverse to L; it is obvious that G_L is an open, dense subset of G_p. Each subspace $h\in G_L$ is the graph of a linear mapping $\alpha: M\to L$, i.e., it consists of points $(\alpha z, z)$, $z\in M$. Therefore, it is possible to take the elements of the matrix of the linear mapping α in the selected basis as coordinates on G_p.

In each subspace $h(\alpha)\in G_L$, we prescribe a basis of vectors $u_i(\alpha)=f_i+\alpha f_i$ $(i=1,\ldots,p+1)$. Then the mapping

$$\alpha\mapsto u(\alpha)=(u_1(\alpha),\ldots,u_{p+1}(\alpha))$$

is a local section of the bundle $E_p\to G_p$.

We shall now define a local section of the bundle $F_p\to G_p$. Let $e^1,\ldots,e^{n-p}, f^1,\ldots,f^{p+1}$ be the basis in $(\mathbf{R}^{n+1})'$ dual to the basis $e_1,\ldots,e_{n-p}, f_1,\ldots,f_{p+1}$, and let L' and M' be the subspaces spanned by $e^1,\ldots,e^{n-p}$ and $f^1,\ldots,f^{p+1}$, respectively. To each $\alpha: M\to L$ we assign the $(n-p)$-frame $\xi(\alpha)=(\xi^1(\alpha),\ldots,\xi^{n-p}(\alpha))$, where $\xi^i(\alpha)=e^i-\alpha' e^i$, $\alpha': L'\to M'$ is the mapping adjoint to $\alpha: M\to L$. It is obvious that the frame $\xi(\alpha)$ is orthogonal to the frame $u(\alpha)$, and it therefore defines the subspace $h(\alpha)\in G_L$ corresponding to α; hence, the mapping

$$\alpha\mapsto\xi(\alpha)$$

is a local section of the bundle $F_p\to G_p$.

We shall write out the formula for the restrictions of the functions $(\mathcal{J}_p f)(u)$ and $(\mathcal{J}'_p f)(\xi)$ to the respective sections. It is not hard to see that the polyvectors $\hat{u}(\alpha)$ and $\hat{\xi}(\alpha)$ are dual in the sense of the definition given in part 3. Therefore, by Proposition 1 we have

$$(\mathcal{J}'_p f)(\xi(\alpha))=(\mathcal{J}_p f)(u(\alpha))=\int_{\mathbf{R}^p} f(\alpha z, z)\,\alpha_p(z, dz).$$

7. Inversion Formula and Description of the Image for the Radon Transform in Projective Space. Before taking up the transform $\mathcal{J}_p$ for arbitrary $p\leqslant n-1$, we consider in detail the

transform $\mathcal{I}_{n-1}$, which we shall call the Radon transform. Thus, the projective Radon transform is defined as an operator

$$\mathcal{I}_{n-1}:\Phi_{n-1}(\mathbf{RP}^n)\to\Phi_0((\mathbf{RP}^n)'),$$

where $\Phi_{n-1}(\mathbf{RP}^n)$ and $\Phi_0((\mathbf{RP}^n)')$ are the spaces of even homogeneous C^∞ functions, respectively, on $\mathbf{R}^{n+1}\setminus 0$ and $(\mathbf{R}^{n+1})'\setminus 0$ of degrees of homogeneity $-n$ and -1. This operator is given by the following formula:

$$(\mathcal{I}_{n-1}f)(\xi)=\frac{1}{2}\int_K f(x)\,\delta(\langle\xi,x\rangle)\,\sigma_n(x,dx), \tag{2.14}$$

where $K\subset\mathbf{R}^{n+1}$ is any surface intersecting once each ray issuing from the point $x=0$.

<u>THEOREM 1.</u> The operator $\mathcal{I}_{n-1}$ realizes an isomorphism of the spaces $\Phi_{n-1}(\mathbf{RP}^n)$ and $\Phi_0((\mathbf{RP}^n)')$. The operator inverse to $\mathcal{I}_{n-1}$ is given by the formula

$$(\mathcal{I}_{0,n-1}\varphi)(x)=\frac{(-1)^{(n-1)/2}}{2(2\pi)^{n-1}}\int_K\varphi(\xi)\,\delta^{(n-1)}(\langle\xi,x\rangle)\,\sigma_n(\xi,d\xi) \tag{2.15}$$

for odd n and by

$$(\mathcal{I}_{0,n-1}\varphi)(x)=\frac{(-1)^{n/2-1}(n-1)!}{2(2\pi)^n}\int_K\varphi(\xi)\,\langle\xi,x\rangle^{-n}\sigma_n(\xi,d\xi) \tag{2.15'}$$

for even n.*

<u>Proof.</u> We first recall that the affine Radon transform was defined on the Schwartz space $S(\mathbf{R}^n)$; this space is imbedded in $\Phi_{n-1}(\mathbf{RP}^n)$ as the subspace $\Phi^{\eta}_{n-1}(\mathbf{RP}^n)$, where η is any vector of $(\mathbf{R}^{n+1})'\setminus 0$ (see part 5). Therefore, to construct the inverse operator for $\mathcal{I}_{n-1}$ in projective space we can use the inversion formula for the affine Radon transform.

It is obvious that the operator $\mathcal{I}_{0,n-1}$ acts from $\Phi_0((\mathbf{RP}^n)')$ to $\Phi_{n-1}(\mathbf{RP}^n)$. We fix any $\eta\in(\mathbf{R}^{n+1})'\setminus 0$ and show that on the subspace $\mathcal{I}_{n-1}\Phi^{\eta}_{n-1}(\mathbf{RP}^n)$ the operator $\mathcal{I}_{0,n-1}$ agrees with the operator inverse to the affine Radon transform, i.e., it goes over into it under the isomorphism $\Phi^{\eta}_{n-1}(\mathbf{RP}^n)\simeq S(\mathbf{R}^n)$.

It may be assumed with no loss of generality that $\eta=(0,\ldots,0,1)$. We write out expressions for $\mathcal{I}_{0,n-1}$ by taking as K a cylindrical surface with generator parallel to η:

$$(\mathcal{I}_{0,n-1}\varphi)(x)=\frac{(-1)^{\frac{n-1}{2}}}{2(2\pi)^{n-1}}\int_{K'}\int_{-\infty}^{+\infty}\varphi(\xi)\,\delta^{(n-1)}(\langle\xi,x\rangle)\,d\xi_{n+1}\wedge\sigma_{n-1}(\xi',d\xi') \tag{2.16}$$

for odd n;

$$(\mathcal{I}_{0,n-1}\varphi)(x)=\frac{(-1)^{n/2-1}(n-1)!}{2(2\pi)^n}\int_{K'}\int_{-\infty}^{+\infty}\varphi(\xi)\,\langle\xi,x\rangle^{-n}d\xi_{n+1}\wedge\sigma_{n-1}(\xi',d\xi') \tag{2.16'}$$

for even n. Here $\xi'=(\xi_1,\ldots,\xi_n)$, $\sigma_{n-1}(\xi',d\xi')=\sum_{k=1}^{n}(-1)^{k-1}\xi_k d\xi_1\wedge\ldots\wedge\widehat{d\xi_k}\wedge\ldots\wedge d\xi_n$; K' is an arbitrary surface on the plane $\xi_{n+1}=0$ enclosing the point $\xi'=0$.

Under the isomorphism $\Phi^{\eta}_{n-1}(\mathbf{RP}^n)\simeq S(\mathbf{R}^n)$ to a function $f\in\Phi^{\eta}_{n-1}(\mathbf{RP}^n)$ there corresponds the function $F(x^1,\ldots,x^n)=f(x^1,\ldots,x^n,1)$. From (2.14), where we take as K the pair of hyperplanes $x^{n+1}=\pm 1$, it follows that $\varphi=\mathcal{I}_{n-1}f$ is the affine Radon transform of the function

---

*In this expression the factor $\langle\xi,x\rangle^{-n}$ must be interpreted as a generalized function (see the definition of the generalized function P^{-n} in [24]; see also Chap. I).

$F\in S(\mathbf{R}^n)$. Then the integrals (2.16) and (2.16') for $x^{n+1}=1$ coincide with the inversion formulas for the affine Radon transform (see Sec. 1), and they are thus equal to $F(x)$. Hence, passing anew to homogeneous coordinates $(x^i \to x^i/x^{n+1})$, we find that $\mathcal{I}_{0,n-1}\mathcal{I}_{n-1}f=f$ for functions $f\in\Phi_p^{\eta}(\mathbf{RP}^n)$.

Since the space $\Phi_p(\mathbf{RP}^n)$ is the linear hull of a finite number of spaces $\Phi_p^{\eta}(\mathbf{RP}^n)$ (e.g., it is possible to take the spaces corresponding to the vectors $\eta^i=x^i$, $i=1,\ldots,n+1$), we have thus proved that $\mathcal{I}_{0,n-1}\mathcal{I}_{n-1}f=f$ on all of $\Phi_p(\mathbf{RP}^n)$. This implies that the operator $\mathcal{I}_{n-1}$ is injective, $\mathcal{I}_{0,n-1}$ is surjective, and the operator $\mathcal{I}_{n-1}\mathcal{I}_{0,n-1}$ is the identity operator on the image of $\mathcal{I}_{n-1}$. To complete the proof it suffices to show that the operator $\mathcal{I}_{0,n-1}$ is injective. Thus, suppose that $\mathcal{I}_{0,n-1}\varphi=0$, where $\varphi\in\Phi_0((\mathbf{RP}^n)')$; we shall prove that $\varphi=0$.

Suppose first that n is an odd number. Then $\frac{\partial^{n-1}\varphi}{\partial\xi_{i_1}\ldots\partial\xi_{i_{n-1}}}\in\Phi_{n-1}$, and by integration by parts we obtain

$$\mathcal{I}_{n-1}\frac{\partial^{n-1}\varphi}{\partial\xi_{i_1}\ldots\partial\xi_{i_{n-1}}}=c\int_K\frac{\partial^{n-1}\varphi(\xi)}{\partial\xi_{i_1}\ldots\partial\xi_{i_{n-1}}}\delta(\langle\xi,x\rangle)\sigma_n(\xi,d\xi)=cx^{i_1}\ldots x^{i_{n-1}}\mathcal{I}_{0,n-1}\varphi=0.$$

Since $\mathcal{I}_{n-1}$ is injective, this implies that $\frac{\partial^{n-1}\varphi}{\partial\xi_{i_1}\ldots\partial\xi_{i_{n-1}}}=0$ for any $i_1,\ldots,i_{n-1}$, i.e., φ is a polynomial. On the other hand, since φ has degree of homogeneity -1, it follows that $\varphi\equiv0$.

In the case of even n we represent the operator $\mathcal{I}_{0,n-1}$ in the form

$$(\mathcal{I}_{0,n-1}\varphi)(x)|_{x^{n+1}=1}=c\int\psi(\xi)\delta^{(n-1)}(\xi_1x^1+\ldots+\xi_nx^n+\xi_{n+1})d\xi_{n+1}\wedge\sigma_{n-1}(\xi',d\xi'),\tag{2.17}$$

where

$$\psi(\xi',\xi_{n+1})=\int_{-\infty}^{+\infty}\varphi(\xi',p)(p-\xi_{n+1})^{-1}dp\tag{2.18}$$

(the integral (2.18) is understood in the sense of principal value). We observe that the transform (2.18) preserves the degree of homogeneity but it changes the parity. On the basis of (2.17), just as in the case of odd n, we see that $\psi(\xi)\equiv0$. Since the transform (2.18) is injective (this can be seen by going over to the Fourier transform), this implies that $\varphi\equiv0$.*

Remark. Theorem 1 admits the following generalization. We denote by $\Phi_{k,\varepsilon}(\mathbf{RP}^n)$, where $0\leqslant k\leqslant n-1$, $\varepsilon=0,1$, the space of C^∞ functions on $\mathbf{R}^{n+1}\setminus0$ satisfying the homogeneity condition $f(\lambda x)=\lambda^{-k-1}(\operatorname{sgn}\lambda)^{\varepsilon}f(x)$; we define the space $\Phi_{k,\varepsilon}((\mathbf{RP}^n)')$ similarly. We introduce the operators $\mathcal{I}^{\varepsilon}_{k,n-k-1}$ by the formulas

$$(\mathcal{I}^0_{k,n-k-1}f)(\xi)=\int_K\delta^{(n-k-1)}(\langle\xi,x\rangle)f(x)\sigma_n(x,dx),$$

$$(\mathcal{I}^1_{k,n-k-1}f)(\xi)=\int_K\langle\xi,x\rangle^{-n+k}f(x)\sigma_n(x,dx).$$

THEOREM 1'. The operators $\mathcal{I}^{\varepsilon}_{k,n-k-1}$ realize isomorphisms of the following spaces:

$$\mathcal{I}^{\varepsilon}_{k,n-k-1}:\Phi_{k,\varepsilon}(\mathbf{RP}^n)\to\Phi_{n-k-1,\bar\varepsilon}((\mathbf{RP}^n)')$$

*We note that the spaces $\Phi_p(\mathbf{RP}^n)$ and $\Phi_0((\mathbf{RP}^n)')$ are canonically isomorphic to the space $\Phi(S^n)$ of C^∞ functions on the sphere S^n; the operators $\mathcal{I}_{n-1}$ and $\mathcal{I}_{0,n-1}$, considered as operators from $\Phi(S^n)$ to $\Phi(S^n)$, commute with rotations of the sphere and with one another.

for even n, where ε and $\bar{\varepsilon}$ have different parity;

$$\mathcal{I}^{\varepsilon}_{k,n-k-1}:\Phi_{k,\bar{\varepsilon}}(\mathrm{RP}^n)\to\Phi_{n-k-1,\bar{\varepsilon}}((\mathrm{RP}^n)')$$

for odd n. Here the operator inverse to $\mathcal{I}^{\varepsilon}_{k,n-k-1}$ up to a factor is $\mathcal{I}^{\bar{\varepsilon}}_{n-k-1,k}$ for even n and $\mathcal{I}^{\varepsilon}_{n-k-1,k}$ for odd n.

8. Description of the Image of the Operator $\mathcal{I}_p$ for $p<n-1$. The problem of describing the image of the operator $\mathcal{I}_p$ for $p<n-1$ is closely related to the construction of the inverse operator (the inversion formula). Usually the operator inverse to $\mathcal{I}_p$ is determined as the restriction of an operator defined on a larger space than the image of $\mathcal{I}_p$; the nonuniqueness of the inversion formula is connected with this. We discuss this question in detail for even p in parts 10 and 11. We shall temporarily use an inversion formula of special form which suffices to describe the image of $\mathcal{I}_p$. This formula is derived on the basis of the inversion formula for the case $p=n-1$ (the Radon transform) obtained in part 7.

As we know, for $p<n-1$ the operator $\mathcal{I}_p$ is given in the coordinates ξ by the formula

$$(\mathcal{I}_p f)(\xi)=\frac{1}{2}\int_K f(x)\prod_{i=1}^{n-p}\delta(\langle\xi^i,x\rangle)\sigma_n(x,dx), \tag{2.19}$$

where $\xi=(\xi^1,\ldots,\xi^{n-p})\in F_p$. It defines a mapping

$$\mathcal{I}_p:\Phi_p(\mathrm{RP}^n)\to\Phi_0(F_p),$$

where $\Phi_0(F_p)$ is the space of C^∞ functions on the manifold $F_p=F_{p,n}$ of $(n-p)$-frames ξ in $(\mathrm{R}^{n+1})'$ which for any $g\in GL(n-p,\mathrm{R})$ satisfy the condition $\varphi(\xi g)=|\det g|^{-1}\varphi(\xi)$.

THEOREM 2. In order that the C^∞ function $\varphi(\xi)$ on F_p belong to the image of the operator $\mathcal{I}_p$, it is necessary and sufficient that it satisfy the following conditions:

$$1)\quad \varphi(\lambda_1\xi^1,\ldots,\lambda_{n-p}\xi^{n-p})=|\lambda_1\ldots\lambda_{n-p}|^{-1}\varphi(\xi^1,\ldots,\xi^{n-p}),$$

$$2)\quad \frac{\partial^2\varphi(\xi)}{\partial\xi_i^k\,\partial\xi_{i'}^{k'}}=\frac{\partial^2\varphi(\xi)}{\partial\xi_i^{k'}\,\partial\xi_{i'}^{k}}\ (1\leqslant i,i'\leqslant n+1,\ 1\leqslant k,k'\leqslant n-p). \tag{2.20}$$

Remark. We do not assume beforehand that $\varphi\in\Phi_0(F_p)$; this will follow from conditions 1 and 2.

Proof. If $\varphi=\mathcal{I}_p f$, then conditions 1 and 2 follow directly from formula (2.19). We shall prove that they are sufficient.

If $\varphi=\mathcal{I}_p f$, then the inverse operator is constructed as follows. Through any point $x\in\mathrm{R}^{n+1}\setminus 0$ we pass an arbitrary $(p+2)$-dimensional subspace L and we restrict φ to the set of $(p+1)$-dimensional subspaces belonging to L. Then $f(x)$ is obtained by means of the inversion formula for the Radon transform in L. We apply this inversion formula to any function φ satifying the conditions of the theorem and prove that the result does not depend on the choice of subspace L passing through x. We thus define a function f on $\mathrm{R}^{n+1}\setminus 0$. Now it is clear that $\mathcal{I}_p f=\varphi$, since this equality is valid for the restriction of φ to a set of subspaces belonging to an arbitrary $(p+2)$-dimensional subspace.

Thus, let $\varphi(\xi)$ satisfy the conditions of the theorem. We introduce the manifold A of pairs (x,ξ'), where $x\in\mathrm{R}^{n+1}\setminus 0$, and $\xi'=(\xi^2,\ldots,\xi^{n-p})$ is an $(n-p-1)$-frame in $(\mathrm{R}^{n+1})'$ orthogonal to x (i.e., $\langle\xi^i,x\rangle=0$, $i=2,\ldots n-p$).

Let $(x_0,\xi')\in A$. We denote by $\hat{L}_{\xi'}\subset(\mathrm{R}^{n+1})'$ the $(n-p-1)$-dimensional subspace spanned by ξ', and by $L_{\xi'}\subset\mathrm{R}^{n+1}$ the subspace of dimension $p+2$ orthogonal to $\hat{L}_{\xi'}$, i.e., the subspace defined by the equations $\langle \xi^i, x \rangle = 0$, ($i=2,\dots,n-p$). By definition $x_0 \in L_{\xi'}$. For any $\xi^1 \in (\mathbf{R}^{n+1})' \setminus \hat{L}_{\xi'}$ we set

$$\varphi_{\xi'}(\xi^1) = \varphi(\xi^1, \xi').$$

LEMMA 1. If $\xi' = (\xi^2, \dots, \xi^{n-p})$, then

$$\varphi_{\xi'}(\xi^1 + \lambda \xi^i) = \varphi_{\xi'}(\xi^1) \tag{2.21}$$

for any $i=2,\dots,n-p$ and $\lambda \in \mathbf{R}$.

Proof. The equality (2.21) is equivalent to the relation

$$\sum_{s=1}^{n+1} \frac{\partial \varphi}{\partial \xi_s^1} \xi_s^i = 0 \quad (i=2,\dots,n-p),$$

which we shall prove. Applying conditions 1 and 2, we find that for any r:

$$\frac{\partial}{\partial \xi_r^i}\left(\sum_s \frac{\partial \varphi}{\partial \xi_s^1} \xi_s^i\right) = \frac{\partial \varphi}{\partial \xi_r^1} + \sum_s \frac{\partial^2 \varphi}{\partial \xi_r^i \partial \xi_s^1} \xi_s^i = \frac{\partial \varphi}{\partial \xi_r^1} + \sum_s \frac{\partial^2 \varphi}{\partial \xi_r^1 \partial \xi_s^i} \xi_s^i = \frac{\partial \varphi}{\partial \xi_r^1} - \frac{\partial \varphi}{\partial \xi_r^1} = 0.$$

This proves that the function $\sum_s \frac{\partial \varphi}{\partial \xi_s^1} \xi_s^i$ does not depend on ξ^i. On the other hand, by condition 1 it is an odd function of ξ^i. Hence $\sum_s \frac{\partial \varphi}{\partial \xi_s^1} \xi_s^i = 0$.

COROLLARY. $\varphi_{\xi'}$ is a function on $(L_{\xi'})' \setminus 0$, where $(L_{\xi'})' = (\mathbf{R}^{n+1})' / \hat{L}_{\xi'}$ is the factor space dual to $L_{\xi'}$.

It is obvious that the function $\varphi_{\xi'}$ is infinitely differentiable on $(L_{\xi'})' \setminus 0$, and by condition 1 it satisfies the condition $\varphi_{\xi'}(\lambda \xi^1) = |\lambda|^{-1} \varphi_{\xi'}(\xi^1)$. Hence, by Theorem 1 $\varphi_{\xi'}$ is the Radon transform of a C^∞ function $f_{\xi'}(x)$ on $L_{\xi'} \setminus 0$ which satisfies the condition $f_{\xi'}(\lambda x) = |\lambda|^{-p-1} f(x)$. Thus, on the manifold $\mathring{A}$ the following function is defined:

$$F(x, \xi') = f_{\xi'}(x).$$

LEMMA 2. For any fixed $x \in \mathbf{R}^{n+1} \setminus 0$ the function $F(x, \xi')$ does not depend on the frame ξ' orthogonal to x.

Proof. It suffices to prove the assertion for any fixed point x, e.g., for $x = x_0 = (1, 0, \dots, 0)$. In this case the condition of the orthogonality of the frame ξ' to x is $\xi_1^2 = \dots = \xi_1^{n-p} = 0$.

Using the inversion formula for the Radon transform, we write out an explicit expression for the function $F(x_0, \xi') = f_{\xi'}(x_0)$ in terms of $\varphi_{\xi'}$. For this in $(\mathbf{R}^{n+1})'$ we fix a $(p+2)$-dimensional coordinate subspace $\{\xi_1^1, \xi_{i_1}^1, \dots, \xi_{i_{p+1}}^1\}$ transversal to $\hat{L}_{\xi'}$; it may then be assumed that $\varphi_{\xi'}$ is a function on this subspace. To be specific we fix the subspace $\{\xi_1^1, \xi_2^1, \dots, \xi_{p+2}^1\}$. Since $\langle \xi^1, x_0 \rangle = \xi_1^1$, by the inversion formula for the Radon transform we have

$$F(x_0, \xi') = c \int_K \varphi_{\xi'}(\xi') \delta^{(p)}(\xi_1^1) \sigma_{p+1}(\xi^1, d\xi^1)$$

for even p;

$$F(x_0, \xi') = c \int_K \varphi_{\xi'}(\xi^1) (\xi_1^1)^{-p-1} \sigma_{p+1}(\xi^1, d\xi^1)$$

for odd p. To prove the lemma it suffices to show that

$$\frac{\partial F}{\partial \xi_i^k} = 0 \text{ for } 2 < i < p+2, \quad 2 < k < n-p. \tag{2.22}$$

We fix i and choose for K in the inversion formulas the planes $\xi_j^1=\pm 1$, where $j\neq 1, i$. Because of the parity of $\varphi_{\xi'}$, the integral over K is equal to a double integral over the plane $\xi_j^1=0$. As a result, in the case of even p we obtain

$$F(x_0,\xi')=c\int_{\xi_j^1=1}\varphi_{\xi'}(\xi^1)\delta^{(p)}(\xi_1^1)\,d\xi_1^1\wedge\ldots\wedge\widehat{d\xi_j^1}\wedge\ldots\wedge d\xi_{p+2}^1=c\int\frac{\partial^p\varphi_{\xi'}(\xi^1)}{(\partial\xi_1^1)^p}\Bigg|_{\substack{\xi_1^1=0\\ \xi_j^1=1}}d\xi_2^1\wedge\ldots\wedge\widehat{d\xi_j^1}\wedge\ldots\wedge d\xi_{p+2}^1.$$

By differentiating with respect to ξ_i^k and using the relation $\frac{\partial^2\varphi}{\partial\xi_i^k\partial\xi_1^1}=\frac{\partial^2\varphi}{\partial\xi_i^1\partial\xi_1^k}$, from this we obtain

$$\frac{\partial F}{\partial\xi_i^k}=c\int\frac{\partial}{\partial\xi_i^1}\left(\frac{\partial^p\varphi(\xi)}{(\partial\xi_1^1)^{p-1}\partial\xi_1^k}\right)_{\substack{\xi_1^1=0\\ \xi_j^1=1}}d\xi_2^1\wedge\ldots\wedge\widehat{d\xi_j^1}\wedge\ldots\wedge d\xi_{p+2}^1=0$$

(as the integral of an exact form).

In the case of odd p the proof of Eq. (2.22) is analogous. The proof of the lemma is complete.

On the basis of this lemma we define the function $f(x)$ on $\mathbf{R}^{n+1}\setminus 0$ by the formula

$$f(x)=f_{\xi'}(x),$$

where $\xi'=(\xi^2,\ldots,\xi^{n-p})$ is an arbitrary frame in $(\mathbf{R}^{n+1})'$ orthogonal to x. It is infinitely differentiable on $L_{\xi'}\setminus 0$, where $L_{\xi'}$ is any $(p+2)$-dimensional subspace of $\mathbf{R}^{n+1}$; hence, it is infinitely differentiable on $\mathbf{R}^{n+1}\setminus 0$. Further, f satisfies the condition $f(\lambda x)=|\lambda|^{-p-1}f(x)$; hence, $f\in\Phi_p(\mathbf{RP}^n)$. As noted at the beginning of the proof, we have $\mathcal{J}_p f=\varphi$. The proof of the theorem is complete.

Remark. We see that Eqs. (2.20) are partially responsible for the fact that $\varphi\in\Phi_0(F_p)$, i.e., it drops onto the one-dimensional bundle over G_p, and they are partially responsible for the fact that $\varphi\in\Phi_0(F_p)$ belongs to the image of $\mathcal{J}_p$. In order to distinguish equations which precisely describe the image of $\mathcal{J}_p$ in $\Phi_0(F_p)$, it is convenient to select local coordinates α on G_p and give the section $\alpha\mapsto\xi(\alpha)$ (see part 6). Setting $\varphi(\xi(\alpha))=\psi(\alpha)$, we then obtain the following system of equations:

$$\frac{\partial^2\psi(\alpha)}{\partial\alpha_j^i\partial\alpha_{j'}^{i'}}=\frac{\partial^2\psi(\alpha)}{\partial\alpha_{j'}^{i}\partial\alpha_{j}^{i'}}.$$

9. Description of the Image of the Operator $\mathcal{J}_p$ in the Coordinates u. We recall that according to the first definition of the operator $\mathcal{J}_p$, it acts from $\Phi_p(\mathbf{RP}^n)$ to $\Phi_0(E_p)$, where $\Phi_0(E_p)$ is the space of C^∞ functions on the manifold E_p of $(p+1)$-frames $u=(u_1,\ldots,u_{p+1})$ in $\mathbf{R}^{n+1}$ which for any $g\in GL(p+1,\mathbf{R})$ satisfy the condition (2.6): $\varphi(ug)=|\det g|^{-1}\varphi(u)$.

THEOREM 3. In order that a C^∞ function $\varphi(u)$ on E_p belong to the image of the operator $\mathcal{J}_p$, it is necessary and sufficient that it satisfy the following conditions:

$$1)\quad \varphi(\lambda^1u_1,\ldots,\lambda^{p+1}u_{p+1})=|\lambda^1\ldots\lambda^{p+1}|^{-1}\varphi(u_1,\ldots,u_{p+1}),$$

$$2)\quad \frac{\partial^2\varphi(u)}{\partial u_k^i\partial u_{k'}^{i'}}=\frac{\partial^2\varphi(u)}{\partial u_k^{i'}\partial u_{k'}^{i}}\quad(1\leqslant i,\ i'\leqslant n+1,\ 1\leqslant k,\ k'\leqslant p+1). \qquad (2.23)$$

Proof. If $\varphi=\mathcal{J}_p f$, then conditions 1 and 2 follow directly from formula (2.5) for $\mathcal{J}_p f$. We shall prove that they are sufficient. Thus, suppose that φ satisfies the conditions of the theorem.

We shall first prove that φ satisfies the relation (2.6). In view of condition 1, it suffices to show that φ is preserved under any transformation $u_k \to u_k + \lambda u_l$ $(k \neq l)$ (since these transformations generate the group $SL(p+1, \mathbf{R})$). The latter condition is equivalent to the relation

$$\sum_{s=1}^{n+1} \frac{\partial \varphi}{\partial u_k^s} u_l^s = 0 \quad (k \neq l)$$

which we shall prove. Applying conditions 1 and 2, for any r we obtain

$$\frac{\partial}{\partial u_l^r}\left(\sum_s \frac{\partial \varphi}{\partial u_k^s} u_l^s\right) = \frac{\partial \varphi}{\partial u_k^r} + \sum_s \frac{\partial^2 \varphi}{\partial u_k^r \partial u_l^s} u_l^s = \frac{\partial \varphi}{\partial u_k^r} - \frac{\partial \varphi}{\partial u_k^r} = 0.$$

The function $\sum_s \frac{\partial \varphi}{\partial u_k^s} u_l^s$ thus does not depend on u_l. Since, on the other hand, by condition 1, it is an odd function of u_l, this function is equal to zero.

We now define the function $\tilde{\varphi}(\xi)$ on F_p from the condition $\tilde{\varphi}(\xi) = \varphi(u)$, if the polyvectors $\hat{\xi} = \xi^1 \wedge \ldots \wedge \xi^{n-p}$ and $\hat{u} = u_1 \wedge \ldots \wedge u_{p+1}$ are dual (see part 4). Because of relation (2.6), this definition is correct, and $\tilde{\varphi}(\xi) \in \Phi_0(F_p)$. In particular, if $\alpha \mapsto \xi(\alpha)$ and $u \mapsto u(\alpha)$ are local sections, respectively, of the bundles $F_p \to G_p$ and $E_p \to G_p$, defined in part 6, then $\tilde{\varphi}(\xi(\alpha)) = \varphi(u(\alpha)) = \psi(\alpha)$.

It is not hard to see that for the functions $\varphi \in \Phi_0(E_p)$ and $\tilde{\varphi} \in \Phi_0(F_p)$ each system of equations $\frac{\partial^2 \varphi(u)}{\partial u_k^i \partial u_{k'}^{i'}} = \frac{\partial^2 \varphi(u)}{\partial u_k^{i'} \partial u_{k'}^i}$ and $\frac{\partial^2 \tilde{\varphi}(\xi)}{\partial \xi_i^k \partial \xi_{i'}^{k'}} = \frac{\partial^2 \tilde{\varphi}(\xi)}{\partial \xi_i^{k'} \partial \xi_{i'}^k}$ is equivalent to the system of equations on the bundle $\frac{\partial^2 \psi(\alpha)}{\partial \alpha_j^i \partial \alpha_{j'}^{i'}} = \frac{\partial^2 \psi(\alpha)}{\partial \alpha_j^{i'} \partial \alpha_{j'}^i}$. Therefore, it follows from the hypotheses of the theorem that the function $\tilde{\varphi}$ satisfies the conditions of Theorem 2 and hence lies in the image of $\mathcal{I}_p'$. Therefore, the function $\varphi(u)$ belongs to the image of $\mathcal{I}_p$.

10. Inversion Formula for Even p in the Coordinates u. We recall that the function $\varphi = \mathcal{I}_p f$ has been defined in two ways: as a function on the manifold E_p of $(p+1)$-frames in $\mathbf{R}^{n+1}$ and as a function on the manifold F_p of $(n-p)$-frames in the dual space $(\mathbf{R}^{n+1})'$. We shall consider both definitions of $\varphi = \mathcal{I}_p f$ and for each of them in the case of even p we shall obtain an inversion formula for the transformation $\mathcal{I}_p$. In this part we consider the case where $\hat{f} = \mathcal{I}_p f$ is defined as a function on the manifold of frames $u = (u_1, \ldots, u_{p+1})$ in $\mathbf{R}^{n+1}$ by formula (2.5).

We set $u = (v, x)$, where $v = (v_1, \ldots, v_p)$, and we introduce the following differential p-form:

$$\varkappa_x \hat{f} = \sum_{i_1, \ldots, i_p = 1}^{n+1} \frac{\partial^p \hat{f}(v_1, \ldots, v_p, x)}{\partial x^{i_1} \ldots \partial x^{i_p}} dv_1^{i_1} \wedge \ldots \wedge dv_p^{i_p}. \tag{2.24}$$

For fixed x $\varkappa_x \hat{f}$ is a differential form on the manifold E^x of all p-frames v such that the subspace spanned by v does not contain x. We shall study the properties of this form.

We denote by $\tilde{G}^x$ the manifold of oriented p-dimensional subspaces not passing through x, and we denote by $\pi: E^x \to \tilde{G}^x$ the canonical projection (which assigns to each $v \in E^x$ the oriented subspace spanned by v; the orientation is determined by the frame v). It is obvious that $\pi v = \pi v'$ if and only if $v' = vg$, $g \in GL_+(p, \mathbf{R})$. Thus, $\pi: E^x \to \tilde{G}^x$ is a principal bundle with structural group $GL_+(p, \mathbf{R})$.

Proposition 2. The form $\varkappa_x \hat{f}$, where $\hat{f} = \mathcal{I}_p \hat{f}$, drops from E^x to $\tilde{G}^x$.

For the proof it suffices to show that 1) the form $\varkappa_x\hat{f}$ is invariant under $GL_+(p, \mathbf{R})$ and 2) the form $\varkappa_x\hat{f}$ is preserved under the substitution $dv_i \to dv_i + \sum_k v_k dg_i^k$. Both facts follow directly from the relation $\hat{f}(ug) = |\det g|^{-1}\hat{f}(u)$ and the explicit expression for $\varkappa_x\hat{f}$.

Proposition 3. The form $\varkappa_x\hat{f}$ is closed on $\tilde{G}^x$.

The assertion follows from the relations

$$\frac{\partial^2\hat{f}}{\partial x^i \partial v_k^j} = \frac{\partial^2\hat{f}}{\partial x^j \partial v_k^i} \quad (1 \leqslant i,\ j \leqslant n+1,\ 1 \leqslant k \leqslant p).$$

Since it is closed, the form $\varkappa_x\hat{f}$ defines a linear functional on the space $H_p(\tilde{G}^x)$ of p-dimensional homologies. We shall describe this space and compute the corresponding functional.

We denote by G^x the Grassmann manifold of oriented $(p+1)$-dimensional subspaces $h \in G_p$, passing through x; $G^x \simeq G_{p-1,n-1}$. We consider the canonical projection $\tilde{G}^x \to G^x$ (which assigns to a subspace $h \in \tilde{G}^x$ the oriented space spanned by h and x). In the bundle $\tilde{G}^x \to G^x$ the fibers are contractible (they are isomorphic to $\mathbf{R}^{n-p}$), and therefore the homology of $\tilde{G}^x$ is obtained by lifting the homology of G^x; in particular, $H_p(\tilde{G}^x) = H_p(G^x)$. Cycles in G^x consisting of p-dimensional planes contained in a fixed $(p+1)$-dimensional plane and also any lift of such a cycle to $\tilde{G}^x$ we call Euler cycles.

THEOREM 4. If $\hat{f} = \mathcal{I}_p f$, where p is an even number, and $\varkappa_x\hat{f}$ is the differential p-form on $\tilde{G}^x$ defined by formula (2.24), then there is the following inversion formula:

$$\int_\gamma \varkappa_x\hat{f} = c\,d(\gamma)\,f(x), \quad (c \neq 0), \tag{2.25}$$

where $\gamma \subset \tilde{G}^x$ is an arbitrary p-dimensional cycle, and $d(\gamma)$ is the index of intersection of with the Euler cycle.

Proof. We determine a basis of the p-dimensional homologies in $\tilde{G}^x$. To this end we fix in $\mathbf{RP}^n$ an $(n-1)$-dimensional plane $\mathbf{RP}^{n-1}$ not passing through x, and we consider the submanifold $G_{p-1,n-1} \subset \tilde{G}^x$ of all $(p-1)$-dimensional planes in $\mathbf{RP}^{n-1}$. Since there is a natural isomorphism between $G_{p-1,n-1}$ and G^x and $H_p(\tilde{G}^x) = H_p(G^x)$, a basis of the p-dimensional homologies in $G_{p-1,n-1}$ will also be a basis of the p-dimensional homologies in $\tilde{G}^x$. As such a basis we choose the p-dimensional Schubert cells which include the Euler cycle γ_0. The proof of the theorem reduces to verifying the following assertions:

$$\int_\gamma \varkappa_x\hat{f} = 0, \ \text{if} \quad \gamma \neq \gamma_0; \tag{2.26}$$

$$\int_{\gamma_0} \varkappa_x\hat{f} = c f(x), \ c \neq 0. \tag{2.27}$$

We shall first prove relation (2.26). It is known that for any p-dimensional Schubert cell $\gamma \subset G_{p-1,n-1}$ distinct from the Euler cycle γ_0, all planes $h \in \gamma$ pass through a fixed point x_0. Therefore, in the frames $v = (v_1, \ldots, v_p)$ defining the planes $h \in \gamma$, the vector v_1 may be assumed fixed. This implies that $\varkappa_x\hat{f}|_\gamma = 0$, and hence $\int_\gamma \varkappa_x\hat{f} = 0$.

We shall now prove relation (2.27). To be specific, suppose that $x = x_0 = (1, 0, \ldots, 0)$. We consider the $(p+2)$-dimensional coordinate subspace $L \subset \mathbf{R}^{n+1}$ containing x_0: $x^{p+3} = \ldots = x^{n+1} = 0$. We denote by f_0 the restriction of the function f to the subspace L and by $\hat{f}_0$ the restriction of $\hat{f}$ to the submanifold of frames belonging to L. It is obvious that $\hat{f}_0$ is the Radon transform of the function f_0. We take the Euler cycle γ_0 in $\tilde{G}^x$ consisting of subspaces belonging to L. On the basis of the inversion formula for the Radon transform in L, we then obtain

$$\int_{\gamma_0} \varkappa_x \hat{f} = \int_{\gamma_0} \varkappa_x \hat{f}_0 = c f_0(x) = c f(x),\ c \neq 0,$$

as was required to prove.

11. Inversion Formula for Even p in the Coordinates ξ. In the coordinates ξ we wish to construct a differential form analogous to $\varkappa_x f$. The form $\varkappa_x f$ drops to a flag manifold: a p-dimensional plane h passing through x, and a $(p-1)$-dimensional plane belonging to h but not passing through x. Considering this, in the case of the coordinates ξ, it is natural to construct the differential form on the manifold F^x of frames (ξ, η), where $\xi = (\xi^1, \ldots, \xi^{n-p})$ defines a p-dimensional plane passing through x (i.e., $\langle \xi^i, x \rangle = 0$, $i = 1, \ldots, n-p$), while the intersection of this plane with the plane $\langle \eta, y \rangle = 0$ defines a $(p-1)$-dimensional plane not passing through x (i.e., $\langle \eta, x \rangle \neq 0$); to be specific, we assume that $\langle \eta, x \rangle = 1$.

To describe this form we introduce some notation. Suppose there are given the collections $f^i = (f_1^i, \ldots, f_{n+1}^i)$, $i = 1, \ldots, n-p+1$ and $\omega^i = (\omega_1^i, \ldots, \omega_{n+1}^i)$, $i = 1, \ldots, p$, where f_j^i are functions, and ω_j^i are differential 1-forms with function or operator coefficients. We define

$$[f^1, \ldots, f^{n-p+1}, \omega^1, \ldots, \omega^p]\varphi = \left(\sum \operatorname{sgn}(j_1, \ldots, j_{n+1}) f_{j_1}^1 \ldots f_{j_{n-p+1}}^{n-p+1} \omega_{j_{n-p+2}}^1 \wedge \ldots \wedge \omega_{j_{n+1}}^p\right)\varphi.$$

It is here assumed that the coefficients of the form ω_j^i act as operators only on the function φ; the summation goes over all permutations of the indices $1, \ldots, n+1$. This expression can also be written in determinant form:

$$[f^1, \ldots, f^{n-p+1}, \omega^1, \ldots, \omega^p] = \begin{vmatrix} f_1^1 \ldots f_1^{n-p+1} & \omega_1^1 \ldots \omega_1^p \\ \vdots & \vdots \\ f_{n+1}^1 \ldots f_{n+1}^{n-p+1} & \omega_{n+1}^1 \ldots \omega_{n+1}^p \end{vmatrix},$$

where in the expansion of the determinant the multiplication of the differential forms is exterior multiplication. It follows from the definition that this determinant is antisymmetric with respect to any two rows and the columns $f^1, \ldots, f^{n-p+1}$, and it is symmetric with respect to the columns $\omega^1, \ldots, \omega^p$.

We introduce the differential 1-forms with operator coefficients

$$\omega_k = \sum_{j=1}^{n-p} \langle \eta, \frac{\partial}{\partial \xi^j} \rangle d\xi_k^j = \sum_{j=1}^{n-p} \sum_{s=1}^{n+1} \eta_s \frac{\partial}{\partial \xi_s^j} d\xi_k^j$$

and we set $\omega = (\omega_1, \ldots, \omega_{n+1})$.

Definition. Using the notation introduced, we define the differential form $\tilde{\varkappa}_x \varphi$ on the manifold F^x of frames (ξ, η) by the formula

$$\tilde{\varkappa}_x \varphi = \sum_{k+l=p} \frac{p!}{k!(l!)^2} [\xi^1, \ldots, \xi^{n-p}, \zeta, \underbrace{d\eta, \ldots, d\eta}_{k}, \underbrace{\omega, \ldots, \omega}_{l}]\varphi, \tag{2.28}$$

where $\varphi \in \Phi_0(F_p)$; ζ is any vector satisfying the condition $\langle \zeta, x \rangle = 1$; the form does not depend on the choice of this vector. We shall present several other expressions for the form $\tilde{\varkappa}_x \varphi$. We have

$$\tilde{\varkappa}_x \varphi = \sum_{k+l=p} \frac{p!}{k!(l!)^2} \tau_{k,l}\varphi,$$

where $\tau_{k,l}=[\xi^1,\ldots,\xi^{n-p},\zeta,\underbrace{d\eta,\ldots,d\eta}_{k},\underbrace{\omega,\ldots,\omega}_{l}]$.

We further define the forms $\rho_{k,l}^{j_1,\ldots,j_l}$ as the interior products

$$\rho_{k,l}^{j_1,\ldots,j_l}(x;\xi,d\eta,d\xi)=\langle x,d\zeta\rangle \lrcorner [\xi^1,\ldots,\xi^{n-p},d\zeta,\underbrace{d\eta,\ldots,d\eta}_{k},d\xi^{j_1},\ldots,d\xi^{j_l}].$$

We then have

$$\tilde{\varkappa}_x\varphi=\sum_{k+l=p}\sum_{j_1,\ldots,j_l=1}^{n-p}\frac{p!}{k!\,(l!)^2}\left(\prod_{s=1}^{l}\langle\eta,\frac{\partial}{\partial\xi^{j_s}}\rangle\right)\varphi\rho_{k,l}^{j_1,\ldots,j_l}(x;\xi,d\eta,d\xi). \qquad (2.28')$$

We shall study the properties of the differential form $\tilde{\varkappa}_x\varphi$.

We denote by $\tilde{G}^x$ the manifold of oriented p-dimensional subspaces of $\mathbf{R}^{n+1}$ not passing through x, and we define the projection $\pi:F^x\to\tilde{G}^x$ by assigning to each $(n-p+1)$-frame $(\xi,\eta)\in F^x$ the oriented subspace in $\mathbf{R}^{n+1}$ orthogonal to it (the orientation is determined by the orientation of the frame). It is obvious that $\pi(\xi,\eta)=\pi(\xi',\eta')$ if and only if $\xi'=\xi g$, $g\in GL_+(n-p,\mathbf{R})$ and $\eta'=\eta+\lambda_1\xi^1+\ldots+\lambda_{n-p}\xi^{n-p}$. Thus, $\pi:F^x\to\tilde{G}^x$ is a principal bundle with structural group which consists of matrices of the form $\begin{pmatrix} g & * \\ 0 & 1\end{pmatrix}$, where $g\in GL_+(n-p,\mathbf{R})$.

Proposition 4. The differential form $\tilde{\varkappa}_x\varphi$ drops from F^x to $\tilde{G}^x$.

Proof. It suffices to show that the form $\tilde{\varkappa}_x\varphi$ is 1) invariant under the structural group and 2) is preserved by the substitutions $d\xi\to d\xi+\xi dg$ and $d\eta\to d\eta+\sum\xi^i d\lambda_i$. It is obvious, first of all, that under the substitutions $d\xi\to d\xi+\xi dg$, $d\eta\to d\eta+\sum\xi^i d\lambda_i$ each of the operators τ_{kl} and hence also $\tilde{\varkappa}_x\varphi$ are preserved. We consider further the transformation $\xi\to\xi g$, $g\in GL_+(n-p,\mathbf{R})$. Under this transformation each of the forms ω_k is preserved, and therefore the operator τ_{kl} is multiplied by g. Since the function itself is multiplied by $(\det g)^{-1}$ under this transformation, each of the forms $\tau_{kl}\varphi$ and hence also the form $\tilde{\varkappa}_x\varphi$ are preserved. It remains to verify that the form $\tilde{\varkappa}_x\varphi$ is invariant under the transformations $\eta\to\eta+\lambda_i\xi^i$. This fact is established by a somewhat more lengthy computation (see [10]) which we shall omit here.

Proposition 5. If $\varphi=\mathcal{J}_p f$, then the form $\tilde{\varkappa}_x\varphi$ is closed on $\tilde{G}^x$.

For the proof see [10]; it requires simple but somewhat involved computations which we shall omit here.

THEOREM 5. If $\varphi=\mathcal{J}_p f$, where p is an even number and $\tilde{\varkappa}_x\varphi$ is the differential p-form on $\tilde{G}^x$ defined by formula (2.28), then there is the following inversion formula:

$$\int_\gamma\tilde{\varkappa}_x\varphi=cd(\gamma)f(x),\quad (c\neq 0), \qquad (2.29)$$

where $\gamma\subset\tilde{G}^x$ is any p-dimensional cycle, and $d(\gamma)$ is the index of intersection of γ with the Euler cycle.

Proof. As in the proof of Theorem 4, we fix a plane $\langle\eta,y\rangle=0$ in $\mathbf{RP}^n$ not passing through x ($\langle\eta,x\rangle\neq 0$). In this plane we consider the submanifold $G_{p-1,n-1}\subset\tilde{G}^x$ of $(p-1)$-dimensional planes. The p-dimensional Schubert cells $\gamma\subset G_{p-1,n-1}$ then form a basis of the p-dimensional homology on $\tilde{G}^x$. Since $\tilde{\varkappa}_x\varphi$ is closed, the proof of the theorem reduces to verifying the two relations

$$\int_{\gamma} \tilde{\varkappa}_x \varphi = 0$$

if $\gamma \subset G_{p-1,n-1}$ is a Schubert cell distinct from the Euler cycle;

$$\int_{\gamma_0} \tilde{\varkappa}_x \varphi = c f(x) \quad (c \neq 0) \tag{2.30}$$

if γ_0 is the Euler cycle.

Suppose first that the Schubert cell γ is distinct from the Euler cycle. It is known that then all the subspaces $h \in \gamma$ contain a fixed, one-dimensional subspace . It may therefore be assumed that all $(n-p-1)$-frames (ξ, η) in $(\mathbf{R}^{n+1})'$, defining the subspaces are contained in the n-dimensional subspace orthogonal to L^1. In this case the rows of all determinants contained in the expression for $\tilde{\varkappa}_x \varphi$ are linearly dependent. Hence $\tilde{\varkappa}_x \varphi|_{\gamma} = 0$, and therefore $\int_{\gamma} \tilde{\varkappa}_x \varphi = 0$.

We shall now prove equality (2.30). To be specific, suppose that $x = x_0 = (0, \ldots, 0, x_0^{p+2}, 0, \ldots, 0)$, and let $L_{p+2} \subset \mathbf{R}^{n+1}$ be the subspace of points $x = (x^1, \ldots, x^{p+2}, 0, \ldots, 0)$. In $\mathbf{R}^{n+1)'}$ we fix the vector $\eta = e^{p+2}$ and an $(n-p-1)$-frame orthogonal to e^{p+2}: $\xi = (\xi^1, \ldots, \xi^{n-p-1})$, where $\xi^i = e^{p+2+i}$, $i = 1, \ldots, n-p-1$ ($e^1, \ldots, e^{n+1}$ are basis vectors). The frame $(\xi^1, \ldots, \xi^{n-p-1}, \eta)$ defines a (p + 1)-dimensional subspace $L_{p+1} \subset L_{p+2}$ not containing x_0. For γ_0 we take the manifold of all p-dimensional subspaces in L_{p+1}. Each $h \in \gamma_0$ is defined by an $(n-p+1)$-frame $(\xi^1, \ldots, \xi^{n-p-1}, \xi', \eta)$, where $\xi^1, \ldots, \xi^{n-p-1}$, η are the fixed vectors defined above and $\xi' = (\xi_1', \ldots, \xi_{p+2}', 0, \ldots, 0)$.

We denote by f_0 the restriction of the function f to the subspace L_{p+2}, and we set $\psi(\xi') = \varphi(\xi^1, \ldots, \xi^{n-p-1}, \xi')$. It is obvious that $\psi(\xi')$ is the Radon transform of the function f_0. From the definition of $\tilde{\varkappa}_x \varphi$ it follows that

$$\tilde{\varkappa}_{x_0} \varphi|_{\gamma_0} = \frac{1}{p!} |x_0^{p+2}|^{-p-1} \frac{\partial^p \psi(\xi')}{(\partial \xi_{p+2}')^p} \bigg|_{\xi_{p+2}'=0} \sum_{k=1}^{p+1} (-1)^{k-1} \xi_k' d\xi_1' \wedge \ldots \wedge \widehat{d\xi_k'} \wedge \ldots \wedge d\xi_{p+1}'.$$

We therefore have

$$\int_{\gamma_0} \tilde{\varkappa}_{x_0} \varphi = c \int \psi(\xi') \delta^{(p)}(\langle \xi', x_0 \rangle) \sigma_{p+1}(\xi', d\xi').$$

By the inversion formula for the Radon transform (see part 7) the integral on the right is equal to $c_1 f_0(x_0) = c_1 f(x_0)$ $(c_1 \neq 0)$. The proof of the theorem is complete.

12. Remarks on the Forms $\varkappa_x \varphi$ and $\tilde{\varkappa}_x \varphi$. 1°. Since the forms $\varkappa_x \varphi$ and $\tilde{\varkappa}_x \varphi$ constructed above are closed on the manifold $\tilde{G}^x$ of p-dimensional, oriented subspaces in $\mathbf{R}^{n+1}$ not passing through x, and the difference $\varkappa_x \varphi - c\tilde{\varkappa}_x \varphi$ for suitable choice of the coefficient c is orthogonal to all p-dimensional Schubert cells, this difference is an exact differential form on $\tilde{G}^x$.

2°. We denote by G^x the manifold of $(p+1)$-dimensional, oriented subspaces in $\mathbf{R}^{n+1}$ containing x, and we let $\pi: \tilde{G}^x \to G^x$ denote the canonical projection. The restriction $s^* \tilde{\varkappa}_x \varphi$ of the form $\tilde{\varkappa}_x \varphi$ to an arbitrary section s of the bundle $\tilde{G}^x \to G^x$ may be considered a differential form on G^x. Thus, to the form $\tilde{\varkappa}_x \varphi$ on $\tilde{G}^x$ there corresponds a family of differential p-forms $s^* \tilde{\varkappa}_x \varphi$ on G^x. If $\varphi = \mathcal{J}_p f$, then the difference of any two such forms is an exact form on G^x, while the integral of each of them over a p-dimensional cycle in G^x is equal to $f(x)$, up to a factor not depending on f.

Of course, if a differential p-form on G^x gives an inversion formula, then a differential form of the form $d(A\varphi)$ can be added to it; here A is an arbitrary operator from the space of differential $(p-1)$-forms on G^x. We emphasize that the operator taking φ into the difference of the restrictions of $\tilde{\varkappa}_x\varphi$ to various sections is not an operator of this form: its image is an exact form only for functions φ satisfying the system of equations (2.20), i.e., functions belonging to the image of the operator $\mathcal{I}_p$.

3°. We consider the restriction of the form $\tilde{\varkappa}_x\varphi$ to a section corresponding to a fixed plane $\eta=\text{const}$. This restriction coincides with the principal term of the expression for $\tilde{\varkappa}_x\varphi$ (see formula (2.28')):

$$\tilde{\varkappa}_x\varphi\,|_{\eta=\text{const}}=\frac{1}{p!}\sum_{j_1,\ldots,j_p=1}^{n-p}\left(\prod_{s=1}^{p}\langle\,\eta,\frac{\partial}{\partial\xi^{j_s}}\,\rangle\right)\varphi\rho_{0,p}^{j_1,\ldots,j_p}(x,\xi,d\xi). \tag{2.31}$$

The coefficients of this form depend on η as a parameter, and the forms corresponding to distinct η differ by an exact form.

For the case of the Radon transform in part 6 inversion formulas were constructed in which differential forms not containing η are integrated. The reason for this is that there differential forms were constructed on the entire manifold of n-dimensional subspaces in $\mathbf{R}^{n+1}$, while $\varkappa_x\varphi|_{\eta=\text{const}}$ is a form on the submanifold G^x; this form depends in an essential way on η. It can naturally be interpreted as the residual form of a differential form on $(\mathbf{R}^{n+1})'$ on the submanifold $\langle\xi,x\rangle=0$. A residual form is not determined uniquely, and this occasions the dependence on the parameter η.

4°. In Chap. I for the case of even p inversion formulas were given for the affine transform $\mathcal{I}_p$. We shall show how they are obtained from the projective inversion formulas. Suppose an affinization of $\mathbf{RP}^n$ is fixed, i.e., a subspace $\langle\,\eta,x\,\rangle=0$ in $\mathbf{R}^{n+1}$ is fixed. In $\mathbf{R}^{n+1}$ we introduce coordinates so that $\langle\,\eta,x\,\rangle=x^{n+1}$. Then $x^1/x^{n+1},\ldots,x^n/x^{n+1}$ are the corresponding inhomogeneous coordinates. As noted in part 5, we obtain the affine transform $\mathcal{I}_p$ if we take for the vectors v_i the vectors $v_i=(\alpha_i,0)$, where $\alpha_i=(\alpha_i^1,\ldots,\alpha_i^n)$. Suppose further that $x\in\mathbf{R}^{n+1}\setminus 0$, where $x^{n+1}=1$. Restricting the form $\varkappa_x\hat{f}$ (see formula (2.24)) to frames of this type, we then obtain

$$\varkappa_x\varphi=\sum_{i_1,\ldots,i_p=1}^{n}\frac{\partial^p\varphi(\alpha,x)}{\partial x^{i_1}\ldots\partial x^{i_p}}\,d\alpha_1^{i_1}\wedge\ldots\wedge d\alpha_p^{i_p}. \tag{2.32}$$

This form coincides with the affine version of $\varkappa_x\varphi$ introduced in Chap. I. Similarly, setting, in formula (2.28) for $\tilde{\varkappa}_x\varphi$, $\xi_{n+1}^i=1$ $(i=1,\ldots,n-p)$ and restricting it to the section $\eta=(0,\ldots,0,1)$, we obtain the affine formula (1.37) for $\tilde{\varkappa}_x\varphi$ of Chap. I.

In order to clarify the extent to which the projective inversion formula is broader than the affine formula, we go over from frames (v,x), where $x^{n+1}=1$, to the manifold of triples (x,α,c), where $c=(c_1,\ldots,c_p)$, $c_j=v_j^{n+1}$, $\alpha_j=v_j-c_jx$, $j=1,\ldots,p$. We obtain a vector bundle over the space of frames (α,x) of the affine problem. Passing in (2.24) to the coordinates α, c, x, we obtain

$$\varkappa_x\varphi=\sum_{i_1,\ldots,i_p=1}^{n+1} L_{i_1}\ldots L_{i_p}(t\varphi(\alpha,x))|_{t=1}\,\omega_1^{i_1}\wedge\ldots\wedge\omega_p^{i_p}, \tag{2.33}$$

where $L_i=t\left(\frac{\partial}{\partial x^i}-\sum_{j=1}^{p}c_j\frac{\partial}{\partial\alpha_j^i}\right)(1\leq i\leq n)$, $L_{n+1}=-t^2\frac{\partial}{\partial t}-\sum_{k=1}^{n}x^kL_k$, $\omega_j^i=d\alpha_j^i+x^jdc_j$ $(1\leq i\leq n)$, $\omega_j^{n+1}=dc_j$. We observe that the affine form (2.32) is obtained from (2.33) for $c\equiv0$. By choosing distinct $c=c(\alpha)$, we obtain p-forms which differ from (2.32) by an exact form.

2. Paley–Wiener Theorem for the Integral Transform in Affine Space

In Chap. I a description was obtained of the image of the space $S(\mathbf{R}^n)$ of rapidly decreasing functions on $\mathbf{R}^n$ under the transform $\mathcal{I}_p$. Here this result will be obtained on the basis of the projective problem considered in Sec. 1. To this end we fix an arbitrary vector $\eta\in(\mathbf{R}^{n+1})'\setminus0$ and imbed $S(\mathbf{R}^n)$ in $\Phi_p(\mathbf{RP}^n)$ as the subspace $\Phi_p^\eta(\mathbf{RP}^n)$ of functions $f\in\Phi_p(\mathbf{RP}^n)$ which vanish together with all partial derivatives on the plane $\langle\eta,x\rangle=0$ (see Sec. 1, part 5). We remark that a function $f\in\Phi_p(\mathbf{RP}^n)$ belongs to the subspace $\Phi_p^\eta(\mathbf{RP}^n)$ if and only if $f(x)\langle\eta,x\rangle^{-k}$ for any $k=1,2,\ldots$ is a C^∞ function on $\mathbf{R}^{n+1}\setminus0$.

Thus, the problem of describing the image of the space $S(\mathbf{R}^n)$ under the affine transform $\mathcal{I}_p$ reduces to a description of the image of the subspace $\Phi_p^\eta(\mathbf{RP}^n)\subset\Phi_p(\mathbf{RP}^n)$ under the projective transform $\mathcal{I}_p$. The latter problem we solve in parts 2 and 3 using the fact that the description of the image of the entire space $\Phi_p(\mathbf{RP}^n)$ has already been obtained in Sec. 1. We recall that the cases $p<n-1$ and $p=n-1$ are essentially different.

The next part is of auxiliary character.

1. Elements of Operational Calculus. In the theory of integral transforms operators whose images under these transforms have a simple form play an important role (e.g., in the case of the Fourier transform these are linear differential operators with constant coefficients and operators of multiplication by polynomials). We now consider certain analogous operators connected with the integral transforms $\mathcal{I}_p$ and $\mathcal{I}_p'$.

Proposition 6. Let $y\in\mathbf{R}^{n+1}$ and $\zeta\in(\mathbf{R}^{n+1})'$, with $\langle\zeta,y\rangle=0$. Then the differential operators

$$(D_{y,\zeta}f)(x)=\langle\zeta,x\rangle\langle y,\frac{\partial}{\partial x}\rangle f(x)\quad(x\in\mathbf{R}^{n+1}\setminus0),$$

$$(\hat{D}_{y,\zeta}\varphi)(u)=\sum_{j=1}^{p+1}\langle\zeta,u_j\rangle\langle y,\frac{\partial}{\partial u_j}\rangle\varphi(u)\quad(u=(u_1,\ldots,u_{p+1})\in E_p),$$

$$(\tilde{D}_{y,\zeta}\varphi)(\xi)=\sum_{j=1}^{n-p}\langle y,\xi^j\rangle\langle\zeta,\frac{\partial}{\partial\xi^j}\rangle\varphi(\xi)\quad(\xi=(\xi^1,\ldots,\xi^{n-p})\in F_p),$$

where $\langle y,\frac{\partial}{\partial u_j}\rangle=\sum_{i=1}^{n+1}y^i\frac{\partial}{\partial u_j^i}$, $\langle\zeta,\frac{\partial}{\partial\xi^j}\rangle=\sum_{i=1}^{n+1}\zeta_i\frac{\partial}{\partial\xi_i^j}$, are defined as bounded operators on the spaces $\Phi_p(\mathbf{RP}^n)$, $\Phi_0(E_p)$ and $\Phi_0(F_p)$, respectively (for the definition of these spaces see Sec. 1, parts 2 and 3).

Proof. For $D_{y,\zeta}$ the assertion is obvious. It is also obvious that the operators $\hat{D}_{y,\zeta}$ and $\tilde{D}_{y,\zeta}$ are defined on the spaces of all smooth functions on E_p and F_p, respectively. It therefore remains only to check that they preserve the respective subspaces $\Phi_0(E_p)$ and $\Phi_0(F_p)$. The fact that the subspaces $\Phi_0(E_p)$ and $\Phi_0(F_p)$ are preserved follows from the fact that the operators $\hat{D}_{y,\zeta}$ and $\tilde{D}_{y,\zeta}$ commute with the action of the respective groups $GL(p+1,\mathbf{R})$ and

$GL(n-p, \mathbf{R})$ on E_p and F_p; the last fact is easily checked directly.

Proposition 7. For any $f\in\Phi_p(\mathbf{RP}^n)$ we have

$$\mathcal{I}_p D_{y,\zeta} f = \hat{D}_{y,\zeta}\mathcal{I}_p f, \tag{2.34}$$

$$\mathcal{I}'_p D_{y,\zeta} f = -\tilde{D}_{y,\zeta}\mathcal{I}'_p f, \tag{2.34'}$$

where $\mathcal{I}_p:\Phi_p(\mathbf{RP}^n)\to\Phi_0(E_p)$ and $\mathcal{I}'_p:\Phi_p(\mathbf{RP}^n)\to\Phi_0(F_p)$ are the operators of integration over p-dimensional planes introduced in Sec. 1.

Proof. According to Sec. 1, for any $f\in\Phi_p(\mathbf{RP}^n)$ we have

$$(\mathcal{I}_p f)(u) = \int_{\mathbf{RP}^p} f(ut)\sigma_p(t, dt),$$

where $ut = u_1 t^1 + \ldots + u_{p+1} t^{p+1}$. Hence,

$$(\hat{D}_{y,\zeta}\mathcal{I}_p f)(u) = \sum_{j=1}^{p+1} \langle \zeta, u_j \rangle \int_{\mathbf{RP}^p} \left(\langle y, \frac{\partial}{\partial x} \rangle f \right)(ut)\, t^j \sigma_p(t, dt) =$$

$$= \int_{\mathbf{RP}^p} \left(\langle \zeta, x \rangle \langle y, \frac{\partial}{\partial x} \rangle f(x) \right)_{x=ut} \sigma_p(t, dt) = (\mathcal{I}_p D_{y,\zeta} f)(u).$$

Similarly, the equality (2.34') follows from the explicit formula for $\mathcal{I}'_p$ (see Sec. 1, part 3).

COROLLARY. The subspaces $\mathcal{I}_p\Phi_p(\mathbf{RP}^n)\subset\Phi_0(E_p)$ and $\mathcal{I}'_p\Phi_0(\mathbf{RP}^n)\subset\Phi_0(F_p)$ are invariant under the operators $\hat{D}_{y,\zeta}$ and $\tilde{D}_{y,\zeta}$, respectively.

Remark. Because of the condition $\langle \zeta, y \rangle = 0$, the operator $D_{y,\zeta}$ can also be written in the form $(D_{y,\zeta} f)(x) = \langle y, \frac{\partial}{\partial x} \rangle \langle \zeta, x \rangle f(x)$; a similar statement holds for the operators $\hat{D}_{y,\zeta}$ and $\tilde{D}_{y,\zeta}$. It is possible to not impose the condition $\langle \zeta, y \rangle = 0$ and to consider two versions of the definition of the operators $D_{y,\zeta}$, $\hat{D}_{y,\zeta}$, $\tilde{D}_{y,\zeta}$. The connection between the operators becomes somewhat more complicated.

We fix the form $\sigma_n(x, dx)$ in the space $\mathbf{R}^{n+1}$ and define an isomorphism of the spaces $\tau:\Phi_0(E_p)\to\Phi_0(F_p)$ by setting $(\tau\varphi)(\hat{\xi}) = \varphi(\hat{u})$, where $\hat{u} = u_1\wedge\ldots\wedge u_{p+1}$ is the polyvector dual to $\hat{\xi} = \xi^1\wedge\ldots\wedge\xi^{n-p}$ (see the definition in Sec. 1, part 4).

Proposition 8. Under the isomorphism $\tau:\Phi_0(E_p)\to\Phi_0(F_p)$ the operator $\hat{D}_{y,\zeta}$ goes over into the operator $\tilde{D}_{y,\zeta}$, i.e., for any $\varphi\in\Phi_0(E_p)$ we have

$$\tau\hat{D}_{y,\zeta}\varphi = -\tilde{D}_{y,\zeta}\tau\varphi.$$

Proof. We choose systems of dual coordinates in $\mathbf{R}^{n+1}$ and $(\mathbf{R}^{n+1})'$, in which $y = (1, 0, \ldots, 0)$, $\zeta = (0, \ldots, 0, 1)$. In these coordinates the expressions for $\hat{D}_{y,\zeta}$ and $\tilde{D}_{y,\zeta}$ take the form

$$\hat{D}_{y,\zeta}\varphi = \sum_{k=1}^{p+1} u_k^{n+1} \frac{\partial\varphi}{\partial u_k^1}, \quad \tilde{D}_{y,\zeta}\tilde{\varphi} = \sum_{k=1}^{n-p} \xi_1^k \frac{\partial\tilde{\varphi}}{\partial \xi_{n+1}^k}. \tag{2.35}$$

Let $\alpha\mapsto u(\alpha)$ and $\alpha\mapsto\xi(\alpha)$ be sections of the bundles $E_p\to G_p$ and $F_p\to G_p$, introduced in Sec. 1, part 6. By what was proved there, for any $\varphi\in\Phi_0(E_p)$ and $\tilde{\varphi}\in\Phi_0(F_p)$ the condition $\tilde{\varphi} = \tau\varphi$ is equivalent to the condition $\tilde{\varphi}(\xi(\alpha)) = \varphi(u(\alpha))$ for any α. Therefore, it suffices for us to prove that if $\tilde{\varphi}(\xi(\alpha)) = \varphi(u(\alpha)) = \psi(\alpha)$, then

$$(\tilde{D}_{y,\zeta}\tilde{\varphi})(\xi(\alpha)) = -(\hat{D}_{y,\zeta}\varphi)(u(\alpha)). \tag{2.36}$$

Equality (2.36) follows directly from formulas (2.35) for the operators $\hat{D}_{y,\zeta}$, $\tilde{D}_{y,\zeta}$ and the definition of the sections $u(\alpha)$ and $\xi(\alpha)$. Namely, we have $(\hat{D}_{y,\zeta}\varphi)(u(\alpha)) = -(\tilde{D}_{y,\zeta}\tilde{\varphi})(\xi(\alpha)) = \frac{\partial\psi(\alpha)}{\partial\alpha_{p+1}^1}$.

2. Paley–Wiener Theorem for the Case $p<n-1$. We fix an arbitrary vector $\eta\in(\mathbf{R}^{n+1})'\setminus 0$ and denote by G^η the submanifold of ("infinitely distant") $(p+1)$-dimensional subspaces in $\mathbf{R}^{n+1}$ belonging to the subspace $\langle \eta, x\rangle=0$. We further fix an arbitrary vector $y\in\mathbf{R}^{n+1}$ such that $\langle \eta, y\rangle\neq 0$.

THEOREM 6 (Paley–Wiener). Let $\varphi(\xi)$ be a C^∞ function on F_p $(p<n-1)$. In order that the function φ belong to the image of the space $\Phi_p^\eta(\mathbf{RP}^n)$ under the mapping $\mathcal{I}_p$, it is necessary and sufficient that it satisfy the following conditions:

1) $\varphi(\lambda_1\xi^1, \ldots, \lambda_{n-p}\xi^{n-p})=|\lambda_1 \ldots \lambda_{n-p}|^{-1}\varphi(\xi^1, \ldots, \xi^{n-p})$,

2) $\dfrac{\partial^2\varphi(\xi)}{\partial\xi_i^k\partial\xi_{i'}^{k'}}=\dfrac{\partial^2\varphi(\xi)}{\partial\xi_i^{k'}\partial\xi_{i'}^k}$ $(1\leqslant i, i'\leqslant n+1,\ 1\leqslant k, k'\leqslant n-p)$,

3) $(\tilde{D}_{y,\zeta})^s\varphi|_{G^\eta}=0$ for any $s=0, 1, \ldots$ and any $\zeta\in(\mathbf{R}^{n+1})'$ orthogonal to y.

Remark. Conditions 1 and 2 ensure the existence of a function $f\in\Phi_p(\mathbf{RP}^n)$ such that $\varphi=\mathcal{I}_pf$; condition 3 ensures that the function f belong to the subspace $\Phi_p^\eta(\mathbf{RP}^n)$.

We obtain this theorem as a corollary of a stronger assertion. Namely, a description is given of functions on F_p which can be represented in the form $\varphi=\mathcal{I}_pf$, where f belongs to $\Phi_\eta^p(\mathbf{RP}^n)$ and has on the subspace $\langle \eta, x\rangle=0$ a zero of order $\geqslant k$. The Paley–Wiener theorem is obtained when the corresponding conditions are satisfied for all k.

THEOREM 6'. Let $\varphi(\xi)$ be a C^∞ function on F_p. In order that the function φ be representable in the form $\varphi=\mathcal{I}_pf$, $f\in\Phi_p(\mathbf{RP}^n)$, where all partial derivatives of the function f through order $k-1$ are equal to zero on the subspace $\langle \eta, x\rangle=0$, it is necessary and sufficient that the function φ satisfy conditions 1 and 2 of Theorem 6 and also the following condition:

3') $(\tilde{D}_{y,\zeta})^s\varphi|_{G^\eta}=0$ for any $s=0, 1, \ldots, k-1$ and any vector $\zeta\in(\mathbf{R}^{n+1})'$ orthogonal to y.

Proof. By Theorem 2, conditions 1 and 2 are necessary and sufficient for the function φ to be representable in the form $\varphi=\mathcal{I}_pf$, where $f\in\Phi_p(\mathbf{RP}^n)$. Further, since $\langle \eta, y\rangle\neq 0$, it follows that for any function $f\in\Phi_p(\mathbf{RP}^n)$ the vanishing on the subspaces $\langle \eta, x\rangle=0$ of all its partial derivatives through order $k-1$ is equivalent to the condition

$$\langle y, \frac{\partial}{\partial x}\rangle^s f(x)\Big|_{\langle \eta, x\rangle=0}=0,\quad s=0, \ldots, k-1. \tag{2.37}$$

Condition (2.37), in turn, is equivalent to the condition

$$(D_{y,\zeta})^s f(x)|_{\langle \eta, x\rangle=0}\equiv\langle \zeta, x\rangle^s\langle y, \frac{\partial}{\partial x}\rangle^s f(x)\Big|_{\langle \eta, x\rangle=0}=0 \tag{2.38}$$

for any $s=0, \ldots, k-1$ and any vector $\zeta\in(\mathbf{R}^{n+1})'$ orthogonal to y. Indeed, for any x satisfying the condition $\langle \eta, x\rangle=0$, because of the inequality $\langle \eta, y\rangle\neq 0$, there is a vector ζ such that $\langle \zeta, y\rangle=0$ and $\langle \zeta, x\rangle\neq 0$; therefore, (2.37) follows from (2.38). It thus suffices to prove the equivalence of the conditions $(D_{y,\zeta})^s f|_{\langle \eta, x\rangle=0}$ and $(\tilde{D}_{y,\zeta})^s\varphi|_{G^\eta}=0$. For this we use the equality

$$(\tilde{D}_{y,\zeta})^s\varphi=(-1)^s\mathcal{I}_p(D_{y,\zeta})^s f. \tag{2.39}$$

From (2.39) it is obvious that if $(D_{y,\zeta})^s f|_{\langle \eta, x\rangle=0}=0$, then $(\tilde{D}_{y,\zeta})^s\varphi|_{G^\eta}=0$. Conversely, suppose that $(\tilde{D}_{y,\zeta})^s\varphi|_{G^\eta}=(-1)^s\mathcal{I}_p(D_{y,\zeta})^s f|_{G^\eta}=0$. The values of the function $(D_{y,\zeta})^s f$ at points of the subspace $\langle \eta, x\rangle=0$ $(x\neq 0)$ can be recovered on the basis of the inversion formulas for the transform $\mathcal{I}_p$. Here, since $p<n-1$, it is possible to choose a version of the inversion

formula such that it contains only planes of G^{η} (see Sec. 1, part 7). Hence, since $\mathcal{J}_p(D_{y,\zeta})^s f|_{G^{\eta}}=0$, it follows that $(D_{y,\zeta})^s f|_{\langle\eta,x\rangle=0}=0$. The proof of the theorem is complete.

COROLLARY. Let $\eta^1,\ldots,\eta^s$ be fixed, pairwise noncollinear vectors in $(\mathbf{R}^{n+1})'$, and let $k_1,\ldots,k_s$ be natural numbers. The function $f(x)\prod_{i=1}\langle\eta^i,x\rangle^{-k_i}$, where $f\in\Phi_p(\mathbf{RP}^n)$, $p<n-1$, is infinitely differentiable on $\mathbf{R}^{n+1}\setminus 0$ if and only if the function $\varphi=\mathcal{J}_p f$ satisfies for each pair $\eta=\eta^i$, $k=k_i$ $(i=1,\ldots,s)$ condition 3' of the theorem.

3. Paley–Wiener Theorem for the Case $p=n-1$. We recall that for $p=n-1$ the image of a function $f\in\Phi_{n-1}(\mathbf{RP}^n)$ belongs to the space $\Phi_0((\mathbf{RP}^n)')$ and is given by the formula

$$(\mathcal{J}_{n-1}f)(\xi)=\frac{1}{2}\int_K f(x)\delta(\langle\xi,x\rangle)\sigma_n(x,dx), \tag{2.40}$$

where $\xi\in(\mathbf{R}^{n+1})'\setminus 0$, and $K\subset\mathbf{R}^{n+1}$ is any surface intersecting once each ray issuing from the point x = 0.

THEOREM 7 (Paley–Wiener). In order that a function $\varphi(\xi)\in\Phi_0((\mathbf{RP}^n)')$ belong to the image of the subspace $\Phi^{\eta}_{n-1}(\mathbf{RP}^n)$ under the mapping $\mathcal{J}_{n-1}$, it is necessary and sufficient that it satisfy the following conditions:

1) $\langle\zeta,\frac{\partial}{\partial\xi}\rangle^s\varphi\big|_{\xi=\eta}=0$ for any $s=0,1,\ldots$ and any $\zeta\in(\mathbf{R}^{n+1})'$;

2) the integral $\psi_s(\zeta;\eta)=\int_{-\infty}^{+\infty}\varphi(\zeta+t\eta)t^s dt$ converges for any $s=0,1,\ldots$ and any $\zeta\in(\mathbf{R}^{n+1})'$ and is a homogeneous polynomial of degree s in ζ.

As in part 2, we shall find conditions under which a function $\varphi\in\Phi_0((\mathbf{RP}^n)')$ can be represented in the form $\varphi=\mathcal{J}_{n-1}f$, where f belongs to $\Phi_{n-1}(\mathbf{RP}^n)$ and has on the plane $\langle\eta,x\rangle=0$ a zero of order $\geqslant k$; the Paley–Wiener theorem is obtained when these conditions are satisfied for all k.

THEOREM 7'. In order that a function $\varphi(\xi)\in\Phi_0((\mathbf{RP}^n)')$ be representable in the form $\varphi=\mathcal{J}_{n-1}f$, $f\in\Phi_{n-1}(\mathbf{RP}^n)$, where all partial derivatives of the function f through order $k-1$ vanish on the subspace $\langle\eta,x\rangle=0$, it is necessary and sufficient that it satisfy conditions 1 and 2 of Theorem 7 for any $\zeta\in(\mathbf{R}^{n+1})'$ and $s=0,1,\ldots,k-1$.

Proof. We note that it follows from condition 1 and the definition of ψ_s that $\psi_s(\eta;\eta)=0$ and $\psi_s(\zeta-\lambda\eta;\eta)=\sum_{i=0}^{s}\binom{s}{i}\lambda^{s-i}\psi_i(\zeta;\eta)$. It may thus be assumed in condition 2 that ζ is a vector of a fixed subspace $H_\eta\subset(\mathbf{R}^{n+1})'$ complementary to the vector η. We choose a coordinate system in which $\eta=(0,\ldots,0,1)$, and H_η is the coordinate subspace $\xi_{n+1}=0$. It is then obvious that the condition on the function f is equivalent to the following: the function $f(x)(x^{n+1})^{-k}$ is a C^∞ function on $\mathbf{R}^{n+1}\setminus 0$ or, equivalently,

$$\frac{\partial^s f(x',x^{n+1})}{(\partial x^{n+1})^s}\Big|_{x^{n+1}=0}=0\quad(s=0,\ldots,k-1) \tag{2.41}$$

for any $x'=(x^1,\ldots,x^n)\neq 0$. The conditions on the function φ assume the form

1) $\left(\zeta_1\frac{\partial}{\partial\xi_1}+\ldots+\zeta_n\frac{\partial}{\partial\xi_n}\right)^s\varphi|_{\xi=(0,\ldots,0,1)}=0$ for any $s=0,\ldots,k-1$ and any $\zeta'=(\zeta_1,\ldots,\zeta_n)$;

2) the integral

$$\psi_s(\xi')=\int_{-\infty}^{+\infty}\varphi(\xi',\xi_{n+1})(\xi_{n+1})^s d\xi_{n+1}\quad(s=0,\ldots,k-1) \tag{2.42}$$

converges and is a homogeneous polynomial of degree s in $\xi' = (\xi_1, \ldots, \xi_n)$.

We first show that condition (2.41) on f implies conditions 1 and 2 on $\varphi = \mathcal{I}_{n-1}f$. The validity of condition 1 is verified in the same way as for $p < n-1$. Further, it is not hard to see that condition 1 implies the convergence of the integrals $\psi_s(\xi')$. In order to prove that these integrals are polynomials in ξ', it suffices to substitute in (2.42) the expression (2.40) for φ, where for K we take the pair of hyperplanes $x^{n+1} = \pm 1$. Integrating the expression obtained on ξ_{n+1}, we find that

$$\psi_s(\xi') = \int f(x^1, \ldots, x^n, 1)(\xi_1 x^1 + \ldots + \xi_n x^n)^s dx^1 \wedge \ldots \wedge dx^n.$$

This implies that $\psi_s(\xi')$ is a homogeneous polynomial of degree s.

We shall now prove that if φ satisfies conditions 1 and 2, then $\varphi = \mathcal{I}_{n-1}f$, where f belongs to $\Phi_{n-1}(\mathbf{RP}^n)$ and satisfies condition (2.41).

Since $\varphi \in \Phi_0((\mathbf{RP}^n)')$, by Theorem 1 we have $\varphi = \mathcal{I}_{n-1}f$, where $f \in \Phi_{n-1}(\mathbf{RP}^n)$. It remains to check that f satisfies condition (2.41). For this we use the inversion formulas for the Radon transform (see Sec. 1):

$$f(x) = c \int_{K'} \int_{-\infty}^{+\infty} \varphi(\xi', \xi_{n+1}) \delta^{(n-1)}(\langle \xi, x \rangle) d\xi_{n+1} \wedge \sigma_{n-1}(\xi', d\xi') \tag{2.43}$$

for odd n;

$$f(x) = c \int_{K'} \int_{-\infty}^{+\infty} \varphi(\xi', \xi_{n+1}) \langle \xi, x \rangle^{-n} d\xi_{n+1} \wedge \sigma_{n-1}(\xi', d\xi') \tag{2.43'}$$

for even n. For odd n it follows from (2.43) that

$$\frac{\partial^s f(x', x^{n+1})}{(\partial x^{n+1})^s}\bigg|_{x^{n+1}=0} = -c \int_{K'} \int_{-\infty}^{+\infty} \varphi(\xi', \xi_{n+1})(\xi_{n+1})^s \delta^{(n+s-1)}(\langle \xi', x' \rangle) d\xi_{n+1} \wedge \sigma_{n-1}(\xi', d\xi') = c \int_{K'} \psi_s(\xi') \delta^{(n+s-1)}(\langle \xi', x' \rangle) \sigma_{n-1}(\xi', d\xi'), \tag{2.44}$$

where $\psi_s(\xi')$ is a polynomial of degree s $(s = 0, \ldots, k-1)$.

Similarly, for even n it follows from (2.43') that

$$\frac{\partial^s f(x', x^{n+1})}{(\partial x^{n+1})^s}\bigg|_{x^{n+1}=0} = c \int_{K'} \psi_s(\xi') \langle \xi', x' \rangle^{-n-s} \sigma_{n-1}(\xi', d\xi'). \tag{2.44'}$$

We now observe that the following identities hold:

$$\int_{K'} \delta^{(n-1)}(\xi_1 x^1 + \ldots + \xi_n x^n) \sigma_{n-1}(\xi', d\xi') = 0 \tag{2.45}$$

for odd n;

$$\int_{K'} (\xi_1 x^1 + \ldots + \xi_n x^n)^{-n} \sigma_{n-1}(\xi', d\xi') = 0 \tag{2.45'}$$

for even n; K' is any surface in $(\mathbf{R}^n)'$ which intersects once each ray issuing from the point $\xi' = 0$. Indeed, it is obvious that the integrals (2.45) and (2.45') do not depend on the choice of the surface K', and therefore for K' it is possible to take the pair of planes $\xi_n = \pm 1$; with this choice of K' the formulas (2.45) and (2.45') become trivial.

Differentiating the identities (2.45) and (2.45') s times with respect to $x^1, \dots, x^n$, we see directly that the integrals (2.44) and (2.44') are equal to zero for any homogeneous polynomial $\psi_s(\xi')$ of degree s.

COROLLARY. Suppose that $\eta^1, \dots, \eta^s$ are any pairwise noncollinear vectors of $(\mathbf{R}^{n+1})'$ and $k_1, \dots, k_s$ are natural numbers. The function $f(x)\prod_{i=1}^{s} \langle \eta^i, x \rangle^{-k_i}$, where $f \in \Phi_{n-1}(\mathbf{RP}^n)$, is infinitely differentiable on $\mathbf{R}^{n+1} \setminus 0$ if and only if the function $\varphi = I_{n-1}f$ satisfies for each pair $\eta = \eta^i$, $k = k_i$ the hypotheses of Theorem 7'.

3. Integral Geometry for Arbitrary One-Dimensional Bundles over Projective Space

In this section we consider a generalization of the integral transform $\mathcal{I}_p$ of Sec. 1 to other one-dimensional bundles over $\mathbf{RP}^n$. The basic object here is the space $\Phi_{p,s}(\mathbf{RP}^n)$ of all C^∞ functions on $\mathbf{R}^{n+1} \setminus 0$ satisfying the homogeneity condition

$$f(\lambda x) = |\lambda|^{-p-1}\lambda^{-s} f(x), \tag{2.46}$$

$s = 1, 2, \dots$. This space can be interpreted as the space of smooth sections of a certain one-dimensional vector bundle over $\mathbf{RP}^n$. On the space $\Phi_{p,s}(\mathbf{RP}^n)$ we define an operator $\mathcal{I}_{p,s}$ of integration over p-dimensional planes. The case $s=0$ was treated in Sec. 1; regarding the case $s = -1, -2, \dots$, see the remark in part 2 of this section.

1. Definition of the Operator $\mathcal{I}_{p,s}$. Here we retain the basic notation of Sec. 1. As in Sec. 1, we introduce the operation of integrating over p-dimensional planes in two ways; the first way is based on defining p-dimensional planes by means of a $(p+1)$-frame $u = (u_1, \dots, u_{p+1})$ in $\mathbf{R}^{n+1}$, while the second uses an $(n-p)$-frame $\xi = (\xi^1, \dots, \xi^{n-p})$ in the dual space $(\mathbf{R}^{n+1})'$. In this part we present the first approach.

Let $f \in \Phi_{p,s}(\mathbf{RP}^n)$, let h_u be a $(p+1)$-dimensional oriented subspace in $\mathbf{R}^{n+1}$ spanned by the frame $u = (u_1, \dots, u_{p+1}) \in E_p$, and let $t = (t^1, \dots, t^{p+1})$ be the coordinates on h_u relative to the basis $u_1, \dots, u_{p+1}$. We introduce the following differential forms on h_u:

$$f(ut)\, t^\alpha \sigma_p(t, dt) \equiv f(u_1 t^1 + \dots + u_{p+1} t^{p+1})\, t^{\alpha_1} \dots t^{\alpha_s} \sigma_p(t, dt),$$

where $\alpha = (\alpha_1, \dots, \alpha_s)$ is any collection of s indices, $1 \leqslant \alpha_k \leqslant p+1$. As in the case $s=0$ (see Sec. 1), it is easy to see that this form drops to the manifold of rays in h_u issuing from the point 0.

Definition. Let $\mathcal{I}_{p,s}$ be the operator assigning to each function $f \in \Phi_{p,s}(\mathbf{RP}^n)$ the collection of functions $\hat{f}^\alpha(u), \alpha = (\alpha_1, \dots, \alpha_s)$, $1 \leqslant \alpha_k \leqslant p+1$ on E_p, where

$$\hat{f}^\alpha(u) = \frac{1}{2}\int_K f(ut)\, t^\alpha \sigma_p(t, dt), \tag{2.47}$$

and K is an arbitrary surface on h_u intersecting once each ray issuing from the point 0.

As in Sec. 1 for the case $s=0$, the integral (2.47) can be interpreted as follows:

$$\hat{f}^\alpha(u) = \int_{\mathbf{RP}^p} f(ut)\, t^\alpha \sigma_p(t, dt).$$

Remark. If in place of the homogeneity condition (2.46) it is assumed that $f(\lambda x) = \operatorname{sgn} \lambda |\lambda|^{-p-1}\lambda^{-s} f(x)$, then all the integrals (2.47) are equal to zero.

We note the basic properties of the functions $\hat{f}^\alpha$ (they follow directly from the definition of the operator $\mathcal{I}_{p,s}$):

(i) The functions $\hat{f}^{\alpha}$ are symmetric in $\alpha=(\alpha_1, \ldots, \alpha_s)$ $(1\leqslant\alpha_l\leqslant p+1)$.

(ii) For any $g\in GL(p+1, \mathbf{R})$ we have

$$\hat{f}^{\alpha}(ug)=|\det g|^{-1}\sum_{\beta}(g^{-1})^{\alpha}_{\beta}\hat{f}^{\beta}(u), \tag{2.48}$$

where $(g^{-1})^{\alpha}_{\beta}=(g^{-1})^{\alpha_1}_{\beta_1}\ldots(g^{-1})^{\alpha_s}_{\beta_s}$, and $(g^{-1})^{\alpha_l}_{\beta_l}$ are the elements of the matrix g^{-1}.

Thus, the operator $\mathcal{I}_{p,s}$ defines a mapping

$$\mathcal{I}_{p,s}:\Phi_{p,s}(\mathbf{RP}^n)\to\Phi_{0,s}(E_p),$$

where $\Phi_{0,s}(E_p)$ is the space of all collections $\{\hat{f}^{\alpha}\}$, $\alpha=(\alpha_1, \ldots, \alpha_s)$, $1\leqslant\alpha_k\leqslant p+1$, of C^∞ functions on E_p which are symmetric in α and satisfy condition (2.48). We note that the elements of this space can be interpreted as symmetric tensor fields of rank s on G_p relative to the group $SL(p+1, \mathbf{R})$ which satisfy the homogeneity condition $\hat{f}^{\alpha}(\lambda u)=|\lambda|^{-p-1}\lambda^{-s}\hat{f}^{\alpha}(u)$.

(iii) For any $\alpha'=(\alpha_1, \ldots, \alpha_{s-1})$, $k, l=1, \ldots, p+1$ and for any $y\in\mathbf{R}^{n+1}$ the following relation holds:

$$\langle y, \frac{\partial}{\partial u_k}\rangle\hat{f}^{\alpha', l}=\langle y, \frac{\partial}{\partial u_l}\rangle\hat{f}^{\alpha', k}.$$

<u>Remark.</u> Instead of the functions $\hat{f}^{\alpha}$ defined by formula (2.47) we could introduce the integrals

$$\hat{f}(u; \chi)=\int_{\mathbf{RP}^p}f(ut)\chi(t)\sigma_p(t, dt),$$

where $\chi(t)$ is any homogeneous polynomial of degree s. By expanding $\chi(t)$ into a sum of monomials, we find that $\hat{f}(u; \chi)$ is a linear combination of the functions $\hat{f}^{\alpha}$.

<u>2. Another Method of Defining the Operator $\mathcal{I}_{p,s}$.</u> We now construct an integral transform $\mathcal{I}'_{p,s}$ assigning to functions $f\in\Phi_{p,s}(\mathbf{RP}^n)$ functions of $(n-p)$-frames $\xi=(\xi^1, \ldots, \xi^{n-p})$ in the space $(\mathbf{R}^{n+1})'$ dual to $\mathbf{R}^{n+1}$. We shall see that the transform $\mathcal{I}'_{p,s}$ is essentially another way of defining the transform $\mathcal{I}_{p,s}$.

Let $f\in\Phi_{p,s}(\mathbf{RP}^n)$. In $(\mathbf{R}^{n+1})'$ we fix any s vectors $\zeta^1, \ldots, \zeta^s$ and consider the function

$$f(x)\langle\zeta^1, x\rangle\ldots\langle\zeta^s, x\rangle.$$

This function belongs to $\Phi_p(\mathbf{RP}^n)$, and therefore the operator $\mathcal{I}'_p$ introduced in Sec. 1, part 3 can be applied to it.

<u>Definition.</u> We denote by $\mathcal{I}'_{p,s}$ the operator assigning to each function $f\in\Phi_{p,s}(\mathbf{RP}^n)$ the function $\varphi(\xi, \zeta)$, where $\xi=(\xi^1, \ldots, \xi^{n-p})\in F_p$, $\zeta=(\zeta^1, \ldots, \zeta^s)$, $\zeta^i\in(\mathbf{R}^{n+1})'$, defined by the formula

$$\varphi(\xi, \zeta)=\mathcal{I}'_p(f(x)\langle\zeta^1, x\rangle\ldots\langle\zeta^s, x\rangle)=\frac{1}{2}\int_{K'}f(x)\langle\zeta^1, x\rangle\ldots\langle\zeta^s, x\rangle\,\omega_{\xi}(x, dx); \tag{2.49}$$

the integral is taken over any surface $K'\subset h_{\xi}$ intersecting once each ray in h_{ξ} issuing from the point 0. In another notation

$$\varphi(\xi, \zeta)=\frac{1}{2}\int_K f(x)\prod_{i=1}^{s}\langle\zeta^i, x\rangle\prod_{j=1}^{n-p}\delta(\langle\xi^j, x\rangle)\sigma_n(x, dx), \tag{2.49'}$$

where $K\subset\mathbf{R}^{n+1}$ is an arbitrary surface intersecting once each ray in $\mathbf{R}^{n+1}$ issuing from the point 0.

<u>Remark.</u> In analogy with (2.49) it is possible to define the transform $\mathcal{I}'_{p,s}$ also for negative s, viz.,

$$(\mathcal{I}'_{p,s}f)(\xi, y) = \mathcal{I}'_p\left(\langle y_1, \frac{\partial}{\partial x}\rangle \ldots \langle y_{-s}, \frac{\partial}{\partial x}\rangle f(x)\right),$$

where $y = (y_1, \ldots, y_{-s})$, $y_i \in \mathbf{R}^{n+1}$.

We note the basic properties of the functions $\varphi = \mathcal{I}'_{p,s}f$ (they follow directly from the definition of the operator $\mathcal{I}'_{p,s}$).

(i) For any $g \in GL(n-p, \mathbf{R})$ we have

$$\varphi(\xi g, \zeta) = |\det g|^{-1}\varphi(\xi, \zeta).$$

(ii) For any fixed ξ, $\varphi(\xi, \zeta)$ is a symmetric polylinear function of $\zeta^1, \ldots, \zeta^s$, and $\varphi(\xi, \zeta) = 0$ if at least one of the vectors ζ^i is a linear combination of the vectors $\xi^1, \ldots, \xi^{n-p}$. Thus, $\varphi(\xi, \zeta)$ is a symmetric polylinear function of the vectors $\zeta^1, \ldots, \zeta^s$ on the factor space $(\mathbf{R}^{n+1})'/\{\xi^1, \ldots, \xi^{n-p}\}$.

(iii) For any $\zeta' = (\zeta^1, \ldots, \zeta^{s-1})$, $\eta^1, \eta^2 \in (\mathbf{R}^{n+1})'$ and any $k = 1, \ldots, n-p$ we have the relation

$$\langle \eta^1, \frac{\partial}{\partial \xi^k}\rangle \varphi(\xi; \zeta', \eta^2) = \langle \eta^2, \frac{\partial}{\partial \xi^k}\rangle \varphi(\xi; \zeta', \eta^1).$$

The definition of the operator $\mathcal{I}'_{p,s}$ introduced here admits the following specializations.

1°. Let $\varphi(\xi, \zeta) = \mathcal{I}'_{p,s}f$. We fix an arbitrary basis $e^1, \ldots, e^{n+1}$ in $(\mathbf{R}^{n+1})'$, and as the vectors ζ^k we take only vectors of this basis. Thus, in place of the function $\varphi(\xi, \zeta)$ we define a collection of functions $\hat{\varphi}^{i_1, \ldots, i_s}(\xi) = \varphi(\xi; e^{i_1}, \ldots, e^{i_s})$ $(1 \leqslant i_k \leqslant n+1)$. These functions $\hat{\varphi}^{i_1, \ldots, i_s}$ are the components of a symmetric tensor field relative to the group $SL(n+1, \mathbf{R})$. It is clear that the original function φ can be uniquely expressed in terms of them.

2°. Let $\eta = (\eta^1, \ldots, \eta^{p+1})$ be any collection of $p+1$ vectors in $(\mathbf{R}^{n+1})'$. Instead of the function $\varphi(\xi, \zeta)$ we introduce the collection of functions $\tilde{\varphi}^{\alpha_1, \ldots, \alpha_s}(\xi, \eta) = \varphi(\xi; \eta^{\alpha_1}, \ldots, \eta^{\alpha_s})$, where $1 \leqslant \alpha_k \leqslant p+1$. These functions are the components of a symmetric tensor field relative to the group $SL(p+1, \mathbf{R})$. The original function $\varphi(\xi, \zeta) = \mathcal{I}'_{p,s}f$ can be uniquely expressed in terms of the functions $\tilde{\varphi}^{\alpha_1, \ldots, \alpha_s}(\xi, \eta)$. Namely, if the vectors $\xi^1, \ldots, \xi^{n-p}, \eta^1, \ldots, \eta^{p+1}$ are linearly independent and $\zeta^i = \sum_k a_k^i \xi^k + \sum_k b_k^i \eta^k$ is the expansion of the vectors $\zeta^i \in (\mathbf{R}^{n+1})'$ in terms of these vectors, then by property (ii) of the function φ, we have

$$\varphi(\xi, \zeta) = \sum_{\alpha_1, \ldots, \alpha_s} b^1_{\alpha_1} \ldots b^s_{\alpha_s} \tilde{\varphi}^{\alpha_1, \ldots, \alpha_s}(\xi, \eta).$$

3. Connection between the Operators $\mathcal{I}_{p,s}$ and $\mathcal{I}'_{p,s}$. Proposition 9. Let $\{\hat{f}^\alpha\} = \mathcal{I}_{p,s}f$ and $\varphi = \mathcal{I}'_{p,s}f$. If the polyvectors $\hat{u} = u_1 \wedge \ldots \wedge u_{p+1}$ and $\hat{\xi} = \xi^1 \wedge \ldots \wedge \xi^{n-p}$ are dual in the sense of the definition of part 4, Sec. 1, then

$$\varphi(\xi, \zeta) = \sum_\alpha \hat{f}^\alpha(u) \langle \zeta^1, u_{\alpha_1}\rangle \ldots \langle \zeta^s, u_{\alpha_s}\rangle.$$

In particular, if the vectors ζ^i are chosen so that $\langle \zeta^i, u_j\rangle = \delta_j^{\alpha_i}$ $(i = 1, \ldots, s,\ j = 1, \ldots, p+1)$, then $f^\alpha(u) = \varphi(\xi, \zeta)$.

For the proof it suffices to write out the expression for $\mathcal{I}'_p(f(x)\langle \zeta^1, x\rangle \ldots \langle \zeta^s, x\rangle)$ in the coordinates u.

4. Description of the Image of the Operator $\mathcal{I}'_{p,s}$ for a Fixed Collection $\zeta = (\zeta^1, \ldots, \zeta^s)$. On the basis of the corollaries of Theorems 6' and 7' (Sec. 2) we immediately obtain a description of the functions $\varphi(\xi, \zeta) = \mathcal{I}'_{p,s}f$ for any fixed ζ.

THEOREM 8. Suppose the collection of vectors $\eta=(\underbrace{\eta^1,\ldots,\eta^1}_{k_1},\ldots,\underbrace{\eta^r,\ldots,\eta^r}_{k_r})$ is fixed, where the vectors $\eta^i\in(\mathbf{R}^{n+1})'$ are pairwise noncollinear and $k_1+\ldots+k_r=s$. In order that the C^∞ function $\psi(\xi)$ on F_p be representable in the form $\psi(\xi)=\varphi(\xi,\eta)$, where $\varphi=\mathcal{J}'_{p,s}f$, $f\in\Phi_{p,s}(\mathbf{RP}^n)$, it is necessary and sufficient that it belong to the image of the operator $\mathcal{J}_p$ and satisfy the following conditions:

a) The case $p<n-1$. For any vectors $\zeta^i\in(\mathbf{R}^{n+1})'$ and $y_i\in\mathbf{R}^{n+1}$ such that $\langle\zeta^i,y_i\rangle=0$ and $\langle\eta^i,y_i\rangle\neq 0$ $(1\leq i\leq r)$, the following equalities are satisfied: $(\tilde D_{y_i,\zeta^i})^s\psi|_{G^i}=0$, $s=0,\ldots,k_i-1$, where G^i is the manifold of $(p+1)$-dimensional subspaces of $\mathbf{R}^{n+1}$ belonging to the subspace $\langle\eta^i,x\rangle=0$ (for the definition of the operator $\tilde D_{y,\zeta}$ see Sec. 2, part 1).

b) The case $p=n-1$. For any $s=0,\ldots,k_i-1$ and $\zeta^i\in(\mathbf{R}^{n+1})'$ the following conditions are satisfied: $\langle\zeta^i,\frac{\partial}{\partial\xi}\rangle^s\psi|_{\xi=\eta^i}=0$, and the integral $\psi_s(\zeta^i,\eta^i)=\int_{-\infty}^{+\infty}\psi(\zeta^i+t\eta^i)t^s dt$ converges and is a homogeneous polynomial of degree s in ζ^i.

5. Description of the Image of the Operator $\mathcal{J}'_{p,s}$. THEOREM 9. A function $\varphi(\xi,\zeta)$, where $\xi=(\xi^1,\ldots,\xi^{n-p})\in F_p$, $\zeta=(\zeta^1,\ldots,\zeta^s)$, $\zeta^i\in(\mathbf{R}^{n+1})'$, belongs to the image of the operator $\mathcal{J}'_{p,s}$ if and only if the following conditions are satisfied:

1) the function φ is infinitely differentiable with respect to $\xi\in F_p$ and is a symmetric polylinear function in $\zeta^1,\ldots,\zeta^s$;

2) $\varphi(\lambda_1\xi^1,\ldots,\lambda_{n-p}\xi^{n-p};\zeta)=|\lambda_1\ldots\lambda_{n-p}|^{-1}\varphi(\xi^1,\ldots,\xi^{n-p};\zeta)$;

3) for any $\zeta'=(\zeta^1,\ldots,\zeta^{s-1})$, $\eta^1,\eta^2\in(\mathbf{R}^{n+1})'$ and any $k=1,\ldots,n-p$ the following relation holds:

$$\langle\eta^1,\frac{\partial}{\partial\xi^k}\rangle\varphi(\xi;\zeta',\eta^2)=\langle\eta^2,\frac{\partial}{\partial\xi^k}\rangle\varphi(\xi;\zeta',\eta^1). \tag{2.50}$$

The necessity of these conditions has been noted in part 2. We break the proof of their sufficiency into several lemmas.

LEMMA 1. If the function $\varphi(\xi,\zeta)$ satisfies condition (2.50), then for any vectors η^1, $\eta^2\in(\mathbf{R}^{n+1})'$ and any k, $l=1,\ldots,n-p$ we have

$$\langle\eta^1,\frac{\partial}{\partial\xi^k}\rangle\langle\eta^2,\frac{\partial}{\partial\xi^l}\rangle\varphi(\xi,\zeta)=\langle\eta^2,\frac{\partial}{\partial\xi^k}\rangle\langle\eta^1,\frac{\partial}{\partial\xi^l}\rangle\varphi(\xi,\zeta).$$

In particular,

$$\frac{\partial^2\varphi(\xi,\zeta)}{\partial\xi_i^k\,\partial\xi_{i'}^{k'}}=\frac{\partial^2\varphi(\xi,\zeta)}{\partial\xi_i^{k'}\,\partial\xi_{i'}^{k}}\ (1\leq i,i'\leq n+1,\ 1\leq k,k'\leq n-p). \tag{2.51}$$

Indeed, setting $\zeta=(\zeta',\eta)$, $\zeta'=(\zeta^1,\ldots,\zeta^{s-1})$, $\eta\in(\mathbf{R}^{n+1})'$ and applying relation (2.50) several times, we obtain

$$\langle\eta^1,\frac{\partial}{\partial\xi^k}\rangle\langle\eta^2,\frac{\partial}{\partial\xi^l}\rangle\varphi(\xi;\zeta',\eta)=\langle\eta,\frac{\partial}{\partial\xi^k}\rangle\langle\eta^2,\frac{\partial}{\partial\xi^l}\rangle\varphi(\xi;\zeta',\eta^1)=$$
$$=\langle\eta,\frac{\partial}{\partial\xi^k}\rangle\langle\eta^1,\frac{\partial}{\partial\xi^l}\rangle\varphi(\xi;\zeta',\eta^2)=\langle\eta^2,\frac{\partial}{\partial\xi^k}\rangle\langle\eta^1,\frac{\partial}{\partial\xi^l}\rangle\varphi(\xi;\zeta',\eta).$$

Suppose that the function φ satisfies the hypotheses of the theorem. By Theorem 2 (Sec. 1) from these conditions and the relation (2.51) it then follows that for each fixed ζ the function $\varphi(\xi,\zeta)$ belongs to the image of the operator $\mathcal{J}'_p$. Thus, there exists a function $f(x,\zeta)$ that for any fixed ζ belongs to the space $\Phi_p(\mathbf{RP}^n)$ and is such that

$$\varphi(\xi,\zeta)=\mathcal{J}'_p f=\frac{1}{2}\int_K f(x,\zeta)\prod_{i=1}^{n-p}\delta(\langle\xi^i,x\rangle)\sigma_n(x,dx).$$

Since the operator $\mathcal{J}'_p$, is injective, this function is unique and is a symmetric polylinear function with respect to $\zeta=(\zeta^1,\ldots,\zeta^s)$.

LEMMA 2. For any $\zeta'=(\zeta^1,\ldots,\zeta^{s-1})$ and $\eta^1,\eta^2\in(\mathbf{R}^{n+1})'$ there is the equality

$$\langle \eta^1, x\rangle f(x;\zeta',\eta^2)=\langle \eta^2, x\rangle f(x;\zeta',\eta^1). \tag{2.52}$$

Proof. From the definition of the operator $\mathcal{J}'_p$ it follows easily that for any $y\in\mathbf{R}^{n+1}$ we have

$$\mathcal{J}'_p\left\{\langle y,\frac{\partial}{\partial x}\rangle(\langle \eta^1, x\rangle f(x;\zeta',\eta^2))\right\}=-\sum_{k=1}^{n-p}\langle y,\xi^k\rangle\langle \eta^1,\frac{\partial}{\partial\xi^k}\rangle\varphi(\xi;\zeta',\eta^2).$$

Hence, by condition 3)

$$\mathcal{J}'_p\left\{\langle y,\frac{\partial}{\partial x}\rangle(\langle \eta^1, x\rangle f(x;\zeta',\eta^2))\right\}=\mathcal{J}'_p\left\{\langle y,\frac{\partial}{\partial x}\rangle(\langle \eta^2, x\rangle f(x;\zeta',\eta^1))\right\}$$

and, since $\mathcal{J}'_p$ is injective,

$$\langle y,\frac{\partial}{\partial x}\rangle(\langle \eta^1, x\rangle f(x;\zeta',\eta^2)-\langle \eta^2, x\rangle f(x;\zeta',\eta^1))=0$$

for any y. This implies (2.52).

COROLLARY. For any $\zeta=(\zeta^1,\ldots,\zeta^s)$ and $\eta=(\eta^1,\ldots,\eta^s)$ we have

$$f(x,\zeta)\prod_{i=1}^{s}\langle \eta^i, x\rangle=f(x,\eta)\prod_{i=1}^{s}\langle \zeta^i, x\rangle;$$

hence, the function

$$f(x)=f(x,\zeta)\prod_{i=1}^{s}\langle \zeta^i, x\rangle^{-1} \tag{2.53}$$

does not depend on ζ.

LEMMA 3. Function $f(x)$ defined by formula (2.53) belongs to the space $\Phi_{p,s}(\mathbf{RP}^n)$.

Proof. In (2.53) we set $\zeta=(\zeta^1,\ldots,\zeta^1)$, where $\zeta^1\neq 0$ is a fixed vector. From equality (2.53) it then follows that the function $f(x)$ is C^∞ everywhere with the possible exception of points of the subspace $\langle \zeta^1, x\rangle=0$. Hence, since ζ^1 is arbitrary, the function $f(x)$ is C^∞ on $\mathbf{R}^{n+1}\setminus 0$. It follows also from equality (2.53) that $f(\lambda x)=|\lambda|^{-p-1}\lambda^{-s}f(x)$; hence $f\in\Phi_{p,s}(\mathbf{RP}^n)$.

It has thus been proved that $\varphi(\xi,\zeta)=\mathcal{J}'_p(f(x)\langle \zeta^1, x\rangle\ldots\langle \zeta^s, x\rangle)$, where $f\in\Phi_{p,s}(\mathbf{RP}^n)$, i.e., $\varphi=\mathcal{J}'_{p,s}f$. The proof of the theorem is complete.

6. Description of the Image of the Operator $\mathcal{J}_{p,s}$. THEOREM 10. In order that a collection of C^∞ functions $\{\varphi^\alpha(u)\}$ on E_p, where $\alpha=(\alpha_1,\ldots,\alpha_s)$, $1\leqslant\alpha_k\leqslant p+1$, belong to the image of the operator $\mathcal{J}_{p,s}$, it is necessary and sufficient that the following conditions be satisfied:

1) the functions φ^α are symmetric with respect to $\alpha=(\alpha_1,\ldots,\alpha_s)$;

2) $\varphi^\alpha(\lambda^1u_1,\ldots,\lambda^{p+1}u_{p+1})=|\lambda^1\ldots\lambda^{p+1}|^{-1}(\lambda^{\alpha_1}\ldots\lambda^{\alpha_s})^{-1}\varphi^\alpha(u_1,\ldots,u_{p+1})$;

3) for any $\alpha'=(\alpha_1,\ldots,\alpha_{s-1})$, $k,l=1,\ldots,p+1$ and any $y\in\mathbf{R}^{n+1}$ the following relations hold:

$$\langle y,\frac{\partial}{\partial u_k}\rangle\varphi^{\alpha',l}=\langle y,\frac{\partial}{\partial u_l}\rangle\varphi^{\alpha',k}.$$

The necessity of the conditions has already been noted in part 1. We shall prove that they are sufficient. Thus, suppose that the collection $\{\varphi^\alpha\}$ satisfies the conditions of the theorem. We define the function $\psi(u,\zeta)$, where $\zeta=(\zeta^1,\ldots,\zeta^s)$, $\zeta^i\in(\mathbf{R}^{n+1})'$, by the following formula:

$$\psi(u,\zeta)=\sum_{\alpha_1,\dots,\alpha_s}\varphi^{\alpha_1,\dots,\alpha_s}(u)\langle\zeta^1,u_{\alpha_1}\rangle\dots\langle\zeta^s,u_{\alpha_s}\rangle.$$

It is obvious that $\psi(u,\zeta)$ is a symmetric polylinear function with respect to $\zeta^1,\dots,\zeta^s$.

LEMMA 1. The function $\psi(u,\zeta)$ satisfies the following relations:

a) $\psi(\lambda^1u_1,\dots,\lambda^{p+1}u_{p+1};\zeta)=|\lambda^1\dots\lambda^{p+1}|^{-1}\psi(u_1,\dots,u_{p+1};\zeta)$;

b) for any $\zeta'=(\zeta^1,\dots,\zeta^{s-1})$, $\eta^1,\eta^2\in(\mathbf{R}^{n+1})'$ and any $y\in\mathbf{R}^{n+1}$ there are the equalities

$$\left(\langle\eta^1,y\rangle+\sum_{k=1}^{p+1}\langle\eta^1,u_k\rangle\langle y,\frac{\partial}{\partial u_k}\rangle\right)\psi(u;\zeta',\eta^2)=\left(\langle\eta^2,y\rangle+\sum_{k=1}^{p+1}\langle\eta^2,u_k\rangle\langle y,\frac{\partial}{\partial u_k}\rangle\right)\psi(u;\zeta',\eta^1);$$

c) $\langle y_1,\frac{\partial}{\partial u_k}\rangle\langle y_2,\frac{\partial}{\partial u_l}\rangle\psi(u,\zeta)=\langle y_2,\frac{\partial}{\partial u_k}\rangle\langle y_1,\frac{\partial}{\partial u_l}\rangle\psi(u,\zeta)$ for any vectors $y_1,y_2\in\mathbf{R}^{n+1}$ and any $k,l=1,\dots,p+1$.

These relations can be established directly on the basis of the definition of the function $\psi(u,\zeta)$ and the conditions on $\varphi^\alpha(u)$.

Remark. From relations a) and c) it follows (see Sec. 1) that for fixed ζ the function $\psi(u,\zeta)$ depends only on the polyvector $\hat{u}=u_1\wedge\dots\wedge u_{p+1}$. It is therefore possible to go over from the coordinates $u=(u_1,\dots,u_{p+1})\in E_p$ to the coordinates $\xi=(\xi^1,\dots,\xi^{n-p})\in F_p$ by setting $\tilde{\psi}(\xi,\zeta)=\psi(u,\zeta)$, where $\hat{u}=u_1\wedge\dots\wedge u_{p+1}$ is the polyvector dual to $\hat{\xi}=\xi^1\wedge\dots\wedge\xi^{n-p}$ (see Sec. 1, part 4). If we show that $\tilde{\psi}(\xi,\zeta)$ satisfies the hypotheses of Theorem 9, then our theorem will follow from Theorem 9. We observe that only condition 3 of Theorem 9 needs verification, since the first conditions are obvious. However, we shall prove Theorem 10 differently; we will thus obtain an indirect proof of the assertion regarding the functions $\tilde{\psi}(\xi,\zeta)$.

By Theorem 3 (Sec. 1) it follows from relations a) and c) that for any fixed ζ the function $\psi(u,\zeta)$ belongs to the image of the operator $\mathcal{I}_p$. Thus, there exists a function $f(x,\zeta)$ that for any fixed ζ belongs to the space $\Phi_p(\mathbf{RP}^n)$ and is such that

$$\psi(u,\zeta)=\mathcal{I}_pf=\int_{\mathbf{RP}^p}f(ut,\zeta)\sigma_p(t,dt).$$

Since the operator $\mathcal{I}_p$ is injective, this function is unique and is a symmetric polylinear function with respect to $\zeta=(\zeta^1,\dots,\zeta^s)$.

LEMMA 2. For any $\zeta'=(\zeta^1,\dots,\zeta^{s-1})$ and $\eta^1,\eta^2\in(\mathbf{R}^{n+1})'$ the following equality holds:

$$\langle\eta^1,x\rangle f(x;\zeta',\eta^2)=\langle\eta^2,x\rangle f(x;\zeta',\eta^1).\tag{2.54}$$

Proof. From the definition of the operator $\mathcal{I}_p$ it follows immediately that for any $y\in\mathbf{R}^{n+1}$ we have

$$\mathcal{I}_p\left\{\langle y,\frac{\partial}{\partial x}\rangle(\langle\eta^1,x\rangle f(x;\zeta',\eta^2))\right\}=\left(\langle\eta^1,y\rangle+\sum_{k=1}^{p+1}\langle\eta^1,u_k\rangle\langle y,\frac{\partial}{\partial u_k}\rangle\right)\psi(u;\zeta',\eta^2).$$

Therefore, by relation b) we obtain

$$\mathcal{I}_p\left\{\langle y,\frac{\partial}{\partial x}\rangle(\langle\eta^1,x\rangle f(x;\zeta',\eta^2)-\langle\eta^2,x\rangle f(x;\zeta',\eta^1))\right\}=0,$$

whence equality (2.54) follows (see Lemma 2 in part 5). The proof of the lemma is complete.

On the basis of relation (2.54), just as in the proof of Theorem 9, we find that

$$\psi(u,\zeta)=\mathcal{I}_p\left(f(x)\prod_{i=1}^{s}\langle\zeta^i,x\rangle\right),$$

where $f\in\Phi_{p,s}(\mathbf{RP}^n)$, i.e.,

$$\sum_\alpha \varphi^\alpha(u)\langle \zeta^1, u_{\alpha_1}\rangle\dots\langle \zeta^s, u_{\alpha_s}\rangle=\sum_\alpha\Big(\int_{\mathbf{RP}^p} f(ut)\,t^\alpha\sigma_p(t, dt)\Big)\langle \zeta^1, u_{\alpha_1}\rangle\dots\langle \zeta^s, u_{\alpha_s}\rangle.$$

This implies that

$$\varphi^\alpha(u)=\int_{\mathbf{RP}^p} f(ut)\,t^\alpha\sigma_p(t, dt)$$

for any $\alpha=(\alpha_1, \dots, \alpha_s)$. The proof of the theorem is complete.

7. Integral Transform $\mathcal{I}^0_{p,s}$. In part 1 to each function $f\in\Phi_{p,s}(\mathbf{RP}^n)$ there was assigned a tensor field $\{\hat f^\alpha\}$ on G_p. Since all components of this field can be uniquely expressed in terms of one of them, it is natural to consider for each fixed $\alpha=(\alpha_1, \dots, \alpha_s)$ the integral transform $\mathcal{I}^\alpha_{p,s}: f\mapsto \hat f^\alpha$ which assigns to each function $f\in\Phi_{p,s}(\mathbf{RP}^n)$ a fixed component of the tensor field $\{\hat f^\alpha\}$. Here we consider the case $\alpha=(p+1, \dots, p+1)$ and for simplicity we denote the corresponding operator by $\mathcal{I}^0_{p,s}$. Thus, by definition we have

$$(\mathcal{I}^0_{p,s}f)(u)=\int_{\mathbf{RP}^p} f(ut)\,(t^{p+1})^s\,\sigma_p(t, dt). \tag{2.55}$$

Our task is to describe the image of this operator.

THEOREM 11. In order that a C^∞ function $\varphi(u)$ on E_p belong to the image of the operator $\mathcal{I}^0_{p,s}$, it is necessary and sufficient that the following conditions be satisfied:

1) $\varphi(\lambda^1 u_1, \dots, \lambda^{p+1}u_{p+1})=|\lambda^1\dots\lambda^{p+1}|^{-1}(\lambda^{p+1})^{-s}\varphi(u_1, \dots, u_{p+1})$;

2) $\dfrac{\partial^2\varphi(u)}{\partial u^i_k\partial u^{i'}_{k'}}=\dfrac{\partial^2\varphi(u)}{\partial u^{i'}_k\partial u^i_{k'}}$ $(1\leq i, i'\leq n+1,\ 1\leq k, k'\leq p+1)$.

The necessity of the conditions of the theorem follows from formula (2.55) for $\hat f=\mathcal{I}^0_{p,s}f$. We shall prove that they are sufficient. Thus, suppose that the function φ satisfies the conditions of the theorem.

LEMMA 1. The function φ satisfies the relations

$$\sum_{i=1}^{n+1}\frac{\partial\varphi}{\partial u^i_k}u^i_l=0 \tag{2.56}$$

for any $k=1, \dots, p+1$, $l=1, \dots, p$; $k\neq l$.

Proof. We denote the left side of equality (2.56) by $\psi_{k,l}$, and we prove that $\psi_{k,l}$ does not depend on u_l. Indeed, on the basis of conditions 1 and 2 we have for any $j=1, \dots, n+1$:

$$\frac{\partial\psi_{k,l}}{\partial u^j_l}=\sum_{i=1}^{n+1}\frac{\partial^2\varphi}{\partial u^j_l\partial u^i_k}u^i_l+\frac{\partial\varphi}{\partial u^j_k}=\sum_{i=1}^{n+1}\frac{\partial^2\varphi}{\partial u^j_k\partial u^i_l}u^i_l+\frac{\partial\varphi}{\partial u^j_k}=-\frac{\partial\varphi}{\partial u^j_k}+\frac{\partial\varphi}{\partial u^j_k}=0.$$

Since, on the other hand, by condition 1, $\psi_{k,l}$ is an odd function of u_l, it follows that $\psi_{k,l}=0$.

We shall further write $u=(v, x)$, where $v=(v_1, \dots, v_p)$. Then by Lemma 1 and condition 1 we have the following result.

COROLLARY. For any $g\in GL(p, \mathbf{R})$ and any numbers $\lambda, \lambda^1, \dots, \lambda^p$ there is the equality

$$\varphi\Big(vg, \lambda x+\sum_l\lambda^l v_l\Big)=|\det g|^{-1}|\lambda|^{-1}\lambda^{-s}\varphi(v, x).$$

LEMMA 2. For any $k_1, \dots, k_{s+1}=1, \dots, p$ we have

$$\sum_{i_1,\dots,i_{s+1}}\frac{\partial^{s+1}\varphi(v, x)}{\partial v^{i_1}_{k_1}\dots\partial v^{i_{s+1}}_{k_{s+1}}}x^{i_1}\dots x^{i_{s+1}}=0. \tag{2.57}$$

Proof. We denote the left side of Eq. (2.57) by ψ and prove that ψ does not depend on x. Indeed, applying condition 2, we obtain for any $j=1, \ldots, n+1$:

$$\frac{\partial\psi}{\partial x^j}=\frac{1}{s+1}\times$$

$$\times\sum_{\beta=1}^{s+1}\frac{\partial}{\partial v^j_{k_\beta}}\left(\sum_{i_1,\ldots,i_{s+1}}\frac{\partial^{s+1}\varphi(v, x)}{\partial v^{i_1}_{k_1}\ldots\partial x^{i_\beta}\ldots\partial v^{i_{s+1}}_{k_{s+1}}}x^{i_1}\ldots x^{i_{s+1}}\right)+\sum_{\beta=1}^{s+1}\sum_{i_1,\ldots,\hat{i}_\beta,\ldots,i_{s+1}}\frac{\partial^{s+1}\varphi(v, x)}{\partial v^{i_1}_{k_1}\ldots\partial v^j_{k_\beta}\ldots\partial v^{i_{s+1}}_{k_{s+1}}}x^{i_1}\ldots\widehat{x^{i_\beta}}\ldots x^{i_{s+1}}.$$

Since by the homogeneity conditions for φ, $\sum_{i_3}\frac{\partial\varphi(v, x)}{\partial x^{i_\beta}}x^{i_\beta}=-(s+1)\varphi(v, x)$, the expression obtained is equal to zero. Thus, the function ψ does not depend on x. Since, on the other hand, ψ is an odd function of x, it follows that $\psi=0$.

COROLLARY. The function $\varphi(v_1-\lambda_1 x, \ldots, v_p-\lambda_p x, x)$ is a polynomial in $\lambda_1, \ldots, \lambda_p$ of degree $\leqslant s$. Thus, setting $\lambda_{p+1}=1$, we have

$$\varphi(v_1-\lambda_1 x, \ldots, v_p-\lambda_p x, x)=\sum_\alpha\lambda_\alpha\varphi^\alpha(v, x), \tag{2.58}$$

where $\alpha=(\alpha_1, \ldots, \alpha_s)$ $(1\leqslant\alpha_i\leqslant p+1)$, $\lambda_\alpha=\lambda_{\alpha_1}\ldots\lambda_{\alpha_s}$; $\varphi^\alpha(v, x)$ is a C^∞ function on E_p, and, in particular, $\varphi^{p+1,\ldots,p+1}=\varphi$.

From the conditions on the function φ we easily find the following result.

LEMMA 3. The collection of functions $\varphi^\alpha(u)$ defined by Eq. (2.58) satisfies the conditions of Theorem 10.

Thus, by this theorem $\{\varphi^\alpha\}=\mathcal{I}_{p,s}f$, where $f\in\Phi_{p,s}(\mathbf{RP}^n)$. In particular,

$$\varphi(u)=\varphi^{p+1,\ldots,p+1}(u)=\int_{\mathbf{RP}^p}f(ut)(t^{p+1})^s\sigma_p(t, dt),$$

i.e., φ belongs to the image of the operator $\mathcal{I}^0_{p,s}$. The proof of the theorem is complete.

8. Inversion Formula for the Case of Even p. We first write out an inversion formula for the integral transform $\varphi(\xi, \zeta)=\mathcal{I}'_{p,s}f$, where $f\in\Phi_{p,s}(\mathbf{RP}^n)$. Since by definition $\varphi(\xi, \zeta)=\mathcal{I}'_p(f(x)\langle\zeta^1, x\rangle\ldots\langle\zeta^s, x\rangle)$, the inversion formula for $\mathcal{I}'_{p,s}$ follows directly from the inversion formula for the operator $\mathcal{I}'_p$ presented in Sec. 1, part 11. Namely, let $x\in\mathbf{R}^{n+1}\setminus 0$ and suppose that $\zeta=(\zeta^1, \ldots, \zeta^s)$ are any vectors in $(\mathbf{R}^{n+1})'$ such that $\langle\zeta^i, x\rangle\neq 0$ $(i=1, \ldots, s)$. We set

$$\tilde{\varkappa}^\zeta_x=\langle\zeta^1, x\rangle^{-1}\ldots\langle\zeta^s, x\rangle^{-1}\tilde{\varkappa}_x,$$

where $\tilde{\varkappa}_x$ is the operator defined in Sec. 1, part 11. Retaining the notation introduced there, we have the following result.

Proposition 10. If $\varphi=\mathcal{I}'_{p,s}f$, $f\in\Phi_{p,s}(\mathbf{RP}^n)$, where p is an even number, then there is the inversion formula

$$\int_\gamma\tilde{\varkappa}^\zeta_x(\varphi(\xi, \zeta))=cd(\gamma)f(x) \quad (c\neq 0).$$

We now write out an inversion formula for the integral transform $\varphi(u)=\mathcal{I}^0_{p,s}f$. We consider the differential form $\varkappa_x\varphi$, where $\varkappa_x$ is the operator introduced in Sec. 1, part 10. In Sec. 1 this operator was applied to functions $\varphi=\mathcal{I}_p f$, where $f\in\Phi_p(\mathbf{RP}^n)$. However, it is not hard to see that the basic assertions regarding the differential form $\varkappa_x\varphi$ in Sec. 1 remain valid in our case as well with $\varphi=\mathcal{I}^0_{p,s}f$, $f\in\Phi_{p,s}(\mathbf{RP}^n)$. Thus, just as in Sec. 1, we obtain the following result.

Proposition 11. If $\varphi=\mathcal{J}^0_{p,s}f$, $f\in\Phi_{p,s}(\mathbf{RP}^n)$, where p is an even number, then there is the inversion formula

$$\int_\gamma \varkappa_x\varphi=cd(\gamma)f(x)\ (c\neq 0)$$

(the notation is the same as in formula (2.25)).

APPENDIX. SUMMARY OF FORMULAS FOR THE COMPLEX SPACE $\mathbf{C}P^n$

1°. Definition of the Operators $\mathcal{J}_p$ and $\mathcal{J}'_p$.

$$(\mathcal{J}_pf)(u)=\left(\frac{i}{2}\right)^p\int_{\mathbf{C}P^p} f(ut)\,\sigma_p(t,dt)\wedge\sigma_p(\bar{t},d\bar{t}),$$

where $\sigma_p(t,dt)$ is defined by formula (2.2) and $u=(u_1,\ldots,u_{p+1})$ is a $(p+1)$-frame in $\mathbf{C}^{n+1}$;

$$(\mathcal{J}'_pf)(\xi)=\left(\frac{i}{2}\right)^n\int_K f(z)\prod_{i=1}^{n-p}\delta(\langle\xi^i,z\rangle)\,\sigma_n(z,dz)\wedge\sigma_n(\bar{z},d\bar{z}),$$

where $\xi=(\xi^1,\ldots,\xi^{n-p})$ is an $(n-p)$-frame in $(\mathbf{C}^{n+1})'$. The operators act on C^∞ functions on $\mathbf{C}^{n+1}\setminus 0$ which satisfy the homogeneity condition $f(\lambda z)=|\lambda|^{-2p-2}f(z)$.

2°. Description of the Image of the Operator $\mathcal{J}_p$. A C^∞ function $\varphi(u)$ on E_p belongs to the image of the operator $\mathcal{J}_p$ if and only if it satisfies the relations

1) $\varphi(\lambda^1u_1,\ldots,\lambda^{p+1}u_{p+1})=|\lambda^1\ldots\lambda^{p+1}|^{-2}\varphi(u_1,\ldots,u_{p+1})$;

2) $\dfrac{\partial^2\varphi(u)}{\partial u_k^i\,\partial u_{k'}^{i'}}=\dfrac{\partial^2\varphi(u)}{\partial u_k^{i'}\,\partial u_{k'}^{i}},\quad \dfrac{\partial^2\varphi(u)}{\partial\overline{u_k^i}\,\partial\overline{u_{k'}^{i'}}}=\dfrac{\partial^2\varphi(u)}{\partial\overline{u_k^{i'}}\,\partial\overline{u_{k'}^{i}}}$

for any $i,i'=1,\ldots,n+1$, $k,k'=1,\ldots,p+1$. There is an analogous result for the operator $\mathcal{J}'_p$.

3°. Inversion Formula for the Radon Transform ($p=n-1$). If $\hat{f}=\mathcal{J}'_{n-1}f$, then

$$f(z)=(-1)^{n-1}\pi^{-2n+2}\left(\frac{i}{2}\right)^n\int_\Gamma\hat{f}(\xi)\,\delta^{(n-1,n-1)}(\langle\xi,z\rangle)\sigma_n(\xi,d\xi)\wedge\sigma_n(\bar{\xi},d\bar{\xi});$$

the integral is taken over any surface Γ in $\mathbf{C}^{n+1}$ which intersects once almost every line passing through the point $\xi=0$.

4°. Inversion Formula for Arbitrary $p\leqslant n-1$.

$$\int_\gamma(\varkappa_z\wedge\bar{\varkappa}_z)\,\hat{f}=(-1)^p\pi^{2p}d(\gamma)f(z), \tag{2.59}$$

where $\bar{\varkappa}_z$ is defined in the coordinates u and ξ, respectively, by formulas (2.24) and (2.28) in which the partial derivatives with respect to real directions are replaced by partial derivatives with respect to the complex directions $\frac{\partial}{\partial z^i}$, $\frac{\partial}{\partial\xi_i}$; correspondingly $\bar{\varkappa}_z$ contains partial derivatives $\frac{\partial}{\partial\bar{z}^i}\frac{\partial}{\partial\bar{\xi}_i}$. Thus, $(\varkappa_z\wedge\bar{\varkappa}_z)\hat{f}$ is a differential form of type (p, p). The integral is taken over any cycle γ of real dimension $2p$ in the manifold of complex p-dimensional subspaces not passing through the point z; $d(\gamma)$ is the index of intersection γ with the Euler cycle.

We emphasize that in the complex case the local inversion formula (2.59) holds for any $p\leqslant n-1$.

5°. Definition of the Operators $\mathcal{J}_{p,s}$, $\mathcal{J}'_{p,s}$, $s=(s',s'')$, Where $s',s''=0,1,\ldots$.

$$(\mathcal{J}_{p,s}f)^{\alpha,\beta}(u)=\left(\frac{i}{2}\right)^p\int_{\mathbf{C}R^p} f(ut)\,t^\alpha\bar{t}^\beta\sigma_p(t,dt)\wedge\sigma_p(\bar{t},d\bar{t}),$$

where $\alpha=(\alpha_1,\ldots,\alpha_{s'})$, $\beta=(\beta_1,\ldots,\beta_{s''})$, $1\leqslant\alpha_i$, $\beta_j\leqslant p+1$.

$$(\mathcal{I}'_{p,s}f)(\xi;\zeta,\eta)=\left(\frac{i}{2}\right)^n\int\limits_K f(z)\prod_{i=1}^{s'}\langle\zeta^i,z\rangle\prod_{j=1}^{s''}\langle\bar{\eta}^j,\bar{z}\rangle\prod_{k=1}^{n-p}\delta(\langle\xi^k,z\rangle)\,\sigma_n(z,dz)\wedge\sigma_n(\bar{z},d\bar{z}),$$

where $\xi=(\xi^1,\ldots,\xi^{n-p})$ is an $(n-p)$-frame in $(\mathbf{C}^{n+1})'$, $\zeta=(\zeta^1,\ldots,\zeta^{s'})$, $\eta=(\eta^1,\ldots,\eta^{s''})$, $\zeta^i,\eta^j\in(\mathbf{C}^{n+1})'$. The operators act on C^∞ functions on $\mathbf{C}^{n+1}\setminus 0$, which satisfy the condition: $f(\lambda z)=\lambda^{-p-s'-1}\cdot\bar{\lambda}^{-p-s''-1}f(z)$.

<u>6°. Inversion Formulas for the Operators $\mathcal{I}_{p,s}$, $\mathcal{I}'_{p,s}$.</u> If $\varphi(\xi;\zeta,\eta)=\mathcal{I}'_{p,s}f$, then

$$\int\limits_\gamma(\tilde{\varkappa}_z\wedge\bar{\tilde{\varkappa}}_z)\varphi(\xi;\zeta,\eta)=(-1)^p\pi^{2p}d(\gamma)\prod_{i=1}^{s'}\langle\zeta^i,z\rangle\prod_{j=1}^{s''}\langle\bar{\eta}^j,\bar{z}\rangle f(z).$$

If $\{\hat{f}^{\alpha,\beta}\}=\mathcal{I}_{p,s}f$, then setting $\hat{f}=\hat{f}^{p+1,\ldots,p+1}$, we have

$$\int\limits_\gamma(\varkappa_z\wedge\bar{\varkappa}_z)\hat{f}=(-1)^p\pi^{2p}d(\gamma)f(z).$$

(Here the notation is the same as in 4°.)

CHAPTER III

PROBLEMS OF INTEGRAL GEOMETRY FOR DIFFERENTIAL FORMS IN THREE-DIMENSIONAL AFFINE AND PROJECTIVE SPACES

Chapters III and IV are devoted to problems connected with the integration of differential forms. The basic idea is that if on a manifold M there is given some manifold N of p-dimensional submanifolds (e.g., in $\mathbf{R}^n$ there is given a manifold of p-dimensional planes), then over the elements of N it is possible to integrate not only differential p-forms but also any -forms with $r>p$. The result of the integration is $(r-p)$-forms on N (in particular, if $r=p$, functions on N are obtained). In Chaps. III and IV this idea is realized for the integration of r-forms on p planes in $\mathbf{R}^n$ or $\mathbf{RP}^n$.

We observe that the space of n-forms on $\mathbf{RP}^n$ is canonically isomorphic to one of the spaces considered in Sec. 3 of Chap. II (viz., the space $\Phi_{p,\,n-p-1}(\mathbf{RP}^n)$), and problems connected with the integration of these forms over p planes have essentially already been treated there. In the case of differential forms of degree $r<n$ we deal with sections of certain multidimensional bundles over $\mathbf{RP}^n$, and the problems that arise here are essentially different from those considered in the preceding chapters. However, we also consider the case $r=n$, since the language of integrating differential forms makes it possible to give a different interpretation of a number of results (in particular, the description of the image).

In Chap. III we treat problems connected with the integration of differential forms over lines in three-dimensional space. The cases of 3-forms, 2-forms, and 1-forms each has its special features, and these cases are considered individually. The content of the chapter may be considered as an extended introduction to the entire theory. It is found, moreover, that the three-dimensional case is contained in a certain sense as the most important component in the multidimensional case.

The works of Penrose on the theory of twistors [62, 63] and especially the subsequent works on the theory of self-dual solutions of the Yang–Mills equations (instantons) [46] have had a major influence on the exposition.

In conclusion, as in the preceding chapters, a summary is given of the formulas for the complex case. Here the operation of integrating differential (r,s)-forms, $r\geqslant p$, $s\geqslant p$, over (analytic) planes is considered. We note that in the case $s=p$ the corresponding mapping has a kernel. However, it was shown in [27] that due to this it is possible to go over from integral geometry on all of $\mathbf{CP}^n$ to integral geometry in q-linear concave domains (i.e., domains which are the union of the analytic p-planes contained in them). Here the isomorphic mapping does not occur on all p-forms but on the $\bar{\partial}$-cohomology.

1. Definition of the Integral of an r-Form Over Lines in $\mathbf{R}^3$

In Secs. 1-3 we shall integrate differential r-forms over lines in $\mathbf{R}^3$ ($r=1, 2, 3$). The basic problem is to describe the $(r-1)$-forms on the manifold of lines H that are hereby obtained. A special feature of the exposition is that we endeavor to write explicit formulas in concrete, simple situations of maximal simplicity. Therefore, many definitions and formulas are given in the simplest coordinate systems although they actually have an invariant meaning which we shall clarify later in Chap. IV.

1. Coordinates in H and the Spaces of Differential Forms $\Omega^r(\mathbf{R}^3)$ and $\Omega^r(H)$. We choose the simplest coordinates on the manifold H of lines in $\mathbf{R}^3$. Namely, if (x^1, x^2, x^3) are coordinates in $\mathbf{R}^3$, then we restrict attention to lines intersecting the plane $x^3=0$; in the set $H_0 \subset H$ of such lines we introduce the single coordinate system $(\alpha^1, \alpha^2, \beta^1, \beta^2)$ defining lines by the equations

$$x^i=\alpha^i x^3+\beta^i, \quad i=1, 2.$$

Over such lines we shall integrate differential r-forms on $\mathbf{R}^3$ with coefficients in the Schwartz space the space of these forms we denote by $\Omega^r(\mathbf{R}^3)$.

On the manifold of lines H we consider spaces of differential r-forms with infinitely differentiable, rapidly decreasing coefficients which we denote by $\Omega^r(H)$. Due to the fact that H_0 is dense in H, the conditions of rapid decrease of a function f on H can be formulated in the coordinates $(\alpha^1, \alpha^2, \beta^1, \beta^2)$. They consist, first of all, in the condition that must be infinitely differentiable and rapidly decreasing with respect to β together with all partial derivatives uniformly with respect to α in any compact region of variation of α. Secondly, the function f must extend from H_0 to H with preservation of smoothness and rapid decay. In order to write down this last condition, it is necessary to go over to other local coordinates on H by defining lines by the equations $x^i=\alpha^i x^1+\beta^i$ ($i=2, 3$) or the equations $x^i=\alpha^i x^2+\beta^i$ ($i=1, 3$), and to represent f as a function of these new coordinates. We obtain, respectively, the following functions:

$$f_1(\alpha^2, \alpha^3, \beta^2, \beta^3)=f\left(\frac{1}{\alpha^3}, \frac{\alpha^2}{\alpha^3}, -\frac{\beta^3}{\alpha^3}, \beta^2-\frac{\alpha^2\beta^3}{\alpha^3}\right),$$

$$f_2(\alpha^1, \alpha^3, \beta^1, \beta^3)=f\left(\frac{\alpha^1}{\alpha^3}, \frac{1}{\alpha^3}, \beta^1-\frac{\alpha^1\beta^3}{\alpha^3}, -\frac{\beta^3}{\alpha^3}\right).$$

The second condition is that the functions f_1, f_2 defined for $\alpha^3\neq 0$ must extend to C^∞ functions on the entire coordinate space and be rapidly decreasing in β.

2. Definition of the Form $\mathcal{I}\omega$. It is well known how to integrate 1-forms over lines; the result of the integration is a function on the manifold of lines. It is found that it is also possible to define the integral of r-forms for $r>1$ over lines. The result is an $(r-1)$-form on the manifold of lines. In the next chapter we give a relatively general definition of the integral of an r-form on a manifold over p-dimensional manifolds for $p<r$. Here

we give a definition for the particular case associated with fixed coordinate systems in and the space of lines

Thus, suppose there is given an r-form on $\mathbf{R}^3$,

$$\omega=\sum_{i_1<\ldots<i_r} f_{i_1,\ldots,i_r}(x)\,dx^{i_1}\wedge\ldots\wedge dx^{i_r}$$

($r=1, 2, 3$). We set $x^i=\alpha^i x^3+\beta^i$ ($i=1, 2$) and we go over from the coordinates x^1, x^2, x^3 to the coordinates α, β, x^3*, i.e., in the expression for ω we substitute: $x^i=\alpha x^3+\beta^i$, $dx^i=\alpha^i dx^3+x^3 d\alpha^i+d\beta^i$ ($i=1, 2$). In the expression obtained we drop all monomials not containing dx^3. We integrate the expression obtained on x^3 for fixed $\alpha, \beta, d\alpha, d\beta$. As a result, we obtain an $(r-1)$-form on the manifold of lines which is given in the coordinates (α, β). We call this form the integral of the original r-form ω over lines in $\mathbf{R}^3$ and denote it by $\mathcal{I}\omega$. It is obvious that for $r=1$ the present definition coincides with the usual definition.

By applying this definition to forms ω of different degrees on $\mathbf{R}^3$, we obtain explicit expressions for $\mathcal{I}\omega$. Thus, if $\deg\omega=3$, i.e.,

$$\omega=f(x)\,dx^1\wedge dx^2\wedge dx^3,$$

then

$$\mathcal{I}\omega=\hat{f}^{(2)}(\alpha,\beta)\,d\alpha^1\wedge d\alpha^2+\hat{f}^{(1)}(\alpha,\beta)\,(d\alpha^1\wedge d\beta^2-d\alpha^2\wedge d\beta^1)+\hat{f}^{(0)}(\alpha,\beta)\,d\beta^1\wedge d\beta^2,$$

where

$$\hat{f}^{(l)}(\alpha,\beta)=\int_{-\infty}^{+\infty} f(\alpha^1x^3+\beta^1,\ \alpha^2x^3+\beta^2,\ x^3)(x^3)^l dx^3 \quad (l=0, 1, 2).$$

For the formula for $\mathcal{I}\omega$ in the case $\deg\omega=2$, see Sec. 3.

The problem now consists in describing the space of forms $\mathcal{I}\omega$ for $\omega\in\Omega^r(\mathbf{R}^3)$, $r=1, 2, 3$. In addition, we shall show that for $r=2, 3$ the mapping $\mathcal{I}$ has no kernel (for $r=1$ any exact form belongs to the kernel).

3. Motivation for the Definition Given. We wish to define the integral $\mathcal{I}\omega$ over lines of a differential r-form ω on $\mathbf{R}^3$, where $r=2, 3$, so that the following condition is satisfied. Let S be an $(r-1)$-dimensional surface in the manifold of lines. We denote by $\Sigma(S)$ the manifold of points in $\mathbf{R}^3$ lying on lines $h\in S$. As a rule, in general position $\dim\Sigma(S)=r$. We require that

$$\int_S \mathcal{I}\omega=\int_{\Sigma(S)}\omega. \tag{3.1}$$

We shall treat the cases of 2-forms and 3-forms on $\mathbf{R}^3$ in detail.

Let ω be a 2-form on $\mathbf{R}^3$, let h be a point in the manifold of lines H, and let ξ be a vector in the space T_h tangent to H at the point h. We consider a curve $s(t)$ on H such that $s(0)=h$, $\dot{s}(0)=\xi$. We set $S_\varepsilon=\{s(t);\ 0\leqslant t\leqslant\varepsilon\}$ and denote by $\Sigma(S_\varepsilon)$ the set of points in $\mathbf{R}^3$ lying on lines $h\in S_\varepsilon$. Condition (3.1) leads naturally to the following definition:

$$\mathcal{I}\omega(h;\xi)=\lim_{\varepsilon\to 0}\frac{1}{\varepsilon}\int_{\Sigma(S_\varepsilon)}\omega.$$

*Actually (α, β, x^3) are coordinates in the set of pairs: a point $x\in\mathbf{R}^3$, a line passing through it.

This limit depends only on h and ξ, and $\mathcal{I}\omega(h;\xi)$ depends linearly on ξ, i.e., $\mathcal{I}\omega(h;\xi)$ is a differential 1-form on the manifold of lines. It can be verified that we thus arrive at the definition of $\mathcal{I}\omega$ given above for the case $r=2$.

Suppose now that ω is a 3-form on $\mathbf{R}^3$, $h\in H$, and $\xi_1, \xi_2 \in T_h$. We consider the two-dimensional surface $s(t_1, t_2)$ in H such that $s(0,0)=h$, $s'_{t_1}(0,0)=\xi_1$ and $s'_{t_2}(0,0)=\xi_2$. We set $S_\varepsilon=\{s(t_1, t_2);\ 0\leqslant t_1, t_2\leqslant\varepsilon\}$ and we denote by $\Sigma(S_\varepsilon)$ the manifold of points in $\mathbf{R}^3$ lying on lines in $h\in S_\varepsilon$. We then set

$$\mathcal{I}\omega(h;\xi_1,\xi_2)=\lim_{\varepsilon\to 0}\frac{1}{\varepsilon^2}\int\limits_{\Sigma(S_\varepsilon)}\omega.$$

On evaluating this limit, we obtain the expression for $\mathcal{I}\omega$ for r = 3 given in the preceding part.

2. Study of $\mathcal{I}\omega$ for the Case Where ω Is a 3-Form on $\mathbf{R}^3$

Let $\omega\in\Omega^3(\mathbf{R}^3)$, i.e., $\omega=f(x)\,dx^1\wedge dx^2\wedge dx^3$, where f is a function in the Schwartz space $S(\mathbf{R}^3)$. As we have already mentioned, in the local coordinates (α,β) on H $\mathcal{I}\omega$ has the following form:

$$\mathcal{I}\omega=\hat{f}^{(2)}(\alpha,\beta)\,d\alpha^1\wedge d\alpha^2+\hat{f}^{(1)}(\alpha,\beta)\,(d\alpha^1\wedge d\beta^2-d\alpha^2\wedge d\beta^1)+\hat{f}^{(0)}(\alpha,\beta)\,d\beta^1\wedge d\beta^2, \tag{3.2}$$

where

$$\hat{f}^{(i)}(\alpha,\beta)=\int\limits_{-\infty}^{+\infty}f(\alpha^1x^3=\beta^1,\ \alpha^2x^3+\beta^2,\ x^3)(x^3)^i\,dx^3\ (i=0,1,2). \tag{3.3}$$

From the formulas for the coefficients $\hat{f}^{(i)}$ it follows that $\mathcal{I}\omega\in\Omega^2(H)$, where $\Omega^2(H)$ is the space of differential 2-forms on H with rapidly decreasing coefficients introduced in part 1, Sec. 1.

The basic results of this section are as follows: 1) if $\omega\in\Omega^3(\mathbf{R}^3)$, then $\mathcal{I}\omega=0$ only in the case $\omega=0$; 2) a form $\sigma\in\Omega^2(H)$ can be represented in the form $\sigma=\mathcal{I}\omega$ if and only if it is closed and self-dual, i.e., $*\sigma=\sigma$, where the $*$ operation is invariantly defined in the space of differential forms on H and is constructed on the basis of a certain nondegenerate metric on H. The question of recovering ω on the basis of $\mathcal{I}\omega$ reduces to the corresponding problem for functions.

1. Properties of the Form $\mathcal{I}\omega$. It is evident from (3.2) that $\mathcal{I}\omega$ cannot be an arbitrary 2-form on H_0. First of all, it satisfies the condition on the coefficients formulated in the following definition.

Definition. A 2-form on the manifold H_0

$$\sigma=a\,d\alpha^1\wedge d\alpha^2+b_1\,d\alpha^1\wedge d\beta^2+b_2\,d\alpha^2\wedge d\beta^1+c\,d\beta^1\wedge d\beta^2+c_1\,d\alpha^1\wedge d\beta^1+c_2\,d\alpha^2\wedge d\beta^2$$

is called admissible if $c_1=c_2=0$, $b_2=-b_1$.

We thus have the following assertion.

Proposition 1. The differential form $\mathcal{I}\omega$ is admissible.

The geometric meaning of admissibility will be clarified in part 2.

Proposition 2. The differential form $\mathcal{I}\omega$ is closed.

Proof. From formulas (3.3) for the coefficients $\hat{f}^{(i)}$ of the form $\mathcal{I}\omega$ it follows that

$$\frac{\partial\hat{f}^{(2)}}{\partial\beta^i}=\frac{\partial\hat{f}^{(1)}}{\partial\alpha^i},\quad \frac{\partial\hat{f}^{(1)}}{\partial\beta^i}=\frac{\partial\hat{f}^{(0)}}{\partial\alpha^i},\quad i=1,2.$$

These relations are equivalent to the closedness of the form $\mathcal{I}\omega$.

COROLLARY. The coefficients of the form $\mathcal{I}\omega$ satisfy the ultrahyperbolic equation

$$\Delta\hat{f}^{(i)}\equiv\frac{\partial^2\hat{f}^{(i)}}{\partial\alpha^1\partial\beta^2}-\frac{\partial^2\hat{f}^{(i)}}{\partial\alpha^2\partial\beta^1}=0,\ i=0,1,2.$$

Remark. The closedness of $\mathcal{I}\omega$ is a consequence of the more general fact regarding the commutativity of the operator $\mathcal{I}$ and the operator d of exterior differentiation if it is noted that the form ω is closed as a form of maximal degree. The commutativity of d and $\mathcal{I}$ can be obtained from (3.1) on the basis of Stokes' formula. Namely, we observe that $\partial\Sigma(S)=\Sigma(\partial S)$, and therefore by Stokes' formula $\int_{\Sigma(\partial S)}\omega=\int_{\Sigma(S)}d\omega$. Hence, by (3.1) $\int_{\partial S}\mathcal{I}\omega=\int_{S}\mathcal{I}(d\omega)$ for any surface $S\subset H(\dim S=\deg\omega)$, where $d\mathcal{I}\omega=\mathcal{I}d\omega$.

2. An Invariant Definition of the Admissibility of 2-Forms. Our definition of admissibility is bound to a system of coordinates in H. Its geometric meaning is more simply related to the motivation for the integral over lines $\mathcal{I}\omega$ given in part 3, Sec. 1. In the definition given there, to each two-dimensional surface $S\subset H$ we assign a surface $\Sigma(S)\subset\mathbf{R}^3$. As a rule, $\dim\Sigma(S)=3$. However, if S is a manifold of lines lying on a fixed plane, then $\Sigma(S)$ degenerates into a two-dimensional surface (a plane). As a result, it follows from (3.1) that the restriction of $\mathcal{I}\omega$ to such a surface S must be equal to zero. We proceed to precise definitions.

Definition. We call a pair of vectors $\xi_1,\xi_2\in T_h$ degenerate if the subspace they span is tangent to a submanifold of lines lying in some plane passing through h.

Proposition 3. A differential 2-form on H is admissible if and only if it vanishes on any degenerate pair of vectors $\xi_1,\xi_2\in T_h$.

It is obvious that Proposition 3 is equivalent to the following assertion.

Proposition 4. A differential 2-form on H is admissible if and only if its restriction to the submanifold of lines lying in any fixed plane is equal to zero.

Proof of Proposition 4. Let σ be a 2-form on H

$$\sigma=a\,d\alpha^1\wedge d\alpha^2+b_1d\alpha^1\wedge d\beta^2+b_2d\alpha^2\wedge d\beta^1+c\,d\beta^1\wedge d\beta^2+c_1d\alpha^1\wedge d\beta^1+c_2d\alpha^2\wedge d\beta^2$$

and let $h_0=(\alpha_0,\beta_0)$ be any line. We observe that the manifold M of lines lying in any fixed plane passing through h_0 is given by the equations: $\lambda_1(\alpha^1-\alpha_0^1)+\lambda_2(\alpha^2-\alpha_0^2)=0$, $\lambda_1(\beta^1-\beta_0^1)+\lambda_2(\beta^2-\beta_0^2)=0$. Therefore, the condition that the line $h=(\alpha_0+d\alpha,\ \beta_0+d\beta)$ belong to M has the form $\lambda_1d\alpha^1+\lambda_2d\alpha^2=0$, $\lambda_1d\beta^1+\lambda_2d\beta^2=0$, whence $d\alpha^2=\lambda d\alpha^1$, $d\beta^2=\lambda d\beta^1$ $(\lambda=-\lambda_1/\lambda_2)$. Thus, in the coordinates α^1, β^1 on M the form $\sigma|_M$ has at the point $h_0\in M$ the form

$$\sigma|_M=(\lambda^2c_2+\lambda(b_1+b_2)+c_1)\,d\alpha^1\wedge d\beta^1.$$

Hence, if $c_1=c_2=b_1+b_2=0$, then $\sigma|_M=0$; conversely, if $\sigma|_M=0$ for any M, then, since λ is arbitrary, we have $c_1=c_2=b_1+b_2=0$. Thus, the condition that $\sigma_M|=0$ for any M is equivalent to the condition $c_1=c_2=b_1+b_2=0$, as was required to prove.

Remark. Proposition 3 shows that the concept of an admissible form is invariant under all automorphisms in the manifold of lines, since in the definition of a degenerate pair of vectors only the concept of incidence of points, lines, and planes is used.

3. Admissibility and Self-Duality. We call two lines incident if they have a common point, and we denote by H_h the set of all lines incident with a fixed line $h\in H$. It is obvious that H_h is a three-dimensional surface in H with a singularity at h. We note that under an automorphism of H taking h_1 into h_2, the surface H_{h_1} goes over into H_{h_2}.

In the tangent space T_h to H we consider the set of lines tangent to H_h. The conical surface in T_h formed by these lines we call the cone of incidence and denote it by K_h.

If $h=(\alpha_0^1, \alpha_0^2, \beta_0^1, \beta_0^2)$, then the surface H_h is given by the equation

$$(\alpha^1-\alpha_0^1)(\beta^2-\beta_0^2)-(\alpha^2-\alpha_0^2)(\beta^1-\beta_0^1)=0,$$

while the equation of the cone K_h has the form

$$d\alpha^1 d\beta^2-d\alpha^2 d\beta^1=0.$$

Thus, in the coordinates (α, β) the cone K_h is naturally identified with H_h.

From the equations of the cone K_h it is clear that there are two families Π_1, Π_2 of two-dimensional planes on it:

$$\Pi_1: d\alpha^2=\lambda d\alpha^1, \quad d\beta^2=\lambda d\beta^1;$$
$$\Pi_2: d\beta^1=\lambda d\alpha^1, \quad d\beta^2=\lambda d\alpha^2.$$

Moreover, through each point of the cone distinct from the vertex there passes exactly one plane of each family. These families of planes in T_h have a simple geometric interpretation: the planes of the family Π_1 are tangent to submanifolds of lines lying in one plane of the space $\mathbf{R}^3$ (passing through h); the planes of the family Π_2 are tangent to submanifolds of lines intersecting h in one point.

From what has been said in part 2 (Proposition 4) it follows that a 2-form on H is admissible if and only if it is equal to zero on planes of the family Π_1.

The cone of incidence K_h, up to a constant multiple, determines an indefinite metric in the tangent space T_h, of which it is the isotropic cone. Up to an arbitrary functional multiple in H there is thus defined an indefinite metric ds^2, viz.,

$$ds^2=c(\alpha, \beta)(d\alpha^1 d\beta^2-d\alpha^2 d\beta^1). \tag{3.4}$$

On differential forms on H we introduce the operation $*$ corresponding to this metric ds^2 (see, e.g., [37]). We recall that for this we first introduce an indefinite scalar product (σ_1, σ_2) in the space of differential forms on H. For example, if $\sigma_1=\frac{1}{2}\sum a_{i_1 i_2} d\xi^{i_1}\wedge d\xi^{i_2}$, $\sigma_2=\frac{1}{2}\sum b_{i_1 i_2} d\xi^{i_1}\wedge d\xi^{i_2}$ are two 2-forms on H, where $(\xi^1, \xi^2, \xi^3, \xi^4)=(\alpha^1, \alpha^2, \beta^1, \beta^2)$, then $(\sigma_1, \sigma_2)=\sum g^{i_1 j_1} g^{i_2 j_2} a_{i_1 i_2} b_{j_1 j_2}$, where g^{ij} is the matrix inverse to the matrix g_{ij} of the quadratic form ds^2; if σ_1, σ_2 are forms of different degrees, then $(\sigma_1, \sigma_2)=0$. The differential form $*\sigma_1$ is defined from the condition that

$$\sigma\wedge(*\sigma_1)=(\sigma, \sigma_1)c^2 d\alpha^1\wedge d\alpha^2\wedge d\beta^1\wedge d\beta^2$$

for any differential form σ. In particular, if σ_1 is a 2-form, then $*\sigma_1$ is also a 2-form. We observe that on 2-forms σ the operator $*$ does not depend on the factor c in the metric ds^2 and is hence determined only by the cone of incidence K_h, and is hence invariantly defined.

<u>Proposition 5.</u> The differential 2-form σ is admissible if and only if it is self-dual, i.e., $*\sigma=\sigma$.

<u>Proof.</u> It is obvious that for $c=1$ the matrix g_{ij} of the quadratic form (3.4) coincides with its inverse g^{ij}: $g^{ij}=g_{ij}$, where $g_{14}=g_{41}=-g_{23}=-g_{32}=1$, while the remaining elements are equal to zero. From this we find the scalar products on simple 2-forms: $(d\alpha^1\wedge d\alpha^2, d\beta^1\wedge d\beta^2)=(d\alpha^1\wedge d\beta^1, d\alpha^2\wedge d\beta^2)=-(d\alpha^1\wedge d\beta^2, d\alpha^1\wedge d\beta^2)=-(d\alpha^2\wedge d\beta^1, d\alpha^2\wedge d\beta^1)=1$, while the remaining scalar products are equal to zero. From these equalities it follows that the operation $*$

preserves the forms $d\alpha^1\wedge d\alpha^2$ and $d\beta^1\wedge d\beta^2$, changes the sign of the forms $d\alpha^1\wedge d\beta^1$ and $d\alpha^2\wedge d\beta^2$, while the forms $d\alpha^1\wedge d\beta^2$ and $d\beta^1\wedge d\alpha^2$ go over into one another. From this it is obvious that the condition of admissibility of the form σ is equivalent to the condition $*\sigma=\sigma$.

COROLLARY. A 2-form on H is self-dual if and only if for any $h\in H$ it is equal to zero on planes of the family Π_1 of the cone of incidence K_h.

Remark 1. Similarly, it can be shown that a 2-form σ on H is antiself-dual (i.e., $*\sigma=-\sigma$) if and only if for any $h\in H$ it is equal to zero on planes of the second family Π_2 of the cone K_h.

Remark 2. If on H we consider closed nondegenerate 2-forms, then each of them defines on H the structure of a symplectic manifold, and the condition of admissibility or, equivalently, of self-duality of such a form means that all the planes of the family Π_1 must be Lagrangian planes. Similarly, the condition of antiself-duality means that the planes of the family Π_2 are Lagrangian planes.

Remark 3. In the space H it is possible to introduce in the usual manner the Laplace operator Δ by the formula $\Delta=d\delta+\delta d$, $\delta=-*d*$. We observe that up to a factor this operator coincides with an ultrahyperbolic operator, viz., $\Delta=-4\left(\frac{\partial^2}{\partial\alpha^1\partial\beta^2}-\frac{\partial^2}{\partial\alpha^2\partial\beta^1}\right)$.

4. Regularization of the Form $\mathcal{I}\omega$. We denote by A the manifold of pairs (h, x), where h is a line in $\mathbf{R}^3$, and x is a point of the line h. The manifold A has canonical projections onto the original manifolds $\mathbf{R}^3$ and H, viz., $(h, x)\mapsto x$ and $(h, x)\mapsto h$. In A we introduce two coordinates systems connected with these projections. The first system is (x, α), where $x=(x^1, x^2, x^3)$ is a point of $\mathbf{R}^3$, and $\alpha=(\alpha^1, \alpha^2)$ gives the line passing through the point x. The second system is (α, β, x^3), where (α, β) are the coordinates of a line in H, and x^3 is a coordinate of a point on this line. These coordinate systems are connected by the relations $x_1=\alpha^1x^3+\beta^1$, $x^2=\alpha^3x^3+\beta^2$. The mapping $A\to\mathbf{R}^3$ is conveniently considered in the coordinates (x, α), viz., $(x, \alpha)\mapsto x$, while the mapping $A\to H$ is conveniently considered in the coordinates (α, β, x^3), viz., $(\alpha, \beta, x^3)\mapsto(\alpha, \beta)$.

We now introduce the operator of regularization R. We recall that the mapping $\omega\mapsto\mathcal{I}\omega$ defined in Sec. 1 is carried out in two stages: we first lift the form ω from $\mathbf{R}^3$ to A (i.e., we go over from the coordinates x^1, x^2, x^3 to the "redundant" coordinates $\alpha^1, \alpha^2, \beta^1, \beta^2, x^3$), and we then integrate the form obtained on A over fibers of the bundle $A\to H$ (i.e., on x^3). We shall see that in order to recover the form ω from the form $\sigma=\mathcal{I}\omega$ it is also necessary to proceed in two steps. The first step is the regularization of the form σ, which will now be defined.

Suppose there is given an admissible 2-form on H

$$\sigma=\varphi_2(\alpha, \beta)\,d\alpha^1\wedge d\alpha^2+\varphi_1(\alpha, \beta)(d\alpha^1\wedge d\beta^2-d\alpha^2\wedge d\beta^1)+\varphi_0(\alpha, \beta)\,d\beta^1\wedge d\beta^2.$$

We lift the form σ from H to A and take the exterior product with dx^3. We write the form obtained $\sigma\wedge dx^3$ on A in the coordinates (x, α) on A (for this, in the expression for $\sigma\wedge dx^3$ it is necessary to substitute $\beta^i=x^i-\alpha^ix^3$, $d\beta^i=dx^i-\alpha^idx^3-x^3d\alpha^i$ $(i=1, 2)$). We obtain

$$\sigma\wedge dx^3=\varphi_2 d\alpha^1\wedge d\alpha^2\wedge dx^3+\varphi_1(d\alpha^1\wedge(dx^2-x^3d\alpha^2)-$$
$$d\alpha^2\wedge(dx^1-x^3d\alpha^1))\wedge dx^3+\varphi_0(dx^1-x^3d\alpha_1)\wedge(dx^2-x^3d\alpha^2)\wedge dx^3.$$

Definition. The regularization $R\sigma$ of the form σ is the homogeneous component of the form $\sigma\wedge dx^3$ of bi-degree (3, 0) relative to dx and $d\alpha$; in other words, $R\sigma$ is the 3-form on A obtained from $\sigma\wedge dx^3$ by dropping all terms with $d\alpha$.

From the definition it follows that

$$R\sigma=\varphi_0(\alpha,\beta)\,dx^1\wedge dx^2\wedge dx^3, \tag{3.5}$$

where $\varphi_0(\alpha,\beta)$ is the coefficient of the form σ preceding $d\beta^1\wedge d\beta^2$. In particular, if $\sigma=\mathcal{I}\omega$, where $\omega=f(x)\,dx^1\wedge dx^2\wedge dx^3$, then

$$\varphi_0(\alpha,\beta)=\int_{-\infty}^{+\infty} f(\alpha^1x^3+\beta^1,\ \alpha^2x^3+\beta^2,\ x^3)\,dx^3. \tag{3.6}$$

The operator R is important for the inversion of the transform $\mathcal{I}$. Namely, in order to recover the form ω on the basis of the form $\sigma=\mathcal{I}\omega$, we first construct the form $R\sigma$. Its coefficient φ_0 is related to the coefficient f of the original form ω by relation (3.6). As a result, the problem of inverting the transform $\mathcal{I}$ for forms reduces to the problem of inverting the transform (3.6) for functions, which has already been considered in Chap. I.

We shall establish some properties of the operation R.

Definition. We call a form of the form (3.5) on A harmonic if its coefficient φ_0 is a harmonic function, i.e.,

$$\Delta\varphi_0\equiv\frac{\partial^2\varphi_0}{\partial\alpha^1\partial\beta^2}-\frac{\partial^2\varphi_0}{\partial\alpha^2\partial\beta^1}=0.$$

Proposition 6. If σ is a closed admissible 2-form on H, then $R\sigma$ is a harmonic form on A. Conversely, any harmonic form $\tilde\sigma$ on A can be represented in the form $\tilde\sigma=R\sigma$, where σ is a closed admissible 2-form.

Proof. The first assertion follows immediately from formula (3.5) and the fact that the coefficients of a closed admissible 2-form on H are harmonic functions. Conversely, any harmonic form $\tilde\sigma=\varphi_0(\alpha,\beta)\times dx^1\wedge dx^2\wedge dx^3$ can be represented in the form $\tilde\sigma=R\sigma$, where $\sigma=\varphi_2(\alpha,\beta)\,d\alpha^1\wedge d\alpha^2+\varphi_1(\alpha,\beta)(d\alpha^1\wedge d\beta^2-d\alpha^2\wedge d\beta^1)+\varphi_0(\alpha,\beta)\,d\beta^1\wedge d\beta^2$, and the coefficients φ_2, φ_1 are determined from the equations

$$\frac{\partial\varphi_1}{\partial\beta^i}=\frac{\partial\varphi_0}{\partial\alpha^i},\ \frac{\partial\varphi_2}{\partial\beta^i}=\frac{\partial\varphi_1}{\partial\alpha^i},\ i=1,2; \tag{3.7}$$

the compatibility of these equations follows from the fact that φ_0 is harmonic.

COROLLARY. The differential form $R\mathcal{I}\omega$ is harmonic.

Remark. The condition that the differential form $R\sigma$ is harmonic can be formulated in a somewhat different form. Namely, on forms in A we introduce the following operator $\varkappa$:

$$\varkappa\tilde\sigma=d\alpha^1\wedge\frac{\partial\tilde\sigma}{\partial\beta^1}+d\alpha^2\wedge\frac{\partial\tilde\sigma}{\partial\beta^2}=d\alpha^1\wedge\frac{\partial\tilde\sigma}{\partial x^1}+d\alpha^2\wedge\frac{\partial\tilde\sigma}{\partial x^2}.$$

The next assertion is obvious.

Proposition 7. Harmonicity of the form $R\sigma$ is equivalent to closedness of the form $\varkappa R\sigma$.

In the complex case analogous considerations make it possible on the basis of the form $\varkappa R\mathcal{I}\omega$ to construct the operator inverse to $\mathcal{I}$.

Proposition 8. If σ is a closed admissible 2-form on H and $\sigma\in\Omega^2(H)$, then $R\sigma=0$ implies $\sigma=0$.

Proof. Let $\sigma=\varphi_2 d\alpha^1\wedge d\alpha^2+\varphi_1(d\alpha^1\wedge d\beta^2-d\alpha^2\wedge d\beta^1)+\varphi_0 d\beta^1\wedge d\beta^2$; then $R\sigma=\varphi_0 dx^1\wedge dx^2\wedge dx^3$. If $R\sigma=0$, then $\varphi_0=0$; but then from Eqs. (3.7) and the rapid decrease of the functions φ_i in β it follows that also $\varphi_1=\varphi_2=0$.

5. Description of the Image of the Mapping $\mathcal{J}$. THEOREM 1. A differential 2-form $\sigma\in\Omega^2(H)$ belongs to the image of the mapping $\mathcal{J}:\Omega^3(R^3)\to\Omega^2(H)$ if and only if it is admissible and closed, i.e., $*\sigma=\sigma$ and $d\sigma=0$.

Proof. The necessity of the conditions has already been established. We prove that they are sufficient. Suppose that the differential form σ satisfies the conditions of the theorem. We construct the form $R\sigma=\varphi dx^1\wedge dx^2\wedge dx^3$. Its coefficient $\varphi(\alpha, \beta)$ is a function in the space S on H which satisfies the condition of harmonicity. According to Chap. I, such a function belongs to the image of the transform $\mathcal{J}_1$ defined on $S(R^3)$, i.e., it can be represented in the form

$$\varphi(\alpha, \beta)=\int_{-\infty}^{+\infty} f(\alpha^1 x^3+\beta^1, \alpha^2 x^3+\beta^2, x^3)\,dx^3,$$

where $f\in S(R^3)$. We consider the differential form $\omega=f(x)\,dx^1\wedge dx^2\wedge dx^3$. From equalities (3.2) and (3.3) it follows that $R\mathcal{J}\omega=R\sigma$. Hence, by Proposition 8 $\sigma=\mathcal{J}\omega$.

3. Study of $\mathcal{J}\omega$ for the Case Where ω Is a 2-Form on R^3

Let

$$\omega=f_1(x)\,dx^2\wedge dx^3+f_2(x)\,dx^3\wedge dx^1+f_3(x)\,dx^1\wedge dx^2$$

be any form of $\Omega^2(R^3)$. On computing its integral over lines $\mathcal{J}\omega$ according to the rule set forth in Sec. 1, we obtain

$$\mathcal{J}\omega=a_1 d\alpha^1+a_2 d\alpha^2+b_1 d\beta^1+b_2 d\beta^2, \tag{3.8}$$

where

$$a_1=\alpha^2\psi_3-\psi_2,\quad a_2=-\alpha^1\psi_3+\psi_1,\quad b_1=\alpha^2\varphi_3-\varphi_2,\quad b_2=-\alpha^1\varphi_3+\varphi_1, \tag{3.9}$$

$$\varphi_i(\alpha, \beta)=\int_{-\infty}^{+\infty} f_i(\alpha^1 x^3+\beta^1, \alpha^2 x^3+\beta^2, x^3)\,dx^3, \tag{3.10}$$

$$\psi_i(\alpha, \beta)=\int_{-\infty}^{+\infty} f_i(\alpha^1 x^3+\beta^1, \alpha^2 x^3+\beta^2, x^3)\,x^3 dx^3.$$

We note that $\mathcal{J}\omega\in\Omega^1(H)$.

1. Properties of the Form $\mathcal{J}\omega$. From expressions (3.9) and (3.10) for the coefficients of the differential form $\mathcal{J}\omega$ we immediately obtain the following relations between them:

$$\frac{\partial a_1}{\partial\beta^1}=\frac{\partial b_1}{\partial\alpha^1},\quad \frac{\partial a_2}{\partial\beta^2}=\frac{\partial b_2}{\partial\alpha^2},\quad \frac{\partial a_1}{\partial\beta^2}-\frac{\partial b_1}{\partial\alpha^2}=-\left(\frac{\partial a_2}{\partial\beta^1}-\frac{\partial b_2}{\partial\alpha^1}\right). \tag{3.11}$$

We shall clarify the meaning of these relations.

Proposition 9. Conditions (3.11) are equivalent to the condition of admissibility of the differential $d\sigma$ of the 1-form σ.

The assertion follows from the explicit expression for $d\sigma$:

$$d\sigma=\left(\frac{\partial a_2}{\partial\alpha^1}-\frac{\partial a_1}{\partial\alpha^2}\right)d\alpha^1\wedge d\alpha^2+\left(\frac{\partial b_2}{\partial\alpha^1}-\frac{\partial a_1}{\partial\beta^2}\right)d\alpha^1\wedge d\beta^2+\left(\frac{\partial b_1}{\partial\alpha^2}-\frac{\partial a_2}{\partial\beta^1}\right)d\alpha^2\wedge$$
$$\wedge d\beta^1+\left(\frac{\partial b_2}{\partial\beta^1}-\frac{\partial b_1}{\partial\beta^2}\right)d\beta^1\wedge d\beta^2+\left(\frac{\partial b_1}{\partial\alpha^1}-\frac{\partial a_1}{\partial\beta^1}\right)d\alpha^1\wedge d\beta^1+\left(\frac{\partial b_2}{\partial\alpha^2}-\frac{\partial a_2}{\partial\beta^2}\right)d\alpha^2\wedge d\beta^2.$$

COROLLARY. Conditions (3.11) are equivalent to the condition $*d\sigma = d\sigma$.

Remark. We emphasize that the admissibility of the differentials $d\mathcal{I}\omega$, where ω is a 2-form, follows from the admissibility of the integrals of 3-forms over lines and the fact that the operations $\mathcal{I}$ and d commute.

Definition. We call a differential 1-form σ on H partially closed if the restriction of σ to any submanifold of lines lying in an arbitrary two-dimensional plane is a closed form on this submanifold.

Proposition 10. Conditions (3.11) are equivalent to the condition that the form σ be partially closed.

Indeed, by Proposition 4 the condition of partial closedness of the 1-form σ is equivalent to the condition of admissibility of its differential $d\sigma$ and hence to the conditions (3.11).

Remark. The necessity of the condition of partial closedness of $\mathcal{I}\omega$ can be explained as follows. To compute the restriction of $\mathcal{I}\omega$ to a manifold of lines lying in a plane L, it is possible to restrict ω to L; now $\omega|_L$ is a closed form as a form of maximal degree on L, and hence the restriction of the form $\mathcal{I}\omega$ is also closed (see part 1, Sec. 2).

Proposition 11. From the partial closedness of $\mathcal{I}\omega$ there follow the following equations of third order for the coefficients of the form $\mathcal{I}\omega$:

$$\Delta\left(\frac{\partial a_1}{\partial\alpha^2}-\frac{\partial a_2}{\partial\alpha^1}\right)=0,\quad \Delta\left(\frac{\partial b_1}{\partial\beta^2}-\frac{\partial b_2}{\partial\beta^1}\right)=0, \tag{3.12}$$

$$\Delta\left(\frac{\partial b_1}{\partial\alpha^2}-\frac{\partial a_2}{\partial\beta^1}\right)=0,\quad \Delta\left(\frac{\partial a_1}{\partial\beta^2}-\frac{\partial b_2}{\partial\alpha^1}\right)=0. \tag{3.12'}$$

Proof. Differentiating the third of Eqs. (3.11) with respect to α^1, α^2 and using the first two equations, we obtain $\Delta\left(\frac{\partial a_1}{\partial\alpha^2}-\frac{\partial a_2}{\partial\alpha^1}\right)=0$. We obtain the equation $\Delta\left(\frac{\partial b_1}{\partial\beta^2}-\frac{\partial b_2}{\partial\beta^1}\right)=0$ in a similar way. Further, using the relations (3.11), we find

$$\left(\frac{\partial^2}{\partial\alpha^1\partial\beta^2}-\frac{\partial^2}{\partial\alpha^2\partial\beta^1}\right)\frac{\partial b_1}{\partial\alpha^2}=\frac{\partial^3 b_1}{\partial\alpha^1\partial\alpha^2\partial\beta^2}-\frac{\partial^2}{\partial\alpha^2\partial\beta^1}\left(-\frac{\partial b_2}{\partial\alpha^1}+\frac{\partial a_1}{\partial\beta^2}+\frac{\partial a_2}{\partial\beta^1}\right)=\frac{\partial^3 a_1}{\partial\alpha^2\partial\beta^1\partial\beta^2}+\frac{\partial^3 a_2}{\partial\alpha^1\partial\beta^1\partial\beta^2}-$$
$$-\frac{\partial^3 a_1}{\partial\alpha^2\partial\beta^1\partial\beta^2}-\frac{\partial^3 a_2}{\partial\alpha^2(\partial\beta^1)^2}=\left(\frac{\partial^2}{\partial\alpha^1\partial\beta^2}-\frac{\partial^2}{\partial\alpha^2\partial\beta^1}\right)\frac{\partial a_2}{\partial\beta^1},$$

whence $\Delta\left(\frac{\partial b_1}{\partial\alpha^2}-\frac{\partial a_2}{\partial\beta^1}\right)=0$.

2. Additional Differential Equations for the Coefficients of the Form $\mathcal{I}\omega$. Proposition 1.2. The coefficients of the differential form $\mathcal{I}\omega$ satisfy the following equation of third order:

$$\Delta\left(\frac{\partial a_1}{\partial\beta^2}-\frac{\partial b_1}{\partial\alpha^2}\right)=0. \tag{3.13}$$

Proof. From Eqs. (3.10) it follows that $\frac{\partial\varphi_i}{\partial\beta^j}=\frac{\partial\varphi_i}{\partial\alpha^j}$, $i=1, 2, 3;\ j=1, 2$. Expressions (3.9) for a_1 and b_1 therefore imply that $\frac{\partial a_1}{\partial\beta^2}-\frac{\partial b_1}{\partial\alpha^2}=-\varphi_3$. Equation (3.13) is a consequence of the harmonicity of the function φ_3.

Remark. Equation (3.13) is not a consequence of partial closedness of the form or even of closedness. Indeed, we consider the exact form $\sigma=d\varphi(\alpha, \beta)$, where φ is an arbitrary function on H. For this form condition (3.13) becomes $\Delta^2\varphi=0$, which, of course, need not be satisfied for an arbitrary function φ.

Proposition 13. Under the condition that the form σ be partially closed, Eq. (3.13) is equivalent to each of the following equations:

$$\Delta\left(\frac{\partial a_2}{\partial\beta^1}-\frac{\partial b_2}{\partial\alpha^1}\right)=0,\quad \Delta\left(\frac{\partial a_1}{\partial\beta^2}-\frac{\partial a_2}{\partial\beta^1}\right)=0,\quad \Delta\left(\frac{\partial b_1}{\partial\alpha^2}-\frac{\partial b_2}{\partial\alpha^1}\right)=0. \tag{3.14}$$

Proof. The assertion follows from relations (3.11) and (3.12).

Proposition 14. For a partially closed form σ relation (3.13) is equivalent to the following condition: $\delta d\delta\sigma=0$.

Proof. From the definition of the operation $*$ it follows that

$$*(a_1d\alpha^1+a_2d\alpha^2+b_1d\beta^1+b_2d\beta^2)=-a_1d\alpha^1\wedge d\alpha^2\wedge d\beta^1-a_2d\alpha^1\wedge d\alpha^2\wedge d\beta^2+b_1d\alpha^1\wedge d\beta^1\wedge d\beta^2+b_2d\alpha^2\wedge d\beta^1\wedge d\beta^2.$$

Hence,

$$d*\sigma=\left(\frac{\partial a_1}{\partial\beta^2}-\frac{\partial a_2}{\partial\beta^1}-\frac{\partial b_1}{\partial\alpha^2}+\frac{\partial b_2}{\partial\alpha^1}\right)d\alpha^1\wedge d\alpha^2\wedge d\beta^1\wedge d\beta^2=2\left(\frac{\partial a_1}{\partial\beta^2}-\frac{\partial b_1}{\partial\alpha^2}\right)d\alpha^1\wedge d\alpha^2\wedge d\beta^1\wedge d\beta^2,$$

whence

$$\delta\sigma=-*d*\sigma=-2\left(\frac{\partial a_1}{\partial\beta^2}-\frac{\partial b_1}{\partial\alpha^2}\right).$$

Therefore, since $\Delta=d\delta+\delta d$ and $\delta^2=0$, it follows that

$$\delta d\delta\sigma=\Delta\delta\sigma=-2\Delta\left(\frac{\partial a_1}{\partial\beta^2}-\frac{\partial b_1}{\partial\alpha^2}\right),$$

which proves our assertion.

Remark. Condition (3.13) on the form $\mathcal{I}\omega$ can be clarified as follows. If $\sigma=a_1d\alpha^1+a_2d\alpha^2+b_1d\beta^1+b_2d\beta^2$, then we set $\sigma_\alpha=a_1d\alpha^1+a_2d\alpha^2$, $\sigma_\beta=b_1d\beta^1+b_2d\beta^2$.

LEMMA. For any differential 2-form $\omega\in\Omega^2(\mathbf{R}^3)$ there exists a form $\omega'\in\Omega^2(\mathbf{R}^3)$ such that $(\mathcal{I}\omega')_\beta=(\mathcal{I}\omega)_\alpha$.

Indeed, $\omega'=x^3\omega$ is such a form; see expressions (3.9) and (3.10) for the coefficients of the form $\mathcal{I}\omega$.

Equation (3.13) for the coefficients of the form $\mathcal{I}\omega$ is a consequence of the partial closedness of the differential form $\mathcal{I}\omega'$. Indeed, the partial closedness of $\mathcal{I}\omega'$ implies (Proposition 11) that its coefficients b_1', b_2' satisfy the equation $\Delta\left(\frac{\partial b_1'}{\partial\beta^2}-\frac{\partial b_2'}{\partial\beta^1}\right)=0$. Hence, since $b_i'=a_i$, we obtain $\Delta\left(\frac{\partial a_1}{\partial\beta^2}-\frac{\partial a_2}{\partial\beta^1}\right)=0$, which is equivalent to Eq. (3.13).

We note that actually on $\mathbf{R}^3$ there also exists a 2-form ω'' such that $(\mathcal{I}\omega'')_\alpha=(\mathcal{I}\omega)_\beta$, viz., $\omega''=(x^3)^{-1}\omega$. As we see, here it is necessary to extend the class of differential forms by admitting forms with singularities. However, this fact cannot be reflected by the differential relations which hereby arise.

3. Regularization of $\mathcal{I}\omega$ and Trivial Forms. As in the preceding section, we reduce the problem of inverting the operator $\mathcal{I}$ to the corresponding problem for functions.

We note that since the coefficients a_1, a_2 and the coefficients b_1, b_2 of the form $\mathcal{I}\omega$ are connected by relations (3.11), we are actually given two functions b_1, b_2 of four variables. The problem consists in recovering the form ω on the basis of them, i.e., three unknown functions of three variables. At first glance this problem may seem underdetermined. However, as we shall see, the shortage of functions is compensated by the number of variables.

Intuitively it is natural to call the regularization of the differential form $\mathcal{I}\omega$ a form of the type

$$R\mathcal{I}\omega=\varphi_1(\alpha,\beta)\,dx^2\wedge dx^3+\varphi_2(\alpha,\beta)\,dx^3\wedge dx^1+\varphi_3(\alpha,\beta)\,dx^1\wedge dx^2,$$

where φ_i are expressed in terms of the coefficients f_i of the form ω by formula (3.10), i.e., they are integrals of the functions f_{ii} over lines in $\mathbf{R}^3$. Here, as in the case of 3-forms, this form is harmonic, and, using the results of the integral geometry of functions in $\mathbf{R}^n$ (see Chap. I), the form ω can be recovered from the form $R\mathcal{I}\omega$.

<u>Proposition 15.</u> The coefficients of the form $R\mathcal{I}\omega$ are expressed in terms of the coefficients of the form $\mathcal{I}\omega$ by the formulas

$$\varphi_3=\frac{\partial b_1}{\partial\alpha^2}-\frac{\partial a_1}{\partial\beta^2},\quad \varphi_1=\alpha^1\varphi_3+b_2,\quad \varphi_2=\alpha^2\varphi_3-b_1.$$

The assertion follows immediately from the expressions (3.9) and (3.10) for the coefficients of the form $\mathcal{I}\omega$.

<u>COROLLARY.</u> The differential form $\omega\in\Omega^2(\mathbf{R}^3)$ can be uniquely recovered from its image $\mathcal{I}\omega$.

We shall now give another definition of the operator R. We first recall how the operator $\omega\mapsto\mathcal{I}\omega$ is defined for forms of degree 2. The form ω is lifted to A, written in the coordinates (α,β,x^3), and the terms of degree 0 in dx^3 are dropped. The form obtained is integrated on x^3 for fixed $\alpha,\beta,d\alpha,d\beta$; as a result, we obtain $\mathcal{I}\omega$. We emphasize that while for forms of degree 3 in the forms lifted to A there were no components of degree 0 in dx^3, in the present situation there may, in general, be such components.

<u>Definition 6.</u> We call a differential form of degree 2 on A trivial if it has degree zero in $d\alpha$ in the coordinates (x,α) and degree zero in dx^3 in the coordinates (α,β,x^3).

It is clear that by dropping the terms of degree 0 in dx^3 in the lift of the form ω, we thus neglect trivial terms. (We have included in the definition the condition that the form have degree 0 in $d\alpha$ in the coordinates (x,α), since the form ω itself possesses this property.)

<u>Proposition 16.</u> Any trivial form has the form

$$\psi(dx^1-\alpha^1dx^3)\wedge(dx^2-\alpha^2dx^3)=\psi(dx^1\wedge dx^2-\alpha^1dx^3\wedge dx^2-\alpha^2dx^1\wedge dx^2).$$

<u>Proof.</u> It is obvious that any differential 2-form on A which in the coordinates (x,α) has degree 0 relative to $d\alpha$ can be represented as a sum of monomials of the following forms:

$$\psi(dx^1-\alpha^1dx^3)\wedge(dx^2-\alpha^2dx^3),$$
$$\psi(dx^1-\alpha^1dx^3)\wedge dx^3\equiv\psi dx^1\wedge dx^3,$$
$$\psi(dx^2-\alpha^2dx^3)\wedge dx^3\equiv\psi dx^2\wedge dx^3.$$

Since $dx^i-\alpha^i dx^3=x^3d\alpha^i+d\beta^i$ $(i=1,2)$, in the coordinates (α,β,x^3) only the monomials of the first type have degree 0 relative to dx^3.

We now proceed to the definition of the operator R. Suppose that there is given the 1-form on H

$$\sigma=a_1d\alpha^1+a_2d\alpha^2+b_1d\beta^1+b_2d\beta^2.$$

We first perform on σ the same operations as in the case $r=3$ (see Sec. 2), viz., we lift σ from H to the manifold A, we take the exterior product (from the right) with dx^3, we write the form $\sigma\wedge dx^3$ obtained in the coordinates (x,α) and we drop terms containing $d\alpha^1$, $d\alpha^2$. We denote the 2-form on A obtained by $(\sigma\wedge dx^3)_0$. It is not hard to see that

$$(\sigma\wedge dx^3)_0=b_1dx^1\wedge dx^3+b_2dx^2\wedge dx^3.$$

We define the operation R in correspondence with the following proposition.

Proposition 17. Let σ be a partially closed 1-form on H which satisfies the condition (3.13). Then there exists a unique differential 2-form $R\sigma$ on A which satisfies the following conditions: 1) $R\sigma$ is a harmonic form, i.e., $\Delta R\sigma=0$; 2) the difference $R\sigma-(\sigma\wedge dx^3)_0$ is a trivial form.

Proof. We consider the differential form

$$R\sigma=\varphi_1(\alpha,\beta)\,dx^2\wedge dx^3+\varphi_2(\alpha,\beta)\,dx^3\wedge dx^1+\varphi_3(\alpha,\beta)\,dx^1\wedge dx^2,$$

where

$$\varphi_3=\frac{\partial b_1}{\partial\alpha^2}-\frac{\partial a_1}{\partial\beta^2}=\frac{\partial a_2}{\partial\beta^1}-\frac{\partial b_2}{\partial\alpha^1},\quad \varphi_1=\alpha^1\varphi_3+b_2,\quad \varphi_2=\alpha^2\varphi_3-b_1.$$

We shall show that this form satisfies the conditions of the proposition. First of all, it is harmonic. Indeed, $\Delta\varphi_3=0$, by condition (3.13). This implies that

$$\Delta\varphi_1=\Delta b_2+\frac{\partial\varphi_3}{\partial\beta^2}=\frac{\partial^2 b_2}{\partial\alpha^1\partial\beta^2}-\frac{\partial^2 b_2}{\partial\alpha^2\partial\beta^1}+\frac{\partial^2 a_2}{\partial\beta^1\partial\beta^2}-\frac{\partial^2 b_2}{\partial\alpha^1\partial\beta^2}=\frac{\partial}{\partial\beta^1}\left(\frac{\partial a_2}{\partial\beta^2}-\frac{\partial\beta_2}{\partial\alpha^2}\right)=0.$$

It can be verified similarly that $\Delta\varphi_2=0$. Further, substituting into $R\sigma$ the expressions for φ_1 and φ_2, we obtain

$$R\sigma=b_2dx^2\wedge dx^3-b_1dx^3\wedge dx^1+\varphi_3(dx^1\wedge dx^2+\alpha^1dx^2\wedge dx^3+\alpha^2dx^3\wedge dx^1).$$

Hence,

$$R\sigma-(\sigma\wedge dx^3)_0=\varphi_3(dx^1-\alpha^1dx^3)\wedge(dx^2-\alpha^2dx^3),$$

i.e., this difference is the trivial form. The uniqueness of $R\sigma$ follows from the conditions of rapid decay imposed on the coefficients of the forms considered.

Proposition 18. Any harmonic 2-form on A of the form

$$\tilde\sigma=\varphi_1(\alpha,\beta)\,dx^2\wedge dx^3+\varphi_2(\alpha,\beta)\,dx^3\wedge dx^1+\varphi_3(\alpha,\beta)\,dx^1\wedge dx^2$$

can be represented in the form $\tilde\sigma=R\sigma$, where σ is a partially closed form on H satisfying the condition (3.13).

Proof. We consider the differential form

$$\sigma=a_1d\alpha^1+a_2d\alpha^2+b_1d\beta^1+b_2d\beta^2,$$

where $b_1=\alpha^2\varphi_3-\varphi_2$, $b_2=-\alpha^1\varphi_3+\varphi_1$, and the coefficients a_1, a_2 are determined from Eqs. (3.11). It follows from the definition that σ is a partially closed form. The harmonicity of $\tilde\sigma$ implies that $\Delta\varphi_3=\Delta(b_2+\alpha^1\varphi_3)=\Delta(-b_1+\alpha^2\varphi_3)=0$, whence $-\frac{\partial\varphi_3}{\partial\beta^1}=\Delta b_1$, $-\frac{\partial\varphi_3}{\partial\beta^2}=\Delta b_2$. This implies that $\Delta\left(\frac{\partial b_1}{\partial\alpha^2}-\frac{\partial b_2}{\partial\alpha^1}\right)=0$; hence σ satisfies the Eq. (3.13). It remains to note that $\tilde\sigma-(\sigma\wedge dx^3)_0=\varphi_3(dx^1-\alpha^1dx^3)\wedge(dx^2-\alpha^2dx^3)$, i.e., this difference is a trivial form.

Proposition 19. The mapping $\sigma\mapsto R\sigma$ defined on partially closed forms σ satisfying condition (3.13) is injective.

Proof. Suppose that $R\sigma=(\sigma\wedge dx^3)_0+\tilde\sigma_1=0$. Then $(\sigma\wedge dx^3)_0=b_1dx^1\wedge dx^3+b_2dx^2\wedge dx^3$ is a trivial form, which is possible only in the case $b_1=b_2=0$. From the conditions of partial closedness (3.11) it follows that then also $a_1=a_2=0$, i.e., $\sigma=0$.

4. Description of the Image of the Mapping $\mathcal{I}$. THEOREM 2. The differential 1-form $\sigma\in\Omega^1(H)$ belongs to the image of the mapping $\mathcal{I}:\Omega^2(\mathbb{R}^3)\to\Omega^1(H)$ if and only if it is partially closed and its codifferential is harmonic, i.e., the relations $*d\sigma=d\sigma$ and $\delta d\delta\sigma=0$ are satisfied.

Proof. The necessity of the conditions was established earlier. We shall prove that they are sufficient. Let the differential form σ satisfy the conditions of the theorem. In correspondence with Proposition 17 we construct the harmonic form $R\sigma=\varphi_1 dx^2\wedge dx^3+\varphi_2 dx^3\wedge dx^1+\varphi_3 dx^1\wedge dx^2$. Its coefficients φ_i are harmonic functions in the space S on H. Hence, they can be represented as integrals over lines of functions in the space $S(\mathbf{R}^3)$ (see part 5, Sec. 2):

$$\varphi_i(\alpha,\beta)=\int_{-\infty}^{+\infty} f_i(\alpha^1 x^3+\beta^1, \alpha^2 x^3+\beta^2, x^3)\,dx^3, \quad i=1,2,3.$$

We consider the differential form

$$\omega=f_1(x)\,dx^2\wedge dx^3+f_2(x)\,dx^3\wedge dx^1+f_3(x)\,dx^1\wedge dx^2.$$

From the definition of the operations $\mathcal{I}$ and R it follows that $R\mathcal{I}\omega=R\sigma$. Hence, by Proposition 19 it follows that $\sigma=\mathcal{I}\omega$.

COROLLARY. If $\sigma\in\Omega^1(H)$, $*d\sigma=d\sigma$ and $d*\sigma=0$, then σ belongs to the image of the mapping $\mathcal{I}$.

5. A Remark Regarding the Conditions of Partial Closedness. The added condition $\delta d\delta\sigma=0$ imposed on the form $\sigma=\mathcal{I}\omega$ has still another important interpretation.

Definition 7. Let M be a manifold, and let T_pM be the tangent space at the point p. An (odd) r-density on M is an infinitely differentiable mapping which assigns to each point $p\in M$ and each r-dimensional subspace $L\subset T_pM$ a skew r-form $\Psi(p;\xi_1,\dots,\xi_r)$ on L.

In particular, any differential r-form on M is an r-density. The converse is, in general, not true with the exception of the case $\dim M=r$.

The concept of the integral of a 2-density over an oriented line in $\mathbf{R}^3$ is defined according to the same scheme as the concept of the integral of a 2-form. As the result of the integration a 1-density on the manifold of lines is obtained.

It is not hard to see that any partially closed differential 1-form σ on the manifold of lines is the integral over lines of some 2-density on $\mathbf{R}^3$. Indeed, let L be any plane in $\mathbf{R}^3$. The restriction $\sigma|_L$ of a form σ to the submanifold of lines lying in L is a closed 1-form by hypothesis. Hence, in L there exists a well defined differential 2-form $\omega(x;\xi_1,\xi_2)$, whose integral over lines in L is equal to $\sigma|_L$. Since L is arbitrary, the function $\omega(x;\xi_1,\xi_2)$ is defined for all $x\in\mathbf{R}^3$ and $\xi_1,\xi_2\in T_x$ and is a 2-density. The added condition $\delta d\delta\sigma=0$ imposed on the partially closed form σ is thus equivalent to the condition that a 2-density on $\mathbf{R}^3$ which yields the form σ on integration over lines is a differential form.

6. Integration of 1-Forms on $\mathbf{R}^3$. Let $\omega\in\Omega^1(\mathbf{R}^3)$:

$$\omega=f_1(x)\,dx^1+f_2(x)\,dx^2+f_3(x)\,dx^3.$$

Integrating ω over lines in $\mathbf{R}^3$, we obtain a function φ in the space S on the manifold of lines. In contrast to the case of 2-forms and 3-forms, the mapping $\mathcal{I}:\omega\mapsto\varphi$ has a nonzero kernel, since exact forms map into zero under this mapping.

Proposition 20. $\operatorname{Ker}\mathcal{I}$ coincides with the space of all exact 1-forms in $\Omega^1(\mathbf{R}^3)$.

Proof. Any exact 1-form ω on $\mathbf{R}^3$ belongs to $\operatorname{Ker}\mathcal{I}$. Conversely, let $\omega\in\operatorname{Ker}\mathcal{I}$, i.e., $\mathcal{I}\omega=0$. Then $\mathcal{I}d\omega=d\mathcal{I}\omega=0$, and since the mapping $\mathcal{I}$ is a monomorphism on 2-forms, it follows that $d\omega=0$. Closedness of the 1-form ω implies its exactness.

Proposition 21. A function φ in the space S on the manifold of lines in $\mathbf{R}^3$ belongs to Im $\mathcal{I}$ if and only if its differential $d\varphi$ is an integral along lines of some 2-form in $\Omega^2(\mathbf{R}^3)$.

Proof. Let $\varphi=\mathcal{I}\omega_1$; then $d\varphi=d\mathcal{I}\omega_1=\mathcal{I}d\omega_1$. Conversely, let $d\varphi=\mathcal{I}\omega_2$; then $\mathcal{I}d\omega_2=d\mathcal{I}\omega_2=0$, and since the mapping $\mathcal{I}$ is a monomorphism on 3-forms, it follows that $d\omega_2=0$. Since any closed 2-form in $\Omega^2(\mathbf{R}^3)$ is an exact form, it follows that $\omega_2=d\omega_1$, where $\omega_1\in\Omega^1(\mathbf{R}^3)$. Thus, $d\varphi=\mathcal{I}d\omega_1$, whence $d(\varphi-\mathcal{I}\omega_1)=0$; hence, $\varphi=\mathcal{I}\omega_1$.

COROLLARY. A function φ in the space S on the manifold of lines is the integral over lines of some 1-form in $\Omega^1(\mathbf{R}^3)$ if and only if it satisfies the equation

$$\Delta^2\varphi=0\quad\left(\Delta=\frac{\partial^2}{\partial\alpha^1\partial\beta^2}-\frac{\partial^2}{\partial\alpha^2\partial\beta^1}\right).$$

Indeed, it suffices to apply Theorem 2 to the form $d\varphi$.

4. Integration of Differential Forms on the Manifold of Planes in $\mathbf{R}^3$

We have shown that the problem of integrating 3-forms on $\mathbf{R}^3$ over lines leads to closed self-dual forms on the manifold of lines. It is natural to expect that there is a problem of integral geometry connected with anti-self-dual forms. This problem we consider in the present section. We shall consider differential forms defined on the manifold $(\mathbf{R}^3)'$ of all planes in $\mathbf{R}^3$. Since there is an incidence relation between elements of $(\mathbf{R}^3)'$ and H (a line lies in a plane), we can integrate differential forms on $(\mathbf{R}^3)'$ over sets of planes incident with an arbitrary fixed line. It is found that in the integration of 3-forms on $(\mathbf{R}^3)'$ closed anti-self-dual forms are obtained. We proceed to precise definitions in coordinate form.

We adopt the convention of defining planes in $\mathbf{R}^3$ by equations solved for x^2:

$$x^2=\xi_1x^3-\xi_3x^1+\xi_2.$$

The coefficients (ξ_1,ξ_2,ξ_3) of these equations we adopt as coordinates in $(\mathbf{R}^3)'$.

We choose any line $h\in H$ with coordinates $(\alpha^1,\alpha^2,\beta^1,\beta^2)$:

$$x^i=\alpha^ix^3+\beta^i,\quad i=1,2.$$

The sheaf of planes M_h incident with h is then given by the following equations:

$$\xi_1=\alpha^1\xi_3+\alpha^2,\quad \xi_2=\beta^1\xi_3+\beta^2. \tag{3.15}$$

If the space of planes $(\mathbf{R}^3_x)'$ is interpreted as the point space $\mathbf{R}^3_\xi$ with coordinates (ξ_1,ξ_2,ξ_3), then Eqs. (3.15) can be written in the form

$$\xi_1=\tilde\alpha^1\xi_3+\tilde\beta^1,\quad \xi_2=\tilde\alpha^2\xi_3+\tilde\beta^2,$$

where

$$\tilde\alpha^1=\alpha^1,\ \tilde\alpha^2=\beta^1,\ \tilde\beta^1=\alpha^2,\ \tilde\beta^2=\beta^2. \tag{3.16}$$

Thus, elements of the space H can be interpreted both as lines in the space $\mathbf{R}^3_x$ and as lines in the space $\mathbf{R}^3_\xi\cong(\mathbf{R}^3_x)'$. In correspondence with this, on H there are two systems of coordinates (α,β) and $(\tilde\alpha,\tilde\beta)$ connected by the relation (3.16).

In the coordinates $(\tilde\alpha,\tilde\beta)$ the equation of the cone of incidence has the same form as in the coordinates (α,β), while the equations of the family of planes Π_1 go over into the equations of the family Π_2. Therefore, in the system of coordinates $(\tilde\alpha,\tilde\beta)$ connected with the interpretation of the elements of H as lines in $\mathbf{R}^3_\xi$, it is necessary to denote the family Π_1 by $\tilde\Pi_2$ and the family Π_2 by $\tilde\Pi_1$.

The system of coordinates $(\tilde{\alpha}, \tilde{\beta})$ defines an orientation on H opposite to the orientation connected with the coordinate system (α, β), while the volume element $d\tilde{\alpha}^1\wedge d\tilde{\alpha}^2\wedge d\tilde{\beta}^1\wedge d\tilde{\beta}^2$ differs in sign from the volume element $d\alpha^1\wedge d\alpha^2\wedge d\beta^1\wedge d\beta^2$. Therefore, the operation $*$ connected with the coordinates $(\tilde{\alpha}, \tilde{\beta})$ differs in sign from the operation $*$ introduced earlier, and hence self-duality relative to (α, β) goes over into anti-self-duality relative to $(\tilde{\alpha}, \tilde{\beta})$. This is consistent with the relations $\tilde{\Pi}_1=\Pi_2$ and $\tilde{\Pi}_2=\Pi_1$. We shall not introduce a new notation for the operation $*$ connected with the new interpretation of the elements of H, but write it as $*$, where $*$ is the operation on forms on H introduced earlier.

The operation $\mathcal{I}'$ of integrating forms on $(\mathbf{R}^3)'$ over sheaves of planes could be defined directly; however, it is simpler to interpret sheaves of planes in $\mathbf{R}^3_x$ as lines in $\mathbf{R}^3_\xi\cong(\mathbf{R}^3_x)'$ and to use the definition given earlier of the integral of a differential form over lines. Here all assertions presented earlier for the form $\mathcal{I}'\omega$, written in the coordinates $(\tilde{\alpha}, \tilde{\beta})$, are valid. After this it is possible to rephrase all assertions in the coordinates (α, β). As a result, we obtain the following assertions.

Proposition 22. A differential 2-form $\sigma\in\Omega^2(H)$ can be represented in the form $\sigma=\mathcal{I}'\omega$, $\omega\in\Omega^3((\mathbf{R}^3)')$, if and only if $d\sigma=0$ and $*\sigma=-\sigma$.

COROLLARY. Any closed 2-form $\sigma\in\Omega^2(H)$ can be represented uniquely in the form $\sigma=\mathcal{I}\omega+\mathcal{I}'\omega'$, where $\omega\in\Omega^3(\mathbf{R}^3)$, $\omega'\in\Omega^3((\mathbf{R}^3)')$.

Proposition 23. A differential 1-form $\sigma\in\Omega^1(H)$ can be represented in the form $\sigma=\mathcal{I}'\omega$, $\omega\in\Omega^2((\mathbf{R}^3)')$, if and only if $*d\sigma=-d\sigma$ and $\delta d\delta\sigma=0$.

Proposition 24. A function φ in the space S on H can be represented in the form $\varphi=\mathcal{I}'\omega$, $\omega\in\Omega^1((\mathbf{R}^3)')$, if and only if $\Delta^2\varphi=0$.

Remark. Instead of going over from the coordinates (α, β) to the coordinates $(\tilde{\alpha}, \tilde{\beta})$, it is possible to consider the involution on H induced by any affine correspondence between $\mathbf{R}^3$ and $(\mathbf{R}^3)'$. Under this correspondence, to lines in $\mathbf{R}^3$ there correspond sheaves of planes, the incidence relation is preserved, and the families of planes Π_1 and Π_2 change places. In particular, if x and ξ are the coordinates in $\mathbf{R}^3$ and $(\mathbf{R}^3)'$ introduced above and $x\mapsto\xi=x$, then lines with coordinates (α, β) go over into lines with coordinates $(\tilde{\alpha}, \tilde{\beta})$ defined by Eqs. (3.16).

5. Differential Forms on Projective Space and on the Manifold of Lines in $\mathbf{RP}^3$

In Secs. 5-8 the results on integrating differential forms on $\mathbf{R}^3$ are carried over to projective space $\mathbf{RP}^3$. Passage from affine space to projective space has two advantages. First of all, since the space $\mathbf{RP}^3$ itself and the Grassmann manifold G of oriented lines in $\mathbf{RP}^3$ are compact there are natural spaces of differential forms on them — forms with infinitely differentiable coefficients. This eliminates the necessity of investigating the behavior of forms at infinity. Further, all the formulas that arise in the projective case, although they are more involved, are characterized by greater invariance.

In this section we present some preliminary facts.

1. Coordinates in $\mathbf{RP}^3$ and G. Elements of $\mathbf{RP}^3$ we write in homogeneous coordinates $x=(x^1, x^2, x^3, x^4)$, i.e., as points in $\mathbf{R}^4\setminus 0$; two points $x, x'\in\mathbf{R}^4\setminus 0$ give the same element of $\mathbf{RP}^3$

if and only if $x'=\lambda x$. This means that we introduce the principal bundle $\mathbf{R}^4\setminus 0\to\mathbf{RP}^3$ with structural group $\mathbf{R}^\times$.

We introduce coordinates on the manifold G of oriented lines in $\mathbf{RP}^3$ or, equivalently, on the manifold of oriented two-dimensional subspaces of $\mathbf{R}^4$. We denote by E the manifold of 2-frames $z=(u,v)$ in $\mathbf{R}^4$ (a Stiefel manifold), i.e., the manifold of matrices $z=\begin{pmatrix}u^1 & u^2 & u^3 & u^4\\ v^1 & v^2 & v^3 & v^4\end{pmatrix}$ of rank 2. To each $z=(u,v)$ we assign the oriented subspace $h\in G$ spanned by u and v, and we obtain the mapping $E\to G$. It is obvious that two matrices z and z' define the same element $h\in G$ if and only if $z'=gz$, $g\in GL_+(2,\mathbf{R})$. Thus, $E\to G$ is a principal bundle with structural group $GL_+(2,\mathbf{R})$. The elements of any matrix z in the preimage of $h\in G$ we call the homogeneous coordinates of h.

Another classical method of defining elements in G is provided by Plücker coordinates. We recall that the Plücker coordinates of a line $h\in G$ form a system of numbers $p_{ij}=u^iv^j-u^jv^i$, where u^i, v^j are the homogeneous coordinates of the line. Since $p_{ij}=-p_{ji}$, among the numbers p_{ij} there are only 6 essential coordinates. We present two elementary assertions regarding Plücker coordinates.

1) The Plücker coordinates are connected by the relation

$$p_{12}p_{34}-p_{13}p_{24}+p_{14}p_{23}=0. \tag{3.17}$$

Conversely, if the numbers p_{ij}, $i<j$, satisfy relation (3.17), then they are the Plücker coordinates of some line $h\in G$. 2) Two elements of G with Plücker coordinates p_{ij} and p'_{ij} coincide if and only if $p'_{ij}=\lambda p_{ij}$, $\lambda>0$.

We denote by K the cone in the space $\mathbf{R}^6$ with coordinates p_{ij}, $i<j$, given by Eq. (3.17). From 1) and 2) it follows that $K\setminus 0$ is a principal bundle over G with structural group $(\mathbf{R}_+)^\times$ (the multiplicative group of positive numbers).

2. Geometry on the Grassman manifold G. a) The action of the group $GL(4,\mathbf{R})$. The manifold of lines G, just as the projective space $\mathbf{RP}^3$, is a homogeneous space of the group $GL(4,\mathbf{R})$. In homogeneous coordinates z on G the action of an element $g\in GL(4,\mathbf{R})$ reduces to the multiplication of z on the right by g: $z\mapsto zg$. In Plücker coordinates the transformation g is given by the following formula: $p'_{ij}=\sum_{k,l}p_{kl}g_{ki}g_{lj}$.

b) Involution in G. Any projective mapping $x\mapsto\xi_x$ of the space $\mathbf{RP}^3$ to the dual space $(\mathbf{RP}^3)'$ induces an involution in the manifold of lines G. Namely, to each line $h\in G$ there corresponds under this involution a line h' lying on the intersection of the planes ξ_x, where x runs through the points of the original line. In particular, we consider the mapping $\mathbf{RP}^3\to(\mathbf{RP}^3)'$ assigning to a line with homogeneous coordinates $x=(x^1,x^2,x^3,x^4)$ the two-dimensional plane $\xi_x=x$, i.e., the plane with the same homogeneous coordinates (x^1,x^2,x^3,x^4). The corresponding involution in G is then given in Plücker coordinates on G as follows: $(p_{ij})\mapsto(q_{ij})$, where q_{ij} $(i\neq j)$ are defined from the relation $q_{ij}=p_{kl}$, where (i,j,k,l) is an even permutation of the indices $(1,2,3,4)$. In homogeneous coordinates on G this involution assigns to an element with homogeneous coordinates z the element with homogeneous coordinates w, where z and w are connected by the relation

$$zw'=0,\quad \det\begin{pmatrix}z\\ w\end{pmatrix}>0.$$

It is easy to verify that this relation uniquely determines a (2×4)-matrix w of rank 2 up to left multiplication by an element $g\in GL_+(2,\mathbf{R})$.

c) The cone of incidence. In Sec. 2, part 3 we introduced the manifold $G_h\subset G$ of lines incident with $h\in G$ and the cone of incidence $K_h\subset T_h(G)$ tangent to it. We shall write the equation of these manifolds in homogeneous coordinates. We note, first of all, that the condition of incidence of two lines with homogeneous coordinates z', z'' has the form $\det\binom{z'}{z''}=0$. By expanding this determinant with respect to the first two rows, we obtain the incidence condition in Plücker coordinates: $\sum p'_{ij}p''_{kl}=0$, where the sum is taken over all even permutations of the indices 1, 2, 3, 4. This implies that the equation of the submanifold G_h and the equation of the cone of incidence in homogeneous and Plücker coordinates have the respective forms:

$$G_h\colon\ \det\binom{z_0}{z}=0;\quad \sum p^0_{ij}p_{kl}=0,$$

where the sum is taken over even permutations (i, j, k, l), and z_0, (p^0_{ij}) are the coordinates of h;

$$K_h\colon\ \det\binom{z_0}{dz}=0;\quad \sum dp_{ij}dp_{kl}=0,$$

where the differentials dp_{ij} are connected by the relation $\sum p^0_{ij}dp_{kl}=0$.

We recall that on the cone of incidence K_h there are two families of planes: 1) the family Π_1 of planes tangent to the manifolds of lines belonging to a fixed plane in $\mathbf{RP}^3$ containing h and 2) the family Π_2 of planes tangent to the manifolds of lines passing through a fixed point $x\in h$. It is obvious that under the involution the families Π_1 and Π_2 change roles.

LEMMA. The planes of the family Π_1 are given in homogeneous coordinates $z=(u, v)$ by the equations

$$(\xi, du)=0,\quad (\xi, dv)=0,$$

where $\xi\in(\mathbf{R}^4)'\setminus 0$ satisfies the relations $(\xi, u_0)=0$, $(\xi, v_0)=0$.

Indeed, it suffices to note that any plane in $\mathbf{RP}^3$ is given by a point $\xi\in(\mathbf{R}^4)'\setminus 0$, while the condition that a line $h(u, v)$ belong to this plane has the form $(\xi, u)=0$, $(\xi, v)=0$.

COROLLARY. In Plücker coordinates the equations of a plane of the family Π_1 have the form

$$\sum_i \xi_i dp_{ij}=0,\quad j=1, 2, 3, 4,$$

where ξ satisfies the relations $\sum_i \xi_i p^0_{ij}=0$, $j=1, 2, 3, 4$.

The planes of the second family Π_2 are given by analogous equations in dual coordinates. In particular, in Plücker coordinates their equations are

$$\sum_i x^i dq_{ij}=0,\ \text{where} \sum_i x^i q^0_{ij}=0\quad (j=1, 2, 3, 4).$$

3. Differential Forms in Homogeneous Coordinates on $\mathbf{RP}^3$ and G. We write differential forms on $\mathbf{RP}^3$ in homogeneous coordinates $x=(x^1, x^2, x^3, x^4)$, i.e., in the lift to $\mathbf{R}^4\setminus 0$. For example, any differential 2-form on $\mathbf{R}^4$ has the form $\omega=\sum_{i,j=1}^4 f_{ij}(x)dx^i\wedge dx^j$, where the coefficients f_{ij} are antisymmetric in i and j. In order to find a general expression for such

forms we use the following criterion. In order that a differential form on $\mathbf{R}^4\setminus 0$ be the lift of a form on $\mathbf{RP}^3$, it is necessary and sufficient that it be preserved under the transformation $x\to\lambda x$, $dx\to\lambda dx+x d\lambda$ (a general criterion for dropping forms defined on principal bundles can be found in Sec. 1, Chap. IV). On the basis of this criterion it is easy to find the general form of differential r-forms on $\mathbf{RP}^3$, written in homogeneous coordinates, viz.:

1) The case $r=3$. $\omega=f(x)(x^1dx^2\wedge dx^3\wedge dx^4 - x^2dx^1\wedge dx^3\wedge dx^4+x^3dx^1\wedge dx^2\wedge dx^4-x^4dx^1\wedge dx^2\wedge dx^3)$, where f satisfies the condition $f(\lambda x)=\lambda^{-4}f(x)$.

2) The case $r=2$. $\omega=\sum_{i,j=1}^{4} f_{ij}(x)dx^i\wedge dx^j$, where the functions f_{ij} are antisymmetric in i and j and satisfy the relations $f_{ij}(\lambda x)=\lambda^{-2}f_{ij}(x)$; $\sum_i f_{ij}(x)x^i=0$, $j=1, 2, 3, 4$.

3) The case $r=1$. $\omega=\sum_{i=1}^{4} f_i(x)dx^i$, where the functions f_i satisfy the relations $f_i(\lambda x)=\lambda^{-1}f_i(x)$, $\sum_i f_i(x)x^i=0$.

We usually write differential forms on G in homogeneous coordinates $z=(u, v)$, i.e., in the lift to the manifold E of 2-frames in $\mathbf{R}^4$. For example, any 1-form on E has the form

$$\sigma=\sum_{i=1}^{4}(a_i(u, v)\,du^i+b_i(u, v)\,dv^i).$$

These forms on E are not arbitrary, since they drop to G. In order to find their general form, it is necessary to use the following criterion. In order that a form σ on E be the lift of a form on G, it is necessary and sufficient that it be invariant under the group $GL_+(2,\mathbf{R})$ and be preserved under the substitutions $du\to du+ud\alpha+vd\gamma$, $dv\to dv+ud\beta+vd\delta$.

By applying this criterion, we obtain conditions for dropping a form from E onto G in the form of relations on its coefficients. These relations will be presented later separately for forms of different degrees.

Remark. Any transformation of the group $GL(2, \mathbf{R})$ can be represented as the composition of the following transformations: $u\to\lambda u$, $v\to\mu v$ $(\lambda\mu>0)$; $u\to u+\lambda v$, $v\to v$ and $u\to u$, $v\to v+\lambda u$. Therefore, if a form is invariant under these transformations, then it is invariant under the entire group $GL_+(2, \mathbf{R})$. Similarly, it is possible to use the fact that any transformation of $GL_+(2, \mathbf{R})$ can be represented as the composition of the following transformations: $u\to\lambda u$, $v\to\mu v$ $(\lambda\mu>0)$; $u\to u$, $v\to v+\lambda u$ and $u\to x$, $v\to -u$.

4. Connection between the Affine and Projective Problems. In order to relate the problems of integral geometry for affine and projective space, we prescribe an affinization of $\mathbf{RP}^3$, i.e., we choose an "infinitely distant" two-dimensional plane $L\subset\mathbf{RP}^3$. To be specific, suppose that L is given by the equation $x^4=0$. Then in the affine space $\mathbf{R}^3=\mathbf{RP}^3\setminus L$ it is possible to go over to the inhomogeneous coordinates $(x^1, x^2, x^3, x^4)\mapsto(x^1/x^4,\ x^2/x^4,\ x^3/x^4)$. Each differential form on $\mathbf{R}^3$ given in inhomogeneous coordinates can be rewritten in homogeneous coordinates by using the expression for the inhomogeneous coordinates in terms of homogeneous coordinates. The differential form on $\mathbf{R}^4\setminus 0$ so obtained will satisfy the condition for dropping onto $\mathbf{RP}^3$, while if the original form satisfies natural conditions of decay at infinity, then the form obtained $\mathbf{R}^4\setminus 0$ will be smooth. Thus, any smooth differential form on $\mathbf{RP}^3$ can be considered as a form on the affine space $\mathbf{R}^3=\mathbf{RP}^3\setminus L$ which satisfies certain auxiliary conditions at infinity.

We proceed now to the manifold of lines G. We denote by H_0 the submanifold of lines with homogeneous coordinates

$$z=\begin{pmatrix} u_1 & u_2 & u_3 & u_4 \\ v_1 & v_2 & v_3 & v_4 \end{pmatrix},$$

where

$$\det\begin{pmatrix} u_3 & u_4 \\ v_3 & v_4 \end{pmatrix}>0.$$

Using the fact that the matrices z and gz give the same line, we introduce in H_0 the inhomogeneous coordinates $(\alpha^1, \alpha^2, \beta^1, \beta^2)$ defined from the relation

$$\begin{pmatrix} \alpha^1 & \alpha^2 & 1 & 0 \\ \beta^1 & \beta^2 & 0 & 1 \end{pmatrix}=\begin{pmatrix} u^3 & u^4 \\ v^3 & v^4 \end{pmatrix}^{-1}\begin{pmatrix} u^1 & u^2 & u^3 & u^4 \\ v^1 & v^2 & v^3 & v^4 \end{pmatrix}.$$

Since each line in H_0 with coordinates $(\alpha^1, \alpha^2, \beta^1, \beta^2)$ is given by the equations $x^1=\alpha^1 t^1+\beta^1 t^2$, $x^2=\alpha^2 t^1+\beta^2 t^2$, $x^3=t^1, x^4=t^2$, it follows that in inhomogeneous coordinates (x^1, x^2, x^3) on $\mathbf{R}^3$ its equations take the form $x^i=\alpha^i x^3+\beta^i$ $(i=1, 2)$. This implies that H_0 coincides with the submanifold of lines in $\mathbf{R}^3$ introduced in Sec. 1, while the coordinates $(\alpha^1, \alpha^2, \beta^1, \beta^2)$ coincide with the coordinates introduced there.

The inhomogeneous coordinates $(\alpha^1, \alpha^2, \beta^1, \beta^2)$ on G can be expressed in terms of the homogeneous coordinates $p_{ij}=u^i v^j-u^j v^i$ or z by the formulas $\alpha^i=\frac{p_{i4}}{p_{34}}$, $\beta^i=\frac{p_{i3}}{p_{43}}$ $(i=1, 2)$. Using these relations, we can write each differential form on the manifold of lines in $\mathbf{R}^3$, given in the coordinates $(\alpha^1, \alpha^2, \beta^1, \beta^2)$, in homogeneous coordinates; as a result, a differential form on the submanifold H_0 of lines in $\mathbf{RP}^3$ is obtained. If here the original form satisfies certain natural smoothness conditions at infinity, then the form obtained extends uniquely to a smooth form on the entire manifold of oriented lines G.

In view of what has been said, all definitions and all formulas for the case $\mathbf{R}^3$ of the preceding sections carry over to the case of projective space; in particular, in order to obtain the formulas for projective space it suffices to go over from inhomogeneous coordinates to homogeneous coordinates. We observe, however, that it is technically more convenient to obtain these formulas independently by repeating the entire calculation in homogeneous coordinates.

6. Integration of Differential Forms Over Lines Given in Homogeneous Coordinates

1. Definition of $\mathcal{I}\omega$. Let ω be a differential r-form on $\mathbf{RP}^3$ given in homogeneous coordinates $x=(x^1, x^2, x^3, x^4)$, and let $z=(u, v)$ be homogeneous coordinates on the manifold of lines G. We observe that points x of any line $z=(u, v)$ are given by homogeneous coordinates t^1, t^2 defined from the relation $x=t^1u+t^2v$. We write the form ω in the new coordinates u, v and t connected with x by the relation $x=t^1u+t^2v$; for this in the expression for ω it is necessary to substitute $x=t^1u+t^2v$, $dx=t^1du+t^2dv+udt^1+vdt^2$. In the expression obtained we separate out terms of degree 1 in dt^1, dt^2. In the space of the variables (t^1, t^2) we then choose any cycle homologous to the circle $S^1: (t^1)^2+(t^2)^2=1$, and we integrate the form obtained on t over this cycle. (We may take for such a cycle, e.g., the circle S^1 or any other curve enclosing the point 0 and intersecting once each ray issuing from the point 0.) It is easy to see that the integral does not depend on the choice of cycle. One half of the integral obtained we call the integral of the form ω over lines in $\mathbf{RP}^3$ and denote it by $\mathcal{I}\omega$. Thus, $\mathcal{I}\omega$ is a differential $(r-1)$-form defined on the manifold E of 2-frames (u, v) in $\mathbf{R}^4$.

We need to show that the definition given here is consistent with the definition of the integral over lines in $\mathbf{R}^3$ presented in Sec. 1. For this it is necessary to show that 1) the differential form $\mathcal{I}\omega$ drops from E to G and hence does not depend on the choice of inhomogeneous coordinates on G; 2) in inhomogeneous coordinates $(\alpha^1, \alpha^2, \beta^1, \beta^2)$ the definition of $\mathcal{I}\omega$ coincides with the definition given in Sec. 1.

The second assertion is obvious (the integral over the circle S^1 in inhomogeneous coordinates on $\mathbf{R}^3$ reduces to a double integral over x^3). The first assertion follows easily from the explicit expression for $\mathcal{I}\omega$ to be given later.

2. An Expression for $\mathcal{I}\omega$ in the Case $r=3$. We have already noted that in homogeneous coordinates any 3-form on $\mathbf{RP}^3$ has the following form:

$$\omega=f(x)(x^1dx^2\wedge dx^3\wedge dx^4-x^2dx^1\wedge dx^3\wedge dx^4+x^3dx^1\wedge dx^2\wedge dx^4-x^4dx^1\wedge dx^2\wedge dx^3),$$

where f is a homogeneous function of degree of homogeneity -4. In more compact form this expression can also be written

$$\omega=\frac{1}{3!}f(x)[x, dx, dx, dx], \tag{3.18}$$

where $[x, dx, dx, dx]=\sum \operatorname{sgn}(i_1, i_2, i_3, i_4)\, x^{i_1}dx^{i_2}\wedge dx^{i_3}\wedge dx^{i_4}$ (the sum goes over permutations of the indices 1, 2, 3, 4).

On the basis of the definition of the form $\mathcal{I}\omega$ given in part 1 we easily obtain the following result.

Proposition 25. The integral of the differential form ω over lines in $\mathbf{RP}^3$ is given in homogeneous coordinates (u, v) by the following formula:

$$\mathcal{I}\omega=\frac{1}{2}\hat{f}^{(2)}(u, v)[u, v, du, du]+\hat{f}^{(1)}(u, v)[u, v, du, dv]+\frac{1}{2}\hat{f}^{(0)}(u, v)[u, v, dv, dv], \tag{3.19}$$

where

$$\hat{f}^{(i)}(u, v)=\frac{1}{2}\int_{S^1} f(t^1u+t^2v)(t^1)^i(t^2)^{2-i}(t^1dt^2-t^2dt^1). \tag{3.20}$$

Proposition 26. The form $\mathcal{I}\omega$ drops from E to G.

According to Sec. 5, for the proof it is necessary to show that the form $\mathcal{I}\omega$ is invariant under transformations of the group $GL_+(2, \mathbf{R})$ and that it is preserved by the substitutions $du\to du+ud\alpha+vd\gamma$, $dv\to dv+ud\beta+vd\delta$. Both facts are established by direct verification.

Remark. We observe that under the action of the full linear group the form $\mathcal{I}\omega$ is multiplied by sgn(det g). Hence, it changes sign with change of orientation of the lines $h\in G$ (and hence does not drop onto the manifold of nonoriented lines). In connection with this only such forms on G will be considered, with no special mention.

We shall establish the connection between formulas in homogeneous and inhomogeneous coordinates. Suppose that the form ω in homogeneous coordinates has the form (3.18), and let

$$\omega'=f_0(x^1, x^2, x^3)\,dx^1\wedge dx^2\wedge dx^3$$

be its expression in inhomogeneous coordinates on $\mathbf{R}^3$. The form ω' is obtained by restricting the form ω, considered as a form on $\mathbf{R}^4$, to the plane $x^4=1$. It is obvious that this restriction is equal to $\omega'=-f(x^1, x^2, x^3, 1)\,dx^1\wedge dx^2\wedge dx^3$; hence $f_0(x^1, x^2, x^3)=-f(x^1, x^2, x^3, 1)$. Conversely, the form ω is obtained from ω' by the passage to homogeneous coordinates $x^i\to x^i/x^4$, $dx^i\to d(x^i/x^4)$; as a result, we obtain $f(x^1, x^2, x^3, x^4)=-(x^4)^{-4}f_0(x^1/x^4, x^2/x^4, x^3/x^4)$.

Remark. In order that the extension of a differential form ω' to the space $\mathbf{RP}^3$ belong to the class C^∞, the function f_0 must satisfy an additional condition, viz., the function $(x^4)^{-4}f_0(x^1/x^4, x^2/x^4, x^3/x^4)$ must be of class C^∞ on $\mathbf{R}^4\setminus 0$.

We present the relation between the coefficients of the form $\mathcal{J}\omega$ given in homogeneous coordinates (u, v) by formula (3.19) and the coefficients of this form given in the inhomogeneous coordinates (α, β): $\mathcal{J}\omega' = \hat{f}_0^{(2)}(\alpha, \beta)\, d\alpha^1\wedge d\alpha^2 + \hat{f}_0^{(1)}(\alpha, \beta)(d\alpha^1\wedge d\beta^2 - d\alpha^2\wedge d\beta^1) + \hat{f}_0^{(0)}(\alpha, \beta)\, d\beta^1\wedge d\beta^2$. Here $\hat{f}_0^{(i)}(\alpha, \beta)$ are expressed in terms of the function $f_0(x^1, x^2, x^3)$ by formulas (3.3). We have $\hat{f}_0^{(i)}(\alpha, \beta) = \hat{f}^{(i)}(\alpha^1, \alpha^2, 1, 0; \beta^1, \beta^2, 0, 1)$. Conversely,

$$\hat{f}^{(2)}(u, v) = p_{34}^{-3}\left((v^4)^2\hat{f}_0^{(2)}(\alpha, \beta) - 2v^3v^4\hat{f}_0^{(1)}(\alpha, \beta) + (v^3)^2\hat{f}_0^{(0)}(\alpha, \beta)\right),$$

$$\hat{f}^{(1)}(u, v) = -p_{34}^{-3}\left(u^4v^4\hat{f}_0^{(2)}(\alpha, \beta) - (u^3v^4 + u^4v^3)\hat{f}_0^{(1)}(\alpha, \beta) + u^3v^3\hat{f}_0^{(0)}(\alpha, \beta)\right),$$

$$\hat{f}^{(0)}(u, v) = p_{34}^{-3}\left((u^4)^2\hat{f}_0^{(2)}(\alpha, \beta) - 2u^3u^4\hat{f}_0^{(1)}(\alpha, \beta) + (u^3)^2\hat{f}_0^{(0)}(\alpha, \beta)\right),$$

where $\alpha^i = \frac{p_{i4}}{p_{34}}$, $\beta^i = \frac{p_{i3}}{p_{43}}$, $p_{ij} = u^iv^j - u^jv^i$.

3. An Expression for $\mathcal{J}\omega$ in the Case $r=2$. As already noted, any differential 2-form on $\mathbf{RP}^3$ has in homogeneous coordinates $x = (x^1, x^2, x^3, x^4)$ the form

$$\omega = \sum_{i_1, i_2=1}^{4} f_{i_1 i_2}(x)\, dx^{i_1}\wedge dx^{i_2}, \tag{3.21}$$

where $f_{i_1 i_2}$ are antisymmetric in i_1, i_2 and satisfy the following conditions for dropping to $\mathbf{RP}^3$: $f_{i_1 i_2}(\lambda x) = \lambda^{-2} f_{i_1 i_2}(x)$, $\sum_i f_{i_1 i_2}(x)\, x^{i_1} = 0$. On the basis of the definition of the operation we obtain the following result by a simple computation.

Proposition 27. The integral of the differential form $\mathcal{J}$ over lines in $\mathbf{RP}^3$ is given by the following formula:

$$\mathcal{J}\omega = \sum_{i=1}^{4} a_i(u, v)\, du^i + \sum_{i=1}^{4} b_i(u, v)\, dv^i, \tag{3.22}$$

where

$$a_i(u, v) = \sum_j \varphi_{ij}(u, v)\, v^j,\quad b_i(u, v) = -\sum_j \varphi_{ij}(u, v)\, u^j, \tag{3.23}$$

$$\varphi_{ij}(u, v) = \int_{S^1} f_{ij}(t^1u + t^2v)(t^1dt^2 - t^2dt^1). \tag{3.24}$$

It follows from (3.24) that under transformations $g\in GL_+(2, \mathbf{R})$ the functions $\varphi_{ij}(u, v)$ are multiplied by $(\det g)^{-1}$. Therefore, they are homogeneous functions of the Plücker coordinates $p_{ij} = u^iv^j - u^jv^i$ of degree of homogeneity -1, and we can write $\varphi_{ij}(u, v) = \varphi_{ij}(p)$. We note the equality

$$\sum_{i,j} \varphi_{ij}(p)\, p_{ij} = 0. \tag{3.25}$$

COROLLARY. The form $\mathcal{J}\omega$ can be written in Plücker coordinates p_{ij} in the following manner:

$$\mathcal{J}\omega = \frac{1}{2}\sum_{i,j} \varphi_{ij}(p)\, dp_{ij}, \tag{3.26}$$

where the functions $\varphi_{ij}(p)$ are given by formula (3.24).

Proposition 28. The form $\mathcal{J}\omega$ drops from E to the manifold G.

Proof. From what has been said above it follows that the form $\mathcal{I}\omega$ drops from E to $K\setminus 0$, where K is the cone in the space $\mathbf{R}^6$ with coordinates p_{ij}, $i<j$ $(i, j=1,2,3,4)$, given by the equation $p_{12}p_{34}-p_{13}p_{24}+p_{14}p_{23}=0$. To prove that $\mathcal{I}\omega$ drops from $K\setminus 0$ to G, it is necessary to show that this form is preserved under the transformations $p_{ij}\to\lambda p_{ij}$ $(\lambda>0)$ and the substitutions $dp_{ij}\to dp_{ij}+p_{ij}d\lambda$. The first follows from the homogeneity of the functions φ_{ij}, and the second from inequality (3.25).

We shall establish the connection between the formulas in homogeneous and inhomogeneous coordinates. Let $\omega=\sum_{i,j=1}^{3} f_{ij}^0(x)\,dx^i\wedge dx^j$ be the form (3.21) written in inhomogeneous coordinates (x^1, x^2, x^3) on $\mathbf{R}^3$. We then have $f_{ij}^0(x^1, x^2, x^3)=f_{ij}(x^1, x^2, x^3, 1)$, $i, j=1,2,3$. Conversely, $f_{ij}(x^1, x^2, x^3, x^4)=(x^4)^{-2} f_{ij}^0(x^1/x^4, x^2/x^4, x^3/x^4)$ for $i, j=1,2,3$; $f_{i4}(x^1, x^2, x^3, x^4)=-(x^4)^{-3}\sum_{j=1}^{3} f_{ij}^0(x^1/x^4, x^2/x^4, x^3/x^4)x^j$.

We now present relations between the coefficients of the form $\mathcal{I}\omega$ given in homogeneous coordinates by formula (3.22) and the coefficients of this form given in inhomogeneous coordinates $\mathcal{I}\omega=\sum_{i=1}^{3} a_i^0(\alpha,\beta)\,d\alpha^i+\sum_{i=1}^{3} b_i^0(\alpha,\beta)\,d\beta^i$. We have

$$a_i^0(\alpha,\beta)=a_i(\alpha^1, \alpha^2, 1, 0;\ \beta^1, \beta^2, 0, 1),$$
$$b_i^0(\alpha,\beta)=b_i(\alpha^1, \alpha^2, 1, 0;\ \beta^1, \beta^2, 0, 1).$$

Conversely, $a_i(u, v)=p_{34}^{-1}(a_i^0(\alpha,\beta)v^4-b_i^0(\alpha,\beta)v^3)$, $b_i(u, v)=p_{34}^{-1}(-a_i^0(\alpha,\beta)u^4+b_i^0(\alpha,\beta)u^3)$, $i=1, 2$. The remaining coefficients are determined from the relations $\sum_i a_ip_{ij}=0$, $\sum_i b_ip_{ij}=0$, $i,j=1, 2, 3, 4$, where $\alpha^i=p_{i3}/p_{34}$, $\beta^i=p_{i4}/p_{43}$, $p_{ij}=u^iv^j-u^jv^i$.

7. Description of the Forms $\mathcal{I}\omega$ in Homogeneous Coordinates: The Case $r=3$

In Sec. 1 it was proved that the form $\mathcal{I}\omega$ is closed and admissible. We recall that the condition of admissibility was first introduced in coordinate form and was then obtained by several equivalent invariant definitions. Here we shall proceed from the following two equivalent definitions. 1) A form is called admissible if its restriction to a submanifold of lines lying in any two-dimensional plane is equal to zero. 2) A form is called admissible if it vanishes on planes of the family Π_1 of the cone of incidence (see part 2, Sec. 5). We shall establish how the admissibility and closedness conditions are written in homogeneous coordinates.

1. Condition of Admissibility. Let σ be any differential 2-form on E,

$$\sigma=\sum_{i,j} a_{ij}(u, v)\,du^i\wedge du^j+\sum_{i,j} b_{ij}(u, v)\,du^i\wedge dv^j+\sum_{i,j} c_{ij}(u, v)\,dv^i\wedge dv^j,$$

where a_{ij} and c_{ij} are antisymmetric in i and j.

Proposition 29. If a form σ drops to G, then it is admissible if and only if it can be written in the form

$$\sigma=\frac{1}{2}\hat{f}^{(2)}(u, v)[u, v, du, du]+\hat{f}^{(1)}(u, v)[u, v, du, dv]+\frac{1}{2}\hat{f}^{(0)}(u, v)[u, v, dv, dv]. \tag{3.27}$$

Proof. Suppose that the form σ can be represented in the form (3.27). By going over to the inhomogeneous coordinates (α, β), we then obtain

$$\sigma=\hat{f}_0^{(2)}(\alpha,\beta)\,d\alpha^1\wedge d\alpha^2+\hat{f}_0^{(1)}(\alpha,\beta)(d\alpha^1\wedge d\beta^2-d\alpha^2\wedge d\beta^1)+\hat{f}_0^{(0)}(\alpha,\beta)\,d\beta^1\wedge d\beta^2. \tag{3.28}$$

Hence, the form σ is admissible (see the definition of admissibility of a form in Sec. 2). Conversely, if the form σ is admissible, then in the inhomogeneous coordinates (α, β) it has the form (3.28). Going over in (3.28) to the homogeneous coordinates $\alpha^i = p_{i3}/p_{34}$, $\beta^i = p_{i4}/p_{43}$, where $p_{ij} = u^i v^j - u^j v^i$, it is not hard to see that the form obtained reduces to the form (3.28).

We emphasize that the representation of an admissible form in the form (3.27) has been done under the assumption that it drops to G. However, not every form of this type drops to G.

Proposition 30. The conditions that a differential form on E of the type (3.27) drop to G, have the following form:

$$1)\ \hat{f}^{(i)}(\lambda u, \mu v) = \lambda^{-i-1}\mu^{i-3}\hat{f}^{(i)}(u, v) \text{ for } \lambda\mu > 0; \tag{3.29}$$

$$2)\ \sum_s \frac{\partial \hat{f}^{(2)}}{\partial u^s} v^s = 0,\ \sum_s \frac{\partial \hat{f}^{(0)}}{\partial v^s} u^s = 0; \tag{3.30}$$

$$3)\ \hat{f}^{(2)} + \sum_s \frac{\partial \hat{f}^{(1)}}{\partial u^s} v^s = 0,\ 2\hat{f}^{(1)} + \sum_s \frac{\partial \hat{f}^{(0)}}{\partial u^s} v^s = 0; \tag{3.31}$$

$$4)\ \hat{f}^{(0)} + \sum_s \frac{\partial \hat{f}^{(1)}}{\partial v^s} u^s = 0,\ 2\hat{f}^{(1)} + \sum_s \frac{\partial \hat{f}^{(2)}}{\partial v^s} u^s = 0.^* \tag{3.32}$$

Proof. First of all, we note that any form of the type (3.27) is preserved under the substitutions $du \to du + u\,d\alpha + v\,d\gamma$, $dv \to dv + u\,d\beta + v\,d\delta$. Further, it is not hard to see that the condition (3.29) is equivalent to the condition of invariance of the form σ under the transformations $u \to \lambda u$, $v \to \mu v$; the condition (3.31) together with the first of the conditions (3.30) are equivalent to the condition of invariance of σ relative to the transformation $u \to u + \lambda v$; the condition (3.32) together with the second of conditions (3.30) are equivalent to the invariance of σ under the transformations $v \to v + \lambda u$. Therefore, by the criterion formulated in Sec. 5, part 3, conditions (3.29)-(3.32) are equivalent to the condition that σ drop from E to G.

Remark 1. If $\sigma = \mathcal{I}\omega$, then relations (3.29)-(3.32) follow easily from the explicit expressions for the coefficients of the form $\mathcal{I}\omega$.

Remark 2. Using the remark at the end of part 3, Sec. 5, it is possible to formulate necessary and sufficient conditions that a form σ drop to G in a different form, viz.,

$$1)\ \hat{f}^{(i)}(\lambda u, \mu v) = \lambda^{-i-1}\mu^{i-3}\hat{f}^{(i)}(u, v) \text{ for } \lambda\mu > 0;$$

$$2)\ \sum_s \frac{\partial \hat{f}^{(2)}}{\partial u^s} v^s = 0,\ \hat{f}^{(2)} + \sum_s \frac{\partial \hat{f}^{(1)}}{\partial u^s} v^s = 0,\ 2\hat{f}^{(1)} + \sum_s \frac{\partial \hat{f}^{(0)}}{\partial u^s} v^s = 0;$$

$$3)\ \hat{f}^{(0)}(u, v) = \hat{f}^{(2)}(v, u),\quad \hat{f}^{(1)}(v, u) = -\hat{f}^{(1)}(u, v).$$

The difference between the affine and projective cases merits attention. In the projective case the condition that a form drop to G implies that it can be recovered from one coefficient, e.g., $\hat{f}^{(0)}$. In the affine case this is not so. On the other hand, a projective form can be recovered from its affine section. The reason for this seeming contradiction is that the coefficients of a projective form depend on 6 rather than 4 variables, and all coefficients of this form are contained in the expression of each of them in terms of the coefficients of the affine form (see part 2, Sec. 5).

*The conditions that an arbitrary 2-form σ drop to G are more involved; we shall not need them.

2. Condition of Self-Duality. In Sec. 2, part 3 the operator $*$ was introduced in the space of differential forms on a manifold of lines, and it was proved that the conditions of admissibility and self-duality of 2-forms are equivalent. We introduce the concept of a self-dual form on the manifold of frames E. Let $s: G\to E$ be any section of the bundle $E\to G$. The section s induces an isomorphism of the spaces of differential forms on G and sG. We denote by $*_s$ the image of the operator $*$ under this mapping.

Remark. We thus define $*$ at each point $z\in E$ on each four-dimensional subspace of the tangent space T_z transversal to the fiber of the bundle $E\to G$.

Definition. We call a differential form σ on E self-dual if its restriction $\sigma|_{sG}$ to any section s of the bundle $E\to G$ is a self-dual form, i.e., $*_s(\sigma|_{sG})=\sigma|_{sG}$.

It is clear that if a form σ drops to G, then the conditions $*_s(\sigma|_{sG})=\sigma|_{sG}$ for various sections s are equivalent to one another and are equivalent to the self-duality of the form on G.

Proposition 31. If the differential 2-form

$$\sigma=\sum_{i,j}a_{ij}du^i\wedge du^j+\sum_{i,j}b_{ij}du^i\wedge dv^j+\sum_{i,j}c_{ij}dv^i\wedge dv^j$$

drops to G, then it is self-dual if and only if $b_{ii}=b_{jj}=0$, $b_{ij}=-b_{ji}$ for any fixed indices $i\neq j$.

Proof. As a local section sG we choose a manifold of frames for which the last two coordinates are fixed and $\det\begin{pmatrix}u^3 & u^4\\ v^3 & v^4\end{pmatrix}\neq 0$. We then have

$$\sigma|_{sG}=2a_{12}du^1\wedge du^2+\sum_{i,j=1}^{2}b_{ij}du^i\wedge dv^j+2c_{12}dv^1\wedge dv^2.$$

Going over from the coordinates (u^1, u^2, v^1, v^2) to the local coordinates $(\alpha^1, \alpha^2, \beta^1, \beta^2)$ and using the condition of self-duality in these coordinates, we easily find that the self-duality of $\sigma|_{sG}$ is equivalent to the condition $b_{ij}=-b_{ji}$, $i,j=1,2$. Since the property of self-duality does not depend on the choice of section s, the condition $b_{11}=b_{22}=0$, $b_{12}=-b_{21}$ is equivalent to the condition $b_{ii}=b_{jj}=0$, $b_{ij}=-b_{ji}$ for any other fixed indices $i\neq j$.

Conversely, suppose that $b_{ij}=-b_{ji}$ for any i, j. From the condition that the form σ drops to G it follows, in particular, that

$$\sum_i a_{ij}u^i=\sum_i a_{ij}v^i=\sum_i b_{ij}u^i=\sum_i b_{ij}v^i=\sum_i c_{ij}u^i=\sum_i c_{ij}v^i=0.$$

The equations $\sum_i b_{ij}u^i=0$, $\sum_i b_{ij}v^i=0$ and the fact that b_{ij} are antisymmetric then imply that

$$\frac{b_{12}}{p_{34}}=\frac{b_{13}}{p_{42}}=\frac{b_{14}}{p_{23}}=\frac{b_{23}}{p_{14}}=\frac{b_{24}}{p_{31}}=\frac{b_{34}}{p_{12}}=\hat{f}^{(1)},$$

where $p_{ij}=u^iv^j-u^jv^i$. Hence, $\sum_{i,j}b_{ij}du^i\wedge dv^j=\hat{f}^{(1)}[u, v, du, dv]$. Similarly, since a_{ij} and c_{ij} are antisymmetric, $\sum_{i,j}a_{ij}du^i\wedge du^j=\frac{1}{2}\hat{f}^{(2)}[u, v, du, du]$, $\sum_{i,j}c_{ij}dv^i\wedge dv^j=\frac{1}{2}\hat{f}^{(0)}[u, v, dv, dv]$, i.e., the form σ can be represented in the form (3.27). Hence, the form σ is admissible and thus self-dual.

We denote by Δ_s the image of the Laplace operator on G, $\Delta=\delta d+d\delta$, under the isomorphism $G\to sG$. We call a differential form σ on E harmonic if its restriction to any section s of the bundle $E\to G$ is a harmonic form, i.e., $\Delta_s(\sigma|_{sG})=0$. It is clear that if the form σ drops to G, then its harmonicity on one section s implies its harmonicity on other sections and is equivalent to the condition of its harmonicity on G.

It is not hard to see that in the special case where sG is the manifold of frames $z=(u, v)$ for which the i-th and j-th coordinates are fixed the operator Δ_s coincides up to a numerical factor with the ultrahyperbolic operator

$$\Delta_{ij}=\frac{\partial^2}{\partial u^i \partial v^j}-\frac{\partial^2}{\partial u^j \partial v^i}.$$

3. Reconstruction of an Admissible Form on G from One Coefficient. Proposition 32. In order that a function $\hat{f}^{(0)}(u, v)$ on E be the coefficient of $[u, v, dv, dv]$ for some admissible form on G, it is necessary and sufficient that it be a homogeneous function of u and v of bidegree $(-1, -3)$ and satisfy the following conditions:

$$1) \sum_s \frac{\partial \hat{f}^{(0)}}{\partial v^s} u^s=0; \tag{3.33}$$

$$2) \sum_{i,j,k} \frac{\partial^3 \hat{f}^{(0)}}{\partial u^i \partial u^j \partial u^k} v^i v^j v^k=0. \tag{3.34}$$

Proof. Necessity follows from Proposition 30. We shall prove sufficiency. Suppose that the conditions of the proposition are satisfied. We consider the differential form (3.27), where

$$\hat{f}^{(1)}=-\frac{1}{2}\sum_i \frac{\partial \hat{f}^{(0)}}{\partial u^i} v^i,$$

$$\hat{f}^{(2)}=-\sum_i \frac{\partial \hat{f}^{(1)}}{\partial u^i} v^i=\frac{1}{2}\sum_{i,j} \frac{\partial^2 \hat{f}^{(0)}}{\partial u^i \partial u^j} v^i v^j.$$

We must show that this form drops to G, i.e., that the conditions of Proposition 30 are satisfied. It is obvious that the functions $\hat{f}^{(i)}$ satisfy the conditions (3.29) and (3.31) of Proposition 30. It is further obvious that conditions (3.33) and (3.34) are equivalent to condition (3.30) of Proposition 30. It remains to verify condition (3.32). We first prove that

$$\sum_i u^i \Delta_{ij} \hat{f}^{(0)}=0 \quad \left(\Delta_{ij}=\frac{\partial^2}{\partial u^i \partial v^j}-\frac{\partial^2}{\partial u^j \partial v^i}\right). \tag{3.35}$$

Indeed, differentiating $\sum_i \frac{\partial \hat{f}^{(0)}}{\partial v^i} u^i=0$ with respect to u^j, we obtain $\sum_i \frac{\partial^2 \hat{f}^{(0)}}{\partial v^i \partial u^j} u^i+\frac{\partial \hat{f}^{(0)}}{\partial v^j}=0$. On the other hand, from the conditions of homogeneity it follows that $\sum_i \frac{\partial^2 \hat{f}^{(0)}}{\partial v^j \partial u^i} u^i+\frac{\partial \hat{f}^{(0)}}{\partial v^j}=0$. Subtracting the second equality from the first, we obtain Eq. (3.35). Applying this equality, we have

$$\hat{f}^{(0)}+\sum_s \frac{\partial \hat{f}^{(1)}}{\partial v^s} u^s=\hat{f}^{(0)}-\frac{1}{2}\sum_{i,s} \frac{\partial^2 \hat{f}^{(0)}}{\partial v^s \partial u^i} u^s v^i-\frac{1}{2}\sum_s \frac{\partial \hat{f}^{(0)}}{\partial u^s} u^s=$$
$$=\hat{f}^{(0)}-\frac{1}{2}\sum_{i,s} \frac{\partial^2 \hat{f}^{(0)}}{\partial u^s \partial v^i} u^s v^i-\frac{1}{2}\sum_s \frac{\partial \hat{f}^{(0)}}{\partial u^s} u^s=\hat{f}^{(0)}-\frac{3}{2}\hat{f}^{(0)}+\frac{1}{2}\hat{f}^{(0)}=0;$$

$$2\hat{f}^{(1)}+\sum_s \frac{\partial \hat{f}^{(2)}}{\partial v^s} u^s=-\sum_i \frac{\partial \hat{f}^{(0)}}{\partial u^i} v^i+\frac{1}{2}\sum_{i,j,s} \frac{\partial^3 \hat{f}^{(0)}}{\partial u^i \partial u^j \partial v^s} v^i v^j u^s+\sum_{i,j} \frac{\partial^2 \hat{f}^{(0)}}{\partial u^i \partial u^j} v^i u^j=-3\sum_i \frac{\partial \hat{f}^{(0)}}{\partial u^i} v^i+$$
$$+\frac{1}{2}\sum_i \frac{\partial}{\partial u^i}\left(\sum_{j,s} \frac{\partial^2 \hat{f}^{(0)}}{\partial u^j \partial v^s} v^i v^j u^s\right)-\frac{1}{2}\sum_{i,j} \frac{\partial^2 \hat{f}^{(0)}}{\partial u^j \partial v^i} v^i v^j=-\frac{3}{2}\sum_i \frac{\partial \hat{f}^{(0)}}{\partial u^i} v^i+\frac{1}{2}\sum_i \frac{\partial}{\partial u^i}\left(\sum_{j,s} \frac{\partial^2 \hat{f}^{(0)}}{\partial u^s \partial v^j} v^i v^j u^s\right)=0.$$

4. Condition of Closedness. Proposition 33. If a form σ of the form (3.27) drops to G, then the condition that it be closed is equivalent to the relations

$$\frac{\partial \hat{f}^{(2)}}{\partial v^i}=\frac{\partial \hat{f}^{(1)}}{\partial u^i},\quad \frac{\partial \hat{f}^{(1)}}{\partial v^i}=\frac{\partial \hat{f}^{(0)}}{\partial u^i},\quad i=1,2,3,4. \tag{3.36}$$

Proof. Since σ drops to G, the condition that it be closed is equivalent to the closedness of the restriction of σ to any section of the bundle $E\to G$. We take for this section the submanifold of frames for which two coordinates, e.g., the first and second, are fixed. The restriction of σ to this submanifold has the form

$$\sigma=p_{12}(\hat{f}^{(2)}du^3\wedge du^4+\hat{f}^{(1)}(du^3\wedge dv^4-du^4\wedge dv^3)+\hat{f}^{(0)}dv^3\wedge dv^4).$$

It is obvious that the condition that this form be closed is given by Eqs. (3.36) for $i=3, 4$, as required to prove.

Remark 1. From the argument presented it follows that for the closedness of a 2-form it is sufficient that the relations (3.36) be satisfied for any two indices i, j.

Remark 2. We know that in the affine case the coefficients of the form $\mathcal{I}\omega$ are related by the conditions of closedness. In the projective case they are further related by the conditions (3.29)-(3.32) of dropping to G. We note that these conditions partially overlap. Namely, we shall show that conditions (3.36) and (3.29) imply conditions (3.31) and (3.32).

Indeed, multiplying the first of Eqs. (3.36) by v and summing on i, we obtain $-\hat{f}^{(2)}=\sum_i \frac{\partial \hat{f}^{(1)}}{\partial u^i}v^i$. The remaining relations of (3.31) and (3.32) are obtained in a similar way.

5. Closedness and Conditions of Harmonicity. From the conditions of closedness (3.36) it follows that the coefficients of a form σ of the type (3.27) satisfy the condition of harmonicity (see part 2), in particular, they satisfy the system of ultrahyperbolic equations

$$\Delta_{ij}\hat{f}^{(k)}\equiv\frac{\partial^2 \hat{f}^{(k)}}{\partial u^i\partial v^j}-\frac{\partial^2 \hat{f}^{(k)}}{\partial u^j\partial v^i}=0. \tag{3.37}$$

We note that although the form σ itself drops to G, its coefficients do not drop to G; in the harmonicity condition it is thus not possible to restrict attention to one equation. However, it is found that Eqs. (3.37) for the harmonicity of a function are already sufficient. We have the following assertion.

Proposition 34. If the form (3.27) drops to G and its coefficient $\hat{f}^{(0)}$ satisfies Eqs. (3.37), then this form is closed.

Proof. From the condition that a form drop to G it follows that $2\hat{f}^{(1)}+\sum_s \frac{\partial \hat{f}^{(0)}}{\partial u^s}v^s=0$, $\hat{f}^{(2)}+\sum_s \frac{\partial \hat{f}^{(1)}}{\partial u^s}v^s=0$. Differentiating the first equality with respect to v^i and using the fact that $\frac{\partial^2 \hat{f}^{(0)}}{\partial u^s\partial v^i}=\frac{\partial^2 \hat{f}^{(0)}}{\partial u^i\partial v^s}$ and $\hat{f}^{(0)}$ has degree of homogeneity -3 in v, we obtain $2\frac{\partial \hat{f}^{(1)}}{\partial v^i}+\sum_s \frac{\partial^2 \hat{f}^{(0)}}{\partial v^s\partial u^i}v^s+\frac{\partial \hat{f}^{(0)}}{\partial u^i}=0$, whence $2\frac{\partial \hat{f}^{(1)}}{\partial v^i}-2\frac{\partial \hat{f}^{(0)}}{\partial u^i}=0$. From this relation it also follows that $\Delta_{ij}\hat{f}^{(1)}=0$. Similarly, from the equality $f^{(2)}+\sum_s \frac{\partial \hat{f}^{(1)}}{\partial u^s}v^s=0$ it follows that $\frac{\partial \hat{f}^{(2)}}{\partial v^i}-\frac{\partial \hat{f}^{(1)}}{\partial u^i}=0$.

6. Reconstruction of a Closed Form on G from One Coefficient. Proposition 35. In order that a function $\hat{f}^{(0)}(u, v)$ on E be the coefficient of $[u, v, dv, dv]$ in a closed form on G,

it is necessary and sufficient that it be harmonic, i.e., $\Delta_{ij}\hat{f}^{(0)}(u, v)=0$, and satisfy the conditions of Proposition 32.

The assertion follows from Propositions 22 and 24.

Regularization of the Form $\mathcal{I}\omega$. We denote by $\tilde{E}$ the manifold of triples (x, u, v) of pairwise noncollinear vectors in $\mathbf{R}^4$ such that $x=ut+v$. We define the projection

$$\tilde{E}\to E\colon (x, u, v)\mapsto(u, v);$$

we introduce in $\tilde{E}$ two systems of coordinates: (x, u, t) and (u, v, t).

Let σ be an arbitrary 2-form on E. We perform the following operations on σ: 1) we lift σ from E to $\tilde{E}$; 2) we separate out the terms of degree 2 in dv; 3) we multiply the expression obtained exteriorly by dt. We denote the 3-form on $\tilde{E}$ thus obtained by $(\sigma\wedge dt)_0$. We define the form $R\sigma$ according to the following proposition.

Proposition 36. If σ is an admissible form of the type (3.27), then on $\tilde{E}$ there exists a unique differential form $R\sigma$ of the form

$$R\sigma=\psi(u, x)[x, dx, dx, dx]$$

such that in the coordinates (u, v, t) on $\tilde{E}$ its homogeneous component of bidegree (2.1) in dv and dt coincides with $(\sigma\wedge dt)_0$.

Proof. Carrying out the operations indicated on the form σ, we obtain the form

$$(\sigma\wedge dt)_0=\frac{1}{2}\hat{f}^{(0)}(u, x)[udt, v, dv, dv].$$

It is not hard to see that the following form satisfies the condition of the proposition:

$$R\sigma=-\frac{1}{3!}\hat{f}^{(0)}(u, x)[x, dx, dx, dx]. \tag{3.38}$$

This form is unique, since the mapping assigning to forms $\psi(u, x)[x, dx, dx, dx]$ their homogeneous components of bidegree (2, 1) in dv and dt is injective.

Thus, according to the definition, the 3-form (3.38) is a regularization of the admissible form (3.27).

The operator R is important for the inversion of the transform $\mathcal{I}$. Namely, in order to recover the form ω from the form $\sigma=\mathcal{I}\omega$, we first construct the form $R\sigma$. Its coefficient $\hat{f}^{(0)}(u, x)$ is connected with the coefficient of the original form ω by the relation

$$\hat{f}^{(0)}(u, x)=\frac{1}{2}\int_{S^1} f(t^1u+t^2x)(t^2)^2(t^1dt^2-t^2dt^1). \tag{3.39}$$

Thus, the problem of inverting the transform $\mathcal{I}$ for forms reduces to the problem of inverting the transform (3.39) for functions already considered in Sec. 3, Chap. II.

Proposition 37. If an admissible form σ is closed, then the coefficient $\hat{f}^{(0)}(u, x)$ of the form $R\sigma$ satisfies the ultrahyperbolic equations $\Delta_{ij}\hat{f}^{(0)}=0$. Conversely, if $\tilde{\sigma}$ is a form of the type (3.38) where the function $\hat{f}^{(0)}$ satisfies the conditions of Proposition 30 and the equations $\Delta_{ij}\hat{f}^{(0)}=0$, then $\tilde{\sigma}=R\sigma$, where σ is an admissible closed form on G.

Indeed, the first assertion is obvious, and the second follows from Proposition 35.

8. Description of the Image of the Mapping $\mathcal{I}$. THEOREM 3. A differential 2-form σ on G of class C^∞ can be represented in the form $\sigma=\mathcal{I}\omega$, where ω is a 3-form on $\mathbf{RP}^3$ of class C^∞, if and only if it is admissible and closed, i.e., $*\sigma=\sigma$ and $d\sigma=0$.

Proof. The necessity of the conditions was proved earlier. We shall prove that they are sufficient. Let σ satisfy the conditions of the theorem. It can then be represented in the form (3.27), where f^i are C^∞ functions on the manifold of frames E which satisfy relations (3.29)-(3.32).

The coefficient $\hat{f}^{(0)}(u, v)$ of this form is a harmonic function and satisfies the condition of homogeneity $\hat{f}^{(0)}(\lambda u, \mu v) = |\lambda\mu|^{-1}\mu^{-2}\hat{f}^{(0)}(u, v)$. Therefore, according to Theorem 11 of Sec. 3, Chap. II, the function $\hat{f}^{(0)}(u, v)$ can be represented as the integral

$$\hat{f}^{(0)}(u, v) = \frac{1}{2}\int_{S^1} f(t^1u + t^2v)(t^2)^2(t^1dt^2 - t^2dt^1),$$

where $f(x)$ is a C^∞ function on $\mathbf{R}^4\setminus 0$ which satisfies the homogeneity condition $f(\lambda x) = \lambda^{-4}f(x)$.

We consider the differential form on G

$$\omega = \frac{1}{3!} f(x)[x, dx, dx, dx].$$

From the definition of the operations $\mathcal{I}$ and R it follows that $R\mathcal{I}\omega = R\sigma = \frac{1}{3!}\hat{f}^{(0)}(u, x)[x, dx, dx, dx]$. Hence, since the mapping R is injective, $\sigma = \mathcal{I}\omega$.

Remark. In the proof we have used only the harmonicity of the coefficient $\hat{f}^{(0)}$ of the form σ $(\Delta_{ij}\hat{f}^{(0)} = 0)$. It has thus been established that the harmonicity of $\hat{f}^{(0)}$ implies the closedness of the form σ.

We shall give another formulation of this theorem. If σ is a form of the type (3.27), then we set

$$\varkappa\sigma = \left(\sum_s \frac{\partial \hat{f}^{(0)}(u, x)}{\partial x^s} du^s\right)\wedge[x, dx, dx, dx].$$

It is easy to see that the differential form $\varkappa\sigma$ drops from the manifold E to the manifold of pairs $(\tilde{u}, \tilde{x})$ of rays in $\mathbf{R}^4$, where $\tilde{x} \neq \tilde{u}$.

THEOREM 3'. In order that a differential 2-form C^∞ of class σ on G be representable in the form $\sigma = \mathcal{I}\omega$, where ω is a differential 3-form of class C^∞ on $\mathbf{RP}^3$, it is necessary and sufficient that it be admissible and the form $\varkappa\sigma$ be closed.

For the proof it suffices to note that the condition that the form $\varkappa\sigma$ be closed is equivalent to the relations $\Delta_{ij}\hat{f}^{(0)} = 0$, $i, j = 1, 2, 3, 4$. Therefore, the theorem follows from the previous theorem and the remark made above.

8. Description of the Form $\mathcal{I}\omega$ in Homogeneous Coordinates: The Case $r=2$

1. Conditions for Dropping to G. Proposition 38. The conditions that a differential 1-form σ on E,

$$\sigma = \sum_i a_i(u, v)\,du^i + \sum_i b_i(u, v)\,dv^i, \tag{3.40}$$

drops to G are as follows:

$$1)\ a_i(\lambda u, \mu v) = \lambda^{-1}a_i(u, v), \quad b_i(\lambda u, \mu v) = \mu^{-1}b_i(u, v); \tag{3.41}$$

$$2)\ a_i + \sum_s \frac{\partial b_i}{\partial u^s} v^s = 0, \quad \sum_s \frac{\partial a_i}{\partial u^s} v^s = 0; \tag{3.42}$$

$$3)\ b_i + \sum_s \frac{\partial a_i}{\partial v^s} u^s = 0, \quad \sum_s \frac{\partial b_i}{\partial v^s} u^s = 0; \tag{3.43}$$

$$4)\ \sum_i a_i u^i = \sum_i a_i v^i = \sum_i b_i u^i = \sum_i b_i v^i = 0. \tag{3.44}$$

For the proof it is necessary to use the criterion for dropping a form from E to G formulated in Sec. 5. It is not hard to see that the condition of invariance of the form σ under the transformations $u \to \lambda u$ and $v \to \mu v$ is expressed by relations (3.41), invariance under the transformations $u \to u + \lambda v$, $v \to v$ by relations (3.42), and invariance under $u \to u$ and $v \to v + \lambda u$ by relations (3.43). Finally, the condition of invariance of σ under the substitutions $du \to du + ud\lambda$, $du \to du + vd\lambda$, $dv \to dv + ud\lambda$ and $dv \to dv + vd\lambda$ is equivalent to relations (3.44).

Remark. By using the remark at the end of part 3, Sec. 2, it is possible to express the condition for dropping σ from E to G in another form, viz.: 1) $b_i(\lambda u, \mu v) = \mu^{-1} b_i(u, v)$, 2) $\sum_s \frac{\partial b_i}{\partial v^s} u^s = 0$, $b_i + \sum_s \frac{\partial a_i}{\partial v^s} u^s = 0$, 3) $\sum_i b_i u^i = \sum_i b_i v^i = 0$, 4) $a_i(u, v) = -b_i(v, u)$. In particular, this implies the following result.

In order that a differential form $\sum_i b_i dv^i$ on E may be extended to a form $\sum_i a_i du^i + \sum_i b_i dv^i$, which drops to G, it is necessary and sufficient that its coefficients satisfy the relations 1) and 3) and also the relations $\sum_s \frac{\partial b_i}{\partial v^s} u^s = 0$ and $\sum_s \frac{\partial b_i}{\partial u^s} v^s = b_i(v, u)$. Under these conditions the coefficients a_i of the extended form are uniquely determined by the equalities $a_i(u, v) = -b_i(v, u)$.

2. Condition of Partial Closedness. In Sec. 3 it was proved that the form $\mathcal{I}\omega$ is partially closed. We formulate the condition of partial closedness of a form given in homogeneous coordinates.

Proposition 39. If a form σ of type (3.40) drops to G, then it is partially closed if and only if its coefficients a_i, b_i satisfy the relations

$$\frac{\partial a_i}{\partial v^j} - \frac{\partial b_i}{\partial u^j} = -\frac{\partial a_j}{\partial v^i} + \frac{\partial b_j}{\partial u^i}, \quad i, j = 1, 2, 3, 4. \tag{3.45}$$

In particular, $\frac{\partial a_i}{\partial v^i} = \frac{\partial b_i}{\partial u^i}$.

Proof. It is obvious that equality (3.45) is equivalent to the condition that the coefficients of the form $d\sigma$ in front of $du^i \wedge dv^j$ are antisymmetric in i and j. In turn, the fact that these coefficients are antisymmetric is equivalent to the condition of admissibility of the form $d\sigma$ and hence to the condition of partial closedness of the form σ.

We note that relations (3.45) for one pair of indices i, j imply the relations for any i, j.

3. Additional Conditions on the Coefficients of the Form $\mathcal{I}\omega$. Let $\sigma = \mathcal{I}\omega$, where $\omega = \sum f_{ij}(x)\, dx^i \wedge dx^j$. According to Proposition 27, the coefficients a_i, b_i of the form σ can then be expressed in terms of the coefficients f_{ij} of the form ω: $a_i = \sum_j \varphi_{ij} v^j$, $b_i = -\sum_j \varphi_{ij} u^j$, where

$$\varphi_{ij}(u, v) = \int_{S^1} f_{ij}(t^1 u + t^2 v)(t^1 dt^2 - t^2 dt^1). \tag{3.46}$$

LEMMA. The functions φ_{ij} defined by Eq. (3.46) can be expressed in terms of the coefficients of the form $\sigma = \mathcal{I}\omega$ in the following manner:

$$\varphi_{ij} = \frac{\partial a_i}{\partial v^j} - \frac{\partial b_i}{\partial u^j}. \tag{3.47}$$

Proof. From the expressions for the coefficients a_i, b_i it follows that $\frac{\partial a_i}{\partial v^j}=\sum_s \frac{\partial \varphi_{is}}{\partial v^j} v^s+\varphi_{ij}$, $\frac{\partial b_i}{\partial u^j}=-\sum \frac{\partial \varphi_{is}}{\partial u^j} u^s-\varphi_{ij}$; hence, $\frac{\partial a_i}{\partial v^j}-\frac{\partial b_i}{\partial u^j}=\sum_s\left(\frac{\partial \varphi_{is}}{\partial u^j} u^s+\frac{\partial \varphi_{is}}{\partial v^j} v^s\right)+2\varphi_{ij}$. Further, from expression (3.46) for φ_{ij} and the relation $\sum_s f_{is}(x)x^s=0$ it follows that

$$\sum_s\left(\frac{\partial \varphi_{is}}{\partial u^j} u^s+\frac{\partial \varphi_{is}}{\partial v^j} v^s\right)=\int_{S^1}\left(\sum \frac{\partial f_{is}}{\partial x^j} x^s\right)_{x=t^1u+t^2v}(t^1dt^2-t^2dt^1)=-\int_{S^1} f_{ij}(t^1dt^2-t^2dt^1)=-\varphi_{ij}.$$

The proof of the lemma is complete.

Proposition 40. The coefficients a_i, b_i of the differential form $\sigma=\mathcal{I}\omega$ satisfy the following equations of third order:

$$\Delta_{kl}\left(\frac{\partial a_i}{\partial v^j}-\frac{\partial b_i}{\partial u^j}\right)=0 \quad \left(\Delta_{kl}=\frac{\partial^2}{\partial u^l\partial v^j}-\frac{\partial^2}{\partial u^j\partial v^l}\right). \tag{3.48}$$

Proof. The functions φ_{ij} defined by Eqs. (4.46) satisfy the equations $\Delta_{kl}\varphi_{ij}=0$. Therefore, the assertion follows from the preceding lemma.

4. Description of the Image of the Mapping $\mathcal{I}$. THEOREM 4. In order that a differential 1-form σ of class C^∞ on G be representable in the form $\sigma=\mathcal{I}\omega$, where ω is a 2-form of class C^∞ on $\mathbf{R}P^3$, it is necessary and sufficient that it be partially closed and that its coefficients a_i, b_i satisfy the additional differential relations (3.48).

Proof. We set

$$\varphi_{ij}(u, v)=\frac{\partial a_i}{\partial v^j}-\frac{\partial b_i}{\partial u^j}=-\frac{\partial a_j}{\partial v^i}+\frac{\partial b_j}{\partial u^i}.$$

The functions φ_{ij} are harmonic and satisfy the homogeneity condition $\varphi_{ij}(\lambda u, \mu v)=|\lambda\mu|^{-1}\varphi_{ij}(u, v)$. Hence, by Theorem 3 of Sec. 1, Chap. II they can be represented in the form

$$\varphi_{ij}(u, v)=\int_{S^1} f_{ij}(t^1u+t^2v)(t^1dt^2-t^2dt^1),$$

where f_{ij} are homogeneous C^∞ functions on $\mathbf{R}^4\setminus 0$ of degree of homogeneity -2. We consider the differential form on $\mathbf{R}^4\setminus 0$:

$$\omega=\sum f_{ij}(x)\,dx^i\wedge dx^j.$$

It is not hard to verify that it drops onto G, i.e., that $\sum_j f_{ij}(x)x^j=0$ (see [4]).

We shall show that $\mathcal{I}\omega=\sigma$. Let $\mathcal{I}\omega=\sum a_i' du^i+\sum b_i' dv^i$. From the formulas for the coefficients of the form ω it follows that $a_i'=\sum_j \varphi_{ij}v^j=a_i$, $b_i'=-\sum_j \varphi_{ij}u^j=b_i$, and therefore $\mathcal{I}\omega=\sigma$.

5. Regularization of the Form $\mathcal{I}\omega$ and Trivial Forms. We shall show that the problem of inverting the operator $\mathcal{I}$ can be reduced to the corresponding problem for functions. We call the following differential form a regularization of the differential form $\sigma=\mathcal{I}\omega$, $\omega\in\Omega^2(\mathbf{R}P^3)$:

$$R\sigma=\frac{1}{2}\sum_{i,j}\varphi_{ij}(u, x)\,dx^i\wedge dx^j,$$

where φ_{ij} are expressed in terms of the coefficients f_{ij} of the form ω by formulas (3.46), i.e., they are integrals of the functions f_{ij} over lines in $\mathbf{R}P^3$. This form is harmonic, i.e., $\Delta_{kl}(R\sigma)=0$ for any k, l, and its coefficients are homogeneous functions of u and x of bidegree $(-1, -1)$. Using the results of integral geometry for functions (see Sec. 1 of

Chap. II), on the basis of the functions φ_{ij} it is possible to recover f_{ij} and thus on the basis of the form $R\sigma$ to recover the form ω. On the other hand, by the lemma of part 3 the coefficients of the form $R\sigma$ can be expressed in terms of the coefficients of the form $\sigma=\mathcal{I}\omega$ by the formulas

$$\varphi_{ij}=\frac{\partial a_i}{\partial v^j}-\frac{\partial b_i}{\partial u^j}.$$

Therefore, the form ω can be uniquely recovered from its image $\mathcal{I}\omega$.

Another definition of the operator R is given below. Let $\tilde{E}$ be the manifold of triples (x, u, v) of pairwise noncollinear vectors in $\mathbf{R}^4$ such that $x=ut+v$ $(t\neq 0)$ which was already introduced in Sec. 7. In the manifold there are two coordinate systems: the system (x, u, t) and the system (u, v, t).

Definition. We call a differential 2-form on $\tilde{E}$ trivial if it has degree zero in du and dt in the coordinates (u, v, t).

Proposition 41. Any trivial form on $\tilde{E}$ is a sum of monomials of the form

$$\psi_{ij}(u^k dx^i-u^i dx^k)\wedge(u^k dx^j-u^j dx^k),$$

where k is any fixed index.

Proof. For fixed u and k the differential forms $u^k dx^i-u^i dx^k$ $(i\neq k)$ and dx^k form a basis in the space of 1-forms on $\mathbf{R}^4$. Hence, any 2-form on $\tilde{E}$ of degree zero in du and dt in the coordinates (x, u, t), can be represented as a sum of monomials of the form $\psi_{ij}(u^k dx^i-u^i dx^k)\wedge(u^k dx^j-u^j dx^k)$ and $\psi_i(u^k dx^i-u^i dx^k)\wedge dx^k=u^k\psi_i dx^i\wedge dx^k$.

In each monomial we go over from the coordinates (x, u, t) to the coordinates (u, v, t). Since $u^k dx^i-u^i dx^k=(u^k dv^i-u^i dv^k)+t(u^k du^i-u^i du^k)$, the monomials of the first type in coordinates (u, v, t) have degree zero relative to dt. On the other hand, monomials of the second type and sums of them have a nonzero component of degree 1 in dt.

We proceed to the construction of the operator R. Let σ be a 1-form on E. We carry out the following operations: 1) We define the mapping $\tilde{E}\to E:(x, u, v)\mapsto(u, v)$ and we lift σ to $\tilde{E}$. 2) We multiply σ exteriorly by an arbitrary 1-form of the type

$$\tau=\sum_i \rho_i(u, v, t)\,dx^i, \tag{3.49}$$

normalized by the condition $\sum_i \rho_i u^i=1$. 3) We write the form $\tau\wedge\sigma$ in coordinates (x, u, t), and in the expression obtained we separate out the component of degree zero in du and dt. The differential 2-form on dx so obtained we denote by $(\tau\wedge\sigma)_0$. We now define the form $R\sigma$ according to the following proposition.

Proposition 42. Let σ be a partially closed form on G which satisfies condition (3.48). Then there exists a unique 2-form $R\sigma$ on $\tilde{E}$ which satisfies the following conditions:

a) the form $R\sigma$ in coordinates (x, u, t) has degree 0 in du and dt; its coefficients do not depend on t and are homogeneous functions of x and u of bidegree $(-1, -1)$;

b) the form $R\sigma$ is harmonic, i.e., $\Delta_{ij}R\sigma=0$, $i, j=1, 2, 3, 4$;

c) the difference $R\sigma-(\tau\wedge\sigma)_0$ is a trivial form for any form τ.

Proof. Let σ satisfy the conditions of the proposition. We consider the differential form

$$R\sigma=\frac{1}{2}\sum_{i,j}\varphi_{ij}(u, x)\,dx^i\wedge dx^j,$$

where the functions φ_{ij} are expressed in terms of the coefficients a_i, b_i of the form σ by formulas (3.47), i.e., $\varphi_{ij}=\frac{\partial a_i}{\partial x^j}-\frac{\partial b_i}{\partial u^j}$. It is obvious that the form $R\sigma$ satisfies conditions a) and b). We shall prove that it also satisfies condition c). From the condition that σ drop to G it follows that $\sum_j\varphi_{ij}u^j=\sum_j\frac{\partial a_i}{\partial x^j}u^j-\sum_j\frac{\partial b_i}{\partial u^j}u^j=-b_i$. Hence, for any $k=1, 2, 3, 4$ we have

$$(dx^k\wedge\sigma)_0=\sum_j b_j dx^k\wedge dx^j=\sum_{i,j}\varphi_{ij}u^i dx^k\wedge dx^j.$$

We represent $R\sigma$ in the form

$$R\sigma=\sum_{j\neq k}\varphi_{kj}dx^k\wedge dx^j+\frac{1}{2}\sum_{i,j\neq k}\varphi_{ij}dx^i\wedge dx^j.$$

This implies that

$$u^kR\sigma-(dx^k\wedge\sigma)_0=\frac{1}{2}\sum_{i,j\neq k}\varphi_{ij}u^k dx^i\wedge dx^j-\sum_{i,j\neq k}\varphi_{ij}u^i dx^k\wedge dx^j=\frac{1}{2u^k}\sum_{i,j\neq k}\varphi_{ij}(u^k dx^i-u^i dx^k)\wedge(u^k dx^j-u^j dx^k),$$

i.e., this difference is a trivial form. Since k is arbitrary, this implies that the difference $R\sigma-(\tau\wedge\sigma)_0$, where τ is any 1-form of the type (3.49), is also a trivial form. Thus, the existence of a form $R\sigma$ satisfying the conditions of the proposition has been proved. Its uniqueness follows from the fact that any trivial form satisfying the conditions of Proposition 42 is equal to zero.

We note that if $\sigma=\mathcal{I}\omega$, then both definitions of the operator R presented above coincide.

The condition that a form σ belong to the image of $\mathcal{I}$ can now be formulated in another way.

THEOREM 5. A differential 1-form σ on G of class C^∞ can be represented in the form $\sigma=\mathcal{I}\omega$, where ω is a 2-form of class C^∞ on $\mathbf{RP}^3$, if and only if it is partially closed and for it there exists a regularization $R\sigma$, i.e., a 2-form satisfying conditions a)-c) of Proposition 42.

Proof. The necessity of the conditions has already been established. We shall prove that they are sufficient. Suppose that the form $\sigma=\sum_i a_i du^i+\sum_i b_i dv^i$ satisfies the conditions of the theorem, and let $R\sigma=\frac{1}{2}\sum_{i,j}\varphi_{ij}dx^i\wedge dx^j$ be its regularization. We shall prove that the coefficients of the form σ satisfy Eqs. (3.48). This will imply that σ satisfies the hypotheses of Theorem 4 and hence belongs to the image of $\mathcal{I}$.

It follows from the conditions of the theorem that $R\sigma-\left(\frac{dx^1}{u^1}\wedge\sigma\right)_0=\sigma_1$ is a trivial form. We represent σ_1 in the form

$$\sigma_1=\frac{1}{2(u^1)^2}\sum_{i,j\neq 1}\psi_{ij}(u^1dx^i-u^idx^1)\wedge(u^1dx^j-u^jdx^1).$$

Then

$$\left(\frac{dx^1}{u^1}\wedge\sigma\right)_0+\sigma_1=\frac{1}{2}\sum_{i,j\neq 1}\psi_{ij}dx^i\wedge dx^j+\frac{1}{u^1}\sum_{j\neq 1}\Big(b_j-\sum_{i\neq 1}\psi_{ij}u^i\Big)dx^1\wedge dx^j.$$

Therefore, $\psi_{ij}=\varphi_{ij}$ for $i, j\neq 1$. By hypothesis the coefficients of this form satisfy the equations $\Delta_{kl}\varphi=0$. This implies that

$$\Delta_{kl}b_j=\frac{\partial\varphi_{kj}}{\partial v^l}-\frac{\partial\varphi_{lj}}{\partial v^k} \tag{3.50}$$

for $j, k, l\neq 1$. Since the index 1 was selected arbitrarily, (3.50) is valid for any j, k, l. Further, we have

$$\Delta_{kl}a_j=-\Delta_{kl}\Big(\sum_s\frac{\partial b_j}{\partial u^s}v^s\Big)=-\sum_s\frac{\partial}{\partial u^s}(\Delta_{kl}b_j)v^s=-\sum_s\Big(\frac{\partial^2\varphi_{kj}}{\partial v^l\partial u^s}-\frac{\partial^2\varphi_{lj}}{\partial v^k\partial u^s}\Big)v^s=-\sum_s\Big(\frac{\partial^2\varphi_{kj}}{\partial u^l\partial v^s}-\frac{\partial^2\varphi_{lj}}{\partial u^k\partial v^s}\Big)v^s,$$

whence

$$\Delta_{kl}a_j=\frac{\partial\varphi_{kj}}{\partial u^l}-\frac{\partial\varphi_{lj}}{\partial u^k}. \tag{3.51}$$

It follows from Eqs. (3.50) and (3.51) and the harmonicity of φ_{ij} that $\Delta_{kl}\Big(\frac{\partial a_j}{\partial v^l}-\frac{\partial b_j}{\partial u^l}\Big)=0,$

as was required to prove.

APPENDIX. SUMMARY OF THE BASIC DEFINITIONS AND RESULTS FOR THE SPACE C^3

1°. The operation $\mathcal{J}$ (of integration over complex lines in C^3) is defined in analogy with the real case and defines a mapping $\mathcal{J}:\Omega^{r,s}(C^3)\to\Omega^{r-1,s-1}(H)$ ($\Omega^{r,s}$ is the space of forms of type (r,s)).

2°. The Condition of Admissibility. A differential form σ on H of type (2, 2) is called admissible if $\sigma(h;\xi_1,\xi_2;\bar\eta_1,\bar\eta_2)=0$ whenever (ξ_1,ξ_2) or (η_1,η_2) is a degenerate pair of vectors (the definition of a degenerate pair is exactly the same as in the real case). Admissible (2, 1)- and (1, 2)-forms are defined similarly.

3°. Description of the Image of the Mapping $\mathcal{J}$. 1) A differential form $\sigma\in\Omega^{2,2}(H)$ can be represented in the form $\sigma=\mathcal{J}\omega$, $\omega\in\Omega^{3,3}(C^3)$ if and only if it is admissible and closed.

2) A differential form $\sigma\in\Omega^{r,s}(H)$, where $1\leqslant r,\ s\leqslant 2,\ r+s<4$, can be represented in the form $\sigma=\mathcal{J}\omega$, $\omega\in\Omega^{r+1,s+1}(C^3)$ if and only if 1) the forms σ, $\partial\sigma$ and $\bar\partial\sigma$ are admissible and 2) the coefficients of the form σ satisfy additional equations of third order analogous to equations (3.13); see [3].

4°. The Inversion Formula. If $\sigma=\mathcal{J}\omega$, where $\omega\in\Omega^{3,3}(C^3)$, then we set

$$\varkappa\sigma=\sum_{i,j=1}^{2}\frac{\partial^2\varphi(\alpha,x'-\alpha x^3)}{\partial\beta^i\partial\bar\beta^j}\,d\alpha^i\wedge d\bar\alpha^j\wedge dx\wedge d\bar x,$$

where $\varphi(\alpha,\beta)$ is the coefficient of the form σ for $d\beta^1\wedge d\beta^2\wedge d\bar\beta^1\wedge d\bar\beta^2$; $dx=dx^1\wedge dx^2\wedge dx^3$. Then

$$\int_{\gamma_x}\varkappa\sigma=c_{\gamma_x}\omega,$$

where the integral is taken over any cycle γ_x of real dimension 2 in the manifold CP^2 of complex lines in C^3 which pass through x; c_{γ_x} depends only on the homology class of the cycle γ_x (if γ_x is a line in CP^2, then $c_{\gamma_x}=8\pi^2 i$).

Similar inversion formulas hold for any $1\leqslant r,s\leqslant 2$.

For the space CP^3 the definitions and results are analogous; see [4].

CHAPTER IV

BINARY BUNDLES AND INTEGRAL GEOMETRY FOR DIFFERENTIAL FORMS IN n-DIMENSIONAL PROJECTIVE SPACE

The principal content of this chapter is the investigation of integral transforms consisting in the integration of differential r-forms on RP^n over p-dimensional planes for $r\geqslant p$.

The beginning of the chapter (Secs. 1 and 2) is devoted to a discussion of the general problem of integrating differential forms on the binary bundles introduced in [21]. Essentially, the language of binary bundles was already present in our preceding considerations, and here we explain how the concepts introduced in Chap. III carry over to the general situation.

As we have seen in Chap. III, for $n=3$, $p=1$ the case of forms of degree $r=3$ was considerably simpler than the case $r=2$. For arbitrary n and p the situation is analogous: for $r-p>1$ the results are quite simple, and the description of the image of the integral transform is given in the language of the invariant concepts of binary bundles (admissibility, partial closedness). For $r-p=1$ the description of the image is more complicated. In Sec. 7 it is shown that the problem considered in this chapter can in a certain sense be decomposed into a series of problems equivalent to the problem for lines in three-dimensional space treated in Chap. III.

1. Differential Forms on a Fiber Space

The basic object of integral geometry is a binary bundle which we define in Sec. 2. However, all the basic concepts are formed from concepts connected with ordinary bundles. Therefore, we first recall certain facts concerning differential forms on a fiber space.

1. Operator π^* of the Lift (Inverse Image) of a Differential Form. We consider a C^∞ smooth fibration $\pi: A\to B$ of the manifold A. The projection π induces mappings of the tangent spaces $d\pi: T_a\to T_{\pi a}$. We denote by $\Omega(A)$ and $\Omega(B)$, respectively, the spaces of differential forms on A and B.

Definition. We define the mapping $\pi^*:\Omega(B)\to\Omega(A)$ by the formula

$$\pi^*\omega(a;\eta_1,\ldots,\eta_r)=\omega(\pi a;\pi\eta_1,\ldots,\pi\eta_r),$$

where $\eta_1,\ldots,\eta_r\in T_a$ (for brevity we write $\pi\eta_i$ instead of $(d\pi)\eta_i$).

We call the form $\pi^*\omega$ the lift of the form ω from B to A or the inverse image of the form ω. If $\omega_1=\pi^*\omega$, then we also say that the form ω_1 drops from A to B, and we call the form ω the drop onto B of the form ω_1.

Because of this definition, a form $\omega_1(a;\eta_1,\ldots,\eta_r)$ on A can be dropped to B if and only if it is constant on the fibers of the bundle $\pi: A\to B$, i.e.,

$$\omega_1(a;\eta_1,\ldots,\eta_r)=\omega_1(a',\eta_1',\ldots,\eta_r')$$

whenever $\pi a=\pi a'$, $\pi\eta_i=\pi\eta_i'$, $i=1,\ldots,r$.

If $\pi: A\to B$ is a principal bundle with structural group G, then the criterion for dropping a form from A to B can be formulated in another way which is more convenient for applications.

Proposition 1. In order that a differential form $\omega_1(a;\eta_1,\ldots,\eta_r)$ on A drop from A to B, the following conditions are necessary and sufficient: 1) the form ω_1 is invariant under G, i.e.,

$$\omega_1(ag;\eta_1 g,\ldots,\eta_r g)=\omega_1(a;\eta_1,\ldots,\eta_r)$$

for any $g\in G$;

2) $\omega_1(a;\eta_1+\eta_1',\ldots,\eta_r+\eta_r')=\omega_1(a;\eta_1,\ldots,\eta_r)$ if the vectors $\eta_i'\in T_a$ are tangent to the fiber of the bundle π.

2. Filtration of Forms in $\Omega(A)$. We denote by $\Omega^q(A)$ the subspace of forms $\omega(a; \eta_1, \ldots, \eta_r)$ which vanish when more than q vectors η_i lie in the subspace tangent to the fiber, i.e., $\omega(a; \eta_1, \ldots, \eta_r)=0$ if $\eta_1, \ldots, \eta_{q+1}$ are tangent to the fiber at the point $a \in A$. The spaces $\Omega^q(A)$ define an ascending filtration of $\Omega(A)$, i.e.,

$$0 \subset \Omega^0(A) \subset \ldots \subset \Omega^N(A) = \Omega(A),$$

where N is the dimension of the fiber.

Definition. If $\omega \in \Omega^q(A) \setminus \Omega^{q-1}(A)$, then we say that the form ω has relative degree q in the bundle $\pi: A \to B$, and write $\mathrm{Deg}_B \omega = q$.

Remark. The filtration of forms and the lift of the relative degree $\deg_b \omega$ of a form ω can be defined independently for each point $b \in B$. It is obvious that $\mathrm{Deg}_B \omega = \max_{b \in B} \deg_b \omega$. In cases of interest to us $\deg_b \omega = \mathrm{Deg}_B \omega$ almost everywhere on B.

We shall show how to define the filtration and relative degree of a form given in coordinates on A. Since A locally has the structure of a direct product, in a neighborhood of any point $a \in A$ it is possible to introduce a local coordinate system $\alpha^1, \ldots, \alpha^N, x^{N+1}, \ldots, x^{N+M}$ such that α^i are the coordinates in the fiber and x^j are the coordinates in the projection. In these coordinates any form $\omega \in \Omega(A)$ can be written

$$\omega = \sum_{\nu_1 < \ldots < \nu_r} \varphi_{\nu_1, \ldots, \nu_r}(\alpha, x)\, d\alpha^{\nu_1} \wedge \ldots \wedge d\alpha^{\nu_t} \wedge dx^{\nu_{t+1}} \wedge \ldots \wedge dx^{\nu_r}. \tag{4.1}$$

We associate with the coordinate system (α, x) a gradation in the space of differential forms: $\omega = \omega_0 + \ldots + \omega_N$ (N is the dimension of the fiber), where ω_q $(q=0, \ldots, N)$ is a differential form of the type (4.1) which in each term contains precisely q differentials $d\alpha^i$. The next assertion is then obvious.

LEMMA. 1) $\omega \in \Omega^q(A)$ if and only if $\omega_{q+1} = \ldots = \omega_N = 0$, i.e., the number of differentials $d\alpha^i$ in each term of expression (4.1) does not exceed q;

2) a form ω has relative degree q if and only if $\omega_{q+1} = \ldots \omega_N = 0$ and $\omega_q \neq 0$.

COROLLARY. Exterior differentiation raises the relative degree of a form ω by no more than one:

$$d: \Omega^q(A) \to \Omega^{q+1}(A).$$

Definition. We call a differential form $\omega \in \Omega(A)$ of relative degree q in the bundle $A \to B$ relatively closed if $d\omega \in \Omega^q(A)$, i.e., exterior differentiation does not raise the relative degree of the form.

In the coordinate system (α, x) we represent the operator d as the sum $d = d_\alpha + d_x$. The condition of relative closedness of the form $\omega = \omega_0 + \ldots + \omega_q$ $(\omega_q \neq 0)$ is then equivalent to the condition that $d_\alpha \omega_q = 0$.

Remark. The concepts introduced admit the following generalization. Let A be any smooth manifold, and suppose there is given an N-dimensional distribution on A, i.e., to each point $a \in A$ there is assigned an N-dimensional subspace $L_a \subset T_a$. We denote by $\Omega^q(A)$ the subspace of differential forms ω on A such that $\omega(a; \eta_1, \ldots, \eta_r) = 0$ whenever more than q vectors $\eta_i \in T_a$ belong to the subspace L_a. It is obvious that the subspaces $\Omega^q(A)$ form an ascending filtration of the space of differential forms; on the basis of this filtration, it is possible to introduce the concept of the relative degree of a form and a relatively closed form for the given distribution on A.

3. Operation ρ of Restricting Forms to Fibers. On forms on A we introduce the operation ρ, which we call the restriction to fibers of the bundle $\pi: A\to B$. Let ω be an r-form on A of relative degree $\mathrm{Deg}_B\omega=k$.

We first consider the case $k=r$; in this case the degree of the form does not exceed the dimension of the fiber. We define $(\rho\omega)_b$ as the restriction (in the usual sense) of the form ω to the fiber $\pi^{-1}b$. Thus, $\rho\omega$ is a function on B, the values of which are r-forms on the fibers $\pi^{-1}b$.

We now define $\rho\omega$ for the case $k<r$. Let $b\in B$ and $\eta_1, \ldots, \eta_{r-k}\in T_b$. We consider the fiber $\pi^{-1}b$ and for each point $a\in\pi^{-1}b$ we define vectors $\zeta_i(a, \eta_i)\in T_a$ such that $\pi\zeta_i=\eta_i$, $i=1, \ldots, r-k$. For any $\xi_1, \ldots, \xi_k\in T_a(\pi^{-1}b)$ we set

$$(\rho\omega)_{b;\eta_1,\ldots,\eta_{r-k}}(a;\ \xi_1, \ldots, \xi_k)=\omega(\alpha;\ \zeta_1(a, \eta_1), \ldots, \zeta_{r-k}(a, \eta_{r-k}), \xi_1, \ldots, \xi_k). \tag{4.2}$$

LEMMA. Expression (4.2) does not depend on the manner of lifting the vectors η_i, i.e., on the choice of the functions $\zeta_i(a;\ \eta_i)$. Thus, $\rho\omega$ is a differential $(r-k)$-form on B, the value of which for any $b\in B$ and $\eta_1, \ldots, \eta_{r-k}\in T_b$ is a k-form on the fiber $\pi^{-1}b$.

Indeed, if $\zeta_i'(a;\ \eta_i)$ are other vectors such that $\pi\zeta_i'=\eta_i$, then the vectors $\zeta_i'-\zeta_i=\xi_i'$ are tangent to the fiber at the point a. Since the vectors $\xi_1, \ldots, \xi_k$ are tangent to the fiber and $\mathrm{Deg}_B\omega=k$, it follows that $\omega(a;\ \zeta'_1, \ldots, \zeta'_{r-k}, \xi_1, \ldots, \xi_k)=\omega(a;\ \zeta_1, \ldots, \zeta_{r-k}, \xi_1, \ldots, \xi_k)$.

We note that $\rho\omega=0$ at all points $b\in B$, where $\deg_b\omega<\mathrm{Deg}_B\omega$.

Proposition 2. In the coordinate system (α, x) on A introduced above the form $\rho\omega$ is the homogeneous component of the form ω of maximal degree k relative to $d\alpha^i$.

Proof. In the coordinate system (α, x) the lifts ζ_i of the vectors $\eta_i\ni T_b$ to T_a, $a\in\pi^{-1}b$, are canonically defined, and it is thus possible to assume that the vectors ζ_i do not depend on a. Therefore, $\rho\omega$ is obtained from ω by separating out terms of degree k in $d\alpha^i$, where $k=\mathrm{Deg}_B\omega$.

4. Operation π_* of Integrating a Differential Form along Fibers (the Direct Image). We consider the C^∞ fibration $\pi: A\to B$ of the manifold A with fibers which are assumed to be compact oriented manifolds. We define the operation

$$\pi_*: \Omega(A)\to\Omega(B)$$

of integrating differential forms on A along the fibers of this bundle.

Definition (of the direct image). Let ω be an r-form on A of relative degree k. If $k<N$, where N is the dimension of the fiber, then we set $\pi_*\omega=0$. If $k=N$, then for any $b\in B$ and $\eta_1, \ldots, \eta_{r-N}\in T_b$ we define

$$(\pi_*\omega)(b;\ \eta_1, \ldots, \eta_{r-N})=\int_{\pi^{-1}b}(\rho\omega)_{b;\eta_1,\ldots,\eta_{r-N}}, \tag{4.3}$$

where $(\rho\omega)_{b;\eta_1,\ldots,\eta_{r-N}}$ is the N-form on $\pi^{-1}b$ defined by Eq. (4.2).

Thus, $\pi_*\omega$ is an $(r-N)$-form on B.

We present an explicit expression for $\pi_*\omega$ in coordinates on B. As before, we prescribe the coordinates $(\alpha^1, \ldots, \alpha^N, x^{N+1}, \ldots, x^{N+M})$ on A, where α^i are coordinates in the fiber and x^j are coordinates in the base. In these coordinates the differential form ω has the form

$$\omega=\sum_{i_1<\ldots<i_{r-N}} f_{i_1,\ldots,i_{r-N}}(\alpha, x)\,dx^{i_1}\wedge\ldots\wedge dx^{i_{r-N}}\wedge d\alpha^1\wedge\ldots\wedge d\alpha^N+\ldots,$$

where the dots denote terms of relative degree less than N.

Proposition 3. We have

$$\pi_*\omega=\sum_{i_1<\ldots<i_{r-N}} \varphi_{i_1,\ldots,i_{r-N}}(x)\,dx^{i_1}\wedge\ldots\wedge dx^{i_{r-N}},$$

where

$$\varphi_{i_1,\ldots,i_{r-N}}(x)=\int f_{i_1,\ldots,i_{r-N}}(\alpha, x)\,d\alpha^1\wedge\ldots\wedge d\alpha^N.$$

The assertion follows immediately from Proposition 2 and the formula (4.3) for $\pi_*\omega$.

We shall prove an assertion analogous to the assertion regarding the integration of exact forms.

LEMMA. If the relative degree of ω is less than N, where N is the dimension of the fiber, then $\pi_*d\omega=0$.

Proof. The assertion is obvious if the relative degree of ω is less than $N-1$. Suppose that the relative degree of ω is equal to $N-1$. We introduce coordinates (α, x) on A and represent the operator of exterior differentiation d as the sum $d=d_\alpha+d_x$. Then, on the one hand, $\pi_*d_x\omega=0$, since the operator d_x does not raise the relative degree of ω; hence, the relative degree of $d_x\omega$ is less than N. On the other hand, $\pi_*d_\alpha\omega=0$ as the integral along the fiber of an exact form in the fiber. Hence, $\pi_*d\omega=0$.

Proposition 4. The operators π^* and π_* commute with the operator of exterior differentiation, viz.,

$$\pi^*\circ d=d\circ\pi^*,\quad \pi_*\circ d=d\circ\pi_*.$$

Proof. The first equality is obvious; we shall prove the second equality. In A we introduce the coordinates (α, x) as above and consider the corresponding gradation of differential forms

$$\omega=\omega_0+\ldots+\omega_N,$$

where ω_q is the component of relative degree q. According to the lemma, for $q<N$ we have $\pi_*d\omega_q=d\pi_*\omega_q=0$. Suppose now that $\omega=\omega_N$. It is obvious that $\pi_*\circ d_x=d_x\circ\pi_*$. Hence, since $d_\alpha\omega_N=0$, $\pi_*d\omega_N=\pi_*d_x\omega_N=d_x\pi_*\omega_N=d\pi_*\omega_N$.

We make a remark which we shall need in Sec. 7. Suppose there is given the commutative diagram

where A, B, C, D are manifolds and π_i are smooth fibrations of the corresponding manifolds. In view of the commutativity of the diagram, π_1 maps each fiber of the bundle $A\to C$ into a fiber of the bundle $B\to D$.

Proposition 5. If the preimage of each fiber of the bundle $B\to D$ is a fiber of the bundle $A\to C$ (i.e., π_1 realizes a homeomorphism between fibers of the bundle $A\to C$ and fibers of the bundle $B\to D$), then

$$(\pi_2)_*\pi_1^*=\pi_4^*(\pi_3)_*. \tag{4.4}$$

Proof. It may be assumed with no loss of generality that the bundle $B \to D$ is trivial, i.e., $B = D \times X$. By hypothesis the trivialization of B induces a trivialization of $A = C \times X$, viz., $a \mapsto (\pi_2 a, x)$, where x is determined from the equality $\pi_1 a = (d, x)$. In this case Eq. (4.4) follows from the definition of the operations * and $_*$.

Remark. Below we shall consider bundles with noncompact fibers. For these bundles the definitions and results presented above remain in force if additional conditions are imposed on the class of differential forms which ensure, first of all, the convergence of the integrals along the fibers and, secondly, the equality $\pi_* d\omega = 0$ for forms ω of relative degree $N-1$.

In many interesting examples there exist compactifications $\overline{A}$, $\overline{B}$ of the manifolds A, B, and the bundle $A \to B$ extends to a bundle $\overline{A} \to \overline{B}$. In this case differential forms on A which extend to smooth forms on $\overline{A}$ satisfy the conditions formulated.

5. Integration of Relatively Closed Forms. The operation of integrating forms of maximal relative degree along fibers can be generalized to relatively closed forms of any relative degree.

Let ω be a relatively closed r-form on A of relative degree k. Then, as in the definition of part 4, we consider for any point $b \in B$ and any vectors $\eta_1, \ldots, \eta_{r-k} \in T_b$ the differential k-form $(\rho\omega)_{b;\eta_1,\ldots,\eta_{r-k}}$ on the fiber $\pi^{-1}b$ defined by Eq. (4.2). Since ω is relatively closed, the form $(\rho\omega)_{b;\eta_1,\ldots,\eta_{r-k}}$ is closed on the fiber $\pi^{-1}b$. Therefore, to it there corresponds an element $q(\omega; b; \eta_1, \ldots, \eta_{r-k})$ of the cohomology group $H^k(\pi^{-1}b)$ of the fiber.

The operator $(b; \eta_1, \ldots, \eta_{r-k}) \mapsto q(\omega; b; \eta_1, \ldots, \eta_{r-k})$ which arises can naturally be interpreted as a differential form of degree $r-k$ on B with values in a bundle over B, with fiber over $b \in B$ being the group $H^k(\pi^{-1}b)$. We note that this bundle over B has a canonical flat connection, and in the case where the base of the bundle $A \to B$ is simply connected this bundle is trivial.

Thus, to each relatively closed r-form on A of relative degree k we have assigned an $(r-k)$-form $\pi_*\omega$ on B with values in the k-dimensional cohomology groups of the fibers of the bundle $\pi: A \to B$. In the case $k = N$, where N is the dimension of the fiber, the cohomology group is one-dimensional, and it may be assumed that $\pi_*\omega$ assumes numerical values. Observing this remark, the definition of $\pi_*\omega$ in this part agrees for $k = N$ with the definition given in part 4.

Remark. The operation π_* was defined under the assumption that the fibers of the bundle $A \to B$ are oriented manifolds. However, in many examples these fibers are either nonoriented or they are orientable but there is no global way of fixing an orientation in each fiber. We shall show how to carry over the definition of the operation π_* to these cases.

a) Orientable Fibers. Since for the integration of a form along a fiber it is necessary to fix its orientation, as a result of the integration a form is obtained not on B, but on the manifold $\tilde{B}$ of oriented fibers which forms a two-sheeted covering of B. This form changes sign under the involution of $\tilde{B}$ corresponding to a change of orientation, and in this sense it can be considered as an odd differential form on B (see de Rham [37]).

b) Nonorientable Fibers. Let $\tilde{A}$ be a two-sheeted covering of A for which the corresponding coverings of all fibers of the bundle $A \to B$ are orientable. Then in place of bundle $A \to B$ we take the bundle $\tilde{A} \to B$ induced by it and apply the previous construction to the latter. As a result, differential forms are obtained on the manifold $\tilde{B}$ of oriented fibers of the bundle $\tilde{A} \to B$ ($\tilde{B}$ is a two-sheeted covering of B) which are odd with respect to the involution of $\tilde{B}$. We note that here on $\tilde{A}$ differential forms should be considered which do not drop to A (they are odd relative to the involution of $\tilde{A}$).

2. Binary Bundles and the Formulation of Problems of Integral Geometry

1. Binary Bundles. We suppose there are given two infinitely differentiable manifolds B and C and that between certain pairs $b \in B$ and $c \in C$ there is given an incidence relation which satisfies the following axioms.

We denote by A the set of all incident pairs $(b, c) \in B \times C$ and let

$$\begin{array}{ccc} & A & \\ {\scriptstyle \pi_1} \swarrow & & \searrow {\scriptstyle \pi_2} \\ B & & C \end{array}$$

be the natural projections of A onto B and C. For each $c \in C$ we denote by B_c the subset of points $b \in B$ incident with c; similarly, for each $b \in B$ we denote by C_b the set of points $c \in C$ incident with b. We note that $B_c = \pi_1 \pi_2^{-1} c$, $C_b = \pi_2 \pi_1^{-1} b$.

Axiom 1. If $B_{c_1} = B_{c_2}$, then $c_1 = c_2$; similarly, if $C_{b_1} = C_{b_2}$, then $b_1 = b_2$.

Axiom 2. A is an infinitely differentiable submanifold of $B \times C$.

Axiom 3. The projections π_1, π_2 are infinitely differentiable fibrations of the manifold A over B and C, respectively.

Thus, the incidence relation between elements of the manifolds B and C is a prescription of a manifold $A \subset B \times C$ satisfying axioms 1-3.

Definition. The triple of manifolds B, C and $A \subset B \times C$ together with the projections $\pi_1 : A \to B$ and $\pi_2 : A \to C$ satisfying axioms 1-3 we call the binary bundle connected with the given incidence relation.

2. Formulation of the Problem of Integral Geometry. Suppose there is given a binary bundle

$$\begin{array}{ccc} & A & \\ {\scriptstyle \pi_1} \swarrow & & \searrow {\scriptstyle \pi_2} \\ B & & C \end{array},$$

where it is additionally assumed that the fibers of π_2 are compact oriented manifolds, and let N be the dimension of the fiber of π_2. We define the operation $\mathcal{I}$ on differential forms on B as the composition of the mappings π_1^* and $(\pi_2)_*$, i.e., $\mathcal{I} = (\pi_2)_* \pi_1^*$. Because of the definition, this operation provides for any $r \geq N$ a mapping

$$\mathcal{I} : \Omega^r(B) \to \Omega^{r-N}(C),$$

where $\Omega^r(B)$, $\Omega^{r-N}(C)$ are, respectively, the space of C^∞ smooth forms on B and C of degrees r and $r-N$.

In correspondence with the remark at the end of Sec. 1 this definition carries over also to the case of nonoriented fibers. Namely, if the fibers are orientable, but their orientation is not fixed, then $\mathcal{I}$ defines a mapping

$$\mathcal{I}:\Omega^r(B)\to\Omega^{r-N}(\tilde{C}),$$

where $\tilde{C}$ is the manifold of oriented fibers of the bundle $A\to C$ (i.e., a two-sheeted covering of C), and $\Omega^{r-N}(\tilde{C})$ is the space of differential forms on $\tilde{C}$ which are odd under the involution in $\tilde{C}$.

If the fibers of the bundle $A\to C$ are not orientable, then we choose a two-sheeted covering $\tilde{B}$ of the manifold B such that for the binary bundle induced by it

the fibers of the bundle $\tilde{A}\to C$ are orientable manifolds. On $\tilde{B}$ we choose the space $\Omega^r(\tilde{B})$ of differential forms which are odd under involution. The operation $\mathcal{I}$ in this case defines a mapping

$$\mathcal{I}:\Omega^r(\tilde{B})\to\Omega^{r-N}(\tilde{C}),$$

where $\tilde{C}$ is the manifold of oriented fibers of the bundle $\tilde{A}\to C$ (i.e., a two-sheeted covering of C), and $\Omega^{r-N}(\tilde{C})$ is the space of differential forms which are odd under the involution of $\tilde{C}$.

In the case of complex manifolds a canonical orientation is introduced on the fibers, and all considerations are carried out in the language of this orientation.

3. Admissibility of the Forms $\mathcal{I}\omega$. We shall establish some general properties of differential forms $\sigma\in\Omega(C)$ which can be represented in the form $\sigma=\mathcal{I}\omega$. First of all, from Proposition 4 we obtain the following result.

Proposition 6.

$$\mathcal{I}\circ d = d\circ\mathcal{I},$$

where d is the operator of exterior differentiation.

COROLLARY. The image of an exact form is an exact form; the image of a closed form is a closed form.

Remark. The converse is not true in general. However, if the mapping $\mathcal{I}$ is injective on forms of degree one greater, then the preimage of a closed form (if it exists) is a closed form. Indeed, if $d\mathcal{I}\omega=0$, then $\mathcal{I}d\omega=0$, and therefore $d\omega=0$.

Definition. We call an $(r-N)$-dimensional subspace $L\subset T_cC$, $c\in C$, degenerate if for any point $a\in\pi_2^{-1}$ the preimage $L_a\subset T_a$ of the space L is not transversal to the fibers of the bundle $\pi_1: A\to B$ (i.e., it contains at least one nonzero vector tangent to a fiber of this bundle).

Since the fiber of the bundle $A\to B$ passing through the point $a=(b, c)$ is (b, C_b), the present definition is equivalent to the following definition.

Definition. An $(r-N)$-dimensional subspace $L\subset T_c$ is called degenerate if for any point $b\in B$ incident with c it contains a nonzero vector tangent to the submanifold $C_b\subset C$.

Definition. We call a differential $(r-N)$-form σ on C admissible if $\sigma(c;\eta_1,\ldots,\eta_{r-N})=0$ whenever the subspace $L\subset T_c$ spanned by $\eta_1,\ldots,\eta_{r-N}$ is degenerate.

Proposition 6. The differential $(r-N)$-form $\sigma=\mathcal{I}\omega$, $\omega\in\Omega^r(B)$, is admissible.

Proof. Let $c \in C$, and suppose that the subspace L spanned by the vectors $\eta_1, \ldots, \eta_{r-N}$ is degenerate. According to the definition, $\sigma(c; \eta_1, \ldots, \eta_{r-N})$ is obtained by integrating along $\pi_2^{-1} c$ the differential N-form

$$(\pi_1{}^*\omega)\ (a; \zeta_1, \ldots, \zeta_{r-N}, \xi_1, \ldots, \xi_N), \tag{4.5}$$

where $a \in \pi_2^{-1} c$, $\xi_1, \ldots, \xi_N$ are vectors tangent to the fiber of the fibration π_2 and $\pi_2 \zeta_i = \eta_i$, $i = 1, \ldots, r-N$ (expression (4.5) does not depend on the choice of the vectors ζ_i). If the vectors ζ_i, ξ_j are linearly independent, then the subspace L_a they generate is the full pre-image of the space L in T_a. Hence, by assumption, it contains a nonzero vector tangent to the fiber of the projection π_1 at the point a. Then since the differential form $\pi^*{}_1\omega$ has relative degree 0 in the fibration π_1, expression (4.5) is equal to zero. Hence, the integral of the differential form (4.5) along the fiber $\pi_2^{-1}c$ is also equal to zero.

COROLLARY. If $\sigma = \mathscr{I}\omega$, then $d\sigma$ is an admissible form.

Remark. The admissibility of σ does not, in general, imply the admissibility of $d\sigma$. For example (see Chap. III), the condition of admissibility of a 2-form σ on the manifold of lines in $\mathbf{R}^3$ is expressed by algebraic relations among its coefficients, while the admissibility of $d\sigma$ involves differential expressions.

4. Partially Closed Forms. Let $B_1 \subset B$ be an arbitrary submanifold. To B_1 we assign the subset C_1 of points $c \in C$ such that $B_c \subset B_1$. We call the submanifold B_1 regular if $C_1 \subset C$ is a submanifold in C, and $\dim C_1 \geqslant \dim B_1 - N$ (N is the dimension of a fiber of the bundle $A \to C$).

Definition. We call a differential form $\sigma \in \Omega^{r-N}(C)$ partially closed if its restriction to any submanifold $C_1 \subset C$ corresponding to a regular r-dimensional submanifold $B_1 \subset B$ is a closed form on C_1.

We note that in the case $r = \dim B$ the condition of partial closedness coincides with the condition of closedness.

Remark. If $\dim C_1 = r - N$, then the form $\sigma|_{C_1}$ is closed on C_1 as a form of maximal degree. Therefore, in the definition it is possible to restrict attention to submanifolds C_1 of dimension greater than $r - N$.

Proposition 7. If the form $d\sigma$ is admissible, then the form σ is partially closed.

Proof. Let $\deg \sigma = r$, and suppose that the submanifold $C_1 \subset C$ corresponds to a regular r-dimensional manifold $B_1 \subset B$, and $\dim C_1 = r - N + s$, $s > 0$. We must prove that $d\sigma|_{C_1} = 0$, i.e., for any $c \in C_1$ and any linearly independent vectors $\eta_1, \ldots, \eta_{r-N+1}$ in the tangent T_c to C_1 we have

$$d\sigma(c; \eta_1, \ldots, \eta_{r-N+1}) = 0. \tag{4.6}$$

We set $A_1 = \pi_2^{-1} C_1$, $B'_1 = \pi_1 A_1$, and we consider the binary bundle

$$\begin{array}{ccc} & A_1 & \\ {\scriptstyle\pi_1}\swarrow & & \searrow{\scriptstyle\pi_2} \\ B_1 & & C_1 \end{array}$$

induced by the original binary bundle

$$\begin{array}{ccc} & A & \\ {\scriptstyle\pi_1}\swarrow & & \searrow{\scriptstyle\pi_2} \\ B & & C \end{array}.$$

It is obvious that $\dim A_1=r+s$ and that $B_1'\subset B_1$; therefore, $B_1'=r'\leqslant r$. We consider the subspace $L\subset T_c$ spanned by the vectors $\eta_1,\ldots,\eta_{r-N+1}$, and we shall prove that L is degenerate. Let $a\in\pi_2^{-1}c$, and let T_a be the tangent space to A_1 at the point a, and let $L_a\subset T_a$ be the full preimage of L in the space T_a. We then have $\operatorname{codim} L_a=\operatorname{codim} L=s-1$, and hence $\dim L_a=r+1$. On the other hand, suppose that $M\subset T_a$ is the subspace tangent to a fiber of the bundle $A_1\to B_1'$. Since $\dim A_1=r+s$, $\dim B_1'\leqslant r$, it follows that $\dim M\geqslant s$. Hence, $L_a\cap M\neq 0$, i.e., the space L is degenerate. The admissibility of the form $d\sigma$ then implies equality (4.6).

COROLLARY. The differential forms $\sigma=\mathcal{I}\omega$ are partially closed.

3. Differential Forms on $\mathbf{RP}^n$ and the Grassmann Manifold $G_{p,n}$

In subsequent sections the operation $\mathcal{I}$ of integrating differential forms introduced in Sec. 2 for any binary bundles is studied in detail for the case where $B=\mathbf{RP}^n$ and $C=G_{p,n}$ — the manifold of p-dimensional planes in $\mathbf{RP}^n$; in this case we are dealing with the binary bundle

$$\mathbf{RP}^n \xleftarrow{\ \pi_1\ } A \xrightarrow{\ \pi_2\ } G_{p,n},$$

where A is the manifold of incident pairs (h, x), $h\in G_{p,n}$, $x\in\mathbf{RP}^n$, and π_1,π_2 are the canonical projections.

It is known that p-dimensional planes in $\mathbf{RP}^n$, i.e., the spaces $\mathbf{RP}^p$, are orientable for odd p and nonorientable for even p. Therefore, in order to define the operation $\mathcal{I}$ in the case of even p, we must replace $B=\mathbf{RP}^n$ by a two-sheeted covering $\tilde{B}$ such that for the induced binary bundle $\tilde{B}\leftarrow\tilde{A}\to G_{p,n}$ the fibers of the bundle $\tilde{A}\to G_{p,n}$ are orientable. The sphere S^n is such a two-sheeted covering.

It is convenient to combine the even- and odd-dimensional cases by considering differential forms on S^n in both cases. The only difference is that for odd p forms are taken which are even with respect to the involution $x\mapsto -x$ (and they therefore drop onto $\mathbf{RP}^n$), while for even p odd forms are taken. The corresponding spaces of C^∞ differential forms we denote by $\Omega^r(\mathbf{RP}^n)$. The operators $\mathcal{I}=\mathcal{I}_p$ realize the mappings

$$\mathcal{I}_p:\Omega^r(\mathbf{RP}^n)\to\Omega^{r-p}(G_{p,n}).$$

Here $\Omega^{r-p}(G_{p,n})$ is the space of odd C^∞ differential forms on a two-sheeted covering of $G_{p,n}$ (the latter consists of oriented sections of the sphere S^n by $(p+1)$-dimensional subspaces or, equivalently, of oriented $(p+1)$-subspaces in $\mathbf{R}^{n+1}$).

Differential forms on $\mathbf{RP}^n$ and $G_{p,n}$ are conveniently written in homogeneous coordinates, i.e., in the lifts, respectively, to $\mathbf{R}^{n+1}\setminus 0$ and to the manifold E_p of $(p+1)$-frames $u=(u_1,\ldots,u_{p+1})$ in $\mathbf{R}^{n+1}$. Thus, any form $\omega\in\Omega^r(\mathbf{RP}^n)$ has the form

$$\omega=\sum_{|I|=r} f_I(x)\,dx^I,$$

where $I=(i_1,\ldots,i_r)$, $dx^I=dx^{i_1}\wedge\ldots\wedge dx^{i_r}$, and the coefficients $f_I=f_{i_1,\ldots,i_r}$ are assumed to be antisymmetric with respect to $i_1,\ldots,i_r$. The conditions that $\omega\in\Omega^r(\mathbf{RP}^n)$, i.e., that ω drops to S^n and has the appropriate parity, are

1) $f_I(\lambda x)=\lambda^{-r}f_I(x)$ for $\lambda>0$, $f_I(-x)=(-1)^{r-p+1}f_I(x)$;

2) $\sum_{i_1} f_{i_1,\ldots,i_r}(x)\, x^{i_1}=0.$

Any s-form on $G_{p,n}$, lifted to E_p, has the form

$$\sigma=\sum_{|I|=|J|=s}\varphi_I^J(u)\, du_J^I,$$

where $du_J^I=du_{j_1}^{i_1}\wedge\ldots\wedge du_{j_s}^{i_s}$, and $\varphi_I^J=\varphi_{i_1,\ldots,i_s}^{j_1,\ldots,j_s}$ are antisymmetric with respect to $(i_1, j_1),\ldots,(i_s, j_s)$. The conditions that σ drop to $G_{p,n}$ are involved; they will be formulated below only for forms of special type.

Sometimes in the notation for the coefficients of forms it will be useful to break the multiindices into parts. For example, in place of f_I, $|I|=r$ it is possible to write f_{iI}, $|I|=r-1$ or $f_{I',I''}$, where $|I'|=r'$, $|I''|=r''$, $r'+r''=r$. In particular, condition 2) for dropping the form ω can conviently be written in the form $\sum_i f_{iI}(x)\, x^i=0$.

1. The Condition of Admissibility. According to Sec. 2, a linear subspace $L\subset T_h$ $(h\in G_{p,n})$ is called degenerate if for any point $x\in h$ it is not transversal to the submanifold of planes $h'\in G_{p,n}$ incident with x, i.e., it contains a nonzero vector tangent to this submanifold. Roughly speaking, L is degenerate if infinitesimal shifts of h in $\mathbf{RP}^n$ in the directions of L fill out a plane of dimension less than $\dim L+p$, while for a subspace L in general position this dimension is equal to $\dim L+p$. A differential form of degree s on $G_{p,n}$ is called admissible if it is equal to zero on any degenerate s-dimensional subspace.

THEOREM 1. Suppose that the differential s-form σ on E_p drops to $G_{p,n}$. Then the following conditions are equivalent:

1) the form σ is admissible;

2) the restriction of σ to a submanifold of $(p+1)$-dimensional subspaces lying in any fixed $(s+p)$-dimensional space is equal to zero;

3) the coefficients $\varphi_I^J=\varphi_{i_1,\ldots,i_s}^{j_1,\ldots,j_s}$ of the form σ are antisymmetric with respect to $i_1,\ldots,i_s$.

Proof. Since the elements of E_p are given by $(n+1)\times(p+1)$ matrices of rank $p+1$, the tangent space at each point of E_p is naturally identified with the space T of all $(n+1)\times(p+1)$ matrices. We choose in T the basis $\{e_j^i\}_{i=1,\ldots,n+1,\; j=1,\ldots,p+1}$, where e_j^i is a matrix with a 1 at the intersection of the j-th column and i-th row and zeros elsewhere. It is obvious that if $\sigma=\sum_{|I|=|J|=s}\varphi_I^J(u)\, du_J^I$, then

$$\varphi_{i_1,\ldots,i_s}^{j_1,\ldots,j_s}(u)=\frac{1}{s!}\,\sigma(u; e_{j_1}^{i_1},\ldots,e_{j_s}^{i_s}).$$

LEMMA. If the form σ is equal to zero on any system of vectors of the form

$$e_{j_1}^{i_1}+e_{j_2}^{i_1},\quad e_{j_2}^{i_2}+e_{j_1}^{i_2},\quad e_{j_3}^{i_3},\ldots,e_{j_s}^{i_s}, \tag{4.7}$$

then its coefficients φ_I^J are antisymmetric with respect to I.

Indeed, we have

$$0=\sigma(u; e_{j_1}^{i_1}+e_{j_2}^{i_1}, e_{j_2}^{i_2}+e_{j_1}^{i_2}, e_{j_3}^{i_3},\ldots,e_{j_s}^{i_s})=s!\,(\varphi_{i_1i_2I'}^J+\varphi_{i_1i_1I'}^J+\varphi_{i_2i_2I'}^J+\varphi_{i_2i_1I'}^J),$$

where $I'=(i_3,\ldots,i_s)$, $J=(j_1,\ldots,j_s)$. From this it follows immediately that $\varphi_{i_1i_2I'}^J=0$ for $i_1=i_2$ and $\varphi_{i_1i_2I'}^J+\varphi_{i_2i_1I'}^J=0$ for any i_1, i_2. Thus, the coefficients φ_I^J are antisymmetric in i_1, i_2 and hence with respect to any pair i_k, i_l, $k, l=1,\ldots,s$.

We return to the proof of the theorem. 1) 1⇒2. Let $L\subset T_h$ be the linear subspace tangent to the submanifold of $(p+1)$-dimensional subspaces belonging to a fixed $(s+p)$-dimensional subspace (containing h); let $L_x\subset T_h$ be the subspace tangent to the manifold of $(p+1)$-dimensional subspaces incident with $x\in h$. It is obvious that $\dim L=(p+1)(s-1)$, $\dim(L\cap L_x)=p(s-1)$, whence $\dim(L\cap L_x)+s=\dim L+1$. Therefore, any s-dimensional subspace $L'\subset L$ has nonzero intersection with L_x for any $x\in h$, and it is thus degenerate. Thus, if an s-form σ is admissible, then it is equal to zero on L.

2) 2⇒3. Let $h=(u_1,\dots,u_{p+1})$ and let $\zeta_1,\dots,\zeta_{s-1}$ be any vectors in R^{n+1}. We consider the $(s+p)$-dimensional subspace $L_{s+p}\subset R^{n+1}$ containing the vectors $u_1,\dots,u_{p+1},\zeta_1,\dots,\zeta_{s-1}$. Let M be the manifold of $(p+1)$-dimensional subspaces belonging to L_{s+p}, let $L\subset T_h$ be a subspace tangent to M, and let $\tilde{L}\subset T$ be the preimage of L in the standard tangent space to E_p. Since M contains any subspace with homogeneous coordinates

$$(u_1,\dots,u_{k-1},u_k+t\zeta_i,u_{k+1},\dots,u_{p+1}),\ 1\leqslant i\leqslant s-1,\ 1\leqslant k\leqslant p+1,$$

the space $\tilde{L}$ contains matrices of the form

$$(0,\dots,0,\zeta_i,0,\dots,0), \tag{4.8}$$

where ζ_i stands at the k-th place. Hence, if the form satisfies condition 2, then it is equal to zero on any system of s matrices of the form (4.8). In particular, it is equal to zero on matrices of the form $e^{i_1}_{j_1}+e^{i_1}_{j_1'}, e^{i_1}_{j_2}+e^{i_1}_{j_2'}, e^{i_2}_{j_3},\dots,e^{i_s}_{j_s}$. But then, according to the lemma, it satisfies condition 3.

3) 3⇒1. We consider a coordinate neighborhood in $G_{p,n}$ in which, e.g., $\det\|u^i_j\|_{i=n-p+1,\dots,n+1;\ j=1,\dots,p+1}\neq 0$. The elements of this neighborhood may be given by matrices of the special form $u=\binom{v}{e}$, where e is the identity $(p+1)$ matrix. We choose the elements of the $(n-p)\times(p+1)$ matrix v as coordinates on $G_{p,n}$, and we write the form in these coordinates: $\sigma=\sum\varphi^J_I(v)\,dv^I_J$. It follows from condition 3 that the coefficients φ^J_I are symmetric in J and antisymmetric in I. To prove the admissibility of the form σ it suffices to show that in general between the coefficients φ^J_I of the admissible form there do not exist other linear relations aside from the conditions of antisymmetry. To this end we consider the admissible forms $\sigma=\mathcal{J}_p\omega$, where the form ω in the inhomogeneous coordinates $x^1,\dots,x^n$ $(x^{n+1}=1)$ has the following form: $\omega=\sum f_{I,n-p+1,\dots,n}(x)\,dx^I\wedge dx^{n-p+1}\wedge\dots\wedge dx^n$. In this case the coefficients φ^J_I of the form $\sigma=\mathcal{J}_p\omega$ are expressed by the formulas

$$\varphi_I^{j_1,\dots,j_{r-p}}(v)=\int f_{I,n-p+1,\dots,n}(x)\,x^{j'_1}\dots x^{j'_{r-p}}dx^{n-p+1}\wedge\dots\wedge dx^n,$$

where $j'_k=j_k+n-p$, and the integral is taken over the plane with coordinates v. From these explicit formulas for φ^J_I the assertion follows easily.

<u>2. Cone of Incidence and Admissibility.</u> We say that two p-dimensional planes h_1, h_2 in RP^n are incident if $\dim(h_1\cap h_2)=p-1$. With each p-dimensional plane $h\in G_{p,n}$ we associate the set G_h of all planes $h'\in G_{p,n}$ incident with it. It is not hard to verify that $\dim G_h=n$.

We denote by K_h the set of all vectors of the tangent space T_h which are tangent to G_h, and we call K_h the cone of incidence. On the cone K_h there are two families Π_1 and Π_2 of generators. Each generator of the family Π_1 is given by a $(p+1)$-dimensional plane

$h_{p+1} \supset h$. Namely, we must take the manifold $\mathbf{RP}^{p+1}$ of all p-dimensional planes in $\mathbf{RP}^n$ belonging to h_{p+1}. The tangent space to this manifold at the point h is by definition the corresponding generator of the family Π_1. Thus, Π_1 is an $(n-p-1)$-dimensional family of $(p+1)$-dimensional planes. Each generator of the family Π_2 is given by a $(p-1)$-dimensional plane $h_{p-1} \subset h$. Namely, we must take the manifold $\mathbf{RP}^{n-p}$ of all p-dimensional planes containing h_{p-1}. The tangent space to this manifold is by definition the corresponding generator of the family Π_2. Thus, Π_2 is a p-dimensional family of $(n-p)$-dimensional planes. It is obvious that two planes of one family intersect only at the vertex of the cone of incidence K_h, and any two planes of different families intersect along a line. Moreover, through each point of the cone K_h distinct from the vertex there passes precisely one plane of each family.

Proposition 8. A differential s-form on $G_{p,n}$ is admissible if and only if it is equal to zero on any s-dimensional subspace $L \subset T_h$ such that the dimension of its intersection with at least one plane of the family Π_1 is not less than 2.

Proof. Let L_1 be a generator of the family Π_1, and let $L_2 \subset L_1$ be any linear subspace with $\dim L_2 \geqslant 2$. We shall prove that the subspace L_2 is degenerate. Indeed, let $L^x \subset T_h$ be the subspace tangent to the manifold of p-dimensional planes in $\mathbf{RP}^n$ passing through the point $x \in h$. It is not hard to see that $\dim (L^x \cap L_1) = p$. Thus, since $\dim L_1 = p+1$, it follows that $\dim (L^x \cap L_2) \geqslant 1$.

Now suppose that the subspace $L \subset T_h$ satisfies the condition of the proposition. By what has already been proved, L is degenerate. Hence, if a form is admissible, then it is equal to zero on L. Conversely, suppose that a form σ is equal to zero on each subspace satisfying the conditions of the proposition. We consider a system of vectors of form (4.7). It is obvious that the first two of these vectors define a subspace in T_h belonging to one of the generators of the family Π_1. Hence, the subspace L spanned by the vectors (4.7) satisfies the conditions of the proposition, and therefore the form σ is equal to zero on these vectors. By the lemma and Theorem 1, the form σ is admissible.

3. A Remark on the Dual Problem. There exists a natural identification of $G_{p,n}$, realized as a manifold of p-dimensional planes in $\mathbf{RP}^n$, with $G_{n-p-1,n}$, realized as a manifold of $(n-p-1)$-dimensional planes in the dual space $(\mathbf{RP}^n)'$ (the points of $(\mathbf{RP}^n)'$ are hyperplanes in $\mathbf{RP}^n$). Under this identification the cone of incidence is preserved, while the families of generators Π_1, Π_2 change places. We consider the integral transform consisting in integrating differential forms on $(\mathbf{RP}^n)'$ over $(n-p-1)$-dimensional planes. The differential forms on $G_{p,n} \cong G_{n-p-1,n}$ hereby obtained satisfy the condition of admissibility formulated in Proposition 8 with the difference that the family Π_1 must be replaced by Π_2.

4. Condition of Partial Closedness. We call an s-form on $G_{p,n}$ partially closed ($s \leqslant n-p$) if its restriction to the submanifold of p-dimensional planes belonging to any $(s+p)$-dimensional plane is closed in this submanifold. By Theorem 1, the partial closedness of σ is equivalent to the admissibility of the form $d\sigma$. We note that for $s=n-p$ the condition of partial closedness coincides with the condition of closedness.

We introduce the notation $\partial^k_i = \partial/\partial u^i_k$.

<u>Proposition 9.</u> If an s-form σ on E_p

$$\sigma=\sum \varphi_I^J(u)\,du_J^I$$

drops to $G_{p,n}$ and is admissible, then the condition that it be partially closed is equivalent to the following relation between its coefficients:

$$\partial_i^k\varphi_{lI'}^{jJ'}-\partial_i^j\varphi_{lI'}^{kJ'}=\partial_l^j\varphi_{iI'}^{kJ'}-\partial_l^k\varphi_{iI'}^{jJ'}$$
$$(1\leqslant j,k\leqslant p+1,\ 1\leqslant i,l\leqslant n+1). \tag{4.9}$$

<u>Proof.</u> It is obvious that

$$d\sigma=\frac{1}{s+1}\sum_{|I|=|J|=s+1}\psi_I^J(u)\,du_J^I,$$

where

$$\psi_I^J=\psi_{i_0,\ldots,i_s}^{j_0,\ldots,j_s}=\sum_{k=0}^{p}(-1)^k\,\partial_{i_k}^{j_k}\varphi_{i_0,\ldots,\hat{i}_k,\ldots,i_s}^{j_0,\ldots,\hat{j}_k,\ldots,j_s}.$$

By Theorem 1, the condition of closedness of $d\sigma$ is equivalent to the condition that the coefficients ψ_I^J are symmetric in J (are antisymmetric in I). Since the form σ is admissible, its coefficients φ_I^J are symmetric in J; therefore, the condition of admissibility of $d\sigma$ is equivalent to ψ_I^J being symmetric with respect to any two fixed indices of J, e.g., j_0, j_1, i.e.,

$$\partial_{i_0}^{j_0}\varphi_{i_1,\ldots,i_s}^{j_1,\ldots,j_s}-\partial_{i_1}^{j_1}\varphi_{i_0,i_2,\ldots,i_s}^{j_0,j_2,\ldots,j_s}+\sum_{k=2}^{s}(-1)^k\,\partial_{i_k}^{j_k}\varphi_{i_0,\ldots,\hat{i}_k,\ldots,i_s}^{j_0,\ldots,\hat{j}_k,\ldots,j_s}=\partial_{i_0}^{j_1}\varphi_{i_1,\ldots,i_s}^{j_0,j_2,\ldots,j_s}-\partial_{i_1}^{j_0}\varphi_{i_0,i_2,\ldots,i_s}^{j_1,\ldots,j_s}+\sum_{k=2}^{s}(-1)^k\,\partial_{i_k}^{j_k}\varphi_{i_0,\ldots,\hat{i}_k,\ldots,i_s}^{j_1,j_0,j_2,\ldots,\hat{j}_k,\ldots,j_s}.$$

In view of the symmetry of φ_I^J in J, the corresponding terms following the sums on the left and right sides of the equality coincide, and we obtain the relation (4.9).

5. <u>Conditions for Dropping to $G_{p,n}$.</u> <u>Proposition 10.</u> If the coefficients of the s-form σ are antisymmetric in the lower incides, then the condition that this form drop to $G_{p,n}$ is equivalent to the following relations:

$$1)\ \sum_{k=1}^{n+1}u_{j'}^k\partial_k^l\varphi_I^J=0,\quad \text{if}\quad j'\notin J; \tag{4.10}$$

$$\sum_{k=1}^{n+1}u_{j'}^k\partial_k^l\varphi_I^{j'J'}+l_{j'J'}\varphi_I^{lJ'}=0,$$

where $l_{j'J'}$ is the multiplicity with which j' enters the multiindex (j', J');

$$2)\ \sum_{i=1}^{n+1}u_j^i\varphi_{iI'}^J=0\ \text{ for any }\quad J,\ I'\text{ and }j. \tag{4.11}$$

<u>Proof.</u> The condition that the form σ drop to $G_{p,n}$ is equivalent to the conditions that 1) the form is invariant under $GL_+(p+1,\mathrm{R})$ and 2) the form is preserved by the substitutions $du_j^i\to du_j^i+\sum_k u_k^i dg_j^k$. It is obvious that the second condition is equivalent to relation (4.11). Further, the invariance of σ under $GL_+(p+1,\mathrm{R})$ is equivalent to invariance under any transformation of the form $u_j\to u_j+\lambda u_{j'}$ $(1\leqslant j,\ j'\leqslant p+1)$. It is not hard to see that the condition of invariance of σ under the transformations $u_j\to u_j+\lambda u_{j'}$ is expressed by the relations (4.10).

COROLLARY. All the coefficients φ_I^J of an admissible form σ on $G_{p,n}$ can be uniquely expressed in terms of the coefficients $\varphi_I^{p+1,\dots,p+1}$.

Remark. We recall that the differential forms σ on $G_{p,n}$ are assumed to be odd. This added condition is equivalent to the condition that σ change sign under the transformation $u_1 \to -u_1$.

4. Some Properties of the Form $\mathcal{I}_p\omega$

The problem of this section is to find an explicit form for the coefficients of the form $\mathcal{I}_p\omega$ and to obtain relations for these coefficients. The extent to which these conditions follow from the conditions of admissibility and partial closedness will be established below.

1. An Expression for the Coefficients of the Form $\mathcal{I}_p\omega$. Suppose there is given an r-form $\omega \in \Omega^r(\mathbf{RP}^n)$, where $r > p$,

$$\omega = \sum_{|J|=r} f_I(x)\, dx^I.$$

Then

$$\mathcal{I}_p\omega = \sum_{|I|=|J|=r-p} \varphi_I^J(u)\, du_J^I, \quad \mathcal{I}_p\omega \in \Omega^{r-p}(G_{p,n}).$$

The form $\mathcal{I}_p\omega$ is admissible and partially closed (see Sec. 2). Therefore, its coefficients φ_I^J are antisymmetric in I (are symmetric in J) and satisfy relations (4.9). Moreover, since $\mathcal{I}_p\omega$ is a form on $G_{p,n}$, the coefficients φ_I^J satisfy the conditions of dropping to $G_{p,n}$ given in part 5, Sec. 3.

We introduce the following functions on E_p:

$$\hat{f}_I^J(u) = \frac{1}{2}\int_{S^p} f_I(tu)\, t^J \sigma_p(t, dt), \tag{4.12}$$

where $J = (j_1, \dots, j_{r-p-1})$, $1 \leqslant j_k \leqslant p+1$; $t^J = t^{j_1} \dots t^{j_{r-p-1}}$, $tu = t^1u_1 + \dots + t^{p+1}u_{p+1}$; $\sigma_p(t, dt) = \sum_{k=1}^{p+1}(-1)^{k-1} t^k dt^1 \wedge \dots \wedge \widehat{dt^k} \wedge \dots \wedge dt^{p+1}$ (the integral is taken over the sphere S^p or over any cycle in $\mathbf{R}^{p+1}_t \setminus 0$ homologous to S^p).

We observe that if the parity of ω is equal to the parity of p, then $\hat{f}_I^J(u) \equiv 0$.

Proposition 11. For any k and J', where $1 \leqslant k \leqslant p+1$, $|J'| = r-p-1$, we have

$$\varphi_I^{kJ'}(u) = (-1)^{k-1} \frac{r!}{(r-p)!} \sum_{|I'|=p} \hat{f}_{I,I'}^{J'}(u)\, u_{\hat{k}}^{I'}, \tag{4.13}$$

where $\hat{k} = (1, \dots, \hat{k}, \dots, p+1)$.

Proof. In order to obtain $\mathcal{I}_p\omega$, it is first necessary to substitute, into ω, $x = \sum_j t^j u_j$, $dx = \sum_j (dt^j \cdot u_j + t^j du_j)$ and in the expression obtained to separate out terms of degree p in dt. We obtain

$$\omega = C_r^p \sum f_{I,I'}(tu)\, u_{J'}^{I'} t^J du_J^I \wedge dt^{J'} = C_r^p \sum f_{II'}(tu)\, u_{J'}^{I'} t^{j} t^J du_j^i \wedge du_J^I \wedge dt^{J'},$$

where in the last expression $|I| = |J| = r-p-1$, $|I'| = |J'| = p$. We collect terms with fixed I, J, i.e., we consider the expressions

$$\omega_I^J = \sum f_{II'}(tu)\, u_{J'}^{I'} t^j du_j^i \wedge dt^{J'},$$

where the summation is taken only over i, j, I', J'. Here it may be assumed that the indices $J'=(j'_1, \ldots, j'_p)$ are pairwise distinct. We denote by j'_{p+1} an index distinct from $j'_1, \ldots, j'_p$, $1 \leq j'_{p+1} \leq p+1$, so that $j'_1, \ldots, j'_{p+1}$ form a permutation of the numbers $1, \ldots, p+1$. We then have

$$\omega_I^J = \sum_{i,I',J'} f_{iII'}(tu)\, u_{J'}^{I'} t^{j'_{p+1}} du^i_{j'_{p+1}} \wedge dt^{J'} + \sum_{i,I',J'} \sum_{k=1}^{p} f_{iII'}(tu)\, u_{J'}^{I'} t^{j'_k} du^i_{j'_k} \wedge dt^{J'}. \tag{4.14}$$

We make use of the equality $\sum_i f_{iI}(x)x^i = 0$; from this it follows that for any $k=1, \ldots, p$ we have

$$\sum_{i'_k} f_{iII'} t^{j'_k} u^{i'_k}_{j'_k} = -\sum_{i'_k} \sum_{l \neq k} f_{iII'} t^{j'_l} u^{i'_k}_{j'_l} = -\sum_{i'_k} f_{iII'} t^{j'_{p+1}} u^{i'_k}_{j'_{p+1}} - \sum_{i'_k} \sum_{\substack{l=1 \\ (l \neq k)}}^{p} f_{iII'} t^{j'_l} u^{i'_k}_{j'_l}.$$

Therefore, the second term in (4.14) is equal to

$$-\sum_{i,I',J'} \sum_{k=1}^{p} f_{iII'} u^{i'_1}_{j'_1} \ldots u^{i'_k}_{j'_{p+1}} \ldots u^{i'_p}_{j'_p} t^{j'_{p+1}} du^i_{j'_k} \wedge dt^{J'} - \sum_{i,I',J'} \sum_{k=1}^{p} \sum_{\substack{l=1 \\ (l \neq k)}}^{p} f_{iII'} u^{i'_1}_{j'_1} \ldots u^{i'_k}_{j'_l} \ldots u^{i'_p}_{j'_p} t^{j'_l} du^i_{j'_k} \wedge dt^{J'}. \tag{4.15}$$

Since $f_{iII'}$ is antisymmetric, the second term in (4.15) is equal to zero (since, on the one hand, it is preserved by permutation of the indices i'_k and i'_l and, on the other hand, it changes sign under this permutation). In the first term in (4.15) we change notation: $j'_{p+1} \leftrightarrow j'_k$. It then has the form

$$\sum_{i,I',J'} \sum_{k=1}^{p} (-1)^{p-k+1} f_{iII'} u^{I'}_{J'} du^i_{j'_{p+1}} \wedge \left(t^{j'_k} dt^{j'_1} \wedge \ldots \wedge \widehat{dt^{j'_k}} \wedge \ldots \wedge dt^{j'_{p+1}}\right).$$

As a result, we obtain

$$\omega_I^J = (-1)^p \sum_{i,I',J'} \operatorname{sgn}(j'_1, \ldots, j'_{p+1})\, f_{iII'}(tu)\, u^{I'}_{J'} du^i_{j'_{p+1}} \wedge \sigma_p(t, dt).$$

Hence,

$$\omega = (-1)^p C^p_r \sum f_{iII'}(tu)\, u^{I'}_{J'} t^J \operatorname{sgn}(j'_1, \ldots, j'_{p+1})\, du^i_{j'_{p+1}} \wedge du^I_J \wedge \sigma_p(t, dt).$$

Integrating on t, we thus obtain

$$\mathcal{I}_p \omega = (-1)^p C^p_r \sum \hat{f}^J_{iII'}(u) \operatorname{sgn}(j'_1, \ldots, j'_{p+1})\, u^{I'}_{J'} du^i_{j'_{p+1}} \wedge du^I_J,$$

where $\hat{f}^J_{iII'}$ is given by formula (4.12). Using the antisymmetry of the coefficients in the lower indices, this expression can be represented as follows:

$$\mathcal{I}_p \omega = \frac{r!}{(r-p)!} \sum \hat{f}^J_{iII'}(u) (-1)^{k-1} u^{I'}_{\hat{k}} du^i_k \wedge du^I_J,$$

where $\hat{k} = (1, \ldots, \hat{k}, \ldots, p+1)$, i.e.,

$$\mathcal{I}_p \omega = \sum \varphi^{k,J}_{iI}(u)\, du^i_k \wedge du^I_J,$$

where

$$\varphi^{k,J}_I(u) = (-1)^{k-1} \frac{r!}{(r-p)!} \sum \hat{f}^J_{I,I'}(u)\, u^{I'}_{\hat{k}}. \tag{4.16}$$

It remains to verify that the coefficients φ^{kJ}_I are antisymmetric with respect to simultaneous permutations of the upper and lower indices. First of all, it follows from (4.16) that they are antisymmetric relative to I and symmetric relative to J. It therefore suffices to prove that

$$(-1)^{k-1}\sum \hat{f}^{l,J'}_{I,I'}(u)\,u^{l'}_{k} = (-1)^{l-1}\sum \hat{f}^{k,J'}_{I,I'}(u)\,u^{l'}_{l}.$$

To this end we make use of the relation

$$\sum_{i_l} \hat{f}^{l,J'}_{I,I'} u^{i}_{l} = -\sum_{s\neq l}\sum_{i_l} \hat{f}^{s,J'}_{I,I'} u^{i}_{s}.$$

It implies that for $l<k$

$$\sum_{I'} \hat{f}^{l,J'}_{I,I'}(u)\,u^{i'_1}_{1}\ldots u^{i'_{k-1}}_{k-1}u^{i'_k}_{k+1}\ldots u^{i'_p}_{p+1} = -\sum_{s\neq l}\sum_{I'} \hat{f}^{s,J'}_{I,I'} u^{i'_1}_{1}\ldots u^{i'_l}_{s}\ldots u^{i'_{k-1}}_{k-1}u^{i'_k}_{k+1}\ldots u^{i'_p}_{p+1}.$$

In view of the antisymmetry of $\hat{f}^{s,J'}_{I,I'}$ in I', the inner sum is equal to zero for $s\neq k$. Hence,

$$\sum_{I'} \hat{f}^{l,J'}_{I,I'}(u)\,u^{I'}_{\hat{k}} = -\sum \hat{f}^{k,J'}_{I,I'}(u)\,u^{i'_1}_{1}\ldots u^{i'_l}_{k}\ldots u^{i'_{k-1}}_{k-1}u^{i'_k}_{k+1}\ldots u^{i'_p}_{p+1} = (-1)^{k-l}\sum \hat{f}^{k,J'}_{I,I'}(u)\,u^{I'}_{\hat{l}},$$

as required.

2. Reconstruction of the Functions $\hat{f}^{J}_{I}(u)$ on the Basis of the Coefficients of the Form $\mathcal{J}_p\omega$.

Proposition 12. The functions $\hat{f}^{J}_{I}(u)$ defined by Eq. (4.12) can be expressed in terms of the coefficients φ^{J}_{I} of the form $\mathcal{J}_p\omega$ as follows:

$$\hat{f}^{J}_{I,I'}(u) = \frac{(r-p)!}{r!\,p!}\sum_{k=1}^{p+1}(-1)^{k-1}\Delta^{\hat{k}}_{I'}\varphi^{kJ}_{I}(u), \tag{4.17}$$

where $\Delta^{J}_{I} \equiv \Delta^{j_1,\ldots,j_p}_{i_1,\ldots,i_p} = \det\left\|\partial^{j_\beta}_{i_\alpha}\right\|_{\alpha,\beta=1,\ldots,p}$; in somewhat different notation

$$\hat{f}^{J}_{I,i'_1,\ldots,i'_p}(u) = \frac{(r-p)!}{r!\,p!}(-1)^{p}\sum \operatorname{sgn}(j_1,\ldots,j_{p+1})\,\partial^{j_1}_{i'_1}\ldots\partial^{j_p}_{i'_p}\varphi^{j_{p+1},J}_{I}(u),$$

where the sum is taken over all permutations of the indices $1,\ldots,p+1$.

Proof. According to Proposition 11,

$$\varphi^{k} = (-1)^{k-1}\frac{r!}{(r-p)!}\sum \hat{f}_{i'_1,\ldots,i'_p}\,u^{i'_1}_{1}\ldots u^{i'_{k-1}}_{k-1}u^{i'_k}_{k+1}\ldots u^{i'_p}_{p+1}$$

(to simplify notation we write φ^{k} in place of φ^{kJ}_{I} and $\hat{f}_{i'_1,\ldots,i'_p}$ in place of $\hat{f}^{J}_{I,i'_1,\ldots,i'_p}$). To this equality we apply the operator

$$\Delta^{1,\ldots,\hat{k},\ldots,p+1}_{i_1,\ldots,i_p} = \sum \operatorname{sgn}(\alpha_1,\ldots,\alpha_p)\,\partial^{1}_{i_{\alpha_1}}\ldots\partial^{k-1}_{i_{\alpha_{k-1}}}\partial^{k+1}_{i_{\alpha_k}}\ldots\partial^{p+1}_{i_{\alpha_p}}.$$

Since the functions $\hat{f}_I$ satisfy the relations $\partial^{i_1}_{k_1}\partial^{i_2}_{k_2}\hat{f}_I = \partial^{i_2}_{k_1}\partial^{i_1}_{k_2}\hat{f}_I$, the right side contains derivatives of the function $\hat{f}_I$ of order no higher than first. As a result, we obtain

$$(-1)^{k-1}\Delta^{1,\ldots,\hat{k},\ldots,p+1}_{i_1,\ldots,i_p}\varphi^{k} = \frac{r!\,p!}{(r-p)!}\hat{f}_{i_1,\ldots,i_p} +$$

$$+\frac{r!}{(r-p)!}\sum_{l=1}^{p}\sum_{(\alpha_1,\ldots,\alpha_p)}\sum_{i'_l}\operatorname{sgn}(\alpha_1,\ldots,\alpha_p)\,\partial^{r_{l,k}}_{i_{\alpha_l}}\hat{f}_{i_{\alpha_1},\ldots,i_{\alpha_{l-1}},i'_l,i_{\alpha_{l+1}},\ldots,i_{\alpha_p}}\,u^{i'_l}_{r_{l,k}}, \tag{4.18}$$

where $r_{l,k}=l$ for $l\leqslant k-1$, $r_{l,k}=l+1$ for $l\geqslant k$. We transform the second term separately:

$$U^{k} = \sum_{(\alpha_1,\ldots,\alpha_p)}\sum_{l=1}^{p}\sum_{i'_l}\operatorname{sgn}(\alpha_1,\ldots,\alpha_p)\,u^{i'_l}_{r_{l,k}}\partial^{r_{l,k}}_{i_{\alpha_l}}\hat{f}_{i_{\alpha_1},\ldots,i_{\alpha_{l-1}},i'_l,i_{\alpha_{l+1}},\ldots,i_{\alpha_p}} =$$

$$=(p-1)!\sum_{l=1}^{p}\sum_{s=1}^{p}\sum_{i'}(-1)^{s-1}u^{i'}_{r_{l,k}}\partial^{r_{l,k}}_{i_s}\hat{f}_{i',i_1,\ldots,i_{s-1},i_{s+1},\ldots,i_p} = (p-1)!\sum_{r\neq k}\sum_{s=1}^{p}\sum_{i'}(-1)^{s-1}u^{i'}_{r}\partial^{r}_{i_s}\hat{f}_{i',i_1,\ldots,i_{s-1},i_{s+1},\ldots,i_p}.$$

Summing on k, we obtain

$$\sum_{k=1}^{p+1}U^{k} = p!\sum_{r=1}^{p+1}\sum_{s=1}^{p}\sum_{i'}(-1)^{s-1}u^{i'}_{r}\partial^{r}_{i_s}\hat{f}_{i',i_1,\ldots,i_{s-1},i_{s+1},\ldots,i_p}.$$

We shall prove that

$$\sum_{l'}\sum_{r} u_r^{l'}\partial_l^r \hat{f}_{l'I} = -\hat{f}_{lI}. \tag{4.19}$$

Indeed, since $\sum_{l'} f_{l'I}x^{l'}=0$, it follows that $\sum_{l'} \frac{\partial f_{l'I}}{\partial x^l}x^{l'} = -f_{lI}$. Hence,

$$\sum_{l'}\sum_{r} u_r^{l'}\partial_l^r \hat{f}_{l'I} = \frac{1}{2}\sum_{l'}\int_{S^p}\frac{\partial}{\partial x^l} f_{l'I}(tu)(tu)^{l'}\, t^J\sigma_p(t, dt) = -\frac{1}{2}\int_{S^p} f_{lI}(tu)\,t^J\,\sigma_p(t, dt) = -\hat{f}_{lI}.$$

On the basis of Eq. (4.19) we obtain

$$\sum_{l'}\sum_{r} u_r^{l'}\partial_{l_s}^r \hat{f}_{l', l_1,\dots,l_{s-1}, l_{s+1},\dots,l_p} = -\hat{f}_{l_s, l_1,\dots,l_{s-1}, l_{s+1},\dots,l_p} = (-1)^s\hat{f}_{l_1,\dots,l_p}.$$

Hence,

$$\sum_{k=1}^{p+1} U^k = -p!\,p\hat{f}_{l_1,\dots,l_p}. \tag{4.20}$$

On the basis of Eqs. (4.18) and (4.20) we obtain

$$\sum_k (-1)^{k-1}\Delta_{l_1,\dots,l_p}^{1,\dots,\hat{k},\dots,p+1}\varphi^k = \frac{r!}{(r-p)!}((p+1)!\,\hat{f}_{l_1,\dots,l_p} - p!\,p\hat{f}_{l_1,\dots,l_p}) = \frac{r!\,p!}{(r-p)!}\hat{f}_{l_1,\dots,l_p},$$

as required.

3. Relations between the Coefficients of the Form $\mathcal{I}_p\omega$. From the definition of the functions $\hat{f}_I^J$ we immediately obtain the following result.

Proposition 13. The functions $\hat{f}_I^J(u)$, defined by Eq. (4.12) satisfy the following relations:

$$\partial_i^{l_1}\hat{f}_I^{l_2 J} = \partial_i^{l_2}\hat{f}_I^{l_1 J}, \tag{4.21}$$

$$(\partial_{i_1}^{l_1}\partial_{i_2}^{l_2} - \partial_{i_1}^{l_2}\partial_{i_2}^{l_1})\hat{f}_I^J \tag{4.22}$$

for any $l_1, l_2=1,\dots,p+1$, $i, i_1, i_2=1,\dots,n+1$.

Since the functions $\hat{f}_I^J$ can be expressed in terms of the coefficients φ_I^J of the form $\mathcal{I}_p\omega$ by formulas (4.17), from this we obtain the following conditions on the coefficients φ_I^J.

Proposition 14. The expressions $\sum_{k=1}^{p+1}(-1)^{k-1}\Delta_{I'}^{\hat{k}}\varphi_I^{kJ}$ are antisymmetric with respect to the set of indices I, I' and satisfy the relations

$$\partial_i^{l_1}\left(\sum_{k=1}^{p+1}(-1)^{k-1}\Delta_{I'}^{\hat{k}}\varphi_I^{kl_2J'}\right) = \partial_i^{l_2}\left(\sum_{k=1}^{p+1}(-1)^{k-1}\Delta_{I'}^{\hat{k}}\varphi_I^{kl_1J'}\right);$$

$$(\partial_{i_1}^{l_1}\partial_{i_2}^{l_2} - \partial_{i_1}^{l_2}\partial_{i_2}^{l_1})\left(\sum_{k=1}^{p+1}(-1)^{k-1}\Delta_{I'}^{\hat{k}}\varphi_I^{kJ}\right) = 0.$$

Proposition 15. If $r>p+1$, then relations (4.22) are a consequence of relations (4.21).

Proof. Since in the expression for $\hat{f}_I^J$ $|J|=r-p-1$, it follows that in the case $r>p+1$ we can write $J=(l, J')$, $|J'|=r-p-2$. Applying relation (4.21) several times, we have $\partial_{i_1}^{l_1}\partial_{i_2}^{l_2}\hat{f}_I^{lJ'} = \partial_{i_1}^{l_1}\partial_{i_2}^{l}\hat{f}_I^{l_2J'} = \partial_{i_1}^{l_2}\partial_{i_2}^{l}\hat{f}_I^{l_1J'} = \partial_{i_1}^{l_2}\partial_{i_2}^{l_1}\hat{f}_I^{lJ'}$, as required.

In the next section we shall prove that the conditions of antisymmetry and relations (4.21) follow from admissibility and partial closedness of the form on $G_{p,n}$. Therefore, for $r>p+1$ admissibility and partial closedness imply conditions (4.22) also. In the special case $r=p+1$ relations (4.21) are absent, while relations (4.22) are added conditions on forms belonging to the image of the transform $\mathcal{I}_p$.

4. Regularization of the Form $\mathcal{I}_p\omega$ and Trivial Forms. We pose the problem of recovering the initial form ω on the basis of the form $\sigma=\mathcal{I}_p\omega$. In this part we show how this problem reduces to an analogous problem for sections f_I of a one-dimensional bundle over $\mathbf{RP}^n$. We set

$$\hat{f}_I(u_1, \ldots, u_{p+1}) = \frac{1}{2}\int_{S^p} f_I(t^1u_1+\ldots+t^{p+1}u_{p+1})(t^{p+1})^{r-p-1}\sigma_p(t, dt), \tag{4.23}$$

where f_I are the coefficients of ω, i.e., in the notation of part 1, $\hat{f}_I=\hat{f}_I^{p+1,\ldots,p+1}$. According to part 2, for $r>p$ we have

$$\hat{f}_{I,I'}(u)=\frac{(r-p)!}{r!p!}\sum_{k=1}^{p+1}(-1)^{k-1}\Delta_{I'}^{\hat{k}}\varphi_I^{k,p+1,\ldots,p+1}(u), \tag{4.24}$$

where $|I|=r-p$, $|I'|=p$; φ_I^J are the coefficients of the form $\mathcal{I}_p\omega$.

We call the following differential r-form on E_p a regularization of the form $\sigma=\mathcal{I}_p\omega$

$$R\sigma=\sum \hat{f}_I(u_1, \ldots, u_p, x)\,dx^I. \tag{4.24'}$$

The problem of recovering ω on the basis of the form $\mathcal{I}_p\omega$ reduces to the problem of recovering ω on the basis of the form $R\sigma$, i.e., to the inversion of the integral transform (4.23). It was proved in Chap. II that the latter always has a unique solution, and for the case of even p an explicit inversion formula was obtained.

Proposition 16. If $r>p$, then on the space of forms $\Omega^r(\mathbf{RP}^n)$ the mapping $\mathcal{I}_p$ is injective.

We now give another definition of the operation R. Let $\Omega(E_p)$ be the space of differential forms on E_p. In each frame u we distinguish the last vector $u_{p+1}=x$, and we set $u=(v, x)$, where v is a p-frame; we denote by $\Omega_0(E_p)$ the subspace of forms of degree 0 in dv. The natural projection of a form $\sigma\in\Omega(E_p)$ onto the subspace $\Omega_0(E_p)$ we denote by σ_0: if

$$\sigma=\sum\varphi_I^J(u)\,du_J^I, \text{ then } \sigma_0=\sum\varphi_I^{p+1,\ldots,p+1}(v, x)\,dx^I.$$

Definition. We call a differential form $\sigma\in\Omega_0(E_p)$ trivial if under the substitution $x\to x+tv$, $dx\to dx+dtv+tdv$ a form is obtained which has degree less than p relative to dt.

Proposition 17. An r-form $\sigma\in\Omega_0(E_p)$ $(r\geqslant p)$

$$\sigma=\sum_{|I|=r}a_I\,dx^I$$

is trivial if and only if its coefficients satisfy the following relation:

$$\sum_{i_1,\ldots,i_p}a_{I',i_1,\ldots,i_p}v_1^{i_1}\ldots v_p^{i_p}=0 \quad (|I'|=r-p). \tag{4.25}$$

Proof. It suffices to note that under the substitution $dx\to dx+dtv+tdv$ the following form is obtained:

$$\sum_{I'}\sum_{i_1,\ldots,i_p}a_{I',i_1,\ldots,i_p}v_1^{i_1}\ldots v_p^{i_p}dx^{I'}\wedge dt^1\wedge\ldots\wedge dt^p+\ldots,$$

where the dots denote terms of lower degree in dt.

LEMMA. If a form $\tilde{\sigma}\in\Omega_0(E_p)$ is trivial and harmonic (i.e., $(\partial^2/\partial v_i^k\partial x^l-\partial^2/\partial v_i^l\partial x^k)\tilde{\sigma}=0$), then $\tilde{\sigma}=0$.

Proof. We use the following simple assertion: if $a_{ij}(x)$ are smooth functions of x on $\mathbf{R}^{n+1}\setminus\{v\}$, are antisymmetric in i, j and $\frac{\partial a_{ij}}{\partial x^k}-\frac{\partial a_{kj}}{\partial x^i}=0$ for $i, j, k=1, \ldots, n+1$, then $a_{ij}\equiv 0$. We set $b_I=\sum_{i_1,\ldots,i_{p-1}} a_{I,i_1,\ldots,i_{p-1}}v_1^{i_1}\ldots v_{p-1}^{i_{p-1}}$; from (4.25) it then follows that $\sum_{i_p} b_{I,i_p}v_p^{i_p}=0$. To this equality we apply the operator $\partial^2/\partial v_p^k\partial x^l-\partial^2/\partial v_p^l\partial x^k$. Using the harmonicity of the coefficients a_I, we obtain $\partial b_{I,k}/\partial x^l-\partial b_{I,l}/\partial x^k=0$. Hence, by the assertion made at the start, $b_{I,k}=0$, i.e., $\sum_{i_1,\ldots,i_{p-1}} a_{I,i_1,\ldots,i_{p-1}}v_1^{i_1}\ldots v_{p-1}^{i_{p-1}}=0$ $(|I|=r-p+1)$. Repeating this process p times, we obtain finally $a_I=0$.

Definition. We call a form $\tau\in\Omega_0^p(E_p)$, $\tau=\sum\tau_I(v, x)\,dx^I$ p-flat if $p!\sum\tau_{i_1,\ldots,i_p}v_1^{i_1}\ldots v_p^{i_p}=1$ (by the agreement adopted, τ_I are antisymmetric in I).

An example of a p-flat form in the region $\delta^{k_1,\ldots,k_p}\equiv\det\|v_j^{k_i}\|_{i,j=1,\ldots,p}\neq 0$ is the form $(\delta^{k_1,\ldots,k_p})^{-1}dx^{k_1}\wedge\ldots\wedge dx^{k_p}$.

We proceed to the construction of the operator R. Let $\sigma=\mathcal{I}_p\omega$, $\omega\in\Omega^r(\mathbf{RP}^n)$. We lift σ to E_p, and we denote by σ_0 the projection of σ onto the subspace $\Omega_0(E_p)$, i.e.,

$$\sigma_0=\sum\varphi_I^{p+1,\ldots,p+1}(v, x)\,dx^I=(-1)^p\frac{r!}{(r-p)!}\sum\hat{f}_{I,i_1,\ldots,i_p}(v, x)\,v_1^{i_1}\ldots v_p^{i_p}dx^I. \tag{4.26}$$

We define the form $R\sigma$ according to the following proposition.

Proposition 18. If $\sigma=\mathcal{I}_p\omega$, then there exists a unique r-form $R\sigma\in\Omega_0^r(E_p)$ which satisfies the following conditions:

a) the form $R\sigma$ is harmonic, i.e.,

$$\left(\frac{\partial^2}{\partial v_i^k\partial x^l}-\frac{\partial^2}{\partial v_i^l\partial x^k}\right)R\sigma=0\quad\text{for any}\quad i, k, l;$$

b) the difference $R\sigma-(-1)^p\sigma_0\wedge\tau$ is a trivial form for any p-flat form τ.

Proof. We shall show that the form $R\sigma$ defined by Eqs. (4.24), (4.24') satisfies the conditions of the proposition. The validity of condition a) is obvious; we shall verify condition b). In the expression for $R\sigma$ we make the substitution $dx\to dx+dt\cdot v+t\,dv$ and we separate out the homogeneous component $(R\sigma)^p$ of degree p in dt. We obtain

$$(R\sigma)^p=\frac{r!}{(r-p)!}\sum\hat{f}_{I,i_1,\ldots,i_p}(v, x)\,(t^1dv_1^I+\ldots+t^pdv_p^I)\wedge v_1^{i_1}\ldots v_p^{i_p}dt^1\wedge\ldots\wedge dt^p. \tag{4.27}$$

We do the same with the form $\sigma_0\wedge dx^{k_1}\wedge\ldots\wedge dx^{k_p}$, where $k_1, \ldots, k_p$ are any fixed indices; we obtain

$$(\sigma_0\wedge dx^{k_1}\wedge\ldots\wedge dx^{k_p})^p=\delta^{k_1,\ldots,k_p}(v)\frac{r!}{(r-p)!}(-1)^p\times$$
$$\times\sum\hat{f}_{I,i_1,\ldots,i_p}(v, x)\,v_1^{i_1}\ldots v_p^{i_p}(t^1dv_1^I+\ldots+t^pdv_p^I)\wedge dt^1\wedge\ldots\wedge dt^p, \tag{4.28}$$

where $\delta^{k_1,\ldots,k_p}(v)=\det\|v_j^{k_i}\|_{i,j=1,\ldots,p}$. From equalities (4.27) and (4.28) it follows that the form

$$\delta^{k_1,\ldots,k_p}(v)\,R\sigma-(-1)^p\sigma_0\wedge dx^{k_1}\wedge\ldots\wedge dx^{k_p} \tag{4.29}$$

is trivial for any $k_1, \ldots, k_p$. Multiplying by $\tau_{k_1,\ldots,k_p}$ and summing on $k_1, \ldots, k_p$, we obtain the form $R\sigma-(-1)^p\sigma_0\wedge\tau$; hence this form is trivial. The uniqueness of the form $R\sigma$ follows from the lemma proved above.

5. Regularization of Admissible, Partially Closed Forms; Description of the Image of the Mapping $\mathcal{I}_p$ and the Inversion Formula

1. Functions ψ^J_I. Suppose there is given an admissible, partially closed form $\sigma\in\Omega^{r-p}(G_{p,n})$, where $r-p>0$,

$$\sigma=\sum_{|I|=|J|=r-p}\varphi^J_I(u)\,du^I_J.$$

We define the functions $\psi^J_{I,I'}(u)$, $|I|=r-p$, $|I'|=p$, $|J|=r-p-1$ by the formula

$$\psi^J_{I,I'}(u)=\frac{(r-p)!}{r!\,p!}\sum_{k=1}^{p+1}(-1)^{k-1}\Delta^{\hat{k}}_{I'}\varphi^{kJ}_I(u), \tag{4.30}$$

where $\hat{k}=(1,\ldots,\hat{k},\ldots,p+1)$, $\Delta^{j_1,\ldots,j_s}_{i_1,\ldots,i_s}=\det\|\partial^{j_\beta}_{i_\alpha}\|_{\alpha,\beta=1,\ldots,s}$. The present definition is motivated by the fact that in the case $\sigma=\mathcal{I}_p\omega$ the functions $\psi^J_{I,I'}(u)$ coincide with the integrals $\hat{f}^J_{I,I'}$ of the coefficients of the form ω over p-dimensional planes (see formula (4.17)). We shall establish the properties of the functions $\psi^J_{I,I'}$.

From the condition of admissibility it follows that the functions $\psi^J_{I,I'}$ are symmetric in J and antisymmetric in I and I'. This follows immediately from the following assertion.

Proposition 19. The sum

$$F_l=\sum_{k=1}^{p+1}(-1)^{k+l}\Delta^{\hat{k}}_{i_1,\ldots,\hat{i}_l,\ldots,i_{p+1}}\varphi^{kJ}_{i_lI'}, \tag{4.31}$$

where $|I'|=|J|=r-p-1$, does not depend on l $(1\leqslant l\leqslant p+1)$.

Proof. Expanding for $l\neq p+1$ the determinants $\Delta^{\hat{k}}_{i_1,\ldots,\hat{i}_l,\ldots,i_{p+1}}$ in terms of elements of the last row, we obtain

$$F_l=(-1)^{p+l}\sum_{s<k}(-1)^{k+s}\Delta^{1,\ldots,\hat{s},\ldots,\hat{k},\ldots,p+1}_{i_1,\ldots,\hat{i}_l,\ldots,i_p}\partial^s_{i_{p+1}}\varphi^{kJ}_{i_lI'}-(-1)^{p+l}\sum_{s>k}(-1)^{k+s}\Delta^{1,\ldots,\hat{k},\ldots,\hat{s},\ldots,p+1}_{i_1,\ldots,\hat{i}_l,\ldots,i_p}\partial^s_{i_{p+1}}\varphi^{kJ}_{i_lI'}$$

(the summation goes over k and s). In the first term we substitute $\partial^s_{i_{p+1}}\varphi^{kJ}_{i_lI'}=\partial^k_{i_l}\varphi^{sJ}_{i_{p+1}I'}+\partial^k_{i_{p+1}}\varphi^{sJ}_{i_lI'}-\partial^s_{i_l}\varphi^{kJ}_{i_{p+1}I'}$; we obtain

$$F_l=(-1)^{p+l}\sum_{s<k}(-1)^{k+s}\Delta^{1,\ldots,\hat{s},\ldots,\hat{k},\ldots,p+1}_{i_1,\ldots,\hat{i}_l,\ldots,i_p}\partial^k_{i_l}\varphi^{sJ}_{i_{p+1}I'}-(-1)^{p+l}\sum_{s<k}(-1)^{k+s}\Delta^{1,\ldots,\hat{s},\ldots,\hat{k},\ldots,p+1}_{i_1,\ldots,\hat{i}_l,\ldots,i_p}\partial^s_{i_l}\varphi^{kJ}_{i_{p+1}I'}+$$

$$+(-1)^{p+l}\sum_{s<k}(-1)^{k+s}\Delta^{1,\ldots,\hat{s},\ldots,\hat{k},\ldots,p+1}_{i_1,\ldots,\hat{i}_l,\ldots,i_p}\partial^k_{i_{p+1}}\varphi^{sJ}_{i_lI'}-(-1)^{p+l}\sum_{s>k}(-1)^{k+s}\Delta^{1,\ldots,\hat{s},\ldots,\hat{k},\ldots,p+1}_{i_1,\ldots,\hat{i}_l,\ldots,i_p}\partial^s_{i_{p+1}}\varphi^{kJ}_{i_lI'}.$$

It is obvious that the last two terms cancel. If in the first sum we change the notation $k\leftrightarrow s$ and then sum on s, then we obtain

$$F_l=\sum_{k=1}^{p+1}(-1)^{p+k+1}\Delta^{1,\ldots,\hat{k},\ldots,p+1}_{i_1,\ldots,i_p}\varphi^{kJ}_{i_{p+1}I'}=F_{p+1}.$$

Proposition 20. The functions $\partial^l_k\psi^J_I$ are symmetric with respect to the collection of indices (l, J).

Proof. We shall prove the following relation:

$$\partial^l_k\psi^J_{I',I}=\frac{(r-p)!}{r!\,(p+1)!}(-1)^{r-p-1}\left(pD_I\varphi^{lJ}_{kI'}+\sum_{s=1}^{p+1}D_{I_s}\varphi^{lJ}_{i_sI'}\right), \tag{4.32}$$

where $I=(i_1,\ldots,i_{p+1})$, and I_s is obtained from I by replacing i_s by k; $D_I=\Delta^{1,\ldots,p+1}_{i_1,\ldots,i_{p+1}}$. Our assertion follows immediately from this relation.

Expanding the determinant D_{I_s}, in terms of elements of the s-th row, we obtain

$$D_{I_s}\varphi^{lJ}_{i_s I'} = \sum_{\alpha\neq l}(-1)^{s+\alpha}\Delta^{1,\dots,\hat{\alpha},\dots,p+1}_{i_1,\dots,\hat{i}_s,\dots,i_{p+1}}\partial^{\alpha}_{k}\varphi^{lJ}_{i_s I'} + (-1)^{s+1}\Delta^{1,\dots,\hat{l},\dots,p+1}_{i_1,\dots,\hat{i}_s,\dots,i_{p+1}}\partial^{l}_{k}\varphi^{lJ}_{i_s I'}.$$

By the condition of partial closedness (4.9), we have

$$\partial^{\alpha}_{k}\varphi^{lJ}_{i_s I'} = \partial^{l}_{i_s}\varphi^{\alpha J}_{kI'} + \partial^{l}_{k}\varphi^{\alpha J}_{i_s I'} - \partial^{\alpha}_{i_s}\varphi^{lJ}_{kI'}.$$

Hence,

$$D_{I_s}\varphi^{lJ}_{i_s I'} = -\sum_{\alpha\neq l}(-1)^{s+\alpha}\Delta^{1,\dots,\hat{\alpha},\dots,p+1}_{i_1,\dots,\hat{i}_s,\dots,i_{p+1}}\partial^{\alpha}_{i_s}\varphi^{lJ}_{kI'} +$$

$$+ \sum_{\alpha\neq l}(-1)^{s+\alpha}\Delta^{1,\dots,\hat{\alpha},\dots,p+1}_{i_1,\dots,\hat{i}_s,\dots,i_{p+1}}\partial^{l}_{i_s}\varphi^{\alpha J}_{kI'} + \partial^{l}_{k}\sum_{\alpha}(-1)^{s+\alpha}\Delta^{1,\dots,\hat{\alpha},\dots,p+1}_{i_1,\dots,\hat{i}_s,\dots,i_{p+1}}\varphi^{\alpha J}_{i_s I'}.$$

We sum this expression on s and note that

$$\sum_{s}(-1)^{s+\alpha}\Delta^{1,\dots,\hat{\alpha},\dots,p+1}_{i_1,\dots,\hat{i}_s,\dots,i_{p+1}}\partial^{\alpha}_{i_s} = D_I, \quad \sum_{s}(-1)^{s+\alpha}\Delta^{1,\dots,\hat{\alpha},\dots,p+1}_{i_1,\dots,\hat{i}_s,\dots,i_{p+1}}\partial^{l}_{i_s} = 0$$

for $l\neq\alpha$. As a result, we obtain

$$\sum_{s} D_{I_s}\varphi^{lJ}_{i_s I'} = -pD_I\varphi^{lJ}_{kI'} + \partial^{l}_{k}\sum_{\alpha,s}(-1)^{s+\alpha}\Delta^{1,\dots,\hat{\alpha},\dots,p+1}_{i_1,\dots,\hat{i}_s,\dots,i_{p+1}}\varphi^{\alpha J}_{i_s I'} =$$

$$= -pD_I\varphi^{lJ}_{kI'} + \frac{r!\,p!}{(r-p)!}(p+1)(-1)^p\partial^{l}_{k}\psi^{J}_{i_{p+1},I',i_1,\dots,i_p} = -pD_I\varphi^{lJ}_{kI'} + \frac{r!\,(p+1)!}{(r-p)!}(-1)^{r-p-1}\partial^{l}_{k}\psi^{J}_{I',i_1,\dots,i_{p+1}},$$

as required.

<u>COROLLARY.</u> In the case $r>p+1$ the functions ψ^{J}_{I} satisfy the relation

$$\partial^{l'}_{k}\psi^{lJ'}_{I} = \partial^{l}_{k}\psi^{l'J'}_{I}. \tag{4.33}$$

<u>Proposition 21.</u> If $r>p+1$, then the functions ψ^{J}_{I} are harmonic, i.e., for any $1\leqslant l_1,$ $l_2\leqslant p+1$ and $1\leqslant k_1, k_2\leqslant n+1$ we have

$$(\partial^{l_1}_{k_1}\partial^{l_2}_{k_2} - \partial^{l_2}_{k_1}\partial^{l_1}_{k_2})\psi^{J}_{I} = 0.$$

<u>Proof.</u> Applying relation (4.33) several times, we obtain

$$\partial^{l_1}_{k_1}\partial^{l_2}_{k_2}\psi^{J}_{I} = \partial^{l_1}_{k_1}\partial^{l_2}_{k_2}\psi^{lJ'}_{I} = \partial^{l_1}_{k_1}\partial^{l}_{k_2}\psi^{l_2J'}_{I} = \partial^{l_2}_{k_1}\partial^{l}_{k_2}\psi^{l_1J'}_{I} = \partial^{l_2}_{k_1}\partial^{l_1}_{k_2}\psi^{lJ'}_{I},$$

as required.

<u>2. Operator R.</u> Let $\sigma\in\Omega^{r-p}(G_{p,n})$ be an admissible, partially closed form. We define the r-form $R\sigma\in\Omega^{r}_{0}(E_p)$ by the formula

$$R\sigma = \sum_{|I|=r}\psi_I(v, x)\,dx^I, \tag{4.34}$$

where $\psi_I = \psi^{p+1,\dots,p+1}_{I}$. This definition is motivated by the fact that in the case $\sigma=\mathcal{J}_p\omega$ the form $R\sigma$ is the regularization of σ in the sense of Proposition 18.

<u>Proposition 22.</u> The coefficients $\varphi^{p+1,\dots,p+1}_{I}$ of the form σ can be expressed in terms of the coefficients of the form $R\sigma$ by the following relations:

$$\varphi^{p+1,\dots,p+1}_{I'} = (-1)^p\frac{r!}{(r-p)!}\sum_{i_1,\dots,i_p}\psi_{I',i_1,\dots,i_p}v^{i_1}_{1}\dots v^{i_p}_{p}. \tag{4.35}$$

<u>Proof.</u> For brevity we write φ^k and $\psi_{i_1,\dots,i_p}$ in place of $\varphi^{k,p+1,\dots,p+1}_{I'}$ and $\psi_{I',i_1,\dots,i_p}$. According to (4.30), we then have

$$\psi_{i_1,\dots,i_p}(u) = C_p\sum_{k=1}^{p+1}(-1)^{k-1}\Delta^{1,\dots,\hat{k},\dots,p+1}_{i_1,\dots,i_p}\varphi^k(u) = (-1)^pC_p\sum\operatorname{sgn}(j_1,\dots,j_{p+1})\,\partial^{j_1}_{i_1}\dots\partial^{j_p}_{i_p}\varphi^{j_{p+1}}(u),$$

where $C_p=\frac{(r-p)!}{r!\,p!}$, and the last sum is taken over all permutations $(j_1,\dots,j_{p+1})$ of the indices $1,\dots,p+1$. We compute the sum

$$S=\sum \psi_{i_1,\dots,i_p} v_1^{i_1}\dots v_p^{i_p}=(-1)^p C_p \sum\sum \operatorname{sgn}(j_1,\dots,j_{p+1}) v_1^{i_1}\dots v_p^{i_p}\partial_{i_1}^{j_1}\dots\partial_{i_p}^{j_p}\varphi^{j_{p+1}}.$$

To this end we permute $v_p^{i_p}$ successively with the operators $\partial_{i_1}^{j_1},\dots,\partial_{i_{p-1}}^{j_{p-1}}$ in correspondence with the following equality: $v_p^{i_p}\partial_{i_k}^{j_k}=\partial_{i_k}^{j_k}v_p^{i_p}-\delta_p^{j_k}\delta_{i_k}^{i_p}$, where δ_j^i is the Kronecker symbol. We obtain

$$S=(-1)^p C_p \sum\sum \operatorname{sgn}(j_1,\dots,j_{p+1}) v_1^{i_1}\dots v_{p-1}^{i_{p-1}}\times$$
$$\times\left(\partial_{i_1}^{j_1}\dots\partial_{i_{p-1}}^{j_{p-1}}v_p^{i_p}\partial_{i_p}^{j_p}\varphi^{j_{p+1}}-\sum_{k=1}^{p-1}\delta_p^{j_k}\delta_{i_k}^{i_p}\partial_{i_1}^{j_1}\dots\widehat{\partial_{i_k}^{j_k}}\dots\partial_{i_p}^{j_p}\varphi^{j_{p+1}}\right).$$

From this, noting that $\sum_{i_p} v_p^{i_p}\partial_{i_p}^{j_p}\varphi^{j_{p+1}}=-\delta_p^{j_{p+1}}\varphi^{j_p}$, after elementary transformations we obtain

$$S=(-1)^{p-1}pC_p\sum\sum \operatorname{sgn}(j_1,\dots,j_p,p)\, v_1^{i_1}\dots v_{p-1}^{i_{p-1}}\partial_{i_1}^{j_1}\dots\partial_{i_{p-1}}^{j_{p-1}}\varphi^{j_p},$$

where the inner sum is taken over permutations of the indices $1,\dots,p-1,\ p+1$. Repeating this procedure with each of the factors $v_{p-1}^{i_{p-1}},\dots,v_1^{i_1}$, we finally obtain

$$S=p!\,C_p\operatorname{sgn}(p+1,1,\dots,p)\,\varphi^{p+1}=(-1)^p\frac{(r-p)!}{r!}\varphi^{p+1},$$

as required.

COROLLARY. The mapping $\sigma\mapsto R\sigma$ is injective.

3. Regularization of the Form σ. Proposition 23. If $\sigma\in\Omega^{r-p}(G_{p,n})$, $r>p+1$, is an admissible, partially closed form, then there exists a unique r-form $R\sigma\in\Omega^r{}_0(E_p)$ satisfying the conditions of Proposition 18. This is just the form $R\sigma$ defined in part 2.

Proof. Condition a) follows from Proposition 21. The verification of condition b) is carried out on the basis of relation (4.35) in exactly the same way as in Proposition 18.

We call the form $R\sigma$ the regularization of the admissible, partially closed form σ.

4. Description of the Image of the Mapping $\mathcal{I}_p$ for the Case $r>p+1$. THEOREM 2. In order that a differential form $\sigma\in\Omega^{r-p}(G_{p,n}), r>p+1$, be representable in the form $\sigma=\mathcal{I}_p\omega$, where $\omega\in\Omega^r(\mathbf{RP}^n)$, it is necessary and sufficient that it be admissible and partially closed.

The necessity of the conditions follows from general theorems on the operation $\mathcal{I}_p$ (see Sec. 2). We shall prove that they are sufficient. Let $\sigma=\sum\varphi_J^I(u)\,du_J^I$, and let $R\sigma=\sum\psi_I(v,x)\,dx^I$ be the regularization of the form σ.

Proposition 24. The coefficients ψ_I of the form $R\sigma$ satisfy the following relations:

1) $$\psi_I(\lambda^1u_1,\dots,\lambda^{p+1}u_{p+1})=|\lambda^1\dots\lambda^{p+1}|^{-1}(\lambda^{p+1})^{-r+p+1}\psi_I(u_1,\dots,u_{p+1});\qquad(4.36)$$

2) $$(\partial_{k_1}^{i_1}\partial_{k_2}^{i_2}-\partial_{k_1}^{i_2}\partial_{k_2}^{i_1})\psi_I=0.$$

Indeed, condition 1) follows directly from the definition of ψ_I, and the relation 2) was proved in Proposition 21.

On the basis of Theorem 11 of Chap. II, from this it follows that the functions ψ_I can be represented in the form

$$\psi_I(u)=\frac{1}{2}\int_{S^p} f_I(t^1u_1+\dots+t^{p+1}u_{p+1})(t^{p+1})^{r-p-1}\sigma_p(t,dt),\qquad(4.37)$$

where f_I are C^∞ functions on $\mathbf{R}^{n+1}\setminus 0$ which satisfy the homogeneity condition

$$f_I(\lambda x)=|\lambda|^{-p-1}\lambda^{-r+p+1}f_I(x).\qquad(4.38)$$

We consider the differential r-form on $\mathbf{R}^{n+1}\setminus 0$, $\omega=\sum_{|I|=r} f_I(x)\,dx^I$, where f_I are functions on $\mathbf{R}^{n+1}\setminus 0$ defined in terms of the coefficients ψ_I of the form $R\sigma$ in correspondence with formula (4.37).

We shall show that the form ω drops from $\mathbf{R}^{n+1}\setminus 0$ to $\mathbf{RP}^n$, and $\omega\in\Omega^r(\mathbf{RP}^n)$, i.e., ω has the appropriate parity on $\mathbf{RP}^n$. We must show that $\sum_i f_{iI}(x)\,x^i=0$ or, equivalently,

$$F_k(x)\equiv\frac{\partial}{\partial x^k}\Big(\sum_i f_{iI}(x)\,x^i\Big)=0 \quad \text{for} \quad k=1,\,\dots,\,n+1. \tag{4.39}$$

We use the fact that the mapping $f_I\mapsto\psi_I$ given by Eq. (4.37) is injective. Hence, the relation (4.39) is equivalent to the equality

$$\frac{1}{2}\int_{S^p}\frac{\partial}{\partial x^k}\Big(\sum f_{iI}(x)\,x^i\Big)\Big|_{x=tu}(t^{p+1})^{r-p-1}\,\sigma_p(t,\,dt)=0,$$

or, equivalently,

$$\sum_{i,s} u^i_s\partial^s_k\psi_{iI}+\psi_{kI}=0. \tag{4.40}$$

Thus, it suffices to prove the relation (4.40) for any $k=1,\dots,n+1$. We have

$$\sum_{i,s} u^i_s\partial^s_k\psi_{iI}=\sum_{i,s}\partial^s_k(u^i_s\psi_{iI})-(p+1)\,\psi_{kI}.$$

Further,

$$\sum_{i,s}\partial^s_k(u^i_s\psi_{iI'I})=C_p\sum_{i,s}\partial^s_k\Big(u^i_s\sum_{(j_1,\dots,j_{p+1})}\operatorname{sgn}(j_1,\,\dots,\,j_{p+1})\,\partial^{j_1}_{i_1}\dots\partial^{j_p}_{i_p}\varphi^{j_{p+1}}_{iI'}\Big)=$$

$$=-C_p\sum_{i,s}\sum_{\alpha=1}^{p}\sum_{(j_1,\dots,j_{p+1})}\operatorname{sgn}(j_1,\,\dots,\,j_{p+1})\,\delta^i_{i_\alpha}\delta^{j_\alpha}_s\partial^s_k\partial^{j_1}_{i_1}\dots\widehat{\partial^{j_\alpha}_{i_\alpha}}\dots\partial^{j_p}_{i_p}\varphi^{j_{p+1}}_{iI'}=$$

$$=-C_p\sum_{\alpha=1}^{p}\sum_{(j_1,\dots,j_{p+1})}\operatorname{sgn}(j_1,\,\dots,\,j_{p+1})\,\partial^{j_1}_{i_1}\dots\partial^{j_\alpha}_{k}\dots\partial^{j_p}_{i_p}\varphi^{j_{p+1}}_{i_\alpha I'}=$$

$$=C_p\sum_{\alpha=1}^{p}\sum_{(j_1,\dots,j_{p+1})}\operatorname{sgn}(j_1,\,\dots,\,j_{p+1})\,\partial^{j_1}_{i_1}\dots\partial^{j_\alpha}_{i_\alpha}\dots\partial^{j_p}_{i_p}\varphi^{j_{p+1}}_{kI'}=p\psi_{kI'I}.$$

Thus, $\sum_{i,s} u^i_s\partial^s_k\psi_{iI}=p\psi_{kI}-(p+1)\,\psi_{kI}=-\psi_{kI}$. The fact that ω has the required parity on $\mathbf{RP}^n$ follows from the homogeneity condition (4.35).

Finally, we shall prove that $\mathcal{J}_p\omega=\sigma$. Indeed, suppose that $\mathcal{J}_p\omega=\sum\tilde{\varphi}^J_I(u)\,du^I_J$, and let

$$\tilde{\psi}_{I,I'}(u)=\frac{(r-p)!}{r!\,p!}\sum_{k=1}^{p+1}(-1)^{k-1}\Delta^{\hat{k}}_{I'}\tilde{\varphi}^{kJ}_I(u).$$

According to (4.17), we have $\tilde{\psi}_I=\psi_I$. Hence, by Proposition 22, $\tilde{\varphi}^{p+1,\dots,p+1}_I=\varphi^{p+1,\dots,p+1}_I$. But then (see the corollary of Proposition 10), $\tilde{\varphi}^J_I=\varphi^J_I$ for any J and I, i.e., $\mathcal{J}_p\omega=\sigma$. The proof of the theorem is complete.

<u>COROLLARY.</u> If $\sigma=\sum\varphi^J_I(u)\,du^I_J$ is an admissible, partially closed form on $G_{p,n}$, and

$$\psi^J_{I,I'}(u)=\frac{(r-p)!}{r!\,p!}\sum_{k=1}^{p+1}(-1)^{k-1}\Delta^{\hat{k}}_{I'}\varphi^{kJ}_I(u)=\frac{(r-p)!}{r!\,p!}(-1)^p\sum_{(j_1,\dots,j_{p+1})}\operatorname{sgn}(j_1,\,\dots,\,j_{p+1})\,\partial^{j_1}_{i_1}\dots\partial^{j_p}_{i_p}\varphi^{j_{p+1}J}_I, \tag{4.41}$$

then the following inversion of Eq. (4.41) holds:

$$\varphi^{kJ}_I(u)=(-1)^{k-1}\frac{r!}{(r-p)!}\sum_{i_1,\dots,i_p}\psi^J_{I,\,i_1,\dots,i_p}(u)\,u^{i_1}_1\dots u^{i_{k-1}}_{k-1}u^{i_k}_{k+1}\dots u^{i_p}_{p+1}. \tag{4.42}$$

Remark. In Proposition 22 formula (4.22) was established for the special case $J=(p+1, \ldots, p+1)$, $k=p+1$.

5. Description of the Image of the Mapping $\mathcal{I}_p$ for the Case $r=p+1$. THEOREM 3. In order that a C^∞ 1-form $\sigma=\Sigma\varphi^j_i(u)du^i_j$ on $G_{p,n}$, $\sigma\in\Omega^1(G_{p,n})$, be representable in the form $\sigma=\mathcal{I}_p\omega$, where ω is a C^∞ $(p+1)$-form on $\mathbf{RP}^n$, it is necessary and sufficient that it be partially closed and its coefficients satisfy the additional relations

$$(\partial^{l_1}_{k_1}\partial^{l_2}_{k_2}-\partial^{l_2}_{k_1}\partial^{l_1}_{k_2})\sum_{k=1}^{p+1}(-1)^{k-1}\Delta^{\hat{k}}_{i_1,\ldots,i_p}\varphi^k_i(u)=0 \tag{4.43}$$

for any indices contained in this expression.

Proof. We set

$$\psi_{i,i_1,\ldots,i_p}(u)=\frac{1}{p!\,(p+1)!}\sum_{k=1}^{p+1}(-1)^{k-1}\Delta^{\hat{k}}_{i_1,\ldots,i_p}\varphi^k_i(u). \tag{4.44}$$

If $\sigma=\mathcal{I}_p\omega$, where $\omega=\sum f_I(x)dx^I$, then the form σ is partially closed, and by Proposition 12

$$\psi_I(u)=\frac{1}{2}\int_{S^p} f_I(tu)\,\sigma_p(t,dt). \tag{4.45}$$

It follows from (4.45) that $(\partial^{l_1}_{k_1}\partial^{l_2}_{k_2}-\partial^{l_2}_{k_1}\partial^{l_1}_{k_2})\psi_I(u)=0$, i.e., relation (4.43) is satisfied. The necessity of the conditions has thus been proved. Let σ be an admissible form, and let ψ_I be the function defined by Eq. (4.44). As was proved earlier, the functions ψ_I are antisymmetric in I and satisfy the conditions of Proposition 24. Therefore, by Theorem 11 of Chap. II they can be represented in the form

$$\psi_I(u)=\frac{1}{2}\int_{S^p} f_I(tu)\,\sigma_p(t,dt),$$

where f_I is a C^∞, even, homogeneous function on $\mathbf{R}^{n+1}\setminus 0$ of degree of homogeneity r. We consider the differential form on $\mathbf{R}^{n+1}\setminus 0$

$$\omega=\sum f_I(x)\,dx^I.$$

Just as in the proof of Theorem 2 for forms of degree $r>p+1$, we see that this form drops to $G_{p,n}$, and $\mathcal{I}_p\omega=\sigma$.

Remark. Thus, for any $r>p+1$ the necessary and sufficient conditions that σ belong to the image of the mapping $\mathcal{I}_p$ are: 1) admissibility and partial closedness of σ and 2) harmonocity of the functions ψ_I. The difference between the case $r>p+1$ and the case $r=p+1$ is that in the first case condition 2) follows from 1), while in the second case these conditions are independent.

6. Inversion Formula for Even p. For the case of even p, we shall construct an operator inverse to the operator $\mathcal{I}_p$. Let $\sigma=\mathcal{I}_p\omega$, where $\omega\in\Omega^r(\mathbf{RP}^n)$, $r>p$, and let

$$R\sigma=\sum \hat{f}_I(v,x)\,dx^I$$

be the regularization of the form σ. We recall that the coefficients $\hat{f}_I$ of the form $R\sigma$ can be expressed in terms of the coefficients of the form σ by formulas (4.24), and they are related to the coefficients f_I of the original form ω by the relation (4.23):

$$\hat{f}_I(v,x)=\int_{\mathbf{RP}^p} f_I(t^1v_1+\ldots+t^pv_p+t^{p+1}x)(t^{p+1})^{r-p-1}\sigma_p(t,dt).$$

The problem of inverting the operator $\mathcal{I}_p$ thus reduces to the problem of inverting the integral transform (4.23) for functions which was treated in Sec. 3, Chap. II. On the basis of the result obtained there, the inverse of the operator $\mathcal{I}_p$ is constructed in the following manner.

On differential forms in E_p we introduce the operator

$$\varkappa_x\theta=\sum_{i_1,\dots,i_p}\frac{\partial^p\theta}{\partial x^{i_1}\dots\partial x^{i_p}}\wedge dv_1^{i_1}\wedge\dots\wedge dv_p^{i_p}.$$

This operator is applied to the form $R\sigma$. As a result, we obtain a differential form $\varkappa_x R\sigma$ of degree $r+p$ on E_p. This form can be dropped from E_p to the manifold F of pairs (h, x'), where $x'\in \mathbf{RP}^n$, and h is a p-dimensional subspace of $\mathbf{R}^{n+1}$ not incident with x' (the projection $E_p\to F$ is defined by the formula $\pi(v, x) = (h, x')$, where h is the subspace spanned by the frame v, and x' is the image of the point x under the canonical projection $\mathbf{R}^{n+1}\setminus 0\to \mathbf{RP}^n$).

We consider the canonical projection $F\to\mathbf{RP}^n$. It is not hard to show that the form $\varkappa_x R\sigma$ is relatively closed in the bundle $F\to\mathbf{RP}^n$ and has relative degree p. Let γ be any p-dimensional cycle in the standard fiber of the bundle $F\to\mathbf{RP}^n$, and let $\gamma^x\subset F^x$ be any cycle homologous to γ in the fiber over x.

THEOREM 4. Let $\sigma=\mathcal{I}_p\omega$, where $\omega\in\Omega^r(\mathbf{RP}^n)$, $r>p$, and p is an even number. Then

$$\int_{\gamma^x}\varkappa_x R\sigma=c\,d(\gamma)\,\omega\quad(c\neq 0), \tag{4.46}$$

where $d(\gamma)$ is the index of intersection of γ with the Euler cycle.

The assertion follows from Proposition 11 of Chap. II.

6. Affine Problem as a Special Case of the Projective Problem

1. Spaces of Differential Forms. Here we study the operation $\mathcal{I}_p$ on forms in affine space $\mathbf{R}^n$, where it is defined as in the case of projective space. The basic results regarding this operation can be obtained as corollaries from the results for the projective case presented in Secs. 3-5. For this we prescribe an affinization of $\mathbf{RP}^n$, i.e., we choose an "infinitely distant" $(n-1)$-dimensional plane $L\subset\mathbf{RP}^n$, e.g., the plane $x^{n+1}=0$. In the affine space $\mathbf{R}^n=\mathbf{RP}^n\setminus L$ it is possible to go over to inhomogeneous coordinates $(x^1,\dots, x^{n+1})\to(x^1/x^{n+1}, \dots, x^n/x^{n+1})$. Using the relation between homogeneous and inhomogeneous coordinates, we can consider each smooth differential form on $\mathbf{RP}^n$ as a smooth differential form on $\mathbf{R}^n=\mathbf{RP}^n\setminus L$ which satisfies some natural conditions at infinity (see Chap. III, Sec. 5).

In the case of affine space $\mathbf{R}^n$ it is convenient to introduce the space $\Omega^r(\mathbf{R}^n)$ of differential (even or odd) r-forms on $\mathbf{R}^n$ with coefficients in the Schwartz space $S(\mathbf{R}^n)$. It is clear that forms in $\Omega^r(\mathbf{R}^n)$ extend to smooth forms on $\mathbf{RP}^n$ of the corresponding parity, i.e., there is the imbedding $\Omega^r(\mathbf{R}^n)\subset\Omega^r(\mathbf{RP}^n)$.

Other objects in the affine problem are the manifold $H_{p,n}$ of p-dimensional planes in $\mathbf{R}^n$ $(\mathbf{R}^n=\mathbf{RP}^n\setminus L)$ and its two-sheeted covering — the manifold $\tilde{H}_{p,n}$ of oriented p-planes. The manifold $H_{p,n}$ is naturally imbedded in $G_{p,n}$ as an open, dense subset: the image of $H_{p,n}$ in $G_{p,n}$ consists of all $h\in G_{p,n}$ such that $h\not\subset L$. By virtue of this imbedding, any smooth differential form on $G_{p,n}$ can be considered as a smooth form on $H_{p,n}$ which satisfies additional conditions "at infinity" (cf. Chap. III, Sec. 5).

Since $H_{p,n}$ and $\tilde{H}_{p,n}$ are fiber spaces with a compact base $G_{p-1,n-1}$ and fiber $\mathbf{R}^{n-p}$ (see Chap. I, Sec. 1), the concept of differential r-forms on them with infinitely differentiable, rapidly decreasing coefficients can be introduced in a natural way. The subspace of all such forms on $\tilde{H}_{p,n}$ which satisfy the added condition of being odd we denote by $\Omega^r(H_{p,n})$. Any differential form in $\Omega^r(H_{p,n})$ extends to a smooth, odd differential form on $\tilde{G}_{p,n}$, i.e., there is the imbedding $\Omega^r(H_{p,n})\subset\Omega^r(G_{p,n})$.

2. Description of the Image of the Mapping $\mathcal{J}_p$. If in going over to the affine problem we take precisely the spaces of differential forms on $\mathbf{R}^n$ and $H_{p,n}$ which extend to smooth forms on $\mathbf{RP}^n$ and $G_{p,n}$, respectively, then the affine problem in question would require no new proofs. However, we have gone over to other spaces of differential forms, viz., to the spaces $\Omega^r(\mathbf{R}^n)$ and $\Omega^r(H_{p,n})$ of forms with rapidly decreasing coefficients on $\mathbf{R}^n$ and $H_{p,n}$, respectively.

The space $\Omega^r(\mathbf{R}^n)$ is a subspace of $\Omega^r(\mathbf{RP}^n)$, and the affine operator $\mathcal{J}_p$ as an operator from $\Omega^r(\mathbf{R}^n)$ to $\Omega^{r-p}(G_{p,n})$ is obtained as the restriction of the projective operator $\mathcal{J}_p:\Omega^r(\mathbf{RP}^n)\to\Omega^{r-p}(G_{p,n})$. Here it is not hard to show that its image lies in the subspace $\Omega^{r-p}(H_{p,n})\subset\Omega^{r-p}(G_{p,n})$. It is somewhat more difficult to prove that the mapping $\mathcal{J}_p:\Omega^r(\mathbf{R}^n)\to(\mathcal{J}_p\Omega^r(\mathbf{RP}^n))\cap\Omega^{r-p}(H_{p,n})$ for $p<n-1$ is surjective (for even p this follows from the inversion formula). We formulate these assertions as a proposition.

Proposition 25. If $p<n-1$, then

$$\mathcal{J}_p\Omega^r(\mathbf{R}^n)=(\mathcal{J}_p\Omega^r(\mathbf{RP}^n))\cap\Omega^{r-p}(H_{p,n}).$$

In particular, $\mathcal{J}_p\Omega^r(\mathbf{R}^n)\subset\Omega^{r-p}(H_{p,n})$ and the following diagram is commutative:

$$\begin{array}{ccc}\Omega^r(\mathbf{R}^n) & \xrightarrow{\mathcal{J}_p} & \Omega^{r-p}(H_{p,n})\\ \downarrow & & \downarrow\\ \Omega^r(\mathbf{RP}^n) & \xrightarrow{\mathcal{J}_p} & \Omega^{r-p}(G_{p,n})\end{array}$$

(the vertical arrows are imbedding mappings).

Because of Proposition 25, the description of the image of $\mathcal{J}_p$ for $\mathbf{R}^n$ for $p<n-1$ reduces to the description of the image of $\mathcal{J}_p$ for $\mathbf{RP}^n$ obtained in Sec. 5. Thus, from Theorem 2 we obtain the following result.

THEOREM 2'. In order that the differential form $\sigma\in\Omega^{r-p}(H_{p,n})$, where $p+1<r\leqslant n$, be representable in the form $\sigma=\mathcal{J}_p\omega$, $\omega\in\Omega^r(\mathbf{R}^n)$, it is necessary and sufficient that the forms σ and $d\sigma$ be admissible.

To describe the image of $\Omega^{p+1}(\mathbf{R}^n)$ we need coordinates. Let $x=(y,z)$, $y=(y^1,\dots,y^{n-p})$, $z=(z^1,\dots,z^p)$ be inhomogeneous coordinates on $\mathbf{R}^n$. Any p-dimensional plane $h\in H_{p,n}$ transversal to the $(n-p)$-dimensional plane $z=0$ is then given by the equation $y=\alpha z+\beta$, where $\beta\in\mathbf{R}^{n-p}$ and α is an $(n-p)\times p$ matrix. We take the coordinates of the vector β and the elements of the matrix α as local coordinates on $H_{p,n}$. (In other words, this means that we define a plane $h\in H_{p,n}\subset G_{p,n}$ by matrices $u\in E_p$ of the special form $u=\left(\frac{v}{e}\right)$, where e is the identity $(p+1)$ matrix and $v=(\alpha,\beta)$.) From the results of Sec. 5 for $\mathbf{RP}^n$ (Theorem 3) we obtain the following assertion.

THEOREM 3'. In order that a 1-form $\sigma\in\Omega^1(H_{p,n})$, $p<n-1$,

$$\sigma=\sum\varphi_i^j(\alpha,\beta)\,d\alpha_j^i+\sum\psi_i(\alpha,\beta)\,d\beta^i$$

be representable in the form $\sigma=\mathcal{I}_p\omega$, $\omega\in\Omega^{p+1}(\mathbf{R}^n)$, it is necessary and sufficient that the form $d\sigma$ be admissible and that the relations

$$\left(\frac{\partial^2}{\partial\alpha_k^{i_1}\partial\beta^{l_2}}-\frac{\partial^2}{\partial\alpha_k^{i_2}\partial\beta^{l_1}}\right)\left(\Delta_{i_1,\ldots,i_p}\psi_l-\sum_{j=1}^{p}\Delta^j_{i_1,\ldots,i_p}\varphi^j_l\right)=0$$

be satisfied for any indices contained in this expression. Here $\Delta_{i_1,\ldots,i_p}=\det\|\partial/\partial\alpha_j^{i_k}\|_{k,j=1,\ldots,p}$, and $\Delta^j_{i_1,\ldots,i_p}$ is obtained from $\Delta_{i_1,\ldots,i_p}$ by replacing $(\partial/\partial\alpha_j^{i_1},\ldots,\partial/\partial\alpha_j^{i_p})$ by $(\partial/\partial\beta^{i_1},\ldots,\partial/\partial\beta^{i_p})$.

Remark. In the case $p=n-1$, $r=n$ the transformation $\mathcal{I}_p$ is essentially the Radon transform for functions which was considered in Chap. II. In this case additional integral relations (the Cavalieri conditions) arise in describing the image of $\Omega^r(\mathbf{R}^n)$.

3. Inversion Formula for Even p. On the basis of the results of Sec. 5 we shall find the operator inverse to $\mathcal{I}_p$, where p is an even number. Here it is essentially a question of writing the inversion formula (4.46) of Sec. 5 in the coordinates (α,β) on $H_{p,n}$.

Let $\omega=\sum f_I(x)\,dx^I\in\Omega^r(\mathbf{R}^n)$, and let

$$\sigma=\mathcal{I}_p\omega=\sum\varphi^J_{I,I'}(\alpha,\beta)\,d\alpha^I_J\wedge d\beta^{I'}.$$

We note that the coefficients of the form σ are expressed in terms of the coefficients of the form ω in the following manner:

$$\varphi^J_{I,I'}(\alpha,\beta)=\frac{r!}{k!l!}\sum_{i_1,\ldots,i_p=1}^{n}\hat{f}^J_{I,I',i_1,\ldots,i_p}(\alpha,\beta)\,\alpha_1^{i_1}\ldots\alpha_p^{i_p},\tag{4.47}$$

where $k=|I|$, $l=r-p-k$, $\alpha_j^{n-p+i}=\delta^i_j$ for $i,j=1,\ldots,p$,

$$\hat{f}^J_I(\alpha,\beta)=\int_{\mathbf{R}^p}f_I(\alpha z+\beta,z)\,z^J dz$$

$$(z^J=z^{j_1}\ldots z^{j_k},\ dz=dz^1\wedge\ldots\wedge dz^p).$$

In the coordinates introduced on $\mathbf{R}^n$ and $H_{p,n}$ the regularization $R\sigma$ of the form σ has the following form:

$$R\sigma=\sum\hat{f}_I(\alpha,\beta)\,dx^I,$$

where

$$\hat{f}_I(\alpha,\beta)=\int_{\mathbf{R}^p}f_I(\alpha z+\beta,z)\,dz.\tag{4.48}$$

The coefficients $\hat{f}_I$ of the form $R\sigma$ can be expressed directly in terms of part of the coefficients of the form σ, viz., in terms of the coefficients φ_I of $d\beta^I$ ($|I|=r-p$) and the coefficients $\varphi^j_{i,I}$ of $d\alpha^i_j\wedge d\beta^I$ ($|I|=r-p-1$). These expressions can be obtained from the relations (4.47) as in Sec. 5 for forms written in homogeneous coordinates on $\mathbf{RP}^n$ and $G_{p,n}$; however, in inhomogeneous coordinates they are more complicated. (The expression for f_I depends on how many indices contained in I are greater than $n-p$.) We present them only for the cases $p=1$ and $p=2$.

1) The case $p=1$. Let $I=(i_1,\ldots,i_{r-2})$, where $1\leqslant i_k\leqslant n-1$; then

$$\hat{f}_{I,i}=\frac{1}{r}\frac{\partial\varphi_I}{\partial\alpha_1^i}-\frac{1}{r(r-1)}\frac{\partial\varphi_I^1}{\partial\beta^i}\ \text{for}\quad i<n,$$

$$\hat{f}_{I,n}=\frac{1}{r}\varphi_I-\sum_{i=1}^{n-1}\alpha_1^i\hat{f}_{I,i}.$$

2) The case $p=2$.

$$\hat{f}_{I,i_1,i_2}=\frac{1}{2r(r-1)(r-2)}\left((r-2)\Delta_{i_1,i_2}\varphi_I-\Delta^1_{i_1,i_2}\varphi^1_I-\Delta^2_{i_1,i_2}\varphi^2_I\right)$$

for $1\leqslant i_1, i_2\leqslant n-2$; here

$$\Delta_{i_1,i_2}=\begin{vmatrix}\partial/\partial\alpha_1^{i_1} & \partial/\partial\alpha_2^{i_1}\\ \partial/\partial\alpha_1^{i_2} & \partial/\partial\alpha_2^{i_2}\end{vmatrix},$$

and $\Delta^j_{i_1,i_2}$ is obtained from Δ_{i_1,i_2} by replacing the j-th column by $(\partial/\partial\beta^{i_1}, \partial/\partial\beta^{i_2})$;

$$\hat{f}_{I,i,n-1}=-\frac{1}{r(r-1)(r-2)}\left((r-2)\frac{\partial\varphi_I}{\partial\alpha_2^i}-\frac{\partial\varphi_I^2}{\partial\beta^i}\right)-\sum_{i'=1}^{n-2}\alpha_1^{i'}\hat{f}_{I,i,i'},$$

$$\hat{f}_{I,i,n}=\frac{1}{r(r-1)(r-2)}\left((r-2)\frac{\partial\varphi_I}{\partial\alpha_1^i}-\frac{\partial\varphi_I^1}{\partial\beta^i}\right)-\sum_{i'=1}^{n-2}\alpha_2^{i'}\hat{f}_{I,i,i'}$$

$$\text{for}\quad 1\leqslant i\leqslant n-2;$$

$$\hat{f}_{I,n-1,n}=\frac{1}{r(r-1)}\varphi_I-\sum_{i=1}^{n-2}\alpha_1^i\hat{f}_{I,i,n}+\sum_{i=1}^{n-2}\alpha_2^i\hat{f}_{I,i,n-1}-\sum_{i,i'=1}^{n-2}\alpha_1^i\alpha_2^{i'}\hat{f}_{I,i,i'}.$$

We set

$$\varkappa R\sigma=\sum_{I,i_1,\ldots,i_p}\frac{\partial^p\hat{f}_I(\alpha, y-\alpha z)}{\partial\beta^{i_1}\ldots\partial\beta^{i_p}}dx^I\wedge d\alpha_1^{i_1}\wedge\ldots\wedge d\alpha_p^{i_p}.$$

For fixed $x=(y, z)$ and dx this expression is a closed differential form on the manifold of -dimensional subspaces in $\mathbf{R}^n$ given by the matrices α. On the basis of Theorem 4 we obtain the following assertion.

THEOREM 4'. If γ is any p-dimensional cycle on the manifold of p-dimensional subspaces in $\mathbf{R}^n$ and $d(\gamma)$ is its incidence index with the Euler cycle, then

$$\int_\gamma\varkappa R\sigma=cd(\gamma)\,\omega, \tag{4.49}$$

where c is a constant not depending on γ, while $c\neq0$ for even p and $c=0$ for odd p.

Remark. Formula (4.49) can be obtained directly by using the inversion formula for the integral transform (4.48) obtained in Chap. I.

7. Connection with the Problem of Integral Geometry for Lines in Three-Dimensional Space

1. Here we shall show that the problem of integral geometry for -dimensional planes in $\mathbf{RP}^n$ in a certain sense reduces to the analogous problem for lines in $\mathbf{RP}^3$. Namely, we shall establish a connection between the conditions of admissibility for forms on $G_{p,n}$ and the conditions of admissibility for forms on the manifold $G_{1,3}$ of lines in $\mathbf{RP}^3$.

We introduce the following flag manifolds: the manifold B of incident pairs (h_1, h_2), $h_1\in G_{p-2,n}$, $h_2\in G_{p+2,n}$, $h_1\subset h_2$, and the manifold A of incident triples (h_1, h, h_2), $h_1\in G_{p-2,n}$, $h\in G_{p,n}$, $h_2\in G_{p+2,n}$, $h_1\subset h\subset h_2$. In the case $p=1$, it is assumed that $G_{p-2,n}=\varnothing$ and hence $B=G_{3,n}$, $A=G_{p,p+2,n}$. The manifold A is equipped with a natural structure of a binary bundle:

$$G_{p,n}\xleftarrow{\ \pi_1\ }A\xrightarrow{\ \pi_2\ }B.$$

Here each fiber of the bundle π_2 is isomorphic to the manifold $G_{1,3}$ of two-dimensional subspaces in $\mathbf{R}^4$ or, equivalently, to the manifold of lines in $\mathbf{RP}^3$.

Let σ be a differential form of degree $s=r-p$ on $G_{p,n}$. We lift σ from $G_{p,n}$ to A, i.e., we consider the differential form $\tilde{\sigma}=\pi_1^*\sigma$ on A. It is not hard to see that if $s\leqslant 2$, then $\mathrm{Deg}_{\pi_2}\tilde{\sigma}=\deg\sigma$, where $\mathrm{Deg}_{\pi_2}\tilde{\sigma}$ is the relative degree of $\tilde{\sigma}$ in the bundle $\pi_2: A\to B$ (for the definition of relative degree see Sec. 1); if $s>2$, then $\mathrm{Deg}_{\pi_2}\tilde{\sigma}\geqslant 2$. For $s>2$ we shall henceforth consider only forms σ for which $\mathrm{Deg}_{\pi_2}\tilde{\sigma}=2$. This condition is motivated by the following assertion.

Proposition 26. If σ is an admissible form of degree $s>2$ on $G_{p,n}$, then $\mathrm{Deg}_{\pi_2}(\pi_1^*\sigma)=2$.

Proof. Let $(h, b)\in A$. We denote by $G_p(b)$ the manifold of planes $h'\in G_{p,n}$ incident with b; $\dim G_p(b)=4$. We must show that $\sigma(h; \xi_1,\dots, \xi_s)=0$ if three of the vectors $\xi_i\in T_hG_{p,n}$ are tangent to $G_p(b)$.

We denote by L^x the submanifold of planes $h'\in G_{p,n}$ incident with $x\in\mathbf{RP}^n$. We recall that the form σ is called admissible if $\sigma(h; \xi_1,\dots, \xi_s)=0$ whenever for any point $x\in h$ the space spanned by the vectors $\xi_1,\dots, \xi_s$ contains a nonzero vector tangent to L^x. It is not hard to see that $\dim (L^x\cap G_p(b))\geqslant 2$, and therefore the codimension of $L^x\cap G_p(b)$ in $G_p(b)$ is no greater than 2. Hence, if three of the vectors $\xi_1,\dots, \xi_s$ are linearly independent and tangent to $G_p(b)$, then the three-dimensional subspace in $T_hG_{p,n}$ which they span contains a nonzero vector tangent to L^x for any $x\in h$; hence $\sigma(h; \xi_1, \dots, \xi_s)=0$, as required.

Remark. In the case $p=1$, the converse is also true: if $s>2$ and $\mathrm{Deg}_{\pi_2}(\pi_1^*\sigma)=2$, then σ is an admissible form on $G_{p,n}$. Indeed, suppose that $\mathrm{Deg}_{\pi_2}(\pi_1^*\sigma)=2$. We consider the family Π_1 of two-dimensional generators of the cone of incidence $K_h\subset T_hG_{1,n}$. Let $L_0\in\Pi_1$ and let $\eta\in T_h$ be an arbitrary vector. It is obvious that the planes of Π_1 generate the entire space $T_hG_{1,n}$; therefore, there exist planes $L_i\in\Pi_1$, $i=1, \dots, k$, and vectors $\eta_i\in L_i$ such that $\eta=\eta_1+\dots+\eta_k$. It is further obvious that each of the subspaces L_0+L_i is tangent to some $G_1(b)$, $b\in B=G_{3,n}$. Hence, if $\xi_0, \xi_1\in L_0$, then $\sigma(h; \xi_0, \xi_1, \eta_i, \dots)=0$ for any $i=1, \dots, k$ (since $\mathrm{Deg}_{\pi_2}(\pi_1^*\sigma)=2$). This implies that $\sigma(h; \xi_0, \xi_1, \eta\dots)=0$; thus, by Proposition 8 the form σ is admissible.

2. The restriction of the s-form σ to the submanifolds of planes $G_p(b)\subset G_{p,n}$ is the following form:

$$\tau\sigma=\rho(\pi_1^*\sigma),$$

where ρ is the operation on forms in the bundle $A\to B$ introduced in Sec. 1, part 3. According to the definition given there, in the case $s\leqslant 2$, $\tau\sigma$ is a function on B whose value at a point $b\in B$ is the restriction (in the usual sense) of the form σ to the submanifold $G_p(b)\subset G_{p,n}$. In the case $s>2$ $\tau\sigma$ is the differential $(s-2)$-form on B whose value for any $b\in B$ and $\eta_1, \dots, \eta_{s-2}\in T_bB$ is the 2-form $(\tau\sigma)_{b;\eta_1,\dots,\eta_{s-2}}$ on the fiber $\pi_2^{-1}b=G_p(b)$ defined by the following formula:

$$(\tau\sigma)_{b;\eta_1,\dots,\eta_{s-2}}(h; \xi_1, \xi_2)=(\pi_1^*\sigma)(h, b; \eta_1', \dots, \eta_{s-2}', \xi_1, \xi_2).$$

Here $\eta_i'\in T_{h,b}A$ are vectors such that $\pi_2\eta_i'=\eta_i$, $1\leqslant i\leqslant s-2$; for fixed $\eta_1, \dots, \eta_{s-2}$ the expression does not depend on the particular choice of them.

Since all 1-forms on $G_{p,n}$ are admissible, it is possible to restrict attention to forms of degree $s>1$. The connection between the conditions of admissibility of an s-form σ on $G_{p,n}$ and of the corresponding 2-form $(\tau\sigma)_{b;\eta_1,\dots,\eta_{s-2}}$ on $G_p(b)\cong G_{1,3}$ is established by the following theorem.

THEOREM 5. A differential s-form σ on $G_{p,n}$ is admissible if and only if for any $b\in B$ and $\eta_1, \ldots, \eta_{s-2}\in T_bB$ the 2-form $(\tau\sigma)_{b;\eta_1,\ldots,\eta_{s-2}}$ on $G_p(b)\cong G_{1,3}$ is admissible.

It is obvious that the assertion of the theorem is equivalent to the following assertion (a stronger version of Proposition 8 of Sec. 3).

Proposition 27. The following assertions are equivalent: (i) the form σ is admissible; (ii) $\sigma(h; \xi_1, \xi_2, \ldots, \xi_s)=0$ whenever the vectors $\xi_1, \xi_2\in T_hG_{p,n}$ are tangent to the submanifold $G_p(h_1, h_2)\cong G_{1,2}$ of planes $h'\in G_{p,n}$ such that $h_1\subset h'\subset h_2$, where $h_1\in G_{p-2,n}$, $h_2\in G_{p+1,n}$.

Proof. It is easy to see that a subspace $L\subset T_hG_{p,n}$ tangent to $G_p(h_1, h_2)$ is degenerate (see the proof of Proposition 8). Hence (i)⇒(ii). Conversely, suppose that condition (ii) is satisfied, and let $h=(u_1, \ldots, u_{p+1})$. We fix arbitrary indices $i_1, i_2 (1\leqslant i_1, i_2\leqslant n+1)$ and j_1, j_2 $(1\leqslant j_1, j_2\leqslant p+1)$, and we set $u_0=(\varepsilon^1, \ldots, \varepsilon^{n+1})\in\mathbf{R}^{n+1}$, where $\varepsilon^{i_1}=\varepsilon^{i_2}=1$, $\varepsilon^i=0$ for $i\neq i_1, i_2$. We consider the curves on the Grassmann manifold $G_{p,n}$: $h_t'=(u_1, \ldots, u_{j_1}+tu_0, \ldots, u_{p+1})$ and $h_t''=(u_1, \ldots, u_{j_2}+tu_0, \ldots, u_{p+1})$. Both curves belong to the manifold $G_p(h_1, h_2)\subset G_{p,n}$, where $h_1\in G_{p-2,n}$ and $h_2\in G_{p+1,n}$ are generated, respectively, by the frames $(u_1, \ldots, \widehat{u}_{j_1}, \ldots, \widehat{u}_{j_2}, \ldots, u_{p+1})$ and $(u_0, u_1, \ldots, u_{p+1})$. Hence, if ξ_1, ξ_2 are tangent vectors to these curves at the point h, then by condition (ii) $\sigma(h; \xi_1, \xi_2, \xi_3, \ldots, \xi_s)=0$ for any $\xi_3, \ldots, \xi_s$. In the notation of Sec. 3 we have $\xi_1=e^{i_1}_{j_1}+e^{i_2}_{j_1}$, $\xi_2=e^{i_1}_{j_2}+e^{i_2}_{j_2}$; therefore, the equality $\sigma(h; \xi_1, \ldots, \xi_s)=0$ implies the admissibility of the form σ according to what has been proved in Sec. 3, part 1.

Remark. If the forms σ and $d\sigma$ are admissible, then the 2-form $(\tau\sigma)_{b;\eta_1,\ldots,\eta_{s-2}}$ is closed on $G_p(b)$.

Indeed, since $\mathrm{Deg}_{\pi_2}(d\pi_1^*\sigma)=\mathrm{Deg}_{\pi_2}(\pi_1^*d\sigma)=\mathrm{Deg}_{\pi_2}(\pi_1^*\sigma)=2$, the form $\pi_1^*\sigma$ is relatively closed in the bundle $A\to B$.

3. We shall find an expression for $\tau\sigma$ in coordinate form. To this end we shall consider the form $\tau\sigma$ not on the entire manifold B, but on certain of its submanifolds $B^{i_1,i_2}_{j_1,j_2}$, which are defined below.

We first introduce local coordinates on $G_{p,n}$. We decompose $\mathbf{R}^{n+1}$ into the direct sum of subspaces $\mathbf{R}^{n+1}=\mathbf{R}^{n-p}\oplus\mathbf{R}^{p+1}$. Any linear mapping α: $\mathbf{R}^{p+1}\to\mathbf{R}^{n-p}$ defines a $(p+1)$-dimensional subspace h in $\mathbf{R}^{n+1}$: $h=\{(\alpha z, z), z\in\mathbf{R}^{p+1}\}$; all $(p+1)$-dimensional subspaces h such that $h\cap\mathbf{R}^{n-p}=0$ are hereby obtained. If coordinates are introduced in $\mathbf{R}^{p+1}$ and $\mathbf{R}^{n-p}$, then α is given by a $(p+1)\times(n-p)$ matrix whose elements we take as local coordinates on $G_{p,n}$. Thus, if $y=(y^1, \ldots, y^{n-p})$, $z=(z_1, \ldots, z^{p+1})$ are coordinates on $\mathbf{R}^{n-p}$ and $\mathbf{R}^{p+1}$, respectively, then the plane $h\in G_{p,n}$ with coordinates $\|\alpha^i{}_j\|$ is given by the equations

$$y^i=\sum_{j=1}^{p+1}\alpha^i_jz^j \quad (1\leqslant i\leqslant n-p). \tag{4.50}$$

In the coordinates α^i_j any differential s-form on $G_{p,n}$ has the form

$$\sigma=\sum_{|I|=|J|=s}\sigma^J_I(\alpha)\,d\alpha^I_J,$$

where $I=(i_1, \ldots, i_s)$, $J=(j_1, \ldots, j_s)$, $d\alpha^I_J=d\alpha^{i_1}_{j_1}\wedge\ldots\wedge d\alpha^{i_s}_{j_s}$, and the coefficients $\sigma^J_I=\sigma^{j_1,\ldots,j_s}_{i_1,\ldots,i_s}$ are antisymmetric with respect to permutations of the pairs $(i_1, j_1), \ldots, (i_s, j_s)$. Here the form σ is admissible if and only if its coefficients $\sigma^{j_1,\ldots,j_s}_{i_1,\ldots,i_s}$ are antisymmetric in $i_1, \ldots, i_s$.

We fix indices $i_1, \ i_2 \ (1 \leqslant i_1, \ i_2 \leqslant n-p; \ i_1 \neq i_2)$ and $j_1, \ j_2 \ (1 \leqslant j_1, \ j_2 \leqslant p+1; \ j_1 \neq j_2)$, and and we denote by $B^{i_1, i_2}_{j_1, j_2}$ the submanifold of flags $b=(h_1, h_2) \in B$, where $h_1 \in G_{p-2, n}$ is given by the system of equations

$$y^i = \sum_{j=1}^{p+1} a^i_j z^j \quad (i=1, \ldots, n-p); \tag{4.51}$$

and

$$z^{j_1}=0, \quad z^{j_2}=0,$$

while $h_2 \in G_{p+2, n} (h_2 \supset h_1)$ is given by the system of equations (4.51) for $i \neq i_1, i_2$. The coefficients a^i_j of Eqs. (4.51), where $(i, j) \neq (i_k, j_l)$, $k, l=1, 2$, we take as coordinates on $B^{i_1, i_2}_{j_1, j_2}$.

Proposition 28. For $s \geqslant 2$ on the submanifold $B^{i_1, i_2}_{j_1, j_2} \subset B$ the form $\tau\sigma$ coincides with the homogeneous component of the form σ of degree 2 in $da^{i_k}_{j_l}$, $k, l=1, 2$.

Proof. The condition that a plane $h \in G_{p, n}$ given by Eqs. (4.50) is incident with $(h_1, h_2) \in B^{i_1, i_2}_{j_1, j_2}$, i.e., $h_1 \subset h \subset h_2$, can be written in coordinate form as follows:

$$\alpha^i_j = a^i_j \quad \text{for} \quad (i, j) \neq (i_k, j_l), \ k, l=1, 2.$$

Therefore, in order to obtain an expression for $\tau\sigma$ on the submanifold $B^{i_1, i_2}_{j_1, j_2}$ in the expression for σ it is necessary to substitute da^i_j in place of $d\alpha^i_j$ for $(i, j) \neq (i_k, j_l)$ and discard all terms of degree less than two in $d\alpha^{i_k}_{j_l}$, $k, l=1, 2$.

COROLLARY. If the form $\tau\sigma$ is admissible on the fibers $G_p(b)$ for $b \in B^{i_1, i_2}_{j_1, j_2} (1 \leqslant i_1, \ i_2 \leqslant n-p, \ 1 \leqslant j_1, \ j_2 \leqslant p+1)$, then the form σ is admissible on $G_{p, n}$.

8. Connection between the Operators $\mathcal{I}_p$ for Different p

Here we establish the connection between the differential forms $\mathcal{I}_p\omega$ and $\mathcal{I}_q\omega$, where $\omega \in \Omega^r(\mathbf{RP}^n)$, $p<q \leqslant r$, and p and q have the same parity. (We recall that the space $\Omega^r(\mathbf{RP}^n)$ depends on the parity of p.) We shall show that

$$\mathcal{I}_q\omega = \mathcal{I}_{q, p} \circ \mathcal{I}_p \omega,$$

where $\mathcal{I}_{q, p}$ is an operator which will be constructed below.

We introduce the manifold $G_{p, q, n}$ of flags $h_1 \subset h_2$, $h_1 \in G_{p, n}$, $h_2 \in G_{q, n}$, and we consider the natural binary bundle

$$G_{p,n} \xleftarrow{\ \pi_1\ } G_{p,q,n} \xrightarrow{\ \pi_2\ } G_{q,n}.$$

We note that the fiber of the fibration π_2 is $G_{p,q}$.

Proposition 29. If a differential form $\sigma \in \Omega^{r-p}(G_{p,n})$ is admissible, then $\operatorname{Deg}_{\pi_2}(\pi_1^*\sigma)=q-p$.

Proof (cf. the Proof of Proposition 26). We denote by $G_p(h_2)$ the submanifold of planes $h_1 \in G_{p,n}$ incident with the plane $h_2 \in G_{q,n}$. Let $(h_1, h_2) \in G_{p,q,n}$, $x \in h_1$ and let L^x be the manifold of planes $h \in G_{p,n}$ incident with x. It is not hard to see that $\dim G_p(h_2)=(p+1)(q-p)$ and $\dim (L^x \cap G_p(h_2)) = p(q-p)$, and therefore the codimension of $L^x \cap G_p(h_2)$ in $G_p(h_2)$ is equal to $q-p$. This implies that any subspace in $T_{h_1}G_p(h_2)$ of dimension greater than $q-p$ contains a vector tangent to L^x, i.e., it is degenerate. Therefore, $(\pi_1^*\sigma)(h_1, h_2; \xi_1, \ldots, \xi_{r-p})=0$ if more than $q-p$ vectors ξ_i are tangent to a fiber of the bundle π_2.

Proposition 30. The differential form $\sigma=\mathcal{I}_p\omega$, where $\omega \in \Omega^r(\mathbf{RP}^n)$, satisfies the following conditions: (i) $\operatorname{Deg}_{\pi_2}(\pi_1^*\sigma)=q-p$, and the form $\pi_1^*\sigma$ is relatively closed in the bundle

$\pi_2: G_{p,q,n} \to G_{q,n}$; (ii) $\sigma(h; \xi_1, \ldots, \xi_{r-p}) = 0$ whenever $q-p$ vectors ξ_i are tangent to the submanifold of planes $h' \in G_{p,n}$ belonging to a fixed $(q-1)$-dimensional plane.

Proof. (i) If $\sigma = \mathcal{I}_p \omega$, then the forms σ and $d\sigma$ are admissible. Hence, by Proposition 29 $\mathrm{Deg}_{\pi_2}(\pi_1^* \sigma) = q-p$ and $\mathrm{Deg}_{\pi_2}(d\pi_1^* \sigma) = \mathrm{Deg}_{\pi_2}(\pi_1^* d\sigma) = q-p$. Since the relative degree of the form $\pi_1^* \sigma$ does not increase on exterior differentiation, this form is relatively closed.

(ii) The equality $\mathrm{Deg}_{\pi_2}(\pi_1^* \sigma) = q-p$ means that $\sigma(h; \xi_1, \ldots, \xi_{r-p}) = 0$ whenever $q-p+1$ of the vectors ξ_i are tangent to the submanifold of planes $h' \in G_{p,n}$ belonging to a fixed q-dimensional plane. Since in this condition $q > p$ is arbitrary, replacing q by $q-1$, we obtain condition (ii).

We denote by $\Omega_q^{r-p}(G_{p,n})$ the subspace of all differential forms $\sigma \in \Omega^{r-p}(G_{p,n})$ satisfying the conditions (i) and (ii). Below we define an operator

$$\mathcal{I}_{q,p}: \Omega_q^{r-p}(G_{p,n}) \to \Omega^{r-q}(G_{q,n}).$$

We consider the bundle $(G_{q,n}, H^{q-p}(G_{p,q}))$ with base $G_{q,n}$ and fiber $H^{q-p}(G_{p,q})$, which corresponds canonically to the bundle $\pi_2: G_{p,q,n} \to G_{q,n}$. To each $(r-p)$-form on $G_{p,q,n}$ of relative degree $q-p$ in the bundle $G_{p,q,n} \to G_{q,n}$ and relatively closed in this bundle there canonically corresponds an $(r-q)$-form on $G_{q,n}$ with coefficients in the bundle $(G_{q,n}, H^{q-p}(G_{p,q}))$. In particular, this holds for forms $\sigma \in \Omega_q^{r-p}(G_{p,n})$. We shall study this mapping in more detail.

We consider the standard fiber $G_{p,q} \cong G_p(h)$ of the bundle $G_{p,q,n} \to G_{q,n}$. To each $(q-p)$-dimensional cycle $\gamma \subset G_{p,q}$ we assign a collection of cycles $\gamma_h \subset G_p(h)$, $h \in G_{q,n}$, homologous to γ. We call Euler cycles in $G_{p,q}$ the $(q-p)$-dimensional cycles consisting of all planes $h' \in G_{p,q}$ containing a fixed $(p-1)$-dimensional plane.

Proposition 31. If $\sigma \in \Omega_q^{r-p}(G_{p,n})$, then for any $(q-p)$-dimensional cycle $\gamma \subset G_{p,q}$ we have

$$\int_{\gamma_h} \pi_1^* \sigma = d(\gamma) \tilde{\sigma}, \tag{4.52}$$

where $d(\gamma)$ is the index of intersection of γ with an Eucler cycle, and $\tilde{\sigma} \in \Omega^{r-q}(G_{q,n})$ does not depend on the choice of the cycle γ.

Proof. We consider a basis $\{\gamma\}$ of the $(q-p)$-dimensional homologies in $G_{p,q}$ which consists of $(q-p)$-dimensional Schubert cells. Among the elements of this basis there is an Euler cycle γ_0. To prove relation (4.52) it suffices to show that $\int_{\gamma_h} \pi_1^* \sigma = 0$ if γ is a $(q-p)$-dimensional Schubert cell distinct from the Euler cycle.

We use the following well known fact: if $\gamma \subset G_{p,q}$ is a $(q-p)$-dimensional Schubert cell distinct from the Euler cycle, then all planes $h \in \gamma$ are contained in a fixed $(q-1)$-dimensional plane. Hence, by condition (ii) it follows that $(\pi_1^* \sigma)(h; \xi_1, \ldots, \xi_{r-p}) = 0$ if $q-p$ of the vectors ξ_i are tangent to γ_h; hence $\int_{\gamma_h} \pi_1^* \sigma = 0$.

In correspondence with Proposition 31 we define the operator

$$\mathcal{I}_{q,p}: \Omega_q^{r-p}(G_{p,n}) \to \Omega^{r-q}(G_{q,n})$$

by the formula

$$\mathcal{I}_{q,p} \sigma = \frac{1}{d(\gamma)} \int_{\gamma_h} \pi_1^* \sigma, \tag{4.53}$$

where $\gamma \subset G_{p,q}$ is any $(q-p)$-dimensional cycle whose index of intersection $d(\gamma)$ with an Euler cycle is equal to zero.

THEOREM 6. For any form $\omega \in \Omega^r(\mathbf{RP}^n)$ and any numbers p, q of the same parity $(p < q \leq r)$ there is the equality

$$\mathcal{I}_q \omega = \mathcal{I}_{q,p} \circ \mathcal{I}_p \omega,$$

where $\mathcal{I}_{q,p}$ is the operator on $\Omega_q^{r-p}(G_{p,n})$ defined by formula (4.53).

Proof. It suffices to show that

$$\int_{\gamma_h} \pi_1^*(\mathcal{I}_p \omega) = \mathcal{I}_q \omega,$$

where γ is an Euler cycle. For the proof we use the definition of $\mathcal{I}_p \omega$. To compute $\mathcal{I}_p \omega$ it is necessary to substitute, in the form ω, $x = t^1 u_1 + \ldots + t^p u_p + u_{p+1}$, $dx = dt^1 u_1 + \ldots + dt^p u_p + t^1 du_1 + \ldots + t^p du_p + du_{p+1}$, separate out the homogeneous component of maximal degree p in $dt^1, \ldots, dt^p$, and integrate this component over $t^1, \ldots, t^p$ for fixed u_i, du_i. The $(r-p)$-form obtained on the manifold E_p of frames $u = (u_1, \ldots, u_{p+1})$ drops to $G_{p,n}$ under the natural projection $E_p \to G_{p,n}$; this is the form $\mathcal{I}_p \omega$.

Now let $h = (u_1, \ldots, u_{q+1}) \in G_{q,n}$. We choose for γ_h the $(q-p)$-dimensional family of p-dimensional planes in h containing the $(p-1)$-dimensional plane $h' = (u_1, \ldots, u_p)$. The operation $\sigma \mapsto \int_{\gamma_h} \pi_1^* \sigma$ then consists in the following. We replace u_{p+1} by $t^{p+1} u_{p+1} + \ldots + t^q u_q + u_{q+1}$ and du_{p+1} by $dt^{p+1} u_{p+1} + \ldots + dt^q u_q + t^{p+1} du_{p+1} + \ldots + t^q du_q + du_{q+1}$, we separate out the homogeneous component of maximal degree $q-p$ in $dt^{p+1}, \ldots, dt^q$ in the expression obtained, and we integrate this component on $t^{p+1}, \ldots, t^p$ for fixed u_i and du_i. From this it is clear that the composition of the mappings $\omega \mapsto \mathcal{I}_p \omega$ and $\mathcal{I}_p \omega \mapsto \int_{\gamma_h} \pi_1^*(\mathcal{I}_p \omega)$ is $\omega \mapsto \mathcal{I}_q \omega$. The proof of the theorem is complete.

APPENDIX. INTEGRATION OF DIFFERENTIAL FORMS OVER p-DIMENSIONAL PLANES IN THE SPACE $\mathbf{CP}^n$

All the assertions and formulas for $\mathbf{RP}^n$ presented in Secs. 3-8 carry over automatically to the complex space $\mathbf{CP}^n$. Here the number of admissibility conditions, conditions for dropping forms from E_p to $G_{p,n}$, etc. is correspondingly doubled.

1°. The operator $\mathcal{I}_p$ of integrating differential forms on $\mathbf{CP}^n$ over complex p-dimensional planes takes C^∞ forms of type (r', r'') on $\mathbf{CP}^n$, where $r' \geq p$, $r'' \geq p$, into forms of type $(r'-p, r''-p)$ on the complex Grassmann manifold $G_{p,n}$. The mapping $\mathcal{I}_p$ is injective for $r' > p$, $r'' > p$. If

$$\omega = \sum_{|I'| = r',\, |I''| = r''} f_{I',I''}(z)\, dz^{I'} \wedge d\bar{z}^{I''}, \quad z = (z^1, \ldots, z^{n+1}) \in \mathbf{C}^{n+1},$$

then

$$\mathcal{I}_p \omega = \sum_{\substack{|I'| = |J'| = r'-p \\ |I''| = |J''| = r''-p}} \varphi^{J',J''}_{I',I''}(u)\, du^{I'}_{J'} \wedge d\bar{u}^{I''}_{J''}, \quad u = (u_1, \ldots, u_{p+1}) \in E_p,$$

where

$$\varphi^{I'_1 J', I''_1 J''}_{I'_1, I''_1}(u) = (-1)^{J'+J''} \frac{r'!\, r''!}{(r'-p)!\, (r''-p)!} \sum_{|I'_2| = |I''_2| = p} \hat{f}^{J',J''}_{I'_1, I'_2;\, I''_1, I''_2}(u)\, u^{I'_2}_{\hat{j}} \bar{u}^{I''_2}_{\hat{j}}, \quad \hat{j} = (1, \ldots, \hat{j}, \ldots, p+1),$$

$$\hat{f}^{J',J''}_{I',I''}(u) = \int_{|t|=1} f_{I',I''}(tu)\, t^{J'} \bar{t}^{J''} \sigma_p(t, dt) \wedge \sigma_p(\bar{t}, d\bar{t}).$$

2°. A differential form σ on $G_{p,n}$ of type (s', s'') is called admissible if $\sigma(h; \xi_1, \ldots, \xi_{s'}; \bar{\eta}_1, \ldots, \bar{\eta}_{s''})=0$ $(\xi_i, \eta_j \in T_h G_{p,n})$ whenever either $(\xi_1, \ldots, \xi_{s'})$ or $(\eta_1, \ldots, \eta_{s''})$ generate a degenerate subspace (the definition of a degenerate subspace is the same as in the real case).

3°. Description of the Image of $\mathcal{I}_p$. A differential C^∞ form σ on $G_{p,n}$ of type $(r'-p, r''-p)$, where $r'-p>1$, $r''-p>1$, belongs to the image of $\mathcal{I}_p$ if and only if the forms σ, $\partial\sigma$ and $\bar{\partial}\sigma$ are admissible. In the special case where at least one of the indices r', r'' is equal to $p+1$, additional relations analogous to those presented in Theorem 3 for the real case must be satisfied in addition to the admissibility conditions.

4°. The Inversion Formula. Let $\sigma=\mathcal{I}_p\omega$, where ω is a form of type (r', r'') on $\mathbf{CP}^n$. We set

$$R\sigma=\sum_{|I'|=r', |I''|=r''} \hat{f}_{I', I''}(u)\, dz^{I'} \wedge d\bar{z}^{I''},$$

where

$$\hat{f}_{I_1', I_2', I_1'', I_2''}(u)=\frac{(r'-p)!(r''-p)!}{r'!r''!(p!)^2} \sum_{j', j''=1}^{p+1} (-1)^{j'+j''} \Delta_{I_2'}^{\hat{j}'} \bar{\Delta}_{I_2''}^{\hat{j}''} \varphi_{I_1', I_1''}^{J'j', J''j''}(u).$$

Here φ_I^J are the coefficients of the form σ; $J'=(p+1, \ldots, p+1)$, $J''=(p+1, \ldots, p+1)$; for the definition of the operators $\Delta_I^{\hat{j}}$ see Sec. 4.

We introduce the differential form of type $(r'+p, r''+p)$ on $A:(\varkappa_z \wedge \bar{\varkappa}_z) R\sigma$, where the operator $\varkappa_z$ is given by a formula analogous to that in Sec. 5, part 6. For this form there are assertions analogous to those presented in Sec. 5, part 6 for the form $\varkappa_x R\sigma$ in the real case. In particular, in the notation used there we have

$$\int_{\gamma^z} (\varkappa_z \wedge \bar{\varkappa}_z) R\sigma=c d(\gamma)\, \omega \quad (c \neq 0),$$

where γ^z is any cycle in F^z of real dimension 2p. We emphasize that the inversion formula holds for any p.

LITERATURE CITED

1. I. M. Gel'fand, "Integral geometry and its connection with representation theory," Usp. Mat. Nauk, 15, No. 2, 155-164 (1960).
2. I. M. Gel'fand and S. G. Gindikin, "Nonlocal inversion formulas in real integral geometry," Funkts. Anal. Ego Prilozhen., 11, No. 3, 12-19 (1977).
3. I. M. Gel'fand, S. G. Gindikin, and M. I. Graev, "Problems of integral geometry related to the integration of differential forms along lines in $\mathbf{R}^3$ and $\mathbf{C}^3$," Inst. Prikl. Mat. Akad. Nauk SSSR, Moscow, Preprint No. 24 (1979).
4. I. M. Gel'fand, S. G. Gindikin, and M. I. Graev, "Problems of integral geometry related to the integration of differential forms along lines in three-dimensional projective space," Inst. Prikl. Mat. Akad. Nauk SSSR, Moscow, Preprint No. 41 (1979).
5. I. M. Gel'fand, S. G. Gindikin, and M. I. Graev, "Binary bundles and problems of integral geometry related to the integration of differential forms along lines," Inst. Prikl. Mat. Akad. Nauk SSSR, Moscow, Preprint No. 60 (1979).
6. I. M. Gel'fand, S. G. Gindikin, and M. I. Graev, "Problems of integral geometry in projective space related to the integration of differential forms," Inst. Prikl. Mat. Akad. Nauk SSSR, Moscow, Preprint No. 79 (1979).
7. I. M. Gel'fand, S. G. Gindikin, and M. I. Graev, "Some questions related to the integration of differential forms over planes in $\mathbf{RP}^n$ and $\mathbf{CP}^n$," Inst. Prikl. Mat. Akad. Nauk SSSR, Moscow, Preprint No. 126 (1979).
8. I. M. Gel'fand, S. G. Gindikin, and M. I. Graev, "A problem of integral geometry in $\mathbf{RP}^n$ related to the integration of differential forms," Funkts. Anal. Prilozhen., 13, No. 4, 64-66 (1979).

9. I. M. Gel'fand, S. G. Gindikin, and M. I. Graev, "Problems of integral geometry in affine space related to the integration of functions over planes," Inst. Prikl. Mat. Akad. Nauk SSSR, Moscow, Preprint No. 152 (1979).
.0. I. M. Gel'fand, S. G. Gindikin, and M. I. Graev, "Integral geometry for some one-dimensional bundles over projective space," Inst. Prikl. Mat. Akad. Nauk SSSR, Moscow, Preprint No. 24 (1980).
.1. I. M. Gel'fand, S. G. Gindikin, and Z. Ya. Shapiro, "The local problem of integral geometry in a space of curves," Funkts. Anal. Prilozhen., 13, No. 2, 11-31 (1979).
2. I. M. Gel'fand and M. I. Graev, "The geometry of homogeneous spaces, group representations in homogeneous spaces, and related questions of integral geometry. I," Tr. Mosk. Mat. Obshch., 8, 321-390 (1959)
3. I. M. Gel'fand and M. I. Graev, "Integrals over hyperplanes of test functions and generalized functions," Dokl. Akad. Nauk SSSR, 135, No. 6, 1307-1310 (1960).
.4. I. M. Gel'fand and M. I. Graev, "Integral transforms connected with complexes of lines in complex affine space," Dokl. Akad. Nauk SSSR, 138, No. 6, 1266-1269 (1961).
5. I. M. Gel'fand and M. I. Graev, "Complexes of lines in the space C^n," Funkts. Anal. Prilozhen., 2, No. 3, 39-52 (1968).
.6. I. M. Gel'fand and M. I. Graev, "Complexes of k-dimensional planes in the space C^n and the Plancherel formula for the group $GL(n, C)$," Dokl. Akad. Nauk SSSR, 179, No. 3, 522-525 (1968).
7. I. M. Gel'fand, M. I. Graev, and N. Ya. Vilenkin, Integral Geometry and Representation Theory, Academic Press (1966).
.8. I. M. Gel'fand, M. I. Graev, and I. I. Pyatetskii-Shapiro, Theory of Representations and Automorphic Functions [in Russian], Nauka, Moscow (1966).
.9. I. M. Gel'fand, M. I. Graev, and Z. Ya. Shapiro, "Integral geometry on the manifold of k-dimensional planes," Dokl. Akad. Nauk SSSR, 168, No. 6, 1236-1238 (1966).
•0. I. M. Gel'fand, M. I. Graev, and Z. Ya. Shapiro, "Integral geometry on k-dimensional planes," Funkts. Anal. Prilozhen., 1, No. 1, 15-31 (1967).
1. I. M. Gel'fand, M. I. Graev, and Z. Ya. Shapiro, "Differential forms and integral geometry," Funkts. Anal. Prilozhen., 3, No. 2, 24-40 (1969).
2. I. M. Gel'fand, M. I. Graev, and Z. Ya. Shapiro, "Integral geometry in projective space," Funkts. Anal. Prilozhen., 4, No. 1, 14-32 (1970).
'3. I. M. Gel'fand, M. I. Graev, and Z. Ya. Shapiro, "A problem of integral geometry related to a pair of Grassmann manifolds," Dokl. Akad. Nauk SSSR, 193, No. 2, 259-262 (1970).
4. I. M. Gel'fand and G. E. Shilov, Generalized Functions, Vol. 1, Properties and Operations, Academic Press (1964).
'5. S. G. Gindikin, "Unitary representations of groups of automorphisms of symmetric Riemannian spaces of zero curvature," Funkts. Anal. Prilozhen., 1, No. 1, 32-37 (1967).
6. S. G. Gindikin and F. I. Karpelevich, "On a problem of integral geometry," in: In Honor of the Memory of N. G. Chebotarev, 1894-1947 [in Russian], Kazansk. Univ., Kazan, 1964, pp. 30-43.
:7. S. G. Gindikin and G. M. Khenkin, "Integral geometry for $\bar{\partial}$-cohomologies in q-linearly concave domains in CP^n," Funkts. Anal. Prilozhen., 12, No. 4, 6-20 (1978).
8. S. G. Gindikin and G. M. Khenkin, "The Penrose transformation and complex integral geometry," in: Sovrem. Probl. Mat., Vol. 11 (Itogi Nauki i Tekhn. VINITI AN SSSR), Moscow, 1980 (in press).
:9. F. John, Plane Waves and Spherical Means in Application to Partial Differential Equations [Russian translation], IL (1958).
•0. A. A. Kirillov, "On a problem of I. M. Gel'fand," Dokl. Akad. Nauk SSSR, 137, No. 2, 276-277 (1961).
•1. K. Maius, "The structure of admissible complexes of lines in C^n," Funkts. Anal. Prilozhen., 7, No. 1, 79-81 (1973).
•2. K. Maius, "Admissible complexes of lines with one critical point," Funkts. Anal. Prilozhen., 9, No. 2, 81-82 (1975).
•3. K. Maius, "The structure of admissible complexes of lines in CP^n," Tr. Mosk. Mat. Obshch., 39, 181-211 (1979).
•4. E. E. Petrov, "The Radon transform in spaces of matrices and Grassmann manifolds," Dokl. Akad. Nauk SSSR, 177, No. 4, 782-785 (1967).
•5. E. E. Petrov, "The Radon transform in a space of matrices," Proc. of the Seminar on Vector and Tensor Analysis and Their Applications to Geometry, Mechanics, and Physics, Moscow State Univ., No. 15, 299-315 (1970).

36. K. Maius, "The Gel'fand–Graev–Shapiro formula for Radon complexes," Sb. Nauch. Tr. MGZPI, No. 39, 237-245 (1974).
37. G. de Rham, Differentiable Manifolds [Russian translation], IL, Moscow (1956).
38. V. I. Semyanistyi, "On some integral transforms in Euclidean space," Dokl. Akad. Nauk SSSR, 134, No. 3, 536-539 (1960).
39. V. I. Semyanistyi, "Homogeneous functions and some problems of integral geometry in spaces of constant curvature," Dokl. Akad. Nauk SSSR, 136, No. 2, 288-291 (1961).
40. V. I. Semyanistyi, "Some integral transforms and integral geometry in an elliptic space," Proc. of the Seminar on Vector and Tensor Analysis and Their Application to Geometry, Mechanics, and Physcs, Moscow State Univ., No. 12, 397-441 (1963).
41. V. I. Semyanistyi, "Some problems of integral geometry in pseudo-Euclidean and non-Euclidean spaces," Proc. of the Seminar on Vector and Tensor Analysis and Their Application to Geometry, Mechanics, and Physics, Moscow State Univ., No. 13, 244-302 (1966).
42. V. I. Semyanistyi and Z. F. Shibasova, "Integral geometry in Euclidean space and its relation to boundary problems," Dokl. Akad. Nauk SSSR, 176, No. 2, 269-272 (1967).
43. A. A. Khachaturov, "Determination of the value of a measure for a domain of n-dimensional Euclidean space in terms of its values for all half spaces," Usp. Mat. Nauk, 9, No. 3, 205-212 (1954).
44. W. Hodge and D. Pidoe, Methods of Algebraic Geometry [Russian translation], Vol. I, IL, Moscow (1954).
45. J. Schwartz, Differential Geometry and Topology [Russian translation], Mir, Moscow (1970).
46. M. F. Atiyah and R. S. Ward, "Instantons and algebraic geometry," Commun. Math. Phys., 55, No. 2, 117-124 (1977).
47. P. A. Griffiths, "Complex differential and integral geometry and curvature integrals associated to singularities of complex analytic varieties," Duke Math. J., 45, No. 3, 427-512 (1978).
48. V. Guillemin and S. Sternberg, Geometric Asymptotics, Math. Surveys, No. 14, AMS, Providence, Rhode Island (1978).
49. V. Guillemin and S. Sternberg, "Some problems in integral geometry," Am. J. Math., 101, No. 1, 915-955 (1979).
50. S. Helgason, "Differential operators on homogeneous spaces," Acta Math., 102, 239-299 (1959).
51. S. Helgason, "Duality and Radon transform for symmetric spaces," Am. J. Math., 85, No. 4, 667-692 (1963).
52. S. Helgason, "A duality in integral geometry; some generalizations of the Radon transform," Bull. Am. Math. Soc., 70, No. 4, 435-446 (1964).
53. S. Helgason, "The Radon transform on Euclidean spaces, compact two-point homogeneous spaces, and Grassmann manifolds," Acta Math., 113, Nos. 3-4, 153-180 (1965).
54. S. Helgason, "Radon–Fourier transforms on symmetric spaces and related group representations," Bull. Am. Math. Soc., 71, No. 5, 757-763 (1965).
55. S. Helgason, "Applications of the Radon transform to representations of semisimple Lie groups," Proc. Nat. Acad. Sci. USA, 63, No. 3, 643-647 (1969).
56. S. Helgason, "A duality for symmetric spaces with applications to group representations," Adv. Math., 5, No. 1, 1-154 (1970).
57. S. Helgason, The Radon Transform, Progress in Math., Vol. 5, Birkhäuser (1980).
58. F. John, "Bestimmung einer Funktion aus ihren Integralen über gewisse Mannigfaltigkeiten," Math. Ann., 100, 488-520 (1934).
59. F. John, "Abhängigkeiten zwischen den Flächenintegralen einer stetigen Funktion," Math. Ann., 111, 541-559 (1935).
60. F. John, "The ultrahyperbolic differential equation with four independent variables," Duke Math. J., 4, 300-322 (1938).
61. D. Ludwig, "The Radon transform in Euclidean space," Commun. Pure Appl. Math., 19, 49-81 (1966).
62. R. Penrose, Twistor Theory, Its Aims and Achievements, Quantum Gravity, Oxford Symp., Oxford (1975).
63. R. Penrose, Massless Fields and Sheaf Cohomology, Twistor Newsletter, Oxford (1977).
64. J. Radon, "Über die Bestimmung von Funktionen durch ihre Integralwärte längs gewisser Mannigfaltigkeiten," Ber. Verh. Sachs. Acad., 69, 262-277 (1917).
65. K. T. Smith, D. C. Solmon, and S. L. Wagner, "Practical and mathematical aspects of the problem of reconstructing objects from radiographs," Bull. Am. Math. Soc., 83, 1227-1270 (1977).

66. T. Tsujishita, "Integral transformations associated with double fiberings," Osaka J. Math., 15, 391-418 (1978).
67. R. O. Wells, "Complex manifolds and mathematical physics," Bull. Am. Math. Soc., New Ser., 1, No. 2, 296-336 (1979).

12.

(with G. S. Shmelev)

Geometric structures of double bundles and their relation to certain problems in integral geometry

Funkts. Anal. Prilozh. **17** (2) (1983) 7-22 [Funct. Anal. Appl. **17** (1983) 84-96].
Zbl. **519**:53058

INTRODUCTION

By definition, a double bundle (DB) is a diagram (see also [9])

$$\begin{array}{ccc} & A & \\ \pi_1 \swarrow & & \searrow \pi_2, \\ B & & \Gamma \end{array}$$

and A, B, and Γ are smooth manifolds, and: 1) A is a smooth bundle with respect to both projections π_1 and π_2; 2) the mapping $\pi_1 \times \pi_2: A \to B \times \Gamma$ is a nondegenerate injective diffeomorphism; 3) for any two points x_1, $x_2 \in B$, the surfaces $\pi_2 \circ \pi_1^{-1} x_1$ and $\pi_2 \circ \pi_1^{-1} x_2$ do not coincide, and a similar condition holds for any two points $\xi_1, \xi_2 \in \Gamma$. The surfaces $\pi_2 \circ \pi_1^{-1} x$ and $\pi_1 \circ \pi_2^{-1} \xi$ will be denoted by Γ_x and B_ξ, respectively.

An example: $B = P^3$, $\Gamma = G(4, 2)$ is the manifold of lines in P^3, and A is the manifold of pairs (line, point on the line).

The mappings π_1 and π_2 enable us to carry the various analytic objects (functions, forms, etc.) from B to Γ, by first lifting them to A in some way or another, and subsequently descending them to Γ. To see an example, suppose that there are given measures χ_ξ on the surfaces B_ξ. Define the operator of integration over the B_ξ's as $If(\xi) = \int f\chi_\xi$. Then we are presented with a nice and profound problem: Recover the function f given its image $\varphi = If$. In the case $\dim B_\xi = 1$, a complete solution to this problem for the so-called local inversion formulas has been obtained in the work of Gel'fand, Graev, Shapiro, Gindikin, Bernshtein, and Maius (see [2-5, 13]). The papers [9-12] and [14] are devoted to the case $\dim B_\xi > 1$. Nonlocal inversion formulas are discussed in the work of Gel'fand and Gindikin [15], and of Gel'fand, Graev, and Rosu [12]. All the local inversion formulas are given by an integral over Γ_x or over some cycle in Γ_x and are constructed from a form whose coefficients are partial derivatives of φ. The formulas arising in this way have a lot in common: there always exists a larger DB: $B \overset{\pi_1'}{\leftarrow} A' \overset{\pi_2'}{\rightarrow} \Gamma'$ (which may possibly coincide with the original one), such that $A \subset A'$, $\Gamma \subset \Gamma'$, $(\pi_2')^{-1}\Gamma = A$, and the form $\varkappa$, giving the local inversion formula on the original DB is the restriction of the form $\varkappa'$ giving on the new DB (moreover, the partial derivatives appearing in $\varkappa$, must be computed in terms of the original DB; see [2, 10]). Even more, the new form $\varkappa'$ always has the following structure. For each point $\xi \in \Gamma'$ and some linear spaces V_1 and V_2, with $\dim V_1 = \dim B_3 + 1$. and $\dim V_2 = \dim \Gamma_x$, a linear map $\Psi'_\xi: V_1 \otimes V_2 \to T_\xi \Gamma$ is fixed, with the property that given any $x \in B_\xi$, there is $v \in PV_1$ such that $\Psi_\xi(v \otimes V_2) = T_\xi \Gamma_x$. Then one identifies B_ξ with a domain in PV_1. Next, a hyperplane H_ξ and an affine chart are fixed on V_1, so that H_ξ is the plane at infinity in this chart; the measure χ_ξ coincides with the standard volume in this chart. It turns out that $\varkappa'$ is a k-form with $k = \dim B_\xi$, whose values are k-th order differential operators. Moreover, by taking the quotient with respect to $T_\xi \Gamma_x$, the symbol of the operator $\varkappa(\tau_1, \ldots, {}_k)$, where $\tau_1, \ldots, \tau_k \in T_\xi \Gamma_x$, is transformed into a polynomial $\Delta(\tau_1, \ldots, \tau_k)$, which may be described as follows. Identity $T_\xi \Gamma_x$ and V_2 via the mapping $\Psi|_{v \otimes V_2}$, for a suitable $v \in PV_1$; then, for some basis $e_1, \ldots, e_k$ in H_ξ, which does not depend on $\tau_1, \ldots, \tau_k$, we shall have $\Delta(\tau_1, \ldots, \tau_k) = \sum_\sigma (-1)^\sigma \times \prod_{i=1} \Psi'(e_i \otimes \tau_{\sigma(i)})$, where the sum is taken over all permutations σ, and Ψ_ξ' is the composition of Ψ_ξ and the factorization with respect to $T_\xi \Gamma_x$.

Institute of Applied Mathematics, Academy of Sciences of the USSR. Moscow State University. Translated from Funktsional'nyi Analiz i Ego Prilozheniya, Vol. 17, No. 2, pp. 7-22, April-June, 1983. Original article submitted December 29, 1982.

In the present work we show that if the double bundle is, in some sense, nondegenerate, then the form $\varkappa$ necessarily satisfies the conditions described above. Here we formulate the nondegeneracy condition in terms of the characteristic mapping (see Sec. 1, No. 2 for the definition), which is a geometric invariant of the DB. The corresponding relation among the dimensions is $\dim A \leqslant \frac{2}{3}(\dim B + \dim \Gamma) - \frac{1}{3}$ (for an arbitrary DB one simply has $\dim A < \dim B + \dim \Gamma$). In other words, we find here the very conditions necessary for the existence of a local inversion formula for a "general position" double bundle satisfying this inequality.

The material is organized as follows. In Sec. 1 we give the definitions and the required geometric constructions. In Sec. 2 we define the concept of a pseudotensorial structure, which generalizes in some sense the classical almost-Grassmanian structures, without being a G-structure itself. We then introduce the notion of a right semi-Grassmanian double bundle, as a double bundle for which the Γ_x's are integral surfaces of the pseudotensorial structure on Γ. In Sec. 3 the admissible double bundles are introduced and one proves the basic result concerning the relation between admissible and right semi-Grassmanian double bundles. Finally, Sec. 4 is devoted to the proof of this result.

In conclusion, we express our gratitude to M. I. Graev and A. B. Goncharova for fruitful and stimulating discussions.

1. Double Bundles

1. Duality of the Structure of a Double Bundle. The definition of a double bundle (DB) was given in the Introduction.

Let $a \in A$, $x = \pi_1 a$, and $\xi = \pi_2 a$. We denote the fibers over the points x and ξ by F_x and Φ_ξ, respectively. From the definition of a DB it follows that $T_a F_x \cap T_a \Phi_\xi = 0$. Set $L_a = T_a F_x \oplus T_a \Phi_\xi$. The projections $\pi_{1*} T_a A \to T_x B$ and $\pi_{2*}: T_a A \to T_\xi \Gamma$ induce isomorphisms $T_x B / T_x B_\xi \simeq T_a A / L_a \simeq T_\xi \Gamma / T_\xi \Gamma_x$. (Note that $B_\xi = \pi_1 \Phi_\xi$ and $\Gamma_x = \pi_2 F_x$.) We shall denote the isomorphism $T_x B / T_x B_\xi \simeq T_\xi \Gamma / T_\xi \Gamma_x$ by $n(x, \xi)$.

Fix a point $\xi_0 \in \Gamma$. Let $\mathscr{D}$ be a differential operator at the point ξ_0. For all ξ belonging to some neighborhood of ξ_0, pick parametrizations of the surfaces B_ξ depending smoothly on ξ. Then for each point $x \in B_\xi$, we have a mapping $\xi \to y(\xi, x)$, where $y(\xi, x)$ is the point on B_ξ having the same value of the parameter as x has on B_{ξ_0}. At the same time, the operator $\mathscr{D}$ is transformed into an operator $\mathscr{D}_x$ at the point x.

Let f_1 be a function such that $df_1 = 0$ at the point x_0 on B_{ξ_0}, and let $f = (f_1)^k$, with $k = \operatorname{ord} \mathscr{L}$. The following statement is readily verified.

Proposition. The value $\mathscr{D}_x f$ does not depend upon the choice of parametrization on B_ξ and equals $\sigma\mathscr{D}(n^*(x, \xi_0) df_1)$, where $\sigma\mathscr{D}$ is considered as a polynomial of degree k on $T^*_{\xi_0}\Gamma$.

This proposition clearly remains valid when one replaces B by Γ.

COROLLARY. To every vector $v \in T_\xi \Gamma$ (or $u \in T_x B$) corresponds a section γ_v (respectively, δ_u) of the normal bundle NB_ξ (respectively, $N\Gamma_x$).

2. Conjugate Points and Characteristic Vectors. Definition. Let $\xi \in \Gamma$ be a fixed point. Two points x_1, x_2 on B_ξ are said to be conjugate on B_ξ relative to $v \in T_\xi \Gamma$, if $\gamma_v \times (x_1) = \gamma_v(x_2) = 0$.

The infinitesimal analog of the idea of conjugate points is the notion of characteristic vector.

Definition. Set $V_a = T_a A / L_a$, $\Pi_x = \pi_2^{-1} \Gamma_x$, and $P_\xi = \pi_1^{-1} B_\xi$. We now define a linear map Θ: $S^2 L_a \to V_a$. Suppose that $\sigma \in S^2 L_a$ and $\mathscr{D}_1$ and $\mathscr{D}_2$ are second-order differential operators such that $\sigma(\mathscr{D}_1) = \sigma(\mathscr{D}_2) = \sigma$ and $\mathscr{D}_1$ is tangent to P_ξ, and $\mathscr{D}_2$ to Γ_x (an operator $\mathscr{D}$ is said to be tangent to a submanifold if for any function f vanishes on the submanifold, the function $\mathscr{D} f$ also vanishes on the submanifold). The difference $\mathscr{D}_1 - \mathscr{D}_2$ is a first-order operator, i.e., a vector. Set $\Theta(\sigma) = \pi(\mathscr{D}_1 - \mathscr{D}_2)$, where $\pi: T_a A \to V_a$ is the projection. Θ is well defined as a map from $S^2 L_a$ into V_a. Since $\Pi_x \cap P_\xi \supset F_x \cup \Phi_\xi$, we have $\Theta(\sigma) = 0$, whenever σ is generated either only by vectors from $T_a F_x$, or only by vectors from $T_a \Phi_\xi$. That is to say, Θ is a map from $T_aF_x \otimes T_a\Phi_\xi$ into V_a. We say that the vectors $u \in T_aF_x$ and $v \in T_a\Phi_\xi$ are conjugate characteristic if $\Theta(u \otimes v) = 0$. The absence of conjugate characteristic vectors implies that $\operatorname{Ker}\Theta$ does not intersect the vectors of rank 1. The map Θ is referred to as characteristic.

Let us give an alternative definition of the conjugate characteristic vectors. Suppose that a_0, $x_0 = \pi_1 a_0$, and $\xi_0 = \pi_2 a_0$ are fixed. We shall consider that B_ξ is embedded in NB_ξ as the zero section. Then $T_{x_0}NB_\xi = N_{x_0} \oplus T_{x_0}B_\xi$, where N_{x_0} is the fiber over x_0. Let $v \in T_\xi\Gamma_{x_0}$. Then $\gamma_v(x_0) = 0$ and the differential $d\gamma_v$ at the point x_0 can be interpreted as the graph of a linear map $\Theta(v)$ from $T_{x_0}B_{\xi_0}$ into $N_{x_0} \simeq V_{a_0}$. Two vectors $v \in T_{\xi_0}\Gamma_{x_0}$ and $u \in T_{x_0}B_{\xi_0}$ are said to be conjugate characteristic if $\Theta(v)(u) = 0$. In the sequel we shall write $\Theta(v)(u)$. as well as $\Theta(v \otimes u)$.

This definition is equivalent to the previous one and it becomes evident that characteristic vectors are indeed the infinitesimal analog of the conjugate points.

2. Right Semi-Grassmanian Double Bundles

1. Manifolds with Pseudotensorial Structure. Definition. A manifold Γ is said to be (k, l)-pseudotensorial [or a manifold with a (nondegenerate) (k, l)-pseudotensorial structure] if for each point $\xi \in \Gamma$ there is given a linear map $\Psi_\xi\colon V_1 \otimes V_2 \to T_\xi\Gamma$, where $\dim V_1 = k$, $\dim V_2 = 1$, such that for the points $v \in PV_1$ in general position the intersection $v \otimes V_2 \cap \operatorname{Ker}\Psi_\xi$ is empty.

This sort of concept was first introduced in [2] (see also [4], where it appears as one of the admissibility conditions). Then it naturally occurred in [16, 13] for k = 2 and arbitrary l in the case of the admissible complex of linear. We owe to I. N. Bernshtein and S. G. Gindikin its general formulation for the case of curves. They were the first to underscore the close relationship between this notion and the canonical rational structure on a curve (see [5], in particular the definition of the $\mathcal{P}$- and $\mathcal{P}_1$-structures). Since these authors considered admissible complexes only for $\dim B_\xi = 1$, they formulated the notion of (k, l)-pseudotensorial structure only for k = 2, but there is no doubt that they fully understood its generalization to arbitrary k. In our work we demonstrate the intimate connection between the general definition and the theory of admissible double bundles in the case of surfaces. We should also emphasize that this structure is of independent differential-geometric interest, aside from its connection with the theory of double bundles.

Examples. 1. A Pfaff system of k-dimensional planes in Γ defines on Γ a (k, 1)- [or (1, k)-] pseudotensorial structure.

2. Suppose that $\dim\Gamma = 3$ and at each point $\xi \in \Gamma$ there is given a quadratic cone in $T_\xi\Gamma$. The system of planes tangent to it defines a (2, 2)-pseudotensorial structure in $T_\xi\Gamma$.

3. Suppose that $\dim\Gamma = 4$ and at each point $\xi \in \Gamma$ there is given a quadratic cone of signature (+, +, −, −) in $T_\xi\Gamma$. The system of isotropic two-dimensional planes of this cone defines a (2, 2)-pseudotensorial structure in $T_\xi\Gamma$.

4. Let Γ be the manifold $G(k + 1, k)$ of k-dimensional planes in the $(k + l)$-dimensional space. Then there is a pseudotensorial structure on Γ which coincides with the pseudo-Grassmanian G-structure (see [6]).

The last example shows that the pseudo-Grassmanian structures are particular cases of pseudotensorial structures. This is the place to mention that pseudo-Grassmanian structures have been studied until now in the frame of the theory of G-structures (see [6–8]). In contrast, a pseudotensorial structure is not a G-structure, because its structure group is, generally speaking, trivial.

Remark. To give the map $\Psi_\xi\colon V_1 \otimes V_2 \to T_\xi\Gamma$ is the same as to give a linear complex of planes in $T_\xi\Gamma$. The fact that there are no characteristic vectors says that the planes of this complex have no critical points (for the definition of critical points, see [2, 13]). It seems likely that, using the same arguments as in the proof of our main theorem, one can prove under the assumption that there are no critical points, that a complex is linear if given any two of its planes one can find a one-parameter family of planes of the complex in the space they span.

Let V_1' and V_2' be subspaces of V_1 and V_2 with $\dim V_1' = \dim V_2' = s$, and let μ_1 and μ_2 be fixed volume elements on V_1' and V_2'. Further, let $e_1, \ldots, e_s$ be based in V_1' and ε_1,

$\ldots, \varepsilon_s$ be based in $V_2^!$, satisfying $\mu_1(e_1,\ldots,e_s) = \mu_2(\varepsilon_1,\ldots,\varepsilon_s) = 1$. Set $\Delta_\Psi(V_1', V_2', \mu_1, \mu_2) = \sum_\sigma \times (-1)^\sigma \prod_{i=1}^{s} \Psi_\xi(e_i \otimes \varepsilon_{\sigma(i)})$, where the product is understood as an element from $S^k(T_\xi\Gamma)$, and the sum is taken over all permutations σ. Note that $\Delta_\Psi(V_1', V_2', \mu_1, \mu_2)$ depends only upon the choice of the measures μ_1 and μ_2.

Set $P_u = \Psi_\xi(u \otimes V_2)$, $u \in PV_1$; and $R_v = \Psi_\xi(V_1 \otimes v)$, $v \in PV_2$. A surface Γ' in Γ is said to be a right (left) integral surface of the pseudotensorial structure if at each point of Γ' its tangent planes coincide with the planes P_u (respectively, R_v) for some u (respectively, for some v).

2. Semi-Grassmanian Double Bundles. Definition. A double bundle

$$B \swarrow A \searrow \Gamma'$$

with $\dim B_\xi = k$, $\dim \Gamma = l$, is called right $(k+1, l)$-semi-Grassmanian if there is a $(k+1, l)$-pseudotensorial structure $\Psi_\xi: V_1 \otimes V_2 \to T_\xi\Gamma$, on Γ for which the surfaces Γ_x are right integral surfaces and, in addition, for general position $\xi \in \Gamma$, the set of all $v \in PV_1$, such that $\Psi_\xi(v \otimes V_2) = T_\xi\Gamma_x$ for some $x \in B_\xi$ is an open domain in PV_1. In this case we shall say also that the surfaces Γ_ξ define a semi-Grassmanian structure in Γ.

Examples of Right Semi-Grassmanian DB's. 1. Example 2 in No. 2 applies here too.

2. In P^n fix manifolds $M_1,\ldots,M_s$, $s \leqslant n-2$, such that $\operatorname{codim} M_i \leqslant 2$. Let the family of surfaces B_ξ be precisely the family of all lines in P^n which have a contact of order 2-$\operatorname{codim} M_i$ with M_i. Then, if $M_1,\ldots,M_s$ are in general position (for more precision see [2, 13]), oru DB is right semi-Grassmanian (for a generalization of this example, see [5, 13]).

3. Let us generalize the previous example. In P^n fix surfaces $M_1,\ldots,M_s$, $s \leqslant n-d-1$, such that $\operatorname{codim} M_i \leqslant d+1$.

Define the surfaces B_ξ as the family Γ of all d-dimensional planes in P^n which are tangent to M_i along a subspace of dimension no less than $d - \operatorname{codim} M_i + 1$. Denote by B and A the manifold of all $(d-1)$-dimensional planes in P^n, and the manifold of all pairs [d-dimensional plane from Γ, $(d-1)$-dimensional subplane], respectively. Then the corresponding DB is right semi-Grassmanian.

4. Let C be a complex of $(d-1)$-dimensional planes in P^n, and let Γ be the family of all d-dimensional planes in P^n which each contain at least one plane from C. Then the corresponding DB is right semi-Grassmanian (this DB was first considered in [17]).

3. Admissible Double Bundles

1. Suppose that on the surfaces B_ξ there are given measures χ_ξ depending smoothly on ξ. Denote by $I: C_0^\infty(B) \to C^\infty(\Gamma)$ the integral transform

$$(If)(\xi) = \int_{B_\xi} f\chi_\xi.$$

Further, let $\Omega^k \to B$ be the smooth bundle over B whose fiber over each point $x \in B$ is the space of exterior (differential) k-forms with smooth coefficients on the manifold F_x. We denote the space of sections of this bundle by $\Gamma(\Omega^k)$.

We shall say that I admits a local inversion formula if on A there is a differential operator $\varkappa$ acting from $C^\infty(A)$ into $\Gamma(\Omega k)$, and such that $\varkappa(\pi_2^{-1}If)$ is a closed form on F_x for any function f on B and any $x \in B$; in addition, we require that this form be exact whenever $\operatorname{supp} f \not\ni x$. A DB with a fixed system of measures χ_ξ for which this holds will be referred to as admissible.

We say that the operator $\varkappa$ is trivial if there exists a differential operator $\varkappa_1: C^\infty(A) \to \Gamma(\Omega^{k-1})$ such that $\varkappa(\pi_2^{-1}If) = d\varkappa_1(\pi_2^{-1}If)$ for any $f \in C_0^\infty(B)$. Two operators $\varkappa_1$ and $\varkappa_2$ are regarded as equivalent if $\varkappa_1 - \varkappa_2$ is the trivial operator.

Since we shall consider only the action of $\varkappa$ on functions of the form $\pi_2^{-1}\varphi$, with $\varphi \in C^\infty(\Gamma)$, it is convenient to regard $\varkappa$ as a form given on the manifolds Γ_x. If $\xi \in \Gamma_x$ and

$\tau_1,\ldots,\tau_k \in T_\xi\Gamma_x$ are fixed vectors, then $\varkappa$ defines a differential operator $\varkappa(\tau_1,\ldots,\tau_k)$ at the point ξ on Γ. We denote its symbol by $\sigma(\tau_1,\ldots,\tau_k)$.

For examples of admissible DB's, see [1-5, 9-11], as well as the examples in Sec. 2, No. 2.

2. The Fundamental Lemma. Let $\tau_1,\ldots,\tau_k \in T_\xi\Gamma$. We let $l(\tau_1,\ldots,\tau_k)$ denote the space spanned by these vectors. For $V \subset T_\xi\Gamma$, we let Ann V denote its annihilator in $T_\xi^*\Gamma$.

Fundamental Lemma. Suppose that the points ξ, x_0, x are such that $x_0, x \in B_\xi$, and x_0 and x are not conjugate in B, for any $V \in T_\xi\Gamma$. Then given any $\tau_1,\ldots,\tau_k \in T_\xi\Gamma_{x_0}$, the polynomial $\sigma(\tau_1,\ldots,\tau_k)$ vanishes on the intersection $\operatorname{Ann} l(\tau_1,\ldots,\tau_k) \cap \operatorname{Ann} T_\xi\Gamma_x$.

Proof. Let C be a smooth k-dimensional cycle in Γ_{x_0}. Set $\Omega^C_{x_0} = \bigcup B_\xi$, $\xi \in C$. Since the points x and x_0 are not conjugate in B_ξ, the set $\Omega^C_{x_0}$ is a smooth 2k-dimensional manifold in some neighborhood U of the point x. Moreover, one can choose coordinates on U so that the surfaces B_ξ are given by the equation $x_{k+1} = c_1,\ldots,x_{2k} = c_k$, $x_{2k+1} = \ldots = x_n$. Then if supp f $\subset$ U,

$$\int_C \varkappa\varphi = \int_C \varkappa\Big(\int_{B_\xi,\ \xi\in C} f(x_1,\ldots,x_k,c_1,\ldots,c_k)\,\chi(x_1,\ldots,x_k,c_1,\ldots,c_k)\Big)\,dx_1\ldots dx_k = \int_{\Omega^C_{x_0}} \mathcal{D}f\,dx_1\ldots dx_{2k},$$

where $\mathcal{D}$ is a differential operator on B with $\operatorname{ord}\mathcal{D} = \operatorname{ord}\varkappa$.

Suppose now that $f = f_1^s$, where f_1 vanishes on $\Omega^C_{x_0}$ and $s = \operatorname{ord}\mathcal{D} = \operatorname{ord}\varkappa$. Then the integral equals $\int_{\Omega^C_{x_0}} \sigma\mathcal{D}(df_1)\,dx_1\ldots dx_{2k}$, which, according to the proposition in Sec. 1, No. 1 is just

$$\int_{\Omega^C_{x_0}} \chi'(x_1,\ldots,x_{2k})\,[\sigma(\tau_1,\ldots,\tau_k)(n^*(x,\xi)\,df_1)]\,dx_1\ldots dx_{2k}, \qquad (*)$$

where the vectors $\tau_1,\ldots,\tau_k$ are tangent to C at the point $\xi \in C \cap \Gamma_x$, and χ' does not depend on f_1.

Since the function f_1 can be selected so that the differential df_1 is any preassigned vector from $\operatorname{Ann} T_x\Omega^C_{x_0}$ at every point $x \in U$, (*) implies that $\sigma(\tau_1,\ldots,\tau_k)(n^*(x,\xi)df_1) = 0$. The map $n^*(x,\xi)$ induces an isomorphism $\operatorname{Ann} T_x\Omega^C_{x_0} \to \operatorname{Ann} T_\xi\Gamma_x \cap \operatorname{Ann} l(\tau_1,\ldots,\tau_k)$, which completes the proof of the lemma.

Remark. The same arguments lead to the following statement. Let $C_\xi = \bigcup_{x\in B_\xi} \operatorname{Ann} T_\xi\Gamma_x$. If $\mathcal{D}$ is a differential operator such that $\mathcal{D}(If) = 0$ for all $f \in C_0^\infty(B)$, then $\sigma\mathcal{D}$ vanishes identically on C_ξ.

3. Formulation of the Main Theorem. Henceforth, the manifolds A, B, and Γ, and the mappings π_1 and π_2 are assumed to be analytic. In addition, we assume that $\dim B \leqslant \dim \Gamma$, implying $\dim B_\xi \leqslant \dim \Gamma_x$ for any x, ξ.

We let $\sigma'(\tau_1,\ldots,\tau_k)$ denote the image of the symbol $\sigma(\tau_1,\ldots,\tau_k)$ under the projection $T_\xi\Gamma \to T_\xi\Gamma/T_\xi\Gamma_x$.

THEOREM. Suppose that the given DB is admissible for some family of measures on B_ξ. If in the tangent space of a generic point $a \in A$ there are no conjugate characteristic vectors, then: 1) $\deg\varkappa \geqslant \dim B_\xi$; 2) if $\deg\varkappa = \dim B_\xi$, then the surfaces Γ_x define a semi-Grassmanian structure on Γ and, in particular, for every ξ, the surface B_ξ is canonically identified with a domain in the projective space PV_ξ for some V_ξ; 3) given arbitrary vectors $\tau_1,\ldots,\tau_k \in T_\xi\Gamma_x$, the symbol $\sigma'(\tau_1,\ldots,\tau_k) = \Delta_\Theta(\tau_1,\ldots,\tau_k;\mu)$, where $\Theta\colon T_xB_\xi \otimes T_\xi\Gamma_x \to V_a$ is the characteristic map, and μ is a measure on T_xB_ξ.

4. Proof of the Theorem

1. Let $\Psi\colon V_1 \otimes V_2 \to V_3$ be a linear map, where $\dim V_1 = k$, $\dim V_2 = l$, and $k \leqslant l$. We let Ψ^* denote the dual map $V_3^* \to V_1^* \otimes V_2^*$. The vectors of $V_1^* \otimes V_2^*$ may be thought of as linear maps $V_2 \to V_1^*$. Let C^s_Ψ and C^s denote the cone of vectors of rank $k - s$ in $V_1^* \otimes V_2^*$ and its preimage in V_3^*, respectively. Set

$$R_v = V_1 \otimes v, \quad v \in PV_2, \quad P_u = u \otimes V_2, \quad u \in PV_1,$$
$$S_u = \mathrm{Ann}\, P_u, \quad T_v = \mathrm{Ann}\, R_v; \; S_u, \; T_v \subset V_1^* \otimes V_2^*.$$

Further, let $R_{v,\psi}$ and $P_{u,\psi}$ ($S_{u,\psi}$ and $T_{v,\psi}$) denote the images (respectively, preimages) of these planes in V_3 (respectively, in V_3^*). Let $x \in C^s \setminus C^{s+1}$. The tangent space $T_x C^s$ is precisely the space of those maps $V_2 \to V_1^*$ which take Ker x into Im x. The quotient $V_1^* \otimes V_2^* / T_x C^s$ has a natural tensor product structure, $W_1^* \otimes W_2^*$, with $W_1^* = V_1^*/\mathrm{Im}\, x$ and $W_2 = \mathrm{Ker}\, x$. Recall that given $L_1 \subset V_1$, and $L_2 \subset V_2$, with $\dim L_1 = \dim L_2$, $\Delta_\Psi(L_1, L_2, \mu_1, \mu_2)$ is a polynomial constructed from the restriction of Ψ to $L_1 \otimes L_2$ (see Sec. 2, No. 1). When $\dim V_1 = \dim V_2$, we shall denote the polynomial $\Delta_\Psi(V_1, V_2, \mu_1, \mu_2)$ simply by Δ_Ψ, if we are concerned with it only up to a proportionality. If the k vectors $\tau_1, \ldots, \tau_k$ have been fixed in V_2, then we shall also write $\Delta_\Psi(\tau_1, \ldots, \tau_k; \mu)$ instead of $\Delta_\Psi(V_1, L, \mu, \mu_0)$, where L is spanned by $\tau_1, \ldots, \tau_k$, and μ is normalized on $\tau_1, \ldots, \tau_k$. If $\dim L < k$, then we put $\Delta_\Psi(\tau_1, \ldots, \tau_k; \mu) = 0$.

A point on C_ψ^1 will be said to be in general position if one can find a polynomial $\Delta_\Psi(L_1, L_2, \mu_1, \mu_2)$. $\dim L_1 = \dim L_2 = k - 1$, which does not vanish at this point.

Proposition 1. 1) If $k \leqslant l$, then C_ψ^1 coincides with the set of common zeros of the polynomials $\Delta_\Psi(\tau_1, \ldots, \tau_k; \mu)$; 2) if $k \leqslant l$, then through every point in general position on C_ψ^1 passes exactly one plane $S_{u,\psi}$ and a $(k - l)$-parameter family of planes $T_{v,\psi}$. Conversely, if through a given point on C_ψ^1 passes exactly one plane $S_{u,\psi}$, then it is in general position.

Proposition 2. If Ker Ψ does not intersect the cone of vectors of rank 1, then: 1) the dimensions of the plane $S_{u,\psi}$ and $P_{u,\psi}$ do not depend on u or v; 2) every vector $x \in V_3$ is contained in no more than a finite set of planes $P_{u,\psi}$; 3) $\dim V_3 \geqslant k + l - 1$, and $\dim \times S_{u,\psi} \geqslant k - 1$.

LEMMA 1. If Ker Ψ does not intersect the cone of vectors of rank 1, then C_ψ^1 is irreducible and for $k = l$, Δ_ψ is the generator of the ideal corresponding to C_ψ^1.

Proof. Since all the planes $S_{u,\psi}$ have one and the same dimension and u runs over all PV_1, we see that all these planes must lie in one and the same irreducible component, i.e., C_ψ^1 is irreducible. The cone of tangent spaces to C_ψ^1 coincides with the cone of vectors of rank 1 in $V_1 \otimes V_2$. Since Ker Ψ does not intersect the latter, Im Ψ^* is in general position with C^1. Therefore, if h is a line in Im Ψ^* which is in general position with $C_\Psi^1 = \mathrm{Im}\, \Psi^* \cap C^1$, then h is in general position with C^1 too, i.e., C_ψ^1, together with C^1, is a hypersurface of degree k, which proves the second part of the lemma.

Proposition 3. Suppose that Ker Ψ does not intersect the cone of vectors of rank 1. Then there is an open set of general position points on the cone C_ψ^1.

Proof. If there are no such points, then through every point passes at least a one-parameter family of planes $S_{u,\psi}$, and codim C_ψ^1 in V_3^* is $l - k + 2$, which is larger than codim C^1 in V_3; contradiction.

Let Ψ: $V_1 \otimes V_2 \to V_3$ be a linear map, $\dim V_2 = 2$, and let v_1, v_2 be a basis in V_2. We let Ψ_1 and Ψ_2 denote the restrictions of Ψ to $V_1 \otimes v_1$ and $V_1 \otimes v_2$, respectively. Let V_{12} be a maximal subspace of V_1 with the property $\Psi_2(V_{12}) \subset \Psi_1(V_{12})$.

LEMMA 2. Suppose that there are given linear maps Ψ: $V_1 \otimes V_2 \to V_3$ and L: $V_2 \to V_3$, $\dim V_2 = 2$, and let v_1, v_2 be a basis in V_2. Assume that: 1) Ψ_1 is nondegenerate: 2) the operator $A = \Psi_1^{-1} \circ \Psi_2|_{V_{12}}$, acting from V_{12} into V_{12}, has pairwise distinct eigenvalues: 3) $L(v) \in \Psi(V_1 \otimes v)$ for all $v \in V_2$. Then there exists $u \in V_1$, such that $L(v) = \Psi(u \otimes v)$ for all $v \in V_2$.

Proof. a) Suppose that $\dim V_1 = \dim V_3$. We proceed by induction on $\dim V_1$. For $\dim V_1 = 1$ the claim is obvious. Note that in this case V_{12} is just V_1. Let u be an eigenvector of A and set $x = \Psi_1 u$. Then for any v one has $\Psi(u \otimes v) = \lambda(v) x$, where λ is linear in v. Consequently, the map Ψ': $V_1/u \otimes V_2 \to V_3/x$ is well-defined. Let L' denote the map $V_2 \to V_3/x$. Then, using the inductive hypothesis, $L'(v) = \Psi'(u' \otimes v)$ for some $u' \in V_1/u$ and any $v \in V_2$. Let u" be the preimage of u' in V , and put $L'' = L - \Psi(u'' \otimes \cdot)$. Then $L''(v) = \mu(v)x$ for all v, where μ is linear in v. Now let $\lambda(v_0) = 0$. Then the conditions of the lemma imply that $x \notin \Psi(V_1 \otimes v_0)$, whence $\mu(v_0) = 0$ too. Consequently, $\lambda = \alpha\mu$ and $L''(v) = \Psi(\alpha u \otimes v)$.

b) We next prove the following auxiliary statement. If in V_1 there is no subspace V such that $\Psi_2(V) \subset \Psi_1(V)$, then there is a subspace $M \subset V_3$, such that $V_3 = M \oplus \mathrm{Im}\, \Psi_1$, and for the map $\Psi'\colon V_1\otimes V_2\to V_3/M$, the allied operator $A'=(\Psi_1')^{-1}\circ\Psi_2'$ has an arbitrarily prescribed spectrum. We proceed again by induction on $\dim V_1$. For $\dim V_1 = 1$, the statement is obvious. Set $V_1^1=\Psi_1^{-1}(\operatorname{Im}\Psi_1\cap\operatorname{Im}\Psi_2)$, and $V_3^1=\operatorname{Im}\Psi_1$. Let $\Psi^1\colon V_1^1\otimes V_2\to V_3^1$ be the restriction of Ψ to $V_1^1\otimes V_2$. Then the induction hypothesis applies to Ψ^1. Choose M^1 to be a complement to $\operatorname{Im}\Psi_1^1$, such that for $\Psi^{1'}\colon V_1^1\otimes V_2\to V_3^1/M^1$, the allied operator $A_1^1=(\Psi_1^{1'})^{-1}\circ\Psi_2^{1'}$ has an arbitrarily prescribed spectrum. Set $M_1^1=\Psi_1^{-1}M^1$. Then $\Psi_2'(M_1^1)\cap\Psi_1(V_1)=0$. Picking a suitable projection $\pi\colon V_3\to\operatorname{Im}\Psi_1$, we can arrange it that the operator $A_2^1=\Psi_1^{-1}\circ\pi\circ\Psi_2|_{M_1^1}$ acting from M_1^1 into M_1^1 also has an arbitrarily prescribed. Finally, note that the spectrum of A' is the union of the spectra of A_1^1 and A_2^1.

c) Now let $\dim V_3 > \dim V_1$. We use induction on both $\dim V_1$ and $\dim V_3$. Let $V=\Psi_1(V_{12})$, $V_1'=V_1/V_{12}$, $V_3'=V_3/\Psi_1(V_{12})$. Then there are well-defined maps $\Psi'\colon V_1'\otimes V_2\to V_3'$ and $L'\colon V_2\to V_3'$, and Ψ' satisfies the conditions of b). Therefore, there exists $M\subset V_3'$, complementary to $\operatorname{Im}\Psi_1'$, and such that the maps $V_1'\otimes V_2\to V_3'/M$ and $L''\colon V_2\to V_3'/M$ satisfy the conditions of the lemma. The inductive hypothesis yields $L''=\Psi''(u\otimes\cdot)$. Then $L_1 = L' - \Psi'(u\otimes\cdot)$ takes values only in M. But M does not intersect $\Psi(V_1\otimes v)$ for almost all $v\in V_2$; hence $L_1\equiv 0$. Put $L_1'=L-\Psi(\tilde u\otimes\cdot)$. Then L_1' takes its values in V. Since due to the choice of V_{12}, $\Psi(V_{12}\otimes v)\subset V$, for all $v\in V_2$, the conditions of the lemma are satisfied for the maps L_1' and $\Psi'''\colon V_{12}\otimes V_2\to V$, where Ψ''' is the restriction of Ψ. By the inductive hypothesis, $L_1'=\Psi'''(u_1\otimes\cdot)$, and the proof is complete.

Recall that if one is given a family of planes $L(t)$ in V, where $t\in U$. $\dim U=r$, and $L(t)$ depends smoothly on t, then for every point we have the tangent map $\Lambda(t,T_tU)\colon L(t)\to V/L(t)$, or, equivalently, $\Lambda(t)\colon T_t^*U\otimes L(t)\to V/L(t)$.

Let Σ denote the tangent map to the family $S_{u,\Psi}$.

LEMMA 3. Suppose that the one-parameter family of planes $L(t)$ is such that $L(0) = S_{v_0,\Psi}$ for some v_0, and let $L(t)\subset C_\Psi^1$ for all t. If v_0 is in general position, then the tangent map $\Lambda(t,\partial t)\colon L(0)\to V_3^*/L(0)$ is identical with the restriction of the tangent map $\Sigma(\tau\otimes L(0))$ for some $\tau\in T_{v_0}PV_1$.

Proof. Denote $T_{v_0}PV_1$ by V'. From Lemma 1 it follows that for v_0 in general position the plane $S_{v_0,\Psi}$ contains a point in general position. Now if x is a general position, then $\Lambda(0,\partial t)(x)\subset T_xC_\Psi^1/L(0)=\Sigma(V'\otimes x)$. Note that since $k\leqslant l$, at a point x in general position $\Sigma(v\otimes x)\neq 0$ for all v.

a) Suppose that the singular points form on $S_{v_0,\Psi}$ a cone $M=S_{v_0,\Psi}\cap C_\Psi^2$ of codimension 2. Let H be a two-dimensional subspace of $S_{v_0,\Psi}$ that intersects M only at zero. Then the restrictions of Λ and Σ to H satisfy the conditions of Lemma 2, and hence $\Lambda|_H=\Sigma(\tau\otimes\cdot)|_H$ for some $\tau\in V'$. Since H is in general position, the same is true throughout $S_{v_0,\Psi}$.

b) Suppose that $\operatorname{codim} M$ in $S_{v_0,\Psi}$ equals 1 for v_0 in general position.

Let $N=S_{v_0,\Psi}\cap C_\Psi^3$, and suppose that the complement $M\setminus N$ is nonempty. Let $x\in M\setminus V$; then through x passes a one-parameter family of planes $S_{v,\Psi}$. For each such plane we let $S_{v,\Psi}'$ denote the sum of $S_{v,\Psi}$ and $T_xC_\Psi^2$. Fix in $L(t)$ bases depending smoothly on t. Let $\gamma(t)$ be the point in $L(t)$ having the same coordinates as x. Since $\gamma(t)\in c_\Psi^1$, the velocity vector of the curve γ at the point x belongs to one of the spaces $S_{v,\Psi}'$. While choosing bases in $L(t)$ in all the possible ways, the velocity vector of γ describes the space $L_x = \Lambda^{-1}(\Lambda(0,\partial t)(x))$. This subspace should be entirely contained in the union of the spaces $S_{v,\Psi}'$. The map $\Psi^*\colon V_3^*\to V_1^*\otimes V_2^*$ takes $S_{v,\Psi}'$ into S_v'. The quotient of $V_1^*\otimes V_2^*$ by T_xC^2 carries a canonical tensorial structure $C^2\otimes C^{2+l-k}$, and the images S_v' coincide with spaces of the form $v\otimes C^{2+l-k}$, while $S_{v,\Psi}'$ is taken into the intersection of S_v' with the plane H_Ψ, representing the image of $\operatorname{Im}\Psi^*$ in the above quotient space. Since H_Ψ is transverse to all the planes of the form $v\otimes C^{2+l-k}$ [this follows from the condition: $\operatorname{Ker}\Psi\cap$ (the cone of vectors of rank 1 in $V_1\otimes V_2$) $= 0$], its intersection with all the $v\otimes C^{2+l-k}$ is at least one-dimensional. Moreover, since the set of singular points in $S_{v,\psi}$ has codimension 1 for v in general position, then this intersection is exactly one-dimensional.

We conclude that H_ψ cannot contain a plane of the form $C^2 \otimes u$: otherwise, H_ψ would coincide with this plane, contradicting the fact that H_ψ is transverse to all $v \otimes C^{3+l-k}$. Consequently, the image L_x lies in a space $v \otimes C^{3+l-k}$ for some v; indeed, according to [8], if L_x lies in the cone of vectors of rank 1, then it lies either in $C^2 \otimes u$, or in $v \otimes C^{3+l-k}$, for some u, v. Hence L_x lies in $S'_{v_0,\Psi}$, i.e., $\Lambda(0, \partial t)(x) \in \Sigma(V' \otimes x)$. Since $S_{v_0,\Psi} \not\subset T_xC^2$, one can pick in $S_{v_0,\Psi}$ a two-dimensional subspace K which is transverse to $T_xC^2 \cap S_{v_0,\Psi}$ in $S_{v_0,\Psi}$. Then the restrictions of Λ and Σ to K satisfy the conditions of Lemma 1, and hence $\Lambda|_K = \Sigma(\tau \otimes \cdot)|_K$ for some $\tau \in V'$. Since K is in general position, this holds throughout $S_{v_0,\Psi}$.

c) Suppose that codim M in $S_{v_0,\Psi}$ is 1 and N = M. Suppose $x \in M$. The images of the planes $\Psi^*(S'_{v,\Psi})$ in the quotient of $V_3^!$ by T_xC^3 are intersections of H_ψ (i.e., of $\mathrm{Im}\Psi^*$), with planes of the form $P \otimes C^{3+l-k}$, where dimP = 2. Recalling that H_ψ is transverse to all such planes, we conclude that these intersections have codimension no less than 2 (see Proposition 2), which contradicts the condition codimM = 1.

If we now allow for L(t) only parametrizations for which the Taylor series at zero starts with the (k + 1)-th term, then we may identify the (k + 1)-th differential $\Lambda(t_0, (\partial t)^{k+1})$ with a map $L(t_0) \to V_3^*/L(t_0)$, and prove for it the following analog of Lemma 3.

<u>LEMMA 3'.</u> Suppose that: 1) $L(t) \subset C^1_\Psi$ for all t; and 2) for t_0, $L(t_0) = S_{v_0,\psi}$, where v_0 is in general position. Then $\Lambda(t_0, (\partial t)^{k+1}) = \Sigma(\tau \otimes L(t_0)) \otimes (\partial t)^{k+1}$ for some $\tau \in T_{v_0}PV_1$, provided that all the preceding terms in the Taylor series vanish.

2. Fix a point $a_0 \in A$, and let $x_0 = \pi_1 a_0$, $\xi_0 = \pi_2 a_0$. Denote $\mathrm{Ann}\, T_\xi \Gamma_{x_0} = V_{a_0}$. Given any k linearly independent vectors $\tau_1, \ldots, \tau_k \in T_\xi \Gamma_{x_0}$ and any vector $v \in T_{x_0} B_{\xi_0}$, we let Ann × $(\tau_1, \ldots, \tau_k; v)$ denote the annihilator of the subspace $\Theta(v)(l(\tau_1, \ldots, \tau_k))$, where $l(\tau_1, \ldots, \tau_k)$ is the subspace of $T_\xi \Gamma_{x_0}$ spanned by $\tau_1, \ldots, \tau_k$. Set $C_\Theta(\tau_1, \ldots, \tau_k) = \bigcup_{v \in B_\xi} \mathrm{Ann}(\tau_1, \ldots, \tau_k; v)$.

<u>LEMMA 4.</u> Suppose that Ker Θ does not intersect the cone of vectors of rank 1. Then $\sigma'(\tau_1, \ldots, \tau_k)$ vanishes identically on $C_\Theta(\tau_1, \ldots, \tau_k)$.

<u>Proof.</u> According to the fundamental lemma, $\sigma(\tau_1, \ldots, \tau_k)$ vanishes on all the planes $\mathrm{Ann}(\pi T_\xi \Gamma_x) = \mathrm{Ann}\, T_\xi \Gamma_{x_0} \cap \mathrm{Ann}\, l(\tau_1, \ldots \tau_k)$, where $\pi: T_\xi \Gamma \to T_\xi \Gamma / l(\tau_1, \ldots, \tau_k)$ is the projection. Pick in each plane $T_{\xi_0}\Gamma_x$ a basis $e_1(x), \ldots, e_l(x)$ depending smoothly on x, and such that $e_1 \times (x_0) = \tau_1, \ldots, e_k(x_0) = \tau_k$. Let $e'_1(x), \ldots, e'_l(x)$ be the images of this basis in $\pi T_\xi \Gamma_x$. Now let x approach x_0 along a smooth curve in B_ξ with parameter t and initial velocity vector v. Then the vectors $(1/t)e'_1(x), \ldots, (1/t)e'_k(x)$ tend to some limits $g_1(v), \ldots, g_k(v)$, and clearly $\pi_{x_0}(g_i(v)) = \Theta(v)(\tau_i)$, where π_{x_0} is the projection $T_\xi \Gamma \to T_\xi \Gamma / T_\xi \Gamma_{x_0}$. Since $\Theta(v)$ is nondegenerate on $l(\tau_1, \ldots, \tau_k)$, the vectors $g_1(v), \ldots, g_k(v), e_{k+1}(x_0), \ldots, e_l(x_0)$ are linearly independent. Therefore, the plane they span is the limit of the planes $\pi T_\xi \Gamma_x$, and its annihilator lies in the zero-set of the symbol $\sigma(\tau_1, \ldots, \tau_k)$. It follows that the intersection of this annihilator with $\mathrm{Ann}\, T_\xi \Gamma_{x_0}$, that is, the space $\mathrm{Ann}(\tau_1, \ldots, \tau_k; v)$, lies in the zero-set of $\sigma'(\tau_1, \ldots, \tau_k)$.

<u>LEMMA 5.</u> The degree of any polynomial that vanishes identically on C_Θ, is no less than k.

The proof below is due to S. L. Tregub. Notation. Set $X_0 = C^1$. Since $C^1_\Psi = (\Psi^*)^{-1}(C^1 \cap H)$ for $H = \mathrm{Im}\Psi^*$, we may consider a sequence of planes H_s, s = 0,...,r, r = codimH, such that $H_0 = V_1^* \otimes V_2^*$, $H_s \subset H_{s-1}$, $H_r = H$, $\mathrm{codim}\, H_s$ in H_{s-1} is equal to 1, and for each s the plane H_s is transverse to all S_v. Set $X_s = X_{s-1}$, $P^{k-1} = P(V_1)$, and let $\overline{X}_s$ be the desingularization of X_s; $\overline{X}_s$ is a bundle over P^{k-1} whose fiber over the point v is $S_v \cap H_{s-1}$. Let $\mathcal{H}_s = \mathcal{O}_{H_s}(1)$; then H_{s+1} and X_{s+1} are recognized as the divisor of zeros of this sheaf on H_s, and, respectively, as the divisor of its restriction to X_s. (For all these notions we refer the reader to [19].) Finally, we let $\mathcal{L}_s$ and L_s denote the sheaf $\mathcal{O}(1)$ on X_s and, respectively, its divisor on $\overline{X}_s$; then $L_s = \overline{X}_{s+1}$.

Let R be the kernel of the map $H^0(H_r, \mathcal{H}_r^t) \to H^0((X_r)_{\mathrm{red}}, \mathcal{H}_r^t)$, where $(X_r)_{\mathrm{red}}$ is the reduced variety of X_r. We must show that $R = 0$. Set $L_{-1} = \overline{X}_0$. We proceed by induction on s and verify that the following is true. 1°. The sequence $0 \to H^0(H_s, \mathcal{H}_s^t) \to H^0(L_{s-1}, \mathcal{L}_s^t) \to 0$ is exact for $0 \leqslant t < k$; 2°. $H^i(\overline{X}_s, \mathcal{L}_s^t) = 0$ for $i > 0$, $0 \leqslant t < k$. We denote $\dim H^i(\cdot, \cdot)$ by $h^i(\cdot, \cdot)$. With the aid of Leray's spectral sequence (see [19]), one can show that

$$h^0(\overline{X}_0, \mathcal{L}_0^t) = \binom{N+t}{t} = h^0(H_0, \mathcal{H}_0^t), h^i(\overline{X}_0, \mathcal{L}_0^t) = 0$$

for $i > 0$, where $N = \dim(V_1^* \otimes V_2^*)$, which implies 2° for $s = 0$. The proof of 1° follows from the fact that the ideal defining X_0 is generated by polynomials of degree k. Now use the induction. Assertion 2° follows from the exact sequence $0 \to \mathcal{L}_{s-1}^{t-1} \to \mathcal{L}_{s-1}^t \to \mathcal{L}_{s-1}^t|_{L_{s-1}} \to 0$, because $\mathcal{L}_{s-1}^t|_{L_{s-1}} = \mathcal{L}_s^t$ and $h^i(\mathcal{L}_{s-1}^t) = 0$ for $t < k$ and $i > 0$. To prove 1°, note that the sequence

$$0 \to H^0(H_{s-1}, \mathcal{H}_{s-1}^{t-1}) \to H^0(H_{s-1}, \mathcal{H}_{s-1}^t) \to H^0(H_s, \mathcal{H}_s^t) \to 0. \tag{1}$$

and

$$0 \to H^0(\overline{X}_{s-1}, \mathcal{L}_{s-1}^{t-1}) \to H^0(\overline{X}_{s-1}, \mathcal{L}_{s-1}^t) \to H^0(\overline{X}_s, \mathcal{L}_s^t) \to 0 \tag{2}$$

are exact, because $\overline{X}_s = L_{s-1}$ and $H^1(\overline{X}_{s-1}, \mathcal{L}_{s-1}^t) = 0$ for $t < k$. Whence $h^0(H_s, \mathcal{H}_s^t) = h^0(\overline{X}_s, \mathcal{L}_s^t)$, because $h^0(H_{s-1}, \mathcal{H}_{s-1}^{t-1}) = h^0(\overline{X}_{s-1}, \mathcal{L}_{s-1}^{t-1})$, and $h^0(H_{s-1}, \mathcal{H}_{s-1}^t) = h^0(\overline{X}_{s-1}, \mathcal{L}_{s-1}^t)$. Finally, the fact that the map $H^0(H_s, \mathcal{H}_s^t) \to H^0(\overline{X}_s, \mathcal{L}_s^t)$ is an epimorphism is seen from the diagram

$$\begin{array}{ccccc} H^0(H_{s-1}, \mathcal{H}_{s-1}^t) & \to & H^0(\overline{X}_{s-1}, \mathcal{L}_{s-1}^t) & \to & 0, \\ \downarrow & & \downarrow & & \\ H^0(H_s, \mathcal{H}_s^t) & \to & H^0(\overline{X}_s, \mathcal{L}_s^t) & \to & 0, \\ \downarrow & & \downarrow & & \\ 0 & & 0 & & \end{array}$$

where the upper row is exact by the induction hypothesis, the right column follows from (2), and the left one is well known. To complete the proof, note that since $(X_r)_{\mathrm{red}} = X_r$ (due to the properties of Ψ), R is contained in the kernel of the map $H^0(H_r, \mathcal{H}_r^t) \to H^0(L_{r-1}, \mathcal{L}_r^t)$, i.e., $R = 0$.

<u>LEMMA 6.</u> $\deg \varkappa \geqslant k$, and if $\deg \varkappa = k$, then for any $\tau_1, \ldots, \tau_k \in T_\xi\Gamma_{x_0}$ we have $\sigma'(\tau_1, \ldots, \tau_k) = \Delta_\Psi(\tau_1, \ldots, \tau_k; \mu)$ for some volume element μ.

<u>Proof.</u> Recall that $\sigma(\tau_1, \ldots, \tau_k)$ is a polynomial on $\mathrm{Ann}\, l(\tau_1, \ldots, \tau_k)$. If $\sigma'(\tau_1, \ldots, \tau_k) = 0$, then $\sigma(\tau_1, \ldots, \tau_k)$ vanishes on $\mathrm{Ann}\, T_\xi\Gamma_{x_0}$. Let s be such that for all $i < s_i$ and arbitrary $v_1, \ldots, v_i \in \mathrm{Ann}\, l(\tau_1, \ldots, \tau_k)$, all the polynomials $\frac{\partial^i}{\partial v_1 \ldots \partial v_i} \sigma(\tau_1, \ldots, \tau_k)$ vanish on $\mathrm{Ann}\, T_\xi\Gamma_{x_0}$, whereas for $i = s$, one can find a collection $v_1, \ldots, v_s$ such that $\frac{\partial^s}{\partial v_1 \ldots \partial v_s} \sigma(\tau_1, \ldots, \tau_k)$ does not vanish identically on $\mathrm{Ann}\, T_\xi\Gamma_{x_0}$. Let $\zeta \in \mathrm{Ann}\, T_\xi\Gamma_{x_0} \cap \mathrm{Ann}\, T_\xi\Gamma_x$. If x is close enough to x_0, then the planes $\mathrm{Ann}\, T_\xi\Gamma_x \cap l(\tau_1, \ldots, \tau_k)$ and $\mathrm{Ann}\, T_\xi\Gamma_{x_0}$ are transverse in $l(\tau_1, \ldots, \tau_k)$. Now decompose each vector v_j into a sum of two components: $v_j = u_j + w_j$, where $u_j \in \mathrm{Ann}\, T_\xi\Gamma_{x_0}$, and $w_j \in \mathrm{Ann}\, T_\xi\Gamma_x \cap \mathrm{Ann}\, l(\tau_1, \ldots, \tau_k)$, and express $\frac{\partial^s}{\partial v_1 \ldots \partial v_s} \sigma(\tau_1, \ldots, \tau_k)$ as a sum of two terms, one containing the derivatives with respect to u_j, and the other the derivatives with respect to w_j. Then each term in which for j appears a derivative with respect to u_j, vanishes, by our assumption on derivatives. The term $\frac{\partial^s}{\partial u_1 \ldots \partial u_s} \sigma(\tau_1, \ldots, \tau_k)$ also vanishes at the point ζ, because $\sigma(\tau_1, \ldots, \tau_k)$ vanishes identically on $\mathrm{Ann}\, T_\xi\Gamma_x \cap \mathrm{Ann}\, l(\tau_1, \ldots, \tau_k)$. Consequently, $\frac{\partial^s}{\partial v_1 \ldots \partial v_s} \sigma(\tau_1, \ldots, \tau_k)$ vanishes identically on the intersections $\mathrm{Ann}\, T_\xi\Gamma_{x_0} \cap \mathrm{Ann}\, T_\xi\Gamma_{x_0}$. Hence it vanishes also on

C_Θ (this is verified as in Lemma 4). Now Lemma 5 implies $\deg \frac{\partial^s}{\partial v_1 \ldots \partial v_s} \sigma(\tau_1,\ldots,\tau_k) \geqslant k$. Therefore, if $\sigma'(\tau_1,\ldots,\tau_k) = 0$, then $\deg \sigma(\tau_1,\ldots,\tau_k) > k$.

Suppose that $\sigma'(\tau_1,\ldots,\tau_k) \neq 0$. Then, according to Lemma 4 and Proposition 1, $\sigma'(\tau_1,\ldots,\tau_k)$ belongs to the ideal generated by $\Delta_\Psi(\tau_1,\ldots,\tau_k)$, i.e., $\deg \sigma(\tau_1,\ldots,\tau_k) = \deg \sigma'(\tau_1,\ldots,\tau_k) \geqslant k$. If $\deg \sigma'(\tau_1,\ldots,\tau_k) = k$, then $\sigma'(\tau_1,\ldots,\tau_k) = \Delta_\Psi(\tau_1,\ldots,\tau_k;\ \mu(\tau_1,\ldots,\tau_k))$ for some $\mu(\tau_1,\ldots,\tau_k)$. Since $\Delta_\Psi(\tau_1,\ldots,\tau_k;\mu)$, for fixed μ, and $\sigma'(\tau_1,\ldots,\tau_k)$ are multilinear in τ, we see that actually $\mu(\tau_1,\ldots,\tau_2)$ does not depend on $\tau_1,\ldots,\tau_k$.

Let M_Θ be the family of planes $\operatorname{Ann}\Theta(v \otimes T_\xi\Gamma_{x_0})$, $v \in T_{x_0}B_\xi$, and let M be the family of planes $\operatorname{Ann}(\pi_{x_0}T_{\xi_0}\Gamma_x)$.

LEMMA 7. If $\deg \varkappa = k$ and the conditions of the theorem are fulfilled, then $M_\Theta \supset \mathcal{M}$.

Proof. Let us show that for arbitrary r one can choose analytic parameters $t_1,\ldots,t_k$ on B_{ξ_0} in a neighborhood of x_0, and a mapping E(t) from $H_0 = T_\xi\Gamma_{x_0}$ into the plane $\pi_{x_0}T_\xi\Gamma_x$, such that the Taylor series expansion of E(t) at the point x_0 will have the form $E(t) = \sum\Theta_i t_i + S$, where S contains only terms of order higher than r, Θ_i is the restriction of Θ to $e_i \otimes T_\xi\Gamma_{x_0}$, and $e_1,\ldots,e_k$ is the basis of $T_{x_0}B_{\xi_0}$, dual to dt_i. We proceed by induction on r. For r = 1, the assertion follows immediately from the definition of Θ. Suppose that it is true for $r' > 1$. Let γ be a curve of the form $t_1 = \lambda_1\tau,\ldots,t_k = \lambda_k\tau$. Set $E_1(\tau) = E(\tau)/\tau$, and denote by $M(\tau)$ and $L(\tau)$ the plane spanned by $E_1(\tau)$ and, respectively, its annihilator in V_3^*; note that $E_1(0) = \sum_i \lambda_i\Theta_i$. The first $r' - 1$ derivatives of $E_1(\tau)$ at $\tau = 0$ vanish; hence Lemma 3' applies to the family $L(\tau)$, i.e., the value of the r'-th differential on the vector $\lambda = \lambda_1 e_1 + \ldots + \lambda_k e_k$ coincides with the restriction of Σ to $v(\lambda) \otimes L(0)$ for some vector $v(\lambda) \in V'$, where $V' = T_{x_0}B_{\xi_0}$. The map $\lambda \to v(\lambda)$ is analytic in λ and homogeneous of degree $r' + 1$. From the exact sequence $0 \to \mathcal{O}(r') \to V' \otimes \mathcal{O}(r'+1) \to TPV' \otimes \mathcal{O}(r') \to 0$ and from the fact that $H^1(\mathcal{O}(r')) = 0$, it follows that $v(\lambda) = A_\lambda(\lambda)$, where A is a linear map $S^{r'+1}V' \to V'$, and $A_\lambda(\lambda)$ denotes the projection of the value it takes on the monomial $\lambda^{r'+1}$ in the quotient with respect to λ. Write $A = \sum e_i P_i(\lambda)$, with polynomials $P_i(\lambda)$ of degree $r' + 1$ on V'. Perform the change of variables $t_i' = t_i + P_i$. Then for any curve $t_1' = \lambda_1\tau,\ldots,t_k' = \lambda_k\tau$, the $(r'+1)$-th differential of the family $L(\tau)$ at $\tau = 0$ vanishes. This means that in the expansion $E(t') = \sum\Theta_i t_i' + E^{r'+1}(t') +$ (terms of order higher than $r' + 1$), the map $E^{r'+1}$ is such that $\operatorname{Im} E^{r'+1}(\lambda) \subset \operatorname{Im}\Theta(\lambda)$, for all $\lambda = (\lambda_1,\ldots,\lambda_k)$, where $\Theta(\lambda) = \sum\lambda_i\Theta_i$. Set $C(\lambda) = \Theta^{-1}(\lambda)\circ E^{r'+1}(\lambda)$; then $C(\lambda)$ is an analytic mapping from V' into $\operatorname{Hom}(H_0, H_0)$, homogeneous of order r'. Consequently, C is a linear mapping from $S^{r'}V'$ into $\operatorname{Hom}(H_0, H_0)$ and $E^{r'+1}(\lambda) = \Theta(\lambda)\circ C(\lambda)$. Now change the basis: $E'(\lambda) = E(\lambda)\circ(1 - C(\lambda))$. Then in the Taylor series expansion of $E'(\lambda)$ the $(r'+1)$-th terms vanishes, as claimed. Under the analyticity assumption, this yields $\mathcal{M} \subset M_\Theta$. Moreover, the analytic coordinates $t_1,\ldots,t_k$ in a neighborhood of $x_0 \in B_{\xi_0}$ can be choosen to ensure that $\operatorname{Im} E(t) = \operatorname{Im}(\sum t_i\Theta_i)$. In fact, let $t_1',\ldots,t_k'$ be some coordinates. To each point t we associate a vector v(t) such that $\operatorname{Im} E(t) = \operatorname{Im}\Theta(v \otimes H_0)$, and the Plücker coordinates of the bases $e_1(t),\ldots,e_k(t)$ and $\Theta(v \otimes e_i^0)$ are equal. The mapping $t \to v(t)$ is analytic everywhere, except at zero. Since for $|v| \to \infty$ the Plücker coordinates of the base $\Theta(v \otimes e_i^0)$, $i = 1,\ldots,k$, tend to ∞, this mapping is analytic at the point x_0, too, as required.

3. Proposition 4. Suppose that we are given two linear maps, $\Psi_1 \colon V_1 \otimes V_2 \to V$ and $\Psi_2 \colon W_1 \otimes W_2 \to V$, such that $\dim V_1 = \dim W_1 = 2$, $\dim V_2 = \dim W_2 = l$, and $\operatorname{Ker}\Psi_1$ and $\operatorname{Ker}\Psi_2$ do not intersect the cone of vectors of rank 1. If for every $v \in V_1$ one can find $w(v) \in W_1$, such that $\operatorname{Im}\Psi_1(v \otimes V_2) = \operatorname{Im}\Psi_2(w(v) \otimes W_2)$, then $\Psi_1 = (i_1 \otimes i_2)\circ\Psi_2$, where $i_1 \colon W_1 \to V_1$ and $i_2 \colon W_2 \to V_2$ are isomorphisms.

Proof. Let v_1, v_2 and $e_1,\ldots,e_l$ be bases in V_1 and V_2, respectively. Then $w_1 = w(v_1)$, $w_2 = w(v_2)$ is a base in W_1. Let $\varepsilon_1,\ldots,\varepsilon_l$ be a basis in W_2 such that $\Psi_2(w_1 \otimes \varepsilon_i) = \Psi_1(v_1 \otimes e_i)$.

Then the linear map $\Psi_2^{-1}(w_2\otimes W_2)\circ\Psi_1(v_2\otimes V_2)$ from V_2 into W_2 is well defined and is given in the above bases by a matrix A. Raplacing w_2 by $w_2' = \lambda w_2$, we may assume that A has the eigenvalue 1 with some corresponding eigenvector $x = x_1e_1 + \ldots + x_le_l$. Let $x' = x_1\varepsilon_1 - \ldots + x_le_l$. Then $\Psi_1((\lambda_1v_1+\lambda_2v_2)\otimes x) = \Psi_2((\lambda_1w_1-\lambda_2w_2)\otimes x')$ for all λ_1, λ_2. We let L denote the linear space spanned by $\chi_1 = \Psi_1(v_1\otimes x) = \Psi_2(w_1\otimes x')$ and $\chi_2 = \Psi_1(v_2\otimes x) = \Psi_2(w_2\otimes x')$. Now let w(v) = $\mu_1'w_1 + \mu_2'w_2$ for $v = \mu_1v_1 + \mu_2v_2$. If $(\mu_1, \mu_2) \neq \alpha(\mu_1', \mu_2')$, then $\operatorname{Im}\Psi_1(v\otimes V_2)$, being equal to $\operatorname{Im}\Psi_2(w(v)\otimes W_2)$, contains the vectors $\mu_1\chi_1 + \mu_2\chi_2$ and $\mu_1'\chi_1 + \mu_2'\chi_2$, i.e., it contains the entire space L. However, since $\operatorname{Ker}\Psi_i$ does not intersect the cone of vectors of rank 1, $L\subset \operatorname{Im}\Psi_i(v\otimes V_2)$ only for a finite number of vectors v.

Therefore, we may assume that w(v) has the same coordinates as v, for any vector v, and hence identity V_1 and $W_1 : V_1 = W_1 = V$. To each vector $v\in V$ we associate the matrix A(v) of the linear map $\Psi_1^{-1}(v\otimes V_2)\circ\Psi_2(v\otimes W_2)$ with respect to the bases $e_1,\ldots,e_l$ and $\varepsilon_1,\ldots,\varepsilon_l$. The function A(v) is analytic on V and homogeneous of degree 0, hence it is constant and equals $A(v_1) = E$, as claimed.

<u>Proposition 5.</u> Suppose that in the space V there is given a one-parameter family of planes L(t), $\dim L(t) = l$, such that: 1) $L(t_1)\cap L(t_2) = 0$ for t_1 and t_2 close enough; 2) in $L(t_0)$ there is a cone Γ, lying in no hyperplane, and so that through every line in Γ passes a two-dimensional plane which intersects all the planes sufficiently close to L(t). Then for some t_1 there exist in $L(t_0)$ and $L(t_1)$ bases $e_1,\ldots,e_l$ and $\varepsilon_1,\ldots,\varepsilon_l$ such that for sufficiently close t and some λ_1, λ_2, the vectors $\lambda_1e_i + \lambda_2\varepsilon_i$ yields a basis in L(t), and, in addition, if the basis $e_1,\ldots,e_l$ is fixed, then the basis $\varepsilon_1,\ldots,\varepsilon_l$ is uniquely chosen up to proportionality.

<u>Proof.</u> Let L be the complement of $L(t_0)$ in V. Then to every plane L(t) which is sufficiently close to $L(t_0)$ corresponds to nondegenerate linear operator $A(t):L(t_0)\to L$. Moreover, for distinct t_1 and t_2 the operators $A_1(t_1)$ and $A(t_2)$ are proportional on the vectors of Γ. Hence all A(t) are proportional, which completes the proof of the proposition.

<u>LEMMA 8.</u> Let M(v) be a family of l-dimensional planes in the linear space V, parametrized by the points v of the k-dimensional domain U, and satisfying: 1) the kernel of the tangent map $\Theta: T_vU\otimes M(v)\to V/M(v)$ does not intersect the cone of vectors of rank 1 for almost all v; 2) for almost any v_0, the images of the planes M(v) under the projection $\pi_{v_0}: V\to V/M(v_0)$ are precisely the planes $\Theta(\tau\otimes M(v_0))$, where $\tau\in T_{v_0}U$. Then there exists a linear map $\Psi: V_1\otimes V_2\to V$, $\dim V_2 = l$, $\dim V_1 = k+1$, such that U is identified with a domain in PV_1 and $\operatorname{Im}\Psi(v\otimes V_2) = M(v)$.

<u>Proof.</u> a) From the assumptions of the lemma it follows that for v_0 and v_1 in general position, $M(v_0)\cap M(v_1) = 0$. Since the images of the planes M(v) under the projection π_{v_0} depend on k − 1 parameters, there exists an (analytic) curve $\gamma_{v_0v_1}$ in U with an analytic parameter t on it, such that $M(t)\subset M(v_0)\otimes M(v_1)$ for all $t\in\gamma_{v_0v_1}$. Let us verify that one can choose bases $e_1,\ldots,e_l$ in $M(v_0)$, and $\varepsilon_1,\ldots,\varepsilon_l$ in $M(v_1)$, so that for some λ_1, λ_2 the vectors $\lambda_1e_i + \lambda_2\varepsilon_i$, $i = 1,\ldots,l$, will form a basis in M(t) (we shall refer to such a family of bases as compatible).

Let $\pi_{v_0}M(t) = \operatorname{Im}\Theta(\tau_1)$ for some τ_1. For τ in general position, $\operatorname{Im}\Theta(\tau)\cap\operatorname{Im}\Theta(\tau_1)$ has codimension $r\geqslant k-1$ in $\operatorname{Im}\Theta(\tau_1)$. Let γ' be the curve in U constituted from the vectors v for which $\pi_{v_0}M(v) = \operatorname{Im}\Theta(\tau)$, and let s be an analytic parameter on γ'. Since from now on we shall consider only M(t) for $t\in\gamma_{v_0v_1}$, then one can replace π_V by the projection of the space $V_1 = M(v_0)\oplus M(v_1)$ on $H(v) = V_1\cap M(v)$. If $v\in\gamma'$, then, as $v\to v_0$, H(v) approaches $S(\tau) = \Theta^{-1}(\tau)(\operatorname{Im}\Theta(\tau_1)\cap\operatorname{Im}\Theta(\tau))$, $S(\tau)\subset M(v_0)$. We shall identify $V_1/M(v_0)$ with $M(v_1)$. Let $S'(\tau)$ be an arbitrary hyperplane in $M(v_0)$, with $S'(\tau)\supset S(\tau)$. Further, let $e_1,\ldots,e_{l-1}$ be a basis in $M(v_0)$ such that $e_0\in M(v_0)\setminus S'(\tau)$, $e_1,\ldots,e_{r-1}\in S(\tau)\setminus S(\tau_1)$ $e_r,\ldots,e_{l-1}\in S(\tau)$.

Now set $T_1 = \Theta(\tau_1)(S'(\tau))$, and pick an $(l-1)$-dimensional subspace T_2 of $V_1/M(v_0)$ with $T_2\supset\operatorname{Im}\Theta(\tau)\cap\operatorname{Im}\Theta(\tau_1)$. Next choose a basis $a_1,\ldots,a_{l-1}$ in T_2 such that: 1) $a_r,\ldots,a_{l-1}\in\operatorname{Im}\Theta(\tau)\cap\operatorname{Im}\Theta(\tau_1)$, and 2) the vectors $a_i(\lambda) = a_i + \lambda\Theta(\tau_1\otimes e_i)$, $i = 1,\ldots,l-1$, are linearly independent for all λ. Further, for each $H(s)$, $s \in \gamma'$, pick an $(l-1)$-dimensional subspace $H'(s) \supset H(s)$ and a basis $f_i(s)$ in $H'(s)$ such that $f_r(s), \ldots, f_{l-1}(s) \in H(s)$ and $f_i(s) = e_i + sa_i + \mu_i e_0 + O(s \to v_0)$. We shall consider that the vectors $\Theta(\tau \otimes e_i)$ lie in $M(v_1)$.

Given any $s \in \gamma'$ in general position, one can find a basis $\varepsilon_0(s), \ldots, \varepsilon_{l-1}(s)$ in $M(v_1)$ such that for any $t \in \gamma_{v_0v_1}$ and for some λ_1 and λ_2, the vectors $\varepsilon_i(t, s) = \lambda_1\pi_s e_i + \lambda_2\pi_s\varepsilon_i(s)$, $i = 0, \ldots, l-1$, constitute a basis in $\pi_s M(t)$. We let $Q_0(s)$ the linear subspace of S^2V_1 spanned by the quadratics $\sigma_i = e_0\varepsilon_i(s) - e_i\varepsilon_0(s)$, $i = 1, \ldots, l-1$; $\dim Q_0(s) = l-1$. Set $L(t) = \operatorname{Ann} M(t)$ in V_1^* and $\Omega = \bigcup L(t)$, where the union is taken over all $t \in \gamma_{v_0v_1}$. Then the intersection $\alpha_s = \operatorname{Ann} H'(s) \cap \Omega$ is a curve in PV_1^*. Let σ_i' be the image of σ_i in the quotient space $V_1/H'(s)$. Then σ_i' are polynomials on $\operatorname{Ann} H'(s)$ and vanish on α_s. Let Λ be the subspace of V_1 spanned by $M(v_1)$ and e_0; for s close enough to v_0, Λ and $H'(s)$ are transverse. Henceforth, we shall identify Λ with $V_1/H'(s)$ and regard $Q_0(s)$ as a subspace of $S^2\Lambda$. Let $\varepsilon_0'(s)$ be the line spanned by $\varepsilon_0(s)$. We can choose a sequence $s_n \to v_0$ such that the pair $(Q_0(s_n), \varepsilon_0'(s)) \in G(S^2V, l-1) \times P\Lambda$ has a limit (Q_0, ε_0'). Every quadric belonging to Q_0 may be written as $\sigma = e_0v' + \varepsilon_0'v''$, with $v', v'' \in \Lambda$.

Let us find the limit points for α_{s_n}, $n \to \infty$. The kernel of any functional, which is a point on α_{s_n}, is the intersection $K(t, s_n) = (M(t) \oplus H'(s_n)) \cap \Lambda$, provided t and s_n are in general position and $t \in \gamma_{v_0v_1}$. Pick a parametrization z of $\gamma_{v_0v_1}$ with initial velocity vector $\lambda_0\tau_1$. In $M(z)$ one can produce a basis $g_i(z)$ satisfying $g_i(z) = e_i + \lambda_0 z \cdot \Theta(\tau_1 \otimes e_i) + 0(z)$. Then we find in $K(t, s_n)$ the vectors $\lambda_0 z\Theta(\tau_1 \otimes e_i) - s_n a_i + 0(z) + 0(s_n \to v_0)$, $i=1, \ldots, l-1$, and $e_0 + 0(z)$. As $s_n \to v_0$, $K(s_n, s_n)$ becomes the plane spanned by the vectors e_0 and $a_i(\lambda_0)$, $i = 1, \ldots, l-1$; indeed, these vectors are linearly independent. This plane corresponds to a point on $L(v\ )$.

As λ varies, this point on $L(v_0)$ describes a certain curve δ. Moreover, δ lies in no hyperplane, because there is no vector linearly independent from $a_i(\lambda)$ for an infinite number of values of λ [if not, the vectors $a_i(\lambda)$ will be themselves linearly dependent for some λ]. But every quadric σ from Q_0 vanishes on δ; hence its restriction to $L(v_0)$ cannot have rank 2, i.e., $\sigma = e_0v_0'$ and $\sigma|_{L(v_0)} \equiv 0$. Since $\dim Q_0 = l-1$, we can choose in Q_0 a basis $\rho_i = e_0v_i$, $i = 1, \ldots, l-1$, such that the vectors v_i are linearly independent in Λ. Consequently, the set of common zeros of ρ_i consists of $L(v_0)$ and, possibly, of a line $l(\tau)$ (in PV_1^*) which intersects $L(v_0)$ at one point. Finally, for t = const and $s_n \to v_0$, the limit of $K(t, s_n)$ does not contain e_0, i.e., among the limit points of α_{s_n} there is a curve which does not lie in $L(v_0)$, and hence coincides with $l(\tau)$. Moreover, $l(\tau) \subset \Omega$. The points of $l(\tau)$ are functionals on V_1 with kernel $M(t) + H'(v_0)$. As $t \to v_0$, this kernel approaches $M(v_0) \oplus \Theta(\tau_1 \otimes S'(\tau))$; the corresponding point $\beta(\tau, S'(\tau))$ lies in $L(v_0)$.

When $S'(\tau)$ varies for fixed τ, $\beta(\tau, S'(\tau))$ runs over the plane $N(\tau) = \operatorname{Ann} S(\tau) \cap L(v_0)$. When τ varies, the planes $N(\tau)$ describe a certain surface Γ, which is contained in no hyperplane: assuming the contrary, one would find a vector lying in $\operatorname{Im}\Theta(\tau) \cap \operatorname{Im}\Theta(\tau_1)$ for an infinity of values of τ. Through every point of Γ (in PV_1^*) passes a line which intersects all $L(t)$ for t sufficiently close to v_0. Therefore, in virtue of the analyticity and of Definition 4, we conclude that in $L(t)$ there exists a compatible system of bases. Hence such a system exists in $M(t)$, too.

b) Fix $v_0 \in U$ and pick a basis $e_1, \ldots, e_l$ in $M(v_0)$. For each v pick in $M(v)$ a basis $E(v) = (e_1(v)), \ldots, e_l(v))$ such that for every curve $\gamma_{v_0v_1}$ these bases are compatible. Let us show that the bases $E(v)$ are then compatible for all curves $\gamma_{v_1v_2}$. According to step a), for $t \in \gamma_{v_1v_2}$, there exists in $M(t)$ a family of compatible bases $E_1(t)$. Let v_3 be distinct from v_0, v_1, and v_2. The projection π_{v_0} takes the bases $E(v)$ into bases $E'(v)$ for the planes $M(v)$, which are compatible on each curve γ_{v_0v}. Whence. When, according to Proposition 4, $E'(v)$ is proportional to $\Theta(\tau \otimes E(v_3))$ the corresponding τ. In particular, $E(v)$ are compatible on $\gamma_{v_1v_2}$, too. Applying once more Proposition 4, we deduce that $E'(v)$ are proportional to $\pi_{v_3}E_1(v)$, i.e., $E(v)$ and $E_1(v)$ are proportional.

At last, let us note that the existence of a family of compatible bases is equivalent to the statement of the lemma.

The theorem is now seen to be a corollary of Lemmas 6, 7, and 8.

LITERATURE CITED

1. I. M. Gel'fand, M. I. Graev, and N. Ya. Vilenkin, Integral Geometry and Related Problems in Representation Theory [in Russian], Fizmatgiz, Moscow (1962).
2. I. M. Gel'fand and M. I. Graev, "Admissible complexes of lines in C^n," Funkts. Anal. Prilozhen., 2, No. 3, 39-52 (1968).
3. I. M. Gel'fand, S. G. Gindikin, and Z. Ya. Shapiro, "The local problem of integral geometry in the space of curves," Funkts. Anal. Prilozhen., 13, No. 2, 11-31 (1979).
4. I. M. Gel'fand and M. I. Graev, "Admissible n-dimensional complexes of curves in R^n," Funkts. Anal. Prilozhen., 14, No. 4, 36-44 (1980).
5. S. G. Gindikin, "Twistors and integral geometry," in: Lecture Notes Math., Springer-Verlag, New York (1982).
6. M. A. Akivis, "Textures and almost-Grassmanian manifolds," Dokl. Akad. Nauk SSSR, 252, No. 2, 267-270 (1980).
7. Th. Hangan, "Geometrie differentielle grassmannienne," Rev. Roum. Math. Pures App., 11, No. 5, 519-531 (1966).
8. A. B. Goncharov, "Infinitesimal structures connected to Hermitian symmetric spaces," Funkts. Anal. Prilozhen., 15, No. 3, 83-84 (1981).
9. I. M. Gel'fand, M. I. Graev, and Z. Ya. Shapiro, "Integral geometry on k-dimensional planes," Funkts. Anal. Prilozhen., 1, No. 1, 15-31 (1967).
10. I. M. Gel'fand and M. I. Graev, "Complexes of k-dimensional planes and the Plancherel formula for SL(n, C)," Dokl. Akad. Nauk SSSR, 179, No. 3, 522-525 (1968).
11. I. M. Gel'fand and M. I. Graev, "Geometry of homogeneous spaces, rpr group representations in homogeneous spaces, and related problems of integral geometry," Tr. Mosk. Mat. Ob., 8, 321-390 (1959).
12. I. M. Gel'fand, M. I. Graev, and R. Rosu, "Nonlocal inversion formulas in a problem of integral geometry related with p-dimensional planes in real projective space," Funkts. Anal. Prilozhen., 16, No. 3, 49-51 (1982).
13. K. Maius, "The structure of admissible complexes of lines in CP^n," Tr. Mosk. Mat. Ob., 39, 181-211 (1979).
14. I. M. Gel'fand, M. I. Graev, and Z. Ya. Shapiro, "A problem in integral geometry connected with a pair of Grassman manifolds," Dokl. Akad. Nauk SSSR, 193, No. 2, 259-262 (1970).
15. M. I. Gel'fand and S. G. Gindinkin, "Nonlocal inversion formulas in real integral geometry," Funkts. Anal. Prilozhen., 11, No. 3, 12-19 (1977).
16. K. Maius, "The structure of admissible complexes of lines in C^n," Funkts. Anal. Prilozhen., 7, No. 1, 78-81 (1973).
17. S. G. Gindikin, "Unitary representations of the automorphism groups of Riemannian symmetric spaces of constant curvature," Funkts. Anal. Prilozhen., 1, No. 1, 32-37 (1967).
18. A. A. Kirillov, "On a problem raised by I. M. Gel'fand," Dokl. Akad. Nauk SSSR, 137, No. 2, 276-277 (1961).
19. R. Harthsorne, Algebraic Geometry, Graduate Texts in Math., 52, Springer-Verlag, New York (1977).

13.
(with M. I. Graev and R. Rosu)

The problem of integral geometry and intertwining operators for a pair of real Grassmannian manifolds

J. Oper. Theory **12** (2) (1984) 359–383. MR **86c**:22016. Zbl. **551**:53034

INTRODUCTION

Let G_p, G_q be the Grassmannian manifolds of the p respective q dimensional subspaces of the real linear space $V\,(p<q)$. This paper is concerned with the remarkable integral transform associated with G_p and G_q. It can be defined as the transform which associates with each function on G_p, a function on G_q; namely, if f is a function on G_p and $a \in G_q$, then $(\mathscr{I}f)(a)$ is the value of the integral of the function f on $G_p(a)$—the set of the subspaces $b \in G_p$ which are contained in a. When so defining it, we already assume that on each $G_p(a)$ there is a given measure and, by this, a supplementary structure is being introduced on G_p. In order to avoid this we shall define the operator $\mathscr{I}$ slightly different: we introduce the function spaces F_p^λ on $\tilde{G}_p$ — the manifold of the pairs (b, β) where $b \in G_p$ and β is a non-oriented volume element in b, which satisfy the homogeneity condition $f(b, t\beta) = t^\lambda f(b, \beta)$ for any $t > 0$. We shall define (see § 1, p. 3) $\mathscr{I}$ as an operator

$$\mathscr{I}: F_p^q \to F_q^p$$

which, for the natural representation of $\mathrm{SL}(V)$ on F_q^p and F_p^q, is an intertwining operator (that means, by definition, that it commutes with the operators of the representation). If an Euclidian structure is introduced on V and, by this, we have a measure on each $G_p(a)$ then the definition of the operator coincides with the one at the beginning.

The main purpose of this paper is to explicitly construct an operator $F_q^p \to F_p^q$ which on $\mathrm{Im}\,\mathscr{I}$, the image of $\mathscr{I}$, coincides with the inverse of $\mathscr{I}$ — that is obtaining an inversion formula for $\mathscr{I}$. (We shall consider the case $p+q \leqslant n$, when $\mathscr{I}$ is injective.)

Several approaches on the subject are already known: in [6] it was constructed the operator $\varkappa$ which associates with every function $\varphi \in F_q^p$ and every $b \in G_p$ the $p(q-p)$ differential form $\varkappa_b \varphi$ on the manifold G_{q-p}. It was proved that if $\varphi = \mathscr{I}f$

then $\varkappa_b\varphi$ is a closed form on G_{q-p} and, moreover, for any $p(q-p)$-dimensional cycle $\gamma \subset G_{q-p}$

$$(1) \qquad \int_\gamma \varkappa_b\varphi = C_\gamma f(b)$$

where C_γ is depending only on the homology class of the cycle γ. For even $q-p$ there are cycles γ with $C_\gamma \neq 0$ and then (1) is an inversion formula for $\mathscr{I}$; for odd $q-p$ there are no such cycles. The affine variant of the problem was solved in [1] for $p=1$ and arbitrary q, and was considered in [12] for arbitrary p and q. In the projective case the problem was solved in [4] for $p=1$ and arbitrary q.

In this paper the inversion formula for $\mathscr{I}$ is obtained by means of simple standard constructions. We start by defining (see § 2) a sequence of functions P_k $(1 \leqslant k < n-p)$ which by their geometrical intrinsic sense are in themselves important. The function P_{q-p} is crucial in constructing the inversion formula. We then consider the manifold $\mathscr{E}$ of the pairs $(c, \tilde{b})$ where $c \in G_{q-p}$, $\tilde{b} = (b, \beta) \in \tilde{G}_p$ and $b \cap c = 0$. $\mathscr{E}$ is a bundle over $\tilde{G}_p$ having as fibre at $\tilde{b}$ the set $\pi^{-1}(\tilde{b}) = G^b_{q-p}$ consisting of those $c \in G_{q-p}$ with $c \cap b = 0$.

A central achievement of this paper is the construction of the intertwining operator

$$\chi : F^p_q \to \Omega^N(\mathscr{E})$$

where Ω^N is the space of N-densities on $\mathscr{E}$ $(N = p(q-p))$ with the natural action of the representation of $\mathrm{SL}(V)$. This operator is explicitly constructed in §3 by composing several mappings which, on the respective manifolds, commute with the action of the group $\mathrm{SL}(V)$.

Afterwards we obtain the inversion formula for $\mathscr{I}$ in §4 by means of the operator χ. Namely we introduce a class of N-dimensional manifolds $C \subset G_{q-p}$, the harmonic manifolds, which play for the densities the same role the N-dimensional cycles play for differential forms. It is proved that if $C \subset \pi^{-1}(b)$ is a harmonic manifold, then the following inversion formula is true:

$$(2) \qquad \int_C (\chi\varphi)(\cdot|\tilde{b}) = \mathrm{Cr}(C) f(\tilde{b})$$

where the cofficient $\mathrm{Cr}(C)$ is depending only on C. We like to point out that, in the same manner, the operator $\varkappa$ can be defined as an intertwining operator (in [6] the operator $\varkappa$ was defined in a different way).

The inversion formula (2) naturally leads us to the notion of permissible complex in G_q. Namely, since for $p < q < n-p$ we have $\dim G_p < \dim G_q$, in this case, the knowledge of the function $\mathscr{I}f$ on the whole manifold G_q is over-sufficient for recovering the function f. We can assume that the function $\mathscr{I}f$ is given only on a certain submanifold $K \subset G_q$, whose dimension is $k \geqslant \dim G_p$, and recover the function f by means of the function $\mathscr{I}_K f = \mathscr{I}f|_K$. We call the submanifold $K \subset G_q$ a permissible complex if for almost every $b \in G_p$ there is a harmonic manifold C_b, such that for any $\varphi \in F_q^p$, $(\chi\varphi)(\cdot|\tilde{b})$ restricted at C_b is uniquely defined by means of $\varphi|_K$. It is true that for permissible complexes, the associated operator $\mathscr{I}_K$ is one to one, but remarkable for them is the fact that the inversion formula for $\mathscr{I}_K$ has the explicit form given by (2) for $C = C_b$.

Using the structure of the operator χ we define in §5 a large class of permissible complexes. The problem of describing all the permissible complexes is of great interest.

1. THE INTERTWINING OPERATOR $\mathscr{I}$

1. PRELIMINARIES. 1° We define the (non-oriented) volume element ω in the real, n-dimensional vector space V, as a non-negative function $\omega(v_1, \ldots, v_n)$, $v_i \in V$, which for any change $v_i \to \sum_{j=1}^{n} g_{ij} v_j$ $(i = 1, \ldots, n)$ is multiplied by $|\det g|$; in particular $\omega(v_1, \ldots, v_n) = 0$ for $v_1, \ldots, v_n$ linearly dependent vectors. It is obvious that the volume element ω is uniquely determined by the number $\omega(e_1, \ldots, e_n)$, where $\{e_1, \ldots, e_n\}$ is a basis in V.

We introduce some notations. Let ω be a volume element in V. We then denote by ω' the volume element in the dual space V' given by the equality

$$\omega'(e^1, \ldots, e^n)\omega(e_1, \ldots, e_n) = 1$$

where $\{e^i\}$ and $\{e_j\}$ are dual bases respectively in V' and V. Next let L be a subspace in V and α a volume element in L. We denote by ω/α the volume element in the factor space V/L, determined by the pair ω, α. That is if $\{e_1, \ldots, e_k\}$ is a basis in L and $\{e_{k+1}, \ldots, e_n\}$ completes it up to a basis in V, then let $\tilde{e}_i$ $(k+1 \leqslant i \leqslant n)$ be the projection of e_i on the factor space V/L. Then

$$\omega/\alpha(\tilde{e}_{k+1}, \ldots, \tilde{e}_n) = \frac{\omega(e_1, \ldots, e_n)}{\alpha(e_1, \ldots, e_k)}.$$

Finally, if in the vector spaces L_1, L_2 there are respectively given the volume elements ν_1, ν_2, then in the tensorial product space $L_1 \otimes L_2$ we canonically define a volume element which is denoted by $\nu_1 \otimes \nu_2$. That is, if $\{e_i\}$, $\{f_j\}$ are bases respectively in L_1 and L_2 then

$$(\nu_1 \otimes \nu_2)(\{e_i \otimes f_j\}) = \nu_1(\{e_i\})\nu_2(\{f_j\}).$$

It is easy to verify that the introduced definitions are correct, that is, they are independent of the choice of the bases in the respective vector spaces. The next lemma follows from the definitions.

LEMMA. *For any $t > 0$, $s > 0$, the following are true:*

$$(1.1') \qquad (t\omega)' = t^{-1}\omega$$

$$(1.1'') \qquad s\omega/t\alpha = s/t\ \omega/\alpha$$

$$(1.1''') \qquad (sv_1) \otimes (tv_2) = s^{\dim L_2}\, t^{\dim L_1}(v_1 \otimes v_2).$$

2° We mean by k (even) density on the smooth manifold X, $\dim X = n > k$, a smooth function σ which associates with each pair (x, h) where $x \in X$ and h is a k-dimensional subspace of T_xX, a volume element in h. In other words, σ is a function of $x \in X$ and of the vectors $\xi_1, \ldots, \xi_k \in T_xX$, which for any change $\xi_i \to \sum_{j=1}^{k} g_{ij}\xi_j$ $(i = 1, \ldots, k)$ is multiplied by $|\det g|$.

For k-densities we can define by the same methods used for k differential forms, the following operations:

1) integration on k dimensional submanifolds in X,

2) the pullback π^* from X to Y where $\pi : Y \to X$ is a bundle,

3) integration of the k-densities on Y denoted by π_* on the fibres of the bundle $\pi : Y \to X$ (π_* is defined for $k \geqslant l$, l — the fibre's dimension, and it transforms the k-densities on Y into $k - l$ densities on X).

2. THE SPACES F_p^λ. Let $G_p(V)$ be the manifold of the p dimensional subspaces of the vector space V, $\dim V = n$. We denote by $\tilde{G}_p(V)$ the manifold of the pairs $\tilde{b} = (b, \beta)$ where $b \in G_p(V)$, β is a volume element in b (clearly $\tilde{G}_p(V)$ can be viewed as line bundle over $G_p(V)$). We introduce the spaces $F_p^\lambda = F_p^\lambda(V)$ $(p = 1, \ldots, n-1, \lambda \in \mathbf{C})$ of C^∞ functions on $\tilde{G}_p(V)$ which satisfy the homogeneity condition

$$(1.2) \qquad f(b, t\beta) = t^\lambda f(b, \beta)$$

for any $t > 0$. We define a representation of the group $\mathrm{SL}(V)$ on each space F_p^λ by:

$$(T(g)f)(\tilde{b}) = f(\tilde{b}g)$$

where $\tilde{b} \to \tilde{b}g$ is the natural action of the element $g \in \mathrm{SL}(V)$ on $\tilde{G}_p(V)$.

Next, we shall give several interpretations for the spaces F_p^λ that shall be used later on.

1) $F_p^\lambda(V) \simeq F_{n-p}^\lambda(V')$ where V' is the dual space of V. This isomorphism is induced by the one to one and onto mapping $\tilde{G}_p(V) \to \tilde{G}_{n-p}(V')$ which carries $(b, \beta) \in \tilde{G}_p(V)$ to $(\operatorname{Ann} b, (\omega/\beta)')$ in $\tilde{G}_{n-p}(V')$, ω — a fixed volume element in V.

2) We consider $F_p^\lambda(V)$ as a C^∞ function space on $G_p(V)$. In order to do this we give an Euclidian structure on V and denote by β_b the Euclidian volume element in $b \in G_p(V)$. We obtain the desired interpretation by associating with each function $f \in F_p^\lambda(V)$ the function $\varphi(b) = f(b, \beta_b)$ on $\tilde{G}_p(V)$.

3) Notice that $\tilde{G}_p(V)$ can be interpreted as the manifold of p vectors $u = u_1 \wedge \ldots \wedge u_p \neq 0$ of V, where we identified u with $-u$. That is, for each pair $(b, \beta) \in \tilde{G}_p(V)$ there is a corresponding p-vector $u = u_1 \wedge \ldots \wedge u_p$ given, up to multiplication by -1, by the conditions: the space spanned by $u_1, \ldots, u_p$ is b and $\beta(u_1, \ldots, u_p) = 1$. By this, F_p^λ can be regarded as the space of C^∞ functions on the manifold of the p-vectors $u \neq 0$ which satisfy

$$f(tu) = |t|^{-\lambda} f(u) \quad \text{for any } t \neq 0.$$

From this interpretation we get two more others.

4) We regard F_p^λ as the space of C^∞ functions on the manifold $E_{p,n}$ of the p-frames in V, which satisfy the condition

$$f(gx) = |\det g|^{-\lambda} f(x) \tag{1.3}$$

for any $g \in \mathrm{GL}(p, \mathbf{R})$. That is, as $E_{p,n}$ is a bundle on the manifold of non-zero p-vectors, we obtain the desired interpretation associating with each function defined on the manifold of p vectors, its pullback on $E_{p,n}$.

5) In a coordinate system for V, any p-frame is given by a $p \times n$ matrix of rank p. So, F_p^λ is interpreted as the C^∞ function space on the manifold of $p \times n$ matrices of rank p, which satisfy the condition (1.3), or, equivalently, as the space of C^∞ functions that are even, homogenous, of $-\lambda$ homogeneity on the p-minors of the matrix x.

6) F_p^λ interpreted as the C^∞ function space on the manifold $M_{p,n-p}$ of $p \times (n-p)$ matrices *). That is done by associating with each matrix $u = \|u_i^j\|$ in $M_{p,n-p}$, the p-vector $\tau u = v_1 \wedge \ldots \wedge v_p$ where

$$v_i = (\delta_i^1, \ldots, \delta_i^p, u_i^1, \ldots, u_i^{n-p}) \quad i = 1, \ldots, p$$

(δ_j^i - Kronecker's symbol). It is obvious that the projection of the set $\tau M_{p,n-p}$ on $G_p(V)$ is an open and dense subset of $G_p(V)$. We obtain the desired interpretation, associating with each function f defined on $\tilde{G}_p(V)$, the function $(\tau^* f)(u) = f(\tau u)$ defined on $M_{p,n-p}$.

NOTE. There is a purely group interpretation of the spaces $F_p^\lambda(V)$. They are interpreted as the C^∞ function spaces on the group $\mathrm{SL}(n, \mathbf{R})$, satisfying the condition $f(ug) = |\det a|^\lambda f(u)$ for any block triangular matrix $g = \begin{pmatrix} a & 0 \\ b & c \end{pmatrix}$ where a is a p-matrix. In this interpretation, the operators of the representation act by right translations.

*) These functions satisfy supplementary conditions of decreasing at infinity, which are not presented here.

3. The Definition of the Intertwining Operator $\mathcal{J}$. The main subject of this paper is the intertwining operator

$$\mathcal{J} : F_p^q \to F_q^p \quad (1 \leqslant p < q < n)$$

that is, an operator which commutes with the operators of the representation. This operator has a simple geometrical sense: roughly speaking, $(\mathcal{J}f)(a)$, $a \in G_q(V)$ is equal to the value of the integral of the function f on the set of the p subspaces that are contained in a; we shall now give the exact definition of the operator $\mathcal{J}$.

Let $f \in F_p^q$, $\tilde{a} = (a, \alpha) \in \tilde{G}_q(V)$, $\tilde{b} = (b, \beta) \in \tilde{G}_p(a)$. We denote by $\sigma_{\tilde{a}}(\tilde{b})$ the volume element in the tangent space $T_bG_p(a) = b' \otimes a/b$ canonically defined by the volume elements α and β:

$$\sigma_{\tilde{a}}(\tilde{b}) = \beta' \otimes \alpha/\beta. \tag{1.4}$$

By the homogeneity conditions (1.1) and (1.2) it follows that the product $f(\tilde{b})\sigma_{\tilde{a}}(\tilde{b})$ is independent of the choice of the volume element β and, therefore, it defines for any fixed a a N-density on $G_p(a)$, where $N = p(q-p)$. We define

$$(\mathcal{J}f)(\tilde{a}) = \int_{G_p(a)} f(\tilde{b})\sigma_{\tilde{a}}(\tilde{b}). \tag{1.5}$$

It follows immediately from the definition that $\mathcal{J}f \in F_q^p$ and that $\mathcal{J}$ is an intertwining operator.

We give now the formula of $\mathcal{J}$ in coordinates when F_p^q and F_q^p are interpreted as function spaces respectively on the manifold of $p \times (n-p)$ matrices and on the manifold of $q \times (n-q)$ matrices. We write $u \in M_{p,n-p}$ and $v \in M_{q,n-q}$ as block matrices: $u = (u_1, u_2)$ where $u_1 \in M_{p,q-p}$, $u_2 \in M_{p,n-q}$ and $v = \begin{pmatrix} v_1 \\ v_2 \end{pmatrix}$ where $v_1 \in M_{p,n-q}$, $v_2 \in M_{q-p,n-q}$. Then

$$(\mathcal{J}f)(v) = \int_{M_{p,q-p}} f(t, v_1 + tv_2)\mathrm{d}t \tag{1.6}$$

where $\mathrm{d}t = \prod \mathrm{d}t_i^j$.

Our purpose is to obtain inversion formulas for $\mathcal{J}$, that is to construct an inverse operator $\mathcal{J}^{-1}$ on $\operatorname{Im}\mathcal{J}$. As for $q > n-p$ implies $\dim G_p > \dim G_q$, $\operatorname{Ker}\mathcal{J}$, the kernel of the operator $\mathcal{J}$, is non-zero in this case. That is why we shall from now on assume $q \leqslant n-p$.

2. THE SEQUENCE OF FUNCTIONS P_k

1. Definition of the Functions P_k in Geometric Manner. By means of simple geometric constructions we shall define a sequence of functions $P_k(\tilde{c}, A|\tilde{h})$, $1 \leqslant k < n-p$, where $\tilde{c} = (c, \gamma) \in \tilde{G}_k(V)$, $\tilde{h} = (h, \eta) \in \tilde{G}_p(V')$, $c \perp h$ and $A = \{A_i\}$, $i = 1, \ldots, p \cdot k$, vectors in the tangent space $T_cG_k(V)$.

The function P_{q-p} plays a central role in the construction of the operator χ.

Let k be fixed. We consider $\mathscr{B}$ the manifold of the pairs (c, h) where $c \in G_k(V)$, $h \in G_p(V')$ and $c \perp h$. We define the mappings $\pi_1 : \mathscr{B} \to G_p(V')$ and $\pi_2 : \mathscr{B} \to G_k(V)$ by $\pi_1(c, h) = h$, $\pi_2(c, h) = c$. We obtain a double fibration

$$\begin{array}{ccc} & \mathscr{B} & \\ {\scriptstyle \pi_1} \swarrow & & \searrow {\scriptstyle \pi_2} \\ G_p(V') & & G_k(V)\, . \end{array}$$

We notice that $\pi_2^{-1}(c) \simeq G_p((V/c)') = G_p(\mathrm{Ann}\, c)$.

Let $\psi(\tilde{h})$ be an arbitrary smooth function on $\tilde{G}_p(V')$ which satisfies the condition $\psi(h, t\eta) = t^n\psi(h, \eta)$ for any $t > 0$. Then $\psi(\tilde{h})\sigma_{(\tilde{V}')}(\tilde{h})$ is independent of the choice of the volume element and defines a $p(n-p)$ density on $G_p(V')$. Consider on $\mathscr{B}$ the pullback of this density, that is, the $p(n-p)$ density $\tau = \pi_1^*(\psi\sigma_{\tilde{V}'})$ given by the following equality:

$$\tau(c, h, B_1, \ldots, B_r) = (\psi(\tilde{h})\sigma_{\tilde{V}'}(\tilde{h})((D\pi_1)B_1, \ldots, (D\pi_1)B_r)$$

where $B_i \in T_{c,h}\mathscr{B}$ and $r = p(n-p)$.

We fix $A = \{A_i\}$ an arbitrary system of $p \cdot k$ vectors in the tangent space $T_cG_k(V)$ and let $\bar{A} = \{\bar{A}_i\}$ where $\bar{A}_i \in T_{c,h}\mathscr{B}$ are vectors in the preimage of A_i. Then in $T_{c,h}\mathscr{B}$ we choose $B = \{B_i\}$ a system of $p(n-k-p)$ tangent vectors to the fibre of the bundle π_2, that is $B_i \in T_hG_p((V/c)')$. Then, for fixed c, $\tilde{h}$ and A, $\tau(c, h, A, B)$ is independent of the choice of the vectors $\bar{A}_i$ and defines a volume element in $T_hG_p(\mathrm{Ann}\, c)$. On the other hand $\psi(\tilde{h})\sigma_{\mathrm{Ann}\, c, (\omega/\gamma)'}(\tilde{h})$ is also a volume element in $T_hG_p(\mathrm{Ann}\, c)$ hence $\tau(c, h, \bar{A}, B)$ and $(\psi(\tilde{h})\sigma_{\mathrm{Ann}\, c, (\omega/\gamma)'}(\tilde{h}))(B)$ differ by a multiplicative factor which is independent of B.

DEFINITION. We define $P_k(\tilde{c}, A|\tilde{h})$ by the equality

$$\pi_1^*(\psi(\tilde{h})\sigma_{\tilde{V}'}(\tilde{h}))_{(c, h, \bar{A}, B)} = P_k(\tilde{c}, A|\tilde{h})(\psi(\tilde{h})\sigma_{\mathrm{Ann}\, c, (\omega/\gamma)'}(\tilde{h}))(B). \tag{2.1}$$

It is obvious that P_k is independent of the choice of ψ. It follows from the definition that the function P_k is invariant for the action of the group $\mathrm{SL}(V)$ on the manifold of the triplets $(\tilde{c}, A|\tilde{h})$. From (2.1) we have

$$P_k(\tilde{c}, A|\cdot) \in F_p^{-k}((V/c)').$$

(Rigourously $P_k(\tilde{c}, A|\cdot)$ belongs to some extension of $F_p^{-k}((V/c)')$ consisting of piecewise smooth functions.)

2. Explicit Formula for the Functions $P_k(\tilde{c}, A|\tilde{h})$.

Proposition. *Let* $\tilde{c} = (c, \gamma)$, $\tilde{h} = (h, \eta)$, $A = \{A_1, \ldots, A_{p \cdot k}\}$ *where* $A_i \in T_c G_k(V) =$ $= \mathrm{Hom}(c, V/c)$. *Then*

$$P_k(\tilde{c}, A|\tilde{h}) = |\det\|\langle v^i, A_s v_j\rangle\|| \tag{2.2}$$

where $\{v_j\}$, $\{v^i\}$ *are some unitary bases respectively in* c *and* $h \in G_p((V/c)')$, *that is* $\eta(\{v^i\}) = \gamma(\{v_j\}) = 1$ *(the right side is independent of their choice).*

We point out that if in $\tilde{c}$ and $\tilde{h}$ there are respectively given the frames $z =$ $= (z_1, \ldots, z_n)$ and $\xi = (\xi^1, \ldots, \xi^p)$ then (2.2) can be written as

$$P_k(z, A|\xi) = |\det\|\langle \xi^i, A_s z_j\rangle\|| \quad (z\xi' = 0). \tag{2.3}$$

Proof. Both sides of (2.2) are invariant for the action of the group SL(V). Therefore, as the group acts transitively on the manifold of pairs $(\tilde{c}, \tilde{h})$ it suffices to prove (2.2) for a certain pair $\tilde{c} = \tilde{c}_0$, $\tilde{h} = \tilde{h}_0$. We assume that there are given dual coordinate systems in V and V'. On $G_p(V')$ we define local coordinates (ξ, η)

$$\xi = \|\xi_i^j\|_{\substack{i=1,\ldots,p,\\ j=1,\ldots,k}} \qquad \eta = \|\eta_i^j\|_{\substack{i=1,\ldots,p\\ j=1,\ldots,n-p-k}} \quad ;$$

for each pair (ξ, η) there is a corresponding p-subspace spanned by the row vectors of the $p \times k$ matrix $h = (\xi, I_p, \eta)$. At the same time we consider $F_p^\lambda(V')$ as space of functions of (ξ, η). Analogously we can define on $G_k(V)$ local coordinates

$$a = \|a_i^j\|_{\substack{i=1,\ldots,k,\\ j=1,\ldots,p}} \qquad b = \|b_i^j\|_{\substack{i=1,\ldots,k\\ j=1,\ldots,n-k-p}} \quad ;$$

for each pair (a, b) there is a corresponding k-subspace spanned by the row vectors of the matrix (I_k, a, b). As the ortogonality condition of $c = (I_k, a, b)$ and $h = (\xi, I_p, \eta)$ is written $a + \xi' + b\eta' = 0$, we can choose as coordinates on $\mathcal{B}$ the triplet (a, b, η). In these coordinates, we have

$$\pi_1(a, b, \eta) = (-a' - \eta b', \eta).$$

Then if we fix $c_0 = (I_k, 0, 0)$ and $h_0 = (0, I_p, 0)$ we find that in (c_0, h_0) the differential of the mapping π_1 is written

$$(D\pi_1)(A, B, \eta) = (-A', \eta).$$

By standard computation we get the equality (2.2) for $(c, h) = (c_0, h_0)$.

3. The Intertwining Operator χ

1. The Space $\Omega^N(\mathcal{E})$. We shall denote by $\mathcal{E} = \mathcal{E}_{q-p}$ the manifold of the pairs $(c, \tilde{b})$ where $c \in G_{q-p}(V)$, $\tilde{b} = (b, \beta) \in \tilde{G}_p(V)$ and $b \cap c = 0$; and by $\Omega^N(\mathcal{E})$, $N = p(q-p)$, the space of N-densities $\tau(c, b)$ on $\mathcal{E}$ satisfying the condition

$\tau(c, (b, t\beta)) = t^q\tau(c, (b, \beta))$ for any $t > 0$. As SL(V) acts naturally on $\mathscr{E}$, we can define on $\Omega^N(\mathscr{E})$ a representation of SL(V) by translation operators.

In this paragraph we shall construct an intertwining operator

$$\chi : F_q^p \to \Omega^N(\mathscr{E})$$

which is crucial in the inversion formula. We shall obtain the inversion formula for $\mathscr{I}$ in §4.

2. THE INTERTWINING OPERATORS R_p^λ. The construction of χ will be done in several steps. We start by constructing the intertwining operators

$$R_p^\lambda(V, \omega) : F_p^\lambda(V) \to F_q^{n-\lambda}(V').$$

(Recall that $F_p^{n-\lambda}(V') \simeq F_{n-p}^{n-\lambda}(V)$.) In order to do this we introduce the function $U^{(\lambda)}(\tilde{h}, \tilde{b})$ on the manifold of the pairs $(\tilde{h}, \tilde{b})$ where $\tilde{h} = (h, \eta) \in \tilde{G}_p(V')$, $\tilde{b} = (b, \beta) \in \tilde{G}_p(V)$, defined by the equality

$$U^{(\lambda)}(\tilde{h}, \tilde{b}) = C_{p,n}(\lambda) |\det \| \langle f^i, e_j \rangle \| | \quad (\lambda \in \mathbf{C}). \tag{3.1}$$

Here $\{f^i\}$, $\{e_j\}$ are unitary bases respectively in h and b (the right side is independent of their choice),

$$C_{p,n}(\lambda) = \pi^{-p(n-p)} \Gamma_p\left(\frac{\lambda + n}{2}\right) \Gamma_p\left(\frac{\lambda + p}{2}\right) \tag{3.2}$$

where $\Gamma_p(s) = \prod_{k=0}^{p-1} \Gamma\left(s - \frac{k}{2}\right)$. We point out that

$$U^{(\lambda)}(h, t\eta; b, s\beta) = (ts)^{-\lambda} U^{(\lambda)}(h, \eta; b, \beta) \tag{3.3}$$

for any $t > 0$, $s > 0$.

Let $f \in F_p^\lambda$. From (1.1), (1.2) and (3.3) it follows that the product $U^{(\lambda-n)}(\tilde{h}, \tilde{b}) f(\tilde{b}) \sigma_{\tilde{V}}(b)$, is independent of β. Here $\tilde{V} = (V, \omega)$ and $\sigma_{\tilde{V}}(\tilde{b}) = \beta' \otimes \omega/\beta$. Therefore for any fixed $\tilde{h}$, it defines a $p(n-p)$ density on $G_p(V)$. By definition:

$$(R_p^\lambda(\tilde{V})f)_{(\tilde{h})} = \int_{G_p(V)} U^{(\lambda-n)}(\tilde{h}, \tilde{b}) f(\tilde{b}) \sigma_{\tilde{V}}(\tilde{b}). \tag{3.4}$$

The integral (3.4) is convergent for $\operatorname{Re}\lambda > n - 1$. In order to make it have sense for $\operatorname{Re}\lambda \leqslant n - 1$, we give an Euclidian structure on V and regard F_p^λ as C^∞ function space on $G_p(V)$ (see §1 p. 2). Then the operator R_p^λ is written:

$$(R_p^\lambda f)_{(h)} = \int_{G_p(V)} U^{(\lambda-n)}(h, b) f(b) \sigma(b) \tag{3.5}$$

where $U_0^\lambda(h, b) = U^{(\lambda)}(h, \eta_h; b, \beta_b)$, $\sigma(b) = \sigma_{\tilde{V}}(b, \beta_b)$ and η_h, β_b are the Euclidian volume elements respectively in h and b.

For a fixed f, the integral (3.5) is convergent in the domain $\operatorname{Re}\lambda > n-1$ and is λ-analytic in this domain. For $\operatorname{Re}\lambda \leqslant n-1$ we define $R_p^\lambda f$ as the analytic continuation in λ. It is easy to verify that this definition is independent of the choice of the Euclidian structure on V.

It follows from the definition that in any point in whose neighbourhood $R_p^\lambda(\tilde{V})$ is analytic (such points will be called regular for the operators $R_p^\lambda(\tilde{V})$), it carries the spaces $F_p^\lambda(V)$ into $F_p^{n-\lambda}(V')$ and is an intertwining operator.

NOTE. If we change ω by $t\omega$ $(t > 0)$, then, $\sigma_{V,\omega}(\tilde{b})$ is multiplied by t^p, so from (3.4) we have

$$R_p^\lambda(V, t\omega) = t^p R_p^\lambda(V, \omega) \quad \text{for any } t > 0.$$

Elementary computation leads us to the formula of $R_p^\lambda(V, \omega)$ in coordinates, when $F_p^\lambda(V)$ is interpreted as the C^∞ function space on the $p \times (n-p)$ matrices $u = \|u_i^j\|$ and analogously $F_p^{n-\lambda}(V')$ is interpreted as the C^∞ function space on the matrices $(n-p)\times p$, $\xi = \|\xi_j^i\|$. Namely

$$(R_p^\lambda(\tilde{V})f)_{(\xi)} = C_{p,n}(\lambda - n) \int\limits_{M_{p,\, n-p}} f(u)|I_p + u\xi|^{\lambda-n}\mathrm{d}u \tag{3.6}$$

where $\mathrm{d}u = \prod \mathrm{d}u_i^j$, I_p being the p identity matrix and we denote by $|\cdot|$, the absolute value of the determinant of the respective matrix.

3. Regularity Conditions for the Operators $R_p^\lambda(\tilde{V})$ and the Inversion Formula.

PROPOSITION 1. *For $p > 1$, $R_p^\lambda(\tilde{V})$ is λ regular whenever $\lambda \neq (p-1) - k$, $k = 0, 1, \ldots$; for $p = 1$, $R_p^\lambda(\tilde{V})$ is λ regular whenever $\lambda \neq 2k$, $k = 0, 1, \ldots$.*

Proof. It is known (see for instance [10], [11]) that the generalized function $\dfrac{|\det x|^\lambda}{\Gamma_p\left(\frac{\lambda+2}{2}\right)}$ on the manifold of p-matrices is a λ-entire function. Therefore, the singular points of R_p^λ as function of λ, coincide with the singular points of the function $\Gamma_p\left(\dfrac{\lambda}{2}\right)$. Hence our statement follows immediately.

PROPOSITION 2. *If λ and $n - \lambda$ are regular points for the operators R_p^λ then $R_p^{n-\lambda}(\tilde{V}')\circ R_p^\lambda(\tilde{V})$ is the identity operator on $F_p^\lambda(V)$.*

COROLLARY. *With the same assumption on λ, the mappings*

$$R_p^\lambda(V) : F_p^\lambda(V) \to F_p^{n-\lambda}(V')$$

and

$$R_p^{n-\lambda}(\tilde{V}') : F_p^{n-\lambda}(V') \to F_p^{\lambda}(V)$$

are isomorphisms.

Proof. We shall make use of the following statement: any intertwining operator which maps $F_p^{\lambda}(V)$ on itself is proportional to the identity operator E. By this statement we have $R_p^{n-\lambda}(\tilde{V}')\circ R_p^{\lambda}(V) = cE$. So we have to show that $c = 1$. In order to do this we give respectively on V and V', dual inner products $(\ ,\ )$ and denote by $SO(n)$ the subgroup of the group $SL(V)$ which consists of those transformations which preserve the inner product. In $F_p^{\lambda}(V)$ and $F_p^{n-\lambda}(V')$ there is respectively a unique vector, up to a multiplicative constant, which is invariant for $SO(n)$. Namely the functions $f(b, \beta) = |\det\|(v_i, v_j)\||^{-\lambda/2}$ respectively $\varphi(h, \eta) = |\det\|(v^i, v^j)\||^{-((n-\lambda)/2)}$ where $\{v_i\}$ and $\{v^j\}$ are unitary basis respectively in $b \in G_p(V)$ and $h \in G_p(V')$. Hence $R_p^{\lambda}(\tilde{V})f = S_\lambda \varphi$, $R_p^{n-\lambda}(\tilde{V}')\varphi = S_{n-\lambda} f$ and it suffices to show that $S_\lambda = S_{n-\lambda} = 1$. For that we use the matriceal interpretation of $F_p^{\lambda}(V)$ and $F_p^{n-\lambda}(V')$ where we have

$$f(u) = |I_p + uu'|^{-\lambda/2}, \quad \varphi(\xi) = |I_p + \xi'\xi|^{-\frac{n-\lambda}{2}}.$$

Thus the equality $R_p^{\lambda} f = S_\lambda \varphi$ becomes

$$S_\lambda |I_p + \xi'\xi|^{-\frac{n-\lambda}{2}} = C_{p,n}(\lambda - n) \int\limits_{M_{p,n-p}} |I_p + uu'|^{-\lambda/2} |I_p + u\xi|^{\lambda-n} du.$$

Computing for $\xi = 0$, we get

$$S_\lambda = C_{p,n}(\lambda - n) \int\limits_{M_{p,n-p}} |I_p + uu'|^{-\lambda/2} du.$$

As $\int\limits_{M_{p,n-p}} |I_p + uu'|^{-\lambda/2} du = [C_{p,n}(\lambda - n)]^{-1}$ (see Appendix 2 following this point), we have $S_\lambda = 1$.

APPENDIX 1. The explicit formula of the generalized function $\dfrac{|\det x|^\lambda}{\Gamma_p\left(\dfrac{\lambda+p}{2}\right)}$ for $\lambda = -p - k$, $k = 0, 1, \ldots$ (see [10], [11]).

Let $\delta(x)$ be the delta function on M_p-the manifold of p-matrices $(\delta, f) = f(0)$ for any test function f. As it is known $\delta(x) = (2\pi)^{-p^2}\mathscr{F}(\mathbf{1})$, where $\mathscr{F}$ is the Fourier transform defined on the test functions by the equality

$$(\mathscr{F}f)(x) = \int f(x) e^{i \cdot \mathrm{tr}(y'x)} dy \qquad dy = \prod dy_{ij},$$

1 being the function identically equal to 1.

Motivated by the definition of the delta function $\delta(x)$, we introduce the distribution $\gamma(x)$ by the equality

$$\gamma(x) = (2\pi)^{-p^2}\mathscr{F}(\operatorname{sgn}(\det x)).$$

Unlike for $\delta(x)$, the support of the distribution γ is the whole manifold M_p. Considering $\delta^{(k)}(x) = \mathscr{D}^k\delta(x)$, $\gamma^{(k)}(x) = \mathscr{D}^k\gamma(x)$, $k = 0, 1, \ldots$ where $\mathscr{D} = \det\dfrac{\partial}{\partial x} = \det\left\|\dfrac{\partial}{\partial x_{ij}}\right\|$, we have with these notations:

$$\left.\frac{|\det x|^\lambda}{\Gamma_p\left(\frac{\lambda+p}{2}\right)}\right|_{\lambda=-p-2k} = \frac{(-1)^{p\cdot k}\cdot\pi^{p^2/2}}{2^{2kp}\Gamma_p\left(k+\frac{p}{2}\right)}\,\delta^{(2k)}(x)$$

$$\left.\frac{|\det x|^\lambda}{\Gamma_p\left(\frac{\lambda+p}{2}\right)}\right|_{\lambda=-p-2k-1} = \frac{(-\mathrm{i})^{(2k+1)p}\cdot\pi^{p^2/2}}{2^{(2k+1)}\Gamma_p\left(k+\frac{p+1}{2}\right)}\,\gamma^{(2k+1)}(x).$$

APPENDIX 2. Let us make the computations for the formulas

$$I^\lambda_{p,m} = \int\limits_{M_{p,m}} |I_p + xx'|^{-\lambda/2}\mathrm{d}x = \pi^{\frac{pm}{2}}\frac{\Gamma_p\left(\frac{\lambda-m}{2}\right)}{\Gamma_p\left(\frac{\lambda}{2}\right)}.$$

For $p = 1$ we have $I^\lambda_{1,m} = \int(1 + x_1^2 + \ldots + x_m^2)^{-\lambda/2}\,\mathrm{d}x_1\ldots\mathrm{d}x_m$. This integral is easily computed by passing to spherical coordinates:

$$I^\lambda_{1,m} = \pi^{m/2}\frac{\Gamma\left(\frac{\lambda-m}{2}\right)}{\Gamma\left(\frac{\lambda}{2}\right)}.$$

Suppose now that $p > 1$. We write x as $x = \begin{pmatrix} y \\ z\end{pmatrix}$ where z is the last row of the matrix x. The matrix $I_m + xx'$ is positively defined; we consider $u = (I_m + y'y)^{1/2}$, $v = zu^{-1}$. Then we have $I_m + x'x = I_m + y'y + z'z = u(I_m + v'v)u$ and it follows that $|I_m + x'x| = |I_m + v'v|\cdot|I_m + y'y|$. Therefore, using the evident equality $|I_p + xx'| = |I_m + x'x|$ we have $|I_p + x'x| = (1 + vv')|I_{p-1} + yy'|$. And as $\mathrm{d}x =$

$= dy\,dz = |I_{p-1} + yy'|^{1/2}\,dy\,dv$ we conclude that

$$I_{p,m} = \int (1+vv')^{-\lambda/2}\,dv \int |I_{p-1}+yy'|^{-\frac{\lambda-1}{2}}\,dy = I^{\lambda}_{1,m} I^{\lambda-1}_{p-1,m}.$$

It follows that

$$I^{\lambda}_{p,m} = I^{\lambda}_{1,m}\cdot I^{\lambda-1}_{1,m}\,\ldots\, I^{\lambda-p-1}_{1,m} = \pi^{\frac{pm}{2}} \frac{\Gamma_p\left(\frac{\lambda-m}{2}\right)}{\Gamma_p\left(\frac{\lambda}{2}\right)}.$$

REMARK. By $|I_p + xx'| = |I_m + x'x|$ we have that $I^{\lambda}_{p,m} = I^{\lambda}_{m,p}$. Therefore we find the following interesting formula for Γ_p

$$\frac{\Gamma_p\left(\frac{\lambda-m}{2}\right)}{\Gamma_p\left(\frac{\lambda}{2}\right)} = \frac{\Gamma_m\left(\frac{\lambda-p}{2}\right)}{\Gamma_m\left(\frac{\lambda}{2}\right)}.$$

4. THE OPERATORS $\hat{u} : F_p(V) \to F_p^{\lambda+\mu}(V)$. If $f \in F_p(V)$, $u \in F_p^{-\mu}(V')$ then $u\cdot(R_p^{\lambda}(\tilde{V})f) \in F_p^{n-\lambda-\mu}(V')$. We associate with the function u the operator $\hat{u} : F_p^{\lambda}(V) \to$ $\to F_p^{\lambda+\mu}(V)$ defined by the equality

$$\hat{u}f = R_p^{n-\lambda-\mu}(\tilde{V}')(u\cdot R_p^{\lambda}(\tilde{V})f). \tag{3.7}$$

We shall call u the symbol of the operator $\hat{u}$ and μ its order. The given definition still makes sense when u is a piecewise smooth function. In this case $\hat{u}f$ will be a function from a certain extension of the space $F_p^{\lambda+\mu}(V)$.

We now describe the case when $\hat{u}$ is a differential operator. Let $F_p^{\lambda}(V)$ be interpreted as the function space on the $p\times(n-p)$ matrices of rank p and respectively $F_p^{\lambda}(V')$ as function space on the $p\times(n-p)$ matrices ξ. If μ is a non-negative even number then there is an invariant finite dimensional subspace $\Phi_p^{-\mu}(V')$ which consists of the homogenous polynomials of order μ of the p-minors of the matrix ξ.

PROPOSITION 3. *If $u(\xi) \in \Phi_p^{-\mu}(V')$ where μ is an arbitrary non-negative even number, then $\hat{u}$ is a differential operator of order $p\cdot\mu$, given, up to a multiplicative constant, by the equality*

$$\hat{u} = u\left(\frac{\partial}{\partial x}\right)$$

(that is $\hat{u}$ is determined by changing, in the formula of u, the elements ξ_{ij} with the operators $\frac{\partial}{\partial x_{ij}}$).

Proof. If $f \in F_p^\lambda(V)$ and $\varphi(\xi) = (R_p(\tilde{V})f)(\xi)$, then considering the chosen interpretation for the spaces F_p^λ, $\hat{u}$ is written like this:

$$(3.8) \qquad (\hat{u}f)(x) = C_{p,n}(-\lambda-\mu)\int_\Omega |x\xi'|^{-\lambda-\mu}u(\xi)\varphi(\xi)\sigma(\xi)$$

where $\sigma(\xi)$ is a $p\times(n-p)$ density in whose explicit formula we take no interest, and the integral is computed on an arbitrary section of the boundle $E_{p,n} \to G_p(V')$. From Proposition 2 we obtain

$$(3.9) \qquad f(x) = C_{p,n}(-\lambda)\int_\Omega |x\xi'|^{-\lambda}\varphi(\xi)\sigma(\xi).$$

We apply the operator $u\left(\frac{\partial}{\partial x}\right)$ to both sides of the equality (3.9). To write explicitly the obtained formula, we define on the manifold of p-matrices $z = \|z_{ij}\|$ the operator $\frac{\partial}{\partial z} = \det\left\|\frac{\partial}{\partial z_{i,j}}\right\|$. It is obvious that

$$u\left(\frac{\partial}{\partial x}\right)|x\xi'|^{-\lambda} = u(\xi)\left(\frac{\partial}{\partial z}\right)^\mu |z|^{-\lambda}\Big|_{z=u\xi'}.$$

On the other hand, it is known (see for instance [10]) that

$$\left(\frac{\partial}{\partial z}\right)^\mu |z|^{-\lambda} = C_{\lambda,\mu}|z|^{-\lambda-\mu}$$

where $C_{\lambda,\mu}$ is a numerical constant. This is how we obtain

$$(3.10) \qquad \left(u\left(\frac{\partial}{\partial x}\right)f\right)(x) = C_{p,n}(-\lambda)C_{\lambda,\mu}\int_\Omega |x\xi'|^{-\lambda-\mu}u(\xi)\varphi(\xi)\sigma(\xi).$$

Combining (3.9) and (3.10) we get the statement of the proposition.

For us, the main example of an $\hat{u}$ operator is given by the following.

DEFINITION 1. We shall denote by $P(c, A)$ the operator

$$\hat{P}(\tilde{c}, A) : F_p^p(V/c) \to F_p^q(V/c)$$

given by the symbol $\hat{P}(c, A|\tilde{h}) = P_{q-p}(\tilde{c}, A\tilde{h})$ (see §2). If $c \in G_{q-p}(V)$ is given by the frame $z = (z_1, \ldots, z_{q-p})$, then we write $\hat{P}(z, A)$ instead of $\hat{P}(\tilde{c}, A)$.

By the proof of the previous proposition, and by Proposition 1, §2 we have (with the obvious notation):

PROPOSITION 4. *If $z = (z_1, \ldots, z_{q-p})$ and A are such that*

$$(3.11) \qquad \det\|\langle\xi^i, A_s z_j\rangle\| \geqslant 0$$

for any p-vector $\xi = (\xi^1, \dots, \xi^p)$ *then* $\hat{P}(z, A)$ *is a differential operator of order* $N = p(q-p)$.

REMARK. The function $\det\|\langle \xi^i, A_s z_j\rangle\|$ is a homogenous polynomial of degree $q-p$ in each of the vectors ξ^i and, consequently, the inequality (3.11) is possible only for even $q-p$.

5. DEFINITION OF THE OPERATOR χ. Let $\tilde{c} = (c, \gamma) \in \tilde{G}_{q-p}(V)$. We define the injective mapping

$$\theta_{\tilde{c}}: \tilde{G}_p(V/c) \to \tilde{G}_q(V)$$

given by the equality $\theta_{\tilde{c}}(a, \alpha) = (a_1, \alpha_1)$ where $a_1 \in G_q(V)$ is the preimage of $a \in G_p(V/c)$ and α_1 is defined by the equality $\alpha_1/\gamma = \alpha$. We shall associate with every function $\varphi \in F_q^p(V)$ and every $\tilde{c} \in \tilde{G}_{q-p}(V)$, the function $\varphi_{\tilde{c}} \in F_p^p(V/c)$ given by the equality

$$\varphi_{\tilde{c}}(\tilde{a}) = \varphi(\theta_{\tilde{c}}\tilde{a}) \quad \tilde{a} \in \tilde{G}_p(V/c). \tag{3.12}$$

Let $\mathscr{E}$ be the manifold of the pairs $(c, \tilde{b})$ we have introduced in Subsection 1, $c \in G_{q-p}(V)$, $\tilde{b} = (b, \beta) \in \tilde{G}_p(V)$ and $b \cap c = 0$. We construct the mappings $\rho: \mathscr{E} \to G_{q-p}(V)$ and $\pi: \mathscr{E} \to G_p(V)$ by the equalities $\rho(c, \tilde{b}) = c$, $\pi(c, \tilde{b}) = \tilde{b}$.

DEFINITION 2. Let $\varphi \in F_q^p(V)$. For any $\tilde{c} = (c, \gamma) \in \tilde{G}_{q-p}(V)$, $\tilde{b} = (b, \beta) \in \tilde{G}_p(V)$ with $b \cap c = 0$ and any $A = \{A_1, \dots, A_N\}$, $A_i \in T_{c,b}\mathscr{E}$, $N = p(q-p)$, we define:

$$(\chi\varphi)_{(c, \tilde{b}; A)} = (\hat{P}(\tilde{c}, \rho A)\, \varphi_{\tilde{c}})(b \oplus c/c, \beta \oplus \gamma/\gamma) \tag{3.13}$$

where $\rho A = \{\rho A_i\}$, $\rho A_i \in T_c G_{q-p}(V)$.

We point out that the right side is independent of the choice of the volume element γ in c, as when changing γ in $t\gamma$ $(t > 0)$, $\varphi_{\tilde{c}}$ is multiplied by t^p and $\hat{P}(\tilde{c}, \rho A)$ is multiplied by t^{-p}. It follows from the definition that $\chi\varphi$ is an N-density on $\mathscr{E}$. It is easy to verify that $(\chi\varphi)(c, (b, t\beta), A) = t^q(\chi\varphi)(c, (b, \beta), A)$; so χ determines a mapping

$$\chi: F_q^p(V) \to \Omega^N(\mathscr{E})$$

where $\Omega^N(\mathscr{E})$ is the space of N densities on $\mathscr{E}$ we defined in Subsection 1.

PROPOSITION 5. *χ is an intertwining operator.*

12 – 2294

Proof. We obtain χ as a composition of several mappings

$$\varphi \to \varphi_{\tilde{c}} \to \psi \to P_\psi \to \chi\varphi$$

where $\psi = R_p^\lambda(\tilde{V}/c)\varphi_{\tilde{c}}$. In the above sequence, each element is a function whose domain is a manifold on which the group $SL(V)$ acts naturally, for instance $\varphi_{\tilde{c}}$ is a function defined on the manifold of the pairs $(\tilde{c}, \tilde{a})$ where $\tilde{c} = (c, \gamma) \in \tilde{G}_{q-p}(V)$, $\tilde{a} = (a, \alpha) \in \tilde{G}_p(V/c)$, ψ is a function on the manifold of the pairs $(\tilde{c}, \tilde{h})$ where $\tilde{c} = (c, \gamma) \in \tilde{G}_{q-p}(V)$, $\tilde{h} = (h, \eta) \in \tilde{G}_p(\mathrm{Ann}\, c)$, and so on. From the definition of these mappings we see that on the respective manifold, they commute with the action of the group $SL(V)$.

DEFINITION 3. We shall denote by $(\chi\varphi)(c, A|\tilde{b})$ the restriction of the N-density φ at the fibre $\pi^{-1}(\tilde{b}) = G^b_{q-p}(V)$ of the bundle $\pi: \mathscr{E} \to G_p(V)$

$$(\chi\varphi)_{(c,A|\tilde{b})} = (\chi\varphi)_{(c,\tilde{b},A)}\Big|_{\pi^{-1}(b)} \tag{3.14}$$

(where $G^b_{q-p}(V)$ is the manifold of the subspaces $c \in G_{q-p}(V)$ which satisfy $c \cap b = 0$).

For a fixed $\tilde{b} \in \tilde{G}_p(V)$, $(\chi\varphi)_{(c,A,\tilde{b})}$ is a N-density on $G^b_{q-p}(V)$.

6. LEMMA FOR PASSING TO SUBSPACES $W \subset V$. In the definition of the operator χ, the framework was the vector space V. In order to stress this, hereafter we shall use χ^V instead of χ. Recall that the dimension of V satisfies $n = \dim V \geqslant p + q$.

LEMMA. *Let $W \subset V$ be an arbitrary subspace of V, $\dim W = m \geqslant p + q$ and et φ^W be the restriction of the function $\varphi \in F^p_q(V)$ at $G_q(W)$ (that means $\varphi^W \in F^p_q(W)$). Then for any triplet $(c, A, \tilde{b})$ where $c \in G_{q-p}(W)$, $A = \{A_i\}$, $A_i \in T_cG_{q-p}(W)$, $b \in \tilde{G}_p(W)$ we have*

$$(\chi^V\varphi)(c, A|\tilde{b}) = (\chi^W\varphi^W)(c, A|\tilde{b}). \tag{3.15}$$

Proof. We can assume without loss of generality that in the coordinate system of V, the space W is given by the equations $x^{m+1} = \ldots = x^n = 0$ and c is the space spanned by the vectors $v_1, \ldots, v_{q-p}$ where $v_i = (\delta^1_{i+p}, \ldots, \delta^n_{i+p})$ (δ^j_i — Kronecker's symbol). We identify V/c with the subspace spanned by $(x^1, \ldots, x^n)$ where $x^{p+1} = \ldots = x^q = 0$. We regard $F^q_p(V/c)$ and $F^{n-q}_p((V/c)')$ as function spaces respectively on the manifold of $p \times (n-q)$ matrices u and on the manifold of $(n-q) \times p$ matrices ξ (where u and ξ are local coordinates respectively for the Grassmannians $G_p(V/c)$ and $G_p((V/c)')$. We write them as block matrices:

$$u = (u_1, u_2), \quad \xi = \begin{pmatrix} \xi_1 \\ \xi_2 \end{pmatrix} \quad \text{where } u_1, \xi_1' \in M_{p,n-q}\,;\ u_2, \xi_2' \in M_{p,n-m}\,.$$

Considering this the Grassmannian $G_p(W/c) \subset G_p(V/c)$ is given by the equation $u_2 = 0$; thus $\varphi^W_{\tilde{c}}$ is a function of u_1, given by the equation

$$\varphi^W_{\tilde{c}}(u_1) = \varphi_{\tilde{c}}(u_1, 0).$$

Using the notation $\psi(\xi_1, \xi_2) = (R^p_p(\widetilde{V/c})\varphi_{\tilde{c}})(\xi_1, \xi_2)$, $\varphi^W(\xi_1) = (R^p_p(\widetilde{W/c})\varphi^W_{\tilde{c}})(\xi_1)$ we prove that

$$(3.16) \qquad \varphi^W(\xi_1) = \alpha \int \psi(\xi_1, \xi_2) d\xi_2 \quad \text{where } \alpha = \left. \frac{C_{p,m-q+p}(\lambda)}{C_{p,n-q+p}(\lambda)} \right|_{\lambda=-q}$$

and as usually $d\xi_2$ is $\prod d(\xi_2)_{ij}$. Indeed we have

$$\varphi_{\tilde{c}}(u_1, u_2) = C_{p,n-q+p}(\lambda) \int |I_p + u_1\xi_1 + u_2\xi_2|^\lambda \psi(\xi_1, \xi_2) d\xi_1 d\xi_2 \Big|_{\lambda=-q}$$

which gives us for $u_2 = 0$:

$$(3.17) \qquad \varphi^W_{\tilde{c}}(u_1) = C_{p,n-q+p}(\lambda) \int |I_p + u_1\xi_1|^\lambda \left(\int \psi(\xi_1, \xi_2) d\xi_2 \right) d\xi_1 \Big|_{\lambda=-q}.$$

On the other hand we have

$$(3.18) \qquad \psi^W_{\tilde{c}}(u_1) = C_{p,m-q+p}(\lambda) \int |I_p + u_1\xi_1|^\lambda \psi(\xi_1) d\xi_1 |_{\lambda=-q}.$$

Combining (3.17) and (3.18) we get the equality (3.16). We shall now proceed to prove the equality (3.15). Presuming that the assumptions of the lemma are fulfilled, it is easy to verify that $P(c, A|\xi_1, \xi_2)$ is independent of ξ_2 and $P(c, A|\xi_1, \xi_2) = P^W(c, A, \xi_1)$. Next, $b \in G_p(V)$ is given by the matrix $v = (v_1, v_2)$ and the condition $b \in G_p(W)$ becomes $v_2 = 0$. We find that

$$(\chi^V\varphi)(c, A|\tilde{b}) = C_{p,n-q+p}(\lambda) \int |I_p + u_1\xi_1|^\lambda P^W(c, A|\xi_1)\psi(\xi_1, \xi_2) d\xi_1 d\xi_2 \Big|_{\lambda=-q}.$$

(3.16) shows us that the right side is equal to $(\chi^W\varphi^W)(c, A|\tilde{b})$ and the proof is ended.

4. THE INVERSION FORMULA FOR THE OPERATOR $\mathscr{I}$

1. Main Lemma. We shall denote by $\tilde{\mathscr{B}}$ the manifold of the pairs $(c, \tilde{h})$ where $c \in G_{q-p}(V)$, $\tilde{h} = (h, \eta) \in \tilde{G}_p(V')$ and $c \perp h$. We shall associate with each

function $\varphi \in F_q(V)$ the function $\tilde{\varphi}$ on $\tilde{\mathscr{B}}$ given yb the following equality

$$\tilde{\varphi}(c, \tilde{h}) = (R_p^p(V/c, \omega/\gamma)\varphi_{c,\gamma})(\tilde{h}) . \tag{4.1}$$

Here γ is a fixed volume element in c and $\varphi_{c,\gamma} \in F_p^p(V/c)$ is given by the equality (3.12) (easy computation shows us that the right side is independent of the choice of γ).

LEMMA. *If $\varphi = \mathscr{I}f$ where $f \in F_p^q(V)$, then for any $(c, \tilde{h}) \in \tilde{\mathscr{B}}$ we have*

$$\tilde{\varphi}(c, \tilde{h}) = \alpha_{p,q}(R_p^p(V, \omega)f)(\tilde{h}) \tag{4.2}$$

where $\alpha_{p,q} = \pi^{p(q-p)}\Gamma_p(p/2)/\Gamma_p(q/2)$.

Proof. It suffices to show that the statement of the lemma is true for a certain fixed $c = c_0$. On the manifolds $G_p(V)$, $G_q(V)$ and $G_p(V')$ we respectively introduce local coordinates $u = (u_1, u_2)$, $v = \begin{pmatrix} v_1 \\ v_2 \end{pmatrix}$ and $\xi = \begin{pmatrix} \xi_1 \\ \xi_2 \end{pmatrix}$ where $u_1, \xi_1' \in M_{p,q-p}$; $u_2, \xi_2' \in M_{p,n-q}$, $v_1 \in M_{p,n-q}$, $v_2 \in M_{q-p,n-q}$ and we regard $F_p^q(V)$, $F_q^p(V)$ and $F_p^{n-q}(V')$ as function spaces respectively on the manifolds of the matrices u, v, ξ (see Subsection 2 of Section 1).

In this interpretation the function $\tilde{f} = R_p^q(\tilde{V})f$ is given by the equality

$$\tilde{f}(\xi_1, \xi_2) = C_{p,n}(\lambda - n)\int |I_p + u_1\xi_1 + u_2\xi_2|^{\lambda-n} f(u_1, u_2)\, du_1\, du_2\Big|_{\lambda=q} \tag{4.3}$$

(see Subsection 2 of Section 2). Next we give $c \in G_{q-p}(V)$ using the $(q-p)\times n$ matrix $c = (a, I_{q-p}, b)$ where $a \in M_{q-p,p}$, $b \in M_{q-p,n-q}$. Suppose $c_0 = (0, I_{q-p}, 0)$; then the orthogonality condition for $\xi = \begin{pmatrix} \xi_1 \\ \xi_2 \end{pmatrix}$ and c_0 is $\xi_1 = 0$. It follows from the definition of $\tilde{\varphi}(c, \tilde{h})$ that for $c = c_0$, $\tilde{\varphi}$ is a function of ξ_2 given by

$$\tilde{\varphi}(b, \xi_2) = C_{p,n-(q-p)}(\lambda - n)\int |I_p + v_1\xi_2|^{\lambda-n} \varphi\begin{pmatrix} v_1 \\ 0 \end{pmatrix} dv_1\Big|_{\lambda=q} . \tag{4.4}$$

Combining this with the formula of φ as function of f (see 1.6) we get

$$\tilde{\varphi}(c_0, \xi_2) = C_{p,n-(q-p)}(\lambda - n)\int |I_p + v_1\xi_2|^{\lambda-n} f(t, v_1)\, dt\, dv_1\Big|_{\lambda=q} . \tag{4.5}$$

Therefore $\tilde{\varphi}(c_0, \xi_2) = \alpha_{p,q}\tilde{f}(0, \xi_2)$ where

$$\alpha_{p,q} = \frac{C_{p,n-(q-p)}(\lambda - n)}{C_{p,n}(\lambda - n)}\Big|_{\lambda=q} = \pi^{p(q-p)}\frac{\Gamma_p(p/2)}{\Gamma_p(q/2)} .$$

Hence our lemma is proved.

COROLLARY. *The image of the operator $\mathscr{I}$, $\operatorname{Im}\mathscr{I}$, is the set of those functions $\varphi \in F_q^p(V)$ which satisfy the condition $\tilde{\varphi}(c_1, \tilde{h}) = \tilde{\varphi}(c_2, \tilde{h})$ for any $(c_1, \tilde{h})$ and $(c_2, \tilde{h})$ of $\tilde{\mathscr{B}}$.*

NOTE. An equivalent characterization for $\operatorname{Im}\mathscr{I}$ was given in [6].

2. THE INVERSION FORMULA. Let $C \subset G_{q-p}(V)$ be a $N = p(q-p)$ dimensional submanifold. By the Crofton symbol of C we shall mean the function $\mathrm{Cr}_C(h)$ on $G_p(V')$ which assigns to h the number of subspaces $c \in C$ which are orthogonal to h. It is easy to see that Cr_C is almost everywhere finite. C will be called non-degenerate if supp Cr_C contains an open subset and will be called non-singular if $\operatorname{supp}\mathrm{Cr}_C = G_p(V')$; we say C is harmonic if $\mathrm{Cr}_C(h) = \text{const} \neq 0$ almost everywhere. We shall denote this contains by Cr_C and shall call it the Crofton number. We shall denote by $\widehat{\mathrm{Cr}}_C$ the operator $\widehat{\mathrm{Cr}}_C : F_p^q(V) \to F_p^q(V)$ given by the symbol $\mathrm{Cr}_C(h)$ (see Subsection 4 of Section 3). We notice that if C is harmonic then $\widehat{\mathrm{Cr}}_C = \mathrm{Cr}(C) \cdot E$ where E is the identity operator.

THEOREM. *Let $\tilde{b} = (b, \beta) \in \tilde{G}_p(V)$ and C a non-singular submanifold of $G_{q-p}^b(V)$ whose dimension is $N = p(q-p)$. Then if $f \in F_p^q(V)$ and $\varphi = \mathscr{I}f$*

$$\int_C (\chi\varphi)(\cdot\,|\tilde{b}) = (\widehat{\mathrm{Cr}}_C f)(\tilde{b}). \tag{4.6}$$

In particular, if C is a harmonic submanifold, then

$$\int_C (\chi\varphi)(\cdot\,|\tilde{b}) = \mathrm{Cr}(C)\cdot f(\tilde{b}). \tag{4.7}$$

3. PROOF OF THE THEOREM. According to the definition of $\widehat{\mathrm{Cr}}_C$, we have $\widehat{\mathrm{Cr}}_C f = R_p^{n-q}(\tilde{V}')(\mathrm{Cr}_C \tilde{f})$ where $\tilde{f} = R_p^q(\tilde{V})f$, that is

$$(\widehat{\mathrm{Cr}}_C f)(\tilde{b}) = \int_{G_p(V')} U^{(-q)}(\tilde{b}, \tilde{h})\mathrm{Cr}_C(h)\tilde{f}(\tilde{h})\sigma_{\tilde{V}'}(\tilde{h}). \tag{4.8}$$

We shall consider the submanifold $\mathscr{B}_c \subset \mathscr{B}$ consisting of the pairs (c, h) where $c \in C$, $h \in G_p(V')$ and $h \perp c$. We define the projections $\pi_1 : \mathscr{B}_c \to G_p(V')$ and $\pi_2 : \mathscr{B}_c \to C$ by $\pi_1(c, h) = h$ and $\pi_2(c, h) = c$. Let $U = \pi_1(\mathscr{B}_c)$; then we can consider that in (4.8) the integration is done on U. For almost every point $h \in U$, the preimage $\pi_1^{-1}(h)$ consists of a finite number of points $\mathrm{Cr}_C(h) \neq 0$; therefore (4.8) can be represented as an integral on the manifold $\mathscr{B}_c$:

$$(\widehat{\mathrm{Cr}}_C f)(\tilde{b}) = \int_{\mathscr{B}_c} \pi_1^*(U^{(-q)}(\tilde{b}, \tilde{h})\tilde{f}(\tilde{h})\sigma_{\tilde{V}'}(\tilde{h})). \tag{4.9}$$

According to the Main Lemma we have $\tilde{f}(\tilde{h}) = \alpha_{p,q}^{-1}\tilde{\varphi}(c, h)$, where $c \in C$ is an arbitrary element orthogonal to h. Hence

$$(4.10) \qquad (\mathrm{Cr}_C f)(\tilde{b}) = \alpha_{p,q}^{-1} \int_{\mathscr{B}_c} \pi_1^*(U^{(-q)}(\tilde{b}, \tilde{h})\tilde{\varphi}(c, h)\sigma_{\tilde{V}'}(\tilde{h})).$$

We shall compute this integral as an iterated integral in which we first integrate on the fiber and then on the basis of the bundle $\pi_2 : \mathscr{B}_c \to C$. According to (2.1) we have:

$$(\widehat{\mathrm{Cr}}_C f)(\tilde{b}) = \alpha_{p,q}^{-1} \int_C \Big(\int_{\pi_2^{-1}(c)} U^{(-q)}(\tilde{b}, \tilde{h}) P(c, A|\tilde{h})\tilde{\varphi}(c, \tilde{h})\sigma_{(\tilde{V}/c)'}(\tilde{h}) \Big).$$

It follows from the definition of the operator χ that the first integral is equal to $\alpha_{p,q}(\chi\varphi)(c, A|\tilde{b})$. So

$$(\widehat{\mathrm{Cr}}_C f)(\tilde{b}) = \int_C (\chi\varphi)(\cdot|\tilde{b}).$$

NOTE. For each harmonic submanifold there is a corresponding application

$$\sigma : \Omega^N(\mathscr{E}) \to F_p^q(V)$$

(where $F_p^q(V)$ is a certain extension of $F_p^q(V)$), defined by the equality

$$\sigma(\tau)(\tilde{b}) = \int_C \tau(\cdot|\tilde{b}).$$

This application is not an intertwining operator on the whole $\Omega^N(\mathscr{E})$ but according to the proof, its restriction at the subspace $\chi\mathscr{I}(F_p^q) \subset \Omega^N(\mathscr{E})$ is an intertwining operator.

REMARK. In the same manner, for any $k \leqslant q - p$ we can define the operator

$$\chi_k : \mathrm{Im}\,\mathscr{I} \to \Omega^{p \cdot k}(\mathscr{E}_k)$$

where $\mathscr{E}_k$ — the manifold of the pairs $(c_k, \tilde{b})$, $c_k \in G_k(V)$ and $\tilde{b} \in G_p(V)$.

That is, let $\varphi \in \mathrm{Im}\,\mathscr{I}$ and $\tilde{\varphi}(c, \tilde{h})$ the function on $\mathscr{B}$ given by (4.1). We define the function $\tilde{\varphi}_k(c_k, \tilde{h})$ on the manifold of the pairs $(c_k, \tilde{h})$ where $c \in G_k(V)$, $\tilde{h} \in \tilde{G}_p(V')$ and $c_k \perp h$, by the equality

$$\tilde{\varphi}_k(c_k, \tilde{h}) = \tilde{\varphi}(c, \tilde{h})$$

where $c \in G_{q-p}(V)$ is a certain subspace satisfying $c_k \subset c$, $c \perp h$. According to the Main Lemma the given definition is correct. By definition

$$\chi_k \varphi = R_p^{q-n-k}(\mathrm{Ann}\, c_k, (\omega/\gamma_k)')(P_k(\tilde{c}_k, A|\tilde{h})\tilde{\varphi}(c_k, \tilde{h}))$$

where P_k is given by (2.1).

We underline that for $k < q-p$ the operators are defined on the subspace $\mathrm{Im}\,\mathscr{I} \subset F_q^p$.

Using the operators χ_k we can construct other inversion formulas for the operator $\mathscr{I}$.

APPENDIX. ON THE OPERATOR $\varkappa$. For any $\tilde{b} \in \tilde{G}_p(V)$ there was defined in [6] the operator $\varkappa_{\tilde{b}}$ which associates with a function $\varphi \in F_q^p(V)$ a N-differential form on $G_{q-p}^b(V)$, $(N = p(q-p))$. It was proved that if $\varphi = \mathscr{I}f$, $f \in F_p^q(V)$, then $\varkappa_{\tilde{b}}\varphi$ is a closed form on $G_{q-p}^b(V)$ and we have for any N-dimensional cycle $\gamma \subset G_{q-p}^b(V)$:

$$\int_\gamma \varkappa_b \varphi = C_\gamma f(\tilde{b}) \tag{4.11}$$

where C_γ is depending only on the homology class of γ. Moreover, for even $q-p$ there are cycles γ for which $C_\gamma \neq 0$ and for these cycles (4.11) gives the inversion formula. There are no such cycles for odd $q-p$.

The operator $\varkappa$ can be defined analogously to χ as intertwining operator. In order to do that, we must replace, in the definitions given in this paper, the non--oriented volume element by an oriented volume element. Therefore we shall regard $\tilde{G}_p(V)$ as the manifold of the pairs (b, β) where $b \in G_p(V)$ and β is an oriented volume element in b. Instead of $F_p^\lambda(V)$ we must introduce a larger class of spaces, $F_p^{\lambda,\varepsilon}(V)$ $(\varepsilon = 0, 1)$ the spaces of C^∞ functions on $\tilde{G}_p(V)$ satisfying the condition $f(b, t\beta) = |t|^\lambda \operatorname{sgn}^\varepsilon t \cdot f(b, \beta)$ for any $t \neq 0$ (we notice that $F_p^{\lambda,0} \simeq F_p^\lambda$). If in the definition of χ we change $P(\tilde{c}, A|\tilde{h}) = |\det\|\langle \eta^i, A_s v_j\rangle\||$ with $Q(\tilde{c}, A|h) = \det\|\langle \eta^i, A_s v_j\rangle\|$ then instead of χ we get the intertwining operator

$$\varkappa\colon F_q^p(V) \to \Omega_0^N(\mathscr{E})$$

where $\mathscr{E}$ is a bundle over $\tilde{G}_p(V)$ with the fibre at $\tilde{b} = (b, \beta) \in \tilde{G}_p(V)$ equal to $G_{q-p}^b(V)$, and $\Omega_0^N(\mathscr{E})$ is the space of the N-differential forms which satisfy the condition $\tau(c, (b, t\beta)) = |t|^q\, \tau(c, (b, \beta))$ for any $t \neq 0$.

It can be shown, by passing to coordinates, that the restriction of $\varkappa_{\tilde{b}}\varphi$ of the differential form $\varkappa\varphi$, $\varphi \in F_q^p(V)$ at the fiber $\pi^{-1}(b)$, $\pi\colon \mathscr{E} \to \tilde{G}_p(V)$ coincides with the differential form introduced in [6].

5. PERMISSIBLE COMPLEXES

1. Definition of the Permissible Complex. As for $p < q < n - p$, $\dim G_p(V) < \dim G_q(V)$, in this case, knowing the values of $\mathscr{I}f$ on the whole manifold $G_q(V)$ is over-sufficient for recovering the function f; we can assume that the function $\mathscr{I}f$ is given only on a certain submanifold $K \subset G_q(V)$ whose dimension is $k \geqslant \dim G_p(V)$. There arises the problem of recovering the function f by means of the function $\mathscr{I}_K f = \mathscr{I}f \mid K$.

We introduce a class of submanifolds $K \subset G_q(V)$ for which $\mathscr{I}_K$ is one to one and the inversion formula for $\mathscr{I}_K$ follows immediately from the inversion formula for $\mathscr{I}$ we obtained in Section 4.

Definition. The submanifold $K \subset G_q(V)$ will be called a *permissible complex* (more exactly : a *p-permissible complex*) if for almost every $b \in G_p(V)$ there is a harmonic submanifold $C_b \subset G^b_{q-p}(V)$ such as if $\varphi|K = 0$ then $(\chi\varphi)(\cdot\,|\,\tilde{b})|C_b = 0$ for any $\varphi \in F^p_q(V)$.

In other words the restriction of the density $(\chi\varphi)(\cdot\,|\tilde{b})$ at C_b is uniquely determined by the restriction of φ at K. It is obvious that in the case of a permissible complex K, the function $f \in F^q_p(V)$ is computed from its image $\varphi = \mathscr{I}_K f$ by means of the inversion formula (4.7) for $C = C_b$.

Note. By replacing in the definition the operator χ by the operator $\varkappa$ and the harmonic manifold by the $p(q-p)$ cycle in $G^b_{q-p}(V)$ we obtain the notion of permissible complex earlier introduced for complex vector spaces and whose complete characterization is given in [3] and [9] for $p = 1$, $q = 2$.

2. Examples of Permissible Complexes. 1°. *The complex* K_C. We associate with each harmonic submanifold $C \subset G_{q-p}(V)$ the complex $K_C \subset G_q(V)$ consisting of the subspaces $a \in G_q(V)$ which contain at least one subspace $c \in C$. We notice that $\dim K_C = \dim G_p(V) = p(n-p)$. Indeed the set of the subspaces $a \in G_q(V)$ which contain a fixed subspace $c \in C$, is equivalent to $G_p(V/c)$ and therefore its dimension is $p(n-q)$. It follows that $\dim K_C = p(q-p) + p(n-q) = p(n-p)$. Obviously, the complex K_C is permissible.

We shall give an example of a K_C complex for $p = 1$, $q = 3$. We consider a pair of lines l_1, l_2 in $\mathbf{P}^{n-1}$ situated in general position. We denote by C the set of the lines in $\mathbf{P}^{n-1}$ which intersect the given lines. It is obvious that C is a harmonic manifold with the Crofton symbol 1. The complex K_C consists of the set of 2-dimensional planes of $\mathbf{P}^{n-1}$ which intersect the two lines l_1, l_2.

2° *Radon complexes*. The manifold $K \subset G_q(V)$ will be called a Radon complex (more exactely: Radon p-complex) if for almost every $b \in G_p(V)$ there is a linear subspace $W_b \subset V$ such that $b \subset W_b$, $\dim W_b \geqslant p + q$ and $G_q(W_b) \subset K$. It follows directly from the lemma in Subsection 6 of Section 3, that any Radon complex is permissible.

Remark. In the definition of the Radon complexes, we can assume that $\dim W_b = p + q$. Then the inversion formula for $\mathscr{I}_K$ is reduced at the inversion formula for the Radon transform in the projective space.

We shall give an example of Radon complex for $p = 1$, $q = 2$. Consider in $\mathbf{P}^{n-1}$ an arbitrary $(n - 3)$-dimensional family of 2-dimensional planes which contains all the points of $\mathbf{P}^{n-1}$. The complex K consists of the set of all the lines which are contained in at least one of the planes of the considered family. It is obvious that, generally, this complex K is not a K_C complex. At the same time, the complex of 2-dimensional planes in Example 1°, is not a Radon complex.

3. $\mathscr{I}$-complexes. We shall introduce a class of permissible complexes $K \subset G_q(V)$ which contains both the K_C complexes and the Radon complexes.

Definition. The submanifold $K \subset G_q(V)$ will be called a *$\mathscr{I}$-complex* if for almost every $b \in G_p(V)$ there is a subspace $W_b \subset V$ of dimension $\dim W_b \geqslant p + q$ and a harmonic manifold $C_b \subset G_{q-p}(W_b)$ such that $b \subset W_b$ and $K_{C_b} \cap G_q(W_b) \subset K$.

It follows immediately from the lemma in Subsection 6 of Section 3 that any $\mathscr{I}$-complex is permissible and, moreover, the inversion formula is (4.7) for $C = C_b$. It is also obvious that both the K_C complexes and the Radon complexes are $\mathscr{I}$-complexes.

We shall now construct an example of a $\mathscr{I}$-complex which is neither a K_C complex nor a Radon complex. Let $p = 1$, $q = 3$. Consider in $\mathbf{P}^{n-1}$ an arbitrary one-dimensional family π_t of $(n - 2)$-dimensional planes which contain all the points of $\mathbf{P}^{n-1}$. In each plane π_t we fix an arbitrary pair of lines l_t^1, l_t^2 situated in general position. The complex K consists of all the 2-dimensional planes in $\mathbf{P}^{n-1}$ contained in at least one of the planes π_t and which intersect each of the corresponding lines l_t^1 and l_t^2. We notice that $\dim K = n - 1$.

4. The Integral Transform $\mathscr{I}_K$ for an Arbitrary Complex K_C. Let $C \subset G_{q-p}(V)$ be an arbitrary submanifold, not necessary harmonic, K_C the manifold of those $a \in G_q(V)$ which contain at least one subspace $c \in C$. Generally, the complex $K = K_C$ is not permissible and $\operatorname{Ker} \mathscr{I}_K$, the kernel of the associated integral transform $\mathscr{I}_K$, can be non-zero. We shall compute this kernel. Let $M_C = \{h \in G_p(V') \mid \mathrm{Cr}_C(h) = 0\}$.

Proposition 1. $\operatorname{Ker} \mathscr{I}_K$, $K = K_C$, *consists of the set of those functions* $f \in F_p^q(V)$ *for which* $\operatorname{supp}(R_p^q(\tilde{V})f) \subset \tilde{M}_C$. *In particular, the mapping* $\mathscr{I}_K$ *is one to one if and only if C is non-degenerated, that is,* $\mathrm{Cr}_C(h) \neq 0$ *almost everywhere.*

Proof. Let $f \in F_p^q(V)$, $\varphi = \mathscr{I}_K f$ and $\tilde{f} = R_p^q(V)f$. We shall assume without any confusion, that the function $\tilde{f}$ is given as a function on $G_p(V')$ and not on the bundle $\tilde{G}_p(V')$ over $G_p(V')$. We denote by G_p^1 the set of the subspaces $h \in G_p(V)$ for which $\mathrm{Cr}_C(h) \neq 0$. By the assumptions in the lemma and by the fact that the mapping

$\varphi \to \tilde{\varphi}(c, \tilde{h}) = (R_p^p(\widetilde{V}/c)\varphi_{\tilde{c}})(h)$ is one to one, it follows that

$$\varphi \equiv 0 \Leftrightarrow \tilde{f}|\overline{G_p^1} \equiv 0$$

where $\overline{G_p^1}$ is the closure of G_p^1. Hence our statement follows immediately.

Let now $\mathscr{B}_c \subset \mathscr{B}$ be the manifold of the pairs $(c, \tilde{h}) \in \mathscr{B}$ where $c \in C$. We notice that for $c \in C$, $\tilde{\varphi}(c, \tilde{h})$ is determined only by the restriction $\varphi = \mathscr{I}_K f$ of the function $\mathscr{I} f$ at the complex K. Therefore by the results in Subsection 1 of Section 4 we have:

Proposition 2. Im $\mathscr{I}_K$, *the image of the application* $\mathscr{I}_K$, *consists of the set of those* C^∞ *functions* φ *on* K *which satisfy* $\tilde{\varphi}(c_1, \tilde{h}) = \tilde{\varphi}(c_2, \tilde{h})$ *for any* $(c_1, \tilde{h})$ *and* $(c_2, \tilde{h})$ *in* $\mathscr{B}_c$.

REFERENCES

1. Gelfand, I. M.; Gindikin, S. G., Non-local inversion formulas in integral geometry in real spaces (Russian), *Funktsional. Anal. i Priložen.*, **11**:3(1977), 12–19.
2. Gelfand, I. M.; Gindikin, S. G.; Graev, M. I., Integral geometry in affine and projective spaces (Russian), *Actual problems in mathematics, VINITI Moscow*, **16**(1980), 53–226.
3. Gelfand, I. M.; Graev, M. I., Line complexes in the spaces $\mathbf{C}^n$ (Russian), *Funktsional. Anal. i Priložen.*, **2**:3(1968), 39–52.
4. Gelfand, I. M.; Graev, M. I.; Roşu, R., The problem of integral geometry for p-dimensional planes in real projective space (the nonlocal variant), in *Operator algebras and group representations*, vol. I, Pitman, 1983, 192–207.
5. Gelfand, I. M.; Graev, M. I.; Šapiro, Z. Ja., Integral geometry in a projective space (Russian), *Funktsional. Anal. i Priložen.*, **4**:1(1970), 14–32.
6. Gelfand, I. M.; Graev, M. I.; Šapiro, Z. Ja., A problem of integral geometry for a pair of Grassmann manifolds (Russian), *Dokl. Akad. Nauk SSSR*, **193**:2(1970), 892–895.
7. Gelfand, I. M.; Šmelov, G. S., Geometrical structures of double fibrations and their connections with problems of integral geometry (Russian), *Funktsional. Anal. i Priložen.*, **17**:2(1983), 7–23.
8. Gindikin, S. G., Unitary representations of the group of automorphisms of the symmetric Riemannian spaces with null curvature (Russian), *Funktsional. Anal. i Priložen.*, **1**:1(1967), 32–37.
9. Mályusz, K., The structure of permissible line complexes in $\mathbf{CP}^n$ (Russian), *Trudy Moscow Mat. Obšč.*, **39**(1979), 181–211.
10. Petrov, E. E., The Radon transform in matrix spaces (Russian), *Proceedings of Vectorial and Tensorial Analysis, Seminar M.G.U.*, Moscow, vol. 15 (1970), 299–315.
11. Raïs, M., Distributions homogènes des espaces de matrices, *Bull. Soc. Math. France Mem.*, **30**(1972).

12. Šifrin, M. A., Non-local inversion formula for the problem of integral geometry for a pair of real grassmannian manifolds (affine variant) (Russian), Preprint I.P.M., Moscow, no. **91**(1981).

I. M. GELFAND
Laboratory of Mathematical Methods in Biology, Corpus A, State University of Moscow, Moscow V—234 U.R.S.S.

M. I. GRAEV
Keldysh Institute of Applied Mathematics, Minsskaya sq. 4, Moscow A — 47, U.R.S.S.

R. ROŞU
Institute of Mathematics, University of Bucharest, Academiei 14, Bucharest, Romania

Received December 12 1983.

14.

(with A. B. Goncharov)

On a characterization of Grassmann manifolds

Dokl. Akad. Nauk SSSR **289** (5) (1986) 1047–1052
[Sov. Math., Dokl. **34** (1987) 189–193]

1. Statement and discussion of results. We will say that a family of k-dimensional submanifolds of an n-dimensional manifold M *satisfies condition* A if for every point $x \in M$ each K-dimensional subspace of T_xM is the tangent space to exactly one of the submanifolds of the family.

THEOREM A. *In the category of complex-analytic manifolds a family of k-dimensional submanifolds with $k > 1$ of an n-dimensional manifold is locally isomorphic to the family of all k-dimensional planes in $\mathbf{CP}^n$ if and only if condition* A *holds for this family.*

Of course, the necessity of condition A is obvious.

The analogous theorem in the category of real-analytic manifolds is false. We will give a counterexample at the end of §2.

Already the special case of this theorem for surfaces in three-dimensional space was a surprise to us. Namely for $k = 2$ and $n = 3$ Theorem A reduces to the following

ASSERTION. *A three-parameter family of holomorphic surfaces in a three-dimensional domain U is (locally) biholomorphic to the family of all planes in $\mathbf{CP}^3$ if and only if at every point $x \in U$ each two-dimensional subspace of T_xU is tangent to exactly one of the surfaces.*

The analogous result for $k = 1$ is false, because condition A holds for families of geodesics of any projective connection (and only for them [1]).

We note that in the case $k = 1$, $n = 2$ there is a classical criterion going back to Tresse [2] for determining whether a family of curves in the plane can be straightened (see also [1] and [3]). To formulate it we need the concept of a dual family. Namely, if a family of curves in the domain of (x_1, x_2) depending on parameters ξ_1, ξ_2 is given by an equation $\Phi(x_1, x_2; \xi_1, \xi_2) = 0$, then by viewing x_1 and x_2 as parameters we obtain in the domain of (ξ_1, ξ_2) a family of curves dual to the original one.

THEOREM [1]–[4]. *A two-parameter family of holomorphic curves in a two-dimensional manifold is (locally) isomorphic to the family of lines in the plane if and only if condition* A *holds both for the original and for the dual family.*

Below we will show that the difference between the statement of this theorem and the above Assertion is due to the fact that the "theorem" of Desargues is one of the axioms of projective geometry in the plane and is a corollary of them in space.

2. The main theorem in the language of differential equations. Let $\mathrm{Gr}_k(T_xM)$ be the manifold of k-dimensional subspaces of T_xM. We remark that $\mathrm{Gr}_k(T_xM)$ is the manifold of 1-jets of k-dimensional submanifolds at x. We denote by $E^{(x)}$ the manifold of 2-jets of k-dimensional submanifolds at x. By associating to a 2-jet its 1-jet we obtain

1980 *Mathematics Subject Classification* (1985 *Revision*). Primary 58A99; Secondary 43A85, 51A05, 32C25.

a (vector) bundle $\pi\colon E^{(x)} \to \mathrm{Gr}_k(T_xM)$. Its fiber over the subspace $h \in \mathrm{Gr}_k(T_xM)$ is canonically isomorphic to $S^2(h)^* \otimes T_xM/h$.

We denote by Γ_x the collection of manifolds of a given family passing through the point x. We define a mapping $s\colon \Gamma_x \to E^{(x)}$ by associating to a submanifold of the family its 2-jet at the point x. Condition A means precisely that $\pi \circ s\colon \Gamma_x \to \mathrm{Gr}_k(T_xM)$ is an isomorphism; that is, $s(\Gamma_x) \subset (\mathrm{Gr}_k(T_xM), E^{(x)})$. The $(k+1)(n-k)$-parameter properties of k-dimensional submanifolds in general position are conveniently described in the language of differential equations. Namely, suppose a family is given by the equations $x^{k+i} = \Phi^i(x^\alpha, \xi_s)$, where $1 \le s \le (k+1)(n-k)$, $1 \le i \le n-k$, and $1 \le \alpha \le k$. We add to them the equations

$$\frac{\partial x^{k+i}}{\partial x^\beta} = \frac{\partial \Phi^i}{\partial x^\beta}(x^\alpha, \xi_x)$$

and from the resulting system we find the ξ_s as functions of $p^j_\beta = \partial x^{k+j}/\partial x^\beta$. Substituting them in the equation

$$\frac{\partial^2 x^{k+i}}{\partial x^{\beta_1} \partial x^{\beta_2}} = \frac{\partial^2 \Phi^i}{\partial x^{\beta_1} \partial x^{\beta_2}}(x^\alpha, \xi_s),$$

we obtain a consistent system of differential equations of the form

$$\text{(1)} \qquad \frac{\partial^2 x^{k+i}}{\partial x^{\beta_1} \partial x^{\beta_2}} = f^i_{\beta_1\beta_2}(x^\alpha, p^j_\beta).$$

Conversely, the solutions of a consistent system of differential equations (1) of the form constitute a $(k+1)(n-k)$-parameter family of k-dimensional manifolds.

LEMMA 1. *In the category of complex-analytic manifolds, if a family of k-dimensional submanifolds satisfies condition* A, *then the system of the form* (1) *whose solutions are the manifolds of the family satisfies the following*

CONDITION B. *In any system of coordinates on M the functions $F^i_{\beta_1\beta_2}$ are polynomials with respect to p^j_β.*

PROOF. By the GAGA principle,* $\Gamma(\mathrm{Gr}_k(T_xM), E^{(x)})$ consists of algebraic sections.

Theorem A is a consequence of the following theorem, which holds even in the category of C^∞-manifolds.

THEOREM B. *When $k > 1$ a consistent system of differential equations* (1) *for which condition* B *holds can be reduced, by a change of dependent and independent variables, to the form*

$$\frac{\partial^2 x^{k+i}}{\partial x^\alpha \partial x^\beta} = 0.$$

LEMMA 2. $\Gamma(\mathrm{Gr}_k(T_xM), E^{(x)}) \cong \ker f$, *where $f\colon S^2(T_xM)^* \otimes T_xM \to T_xM^*$ is the natural mapping.*

Using Lemma 2 we can find an explicit form of systems (1) satisfying condition B. For $k = 1$ and $n = 2$ we obtain the equation

$$\text{(2)} \qquad y''_{xx} = a_0 + a_1 y'_x + a_2 (y'_x)^2 + a_3 (y'_x)^3, \qquad a_i = a_i(x, y).$$

For $k = 2$ and $n = 3$ we obtain the system

$$\text{(3)} \qquad \begin{aligned} z''_{xx} &= p^2 L_0 + 2pL_1 + L_3, \quad p = z'_x, \quad q = z'^{*}_y, \\ z''_{xy} &= pqL_0 + pL_2 + qL_1 + L_4, \qquad L_i = a^0_i = a'_i p + a^2_i q, \\ z''_{yy} &= q^2 L_0 + 2qL_2 + L_5, \qquad a^j_i = a^j_i(x, y, z). \end{aligned}$$

---

* *Translator's note.* This principle, named for Serre's paper "Géométrie Algébrique et Géométrie Analytique," says that a global analytic object on an algebraic variety is algebraic.

COROLLARY 1. *A three-parameter family of surfaces over* $\mathbf{R}$ *or* $\mathbf{C}$ *is locally diffeomorphic to the family of all planes in the space* $\mathbf{RP}^3$ *or* $\mathbf{CP}^3$ *respectively if and only if the corresponding system of differential equations in some (and hence in every) coordinate system has the form* (3).

We now give an example of a three-parameter family of real surfaces in $\mathbf{R}^3$ satisfying condition A for which the corresponding system of differential equations of the form (1) does not satisfy condition B. Let f be a nonconstant analytic function on $\mathbf{RP}^2$. We consider the system $z''_{xx} = z''_{xy} = z''_{yy} = f(p,q)$ and also the two systems obtained from it by considering x as a function of y and z or y as a function of x and z. Obviously each of these three systems is consistent. We define a three-parameter family of surfaces in the space of (x,y,z) as the family of surfaces each of which is the graph of a solution of at least one of these systems. Thus this family sataisfies condition A, but it does not satisfy condition B because $f(p,q)$ is not a polynomial with respect to p and q. Therefore it is not even locally diffeomorphic to the family of all planes in $\mathbf{RP}^3$. Consequently Theorem A does not hold in the category of real manifolds.

3. The inequivalence of dependent and independent variables in the machinery of partial derivatives does not allow one to see why equation (2) and system (3) (and the corresponding systems for all k and n) preserve their form under diffeomorphisms of M.

Below we will describe system (1) in a more symmetric, parametric form. Passing to this form has the same advantages as passing from affine to homogeneous coordinates. To avoid encumbering the exposition with superflous details, we consider the typical example $k=2, n=3$. At the end we give the formulas in general form.

Let x^1, x^2, and x^3 be functions of the parameters t_1 and t_2. Let

$$\Delta_k(x) = (-1)^\sigma D\begin{pmatrix} x^i, & x^j \\ t_1, & t_2 \end{pmatrix},$$

where σ is the sign of the permutation

$$\begin{pmatrix} 1 & 2 & 3 \\ k & i & j \end{pmatrix}$$

and

$$D\begin{pmatrix} x^i, & x^j \\ t_1, & t_2 \end{pmatrix}$$

is the Jacobian. We set

$$A_{\alpha\beta}(x) = x^s_{,\alpha\beta}\Delta_s(x),$$

where $1 \le \alpha, \beta \le 2$ and $x^s_{,\alpha} = \partial x^s/\partial t_\alpha$ and so on.

When $x^1 = t_1$ and $x^2 = t_2$ the quantities $A_{\alpha\beta}(x)$ reduce to $\partial^2 x^3/\partial x^\alpha \partial x^\beta$. Their advantage is that under diffeomorphisms of the domain of the parameters t_1 and t_2 they transform like tensor fields.

It is easy to show that

$$A_{\alpha,\beta}(x) = \Delta_3(x)\frac{\partial^2 x^3}{\partial x^{s_1}\partial x^{s_2}} x^{s_1}_{\ ,\alpha}\cdot x^{s_2}_{\ ,\beta}, \qquad 1 \le s_1, s_2, \le 2.$$

Therefore system (1) can be written in the form

$$A_{\alpha\beta}(x) = \Phi_{\alpha\beta}(x; x^i_{\ ,\gamma}), \qquad 1 \le i \le 3,\ 1 \le \gamma \le 2.$$

In particular, system (3) in parametric form looks like this:

(4) $$A_{\alpha\beta}(x) = c^k_{ij}(x)\cdot\Delta_k(x)\cdot x^j_{\ ,\alpha}\cdot x^j_{\ ,\beta}, \qquad c^k_{ij} = c^k_{ji}.$$

The quantities $A_{\alpha\beta}(x)$ transform in the following way under the coordinate change $y^i = y^i(x^1, x^2, x^3)$:

$$
(5) \quad \begin{aligned} A_{\alpha\beta}(x) &= D\begin{pmatrix} y^1, & y^2, & y^3 \\ x^1, & x^2, & x^3 \end{pmatrix} \cdot A_{\alpha\beta}(x) \\ &+ \frac{1}{2}\sum (-1)^{\sigma} \frac{\partial^2 y^k}{\partial x^{\alpha_1}\partial x^{\alpha_2}} \cdot D\begin{pmatrix} y^i, & y^j \\ x^{\alpha_3}, & x^{\alpha_4} \end{pmatrix} \cdot \frac{\partial x^{\alpha_3}}{\partial t_1} \cdot \frac{\partial x^{\alpha_4}}{\partial t_2} \cdot \frac{\partial x^{\alpha_1}}{\partial t_\alpha} \cdot \frac{\partial x^{\alpha_2}}{\partial t_\beta}, \end{aligned}
$$

$$
\sigma = \operatorname{sgn}\begin{pmatrix} 1 & 2 & 3 \\ k & i & j \end{pmatrix}.
$$

Since $\partial x^i/\partial t = (\partial x^i/\partial y^j) \cdot (\partial y^j/\partial t_\alpha)$, we obtain

$$
(6) \qquad A_{\alpha\beta}(y) = D\begin{pmatrix} y^1, & y^2, & y^3 \\ x^1, & x^2, & x^3 \end{pmatrix} \cdot A_{\alpha\beta}(x) + C^k_{\alpha_1\alpha_2}(y) \cdot \Delta_k(y) \cdot \frac{\partial y^{\alpha_1}}{\partial t_\alpha} \frac{\partial y^{\alpha_1}}{\partial t_\beta},
$$

where $C^k_{\alpha_1\alpha_2}(y) = C^k_{\alpha_2\alpha_1}(y)$. Hence we see that system (3) preserves its form under the changes $y^i = y^i(x^j)$. If we set $x^1 = t_1, x^2 = t_2$, and $x^3 = x^3(t_1, t_2)$, then system (4) takes the form (3).

4. Sketch of the proof of Theorem B. We will say that a family of k-submanifolds in the domain U *can be straightened to mth order* if for every point $x \in U$ there is a diffeomorphism f of a neighborhood of x onto a domain in $\mathbf{R}^n$ identifying the manifold of m-jets of submanifolds of the family at x with the manifold of m-jets of k-planes at $f(x)$.

Condition A means precisely that the family under consideration can be straightened to first order.

LEMMA 3. *The family of solutions of a consistent system of equations* (1) *satisfying condition* B *can be straightened to second order (for any* k*).*

PROOF. Let $x = 0, k = 2$, and $n = 3$. It follows from (5) that after the change $y^i = x^i + a^i_{\alpha_1,\alpha_2}x^{\alpha_1}x^{\alpha_2}$ the ststem $A_{\alpha\beta}(x) = 0$ at the point $y = 0$ takes the form

$$
A_{\alpha\beta}(y)|_{y=0} = \frac{1}{2} a^i_{\alpha_1\alpha_2} \cdot \frac{\partial y^{\alpha_1}}{\partial t_\alpha} \cdot \frac{\partial y^{\alpha_2}}{\partial t_\beta}.
$$

This means that after a suitable diffeomorphism the coefficients $c^k_{ij}(x)$ in (4) can be made equal to zero at any point $x = x_0$.

LEMMA 4. *If a three-parameter family of surfaces in space can be straightened to second order, then on each surface of the family the remaining surfaces cut out a two-parameter family of curves.*

PROOF. It follows from Lemma 3 that the family of curves on a surface (cut out by the other surfaces) can be straightened to second order. Therefore it is the family of solutions of some differential equation of second order, and consequently depends on two parameters.

Thus it follows from Lemmas 3 and 4 that the family of solutions of a consistent system of differential equations (3) determines in space a four-parameter family of curves (along which the surfaces of the family intersect). Moreover, points, these curves, and the surfaces satisfy all the axioms of projective geometry in space (see [5]). Using the axioms and elementary geometric constructions of projective geometry we can construct in space a system of coordinates in which the surfaces under consideration are planes (cf. [5], Chapter 5). Elsewhere we will give another proof of the existence of such a system of coordinates not using the axioms of projective space. Theorem B is proved for

$k = 2$ and $n = 3$. It can be proved analogously in the general case $k > 1$. It is convenient to consider, in place of the partial derivatives of second order, the quantities

$$(7)\qquad A_{\alpha\beta}^{i_1,\dots,i_{k+1}}(x^1,\dots,x^n) = \sum_{1\le s\le n} (-1)^{s+1} \frac{\partial^2 x^{k+i}}{\partial t_\alpha \partial t_\beta} D\begin{pmatrix} x^{i_1},\dots,\widehat{x^{i_s}},\dots,x^{i_{k+1}} \\ t_1,\dots,t_k \end{pmatrix},$$

where $1 \le i_1 < \cdots < i_{k+1} \le n$. A system of the form (1) satisfying condition B can be written as

$$(8)\qquad A_{\alpha\beta}^{i_1,\dots,i_{k+1}}(x^1,\dots,x^n) = C^{i_1,\dots,i_{k+1}}_{\alpha_1,\dots,\alpha_{k-1};\alpha_k,\alpha_{k+1}}(x) \cdot D\begin{pmatrix} x^{\alpha_1},\dots,x^{\alpha_{k-1}} \\ t_1,\dots,t_{k-1} \end{pmatrix} \frac{\partial x^{\alpha_k}}{\partial t_{\alpha_k}} \frac{\partial x^{\alpha_{k+1}}}{\partial t_{\alpha_{k+1}}},$$

where the functions $C^{i_1,\dots,i_{k+1}}_{\alpha_1,\dots,\alpha_{k-1};\alpha_k,\alpha_{k+1}}$ are antisymmetric in the first set of lower indices and symmetric in the last two.

5. Concluding remarks. a) *Connection with integral geometry.* A family of submanifolds B_ξ in B, where $\xi \in \Gamma$, is called *admissible* if there are densities $d\chi_\xi$ on B_ξ such that the integral transform

$$I\colon C_0^\infty(B) \to C^\infty(\Gamma); \qquad I\colon f(x) \mapsto \int_{B_\xi} f(x)\, d\chi_\xi$$

admits a local inversion forumla ($\dim B = \dim \Gamma$).

THEOREM [6]. *A two-parameter family of holomorphic curves in a complex two-dimensional manifold is admissible if and only if the dual family of curves satisfies condition* A.

It follows from the main result of work of A. B. Goncharov, to appear, that an n-parameter family of complex hypersurfaces in Γ dual to an admissible family of hypersurfaces in B satisfies a weaker version of condition A: for every point $\xi \in \Gamma$ almost every subspace $h \subset T_\xi\Gamma$ with $\operatorname{codim} h = 1$ is the tangent space to exactly one submanifold of the family. (The subspaces h for which this condition is not fulfilled form a variety of nonzero codimension in $P(T_\xi^*\Gamma)$.)

b) For a system of equations (1) to have solutions for $k > 1$, consistency conditions must be satisfied. For a system of equations of the form (8) these conditions are a system of nonlinear first-order differential equations in the coefficients $C^{i_1,\dots,i_{k+1}}_{\alpha_1,\dots,\alpha_{k-1};\alpha_k,\alpha_{k+1}}$. For example, when $k = 2$ and $n = 3$ we obtain a system of 30 equations in 15 unknown functions. Theorem B makes it possible to solve the system. It would be very interesting to find a proof of Theorem B via a direct integaration of this system.

The authors thank M. L. Kontsevich for helpful discussions.

Scientific Council on the Complex Problem "Cybernetics"
Academy of Sciences of the USSR
Moscow
Received 22/MAY/86

BIBLIOGRAPHY

1. É. Cartan, Bull. Soc. Math. France **52** (1924), 205–241.

2. Ar. Tresse, Acta Math. **18** (1894), 1–88.

3. V. I. Arnol'd, *Supplementary chapters to the theory of ordinary differential equations*, "Nauka", 1978; English transl. *Geometrical methods in the theory of ordinary differential equations*, Springer-Verlag, 1982.

4. S. G. Gindikin, Twistor Geometry and Nonlinear Systems (Primorsko, 1980), Lecture Notes in Math., vol. 970, Springer-Verlag, 1982, pp. 2–42.

5. David Hilbert, *Grundlagen der Geometrie*, 7th ed., Teubner, Leipzig, 1930; English transl. of 1st ed., The Open Court, LaSalle, Ill., 1902; reprint, 1959.

6. I. M. Gel'fand, S. G. Gindikin, and Z. Ya. Shapiro, Funktsional. Anal. i Prilozhen. **13** (1979), no. 2, 11–31; English transl. in Functional Anal. Appl. **13** (1979).

Translated by HAROLD P. BOAS

Part II

Cohomology and characteristic classes

1.
(with D. B. Fuks)
On classifying spaces for principal fiberings with Hausdorff bases

Dokl. Akad. Nauk SSSR **181** (1968) 515–518 [Sov. Math., Dokl. **9** (1968) 851–854].
MR **38**:716. Zbl. **181**:266

In a recent work [1] (see also [2, 3]) we defined the analog of the universal fibering for principal G-fiberings with Hausdorff bases, where G is a closed Lie subgroup of the group $GL(n, \mathbf{R})$. The present note contains a generalization of this construction which enables us to construct a universal G-fibering for every topological group G. This construction contains the construction of [1] (with a trivial modification) as well as an analog of the well-known construction of Milnor [6] as particular cases. By means of this we are able to extend the results of [1] to the case of an arbitrary connected Lie group and establish that the group $H^q_{alg}(G; V)$ of characteristic classes of principal G-fiberings (with Hausdorff bases) with coefficents in the topological G-module V coincides with the group $H^q_c(G; V)$ of "continuous Eilenberg-MacLane cohomologies" [4] for every paracompact group G.

1. Let G be a topological group. By a principal G-fibering we mean a triple consisting of a topological space E on which the group G operates from the right without fixed points (i.e., if $yg = y$ for some $g \in G$, $y \in E$, then $g = e$), a factor-space X and a projection $p: E \to X$. The principal G-fibering $\xi = (E, P, X)$ is said to be locally trivial in the classical sense (or just locally trivial) if every point $x \in X$ has a neighborhood U such that the complete preimage $p^{-1}(U) \subset E$ as a G-space is isomorphic to the product $U \times G$.

The elementary G-object $\mathbf{T}$ is the triple $\mathbf{T} = (T, A, \mu)$, where T is a Hausdorff space with a distinguished point $*$; A is a topological semigroup with unit containing the group G; μ: $T \times A \to T$ is the continuous action of the semigroup A on T such that:

1^0. Every transformation $a \in A$ of the space T leaves the point $*$ fixed.

2^0. Different elements of the semigroup A determine different transformations of the space T.

3^0. There exists a homotopy ϕ_t: $T \to T$ such that $\phi_t \in A$ for all t, ϕ_0 is the identity transformation, $\phi_1(z) = *$ for all $z \in T$.

4^0. There exists a neighborhood W of the group G in the semigroup A and a mapping λ: $W \to G$ such that: (a) every element of W is invertible in the semigroup A: (b) $ag \in W$ for all $a \in W$, $g \in G$; (c) $\lambda(ag) = \lambda(a)g$ for all $a \in A$, $g \in G$; (d) $\lambda(g) = g$ for every $g \in G$.

We note that the group G acts both on the space T and on the semigroup A (multiplication from the right); here the invertible elements of the semigroup A form a G-invariant subset on which the group A acts without fixed points.

The following two examples are our basic examples of elementary objects.

Example 1. G is a closed Lie subgroup of the group $GL(n, \mathbf{R})$, $T = \mathbf{R}^n$, A is the semigroup of all linear mappings of $\mathbf{R}^n$ into itself.

Example 2. G is any topological group, T is a cone CG over G (i.e., the collection of pairs (g, t) where $g \in G$ and t is a nonnegative number where the pairs $(g', 0)$ and $(g'', 0)$ are identified for all $g', g'' \in G$). The semigroup A as a topological space coincides with T; multiplication in A (and the action of A on T) is defined by the formula $(g_1, t_1)(g_2, t_2) = (g_1 g_2, t_1 t_2)$. The inclusion

$G \subset A$ associates the element $g \in G$ with the point $(g, 1) \in A$. The element ϕ_t is chosen to be $(e, 1-t)$.

We fix the group G and the elementary G-object $\mathbf{T} = (T, A, \mu)$.

For every principal G-fibering $\xi = (E, p, X)$ we can construct a fibering $\xi_T(E_T, p_T, X)$ with fiber T by setting $E_T = (E \times T)/G$ (the action of the group G on $E \times T$ is determined coordinatewise). To each element $y \in E$ corresponds a mapping $\eta_y: T \to E_T$ (the composition $T = y \times T \subset E \times T \to E_T$) which maps T homeomorphically onto a fiber of the fibering ξ_T over the point $p(y)$. Obviously $\eta_{yg} = \eta_y g$ for all $y \in E$, $g \in G$, and different mappings η_{y_1}, η_{y_2} correspond to different points $y_1, y_2 \in E$. It is also obvious that the mapping η_y depends continuously on the point y.

Definition. The principal G-fibering $\xi = (E, p, X)$ is said to *locally* **T**-*trivial* if for each point $x \in X$ there exists a continuous mapping $\pi_x: E_T \to T$ such that for all $y \in E$ the mapping $\pi_x\eta_y : T \to T$ belongs to the semigroup A and when $py = x$ is its invertible element.

Remark. The definition would not be changed if we required that when $py = x$ the element $\pi_x\eta_y \in A$ belong to $G \subset A$: for this condition to be satisfied it suffices to take the mapping $(\pi_x\eta_y)^{-1}\pi_x$, where $py = x$, instead of the mapping π_x.

Obviously the fibering induced by the locally **T**-trivial fibering $\xi = (E, p, X)$ for any mapping $f: X' \to X$ is also locally **T**-trivial.

Proposition 1. *Every locally* **T**-*trivial fibering is locally trivial (in the classical sense).*

Proof. Let x be any point of the basis of the locally **T**-trivial fibering $\xi = (E, p, X)$. As has been noted, we can assume that the mapping $\pi_x\eta_y$ is an element of the group G when $py = x$. The set of points $y \in E$ for which $\pi_x\eta_y \in W$ is open in E. Since $\pi_{yg} = \eta_y g$, this set is invariant with respect to the action of the group G on E, i.e., it is a complete preimage of some open set $U \subset X$ for the projection p. The mapping $p^{-1}(U) \to U \times G$, associating the element $(p(y), \lambda(\pi_x\eta_y)) \in U \times G$ with the element $y \in p^{-1}(U)$, is compatible with the action of the group G. Consequently the G-spaces $p^{-1}(U)$ and $U \times G$ are isomorphic.

Proposition 2. *If the basis X of a locally trivial fibering ξ is completely regular, then ξ is locally* **T**-*trivial.*

Proof. Let $x \in X$ be a point; let $\sigma: U \to E$ be a cross-section of the fibering ξ over the neighborhood $U \ni x$; let h be a continuous real function on X taking on values from 0 to 1, at the point x and 1 outside U. Set $\pi_x(z) = \phi_{h(z)}\eta^{-1}_{\sigma(p_T(z))}(z)$ for $p_T(z) \in U$ and $\pi_x(z) = *$ for the remaining z.

2. We proceed to a presentation of the construction of the universal fibering. For the topological space X and the set I we let X^I denote the topological space whose points are collections $\{x_i\}$ of points of the space X enumerated by elements of I. The basis of open sets of the space X^I is made up of sets $\Gamma(i_1, \cdots, i_k; U_1, \cdots, U_k)$, where $i_1, \cdots, i_k \in I$, $U_1, \cdots, U_k$ are open sets in X, $\Gamma(i_1, \cdots, i_k; U_1, \cdots, U_k) = \{\{x_i\} \mid x_{i_1} \in U_1, \cdots, x_{i_k} \in U_k\}$.

Let $\mathfrak{E}^I_G$ denote the subspace of the space A^I consisting of collections $\{a_i\}$ in which at least one element a_i is invertible in the semigroup A. The action of the group G on $\mathfrak{E}^I_G$ is defined by the formula $\{a_i\}g = \{a_i g\}$. We denote the factor-space $\mathfrak{E}^I_G/G$ by $\mathfrak{S}^I_G$ and the projection $\mathfrak{E}^I_G \to \mathfrak{S}^I_G$ by p^I_G or briefly by p.

Proposition 3. *The space $\mathfrak{S}^I_G$ is Hausdorff.*

The proof is obvious.

Proposition 4. *The fibering* $(\mathcal{E}^I_G, p^I_G, \mathcal{S}^I_G)$ *is locally* **T**-*trivial.*

Proof. The points of the space $\mathcal{E}^I_G$ can be identified with imbeddings $T \to T^I$ whose composition with the projection $\rho_i : T^I \to T$ given by the formula $\rho_i(\{z_i\}) = z_i$ for every $i \in I$ belongs to the semigroup A. The points of the space $(\mathcal{E}^I_G)_T$ can be identified with pairs (ϕ, z) where ϕ is such an imbedding, $z \in \phi(T)$; here $(\phi', z) = (\phi'', z)$ if $\phi' = \phi'' g$, $g \in G$. The projection $(\mathcal{E}^I_G)_T \to \mathcal{S}^I_G$ consists of discarding the second element of the pair. Let $\pi_i : (\mathcal{E}^I_G)_T \to T$ $(i \in I)$ denote the mapping given by the formula $\pi_i(\phi, z) = \rho_i(z)$. The requirements of the definition of local **T**-triviality are satisfied by the mapping $\pi_x : (\mathcal{E}^I_G)_T \to T$ which is defined for each point $x \in \mathcal{S}^I_G$ as the mapping coinciding with the mapping π_i where $i \in I$ is an element such that there exists a point $y = \{a_i\} \in \mathcal{E}^I_G$ for which $p(y) = x$ and the element a_i is invertible in A.

Theorem 1. *For every locally* **T**-*trivial fibering* $\xi = (E, p, X)$ *such that the cardinality of the basis of open sets of the space* X *does not exceed the cardinality of the set* I, *there exists a continuous mapping* $\phi: X \to \mathcal{S}^I_G$ *such that the fibering* ξ *is equivalent to the fibering induced by the fibering* $(\mathcal{E}^I_G, p, \mathcal{S}^I_G)$ *by the map* ϕ.

Proof. It suffices to construct a continuous mapping of the space E into the space $\mathcal{E}^I_G$ which commutes with the action of the group G. For every point $y \in E$ a mapping $\eta_y : T \to E_T$ is defined such that for every $x \in X$ such that $p(y) = x$, the mapping $\pi_x \eta_y : T \to T$ is an invertible element of the semigroup A. From 4^0 of the definition of elementary G-object it follows that elements $\pi_x \eta_y$ for those $y \in E$ for which $p(y) \in U_x$ where U_x is some neighborhood of x, are also invertible in the semigroup A. The neighborhoods U_x constitute coverings of a space in which, by hypothesis, we can inscribe a covering whose cardinality does not exceed the cardinality of the set I. Therefore from all the mappings π_x we can choose a collection of mappings $\{\pi_i\}$, numbered by elements of i, such that for every point $y \in E$ at least one of the mappings $\pi_i \eta_y$ is invertible in the semigroup A. It remains to define the mapping $\Phi : E \to \mathcal{E}^I_G$ by $\Phi(y) = \{\pi_i \eta_y\}$.

Thus the fibering $\mathcal{E}^I_G = (\mathcal{E}^I_G, p, \mathcal{S}^I_G)$ is universal for locally **T**-trivial principal G-fiberings whose bases have a basis of neighborhoods whose cardinality does not exceed the cardinality of the set I.

We investigate the two basic examples of elementary G-objects in more detail. If G is a Lie group, $T = \mathbf{R}^n$, A is the semigroup of all linear mappings, then $\mathcal{E}^I_G$ is the space of all linear imbeddings $\mathbf{R}^n \to (\mathbf{R}^n)^I$ whose composition with one of the projections $(\mathbf{R}^n)^I \to \mathbf{R}^n$ is nondegenerate. We recall that in [1] we considered the space of all linear imbeddings $\mathbf{R}^n \to (\mathbf{R}^n)^I$ for the space $\mathcal{E}^I_G$. Thus the universal fibering obtained from the proposed general construction is a subfibering of the universal fibering of [1], and the basis of the first is an open, everywhere dense set in the basis of the second.

If $T = CG$, then the space $\mathcal{E}^I_G$ can be obtained from $(CG)^I$ by removing the unique point $\{a_i\}$, where $a_i = *$ (vertex of the cone) for all I. We note that if T is a set of n elements, then the space $\mathcal{E}^I_G$ is homeomorphic to the direct product of an $(n-1)$-fold join $G * \cdots * G$ of the group G by the straight line $\mathbf{R}$, where the action of the group does not change the coordinates of the point on $\mathbf{R}$. Recall that the classical construction of Milnor of a universal fibering for the group G consists of defining the action of the group on the space $E_n = \underbrace{G * \cdots * G}_{n}$ and then taking (E_G, p, B_G), where $E_G = \varinjlim E_n$ and $B_G = E_G/G$ as the universal fibering for the group G. Thus in the case $T = CG$ our construction yields an analog of Milnor's universal fibering.

3. **Definition.** We say that we have a q-dimensional *characteristic class* of locally **T**-trivial principal G-fiberings with coefficients in the topological G-module V if with each locally **T**-trivial principal G-fibering $\xi = (E, p, X)$ is associated an element $\alpha(\xi) \in H^q(X; \mathbf{V})$, where **V** is a bundle of germs of sections of a fibering with fiber V induced by the fiber ξ, and for every mapping $\phi : \xi' = (E', p', X') \to \xi'' = (E'', p'', X'')$ we have the equality $\tilde{\phi}^* \alpha(\xi'') = \alpha(\xi')$. Here $\tilde{\phi} : X' \to X''$ is the mapping of bases corresponding to the mapping ϕ.

The characteristic classes form an abelian group which we denote by $H^q_{alg}(G; V)$

Definition. The qth homology group of the complex

$$F_0 \xrightarrow{\partial_0} F_1 \xrightarrow{\partial_1} \cdots,$$

where F_q is a group of continuous mappings of the product $\underbrace{G \times \cdots \times G}_{q}$ into V and

$$\partial_a f(g_1, \ldots, g_{q+1}) = g_1 f(g_2, \ldots, g_{q+1})$$

$$\sum_i (-1)^i f(g_1, \ldots, g_i g_{i+1}, \ldots, g_{q+1}) + (-1)^{q+1} f(g_1, \ldots, q_q)$$

is called a continuous Eilenberg-MacLane cohomology group $H^q_c(G; V)$ of the group G with coefficients in the topological G-modulus V.

Theorem 2. *For every paracompact group G the equality $H^q_{alg}(G; V) = H^q_c(G; V)$ holds. (In particular, the group $H^q_{alg}(G; \mathbf{V})$ does not depend on T.)*

The proof of this theorem is based on the fact that if the characteristic classes α' and α'' coincide for all fiberings ξ^l_G, then they are equal. A decisive role is played by the following lemma.

Lemma. *For every continuous real function on the space $\mathfrak{E}^l_G$ there is a subset $l' \subset l$ such that the difference $l \setminus l'$ is not more than countable, and the restriction $f|_{\mathfrak{E}^{l'}_G}$ is constant.*

The proof is the same as the proof of the analogous lemma in [1] (see Lemma 2.4).

We recall that in [1] we proved for closed Lie subgroups G of the group $GL(n; \mathbf{R})$ that the group $H^q_{alg}(G; \mathbf{R})$ is isomorphic to the qth homology group of the complex of G-invariant differential froms on the space $G/\hat{G}$, where $\hat{G} \subset G$ is a maximal compact subgroup. That this group is isomorphic to the group $H^q_c(G; V)$ was established by Hochschild and Mostow [5].

In conclusion we express our gratitude to Professor A. Borel for his interest in this work and for calling the pertinent literature to our attention.

Moscow State University
Received 17/APR/68

BIBLIOGRAPHY

[1] I. M. Gel'fand and D. B. Fuks, Funkcional. Anal. i Priložen. 1 (1967), no. 4.

[2] ———, Dokl. Akad. Nauk SSSR 176 (1967), 24 = Soviet Math. Dokl. 8 (1967), 1031.

[3] ———, Dokl. Akad. Nauk SSSR 177 (1967), 763 = Soviet Math. Dokl. 8 (1967), 1483.

[4] W. T. van Est, Nederl. Akad. Wetensch. Proc. Ser. A. 56 = Indag. Math. 15 (1953), 484. MR 15, 505.

[5] G. Hochschild and G. D. Mostow, Illinois J. Math. 6 (1962), 367. MR 26 #5092.

[6] J. W. Milnor, Ann. of Math (2) 63 (1956), 272. MR 17, 994.

Translated by Lisa Rosenblatt

2.
(with D. B. Fuks)

Topology of noncompact Lie groups

Funkts. Anal. Prilozh. 1 (4) (1967) 33–45 [Funct. Anal. Appl. 1 (1967) 285–295].
MR 37:2253. Zbl. 169:547

INTRODUCTION

1. Every topological group has associated with it, in addition to the customary algebraic topological invariants (cohomology, homotopy, the K-functor, etc.) describing its purely topological structure, another series of invariants describing its group structure. The starting point in this case is not the group G per se, but the category of principal G-foliations. Let h be a contravariant functor of a certain category of topological spaces in the category of Abelian groups (or in the category of sets). We say that the characteristic class α of principal G-foliations with values in h is specified when to every principal G-foliation $(\xi)p : E \to X$ is referred an element $\alpha(\xi) \in h(X)$, such that, if

$$(\varphi)\quad \begin{array}{ccc} E_1 & \xrightarrow{\varphi''} & E_2 \\ \downarrow p_1 & & \downarrow p_2 \\ X_1 & \xrightarrow{\varphi'} & X_2 \end{array}$$

is a mapping of the foliation $(\xi_1)p_1 : E_1 \to X_1$ into the foliation $(\xi_2)p_2 : E_2 \to X_2$, then $[h(\varphi')]\alpha(\xi_2) = \alpha(\xi_1)$. The set of all such characteristic classes is designated $h_{alg}(G)$. If h is a functor with values in the category of Abelian groups, the set $h_{alg}(G)$ itself comprises a group in the obvious sense.

2. Now it is time to specify more precisely what is meant by the principal G-foliation. Usually the concept is narrowed to locally trivial principal G-foliations, the bases of which are cell complexes. In this case, as we know, there exists a so-called universal principal G-foliation $(\Xi)p_G : E_G \to B_G$, such that for every locally trivial principal foliation $(\xi)p : E \to X$, the base of which is a cell complex, there exists a mapping φ of the foliation ξ into the foliation Ξ, which is unique in the sense that the mapping $\varphi' : X \to B_G$ is uniquely defined correct to a homotopy. Principal foliations $(\xi)p : E \to X$ with fixed base X have a one-to-one correspondence with homotopic classes of mappings of X into B_G. Clearly, the characteristic class $\alpha \in h_{alg}(G)$, given this definition of the category, is completely defined by its value on the universal foliation, i.e., by the element $\alpha(\Xi) \in h(B_G)$; if, however, the functor h is homotopic, the equation $h_{alg}(G) = h(B_G)$ holds.

The space B_G is called the classification space of the group G. It is constructed from the group G correct to homotopic equivalence. Here the classification space of a Lie group and its maximal compact subgroup are identical. Consequently, not only the characteristic classes, but also the categories of principal foliations themselves, are the same for any Lie group and its maximal compact subgroup. But the algebraic structure of these groups, of course, do not by any means necessarily reduce one into the other.

This is especially clearly evinced by the following example. Let EU(X) be the set of classes of stable equivalent complex vector sheaves with base X. This is a contravariant functor with values in the category of sets, and we can define for any group G a set $EU_{alg}(G)$ (we interpret this set in the sense of a category of principal foliations with cell bases). Atiyah [1] has shown that if G is a compact Lie group, $EU_{alg}(G)$ is congruent with the set of classes of equivalent unitary representations of the group G. But if G is a noncompact group, then it follows from the foregoing that $EU_{alg}(G) = EU_{alg}(\hat{G})$, where $\hat{G}$ is the maximal compact subgroup of the group G, i.e., $EU_{alg}(G)$ is expressed in terms of representations of the group $\hat{G}$, which do not have anything in common with the representations of the group G.

3. The incorporation of noncompact Lie groups into the general theory can be achieved by expanding the given category of principal G-foliations. Specifically, we require nothing of the foliation base except

---

Moscow State University. Translated from Funktsional'nyi Analiz i Ego Prilozheniya, Vol. 1, No. 4, pp. 33–45, October- December, 1967. Original article submitted July 15, 1967.

that it be a Hausdorff base. The resulting change in the category of principal G-foliations turns out to be inconsequential if the group G is compact, and very significant if G is noncompact.

This is due to the following. The reason the principal G-foliation whose layer is a compact group G can be nontrivial is purely algebraic: There is no transversal surface on a cycle of a particular dimensionality. This fact shows up in full measure even in foliations on cell complexes. If, on the other hand, the group G is not compact, the principal G-foliations can still be nontrivial, insofar as the portion of the layers going to infinity come too close together. The base of this foliation possesses poor topological properties (in the sense of separability) and, of course, cannot be a cellular complex.

For example, if G = R is the group of real numbers, then any principal G-foliations on a cell complex and on any paracompact space in general is trivial. In this sense R does not differ from $\hat{R} = \{0\}$. But nontrivial principal R-foliations do nevertheless occur. We give an example. Let $\overline{\mathscr{E}}$ be the space of all real sequences $(x_1, x_2, \ldots)$ in a Tychonoff topology, and let $\mathscr{E} = \overline{\mathscr{E}} \setminus \{0\}$ be the same space with the null sequence discarded. The group R operates on $\mathscr{E}$ like a group of coordinate multiplications by positive real numbers. The trajectories of this group are rays emanating from the coordinate origin. The space $\mathscr{S}$ of trajectories (sphere analog) is Hausdorffian, but not regular. (Moreover, as is easily shown, every function continuous on $\mathscr{S}$ is a constant!) The principal R-foliation $(\xi)p : \mathscr{E} \to \mathscr{S}$ is nontrivial. In fact, the transversal surface $\mathscr{S} \to \mathscr{E}$ does not exist, because if the mapping $\pi_n : \mathscr{E} \to R$ is the n-th coordinate, the superposition $\mathscr{S} \to \mathscr{E} \xrightarrow{\pi_n} R$ is a continuous function on $\mathscr{S}$, i.e., a constant. Therefore, every continuous mapping $\mathscr{S} \to \mathscr{E}$ is a mapping into a point.

4. In this paper, we investigate certain functors of the type $h_{alg}(G)$, defined by means of such an expanded category. The group G is assumed to be a closed Lie subgroup of the group GL(n; R) of matrices of order n.

In the first section we investigate the category itself. Here, we construct an analog of the universal foliation and expound some of its properties. The formulation is founded on the same notion as the example given above. The second section presents a calculation of the groups $H^q_{alg}(G; R)$. We interpret $H^q_{alg}(G; R)$ as $h_{alg}(G)$ for the case when the functor h is $X \to H^q(X; R)$, where $H^q(X; R)$ is a group of q-dimensional cohomologies of the space X with coefficients in the sheaf R of germs of real functions.* We recall that if the space X is paracompact, $H^q(X; R) = 0$ for $q > 0$, so that the extension of the category to the category of all Hausdorff spaces is significant here. Accordingly, if G is a compact group, $H^q_{alg}(G; R) = 0$ for $q > 0$.

As a whole, the answer turns out to be exceedingly simple. The groups $H^q_{alg}(G; R)$ are always trivial, beginning with a certain dimensionality. For example, if G is a semisimple group, $\overline{G}$ is its compact form, and $\hat{G}$ is the maximal compact subgroup, then the group $H^q_{alg}(G; R)$ is isomorphic to the usual group of q-dimensional real cohomologies of the space $\overline{G}/\hat{G}$.

In the third section, we define and calculate the groups $H^q_{alg}(G; M)$, where M is any representation of the group G. Here also we show how $H^q_{alg}(G; M) = \mathrm{Ext}^q_G(1; M)$, where 1 is the unity representation.

In the fourth section, we calculate the functor $EH_{alg}(G)$ (analogous to the functor EU_{alg} mentioned above), where H is any topological group. It turns out that the set $EH_{alg}(G)$ is naturally equivalent to the set of topological homomorphisms of G into H.

In this sense the ideology of our paper is complementary to the ideology of Dixmier and Douady in [2, 3]. It is shown in the latter that it is impossible within the framework of paracompact spaces to formulate a meaningful theory of foliations with an infinite-dimensional layer. Conversely, setting H in our

* We have everywhere in mind Czech cohomologies.

work equal to a group of transformations of an infinite-dimensional representation of a Lie group G, we obtain nontrivial infinite-dimensional vector foliations on certain nonparacompact spaces. The present paper antedates the papers published in the Proceedings of the Academy of Sciences of the USSR [Doklady AN SSSR] [4]. In the course of writing the article, we had useful consultations with A. V. Arkhangel'skii, A. L. Onishchik, and E. G. Sklyarenko, to whom we convey our utmost gratitude.

1. The Category

1. Let G be a closed Lie subgroup of a group GL(n, R) of nondegenerate real matrices of order n. We may say that a principal G-foliation is specified when a Hausdorff-topological space E is given on which the group G acts on the right without stationary points, where the space of orbits X is also Hausdorffian. Notation: $(\xi)p : E \to X$ (here p denotes a natural projection, ξ the complete structure in the large). The entire discussion that follows would be valid if we included in this definition the requirement of complete regularity on the part of the space E (but not X).

We interpret the mapping $\varphi : \xi_1 \to \xi_2$ of a principal G-foliation $(\xi_1)p_1 : E_1 \to X_1$ into the principal G-foliation $(\xi_2)p_2 : E_2 \to X_2$ as the commutative mapping diagram

$$(\varphi)\quad \begin{array}{ccc} E_1 & \xrightarrow{\varphi''} & E_2 \\ \downarrow p_1 & & \downarrow p_2 \\ X_1 & \xrightarrow{\varphi'} & X_2 \end{array},$$

in which φ'' is matched with the action of the group.

Let $(\xi)p : E \to X$ be a principal G-foliation, R^n an n-dimensional real vector space. The group G acts on R^n on the left as a subgroup of the group of matrices. We consider the direct product $E \times R^n$, on which the group G acts according to the relation $(\alpha, \beta)g = (\alpha g, g^{-1}\beta)$. The space of orbits $E^{\#} = E \times R^n/G$, clearly, is Hausdorffian. Together with the natural mapping $p^{\#} : E^{\#} \to X$, it is called a vector foliation induced by the principal foliation ξ. It is apparent that the complete image of the point $x \in X$ in the mapping $p^{\#}$ is a vector space.

The mapping $\varphi : \xi_1 \to \xi_2$ corresponds to a mapping of the induced vector foliations, i.e., to the commutative diagram

$$(\varphi^{\#})\quad \begin{array}{ccc} E_1^{\#} & \xrightarrow{\varphi^{\#}} & E_2^{\#} \\ \downarrow p_1^{\#} & & \downarrow p_2^{\#} \\ X_1 & \xrightarrow{\varphi'} & X_2 \end{array}.$$

Every point $\bar{x} \in E$ corresponds to an isomorphism $R^n \to E \times R^n \to E^{\#}$ of the standard copy of the space R^n onto the layer $(p^{\#})^{-1}p(\bar{x})$ of the space $E^{\#}$. Consequently, we have juxtaposed to every point $\bar{x} \in E$ a basis $P_{\bar{x}}$ in a layer of the induced foliation over the point $x = p(\bar{x})$, where juxtaposed to the point $\bar{x}g$ is a basis occurring in the same layer and differing from $P_{\bar{x}}$ by the matrix g.

2. We call a vector foliation $(\xi^{\#})p^{\#} : E^{\#} \to X$ locally trivial if for every point $x \in X$ there exists a mapping $\eta_x : E^{\#} \to R^n_x$ of the space $E^{\#}$ onto layer R^n_x over the point x, where this mapping is identical on R^n_x and linear on every layer of the foliation $\xi^{\#}$. A principal G-foliation is called locally trivial if the vector foliation induced by it is locally trivial. Our definition of local triviality differs from the conventional one. A foliation that is locally trivial in our sense is also locally trivial in the conventional sense. Thus, there exists about any point $x \in X$ a neighborhood U such that the mapping η_x is nondegenerate on the layer R^n_y for $y \in U$. Clearly, the restriction of the foliation ξ (and the foliation $\xi^{\#}$) over U is trivial. Conversely, the local triviality of a principal foliation in the conventional sense implies its local triviality in our sense under the condition that its base is completely regular.

Thus, we are investigating the category of locally trivial principal G-foliations, which throughout the entire ensuring discussion we refer to simply as G-foliations.

3. Let I be any set. We denote by $\overline{\mathscr{E}}^I$ the Tychonoff product of lines indexed by elements of the set I. Thus, a point $x \in \overline{\mathscr{E}}^I$ comprises a set $\{x_i | i \in I\}$ of real numbers labeled with indices from I. We denote by $\mathscr{E}^I$ the set of all nondegenerate n-hedra in the linear space $\overline{\mathscr{E}}^I$. The space $\mathscr{E}^I$ is acted upon in an obvious manner by the group GL(n, R), and its subgroup G. The factor space $\mathscr{E}^I/G = \mathscr{E}^I_G$, clearly, is Hausdorffian, and the projection $p^I : \mathscr{E}^I \to \mathscr{E}^I_G$ defines a foliation ξ^I. Obviously, the embedding $j : I_1 \to I_2$ of a set I_1 into a set I_2 defines the mapping $j_* : \xi^{I_1} \to \xi^{I_2}$.

Consequently, in our category we have picked out a subcategory comprising the foliations ξ^I and the mappings j_*. This subcategory plays a role analogous to that of the universal foliation.†

PROPOSITION 1.1. For any G-foliation ξ there exists a mapping $\varphi : \xi \to \xi^I$, where I is a certain set.

Proof. Mappings $\eta_x : E^\# \to R^n_x$ existing according to the definition of local triviality for all $x \in X$ collectively define a mapping $\psi : E^\# \to \prod_{x \in X} R^n_x = \overline{\mathscr{E}}^{X \times \{1, 2, \ldots, n\}}$. In this mapping every layer of the foliation $\pi : E^\# \to X$ is mapped linearly and nondegenerately into the space $\overline{\mathscr{E}}^{X \times \{1, 2, \ldots, n\}}$. We introduce the following notation: $I = X \times \{1, 2, \ldots, n\}$. As remarked above, every point $\bar{x} \in E$ corresponds to an n-hedron in the layer $p^{\# -1} p(\bar{x})$. In the mapping ψ this n-hedron goes over into an n-hedron and defines a point $\varphi''(\bar{x}) \in \mathscr{E}^I$. The mapping φ'' commutes with the action of the group and defines the mapping φ of the foliation ξ into the foliation ξ^I.

4. We wish to point out two additional properties of the spaces $\mathscr{E}^I$.

PROPOSITION 1.2. The space $\mathscr{E}^I$ is completely regular for any set I, and if the set I is denumerable, the space $\mathscr{E}^I$ is paracompact.

PROPOSITION 1.3. For any continuous function f on $\mathscr{E}^I$ there exists a subset $I' \subset I$, such that the difference $I \setminus I'$ is at most denumerable, and the function f is constant on $\mathscr{E}^{I'} \subset \mathscr{E}^I$.

Proof. We pick out n arbitrary elements in I. In the space $\overline{\mathscr{E}}^I$ we separate a corresponding n-dimensional space and investigate the set $K \subset \mathscr{E}^I$ of its bases. We pick an $\varepsilon > 0$. For every point $y_0 = (e_1, e_2, \ldots, e_n) \in K$ there exists a neighborhood $U \subset \mathscr{E}^I$, in which $|f(y) - f(y_0)| < \varepsilon$. According to the definition of the Tychonoff topology, this neighborhood includes all bases of the form $(e_1 + \tilde{e}_1, e_2 + \tilde{e}_2, \ldots, e_n + \tilde{e}_n)$, where $\tilde{e}_1, \tilde{e}_2, \ldots, \tilde{e}_n$ are arbitrary vectors from $\overline{\mathscr{E}}^{I(y_0, \varepsilon)} \subset \mathscr{E}^I$, and the difference $I \setminus I_0(y_0, \varepsilon)$ is finite. Let $K_0 \subset K$ be a denumerable everywhere dense subset. We introduce the following notation: $I' = \bigcap_{y_0 \in K_0} \bigcap_{n=1}^{\infty} I(y_0, 1/n)$. It is clear that the set $I \setminus I'$ is at most denumerable. For every point $y_0 \in K_0$ and, hence, for every point $y = (e_1, e_2, \ldots, e_n) \in K$ the value of the function $f(\tilde{y})$, where $\tilde{y} = (e_1 + \tilde{e}_1, \tilde{e}_2 + \tilde{e}_2, \ldots, e_n + \tilde{e}_n)$, does not depend on $\tilde{e}_1, \tilde{e}_2, \ldots, \tilde{e}_n$ for $\tilde{e}_1, \tilde{e}_2, \ldots, \tilde{e}_n \in \overline{\mathscr{E}}^{I'}$. Hence, multiplying $e_1, e_2, \ldots, e_n$ by t and letting t tend to zero, we deduce that our function is constant on $\mathscr{E}^{I'} \subset \mathscr{E}^I$.

The following stronger version of this proposition is proved analogously.

PROPOSITION 1.3'. For any continuous mapping $f : \mathscr{E}^I \to X$, where X is any topological space, there exists a subset $I' \subset I$, such that the difference $I \setminus I'$ has a power no greater than the base of the neighborhoods of the space X, and the image of the subspace $\mathscr{E}^{I'} \subset \mathscr{E}^I$ in the mapping f comprises one point.

2. Cohomologies with Real Coefficients

1. Throughout this and the next section the group G is assumed to be connected or to have a finite set of connected components.

Let R be a sheaf of germs of continuous real functions. The contravariant functor $X \to H^q(X; R)$ defines a group $H^q_{alg}(G; R)$ in the sense of the general definition given in the Introduction.

† If we alter the category by introducing the additional requirement of denumerability on the part of the base of neighborhoods of the given topological spaces, we could replace this subcategory by a single foliation, regarding I as a standard denumerable set. For such a category, however, we are unable to prove any of the theorems, because Proposition 1.3 plays an important part in our later discussions. We are hopeful that the results of the paper can be carried over to this case as well.

Let $\hat{G} \subset G$ be a maximal compact subgroup. We denote by G^* a homogeneous space $G/\hat{G}$. We denote by $\Omega_s^q(G^*)$ a group of differential smooth (the word "smooth" for us always denotes the "class C^∞") forms of degree q, invariant with respect to the action of the group G on G^*. We investigate the complex

$$(\Omega_s(G^*))\ 0 \xrightarrow{d_{-1}} \Omega_s^0(G^*) \xrightarrow{d_0} \Omega_s^1(G^*) \xrightarrow{d_1} \dots \xrightarrow{d_{r-1}} \Omega_s^r(G^*) \xrightarrow{d_r} 0 \longrightarrow \dots,$$

where d_k is a differential, and r is the dimensionality of the space G^* (clearly, the differential of an invariant form is again an invariant form).

Following is the principal result of this section:

THEOREM 2.1. The groups $H^q_{alg}(G; R)$ are congruent with homologies of the complex $\Omega_s(G^*)$, i.e.,

$$H^q_{alg}(G; \mathbf{R}) = \operatorname{Ker} d_q / \operatorname{Im} d_{q-1}.$$

2. Let $(\xi)p : E \to X$ be some locally trivial (in general, not smooth) foliation, the layer of which is a smooth manifold (with matched smooth structures being specified on all the layers). At every point $x \in E$ it is possible to analyze a tangential plane F_x to the layer at this point. We denote by F_x^q the space of piecewise-symmetric q-linear forms in F_x. We say that a smooth differential form φ of degree q of the foliation ξ is specified in an open set $U \subset E$ (and, accordingly, on the open set $V \subset X$), if to every point x of the set U [and, accordingly, the set $p^{-1}(V)$] is referred an element $\varphi_x \in F_x^q$, where φ_x depends continuously on x within the limits of U [and, accordingly, $\pi^{-1}(V)$], and depends smoothly on x within the limits of each layer. In other words, a form on $U \subset E$ (and, accordingly, on $V \subset X$) is a family of differential forms specified in the intersections of the layers with U (and, accordingly, in the layers lying over V). Correspondingly, it is possible to construct two sheaves of germs of such forms: a sheaf $\Omega^q(\xi)$ with base X and $\tilde{\Omega}^q(\xi)$ with base E. Here the group of cross sections of the sheaf $\Omega^q(\xi)$ over the open set $V \subset X$ is isomorphic to the group of cross sections of the sheaf $\tilde{\Omega}^q(\xi)$ over the set $p^{-1}(V) \subset X$.

The following is a consequence of Theorem 4.17.1 in Godement's book [6]:

PROPOSITION 2.2. $H^p(X; \Omega^q(\xi)) \approx H^p(E; \tilde{\Omega}^q(\xi))$ for $p \geq 0$, $q \geq 0$.

We consider separately the case when $(\xi)p : E \to X$ is a G-foliation. Clearly, once we pick the basis of the tangential space at unity to the group G, i.e., an algebra $\mathfrak{G}$ of the group G, we can then choose canonically the basis in all spaces F_x tangential to the layers of all G-foliations. Then we can replace the piecewise-symmetric q-linear form φ_x by its coordinates φ_x^i ($i = 1, 2, \dots, C_m^q$; $m = \dim G$) in this basis. Consequently, to specify the differential form φ of the foliation $(\xi)p : E \to B$ in a neighborhood $U_i \subset E$ is the same as to specify in this neighborhood C_m^q continuous functions φ_x^i ($i = 1, 2, \dots, C_m^q$) smooth on the layers. In other words, the following statement holds true:

PROPOSITION 2.3. For every G-foliation there occurs an isomorphism $\tilde{\Omega}^q(\xi) \approx \underbrace{\tilde{R} + \tilde{R} + \dots + \tilde{R}}_{C_m^q}$,

where $\tilde{R}$ is a sheaf of germs of continuous real layerwise-smooth functions. This isomorphism is constructed on the basis of a Lie algebra in the group G and is natural in the category of G-foliations.

3. Let $(\xi)p : E \to X$ be a G-foliation. The group $\hat{G} \subset G$ acts on a space E without stationary points. We denote by E^* the space of trajectories of this group. We can represent the mapping p in the form of a composition of two mappings $E \xrightarrow{\hat{p}} E^* \xrightarrow{p^*} X$, the first of which defines a $\hat{G}$-foliation $(\hat{\xi})\hat{p} : E \to E^*$, the second of which defines a locally trivial foliation $(\xi^*)p^* : E^* \to X$.

Let us examine the sequence of sheaves on the space X

$$0 \to R \xrightarrow{j} \Omega^0(\xi^*) \xrightarrow{d_0} \Omega^1(\xi^*) \xrightarrow{d_1} \dots \tag{2.4}$$

Here the mappings d_k are induced by differentials, and the embedding $j : R \to \Omega^0(\xi^*)$ is induced by the "ascent" of continuous functions from X onto E^*. From the acyclicity of the space G^* (here only once do we use the finiteness of the set of connected components of the group G) it follows that this sequence is exact.

4. For brevity we write Ω^q in place of $\Omega^q(\xi^*)$, until subsection 5.

PROPOSITION 2.5. There exists a spectral sequence $\{E_r^{p,q}; d_r^{p,q}: E_r^{p,q} \to E_r^{p-r+1,q+r}\}$, such that

1°. $E_1^{p,q} = H^p(X; \Omega^q)$;

2°. the differential $d_1^{p,q}: H^p(X; \Omega^q) \to H^p(X; \Omega^{q+1})$ is induced by the homomorphism $d_q: \Omega^q \to \Omega^{q+1}$;

3°. the group $E_\infty^m = \sum_{p+q=m} E_\infty^{p,q}$ is associated with $H^m(X; R)$ relative to a certain filtration.

Proof. We denote by K^q the kernel of the mapping $d_q: \Omega^q \to \Omega^{q+1}$. The exact sequence (2.4) may be represented in the form of a set of exact sequences

$$\begin{array}{l} 0 \to R \xrightarrow{i} \Omega^0 \to K^1 \to 0, \\ 0 \to K^q \to \Omega^q \to K^{q+1} \to 0 \qquad (q = 1, 2, \ldots). \end{array} \tag{2.6}$$

We introduce the following notation: $\Omega^{p,q} = H^p(X; \Omega^q)$, $K^{p,q} = H^q(X; K^q)$. The exact cohomological sequences induced by the exact sequences of coefficients (2.6) are conveniently written in the following commutative diagram form:

$$\begin{array}{cccccc}
 & 0 & & 0 & & \\
 & \downarrow & & \downarrow & & \\
\cdots \Omega^{0,p+q-1} & \xrightarrow{\pi} K^{0,p+q} & \xrightarrow{\eta} \Omega^{0,p+q} & \xrightarrow{\pi} K^{0,p+q+1} & \xrightarrow{\eta} \Omega^{0,p+q+1} & \cdots \\
 & \downarrow \partial & & \downarrow \partial & & \\
 & \cdots & & \cdots & & \\
 & \downarrow \partial & & \downarrow \partial & & \\
\cdots \Omega^{p-1,q} & \xrightarrow{\pi} K^{p-1,q+1} & \xrightarrow{\eta} \Omega^{p-1,q+1} & \xrightarrow{\pi} K^{p-1,q+2} & \xrightarrow{\eta} \Omega^{p-1,q+2} & \cdots \\
 & \downarrow \partial & & \downarrow \partial & & \\
\cdots \Omega^{p,q-1} & \xrightarrow{\pi} K^{p,q} & \xrightarrow{\eta} \Omega^{p,q} & \xrightarrow{\pi} K^{p,q+1} & \xrightarrow{\eta} \Omega^{p,q+1} & \cdots \\
 & \downarrow \partial & & \downarrow \partial & & \\
\cdots \Omega^{p+1,q-2} & \xrightarrow{\pi} K^{p+1,q-1} & \xrightarrow{\eta} \Omega^{p+1,q-1} & \xrightarrow{\pi} K^{p+1,q} & \xrightarrow{\eta} \Omega^{p+1,q} & \cdots \\
 & \downarrow \partial & & \downarrow \partial & & \\
 & \cdots & & \cdots & & \\
 & \downarrow \partial & & \downarrow \partial & & \\
 & H^{p+q}(X; R) & \xrightarrow{\eta} \Omega^{p+q,0} & \xrightarrow{\pi} K^{p+q,1} & \xrightarrow{\eta} \Omega^{p+q,1} & \cdots \\
 & & & \downarrow \partial & & \\
 & & & H^{p+q+1}(X; R) & \xrightarrow{\eta} \Omega^{p+q+1,0}. &
\end{array}$$

Here the sequences of homomorphisms $(\pi, \partial, \eta, \pi, \partial, \eta, \ldots)$ are exact. Let $r \geq 1$. We analyze two subgroups in $\Omega^{p,q}$: $Z_r^{p,q} = (\pi)^{-1}[\partial \ldots \partial (K^{p-r+1,q+r})]$ and $B_r^{p,q} = \eta(\underbrace{\mathrm{Ker}\, \partial \ldots \partial}_{r-1})$. We introduce the following notation: $E_r^{p,q} = Z_r^{p,q}/B_r^{p,q}$. The differential $d_r^{p,q}$ is defined in the following manner. Let $\xi \in E_r^{p,q}$. We pick the representative $z \in Z_r^{p,q}$. By definition, $\pi^{p,q} z = \underbrace{\partial \ldots \partial}_{r-1} w$, where $w \in K^{p-r+1,q+r}$. It is clear that $w \in Z_s^{p-r+1,q+r}$ for all s and defines an element $\eta \in E_r^{p-r+1,q+r}$, which is denoted $d_r^{p,q}\xi$. The validity of the definitions of the differential and the equation $E_{r+1}^{p,q} = \mathrm{Ker}\, d_r^{p,q}/\mathrm{Im}\, d_r^{p+r-1,q-r}$ are tested in the usual manner.

Obviously, $E_\infty^{p,q} = \mathrm{Im}\,[K^{p,q} \to H^{p+q}(X; R)]/\mathrm{Im}\,[K^{p-1,q+1} \to H^{p+q}(X; R)]$.

This proves the proposition.

5. Thus, for every G-foliation $(\xi)p : E \to B$ we have specified a spectral sequence. More precisely, we have constructed a contravariant functor from the category of G-foliations into the category of such sequences, since a mapping of G-foliations corresponds to a mapping of these spectral sequences in the opposite direction. Our next step is analogous in some degree to the transition $h \mapsto h_{alg}$.

Let $E_r^{p,q}(\xi)$ be a group entering into the constructed spectral sequence for the G-foliation ξ. We say that an element $\alpha \in \widetilde{E}_r^{p,q}$ is specified if for every G-foliation ξ an element $\alpha(\xi) \in E_r^{p,q}(\xi)$ is specified; if $\varphi : \xi_1 \to \xi_2$ is a mapping of the foliations and $\varphi_r^{p,q} : E_r^{p,q}(\xi_2) \to E_r^{p,q}(\xi_1)$ is a homomorphism entering into the induced mapping of the spectral sequences, then $\varphi_r^{p,q}\alpha(\xi_2) = \alpha(\xi_1)$. The homomorphisms $d_r^{p,q} : E_r^{p,q}(\xi) \to E_r^{p-r+1,q+r}(\xi)$ defined for any G-foliation ξ, define a homomorphism $\widetilde{d}_p^{p,q} : \widetilde{E}_r^{p,q} \to \widetilde{E}_p^{p-r+,q+r}$.

The truth of the following is easily verified:

<u>PROPOSITION 2.7.</u> The groups $\widetilde{E}_r^{p,q}$ and the mappings $\widetilde{d}_r^{p,q}$ for a spectral sequence whose limiting term $\widetilde{E}_\infty^m = \sum \widetilde{E}_\infty^{p,q}$ is joined to $H_{alg}^m(G; R)$.

It turns out that the initial term of the spectral sequence $\{\widetilde{E}_r^{p,q}, \widetilde{d}_r^{p,q}\}$ is constructed maximally in simple fashion. Specifically, we have the following:

<u>PROPOSITION 2.8.</u> $\widetilde{E}_1^{0,q} = \Omega_s^q(G^*)$ for all $q \geq 0$; $\widetilde{E}_1^{p,q} = 0$ for $p > 0$ and all $q \geq 0$. The differential $d_1^{0,q}$ coincides with differentiation of the forms.

Theorem 2.1 is an obvious consequence of Propositions 2.7 and 2.8.

<u>Proof of Proposition 2.8.</u> By definition, the element $\alpha \in \widetilde{E}_1^{p,q}$ is a function referring to each G-foliation $(\xi)p : E \to B$ an element $\alpha(\xi) \in \widetilde{E}_1^{p,q}(\xi) = H^0(X; \Omega^q(\xi^*))$, where if $\varphi : \xi_1 \to \xi_2$ is a mapping of the foliations, then $\varphi^*\alpha(\xi_2) = \alpha(\xi_1)$. Here φ^* is the mapping of corresponding cohomology groups induced by the mapping φ.

We first consider the case $p = 0$. Then $\alpha(\xi) \in H^0(X; \Omega^q(\xi^*))$ is the cross section of the sheaf $\Omega^q(\xi^*)$ over X, i.e., a smooth differential form of degree q of the foliation ξ^* over all X. Applying to this form the homomorphism induced by the mapping $\hat{p} : E \to E^*$, we obtain the form $\widetilde{\alpha}(\xi)$ of the foliation $(\xi)p : E \to X$, i.e., as proved in Subsection 2, a set of C_m^q layerwise smooth continuous functions $\widetilde{\alpha}^i(\xi)$ on E. Such functions are specified on the space of any G-foliation and in the mapping of one G-foliation into another one maps into another.

It follows from Propositions 1.1 and 1.3 that these functions are constants identical for all foliations. In fact, in the mapping induced by a continuous mapping constants are mapped into constants. Inasmuch as, according to Proposition 1.1, any G-foliation can be mapped into a foliation ξ^I, it is sufficient to prove the constancy of these functions for the foliations ξ^I. According to Proposition 1.3, the foliation ξ^I can be mapped into a foliation ξ^J, where $J = I \cup \{1, 2, \dots\}$, such that in the induced mapping the functions $\widetilde{\alpha}^i(\xi^J)$ are mapped into constants.

Thus, our form has the same coordinates for all foliations in all bases obtained by left displacements from some fixed basis in the Lie algebra of the group G. This has the precise meaning that the form $\widetilde{\alpha}(\xi)$ in each layer of the foliation ξ represents a left-invariant form, which is the same in all layers of the foliation ξ and is the same for all foliations ξ. This means that the form $\alpha(\xi)$, specified on the foliation ξ, was also invariant on every layer.

It is clear, on the other hand, that it is possible by means of every invariant form of the homogeneous space G^* to define a form on all foliations $(\xi^*)p^* : E^* \to X$ in a manner that is natural in the category of G-foliations.

Thus, we have established an isomorphism between the group $\widetilde{E}_1^{0,q}$ and the group $\Omega_s^q(G^*)$ of invariant forms of degree q on G^*. This completes the proof of Proposition 2.8 for the case p = 0.

7. We now consider the case $p > 0$. Let $\alpha \in \widetilde{E}_1^{p,q}$, i.e., let $\alpha(\xi) \in H^p(X; \Omega^q(\xi^*))$ be specified for all G-foliations ξ. The homomorphism $H^p(X; \Omega^q(\xi^*)) = H^p(E^*; \widetilde{\Omega}^q(\xi^*)) \to H^p(E; \widetilde{\Omega}^q(\xi))$ is constructed in obvious fashion. It is a monomorphism. Thus, we consider a sheaf $\widetilde{\widetilde{\Omega}}^q(\xi)$ with base E^*. According to Proposition 2.2 $H^p(E; \widetilde{\Omega}^q(\xi)) = H^p(E^*; \widetilde{\widetilde{\Omega}}^q(\xi))$. In the sense of this equation our homomorphism $H^p(E^*; \widetilde{\Omega}^q(\xi^*)) \to H^p(E; \widetilde{\Omega}^q(\xi))$ induces the obvious embedding of sheaves $\Omega^q(\xi^*) \to \widetilde{\widetilde{\Omega}}^q(\xi)$. But the subsheaf $\Omega^q(\xi^*) \subset \widetilde{\widetilde{\Omega}}^q(\xi)$ is separated by a direct term; the mapping $\widetilde{\widetilde{\Omega}}^q(\xi) \to \Omega^q(\xi^*)$ is constructed by averaging the forms on compact layers of the foliation $\hat{\xi}$. Therefore, $H^p(X; \Omega^q(\xi^*)) \subset H^p(E; \widetilde{\Omega}^q(\xi))$. Furthermore, as shown above, $H^p(E; \widetilde{\Omega}^q(\xi)) = (H^p(E; \widetilde{R}))^{C^q_m}$. Finally, it is seen that $H^p(E; \widetilde{R}) = H^p(E; R)$ (we recall that the difference between the sheaves R and $\widetilde{R}$ lies in the fact that all continuous functions are considered in the construction of the sheaf R, while only layerwise smooth functions are considered in the construction of $\widetilde{R}$).

Thus, there occurs the monomorphism $H^p(X; \Omega^q(\xi^*)) \to [H^p(E; R)]^{C^q_m}$, which is unique in the category of G-foliations. Therefore, in order to show that $\alpha(\xi) = 0$, it is sufficient to prove the following. Let every G-foliation $(\xi)p: E \to X$ have referred to it an element $\beta(\xi) \in H^p(E; R)$, and if $\varphi: \xi_1 \to \xi_2$ is a mapping of foliations, let $\varphi^* \beta(\xi_2) = \beta(\xi_1)$. Then $\beta(\xi) \equiv 0$.

We note first of all that the equation $\beta(\varepsilon) = 0$ is obvious for the foliation $(\varepsilon): G \to *$, because $H^p(G; R) = 0$ for $p > 0$. Consequently, it is valid also for any trivial foliation, since a trivial foliation can be mapped into ε. We denote by $(p^*\xi)\widetilde{p}: \widetilde{E} \to E$ the foliation induced by a certain G-foliation $(\xi)p: E \to X$ in the mapping p. Clearly, $p^*\xi$ is a trivial foliation, therefore, $\widetilde{E} = E \times G$. The obvious mapping $\varphi: p^*\xi \to \xi$ has the property that $(p^*\xi)^*: H^p(E; R) \to H^p(\widetilde{E}; R)$ is an isomorphism. Therefore, $\beta(\xi) = 0$.

Proposition 2.8 and, along with it, Theorem 2.1 have been completely proved.

8. The result obtained above expresses our new invariants in terms of a well-studied object, viz., invariant forms on a homogeneous space. The classical theory of Lie groups makes it possible in certain cases to give another formulation of the result. We present a few such formulations and examples. All the facts required for the proof may be found in the Sophus Lie Seminar [5].

<u>THEOREM 2.9.</u> Let G be a semisimple Lie group, $\overline{G}$ its compact form, and $\hat{G}$ the maximal compact subgroup. Then $H^q_{alg}(G; R) = H^q(\overline{G}/\hat{G})$, where $H^q(Y)$ is a group of cohomologies of the space Y with real coefficients.

<u>Example.</u> Let G = SL(n; R) be a group of matrices of order n with determinant equal to one. Then $\overline{G} = SU(n)$, $\hat{G} = SO(n)$, $H^q_{alg}(SL(n; R); R) = H^q(SU(n)/SO(n))$. For example, $H^*_{alg}(SL(2; R)) = H^*(S^2)$.

<u>THEOREM 2.10.</u> Let G be a solvable Lie group, $\mathfrak{G}$ its algebra. Then $H^q_{alg}(G; R) = H^q(\mathfrak{G}; 1)$ (homologies of a Lie algebra with coefficients in the unity representation).

<u>Example.</u> Let G = R be the group of real numbers. Then $H^q_{alg}(G; R) = H^1_{alg}(G; R) = R$; $H^q_{alg}(G; R) = 0$ for $q > 0$. The nontriviality of a one-dimensional group of cohomologies is already apparent from the Introduction; the group $H^1(X; R)$ is isomorphic to the group of classes of equivalent principal R-foliations over the space X.

<u>THEOREM 2.11.</u> There exists a cohomological spectral sequence $\{E_r^{p,q}, d_r^{p,q}\}$ for which $E_2^{p,q} = H^p_{alg}(G; R) \otimes H^q(G)$, E^m_∞ is joined to the m-th group of cohomologies of the Lie algebra of a group G.

Consequently, in a certain sense the groups $H^q_{alg}(G; R)$ "measure the distance" between the groups of cohomologies of a group itself and its algebras, which, as we know [5], coincide for any compact Lie group.

3. Cohomologies with Coefficients in a Representation

1. Let M be a certain representation of a group G, i.e., a linear topological space (finite-dimensional or infinite-dimensional), and the continuous action $G \times M \to M$ of the group G on this space as a group of linear operators. For any principal G-foliation $(\xi)p : E \to B$ we define an induced foliation with layer M; the space of this foliation is the space of orbits $M \times E/G$, where the group G acts on $M \times E$ coordinatewise, and the projection is uniquely defined. It is clear that the mapping of G-foliation $\varphi : \xi_1 \to \xi_2$ corresponds to a mapping of such induced foliations. We also denote the sheaf of germs of continuous cross sections of the induced foliation over X by the letter M. An element $\alpha \in H^q_{alg}(G, M)$ is, by definition, a function referring to the G-foliations $(\xi)p : E \to X$ an element $\alpha(\xi) \in H^q(X; M)$ in a manner natural with regard to the mapping of foliations.

The calculation of the groups $H^q_{alg}(G; M)$ follows verbatim the calculation of the groups $H^q_{alg}(G; R)$ in Sec. 2. We present herein only the final formulations, viz., the analogs of Theorem 2.1 and Proposition 2.7.

Let Y be a homogeneous space of the group G. We denote by $\Omega^q(Y; M)$ a space of smooth differential forms of degree q on Y with values in M. The action of the group G on the space $\Omega^q(Y; M)$ is defined as follows. Let $\varphi \in \Omega^q(Y; M)$ be a form, and let $\varepsilon_1, \ldots, \varepsilon_q$ be vectors from the tangential space at any point $x \in Y$, $g \in G$. Then $[g\varphi](\varepsilon_1, \ldots, \varepsilon_2) = g(\varphi(g^{-1}\varepsilon_q))$ $(g^{-1}\varepsilon_1, \ldots, g^{-1}\varepsilon_q$ are vectors planted at the point $g^{-1}x \in Y)$. If any element of the group G keeps the form φ in place, then such a form is called invariant. We denote by $\Omega^\alpha_S(Y; M)$ a subgroup of invariant forms.

THEOREM 3.1. The groups $H^q_{alg}(G; M)$ coincide with the groups of homologies of the complex

$$\Omega_S(Y; M) \quad \Omega^0_S(Y; M) \xrightarrow{d_0} \Omega^1_S(Y; M) \xrightarrow{d_1} \ldots .$$

THEOREM 3.2. There exists a cohomological spectral sequence $\{E^{p,q}_r, d^{p,q}_r\}$, such that $E^{p,q}_2 = H^p_{alg}(G; M) \otimes H^q(\hat{\mathfrak{G}}; M)$, and E^m_∞ is joined to $H^m(\mathfrak{G}; 1)$. Here $\hat{\mathfrak{G}}$ is the subalgebra corresponding to the maximal compact subgroup, and the groups of cohomologies of an algebra with coefficients in the representation are defined as in [5].

2. Another interpretation of the foregoing results may be given in terms of the functor Ext in the category of representations (in the sense of the definition given above).

THEOREM 3.3. $H^q_{alg}(G; M) = \mathrm{Ext}^q(1, M)$.

Proof. Let us examine the exact sequence

$$0 \to M \xrightarrow{j} \Omega^0(G^*; M) \xrightarrow{d_0} \Omega^1(G^*; M) \xrightarrow{d_1} \ldots$$

of representations of the group G (its exactness follows from the acyclicity of the space G*). Here the first homomorphism j is an embedding of the group of constants into the group of all smooth functions.

The representations $\Omega^q(G^*; M)$, $q \geq 0$, are injective in the category of all representations. We prove this, for example, for $\Omega^0(G^*; M)$.

Let us consider the exact sequence $0 \to A \to B$ and the homomorphism $A \to \Omega^0(G^*; M)$. This homomorphism refers to every $a \in A$, a function α on G*. We denote by $\alpha_0 \in M$ the value of this function at the reference point * (i.e., at the point whose stationary subgroup is the group $\hat{G}$). We obtain the homomorphism $A \to M$, which is commutative, in general, with the action, not of the whole group G, but only of the subgroup $\hat{G}$. On the other hand, the homomorphism $A \to M$ is commutative with the action of the group $\hat{G}$, whose definition is uniquely extended to the homomorphism $A \to \Omega^0(G^*; M)$, which is commutative with the action of the whole group G. Our problem – to continue the homomorphism $A \to \Omega^0(G^*; M)$ to the homomorphism $B \to \Omega^0(G^*; M)$ – is therefore equivalent to the problem of continuing the homomorphism $A \to M$ to the homomorphism $B \to M$, which is commutative with the action of the group B. But every representation of a compact group is completely reducible, so that $B = A + C$, where C is a representation of the group $\hat{G}$ (A and B in this equation are also interpreted as representations of the group $\hat{G}$). Consequently, the desired continuation is possible. The injectivity of all the remaining moduli is proved analogously.

By definition, $\mathrm{Ext}^q(1; M)$ represents the q-th group of homologies of the complex

$$\mathrm{Hom}(1, \Omega^0(G^*; M)) \to \mathrm{Hom}(1, \Omega^1(G^*; M)) \to \ldots.$$

However, to map 1 into $\Omega^q(G^*; M)$ is the same as picking an invariant element in $\Omega^q(G^*; M)$; therefore, $\mathrm{Hom}(1, \Omega^q(G^*; M)) = \Omega^q_s(G^*; M)$, and $\mathrm{Ext}^q(1; M) = H^q_{\mathrm{alg}}(G; M)$.

This completes the proof of the theorem.

4. Principal H-Foliations

1. Let H be a topological group. In using the term H-foliation, we mean a locally trivial principal H-foliation, where local triviality in this case is interpreted both in the conventional sense and (in the event of a Lie group H) in the sense of the definition given in Sec. 1; this does not affect the end result.

We denote by EH(X) the set of all principal H-foliations with base X. The symbol EH represents a contravariant functor, which we use to construct the set $EH_{\mathrm{alg}}(G)$. The element $\alpha \in EH_{\mathrm{alg}}(G)$ is a function referring a certain H-foliation with the same base to every G-foliation. This function is easily formulated if we specify the homormorphism $\varphi: G \to H$; the G-foliation $(\xi)p: E \to X$ corresponds to an H-foliation $E \times H/G \to X$. We will show that this comprises the complete set of examples.

THEOREM 4.1. The following one-to-one correspondence, natural to G and H, holds: $EH_{\mathrm{alg}}(G) = \widetilde{\mathrm{Hom}}(G, H)$. Here $\widetilde{\mathrm{Hom}}(G, H)$ is a set of classes of homomorphisms of G into H relative to the conventional relation; the homomorphisms $\varphi_1, \varphi_2: G \to H$ are equivalent if there exists an $h \in H$, such that $h\varphi_1(g)h^{-1} = \varphi_2(g)$ for all $g \in G$.

Proof. We have set in juxtaposition to every homomorphism $\varphi: G \to H$ an element $\varphi^* \in EH_{\mathrm{alg}}(G)$. It remains for us to show that every element of $EH_{\mathrm{alg}}(G)$ is an element of the same form and that $\varphi_1^* = \varphi_2^*$ when and only when $\varphi_1 \sim \varphi_2$.

Let $\alpha \in EH_{\mathrm{alg}}(G)$. We note first of all that since any H-foliation whose base is one point is trivial and since a trivial G-foliation can be mapped into a foliation $(\xi)p: G \to *$, the foliation $\alpha(\xi)$ is, therefore, trivial if the G-foliation ξ is trivial.

Let $(\xi)p: E \to X$ be a certain G-foliation. The foliation $p^*\xi$ with base E, as noted above, is trivial; consequently, the foliation $\alpha(p^*\xi)$ is also trivial, i.e., the points of its space can somehow be set in one-to-one correspondence with couples (x,h), where $x \in E$, $h \in H$. We pick an element $g \in G$. By the construction of the foliation $p^*\xi$, there exists a canonical isomorphism between the layers of this foliation that lie over the points $x \in E$ and xg. It maps the point (x, e) (e is the identity element of the group H) into some point (xg, h). Consequently, we have referred an element $h \in H$ to every point $x \in E$, i.e., we have obtained a continuous mapping $E \to H$. This continuous mapping is not uniquely renewed in the foliation ξ, but is constructed in it with a certain arbitrariness, depending on the choice of isomorphism between the foliation $p^*\xi$ and the trivial foliation. But, within the span of this arbitrariness, the given mapping can always be chosen for every foliation as a mapping that takes all of E into a single point $h \in H$ in such fashion that the correspondence $g \to h$ defines a homomorphism $G \to H$ (possibly, a layer for every foliation). This is done as follows. If $\varphi: \xi_1 \to \xi_2$ is a mapping of foliations and $\alpha_2: E_2 \to H$ is a mapping constructed on the element $g \in G$ for the foliation ξ_2, then for the foliation ξ_1 it is possible to adopt the mapping $\varphi'' \circ \alpha_2$ as the mapping corresponding to the element g. Therefore, according to Proposition 1.1, we need only regard the foliations ξI. According to Proposition 1.3, it is possible for every set I to indicate a set J and an embedding $I \subset J$, such that all mappings $\mathscr{E}^J \to H$ constructed on all $g \in G$ will be constant on $\mathscr{E}^I \subset \mathscr{E}^J$.

We are now left with the task of finding out to what extent the constructed homomorphisms $G \to H$ differ for different foliations. We note first of all that the trivialization of the foliation $p^*\xi$ is most easily changed by setting $[x, h] = (x, hh_0)$, where h_0 is a fixed element. Then the homomorphism $G \to H$ is replaced by an equivalent. Hence, this homomorphism $G \to H$ is defined at least correct to equivalence. Conversely, if two trivializations are chosen for a certain foliation over $\mathscr{E}^J$ and $[x, h_1] = (x, h_2)$, then the mapping $x \to h_1 h_2^{-1}$ is constant on $\mathscr{E}' \subset \mathscr{E}^J$, according to Proposition 1.3, and, therefore, the homomorphisms $G \to H$ constructed on the basis of these trivializations are equivalent.

This completes the proof of Theorem 4.1.

2. We point out that this proof did not rely on the connectedness of the group G. Inasmuch as for any space Y there exists a known isomorphism between the group $H^1(Y; R)$ and the group of principal R-foliations over Y, we can make use of Theorem 4.1 to extend all the results of Sec. 2 with $q = 1$ to the case when the group G is not connected. For example, as the reader can verify for himself, the following equation holds for the group Z of integers:

$$H^q_{\mathrm{alg}}(\mathbf{Z}; \mathbf{R}) = \begin{cases} \mathbf{R} & \text{for} \quad q = 0, 1, \\ 0 & \text{for} \quad q > 1. \end{cases}$$

We anticipate the results of the article to be valid for all Lie groups. The only place in the article where it would not be permissible to regard the group G as any group whatsoever was in the construction of the foliation $\mathscr{E}'$ (connectedness of the group was also used in Sec. 2; we treated the space G^* as acyclic). For a generalization of the results of the article, therefore, it is sufficient to construct certain G-foliations analogous in their properties to the foliations $\mathscr{E}'$ for any group G.

LITERATURE CITED

1. M. Atiyah, Lectures on the Equivariant K-Theory, University of Warwick (1965).
2. I. Dixmier and A. Douady, "Champs continus d'espaces hilbertiens et de C* -algebres [Continuous fields of Hilbert spaces and of C*-algebras]," Bull. Soc. Math. France, 91 (1963), 227-284.
3. I. Dixmier, "Champs continus d'espaces hilbertiens et de C* -algebres [Continuous fields of Hilbert spaces and of C* -albegras]," J. Math. Pure et Appl., 128, (1963), 1-20.
4. I. M. Gel'fand and D. B. Fuks, "Cohomologies of Lie groups with real coefficients," Dokl. Akad. Nauk SSSR, 176, No. 1 (1967).
5. Group Topology and Lie Algebra (collected papers) [Russian translation], IL, Moscow (1965).
6. R. Godement, Algebraic Topology and Sheaf Theory [Russian translation], IL, Moscow (1964).

3.

(with D. B. Fuks)

The cohomologies of the Lie algebra of the vector fields in a circle

Funkts. Anal. Prilozh. **2** (4) (1968) 92-93 [Funct. Anal. Appl. **2** (1968) 342-343].
MR 39:6348a. Zbl. **176**:115

In any manifold M, the space of tangential vector fields of class C^∞ is a Lie algebra with respect to the operation of commutation of differential operators of the first order. Let us denote this algebra by $\mathfrak{A}(M)$.

We recall that the cohomologies of the Lie algebra $\mathfrak{A}$ with coefficients in the trivial one-dimensional representation are the cohomologies of the complex (C_q, ∂_q), where C_q is the space of real, skew-symmetric, q-linear systems in $\mathfrak{A}$, $\partial_q \varphi(x_1, \ldots, x_{q+1}) = \sum\limits_{1 \leqslant i < j \leqslant q+1} (-1)^{i+j-1} \varphi([x_i, x_j], x_1, \ldots \hat{i}\, \hat{j} \ldots, x_{q+1})$, and multiplication μ: $C_q \otimes C_{q'} \to C_{q+q'}$ is given by the formula

$$\varphi \cdot \psi(x_1, \ldots, x_{q+q'}) = \sum_{1 < i_1 < \ldots < i_q \leqslant q+q'} (-1)^{i_1 + \ldots + i_q - \frac{1}{2} q(q+q)} \times \varphi(x_{i_1}, \ldots, x_{i_q}) \cdot \psi(x_1, \ldots \hat{i}_1\, \hat{i}_2\, \hat{i}_q \ldots, x_{q+q'}).$$

The ring of the cohomologies of the Lie algebra $\mathfrak{A}(M)$ depends only on the manifold M. Let us denote this ring by $\mathfrak{H}^*(M)$. As far as we can judge, neither the covariant, nor the contravariant of the functor of the function $\mathfrak{H}$ has a natural structure. Evidently, $\mathfrak{H}^*(M)$ is an algebra over the field **R**.

Let us denote a circle by S.

THEOREM 1. The algebra $\mathfrak{H}(S)$ is a tensor product of the algebra of polynomials from a two-dimensional generator multiplied by an external algebra from a three-dimensional generator.

The scheme of the proof of this theorem is the following: a smooth vector field in a circle is a smooth function in it. Thus, a skew-symmetric, q-linear system in an algebra of vector fields is a skew-symmetric, generalized function of q variables having values in the circle. In terms of generalized functions, the differential ∂_q is written by the formula

$$\partial_q \varphi(s_1, \ldots, s_{q+1}) = \frac{1}{2} \sum_{1 \leqslant i < j \leqslant q+1} (-1)^{i+j-1} [\varphi(s_i, s_1, \ldots \hat{i}\, \hat{j} \ldots, s_{q+1}) + \varphi(s_j, s_1, \ldots \hat{i}\, \hat{j} \ldots, s_{q+1})]\, \delta'(s_i - s_j). \tag{1}$$

For convenience, we shall call a generalized function φ such that $\partial_q \varphi = 0$, a cycle; similarly, we shall call a generalized function from $\operatorname{Im} \partial_{q-1}$, a boundary; in speaking about the product of cycles, we shall regard it as multiplication μ.

From Formula (1), it is easy to deduce that each cycle is a generalized function concentrated into a set of points $(s_1, \ldots, s_q)$ such that $s_i = s_j$ for some i and j, $i \neq j$. Each such generalized function φ is in the form of an infinite sum

$$\sum_{1 \leqslant i < j \leqslant q,\ k = 0, 1, \ldots} \varphi_{ij}^k (s_i, s_1, \ldots \hat{i}\, \hat{j} \ldots, s_q)\, \delta^{(k)}(s_i - s_j),$$

where φ_{ij}^k are skew-symmetric, generalized functions in all of the variables except the first. Applying Formula (1) to this sum and examining the resulting expression, we can observe that each function φ_{ij}^k is concentrated into a set of points $(\sigma_1, \ldots, \sigma_{q-1})$, such that $\sigma_l = \sigma_m$ for some l and m, $1 < l < m \leq q-1$. This circumstance permits us to write the function φ in the form of a sum, in which each summand contains a product of two δ functions or of their derivatives.

Analogously, we arrive at the following result: each cycle, correct to within the boundary and the factorable cycles, is equal to the generalized function φ_0 concentrated into the set $s_1 = \ldots = s_q$. We can also show that $\varphi_0(s_1, \ldots, s_q = \varphi_0(s_1 + s, \ldots, s_q + s)$ for any $s \in S$.

Moscow State University. Translated from Funktsional'nyi Analiz i Ego Prilozheniya, Vol. 2, No. 4, pp. 92-93, October-December, 1968. Original article submitted July 5, 1968.

Every such function φ_0 is presented in the form of a linear combination with constant coefficients of the functions

$$\delta_P = P(\partial/\partial s_1, \ldots, \partial/\partial s_q)[\delta(s_1 - s_2) \ldots \delta(s_1 - s_q)],$$

where P is a skew-symmetric polynomial. In this connection, $\delta P = 0$ when and only when $P(\xi_1, \ldots, \xi_q)$ is divisible by $\xi_1 + \ldots + \xi_q$. The simple computation yields

$$\partial_q \delta_P = \delta_{\nabla P},$$

where

$$\nabla P(\xi_1, \ldots, \xi_{q+1}) = \sum_{1 \leq i < j \leq q+1} (-1)^{i+j-1} P(\xi_i + \xi_j, \xi_1, \ldots \hat{i}\, \hat{j} \ldots, \xi_{q+1}).$$

Let us denote by F/q the quotient space of skew-symmetric polynomials from the q variables in the subspace of polynomials of the form $P \cdot \Sigma$, where $\Sigma(\xi_1, \ldots, \xi_q) = \xi_1 + \ldots + \xi_q$. From what has been said above, it follows that for each multiplicative generator of the ring $\mathfrak{H}^*(S)$, there corresponds an additive generator of the group of cohomologies of the complex $\{F_q, \nabla\}$.

<u>LEMMA</u>. The cohomologies of the complex $\{F_q, \nabla\}$ have two generators; their dimensionalities are two and three.

Although its proof is elementary, the lemma is not completely evident. The generators of the groups of two-dimensional and three-dimensional cohomologies are in correspondence with the polynomials $P_1(\xi_1, \xi_2) = \xi_1^3 - \xi_2^3$ and $P_2(\xi_1, \xi_2, \xi_3) = (\xi_1 - \xi_2)(\xi_1 - \xi_3)(\xi_2 - \xi_3)$. In this connection, $\nabla P_1 = \Sigma P_2$, and $\nabla P_2 = 0$.

From the Lemma, it follows that the ring $\mathfrak{H}^*(S)$ is multiplicatively generated by the cohomological classes of generalized functions $\varphi_1(s_1, s_2) = \delta'''(s_1 - s_2)$ and $\varphi_2(s_1, s_2, s_3) = (\partial/\partial s_1 - \partial/\partial s_2)(\partial/\partial s_1 - \partial/\partial s_3)(\partial/\partial s_2 - \partial/\partial s_3)[\delta(s_1 - s_2)\delta(s_1 - s_3)]$. To prove the Theorem, we must still verify that among these classes there are no other relations except the relation of skew commutation.

The result can be used to compute the groups $H^q_{alg}(G; \mathbb{R})$ (see [1]), where G is the group of all the diffeomorphisms of the circle. (We recall that according to Theorem 2 of our note [2], the groups $H^q_{alg}(G; \mathbb{R})$ are isomorphic to the groups $H^q_C(G; \mathbb{R})$ considered by van Est [3] "of continuous cohomologies of the group G in the sense of Eilenberg-MacLane." Besides, we have not yet published the proof of this Theorem.) The computation of the groups $H^q_{alg}(G; \mathbb{R})$ basically repeats their computation for the Lie groups in [1]. We thus obtain

<u>THEOREM 2.</u> The algebra $H^*_{alg}(G; \mathbb{R}) = \sum_q H^q_{alg}(G; \mathbb{R})$ is a tensor product of the algebra of polynomials from two two-dimensional generators multiplied by an external algebra from a three-dimensional generator.

LITERATURE CITED

1. I. M. Gel'fand and D. B. Fuks, Funksional'. Analiz. i Ego Prilozhen., 1, No. 4, 33-35 (1967).
2. I. M. Gel'fand and D. B. Fuks, Dokl. Akad. Nauk SSSR, 181, No. 3, 515-518 (1968).
3. W. T. van Est, Indag. Math., 15, No. 5, 484-492 (1953).

4.

(with D. B. Fuks)

Cohomologies of the Lie algebra of tangential vector fields of a smooth manifold

Funkts. Anal. Prilozh. **3** (3) (1969) 32–52 [Funct. Anal. Appl. **3** (1969) 194–210].
MR **41**:1067. Zbl. **216**:203

0.1. With any vector field there is associated a whole series of distinct differential objects: smooth functions, differential forms, differential operators, smooth vector and tensor fields, and so on. From these differential objects are constructed spaces naturally equipped with further algebraic structures. In this way associative algebras of smooth functions and differential operators, the Lie algebra of vector fields, and of complex exterior differential forms are obtained. Each of the enumerated algebraic structures defines uniquely to within a diffeomorphism the original manifold, although in practice it is difficult to use this fact since all the mentioned spaces are infinite dimensional.

0.2. Nevertheless the mentioned fact in principle creates the possibility of expressing the usual invariants of the smooth manifolds (both homotopic and differential topologic) in terms of the differential objects enumerated in (0.1). There exist a whole series of relations of such a type: first the classical theorem of de Rahm, and also the formulas expressing the characteristic classes of manifolds in terms of the elements of a Riemannian structure [3], the formula of Atiyah–Singer for the index of an elliptic operator, and the recent results of Singer on the Whitehead invariant and so on. A common characteristic in the proof of similar theorems consists in the fact that a certain expression containing as parameters the elements of an infinite dimensional space in fact depends on a finite parameter (for example the integral of a closed form over a circle depends only on the cohomology class of this form; the index of the elliptic operator $d: \Gamma(E_1) \to \Gamma(E_2)$ depends on the element $\varkappa(\sigma(d)) \in K(T(E_1))$.

0.3. Another approach to the study of differential objects on a smooth manifold is possible. In the homological algebra there are definitions of cohomologies of associative algebras and the Lie algebras, and these cohomologies can be shown to be finite dimensional even if the original algebras are infinite dimensional. There arises the problem of finding those invariants connected with the algebraic objects enumerated in (0.1), which are finite dimensional for all smooth manifolds or for a sufficiently wide class of them.

0.4. In the present work we study the cohomologies of the Lie algebra of smooth vector fields on a closed smooth manifold with coefficients in a trivial representation. The principal result of this paper is Theorem 8.7 on the finite dimensionality of the considered cohomologies. A number of concrete results are also obtained; namely, in a standard complex for the calculation of the homologies of the Lie algebra of vector fields a subcomplex, designated diagonal by us, whose homologies are also differential invariants of the manifold and therefore are of intrinsic interest. For the homologies of this complex a spectral sequence is constructed whose second term is a tensor product of the homologies of the manifold on a calibrated finite dimensional space D, depending on the dimensionality of the manifold (for one dimensional manifolds D is one-dimensional, for two-dimensional ones, it is five-dimensional). The homologies of the diagonal of a complex are connected in a definite manner with the cohomologies of the algebra of vector fields under examination here. This relation allows us to prove the already-mentioned Theorem 8.7 and also to calculate in § 9 the cohomologies of the Lie algebra of vector fields on an n-dimensional torus.

0.5. Our method is also applicable to the solution of other similar problems. We formulate some of them:

Moscow State University. Translated from Funktsional'nyi Analyiz i Ego Prilozheniya, Vol. 3, pp. 32–52, July–September, 1969. Original article submitted March 24, 1969.

1. Vector fields of the manifold M vanishing on its subset A constitute a Lie algebra. To find its cohomologies.

2. If the manifold M is equipped with a Riemannian contact or symplectic structures then the vector fields coinciding with these structures constitute a Lie algebra. To find their cohomologies.*

3. To find the cohomologies of the algebra of vector fields with coefficients in non-trivial representations of this algebra (in the space of smooth functions on a manifold, in the spaces of vector and tensor fields and so on).

We propose to devote subsequent publications to these questions.

0.6. The present article was preceded by two short publications [1, 2]. Some of the results announced there are proved here.

§ 1. BASIC DEFINITIONS

1.1. Let M be a smooth† n-dimensional manifold without an edge. We denote through $\mathfrak{A}(M)$ the Lie algebra of smooth finite vector fields on M in a C^∞ topology with an intrinsic commutation operator, and by $\mathfrak{H}^*(M)$ the calibrated algebra of the cohomologies of the algebra $\mathfrak{A}(M)$ with coefficients in a trivial real representation (i.e. in the field $\mathbf{R}$ with a trivial action of the algebra $\mathfrak{A}(M)$). We recall that the groups $\mathfrak{H}^q(M)$ are defined as groups of homologies of the complex $\mathscr{C}(M) = \{C^q, d^q\}$ where C^q is the space of skew-symmetric q-liner continuous real forms on $\mathfrak{A}(M)$, the homomorphism $d = d^q: C^q \to C^{q+1}$ is given by the formula

$$dL(\xi_1, \ldots, \xi_{q+1}) = \sum_{1 \leqslant s < t \leqslant q+1} (-1)^{s+t-1} L([\xi_s, \xi_t], \xi_1, \ldots \hat{\xi}_s \ldots \hat{\xi}_t \ldots, \xi_{q+1}),$$

where $\xi_1, \ldots, \xi_{q+1} \in \mathfrak{A}(M)$ and $[\eta, \zeta]$ denotes the commutator of the vector fields η, ζ. The multiplication $C^q \otimes C^r \to C^{q+r}$ is defined by the formula

$$L_1 \cdot L_2(\xi_1, \ldots, \xi_{q+r}) = \sum_{1 \leqslant i_1 < \ldots < i_q \leqslant q+r} (-1)^{i_1 + \ldots + i_q - (q(q+1))/2} \times L_1(\xi_{i_1}, \ldots, \xi_{i_q}) \cdot L_2(\xi_1, \ldots \hat{\xi}_{i_1} \ldots \hat{\xi}_{i_2} \ldots \hat{\xi}_{i_q}, \ldots, \xi_{q+r}).$$

We shall denote the elements of the space C^q as q-dimensional cochains; the elements of the space $\operatorname{Ker} d^q \subset C^q$ as cocycles; cocycles whose difference lies in $\operatorname{Im} d^{q-1}$ we call cohomologies.

1.2. Our immediate objective is to define a filtration in the space C^q. First, we shall give the formal definition and after that we shall clarify it by a number of examples. We say that cochain $L \in C^q$ has a filtration $\leq k$ if, from $L(\xi_1, \ldots, \xi_q) \neq 0$, it follows that there exist k points $x_1, \ldots, x_k \in M$ such that each of the vector fields $\xi_1, \ldots, \xi_q$ is not identically zero in the neighborhood of at least one of the points $x_1, \ldots, x_k$. A filtration ≤ 0 has only a zero cochain: the condition which must be satisfied by the fields $\xi_1, \ldots, \xi_q$ with $L(\xi_1, \ldots, \xi_q) \neq 0$ in this case is not fulfilled. Conversely, any cochain has a filtration $\leq q$: if $L(\xi_1, \ldots, \xi_q) \neq 0$ then all the fields $\xi_1, \ldots, \xi_q$ are nonzero, i.e. there exist points $x_1, \ldots, x_q$ such that the field ξ_i is different from zero at the point x_i (this means neither identically equal to zero at the point nor in any neighborhood of the point) for $i = 1, \ldots, q$. The cochain $L \in C^q$ has a filtration ≤ 1 if $L(\xi_1, \ldots, \xi_q)$ can be nonzero only when the intersection $\bigcap_{i=1}^{q} S(\xi_i)$ is nonempty, where $S(\xi)$ is the closure of the set of those points of the manifold M in which the vector field ξ is different from zero. Further, the cochain $L \in C^q$ has a filtration $\leq q-1$ if, from $L(\xi_1, \ldots, \xi_q) \neq 0$, it follows that at least one of the pairwise intersections of the sets $S(\xi_1)$, $\ldots$, $S(\xi_q)$ is nonempty.

We denote by C^q_k the subspace of the space C^q consisting of the elements of the filtration $\leq k$. It is clear that $C^q_k = 0$ for $k \leq 0$, $C^q_k = C^q$ for $k \geq q$ and $0 = C^q_0 \subset C^q_1 \subset \ldots \subset C^q_q = C^q$.

*One-dimensional cohomologies of some of these algebras were found recently by V. I. Arnold. (This work is, as yet, unpublished.)

†The smoothness of manifolds, reflections, vector fields, etc., are everywhere understood as pertaining to the class.

1.3. The image of the space C_k^q under a homomorphism d^q is contained in C_k^{q+1}.

Proof. Let $L \in C_k^q$ and let $dL(\xi_1, \ldots, \xi_{q+1}) \neq 0$. In view of the definition of the homomorphism d this means that $L([\xi_s, \xi_t], \xi_1, \ldots \hat{\xi}_s \ldots \hat{\xi}_t \ldots, \xi_{q+1}) \neq 0$ for some s, t. This in turn means that there exist such points $x_1, \ldots, x_k$ that each of the fields $[\xi_s, \xi_t], \xi_1, \ldots \xi_s \ldots \xi_t \ldots, \xi_{q+1}$ is different from zero in the neighborhood of one of these points. It is clear that the two fields ξ_s, ξ_t are different from zero in any neighborhood in which the field $[\xi_s, \xi_t]$ is different from zero. Therefore each of the fields $\xi_1, \ldots, \xi_{q+1}$ is different from zero in the neighborhood of one of the points. Hence $dL \in C_k^{q+1}$.

1.4. From the proposition proved it follows that the subspaces $C_1^q \subset C^q$ consist of a subcomplex of of the complex $\mathscr{C}$ (M) and this subcomplex is denoted by $\mathscr{C}_1$ (M) and is called diagonal (the reason for this statement will become clear in § 2). We emphasize that a diagonal complex, in general, is not closed under multiplication.

1.5. If $dL \in C_{q-1}^{q+1}$ then $L \in C_{q-1}^q$. In particular, any q-dimensional cochain belongs to the space C_{q-1}^q.

Proof. We must establish that if $dL \in C_{q-1}^{q+1}$, then for any $\xi_1, \ldots, \xi_q \in \mathfrak{A}(M)$ such that the sets $S(\xi_i)$ do not intersect pairwise, the equality $L(\xi_1, \ldots, \xi_q) = 0$ is valid. Representing each of the fields ξ_i in the form of a sum $\sum_k \varphi_k \cdot \xi_i$, where $\{\varphi_k\}$ is the partition of unity subjected to a sufficiently small local finite coordinate covering of the manifold M, we reduce this statement to the case when each of the sets $S(\xi_i)$ is contained in the open set $U_i \subset M$ which is diffeomorphic to $\mathbf{R}^n$ and the sets U_i do not intersect pairwise.

LEMMA. It is possible to represent any finite, smooth vector field on $\mathbf{R}^n$ in the form of a finite sum $\sum_k [\eta_k, \zeta_k]$, where η_k, ζ_k are smooth finite vector fields on $\mathbf{R}^n$.

Before proving the lemma we show how to complete the proof of our statement with its help. Since $S(\xi_1) \subset U_1$ and U_1 is diffeomorphic to $\mathbf{R}^n$, then $\xi_1 = \sum_k [\eta_k, \zeta_k]$ where $S(\eta_k), S(\zeta_k) \subset U_1$. Since for every k among the sets $S(\eta_k), S(\zeta_k), S(\xi_2), \ldots, S(\xi_q)$ the only intersecting pair is $S(\eta_k), S(\zeta_k)$ then $dL(\eta_k, \zeta_k, \xi_2, \ldots, \xi_q) = L([\eta_k, \zeta_k], \xi_2, \ldots, \xi_q)$ and hence, $\sum_k dL(\eta_k, \zeta_k, \xi_2, \ldots, \xi_q) = L \sum_k [\eta_k, \zeta_k], \xi_2, \ldots, \xi_q = L(\xi_1, \ldots, \xi_q)$. Finally, since $dL \in C_{q-1}^{q+1}$, therefore $dL(\eta_k, \zeta_k, \xi_2, \ldots, \xi_q) = 0$ for all k: in the opposite case q + 1 points can be found such that each of the q + 1 vector fields η_k, ζ_k, ξ_k is different from zero in the neighborhood of one of them which is only possible if, among the sets $S(\eta_k), S(\zeta_k), S(\xi_2), \ldots, S(\xi_q)$ there are at least two intersecting pairs. Hence follows the derived equality $L(\xi_1, \ldots, \xi_q) = 0$.

Proof of the Lemma. We denote by $\varepsilon_k (k = 1, \ldots, n)$ a vector field, on $\mathbf{R}^n$ consisting of unit vectors parallel to the k-th coordinate axis. Any smooth finite vector field ξ on $\mathbf{R}^n$ can be represented in the form $\sum \varphi_k \varepsilon_k$, where φ_k $(k = 1, \ldots, n)$ are smooth finite functions on $\mathbf{R}^n$. It is easy to verify that for any k any smooth finite function φ_k can be represented in the form of a finite sum $\sum_i \left(\psi'_{i,k} \frac{\partial \psi''_{i,k}}{\partial x_k} - \psi''_{ik} \frac{\partial \psi'_{ik}}{\partial x_k} \right)$, where ψ'_{ik}, $\psi''_{i,k}$ are smooth finite functions (in fact, two terms are sufficient; moreover for the functions $\psi'_{1,k}, \psi'_{2,k}$ it is possible to take finite functions coinciding in a sphere of sufficiently large radii with 1 and x_k, respectively). We assume $\eta_{i,k} = \psi'_{i,k}\varepsilon_k$, $\zeta_{i,k} = \psi''_{i,k}\varepsilon_k$. It is clear that $\xi = \sum_{i,k} [\eta_{i,k}, \zeta_{i,k}]$.

§ 2. GENERALIZED SECTIONS

2.1. Definition. By a generalized section of a smooth vector fiber ν over a smooth manifold N is meant a continuous linear functional on the space $\Gamma(\nu')$ of finite smooth sections of a fiber ν' associated with the fiber ν.*

*The title "Generalized Section" is surely not entirely relevant here: the ordinary smooth sections of fiber ν are not, in the sense of our definition, generalized sections of this fiber. This defect disappears for fibers equipped with definite additional structures which will only be discussed starting from § 3.

In particular, for a generalized function on N it is a generalized section of the trivial one-dimensional fiber.

2.2. If the fiber ν is trivial and $F_1, \ldots, F_n$ are its sections, then every generalized section of the fiber ν can evidently be written in the form $\sum_{i=1}^{n} \varphi_i F_i$, where φ_i are generalized functions on the manifold N.

2.3. If ν_1, ν_2 are smooth vector fibers over smooth manifolds N_1, N_2, then over the product $N_1 \times N_2$ are defined smooth vector fibers $\pi_1^*\nu_1, \pi_2^*\nu_2$, where $\pi_1 : N_1 \times N_2 \to N_1$, $\pi_2 : N_1 \times N_2 \to N_2$ are the projections. The tensor product $\Gamma(\nu_1) \otimes \Gamma(\nu_2)$ over R of the spaces of finite smooth sections of the fibers ν_1, ν_2 are naturally embedded in the space $\Gamma(\pi_1^*\nu_1 \otimes \pi_2^*\nu_2)$ of finite smooth sections of the fiber $\pi_1^*\nu_1 \otimes \pi_2^*\nu_2$ with the basis $N_1 \times N_2$.

Every continuous linear functional on the space $\Gamma(\nu_1) \otimes \Gamma(\nu_2)$ is extended up to a continuous linear functional on the space $\Gamma(\pi_1^*\nu_1 \otimes \pi_2^*\nu_2)$, and such an extension is unique.

This proposition is a generalization of the classical nuclear theorem of L. Schwartz (see [4]) and obviously reduces to this theorem with the help of the partition of unity in a local coordinate system.

2.4. A generalized section F of the fiber ν over N is called concentrated on a closed set $A \subset N$ if $F(\sigma) = 0$ for any sections σ of the fiber ν' coinciding with the zero section in the neighborhood of the set A.

For any generalized section F of the fiber ν concentrated on a compact set $A \subset N$ is to be found a number s such that if a finite smooth section σ of the fiber ν' has, at every point of the set A, a tangent of order $\geq s$ with a zero section, then $F(\sigma) = 0$.

This proposition is an extension of the well known theorem of the theory of generalized functions [5] and in a standard manner reduces to this theorem.

The smallest of the numbers s fulfilling the condition of last assumption for a given generalized section F is called the order of the generalized section F with respect to the set A.

2.5. We return to our manifold M. As is known, a finite smooth vector field on M is a finite section of its tangential fiber $\tau = \tau(M)$. A continuous q-linear real form on $\mathfrak{A}(M)$ is thus a continuous linear functional on the space $\underbrace{\Gamma(\tau) \otimes \ldots \otimes \Gamma(\tau)}_{q}$. In view of 2.3 this functional is uniquely extended up to a linear functional on the space $\Gamma(\tau_q)$ where $\tau_q = \pi_1^*\tau \otimes \ldots \otimes \pi_q^*\tau$, $\pi_i^*\tau$ is a vector fiber with the basis $M^q = \underbrace{M \times \ldots \times M}_{q}$, induced by the fiber τ under the projection $\pi_i : M^q \to M$ on the i-th factor. Thus, continuous q-linear real forms on $\mathfrak{A}(M)$ may be identified with generalized sections of the fiber $\tau_q' = \pi_1^*\tau' \otimes \ldots \otimes \pi_q^*\tau'$ with the basis M^q. The tensor product of spaces cotangent to the manifold M at the points $(x_1, \ldots, x_q)$ serves as a layer of the fiber τ_q' over the point $(x_1, \ldots, x_q)$.

2.6. The mapping $p_{ij} : M^q \to M^q$ permuting the i-th coordinate with the j-th is naturally covered by an automorphism of the fiber τ_q'. A generalized section L of the fiber τ_q' is called skew-symmetric if any automorphism p_{ij} converts it into $(-L)$. It is clear that the space of cochains C^q is not so different from the space of generalized skew-symmetric sections of the fiber τ_q'.

2.7. We denote by M_k^q the subset of the product M^q consisting of points $(x_1, \ldots, x_q)$ such that among the points $x_1, \ldots, x_q \in M$ no more than k are distinct. From the definition of a filtration in the space C^q it is easy to deduce that C_k^q is none other than the space of generalized skew-symmetric sections of the fiber τ_q' concentrated on the set M_k^q. In particular, the cochains of diagonal subcomplex $\mathscr{C}_1(M)$ of the complex $\mathscr{C}(M)$ are the generalized skew-symmetric sections of the fiber τ_q' concentrated on the diagonal $\Delta(M) \subset M^q$. Thus the title "diagonal" is explained.

2.8. We shall now assume that the manifold M is compact and denote the space of generalized sections of the fiber τ_q' by $C_{k,m}^q$, concentrated on M_k^q and having, with respect to M_k^q, an order $\geq m$. It is clear that $0 = C_{k-1}^q \subset C_{k,0}^q \ldots \subset C_k^q$. From the proposition 2.4 it follows that $\bigcup_m C_{k,m}^q = C_k^q$, i.e. that the spaces $C_{k,m}^q$ arise from an exhaustive filtration of the space C_k^q.

We note that the filtration $\{C^q_{k,m}\}$ is empty for q = k: since $M^q_q = M^q$, therefore any generalized section is concentrated on M^q_q and has the order 0 with respect to M^q_q.

2.9. We assume that $\xi_i : M^q \to \tau(M)$ is a smooth function connecting at the point $(x_1, \ldots, x_q) \in M^q$ the tangential vector $\xi_i(x_1, \ldots, x_q)$ to the manifold to the point M at the point x_i. The expression $\xi_1 \otimes \ldots \otimes \xi_q$ may be regarded as the cross section of the fiber τ_q over M^q, i.e. as an element of that space the functionals on which we have identified with the cochains from C^q. Thus, if $L \in C^{q-1}$, then

$$(d^{q-1}L)(\xi_1 \otimes \ldots \otimes \xi_q) = \sum_{1 \leq i < j \leq q} (-1)^j L(\eta^{(i,j)} \otimes \eta_1^{(i,j)} \otimes \ldots \hat{\eta}_i^{(i,j)} \ldots \hat{\eta}_j^{(i,j)} \ldots \otimes \eta_q^{(i,j)}),$$

where $\eta_k^{(i,j)}(x_1, \ldots, x_{q-1}) = \xi_k(x_2, \ldots, x_{i-1}, x_1, x_{i+1}, \ldots, x_{j-1}, x_1, x_{j+1}, \ldots, x_{q-1})$ and $\eta^{(i,j)}(x_1, \ldots, x_{q-1})$ are for every $x_2, \ldots, x_{q-1}$ the values at the point $x_1 \in M$ of a commutator of the vector fields $\zeta_1(x) = \xi_i(x_2, \ldots, x_{i-1}, x_{i+1}, \ldots, x_{j-1}, x, x_{j+1}, \ldots, x_{q-1})$, $\zeta_2(x) = \xi_j(x_2, \ldots, x_{i-1}, x, x_{i+1}, \ldots, x_{j-1}, x, x_{j+1}, \ldots, x_{q-1})$ on the manifold M. In particular, if q = 2, then $d^1L(\xi_1 \otimes \xi_2) = L(\eta)$, where $\eta = [\xi_1, \xi_2]$.

The condition for the functional $L \in C^q$ to be a member of the space $C^q_{k,m}$, formulated in 2.8, is evidently equivalent to this condition: if for the points $x_1, \ldots, x_q \in M$, among which no more than k are distinct, the functions $\xi_1, \ldots, \xi_q$ possess at the point $(x_1, \ldots, x_q)$ zero of multiplicities $m_1, \ldots, m_q$ such that $m_1 + \ldots + m_q > m$, then $L(\xi_1 \otimes \ldots \otimes \xi_q) = 0$. We say that a function $\xi : M^q \to \tau(M)$ has a zero of multiplicity m_0 at the point $(x_1, \ldots, x_q)$ if all the scalar functions, being given in a local coordinate, together with derivatives up to the order $m_0 - 1$ inclusive, vanish at the point $(x_1, \ldots, x_q)$; in particular, the function ξ has a zero of multiplicity 0 at any point, a zero of multiplicity 1 at any point $(x_1, \ldots, x_q)$ for which $\xi(x_1, \ldots, x_q) = 0$.

2.10. The image of the space $C^q_{k,m}$ under the mapping d^q is contained in $C^{q+1}_{k,m+1}$.

This follows from the following elementary fact. If the point x is with a zero of multiplicity m' of the field ξ', and with a zero of multiplicity m" for the field ξ'', then it is with a zero of m' + m" − 1 of the field $[\xi', \xi'']$.

2.11. If $L \in C^q_{k,m}$, $L' \in C^{q'}_{k',m'}$, then $L \cdot L' \in C^{q+q'}_{k+k',m+m'}$.

2.12. The proof is obvious. In view of 1.3 the homomorphism d^q induces a homomorphism $d^q_k : C^q_k / C^{q-1}_{k-1} \to C^{q+1}_k / C^{q+1}_{k-1}$ and in view of 2.10 the latter coincides with a filtration of the groups C^q/C^{q-1}_{k-1} with the subgroups $C^q_{k,m}/(C^q_{k,m} \cap C^q_{k-1})$. Therefore, for any k there arises a spectral sequence connecting $\bigoplus_{q,m} (C^q_{k,m}/(C^q_{k,m-1} + (C^q_{k,m} \cap C^q_{k-1})))$ with the homologies of the complex $\{C^q_k/C^q_{k-1}, d^q_k\}$. In § 3-7 we shall study in detail this spectral sequence for k = 1, and in § 8 obtain some results for k > 1.

We fix some definitions. Let $E_0^{q-m,m} = C^q_{1,m}/C^q_{1,m-1}$ and let $\delta_0^{q-m,m} : E_0^{q-m,m} \to E_0^{q-m,m+1}$ be a homomorphism obtained from $d^q|_{C^q_{1,m}} : C^q_{1,m} \to C^{q+1}_{1,m+1}$ by factorization. The complex $\mathscr{E}_0 = \{E_0^{u,v}, \delta_0^{u,v}\}$ is the zeroth term of the spectral sequence $\{E_r^{u,v}, \delta_r^{u,v} : E_r^{u,v} \to E_r^{u+r,v-r+1}\}$ and converges to the homologies of the diagonal complex $\mathscr{E}_1 = \{C^q, d^q|_{C^q_1}\}$.

§ 3. THE SPACE $E_0^{u,v}$ (DESCRIPTION IN AN INVARIANT TERMINOLOGY)

3.1. According to definition $E_0^{q-m,m} = C^q_{1,m}/C^q_{1,m-1}$. The space $C^q_{1,m}$ is the space of skew-symmetric functionals L on the space of smooth sections of the fiber τ_q such that $L(\sigma) = 0$ if the section σ, together with derivatives up to the order m in all directions, vanishes on $\Delta(M) \subset M^q$. We denote by A^q_m the space of these sections and set $B^q_m = A^q_{m-1}/A^q_m$. It is clear that $E^{q,m}$ is the space of functionals on B^q_m subjected to the additional condition of skew-symmetry.

3.2. We shall fix on the manifold M a Riemannian metric. This metric induces, on the one hand, a connection in the fiber τ_q, and on the other hand it induces the structure of a vector of O-fibers in a tubular neighborhood of the submanifold $\Delta(M)$ in M^q. Thanks to this for arbitrary vectors $\gamma_1, \ldots, \gamma_m$, normal to A(M) at some point $x \in \Delta(M)$ and for any smooth section σ of the fiber τ_q is defined the derivative $(\partial^m / \partial\gamma_1 \ldots \partial\gamma_m)\sigma$ whose values at the point x is a vector from a layer of the fiber τ_q over the point x. Relating this vector to the vectors $\gamma_1, \ldots, \gamma_m$ we evidently obtain a linear mapping $[d_m\sigma](x) : S^m\nu_q(x) \to \tau_q(x) = \otimes^q\tau(x)$ where $\nu_q(x)$ is a space, normal at the point x to $\Delta(M)$ (in the manifold M^q), $\tau(x)$ is the tangential space to $\Delta(M)$ at the point x, and by S^m and $\otimes^q$ we have denoted the operators of taking the m-th symmetry and q-th tensor class. We note that the space $\nu_q(x)$ is a factor-space of the space $\oplus\tau(x) = \underbrace{\tau(x) + \ldots + \tau(x)}_{q}$ 'diagonal' to the subspace $\Delta\tau(x)$ consisting of vectors of the form $(\xi, \ldots, \xi) \in \underset{q}{\oplus}\tau(x)$ with $\xi \in \tau(x)$.

The totality of the mappings $d_m\sigma(x)$ for all $x \in \Delta(M)$ defines a smooth section $d_m\sigma$ of a vector of the fiber $\mathcal{S}^q_m = \mathrm{Hom}(S^m\nu_q, \otimes^q\tau)$ with the basis $\Delta(M)$ canonically diffeomorphic to M (here ν_q and τ are tangential and normal fibers of the manifold $\Delta(M) \subset M^q$; from the remarks made above, it follows that ν_q is a factor-fiber of the fiber $\oplus\tau = \tau\oplus \ldots \oplus\tau$ with respect to the subfiber $\Delta\tau \subset \underset{q}{\oplus}\tau$).

The juxtaposition $\sigma \to d_m\sigma$ defines an isomorphism of the space B^q_m on the space of all smooth sections of the fibers $\mathcal{S}^q_m = \mathrm{Hom}(S^m\nu_q, \otimes^q\tau)$.

The proof is evident.

3.3. Since the elements of the space $E_0^{q-m,m}$ can be represented as functionals on B^q_m, they may also be interpreted as generalized sections of the fiber associated with $\mathcal{S}^q_m$. In addition, not every generalized section of this kind gives an element of $E_0^{q-m,m}$: in order for this it is still necessary to make use of the condition of skew symmetry. This condition in the new language signifies the following. The mapping $p_{ij} : M^q \to M^q$ transposes the i-th coordinate with the j-th identically on the diagonal $\Delta(M)$, but generates a non-trivial automorphism of the fibers ν_q and $\tau_q|_{\Delta(M)} = \otimes^q\tau$. Namely, in the fiber $\otimes^q\tau$ a permutation of the i-th and the j-th factors is induced, in the fiber ν_q also an automorphism, obtained from the permutation of the i-th and j-th terms in $\underset{q}{\oplus}\tau$, is induced. The obtained automorphisms of the fibers ν_q and $\otimes^q\tau$ generate an automorphism of the fiber $\mathcal{S}^q_m = \mathrm{Hom}(S^m\nu_q, \otimes^q\tau)$ which we denote through p^*_{ij}. It is clear that the space $E_0^{q-m,m}$ consists of such functionals F on the space of smooth sections of the fiber $\mathcal{S}^q_m$ for which $F \circ p^*_{ij} = -p^*_{ij} \circ F$ for $1 \le i < j \le q$. Such functionals may be regarded as generalized sections of some subfiber of the fiber associated with $\mathcal{S}^q_m$. Namely, for every point $x \in \Delta(M)$ there is defined a mapping $p^*_{ij}(x) : S^q_m(x) \to S^q_m(x)$ of the layer $S^q_m(x) = \mathrm{Hom}(S^q_m\nu_q(x), \otimes^q\tau(x))$ into itself. We denote through $P^q_m(x)$ the subspace of the space $(S^q_m(x))'$ consisting of those functionals $F : S^q_m(x) \to R$ for which $F \circ p^*_{ij}(x) = -p^*_{ij}(x) \circ F$ for $1 \le i < j \le q$. The union $\bigcup_{x \in \Delta(M)} P^q_m(x)$ is the space of vector fibers over $\Delta(M)$ which serves as the subfiber associated with $\mathcal{S}^q_m$. We denote this fiber by $\mathcal{P}^q_m$. Combining all the statements we obtain the interpretation of the space $E^{q-m,m}$ as the spaces of generalized sections of the fiber $\mathcal{P}^q_m$.

§ 4. THE SPACE $E_0^{u,v}$ (DESCRIPTION IN LOCAL COORDINATES)

In this paragraph we shall give a description of the fiber $\mathcal{P}^q_m$ (by the same token of the spaces $E_0^{q-m,m}$) in terms of local coordinates on the manifold M.

4.1. To begin with we describe two abstract linear spaces $\tilde{P}^q_m$ and P^q_m, the second of which, as we shall see further, is a standard example of the layer of a fiber $\mathcal{P}^q_m$.

An element of the space $\tilde{P}^q_m$ is the set $\{\lambda^\tau\}$ of homogeneous polynomials $\lambda^\tau(y_{11}, \ldots, y_{1n}; \ldots; y_{21}, \ldots, y_{2n}; \ldots; y_{q1}, \ldots, y_{qn})$ of degree m in nq variables $y_{11}, \ldots, y_{1n}; y_{21}, \ldots, y_{2n}; \ldots; y_{q1}, \ldots, y_{qn}$. These polynomials are enumerated by the sequences $\tau = \{\tau_1, \ldots, \tau_n\}$ of nonnegative integers with $\tau_1 + \ldots + \tau_n = q$. It is assumed that every polynomical λ^τ is skew-symmetric in the first τ_1 groups of n variables, in second τ_2 groups of n variables, and so on, i.e. if $\tau_1 + \ldots + \tau_s < r < \tau_1 + \ldots + \tau_{s+1}$ for some s, then

$$\lambda^\tau(y_{11}, \ldots; y_{1n}; \ldots; y_{q1}, \ldots, y_{qn}) = -\lambda^\tau(y_{11}, \ldots, y_{1n}; \ldots; y_{r+1,1}, \ldots, y_{r+1,n}; y_{r1}, \ldots, y_{rn}; \ldots; y_{q1}, \ldots, y_{qn}).$$

We denote through R^q_m a subspace of the space $\tilde{P}^q_m$ which is defined by the condition: the set $\{\lambda^\tau\}$ belongs to R^q_m if for any τ the polynomial λ^τ belongs to an ideal which is generated by the polynomials $\sigma_1 = y_{11} + \ldots + y_{q1}, \ldots, \sigma_n = y_{1n} + \ldots + y_{qn}$. We set $P^q_m = \tilde{P}^q_m / R^q_m$.

4.2. Now we demonstrate that the layer $P^q_m(x)$ of the fiber $\mathcal{F}^q_m$ over the point $x \in \Delta(M)$ is isomorphic to P^q_m and that the isomorphism $P^q_m(x) \to P^q_m$ is determined canonically if, $\tau(x)$ has a fixed basis in the space. It is possible to identify that space which is associated with $\mathrm{Hom}(S^m \nu_q(x), \otimes^q \tau(x))$ with $\mathrm{Hom} \cdot (\otimes^q \tau(x), S^m \nu_q(x))$.* Moreover it is possible to interpret $S^m \nu_q(x)$ as the space of homogeneous polynomials (forms) of degree m, given on the space $\nu_q'(x)$ associated with $\nu_q(x)$. The space $\nu_q'(x)$ is a subspace of the space $\bigoplus_q \tau'(x) = \tau'(x) \oplus \ldots \oplus \tau'(x)$. It is clear that every polynomial which is specified on $\nu_q'(x)$ can be extended up to a polynomial specified on $\bigoplus_q \tau'(x)$; and moreover the latter, which is not defined uniquely but to within the addition of a polynomial, vanishes on $\Delta\tau'(x) \subset \bigoplus_q \tau'(x)$.

We choose a basis $\{e_1, \ldots, e_n\}$ in the space $\tau(x)$. The dual basis in $\tau'(x)$ is denoted by $\{y_1, \ldots, y_n\}$. Then there arises a basis in $\bigoplus_q \tau'(x)$; we denote it by $\{y_{11}, \ldots, y_{1n}; y_{21}, \ldots, y_{2n}; \ldots; y_{q1}, \ldots, y_{qn}\}$. The subspace $\Delta\tau'(x) \subset \bigoplus_q \tau'(x)$ is spanned by the elements $y_{11} + \ldots + y_{q1}, \ldots, y_{1n} + \ldots + y_{qn}$. A basis in the space $\otimes^q \tau(x)$ is composed of elements $e_{i_1 \ldots i_q} = e_{i_1} \otimes \ldots \otimes e_{i_q}$. The homomorphism $\otimes^q \tau(x) \to S^m \nu_q(x)$ is an element of the space $P^q_m(x)$; the representation of such a homomorphism is equivalent to fixing for each set $\{i_1, \ldots, i_q\}$ of homogeneous polynomial $\lambda_{i_1 \ldots i_q}(y_{11}, \ldots, y_{1n}; \ldots; y_{q1}, \ldots, y_{qn})$ of degree m, specified to within a polynomial of the form: $\mu_1\sigma_1 + \ldots + \mu_n\sigma_n$. Thus

$$\lambda_{i_1 \ldots i_q}(y_{11}, \ldots, y_{1n}; \ldots; y_{q1}, \ldots, y_{qn}) = -\lambda_{i_1 \ldots i_{r+1} i_r \ldots i_q}(y_{11}, \ldots, y_{1n}; \ldots; y_{r+1,1}, \ldots, y_{r+1,n}; y_{r1}, \ldots, y_{rn}; \ldots; y_{q1}, \ldots, y_{qn}).$$

It is clear that all the polynomials $\lambda_{i_1 \ldots i_q}$ would be determined if one were given only polynomials $\lambda_{i_1 \ldots i_q}$ with $i_1 \le \ldots \le i_q$ satisfying the requirement that if $i_r = i_{r+1}$, then

$$\lambda_{i_1 \ldots i_q}(y_{11}, \ldots, y_{1n}; \ldots; y_{q1}, \ldots, y_{qn}) = -\lambda_{i_1 \ldots i_q}(y_{11}, \ldots, y_{1n}; \ldots; y_{r+1,1}, \ldots, y_{r+1,n}; y_{r1}, \ldots, y_{rn}; \ldots; y_{q1}, \ldots, y_{qn}).$$

Let $\tau = \{\tau_1, \ldots, \tau_n\}$ be a sequence of nonnegative integers with $\tau_1 + \ldots + \tau_n = q$. We set $\lambda^\tau = \lambda_{\underbrace{1\ldots1}_{\tau_1}\underbrace{2\ldots2}_{\tau_2}\ldots\underbrace{n\ldots n}_{\tau_n}}$. Evidently, the set $\{\lambda^\tau\}$ is an element of the space $\tilde{P}^q_m$, and consequently defines an element of the space P^q_m. It is clear that there arises the mapping $P^q_m(x) \to P^q_m$ and that this mapping is an isomorphism.

4.3. Fixing a local system of coordinates on the manifold in the neighborhood U of the point x, we also obtain the isomorphism $\mathcal{F}^q_m|_U = U \times P^q_m$. Therefore, every generalized section of the fiber $\mathcal{F}^q_m$ concentrated on the set U can be written in local coordinates in the form $\sum_{i=1}^N \varphi_i \lambda_i$, where φ_i is a generalized

* For any finite linear spaces, A, B there exists a canonical isomorphism $\mathrm{Hom}(B, A) \to (\mathrm{Hom}(A, B))'$. It relates an element $\varphi \in \mathrm{Hom}(B, A)$ to the functional $\psi \to \mathrm{Tr}(\psi \circ \varphi)$ on the space $\mathrm{Hom}(A, B)$.

function on M, concentrated on U, and $\{\lambda_i\}$ is an element of the space P^q_m (see 2.2). Using the partition of unity we can represent any generalized section of the fiber $\mathscr{P}^q_m$ in the form of a linear combination of such generalized sections.

4.4. It is possible to construct the isomorphism $\mathscr{P}^q_m|_U = U \times P^q_m$ not only if there is given, in U a local coordinate system but also if, at each point of U there is chosen the basis of a tangent space depending continuously on the point. For example, if M is a parallelizable n-dimensional manifold then defining on it n smooth tangent fields, linearly independent at each point, we obtain an isomorphism of the entire fiber $\mathscr{P}^q_m$ onto a trivial fiber $M \times P^q_m$, and an isomorphism of the space $E_0^{q-m,m}$, on a tensor product of the space P^q_m onto the space of generalized functions on the manifold M.

4.5. An element of the space $E_0^{q-m,m}$, corresponding to the generalized section $\varphi \cdot \lambda$, where φ is a generalized function on M, concentrated in the neighborhood U with a local system of coordinates, and λ is an element of the space P^q_m, can be immediately described in terms of local coordinates in the following manner. Let $\xi_1, \ldots, \xi_q$ be vector fields on the manifold M and let $\xi_i^{(1)}(x), \ldots, \xi_i^{(n)}(x)$ be functions defined on the set U, and let there be given the coordinates of a vector ξ_i located in U in a local system of coordinates. Let $\{\lambda^\tau\}$ be a representative of the element λ in the space $\tilde{P}^q_m$. We take any sequence $\tau = \{\tau_1, \ldots, \tau_n\}$ with $\tau_1 + \ldots + \tau_n = q$. We replace in the polynomial $\lambda^\tau(y_{11}, \ldots, y_{qn})$ the monomial $\prod_{i,j} y_{ij}^{m_{ij}}$ by the function

$$\prod_{s=1}^{n} \prod_{\tau_1+\ldots+\tau_{s-1} < i \leqslant \tau_1+\ldots+\tau_s} \frac{\partial^{m_{i1}+\ldots+m_{in}} \xi_i^{(s)}}{\partial x_1^{m_{i1}} \ldots \partial x_n^{m_{in}}},$$

where $x_1, \ldots, x_n$ are the local coordinates of a point $x \in U$. As a result of this substitution we obtain a function $f^\tau_{\xi_1,\ldots,\xi_q}(x)$ defined on U. We make this function skew-symmetric in $\xi_1, \ldots, \xi_q$ by setting

$$g^\tau_{\xi_1,\ldots,\xi_q} = \sum_{(i_1,\ldots,i_q)} \varepsilon(i_1, \ldots, i_q) \cdot f^\tau_{\xi_{i_1},\ldots,\xi_{i_q}},$$

where $\varepsilon(i_1, \ldots, i_q) = +$ or -1 depending on the parity of the permutation $(i_1, \ldots, i_q)$. Finally, we set

$$L(\xi_1, \ldots, \xi_q) = \varphi\Big(\sum_\tau g^\tau_{\xi_1,\ldots,\xi_q}\Big),$$

where the summation of the right-hand side is over the whole sequence $\tau = \{\tau_1, \ldots, \tau_n\}$ with $\tau_1 + \ldots + \tau_n = q$.

The obtained cochain L depends on the generalized function φ and on the element $\{\lambda^\tau\} \in \tilde{P}^q_m$. We denote it by $\varphi \cdot \{\lambda^\tau\}$.

It is easy to show that the cochain $\varphi \cdot \{\lambda^\tau\}$ belongs to the space $C^q_{1,m}$, but if $\{\lambda^\tau\} \in R^q_m$, then $\varphi \cdot \{\lambda^\tau\} \in C^q_{1,m-1}$. Hence the chain L defines an element $\tilde{L} \in E_0^{q-m,m}$, depending only on λ and φ. From the definitions given in the previous paragraph it is immediately deduced that $\tilde{L} = \varphi \cdot \lambda$.

4.6. If the local coordinates $x_1, \ldots, x_n$ given in U have the same dimensions as in a Riemann metric then the identity

$$\varphi \cdot \{\sigma_s \lambda^\tau\} = -\left(\frac{\partial \varphi}{\partial x_s}\right) \cdot \{\lambda^\tau\}$$

is valid.

The proof is evident (for definition of differentiation of generalized functions see [5]).

§ 5. THE HOMOMORPHISMS $\delta_0^{u,v}$

In this paragraph we shall establish that a homomorphism $\delta_0^{q-m,m}: E_0^{q-m,m} \to E_0^{q-m,m+1}$ is induced by some linear mapping of the fiber $\mathscr{P}^q_m$ in the fiber $\mathscr{P}^{q+1}_{m+1}$ and study this linear mapping.

5.1. To begin with, we shall describe some linear mapping $\widetilde{P}^q_m \to \widetilde{P}^{q+1}_{m+1}$. Let $\lambda = \{\lambda^\tau\} \in \widetilde{P}^q_m$ be any element, $\tau = \{\tau_1, \ldots, \tau_n\}$ be a sequence of nonnegative terms $\tau_1 + \ldots + \tau_n = q + 1$. We assume

$$\mu^\tau(y_{11}, \ldots, y_{1n}; \ldots; y_{q+1,1}, \ldots y_{q+1,n}) = \sum_{1 \leqslant r \leqslant n} \sum_{\tau_1+\ldots+\tau_{r-1} < i < j \leqslant \tau_1+\ldots+\tau_r} (-1)^{\tau_1+\ldots+\tau_{r-1}+i+j-1}$$
$$\times (y_{jr} - y_{ir}) \lambda^{\tau_-(r)} (y_{11}, \ldots, y_{\tau_1+\ldots+\tau_{r-1},n}; y_{i1} + y_{j1}, \ldots, y_{in} + y_{jn};$$
$$y_{\tau_1+\ldots+\tau_{r-1}+1,1}, \ldots; \hat{y}_{i1}, \ldots, \hat{y}_{in}; \ldots; \hat{y}_{j1}, \ldots, \hat{y}_{jn}; \ldots y_{q+1,n})$$
$$+ \sum_{1 \leqslant s < t \leqslant n} \sum_{\substack{\tau_1+\ldots+\tau_{s-1} < i \leqslant \tau_1+\ldots+\tau_s \\ \tau_1+\ldots+\tau_{t-1} < j \leqslant \tau_1+\ldots+\tau_t}} [(-1)^{\tau_1+\ldots+\tau_{t-1}+i+j-1}$$
$$\times y_{js} \lambda^{\tau_-(s)} (y_{11}, \ldots; \hat{y}_{i1}, \ldots, \hat{y}_{in}; \ldots, y_{\tau_1+\ldots+\tau_{t-1},n};$$
$$y_{i1} + y_{j1}, \ldots, y_{in} + y_{jn}; y_{\tau_1+\ldots+\tau_{t-1}+1,1}, \ldots; \hat{y}_{j1}, \ldots, \hat{y}_{jn}; \ldots, y_{q+1,n})$$
$$+ (-1)^{\tau_1+\ldots+\tau_{s-1}+i+j-1} y_{it} \cdot \lambda^{\tau_-(t)} (y_{11}, \ldots, y_{\tau_1+\ldots+\tau_{s-1},n};$$
$$y_{i1} + y_{j1}, \ldots, y_{in} + y_{jn}; y_{\tau_1+\ldots+\tau_{s-1}+1,1}, \ldots; \hat{y}_{i1}, \ldots, \hat{y}_{in}; \ldots; \hat{y}_{j1}, \ldots, \hat{y}_{jn}; \ldots y_{q+1,n})],$$

where $\tau_-(u)$ is the set obtained from τ by decreasing τ_u by unity and $\lambda^{\tau_-(u)}$ is a polynomial entering within the set λ. Thus if $\tau_u = 0$, then $\tau_-(u)$ is not defined, but then, as can be noted, in this case $\lambda^{\tau_-(u)}$ does not enter in the right-hand side of the formula.

We leave it to the reader to prove that the set $\mu = \{\mu^\tau\}$ is an element of the space $\widetilde{P}^{q+1}_{m+1}$ (in particular, that the polynomial μ^τ satisfies the condition of skew-symmetry of 4.1), that the function $\lambda \to \mu$ is linear and that if $\lambda \in R^q_m$ (see 4.1), then $\mu \in R^{q+1}_{m+1}$.

The constructed linear mapping $\widetilde{P}^q_m \to \widetilde{P}^{q+1}_{m+1}$ is denoted by $\widetilde{\nabla}$, and the mapping $P^q_m \to P^{q+1}_{m+1}$, obtained from $\widetilde{\nabla}$ by factorization, by ∇.

5.2. For any generalized function φ concentrated in a coordinate neighborhood and for any element $\lambda \in \widetilde{P}^q_m$ the equality

$$d(\varphi \cdot \lambda) = \varphi \cdot \widetilde{\nabla}\lambda$$

is valid. (For the definition of the differential d, see in 1.1, and of the product $\varphi \cdot \lambda$ in 4.5.) The statement follows immediately from 4.5 and 5.1.

5.3. There exists a mapping $\partial = \partial^q_m$ of a vector of the fiber $\mathcal{P}^q_m$ into vector of fiber $\mathcal{P}^{q+1}_{m+1}$ which on each layer induces a mapping $\nabla: P^q_m \to P^{q+1}_{m+1}$ (more accurately, the diagram

$$\begin{array}{ccc} P^q_m(x) & \xrightarrow{\partial(x)} & P^{q+1}_{m+1}(x) \\ \downarrow & & \downarrow \\ P^q_m & \xrightarrow{\nabla} & P^{q+1}_{m+1}, \end{array}$$

in which vertical arrows denote isomorphisms induced in some local system of coordinates in the neighborhood of the point $x \in M$ (see 4.2), and which commute for any point x and for any local system of coordinates).

2. The homomorphism of the space of generalized sections of the fiber $\mathcal{P}^q_m$ into the space of generalized sections of the fiber $\mathcal{P}^{q+1}_{m+1}$, induced by the mapping $\delta_0^{q-m,m}$, coincides with the homomorphism $E_0^{q-m,m} \to E_0^{q-m,m+1}$.

For proof it is sufficient to establish that for any coordinate neighborhood U in the manifold M of any generalized function φ on M concentrated on U and for any element $\lambda \in P^q_m$ there holds the equality $\delta_0^{q-m,m} \cdot (\varphi \cdot \lambda) = \varphi \cdot \nabla\lambda$. But this equality follows from 5.2.

5.4. (COROLLARY) The homomorphism $\delta_0^{u,v}$ permutes with multiplication by a smooth function.

5.5. From the proposition 5.3 it follows that the composition $\nabla \circ \nabla : P^{q-1}_{m-1} \to P^{q+1}_{m+1}$ is a null homomorphism. Moreover from the definition of an homomorphism it is easy to deduce the triviality of the composition $\tilde{\nabla} \circ \tilde{\nabla} : \tilde{P}^{q-1}_{m-1} \to \tilde{P}^{q+1}_{m+1}$. Thus arise the homology spaces $D^q_m = (\operatorname{Ker} \nabla|_{P^q_m})/(P^{q-1}_{m-1})$ and $\tilde{D}^q_m = (\operatorname{Ker} \tilde{\nabla}|_{\tilde{P}^q_m})/\tilde{\nabla}(\tilde{P}^{q-1}_{m-1})$. The composition $\partial^q_m \circ \partial^{q-1}_{m-1} : \mathscr{P}^{q-1}_{m-1} \to \mathscr{P}^{q+1}_{m+1}$ is also a null mapping due to which is generated a vector of the fiber $\mathscr{D}^q_m = \operatorname{Ker} \partial^q_m / \operatorname{Im} \partial^{q-1}_{m-1}$ with the base M. With a layer of the fiber $\mathscr{D}^q_m$ there is a space D^q_m.

The homologies of the complex $\mathscr{E}_0 = \{E^{u,v}_0, \delta^{u,v}_0\}$ are a bigraded space $E_1 \oplus E^{u,v}_1$, where $E^{u,v}_1 = \operatorname{Ker} \delta^{u,v}_0 / \operatorname{Im} \delta^{u,v-1}_0$. Proposition 5.3 permits us to reach the following conclusion about this space.

The space $E^{q-m,m}_1$ is a space of generalized sections of the fiber $\mathscr{D}^q_m$.

5.6. In the case when the manifold M is parallelizable, the fiber $\mathscr{D}^q_m$ as well as the fiber $\mathscr{P}^q_m$ are trivial. Therefore, the space $E^{q-m,m}_1$ in this case is isomorphic to the tensor product of the space D^q_m on the space of generalized functions on the manifold M. Such an isomorphism can be constructed if there is specified a system of linearly independent smooth tangential vector fields on the manifold.

§ 6. HOMOLOGIES OF THE COMPLEXES $\{P^q_m, \nabla\}$ AND $\{\tilde{P}^q_m, \tilde{\nabla}\}$

To begin with, we investigate the homologies $\tilde{D}^q_m = (\operatorname{Ker} \tilde{\nabla}|_{P^q_m})/\tilde{\nabla}(\tilde{P}^{q-1}_{m-1})$ of the complex $\{\tilde{P}^q_m, \tilde{\nabla}\}$. As we shall see, the homologies of the complex $\{P^q_m, \nabla\}$ are simply expressed through them.

6.1. We call an element $\lambda = \{\lambda^\tau\} \in \tilde{P}^q_m$ of uniform multiplicity $(m_1, \ldots, m_n)$ where $m_1 + \ldots + m_n = m - q$ if for any sequence $\tau = \{\tau_1, \ldots, \tau_n\}$ with $\tau_1 + \ldots + \tau_n = q$ the polynomial λ^τ is homogeneous for $s = 1, \ldots, n$ in the variables $y_{1s}, \ldots, y_{qs}$ and is of degree $m_s + \tau_s$. It is clear that any element of P^q_m is uniquely represented in the form of a finite sum of the homogenous.

From formula 5.1 it is clear that if $\lambda \in \tilde{P}^q_m$ is a homogeneous element, then $\tilde{\nabla}\lambda$ is also a homogeneous element and the multiplicity of the element λ and $\tilde{\nabla}\lambda$ is identical.

6.2. For any element $\lambda = \{\lambda^\tau\} \in \tilde{P}^q_m$ and $s = 1, \ldots, n$ we define an element $\lambda_s = \{\lambda^\tau_s\} \in \tilde{P}^{q-1}_{m-1}$ by the equality

$$\lambda^\tau_s(y_{11}, \ldots, y_{q-1,n}) = \frac{\partial}{\partial z_s} \lambda^{\tau+(s)}(y_{11}, \ldots, y_{\tau_1+\ldots+\tau_{s-1},n}; z_1, \ldots, z_n; y_{\tau_1+\ldots+\tau_{s-1}+1,1}, \ldots, y_{q-1,n})\big|_{z_1=\ldots=z_n=0}$$

for $\tau = \{\tau_1, \ldots, \tau_n\}$, $\tau_1 + \ldots + \tau_n = q - 1$, where $\tau_+(s) = \{\tau_1, \ldots, \tau_s + 1, \ldots, \tau_n\}$.

If $\lambda = \tilde{P}^q_m$ is a homogeneous element of multiplicity $(m_1, \ldots, m_n)$ and $\tilde{\nabla}\lambda = 0$, then for $s = 1, \ldots, n$ we have the equality

$$(-1)^q \cdot m_s \lambda = \tilde{\nabla}\lambda_s.$$

It is obtained from the equality $\tilde{\nabla}\lambda = 0$ if, for every $\tau = \{\tau_1, \ldots, \tau_n\}$ with $\tau_1 + \ldots + \tau_n = q + 1$, τ_s we differentiate the equality $\mu^\tau = 0$, where $\mu = \tilde{\nabla}\lambda$ (see 5.1), with respect to $y_{\tau_1+\ldots+\tau_{s-1}+1,s}$ and put $y_{\tau_1+\ldots+\tau_{s-1}+1,1} = \ldots = y_{\tau_1+\ldots+\tau_{s-1}+1,n} = 0$.

6.3. The space $D = \bigoplus_{q,m} D^q_m$. If $D^q_m \neq 0$, then $q = m \leq n(n+2)$.

Proof. It is clear that if $\lambda = \{\lambda^\tau\} \in \tilde{P}^q_m$, $\tilde{\nabla}\lambda = 0$, then for any homogeneous component μ of the element λ (see 6.1) we have the equality $\tilde{\nabla}\mu = 0$. If the multiplicity of the element μ is not equal to $(0, \ldots, 0)$, then according to 6.2, from $\mu \in \operatorname{Ker} \tilde{\nabla}$, it follows that $\mu \in \operatorname{Im} \tilde{\nabla}$. Thus on any coset of $(\operatorname{Ker} \tilde{\nabla}|_{\tilde{P}^q_m})/\tilde{\nabla}(\tilde{P}^{q-1}_{m-1})$

can be found a homogeneous element of multiplicity (0, ..., 0). Hence it is seen that if $D^q_m \neq 0$, then $q = m$. In particular, the condition of skew-symmetry means that the polynomial $\prod_{i,k} y_{ik}^{m_{ik}}$ can enter into λ^τ only under the condition that for any s among the rows $(m_{i1}, \ldots, M_{in})$ with $\tau_1 + \ldots + \tau_{s-1} < i \leq \tau_1 + \ldots + \tau_s$ none are congruent. In particular, among them none is zero and none equal to n with the sum of the elements equal to unity. Therefore

$$N(s) = \sum_{\tau_1+\ldots+\tau_s < i \leqslant \tau_1+\ldots+\tau_{s+1}} \sum_{k=1}^{n} m_{ik} \geqslant \begin{cases} \tau_s - 1, & \text{if } \tau_s \leqslant n+1, \\ 2\tau_s - (n+2), & \text{if } \tau_s \geqslant n+1. \end{cases}$$

Hence it is seen that if the degree of our polynomial, equal to $\sum_s N(s)$, does not exceed $q = \sum_s \tau_s$, then $q \leq n(n+2)$. But there exists only a finite number of monomials of qn variables of degree $\leq n(n+2)$. Therefore, the space $\widetilde{D}$ is finite dimensional.

6.4. We proceed to the calculation of the homologies $D = \bigoplus_{q,m} D^q_m$ of the complex $\{P^q_m, \nabla\}$.

If $\lambda = \{\lambda^\tau\} \in \widetilde{P}^q_m$ and $1 \leq s \leq n$, then the set $\{\sigma_s \lambda^\tau\} \in \widetilde{P}^q_{m+1}$ (here as well as before $\sigma_s = y_{1s} + \ldots + y_{qs}$) is denoted by $\sigma_s \lambda$. From formula 5.1 it is clear that $\widetilde{\nabla}(\sigma_s \cdot \lambda) = \sigma_s \cdot \widetilde{\nabla}\lambda$ (we remark that in the latter equality σ_s on the left means $y_{1s} + \ldots + y_{qs}$ and on the right $y_{1s} + \ldots + y_{q+1,s}$).

Let $\widetilde{\alpha} \in \widetilde{D}^q = \widetilde{D}^q_q$ and let $\lambda \in \widetilde{P}^q_q$ be a homogeneous element of multiplicity (0, ..., 0) representing α. For any s the product $\sigma_s \cdot \lambda$ is a homogeneous element of nonzero multiplicity and $\nabla(\sigma_s \cdot \lambda) = \sigma_s \cdot (\nabla\lambda) = 0$. In view of 6.2 there exists an element $\lambda_{(s)} \in P^{q-1}_q$ with $\widetilde{\nabla}\lambda_s = \sigma_s \cdot \lambda$. Further for any s, t we have $\widetilde{\nabla}(\sigma_s \lambda_{(t)} - \sigma_t \lambda_{(s)}) = 0$ and, in view of 6.2, $\sigma_s \lambda_{(t)} - \sigma_t \lambda_{(s)} = \widetilde{\nabla}\lambda_{(s,t)}$, where $\lambda_{(st)} \in \widetilde{P}^{q-2}_q$. The elements $\lambda_{(s_1 \ldots s_m)} \in \widetilde{P}^{q-m}_m$ with $\widetilde{\nabla}\lambda_{(s_1 \ldots s_m)} = \sum_{i=1}^{m} (-1)^{i-1} \sigma_{s_i} \times \lambda_{(s_1 \ldots \hat{s}_i \ldots s_n)}$ are analogously constructed. It is easy to verify that the elements $\lambda;\ \lambda_{(1)}, \ldots, \lambda_{(n)};\ \ldots,\ \lambda_{(s_1 \ldots s_m)} \ldots$ with $s_1 < \ldots < s_m$ specify the elements $\alpha \in D^q_q;\ \alpha_{(1)}, \ldots, \alpha_{(n)} \in D^{q-1}_q;\ \ldots,\ \alpha_{(s_1 \ldots s_m)} \in D^{q-m}_q, \ldots$ depending only on α. Moreover, if the elements $\alpha^{(i)}$ $(1 \leq i \leq N)$ form a basis in $\widetilde{D}$, then the elements corresponding to them $\alpha^{(i)}_{(s_1 \ldots s_m)}$ $(1 \leq i \leq N,\ 0 \leq m \leq n,\ s_1 < \ldots < s_m)$ form a basis in D.

Thus if the space $\widetilde{D}$ has dimension N then the space D has dimension $N \cdot 2^n$.

6.5. For any m the equality $D^1_m = 0$ holds.

Proof. For $q = 1$ we have $\sigma_1 = y_{11}, \ldots, \sigma_n = y_{1n}$. Therefore, in order for the element $\{\lambda^\tau\} \in \widetilde{P}^1_m$ to belong to R^1_m it is necessary and sufficient that the free terms of all the polynomials $\lambda^\tau(y_{11}, \ldots, y_m)$ vanish. Thus, any element of P^1_m is represented in $\widetilde{P}^1_m$ by such an element $\{\lambda^\tau\}$ that all λ^τ are constants. The proposition that such an element is a cocycle then and only then when it is equal to zero is then immediately proved.

6.6. If $q \leq n + 1$, then $\widetilde{D}^q = 0$.

In fact, if $\widetilde{D}^q \neq 0$ then, in view of 6.4, $D^{q-n}_q \neq 0$, from which according to 6.5, $q - n > 1$.

6.7. From 6.3-6.6 it follows that D^q_m may differ from zero only if the inequalities $n + 1 < m \leq n \cdot (n+2)$, $0 \leq m - q \leq n$ are simultaneously obeyed. Consequently, the space $E^{u,v}_1$ can differ from zero only if $-n \leq u \leq 0$, $n + 1 \leq v \leq n(n+2)$.

6.8. In all the previous considerations the dimentionality n of the manifold M was fixed by us and the problem of finding the dimentionality of the spaces D^q_m, $\widetilde{D}^q_m$ was therefore posed separately for each [n]. In view of 6.2-6.3 this problem is reduced to the calculation of the homologies of a finite complex consisting of the elements spaces P^q_m, $\widetilde{P}^q_m$ of zero multiplicity. However, the dimensionality of this complex increases rapidly with increasing n and it is possible to determine its homologies only with the help of extremely cumbersome computations. We limit ourselves to the cases n = 1, 2; moreover, for the second case we only mention the final result.

6.9. If n = 1 then $\widetilde{P}^q_m$ is a space of homogenous skew-symmetric polynomials of degree m in q-variables and the homomorphism $\widetilde{\nabla}: P^q_m \to P^{q+1}_{m+1}$ in this case acts according to the formula

$$(\widetilde{\nabla}P)(y_1, \ldots, y_{q+1}) = \sum_{1 \leqslant i < j \leqslant q+1} (-1)^{i+j-1} P(y_i + y_j, y_1, \ldots \hat{y}_i \ldots \hat{y}_j \ldots, y_{q+1}).$$

A uniform element of zero multiplicity, in the given case, is a skew-symmetric manifold whose degree is equal to the number of variables. It is easy to see that the space of such polynomials is three-dimensional and that a basis in it consists of polynomials $P_1(y) = y \in \widetilde{P}^1_1$, $P_2(y_1, y_2) = y_1^2 - y_2^2 \in \widetilde{P}^2_2$, $P_3(y_1, y_2, y_3) = (y_1 - y_2) \cdot (y_2 - y_3)(y_3 - y_1) \in \widetilde{P}^2_3$. From the formula derived above we obtain that $\widetilde{\nabla}P_1 = P_2$, $\widetilde{\nabla}P_2 = \widetilde{\nabla}P_3 = 0$, and therefore the space D is one-dimensional and its generator is an element of $\widetilde{D}^3 = \widetilde{D}^3_3$ represented by the polynomial P_3.

According to 6.4 the space D when n = 1 is two-dimensional and its generators lie in the spaces D^3_3, D^2_3. It is easy to see that they are represented by the polynomials P_3, $P_4(y_1, y_2) = y_1^2 y_2 - y_1 y_2^2$.

6.10. If n = 2, then the dimensionality of the space of homogeneous elements of multiplicity (0, 0) lying in $\widetilde{P}^q_q$ is equal for q = 1, 2, 3, 4, 5, 6, 7, 8, respectively 2, 6, 18, 13, 42, 34, 18, 5. If we choose in the enumerated space suitable bases and write the matrices of the operators $\widetilde{\nabla}$, it is possible to determine the dimensionality of the spaces of the homologies. It turns out that for n = 1, 2, 3, 4, 5, 6, 7, 8 the dimensionality of the space $\widetilde{D}^q$ is respectively equal to 0, 0, 0, 0, 2, 1, 0, 2, and hence, in view of 6.4, $\dim(\bigoplus_m D^q_m) = 0$, 0, 2, 5, 4, 3, 4, 2.

§ 7. THE DIFFERENTIAL $\delta^{u,v}_1$ AND THE SPACES $E^{u,v}_2$

7.1. The connection between the spaces D and $\widetilde{D}$, obtained in 6.4, permits us to represent the space D in the form $\widetilde{D} \otimes \Lambda(\zeta_1, \ldots, \zeta_n)$, where $\Lambda(\zeta_1, \ldots, \zeta_n)$ is the exterior algebra of formal variables $\zeta_1, \ldots, \zeta_n$; it is sufficient to put $\widetilde{\alpha} \otimes (\zeta_{j_1} \wedge \ldots \wedge \zeta_{s_m}) = \alpha_{(s_1 \ldots s_m)}$ for $\widetilde{\alpha} \in D$, $s_1 < \ldots < s_m$ (see 6.4). We remark that under an isomorphism $\widetilde{D} \otimes \Lambda(\zeta_1, \ldots, \zeta_n) \to D$ the space $\widetilde{D}^q \otimes \Lambda^r(\zeta_1, \ldots, \zeta_n)$ goes over into D^{q-r}_q.

Fixing the basis $e_1, \ldots, e_n$ in space $\tau(x)$ to the tangent manifold M at the point $x \in M$ we can construct an isomorphism, $J_{e_1 \ldots e_n}$ between the layer D(x) of the fiber $\mathscr{D}$ over point x and the space D. The inverse image $\widetilde{D}(x) \subset D(x)$ of the space $\widetilde{D} \subset D$ under the isomorphism does not depend on the choice of the basis: this follows from the fact that $\widetilde{D} = \bigoplus_q D^q_q \subset D$. We define an isomorphism $I_{e_1 \ldots e_n} \circ D = \widetilde{D} \otimes \Lambda(\zeta_1, \ldots, \zeta_n) \to \widetilde{D} \cdot (x) \otimes \Lambda(\tau(x)))$ where $\Lambda(\tau(x))$ is an exterior algebra over the space $\tau(x)$, as coinciding with $J^{-1}_{e_1 \ldots e_n}$ on $\widetilde{D}$ and mapping $\zeta_1, \ldots, \zeta_n$ respectively, into $e_1, \ldots, e_n$. It is easy to see that the composition $I_{e_1 \ldots e_n} \circ J_{e_1 \ldots e_n}$: $D(x) \to \widetilde{D}(x) \otimes \Lambda(\tau(x))$ does not depend on the basis $e_1, \ldots, e_n$. Thus we obtain an isomorphism of the fiber $\mathscr{D}$ on the tensor product of the fibering $\widetilde{\mathscr{D}}$, whose layer is isomorphic to $\widetilde{D}$, on the fiber $\Lambda(\tau) = \bigoplus_r \Lambda^r(\tau)$.

7.2. Let $U \subset M$ be a coordinate neighborhood in which a local coordinate system $x_1, \ldots, x_n$ is specified, and which determines in U the same Riemannian structure as that given in M. Further, let φ be a generalized function on M, concentrated in U; $\alpha \in \widetilde{D}^q$ be any element; $s_1, \ldots, s_m$ are integers such that $1 \leq s_1 < \ldots < s_m \leq q$. Then it is possible to interpret the product $\varphi \cdot (\alpha \otimes (\zeta_{s_1} \wedge \ldots \wedge \zeta_{s_m}))$ as a generalized section of the fiber $\mathscr{D}^{q-m}_q$, i.e. as an element of $E^{-m,q}_1$.

The following equality is valid

$$\delta_1^{-m,q}(\varphi \cdot (\alpha \otimes (\zeta_{s_1} \wedge \cdots \wedge \zeta_{s_m}))) = \sum_{l=1}^m (-1)^l \cdot \frac{\partial \varphi}{\partial x_{s_l}} \cdot (\alpha \otimes (\zeta_{s_1} \wedge \cdots \hat{\zeta}_{s_l} \cdots \wedge \zeta_{s_m})).$$

Proof. As we know from 6.4, the element $\alpha \otimes (\zeta_{s_1} \wedge \ldots \wedge \zeta_{s_m})$ is represented in $\tilde{P}^{q-m}_m$ by an element $\lambda_{(s_1 \ldots s_m)}$, where λ is a representative of the element α. The element $\varphi \cdot (\alpha \otimes (\zeta_{s_1} \wedge \ldots \wedge \zeta_{s_m}))$ has the representative $\varphi \cdot \lambda_{(s_1 \ldots s_m)}$ (see 4.5) in $C^{q-m}_{1,q}$. We have

$$d(\varphi \cdot \lambda_{(s_1 \ldots s_m)}) = \varphi \cdot \tilde{\nabla} \lambda_{(s_1 \ldots s_m)} \qquad \text{(see 5.2)}$$

$$= \varphi \cdot \sum_{i=1}^{m} (-1)^{i-1} \sigma_{s_i} \lambda_{(s_1 \ldots \hat{s}_i \ldots s_m)} \qquad \text{(see 6.4)}$$

$$= \sum_{i=1}^{m} (-1)^{i-1} \left(-\frac{\partial \varphi}{\partial x_{s_i}} \right) \lambda_{(s_1 \ldots \hat{s}_i \ldots s_m)} \qquad \text{(see 4.6)}$$

and hence follows the proposition to be proved.

7.3. If the manifold M is orientable then the fiber $\tilde{\mathcal{D}}$ is trivial.

Proof. It is clear that the fiber $\tilde{\mathcal{D}}$ over an oriented Riemannian manifold M is induced by a tangent fiber of the manifold M and by some action of the group SL(n, R) on $\tilde{D}$. It suffices for us to prove that this action is trivial. Let $\alpha_1, \ldots, \alpha_N$ be a basis in $\tilde{D}^q$, U a coordinate neighborhood which was considered in 7.2, V, W $\subset$ U open subsets, with $\bar{V} \subset U$, $W \subset V$. We denote through φ a smooth function on M vanishing outside of U and equal to unity in V. According to 7.2, $\delta_1^{-n,q}(\varphi \cdot (\alpha_i \otimes (\zeta_1 \wedge \ldots \wedge \zeta_n)))$ is equal to zero on V. We take any measure-preserving diffeomorphism $f: M \to M$ equal to identity on W and outside of V. The diffeomorphism f induces a mapping $E_1^{-n,q} \to E_1^{-n,q}$, transforming $\varphi \cdot (\alpha_i \otimes (\zeta_1 \wedge \ldots \wedge \zeta_n))$ into $\varphi \cdot \left(\sum_j \psi_{ij} \alpha_j \otimes \cdot (\zeta_1 \wedge \ldots \wedge \zeta_n) \right)$, where $\psi_{ij}(x)$ is the transformation matrix (in the basis $\alpha_1, \ldots, \alpha_N$) induced by the differential of the diffeomorphism f at the point $x \in M$. We have $\delta_1^{-n,q}(\varphi \cdot \left(\sum \psi_{ij} \alpha_j \otimes (\zeta_1 \wedge \ldots \wedge \zeta_n) \right)) = 0$, hence it follows that all functions ψ_{ij} are constants, and since $\psi_{ij}(x) = \delta_{ij}$ for $x \in W$, therefore, $\psi_{ij}(x) = \delta_{ij}$ for all x. Thus the differential of the diffeomorphism f at any point x induces an identity mapping of the space $\tilde{D}^q$, but this differential can be any element from SL(n; R). The proposition is proved.

7.4. Thus the fiber $\mathcal{D}$ is the tensor product of the fiber $\Lambda(\tau)$ on the space $\tilde{D}$. Consequently, the fiber $\mathcal{D}'$ associated with $\mathcal{D}$ is a tensor product $\Lambda(\tau') \otimes \tilde{D}'$ and the space $\Gamma(\mathcal{D}')$ of smooth sections of the fiber D' is therefore not that different from the tensor product of the space $\tilde{D}'$ on the space $\Omega = \sum_s \Omega^s$ of exterior differential forms of the manifold M. Thus E_1 is associated with $\Omega \otimes \tilde{D}'$; in particular, $E_1^{u,v}$ is associated with $\Omega^{-u} \otimes (\tilde{D}^v)'$.

From 7.2 it follows that the differential $\delta_1^{u,v}: E_1^{u,v} \to E_1^{u+1,v}$ is associated with $d \otimes \mathrm{id}: \Omega^{-u-1} \otimes (\tilde{D}^v)' \to \Omega^{-u} \otimes (\tilde{D}^v)'$ where d is the exterior differential.

7.5. The equality $E_2^{u,v} = H_{-u}(M; R) \otimes \tilde{D}^v$ is valid.

This follows from 7.4.

§ 8. HOMOLOGIES OF THE COMPLEXES $\{C^q_k / C^q_{k-1}, d^q_k\}$

The subspaces $C^q_{k,m}/(C^q_{k,m} \cap C^q_{k-1})$ of the space C^q_k/C^q_{k-1} define in it a filtration with which the differential d^q_k (see 2.12) coincides. The spectral sequence connecting the space $\underset{q,m}{\otimes} (C^q_{k,m}/(C^q_{k,m-1} + (C^q_{k,m} \cap C^q_{k-1})))$ with the homologies of the complex $\{C^q_k/C^q_{k-1}, d^q_k\}$ thus arises. We denote this spectral sequence by $\{{}^{(k)}E_r^{u,v}, {}^{(k)}\delta_r^{u,v}\}$ (in particular, $E_0^{q-m,m} = C^q_{k,m}/(C^q_{k,m-1} + (C^q_{k,m} \cap C^q_{k-1}))$). In this paragraph we shall communicate some results dealing with this sequence.

8.1. Let $S = \{S_1, \ldots, S_k\}$ be a partition of the set $\{1, \ldots, q\}$ on k nonempty subsets. We denote by M^k_S a subset of the manifold M^q consisting of such points $(x_1, \ldots, x_q)$ that $x_i = x_j$, if i and j belong to one and the same set from the partitioning S. It is clear that each of M^k_S is canonically diffeomorphic to M^k, that the union of these sets with the whole of S is M^q_k and that their pair intersections lie in M^q_{k-1}. The set M^k_S can also be obtained by fixing a representation of the manifold M^q in the form $M^{s_1}\times\ldots\times M^{s_k}$, taking in M^{s_i} a diagonal and multiplying these diagonals. The diagonal of the product M^{s_i} serves as the base of the fiber $\mathscr{P}^{s_i}_m$, m = 0, 1, ... (see 3.3); the product M^k_S of these diagonals therefore serves as basis of the fiber $\mathscr{P}^{s_1}_{m_1}\otimes\ldots\otimes\mathscr{P}^{s_k}_{m_k}$ and, consequently, of the fiber $\bigoplus_{m_1+\ldots+m_k+m}\mathscr{P}^{s_1}_{m_1}\otimes\ldots\otimes\mathscr{P}^{s_k}_{m_k}$, which we denote by $\mathscr{P}^q_{S,m}$.

Since $\cup M^k_S = M^q_k$ and $M^k_{S'}\cap M^k_{S''}\subset M^q_{k-1}$ for any S', S", therefore from the fibers $\mathscr{P}^q_{S,m}$ it is possible to construct a fiber vector over $M^q_k\setminus M^q_{k-1}$ (in general the dimensions of the layers which connect different components of the space $M^q_k\setminus M^q_{k-1}$ are not identical). We denote this fiber by $\mathscr{P}^q_{k,m}$. The layer of the $\mathscr{P}^q_{k,m}$ over a point $x\in M^k_S\subset M^q_k$ is $\bigoplus_{m_1+\ldots_k m=m}(P^{s_1}_{m_1}\otimes\ldots\otimes P^{s_k}_{m_k})$.

Each of mappings $p_{ij}: M^q\to M^q$ induces an automorphism of the fiber $\mathscr{P}^q_{k,m}$.

8.2. The space of generalized sections of the fiber $\mathscr{P}^q_{S,m}$ can be identified with the factor-space of the space of generalized sections of the fiber τ'_q, which is concentrated on M^k_S, having an order m with respect to M^k_S and is multiplied by −1 under the mapping p_{ij}, preserving M^k_S in the subspace consisting of sections of order m − 1 with respect to M^k_S.

Any generalized section of the fiber τ'_q concentrated on M^q_k and of order m with respect to M^q_k can be represented in the form of a sum of generalized sections, concentrated on the sets M^q_k, and having an order m with respect to M^k_S, where these terms are defined uniquely up to the addition of generalized sections, concentrated on M^q_{k-1}. Therefore, to every element of the space ${}^{(k)}E_0^{q-m,m}$ can be associated a generalized section of the fiber $\mathscr{P}^q_{k,m}$. This generalized section satisfies two conditions: first, it can be extended up to a generalized section of each of the fibers $\mathscr{P}^q_{k,S}$, and secondly, that it is multiplied by −1 under the action of each of the mappings p_{ij}. It is possible to verify that this correspondence defines an isomorphism ${}^{(k)}E_0^{q-m,m}$ on the space of generalized sections of the fiber $\mathscr{P}'^q_{k,m}$ possessing the mentioned properties.

Let $x\in M^k_S\setminus(M^k_S\cap M^q_{k-1})$ be any point and U be such a coordinate neighborhood of the point x that $U\cap M^q_k\subset M^k_S\setminus(M^k_S\cap M^q_{k-1})$. We denote by $\widetilde{U}$ the union of sets obtained from U under the action of the group of permutations. From what was said above it follows that any element of the space ${}^{(k)}E_0^{q-m,m}$, concentrated on $\widetilde{U}$, can be represented in the form of a finite sum of expressions of the form $\varphi\cdot(\lambda_1\otimes\ldots\otimes\lambda_k)$ where φ is symmetric generalized function on $M^q_k\setminus M^q_{k-1}$, concentrated on $\widetilde{U}\cap M^q_k$, $\lambda_i\in P^{s_i}_{m_i}$, $m_1+\ldots+m_k = m$. It is clear that it is sufficient to give such a generalized function on the set $U\cap M^q_k$.

8.3. We shall now describe the action of the homomorphism ${}^{(k)}\delta_0^{q-m,m}$ on the elements of the form $\varphi\cdot(\lambda_1\otimes\ldots\otimes\lambda_k)$. We denote by A_i the image of the set $U\cap M^q_k$ under the mapping $\gamma_i: M^q\to M^{q+1}$ relating the point $(x_1, \ldots, x_q)$ to the point $(x_i, x_1, \ldots, x_q)$. It is clear that $A_i\subset M^{q+1}$, that sets A_i and even their images $\widetilde{A}_i$ under the action of the group of permutations are mutually disjointed and that $\gamma_i|_{U\cap M^q_k}: U\cap M^q_k \to A_i$ is a diffeomorphism. Therefore, the generalized function φ defines generalized functions φ_i on M^q_{k+1}, concentrated on A_i.

The formula

$$ {}^{(k)}\delta_0^{q-m,m}(\varphi\cdot(\lambda_1\otimes\ldots\otimes\lambda_k)) = \sum_{i=1}^{k}(-1)^{s_1+\ldots+s_{i-1}}\varphi_i\cdot\lambda_1\otimes\ldots\otimes\nabla\lambda_i\otimes\ldots\otimes\lambda_k. $$

is valid. It is proved with the help of an obvious computation.

8.4. With the help of this formula it is possible to obtain exhaustive information about the homologies ${}^{(k)}E_1^{u,v}$ of the complex $\{{}^{(k)}E_0^{u,v}, {}^{(k)}\delta_0^{u,v}\}$.

We define the fiber $\mathcal{D}^{q}_{S,m}$ with the basis M^k_S as $\bigoplus_{m_1+\ldots+m_k=m}\left(\mathcal{D}^{s_1}_{m_1}\otimes\ldots\otimes\mathcal{D}^{s_k}_{m_k}\right)$ and the fiber $\mathcal{T}^{q}_{k,m}$ with the basis $M^q_k\setminus M^q_{k-1}$ as coinciding with $\mathcal{D}^{q}_{S,m}$ on $M^k_S\setminus(M^k_S\cap M^q_k)$. The space ${}^{(k)}E_1^{q-m,m}$ is the space of generalized sections of the fiber $\mathcal{D}^{q}_{k,m}$ which is extended with respect to generalized sections of each of the fibers $\mathcal{D}^{q}_{S,m}$ and satisfies the condition of symmetry. This is proved in the same manner as the analogous proposition of § 5 (see 5.5).

Using the information about the structure of the fibers $\mathcal{D}^{s}_{m}$ contained in § 7 we obtain that the space of smooth sections of the fiber associated with $\mathcal{D}^{s_1}_{m_1}\otimes\ldots\otimes\mathcal{D}^{s_k}_{m_k}$ is $\Omega^{m_1-s_1,\ldots,m_k-s_k}(M^k)\otimes(\tilde{D}^{m_1}\otimes\ldots\otimes\tilde{D}^{m_k})'$, where $\Omega^{u_1,\ldots,u_k}(M^k)$ is the space of smooth differential forms on M^k of degree u_i along the i-th factor (i.e. the space of smooth sections of the fiber $\bigoplus_i\Lambda^{u_i}(\pi_i^*\tau')$). In order to give an element of the space ${}^{(k)}E_1^{q-m,m}$, it is sufficient to give for all $m_1,\ldots,m_k$, $s_1,\ldots,s_k$ with $m_1+\ldots+m_k=m$, $s_1+\ldots+s_k=q$ a functional on the space $\Omega^{m_1-s_1,\ldots,m_k-s_k}(M^k)\otimes(\tilde{D}^{m_1}\otimes\ldots\otimes\tilde{D}^{m_k})'$ (using the condition of skew symmetry it might be possible to define it over all spaces $\Gamma((\mathcal{D}^{q}_{S,m})'))$. But

$$ \bigoplus_{\substack{m_1+\ldots+m_k=m\\ s_1+\ldots+s_k=q}}\Omega^{m_1-s_1,\ldots,m_k-s_k}(M^k)\otimes(\tilde{D}^{m_1}\otimes\ldots\otimes\tilde{D}^{m_k})' = \Omega^{m-q}(M^k)\otimes\Big(\bigoplus_{m_1+\ldots+m_k=m}(\tilde{D}^{m_1}\otimes\ldots\otimes\tilde{D}^{m_k})'\Big), $$

and therefore an element of ${}^{(k)}E_1^{q-m,m}$ is defined by a functional on $\Omega^{m-q}(M^k)\otimes({}^{(k)}\tilde{D}^m)'$ where ${}^{(k)}\tilde{D}^m = \bigoplus_{m_1+\ldots+m_k=m}(\tilde{D}^{m_1}\otimes\ldots\otimes\tilde{D}^{m_k})$. However, the assignment of such a functional is not equivalent to the assignment of an element of ${}^{(k)}E_1^{q-m,m}$: firstly, this element must still satisfy some condition of skew symmetry; secondly, it does not uniquely define an element of ${}^{(k)}E_1^{q-m,m}$ but only up to addition of a functional $F:\Omega^{m-q}(M^k)\otimes({}^{(k)}\tilde{D}^m)'\to R$ such that $F(\alpha\otimes\beta)=0$ for all forms $\alpha\in\Omega^{m-q}(M^k)$, which vanish in the neighborhood of the set $M^k_{k-1}\subset M^k$.

8.5. In the case $k>1$ propositions 7.4 and 7.5 are directly generalized, and namely, it is found that the homomorphism associated with $d\otimes \mathrm{id}:\Omega^{m-q-1}(M^k)\otimes({}^{(k)}\tilde{D}^m)'\to\Omega^{m-q}(M^k)\otimes({}^{(k)}\tilde{D}^m)'$, where $d:\Omega^{m-q-1}(M^k)\to\Omega^{m-q}(M^k)$ is an exterior differential, induces the homomorphism ${}^{(k)}E_1^{q-m,m}\to{}^{(k)}E_1^{q-m+1,m}$, and coincides with ${}^{(k)}\delta_1^{q-m,m}$. The space ${}^{(k)}E_2^{q-m,m}$ is isomorphic to the space, obtained from $H_{m-q}(M^k)\otimes{}^k\tilde{D}^m$ by the following identification:

$$ \alpha\otimes\lambda_1\otimes\ldots\otimes\lambda_k=(-1)^{m_i m_j}(p_{ij})_*\alpha\otimes\lambda_1\otimes\ldots\otimes\lambda_{i-1}\otimes\lambda_j\otimes\lambda_{i+1}\otimes\ldots\otimes\lambda_{j-1}\otimes\lambda_i\otimes\lambda_{j+1}\otimes\ldots\otimes\lambda_k $$

for any $\alpha\in H_{m-q}(M^k)$, $\lambda_1\in\tilde{D}^{m_1},\ldots,\lambda_k\in\tilde{D}^{m_k}$.

In particular the short form of such a proposition is:

The spaces ${}^{(k)}E_2^{u,v}$ are all finite dimensional and can be different from zero only if $-kn\le u\le 0$, $kn+k+1<v\le k\cdot n(n+2)$.

8.6. From the finite dimensionality of the spaces ${}^{(k)}E_2^{u,v}$ follows the finite dimensionality of the spaces ${}^{(k)}E_\infty^{u,v}$ and consequently, the finite dimensionality of the homologies of the complex $\{C^q_k/C^q_{k-1}, d^q\}$.

For finding the homologies of the original complex $\mathscr{C} = \{C^q, d^q\}$ we still need to investigate the spectral sequence corresponding to the filtrations of the group C^q by the subgroups C^q_k. The first term of this spectral sequence is $\bigoplus_{q,k} \operatorname{Ker} d^q_k / \operatorname{Im} d^{q-1}_k$. For this purpose it is sufficient to take the spaces $\operatorname{Ker} d^q_k/\operatorname{Im} d^{q-1}_k$ with $k \le q-1$: all cocycles of the complex $\mathscr{C}$ lie in C^q_{q-1} (see 1.5), those of them which do not lie in C^q_{q-2} (and even in $C^q_{[\frac{q+1}{2}]}$) are cohomologically zero (this follows from 6.5), but if a cocycle, lying in C^q_{q-2}, is cohomologically zero then it serves as a coboundary of an element of C^q_{q-1} (see 1.5).

8.7. The groups of cohomologies of the complex $\mathscr{C}$ are finite dimensional.

This follows from 8.5 and 8.6.

§9. THE CASE OF A TORUS

9.1. If M is an n-dimensional torus T^n, then the spectral sequence $\{E^{u,v}_r, \delta^{u,v}_r\}$, converging to the homologies of the diagonal of the complex, is trivial beginning with the first term (i.e. $\delta^{u,v}_r = 0$ for $r \ge 2$).

Proof. Since the torus is parallelizable, every cochain from the diagonal of the complex can be represented in the form $\sum_i \varphi_i\lambda_i$, where φ_i are generalized functions on the torus and $\lambda_i \in \oplus \tilde{P}^q_m$. Since on a torus there exist a complete system of linearly independent pairs of commuting vector fields, a straightforward calculation shows that $d\left(\sum_i \varphi_i\lambda_i\right) = \sum_i \varphi_i\tilde{\nabla}\lambda_i$. Let $\rho \otimes \alpha \in H_{-u}(T^n) \otimes \tilde{D}^v = E^{u,v}_2$, be any element, $\varphi = \sum \varphi_{i_1 \dots i_{n-q}} dx_{i_1} \wedge \dots \wedge dx_{i_{n-q}}$ is a closed differential form on torus, representing the class of cohomologies $D\rho \in H^{n+u}(T^n)$, dual to the class ρ, and λ is the representative of the class α in $\tilde{P}^v$. Then the cochain $\sum \varphi_{i_1\dots i_{n-q}}\lambda_{(i_1\dots i_{n-q})}$ of the diagonal of the complex is a cocycle representing the element $\rho \otimes \alpha \in E^{-u,v}_2$. Thus, every element of $E^{-u,v}_2$ is represented by a cocycle of the diagonal of complex; hence it follows that all differentials of the spectral sequence $\{E^{u,v}_r, \delta^{u,v}_r\}$ starting from the second are trivial.

From the proved proposition it follows that the homologies of the complex $\mathscr{C}_1(T^n)$ are isomorphic to $H_*(T^n) \otimes \tilde{D}$, i.e. isomorphic to D.

9.2. If M is a circumference, then the cohomologies of the Lie algebra of vector fields are a tensor product of the algebra of polynomials from a two-dimensional generator onto the exterior algebra of a three-dimensional generator.

Proof. Since $\tilde{D}$ is generated from one-dimensional manifolds by one of the three-dimensional generators (see 6.9) then, according to 8.5, ${}^{(k)}E_2 = \oplus\, {}^{(k)}E^{u,v}_2$ consists of those elements of the space $H_*((S^1)^k, (S^1)^k_{k-1}) \otimes D \otimes \dots \otimes D = H_*((S^1)^k, (S^1)^k_{k-1})$ which are multiplied by -1 under the mapping $H_*((S^1)^k, (S^1)^k_{k-1}) \to H_*((S^1)^k, (S^1)^k_{k-1})$ induced by a transposition of the factors.

The space $(S^1)^k \setminus (S^1)^k_{k-1}$ for $k \ge 2$ consists of $k(k-1)/2$ connected components each of which is homotopically equivalent to the circumference. The space $H^*((S^1)^k, (S^1)^k_{k-1}) = H^*((S^1)^k \setminus (S^1)^k_{k-1})$ has therefore $k(k-1)/2$ generators of dimensionality k and $k(k-1)/2$ generators of dimensionality $(k-1)$. It is easy to show that, after the identifications mentioned in 8.5, there remains a single element of each these groups, and hence ${}^{(k)}E^{u,v}_2 = 0$ for $(u, v) \ne (-k, 3k)$, $(-k+1, 3k)$ and ${}^{(k)}E^{-k,k}_2 = {}^{(k)}E^{-k+1,3k}_2 = \mathbf{R}$.

Using the proposition 2.11 it is possible to define the multiplication ${}^{(k)}E^{u,v}_r \otimes {}^{(l)}E^{u',v'}_r \to {}^{(k+l)}E^{u+u',v+v'}_r$. It is clear that it is connected by the usual formulas with the differentials ${}^{(k)}\delta^{u,v}_r$. It is easy to verify that the spaces ${}^{(k)}E^{-k,3k}_2$, ${}^{(k)}E^{-k+1,3k}_2$ are generated by the elements α^k, $\beta\alpha^{k-1}$, where $\alpha \in E^{-1,3}_1 = H_1(S^1) \otimes \tilde{D}^3$, $\beta \in E^{0,3}_2 = H_0(S^1) \otimes \tilde{D}^3$ are the generators.

It follows from 9.1 that elements α, β belong to the nucleus of all differentials of the spectral sequence $\{E_r^{u,v}, \delta_r^{u,v}\}$. Therefore this is also true for elements α^k, $\beta\alpha^{k-1}$ in the spectral sequences $\{{}^{(k)}E^{u,v}, {}^{(k)}\delta^{u,v}\}$. Moreover, since α and β are represented by cocycles of the diagonal of the complex, therefore α^k, $\beta\alpha^{k-1}$ are represented by cocycles of the complex $\mathscr{C}$, lying in C_k^q. Therefore the spectral sequence connecting the space $\bigotimes_{q,k} \operatorname{Ker} d_k^q / \operatorname{Im} d_k^{q-1}$ with the homologies of the complex $\mathscr{C}$ (see 8.6) are trivial, starting with the first term. Thus the cohomologies of the complex $\mathscr{C}$ are generated by adding the elements a^k, $b \cdot a^{k-1}$, where a, b are the homology classes of the cochains of the diagonal of the complex representing α, β. Since, evidently $ab = ba$ and $b^2 = 0$, therefore, from here the proof of the proposition follows.

LITERATURE CITED

1. I. M. Gel'fand and V. B. Fuks, "Cohomologies of the Lie algebra of vector fields on a circumference," Funkts. Analiz., 3, No. 4, 92–93 (1968).
2. I. M. Gel'fand and V. B. Fuks, "Cohomologies of the Lie algebra of vector fields on a manifold," Funkts. Analiz., 3, No. 2, 97 (1969).
3. J. Schwartz, "Differential geometry and topology (1967) (mimeograph).
4. I. M. Gel'fand and C. E. Shilov, Generalized Functions, New York (1961–68).
5. L. Schwartz, Théorie des distributions, Paris (1952).

5.

(with D. B. Fuks)

Cohomology of the Lie algebra of formal vector fields

Izv. Akad. Nauk SSSR, Ser. Mat **34** (1970) 322–337 [Math. USSR, Izv. **34** (1970) 327–342]. MR **44**:1103. Zbl. **216**:203

Abstract. We calculate the cohomology of the Lie algebra of formal vector fields at the origin in a euclidean space. The results are applied to the investigation of the Lie algebra of tangent vector fields on a smooth manifold.

Introduction

0.1. By a *formal vector field* at the point $0 = (0, \cdots, 0)$ of the space $\mathbf{R}^n$ we mean a linear combination of the form $\Sigma_{i=1}^n p_i(x_1, \cdots, x_n)e_i$, where $e_1, \cdots, e_n$ are the vectors of the standard basis of $\mathbf{R}^n$ and the $p_i(x_1, \cdots, x_n)$ are formal power series with real coefficients in the coordinates $x_1, \cdots, x_n$ of the space. The set of formal vector fields can also be defined as the inverse limit of the system $\{S_r, \pi_r\}$, where S_r is the space of r-sets of vector fields of class C^r at the point $0 = (0, \cdots, 0)$ and π_r: $S_r \to S_{r-1}$ is the natural projection. From this definition we see that the formal vector fields constitute a linear topological space.

The commutator $[\xi', \xi'']$ of the formal vector fields

$$\xi' = \sum_{i=1}^{n} p'_i(x_1, \ldots, x_n) e_i, \quad \xi'' = \sum_{i=1}^{n} p''_i(x_1, \ldots, x_n) e_i$$

is defined by the formula

$$[\xi', \xi''] = \sum_{i=1}^{n} \left[\sum_{j=1}^{n} \left(p'_j \frac{\partial p''_i}{\partial x_j} - p''_j \frac{\partial p'_i}{\partial x_j} \right) \right] e_i.$$

With respect to commutation, the formal vector fields at 0 constitute a topological Lie algebra, denoted by W_n. The subject of this paper is the calculation of the cohomology of this topological Lie algebra with coefficients in the identity real representation (i.e., in the field $\mathbf{R}$ with trivial action of the algebra W_n).

0.2. In order to formulate the final result, we have to describe certain auxiliary topological spaces X_n $(n = 1, 2, \cdots)$. Suppose $N \geq 2n$, and let $(E(N, n), p, G(N, n))$ be the canonical $U(n)$-bundle over the (complex) Grassmann manifold $G(N, n)$. The usual cell decomposition of the manifold $G(N, n)$ (cf., e.g., [1], Russian p. 89) has the property that the $2n$th skeleton $(G(N, n))_{2n}$ for $N \geq 2n$ is independent of N. The inverse image of the set $(G(N, n))_{2n}$ under the mapping p we denote by X_n.

The space X_1 is a three-dimensional sphere. The remaining spaces X_i do not have such a simple geometric interpretation.

0.3. The central result of the paper is the following:

Theorem. *For all q we have an isomorphism:*

$$H^q(W_n; \mathbf{R}) = H^q(X_n; \mathbf{R}).$$

Multiplication in the ring $H^*(W_n; \mathbf{R})$ and in the ring $H^*(X_n; \mathbf{R})$ is trivial, i.e., the product of any two elements of positive dimension is zero.

The cohomology of the space X_n can be found by using standard methods of topology. Considerable information is obtained about this cohomology in the process of proving Theorem 0.3 (see §5). For example, it is trivial for $0 < q \leq 2n$ and for $q > n(n+2)$.

0.4. The results of this paper are closely connected with our results on the cohomology of Lie algebras of vector fields on manifolds [2]. Specifically, the standard cochain complex of the algebra W_n is, as we shall see in §1 (cf. 1.6), just the complex $\{\oplus_m \tilde{P}^q_m, \tilde{\nabla}\}$ considered in [2] (cf. [2], §§4–6), and so the spaces $H^q(W_n; \mathbf{R})$ coincides with the spaces $\tilde{D}^q$ of homology of this complex. This makes it possible to obtain some new information from Propositions 7.5 and 8.5 of [2]. For example, we prove that the spaces $\mathfrak{H}^q(M)$ of cohomology of dimension q of the Lie algebra of smooth vector fields on an n-dimensional orientable compact manifold M with coefficients in $\mathbf{R}$ are trivial for $0 < q \leq n$ (Theorem 6.3). Of other applications of Theorem 0.3, we note the recently published calculation [3] of the rings $\mathfrak{H}^*(T^n) = \Sigma_q \mathfrak{H}^q(T^n)$, where T^n is a torus of dimension n.

0.5. With the exception of §6, this paper can be read independently of [2]. However, §1.6 would have to be omitted and Proposition 1.8 taken for granted (or re-proved).

0.6. An important point in the proof of Theorem 0.3 is the fact that the cohomology of the unitary group $U(n)$ occurs in it as the cohomology of the Lie algebra $\mathfrak{gl}(n; \mathbf{R})$ of real matrices of order n. For this observation we are indebted to M. V. Losik, to whom we are glad to express our sincere gratitude.

§1. The standard complex

1.1. We recall that the cohomology of a topological Lie algebra $\mathfrak{g}$ with coefficients in a $\mathfrak{g}$-module M is defined as the homology of the complex $\{C^q(\mathfrak{g}; M), d^q(\mathfrak{g}; M)\}$, where $C^q(\mathfrak{g}; M)$ is the space of continuous skew-symmetric q-linear functionals on $\mathfrak{g}$ with values in M, and the differential $d^q(\mathfrak{g}; M)$: $C^q(\mathfrak{g}; M) \to C^{q+1}(\mathfrak{g}; M)$ is defined by the formula

$$[d^q(\mathfrak{g}; M)P](\xi_1, \ldots, \xi_{q+1}) =$$

$$= \sum_{1\leqslant s<t\leqslant q+1} (-1)^{s+t-1} P([\xi_s, \xi_t], \xi_1, \ldots, \hat{\xi}_s, \ldots, \hat{\xi}_t, \ldots, \xi_{q+1}) + \sum_{1\leqslant s\leqslant q+1} (-1)^s \xi_s P(\xi_1, \ldots, \hat{\xi}_s, \ldots, \xi_{q+1}).$$

Here $\xi_i \in \mathfrak{g}$, and $\xi_s P(\xi_1, \cdots, \hat{\xi}_s, \cdots, \xi_{q+1})$ denotes the result of the action of the element $\xi_s \in \mathfrak{g}$ on $P(\xi_1, \cdots, \hat{\xi}_s, \cdots, \xi_{q+1})$ as on an element of the $\mathfrak{g}$-module M.

The complex $\{C^q(\mathfrak{g}; M), d^q(\mathfrak{g}; M)\}$ is called the *standard cochain complex of the algebra* $\mathfrak{g}$ *with values in* M.

1.2. For brevity we denote the space $\mathbf{R}^n$ by T and the Lie algebra W_n by W. The standard cochain complex of W with values in $\mathbf{R}$ we shall denote simply by $\{C^q, d^q\}$. Thus, C^q is the space of continuous skew-symmetric q-linear real functionals on W, and $d^q\colon C^q \to C^{q+1}$ is the homomorphism defined by

$$(d^q P)(\xi_1, \ldots, \xi_{q+1}) = \sum_{1\leqslant s<t\leqslant q+1} (-1)^{s+t-1} P([\xi_s, \xi_t], \xi_1, \ldots, \hat{\xi}_s, \ldots, \hat{\xi}_t, \ldots, \xi_{q+1}). \tag{1}$$

Every continuous linear functional on W can be represented uniquely as a finite sum of functionals of the form

$$\Sigma_i p_i e_i \to \nu_0 (D(\Sigma_i \alpha(e_i) p_i)),$$

where α is a functional on T, D is a differential operator, and ν_0 is the functional assigning to every power series its free term. The space of differential operators in the space of power series is just $S^* T = \bigoplus_m S^m T$—the sum of all symmetric powers of the space T. Thus, the space W' conjugate to W is canonically isomorphic with $S^* T \otimes T'$. Finally, the space of continuous skew-symmetric q-linear functionals on W is obviously $\Lambda^q(W')$—the qth exterior power of the space W'. In short, we have:

$$C^q = \Lambda^q (S^* T \otimes T'). \tag{2}$$

1.3. We shall sometimes regard the elements of the space C^q as functionals on the space conjugate to C^q, i.e., as functions of $2q$ vector arguments $\alpha_1, \cdots, \alpha_q \in T'$, $\beta_1, \cdots, \beta_q \in T$, polynomially dependent on the components of the covectors $\alpha_1, \cdots, \alpha_q$, multilinear in $\beta_1, \cdots, \beta_q$, and changing sign under simultaneous interchange of α_i with α_j and β_i with β_j. In this notation the differential $d = d^q\colon C^q \to C^{q+1}$ is given by the formula

$$(dP)(\alpha_1, \ldots, \alpha_{q+1}, \beta_1, \ldots, \beta_{q+1}) = \sum_{1\leqslant s<t\leqslant q+1} (-1)^{s+t-1} P(\alpha_s + \alpha_t, \alpha_1, \ldots, \hat{\alpha}_s, \ldots, \hat{\alpha}_t, \ldots, \alpha_{q+1}; \beta(s,t), \beta_1, \ldots, \hat{\beta}_s, \ldots, \hat{\beta}_t, \ldots, \beta_{q+1}), \tag{3}$$

where $\beta(s, t) = (\alpha_t, \beta_s)\beta_t - (\alpha_s, \beta_t)\beta_s$.

1.4. The standard cochain complex of a Lie algebra $\mathfrak{g}$ with values in $\mathbf{R}$ has a canonical multiplicative structure: the multiplication $C^q(\mathfrak{g}; \mathbf{R}) \otimes C^r(\mathfrak{g}; \mathbf{R}) \to C^{q+r}(\mathfrak{g}; \mathbf{R})$ is defined by the formula

$$(PQ)(\xi_1, \ldots, \xi_{q+r})$$
$$= \sum (-1)^{i_1+\ldots+i_q-q(q+1)/2} P(\xi_{i_1}, \ldots, \xi_{i_q}) Q(\xi_{j_1}, \ldots, \xi_{j_r}),$$

where the summation is over all partitions of the set $\{1, \cdots, q+r\}$ into disjoint subsets $\{i_1, \cdots, i_q\}$, $\{j_1, \cdots, j_r\}$ with $i_1 < \cdots < i_q$, $j_1 < \cdots < j_r$.

This multiplication is connected with the differential $d = d(\mathfrak{g}; \mathbf{R})$ through the usual formula $d(PQ) = (dP)Q + (-1)^{\dim P} P(dQ)$ and defines in the space $H^*(\mathfrak{g}; \mathbf{R}) = \Sigma_q H^q(\mathfrak{g}; \mathbf{R})$ the structure of an algebra over the field $\mathbf{R}$.

When $\mathfrak{g} = W$, cochain multiplication is expressed in the language of 1.3 by the formula

$$(PQ)(\alpha_1, \ldots, \alpha_{q+r}; \beta_1, \ldots, \beta_{q+r})$$
$$= \sum (-1)^{i_1+\ldots+i_q-q(q+1)/2} P(\alpha_{i_1}, \ldots, \alpha_{i_q}; \beta_{i_1}, \ldots, \beta_{i_q})$$
$$\times Q(\alpha_{j_1}, \ldots, \alpha_{j_r}; \beta_{j_1}, \ldots, \beta_{j_r}). \tag{4}$$

1.5. As is clear from (2), there are canonical representations of the group $GL(n, \mathbf{R})$ in the spaces C^q. In the language of 1.3 these representations are given by the formula

$$(gP)(\alpha_1, \ldots, \alpha_q; \beta_1, \ldots, \beta_q) = P(\alpha_1 g', \ldots, \alpha_q g'; g\beta_1, \ldots, g\beta_q). \tag{5}$$

We see from (3) and (4) that the differential d and multiplication $C^q \otimes C^r \to C^{q+r}$ are compatible with these representations.

The above representations induce representations of the Lie algebra $\mathfrak{gl}(n, \mathbf{R})$ in the spaces C^q. The formulas describing these representations can be obtained by using the standard basis $\{A_{kl}\}$ in $\mathfrak{gl}(n, \mathbf{R})$ (here $A_{kl} = \|a_{ij}(k, l)\|$, where $a_{ij}(k, l) = \delta_{ik}\delta_{jl}$). The element $A_{kl} \in \mathfrak{gl}(n, \mathbf{R})$ acts in C^q according to the formula

$$(A_{kl}P)(\alpha_1, \ldots, \alpha_q; \beta_1, \ldots, \beta_q)$$
$$= \sum_{i=1}^{q} \Big[P(\alpha_1, \ldots, \alpha_q; \beta_1, \ldots, \beta_{i-1}, A_{kl}\beta_i, \beta_{i+1}, \ldots, \beta_q)$$
$$- x_{ik} \frac{\partial P}{\partial x_{il}} (\alpha_1, \ldots, \alpha_q; \beta_1, \ldots, \beta_q) \Big]; \tag{6}$$

here $x_{ij} = (\alpha_i, e_j)$ is the jth component of the covector α_i.

1.6. Let F be an element of C^q, and τ a set of nonnegative integers $\tau_1, \cdots, \tau_n$ with $\tau_1 + \cdots + \tau_n = q$ The formula

$$P^\tau(\alpha_1, \ldots, \alpha_q) = P(\alpha_1, \ldots, \alpha_q; \underbrace{e_1, \ldots, e_1}_{\tau_1}, \ldots, \underbrace{e_n, \ldots, e_n}_{\tau_n})$$

defines a polynomial P^τ in the components of the covectors $\alpha_1, \cdots, \alpha_q$ such that if $\tau_1 + \cdots + \tau_{s-1} < i < j \leq \tau_1 + \cdots + \tau_s$, where $1 \leq s \leq n$, then

$$P^{\tau}(\alpha_1, \ldots, \alpha_q) = -P^{\tau}(\alpha_1, \ldots, \alpha_{i-1}, \alpha_j, \alpha_{i+1}, \ldots, \alpha_{j-1}, \alpha_i, \alpha_{j+1}, \ldots, \alpha_q).$$

Clearly, specifying an element $P \in C^q$ is equivalent to specifying for all τ polynomials P^{τ} with the indicated property, i.e., the space C^q coincides with the space $\oplus_m \widetilde{P}^q_m$ of [2] (cf. [2], 4.1). It is easy to see that the differential $d: C^q \to C^{q+1}$ also coincides with the mapping $\widetilde{\nabla}: \oplus_m \widetilde{P}^q_m \to \oplus_m \widetilde{P}^{q+1}_m$ defined in §5.1 of [2] and so the spaces $\widetilde{D}^q = \oplus_m \widetilde{D}^q_m$ of [2] coincide with the spaces $H^q(W_n; \mathrm{R})$ under investigation.

1.7. The element $P = P(\alpha_1, \cdots, \alpha_q; \beta_1, \cdots, \beta_q)$ is a polynomial in the variables x_{ij}, y_{ij} $(1 \le i \le q, 1 \le j \le n)$, where $x_{i1}, \cdots, x_{in}$ are the components of the covector α_i and $y_{i1}, \cdots, y_{in}$ components of the vector β_i. If P is homogeneous of degree m in the variables x_{ij}, then, as is clear from (3), dP is homogeneous of degree $m+1$ in the variables x_{ij}. Hence the complex $\{C^q, d^q\}$ splits into the direct sum of subcomplexes $\{C^q_k, d^q\}$ $(k = \cdots, -1, 0, 1, \cdots)$, where C^q_k is the subspace of C^q composed of the polynomials $P(x_{ij}, y_{ij})$ homogeneous of degree $q+k$ in the x_{ij}. Clearly this decomposition is invariant with respect to the action of the group $GL(n, \mathrm{R})$, and if $\alpha \in C^q_k$, $\beta \in C^r_l$, then $\alpha\beta \in C^{q+r}_{k+l}$. In particular, $\{C^q_0, d^q\}$ is a multiplicative subcomplex of the complex $\{C^q, d^q\}$, invariant with respect to the action of $GL(n, \mathrm{R})$.

1.8. *The inclusion of the complex* $\{C^q_0, d^q\}$ *in the complex* $\{C^q, d^q\}$ *induces an isomorphism of homology.*

To prove this it is sufficient to establish that the complexes $\{C^q_k, d^q\}$ with $k \neq 0$ have zero homology. But this follows from Proposition 6.2 of [2].

Remark. The cited Proposition 6.2 lets us select from the complex $\{C^q, d^q\}$ a smaller subcomplex with the same homology (the subcomplex of "elements of zero polydegree"), as done in [2]. But this subcomplex is inconvenient here since it is not invariant with respect to the action of $GL(n, \mathrm{R})$.

1.9. *If* $0 < q < n$, *then* $H^q(W_n; \mathrm{R}) = 0$.

This follows from Proposition 6.6 of [2]. Nevertheless, we give the proof.

Proof. For each element $Q \in C^q$ define the elements $\sigma_i Q \in C^q$ $(i = 1, \cdots, n)$ by the equality

$$\sigma_i Q(\alpha_1, \ldots, \alpha_q; \beta_1, \ldots, \beta_q)$$
$$= [Q(\alpha_1, \ldots, \alpha_q; \beta_1, \ldots, \beta_q)](\alpha_1 + \ldots + \alpha_q, e_i).$$

It is easy to show that if $Q_1, \cdots, Q_k \in C^q$ and $\sigma_{i_1} Q_1 + \cdots + \sigma_{i_k} Q_k = 0$ $(i_1, \cdots, i_k$ distinct), then Q_k can be put in the form $\sigma_{i_1} R_1 + \cdots + \sigma_{i_{k-1}} R_{k-1}$, where $R_1, \cdots, R_{k-1} \in C^q$ (this follows from the independence of the elements $x_{1i} + \cdots + x_{qi}$ $(i = 1, \cdots, n)$ in the ring of polynomials in $\alpha_1, \cdots, \alpha_q; \beta_1, \cdots, \beta_q$). It is also clear that $d(\sigma_i Q) = \sigma_i dQ$, and if $Q \in C^q_k$, then $\sigma_i Q \in C^q_{k+1}$.

Let $P \in C^q_0$, $q < n$ and $dP = 0$. We show that $P \in \mathrm{Im}\, d$.

Since $d(\sigma_i P) = 0$ for all i and $\sigma_i P \in C^q_1$, there are elements $P_i \in C^{q-1}_1$ with

$dP_i = \sigma_i P$ (cf. 1.8). Similarly, for $s > 1$, elements $P_{i_1 \cdots i_s} \in C_s^{q+s}$ with $1 \le i_1 < \cdots$ $\cdots < i_s \le n$ are defined (inductively with respect to s) by the equalities

$$dP_{i_1 \dots i_s} = \sum_{k=1}^{s} (-1)^{k-1} \sigma_{i_k} P_{i_1 \dots \hat{i}_k \dots i_s}. \tag{7}$$

Of course, these elements are not defined uniquely: if the elements $P_{j_1 \cdots j_{s-1}}$ for all sets $j_1 \cdots j_{s-1}$, have already been chosen, then $P_{i_1 \cdots i_s}$ is defined up to terms of the form dR with $R \in C_s^{q-s-1}$. Since $P_{1 \cdots q+1} \in C^{-1} = 0$, we have $P_{1 \cdots q+1} = 0$, and so

$$\sum_{k=1}^{q+1} (-1)^{k-1} \sigma_k P_{1 \dots \hat{k} \dots q+1} = 0.$$

Hence $P_{1 \cdots q} = \Sigma_{k=1}^{q} \sigma_k R_k$, where $R_k \in C^0$. Replace $P_{1 \dots \hat{k} \dots q}$ by $P_{1 \dots \hat{k} \dots q} +$ $(-1)^k dR_k$. Since

$$\sum_{k=1}^{q} (-1)^{k-1} \sigma_k [P_{1 \dots \hat{k} \dots q} + (-1)^k dR_k] = d\left(P_{1 \dots q} - \sum_{k=1}^{q} \sigma_k R_k\right) = 0,$$

we can then take $P_{1 \cdots q}$ to be 0. Thus, the elements $P_{i_1 \cdots i_s}$ satisfying (7) can be chosen so that $P_{1 \cdots q} = 0$. Repeating the same argument we arrive at a system of elements $P_{i_1 \cdots i_s}$ satisfying conditions (7) and such that $P_{12} = 0$. Then $\sigma_1 P_2 - \sigma_2 P_1 = dP_{12} = 0$, so that $P_1 = \sigma_1 R$ for some $R \in C^q$ and $\sigma_1 (P - dR) = dP_1 - \sigma_1 dR = 0$. Hence $P = dR$.

§2. The filtration

The aim of this section is to define in the complex $\{C_0^q, d^q\}$ a filtration of which the corresponding spectral sequence will be used in what follows to compute the homology of the complex.[1)]

2.1. Each monomial in the polynomial $P = P(\alpha_1, \cdots, \alpha_q; \beta_1, \cdots, \beta_q) = P(x_{ij}, y_{ij}) \in C_0^q$ has degrees m_{ij} with respect to the variables x_{ij}. The numbers $m_i = \Sigma_{j=1}^n m_{ij}$ $(i = 1, \cdots, q)$ we shall call simply the *degrees* of the monomial. Clearly, $m_1, \cdots, m_q$ are nonnegative integers, and $m_1 + \cdots + m_q = q$.

We shall say that the element P has filtration $\ge u$ if among the degrees of each of its monomials there are at least u different from unity. The elements of filtration $\ge u$ form a subspace of the space C_0^q, denoted by $F^u C_0^q$. Clearly, $F^0 C_0^q = C_0^q$,

1) The filtration defined here does not differ essentially from the Serre-Hochschild filtration [5], constructed relative to the subalgebra $\mathfrak{gl}(n, \mathbf{R})$ of the algebra W_n (the imbedding of $\mathfrak{gl}(n, \mathbf{R})$ into W_n is induced in a natural manner by the imbedding of the group of linear transformations into the group of diffeomorphisms).

$F^u C_0^q \supset F^{u'} C_0^q$ for $u \le u'$, and the intersection $\bigcap_{u=0}^{\infty} F^u C_0^q$ consists of zero alone.

2.2. *The filtration F is compatible with the differential d.*

Indeed, in passing from the element P to the element dP, a monomial with degrees $m_1, \cdots, m_q$ is transformed, as is clear from (3), into a sum of monomials with degrees $m_2, \cdots, m_s, \alpha, m_{s+1}, \cdots, m_{t-1}, \beta, m_t, \cdots, m_q$, where $1 \le s < t \le q+1$, $\alpha + \beta = m_1 + 1$. The number of degrees different from unity is not reduced by this transformation: if $m_1 \ne 1$, then $m_1 + 1 \ne 2$, and so either $\alpha \ne 1$ or $\beta \ne 1$. Hence $d(F^u C_0^q) \subset F^u C_0^{q+1}$.

2.3. *The filtration F is multiplicative, i.e. if $P \in F^u C_0^q$, $Q \in F^{u'} C_0^{q'}$, then $PQ \in F^{u+u'} C_0^{q+q'}$.*

2.4. *The filtration F is compatible with the action of the group $GL(n, \mathbf{R})$, i.e., $g(F^u C_0^q) \subset F^u C_0^q$ for $g \in GL(n, \mathbf{R})$, $0 \le u < \infty$.*

2.5. Put $E_0^{u,v} = (F^u C_0^{u+v})/(F^{u+1} C_0^{u+v})$ and denote by $d_0^{u,v}$ the homomorphism $E_0^{u,v} \to E_0^{u,v+1}$ obtained by factorization from d^{u+v}. The bigraded complex $\{E_0^{u,v}, d_0^{u,v}\}$ is the zero term of the multiplicative spectral sequence $\{E_r^{u,v}, d_r^{u,v}: E_r^{u,v} \to E_r^{u+r,v-r+1}\}$, whose limit term is the ring associated with the homology ring of the complex $\{C_0^q, d^q\}$, i.e., with $H^*(W_n; \mathbf{R})$. The group $GL(n, \mathbf{R})$ and the algebra $\mathfrak{gl}(n, \mathbf{R})$ act on each of the spaces $E_r^{u,v}$, and the differentials $d_r^{u,v}$ are compatible with these actions (cf. 1.4, 2.4).

We turn now to the study of the spectral sequence $\{E_r^{u,v}, d_r^{u,v}\}$.

§3. The zero term in the spectral sequence

The results of this section can also be derived from §2 of [5]. We prefer, however, an independent derivation, since direct references would be difficult.

3.1. The space $E_0^{u,v}$ can be identified with the space of polynomials in $C_0^{u,v}$ comprised of those monomials that have v degrees equal to unity and u different from unity. Each such monomial $p(\alpha_1, \cdots, \alpha_q; \beta_1, \cdots, \beta_q)$ can be represented uniquely as a product $p'(\alpha_{i_1}, \cdots, \alpha_{i_v}; \beta_{i_1}, \cdots, \beta_{i_v}) \cdot p'(\alpha_{j_1}, \cdots, \alpha_{j_u}; \beta_{j_1}, \cdots, \beta_{j_u})$, in which the first factor is a multilinear function of all its arguments. If we split up in this way each of the monomials comprising the polynomial $P \in E_0^{u,v}$, and group terms obtained from each other by simultaneous interchange of α_i with α_j and β_i with β_j, we obtain a representation of the element P as a finite sum $\Sigma_k P_k' P_k''$, where $P_k' \in E_0^{0,v}$, $P_k'' \in E_0^{u,0}$ (multiplication is in the sense of 1.4). Clearly, this representation defines an isomorphism

$$E_0^{u,v} = E_0^{0,v} \otimes E_0^{u,0}.$$

We emphasize that the ring structure in E_0 is compatible with this decomposition, i.e., if $P \in E_0^{0,v}$, $Q \in E_0^{u,0}$, then the product PQ is equal to the element $P \otimes Q \in E_0^{u,v}$.

3.2. The following proposition describes the "zero column" of the term E_0.

The complex $\{E_0^{0,q}, d_0^{0,q}\}$ is isomorphic with the standard cochain complex of the algebra $\mathfrak{gl}(n, \mathbf{R})$ with values in $\mathbf{R}$ and trivial action of the algebra.

Proof. By 3.1, $E_0^{0,q}$ is the space of functions of covectors $\alpha_1, \cdots, \alpha_q$ and vectors $\beta_1, \cdots, \beta_q$, which are linear in all their arguments and change sign under simultaneous interchange of α_i with α_j and β_i with β_j, i.e., $E_0^{0,q} = \Lambda^q(T \otimes T')$. The space $(T \otimes T')' = T' \otimes T$ can be identified with the space $\mathfrak{gl}(n, \mathbf{R})$ of linear operators in the space T (the element $\alpha \otimes \beta$ acts in T according to the formula $x \to \alpha(x)\beta$), and so $E_0^{0,q} = \Lambda^q(\mathfrak{gl}(n, \mathbf{R})')$. Let $P = P(\alpha_1, \cdots, \alpha_q; \beta_1, \cdots, \beta_q) \in E_0^{0,q}$; using the multilinearity of F, we can rewrite (3) as

$$dP(\alpha_1, \ldots, \alpha_{q+1}; \beta_1, \ldots, \beta_{q+1})$$
$$= \sum_{1\leqslant s<t\leqslant q+1} (-1)^{s+t-1}[P(\alpha_s, \alpha_1, \ldots, \hat{\alpha}_s, \ldots, \hat{\alpha}_t, \ldots, \alpha_{q+1}; (\alpha_t, \beta_s)\beta_t, \beta_1, \ldots$$
$$\ldots, \hat{\beta}_s, \ldots, \hat{\beta}_t, \ldots, \beta_{q+1}) - P(\alpha_t, \alpha_1, \ldots, \hat{\alpha}_s, \ldots, \hat{\alpha}_t, \ldots, \alpha_{q+1};$$
$$(\alpha_s, \beta_t)\beta_s, \beta_1, \ldots, \hat{\beta}_s, \ldots, \hat{\beta}_t, \ldots, \beta_{q+1})]$$
$$+ \sum_{1\leqslant s<t\leqslant q+1} (-1)^{s+t-1}[P(\alpha_t, \alpha_1, \ldots, \hat{\alpha}_s, \ldots, \hat{\alpha}_t, \ldots, \alpha_{q+1};$$
$$(\alpha_t, \beta_s)\beta_t, \beta_1, \ldots, \hat{\beta}_s, \ldots, \hat{\beta}_t, \ldots, \beta_{q+1}) - P(\alpha_s, \alpha_1, \ldots, \hat{\alpha}_s, \ldots, \hat{\alpha}_t, \ldots, \alpha_{q+1};$$
$$(\alpha_s, \beta_t)\beta_s, \beta_1, \ldots, \hat{\beta}_s, \ldots, \hat{\beta}_t, \ldots, \beta_{q+1})].$$

The first sum on the right-hand side is a linear function of all its arguments, and the second sum, by contrast, consists of terms, each of which is not linearly dependent on each of the covectors $\alpha_1, \cdots, \alpha_q$. Hence the second sum belongs to $F^1C_0^q$, and $d_0^{0,q}P$ coincides with the first sum. Since the commutator of the elements $\alpha \otimes \beta, \alpha' \otimes \beta' \in \mathfrak{gl}(n, \mathbf{R})$, where $\alpha, \alpha' \in T'$, $\beta, \beta' \in T$, is obviously equal to $(\alpha, \beta')\alpha' \otimes \beta - (\alpha', \beta)\alpha \otimes \beta'$, the differential $d_0^{0,q}$ is just the homomorphism $d\colon \Lambda^q(\mathfrak{gl}(n, \mathbf{R})') \to \Lambda^{q+1}(\mathfrak{gl}(n, \mathbf{R})')$, acting according to the formula

$$dP(\xi_1, \ldots, \xi_{q+1})$$
$$= \sum_{1\leqslant s<t\leqslant q+1} (-1)^{s+t-1}P([\xi_s, \xi_t], \xi_1, \ldots, \hat{\xi}_s, \ldots, \hat{\xi}_t, \ldots, \xi_{q+1}),$$

i.e., it coincides with the differential of the standard complex of the algebra $\mathfrak{gl}(n, \mathbf{R})$.

3.3. Now we examine the action of the differential d_0 on the elements of the "zero row", i.e., on the elements of the spaces $E_0^{q,0}$. Let $\{A'_{kl}\}$ be the basis in $\mathfrak{gl}(n, \mathbf{R})' = T \otimes T' = E_0^{0,1}$ conjugate to the basis $\{A_{kl}\}$ in $\mathfrak{gl}(n, \mathbf{R}) = T' \otimes T$ (note that $A_{kl} = e'_l \otimes e_k$, $A'_{kl} = e_l \otimes e'_k$, where $\{e'_i\}$ is the basis in T' conjugate to the basis $\{e_i\}$ in T).

For any element $P \in E_0^{q,0}$ we have the equality

$$d_0^{q,0}P = -\sum\nolimits_{k,l} A'_{kl} \otimes (A_{kl}P) \tag{8}$$

(where $A_{kl}P$ is the image of P under the action of the element $A_{kl} \in \mathfrak{gl}(n, \mathbf{R})$; cf. 1.5).

Proof. By definition, $d_0^{q,0}P$ is the part of the polynomial dP comprised of monomials linearly dependent on one of the covectors α, i.e.,

$$(d_0^{q,0}P)(x_{ij}, y_{ij})$$

$$= \sum_{i=1}^{q+1}\sum_{l=1}^{n} x_{il}\left[\frac{\partial}{\partial x_{il}} dP(\alpha_1, \ldots, \alpha_{q+1}; \beta_1, \ldots, \beta_{q+1})\right]_{\alpha_i=0}$$

$$= \sum_{i=1}^{q+1}\sum_{l=1}^{n}\sum_{k=1}^{n} x_{il}y_{ik}\left[\frac{\partial}{\partial x_{il}} dP(\alpha_1, \ldots, \alpha_{q+1}; \beta_1, \ldots, \beta_{q+1})\right]_{\alpha_i=0,\ \beta_i=e_k}.$$

From this we see that

$$d_0^{q,0}P = \textstyle\sum_{k,l} A'_{k,l} \otimes P_{k,l},$$

where

$$P_{k,l} = (-1)^q\left[\frac{\partial}{\partial x_{q+1,l}} dP(\alpha_1, \ldots, \alpha_{q+1}; \beta_1, \ldots, \beta_q, e_k)\right]_{\alpha_{q+1}=0}.$$

Replace dP in accordance with (3); in doing so we can drop the terms with $t \neq q+1$, since the degrees of the monomials in each of these terms in the components of the covector α_{q+1} are different from unity, and after differentiating with respect to $x_{q+1,l}$ and equating α_{q+1} to zero they become zero. We obtain:

$$P_{k,l} = \frac{\partial}{\partial x_{q+1,l}}\Big[\sum_{1\leqslant s\leqslant q} (-1)^s P(\alpha_s + \alpha_{q+1}, \alpha_1, \ldots, \hat{\alpha}_s, \ldots$$

$$\ldots, \alpha_q; (\alpha_{q+1}, \beta_s)e_k - x_{s\kappa}\beta_s, \beta_1, \ldots, \hat{\beta}_s, \ldots, \beta_q)\Big]_{\alpha_{q+1}=0}$$

$$= \sum_{1\leqslant s\leqslant q}\Big[x_{sq}\frac{\partial}{\partial x_{sl}} P(\alpha_1, \ldots, \alpha_q; \beta_1, \ldots, \beta_q)$$

$$- P(\alpha_1, \ldots, \alpha_q; \beta_1, \ldots, \beta_{s-1}, y_{sl}e_k, \beta_{s+1}, \ldots, \beta_q)\Big].$$

It remains only to observe that $y_{sl}e_k = A_{kl}\beta_s$ and use formula (6) of 1.4.

3.4. From Propositions 3.2 and 3.3 we can obtain a complete description of the term E_0.

For each u and v there is a canonical isomorphism of the space $E_0^{u,v} = E_0^{0,v} \otimes E_0^{u,0}$ onto the space $\mathrm{Hom}((E_0^{0,v})', E_0^{u,0}) = \mathrm{Hom}(\Lambda^v(\mathfrak{gl}(n, \mathbf{R})), E_0^{u,0})$, i.e., a canonical isomorphism $E_0^{u,v} \to C^v(\mathfrak{gl}(n, \mathbf{R}); E_0^{u,0})$ (cf. 1.1). We denote it by $\eta^{u,v}$.

The isomorphisms $\eta^{u,v}$ define for each u isomorphisms of the complex $\{E_0^{u,v}, d_0^{u,v}\}_{0\leq v<\infty}$ onto the complex $\{C^v(\mathfrak{gl}(n, \mathbf{R}); E_0^{u,0}), d^v(\mathfrak{gl}(n, \mathbf{R}); E_0^{u,0})\}$.

Proof. We must prove that for any elements $P \in E_0^{0,v}$, $Q \in E_0^{u,0}$ the element $P \otimes Q \in E_0^{u,v}$ has identical images under the homomorphisms

$$\eta^{u,v+1}\circ d_0^{u,v}, \quad [d^v(\mathfrak{gl}(n, \mathbf{R}); E_0^{u,0})]\circ\eta^{u,v}: E_0^{u,v} \to C^{v+1}(\mathfrak{gl}(n, \mathbf{R}); E_0^{u,0})$$

(for $u = 0$ this has already been proved; cf. 3.2). By the definition of the homomorphism $\eta^{u,v}$, the value of the cochain $\eta^{u,v}(P \otimes Q) \in C^v(\mathfrak{gl}(n, \mathbf{R}); E_0^{u,})$ on the elements $\xi_1, \cdots, \xi_v \in \mathfrak{gl}(n, \mathbf{R})$ is equal to $[(\eta^{0,v}P)(\xi_1, \cdots, \xi_v)]Q \in E_0^{u,0}$. Hence for any $\xi_1, \cdots, \xi_{v+1} \in \mathfrak{gl}(n, \mathbf{R})$ we have:

$$\{[d^v(\mathfrak{gl}(n, \mathbf{R}); E_0^{u,0})](\eta^{u,v}(P\otimes Q))\}(\xi_1, \ldots, \xi_{v+1})$$

$$= \sum_{1\leqslant s<t\leqslant v+1} (-1)^{s+t-1}[(\eta^{0,v}P)([\xi_s, \xi_t,]\ \xi_1, \ldots, \hat{\xi}_s, \ldots, \hat{\xi}_t, \ldots, \xi_{v+1})]Q$$

$$+ \sum_{1\leqslant s\leqslant v+1} (-1)^s[(\eta^{0,v}P)(\xi_1, \ldots, \hat{\xi}_s, \ldots, \xi_{v+1})](\xi_s Q);$$

$$[\eta^{u,v+1}(d_0^{u,v}(P\otimes Q))](\xi_1, \ldots, \xi_{v+1})$$

$$= [\eta^{u,v+1}(d_0^{0,v}P\otimes Q + (-1)^v P\otimes d_0^{u,0}Q)](\xi_1, \ldots, \xi_{v+1})$$

$$= \sum_{1\leqslant s<t\leqslant v+1} (-1)^{s+t-1}[(\eta^{0,v}P)([\xi_s, \xi_t], \xi_1, \ldots, \hat{\xi}_s, \ldots, \hat{\xi}_t, \ldots, \xi_{v+1})]Q$$

$$- \sum_{1\leqslant s\leqslant v+1} (-1)^v(-1)^{v-s}\sum\nolimits_{k,l}(\eta^{0,v}P)(\xi_1, \ldots, \hat{\xi}_s, \ldots, \xi_{v+1})A'_{kl}(\xi_s)A_{kl}Q.$$

It remains only to observe that since the bases $\{A'_{kl}\}$, $\{A_{kl}\}$ are conjugate, we have

$$\sum_{k,l} A'_{k,l}(\xi_s)A_{k,l} = \xi_s.$$

3.5. *For all u and v the space $E_1^{u,v}$ is canonically isomorphic to the space $H^v(\mathfrak{gl}(n, \mathbf{R}); E_0^{u,0})$ of cohomology of dimension v of the Lie algebra $\mathfrak{gl}(n, \mathbf{R})$ with coefficients in the $\mathfrak{gl}(n, \mathbf{R})$-module $E_0^{u,0}$.*

This follows from 3.4.

§4. The first term of the spectral sequence

The problem of describing the term E_1 of our spectral sequence has thus been reduced to that of examining the cohomology of the Lie algebra $\mathfrak{gl}(n, \mathbf{R})$ with coefficients in various representations; this is the subject of the present section.

We start with two well-known facts from the theory of Lie algebras.

4.1. Let M be an arbitrary finite-dimensional $\mathfrak{gl}(n, \mathbf{R})$-module. Denote by M_0 the subspace of M consisting of those elements m such that $Am = 0$ for any matrix $A \in \mathfrak{gl}(n, \mathbf{R})$ with $\operatorname{Tr} A = 0$. Obviously M_0 is a submodule of the $\mathfrak{gl}(n, \mathbf{R})$-module M: if $m \in M_0$ and $B \in \mathfrak{gl}(n, \mathbf{R})$, then for any matrix $A \in \mathfrak{gl}(n, \mathbf{R})$ with $\operatorname{Tr} A = 0$ we have the equality $ABm = A(B - (\operatorname{Tr} B)E)m + (\operatorname{Tr} B)Am = 0$, so that $Bm \in M_0$.

The inclusion $M_0 \to M$ induces an isomorphism $H^*(\mathfrak{gl}(n, \mathbf{R}), M_0) = H^*(\mathfrak{gl}(n, \mathbf{R}); M)$.

Proof. The direct decomposition $\mathfrak{gl}(n, \mathbf{R}) = \mathbf{R} \oplus \mathfrak{sl}(n, \mathbf{R})$ induces canonical isomorphisms $H^*(\mathfrak{gl}(n, \mathbf{R}); M_0) = H^*(\mathbf{R}; H^*(\mathfrak{sl}(n, \mathbf{R}); M_0))$, and $H^*(\mathfrak{gl}(n, \mathbf{R}); M) = H^*(\mathbf{R}; H^*(\mathfrak{sl}(n, \mathbf{R}; M))$; the homomorphism

$$H^*(\mathfrak{sl}(n, \mathbf{R}); M_0) \to H^*(\mathfrak{sl}(n, \mathbf{R}); M),$$

induced by the inclusion $M_0 \to M$ is likewise an isomorphism, in view of the simplicity of the algebra $\mathfrak{sl}(n, \mathbf{R})$ ([4], Russian p. 84).

4.2. *There exists a ring isomorphism* $H^*(\mathfrak{gl}(n, \mathbf{R}); \mathbf{R}) = H^*(U(n); \mathbf{R})$. *In other words, there exist elements* $\phi_i = \phi_i(n) \in H^*(\mathfrak{gl}(n, \mathbf{R}); \mathbf{R})$, $i = 1, \cdots, n$, *such that* $\dim \phi_i(n) = 2i - 1$ *and* $H^*(\mathfrak{gl}(n, \mathbf{R}); \mathbf{R})$ *is the exterior algebra in* $\phi_1(n), \cdots, \phi_n(n)$ *over* $\mathbf{R}$ ([6], *Russian* p. 191). *The generators* $\phi_i(n)$ *can be chosen so that the cohomology isomorphism induced by the usual inclusion* $\mathfrak{gl}(n, \mathbf{R}) \to \mathfrak{gl}(m, \mathbf{R})$ $(m \geq n)$ *takes* $\phi_i(m)$ *into* $\phi_i(n)$ *for* $i \leq n$ *and into zero for* $i > n$.

Proof. The complexifications of the Lie algebras $\mathfrak{gl}(n, \mathbf{R})$ and $\mathfrak{u}(n)$ are canonically isomorphic (both coincide with $\mathfrak{gl}(n, \mathbf{C})$), so that $H^*(\mathfrak{gl}(n, \mathbf{R}); \mathbf{R}) = H^*(\mathfrak{u}(n); \mathbf{R})$. In turn, a canonical isomorphism $H^*(\mathfrak{u}(n); \mathbf{R}) \to H^*(U(n); \mathbf{R})$ is induced, since $U(n)$ is compact, by the natural imbedding of the standard complex of the algebra $\mathfrak{u}(n)$ into the de Rham complex of the group $U(n)$. The last assertion of the proposition is obvious.

Remark. For the generators $\phi_i(n)$ of the exterior algebra $H^*(\mathfrak{gl}(n, \mathbf{R}); \mathbf{R})$ we can take the cohomology classes of the cocycles $\Phi_i \in C^{2i-1}(\mathfrak{gl}(n, \mathbf{R}); \mathbf{R})$, where

$$\Phi_i(\xi_1, \ldots, \xi_{2i-1}) = \sum \varepsilon(k_1, \ldots, k_{2i-1}) \operatorname{Tr}(\xi_{k_1}, \ldots, \xi_{k_{2i-1}}),$$

with the summation over all permutations $(k_1, \cdots, k_{2i-1})$ of the numbers $1, \cdots \cdots, 2i-1$, and $\epsilon(k_1, \cdots, k_{2i-1}) = 1$ or -1 according to the parity of the permutation $(k_1, \cdots, k_{2i-1})$. This observation will not be used, however, in what follows.

4.3. We proceed to examination of the term E_1. Let K^u be the subspace of $E_0^{u,0}$ consisting of those elements on which the action of the algebra $\mathfrak{sl}(n, \mathbf{R})$ is trivial. Clearly $K^* = \Sigma_{u=0}^{\infty} K^u$ is a subring of the ring $E_0^{*,0} = \Sigma_{u=0}^{\infty} E_0^{u,0}$.

We construct *a canonical isomorphism of the bigraded rings* $E_1 = \Sigma_{u,v} E_1^{u,v}$ *and* $K^* \otimes H^*(U(n); \mathbf{R})$.

Observe first of all that the algebra $\mathfrak{gl}(n, \mathbf{R})$ also acts trivially on K^*. This follows from the fact that the equality $e(P) = 0$, where $e \in \mathfrak{gl}(n, \mathbf{R})$ is the identity matrix, holds for all $P \in C_0^q$ (which in turn follows easily from (6), using the fact that the degree of each monomial in the polynomial P for the totality of the variables x_{ij} is equal to q). By 4.1, the inclusion $K^u \to E_0^{u,0}$ induces an isomorphism $H^*(\mathfrak{gl}(n, \mathbf{R}); K^u) = H^*(\mathfrak{gl}(n, \mathbf{R}); E_0^{u,0})$. As for the equality $H^*(\mathfrak{gl}(n, \mathbf{R}); K^*) = K^* \otimes H^*(U(n); \mathbf{R})$, it follows, since $\mathfrak{gl}(n, \mathbf{R})$ acts trivially on K^*, from 4.2.

4.4. *Every element of* C_0^q *invariant with respect to the action of the group* $SL(n, \mathbf{R})$ *is a linear combination of polynomials in* $\alpha_1, \cdots, \alpha_q; \beta_1, \cdots, \beta_q$ *of the form*

$$(\alpha_{s_1}, \beta_1) \ldots (\alpha_{s_q}, \beta_q) \tag{9}$$

(*equality allowed among the indices* $s_1, \cdots, s_q$). *In particular, this holds for the elements of the space* K^*.

Proof. Since the degrees of the monomials (see 2.1) are invariant with respect to the action of the group $SL(n, \mathbf{R})$, it suffices to prove this for invariant cochains $P \in C_0^q$ that are homogeneous in the sense that all the monomials of the polynomial P have, up to order, the same degree $m_1, \cdots, m_q$. More than that, it suffices to prove that we already have a sum of polynomials of the form (9) in the polynomial $P_0(\alpha_1, \cdots, \alpha_q; \beta_1, \cdots, \beta_q)$, comprised of those monomials of P that have degree m_1 in α_1, degree m_2 in $\alpha_2, \cdots$, degree m_q in α_q, where $m_1 \geq m_2 \geq \cdots \geq m_q$.

The polynomial P_0 can be regarded as a tensor. More precisely, there exists a unique tensor

$$\tau = \tau(\gamma_1, \ldots, \gamma_q; \beta_1, \ldots, \beta_q) \in [(\otimes^q T') \otimes (\otimes^q T)]'$$

such that

$$P(\alpha_1, \ldots, \alpha_q; \beta_1, \ldots, \beta_q) = \tau(\underbrace{\alpha_1, \ldots, \alpha_1}_{m_1}, \ldots, \underbrace{\alpha_q, \ldots, \alpha_q}_{m_q}; \beta_1, \ldots, \beta_q) \tag{10}$$

and symmetric with respect to the arguments $\gamma_{m_1+\cdots+m_{s-1}+1}, \cdots, \gamma_{m_1+\cdots+m_s}$ for $1 \leq s \leq q$. Clearly the tensor τ is also invariant with respect to the group $SL(n, \mathbf{R})$.

By a classical theorem of the theory of invariants, the algebra of tensors invariant with respect to the group $SL(n, \mathbf{R})$ is generated by tensors of the form:

1) (α, β), where $\alpha \in T'$, $\beta \in T$;

2) $\det(\alpha_1, \cdots, \alpha_n)$, where $\alpha_1, \cdots, \alpha_n \in T'$;

3) $\det(\beta_1, \cdots, \beta_n)$, where $\beta_1, \cdots, \beta_n \in T$ ([7], Theorem II 6.A).

Since $\det(\alpha_1, \cdots, \alpha_n) \cdot \det(\beta_1, \cdots, \beta_n) = \det \|(\alpha_i, \beta_j)\|$, invariant tensors depending on the same number of vectors and covectors are generated by generators of only the first form. It follows that τ is a linear combination of tensors of the form $(\alpha_{i_1}, \beta_1) \cdots (\alpha_{i_q}, \beta_q)$ (where the indices $i_1, \cdots, i_q$ are distinct), and application of (10) yields our assertion.

4.5. *Let ψ_k be the element of $E_0^{2k,0}$ defined by the equality*

$$\begin{aligned}&\psi_k(\alpha_1, \ldots, \alpha_{2k}; \beta_1, \ldots, \beta_{2k})\\ &= \Sigma_{(i_1,\ldots,i_{2k})} \varepsilon(i_1, \ldots, i_{2k}) (\alpha_{i_k}, \beta_{i_1})(\alpha_{i_1}, \beta_{i_2}) \ldots (\alpha_{i_{k-1}}, \beta_{i_k})\\ &\quad \times (\alpha_{i_1}, \beta_{i_{k+1}}) \ldots (\alpha_{i_k}, \beta_{i_{2k}})\end{aligned}$$

(summation over all permutations $(i_1, \cdots, i_{2k})$ of the numbers 1, 2, $\cdots$, 2k). Then the algebra K^* is generated by the elements $\psi_1, \cdots, \psi_n$. If $s_1 + 2s_2 + \cdots + ns_n > n$, then $\psi_1^{s_1} \psi_2^{s_2} \cdots \psi_n^{s_n} = 0$; and these relations constitute a basis in the space of relations between monomials in $\psi_1, \cdots, \psi_n$.

Proof. *Take $P \in K^q$.* By 4.4 we have:

$$P(\alpha_1, \ldots, \alpha_q; \beta_1, \ldots, \beta_q) = \sum_{s_1,\ldots,s_q} a_{s_1,\ldots,s_q} (\alpha_{s_1}, \beta_1) \ldots (\alpha_{s_q}, \beta_q).$$

Suppose a coefficient $a_{s_1,\cdots,s_q}$ is different from zero. Since $P \in E_0^{q,0}$, i.e., since the filtration of P is equal to q, the expression $a_{s_1,\cdots,s_q}(\alpha_{s_1}, \beta_1)\cdots(\alpha_{s_q}, \beta_q)$ is not linear in any α, and therefore among the numbers $s_1, \cdots, s_q$ none is encountered only once; in particular, among them at most $[q/2]$ are distinct. Let $j_1, \cdots, j_r$ be all the natural numbers not exceeding q and not encountered among $s_1, \cdots, s_q$. By the preceding remark we have $r \geq q/2$. But the numbers $s_{j_1}, \cdots, s_{j_r}$ are distinct: if $s_{jk} = s_{jl}$, then the product $(\alpha_{s_1}, \beta_1)\cdots(\alpha_{s_q}, \beta_q)$ goes into itself under simultaneous interchange of α_{jk} with α_{jl} and β_{jk} with β_{jl}, so that $a_{s_1\cdots s_q} = -a_{s_1\cdots s_q}$; consequently, we have also $r \leq q/2$; i.e., $r = q/2$. We conclude that q is even, and that among the indices $s_1, \cdots, s_q$ exactly half the numbers $1, \cdots, q$ are not encountered at all, and the rest exactly twice each. Using skew-symmetry, we find that P is a linear combination of polynomials

$$P_\tau(\alpha_1, \ldots, \alpha_q; \beta_1, \ldots; \beta_q) \\ = \Sigma_{(i_1\ldots i_q)}\varepsilon(i_1\ldots i_q)(\alpha_{\tau(i_1)}, \beta_{i_1})\ldots(\alpha_{\tau(i_r)}, \beta_{i_r})(\alpha_{i_1}, \beta_{i_{r+1}})\ldots(\alpha_{i_r}, \beta_{i_{2r}}), \tag{11}$$

where $r = q/2$ and τ is a permutation of the indices $i_1, \cdots, i_r$. Since every permutation factors into a product of cyclic permutations, the element P_τ is representable as a sum of products of the elements ψ_k.

The part of the sum (11) corresponding to fixed values of the indices $i_1, \cdots, i_r$ is multilinear and skew-symmetric in $\beta_{i_{r+1}}, \cdots, \beta_{i_{2r}}$; that is to say, in r vectors of an n-dimensional space. Hence if $r = q/2 > n$, then $P_\tau = 0$. This means that $\psi_1^{s_1}\psi_2^{s_2}\cdots \cdots\psi_N^{s_N} = 0$ for $s_1 + 2s_2 + \cdots + Ns_N > n$, and in particular, $\psi_i = 0$ for $i > n$.

Finally, for $r \leq n$ the elements P_τ are linearly independent. Indeed, if $\Sigma a_\tau P_\tau = 0$, then putting $\alpha_1 = e_1, \cdots, \alpha_r = e_r, \alpha_{r+1} = \cdots = \alpha_{2r} = 0, \beta_1 = e_{\tau(1)}, \cdots, \beta_r = e_{\tau(r)}, \beta_{r+1} = e_1, \cdots, \beta_{2r} = e_r$, we obtain that $r! \cdot a_\tau = 0$.

4.6. Proposition 4.5 together with the isomorphism 4.3 allows us to give a complete description of the term E_1.

The bigraded ring E_1 *is isomorphic to the tensor product* $\Lambda(\phi_1, \cdots, \phi_n) \otimes \{R[\psi_1, \cdots, \psi_n]/I\}$, *where* $\phi_k \in E_1^{0,2k-1}$, $\psi_k \in E_1^{2k,0}$, *and* I *is the ideal in the ring* $R[\psi_1, \cdots, \psi_n]$ *generated by the monomials* $\psi_1^{s_1}\psi_2^{s_2}\cdots\psi_n^{s_n}$ *with* $s_1 + 2s_1 + \cdots \cdots + ns_n > n$.

4.7. As follows from 4.6, $E_1^{u,v} = 0$ for odd u. Since $d_1^{u,v}$ is a homomorphism from the space $E_1^{u,v}$ to the space $E_1^{u+1,v}$, it follows that $d_1 \equiv 0$. We arrive therefore at the following proposition.

The term E_2 *of our spectral sequence is isomorphic to the term* E_1.

§5. The remaining terms of the spectral sequence

Throughout this section we shall denote by the same letter an element of $\operatorname{Ker} d_r \subset E_r$ and the element it determines in E_{r+1}. For example, if $\alpha \in E_2$ and $d_2\alpha = 0$, there is defined an element $d_3\alpha \in E_3$; if $d_3\alpha = 0$, there is defined an element $d_4\alpha \in E_4$; etc.

5.1. *If* $1 \le k \le n$, *then* $d_i^{0,2k-1}\phi_k = 0$ *for* $i = 2, \cdots, 2k-1$, *and* $d_{2k}^{0,2k-1}\phi_k = \psi_k$.

Proof. The inclusion $W_n \to W_m$ induces a homomorphism of the spectral sequence $\{'E_r^{u,v}, 'd_r^{u,v}\}$, corresponding to the algebra W_m into the given spectral sequence $\{E_r^{u,v}, d_r^{u,v}\}$. The ring $'E_r$ is generated by generators $\phi_1', \cdots, \phi_m', \psi_1', \cdots, \psi_m'$ analogous to $\phi_1, \cdots, \phi_n, \psi_1, \cdots, \psi_n$, and the homomorphism $'E_r \to E_r$ takes $\phi_1', \cdots, \phi_n', \psi_1', \cdots, \psi_n'$ into $\phi_1, \cdots, \phi_n, \psi_1, \cdots, \psi_n$ (for ψ this is obvious from the definition in 4.5, and for ϕ it follows from Proposition 4.2). By Proposition 1.9, $H^q(W_m; \mathrm{R}) = 0$ for $0 < q < m$ and therefore $'E_\infty^{u,v} = 0$ for $0 < u + v < m$. Now apply to the spectral sequence $\{'E, 'd\}$ the theorem of Borel ([6], Theorem 13.1). We obtain that if m is sufficiently large, then $'d_i^{0,2k-1}\phi_k' = 0$ and $'d_{2k}^{0,2k-1}\phi_k' = \psi_k'$ for $i = 2, \cdots, 2k-1$ and $k = 1, \cdots, n$. This implies our assertion.

5.2. Proposition 5.1 and the formulas connecting the differential of the spectral sequence with multiplication allow us to give a complete description of our spectral sequence. We obtain the following.

Consider an element $P = \phi_{i_1}, \phi_{i_k}\psi_{j_1}^{s_1} \cdots \psi_{j_l}^{s_l} \in E_2^{2(j_1 s_1 + \cdots + j_l s_l),\, 2(i_1 + \cdots + i_k) - k}$, *where* $i_1 < \cdots < i_k$, $j_1 < \cdots < j_l$ *and all the* s_i *are different from zero. If* $k = 0$, *or* $i_1 > j_1$, *or* $i_1 + j_1 s_1 + \cdots + j_l s_l > n$, *then* $d_r P = 0$ *for all* r. *Otherwise,* $d_r P = 0$ *for* $r > 2i_1$, *and* $d_{2i_1} P = \phi_{i_2} \cdots \phi_{i_k}\psi_{i_1}\psi_{j_1}^{s_1} \cdots \psi_{j_l}^{s_l}$.

This proposition yields an algorithm for determining the spaces $E_\infty^{u,v}$, and thereby the dimensions of the spaces $H^q(W_n; \mathrm{R})$.

5.3. *If* $u \le n$ *and* $u + v > 0$, *then* $E_\infty^{u,v} = 0$. *If* $0 < u + v \le 2n$, *then* $E_\infty^{u,v} = 0$.

Proof. Take $P = \phi_{i_1} \cdots \phi_{i_k}\psi_{j_1}^{s_1} \cdots \psi_{j_l}^{s_l} \in E_2^{u,v}$, and suppose $u + v > 0$ and either $u \le n$ or $u + v \le 2n$. There are two possibilities. The first is that $i_1 > j_1$ (or $k = 0$), and then $2j_1 \le n$; in this case P is the image of the element $\phi_{j_1}\phi_{i_1} \cdots \cdots \phi_{i_k}\psi_{j_1}^{s_1 - 1}\psi_{j_2}^{s_2} \cdots \psi_{j_l}^{s_l} \in E_2$ under the differential $d_2^{u - 2j_1, v + 2j_1 - 1}$. The second possibility is that $i_1 \le j_1$ (or $l = 0$), and then $i_1 + u/2 \le n$; in this case $d_{2i_1} P \neq 0$. In either case, P does not remain in E_∞.

5.4. **Corollary.** *If* $0 < q \le 2n$, *then* $H^q(W_n; \mathrm{R}) = 0$.

5.5. It is easy to find the first nontrivial cohomology space for the algebra W_n. *This is the space* $H^{2n+1}(W_n; \mathrm{R})$; *its dimension* ρ_n *is one less than the number of representations of the integer* $n + 1$ *as a sum of nonnegative integers (representations differing only in the order of summands being regarded as the same).* The proof is left to the reader.

5.6. *Multiplication in the ring* $H^*(W_n; \mathrm{R})$ *is trivial.*

Proof. Proposition 5.3 implies that every element of $H^*(W_n, \mathrm{R})$, of positive dimension is represented by a cocycle of filtration $> n$, and therefore the product of two such elements is represented by a cocycle of filtration $> 2n$. Since $E_\infty^{u,v} = 0$ for $u > 2n$, the product is zero.

5.7. We can now prove Theorem 0.3 of the Introduction. The second term of the real cohomology spectral sequence of the bundle $(X_n, p, (G(N, n))_{2n})$ described in the Introduction is $H^*((G(N, n))_{2n}; \mathbf{R}) \otimes H^*(U(n); \mathbf{R})$. From Theorem 19.6 of [6] it follows that this tensor product is isomorphic as a bigraded algebra to the term E_2 of our spectral sequence, and also that in the two spectral sequences the actions of the differentials are the same. This means that the limits of the spectral sequences are the same, and thus $H^q(W_n; \mathbf{R}) = H^q(X_n, \mathbf{R})$ for all q. This equality together with 5.6 is the content of Theorem 0.3.

5.8. We note in conclusion that Proposition 5.1 can be proved by direct computation of differentials, without reference to Proposition 1.9. For this one can use the Remark in §4.2.

§6. Applications to the Lie algebras of smooth vector fields on manifolds

6.1. Let $\mathfrak{A}(M)$ be the Lie algebra of vector fields of class C^∞ on a compact orientable n-dimensional manifold M of class C^∞. In the cochain complex $\mathfrak{C} = \mathfrak{C}(M)$ of this algebra with values in $\mathbf{R}$ consider the subcomplex $\mathfrak{C}_d = \mathfrak{C}_d(M)$ consisting of those cochains $P \in \mathfrak{C}^q$ such that whenever the supports of the vector fields $\xi_1, \cdots, \xi_q$ have (taken together) void intersection, then $P(\xi_1, \cdots, \xi_q) = 0$. In [2] we called this subcomplex diagonal. To compute its cohomology (to which in a certain sense we reduced the problem of finding the cohomology of the algebra $\mathfrak{A}(M)$) a spectral sequence was constructed. The main proposition concerning this spectral sequence ([2], Proposition 7.5) can be formulated as follows, using the results of the present paper (viz., Theorem 0.3 and Propositions 1.6 and 5.4):

There exists a spectral sequence $\{E_r^{p,q}, d_r^{p,q}\}$ *which converges to the cohomology of the complex* $\mathfrak{C}_d$, *such that* $E_2^{p,q} = H_{-p}(M; \mathbf{R}) \otimes H^q(X_n; \mathbf{R})$. *In particular,* $E_2^{p,q} = 0$ *for* $0 < p + q \leq n$.

A similar sharpening can be made of Proposition 8.5 of [2]. Without recalling the relevent definitions, we remark only that for the spaces ${}^{(k)}E_2^{p,q}$ considered there it follows from the results of the present paper that ${}^{(k)}E_2^{p,q} = 0$ for $0 < u + v \leq kn + k - 1$.

6.2. *The first nontrivial cohomology space for the diagonal complex is* $H_{n+1}(\mathfrak{C}_d)$. *Its dimension (over* $\mathbf{R}$*) is* ρ_n (see 5.5).

This follows from the preceding remarks and §5.5.

6.3. *The q-dimensional cohomology space of the Lie algebra of smooth vector fields of a compact orientable n-dimensional manifold M coincides for* $q \leq 2n$ *with the q-dimensional cohomology space of the complex* $\mathfrak{C}_d(M)$. *In particular,* $H^q(\mathfrak{A}(M); \mathbf{R}) = 0$ *for* $q \leq n$, *and* $\dim H^{n+1}(\mathfrak{A}(M); \mathbf{R}) = \rho_n$.

This follows from 6.2 and Proposition 8.5 of [2].

Received 4 NOV 69

BIBLIOGRAPHY

[1] S. S. Chern, *Complex manifolds*, Textos de Matemática, no. 5, Instituto de Física e Matemática, Universidade do Recife, 1959; Russian transl., IL, Moscow, 1961. MR 22 #1920.

[2] I. M. Gel'fand and D. B. Fuks, *Cohomologies of the Lie algebra of tangent vector fields of a smooth manifold*, Funkcional. Anal. i Priložen **3** (1969), no. 3, 32–52 = Functional Anal. Appl. 3 (1969), 194–210. MR 41 #1067.

[3] ———, *On cohomologies of a Lie algebra of smooth vector fields*, Dokl. Akad. Nauk SSSR **190** (1970), 1267–1270 = Soviet Math. Dokl. 11 (1970), 268–271.

[4] *Séminaire "Sophus Lie" de l'École Normale Supérieure*, 1954/55. *Théorie des algèbres de Lie. Topologie des groupes de Lie*, Secrétariat mathématique, Paris, 1955; Russian transl., IL, Moscow, 1962. MR 17, 384.

[5] G. Hochschild and J. P. Serre, *Cohomology of Lie algebras*, Ann. of Math. (2) 57 (1953), 591–603. MR 14, 943.

[6] A. Borel, *Sur la cohomologie des espaces fibrés principaux et des espaces homogènes de groupes de Lie compacts*, Ann. of Math. (2) 57 (1953), 115–207; Russian transl., Chapter IV in *Fiber spaces and their applications*, II, Moscow, 1958.

[7] H. Weyl, *The classical groups. Their invariants and representations*, Princeton, Univ. Press, Princeton, N. J., 1939; 2nd ed., 1946; Russian transl., IL, Moscow, 1947. MR 1, 42.

6.

(with D. B. Fuks)

Cohomologies of Lie algebra of tangential vector fields II

Funkts. Anal. Prilozh. 4 (4) (1970) 23–31 [Funct. Anal. Appl. 4 (1970) 110–116].
MR 44:2248. Zbl. **208**:514

This paper is a continuation of an earlier paper by us which we shall cite as [1]. We recall that in [1] we studied the cohomologies of the Lie algebra $\mathfrak{A}(M)$ of smooth tangential vector fields of a smooth, compact orientable manifold M with coefficients in a trivial real representation. The main result of [1] was a theorem about the finite dimensionality of these cohomologies (in every dimension). In the course of the proof we introduced in the standard complex $\mathscr{C}(M)=\{C^q(M), d^q\}$ of the Lie algebra $\mathfrak{A}(M)$ a subcomplex $\mathscr{C}_1(M)=\{C_1^q(M), d^q\}$, designated "diagonal" by us, constructed from the spectral sequence $\mathscr{E}=\{E_r^{u,v}, \delta_r^{u,v}=E_r^{u,v}\to E_r^{u+r,v-r+1}\}$ which converged to the homologies of the diagonal complex, and an expression for its first (second) term was derived.

The present work consists of two parts. In the first a new interpretation of the second term of the spectral sequence $\mathscr{E}$ is given which permits, in particular, proof of the triviality of some of its differentials. In the second part the relation between the passage to the limit of the spectral sequence $\mathscr{E}$ (i.e., between the homologies of the diagonal complex) and the cohomologies of the algebra $\mathfrak{A}(M)$ is discussed. The most complete information is obtained for the case when the spectral sequence $\mathscr{E}$ is trivial (i.e., when $E_2 = E_\infty$). Making use of the results of both the parts we obtain, for some of the manifolds, in particular for toruses of any dimension and for all orientable two-dimensional manifolds, a description of the ring, adjoined (with respect to the filtrations introduced in 1.2 of [1]) to the ring of cohomologies of the Lie algebra of tangential vector fields. In the present work we shall follow the notation introduced in [1] and will not repeat the definitions given there. For ease in reading we shall mention at the appropriate places the number of section of [1] containing the necessary explanations.

The present work was preceded by a short note [3].

§1. SPECTRAL SEQUENCE OF A DIAGONAL COMPLEX

1.1. We shall begin with the observation that the complex $\{\bigoplus_m P_m^q, \tilde{\nabla}\}$ (see [1], pp. 4.1, 5.1) is none other than the standard complex of the topological Lie algebra W_n of formal vector fields at the point $0 \in R^n$ with coefficients in R (i.e., in a trivial real representation).

We recall that a formal vector field at the point $0 \in R^n$ is an expression of the form $\sum_{i=1}^n f_i(x_1, \ldots, x_n)e_i$, where $f_1, \ldots, f_n$ are formal power series in powers of the coordinates $x_1, \ldots, x_n$ of the space R^n, and $e_1, \ldots, e_n$ is a standard basis in R^n. The operation of commutation transforming the space W_n of formal vector fields into a topological Lie algebra is given by the formula

$$\Big[\sum_i f_ie_i,\ \sum_j g_je_j\Big]=\sum_k\Big(\sum_i f_i\frac{\partial g_k}{\partial x_i}-\sum_j g_j\frac{\partial f_k}{\partial x_j}\Big)e_k.$$

The standard complex of the topological algebra W_n is a complex $\{C^q(W_n;\ R), d^q\}$ where $C^q(W_n;\ R)$ is the space of continuous skew-symmetric q-linear real functionals on W_n, and the differential $d^q : C^q(W_n;\ R)\to C^{q+1}(W_n;\ R)$ defined by the formula $(d^q\alpha)(\eta_1, \ldots, \eta_{q+1}) = \sum_{1\le s<t\le q+1}(-1)^{s+t-1}\alpha([\eta_s, \eta_t], \eta_1, \ldots \hat{\eta}_s \ldots \hat{\eta}_t, \ldots, \eta_{q+1})$.

Moscow State University. Translated from Funktsional'nyi Analiz i Ego Prilozheniya, Vol. 4, No. 2, pp. 23-31, April-June, 1970. Original article submitted December 17, 1969.

Every element $\{\lambda^\tau\} \in \tilde{P}^q_m$ can be included in the space $C^q(W_n;\ R)$, by identifying it with a q-linear functional on W_n, acting in accordance with the formula

$$\left(\sum_i f_{1,i}e_i,\ \ldots,\ \sum_i f_{q,i}e_i\right) \mapsto \sum_\tau v_0\left[\lambda^\tau\left(\frac{\partial}{\partial x_{11}},\ldots,\frac{\partial}{\partial x_{1n}};\ldots;\frac{\partial}{\partial x_{q1}},\ldots,\frac{\partial}{\partial x_{qn}}\right)\prod_{r=1}^{n} f_{r,s_r}(x_{r1},\ \ldots,\ x_{rn})\right]. \quad (1)$$

Here S_r is a natural number such that $\tau_1 + \ldots + \tau_{S_r - 1} < r \le \tau_1 + \ldots + \tau_{S_r}$, and $\nu_0[\ldots]$ is a free term indicated in the parentheses of the power series.

Such an identification defines, as can be easily proved, an isomorphism of the spaces $\oplus \tilde{P}^q_m$ and $C^q(W_n;\ R)$; the homomorphism $\tilde{\nabla}$ goes over into a differential d^q of the standard complex $\{C^q(W_n;\ R), d^q\}$ of the algebra W_n. Thus the homologies $\tilde{D}^q$ of the complex $\{\underset{m}{\oplus}\tilde{P}^q_m,\ \tilde{\nabla}\}$ coincide with the cohomologies $H^q(W_n;\ R)$ of the topological Lie algebra W_n with coefficients in R.

More details and other viewpoints than that asserted here are presented in our paper [2].

1.2. The first term $E_2 = \underset{u,v}{\oplus} E_2^{u,v}$ of the spectral sequence $\mathscr{E} = \{E_r^{u,v},\ \delta_r^{u,v}\}$ (see [1], p. 2.12) is isomorphic (as the bigraded space) to the sum $\underset{u,v}{\oplus}(H_{-u}(M;\ R) \otimes \tilde{D}^v) = \underset{u,v}{\oplus}(H^{n+u}(M;\ R) \otimes H^v(W_n;\ R))$ (see [1], p. 7.5). In paragraphs 1.3-1.4 below we shall give a new description of the isomorphism $H^{n+u}(M;\ R) \otimes H^v(W_n;\ R) \to E_2^{u,v}$ (we shall not prove again that this is an isomorphism).

1.3. We begin with the construction of certain spectral cochains of the diagonal complex (see [1], p. 1.4).

Let U be a coordinate neighborhood on the manifold M with local coordinates $y_1, \ldots, y_n$. Every vector field $\xi \in \mathfrak{A}(M)$ can be represented in U in the form $\sum_{i=1}^{n} g_i(y_1,\ \ldots,\ y_n)\frac{\partial}{\partial y_i}$; it defines at every point $z = (y_1^0, \ldots, y_n^0) \in U$ a formal vector field $\xi(z,\ U) = \sum_{i=1}^{n} f_i(z)e_i$, where

$$[f_i(z)](x_1,\ \ldots,\ x_n) = \sum a_{i,k_1,\ldots,k_n} x_1^{k_1} \ldots x_n^{k_n},$$

$$a_{.,k_1,\ldots,k_n} = \frac{1}{k_1! \ldots k_n!}\left[\frac{\partial^{k_1+\ldots+k_n} g_i}{\partial y_1^{k_1} \ldots \partial y_n^{k_n}}\right]_{y_1=y_1^0,\ldots,y_n=y_n^0}$$

For any of the q vector fields $\xi_1, \ldots, \xi_q \in \mathfrak{A}(M)$ and any cochain $\alpha \in C^{q+r}(W_n;\ R) = \underset{m}{\oplus}\tilde{P}^{q+r}_m$, we set

$$\varphi(\alpha,\ U;\ \xi_1,\ \ldots,\ \xi_q) = \sum_{1 \le i_1 < \ldots < i_r \le n} \alpha(\xi_1(z,\ U), \ldots,\ \xi_q(z,\ U),\ e_{i_1}, \ldots,\ e_{i_r})\, dy_{i_1} \wedge \ldots \wedge dy_{i_r}.$$

As can be seen from this formula $\varphi(\alpha,\ U;\ \xi_1, \ldots, \xi_q)$ is a smooth, real differential form of degree r which is defined in U.

We shall now concentrate on the following objects: 1) the element $\alpha \in C^{q+r}(W_n;\ R)$; 2) smooth differential form ψ on M of degree $n-r$; 3) the coordinate covering $\Gamma = \{U_1, \ldots, U_N\}$ with local coordinates $y_{i1}, \ldots, y_{in}$ in U_i; 4) the partition of unity $\rho = \{\rho_1, \ldots, \rho_N\}$ subordinated to this. For any of the vector fields $\xi_1, \ldots, \xi_q \in \mathfrak{A}(M)$ we set

$$J(\alpha,\ \psi,\ \Gamma,\ \rho;\ \xi_1,\ \ldots,\ \xi_q) = \int_M \psi \wedge \left[\sum_{i=1}^{N} \rho_i \varphi(\alpha,\ U_i;\ \xi_1,\ \ldots,\ \xi_q)\right]. \quad (2)$$

It is clear that $J(\alpha, \psi, \Gamma, \rho)$ is a continuous skew-symmetric q-linear real functional on $\mathfrak{A}(M)$, i.e., $J(\alpha, \psi, \Gamma, \rho) \in C^q(M)$ belongs to C_1^q, i.e., it is a cochain of the diagonal complex. Indeed, if the carriers of the vector fields $\xi_1, \ldots, \xi_q$ do not intersect in a union, i.e., if in the neighborhood of every point of the manifold M one of the fields $\xi_1, \ldots, \xi_q$ vanishes, then each of the forms $\varphi(\alpha,\ U;\ \xi_1, \ldots, \xi_q)$, as can be easily seen from the definition, identically vanishes, and the number $J(\alpha, \psi, \Gamma, \rho;\ \xi_1, \ldots, \xi_q)$ denotes the functional $J(\alpha, \psi, \Gamma, \rho)$ on the fields $\xi_1, \ldots, \xi_q$.

We emphasize that the symbol $J(\alpha, \psi, \Gamma, \rho)$ makes sense for any $\alpha, \psi, \Gamma, \rho$ such that $\deg \alpha + \deg \psi \geq n$ It defines an element of $C_1^{\deg\alpha+\deg\psi-n}(M)$.

1.4. Let $a \in \widetilde{D}^{q+r} = H^{q+r}(W_n;\ R)$, $\Psi \in H^{n-r}(M, R)$ be an arbitrary element. Then, whatever the following may be:

(1) the ring $\alpha \in \widetilde{P}^{q+r}_{q+r} \subset C^{q+r}(W_n;\ R)$, representing the element a;

(2) the smooth, empty, exterior differential form ψ on M of degree $n-r$ representing the class of cohomologies a;

(3) the coordinate covering $\Gamma = \{U_1, \ldots, U_N\}$ of the manifold M;

(4) the partition of unity subordinated to the covering constructed according to (1)-(4), the cochain $J(\alpha, \psi, \Gamma, \rho) \in C^q_1(M)$ is contained in $C^q_{1,q+r}$ and defines in $E_2^{-r,q+r} = H^{n-r}(M;\ R) \otimes H^{q+r}(W_n;\ R)$ an element equal to $\Psi \otimes a$.

Proof. First of all we notice that the functional relating the q formal vector fields $\eta_1, \ldots, \eta_q$ to the number $\alpha(\eta_1, \ldots, \eta_q, e_{i_1}, \ldots, e_{i_r})$ is none other than $\alpha_{(i_1 \ldots i_r)}(\eta_1, \ldots, \eta_q)$ (see [1], p. 6.4). Indeed as follows immediately from a calculation, if $\beta \in \widetilde{P}^{s+1}_m = C^{s+1}(W_n;\ R)$ and $\gamma(\eta_1, \ldots, \eta_s) = \beta(\eta_1, \ldots, \eta_s, e_k)$, then $\widetilde{\nabla}\gamma = \sigma_k \beta$.

The cochain $J(\alpha, \psi, \Gamma, \rho)$ can be decomposed into a sum of cochains of the form

$$(\xi_1, \ldots, \xi_q) \longmapsto \int_{U_i} \alpha(\xi_1(z, U), \ldots, \xi_q(z, U), e_{i_1}, \ldots, e_{i_r})\, \rho_i \psi \wedge dy_{i_1} \wedge \ldots \wedge dy_{i_r} \tag{3}$$

$(1 \leq i \leq N, 1 \leq i_1 < \ldots < i_r \leq n)$, where $y_1, \ldots, y_n$ are local coordinates in U_i. The form $\rho_i\psi$ can be represented in the form $\sum \psi_{i_1 \ldots i_r}\, dy_1 \wedge \ldots dy_{j_1} \ldots dy_{j_r} \ldots \wedge dy_n$, where $\psi_{j_1 \ldots j_r}$ are smooth finite functions on U_i. The cochain (3) is evidently none other than $\psi_{i_1 \ldots i_r}\alpha_{(i_1 \ldots i_r)}$, where $\psi_{i_1 \ldots i_r}$ can be regarded as a generalized function on U_i (see [1], 4.5). Consequently, the cochain (3) is contained in $C^q_{1,q+r}$ (see [1], 4.5), and is definable through its element in $E_0^{-r,q+r}$ and belongs to the kernel of the differential $\delta_0^{-r,q+r}$ (see [1], 5.2-5.3). In $E_1^{-r,q+r}$ the cochain (3) defines an element $\psi_{i_1 \ldots i_r}(a \otimes (\zeta_{i_1} \wedge \ldots \wedge \zeta_{i_r}))$ (see [1], 7.2) which, after identification of $E_1^{-r,q+r}$ with $(\Omega^r(M))' \otimes \widetilde{D}^{q+r}$ (see [1], 7.4) is transformed into the element $(\psi_{i_1 \ldots i_r}\, dy_1 \wedge \ldots dy_{i_1} \ldots dy_{i_r} \ldots \wedge dy_n) \otimes a$ (a form of degree $n-r$ on M is a functional on forms of power r, i.e., is an element of space $(\Omega^r(M))'$). Summing all the cochains (3) we arrive at the fact that the cochain $J(\alpha, \psi, \Gamma, \rho)$ belongs to $C^q_{1,q+r}$ and defines in $E_1^{-r,q+r} = (\Omega^r(M))' \otimes \widetilde{D}^{q+r}$ an element $\psi \otimes a$. So long as ψ is a closed form, this element belongs to the kernel of the differential $\delta_1^{-r,q+r}$ (see [1], 7.4) and defines in $E_2^{-r,q+r} = (H^r(M;\ R))' \otimes \widetilde{D}^{q+r} = H^{n-r}(M;\ R) \otimes H^{q+r}(W_n;\ R)$ an element $\Psi \otimes a$.

1.5. Our next purpose is to describe in a suitable terminology the differential of the element $J(\alpha, \psi, \Gamma, \rho)$ (i.e., its image under the homomorphism $d^q : C^q(M) \to C^{q+1}(M)$). The following proposition exists:

The element $dJ(\alpha, \psi, \Gamma, \rho) \in C^{q+1}(M)$ is equal to the sum

$$J(d^{q+r}\alpha, \psi, \Gamma, \rho) + (-1)^q \widetilde{J}(\alpha, \psi, \Gamma, \rho),$$

where $\widetilde{J}(\alpha, \psi, \Gamma, \rho)$ is an element of the space definable through the formula

$$\widetilde{J}(\alpha, \psi, \Gamma, \rho; \xi_1, \ldots, \xi_{q+1}) = \int_M \psi \wedge \left\{\sum_{i=1}^N \rho_i d[\varphi(\alpha, U_i; \xi_1, \ldots, \xi_{q+1})]\right\}.$$

Proof. According to the definition,

$$\begin{aligned}
&[dJ(\alpha, \psi, \Gamma, \rho)](\xi_1, \ldots, \xi_{q+1}) \\
&= \sum_{1 \leqslant s < t \leqslant q+1} (-1)^{s+t-1} J(\alpha, \psi, \Gamma, \rho; [\xi_s, \xi_t], \xi_1, \ldots \hat{\xi}_s \ldots \hat{\xi}_t \ldots, \xi_{q+1}) \\
&= \int_M \psi \wedge \left[\sum_{i=1}^N \rho_i \sum_{1 \leqslant s < t \leqslant q+1} (-1)^{s+t-1} \varphi(\alpha, U_i; [\xi_s, \xi_t], \xi_1, \ldots \hat{\xi}_s \ldots \hat{\xi}_t \ldots, \xi_{q+1}\right].
\end{aligned}$$

We investigate the integrand of the expression. We have

$$\sum_{1\le s<t\le q+1} (-1)^{s+t-1}\varphi(\alpha, U_i; [\xi_s, \xi_t], \xi_1, \ldots \hat{\xi}_s \ldots \hat{\xi}_t \ldots, \xi_{q+1})$$
$$= \sum_{\substack{1\le s<t\le q+1\\ 1\le i_1<\ldots<i_r\le n}} (-1)^{s+t-1}\alpha([\xi_s(z, U_i), \xi_t(z, U_i)], \xi_1(z, U_i),$$
$$\ldots \hat{\xi}_s(z, U_i) \ldots \hat{\xi}_t(z, U_i) \ldots, \xi_{q+1}(z, U_i), e_{i_1}, \ldots, e_{i_r})\, dy_{i_1}\wedge\ldots\wedge dy_{i_r}$$
$$= \sum_{1\le i_1<\ldots<i_r\le n} [(d^{q+r}\alpha)(\xi_1(z, U_i), \ldots, \xi_{q+1}(z, U_i), e_{i_1}, \ldots, e_{i_r})]\, dy_{i_1}\wedge\ldots\wedge dy_{i_r}$$
$$- \sum_{\substack{1\le s\le q+1\\ 1\le t\le r}} (-1)^{q+s+t}\alpha([\xi_s(z, U_i), e_{i_t}], \xi_1(z, U_i), \ldots \widehat{\xi_s(z, U_i)}$$
$$\ldots, \xi_{q+1}(z, U_i), e_{i_1}, \ldots \hat{e}_{i_t} \ldots, e_{i_r})\, dy_{i_1}\wedge\ldots\wedge dy_{i_r}.$$

Here we have made use of the fact that $[e_k, e_l] = 0$. In the obtained expression the first part is none other than $\varphi(d^{q+r}\alpha, U_i; \xi_1, \ldots, \xi_{q+1})$, and the second part we transform, making use of the fact that $[\xi, e_k] = -\frac{\partial \xi}{\partial x_k}$ (differentiation of the formal vector field leads properly to $\frac{\partial}{\partial x_k}\left[\sum_{i=1}^{n} f_i e_i\right] = \sum_{i=1}^{n}\frac{\partial f_i}{\partial x_k} e_i$). We obtain

$$\sum_{s,t}(-1)^{q+s+t}\alpha([\xi_s(z, U_i), e_{i_t}], \xi_1(z, U_i), \ldots \hat{\xi}_s(z, U_i)$$
$$\ldots, \xi_{q+1}(z, U_i), e_{i_1}, \ldots \hat{e}_{i_t} \ldots, e_{i_r})\, dy_{i_1}\wedge\ldots\wedge dy_{i_r}$$
$$= \sum_{s,t}(-1)^{q+s+t-1}\alpha\left(\frac{\partial}{\partial x_{i_t}}\xi_s(z, U_i), \xi_1(z, U_i), \ldots \hat{\xi}_s(z, U_i)\right.$$
$$\left.\ldots, \xi_{q+1}(z, U_i), e_{i_1}, \ldots \hat{e}_{i_t} \ldots, e_{i_r}\right) dy_{i_1}\wedge\ldots\wedge dy_{i_r}$$
$$= \sum_t (-1)^{q+1}\left[\frac{\partial}{\partial y_{i_t}}\alpha(\xi_1(z, U_i), \ldots, \xi_{q+1}(z, U_i), e_{i_1}, \ldots \hat{e}_{i_t} \ldots, e_{i_r})\right] dy_{i_t}$$
$$\wedge dy_{i_1}\wedge\ldots \widehat{dy_{i_t}} \ldots \wedge dy_{i_r} = (-1)^{q+1} d\varphi(\alpha, U_i; \xi_1, \ldots, \xi_{q+1})$$

(in the last expression d denotes the ordinary exterior differentiation of a form). Thus,

$$[dJ(\alpha, \psi, \Gamma, \rho)](\xi_1, \ldots, \xi_{q+1})$$
$$= \int_M \psi\wedge\sum_{i=1}^{N}\rho_i[\varphi(d^{q+r}\alpha, U_i; \xi_1, \ldots, \xi_{q+1}) + (-1)^q d\varphi(\alpha, U_i; \xi_1, \ldots, \xi_{q+1})],$$

i.e.,

$$dJ(\alpha, \psi, \Gamma, \rho) = J(d^{q+r}\alpha, \psi, \Gamma, \rho) + (-1)^q \tilde{J}(\alpha, \psi, \Gamma, \rho).$$

1.6. Let $a \in H^{q+r}(W_n; R)$. We assume that a can be represented by such a cocycle $\alpha \in \tilde{F}^{q+r}_{q+r}$, that for some coordinate covering $\Gamma = \{U_1, \ldots, U_N\}$ of the manifold M the forms $\varphi(\alpha, U_i; \xi_1, \ldots, \xi_q)$ are congruent in the intersections $U_i \cap U_j$ (i.e., through contractions on the neighborhood U_i of a form φ) for any $\xi_1, \ldots, \xi_q$. Then the cochain $J(\alpha, \psi, \Gamma, \rho)$ is a cocycle whatever may be the closed form ψ of degree $n-r$ and the partition of unity ρ, subordinated to the covering Γ. Indeed in this case, according to 1.5,

$$dJ(\alpha, \psi, \Gamma, \rho) = J(d^{q+r}\alpha, \psi, \Gamma, \rho) + (-1)^q\tilde{J}(\alpha, \psi, \Gamma, \rho) = (-1)^q\tilde{J}(\alpha, \psi, \Gamma, \rho);$$
$$\tilde{J}(\alpha, \psi, \Gamma, \rho; \xi_1, \ldots, \xi_{q+1})$$
$$= \int_M \psi\wedge\sum_{i=1}^{N}\rho_i d\varphi(\alpha, U_i; \xi_1, \ldots, \xi_q) = \int_M \psi\wedge d\varphi = (-1)^{n-r+1}\int_M d\psi\wedge\varphi = 0.$$

From that which has been proved and from the proposition 1.4 it follows that, in the stated case, for any $\Psi \in H^{n-r}(M; R)$ the element $\Psi\otimes\alpha \in E_2^{-r,q+r}$ belongs to the kernel of all the differentials (i.e., $d_2^{-r,q+r}(\Psi\otimes a) = d_3^{-r,q+r}(\Psi\otimes a) = \ldots = 0$).

1.7. There arises the question: how does the form change with the variation of local coordinates?

Let U and V be coordinate neighborhoods overlapping the point $z \in M$. We displace the origin of both the coordinate systems to the point z and denote by τ a diffeomorphism of the neighborhood of the point $0 \in R^n$ into another neighborhood of the same point connecting the two obtained systems of local coordinates. The diffeomorphism τ can be applied to any formal vector field from W_n. In particular, it is clear that $\tau\xi(z, U) = \xi(z, V)$ for $\xi \in \mathfrak{A}(M)$. Hence it follows that

$$\varphi(\alpha, U; \xi_1, \ldots, \xi_q) = \varphi(\tau^* \alpha, V; \xi_1, \ldots, \xi_q),$$

where

$$\tau^* \alpha(\eta_1, \ldots, \eta_{q+r}) = \alpha(\tau\eta_1, \ldots, \tau\eta_{q+r}).$$

1.8. If the element $a \in H^v(W_n; R)$ can be represented in $\tilde{P}_v^v$ by a cocycle invariant under any diffeomorphism τ connecting two overlapping coordinate systems on M then the element $\Psi \otimes a \in E_2$ belongs to the kernel of all the differentials for every $\Psi \in H^*(M; R)$. In particular, the latter is true if a can be represented by a cocycle invariant under any orientation preserving diffeomorphism of the neighborhood point $0 \in R^n$, translating 0 to 0.

This follows from 1.6 and 1.7.

1.9. Applications of proposition 1.8 are based on the following lemma.

LEMMA. Any element $H^v(W_n; R)$ can be represented in $\tilde{P}_v^v$ by a cocycle invariant under the group $G_+L(n, R)$ of orientation preserving linear transformations of the space R^n.

Every element of $H^{n(n+2)}(W_n; R)$ can be represented in $\tilde{P}_{n(n+2)}^{n(n+2)}$ by a cocycle, invariant under all the orientation-preserving diffeomorphisms of the space R^n, translating 0 to 0.

Explanation. We recall that $n(n + 2)$ is the maximum dimension for nontrivial cohomologies of the algebra W_n (see [1], 6.3).

Proof. It is clear that every element of $\tilde{P}_v^v$ is invariant under similarity transformations (this is seen from formula (1) of 1.1). Therefore for the proof of the first of the two assertions it is sufficient to represent $a \in H^v(W_n; R)$ by the cocycle α, invariant under the group $SL(n, R)$. The class a is invariant under $SL(n, R)$ (see [1], 7.3). We denote through V the subspace (finite dimensional) of the space $\tilde{P}_v^v$ consisting of cocycles belonging to the class of cohomologies of the form μa, where $\mu \in R$. For every such cocycle number μ, we obtain a linear mapping $V \to R$, commuting with the action of the group $SL(n, R)$ in V and in R. From Schur's lemma it follows that there exists an $SL(n, R)$-invariant element $\alpha \in V$ going over under this mapping into $1 \in R$. The first of the assertions is proved.

For proof of the second assertion we represent any element $a \in H^{n(n+2)}(W_n; R)$ through the cocycle $\alpha \in \tilde{P}_{n(n+2)}^{n(n+2)}$, invariant under the group $G_+L(n, R)$, and prove that the latter goes over into itself under the action of a diffeomorphism whose differentials are identical. But such a diffeomorphism evidently transforms α into $\alpha + \sum_{m<n(n+2)} \alpha_m$, where $\alpha_m \in \tilde{P}_m^{n(n+2)}$. Also the space $\tilde{P}_m^{n(n+2)}$, as can be easily perceived from the proof of Proposition 6.3 of [1], is trivial for $m < n(n + 2)$. The lemma is proved.

1.10. If all the functions are transformed from one local coordinate on the manifold M to another, linearly, (for example, if M is a torus or a compact factor manifold R^n under a discrete subgroup of the group $G_+L(n, R)$), then the spectral sequence $\mathscr{E}$ is trivial (starting with E_2). This follows from 1.8 and 1.9.

1.11. If M is a two-dimensional orientable manifold then the spectral sequence $\mathscr{E}$ is trivial.

Proof. The space $H^q(W_2; R)$ is not trivial only for $q = 0, 5, 7, 8$ (see [2]*). Therefore the differential $\delta_r^{u,v}$ for $r \geq 2$ is trivial from considerations of dimensionality since we shall exclude the differential $\delta_2^{-2,8} : E_2^{-2,8} \to E_2^{0,7}$, which is trivial according to 1.8 and 1.9.

§2. TORUSES AND TWO-DIMENSIONAL MANIFOLDS

2.1. Together with the spectral sequence $\mathscr{E}$, converging to the homologies of the diagonal complex, we investigated in [1] spectral sequences ${}^{(k)}\mathscr{E} = \{{}^{(k)}E_r^{u,v}, {}^{(k)}\delta_r^{u,v}\}$, which converge to the homologies of the

*The spaces $\tilde{D}^q$ for n = 2 were calculated in [1] (see [1], 6.10). However, there was a misprint which distorted the answer.

factor complexes $\mathscr{C}_k/\mathscr{C}_{k-1}$ (see [1], 1.2, 1.3, 2.12, and the introduction to §8). The first term ${}^{(k)}E_2 = \sum_{u,v} {}^{(k)}E_2^{u,v}$ of such a spectral sequence is described as follows. We set

$$E(k, q) = \oplus [H^{m_1}(W_n; \mathbf{R}) \otimes \ldots \otimes H^{m_k}(W_n; \mathbf{R})],$$

where the summation is to be carried out over all the integer positive numbers $m_1, \ldots, m_k$ with $m_1 + \ldots + m_k = q$. Further we set

$$H(k, p, q) = H_p(M^k, M^k_{k-1}; E(k, q)) = H^{kn-p}(M^k \setminus M^k_{k-1}; \mathbf{R}) \otimes E(k, q),$$

where $M^k = \underbrace{M \times \ldots \times M}_{k}$, M^k_{k-1} is a part of the product M^k, consisting of points of these products, the coordinates of each of which are mutually disjoint. The group S(k) of permutations of k elements acts on H(k, p, q): it acts on M^k, transposing the coordinates, and on E(k, q), giving the same product of rearranged elements but multiplied by a factor which is a multiple of -1.

The space ${}^{(k)}E_2^{u,v}$ is isomorphic to the space of S(k)-invariant elements of H(k, −u, v).

This assumption in another form was proved in §8 of [1].

For k = 1 the spectral sequence ${}^{(k)}\mathscr{E}$ coincides with $\mathscr{E}$.

2.2. The multiplication $C^p_k(M) \otimes C^q_l(M) \to C^{p+q}_{k+l}(M)$ (see [1], 2.11) for every r induces a multiplication ${}^{(k)}E_r^{u_1,v_1} \otimes {}^{(l)}E_r^{u_2,v_2} \to {}^{(k+l)}E_r^{u_1+u_2,v_1+v_2}$, connected with a diffeomorphism through the formula

$${}^{(k+l)}\delta_r^{u_1+u_2,v_1+v_2}\eta_1\eta_2 = ({}^{(k)}\delta_r^{u_1,v_1}\eta_1)\,\eta_2 + (-1)^{u_1+v_1}\,\eta_1\,{}^{(l)}\delta_r^{u_2,v_2}\eta_2.$$

In particular there arises the product $E_r^{u_1,v_1} \otimes \ldots \otimes E_r^{u_k,v_k} \to {}^{(k)}E_r^{u_1+\ldots+u_k,v_1+\ldots+v_k}$.

The product of elements $\Psi_i \otimes \alpha_i \in H^{n-r_i}(M; \mathbf{R}) \otimes H^{q_i+r_i}(M; \mathbf{R}) = E_2^{-r_i,q_i+r_i}$ (i = 1, ..., k) can be equated to

$$\sum \varepsilon(i_1, \ldots, i_k)\, \pi^*(\Psi_{i_1} \otimes \ldots \otimes \Psi_{i_k}) \otimes \alpha_{i_1} \otimes \ldots \otimes \alpha_{i_k} \in {}^{(k)}E_2^{-r_1-\ldots-r_k,\, q_1+\ldots+q_k+r_1+\ldots+r_k}.$$

Here summation is carried out over all the permutations $(i_1, \ldots, i_k)$ of the numbers 1, ..., k; π is naturally the enclosure of the space $M^k \setminus M^k_{k-1}$ in M^k; $\varepsilon(i_1, \ldots, i_k) = (-1)^{\sum_{s<t} \deg \alpha_{i_s} \deg \Psi_{i_t}}$.

This assertion follows from 8.4 of [1].

2.3. Every element of ${}^{(k)}E_2$ can be represented in the form of a sum of products of the elements from E_2.

Proof. It is sufficient to show that every S(k)-invariant element of $H_p(M^k, M^k_{k-1}; E(k, q))$ belongs to the generator of a homomorphism π_*: $H_p(M^k; E(k, q)) \to H_p(M^k, M^k_{k-1}; E(k, q))$. For this reason it is sufficient to establish that no invariant element of $H^p(M^k, M^k_{k-1}; E(k, q))$ under a homomorphism π^* associated with π_* fails to vanish. In view of the exactness of the sequence pair (M^k, M^k_{k-1}) and the finiteness of the group S(k) the last is equivalent to the fact that any invariant element of $H^p(M^k_{k-1}; E(k, q))$ is the image of some element of $H^p(M^k, E(k, q))$. But this is evident: if f: $M^{k-1} \to M^k_{k-1}$ is a mapping given by the formula $f(x_1, \ldots, x_{k-1}) = (x_1, \ldots, x_{k-1}, x_{k-1})$, then the invariant element $\alpha \in H^p(M^k_{k-1}; E(k, q))$ serves as the image of an element obtained by symmetrization of $f^*\alpha \otimes 1 \in H^p(M^k; E(p, q))$.

2.4. Proposition 2.3 creates the possibility of regenerating the spectral sequences ${}^{(k)}\mathscr{E}$ through the spectral sequence $\mathscr{E}$. In the subsequent we shall exclusively consider the case when the spectral sequence $\mathscr{E}$ is trivial. (We recall that the sufficiency condition for triviality of a spectral sequence was formulated in 1.8-1.11.)

2.5. If the spectral sequence $\mathscr{E}$ is trivial then there exist the following assertions.

1°. All the spectral sequences ${}^{(k)}\mathscr{E}$ are trivial, and this means that the homologies of the complex $\mathscr{C}_k(M)/\mathscr{C}_{k-1}(M)$ coincide with ${}^{(k)}E_2$ (see 2.1).

2°. The space $\mathfrak{H}^q(M)$ is isomorphic to the sum $\sum_k \sum_{u+v=q} {}^{(k)}E_2^{u,v}$ (in other words $\mathfrak{H}^*(M) = \sum_k {}^{(k)}E_2$).

3°. The ring $\mathfrak{H}^*(M)$ is multiplicatively generated through additive transformations of the homologies of the complex $\mathscr{C}_1(M)$ (this means that the ring which is generated is multiplicatively

Proof. Assertion 1° follows from prop. 2.3 in an obvious manner. Furthermore every class of homologies of the complex $\mathscr{C}_k(M)/\mathscr{C}_{k-1}(M)$ is represented by a cycle of the complex $\mathscr{C}_k(M)$, i.e. by a cycle of the entire complex $\mathscr{C}(M)$. Hence it follows that the homologies of the complex $\mathscr{C}(M)$ are isomorphic to the sum of the homologies of all the complexes $\mathscr{C}_k(M)/\mathscr{C}_{k-1}(M)$ and are generated by the products of homology classes of the complex $\mathscr{C}_1(M)$.

2.6. With the help of proposition 2.5 it is possible to obtain for any manifold M with a trivial spectral sequence $\mathscr{E}$ a description of the ring adjoined to the ring $\mathfrak{H}^*(M)$ with respect to the filtrations $\mathscr{C}_1(M) \subset \mathscr{C}_2(M) \subset \ldots \subset \mathscr{C}(M)$. This adjoined ring is a sum $\sum_k {}^{(k)}E_2$ with a multiplication which is generated through the multiplications ${}^{(k)}E_2 \otimes {}^{(l)}E_2 \to {}^{(k+l)}E_2$. A description of the space ${}^{(k)}E_2$ and the mentioned multiplication, carried out in 2.1 and 2.2, enables us to determine this ring, knowing the ring of cohomologies of the manifold M and the rings $H^*(W_n; R)$.

2.7. From Proposition 2.5 it follows that if M is any of the manifolds satisfying the conditions of one of the Propositions 1.9–1.11, then the ring $\mathfrak{H}^*(M)$ is multiplicatively finitely generated. In particular this is true for the toruses and for two-dimensional orientable manifolds. Among the generators of this ring there is an additive generator of the homologies of the diagonal complex, i.e., the additive generator of the space $H^*(M; R) \otimes H^*(W_n; R)$. For example $\mathfrak{H}^*(S^2)$ is multiplicatively generated by ten generators and $\mathfrak{H}^*(S^1 \times S^1)$ by twenty. Between these generators there exist relations different from the obvious relations of skew-symmetricity; this arises due to the fact that the homomorphism π_* (see 2.2) is not a monomorphism. For example, if a is a nontrivial element of $H^8(W_2; R)$ and 1 is the generator of the group $H^0(S^2; R)$, then the square of an element of $\mathfrak{H}^8(S^2)$, serving as an element of $1 \otimes a \in E_2^{0,8}$, vanishes: according to 2.2 it vanishes in the homologies of the complex $\mathscr{C}_2(M)/\mathscr{C}_1(M)$, i.e., it can be represented by a cycle lying in $\mathscr{C}_1(M)$; however, the dimensionality of a nontrivial cycle of $\mathscr{C}_1(M)$ cannot exceed eight.

It is not known whether there exists a finitely generated ring for any compact orientable manifold M.

Remark. The rings $H^*(W_n; R)$ have been calculated by us in [2]. There it was shown, in particular, that multiplication within them is trivial, i.e., the product of elements of positive dimensionality vanished.

In conclusion we mention some misprints in [1].

1. On p. 41 in line 12 from the bottom instead of $f^T_{\xi_1,\ldots,\xi_q}$ there should be $f^T_{\xi_{i_1},\ldots,\xi_{i_q}}$.

2: On p. 46 the last sentence of paragraph 6.10 ought to read as: it is proved that for q = 1, 2, 3, 4, 5, 6, 7, 8 the dimensionality of $\tilde{D}^q$ is equal to 0, 0, 0, 0, 2, 0, 1, 2, respectively; hence, in view of 1.4, $\dim(\bigoplus_m D^q_m) = 0, 0, 2, 5, 3, 4, 5, 2$.

3. On p. 50, in lines 17 and 20 from the bottom instead of $H_{m-q}(M^k)$ there ought to stand $H_{m-q}(M^k, M^k_{k-1})$.

LITERATURE CITED

1. I. M. Gel'fand and D. B. Fuks, Funkts. Analiz, 3, No. 3, 32-52 (1969).
2. I. M. Gel'fand and D. B. Fuks, Izv. Akad. Nauk SSSR, Seriya Matem., 34 (1970).
3. I. M. Gel'fand and D. B. Fuks, Dokl. Akad. Nauk SSSR, 190, No. 6 (1970).

7.

(with D. B. Fuks)

Cohomologies of Lie algebra of vector fields with nontrivial coefficients

Funkts. Anal. Prilozh. 4 (3) (1970) 10–25 [Funct. Anal. Appl. 4 (1970) 181–192].
MR 44:7752. Zbl. 222:58001

In this paper we continue the investigation which we began in [1-3] of the cohomologies of the Lie algebra of formal as well as smooth vector fields (on a smooth manifold). Specifically we study the cohomologies of the algebras with coefficients in various representations, mainly with coefficients in the spaces of exterior differential forms, formal or smooth, respectively. The results obtained in these papers involving cohomologies with coefficients in spaces of forms of degree 0 (i.e., in the spaces of smooth functions or of formal power series), were also contained in the work of M. V. Losik [4]. It is true that theorems about the algebra of formal vector fields were not singled out, but were essentially contained in them (see [4], §2). We shall not rely here on the results of M. V. Losik since we will prove them again. Our proof in the corresponding part is not, in principle, different from the proof of M. V. Losik, although it is considerably shorter. The investigation of cohomologies with coefficients in forms of degree greater than zero encounters a series of new difficulties which are overcome with the application of the results of [3].

The first part of the present work is devoted to the algebra of formal vector fields. Using the concept of induced representation we reduce the investigation of the cohomologies of this algebra with coefficients in the spaces of formal exterior differential forms to the study of the cohomologies of the subalgebra of an algebra originating from the fields having a singularity at the origin of coordinates with coefficients now in a finite dimensional module. The latter problem is connected with this problem, now solved in [3]: the desired cohomologies are found through isomorphisms with the displacement of the dimensions over the columns of the initial term of the spectral sequence to be investigated in this paper. In particular, the cohomologies of the Lie algebra of formal vector fields with coefficients in formal exterior differential forms turned out to be finite-dimensional.

In the second part the Lie algebras of smooth vector fields on smooth manifolds are investigated. In the standard complex of cochains with coefficients in the space of smooth exterior differential forms (in general in the space of sections of a finite-dimensional vector fiber induced by a tangent fiber) we select a subcomplex which, in analogy with [1], we shall call diagonal. We emphasize that the diagonal complex to be studied here and the diagonal complex of [1], although they are defined similarly, play quite different roles. If, in the case of constant coefficients, a diagonal complex, in some sense, multiplicatively generates the entire complex of cochains then, for example, in the case of coefficients in the space of exterior differential forms, the diagonal subcomplex is the subring of the whole standard complex. Thus the ring of cohomologies of the algebra of smooth vector fields with coefficients in the full ring of smooth exterior differential forms is a module over the ring of cohomologies of the corresponding diagonal complex.*

The main result of the second part is the calculation of the ring of cohomologies of the diagonal complex with coefficients in the ring of smooth exterior differential forms. These cohomologies are found to be finite-dimensional: they depend only on the rational cohomologies of the manifold and on the rational classes of Pontryagin. In the general case (i.e., in the case when the coefficients are taken in he space of smooth sections of a fiber induced by tangent) the constructed spectral sequence converges to the homologies of the diagonal complex. The second term of this spectral sequence is a tensor product of real

*This ring was designated by M. V. Losik as the ring of differential cohomologies of the algebra of vector fields.

Moscow State University. Translated from Funktional'nyi Analiz i Ego Prilozheniya, Vol. 4, No. 3, pp. 10-25, July-September, 1970. Original article submitted March 31, 1970.

cohomologies of the manifold and the cohomologies of the algebra of formal vector fields with appropriate coefficients. In particular if the latter cohomologies are finite-dimensional, then the homologies of the diagonal complex are also finite-dimensional.

The question over the arrangement of homologies of a completely standard complex of the algebra of smooth vector fields with nontrivial coefficients remains completely open.

I. ALGEBRAS OF FORMAL FIELDS

§1. Cohomologies of Algebra L_0 with Coefficients in Skew-Symmetric Representations

1.1. We shall denote by W the Lie algebra of formal vector fields at the point 0 of the point R^n. The subalgebra of the algebra W is denoted by L_k $k = 0, 1, \ldots$), consisting of those fields $\sum_{i=1}^{n} a_i(x_1, \ldots, x_n) e_i$ (where $e_1, \ldots, e_n$ are vectors of a standard basis of space R^n) for which in the power series $a_i(x_1, \ldots, x_n)$ terms of degree less than k are totally absent. It is evident that the space W/L_k is finite-dimensional and that $L_1, L_2, \ldots$ is the normal subalgebra of the algebra L_0. The factor algebra L_0/L_1 is isomorphic to $\mathfrak{gl}(n, R)$, moreover, there exists an imbedding of the algebra $\mathfrak{gl}(n, R)$ into L_0 whose superposition is identical with the projection $L_0 \to L_0/L_1 = \mathfrak{gl}(n, R)$: this imbedding relates the element $A_{ij} \in \mathfrak{gl}(n, R)$ to the element $x_i e_j \in L_0$ (here A_{ij} is a matrix in which the only nonvanishing element is at the intersection of the i-th row with the j-th column and equals unity).

Any representation of the algebra $\mathfrak{gl}(n, R)$ induces, owing to the existence of the projection $L_0 \to \mathfrak{gl}(n, R)$, a representation of the algebra L_0. In particular we will consider as the L_0-modules the spaces $\Lambda^r(R^n)'$ of exterior forms of degree r of the space R^n (in which the algebra $\mathfrak{gl}(n, R)$ acts in a natural manner). It is evident that the action of the algebra L_0 in $\Lambda^*(R^n)' = \sum_{r=0}^{n} \Lambda^r(R^n)'$ is congruent with the ring structure in $\Lambda^*(R^n)'$ since the cohomologies of the algebra L_0 with coefficients in $\Lambda^*(R^n)'$ represent the same algebra over R. In what follows we shall write $\Lambda^r(R^n)', \Lambda^*(R^n)'$ as Λ^r, Λ^*.

1.2. We formulate the main results of this paragraph.

THEOREM. A bigraded ring $H^*(L_0; \Lambda^*)$ is multiplicatively generated by 2n generators

$$\rho_i \in H^{2i-1}(L_0; \Lambda^0) \quad (i = 1, \ldots, n), \qquad \tau_j \in H^j(L_0; \Lambda^j) \quad (j = 1, \ldots, n).$$

These generators are connected only through such relations:

$$\rho_k\rho_l = -\rho_l\rho_k; \quad \rho_k\tau_l = \tau_l\rho_k; \quad \tau_k\tau_l = \tau_l\tau_k; \quad \tau_1^{j_1}\tau_2^{j_2}\ldots\tau_n^{j_n} = 0 \text{ for } j_1 + 2j_2 + \ldots + nj_n > n.$$

In particular the ring $H^*(L_0; R) = H^*(L_0; \Lambda^0)$ is an exterior algebra from the generators with dimensions of 1, 3, 5, ..., 2n−1; i.e., $H^*(L_0; R) = H^*(\mathfrak{gl}(n, R); R)$. Moreover,

$$H^q(L_0; \Lambda^r) = \begin{cases} 0 & \text{for } q < r, \\ H^r(L_0; \Lambda^r) \otimes H^{q-r}(\mathfrak{gl}(n, R); R) & \text{for } q \geq r, \end{cases}$$

and the dimension of the space $H^r(L_0; \Lambda^r)$ is equal to the number of different representations of the number r in the form of natural terms.

Proof of this theorem is contained in §1.3-1.7.

The plan of the proof is as follows. We compare two spectral sequences which converge to the cohomologies of the algebra W with coefficients in R. Essentially, this is the spectral sequence of Serre-Hochshild [6] corresponding to subalgebras $\mathfrak{gl}(n, R)$ and L_0 of the algebra W. Next we shall prove that E_1 terms of these spectral sequences are isomorphic (up to a permutation of the indices) and at the same time one of these is $H^*(L_0; \Lambda^*)$, and the other was calculated by us in [3].

1.3. We shall begin by recapitulating some facts from [3].

The elements of the space $C^q(W; R)$, i.e., the continuous skew-symmetric q-linear real functionals on W, can be identified with the functions depending on q vectors $\beta_1, \ldots, \beta_q \in R^n$ and on q covectors $\alpha_1, \ldots, \alpha_q \in (R^n)'$, multilinear in $\beta_1, \ldots, \beta_q$, polynomials in the coordinates of the covectors $\alpha_1, \ldots, \alpha_q$, and change sign under the simultaneous interchange of α_i with α_j, β_i with β_j. Namely, to the function

$$P(\alpha_1, \ldots, \alpha_q; \beta_1, \ldots, \beta_q) = \sum A_{m_{11},\ldots,m_{qn}; i_1,\ldots,i_q} \alpha_{11}^{m_{11}} \ldots \alpha_{qn}^{m_{qn}} \beta_{1i_1} \ldots \beta_{qi_q},$$

where $\alpha_{i1}, \ldots, \alpha_{in}$ are the coordinates of the covector α_i, and $\beta_{i1}, \ldots, \beta_{in}$ are the coordinates of the vector β_i, there corresponds a cochain $p \in C^q$ (W; **R**) definable by the formula

$$p\left(\sum a_{1i} e_i, \ldots, \sum a_{qi} e_i\right) = \sum A_{m_{11},\ldots,m_{qn}; i_1,\ldots,i_q} \left[\prod_{s=1}^{n} \frac{\partial^{m_{s1}+\cdots+m_{sn}} a_{si_s}}{\partial^{m_{s1}} x_1 \ldots \partial^{m_{sn}} x_n}\right]_{\substack{x_1=0\\ \cdots \\ x_n=0}} \tag{1}$$

The number $m_i = m_{i1} + \ldots + m_{in}$ is called the degree of the monomial $\alpha_{11}^{m_{11}} \ldots \alpha_{qn}^{m_{qn}} \beta_{1i_1} \ldots \beta_{qi_q}$ in α_i. We denote through C_0^*(W; **R**) a subcomplex of the complex C^*(W; **R**) consisting of monomials in which $m_1 + \ldots + m_q = q$. It was shown (see [3], § 1.4; [1] § 6.2) that the imbedding induces an isomorphism of the homologies.

The subspace F_s of the space C_0^*(W; **R**) is defined as the set of those polynomials, the mean degree of each of their monomials in $\alpha_1, \ldots, \alpha_q$ is not less than s, different from 1. The spaces F_s make up a filtration

$$C_0^*(W; \mathbf{R}) = F_0 \supset F_1 \supset F_2 \supset \ldots$$

This filtration is identical with the differential [see [3], § 2.2] and generates therefore a spectral sequence converging to the homologies of the complex C_0^*(W; **R**), i.e., to H^*(W; **R**). We denote this spectral sequence by $\mathscr{E} = \{E_r^{s,t}, d_r^{s,t}\}$.

The ring $E_1 = \sum E_1^{s,t}$ is generated by the generators $\varphi_i \in E_1^{0,2i-1}$ (i = 1, . . ., n), $\psi_j \in E_1^{2j,0}$ (j = 1, . . ., n), which are connected by the relations

$$\varphi_k\varphi_l = -\varphi_l\varphi_k;\ \varphi_k\psi_i = \psi_i\varphi_k;\quad \psi_i\psi_k = \psi_k\psi_i;\ \psi_1^{j_1} \ldots \psi_n^{j_n} = 0 \text{ for } j_1 + 2j_2 + \ldots + nj_n > n.$$

Proof is carried out in § 5 of [3].

1.4. Alongside the filtration $\{F_s\}$ we shall investigate in the complex C_0^*(W; **R**) the filtration $\{F_s'\}$ defined as follows. We denote by F_{2r}' the subspace of the space C_0^*(W; **R**) consisting of those polynomials, among each of whose monomials at least r vanish, and we put $F_{2r-1}' = F_{2r}'$. The filtration

$$C_0^*(W; \mathbf{R}) = F_0' \supset F_1' \overset{(=)}{\supset} F_2' \supset F_3' \overset{(=)}{\supset} F_4' \supset \ldots$$

also agrees with the differential in the complex C_0^*(W; **R**) [this follows directly from the formula (1)] and therefore generates a spectral sequence which converges to H^*(W; **R**). We denote this spectral sequence by $\mathscr{E}' = \{'E_r^{s,t}, 'd_r^{s,t}\}$. It is clear that $'E_r^{s,t} = 0$ if s is odd.

There exists an imbedding $F_s \supset F_s'$; indeed if among nonnegative integers $m_1, \ldots, m_q$ with $m_1 + \ldots + m_q = q$ at least s are different from unity then among these at least $[(s+1)/2]$ are zero. This imbedding induces a homomorphism of the spectral sequence $\mathscr{E}$ in the spectral sequence $\mathscr{E}'$; we denote it by $\varkappa = \{\varkappa_r : E_r \to 'E_r\}$.

1.5. The homomorphism $\varkappa_r : E_r \to 'E_r$ is an isomorphism for $r \geq 1$.

Proof. In each of the spaces C_0^*(W; **R**), E_0, $'E_0$ it is possible to take as basis the set of skew-symmetric monomials of the form

$$\alpha_{11}^{m_{11}} \ldots \alpha_{qn}^{m_{qn}} \beta_{1i_1} \ldots \beta_{qi_q}. \tag{2}$$

In the differential dP of such a skew-symmetric monomial P of degrees $m_1, \ldots, m_q$ in $\alpha_1, \ldots, \alpha_q$ there enter monomials whose degrees in $\alpha_1, \ldots, \alpha_{q+1}$ (taken in some order) are equal to $m_1, \ldots, \hat{m}_i, \ldots, m_q, m_i', m_i''$, where $m_i' + m_i'' = m_i + 1$, $m_i' \geq m_i''$ (this follows from p. 1.3 of [3]). The differential $'d_0P$ of this same monomial in $'E_0$ is obtained if we choose from dP those monomials in which $m_i'' = 0$, $m_i \neq 0$; the differential d_0P of the element P in E_0 is obtained if we leave in dP only those terms in which one of the numbers m_i', m_i'' is equal to m_i, and the remaining are unity.

The complex E_0 can be decomposed into a direct sum of two subcomplexes: $E_0^{(1)}$ and $E_0^{(2)}$; the first is generated by skew-symmetric monomials, each of whose degree is less than or equal to 2, the second by the remaining. From §4.6 of [3] it can be seen that all the elements from E_1, i.e., all classes of homologies of the complex E_0 can be represented by polynomials consisting of monomials whose degree is less than or equal to 2. Hence it follows that the imbedding $E_0^{(1)} \to E_0$ induces a cohomological isomorphism and the complex $E_0^{(2)}$ is acyclic.

The homomorphism $\varkappa_0$ isomorphically maps $E_0^{(1)}$ onto some subcomplex $'E_0^{(1)}$ of the complex E_0 and identically vanishes on $E_0^{(1)}$. Since $\varkappa_1$ is a homological homomorphism induced by the homomorphism $\varkappa_0$, our statement is reduced to the fact that the imbedding $'E_0^{(1)} \to 'E_0$ induces an isomorphism of the homologies. The latter, in view of the exactness of the pair sequence $('E_0, 'E_0^{(1)})$ is equivalent to the acyclicity of the complex $'E_0^{(2)} = 'E_0/'E_0^{(1)}$.

Let us compare the complexes $E_0^{(2)}$ and $'E_0^{(2)}$. They have the same basis – the set of skew-symmetric monomials of the form (2), among which there is none with degree less than 3. We denote a filtration of a skew-symmetric monomial P with degrees $m_1, \ldots, m_q$ by the quantity $\varphi(P) = m_1^2 + \ldots + m_q^2 - q$. It is clear that $'d_0P$ is a linear combination of monomials of a filtration less than equal to $\varphi(P)$ and d_0P is obtained from $'d_0P$ by the crossing out of monomials of a filtration smaller than $\varphi(P)$. (Both the statements follow from the fact that for any natural m', m", and also for m' = 0, m" = 1, there exists the inequality $(m')^2 + (m'')^2 - 1 \geq (m' + m'' - 1)^2$, where the equality is attained only when one of the numbers m', m" equals unity.) Hence it can be seen that $E_0^{(2)}$ is none other than the initial term of the spectral sequence corresponding to this filtration in the complex $'E_0^{(2)}$. Hence, from the acyclicity of the complex $E_0^{(2)}$ follows the acyclicity of the complex $'E_0^{(2)}$. The proposition is proved.

1.6. From §1.5, by having taken $F'_{2r-1} = F'_{2r}$, we have naturally stretched the enumeration in the definition of the filtration $\{F'_s\}$. We will however consider the spectral sequence $\mathscr{E}'' = \{''E_r^{s,t}, ''d_r^{s,t}\}$, corresponding to the "nonexpanding" of the filtration

$$C_0^*(W; \mathbf{R}) = F''_0 \supset F''_1 \supset F''_2 \supset \ldots,$$

where $F''_s = F'_{2s}$. It is evident that the spectral sequences $\mathscr{E}'$ and $\mathscr{E}''$ differ only in enumerations (see Fig. 1):

$$'E_r^{s,t} = \begin{cases} 0, & \text{if } s \text{ is odd,} \\ ''E_{[(r+1)/2]}^{s/2,\, t+(s/2)}, & \text{if } s \text{ is even} \end{cases}$$

$$'d_r^{s,t} = \begin{cases} 0, & \text{if } s \text{ is odd,} \\ ''d_{[(r+1)/2]}^{s/2,\, t+(s/2)}, & \text{if } s \text{ is even.} \end{cases}$$

Thus the isomorphism $''E_{[(r+1)/2]} \to 'E_r$, translating $''E_{[(r+1)/2]}^{u,v}$ into $'E_r^{2u,v-u}$, is multiplicative.

1.7. There exists an isomorphism of bigraded rings

$$H^*(L_0; \Lambda^*) = \sum_{q,r} H^q(L_0; \Lambda^r) \quad u \quad ''E_1 = \sum_{q,r} {''E_1^{q,r}}.$$

To prove this proposition we shall complete, in view of §1.3, 1.5, 1.6, the proof of Theorem 1.2. We denote through $\widetilde{F}''_r$ a subcomplex of the complex $C^*(W; \mathbf{R})$ consisting of such polynomials that among the powers of each of these monomials at least r vanish. Evidently $F''_r = \widetilde{F}''_r \cap C_0^*(W; \mathbf{R})$. The imbedding $F''_r \to \widetilde{F}''_r$ induces an isomorphism of the homologies (this follows from § 6.2 of [1] and §1.6 of [3]). From the lemma over the five homologies it follows that an isomorphism of homologies also induces the imbedding $F''_r/F''_{r+1} \to \widetilde{F}''_r/\widetilde{F}''_{r+1}$.

The subcomplex $\widetilde{F}''_r$ of the complex $C^*(W; \mathbf{R})$ can be defined also as consisting of such rings $p \in C^*(W; \mathbf{R})$ that $p(\xi_1, \ldots, \xi_q) = 0$ for $\xi_r, \ldots, \xi_q \in L_0$ (here q = dim p). The equivalence of the two definitions of the subcomplex $\widetilde{F}''_r$ immediately follows from the formula (1).

For all the formal vector fields $\xi = \sum a_i(x_1, \ldots, x_n) e_i \in W$ we denote by ξ^0 its free term (i.e., the vector $\sum a_i(0, \ldots, 0) e_i$ and by $\widetilde{\xi}$ the difference $\xi - \xi^0 \in L_0$. We define the mapping $g : C^q(L_0; \Lambda^r) \to C^{q+r}(W; \mathbf{R})$, by putting

$$[g(P)](\xi_1, \ldots, \xi_{q+r}) = \sum_{1 \leq i_1 < \ldots < i_q \leq q+r} (-1)^{i_1 + \ldots + i_q - q(q+1)/2} [P(\widetilde{\xi}_{i_1}, \ldots, \widetilde{\xi}_{i_q})](\xi^0_{j_1}, \ldots, \xi^0_{j_r}),$$

where $P \in C^q(L_0; \Lambda^r)$, $\xi_1, \ldots, \xi_{q+r} \in W$, and $\{j_1, \ldots, j_r\}$ is the complement of the set $\{i_1, \ldots, i_q\}$ in the set $\{1, \ldots, q+r\}$, where $j_1 < \ldots < j_r$.

It is clear that $g(C^*(L_0;\ \Lambda^r)) \subset \widetilde{F}''_r$. The composition g_0 of the mapping g with the projection $\widetilde{F}''_r \to \widetilde{F}''_r/\widetilde{F}''_{r+1}$ is the isomorphism $C^*(L_0;\ \Lambda^r) \to \widetilde{F}''_r/\widetilde{F}''_{r+1}$. Indeed the mapping $h : \widetilde{F}''_r \to C^*(L_0;\ \Lambda^r)$, defined by the formula

$$[h(P')(\xi_1, \ldots, \xi_q)](\varepsilon_1, \ldots, \varepsilon_r) = P'(\xi_1, \ldots, \xi_q, \varepsilon_1, \ldots, \varepsilon_r),$$

where $\xi_1, \ldots, \xi_q \in L_0 \subset W$, $\varepsilon_1, \ldots, \varepsilon_r \in R^n \subset W$, is trivial on $\widetilde{F}''_{r+1}$ and thus $h \circ g = 1$, $\mathrm{Im}[(g \circ h) - 1] \subset \widetilde{F}''_{r+1}$. It is easy to verify that the mapping h is permutable with bounded operators so that g_0 is an isomorphism of the complexes. Consequently the mapping g_0 induces an isomorphism of q-dimensional cohomologies of the algebra L_0 with coefficients in Λ^r and (q + r)-dimensional homologies of the complex $\widetilde{F}''_r/\widetilde{F}''_{r+1}$, i.e., an isomorphism between $H^q(L_0;\ \Lambda^r)$ and $E_1^{q,r}$. The multiplicativity resulting from the isomorphism $H^*(L_0;\ \Lambda^*) \to E_1$ is seen, for example, from the definition of the mapping h.

§1.7 is proved and with it Theorem 1.2 is also proved.

Remark. The coincidence of the homologies of the complex $\widetilde{F}''_r/\widetilde{F}''_{r+1}$ with $H^*(L_0;\ \Lambda^r)$ can be derived also from the well-known Serre-Hochshild theorem [6].

1.8. The imbedding $\mathfrak{gl}(n, R) \to L_0$ induces an isomorphism $H^*(\mathfrak{gl}(n, R)) \to H^*(L_0;\ R)$.

The correctness of this version of Theorem 1.2 can be discerned from the proof of the latter.

2. Relations Between the Cohomologies of the Lie Algebras L_0 and W

In this paragraph the results of §1 are processed into information over the cohomologies of the algebra W. All the propositions proved here are corollaries of the corresponding assertions of §1 and the classical theorem over the cohomologies of the Lie algebra. However, since the general theory of the cohomologies of the topological Lie algebras has not been set forth anywhere, we shall briefly give a straightforward proof.

2.1. Let M be a finite-dimensional module over the algebra L_0. We denote by $\widetilde{M}$ the space of formal power series of the variables $x_1, \ldots, x_n$ (identifiable with the coordinates in the space R^n) with coefficients from M. The action of the algebra W on $\widetilde{M}$ is defined as follows. If $a(x_1, \ldots, x_n)$ is a power series (with real coefficients) and m is an element of M then the element $ma \in \widetilde{M}$ under the action of the vector field $\xi \in W$ goes over into the element

$$m\xi(a) + \sum_{q_1\geq 0,\ldots,q_n\geq 0} \frac{1}{q_1!\ldots q_n!}\left[\left(\frac{\partial^{q_1+\ldots+q_n}\xi}{\partial x_1^{q_1}\ldots\partial x_n^{q_n}}\right)m\right] x_1^{q_1}\ldots x_n^{q_n}$$

of the module $\widetilde{M}$. Here $\xi(a)$ is the result of the application of the differential operator ξ to the power series a, the differentiation of the vector field is with respect to the coordinates, the tilde means the same as in §1.7.

It is immediately verified that the mentioned action of the algebra W on $\widetilde{M}$ agrees with the operation of commutation in W. Thus $\widetilde{M}$ is equipped with the structure of the W-module. One says that the W-module of $\widetilde{M}$ is induced by an L_0-module of M.

Evidently if M is a ring and mappings of the algebra L_0 are through its differentiations then $\widetilde{M}$ is a ring and the mappings of the algebra W are also through its differentiations.

Example. The module Λ^r over the algebra L_0 induces a W-module Ω^r of the formal exterior differential forms of degree r with the ordinary action of the algebra W; in particular the trivial L_0-module of W induces a W-module of the power series.

Remark. We leave it to the reader to prove that our definition of the W-module agrees with the ordinary definition of the induced module which is as follows. If A is a Lie algebra, and B its subalgebra, then the A-module, induced by the B-module of M is defined as $\mathrm{Hom}_{[B]}([A], M)$, where [A], [B] are the enveloping algebras for A, B.

2.2. For every finite dimensional L_0-module of M there exists an isomorphism between the cohomologies of the algebra L_0 with coefficients in M and the cohomologies of the algebra W with coefficients in $\widetilde{M}$. If M is a ring and the action of the algebra L_0 is multiplicative then the mentioned isomorphism is annular.

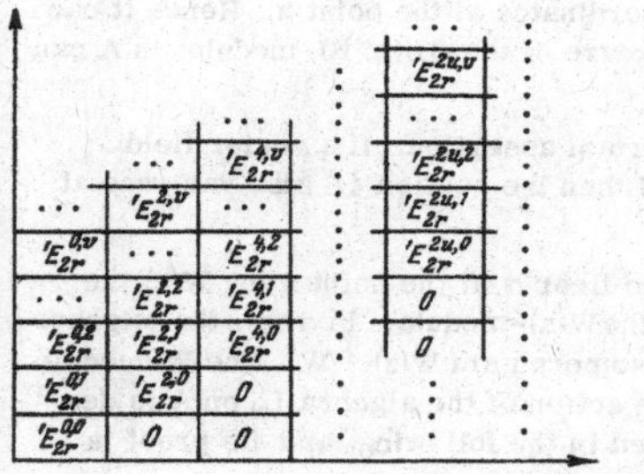

Fig. 1

In particular the bigraded rings $H^*(W;\ \Omega^r)$ and $H^*(L_0;\ \Lambda^r)$ are isomorphic. Thus Theorem 1.2 gives a description of the ring $H^*(W;\ \Omega^r)$.

The reciprocally inverse homomorphisms $H^q(W;\ \Omega^r) \to H^q(L_0;\ \Lambda^r)$, $H^q(L_0;\ \Lambda^r) \to H^r(W;\ \Omega_r)$ are induced by the mappings

$$f^*: C^q(W;\ \Omega^r) \to C^q(L_0;\ \Lambda^r), \qquad g: C^q(L_0;\ \Lambda^r) \to C^q(W;\ \Omega^r),$$

acting according to the formula

$$[f(P)](\xi_1, \ldots, \xi_q) = \nu_0(P(\xi_1, \ldots, \xi_q))$$

(here $P \in C^q(W;\ \tilde{M})$; $\xi_1, \ldots, \xi_q \in L_0 \subset W$; ν_0 is a function relating a power series to its free term);

$$[g(Q)](\xi_1, \ldots, \xi_q) = \sum \frac{1}{r_{11}! \ldots r_{qn}!} Q\left(\frac{\partial^{r_{11}+\cdots+r_{1n}}\xi_1}{\partial^{r_{11}}x_1 \ldots \partial^{r_{1n}}x_n}, \ldots, \frac{\partial^{r_{q1}+\cdots+r_{qn}}\xi_q}{\partial^{r_{q1}}x_1 \ldots \partial^{r_{qn}}x_n}\right) x_1^{q_1} \ldots x_n^{q_n}$$

(here $Q \in C^q(L_0;\ M)$; $\xi_1, \ldots, \xi_p \in W$; the summation is carried out over all nonnegative integers $r_{11}, \ldots, r_{1n}, \ldots, r_{q1}, \ldots, r_{qn}, q_1, \ldots, q_n$ such that $r_{11} + \ldots + r_{q1} = q_1, \ldots, r_{1n} + \ldots + r_{qn} = q_n$).

Remark. The equality $\bar{H}^*(B;\ M) = H^*(A;\ \mathrm{Hom}_{[B]}([A], M))$ (the notations are borrowed from the remark to §2.1) in the classical theory of the cohomologies of the Lie algebra is well known (see for example [5], § XIII, 4.2).

2.3. The spectral sequence $\mathscr{E}''$ (see, §1.6) can be interpreted as a spectral sequence, corresponding to no acyclic resolvent

$$0 \to R \xrightarrow{d} \Omega^1 \xrightarrow{d} \Omega^2 \to \ldots \to \Omega^n \to 0$$

of the W-module of R (grossly speaking as the second spectral sequence of the double complex $C^*(W;\ \Omega^*)$; see [5], XV, § 6). Proof is left to the reader.

II. ALGEBRAS OF SMOOTH FIELDS

§3. The Diagonal Complex

3.1. Let A be a finite-dimensional GL(n, R)-module (i.e., a finite-dimensional real linear space in which a representation of the group GL(n, R) is given) and let M be a smooth (class C^∞) connected manifold. We emphasize that we assume M to be neither orientable nor compact, nor having an edge. By α we denote a vector fiber over M with a layer isomorphic to A, induced through the tangent fiber by means of a representation of the group GL(n, R) in A. By $\mathcal{A}$ we denote the space of smooth sections of the fiber α. Every diffeomorphism of the manifold M induces a layerwise diffeomorphism of the space of tangent fiber into itself, and hence a diffeomorphism of the space of fiber α into itself. As a result the space $\mathcal{A}$ is equipped with the structure of a module over the group of diffeomorphisms of the manifold M and hence over the Lie algebra $\mathfrak{A}(M)$ of smooth vector fields of the manifold M.

Our purpose is to study the cohomologies of the algebra $\mathfrak{A}(M)$ with coefficients in the $\mathfrak{A}(M)$-module of $\mathcal{A}$.

We note that if A is an algebra over R and operators of GL(n, R) are automorphisms of this algebra (we shall call this case in the subsequent as simply multiplicative) then $\mathcal{A}$ is also an algebra over R and operators from $\mathfrak{A}(M)$ are diffeomorphisms of this algebra. In this case the spaces $C^*(\mathfrak{A}(M);\ \mathcal{A})$ and $H^*(\mathfrak{A}(M);\ \mathcal{A})$ are equipped with natural structures of graded real algebras.

3.2. We mention an important reciprocal connection existing between the $\mathfrak{A}(M)$-module $\mathcal{A}$ and the GL(n, R)-module of A. We choose in the neighborhood U of a point $x \in M$ a local coordinate system. We obtain a canonical layerwise diffeomorphism of the total image of the neighborhood U in the space of the fiber α onto the product $U \times A$. We identify, with the help of this diffeomorphism, the layer of the fiber α at the point x with A. Let ξ be a vector field on M vanishing at the point x and let f be a smooth section of the fiber α. Since $\mathcal{A}$ is a $\mathfrak{A}(M)$-module, the element $\xi f \in \mathcal{A}$ is defined. The points $f(x)$ and $[\xi f](x)$ are elements of the space A. As can be verified $[\xi f](x) = \Xi f(x)$, where $\Xi \in \mathfrak{gl}(n, R)$ is a matrix consisting of products of coordinate functions of the field ξ in terms of local coordinates at the point x. Hence it can be seen that the structure of the $\mathfrak{gl}(n, R)$-module (and hence the structure of the GL(n, R)-module) in A can be reconstructed through the structure of the $\mathfrak{A}(M)$-module in $\mathcal{A}$.

3.3. We single out from what has been said the following important assertion. If a vector field $\xi \in \mathfrak{A}(M)$ and the section $f \in \mathcal{A}$ of the fiber α vanish at the point $x \in M$ then the section ξf also vanishes at the point x.

3.4. We denote through A(x) the space of formal sections of the fiber α at the point x by W(x) the space of formal vector fields at this point. Evidently A(x) is within the W(x)-module. Fixing in the neighborhood of the point x a local coordinate system we obtain a canonical isomorphism W(x) = W. The W-module A(x) is none other than W-module induced by the L_0-module of A (the action of the algebra L_0 on A is defined by the projection $L_0 \to \mathfrak{gl}(n, R)$). This assertion is not employed in the following, and its proof is left to the reader.

3.5. All of what follows can, without difficulty, be carried over to the case when $\mathcal{A}$ is the space of smooth sections of a vector fiber over M in which the connectedness is specified. It is clear that such a fiber can not be induced by tangents, so that this case is more general than that described in §3.1 although it requires the introduction of additional structure, namely, connectedness.

3.6. We proceed to the definition in $C^*(\mathfrak{A}(M); \mathcal{A})$ of a filtration, analogous to that considered in [1] (see [1], §1.2). We recall that $C^q(\mathfrak{A}(M); \mathcal{A})$ is the space of q-linear continuous functionals defined on $\mathfrak{A}(M)$ and with starting values in $\mathcal{A}$.

We shall say that the ring $P \in C^q(\mathfrak{A}(M); \mathcal{A})$ has a filtration no larger than k, if, for arbitrary vector fields $\xi_1, \ldots, \xi_q \in \mathfrak{A}(M)$ such that the section $P(\xi_1, \ldots, \xi_q)$ of the fiber α is different from zero at some point $x \in M$, it is required to find points $x_1, \ldots, x_k \in M$ such that each of the fields $\xi_1, \ldots, \xi_q$ does not identically vanish in an arbitrarily small neighborhood of one of the points $x_1, \ldots, x_k$, x.

For example a filtration, not greater than −1, has only a zero ring; a filtration not greater than 0 has a ring P if and only if the section $P(\xi_1, \ldots, \xi_q)$ vanishes at every point of the manifold M in a neighborhood of which one of the fields $\xi_1, \ldots, \xi_q$ vanishes.

The set of q-dimensional rings of a filtration, not exceeding k, will be denoted by $C^q_k(\mathfrak{A}(M); \mathcal{A})$. It is clear that $C^q_k(\mathfrak{A}(M); \mathcal{A})$ is the subspace of the space $C^q(\mathfrak{A}(M); \mathcal{A})$ and that $d(C^q_k(\mathfrak{A}(M); \mathcal{A})) \subset C^{q+1}_k(\mathfrak{A}(M); \mathcal{A})$.

The subcomplex $C^*_0 = \bigoplus_q C^q_0(\mathfrak{A}(M); A)$ of the complex $C^*(\mathfrak{A}(M); \mathcal{A})$ is called <u>diagonal</u> and is denoted by C^q_Δ. Evidently the ring P belongs to the diagonal complex if and only if the value of the section $P(\xi_1, \ldots, \xi_q)$ of the fiber α at any point $x \in M$ depends only on the germs of the fields $\xi_1, \ldots, \xi_q$ at the point x.

In the multiplicative case the introduced filtration is multiplicative: if $\alpha \in C^q_k(\mathfrak{A}(M); \mathcal{A})$, $\beta \in C^{q'}_{k'}(\mathfrak{A}(M); \mathcal{A})$, then $\alpha\beta \in C^{q+q'}_{k+k'}(\mathfrak{A}(M); \mathcal{A})$. In particular, the diagonal complex is a multiplicative subcomplex of the whole complex $C^*(\mathfrak{A}(M); \mathcal{A})$. The main difference of the considered case from the case of constant coefficients was investigated in [1]: here the filtration was also multiplicative but for the diagonal complex the complex $C^*_1(\mathfrak{A}(M); R)$, which was not multiplicative, was taken.

3.7. In this paper we limit ourselves to the investigation of the ring $H^*_\Delta(\mathfrak{A}(M); \mathcal{A})$ of cohomologies of the diagonal complex $C^*_\Delta(\mathfrak{A}(M); \mathcal{A})$. We shall leave open the question of the relation of the ring $H^*_\Delta(\mathfrak{A}(M); \mathcal{A})$ with the ring $H^*(\mathfrak{A}(M); \mathcal{A})$ of homologies of the whole complex $C^*(\mathfrak{A}(M); \mathcal{A})$. We shall only remark that $H^*(\mathfrak{A}(M); \mathcal{A})$ is a $H^*_\Delta(\mathfrak{A}(M); \mathcal{A})$-module.

4. The Spectral Sequence

We shall begin the study of the homologies of the complex $C^*_\Delta(\mathfrak{A}(M); \mathcal{A})$ by defining in this complex a filtration and investigate the spectral sequence arising therefrom.

4.1. We shall denote by Φ^q_r the subsequence of the space $C^q_\Delta(\mathfrak{A}(M); \mathcal{A})$, consisting of rings P such that if q + 1 − r of vector fields $\xi_1, \ldots, \xi_q$ vanish at some point $x \in M$, then the section $P(\xi_1, \ldots, \xi_q)$ of the fiber α also vanishes at the point x. We also put $\Phi_r = \bigoplus_q \Phi^q_r$. It is evident that

$$C^*_\Delta(\mathfrak{A}(M); \mathcal{A}) = \Phi_0 \supset \Phi_1 \supset \Phi_2 \supset \ldots$$

and that Φ_r^q = for q < r.

4.2. The filtration $\{\Phi_r\}$ agrees with a diffeomorphism.

Proof. Let $P \in \Phi_r^q$ and let $\xi_1, \ldots, \xi_{q+1} \in \mathfrak{A}(M)$ be vector fields where $\xi_1, \ldots, \xi_{q+2-r}$ vanish at the point $x \in M$. Then

$$(dP)(\xi_1, \ldots, \xi_{q+1}) = \sum_{1\leq s<t\leq q+1} (-1)^{s+t-1} P([\xi_s, \xi_t], \xi_1, \ldots \hat{\xi}_s \ldots \hat{\xi}_t \ldots, \xi_{q+1})$$

$$+ \sum_{1\leq s\leq q+1} (-1)^{s-1} \xi_s P(\xi_1, \ldots \hat{\xi}_s \ldots, \xi_{q+1}).$$

The first sum on the right-hand order vanishes at the point x by virtue of the fact that if two vector fields vanish at some point, then their commutator also vanishes at the same point. From Proposition 3.3 it follows the vanishing at the point x of terms of the second sum corresponding to the values of s less than or equal to q + r − 2. Finally the vanishing of the remaining terms at the point x follows from the following lemma.

LEMMA. If $P \in \Phi_r^m$ and $\xi \in \mathfrak{A}(M)$, then $\xi P \in \Phi_{r-1}^m$ (here ξP is defined by the formula $(\xi P)(\xi_1, \ldots, \xi_m) = \xi[P(\xi_1, \ldots, \xi_m)]$).

Proof of the Lemma. Evidently it is possible to limit ourselves to the case when ξ has a compact carrier.

If φ is a diffeomorphism of the manifold into itself and $P \in C_\Delta^m(\mathfrak{A}(M); \mathcal{A})$, then we denote by φP the ring given by the formula

$$(\varphi P)(\xi_1, \ldots, \xi_m) = \varphi[P(\varphi^{-1}\xi_1, \ldots, \varphi^{-1}\xi_m)].$$

It is clear that if $P \in \Phi_r$, then also $\varphi P \in \Phi_r$.

Let now $\xi, \xi_1, \ldots, \xi_q \in \mathfrak{A}(M)$, where the fields $\xi_1, \ldots, \xi_{q+2-r}$ vanish at the point x. The vector field ξ defines a set of diffeomorphisms φ_t: M → M. We have

$$[\xi P(\xi_1, \ldots, \xi_q)](x) = \frac{d}{dt}[\varphi_t(P(\xi_1, \ldots, \xi_q))](x),$$

$$\varphi_t[P(\xi_1, \ldots, \xi_q)](x) = [(\varphi_t P)(\varphi_t \xi_1, \ldots, \varphi_t \xi_q)](x) = (\varphi_t P)(\xi_1, \ldots, \xi_q)(x)$$

$$+ t\sum_s (\varphi_t P)\left(\xi_1, \ldots, \xi_{s-1}, \frac{\partial \varphi_t \xi_s}{\partial t}, \xi_{s+1}, \ldots, \xi_q\right)(x) + o(t).$$

Since $\varphi_t P \in \Phi_r$, the first two terms of the last sum vanish. Thus $\xi P \in \Phi_{r-1}$.

4.3. If A is the algebra and the operators from GL(n, R) are automorphisms then the filtration $\{\Phi_r\}$ is multiplicative. This is evident.

4.4. In view of §4.2, 4.3 the filtration $\{\Phi_r\}$ generates a spectral sequence which is multiplicative in the multiplicative case converging to $H_\Delta^*(\mathfrak{A}(M); \mathcal{A})$. We denote this spectral sequence by $\{E_r^{s,t}, d_r^{s,t}\}$.

We begin with the zeroth term of this spectral sequence. We denote through A(x) a layer of the fiber α over the point x, through $L_0(x)$ the algebra of formal vector fields at the point x vanishing at this point. In agreement with 3.2 A(x) is an $L_0(x)$-module where, if we fix a local coordinate system in the neighborhood of the point x there arises the isomorphisms A(x) = A, $L_0(x) = L_0$ wherefrom is obtained the structure of the L_0-module on A induced by the structure of $\mathfrak{gl}(n, R)$-module on A by means of the projection $L_0 \to \mathfrak{gl}(n, R)$. We denote through $G^{s,t}(x)$ the space of linear functionals acting from the s-th exterior series $\Lambda^s \tau(x)$ of the tangent space to M at the point x in the space $C^t(L_0(x);\ A(x))$ of rings of dimensions t of the algebra $L_0(x)$ with coefficients in A(x). The union of the spaces $G^{s,t}(x)$, for all $x \in M$ in an obvious fashion, is equipped with the structure of a smooth vector fiber with the basis M. We denote it through $\mathcal{G}^{s,t}$.

We shall construct a canonical isomorphism of the space $E_0^{s,t}$ on the space $\Gamma(\mathcal{G}^{s,t})$ of smooth sections of the fiber $\mathcal{G}^{s,t}$.

We fix on M a Riemannian metric. Let h: M → R be such a positive continuous function on M that in the h(x)-neighborhood of the point x any two points are connected by a single geodesic; further, let g: M × M → R be a smooth function such that $0 \leq g(x, y) \leq 1$ for any x, y; g(x, y) = 1 for $\rho(x, y) < (h(x)/2)$;

$g(x, y) = 0$ for $\rho(x, y) > h(x)$. If ε is a tangent vector to the manifold M at the point x then we define a vector field $\widetilde{\varepsilon}$ in the following manner. The exponential mapping exp: $\tau(x) \to M$ is a diffeomorphism on the original $h(x)$-neighborhood U of the point x. We choose in $\exp^{-1}(U) \subset \tau(x)$ a vector field obtained by a parallel displacement of the vector $\exp^{-1}\varepsilon$. We translate this vector field in U by means of an exp mapping, we multiply a vector of this field at every point $y \in U$ by $g(x, y)$ and in the obtained vector field in U we define $M \setminus U$ through the zero.

If ξ is a smooth vector field on M and $x \in M$ is a point then we define the element $\xi^*(x) \in L_0(x)$ as a formal vector field at the point x corresponding to a smooth vector field $\xi - \widetilde{\xi(x)}$.

We now define a homomorphism $F: \Gamma(\mathcal{G}^{s,t}) \to E_0^{s,t}$. Let $\varphi \in \Gamma(\mathcal{G}^{s,t})$. We denote through $\widetilde{\varphi}$ the ring of $C^{s+t}(\mathfrak{A}(M); \mathcal{A})$, corresponding to vector fields $\xi_1, \ldots, \xi_q \in \mathfrak{A}(M)$ of the fiber section α assuming at the point $x \in M$ the value

$$\sum (-1)^{i_1+\ldots+i_s - s(s+1)/2} \{[(\varphi(x)](\xi_{i_1}(x), \ldots, \xi_{i_s}(x))\} (\xi^*_{j_1}(x), \ldots, \xi^*_{j_t}(x)) \in A(x).$$

Here $\varphi(x) \in \mathrm{Hom}(\Lambda^s \tau(x), C^t(L_0(x), A(x)))$ is the value of the section φ at the point x; summation is carried out over all the partitions of the set $\{1, \ldots, s+t\}$ on the subsets $\{i_1, \ldots, i_s\}$, $\{j_1, \ldots, j_t\}$, where $i_1 < \ldots < i_s$, $j_1 < \ldots < j_t$. The ring φ belongs to Φ_s^{s+t}: if $q+1-s$ of the fields $\xi_1, \ldots, \xi_q$ vanish at the point x then one of the vectors $\xi_{i_1}(x), \ldots, \xi_{i_s}(x)$ must necessarily vanish.

The residue class of the ring $\widetilde{\varphi}$ with respect to Φ_{s+1}^{s+t} is an element of $\Phi_s^{s+t}/\Phi_{s+1}^{s+t} = E_0^{s,t}$. We denote it through $F(\varphi)$.

The monomorphicity of the mapping F is almost evident: if the section φ differs from zero, i.e., if there is a point $x \in X$, vectors $\varepsilon_1, \ldots, \varepsilon_s \in \tau(x)$ and elements $\eta_1, \ldots, \eta_t \in L_0(x)$ such that

$$\{[\varphi(x)](\varepsilon_1, \ldots, \varepsilon_s)\}(\eta_1, \ldots, \eta_t) \neq 0,$$

then the ring $\widetilde{\varphi}$ takes a value different from zero at the point x on the vector fields $\widetilde{\varepsilon}_1, \ldots, \widetilde{\varepsilon}_s, \zeta_1, \ldots, \zeta_t$, where $\zeta_1, \ldots, \zeta_t$ are smooth vector fields defining at the point x formal vector fields $\eta_1, \ldots, \eta_t$. Hence it follows the ring $\widetilde{\varphi}$ not only is different from zero but does not belong to Φ_{s+1}, i.e., $F(\varphi) \neq 0$.

Finally, we prove that the mapping F is epimorphic. Let $\psi \in \Phi_s^{s+t}$. We denote through φ the section of the fiber $\mathcal{G}^{s,t}$, defined by the formula

$$\{[\varphi(x)](\varepsilon_1, \ldots, \varepsilon_s)\}(\eta_1, \ldots, \eta_t) = \psi(\widetilde{\varepsilon}_1, \ldots, \widetilde{\varepsilon}_s, \zeta_1, \ldots, \zeta_t)(x),$$

where $\zeta_1, \ldots, \zeta_t$ are vector fields which define at the point x formal vector fields $\eta_1, \ldots, \eta_t$ (in view of the continuity of the functional ψ, $\psi(\varepsilon_1, \ldots, \varepsilon_s, \zeta_1, \ldots, \zeta_t)(x)$ does not depend on the choice of these fields $\zeta_1, \ldots, \zeta_t$). The difference $\psi - \widetilde{\varphi}$ belongs to Φ_{r+1}: if $q-s$ of these fields $\xi_1, \ldots, \xi_q$, say, $\xi_{s+1}, \ldots, \xi_q$ vanish at the point x, then

$$]\psi - \widetilde{\varphi}](\xi_1, \ldots, \xi_q)(x) = [\psi(\xi_1, \ldots, \xi_q)]x - \sum \{[\varphi(x)](\xi_{i_1}(x), \ldots, \xi_{i_s}(x))\}(\xi^*_{j_1}(x), \ldots, \xi^*_{j_{q-s}}(x))$$

$$= [\psi((\xi_1 - \widetilde{\xi_1(x)}) + \widetilde{\xi_1(x)}, \ldots, (\xi_s - \widetilde{\xi_s(x)}) + \widetilde{\xi_s(x)}, \xi_{s+1}, \ldots, \xi_q) - \psi(\widetilde{\xi_1(x)}, \ldots, \widetilde{\xi_s(x)}, \xi_{s+1}, \ldots, \xi_q)](x).$$

But since $\psi \in \Phi_r$ and the vector fields $\xi_1 - \widetilde{\xi_1(x)}, \ldots, \xi_s - \widetilde{\xi_s(x)}, \xi_{s+1}, \ldots, \xi_q$ vanish at the point x, then

$$[\psi(\xi_1, \ldots, \xi_q)](x) = [\psi(\widetilde{\xi_1(x)}, \ldots, \widetilde{\xi_s(x)}, \xi_{s+1}, \ldots, \xi_q)](x).$$

Thus the class of the element ψ in $E_r^{s,t}$ is equal to $F(\varphi)$. The isomorphicity of the mapping F is proved.

4.5. The differential $d_0^{s,t}: E_0^{s,t} \to E_0^{s,t+1}$ coincides up to a sign with the homomorphism $\Gamma(\mathcal{G}^{s,t}) \to \Gamma(\mathcal{G}^{s,t+1})$, induced by the differentials $d: C^t(L_0(x); A(x)) \to C^{t+1}(L_0(x); A(x))$.

<u>Proof.</u> We choose in the section $\varphi \in \Gamma(\mathcal{G}^{s,t})$, the point $x \in X$, vectors $\varepsilon_1, \ldots, \varepsilon_s \in \tau(x)$ and the elements $\eta_1, \ldots, \eta_{t+1} \in L_0(x)$. We have

$$\begin{aligned} \{[d_0^{s,t}\varphi](\varepsilon_1, \ldots, \varepsilon_s)\}(\eta_1, \ldots, \eta_{t+1})(x) &= [d\widetilde{\varphi}](\widetilde{\varepsilon}_1, \ldots, \widetilde{\varepsilon}_s, \zeta_1, \ldots, \zeta_{t+1})(x) \\ &= \sum_{1 \leqslant i < j \leqslant t+1} (-1)^{i+j+s-1} \widetilde{\varphi}(\widetilde{\varepsilon}_1, \ldots, \widetilde{\varepsilon}_s, [\zeta_i, \zeta_j], \zeta_1, \ldots \hat{\zeta}_i \ldots \hat{\xi}_j \ldots, \zeta_{t+1})(x) \\ &+ \sum_{1 \leqslant i \leqslant s,\ 1 \leqslant j \leqslant t+1} (-1)^{i+j+s-1} \widetilde{\varphi}([\widetilde{\varepsilon}_i, \zeta_j], \widetilde{\varepsilon}_1, \ldots \hat{\widetilde{\varepsilon}}_i \ldots, \widetilde{\varepsilon}_s, \zeta_1, \ldots \hat{\zeta}_j \ldots, \zeta_{t+1})(x) \end{aligned}$$

$$+ \sum_{1 \leqslant i \leqslant s} (-1)^{i-1} \tilde{\varepsilon}_i [\tilde{\varphi}(\tilde{\varepsilon}_1, \ldots \hat{\tilde{\varepsilon}}_i \ldots, \tilde{\varepsilon}_s, \zeta_1, \ldots, \zeta_{t+1})](x) +$$
$$+ \sum_{1 \leqslant j \leqslant t+1} (-1)^{j+s-1} \zeta_j [\tilde{\varphi}(\tilde{\varepsilon}_1, \ldots, \tilde{\varepsilon}_s, \zeta_1, \ldots \hat{\zeta}_j \ldots, \zeta_{t+1})](x)$$

(here $\zeta_1, \ldots, \zeta_{t+1}$ are smooth vector fields which define at the point x formal vector fields $\eta_1, \ldots, \eta_{t+1}$).

The second and the third terms in the sum vanish at the point x (proof is analogous to the proof of lemma of § 4.2); the value of the remaining part of the sum at the point x is equal to

$$(-1)^s \Big[\sum_{1 \leqslant i < j \leqslant t+1} (-1)^{i+j-1} \{[\varphi(x)](\varepsilon_1, \ldots, \varepsilon_s)\} ([\eta_i, \eta_j], \eta_1, \ldots \hat{\eta}_i \ldots \hat{\eta}_j \ldots, \eta_{t+1})$$
$$+ \sum_{1 \leqslant i \leqslant t+1} (-1)^{i-1} \eta_i \{[\varphi(x)](\varepsilon_1, \ldots, \varepsilon_s)\} (\eta_1, \ldots \hat{\eta}_i \ldots, \eta_{t+1}).$$

i.e., it is equal to the value of the differential of the ring $[\varphi(x)](\varepsilon_1, \ldots, \varepsilon_s) \in C^t(L_0(x); A(x))$ on the fields $\eta_1, \ldots, \eta_{t+1}$. The proposition is proved.

4.6. From Proposition 4.5 it follows that $E_1^{s,t}$ in the space of smooth sections of the fiber is composed of spaces $\mathrm{Hom}(\Lambda^s \tau(x), H^t(L_0(x); A(x)))$ for all $x \in M$. Since the spaces $H^t(L_0(x); A(x))$ are canonically isomorphic to the space $H^t(L_0; A)$ (this follows from the fact that every Lie algebra acts trivially on its cohomologies with arbitrary coefficients) there exists the canonical isomorphism,

$$E_1^{s,t} = \Omega^s(M) \otimes H^t(L_0; A),$$

where $\Omega^s(M)$ is the space of smooth exterior differential forms of degree s on M.

4.7. The differential $d_1^{s,t} : E_1^{s,t} \to E_1^{s+1,t}$ is induced by the exterior differential form $d : \Omega^s(M) \to \Omega^{s+1}(M)$.

<u>Proof.</u> We take the elements $\beta \in \Omega^s(M)$, $\gamma \in H^t(L_0; A)$ and choose a cocycle $P \in C^t(L_0; A)$ representing γ. We fix a point $x \in M$ and choose in its neighborhood U the same coordinate system as in 4.4. There arises a layerwise diffeomorphism of the original neighborhood U in the fiber α on $U \times A$. The element $\beta \otimes \gamma \in \Omega^s(M) \otimes H^t(L_0; A) = E_1^{s,t}$ represents in $E_0^{s,t}$ the section φ of the fiber $\mathcal{G}^{s,t}$, which, in U, is specified by the formula

$$\{[\varphi(y)](\varepsilon_1, \ldots, \varepsilon_s)\}(\eta_1, \ldots, \eta_t) = \beta(\varepsilon_1, \ldots, \varepsilon_s) P(\eta_1, \ldots, \eta_t),$$

where $y \in U$; $\varepsilon_1, \ldots, \varepsilon_s \in \tau(y)$. We must find out which section defines the ring $d\tilde{\varphi}$ which is no longer in $\mathcal{G}^{s,t+1}$, but in $\mathcal{G}^{s+1,t}$, i.e., must define

$$d\tilde{\varphi}(\tilde{\nu}_1, \ldots, \tilde{\nu}_{s+1}, \xi_1, \ldots, \xi_t)(x),$$

where $\tilde{\nu}_1, \ldots, \tilde{\nu}_{s+1} \in \tau(x)$; $\xi_1, \ldots, \xi_t \in \mathfrak{A}(M)$ are fields vanishing at the point x. We have

$$d\tilde{\varphi}(\tilde{\nu}_1, \ldots, \tilde{\nu}_{s+1}, \xi_1, \ldots, \xi_t)(x) = \sum_{\substack{1 \leqslant i \leqslant s+1 \\ 1 \leqslant j \leqslant t}} (-1)^{i+j-1} \tilde{\varphi}(\tilde{\nu}_1, \ldots \hat{\tilde{\nu}}_i \ldots, \tilde{\nu}_{s+1}, [\tilde{\nu}_i, \xi_j], \xi_1, \ldots \hat{\xi}_j \ldots, \xi_t)(x)$$
$$+ \sum_{1 \leqslant i \leqslant s+1} (-1)^{i-1} \tilde{\nu}_i [\varphi(\tilde{\nu}_1, \ldots \hat{\tilde{\nu}}_i \ldots, \tilde{\nu}_{s+1}, \xi_1, \ldots, \xi_t)](x)$$

(the remaining terms vanish at the point x for obvious reasons)

$$= \sum_{1 \leqslant i \leqslant s+1} (-1)^{i-1} \{\tilde{\nu}_i [\beta(\tilde{\nu}_1, \ldots \hat{\tilde{\nu}}_i \ldots, \tilde{\nu}_{s+1}) P(\xi_1, \ldots, \xi_t)]$$
$$- \beta(\tilde{\nu}_1, \ldots \hat{\tilde{\nu}}_i \ldots, \tilde{\nu}_{s-1}) \tilde{\nu}_i P(\xi_1, \ldots, \xi_t)\}(x) =$$

(the application of the form β and the cocycle P to the smooth vector fields gives smooth functions in U; more explicitly the function $P(\xi_1, \ldots, \xi_t)$ assumes at the point $y \in U$ a value equal to

$$= \sum_{1 \leqslant i \leqslant s+1} (-1)^{i-1} \{[\tilde{\nu}_i \beta(\tilde{\nu}_1, \ldots \hat{\tilde{\nu}}_i \ldots, \tilde{\nu}_{s+1})] P(\xi_1^{\bullet}, \ldots, \xi_t^{\bullet})\}(x)$$
$$= d\beta(\nu_1, \ldots, \nu_{s+1}) P(\xi_1^{\bullet}, \ldots, \xi_t^{\bullet}).$$

The proposition is proved.

4.8. There exists the canonical isomorphism $E_2^{s,t} = H^s(M; R) \otimes H^t(L_0; A)$.

This isomorphism is obtained from §4.6 and 4.7. We emphasize that in the multiplicative case this isomorphism is annular.

With this we complete investigation of the general case. From Theorem 4.8 it follows that if the spaces $H^*(L_0; A)$, $H^*(M; R)$ are finite-dimensional in the large or in every dimension, then the analogous assertion is valid also for the spaces $H^*_\Delta(\mathfrak{A}(M); \mathcal{A})$. Applying then 1.2 we can reach a conclusion about the finite dimensionality of the cohomologies of the diagonal complex with coefficients in the space of smooth exterior differential forms. This result, however, will be covered in §5 where the mentioned cohomologies will be calculated.

§5. Cohomologies with Coefficients in Exterior Forms

5.1. Theorem. Let M be a smooth connected n-dimensional manifold, $\Omega^\cdot(M) = \sum \Omega^q(M)$ is the algebra of smooth exterior differential forms on M. A bigraded ring $H^*_\Delta(\mathfrak{A}(M); \Omega^*(M))$ is isomorphic to the tensor product of the two of its subrings: the ring $H^*_\Delta(\mathfrak{A}(M); \Omega^0)$ and the ring $\sum_q H^q_\Delta(\mathfrak{A}(M); \Omega^q)$.

A graded ring $H^*_\Delta(\mathfrak{A}(M); \Omega^0)$ is isomorphic to the ring of real cohomologies of the space T of a smooth U(n)-fiber induced by the complexification of the tangent fiber.*

A graded ring $\sum_q H^q_\Delta(\mathfrak{A}(M); \Omega^q(M))$ is isomorphic to the ring $\sum_q H^q(L_0; \Lambda^q)$, i.e., is generated by the commutations of the generators $\lambda_q \in H^q(L_0; \Lambda^q)$ $(q = 1, \ldots, n)$, connected by the relations $\lambda_1^{s_1} \ldots \lambda_n^{s_n} = 0$ for $s_1 + 2s_2 + \ldots + ns_n > n$. Thus

$$H^r_\Delta(\mathfrak{A}(M); \Omega^q(M)) = \begin{cases} 0 & \text{for } r < q, \\ H^q_\Delta(\mathfrak{A}(M); \Omega^q(M)) \otimes H^{r-q}(T; R) & \text{for } r \geq q, \end{cases}$$

where the dimension of the space $H^q_\Delta(\mathfrak{A}(M); \Omega^q(M))$ is equal to the number of partitions of the number q into a sum of natural terms.

Proof. According to §4 there exists a spectral sequence $\{E_r^{s,t}, d_r^{s,t}\}$, converging to the ring $H_0^*(\mathfrak{A}(M); \Omega^*(M))$ with $E_2^{s,t} = H^s(M; R) \otimes H^t(L_0; \Lambda^*)$. Under imbedding the spectral sequence corresponding to the subring $C^\infty(M) = \Omega^0(M) \subset \Omega^*(M)$. To begin with we study this subsequence.

The space T is a manifold on which the group U(n) acts. Since the group U(n) is compact the cohomologies of the space T are isomorphic to the homologies of the subcomplex $\tilde{\Omega}^*(T)$ of the de Rahm complex of this manifold composed of U(n)-invariant forms. We denote through τ^* the vector fiber over M with the layer $\mathfrak{gl}(n, R)$, induced by the tangent fiber. Since the space of left-invariant exterior forms on U(n) is canonically isomorphic to the space of linear skew-symmetric forms on $\mathfrak{gl}(n, R)$, the elements of the space $\tilde{\Omega}^*(T)$ can be regarded as smooth functions corresponding to each point x of the manifold M with a linear skew-symmetric form on the space $\tau(x) \oplus \tau^*(x)$. We remark now that a smooth vector field on M (more accurately its 1-jet) defines a smooth section of the fiber $\tau \oplus \tau^*$. Therefore, if q vector fields $\xi_1, \ldots, \xi_q$ are specified on M, and the element $\alpha \in \tilde{\Omega}^q(t)$, then by calculating the value of the form α on the 1-jets of fields $\xi_1, \ldots, \xi_q$ at every point of the manifold we obtain a smooth function on M. Thus we construct the homomorphism

$$\chi : \tilde{\Omega}^q(T) \to C^q_\Delta(\mathfrak{A}(M); C^\infty(M)).$$

It is easy to verify that it permutes with the differentials.

In the complex $\tilde{\Omega}^*(T)$ there is a filtration due to the presence in T of the structure of a fiber space (see for example [7], p. 178-180). The homomorphism χ coincides with filtrations existing in both the complexes (this follows from the definition of filtrations). The first terms induced by these filtrations (of the spectral sequences) are isomorphic to $\Omega^*(M) \otimes H^*(\mathfrak{gl}(n, R); R)$, $\Omega^*(M) \otimes H^*(L_0; R)$. respectively. From the construction of these isomorphisms it can be seen that the homomorphism, induced by the homomorphism χ on $\Omega^*(M)$, is an identity but on $H^*(\mathfrak{gl}(n, R); R)$ defines a mapping into $H^*(L_0; R)$ coinciding with the homomorphism induced by the imbedding $\mathfrak{gl}(n, R) \to L_0$. Thus from §1.8 it follows that the homomorphism χ

*This result is due to M. V. Losik [4].

induces an isomorphism of the first terms of the spectral sequences and hence it induces an isomorphism of the rings of cohomologies. The part of the theorem relating to $H^*(\mathfrak{A}(M);\ \Omega^0(M))$ is proved.

The spectral sequence $\{E_r^{s,t}, d_r^{s,t}\}$ is decomposed into a sum of spectral sequences $\{{}^{(q)}E_r^{s,t}, {}^{(q)}d_r^{s,t}\}$, corresponding to the submodules $\Omega^q(M) \subset \Omega^*(M)$. In accordance with § 4.8 ${}^{(q)}E_2^{s,t} = H^s(M;\ R) \otimes H^t(L_0;\ \Lambda^q)$. As is known (Theorem 1.2), $H^*(L_0;\ \Lambda^q)$ is a free $H^*(L_0;\ R)$-module generated by the additive basis of the space $H^q(L_0;\ \Lambda^q)$.† Using the multiplicative structure in the spectral sequence $\{E_r^{s,t}, d_r^{s,t}\}$, we may reconstruct through the differentials within the spectral sequence $\{{}^{(0)}E_r^{s,t}, {}^{(0)}d_r^{s,t}\}$ (but this spectral sequence coincides with the spectral sequence of the fiber $T \to M$) the differentials in the spectral sequence $\{{}^{(q)}E_r^{s,t}, {}^{(q)}d_r^{s,t}\}$. We obtain that ${}^{(q)}d_r^{0,q} = 0$ for $r \geq 2$, that $H^q_\Delta(\mathfrak{A}(M);\ \Omega^q) = {}^{(q)}E_\infty^{0,q} = {}^{(q)}E_2^{0,q}) = H^q(L_0;\ \Lambda^q)$, and that $H^*_\Delta(\mathfrak{A}(M);\ \Omega^q(M))$ is a free $H^*_\Delta(\mathfrak{A}(M), \Omega^0(M))$-module, generated by an additive basis of the space $H^q_\Delta(\mathfrak{A}(M);\ \Omega^q(M))$. Thus $H^*_\Delta(\mathfrak{A}(M);\ \Omega^*(M))$ is a tensor product of the subrings $H^*_\Delta(\mathfrak{A}(M);\ \Omega^0(M)), \sum_q H^q_\Delta(\mathfrak{A}(M); \Omega^q(M))$.

It remains to remark that the isomorphism $\sum_q H^q_\Delta(\mathfrak{A}(M); \Omega^q(M)) = \sum_q H^q(L_0; \Lambda^q)$ is evidently multiplicative.

LITERATURE CITED

1. I. M. Gel'fand and D. B. Fuks, "Cohomologies of Lie algebra of tangential vector fields of a smooth manifold," Funktsional'. Analiz i Ego Prilozhen., 3, No. 3, 32-52 (1969).
2. I. M. Gel'fand and D. B. Fuks, "Cohomologies of Lie algebra of tangential vector fields of a smooth manifold. II," Funktsional'. Analiz i Ego Prilozhen., 4, No. 2, 23-31 (1970).
3. I. M. Gel'fand and D. B. Fuks, "Cohomologies of Lie algebra of formal vector fields," Izv. AN SSSR, seriya matem., 34, No. 2, 322-337 (1970).
4. M. V. Losik, "On the cohomologies of infinite dimensional Lie algebras of formal vector fields," Funktsional'. Analiz i Ego Prilozhen., 4, No. 2, 43-53 (1970).
5. H. Cartan and S. Eilenberg, Cohomological Algebra [Russian translation], IL, Moscow (1960).
6. J. P. Serre and C. Hochshild, Cohomology of group extensions, Trans. Amer. Math. Soc., 74 (1953), 110-134.
7. D. B. Fuks, "Spectral Sequences of Fibers," Uspekhi Matem. Nauk, 21, No. 5, 149-181 (1966).

†Hence it follows that ${}^{(q)}E_2$ is the free ${}^{(0)}E_2$-module generated by the additive basis of the space ${}^{(q)}E^{0q}$.

8.

(with A. S. Mishchenko)

Quadratic forms over commutative group rings and the K-theory

Funkts. Anal. Prilozh. **3** (4) (1969) 28-33 [Funct. Anal. Appl. **3** (1969) 277-281].
MR **41**:9243. Zbl. **239**:55004

This work originated from the solution of a problem stated by S. P. Novikov [1]. Namely, he noticed that in classification of smooth manifolds and also in other problems of smooth topology of not simply connected manifolds, an algebraic problem about classification of quadratic forms with coefficients in group rings arises. This problem is considered in the second section of this work for the commutative case. It is very interesting that the invariants obtained can be easily described in terms of the K-functor. It is better to consider the given problem as a particular case of a more general problem, namely the equivalence problem of quadratic forms, the coefficients of which depend on a point of the manifold.

We consider, for example, the simplest setting up of the problem. We consider a commutative discrete group G, for example, a free abelian group with a finite number of generators. Let $\|a_{ik}(g)\| = A$ be a matrix the elements of which are complex valued (for example, finite) functions over the group (elements of the group ring). We assume that A is an Hermitian matrix, i.e., that $A^* = A$, where $A^* = \|\bar{a}_{ki}(g^{-1})\|$. We say that A and B are equivalent if an invertible matrix $X = \|x_{ik}(g)\|$ exists such that $B = X^*AX$. The problem is to determine conditions of equivalence of matrices, among other conditions for a matrix to be equivalent to

$$A \sim \begin{pmatrix} E_p & 0 \\ 0 & -E_q \end{pmatrix},$$

where E_p and E_q are identity matrices of orders p and q, respectively.

We discuss in brief the initial considerations in solution of this problem. Passing from the group G to the character group G*, we obtain that a quadratic form $a_{ik}(\chi)$, depending on the point $\chi \in G^*$ is given; here $a_{ki}(\chi) = \overline{a_{ik}(\chi)}$. There arises then the equivalence problem of quadratic forms $A(\chi) = \|a_{ik}(\chi)\|$, where χ runs through the character group. The idea of description of the invariants of the quadratic form $A(\chi)$ is to consider a stratification, the layer of which at every point χ consists of all methods of reduction of the quadratic form at the same point χ to a canonical form.

It turns out that the answer is formulated in terms of a K-functor of the character group of the free abelian group G. The majority of these elementary considerations do not depend on the fact that G is a character group. Thus, the classification of quadratic forms over group rings is reduced to the problem of classification of quadratic forms depending continuously on the point of a complex.

Analogous results hold for real symmetric and antisymmetric forms. They lead to a KR-functor and to a new functor, which we denote by SK.

We bring an example of a theorem. Assume that we have a matrix $\|a_{ij}(t)\|$, where $t = (t_1, \ldots, t_n)$ runs through a n-dimensional torus. Two forms $A(t) = \|a_{ij}(t)\|$ and $B(t) = \|b_{ij}(t)\|$ are called equivalent if a matrix $X(t) = \|x_{ik}(t)\|$ (Det $X(t) \neq 0$ for any t) exists such that $B(t) = X^*(t)A(t)X(t)$. Assume that Det $A(t) \neq 0$. The invariants of A(t) will then be, first of all, the signature p-q of the matrix A(t) at each point t. This signature is constant as a result of continuity. It turns out that in addition to the signature we have $2^{n-1}-1$ integral invariants of the form $\|a_{ik}(t)\|$ (if p and q are not simultaneously smaller than n). For example, on the circle (n = 1) the signature is the unique invariant.

Moscow State University. Translated from Funktsional'nyi Analiz i Ego Prilozheniya, Vol. 3, No.4, pp. 28-33, October-December, 1969. Original article submitted June 16, 1969.

This work allows us also to compute simply the rank of the group of quadratic forms over group rings, obtained earlier by Shaneson from considerations of smooth topology.

1. A Useful Construction of a Classifying Space of the Group U(n)

Let X be a finite cell complex, and let A(x) be a function on the complex X assuming values in the spaces S_n of Hermitian non-degenerate matrices of order n. We put in correspondence to each such function A(x) an element of the group K(X).

1.1. Equivalence of Hermitian forms Depending on a Point of the Space. Let S(p, q), p + q = n be a subspace of the S_n, consisting of all matrices with signature equal p−q. Every matrix A ∈ S(p, q), hence, it can be brought by transformations of the form $A \to C^*AC$ to the canonical form

$$I_{p,q} = \begin{pmatrix} E_p & 0 \\ 0 & -E_q \end{pmatrix},$$

where E_p, E_q are identity matrices of orders p, q respectively. It is therefore possible to construct a mapping π: GL(n, C) → S(p, q), which puts in correspondence to each non-degenerate matrix C ∈ L(n, C) the matrix $\pi(C) = C^*I_{p,q}C \in S(p, q)$. Let U(p, q) be the subgroup of the group GL(n, C), which leaves fixed the matrix $I_{p,q}$, i.e., $C^*I_{p,q}C = I_{p,q}$ for all C ∈ U(p, q). It is easy to see that the mapping is a principal stratification with the layer U(p, q). Insofar as the group U(p, q) contracts to the group U(p) × U(q), and the group L(n, C) to the group U(n), it follows that the space S(p, q) is homotopically equivalent to the space U(n)/U(p) × U(q). Introducing the stabilization of the matrices S(p, q)→S(p+1, q+1) according to the formula $A \to \begin{pmatrix} A & 0 & 0 \\ 0 & 1 & 0 \\ 0 & 0 & -1 \end{pmatrix}$, we obtain that the space $S = \varinjlim S_n$ is homotopically equivalent to the classifying space Z × BU of an infinite dimensional unitary group U (U is understood to be $\varinjlim U(n)$). Thus, the space S of Hermitian non-degenerate matrices is a classifying space of a unitary K-theory. An element of the group K(X) determined by the function A(x) will be denoted by [A].

If two functions A(x) and B(x) are homotopic, then they define the same element of the group K(X). We shall, however, be interested in what follows in another equivalence relation of functions. We say that two functions A(x), B(x): X → S_n are equivalent if there exists a function C(x): X → L(n, C), such that $C^*(x)A(x)C(x) = B(x)$, x ∈ X.

PROPOSITION 1.1. Two functions A(x), B(x): X → S(p, q) are equivalent if and only if they are homotopic.

Proof. To every function A(x): X → S(p, q) there corresponds a principal U(p) × U(q)-stratification ξ_A, induced by the stratification π: L(n, C) → S(p, q). If the functions A(x) and B(x) are equivalent then the principal U(p) × U(q)-stratifications corresponding to them are also equivalent. Indeed, let C(x): X → L(n, C) be a function such that $C^*(x)A(x)C(x) = B(x)$. Let a: ξ_A → L(n, C), b: ξ_B → L(n, C) be mappings of the principal stratifications induced by the functions A, B, respectively. We construct a mapping f: $\xi_A \to \xi_B$. If $y \in \xi_A$ is a point covering the point x ∈ X, then $f(y) \in \xi_B$ is determined as a unique point covering the point x for which $b(f(y)) = a(y)C(x)$.

It follows from here that the principal U(p)-stratifications which are projections, say, on the first factor of the group U(p) × U(q) are also equivalent. Insofar as the space S is a classifying space for U(p)-stratifications, the mappings A(x) and B(x) are homotopic.

Conversely, if the mappings A(x) and B(x) are homotopic, then there exists a principal U(p) × U(q)-stratification η on the complex X × I, and $\eta|X \times \{0\} = \xi_A$, $\eta|X \times \{1\} = \xi_B$. Hence, the stratifications ξ_A and ξ_B are equivalent. Let, as before, $\xi_A \to \xi_B$ be a mapping of the principal stratifications. Let the point $y \in \xi_A$ cover the point x ∈ X. We set $C(x) = a(y)^{-1}b(f(y))$. The value of C(x) does not depend on the choice of the point y covering the point x, and, as it is easy to see, the relation $B(x) = C^*(x)A(x)C(x)$ holds. The functions A(x) and B(x) are consequently equivalent.

Thus, if we form the Grothendieck group, generated by all classes of equivalent functions A(x), the addition in which is defined as a direct sum of matrices, then this group is isomorphic to the group K(X). We notice the following property of the functions A(x).

PROPOSITION 1.2. Let ξ_A^1, ξ_A^2 be projections of the principal U(p) × U(q)-stratification ξ_A on the first and second factors of the group U(p) × U(q). Then $[\xi_A^1] = -[\xi_A^2]$, where $[\xi_A]$ is an element of the group $\widetilde{K}(X)$, corresponding to the principal U(p) × U(q)-stratification.

Proof. We can assume, without loss of generality, that the signature of the quadratic form A(x) equals zero. Let π: U(2n) → U(2n)/U(n) × U(n) be the principal U(n) × U(n)-stratification. We want to prove that it is trivial over the group U(2n). We associate with the stratification π the stratification with the group U (2n) over the basis X = U(2n)/U(n) × U(n). The space of this stratification E can be constructed as a factor-space of the space U(2n) × U(2n) with respect to the equivalence relation

$$(ag,\ b) = (a,\ gb),\quad g \in U(n) \times U(n) \subset U(2n).$$

We construct a section in this stratification: f: X → E. Let x ∈ X, π(a) = x. Then $f(x) = (a, a^{-1})$. The definition of the function f is correct, since if $\pi(a') = x$, then a' = ag; hence $(a', a'^{-1}) = (a, a^{-1})$.

COROLLARY 1.3. If C(x) = A(x) ⊗ B(x), then we have in the group $\widetilde{K}(X)$ the equality |C| = 2[A][B].

1.2. Real Quadratic Forms. An analog of a real K-theory for complexes with involutions has been considered in [3]. We consider for every complex X with involution a complex vector stratification ξ, with an anticomplex involution τ, i.e., if x ∈ ξ, then $\tau(\lambda x) = \bar{\lambda}\tau(x)$. Grothendieck's group generated by the above stratifications is denoted by KR(X). For the theory of cohomologies constructed over the functor KR(X) cohomologies of a point are computed, which allows us to compute the groups KR(X) for various spaces.

The construction done in point 1.1 can be extended also to the case of the functor KR(X). We consider the functions A(x): X → S(p, q), satisfying the condition $\overline{A(x)} = A(\tau(x))$. Two functions A(x) and B(x) are called equivalent if a function C(x): X → L(n, C) exists such that $\overline{C}(x) = C(\tau(x))$, and $C^*(x)A(x)C(x) = B(x)$. It turns out that the classes of equivalent functions obtained in this way generate a group isomorphic to the group KR(X). Indeed, we have

PROPOSITION 1.4. A representing object for the functor KR(X) is the space U(n)/U(p) × U(q), p + q = n, on which the involution τ acts according to the formula

$$\tau(C) = \overline{C},\quad C \in U(n).$$

Proof. According to the results of [4], if π: E → X is a principal G-stratification in the category of spaces with involution, where G is a group with involution and the functor of the homotopic groups $\omega_q(E)$ is trivial, $0 \le q \le N$, N sufficiently large, then the space X is a representing object of the category of principal G-stratifications over an arbitrary complex with involution Y. We verify first that the mapping π: U(n) → U(n)/U(p) × U(q) is a principal stratification in the category of spaces with involution. Let O(n) be the space of fixed points of the space U(n); let X_0 be the space of fixed points of the space X. Then $\pi(O(n)) = X_0$. Indeed, a point x ∈ X can be represented as a p-dimensional complex subspace L of the space C^n. If $\tau(x) = x$, then the subspace L is invariant with respect to the involution of complex conjugacy, i.e., it is determined by equations with real coefficients. Hence, there exists a y ∈ O(n), such that $\pi(y) = x$. Further, if $y_1, y_2 \in O(n)$ and $\pi(y_1) = \pi(y_2) \in X_0$, then $y_1 y_2^{-1} \in O(p) \times O(q)$. Thus, the fixed points form a principal O(p) × O(q)-stratification:

$$\pi': O(n) \to O(n)/O(p) \times O(q).$$

Similarly, the fixed points form, in the space U(n)/U(p), a set isomorphic to O(n)/O(p). The mapping U(n)/U(p) → U(n)/U(p) × U(q) is, therefore, a principal U(q)-stratification in the category of spaces with involution, and the functor of homotopic groups $\omega_q(U(n)/U(p))$ is trivial up to dimension p, which was what we were required to prove.

PROPOSITION 1.5. Two functions A(x) and B(x) are equivalent if and only if they are homotopic in the category of spaces with involution.

The proof is analogous to the proof of Proposition 1.1.

1.3. Real Antisymmetric Forms. A second analog of K-theory for spaces with involution is the group, generated by all functions A(x): X → S_n, for which equality $\overline{A(x)} = -A(\tau x)$, x ∈ X holds. Two functions A(x) and B(x) are said to be equivalent if a function C(x): X → L(n, C) exists such that $C(x) = C(\tau(x))$ and $C^*(x)A(x)C(x) = B(x)$. We denote Grothendieck's group generated by the above functions by SK(X).

PROPOSITION 1.6. There exist natural homomorphisms α and β

$$\alpha: K(X)\to SK(X),\quad \beta: SK(X)\to K(X)$$

such that $\beta\circ\alpha(x) = x-\bar{x}$ $\alpha\circ\beta(x) = 2x$. $\bar{x}$ denotes here a complex-conjugate stratification.

Proof. We consider the matrix $J=\frac{1}{\sqrt{2}}\begin{pmatrix}-i & -1\\ 1 & i\end{pmatrix}$. The following relations:

$$J^{*}=J^{-1},$$

$$C(x)=J^{*}\begin{pmatrix}A(x) & 0\\ 0 & -\bar{A}(\tau x)\end{pmatrix}J=\frac{1}{2}\begin{pmatrix}A(x)-\bar{A}(\tau x) & i(A(x)+\bar{A}(\tau x))\\ i(A(x)+\bar{A}(\tau x)) & A(x)-\bar{A}(\tau x)\end{pmatrix}$$

hold for this matrix. It is clear that $C(\tau x) = -\overline{C(x)}$. The mapping α puts in correspondence to the function A(x) the function C(x). To equivalent matrices correspond equivalent ones. The mapping β reduces, in fact, to "forgetting" of the involution on the complex X.

COROLLARY 1.7. $SK(S^{4k+2})\otimes Q = Q$, $SK(S^{4k})\otimes Q = 0$. Q is here the field of rational numbers.

2. Quadratic Forms Over Commutative Group Rings

The results of the previous section allow us to determine the invariants of quadratic forms,the coefficients of which belong to a commutative group ring, in homotopic terms. Let G be an abelian finitely generated group, and Z(G) its group ring. We consider a nondegenerate symmetric or antisymmetric scalar product in a finitely generated free Z(G)-module M. The scalar product will be assumed trivial if it can be represented in some basis by the matrix $\begin{pmatrix}0 & E\\ \pm E & 0\end{pmatrix}$ where E is the identity matrix. All such Z(G)-modules with a scalar product generate a Grothendieck group L(G). The problem consists of describing the group L(G) for various groups G. We fix in the module M an arbitrary Z(G)-basis and obtain by the scalar product a matrix $A = \|a_{ik}\|$, the elements of which belong to the ring Z(G) and satisfy the condition $a_{ik}^{*} = \pm a_{ki}$, where the involution * is generated by the correspondence $g\to g^{-1}$, $g\in G$.

In order to pass to homotopic terms of description of invariants of the group L(G) we replace the algebraic problem by an analytic one. First we replace the ring of integers Z by the field of real or complex numbers. Secondly, we replace the group ring of the abelian group G by a ring consisting of continuous functions on the character group G* of the group G. Distinct completions of the group ring lead to distinct subrings of the ring of continuous functions on the character group G*; however, this probably influences only slightly the homotopic invariants determined by those subrings.

We shall concentrate in the case of a free abelian group G, for example, on the ring of series $\sum_{g\in G} a(g)g$, the coefficients a(g) of which are such that $P(g)\cdot a(g)$ tends to zero for any polynomial P(g) (the elements g are considered as elements of an integral lattice of an Euclidean space). The obtained ring $\hat{C}(G)$ is isomorphic to a ring of functions of the class C^∞ on the character group G* (G* is a torus in this case). We denote the corresponding group of scalar products by $L_C(G)$. We consider the subring $\hat{R}(G)$ of the ring $\hat{C}(G)$, consisting from series with real coefficients. We denote the group of scalar products corresponding to the ring $\hat{R}(G)$ by $L_{R,0}(G)$ in the symmetric case and by $L_{R,1}(G)$ in the antisymmetric case.

We are now in the position to formulate the basic theorem on homotopic description of invariants of a group of scalar products over a commutative group ring. Let an involution $\tau(\chi) = \chi^{-1}$ be defined on the character group G*.

THEOREM. There exist the isomorphisms:

$$L_C(G)\approx K(G^{*}),\quad L_{R,0}(G)\approx KR(G^{*}),\quad L_{R,1}(G)\approx SK(G^{*}).$$

This theorem is a particular case of the assertions of §1, in points 1.1, 1.2, and 1.3, respectively.

Remark. If one is interested only in the ranks of the groups $L_C(G)$, $L_{R,0}(G)$, $L_{R,1}(G)$, then the group $L_C(G)$ decomposes into a direct sum of groups $L_{R,0}(G)$ and $L_{R,1}(G)$: $L_C(G)\otimes Q\approx L_{R,0}(G)\otimes Q + L_{R,1}(G)\otimes Q$. Groups of scalar products over the integer ring Z(G) are computed up to elements of finite order in [2]. Not complicated computations show that the groups $L_{R,0}(G)\otimes Q$ and $L_{R,1}(G)\otimes Q$ are isomorphic to the corresponding groups of scalar products over the ring Z(G). This shows that the method described in the present article does not lead to an essential loss of information on groups of scalar products over group rings.

LITERATURE CITED

1. S. P. Novikov, "Characteristic classes of Pontryagin," Lecture at the International Mathematical Congress, Moscow (1966).
2. I. Shaneson, "Wall's surgery obstruction groups for Z × G, for suitable groups G," Bull. Amer. Math. Soc., 74, 467-471 (1968).
3. M. At'ya, Lectures about K-theory, K-theory and Reality [in Russian], Supplement II, "Mir," Moscow (1967).
4. G. E. Bredon, Equivariant Cohomology Theories, Springer-Verlag, Berlin (1967).

9.

(with D. B. Fuks)

Upper bounds for cohomology of infinite-dimensional Lie algebras

Funkts. Anal. Prilozh. 4 (4) (1970) 70–71 [Funct. Anal. Appl. 4 (1970) 323-324].
MR 44:4792. Zbl. 224:18013

In [1] we studied cohomology with real coefficients of Lie algebras of smooth vector fields on smooth manifolds. Under certain conditions we proved the finite-dimensionality of this cohomology. One of the intermediate assertions was the assertion of the finite-dimensionality of the complete cohomology space of a Lie algebra of formal vector fields (see [1], Proposition 6.2; the treatment of this result via formal vector fields is obtained in §1 of [2]). The purpose of the present note is the formalization and generalization of the method we used to prove this assertion. With the aid of similar methods, as remarked by B. I. Rosenfeld, it may be shown that the cohomology space of the Lie algebra of formal tangent vector fields is finite-dimensional. At the same time, for algebras, for example, this method does not succeed in obtaining a proof of the same finite-dimensionality for Hamiltonian formal vector fields, although a certain reduction of the standard cochain complexes is obtained. We do not repeat here the method in question: it is not necessary for understanding what follows. The reader can observe for himself the direct connection between the material set forth below and the proof of Proposition 6.2 of [1].

1. Let $W_n = \left\{\sum P_i\ (\partial/\partial x_i) \mid P_i \in K[[x_1, \dots, x_n]]\right\}$ be the algebra of formal vector fields in K^n, where K is the ground field. Further, let $a_1, \dots, a_n$ be elements of the field K and let A be the additive group they generate. Denote by S_a, where $a \in A$, the subspace of the algebra W_n spanned by the monomials $x_1^{k_1} \dots x_n^{k_n}$ $(\partial/\partial x_i)$ with $k_1 a_1 + \dots + k_n a_n - a_i = a$. It is clear that $W_n = \bigoplus_{a \in A} S_a$ and that $[S_{a'}, S_{a''}] = S_{a' + a''}$. Thus the algebra W_n is graded by the elements of A.

<u>Example.</u> If $a_1 = \dots = a_n = 1$, then the grading just constructed coincides with the ordinary grading $\bigoplus_{k \in Z} S_k$ of the algebra W_n, where $S_k = \left\{\sum P_i\,(\partial/\partial x_i) \mid \deg P_i = k+1\right\}$.

2. Set $\alpha = \sum a_i x_i\,(\partial/\partial x_i)$. Clearly, if $\xi \in S_a$, then $[\alpha, \xi] = a\xi$.

3. Let $\mathfrak{g}$ be a subalgebra of the algebra W_n containing the element α. Set $T_a = \mathfrak{g} \cap S_a$.

<u>PROPOSITION.</u> The algebra $\mathfrak{g}$ is the direct sum of its subspaces $T_a (a \in A)$.

<u>Proof.</u> It is clear that vectors belonging to subspaces T_a with distinct a are linearly independent. Therefore it suffices to show that if $\xi_1 \in S_{a(1)}, \dots, \xi_m \in S_{a(m)}$, where $a(1), \dots, a(m) \in A$, are unequal in pairs, and if $\xi = \xi_1 + \dots + \xi_m \in \mathfrak{g}$, then $\xi_1 \in \mathfrak{g}, \dots, \xi_m \in \mathfrak{g}$. The latter follows from the fact that together with ξ the elements $\xi^{(i)} = [\alpha, [\alpha, \dots [\alpha, \xi]] \dots] = a(1)^i \xi_1 + \dots + a(m)^i \xi_m$ belong to $\mathfrak{g}$, and $\det \| a(k)^l \|_{\substack{1 \leq k \leq m \\ 0 \leq l \leq m-1}} \neq 0$.

4. The standard cochain complex $\mathscr{C} = \{C^q, d^q\}$ of the algebra $\mathfrak{g}$ with coefficients in K decomposes into a direct sum of the subcomplexes $\mathscr{C}_a = \{C_a^q, d^q\}\ (a \in A)$, where C_a^q consists of cochains $c : \Lambda^a \mathfrak{g} \to K$, such that if $\xi_1 \in T_{a(1)}, \dots, \xi_q \in T_{a(q)}$ and $a(1) + \dots + a(q) \neq a$, then $c(\xi_1, \dots, \xi_q) = 0$ (we assume also that $C_0^0 = C^0$, $C_a^0 = 0$ for $a \neq 0$).

<u>THEOREM 1.</u> The complex $\mathscr{C}_a$ is acyclic for $a \neq 0$. Hence the inclusion $\mathscr{C}_0 \to \mathscr{C}$ induces a cohomology isomorphism.

It follows from this that each Lie algebra acts trivially on its cohomology, and at the same time, if $\eta \in H^q(\mathscr{C}_a)$, then $\alpha\eta = a\eta$.

Moscow State University. Translated from Funktsional'nyi Analiz i Ego Prilozheniya, Vol. 4, No. 4, pp. 70–71, October–December, 1970. Original article submitted July 6, 1970.

5. Theorem 1 can be strengthened by considering all elements of the form $\sum a_i x_i (\partial/\partial x_i)$, contained in $\mathfrak{g}$. Namely, the cohomology of the complex $\mathscr{C}$ reduces to the cohomology of the intersection of all complexes $\mathscr{C}_0$, corresponding to these elements.

6. An analog of Theorem 1 holds in the case of nontrivial coefficients. Namely, we have

THEOREM 2. Assume given a $\mathfrak{g}$-module $M = \bigoplus_{\alpha \in \tilde{A}} M_\alpha$, where $\tilde{A} \supset A$ is a subgroup of the additive group of the field K, and M_α is a linear space over K. Assume further that $\xi(M_\alpha) \subset M_{\alpha+\alpha'}$ for $\xi \in T_{\alpha'}$, $\alpha' \in A$, and that the operator $M_\alpha \to M_\alpha$, taking $x \in M_\alpha$ to $\alpha x - ax$, is nilpotent for $\alpha \in \tilde{A}$. For $\alpha \in \tilde{A}$, denote by $\mathscr{C}_\alpha$ the subcomplex of the standard complex $\mathscr{C} = \{C^q(\mathfrak{g}; M); d^q\}$, consisting of the cochains $\Lambda^\alpha(\mathfrak{g}) \to M$, assigning its homogeneous mappings degree α. Then the complex $\mathscr{C}_\alpha$ is acyclic for $\alpha \neq 0$, and hence the inclusion $\mathscr{C}_0 \to \mathscr{C}$ induces a cohomology isomorphism.

Remarks. 1°. The conditions of the theorem are satisfied if M is the identity representation or the adjoint representation of the algebra $\mathfrak{g}$.

2°. If the $\mathfrak{g}$-module M is finite-dimensional over K, then as is easily verified, the $(\mathfrak{g} \otimes_K K^*)$-module $M \otimes_K K^*$, where K^* is some finite algebraic extension of the field K, has a grading which satisfies the stated conditions.

7. THEOREM 3. If K = **R** or K = C and the algebra $\mathfrak{g}$ contains the element $\Sigma a_i x_i(\partial/\partial x_i)$, where $a_1, \ldots, a_n$ are positive real numbers, and if M is a finite-dimensional $\mathfrak{g}$-module then the complete cohomology space $H^*(\mathfrak{g}; M)$ is finite-dimensional. This is true if K is an arbitrary field of zero characteristic and $a_1, \ldots, a_n$ are natural numbers.

Remark. As well as $\mathfrak{g}$, the algebra $\mathfrak{g}_0 = \mathfrak{g} \cap L_0$ (where $L_0 = \left\{\sum P_i (\partial/\partial x_i) \mid P_1(0) = \ldots = P_n(0) = 0\right\}$) clearly also satisfies the conditions of the theorem.

8. COROLLARY 1. Of the six types of primitive infinite-dimensional filtered algebras over C (see [3]), four are: 1) the algebra of all formal vector fields in C^n; 2) the algebra of formal vector fields in C^n with a fixed differentiation; 3) the algebra of formal vector fields in C^n preserving Hamiltonian forms up to a constant factor; 4) the algebra of tangent vector fields consisting of algebras having finite-dimensional cohomology with coefficients in an arbitrary finite-dimensional module.

This follows from Theorem 3; in the first three cases we may choose all a_i equal to the identity, and in the last case we may take $a_1 = \ldots = a_{2n} = 1, a_{2n+1} = 2$.

Remark. Algebras of two other types (algebras without divergence and Hamiltonian formal vector fields), on the other hand, do not satisfy the conditions of Theorem 3. It is not known whether they have finite- or infinite-dimensional cohomology. The former seems most likely.

9. By the remark after Theorem 3, the complete cohomology space with coefficients in a finite-dimensional module of the intersection of the algebra L_0 with any of the algebras listed in Corollary 1 is finite-dimensional. As is well known, if $\mathfrak{h}$ is a subalgebra of an algebra $\mathfrak{g}$, then the cohomology of the algebra $\mathfrak{h}$ with coefficients in an arbitrary $\mathfrak{h}$-module M coincides with the cohomology of the algebra $\mathfrak{g}$ with coefficients in the induced $\mathfrak{g}$-module $\hat{M}$ (see [4], Proposition XIII, 4.3; see also [5], §3). Hence we have

COROLLARY 2. If $\mathfrak{g}$ is one of the algebras listed in Corollary 1, the cohomology of the algebra $\mathfrak{g}$ with coefficients in an arbitrary representation induced by the finite-dimensional representation of the algebra $\mathfrak{g} \cap L_0$ is finite-dimensional. It is related, for example, to the cohomology with coefficients in the space of formal tensor fields of an arbitrary form (in particular, with coefficients in the adjoint representation) with coefficients in the space of formal fields arising from an arbitrary form, etc.

LITERATURE CITED

1. I. M. Gel'fand and D. B. Fuks, Funkts. Analiz, 3, No. 3, 32-52 (1969).
2. I. M. Gel'fand and D. B. Fuks, Izv. Akad. Nauk SSSR, Ser. Mat., 34, 322-337 (1970).
3. V. Guillemin, D. Quillen, and S. Sternberg, Proc. Nat. Acad. Sci. USA, 55, 687-690 (1966).
4. A. Cartan and S. Eilenberg, Homological Algebra [Russian translation], IL, Moscow (1960).
5. I. M. Gel'fand and D. B. Fuks, Funkts. Analiz, 4, No. 3, 10-23 (1970).

10.
(with D. B. Fuks and D. A. Kazhdan)
The actions of infinite-dimensional Lie algebras

Funkts. Anal. Prilozh. 6 (1) (1972) 10–15 [Funct. Anal. Appl. 6 (1972) 9–13].
MR 46:922. Zbl. 267:18023

The actions of Lie algebras on smooth manifolds are the subject of a classical and far advanced theory. It is not surprising that the main concepts of this theory may be extended to the infinite-dimensional case; however, in the latter, the theory acquires in an unexpected way what appears to us to be a new and very interesting content. We discuss this infinite-dimensional theory here, and at the end we indicate the connection with problems of calculating the cohomologies of infinite-dimensional Lie algebras.

§1. Definition of the Action. Principal Homogeneous Spaces

1. First we give the main definitions in the most general form. Let A be a nuclear commutative associative **C**-algebra with a unit element, and let S denote the set of closed maximal ideals of A, that is, the set of continuous ring homomorphisms $A \to \mathbf{C}$. By a tangent vector to S at the point $s \in S$ we mean a continuous linear mapping d: $A \to \mathbf{C}$, such that $d(a_1a_2) = s(a_1)\,d(a_2) + s(a_2)\,d(a_1)$. The set $\tau_s S$ of these tangent vectors is, in an obvious sense, a linear topological space over **C**; it is called the tangent space to S at s. By a vector field on S we mean a derivation of A, that is, a continuous linear mapping D: $A \to A$, such that $D(a_1a_2) = a_1D(a_2) + a_2D(a_1)$. The set $\mathfrak{A}(S)$ of these vector fields is, in an obvious sense, a topological Lie algebra and a topological A-module. For every $s \in S$ the composition $D \mapsto s \circ D$ defines a homomorphism of the A-module $\mathfrak{A}(S)$ into the **C**-module $\tau_s S$, compatible with the homomorphism s: $A \to \mathbf{C}$. The image of a vector field under this homomorphism v(s): $\mathfrak{A}(S) \to \tau_s S$ is called the value of the field at the point $s \in S$.

Let $\mathfrak{g}$ be a nuclear Lie algebra. We say that $\mathfrak{g}$ acts on S, or that S is a $\mathfrak{g}$-space, if there is a continuous homomorphism of $\mathfrak{g}$ into the Lie algebra $\mathfrak{A}(S)$. If the action φ: $\mathfrak{g} \mapsto \mathfrak{A}(S)$ has the property that the mapping of the A-module $A \otimes_{\mathbf{C}} \mathfrak{g}$ into the A-module $\mathfrak{A}(S)$ defined by $a \otimes g \mapsto a\varphi(g)$ is an isomorphism, then S is called a principal homogeneous $\mathfrak{g}$-space.

2. Next we consider the significance of these definitions in a geometrical situation. The simplest geometrical situation is the case when S is a smooth manifold (finite- or infinite-dimensional) and A is the algebra of complex-valued smooth functions on S. In this case the tangent vectors, the vector fields, and the action of the Lie algebra obviously have the usual meaning, and we need interpret only the concept of a principal homogeneous $\mathfrak{g}$-space.

PROPOSITION 1.1. Let φ: $\mathfrak{g} \to \mathfrak{A}(S)$ denote the action of the Lie algebra $\mathfrak{g}$ on the smooth manifold S. For S to be a principal homogeneous space relative to this action, it is necessary and sufficient that, for each $s \in S$ the through homomorphism

$$\mathfrak{g} \xrightarrow{\varphi} \mathfrak{A}(S) \xrightarrow{v(s)} \tau_s S$$

is an isomorphism.

We do not use the necessity of the condition in what follows; therefore we leave the proof of it to the reader; we prove the sufficiency of the condition.

Moscow State University. Translated from Fuktsional'nyi Analiz i Ego Prilozheniya, Vol. 6, No. 1, pp. 10–15, January–March, 1972. Original article submitted October 1, 1971.

We must prove that the formula $a \otimes g \mapsto a\varphi(g)$ determines an isomorphism $A \otimes_C \mathfrak{g} \to \mathfrak{A}(S)$. It is obvious that this is a monomorphism, since if $\sum a_i \otimes g_i \mapsto 0$, then for every $s \in S$, under the homomorphism $v(s) \circ \varphi$, the element $\sum s(a_i) g_i \in \mathfrak{g}$ goes into the zero element, and this contradicts our assumption. To prove that the mapping is an epimorphism it is sufficient to note that $A \otimes \mathfrak{g}$ is the space of smooth functions on S with values in $\mathfrak{g}$. The element of $A \otimes \mathfrak{g}$ that goes into a given element $\alpha \in \mathfrak{A}(S)$ is the function $s \mapsto (\varphi \circ v(s))^{-1} v(s) \xi$.

Comment. Proposition 1.1 reveals the extreme poorness of the contact of a principal homogeneous space in the finite-dimensional case. For, if $\mathfrak{g}$ is the Lie algebra of a finite-dimensional simply connected Lie group G, then the connected components of the principal homogeneous $\mathfrak{g}$ -space S must be factor spaces of G with respect to discrete subgroups. On the contrary, in the infinite-dimensional case this concept is meaningful, as is confirmed by the following example.

3. A basic example. Let M be a smooth n-dimensional manifold. Let S(M) denote the space of ∞ -jets of the mappings $\mathbf{R}^n \to M$ which are regular at the point $0 \in \mathbf{R}^n$; more precisely, S(M) is the inverse limit of the sequence

$$M = S_0(M) \leftarrow S_1(M) \leftarrow S_2(M) \leftarrow \dots,$$

where $S_k(M)$ denotes the space of k-jets of the mappings $\mathbf{R}^n \to M$ which are regular at $0 \in \mathbf{R}^n$, and the arrows denote the obvious submersion. It is clear that S(M) is an infinite-dimensional smooth manifold, and that the algebra A of smooth complex-valued functions on S(M) is the direct limit of the sequence of algebras A_k of smooth functions on $S_k(M)$.

We make the Lie algebra W_n of formal vector fields at $0 \in \mathbf{R}^n$ act on S(M); the elements of W_n are interpreted as ∞ -jets of one-parameter families of diffeomorphisms $\mathbf{R}^n \to \mathbf{R}^n$, that is, tangent vectors to $S(\mathbf{R}^n)$ at the point corresponding to the identity mapping $\mathbf{R}^n \to \mathbf{R}^n$. Moreover, this interpretation enables us to determine an isomorphism of the space W_n onto this tangent space. Since a point of S(M) is an ∞ -jet of the mapping $\mathbf{R}^n \to M$, it also determines a mapping $S(\mathbf{R}^n) \to S(M)$ which induces an isomorphism of the tangent spaces. Thus, all tangent spaces of the manifold S(M) may be canonically identified with the tangent space to $S(\mathbf{R}^n)$, that is, with W_n. This also determines a homomorphism $W_n \to \mathfrak{A}(S(M))$ (its permutability with commutation may be verified trivially), that is, the action of W_n on S(M).

By Proposition 1.1, S(M) is a principal homogeneous W_n-space.

4. We return to the general theory developed in Sec. 1. Let $A, S, \mathfrak{A}(S)$ have the same meaning as in Sec. 1, and let B be a nuclear space. By an exterior differential form of degree q on S with values in B we mean a continuous A-linear mapping $\Lambda^q_A \mathfrak{A}(S) \to A \otimes_C B$ (here $\Lambda^q_A \mathfrak{A}(S)$ is the skew-symmetrized q-th power of the A-module $\mathfrak{A}(S)$ relative to tensor multiplication over A). Let $\Omega^q = \Omega^q(S, B)$ denote the space of these forms and let $\Omega^* = \bigoplus_q \Omega^q$. Exterior differentiation d: $\Omega^q \to \Omega^{q+1}$ is defined by the usual formula:

$$(d\alpha)(D_1, \dots, D_{q+1}) = \sum_{1 \leqslant s < t \leqslant q+1} (-1)^{s+t-1} \alpha([D_s, D_t], D_1, \dots \hat{D}_s \dots \hat{D}_t \dots, D_{q+1}) + \sum_{1 \leqslant s \leqslant q+1} (-1)^s D_s \alpha(D_1, \dots \hat{D}_s \dots, D_{q+1}).$$

It is clear from the A-linearity of the differential form $\Lambda^q_A \mathfrak{A}(S) \to A \otimes_C B$, that, for every $s \in S$ it induces a linear (over C) mapping $\Lambda^q_C \tau_s S \to B$ (this accounts for the name).

The complex $\{\Omega^*(S, B), d\}$ is called the de Rham complex of the manifold S with coefficients in B.

We assume now that S is a principal homogeneous $\mathfrak{g}$ -space. Then the mapping $\omega : \mathfrak{A}(S) \to A \otimes \mathfrak{g}$, inverse to $\mathrm{id} \otimes \varphi$ (see Sec. 1) is a 1-form on S with values in $\mathfrak{g}$.

PROPOSITION 1.2. (The Maurer-Cartan formula).

$$d\omega = -\frac{1}{2}[\omega, \omega],$$

that is, $d\,\omega(D_1, D_2) = -\frac{1}{2}[\omega(D_1), \omega(D_2)]$ for any $D_1, D_2 \in \mathfrak{A}(S)$. (The commutator on the right-hand side of the last equality is taken in $A \otimes \mathfrak{g}$; it is defined by the formula $[a_1 \otimes g_1, a_2 \otimes g_2] = a_1 a_2 \otimes [g_1, g_2]$.)

To prove the proposition it is sufficient to verify the equality $\omega([D_1, D_2]) - D_1\omega(D_2) + D_2\omega(D_1) = [\omega(D_1), \omega(D_2)]$ for elements D_i of the form $a_i\varphi(g_i)$, where $a_i \in A$, and $g_i \in \mathfrak{g}$. By taking into account that $\omega = (\mathrm{id} \otimes \varphi)^{-1}$ and that $[a_1D_1, a_2D_2] = D_1(a_2)D_2 - D_2(a_1)D_1 + a_1a_2[D_1, D_2]$ for any $D_1, D_2 \in \mathfrak{A}(S)$, we obtain that $\omega([D_1, D_2]) = a_1a_2 \otimes [g_1, g_2] + D_1(a_2) \otimes g_2 - D_2(a_1) \otimes g_1$, whence follows the desired equality.

§ 2. Geometric $\mathfrak{g}$-Complexes

1. The concept of a geometric $\mathfrak{g}$-complex, defined below, is not essentially new (see, for example, [6]). We are interested in the interaction of this concept with those introduced in Sec. 1.

Let $\mathfrak{g}$ be a Lie algebra and let $\{K = \bigoplus_q K^q, d\}$ be a complex for which the differential d has degree +1. We say that the geometric action of $\mathfrak{g}$ in K is defined (or K is a geometric $\mathfrak{g}$-complex, or simply, a $\mathfrak{g}$-complex) if, for each $g \in \mathfrak{g}$ there is a linear mapping $\bar{g}: K \to K$ of degree -1 such that the following equalities hold for any $g_1, g_2 \in \mathfrak{g}$

1°. $\overline{(\alpha g_1 + \beta g_2)} = \alpha\bar{g}_1 + \beta\bar{g}_2 \qquad (\alpha, \beta \in \mathbf{C}),$

2°. $\bar{g}_1\bar{g}_2 = -\bar{g}_2\bar{g}_1,$

3°. $\overline{[g_1, g_2]} + \bar{g}_1 d\bar{g}_2 - \bar{g}_2 d\bar{g}_1 + d\bar{g}_1\bar{g}_2 - \bar{g}_2\bar{g}_1 d = 0.$

When we put $a(g) = \bar{g}d + d\bar{g}$, we obtain a representation of $\mathfrak{g}$ in the endomorphisms of the complex K. It is easy to appreciate, however, that not every such representation is obtained in this way from a geometric action (for example, the operators $a(g)$ arising from the geometric action induce trivial homological endomorphisms).

If K is a geometric $\mathfrak{g}$-complex, we denote by $K^{\mathfrak{g}}$ the subcomplex of K consisting of those $k \in K$ such that $\bar{g}k = a(g)k = 0$ (or, $\bar{g}k = \bar{g}\,dk = 0$) for all $g \in \mathfrak{g}$.

2. The fundamental examples. A. The de Rham complex, with arbitrary coefficients, of the $\mathfrak{g}$-space S (see Sec. 1.1) has the natural structure of a geometric $\mathfrak{g}$-complex. The geometric action is defined by the formula $(g\alpha)(D_1, \ldots, D_{q-1}) = \alpha(\varphi(g), D_1, \ldots, D_{q-1})$, where $g \in \mathfrak{g}; \alpha \in \Omega^{q-1}$ (S; the space of coefficients), $D_1, \ldots, D_{q-1} \in \mathfrak{A}(S)$ and φ denotes the action $\mathfrak{g} \to \mathfrak{A}(S)$ of $\mathfrak{g}$ in S.

B. Let $\mathfrak{g}$ be a subalgebra of the Lie algebra $\mathfrak{h}$. The standard complex of cochains of $\mathfrak{h}$ (with arbitrary coefficients) has the natural structure of a geometric $\mathfrak{g}$-complex (here it would be appropriate to recall the definition of a standard complex, but we delay this slightly since in Sec. 3 we shall require more extensive information about cochains and cohomlogies of Lie algebras; if necessary, the reader should now consult the beginning of Sec. 3). The geometric action is given by the formula $(gL)(h_1, \ldots, h_{q-1}) = L(g, h_1, \ldots, h_{q-1})$, where $g \in \mathfrak{g}$, $L \in C^{q-1}$ ($\mathfrak{g}$; the module of coefficients), and $h_1, \ldots, h_{q-1} \in \mathfrak{h}$.

3. We clarify Example A in the finite-dimensional case. Let S be a finite-dimensional manifold, and suppose that a finite-dimensional Lie group G, whose Lie algebra is $\mathfrak{g}$, acts on S smoothly and without fixed points. In this case it is obvious that the operator $a(g)$: $\Omega^q(S) \to \Omega^{q-1}(S)$ (coefficients in $\mathbf{C}$), determined by an element $g \in \mathfrak{g}$ coincides with the orginary Lie differentiation along the vector field corresponding to g under the action $\mathfrak{g} \to \mathfrak{A}(S)$.

PROPOSITION 2.1. The subcomplex $(\Omega^*(S))^{\mathfrak{g}}$ of the complex $\Omega^*(S)$ coincides with the image of the monomorphism p^*: $\Omega^*(S/G) \to \Omega^*(S)$, where S/G denotes the factor manifold, and p is a projection.

Proof. If the form $\alpha \in \Omega^*(S)$ belongs to $(\Omega^*(S))^{\mathfrak{g}}$ it means that $a(g)\alpha = 0$ and that $\bar{g}\alpha = 0$ for $g \in \mathfrak{g}$; the first condition is equivalent to the condition that α is a G-invariant form, while the second condition is equivalent to the condition that α is annihilated by the substitution of a vector tangential to the fiber. Together these conditions mean that α is a form lifted from S/G.

Remark. A similar result holds for infinite-dimensional manifolds if the Lie algebra $\mathfrak{g}$ corresponds to a Lie group G, and if S and S/G, are smooth manifolds.

4. If $K_1 = \bigoplus_q K_1^q$, $K_2 = \bigoplus_q K_2^q$ are two $\mathfrak{g}$-complexes, we may canonically form the $\mathfrak{g}$-complex $\mathrm{Hom}(K_1, K_2) = \bigoplus_{-\infty<q<\infty} \mathrm{Hom}^q(K_1, K_2)$, where $\mathrm{Hom}^q(K_1, K_2) = \bigoplus_r \mathrm{Hom}(K_1^r, K_2^{r+q})$, the boundary operator is given by the formula $(d\rho)(k) = d(\rho(k)) - \rho(dk)$, and the geometric action of the algebra $\mathfrak{g}$ is defined by the formula $(\bar{g}\rho)(k) = \bar{g}(\rho(k)) - \rho(\bar{g}k)$.

§3. Application: The Homologies of the Diagonal Complex

1. In Sec. 3 we shall be concerned with things connected with cohomologies of infinite-dimensional Lie algebras. We recall certain relevant definitions and results (for the details, see [1-3]).

By the standard complex of cochains of the topological Lie algebra $\mathfrak{g}$ (over C) with coefficients in the $\mathfrak{g}$-module B we mean the complex $\{C^q, d^q\}$, where C^q denotes the space of continuous skew-symmetric q-linear functionals over C^q with values in B, and the differential $d^q : C^q \to C^{q+1}$ is defined by the formula

$$(d^qL)(g_1, \ldots, g_{q+1}) = \sum_{1 \le s < t \le q+1} (-1)^{s+t-1} L([g_s, g_t], g_1, \ldots \hat{g}_s \ldots \hat{g}_t \ldots, g_{q+1}) + \sum_{1 \le s \le q+1} (-1)^s g_s L(g_1, \ldots \hat{g}_s \ldots, g_{q+1}).$$

The homologies of this standard complex are called the cohomologies of the algebra $\mathfrak{g}$ with coefficients in B, and we use the notation $H^*(\mathfrak{g}; B) = \bigoplus_q H^q(\mathfrak{g}; B)$.

When B is a ring, in particular, when B = C, the standard complex is multiplicative, and $H^*(\mathfrak{g}; B)$ is a graded ring.

The cohomologies of W_n with coefficients in C were found in [3]† namely, the authors constructed the isomorphism $H^*(W_n; C) \to H^*(X_n; C)$, where X_n is the inverse-image of the 2n-skeleton of the base BU(n) of the universal U(n)-fibering in the total space of this fibering (it is assumed that the space BU(n) is split into cells of even dimensions).

It was shown in [1] that, for a compact orientable manifold M, the cohomologies of the Lie algebra of the smooth vector fields on M with coefficients in C are finite-dimensional in each dimension.‡ In a certain sense the calculation of these cohomologies is reduced to the calculation of the homologies of the "diagonal complex," that is, the subcomplex of $\{C^q(\mathfrak{A}(M); C); d^q\}$ defined in the following way. The chain $L \in C^q(\mathfrak{A}(M); C)$ is said to be diagonal if, for any vector fields $D_1, \ldots, D_q \in \mathfrak{A}(M)$ whose supports are nonintersecting in aggregate, the value $L(D_1, \ldots, D_q)$ is zero. The diagonal chains form a subcomplex of the diagonal complex; we call this subcomplex diagonal also and denote it by $C_1^{\bullet}$.

To calculate the homologies of the diagonal complex the authors of [1] constructed a spectral sequence, the second term of which is calculated from the formula $E_2^{s,t} = H_{-s}(M) \otimes H^t(W_n; C)$.

2. The subsequent reasoning is based on a study of the geometric W_n-complex $K = \mathrm{Hom}(C^*(W_n; C), \Omega^*(S(M)))$ (see Secs. 2.2 and 2.4).

PROPOSITION 3.1. The canonical isomorphism $[(K^{-q})^{W_n}]' = C_1^q(\mathfrak{A}(M))$ holds for all q (the prime denotes conjugation). This isomorphism permutes with the boundary operators d': $[(K^{-q})^{W_n}]' \to [(K^{-q-1})^{W_n}]'$ and $d: C_1^q(\mathfrak{A}(M)) \to C_1^{q+1}(\mathfrak{A}(M))$.

Proof. We confine ourselves to the construction of the mapping $[K^{-q})^{W_n}]' \to C_1^q(\mathfrak{A}(M))$. By using the fact that S(M) is a principal homogeneous W_n-space, in particular, by using the Maurer-Cartan formula, one may verify automatically that this mapping is an isomorphism and that it permutes with the differentials.

We assume that there is a functional on the space $(K^{-q})^{W_n} = \bigoplus_r \mathrm{Hom}(C^r(W_n; C), \Omega^{r-q}(S(M)))$ and, we construct, in terms of it, a chain of the diagonal complex. We take q vector fields $D_1, \ldots, D_q \in \mathfrak{A}(M)$. For each r they determine a homomorphism $C^r(W_n; C) \to \Omega^{r-q}(S(M))$, since to the chain $L \in C^r(W_n; C)$ corresponds the form which relates to the vectors $a_1, \ldots, a_{r-q} \in \tau_s\mathfrak{A}(M)$ tangential to $\mathfrak{A}(M)$ at some point, the number $L(D_1, \ldots, D_q, a_1, \ldots, a_{r-q})$ (we may regard the fields $D_1, \ldots, D_q$ as elements of W_n, because the fixing of a point of S(M) corresponds to the fixing of a local coordinate system of M; we may regard the vectors $a_1, \ldots, a_{r-q}$ as elements of W_n, since $\tau_s S(M)$ is identified with W_n). It is obvious that the element of $\mathrm{Hom}(C^r(W_n; C), \Omega^{r-q}(S(M)))$ we have obtained is contained in K^{-q}, so that our functional has a certain value

† Gel'fand and Fuks [3] found the cohomologies of the real algebra W_n with real coefficients; however, the result is not affected by replacing R by C.
‡ An attentive reader of [1] will observe that this result holds without the assumptions of compactness and orientability.

at this element. We take it to be $L(D_1, \ldots, D_q)$. It is obvious that the chain L_1 lies in the diagonal complex. The required homomorphism is given by the juxtaposition $L \mapsto L_1$.

3. The complex K is a geometric L_0-complex, where L_0 is the algebra of formal vector fields vanishing at $0 \in R^n$. We study the complex K^{L_0}.

PROPOSITION 3.2. The inclusion $K^{W_n} \to K^{L_0}$ induces a homological isomorphism.

Proof. This is obvious when $M = R^n$, since the homologies of both complexes are isomorphic (with a change of enumeration) to the cohomologies of the algebra W_n. To go from here to the general case it is sufficient to fix a triangulation of M and use it to define a filtration in the complex K (it is defined with the help of the usual filtration in the de Rham complex of S(M), fibered over M), and hence in the complexes K^{L_0} and K^{W_n}. This filtration determines spectral sequences, which are convergent to homologies of the complexes K^{L_0} and K^{W_n}, and the established result for the case $M = R^n$ means that, for these spectral sequences the second terms are mapped isomorphically under the homomorphism induced by the inclusion $K^{W_n} \to K^{L_0}$.

4. By using Propositions 3.1 and 3.2 the problem of calculating the homologies of the diagonal complex is reduced to the problem of calculating the homologies of K^{L_0}. It is clear from here that the homologies of the diagonal complex depend only on topological properties of the tangent, fibering of the manifold M. For, let $S^0(M)$ denote the space of ∞-jets of the regular mappings of R^n into the tangent spaces of M that map 0 into 0. It is clear that $S^0(M)$ is the total space of the fibering over M induced by the tangent space; it is also clear that $S^0(M)$ is an L_0-space (but not a W_n-space), and as an L_0-space it is defined by a tangent fibering. At the same time, by using any Riemannian metric on M we may construct an L_0-diffeomorphism $S^0(M) \to S(M)$, so that $\mathrm{Hom}(C^r(W_n; C), \Lambda^{r-q}(S(M)))^{L_0} = \mathrm{Hom}(C^r(W_n; C), \Omega^{r-q}(S^0(M)))^{L_0}$.

We may also use Propositions 3.1 and 3.2 when calculating the cohomologies of the diagonal complex. For example, the following theorem holds.

THEOREM. If the manifold M is parallelizable, then

$$H_1^q(\mathfrak{A}(M)) = \bigoplus_{t-s=q} H_s(M) \otimes H_c^t(W_n; C).$$

Indeed, in this case the fibering $S(M) \to M$ has a cross section. Hence, there is a mapping $\Omega^*(S(M)) \to \Omega^*(M)$, which induces a mapping $K \to \mathrm{Hom}(C^*(W_n; C), \Omega^*(M))$ under which the complex K^{L_0} is mapped isomorphically onto $\mathrm{Hom}(C^*(W_n; C), \Omega^*(M))$. It remains to apply Propositions 3.1 and 3.2, and the Kunneth formula.

CONCLUSION

The short paper [4] preceded the present article. M. V. Losik [5] recently obtained new results about the homologies of the diagonal complex; the results in Sec. 3.4 follow from his work.

LITERATURE CITED

1. I. M. Gel'fand and D. B. Fuks, "Cohomologies of the Lie algebra of tangent vector fields of a smooth manifold," Funktsional. Analiz i Ego Prilozhen., 3, No. 3, 32-52 (1969).
2. I. M. Gel'fand and D. B. Fuks, "Cohomologies of the Lie algebra of tangent vector fields of a smooth manifold. II," Funktsional. Analiz. i Ego Prilozhen., 4, No. 2, 23-32 (1970).
3. I. M. Gel'fand and D. B. Fuks, "Cohomologies of the Lie algebra of formal vector fields," Izv. Akad. Nauk SSSR. Ser. Matem., 34, 322-337 (1970).
4. I. M. Gel'fand and D. A. Kazhdan, "Some questions on differential geometry and the calculation of the cohomologies of the Lie algebras of vector fields," Dokl. Akad. Nauk SSSR, 200, No. 2, 269-272 (1971).
5. M. V. Losik, "On the cohomologies of the Lie algebra of vector fields with coefficients in the trivial identity representation," Funksional. Analiz i Ego Prilozhen., 6, No. 1, 24-36 (1972).
6. H. Cartan, Notions d'algèbre differentielle; application aux groups de Lie et aux varietés oû opére un group de Lie, Coll. de Toplogie, Bruxelles, 1950, Georges Thone, Liege, Masson et Cie., Paris (1951), 15-27.

11.

(with D. B. Fuks and D. I. Kalinin)

Cohomology of the Lie algebra of Hamiltonian formal vector fields

Funkts. Anal. Prilozh. **6** (63) (1972) 25–29 [Funct. Anal. Appl. **6** (1972) 193–196].
MR **47**:1088. Zbl. **259**:57023

As noted in [2], computing the cohomology of the Lie algebra of Hamiltonian formal vector fields is an essentially more difficult problem than, say, computing the cohomology of the Lie algebra of all formed vector fields, which was done in [1]. The methods used in [1] allow one to find the homology of a certain direct summand of the cochain complex of the algebra of Hamiltonian fields without great difficulty, but they yield no information about the complementary summand. In order to test the hypothesis that this complementary summand is acyclic, we have made some computations on an electronic computer. As a result, the above hypothesis has been rejected: we have discovered new and nontrivial cohomology classes of the algebra of Hamiltonian formal vector fields in R^2. The important difference between these classes and the cohomology classes of the algebra of all formal vector fields found in [1] is that the former cannot be represented by cocycles depending only on the 2-sets of their arguments (see [3]).

This partial result seems interesting to us for two reasons. The first is methodological: it turns out that the difficulties encountered in computing the cohomology of the algebra of Hamiltonian fields are fundamental in origin. The second is as follows. The construction of Godbillon, Vey, Bernshtein, and Rozenfel'd (see [4, 5]) can be carried over to the Hamiltonian case, making it possible to construct, for each cohomology class of the algebra of Hamiltonian formal vector fields, a characteristic class of Hamiltonian fibers of corresponding codimension. In particular, the classes which we will point out here furnish characteristic classes of Hamiltonian fibers of codimension 2. It would be interesting to determine whether or not these characteristic classes are nontrivial.

1. GENERAL DEFINITIONS

1. A formal vector field $\xi = \sum_{i=1}^{n} P_i(x_1, \ldots, x_n, y_1, \ldots, y_n) \frac{\partial}{\partial x_i} + \sum_{i=1}^{n} Q_i(x_1, \ldots, x_n, y_1, \ldots, y_n) \frac{\partial}{\partial y_i}$ in R^{2n} is called Hamiltonian if there exists a formal power series $H(x_1, \ldots, x_n, y_1, \ldots, y_n)$ with $\frac{\partial H}{\partial x_i} = Q_i$, $\frac{\partial H}{\partial y_i} = -P_i$ $(i = 1, \ldots, n)$. The series H determines the field up to a constant summand; it is called the Hamiltonian of the field ξ and is denoted by $H(\xi)$. The set V_n of all Hamiltonian formal vector fields is closed with respect to Poisson brackets and therefore inherits the structure of a topological Lie algebra from the algebra W_{2n} of all formal vector fields in R^{2n}. The cohomologies $H^*(V_n; 1) = H^*(V_n; 1) = \bigoplus_q H^q(V_n; 1) = \bigoplus_q H^q_n$ of the algebra V_n (with constant coefficients) are defined to be the homologies of the complex $\mathscr{C}_n = \{C^q_n, d^q_n\}$, where C^q_n is the space of continuous skew-symmetric real q-linear functionals on V_n, and d^q_n acts as follows:

$$(d^q_n L)(\xi_1, \ldots, \xi_{q+1}) = \sum_{1 \le s < t \le q+1} (-1)^{s+t-1} L([\xi_s, \xi_t], \xi_1, \ldots \hat{\xi}_s \ldots \hat{\xi}_t \ldots, \xi_{q+1}).$$

Multiplication is defined as usual in $\mathscr{C}_n$, turning C_n and $H^*(V_n; 1)$ into graded rings.

2. We identify the elements of the space C^q_n with real polynomials $p(\alpha_1, \ldots, \alpha_q, \beta_1, \ldots, \beta_q)$ of vector arguments $\alpha_i = (\alpha_{i1}, \ldots, \alpha_{in})$, and $\beta_i = (\beta_{i1}, \ldots, \beta_{in})$ such that $p(\alpha_1, \ldots, \alpha_{i-1}, 0, \alpha_{i+1}, \ldots, \alpha_n, \beta_1, \ldots, \beta_{i-1}, 0,$

Moscow State University. Translated from Funktsional'nyi Analiz i Ego Prilozheniya, Vol. 6, No. 3, pp. 25–29, July–September, 1972. Original article submitted April 4, 1972.

$\beta_{i+1}, \ldots, \beta_n) = 0$ for every i and $p(\alpha_{i_1}, \ldots, \alpha_{i_n}, \beta_{i_1}, \ldots, \beta_{i_n}) = \operatorname{sgn}(i_1, \ldots, i_n)\; p(\alpha_1, \ldots, \alpha_n, \beta_1, \ldots, \beta_n)$ for any permutation $(i_1, \ldots, i_n)$. Specifically, the polynomial

$$\sum A(a_{11}, \ldots, a_{qn}, b_{11}, \ldots, b_{qn}) \prod_{i=1}^{q} \prod_{j=1}^{n} \alpha_{ij}^{a_{ij}} \beta_{ij}^{b_{ij}} \tag{1}$$

is identified with the functional

$$(\xi_1, \ldots, \xi_q) \mapsto \sum A(a_{11}, \ldots, a_{qn}, b_{11}, \ldots, b_{qn}) \prod_{i=1}^{q} \frac{\partial^{a_{i1}+\ldots+a_{in}+b_{i1}+\ldots+b_{in}} H(\xi_i)}{\partial x_1^{a_{i1}} \ldots \partial x_n^{a_{in}} \partial y_1^{b_{i1}} \ldots \partial y_n^{b_{in}}}.$$

In these terms the differential d^q_n can be expressed by the formula

$$(d^q_n p)(\alpha_1, \ldots, \alpha_{q+1}, \beta_1, \ldots, \beta_{q+1}) = \sum_{1 \leq s < t \leq q+1} (-1)^{s+t-1} [(\alpha_s, \beta_t) - (\alpha_t, \beta_s)]$$
$$\times\, p(\alpha_s + \alpha_t, \alpha_1, \ldots \hat{\alpha}_s \ldots \hat{\alpha}_t \ldots, \alpha_{q+1}, \beta_s + \beta_t, \beta_1, \ldots \hat{\beta}_s \ldots \hat{\beta}_t \ldots, \beta_{q+1})$$

(here $(\alpha_i, \beta_j) = \alpha_{i1}\beta_{j1} + \ldots + \alpha_{in}\beta_{jn}$).

3. For any integers N_1 and N_2, let C^{q,N_1,N_2}_n denote the subspace of C^q_n consisting of polynomials (1) with the following properties: if $A(a_{11}, \ldots, a_{qn}, b_{11}, \ldots, b_{qn}) \neq 0$, then $a_{11} + a_{qn} = N_1 + q$ and $b_{11} + \ldots + b_{qn} = N_2 + q$. It is clear from (2) that $d^q_n(C^{q,N_1,N_2}_n) \subset C^{q+1,N_1,N_2}_n$. Therefore the spaces C^{q,N_1,N_2}_n $(q = 0, 1, \ldots)$ form a subcomplex of $\mathscr{C}_n$ for any N_1, N_2, and n. We will denote this subcomplex by $\mathscr{C}^{N_1,N_2}_n$, and the q-th space of its homologies by H^{q,N_1,N_2}_n. It is clear that $\mathscr{C}_n = \bigoplus_{N_1,N_2} \mathscr{C}^{N_1,N_2}_n$ and $\dim \mathscr{C}^{N_1,N_2}_n < \infty$ for any N_1 and N_2, and therefore $H^q_n = \bigoplus_{N_1,N_2} H^{q,N_1,N_2}_n$. It is also clear that the product of an element in $\mathscr{C}^{q,N_1,N_2}_n$ by an element in $C^{q',N_1',N_2'}_n$ lies in $C^{q+q',N_1+N_1',N_2+N_2'}_n$, and that the product of an element in H^{q,N_1,N_2}_n by one in $H^{q',N_1',N_2'}_n$ lies in $H^{q+q',N_1+N_1',N_2+N_2'}_n$.

4. V_n contains a finite-dimensional subalgebra which is usually identified with the canonically isomorphic algebra sp(n); this subalgebra consists of fields $\sum P_i \frac{\partial}{\partial x_i} + \sum Q_i \frac{\partial}{\partial y_i}$, where P_i and Q_i are homogeneous linear functions. Let $\widetilde{\mathscr{C}}_n = \{\widetilde{C}^q_n, d^q_n\}$ be the subcomplex of $\mathscr{C}_n$, consisting of those $L \in C^q_n$ such that $L(\xi_1, \ldots, \xi_{q-1}, \xi) = 0$ for any $\xi_1, \ldots, \xi_{q-1} \in V_n$, $\xi \in \mathrm{sp}(n)$ and $dL(\xi_1, \ldots, \xi_q, \xi) = 0$ for any $\xi_1, \ldots, \xi_q \in V_n$ and $\xi \in \mathrm{sp}(n)$. The homologies of the complex $\widetilde{\mathscr{C}}_n$ are called the relative cohomologies of V_n modulo sp(n), and are denoted by $H^*(V_n, \mathrm{sp}(n); 1)$. We will use the abbreviated notation $\widetilde{H}^*_n = H^*(V_n, \mathrm{sp}(n); 1)$.

In terms of the polynomials, the complex $\widetilde{\mathscr{C}}_n$ can be described as follows. A polynomial $p(\alpha_1, \ldots, \alpha_q, \beta_1, \ldots, \beta_q)$ belongs to $\widetilde{\mathscr{C}}_n$ if and only if 1) for any i, the total degree of each monomial in $\alpha_{i1}, \ldots, \alpha_{in}, \beta_{i1}, \ldots, \beta_{in}$ in p is not equal to two, and 2) the same holds for the polynomial dp. Obviously, $\widetilde{C}^q_n = \bigoplus_{N_1,N_2} \widetilde{C}^{q,N_1,N_2}_n$, where

$$\widetilde{C}^{q,N_1,N_2}_n = \widetilde{C}^q_n \cap C^{q,N_1,N_2}_n, \quad \text{and} \quad \widetilde{H}^q_n = \bigoplus_{N_1,N_2} \widetilde{H}^{q,N_1,N_2}_n,$$

where $\widetilde{H}^{q,N_1,N_2}_n$ is the q-th space of the cohomologies of the complex $\widetilde{\mathscr{C}}^{N_1,N_2}_n = \widetilde{\mathscr{C}}_n \cap \mathscr{C}^{N_1,N_2}_n$.

5. With the pair $(V_n, \mathrm{sp}(n))$ we can associate a Serre-Hochschild spectral sequence in the usual way. Since the algebra sp(n) is simple, the second term of this spectral sequence is $\widetilde{H}^*_n \otimes H^*(\mathrm{sp}(n); 1)$, and the limit term is associated with $H^*(V_n; 1)$. The decompositions of the complexes $\mathscr{C}_n$ and $\widetilde{\mathscr{C}}_n$ into direct sums induce a decomposition of the Serre-Hochschild spectral sequence into a direct sum, giving rise to the spectral sequences

$$\widetilde{H}^{\bullet,N_1,N_2}_n \otimes H^*(\mathrm{sp}(n); 1) \Rightarrow H^{\bullet,N_1,N_2}_n.$$

2. A PARTIAL COMPUTATION

1. If $N_1 \neq N_2$, then $C^{q,N_1,N_2}_n = 0$ and $H^{q,N_1,N_2}_n = 0$ for any q and n.†

<u>Proof.</u> Define a homomorphism $D_q : C^{q+1,N_1,N_2}_n \to C^{q,N_1,N_2}_n$ by the formula

†This proposition, like the decomposition $\mathscr{C}_n = \oplus\, \mathscr{C}^{N_1,N_2}_n$ itself, is essentially contained in [2].

$$p \mapsto \left[\sum_{i=1}^{n} \frac{\partial^2 p}{\partial \alpha_{q+1,\,i}\, \partial \beta_{q+1,\,i}}\right]_{\substack{\alpha_{q+1}=0,\\ \beta_{q+1}=0}}.$$

It is obvious that $D_q(\widetilde{C}_n^{q+1,N_1,N_2}) = 0$. Applying D_q to the left- and right-hand sides of (2), we get

$$D_q(dp) = (-1)^q (N_2 - N_1)p + d(D_{q-1}(p)).$$

Therefore, 1) if dp = 0, then p = $\pm d[D_{q-1}(p)/(N_2-N_1)]$, and thus $H_n^{q,N_1,N_2} = 0$; 2) if $p \in C_n^{N_1,N_2}$, then $D_{q-1}(p) = 0$ and $D_q(dp) = 0$, so p = 0.

COROLLARY. $H_n^{\bullet} = \bigoplus_N H_n^{\bullet,N,N}$; $\widetilde{H}_N^{\bullet} = \bigoplus_N \widetilde{H}_n^{\bullet,N,N}$.

Below we will write $H_n^{\bullet,N}$, $\widetilde{H}_n^{\bullet,N}$, and $\mathscr{C}_n^N$, $\widetilde{\mathscr{C}}_n^N$ instead of $H_n^{\bullet,N,N}$, $\widetilde{H}_n^{\bullet,N,N}$, and $\mathscr{C}_n^{N,N}$, $\widetilde{\mathscr{C}}_n^{N,N}$.

2. The ring $\bigoplus_{N\le 0} \widetilde{H}_n^{\bullet,N}$ is generated (multiplicatively) by the cohomology classes $\Gamma \in \widetilde{H}_n^{2n-1}$ and $\Psi_i \in \widetilde{H}_n^{4i,0}$ $(i = 0, \ldots, n)$, represented by the polynomials

$$\gamma(\alpha_1, \alpha_2, \beta_1, \beta_2) = \gamma_{12},$$

$$\psi_i(\alpha_1, \ldots, \alpha_{4i}, \beta_1, \ldots, \beta_{4i}) = \sum \operatorname{sgn}(j_1, \ldots, j_{4i})(\gamma_{j_1 j_2}\gamma_{j_2 j_3} \cdots \gamma_{j_{2i-1} j_2}\gamma_{j_{2i} j_1})(\gamma_{j_1 j_{2i+1}} \cdots \gamma_{j_{2i} j_{4i}}),$$

where $\gamma_{st} = (\alpha_s, \beta_t) - (\alpha_t, \beta_s)$ and the summation in the second equality is over all permutations $(j_1, \ldots, j_{4i})$ of the numbers 1, ..., 4i. The module of relations among the classes Γ and ψ_i is generated by the relations

$$\Gamma^k \Psi_1^{k_1} \ldots \Psi_n^{k_n} = 0,$$

where $k + k_1 + 2k_2 + \ldots + nk_n > n$.

In the Serre-Hochschild spectral sequence,

$$\left[\bigoplus_{N\le 0} \widetilde{H}_n^{\bullet,N}\right] \otimes H^*(\mathrm{sp}(n); 1) \Rightarrow \bigoplus_{N\le 0} H_n^{\bullet,N},$$

the standard generators of the ring $H^*(\mathrm{sp}(n); 1)$ (of dimension 3, 7, ..., 4n−1) are transgressive and are transformed by transgressions into $\Psi_1, \Psi_2, \ldots, \Psi_n$.†

The proof of this theorem is a repetition, with the obvious changes, of the proof of Theorem 0.3 in [1].

Theorem 2 gives us complete information about the cohomologies $\widetilde{H}_n^{*,N}$ and $H_n^{*,N}$ with $N \le 0$. Thus only $\widetilde{H}_n^{*,N}$ and $H_n^{*,N}$ for N > 0 have not been computed.

3. HOMOLOGIES OF THE COMPLEXES $\widetilde{\mathscr{C}}_1^N$ AND $\mathscr{C}_1^N$ FOR N = 1, 2, 3, 4

1. The complexes $\widetilde{\mathscr{C}}_1^1$, $\widetilde{\mathscr{C}}_1^2$, and $\widetilde{\mathscr{C}}_1^3$ are acyclic. The homologies of $\widetilde{\mathscr{C}}_1^{4'}$ are generated (additively) by one generator lying in $H_1^{7,4}$ ($\subset H^7(V_1, \mathrm{sp}(1); 1)$). This generator is represented by the polynomial

$$p(\alpha_1, \ldots, \alpha_7, \beta_1, \ldots, \beta_7) = \sum \operatorname{sgn}(j_1, \ldots, j_7)\Big[8\gamma_{j_1 j_2} \prod_{3\le s<t\le 7} \gamma_{j_s j_t} - 5\gamma_{j_1 j_2}\gamma_{j_3 j_4}^3 \gamma_{j_5 j_6}^2 \gamma_{j_5 j_7}^2 \gamma_{j_6 j_7}^3$$
$$- 40\gamma_{j_1 j_2}\gamma_{j_2 j_3}^2 \gamma_{j_3 j_4}\gamma_{j_4 j_5}^2 \gamma_{j_5 j_6}\gamma_{j_5 j_7}\gamma_{j_6 j_7}^3 - 10\gamma_{j_1 j_2}\gamma_{j_3 j_4}^2 \gamma_{j_3 j_5}\gamma_{j_4 j_5}\gamma_{j_4 j_7}\gamma_{j_5 j_6}\gamma_{j_5 j_7}\gamma_{j_6 j_7}^3 + 40\gamma_{j_1 j_2}\gamma_{j_3 j_4}^2 \gamma_{j_3 j_5}\gamma_{j_4 j_5}^2 \gamma_{j_5 j_7}\gamma_{j_6 j_7}^4\Big],$$

where γ_{uv} denotes the expression $\alpha_u\beta_v - \alpha_v\beta_u$ and the sum is carried out over all permutations $(j_1, \ldots, j_7)$ of the numbers 1 through 7.

The proof consists of a direct calculation. This calculation was carried out with the help of a computer.

We present an intermediate result of the computation: the dimensions of the spaces $C_1^{q,N}$.

The spaces $C_1^{q,N}$, $1 \le N \le 4$ and $q \ge 9$ are trivial.

2. The complexes $\mathscr{C}_1^1$, $\mathscr{C}_1^2$, and $\mathscr{C}_1^3$ are acyclic. The homologies of the complex $\mathscr{C}_1^4$ are generated (additively) by two generators lying in $H_1^{7,4}$ ($\subset H^7(V_1; 1)$) and $H_1^{10,4}$ ($\subset H^{10}(V_1; 1)$) respectively.

†The author thanks V. I. Rozenfel'd for his discussion of this theorem. The part of the theorem dealing with $\mathscr{C}_n^*$ can be deduced from his results in [6].

TABLE 1

q	1	2	3	4	5	6	7	8
dim $\tilde{C}_1^{q,1}$	0	1	1	0	1	1	0	0
dim $\tilde{C}_1^{q,2}$	0	0	2	4	2	0	0	0
dim $\tilde{C}_1^{q,3}$	0	1	2	4	8	6	1	0
dim $\tilde{C}_1^{q,4}$	0	0	5	13	17	18	14	4

This theorem can be deduced from the preceding one by means of the Serre-Hochschild spectral sequence.

3. In conclusion, we note that V_1 coincides with the algebra of formal vector fields on the plane without divergence, since the elements of nonstandard form are present in the cohomologies of this Cartan series also.

LITERATURE CITED

1. I. M. Gel'fand and D. B. Fuks, "Cohomology of the Lie algebra of formal vector fields," Izv. Akad. Nauk SSSR, Ser. Matem., 34, 322-337 (1970).
2. I. M. Gel'fand and D. B. Fuks, "An upper bound for the cohomologies of infinite-dimensional Lie algebras," Funktsional. Analiz i Ego Prilozhen., 4, No. 4, 70-71 (1970).
3. I. M. Gel'fand and D. B. Fuks, "On cycles representing the cohomology classes of the Lie algebra of formal vector fields," Uspekhi Matem. Nauk, 25, No. 5, 239-240 (1970).
4. C. Godbillon and J. Vey, "Un invariant des feuilletages de codimension 1," Comptes Rendus Acad. Sci. Paris, 273, No. 2, 273 (1971).
5. I. N. Bernshtein and B. I. Rozenfel'd, "On characteristic fiber classes," Funktsional. Analiz i Ego Prilozhen., 6, No. 1, 68-69 (1972).
6. V. I. Rozenfel'd, "Cohomologies of some infinite-dimensional Lie algebras," Funktsional. Analiz i Ego Prilozhen., 5, No. 4, 84-85 (1971).

12.
(with D. B. Fuks)
PL Foliations

Funkts. Anal. Prilozh. 7 (4) (1973) 29–37 [Funct. Anal. Appl. 7 (1973) 278–284].
MR 49:3958. Zbl. 294:57016

The present study grew out of an attempt to comprehend the local geometric nature of some recently discovered characteristic classes of foliations. The analogy between smooth and piecewise-linear (PL) topologies suggests that such characteristic classes can be readily constructed for foliations on combinatorial manifolds reducible in every simplex to a family of parallel planes. Such foliations (which we call affine PL foliations) are defined in §1, subsection 2; they have a relatively uncomplicated classifying space (subsections 6 and 9) and are canonically smoothed (subsections 14 and 15). It turns out, however, that even the first of the characteristic classes, namely the Godbillon–Vey class, is identically equal to zero on these foliations (subsection 16). To render it nontrivial we extend the class of PL foliations by allowing the leaves in the simplexes to be planes parallel in the projective sense ("projective PL foliations;" see subsection 3). On these foliations the Godbillon–Vey class is no longer trivial (subsection 17).

Like the affine,PL foliations have a straightforward classifying space (subsections 7 and 10).

The present article is limited to foliations of codimension 1, but some of the results, including all those of §§ 1 and 2, are easily translated to arbitrary codimensions.

Note on Literature References. The general theory of foliations and their classifying spaces is described in the form necessary for our intentions in Haefliger's book [4]. The Godbillon–Vey class is introduced in [3], and various extensions of this class are given in [1, 2, 5]. An adequate familiarity with all of these matters can be gained in the survey paper [7]. We also mention the felicitous conceptual affinity of the latter article with a recent preprint by R. Bott and A. Haefliger: "Some Remarks on Continuous Cohomology."

§1. Fundamental Definitions

1. The most useful category of topological spaces for our objectives is the category of semisimplicial spaces and simplicial mappings. Following are the requisite precise definitions.

A standard simplex of dimension q is a set

$$T^q = \{(x_1, \ldots, x_{q+1}) \in \mathbf{R}^{q+1} \mid x_1 \geqslant 0, \ldots, x_{q+1} \geqslant 0, \textstyle\sum x_i = 1\}.$$

For $r \leq q$ we consider T^r to be embedded in T^q, identifying $(x_1, \ldots, x_{r+1})$ with $(x_1, \ldots, x_{r+1}, 0, \ldots, 0)$. It is always understood with reference to an affine mapping of a standard simplex into a standard simplex that vertices map into vertices.

A semisimplicial space is a geometric realization of a semisimplicial complex. More precisely, a semisimplicial space is a cellular space (CW complex) for which each cell e is labeled by $(q + 1)!$, where $q = \dim e$, pairwise-distinct continuous mappings of a standard simplex T^q into that space which differ from one another by affine automorphisms $T^q \to T^q$ and have the following properties: 1) the mapping labeled for cell e maps the interior of the simplex T^q onto e and takes the boundary of T^q into a union of cells of dimension less than q; 2) if e' is a cell contained in the closure of cell e, then each of the mappings labeled for e' acts as a restriction of a certain mapping labeled for e. We emphasize that the labeled mappings are not necessarily embedded; they can glue different faces and even map faces onto cells of lower dimensions.

Moscow State University. Translated from Funktsional'nyi Analiz i Ego Prilozheniya, Vol. 7, No. 4, pp. 29–37, October-December, 1973. Original article submitted June 6, 1973.

A continuous mapping f of a semisimplicial space X_1 into a semisimplicial space X_2 is said to be simplicial if for every labeled mapping φ: $T^q \to X_1$ there are a labeled mapping φ_2: $T^r \to X_2$ and an affine mapping h: $T^q \to T^r$ such that $f \circ \varphi_1 = \varphi_2 \circ h$.

A semisimplicial space X' is called a subdivision of a semisimplicial space X if X' is congruent with X as a topological space and for every labeled mapping φ: $T^r \to X'$ there is a labeled mapping ψ: $T^q \to X$ and a linear mapping h of the simplex T^r onto part of the simplex T^q such that $\psi \circ h = \varphi$.

2. An oriented affine PL foliation of codimension 1 (or, simply, an affine PL foliation) on a standard simplex T^q is a class of linear functions on T^q equivalent in the following sense: $f \sim g$ if $f = ag + b$, where $a \in R$, $b \in R$, and $a > 0$. If $r \leq q$ and $\mathcal{F}$ is an affine PL foliation on T^q, then the restrictions of functions from $\mathcal{F}$ onto T^r form an affine PL foliation on T^r, which we call the restriction of the foliation $\mathcal{F}$. An affine mapping h: $T^q \to T^r$ is said to be consistent with affine PL foliations $\mathcal{F}_1$, and $\mathcal{F}_2$ on T^q and T^r if the membership $f \in \mathcal{F}_2$ implies $f \circ h \in \mathcal{F}_1$. An affine PL foliation $\mathcal{F}$ on T^q is said to be degenerate if there are an affine PL foliation $\mathcal{F}'$ on T^r with $r < q$ and an affine mapping $T^q \to T^r$ consistent with $\mathcal{F}$ and $\mathcal{F}'$, and otherwise it is said to be nondegenerate.

An oriented affine PL foliation of codimension 1 (or, simply, an affine PL foliation) on a semisimplicial space X is a function that associates with every labeled mapping φ: $T^q \to X$ a particular affine PL foliation $\mathcal{F}(\varphi)$ on T^q such that if the labeled mapping ψ: $T^r \to X$ is the restriction of the labeled mapping φ: $T^q \to X$, then the foliation $\mathcal{F}(\psi)$ is the restriction of the foliation $\mathcal{F}(\varphi)$. A simplicial mapping f of a semisimplicial space X_1 with an affine PL foliation $\mathcal{F}_1$ into a semisimplicial space X_2 with an affine PL foliation $\mathcal{F}_2$ is said to be consistent with $\mathcal{F}_1$, and $\mathcal{F}_2$, if for any labeled mappings φ_1: $T^q \to X_1$ and φ_2: $T^r \to X_2$ every affine mapping h: $T^q \to T^r$ satisfying the relation $f \circ \varphi_1 = \varphi_2 \circ h$ is consistent with $\mathcal{F}_1(\varphi_1)$, and $\mathcal{F}_2(\varphi_2)$. It is clear that if f is a simplicial mapping of a semisimplicial space X_1 into a semisimplicial space X and $\mathcal{F}$ is an affine PL foliation on X, then there exists on X_1 a unique affine PL foliation $\mathcal{F}_1$ such that f is consistent with $\mathcal{F}_1$, and $\mathcal{F}$; the foliation $\mathcal{F}_1$ is said to be induced by $\mathcal{F}$ through f and is denoted by $f^*\mathcal{F}$.

An affine PL foliation $\mathcal{F}$ on a semisimplicial space is said to be nondegenerate if all foliations $\mathcal{F}(\varphi)$ are nondegenerate. It is clear that every affine PL foliation is induced by a nondegenerate PL foliation through a simplicial mapping.

An affine PL foliation on a semisimplicial space X determines in an obvious way an affine PL foliation on every subdivision X' of the space X. In this case the nondegeneracy of the second foliation implies the nondegeneracy of the first foliation.

3. An oriented projective PL foliation of codimension 1 (or, simply, a projective PL foliation) on a standard simplex T^q is a class of pairs of simultaneously nonvanishing linear functions on T^q equivalent in the following sense: $(f_1, f_2) \sim (g_1, g_2)$ if either the functions f_1 and f_2 are proportional and the functions g_1 and g_2 are proportional or $g_1 = a_{11}f_1 + a_{12}f_2$ and $g_2 = a_{21}f_1 + a_{22}f_2$, where $a_{11} \in R$, $a_{12} \in R$, $a_{21} \in R$, $a_{22} \in R$, and $a_{11}a_{22} - a_{12}a_{21} > 0$. All the definitions given in subsection 2 are repeated verbatim with the replacement of affine PL foliations by projective PL foliations. In this way we define oriented projective PL foliations of codimension 1 (or, simply, projective PL foliations) on semisimplicial spaces, induced projective PL foliations, and nondegenerate projective PL foliations; a projective PL foliation on a semisimplicial space X determines a projective PL foliation on any subdivision of X, and the nondegeneracy of the second foliation implies the nondegeneracy of the first foliation.

4. Affine PL foliations constitute a special case of projective PL foliations; the function f is identified with the pair $(f, 1)$. Neither projective nor affine PL foliations constitute a special case of ordinary foliations (including even C^0 foliations with singularities in the sense of Haefliger; see Fig. 1); the relationship of PL foliations to ordinary foliations is discussed in § 4.

§ 2. Classifying Spaces

5. Let X be a semisimplicial space, X_0 and X_1 subdivisions thereof, $\mathcal{F}_0$ an affine (projective) PL foliation on X_0, and $\mathcal{F}_1$ an affine (projective) PL-foliation on X_1. We say that $\mathcal{F}_0$ and $\mathcal{F}_1$ are homotopic in the class of affine (projective) PL foliations or, simply, homotopic if there are a semisimplicial space Z, congruent as a topological space with $X \times I$, and an affine (projective) PL foliation $\mathcal{F}$ on Z such that the formulas $i_0(x) = (x, 0)$ and $i_1(x) = (x, 1)$ determine simplicial mappings i_0: $X_0 \to Z$ and i_1: $X_1 \to Z$ and that $i_0^*\mathcal{F} = \mathcal{F}_0$, $i_1^*\mathcal{F} = \mathcal{F}_1$. It is clear that if $\mathcal{F}$ is an affine or a projective PL foliation on X, X' is a subdivision of X, and $\mathcal{F}'$ is the foliation determined on X' by the foliation $\mathcal{F}$, then $\mathcal{F}$ and $\mathcal{F}'$ are homotopic.

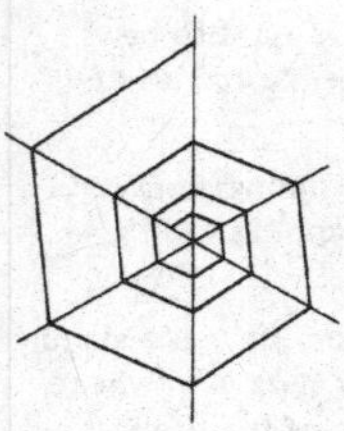

Fig. 1

We denote by Af(X) [Pro(X)] the set of homotopic classes of affine (projective) PL foliations on all possible subdivisions of the semisimplicial space X. Associated with a simplicial mapping f: X → Y in an obvious way are mappings AF(f): Af(Y) → Af(X), Pro(f): Pro(Y) → Pro(X), and this correspondence determines contravariant functors Af and Pro from the category of semisimplicial spaces into the category of sets. By the simplicial approximation theorem (which is applicable to semisimplicial spaces) these functors are homotopic; i.e., they take homotopic mappings into congruent mappings. Consequently, Af and Pro can be regarded as functions from the category of weak homotopic types into the category of sets.

Our most immediate objective is to construct representing spaces for the functors Af and Pro.

6. We denote by A_q the set of all nondegenerate affine PL foliations on T^q. In the space $\mathcal{A} = \coprod_{q=0}^{\infty}(A_q \times T^q)$ we make the following identifications: for every triple $(x, \mathcal{F}, \varphi)$, where $x \in T^r$, $\mathcal{F} \in A_q$, and φ is the affine embedding $T^r \to T^q$, we glue the point $(\varphi^* \mathcal{F}, x) \in A_r \times T^r \subset \mathcal{A}$ to the point $(\mathcal{F}, \varphi(x)) \in A_q \times T^q \subset \mathcal{A}$. The resulting space BAf has a natural semisimplicial structure: the following direct mappings act as labeled mappings:

$$\Phi(\mathcal{F}): T^q \xrightarrow{j(\mathcal{F})} A^q \times T^q \xrightarrow{\text{incl}} \mathcal{A} \xrightarrow{\text{proj}} B\,\text{Af}$$

with $\mathcal{F} \in A_q$, where $j(\mathcal{F})$ is an embedding taking $x \in T^q$ into $(\mathcal{F}, x) \in A_q \times T^q$. The function $\Phi(\mathcal{F}) \mapsto \mathcal{F}$ determines on BAf an affine PL foliation, which we denote by UAf. Clearly, this foliation is nondegenerate.

<u>THEOREM.</u> The formula $f \mapsto f^*\,U\text{Af}$ determines for every semisimplicial space X a one-one correspondence between simplicial mappings X → BAf and affine PL foliations on X. This correspondence, in turn, determines a one-one correspondence between the set π(X, BAf) of homotopic classes of continuous mappings X → BAf and the set Af(X), i.e., the equivalence of the functors Af and π(, BAf).

The first statement is obvious, and the second follows from the first and the simplicial approximation theorem.

7. We denote by P_q the set of all nondegenerate projective PL foliations on T^q. In the space $\mathcal{P} = \coprod_{q=0}^{\infty}$ $(P_q \times T^q)$ we make the following identification: For every triple $(x, \mathcal{F}, \varphi)$, where $x \in T^r$, $\mathcal{F} \in P_q$, and φ is the affine embedding $T^r \to T^q$, we glue the point $(\varphi^* \mathcal{F}, x) \in P_r \times T^r \subset \mathcal{P}$ to the point $(\mathcal{F}, \varphi(x)) \in P_q \times T^q \subset \mathcal{P}$. The resulting space BPro has a natural semisimplicial structure: the following direct mappings act as labeled mappings:

$$\Psi(\mathcal{F}): T^q \xrightarrow{j(\mathcal{F})} P_q \times T^q \xrightarrow{\text{incl}} \mathcal{P} \xrightarrow{\text{proj}} B\,\text{Pro}$$

with $\mathcal{F} \in P_q$. The function $\Psi(\mathcal{F}) \mapsto \mathcal{F}$ determines on BPro a projective PL foliation, which we denote by UPro. Clearly, this foliation is nondegenerate.

<u>THEOREM.</u> The formula $f \mapsto f^*\,U\text{Pro}$ determines for every semisimplicial space X a one-one correspondence between simplicial mappings X → BPro and projective PL foliations on X. This correspondence, in turn, determines a natural one-one correspondence between the set π(X, BPro) and the set Pro(X), i.e., the equivalence of the functors Pro and π(, BPro).

8. Inasmuch as affine PL foliations are projective PL foliations, the space BAf is a subspace of the space BPro. It is clear that UAf is the restriction of the foliation UPro onto BAf.

<u>§3. A More Explicit Description of the Classifying Spaces</u>

9. Let R^δ be the group of real numbers in discrete topology, and let W be the discrete group of orientation-preserving affine transformations of a straight line. Below we define homotopic equivalence between the space BAf and the factor space $K(W, 1)/K(R^\delta, 1)$; the embedding $K(R^\delta, 1) \to K(W, 1)$ is induced by the embedding $R^\delta \to W$, which associates with the number r a transformation $t \to te^r$ [as usual, we denote by $K(\pi, 1)$ the cellular space with fundamental group π and a contractible universal covering group].

We denote by C a simplex spanned by R^δ, i.e., a simplicial space with vertex set R^δ such that every finite subset of R^δ acts as the vertex set of a certain simplex. Clearly, the space C is contractible. The action of the group W on R determines the action of that group on C by simplicial automorphisms. It is clear that the vertices of C form an orbit of W, with R^{δ} as the stationary subgroup of the vertex 0; on the complement of this orbit W acts freely. It is also obvious that the space of orbits C/W has a natural semisimplicial structure and is congruent as a semisimplicial space with BAf.

There corresponds to every $r \in R$ a representation of the space C as a simplicial cone over a certain subspace C_r of C with vertex r. We denote by D_r and E_r the images of the spaces C and C_r under contraction to that vertex with coefficient 1/3. It is clear that the set $C \setminus \bigcup_r D_r$ is invariant under W, that the sets D_r are pairwise nonintersecting, and that

$$C/W = [(C \setminus \bigcup_r D_r)/W] \cup [(\bigcup_r D_r)/W] =$$
$$= [(C \setminus \bigcup_r D_r)/W] \cup (D_0/R^{\delta}) = [(C \setminus \bigcup_r D_r)/W] \cup \text{con}(E_0/R^{\delta}),$$

where the symbol "con" denotes a cone. The group W acts freely on $C \setminus \bigcup_r D_r$, the group R^{δ} acts freely on E_0, and the spaces $C \setminus \bigcup_r D_r$, E_0 are contractible; the contractibility of E_0 is obvious, while the contractibility of $C \setminus \bigcup_r D_r$ follows from the contractibility of C and the deformation retractibility of C on $C \setminus \bigcup_r D_r$. Consequently, $(C \setminus \bigcup_r D_r)/W = K(W, 1)$, $E_0/R^{\delta} = K(R^{\delta}, 1)$, and we obtain a representation of the space BAf = C/W as the result of gluing to K(W, 1) a cone over $K(R^{\delta}, 1)$. In this way we arrive at the anticipated homotopic equivalence.

10. Let L be the discretized (topology-free) universal covering connectivity component $GL_+(2, R)$ of the group GL(2, R), and let V be the discretized subgroup of $GL_+(2, R)$ formed by the matrices $\begin{pmatrix} a & b \\ 0 & a^{-1} \end{pmatrix}$ with $a \in R$, $b \in R$, and $a > 0$. The inclusion relation $V \to GL_+(2, R)$ is covered by the unique embedding $V \to L$. Below we define homotopic equivalence between the space BPro and the factor space K(L, 1)/ K(V, 1).

We denote by F a simplicial space with vertex set $(R^{\delta} \times R^{\delta}) \setminus (0, 0)$ such that a finite subset A of the set $(R^{\delta} \times R^{\delta}) \setminus (0, 0)$ spans a simplex if and only if the convex hull of the set A in R^2 does not contain (0, 0). It is clear that the natural mapping $\gamma: F \to R^2 \setminus (0, 0)$ is a homotopic equivalence. We denote by F_{θ}, where $\theta \in S^1$, the subspace of F formed by simplexes φ: such that $\gamma(\varphi) \subset \{t\theta \mid t \in R, t > 0\}$. The action of the group $GL_+(2, R)$ on R^2 determines the action of that group (without topology) on F by simplicial automorphisms. Clearly, the set $\bigcup_{\theta} F_{\theta}$ is invariant, where the maximal subgroup under which the set $F_{(0,1)}$ is invariant is V; on $F \setminus \bigcup_{\theta} F_{\theta}$ the group $GL_+(2, R)$ acts freely. It is also evident that the space $[F/GL_+(2, R)] / [(\bigcup_{\theta} F_{\theta}) / GL_+(2, R)]$ has a natural semisimplicial structure and is congruent as a semisimplicial space with BPro.

We denote by I_{θ} the union of all simplexes of F that intersect with F_{θ}, and by J_{θ} the union of all simplexes in I_{θ} that do not intersect with F_{θ}. Obviously, J_{θ} includes precisely the simplexes φ in F for which γ maps the vertices into $R^2 \setminus \{t\theta \mid t \in R\}$, and the entire simplex T^q into $R^2 \setminus \{t\theta \mid t \in R, t \leqslant 0\}$. It is also apparent that I_{θ} is the join $F_{\theta} * J_{\theta}$, i.e., the space obtained from the product $F_{\theta} \times J_{\theta} \times I$ by the identification of the point (x, y', 0) with the point (x, y", 0) for $x \in F_{\theta}$, $y' \in J_{\theta}$, $y'' \in J_{\theta}$ and of the point (x', y, 1) with the point (x", y, 1) for $x' \in F_{\theta}$, $x'' \in F_{\theta}$, $y \in J_{\theta}$. The part of I_{θ} obtained by this identification from $F_{\theta} \times J_{\theta} \times [0, 1/3]$ is denoted by G_{θ}, and the part obtained from $F_{\theta} \times J_{\theta} \times \{1/3\}$ is denoted by H_{θ}. It is clear that the set $F \setminus \bigcup_{\theta} G_{\theta}$ is invariant under $GL_+(2, R)$, that the sets G_{θ} are nonintersecting, and that

$$[F/GL_+(2, R)] / [(\bigcup_{\theta} F_{\theta})/GL_+(2, R)] =$$
$$= [(F \setminus \bigcup_{\theta} G_{\theta})/GL_+(2, R)] \cup \{[(\bigcup_{\theta} G_{\theta})/GL_+(2, R)] / [(\bigcup_{\theta} F_{\theta})/GL_+(2, R)]\} =$$
$$= [(F / \bigcup_{\theta} G_{\theta}) / GL_+(2, R)] \cup [(G_{(0,1)}/V) / (F_{(0,1)}/V)] =$$
$$= [(F \setminus \bigcup_{\theta} G_{\theta}) / GL_+(2, R)] \cup \text{con}(H_{(0,1)}/V).$$

The group $GL_+(2, R)$ acts freely on $F \setminus \bigcup_{\theta} G_{\theta}$, the group V acts freely on $H_{(0,1)}$, and the space $H_{(0,1)}$, represented as the product of the contractible spaces $F_{(0,1)}$ and $J_{(0,1)}$, is contractible. But the space $F \setminus \bigcup_{\theta} G_{\theta}$ is the deformation retract of the space F, and F is homotopically equivalent to a circle. The action of the group $GL_+(2, R)$ on F is elevated to the action of the group L on the universal covering group over F, which is contractible. Consequently, $(F \setminus \bigcup_{0} G_\theta) / GL_+(2, R) = K(L, 1)$, $H_{(0,1)}/V = K(V, 1)$, and we obtain a representation of the space BPro as the result of gluing to K(L, 1) a cone over K(V, 1). In this way we arrive at the desired homotopic equivalence.

11. We now augment and refine somewhat the discussions of subsections 9 and 10. Each of the semisimplicial spaces BAf and BPro has a single vertex. The complement of the vertex in BAf is a space of the type K(W, 1), and the same complement in BPro is a space of the type K(L, 1). The embedding BAf → BPro takes the complement of the vertex into the complement of the vertex, and the resulting mapping K(W, 1) → K(L, 1) is induced by the natural embedding W → L. Inasmuch as the image of this embedding is contained in the image of the embedding V → L, the embedding (BAf\vertex) → BPro is homotopically trivial.

The pair (BAf, vertex) is locally homeomorphic to the pair (con $K(R^\delta, 1)$, vertex), and the pair (BPro, vertex) is locally homeomorphic to the pair (con K(V, 1), vertex). This fact yields the following unexpected consequence:

If a semisimplicial space X is simply connected and locally simply connected (for example, if X is a simply connected manifold of dimension not less than 3), then every nondegenerate (affine or projective) PL foliation is homotopic to zero; if a semisimplicial space X is locally simply connected (for example, if X is an arbitrary manifold of dimension not less than 3), then every nondegenerate affine PL foliation is homotopic to zero in the class of projective PL foliations.

Indeed, the simplicial mapping f: X → BAf (BPro) corresponding to a nondegenerate affine (projective) PL foliation takes the complement of the vertex into the complement of the vertex. It suffices now to show that if X is locally simply connected, then f is homotopic to a mapping whose image does not contain a vertex. This result is obvious, because the restriction of f onto the boundary of a neighborhood of every vertex of X can be connected to a constant mapping of a homotopy passing near the vertex of the space BAf(BPro) but not touching it.

Therefore, it is inadmissible to limit the investigation of PL foliations on manifolds to the nondegenerate case.

§ 4. Smoothing of PL Foliations

12. We define a smooth foliation as an oriented C^∞ foliation of codimension 1 with singularities in the sense of Haefliger. We now give some precise definitions for the reader's convenience.

An atlas of a smooth foliation on a topological space X is a set comprising an open covering $\mathcal{U}$ of X and the sets of continuous mappings $\{\alpha_U : U \to R \mid U \in \mathcal{U}\}$, $\{\beta_{U,V} : U \cap V \to \Gamma \mid U, V \in \mathcal{U}\}$, where Γ is the topological groupoid of germs of orientation-preserving diffeomorphisms of a straight line, such that: 1) if $x \in U \cap V (U, V \in \mathcal{U})$, then $\beta_{U,V}(x) \circ g_x(\alpha_U) = g_x(\alpha_V)$ (g_x is the germ at the point x); 2) if $x \in U \cap V \cap W$ $(U, V, W \in \mathcal{U})$, then $\beta_{W,U}(x)\beta_{V,W}(x)\beta_{W,U}(x) = g_{\alpha_U(x)}(\mathrm{id})$. Two atlases are said to be equivalent if they act as subatlases of a third atlas, and a class of equivalent atlases is a smooth foliation. A smooth foliation induced by another foliation through a continuous mapping is defined in a natural way, as is the homotopicity of smooth foliations.

As we know, the functor associating with a cellular space X a set of homotopic classes of smooth foliations on X is representable. The representing space is called the Haefliger classifying space and is denoted by BΓ.

13. A smooth foliation $\mathcal{D}$ on a semisimplicial space X is said to be compatible with a projective PL foliation $\mathcal{F}$ on X if $\mathcal{D}$ has an atlas $\mathcal{U}$, $\{\alpha_U\}$, $\{\beta_{U,V}\}$, such that for every labeled mapping φ: $T^q \to X$ there are a $U \in \mathcal{U}$ and a pair $(f_1, f_2) \in \mathcal{F}(\varphi)$ such that $\varphi(T^q) \subset U$ and $f_1(x) = \alpha_U(\varphi(x)) f_2(x)$ for $x \in T^q$.

14. In subsection 15 below we define a certain smooth foliation $\mathcal{S}$ on BPro compatible with a certain projective PL foliation on BPro homotopic to the universal foliation UPro. The role of this smooth foliation is that of defining a canonic smoothing operation for projective (and so for affine) PL foliations. Specifically, the mapping BPro → BΓ corresponding to the foliation $\mathcal{S}$ induces a mapping of the functor $\pi($, BPro) into the functor $\pi($, BΓ), i.e., associates with every homotopic class of projective PL foliations on a semisimplicial space X a definite homotopic class of smooth foliations on X. The latter class includes a foliation compatible with a representative of the initial class of projective PL foliations.

Referring back to subsection 4, we see that an individual PL foliation can turn out not to be compatible with any smooth foliation. But homotopically every PL foliation is canonically smoothed, making it possible to transfer characteristic classes to PL foliations. We discuss this topic in further detail in §5.

15. Below we use the notation of subsection 10. Being a part of a singular complex of the perforated plane $\mathbf{R}^2\setminus(0, 0)$, the space F has a natural mapping into $\mathbf{R}^2\setminus(0, 0)$, i.e., a pair of simultaneously nonvanishing continuous functions is defined on F; these functions are linear on every simplex and determine a certain projective PL foliation $\mathscr{H}$. It is clear that $\mathscr{H}$ is compatible with a certain smooth foliation $\mathscr{R}$ and coincides with the foliation induced by UPro through the projection F → BPro. We denote by $\widetilde{F}$ the space obtained from F by the substitution of $F_\theta \times J_\theta \times [0, 1/3]$ for each of the sets G_θ:

$$\widetilde{F} = (F \setminus \bigcup_\theta G_\theta) \cup [\bigcup_\theta (F_\theta \times J_\theta \times [0,1/3])];$$

$$(F \setminus \bigcup_\theta G_\theta) \cap (F_\theta \times J_\theta \times [0, 1/3]) = F_\theta \times J_\theta \times \{1/3\} = H_\theta.$$

The action of the group L on F is elevated in an obvious way to the action of the group L on $\widetilde{F}$, and it is clear that the latter action is free. The foliations $\widetilde{\mathscr{H}}$ and $\widetilde{\mathscr{R}}$, induced by the foliations $\mathscr{H}$ and $\mathscr{R}$ through the projection $\widetilde{F} \to F$ are compatible and invariant under this action and therefore determine on $\widetilde{F}/L = K(L, 1)$ a projective PL foliation $\mathscr{H}_0$ and a smooth foliation $\mathscr{R}_0$, which are compatible. The space $[\bigcup_0 (F_\theta \times J_\theta \times 0)]/L = (F_{(0,1)} \times J_{(0,1)} \times 0)/V = K(V, 1)$ lies in one leaf of each of these foliations (i.e., the restrictions of these foliations on K(V, 1) coincide with the foliations induced by projection to a point). The factorization K(L, 1) → K(L, 1)/K(V, 1) transforms the foliation $\mathscr{H}_0$ into UPro and annihilates the foliation $\mathscr{R}_0$. We replace this factorization by gluing of a cone over K(V, 1); attaching the glued cone to the layer containing K(V, 1), we transform $\mathscr{H}_0$ and $\mathscr{R}_0$ into a compatible projective PL foliation $\mathscr{K}$ and smooth foliation $\mathscr{S}$ on K(L, 1) ∪ con K(V, 1), whereupon it merely remains to note the fact that the foliation $\mathscr{K}$ is homotopic to the foliation induced by UPro through the homotopic equivalence K(L, 1) ∪ con K(V, 1) → K(L, 1)/K(V, 1) = BPro.

§5. Characteristic Classes

16. THEOREM. The homomorphic inclusion H*(BPro; A) → H*(BAf; A) is trivial for any coefficient of the group A.

Proof. Inasmuch as the embedding $\mathbf{R}^\delta \to W$ is semi-invertible, the homomorphism $H^*(\text{BAf}; A) = H^*(K(W, 1)/(K(\mathbf{R}^\delta, 1); A) \to H^*(K(W, 1); A)$ is a monomorphism. On the other hand, the direct mapping K(W, 1) → K(L, 1) → K(L, 1)/K(V, 1) is homotopically trivial (subsection 11) and induces a trivial cohomological homomorphism. All we have to do now is to use the commutativity of the diagram

$$\begin{array}{ccc} K(W,1) & \xrightarrow{\text{incl}} & K(L,1) \\ \downarrow \text{proj} & & \downarrow \text{proj} \\ K(W,1)/K(\mathbf{R}^\delta,1) & & K(L,1)/K(V,1) \\ \| & & \| \\ BAf & \xrightarrow{\text{incl}} & BPro. \end{array}$$

17. It follows from the theorem of subsection 16 that the inclusion relation H*(BΓ; A) → H*(BAf; A) is also trivial. We therefore have the proposition: any characteristic class of smooth foliations, including the Godbillon-Vey class, acts to prevent a given smooth foliation from becoming a smoothing of an affine PL foliation.

18. THEOREM. The homomorphic inclusion H*(BI; **R**) → H* (BPro; **R**) is nontrivial on the Godbillon–Vey class.

For the proof it is sufficient to show that a certain foliation with a nontrivial Godbillon–Vey class is a smoothing of a projective PL foliation. However, this property is already inherent in the foliation used by Godbillon and Vey to prove the nontriviality of their class.

We recall the structure of this foliation. We pick the following two subgroups in the group SL(2, **R**): the subgroup H composed of the matrices $\begin{pmatrix} a & b \\ 0 & a^{-1} \end{pmatrix}$ with $a \in \mathbf{R}$, $b \in \mathbf{R}$, and $a > 0$, as well as an arbitrary discrete subgroup Δ with the compact factor space SL(2, **R**)/Δ. The left adjacent classes of the group H form a certain foliation on SL(2, **R**). This foliation is invariant under right translations and therefore determines a certain foliation $\mathscr{F}$ on the right factor space SL(2, **R**)/Δ, and that foliation has precisely a nontrivial Godbillon–Vey class.

We construct the projective PL foliation for which $\mathcal{F}$ acts as a smoothing as follows. The fundamental group of the manifold SL(2, R)/Δ is the inverse image Δ' of the group Δ under the projection L → GL+(2, R). The inclusion Δ' → L determines a homotopic class of mappings SL(2, R)/Δ → K(L, 1), and the composition with the projection K(L, 1) → K(L, 1)/K(V, 1) = BPro transforms it into a homotopic class of mappings SL(2, R)/Δ → BPro, i.e., into a homotopic class of projective PL foliations on SL(2, R)/Δ. It is clear that the smoothing of the foliations in this class at once yields the homotopic class of the foliation $\mathcal{F}$.

19. Concluding Remarks. 1°. The cohomologies of the space BPro can be bound from the cohomological sequence of the pair (K(L, 1), K(V, 1)). Since the cohomologies of the space K(π, 1) coincide with the cohomologies of the group π, the indicated sequence assumes the form

$$\ldots \to H^q(L; A) \to H^q(V; A) \to H^{q+1}(B\mathrm{Pro}; A) \to H^{q+1}(L; A) \to H^{q+1}(V; A) \to \ldots.$$

The cohomologies of the groups L and V are not completely known; however, the "continuous" cohomologies of these groups with coefficients in R are known [6]; they are as follows:

$$\dim H_c^q(L; \mathbf{R}) = \begin{cases} 1 & \text{for } q = 0, 1, 3, 4, \\ 0 & \text{for } q \neq 0, 1, 3, 4; \end{cases} \quad \dim H_c^q(V; \mathbf{R}) = \begin{cases} 1 & \text{for } q = 0, 1, \\ 0 & \text{for } q \neq 0, 1. \end{cases}$$

Hence it follows that the groups H^3(BPro; R) and H^4(BPro, R) have at least one generator each. In H_3 the latter is the Godbillon-Vey class; the nature of the four-dimensional class is still not clear.

2°. A similar remark applies to BAf, but nonnull cohomology classes are not obtained in this case.

3°. It can be shown that the spaces BAf and BPro are 2-connected; in this respect they are similar to BΓ.

LITERATURE CITED

1. I. N. Bernshtein and B. I. Rozenfel'd, "Characteristic classes of foliations," Funktsional. Analiz i Ego Prilozhen., 6, No. 1, 68-69 (1972).
2. R. Bott and A. Haefliger, "On characteristic classes of Γ foliations," Bull. Amer. Math. Soc., 78, No. 6, 1039-1044 (1972).
3. C. Godbillon and J. Vey, "Un invariant des feuilletages de codimension 1," Compt. Rend., 273, No. 2, A92-A95 (1971).
4. A. Haefliger, Homotopy and Integrability, Amsterdam (1971); Springer Lecture Notes, Vol. 197, pp. 133-163.
5. A. Haefliger, "Sur les classes caracteristiques des feuilletages," Seminaire Bourbaki, No. 412 (June, 1972).
6. G. Hochshild and G. D. Mostow, "Cohomology of Lie groups," Illinois J. Math., 6, No. 3, 367-401 (1962).
7. D. B. Fuks, "Characteristic classes of foliations," Usp. Matem. Nauk, 28, No. 2, 3-17 (1973).

13.

(with B. L. Feigin and D. B. Fuks)

Cohomologies of the Lie algebra of formal vector fields with coefficients in its adjoint space and variations of characteristic classes of foliations

Funkts. Anal. Prilozh. **8** (2) (1974) 13–29 [Funct. Anal. Appl. **8** (1974) 99–112].
MR **50**:8553. Zbl. **298**:57011

Introduction

0.1. The present article deals with the characteristic classes of foliations (see [9, 1, and 7]).

There are two constructions of such classes. The first relates to a generalized foliation (foliation in the sense of Haeflinger, see [10]) of codimension n with a trivialized normal stratification on a topological space X the homomorphism

$$H^*(W_n;\ \mathbf{R}) \to H^*(X;\ \mathbf{R}),$$

where W_n is a (topological) Lie algebra of formal vector fields in $\mathbf{R}^n$. This homomorphism is natural relative to continuous mappings and, therefore, relates to each element of the ring $H^*(W_n; \mathbf{R})$ a characteristic class of (generalized) foliations of codimension n with a trivialized normal stratification. The second construction starts from an arbitrary oriented generalized foliation of codimension n on X and gives the homomorphism

$$H^*(W_n,\ \mathfrak{o}_n;\ \mathbf{R}) \to H^*(X;\ \mathbf{R}),$$

where $\mathfrak{o}_n$ is the Lie algebra of the group of orthogonal matrices of order n naturally imbedded in W_n. This homomorphism allows us to interpret elements of the ring $H^*(W_n, \mathfrak{o}_n; \mathbf{R})$ as the characteristic classes of oriented foliations of codimension n.

Both the constructions are C^∞-continuous; i.e., the characteristic classes of C^∞-near foliations on X are near in $H^*(X; \mathbf{R})$. Moreover, a smooth deformation of a foliation induces a smooth deformation of its characteristic classes. Naturally the question arises as to how much the characteristic classes of foliations are variable. In this context two results are known. The first result is due to Thurston [13] who constructed a smooth one-parameter family of oriented (even nonsingular) foliations of codimension 1 on the three-dimensional sphere with a nonconstant Godbillon–Vey class [a Godbillon–Vey class is the characteristic class corresponding to a nonzero element of the group $H^3(W_1;\ \mathbf{R}) = H^3(W_1,\ \mathfrak{o}_1;\ \mathbf{R}) \cong \mathbf{R}$]; as mentioned in [8], Thurston also varied in a similar manner one of the generators of the group $H^5(W_2; \mathbf{R})$. On the other hand, Heitsch [11] proved rigidity for a (small) part of characteristic classes of foliations, i.e., invariability for smooth variations of foliations.

0.2. In the present article we offer an algebraic approach to the variations of characteristic classes of foliations. Side by side with the characteristic classes of foliations we consider the characteristic classes of variations of foliations, understanding by a variation of foliation a 1-jet relative to the parameter of a smooth one-parameter family of foliations (on the same space). Two sorts of such classes are defined: the characteristic classes of variations of foliations with a trivialized normal stratification and the characteristic classes of variations of oriented foliations (of course, we can define the characteristic classes for an arbitrary variety of "Γ-foliations," see [10]). The characteristic classes of foliations serve as the

---

Moscow State University. Translated from Funktsional'nyi Analiz i Ego Prilozheniya, Vol. 18, No. 2, pp. 13–29, April–June, 1974. Original article submitted December 10, 1973.

simplest examples of the characteristic classes of the variations of foliations: the value of such a characteristic class on the variation of a foliation is defined simply as its value on the variable foliation. The derivatives relative to the parameter of the characteristic classes of foliations give a more important class of examples; these are also essential, of course, for the consideration of the question concerning the variability of the characteristic classes of foliations.

For example, with the variations of oriented foliations of codimension 1 are connected at least two characteristic classes: the Godbillon–Vey class and its derivative relative to the parameter. Thurston's theorem shows that the second class is nontrivial (the nontriviality of the first class is established in [9]).

0.3. Let us denote by $\hat{W}_n$ the tensor product of the Lie algebra W_n with the associative and commutative algebra $\mathbf{R}[t]/t^2\mathbf{R}[t]$. This is a Lie algebra; we can write its elements in the form $\xi+t\eta$, $(\xi, \eta \in W_n)$; the commutation operation is given by the equation

$$[\xi + t\eta, \xi' + t\eta'] = [\xi, \xi'] + t([\xi, \eta'] + [\eta, \xi']).$$

The functions $\xi \mapsto \hat{\xi} = \xi + t0$, $\xi + t\eta \mapsto \xi$ define the imbedding $W_n \to \hat{W}_n$ and the projection $\hat{W}_n \to W_n$ with the identical composition $W_n \to \hat{W}_n \to W_n$. In what follows we shall regard W_n as a subalgebra of the algebra $\hat{W}_n$.

The algebra $\hat{W}_n$ plays for the characteristic classes of the variations of foliations the same role as the algebra W_n plays for the characteristic classes of foliations: the elements of the ring $H^*(\hat{W}_n; \mathbf{R})$ supply the characteristic classes of variations of foliations with a trivialized normal stratification, and the elements of the ring $H^*(\hat{W}_n, \mathfrak{o}_n; \mathbf{R})$ supply the characteristic classes of the variations of oriented foliations. The natural imbeddings

$$H^*(W_n; \mathbf{R}) \to H^*(\hat{W}_n; \mathbf{R}), \quad H^*(W_n, \mathfrak{o}_n; \mathbf{R}) \to H^*(\hat{W}_n, \mathfrak{o}_n; \mathbf{R})$$

correspond to the above-mentioned interpretations of the characteristic classes of foliations as the characteristic classes of the variations of foliations. Moreover, there exist natural homomorphisms (additive)

$$\text{var}: H^*(W_n; \mathbf{R}) \to H^*(\hat{W}_n; \mathbf{R}), \quad \text{var}: H^*(W_n, \mathfrak{o}_n; \mathbf{R}) \to H^*(\hat{W}_n, \mathfrak{o}_n; \mathbf{R}),$$

possessing the property that the characteristic class of the variations of foliations determined by the element var (α) is the derivative relative to the parameter of the characteristic class of foliations determined by the element α. This homomorphism is important for the study of the variability of the characteristic classes of foliations; e.g., if var $(\alpha) = 0$, then the characteristic class corresponding to α is rigid.

0.4. For the cohomologies of the Lie algebra $\hat{W}_n$ we construct the direct factorizations

$$H^q(\hat{W}_n; \mathbf{R}) = \bigoplus_{p+r=q} H^p(W_n; \Lambda^r W_n'),$$
$$H^q(\hat{W}_n, \mathfrak{o}_n; \mathbf{R}) = \bigoplus_{p+r=q} H^p(W_n, \mathfrak{o}_n; \Lambda^r W_n'),$$

where W_n' is the adjoint space of W_n (with the natural structure of a W_n-module), and Λ^r denotes the r-th exterior degree. In addition, the canonical isomorphisms

$$H^*(\hat{W}_n; \mathbf{R}) \cong H^*(W_n; \Lambda^* W_n'), \quad H^*(\hat{W}_n, \mathfrak{o}_n; \mathbf{R}) \cong H^*(W_n, \mathfrak{o}_n; \Lambda^* W_n')$$

are annular.

The imbeddings of the terms $H^q(W_n; \Lambda^0 W_n') = H^q(W_n; \mathbf{R})$, $H^q(W_n, \mathfrak{o}_n; \Lambda^0 W_n') = H^q(W_n, \mathfrak{o}_n; \mathbf{R})$ in the sums $\bigoplus H^p(W_n, \Lambda^r W_n') = H^q(\hat{W}_n; \mathbf{R})$, $\bigoplus H^p(W_n, \mathfrak{o}_n; \Lambda^r W_n') = H^q(\hat{W}_n, \mathfrak{o}_n; \mathbf{R})$ coincide, as it turns out, with the natural imbeddings $H^q(W_n; \mathbf{R}) \to H^q(\hat{W}_n; \mathbf{R})$, $H^q(W_n, \mathfrak{o}_n; \mathbf{R}) \to H^q(\hat{W}_n, \mathfrak{o}_n; \mathbf{R})$ (see Sec. 0.3). Further, the images under the homomorphisms var: $H^q(W_n; \mathbf{R}) \to \hat{H}^q(W_n; \mathbf{R})$ and var: $H^q(W_n, \mathfrak{o}_n; \mathbf{R}) \to H^q(\hat{W}_n, \mathfrak{o}_n; \mathbf{R})$ lie in $H^{q-1}(W_n; W_n')$ and $H^{q-1}(W_n, \mathfrak{o}_n; W_n')$ because the homomorphisms var can be considered as the homomorphisms $H^q(W_n; \mathbf{R}) \to H^{q-1}(W_n; W_n')$ and $H^q(W_n, \mathfrak{o}_n; \mathbf{R}) \to H^{q-1}(W_n, \mathfrak{o}_n; W_n')$; in this capacity they also have a direct description.

0.5. The principal technical results of this article are concerned with the cohomologies $H^*(W_n; W_n')$ and $H^*(W_n, \mathfrak{o}_n; W_n')$. We give independent significance to these cohomologies and investigate them in more detail than is necessary for the characteristic classes of foliations.

Let us observe that $H^*(W_n; W_n')$ is a $H^*(W_n; \mathbf{R})$-module and $H^*(W_n, \mathfrak{o}_n; W_n')$ is a $H^*(W_n, \mathfrak{o}_n; \mathbf{R})$-module.

We prove the following theorems.

THEOREM I. $H^q(W_n; W_n') \cong H^{2n+1}(W_n; \mathbf{R}) \otimes H^{q-2n}(\mathfrak{gl}(n); \mathbf{R})$; the structure of a $H^*(W_n; \mathbf{R})$-module in $H^*(W_n; W_n')$ is trivial [i.e., if $y \in H^p(W_n; \mathbf{R})$, $x \in H^q(W_n; W_n)$ and $p > 0$, $q > 0$, then $xy = 0$].

THEOREM II.

$$H^q(W_n, \mathfrak{o}_n; W_n') \cong H^{2n+1}(W_n; \mathbf{R}) \otimes H^{q-2n}(\mathfrak{gl}(n), \mathfrak{o}_n; \mathbf{R});$$

the structure of a $H^*(W_n, \mathfrak{o}_n; \mathbf{R})$-module in $H^*(W_n; \mathfrak{o}_n; W_n')$ is trivial.

The field $\mathbf{R}$ in both these theorems can be replaced by an arbitrary field of characteristic zero.

The homomorphisms var are further described; in particular, their kernels are found.

0.6. As mentioned in Sec. 0.3, the kernels of the homomorphisms var consist of rigid characteristic classes of foliations. However, it turns out that our computation of the kernels gives precisely Heitsch's theorem. (Without claiming precision we can say that this means that "in the domain of formulas" it is not necessary to prove a stronger theorem of rigidity than Heitsch's theorem.) On the other hand, the statements about the triviality of a modular structure have a curious corollary about the variations of the characteristic classes of foliations. It follows from them that whatever a foliation or its variation may be, the product of an arbitrary characteristic class with the derivative relative to the parameter of another arbitrary (possibly, the same) characteristic class equals zero. Let us illustrate this statement by an example. Let x_1, x_2 be a natural basis of the space $H^3(S^3 \times S^3; \mathbf{R})$ and let F be an oriented foliation of codimension 1 on $S^3 \times S^3$ whose Godbillon–Vey class is proportional to x_1 (such a foliation may be obtained by multiplication of an arbitrary foliation of the Thurston's family by S^3). Then for the variation of the foliation F the Godbillon–Vey class can be varied (this can be seen from Thurston's example) but necessarily remains proportional to x_1.

0.7. The structure of the article is as follows. The first section is devoted to the computation of the cohomologies $H^*(W_n; W_n')$ and $H^*(W_n, \mathfrak{o}_n; W_n')$. In the second section are computed the homomorphisms var, the third section is devoted to the cohomologies of the algebra $\hat{W}_n$, and finally, the fourth section is devoted to the characteristic classes of foliations and their variations.

§ 1. $H^*(W_n; W_n')$ and $H^*(W_n, \mathfrak{o}_n; W_n')$

1.1. Let us begin by recalling the fundamental definitions.

Let $\mathfrak{g}$ be a topological Lie algebra over a field k of characteristic 0, and let M be an arbitrary $\mathfrak{g}$-module. By a q-dimensional cochain of the algebra $\mathfrak{g}$ with coefficient in M we understand a continuous skew-symmetric q-linear function on $\mathfrak{g}$ with values in M, i.e., a continuous homomorphism $\Lambda^q\mathfrak{g} \to M$. The space of these cochains is denoted by $C^q(\mathfrak{g}; M)$. The $(q+1)$-dimensional cochain dL, defined by the equation

$$dL(\xi_1, \ldots, \xi_{q+1}) = \sum_{1 \leq s < t \leq q+1} (-1)^{s+t-1} L([\xi_s, \xi_t], \xi_1, \ldots \hat{\xi}_s \ldots \hat{\xi}_t \ldots, \xi_{q+1}) + \sum_{1 \leq s \leq q+1} (-1)^s \xi_s L(\xi_1, \ldots \hat{\xi}_s \ldots, \xi_{q+1}),$$

is called the differential of the q-dimensional cochain L. It is clear that the function $L \mapsto dL$ defines a linear mapping $d: C^q(\mathfrak{g}; M) \to C^{q+1}(\mathfrak{g}; M)$ and that the composition $d \circ d: C^q(\mathfrak{g}; M) \to C^{q+2}(\mathfrak{g}; M)$ is trivial. Thus, the spaces $C^q(\mathfrak{g}; M)$ and the homomorphisms d make up a complex; the q-th homology group of this complex is called the q-th cohomology group of the algebra $\mathfrak{g}$ with coefficients in M and is denoted by $H^q(\mathfrak{g}; M)$.

Let $\mathfrak{h}$ be a subalgebra of the algebra $\mathfrak{g}$. The subspace of the space $C^q(\mathfrak{g}; M)$ chosen by the conditions $L(\xi_1, \ldots \xi_q) = 0$ and $dL(\xi_1, \ldots, \xi_{q+1}) = 0$ for $\xi_1 \in \mathfrak{h}$ is denoted by $C^q(\mathfrak{g}, \mathfrak{h}; M)$. It is obvious that the spaces $C^q(\mathfrak{g}, \mathfrak{h}; M)$ constitute a subcomplex of the complex $\{C^q(\mathfrak{g}; M), d\}$; the q-th homology group of this subcomplex is called the q-th cohomology group of the algebra $\mathfrak{g}$ modulo $\mathfrak{h}$ with coefficients in M and is denoted by $H^q(\mathfrak{g}, \mathfrak{h}; M)$.

Let $L \in C^q(\mathfrak{g}; M)$ and $K \in C^p(\mathfrak{g}; k)$ (unless stated otherwise, k is considered to be a trivial $\mathfrak{g}$-module). The equation

$$KL(\xi_1, \ldots, \xi_{p+q}) = \sum_{1 \leq i_1 < \ldots < i_p \leq p+q} (-1)^{i_1 + \ldots + i_p - [p(p+1)/2]} K(\xi_{i_1}, \ldots, \xi_{i_p}) \, L(\xi_1, \ldots \hat{\xi}_{i_1} \ldots \hat{\xi}_{i_p} \ldots, \xi_{p+q})$$

defines the cochain $KL \in C^{p+q}(\mathfrak{g}; M)$ and the product $C^p(\mathfrak{g}; k) \otimes C^q(\mathfrak{g}; M) \to C^{p+q}(\mathfrak{g}; M)$. It is clear that $d(KL) = (dK)L + (-1)^p K(dL)$, due to which the product $H^p(\mathfrak{g}; k) \otimes H^q(\mathfrak{g}; M) \to H^{p+q}(\mathfrak{g}; M)$ is also defined. This product for $M = k$ defines in $H^*(\mathfrak{g}; k) = \bigoplus H^q(\mathfrak{g}; k)$ the structure of a graded ring and, for arbitrary M, defines in $H^*(\mathfrak{g}; M) = \bigoplus_q H^q(\mathfrak{g}; M)$ the structure of a graded $H^*(\mathfrak{g}; k)$-module. In the same manner, $H^*(\mathfrak{g}, \mathfrak{h}; k)$ is a graded ring and $H^*(\mathfrak{g}, \mathfrak{h}; M)$ is a graded $H^*(\mathfrak{g}, \mathfrak{h}; k)$-module.

In what follows we shall denote by W_n the topological Lie algebra of formal vector fields in the n-dimensional space over k, by W_n' is denoted the space of continuous functionals on W_n (with the natural structure of a W_n-module), by $\mathfrak{gl}(n)$ is denoted that subalgebra of the algebra W_n which consists of fields of the form $\sum a_{ij}x_i\, \partial/\partial x_j$ (isomorphic in an obvious manner to the usual algebra of matrices), and by $\mathfrak{o}_n$ is denoted that subalgebra of the algebra $\mathfrak{gl}(n)$ which is selected by the condition $a_{ij} = a_{ji}$.

The present section is devoted to the computation of a $H^*(W_n; k)$-module of $H^*(W_n; W_n')$ and the computation of a $H^*(W_n, \mathfrak{o}_n; k)$-module of $H^*(W_n, \mathfrak{o}_n; W_n')$. These computations differ little from each other, and we shall carry out only the first computation in detail, confining ourselves for the second computation to the formulation of the final result and pointing out of the differences from the first computation.

1.2. The space $C^q(W_n; W_n')$ is $(\Lambda^q W_n)' \otimes W_n' = \Lambda^q(W_n') \otimes W_n'$. In its turn W_n' is naturally identified with $S^*T \otimes T'$, where $T = k^n$ and S^* denotes a symmetric algebra (cf. [4], Sec. 1.2). Thus,

$$C^q(W_n; W_n') = \Lambda^q(S^*T \otimes T') \otimes (S^*T \otimes T'),$$

which allows us to write the elements of the space $C^q(W_n; W_n')$ as functions of the arguments $\alpha_1, \ldots, \alpha_{q+1} \in T'$; $\beta_1, \ldots, \beta_{q+1} \in T$, depending polynomially on the coordinates of the covectors $\alpha_1, \ldots, \alpha_{q+1}$, multilinear relative to $\beta_1, \ldots, \beta_{q+1}$, and changing sign for the simultaneous interchanges of α_i with α_j and of β_i with β_j if $i < q+1$, $j < q+1$. The differential d is given for such a representation by the equation

$$\begin{aligned} dP(\alpha_1, \ldots, \alpha_{q+2}; \beta_1, \ldots, \beta_{q+2}) = &\sum_{1 \leqslant s < t \leqslant q+1} (-1)^{s+t-1} P(\alpha_s + \alpha_t, \alpha_1, \ldots \\ &\ldots \hat{\alpha}_s \ldots \hat{\alpha}_t \ldots, \alpha_{q+2}; \beta(s,t), \beta_1, \ldots \hat{\beta}_s \ldots \hat{\beta}_t \ldots, \beta_{q+1}) + \\ &+ \sum_{1 \leqslant s \leqslant q} (-1)^{s+1} P(\alpha_1, \ldots \hat{\alpha}_s \ldots, \alpha_{q+1}, \alpha_s + \alpha_{q+2}; \beta_1, \ldots \hat{\beta}_s \ldots, \beta_{q+1}, \beta(s, q+2)), \end{aligned}$$

where $\beta(s,t) = (\alpha_t, \beta_s)\beta_t - (\alpha_s, \beta_t)\beta_s$. In the same manner, cochains of $C^q(W_n; k)$ are identified with functions of $\alpha_1, \ldots, \alpha_q \in T'$; $\beta_1, \ldots, \beta_q \in T$, polynomial relative to α, multilinear relative to β, and changing sign for the simultaneous interchanges of α_i with α_j and of β_i with β_j (cf. [4], Sec. 1.3). The product $C^p(W_n; k) \otimes C^q(W_n; W_n') \to C^{p+q}(W_n; W_n')$ acts according to the equation

$$\begin{aligned} PQ(\alpha_1, \ldots, \alpha_{p+q+1}; \beta_1, \ldots, \beta_{p+q+1}) = &\sum_{1 \leqslant i_1 < \ldots < i_p \leqslant p+q} (-1)^{i_1 + \ldots + i_p - [p(p+1)/2]} \times \\ \times P(\alpha_{i_1}, \ldots, \alpha_{i_p}; \beta_{i_1}, \ldots, \beta_{i_p})\, Q(\alpha_1, \ldots \hat{\alpha}_{i_1} \ldots \hat{\alpha}_{i_p} \ldots, \alpha_{p+q+1}; \beta_1 \ldots \hat{\beta}_{i_1} \ldots \hat{\beta}_{i_p} \ldots, \beta_{p+q+1}) \end{aligned}$$

(cf. [4], Sec. 1.4).

1.3. We use the Serre–Hochschild spectral sequence (see [12]) corresponding to the subalgebra $\mathfrak{gl}(n)$ of the algebra W_n. This spectral sequence converges to $H^*(W_n; W_n')$, and its first term is defined by the equation

$$E_1^{qr} = H^r(\mathfrak{gl}(n); C^q(W_n, \mathfrak{gl}(n); W_n')).$$

The space $C^q(W_n, \mathfrak{gl}(n); W_n')$ is a subspace of the space $C^q(W_n; W_n')$ consisting of the polynomials $P(\alpha_1, \ldots, \alpha_{q+1}; \beta_1, \ldots, \beta_{q+1})$ satisfying the condition: the total degree relative to the coordinates of the covectors $\alpha_1, \ldots, \alpha_q$ of each monomial occuring in P is different from 1. Although the $\mathfrak{gl}(n)$-module $C^q(W_n, \mathfrak{gl}(n); W_n')$ is infinite-dimensional, it splits up into the sum of finite-dimensional modules (the total degree relative to $\alpha_1, \ldots, \alpha_{q+1}$ is a $\mathfrak{gl}(n)$-invariant), and therefore

$$H^r(\mathfrak{gl}(n);\ C^q(W_n, \mathfrak{gl}(n);\ W_n')) = [C^q(W_n, \mathfrak{gl}(n);\ W_n')]^{\mathfrak{gl}(n)} \otimes H^r(\mathfrak{gl}(n); \mathrm{k}),$$

where $[\ ]^{\mathfrak{gl}(n)}$ denotes the space of $\mathfrak{gl}(n)$-invariant elements (see [6], Chap. 5, Theorem 1).

The differential $d_1^{qr}: E_1^{qr} \to E_1^{q+1,r}$ is induced by the differential $d: C^q(W_n, \mathfrak{gl}(n); W_n') \to C^{q+1}(W_n, \mathfrak{gl}(n); W_n')$; because of this E_2^{qr} is the tensor product of the q-th homology space of the complex

$$\{[C^q(W_n,\ \mathfrak{gl}(n);\ W_n')]^{\mathfrak{gl}(n)},\ d\} \tag{1}$$

with $H^r(\mathfrak{gl}(n); \mathrm{k})$. Thus, our next problem is to compute the homology of the complex (1). This is done below: Propositions 1 and 2 give a clearer description of the spaces constituting the complex (1), Proposition 3 describes its differential, and, finally, Proposition 4 describes its homology.

1.4. It follows from a basic theorem of the theory of invariants (see [3], Theorem II 6A) that every $\mathfrak{gl}(n)$-invariant cochain has the form

$$\sum_{i_1=1}^{q} \cdots \sum_{i_{q+1}=1}^{q} k_{i_1 \ldots i_{q+1}} (\alpha_{i_1}, \beta_1) \ldots (\alpha_{i_{q+1}}, \beta_{q+1}), \tag{2}$$

where $k_{i_1 \ldots i_{q+1}} \in \mathrm{k}$ (cf. [4], Proposition 4.4). First of all, let us clarify which expressions of the form (2) satisfy the condition of skew symmetry and the condition that the degree relative to $\alpha_1, \ldots, \alpha_q$ is different from unity.

Let us define the cochains

$$\rho_{st}^{\varepsilon_1\varepsilon_2} \in C^{2s+2t+\varepsilon_1+\varepsilon_2}(W_n, \mathfrak{gl}(n);\ W_n'),$$
$$\sigma_u^{\varepsilon} \in C^{2u+\varepsilon}(W_n, \mathfrak{gl}(n); W_n'),\quad \psi_v \in C^{2v}(W_n; \mathrm{k}),$$

where s, t, u, $v \geq 0$ and ε_1, ε_2, $\varepsilon = 0$ or 1, by the equations

$$\rho_{st}^{\varepsilon_1\varepsilon_2}(\alpha_1, \ldots, \alpha_{2s+2t+\varepsilon_1+\varepsilon_2+2};\ \beta_1, \ldots, \beta_{2s+2t+\varepsilon_1+\varepsilon_2+2}) =$$
$$= \sum_{(i_1, \ldots, i_{2s+2t+\varepsilon_1+\varepsilon_2+1})} \operatorname{sign}(i_1, \ldots, i_{2s+2t+\varepsilon_1+\varepsilon_2+1}) (\alpha_{i_2}, \beta_{i_1})(\alpha_{i_3}, \beta_{i_2}) \ldots (\alpha_{i_s}, \beta_{i_{s-1}}) \times$$
$$\times (\alpha_{i_{s+t+1}}, \beta_{i_s})(\alpha_{i_{s+2}}, \beta_{i_{s+1}}') \ldots (\alpha_{i_{s+t+1}}, \beta_{i_{s+t}}^-)(\alpha_{i_{s+1}}, \beta_{i_{s+t+1}}) \times$$
$$\times (\alpha_{i_1}, \beta_{i_{s+t+2}})(\alpha_{i_2}, \beta_{i_{s+t+3}}) \ldots (\alpha_{i_{s+t}}, \beta_{i_{2s+2t+1}}) \times$$
$$\times \begin{cases} (\alpha_{i_1}, \beta_{2s+2t+2}), & \text{if } \varepsilon_1 = \varepsilon_2 = 0, \\ (\alpha_{i_{s+t+1}}, \beta_{i_{2s+2t+2}})(\alpha_{i_1}, \beta_{2s+2t+3}), & \text{if } \varepsilon_1 = 0,\ \varepsilon_2 = 1, \\ (\alpha_{2s+2t+3}, \beta_{i_{2s+2t+2}})(\alpha_{i_1}, \beta_{2s+2t+3}), & \text{if } \varepsilon_1 = 1,\ \varepsilon_2 = 0, \\ (\alpha_{i_{s+t+1}}, \beta_{i_{2s+2t+2}})(\alpha_{2s+2t+4}, \beta_{i_{2s+2t+3}})(\alpha_{i_1}, \beta_{2s+2t+4}), & \text{if } \varepsilon_1 = \varepsilon_2 = 1; \end{cases}$$

$$\sigma_u^{\varepsilon}(\alpha_1, \ldots, \alpha_{2u+\varepsilon+1};\ \beta_1, \ldots, \beta_{2u+\varepsilon+1}) = \sum_{(i_1, \ldots, i_{2u+\varepsilon})} \operatorname{sign}(i_1, \ldots, i_{2u+\varepsilon}) \times$$
$$\times (\alpha_{i_2}, \beta_{i_1})(\alpha_{i_3}, \beta_{i_2}) \ldots (\alpha_{i_u} \beta_{i_{u-1}})(\alpha_{i_1}, \beta_{i_{u+1}})(\alpha_{i_2}, \beta_{i_{u+2}}) \ldots (\alpha_{i_u} \beta_{i_{2u}}) \times$$
$$\times \begin{cases} (\alpha_{2u+1}, \beta_{i_u})(\alpha_{i_1}, \beta_{2u+1}), & \text{if } \varepsilon = 0, \\ (\alpha_{2u+2}, \beta_{i_u})(\alpha_{2u+2}, \beta_{i_{2u+1}})(\alpha_{i_1}, \beta_{2u+2}), & \text{if } \varepsilon = 1; \end{cases}$$

$$\psi_v(\alpha_1, \ldots, \alpha_{2v};\ \beta_1, \ldots, \beta_{2v}) = \sum_{(i_1, \ldots, i_{2v})} \operatorname{sign}(i_1, \ldots, i_{2v})(\alpha_{i_2}, \beta_{i_1})(\alpha_{i_3}, \beta_{i_2}) \ldots$$
$$\ldots (\alpha_{i_v}, \beta_{i_{v-1}})(\alpha_{i_1}, \beta_{i_v})(\alpha_{i_1}, \beta_{i_{v+1}})(\alpha_{i_2}, \beta_{i_{v+2}}) \ldots (\alpha_{i_v}, \beta_{i_{2v}}).$$

PROPOSITION 1. The space $[C^*(W_n,\ \mathfrak{gl}(n);\ W_n')]^{\mathfrak{gl}(n)}$ is additively generated by cochains of the form $\psi_{v_1} \ldots \psi_{v_m} \rho_{st}^{\varepsilon_1\varepsilon_2}$ and $\psi_{v_1} \ldots \psi_{v_m} \sigma_u^{\varepsilon}$.

Proof. Every cochain of $[C^q(W_n, \mathfrak{gl}(n); W_n')]^{\mathfrak{gl}(n)}$ is the sum of polynomials each of which is obtained by skew-symmetrization from a monomial, and we can restrict ourselves to the consideration of such polynomials. Let

$$(\alpha_{k_1}, \beta_1) \ldots (\alpha_{k_{q+1}}, \beta_{q+1})$$

be a monomial whose skew-symmetrization gives the proposed cochain P. It follows from the conditions imposed upon the polynomials that

(i) each of the numbers 1, ..., q either does not occur at all among $k_1, \dots, k_{q+1}$ or occurs at least twice;

(ii) if $1 \le \mu < \nu \le q$ and the numbers μ, ν do not occur among $k_1, \dots, k_{q+1}$, then $k_\mu \neq k_\nu$ (or else the simultaneous interchange of α_μ with α_ν and of β_μ with β_ν would not change the monomial and its skew-symmetrization would be equal to zero).

Case 1: $q = 2r$. In this case exactly r of the numbers 1, ..., q are not contained among $k_1, \dots, k_{q+1}$. Indeed, if, let us say, r, ..., q do not occur among $k_1, \dots, k_{q+1}$, then $k_r, \dots, k_q$ do not contain more than r different numbers, i.e., at least two of these numbers coincide, which contradicts (i). On the other hand, if, let us say, the numbers 1, ..., r + 1 are contained among $k_1, \dots, k_{q+1}$, then by virtue of (ii) each of the former numbers occurs among $k_1, \dots, k_{q+1}$ at least twice, whence $q + 1 \ge 2(r + 1)$.

We assume that 1, ..., r (and perhaps q + 1) occur and r + 1, ..., q do not occur among the numbers $k_1, \dots, k_{q+1}$. In addition the numbers $k_{r+1}, \dots, k_q$ are pairwise distinct by virtue of (ii).

Subcase 11: q + 1 does not occur among $k_{r+1}, \dots, k_q$. Then we may assume that $k_{r+1} = 1, \dots, k_q = r$. Then the numbers 1, ..., r must occur among $k_1, \dots, k_r, k_{q+1}$ at least once. Let us construct the sequence

$$l_0 = q + 1,\ l_1 = k_{l_0},\ l_2 = k_{l_1}, \dots .$$

Let l_u be the first element of this sequence which is equal to one of the preceding elements. Then among the u indices $k_{l_0}, \dots, k_{l_u}$ occur exactly u−1 of the numbers r + 1, ..., q (namely, $l_1, \dots, l_{u-1}$). This means that the indices k_i with $i \in I = \{1, \dots, \hat{l}_1 \dots \hat{l}_{u-1} \dots, q\}$ are pairwise distinct, and the function $i \mapsto k_i$ defines a permutation $I \to I$. The latter splits up into a product of cycles; let $v_1, \dots, v_m$ be the lengths of these cycles.

The monomial is completely described; if $l_u = q + 1$, then $P = \pm \psi_{v_1} \dots \psi_{v_m} \sigma_u^0$, and if $l_u = l_s$ with $0 < s < u$, then $P = \pm \psi_{v_1} \dots \psi_{v_m} \rho^{01}_{r-1,\,u-s}$.

Subcase 12: q + 1 occurs among $k_{r+1}, \dots, k_q$. Then we may assume that $k_{r+1} = 1, \dots, k_{q-1} = r-1$, $k_q = q+1$. The number r must occur twice, and the numbers 1, ..., r−1 must occur once, among the numbers $k_1, \dots, k_r, k_{q+1}$. Let us again construct the sequence $l_0 = q + 1$, $l_1 = k_{l_0}, \dots$. If l_u is the first element of this sequence which is equal to one of the preceding elements l_s, then $0 < s < u$ and $l_s = l_u = r$. The function $i \mapsto k_i$ again defines a permutation of the set $\{1, \dots, \hat{l}_1 \dots \hat{l}_{u-1} \dots, q\}$; the latter splits up into a product of cycles, and if $v_1, \dots, v_m$ are the lengths of these cycles, then $P = \pm \psi_{v_1} \dots \psi_{v_m} \rho^{10}_{s-1,\,u-s}$.

Case 2. $q = 2r-1$. In this case either r or r−1 of the numbers 1, ..., q do not occur among $k_1, \dots, k_{q+1}$. Indeed, if, let us say, r−1, ..., q do not occur among $k_1, \dots, k_{q+1}$, then $k_{r-1}, \dots, k_q$ do not contain more than r−2 distinct numbers, i.e., at least two of these numbers coincide, which contradicts the condition (ii). On the other hand, if, let us say, the numbers 1, ..., r + 1 occur among the numbers $k_1, \dots, k_{q+1}$, then by virtue of (i) each of these numbers occurs among $k_1, \dots, k_{q+1}$ at least twice, whence $q + 1 \ge 2(r + 1)$.

Subcase 21. r of the numbers 1, ..., q occur among $k_1, \dots, k_{q+1}$.

Subcase 22. r−1 of the numbers 1, ..., q occur among $k_1, \dots, k_{q+1}$.

We omit the detailed consideration of these subcases, which is analogous to the preceding. In the subcase 21 the cochain coincides, to within sign, with a cochain of the form $\psi_{v_1} \dots \psi_{v_m} \sigma_u^1$ or $\psi_{v_1} \dots \psi_{v_m} \rho^{11}_{st}$, and in the subcase 22 it coincides with a cochain of the form $\psi_{v_1} \dots \psi_{v_m} \rho^{00}_{st}$.

1.5. Our next problem is to compute the relations between the generators obtained above. Let us observe that the system of generators indicated by us obviously does not depend on n so that the dependence on n of the space $[C^q(W_n, \mathfrak{gl}(n); W_n')]^{\mathfrak{gl}(n)}$ must appear only on the relations. For convenience in further formulations let us choose in $[C^*(W_n, \mathfrak{gl}(n); W_n')]^{\mathfrak{gl}(n)}$ five sequences of subspaces A_r, B_r, C_r, D_r, F_r.

The systems of cochains generating these spaces are shown in the second column of Table 1. Each of these subspaces consists of cochains of a single dimension given in the third column of the table. In the last two columns are mentioned the degrees relative to the elements of the corresponding spaces (more precisely: an arbitrary cochain of the space under consideration can be written down in the form of a polynomial in α, β, and the degree relative to $\alpha_1, \dots, \alpha_{q+1}$ of an arbitrary monomial occuring in this polynomial is computed; it is easy to see that the degrees relative to $\alpha_1, \dots, \alpha_q$, to within order, do not depend on the choice of the cochain and the monomial, and the degree relative to α_{q+1} does not depend at all on this choice). Proposition 1 means that

TABLE 1

Space	Generators	Dimension (q)	Degrees relative to $\alpha_1, \ldots, \alpha_q$	Degree relative to α_{q+1}
A_r	$\psi_{v_1} \ldots \psi_{v_m} \rho_{st}^{00}$ with $v_1 + \ldots + v_m + s + t = r - 1$	$2r-1$	$\underbrace{0\ldots0}_{r-1}\ \underbrace{2\ldots2}_{r}$	0
B_r	$\psi_{v_1} \ldots \psi_{v_m} \rho_{st}^{11}$ with $v_1 + \ldots + v_m + s + t = r - 2$	$2r-1$	$\underbrace{0\ldots0}_{r}\ \underbrace{2\ldots2}_{r-2}3$	1
C_r	$\psi_{v_1} \ldots \psi_{v_m} \sigma_u^1$ with $v_1 + \ldots + v_m + u = r - 1$	$2r-1$	$\underbrace{0\ldots0}_{r}\ \underbrace{2\ldots2}_{r-1}$	2
D_r	$\psi_{v_1} \ldots \psi_{v_m} \rho_{st}^{01}$ with $v_1 + \ldots + v_m + s + t = r - 1$	$2r$	$\underbrace{0\ldots0}_{r}\ \underbrace{2\ldots2}_{r-1}3$	0
F_r	$\psi_{v_1} \ldots \psi_{v_m} \rho_{st}^{10}$ with $v_1 + \ldots + v_m + s + t = r - 1$; $\psi_{v_1} \ldots \psi_{v_m} \sigma_u^0$ with $v_1 + \ldots + v_m + u = r$	$2r$	$\underbrace{0\ldots0}_{r}\ \underbrace{2\ldots2}_{r}$	1

$$[C^{2r-1}(W_n, \mathfrak{gl}(n);\ W_n')]^{\mathfrak{gl}(n)} = A_r + B_r + C_r,$$
$$[C^{2r}(W_n, \mathfrak{gl}(n);\ W_n')]^{\mathfrak{gl}(n)} = D_r + F_r.$$

Since all the collections of degrees in the table are pairwise distinct, our spaces are pairwise disjoint and the problem of determination of the relations splits up into the problem of determination of the relations in each of the spaces A_r, B_r, C_r, D_r, F_r.

PROPOSITION 2. (i) The spaces A_r with $r \geq n+2$, B_r with $r \geq n+1$, C_r with $r \geq n+1$, D_r with $r \geq n+1$, and F_r with $r \geq n+1$ are trivial.

(ii) In the spaces A_r with $r \leq n$, B_r with $r \leq n$, C_r with $r \leq n$, D_r with $r \leq n$, and F_r with $r \leq n-1$ the generators indicated in Table 1 are linearly independent.

(iii) The equations

$$\psi_{v_1} \ldots \psi_{v_m} \sigma_u^0 = \sum_{s+t=u-1} \psi_{v_1} \ldots \psi_{v_m} \rho_{st}^{10} + \sum_{i=1}^{m} (-1)^{uv_i} v_i \psi_{v_1} \ldots \hat{\psi}_{v_i} \ldots \psi_{v_m} \rho_{u, v_i-1}^{10} \tag{3}$$

make up the defining system of relations in the space F_n.

(iv) The equations

$$\psi_{v_1} \ldots \psi_{v_m} \rho_{0u}^{00} = \sum_{s+t=u-1} \psi_{v_1} \ldots \psi_{v_m} \rho_{s+1, t}^{00} + \sum_{i=1}^{m} (-1)^{(u+1)v_i} v_i \psi_{v_1} \ldots \hat{\psi}_{v_i} \ldots \psi_{v_m} \rho_{u+1, v_i-1}^{00};$$

$$\sum_{i=1}^{m} v_i \psi_{v_1} \ldots \hat{\psi}_{v_i} \ldots \psi_{v_m} \rho_{0 v_i-1}^{00} = 0$$

make up the defining system of relations in the space A_{n+1}.

The statements (i) and (ii) follow from the following statements (i') and (ii') in an obvious manner.

(i') If the monomial $(\alpha_{k_1}, \beta_1) \ldots (\alpha_{k_{q+1}}, \beta_{q+1})$ does not depend on more than n of the covectors $\alpha_1, \ldots, \alpha_q$, then its skew-symmetrization (relative to $\alpha_1, \ldots, \alpha_q$; $\beta_1, \ldots, \beta_q$) is equal to zero.

(ii') The monomials of the form $(\alpha_{k_1}, \beta_1) \ldots (\alpha_{k_{q+1}}, \beta_{q+1})$ having a positive degree relative to not more than n of the covectors $\alpha_1, \ldots, \alpha_{q+1}$ are linearly independent.

Proof of the Statement (i'). Let us choose from the skew-symmetrization under consideration the sum of monomials which do not depend upon $\alpha_{j_1}, \ldots, \alpha_{j_{n+1}}$, where $1 \le j_1 < \ldots < j_{n+1} \le q$. This sum is multilinear and skew-symmetric relative to $\beta_{j_1}, \ldots, \beta_{j_{n+1}}$ and is therefore equal to zero.

Proof of the Statement (ii'). Let $e_1, \ldots, e_n$ be a basis in T and $e_1', \ldots, e_n'$ be the adjoint basis in T', and let $(\alpha_{k_1}, \beta_1) \ldots (\alpha_{k_{q+1}}, \beta_{q+1})$ be a monomial having a positive degree relative to $\alpha_{j_1}, \ldots, \alpha_{j_r}$ $(r \le n$, $1 \le j_1 < \ldots < j_r \le q+1)$ and zero degree relative to the remaining j_r. Let us put $\alpha_{j_1} = e_1', \ldots, \alpha_{j_r} = e_r'$; $\alpha_j = 0$ for $j \ne j_1, \ldots, j_r$ and let us define β_i by the condition: $\alpha_{k_i} = \beta_i'$. The substitution of these α and β in our monomial gives unity, but the substitution of these α and β in any other monomial having a positive degree relative to not more than n of the α gives zero.

Proof of Statement (iii). First of all, let us observe that for the proof of the nontriviality of a cochain $P \in P_n$ it is sufficient to prove that not all the numbers (elements of the field k)

$$\gamma_{i_1 \ldots i_{n+1}}(P) = P(\underbrace{0, \ldots, 0}_{n}, e_1', \ldots, e_n', e_1'; e_1, \ldots, e_n, e_{i_1}, \ldots, e_{i_{n+1}})$$

with $1 \le i_1 \le n, \ldots, 1 \le i_{n+1} \le n$ are zero.

Indeed, since

$$(\beta_1, \ldots, \beta_{2n+1}) \mapsto P(0, \ldots, 0, e_1', \ldots, e_n', e_1'; \beta_1, \ldots, \beta_{2n+1})$$

is a multilinear function, skew-symmetric relative to $\beta_1, \ldots, \beta_n$, it follows from the vanishing of the numbers γ that

$$P(0, \ldots, 0, e_1', \ldots, e_n', e_1'; \beta_1, \ldots, \beta_{2n+1}) = 0$$

for arbitrary $\beta_1, \ldots, \beta_{2n+1}$. Using the invariance of P relative to the permutation of the coordinates and by the condition of skew-symmetry, it follows that

$$P(0, \ldots, 0, e_1', \ldots, e_n', e_i'; \beta_1, \ldots, \beta_{2n+1}) = 0$$

for arbitrary $\beta_1, \ldots, \beta_{2n+1}$ and every i. Since, further,

$$\alpha_{2n+1} \mapsto P(0, \ldots, 0, e_1', \ldots, e_n', \alpha_{2n+1}; \beta_1, \ldots, \beta_{2n+1})$$

is a linear function, the latter means that

$$P(0, \ldots, 0, e_1', \ldots, e_n', \alpha_{2n+1}; \beta_1, \ldots, \beta_{2n+1}) = 0$$

for arbitrary $\beta_1, \ldots, \beta_{2n+1}, \alpha_{2n+1}$. It follows from this, again in view of the invariance of P, that

$$P(0, \ldots, 0, \alpha_{n+1}, \ldots, \alpha_{2n+1}; \beta_1, \ldots, \beta_{2n+1}) = 0$$

for arbitrary $\alpha_{n+1}, \ldots, \alpha_{2n+1}; \beta_1, \ldots, \beta_{2n+1}$ satisfying the additional condition of linear independence of $\alpha_{n+1}, \ldots, \alpha_{2n}$, and this additional condition may be discarded by virtue of the continuity of P. Finally it follows clearly from the last equation that $P = 0$ since

$$P = \sum_{1 \leqslant i_1 < \ldots < i_n \leqslant 2n} P|_{\alpha_{i_1} = \ldots = \alpha_{i_n} = 0},$$

and the terms of this sum are obtained from each other by the permutations $(\alpha_i, \beta_i) \leftrightarrow (\alpha_j, \beta_j)$ $(1 \leqslant i < j \leqslant 2n)$.

Further, since the degree of the polynomial $P \in F_n$ relative to α_{n+1} is equal to 1, and is zero or 2 relative to the remaining α, we can have $\gamma_{i_1 \ldots i_{n+1}}(P)$ different from zero only under the condition that 1 occurs twice and 2, ..., n occur once among $i_1, \ldots, i_{n+1}$. Let $i_1, \ldots, i_{n+1}$ be such a collection. We shall consider two cases.

First Case: $i_{n+1} = 1$. Then $(i_1, \ldots, i_n)$ is a permutation of the numbers 1, ..., n. It splits up into a product of cycles; let u be the length of the cycle containing 1, and $v_1, \ldots, v_m$ be the lengths of the remaining cycles. It is easy to see that $\gamma_{i_1 \ldots i_{n+1}}(P)$ is equal to zero for all the monomials P of our system of generators except three:

$$\psi_{v_1}\ldots\psi_{v_m}\rho^{10}_{0,\,u-1},\quad \psi_{v_1}\ldots\psi_{v_m}\sigma^0_u,\quad \psi_{v_1}\ldots\psi_{v_m}\psi_u\sigma^0_0,$$

for which $\gamma_{i_1\ldots i_{n+1}}$ is different from zero (and is easily computed).

Second Case. $i_{n+1} \neq 1$. Let us construct the sequence $l_0 = i_{n+1}$, $l_1 = i_{l_0}$, $l_2 = i_{l_1}, \ldots$, Let l_u be the first element of this sequence which is equal to one of the preceding elements l_s. It is clear that $0 < s < u$ and $l_s = l_u = 1$ and that on the set $\{1, \ldots, n\} \setminus \{l_0, \ldots, l_{u-1}\}$ the function $k \mapsto i_k$ defines a permutation. This permutation splits up into a product of cycles. Let $v_1, \ldots, v_m$ be the lengths of these cycles. Then $\gamma_{i_1\ldots i_{n+1}}(P)$ is equal to zero for all the monomials P of our system of generators, except three:

$$\psi_{v_1}\ldots\psi_{v_m}\rho^{10}_{s,\,u-s-1},\quad \psi_{v_1}\ldots\psi_{v_m}\sigma^0_u,\quad \psi_{v_1}\ldots\psi_{v_m}\psi_{u-s}\sigma^0_s,$$

for which $\gamma_{i_1\ldots i_{n+1}}$ is different from zero (and is again easily computed).

From what we have said it follows that the cochains $\psi_{v_1}\ldots\psi_{v_m}\rho^{10}_{st}$ are linearly independent in E_n, and the cochains $\psi_{v_1}\ldots\psi_{v_m}\sigma^0_u$ are linearly expressed through them (since each γ is different from zero exactly for one cochain $\psi_{v_1}\ldots\psi_{v_m}\rho^{10}_{st}$ and a certain γ is different from zero on every cochain $\psi_{v_1}\ldots\psi_{v_m}\rho^{10}_{st}$). In order to prove Eq. (3), it remains to compute the values of γ on the cochains indicated above and convince ourselves that the results do not contradict this equation. This is automatic, but we omit the rather cumbersome computation.

The proof of the statement (iv) is analogous to the preceding one. We restrict ourselves to the indication that for the triviality of the cochain it is sufficient that the numbers

$$P(0, \ldots, 0, e_1', \ldots, e_n', e_1', 0; e_1, \ldots, e_n, e_{i_1}, \ldots, e_{i_{n+2}}),$$
$$P(0, \ldots, 0, e_1', \ldots, e_n', e_1' + e_2', 0; e_1, \ldots, e_n, e_{i_1}, \ldots, e_{i_{n+2}})$$

vanish.

1.6. The following automatically proved proposition describes the action of the differential d in $[C^*(W_n, \mathfrak{gl}(n); W_n')]^{\mathfrak{gl}(n)}$.

PROPOSITION 3. $d\rho^{00}_{st} = -\rho^{01}_{st} + (-1)^{s+t+1}\rho^{01}_{st}$; $d\rho^{01}_{st} = -\rho^{00}_{s+1,\,t} + \rho^{11}_{st}$; $d\rho^{10}_{st} = (-1)^{s+t}\rho^{00}_{s+1,\,t} + (-1)^{s+t-1}\rho^{11}_{st}$; $d\rho^{11}_{st} = (-1)^{s+t}\rho^{01}_{s+1,\,t} - \rho^{10}_{s+1,\,t}$; $d\sigma^0_u = (-1)^{u+1}\rho^{00}_{0u} + (-1)^u\sigma^1_u$; $d\sigma^1_u = (-1)^{u+1}\rho^{01}_{0u} - \rho^{10}_{0u}$.

1.7. Propositions 1-3 permit us to compute the homologies of the complex $\{[C^q(W_n, \mathfrak{gl}(n); W_n')]^{\mathfrak{gl}(n)}, d\}$ without any difficulty. The following result is obtained.

PROPOSITION 4. If $r \neq 2n$, then the r-th homological space H_r of the complex $\{[C^q(W_n, \mathfrak{gl}(n); W_n')]^{\mathfrak{gl}(n)}\}$ is trivial; the dimension of the space H_{2n} is one less than the number of representations of the number $n+1$ as the sum of natural numbers (representations differing only in the order of their terms are regarded as the same).

More precisely: let us denote by $x_{v_1\ldots v_m}$, where $0 < v_1 \le \ldots \le v_m$ and $v_1 + \ldots + v_m = n+1$, the homology class of the cycle

$$\xi_{v_1\ldots v_m} = \sum_{i=1}^{m} (-1)^{v_i} v_i \psi_{v_1}\ldots\hat{\psi}_{v_i}\ldots\psi_{v_m}\sigma^0_{v_i-1} \in [C^{2n}(W_n, \mathfrak{gl}(n); W_n')]^{\mathfrak{gl}(n)};$$

the space H_{2n} is generated by its elements $x_{v_1\ldots v_m}$ connected by the single relation

$$\sum_{v_1\ldots v_m} \frac{(-1)^{\frac{1}{2}\sum_{i=1}^{m}(v_i-1)(v_i-2)}}{\prod_{j=1}^{n+1}(k_j!)\prod_{i=1}^{m} v_i}\, x_{v_1\ldots v_m} = 0,$$

where k_j is the number of times j occurs among $v_1, \ldots, v_m$.

1.8. Now we can prove those parts of Theorems I and II of the introduction which are related to $H^*(W_n; W_n')$. For this let us return to the Serre–Hochschild spectral sequence (see Sec. 1.3). Proposition 4 shows that this spectral sequence has

$$E_2^{qr} = \begin{cases} 0, & \text{if } q \neq 2n, \\ E_2^{2n,0} \otimes H^r(\mathfrak{gl}(n); k), & \text{if } q = 2n, \end{cases}$$

and dim $E_2^{2n,0}$ is one less than the number of representations of the number $n+1$ as the sum of natural numbers. All further differentials of the spectral sequence are trivial by dimensional arguments, and we get the canonical isomorphism

$$H^q(W_n; W_n') = E_2^{2n,0} \otimes H^{q-2n}(\mathfrak{gl}(n); k).$$

Combining the information obtained above about dim $E_2^{2n,0}$ with Theorem 5.5 of [4], we get the (this time, a noncanonical) isomorphism

$$H^q(W_n; W_n') \cong H^{2n+1}(W_n; k) \otimes H^{q-2n}(\mathfrak{gl}(n); k).$$

It remains to prove that for every $y \in H^p(W_n; k)$ and $x \in H^q(W_n; W_n')$ with $p > 0$, $q > 0$ the product xy is equal to zero. Proof: the Serre–Hochschild filtration of the class y relative to the subalgebra $\mathfrak{gl}(n)$ is not less than n (see [4], Proposition 5.3), and the filtration of the class x is equal to $2n$. By virtue of the multiplicativity of this filtration the product xy has a filtration which is not less than $3n$; consequently, $xy = 0$.

1.9. The statements of Theorems I and II related to $H^*(W_n, \mathfrak{o}_n; W_n')$ are proved in exactly the same way. The Serre–Hochschild spectral sequence with

$$E_1^{q,r} = H^r(\mathfrak{gl}(n), \mathfrak{o}_n; C^q(W_n, \mathfrak{gl}(n); W_n')) = H^r(\mathfrak{gl}(n), \mathfrak{o}_n; k) \otimes [C^q(W_n, \mathfrak{gl}(n); W_n')]^{\mathfrak{gl}(n)},$$

converges to $H^*(W_n, \mathfrak{o}_n; W_n')$, and on repeating with suitable changes everything we said in Sec. 1.8, we deduce from Proposition 4 that

$$H^q(W_n, \mathfrak{o}_n; W_n') \cong H^{2n+1}(W_n; k) \otimes H^{q-2n}(\mathfrak{gl}(n), \mathfrak{o}_n; k).$$

The triviality of $H^*(W_n, \mathfrak{o}_n; W_n')$ as a $H^*(W_n, \mathfrak{o}_n; k)$-module is again proved on the basis of the multiplicativity of the Serre–Hochschild filtration, although with regard to the filtration of the elements of a positive dimension of the space $H^*(W_n, \mathfrak{o}_n; k)$ we can state only that it is not less than 2.

§ 2. Homomorphisms var

2.1. The space $C^{q+1}(W_n; k) = \Lambda^{q+1} W_n'$ is naturally imbedded in $C^q(W_n; W_n') = (\Lambda^q W_n') \otimes W_n'$, and this imbedding is obviously permutable with the differential. The resulting homomorphism $H^{q+1}(W_n; k) \to H^q(W_n; W_n')$ is denoted by var. This is the homomorphism which was promised in the introduction. We shall discuss its connection with the characteristic classes of foliations in § 4, and in this section we study it as an independent object.

2.2. We know that the cocycles $\psi_v \in C^{2v}(W_n, \mathfrak{gl}(n); k)$ are cohomologous to zero in $C^*(W_n; k)$: the cochains $\varphi_v \in C^{2v-1}(W_n; k)$ with $d\varphi_v = \psi_v$ are naturally defined (see [4, 5]). It turns out that the cochains $\varphi_{u_1} \dots \varphi_{u_l} \psi_{v_1} \dots \psi_{v_m}$ with $1 \le u_1 < \dots < u_l \le n$, $1 \le v_1 \le \dots \le v_m \le n$, $u_1 \le v_1$, $v_1 + \dots + v_m \le n$, $u_1 + v_1 + \dots + v_m > n$ are cocycles, and the cohomology classes $C_{u_1 \dots u_l; v_1 \dots v_m}$ of these cocycles form an (additive) basis of the space $H^*(W_n; k)$.

In what follows $f_1, \dots, f_n$ denote the natural multiplicative generators of the algebra $H^*(\mathfrak{gl}(n); k)$ [dim $f_i = 2i - 1$].

THEOREM III.

$$\operatorname{var} C_{u_1 \dots u_l; v_1 \dots v_m} = \begin{cases} 0 & \text{for } u_1 + v_1 + \dots + v_m > n+1, \\ x_{u_1 v_1 \dots v_m} \otimes f_{u_2} \dots f_{u_l} & \text{for } u_1 + v_1 + \dots + v_m = n+1. \end{cases}$$

Proof. Since the class $x_{u_1 v_1 \dots v_m} \otimes f_{u_2} \dots f_{u_l}$ is represented by the cocycle $\varphi_{u_2} \dots \varphi_{u_l} \xi_{u_1 v_1 \dots v_m}$, it is sufficient to show that

$$\operatorname{var} C_{v_0; v_1 \dots v_m} = \begin{cases} 0 & \text{for } v_0 + v_1 + \dots + v_m > n+1, \\ x_{v_0 v_1 \dots v_m} & \text{for } v_0 + v_1 + \dots + v_m = n+1. \end{cases}$$

Let us consider the cochain $\varphi_{v_0}\psi_{v_1}\dots\psi_{v_m}$ in $C^*(W_N; k)$, where $N \gg n$. This cochain is not yet a cocycle: its differential is equal to $\psi_{v_0}\psi_{v_1}\dots\psi_{v_m}$. It is easy to see that the cochain homomorphism $C^*(W_N; k) \to C^*(W_N; W'_N)$ transforms $\psi_{v_0}\psi_{v_1}\dots\psi_{v_m}$ into

$$\sum_{i=0}^{m} v_i\psi_{v_0}\dots\hat{\psi}_{v_i}\dots\psi_{v_m}(\rho^{00}_{0v_i-1}-\sigma^1_{v_i-1}) \tag{4}$$

and, consequently, transforms $\varphi_{v_0}\psi_{v_1}\dots\psi_{v_m}$ into some gl(n)-invariant cochain whose differential is equal to (4). The cochain

$$\sum_{i=0}^{m}(-1)^{v_i}v_i\psi_{v_0}\dots\hat{\psi}_{v_i}\dots\psi_{v_m}\sigma^0_{v_i-1} \tag{5}$$

is an invariant cochain with the differential (4). Consequently, the homomorphism $C^*(W_N; k) \to C^*(W_N; W'_N)$ transforms the cochain $\varphi_{v_0}\psi_{v_1}\dots\psi_{v_m}$ into the sum of the cochain (5) and a certain invariant cocycle. Since N is large, the latter is the differential of some invariant cochain $\omega \in C^*(W_N; W'_N)$. The cochain ω can be uniquely (N is large!) written down in terms of the scalar products (α_μ, β_ν) (see Sec. 1.4). Let ω_n be a cochain of $C^*(W_n; W'_n)$ expressible by these scalar products, just as ω; then the image of $\varphi_{v_0}\psi_{v_1}\dots\psi_{v_m} \in C^*(W_n; k)$ under the homomorphism $C^*(W_n; k) \to C^*(W_n; W'_n)$ is

$$\sum(-1)^{v_i}v_i\psi_{v_0}\dots\hat{\psi}_{v_i}\dots\psi_{v_m}\sigma^1_{v_i-1}+d\omega_n,$$

i.e., $\xi_{v_0\dots v_m}+d\omega_n$ for $v_0+\dots+v_m = n+1$ and $d\omega_n$ for $v_0+\dots+v_m > n+1$. The theorem is proved.

COROLLARY. The sequence

$$H^*(W_{n+1}; k) \to H^*(W_n; k) \xrightarrow{\text{var}} H^*(W_n; W'_n),$$

in which the first homomorphism is induced by the inclusion $W_n \to W_{n+1}$, is exact.

2.4. The homomorphism var: $H^{q+1}(W_n, \mathfrak{o}_n; k) \to H^q(W_n, \mathfrak{o}_n; W'_n)$ is defined verbatim as its absolute analog in Sec. 2.1. We have the commutative diagram

$$\begin{array}{ccc} H^{q+1}(W_n, \mathfrak{o}_n; k) & \xrightarrow{\text{var}} & H^q(W_n, \mathfrak{o}_n; W'_n) \\ \downarrow & & \downarrow \\ H^{q+1}(W_n; k) & \xrightarrow{\text{var}} & H^q(W_n; W'_n), \end{array}$$

whose vertical arrows are induced by the inclusions of complexes. It follows from Theorems I and II that the vertical homomorphism on the right-hand side is a monomorphism, by virtue of which the diagram gives a complete description of the upper homomorphism var in terms of the lower. We restrict ourselves to the formulation of the relative variant of the corollary from Theorem III.

The sequence

$$H^*(W_{n+1}, \mathfrak{o}_{n+1}; k) \to H^*(W_n, \mathfrak{o}_n; k) \xrightarrow{\text{var}} H^*(W_n, \mathfrak{o}_n; W'_n),$$

in which the first homomorphism is induced by inclusion, is exact.

§ 3. Cohomologies of the Lie Algebra $\hat{W}_n$

3.1. Let us put $\hat{W}_n = W_n \otimes (k[t]/t^2k[t])$. As in the case $k = \mathbf{R}$ (see Sec. 0.3), $\hat{W}_n$ is a Lie algebra; as in the case $k = \mathbf{R}$ its elements can be written in the form $\xi + t\eta$ with $\xi, \eta \in W_n$. The algebra W_n is naturally imbedded into $\hat{W}_n$ ($\xi \mapsto \xi + t0$); the function $\xi + t\eta \mapsto \xi$ defines a retraction $\hat{W}_n \to W_n$.

Let us define $\gamma_s: C^{q-s}(W_n; \Lambda^s W'_n) \to C^q(\hat{W}_n; k)$ by the equation

$$(\gamma_s L)(\xi_1+t\eta_1,\dots,\xi_q+t\eta_q) = \sum_{1\le i_1<\dots<i_s\le q}(-1)^{i_1+\dots+i_s-s(s+1)/2}[L(\xi_1,\dots\hat{\xi}_{i_1}\dots\hat{\xi}_{i_s}\dots,\xi_q)](\xi_{i_1}\wedge\dots\wedge\xi_{i_s}).$$

The homomorphisms γ_s are obviously permutable with the differential and define certain homomorphisms $g_s: H^{q-s}(W_n; \Lambda^s W'_n) \to H^q(\hat{W}_n; k)$. The sum of the homomorphisms g_s gives the homomorphism

$$g = \bigoplus_s H^{q-s}(W_n;\, \Lambda^s W_n') \to H^q(\widehat{W}_n;\, k).$$

PROPOSITION 5. The homomorphism g is an isomorphism.

Proof. The sum of the homomorphisms $H^q(\widehat{W}_n; k) \to H^{q-s}(W_n; \Lambda^s W_n')$ induced by the homomorphisms $\beta_s : C^q(\widehat{W}_n; k) \to C^{q-s}(W_n; \Lambda^s W_n')$, defined by the equation

$$[(\beta_s L)(\xi_1, \ldots, \xi_{q-s})](\eta_1 \wedge \ldots \wedge \eta_s) = L(\xi_1, \ldots, \xi_{q-s}, t\eta_1, \ldots, t\eta_s),$$

is inverse to g.

Since $\Lambda^* W_n'$ is a ring (with the usual inner product), $H^*(\widehat{W}_n; \Lambda^* W_n')$ is also a ring. The following proposition is an obvious supplement to Proposition 5.

PROPOSITION 6. The isomorphism $H^*(\widehat{W}_n; \Lambda^* W_n') \to H^*(\widehat{W}_n; k)$ defined by the isomorphisms of Proposition 5 is annular.

Proposition 5 interprets the computations of § 1 as the partial computation of the ring $H^*(\widehat{W}_n; k)$. We shall not have the complete computation of this ring at present. Let us restrict ourselves to the remark that, as a simple calculation shows, $H^q(\widehat{W}_1; k) = H^q(W_1; k) \oplus H^{q-1}(W_1; W_1')$.

3.2. Let us define one more homomorphism $\delta : C^q(W_n; k) \to C^q(\widehat{W}_n; k)$ by the equation

$$\delta L(\xi_1 + t\eta_1, \ldots, \xi_q + t\eta_q) = \sum_{i=1}^{q} L(\xi_1, \ldots, \xi_{i-1}, \eta_i, \xi_{i+1}, \ldots, \xi_q).$$

This homomorphism also permutes with the differential and defines a certain homomorphism

$$\mathrm{var} : H^q(W_n; k) \to H^q(\widehat{W}_n; k).$$

This notation is motivated by the easily verified fact that the diagram

$$\begin{array}{ccc} & \overset{\mathrm{var}}{\nearrow} & H^{q-1}(W_n;\, W_n') \\ H^q(W_n;\, k) & & \downarrow g_1 \\ & \underset{\mathrm{var}}{\searrow} & H^q(\widehat{W}_n;\, k) \end{array}$$

is commutative.

3.3. $\mathfrak{o}_n$, together with W_n, is also a subalgebra of the algebra $\widehat{W}_n$. The verbatim repetition of the preceding part of the section gives the isomorphism

$$\bigoplus_s H^{q-s}(W_n,\, \mathfrak{o}_n;\, \Lambda^s W_n') \to H^q(\widehat{W}_n,\, \mathfrak{o}_n;\, k),$$

constituting the annular isomorphism $H^*(W_n, \mathfrak{o}_n; \Lambda^* W_n') \to H^*(\widehat{W}_n; k)$, and the homomorphism var: $H^q(W_n, \mathfrak{o}_n; k) \to H^q(\widehat{W}_n, \mathfrak{o}_n; k)$ with the commutative diagram

$$\begin{array}{ccc} & \overset{\mathrm{var}}{\nearrow} & H^{q-1}(W_n,\, \mathfrak{o}_n;\, W_n') \\ H^q(W_n,\, \mathfrak{o}_n;\, k) & & \downarrow g_1 \\ & \underset{\mathrm{var}}{\searrow} & H^q(\widehat{W}_n,\, \mathfrak{o}_n;\, k). \end{array}$$

§ 4. Characteristic Classes of Variations of Foliations

4.1. The interpretation of the elements of the rings $H^*(\widehat{W}_n; R)$ and $H^*(\widehat{W}_n, \mathfrak{o}_n; R)$ as characteristic classes of variations of foliations repeats almost verbatim the well-known interpretation of the elements of the rings $H^*(W_n; R)$ and $H^*(W_n, \mathfrak{o}_n; R)$ as characteristic classes of foliations (see, e.g., [2]). Therefore we set forth this interpretation in a sketchy form.

We restrict ourselves to nonsingular foliations on smooth manifolds since from the homotopical point of view this is the general case.

Let F be an oriented foliation of the class C^∞ of codimension n on a smooth manifold X. By a deformation of this foliation we understand a C^∞-foliation of codimension n+1 on $X \times \mathbf{R}$ whose fibres lie in sets of the form $X \times t$ and whose intersection with $X \times 0$ coincides with F (for the natural identification $X = X \times 0$). Two deformations are said to be tangential to each other if at each point of the set $X \times 0 \subset X \times \mathbf{R}$ the 1-jets of these deformations coincide. The class of tangential deformations of a foliation F is called the variation of the foliation F.

Let Φ be the variation of the foliation F. Let us fix the deformation φ representing the variation Φ, and let us denote by $S(\varphi)$ the manifold of ∞-jets of the submersions $X \times \mathbf{R} \to \mathbf{R}^{n+1}$, with the source lying in the set $X \times 0$ and with the orifice $0 \in \mathbf{R}^{n+1}$, constant on the fibres of the foliation φ, compatible with the orientations of this foliation, and such that the diagram

$$\begin{array}{ccc} X \times \mathbf{R} & \to & \mathbf{R}^{n+1} \\ \text{projection} \searrow & & \swarrow (x_1, \dots, x_{n+1}) \mapsto x_{n+1} \\ & \mathbf{R} & \end{array}$$

is commutative. Further, the elements $\xi, \eta \in S(\varphi)$ are called tangential if (i) the restrictions $\xi \mid_{X \times 0}$, $\eta \mid_{X \times 0}$ and (ii) the 2-jet submersions $X \times \mathbf{R} \to \mathbf{R}^{n+1}$ determined by ξ, η coincide. Identifying the tangential elements of the manifold $S(\varphi)$ we get a new manifold; the latter is determined by the variation Φ in an obvious sense, and we denote it by $S(\Phi)$.

The usual differential notions, in particular, the tangent spaces and de Rham complex, have meaning for $S(\Phi)$. The homologies of the de Rham complex coincide with the cohomologies of the manifold $S(\Phi)$. Let us observe that the natural projection of the manifold $S(\Phi)$ onto the manifold of the tangential k-hedrons of $X \times 0$, normal to the foliation and compatible with the orientation of the latter, is a homotopy equivalence. Moreover, the group SO(n) acts freely on $S(\Phi)$ (as the group of motions of the space $\mathbf{R}^{n+1}$, stationary on the x_{n+1} axis), and the factor space $S(\Phi)/SO(n)$ is naturally homotopically equivalent to X.

The construction, usual in the theory of the characteristic classes of foliations, defines the canonical epimorphism of the tangent space $T_\xi S(\Phi)$ at an arbitrary point $\xi \in S(\Phi)$ onto $\hat{W}_n = W_n \otimes (\mathbf{R}[t]/t^2 \mathbf{R}[t])$. Namely, let us fix a smooth curve on $S(\Phi)$ with origin ξ touching the proposed vector of $T_\xi S(\Phi)$. This curve can be represented as a one-parameter family of submersions φ_t of some open set $U \subset X \times I$ into $\mathbf{R}^{n+1}$. Further, these submersions can be obtained from a single (φ_0), by composing with the diffeomorphisms $\rho_t : \mathbf{R}^{n+1} \to \mathbf{R}^{n+1}$, not changing the last coordinate and depending smoothly on t. The family ρ_t determines a smooth vector field in $\mathbf{R}^{n+1}$ with zero (n+1)-th components, and the arbitrariness of the construction makes it definite only up to the addition of a planar vector field and a vector field divisible by the square of the last coordinate, i.e., as an element of the algebra $\hat{W}_n$.

The constructed homomorphism $T_\xi S(\Phi)$ determines the homomorphism $\Lambda^q(\hat{W}_n) \to \Lambda^q(T_\xi S(\Phi))'$ and, since the latter is determined for an arbitrary ξ, also the homomorphism

$$C^*(\hat{W}_n; \mathbf{R}) \to \Omega^*(S(\Phi)), \tag{6}$$

where Ω^* denotes the de Rham complex. The homomorphism (6) permutes with the action of the group SO(n) and with the differential, and therefore it determines the homomorphisms

$$C^*(\hat{W}_n, \mathfrak{o}_n; \mathbf{R}) \to \Omega^*(S(\Phi)/SO(n)),$$

$$H^*(\hat{W}_n; \mathbf{R}) \to H^*(S(\Phi); \mathbf{R}), \tag{7}$$

$$H^*(\hat{W}_n, \mathfrak{o}_n; \mathbf{R}) \to H^*(S(\Phi)/SO(n); \mathbf{R}) = H^*(X; \mathbf{R}), \tag{8}$$

and if the normal stratification of the foliation F is trivialized, then there exists a section $X \to S(\Phi)$ defining a homomorphism $H^*(S(\Phi); \mathbf{R}) \to H^*(X; \mathbf{R})$ whose composition with (7) gives the homomorphism

$$H^*(\hat{W}_n; \mathbf{R}) \to H^*(X; \mathbf{R}). \tag{9}$$

The homomorphisms (8) and (9) also permit us to interpret the elements of the ring $H^*(\hat{W}_n, \mathfrak{o}_n; \mathbf{R})$ as characteristic classes of the variations of foliations, and the elements of the ring $H^*(\hat{W}_n; \mathbf{R})$ as characteristic classes of the variations of foliations with a trivialized normal stratification.

4.2. Let F be a smooth foliation of codimension n on X and φ be its deformation. Let us denote by F_t ($t \in \mathbf{R}$) the foliation on X determined by the foliation φ under the identification $X = X \times t$, and by Φ the variation of the foliation F determined by φ. The foliation F_t determines a certain homomorphism

$$\alpha_t : H^*(W_n, \mathfrak{o}_n; \mathbf{R}) \to H^*(X; \mathbf{R}),$$

and the variation Φ, by virtue of the construction of 4.1, determines a certain homomorphism

$$\beta : H^*(\widehat{W}_n, \mathfrak{o}_n; \mathbf{R}) \to H^*(X; \mathbf{R}),$$

and it is clear that the diagrams

$$\begin{array}{ccc} H^*(W_n, \mathfrak{o}_n; \mathbf{R}) & \overset{\alpha_0}{\searrow} & \\ \text{inclusion} \downarrow & & H^*(X; \mathbf{R}), \\ H^*(\widehat{W}_n, \mathfrak{o}_n \mathbf{R}) & \nearrow_{\beta} & \end{array} \qquad \begin{array}{ccc} H^*(W_n, \mathfrak{o}_n; \mathbf{R}) & \searrow \left.\frac{\partial\alpha}{\partial t}\right|_{t=0} & \\ \text{var} \downarrow & & H^*(X; \mathbf{R}) \\ H^*(\widehat{W}_n, \mathfrak{o}_n; \mathbf{R}) & \nearrow_{\beta} & \end{array}$$

are commutative. If the normal stratification of the foliation F is trivialized, then this is also true for the cohomologies $H^*(W_n; \mathbf{R})$ and $H^*(\widehat{W}_n; \mathbf{R})$ (we retain in this case the symbols α_t and β).

PROPOSITION 7. If a class $x \in H^*(W_n, \mathfrak{o}_n; \mathbf{R})$ lies in the image of the inclusion homomorphism $H^*(W_{n+1}, \mathfrak{o}_{n+1}; \mathbf{R}) \to H^*(W_n, \mathfrak{o}_n; \mathbf{R})$, then $\alpha_t(x)$ does not depend on t. If the normal stratification of the foliation F is trivialized and the class $x \in H^*(W_n; \mathbf{R})$ lies in the image of the inclusion homomorphism $H^*(W_{n+1}; \mathbf{R}) \to H^*(W_n; \mathbf{R})$, then $\alpha_t(x)$ does not depend on t.

This proposition coincides with the Heitsch's theorem [11]. It follows from the corollary to Theorem (III) (Sec. 2.3) and the proposition of Sec. 2.4. Let us observe that an entirely simple proof of the Heitsch's theorem is given in [2].

PROPOSITION 8. For arbitrary $x, y \in H^*(W_n, \mathfrak{o}_n; \mathbf{R})$ the class $\alpha_t(x)[\partial\alpha_t(y)/\partial t]_{t=0}$ is equal to zero. If the normal stratification of a foliation is trivialized, then the same is true for arbitrary $x, y \in H^*(W_n; \mathbf{R})$.

This follows from Theorems I and II and the multiplicativity of the homomorphisms (8) and (9).

LITERATURE CITED

1. I. N. Bernshtein and B. I. Rosenfel'd, "On characteristic classes of foliations," Funktsional. Analiz i Ego Prilozhen., 6, No. 1, 68-69 (1972).
2. I. N. Bernshtein and B. I. Rosenfel'd, "Homogeneous spaces of infinite-dimensional Lie algebras and characteristic classes of foliations," Uspekhi Matem. Nauk, 28, No. 4, 103-138 (1973).
3. H. Weil, Classical Groups, Their Invariants and Representations, Princeton University Press, Princeton (1946).
4. I. M. Gel'fand and D. B. Fuks, "Cohomologies of Lie algebras of formal vector fields," Izv. Akad. Nauk SSSR, Ser. Matem., 34, 322-337 (1970).
5. C. Godbillon, "Cohomologies of Lie algebras of formal vector fields," Uspekhi Matem. Nauk, 28, No. 4, 139-152 (1973).
6. Theory of Lie Algebras. Topology of Lie Groups [Russian translation], Izd. Inost. Lit., Moscow (1962).
7. R. Bott and A. Haeflinger, "On characteristic classes of Γ-foliations," Bull. Amer. Math. Soc., 78, No. 6, 1039-1044 (1972).
8. R. Bott, "Some remarks on continuous cohomology," Symposium "Manifolds," Tokyo (1973).
9. C. Godbillon and J. Vey, "Un invariant des feuilletages de codimension 1," Compt. Rend. Acad. Sci. (Paris), 273, No. 2, A92-A95 (1971).
10. A. Haeflinger, Homotopy and Integrability, Springer Lecture Notes in Mathematics, No. 197, Springer-Verlag, New York (1971), pp. 133-163.
11. J. Heitsch, "On deformations of secondary characteristic classes, Topology, 12, No. 4, 381-388 (1973).
12. G. Hochschild and J.-P. Serre, "Cohomology of Lie algebras," Ann. Math., 57, 591-603 (1953).
13. W. Thurston, "Non-cobordant foliations of S^3," Bull. Amer. Math. Soc., 78, No. 4, 511-514 (1972).

14.

(with D. B. Fuks and A. M. Gabrielov)

The Gauss-Bonnet theorem and the Atiyah-Patodi-Singer functionals for the characteristic classes of foliations

Topology **15** (1976) 165–188. MR **55**:4201. Zbl. **347**:57009

WE PROVE in this article some formulae of Gauss–Bonnet kind for the characteristic classes of foliations (see [2], [5]). Namely, fixing a Riemannian metric on a manifold with a foliation allows one to determine explicitly for each of these characteristic classes a representing differential form (see, for example, [4]). If a domain X with a piecewise smooth boundary transversal to the foliation is given in the manifold, then the integral of such a form over X, corrected by adding the integrals over faces of different dimensions of certain forms depending on the foliation and the metric near ∂X, depends only on the induced metric on ∂X (and, of course, on the foliation on X). By using these formulae one can, in particular, extend the definition of the characteristic classes of foliations to the piecewise smooth case.

The general idea of making topological invariants local by means of explicit formulae depending on a metric or another auxiliary structure was suggested by I. M. Gel'fand at the Nice congress [6]. This idea is partially realized here for the characteristic classes of foliations (though the general theory developed in §§2, 4 and 5 will possibly have wider applications). Remarkable progress in this direction has been achieved in the recent work of Atiyah–Patodi–Singer [1] where a formula of Gauss–Bonnet kind is obtained for the Hirzebruch polynomials (of the Pontryagin forms) and the signature. The essentially new aspect of the Atiyah–Patodi–Singer formula is that its right-hand-side—a quantity depending on the induced metric of the boundary—is not expressible as the integral over the boundary of a form locally determined by the metric. (It measures the degree of the asymmetry of the spectrum of the elliptic operator in the space of forms which takes φ into $(-1)^p(d*\varphi - *d\varphi)$, where $p = (\deg\varphi)/2)$. This property of the 'boundary functional'—its 'non-localizability' is still valid in our formula.

In further comparing the Atiyah–Patodi–Singer formula with ours we should point out an important advantage of the former. Our boundary functional depends on the auxiliary structure on the boundary and on the whole foliation. The corresponding part of the Atiyah–Patodi–Singer formula is divided into two parts: a quantity depending only on the metric properties of the boundary, and a quantity depending only on the topological properties of the whole manifold. We do not see how to obtain such a decomposition of our boundary functional.

More detailed discussion of the interrelations between our formula, the Atiyah–Patodi–Singer formula, and the classical Gauss–Bonnet formula will be found in #4 of §1.

The first section is devoted to codimension one foliations; in its plan it is a reduced copy of the remainder of the article. We have tried to write it more concretely and without omitting calculations, especially so because the method of calculation is often of more importance for us than the result.

The contents of the remaining sections will be briefly described at the end of §1.

§1. THE GAUSS–BONNET THEOREM FOR THE GODBILLON–VEY CLASS

1. The Godbillon–Vey form and the form Γ^1

Definition 1.1. Let $\mathcal{F}$ be a smooth (C^∞–) orientable foliation of codimension 1 on a smooth manifold X. A pair (ω, η) of differential 1-forms on X is called associated with $\mathcal{F}$ if ω is a nondegenerate (nowhere vanishing) form determining $\mathcal{F}$ and η satisfies the condition $\eta \wedge \omega = d\omega$.

Note that one can construct an associated pair of forms for any orientable foliation of codimension 1: ω exists because of orientability, and the existence of η with $\eta \wedge \omega = d\omega$ is the classical integrability condition for the Pfaffian system $\omega = 0$. Moreover ω is unique up to multiplication by a nowhere-vanishing smooth function.

An associated pair of forms can be constructed explicitly if X is furnished not only with $\mathcal{F}$ but also with a smooth Riemannian metric, say g. In this case one can put two extra conditions on ω and η: $\|\omega\| = 1$, and $(\eta, \omega) = 0$. It is easy to see that associated pairs ω, η with these properties do exist. Moreover if X is connected, there are precisely two such pairs of which the second can be obtained from the first by changing ω to $-\omega$. Choosing one of these pairs is equivalent to choosing an orientation of the foliation. The pairs are called associated with $\mathcal{F}$ and g.

Definition 1.2. If (ω, η) is associated with a foliation $\mathcal{F}$, then the 3-form $\Gamma^0(\eta) = \eta \wedge d\eta$ is called the Godbillon–Vey form of $\mathcal{F}$ attached to (ω, η). If (ω, η) is associated with $\mathcal{F}$ and with a metric g, then $\Gamma^0(\eta)$ is denoted also by $\Gamma^0(g)$, and is called the Godbillon–Vey form of $\mathcal{F}$ attached to g.

The following properties of the Godbillon–Vey form are well known (see [11]).

PROPOSITION 1.1. (*i*) *The Godbillon–Vey form is closed.*

(*ii*) *The cohomology class of the Godbillon–Vey form of a given foliation does not depend on the associated pair of forms.*

Proof. Let (ω, η) be a pair of forms associated with the foliation. The relation $\eta \wedge \omega = d\omega$ implies that $d(\eta \wedge \omega) = 0$ and $d(\eta \wedge \omega) = d\eta \wedge \omega - \eta \wedge d\omega = d\eta \wedge \omega - \eta \wedge \eta \wedge \omega = d\eta \wedge \omega$. Hence $d\eta \wedge \omega = 0$, and therefore $d\eta$ is the product of a form ζ and ω. Thus

$$d\Gamma^0 = d\eta \wedge d\eta = \zeta \wedge \omega \wedge \zeta \wedge \omega = 0.$$

Now let (ω', η') be another pair of forms associated with the same foliation. Then define

$$\Gamma^1(\omega, \eta; \omega', \eta') = \eta' \wedge \eta - d \log \left|\frac{\omega'}{\omega}\right| \wedge (\eta' + \eta)$$

(here ω'/ω is the function with which one has to multiply ω to obtain ω'). Clearly

$$d \log \left|\frac{\omega'}{\omega}\right| \wedge \omega' = \frac{\omega}{\omega'} d\left(\frac{\omega'}{\omega}\right) \wedge \omega' = d\left(\frac{\omega'}{\omega}\right) \wedge \omega = d\omega' - \frac{\omega'}{\omega} d\omega = \eta' \wedge \omega' - \frac{\omega'}{\omega} \eta \wedge \omega = (\eta' - \eta) \wedge \omega'.$$

Hence $d \log |\omega'/\omega| = \eta' - \eta + \varphi\omega'$ where φ is a smooth function, and

$$\begin{aligned} d\Gamma^1(\omega, \eta; \omega', \eta') &= d\eta' \wedge \eta - \eta' \wedge d\eta + d \log \left|\frac{\omega'}{\omega}\right| \wedge (d\eta' + d\eta) \\ &= d\eta' \wedge \eta - \eta' \wedge d\eta + (\eta' - \eta) \wedge (d\eta' + d\eta) + \varphi\omega' \wedge d\eta' \\ &\quad + \varphi \frac{\omega'}{\omega} \omega \wedge d\eta = \eta' \wedge d\eta' - \eta \wedge d\eta = \Gamma^0(\eta') - \Gamma^0(\eta) \end{aligned}$$

(we used the above relation $\omega \wedge d\eta = 0$ and the similar relation $\omega' \wedge d\eta' = 0$). Thus the difference $\Gamma^0(\eta') - \Gamma^0(\eta)$ is an exact form.

Q.E.D.

The explicit formulae obtained in the last proof will be of no less importance for us than the proposition itself. We shall be particularly interested in the form $\Gamma^1(\omega, \eta; \omega', \eta')$ and in the equality $d\Gamma^1(\omega, \eta; \omega', \eta') = \Gamma^0(\eta') - \Gamma^0(\eta)$.

Note that $\Gamma^1(\omega', \eta'; \omega, \eta) = -\Gamma^1(\omega, \eta; \omega', \eta')$.

Note also that changing the sign of ω or ω' does not change $\Gamma^1(\omega, \eta; \omega', \eta')$ so the last is well defined by a pair of metrics g, g': one takes for (ω, η) and (ω', η') arbitrary pairs of forms associated with $\mathcal{F}$, g and with $\mathcal{F}$, g'. The resulting form is denoted by $\Gamma^1(g, g')$; the relations $d\Gamma^1(g, g') = \Gamma^0(g') - \Gamma^0(g)$ and $\Gamma^1(g', g) = -\Gamma^1(g, g')$ evidently hold.

2. The form Γ^2

Let (ω, η), (ω', η'), (ω'', η'') be three pairs of forms associated with a certain foliation. Then define

$$\Gamma^2(\omega, \omega', \omega'') = \frac{1}{2}\left(\log \left|\frac{\omega'}{\omega}\right| d \log \left|\frac{\omega''}{\omega}\right| - \log \left|\frac{\omega''}{\omega}\right| d \log \left|\frac{\omega'}{\omega}\right|\right).$$

Clearly $\Gamma^2(\omega, \omega', \omega'')$ is skew-symmetric in $\omega, \omega', \omega''$. It is clear also that if the pairs (ω, η),

$(\omega', \eta'), (\omega'', \eta'')$ are attached to metrics g, g', g'', then $\Gamma^2(\omega, \omega', \omega'')$ is well defined by the metrics; it is denoted then by $\Gamma^2(g, g', g'')$.

PROPOSITION 1.2. $d\Gamma^2(\omega, \omega', \omega'') = \Gamma^1(\omega', \eta'; \omega'', \eta'') - \Gamma^1(\omega, \eta; \omega'', \eta'') + \Gamma^1(\omega, \eta; \omega', \eta')$.

Proof. Since $d \log |\omega'/\omega| - \eta' + \eta$ is divisible by ω' (see the proof of Proposition 1.1), and $d \log |\omega''/\omega| - \eta'' + \eta$ is divisible by ω'' and since $\omega' \wedge \omega'' = 0$, we have

$$\left(d \log \left|\frac{\omega'}{\omega}\right| - \eta' + \eta\right) \wedge \left(d \log \left|\frac{\omega''}{\omega}\right| - \eta'' + \eta\right) = 0.$$

On the other hand we have

$$\begin{aligned}
&\left(d \log \left|\frac{\omega'}{\omega}\right| - \eta' + \eta\right) \wedge \left(d \log \left|\frac{\omega''}{\omega}\right| - \eta'' + \eta\right) \\
&= d \log \left|\frac{\omega'}{\omega}\right| \wedge d \log \left|\frac{\omega''}{\omega}\right| + \eta' \wedge \eta'' - \eta \wedge \eta'' + \eta \wedge \eta' \\
&\quad + d \log \left|\frac{\omega'}{\omega}\right| \wedge (\eta' - \eta'') + d \log \left|\frac{\omega''}{\omega}\right| \wedge (\eta' - \eta) \\
&= d \log \left|\frac{\omega'}{\omega}\right| \wedge d \log \left|\frac{\omega''}{\omega}\right| - \eta'' \wedge \eta' + \eta'' \wedge \eta - \eta' \wedge \eta \\
&\quad + d \log \left|\frac{\omega'}{\omega}\right| \wedge (\eta'' + \eta') - d \log \left|\frac{\omega''}{\omega}\right| \wedge (\eta'' + \eta) + d \log \left|\frac{\omega'}{\omega}\right| \wedge (\eta' + \eta) \\
&= d\Gamma^2(\omega, \omega', \omega'') - \Gamma^1(\omega', \eta'; \omega'', \eta'') + \Gamma^1(\omega, \eta; \omega'', \eta'') \\
&\quad - \Gamma^1(\omega, \eta; \omega', \eta').
\end{aligned}$$

PROPOSITION 1.3. *If (ω''', η''') is a fourth form associated with our foliation, then*

$$\Gamma^2(\omega', \omega'', \omega''') - \Gamma^2(\omega, \omega'', \omega''') + \Gamma^2(\omega, \omega', \omega''') - \Gamma^2(\omega, \omega', \omega'') = 0.$$

Proof.

$$\begin{aligned}
\Gamma^2(\omega, \omega', \omega''') - \Gamma^2(\omega, \omega', \omega'') &= \frac{1}{2} \log \left|\frac{\omega'}{\omega}\right| \left(d \log \left|\frac{\omega'''}{\omega}\right| - d \log \left|\frac{\omega''}{\omega}\right|\right) \\
&\quad - \frac{1}{2}\left(\log \left|\frac{\omega'''}{\omega}\right| - \log \left|\frac{\omega''}{\omega}\right|\right) d \log \left|\frac{\omega'}{\omega}\right| \\
&= \frac{1}{2}\left(\log \left|\frac{\omega''}{\omega'''}\right| d \log \left|\frac{\omega'}{\omega}\right| - \log \left|\frac{\omega'}{\omega}\right| d \log \left|\frac{\omega''}{\omega'''}\right|\right); \\
\Gamma^2(\omega', \omega'', \omega''') - \Gamma^2(\omega, \omega'', \omega''') &= \Gamma^2(\omega'', \omega''', \omega') - \Gamma^2(\omega'', \omega''', \omega) \\
&= \frac{1}{2}\left(\log \left|\frac{\omega'}{\omega}\right| d \log \left|\frac{\omega''}{\omega'''}\right| - \log \left|\frac{\omega''}{\omega'''}\right| d \log \left|\frac{\omega'}{\omega}\right|\right).
\end{aligned}$$

For convenience of reference we collect here the relations

$$d\Gamma^0(\eta) = 0 \tag{1}$$

$$d\Gamma^1(\omega, \eta; \omega', \eta') = \Gamma^0(\eta') - \Gamma^0(\eta); \tag{2}$$

$$d\Gamma^2(\omega, \omega', \omega'') = \Gamma^1(\omega', \eta'; \omega'', \eta'') - \Gamma^1(\omega, \eta; \omega'', \eta'') + \Gamma^1(\omega, \eta; \omega', \eta'); \tag{3}$$

$$0 = \Gamma^2(\omega', \omega'', \omega''') - \Gamma^2(\omega, \omega'', \omega''') + \Gamma^2(\omega, \omega', \omega''') - \Gamma^2(\omega, \omega', \omega''), \tag{4}$$

where (ω, η), (ω', η'), (ω'', η''), (ω''', η''') are pairs of forms associated with a certain foliation. The metric version is

$$d\Gamma^0(g) = 0; \tag{1'}$$

$$d\Gamma^1(g, g') = \Gamma^0(g') - \Gamma^0(g); \tag{2'}$$

$$d\Gamma^2(g, g', g'') = \Gamma^1(g', g'') - \Gamma^1(g, g'') + \Gamma^1(g, g'); \tag{3'}$$

$$0 = \Gamma^2(g', g'', g''') - \Gamma^2(g, g', g''') + \Gamma^2(g, g', g'') - \Gamma^2(g, g', g''), \tag{4'}$$

where g, g', g'', g''' are metrics.

3. The Gauss–Bonnet formula

Let X be a connected oriented closed domain in a smooth three-dimensional manifold W without a boundary. Suppose X is the intersection of a finite set of compact smooth three-dimensional submanifolds of W whose boundaries are in general position. There is a natural decomposition of X into the disjoint sum of four smooth manifolds ('open skeleta'): a three-dimensional one, int X, a two-dimensional one, Y, a one-dimensional one, Z, and a null-dimensional one. Let $Y_1, \ldots, Y_k$ be the components of Y, and $Z_1, \ldots, Z_l$ the components of Z. It is clear that $X \cup Y$ is a smooth 3-manifold whose boundary is Y, and $Y_i \cup (\bar{Y}_i \cap Z)$ is a smooth 2-manifold whose boundary is $\bar{Y}_i \cap Z$.

Let W have an orientable codimension-one foliation $\mathcal{F}$, transversal to Y and Z. We fix three pairs of forms: a pair (ω, η) associated with F on W, a pair (ω_1, η_1) associated with $\mathcal{F}|_Y$ on Y, and a pair (ω_2, η_2) associated with $\mathcal{F}|_Z$ on Z (the last means only that ω_2 is non-degenerate). We assume also that the forms $\omega_1|_{Y_2}$, $\eta_1|_{Y_i}$ can be extended to $\bar{Y}_i$ and the forms $\omega_2|_{Z_i}$, $\eta_2|_{Z_i}$ can be extended to $\bar{Z}_i$.

Define

$$H(\omega, \eta; \omega_1, \eta_1; \omega_2) = \int_X \Gamma^0(\eta) + \int_Y \Gamma^1(\omega, \eta; \omega_1, \eta_1; \omega_2) - \sum_{i=1}^{k} \int_{\bar{Y}_i \cap Z} \Gamma^2(\omega, \omega_1|_{Y_i}, \omega_2) \qquad (5)$$

(the orientation of Y is induced by that of Z, the orientation of $\bar{Y}_i \cap Z$ is induced by that of Y, and the form $\omega_1|_{Y_i}$ is considered as defined on $\bar{Y}_i \cap Z$).

The most important case of the above situation is that when all forms $\omega, \eta; \omega_1, \eta_1; \omega_2$ are defined by a Riemannian metric g in W: this metric induces Riemannian metrics in int X, Y, and Z, and they interact with $\mathcal{F}$ and give forms $\omega, \eta; \omega_1, \eta_1; \omega_2, \eta_2$ satisfying the above extendability condition. Here $\omega, \omega_1, \omega_2$ are defined only up to sign; nevertheless the forms $\Gamma^0, \Gamma^1, \Gamma^2$ are well defined. Substituting them into the right-hand-side of (5) we obtain a value depending (when $\mathcal{F}$ is fixed) only on g; we denote it by $H(g)$.

Note that unlike the metrics in X, Y, Z the pairs (ω, η), (ω_1, η_1), (ω_2, η_2) are not induced by a single pair of forms in W. This is our motivation for choosing (ω, η), (ω_1, η_1), (ω_2, η_2) independently in the above construction.

THEOREM 1.1. *$H(\omega, \eta; \omega_1, \eta_1; \omega_2)$ does not depend (when $\mathcal{F}$, $\omega_1, \eta_1, \omega_2$ are fixed) on ω and η. In particular, if g, g' are two Riemannian metrics in W inducing the same metric in Y then $H(g) = H(g')$.*

It is essential that the integrands in the right-hand-side of (5) do depend on ω and η and that, moreover, $H(\omega, \eta; \omega_1, \eta_1; \omega_2)$ cannot be represented as a combination of integrals of forms universally expressible in terms of $\omega_1, \eta_1, \omega_2$. The precise statement and the proof of this property of H (the 'non-localizability' of H) will be given in subsection 5.

Instead of Theorem 1.1 we prove the following more precise assertion.

THEOREM 1.2. *Let (ω', η'), (ω_1', η_1'), (ω_2', η_2') be other pairs of forms associated with F, $F|_Y$, $F|_Z$, and satisfying the extendability conditions. Then*

$$H(\omega', \eta'; \omega_1', \eta_1'; \omega_2') - H(\omega, \eta; \omega_1, \eta_1; \omega_2)$$
$$= \oint_Y \Gamma^1(\omega_1, \eta_1; \omega_1', \eta_1') + \sum_{i=1}^{k} \int_{\bar{Y}_i \cap Z} [\Gamma^2(\omega_1|_{Y_i}, \omega_2, \omega_2') - \Gamma^2(\omega_1|_{Y_i}, \omega_1'|_{Y_i}, \omega_2')]. \qquad (6)$$

Proof. The formula (2) and Stokes theorem imply that

$$\int_X \Gamma^0(\eta') - \int_X \Gamma^0(\eta) = \int_Y \Gamma^0(\omega, \eta; \omega', \eta'). \qquad (7)$$

Furthermore, (3) implies that

$$d\Gamma^2(\omega, \omega', \omega_1') = \Gamma^1(\omega', \eta'; \omega_1', \eta_1') - \Gamma^1(\omega, \eta; \omega_1', \eta_1') + \Gamma^1(\omega, \eta; \omega', \eta'),$$

$$d\Gamma^2(\omega, \omega_1, \omega_1') = \Gamma^1(\omega_1, \eta_1; \omega_1', \eta_1') - \Gamma^1(\omega, \eta; \omega_1', \eta_1') + \Gamma^1(\omega, \eta; \omega_1, \eta_1).$$

Subtracting these equations and using Stokes' formula again gives the relation

$$\int_Y \Gamma^1(\omega', \eta'; \omega_1', \eta_1') - \int_Y \Gamma^1(\omega, \eta; \omega_1, \eta_1) = \int_Y \Gamma^1(\omega_1, \eta_1; \omega_1', \eta_1')$$
$$- \int_Y \Gamma^1(\omega, \eta; \omega', \eta') + \sum_{i=1}^{k} \int_{\bar{Y}_i \cap Z} [\Gamma^2(\omega, \omega', \omega_1'|_{Y_i}) - \Gamma^2(\omega, \omega_1|_{Y_i}, \omega_1'|_{Y_i})]. \quad (8)$$

Finally (4) implies that

$$0 = \Gamma^2(\omega_1, \omega_1', \omega_2') - \Gamma^2(\omega, \omega_1', \omega_2') + \Gamma^2(\omega, \omega_1', \omega_2') - \Gamma^2(\omega, \omega_1, \omega_1'),$$
$$0 = \Gamma^2(\omega', \omega_1', \omega_2') - \Gamma^2(\omega, \omega_1', \omega_2') + \Gamma^2(\omega, \omega', \omega_2') - \Gamma^2(\omega, \omega', \omega_1'),$$
$$0 = \Gamma^2(\omega_1, \omega_2, \omega_2') - \Gamma^2(\omega, \omega_2, \omega_2') + \Gamma^2(\omega, \omega_1, \omega_2') - \Gamma^2(\omega, \omega_1, \omega_2).$$

and subtracting the sum of the second and the third equalities from the first one gives us the following:

$$\Gamma^2(\omega, \omega', \omega_1') - \Gamma^2(\omega, \omega_1, \omega_1') = [\Gamma^2(\omega_1, \omega_2, \omega_2') - \Gamma^2(\omega_1, \omega_1', \omega_2')]$$
$$+ [\Gamma^2(\omega', \omega_1', \omega_2') - \Gamma^2(\omega, \omega_1, \omega_2)] - [\Gamma^2(\omega, \omega_2, \omega_2') - \Gamma^2(\omega, \omega', \omega_2')]. \quad (9)$$

Substituting the right-hand-side of (9) into (8) and adding (8) to (7) we obtain the equality

$$H(\omega', \eta'; \omega_1', \eta_1'; \omega_2') - H(\omega, \eta; \omega_1, \eta_1; \omega_2)$$
$$= \int_Y \Gamma^1(\omega_1, \eta_1; \omega_1', \eta_1') + \sum_{i=1}^{k} \int_{\bar{Y}_i \cap Z} [\Gamma^2(\omega_1|_{Y_i}, \omega_2, \omega_2')$$
$$- \Gamma^2(\omega_1|_{Y_i}, \omega_1'|_{Y_i}, \omega_2')] - \sum_{i=1}^{k} \int_{\bar{Y}_i \cap Z} [\Gamma^2(\omega, \omega_2, \omega_2') - \Gamma^2(\omega, \omega', \omega_2')].$$

But the last sum is equal to zero as it contains two integrals over each component of Z and in these two integrals the integrands coincide (this is the difference between the last sum and the previous one) and the orientations of the components are different.

The expression $H(\omega, \eta; \omega_1, \eta_1; \omega_2)$, which by Theorem 1.1 does not depend on ω and η, nevertheless depends on $\mathscr{F}$, and, thus on int X. The following proposition puts an essential restriction on the nature of this dependence.

THEOREM 1.3. *If the pair X, $\mathscr{F}$ is replaced by another pair, $\tilde{X}$, $\tilde{\mathscr{F}}$, with X-int X and $\mathscr{F}|_{X\text{-int}\,X}$ unchanged, then the function*

$$(\omega_1, \eta_1, \omega_2) \mapsto H(\omega, \eta; \omega_1, \eta_1; \omega_2)$$

remains unchanged up to an additive constant.

More precisely: let $\tilde{W}, \tilde{X}, \tilde{Y}, \tilde{Z}, \tilde{\mathscr{F}}$ be similar to $W, X, Y, Z, \mathscr{F}$, and let $\varphi: \tilde{X}\text{-int}\,\tilde{X} \to X\text{-int}\,X$ be a diffeomorphism compatible with $\mathscr{F}|_{X\text{-int}\,X}$ and $\tilde{\mathscr{F}}|\tilde{X}\text{-int}\,\tilde{X}$. Then the difference

$$H(\tilde{\omega}, \tilde{\eta}; \varphi^*\omega_1, \varphi^*\eta_1; \varphi^*\omega_2) - H(\omega, \eta; \omega_1, \eta_1; \omega_2)$$

where $\tilde{\omega}, \tilde{\eta}$ is a pair associated with $\tilde{\mathscr{F}}$, depends only on $\tilde{\mathscr{F}}$, $\mathscr{F}$ and φ, but not on $\tilde{\omega}, \tilde{\eta}, \omega, \eta, \omega_1, \eta_1, \omega_2$.

Metric version: the difference $H(\tilde{g}) - H(g)$, where $\tilde{g}$ and g are metrics in $\tilde{W}$ and W, with respect to which the boundaries $\tilde{X}$-int $\tilde{X}$ and X-int X are isometric, does not depend on $\tilde{g}, g$.

Proof

$$[H(\tilde{\omega}', \tilde{\eta}'; \varphi^*\omega_1', \varphi^*\eta_1'; \varphi^*\omega_2') - H(\omega', \eta'; \omega_1', \eta_1'; \omega_2')]$$
$$- [H(\tilde{\omega}, \tilde{\eta}; \varphi^*\omega_1, \varphi^*\eta_1; \varphi^*\omega_2) - H(\omega, \eta; \omega_1, \eta_1; \omega_2)]$$
$$= [H(\tilde{\omega}', \tilde{\eta}'; \varphi^*\omega_1', \varphi^*\eta_1'; \varphi^*\omega_2') - H(\tilde{\omega}, \tilde{\eta}; \varphi^*\omega_1, \varphi^*\eta_1; \varphi^*\omega_2)]$$
$$- [H(\omega', \eta'; \omega_1', \eta_1'; \omega_2') - H(\omega, \eta; \omega_1, \eta_1; \omega_2)]$$
$$= \left\{\int_{\tilde{Y}} \Gamma^1(\varphi^*\omega_1', \varphi^*\eta_1'; \varphi^*\omega_1, \varphi^*\eta_1)\right.$$
$$\left. + \sum_{i=1}^{k} \int_{\tilde{Y}_i \cap \tilde{Z}} [\Gamma^2(\varphi^*\omega_1|_{\tilde{Y}_i}, \varphi^*\omega_2, \varphi^*\omega_2') - \Gamma^2(\varphi^*\omega_1|_{\tilde{Y}_i}, \varphi^*\omega_1'|Y_i, \varphi^*\omega_2)]\right\}$$
$$- \left\{\int_Y \Gamma^1(\omega_1', \eta_1'; \omega_1, \eta_1) + \sum_{i=1}^{k} \int_{\bar{Y}_i \cap Z} [\Gamma^2(\omega_1|_{Y_i}, \omega_2, \omega_2') - \Gamma^2(\omega_1|_{Y_i}, \omega_1'|_{Y_i}, \omega_2)]\right\} = 0.$$

It is interesting to compare Theorems 1.1 and 1.3 with the classical Gauss–Bonnet formula and the Atiyah–Patodi–Singer theorem[1]. The last represents the integral of the Hirzebruch polynomial L_k in the Pontryagin forms over a Riemannian $4k$-manifold with a boundary as the sum of three terms: (i) something depending only on topological properties of X, (ii) something depending only on metric properties of ∂X, (iii) the integral over ∂X of a form defined locally by the metric in the neighbourhood of ∂X.† Moreover, (ii) cannot be represented as the integral of a form locally defined by the metric of ∂X. The Gauss–Bonnet formula represents in a similar way the integral of the Gaussian curvature of a surface with a boundary, but there is no summand like (ii) in this representation; the sources of this phenomenon are pointed out in [1]. The other difference between the Gauss–Bonnet formula and the Atiyah–Patodi–Singer theorem is that the first holds in a more general situation—when the boundary is piecewise smooth. Then the summand (iii) breaks into two parts: in addition to the integral over the boundary of a 'boundary' form there stands an expression depending on the metric near the vertices of the bounding curve.

The metric version of Theorem 1.1 above represents the integral of the Godbillon–Vey form of a codimension one foliation on a compact Riemann 3-manifold with a piecewise smooth boundary as a sum of two summands: one like (iii) (a sum of integrals over the skeletons of a boundary of forms depending on the metric and the foliation near these skeletons), and some $H(g)$ depending on the metric on the boundary and the whole foliation. By Theorem 1.3, when one changes the foliation in int X (and even int X itself), $H(g)$ varies by an additive constant. The sum (i) + (ii) in the Atiyah–Patodi–Singer formula behaves in the same way. But it is still unclear to us whether it is possible to break $H(g)$ into two parts like (i) and (ii). We shall return to this question in subsection 5 below.

4. The Godbillon–Vey invariant for a piecewise smooth foliation

Assume that a closed oriented 3-manifold X is formed by glueing together a finite set of compact domains with piecewise smooth boundaries (like those we considered in the previous subsection), the closure of each 2-face of each of these domains being attached, by some diffeomorphism, to the closure of a 2-face of another domain. There is a natural decomposition of X into the sum of connected smooth 3-manifolds, $X^1, \ldots, X^j$, connected smooth 2-manifolds, $Y^1, \ldots, Y^k$, connected smooth 1-manifolds, $Z^1, \ldots, Z^l$, and single points. Assume also that the closure of each X^i is furnished with a smooth codimension one foliation $\mathcal{F}^i$ transversal to all Y^α and Z^β contained in $\bar{X}^i$, and that the foliations induced in Y^α by different $\mathcal{F}^i$'s coincide. Thus all the manifolds $X^1, \ldots, X^j, Y^1, \ldots, Y^k; Z^1, \ldots, Z^l$ have foliations of codimension 1. Choose associated pairs of forms: $(\omega^1, \eta^1), \ldots, (\omega^j, \eta^j)$; $(\omega_1{}^1, \eta_1{}^1), \ldots, (\omega_1{}^k, \eta_1{}^k)$, $(\omega_2{}^1, \eta_2{}^1), \ldots, (\omega_2{}^l, \eta_2{}^l)$. These forms are assumed extendable to the closures of their manifolds. The pairs of forms in $Y = \cup Y^i$, $Z = \cup Z^i$ defined by the $(\omega_1{}^i, \eta_1{}^i)$, $(\omega_2{}^i, \eta_2{}^i)$ are denoted by (ω_1, η_1), (ω_2, η_2).

THEOREM 1.4. *The sum*

$$\sum_{i=1}^{j} H(\omega^i, \eta^i; \omega_1|_{Y\cap\bar{X}^i}, \eta_1|_{Y\cap\bar{X}^i}; \omega_2|_{Z\cap\bar{X}^i})$$

does not depend on any of forms $\omega^1, \eta^1, \ldots, \omega_2{}^l$, so that it is well defined by the foliations $\mathcal{F}^i$.

This follows from Theorem 1.2: one must take another set of forms, subtract the corresponding sums and transform it according to (6); all the summands in the resulting sum cancel out.

Thus the Godbillon–Vey invariant can be defined even for piecewise smooth foliations on 3-manifolds. It would be interesting to compare this definition with the construction of the Godbillon–Vey class for piecewise linear foliations[8].

5. The localization and splitting problems

The localization problem is one of representing $H(\omega, \eta; \omega_1, \eta_1; \omega_2)$ as a sum of three terms:

(i) the integral over Y of a form $B(\omega_1, \eta_1)$ universally expressed in terms of ω_1, η_1.

†The formula is given in [1] only under an additional condition implying the vanishing of (iii)—see Theorem 2 of [1]. The general formula is described in the note after Theorem 2.

(ii) the integral over Z of a form defined by the forms ω_2, η_2 and the germs of ω_1, η_1 at points of Z;

(iii) a constant depending on the foliation (in the whole X) but not on any form.

The words 'universally expressed in terms of ω_1, η_1' can be interpreted in different ways. The most natural possibility is the following. For any pair of C^∞ 1-forms ω_1, η_1 in an arbitrary planar domain U such that ω_1 is nondegenerate and $\eta_1 \wedge \omega_1 = \mathrm{d}\omega_1$ a C^∞ 2-form $B(\omega_1, \eta_1)$ is given, and for any diffeomorphism φ of another planar domain into V the relation $B(\varphi^*\omega_1, \varphi^*\eta_1) = \varphi^*B(\omega_1, \eta_1)$ holds. Though the impossibility of localization with this interpretation of universality seems very probable, we have succeeded in proving it only under an extra condition of smoothness put on B: for any vectors u, v at a point of U the function $\Omega^1(U)\times\Omega^1(U)\to\mathbf{R}$ defined by $(\omega_1, \eta_1)\mapsto[B(\omega_1, \eta_1)](u, v)$ belongs to C^1. At first sight this condition seems unimportant, but it is essential for the proof below, and we would like to remove it anyhow.

Lemma. *If the function $(\omega_1, \eta_1)\mapsto B(\omega_1, \eta_1)$ satisfies the above universality and smoothness conditions then $B(\omega_1, \eta_1) = a\,\mathrm{d}\omega_1 + b\,\mathrm{d}\eta_1$ where a, b are constants.*

Proof. In an appropriate coordinate system the pair ω_1, η_1 (wtith nondegenerate ω_1 and with $\eta_1 \wedge \omega_1 = \mathrm{d}\omega_1$) takes the form

$$\omega_1 = e^f\,\mathrm{d}x,$$
$$\eta_1 = \mathrm{d}f + g\,\mathrm{d}x.$$

One can also make $f(0) = g(0) = 0$. For a pair f, g of C^∞-germs with $f(0) = g(0) = 0$ we denote by $B(f, g)$ the value of the form $B(e^f\,\mathrm{d}x, \mathrm{d}f + g\,\mathrm{d}x)$ on the vectors $\partial/\partial x$, $\partial/\partial y$ at 0. We are to prove that $B(f, g) = a(\partial f/\partial y)(0) + b(\partial g/\partial y)(0)$ where a, b are constants.

B is (can be considered as) a C^1-function of

$$f^0_{x\ldots xy\ldots y} = \frac{\partial^{k+l} f}{\partial x^k \partial y^l}(0), \qquad g^0_{x\ldots xy\ldots y} = \frac{\partial^{k+l} g}{\partial x^k \partial y^l}(0)$$

with $k \geq 0$, $l \geq 0$, $(k, l) \neq (0, 0)$, $k + l \leq N$ for some N. The invariance under the diffeomorphism $(x, y)\mapsto(x, \lambda y)$ $[\lambda \neq 0]$ implies that for any $\lambda \neq 0$

$$\lambda B(f^0_x, f^0_y, g^0_x, g^0_y, f^0_{xx}, f^0_{xy}, f^0_{yy}, \ldots) = B(f^0_x, \lambda f^0_y, g^0_x, \lambda g^0_y, f^0_{xx}, \lambda f^0_{xy}, \lambda^2 f^0_{yy}, \ldots);$$

this equality holds also for $\lambda = 0$ (by continuity arguments). This implies, because of $B \in C^1$, that B does not depend on partial derivatives with more than one subscript y. Thus

$$B(f, g) = B(f + y^2 f_1, g + y^2 g_1)$$

for any C^∞-germs f_1, g_1. Using the invariance again we obtain that y^2 can be replaced here with $(\alpha x + y)^2$ for any α and hence that B is a function of four variables f^0_x, f^0_y, g^0_x, g^0_y. Then we apply the invariance under the transformation $(x, y)\mapsto(x + \gamma x^3, \beta x + y)$ and obtain that B does not depend also on f^0_x and g^0_x. Finally, the above equality $\lambda B(\ldots, f^0_y, \ldots, g^0_y, \ldots) = B(\ldots, \lambda f^0_y, \ldots, \lambda g^0_y, \ldots)$ implies, again because of $B \in C^1$, that B is a linear function, i.e. $B(f, g) = a f^0_y + b g^0_y$.

Note. Without the smoothness condition the lemma is false. Counterexample:

$$[B(\omega_1, \eta_1)](u, v) = [\mathrm{d}\omega_1(u, v)]^{2/3}[\mathrm{d}\eta_1(u, v)]^{1/3}.$$

Deducing the nonlocalizability from the lemma is straightforward. If X is a manifold with a smooth boundary then Z is empty and

$$H(\omega, \eta; \omega_1, \eta_1; \omega_2) = \int_{\partial X} \Gamma^1(\omega, \eta; \omega_1, \eta_1).$$

As $\int_{\partial X}(a\,\mathrm{d}\omega_1 + b\,\mathrm{d}\eta_1) = 0$ localizability would mean that $\int_X \Gamma^1(\omega, \eta; \omega_1, \eta_1)$ does not depend on ω_1, η_1. But it does depend on them as the simplest examples show.

Another possible way of interpreting the universality is the following: $B(\omega_1, \eta_1)$ must be expressed in terms of ω_1, η_1 by a formula, in which a pair of 1-forms defined in a space of any dimension can be substituted. More precisely, for any pair of C^∞ 1-forms ω_1, η_1 in an arbitrary open subset U of $\mathbf{R}^n$ with arbitrary n, such that ω_1 is non-degenerate and $\eta_1 \wedge \omega_1 = d\omega_1$ a 2-form $B((\omega_1, \eta_1)$ in U must be given, and for any smooth imbedding of any open subset of $\mathbf{R}^m$ with any $m \leq n$ into U with nondegenerate $\varphi^*\omega_1$ the equality $B(\varphi^*\omega_1, \varphi^*\eta_1) = \varphi^*B(\omega_1, \eta_1)$ must hold.

If universality is interpreted in this way, the nonlocalizability is proved quite easily. The simplest way is just to repeat the above arguments, with the lemma being trivally true without any condition like smoothness: almost all pairs of 1-forms of the above type in a space of dimension ≥ 3 can be reduced to the form $e^y\, dx$, $dy + z\, dx$. There is another proof, which is more interesting from the point of view of generalizations to higher codimensions. We give it below.

We assume again that ∂X is smooth. The localizability would imply that

$$H(\omega', \eta'; \omega_1', \eta_1'; \omega_2') - H(\omega, \eta; \omega_1, \eta_1; \omega_2) = \int_{\partial X} [B(\omega_1', \eta_1') - B(\omega_1, \eta_1)].$$

If $\omega_1 = \omega|_{\partial X}$, $\eta_1 = \eta|_{\partial X}$, $\omega_1' = \omega_1|_{\partial X}$, $\eta_1' = \eta_1|_{\partial X}$ then

$$H(\omega, \eta; \omega_1, \eta_1; \omega_2) = \int_X \Gamma^0(\eta), H(\omega', \eta'; \omega_1', \eta_1'; \omega_2') = \int_X \Gamma^0(\eta'),$$

and we should have:

$$\int_X (\Gamma^0(\eta') - \Gamma^0(\eta)) = \int_{\partial X} [B(\omega_1', \eta_1') - B(\omega_1, \eta_1)],$$

i.e.

$$\int_X \{[\Gamma^0(\eta') - dB(\omega_1', \eta_1')] - [\Gamma^0(\eta) - dB(\omega_1, \eta_1)]\} = 0. \tag{10}$$

This implies that

$$\Gamma^0(\eta') - dB(\omega_1', \eta_1') = \Gamma^0(\eta) - dB(\omega_1, \eta_1). \tag{11}$$

In fact, (10) remains valid if we press ∂X into X, preserving the transversality to the foliation. Hence the integrand of (10) has zero integral over an arbitrarily small piece near the boundary; thus it vanishes near the boundary; and it vanishes everywhere by its universality. Furthermore, (11) means that $\Gamma^0(\eta) - dB(\omega_1, \eta_1)$ [with $\omega_1 = \omega|_{\partial X}$, $\eta_1 = \eta|_{\partial X}$] does not depend on ω, η i.e. it is invariant under diffeomorphisms, and this evidently implies that $\Gamma^0(\eta) - dB(\omega_1, \eta_1)$ is equal to zero. Finally, the equality $\Gamma^0(\eta) = dB(\omega_1, \eta_1)$ shows that the Godbillon–Vey class of an arbitrary foliation is trivial, which is known to be false.

The splitting problem is one of representing $H(\omega, \eta; \omega_1, \eta_1; \omega_2)$ as a sum of two terms:

(i) something depending only on the foliation in the boundary and the forms ω_1, η_1, ω_2, η_2, which does not change when these objects are replaced by diffeomorphic ones.

(ii) a constant depending on the foliation (in the whole of X) but not on any form.

As has been said before, there is such a splitting in the Atiyah–Patodi–Singer formula.

Of course, the solvability of the localization problem implies the solvability of the splitting problem.

In the case of a smooth boundary the splitting problem can be reduced to the following question concerning an individual manifold X with a foliation $\mathcal{F}$. Let φ be a diffeomorphism $\partial X \to \partial X$, compatible with the foliation $\mathcal{F}|_{\partial X}$, and let ω_1, η_1 be a pair of forms associated with $\mathcal{F}|_{\partial X}$. Is the integral

$$\int_{\partial X} \Gamma^1(\omega_1, \eta_1; \varphi^*\omega_1, \varphi^*\eta_1) \tag{12}$$

zero?

If the splitting problem has a solution, then, after extending the pairs (ω_1, η_1), $(\varphi^*\omega_1, \varphi^*\eta_1)$ to pairs (ω, η), (ω', η') associated with whole of $\mathcal{F}$, the difference

$$H(\omega, \eta; \omega_1, \eta_1; \lambda) - H(\omega', \eta'; \varphi^*\omega_1, \varphi^*\eta_1; \lambda)$$

(where λ is an 1-form on the empty manifold) will vanish. But it is equal to the integral (12).

Though we cannot see any general reason for the triviality of the integral (12), it is trivial for all examples considered by us, and we are forced to think that it is always trivial. If this is really true, the splitting problem is solvable at least for manifolds with smooth boundaries. It can be solved by means of the Zermelo axiom: one fixes for each (up to diffeomorphism) connected piecewise smooth 2-manifold with a foliation of codimension 1, transversal to all edges, a foliation bounded by it on a smooth 3-manifold, and a pair $\omega_1{}^0, \eta_1{}^0$ of forms associated to this foliation, and then takes as the summand (i) the expression

$$\int_{\partial X} \Gamma^1(\omega_1{}^0, \eta_1{}^0; \omega_1, \eta_1). \tag{13}$$

The triviality of (12) implies the invariance of (13) under diffeomorphisms. But even in this case we cannot see if it is possible to make the splitting effective as in the Atiyah–Patodi–Singer formula. We cannot even properly state the corresponding problem.

However there exists a class of foliations for which H can be split effectively (but probably still not localized). This class consists of foliations whose restriction to the boundary can be defined by a closed form. In this case there exists a pair of forms $\omega_1{}^0, \eta_1{}^0$ associated to the foliation on ∂X with $\eta_1{}^0 = 0$, and it turns out that the integral

$$\int_{\partial X} \Gamma^1(\omega_1{}^0, 0; \omega_1, \eta_1) \tag{14}$$

is invariant when $\omega_1{}^0$ is replaced by any other closed form $\tilde{\omega}_1{}^0$ defining the foliation (this follows from the evident equality $\Gamma^1(\tilde{\omega}_1{}^0, 0; \omega_1{}^0, 0) = 0$). The integral (14) can be taken as the summand (i).

The following §§2–5 contain the extension of the theory developed in §1 to a more general situation, in particular, to the Bernstein–Rosenfeld–Bott–Haefliger characteristic classes of foliations of arbitrary codimensions (see [2], [5]). In §2 we describe some rather general structures playing the role of a metric or a pair of forms associated with a foliation; in this general situation we define objects having the main properties of $\Gamma^0, \Gamma^1, \Gamma^2$. In §3 the theory of §2 is made explicit in the case of foliations. §4 is independent of the previous sections and contains preliminary material for §5, in which the basic results of §1 are generalized to foliations of arbitrary codimension.

§2. FORMAL STRUCTURES AND RELATED BICOMPLEXES

1. Formal structures

Let X be a smooth manifold, and $p: E \to X$ a smooth fibration whose total space is furnished with smooth action of the groupoid Diff X of diffeomorphisms of X which is compatible with p. In other words, to each diffeomorphism φ of an open set $U \subset X$ onto an open set $V \subset X$ is associated a diffeomorphism $\hat{\varphi}: p^{-1}(U) \to p^{-1}(V)$ covering φ, and the function $\varphi \mapsto \hat{\varphi}$ commutes with composing and inverting diffeomorphisms.

Let then $\mathscr{E}$ be (the total space of) the sheaf of germs of sections of p, $E^{(\infty)}$ the space of ∞-jets of these sections (with C^∞ topology), $p^{(\infty)}: E^{(\infty)} \to X$ the natural projection, and $j: \mathscr{E} \to E^{(\infty)}$ the natural map. We assume a (not necessarily finite-dimensional) submanifold C of $E^{(\infty)}$ chosen such that $p^{(\infty)}|_C: C \to X$ is a smooth fibration and $j^{-1}(C)$ is a subsheaf of $\mathscr{E}$. We denote this subsheaf by $\mathscr{C}$, and call it the structural sheaf, its sections $\mathscr{C}$-structures, and elements of $\mathscr{C}$ formal ($\mathscr{C}$-) structures. Elements of Diff X preserving $\mathscr{C}$ are called admissible. They form a subgroupoid of Diff X which is denoted by $G(\mathscr{C})$.

Note that the fibration $C \to X$ has a canonical flat $G(\mathscr{C})$-invariant connection: its horizontal sections are the images of sections of $\mathscr{C}$.

Examples. 1°. The fibration $p: E \to X$ is the symmetric square of the cotangent bundle, $\mathscr{C}$ is the sheaf of germs of positive quadratic forms (Riemannian metrics). In this case $\mathscr{C}$-structures are Riemannian metrics, and $G(\mathscr{C})$ coincides with Diff X.

2°. $E = X \times \mathbf{R}^n$ (where $n = \dim X$), p is the product projection, and $\mathscr{C}$ is the sheaf of germs of sections $X \to E$ such that the composition $X \to E = X \times \mathbf{R}^n \to \mathbf{R}^n$ is regular. In this case C is the space of formal local coordinate systems on $X (C = S(X)$ in the notation of [9]), and $G(\mathscr{C})$ is Diff X again.

3°. Assume that a codimension one foliation $\mathscr{F}$ is given on X. We take for p the Whitney sum of two copies of the cotangent bundle, and for $\mathscr{C}$ the sheaf of germs of pairs of 1-forms associated with $\mathscr{F}$ (see §1, #1). In this case $G(\mathscr{C})$ is the groupoid of foliation preserving (partial) diffeomorphisms of X.

4°. Assume that a codimension q foliation $\mathscr{F}$ is given on X. We take for p the Whitney sum of $q+q^2$ copies of the cotangent bundle, and for $\mathscr{C}$ the sheaf of germs of sets $\{\omega_i, \eta_{jk} \mid 1 \leq i, j, k \leq q\}$ of 1-forms satisfying the following conditions:

(i) the forms ω_i define $\mathscr{F}$;

(ii) $d\omega_j = \Sigma_k \eta_{jk} \wedge \omega_k$ (for all j).

$G(\mathscr{C})$ is again the groupoid of foliation preserving diffeomorphisms.

Evidently, 3° is a special case of 4°; it is clear also that all stalks of $\mathscr{C}$ are non-empty, and that $\mathscr{C}$ has a global section if and only if the normal bundle of $\mathscr{F}$ is trivial.

We denote below the sheaf $\mathscr{C}$ of this example by $\mathscr{C}(\mathscr{F})$.

Note that specifying forms ω_i, η_{jk} satisfying the above conditions (i), (ii) is equivalent to trivializing the second conormal bundle of $\mathscr{F}$, i.e. the bundle of 2-jets of maps $(X, x) \to (\mathbf{R}^q, 0)$ (with $x \in X$) constant on each leaf of $\mathscr{F}$. Specifying forms ω_i defining $\mathscr{F}$ without η_{jk} is equivalent to trivialing the first (usual) conormal bundle of $\mathscr{F}$. One can also define structures consisting in trivializing the kth conormal bundle of $\mathscr{F}$ with any $k \leq \infty$; if $k = \infty$ the space of formal structures is the space $S(\mathscr{F})$ of [3].

5°. Make a fiberwise factorization of the total space E of the fibration p of the previous example, identifying in each fiber of p pairs $(\alpha_i, \beta_{jk}), (\alpha'_i, \beta'_{jk})$ such that there exists an orthogonal matrix $\|a_{lm}\|$ with

$$\alpha'_i = \sum_s a_{is}\alpha_s, \quad \beta'_{jk} + \beta'_{kj} = \sum_{u,v} a_{ju}a_{kv}(\beta_{jk} + \beta_{kj}).$$

This identification turns the space E into a space E_0, and the fibration $p: E \to X$ into a fibration $p_0: E_0 \to X$. Take for a structural sheaf the sheaf $\mathscr{CO}(\mathscr{F})$ of germs of smooth sections of p_0 which are the images of germs of smooth sections of p belonging to $\mathscr{C}(\mathscr{F})$. The groupoid $G(\mathscr{CO}(\mathscr{F}))$ coincides with $G(\mathscr{C}(\mathscr{F}))$.

When $q = 1$ the sheaf $\mathscr{CO}(\mathscr{F})$ is obtained from $\mathscr{C}(\mathscr{F})$ with the identification of (ω, η) and $(-\omega, \eta)$.

The geometrical meaning of a $\mathscr{CO}(\mathscr{F})$-structure consists in trivializing the second conormal bundle of $\mathscr{F}$ factorized by the natural action of $O(q)$.

The main difference between $\mathscr{CO}(\mathscr{F})$-structures and $\mathscr{C}(\mathscr{F})$-structures is that a $\mathscr{CO}(\mathscr{F})$-structure exists for any $\mathscr{F}$. Moreover, it can be canonically defined if X is furnished not only with a foliation but also with a Riemannian metric. In this case we have only to construct near every point of X a set of forms $\omega_1, \ldots, \omega_q$ defining the foliation and such that $(\omega_i, \omega_j) = \delta_{ij}$, to choose vector fields $f_1, \ldots, f_q$ with $\omega_i(f_j) = \delta_{ij}$, and then to define the forms η_{jk} by the formula $\eta_{jk}(\xi) = \omega_j(\nabla_\xi f_k)$, where ξ is a tangent vector, and ∇_ξ denotes the covariant derivative. The forms ω_i and the vector fields f_j are defined by these condition up to a change

$$\omega_i, f_j \to \sum_s a_{is}\omega_s, \sum_s a_{js} f_s,$$

where $\|a_{is}\|$ is a (variable) orthogonal matrix, and it is easy to check that the section of p_0 defined by the forms ω_i, η_{jk}, is invariant under this change. Thus the $\mathscr{CO}(\mathscr{F})$-structure is well defined by a metric.

The more detailed investigation of $\mathscr{CO}(\mathscr{F})$-structures and $\mathscr{C}(\mathscr{F})$-structures will be undertaken in §3.

2. The variation bicomplex

The definition of the variation bicomplex which is presented below, as well as the definition of the difference bicomplex of the next subsection, was invented, in the case of Riemannian metrics, by I. M. Gel'fand and M. V. Losik. The details of their work will appear soon.

Let $\mathscr{C}$ be a structural sheaf over X, and C be the corresponding manifold of formal structures. We denote by $\Omega = \Omega(C) = \bigoplus \Omega^r(C)$ the (sheaf) de Rham complex of C. As the fibration $C \to X$ is furnished with a flat connection, Ω has a natural bicomplex structure: $\Omega^r(C) = \bigoplus_{p+q=r} \Omega^{p,q}$, where

$\Omega^{p,q}$ is the set of such $\alpha \in \Omega^{p+q}(C)$, that for any $i \neq 0$, any vertical (tangent to a fiber of the fibration $C \to X$) vectors $\xi_1, \ldots, \xi_{p+i}$, and any horizontal vectors $\eta_1, \ldots, \eta_{q-i}$ one has

$$\alpha(\xi_1, \ldots, \xi_{p+i}, \eta_1, \ldots, \eta_{q-i}) = 0.$$

The differential of Ω decomposes into the sum of bihomogeneous differentials of bidegrees (0, 1) and (1, 0); they are denoted by d and δ. The equality $(\mathrm{d}+\delta)^2 = 0$ is equivalent to the equalities $\mathrm{d}^2 = 0$, $\delta^2 = 0$, $\mathrm{d}\delta = -\delta \mathrm{d}$.

The decomposition $\Omega = \bigoplus \Omega^{p,q}$ and the differentials d and δ are compatible with the action of $G(\mathscr{C})$. This implies that the $G(\mathscr{C})$-invariant elements of Ω also form a bicomplex. It is called the variation bicomplex associated with $\mathscr{C}$. The sheaves this bicomplex consists of are denoted by $\Omega^{p,q}_{\mathscr{C}}$, and for its differentials we keep the notations d, δ.

3. The difference bicomplex

Denote by $A^{p,q}$ the sheaf of skew-symmetric smooth maps of the sheaf $\mathscr{C}^{p+1}$—the sum of $p+1$ copies of $\mathscr{C}$—into the sheaf of germs of differential q-forms on X. Thus an element of $A^{p,q}$ is a skew-symmetric function relating to $p+1$ germs of sections of C at a point $x \in X$ the germ of a q-form on X at x. Another interpretation of $A^{p,q}$ is that it is the sheaf of germs of horizontal q-forms on $C^{(p+1)}$ (the total space of the sum of $p+1$ copies of the fibrations $C \to X$) which are skew-symmetric under the natural action of the permutation group on $C^{(p+1)}$.

The de Rham differential defines a map $A^{p,q} \to A^{p,q+1}$; this map, multiplied by $(-1)^p$, is denoted by d. The differential $\Delta: A^{p,q} \to A^{p+1,q}$ is defined by the formula

$$\Delta a(f_0, \ldots, f_{p+1}) = \sum_{i=0}^{p+1} (-1)^i a(f_0, \ldots, \hat{f}_i, \ldots, f_{p+1}).$$

Clearly, $\mathrm{d}^2 = 0$, $\Delta^2 = 0$ and $\mathrm{d}\Delta = -\Delta \mathrm{d}$. It is clear also that the both differentials are compatible with the action of $G(\mathscr{C})$ so there is a bicomplex of invariant elements just as before. It is called the difference bicomplex associated with $\mathscr{C}$. The sheaves of this bicomplex are denoted by $A^{p,q}_{\mathscr{C}}$, and its differentials again by d, Δ.

Note that we dealt with the difference bicomplex implicitly in §1, where the structural sheaf of the example 3°(#1) played the rôle of $\mathscr{C}$ (also implicitly). The expressions $\Gamma^0, \Gamma^1, \Gamma^2$ were actually sections of $A^{0,3}_{\mathscr{C}}$, $A^{1,2}_{\mathscr{C}}$, $A^{2,1}_{\mathscr{C}}$, and the relations (1)–(4) mean that $\mathrm{d}\Gamma^0 = 0$, $\mathrm{d}\Gamma^1 = -\Delta\Gamma^0$, $\mathrm{d}\Gamma^2 = \Delta\Gamma^1$, $0 = \Delta\Gamma^2$ or that the sum $\Gamma^0 + \Gamma^1 - \Gamma^2$ belongs to the kernel of the total differential $\mathrm{d}+\Delta$ (is a $(\mathrm{d}+\Delta)$-cycle).

4. Relations between the variation and difference bicomplexes

First of all we construct a homomorphism from the difference bicomplex into the variation one.

Let $a \in A^{p,q}$. As we noted in the previous subsection, a may be regarded as a horizontal form of $C^{(p+1)}$; apply to a successively the differentials along the $(p+1)$th, pth, ..., second summand C, and restrict the resulting form to the diagonal $C \subset C^{(p+1)}$. We obtain a form in $\Omega^{p,q}$, and associating it to a defines a map $A^{p,q} \to \Omega^{p,q}$. It is easy to see that this map is compatible with the action of $G(\mathscr{C})$ and that the diagrams

$$\begin{array}{ccc} A^{p,q} & \to & \Omega^{p,q} \\ \downarrow \mathrm{d} & & \downarrow \mathrm{d} \\ A^{p,q+1} & \to & \Omega^{p,q+1} \end{array} \qquad \begin{array}{ccc} A^{p,q} & \to & \Omega^{p,q} \\ \downarrow \Delta & & \downarrow \delta \\ A^{p+1,q} & \to & \Omega^{p+1,q} \end{array}$$

commute. Thus we get a homomorphism $\{A^{p,q}_{\mathscr{C}}; \mathrm{d}, \Delta\} \to \{\Omega^{p,q}_{\mathscr{C}}; \mathrm{d}, \delta\}$. We denote this homomorphism by τ.

Example. If $\mathscr{C}$ is as in the example 3° of #1, then τ maps the $(\mathrm{d}+\Delta)$-cycle $\Gamma^0 + \Gamma^1 - \Gamma^2$ (see the end of #3) into the form $\eta \wedge \mathrm{d}\eta + (\delta\eta \wedge \eta + 2\eta \wedge \mathrm{d}\epsilon) + \epsilon \wedge \mathrm{d}\epsilon$ where ϵ is the vertical 1-form defined by the formula $\delta\omega = \epsilon \wedge \omega$, and ω, η are the horizontal 1-forms defined as follows. If $(\omega_0, \eta_0) \in C$, that is (ω_0, η_0) is an ∞-jet of a pair of forms associated with the foliation, and ξ is a tangent vector of C at (ω_0, η_0), then $\omega(\xi)$, $\eta(\xi)$ are equal to the values of ω_0, η_0 at the projection of ξ in X.

As we shall see, in some special situations τ is a homotopy equivalence. In order to describe these situations we make the following definition.

Definition 2.1. A linearization of the manifold C of formal $\mathscr{C}$ structures is a smooth function $\varphi: C^{(2)}\times I\to C$ [where $C^{(2)}$ denotes, as before, the subset of $C\times C$ consisting of pairs (c_0, c_1) with $p^{(\infty)}(c_0)=p^{(\infty)}(c_1)$] satisfying the following axioms:

(i) $\varphi(c_0, c_1, 0)=c_0$, $\varphi(c_0, c_1, 1)=c_1$;

(ii) $\varphi(c_0, c_0, t)=c_0$;

(iii) $f(\varphi(c_0, c_1, t))=\varphi(f(c_0), f(c_1), t)$ for $f\in G(\mathscr{C})$;

(iv) if $\gamma_0, \gamma_1: U\to C$ are horizontal sections over an open subset of X, then $u\mapsto\varphi(\gamma_0(u), \gamma_1(u), t)$ is a horizontal section for any $t\in I$.

The existence of a linearisation evidently implies that the fibers of $p^{(\infty)}|_C: C\to X$ are contractible. Hence there is no linearization in the examples 2° and 4° of #1 (the fibers are homotopy equivalent to GL (dim X, R) in the example 2°, and to GL (q, R) in the example 4°). On the contrary, the manifold of formal structures of the example 1° has a natural linearization: it is defined by the linear paths. As to example 5°, its manifold of formal structures is also linearizable; we shall prove this in §3.

PROPOSITION 2.1. *If the manifold of formal $\mathscr{C}$-structures is linearizable, the homomorphism* τ: $\{A^{p,q}_{\mathscr{C}}; d, \Delta\}\to\{\Omega^{p,q}_{\mathscr{C}}; d, \delta\}$ *is a homotopy equivalence.*

The following proof contains a canonical construction associating a homotopy inverse of τ to a linearization of C.

For any $c_0, \ldots, c_p\in C_x=C\cap[(p^{(\infty)})^{-1}(x)]$ define a map $\varphi_{c_0\ldots c_p}$ of the standard simplex $\Delta^p=\{(t_0, \ldots, t_p)\in R^{p+1}|t_0\geq 0, \ldots, t_p\geq 0;\ t_0+\cdots+t_p=1\}$ into C in the following way: φ_{c_0} maps Δ^0 into c_0; if the maps $\varphi_{c_0\ldots c_k}$ with $k<p$ are defined, we put

$$\begin{aligned}\varphi_{c_0\ldots c_p}(t_0, \ldots, t_p)&=c_p \quad\text{if}\quad t_p=1\\ &=\varphi_{cc_p}(t_p)\quad\text{if}\quad t_p<1,\end{aligned}$$

where $c=\varphi_{c_0\ldots c_{p-1}}(t_0/1-t_p, \ldots, t_{p-1}/1-t_p)$.

Let α be an element of $\Omega^{p,q}_{\mathscr{C}}$ over $x\in X$, and let $c_0, \ldots, c_p$ be points of C_x, and $\xi_1, \ldots, \xi_q$ be tangent vectors to X at x. Substituting in α the horizontal vectors covering $\xi_1, \ldots, \xi_q$ we obtain a p-form $\alpha(\xi_1, \ldots, \xi_q)$ in C_x. Put

$$\sigma\alpha(c_0, \ldots, c_p; \xi_1, \ldots, \xi_q)=\sum_{(i_0,\ldots,i_p)}[\operatorname{sign}(i_0, \ldots, i_p)/p!]\int_{\Delta^p}\varphi^*_{c_0\ldots c_p}\alpha(\xi_1, \ldots, \xi_p),$$

where the sum is taken over all the permutations of $0, \ldots, p$. If $f_0, \ldots, f_p$ are germs of horizontal sections of $p^{(\infty)}|_C$ at x, then the formula

$$(y; \eta_1, \ldots, \eta_q)\mapsto\sigma\alpha(f_0(y), \ldots, f_p(y), \eta_1, \ldots, \eta_q),$$

where $y\in X$ and $\eta_1, \ldots, \eta_q$ are tangent vectors to X at y, defines a germ $\sigma\alpha(f_0, \ldots, f_p)$ of a q-form on X at x, the formula

$$(f_0, \ldots, f_p)\mapsto\sigma\alpha(f_0, \ldots, f_p)$$

defines an element $\sigma\alpha$ of $A_{\mathscr{C}}^{p,q}$, and, finally, the formula $\alpha\mapsto\sigma\alpha$ defines a map $\sigma:\Omega_{\mathscr{C}}^{p,q}\to A_{\mathscr{C}}^{p,q}$. It is easy to check that the maps σ compose a homomorphism of the bicomplex $\{\Omega_{\mathscr{C}}^{p,q}; d, \delta\}$ into the bicomplex $\{A_{\mathscr{C}}^{p,q}; d, \Delta\}$ we denote this homomorphism again by σ. We shall prove that σ is homotopy inverse to τ.

The composition $\tau\circ\sigma$ is the identity. To prove that the composition $\sigma\circ\tau$ is homotopic to the identity one approximates it by the operations of multiple barycentric subdivision. More precisely, let $\sigma_{k,1}, \ldots, \sigma_{k,N_k}$ be the highest dimension simplices of the kth barycentric subdivision of the standard simplex Δ^p; $v_{kj0}, \ldots, v_{kjp}$ be the naturally ordered vertices of σ_{kj}; and $\epsilon(j)$ equal to 1 or -1, according to the orientation of σ_{kj} induced by this order. The simplex σ_{kj} is contained in a certain simplex $\sigma_{k-1,j'}$ of the $(k-1)$th barycentric subdivision of Δ^p and corresponds to a certain order of its vertices (in general, different from the natural one). The vertices of $\sigma_{k-1,j'}$ in this order are denoted by $v'_{kj0}, \ldots, v'_{kjp}$. (Thus v'_{kjr} is a vertex of the $(k-1)$th barycentric subdivision of Δ^p.)

For germs $f_0, \ldots, f_p$ of sections of the fibering $C\to X$ we denote by f_{kjr}, f'_{kjr} the germs

$$y\mapsto\varphi_{f_0(y)\ldots f_p(y)}(v_{kjr}),\ y\mapsto\varphi_{f_0(y)\ldots f_p(y)}(v'_{kjr}).$$

The formulae

$$\mathrm{bar}_k a(f_0, \dots, f_p) = \sum_{(i_0,\dots,i_p)} [\mathrm{sign}\,(i_0, \dots, i_p) p\,!] \sum_{j=1}^{N_k} \epsilon(j) a(f_{kji_0}, \dots, f_{kji_p}),$$

$$Db(f_0, \dots, f_p) = \sum_{(i_0,\dots,i_p)} [\mathrm{sign}\,(i_0, \dots, i_p)/p\,!] \sum_{l=1}^{k} \sum_{j=1}^{N_v} \epsilon(j) \sum_{l=0}^{p} (-1)^r b(f'_{kji_0}, \dots, f'_{kji_r}, f_{kji_r}, \dots, f_{kji_p})$$

define an endomorphism bar_k of $\{A_{\mathscr{C}}^{p,q}\}$ and a homotopy D connecting this endomorphism with the identity. It remains to note that the sequence $\{\mathrm{bar}_k\}$ has a limit and that this limit is equal to $\sigma \circ \tau$.

5. A final remark

For all the examples of # 1, as well as for any structure $\mathscr{C}$ with transitive $G(\mathscr{C})$, the variation and difference bicomplexes are actually independent of X. For instance, one can replace X with any open subset of it, taking the appropriate restriction of the structural sheaf. (It would be still more convenient to take for X a "formal point"—the ring of formal power series—but our definitions would need a slight modification.)

For this reason, all computations for the variation and difference bicomplexes connected with foliations can be carried out in the case when the manifold is Euclidean space and the foliation is a family of parallel planes.

§3. THE STRUCTURES ASSOCIATED WITH A FOLIATION

1. The generalized Godbillon–Vey forms

The generalized Godbillon–Vey classes were introduced by Bernstein–Rosenfeld [2], [3] and Bott–Haefliger [4], [5], [12]. The construction presented below is intermediate between the constructions [3] and [4].

Let $\mathfrak{g}$ be a Lie algebra (over R). Recall that the Weil algebra $W(\mathfrak{g})$ is by definition the tensor product $\Lambda^*\mathfrak{g}' \otimes S^*\mathfrak{g}'$ of the exterior and symmetric algebras over $\mathfrak{g}'$—the dual of $\mathfrak{g}$—, equipped with:

(i) the graduation $W(\mathfrak{g}) = \oplus_q W^q(\mathfrak{g})$, where

$$W^q(\mathfrak{g}) = \bigoplus_{i+2j=q} (\Lambda^i\mathfrak{g}' \otimes S^j\mathfrak{g}');$$

(ii) the filtration $W(\mathfrak{g}) = F_0 W(\mathfrak{g}) \supset F_1 W(\mathfrak{g}) \supset \cdots$ where

$$F_r W(\mathfrak{g}) = \bigoplus_{2j \geq r} (\Lambda^*\mathfrak{g}' \otimes S^j\mathfrak{g}');$$

(iii) the multiplicative homogeneous differential $\mathrm{d}: W(\mathfrak{g}) \to W(\mathfrak{g})$, defined by the relation

$$\mathrm{d}(\gamma \otimes 1) = 1 \otimes \gamma + (\nabla\gamma) \otimes 1,$$

where $\gamma \in \mathfrak{g}' = \Lambda^1\mathfrak{g}' = S^1\mathfrak{g}'$, and $\nabla\gamma \in \Lambda^2\mathfrak{g}'$ is defined by the formula $\nabla\gamma(g_1, g_2) = \gamma([g_1, g_2])$. In the sequel we shall deal with the algebra $W(\mathfrak{gl}(q, R))$, more precisely with the factor algebra

$$W(\mathfrak{gl}(q, R))/F_{2q+1} W(\mathfrak{gl}(q, R)).$$

This is called the truncated Weil algebra of $\mathfrak{gl}(q, R)$, and is denoted simply by $W(q)$.

Assume now that a codimension q foliation F of X is given and that a $\mathscr{C}(\mathscr{F})$-structure is fixed, that is, a set of 1-forms ω_i, η_{jk} on X has been constructed such that the forms ω_i define $\mathscr{F}$ and the forms η_{jk} satisfy the conditions $\mathrm{d}\omega_j = \Sigma_k \eta_{jk} \wedge \omega_k$. (Recall that such a structure exists if and only if the normal bundle of F is trivial.) Denote by e_{jk} the functional on $\mathfrak{gl}(q, R)$ mapping a matrix into its entry with indices j, k, and define a map $W^1(\mathfrak{gl}(q, R)) = \mathfrak{gl}(q, R)' \to \Omega^1 X$ by the formula $e_{jk} \mapsto \eta_{jk}$. This map has a unique extension to a multiplicative, homogeneous map $W(\mathfrak{gl}(q, R)) \to \Omega^* X$ commuting with the differential which maps $1 \otimes e_{jk}$ into the 2-form

$$\zeta_{jk} = \mathrm{d}\eta_{jk} - \Sigma_m \eta_{jm} \wedge \eta_{mk}.$$

As a direct calculation shows, this form belongs to the ideal generated by $\omega_1, \dots, \omega_q$ (one need only check that $\zeta_{jk} \wedge \omega_1 \wedge \cdots \wedge \omega_q = 0$), and therefore all the products

$$\zeta_{j_1 k_1} \wedge \cdots \wedge \zeta_{j_r k_r}$$

with $r > q$ vanish. Hence the homomorphism $W(\mathfrak{gl}(q, R)) \to \Omega^* X$ constructed above defines a homomorphism

$$W(q) \to \Omega^* X. \tag{15}$$

The forms in the image of this homomorphism are called the generalized Godbillon–Vey forms. For example, a codimension one foliation with an associated pair of forms (ω, η) has four generalized Godbillon–Vey forms: 1, η, $d\eta$, $\eta \wedge d\eta$.

One can prove that the homology homomorphism induced by (15) does not depend on η_{jk} and is invariant under replacing the set $\{\omega_i\}$ with a homotopic one. The proof is similar to that of Proposition 1,1; we omit it though it is closely related to our subject.

The homology of $W(q)$ is well known: it coincides with the real cohomology of the inverse image of the $2n$-skeleton of the base of the universal $U(q)$-fibration. For example, the reduced homology of $W(1)$ is generated by a single three-dimensional element, corresponding to the Godbillon–Vey class. The reduced homology of $W(2)$ is generated by five elements, and so five polynomials in the forms η_{11}, η_{12}, η_{21}, η_{22} and their differentials represent five characteristic classes in the cohomology of X. Here are the polynomials.

$$d(\eta_{11} + \eta_{22}) \wedge d(\eta_{11} + \eta_{22}) \wedge (\eta_{11} + \eta_{22});$$
$$(\zeta_{11} \wedge \zeta_{22} - \zeta_{12} \wedge \zeta_{21}) \wedge (\eta_{11} + \eta_{22});$$
$$(\zeta_{11} \wedge \zeta_{22} - \zeta_{12} \wedge \zeta_{21}) \wedge (\eta_{11} - \eta_{22}) \wedge \eta_{12} \wedge \eta_{21};$$
$$d(\eta_{11} + \eta_{22}) \wedge d(\eta_{11} + \eta_{22}) \wedge \eta_{11} \wedge \eta_{12} \wedge \eta_{21} \wedge \eta_{22};$$
$$(\zeta_{11} \wedge \zeta_{22} - \zeta_{12} \wedge \zeta_{21}) \wedge \eta_{11} \wedge \eta_{12} \wedge \eta_{21} \wedge \eta_{22}$$

(where, as before, $\zeta_{jk} = d\eta_{jk} - \Sigma_m \eta_{jm} \wedge \eta_{mk}$).

Recall now the connection between the homology of $W(q)$ and the continuous cohomology of the Lie algebra W_q for formal vector fields in R^q. This cohomology is by definition the homology of the complex $\{C^*(W_q), \nabla\}$, where $C^r(W_q)$ is the space of continuous linear functionals $\Lambda^r W_q \to R$, and ∇ is defined by the formula

$$\nabla F(\xi_1, \dots, \xi_{r+1}) = \sum_{1 \le s < t \le r+1} (-1)^{s+t-1} F([\xi_s, \xi_t], \xi_1, \dots, \hat{\xi}_s, \dots, \hat{\xi}_t, \dots, \xi_{r+1}).$$

Define an injection

$$W(q) \to C^*(W_q; R) \tag{16}$$

in the following way. The functional $e_{ij} \in \mathfrak{gl}(q, R) = W^1(\mathfrak{gl}(q, R))$ is taken into the 1-cochain $W_q \to R$, given by the formula

$$\sum a_k \frac{\partial}{\partial x_k} + \sum a_{lm} x_l \frac{\partial}{\partial x_m} + \cdots + \mapsto a_{ij},$$

and the images of the remaining elements of $W(q)$ are well defined by linearity, multiplicativity and compatibility with the differential.

PROPOSITION 3.1 (see [7], [10]). *The homomorphism* (16) *is a homotopy equivalence.*

2. The generalized Godbillon–Vey forms in the variation complex associated with $\mathscr{C}(\mathscr{F})$

There is a natural way to define horizontal 1-forms ω_i, η_{jk} in the manifold $C(\mathscr{F})$ of formal $\mathscr{C}(\mathscr{F})$-structures: the value of the form ω_i (the form η_{jk}) on the tangent vector ξ to $C(\mathscr{F})$ at a point $(\omega_i^0, \eta_{jk}^0) \in C(\mathscr{F})$ is defined as the value of the form representing the jet ω_i^0 (the jet η_{jk}^0) on the projection of ξ in X (cf. the example in #4 of §2). These forms satisfy the relation $d\omega_j = \Sigma \eta_{jk} \wedge \omega_k$ (where d is the 'horizontal component' of the exterior differential of $C(\mathscr{F})$—see #2 of §2) and are $G(\mathscr{C}(\mathscr{F}))$-invariant. Therefore the construction of #1 gives a homomorphism

$$W(q) \to \{\Omega^{0,*}_{\mathscr{C}(\mathscr{F})}, d\} \tag{17}$$

which, as we see, is constructed without any additional choices and even without assuming that the normal bundle of F is trivial. Two extension problems arise. The first is to extend (17) to a homomorphism $W(q) \to \{\Omega^{*,*}_{\mathscr{C}(\mathscr{F})}, d + \delta\}$. The second arises because of the equality $\{\Omega^{0,*}_{\mathscr{C}(\mathscr{F})}, d\} = \{A^{0,*}_{\mathscr{C}(\mathscr{F})}, d\}$ and is to extend (17) to a homomorphism $W(q) \to \{A^{*,*}_{\mathscr{C}(\mathscr{F})}, d + \delta\}$. We put the second problem aside for a while ('till #4) and begin solving the first one.

The forms ω_i define an integrable system on $C(\mathscr{F})$: they define the foliation induced by $\mathscr{F}$ via the projection $C(\mathscr{F})\to X$. In virtue of this there exist forms $\hat{\eta}_{jk} \in \Omega^1(C(F))$ such that

$$\sum \hat{\eta}_{jk} \wedge \omega_k = (\mathrm{d}+\delta)\omega_j. \tag{18}$$

Clearly the $(1,0)$-components ϵ_{jk} of the forms $\hat{\eta}_{jk}$ are well defined by (18); they are $G(\mathscr{C}(\mathscr{F}))$-invariant and satisfy the relation $\Sigma \epsilon_{jk} \wedge \omega_k = \delta\omega_j$. (One can take η_{jk} for the $(0,1)$-components of $\hat{\eta}_{jk}$.) By associating $\eta_{jk}+\epsilon_{jk}$ to e_{jk} we obtain the desired extension

$$W(q)\to\{\Omega^{**}_{C(\mathscr{F})}, \mathrm{d}+\delta\} \tag{19}$$

of (17).

PROPOSITION 3.2. *The homomorphism* (19) *induces a monomorphism on homology.*

Proof. By Proposition 3.1 we need only construct a homomorphism $\{\Omega^{**}_{C(\mathscr{F})}, \mathrm{d}+\delta\}\to C^*(W_q; R)$ whose composition with (19) is the injection (16). Take $x \in X$ and choose a local coordinate system $\{t_i\}$ with origin x in a neighbourhood U of x, such that $t_1,\ldots,t_q$ are locally constant on the leaves of the foliation. Take then the point $y \in C$ over x corresponding to the forms $\omega_i = \mathrm{d}t_i (i=1,\ldots,q)$, $\eta_{jk}=0$. The Lie algebra W_q maps naturally into the tangent space of C at y (as well as at any other point of C): any one-parameter family of diffeomorphisms of R^q defines an one-parameter family of foliation preserving diffeomorphisms of U, and—by means of inducing—a one-parameter family of locally defined forms ω, η (the forms η remain zero). This homomorphism $W_q \to \mathrm{tang}_y C$ gives rise to a homomorphism $\Omega^*(C)\to C^*(W_q; R)$ which obviously takes η_{jk} into 0 and ϵ_{jk} into e_{ik}.

3. The generalized Godbillon–Vey forms associated with a $\mathscr{CO}(\mathscr{F})$-structure

Denote by $WO(q)$ the subcomplex of $W(q)$, consisting of '$O(q)$-basic' elements, that is classes of those elements of $\Lambda^*\mathfrak{gl}(q,R)' \otimes S^*\mathfrak{gl}(q,R)'$ which (i) are $O(q)$-invariant, and (ii) belong to the image of $\Lambda\mathfrak{g}(\mathfrak{l}(q,R)/\mathfrak{D}(q))' \otimes S^*\mathfrak{gl}(q,R)'$. These elements can be characterized also by their being taken into the subcomplex $C^*(W_q, O(q); R)$ of $C^*(W_q; R)$ by the injection (16); recall that $C^*(W_q, O(q); R)$ consists of the $O(q)$-invariant elements of $C^*(W_q; R)$ which are annihilated by the elements of $\mathfrak{O}(q) \subset W_q$. The injection $WO(q)\to C^*(W_q, O(q); R)$ arising is a homotopy equivalence again (see [3]).

Let us study the behaviour of elements of $WO(q)$ under the homomorphisms (15) and (19). As to the homomorphism (15), a direct calculation shows that its restriction to $WO(q)$ depends actually not on the $\mathscr{C}(\mathscr{F})$-structure but on the indiced $\mathscr{CO}(\mathscr{F})$-structure. Thus the homomorphism

$$WO(q)\to\Omega^*(X) \tag{20}$$

can be constructed when a $\mathscr{CO}(\mathscr{F})$-structure is given. Furthermore the induced homology homomorphism

$$H_*(WO(q))\to H^*(X; R)$$

does not depend even on the $\mathscr{CO}(\mathscr{F})$-structure, and thus is well defined by the foliation $\mathscr{F}$. As a $\mathscr{CO}(\mathscr{F})$-structure exists for any foliation $\mathscr{F}$ (see #1 of §2), the last homomorphism is defined for any foliation $\mathscr{F}$. It coincides with the characteristic homomorphism of [2] and [5].

When $q=1$, $WO(q)$ coincides with $W(q)$. When $q=2$ the reduced homology of $W(q)$ is generated by three elements, and so three polynomials in the forms $\eta_{11}, \eta_{12}, \eta_{21}, \eta_{22}$ and their differentials represent three characteristic classes. Here are the polynomials:

$$\zeta_{11}\wedge\zeta_{22}-\zeta_{12}\wedge\zeta_{21}$$
$$\mathrm{d}(\eta_{11}+\eta_{22})\wedge\mathrm{d}(\eta_{11}+\eta_{22})\wedge(\eta_{11}+\eta_{22}),$$
$$(\zeta_{11}\wedge\zeta_{22}-\zeta_{12}\wedge\zeta_{21})\wedge(\eta_{11}+\eta_{22})$$

(where $\zeta_{jk} = \mathrm{d}\eta_{jk} - \Sigma_m \eta_{jm}\wedge\eta_{mk}$). They do not vary if the forms η_{jk} are replaced by others representing a $\mathscr{C}(\mathscr{F})$-structure inducing the same $\mathscr{CO}(\mathscr{F})$-structure. The first of these polynomials represents the first Pontryagin class of the normal bundle of $\mathscr{F}$. It may be replaced by $\mathrm{d}\eta_{11}\wedge\mathrm{d}\eta_{22}-\mathrm{d}\eta_{12}\wedge\mathrm{d}\eta_{21}$.

Note that the homology of $WO(q)$ is well-known as that of $W(q)$ (see [12]).

Turn to the homomorphism (19). One can easily check that it takes elements of $WO(q)$ into forms of $C(\mathcal{F})$ lifted from $CO(\mathcal{F})$. Hence we have a homomorphism

$$WO(q)\to\{\Omega^{**}_{\mathcal{CO}(\mathcal{F})}, \mathrm{d}+\delta\}. \qquad (21)$$

PROPOSITION 3.3. *The homomorphism* (21) *induces a monomorphism on homology.*

Proof. Consider the homomorphism

$$WO(q)\to\{\Omega^{**}_{\mathcal{CO}(\mathcal{F})}, \mathrm{d}+\delta\}\to\{\Omega^{**}_{\mathcal{C}(\mathcal{F})}, \mathrm{d}+\delta\}\to C^*(W_q; R)$$

composed of the homomorphism (21), the homomorphism, induced by the projection, and the homomorphism constructed in the proof of Proposition 3.2. Its image is clearly contained in $C^*(W_q, O(q); R)$, and the induced homomorphism $WO(q)\to C^*(W_q, O(q); R)$ clearly coincides with the natural injection. But the latter is a homotopy equivalence.

4. The linearisation of $\mathcal{CO}(\mathcal{F})$-structures

Note first that an ∞-jet of a $\mathcal{CO}(\mathcal{F})$-structure is a class of ∞-jets of $\mathcal{C}(\mathcal{F})$-structures with respect to the following equivalence relation. An ∞-jet of a $\mathcal{C}(\mathcal{F})$-structure (ω_i, η_{jk}) at a certain point is equivalent to an ∞-jet of a $\mathcal{C}(\mathcal{F})$-structure (ω'_i, η'_{jk}) at the same point if there exist ∞-jets of functions $a_{st}(s, t = 1, \dots, q)$ at the point such that the matrix $\|a_{st}\|$ is orthogonal and

$$\omega'_i = \sum_s a_{is}\wedge\omega_s,\ \eta'_{jk} = \sum_{u,v} a_{ji}a_{kv}\eta_{uv} + \sum_w (\mathrm{d}a_{jw})a_{kw}. \qquad (22)$$

The equivalence of this description of a $\mathcal{CO}(\mathcal{F})$-structure to the initial one follows from three remarks. First: if the jets of $\mathcal{C}(\mathcal{F})$-structures (ω_i, η_{jk}), (ω'_i, η'_{jk}) represent the same $\mathcal{CO}(\mathcal{F})$-structure, then there exists a jet of (variable) orthogonal matrices $\|a_{st}\|$ with $\omega'_i = \Sigma_s a_{is}\omega_s$, and this jet is well defined by ω_i, ω'_i. Second: if (ω_i, η_{jk}) is the jet of a $\mathcal{C}(\mathcal{F})$-structure and the forms ω'_i, η'_{jk} are defined by the formula (22) with orthogonal $\|a_{st}\|$ then (ω'_i, η'_{jk}) is also the jet of a $\mathcal{C}(\mathcal{F})$-structure, and (ω_i, η_{jk}), (ω'_i, η'_{jk}) define the same jet of $\mathcal{CO}(\mathcal{F})$-structure. Third: if (ω_i, η_{jk}), (ω'_i, η'_{jk}) are two jets of $\mathcal{C}(\mathcal{F})$-structures and $\omega'_1, \dots, \omega'_q$ coincide with $\omega_1, \dots, \psi_q$ and $\eta'_{jk} - \eta_{jk}$ is skew-symmetric (that is $\eta'_{jk} + \eta'_{kj} = \eta_{jk} + \eta_{kj}$), then the forms η'_{jk} coincide with η_{jk}.

Now we construct a canonical linearisation of $\mathcal{CO}(\mathcal{F})$-structures.

First recall that the homogeneous space $GL(q, R)/O(q)$ has a natural $GL(q, R)$-invariant Riemannian metric of non-positive curvature. In virtue of this any two points of the space are joined by a unique geodesic path, and these paths are taken into each other by $GL(q, R)$.

Let ζ^0, ζ^1 be two ∞-jets of $\mathcal{CO}(\mathcal{F})$-structures at a point $x \in X$. Take jets of $\mathcal{C}(\mathcal{F})$-structures $(\omega_i^0, \eta^0_{jk})$, $(\omega_i^1, \eta^1_{jk})$ over ζ^0, ζ^1. Relating to a jet of a $\mathcal{C}(\mathcal{F})$-structure (ω_i, η_{jk}) at x the jet of the map $X\to GL(q, R)$ taking the vector $\omega_1^0, \dots, \omega_q^0$ into $\omega_1, \dots, \omega_q$ defines a map of the set of jets of $\mathcal{C}(\mathcal{F})$-structures into the set of jets of maps $X\to GL(q, R)$, and hence a map of the set of jets of $\mathcal{CO}(\mathcal{F})$-structures into the set of jets of maps $X\to GL(q, R)/O(q)$. The last map takes ζ^0 into the constant jet with the value $O(q)$ and takes ζ^1 into a certain jet $f: X\to GL(q, R)/O(q)$. The uniqueness of geodesic paths in $GL(q, R)/O(q)$ gives rise to a canonical path $t\mapsto f(t)$ joining these jets. Fix a path $t\mapsto g(t)$ in the space of jets of maps $X\to GL(q, R)$ beginning at the constant jet with the value E and covering f, and denote by h the jet of the map $X\to GL(q, R)$ taking the vector $\omega_1^0, \dots, \omega_q^0$ into $\omega_1, \dots, \omega_q$. The path

$$t\mapsto\{g(t)\omega^0, (1-t)[g(t)\eta^0(g(t))^{-1} + \mathrm{d}g(t)(g(t))^{-1}] + t[g(t)h^{-1}\eta^1 h(g(t))^{-1} + \mathrm{d}(g(t)h^{-1})h(g(t))^{-1}]\}$$

(where ω^0 denotes the vector $\omega_1^0, \dots, \omega_q^0$, and η^0, η^1 denote the matrices $\|\eta^0_{jk}\|$, $\|\eta^1_{jk}\|$) in the space of jets of $\mathcal{C}(\mathcal{F})$-structures defines a path in the space of jets of $\mathcal{CO}(\mathcal{F})$-structures, which does not depend on any intermediate choices and joins ζ^0 and ζ^1.

Such paths give the required linearisation.

By Proposition 2.1 the linearizability of the $\mathcal{CO}(\mathcal{F})$-structure implies that the homomorphism

$$\tau: \{A^{**}_{\mathcal{CO}(\mathcal{F})}; \mathrm{d}, \Delta\}\to\{\Omega^{**}_{\mathcal{CO}(\mathcal{F})}; \mathrm{d}, \delta\}$$

is a homotopy equivalence. Moreover, the canonical linearization gives rise to a canonical homotopy inverse to τ,

$$\sigma: \{\Omega^{**}_{\mathcal{CO}(\mathcal{F})}; \mathrm{d}, \delta\}\to\{A^{**}_{\mathcal{CO}(\mathcal{F})}; \mathrm{d}, \Delta\}.$$

The composition of σ with (21) is a homomorphism

$$WO(q)\to\{A^{**}_{\mathcal{C}O(\mathcal{F})}, d+\Delta\}, \tag{23}$$

and the composition of (23) with the homomorphism induced by the projection $C(\mathcal{F})\to CO(\mathcal{F})$ is a homomorphism

$$WO(q)\to\{A^{**}_{\mathcal{C}O(\mathcal{F})}, d+\Delta\} \tag{24}$$

extending the restriction to $WO(q)$ of (17).

We conjecture that the homomorphisms (19), (21) [and hence (23)], and (24) are homotopy equivalences. (This would imply that the homomorphism (17) cannot be extended to a homomorphism $W(q)\to\{A^{**}_{\mathcal{C}(\mathcal{F})}, d+\Delta\}$.) We can prove it only for $q=1$, $\dim X\neq 2$ so far.

Remarks. 1°. The procedure for constructing $(d+\delta)$-cycles corresponding to the cycles of $W(q)$ is the following. Take the generalized Godbillon–Vey form written in the usual way in terms of the forms η_{jk} and replace everywhere d by $d+\delta$ and η_{jk} by $\eta_{jk}+\epsilon_{jk}$, taking into account that $\delta\epsilon_{jk}=0$.

The procedure for constructing $(d+\Delta)$-cycles corresponding to the cycles of $WO(q)$ is the following. Denote by $E_{st}((\omega_i,\eta_{jk}),(\omega'_i,\eta'_{jk}))$ the (s,t)-element of the matrix $\log F$, where F is the matrix taking $\omega_1,\ldots,\omega_q$ into $\omega'_1,\ldots,\omega'_q$. Of course E_{st} is not a well defined element of $A^{**}_{\mathcal{C}O(\mathcal{F})}$ because only matrices sufficiently close to unity have logarithms. Ignoring this for a while, take the $(d+\delta)$-cycle for the given cycle of $WO(q)$, and replace in it δ by Δ, ϵ_{st} by E_{st} and the exterior product by the skew-symmetrized cup-product of simplicial cochains. General properties of elements of $WO(q)$ automatically imply that: (i) the resulting expressions can be analytically extended to cochains of $A^{**}_{\mathcal{C}(\mathcal{F})}$, and (ii) each of these cochains is the image of a cochain of $A^{**}_{\mathcal{C}O(\mathcal{F})}$.

2°. Applying the above procedures to the Godbillon–Vey form $\eta\wedge d\eta$ we obtain the $(d+\delta)$-cycle

$$\eta\wedge d\eta+(\eta\wedge\delta\eta+\eta\wedge d\epsilon+\epsilon\wedge d\eta)+(\epsilon\wedge d\epsilon+\epsilon\wedge\delta\eta),$$

and the $(d+\Delta)$-cycle $\Gamma^0+\tilde{\Gamma}^1+\tilde{\Gamma}^2$ where

$$\tilde{\Gamma}^1(\omega,\eta;\omega',\eta')=\eta'\wedge\eta+\frac{1}{2}\left[\log\left|\frac{\omega'}{\omega}\right| d(\eta'+\eta)-d\log\left|\frac{\omega'}{\omega}\right|\wedge(\eta'+\eta)\right],$$

$$\tilde{\Gamma}^2(\omega,\eta;\omega',\eta';\omega'',\eta'')=\Gamma^2(\omega,\omega',\omega'')+\frac{1}{2}\left(\log\left|\frac{\omega'}{\omega}\right|\eta''+\log\left|\frac{\omega''}{\omega'}\right|\eta+\log\left|\frac{\omega}{\omega''}\right|\eta'\right).$$

These expressions are different from the $(d+\delta)$ cycle presented in the example in #4 of §2 and the $(d+\Delta)$-cycle $\Gamma^0+\Gamma^1-\Gamma^2$ of §1, though they are related to them by suitable homology relations. In §1 we preferred the expression $\Gamma^0+\Gamma^1-\Gamma^2$ to the expression $\Gamma^0+\tilde{\Gamma}^1+\tilde{\Gamma}^2$ because of the awkwardness of the latter.

3°. When $q=\dim X$ the foliation $\mathcal{F}$ becomes trivial and can be discarded. In this case a $\mathcal{C}(\mathcal{F})$-structure consists in fixing the 2-jet of a local coordinate system at each point $x\in X$ (continuously with respect to x), and a $\mathcal{C}O(\mathcal{F})$-structure is simply an affine connection. Most of the generalized Godbillon–Vey forms associated with this $\mathcal{C}O(\mathcal{F})$-structure are trivial by dimension arguments. The only exceptions are polynomials in the Pontryagin forms. But all the generalized Godbillon–Vey forms are nontrivial in the extended variation or difference complex; moreover the injection of $WO(q)$ into each of these complexes induces a monomorphism on homology (Proposition 3.3).

§4. THE DE RHAM COMPLEX OF A STRATIFIED SET

1. Stratified sets

Definition 4.1. A smooth n-manifold with a piecewise smooth boundary is a connected subset of a smooth n-manifold without boundary which is the intersection of a finite family of n-submanifolds whose boundaries are in general position (cf. #4 of §1).

Manifolds with piecewise smooth boundaries can be described also by means of local models; one can take for them the products of Euclidean spaces and the octants of Euclidean spaces.

By smooth maps of manifolds with piecewise smooth boundaries, differential forms on such manifolds, their tangent vectors, etc., we mean corresponding objects given in the containing manifold.

The boundary of an n-manifold with a piecewise smooth boundary naturally decomposes into a disjoint sum of smooth connected manifolds of dimensions $0, \dots, n-1$; these manifolds, as well as int X, are called strata of X.

Definition 4.2. A Hausdorff space is called a stratified set, if it is furnished with a locally finite decomposition into a sum of strata X_i which are connected sets satisfying the following conditions:

(i) if $\bar{X}_i \cap X_j \neq \emptyset$, then $\bar{X}_i \supset \bar{X}_j$.

(ii) the closure of each X_i is equipped with a structure of a smooth manifold with a piecewise smooth boundary, such that each stratum X_j contained in X_i is a stratum of X_i, and that the smooth structures of X_i, X_j are compatible.

Examples of stratified sets are: a smooth manifold with a piecewise smooth boundary, and a locally finite simplicial complex.

Definition 4.3. A map from a stratified set X into a stratified set Y is called smooth if it takes each stratum of X to a stratum of Y and is smooth on the closure of each stratum of X.

It is obvious that smooth maps are continuous.

2. Differential forms

Let $X = \cup X_i$ be a stratified set.

We denote by $\Omega(\bar{X}_i)$ the sheaf of germs of differential q-forms on X_i, extended onto X by zero. Define

$$\mathbf{\Omega}^q(X) = \bigoplus_{k=0}^{q} \prod_{(X_{i_0},\dots,X_{i_k})} \Omega^{q-k}(\bar{X}_{i_k})$$

where the product under the summation sign is taken over all k-flags, that is sequences of pairwise different strata $X_{i_0}, \dots, X_{i_k}$ with $\bar{X}_{i_{j-1}} \supset X_{i_j}$ $(j = 1, \dots, k)$. The sections of $\mathbf{\Omega}^q(X)$, which are sometimes called q-forms on X, are written as sets of forms $\omega_{i_0\dots i_k}$, where $0 \leq k \leq q$, $i_0, \dots, i_k$ are indices of strata composing a k-flag, and $\omega_{i_0\dots i_k}$ is a $(q-k)$-form on X_{i_k}.

The differential $\mathbf{d}: \mathbf{\Omega}^q(X) \to \mathbf{\Omega}^{q+1}(X)$ is defined by

$$(\mathbf{d}\omega)_{i_0\dots i_k} = (-1)^k \, d\omega_{i_0\dots i_k} + \sum_{j=0}^{k} (-1)^j \omega_{i_0\dots \hat{i}_j \dots i_k} | X_{i_k}.$$

A direct calculation shows that $\mathbf{d} \circ \mathbf{d} = 0$, i.e. that $\{\mathbf{\Omega}^q, \mathbf{d}\}$ is a complex.

PROPOSITION 4.1. *The complex*

$$\mathbf{\Omega}(X) = \{\mathbf{\Omega}^0(X) \xrightarrow{\mathbf{d}} \mathbf{\Omega}^1(X) \xrightarrow{\mathbf{d}} \mathbf{\Omega}^2(X) \to \cdots\}$$

is acyclic, i.e. the homology of this complex is trivial in positive dimensions and coincides in dimension O with the constant sheaf R, naturally embedded in $\mathbf{\Omega}^0(X)$.

Proof. Let x be an arbitrary point of X, and X_i be its stratum. Let then $X_{i_1}, \dots, X_{i_m}$ be strata different from X_i whose closures contain X_i. In the complex

$$\mathbf{\Omega}_x(X) = \{\mathbf{\Omega}_x^0(X) \xrightarrow{\mathbf{d}} \mathbf{\Omega}_x^1(X) \xrightarrow{\mathbf{d}} \mathbf{\Omega}_x^2(X) \to \cdots\}$$

composed of the stalks of the sheaves $\mathbf{\Omega}^q$ at x, we define a filtration F:

$$\omega \in F_k \mathbf{\Omega}_x^q \quad \text{if} \quad \omega_{j_0\dots j_l} = 0 \text{ for } l \leq K.$$

Clearly

$$0 = F_m \mathbf{\Omega}_x^q \subset F_{m-1} \mathbf{\Omega}_x^q \subset \cdots \subset F_{-1} \mathbf{\Omega}_x^q = \mathbf{\Omega}_x^q.$$

This filtration is compatible with the differential and gives rise to a spectral sequence. In this spectral sequence

$$E_0^{p,q} = \bigoplus_{i_0,\dots,i_p} \Omega_x^q(\bar{X}_{i_p}),$$

where the summation is taken over all p-flags composed of the strata X_i, $X_{i_1}, \dots, X_{i_m}$, and $\Omega_x^q(\bar{X}_{i_p})$ is the space of germs of q-forms on X_{i_p} at the point x. The zero differential of this spectral sequence coincides (up to sign) with the usual exterior differential, and since the complexes $\{\Omega_x^q(X_{i_p}), \mathrm{d}\}$ are acyclic, we have

$$E_1^{p,q} = \begin{cases} 0 & \text{if } q > 0, \\ \bigoplus_{j_1,\dots,j_p} R & \text{if } q = 0. \end{cases}$$

Alternatively, $E_1^{p,q}$ can be interpreted as the space of p-cochains of the semisimplicial set corresponding to the (partially) ordered set $i, i_1, \dots, i_m$. The differential d_1 becomes the usual semisimplicial differential in this interpretation. This set has an initial element (namely i) and therefore the semisimplicial set is acyclic. Hence $E_2^{p,q} = 0$ for $(p, q) \neq (0, 0)$ and $E_2^{0,0} = R$, this implies the acyclicity of $\Omega(X)$.

As the sheaves $\Omega^q(X)$ are fine, the acyclicity of $\Omega(X)$ gives rise to a canonical isomorphism between the homology of the global section complex $\{\Gamma\Omega^q, \mathrm{d}\}$ and the real cohomology of X. Thus Ω is a good substitute for the de Rham complex for stratified sets. We shall establish a further analogy between Ω and the de Rham complex in #3, where we define integration for the sections of Ω and prove the Stokes formula.

Note that a smooth map f from a stratified set X into a stratified set Y defines in an obvious way a homomorphism

$$f^*: \Omega(Y) \to \Omega(X),$$

and the induced homomorphism

$$H_*(\{\Gamma\Omega^q(Y), \mathrm{d}\}) \to H_*(\{\Gamma\Omega^q(X), \mathrm{d}\})$$

coincides with

$$f^*: H^*(Y, R) \to H^*(X, R).$$

3. Integration

Being a locally finite simplicial complex, the standard simplex Δ^q has a natural structure of stratified set. Its strata are its open faces. We denote by $\Delta(i_0, \dots, i_k)$ the face opposite to the vertices with indices $i_0, \dots, i_k$. The faces inherit from Δ^q an ordering of their vertices, and hence orientations.

We call a smooth singular q-simplex of a stratified set X a smooth (in the meaning of the Definition 4.3) map of Δ^q into X. A finite linear combination of smooth singular simplices is called a smooth singular chain. The smooth singular chains form a subcomplex of the usual singular complex of X, and it is homotopically equivalent to the whole complex.

If ω is a q-form of X, i.e. a sectionof $\Omega^q(X)$, and φ is a smooth singular q-simplex, then the integral $\int_\varphi \omega$ is defined as

$$\sum_{k=0}^{q} \sum_{i_1,\dots,i_k} (-1)^{i_1+\dots+i_k+I(i_1,\dots,i_k)} \int_{\Delta(i_1,\dots,i_k)} \varphi^*(\omega)_{\Delta^q, \Delta(i_1),\dots,\Delta(i_1,\dots,i_k)}$$

were the second summation is taken over all (ordered) sequences $i_1, \dots, i_k \in \{0, \dots, q\}$ and $I(i_1, \dots, i_k)$ is the number of pairs of indices s, t such that $s < t$, $i_s < i_t$.

The integral $\int_c \omega$ of a form ω over a smooth singular chain $c = \sum n_i \varphi_i$ is defined as $\sum n_i \int_{\varphi_i} \omega$.

It is clear that for any smooth map f of a stratified set X into a stratified set Y, any form ω on Y, and any smooth singular chain c of X we have

$$\int_c f^*\omega = \int_{f_* c} \omega.$$

It is clear also that the integral of a form over a chain does not change if the chain is replaced by its subdivision.

Proposition 4.2. (*The Stokes formula*). *If c is a smooth singular q-chain of a stratified set X and $\omega \in \Gamma\Omega^{q-1}(X)$, then*

$$\int_c \mathbf{d}\omega = \int_{\partial c} \omega$$

where ∂ is the boundary operator.

Proof. One can restrict oneself to the case when $X = \Delta^q$ and c is the identity. One can also take the form ω to be zero for all flags but one, $\Delta_0, \ldots, \Delta_k$. If $\int_{\partial c} \omega \neq 0$, then $(\dim \Delta_0, \ldots, \dim \Delta_k) = (q-1, \ldots, q-k+1)$, if $\int_c \mathbf{d}\omega \neq 0$, then $(\dim \Delta_0, \ldots, \dim \Delta_k)$ is a subsequence of $(q, \ldots, q-k+1)$, i.e. $(\dim \Delta_0, \ldots, \dim \Delta_k) = (q, \ldots, q-r+1, q-r-1, \ldots, q-k-1)$ with $0 \leq r \leq k+1$. In the last case $\Delta_0 = \Delta^q$, $\Delta_1 = \Delta(i_1), \ldots, \Delta_{r-1} = \Delta(i_1, \ldots, i_{r-1})$, $\Delta_r = \Delta(i_1, \ldots, i_{r+1}), \ldots, \Delta_k = \Delta(i_1, \ldots, i_{k+1})$. There are three possibilities.

(i) $r = 0$. In this case

$$\int_c \mathbf{d}\omega = (-1)^{i_1+\cdots+i_{k+1}+I(i_1,\ldots,i_{k+1})} \int_{\Delta(i_1,\ldots,i_{k+1})} \omega_{\Delta_0,\ldots,\Delta_k};$$

$$\int_{\partial c} \omega = (-1)^{i_1}(-1)^{i'_2+\cdots+i'_{k+1}+I(i_2,\ldots,i_{k+1})} \int_{\Delta(i_1,\ldots,i_{k+1})} \omega_{\Delta_0,\ldots,\Delta_k},$$

where $i'_s = \begin{cases} i_s, & \text{if } i_s < i_1, \\ i_s - 1, & \text{if } i_s > i_1, \end{cases}$

and obviously $\int_c \mathbf{d}\omega = \int_{\partial c} \omega$.

(ii) $0 < r \leq k$. In this case

$$\int_c \mathbf{d}\omega = (-1)^{i_1+\cdots+i_{k+1}+I(i_1,\ldots,i_{k+1})} \int_{\Delta(i_1,\ldots,i_k)} (-1)^r \omega_{\Delta_0,\ldots,\Delta_k}$$
$$+ (-1)^{i_1+\cdots+i_{k-1}+I(i_1,\ldots,\widehat{i_r, i_{r+1}},\ldots,i_{k+1}} \int_{(i_1,\ldots,i_k)} (-1)^r \omega_{\Delta_0,\ldots,\Delta_k} = 0,$$

and $\int_{\partial c} \omega = 0$ (the mark $\frown$ means that one should transpose i_r and i_{r+1}).

(iii) $r = k+1$. Then $\Delta_k = \Delta(i_1, \ldots, i_k)$ and the index i_{k+1} is not involved at all. In this case

$$\int_c \mathbf{d}\omega = \sum_{i \notin \{i_1,\ldots,i_k\}} (-1)^{i_1+\cdots+i_k+i+I(i_1,\ldots,i_k)} \int_{\Delta(i_1,\ldots,i_k,i)} (-1)^{k+1} \omega_{\Delta_0,\ldots,\Delta_k}$$
$$+ (-1)^k(-1)^{i_1+\cdots+i_k+I(i_1,\ldots,i_k)} \int_{\Delta(i_1,\ldots,i_k)} \mathbf{d}\omega_{\Delta_0,\ldots,\Delta_k}$$
$$= (-1)^{i_1+\cdots+i_k+I(i_1,\ldots,i_k)+k} \left[\int_{\Delta(i_1,\ldots,i_k)} \mathbf{d}\omega_{\Delta_0,\ldots,\Delta_k} - \int_{\partial\Delta(i_1,\ldots,i_k)} \omega_{\Delta_0,\ldots,\Delta_k} \right] = 0$$

(the last is in virtue of the usual Stokes formula), and $\int_{\partial c} \omega = 0$.

The most important special case of the above situation is when X is a smooth oriented manifold with a piecewise smooth boundary, and the integration is taken over the fundamental cycle of X (that is, over the chain defined by a smooth ordered triangulation compatible with the stratification). In this case we have, after arbitrarily choosing orientations of the strata of ∂X (the fundamental cycle is denoted by the same symbol as the manifold):

$$\int_X \omega = \sum_{(\text{int}\, X, X_{i_1}, \ldots, X_{i_k})} \epsilon(\text{int}\, X, X_{i1}) \ldots \epsilon(X_{i_{k-1}}, X_{i_k}) \int_{X_{i_k}} \omega_{\text{int}\, X, X_{i1},\ldots,X_{ik}},$$

where the summation is taken over all flags beginning with int X and such that codim $X_{i_s} = s$, and $\epsilon(X_{i_s}, X_{i_{s+1}})$ is equal to 1 if the orientations of $X_{i_s}, X_{i_{s+1}}$ agree, and equal to -1 otherwise.

§5. THE GAUSS–BONNET THEOREM FOR FOLIATIONS OF ARBITRARY CODIMENSION

1. Main theorems

Let $X = \cup X_i$ be a stratified set.

Assume that the closure of each X_i is furnished with a structural sheaf $\mathscr{C}_i$ with the formal structure space C_i (we mean that $\mathscr{C}_i$ is the restriction to $\bar{X}_i$ of a structural sheaf $\tilde{\mathscr{C}}_i$ over a smooth manifold containing X as a domain with a piecewise smooth boundary, and C_i is the inverse image of $\bar{X}_i$ in the space of formal $\bar{\mathscr{C}}_i$-structures). These structural sheaves are assumed to agree in the following sense.

(1) If a diffeomorphism $\bar{X}_i \to \bar{X}_i$ belongs to $G(\mathscr{C}_i)$ and takes $\bar{X}_j$ into $\bar{X}_j$, then the induced diffeomorphism $\bar{X}_j \to \bar{X}_j$ belongs to $G(\mathscr{C}_j)$.

(2) For each pair of strata X_i, X_j with $\bar{X}_i \supset X_j$ a map $\mu_{ij}: \mathscr{C}_i|_{\bar{X}_j} \to \mathscr{C}_j$ is fixed. This map covers $\mathrm{id}_{\bar{X}_j}$ and is compatible with the action of the subgroupoid of the groupoid $G(\mathscr{C}_i)$ consisting of diffeomorphisms taking $\bar{X}_j$ into $\bar{X}_j$.

Examples. 1°. Each X_i with $\dim X_i \geqslant q$ is furnished with a foliation of codimension q which is transverse to each stratum of ∂X_i of dimension $\geqslant q$, and whose restrictions to these strata coincide with the foliations given there. The foliation of X_i is denoted by $\mathscr{F}_i$. When $\dim X_i \geqslant q$, one takes $\mathscr{C}(\mathscr{F}_i)$ for $\mathscr{C}_i$ and when $\dim X_i < q$, one defines $\mathscr{C}_i$ as trivial (the sheaf of one-element sets). The maps μ_{ij} are the restriction operations.

2°. The same as 1° but with $\mathscr{C}\mathscr{O}(\mathscr{F}_i)$ in the place of $\mathscr{F}_i$.

3°. $\mathscr{C}_i$ is the sheaf of germs of Riemannian metrics on X_i, and μ_{ij} is taking the induced metric.

Assume now that a (d + Δ)-cycle $a_i \in A^{p-r,r}_{\mathscr{C}_i}$ is fixed for every i and that the cycles a_i satisfy the following compatibility condition: if $\bar{X}_i \supset X_j$ then $a_i|_{\bar{X}_j} = \mu^*_{ij} a_j$.

The fundamental examples of this situation is the case when the sheaves $\mathscr{C}_i$ and the maps μ_{ij} come from the above examples 1°, 2°, and a_i (for $\dim X_i \geqslant q$) is a closed generalized Godbillon–Vey form (the same for all i), i.e. the image of a cycle of the complex $WO(q)$ under the homomorphism (23) and (24). The compatibility condition holds if there is no stratum of dimension $< q$ or if the generalized Godbillon–Vey form has no $(p-r, r)$-components with $r < q$. Note that the second possibility is not entirely excluded: it holds for the (d + Δ)-cycle $\Gamma^0 + \Gamma^1 - \Gamma^2$ of §1, for the first, fourth and fifth forms of the list of #1 of §3, and for the second form from the list of #3 §1. Note also that the (d + Δ)-extensions of the Pontryagin forms do not satisfy the compatibility condition in the situation of example 3° (though the extensions themselves exist by the remark 3° at the end of §3); even the usual Pontryagin forms are not compatible with taking an induced metric.

Return to the general situation. Assume that in every sheaf $\mathscr{C}_i$ a section σ_i is chosen (the sections σ_i are not subject to any compatibility conditions). The sets $\sigma = \{\sigma_i\}$ and $a = \{a_i\}$ give rise to differential forms $\omega_0(\sigma, a)$ of X (i.e. to sections of $\Omega^p(X)$):

$$[\omega_0(\sigma, a)]_{i_0 \ldots i_k}(\xi_1, \ldots, \xi_{p-k}) = [a_{i_k}(\mu_{i_0 i_k}(\sigma_{i_0}|x_{i_k}), \ldots, \mu_{i_{k-1} i_k}(\sigma_{i_{k-1}}|x_{i_k}))](\tilde{\xi}_1, \ldots, \tilde{\xi}_{p-k}),$$

where $\xi_1, \ldots, \xi_{p-k}$ are tangent vectors of X at a certain point x, and $\tilde{\xi}_1, \ldots, \tilde{\xi}_{p-k}$ are the horizontal tangent vectors of $(C_{i_k})^{(k+1)}$ covering $\xi_1, \ldots, \xi_{p-k}$ at the point defined by the jets at x of the sections $\mu_{i_0 i_k}(\sigma_{i_0}|x_{i_k}), \ldots, \mu_{i_{k-1} i_k}(\sigma_{i_{k-1}}|x_{i_k}), \sigma_{i_k}$.

THEOREM 5.1. *Let c be a smooth singular p-chain of X. If ∂c does not touch X_i, then the integral $\int_c \omega_0(\sigma, a)$ does not depend on σ_i. That is, the integral (25) does not change if all sections σ_i with $\partial c \cap X_i = \emptyset$ are replaced by any other sections.*

Theorem 1.1 is a special case of Theorem 5.1.

Instead of Theorem 5.1 we prove the following more precise assertion (generalizing Theorem 1.2).

THEOREM 5.2. *Let $\sigma' = \{\sigma'_i\}$ be another set of sections. Then*

$$\int_c \omega_0(\sigma', a) - \int_c \omega_0(\sigma, a) = \int \omega_1(\sigma, \sigma'; a), \tag{26}$$

where

$$[\omega_1(\sigma, \sigma'; a)]_{i_0 \ldots i_k}(\xi_1, \ldots, \xi_{p-k-1})$$

$$= \sum_{s=0}^{k} (-1)^s [a_{i_k}(\mu_{i_0 i_k}\sigma_{i_0}|x_{i_k}), \ldots, \mu_{i_s i_k}(\sigma_{i_s}|x_{i_k}), \mu_{i_s i_k}(\sigma'_{i_s}|x_{i_k}), \ldots, \mu_{i_{k-1} i_k}(\sigma'_{i_{k-1}}|x_{i_k}), \sigma'_{i_k})](\tilde{\xi}_1, \ldots, \tilde{\xi}_{p-k-1})$$

where, in turn, $\xi_1, \ldots, \xi_{p-k-1}$ are tangent vectors of X, and $\tilde{\xi}_1, \ldots, \tilde{\xi}_{p-k-1}$ are horizontal tangent vectors of $(C_{i_k})^{(k+1)}$ covering $\xi_1, \ldots, \xi_{p-k-1}$ at the point defined by the jets of the sections $\mu_{i_0 i_k}(\sigma_{i_0}/x_{i_k}), \ldots, \sigma'_{i_k}$.

Proof. The equality (26) follows from the Stokes formula and the relation $\mathbf{d}\omega_1(\sigma, \sigma'; a) = \omega_0(\sigma', a) - \omega_0(\sigma, a)$; the latter follows, in turn, because a is a $(\mathrm{d}+\Delta)$-cycle. More precisely:

$$
\begin{aligned}
[\mathbf{d}\omega_1(\sigma,\sigma';a)]_{i_0\ldots i_k}(\xi) &= (-1)^k\, \mathrm{d}[\omega_1(\sigma,\sigma',a)]_{i_0\ldots i_k}(\xi) + \sum_{j=0}^{k}(-1)^j[\omega_1(\sigma,\sigma';a)]_{i_0\ldots \hat{i}_j\ldots i_k}(\xi)\\
&= (-1)^k \sum_{s=0}^{k}(-1)^s\, \mathrm{d}[a_{i_k}(\sigma_{i_0},\ldots,\sigma_{i_s},\sigma'_{i_s},\ldots,\sigma'_{i_k})](\tilde{\xi})\\
&\quad + \sum_{j=0}^{k}(-1)^j\Bigg[\sum_{s=0}^{j-1}(-1)^s a_{i_k}(\sigma_{i_0},\ldots,\sigma_{i_s},\sigma'_{i_s},\ldots,\hat{\sigma}'_{i_j},\ldots,\sigma'_{i_k})\\
&\quad + \sum_{s=j+1}^{k}(-1)^{s-1} a_{i_k}(\sigma_{i_0},\ldots,\hat{\sigma}_{i_j},\ldots,\sigma_{i_s},\sigma'_{i_s},\ldots,\sigma'_{i_k})\Bigg](\tilde{\xi});
\end{aligned}
$$

$$
\begin{aligned}
(-1)^k &\sum_{s=0}^{k}(-1)^s\, \mathrm{d}[a_{i_k}(\sigma_{i_0},\ldots,\sigma_{i_s},\sigma'_{i_s},\ldots,\sigma'_{i_k})]\\
&= -\sum_{s=0}^{k}(-1)^s\,\Delta a_{i_k}(\sigma_{i_0},\ldots,\sigma_{i_s},\sigma'_{i_s},\ldots,\sigma'_{i_k})]\\
&= -\sum_{s=0}^{k}(-1)^s\Bigg[\sum_{j=0}^{s-1}(-1)^j a_{i_k}(\sigma_{i_0},\ldots,\hat{\sigma}_{i_j},\ldots,\sigma_{i_s},\sigma'_{i_s},\ldots,\sigma'_{i_k})\\
&\quad + (-1)^s a_{i_k}(\sigma_{i_0},\ldots,\sigma_{i_{s-1}},\sigma'_{i_s},\ldots,\sigma'_{i_k})\\
&\quad + (-1)^{s+1} a_{i_k}(\sigma_{i_0},\ldots,\sigma_{i_s},\sigma'_{i_s+1},\ldots,\sigma'_{i_k})\\
&\quad + \sum_{j=s+1}^{k}(-1)^{j+1} a_{i_k}(\sigma_{i_0},\ldots,\sigma_{i_s},\sigma'_{i_s},\ldots,\hat{\sigma}_{i_j},\ldots,\sigma_{i_k})\Bigg];
\end{aligned}
$$

thus

$$
\begin{aligned}
[\mathbf{d}\omega_1(\sigma,\sigma';a)]_{i_0\ldots i_k}(\xi) &= -\sum_{s=0}^{k}(-1)^s[(-1)^s a_{i_k}(\sigma_{i_0},\ldots,\sigma_{i_{s-1}},\sigma'_{i_s},\ldots,\sigma'_{i_k})\\
&\quad + (-1)^{s+1} a_{i_k}(\sigma_{i_0},\ldots,\sigma_{i_s},\sigma'_{i_s+1},\ldots,\sigma'_{i_k})](\tilde{\xi})\\
&= [a_{i_k}(\sigma'_{i_0},\ldots,\sigma'_{i_k})](\tilde{\xi}) - [a_{i_k}(\sigma_{i_0},\ldots,\sigma_{i_k})](\tilde{\xi})\\
&= [\omega_0(\sigma',a)]_{i_0\ldots i_k}(\xi) - [\omega_0(\sigma,a)]_{i_0\ldots i_k}(\xi).
\end{aligned}
$$

(We used abbreviated notation: ξ stands for $\xi_1,\ldots,\xi_{p-k-1}$, $\tilde{\xi}$ for $\tilde{\xi}_1,\ldots,\tilde{\xi}_{p-k-1}$, and the symbols μ_{uv}, $|_{X_w}$ are omitted everywhere).

THEOREM 5.3. *If c is a smooth singular p-cycle, then the integral* (25) *does not depend* (*when $\mathscr{C}_i$, a_i are fixed*) *on the sections σ_i.*

This follows from Theorem 5.1 and generalizes Theorem 1.4.

The most important special case of the above situation is when X is a smooth orientable p-manifold ($p = \deg a$) with a piecewise smooth boundary and c is the fundamental cycle of X (see the end of §4). In this case Theorem 5.1 means that the integral $\int_X \omega_0(\sigma, a)$ does not depend (when $\mathscr{C}_i$, a_i are fixed) on the section σ_0 of the sheaf over $X_0 = \operatorname{int} X$. In other words, this integral is a number-valued function of the set $\{\sigma_i\}$ of $\mathscr{C}_i$-structures over the strata of ∂X.

THEOREM 5.4. *This function changes by an additive constant if X and $\{\mathscr{C}_i\}$ are changed, but ∂X and the sheaves over the strata of X do not vary.*

More precisely: let $\tilde{X}$ be another p-manifold with a piecewise smooth boundary, and let $\tilde{\mathscr{C}}_i$, $\tilde{a}_i$ be the same for $\tilde{X}$ as $\mathscr{C}_i$, a_i are for X. Let φ be a diffeomorphism of $\partial\tilde{X}$ onto ∂X such that $\tilde{\mathscr{C}}_i = \varphi^*\mathscr{C}_i$, $a_i = \varphi^* a_i$ when $X_i \subset \partial X$ (we assume that the stratum of X whose image is X_i also has the index i). Then the difference

$$\int_{\tilde{X}} \omega_0(\{\tilde{\sigma}_0\}, \{\varphi^*\sigma_i\}_{i\neq 0}; \tilde{a}) - \int_X \omega_0(\{\sigma_i\}, a),$$

where $\tilde{\sigma}_0$ is a section of $\tilde{\mathscr{C}}_0$ (the structural sheaf over $\tilde{X}$), does not depend (when $\mathscr{C}_0$, $\mathscr{C}_i$, a_i and φ are fixed) on the set $\{\sigma_i\}$.

The proof is straightforward.

Remarks. 1°. So far the only application of Theorems 5.1–5.4 is that to foliations and the generalized Godbillon–Vey forms. In particular Thoeorem 5.3 gives rise to the characteristic classes of piecewise smooth foliations. But do not forget the restriction: either the generalized Godbillon–Vey form must have no $(p-r, r)$-components with $r < q$, or X must have no strata of dimension $< q$. One can satisfy the second by discarding the strata of dimension $< q$; but of course one can do so only if the chain c does not touch these strata.

2°. There is a hint in the notation ω_0, ω_1, that a whole sequence $\{\omega_i\}$ naturally arises. In fact one can construct such a sequence. Moreover the sum $\omega_0 + \omega_1 + \cdots$ is the decomposition into homogeneous components of a cycle of a certain bicomplex which is to the difference bicomplex of §2 as the de Rham complex of §4 is to the usual de Rham complex. One can hope that this rather complicated bicomplex will be of some use anyway. We do not give its definition because we do not need it in such a general form; the reader will be able to reconstruct it quite easily, if necessary.

2. The localisation and splitting problems

The generalization to higher codimensions of the localization problem of #4 of §1 is the following. Let X be a compact manifold with a piecewise smooth boundary furnished with a codimension q foliation $\mathcal{F}$ transversal to strata of dimensions $\geqslant q$. Let then a $\mathcal{CO}(\mathcal{F})$-structure σ_i be fixed on each stratum X_i. Take a cycle $\mathfrak{A}$ of $WO(q)$ and denote by a_i the corresponding $(\mathrm{d}+\Delta)$-cycle of the difference $\mathcal{CO}(\mathcal{F})$-bicomplex over X_i. Assume, finally, that either X has no strata of dimension $< q$ or $\mathfrak{A}$ has no $(p-r, r)$-components with $r < q$. The problem is to represent the functional

$$\{\sigma_i\}_{X_i \subset \partial X} \mapsto \int_X \omega_0(\{\sigma_i\}, \{a_i\})$$

as a sum

$$\sum_{X_i \subset \partial X} \int_{X_i} \alpha_i,$$

where α_i is a form whose germ at $x \in X_i$ depends only on the germs at x of $\mathcal{CO}(\mathcal{F})$-structures over the strata $X_j \subset \partial X$ whose closures contain x.

The impossibility of such a representation seems beyond any doubt, but even for $q = 1$ we have no complete proof of it (see #5 of §1).

It is very probable that after putting on α_i an extra condition of smoothness the nonlocalizability can be proved by a modification of the argument of #5 of §1. On the other hand one can change the statement of the problem also in the spirit of §1 by requiring the forms α_i to be defined by a $\mathcal{CO}(\mathcal{F})$-structure on a manifold of arbitrary dimension and to be natural with respect to maps transversal to foliations. Repeating word for word what was said in #5 of §1 we obtain the implication

the localization exists	$\Rightarrow$	the generalized Godbillon–Vey form is exact

which shows the impossibility of the localization in the cases when the corresponding characteristic classes of foliations are non-trivial. But this non-triviality is proved only for some of the characteristic classes of foliations (see [5]), so the problem remains unsolved here too.

Note also that the non-homologousness to zero of the extended generalized Godbillon–Vey forms in the difference bicomplex (Proposition 3.3) implies the non-localizability in a still weaker form. Namely, there exists no set $\{b_i\}$ of chains in the difference bicomplex, associated with $\mathcal{CO}(\mathcal{F})$-structures over all strata (including $\operatorname{int} X$), satisfying the compatibility condition, and such that the difference

$$\int_X \omega_0(\{\sigma_i\}, \{a_i\}) - \int_{\partial X} \omega_0(\{\sigma_i\}_{X_i \subset \partial X}, \{b_i\}_{X_i \subset \partial X})$$

depends only on the foliation in $\operatorname{int} X$. This statement is a corollary of the following theorem which follows immediately from the Stokes formula.

THEOREM 5.5 (*a supplement to Theorem* 5.1). *If a* $(\mathrm{d}+\Delta)$*-chain* $b_i \in \oplus_r A^{p-1-r,r}_{\sigma_i}$ *with* $b_i|_{X_j} = \mu^*_{ij} b_j$ *is fixed for each* i, *and for arbitrary smooth singular* p*-chain* c *one has*

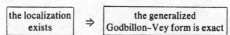

$$\int_c \omega_0(\sigma, a) = \int_{\partial c} \omega_0(\sigma, \{b_i\}) + \text{a quantity not depending on } \sigma$$

then $(d+\Delta)b_i = a_i$.

As to the splitting problem, to solve it for foliations of arbitrary codimension requires the vanishing of the integral generalizing the integral (12); and its vanishing implies (ineffective, in a certain not very clear sense) solvability of the splitting problem for manifolds with smooth boundaries. An effective splitting exists for foliations on manifolds with smooth boundaries which are representable on the boundary by a global system of closed 1-forms.

REFERENCES

1. M. F. ATIYAH, V. K. PATODI and I. M. SINGER: Spectral Asymmetry and Riemannian geometry, *Proc. Lond. Math. Soc.* **5** (1973), 229–234.
2. I. N. BERNSTEIN and B. I. ROSENFELD: On characteristic classes of foliations, *Func. anal. i ego pril.* **6** (1972), 68–69.
3. I. N. BERNSTEIN and B. I. ROSENFELD: Homogeneous spaces of infinite Lie algebras and characteristic classes of foliations, *Uspehi Mat. Nauk.* **28** (1973), 103–138.
4. R. BOTT: Lectures on Characteristic Classes and Foliations, *Springer Lecture Notes in Math* **279** (1972), 1–74.
5. R. BOTT and A. HAEFLIGER: On characteristic classes of Γ-foliations, *Bull. Am. math. Soc.* **78** (1972), 1039–1044.
6. I. M. GEL'FAND: Cohomology of infinite Lie algebras. Some topics in integral geometry, *Actes Congr. Int.* Nice (1970).
7. I. M. GEL'FAND and D. B. FUCHS: Cohomology of the Lie algebra of formal vector fields, *Izv. Akad. Nauk SSSR.* **34** (1970), 332–337.
8. I. M. GEL'FAND and D. B. FUCHS: PL-foliations; *Funct. anal. i ego pril.* **7** (1973), 29–37.
9. I. M. GEL'FAND, D. A. KAZHDAN and D. B. FUCHS: Actions of infinite Lie algebras, *Funct. anal. i ego pril.* **6** (1972), 10–15.
10. C. GODBILLON: Cohomologies d'algebres de Lie de champs de vecteurs formels, *Springer Lect. Notes in Math.* **383** (1974), 69–87.
11. C. GODBILLON and J. VEY: Un invariant des feuilletages de codimension une, *C.r. Acad. sci., Paris* **273** (1971), A92–A95.
12. A. HAEFLIGER: Sur les classes caracteristique des feuilletages, *Seminaire Bourbaki*, Juin 1972, No. 412. *Springer Lect. Notes in Math.* **317** (1973), 239–260).

Lab. of Math. Methods in Biology,
Corpus "A",
Moscow State University,
Moscow 117234,
U.S.S.R.

15.

(with D. B. Fuks)

PL Foliations II

Funkts. Anal. Prilozh. **8** (3) (1974) 7–11 [Funct. Anal. Appl. **8** (1974) 197–200].
MR **54**:6159. Zbl. **316**:57010

In the first part of this work [1] we defined the PL-analog of Haefliger structures (foliations with singularities) of codimension 1. We now investigate PL-foliations of any codimension. The fundamental results pertain to the corresponding classifying spaces, their homologies, and the relationship with the Haefliger classifying spaces $B\Gamma^q$.

Our constructions refer to the case of oriented foliations. The theory can be readily extended, however, to nonoriented foliations and to foliations with additional transversal structures: nondivergent, simplectic, etc.

This part of the study can for the most part be read independently of [1], although the latter contains the geometrical background of the formal constructions that follow.

§1. Fundamental Definitions

1. We start with a description of the semisimplicial sets BAf^q and $BPro^q$.

We put $\tilde{A}^q(n) = R^q \times \ldots \times R^q$ (n + 1 cofactors) and $A^q(n) = \tilde{A}^q(n)/L_+(q)$, where $L_+(q)$ is the group, coordinatewise-real in $\tilde{A}^q(n)$, of orientation-preserving affine transformations of the space R^q. We next define the i-th boundary operator $A^q(n) \to A^{q-1}(n)$ and the i-th degeneracy operator $A^q(n) \to A^{q+1}(n)$ as the factor mappings of the mappings $\tilde{A}^q(n) \to \tilde{A}^{q-1}(n)$ and $\tilde{A}^q(n) \to \tilde{A}^{q+1}(n)$, described by the equations $(x_0, \ldots, x_n) \mapsto (x_0, \ldots \hat{x}_i \ldots x_n)$ and $(x_0, \ldots, x_n) \mapsto (x_0, \ldots, x_i, x_i, \ldots, x_n)$. It is readily verified that the sets $A^q(n)$ with the indicated operators make up a semisimplicial set; the latter is denoted by BAf^q. The sets $\tilde{A}^q(n)$ also form a semisimplicial set, denoted $\widetilde{B}Af^q$; $\widetilde{B}Af^q$ is acted upon by the (discretely topologized) group $L_+(q)$, and $\widetilde{B}Af^q/L_+(q) = BAf^q$.

Next we denote by $\tilde{P}^q(n)$ the subset of degree (n + 1) of R^{q+1} formed by n-tuples $(x_0, \ldots, x_n)$ such that the convex hull of the points $x_0, \ldots, x_n$ does not contain the coordinate origin, and we put $P^q(n) = \tilde{P}^q(n)/GL_+(q+1)$. The same formulas as in the preceding case now define the boundary operators $P^q(n) \to P^{q-1}(n)$ and the degeneracy operators $P^q(n) \to P^{q+1}(n)$, which together with the sets $P^q(n)$ form a semisimplicial set; we denote the latter by $BPro^q$. The sets $\tilde{P}^q(n)$ also form a semisimplicial set, $\widetilde{B}Pro^q$; the latter is acted upon by the (discretely topologized) group $GL_+(q+1)$, and $\widetilde{B}Pro^q/GL_+(q+1) = BPro^q$.

The embedding $R^q \to R^{q+1}$, which is defined by the formula $(y_1, \ldots, y_q) \to (y_1, \ldots, y_q, 1)$, determines a semisimplicial monomorphism $BAf^q \to BPro^q$, which we fix as canonic.

2. Definition. An affine (oriented) PL-foliation of codimension q on a semisimplicial set X is the semisimplicial morphism $X \to BAf^q$. A projective (oriented) PL-foliation of codimension q on X is the semisimplicial morphism $X \to BPro^q$.

The canonic monomorphism $BAf^q \to BPro^q$ permits every affine PL-foliation of codimension q to be treated as a projective PL-foliation of codimension q.

3. For q = 1 these definitions are very close to the corresponding definitions in [1], but they do not reduce to them exactly. The principal difference concerns the main category: we replaced the "semisimplicial spaces" of [1] by semisimplicial sets, which in the terminology of [1] should be called "ordered semisimplicial spaces." The construction of PL-foliations of codimension greater than 1 incurs critical

Moscow State University. Translated from Funktsional'nyi Analiz i Ego Prilozheniya, Vol. 8, No. 3, pp. 7–11, July–September, 1974. Original article submitted April 5, 1974.

difficulties in the category of [1]. The reader of [1] will also note that we have simplified the concept of a projective PL-foliation by somewhat extending the corresponding classifying space.

§ 2. Homotopic Types of Classifying Spaces

The constructions of this section provide, in particular, an explicit description of the homotopic types of spaces realBAf^q and realBPro^q.

4. The following construction defines a covering $\{C_0, \ldots, C_q\}$ of the space realBAf^q such that every intersection $C_{i_1} \cap \ldots \cap C_{i_k}$ $(i_1 < \ldots < i_k)$ is a space of type K(L+ $(i_1, \ldots, i_k)$, 1), where L+ $(i_1, \ldots, i_k)$ is the subgroup formed in the (discretely topologized) group $L_+(i_k)$ by transformations under which the spaces $R^{i_{k-1}}, \ldots, R^{i_1}$ are left invariant. The inclusion relations $C_{i_1} \cap \ldots \cap C_{i_k} \to C_{j_1} \cap \ldots \cap C_{j_l}$ $(\{i_1, \ldots, i_k\} \supset \{j_1, \ldots, j_l\})$ in this case are consistent with the embeddings $L_+(i_1, \ldots, i_k) \to L_+(j_1, \ldots, j_l)$.

We denote by D_i $(0 \le i \le q)$ the part of the semisimplicial set $\tilde{B}$Af^q formed by simplexes $(x_0, \ldots, x_n) \in \tilde{A}^q(n)$, such that the points $x_0, \ldots, x_n$ lie in a single i-dimensional plane, and we define the set $E_i \subset$ real$\tilde{B}$Af^q = realbar$\tilde{B}$Af^q ("real" denotes the geometric realization, and "bar" denotes the baricentric subdivision) as the union of open simplexes in realbar$\tilde{B}$Af^q, whose closure intersects real $D_i \setminus$ real D_{i-1}. It is clear that E_i is an L+(q)-invariant open set containing real $D_i \setminus$real D_{i-1} and that the sets E_i cover real$\tilde{B}$Af^q. We interpret C_i as the image of E_i under the projection real$\tilde{B}$Af^q → realBAf^q.

The components of the set $E_{i_1} \cap \ldots \cap E_{i_k}$ correspond to the flags $L_{i_1} \subset \ldots \subset L_{i_k} \subset R^q$ with dim $L_s = s$. The transformations realized by elements of the group L+(q) represent those components. The maximum subgroup of L+(q) under which the component corresponding to the flag $R^{i_1} \subset \ldots \subset R^{i_k} \subset R^q$, is invariant is L+$(i_1, \ldots, i_k, q)$. The latter component is equivariantly deformation-retracted onto its intersection F with realBAf^{i_k}. The set F is contractible. The group L+$(i_1, \ldots, i_k, q)$ acts on it through its factor group L+ $(i_1, \ldots, i_k)$, whose action is transitive. Thus, $C_{i_1} \cap \ldots \cap C_{i_k} = (E_{i_1} \cap \ldots \cap E_{i_k})/L_+(q)$ is homotopically equivalent to F/L+ $(i_1, \ldots, i_k)$, i.e., is a space of type K(L+ $(i_1, \ldots, i_k)$, 1).

The fulfillment of the consistency condition is obvious.

5. The following construction defines a covering $\{R_0, \ldots, R_q\}$ of the space realBPro^q such that every intersection $R_{i_1} \cap \ldots \cap R_{i_k}$ $(i_1 < \ldots < i_k)$ is the total space of a fibering with basis K (GL+ $(i_1 + 1, \ldots, i_k + 1)$, 1), where GL+ $(i_1 + 1, \ldots, i_k + 1)$ is the subgroup formed in the (discretely topologized) group GL+ $(i_k + 1)$ by transformations under which the spaces $R^{i_1+1}, \ldots, R^{i_{k-1}+1}$ are invariant, and with the fiber $(R^{i_1+1} \setminus 0) \times (R^{i_2-i_1} \setminus 0) \times \ldots \times (R^{i_k-i_{k-1}} \setminus 0)$; the group GL+ $(i_1 + 1, \ldots, i_k + 1)$ acts coordinatewise in the fiber (it acts in $R^{i_s-i_{s-1}} \setminus 0$ as in an invariant subset of the factor space $R^{i_s+1}/R^{i_{s-1}+1}$). The inclusion relation $R_{i_1} \cap \ldots \cap R_{i_k} \to R_{j_1} \cap \ldots \cap R_{j_l}$ $(\{i_1, \ldots, i_k\} \supset \{j_1, \ldots, j_l\})$ in this case is consistent with the natural mapping of the indicated spaces.

The construction itself is analogous to the preceding case. We first introduce semisimplicial sets $S_i \subset \tilde{B}$Pro^q, formed by simplexes $(x_0, \ldots, x_n) \in \tilde{P}^q(n)$, such that the points $x_0, \ldots, x_n$ lie in a single (i + 1)-dimensional subspace of R^{q+1}. Then from that set, just as E_i was defined on the basis of D_i in subsection 4, we define the sets $T_i \subset$ real$\tilde{B}$Pro^q, and adopt the images of the latter in realBPro^q as R_i. By analogy with Sec. 4, the sets T_i are invariant under GL+ (q + 1), and the intersections $T_{i_1} \cap \ldots \cap T_{i_k}$ decompose into components representable by that group, corresponding to the flags of R^{q+1}, and composed of subspaces of dimensions $i_1 + 1, \ldots, i_k + 1$. The stationary subgroup of the component corresponding to the flag $R^{i_1+1} \subset \ldots \subset R^{i_k+1} \subset R^{q+1}$ is GL+$(i_1 + 1, \ldots, i_k + 1, q + 1)$, and this component is equivariantly deformation-retracted onto its intersection U with realBPro^{i_k}. The group GL+ $(i_1 + 1, \ldots, i_k + 1, q + 1)$ acts in U through its factor group GL+ $(i_1 + 1, \ldots, i_k + 1)$, whose action is already transitive. Moreover, we can readily construct a homotopic GL+ $(i_1 + 1, \ldots, i_k + 1)$-equivalence of the set U on $(R^{i_1+1} \setminus 0) \times (R^{i_2-i_1} \setminus 0) \times \ldots \times (R^{i_k-i_{k-1}} \setminus 0)$. Consequently, $R_{i_1} \cap \ldots \cap R_{i_k} = (T_{i_1} \cap \ldots \cap T_{i_k})/GL_+(q + 1)$ is homotopically equivalent to U/GL+$(i_1 + 1, \ldots, i_k + 1)$, i.e., has the required homotopic type.

The fulfillment of the consistency condition is again obvious.

§ 3. Continuous Cohomologies

6. If we knew the cohomologies of the groups L+$(i_1, \ldots, i_k)$ [of the groups GL+ $(i_1, \ldots, i_k)$ in particular], the constructions of § 2 would enable us to determine the cohomologies of the space realBAf^q and realBPro^q by means of exact Mayer–Vietoris sequences. However, the cohomologies of the stated groups are not known. On the other hand, their continuous cohomologies in the sense of Van Est [2] are

known, permitting us to use the same Mayer–Vietoris sequences to determine the continuous cohomologies of the spaces realBAf^q and realBPro^q in the sense of Bott and Haefliger [3].

The latter cohomologies are determined as follows. The sets $A^q(n)$ and $P^q(n)$ of simplexes of the spaces BAf^q and BPro^q have a natural topology, namely, the topology of the space of orbits of the Lie groups $L_+(q)$ and $GL_+(q+1)$. The boundary and degeneracy operators are continuous in this topology, and we can regard BAf^q and BPro^q as semisimplicial objects in the category of topological spaces; in this capacity they are denoted by BAf^q_c and BPro^q_c. The corresponding realizations realBAf^q_c and realBPro^q_c exist, as do the one-to-one continuous mappings ("compactifications")

$$\mathrm{real}BAf^q \to \mathrm{real}BAf^q_c,\quad \mathrm{real}BPro^q \to \mathrm{real}BPro^q_c. \tag{1}$$

The real singular cochains of realBAf^q continuous under the topology of realBAf^q_c are called continuous cochains of the given space, and the cohomologies computed on the basis of those cochains are called continuous cohomologies of the space realBAf^q. Continuous cochains and continuous cohomologies of realBPro^q are defined analogously. The cohomologies are denoted by $H_c(\mathrm{BAf}^q)$ and $H_c(\mathrm{BPro}^q)$.

We point out that the compactifications (1) interact well with the coverings constructed in § 2; the variation in the topology of the elements of those coverings and their intersections is essentially that the spaces $K(L_+(i_1, \ldots, i_k), 1)$, $K(GL_+(i_1, \ldots, i_k), 1)$ are compacted to the classifying spaces of the corresponding topological groups.

7. If $r > 0$, then $H^r_c(\mathrm{BAf}^q) = 0$.

Proof. The inclusion of the Lie group $L_+(0, i_1, \ldots, i_k) = GL_+(i_1, \ldots, i_k)$ in the Lie group $L_+(i_1, \ldots, i_k)$ is a homotopic equivalence and induces isomorphism in the cohomologies of Lie algebras with constant coefficients. According to the results of Hochshild and Mostow [4], this fact implies that the same inclusion induces isomorphism in the continuous cohomologies of those groups as well. This means, in the notation of subsection 4, that the inclusion relation induces an isomorphism of continuous cohomologies for any $i_1, \ldots, i_k$. Applying a Mayer–Vietoris sequence as many times as necessary, we deduce from the foregoing that the inclusion relation $C_0 \to$ realBAf^q also induces an isomorphism of continuous cohomologies, but the continuous cohomologies of C_0 are trivial.

8. Once again, invoking the results of Hochshild and Mostow [4], we can easily show that the continuous cohomologies of the intersection $R_{i_1} \cap \ldots \cap R_{i_k}$ of elements of the covering constructed for realBPro^q in subsection 5 coincide with the real cohomologies of the product

$$[U(i_1+1)/SO(i_1)] \times [U(i_2-i_1)/SO(i_2-i_1-1)] \times \ldots \times [U(i_k-i_{k-1})/SO(i_k-i_{k-1}-1)], \tag{2}$$

and that the homomorphisms of the inclusion relations $R_{i_1} \cap \ldots \cap R_{i_k} \to R_{j_1} \cap \ldots \cap R_{j_l}$ coincide with the homomorphisms induced by the natural embeddings of the corresponding products (2). This fact enables us to compute $H^*_c(\mathrm{BPro}^q)$ by means of Mayer–Vietoris sequences. Following are some of the computational results.

The spaces $H^r_c(\mathrm{BPro}^q)$ with $r > \dim[U(q+1)/SO(q)] = 1 + (q(q+5)/2)$ are trivial. The spaces $H^r_c(\mathrm{BPro}^q)$ with $r = 1 + (q(q+5)/2)$ and $r = q(q+5)/2$ are one-dimensional (for the latter, except in the case $q = 1$); in particular, the ring $\widetilde{H}^*_c(\mathrm{BPro}^q)$ is nontrivial for any $q \geq 1$. The successive dimensions of the spaces $H^r_c(\mathrm{BPro}^q)$ (beginning with $r = 0$) for $q = 1$ are equal to 1, 1, 1, 2, 1, 0, ..., and for $q = 2$ they are equal to 1, 1, 2, 2, 2, 5, 3, 1, 1, 0, ...

§ 4. Smoothing

9. Each of the intersections $R_{i_1} \cap \ldots \cap R_{i_k}$ has the homotopic type of the total space of a fibering for which the fiber is a q-dimensional smooth manifold, while the structural group is discrete and acts in the fiber through orientation-preserving diffeomorphisms. It is well known (see, e.g., [5]) that such a space has a canonic oriented Haefliger structure of codimension q. It is readily perceived that the induced Haefliger structures in the intersections $R_{i_1} \cap \ldots \cap R_{i_k}$ are pairwise consistent and form a Haefliger structure in realBPro^q. This structure makes it possible to use a projective (and, hence, affine) oriented PL-foliation of codimension q in a semisimplicial set X to construct in a natural way an oriented Haefliger structure of codimension q in realX. The latter structure is called a smoothing of the original PL-foliation (see [1], § 4).

We say in the language of classifying spaces that we have constructed a canonic mapping of the space realBPro^q into the Haefliger classifying space $B\Gamma^q_+$.

10. It follows from the theorem of subsection 7 that all the characteristic classes (see [6, 7]) of smoothing of an affine PL-foliation are trivial (these classes are determined by the elements of the space $H_c^*(B\Gamma_+^q)$; see [3]). By contrast, the characteristic classes of a projective PL-foliation can be nontrivial. It is easily shown that the homomorphism $H^{q(q+5)/2}(B\Gamma_+^q) \to H^{q(q+5)/2}(B\mathrm{Pro}^q)$, induced by the smoothing transformation is nontrivial for any q (see [1], § 5).

§ 5. Certain Hypotheses

11. Mather and Thurston have recently proved (see [8]) that the integer-valued homologies of a homotopic fiber of the natural mapping $B\mathrm{Diff}_c^\delta R^q \to B\mathrm{Diff}_c R^q$ coincide with the integer-valued cohomologies of the q-th space of loops of a homotopic fiber of the natural mapping $B\Gamma_+^q \to BSO(q)$; here $\mathrm{Diff}_c R^q$ denotes the group of finite diffeomorphisms of the space R^q, δ denotes discrete topology, and $B\Gamma_+^q$ is the Haefliger classifying space. The same authors have also indicated a natural mapping establishing an isomorphism of integer-valued homologies.

Inasmuch as BAf^q is the PL-analog of the space $B\Gamma_+^q$, we can formulate (as a hypothesis) a PL-analog of the Mather–Thurston theorem. The hypothesis states that the integer-valued homologies of the space $BPL_c^\delta R^q$ (i.e., the group $PL_c R^q$) are isomorphic to the integer-valued homologies of the q-th space of loops of a homotopic fiber of the natural mapping real$BAf^q \to BSO(q)$. The latter can be defined in terms of the operation of smoothing of a PL-foliation (see § 4) and can be described directly in terms of the natural homomorphisms $L_+(i_1, \ldots, i_k) \to GL_+(q)$.

12. We also note that the PL-analog (not yet proved) of the "continuous" version of the Mather–Thurston theorem together with the theorem of subsection 7 above would imply that the continuous cohomologies of the group $PL_c R^q$ coincide with the cohomologies of the q-th space of loops of the group SO(q).

LITERATURE CITED

1. I. M. Gel'fand and D. B. Fuks, "PL-foliations," Funktsional. Analiz i Ego Prilozhen., 7, No. 4, 29-37 (1973).
2. Van Est, "Group cohomology and Lie algebra cohomology in Lie groups," Proc. Nederl. Akad. Wetensch., Ser. A, 56, 484-492, 493-504 (1953).
3. R. Bott, "Some remarks on continuous cohomology," Sympos. Manifolds, Tokyo (1973).
4. G. Hochshild and G. D. Mostow, "Cohomology of Lie groups," Illinois J. Math., 6, No. 3, 367-401 (1962).
5. R. Bott and J. Heitsch, "A remark on the integral cohomology of $B\Gamma_q$," Topology, 11, 141-146 (1972).
6. I. N. Bernshtein and B. I. Rozenfel'd, "Characteristic classes of foliations," Funktsional. Analiz i Ego Prilozhen., 6, No. 1, 68-69 (1972).
7. R. Bott and A. Haefliger, "On characteristic classes of Γ-foliations," Bull. Amer. Math. Soc., 78, No. 6, 1039-1044 (1972).
8. J. N. Mather, "Loops and foliations," Sympos. Manifolds, Tokyo (1973).

16.

(with A. M. Gabrielov and M. V. Losik)

Combinatorial computation of characteristic classes

Funkts. Anal. Prilozh. 9 (2) (1975) 12–28 [Funct. Anal. Appl. 9 (1975) 103–115].
MR 53:14504a. Zbl. 312:57016

INTRODUCTION

We present a method of combinatorial computation of the characteristic classes of a smooth manifold. Technical difficulties increase as one proceeds to higher classes so, until the end, computation in this article is carried out only for the first Pontryagin class. However, we are convinced that, for the most part, the concepts introduced can be used without change for computing the higher classes as well.

In the first section we formulate the basic result of the work (Theorem 1) and introduce what is needed for its notions.

An important role in our construction is played by certain polyhedra, which we call hypersimplexes, in Euclidean space. Barriers to linearizability of cellular decompositions of a smooth manifold are encountered; these come in as corollaries to the main theorem.

In the second section we introduce concepts essential to the proof of the main theorem: variational complex, (d, δ)-co-cycle, and also $S^{k,l}$ forms dual to the hypersimplexes of §1.

A complete proof of the results formulated in §1 is given in the third section.

In the fourth section we construct some special functionals, called Atiyah–Patodi–Zinger functionals, for a smooth manifold with piecewise smooth boundary. These functionals are constructed for any Pontryagin class.

In the Appendix to the first part, we establish a connection of the variational complex with cohomologies of the algebra W_n of formal vector fields. This appendix is not used in the basic text and can be read independently.

§1. COMBINATORIAL CONSTRUCTIONS AND FORMULATION OF RESULTS

1. Cellular Decompositions. Let X be a smooth n-dimensional manifold. A submanifold $Y \subset X$ is called an l-dimensional cell if there exists a neighborhood U of the set Y in X and a diffeomorphic imbedding φ: $U \to R^n$ such that $\varphi(Y)$ is an l-dimensional linear cell (i.e., a convex compact polyhedron) in R^n. The pre-images of the faces of $\varphi(Y)$ are called faces of Y (in particular ϕ and Y are faces of Y).

The cell Y is termed simple if each of its faces of codimension k ($1 \leq k \leq \dim Y$) contains exactly k of its faces of codimension k − 1. The property of simplicity is conserved under products and transversal intersections.

A locally finite decomposition of X into compact subsets X_i is called a cellular decomposition if it fulfills the following conditions:

1) $X_i \neq X_j$ if $i \neq j$;

2) X_i is a cell in X (we shall write $X_i = X_i^l$, where $l = \dim X_i$);

3) if Y is a face of cell X_i then $Y = X_j$ for some j;

Moscow State University. Saratov State University. Translated from Funktsional'nyi Analyz i Ego Prilozheniya, Vol. 9, No. 2, pp. 12-28, April-June, 1975. Original article submitted December 2, 1974.

4) $X_i \cap X_j$ is a face in X_i and in X_j.

2. Sets s(i). Let $\{X_\alpha\}$ be a cellular decomposition of the manifold X and let Y_i be cells of the decomposition $\{X_\alpha\}$. Denote by s(i) the set of indices j such that $X_j \supset X_i$, $\dim X_j = \dim X_i + 1$.

If $j \in s(i)$ and $k \in s(j)$, then there exists a unique cell $X_{j'}$ such that $X_k \supset X_{j'} \supset X_i$, $j' \in s(i)$, $j' \neq j$. We put $j' = \varphi_{ij}(k)$. In this way a mapping φ_{ij}: $s(j) \to s(i)$ is defined.

PROPOSITION 1. Let $\{X_\alpha\}$ be a decomposition in simple cells and let X_i be a cell of the decomposition $\{X_\alpha\}$. If j, $j' \in s(i)$, $k \in s(j) \cap s(j')$, then $\varphi_{ij'} \circ \varphi_{j'k} = \varphi_{ij} \circ \varphi_{jk}$.

This means that the sets s(i) form a bundle under the decomposition $\{X_\alpha\}$. We note that the converse assertion is also true.

PROPOSITION 1'. Let $\{X_\alpha\}$ be a cellular decomposition. If whenever mappings $\varphi_{ij}, \varphi_{jk}, \varphi_{ij'}, \varphi_{j'k}$ are defined we have the equality $\varphi_{ij'} \circ \varphi_{j'k} = \varphi_{ij} \circ \varphi_{jk}$, then all cells of the decomposition $\{X_\alpha\}$ are simple.

Below we shall use only decompositions into simple cells.

3. Complex S_i^{p-1}. Let X_i^{n-p} be a cell of the decomposition $\{X_\alpha\}$. We define a simplicial complex S_i^{p-1} as follows. The set of vertices of S_i^{p-1} coincides with s(i). The family $j_0, \ldots, j_k$ of elements of s(i) is called a k-simplex in S_i^{p-1} if the cells $X_{j_0}^{n-p+1}, \ldots, X_{j_k}^{n-p+1}$ are contained in one cell of dimension $n-p+k+1$. Obviously S_i^{p-1} is a combinatorial sphere.

A geometrical realization of the complex S_i^{p-1} exists in the decomposition induced by the decomposition $\{X_\alpha\}$ on the boundary of a small spherical neighborhood of the point $x \in X_i$ in transversal intersection to the cell X_i drawn through x. It is readily verified that the mapping φ_{ij} for $j \in s(i)$ induces an imbedding of simplicial complexes $\overline{\varphi}_{ij}$: $S_j^{p-2} \to S_i^{p-1}$. From Proposition 1 we easily obtain the following proposition.

PROPOSITION 2. If j, $j' \in s(i)$, $k \in s(j) \cap s(j')$, then $\overline{\varphi}_{ij'} \circ \overline{\varphi}_{j'k} = \overline{\varphi}_{ij} \circ \overline{\varphi}_{jk}$.

4. Co-oriented Chains. We use the term co-orientation of a cell X to refer to an orientation of the complex S_i^{p-1}. In the case p = 1 the complex S_i^0 consists of two points. By orientation of S_i^0 we mean the order of these points (the first, we consider negative, the second, positive). Each n-dimensional cell is considered positively co-oriented.

Consider two cells X_i^{n-p} and X_j^{n-p+1}, $X_j \supset X_i$, of the decomposition $\{X_\alpha\}$, co-oriented in any manner. We define the coefficient of incidence ε_{ij} of cells X_i and X_j. Denote by C_{ij}^{p-1} the subcomplex in S_i^{p-1} which is the closure of the star vertices $j \in S_i^{p-1}$. Then $\overline{\varphi}_{ij}S_j^{p-2}$ coincides with ∂C_{ij}^{p-1}. We put $\varepsilon_{ij} = 1$, if the orientation of $\overline{\varphi}_{ij}\, S_j^{p-2}$ induced from S_j^{p-2} coincides with the orientation of ∂C_{ij}^{p-1} induced from S_i^{p-1}, and we put $\varepsilon_{ij} = -1$ otherwise. If p = 1 and if $S_i^0 = \{j_0, j_1\}$ is oriented in accord with the order j_0, j_1, then we put $\varepsilon_{ij_0} = -1$, $\varepsilon_{ij_1} = 1$.

Let $\hat{C}_q(X)$ be the Abelian group formed by linear combinations (generally speaking, infinite) of co-oriented q-dimensional cells of the decomposition $\{X_\alpha\}$ with integer coefficients, where cells with opposite co-orientations have opposite signs in $\hat{C}_q(X)$. We define the differential $\hat{\partial}$: $\hat{C}_q(X) \to \hat{C}_{q-1}(X)$ by the formula $\hat{\partial} X_j^q = \sum_{i:\, j \in s(i)} \varepsilon_{ij} X_i^{q-1}$. A complex $\hat{C}_*(X) = \oplus \hat{C}_q(X)$ with differential $\hat{\partial}$ is called a complex of co-oriented chains of the decomposition $\{X_\alpha\}$.

Let Λ be an oriented local system in X, i.e., a fibering over X with fiber Z given by means of $\pi_1(X)$ in Z, defined by the formula $\gamma(m) = m$, if a walk by the path γ does not change orientation of X in a neighborhood of the marked point and by $\gamma(m) = -m$ otherwise. Comparing the definitions of co-oriented chains and chains with coefficients in Λ one readily confirms the following assertion.

For any Abelian group G we have $H_q(\hat{C}_*(X) \otimes G) \cong H_q^c(X, G \otimes \Lambda)$, where H_q^c are homologies with closed carriers.

Combining this assertion with Poincaré isomorphism we obtain the following proposition.

PROPOSITION 3. $H_q(\hat{C}_*(X) \otimes G) \cong H^{n-q}(X, G)$.

5. Configuration Spaces. Let X_i^{n-p} be cells of the decomposition $\{X_\alpha\}$. We use the term S_i^{p-1}-configuration for a decomposition σ of space R^p into simplicial cones with vertex at the origin satisfying the following conditions:

a) all one-dimensional edges of the decomposition σ are enumerated by indices from s(i) where for every $j \in s(i)$ there exists a unique edge l_j with number j;

b) indices $j_0, \ldots, j_m$ form an m-simplex in S_i^{p-1} if and only if edges $l_{j_0}, \ldots, l_{j_m}$ belong to one (m+1)-dimensional simplicial cone of the decomposition σ.

We denote by $\Sigma(i)$ a manifold whose elements are S_i^{p-1}-configurations and where two configurations σ and σ' are identified if there exists a linear mapping of space R^p carrying σ to σ' (consistently with the enumeration of edges).

We denote by $\Sigma_0(i)$ a set of configurations σ from $\Sigma(i)$ such that p arbitrary one-dimensional edges of the decomposition σ generate R^p. Suppose further that $\Sigma_1(i)$ is a set of configurations for which there exists not more than one collection of p one-dimensional edges contained in a (p − 1)-dimensional subspace in R^p. We can also define $\Sigma_\nu(i)$ for $\nu = 2, 3, \ldots$ as subsets of configurations from $\Sigma(i)$ with generators of co-dimension ν, but we have no need of this here.

Let $j \in s(i)$. We define a mapping $\varphi_{ij}^{!}: \Sigma(i) \to \Sigma(j)$ as follows. Let $\sigma \in \Sigma(i)$ and let l_j be a one-dimensional edge of the decomposition σ corresponding to index j from s(i). We consider the mapping $R^p \to R^p/l_j \cong R^{p-1}$. Denote by $\varphi_{ij}^{!}(\sigma)$ the decomposition of space R^{p-1} into simplicial cones which are the images of those cones of the decomposition σ which contain l_j. The one-dimensional edges of the decomposition $\varphi_{ij}^{!}(\sigma)$ are enumerated by indices from s(j) in the following way. Let $k \in s(j)$. Then $\varphi_{ij}(k)$ and j are contained in one 1-simplex of the complex S_i^{p-1}. Consequently, the edges $l_{\varphi_{ij}(k)}$ and l_j are contained in one two-dimensional simplicial cone of the decomposition σ. The image of this cone under the projection $R^p \to R^{p-1}$ is a one-dimensional edge of the decomposition $\varphi_{ij}^{!}(\sigma)$ which we denote by l_k. It is easy to verify that the correspondence so defined satisfies conditions a and b of the configuration definition.

We note that $\varphi_{ij}^{\prime}(\Sigma_\nu(i)) \subset \Sigma_\nu(j)$ $(\nu = 0, 1, \ldots)$.

PROPOSITION 4. For $j, j' \in s(i)$ and $k \in s(j) \cap s(j')$ we have $\varphi_{j'k}^{\prime} \circ \varphi_{ij'}^{\prime} = \varphi_{jk}^{\prime} \circ \varphi_{ij}^{\prime}$.

Accordingly, spaces of configurations $\Sigma(i)$ form a co-bundle over the decomposition $\{X_\alpha\}$.

Remark. The space $\Sigma(i)$ is contractible if $\dim X_i = n-2$ and simply connected if $\dim X_i = n-3$. We also demand that the decomposition $\{X_\alpha\}$ fulfill the following condition.

(A) For any cell X_i^{n-4} the space $\Sigma(i)$ is connected.

We do not know whether this is true for an arbitrary decomposition.

6. Geometric Hypersimplexes. Let Δ^p be a standard p-dimensional simplex in Euclidean space, let k be a whole number, $0 \leq p-1$, and let $l = p-k-1$. Denote by $\Delta^{k,l}$ the convex hull of sets of centers of l-dimensional faces of the simplex Δ^p. The cell $\Delta^{k,l}$ we call a (geometric) hypersimplex.

Example. If p = 3, then the cells $\Delta^{0,2}$ and $\Delta^{2,0}$ are ordinary simplexes, and $\Delta^{1,1}$ is an octahedron.

An important role in our construction will be played by hypersimplexes whose vertices are enumerated by various index sets.

Let s be a set of N elements and q a natural number, q < N. Let k, l: $0 \leqslant k \leqslant N-q-1$, $0 \leqslant l \leqslant q-1$ be integers, $p = k + l + 1$. We suppose that the vertices of the simplex Δ^p are enumerated by the numbers 0, . . ., p and that Δ^p is oriented consistently with the order of the vertices. Suppose $j_0, \ldots, j_{k+q}$ is an ordered family of pairwise distinct elements from s. Each vertex Q of the hypersimplex $\Delta^{k,l}$ we place according to a nonordered set of q indices $j_{\alpha_0}, \ldots, j_{\alpha_l}, j_{k+l+2}, \ldots, j_{k+q}$, where $\alpha_0, \ldots, \alpha_l$ are the numbers of vertices of the l-dimensional face of the simplex Δ^p whose center is Q. The cell obtained in this way with indices in the vertices we term an (s,q)-hypersimplex and denote it by $\Delta^{k,l}(j_0, \ldots, j_{k+l+1}; j_{k+l+2}, \ldots, j_{k+q})$. The imbedding $\Delta^{k,l} \to \Delta^p$ defines an orientation of the cell $\Delta^{k,l}$. Clearly, the cells $\Delta^{k,l}(j_0, \ldots, j_{k+l+1}; j_{k+l+2}, \ldots, j_{k+q})$ do not change under arbitrary permutation of the indices $j_{k+l+2}, \ldots, j_{k+q}$, but under permutation of the indices $j_0, \ldots, j_{k+l+1}$ (and with this also the sets of indices in the vertices) a change of orientation takes place depending on the parity of the permutation. A zero-dimensional (s, q)-hypersimplex $\Delta^0(j_0, \ldots, j_{q-1})$ we shall call a point equipped with a non-ordered set of pairwise distinct indices $j_0, \ldots, j_{q-1}$ from s.

PROPOSITION 5.

a) $$\partial\Delta^{k,l}(j_0,\ldots,j_{k+l+1};j_{k+l+2},\ldots,j_{k+q}) = \sum_{\mu=0}^{k+l+1}(-1)^{\mu}\Delta^{k-1,l}(j_0,\ldots\hat{j}_{\mu}\ldots,j_{k+l+1};j_{k+l+2},\ldots,j_{k+q}) -$$
$$-\sum_{\mu=0}^{k+l+1}(-1)^{\mu}\Delta^{k,l-1}(j_0,\ldots\hat{j}_{\mu}\ldots,j_{k+l+1};j_{k+l+2},\ldots,j_{k+q},j_{\mu}), \quad (1)$$

if $k + l \geq 1$, where also in the case $k = 0$ (resp. $l = 0$) the first sum (resp. the second) vanishes.

b) $$\partial\Delta^{0,0}(j_0,j_1;j_2,\ldots,j_q) = \Delta^0(j_1,j_2,\ldots,j_q) - \Delta^0(j_0,j_2,\ldots,j_q). \quad (2)$$

PROPOSITION 6. There exists a one-to-one correspondence between the set of all (s, q)-hypersimplexes and the set of faces of the "maximal" (s, q)-hypersimplex $\Delta^{N-q-1,q-1}(j_0,\ldots,j_{N-1})$, where $\{j_0,\ldots,j_{N-1}\} = s$.

7. Hypersimplicial Complexes. Again let s be a set of N elements, and let q be a natural number, $q < N$. For $p \geq 0$ we define an Abelian group $I_p(s, q)$ as follows. For $p \geq 1$, serving as generators of the group $I_p(s, q)$ are the symbols $\Delta^{k,l}(j_0,\ldots,j_{k+l+1};\ j_{k+l+2},\ldots,j_{k+q})$, where $0 \leqslant k \leqslant N-q-1$, $0 \leqslant l \leqslant q-1$, $k+l+1=p$, and $j_0,\ldots,j_{k+q}$ are an arbitrary ordered collection of pairwise distinct elements of s. These generators are connected by the relations

$$\Delta^{k,l}(\tau j_0,\ldots,\tau j_{k+l+1};\tau' j_{k+l+2},\ldots,\tau' j_{k+q}) = \operatorname{sgn}\tau\Delta^{k,l}(j_0,\ldots,j_{k+l+1};j_{k+l+2},\ldots,j_{k+q}),$$

where τ (resp. τ') is an arbitrary permutation of the indices $j_0,\ldots,j_{k+l+1}$ (resp. $j_{k+l+2},\ldots,j_{k+q}$). The group $I_0(s, q)$ is a free Abelian group with generators $\Delta^0(j_0,\ldots,j_{q-1})$, where $j_0,\ldots,j_{q-1}$ are an arbitrary nonordered collection of q pairwise distinct elements of s.

We use the term (s, q)-hypersimplicial complex for the graduated group $I_*(s, q) = \oplus I_p(s, q)$ with differential ∂: $I_p(s, q) \to I_{p-1}(s, q)$, defined by formulas (1) and (2).

PROPOSITION 7. The homologies of the complex $I_*(s, q)$ are trivial in all positive dimensions, and $H_0(I_*(s, q)) = Z$.

Proof. We consider the geometric hypersimplex $\Delta^{N-q-1,q-1}$. Let K be the cellular complex consisting of the hypersimplex $\Delta^{N-q-1,q-1}$ and all its faces. It is readily seen by Proposition 6 that the complex $I_*(s, q)$ is isomorphic to the complex of chains of the cellular complex K. The assertion now follows from the fact that $\Delta^{N-q-1,q-1}$ is homeomorphic to a sphere.

8. Bundle of Hypersimplicial Complexes. Let X_i^{n-p} be a cell of the decomposition $\{X_\alpha\}$. We put $I_*(i) = I_*(s(i), p)$ if $p \geq 1$. For $p = 0$ we define $I_*(i)$ as the graduated group with the single non-null component $I_0(i) = Z$. If $p > 1$ and $j \in s(i)$, we define a homomorphism of complexes $\widetilde{\varphi}_{ij}$: $I_*(j) \to I_*(i)$ by the formula

$$\widetilde{\varphi}_{ij}(\Delta^{k,l}(\alpha_0,\ldots,\alpha_{k+l+1};\alpha_{k+l+2},\ldots,\alpha_{k+p-1})) = \Delta^{k,l}(\varphi_{ij}\alpha_0,\ldots,\varphi_{ij}\alpha_{k+l+1};\varphi_{ij}\alpha_{k+l+2},\ldots,\varphi_{ij}\alpha_{k+p-1},j);$$
$$\widetilde{\varphi}_{ij}(\Delta^0(\alpha_0,\ldots,\alpha_{p-2})) = \Delta^0(\varphi_{ij}\alpha_0,\ldots,\varphi_{ij}\alpha_{p-2},j).$$

If $p = 1$ and $j \in s(i)$, then the homomorphism $\widetilde{\varphi}_{ij}$: $I_*(j) \to I_*(i)$ is defined by the mapping $Z \to I_0(i)$, carrying 1 to $\Delta^0(j)$.

PROPOSITION 8. For $j, j' \in s(i)$ and $k \in s(j) \cap s(j')$ we have $\widetilde{\varphi}_{ij'} \circ \widetilde{\varphi}_{j'k} = \widetilde{\varphi}_{ij} \circ \widetilde{\varphi}_{jk}$.

Accordingly, the complexes $I_*(i)$ form a bundle over the decomposition $\{X_\alpha\}$.

9. Admissible Hypersimplicial Systems. We shall say that an admissible hypersimplicial system has been given if for each co-oriented cell X_i^{n-p} there is defined an element $b_i \in I_p(i)$ which also fulfills the following conditions:

1) b_i changes sign upon change of co-orientation,

2) if $p = 0$, then $I_0(i) = Z$ and $b_i = 1$,

3) $\partial b_i = \sum_{j\in s(i)} \varepsilon_{ij}\widetilde{\varphi}_{ij}(b_j)$, where ε_{ij} are the incidence coefficients of paragraph 4.

PROPOSITION 9. For any decomposition into simple cells there exists an admissible hypersimplicial system.

Proof. The construction of the elements b_i proceeds by induction on p. Let p = 1. We put $b_i = \Delta^{0,0}(j_0, j_1)$, where j_0 and j_1 are two elements of s(i) and the orientation S_i^0 is defined by the order j_0, j_1. Now, let $p > 1$, and suppose that for $p' < p$ the elements b_α satisfying conditions 1-3 are already constructed. From Proposition 8 and the condition $\partial b_j = \sum_{k\in s(j)} \varepsilon_{jk}\tilde{\varphi}_{jk}(b_k)$ $(j\in s(i))$ it follows that $\partial\left(\sum_{j\in s(i)} \varepsilon_{ij}\tilde{\varphi}_{ij}(b_j)\right) = 0$. According to Proposition 7 there exists an element $b_i \in I_p(i)$ such that $\partial b_i = \sum_{j\in s(i)} \varepsilon_{ij}\tilde{\varphi}_{ij}(b_j)$. The proposition is proved.

Remark 1. Construction of the elements b_i is local relative to the decomposition $\{X_\alpha\}$ in the following sense: for the construction of b_i, already constructed elements b_j are used only for cells X_j containing X_i.

Remark 2. An admissible hypersimplicial system $\{b_\alpha\}$ can be considered as a cycle in the bicomplex of chains of the decomposition $\{X_\alpha\}$ with coefficients in the system of complexes I_*.

10. Construction of the Cycle $\Gamma_{n-4}(X)$. In this paragraph we construct, for each decomposition of the manifold X into simple cells, a certain important cycle $\Gamma_{n-4}(X)$ of codimension 4. In case this decomposition is linearizable, the cycle $\Gamma_{n-4}(X)$ is Poincaré dual to the first Pontraygin class of the manifold X. To construct the cycle $\Gamma_{n-4}(X)$ we first make correspond to each cell X_i^{n-4} a number $\Gamma'(i)$, and then to each cell X_j^{n-3}, containing X_i^{n-4}, we make correspond a number $\Gamma''(i, j)$ and finally we define $\Gamma_{n-4}(X) = \sum \Gamma(i)\, X_i^{n-4}$, where $\Gamma(i) = \Gamma'(i) + \sum_{j\in s(i)} \Gamma''(i, j)$.

Let X_i^{n-4} be a cell of codimension 4 of the decomposition $\{X_\alpha\}$. We define a homomorphism θ_i^0 : $I_4(i) \to H^0(\Sigma_0(i), \mathbf{Q})$ as follows. Let $\sigma \in \Sigma_0(i)$. For each $j \in s(i)$ we choose a vector $e_j \neq 0$ in the one-dimensional edge l_j of the decomposition σ.

As generators of the form $\Delta^{0,3}$ and $\Delta^{3,0}$ for the group $I_4(i)$ we put $\theta_i^0(\Delta^{0,3}) = \theta_i^0(\Delta^{3,0}) = 0$.

Let $\Delta^{1,2} = \Delta^{1,2}(j_0, \ldots, j_4; j_5)$ be a generator for $I_4(i)$. We put $\theta_i^0(\Delta^{1,2})(\sigma) = -\varepsilon/48$, where $\varepsilon = 1$ if among ten of the bases $(e_{j_\alpha}, e_{j_\beta}, e_{j_\gamma}, e_{j_5})$ in $\mathbf{R}^4$ $(0\leqslant\alpha, \beta, \gamma\leqslant 4, \alpha<\beta<\gamma)$ an even number are identically oriented; $\varepsilon = -1$ otherwise.

Let $\Delta^{2,1} = \Delta^{2,1}(j_0, \ldots, j_4; j_5, j_6)$ be a generator for $I_4(i)$. We put $\theta_i^0(\Delta^{2,1})(\sigma) = -\varepsilon/48$, where $\varepsilon = 1$ if among ten of the bases $(e_{j_\alpha}, e_{j_\beta}, e_{j_5}, e_{j_6})$ in $\mathbf{R}^4$ $(0 \leqslant \alpha, \beta \leqslant 4, \alpha < \beta)$ an even number are identically oriented; $\varepsilon = -1$ otherwise.

Now, let X_j^{n-3} be a cell of codimension 3 of the decomposition $\{X_\alpha\}$. We define a homomorphism $\theta_j^1 : I_3(j) \to H^1(\Sigma_1(j), \Sigma_0(j); \mathbf{Q})$ as follows. Let $\sigma(t)$ $(t \in [0, 1])$ be a continuous curve in $\Sigma_1(j)$, $\sigma(0) \in \Sigma_0(j)$, $\sigma(1) \in \Sigma_0(j)$. We choose a vector $e_k(t) \neq 0$ $(k \in s(j))$ on each one-dimensional edge of the decomposition $\sigma(t)$.

As generators of the form $\Delta^{0,2}$ and $\Delta^{2,0}$ for the group $I_3(j)$ we put $\theta_j^1(\Delta^{0,2}) = \theta_j^1(\Delta^{2,0}) = 0$.

Let $\Delta^{1,1} = \Delta^{1,1}(k_0, k_1, k_2, k_3; k_4)$ be a generator of $I_3(j)$. Let $\varkappa(t) = b(t)c(t)/a(t)d(t)$, where

$$e_{k_0}(t) = a(t)e_{k_2}(t) + b(t)e_{k_3}(t) + h_0(t)e_{k_4}(t),$$
$$e_{k_1}(t) = c(t)e_{k_2}(t) + d(t)e_{k_3}(t) + h_1(t)e_{k_4}(t),$$

and, for t varying from 0 to 1, suppose the number $\varkappa(t)$ passes μ_+ times through one of the values 0, 1, or ∞ in the positive direction $(0 \to 1 \to \infty \to 0)$ and μ_- times in the negative direction. Put $\mu = \mu_+ - \mu_-$. Considering $\sigma(t)$ as a relative cycle of the pair $(\Sigma_1(j), \Sigma_0(j))$, we put $\theta_j^1(\Delta^{1,1})(\sigma(t)) = \mu/24$.

Let $\{b_\alpha\}$ be an admissible hypersimplicial system for the decomposition $\{X_\alpha\}$ and assume for each cell X_i^{n-4} a fixed configuration $\sigma_i \in \Sigma_0(i)$ and for each cell X_j^{n-3} a configuration $\sigma_j \in \Sigma_0(j)$. We put $\Gamma'(i) = \theta_i^0 b_i(\sigma_i)$. Let $\sigma_{ij}(t)$ be a continuous curve in $\Sigma_1(j)$ such that $\sigma_{ij}(0) = \sigma_j$, $\sigma_{ij}(1) = \varphi_{ij}(\sigma_i)$.

We put $\Gamma''(i, j) = \varepsilon_{ij}\theta_j^1 b_j(\sigma_{ij}(t))$, where ε_{ij} is the incidence coefficient of the co-oriented cells X_i and X_j. Finally we set $\Gamma(i) = \Gamma'(i) + \sum_{j\in s(i)} \Gamma''(i, j)$.

PROPOSITION 10. Suppose condition (A) is fulfilled. Then the co-oriented chain $\Gamma_{n-4}(X) = \sum_{i:\, \dim X_i = n-4} \Gamma(i)\, X_i$ is a cycle in $\hat{C}_{n-4}(X)\otimes \mathbf{Q}$. The class of homologies $[\Gamma_{n-4}(X)]$ of the cycle $\Gamma_{n-4}(X)$ does not depend on the choice of the admissible hypersimplicial system, the configurations σ_i and σ_j, and curve $\sigma_{ij}(t)$.

The proof of this proposition will be given in the second part of the article.

Remark. The construction of the cycle $\Gamma_{n-4}(X)$ nowhere uses the smooth structure of the manifold X and carries through without changes for any combinatorial manifold.

11. Formulation of the Main Theorem.

Definition. A cellular decomposition is said to be linearizable if an affine structure can be determined in all of the cells of this decomposition such that the structure of any cell will be consistent with the structure of all its faces.

Remark. It is not difficult to show that any decomposition into a product of simplexes, in particular any simplicial decomposition, is linearizable.

THEOREM 1. Let X be a smooth n-dimensional manifold and let $\{X_\alpha\}$ be a linearizable decomposition of X into simple cells satisfying condition (A). Then the class of homologies $[\Gamma_{n-4}(X)]$ under Poincaré isomorphism $H_{n-4}(\hat{C}_* \otimes Q) \cong H^4(X, Q)$ (see Proposition 3) goes over into the first Pontryagin class $P_1(X)$ of the manifold X.

Remark. The assertion of Theorem 1 implies that for an arbitrary decomposition $\{X_\alpha\}$ into simple noncoincident cells of a class $[\Gamma_{n-4}(X)]$ with a Poincaré-dual to $P_1(X)$ there is a barrier to linearizability of the decomposition $\{X_\alpha\}$.

COROLLARY 1. Let $\{X_\alpha\}$ be a linearizable decomposition of a smooth manifold X into simple cells and let Y be a smooth submanifold in X transversal to the decomposition $\{X_\alpha\}$. If the sets $Y_\alpha = Y \cap X_\alpha$ are cells in Y and the decomposition $\{Y_\alpha\}$ is linearizable, then $P_1(X)|_Y = P_1(Y)$.

COROLLARY 2. Let $\{X_\alpha\}$ be a decomposition of a smooth manifold X into simple cells satisfying the condition: every cell of codimension 2 is contained in exactly three cells of codimension 1. If the decomposition $\{X_\alpha\}$ is linearizable then $P_1(X) = 0$.

The following more general assertion is also true.

THEOREM 2. Let $\{X_\alpha\}$ be a linearizable cellular decomposition of a smooth manifold X satisfying the condition: every cell of codimension 2r of the decomposition $\{X_\alpha\}$ is contained in exactly 2r + 1 cells of codimension 2r − 1 of this decomposition. Then the first r Pontryagin classes of the manifold X are equal to zero.

§2. A VARIATIONAL COMPLEX AND $S^{k,l}$ FORMS

12. Space of Connective Flows. Let Y be a smooth variety and let $E \to Y$ be a smooth vector fibering with fibers R^q over Y. Let $\mathcal{F}(E)$ be a bundle (over Y) of connecting jets in the fibering E. Denote by $F^{(k)}(E)$, $F^{(k)}(E) \to Y$, a fibering over Y whose fibers are k-flow sections of the bundle $\mathcal{F}(E)$. The fibering π: $F(E) \to Y$, where $F(E) = \varprojlim F^{(k)}(E)$, is called the connective flow space of the fibering E. Let D(E) be the group of automorphisms of the fibering $E \to Y$, i.e., diffeomorphisms of the space E such as will commute with projections $E \to Y$ and will induce linear mappings of fibers of the fibering E. The group D(E) acts in a natural way in F(E). Associating with each connective set its ∞-flow at points of Y we get a mapping $\mathcal{F}(E) \to F(E)$. The tangent spaces to the images of sections of the bundle $\mathcal{F}(E)$ under this mapping determine in F(E) a planar D(E)-invariant connection ∇.

13. Variational Complex. Let $\bar{\Omega}^* = \oplus \bar{\Omega}^r$ be a bundle of jets of smooth differential forms in the space F(E) (the smooth structure in F(E) is determined by the projective limit $F(E) = \varprojlim F^{(k)}(E)$). The presence in F(E) of the planar connection ∇ permits conversion of $\bar{\Omega}^*$ to the bicomplex $\bar{\Omega}^{s,t}$, considering $\alpha \in \bar{\Omega}^r$ as belonging to $\bar{\Omega}^{s,t}(s + t = r)$ if $\alpha(\eta_1, \ldots, \eta_{s+i}, \eta_{s+i+1}, \ldots, \eta_{s+t}) = 0$ for $i \neq 0$ and for any collection in which $\eta_1, \ldots, \eta_{s+i}$ are vertical vectors (i.e., tangent to the fiber F(E)) and $\eta_{s+i+1}, \ldots, \eta_{s+t}$ are horizontal (with respect to ∇) tangent vectors to F(E). Let $\bar{d}$ be a differential of the complex $\bar{\Omega}^*$. For $\alpha \in \bar{\Omega}^{s,t}$ we put $d\alpha = \Pi^{s,t+1}\bar{d}\alpha$, $\delta\alpha = \Pi^{s+1,t}\bar{d}\alpha$, where $\Pi^{k,l}$ is a projector $\bar{\Omega}^* \to \bar{\Omega}^{k,l}$. Then $d^2 = \delta^2 = 0$, $d + \delta = \bar{d}$. Since connectedness of ∇ is D(E)-invariant the decomposition $\bar{\Omega}^* = \oplus \bar{\Omega}^{s,t}$ and differentials d and δ are consistent with the action of D(E). A subcomplex $\bar{\Omega}^*_{F(E)} = \oplus \bar{\Omega}^{s,t}_{F(E)}$ of forms from $\bar{\Omega}^*$, invariant with respect to the action of D(E), is called a variational complex of connections in E.

14. (d, δ)-Cocycle. Let U be an open set in Y, and let there be a fixed trivialization of a fibering E over U. Then each connection ω in $E|_U$ can be given by a 1-form in U with values in $\mathfrak{gl}(q)$. We define in $F(E)|_U$ a horizontal form $\bar{\omega}$ with values in $\mathfrak{gl}(q)$ as follows. Let $y \in U$, let ω be a connection flow to the point y, and let η be a tangent vector to F(E) at the point $(y, \omega) \in F(E)$. We put $\bar{\omega}(\eta) = \omega(\pi_*\eta)$, where $\pi_*\eta$ is a tangent vector to Y and is a projection of η. Analogously, for two tangent vectors η_1 and η_2 at the point $(y, \omega) \in F(E)$ we put $\bar{R}(\eta_1, \eta_2) = R(\pi_*\eta_1, \pi_*\eta_2)$, where R is the curvature form of the connection ω. Let $P(h_1, \ldots, h_m)$ be an invariant symmetric m-linear form in $\mathfrak{gl}(q)$. If ω is a connection to E and R is the curvature of the connection ω, then $P(R, \ldots, R)$ is a closed 2m-form representing a certain characteristic class P_E of the fibering E. We put $\bar{P}_E = P(\bar{R} + \delta\bar{\omega}, \ldots, \bar{R} + \delta\bar{\omega})$ (the operator δ for forms with vector-valued coefficients is defined just as for ordinary forms in F(E)).

PROPOSITION 11. The form $\bar{P}_E$ does not depend on the choice of the trivial fibering E. Thus, $\bar{P}_E$ is an invariantly determined form in the space F(E). Further, the form $\bar{P}_E$ is D(E)-invariant and $(d + \delta)\cdot(\bar{P}_E) = 0$. Let $[\bar{P}_E]$ be the class of cohomologies of the space F(E) corresponding to the form $\bar{P}_E$. We have $[\bar{P}_E] = \pi^* P_E$.

The form $\bar{P}_E$ is said to be the (d, δ)-cocycle corresponding to the form $P(h_1, \ldots, h_m)$.

15. Difference Cocycle. Let $\bar{P}_E$ be a (d, δ)-cocycle constructed as in paragraph 14 and let $\omega_0, \ldots, \omega_l$ be connections in E. We define a $(2m - l)$-form $P^{l,2m-l} = P^{l,2m-l}(\omega_0, \ldots, \omega_l)$ in Y as follows. Let $y \in Y$, and let $\xi_1, \ldots, \xi_{2m-l}$ be tangent vectors to Y at the point y. If ω is a connection flow to the point y then we denote by $\eta_i(\omega)$ the horizontal tangent vector to F(E) at the point (y, ω), satisfying the condition $\pi_*\eta_i(\omega) = \xi_i$. We define an l-form $Q_{\xi_1, \ldots, \xi_{2m-l}}$ in the fiber (of the fibering F(E)) over the point y according to the formula

$$Q_{\xi_1, \ldots, \xi_{2m-l}}(\zeta_1, \ldots, \zeta_l) = \bar{P}_E(\zeta_1, \ldots, \zeta_l, \eta_1(\omega), \ldots, \eta_{2m-l}(\omega)),$$

where $\zeta_1, \ldots, \zeta_l$ are tangent vectors to the fiber of F(E) at the point (y, ω). We put

$$P^{l,2m-l}(\xi_1, \ldots, \xi_{2m-l}) = \int_{\Delta_y^l(\omega_0, \ldots, \omega_l)} Q_{\xi_1, \ldots, \xi_{2m-l}},$$

where $\Delta_y^l(\omega_0, \ldots, \omega_l)$ is an l-dimensional simplex in the fiber of the fibering F(E) over the point y determined by the mapping

$$\left(t_0, \ldots, t_l \colon t_i \geqslant 0, \sum t_i = 1\right) \mapsto \left(y, \sum t_i (\omega_i)_y\right)$$

(here $(\omega_i)_y$ is a connection flow of ω_i to the point y).

PROPOSITION 12. The forms $P^{l,2m-l}(\omega_0, \ldots, \omega_l)$ are connected by the relations

$$(-1)^{l-1} dP^{l,2m-l}(\omega_0, \ldots, \omega_l) = \sum_{\mu=0}^{l} P^{l-1,2m-l+1}(\omega_0, \ldots \hat{\omega}_\mu \ldots, \omega_l).$$

A collection of forms $P^{l,2m-l}$ is called a difference cocycle corresponding to the form $P(h_1, \ldots, h_m)$.

16. Forms $S^{k,l}$. Let E be a vector fibering with fiber R^q over the manifold Y. A section of the fibering E will be denoted by the letters $m_\alpha = m_\alpha(y)$.

For $k, l\colon k \geqslant 0,\ 0 \leqslant l \leqslant q-1,\ k + l = 1, 2, 3$, we define on Y a differential $(2 - k - l)$-form $S^{k,l}(m_0, \ldots, m_{k+l+1}; m_{k+l+2}, \ldots, m_{k+q})$ (a (−1)-form is a number $c \in R$ while its differential is the same number c considered as constant function in Y).

We put $S^{0,1}(m_0, m_1, m_2; m_3, \ldots, m_q) = \frac{1}{2}(\log|b|\, d\log|a| - \log|a|\, d\log|b|)$, where $a = a(y)$ and $b = b(y)$ are functions in Y defined by the equality $m_0 = am_1 + bm_2 (\mathrm{mod}\ m_3, \ldots, m_q)$;

$S^{1,0}(m_0, m_1, m_2; m_3, \ldots, m_{q+1}) = \frac{1}{2}(\log|a|\, d\log|b| - \log|b|\, d\log|a|)$, where $m_0 = am_1 = bm_2 (\mathrm{mod}\ m_3, \ldots, m_{q+1})$;

$S^{0,2} = S^{2,0} = 0$;

$S^{1,1}(m_0, \ldots, m_3; m_4, \ldots, m_{q+1}) = \Phi(\varkappa)$, where $m_0 = am_2 + bm_3 (\mathrm{mod}\ m_4, \ldots, m_{q+1})$, $m_1 = cm_2 + dm_3 (\mathrm{mod}\ m_4, \ldots, m_{q+1})$, $\varkappa = bc/ad$, $\Phi(\varkappa)$ is a continuous function on the line without points 0 and 1 defined by the conditions $\Phi'(\varkappa) = \log|\varkappa|/(\varkappa - 1) - \log|\varkappa - 1|/\varkappa$, $\Phi(-1) = \Phi(1/2) = \Phi(2) = 0$;

$S^{0,3} = S^{3,0} = 0$;

$S^{1,2}(m_0, \ldots, m_4; m_5, \ldots, m_{q+1}) = -\varepsilon\pi^2/6$, where $\varepsilon = 1$ if among the decimals of the bases $(m_\alpha, m_\beta, m_\gamma, m_5, \ldots, m_{q+1})$ $(0 \leqslant \alpha, \beta, \gamma \leqslant 4, \alpha < \beta < \gamma)$ in the fiber of the fibering E an even number are identically oriented; $\varepsilon = -1$ otherwise;

$S^{2,1}(m_0, \ldots, m_4; m_5, \ldots, m_{q+2}) = -\varepsilon\pi^2/6$, where $\varepsilon = 1$ if among decimals of the bases $(m_\alpha, m_\beta, m_5, \ldots, m_{q+2})$ $(0 \leqslant \alpha, \beta \leqslant 4, \alpha < \beta)$ in the fiber of the fibering E on even number are identically oriented; $\varepsilon = -1$ otherwise.

Remark. Forms $S^{k,l}$ are defined and regular only at those points $y \in Y$ for which the vectors $m_0(y), \ldots, m_{k+q}(y)$ occur in general position.

Proposition 13. The forms $S^{k,l}$ satisfy the following relations:

a) $S^{k,l}(m_0, \ldots, m_{k+l+1}; m_{k+l+2}, \ldots, m_{k+q})$ is cosymmetric with respect to permutations of $m_0, \ldots, m_{k+l+1}$ and symmetric with respect to permutations $m_{k+l+2}, \ldots, m_{k+q}$,

b) for $k + l = 1$

$$dS^{k,l}(m_0, m_1, m_2; m_3, \ldots, m_{k+q}) = -P^{2,2}(\omega_0, \omega_1, \omega_2),$$

where $\omega_0, \omega_1, \omega_2$ connected in the fibering E correspond to the trivializations $(m_0, m_1, m_3, \ldots, m_q)$, $(m_1, m_2, m_3, \ldots, m_q)$, $(m_0, m_2, m_3, \ldots, m_q)$ in case $(k, l) = (0, 1)$ and to the trivializations $(m_0, m_3, \ldots, m_{q+1})$, $(m_1, m_3, \ldots, m_{q+1})$, $(m_2, m_3, \ldots, m_{q+1})$ in case $(k, l) = (1, 0)$, while $P^{2,2}(\omega_0, \omega_1, \omega_2)$ is a component of a difference cocycle corresponding to the form $P(h_1, h_2) = \mathrm{tr}(h_1 \cdot h_2)$,

c) for $k + l = 2, 3$

$$(-1)^{k+l+1} dS^{k,l}(m_0, \ldots, m_{k+l+1}; m_{k+l+2}, \ldots, m_{k+q}) =$$
$$= \sum_{\mu=0}^{k+l+1} (-1)^\mu S^{k-1,l}(m_0, \ldots \hat{m}_\mu \ldots, m_{k+l+1}; m_{k+l+2}, \ldots, m_{k+q}) -$$
$$- \sum_{\mu=0}^{k+l+1} (-1)^\mu S^{k,l-1}(m_0, \ldots \hat{m}_\mu \ldots, m_{k+l+1}; m_{k+l+2}, \ldots, m_{k+q}, m_\mu).$$

We remark that the last formula is dual to formula (1) of paragraph 6 in §1.

APPENDIX

FORMAL VARIATIONAL COMPLEX

Introduced in [2] was the concept of the formal complex of a geometric structure and formal characteristic classes were defined. It turns out to be useful to consider a new concept of formal variational complex of a geometric structure closely connected with the concept of formal complex. In this work, definitions are given of formal complex and formal variational complex for an n-dimensional Riemann structure. It is shown that the given complexes can be defined with the help of a complex of relative cochains of a pair of Lie algebras (W_n, L_0), where W_n is a Lie algebra of formal vectorial fields and L_0 is its subalgebra with coefficients in some W_n-module. In conclusion, generalizations of these concepts are presented with results in arbitrary geometric structures.

§1. PRELIMINARY INFORMATION

In the sequel we consider only manifolds of a C^∞ class. They can be either finite- or infinite-dimensional in the sense of [3]. Smoothness of mappings of manifolds, vector fields, of Riemann metrics, etc. in a manifold are considered throughout as belonging to the C^∞ class.

Let M and N be manifolds (not necessarily finite-dimensional) and let f: M → N be a smooth mapping. We denote by $j_x f$ an ∞-flow of f at the point $x \in M$. In particular f can be a section of a fibering, i.e., a vector field, Riemann metric, etc. We denote by $\Omega^p(M)$ $(p = 0, 1, \ldots)$ the space of smooth differential p-forms on the manifold M and by $d\omega$ the outer differential forms $\omega \in \Omega^p(M)$. If $x \in M$, then we denote by T_x the tangent space to M at the point x, by f_* the induced f mapping of tangent fiberings of the manifolds M and N, and by f^* the induced f mapping $\Omega^p(N) \to \Omega^p(M)$. We denote by $\mathfrak{A}(M)$ the Lie algebra of smooth vector fields in M. If M and N are finite-dimensional, if g is a Riemann metric in N, and f a diffeomorphism of M in an open submanifold of the manifold N, then f^*g denotes the pre-image of g relative to f.

Let W_n be a topological Lie algebra of formal vector fields [1]. Serving as points of W_n are ∞-flows at the point $0 \in R^n$ of vector fields of the space R^n, and a bracket is defined by the equality $[\xi_1 \xi_2] = j_0[\zeta_1, \zeta_2]$, where ζ_1 and ζ_2 are vector fields such that $\xi_1 = j_0\zeta_1$ and $\zeta_2 = j_0\zeta_2$. We recall that the subset L_0 of the Lie algebra W_n, formed by ∞-flows of vector fields vanishing at the point 0 constitutes a subalgebra of the Lie algebra W_n.

Let X be an n-dimensional manifold and let S(X) be an infinite-dimensional manifold of ∞-flows of coordinate systems of the manifold X (see [3]). Serving as points of S(X) are ∞-flows $j_0\alpha$ (at the point $0 \in R^n$) of smooth mappings $\alpha : R^n \to X$ regular at the point 0. The mapping $\pi_1 : S(X) \to X$ prescribed by the condition $\pi_1(s) = \alpha(0)$, where $s = j_0(\alpha)$, determines, in S(X), a structure of a smooth fibering over X. Let $f : Y \to X$ be a diffeomorphism of the manifold Y in an open submanifold of the manifold X and let $f_S : S(Y) \to S(X)$ be the diffeomorphism induced by f. For any vector field $\eta \in \mathfrak{A}(S(X))$ there exists, and uniquely so, a vector field $\varphi_f(\eta) \in \mathfrak{A}(S(Y))$ such that for any point $s \in S(Y)$ we have $(f_S)_*(\varphi_f(\eta)_s) = \eta_{s'}$, where $s' = f(x)$. The correspondence $\eta \to \varphi_f(\eta)$ defines a homomorphism of Lie algebras $\varphi_f : \mathfrak{A}(S(X)) \to \mathfrak{A}(S(Y))$.

The following theorem is known (see [3, 4]).

THEOREM 1. For any n-dimensional manifold X there exists a unique homomorphism of Lie algebras $h_X : W_n \to \mathfrak{A}(S(X))$ admitting the following properties: 1) for each point $s \in S(X)$ h_X induces an isomorphism $h_{X,s}: W_n \to T_S$; 2) $h_{X,s}(L_0)$ is a tangent space at the point s to a fiber of the fibering $\{S(X), \pi_1\}$ passing through that point; 3) a 1-form ω on S(X) with values in W_n, defined at each point $s \in S(X)$ by the equation $\omega_s = h^{-1}_{X,s}$, is smooth; 4) if $f : Y \to X$ is a diffeomorphism of the manifold Y on an open submanifold of the manifold X, then $\varphi f \cdot h_X = h_Y$.

We recall the definition of h_X. With $\xi \in W_n$, let ζ be a vector field in R^n such that $\xi = j_0\zeta$, and $\{\varphi_t\}$ be a local 1-parameter group of mappings of R^n corresponding to ζ. The mapping $\theta_t : S(X) \to S(X)$, defined by the condition $\theta_t(s) = j_0(\alpha \cdot \varphi_t^{-1})$, where $s = j_0\alpha \in S(X)$, does not depend on the choice of α and is a diffeomorphism of S(X). Obviously, $\{\theta_t\}$ is a local 1-parameter group of mappings. We denote by $h_X(\xi)$ the vector field in S(X) corresponding to $\{\theta_t\}$. It is readily seen that $h_X(\xi)$ does not depend on the choice of ζ.

Consider an infinite dimensional manifold $\mathcal{R}$ of ∞-flows (at the point $0 \in R^n$ of Riemann metrics of space R^n. Any point $r \in \mathcal{R}$ is represented in the form $r = j_0 g$, where g is a Riemann metric defined in a neighborhood of the point 0. Let $\xi \in W_n$, ζ be a vector field in R^n such that $\xi = j_0\zeta$, and $\{\varphi_t\}$ corresponding to ζ be a local one-parameter group of mappings of R^n. The mapping $\psi_t : \mathcal{R} \to \mathcal{R}$, defined by the condition $\psi_t(r) = j_0(\varphi_t^* g)$, where $r = j_0 g \in \mathcal{R}$, does not depend on the choice of g and is a diffeomorphism of $\mathcal{R}$. It is obvious that $\{\psi_t\}$ is a local one-parameter group of transformations of $\mathcal{R}$. We denote by $\lambda(\xi)$ the vector field in $\mathcal{R}$ corresponding to $\{\psi_t\}$. It is easy to see that $\lambda(\xi)$ does not depend on the choice of ζ. The correspondence $\xi \to \lambda(\xi)$ defined by the mapping $\lambda : W_n \to \mathfrak{A}(\mathcal{R})$ is a homomorphism of Lie algebras.

We denote by $\delta : \Omega^q(\mathcal{R}) \to \Omega^{q+1}(\mathcal{R})$ $(q = 0, 1, \ldots)$ the outer differential and we shall call it a variation. We note that if $f \in \Omega^0(\mathcal{R})$, then f is a smooth functional on $\mathcal{R}$ and δf is its usual variation. The outer product $\Omega^p(\mathcal{R}) \otimes \Omega^q(\mathcal{R}) \to \Omega^{p+q}(\mathcal{R})$ and the variation δ define in $\Omega^*(\mathcal{R}) = \bigoplus_q \Omega^q(\mathcal{R})$ the structure of a graduated differential algebra. If $\eta \in \mathfrak{A}(\mathcal{R})$, then the usual action of η in $\Omega^0(\mathcal{R})$ uniquely extends to differentiation $\bar{\eta} : \Omega^*(\mathcal{R}) \to \Omega^*(\mathcal{R})$ of the algebra $\Omega^*(\mathcal{R})$, conserving powers and commutativity with δ. It is easy to show that $\bar{\eta}$ is determined by the following formula:

$$(\bar{\eta}\omega)(\zeta_1, \ldots, \zeta_q) = \eta\omega(\zeta_1, \ldots, \zeta_q) + \sum_{1 \leq i \leq q} (-1)^i \omega([\eta, \zeta_i], \ldots \hat{\zeta}_i \ldots, \zeta_q), \tag{1}$$

where $\omega \in \Omega^q(\mathcal{R})$ and $\eta, \zeta_1, \ldots, \zeta_q \in \mathfrak{A}(\mathcal{R})$.

With respect to action $\mathfrak{A}(\mathcal{R})$ thus defined in $\Omega^q(\mathcal{R})$ $(q = 0, 1, \ldots)$ the space $\Omega^q(\mathcal{R})$ is a $\mathfrak{A}(\mathcal{R})$-module. The homomorphism λ induces in $\Omega^q(\mathcal{R})$ the topological structure of a W_n-module with respect to the natural C^∞-topology in $\Omega^q(\mathcal{R})$.

§2. DEFINITIONS OF FORMAL RIEMANN COMPLEX AND FORMAL-VARIATIONAL RIEMANN COMPLEX

We use the term formal Riemann p-cochain φ for a mapping which makes correspond an arbitrary n-dimensional manifold X and a Riemann metric g in its p-form $\varphi(X, g) \in \Omega^p(X)$, possessing the following properties:

1) the coefficients of the form $\varphi(X, g)$ in any system of coordinates in X are smooth functions of the components g_{ij} of a metric tensor in its partial derivatives up to some finite order;

2) if $f: Y \to X$ is a diffeomorphism of the manifold Y on an open submanifold of the manifold X then $f^*\varphi(X, g) = \varphi(Y, f^*g)$.

We denote by $\Phi_r^p (p = 0, 1, \ldots, n)$ the space of formal Riemann p-cochains. We define a mapping $d^p : \Phi_r^p \to \Phi_r^{p+1}$ by means of the formula $(d^p \varphi)(X, g) = d\varphi(X, g)$. It is obvious that the mapping d^p is linear and that $d^{p+1} \cdot d^p = 0$.

The complex $\Phi_r^\cdot = \{\bigoplus_p \Phi_r^p, d^p\}$ is called a formal Riemann complex. Classes of homologies of the complex Φ_r^* are called formal Riemann characteristic classes.

Let $\mathcal{R}(X)$ be an infinite-dimensional manifold of ∞-flows of Riemann metrics of the n-dimensional manifold X. Any point of $\mathcal{R}(X)$ has the form j_xg where $x \in X$ and g is a Riemann metric defined in a neighborhood of the point x. A mapping $\pi_2 : \mathcal{R}(X) \to X$ prescribed by the condition $\pi_2(j_xg) = x$, where $x \in X$ and $j_xg \in \mathcal{R}(X)$, defines in $\mathcal{R}(X)$ the structure of a smooth fibering over X.

We use the term formal-variational Riemann p-cochain ψ for the mapping which places in correspondence with an arbitrary n-dimensional manifold X a p-form $\psi(X) \in \Omega^p(\mathcal{R}(X))$, having the following property: if $f: Y \to X$ is a diffeomorphism of the manifold Y on an open submanifold of the manifold X and if $f_R : \mathcal{R}(Y) \to \mathcal{R}(X)$ is the mapping induced by f, then $f^*\psi(X) = \psi(Y)$.

We denote by $\Psi_r^p (p = 0, 1, \ldots)$ the space of formal Riemann p-cochains. We define a mapping $D^p : \Psi_r^p \to \Psi_r^{p+1}$ by means of the formula $(D^p\psi)(X) = d\psi(X)$, where $\psi \in \Psi_r^p$. It is obvious that the mapping D^p is linear and that $D^{p+1} \cdot D^p = 0$.

We call the complex $\Psi_r^\cdot = \{\bigoplus_p \Psi_r^p, D^p\}$ a formal-variational Riemann complex.

We consider $C_q^\cdot(W_n, \Omega^q(\mathcal{R})) = \{\bigoplus_p C^p(W_n, \Omega^q(\mathcal{R})), d_q^p\}$ $(q = 0, 1, \ldots)$ a standard complex of cochains of a topological Lie algebra W_n with coefficients in a topological W_n-module $\Omega^q(\mathcal{R})$. The space of p-cochains $C^p(W_n, \Omega^q(\mathcal{R}))$ is mapped by continuous p-linear cosymmetric forms in W_n with values in $\Omega^q(\mathcal{R})$. The differential $d_q^p : C^p(W_n, \Omega^q(\mathcal{R})) \to C^{p+1}(W_n, \Omega^q(\mathcal{R}))$ is defined by

$$(d_q^p\omega)(\xi_1, \ldots, \xi_{p+1}) = \sum_{1 \leq i \leq p+1} (-1)^{i-1} \overline{\lambda(\xi_i)}\, \omega(\xi_1, \ldots \hat{\xi}_i \ldots, \xi_{p+1}) + \sum_{1 \leq i < j \leq p+1} (-1)^{i+j} \omega([\xi_i, \xi_j], \xi_1, \ldots \hat{\xi}_i \ldots \hat{\xi}_j \ldots, \xi_{p+1}), \tag{2}$$

where $\omega \in C^p(W_n, \Omega^q(\mathcal{R}))$ and $\xi_1, \ldots, \xi_{p+1} \in W_n$.

Let $C_q^p(\mathcal{R})$ be the subspace of the space $C^p(W_n, \Omega^q(\mathcal{R}))$ formed by p-cochains ω such that for any $\xi \in L_0$ $\xi \lrcorner\, \omega = 0$ and $\xi \lrcorner\, d_q^p\omega = 0$, where $\lrcorner$ denotes interior differentiation. It is obvious that $d_q^p(C_q^p(\mathcal{R})) \subset C_q^{p+1}(\mathcal{R})$.

In this way we obtain a complex $C_q^\cdot(\mathcal{R}) = \{\bigoplus_p C_q^p(\mathcal{R}), d_q^p\}$ $(q = 0, 1, \ldots)$, which is called the complex of continuous relative cochains of a pair of topological Lie algebras (W_n, L_0) with coefficients in the topological W_n-module $\Omega^q(\mathcal{R})$.

We define the mapping $\delta_q^p : C_q^p(\mathcal{R}) \to C_{q+1}^p(\mathcal{R})$ $(p, q = 0, 1, \ldots)$ by the condition

$$(\delta_q^p\omega)(\xi_1, \ldots, \xi_p) = (-1)^p \delta\omega(\xi_1, \ldots, \xi_p), \tag{3}$$

where $\omega \in C_q^p(\mathcal{R})$ and $\xi_1, \ldots, \xi_p \in W_n$. It is obvious that δ_q^p is a linear mapping, that $\delta_{q+1}^p \cdot \delta_q^p = 0$ and that $\delta_q^{p+1} \cdot d_q^p + d_{q+1}^p \cdot \delta_q^p = 0$.

In this way a bicomplex $C^{\cdot\cdot}(\mathcal{R}) = \{\bigoplus_{p,q} C_q^p(\mathcal{R}), d_1, d_2\}$ is defined, where $d_1 = \{d_q^p\}$ and $d_2 = \{\delta_q^p\}$. Denote by $C^\cdot(\mathcal{R})$ the corresponding total complex $\{\bigoplus_k C^k(\mathcal{R}), d_1 + d_2\}$, where $C^k(\mathcal{R}) = \bigoplus_{p+q=k} C_q^p(\mathcal{R})$.

The following theorems hold true.

<u>THEOREM 2.</u> The complexes Ψ_r^* and $C^*(\mathcal{R})$ are canonically isomorphic.

<u>THEOREM 3.</u> The complexes Φ_r^* and $C_0^\cdot(\mathcal{R})$ are canonically isomorphic.

§3. PROOFS OF THEOREMS 2 AND 3

We consider the fiberings $\{S(X), \pi_1\}$ and $\{\mathcal{R}(X), \pi_2\}$. Let E(X) be the subset of points of the manifold $S(X) \times \mathcal{R}(X)$, formed by the points $e = (s, z)$, with $s \in S(X)$ and $z \in \mathcal{R}(X)$, for which $\pi_1(s) = \pi_2(z)$. Obviously E(X) is a smooth submanifold of the manifold $S(X) \times \mathcal{R}(X)$. We define the mappings $p_1 : E(X) \to S(X)$ and $p_2 : E(X) \to \mathcal{R}(X)$ by the conditions $p_1(x, z) = s$ and $p_2(x, z) = z$. Then $\{E(X), p_1\}$ and $\{E(X), p_2\}$ are smooth fiberings over S(X) and $\mathcal{R}(X)$, respectively. Let $e = (s, z) \in E(X)$, $x = \pi_1(s) = \pi_2(z)$, $z = j_0 g$ and $s = j_0 \alpha$. It is readily seen that the ∞-flow $j_0(\alpha^* g)$ belongs to $\mathcal{R}$ and does not depend on the choice of the Riemann metric g and mapping α. Hence, corresponding to $e \to j_0(\alpha^* g)$ there is defined a mapping $p : E(X) \to \mathcal{R}$. The mapping p is smooth and its restriction $p_s : F_s \to \mathcal{R}$ to any fiber F_s of the fibering $\{E(X), p_1\}$ is a diffeomorphism. The pair of projections p_i and p define a canonical representation of E(X) in the form of a direct product $S(X) \times \mathcal{R}$. We place in correspondence with every vector $\xi \in W_n$ a vector field $\widetilde{\xi}$ in E(X) in such a way that $(p_1)_* \widetilde{\xi} = h_x(\xi)$ and $p_* \widetilde{\xi} = \lambda(\xi)$. The field $\widetilde{\xi}$ is uniquely defined and the correspondence $\xi \to \widetilde{\xi}$ defines a homomorphism of Lie algebras $W_n \to \mathfrak{A}(E(X))$.

We denote by ζ_e the value of the vector field $\zeta \in \mathfrak{A}(E(X))$ at the point $e \in E(X)$. Let H_e be a subspace of the tangent space T_e formed by vectors $\widetilde{\xi}_e$ such that $\xi \in W_n$ and let V_e be a subspace of the space T_e formed by vectors $\widetilde{\xi}_e$ such that $\xi \in L_0$. According to Theorem 1 and the definition of $\widetilde{\xi}$ the space V_e is a tangent space to the fiber of the fibering $\{E(X), p_2\}$ passing through the point e. Suppose now $\eta \in \mathfrak{A}(\mathcal{R})$. We denote by $\widetilde{\eta}$ the vector field in E(X) uniquely determined by the conditions $(p_1)_* \widetilde{\eta} = \eta$ and $p_* \widetilde{\eta} = \eta$. The correspondence $\eta \mapsto \widetilde{\eta}$ defines a homomorphism of Lie algebras $\mathfrak{A}(\mathcal{R}) \to \mathfrak{A}(\mathcal{R}(X))$. Let F_e be a subspace of the space T_e formed by vectors $\widetilde{\eta}_e$ where $\eta \in \mathfrak{A}(\mathcal{R})$. It is obvious that $T_e = H_e \oplus F_e$.

Denote by $\Lambda^s(V)$ the s-th outer power of the linear space V and by V' the space conjugate to V. It is known that $\Lambda^k(T'_e) = \bigoplus_{p+q=k} \Lambda^p(H'_e) \otimes \Lambda^q(F'_e)$. Let $\Omega^{p,q}(E(X))$ be the subspace of the space $\Omega^{p+q}(E(X))$, formed by forms ω whose values at any point $e \in E(X)$ belong to $\Lambda^p(H'_e) \otimes \Lambda^q(F'_e)$. Obviously, $\Omega^k(E(X)) = \bigoplus_{p+q=k} \Omega^{p,q}(E(X))$.

Any form $\omega \in \Omega^{p,q}(E(X))$ is uniquely defined by prescribing its values $\omega(\widetilde{\xi}_1, \ldots, \widetilde{\xi}_p, \widetilde{\eta}_1, \ldots, \widetilde{\eta}_q)$, where $\xi_1, \ldots, \xi_p \in W_n$ and $\eta_1, \ldots, \eta_q \in \mathfrak{A}(\mathcal{R})$. Let $\varphi \in C^p(W_n, \Omega^q(\mathcal{R}))$. We consider the (p + q)-form $\sigma_X(\varphi) \in \Omega^{p,q}(E(X))$, defined by the equality $\sigma_X(\varphi)(\widetilde{\xi}_1, \ldots, \widetilde{\xi}_p, \widetilde{\eta}_1, \ldots, \widetilde{\eta}_q) = \varphi(\xi_1, \ldots, \xi_p)(\eta_1, \ldots, \eta_q)$, where $\xi_1, \ldots, \xi_p \in W_n$ and $\eta_1, \ldots, \eta_q \in \mathfrak{A}(\mathcal{R})$. Thus, the correspondence $\varphi \mapsto \sigma_X(\varphi)$ defines a mapping $\sigma_X : C^p(W_n, \Omega^q(\mathcal{R})) \to \Omega^{p,q}(E(X))$.

Let $\omega \in \Omega^{p,q}(E(X))$. It is obvious that $d\omega = d'\omega + d''\omega$, where $d'\omega \in \Omega^{p+1,q}(E(X))$ and $d''\omega \in \Omega^{p,q+1}(E(X))$. Since $d^2 = 0$, then $(d')^2 = 0$, $(d'')^2 = 0$ and $d'd'' + d''d' = 0$. Thus, $\Omega^{**}(E(X)) = \{\bigoplus_{p,q} \Omega^{p,q}(E(X)), d', d''\}$ is a bicomplex with differentials d' and d" while the corresponding total complex $\{\bigoplus_p \Omega^p(E(X)), d\}$ is a de Rama complex of the mapping E(X).

The differentials d' and d" are defined by the following conditions:

$$\begin{aligned} d'\omega(\widetilde{\xi}_1, \ldots, \widetilde{\xi}_{p+1}, \widetilde{\eta}_1, \ldots, \widetilde{\eta}_q) = \sum_{1 \leqslant i \leqslant p+1} (-1)^{i-1} \Big\{ \widetilde{\xi}_i \omega(\widetilde{\xi}_1, \ldots \hat{\widetilde{\xi}}_i \ldots, \widetilde{\xi}_{p+1}, \\ \widetilde{\eta}_1, \ldots, \widetilde{\eta}_q) + \sum_{1 \leqslant j \leqslant q} (-1)^j \omega(\widetilde{\xi}_1, \ldots \hat{\widetilde{\xi}}_i, \ldots, \widetilde{\xi}_{p+1}, \widetilde{[\lambda(\xi_i), \eta_j]}, \widetilde{\eta}_1, \ldots \hat{\widetilde{\eta}}_j \ldots, \widetilde{\eta}_q) \Big\} + \\ + \sum_{1 \leqslant i < j \leqslant p+1} (-1)^{i+j} \omega(\widetilde{[\xi_i, \xi_j]}, \widetilde{\xi}_1, \ldots \hat{\widetilde{\xi}}_i \ldots \hat{\widetilde{\xi}}_j \ldots, \widetilde{\xi}_{p+1}, \widetilde{\eta}_1, \ldots, \widetilde{\eta}_q), \end{aligned} \tag{4}$$

$$\begin{aligned} d''\omega(\widetilde{\xi}_1, \ldots, \widetilde{\xi}_p, \widetilde{\eta}_1, \ldots, \widetilde{\eta}_{q+1}) = (-1)^p \Big\{ \sum_{1 \leqslant i \leqslant q+1} (-1)^i \widetilde{\eta}_i \omega(\widetilde{\xi}_1, \ldots, \widetilde{\xi}_p, \widetilde{\eta}_1, \ldots \hat{\widetilde{\eta}}_i \ldots, \widetilde{\eta}_{q+1}) + \\ + \sum_{1 \leqslant i < j \leqslant q+1} (-1)^{i+j} \omega(\widetilde{\xi}_1, \ldots, \widetilde{\xi}_p, \widetilde{[\eta_i, \eta_j]}, \widetilde{\eta}_1, \ldots \hat{\widetilde{\eta}}_i \ldots \hat{\widetilde{\eta}}_j \ldots, \widetilde{\eta}_{q+1}) \Big\}, \end{aligned} \tag{5}$$

where $\omega \in \Omega^{p,q}(E(X))$, $\xi_1, \ldots, \xi_{p+1} \in W_n$ and $\eta_1, \ldots, \eta_{q+1} \in \mathfrak{A}(\mathcal{R})$. According to (1), (2) and (4), if $\varphi \in C^p(W_n, \Omega^q(\mathcal{R}))$, then $\sigma_X(d_q^p \varphi) = d'(\sigma_X(\varphi))$.

We identify the space $\Omega^k(\mathcal{R}(X))$ $(k = 0, 1, \ldots)$ with a subspace of the space $\Omega^k(E(X))$ by means of an imbedding p_2^*. We recall that at any point e, V_e is a tangent space to a fiber of the fibering $\{E(X), p_2\}$. Hence, the form $\omega \in \Omega^{p,q}(E(X))$ belongs to the space $\Omega^{p,q}(\mathcal{R}(X)) = \Omega^{p+q}(\mathcal{R}(X)) \cap \Omega^{p,q}(E(X))$ if and only if for every $\xi \in L_0$ we have $\widetilde{\xi} \lrcorner \omega = 0$ and $\widetilde{\xi} \lrcorner d\omega = 0$, where $\lrcorner$ denotes interior products. Hence, $\Omega^k(\mathcal{R}$

$(X)) = \bigoplus_{p+q=k} \Omega^{p,q}(\mathcal{R}(X))$ $(k = 0, 1, \ldots)$. Thus $\Omega^{**}(\mathcal{R}(X)) = \{\bigoplus_{p,q} \Omega^{p,q}(\mathcal{R}(X)), d', d''\}$ is a bicomplex and the total complex $\Omega^{**}(\mathcal{R}(X))$ corresponding to $\Omega^{*}(\mathcal{R}(X))$ is a de Rama complex of the manifold $\mathcal{R}(X)$.

If $f: Y \to X$ is a diffeomorphism of the variety Y on an open submanifold of the manifold X, then the mapping $f_{\mathcal{R}}^{*}: \Omega^{**}(\mathcal{R}(X)) \to \Omega^{**}(\mathcal{R}(Y))$ is a homomorphism of bicomplexes. Hence, $\Psi_r^k = \bigoplus_{p+q=k} \Psi_r^{p,q}$ (k = 0, 1, ...), where $\Psi_r^{p,q}$ is the space of formal variational Riemann (p + q)-cochains ψ such that $\psi(X) \in \Omega^{p,q}(\mathcal{R}(X))$ for any variety X. Let $D_1^{p,q}: \Psi_r^{p,q} \to \Psi_r^{p+1,q}$ and $D_2^{p,q}: \Psi_r^{p,q} \to \Psi_r^{p,q+1}$ be mappings defined by the conditions $(D_1^{p,q}\psi)(X) = d'\psi(X)$ and $(D_2^{p,q}\psi)(X) = d''\psi(X)$, where $\psi \in \Psi_r^{p,q}$. It is obvious that $\Psi_r^{**} = \{\bigoplus_{p,q} \Psi_r^{p,q}, D_1^{p,q}, D_2^{p,q}\}$ is a bicomplex and the complex Ψ_r^{*} is the total complex corresponding to Ψ_r^{**}.

By (3) and (4), the conditions $\tilde{\xi} \lrcorner\, \omega = 0$ and $\tilde{\xi} \lrcorner\, d\omega = 0$ are equivalent to the conditions $\tilde{\xi} - \omega = 0$ and $\tilde{\xi} \lrcorner\, d'\omega = 0$. Hence, for any form $\varphi \in C^p(W_n, \Omega^q(\mathcal{R}))$ the form $\sigma_X(\varphi)$ belongs to $\Omega^{p,q}(\mathcal{R}(X))$ if and only if $\varphi \in C_q^p(\mathcal{R})$. Thus, the correspondence $\varphi \to \sigma_X(\varphi)$ for $\varphi \in C_q^p(\mathcal{R})$ $(p, q = 0, 1, \ldots)$ defines a linear mapping $\sigma_X: C^{**}(\mathcal{R}) \to \Omega^{**}(\mathcal{R}(X))$. We already showed that $\sigma_X \cdot d_1 = d' \cdot \sigma_X$. There follows from formulas (3) and (5) that $\sigma_X \cdot d_2 = d'' \cdot \sigma_X$. Thus, for any n-dimensional manifold X the mapping σ_X is a homomorphism of bicomplexes.

If $\varphi \in C_q^p(\mathcal{R})$, then the correspondence $X \to \sigma_X(\varphi)$ defines a (p + q)-cochain of a formal variational Riemann complex Ψ_r^{*}. We define a mapping $\sigma: C^{**}(\mathcal{R}) \to \Psi_r^{**}$ by the following condition: $\sigma(\varphi)(X) = \sigma_X(\varphi)$, where $\varphi \in C_q^p(\mathcal{R})$ $(p, q = 0, 1, \ldots)$ and where X is an arbitrary n-dimensional manifold. Since, for any manifold X, σ_X is a homomorphism of bicomplexes, it follows that σ is a homomorphism of bicomplexes. We show that the homomorphism σ is an isomorphism.

Let $\psi \in \Psi_r^{p,q}$ and let X be an n-dimensional manifold. Consider the restriction ψ_X of the form $\psi(X)$ to the fiber $\mathcal{R}_x$ of the fibering $\{\mathcal{R}(X), \pi_2\}$ at some point $x \in X$. As we showed already, for any point $s \in S(X)$ such that $\pi_1(s) = x$, the form ψ_X is defined by its values $\psi_x(\tilde{\xi}_1, \ldots, \tilde{\xi}_p, \tilde{\eta}_1, \ldots, \tilde{\eta}_q)$, $\xi_1, \ldots, \xi_p \in W_n$ and $\eta_1, \ldots, \eta_q \in \mathfrak{A}(\mathcal{R})$. Since $\psi(X) \in \Omega^{p,q}(\mathcal{R}(X))$, these values do not depend on the choice of the point s. Since for any n-dimensional manifolds X and Y and any points $x \in X$ and $y \in Y$ there exists a diffeomorphism of a neighborhood of point x to a neighborhood of the point y, then according to the definition of formal variational Riemann cochain, the values $\psi_x(\tilde{\xi}_1, \ldots, \tilde{\xi}_p, \tilde{\eta}_1, \ldots, \tilde{\eta}_q)$ do not depend on the choice of manifold X and point x. Denote by φ the p-form on W_n with values in $\Omega^q(\mathcal{R})$, defined by the condition $\varphi(\xi_1, \ldots, \xi_p)\cdot(\eta_1, \ldots, \eta_q) = \psi_x(\tilde{\xi}_1, \ldots, \tilde{\xi}_p, \tilde{\eta}_1, \ldots, \tilde{\eta}_q)$, where $\xi_1, \ldots, \xi_p \in W_n$ and $\eta_1, \ldots, \eta_q \in \mathfrak{A}(\mathcal{R})$. Obviously, $\varphi \in C_q^p(\mathcal{R})$. We set $\tau(\psi) = \varphi$. The correspondence $\psi \to \tau(\psi)$ defines a mapping $\tau: \Psi_r^{**} \to C^{**}(\mathcal{R})$. Verification of the relations $\tau \cdot \sigma = 1$ and $\sigma \cdot \tau = 1$ gives no trouble. Hence, the mapping σ is a canonical isomorphism of the bicomplexes $C^{**}(\mathcal{R})$ and Ψ_r^{**} as well as of the corresponding total complexes $C^{*}(\mathcal{R})$ and Ψ_r^{*}. Thus, Theorem 2 is proved.

Now let $\varphi \in C_0^p(\mathcal{R})$ $(p = 0, 1, \ldots, n)$, let X be an n-dimensional manifold, and let g be a Riemann metric in X. We consider a p-form $\sigma(\varphi)(X)$ in $\mathcal{R}(X)$ and a section $\tilde{g}: X \to \mathcal{R}(X)$ of the fibering $\{\mathcal{R}(X), \pi_2\}$ defined by the metric g. We set $\varphi(X, g) = \tilde{g}^{*}(\sigma(\varphi(X)))$. It is readily seen that $\varphi(X, g) \in \Phi_r^p$. Accordingly, the relation $\varphi \mapsto \varphi(X, g)$ defines a mapping $\sigma_0: C_0^{*}(\mathcal{R}) \to \Phi_r^{*}$. Since $\sigma_X \cdot d_1 = d' \cdot \sigma_X$, σ_0 is a homomorphism of complexes. It follows from the definition of smooth functions on infinite-dimensional manifolds $\mathcal{R}(X)$ that σ_0 is a bijective mapping. Thus, σ_0 is a canonical isomorphism of complexes, proving Theorem 3.

§4. EXTENSIONS AND REMARKS

Let $S_k(X)$ (k = 1, 2, ...) be a manifold of k-flows of coordinate systems of the n-dimensional manifold X. Serving as the points of $S_k(X)$ are the k-flows at the point $0 \in R^n$ of smooth mappings $\alpha: R^n \to X$, regular at the point 0. We consider the Lie group G_k of k-flows at the point 0 of local diffeomorphisms of R^n containing the point 0. It is known [4] that $S_k(X)$ is a smooth principal G_k-fibering over X. Let the group G_k act in a smooth way on some finite-dimensional manifold F. We consider the smooth fibering F(X) with fibers F induced by the principal G_k-fibering $S_k(X)$ and carrying G_k to $S_k(X)$.

We say that a structure of order k and type E is given on the manifold X if a smooth section of the fibering F is given.

Any tensor structure on the manifold (e.g., Riemannian) is a structure of order 1 and type F where F is the space of tensors of the prescribed type in which the group $G_1 = GL(n, R)$ acts in the usual way.

An example of a structure of order 2 is provided by a structure defined in X by a prescribed affine connection.

All definitions and results of §§2 and 3 are conserved if instead of Riemannian structures we take any structure of order k and type F. Here in place of the manifold $\mathscr{R}$ we need to take a manifold of ∞-flows at the point $0 \in \mathbf{R}^n$ of the given structures of order k and type F defined in a neighborhood of the point 0. The proofs of Theorems 2 and 3 are retained without essential changes.

A very important role is played by the formal variational complex $C^*(\Gamma)$, constructed for affine connections. Denote by Γ an infinite-dimensional manifold of ∞-flows of affine connections at the point $0 \in \mathbf{R}^n$. Let g be a Riemann metric in $\mathbf{R}^n$ and let $\gamma(g)$ be the corresponding Riemann connection. The correspondence $g \mapsto \gamma(g)$ induces a smooth mapping $\rho : \mathscr{R} \to \Gamma$ concordant with the action of W_n in $\mathscr{R}$ and Γ. Hence, ρ induces a homomorphism of W_n-modules $\rho^* : \Omega^p(\Gamma) \to \Omega^p(\mathscr{R})$ $(p = 0, 1, \ldots)$ and hence induces a homomorphism of corresponding formal variational complexes $C^*(\Gamma) \to C^*(\mathscr{R})$ and their homologies $H(C^*(\Gamma)) \to H(C^*(\mathscr{R}))$.

We sketch several classes of homologies of the complexes $C^*(\Gamma)$ and $C^*(\mathscr{R})$. Let $\Pi(x_1, \ldots, x_k)$ be an invariant symmetric k-form in the Lie algebra $\mathfrak{gl}(n, \mathbf{R})$ of the complete linear group and let $\Pi(x) = \Pi(x, x, \ldots, x)$. It is easy to verify that $\Pi(R + \delta\gamma)$ (where γ is a form of affine connection and $R = d\gamma + (1/2)[\gamma, \gamma]$ is its curvature form) is a 2k-dimensional cocycle of the complex $C^*(\Gamma)$ (or of the complex $C^*(\mathscr{R})$, if we take for γ only forms of Riemann connections). We note that for even k the form $\Pi(R)$ is one of the Pontryagin forms constructed for Riemann metric g or affine connection γ. Hence, for even k, $\Pi(R)$ can serve as a nontrivial cocycle of the formal complex of affine connections or of the formal Riemann complex.

We say that a p-cochain ψ of the complex Ψ_r^* is rational if for any manifold X the coefficients of the form $\psi(X)$ in any coordinate system in X are polynomials in the components g_{ij} of a metric tensor, in their partial derivatives up to some finite order, and in $\det(g_{ij})^{-1}$. The set $\tilde{\Psi}_r^*$ of all rational cochains of the complex Ψ_r^* forms a subcomplex of the complex Ψ_r^*. All indicated non-trivial classes of homologies of the complex Ψ_r^* are defined by certain rational cochains, i.e., are images of nontrivial classes of homologies of the complex $\tilde{\Psi}_r^*$ under the imbedding $\tilde{\Psi}_r^* \subset \Psi_r^*$. Analogous facts hold also for the formal Riemann complex.

LITERATURE CITED

1. I. M. Gel'fand and D. B. Fuks, "Cohomologies of a Lie algebra of formal vector fields," Izv. Akad. Nauk, Ser. Matem., 34, 322-337 (1970).
2. I. M. Gel'fand, "Cohomologies of infinite-dimensional Lie algebras. Some questions of integral geometry," Report at the International Congress of Mathematicians, Nice, France (1970).
3. I. M. Gel'fand, D. A. Kazhdan, and D. B. Fuks, "Actions of infinite-dimensional Lie algebras," Funktsional. Analiz i Ego Prilozhen., 6, No. 1, 10-15 (1972).
4. I. N. Bernshtein and B. I. Rozenfel'd, "Homogeneous spaces of infinite-dimensional Lie algebras and characteristic classes of fiberings," Uspekh. Matem. Nauk, 28, No. 4, 103-138 (1973).

17.

(with A. M. Gabrielov and M. V. Losik)

Combinatorial computation of characteristic classes

Funkts. Anal. Prilozh. **9** (3) (1975) 5-26 [Funct. Anal. Appl. **9** (1975) 186-202].
MR **53**:14504a. Zbl. **341**:57017

In this paper we give proofs of Theorems 1 and 2 and Proposition 10 formulated in §1 of our previous work of the same title,† which we shall from now on refer to as Part I. At the beginning of Para. 3 we deduce an explicit formula for a characteristic class in the case when a piecewise-smooth connection is given over the manifold (Proposition 16). For this, admissible simplicial systems are introduced in Para. 18. Unlike the hypersimplicial systems of Part I, Para. 9, these, evidently, play an auxiliary role here. From Proposition 16 and simple assertions on the difference cocycle it is easy to deduce Theorem 2 on the triviality of the Pontryagin classes of a manifold with a "simple" complex. Further, in Paras. 23-26, we consider manifolds of the configurations $\Sigma(i)$ and $\tilde{\Sigma}(i)$, define the forms S on $\tilde{\Sigma}(i)$, and investigate the singularities of these forms. In Para. 27, with the aid of forms S, we prove Proposition 10 on the independence of the combinatorially constructed cycle $\Gamma_{n-4}(X)$ from the choice of the hypersimplicial system and configurations. Finally, in Para. 28 we give a proof of Theorem 1, which establishes the coincidence of the first Pontryagin class with the class dual to $\Gamma_{n-4}(X)$ in the case of a linearizable partition.

§3. PROOF OF THE FUNDAMENTAL THEOREMS

17. The Complexes $J_*(i)$

Let $\{X_\alpha\}$ be a partition of the n-dimensional manifold X into simple cells. For each cell X_i we denote by c(i) the set of suffixes of the n-dimensional cells of the partition $\{X_\alpha\}$ containing X_i. Let $J_*(i) = \bigoplus_q J_q(i)$ be a simplicial chain complex of the set c(i): let $J_q(i)$ be an Abelian group with generators $\Delta^q(l_0, \ldots, l_q)$, where $l_0, \ldots, l_q$ is an arbitrary set of suffixes from c(i) and by possessing skew-symmetric correspondences with respect to permutations of the suffixes $l_0, \ldots, l_q$. The differential $\partial: J_q(i) \to J_{q-1}(i)$ is defined by the formula $\partial\Delta^q(l_0, \ldots, l_q) = \sum_{\mu=0}^{q} (-1)^\mu \Delta^{q-1}(l_0, \ldots \hat{l}_\mu \ldots, l_q)$.

If $X_{i_1} \subset X_{i_2}$, then $c(i_2) \subset c(i_1)$, and $J_*(i_2)$ is a subcomplex in $J_*(i_1)$. Thus, the sets c(i) and the complexes $J_*(i)$ form bundles over the partition $\{X_\alpha\}$. We note that if dim $X_i = n$, then $J_0(i) = Z$ is a cyclic group with generator $\Delta^0(i)$.

18. Admissible Simplicial Systems

We recall that the coorientation of the cell X_i^{n-p} (see Part 1, Para. 4) is the orientation of the simplicial complex S_i^{p-1}. The coefficient of identity $\varepsilon_{ij} = \pm 1$ is defined for each pair of coorientated cells

$$X_i^{n-p} \subset X_j^{n-p+1}.$$

We say that an admissible simplicial system $\{D_\alpha\}$ is given if each coorientated cell X_i^{n-p} of the partition $\{X_\alpha\}$ is put into correspondence with the chain $D_i \in J_p(i)$, and the following conditions are satisfied:

†Funktsional'. Analiz i Ego Prilozhen., 9, No. 2, 12-28 (1975).

Moscow State University. Saratov State University. Translated from Funktsional'nyi Analiz i Ego Prilozheniya, Vol. 9, No. 3, pp. 5-26, July-September, 1975. Original article submitted March 31, 1975.

1) D_i changes sign when the coorientation of the cell X_i^{n-p} changes;

2) if $p = 0$, then $D_i = \Delta^0(i)$;

3) if $p \geq 1$, then $\partial D_i = \sum_{j \in s(i)} \varepsilon_{ij} D_j$.

PROPOSITION 14. For any partition into simple cells there exists an admissible simplicial system.

This is easily proved by defining the chains D_i for the cells X_i^{n-p} successively for $p = 1, \ldots, n$ and using the triviality of the homologies of the complexes $J_*(i)$ in positive dimensions.

19. Compatibility of Orientations and Coorientations

Let C be an orientated k-dimensional cell in X, and let X_i^{n-p} ($p \leq k$) be a cell of the partition $\{X_\alpha\}$ such that $C(i) = C \cap X_i^{n-p} \neq \phi$ and the cells C and X_i^{n-p} are transversal in X. Let $x \in C(i)$. Then the tangent space $T_x(C)$ to the cell C at the point x may be expanded as the direct sum $T_x(C) = T^p \oplus T^{k-p}$, where T^{k-p} is the tangent space to C(i), and T^p is some complement to T^{k-p} in $T_x(C)$. The choice of the coorientation of the cell X_i^{n-p} determines the orientation in a fiber of the normal fibering to X_i^{n-p} in X,† and therefore also in T^p. We choose the orientation in the space $T^{k-p} = T_x(C(i))$ in such a way that the orientation of the direct sum $T^p \oplus T^{k-p} = T_x(C)$ coincides with the initial orientation of the cell C. We shall say that the orientation of the manifold C(i) thus defined is compatible with the given coorientation of the cell X_i^{n-p}. Similarly, for any k-dimensional chain $Z = \sum a_\lambda C_\lambda$, transversal to the cell X_i^{n-p} (this means that all cells C_λ are transversal to X_i^{n-p}) we define the (k − p)-dimensional chain $Z(i) = \sum a_\lambda C_\lambda(i)$, taking $C_\lambda(i)$ to be compatibly orientated with the coorientation of the cell X_i^{n-p}.

PROPOSITION 15. Let Z be a cycle in X, transversal to all cells of the partition $\{X_\alpha\}$. Then for every coorientation of the cell X_j^{n-p+1} of the partition $\{X_\alpha\}$, we have

$$\partial Z(j) = (-1)^p \sum_{i: j \in s(i)} \varepsilon_{ij} Z(i).$$

20. Piecewise-Smooth Connections and Characteristic Classes

Let $P(h_1, \ldots, h_r)$ be an invariant symmetric r-linear form on $\mathfrak{gl}(q, \mathbf{R})$. Further, let E be a fibering over X with fiber R^q. If $Y \subset X$, and $\omega_0, \ldots, \omega_k$ are connections in $E|_Y$ then we denote by $P_E^{k, 2r-k}(\omega_0, \ldots, \omega_k)$ the (2r − k)-form on Y which is a component of the difference cocycle constructed with respect to the form $P(h_1, \ldots, h_r)$ (see Part I, Para. 15). We note that $P_E^{0,2r}(\omega) = P(R, \ldots, R)$, where R is the curvature of the connection ω. Thus, the form $P_E^{0,2r}(\omega)$ is the characteristic class P_E, of the fibering E, corresponding to the form $P(h_1, \ldots, h_r)$.

Let us assume that over every cell of higher dimensionality X_l^n a smooth connection ω_l in the fibering $E|_{X_l^n}$ is given. Let X_i^{n-p} be a cell of the partition $\{X_\alpha\}$, and let $\Delta^k = \Delta^k(l_0, \ldots, l_k)$ be a simplex from $J_k(i)$. We put $P_E^{k, 2r-k}(\Delta^k) = P_E^{k, 2r-k}(\omega_{il_0}, \ldots, \omega_{il_k})$, where ω_{il_ν} is the bound of the connection ω_{l_ν} in the fibering E/X_i. If $D = \sum c_\lambda \Delta_\lambda^k \in J_k(i)$, then we put $P_E^{k, 2r-k}(D) = \sum c_\lambda P_E^{k, 2r-k}(\Delta_\lambda^k)$.

PROPOSITION 16. Let $\{D_\alpha\}$ be an admissible simplicial system, and let Z be a 2r-dimensional cycle in X, transversal to the partition $\{X_\alpha\}$. Then

$$\sum_{p=0}^{r} \sum_{i: \dim X_i = n-p} \int_{Z(i)} P_E^{p, 2r-p}(D_i) = (P_E, Z), \tag{1}$$

where (P_E, Z) is the value of the characteristic class P_E on the cycle Z.

Remark. In Proposition 16 the connections ω_l on different cells are given independently of each other. In the case when all these connections are the bounds of some smooth connection ω on X, then $P_E^{p, 2r-p}(D_i) = 0$ for $p > 0$ and Eq. (1) is transformed into the standard definition of the class P_E (see the beginning of the proof of Proposition 16).

†This follows from the fact that the geometrical realization of the complex S_i^{p-1} is a partition induced by the partition $\{X_\alpha\}$ on the boundary of a spherical neighborhood of the cell X_i^{n-p}, in the transversal section to this cell.

Proof. We denote the left-hand side of Eq. (1) by $A(\{\omega_l\})$. It is sufficient to show that $A(\{\omega_l\})$ is, in reality, independent of the choice of connections ω_l. In fact, we assume that all the connections ω_l are bounds on X_l of some smooth connection ω over X. Then for any cell X_i^{n-p} the connections ω_{il} coincide for different $l \in c(i)$. From the skew symmetry of the difference cocycle it follows that $P_E^{p,2r-p}(\omega_{il_0}, \ldots, \omega_{il_p}) = 0$ for any set $l_0, \ldots, l_p$ of suffixes from c(i) if p > 0. Consequently, $P_E^{p,2r-p}(D_i) = 0$, if p > 0. If, however, p = 0, then $P_E^{0,2r}(D_i) = P_E^{0,2r}(\omega)$. Consequently,

$$A(\{\omega_l\}) = \sum_{i:\dim X_i = n} \int_{Z(i)} P_E^{0,2r}(\omega) = \int_Z P_E^{0,2r}(\omega),$$

which by definition is equal to (P_E, Z).

Now let $\{\omega_l^0\}$ and $\{\omega_l^1\}$ be two sets of connections. We shall show that $A(\{\omega_l^1\}) = A(\{\omega_l^0\})$. We denote by $J_*'(i)$ the simplicial chain complex of the set $c(i) \times \{0, 1\}$. The natural embeddings $l \to (l, 0)$ and $l \to (l, 1)$ of c(i) in $c(i) \times \{0, 1\}$ define the embeddings of the complexes $\tau_\nu : J_*(i) \to J_*'(i)$ $(\nu = 0, 1)$. For every simplex $\Delta^k = \Delta^k((l_0, \nu_0), \ldots, (l_k, \nu_k)) \in J_k'(i)$ (here $l_\mu \in c(i)$, $\nu_\mu = 0, 1$) we put

$$P_E^{k,2r-k}(\Delta^k) = P_E^{k,2r-k}(\omega_{il_0}^{\nu_0}, \ldots, \omega_{il_k}^{\nu_k}).$$

If $D = \sum c_\lambda \Delta_\lambda^k \in J_k'(i)$, then we put $P_E^{k,2r-k}(D) = \sum c_\lambda P_E^{k,2r-k}(\Delta_\lambda^k)$. We note that from the correspondences between the forms $P_E^{k,2r-k}$ (Proposition 12), it follows that

$$P_E^{k-1,2r-k-1}(\partial D) = (-1)^{k-1} dP_E^{k,2r-k}(D).$$

Let $\{D_\alpha\}$ be an admissible simplicial system. Then it is possible to put each coorientated cell X_i^{n-p} into correspondence with the chain $D_i' \in J_{p+1}'(i)$, in such a way that the condition

$$\partial D_i' = \sum_{j \in s(i)} \varepsilon_{ij} D_j' + (-1)^p (\tau_1 D_i - \tau_0 D_i)$$

is satisfied. We shall show that for $k \geq 0$ the equality

$$\sum_{p=0}^{k} \sum_{i:\dim X_i = n-p} \int_{Z(i)} (P_E^{p,2r-p}(\tau_1 D_i) - P_E^{p,2r-p}(\tau_0 D_i)) = \sum_{j:\dim X_j = n-k} \int_{Z(j)} dP_E^{k+1,2r-k-1}(D_j'). \tag{2}$$

holds. For the proof, we use induction over k. For k = 0, the assertion follows from the relationships $\tau_1 D_l - \tau_0 D_l = \partial D_l'$, $P_E^{0,2r}(\partial D_l') = dP_E^{1,2r-1}(D_l')$ $(\dim X_l = n)$. Proceeding from k to k + 1, we transform the right-hand side of (2) as follows:

$$\sum_{j:\dim X_j = n-k} \int_{Z(j)} dP_E^{k+1,2r-k-1}(D_j') = \sum_{j:\dim X_j = n-k} \int_{\partial Z(j)} P_E^{k+1,2r-k-1}(D_j') =$$

$$= \sum_{j:\dim X_j = n-k} \sum_{i:j \in s(i)} (-1)^{k+1} \varepsilon_{ij} \int_{Z(i)} P_E^{k+1,2r-k-1}(D_j') = \sum_{i:\dim X_i = n-k-1} (-1)^{k+1} \int_{Z(i)} P_E^{k+1,2r-k-1}\Big(\sum_{j \in s(i)} \varepsilon_{ij} D_j'\Big) =$$

$$= \sum_{i:\dim X_i = n-k-1} (-1)^{k+1} \int_{Z(i)} P_E^{k+1,2r-k-1}(\partial D_i') - \sum_{i:\dim X_i = n-k-1} \int_{Z(i)} P_E^{k+1,2r-k-1}(\tau_1 D_i - \tau_0 D_i) =$$

$$= \sum_{i:\dim X_i = n-k-1} \int_{Z(i)} dP_E^{k+2,2r-k-2}(D_i') - \sum_{i:\dim X_i = n-k-1} \int_{Z(i)} (P_E^{k+1,2r-k-1}(\tau_1 D_i) - P_E^{k+1,2r-k-1}(\tau_0 D_i)).$$

Putting the right-hand sum into the left-hand side of Eq. (2), we obtain the assertion for k + 1.

Now let k = r. Then the left-hand side of Eq. (2) equals $A(\{\omega_l^1\}) - A(\{\omega_l^0\})$, and the right-hand side equals zero, since $P_E^{k',2r-k'} = 0$ for $k' > r$. Consequently, $A(\{\omega_l^0\}) = A(\{\omega_l^1\})$. QED.

21. Some Properties of the Difference Cocycle

We denote by Π_r the invariant symmetric r-linear form on $\mathfrak{gl}(q, \mathbb{R})$ defined by

$$\Pi_r(h_1, \ldots, h_r) = \frac{1}{r!} \sum \mathrm{tr}(h_{\nu_1} \cdot \ldots \cdot h_{\nu_r}),$$

where summation is carried out over all permutations $\nu_1, \ldots, \nu_r$ of the suffixes 1, . . ., r. It is well known that any invariant symmetric r-linear form P_q on $\mathfrak{gl}(q, \mathbb{R})$ [from now on we shall call them simply forms on $\mathfrak{gl}(q, \mathbb{R})$] can be written in the form $P_q = Q(\Pi_1, \ldots, \Pi_r)$, where Q is some polynomial of r variables with real coefficients. We shall call the set of such forms P_q ($q \geq 1$) the stable form corresponding to the polynomial Q, and, dropping the suffix q, we shall denote all these forms by the single letter P.

PROPOSITION 17. Let $R^{q'}$ be a subspace in R^q, and let h_ν ($\nu = 1, \ldots, r$) be elements of $\mathfrak{gl}(q, R)$, i.e., endomorphisms of R^q. Let us assume that $R^{q'}$ is invariant with respect to h_ν. We denote by h'_ν (h''_ν, respectively) the endomorphisms of $R^{q'}$ ($R^q/R^{q'}$) induced by the endomorphisms h_ν. Let P be a stable r-linear form. Then

$$P(h_1, \ldots, h_r) = P(h'_1, \ldots, h'_r) + P(h''_1, \ldots, h''_r).$$

Let E be a vector fibering with fiber R^q over the manifold Y, and let $\omega_0, \ldots, \omega_p$ be connections in E. Let E' be some subfibering in E. Let us assume that E' is invariant with respect to the connections ω_ν (i.e., a parallel translation with respect to these connections carries E' into E'). Then the connections ω_ν induce connections ω'_ν (ω''_ν) in the fibering E' (E/E'). Let P be a stable r-linear form, and let $P_E^{p,2r-p}$, $P_{E'}^{p,2r-p}$, $P_{E/E'}^{p,2r-p}$ be difference cocycles of the fiberings E, E', and E/E', respectively.

PROPOSITION 18. $P_E^{p,2r-p}(\omega_0, \ldots, \omega_p) = P_{E'}^{p,2r-p}(\omega'_0, \ldots, \omega'_p) + P_{E/E'}^{p,2r-p}(\omega''_0, \ldots, \omega''_p)$.*

Let $m_0, \ldots, m_q$ be a set of sections of the fibering E. We denote by ω_ν a connection in the fibering E defined by the trivialization $(m_0, \ldots, \hat{m}_\nu, \ldots, m_q)$.† Let $n_0, \ldots, n_{p+q-1}$ be a set of sections of the fibering E. We denote by υ_ν ($\nu = 0, \ldots, p$) a connection in the fibering E defined by the trivalization $(n_\nu, n_{p+1}, \ldots, n_{p+q-1})$.

PROPOSITION 19'. If $p \neq r$, then $P_E^{p,2r-p}(\omega_0, \ldots, \omega_p) = 0$ and $P_E^{p,2r-p}(\upsilon_0, \ldots, \upsilon_p) = 0$.

First of all, in order to formulate the assertion on the forms $P^{p,2r-p}$ in the mean dimension ($p = r$), we must introduce the forms $S_P^{0,r-1}$ and $S_P^{r-1,0}$. Let $P = \sum c_\lambda \Pi_1^{\lambda_1} \cdot \ldots \cdot \Pi_r^{\lambda_r}$, where $\lambda = (\lambda_1, \ldots, \lambda_r)$, $\lambda_\nu \geqslant 0$, $\sum \nu\lambda_\nu = r$. We put $d(\lambda) = \sum_\nu \lambda_\nu(\nu - 1)$. We define the $(r-1)$-forms $S_P^{0,r-1}$ and $S_P^{r-1,0}$ on Y as follows:

$$S_P^{0,r-1}(m_0, \ldots, m_r; m_{r+1}, \ldots, m_q) = \frac{1}{r}\Big(\sum_\lambda (-1)^{d(\lambda)} c_\lambda\Big) \times$$
$$\times \sum_{\nu=1}^{r} (-1)^{\nu-1} \log|a_\nu|\, d\log|a_1| \wedge \ldots \wedge \widehat{d\log|a_\nu|} \wedge \ldots \wedge d\log|a_r|,$$

where $m_0 = a_1 m_1 + \ldots + a_r m_r \pmod{m_{r+1}, \ldots, m_q}$;

$$S_P^{r-1,0}(n_0, \ldots, n_r; n_{r+1}, \ldots, n_{r+q-1}) = \frac{1}{r}\Big(\sum_\lambda c_\lambda\Big) \sum_{\nu=1}^{r} (-1)^{\nu-1} \log|b_\nu|\, d\log|b_1| \wedge \ldots \wedge \widehat{d\log|b_\nu|} \wedge \ldots \wedge d\log|b_r|,$$

where $n_0 = b_1 n_1 = \ldots = b_r n_r \pmod{n_{r+1}, \ldots, n_{r+q-1}}$.

Remark. In the case $r = 2$ and $P = \Pi_2$ the forms $S_P^{0,1}$ and $S_P^{1,0}$ coincide with the forms $S^{0,1}$ and $S^{1,0}$ defined in Part I, Para. 16. The forms $S_P^{0,r-1}$ are needed for the proof of Theorem 2, and the forms $S_P^{r-1,0}$ for the construction of Atiyah–Patoda–Singer functionals.

PROPOSITION 19". Let $p = r$. Then

$$P_E^{r,r}(\omega_0, \ldots, \omega_r) = (-1)^{\frac{r(r+1)}{2}} dS_P^{0,r-1}(m_0, \ldots, m_r; m_{r+1}, \ldots, m_q),$$
$$P_E^{r,r}(\upsilon_0, \ldots, \upsilon_r) = (-1)^{\frac{r(r-1)}{2}} dS_P^{r-1,0}(n_0, \ldots, n_r; n_{r+1}, \ldots, n_{r+q-1}).$$

Proof. We shall verify the assertions of Propositions 19' and 19" for connections ω_ν (the assertions for connections v_ν are obtained in exactly the same way).

We note that the subfibering E' in E, generated by the sections $m_{p+1}, \ldots, m_q$ is invariant with respect to the connections $\omega_0, \ldots, \omega_p$ and the bounds of all these connections on E' coincide. From Propo-

*As P. Deligne has noted, Propositions 17 and 18 are true only for forms Π_r. Proposition 18 is also true if all the connections ω'_ν coincide and have zero curvature. This is sufficient, however, for the proof of Theorems 1 and 2.

†It is clear that the connections ω_ν are defined only in the neighborhood of those points $y \leq Y$ at which the vectors $m_0(y), \ldots \widehat{m_\nu(y)} \ldots, m_q(y)$ form a basis in the fiber of the fibering E. An analogous remark would be made for the connections v_ν.

sition 18 it follows that $P_E^{p,2r-p}(\omega_0, \ldots, \omega_p) = P_E^{p,2r-p}(\omega_0'', \ldots, \omega_p'')$, where ω_ν'' is the connection in E/E' corresponding to the trivialization $(m_0, \ldots \hat{m}_\nu \ldots, m_p)_{\mathrm{mod}\, E'}$. Hence, replacing E by E/E', we may take q = p.

Let $\Delta^p(\omega_0, \ldots, \omega_p)$ be a p-dimensional simplex in a fiber of the fibering of the streams of connections in E defined by the mapping

$$\left(t = (t_0, \ldots, t_p): t_\nu > 0,\ \sum_{\nu=0}^{p} t_\nu = 1\right) \mapsto \left(\omega(t) = \sum_{\nu=0}^{p} t_\nu \omega_\nu\right).$$

It is easy to verify that the curvature and variation of the connection $\omega(t)$ are given by the expressions

$$R(t) = R_0 + \sum_{\nu=1}^{p} t_\nu (R_\nu - R_0) + \sum_{\nu=1}^{p} t_\nu (\omega_\nu - \omega_0) \wedge (\omega_\nu - \omega_0) - \sum_{\mu,\nu=1}^{p} t_\mu t_\nu (\omega_\mu - \omega_0) \wedge (\omega_\nu - \omega_0),$$

$$\delta\omega(t) = \sum_{\nu=1}^{p} dt_\nu \wedge (\omega_\nu - \omega_0).$$

(These expressions hold for any set of connections ω_ν. Since in our case ω_ν are connections of zero curvature, the terms containing R_ν in the expression for R(t) disappear.)

Let ω_ν be a connection corresponding to the trivialization $(m_0, \ldots \hat{m}_\nu \ldots, m_p)$, and let $m_0 = a_1 m_1 + \ldots + a_p m_p$. We fix the trivialization $(m_1, \ldots, m_p)$ of the fibering E. Then the connection ω_0 is given by the zero matrix and the connection ω_ν ($\nu > 0$), by the matrix $(\omega_\nu)_\alpha^\beta = 0$ if $\alpha \neq \nu$, $(\omega_\nu)_\nu^\beta = -da_\beta / a_\nu$. Then the form

$$P_E^{p,2r-p}(\omega_0, \ldots, \omega_p) = \int_{\Delta^p(\omega_0, \ldots, \omega_p)} P(R(t) + \delta\omega(t), \ldots, R(t) + \delta\omega(t))$$

(for the exact definition see Part I, Para. 15) is a polynomial in the forms da_β / a_ν. Since $P^{p,2r-p}$ is a form of degree 2r − p, and any polynomial in the forms da_β of power greater than r is equal to zero, the assertion of Proposition 19' follows.

Now let p = r. In order to avoid cumbersome notation, let us assume that $P = \Pi_r$. Then

$$P_E^{r,r}(\omega_0, \ldots, \omega_r) = \int_{\Delta^r(\omega_0, \ldots, \omega_r)} P(R(t) + \delta\omega(t), \ldots, R(t) + \delta\omega(t)) =$$

$$= \int_{\Delta^r(\omega_0, \ldots, \omega_r)} P(\delta\omega(t), \ldots, \delta\omega(t)) = (-1)^{\frac{r(r-1)}{2}} \sum_{\nu_1, \ldots, \nu_r} \operatorname{sgn}(\nu_1, \ldots, \nu_r) \times$$

$$\times \int_{\Delta^r(\omega_0, \ldots, \omega_r)} dt_1 \wedge \ldots \wedge dt_r \wedge \operatorname{tr}((\omega_{\nu_1} - \omega_0) \wedge \ldots \wedge (\omega_{\nu_r} - \omega_0)) =$$

$$= \frac{1}{r!}(-1)^{\frac{r(r-1)}{2}} \sum_{\nu_1, \ldots, \nu_r} \operatorname{sgn}(\nu_1, \ldots, \nu_r) \operatorname{tr}((\omega_{\nu_1} - \omega_0) \wedge \ldots \wedge (\omega_{\nu_r} - \omega_0)).$$

Here the summation is carried out over all permutations $\nu_1, \ldots, \nu_r$ of the suffixes 1, . . ., r. Substituting the expressions for $(\omega_\nu)_\alpha^\beta$, we obtain

$$P_E^{r,r}(\omega_0, \ldots, \omega_r) = \frac{1}{r!}(-1)^{\frac{r(r-1)}{2}} \sum_{\nu_1, \ldots, \nu_r} \operatorname{sgn}(\nu_1, \ldots, \nu_r) \left(-\frac{da_{\nu_2}}{a_{\nu_1}}\right) \wedge$$

$$\wedge \left(-\frac{da_{\nu_3}}{a_{\nu_2}}\right) \wedge \ldots \wedge \left(-\frac{da_{\nu_1}}{a_{\nu_r}}\right) = (-1)^{\frac{r(r+1)}{2}} \left(\frac{da_2}{a_1}\right) \wedge \ldots \wedge \left(\frac{da_1}{a_r}\right) =$$

$$= (-1)^{\frac{r(r+1)}{2} + (r-1)} d\log|a_1| \wedge \ldots \wedge d\log|a_r|,$$

whence follows the assertion of Proposition 19". (For the case $P = \Pi_1^{\lambda_1} \cdot \ldots \cdot \Pi_r^{\lambda_r}$ all the computations are carried out in exactly the same way.)

<u>PROPOSITION 20.</u> a) The form $S_P^{0,r-1}(m_0, \ldots, m_r; m_{r+1}, \ldots, m_q)$ is skew-symmetric with respect to the permutations $m_0, \ldots, m_r$; the form $S_P^{r-1,0}(n_0, \ldots, n_r; n_{r+1}, \ldots, n_{r+q-1})$ is skew-symmetric with respect to the permutations $n_0, \ldots, n_r$.

b) $$\sum_{\mu=0}^{r+1}(-1)^{\mu} S_P^{0,r-1}(m_0, \dots \hat{m}_{\mu} \dots, m_{r+1}; m_{r+2}, \dots, m_q, m_{\mu}) = 0;$$

$$\sum_{\mu=0}^{r+1}(-1)^{\mu} S_P^{r-1,0}(n_0, \dots \hat{n}_{\mu} \dots, n_{r+1}; n_{r+2}, \dots, n_{r+q}) = 0.$$

The proof is carried out by direct calculation.

22. Proof of Theorem 2

Let us recapitulate the formulation of Theorem 2 (Part I, Para. 11).

Let $\{X_\alpha\}$ be a linearizable complex of the smooth manifold X, and let every cell of codimensionality 2r of $\{X_\alpha\}$ be contained in exactly 2r + 1 cells of codimensionality 2r − 1 of this complex. Then the first r Pontryagin classes of X are trivial.

Since the Pontryagin forms are expressed in terms of forms Π_r, and for odd r the form Π_r corresponds to a trivial characteristic class in TX, to prove the theorem it is sufficient to verify that the characteristic classes of the fibering TX corresponding to the forms $\Pi_2, \dots, \Pi_{2r}$, are trivial. Since from the conditions of the theorem for some r follows the same condition for all r' < r, it is sufficient to prove the triviality of the class corresponding to the form Π_{2r}. Further, since the form Π_1^{2r} corresponds to a trivial characteristic class, it is sufficient to prove the triviality of the class P(X) corresponding to the form $P = \Pi_{2r} + (-1)^r \Pi_1^{2r}$.

Since $\{X_\alpha\}$ is linearizable, in every cell X_i^{n-p} it is possible to specify a smooth affine connection ω_i such that the following conditions are satisfied:

1) ω_i is a connection of zero curvature.

2) Let $X_i \subset X_k$ and ω_{ik} be the bound of the connection ω_k in $TX_k|X_i$. Then the subfibering TX_i in $TX_k|X_i$ is invariant with respect to the connection ω_{ik}.

3) The connection in TX_i induced by ω_{ik} coincides with ω_i.

For every pair of cells $X_i^{n-p} \subset X_j^{n-p+1}$ we choose a nonzero section m_{ij} of the fibering $(TX_j|X_i)/TX_i$, horizontal with respect to the connection in this fibering induced by ω_{ij}. Due to the embedding $TX_j|X_i \to TX|X_i$, these sections may be also considered as sections of the fibering $(TX|X_i)/TX_i$.

Let X_i^{n-p} be a cell of $\{X_\alpha\}$, $p \le 2r$, and let $X_{j_0(i)}, \dots, X_{j_p(i)}$ be cells of dimensionality n − p + 1 containing X_i^{n-p}. The complex S_i^{p-1} for a cell X_i is the boundary of the simplex $\Delta^p(j_0(i), \dots, j_p(i))$. Let us assume that the orientation of S_i^{p-1}, i.e., the coorientation of X_i^{n-p}, is defined by the order of the suffixes $j_0(i), \dots, j_p(i)$. From the definition of an admissible simplicial system it is not difficult to obtain the following assertion.

<u>PROPOSITION 21.</u> For any admissible simplicial system $\{D_\alpha\}$, we have $D_i = (-1)^p \Delta^p(l_0(i), \dots, l_p(i))$, where $l_\nu(i)$ is the number of the cell $X^n_{l_\nu(i)}$, containing $X_{j_0(i)}, \dots, X_{j_{\nu-1}(i)}, X_{j_{\nu+1}(i)}, \dots, X_{j_p(i)}$.

Let Z by a cycle of dimensionality 4r in X, transversal to $\{X_\alpha\}$. From Propositions 16 and 21 it follows that

$$(P(X), Z) = \sum_{p \le 2r} \sum_{i:\, \dim X_i = n-p} (-1)^p \int_{Z(i)} P_{TX}^{p,\,2r-p}(\omega_{il_0(i)}, \dots, \omega_{il_p(i)}).$$

The assertion of Theorem 2 is thus obtained from the following proposition.

<u>PROPOSITION 22.</u> Let X_i^{n-p} be a cell of $\{X_\alpha\}$, $p \le 2r$. Then $P_{TX}^{p,\,2r-p}(\omega_{il_0(i)}, \dots, \omega_{il_p(i)}) = 0$.

<u>Proof.</u> Since the connections $\omega'_{il_\nu(i)}$ induced by the connections $\omega_{il_\nu(i)}$ in TX_i coincide (and equal ω_i) it follows from Proposition 18 and the skew symmetry of the difference cocycle that

$$P_{TX}^{p,\,2r-p}(\omega_{il_0(i)}, \dots, \omega_{il_p(i)}) = P_{(TX|X_i)/TX_i}^{p,\,2r-p}(\omega''_{il_0(i)}, \dots, \omega''_{il_p(i)}),$$

where $\omega''_{il_\nu(i)}$ are the connections induced by the connections $\omega_{il_\nu(i)}$ in $(TX|X_i)/TX_i$. From the definition of the sections m_{ij} and cells $X^n_{l_\nu(i)}$ it follows that the connection $\omega''_{il_\nu(i)}$ corresponds to the trivialization $(m_{ij_0(i)}, \dots \hat{m}_{ij_\nu(i)} \dots, m_{ij_p(i)})$ of the fibering $(TX|_{X_i})/TX_i$.

Thus, the assertion of Proposition 22 follows from Proposition 19' in the case $p < 2r$ and from 19" in the case $p = 2r$ (since $S_p^{0,2r-1} = 0$).

Thus, the proof of Proposition 22 and, likewise, of Theorem 2 is completed.

23. Manifolds $\widetilde{\Sigma}(i)$

Let X_i^{n-p} be a cell of $\{X_\alpha\}$. By $\Sigma(i)$ (see Part 1, Para. 5) we denote the set of partitions σ of the space R^p by simplicial cones, with vertex at the origin, which induce on the unit sphere in R^p partitions which form a geometrical realization of the complex S_i^{p-1}. The one-dimensional edges of σ are numbered by suffixes $j \in s(i)$, and two such partitions are identical in $\Sigma(i)$ if there exists a linear mapping of R^p which transforms one of them into the other, conserving the numbering of the edges.

We denote by $\widetilde{\sigma} = (\sigma, e_j(\sigma))$ a partition σ in which on each one-dimensional edge l_j ($j \in s(i)$) a nonzero vector $e_j(\sigma)$ is additionally marked. Two such partitions with marked vectors are identical if there exists a linear transformation of R^p which transforms one of them into the other. The set of classes of partitions $\widetilde{\sigma}$ with respect to this identity is denoted by $\widetilde{\Sigma}(i)$. The structure of a smooth manifold on $\widetilde{\Sigma}(i)$ is denoted in the usual way.

The relationship $\widetilde{\sigma} = (\sigma, e_j(\sigma)) \to \sigma$ defines the natural projection $\widetilde{\Sigma}(i) \to \Sigma(i)$.

Let $j \in s(i)$, and let $\widetilde{\sigma} \in \widetilde{\Sigma}(i)$ be a partition with marked vectors. Let l_j be a one-dimensional edge of the partition $\widetilde{\sigma}$ corresponding to the suffix j. We consider the bound of the star of edges l_j in the partition $\widetilde{\sigma}$. Its image under the projection $\pi\colon R^p \to R^p/(l_j) \cong R^{p-1}$ defines a partition of R^{p-1} with marked vectors $e_k = \pi(e_{\varphi_{ij}(k)})$ ($k \in s(j)$) where $e_{j'}$ ($j' \in s(i)$) are the marked vectors of $\widetilde{\sigma}$. Thus, the mapping $\widetilde{\Sigma}(i) \to \widetilde{\Sigma}(j)$ is defined. This mapping is consistent with the mapping $\varphi'_{ij}\colon \Sigma(i) \to \Sigma(j)$ defined in Part I, Para. 5, and we shall denote it by the same letter φ'_{ij}. It is easy to verify that the mapping $\varphi'_{ij}\colon \widetilde{\Sigma}(i) \to \widetilde{\Sigma}(j)$ satisfies the conditions of Proposition 4, i.e., the manifolds $\widetilde{\Sigma}(i)$ form a cobundle over the complex $\{X_\alpha\}$.

We denote by $\widetilde{\Sigma}_\nu(i)$ ($\nu = 0, 1$) the inverse image of the set $\Sigma_\nu(i) \subset \Sigma(i)$ under the projection $\widetilde{\Sigma}(i) \to \Sigma(i)$. We recall that $\Sigma_0(i)$ consists of partitions in which any p one-dimensional edges generate R^p, and $\Sigma_1(i)$ consists of partitions in which there cannot exist more than one set of p one-dimensional edges contained in a $(p-1)$-dimensional subspace in R^p.

Let $E(i) \cong \widetilde{\Sigma}(i) \times R^p$ be a trivial fibering over $\widetilde{\Sigma}(i)$. We define the section m_j ($j \in s(i)$) of $E(i)$ as follows. Let τ be a point of the manifold $\widetilde{\Sigma}(i)$, i.e., the set of partitions $\widetilde{\sigma} = (\sigma, e_j(\sigma))$ with marked vectors which are equivalent with respect to the action of $GL(p, R)$. We choose from the set τ an element $\widetilde{\sigma}_\tau = (\sigma_\tau, e_j(\sigma_\tau))$ which is smoothly dependent on τ. [This may be done, e.g., as follows: we fix the $(p-1)$-simplex $(j_0, \ldots, j_{p-1})$ from S_i^{p-1} and choose $(\sigma_\tau, e_j(\sigma_\tau))$ such that the vectors $e_{j_0}(\sigma_\tau), \ldots, e_{j_p}(\sigma_\tau)$ form a standard basis in R^p.] We put $m_j(\tau) = e_j(\sigma_\tau)$. It is obvious that for different ways of choosing the elements $\widetilde{\sigma}_\tau$ the sets of sections $\{m_j\}_{j \in s(i)}$ are obtained, one from another, by an automorphism of the fibering $E(i)$ over $\widetilde{\Sigma}(i)$.

We need the following assertions on the topology of the manifolds $\Sigma(i)$.

(1) $\Sigma(i)$ is contractible if $\dim X_i = n-2$.

(2) $\Sigma(i)$ is connected and singly connected if $\dim X_i = n-3$.

(3) $\Sigma(i)$ is connected if $\dim X_i = n-4$.

Assertion (1) is trivial. Assertion (2) follows easily from the results of Cairns [Ann. Math., 45, 207-217 (1944)] and Chung-Wu Ho [Trans. Amer. Math. Soc., 181, 213-243 (1973)]. We do not know whether assertion (3) is true for an arbitrary partition. From now on we shall assume that the partition $\{X_\alpha\}$ is such that condition (3) is satisfied for it (see Condition A, Part I, Para. 5).

We note that the manifold $\Sigma(i)$ coincides with the manifold of semilinear homeomorphisms of a $(p-1)$-dimensional sphere with simplicial partition S_i^{p-1} which leave a fixed simplex. Such manifolds have frequently been studied [see, for example, R. Thom, "Des variétés triangulées aux variétés differentiables," Proc. Internat. Congr. Math., Edinburgh, 1958, Cambridge University Press (1960), pp. 248-255].

We also note that the projection $\widetilde{\Sigma}(i) \to \Sigma(i)$ is a homotopic equivalence, hence assertions (1)-(3) also hold for manifolds $\widetilde{\Sigma}(i)$.

24. The Function $\Phi(\varkappa)$

In this paragraph, we investigate the properties of the transcendental function $\Phi(\varkappa)$ defined as follows: $\Phi(\varkappa)$ is a smooth function in $R \setminus \{0, 1\}$,

$$\Phi'(\varkappa) = \log|\varkappa|/(\varkappa-1) - \log|\varkappa-1|/\varkappa, \quad \Phi(-1) = \Phi(1/2) = \Phi(2) = 0.$$

These properties will be used, in particular, to prove the correspondences between the forms $S^{k,l}$ introduced in Part I, Para. 16.

PROPOSITION 23. $\Phi(\varkappa) = -\Phi(1-\varkappa) = -\Phi(1/\varkappa)$.

Proof. It may be verified by direct computation that the form $d\Phi(\varkappa) = (\log|\varkappa|/(\varkappa-1) - \log|\varkappa - 1|/\varkappa)\,dx$ satisfies the relation $d\Phi(\varkappa) = -d\Phi(1-\varkappa)$. Hence, $\Phi(\varkappa) = -\Phi(1-\varkappa) + c(\varkappa)$, where $c(\varkappa)$ is a locally constant function in $R \setminus \{0, 1\}$. From the conditions $\Phi(-1) = \Phi(1/2) = \Phi(2) = 0$ it follows that $c(-1) = c(1/2) = c(2) = 0$. Consequently, $c(\varkappa) \equiv 0$ and $\Phi(\varkappa) = -\Phi(1-\varkappa)$. In exactly the same way, it is proved that $\Phi(\varkappa) = -\Phi(1/\varkappa)$.

COROLLARY. Let $e_0, \ldots, e_3$ be four pairwise-independent vectors R^2, $e_0 = ae_2 + be_3$, $e_1 = ce_2 + de_3$, and let $\varkappa = bc/ad$ be the cross-ratio of the quadruple $(e_0, \ldots, e_3)$. Then $\Phi(\varkappa)$ is skew-symmetric with respect to permutations of the vectors $e_0, \ldots, e_3$.

Proof. It is easy to verify that $\varkappa$ is transformed into $1/\varkappa$ under the permutation $e_0 \leftrightarrow e_1$ and under the permutation $e_2 \leftrightarrow e_3$, and that $\varkappa$ is transformed into $1-\varkappa$ under the permutation $e_1 \leftrightarrow e_2$. Hence, the assertion follows from Proposition 23.

PROPOSITION 24. The limit of $\Phi(\varkappa)$ as $\varkappa \nearrow 0$, $\varkappa \nearrow 1$, $\varkappa \to +\infty$ equals $\pi^2/6$, and the limit as $\varkappa \searrow 0$, $\varkappa \searrow 1$, $\varkappa \to -\infty$ equals $-\pi^2/6$.

Proof. For $|\varkappa| < 1$, we have $\frac{\log(1-\varkappa)}{\varkappa} = -\sum_{k=1}^{\infty} \frac{\varkappa^{k-1}}{k}$. Hence,

$$\int_0^s \frac{\log|1-\varkappa|}{\varkappa}\,d\varkappa = -\sum_{k=1}^{\infty} \frac{s^k}{k^2}, \quad \int_0^1 \frac{\log(1-\varkappa)}{\varkappa}\,d\varkappa = -\sum_{k=1}^{\infty} \frac{1}{k^2} = -\frac{\pi^2}{6}.$$

Consequently,

$$\int_0^1 ((\log|\varkappa|/(\varkappa-1)) - (\log|\varkappa-1|/\varkappa))\,d\varkappa = \pi^2/3.$$

From which, using Proposition 23, it is easy to deduce the assertion of Proposition 24.

PROPOSITION 25. a) Let $e_0, \ldots, e_4$ be five pairwise-independent vectors in R^2, and let $\varkappa_\nu$ ($\nu = 0, \ldots, 4$) be the cross-ratio of the quadruple $(e_0, \ldots \hat{e}_\nu \ldots, e_4)$. Then $\sum_{\nu=0}^{4} (-1)^\nu \Phi(\varkappa_\nu) = -\varepsilon\pi^2/6$ where $\varepsilon = 1$, if among the 10 bases (e_α, e_β) $(\alpha < \beta)$ in R^2 an even number have identical orientation; otherwise, $\varepsilon = -1$.

b) Let $e_0, \ldots, e_4$ be five vectors in R^3, three of which are independent, and let $\varkappa_\nu$ ($\nu = 0, \ldots, 4$) be the cross-ratio of the quadruple $(e_0, \ldots \hat{e}_\nu \ldots, e_4)$ in $R^3/(e_\nu)$. Then $\sum_{\nu=0}^{4} (-1)^\nu \Phi(\varkappa_\nu) = \varepsilon\pi^2/6$, where $\varepsilon = 1$ if among the 10 bases $(e_\alpha, e_\beta, e_\gamma)$ $(\alpha < \beta < \gamma)$ in R^3 an even number have identical orientation; $\varepsilon = -1$, otherwise.

Proof. Let us prove assertion a. It is easy to verify that in the equation $\sum_{\nu=0}^{4} (-1)^\nu \Phi(\varkappa_\nu) = -\varepsilon\pi^2/6$ the left- and right-hand sides are skew-symmetric with respect to permutations of the vectors $e_0, \ldots, e_4$, and do not change under linear transformations nor when any of the vectors is multiplied by a nonzero constant. Hence, it is sufficient to confine ourselves to the case $e_0 = (0, 1)$, $e_1 = (1, 0)$, $e_2 = (1, 1)$, $e_3 = (1, u)$, $e_4 = (1, v)$ $(0 < u < v < 1)$. Further, by direct computation it is possible to verify that $\sum_{\nu=0}^{4} (-1)^\nu d\Phi(\varkappa_\nu) \equiv 0$ as a form on the space of sets of five general-position vectors. (This follows also from the relationship on $dS^{1,1}$ in Proposition 13c.) Hence, $\sum_{\nu=0}^{4} (-1)^\nu \Phi(\varkappa_\nu)$ is a constant function of (u, v) $(0 < u < v < 1)$. In order to determine its value, let $u \to 0$, $v \to 1$. Then $\varkappa_0 \to +\infty$, $\varkappa_1 \searrow 0$, $\varkappa_2 \searrow 0$, $\varkappa_3 \searrow 1$, $\varkappa_4 \to +\infty$, consequently, $\sum_{\nu=0}^{4} (-1)^\nu \Phi(\varkappa_\nu) \to \pi^2/6$ which coincides with the assertion of the proposition. Assertion b is proved in exactly the same way.

We recall that $\Phi(\varkappa)$ was defined in Part I, Para. 16 as one of the forms $S^{k,l}$, namely, $\Phi(\varkappa) = S^{1,1}(m_0, \ldots, m_3; m_4, \ldots, m_{q+1})$, where $\varkappa$ is the cross-ratio of the images of the quadruple $m_0, \ldots, m_3$ (m_ν are sections of the q-dimensional vector fibering E) in the factor $E/(m_4, \ldots, m_{q+1})$. In Proposition 13, relationships were formulated between the forms $S^{k,l}$. We note that assertion b of Proposition 13 is a special case of Proposition 19". The verification of assertions a and c presents no difficulties, except in the following cases: assertion a) for $(k, l) = (1, 1)$ (this assertion was proved in the corollary of Proposition 23), and assertion c) for $k + l = 3$ (this assertion was proved in Proposition 25).

25. The Forms $S_i(b)$

We denote by $\Omega^k = \Omega^k(\tilde{\Sigma}_0(i))$ $(k > 0)$ the space of smooth differential k-forms on $\tilde{\Sigma}_0(i)$. In particular, the 0-forms are smooth functions on $\tilde{\Sigma}_0(i)$. We also define the (−1)-forms on $\tilde{\Sigma}_0(i)$ as sets $\{c_\lambda\}$ of constants from R one for each connected component of $\tilde{\Sigma}_0(i)$. We denote by $\Omega^{-1} = \Omega^{-1}(\tilde{\Sigma}_0(i))$ the space of (−1)-forms on $\tilde{\Sigma}_0(i)$ and define the differential $d\colon \Omega^{-1} \to \Omega^0$, putting into correspondence with the set of numbers $\{c_\lambda\} \in \Omega^{-1}$ the locally constant function on $\tilde{\Sigma}_0(i)$ which takes the value c_λ on the λ-th connected component.

Let $I_*(i)$ be a hypersimplicial complex of cells X_i^{n-p} (see Part 1, Paras. 7 and 8). We recall that the group $I_q(i)$ $(q > 0)$ is generated by generators of form $\Delta^{k,l} = \Delta^{k,l}(j_0, \ldots, j_{k+l+1}; j_{k+l+2}, \ldots, j_{k+p})$ where $j_\nu \in s(i)$, $k + l = q - 1$. If $k + l = 1, 2, 3$, we define the form $S_i(\Delta^{k,l}) \in \Omega^{2-k-l}(\tilde{\Sigma}_0(i))$ using the expression

$$S_i(\Delta^{k,l}) = S^{k,l}(m_{j_0}, \ldots, m_{j_{k+l+1}}; m_{j_{k+l+2}}, \ldots, m_{j_{k+p}}),$$

where m_j are the sections of the fibering E(i) defined in Para. 23. Thus, for any generator of the group $I_q(i)$ (q = 2, 3, 4), and hence for any chain $b \in I_q(i)$, the form $S_i(b) \in \Omega^{3-q}(\tilde{\Sigma}_0(i))$ is defined.

From the properties of the forms $S^{k,l}$ (Proposition 13), it is easy to obtain the following assertion.

PROPOSITION 26. If $b \in I_q(i)$ (q = 3, 4), then $S_i(\partial b) = (-1)^q dS_i(b)$.

Further, from the definition of the mappings $\varphi'_{ij}\colon \tilde{\Sigma}(i) \to \tilde{\Sigma}(j)$ and $\tilde{\varphi}_{ij}\colon I_*(j) \to I_*(i)$, it is easy to obtain the following assertion.

PROPOSITION 27. If $j \in s(i)$ and $b \in I_q(j)$ (q = 2, 3, 4), then $S_i(\tilde{\varphi}_{ij}b) = (\varphi'_{ij})^* S_j(b)$.

26. Singularities of the Forms $S_i(b)$.

Let M be a manifold, the points of which are ordered quadruples of vectors in R^2, no three of which are contained in a one-dimensional subspace. Let M_0 be a submanifold in M, formed by quadruples of pairwise-independent vectors. If $e = (e_0, \ldots, e_3) \in M_0$, $e_0 = ae_2 + be_3$, $e_1 = ce_2 + de_3$, then we put $\varkappa(e) = bc/ad$.

PROPOSITION 28. The mapping $\varkappa\colon M_0 \to R$ is continued to a smooth mapping $M \to RP^1$.

The proof is left to the reader.

Let $b \in I_3(i)$. Then the function $S_i(b)$, smooth on $\tilde{\Sigma}_0(i)$, may have discontinuities at points of $\tilde{\Sigma}(i) \setminus \tilde{\Sigma}_0(i)$. However, the following assertion holds.

PROPOSITION 29. Let $\tau_0 \in \tilde{\Sigma}_1(i)$. Then in some neighborhood U of the point τ_0, $dS_i(b)$ takes the form $d(F|_{\tilde{\Sigma}_0(i)})$, where F is a continuous function in U, smooth in $U \cap \tilde{\Sigma}_0(i)$. If $\dim X_i = n-2$, then the same is true for any point $\tau_0 \in \tilde{\Sigma}(i)$.

Remark. If $\gamma(t)$ $(t \in [0, 1])$ is a curve in U, then we put $\int_{\gamma(t)} dS_i(b) = F(\gamma(1)) - F(\gamma(0))$. Obviously, this definition is independent of the choice of F. It hence follows, by standard means, that for any one-dimensional chain γ in $\tilde{\Sigma}_1(i)$ the integral $\int_\gamma dS_i(b)$ is defined. Obviously, the cochain $\gamma \mapsto \int_\gamma dS_i(b)$ is a cocycle.

Proof of Proposition 29. It is sufficient to verify the assertion for $b = \Delta^{1,1}(j_0, \ldots, j_3;\ j_4, \ldots, j_{p+1})$ ($\dim X_i = n - p$). But in this case, by definition, $S_i(b)(\tau) = \Phi(\varkappa)$, where $\varkappa = \varkappa(e_0(\tau), \ldots, e_3(\tau))$, $e_\nu(\tau)$ are the images of the vectors $m_{j_\nu}(\tau)$ in $R^p/(m_{j_4}(\tau), \ldots, m_{j_{p+1}}(\tau)) \cong R^2$. Since $\tau_0 \in \tilde{\Sigma}_1(i)$, then $(e_0, \ldots, e_3) \in M$. The same is true if $p = 2$ and τ_0 is any point from $\tilde{\Sigma}(i)$. The assertion now follows from Proposition 28, since $d\Phi(\varkappa)$ is a form on RP^1 which is locally the differential of a continuous function.

PROPOSITION 30. Let $\dim X_i = n-2$, $b_i \in I_2(i)$ be an element of an admissible hypersimplicial system (see Part I, Para. 9), and let $\tau_0 \in \tilde{\Sigma}(i)$. Then in some neighborhood U of τ_0, $S_i(b)$ is of the form $\Omega|_{\tilde{\Sigma}(i)} + d(F_{\tilde{\Sigma}(i)})$, where Ω is a smooth 1-form in U, and F is a continuous function in U and smooth in $U \cap \tilde{\Sigma}_0(i)$.

Remark. Unlike Proposition 29, this proposition is not true for any $b \in I_2(i)$, but only for elements b_i of an admissible hypersimplicial system. This is connected with the fact that ∂b_i has a special form (see the proof of the proposition).

Proof of Proposition 30. Let $\tau_0 \in \tilde{\Sigma}(i)$. Then to the sections m_j ($j \in s(i)$) of fibering E(i) it is possible to add a new section $m_{\hat{j}}$ such that it is independent of every m_j in some neighborhood V of τ_0. Let $I_*^{\hat{}}(i) = I_*(s(i) \cup \{\hat{j}\}, 2)$ be a hypersimplicial complex of the set $s(i) \cup \{\hat{j}\}$ (see Para. 7). Then for any chain $b \in I_q^{\hat{}}(i)$ ($q = 2, 3$) it is possible to define the form $S_i(b) \in \Omega^{3-q}(\tilde{\Sigma}_0(i) \cap V)$ exactly as for chains from $I_q(i)$.

Let $s(i) = \{j_0, \ldots, j_N\}$, where the elements j_ν are ordered in such a way that $(j_0, j_1), \ldots, (j_N, j_0)$ are 1-simplices in S_i^1. Let the coorientation of the cell X_i^{n-2}, i.e., the orientation of S_i^1, be chosen such that $S_i^1 = (j_0, j_1) + \cdots + (j_N, j_0)$. Let $b_i \in I_2(i)$ be an element of an admissible hypersimplicial system. It is not difficult to show that

$$\partial b_i = \Delta^{0,0}(j_0, j_2;\ j_1) + \Delta^{0,0}(j_1, j_3;\ j_2) + \cdots + \Delta^{0,0}(j_N, j_1;\ j_0).$$

We define the chain $b^{\hat{}} \in I_2^{\hat{}}(i)$ using the equation

$$\begin{aligned} b^{\hat{}} = {} & \Delta^{1,0}(j_0, j_2, \hat{j};\ j_1) + \Delta^{1,0}(j_1, j_3, \hat{j};\ j_2) + \cdots + \Delta^{1,0}(j_N, j_1, \hat{j};\ j_0) + \\ & + \Delta^{0,1}(j_0, j_1, \hat{j}) + \Delta^{0,1}(j_1, j_2, \hat{j}) + \cdots + \Delta^{0,1}(j_N, j_0, \hat{j}) + \\ & + \Delta^{1,0}(j_0, j_1, j_2;\ \hat{j}) + \Delta^{1,0}(j_0, j_2, j_3;\ \hat{j}) + \cdots + \Delta^{1,0}(j_0, j_{N-1}, j_N;\ \hat{j}). \end{aligned}$$

It is easy to verify that $\partial b^{\hat{}} = \partial b_i$. Since $H_2(I_*^{\hat{}}(i)) = 0$ (Proposition 7), there exists a chain $b' \in I_3^{\hat{}}(i)$, such that $b^{\hat{}} - b_i = \partial b'$. We have $S_i(b_i) = S_i(b^{\hat{}}) + dS_i(b')$. From the fact that the pairs of sections $(m_{j_0}, m_{j_1}), \ldots, (m_{j_N}, m_{j_0})$ define bases in the fiber E(i) over every point $\tau \in \tilde{\Sigma}(i)$, and $m_{\hat{j}}$ is transversal to all m_j ($j \in s(i)$) in V, it follows that $S_i(b^{\hat{}})$ is a smooth form in V. Further, exactly as was done in Proposition 29 for chains from $I_3(i)$, it may be shown that $dS_i(b') = dF$, where F is a continuous function in some neighborhood $U \subset V$ of τ_0, and is smooth in $U \cap \tilde{\Sigma}_0(i)$. The proposition is proved.

Let $\dim X_i = n-3$. If $b \in I_3(i)$, then the class of cohomologies of the spaces $\tilde{\Sigma}_1(i)$, corresponding to the cocycle $dS_i(b)$, is, generally speaking, nontrivial. However, the following assertion holds.

PROPOSITION 31. If $b_i \in I_3(i)$ is an element of an admissible hypersimplicial system, then the cocycle $dS_i(b_i)$ defines a trivial class of cohomologies of the space $\tilde{\Sigma}_1(i)$.

Proof. Since the manifold $\tilde{\Sigma}(i)$ is singly connected (Para. 23, assertion 2), and $\tilde{\Sigma}(i) \setminus \tilde{\Sigma}_1(i)$ is a set of codimensionality two in $\tilde{\Sigma}(i)$, it is sufficient to show that for every point $\tau_0 \in \tilde{\Sigma}(i) \setminus \tilde{\Sigma}_1(i)$ and arbitrarily small neighborhood γ in $\tilde{\Sigma}_1(i)$ with center at τ_0 we have $\int_\gamma dS_i(b_i) = 0$. Since b_i is an element of the admissible hypersimplicial system $\{b_\alpha\}$, then $\partial b_i = \sum_{j \in s(i)} \varepsilon_{ij} \tilde{\varphi}_{ij} b_j$, where b_j are elements of $\{b_\alpha\}$, corresponding to cells X_j^{n-2}. Consequently,

$$dS_i(b_i) = -S_i(\partial b_i) = -\sum_{j \in s(i)} \varepsilon_{ij} S_i(\tilde{\varphi}_{ij} b_j) = -\sum_{j \in s(i)} \varepsilon_{ij} (\varphi_{ij})^* S_j(b_j),$$

$$\int_\gamma dS_i(b_i) = -\sum_{j \in s(i)} \varepsilon_{ij} \int_{\varphi'_{ij}(\gamma)} S_j(b_j).$$

If γ is extended to τ_0, then from Proposition 30 it follows that the integrals on the right-hand side tend to zero. Since the left-hand side remains unchanged, $\int_\gamma dS_i(b_i) = 0$. QED.

27. Proof of Proposition 10

In Part I, Para. 10, the holomorphisms $\theta_i^0: I_4(i) \to H^0(\Sigma_0(i); Q)$ and $\theta_j^1: I_3(j) \to H^1(\Sigma_1(j), \Sigma_0(j); Q)$ were defined for the cells X_i^{n-4} and X_j^{n-3}, respectively. Further, the $(n-4)$-dimensional chain $\Gamma_{n-4}(X) = \sum_{i:\dim X_i = n-4} \Gamma(i) X_i \in \hat{C}_{n-4}(X) \otimes Q$ was constructed. We recall that the numbers $\Gamma(i) \in Q$ were defined as follows. For every cell X_i^{n-4} let the configuration $\sigma_i \in \Sigma_0(i)$ and for every cell X_j^{n-3} the configuration $\sigma_j \in \Sigma_0(j)$ be fixed. Further, if $j \in s(i)$, then the continuous curve $\sigma_{ij}(t) \subset \Sigma_1(j)$, $\sigma_{ij}(0) = \sigma_j$, $\sigma_{ij}(1) = \varphi_{ij}(\sigma_i)$ is given. Finally, let $\{b_\alpha\}$ be an admissible hypersimplicial system. Then $\Gamma(i) = (\theta_i^0 b_i)(\sigma_i) + \sum_{j\in s(i)} \varepsilon_{ij} (\theta_j^1 b_j)(\sigma_{ij}(t))$ [here σ_i is considered as a zero-dimensional cycle in $\Sigma_0(i)$, and $\sigma_{ij}(t)$ is considered as a one-dimensional relative cycle of the pair $(\Sigma_1(j), \Sigma_0(j))$].

Proposition 10 asserts the following:

The chain $\Gamma_{n-4}(X)$ is a cycle in $\hat{C}_{n-4}(X) \otimes Q$ and its class of homologies is independent of the choice of admissible hypersimplicial system $\{b_\alpha\}$, configurations σ_i and σ_j, and curves $\sigma_{ij}(t)$.

The proof of this assertion will be given in Propositions 34-38 of this paragraph.

PROPOSITION 32. a) If $\dim X_i = n-4$ and $b \in I_4(i)$, then the locally constant function $(1/8\pi^2) dS_i(b)$ defines on $\Sigma_0(i)$ a cocycle whose class of homologies coincides with $\theta_i^0(b)$.

b) If $\dim X_j = n-3$, $b \in I_3(j)$ and γ is the relative cycle of the pair $(\Sigma_1(j), \Sigma_0(j))$, then

$$\frac{1}{8\pi^2}\left[\int_\gamma dS_j(b) - S_j(b)(\partial\gamma)\right] = (\theta_j^1 b)(\gamma) \tag{3}$$

[here the function $S_j(b)$ is considered as a 0-cochain on $\Sigma_0(j)$].

Proof. Assertion a is not difficult to obtain by comparing the definition of the homomorphism θ_i^0 and the forms $S^{k,l}$ $(k + l = 3)$. To prove assertion b we note that if $\gamma = \gamma(t)$ is a curve in $\Sigma_1(j)$ with ends in $\Sigma_0(j)$, then the left-hand side of Eq. (3) consists of the sum of the jumps of the function $S_j(b)$ on the curve $\gamma(t)$, multiplied by $-1/8\pi^2$. The assertion is easily obtained by comparing the definition of the homomorphism θ_j^1 and the form $S^{1,1}$ and using Proposition 24.

PROPOSITION 33.

$$\Gamma(i) = \frac{1}{8\pi^2} \sum_{j\in s(i)} \varepsilon_{ij} \left[\int_{\sigma_{ij}(t)} dS_j(b_j) + (S_j(b_j))(\sigma_j)\right]. \tag{4}$$

Proof. We have

$$\theta_i^0(b_i) = \frac{1}{8\pi^2} dS_i(b_i) = \frac{1}{8\pi^2} S_i(\partial b_i) = \frac{1}{8\pi^2} \sum_{j\in s(i)} \varepsilon_{ij} S_i(\tilde{\varphi}_{ij} b_j) = \frac{1}{8\pi^2} \sum_{j\in s(i)} \varepsilon_{ij} (\varphi_{ij}')^* S_j(b_j).$$

Hence,

$$\Gamma(i) = (\theta_i^0 b_i)(\sigma_i) + \sum_{j\in s(i)} \varepsilon_{ij} (\theta_j^1 b_j)(\sigma_{ij}(t)) = \frac{1}{8\pi^2}\Bigg[\sum_{j\in s(i)} \varepsilon_{ij} ((\varphi_{ij}')^* S_j(b_j))(\sigma_i) +$$
$$+ \sum_{j\in s(i)} \varepsilon_{ij} \left[\int_{\sigma_{ij}(t)} dS_j(b_j) - S_j(b_j)(\sigma_{ij}(1)) + S_j(b_j)(\sigma_{ij}(0))\right]\Bigg] =$$
$$= \frac{1}{8\pi^2} \sum_{j\in s(i)} \varepsilon_{ij} \left[S_j(b_j)(\varphi_{ij}'(\sigma_i)) + \int_{\sigma_{ij}(t)} dS_j(b_j) - S_j(b_j)(\varphi_{ij}'(\sigma_i)) + S_j(b_j)(\sigma_j)\right] = \frac{1}{8\pi^2} \sum_{j\in s(i)} \varepsilon_{ij} \left[\int_{\sigma_{ij}(t)} dS_j(b_j) + S_j(b_j)(\sigma_j)\right].$$

PROPOSITION 34. The chain $\Gamma_{n-4}(X)$ is closed.

Proof. We have

$$\partial\Gamma_{n-4}(X) = \sum_{j:\dim X_j = n-4} \sum_{i: j\in s(i)} \varepsilon_{ij} \Gamma(j) X_i = \sum_{i:\dim X_i = n-5} \left(\sum_{j\in s(i)} \varepsilon_{ij}\Gamma(j)\right) X_i.$$

Hence, it is sufficient to show that $\sum_{j\in s(i)} \varepsilon_{ij}\Gamma(j) = 0$ for any cell X_i^{n-5}. From Proposition 33 it follows that

$$\sum_{j\in s(i)} \varepsilon_{ij}\Gamma(j) = \frac{1}{8\pi^2}\Big[\sum_{j\in s(i)}\sum_{k\in s(j)} \varepsilon_{ij}\varepsilon_{jk}S_k(b_k)(\sigma_k)\Big] + \frac{1}{8\pi^2}\Big[\sum_{j\in s(i)}\sum_{k\in s(j)} \varepsilon_{ij}\varepsilon_{jk}\int\limits_{\sigma_{jk}(t)} dS_k(b_k)\Big].$$

Since for every k: $X_k^{n-3} \supset X_i^{n-5}$ there exist exactly two suffixes j_k and j_k' such that $j_k \in s(i)$, $j_k' \in s(i)$, $k \in s(j_k) \cap s(j_k')$ and $\varepsilon_{ij_k}\varepsilon_{j_k k} = -\varepsilon_{ij_k'}\varepsilon_{j_k' k}$, the expression in the first set of square brackets equals zero.

We shall show that the expression in the second set of square brackets also equals zero. For this, we choose some configuration $\sigma_i \in \Sigma_0(X_i)$, and for every $j \in s(i)$ we choose a curve $\sigma_{ij}(t) \subset \Sigma_1(X_j)$ such that $\sigma_{ij}(0) = \sigma_j$ and $\sigma_{ij}(1) = \varphi_{ij}'(\sigma_i)$. (The existence of such a curve follows from Para. 23, assertion 3.) We have

$$\sum_{j\in s(i)}\sum_{k\in s(j)} \varepsilon_{ij}\varepsilon_{jk} \int\limits_{\varphi_{jk}'(\sigma_{ij}(t))} dS_k(b_k) = \sum_{j\in s(i)} \int\limits_{\sigma_{ij}(t)} dS_j\Big(\sum_{k\in s(j)} \varepsilon_{jk}\widetilde{\varphi}_{jk}b_k\Big) = \sum_{j\in s(i)} \varepsilon_{ij} \int\limits_{\sigma_{ij}(t)} dS_j(\partial b_j) = 0$$

as a result of Proposition 26. Hence, it is sufficient to show that the expression

$$\sum_{j\in s(i)}\sum_{k\in s(j)} \varepsilon_{ij}\varepsilon_{jk} \int\limits_{\sigma_{jk}(t)+\varphi_{jk}'(\sigma_{ij}(t))} dS_k(b_k) = \sum_{k:\,X_k^{n-3}\supset X_i^{n-5}} \int\limits_{\gamma_k} dS_k(b_k)$$

equals zero, where $\gamma_k = \varepsilon_{ij_k}\varepsilon_{j_k k}[\sigma_{j_k k}(t) + \varphi_{j_k k}'(\sigma_{ij_k}(t))] + \varepsilon_{ij_k'}\varepsilon_{j_k' k}[\sigma_{j_k' k}(t) + \varphi_{j_k' k}'(\sigma_{ij_k'}(t))]$. The assertion now follows from Proposition 31, since γ_k is a cycle in $\Sigma_1(k)$.

PROPOSITION 35. The cycle $\Gamma_{n-4}(X)$ is independent of the choice a) of the elements b_i, if $\dim X_i \le n-4$; b) of the curves $\sigma_{ij}(t)$; c) of the configuration σ_i, if $\dim X_i = n-4$.

Proof. We make use of Eq. (4) for $\Gamma(i)$ from Proposition 33. This expression does not contain b_i, from which follows assertion a. Further, if we replace $\sigma_{ij}(t)$ by $\sigma_{ij}'(t)$ (with the same ends), Eq. (4) is replaced by $\frac{1}{8\pi^2}\varepsilon_{ij} \int\limits_{\sigma_{ij}'(t)-\sigma_{ij}(t)} dS_j(b_j)$. This expression equals zero as a result of Proposition 31, since $\sigma_{ij}'(t) - \sigma_{ij}(t)$ is a cycle in $\Sigma_1(j)$. This proves assertion b.

To prove assertion c we replace σ_i by σ_i' and choose the curves $\sigma_{ij}'(t)$ as follows. Let $\gamma_i(t) \subset \Sigma_1(i)$, $\gamma_i(0) = \sigma_i$, $\gamma_i(1) = \sigma_i'$ (such a curve exists as a consequence of Para. 23, assertion 3). We put $\sigma_{ij}'(t) = \sigma_{ij}(t) + \varphi_{ij}'(\gamma_i(t))$. If we replace σ_i by σ_i' and $\sigma_{ij}(t)$ by $\sigma_{ij}'(t)$, then Eq. (4) is replaced by

$$\frac{1}{8\pi^2}\sum_{j\in s(i)} \int\limits_{\varphi_{ij}'(\gamma_i(t))} dS_j(b_j) = \frac{1}{8\pi^2}\int\limits_{\gamma_i(t)} dS_i\Big(\sum_{j\in s(i)} \varepsilon_{ij}\widetilde{\varphi}_{ij}b_j\Big) = \frac{1}{8\pi^2}\int\limits_{\gamma_i(t)} dS_i(\partial b_i) = 0$$

by Proposition 26. Hence, Proposition 35 is proved.

PROPOSITION 36. The class of homologies of the cycle $\Gamma_{n-4}(X)$ is independent of the choice of configurations σ_j for X_j^{n-3}.

Proof. We replace σ_j by σ_j' and choose the curves $\sigma_{ij}'(t)$ as follows. Let $\gamma_j(t) \subset \Sigma_1(j)$, $\gamma_j(0) = \sigma_j$, $\gamma_j(1) = \sigma_j'$ (the existence of such a curve follows from Para. 23, assertion 2). We put $\sigma_{ij}'(t) = \sigma_{ij}(t) - \gamma_j(t)$. If we replace σ_j by σ_j' and $\sigma_{ij}(t)$ by $\sigma_{ij}'(t)$, then $\Gamma_{n-4}(X)$ is replaced by

$$\sum_{i:\,j\in s(i)} \varepsilon_{ij}(\theta_j^1 b_j)(\sigma_{ij}'(t) - \sigma_{ij}(t))X_i = -\sum_{i:\,j\in s(i)} \varepsilon_{ij}(\theta_j^1 b_j)(\gamma_j(t))X_i = -(\theta_j^1 b_j)(\gamma_j(t))\,\partial X_j.$$

PROPOSITION 37. The class of homologies of the cycle $\Gamma_{n-4}(X)$ is independent of the choice of b_j for X_j^{n-3}.

Proof. Let X_{j_0} be a cell of dimensionality $n-3$. We replace b_{j_0} by b_{j_0}' and choose for i ($j_0 \in s(i)$), elements b_i' as follows. Let $c_{j_0} \in I_4(j_0)$, $\partial c_{j_0} = b_{j_0}' - b_{j_0}$. Such a chain exists, since $\partial b_{j_0}' = \partial b_{j_0} = \sum_{k\in s(j_0)} \varepsilon_{j_0 k}\widetilde{\varphi}_{j_0 k}b_k$ and $H_3(I_*(j_0)) = 0$. We put $b_i' = b_i + \varepsilon_{ij_0}\widetilde{\varphi}_{ij_0}c_{j_0}$.

Obviously, $\partial b_i' = \sum_{j\in s(i),\, j\ne j_0} \varepsilon_{ij}\widetilde{\varphi}_{ij}b_j + \varepsilon_{ij_0}\widetilde{\varphi}_{ij_0} b_{j_0}'$ and for any cell X_h^{n-5}, $X_h^{n-5} \subset X_{j_0}^{n-3}$, we have

$$\partial b_h = \sum_{i:\, i\in s(h),\, j_0\notin s(i)} \varepsilon_{hi}\tilde{\varphi}_{hi} b_i + \sum_{i:\, i\in s(h),\, j_0\in s(i)} \varepsilon_{hi}\tilde{\varphi}_{hi} b_i'.$$

Hence, replacing b_{j_0} by b_{j_0}' and b_i by b_i' ($j_0 \in s(i)$), we once again obtain an admissible hypersimplicial system. The cycle $\Gamma_{n-4}(X)$ is replaced by the following expression:

$$\sum_{i:\, j_0\in s(i)} [(\theta_i^0(\varepsilon_{ij_0}\tilde{\varphi}_{ij_0}c_{j_0}))(\sigma_i) + \varepsilon_{ij_0}(\theta_{j_0}^1(\partial c_{j_0}))(\sigma_{ij_0}(t))]\, X_i =$$
$$= \frac{1}{8\pi^2} \sum_{i:\, j_0\in s(i)} \varepsilon_{ij_0}\Big[dS_{j_0}(c_{j_0})(\varphi_{ij_0}'\sigma_i) - S_{j_0}(\partial c_{j_0})(\varphi_{ij_0}'\sigma_i) + S_{j_0}(\partial c_{j_0})(\sigma_{j_0}) +$$
$$+ \int\limits_{\sigma_{ij_0}(t)} dS_{j_0}(\partial c_{j_0})\Big] X_i = \sum_{i:\, j_0\in s(i)} \varepsilon_{ij_0}\Big[\frac{1}{8\pi^2} dS_{j_0}(c_{j_0})(\sigma_{j_0})\Big] X_i = \Big(\frac{1}{8\pi^2} dS_{j_0}(c_{j_0})(\sigma_{j_0})\Big)\, \partial X_{j_0}.$$

(Here we use the expressions for θ_i^0 and θ_j^1 from Proposition 32 and, also, Proposition 26.) Obviously, $(1/8\pi^2)dS_{j_0}(c_{j_0})(\sigma_{j_0})$ is a rational number. The proposition is proved.

PROPOSITION 38. The class of homologies of the cycle $\Gamma_{n-4}(X)$ is independent of the choice of b_k for X_k^{n-2}.

Proof. Let X_{k_0} be a cell of dimensionality $n-2$. We replace b_{k_0} by b_{k_0}' and choose for j ($k_0 \in s(j)$) the elements $b_j' \in I_3(j)$ as follows. Let $c_{k_0} \in I_3(k_0)$ be a chain such that $\partial c_{k_0} = b_{k_0}' - b_{k_0}$. We put $b_j' = b_j + \varepsilon_{jk_0}\cdot \tilde{\varphi}_{jk_0} c_{k_0}$. It is easy to verify that by replacing in the admissible hypersimplicial system $\{b_\alpha\}$ the element b_{k_0} by b_{k_0}' and b_i ($k_0 \in s(j)$) by b_j', we once again obtain an admissible hypersimplicial system. The cycle $\Gamma_{n-4}(X)$ is replaced by the following expression:

$$\sum_{j:\, k_0\in s(j)} \sum_{i:\, j\in s(i)} \varepsilon_{ij}\varepsilon_{jk_0}\theta_j^1(\tilde{\varphi}_{jk_0}c_{k_0})(\sigma_{ij}(t))\, X_i = \frac{1}{8\pi^2} \sum_{j:\, k_0\in s(j)} \sum_{i:\, j\in s(i)} \varepsilon_{ij}\varepsilon_{jk_0} \times$$
$$\times \Big[\int\limits_{\sigma_{ij}(t)} dS_j(\tilde{\varphi}_{jk_0}c_{k_0}) - S_j(\tilde{\varphi}_{jk_0}c_{k_0})(\varphi_{ij}'\sigma_i) + S_j(\tilde{\varphi}_{jk_0}c_{k_0})(\sigma_j)\Big] X_i =$$
$$= \frac{1}{8\pi^2} \sum_{j:\, k_0\in s(j)} \sum_{i:\, j\in s(i)} \varepsilon_{ij}\varepsilon_{jk_0}\Big[\int\limits_{\varphi_{jk_0}'\sigma_{ij}(t)} dS_{k_0}(c_{k_0}) - S_{k_0}(c_{k_0})(\varphi_{jk_0}'\varphi_{ij}'\sigma_i) + S_{k_0}(c_{k_0})(\varphi_{jk_0}'\sigma_j)\Big] X_i. \tag{5}$$

We note that for every i: $X_i^{n-4} \subset X_{k_0}^{n-2}$ there exist exactly two suffixes j_i and j_i' such that $j_i \in s(i)$, $j_i' \in s(i)$, $k_0 \in s(j_i) \cap s(j_i')$. Now $\varepsilon_{ij_i}\varepsilon_{j_ik_0} = -\varepsilon_{ij_i'}\varepsilon_{j_i'k_0}$. Since $\varphi_{j_ik_0}'\varphi_{ij_i}' = \varphi_{j_i'k_0}'\varphi_{ij_i'}'$, the second term in the square brackets in Eq. (5) cancels out during summation, and the first term gives, under summation,

$$\frac{1}{8\pi^2} \sum_{i:\, X_i^{n-4}\subset X_{k_0}^{n-2}} \Big[\int\limits_{\gamma_{ik_0}'} dS_{k_0}(c_{k_0})\Big] X_i, \tag{6}$$

where $\gamma_{ik_0}' = \varepsilon_{ij_i}\varepsilon_{j_ik_0}\varphi_{j_ik_0}'\sigma_{ij_i}(t) + \varepsilon_{ij_i'}\varepsilon_{j_i'k_0}\varphi_{j_i'k_0}'\sigma_{ij_i'}(t)$. We choose the configuration $\sigma_{k_0} \in \Sigma_0(k_0)$ and for every j ($k_0 \in s(j)$) we choose the curve $\sigma_{jk_0}(t) \subset \Sigma_1(k_0)$ such that $\sigma_{jk_0}(0) = \sigma_{k_0}$, $\sigma_{jk_0}(1) = \varphi_{jk_0}'\sigma_j$ (such a curve exists by virtue of Para. 23, assertion 1). We put $\gamma_{ik_0}'' = \varepsilon_{ij_i}\varepsilon_{j_ik_0}\sigma_{j_ik_0}(t) + \varepsilon_{ij_i'}\varepsilon_{j_i'k_0}\sigma_{j_i'k_0}(t)$. Then $\gamma_{ik_0} = \gamma_{ik_0}' + \gamma_{ik_0}''$ is a cycle in $\Sigma_1(k_0)$. From Proposition 29 and the contractibility of $\Sigma(k_0)$ it follows that $\int\limits_{\gamma_{ik_0}} dS_{k_0}(c_{k_0}) = 0$. Then Eq. (6) may be rewritten in the form

$$\frac{1}{8\pi^2} \sum_{i:\, X_i^{n-4}\subset X_{k_0}^{n-2}} \Big[\int\limits_{-\gamma_{ik_0}''} S_{k_0}(c_{k_0})\Big] X_i = -\frac{1}{8\pi^2} \sum_{j:\, k_0\in s(j)} \sum_{i:\, j\in s(i)} \varepsilon_{ij}\varepsilon_{jk_0}\Big[S_{k_0}(c_{k_0})(\sigma_{k_0}) + \int\limits_{\sigma_{jk_0}(t)} dS_{k_0}(c_{k_0})\Big] X_i \tag{7}$$

[the term $S_{k_0}(c_{k_0})(\sigma_{k_0})$ which is added in the latter expression cancels out under summation]. Substituting Eq. (7) for the first term in (5), we obtain for the increment of the cycle $\Gamma_{n-4}(X)$ the expression

$$-\frac{1}{8\pi^2} \sum_{j:\, k_0\in s(j)} \varepsilon_{jk_0} \sum_{i:\, j\in s(i)} \varepsilon_{ij}\Big[\int\limits_{\sigma_{jk_0}(t)} dS_{k_0}(c_{k_0}) - S_{k_0}(c_{k_0})(\varphi_{jk_0}'\sigma_j) + S_{k_0}(c_{k_0})(\sigma_{k_0})\Big] X_i = -\sum_{j:\, k_0\in s(j)} \varepsilon_{jk_0}(\theta_{k_0}^1 c_{k_0})(\sigma_{jk_0}(t))\, \partial X_j,$$

where $(\theta_{k_0}^1 c)(\gamma) = \frac{1}{8\pi^2}\Big[\int\limits_{\gamma} dS_{k_0}(c) - S_{k_0}(c)(\partial\gamma)\Big]$. From Proposition 24 it is easy to deduce that $(\theta_{k_0}^1 c)(\gamma)$ is a rational number. Hence, the proposition is proved.

28. Proof of Theorem 1

In this point we shall denote by $P = P(h_1, h_2)$ the form $\Pi_2 = \mathrm{tr}(h_1 \cdot h_2)$. As is well known, this form corresponds to the characteristic class $P_{TX} = -8\pi^2 P_1(X)$, where $P_1(X)$ is the first Pontryagin class of the manifold X.

Theorem 1 (see Part I, Para. 11) states the following:

If $\{X_\alpha\}$ is a linear partition of a smooth n-dimensional manifold X into simple cells, then $P_1(X)$ coincides with a class dual in the Poincaré sense to $\Gamma_{n-4}(X)$.

We assume that in each cell X_i there is given a smooth affine connection ω_i so that conditions 1-3 of Para. 22 are satisfied.

Let $\{D_\alpha\}$ be an admissible simplicial system, and let Z be a four-dimensional cycle in X, transversal to $\{X_\alpha\}$.

<u>PROPOSITION 39.</u> The equality

$$-8\pi^2 (P_1(X), Z) = \sum_{i:\dim X_i = n-2} \int_{Z(i)} P_{TX}^{2,2}(D_i).$$

holds.

<u>Proof.</u> By virtue of Proposition 16 it is sufficient to show that $P_{TX}^{p,4-p}(D_i) = 0$ if $\dim X_i = n-p$, $p < 2$. For $p = 0$, this follows from the fact that ω_i is a connection of zero curvature. Let $p = 1$, $s(i) = \{j, j'\}$, and let the orientation of S_i^0 be defined by the order of j, j'. Then $D_i = \Delta^1(j, j')$ and $P_{TX}^{1,3}(D_i) = P_{TX}^{1,3}(\omega_{ij}, \omega_{ij'})$ (here, as before, ω_{ij} is the bound of ω_j in $TX|X_i$). Since the connections induced by the connections ω_{ij} and $\omega_{ij'}$ in TX_i coincide, it follows from Proposition 18 and the skew symmetry of the difference cocycle that $P_{TX}^{1,3}(\omega_{ij}, \omega_{ij'}) = P_{(TX|X_i)/TX_i}^{1,3}(\omega, \omega')$, where ω and ω' are the connections induced in $(TX|X_i)/TX_i$ by ω_{ij} and $\omega_{ij'}$, respectively. Since ω and ω' are connections of zero curvature in a one-dimensional fibering, it is easy to deduce from Proposition 19' that $P_{(TX|X_i)/TX_i}^{1,3}(\omega, \omega') = 0$. The proposition is proved.

We choose for $j \in s(i)$ the nonzero section m_{ij} of the fibering $(TX_j|X_i)/TX_i$, horizontal with respect to the connection induced by ω_j in $(TX_j|X_i)/TX_i$. We shall assume that at every point $x \in X_i$ the vector $m_{ij}(x)$ is directed within the cell X_j. Using the embedding $TX_j|X_i \to TX|X_i$, we shall consider m_{ij} also as a section of the fibering $E_i = (TX|X_i)/TX_i$. From the compatibility condition on ω_i it is not difficult to deduce the following assertion.

<u>PROPOSITION 40.</u> Let $j \in s(i)$, and let η_{ij}: $E_i \to E_j|X_i$ be a natural mapping of the fibering (factorization with respect to $TX_j|X_i$). Then for every $k \in s(j)$

$$\eta_{ij} m_{i\varphi_{ij}(k)} = c_{ijk} m_{jk},$$

where c_{ijk} is a constant function on X_i.

With respect to the set of sections m_{ij} we define the mappings f_i: $X_i^{n-p} \to \Sigma(i)$ and $\tilde{f}_i$: $X_i^{n-p} \to \tilde{\Sigma}(i)$ as follows. Let $x \in X_i^{n-p}$. For each k-simplex $(j_0, \ldots, j_k)$ from S_i^{p-1} we define in $(E_i)_x \cong R^p$ a simplicial cone of dimensionality k + 1, the one-dimensional edges of which are rays passing through $m_{ij_0}(x), \ldots, m_{ij_k}(x)$. It is easy to verify that the partition σ_x thus defined is an S_i^{p-1}-configuration. The correspondence $x \to \sigma_x$ defines the mapping f_i: $X_i^{n-p} \to \Sigma(i)$. The vectors $m_{ij}(x)$ lie on one-dimensional edges of the partition σ_x. Putting the point x into correspondence with the partition σ_x having the set of marked vectors $e_j = m_{ij}(x)$, we define the mapping $\tilde{f}_i$: $X_i^{n-p} \to \tilde{\Sigma}(i)$.

We wish to replace the sections m_{ij} by some other sections of the fibering E_i such that the new set of sections will also define the mappings f_i and $\tilde{f}_i$ and that the following conditions will be satisfied.

(T_1) The sets $f_i^{-1}(\Sigma(i) \setminus \Sigma_\nu(i))$ ($\nu = 0, 1$) have codimensionality $\nu + 1$ in X_i; the sets $f_i^{-1}(\Sigma(i) \setminus \Sigma_\nu(i)) \cap Z(i)$ have codimensionality $\nu + 1$ in Z(i).

(T_2) If $j \in s(i)$, then $\varphi'_{ij} \tilde{f}_i = \tilde{f}_j|_{X_i}$.

<u>PROPOSITION 41'.</u> Let m_{ij} be an arbitrary set of sections of the fiberings E_i, satisfying the conditions of compatibility from Proposition 40. Then there exists a set of sections m'_{ij} such that $m'_{ij} = a_{ij} \cdot (x) m_{ij}$, where $a_{ij}(x)$ is a function on X_i and $\eta_{ij} m'_{i\varphi_{ij}(k)} = m'_{jk}$ for $k \in s(j)$.

PROPOSITION 41″. Let m'_{ij} be the set of sections from Proposition 41′. Then by means of a small deformation of the sections m'_{ij}, preserving the conditions $\eta_{ij}m'_{i\alpha_{ij}(k)} = m'_{jk}$, it is possible to satisfy the following condition: the set of points $x \in X_i^{n-p}$, at which there exists a set of p vectors $m'_{ij}(x)$, not generating $(E_i)_x$, has codimensionality unity in X_i^{n-p}, and the set of points at which there exists more than one such set has codimensionality two; the intersections of these sets with Z(i) have in Z(i) codimensionalities one and two, respectively.

The proofs of Propositions 41′ and 41″ present no difficulties. The construction of the sections m'_{ij}, and then the deformation of these sections should begin with the zero-dimensional skeleton $\{X_\alpha\}$, and is then continued successively to skeletons of higher dimensionality.

Applying Propositions 41′ and 41″ to the set of sections m_{ij} constructed with respect to connections ω_i, we obtain a set of sections satisfying conditions (T_1) and (T_2) (the mappings f_i and $\tilde{f}_i$ are constructed with respect to this set in the same way as for the set m_{ij}).

We wish to show that for this new set of sections an analog of Proposition 39 holds. For this we require several assertions.

Let $\bar{m}_i = \{m_{ij}\}_{j \in s(i)}$ be an arbitrary set of sections of the fibering E_i, satisfying the condition: if $(j_0, \ldots, j_{p-1})$ form a $(p-1)$-simplex in S_i^{p-1}, then the vectors $m_{ij_0}(x), \ldots, m_{ij_{p-1}}(x)$ are independent at every point $x \in X_i^{n-p}$. Then for every $l\colon X_n^l \supset X_i^{n-p}$ the connection $\omega_l(\bar{m}_i)$ is defined in the fibering E_i. This connection is uniquely defined by the following condition: if $X_i^{n-p} \subset X_j^{n-p+1} \subset X_l^n$, then m_{ij} is a horizontal section with respect to the connection $\omega_l(\bar{m}_i)$. If $\Delta^q = \Delta^q(l_0, \ldots, l_q)$ is a generator in $J_q(i)$, then we put $P_{E_i}^{q,4-q}(\Delta^q, \bar{m}_i) = P_{E_i}^{q,4-q}(\omega_{l_0}(\bar{m}_i), \ldots, \omega_{l_q}(\bar{m}_i))$. If $D = \sum c_\lambda \Delta_\lambda^q$ is a generator in $J_q(i)$, then we put $P_{E_i}^{q,4-q}(D, \bar{m}_i) = \sum c_\lambda P_{E_i}^{q,4-q} \cdot (\Delta_\lambda^q, \bar{m}_i)$.

The following assertion holds.

PROPOSITION 42. If $\{D_\alpha\}$ is an admissible simplicial system, Z is a four-dimensional cycle transversal to $\{X_\alpha\}$, and the sets $\bar{m}_i = \{m_{ij}\}_{j \in s(i)}$ satisfy the conditions of compatibility from Proposition 40, then the expression

$$\sum_{i:\dim X_i = n-2} \int_{Z(i)} P_{E_i}^{2,2}(D_i, \bar{m}_i)$$

is independent of the choice of sections m_{ij}.

The proof of this proposition is analogous to that of Proposition 16 and will not be cited here.

PROPOSITION 43. For any set of sections satisfying the conditions of compatibility from Proposition 40, we have

$$-8\pi^2 (P_1(X), Z) = \sum_{i:\dim X_i = n-2} \int_{Z(i)} P_{E_i}^{2,2}(D_i, \bar{m}_i).$$

Proof. In fact, it is easy to verify that for the set m_{ij} constructed with respect to connections ω_i, we have $P_{E_i}^{2,2}(D_i, \bar{m}_i) = P_{TX}^{2,2}(D_i)$. Hence, the assertion follows from Propositions 39 and 42.

From now on we shall denote by m_{ij} the set of sections satisfying conditions (T_1) and (T_2), the existence of which is proved in Propositions 41′ and 41″, and shall denote by f_i and $\tilde{f}_i$ mappings constructed with respect to this set of sections.

Let $\{b_\alpha\}$ be an admissible hypersimplicial system for $\{X_\alpha\}$.

PROPOSITION 44. For every cell X_i^{n-2}, we have $P_{E_i}^{2,2}(D_i, \bar{m}_i) = -f_i^*(dS_i(b_i))$.†

Proof. We denote by $s^2(i)$ the set of all unordered pairs (j, j') of elements from $s(i)$, $j \neq j'$. Let $\hat{J}_*(i)$ be a simplicial chain complex of the set $s^2(i)$. For each cell X_l^n containing X_i^{n-2} there exist exactly two cells $X_{j_l}^{n-1}$ and $X_{j'_l}^{n-1}$ contained in X_l^n and containing X_i^{n-2}. The correspondence $l \mapsto (j_l, j'_l)$ defines the mapping of the sets $c(i) \to s^2(i)$, and, consequently, the homomorphism of the complexes $\iota'\colon J_*(i) \to \hat{J}_*(i)$.

†This equation holds on the set $f_i^{-1}(\Sigma_0(i))$, since the right-hand side is meaningful only on this set. By condition (T_1) the set $f_i^{-1}(\Sigma_0(i))$ is nonempty. From now on we shall not indicate on what set this or that expression is meaningful.

For $q = 1, 2$, we define the homomorphisms $\iota'': I_q(i) \to \hat{J}_q(i)$ as follows:

$$\Delta^{0,0}(j_0, j_1; j_2) \mapsto \Delta^1((j_0, j_2), (j_1, j_2)); \quad \Delta^{0,1}(j_0, j_1; j_2) \mapsto \Delta^2((j_0, j_1), (j_1, j_2), (j_0, j_2)), \quad \Delta^{1,0}(j_0, j_1, j_2; j_3) \mapsto \Delta^2((j_0, j_3), (j_1, j_3), (j_2, j_3)).$$

The homomorphisms ι'' are permutable with the differentials in $I_*(i)$ and $\hat{J}_*(i)$. For any 2-simplex $\Delta^2 = \Delta^2((j_0, j_0'), (j_1, j_1'), (j_2, j_2')) \in \hat{J}_2(i)$ we put $P^{2,2}(\Delta^2) = P^{2,2}_{E_i}(\omega(j_0, j_0'), \omega(j_1, j_1'), \omega(j_2, j_2'))$ where $\omega(j_\nu, j_\nu')$ is a connection in E_i, corresponding to the trivialization $(m_{ij_\nu}, m_{ij_\nu'})$. If $D = \sum c_\lambda \Delta^2_\lambda$ is an arbitrary chain from $\hat{J}_2(i)$, then we put $P^{2,2}(D) = \sum c_\lambda P^{2,2}(\Delta^2_\lambda)$. It is easy to verify that $P^{2,2}(\iota' D) = P^{2,2}_{E_i}(D, \mathfrak{m}_i)$ for $D \in J_2(i)$. Further, from Proposition 13b, and the definition of the mappings $\tilde{f}_i$, it follows that $P^{2,2}(\iota'' b) = -f_i^*(dS_i(b))$ for $b \in I_2(i)$. If $\{D_\alpha\}$ and $\{b_\alpha\}$ are an admissible simplicial and hypersimplicial system, respectively, then it is easy to verify that $\iota' D_j = \iota'' b_j$ for any cell X_j^{n-1}. Hence, for X_i^{n-2} we have $\partial(\iota' D_i - \iota'' b_i) = 0$ and there exists a chain $c_i \in \hat{J}_3(i)$ such that $\partial c_i = \iota' D_i - \iota'' b_i$. From the properties of the difference cocycle (Proposition 12), it is not difficult to deduce that $P^{2,2}(\partial c) = 0$ for any chain $c \in \hat{J}_3(i)$. Hence, $P^{2,2}(\iota' D_i) = P^{2,2}(\iota'' b_i)$, i.e., $P^{2,2}_{E_i}(D_i, \mathfrak{m}_i) = -f_i^*(dS_i(b_i))$. QED.

<u>PROPOSITION 45.</u> The following equality holds:

$$-8\pi^2(P_1(X), Z) = \sum_{i:\dim X_i = n-3} \sum_{j \in s(i)} \varepsilon_{ij} \int_{Z(i)} f_j^* S_j(b_j).$$

<u>Proof.</u> From Propositions 43 and 44, it follows that

$$-8\pi^2(P_1(X), Z) = -\sum_{j:\dim X_j = n-2} \int_{Z(j)} f_j^* dS_j(b_j).$$

Using Propositions 30 and 15, we obtain

$$-\sum_{j:\dim X_j = n-2} \int_{Z(j)} f_j^* dS_j(b_j) = -\sum_{j:\dim X_j = n-2} \int_{\partial Z(j)} f_j^* S_j(b_j) = \sum_{i:\dim X_i = n-3} \sum_{j \in s(i)} \varepsilon_{ij} \int_{Z(i)} f_j^* S_j(b_j).$$

QED.

<u>PROPOSITION 46.</u>

$$8\pi^2(P_1(X), Z) = \sum_{i:\dim X_i = n-3} \int_{Z(i)} f_i^* dS_i(b_i). \tag{8}$$

<u>Proof.</u> Since the sections m_{ij} satisfy condition (T_2),

$$f_j^* S_j(b_j)|_{X_i} = f_i^*((\varphi_{ij}')^* S_j(b_j)) = f_i^*(S_i(\tilde{\varphi}_{ij} b_j)).$$

Hence,

$$8\pi^2(P_1(X), Z) = \sum_{i:\dim X_i = n-3} \sum_{j \in s(i)} \varepsilon_{ij} \int_{Z(i)} f_j^* S_j(b_j) - \sum_{i:\dim X_i = n-3} \int_{Z(i)} f_i^* S_i\Big(\sum_{j \in s(i)} \varepsilon_{ij} \tilde{\varphi}_{ij} b_j\Big) =$$
$$= -\sum_{i:\dim X_i = n-3} \int_{Z(i)} f_i^* S_i(\partial b_i) = \sum_{i:\dim X_i = n-3} \int_{Z(i)} f_i^* dS_i(b_i).$$

QED.

<u>Remark.</u> Since the form $S_i(b_i)$ on $\tilde{\Sigma}_0(i)$ is a raising of the form from $\Sigma_0(i)$, by considering $S_i(b_i)$ as a form on $\Sigma_0(i)$, we can replace $f_i^* dS_i(b_i)$ by $f_i dS_i(b_i)$ in Eq. (8).

Let us now assume, for simplicity, that the cycle Z intersects every cell X_i^{n-4} in not more than one point $z_i \in f_i^{-1}(\Sigma_0(i))$ (this proposition is not a restriction, since any cycle may be deformed into a cycle for which this condition is satisfied). For every cell X_j^{n-3} we choose a point $z_j \in X_j^{n-3}$ $(z_j \in f_j^{-1}(\Sigma_0(j)))$ and join z_j to each of the points z_i $(j \in s(i))$ of the curve $z_{ij}(t) \subset f_j^{-1}(\Sigma_1(j))$ $(z_{ij}(0) = z_j, z_{ij}(1) = z_i)$. This may be done by virtue of condition (T_1).

We put $Z'(j) = \sum_{i: j \in s(i)} \varepsilon_{ij}(X_i \cdot Z) z_{ij}(t)$. Here $(X_i \cdot Z)$ denotes the index of the intersection of the cooriented cell X_i with the cycle Z, i.e., the coefficient with which the point z_i occurs in the zero-dimensional chain $Z(i)$.

PROPOSITION 47. $\partial Z'(j) = \partial Z(j)$.

Proof. Since Z is a cycle, $\sum_{i:j\in s(i)} \varepsilon_{ij}(X_i \cdot Z) = (\partial X_j \cdot Z) = 0$. Hence, $\partial Z'(j) = \sum_{i:j\in s(i)} \varepsilon_{ij}(X_i \cdot Z) z_i = \partial Z(j)$.

PROPOSITION 48. $(P_1(X), Z) = (\Gamma_{n-4}(X) \cdot Z)$.

Proof. Since $\partial Z(j) = \partial Z'(j)$, from Proposition 31 it follows that

$$\int_{Z(j)} f_j^* \, dS_j(b_j) = \int_{Z'(j)} f_j^* \, dS_j(b_j).$$

Further, from Proposition 32b it follows that

$$\int_{Z'(j)} f_j^* \, dS_j(b_j) = (f_j^* S_j(b_j))(\partial Z'(j)) + 8\pi^2 (\theta_j^1 b_j)(f_j Z'(j)) = \sum_{i:j\in s(i)} \varepsilon_{ij}(X_i \cdot Z)\,[S_j(b_j)(f_j z_i) + 8\pi^2(\theta_j^1 b_j)(f_j z_{ij}(t))].$$

From Proposition 46, we obtain

$$\begin{aligned}(P_1(X), Z) &= \frac{1}{8\pi^2} \sum_{j:\dim X_j=n-3} \int_{Z(j)} f_j^* \, dS_j(b_j) = \frac{1}{8\pi^2} \sum_{j:\dim X_j=n-3} \int_{Z'(j)} f_j^* \, dS_j(b_j) = \\ &= \sum_{i:\dim X_i=n-4} (X_i \cdot Z)\Big[\frac{1}{8\pi^2} \sum_{j\in s(i)} \varepsilon_{ij} S_i(\tilde{\varphi}_{ij} b_j)(f_i z_i) + \sum_{j\in s(i)} \varepsilon_{ij}(\theta_j^1 b_j)(f_j z_{ij}(t))\Big] = \\ &= \sum_{i:\dim X_i=n-4} (X_i \cdot Z)\Big[\frac{1}{8\pi^2} S_i(\partial b_i)(f_i z_i) + \sum_{j\in s(i)} \varepsilon_{ij}(\theta_j^1 b_j)(f_j z_{ij}(t))\Big] = \\ &= \sum_{i:\dim X_i=n-4} (X_i \cdot Z)\Big[(\theta_i^0 b_i)(f_i z_i) + \sum_{j\in s(i)} \varepsilon_{ij}(\theta_j^1 b_j)(f_j z_{ij}(t))\Big] = (\Gamma_{n-4}(X) \cdot Z),\end{aligned}$$

where $\Gamma_{n-4}(X)$ is a cycle constructed with respect to the system $\{b_\alpha\}$, the configurations $\sigma_i = f_i(z_i)$, $\sigma_j = f_j(z_j)$, and the curves

This concludes the proof of Theorem 1.

18.

(with A. M. Gabrielov and M. V. Losik)

A local combinatorial formula for the first class of Pontryagin

Funkts. Anal. Prilozh. **10** (1) (1976) 14–17 [Funct. Anal. Appl. **10** (1976) 12–15].
MR 53:14504b. Zbl. 328:57006

In our paper [1] an obvious combinatorial formula was obtained for the first class of Pontryagin. From this formula simple reasoning yields a local combinatorial formula. In view of the interest in the question of the existence of local formulas we present in this paper the derivation, although the local formula we obtain thereby proves to be more complicated than our original formula.

Let X be a combinatorial n-dimensional manifold, i.e., X is a simplicial complex, and for each of its vertices p there exists a homeomorphism of the star of this vertex onto a neighborhood of the origin in R^n, which takes p into the origin and is linear on each simplex.

We shall further assume that X satisfies the following condition:

(A) For each $(n-4)$-simplex $X_i \subset X$ the space $\Sigma(i)$ is connected. [Definition of the spaces $\Sigma(i)$ will follow below.]

THEOREM 1. Let X be a combinatorial manifold satisfying condition (A). Then there exists a construction which enables one to associate a rational number $\Gamma(i)$ to each cooriented* $(n-4)$-simplex X_i of the manifold X such that the following conditions are satisfied:

1) The number $\Gamma(i)$ is defined by the combinatorial scheme of the simplex X_i;

2) the cooriented chain $\Gamma^L_{n-4} = \sum \Gamma(i) X_i$ is a cycle;

3) if the manifold X is smooth then the homology class of the cycle $\Gamma^L_{n-4}(X)$ is in Poincaré duality with the first Pontryagin class $p_1(X)$ of the manifold X.

Definition of the Spaces $\Sigma(i)$ (also see [1, Sec. 5]). Let X_i be an $(n-p)$-simplex of the combinatorial manifold X. Then there exists a partition σ of the space R^p into simplicial cones with vertices at the origin, having the same combinatorial schemes as the stars of the simplices X_i (i.e., to each $(n-p+k)$-simplex of the manifold X, containing X_i, there corresponds a k-dimensional simplicial cone of the partition σ, this correspondence being one-to-one and consistent with the relations of inclusion of the boundaries). The equivalence classes of all such partitions σ relative to the action of the groups $Gl(p, R)$ form a manifold, which we denote by $\Sigma(i)$. We denote by $\Sigma_0(i)$, [correspondingly $\Sigma_1(i)$], the submanifolds in $\Sigma(i)$ formed by the partitions and having no (respectively having not more than one) set of p one-dimensional edges belonging to a $(p-1)$-dimensional subspace of R^p.

Before approaching the proof of Theorem 1, we formulate some results from [1] in a form we need relating to the first class of Pontryagin.

Let X be a combinatorial manifold with a fixed set of complementary structures. We assume a hypersimplicial system $\{b_\alpha\}$ (for the definition of $\{b_\alpha\}$ see [1, Sec. 9]) of points $\sigma_i \in \Sigma_0(i)$ for dim $X_i = n-4$, $\sigma_j \in \Sigma_0(j)$ for dim $X_j = n-3$ and of curves $\sigma_{ij}(t) \subset \Sigma_1(j)$ such that $\sigma_{ij}(0) = \sigma_j$, $\sigma_{ij}(1) = \varphi'_{ij}(\sigma_i)$ (for the definition of σ_{ij} see [1, Sec. 5]).

*That is, with an oriented normal decomposition.

Moscow State University. Saratov State University. Translated from Funktsional'nyi Analiz i Ego Prilozheniya, Vol. 10, No. 1, pp. 14-17, January-March, 1976. Original article submitted September 29, 1975.

In Sec. 10 of [1] from such a set of structures we constructed a rational cooriented (n − 4)-chain $\Gamma_{n-4}(X)$. In that instance we have the following statements.

THEOREM 2. Let X fulfill condition (A). Then

1) the chain $\Gamma_{n-4}(X)$ is a cycle;

2) the homology class of the cycle $\Gamma_{n-4}(X)$ does not depend on the choice made above of the complementary structures;

3) if X is smooth, then the homology class of cycle $\Gamma_{n-4}(X)$ is in Poincaré duality with the first Pontryagin class $p_1(X)$ of the manifold X.

This theorem is a slightly weakened version of Proposition 10 and Theorem 1 of [1]. The proof of the first of its statements is given in Proposition 34 of [1], of the second in Propositions 35-38, and of the third in Propositions 39-48.

We remark, although we do not need it here, that statement 3) of Theorem 2, independently of the fulfillment of condition (A), is true if the complementary structures on X are chosen in the natural way according to the smooth structure. We formulate this more exactly.

Let X be a smooth manifold with simplicial partitions [without any kind of conditions on $\Sigma(i)$]. Then for each (n − p)-simplex X_i there is defined a mapping f_i: $X_i \to \Sigma(i)$ [for $x \in X_i$ the partition of the manifold X defines a partition $\hat\sigma$ of the space $(TX)_x$, and by the property of $f_i(x)$ it follows that the partition induced by $\hat\sigma$ is taken on $(TX)_x/(TX_i)_x \cong R^p$]. We choose in each (n − 4)-simplex X_i a point $x_i \in X_i$ and in each (n − 3)-simplex X_j a point $x_j \in X_j$. For each pair of simplices $X_i \subset X_j$ (dim X_i = n − 4, dim X_j = n − 3) we choose a curve $x_{ij}(t) \subset X_j$, connecting x_j to x_i. We set $\sigma_i = f_i(x_i) \in \Sigma(i)$, $\sigma_j = f_j(x_j) \in \Sigma(j)$, $\sigma_{ij}(t) = f_j(x_{ij}(t)) \subset \Sigma(j)$. By slightly deforming the points σ_i, σ_j, and the curves $\sigma_{ij}(t)$ we may arrange it so that $\sigma_i \in \Sigma_0(i)$, $\sigma_j \in \Sigma_0(j)$, $\sigma_{ij}(t) \subset \Sigma_1(j)$.

THEOREM 3. The chain $\Gamma_{n-4}(X)$ constructed from σ_i, σ_j, and $\sigma_{ij}(t)$, chosen according to the smooth structure and the arbitrary assumed hypersimplicial system $\{b_\alpha\}$ is a cycle whose homology class is in Poincaré duality with $p_1(X)$.

The proof of this theorem is actually done in the proof of Theorem 1 of [1].

Proof of Theorem 1. Let X satisfy the conditions of Theorem 1. We construct a local combinatorial formula for $p_1(X)$, averaging $\Gamma_{n-4}(X)$ by a different method from the choice of complementary structures. We present the proof through several lemmas.

LEMMA 1. The cycle $\Gamma_{n-4}(X)$ depends only on the choice of the elements b_k (dim X_k = n − 2) and b_j (dimX_j = n − 3) of the assumed hypersimplicial system and on the choice of connected components μ_j of the space $\Sigma_0(j)$ for each (n − 3)-simplex X_j.

Proof. That the cycle $\Gamma_{n-4}(X)$ does not depend on the elements b_α if dim $X_\alpha \le n-4$ on the points σ_i for dim X_i = n − 4, and the curves $\sigma_{ij}(t)$ is clear from Proposition 35 of [1]. Moreover, if dim $X_l \ge n-1$, then the element b_l is defined uniquely. In such a manner $\Gamma_{n-4}(X)$ can only depend on the choice of b_k (dim X_k = n − 2), b_j and σ_j (dim X_j = n − 3). The nondependence of the cycle $\Gamma_{n-4}(X)$ on the choice of the points σ_j inside the fixed connected component μ_j of the space $\Sigma_0(j)$ follows immediately from the definition of $\Gamma_{n-4}(X)$. The lemma is proved.

We set $b^2 = \{b_k\}$ (dim X_k = n − 2), $b^3 = \{b_j\}$, and $\mu = \{\mu_j\}$ (dim X_j = n − 3), and we denote the cycle $\Gamma_{n-4}(X)$ constructed from the given b^2, b^3, and μ by $\Gamma_{n-4}(b^2, b^3, \mu)$. Let

$$\Gamma_{n-4}(b^2, b^3, \mu) = \sum_{i:\, \dim X_i = n-4} \Gamma(i; b^2, b^3, \mu)\, X_i.$$

LEMMA 2. The number $\Gamma(i; b^2, b^3, \mu)$ depends only on the choice of b_k, b_j, and μ_j for the simplices X_k and X_j containing X_i (and on the combinatorial scheme of the star of X_i).

Verification of the lemma comes immediately from the construction of the cycle $\Gamma_{n-4}(X)$ (see [1, Sec. 10]).

LEMMA 3. For each collection $b^2 = \{b_k\}$ (dim X_k = n − 2) continuing along the assumed hypersimplicial system there exists a cycle

$$\Gamma_{n-4}(b^2) = \sum_{i:\ \dim X_i = n-4} \Gamma(i; b^2) X_i,$$

homologous to the cycle $\Gamma_{n-4}(X)$ and such that for each i the number $\Gamma(i; b^2)$ is rational and depends only on the choice of b_k for the simplices X_k containing X_i and on the combinatorial scheme of the star of X_i.

Proof. Let X_j be an $(n-3)$-simplex. Denote by M(j) the number of connected components of the space $\Sigma_0(j)$ [this number is finite since $\Sigma_0(j)$ is a semialgebraic set]. Let b^2 be fixed. We call the chain $b_j \in I_3(j)$ (for definition $I_*(j)$ see [1, Sec. 8]) minimal if:

1. The set $\{b^2, b_j\}$ is part of the assumed hypersimplicial system.

2. The sum of the moduli of the coefficients in the decomposition of b_j according to the basis of the hypersimplices in $I_3(j)$ is minimal, i.e., for any chain b_j' satisfying condition 1 the sum for b_j' is not less than for b_j.

We denote by $N(j; b^2)$ the number of different ways of choosing a minimal chain b_j [this number is finite since $I_3(j)$ is an Abelian group with a finite number of sums].

We set

$$\Gamma_{n-4}(b^2) = \Big(\prod_{j:\ \dim X_j = n-3} \frac{1}{M(j)\, N(j; b^2)}\Big) \sum \Gamma_{n-4}(b^2, b^3, \mu),$$

where the sum is taken over all possible choices of minimal chains b_j and connected components μ_j of the space $\Sigma_0(j)$. The cycle $\Gamma_{n-4}(b^2)$ is homologous to $\Gamma_{n-4}(X)$. Let

$$\Gamma_{n-4}(b^2) = \sum_{i:\ \dim X_i = n-4} \Gamma(i; b^2) X_i.$$

Then using Lemma 2 it is easy to show that

$$\Gamma(i; b^2) = \Big(\prod_{\substack{j:\ \dim X_j = n-3,\\ X_j \supset X_i}} \frac{1}{M(j)\, N(j; b^2)}\Big) \sum \Gamma(i; b^2, b^3, \mu), \tag{1}$$

where the sum is taken over all possible choices of minimal chains b_j and connected components μ_j of the space $\Sigma_0(j)$ for $(n-3)$-simplices X_j containing X_i. The number M(j) obviously depends only on the combinatorial scheme of the star of X_j. Moreover, from the definition of the assumed hypersimplicial system it is easy to deduce that the number $N(j; b^2)$ depends only on the choice of b_k for $X_k \supset X_j$ and on the combinatorial scheme of the star of X_j. If $X_j \supset X_i$, then the combinatorial scheme of the star of X_j defines a combinatorial scheme for the star of X_i. From Lemma 2 it follows that the sum of the numbers $\Gamma(i, b^2, b^3, \mu)$ in formula (1) depends only on the choice of b_k for $X_k \supset X_i$ and on the combinatorial scheme for the star of X_i. Thus, the number $\Gamma(i; b^2)$ depends only on the choice of b_k for $X_k \supset X_i$ and on the combinatorial scheme of the star of X_i, which was to be shown.

For each $(n-2)$-simplex X_k we denote by N(k) the number of different ways to choose the chains $b_k \in I_2(k)$ satisfying the following condition:

3. b_k is an element of the assumed hypersimplicial system, and the sum of the moduli of its coefficients is as small as possible.

LEMMA 4. We set

$$\Gamma^L_{n-4}(X) = \Big(\prod_{k:\ \dim X_k = n-2} \frac{1}{N(k)}\Big) \sum \Gamma_{n-4}(b^2),$$

where the summation is over all possible choices of the chain b_k satisfying condition 3. Then

1) the cycle $\Gamma^L_{n-4}(X)$ is homologous to the cycle $\Gamma_{n-4}(X)$;

2) let $\Gamma^L_{n-4}(X) = \sum_{i:\ \dim X_i = n-4} \Gamma(i) X_i$; then the number $\Gamma(i)$ depends only on the combinatorial scheme of the star of the simplex X_i.

The proof of this lemma is exactly the same as the proof of Lemma 3.

Thus statements 1) and 2) of Theorem 1 are proved.

If X is smooth, then, according to Theorem 2, the homology class of the cycle $\Gamma_{n-4}(X)$ is in Poincaré duality with $p_1(X)$. It follows that this is also true for $\Gamma^{L}_{n-4}(X)$.

The theorem is proved.

Remark. The local combinatorial formula obtained here looks more complicated than the clear formula used for $\Gamma_{n-4}(X)$. Nevertheless, the formula used is sufficiently complicated. It is interesting how very essential the difficulty of this formula is and that it is impossible to get a simpler clear formula for $p_1(X)$. It is interesting also how small the denominators of the coefficients can be in the combinatorial formula for $p_1(X)$. This denominator for $\Gamma_{n-4}(X)$ equals 48, but by slightly modifying the construction we can make it 24. In the meantime the question remains open for the higher classes of Pontryagin. The attempt to push through our construction even in the case of the second class of Pontryagin runs into technical difficulties.

LITERATURE CITED

1. A. M. Gabrielov, I. M. Gel'fand, and M. V. Losik, "Combinatorial computations of characteristic classes," Funktsional'. Analiz i Ego Prilozhen., 9, No. 2, 12-28 (1975); No. 3, 5-26 (1975).

19.
(with A. M. Gabrielov and M. V. Losik)

Atiyah-Patodi-Singer functionals for characteristic functionals for tangent bundles

Funkts. Anal. Prilozh. 10 (2) (1976) 13–28 [Funct. Anal. Appl. 10 (1976) 95–107].
MR 54:1245. Zbl. 344:57008

This article grew out of our attempts to generalize the integral Gauss–Bonnet formula for Hirzebruch polynomials obtained by Atiyah, Patodi, and Singer in [1] to the case of Pontryagin forms on manifolds with smooth boundaries.

We consider the case where the boundary is only piecewise smooth. Interesting new forms occur at the discontinuities in the boundary.

Let X be a Riemannian manifold with piecewise smooth boundary ∂X, and let P_X be some characteristic form on the tangent bundle of the manifold X.

We shall assume for simplicity that P_X is a form of dimension higher than X, although all the results of the article are formulated for forms of arbitrary dimension. The integral of the form P_X on the manifold X depends on the choice of the metric on X in the neighborhood of the boundary. However, the sum of the integrals of certain forms on the various faces of ∂X can be added to this integral so that the number Φ_X so obtained depends only on the intrinsic metric of the boundary and on the topology of X. The number Φ_X will be called the Atiyah–Patodi–Singer functional. The forms on the faces of ∂X which form part of the definition of Φ_X are described by explicit formulas in terms of the metric on X in the neighborhood of the corresponding faces and together with the form P_X give a cocycle in a certain complex, the generalized de Rham complex in the piecewise smooth case.

Now let X be a closed smooth manifold decomposed into pieces X_i with piecewise smooth boundaries. Assume that in each piece X_i a metric has been chosen, and that the restrictions of the metrics on the various pieces agree on their common faces. Then the sum of the fractionals Φ_{X_i} does not depend on the choice of the metrics and is an invariant of the decomposition. Put $\Phi_X = \sum \Phi_{X_i}$. If the decomposition is linearizable, e.g., simplicial, then $\Phi_X = 0$. If the decomposition is simple, i.e., if in the neighborhood of each vertex the decomposition is constructed as the decomposition of the boundary of the positive octant in the neighborhood of the coordinate origin, then Φ_X coincides with the characteristic number of the manifold X corresponding to the form P_X.

Moscow State University. Saratov State University. Translated from Funktsional'nyi Analiz i Ego Prilozheniya, Vol. 10, No. 2, pp. 13-28, April-June, 1976. Original article submitted December 22, 1975.

For an arbitrary decomposition, when P_X is the first Pontryagin form, the difference between the characteristic number and Φ_X is equal to $(X, \Gamma(X))$, where $\Gamma(X)$ is the combinatorial cycle constructed in our article [2].

The results of this article parallel those of [3] for the characteristic classes of foliations, but the construction turns out to be substantially more complicated in the tangent bundle case.

We also note that in the article by Atiyah, Patodi, and Singer there is a remarkable splitting of the functional into two terms: one topological, and the other purely dependent on the boundary. We do not perform such a decomposition here and we do not know if it can be done in the general case.

1. THE DE RHAM COMPLEX FOR A STRATIFIED SET

The definitions and results of this section are taken, almost without change, from §4 of [3]. Proofs of the assertions made here are given there.

Let X be a Hausdorff topological space. An n-dimensional chart with corners is an open subset $U \subset X$ together with a continuous map $\rho: U \to R^n$ which takes U homeomorphically onto a neighborhood of the coordinate origin in $R^k \times (\bar{R}_+)^{n-k} \subset R^n$. Two charts $\rho: U \to R^n$ and $\rho': U' \to R^n$ are said to be compatible if the map $\rho'\rho^{-1}: \rho(U \cap U') \to \rho'(U \cap U')$ is a diffeomorphism (we note that the values of k for the two charts can be different).

Definition 1.1. The space X is said to be an n-dimensional manifold with corners if for some cover $\{U_\nu\}$ of X there is a set of pairwise compatible n-dimensional charts with corners $\{\rho_\nu: U_\nu \to R^n\}$. A chart with corners which is compatible with all the charts of this set is said to be smooth.

Example 1. A simple cell (see [2, Sec. 1]) is a manifold with corners.

Example 2. Let M be a smooth n-dimensional manifold and $\{M_j\}$ a set of n-dimensional submanifolds with boundaries in M, where each boundary ∂M_j lies in general position. Then $X = \bigcap_j M_j$ is a manifold with corners.

Let X be a manifold with corners. Define the tangent bundle TX and the bundle of differentiable forms $\Omega(X) = \oplus \Omega^q(X)$ as follows. If $U \subset X$ and $\rho: U \to R^k \times (\bar{R}_+)^{n-k} \subset R^n$ is a smooth chart, then we put $TU = \rho^*(TR^n)$, $\Omega(U) = \rho^*(\Omega(R^n))$. If ρ and ρ' are two smooth charts defined on U and U' respectively, then the restrictions of the bundles TU and TU' (or the bundles $\Omega(U)$ and $\Omega(U')$) are canonically isomorphic on $U \cap U'$. Hence, the bundles TX and $\Omega(X)$, for which $TX|U = TU$ and $\Omega(X)|U = \Omega(U)$ are well-defined.

Again let X be a manifold with corners. We shall say that the point $x \in X$ belongs to the k-dimensional skeleton X(k) of the manifold X if there is an open set U containing x and a smooth chart with corners $\rho: U \to R^{k'} \times (\bar{R}_+)^{n-k'}$, $k' \leqslant k$, such that $\rho(x) = 0$. The closures of the connected components of $X(k) \setminus X(k-1)$ are called the k-dimensional faces of X.

Definition 1.2. By a smooth map of manifolds with corners we mean a continuous map $f: X \to X'$, satisfying the conditions:

a) if $\rho: U \to R^n$ and $\rho': U' \to R^{n'}$ are smooth charts with corners in X and X', respectively, then the map

$$\rho' \circ f \circ \rho^{-1}: \rho(U \cap f^{-1}(U')) \to R^{n'}$$

is smooth, i.e., it can be extended to a smooth map $V \to R^{n'}$, where V is an open set in R^n, containing $\rho(U \cap f^{-1}(U'))$;

b) if X_i is a face of X and X'_j is a face of X' such that $f(\mathrm{int}\, X_i) \cap X'_j \neq \phi$, then $f(X_i) \subset X'_j$.

A manifold with corners Y is a submanifold of the manifold with corners X if there is a smooth regular inclusion $i: Y \to X$, which is a homeomorphism from Y to i(Y) (regularity means that the natural map $i_*: TY \to TX$ is an inclusion).

Remark. A face of X may not be a submanifold (see Fig. 1, where $Y = \partial X$ is the one-dimensional face of X). For simplicity, we shall consider below only such manifolds with corners all of whose faces are submanifolds.

Definition 1.3. A stratified set is a Hausdorff space X together with a locally finite decomposition into nonintersecting connected subsets X_i (the strata) which satisfies the following conditions:

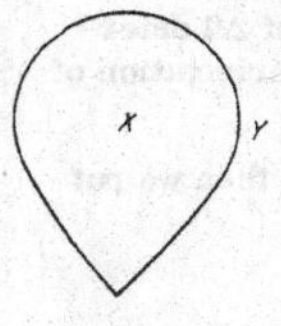

Fig. 1

i) if $\bar{X}_i \cap X_j \neq \phi$, then $\bar{X}_i \supset X_j$;

ii) the closure of each stratum has the structure of a manifold with corners; and

iii) all the strata X_j, contained in $\bar{X}_i$, are the interiors of faces of the manifold with corners X_i, while the smooth structures on $\bar{X}_i$ and $\bar{X}_j$ coincide.

Some examples of stratified sets are: manifolds with corners (the strata are the interiors of the faces), a locally finite simplicial complex, whose decomposition is into its simple cells (see [2, Para. 1]).

Definition 1.4. A map of stratified manifolds $f: X \to X'$ is said to be smooth if the image of each stratum $X_i \subset X$ is contained in some stratum $X'_j \subset X'$ and $f|_{\bar{X}_i}: \bar{X}_i \to \bar{X}'_j$ is smooth as a map of manifolds with corners.

Let X be a stratified set and X_i a stratum of X. Let us denote by $\Omega^q(\bar{X}_i)$ the bundle of q forms on the manifold with corners $\bar{X}_i$, with the bundle extended to X by zero. We then put

$$\Omega^q(X) = \bigoplus_{k=0}^{q} \prod_{X_{i_0},\dots,X_{i_k}} \Omega^{q-k}(\bar{X}_{i_k}),$$

where the product is extended to all k flags, i.e., to sequences of pairwise distinct strata $X_{i_0}, \dots, X_{i_k}$, such that $X_{i_{\nu+1}} \subset \bar{X}_{i_\nu}$.

A section α of the bundle $\Omega^q(X)$ is given as a set of forms $\alpha_{i_0 \dots i_k}$, where $0 \leq k \leq q$, $i_0, \dots, i_k$ being indices of strata of X which form a k flag, and $\alpha_{i_0 \dots i_k}$ is a form of degree $(q-k)$ on $\bar{X}_{i_k}$.

Definition 1.5. The de Rham complex $\Omega(X)$ of a stratified set X is a complex of bundles $\Omega^q(X)$ $(q \geqslant 0)$ with differential $d: \Omega^q(X) \to \Omega^{q+1}(X)$, defined by the formula:

$$(d\alpha)_{i_0 \dots i_k} = (-1)^k d\alpha_{i_0 \dots i_k} + \sum_{\mu=0}^{k} (-1)^{\mu} \alpha_{i_0 \dots \hat{i}_\mu \dots i_k}\Big|_{\bar{X}_{i_k}}$$

(in particular, if k = 0, then $(d\alpha)_{i_0} = d\alpha_{i_0}$).

Proposition 1.6. The complex $\Omega(X)$ is the fine resolvent of the constant bundle R on X. In particular, the homology of the complex $\Gamma(X, \Omega(X))$ is canonically isomorphic to the real cohomology of X.

Let $f: X \to X'$ be a smooth map of stratified sets. Define the homomorphism of complexes $f^*: \Gamma(X', \Omega(X')) \to \Gamma(X, \Omega(X))$ as follows. For each stratum $X_i \subset X$ denote by $X'_{j(i)}$ the stratum of X' containing $f(X_i)$. Let $\alpha = \{\alpha_{j_0 \dots j_l}\}$ be a section of the bundle $\Omega(X')$. For each flag $X_{i_0}, \dots, X_{i_l}$ put

$$(f^*\alpha)_{i_0 \dots i_l} = f^*(\alpha_{j(i_0) \dots j(i_l)}), \quad \text{if all the indices } j(i_\nu) \text{ are distinct,}$$
$$(f^*\alpha)_{i_0 \dots i_l} = 0 \quad \text{otherwise.}$$

It is obvious that the map so constructed is compatible with the differentials. Moreover, the induced homomorphism of homologies coincides with the homomorphism $f^*: H^*(X'; R) \to H^*(X; R)$.

Definition 1.7. Let Δ^q be the standard oriented q simplex with the natural stratification. A smooth singular q simplex of the stratified set X is a smooth map of stratified sets $\varphi: \Delta^q \to X$. A finite combination of smooth singular q simplexes is called a smooth singular q chain. The boundary ∂c of a smooth singular chain c is defined in the usual way.

Let α be a section of the bundle $\Omega^q(\Delta^q)$. Denote by $\Delta^{q-\nu}(i_1, \dots, i_\nu)$ that face of dimension $(q-\nu)$ of the simplex Δ^q which does not contain the vertices $i_1, \dots, i_\nu$.

For each ordered set $i_1, \dots, i_l$ $(0 \leqslant i_\nu \leqslant q,\ 0 \leqslant l \leqslant q)$ let $\alpha(i_1, \dots, i_l)$ denote that component of the form α which corresponds to the l-flag $\Delta^q \supset \Delta^{q-1}(i_1) \supset \dots \supset \Delta^{q-l}(i_1, \dots, i_l)$ (in particular, if $l = 0$ and the set is empty, then $\alpha(\phi)$ is the component of the form α which corresponds to the 0-flag consisting of the single simplex Δ^q). The form $\alpha(i_1, \dots, i_l)$ is defined on $\Delta^{q-l}(i_1, \dots, i_l)$ and has degree $q - l$. Put

$$\int_{\Delta^q} \alpha = \sum_{l=0}^{q} \sum_{i_1,\dots,i_l} \int_{\Delta^{q-l}(i_1,\dots,i_l)} \alpha(i_1, \dots, i_l).$$

Here the orientation of the simplex $\Delta^{q-l}(i_1, \ldots, i_l)$ is determined as follows: The orientation of Δ^q determines an orientation of $\partial\Delta^q$, and so of $\Delta^{q-1}(i_1) \subset \partial\Delta^q$; the orientation of $\Delta^{q-1}(i_1)$ determines an orientation of $\Delta^{q-2}(i_1, i_2)$, in exactly the same way, and we may continue the process to $\Delta^{q-l}(i_1, \ldots, i_l)$.

If X is any stratified set, $\varphi : \Delta^q \to X$ a smooth singular simplex, and α a section of $\Omega^q(X)$, then we put

$$\int_{\varphi(\Delta^q)} \alpha = \int_{\Delta^q} \varphi^*\alpha.$$

If $c = \sum a_i \varphi_i(\Delta^q)$ is a smooth singular q chain on X, then we put

$$\int_c \alpha = \sum a_i \int_{\varphi_i(\Delta^q)} \alpha.$$

<u>Proposition 1.8.</u> If $f : X \to X'$ is a smooth map of stratified sets, c a smooth singular q chain on X, and α a section of the bundle $\Omega^q(X')$, then

$$\int_c f^*\alpha = \int_{f_* c} \alpha.$$

<u>Proposition 1.9</u> (Stokes' Formula). For each section α of the bundle $\Omega^q(X)$ and for each smooth singular (q + 1) chain c we have:

$$\int_c d\alpha = \int_{\partial c} \alpha.$$

<u>Definition 1.10.</u> Let α and β be sections of the bundles $\Omega^p(X)$ and $\Omega^q(X)$, respectively. Then the exterior product $\alpha \wedge \beta \in \Gamma(X, \Omega^{p+q}(X))$ is defined by the formula:

$$(\alpha \wedge \beta)_{i_0 \ldots i_l} = \sum_{\nu=0}^{l} (-1)^{(p-\nu)(l-\nu)} \alpha_{i_0 \ldots i_\nu} \wedge \beta_{i_\nu \ldots i_l}.$$

It is easy to check that the multiplication so defined is associative but not skew-commutative.

<u>Proposition 1.11.</u> The following equality holds: $d(\alpha \wedge \beta)\, d\alpha \wedge \beta + (-1)^p \alpha \wedge d\beta$.

<u>Proof.</u> We have

$$\begin{aligned} d(\alpha \wedge \beta)_{i_0 \ldots i_l} &= (-1)^l d((\alpha \wedge \beta)_{i_0 \ldots i_l}) + \sum_{\mu=0}^{l} (-1)^\mu (\alpha \wedge \beta)_{i_0 \ldots \hat{i}_\mu \ldots i_l} = \\ &= \sum_{\nu=0}^{l} (-1)^{l+(p-\nu)(l-\nu)} d\alpha_{i_0 \ldots i_\nu} \wedge \beta_{i_\nu \ldots i_l} + \sum_{\nu=0}^{l} (-1)^{l+(p-\nu)(l-\nu-1)} \alpha_{i_0 \ldots i_\nu} \wedge d\beta_{i_\nu \ldots i_l} + \\ &+ \sum_{\mu=0}^{l} \Big[\sum_{\nu=0}^{\mu-1} (-1)^{\mu+(p-\nu)(l-1-\nu)} \alpha_{i_0 \ldots i_\nu} \wedge \beta_{i_\nu \ldots \hat{i}_\mu \ldots i_l} + \sum_{\nu=\mu+1}^{l} (-1)^{\mu+(p-\nu-1)(l-\nu)} \alpha_{i_0 \ldots \hat{i}_\mu \ldots i_\nu} \wedge \beta_{i_\nu \ldots i_l} \Big], \end{aligned} \tag{1}$$

$$\begin{aligned} (d\alpha \wedge \beta)_{i_0 \ldots i_l} + (-1)^p (\alpha \wedge d\beta)_{i_0 \ldots i_l} &= \sum_{\nu=0}^{l} (-1)^{(p+1-\nu)(l-\nu)} \Big[(-1)^\nu d\alpha_{i_0 \ldots i_\nu} + \sum_{\mu=0}^{\nu} (-1)^\mu \alpha_{i_0 \ldots \hat{i}_\mu \ldots i_\nu} \Big] \wedge \beta_{i_\nu \ldots i_l} + \\ &+ (-1)^p \sum_{\nu=0}^{l} (-1)^{(p-\nu)(l-\nu)} \alpha_{i_0 \ldots i_\nu} \wedge \Big[(-1)^{l-\nu} d\beta_{i_\nu \ldots i_l} + \sum_{\mu=\nu}^{l} (-1)^{\mu-\nu} \beta_{i_\nu \ldots \hat{i}_\mu \ldots i_l} \Big]. \end{aligned} \tag{2}$$

Equality of the right sides of (1) and (2) comes after a change in the order of summation and a reduction of the corresponding terms on the right side of (2).

<u>Proposition 1.12.</u> The exterior product on the complex $(X, \Omega(X))$ determines a multiplication on its homology groups which coincides with the usual product in $H^*(X; R)$.

2. THE DIFFERENCE COCYCLE AND THE FORM $S^{r-1,0}$

Denote by $\Pi_{r,q}$ the invariant symmetric r-linear form on $\mathfrak{gl}(q, R)$ defined by the formula:

$$\Pi_{r,q}(h_1, \ldots, h_r) = \frac{1}{r!} \sum \operatorname{tr}(h_{\nu_1} \cdot \ldots \cdot h_{\nu_r}),$$

where the summation is carried out over all the permutations $\nu_1, \ldots, \nu_r$ of the indices 1, . . ., r.

Proposition 2.1. Let $R^{q'}$ be a subspace of R^q, and let h_ν ($\nu = 1, \ldots, r$) be elements of $\mathfrak{gl}(q, R)$, i.e., endomorphisms of R^q. Assume that $R^{q'}$ is invariant with respect to the h_ν. Let h'_ν (resp., h''_ν) denote endomorphisms of $R^{q'}$ (resp., $R^q/R^{q'}$), which are induced by the endomorphisms h_ν. Then

$$\Pi_{r,q}(h_1, \ldots, h_r) = \Pi_{r,q'}(h'_1, \ldots, h'_r) + \Pi_{r,q-q'}(h''_1, \ldots, h''_r).$$

Proof. Let the basis $l_1, \ldots, l_q$ for R^q be so chosen that $l_1, \ldots, l_{q'}$ belong to $R^{q'}$. Then the matrices of the endomorphisms h_ν with respect to these bases have the following property: $(h_\nu)^\beta_\alpha = 0$, if $\alpha \leqslant q'$, $\beta > q'$. Therefore,

$$\Pi_{r,q}(h_1, \ldots, h_r) = \frac{1}{r!} \sum_{\nu_1, \ldots, \nu_r} \sum_{\substack{\alpha_1, \ldots, \alpha_r \\ 1 \leqslant \alpha_i \leqslant q}} (h_{\nu_1})^{\alpha_2}_{\alpha_1} (h_{\nu_2})^{\alpha_1}_{\alpha_2} \ldots (h_{\nu_r})^{\alpha_1}_{\alpha_r} =$$

$$= \frac{1}{r!} \sum_{\nu_1, \ldots, \nu_r} \Big[\sum_{\substack{\alpha_1, \ldots, \alpha_r \\ 1 \leqslant \alpha_i \leqslant q'}} (h_{\nu_1})^{\alpha_2}_{\alpha_1} (h_{\nu_2})^{\alpha_1}_{\alpha_2} \ldots (h_{\nu_r})^{\alpha_1}_{\alpha_r} + \sum_{\substack{\beta_1, \ldots, \beta_r \\ q'+1 \leqslant \beta_i \leqslant r}} (h_{\nu_1})^{\beta_2}_{\beta_1} (h_{\nu_2})^{\beta_1}_{\beta_2} \ldots (h_{\nu_r})^{\beta_1}_{\beta_r} \Big] =$$

$$= \Pi_{r,q'}(h'_1, \ldots, h'_r) + \Pi_{r,q-q'}(h''_1, \ldots, h''_r),$$

as was to be proved.

It is well known that any invariant symmetric r-linear form P_q on $\mathfrak{gl}(q, R)$ can be written in the form $P_q = Q(\Pi_{1,q}, \ldots, \Pi_{r,q})$, where Q is some polynomial in r variables with real coefficients (the product of the forms $\Pi_{r,q}$ is taken in the algebra of symmetric forms). The set of forms P_q ($q = 1, 2, \ldots$), constructed from the single polynomial Q, is said to be a stable r-form on $\mathfrak{gl}(\cdot, R)$ and, with the index q omitted, all these forms will be denoted by the one letter P.

We now require a definition for the difference cocycle (see [2, Sec. 15]).

Definition 2.2. Let $E \to X$ be a vector bundle. The difference cocycle in E corresponding to the stable r-form P is a set $\{P^{l,2r-l}\}_{l=0,\ldots,2r}$ (where $P^{l,2r-l} = P^{l,2r-l}(\omega_0, \ldots, \omega_l)$ is a differential $(2r - l)$ form on X depending on the connections $\omega_0, \ldots, \omega_l$ in E) satisfying the conditions:

a) $P^{0,2r}(\omega) = P(R, \ldots, R)$, where R is the curvature form of the connection ω, i.e., $P^{0,2r}(\omega)$ is the characteristic form corresponding to P. In particular, $dP^{0,2r}(\omega) = 0$.

b) The form $P^{l,2r-l}(\omega_0, \ldots, \omega_l)$ is skew-symmetric with respect to permutations of the connections ω_ν.

c) Put $\Delta P^{l-1,2r-l+1}(\omega_0, \ldots, \omega_l) = \sum_{\mu=0}^{l} (-1)^\mu P^{l-1,2r-l+1}(\omega_0, \ldots \hat{\omega}_\mu, \ldots, \omega_l)$. Then

$$(-1)^l dP^{l,2r-l}(\omega_0, \ldots, \omega_l) + \Delta P^{l-1,2r-l+1}(\omega_0, \ldots, \omega_l) = 0$$

for $l = 1, \ldots, 2r$.

d) Let F be a diffeomorphism of E compatible with the structure of the vector bundle, and let f be the corresponding diffeomorphism of X. Let $F^*\omega$ denote the inverse image of the connection ω under the diffeomorphism F. Then $P^{l,2r-l}(F^*_{\omega_0}, \ldots, F^*_{\omega_l}) = f^*P^{l,2r-l}(\omega_0, \ldots, \omega_l)$.

When $P = \Pi_r$ we can write instead of $(\Pi_r)^{l,2r-l}$ the simple expression $\Pi^{l,2r-l}$.

Proposition 2.3. The difference cocycle defined in [2] satisfies the following additional conditions:

e) $P^{l,2r-l} = 0$ for $l > r$.

f) Let $P = \Pi_r$, E' be a subbundle of E, and ω_ν ($\nu = 0, \ldots, l$) connections in E such that parallel transfer with respect to ω_ν takes E' into itself. Let ω'_ν (resp., ω''_ν) denote connections induced in E' (resp., E/E') by the connections ω_ν. Then

$$\Pi^{l,2r-l}(\omega_0, \ldots, \omega_l) = \Pi^{l,2r-l}(\omega'_0, \ldots, \omega'_l) + \Pi^{l,2r-l}(\omega''_0, \ldots, \omega''_l).$$

g) Let $P = \sum_\lambda c_\lambda \Pi_1^{\lambda_1} \cdot \ldots \cdot \Pi_r^{\lambda_r}$ ($\lambda = (\lambda_1, \ldots, \lambda_r)$, $\lambda_i \geqslant 0$, $\sum i\lambda_i = r$), E be a one-dimensional or line bundle, m_ν ($\nu = 0, \ldots, l$) be sections of E which are nowhere zero, and let v_ν be connections on E with respect to which the sections m_ν are horizontal. Then

$$P^{l,2r-l}(v_0, \ldots, v_r) = 0, \quad \text{if} \quad l \neq r,$$

$$P^{r,r}(v_0, \ldots, v_r) = (-1)^{\frac{r(r-1)}{2}} \Big(\sum_\lambda c_\lambda\Big) dS^{r-1,0}(m_0, \ldots, m_r),$$

where

$$S^{r-1,0}(m_0,\ldots,m_r)=\frac{1}{r}\sum_{\mu=1}^{r}(-1)^{\mu-1}\log\left|\frac{m_0}{m_\mu}\right|d\log\left|\frac{m_0}{m_1}\right|\wedge\ldots\widehat{d\log\left|\frac{m_0}{m_\mu}\right|}\ldots\wedge d\log\left|\frac{m_0}{m_r}\right|. \tag{3}$$

Proposition 2.4. a) The form $S^{r-1,0}$ $(m_0, \ldots, m_r)$ is skew-symmetric with respect to permutations of $m_0, \ldots, m_r$;

b) if $m_0, \ldots, m_{r+1}$ is a set of (r + 2) sections, then the expression

$$(\Delta S^{r-1,0})(m_0,\ldots,m_{r+1})=\sum_{\mu=0}^{r+1}(-1)^{\mu}S^{r-1,0}(m_0,\ldots\hat{m}_\mu\ldots,m_{r+1})$$

is zero.

3. CONNECTIONS AND NORMALS

Definition 3.1. Let X be a stratified set. Assume that for each stratum X_i a smooth connection ω_i on $T\overline{X}_i$ is given and that for any two strata X_j and X_k, $X_j \subset \overline{X}_k$, dim X_j = dim $X_k - 1$, a "normal" m_{jk} is chosen, i.e., a section of the bundle $T\overline{X}_k|_{\overline{X}_j}$, which does not intersect with $T\overline{X}_j$. The set $(\omega, m) = (\{\omega_i\}, \{m_{jk}\})$ is called a set of connections and normals for X.

Definition 3.2. Let g be a metric on X which is smooth on the closure of any stratum. Let (ω_g, m_g) denote the set of connections and normals for X constructed from g as follows: ω_i is the connection on $T\overline{X}_i$, determined by the metric $g|_{\overline{X}_i}$, m_{jk} the normal to $\overline{X}_j$ in $\overline{X}_k$ with respect to the metric g.

Let (ω, m) be a set of connections and normals for X. Let X_j and X_l be two strata, $X_j \subset \overline{X}_l$, $\dim X_l = \dim X_j + p$. Then there are exactly p strata $X_{k_1}, \ldots, X_{k_p}$ of dimension dim X_j + 1, which satisfy $X_j \subset \overline{X}_{k_\nu} \subset \overline{X}_l$. Let ω_{jl} denote the connection on $T\overline{X}|_{\overline{X}_j}$, uniquely determined by the conditions: a) $T\overline{X}_j$ is invariant with respect to ω_{jl}, and the connection in TX_j, induced by the connection ω_{jl}, coincides with ω_j; b) m_{jk_ν} is horizontal with respect to ω_{jl} for $\nu = 1, \ldots, p$ (where m_{jk_ν} is considered as a section of $T\overline{X}_l|_{\overline{X}_j}$ through the canonical inclusions $T\overline{X}_{k_\nu}|_{\overline{X}_j} \to T\overline{X}_l|_{\overline{X}_j}$). In particular, $\omega_{jj} = \omega_j$.

Let m_{jl} denote the set $\{m_{jk_\nu}\}_{\nu=1,\ldots,p}$.

Let X_i and X_j be two strata, $X_i \subset \overline{X}_j$, and let X_k be a stratum such that $X_j \subset \overline{X}_k$, $\dim X_k = \dim X_j + 1$. Then there exists a unique stratum $X_{k'}$, satisfying the following condition:

$$\dim X_{k'}=\dim X_i+1,\quad X_i\subset\overline{X}_{k'}\subset\overline{X}_k,\quad X_{k'}\not\subset\overline{X}_j.$$

Let $\varphi_{ij}(k)$ denote k'.

Let $E_{jk}=(T\overline{X}_k|_{\overline{X}_j})/T\overline{X}_j$, $E_{ik'}=(T\overline{X}_{k'}|_{\overline{X}_i})/T\overline{X}_i$. The natural inclusion $T\overline{X}_{k'}|_{\overline{X}_i}\to T\overline{X}_k|_{\overline{X}_i}$ induces the bundle isomorphism $\eta_{ij}: E_{ik'}\xrightarrow{\sim} E_{jk}|_{\overline{X}_i}$.

Let X_i, X_j, and X_k be three strata, $X_i \subset \overline{X}_j \subset \overline{X}_k$, $\dim X_k = \dim X_j + 1$. For each set of strata X_{i_μ}, $0 \leqslant \mu \leqslant r$, $X_i \subset \overline{X}_{i_\mu} \subset \overline{X}_j$, we define the (r − 1) form $S^{r-1,0}_{i,j}(m_{i_0k}, \ldots, m_{i_rk})$ on $\overline{X}_i$ as follows. Put $k_\mu = \varphi_{i_\mu j}(k)$ and let $\widetilde{m}_{i_\mu k_\mu}$ denote the normal form $m_{i_\mu k_\mu}$ under the projection $T\overline{X}_{i_\mu} \mid \overline{x}_{i_\mu} \to E_{i_\mu k_\mu}$. Then

$$S^{r-1,0}_{i,j}(m_{i_0k},\ldots,m_{i_rk})=S^{r-1,0}(\eta_{i,j}(\widetilde{m}_{i_0k_0}),\ldots,\eta_{i,j}(\widetilde{m}_{i_rk_r})), \tag{4}$$

where $S^{r-1,0}$ is the form defined in (3).

If $X_i \subset \overline{X}_j \subset \overline{X}_l$, $\dim X_l = \dim X_j + p$ and X_{k_ν} ($\nu = 1, \ldots p$) are strata of dimension dim X_j + 1, satisfying $X_j \subset \overline{X}_{k_\nu} \subset \overline{X}_l$, then put

$$S^{r-1,0}_{i,j}(m_{i_0l},\ldots,m_{i_rl})=\sum_{\nu=1}^{p}S^{r-1,0}_{i,j}(m_{i_0k_\nu},\ldots,m_{i_rk_\nu}). \tag{5}$$

Proposition 3.3. The following equality holds:

$$dS^{r-1,0}_{i,j}(m_{i_0l},\ldots,m_{i_rl})=(-1)^{\frac{r(r-1)}{2}}[\Pi^{r,r}(\omega_{i_0l},\ldots,\omega_{i_rl})|_{\overline{X}_i}-\Pi^{r,r}(\omega_{i_0j},\ldots,\omega_{i_rj})|_{\overline{X}_i}].$$

Proposition 3.4. If $X_{i_0}, \ldots, X_{i_{r+1}}$ is a set of strata such that $X_i \subset \overline{X}_{i_\mu} \subset \overline{X}_j$ for $\mu = 0, \ldots, r + 1$, then the expression

$$(\Delta S_{i,j}^{r-1,0})(m_{i_0 l},\ldots,m_{i_{r+1}l}) = \sum_{\mu=0}^{r+1}(-1)^{\mu} S_{i,j}^{r-1,0}(m_{i_0 l},\ldots \widehat{m_{i_\mu l}}\ldots,m_{i_{r+1}l})$$

is zero.

The proofs of these assertions are not hard to derive from Propositions 2.3 and 2.4 and the definition of the connections ω_{jl}.

The form $S_{i,k}^{r-1,0}(m_{i_0 l},\ldots,m_{i_r l})$ can be used to derive the following.

Proposition 3.5. a) If $X_{i'} \subset \bar{X}_i$, then

$$S_{i,k}^{r-1,0}(m_{i_0 l},\ldots,m_{i_r l})|_{\bar{X}_{i'}} = S_{i',k}^{r-1,0}(m_{i_0 l},\ldots,m_{i_r l});$$

b) if $X_j \subset \bar{X}_{l'} \subset \bar{X}_l$, then

$$S_{i,j}^{r-1,0}(m_{i_0 l},\ldots,m_{i_r l}) = S_{i,j}^{r-1,0}(m_{i_0 l'},\ldots,m_{i_r l'}) + S_{i,l'}^{r-1,0}(m_{i_0 l},\ldots,m_{i_r l});$$

c) the form $S_{i,k}^{r-1,0}(m_{i_0 l},\ldots,m_{i_r l})$ is skew-symmetric with respect to permutations of the indices $i_0,\ldots,i_r$.

4. CHARACTERISTIC FORMS IN THE COMPLEX $\Omega(X)$

Let P be a stable r form on $gl(\cdot,\mathbf{R})$ (see Sec. 2 above). For each set (ω, m) of connections and normals for the stratified set X we define the section $\Phi^0(P) = \Phi^0(P;\omega,m)$ of the bundle $\Omega^{2r}(X)$ as follows.

If $P = \Pi_r$, then we put

$$\Phi^0(\Pi_r)_{i_0\ldots i_l} = \Pi^{l,2r-l}(\omega_{i_0},\omega_{i_1 i_0},\ldots,\omega_{i_l i_0}); \quad \text{if} \quad l \leqslant r,$$

$$\Phi^0(\Pi_r)_{i_0\ldots i_{r+1}} = (-1)^{\frac{r(r+1)}{2}+1} S_{i_{r+1},i_1}^{r-1,0}(m_{i_1 i_0},\ldots,m_{i_{r+1},i_0}),$$

$$\Phi^0(\Pi_r)_{i_0\ldots i_l} = 0, \quad \text{if} \quad l > r+1.$$

In the general case, if $P = \sum_\lambda c_\lambda \Pi_1^{\lambda_1}\ldots\Pi_r^{\lambda_r}$ $(\lambda = (\lambda_1,\ldots,\lambda_r),\ \lambda_\nu \geqslant 0, \sum \nu\lambda_\nu = r)$ then we put

$$\Phi^0(P) = \sum_\lambda c_\lambda \Phi^0(\Pi_1)^{\lambda_1}\wedge\ldots\wedge\Phi^0(\Pi_r)^{\lambda_r},$$

where the product of the forms from $\Omega(X)$ is as in Definition 1.10.

In particular, the component $\Phi^0(P)_i$, corresponding to the 0-flag X_i, is equal to $P^{0,2r}(\omega_i)$, i.e., it is the characteristic form for the form P constructed from the connection ω_i.

THEOREM 4.1. The section $\Phi^0(P;\omega,m)$ is a cocycle.

Proof. It suffices to prove the theorem for the case $P = \Pi_r$.

Let $l \leq r$. We have

$$(d\Phi^0(\Pi_r))_{i_0\ldots i_l} = (-1)^l d\Pi^{l,2r-l}(\omega_{i_0},\omega_{i_1 i_0},\ldots,\omega_{i_l i_0}) + \Pi^{l-1,2r-l+1}(\omega_{i_1},\omega_{i_2 i_1},\ldots,\omega_{i_l i_1}) +$$

$$+\sum_{\mu=1}^{l}\Pi^{l-1,2r-l+1}(\omega_{i_0},\omega_{i_1 i_0},\ldots\hat{\omega}_{i_\mu i_0}\ldots,\omega_{i_l i_0}) =$$

$$= [(-1)^l d\Pi^{l,2r-l}(\omega_{i_0},\omega_{i_1 i_0},\ldots,\omega_{i_l i_0}) + (\Delta\Pi^{l-1,2r-l+1})(\omega_{i_0},\omega_{i_1 i_0},\ldots,\omega_{i_l i_0})] +$$

$$+ [\Pi^{l-1,2r-l+1}(\omega_{i_1},\omega_{i_2 i_1},\ldots,\omega_{i_l i_1}) - \Pi^{l-1,2r-l+1}(\omega_{i_1 i_0},\omega_{i_2 i_0},\ldots,\omega_{i_l i_0})].$$

The expression in the first set of square brackets is zero from property c) of the difference cocycle, and the expression in the second set of square brackets is zero from properties f) and g).

Let $l = r+1$. We have

$$(d\Phi^0(\Pi_r))_{i_0\ldots i_{r+1}} = (-1)^{r+\frac{r(r+1)}{2}} dS_{i_{r+1},i_1}^{r-1,0}(m_{i_1 i_0},\ldots,m_{i_{r+1}i_0}) +$$

$$+\Pi^{r,r}(\omega_{i_1},\omega_{i_2 i_1},\ldots,\omega_{i_{r+1}i_1}) + \sum_{\mu=1}^{r+1}(-1)^{\mu}\Pi^{r,r}(\omega_{i_0},\omega_{i_1 i_0},\ldots\hat{\omega}_{i_\mu i_0}\ldots,\omega_{i_{r+1}i_0}) =$$

$$= [\Pi^{r,r}(\omega_{i_1 i_0}, \ldots, \omega_{i_{r+1} i_0}) - \Pi^{r,r}(\omega_{i_1}, \omega_{i_2 i_1}, \ldots, \omega_{i_{r+1} i_1})] +$$
$$+ \Pi^{r,r}(\omega_{i_1}, \omega_{i_2 i_1}, \ldots, \omega_{i_{r+1} i_1}) +$$
$$+ \sum_{\mu=1}^{r+1} (-1)^{\mu} \Pi^{r,r}(\omega_{i_0}, \omega_{i_1 i_0}, \ldots \hat{\omega}_{i_\mu i_0} \ldots, \omega_{i_{r+1} i_0}) =$$
$$= (\Delta \Pi^{r,r})(\omega_{i_0}, \omega_{i_1 i_0}, \ldots, \omega_{i_{r+1} i_0}) = 0.$$

(Here Proposition 3.3 has been applied as well as properties c) and e) of the difference cocycle.)

Let $l = r + 2$. We have

$$(-1)^{\frac{r(r+1)}{2}+1} (d\Phi^0(\Pi_r))_{i_0 \ldots i_{r+2}} = [S^{r-1,0}_{i_{r+2}, i_2}(m_{i_2 i_1}, \ldots, m_{i_{r+2} i_1}) -$$
$$- S^{r-1,0}_{i_{r+2}, i_2}(m_{i_2 i_0}, \ldots, m_{i_{r+2} i_0})] + \sum_{\mu=2}^{r+2} (-1)^{\mu} S^{r-1,0}_{i_{r+2}, i_1}(m_{i_1 i_0}, \ldots \widehat{m_{i_\mu i_0}} \ldots, m_{i_{r+2} i_0}) =$$
$$= - S^{r-1,0}_{i_{r+2}, i_1}(m_{i_2 i_0}, \ldots, m_{i_{r+2} i_0}) + \sum_{\mu=2}^{2} (-1)^{\mu} S^{r-1,0}_{i_{r+2}, i_1}(m_{i_1 i_0}, \ldots \widehat{m_{i_\mu i_0}} \ldots, m_{i_{r+2} i_0}) =$$
$$= - (\Delta S^{r-1,0}_{i_{r+2}, i_1})(m_{i_1 i_0}, \ldots, m_{i_{r+2} i_0}) = 0.$$

(Here Propositions 3.5 b) and 3.4 have been used.)

If $l > r + 2$, then $(d\Phi^0(\Pi_r))_{i_0 \ldots i_l} = 0$ follows trivially. And so the proof is complete.

Let (ω, m) and (ω', m') be two sets of connections and normals for X. Define the section $\Phi^1(P) = \Phi^1(P; \omega, m; \omega', m')$ of the bundle $\Omega^{2r-1}(X)$ as follows.

If $P = \Pi_r$, then put

$$\Phi^1(\Pi_r)_{i_0 \ldots i_l} = \sum_{\mu=0}^{l} (-1)^{\mu} \Pi^{l+1, 2r-l-1}(\omega_{i_0}, \omega_{i_1 i_0}, \ldots, \omega_{i_\mu i_0}, \omega'_{i_\mu i_0}, \ldots, \omega'_{i_l i_0})$$

for $l \leq r - 1$;

$$\Phi^1(\Pi_r)_{i_0 \ldots i_r} = \sum_{\mu=1}^{r} (-1)^{\mu + \frac{r(r+1)}{2} + 1} S^{r-1,0}_{i_r, i_1}(m_{i_1 i_0}, \ldots, m_{i_\mu i_0} m'_{i_\mu i_0}, \ldots, m'_{i_r i_0}),$$
$$\Phi^1(\Pi_r)_{i_0 \ldots i_l} = 0 \text{ for } l > r.$$

In the general case, if $P = \sum_\lambda c_\lambda \Pi_1^{\lambda_1} \cdot \ldots \cdot \Pi_r^{\lambda_r}$ $(\lambda = (\lambda_1, \ldots, \lambda_r),\ \lambda_\nu \geqslant 0, \sum \nu\lambda_\nu = r)$ then put

$$\Phi^1(P; \omega, m; \omega', m') = \sum_\lambda c_\lambda \sum_{\nu=1}^{r} \Phi^0(\Pi_1; \omega, m)^{\lambda_1} \wedge \ldots \Phi^0(\Pi_{\nu-1}; \omega, m)^{\lambda_{\nu-1}} \wedge$$
$$\wedge \Phi^1(\Pi_\nu; \omega, m; \omega', m') \wedge [\Phi^0(\Pi_\nu; \omega, m)^{\lambda_\nu - 1} + \Phi^0(\Pi_\nu; \omega, m)^{\lambda_\nu - 2} \Phi^0(\Pi_\nu; \omega', m') + \ldots$$
$$\ldots + \Phi^0(\Pi_\nu; \omega', m')^{\lambda_\nu - 1}] \wedge \Phi^0(\Pi_{\nu+1}; \omega', m')^{\lambda_{\nu+1}} \wedge \ldots \wedge \Phi^0(\Pi_r; \omega', m')^{\lambda_r}.$$

From the definition $\Phi^1(P)$ the following assertion follows easily.

<u>Proposition 4.2.</u> If $(\omega', m') = (\omega, m)$, then $\Phi^1(P; \omega, m; \omega', m') = 0$.

<u>THEOREM 4.3.</u> The following equality holds:

$$\Phi^0(P; \omega', m') - \Phi^0(P; \omega, m) = d\Phi^1(P; \omega, m; \omega', m').$$

<u>Proof.</u> We prove the assertion of the theorem for $P = \Pi_r$ (the assertion for the general case then follows easily).

Let $l \leq r - 1$. We have

$$(d\Phi^1(\Pi_r))_{i_0 \ldots i_l} = (-1)^l \sum_{\mu=0}^{l} (-1)^{\mu} d\Pi^{l+1, 2r-l-1}(\omega_{i_0}, \omega'_{i_1 i_0}, \ldots, \omega_{i_\mu i_0}, \omega'_{i_\mu i_0}, \ldots, \omega'_{i_l i_0}) -$$
$$- \sum_{\mu=1}^{l} (-1)^{\mu} \Pi^{l, 2r-l}(\omega_{i_1}, \omega_{i_2 i_1}, \ldots, \omega_{i_\mu i_1}, \omega'_{i_\mu i_1}, \ldots, \omega'_{i_l i_1}) +$$

$$+\sum_{\nu=1}^{l}(-1)^{\nu}\Big[\sum_{\mu=0}^{\nu-1}(-1)^{\mu}\Pi^{l,2r-l}(\omega_{i_0},\omega_{i_1i_0},\ldots,\omega_{i_\mu i_0},\omega'_{i_\mu i_0},\ldots\hat{\omega}'_{i_\nu i_0}\ldots,\omega'_{i_l i_0})-$$

$$-\sum_{\mu=\nu+1}^{l}(-1)^{\mu}\Pi^{l,2r-l}(\omega_{i_0},\omega_{i_1i_0},\ldots\hat{\omega}_{i_\nu i_0}\ldots,\omega_{i_\mu i_0},\omega'_{i_\mu i_0},\ldots,\omega'_{i_l i_0})\Big]=$$

$$=(-1)^{l}\sum_{\mu=0}^{l}(-1)^{\mu}d\Pi^{l+1,2r-l-1}(\omega_{i_0},\omega_{i_1i_0},\ldots,\omega_{i_\mu i_0},\omega'_{i_\mu i_0},\ldots,\omega'_{i_l i_0})-$$

$$-\sum_{\mu=1}^{l}(-1)^{\mu}\Pi^{l,2r-l}(\omega_{i_1},\omega_{i_2i_1},\ldots,\omega_{i_\mu i_1},\omega'_{i_\mu i_1},\ldots,\omega'_{i_l i_1})+$$

$$+\sum_{\nu=1}^{l}(-1)^{\nu}\Pi^{l,2r-l}(\omega_{i_0},\omega'_{i_0},\omega'_{i_1i_0},\ldots\hat{\omega}'_{i_\nu i_0}\ldots,\omega'_{i_l i_0})+$$

$$+\sum_{\mu=1}^{l}(-1)^{\mu}\Big[-\sum_{\nu=1}^{\mu-1}(-1)^{\nu}\Pi^{l,2r-l}(\omega_{i_0},\omega_{i_1i_0},\ldots\hat{\omega}_{i_\nu i_0}\ldots,\omega_{i_\mu i_0},\omega'_{i_\mu i_0},\ldots,\omega'_{i_l i_0})+$$

$$+\sum_{\nu=\mu+1}^{l}(-1)^{\nu}\Pi^{l,2r-l}(\omega_{i_0},\omega_{i_1i_0},\ldots,\omega_{i_\mu i_0},\omega'_{i_\mu i_0},\ldots\hat{\omega}'_{i_\nu i_0}\ldots,\omega'_{i_l i_0})=$$

$$=\sum_{\mu=0}^{l}(-1)^{\mu}[(-1)^{l}d\Pi^{l+1,2r-l-1}(\omega_{i_0},\omega_{i_1i_0},\ldots,\omega_{i_\mu i_0},\omega'_{i_\mu i_0},\ldots,\omega'_{i_l i_0})-$$

$$-(\Delta\Pi^{l,2r-l})(\omega_{i_0},\omega_{i_1i_0},\ldots,\omega_{i_\mu i_0},\omega'_{i_\mu i_0},\ldots,\omega'_{i_l i_0})]+$$

$$+\Pi^{l,2r-l}(\omega'_{i_0},\omega'_{i_1i_0},\ldots,\omega'_{i_l i_0})-\Pi^{l,2r-l}(\omega_{i_0},\omega'_{i_1i_0},\ldots,\omega'_{i_l i_0})+$$

$$+\sum_{\mu=1}^{l}[\Pi^{l,2r-l}(\omega_{i_0},\omega_{i_1i_0},\ldots,\omega_{i_{\mu-1}i_0},\omega'_{i_\mu i_0},\ldots,\omega'_{i_l i_0})-$$

$$-\Pi^{l,2r-l}(\omega_{i_0},\omega_{i_1i_0},\ldots,\omega_{i_\mu i_0},\omega'_{i_{\mu+1}i_0},\ldots,\omega'_{i_l i_0})]+$$

$$+\sum_{\mu=1}^{l}(-1)^{\mu}\Pi^{l,2r-l}(\omega_{i_1i_0},\ldots,\omega_{i_\mu i_0},\omega'_{i_\mu i_0},\ldots,\omega'_{i_l i_0})-$$

$$-\sum_{\mu=1}^{l}(-1)^{\mu}\Pi^{l,2r-l}(\omega_{i_1},\omega_{i_2i_1},\ldots,\omega_{i_\mu i_1},\omega'_{i_\mu i_1},\ldots,\omega'_{i_l i_1}).$$

In this expression, the term in the first set of square brackets is zero from property c) of the difference cocycle, the next two sums cancel from properties f) and g), and the remaining terms give, after the like terms are reduced,

$$\Pi^{l,2r-1}(\omega'_{i_0},\omega'_{i_1i_0},\ldots,\omega'_{i_l i_0})-\Pi^{l,2r-l}(\omega_{i_0},\omega_{i_1i_0},\ldots,\omega_{i_l i_0})=\Phi^0(\Pi_r;\omega',m')_{i_0\ldots i_l}-\Phi^0(\Pi_r;\omega,m)_{i_0\ldots i_l}.$$

Let $l = r$. We have

$$(d\Phi^1(\Pi_r))_{i_0\ldots i_r}=\sum_{\mu=1}^{r}(-1)^{r+\mu+\frac{r(r+1)}{2}+1}dS^{r-1,0}_{i_r,i_1}(m_{i_1i_0},\ldots,m_{i_\mu i_0},m'_{i_\mu i_0},\ldots,m'_{i_r i_0})-$$

$$-\sum_{\mu=1}^{r}(-1)^{\mu}\Pi^{r,r}(\omega_{i_1},\omega_{i_2i_1},\ldots,\omega_{i_\mu i_1},\omega'_{i_\mu i_1},\ldots,\omega'_{i_r i_1})+$$

$$+\sum_{\nu=1}^{r}(-1)^{\nu}\Big[\sum_{\mu=0}^{\nu-1}(-1)^{\mu}\Pi^{r,r}(\omega_{i_0},\omega_{i_1i_0},\ldots,\omega_{i_\mu i_0},\omega'_{i_\mu i_0},\ldots\hat{\omega}'_{i_\nu i_0}\ldots,\omega'_{i_r i_0})-$$

$$-\sum_{\mu=\nu+1}^{r}(-1)^{\mu}\Pi^{r,r}(\omega_{i_0},\omega_{i_1i_0},\ldots\hat{\omega}_{i_\nu i_0}\ldots,\omega_{i_\mu i_0},\omega'_{i_\mu i_0},\ldots,\omega'_{i_r i_0})\Big]=$$

$$=-\sum_{\mu=1}^{r}(-1)^{\mu}[\Pi^{r,r}(\omega_{i_1i_0},\ldots,\omega_{i_\mu i_0},\omega'_{i_\mu i_0},\ldots,\omega'_{i_r i_0})-$$

$$-\Pi^{r,r}(\omega_{i_1},\omega_{i_2i_1},\ldots,\omega_{i_\mu i_1},\omega'_{i_\mu i_1},\ldots,\omega'_{i_r i_1})]-$$

$$-\sum_{\mu=1}^{r}(-1)^{\mu}\Pi^{r,r}(\omega_{i_1},\omega_{i_2i_1},\ldots,\omega_{i_\mu i_1},\omega'_{i_\mu i_1},\ldots,\omega'_{i_r i_1})+$$

$$+\sum_{\nu=1}^{r}(-1)^{\nu}\Pi^{r,r}(\omega_{i_0},\omega'_{i_0},\omega'_{i_1i_0},\ldots,\hat{\omega}'_{i_\nu i_0}\ldots,\omega'_{i_r i_0})+$$

$$
\begin{aligned}
&+\sum_{\mu=1}^{r}(-1)^{\mu}\Big[-\sum_{\nu=1}^{\mu-1}(-1)^{\nu}\Pi^{r,r}(\omega_{\cdot_0},\omega_{i_1i_0},\dots\hat{\omega}_{i_\nu i_0}\dots,\omega_{i_\mu i_0},\omega'_{i_\mu i_0}\dots,\omega'_{i_r i_0})+\\
&\quad+\sum_{\nu=\mu+1}^{r}(-1)^{\nu}\Pi^{r,r}(\omega_{i_0},\omega_{i_1i_0},\dots,\omega_{i_\mu i_0},\omega'_{i_\mu i_0},\dots\hat{\omega}'_{i_\nu i_0}\dots,\omega'_{i_r i_0})\Big]=\\
&\qquad=-(\Delta\Pi^{r,r})(\omega_{i_0},\omega'_{i_0},\omega'_{i_1i_0},\dots,\omega'_{i_ri_0})+\\
&\qquad+\Pi^{r,r}(\omega'_{i_0},\omega'_{i_1i_0},\dots,\omega'_{i_ri_0})-\Pi^{r,r}(\omega_{i_0}\omega'_{i_1i_0},\dots,\omega'_{i_ri_0})-\\
&\qquad-\sum_{\mu=1}^{r}(-1)^{\mu}(\Delta\Pi^{r,r})(\omega_{i_0},\omega_{i_1i_0},\dots,\omega_{i_\mu i_0},\omega'_{i_\mu i_0},\dots,\omega'_{i_ri_0})+\\
&\qquad+\sum_{\mu=1}^{r}[\Pi^{r,r}(\omega_{i_0},\omega_{i_1i_0},\dots,\omega_{i_{\mu-1}i_0},\omega'_{i_\mu i_0},\dots,\omega'_{i_ri_0})-\\
&\qquad-\Pi^{r,r}(\omega_{i_0},\omega_{i_1i_0},\dots,\omega_{i_\mu i_0},\omega'_{i_{\mu+1}i_0},\dots,\omega'_{i_ri_0})]=\\
&\qquad=\Pi^{r,r}(\omega'_{i_0},\omega'_{i_1i_0},\dots,\omega'_{i_ri_0})-\Pi^{r,r}(\omega_{i_0},\omega_{i_1i_0},\dots,\omega_{i_ri_0})=\\
&\qquad=\Phi^0(\Pi_r;\,\omega',m')_{i_0,\dots,i_r}-\Phi^0(\Pi_r;\,\omega,m)_{i_0,\dots,i_r}.
\end{aligned}
$$

Here, besides changing the order of summation, we have used Proposition 3.3 and properties e) and c) of the difference cocycle.

Let $l = r + 1$. We have

$$
\begin{aligned}
&(-1)^{\frac{r(r+1)}{2}+1}(d\Phi^1(\Pi_r))_{i_0\dots i_{r+1}}=\\
&\quad=-\sum_{\mu=2}^{r+1}(-1)^{\mu}S^{r-1,0}_{i_{r+1},i_1}(m_{i_2i_1},\dots,m_{i_\mu i_1},m'_{i_\mu i_1},\dots,m'_{i_{r+1}i_1})+\\
&\quad+\sum_{\mu=2}^{r+1}(-1)^{\mu}S^{r-1,0}_{i_{r+2},i_1}(m_{i_2i_0},\dots,m_{i_\mu i_0},m'_{i_\mu i_0},\dots,m'_{i_{r+1}i_0})+\\
&+\sum_{\nu=2}^{r+1}(-1)^{\nu}\Big[\sum_{\mu=1}^{\nu-1}(-1)^{\mu}S^{r-1,0}_{i_{r+1},i_0}(m_{i_2i_0},\dots,m_{i_\mu i_0},m'_{i_\mu i_0},\dots\hat{m}'_{i_\nu i_0}\dots,m'_{i_{r+1}i_0})-\\
&\quad-\sum_{\mu=\nu+1}^{r+1}(-1)^{\mu}S^{r-1,0}_{i_{r+1},i_1}(m_{i_1i_0},\dots\hat{m}_{i_\nu i_0}\dots,m_{i_\mu i_0},m'_{i_\mu i_0},\dots,m'_{i_{r+1}i_0})\Big]=\\
&=\sum_{\mu=2}^{r+1}(-1)^{\mu}S^{r-1,0}_{i_{r+1},i_1}(m_{i_2i_0},\dots,m_{i_\mu i_0},m'_{i_\mu i_0},\dots,m'_{i_{r+1}i_0})-\\
&\quad-\sum_{\nu=2}^{r+1}(-1)^{\nu}S^{r-1,0}_{i_{r+1},i_1}(m_{i_1i_0},m'_{i_1i_0},\dots\hat{m}'_{i_\nu i_0}\dots,m'_{i_{r+1}i_0})+\\
&+\sum_{\mu=2}^{r+1}(-1)^{\mu}\Big[-\sum_{\nu=2}^{\mu-1}(-1)^{\nu}S^{r-1,0}_{i_{r+1},i_1}(m_{i_1i_0},\dots\hat{m}_{i_\nu i_0}\dots,m_{i_\mu i_0},m'_{i_\mu i_0},\dots,m'_{i_{r+1}i_0})+\\
&\quad+\sum_{\nu=\mu+1}^{r+1}(-1)^{\nu}S^{r-1,0}_{i_{r+1},i_1}(m_{i_1i_0},\dots,m_{i_\mu i_0},m'_{i_\mu i_0},\dots\hat{m}'_{i_\nu i_0}\dots,m'_{i_{r+1}i_0})\Big]=\\
&\qquad=-(\Delta S^{r-1,0}_{i_{r+1},i_1})(m_{i_1i_0},m'_{i_1i_0},\dots,m'_{i_{r+1}i_0})+\\
&\quad+S^{r-1,0}_{i_{r+1},i_1}(m'_{i_1i_0},\dots,m'_{i_{r+1}i_0})-S^{r-1,0}_{i_{r+1},i_1}(m_{i_1i_0},m'_{i_2i_0},\dots,m'_{i_{r+1}i_0})+\\
&\quad+\sum_{\mu=2}^{r+1}(-1)^{\mu}(\Delta S^{r-1,0}_{i_{r+1},i_1})(m_{i_1i_0},\dots,m_{i_\mu i_0},m'_{i_\mu i_0},\dots,m'_{i_{r+1}i_0})+\\
&\quad+\sum_{\mu=2}^{r+1}[S^{r-1,0}_{i_{r+1},i_1}(m_{i_1i_0},\dots,m_{i_{\mu-1}i_0},m'_{i_\mu i_0},\dots,m'_{i_{r+1}i_0})-\\
&\quad-S^{r-1,0}_{i_{r+1},i_1}(m_{i_1i_0},\dots,m_{i_\mu i_0},m'_{i_{\mu+1}i_0},\dots,m'_{i_{r+1}i_0})\Big]=\\
&\quad=S^{r-1,0}_{i_{r+1},i_1}(m'_{i_1i_0},\dots,m'_{i_{r+1}i_0})-S^{r-1,0}_{i_{r+1},i_1}(m_{i_1i_0},\dots,m_{i_{r+1}i_0})=\\
&\qquad=\Phi^0(\Pi_r;\,\omega',m')_{i_0\dots i_{r+1}}-\Phi^0(\Pi_r;\,\omega,m)_{i_0,\dots,i_{r+1}}.
\end{aligned}
$$

Here we have used Propositions 3.5b) and 3.4.

If $l > r + 1$, then

$$(d\Phi^1(\Pi_r;\omega,m;\omega',m'))_{i_0\dots i_l}=\Phi^0(\Pi_r;\omega',m')_{i_0\dots i_l}-\Phi^0(\Pi_r;\omega,m)_{i_0\dots i_l}=0.$$

Thus, Theorem 4.3 has been proved completely.

Definition 4.4. Let (ω, m) be a set of connections and normals for X, and let X_l be a stratum of X. The connection ω_i (resp., the normal m_{jk}) is interior to X_l, if $X_i \subset \overline{X}_l$ (resp. $X_k \subset \overline{X}_l$.

Remark. If the set (ω_g, m_g) is constructed from the metric g on X (see Definition 3.2), then the connections and normals inferior to X_l are determined by the metric $g|\overline{X}_l$.

The following simple but important assertion is derived immediately from the definitions of $\Phi^0(P)$ and $\Phi^1(P)$.

Proposition 4.5. The forms $\Phi^0(P)_{i_0\ldots i_l}$ and $\Phi^1(P)_{i_0\ldots i_l}$ depend only on the connections and normals inferior to X_{i_0}.

5. FUNCTIONALS FOR CHARACTERISTIC CLASSES

Let X be a stratified set, P a stable r form on $\mathfrak{gl}(\cdot, R)$, and Z a smooth singular 2r chain on X. Put

$$\Phi_Z(P;\omega,m)=\int\limits_Z \Phi^0(P;\omega,m).$$

THEOREM 5.1. The number $\Phi_Z(P; \omega, m)$, as a function of ω_i, m_{jk}, depends only on connections and normals interior to strata having nonempty intersection with ∂Z.

We derive this theorem from the following more precise assertion.

THEOREM 5.2. If (ω, m) and (ω', m') are two sets of connections and normals, then

$$\Phi_Z(P;\omega',m')-\Phi_Z(P;\omega,m)=\int\limits_{\partial Z}\Phi^1(P;\omega,m;\omega',m').$$

Proof. This assertion follows from Theorem 4.3 and the Stokes' formula for the complex $\Omega(X)$.

Proof of Theorem 5.1. Let us assume that, for any stratum X_k whose intersection with ∂Z is nonempty, the connections ω_i and normals m_{ij} are interior to X_k, and coincide with the corresponding connections ω_i' and normals m_{ij}'. From Proposition 4.5 and the definition of the integral on the complex $\Omega(X)$ if follows that

$$\int\limits_{\partial Z}\Phi^1(P;\omega,m;\omega',m')=\int\limits_{\partial Z}\Phi^1(P;\omega,m;\omega,m).$$

On the other hand, $\Phi^1(P; \omega, m; \omega, m) = 0$ according to Proposition 4.2. It then follows from Theorem 5.2 that $\Phi_Z(P; \omega', m') = \Phi_Z(P; \omega, m)$, QED.

Let Y be a closed subspace of X which is the union of several strata of X. Then the space Y with the decomposition $\{X_\alpha : X_\alpha \subset Y\}$, is a stratified set.

THEOREM 5.3. Let Z be a 2r chain on X such that its boundary ∂Z lies in Y. Let g be a metric on Y and $\tilde{g}$ a metric on X such that $\tilde{g}|_Y = g$. Let $(\omega_{\tilde{g}}, m_{\tilde{g}})$ be a set of connections and normals for X constructed from the metric $\tilde{g}$ according to Definition 3.2. Then the number $\Phi_Z(P;g)=\int\limits_Z\Phi^0(P;\omega_{\tilde{g}},m_{\tilde{g}})$ does not depend on any choices made in the selection of $\tilde{g}$ and is therefore a functional of the metric g.

These functionals make their appearance as a result of our attempts to generalize the remarkable Atiyah–Patodi–Singer formula [1] to manifolds with piecewise smooth boundaries, and we shall call them Atiyah–Patodi–Singer functionals.

THEOREM 5.4. Let Z be a cycle on X. Then

a) $\Phi_Z(P; \omega, m)$ does not change when Z is replaced by a cycle homologous to it;

b) $\Phi_Z(P; \omega, m)$ does not depend on the choice of the set (ω, m).

Proof. Assertion a) follows from Theorem 4.1 and the Stokes' formula on the complex $\Omega(X)$. Assertion b) follows from Theorem 5.1.

Thus, $\Phi_Z(P; \omega, m)$ determines a cohomology class $\Phi(P, X) \in H^{2r}(X, R)$, which depends only on the form P and the stratification of the set X.

Definition 5.5. A stratification of X is linearizable if there exists a set of connections $\omega = \{\omega_i\}$ such that

i) ω_i is a flat connection, i.e., its curvature is zero, and

ii) if $X_{i'} \subset \bar{X}_i$, then the subbundle $T\bar{X}_{i'}$ in $T\bar{X}_i|_{\bar{X}_{i'}}$ is invariant with respect to the restriction of the connection ω_i to $T\bar{X}_i|_{\bar{X}_{i'}}$ and the connection, induced by the connection ω_i on $T\bar{X}_{i'}$, coincides with $\omega_{i'}$.

An arbitrary locally finite simplicial complex is an example of a linearizable stratification.

Proposition 5.6. If the stratification of X is linearizable then $\Phi(P, X) = 0$ for any r-form P.

Proof. It suffices to prove the assertion for the case $P = \Pi_r$. Let X be a linearizable stratified set, and let $\omega = \{\omega_i\}$ be a set of connections satisfying the conditions of Definition 5.5. For each pair of strata $X_j \subset \bar{X}_k$, $\dim X_k = \dim X_j + 1$, choose a section m_{jk} of the bundle $T\bar{X}_k|_{\bar{X}_j}$, which is horizontal with respect to the connection ω_k and does not intersect $T\bar{X}_j$. Such a section exists since ω_k is a flat connection. Let $m = \{m_{jk}\}$ denote the chosen set of sections. We now prove that $\Phi^0(\Pi_r; \omega, m) = 0$ for $r > 1$, $\Phi^0(\Pi_1; \omega, m) = d\Psi$, where Ψ is some form from $\Omega^1(X)$.

Let $\Phi^0(\Pi_r) = \Phi^0(\Pi_r; \omega, m)$. We must show that $\Phi^0(\Pi_r)_{i_0 \ldots i_l} = 0$ for $l \geq 0$ and for any l-flag $X_{i_0}, \ldots, X_{i_l}$. Let $l = 0$. Then $\Phi^0(\Pi_r)_{i_0} = 0$ from property a) of the difference cocycle, since ω_{i_0} is a flat connection. If $1 \leq l \leq r$, then $\Phi^0(\Pi_r)_{i_0 \ldots i_l} = \Pi^{l, 2r-l}(\omega_{i_0}, \omega_{i_1 i_0}, \ldots, \omega_{i_l i_0})$. But from the definition of the sections m_{ij} and from condition ii) of Definition 5.5 one may easily deduce that the connection $\omega_{i_\nu i_0}$ coincides with the restriction of the connection ω_{i_0} to $T\bar{X}_{i_0}|_{X_{i_\nu}}$. Therefore

$$\Pi^{l, 2r-l}(\omega_{i_0}, \omega_{i_1 i_0}, \ldots, \omega_{i_l i_0}) = \Pi^{l, 2r-l}(\omega_{i_0}, \ldots, \omega_{i_0}) = 0$$

from property b) of the difference cocycle.

Now let $l = r + 1$. Let X_j and X_i be two strata such that $X_i \subset \bar{X}_j$, and let X_k be a stratum such that $X_j \subset \bar{X}_k$, $\dim X_k = \dim X_j + 1$. Let $k' = \varphi_{ij}(k)$, $E_{ik'} = (T\bar{X}_{k'}|_{\bar{X}_i})/T\bar{X}_i$, $E_{jk} = (T\bar{X}_k|_{\bar{X}_j})/T\bar{X}_j$ and let $\eta_{ij}: E_{ik'} \to E_{jk}|_{\bar{X}_i}$ be the bundle isomorphism of section 3 above. From condition ii) of Definition 5.5 one easily deduces that the connections induced by the connections $\omega_{k'}$ and ω_k on $E_{ik'}$ and E_{jk}, respectively, are transformed into each other under the map η_{ij}. Let $\bar{m}_{ik'}$, $\bar{m}_{jk}$ denote the images of the sections $m_{ik'}$, m_{jk} under the projections $T\bar{X}_{k'}|_{\bar{X}_i} \to E_{ik'}$ and $T\bar{X}_k|_{\bar{X}_j} \to E_{jk}$, respectively. From the definition of the sections m_{jk} it follows that $\eta_{ij}(\bar{m}_{ik'})$ and $m_{jk}|_{\bar{X}_i}$ are horizontal sections with respect to the same connection in the one-dimensional bundle $E_{kj}|_{\bar{X}_i}$. Therefore, their ratio is a constant function on $\bar{X}_i$, and in particular $d \log \left| \frac{\eta_{ij}(\bar{m}_{ik'})}{m_{jk}} \right| = 0$. From this and the definition of the form $S^{r-1,0}$ in (3)-(5) one easily sees that

$$\Phi^0(\Pi_r)_{i_0 \ldots i_{r+1}} = (-1)^{\frac{r(r+1)}{2}+1} S^{r-1,0}_{i_{r+1}, i_1}(m_{i_1 i_0}, \ldots, m_{i_{r+1} i_0}) = 0,$$

if $r > 1$. Thus $\Phi^0(\Pi_r) = 0$ for $r > 1$.

Now let $r = 1$. For each stratum X_k choose a volume form which is invariant with respect to the connection ω_k (this can be done since ω_k is a flat connection). Define the form Ψ from $\Omega^1(X)$ as follows. If $k \neq 1$, then $\Psi_{i_0 \ldots i_k} = 0$. Let X_{i_0} and X_{i_1} be two strata such that $\bar{X}_{i_0} \supset X_{i_1}$, $\dim X_{i_1} = p$. Let $x \in X_{i_1}$ and let $l_1(x), \ldots, l_p(x)$ be some basis in $(TX_{i_1})_x$. Let $V_1(x)$ denote the volume in $(TX_{i_1})_x$ of the parallelepiped generated by the set of vectors $l_1(x), \ldots, l_p(x)$, and let $V_0(x)$ denote the volume in $(T\bar{X}_{i_0})_x$ of the parallelepiped generated by the vectors $l_1(x), \ldots, l_p(x)$ and the set of normals $m_{i_1 i_0}(x)$. Put $\Psi_{i_1 i_0}(x) = \frac{V_0(x)}{V_1(x)}$. It is obvious that $\Psi_{i_1 i_0}(x)$ does not depend on the choice of the vectors $l_1(x), \ldots, l_p(x)$. It is not hard to prove that for the form Ψ so defined the relation $d\Psi = \Phi^0(\Pi_1; \omega, m)$ holds. Thus, the cohomology class $\Phi(\Pi_r, X)$ is trivial for all forms Π_r, as was to be proved.

Let us now assume that X is a smooth n-dimensional manifold equipped with a decomposition $\{X_\alpha\}$ into simple cells (see [2, Sec. 1]), and let the strata of X be the interiors of the cells in the decomposition $\{X_\alpha\}$. Let $P(X) \in H^{2r}(X; \mathbb{R})$ denote the characteristic class of the manifold X corresponding to the form P. Proposition 5.6 implies that $P(X) \neq \Phi(P, X)$ in general (e.g., $\Phi(P, X) = 0$ for any simplicial decomposition). However, the following result is true.

THEOREM 5.7. Assume that the decomposition into simple cells $\{X_\alpha\}$ satisfies the following condition: Each cell of codimension p in this decomposition is contained in exactly $(p + 1)$ cells of codimension $(p - 1)$ (if this condition holds for any value $p \geq 2$, then it holds for all p).

Then $\Phi(P, X) = P(X)$.

Let $\{X_\alpha\}$ be an arbitrary decomposition into simple cells which satisfies condition (A) of Sec. 5 of [2]. Let $P = -(1/8\pi^2)\Pi_2$ be the form which corresponds to the first Pontryagin class of the manifold X. Then the following assertion holds.

THEOREM 5.8. The cohomology class $P(X) - \Phi(P, X)$ coincides with the class which is the dual to the Poincaré combinatorial cycle $\Gamma_{n-4}(X)$, defined in Sec. 10 of [2].

The proofs of Theorems 5.7 and 5.8 are rather long and we do not give them here.

Remark. If the decomposition $\{X_\alpha\}$ is linearizable then Theorem 2 of [2] follows from Theorem 5.7 (for the case where the decomposition is into simple cells); and Theorem 5.8 implies Theorem 1 of [2].

LITERATURE CITED

1. M. F. Atiyah, V. K. Patodi, and I. M. Singer, "Spectral asymmetry and Riemannian geometry," Bull. London Math. Soc., 5, 229-234 (1973).
2. A. M. Gabriélov, I. M. Gel'fand, and M. B. Locik, "Combinatorial calculation of characteristic classes," Funktsional'. Analiz i Ego Prilozhen., 9, No. 2, 12-28 (1975); No. 3, 5-26 (1975).
3. D. B. Fuchs, A. M. Gabriélov, and I. M. Gel'fand, "The Gauss–Bonnet formula and the Atiyah–Patodi–Singer functionals for characteristic classes of foliations," Topology (1976).

20.
(with M. V. Losik)

Computing of characteristic classes of combinatorial vector bundles

Prepr. Inst. Appl. Mat. 99 (1976)

In this paper we introduce the notion of a combinatorial vector bundle and state the problem of the combinatorial computation of characteristic classes of such bundles. This problem is solved only for the first Pontryagin class, but there are reasons to think that the techniques worked out here for this purpose will be useful for the combinatorial computation of other characteristic classes.

The notion of a combinatorial vector bundle arises as a natural generalization of the notion of a smooth manifold with a linearized cellular decomposition. Therefore this work contains, in particular, the main result of the articles [1] and [2] on the combinatorial calculation of the first Pontryagin class of a smooth manifold.

The first chapter contains some necessary preliminary material. The second chapter contains the principal notions and results of the present work. Note that a considerable part of this work is a revision of [1] and [2].

The smoothness of manifolds, mappings, forms, etc. is understood throughout this paper as their belonging to the class C^∞.

The authors are glad to express their sincere gratitude to Pierre Deligne who suggested replacing the "configuration space" of the articles [1] and [2] with a space of "smoothings". The corresponding notion for combinatorial vector bundles is that of the space of fittings (see Section 1 of Chapter II) which is broadly used in this paper.

Chapter I. Preliminary notions and results

1. Cellular decompositions

We shall define a linear l-dimensional cell in the space R^l as a convex compact l-dimensional polyhedron in R^l.

Definition 1.1. Let X be an n-dimensional topological manifold. We shall say that a triple (K, Y, φ), where K is an l-dimensional linear cell in R^l, $Y \subset X$, and $\varphi: K \to Y$ is a homeomorphism, defines an l-dimensional cell in X. Triples (K_1, Y, φ_1) and (K_2, Y, φ_2) define the same l-dimensional cell in X if $\varphi_2^{-1} \circ \varphi_1$ is C^∞-diffeomorphism of K_1 onto K_2.

Thus the set Y is equipped with the structure of a smooth l-dimensional manifold with a piecewise smooth boundary. We shall denote the l-dimensional cell in X simply by Y and assume that the indicated structure is given for Y; the mapping $\varphi: K \to Y$ will be called a chart of the given cell.

The faces of the cell Y are defined in the usual way. It is clear that the faces of the cell Y are again cells in X. Moreover, if Y' is a face of the cell Y then the structures of manifolds with piecewise smooth boundaries on Y and Y' are compatible with each other.

Let Y be an l-dimensional cell in X, and Y' its l'-dimensional face. The number $n-l$ is called the codimension of the cell Y, and the number $l-l'$ is called the codimension of the face Y' of the cell Y. If Y' is the face of codimension k of the cell Y, then there exist at least k different faces of codimension $k-1$ of Y which contain Y'.

Definition 1.2. A cell Y is called simple if any of its faces of codimension k ($k=1,\dots,\dim Y$) is contained precisely in k faces of codimension $k-1$ of the cell Y.

The simplest example of a simple cell is a cell Y in X defined by a triple (K, Y, φ) where the linear cell K is a simplex or a product of simplexes.

Definition 1.3. A family $\mathscr{X} = \{X_i\}_{i\in I}$ of cells of the manifold X is called a cellular decomposition of this manifold if the following conditions are satisfied.

1. The family $\mathscr{X}$ forms a locally finite covering of the manifold X.
2. $X_i \neq X_j$, for $i \neq j$.
3. If Y is a face of a cell X_i then $Y = X_j$ for some $j \in I$.
4. The intersection $X_i \cap X_j$ is a (possibly empty) face of both cells X_i and X_j.

In what follows, an l-dimensional cell X_i of the decomposition $\mathscr{X}$ will sometimes be denoted by X_i^l.

If X is a smooth manifold then we shall consider only those cellular decompositions of X for which the charts of all the cells are regular smooth mappings.

Let $\mathscr{X} = \{X_i\}_{i\in I}$ be a cellular decomposition of a topological manifold X. For $i \in I$ denote by $s(i)$ the set of indices $j \in I$ such that $X_j \supset X_i$ and $\dim X_j = \dim X_i + 1$.

Proposition 1.1. *If $j \in s(i)$ and $k \in s(j)$ then there exists the unique index j' such that $j' \in s(i)$, $k \in s(j')$ and $j' \neq j$.*

Proof. The union of all the faces of the cell X_k, different from X_k, is a topological sphere with a cellular decomposition. The cell X_i has codimension 1 in this sphere, and hence is the common face of two different cells of codimension 0. One of these is X_j, and the other is just $X_{j'}$.

Define the mapping $\varphi_{ij}: s(j) \to s(i) (j \in s(i))$ by the condition $\varphi_{ij}(k) = j'$, where $k \in s(j)$, and j' is defined by Proposition 1.1.

Proposition 1.2. *Let $\mathscr{X}$ be a cellular decomposition of a manifold X with all cells being simple. If $j, j' \in s(i)$ and $k \in s(j) \cap s(j')$, then $\varphi_{ij}\varphi_{jk} = \varphi_{ij'}\varphi_{j'k}$.*

Proof: Let $j' \neq j$ and $l \in s(k)$. Since X_i is a face of codimension 3 of the cell X_l and the cell X_l is simple, then there exist precisely 3 faces of codimension 2 of the cell X_l containing X_i. Two of these are X_j and $X_{j'}$; denote the third by $X_{j''}$. According to the definition of the mappings φ_{ij} we have $\varphi_{ij}\varphi_{jk}(l) = \varphi_{ij'}\varphi_{j'k}(l) = j''$.

Below we consider only cellular decompositions with all cells being simple.

2. *Orientation sheaf and cooriented cells*

Let X be an n-dimensional topological manifold. For any open set $U \subset X$ we put $\Lambda(U) = H_n(X, X-U; \mathbb{Z})$. It is easy to see that the correspondence $U \mapsto \Lambda(U)$ defines a sheaf Λ of groups on X. For any contractible set $Y \subset X$ we have $\Lambda(Y) = \mathbb{Z}$. Hence Λ is a locally constant sheaf whose stalk $\Lambda(x)$ at any point $x \in X$ is isomorphic to $\mathbb{Z}$. We shall call the sheaf Λ on X the orientation sheaf of the manifold X.

The manifold X is called orientable if there exists a global section ε_X of the sheaf Λ whose restriction to any point $x \in X$ is a generator of the group $\Lambda(x)$. If such a global section ε_X is chosen, then the manifold X is called oriented, and the section ε_X is called then orientation of X. The orientation $-\varepsilon_X$ is called opposite to the orientation ε_X.

Now let X be an arbitrary n-dimensional manifold (not necessarily orientable). If Y is a contractible subset of X then $\Lambda(Y) = \mathbb{Z}$, and either of the two sections ε_Y and $-\varepsilon_Y$ of the sheaf Λ on Y, which is restricted for any point $x \in Y$ to the generator of the stalk $\Lambda(x)$, is called a local orientation of the subset Y.

Let Y be an l-dimensional cell on X. If $l > 0$ then the interior $\mathring{Y}$ of the cell Y regarded as an l-dimensional manifold with a piecewise boundary is an l-dimensional manifold without boundary. We shall call any orientation of the interior $\mathring{Y}$ an orientation of the cell Y. In addition to this we shall agree that any 0-dimensional cell in X is orientable and has a unique orientation.

Definition 2.1. One says that there is a coorientation of an l-dimensional cell Y in X if, to any orientation ε of the cell Y, there corresponds a local orientation $o(\varepsilon)$ of Y as a subset of X, and $o(-\varepsilon) = -o(\varepsilon)$. A cell Y with a given coorientation is called cooriented.

If Y is a cooriented cell, then the same cell with the opposite coorientation is denoted by $-Y$. If Y is an n-dimensional cell in X, then any local orientation of the subset Y defines an orientation of the cell Y. This correspondence defines the natural coorientation of an n-dimensional cell Y.

Let Y be an l-dimensional cell in X and let Y' be its face of codimension 1. Then to any orientation of the cell Y corresponds in the usual way some orientation of the cell Y'. One says that coorientations of the cells Y and Y' are compatible if there are mutually corresponding local orientations of the subsets Y and Y' which correspond to mutually corresponding orientations of the cells Y and Y'. If the cells Y and Y are cooriented in an arbitrary way, then we say their incidence number is a number ε which is equal to 1 if the coorientations are compatible and equal to -1 otherwise.

Definition 2.2. Let $\mathscr{X}$ be a cellular decomposition of a manifold X. We denote by $\hat{C}_l(\mathscr{X})$ an Abelian group composed by (possibly infinite) linear combinations of cooriented l-dimensional cells of the decomposition $\mathscr{X}$ with integral coefficients, where the cells with opposite coorientation differ in sign. We define the differential $\hat{\partial}: \hat{C}_l(\mathscr{X}) \to \hat{C}_{l-1}(\mathscr{X})$ by formula $\hat{\partial} X_j^l = \sum_{i; j \in s(i)} \varepsilon_{ij} X_i^{l-1}$, where ε_{ii} is the incidence number of the cooriented cell X_i^{l-1} and X_j^l. The complex $\hat{C}_*(\mathscr{X}) = \{\bigoplus_{0 \leq l \leq n} \hat{C}_l(\tilde{\mathscr{X}}), \hat{\partial}\}$ is called the complex of cooriented chains of the decomposition $\mathscr{X}$ with integral coefficients.

One defines similarly the complex $\hat{C}_*(\mathscr{X}; \mathbb{Q})$ of cooriented chains of the decomposition $\mathscr{X}$ with rational coefficients.

It is known that the group $H_l(\hat{C}_*(\mathscr{X}))$ coincides with the l-th homology group $H_l^c(X; \Lambda)$ of the manifold X with local coefficients Λ and closed supports, while the group $H_l(\hat{C}_*(\mathscr{X}; \mathbb{Q}))$ coincides with the l-th homology group $H_l^c(X; \Lambda \otimes_Z \mathbb{Q})$ of the manifold X with local coefficients $\Lambda \otimes_Z \mathbb{Q}$ and closed supports.

Let $f_X = \sum_\alpha X_\alpha^n$ where X_α^n runs through the whole set of n-dimensional cells of the decomposition $\mathscr{X}$ cooriented in the natural way. Obviously, f_X is a cycle of the complex $\hat{C}_*(\mathscr{X})$ that defines a non-trivial homology class. We call f_X the fundamental cycle of the manifold X.

Let X_1 and X_2 be topological manifolds, Λ_1 and Λ_2 their orientation sheafs, and $\varphi: X_1 \to X_2$ a continuous mapping. An orientation of the mapping φ is defined as a sheaf homomorphism $\bar{\varphi}: \Lambda_1 \to \Lambda_2$ such that for any point $x \in X_1$ it induces an isomorphism of the group $\Lambda_1(x)$ onto the group $\Lambda_2(\varphi(x))$. We shall refer to mapping φ with a given orientation as oriented, for short.

3. *The bundle of connections*

Let G be a Lie group, $\mathfrak{g}$ its Lie algebra and $\mathscr{P}$ a smooth principal G-bundle over a smooth n-dimensional manifold X. Denote by $\hat{g}$ the diffeomorphism of the bundle $\mathscr{P}$ corresponding to $g \in G$. Let $z \in \mathscr{P}$ and let V_z be the vertical space at the point z, that is the tangent space at the point z to the fiber of the bundle $\mathscr{P}$ through z. It is known that as a vector space V_z is canonically isomorphic to the Lie algebra $\mathfrak{g}$.

Definition 3.1. One says that a connection is defined on the principal G-bundle $\mathscr{P}$ if there is a smooth 1-form ω with values in $\mathfrak{g}$ on $\mathscr{P}$ (also called a connection form) which satisfies the following conditions.

1. For any point $z \in \mathscr{P}$ the restriction of the form ω to the vertical space V_z coincides with the canonical isomorphism $V_z \to \mathfrak{g}$.
2. For any $g \in G$ we have $\hat{g}^* \omega = ad\, g^{-1} \omega$.

The curvature form of the connection ω is, by definition, the 2-form R on $\mathscr{P}$ taking values in $\mathfrak{g}$ given by the formula

$$R = d\omega + \tfrac{1}{2}[\omega, \omega].$$

Let $T^*\mathscr{P}$ be the cotangent bundle of the manifold $\mathscr{P}$. Consider the vector bundle $T^*\mathscr{P} \otimes \mathfrak{g}$ over $\mathscr{P}$. Points of the bundle $T^*\mathscr{P} \otimes \mathfrak{g}$ are naturally identified with 1-forms at points of the manifold $\mathscr{P}$ taking values in the Lie algebra $\mathfrak{g}$. Denote by $E(\mathscr{P})$ the subset of the bundle $T^*\mathscr{P}\otimes\mathfrak{g}$ composed of forms ω_z at points $z\in\mathscr{P}$ such that $\omega_z|_{V_z}$ coincides with the canonical isomorphism $V_z\to\mathfrak{g}$. Clearly $E(\mathscr{P})$ is a smooth submanifold of the manifold $T^*\mathscr{P}\otimes\mathfrak{g}$ and the projection of the bundle $T^*\mathscr{P}\otimes\mathfrak{g}$ induces on $E(\mathscr{P})$ the structure of a smooth bundle over $\mathscr{P}$.

Consider the vector subbundle $U(\mathscr{P})$ of the vector bundle $T^*\mathscr{P}\otimes\mathfrak{g}$ composed by forms θ_z at points $z\in\mathscr{P}$ such that $\theta_z(V_z)=0$. Let $z\in\mathscr{P}$, and let $T^*_z\mathscr{P}\otimes\mathfrak{g}$, E_z and U_z be the fibers of the bundles $T^*\mathscr{P}\otimes\mathfrak{g}$, $E(\mathscr{P})$ and $U(\mathscr{P})$ at the point z. It is easy to see that E_z is a plane in the vector space $T^*_z\mathscr{P}\otimes\mathfrak{g}$ not containing 0, and U_z is a vector subspace of the space $T^*_z\mathscr{P}\otimes\mathfrak{g}$ associated with the plane E_z.

The action of the group G on $\mathscr{P}$ and the adjoint representation of G in $\mathfrak{g}$ induce a natural action of the group G on the bundle $T^*\mathscr{P}\otimes\mathfrak{g}$. It is clear that this action preserves the vector bundle structure on $T^*\mathscr{P}\otimes\mathfrak{g}$ and that the bundles $E(\mathscr{P})$ and $U(\mathscr{P})$ are invariant under this action. Using all these notions we give a new definition of connections.

Definition 3.2. A connection on the principal G-bundle $\mathscr{P}$ is a smooth section ω of the bundle $E(\mathscr{P})\to\mathscr{P}$, invariant under the action of the group G on this bundle.

The equivalence of Definitions 3.1 and 3.2 is evident.

Denote the quotient space $E(\mathscr{P})/G$ by $C(\mathscr{P})$. It is easy to see that the space $C(\mathscr{P})$ has the natural structure of a smooth bundle over the manifold X. If $x\in X$, $\mathscr{P}_x$ is the fiber of the bundle $\mathscr{P}$ at the point x and E_z is the fiber of the bundle $E(\mathscr{P})$ at the point z, then the fiber C_x of the bundle $C(\mathscr{P})$ at the point x is equal to $\bigcup_{z\in\mathscr{P}_x}E_z/G$. The factorization of the set $\bigcup_{z\in\mathscr{P}_x}E_z$ by G reduces to identifying the fibers E_z at all points $z\in\mathscr{P}_x$ as preserving their affine structure as planes in vector bundles. Thus the following proposition is proved.

Proposition 3.1. *The space $C(\mathscr{P})$ is a smooth bundle with respect to the natural projection $p\colon C(\mathscr{P})\to X$; the fibers of this bundle have natural structures of affine spaces.*

The following proposition easily follows from Definitions 3.1 and 3.2 and the definition of the bundle $C(\mathscr{P})$.

Proposition 3.2. *Any connection on the principal G-bundle $\mathscr{P}$ defines a smooth section of the bundle $C(\mathscr{P})$, and, vice versa, any smooth section of the bundle $C(\mathscr{P})$ defines a connection on $\mathscr{P}$.*

Further, we call $C(\mathscr{P})$ the bundle of connections on $\mathscr{P}$.

It is easy to see that $E(\mathscr{P})$ is a smooth principal G-bundle over $C(\mathscr{P})$. Moreover, the following proposition holds.

Proposition 3.3. *There exists on the principal G-bundle $E(\mathscr{P})$ a unique connection form χ with the following property: if ω is a connection form on $\mathscr{P}$ and $s\colon\mathscr{P}\to E(\mathscr{P})$ is the section corresponding to ω, then $s^*\chi=\omega$.*

Proof. It is clear that a 1-form χ on $E(\mathscr{P})$ taking values in $\mathfrak{g}$ and satisfying the first condition from Definition 3.1 is uniquely determined by the conditions of Proposition 3.3. The verification of the second of the conditions of Definition 3.1 for χ poses no difficulties.

Call the connection on $E(\mathscr{P})$ defined by the form χ "the canonical connection". We shall need the following proposition below.

Proposition 3.4. *The projection $p: C(\mathscr{P}) \to X$ of the bundle of connections is orientable.*

Proof. For a smooth manifold, a local orientation at a point coincides with an orientation of the tangent space at this point. Therefore, it is sufficient to establish a correspondence between orientations of the tangent space to $C(\mathscr{P})$ at an arbitrary point $\lambda \in C(\mathscr{P})$ and orientations of the tangent space to X at the point $x = p(\lambda)$ continuous with respect to λ. Let the fiber C_x of the bundle $C(\mathscr{P})$ at the point x be equal to $\bigcup_{z \in \mathscr{P}_x} E_z/G$, where $\mathscr{P}_x$ is the fiber of the bundle $\mathscr{P}$ at the point x and E_z is the fiber of the bundle $E(\mathscr{P})$ at the point z. Since E_z is an affine space then tangent spaces to E_z at all points of E_z are canonically isomorphic to the fie($\mathscr{P}$) at the point z. Note that $U_z = T_X^* \otimes \mathfrak{g}$ where T_X^* is the cotangent space to X at the point x. Choose some orientation of the vector space $\mathfrak{g}$. Then the orientation of the vector space U_z is determined by the orientation of the space T_x^*. The tangent space to the affine space C_x at an arbitrary point is canonically isomorphic to the vector space $\bigcup_{z \in \mathscr{P}_x} U_x/G$. Since the action of G is trivial on T_x^* and is induced by the adjoint representation on $\mathfrak{g}$, then the action of G on $\bigcup_{z \in \mathscr{P}_x} U_z/G$ identifies the spaces U_z with each other, preserving their orientations defined by the choice of the orientation of the space T_x^*. Thus, fixing an orientation of the tangent space T_x to the manifold X at the point x determines an orientation of the tangent space to the fiber C_x at any point. It follows from this in an obvious way that there exists an orientation for the projection on $p: C(\mathscr{P}) \to X$.

From now on, we assume that an orientation of the mapping p is fixed, i.e. that p is an oriented mapping.

Let, then, V be a smooth m-dimensional (real or complex) vector bundle over X and let $\mathscr{P}$ be the principal bundle associated with V. One says that a connection is given in the bundle V if a connection is given in the principal bundle $\mathscr{P}$. It is known that a connection in V gives rise to the notion of parallel translation in V. Let $\xi_1, \dots, \xi_m$ be smooth sections of the bundle V which are linearly independent at any point $x \in X$ and hence define a section $s: X \to \mathscr{P}$ of the bundle $\mathscr{P}$. Then there exists a unique connection on $\mathscr{P}$ with a form ω such that all the sections $\xi_1, \dots, \xi_m$ are parallel with respect to this connection, or, equivalently, $s^*\omega = 0$. The curvature form of this connection is obviously equal to zero.

4. *Characteristic classes of bundles*

Let $\mathscr{P}$ be a smooth principal G-bundle over a manifold X, and let R be the curvature form of some connection ω on $\mathscr{P}$. Suppose that $P = P(h_1, \dots, h_k)$ is an invariant symmetric k-linear form on the Lie algebra $\mathfrak{g}$ of the group G. Denote by

$P(R)$ the smooth $2k$-form on X, which is obtained from the form $P(R,\dots,R)$ by alternating. The form $P(R)$ is closed and, according to the well-known theorem of A. Weil, the cohomology class of the manifold X defined by this form does not depend on the choice of a connection form ω. This cohomology class is called the characteristic class of the bundle $\mathscr{P}$ corresponding to the form P. The characteristic classes of the principal bundle associated with the tangent bundle of a manifold X are called the Pontryagin classes of the manifold X.

Let $\mathfrak{g}=\mathfrak{gl}(m,K)$ be the Lie algebra of the general linear group over the field K, which is equal to $\mathbb{R}$ or $\mathbb{C}$. Denote by I_m the algebra of invariant symmetric polylinear forms on the Lie algebra $\mathfrak{g}$ taking values in the field K. Then define the k-form $\pi_k \in I_m$ by the formula

$$\pi_k(h_1,\dots,h_k)=\frac{1}{k!}\sum \operatorname{tr}(h_{v_1}\cdots h_{v_k}),$$

where $h_1,\dots,h_k \in \mathfrak{g}$, and the sum is taken over all permutations $v_1,\dots,v_k$ of the set $\{1,\dots,k\}$.

It is known that the forms $\pi_k (k=1,\dots,m)$ constitute a system of generators of the algebra I_m, that is any form $P \in I_m$ is representable in the form $Q(\pi_1,\dots,\pi_m)$, where $Q(x_1,\dots,x_m)$ is a polynomial in m variables over the field K. On the other hand, if $Q(x_1,\dots,x_k)$ is an arbitrary polynomial in k variables, then for any $m \geqq 1$ and $k \geqq 1$ the form $P_m=Q(\pi_1,\dots,\pi_k)$ belongs to I_m. The set of such forms $P_m (m=1,2,\dots)$ will be called a stable form corresponding to the polynomial Q. We shall usually omit the subscript m and denote all these forms by the same letter P. In particular, the form π_k is the stable form corresponding to the polynomial $Q=x_k$.

Proposition 4.1. *Let the space K^p be naturally embedded into the space $K^m (p<m)$, and the quotient space K^m/K^p be identified with the space K^q where $q=m-p$. Then let h_v $(v=1,\dots,r)$ be elements of the Lie algebra $\mathfrak{g}$, that is endomorphisms of the space K^m, vanishing on K^p. Denote by h'_v the endomorphism of the space K^q induced by h_v. Let P be a stable r-form corresponding to a polynomial Q. Then*

$$P(h_1,\dots,h_r)=P(h'_1,\dots,h'_r).$$

The proof is obvious for the forms $\pi_k (k=1,\dots,r)$. But then the assertion is true for the general form $P=Q(\pi_1,\dots,\pi_r)$.

It was indicated above how characteristic classes of a bundle $\mathscr{P}$ may be constructed by use of a smooth connection ω on $\mathscr{P}$. Now we solve a similar problem in the case when a piecewise smooth connection is given on $\mathscr{P}$. More precisely, let a cellular decomposition on $\mathscr{X}=\{X_i\}_{i\in I}$ of the base space X of the bundle $\mathscr{P}$ be given and for each n-cell X^n_α of this decomposition let a connection ω_α on $\mathscr{P}|_{X^n_\alpha}$ be fixed. We show how characteristic classes of the bundle $\mathscr{P}$ may be constructed with the aid of these data. For this purpose we must recall the theory of even de Rham currents.

5. *Even de Rham currents*

Let X be a smooth n-dimensional manifold and let $A^p(X)$ be a sheaf of germs of smooth p-forms on X. Consider the bundle $\Lambda \otimes_{\mathbb{Z}} A^p(X)$, where Λ is the orientation sheaf of the manifold X. Let us refer to global sections of the sheaf $\Lambda \otimes_Z A^p(X)$ as odd differential p-forms on the manifold X. It is easy to see that a section of the sheaf $\Lambda \otimes_Z A^p(X)$ over a contractible subset Y of X is representable in the form $\varepsilon_Y \otimes \omega$, where ε_Y is a local orientation of the set Y and ω is smooth p-form on Y. Sometimes we shall refer to the usual p-forms as even p-forms, to distinguish them from odd p-forms.

The exterior product and exterior differential for even p-forms induce the exterior product of forms of arbitrary parity and the exterior differential of odd forms. The exterior product of two forms of the same parity is an even form, and the exterior product of two forms of different parity is an odd form.

If $f: X_1 \to X_2$ is a smooth oriented mapping of one manifold into another then the notion of the inverse image $f^*\omega$ of an odd form ω on X_2 is naturally defined, where $f^*\omega$ is an odd form on X_1 and $d(f^*\omega) = f^*d\omega$.

Let Y be a cooriented p-dimensional cell in X and $\varphi: K \to Y$ its chart. Take a smooth, odd p-form λ on Y and let $\lambda = \varepsilon_Y \otimes \omega$ be its representation on Y, where ε_Y is the local orientation of the set Y. Then the integral $\int_Y \lambda$ is defined by the equality $\int_Y \lambda = \int_K \varphi^*\omega$, where the linear cell K is assumed to be furnished with the orientation corresponding to the orientation ε_Y. One defines in a natural way the integral of an odd p-form over a finite cooriented p-dimensional chain and the integral of a compactly supported odd p-form over an infinite cooriented p-dimensional chain. It is easy to see that one has for such integrals the Stokes formula $\int_C d\lambda = \int_{\partial C} \lambda$, where C is a cooriented p-dimensional chain and λ is an odd p-form on X. The integral $\int_X \lambda$, where λ is an odd compactly supported n-form, is defined by the formula $\int_X \lambda = \int_{f_X} \lambda$ where f_X is the fundamental cycle of the manifold X constructed with the use of some cellular decomposition of X.

Consider the space $D^p(X)$ of all compactly supported smooth, odd p-forms on X; it is a topological vector space with respect to the C^∞-topology.

Definition 5.1. An even p-dimensional current on the manifold X is a continuous linear functional on the space $D^p(X)$. The number $n-p$ is called the degree of the p-dimensional current. The space of all even p-dimensional currents on X is denoted by $D^p(X)$.

Examples of even currents

1. A locally integrable even p-form ω on X defines an even current of degree p by means of the condition $\omega(\lambda) = \int_X \omega \wedge \lambda$, where $\lambda \in D^{n-p}(X)$.

2. Define a singular p-dimensional cell in X as a smooth mapping $\varphi: K \to X$, where K is a p-dimensional linear cell in $\mathbb{R}^p$. It is clear that any chart $\varphi: K \to Y$ of a p-dimensional Y of X is a singular p-dimensional cell. A singular p-dimensional cell $\varphi: K \to X$ is called cooriented if an orientation of the mapping φ is given.

Any cooriented p-dimensional singular cell $\varphi: K \to X$ defines an even p-dimensional current φ by means of the formula

$$\varphi(\lambda) = \int_K \varphi^* \lambda, \quad \text{where} \quad \lambda \in D^p(X).$$

If φ is a chart of a cooriented p-dimensional cell Y in X, then the current $\varphi(\lambda)$ does not depend on the choice of the chart φ, and we shall denote this current by Y.

The current $T=\sum k_\alpha \varphi_\alpha$, where $K_\alpha \in \mathbb{R}$ and φ_α is a p-dimensional cooriented singular cell, is called a p-dimensional cooriented singular chain. Note that the current T makes sense when ranges over an infinite set of indices but the chain T is locally finite, that is any compact set in X intersects only with the finite set of supports of cells φ_α. If $\mathscr{X}$ is a cellular decomposition of the manifold X then any cooriented chain of $\mathscr{X}$ is locally finite and hence may be considered as an even p-dimensional current.

3. Let $\varphi: K \to X$ be a cooriented singular cell and ω a smooth even q-form on $X (q \leqq p)$. Define the even $(p-q)$-dimensional current r $\wedge\, \omega$ on X by equality $(\varphi \wedge \omega)(\lambda) = \int_K (\omega \wedge \lambda)$, where $\lambda \in D^{p-q}(X)$.

If $T=\sum k_\alpha \varphi_\alpha$ is a cooriented locally finite singular p-dimensional chain on X and ω is a smooth even q-form on X, then $T \wedge \omega = \sum k_\alpha (\varphi_\alpha \wedge \omega)$ is an even $(p-q)$-dimensional current on X. In particular, if f_X is a fundamental cycle of the manifold X, then the current $f_X \wedge \omega$ coincides with the current ω.

Definition 5.2. Let T be a p-dimensional even current on X. The boundary of the current T is the even $(p-1)$-dimensional current bT on X defined by the equality $bT(\lambda) = T(d\lambda)$, where $\lambda \in D^{p-1}(X)$. The differential of the current T is the current dT equal to $(-1)^{n-p-1} bT$.

The Stokes formula implies immediately that if C is a cooriented p-dimensional chain then the chain $\hat{\partial} C$ regarded as a current coincides with bC. The differential of a smooth form regarded as a current coincides with its exterior differential.

Definition 5.3. Let $f: X_1 \to X_2$ be an oriented smooth mapping of one manifold into another, and T an even p-dimensional current on X_1. The image of the current T under the mapping f is the even p-dimensional current $f_* T$ on X_2 defined by the condition $f_* T(\lambda) = T(f^* \lambda)$, where $\lambda \in D^p(X_2)$.

The definitions above immediately imply the equality $f_*(bT) = b(f_* T)$.

Let $\varphi: K \to X$ be a cooriented p-dimensional singular cell in X. It is easy to see that $b\varphi = \varphi_*(\hat{\partial} K)$, where K is regarded as a p-dimensional cell in $\mathbb{R}^p$ with natural coorientation. Thus $b\varphi$ is a cooriented $(p-1)$-dimensional singular chain in X.

Proposition 5.1. *If $\varphi \wedge \omega$ is the current from Example 3, then*

$$b(\varphi \wedge \omega) = (-1)^q (b\varphi \wedge \omega - \varphi \wedge d\omega).$$

This assertion can be checked without difficulty.

Denote by $D'(X)$ the de Rham complex $\{\bigoplus_{p \geq 0} D^{p'}(X), b\}$ of even currents on the manifold X.

Proposition 5.2 (De Rham). *The p-dimensional homology group of the complex $D'(X)$ is isomorphic to the $(n-p)$-dimensional cohomology group $H^{n-p}(X, \mathbb{R})$ of the manifold X with coefficients in $\mathbb{R}$ or (the Poincaré duality) to the p-dimensional* *homology group $H_p^c(X, \Lambda \otimes_Z \mathbb{R})$ of the manifold X with local coefficients $\Lambda \otimes_Z \mathbb{R}$ and closed supports.*

Let $\mathscr{X}$ be a cellular decomposition of the manifold X and $\hat{C}_*(\mathscr{X})$ the complex of cooriented chains of the decomposition $\mathscr{X}$. If we regard cooriented chains as even currents on X we obtain an inclusion $\hat{C}_*(\mathscr{X}) \subset D'(X)$.

Proposition 5.3. *The homology mapping induced by the inclusion $\hat{C}_*(\mathscr{X}) \subset D'(X)$ coincides with the homology mapping $H_p^c(X, \Lambda) \to H_p^c(X, \Lambda \otimes_Z \mathbb{R})$ induced by the inclusion $\Lambda \subset \Lambda \otimes_Z \mathbb{R}$. A similar statement is true for the complex $\hat{C}_*(\mathscr{X}, \mathbb{Q})$.*

6. *The forms $P^{q,2k-q}$ and their properties*

Let $\mathscr{P}$ be a smooth principal G-bundle over an n-dimensional manifold X and let $\mathscr{X}$ be a cellular decomposition of the manifold X. One says that a connection is defined on an l-dimensional cell X_j if a form ω_j is given on the bundle $\mathscr{P}|_{X_j}$ such that ω_j may be extended to a form ω defining a connection on the bundle $\mathscr{P}|_U$ where U is an open neighborhood of the cell X_j.

Now let X_j be a cooriented l-dimensional cell and $\omega_0, \ldots, \omega_q$ an ordered system of $q+1$ connections on the cell X_j. Denote by S_μ the section of the bundle of connections $C(\mathscr{P})$ over the cell X_j defined by the connection $\omega_\mu (\mu = 0, \ldots, q)$. Take the standard q-dimensional simplex Δ^q and define the mapping $v_j(\omega_0, \ldots, \omega_q): \Delta^q \times X_j \to C(\mathscr{P})$ by the following equality:

$$v_j(\omega_0, \ldots, \omega_q)(t, x) = \sum_{\mu=0}^{q} t_\mu S_\mu(x),$$

where $t \in \Delta^q$, $x \in X_j$ and $t_0, \ldots, t_q$ are the barycentric coordinates of the point t. Note that $\sum_{\mu=0}^q t_\mu S_\mu(x)$ makes sense, for the fibers of the bundle $C(\mathscr{P})$ are affine spaces. It is clear that $v_j(\omega_0, \ldots, \omega_q)$ is a smooth mapping. We equip the mapping $v_j(\omega_0, \ldots, \omega_q)$ with the following orientation. Assume that the simplex Δ^q is oriented in a standard way. Then an orientation of the product $\Delta^q \times X_j$ is determined by the orientation of the cell X_j. Since the cell X_j is cooriented, then an orientation of the cell X_j determines a local orientation of X_j in X, and an orientation of the projection $p: C(\mathscr{P}) \to X$ assigns to a local orientation of X_j a local orientation of the set $v_j(\omega_0, \ldots, \omega_q)(\Delta^q \times X_j) \subset p^{-1}(X_j)$ in $C(\mathscr{P})$. Hence $v_j(\omega_0, \ldots, \omega_q)$ is an oriented mapping, and one may regard it as an even $(q+l)$-dimensional singular cell in $C(\mathscr{P})$. The cell $v_j(\omega_0, \ldots, \omega_q)$ may be viewed geometrically as a film in the space $C(\mathscr{P})$ spanned on the connections $\omega_0, \ldots, \omega_q$.

If $C = \sum k_j X_j$, where $k_j \in \mathbb{Z}$, is an l-dimensional cooriented cell of the decomposition $\mathscr{X}$ and the connections $\omega_0, \ldots, \omega_q$ are defined on all the cells X_j of this chain, then we define the even $(q+l)$-dimensional singular chain $v_C(\omega_0, \ldots, \omega_q)$ by the formula $v_C(\omega_0, \ldots, \omega_q) = \sum k_j v_j(\omega_0, \ldots, \omega_q)$.

Proposition 6.1. *$bv_j(\omega_0, \ldots, \omega_q) = \sum_{\mu=0}^q (-1)^\mu v_j(\omega_0, \ldots, \hat{\omega}_\mu, \ldots, \omega_q) + (-1)^q \sum_{S(i) 8 j} \varepsilon_{ij} v_i(\omega_0, \ldots, \omega_q)$, where ε_{ij} is the incidence number of the cells X_i and X_j.*

Proof is obvious.

Let $P = P(h_1, \dots, h_k)$ be an invariant symmetric k-linear form on the Lie algebra $\mathfrak{g}$, let K be the curvature form of the canonical connection χ on $C(\mathscr{P})$ and let $P(K)$ be the corresponding characteristic $2k$-form on $C(\mathscr{P})$.

Proposition 6.2. *One has*

$$p_*(v_i(\omega_0, \dots, \omega_q)) \wedge P(K) = X_i \wedge P_i^{q,2k-q}(\omega_0, \dots, \omega_q),$$

where $P_i^{q,2k-q}(\omega_0, \dots, \omega_q)$ *is a well-defined smooth* $(2k-q)$*-form on the cell* X_i*, and* $P_1^{q,2k-q}(\omega_0, \dots, \omega_q) = 0$ *if* $q > k$.

Proof. It is easy to see that any section $s: X_j \to \mathscr{P}$ of the bundle $\mathscr{P}$ over the cell X_i induces a section $s: p^{-1}(X_i) \to E(\mathscr{P})$ of the principal G-bundle $E(\mathscr{P})$ over $p^{-1}(X_i)$. For any $\mu = 0, \dots, q$ the form $s^*\omega_\mu = \varphi_\mu$ is a 1-form on X_i taking values in $\mathfrak{g}$. Hence $(v_i(\omega_0, \dots, \omega_q) \circ s')^* = \sum_{\mu=0}^{q} t_\mu \varphi_\mu$. The definition of the curvature form implies the equality

$$(v_i(\omega_0, \dots, \omega_q) \circ s')^* K = \sum_{\mu=0}^{q} t_\mu s^* R_\mu - \frac{1}{2} \sum_{\mu=1}^{q} t_\mu [\varphi_\mu - \varphi_0, \varphi_\mu - \varphi_0]$$
$$+ \frac{1}{2} \sum_{\nu,\mu=1}^{q} \lambda_\nu \lambda_\mu [\varphi_\nu - \varphi_0, \varphi_\mu - \varphi_0] + \sum_{\mu=1}^{q} dt_\mu \wedge (\varphi_\mu - \varphi_0),$$

where R_μ is the curvature form of the connection ω_μ.

It is easy to see that $v_i^*(\omega_0, \dots, \omega_q) P(K) = P((v_i(\omega_0, \dots, \omega_q) \circ s')^* K)$. Therefore $v_i^*(\omega_0, \dots, \omega_q) P(K) = \sum_{j=0}^{k} P_s(\omega_0, \dots, \omega_q)$, where $P_s(\omega_0, \dots, \omega_q)$ is the homogeneous component of $v_i^*(\omega_0, \dots, \omega_q) P(K)$ of degree s with respect to $dt_1, \dots, dt_q$. Clearly $P_q(\omega_0, \dots, \omega_q) = P'_q \wedge P''_q$, where P'_q is a smooth q-form on Δ^q and P''_q is a smooth $(2k-q)$-form on X_i. Put

$$P_i^{q,2k-q}(\omega_0, \dots, \omega_q) = P''_q \int_{\Delta^q} P'_q.$$

Obviously $P_i^{q,2k-q}(\omega_0, \dots, \omega_q) = 0$ if $q > k$.

Let $\lambda \in D^{n+t-2k}(X)$. Then

$$p_*(v_i(\omega_0, \dots, \omega_q) \wedge P(K))(\lambda) = \int_{\Delta^q \times X_i} v_i^*(\omega_0, \dots, \omega_q)(P(K) \wedge p^* \lambda)$$
$$= \int_{X_i} P_i^{q,2k-q}(\omega_0, \dots, \omega_q) \wedge \lambda$$
$$= (X_i \wedge P_i^{q,2k-q}(\omega_0, \dots, \omega_q))(\lambda),$$

and hence $p_*(v_i(\omega_0, \dots, \omega_q) \wedge P(K)) = X_i \wedge P_i^{q,2k-q}(\omega_0, \dots, \omega_q)$. It is clear that the form $P_i^{q,2k-q}(\omega_0, \dots, \omega_q)$ is uniquely determined by this equality.

Suppose now that the connections $\omega_0, \dots, \omega_q$ are defined on the whole of the bundle $\mathscr{P}$. Consider the fundamental cycle $f_X = \sum X_\alpha^n$ of the manifold X and construct on each cell X_α^n the form $P_\alpha^{q,2k-q}(\omega_0, \dots, \omega_q)$. It is easy to see that all the forms $P_\alpha^{q,2k-q}(\omega_0,\ldots,\omega_q)$ are the restrictions of the same smooth $(2k-q)$-form $P^{q,2k-q}(\omega_0,\ldots,\omega_q)$ defined on the whole manifold X. Proposition 6.2 implies the equality

$$p_*(\nu_{fx}(\omega_0,\ldots,\omega_q)\wedge P(K))=P^{q,2k-q}(\omega_0,\ldots,\omega_q).$$

Proposition 6.3. *The forms $P_i^{q,2k-q}(\omega_0,\ldots,\omega_q)$ have the following properties:*
1. *$P_i^{0,2k}(\omega)=P(R)$, where R is the curvature form of the connection ω;*
2. *$P_i^{q,2k-q}(\omega_0,\ldots,\omega_q)=0$ if $q>k$;*
3. *$P_i^{q,2k-q}(\omega_0,\ldots,\omega_q)$ is skew symmetric with respect to permutations of the connections $\omega_0,\ldots,\omega_q$.*

The forms $P^{q,2k-q}(\omega_0,\ldots,\omega_q)$ have the same properties.

The proof of this Proposition follows from the definitions and Proposition 6.2. We shall need the following fact below.

Proposition 6.4. *Let E be a smooth m-dimensional (complex or real) vector bundle, E_1 a smooth p-dimensional vector subbundle of the bundle E, and $E_2=E/E_1$. Take an ordered system $\omega_0,\ldots,\omega_q$ of connections on the bundle E such that the subbundle E_1 is invariant with respect to all these connections and all these connections induce on E_1 the same connection of zero curvature. Denote by ω'_μ the connection on the quotient bundle E_2 by the connection $\omega_\mu(\mu=0,\ldots,q)$. Then the following equality holds:*

$$P^{q,2k-q}(\omega_0,\ldots,\omega_q)=P^{q,2k-q}(\omega'_0,\ldots,\omega'_q).$$

Proof. Let $\mathscr{P}$ be the principal $GL(m,\mathbb{K})$-bundle (where $\mathbb{K}=\mathbb{R}$ or $\mathbb{C}$) associated with the vector bundle E and let G be the subgroup of the group $GL(m,\mathbb{K})$ composed of those automorphisms of the space $\mathbb{K}^m$ which induce the identity mapping on the subspace $\mathbb{K}^p\subset\mathbb{K}^m$. It is easy to see that a subbundle $\mathscr{P}'$ of the bundle $\mathscr{P}$ exists which is a principal *G-bundle and invariant under the connections* $\omega_0,\ldots,\omega_q$. Then the curvature form of the canonical connection on the principal G-bundle $E(\mathscr{P}')$ takes values in the space of such endomorphisms of the space $\mathbb{K}^m$ which vanish on the subspace $\mathbb{K}^p\subset\mathbb{K}^m$. The equality $P^{q,2k-q}(\omega_0,\ldots,\omega_q)=P^{q,2k-q}(\omega'_0,\ldots,\omega'_q)$ now follows from the definition of the connections ω'_μ and Proposition 4.1.

7. *The complex of formal simplicial films and a generalization of A. Weil's theorem*

Let X be an n-dimensional topological manifold with a cellular decomposition $\mathscr{X}=\{X_i\}_{i\in I}$. Denote by $c(i)$ the set of indices of n-dimensional cells of this decomposition which contain the cell X_i. Consider the abstract complex C_i composed of the set $c(i)$ and all its subsets. Let $I_*(i)=\{\bigoplus_{q\geq 0}I_q(i),\partial\}$ be the simplicial chain complex of oriented chains of the complex C_i with integral coefficients. More explicitly, this means that $I_q(i)$ is the Abelian group with generators $\Delta^q(\alpha_0,\ldots,\alpha_q)$, where $\alpha_0,\ldots,\alpha_q\in c(i)$, with the relations of skew symmetry with respect to permutations of the indices $\alpha_0,\ldots,\alpha_q$, and the differential $\partial\colon I_q(i)\to I_{q-1}(i)$ is defined by the formula $\partial\Delta^q(\alpha_0,\ldots,\alpha_q)=\sum_{\mu=0}^q(-1)^\mu\Delta^{q-1}(\alpha_0,\ldots,\hat{\alpha}_\mu,\ldots,\alpha_q)$. If $X_{i_1}\subset X_{i_2}$, then $c(i_1) \supset c(i_2)$, and hence the complex $I_*(i_2)$ is a subcomplex of the complex $I_*(i_1)$.

Definition 7.1. A formal simplicial (p, q)-film of the decomposition is a function assigning to each cooriented p-dimensional cell X_i of the decomposition $\mathscr{X}$ a chain $c(X_i) \in I_q(i)$ such that $c(-X_i) = -c(X_i)$. The set of all formal simplicial (p, q)-films of the decomposition $\mathscr{X}$ forms an Abelian group; it is denoted by the symbol $\hat{C}_{p,q}(\mathscr{X})$.

Let us define the Abelian group homomorphisms $\hat{\partial}_1: \hat{C}_{p,q}(\mathscr{X}) \to \hat{C}_{p-1,q}(\mathscr{X})$ and $\hat{\partial}_2: \hat{C}_{p,q}(\mathscr{X}) \to \hat{C}_{p,q-1}(\mathscr{X})$ by means of the following conditions:

$$\hat{\partial}_1 c(X_i^{p-1}) = (-1)^p \sum_{j \in s(i)} \varepsilon_{ij} c(X_j^p),$$

$$\hat{\partial}_2 c(X_j^p) = \partial c(X_j^p),$$

where $c \in \hat{C}_{p,q}(\mathscr{X})$ and ε_{ij} is the incidence number of the cells X_i^{p-1} and X_j^p.

It is easy to see that $\hat{\partial}_1^2 = 0, \hat{\partial}_2^2 = 0$, and $\hat{\partial}_1 \hat{\partial}_2 + \hat{\partial}_2 \hat{\partial}_1 = 0$. Hence $\{\bigoplus_{p,q \geqq 0} \hat{C}_{p,q}(\mathscr{X}); \hat{\partial}_1, \hat{\partial}_2\}$ is a chain bicomplex. Let $\hat{\partial}_3 = \hat{\partial}_1 + \hat{\partial}_2$ be the total differential of this bicomplex.

Definition 7.2. The chain complex $\hat{C}_{**}(\mathscr{X}) = \{\bigoplus_{p,q \geqq 0} \hat{C}_{p,q}(\mathscr{X}), \hat{\partial}_3\}$ is called the complex of formal simplicial films of the decomposition $\mathscr{X}$.

Let us define the mapping $\lambda: \hat{C}_{**}(\mathscr{X}) \to \hat{C}_*(\mathscr{X})$ by means of the following conditions: (a) $\lambda(\hat{C}_{p,q}(\mathscr{X})) = 0$ if $q > 0$; (b) if $c \in \hat{C}_{p,0}(\mathscr{X})$ and $c(X_i^p) = \sum_{j \in c(i)} k_{ij} \Delta^0(j)$, where $k_{ij} \in \mathbb{Z}$, then $\lambda(c) = \sum_i \sum_{j \in c(j)} k_{ij} X_i^p$ where X_i ranges over the set of all p-dimensional cells of the decomposition $\mathscr{X}$, every one of which is equipped with an arbitrarily chosen coorientation.

Proposition 7.1. *The mapping λ is a complex homomorphism which induces a homology isomorphism.*

Proof. The assertion that λ is a complex homomorphism can be checked without difficulty. Let $A_s = \bigoplus_{p \leqq s} \hat{C}_{p,q}(\mathscr{X})$ and $B_s = \bigoplus_{p \leqq s} \hat{C}_p(\mathscr{X})$. It is clear that A_s and $B_s (s = 0, 1, \ldots)$ are subcomplexes of the complexes $\hat{C}_{**}(\mathscr{X})$ and $\hat{C}_*(\mathscr{X})$ correspondingly, and that $\{A_s\}$ and $\{B_s\}$ are increasing filtrations of the complexes $\hat{C}_{**}(\mathscr{X})$ and $\hat{C}_*(\mathscr{X})$, compatible with the homomorphism λ. Consider the spectral sequences induced by these filtrations. Since $H_q(I_*(i)) = 0$ for $q > 0$ and $H_q(I_*(i)) = \mathbb{Z}$ for $q = 0$, then the term E_1 of the spectral sequence induced by the filtration $\{A_s\}$ is equal to $\bigoplus_{p \geqq 0} (\hat{C}_{p,0}(\mathscr{X}) / \hat{C}_{p-1,0}(\mathscr{X}))$. On the other hand, the differential d_0 of the spectral sequence induced by the filtration $\{B_s\}$ is trivial, and therefore the term E_1 of this spectral sequence is equal to $\bigoplus_{p \geqq 0} (B_p / B_{p-1}) = \bigoplus_{p \geqq 0} \hat{C}_p(\mathscr{X})$. Since the homomorphism λ induces the isomorphism between these terms of the spectral sequences, then λ induces the isomorphism between the homologies of the complexes $\hat{C}_{**}(\mathscr{X})$ and $\hat{C}_*(\mathscr{X})$.

Now let $\mathscr{P}$ be a smooth principal G-bundle over an n-dimensional manifold X

and let $\mathscr{X}$ be a cellular decomposition of the manifold X. Suppose that on each n-dimensional cell X_α of the decomposition a connection is given with the form ω_α.

Let X_i be an arbitrary p-dimensional cooriented cell of the decomposition $\mathscr{X}$. We define a homomorphism v_i of the group $I_q(i)(q=0,1,\ldots)$ into the group of $(p+q)$-dimensional cooriented singular chains of the bundle of connections $C(\mathscr{P})$ by means of the following condition: $v_i(\Delta^q(\alpha_0,\ldots,\alpha_q))=v_i(\omega_{\alpha_0},\ldots,\omega_{\alpha_q})$, where $\alpha_0,\ldots,\alpha_q\in c(i)$. Consider the mapping v of the complex $\hat{C}_{**}(\mathscr{X})$ into the space of cooriented singular chains of the bundle of connections $C(\mathscr{P})$ defined by the formula $v(c)=\sum_{i\in I}v_i(c(X_i))$, where $c\in\hat{C}_{**}(\mathscr{X})$. The chain $v(c)$ gives a geometrical realization of the chain c as a film in the space $C(\mathscr{P})$.

Proposition 7.2. *The mapping v is a homomorphism of the complex $\hat{C}_{**}(\mathscr{X})$ into the complex $D'(X)$.*

This follows easily from the definitions and Proposition 6.1.

Proposition 7.3. *The homology homomorphism $v_*: H_*(\hat{C}_{**}(\mathscr{X}))\to H_*(D'(C(\mathscr{P})))$ induced by the homomorphism v does not depend on the choice of connections ω_α on n-dimensional cells of the decomposition $\mathscr{X}$.*

Proof. Since the projection $p: C(\mathscr{P})\to X$ is a homotopy equivalence, then the mapping $p_*: D'(C(\mathscr{P}))\to D'(X)$ induces a homology isomorphism. Therefore, it is sufficient to prove that the homology homomorphism induced by the mapping $p_*\circ v$ does not depend on the choice of the connections ω_α. It is easy to see that $p_*(v_i(\omega_{\alpha_0},\ldots,\omega_{\alpha_q}))=0$ for $q>0$ and $p_*(v_i(\omega_\alpha))=X_i$, where $\alpha_0,\ldots,\alpha_q\in c(i)$. Hence $p_*\circ v=\rho\circ\lambda$ where $\lambda: \hat{C}_{**}(\mathscr{X})\to\hat{C}_*(\mathscr{X})$ and $\rho: \hat{C}_*(\mathscr{X})\to D'(X)$ is the injection. Since $\rho\circ\lambda$ does not depend on the choice of the connections ω_α, neither does $p_*\circ v$.

Now let P be an invariant symmetric k-linear form on the Lie algebra g. Define the mapping $v_P: \hat{C}_{**}(\mathscr{X})\to D'(X)$ by the following formula: $v_P(c)=p_*(v(c)\wedge P(K))$, where $c\in\hat{C}_{**}(\mathscr{X})$ and K is the curvature form of the canonical connection of the bundle $E(\mathscr{P})$. It follows from Propositions 5.1 and 7.2, the permutability of the mapping p_* with the boundary operator b, and the fact that the form $P(K)$ is closed that the mapping v_P is a complex homomorphism of degree $-2k$.

Proposition 7.4. *The homomorphism v_P induces a homology homomorphism $\zeta: H_p^c(X,\Lambda)\to H_{p-2k}^c(X,\Lambda\otimes_Z\mathbb{R})(p=2k,\ldots,n)$, which does not depend on the choice of connections ω_α on n-dimensional cells of the decomposition $\mathscr{X}$. More precisely, if c is a p-dimensional cycle of the complex $\hat{C}_{**}(\mathscr{X})$ and γ is a homology class of the cycle $\lambda(c)$, then $\zeta(\gamma)$ is the homology class of the cycle $\lambda(c)\wedge P(R)$, where R is the curvature form of some connection on the bundle $\mathscr{P}$. In particular, the mapping ζ assigns to the homology class of the fundamental cycle f_X the Poincaré dual of the characteristic class of the bundle $\mathscr{P}$ corresponding to the form P.*

Proof. Let c be a p-dimensional cycle of the complex $\hat{C}_{**}(\mathscr{X})$. Then, according to Proposition 7.3, a system of connections ω_α being replaced by another system of connections implies the cycle $v(c)$ is replaced by a homologous cycle. Therefore, the cycle $v(c) \wedge P(K)$ is also replaced with a homologous cycle, and hence the homology class of the cycle $v_P(c)$ does not depend on the choice of connections ω_α. Suppose that all the connections ω_α are restrictions of the same connection ω on the bundle $\mathscr{P}$. Then, according to Propositions 6.2 and 6.3 and the definition of the homomorphism $\lambda: \hat{C}_{**}(\mathscr{X}) \to \hat{C}_*(\mathscr{X})$, one has $v_P(c) = \lambda(c) \wedge P(R)$, where R is the curvature form of the connection ω.

Chapter II. The combinatorial computation of characteristic classes of vector bundles

In the articles [1], [2] and [3] the problem of the combinatorial computation of Pontryagin classes of combinatorial manifolds was considered. This problem was completely solved only for the first Pontryagin class, though a considerable number of the techniques developed in the articles cited above may probably be used for combinatorial computation of higher Pontryagin classes. The initial data of this computation are topological manifold and a linearizable cellular decomposition of it, that is a cellular decomposition whose cells can be simultaneously furnished with affine structures such that the structure of each cell is compatible with the structures of all its faces.

In this chapter we consider the problem of combinatorial computation of characteristic classes of an arbitrary (real or complex) vector bundle. The initial datum of this computation is the notion of a combinatorial scheme of a vector bundle which generalizes the notion of a manifold with a linearized cellular decomposition.

1. Combinatorial schemes of vector bundles, and combinatorial vector bundles

Let X be an n-dimensional topological manifold with a cellular decomposition $\mathscr{X} = \{X_i\}_{i \in I}$. Recall that for a cell X_i we denote by $c(i)$ the set of indices of all n-dimensional cells which contain the cell X_i.

Definition 1.1. One says that a scheme of an m-dimensional vector bundle over the field $\mathbb{K}$ (where $\mathbb{K} = \mathbb{R}$ or $\mathbb{C}$) is given on the manifold X with a cellular decomposition $\mathscr{X}$ if for every index $i \in I$ there is a vector space V_i over the field $\mathbb{K}$ whose dimension is equal to $\max\{m - \operatorname{codim} X_i, 0\}$, and the following conditions are satisfied:

1. If $X_i \subset X_j$, then $V_i \subset V_j$.
2. If $\mathbb{K} = \mathbb{R}$, then for each cell X_i a correspondence is given between orientations of all the spaces $V_\alpha (\alpha \in c(i))$.

We define the space of fittings $\sum_i$ for every $i \in I$ in the following way. Consider the family of m-dimensional vector spaces $\{V_\alpha\}_{\alpha \in c(i)}$. A fitting for the cell X_i is, by definition, a family $\{f_{\alpha\beta}\}_{\alpha,\beta \in c(i)}$ of vector space isomorphisms $f_{\alpha\beta}: V_\alpha \to V_\beta$ with the following properties:

1. $f_{\beta\gamma} f_{\alpha\beta} = f_{\alpha\gamma}$, where $\alpha, \beta, \gamma \in c(i)$.
2. If $X_\alpha \cap X_\beta = X_i$, then $f_{\alpha\beta}|_{V_i}$ is the identity mapping.
3. If $\mathbb{K} = \mathbb{R}$, then the isomorphisms $f_{\alpha\beta}$ preserve the given correspondence between the orientations of the spaces V_α and V_β.

The set of all fittings for the cell X_i is called the space of fittings and is denoted by Σ_i. It is clear that the space Σ_i has a natural smooth structure. If $X_i \subset X_j$ then $c(i) \supset c(j)$, and the correspondence $\{f_{\alpha\beta}\}_{\alpha,\beta\in c(i)} \mapsto \{f_{\alpha\beta}\}_{\alpha,\beta\in c(j)}$ defines a smooth mapping $\psi_{ij}: \Sigma_i \to \Sigma_j$.

Definition 1.2. Let a scheme of an m-dimensional combinatorial vector bundle over the field $\mathbb{K}$ be given on a manifold X. Once says that a combinatorial m-dimensional vector bundle $E(\mathscr{X})$ is given on X if for every cell X_i a smooth mapping $f_i: X_i \to \Sigma_i$ is given such that if $X_i \subset X_j$ then $\psi_{ij} f_i = f_j|_{X_i}$.

Let us show that an m-dimensional combinatorial vector bundle $E(\mathscr{X})$ on X defines an m-dimensional vector bundle $E(X)$ on X. Take for each n-dimensional cell X_α the vector bundle $E_\alpha = X_\alpha \times V_\alpha$ on X_α. Let X_i be an arbitrary cell, and $\alpha \in c(i)$. Then $E_\alpha|_{X_i}$ is an m-dimensional vector bundle over the cell X_i. For any $\alpha, \beta \in c(i)$, we paste the bundles $E_\alpha|_{X_i}$ and $E_\beta|_{X_i}$ together by means of the mapping $g_{\alpha\beta}: E_\alpha|_{X_i} \to E_\beta|_{X_i}$ defined by the formula $g_{\alpha\beta}(x, \xi) = (x, f_{\alpha\beta}(x)(\xi))$, where $x \in X_i$, $\xi \in V_\alpha$ and $f_i(x) = \{f_{\alpha\beta}(x)\}_{\alpha,\beta\in c(i)}$. It is evident that these pastings agree with each other and define an m-dimensional vector bundle $E(X)$ on X. Clearly, $E(X)|_{X_\alpha} = E_\alpha$ and a natural connection of zero curvature is given on E_α compatible with the trivialization $E_\alpha = X_\alpha \times V_\alpha$.

Conversely, let a smooth vector bundle E_i of dimension $\max\{m - \operatorname{codim} X_i, 0\}$ be given on each cell X_i of a topological manifold X with a cellular decomposition $\mathscr{X}$, such that if $X_i \subset X_j$ then E_i is a smooth subbundle of the bundle $E_j|_{X_i}$. Suppose that on each bundle $E_\alpha(\alpha \in c(i))$ a connection ω_α of zero curvature is given, and that the following conditions are satisfied.

1. If $X_i \subset X_\alpha$ then the bundle E_i is invariant with respect to the connection ω_α.
2. If $X_i \subset X_\alpha$ and $X_i \subset X_\beta(\alpha, \beta \in c(i))$, then the connections ω_α and ω_β induce the same connection on the bundle E_i.

Choose a point x_α in each n-dimensional cell X_α and denote by V_α the fiber of the bundle E_α at this point. The connection ω_α induces a representation of the bundle E_α in the form $E_\alpha = X_\alpha \times V_\alpha$. If $X_i \subset X_\alpha$, then the bundle E_i is represented similarly in the form $E_i = X_i \times V_{i\alpha}$, where $V_{i\alpha}$ is a subspace of the space V_α. If $X_i \subset X_\alpha$ and $X_i \subset X_\beta$, where $\alpha, \beta \in c(i)$, then the bundle isomorphism $E_i = X_i \times V_{i\alpha} = X_i \times V_{i\beta}$ induces a canonical vector space isomorphism $V_{i\alpha} = V_{i\beta}$ which does not depend on the point $x \in X_i$. We identify the spaces $V_{i\alpha}$ and $V_{i\beta}$ by means of this isomorphism and denote the resulting space by V_i. It is clear that if $X_i \subset X_j$ then $V_i \subset V_j$. If $\mathbb{K} = \mathbb{R}$ we require in addition to this that a correspondence between orientations of all the spaces $V_\alpha(\alpha \in c(i))$ be indicated. Thus, a scheme of a combinatorial m-dimensional vector bundle is defined on the manifold X.

Let the manifold X be smooth, and let a smooth vector bundle $E(X)$ on X be given such that $E_\alpha = E(X)|_{X_\alpha}$ for any n-dimensional cell X_α. Take a cell X_i and let α, $\beta \in c(i)$. Since $E_\alpha|_{X_i} = E_\beta|_{X_i} = E(X)|_{X_i}$, $E_\alpha = X_\alpha \times V_\alpha$ and $E_\beta = X_\beta \times V_\beta$, then an isomorphism $f_{\alpha\beta}(x)\colon V_\alpha \to V_\beta$ is defined for any point $x \in X_i$ and depends smoothly on the point x. Let us define the mapping $f_i\colon X_i \to \sum_i$ by means of the formula $f_i(x) = \{f_{\alpha\beta}(x)\}_{\alpha,\beta\in c(i)}$. It is easy to see that the family of mappings $\{f_i\}_{i\in I}$ defines on X a structure of a combinatorial vector bundle $E(\mathscr{X})$, and the vector bundle on X induced by this structure coincides with $E(X)$.

An example of the situation described above is a topological manifold X with a linearizable cellular decomposition $\mathscr{X}$. For any cell X_i of this decomposition, we take as the bundle E_i the tangent bundle TX_i of this cell, and as the connections ω_α on n-dimensional cells X_α the connections on TX_α defined by affine structures of the cells X_α that are compatible with each other. The correspondence between orientations of the vector spaces $V_\alpha(\alpha \in c(i))$ is defined by the choice of any of the two local orientations of the set $X_i \subset X$.

Let a scheme of m-dimensional combinatorial vector bundle over the field $\mathbb{K}$ on a manifold X with a cellular decomposition $\mathscr{X}$ be given. Take an arbitrary cell X_i of the decomposition $\mathscr{X}$ and the corresponding space of fitting $\sum_i$. Then define in the following way an m-dimensional vector bundle $\mathscr{V}_i$ on the space $\sum_i$. Take the product $\sum_i \times \prod_{\alpha\in c(i)} V_\alpha$ and let $\mathscr{V}_i$ be the subset of this product consisting of points $(\{f_{\alpha\beta}\}_{\alpha,\beta\in c(i)}, \{\xi_\alpha\}_{\alpha\in c(i)})$ such that $f_{\alpha\beta}(\xi_\alpha) = \xi_\beta(\alpha, \beta \in c(i))$. The natural projection $\mathscr{V}_i \to \sum_i$ makes $\mathscr{V}_i$ a smooth m-dimensional vector bundle over the space $\sum_i$.

Proposition 1.1 *For any $\lambda \in c(i)$ there exists a unique connection $\omega_{i\lambda}$ of zero curvature on he bundle $\mathscr{V}_i$, such that all the sections $s_\gamma(\eta)\colon \sum_i \to \mathscr{V}_i(\eta \in V_\lambda)$ defined by the formula*

$$s_\gamma(\eta)(\{f_{\alpha\beta}\}_{\alpha,\beta\in c(i)}) = (\{f_{\alpha\beta}\}_{\alpha,\beta\in c(i)}, \{f_{\gamma\alpha}(\eta)\}_{\alpha\in c(i)})$$

are parallel with respect to this connection.

The proof is evident.

Let $X_i \subset X_j$, let $\sum_i$ and $\sum_j$ be the spaces of fittings for the cells X_i and X_j, and let $\mathscr{V}_i$ and $\mathscr{V}_j$ be the corresponding vector bundles. Define vector bundle mappings $\bar{\psi}_{ij}\colon \mathscr{V}_i \to \mathscr{V}_j$ by means of the formula

$$\bar{\psi}_{ij}(\{f_{\alpha\beta}\}_{\alpha,\beta\in c(i)}, \{\xi_\alpha\}_{\alpha\in c(i)}) = (\{f_{\alpha\beta}\}_{\alpha,\beta\in c(j)}, \{\xi_\alpha\}_{\alpha\in c(j)}),$$

where $(\{f_{\alpha\beta}\}_{\alpha,\beta\in c(i)}, \{\xi_\alpha\}_{\alpha\in c(i)}) \in \mathscr{V}_i$. It is easy to see that $\bar{\psi}_{ij}$ is a smooth mapping and that the base space mapping $\sum_i \to \sum_j$ corresponding to $\bar{\psi}_{ij}$ coincides with the mapping ψ_{ij}.

Proposition 1.2. (a) *The mapping $\bar{\psi}_{ij}$ transforms sections of the bundle $\mathscr{V}_i$ parallel with respect to the connection $\omega_{i\lambda}(\lambda \in c(i))$ into sections of the bundle $\mathscr{V}_j$ parallel with respect to the connection $\omega_{j\lambda}$.*
(b) *If $X_i \subset X_j \subset X_k$, then $\bar{\psi}_{jk} \circ \bar{\psi}_{ij} = \bar{\psi}_{ik}$.*

The *proof* is evident.

In what follows, we consider only schemes of combinatorial vector bundles which have the following property: if X_i is an arbitrary cell of the decomposition $\mathscr{X}$ and $\alpha\in c(i)$ then the vector subspaces V_j of the space V_α which correspond to cells X_j such that $X_i \subset X_j \subset X_\alpha$ are in general position.

Let X_i be some cell of codimension $p \leqq m$, and $\sum_i$ the corresponding space of fittings. Let $\alpha_0 \in c(i)$, $j \in s(i)$, $\alpha \in c(j)$, and $\kappa = \{f_{\alpha\beta}\}_{\alpha,\beta\in c(i)} \in \sum_i$, $V_j(\kappa) = f_{\alpha\alpha_0}(V_j)$. It is clear that $V_j(\kappa)$ is an $(m-p+1)$-dimensional subspace of the space V_{α_0} containing the space V_i. Take some set of indices $j_1,\dots,j_k \in s(i)$ and some nonnegative integer $q \leqq m-p+k$. Denote by $\sum_i(j_1,\dots,j_k;q)$ the subset of the space $\sum_i$ consisting of points κ such that the dimension of the subspace of the space V_{α_0} spanned by the subspaces $V_{j_1}(\kappa),\dots,V_{j_k}(\kappa)$ is no more than q. It is clear that $\sum_i(j_1,\dots,j_k;q)$ is a closed subset of the space $\sum_i$ that does not depend on the choice of the index $\alpha_0 \in c(i)$. A connected component of a finite intersection of subsets of the form $\sum_i(j_1,\dots,j_k;q)$ is called a stratum of the space $\sum_i$. It is easy to see that any subset of the space $\sum_i$ obtained from $\sum_i$ by removing a finite number of strata is an open subset of $\sum_i$. Denote by $\sum_{i,0}$ (by $\sum_{i,1}$, respectively) the subset of the space $\sum_i$ obtained from $\sum_i$ by removing all the strata of codimension 1 (codimension 2). It is clear that the subset $\sum_{i,0}$ consists of points κ such that any system of p mutually different indices $j_1,\dots,j_p \in s(i)$ defines a system of subspaces $V_{j_1}(\kappa),\dots,V_{j_p}(\kappa)$ generating the whole space V_{α_0}. Similarly, the subset $\sum_{i,1}$ consists of points κ such that there exists at most one system of p mutually different indices $j_1,\dots,j_p \in s(i)$ such that the dimension of the linear hull of the subspaces $V_{j_1}(\kappa),\dots,V_{j_p}(\kappa)$ is equal to $m-1$, while all other such systems of indices define systems of subspaces which are in general position.

It is easy to see that if $X_i \subset X_j$ then $\psi_{ij}(\sum_{i,0}) \subset \sum_{j,0}$ and $\psi_{i,j}(\sum_{i,1}) \subset \sum_{j,1}$.

Consider an n-dimensional topological manifold X with a linearized cellular decomposition $\mathscr{X}$ and the corresponding scheme of an n-dimensional combinatorial vector bundle. Let X_α be an n-dimensional cell of the decomposition $\mathscr{X}$ and $x \in X_\alpha$. We identify the vector space V_α with the tangent space to X_α at the point x. Denote by $C_\alpha(x)$ the closed subset of the space V_α consisting of vectors tangent to X_α at the point x and directed inside X_α (in the obvious sense). Take some cell X_i of the decomposition $\mathscr{X}$ and choose a smooth structure in the open neighborhood U of X_i. Then for any point $x \in X_i$ the tangent space $T_x(U)$ is defined and for any $\alpha \in c(i)$ a vector space isomorphism $e_\alpha\colon V_\alpha \to T_x(U)$ is given. It is clear that the sets $e_\alpha(C_\alpha(x))(\alpha \in c(i))$ cover all the space $T_x(U)$. Note that all the sets $e_\alpha(C_\alpha(x))$ contain the tangent space $T_x(X_i)$ to the cell X_i at the point x and the images of these sets under the factorization mapping $T_x(U) \to T_x(U)/T_x(X_i)$ form a decomposition of the quotient space $T_x(U)/T_x(X_i)$ into simplicial cones. All these constructions may be transferred by means of the isomorphism e_α into the space V_α and the quotient space V_α/V_i. The set of mappings $e_\alpha\colon V_\alpha \to T_x(U)(\alpha \in c(i))$ defines a point $y = \{f_{\alpha\beta}\}_{\alpha,\beta\in c(i)}$ of the space $\sum_i$ where $f_{\alpha\beta} = e_\beta^{-1} e_\alpha$. It is evident that the point y does not vary if one multiplies all the mappings e_α $(\alpha \in c(i))$ with the same automorphism of the space $T_x(U)$. The set of all points of the space $\sum_i$ which are related in the way indicated to decompositions of the spaces V_α/V_j into simplicial cones is an open subspace of the space $\sum_i$ which is denoted by $\sum_i'$. It may have empty intersections with some of spectra of $\sum_i$. Note that the configuration space for the cell X_i which was considered in the articles [1] and [2] is homotopically equivalent to the space Σ_i'.

2. *The complex $C_{**}(\sigma)$. Statement of the problem of the combinatorial computation of characteristic classes*

Let a scheme of m-dimensional combinatorial vector bundles over a topological manifold X with a cellular decomposition $\mathscr{X}$ be given. Suppose that in each space of fittings Σ_i an open subset Σ_i' obtained by removing some strata of Σ_i is chosen. Let the system of spaces $\sigma = \{\Sigma_i'\}$ satisfy the following condition: if $X_i \subset X_j$ then $\psi_{ij}(\Sigma_i') \subset \Sigma_j'$. One says that an m-dimensional combinatorial vector bundle determined by the system of mappings $f_i: X_i \to \Sigma_i$ belongs to the type σ if $f_i(X_i) \subset \Sigma_i'$.
Denote by $\Omega_i^q(\sigma)$ the space of smooth q-forms on Σ_i' taking values in $\mathbb{K}$.

Definition 2.1. One says that one has a (p, q)-chain ($p \geqq 0, q \geqq 0$) of type σ of a scheme of an m-dimensional combinatorial vector bundle if, for every cooriented p-dimensional cell X_i of the decomposition $\mathscr{X}$, a form $\omega(X_i) \in \Omega_i^{-q}(\sigma)$ is specified, and $\omega(-X_i) = -\omega(X_i)$. The set of all (p, q)-chains of type σ forms a vector space which is denoted by $C_{p,q}(\sigma)$.

Let us define linear mappings $D_1: C_{p,q}(\sigma) \to C_{p-1,q}(\sigma)$ and $D_2: C_{p,q}(\sigma) \to C_{p,q-1}(\sigma)$ by means of the formulas

$$(D_1\omega)(X_i^{p-1}) = (-1)^q \sum_{j \in s(i)} \varepsilon_{ij} \psi_{ij}^* \omega(X_i^p),$$

$$(D_2\omega)(X_j^p) = (-1)^{q-1} d\omega(X_j^p),$$

where $\omega \in C_{p,q}(\sigma)$ and ε_{ij} is the incidence number of the cooriented cells X_i^{p-1} and X_j^p.
It is easy to see that $D_1^2 = 0$, $D_2^2 = 0$ and $D_1 D_2 + D_2 D_1 = 0$. Hence $\{\bigoplus_{p,q} C_{p,q}(\sigma); D_1, D_2\}$ is a chain bicomplex. Let $D = D_1 + D_2$ be the total differential of this bicomplex.

Definition 2.2. The chain complex $C_{**}(\sigma) = \{\bigoplus_{p,q} C_{p,q}(\sigma), D\}$ is called the complex of type σ of the scheme of an m-dimensional combinatorial vector bundle.

Let P be an invariant symmetric k-linear form on the Lie algebra $\mathfrak{gl}(m. \mathbb{K})$. Take the cell X_i of the decomposition $\mathscr{X}$ and consider the simplicial complex $I_*(i) = \{\bigoplus_{q \geqq 0} I_q(i), \partial\}$ from Section 7 of Chapter I. If $\alpha \in c(i)$ then we denote by $\omega_{i\alpha}'$ the restriction of the connection $\omega_{i\alpha}$ of the bundle V_i to the bundle $V_i | \Sigma_i'$. Define a group homomorphism $P_i: I_q(i) \to \Omega_i^{2k-q}(\sigma)$ by means of the following formula:

$$P_i(\Delta^q(\alpha_0, \ldots, \alpha_q)) = P^{q, 2k-q}(\omega_{i\alpha_0}', \ldots, \omega_{i\alpha_q}'),$$

where $\alpha_0, \ldots, \alpha_q \in c(i)$.

Proposition 2.1. *Let $X_i \subset X_j$ and $\alpha_0, \ldots, \alpha_q \in c(i)$. Then*

$$P_i(\Delta^q(\alpha_0, \ldots, \alpha_q)) = \psi_{ij}^* P_j(\Delta^q(\alpha_0, \ldots, \alpha_q)).$$

The proof of this proposition is of purely technical character. One uses the first part of Proposition 1.2 and the construction of forms $P^{q,2k-q}$ from the proof of Proposition 6.2 of Chapter I.

Proposition 2.2. *If b is the current boundary operator on the space* $\sum_i'$ *then* $bP_i = P_i\partial$.

Proof. It is sufficient to prove that $bP_i(\Delta^q(\alpha_0,\ldots,\alpha_q)) = P_i(\partial\Delta^q(\alpha_0,\ldots,\alpha_q))$, but this follows easily from the definition of the forms $P^{q,2k-q}$ and proposition on 6.1. of Chapter I.

Now define the mapping $\xi_P: \hat{C}_{p,q}(\mathscr{X}) \to C_{p,q-2k}(\sigma)$ by means of the equality $\xi_P(c)(X_i^p) = P_i(c(X_i^p))$, where $c \in \hat{C}_{p,q}(\mathscr{X})$.

Proposition 2.3. *The mapping* ξ_P *is a homomorphism of the bicomplex* $\{\bigoplus_{p,q}\hat{C}_{p,q}(\mathscr{X}); \hat{\partial}_1, \hat{\partial}_2\}$ *into the bicomplex* $\{\bigoplus_{p,q}C_{p,q}(\sigma); D_1, D_2\}$ *of bidegree* $(0, -2k)$. *Hence it defines a homomorphism of the corresponding complete complexes of degree* $-2k$.

This follows from the definitions of the differentials $\hat{\partial}_1, \hat{\partial}_2$, D_1, D_2 and Propositions 2.1 and 2.2.

Suppose now that X is a smooth manifold and that the system of mappings $f_i: X_i \to \sum_i$ defines an m-dimensional combinatorial vector bundle of type $\sigma = \{\sum_i'\}$ on X. Consider the mapping $\eta_\sigma: C_{**}(\sigma) \to D'(X)$ defined by the formula

$$\eta_\sigma(\omega) = \sum X_i \wedge f_i^* \omega(X_i),$$

where $\omega \in C_{**}(\sigma)$ and the summation is taken over all the cells X_i furnished with some coorientation. It follows easily from the definition of the differential D and Proposition 5.1 of Chapter I that the mapping η_σ is a complex homomorphism.

Suppose in addition that the m-dimensional vector bundle induced by the given m-dimensional combinatorial vector bundle is smooth. The following proposition reveals the rôle of the homomorphism ξ_P.

Proposition 2.4. *The complex homomorphism* $\eta_\sigma \circ \xi_P$ *induces the homology homomorphism* $H_p^c(X, \Lambda) \to H_{p-2k}^c(X, \Lambda \otimes_Z \mathbb{R})$ $(p = 2k, \ldots, n)$ *which coincides with the homomorphism* ζ *from Proposition* 7.4. *of Chapter I.*

Proof. We must verify the equality $P^{q,2k-q}(\omega_{\alpha_0}, \ldots, \omega_{\alpha_q}) = f_i^* P^{q,2k-q}(\omega'_{i\alpha_0}, \ldots, \omega'_{i\alpha_q})$, where $\alpha_0, \ldots, \alpha_q \in c(i)$. It is easy to see that the bundle $E_i = E(X)|_{X_i}$ is isomorphic to the bundle $f_i^* V_i$. Let $\bar{f}_i: E_i \to V_i$ be the vector bundle mapping corresponding to the mapping f_i. For any $\alpha \in c(i)$ the connection ω_α on the bundle E_i is made compatible with the connection $\omega_{i\alpha}$ on the bundle V_i by means of the mapping $\bar{f}_i$. Therefore, the desired equality follows from the construction of the forms $P^{q,2k-q}$ given in the proof of Proposition 6.2 of Chapter I.

Note that one may consider the complex $\hat{C}_*(\mathscr{X}, \mathbb{Q})$ of cooriented chains of $\mathscr{X}$

with rational coefficients as a subcomplex of the complex $C_{**}(\sigma)$, namely, a p-dimensional chain $c=\sum k_i X_i^p \in \hat{C}_*(\mathscr{X},\mathbb{Q})$ is identified with the (p,O)-chain ω of the complex $C_{**}(\sigma)$ defined by the equality $\omega(X_i^p)=k_i$. It is easy to see that the resulting embedding $\hat{C}_*(\mathscr{X},\mathbb{Q})\subset C_{**}(\sigma)$ is compatible with the differentials $\hat{\partial}$ and D. Hence $\hat{C}_*(\mathscr{X},\mathbb{Q})$ is a subcomplex of the complex $C_{**}(\sigma)$.

The problem of combinatorial computation of the characteristic class of m-dimensional combinatorial vector bundles of type σ corresponding to the form P may be stated as follows. Let c be a cycle of the complex $\hat{C}_{**}(\mathscr{X})$ such that $\lambda(c)=f_X$ where f_X is the fundamental cycle of the manifold X. Take the cycle $\xi_P(c)$ of the complex $C_{**}(\sigma)$ and suppose that the homology class of the cycle $\xi_P(c)$ is the image of some homology class ρ of the complex $\hat{C}_*(\mathscr{X},\mathbb{Q})$ under the inclusion mapping $\hat{C}_*(\mathscr{X},\mathbb{Q})\subset C_{**}(\sigma)$. Then this homology class ρ depends only on the type σ of the given m-dimensional combinatorial vector bundle. Combinatorial computation of the homology class ρ consists in constructing a cycle Γ of the complex $\hat{C}_*(\mathscr{X},\mathbb{Q})$ which is homologous to the cycle $\xi_P(c)$ in the complex $C_{**}(\sigma)$ and whose coefficients are defined only by the type σ of the m-dimensional combinatorial vector bundle.

3. *Hypersimplexes and hypersimplicial complexes*

Let Δ^p be the standard p-dimensional simplex in the Euclidean space and let k and l be two non-negative integers such that $p=k+l+1$. Denote by $\Delta^{k,l}$ the convex hull of the set of the centers of l-dimensional faces of the simplex Δ^p. The cell $\Delta^{k,l}$ is called a (geometrical) hypersimplex.

Examples. The hypersimplexes $\Delta^{0,p-1}$ and $\Delta^{p-1,0}$ $(p=1,2,\ldots)$ are simplexes. The simplest hypersimplex which is different from simplexes is the hypersimplex $\Delta^{1,1}$ which is an octahedron.

We shall deal below with hypersimplexes whose vertices are labeled with sets of indices.

Take a simplex Δ^p and some hypersimplex $\Delta^{k,l}\subset\Delta^p$ $(k+l+1=p)$. Let S be a set of N elements and q a positive integer such that $l<q$ and $k+q<N$. Take an arbitrary ordered set $j_0,\ldots,j_{k+q}$ of $k+q+1$ mutually different elements of the set S. Label the vertices of the simplex Δ^p with the elements $j_0,\ldots,j_p$ and agree that the orientation on of Δ^p be determined by the ordering $j_0,\ldots,j_p$ of its vertices. Let A be some vertex of the hypersimplex $\Delta^{k,l}$. We assign to it the set of indices $j_{\alpha_0},\ldots,j_{\alpha_l}, j_{p+1},\ldots,j_{k+q}$, where $\alpha_0,\ldots,\alpha_l$ are the numbers of the vertices of the l-dimensional face of the simplex Δ^p whose center is A. We obtain a cell, each vertex of which is labeled with a set of q elements of the set S. We denote this cell by $\Delta^{k,l}(j_0,\ldots,j_p;j_{p+1},\ldots,j_{k+q})$ and call it an (S,q)-hypersimplex. The inclusion $\Delta^{k,l}\subset\Delta^p$ defines an orientation of the cell $\Delta^{k,l}$. It is clear that the cell $\Delta^{k,l}(j_0,\ldots,j_p; j_{p+1},\ldots,j_{k+q})$ remains unchanged under any permutation in each of the groups $j_0,\ldots,j_p$ and $j_{p+1},\ldots,j_{k+q}$, and its orientation changes in accordance with the sign of the permutation of elements of the first group $j_0,\ldots,j_p$.

We call any point equipped with an unordered set of q mutually different elements $j_0,\ldots,j_{q-1}$ of the set S a 0-dimensional hypersimplex and we denote it by $\Delta^0(j_0,\ldots,j_{q-1})$.

Proposition 3.1. *The boundary of an (S,q)-hypersimplex is calculated by means of the following formulas:*

(a) $\partial\Delta^{k,l}(j_0,\ldots,j_p;j_{p+1},\ldots,j_{k+q})$

$$= \sum_{\mu=0}^{p}(-1)^{\mu}\Delta^{k-1,l}(j_0,\ldots,\hat{j}_\mu,\ldots,j_p;j_{p+1},\ldots,j_{k+q})$$

$$-\sum_{\mu=0}^{p}(-1)^{\mu}\Delta^{k,l-1}(j_0,\ldots,\hat{j}_\mu,\ldots,j_p;j_{p+1},\ldots,j_{k+q},j_\mu)$$

if $k+l\geqq 1$, where the first sum disappears in the case $k=0$, and the second sum disappears in the case $l=0$;

(b) $\partial\Delta^{0,0}(j_0,j_1;j_2,\ldots,j_q)=\Delta^0(j_1,j_2,\ldots,j_q)-\Delta^0(j_0,j_1,\ldots,j_q)$.

Proposition 3.2. *There exists a $1-1$ correspondence between the set of all (S,q)-hypersimplexes and the set of all faces of the "maximal" (S,q)-hypersimplex $\Delta^{N-q-1,q-1}(j_0,\ldots,j_{N-1})$, where $S=\{j_0,\ldots,j_{N-1}\}$.*

The proof of these proposition is left to the reader.

Again let S be a set of N elements and q a positive integer less than N. For $p\geqq 0$ we define an Abelian group $J_p(S,q)$ in the following way. If $p=0$ then $J_p(S,q)$ is the free Abelian group with the generators $\Delta^0(j_0,\ldots,j_{q-1})$ where $j_0,\ldots,j_{q-1}$ is an arbitrary unordered set of q mutually different elements of the set S. If $p>0$, then the group $J_p(S,q)$ is generated by symbols $\Delta^{k,l}(j_0,\ldots,j_p;j_{p+1},\ldots,j_{k+q})$, where $k+q<N$, $k+l+1=p$, and $j_0,\ldots,j_{k+q}$ is an arbitrary set of $k+q+1$ mutually different elements of the set S. These generators are tied by relations $\Delta^{k,l}(\tau_{j_0},\ldots,\tau_{j_p};\tau'_{j_{p+1}},\ldots,\tau'_{j_{q+k}})=(\operatorname{sgn}\tau)\Delta^{k,l}(j_0,\ldots,j_p;j_{p+1},\ldots,j_{q+k})$, where τ is an arbitrary permutation of the elements $j_0,\ldots,j_p$ and τ' is an arbitrary permutation of the elements $j_{p+1},\ldots,j_{k+q}$. It is easy to see that the generators $\Delta^0(j_0,\ldots,j_{q-1})$ and $\Delta^{k,l}(j_0,\ldots,j_p;j_{p+1},\ldots,j_{k+q})$ of the groups $J_p(S,q)$ may be interpreted as oriented geometric hypersimplexes.

Definition 3.1. The graded group $J_*(S,q)=\bigoplus_{p\geqq 0}J_p(S,q)$ with the differential $\partial\colon J_p(S,q)\to J_{p-1}(S,q)$ defined by the formulas (a) and (b) from Proposition 3.1 is called the (S,q)-hypersimplicial complex.

Proposition 3.3 $H_p(J_*(S,q))=0$ for $p>0$, and $H_0(J_*(S,q))=\mathbb{Z}$.

Proof. Consider the geometric hypersimplex $\Delta^{N-q-1,q-1}$. Let K be the cellular complex consisting of the hypersimplex $\Delta^{N-q-1,q-1}$ and all its faces. According to Proposition 3.2, the complex $J_*(S,q)$ is isomorphic to the cellular chain complex of the cell complex K. It remains to note that the hypersimplex $\Delta^{N-q-1,q-1}$ is homeomorphic to the ball.

4. *The complex of formal hypersimplicial films*

Let X be an n-dimensional topological space with a cellular decomposition $\mathscr{X}$. Take a cell X_i of codimension p of the decomposition $\mathscr{X}$. For $p \geqq 1$ put $J_*(i) = J_*(s(i), p)$. For $p = 0$ define $J_*(i)$ as the graded group with the only non-trivial component $J_0(i) = \mathbb{Z}$. Then define the graded group homomorphisms $\tilde{\varphi}_{ij}: J_*(j) \to J_*(i)$ ($j \in s(i)$) in the following way. If $p > 1$ then the homomorphism $\tilde{\varphi}_{ij}$ is defined by the formulas

$$\begin{aligned}&\tilde{\varphi}_{ij}(\Delta^{k,l}(\alpha_0, \ldots, \alpha_p; \alpha_{p+1}, \ldots, \alpha_{k+p-1}))\\ &\quad = \Delta^{k,l}(\varphi_{ij}(\alpha_0), \ldots, \varphi_{ij}(\alpha_p); \varphi_{ij}(\alpha_{p+1}), \ldots, \varphi_{ij}(\alpha_{k+p-1}), j),\\ &\quad \cdot \tilde{\varphi}_{ij}(\Delta^0(\alpha_0, \ldots, \alpha_{p-2})) = \Delta^0(\varphi_{ij}(\alpha_0), \ldots, \varphi_{ij}(\alpha_{p-2}), j),\end{aligned}$$

where $\alpha_0, \ldots, \alpha_{k+p-1} \in s(j)$. If $p = 1$ then the homomorphism $\tilde{\varphi}_{ij}$ is defined as the mapping $J_0(j) = \mathbb{Z} \to J_0(i)$ which takes 1 into $\Delta^0(j)$.

Proposition 4.1. *The homomorphism $\tilde{\varphi}_{ij}: J_*(j) \to J_*(i)$ ($j \in s(i)$) is a complex homomorphism. If $j, j' \in s(i)$ and $k \in s(j) \cap s(j')$ then $\tilde{\varphi}_{ij'} \circ \tilde{\varphi}_{j'k} = \tilde{\varphi}_{ij} \circ \tilde{\varphi}_{jk}$.*

Proof. The first assertion can be checked without difficulty. The second assertion follows from the definition of $\tilde{\varphi}_{ij}$ and Proposition 1.2 of Chapter 1.

Now define a complex of formal hypersimplicial films similar to the complex of formal simplicial films $\hat{C}_{**}(\mathscr{X})$.

Definition 4.1. One says that one has a formal hypersimplicial (p, q)-film of the decomposition $\mathscr{X}$ if, to any cooriented p-dimensional cell X_i of the decomposition $\mathscr{X}$, there corresponds a chain $\gamma(X_i) \in J_q(i) = J_q(s(i), n-p)$, such that $\gamma(-X_i) = -\gamma(X_i)$. The set of all formal hypersimplicial (p, q)-films of the decomposition $\mathscr{X}$ forms an Abelian group which is denoted by $\tilde{C}_{pq}(\mathscr{X})$.

Define Abelian group homomorphisms $\tilde{\partial}_1: \tilde{C}_{pq}(\mathscr{X}) \to \tilde{C}_{p-1,q}(\mathscr{X})$ and $\tilde{\partial}_2: \tilde{C}_{pq}(\mathscr{X}) \to C_{p,q-1}(\mathscr{X})$ by means of the following conditions:

$$\tilde{\partial}_1 \gamma(X_i^{p-1}) = (-1)^q \sum_{j \in s(i)} \varepsilon_{ij} \tilde{\varphi}_{ij} \gamma(X_j^p),$$

$$\tilde{\partial} \gamma(X_j^p) = \partial \gamma(X_j^p),$$

where $\gamma \in \tilde{C}_{pq}(\mathscr{X})$ and ε_{ij} is the incidence number of the cells X_i^{p-1} and X_j^p.
It is easy to see that $\tilde{\partial}_1^2 = 0$, $\tilde{\partial}_2^2 = 0$ and $\tilde{\partial}_1 \tilde{\partial}_2 + \tilde{\partial}_2 \tilde{\partial}_1 = 0$. Hence $\{\bigoplus_{p,q \geq 0} \tilde{C}_{pq}(\mathscr{X}); \tilde{\partial}_1, \tilde{\partial}_2\}$ is a chain bicomplex. Let $\tilde{\partial}_3 = \tilde{\partial}_1 + \tilde{\partial}_2$ be the total differential of this complex.

Definition 4.2. The chain complex $\tilde{C}_{**}(\mathscr{X}) = \{\bigoplus_{p,q \geq 0} \tilde{C}_{pq}(\mathscr{X}), \tilde{\partial}_3\}$ is called the complex of formal hypersimplicial films of the decomposition $\mathscr{X}$.

Let us define the mapping $\tilde{\lambda}: \tilde{C}_{**}(\mathscr{X}) \to \hat{C}_*(\mathscr{X})$ by means of the following conditions: (a) $\tilde{\lambda}(\tilde{C}_{pq}(\mathscr{X})) = 0$ if $q > 0$; (b) if $\gamma \in \tilde{C}_{p,0}(\mathscr{X})$, $p < n$ and $\gamma(X_i^p) =$

$\sum k_{j_0,\ldots,j_{n-p-1}} \Delta^0(j_0,\ldots,j_{n-p-1})$, where $j_0,\ldots,j_{n-p-1} \in s(i)$ and $k_{j_0,\ldots,j_{n-p-1}} \in \mathbb{Z}$, then $\tilde{\lambda}(\gamma) = \sum_i \sum_{j_0,\ldots,j_{n-p-1}} k_{j_0,\ldots,j_{n-p-1}} X_i^p$; (c) if $\gamma \in \tilde{C}_{n,0}(\mathscr{X})$, then $\tilde{\lambda}(\gamma) = \sum_\alpha (X_\alpha^n) X^n$.
It is easy to check that $\tilde{\lambda}$ is a complex homomorphism.

Proposition 4.2. *The homomorphism $\tilde{\lambda}$ induces an isomorphism between the homologies of complexes $\tilde{C}_{**}(\mathscr{X})$ and $\hat{C}_*(\mathscr{X})$.*

This proposition is proved in the same way as Proposition 7.1 of Chapter 1.
Let us compare the groups $J_*(i)$ with the groups $I_*(i)$ for cells of codimension $p=0,1$. Let $p=0$. Then after identifying the generator $\Delta^0(i)$ of the group $I_0(i)$ with 1 we get an equality $I_0(i) = J_0(1)$. Let $p=1$. Then $s(i)=c(i)$, and after identifying the generator $\Delta^1(j_0,j_1)$ of the group $I_1(i)$ with the generator $\Delta^{0,0}(j_0,j_1)$ of the group $J_1(i)$ we get an equality $I_1(i) = J_1(i)$.

5. *The forms in the spaces of fittings corresponding to hypersimplexes*

Let X_i be a cell of codimension $p \leqq m$, and $\sum_i$ the corresponding space of fittings. Take an m-dimensional vector bundle $\mathscr{V}_i$ over the space of fittings $\sum_i$, and let $i \in s(i)$, $\alpha \in c(j)$, $y \in \sum_i$, and $\mathscr{V}_i(y)$ be the fibre of the bundle $\mathscr{V}_i$ at the point y. Denote by $\mathscr{V}_{ij}(y)$ the subspace of the space $\mathscr{V}_i(y)$ consisting of points $\{\xi_\beta\}_{\beta \in c(i)}$ such that $\xi_\alpha \in V_j$. It is clear that $\mathscr{V}_{ij} = \bigcup_{y \in \sum_i} \mathscr{V}_{ij}(y)$ is a $(m-p+1)$-dimensional smooth vector subbundle of the bundle $\mathscr{V}_i$ which does not depend on the choice of the index $\alpha \in c(j)$. Any element $\eta \in V_j$ defines a smooth section $s(\eta)$ of the bundle $\mathscr{V}_{ij}$ composed by vectors $\{\xi_\beta\}_{\beta \in c(i)}$ from fibres $\mathscr{V}_{ij}(y)$ such that $\xi_\alpha = \eta$.

Proposition 5.1. *Let $j_1,\ldots,j_p$ be an arbitrary set of pairwise different indices from the set $s(i)$. Then the bundle $\mathscr{V}_{i|\Sigma_{i,0}}$ possesses a unique connection of zero curvature $\omega_{ij_1\ldots j_p}$ such that all the sections $s(\eta_{j_\mu})$ of the bundles $\mathscr{V}_{ij_\mu}$ where $\eta_{j_\mu} \in V_{j_\mu}$ $(\mu=1,\ldots,p)$ are parallel with respect to this connection. If $\alpha \in c(i)$ and $j_1,\ldots,j_p \in c(i)$ are the indices of all faces of codimension $p-1$ of the cell X_α which contain the cell X_i, then $\omega_{ij_1\ldots j_p} = \omega_{i_\alpha}$.*

The *proof* is evident.

Take an oriented geometric hypersimplex $\Delta^{k,l}(j_0,\ldots,j_q;j_{q+1},\ldots,j_{k+p})$ where $j_0,\ldots,j_{k+p} \in s(i)$ and $k+l+1=q$. Each vertex A of this hypersimplex defines p indices of the set $s(i)$ and hence, according to Proposition 5.1, defines a connection ω_A of zero curvature on the bundle $\mathscr{V}_{i|\Sigma_{i,0}}$. Decompose our hypersimplex into p-dimensional simplexes without adding new vertices, and give these p-dimensional simplexes orientation compatible with the orientation of the hypersimplex. Let P be a k-linear symmetric invariant form on the Lie algebra $\mathrm{gl}(m,\mathbb{K})$. Then, to each of the constructed simplexes corresponds a form $P^{p,2k-p}(\omega_0,\ldots,\omega_p)$ on the space $\sum_{i,0}$, where $\omega_0,\ldots,\omega_p$ are the connections corresponding to the vertices of the simplex taken in the order corresponding to the orientation of the simplex. We relate to the hypersimplex $\Delta^{k,l}$ the form $P^{p,2k-p}(\Delta^{k,l})$, which is equal to the sum of all the forms $P^{p,2k-p}$ constructed for the simplexes of our decomposition of $\Delta^{k,l}$.

The form $P^{p,2k-p}(\Delta^{k,l})$ depends, generally, on the decomposition of the hypersimplex $\Delta^{k,l}$ into simplexes. Since the hypersimplexes $\Delta^{0,p-1}$ and $\Delta^{p-1,0}$ are simplexes, then the forms $P^{p,2k-p}(\Delta^{0,p-1})$ and $P^{p,2k-p}(\Delta^{p-1,0})$ are uniquely defined.

Let $\mathscr{V}_{ij_{q+1}\dots j_{q+k}}$ be an $(m-l-1)$-dimensional subbundle of the bundle $\mathscr{V}_{i\|\Sigma_{i,0}}$ spanned by $(m-p-1)$-dimensional subbundles $\mathscr{V}_{ij_{q+1}},\dots,\mathscr{V}_{ij_{q+k}}$ of the bundle $\mathscr{V}_1$. It is clear that the connections corresponding to the vertices of the hypersimplex $\Delta^{k,l}=\Delta^{k,l}(j_0,\dots,j_q;\, j_{q+1},\dots,j_{q+k})$ induce on the bundle $\mathscr{V}_{ij_{q+1}\dots j_{q+k}}$ the same connection of zero curvature. We replace connections corresponding to the vertices of the hypersimplex by the connections induced by them on the quotient bundle $\mathscr{V}_i(\Delta^{k,l})=\mathscr{V}_{i\|\Sigma_{i,0}}/\mathscr{V}_{ij_{q+1}\dots j_{q+k}}$. According to Proposition 6.4 of Chapter I the form $P^{p,2k-p}(\Delta^{k,1})$ does not vary under such a replacement.

6. *The spaces* $\tilde{\Sigma}_i$ *and the forms* $X_{j_1\dots j_p}$

Let X_i be a cell of codimension $p\leqq m$ of the decomposition $\mathscr{X}$, Σ_i the corresponding space of fittings and $\mathscr{V}_i$ an m-dimensional vector bundle over the space Σ_i. Consider the product $\Sigma_i\times V_i$ regarded as a $(m-p)$-dimensional vector bundle over the space Σ_i and identify it with the subbundle of the bundle $\mathscr{V}_i$ by means of the natural embedding $\Sigma_i\times V_i\subset\mathscr{V}_i$ defined by the correspondence $(y,\xi)\to(y,\{\xi_\alpha\}_{\alpha\in c(i)})$, where $y\in\Sigma_i$ and $\xi_\alpha=\xi$.

Let $j\in s(i)$. In Section 5 an $(m-p+1)$-dimensional subbundle $\mathscr{V}_{ij}$ of the bundle $\mathscr{V}_i$ was constructed. It is clear that $\Sigma_i\times V_i\subset\mathscr{V}_{ij}$. Take the quotient bundle $\mathscr{U}_i=\mathscr{V}_i/\Sigma_i\times V_i$. Each index $j\in s(i)$ defines a one-dimensional vector subbundle $\mathscr{U}_{ij}=\mathscr{V}_{ij}/\Sigma_i\times V_j$ of the bundle $\mathscr{U}_i$. Consider the system $(y,\{\eta_j\}_{j\in s(i)})$ composed of the point $y\in\Sigma_i$ and the set of vectors $\{\eta_j\}_{j\in s(i)}$ where η_j is a non-zero vector from the fiber of the bundle $\mathscr{U}_{ij}$ at the point y. Two such system, $(y,\{\eta_j\}_{j\in s(i)})$ and $(y',\{\eta'_j\}_{j\in s(i)})$, will be called equivalent if $y'=y$ and $\eta'_j=\pm\eta_j$ $(j\in s(i))$. Denote the set of equivalence classes of systems $(y,\{\eta_j\}_{j\in s(i)})$ by $\tilde{\Sigma}_i$. It is easy to see that $\tilde{\Sigma}_i$ is a smooth manifold and that the natural projection $\pi_i\colon\tilde{\Sigma}_i\to\Sigma_i$ defines on $\tilde{\Sigma}_i$ the structure of a smooth bundle over the space Σ_i. If $\mathbb{K}=\mathbb{R}$, then the fiber of the bundle $\pi_i\colon\tilde{\Sigma}_i\to\Sigma_i$ is homeomorphic to the space $\mathbb{R}^{N_i}$ where N_i is the cardinality of the set $s(i)$.

Furthermore, we shall use the following.

Proposition 6.1. *Let a scheme of an m-dimensional combinatorial vector bundle on a manifold X with a cellular decomposition* $\mathscr{X}=\{X_i\}_{i\in I}$ *be given. Then there exists metrics on the vector spaces* V_i *(Euclidean if* $\mathbb{K}=\mathbb{R}$ *and Hermitian if* $\mathbb{K}=\mathbb{C}$*) such that if* $X_i\subset X_j$ *then the metric of the space* V_i *is induced by the metric of the space* V_j *and the inclusion* $V_i\subset V_j$.

Proof. It is easy to show that one can embed all the spaces V_i into the space $\mathbb{K}^\infty$ in such a manner that the following conditions hold: (a) for each V_i there exists a natural number n_i such that $V_i\subset\mathbb{K}^{n_i}\subset\mathbb{K}^\infty$; (b) the embeddings $V_i\subset\mathbb{K}^\infty$ are compatible with the given embeddings of the spaces V_i into each other. It is clear that the metric $ds^2=\sum_{s=1}^\infty\xi_s\bar{\xi}_s$ on the space $\mathbb{K}^\infty$, where $\xi_s\in\mathbb{K}$, induces the required metrics on the space V_i.

Assume now that we have mutually compatible metrics on the vector spaces of the given scheme of an m-dimensional combinatorial vector bundle. Then some metric is induced on each fiber of the bundle $\mathscr{U}_i$. Take a system of p pairwise different indices $j_1,\dots,j_p \in s(i)$ and a system $(y,\{\eta_j\}_{j\in s(i)})$, determining a point $\tilde{y}$ of the space $\tilde{\Sigma}_i$. It is clear that the determinant $\delta_{j_1\dots j_p}$ composed of orthonormal coordinates of the vectors $\eta_{j_1},\dots,\eta_{j_p}$ is defined at the point y up to the sign. Hence the ratio $X_{j_1\dots j_p}=(d\delta_{j_1\dots j_p}/\delta_{j_1\dots j_p})$ is a smooth 1-form on the space $\tilde{\Sigma}_{i,0}=\pi_i^{-1}(\Sigma_{i,0})$. It is easy to see that the forms $X_{j_1\dots j_p}$ do not depend on the choice of the mutually compatible metrics or the order of the indices $j_1,\dots,j_p$.

Assume now that $\mathbb{K}=\mathbb{R}$. Then it is evident that $X_{j_1\dots j_p}=d\log\Delta_{j_1\dots j_p}$, where $\Delta_{j_1\dots j_p}=|\delta_{j_1\dots j_p}|$ is a smooth function on the space $\tilde{\Sigma}_{i,0}$. In the case $\mathbb{K}=\mathbb{C}$ the form $X_{j_1\dots j_p}$ is not exact.

Now let $j\in s(i)$. Let us define mappings $\theta_{ij}\colon\tilde{\Sigma}_i\to\tilde{\Sigma}_j$ in the following way. Let $(y_i,\{\eta_j\}_{j\in s(i)})$ be a system which defines a point $\tilde{y}_i$ of the space $\tilde{\Sigma}_i$. The bundle mapping $\psi_{ij}\colon\mathscr{V}_i\to\mathscr{V}_j$ defines the mapping $\mathscr{U}_i\to\mathscr{U}_j$ between the quotient bundles. We define $\theta_{ij}(\tilde{y}_i)$ as the point of the space $\tilde{\Sigma}_j$ which is represented by the system $(y_j,\{\xi_k\}_{k\in s(i)})$, where $y_j=\psi_{ij}(y_i)$ and the vector ζ_k is the image of the vector $\eta_{\varphi_{ij}(k)}$ under the mapping $\mathscr{U}_i\to\mathscr{U}_j$. It is easy to see that θ_{ij} is a smooth mapping. Obviously the following Proposition is valid.

Proposition 6.2. *If $j,j'\in s(i)$ and $k\in s(j)\cap s(j')$ then*

$$\theta_{j'k}\theta_{ij'}=\theta_{jk}\theta_{ij}.$$

Take the section ζ_j of the one-dimensional bundle $\mathscr{U}_{ij}$ induced by the choice of a unit vector in the space V_j/V_i. Then, for any system $(y_i,\{\eta_j\}_{j\in s(i)})$ representing a point $\tilde{y}_i$ of the space $\tilde{\Sigma}_i$, we denote by z_j the ratio η_j/ζ_j. It is clear that z_j is defined up to a multiplicative factor of -1, and that $l_j=dz_j/z_j$ is a smooth 1-form on the space $\tilde{\Sigma}_i$ which does not depend on the choice of the section ζ_j.

Proposition 6.3. *Let X_i be a cell of codimension p and $j\in s(i)$. If $k_1,\dots,k_{p-1}\in s(s(j))$ and $j_\mu=\varphi_{ij}(k_\mu)$ $(\mu=1,\dots,p-1)$ then $\theta_{ij}^*X_{k_1,\dots,k_{p-1}}=X_{j_1\dots j_{p-1}j}-l_j$. If $\mathbb{K}=\mathbb{R}$ then $\Delta_{k_1\dots k_{p-1}}\theta_{ij}=|z_j|^{-1}\Delta_{j_1\dots j_{p-1}j}$.*

This follows in an obvious way from the definitions.

7. *The forms $P^{q,4-q}$ for the stable form π_2*

Take the stable form $P=\pi_2$ on the Lie algebra $\mathrm{gl}(m,\mathbb{K})$. According to Proposition 6.3 of Chapter I we have $p^{0,4}=P(R)$ where R is the connection form ω, $P^{3,1}=0$, and $P^{4,0}=0$. Let us write down the expressions for $P^{1,3}$ and $P^{2,2}$. They may be found by means of direct calculation:

$$P^{1,3}(\omega_0,\omega_1)=-\tfrac{1}{3}(\omega^\alpha_{1,\beta}-\omega^\alpha_{0,\beta})\wedge(\omega^\beta_{1,\gamma}-\omega^\beta_{0,\gamma})\wedge(\omega^\gamma_{1,\alpha}-\omega^\gamma_{0,\alpha})$$
$$+(R^\alpha_{0,\beta}+R^\alpha_{1,\beta})\wedge(\omega^\beta_{1,\alpha}-\omega^\beta_{0,\alpha}),$$
$$P^{2,2}(\omega_0,\omega_1,\omega_2)=-(\omega^\alpha_{0,\beta}\wedge\omega^\beta_{1,\alpha}+\omega^\alpha_{1,\beta}\wedge\omega^\beta_{2,\alpha}+\omega^\alpha_{2,\beta}\wedge\omega^\beta_{0,\alpha}),$$

where $\omega_i = (\omega^{\alpha}_{i,\beta})$ are connection forms ($i = 0, 1, 2; \alpha, \beta, \gamma = 1, \ldots, m$) and the summation is taken over the repeating indices α, β, γ from 1 to m.

Assume that a set of connection forms $\omega_1, \ldots, \omega_N$ on some principal G-bundle $\mathscr{P}$ is given. Let $c = \sum k_i \Delta^2_i$ ($k_i \in \mathbb{Z}$) be a chain of oriented 2-dimensional simplexes Δ^2_i whose vertices are labeled by connections from the set $\omega_1, \ldots, \omega_N$. If $\Delta^2 = \Delta^2(\omega_\lambda, \omega_\mu, \omega_\nu)(\lambda, \mu, \nu = 1, \ldots, N)$ then we put $P^{2,2}(\Delta^2) = P^{2,2}(\omega_\lambda, \omega_\mu, \omega_\nu)$ and define $P^{2,2}(c)$ by the equality $P^{2,2}(c) = \sum k_i P^{2,2}(\Delta^2_i)$. The expression for $P^{2,2}$ above allows the following to be proved easily.

Proposition 7.1. *If $c = \sum k_i \Delta^2_i$ and $c' = \sum k'_j \Delta^2_j$ are two chains of oriented two-dimensional simplexes whose vertices are labeled with connection forms from the set $\omega_1, \ldots, \omega_N$, such that $\partial c = \partial c'$, then $P^{2,2}(c) = P^{2,2}(c')$.*

Let a scheme of m-dimensional combinatorial vector bundles over an n-dimensional topological manifold X with a cellular decomposition $\mathscr{X}$ be given, and let X_i be a cell of codimension $p \leq m$ of this decomposition. Let us take an oriented hypersimplex $\Delta^{0,0} = \Delta^{0,0}(j_0, j_1; j_2, \ldots, j_p)$ where $j_0, \ldots, j_p \in s(i)$. We calculate the form $P^{1,3}(\Delta^{0,0})$ in the way indicated in Section 5. Denote by ω_0 and ω_1 the connections on the one-dimensional bundle $\mathscr{V}_i(\Delta^{0,0})$ induced by the connections $\omega_{j_0 j_2 \cdots j_p}$ and $\omega_{j_1 j_2 \cdots j_p}$, respectively. Then $P^{1,3}(\Delta^{0,0}) = P^{1,3}(\omega_0, \omega_1)$.

Choose in the space V_{j_μ} the vector η_{j_μ} transverse to the subspace $V_i \subset V_{j_\mu}$ and let $s(\eta_{j_\mu})$ be the section of the bundle $\mathscr{V}_{ij_\mu}$ corresponding to the vector η_{j_μ} ($\mu = 0, 1$). Denote by n_μ the image of the section $s(\eta_{j_\mu})$ under the projection $\mathscr{V}_{i|\Sigma_{i,0}} \to \mathscr{V}_i(\Delta^{0,0})$. It is clear that the equality $n_1 = a n_0$ holds, where a is a smooth function on $\Sigma_{i,0}$ which vanishes nowhere. The sections n_0 and n_1 may be regarded as sections of the principal bundle associated with the one-dimensional bundle $\mathscr{V}_i(\Delta^{0,0})$. It is easy to see that $n_0^* \omega_0 = 0$ and $n_0^* \omega_1 = -da/a$. This implies the equality $P^{1,3}(\Delta^{0,0}) = P^{1,3}(\omega_0, \omega_1) = 0$.

Take an oriented hypersimplex $\Delta^{0,1} = \Delta^{0,1}(j_0, j_1, j_2; j_3, \ldots, j_p)$, where $j_0, \ldots, j_p \in s(i)$, and calculate the form $P^{2,2}(\Delta^{0,1})$. Denote by $\omega_0, \omega_1, \omega_2$ the connections on the two-dimensional bundle $\mathscr{V}_i(\Delta^{0,1})$ induced by the connections $\omega_{ij_0 j_1 j_3 \cdots j_p}$, $\omega_{ij_1 j_2 j_3 \cdots j_p}$ and $\omega_{ij_0 j_2 j_3 \cdots j_p}$ respectively. Then $P(\Delta^{0,1}) = P(\omega_0, \omega_1, \omega_2)$. Choose in the space V_{j_μ} a vector n_{j_μ} transverse to the subspace $V_i \subset V_{j_\mu}$, and let $s(\eta_{j_\mu})$ be the section of the bundle $\mathscr{V}_{ij_\mu}$ corresponding to the vector η_{j_μ} ($\mu = 0, 1, 2$). Denote by n_μ the image of the section $s(\eta_{j_\mu})$ under the projection $\mathscr{V}_{i|\Sigma_{i,0}} \to \mathscr{V}_i(\Delta^{0,1})$. Let s be the section of the principal bundle associated with the two-dimensional vector bundle $\mathscr{V}_i(\Delta^{0,1})$, formed by the sections n_0 and n_1. It is clear that the equality $n_2 = a n_0 + b n_1$ holds, where a and b are smooth functions on $\Sigma_{i,0}$ which vanish nowhere. Simple calculations show that the following equalities hold:

$$s^* \omega_0 = 0, \quad s^* \omega_1 = \begin{pmatrix} -\dfrac{da}{a} & 0 \\ -\dfrac{db}{b} & 0 \end{pmatrix}, \quad s^* \omega_2 = \begin{pmatrix} 0 & -\dfrac{da}{a} \\ 0 & -\dfrac{db}{b} \end{pmatrix}.$$

This implies the equality $P^{2,2}(\Delta^{0,1}) = da/a \wedge db/b$.

Note that the functions $a\cdot\pi_i$ and $b\cdot\pi_i$ on the space $\tilde{\Sigma}_i$ coincide up to a constant factor with the ratios $\delta_{j_1\cdots j_p}/\delta_{j_0j_1j_3\cdots j_p}$ and $\delta_{j_0j_2\cdots j_p}/\delta_{j_0j_1j_3\cdots j_p}$, respectively. One easily obtains from this the following equality:

$$\begin{aligned}\pi_i^* P^{2,2}(\Delta^{0,1}) &= X_{j_0j_1j_3\cdots j_p}\wedge X_{j_1j_2j_3\cdots j_p}+X_{j_1j_2j_3\cdots j_p}\\&\wedge X_{j_0j_2j_3\cdots j_p}+X_{j_0j_2j_3\cdots j_p}\wedge X_{j_0j_1j_3\cdots j_p}\\&=3\operatorname{Alt}(0,1,2)(X_{j_0j_1j_3\cdots j_p}\wedge X_{j_1j_2j_3\cdots j_p}),\end{aligned}$$

where Alt (0, 1, 2) denotes the alternation operator over the indices 0, 1, 2.

Take the oriented hypersimplex $\Delta^{1,0}=\Delta^{1,0}(j_0,j_1,j_2;j_3,\ldots,j_{p+1})$, where $j_0,\ldots,j_{p+1}\in s(i)$. Denote by ω_0,ω_1 and ω_2 the connections on the one-dimensional bundle $\mathscr{V}_i(\Delta^{0,1})$ induced by the connections $\omega_{j_0j_3\cdots j_{p+1}}$, $\omega_{j_1j_3\cdots j_{p+1}}$ and $\omega_{j_2j_3\cdots j_{p+1}}$, respectively. Then $P^{2,2}(\Delta^{0,1})=P^{2,2}(\omega_0,\omega_1,\omega_2)$. Choose in the space V_j a vector η_{j_μ} transverse to the subspace $V_i\subset V_{j_\mu}$ and let $s(\eta_{j_\mu})$ be the section of the bundle $\mathscr{V}_{ij_\mu}$ corresponding to the vector η_{j_μ} ($\mu=0,1,2$). Denote by n_μ the image of the section $s(\eta_{j_\mu})$ under the projection $\mathscr{V}_{i|\Sigma_{i,0}}\to\mathscr{V}_i(\Delta^{1,0})$. It is clear that the equalities $n_1=an_0$ and $n_2=bn_0$ hold, where a and b are smooth functions on $\Sigma_{i,0}$ which vanish nowhere. It is easy to see that n_0, n_1 and n_2 may be regarded as sections of the principal bundle associated with the vector bundle $\mathscr{V}_i(\Delta^{1,0})$. Obviously,

$$n_0^*\omega_0=0,\quad n_0^*\omega_1=-\frac{da}{a},\quad n_0^*\omega_2=-\frac{db}{b}.$$

Hence $P^{2,2}(\Delta^{0,1})=db/b\wedge da/a$.

Note that the functions $\pi_i\cdot a$ and $\pi_i\cdot b$ on the space $\tilde{\Sigma}_i$ coincide up to a constant factor with the ratios $\delta_{j_1j_3\cdots j_{p+1}}/\delta_{j_0j_3\cdots j_{p+1}}$ and $\delta_{j_2j_3\cdots j_{p+1}}/\delta_{j_0j_3\cdots j_{p+1}}$, respectively. One easily obtains from this the following equality:

$$\begin{aligned}\pi_i^* P^{2,2}(\Delta^{1,0}) &= -(X_{j_0j_3\cdots j_{p+1}}\wedge X_{j_1j_3\cdots j_{p+1}}+X_{j_1j_3\cdots j_{p+1}}\\&\wedge X_{j_2j_3\cdots j_{p+1}}+X_{j_2j_3\cdots j_{p+1}}\wedge X_{j_0j_3\cdots j_{p+1}})\\&=-3\operatorname{Alt}(0,1,2)(X_{j_0j_3\cdots j_{p+1}}\wedge X_{j_1j_3\cdots j_{p+1}}).\end{aligned}$$

8. *The forms $S(\Delta^{k,l})$ and relations between them*

From this section we shall consider only real m-dimensional combinatorial vector bundles with $m\geqq 4$. Let X_i be a cell X_i of codimension $p\geqq m$ of the decomposition $\mathscr{X}$. For each hypersimplex

$$\Delta^{k,l}=\Delta^{k,l}(j_0,\ldots,j_{k+l+1};j_{k+l+2},\ldots,j_{p+k})\quad(k+1=1,2,3;j_0,\ldots,j_{p+k}\in s(i))$$

we define the forms $S(\Delta^{k,l})$ as indicated below. First of all,

$$\begin{aligned}S(\Delta^{0,1})&=3\operatorname{Alt}(0,1,2)(\log\Delta_{j_0j_1j_3\cdots j_p}X_{j_1j_2\cdots j_p});\\S(\Delta^{1,0})&=3\operatorname{Alt}(0,1,2)(\log\Delta_{j_0j_3\cdots j_{p+1}}X_{j_1j_3\cdots j_{p+1}});\\S(\Delta^{0,2})&=0;\\S(\Delta^{2,0})&=0.\end{aligned}$$

It is clear that $S(\Delta^{0,1})$ and $S(\Delta^{1,0})$ are smooth 1-forms on the space $\tilde{\Sigma}_i$.

Take an oriented hypersimplex $\Delta^{1,1}=\Delta^{1,1}(j_0,j_1,j_2,j_3;j_4,\dots,j_{p+1})$. The ordered quadruple of subbundles $\mathscr{V}_{ij_0}, \mathscr{V}_{ij_1}, \mathscr{V}_{ij_2}; \mathscr{V}_{ij_3}$ of the bundle $\mathscr{V}_i$ defines in the fiber over the point $y\in\sum_{i,0}$ an ordered quadruple of straight lines $l_0(y)$, $l_1(y)$, $l_2(y)$, $l_3(y)$. Denote by $\kappa(y)$ the cross-ratio of the quadruple $l_0(y)$, $l_1(y)$, $l_2(y)$, $l_3(y)$. This means that if n_μ is a non-zero vector belonging to the line $l_\mu(y)$ ($\mu=0,1,2,3$) and $n_2=an_0+bn_1$, $n_3=cn_0+dn_1$, then $\kappa(y)=bc/ad$.

Let $\Phi(\kappa)$ be the smooth function on $\mathbb{R}\backslash\{0,1\}$ defined by the conditions

$$\Phi'(\kappa)=\log|\kappa|/(\kappa-1)-\log|\kappa-1|/\kappa;$$
$$\Phi(-1)=\Phi(1/2)=\Phi(2)=0.$$

We define $S(\Delta^{1,1})$ by the equality

$$S(\Delta^{1,1})=\Phi(\kappa(l_0,l_1,l_2,l_3)).$$

It is clear that $S(\Delta^{1,1})$ is a smooth function on $\sum_{i,0}$.

Furthermore, we put $S(\Delta^{0,3})=S(\Delta^{3,0})=0$.

Now take an oriented hypersimplex $\Delta^{1,2}=\Delta^{1,2}(j_0,j_1,j_2,j_3,j_4;j_5,\dots,j_{p+1})$. The ordered quintuple of subbundles $\mathscr{V}_{ij_0}, \mathscr{V}_{ij_1}, \mathscr{V}_{ij_2}, \mathscr{V}_{ij_3}, \mathscr{V}_{ij_4}$ defines in the fiber over the point $y\in\sum_{i,0}$ of the three-dimensional vector bundle $\mathscr{V}_i(\Delta^{1,2})$ an ordered quintuple of straight lines $l_0(y), l_1(y), l_2(y), l_3(y), l_4(y)$ in general position. Choose on each line $l_\mu(y)$ a non-zero vector $n_\mu(y)$ ($\mu=0,1,2,3,4$) and put

$$S(\Delta^{1,2})=\varepsilon\frac{\pi^2}{6},$$

where $\varepsilon=1$ if, among the ten bases $n_\alpha(y)$, $n_\beta(y)$, $n_\gamma(y)$ ($0\leqq\alpha<\beta<\gamma\leqq 4$) of the three-dimensional space, there are an even number with the same orientation; $\varepsilon=-1$ otherwise. It is easy to verify that $S(\Delta^{1,2})$ does not depend on the choice of the vectors $n_\mu(y)$.

Take, finally, an oriented hypersimplex $\Delta^{2,1}=\Delta^{2,1}(j_0,j_1,j_2,j_3,j_4;j_5,\dots,j_{p+2})$. The ordered quintuple of subbundles $\mathscr{V}_{ij_0}, \mathscr{V}_{ij_1}, \mathscr{V}_{ij_2}, \mathscr{V}_{ij_3}, \mathscr{V}_{ij_4}$ defines in the fiber over the point $y\in\sum_{i,0}$ of the two-dimensional bundle $\mathscr{V}_i(\Delta^{2,1})$ an ordered quintuple of straight lines $l_0(y), l_1(y), l_2(y), l_3(y), l_4(y)$ in general position. Choose on each line $l_\mu(y)$ a non-zero vector $n_\mu(y)$ ($\mu=0,1,2,3,4$) and put

$$S(\Delta^{2,1})=-\varepsilon\frac{\pi^2}{6},$$

where $\varepsilon=1$ if, among the ten bases $n_\alpha(y)$, $n_\beta(y)$ ($0\leqq\alpha<\beta\leqq 4$) of the two-dimensional space, there are an even number with the same orientation; $\varepsilon=-1$ otherwise. It is easy to verify that $S(\Delta^{2,1})$ does not depend on the choice of the vectors $n_\mu(y)$.

It is clear that $S(\Delta^{1,2})$ and $S(\Delta^{2,1})$ are locally constant functions on the space $\sum_{i,0}$.

Extend the forms $S(\Delta^{k,l})$ by linearity to arbitrary linear combinations of hypersimplexes.

Proposition 8.1. *The forms $S(\Delta^{k,l})$ have the following properties:*

(a) *$S(\Delta^{k,l}(j_0,\dots,j_{k+l+1};j_{k+l+2},\dots,j_{p+k}))$ is skew-symmetric with respect to* *permutations of the indices* $j_0, \ldots, j_{k+l+1}$ *and symmetric with respect to permutations of the indices* $j_{k+l+2}, \ldots, j_{p+k}$;

(b) *if* $k+l=1$, *then*

$$dS(\Delta^{k,l}) = \pi_i^* P^{2,2}(\Delta^{k,l});$$

(c) *if* $k+l=2$, *then*

$$dS(\Delta^{k,l}) = -S(\partial\Delta^{k,l});$$

(d) *if* $k+l=3$, *then*

$$S(\Delta^{k,l}) = S(\partial\Delta^{k,l}).$$

Proof. The assertion (a) for $k+l=1$ or 3 is evident, and in the case $k+l=2$ it follows from the properties of the function $\Phi(\kappa)$ proved in the paper [2] (see Corollary from Proposition 23). The assertions (b) and (c) are easily verified directly. The assertion (d) is proved in the paper [2] (see Proposition 25).

Proposition 8.2. *Let* $j \in s(i)$. *Then the following equalities hold:* $\theta_{ij}^* S(\Delta^{k,l}) = S(\tilde{\varphi}_{ij}\Delta^{k,l})$ *if* $k+l=1$; $\psi_{ij}^* S(\Delta^{k,l}) = S(\tilde{\varphi}_{ij}\Delta^{k,l})$ *if* $k+l=2$ *or* 3.

Proof. In the case $k+l=1$ this is proved with the help of Proposition 6.3. In the case $k+l=2$ or 3 the proof is evident.

9. *Combinatorial computation of the first Pontryagin class*

Consider an m-dimensional combinatorial vector bundle $E(\mathscr{X})$ of type σ over a topological manifold X with a cellular decomposition $\mathscr{X}$. We shall assume that the type $\sigma = \{\sum_i'\}$ satisfies the following conditions: (a) if X_k is a cell of codimension 2, then the space $\sum_k$ does not intersect sets of form $\sum_i(l_0, l_1, l_2; 1)$ where $l_0, l_1, l_2 \in s(k)$ and $H^1(\sum_k', \mathbb{R}) = 0$; (b) if X_j is a cell of codimension 3 then the space $\sum_j'$ is connected and $H^1(\sum_j', \mathbb{R}) = 0$; (c) if X_i is a cell of codimension 4 then the space $\sum_i'$ is connected.

Note that all these conditions were used in solving the problem of combinatorial computation of the first Pontryagin class of the manifold X in the papers [1] and [2].

Let f_X be the fundamental cycle of the manifold X, c_X a cycle of the complex $\hat{C}_{**}(\mathscr{X})$ such that $\lambda(c_X) = f_X$, and b_X a cycle of the complex $\tilde{C}_{**}(\mathscr{X})$ such that $\tilde{\lambda}(b_X) = f_X$. Assume that all the cells of the decomposition $\mathscr{X}$ are cooriented in this way or another and for each cooriented cell X_i^{n-p} of codimension p put $c_X(X_i^{n-p}) = c_i^p$ and $b_X(X_i^{n-p}) = b_i^p$. It is easy to see that the following equalities hold:

$$c_i^0 = b_i^0, \quad c_i^1 = b_i^1;$$

$$\partial c_i^{p+1} = (-1)^p \sum_{j \in s(i)} \varepsilon_{ij} c_j^p \quad (p=0, \ldots, n-1);$$

$$\partial b_i^{p+1} = (-1)^p \sum_{j \in s(i)} \varepsilon_{ij} \tilde{\varphi}_{ij} b_j^p \quad (p=0, \ldots, n-1),$$

where ε_{ij} is the incidence number of the cooriented cells X_i and X_j. Note that we have used the identifications $I_0(i)=J_0(i)$ and $I_1(i)=J_1(i)$ indicated in Section 4.

Take the form $P=\pi_2$. Consider the homomorphism ξ_P: $\hat{C}_{**}(\mathscr{X})\to C_{**}(\sigma)$, and let $\beta=\xi_P(c_X)$. It follows from the results of Section 7 that the cycle β vanishes on cells of codimension 0, 1, 3 and 4, that is $\beta\in C_{n-2,-2}(\sigma)$. Let X_k^{n-2} be a cell of codimension 2, $c(k)=\{\alpha_0,\dots,\alpha_N\}$ and $s(k)=\{l_0,\dots,l_N\}$. One may assume the cells X_k^{n-2} and $X_{l_0}^{n-1},\dots,X_{l_N}^{n-1}$ to be cooriented in such way that the following equalities hold:

$$c_{l_0}^1=b_{l_0}^1=\Delta^1(\alpha_N,\alpha_0),\quad c_{l_1}^1=b_{l_1}^1=\Delta^1(\alpha_0,\alpha_1),\dots,\quad c_{l_N}^1=b_{l_N}^1=\Delta^1(\alpha_{N-1},\alpha_N);$$
$$\partial c_k^2=\Delta^1(\alpha_0,\alpha_1)+\Delta^1(\alpha_1,\alpha_2)+\cdots+\Delta^1(\alpha_N,\alpha_0);$$
$$\partial b_k^2=\Delta^{0,0}(l_0,l_2,l_1)+\Delta^{0,0}(l_1,l_3,l_2)+\cdots+\Delta^{0,0}(1_N,l_1,l_0).$$

It is easy to see that the connections corresponding to the vertices of the simplexes $\Delta^1(\alpha_0,\alpha_1),\dots,\Delta^1(\alpha_N,\alpha_0)$ and the hypersimplexes $\Delta^{0,0}(l_0,l_2,l_1),\dots,$ $\Delta^{0,0}(l_N,l_1,l_0)$ coincide with each other. Hence Proposition on 7.1 implies that $\beta(X_k^{n-2})=P^{2,2}(c_k^2)=P^{2,2}(b_k^2)$. Note that $P^{2,2}(b_k^2)$ is the sum of forms defined on the space $\Sigma_{k,0}$, but $P^{2,2}(b_k^2)$ may be regarded as a smooth 2-form defined on the space Σ_k.

Since β is a cycle, the form $\beta(X_k^{n-2})$ is closed. But $H^1(\Sigma_k,\mathbb{R})=0$, and therefore there exists a smooth 1-form $\mu(X_k^{n-2})$ on the space Σ_k' such that $\beta(X_k^{n-2})=d\mu(X_k^{n-2})$. On the other hand, by virtue of Proposition 8.1 we have $\pi_k^*\beta(X_k^{n-2})=\pi_k^*P^{2,2}(b_k^2)=dS(b_k^2)$.

One has the following.

Proposition 9.1. *Let X_k be a cell of codimension 2 and $y\in\pi_k^{-1}(\Sigma_k')$. Then in some neighborhood U of the point y the form $S(b_k^2)$ may be represented as $(\Omega+dF)_{\Sigma_{k,0}\cap U}$, where Ω is a smooth 1-form in U and F is a continuous function in U which is smooth in $U\cap\Sigma_{k,0}$. Hence the form $S(b_k^2)$ is locally integrable on the space $\pi_k^{-1}(\Sigma_k')$.*

This assertion is proved in precisely the same way as Proposition 30 of the article [2].

Thus $\pi_k^*\mu(X_k^{n-2})-S(b_k^2)$ is a locally integrable 1-form on the space $\pi_k^{-1}(\Sigma_k')$. Since $H^1(\pi_k^{-1}(\Sigma_k'),\mathbb{R})=H^1(\Sigma_k',\mathbb{R})=0$, then there exists a current F_k on the space $\pi_k^{-1}(\Sigma_k')$ such that $\pi_k^*\mu(X_k^{n-2})-S(b_k^2)=dF_k$. It follows from Proposition 9.1 that F_k is a continuous function on $\pi_k^{-1}(\Sigma_k')$ and smooth on $\pi_k^{-1}(\Sigma_{k,0}\cap\Sigma_k')$.

Consider the chain $\beta_1\in C_{n-2,-1}(\sigma)$ defined by the equality $\beta_1(X_k^{n-2})=\mu(X_k^{n-2})$. It is easy to show that $\beta-D\beta_1\in C_{n-3,-1}(\sigma)$ is a cycle of the complex $C_{**}(\sigma)$, and in addition to this,

$$\pi_k^*(\beta-D\beta_1)(X_j^{n-3})=\sum_{k\in s(j)}\varepsilon_{jk}\theta_{jk}S(b_k^2)+\sum_{k\in s(j)}\varepsilon_{jk}d(F_k\theta_{jk}),$$

where X_j^{n-3} is a cooriented cell of codimension 3.

With the help of Propositions 8.1 and 8.2 one can reduce the last equality to the form

$$(\beta-D\beta_1)(X_j^{n-3})=-dS(b_j^3)+\sum_{k\in s(j)}\varepsilon_{jk}d(F_k\theta_{jk}).$$

Note that though the functions $F_k \cdot \theta_{jk}$ are defined on $\pi_j^{-1}(\Sigma'_j)$, the sum $\sum_{k \in s(j)} \varepsilon_{jk} d(F_k \cdot \theta_{jk})$ may be regarded as 1-form on the space Σ'_j.

Since $(\beta - D\beta_1)(X_j^{n-3})$ is a closed smooth 1-form on the space Σ'_j and $H^1(\Sigma'_j, \mathbb{R}) = 0$ then there exists a smooth function H_j on Σ'_j such that the following equality holds:

$$dH_j = -dS(b_j^3) + \sum_{k \in s(j)} \varepsilon_{jk} d(F_k \cdot \theta_{jk}).$$

Choose a connected component K_j in the intersection $\Sigma'_j \cap \Sigma_{j,0}$. It is easy to see that it is possible to select the function H_j in such a way that an equality

$$H_j = -S(b_j^3) + \sum_{k \in s(j)} \varepsilon_{jk} F_k \cdot \theta_{jk}$$

holds on K_j.

Consider the function $L_j = -H_j - S(b_j^3) + \sum_{k \in s(j)} \varepsilon_{jk} F_k \theta_{jk}$. It is clear that L_j is a locally constant function on $\Sigma'_j \cap \Sigma_{j,0}$ which is equal to zero on K_j. We shall indicate how to find values of the function L_j on any connected component of the set $\Sigma'_j \cap \Sigma_{j,0}$, using the following assertion (see Proposition 24 of [2]).

Proposition 9.2. *The limit of the function $\Phi(\kappa)$ when $\kappa \uparrow 0$, $\kappa \uparrow 1$ and $\kappa \to +\infty$ is equal to $\pi^2/6$, and the limit of the same function when $\kappa \downarrow 0$, $\kappa \downarrow 1$ and $\kappa \to -\infty$ is equal to $-\pi^2/6$.*

Let $y_0 \in K_j$, $y_1 \in \Sigma'_j \cap \Sigma_{j,0}$ and let $y(t)$ be a continuous curve in the set $\Sigma'_j \cap \Sigma_{j,1}$ such that $y(0) = y_0$ and $y(1) = y_1$. It is clear that the number $-L_j(y_1)$ is equal to the sum of jumps of the function $S(b_j^3)$ along the curve $y(t)$ from the point y_0 to the point y_1. The chain b_j^3 is equal to a linear combination of oriented hypersimplexes of the form $\Delta^{k,l}(k+l=2)$. Of these hypersimplexes, only those with the form $\Delta^{1,1}(k_0, k_1, k_2, k_3; k_4)$ with $k_1, k_2, k_3, k_4 \in s(j)$ make a non-zero contribution to the number $L_j(y_1)$. Consider an ordered quadruple of straight lines $l_0(y(t))$, $l_1(y(t))$, $l_2(y(t))$, $l_3(y(t))$ in the fiber of the two-dimensional vector bundle $\mathscr{V}_j(\Delta^{1,1})$ over the point $y(t)$ and denote their cross-ratio by $\kappa(t)$. Assume that when t changes from 0 to 1 the number $\kappa(t)$ passes μ_+ times through one of the values $0, 1, \infty$ in the positive direction $(0 \to 1 \to \infty \to 0)$ and μ_- times in the negative direction. Then, according to Proposition 9.2, the contribution of the hypersimplex $\Delta^{1,1}(k_0, k_1, k_2, k_3; k_4)$ to the number $L_j(y_1)$ is equal to $(\pi^2/3)(\mu_+ - \mu_-)$.

Let the chain $\beta_2 \in C_{n-3,0}(\sigma)$ be defined by the equality $\beta_2(X_j^{n-3}) = H_j$. Then $\beta - D\beta_1 + D\beta_2 \in C_{n-4,0}(\sigma)$ is a cycle of the complex $C_{**}(\sigma)$ and, in addition to this, $(\beta - D\beta_1 + D\beta_2)(X_i^{n-4}) = \sum_{j \in s(i)} \varepsilon_{ij} H_j \psi_{ij}$ for any cooriented cell X_i^{n-4} of codimension 4.

With the help of Proposition 6.6, it is easy to show that the equality

$$\sum_{j \in s(i)} \sum_{k \in s(j)} \varepsilon_{ij} \varepsilon_{jk} F_k \theta_{jk} \theta_{ij} = 0$$

holds. Hence

$$(\beta - D\beta_1 + D\beta_2)(X_i^{n-4}) = -\sum_{j \in s(i)} \varepsilon_{ij} S(b_j^3) \psi_{ij} - \sum_{j \in s(i)} \varepsilon_{ij} L_j \psi_{ij}.$$

Propositions 8.1 and 8.2 allow one to make the following transformations:

$$\sum_{j\in s(i)} \varepsilon_{ij} S(b_j^3)\psi_{ij} = S\left(\sum_{j\in s(i)} \varepsilon_{ij}\tilde{\varphi}_{ij} b_j^3\right) = -S(\partial b_i^4) = -S(b_i^4).$$

Finally we obtain

$$(\beta - D\beta_1 + D\beta_2)(X_i^{n-4}) = S(b_i^4) - \sum_{j\in s(i)} \varepsilon_{ij} L_j \psi_{ij}.$$

Since $\beta - D\beta_1 + D\beta_2$ is a cycle and the space $\sum_i'$ is connected then $S(b_i^4) - \sum_{j\in s(i)} \varepsilon_{ij} L_j \psi_{ij} = \text{const}$.

It is known that the first Pontryagin class of a smooth vector bundle is determined by the form $P_1 = (1/8\pi^2)\pi_2$. Define a cycle $\Gamma_\sigma \in C_{n-4,0}(\sigma)$ of the complex $C_{**}(\sigma)$ by means of the formula

$$\Gamma_\sigma(X_i^{n-4}) = -\frac{1}{8\pi^2}\left(S(b_i^4) - \sum_{j\in s(i)} \varepsilon_{ij} L_j \psi_{ij}\right),$$

where X_i^{n-4} is an arbitrary cooriented cell of codimension 4.

It is clear that the cycle Γ_σ is homologous to the cycle $\xi_P(c_X)$. In addition, it follows from the definition of the functions $S(\Delta^{k,l})$ $(k+l=3)$ and L_j that $\Gamma_\sigma(X_i^{n-4})$ is a rational number. Hence the cycle Γ_σ belongs to the subcomplex $\hat{C}_*(\mathscr{X}, \mathbb{Q})$ of the complex $C_{**}(\sigma)$ and therefore it defines a homology class in the group $H_{n-4}^c(X, \Lambda \otimes_Z \mathbb{Q})$ or, in view of the Poincaré duality, a cohomology class in $H^4(X, \mathbb{Q})$.

Thus we have the following.

Theorem. *Assume that we are given a real m-dimensional* $(m \geqq 4)$ *combinatorial vector bundle* $E(\mathscr{X})$*, of type* σ *as indicated at the beginning of this section, over an n-dimensional topological manifold X with a cellular decomposition* $\mathscr{X}$*. Then the cycle* Γ_σ *constructed above depends only on the type* σ*. If the vector bundle* $E(X)$ *associated with the combinatorial vector bundle* $E(\mathscr{X})$ *is smooth, then the four-dimensional cohomology class defined by the cycle* Γ_σ *coincides with the first Pontryagin class of the bundle* $E(X)$*.*

Thus the cycle Γ_σ yields a solution to the problem of the combinatorial calculation of the first Pontryagin class for a combinatorial vector bundle of type σ.

When this theorem is compared with Theorem 1 of the paper [1] one should take into account that the incidence numbers ε_{ij} of the cooriented cells X_i^{p-1} and X_j^p as defined in this paper and in the paper [1] differ by the multiple $(-1)^{p-1}$.

References

1. Gabrielov, A.M., Gelfand, I.M., Losik, M.V.: Combinatorial calculation of characteristic classes. Funct. Anal. Appl. **9** (2) (1975) 12–28
2. Gabrielov, A.M., Gelfand, I.M., Losik, M.V.: Combinatorial calculation of characteristic classes. Funct. Anal. Appl. **9** (3) (1975) 5–26
3. Gabrielov, A.M., Gelfand, I.M., Losik, M.V.: Local combinatorial formula for the first Pontryagin class. Funct. Anal. Appl. **10** (1) (1976) 14–17

21.
(with B. L. Feigin and D. B. Fuks)

Cohomology of infinite-dimensional Lie algebras and Laplace operators

Funkts. Anal. Prilozh. **12** (4) (1978) 1-5 [Funct. Anal. Appl. **12** (1978) 243-247].
MR **80i**:58050. Zbl. **396**:17008

Although appreciable progress has been made in the last 10 years in calculating the cohomology of infinite-dimensional Lie algebras, some of the problems of this circle appear to be unapproachable to this day; this relates, in the first place, to the algebras of Hamiltonian and divergence-free vector fields and to the algebras $L_1(n) \subset W_n$. In the solutions of these problems there has been hardly any successful progress without the introduction of new methods.

Such methods could consist of the investigation of the Laplace operators induced by any metrics in the cochain complex. The impression is added that if the metric is introduced in a reasonable way, then the eigenvalues and eigenvectors of the Laplace operator, and hence also the cohomology, will turn out to be calculable. In the following papers we propose to investigate this possibility systematically. In this paper we will analyze one example in which the program indicated can be realized successfully all the way through. It is true that the question concerns a Lie algebra whose cohomology is known: the algebra $L_1(1)$ of formal vector fields on the line having trivial 1-jet. In what follows we will denote this algebra by L_1 and whenever the question is of cohomology, we mean continuous cohomology with trivial coefficients.

The cohomology of the algebra L_1, as well as of the algebras $L_k(1)$ with $k > 1$, was found in [1]. Goncharova's calculation is awkward and does not allow one to find the cohomology of the algebra L_1 without finding the cohomology of the other algebras L_k.

Our paper significantly, as it seems to us, clarifies Goncharova's theorem and contains some new results. In particular, we give an explicit description of cocycles representing the cohomology classes of the algebra L_1 (we also apply this method of describing cocycles to the algebras L_k).

We recall that in the cohomology theory of infinite-dimensional Lie algebras, the cohomology of the algebra L_1 has special significance; the reason is explained in Goncharova's paper. One can add that recently Bukhshtaber and Shokurov discovered a connection between these cohomologies and the Adams—Novikov spectral sequence in complex cobordism theory [2].

Our paper owes much to Goncharova's paper: a whole series of our arguments is implicitly contained in it. It remains to indicate also the connection or in any case the analogy between what is presented below and the theory of Kats—Muda [3].

1. Definitions and General Considerations

The algebra L_1 consists of formal vector fields of the form $\sum_{i\geq 2} a_i x^i (d/dx)$ with $a_i \in \mathbf{R}$ and one imposes the inverse limit topology. The fields $e_i = x^{i+1}(d/dx)$ with $i = 1, 2, \ldots$, constitute a topological basis in L_1. The commutator acts according to the formula $[e_i, e_j] = (j - i)e_{i+j}$.

We denote by ε_i the functional on L_1 acting according to the formula $\varepsilon_i(e_j) = \delta_{ij}$. Obviously $\{\varepsilon_i\}$ is a basis for the space L_1', conjugate with L_1. In L_1' we introduce the metric in which the basis $\{\varepsilon_i\}$ is orthonormal.

The cohomology ring of the algebra L_1 is, by definition, the homology ring of the multiplicative complex $\Lambda^* L_1'$ (Λ^* denotes the exterior algebra), in which the grading and the multiplication are defined in the obvious way, and the differential $\delta: \Lambda^q L_1' \to \Lambda^{q+1} L_1'$ is defined by the formula

---

Moscow State University. Yaroslavl State University. Translated from Funktsional'nyi Analiz i Ego Prilozheniya, Vol. 12, No. 4, pp. 1-5, October-December, 1978. Original article submitted June 14, 1978.

$$\delta\alpha(\xi_1,\ldots,\xi_{q+1})=\sum_{1\leq s<t\leq q+1}(-1)^{s+t-1}\alpha([\xi_s,\xi_t],\xi_1,\ldots\hat{\xi}_s\ldots\hat{\xi}_t\ldots,\xi_{q+1})$$

$(\xi_1, \ldots, \xi_{q+1} \in L_1;\ \alpha \in \Lambda^q L_1')$. The differential is multiplicative, and $\delta e_k = \sum_{\substack{1\leq i<j\\ i+j=k}} (j-i)\,\varepsilon_i \wedge \varepsilon_j$; these properties determine δ completely. Instead of $\Lambda^* L_1'$ and $\Lambda^q L_1'$ we usually write $C^*(L_1)$ and $C^q(L_1)$.

We denote by $C_n^q(L_1)$ the subspace of the space $C^q(L_1)$ generated by the monomials $\varepsilon_{i_1} \wedge \ldots \wedge \varepsilon_{i_q}$ with $i_1 + \ldots + i_q = n$. Obviously, $C^q(L_1) = \bigoplus_n C_n^q(L_1)$ and $\delta(C_n^q(L_1)) \subset C_n^{q+1}(L_1)$. Thus, complex $C^*(L_1)$ decomposes into the sum of its subcomplexes $C_n^*(L_1) = \bigoplus_q C_n^q(L_1)$. Obviously, $\dim C_n^*(L_1) < \infty$ for each n.

The metrization of the space L_1' gives rise to the metrization for each q of the subspace $C^q(L_1)$ in which the monomials $\varepsilon_{i_1} \wedge \ldots \wedge \varepsilon_{i_q}$ constitute an orthonormal basis of this subspace; the decomposition $C^q(L_1) = \bigoplus C_n^q(L_1)$ becomes orthogonal here. (However, neither the differential δ, nor the natural action of the algebra L_1 on $C^q(L_1)$ is compatible with this metric.)

The metric allows one to identify the space $C_n^q(L_1)$ with the conjugate space $C_n^q(L_1)'$ (which is nothing else but the subspace of the space $\Lambda^q L_1$ generated by monomials of the form $e_{i_1} \wedge \cdots \wedge e_{i_q}$ with $i_1 + \ldots + i_q = n$), and the operator $d\colon C_n^{q+1}(L_1)' \to C_n^q(L_1)'$ conjugate with $\delta\colon C_n^q(L_1) \to C_n^{q+1}(L_1)$ turns out to be an operator $C_n^{q+1}(L_1) \to C_n^q(L_1)$. In this role it is defined by the formula

$$d(e_{i_1} \wedge \ldots \wedge e_{i_{q+1}}) = \sum_{1\leq s<t\leq q+1} (-1)^{s+t-1}(i_t - i_s)\, e_{i_s+i_t} \wedge e_{i_1} \ldots \hat{e}_{i_s} \ldots \hat{e}_{i_t} \ldots \wedge e_{i_{q+1}}.$$

Following the classical model, we introduce the Laplace operator

$$\Delta = d\delta + \delta d = (d+\delta)^2\colon C_n^q(L_1) \to C_n^q(L_1).$$

We recall its well-known property. It is a self-adjoint operator all of whose eigenvalues are nonnegative. The operator δ (as well as d) commutes with Δ, thanks to which the complex $C_n^*(L_1)$ decomposes into the sum of orthogonal subcomplexes, corresponding to distinct eigenvalues. Here the complexes corresponding to nonzero eigenvalues are acyclic, and the complex corresponding to eigenvalue zero has trivial differential. Thus, the dimension of the q-th homology space of the complex $C_n^*(L_1)$ is equal to the multiplicity of zero as an eigenvalue of the operator Δ in $C_n^q(L_1)$; moreover, each cohomology class is uniquely represented by an element of the kernel of the operator Δ, a "harmonic cocycle."

It seems likely that not only the kernel of the Laplace operator but also its entire spectrum has homological meaning; for now, however, it is not understood what this meaning consists of.

2. Results

Following Goncharova, we call a collection $(i_1, \ldots, i_q)$ of natural numbers basic, if each of the differences $i_2 - i_1, \ldots, i_q - i_{q-1}$ is greater than or equal to 3. For a basic collection $(i_1, \ldots, i_q)$, we set

$$E(i_1,\ldots,i_q) = \sum_{k=1}^{q} C_{i_k}^3 - \sum_{1\leq l<m\leq q} i_l i_m.$$

THEOREM 1. The set of eigenvalues of the Laplace operator $\Delta\colon C_n^*(L_1) \to C_n^*(L_1)$ coincides with the set of numbers of the form $E(i_1, \ldots, i_q)$, where $(i_1, \ldots, i_q)$ is a basic collection with $i_1 + \ldots + i_q = n$.

For example, for n = 7, the Laplace operator has eigenvalues E(7) = 35, E(1, 6) = 14, E(2, 5) = 0. For any n, the Laplace operator has integral eigenvalues (this would be hard to for see in advance).

The reader will see without difficulty that $E(i_1, \ldots, i_q) > 0$ for any basic collection $(i_1, \ldots, i_q)$ and that $E(i_1, \ldots, i_q) = 0$ in two cases (for fixed q, but not fixed n): when $(i_1, \ldots, i_q) = (1, 4, 7, \ldots, 3q-2)$ or $(i_1, \ldots, i_q) = (2, 5, 8, \ldots, 3q-1)$.

In order to say what the multiplicities of these eigenvalues are, we introduce the following notation. For a basic collection $i = (i_1, \ldots, i_q)$ we set

$$\alpha_1(i) = \begin{cases} 0, & \text{if} \quad i_1 < 3, \\ 1, & \text{if} \quad i_1 \geqslant 3; \end{cases} \quad \alpha_r(i) = \begin{cases} 0, & \text{if} \quad i_r - i_{r-1} = 3, \\ 1, & \text{if} \quad i_r - i_{r-1} > 3 \end{cases} \quad (r = 2, \ldots, q);$$

$$\alpha(i) = \alpha_1(i) + \alpha_2(i) + \ldots + \alpha_q(i).$$

THEOREM 2. Let $i = (i_1 \ldots i_q)$ be a basic collection with $i_1 + \ldots + i_q = n$. The multiplicity of eigenvalue $E(i)$ is equal to 1, if $\alpha(i) = 0$, and equal to $2\alpha(i)$, if $\alpha(i) > 0$).* Here $\alpha(i) = 0$ if and only if $E(i) = 0$; in this case the corresponding eigenvector lies in $C_n^q(L_1)$. Now if $\alpha(i) > 0$, then there exist mutually orthogonal eigenvectors $\gamma_1 \in C_n^{q_1}(L_1), \ldots, \gamma_{2\alpha(i)} \in C_n^{q_{2\alpha(i)}}(L_1)$ corresponding to eigenvalue $E(i)$ with $q_1 = q$, $q_2 = q_3 = q + 1$, $\ldots$, $q_{2\alpha(i)-2} = q_{2\alpha(i)-1} = q + \alpha(i) - 1$, $q_{2\alpha(i)} = q + 2\alpha(i)$.

The "stable cycles" of Goncharova are related to the eigenvectors of the operator Δ. We consider the operator $\sigma\colon C_n^q(L_1) \to C_{n+q}^q(L_1)$ defined by the formula

$$\sigma(e_{i_1} \wedge \ldots \wedge e_{i_q}) = e_{i_1+1} \wedge \ldots \wedge e_{i_q+1}.$$

An element γ of space $C_n^q(L_1)$ is called a stable cycle if $d(\sigma^m\gamma) = 0$ for $m = 0, 1, \ldots$.

THEOREM 3. The space of stable cycles is invariant with respect to Δ. Each of the numbers $E(i)$ is a onefold eigenvalue of the operator Δ on this subspace. The stable cycles which are eigenvectors of the operator with eigenvalue $E(i_1, \ldots, i_q)$, can be written in the form

$$e_{i_1} \wedge \ldots \wedge e_{i_q} + \sum a_{j_1 \ldots j_q} e_{j_1} \wedge \ldots \wedge e_{j_q},$$

where the summation is over collections $(j_1, \ldots, j_q) \neq (i_1, \ldots, i_q)$ with $j_1 \geqslant i_1$, $j_1 + j_2 \geqslant i_1 + i_2$, $\ldots$, $j_1 + \ldots + j_{q-1} \geqslant i_1 + \ldots + i_{q-1}$ $(j_1 + \ldots + j_q = i_1 + \ldots + i_q = n)$.

(This theorem recalls the results of Goncharova which also exploit the correspondence between stable cycles and basic collections.)

Now we indicate an explicit method for constructing stable cycles which is needed to prove our theorems and is important in its own right.

It will sometimes be convenient to identify elements of the space $C_n^q(L_1)$ with skew-symmetric polynomials of degree n in q variables: the monomial $e_{i_1} \wedge \ldots \wedge e_{i_q}$ corresponds to the polynomial $\sum \varepsilon(k_1, \ldots, k_q) z_{k_1}^{i_1} \ldots z_{k_q}^{i_q}$, where the summation is over all permutations $(k_1, \ldots, k_q)$ of the numbers $1, \ldots, q$, and ε denotes the parity of the permutation. This polynomial is obviously divisible by $z_1, \ldots z_q$ and by $\Pi_q = \prod_{1 \leqslant l < m \leqslant q} (z_m - z_l)$.

THEOREM 4. An element of the space $C_n^q(L_1)$ is a stable cycle if and only if the corresponding polynomial is divisible by Π_q^3.

Comparing Theorems 1-4, one can describe the cohomology of the algebra L_1.

THEOREM 5. If $q \geqslant 1$ then $\dim H^q(L_1) = 2$. The cocycles

$$z_1 \ldots z_q \Pi_q^3 \in C^q_{q+3C_q^2}(L_1), \quad z_1^2 \ldots z_q^2 \Pi_q^3 \in C^q_{2q+3C_q^2}(L_1)$$

serve as representatives of the q-dimensional cohomology classes of the algebra L_1.

Aside from the formulas for the cocyles, this theorem is contained in [1].

3. Proofs

We begin with Theorem 4. The polynomial method for writing cochains is convenient for describing the homomorphism d. Namely, if $\gamma \in C_n^q(L_1)$ has the form

$$z_1 \ldots z_q \Pi_q P(z_1, \ldots, z_q)$$

*Of course, the accidental coincidence $E(i) = E(i')$ is possible; in this case the multiplicities are simply added.

(where P is a symmetric polynomial), then $d\gamma$ can be represented by the polynomial

$$z_1 \dots z_{q-1} \Pi_q Q(z_1, \dots, z_{q-1}),$$

where $Q(z_1, \dots, z_{q-1}) = \sum_{i=1}^{q-1} z_i P(z_1, \dots, z_i, z_i, \dots, z_{q-1})(z_i - z_1) \dots (z_i - z_{q-1})$ (the null factor $(z_i - z_i)$ is omitted, naturally).* In order that the cochain γ be a stable cocycle, it is necessary that the right side of the last equation be equal to zero after replacement of the polynomial P by the polynomial $z_1^N \dots z_q^N P$ with arbitrarily large N. But after such a replacement the i-th summand of this sum is multiplied by $z_1^N \dots z_{i-1}^N z_i^{2N} z_{i+1}^N \dots z_{q-1}^N$ and from this it is evident that our requirement is equivalent with the requirement that each summand be equal to zero, i.e., that the polynomial

$$P(z_1, \dots, z_i, z_i, \dots, z_{q-1})$$

be equal to zero for each i. But this means that the polynomial P is divisible by Π_q, which means also by Π_q^2, since this polynomial is symmetric. Thus, γ is divisible by Π_q^3. Thus, stable cycles have the form $z_1 \dots z_q \Pi_q^3 P(z_1, \dots, z_q)$ with symmetric P, and this allows them to be enumerated. In particular, it is evident that in the space of stable cycles one can choose a basis consisting of elements of the form indicated in Theorem 3. We call such a basis triangular.

The proof of the remaining assertions reduces to the explicit enumeration of the eigenvectors of the operator Δ and the calculation of the corresponding eigenvalues. We give this calculation in abbreviated form.

First of all we write the matrix of the operator Δ. Let $I = (i_1, \dots, i_q)$, $J = (j_1, \dots, j_q)$ be ordered collections of natural numbers with $i_1 + \dots + i_q = n$, $j_1 + \dots + j_q = n$. The corresponding matrix element, i.e., the coefficient of $\varepsilon_{j_1} \wedge \dots \wedge \varepsilon_{j_q}$ in the expansion with respect to the standard basis of the cochain $\Delta(\varepsilon_{i_1} \wedge \dots \wedge \varepsilon_{i_q})$ is equal to:

1) $\sum_{r=1} C_{i_r}^3 + 2 \sum_{1 \leq s < t \leq q} i_s i_t - 3 \sum_{u=1}^{q} (q-u) i_u^2$ if $J = I$;

2) $(-1)^{r+s-t-u} 3 i_r (j_u - j_t)$ if there exist r, s, t, u, with $i_r < j_t < j_u < i_s$ and $(i_1, \dots \hat{i}_r \dots \hat{i}_s \dots, i_q) = (j_1, \dots \hat{j}_t \dots \hat{j}_u \dots, j_q)$;

3) $(-1)^{r+s-t-u} 3 j_r (i_u - i_t)$ if there exist r, s, t, u, with $j_r < i_t < i_u < j_s$ and $(j_1, \dots \hat{j}_r \dots \hat{j}_s \dots, j_q) = (i_1, \dots \hat{i}_t \dots \hat{i}_u \dots, i_q)$;

4) 0 in the remaining cases.

Using this description of the matrix of the operator Δ and the previous description of the stable cocycles, it is easy to verify that the space of stable cycles is invariant with respect to Δ and that the restriction of the operator Δ to this space in a triangular basis is determined by a triangular matrix (with respect to the natural order in the triangular basis) with diagonal elements E(i). This completes the proof of Theorem 3.

If $E(i) \neq 0$ then the corresponding stable cycle $\gamma(i)$ cannot be a cocycle (otherwise $\Delta\gamma(i) = 0$), and since Δ and δ commute, $\delta\gamma(i)$ also is an eigenvector of the Laplace operator with eigenvalue E(i).

In order to get the next portion of the eigenvectors one should proceed as follows. It is necessary to take a stable cycle $\gamma \in C_n^q(L_1)$ with $\delta(\gamma) \neq 0$ and to consider in C_{n+q+1}^{q+1} the cochain $\varepsilon_1 \wedge \sigma(\gamma)$ (see Sec. 2 for the definition of σ) and $\sigma(\delta\gamma)$. A short calculation shows that the two-dimensional space generated by these cochains is invariant with respect to Δ, so that it contains two eigenvectors; these eigenvectors can be found without difficulty. Then it is necessary to apply the construction described to these vectors, and so on until we get vectors which do not lie in Ker δ. In a finite number of steps we arrive at a complete system of eigenvectors.

*It should be said that the operator δ cannot be represented by such a compact formula. There is another method for describing the cochains of the algebra L_1 (and W_n) as skew-symmetric polynomials (see [4]), which is convenient for δ, but inconvenient for d. Unfortunately, the Laplace operator is not described by a compact formula in either case.

LITERATURE CITED

1. L. V. Goncharova, "The cohomology of the Lie algebra of formal vector fields on the line," Funkts. Anal. Prilozhen., 7, No. 2, 6-14 (1973); No. 3, 33-44 (1973).
2. V. M. Bukhshtaber and A. A. Shokurov, "The Landweber-Novikov algebra and formal vector fields on the line," Funkts. Anal. Prilozhen., 12, No. 3, 1-11 (1978).
3. V. G. Kats, "Infinite-dimensional Lie algebras and Dedekind's η-function," Funkts. Anal. Prilozhen., 8, No. 1, 77-78 (1974).
4. I. M. Gel'fand and D. B. Fuks, "The cohomology of Lie algebras of formal vector fields," Izv. Akad. Nauk SSSR, Ser. Mat., 34, 322-337 (1970).

22.

(with R. D. MacPherson)

Geometry in Grassmannians and a generalization of the dilogarithm

Adv. Math. **44** (1982) 279–312. Zbl. **504**:57021

In this paper we introduce some polyhedra in Grassman manifolds which we call Grassmannian simplices. We study two aspects of these polyhedra: their combinatorial structure (Section 2) and their relation to harmonic differential forms on the Grassmannian (Section 3). Using this we obtain results about some new differential forms, one of which is the classical dilogarithm (Section 1). The results here unite two threads of mathematics that were much studied in the 19th century. The analytic one, concerning the dilogarithm, goes back to Leibnitz (1696) and Euler (1779) and the geometric one, concerning Grassmannian simplices, can be traced to Binet (1811). In Section 4, we give some of this history along with some recent related results and open problems. In Section 0, we give as an introduction an account in geometric terms of the simplest cases.

Contents. 0. *Introduction.* 0.1. Polyhedra in Grassmann manifolds. 0.2. The first Grassmannian simplices. 0.3. The relation to differential forms. 1. *The generalized dilogarithm forms.* 1.1. Definitions. 1.2. The main results. 1.3. Functional equations. 2. *Grassmannian Simplices.* 2.1. Hypersimplices. 2.2. Projective configurations. 2.3. The closures of H^0 orbits. 2.4. Proof of the differential relation. 3. *Invariant differential forms on Grassman manifolds.* 3.1. Homology of the Grassman manifolds. 3.2. The Pontrjagin character. 3.3. Proof of the vanishing theorem. 4. *Some history, related work, and open problems.* 4.1. The dilogarithm. 4.2. The tetrahedral complex. 4.3. Grassmannian simplices. 4.4. The vanishing theorem.

0. Introduction

0.1. *Polyhedra in Grassmann Manifolds*

0.1.1. Let G_n^m denote the Grassmannian manifold of n dimensional vector subspaces of $\mathbb{R}^{n+m}$. We will think of G_n^m as the manifold of $n-1$ dimensional subspaces of real projective $n+m-1$ space. For example, G_1^m is projective m space itself.

The simplices form a natural class of polyhedra in the projective spaces G_1^m. We want to generalize them to polyhedra in the other Grassmannian manifolds G_n^m.

A key property of the m simplex in G_1^m is that its boundary is a union of $m-1$ simplices in G_1^{m-1} for various embeddings $G_1^{m-1} \subset G_1^m$. The other Grassman manifolds also have similar embeddings according to the pattern:

$$\begin{array}{ccccc} \vdots & & & & \\ G_3^1 & \subset & G_3^2 & & \\ \cup & & \cup & & \vdots \\ G_2^1 & \subset & G_2^2 & \subset & G_2^3 \ldots \\ \cup & & \cup & & \cup \\ G_1^1 & \subset & G_1^2 & \subset & G_1^3 \ldots \end{array}$$

(the horizontal inclusions are the subspaces in a given hyperplane; the vertical inclusions are the subspaces through a given point). We want the face relations of our generalized simplices to similarly respect this embedding pattern.

0.1.2. One tempting route of generalization would be to observe that the submanifolds $G_1^{m-1} \subset G_1^m$ may be constructed from geodesics. In projective 3 space G_1^3, if we are given three geodesic segments forming a triangle, either of these procedures to fill it in with geodesics gives the same 2-simplex:

In G_2^2 already this is no longer true:

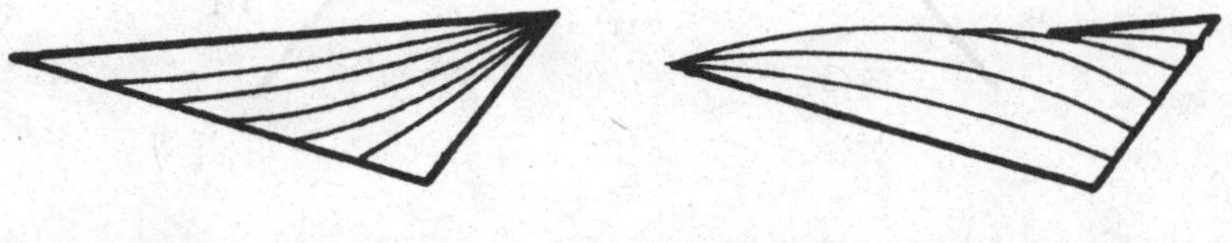

This is a reflection of the fact that G_1^m is a rank one symmetric space (i.e., there is one real invariant of the relative position of two points—the distance), whereas G_2^2 is a rank two symmetric space (there is a two dimensional space of invariants of the relative position of two points).[1]

[1] From the present point of view, this is the reason why finding a combinatorial formula for a Pontyagin class of a polyhedron is so much more difficult a problem than it is for the Euler class.

0.2. *The first Grassmannian simplices*

0.2.1. **Definition.** Let $e_1, e_2, \ldots, e_{n+m}$ be points in real projective $n+m-1$ space $\mathbb{R}P^{n+m-1}$ not all contained in a $n+m-2$ dimensional subspace. Let H^0 be the connected component of the identity of the group of projective automorphisms of $\mathbb{R}P^{n+m-1}$ fixing all the points e_i. Since H^0 takes $n-1$ dimensional planes to $n-1$ dimensional planes, it acts on G_n^m. A *Grassmannian* simplex Γ_n^m is the closure of a generic orbit of H^0.

Grassmannian simplices are our proposed generalization of ordinary simplices in projective space. We will illustrate the idea by looking at the cases Γ_1^2, Γ_2^1, and Γ_2^2.

0.2.2. The generic orbits of H^0 in $G_1^2(=\mathbb{R}P^2)$ are the open regions labelled I, II, III, and IV below. This is because the group of projective automorphisms of $\mathbb{R}P^2$ is transitive on 4-triples of points in general position.

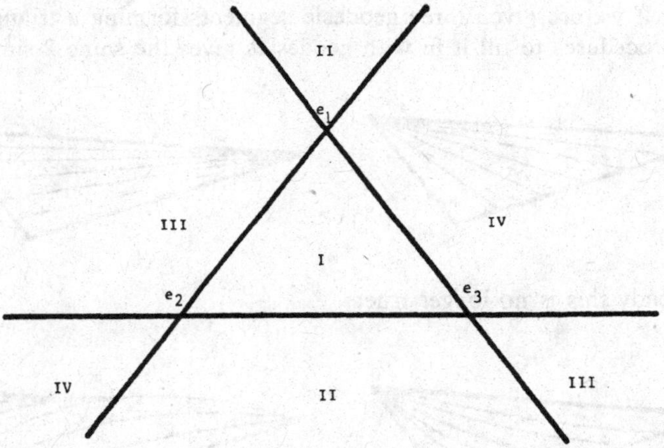

The closure of region I, a Grassmannian simplex Γ_1^2, is an ordinary 2-simplex. Its three edges are closures of non-generic orbits of H^0; they are Grassmannian simplices Γ_1^1. Each is the closure of a generic H^0 orbit in G_1^1 embedded as one of the three lines $\widehat{e_i e_j}$.

0.2.3. The picture in G_2^1, the space of lines in $\mathbb{R}P^2$, is similar by duality. There are four generic H^0 orbits. One is the set of lines not intersecting the closed shaded region in the figure below.

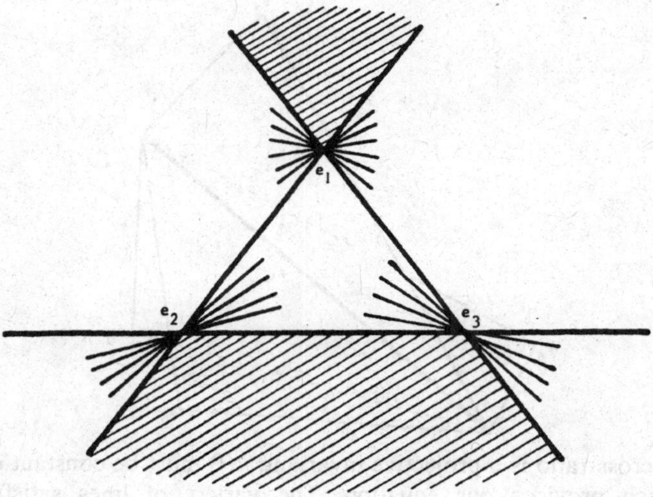

Its closure, a Grassmannian simplex Γ_2^1, has three edges—the three families of lines drawn in the picture. They are Grassmannian simplices Γ_1^1 in G_1^1 embedded as the family of all lines through e_1, e_2, or e_3. The topology of the Γ_2^1 can be viewed this way:

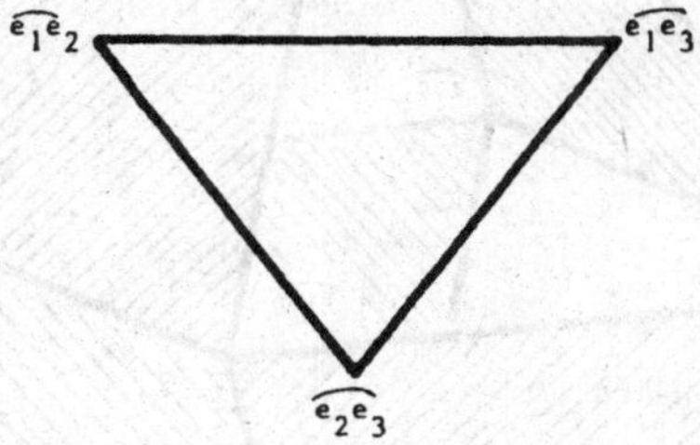

0.2.4. Consider G_2^2, the space of lines l in an $\mathbb{R}P^3$ with four fixed noncoplanar points e_1, e_2, e_3, and e_4. Since H^0 is three dimensional and G_2^2 is four dimensional, we must find an equation satisfied by an H^0 orbit. Let $k(l)$ be the cross-ratio of the four planes $\widehat{e_1 l}, \ldots, \widehat{e_4 l}$ taken in the pencil of planes through l.

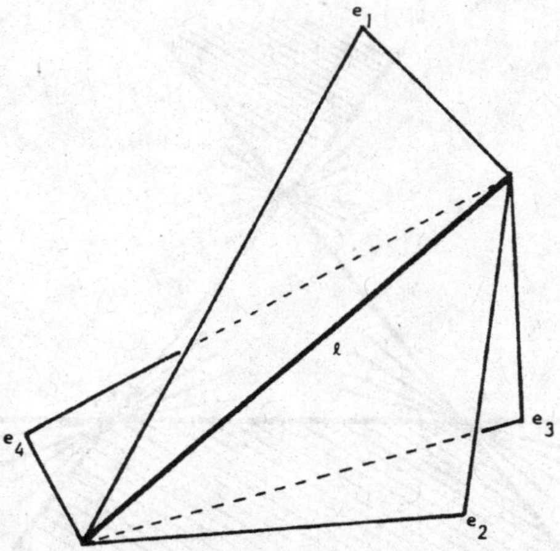

Since the cross-ratio is a projective invariant, $k(l)$ must be constant on an H^0 orbit, which provides our equation. The variety of lines satisfying this equation was classically called the tetrahedral complex (see Subsection 4.2).

We will show (2.3.4) that the topology of a Grassmannian simplex Γ_2^2, the closure of an H^0 orbit, is that of a solid octahedron.

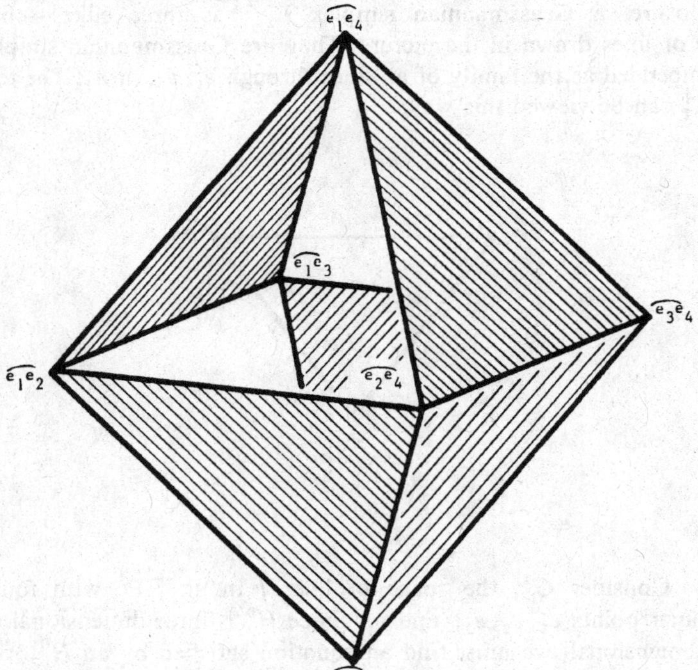

Here the six vertices are the lines $\widehat{e_i e_j}$ as marked.

The four unshaded faces are Grassmannian simplices Γ_2^1. They are closures of generic H^0 orbits in G_2^1 embedded four ways: as the set of lines in the planes $\widehat{e_1 e_2 e_3}$, $\widehat{e_1 e_2 e_4}$, $\widehat{e_1 e_3 e_4}$, and $\widehat{e_2 e_3 e_4}$.

The four shaded faces are Grassmannian simplices Γ_1^2. They are closures of generic H^0 orbits in G_1^2, embedded in four ways: as the set of lines through the point e_1, e_2, e_3, or e_4. The embedding of the Γ_1^2, whose vertices are $\widehat{e_1 e_2}$, $\widehat{e_1 e_3}$ and $\widehat{e_1 e_4}$ in the G_1^2 of lines through e_1, may be pictured as follows:

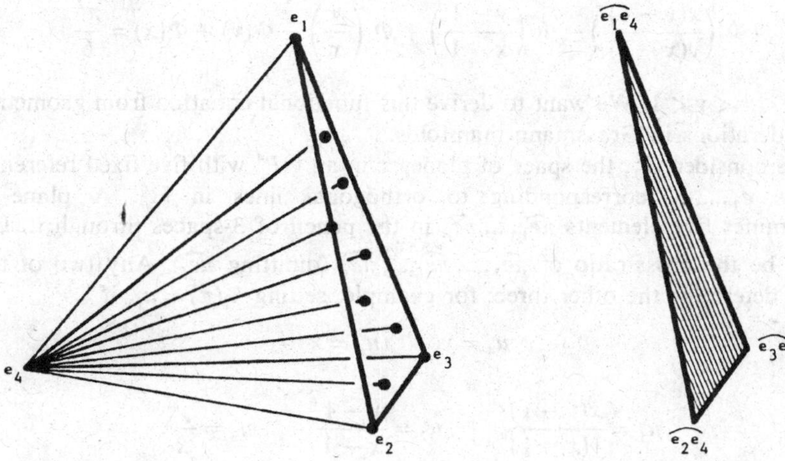

Thus we see that our requirement for the face relations set out in 0.1.1. is satisfied.

0.3. *The Relation to Differential Forms*

0.3.1. Considering G_2^2 as planes through the origin in $\mathbb{R}^4$, we see that the special orthogonal group $SO(4)$ acts on G_2^2. Let P_2^2 be the $SO(4)$ invariant volume form of unit mass on $SO(4)$ and let $e_1,\dots, e_4$ correspond to orthogonal lines in $\mathbb{R}^4$.

Each Grassmannian simplex Γ_2^2 lies in a hypersurface $k(l) = x$. If we let x vary continuously between 0 and 1, we get a continuously varying family of hypersimplices, which we denote by $\Gamma_2^2(x)$. Let $\hat{P}_2^2$ be the integral of P_2^2 over the fibers $\Gamma_2^2(x)$. Since P_2^2 is a 4-form and $\Gamma_2^2(x)$ is three dimensional, $\hat{P}_2^2$ is a 1-form.

Integration.

$$\hat{P}_2^2 = \frac{1}{8\pi^2}\, d\Phi(x),$$

where $\Phi(x) = (Li(x) - Li(1-x))$ and $Li(x) = \sum_{i=1}^{\infty} x^i/i^2$ is the classical dilogarithm function.

From the geometry of 0.2.3, we see that $\Gamma_2^2(x_1)$ and $\Gamma_2^2(x_2)$ have the same boundary, so their union is a clamshell-like closed hypersurface. The above integration shows that the volume it encloses is $\Phi(x_2) - \Phi(x_1)$.

0.3.2. The dilogarithm satisfies a functional equation in two variables. If we extend Φ from $(0, 1)$ to $\mathbb{R} - \{0, 1\}$ by using the symmetry $\Phi(x) = \Phi(1/(1-x))$, then this functional equation may be expressed

$$\Phi\left(\frac{x(y-1)}{y(x-1)}\right) - \Phi\left(\frac{y-1}{x-1}\right) + \Phi\left(\frac{y}{x}\right) - \Phi(y) + \Phi(x) = \frac{\pi^2}{6}$$

for $0 < x < y < 1$. We want to derive this functional equation from geometric considerations in Grassmann manifolds.

We consider G_3^2, the space of planes π in an $\mathbb{R}P^4$ with five fixed reference points $e_1, \ldots, e_5$ corresponding to orthogonal lines in $\mathbb{R}^5$. A plane π determines five elements $\widehat{\pi e}_1, \ldots, \widehat{\pi e}_5$ in the pencil of 3-spaces through it. Let $k_i(\pi)$ be the cross-ratio of $\widehat{\pi e}_1, \ldots, \widehat{\widehat{\pi e}_i}, \ldots, \widehat{\pi e}_5$ (omitting $\widehat{\pi e}_i$). Any two of the $k_i(\pi)$ determine the other three: for example, setting $k_i(\pi) = u_i$, if

$$u_4 = y, \qquad u_5 = x$$

then

$$u_1 = \frac{x(y-1)}{y(x-1)}, \qquad u_2 = \frac{y-1}{x-1}, \qquad u_3 = \frac{y}{x}.$$

Each Grassmannian simplex Γ_3^2 lies in a subvariety $k_5(\pi) = x$ and $k_4(\pi) = y$. If we let x and y vary continuously such that $0 < x < y < 1$, we get a two parameter family of Grassmannian simplices which we denote by $\Gamma_3^2(x, y)$. The Grassmannian simplex $\Gamma_3^2(x, y)$ has five faces Γ_3^1 which do not vary with x and y, and five faces $F_1, \ldots, F_5$ that do. The face F_i is a Grassmannian simplex $\Gamma_2^2(u_i)$ lying in the G_2^2 of planes containing the reference point e_i.

Now we study the 4-form P_3^2 on G_3^2: the harmonic representative of the first Pontrjagin class of the tautological bundle. The form P_3^2 is closed and it restricts to the invariant volume form on the G_2^2 of planes through e_i. We define $\hat{P}_3^2$ to be the integral of P_3^2 over the Grassmannian simplices $\Gamma_3^2(x, y)$. Essentially from Stokes' theorem we find that $d\hat{P}_3^2$ is the sum of the integrals of P_3^2 over the faces F_i. With signs that reflect the induced orientations, we obtain:

Differential Relation (1.2.2).

$$d\hat{P}_3^2 = \sum_{i=1}^{5} (-1)^i \hat{P}_2^2(u_1) = \frac{1}{8\pi^2} \sum_{i=1}^{5} (-1)^i \, d\Phi(u_i).$$

But we also have a rather unexpected result which will be proved in Section 3:

Vanishing Theorem (1.2.4).

P_3^2 is identically zero on $\Gamma_3^2(x, y)$ for any x and y.

Therefore $\hat{P}_3^2$ is identically zero, hence so is $d\hat{P}_3^2$. So we obtain

$$\sum_{i=1}^{5} (-1)^i \, d\Phi(u_i) = 0,$$

so $\sum_{i=1}^{5} (-1)^i(u_i)$ is a constant which we evaluate as $\pi^2/6$ from special values of x and y.

1. The Generalized Dilogarithm Forms

1.1. *Definitions*

1.1.1. Consider the Grassman manifold G_n^m of all n-dimensional (unoriented) planes through the origin in $\mathbb{R}^{n+m}$. By abuse of notation, we will always use the same symbol (usually ξ) for a point $\xi \in G_n^m$ and for the n-plane $\xi \subseteq \mathbb{R}^{n+m}$ corresponding to it.

A coordinate m-plane in $\mathbb{R}^{n+m}$ is one obtained by setting n of the coordinates equal to zero. An n-plane is called *generic* if it intersects each coordinate m-plane only at the origin. The symbol $\tilde{G}_n^m$ denotes the (open) subset of G_n^m consisting of generic n-planes.

The projective linear group $PGL(n+m)$ is the group $GL(n+m)$ of invertible $(n+m) \times (n+m)$ real matrices divided by multiples of the identity. $PGL(n+m)$ acts on G_n^m since multiples of the identity in $GL(n+m)$ act trivially. The Cartan subgroup $H \subset PGL(n+m)$ is the image of the diagonal matrices. It has dimension $n+m-1$.

1.1.2. Proposition. *H preserves $\tilde{G}_n^m$ and acts freely on it.*

This will be proved in 2.2.1.

Definition. *The quotient space $H \backslash \tilde{G}_n^m$ will be denoted by C_n^m and the quotient map by π.*

By the Proposition, C_n^m is a differentiable manifold of dimension $(n-1)(m-1)$. In 2.2.2, it will be shown that it has an interpretation as a space of configurations of $n+m$ points in projective $m-1$ space.

Because of the proposition, the fibres of π identify with the group H so they may be oriented by choosing an ordered basis for the Lie algebra of H.

We choose the usual basis which is the image of $E_{1,1}$, $E_{2,2},\ldots E_{n+m-1,n+m-1}$, where E_{ij} are matrix elements. We denote the induced orientation of π by $[\pi]$.

1.1.3. For each k such that $1 \leqslant k \leqslant n+m$, there are natural maps

$$\begin{array}{ccccc} \tilde{C}^m_{n-1} & \xleftarrow{\tilde{A}_k} & \tilde{C}^m_n & \xrightarrow{\tilde{B}_k} & \tilde{C}^{m-1}_n \\ \downarrow{\scriptstyle \pi} & & \downarrow{\scriptstyle \pi} & & \downarrow{\scriptstyle \pi} \\ C^m_{n-1} & \xleftarrow{\alpha_k} & C^m_n & \xrightarrow{\beta_k} & C^{m-1}_n \end{array}$$

These maps are defined as follows: Let $J_k : \mathbb{R}^{n+m-1} \hookrightarrow \mathbb{R}^{n+m}$ send $(X_1,\ldots, X_{n+m-1})$ to $(X_1,\ldots, X_{k-1}, 0, X_k,\ldots, X_{n+m-1})$. Let $Pr_k : \mathbb{R}^{n+m} \to \mathbb{R}^{n+m-1}$ send $(X_1,\ldots, X_{n+m})$ to $(X_1, X_{k-1}, X_{k+1},\ldots, X_{n+m})$. Then

$$\tilde{A}_k(\xi) = J_k^{-1}\xi,$$
$$\tilde{B}_k(\xi) = \mathrm{Pr}_k \xi$$

(where the expressions on the right define planes in $\mathbb{R}^{n+m-1}$).

The maps α_k and β_k are defined by requiring commutativity of the diagram. This makes sense since π is surjective and constant along the fibers of $\pi \circ \tilde{A}_k$ or $\pi \circ \tilde{B}_k$, as may be easily checked. If the spaces C^m_n are interpreted as projective configurations of points, the corresponding interpretations of α_k and β_k are given in Proposition 2.2.3.

We note that if $n = 1$ then $\tilde{A}_k$ and α_k are not defined; if $m = 1$, $\tilde{B}_k$ and β_k are not defined.

1.2. *The Main Results*

1.2.1. Let us consider a stable characteristic class P on real vector bundles with values in the lth cohomology with real coefficients $H^l(., \mathbb{R})$. (P will be a polynomial in the Pontrjagin classes.) On G^m_n there is a unique $SO(n+m, \mathbb{R})$ invariant closed differential form P^m_n representing the cohomology class $P(\xi)$ where ξ is the tautological bundle over G^m_n. We denote the restriction of P^m_n to $\tilde{G}^m_n$ again by P^m_n.

DEFINITION. The *generalized dilogarithm* form $\hat{P}^m_n$ is given by

$$\hat{P}^m_n = \pi_* P^m_n,$$

where π_* represents integration over the fiber with fiber orientation $[\pi]$. The generalized dilogarithm form $\hat{P}^m_n$ is differential $(l-n-m+1)$-form on C^m_n.

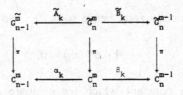

1.2.2. THEOREM (the differential relation). *If $n > 1$ and $m > 1$,*

$$d\hat{P}_n^m = 2 \sum_{k=1}^{n+m} (-1)^k \alpha_k^* \hat{P}_{n-1}^m - 2 \sum_{k=1}^{n+m} (-1)^k \beta_k^* \hat{P}_n^{m-1}.$$

This theorem will be proved in Section 2.

Remark. If $n = 1$ or $m = 1$, then $\hat{P}_n^m$ is identically zero since the tautological vector bundle over G_1^m or G_n^1 has no non-zero stable real characteristic classes.

1.2.3. THEOREM (vanishing). *The generalized dilogarithm form $\hat{P}_n^m$ vanishes identically if it is a 0-form (i.e., if $l = n + m - 1$).*

This follows from the following stronger statement which shows that the integrand is zero in the integration over the fiber $\pi_* P_n^m = \hat{P}_n^m$.

1.2.4. THEOREM. *The restriction of P_n^m to any H orbit vanishes identically.*

This will be proved in 3.3.3.

1.3. *Functional Equations*

1.3.1. DEFINITION. We call a generalized dilogarithm form $\hat{P}_n^m$ *leading* if $\hat{P}_{n+1}^m = 0$ and $\hat{P}_{n+1}^{m-1} = 0$ (for the same characteristic class P) but $\hat{P}_n^m$ is not identically zero.

Dually, we call $\hat{P}_n^m$ *coleading* if $\hat{P}_n^{m+1} = 0$ and $\hat{P}_{n-1}^{m+1} = 0$ but $\hat{P}_n^m$ is not identically zero.

We note that by 1.2.3 a leading generalized dilogarithm form has positive degree.

1.3.2. The following result is an immediate corollary of the differential relation, 1.2.2.

Functional equation. If $\hat{P}_n^m$ is leading, then

$$\sum_{k=1}^{n+m+1} (-1)^k \alpha_k^* \hat{P}_n^m = 0.$$

Dually, if $\hat{P}_n^m$ is coleading then

$$\sum_{k=1}^{n+m+1} (-1)^k \beta_k^* \hat{P}_n^m = 0.$$

1.3.3. Since the leading (or coleading) generalized dilogarithm forms are the ones that satisfy an interesting functional equation, it is important to identify them. This we cannot do in general, but the vanishing theorem 1.2.3 guarantees a rich supply of them.

PROPOSITION. *Consider a stable characteristic class P with values in $H^l(\,;\mathbb{R})$. If for some pair (n, m), $\hat{P}_n^m \neq 0$, then there exists a pair (n, m) so that $\hat{P}_n^m$ is leading and a pair (n, m) so that $\hat{P}_n^m$ is coleading.*

(So far as we know, every nonzero P may satisfy the hypothesis of this proposition. In any case, an open dense set of P satisfies it as may be seen by observing that $\hat{P}_n^m \neq 0$ whenever P_n^m is a non-zero multiple of the invariant volumn form on G_n^m.)

Proof. In the n, m plane, $\hat{P}_n^m$ vanishes if $n+m=l+1$ by the vanishing theorem; $\hat{P}_n^m$ vanishes if $n=1$ or if $m=1$ because then C_n^m is a point. So $\hat{p}_n^m=0$ along the boundary of a triangle. Starting from a point (n, m) where $\hat{P}_n^m \neq 0$, in the interior of the triangle, construct a path through points where $\hat{P}_n^m \neq 0$ such that the step after (n, m) is $(n, m+1)$ or $(n-1, m+1)$. A further step is impossible only when we have reached a leading $\hat{P}_n^m$; but this must happen since we will run into the boundary of the triangle.

1.3.4. Damiano has shown that if P is the ith Pontrjagin class, then $\hat{P}_2^{2i}$ is leading (see [8, 9]).

2. GRASSMANNIAN SIMPLICES

2.1. *Hypersimplices*

2.1.1. DEFINITION. The hypersimplex Δ_n^m is the subset of $\mathbb{R}^{n+m}$ defined by the relations:

$$0 \leqslant t_i \leqslant 1, \qquad \text{for all } i;$$

$$\sum_{i=1}^{n+m} t_i = n,$$

where $t_1, \ldots, t_{n+m}$ are coordinates in $\mathbb{R}^{n+m}$.

The hypersimplex Δ_n^m is a convex $n+m-1$ dimensional polyhedron embedded in the Euclidean space defined by $t_1+t_2+\cdots+t_{n+m}=n$. Its boundary $\partial\Delta_n^m$ is the subset where at least one of the coordinates t_i is either 0 or 1. Its interior denoted by Δ_n^m, is $\Delta_n^m \backslash \partial\Delta_n^m$. See [19] for drawings of the first few hypersimplices.

2.1.2. For all integers k such that $1 \leqslant i \leqslant n+m$, there are natural maps

$$\Delta_{n-1}^{m} \xhookrightarrow{a_k} \Delta_n^m \xhookleftarrow{b_k} \Delta_n^{m-1}$$

defined by

$$a_k(t_1,\dots, t_{n+m-1}) = (t_1,\dots, t_{k-1}, 1, t_k,\dots, t_{n+m-1}),$$
$$b_k(t_1,\dots, t_{n+m-1}) = (t_1,\dots, t_{k-1}, 0, t_k,\dots, t_{n+m-1}).$$

The images of these maps are the codimension one faces of Δ_n^m as a polyhedron; unless $n=m=1$, we have

$$\partial \Delta_n^m = \bigcup_{k=1}^{n+m} \text{image}\,(a_k) \cup \bigcup_{k=1}^{n+m} \text{image}\,(b_k).$$

2.1.3. We choose an orientation $[\Delta_n^m]$ of Δ_n^m given by the ordered coordinates $(t_2-t_1, t_3-t_2,\dots, t_{n+m}-t_{n+m-1})$. $[\Delta_n^m]$ may be interpreted as a generator in $H_{n+m}(\Delta_n^m, \partial\Delta_n^m)$. We have the following homological boundary formula:

PROPOSITION. *Unless $n=m=1$,*

$$\partial[\Delta_n^m] = \sum_{k=1}^{n+m} (-1)^k a_{k*}[\Delta_{n-1}^m] - \sum_{k=1}^{n+m} (-1)^k b_{k*}[\Delta_n^{m-1}]$$

(where if $n=1$ the first summation is regarded as zero, and if $m=1$ the second is regarded as zero).

Proof. See [4, Prop. 5].

This formula may be interpreted as computing the connecting homomorphism for the triple $(\Delta_n^m, \partial\Delta_m^n; c)$ where c is the codimension two-skeleton of Δ_m^n.

2.1.4. Notice that the formula of 2.1.3 for $\partial[\Delta_n^m]$ is similar to the formula of 1.2.2 for $d\hat{p}_n^m$. We say that the Δ_n^m are formally dual to the $\hat{P}_n^m$. In fact Theorem 1.2.2 follows from Proposition 2.1.3 after embedding Δ_n^m in G_n^m as a Grassmannian simplex.

2.2. *Projective Configurations*

2.2.1. Recall that by definition $C_n^m = H\backslash\tilde{G}_n^m$, where $\tilde{G}_n^m$ is the generic part of the Grassmannian of n planes in $\mathbb{R}^{n+m}$ and H is the Cartan subgroup of $PGL(n+m)$, i.e., the diagonal invertible matrices modulo multiples of the identity matrix. The space C_n^m is a differentiable manifold since H acts freely on $\tilde{G}_n^m$ by Proposition 1.1.2.

Proof of Proposition 1.1.2. First, we observe that we may think of G_n^m as the configurations of $n+m$ vectors $V_1,\dots,V_{n+m}$ in $\mathbb{R}^m$ which span $\mathbb{R}^m$ as a vector space, modulo equivalence by $GL(m)$. For given a point in G_n^m, i.e., an n dimensional subspace ζ of $\mathbb{R}^{n+m}$, first choose a vector space epimorphism $\varphi\colon \mathbb{R}^{n+m} \twoheadrightarrow \mathbb{R}^m$ with kernel ζ. Then we obtain a configuration with $V_i=\varphi(e_i)$, where $e_1,\dots,e_{n+m}$ is the standard basis of $\mathbb{R}^{n+m}$. Choosing another φ changes the configuration by a $GL(m)$ element.

From this point of view, $\tilde{G}_n^m$ is the subset of configurations defined by the condition that for all m element subsets S of $\{1,\dots,n+m\}$, the set $\{V_i \mid i\in S\}$ forms a basis of $\mathbb{R}^m$. The H action on G_n^m is expressed this way: the diagonal $GL(n+m)$ element

$$h=\begin{bmatrix} h_1 & & 0 \\ & \ddots & \\ 0 & & h_{n+m}\end{bmatrix}$$

sends $V_1,\dots,V_{n+m}$ to $h_1V_1,\dots,h_{n+m}V_{n+m}$. So if $V_i\neq 0$ the action of H sends V_i into the line l_i through 0 and V_i.

We must prove that any diagonal $GL(n+m)$ element h whose action on a configuration c in G_n^m matches that of a $GL(m)$ element ψ must be a multiple of the identity. Since $n\geqslant 1$, we know that c determines at least $m+1$ lines l_i as above any m of which span $\mathbb{R}^m$. Since ψ fixes these lines, its action on the projectivization $\mathbb{R}P^{m-1}$ of $\mathbb{R}P^m$ must fix the corresponding $m+1$ points, which are in general position. Therefore ψ acts trivially on $\mathbb{R}P^{m-1}$ so ψ must be a multiple k of the identity. Hence $h_1=h_2=\cdots=h_{n+m}=k$ so h is a multiple of the identity.

2.2.2. Let $C_1,\dots,C_{n+m}$ be a configuration of $n+m$ points in real projective $m-1$ space; i.e., $C_i\in\mathbb{R}P^{m-1}$ for each i. The configuration is called *generic* if for no m element subset S of $\{1,2,\dots,n+m\}$ does $\{C_i \mid i\in S\}$ lie in an $m-2$ dimensional projective subspace. The set of generic projective configurations forms an open submanifold of $(\mathbb{R}P^{m-1})^{n+m}=\mathbb{R}P^{m-1}\times\cdots\times\mathbb{R}P^{m-1}$($n+m$ factors). Two such projective configurations are equivalent if there is an element of $PGL(m)$ taking one to the other. The group $PGL(m)$ acts freely on the generic configurations of $n+m$ points since a $PGL(m)$ element is determined by its effect on any $m+1$ independent points. Therefore the space of equivalence classes of generic configurations of $n+m$ points in $\mathbb{R}P^{m-1}$ is a differentiable manifold.

PROPOSITION. *The space C_n^m is naturally diffeomorphic to the space of equivalence classes of generic configurations of $n+m$ points in $\mathbb{R}P^{m-1}$.*

[This proposition says that the generic part of $H\backslash G_n^m$ is naturally diffeomorphic to the generic part of $PGL(n+m)\backslash(\mathbb{R}P^{m-1})^{n+m}$.]

Proof. We us the model for G_n^m presented in the proof in 2.2.1. Consider the space S of $n+m$-tuples of lines $l_1,..., l_{n+m}$ through the origin in $\mathbb{R}^m$ such that any m of them span $\mathbb{R}^m$. We claim that $GL(m)\backslash S = C_n^m$. For, given an element of S, we may choose V_i to be any nonzero point of l_i and any two such choices are related by a diagonal $GL(n+m)$ element. On the other hand, considering each line l_i as a point in $\mathbb{RP}^{m-1}$, we see that $GL(m)$ equivalence classes in S are $PGL(m)$ equivalence classes of generic configurations of $n+m$ points in $\mathbb{RP}^{m-1}$.

2.2.3. Recall that in 1.1.3 the projections $\alpha_k: C_n^m \to C_{n-1}^m$ and $\beta_k: C_n^m \to C_n^{m-1}$ were defined by passing to the quotient space from the projections $\tilde{A}_k: \tilde{G}_n^m \to \tilde{G}_{n-1}^m$ and $\tilde{B}_k: \tilde{G}_n^m \to G_n^{m-1}$, respectively.

PROPOSITION. *Let us interpret C_n^m as projective equivalence classes of generic configurations of $n+m$ points $C_1,..., C_{n+m}$ in $\mathbb{RP}^{m-1}$ as in 2.2.2. Then*

$$\alpha_i(C_1,..., C_{n+m}) = (C_1,..., C_{k-1}, C_{k+1},..., C_{n+m})$$

and

$$\beta_i(C_1,..., C_{n+m}) = (\varphi_k C_1,..., \varphi_k C_{k-1}, \varphi_k C_{k+1},..., \varphi_k C_{n+m})$$

where $\varphi_k: \mathbb{RP}^{m-1} \setminus C_k \to \mathbb{RP}^{m-2}$ is any projective identification of the lines through C_k with $\mathbb{RP}^{m-2}$.

Proof. This proposition results directly from the interpretation of $\tilde{A}_k$ and $\tilde{B}_k$ in the language of the proof in 2.2.1.

$$\tilde{A}_k(V_1,..., V_{n+m}) = (V_1,..., V_{k-1}, V_{k+1},..., V_{1+m})$$

and

$$\tilde{B}_k(V_1,..., V_{n+m}) = (\varphi_k V_1,..., \varphi_k V_{k-1}, \varphi_k V_{k+1},..., \varphi_k V_{n+m}),$$

where $\varphi_k: \mathbb{R}^{n+m} \twoheadrightarrow \mathbb{R}^{n+m-1}$ is any linear epimorphism with V_k in its kernel.

2.2.4. DEFINITION. The space of *enhanced projective configurations*, EC_n^m, is the quotient space $H^0\backslash\tilde{G}_n^m$, where H^0 denotes the connected component of the identity of H.

We have the diagram,

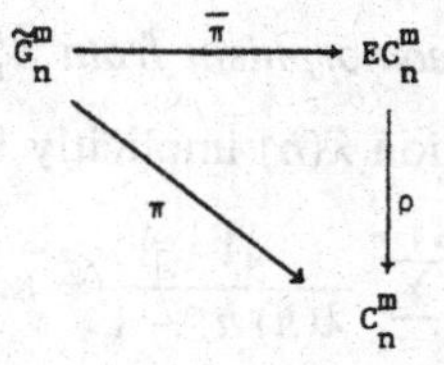

where ρ is the further quotientization. The map ρ is a projection of a 2^{n+m-1} sheeted covering space since $H = H^0 \oplus (\mathbb{Z}/2\mathbb{Z})^{n+m-1}$.

We define the maps $\bar{\alpha}_k$ and $\bar{\beta}_k$ by commutativity of the diagram

$$\begin{array}{ccccc}
\tilde{G}^m_{n-1} & \xleftarrow{\tilde{A}_k} & \tilde{G}^m_n & \xrightarrow{\tilde{B}_k} & \tilde{G}^{m-1}_n \\
\downarrow \bar{\pi} & & \downarrow \bar{\pi} & & \downarrow \bar{\pi} \\
EC^m_{n-1} & \xleftarrow{\bar{\alpha}_k} & EC^m_n & \xrightarrow{\bar{\beta}_k} & EC^{m-1}_n
\end{array}$$

Remark. We may interpret EC^m_n as projective equivalence classes of generic configurations of $n+m$ points in $\mathbb{R}P^{m-1}$ enhanced by a lift of each point to the universal covering space S^{m-1} of $\mathbb{R}P^{m-1}$. Two such lifts that differ by a covering transformation of $S^{m-1} \to \mathbb{R}P^{m-1}$ are deemed to be the same. The maps $\bar{\alpha}_k$ and $\bar{\beta}_k$ have interpretations analogous to those of α_k and β_k in 2.2.3.

2.3. *The Closures of H^0 Orbits*

2.3.1. Recall that $\tilde{\Delta}^m_n$ is the interior of the hypersimplex $\tilde{\Delta}^m_n$; in coordinates $t_1, \ldots, t_{n+m}$ it is the set

$$0 < t_i < 1, \qquad \text{for all } i,$$

$$\sum_{i=1}^{n+m} t_i = n.$$

The group H^0 is the connected component of the identity of the Cartan subgroup H of $PGL(n+m)$; it has coordinates $h_1, \ldots, h_{n+m}$, the entries of the diagonal matrix

$$h = \begin{bmatrix} h_1 & & 0 \\ & \ddots & \\ 0 & & h_{n+m} \end{bmatrix},$$

where $h_i > 0$ for all i and $h_1, \ldots, h_{n+m}$ is equivalent to $\lambda h_1, \ldots, \lambda h_{n+m}$ for all $\lambda > 0$.

We define a map $\eta\colon \tilde{\Delta}^m_n \to H^0$ by the equations

$$h_i = \frac{t_i}{1 - t_i}.$$

PROPOSITION. *η is a diffeomorphism from $\tilde{\Delta}^m_n$ onto H^0.*

Proof. We define a function $\lambda(h)$ implicitly by the equation

$$\sum_{i=1}^{n+m} \frac{1}{\lambda(h)\, h_i + 1} = n.$$

This is everywhere defined, single valued, and always positive because for each h the function of λ,

$$\sum_{i=1}^{n+m} \frac{1}{\lambda \cdot h_i + 1}$$

decreases strictly monotonically from $n+m$ to 0 as λ goes from 0 to ∞.

Now define $\eta^{-1}: H^0 \to \tilde{\Delta}_n^m$ by

$$t_i = \frac{1}{\lambda(h)\, h_i + 1}.$$

The composition $\eta^{-1} \circ \eta$ is the identity by direct calculation, and $\eta \circ \eta^{-1}$ is the identity since it multiplies all the h_i by $\lambda(h)$.

2.3.2. PROPOSITION. *The map*

$$\tilde{\Delta}_n^m \times \tilde{G}_n^m \to G_n^m,$$

which takes (t, ξ) *to* $\eta(t) \cdot \xi$, *has a unique continuous extension to a map*

$$\Delta_n^m \times \tilde{G}_n^m \to G_n^m.$$

(*We denote the image of* (t, ξ) *under this extension by* $t \cdot \xi$. *We think of* $\tilde{\Delta}_n^m$ "*acting*" *on* $\tilde{G}_n^m$ *by a completion of the action of* H^0.)

Proof. Uniqueness is obvious because $\tilde{\Delta}_n^m$ is dense in Δ_n^m. To prove existence, we write a formula for $t \cdot \xi$ in terms of Plücker coordinates.

Let $S \subset \{1,..., n+m\}$ be a subset with n elements and let $\xi \in G_n^m$. Then $P(S, \xi)$ will denote the Plücker coordinate of the point ξ corresponding to the subset S of basis vectors $e_1,..., e_{n+m}$ in $\mathbb{R}^{n+m}$. The action of H^0 on G_n^m may be written as follows

$$P(S, h(\xi)) = \left(\prod_{i \in S} h_i\right) P(S, \xi).$$

Now if $t = (t_1,..., t_{n+m}) \in \tilde{\Delta}_n^m$, we have

$$P(S, \eta(t)\, \xi) = \left(\prod_{i \in S} \frac{t_i}{1 - t_i}\right) P(S, \xi).$$

Since the $P(S, \eta)$ are homogeneous coordinates, we may multiply the expression on th. right by the number

$$\prod_{i=1}^{n+m} 1 - t_i$$

and we obtain

$$P(S, \eta(t)\,\xi) = \Big(\prod_{i \in S} t_i\Big)\Big(\prod_{i \notin S} 1 - t_i\Big)\, P(S, \xi).$$

This last expression clearly extends continuously to all of Δ_n^m (so it gives a formula of $t \cdot \xi$).

2.3.3. We want to express the compatibility of the actions $t \cdot \xi$ for the various G_n^m.

Definition. *For all $k \in \{1,..., n+m\}$, the maps A_k and B_k*

$$G_{n-1}^m \xhookrightarrow{A_k} G_n^m \xhookleftarrow{B_k} G_n^{m-1}$$

are defined as follows: If ξ is a plane of G_{n-1}^m, then $A_k(\xi)$ is the inverse image of ξ by the map $Pr_k\colon \mathbb{R}^{n+m} \to \mathbb{R}^{n+m-1}$ sending *$(X_1,..., X_{n+m})$ to $(X_1,..., X_{k-1}, X_{k+1},..., X_{n+m})$. If ξ is a plane of G_n^{m-1}, then $B_k(\xi)$ is the image of ξ by the map $J_k\colon \mathbb{R}^{n+m-1} \to \mathbb{R}^{n+m}$ sending $(X_1,..., X_{n+m-1})$ to $(X_1,..., X_{k-1}, 0, X_{k+1},..., X_{n+m-1})$.*

Proposition. *If $t \in \Delta_{n-1}^m$ and $\xi \in \tilde{G}_n^m$ then $(a_k t) \cdot \xi = A_k(t \cdot \tilde{A}_k \xi)$. If $t \in \Delta_n^{m-1}$ and $\xi \in G_n^m$ then $(b_k \cdot t) \cdot \xi = B_k(t \cdot \tilde{B}_k \xi)$.*

Proof. The equality follows from calculating the Plücker coordinates of both sides using the formula for $t \cdot \xi$ from the proof in 2.3.2.

2.3.4. Theorem. *Suppose $\xi \in \tilde{G}_n^m$. Then the inclusion*

$$j(\xi)\colon \Delta_n^m \subset G_n^m,$$

$$t \longmapsto t \cdot \xi,$$

is a homeomorphism of the hypersimplex Δ_n^m with the closure of the H^0 orbit of ξ. The map $j(\xi)$ takes the interior Δ_n^m of Δ_n^m diffeomorphically onto the orbit of ξ and it takes interiors of faces (of various dimensions) of Δ_n^m diffeomorphically onto other orbits. The faces of Δ_n^m are mapped into sub-Grassmann manifolds: for example, there is a commutative diagram

$$\begin{array}{ccccc}
\Delta_{n-1}^m & \xrightarrow{a_i} & \Delta_n^m & \xleftarrow{b_i} & \Delta_n^{m-1} \\
\downarrow{\scriptstyle j(A_i\xi)} & & \downarrow{\scriptstyle j(\xi)} & & \downarrow{\scriptstyle j(B_i\xi)} \\
G_{n-1}^m & \xrightarrow{A_i} & G_n^m & \xleftarrow{B_i} & G_n^{m-1}
\end{array}$$

Definition. The image of $j(\xi)$ is called a Grassmannian simplex.

Proof of the theorem. (1) The map $j(\xi)$ gives a diffeomorphism from $\tilde{\Delta}_n^m$ to the H^0 orbit through ξ by Propositions 1.1.2 and 2.3.1.

(2) The commutative diagram follows from Proposition 2.3.3.

(3) This allows us to prove that $j(\xi)$ takes interiors of faces of codimension one diffeomorphically onto orbits by applying (1) above to lower dimensional Grassmann manifolds.

(4) Iterating steps (2) (for similar commutative diagrams in lower dimension) and (3) we prove the interiors of all faces but vertices are mapped diffeomorphically onto orbits.

(5) We may verify directly that the orbits mapped to be different interiors of faces are different and that the vertices are mapped to different coordinate hyperplanes (which are fixed points of the H^0 actions). So $j(\xi)$ is an injection.

(6) The map is continuous by Proposition 2.3.2. Since Δ_n^m is compact, it is a homeomorphism onto the closure of the orbit through ξ.

2.3.5. *Remark.* It follows from Theorem 2.3.4 that there is a natural H^0 action on Δ_n^m. It is determined by continuity by the H^0 action on Δ_n^m given by $h \cdot t = \eta^{-1}(h \cdot \eta(t))$. The closed invariant sets are the unions of faces of Δ_n^m.

2.4. *Proof of the Differential Relation*

We place ourselves in the situation of Subsection 1.2: We have fixed a characteristic class P and therefore differential forms P_n^m on each G_n^m and differential forms $\hat{P}_n^m$ on each C_n^m. We will prove Theorem 1.2.2 relating the forms $\hat{P}_n^m$.

2.4.1. It is more convenient to work on the space of enhanced projective configurations EC_n^m than on C_n^m. Recall from 2.2.4 the diagram

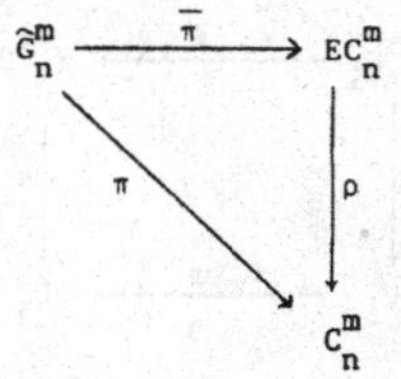

We define $\bar{P}_n^m$ to be $\bar{\pi}_* P_n^m$, where $\bar{\pi}_*$ is computed with respect to a fibre orientation $[\bar{\pi}]$, which is the restriction of $[\pi]$ of 1.1.2.

LEMMA. *Theorem* 1.2.2 *is equivalent to the formula*

$$d\bar{P}^m_n = \sum_{k=1}^{n+m} (-1)^k \bar{\alpha}_k^* \bar{P}^m_{n-1} - \sum_{k-1}^{n+m} (-1)^k \bar{\beta}_k^* \bar{P}^{m-1}_n$$

for $n > 1$ *and* $m > 1$.

Proof. We claim that

$$\bar{P}^m_n = \tfrac{1}{2}^{n+m-1} \rho^* \hat{P}^m_n.$$

This claim and the commutativity of the following diagram imply that the formula of the lemma is ρ^* of the formula of Theorem 1.2.2.

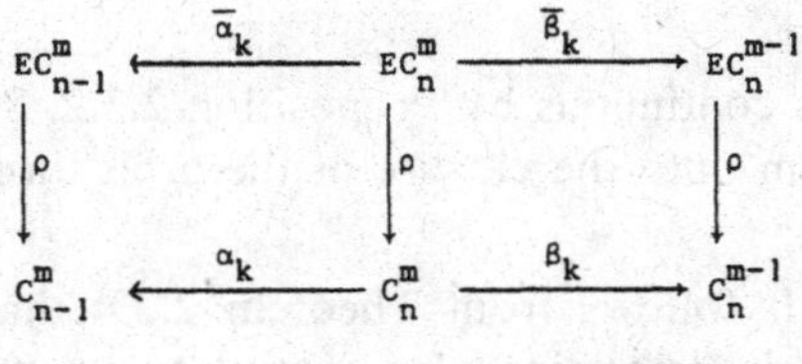

But ρ^* is an injection since ρ is a covering map. Hence the lemma follows.

Now we establish the claim. The form $\omega = \rho^* \tilde{P}^m_n$ is characterized by the following two properties:

(1) $\rho_* \omega = 2^{n+m-1} \tilde{P}^m_n$

(2) Let U be any connected open set in C^m_n such that $\pi^{-1} U$ splits into a disjoint union $U_1 \cup U_2 \cup \cdots$ of copies of U. For each i and j, if $\varphi: U_i \to U_j$ is the diffeomorphism that satisfies $\rho \circ \varphi = \rho$ on U_i, then φ^* takes $\omega | U_j$ to $\omega | U_i$;

The form $2^{n+m-1} \bar{P}^m_n$ satisfies (1) since $\rho_* \bar{P}^m_n = \rho_* \bar{\pi}_* P^m_n = (\rho \circ \bar{\pi})_* P^m_n = \pi_* P^m_n = \tilde{P}^m_n$. To show that it satisfies (2), we write $H = H^0 \oplus F$, where F is the image in $PGL(n+m)$ of the diagonal matrices whose entries are all ± 1. Given i and j there is an element $f \in F$ that takes $\bar{\pi}^{-1} U_i$ diffeomorphically to $\bar{\pi}^{-1} U_j$.

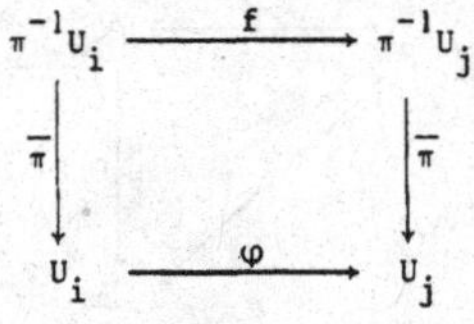

The map f preserves the fiber orientation $[\bar{\pi}]$ since H is Abelian, so $\bar{\pi}_* \circ f_* = \varphi_* \circ \bar{\pi}_*$. And f^* takes $P^m_n | \pi^{-1} U_j$ to $P^m_n | \pi^{-1} U_i$ since f is an orthogonal matrix and P^m_n is invariant under $O(n+m)$. (The $O(n+m)$

invariance of P_n^m, which by its construction is only required to be $SO(n+m)$ invariant, is proved in Section 3—see Corollary 3.2.5). Then

$$\bar{P}_n^m | U_i = \bar{\pi}_* P_n^m | \bar{\pi}^{-1} U_i = \bar{\pi}_* f^* P_n^m | \bar{\pi} U_j$$
$$= \varphi^* \bar{\pi}_* P_n^m | \bar{\pi}^{-1} U_j = \varphi^* \bar{P}_n^m | U_j.$$

2.4.2. We choose a smooth section s of the fibre bundle G_n^m over EC_n^m.

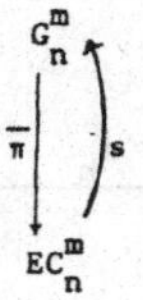

This can be done since the fibers of $\bar{\pi}$ are contractible. (Indeed the fibers are H^0 orbits, H^0 acts freely by Proposition 1.1.2, and H^0 is contractible.)

Using the section s, we define a map

$$r: \Delta_n^m \times EC_n^m \to G_n^m$$

by

$$r(t, c) = t \cdot s(c),$$

where $t \cdot s(c)$ is explained in 2.3.2.

LEMMA. *The map r takes $\tilde{\Delta}_n^m \times EC_n^m \subset \Delta \times EC_n^m$ diffeomorphically to $\tilde{G}_n^m$.*

This follows easily from Proposition 2.3.1.

2.4.3. Let $q: \Delta_n^m \times EC_n^m \to EC_n^m$ be the projection on the second factor.

LEMMA. $\bar{P}_n^m = q_* r^* P_n^m$.

Proof. The following diagram commutes

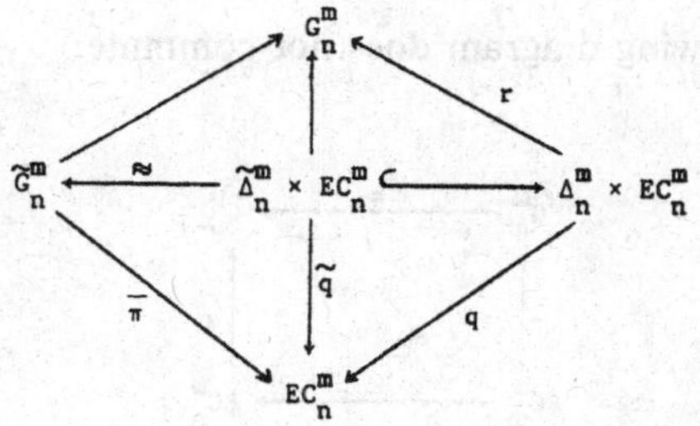

Therefore $\bar{P}_n^m$ is $\tilde{q}_* r^* P_n^m$. But $\Delta_n^m \backslash \tilde{\Delta}_n^m$ is measure zero for the integration over the fibre q_* so $\tilde{q}_*$ and q_* are the same.

2.4.4. Let $Q: \Delta_{n-1}^m \times EC_n^m \to EC_n^m$ be projection onto the second factor and let $Q': \Delta_n^{m-1} \times EC_n^m \to EC_n^m$ also be projection to the second factor.

LEMMA. *For each k such that* $1 \leqslant k \leqslant n+m$, *there is an automorphism* $F_k: \Delta_{n-1}^m \times EC_n^m \to \Delta_{n-1}^m \times EC_n^m$ *such that the following diagram is commutative:*

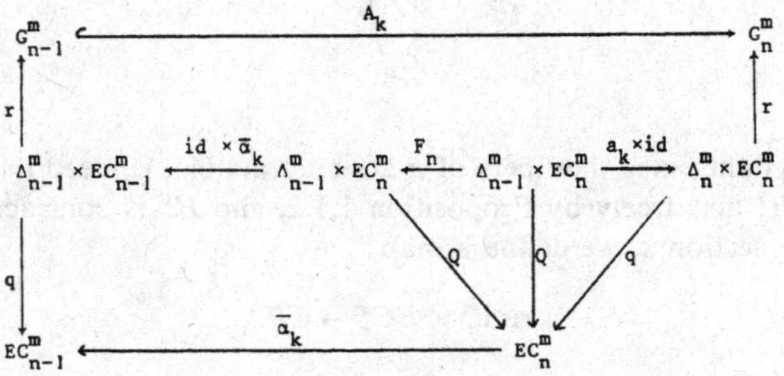

Similarly, for each such k there is an F_k' *such that the following diagram commutes:*

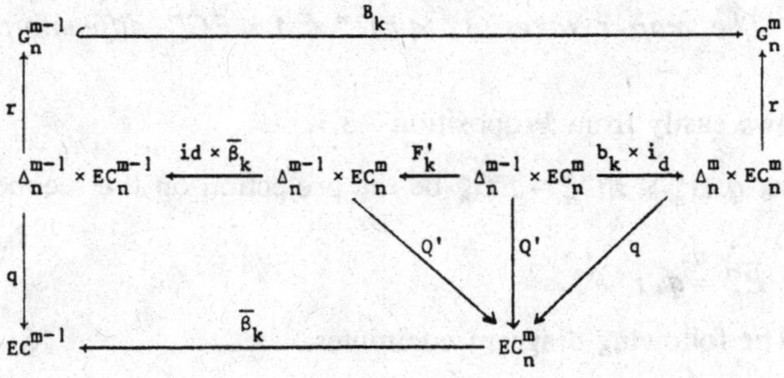

Proof. The following diagram does not commute:

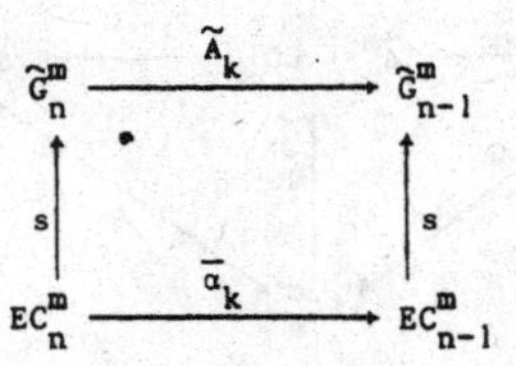

(In fact it is impossible to choose the sections s so that this commutes for all k—a fact of some theoretical significance for combinatorial formulas for Pontrjagin classes, see [19, last sentence of p. 16].) However, for some map $\tilde{s}_k : EC^m_n \to \tilde{\Delta}^m_{n-1}$ we have $s\bar{a}_k(C) = \tilde{s}_k(c) \cdot \tilde{A}_k\, s(c)$. Let $F_k(t, c)$ be $(\tilde{s}_k(c) \cdot t, c)$, where the action of $\tilde{\Delta}^m_{n-1}$ on $\tilde{\Delta}^m_{n-1}$ is obtained from the action of Δ^m_{n-1} on G^m_{n-1} via the embedding of Theorem 2.3.4. Then the commutativity of the first diagram follows from Proposition 2.3.3. Similar remarks apply to the second diagram.

2.4.5. Proof of the differential relation, Theorem 1.2.2: We establish the formula of Lemma 2.4.1, from the differential relation follows.

$$\begin{aligned}
dP^m_n &= d(q_* r^* P^m_n) && \text{by Lemma 2.4.3,}\\
&= q_*\, dr^* P^m_n + (q \,|\, \partial\Delta^m_n \times EC^m_n)_* r^* P^m_n && \text{by Stokes' Theorem,}\\
&= (q \,|\, \partial\Delta^m_n \times EC^m_n)_* r^* P^m_n && \text{since } P^m_n \text{ is closed,}\\
&= \sum_{k=1}^{n+m} (-1)^k Q_*(a_k \times \mathrm{id} \circ r)^* P^m_n - \sum_{k=1}^{n+m} (-1)^k Q'_*(b_k \times \mathrm{id} \circ r)^* P^m_n \\
& && \text{by Proposition 2.1.3,}\\
&= \sum_{k=1}^{n+m} (-1)^k Q_*(\mathrm{id} \times \bar{a}_k \circ r \circ A_k)^* P^m_n \\
&\quad - \sum_{k-1}^{n+m} (-1)^k Q'_*(\mathrm{id} \times \bar{\beta}_k \circ r \circ \beta_k)^* P^m_n \\
& && \text{by Lemma 2.4.4,}\\
&= \sum_{k=1}^{n+m} (-1)^k Q_*(\mathrm{id} \times \bar{a}_k \circ r)^* P^m_{n-1} - \sum_{k=1}^{n+m} (-1)^k Q'_*(\mathrm{id} \times \bar{\beta}_k \circ r)^* P^{m-1}_n \\
& && \text{by stability and naturality of } P,\\
&= \sum_{k=1}^{n+m} (-1)^k \bar{a}^*_k q_* r^* P^m_{n-1} - \sum_{k=1}^{n+m} (-1)^k \bar{\beta}^*_k q_* r^* P^{m-1}_n,\\
&= \sum_{k=1}^{n+m} (-1)^k \bar{a}^*_k \bar{P}^m_{n-1} - \sum_{k=1}^{n+m} (-1)^k \bar{\beta}^*_k P^{m-1}_n. && \text{Q.E.D.}
\end{aligned}$$

3. Invariant Differential Forms on Grassman Manifolds

3.1. *Homology of the Grassman Manifolds*

In this section we recall some well-known facts about the Grassman manifolds G^m_n.

3.1.1. For any $\xi \in G_n^m$, we denote by $SO(n+m)_\xi$ the isotropy subgroup of ξ, i.e., the subgroup of $SO(n+m)$ consisting of elements that stabilize ξ. If $\theta \in SO(n+m)_\xi$, then $d\theta$ maps $T_\xi G_n^m$ (the tangent space to G_n^m at ξ) to itself; this provides a natural action of $SO(n+m)_\xi$ on $T_\xi G_n^m$.

PROPOSITION. *For any n, m, and l the following three real vector spaces are canonically isomorphic:*

(1) *The lth cohomology group of G_n^m with real coefficients, $H^l(G_n^m, \mathbb{R})$.*

(2) *The space of differential l-forms on G_n^m which are invariant under the action of $SO(n+m)$*

(3) *For any $\xi \in G_n^m$, the alternating l-linear functions on $T_\xi G_n^m$ which are invariant under $SO(n+m)_\xi$.*

The maps inducing these isomorphisms are as follows: The map from (2) to (1) takes an i-form to its de Rham cohomology class. The map from (2) to (3) is given by restriction of an i-form to $T_\xi G_n^m$. It inverse is given by this procedure: given an alternating l-linear function φ on $T_\xi(G_n^m)$, to evaluate the corresponding form in $T_{\xi'}(G_n^m)$ apply $d\theta$ where $\theta \in SO(n+m)$ is any element mapping ξ' to ξ. The result is independent of the choice of θ by the $SO(n+m)_\xi$ invariance of φ.

To see the isomorphism of (1) and (2), put an $SO(n+m)$ invariant Riemannian metric on G_n^m. This can be done by averaging since $SO(n+m)$ is compact. The $SO(n+m)$ invariance of an i-form will then be equivalent to the invariance under the connected component of the identity of group of Riemannian isometries of G_n^m. But $H^l(M, \mathbb{R})$ is the space of harmonic forms since G_n^m is compact which is the space of invariant forms since G_n^m is a symmetric space.

3.1.2. Similar constructions to those of the last section give canonical isomorphisms between

(1) a certain subgroup of $H^l(G_n^m; \mathbb{R})$;

(2) the space of differential l-forms on G_n^m which are invariant under the action of $O(n+m)$;

(3) for any $\xi \in G_n^m$, the alternating l-linear functions on $T_\xi G_n^m$ which are invariant under $O(n+m)_\xi$.

Remark. Not all $SO(n+m)$ invariant l-forms are $O(n+m)$ invariant—for example, the $SO(2)$ invariant 1-form on G_1^1.

3.1.3. For any $\xi \in G_n^m$, let $\xi^\perp$ denote the orthogonal complement of ξ (considered as a subspace of $\mathbb{R}^{n+m}$). Then $\mathbb{R}^{n+m}$ identifies with $\xi \oplus \xi^\perp$. We define "graph coordinates" centered at ξ.

$$\mathrm{graph}_\xi : \mathrm{Hom}(\xi, \xi^\perp) \to G_n^m$$

to be the map that takes a linear homomorphism $L: \xi \to \xi^{\perp}$ to its graph $\{(\varepsilon, L(\varepsilon))\}$ in $\xi \oplus \xi^{\perp} = \mathbb{R}^{n+m}$. This is a diffeomorphism onto a neighborhood of ξ.

The differential $d(\text{graph}_{\xi})$ maps $T_0 \operatorname{Hom}(\xi, \xi^{\perp})$ to $T_{\xi} G_n^m$. Since $\operatorname{Hom}(\xi, \xi^{\perp})$ is a vector space, we may identify $T_0 \operatorname{Hom}(\xi, \xi^{\perp})$ with $\operatorname{Hom}(\xi, \xi^{\perp})$ to obtain a cononical isomorphism

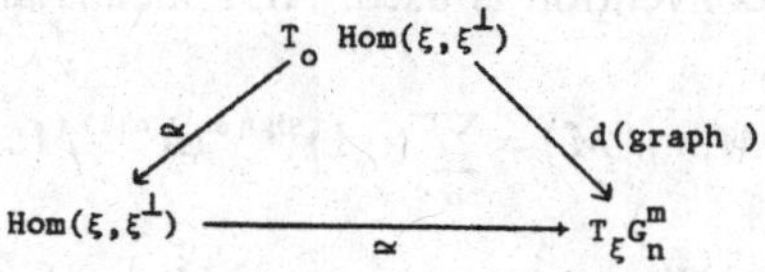

The isotropy subgroup $O(n+m)_{\xi}$ is just the group of orthogonal transformations of $\mathbb{R}^{n+m}$ fixing $\xi \subset \mathbb{R}^{n+m}$ as a set and therefore fixing $\xi^{\perp}$. Any $\bar{\theta} \in O(n+m)_{\xi}$ must restrict to linear maps $\theta: \xi \to \xi$ and $\bar{\theta}: \xi^{\perp} \to \xi^{\perp}$, θ and $\bar{\theta}$ must preserve the restrictions of the inner product on $\mathbb{R}^{n+m}$ to ξ and $\xi^{\perp}$. Any such pair $(\theta, \bar{\theta})$ gives an element of $O(n+m)_{\xi}$.

Proposition. *Choosing orthonormal bases in ξ and in $\xi^{\perp}$, the above construction gives an isomorphism between $T_{\xi} G_n^m$ and $n \times m$ matrices M, between $O(n+m)_{\xi}$ and $O(n) \times O(m)$, and between the action of $O(n+m)_{\xi}$ on $T_{\xi} G_n^m$ and the action*

$$(\theta, \bar{\theta}): M \longmapsto \bar{\theta} \circ M \circ (\theta^{-1}).$$

3.1.4. Summarizing, we have the following result:

To give an $O(n+m)$ invariant differential l-form on G_n^m it suffices to give an alternating i linear function

$$\varphi({}^1M, {}^2M, \ldots, {}^lM),$$

where the ${}^kM$ are matrices ${}^kM_i^j$, $i = 1, \ldots, n$; $j = 1, \ldots, m$, such that

$$\varphi({}^1\bar{M}, {}^2\bar{M}, \ldots, {}^l\bar{M}) = \varphi({}^1M, {}^2M, \ldots, {}^lM) \tag{*}$$

whenever ${}^k\bar{M}_i^j = {}^kM_p^j\,\theta_{p,i}$ for an $n \times n$ orthogonal matrix $\bar{\theta}$ or whenever ${}^k\bar{M}_i^j = {}^kM_i^p\,\bar{\theta}_{pj}$ for an $m \times m$ orthogonal matrix $\bar{\theta}$.

Indeed, by (*) the result will be independent of the choice of orthogonal coordinates in ξ and in $\xi^{\perp}$.

3.2. *The Pontrjagin Character*

In this section we will construct a particularly symmetric expression for the $SO(n+m)$ invariant differential form on G_n^m representing the Pontrjagin character of the tautological bundle over G_n^m.

3.2.1. Following 3.1.3, we define

$$Ph_1({}^1M, {}^2M, {}^3M, {}^4M) = \mathrm{ALT}\ {}^1M^{j(1)}_{i(1)}\ {}^2M^{j(2)}_{i(1)}\ {}^3M^{j(2)}_{i(2)}\ {}^4M^{j(1)}_{i(2)},$$

$$Ph_l({}^1M,..., {}^lM) = \mathrm{ALT}\,{}^1M^{j(1)}_{i(1)}\ {}^2M^{j(2)}_{i(1)}\ {}^3M^{j(2)}_{i(2)} \cdots {}^{4l-1}M^{j(2l)}_{i(2l)}\ {}^{4l}M^{j(1)}_{i(2l)},$$

where the symbols $i(1)$, $i(2)$,..., $j(1)$,... are separate indices and the "repeated index = summation" convention is used. ALT means alternating summation:

$$\mathrm{ALT}\ \psi({}^1M,..., {}^qM) = \sum_\sigma (-1)^{\operatorname{sign}\sigma} \psi({}^{\sigma(1)}M,..., {}^{\sigma(q)}M),$$

where the sum is over all permutations σ of $\{1,..., q\}$. (Note that we do not, as is usual, divide by $q!$ in defining ALT).

PROPOSITION. Ph_l satisfies the condition (*) of 3.1.3 and so defines an $O(n+m)$ invariant $4l$-form on G^m_n (which we also denote by Ph_l).

Proof. If θ is an orthogonal $n \times n$ matrix then

$$({}^1M^{j(1)}_{p(1)}\,\theta_{p(1),i(1)}\ {}^2M^{j(2)}_{q(1)}\,\theta_{q(1),i(1)}) \cdots ({}^{4l-1}M^{j(2l)}_{p(2l)}\,\theta_{p(2l),i(2l)}\ {}^{4l}M^{j(1)}_{q(2l)}\,\theta_{q(2l)i(2l)})$$
$$= ({}^1M^{j(1)}_{i(1)}\ {}^2M^{j(2)}_{i(1)}) \cdots ({}^{4l-1}M^{j(2l)}_{i(2l)}\ {}^{4l}M^{j(1)}_{i(2l)})$$

since θ preserves inner products. A similar manipulation applies to an $m \times m$ orthogonal matrix $\bar{\theta}$.

3.2.2. We now begin to identify Ph_l by applying the standard description of the cohomology of G^m_n in terms of characteristic classes of the tautological bundle. (see [16, 21]).

LEMMA. *Ph_l corresponds (by 3.1.1) to a stable characteristic class in $H^{4l}(G^m_n, \mathbb{R})$ of the tautological bundle over G^m_n (which does not depend on n or m).*

This follows from iterating the following

FACT. *In the diagram (see 2.3.3)*

$$H^{4l}(G^m_{n-1}) \xleftarrow{A_l^*} H^{4l}(G^m_n) \xrightarrow{B_l^*} H^{4l}(G^{m-1}_n),$$

A_l^* maps Ph_l for G^m_n to Ph_l for G^m_{n-1} and B_l^* maps Ph_l for G^m_n to Ph_l for G^{m-1}_n.

Proof. By the invariance property of Ph_l, 3.2.1, we can verify this fact using a special choice of bases in ξ and $\xi^\perp$. With the appropriate choices we need to check that $Ph_l({}^1M,\dots,{}^{4l}M)$ is unchanged if a final row of zeros, or a final column of zeros, is added to each matrix ${}^kM$.

3.2.3 LEMMA. *The characteristic class $\widetilde{Ph}_l$ corresponding to Ph_l is additive; i.e., $\widetilde{Ph}_l(\xi \oplus \xi') = \widetilde{Ph}_l(\xi) + \widetilde{Ph}_l(\xi')$.*

This follows from the

FACT. *Let $\varphi\colon G_n^m \times G_{n'}^{m'} \to G_{n+n'}^{m+m'}$ be the map that sends (ξ, ξ') to $\xi + \xi'$. Then φ^* takes Ph_l for $G_{n+n'}^{m+m'}$ to the sum of the Ph_l for G_n^m and the Ph_l for $G_{n'}^{m'}$ (pulled up by the appropriate projections).*

Proof. Choosing appropriate bases, this amounts to the statement that

$$Ph_l({}^1M + {}^1\bar{M},\dots, {}^{4l}M + {}^{4l}\bar{M}) = Ph_l({}^1M,\dots, 4^lM) + Ph_l({}^1\bar{M},\dots, {}^{4l}\bar{M}),$$

where $M + \bar{M}$ is defined to be the matrix

$$\begin{bmatrix} M & 0 \\ 0 & \bar{M} \end{bmatrix}.$$

This may be verified by observing that every term in the expression for $Ph_l({}^1M + {}^1\bar{M},\dots, {}^{4l}M + {}^{4l}\bar{M})$ is zero unless it is a term in the expression for $Ph({}^1M + 0,\dots, {}^{4l}M + 0)$ or a term in the expression for $Ph(0 + {}^1\bar{M},\dots, 0 + {}^{4l}\bar{M})$.

3.2.4. LEMMA. *Ph_l is nonzero. In*

FACT. *Ph_l is a nonzero multiple of the volume form in G_2^{2l}.*

3.2.5. The only additive stable characteristic classes with values in $H^{4l}(\ ,\mathbb{R})$ are multiples of the $4l$ degree part of the Pontrjagin character (by the splitting principle). Therefore,

THEOREM. *There exist coefficients $C_1, C_2,\dots$ so that the Pontrjagin character is*

$$C_1 Ph_1 + C_2 Ph_2 + \cdots.$$

(It would be interesting to evaluate the C_i.)

COROLLARY. *All $SO(n+m)$ invariant differential forms on G_n^m whose homology class is a stable characteristic class of the tautological bundle are polynomials (by wedge product) in the forms Ph_l. In particular they are all $O(n+m)$ invariant.*

Proof. Any stable characteristic class with values in real cohomology is a polynomial in the pure degree terms of the Pontrjagin character. So an $SO(n+m)$ invariant representative of it, is given by that same polynomial in the $C_i Ph_i$. But there is a unique $SO(n+m)$ invariant representative of it.

3.3. *Proof of the Vanishing Theorem*

3.3.1. First we solve the following problem in elementary geometry: Let $u \in \mathbb{R}^{n+m}$ be a unit vector and let $\mathscr{S}^t_u = \mathbb{R}^{n+m} \to \mathbb{R}^{n+m}$ be a stretch along u by a factor of e^t for small t. (That is, $\mathscr{S}^t_u$ fixes $u^\perp$ pointwise, fixes lines parallel to u as sets, and multiplies distances in these lines by e^t). Suppose we have an orthogonal decomposition $\mathbb{R}^{n+m} = \xi \oplus \xi^\perp$. Then $\mathscr{S}^t_u$ applied to ξ is the graph of what linear function $L^t \colon \xi \to \xi$? The picture is this:

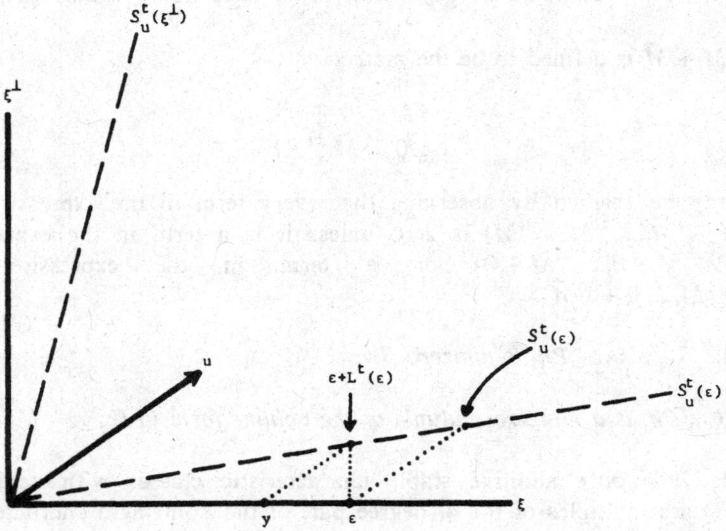

The formula for $\mathscr{S}^t_u(\varepsilon)$ is $\varepsilon + (e^t - 1)(\varepsilon \cdot u)u$. To find the y such that $\mathscr{S}^t_u(y) = \varepsilon + L^t(\varepsilon)$ we need to solve $Pr_\xi \mathscr{S}^t_u y = \varepsilon$, where Pr_ξ is orthogonal projection into ξ. Suppose that $u = v + w$, where $v \in \xi$ and $w \in \xi^\perp$. Then the formula for $Pr_\xi \circ \mathscr{S}^t_u(y)$ is $y + (e^t - 1)(y \cdot v)v$ if $y \in \xi$. We can solve and obtain

$$y = \varepsilon + \frac{1 - e^t}{(e^t - 1)\, v \cdot v + 1} (\varepsilon \cdot v)\, v,$$

as may be checked by direct computation. Therefore

$$L^t(\varepsilon) = \mathscr{S}^t_u y - \varepsilon = (e^t - 1) \left(\left[\varepsilon + \frac{1 - e^t}{(e^t - 1)(v \cdot v) + 1} (\varepsilon \cdot v)\, v \right] \cdot v \right) w. \quad (1)$$

The derivative of this last expression at $t=0$ is

$$\frac{d}{dt} L^t(\varepsilon)\bigg|_{t=0} = (\varepsilon \cdot v)\, w. \tag{2}$$

3.3.2. We want to identify the tangent space to the H orbit $H\xi$ through $\xi \in G_n^m$ as a subspace of $T_\xi G_n^m$ using the model for $T_\xi G_n^m$ as $n \times m$ matrices given in 3.1.3.

Let $e_1, e_2, \ldots, e_{n+m}$ be the standard basis in $\mathbb{R}^{n+m}$. Let $\varepsilon_1, \ldots, \varepsilon_n$ be an orthonormal basis for ξ and let $\bar{\varepsilon}_1, \ldots, \bar{\varepsilon}_m$ be an orthonormal basis for $\xi^\perp$. We define ${}^kV_i$ and ${}^kW^j$ by the relation

$$e_k = \sum_i {}^kV_i \varepsilon_i + \sum_j {}^kW^j \bar{\varepsilon}_j.$$

[That is ${}^kV_i$ (resp. ${}^kW^j$) are the coordinates with respect to the basis $\{\bar{\varepsilon}_i\}$ (resp. $\{\bar{\varepsilon}_j\}$) of the orthogonal projection of e_i into ξ(resp. $\xi^\perp$).]

The ${}^kV_i$ and ${}^kW^j$ satisfy the relation

$$\sum_{i=1}^{n} {}^kV_i\, {}^{k'}V_i = -\sum_{i=1}^{m} {}^kW^i\, {}^{k'}W^i \tag{*}$$

since

$$O + e_k \cdot e_{k'} = \Big(\sum_i {}^kV_i \varepsilon_i + \sum_j {}^kW^j \bar{\varepsilon}_j\Big) \cdot \Big(\sum_i {}^{k'}V_i \varepsilon_i + \sum_j {}^kW^j \bar{\varepsilon}_j\Big)$$

$$= \sum_i {}^kV_i\, {}^{k'}V_i + \sum_j {}^kW^j\, {}^{k'}W^j$$

Let $\mathscr{X}^t_k \subset H^0$ be the element given by $h_i = 1$ for $i \neq k$ and $h_k = e^t$ with respect to coordinates in H from 2.3.1. The diagonal matrix with e^t in the (k, k) position and 1 elsewhere on the diagonal acts on $\mathbb{R}^{n+m}$ by the stretch $\mathscr{S}^t_{e_k}$. Therefore, by the results of the previous section, in graph coordinates centered at ξ, $\mathscr{X}^t_k(\xi)$ is given by Eq. (1), 3.3.1, and its t derivative at $t=0$ is given by (2). Expressing (2) in the basis $\{\varepsilon_i\}$ and $\{\bar{\varepsilon}_j\}$

$$\frac{d}{dt} \mathscr{X}^t_k(\varepsilon_i)\bigg|_{t=0} = (\varepsilon_i \cdot {}^kV)\, {}^kW = {}^kV_i \sum {}^kW^j\, \bar{\varepsilon}_j,$$

so we have

Proposition. *$T_\xi(H_\xi) \subset T_\xi(G_n^m)$ is spanned by the linear transformations*

$${}^kM_i^j = {}^kV_i\, {}^kW^j$$

(no summation over k is implied).

3.3.3. We now prove theorem 1.2.4 which implies the vanishing theorem. We must show that for any ξ and any $O(n+m)$ invariant p-form ω on G_n^m, ω vanishes on any p tangent vectors in $T_\xi(H\xi) \subset T_\xi G_n^m$. Since the Ph_ξ form a basis for the ring of invariant $O(n+m)$ forms (Corollary 2.3.5), it is enough to show this for these forms Ph_ξ. Since the matrices ${}^kM_i^j = {}^kV_i\,{}^hW^j$ form a basis for the vector space $T_\xi(H\xi)$ (Proposition 3.3.2), it is enough to show this for $4l$ of these vectors. By reordering the e_k, we may take these to be ${}^1M_i^j, \ldots, {}^{4l}M_i^j$.

$$\begin{aligned}
&Ph_l({}^1M, \ldots, {}^{4l}M) \\
&= \mathrm{ALT}({}^1V_{i(1)}\,{}^1W^{j(1)})({}^2V_{i(1)}\,{}^2W^{j(2)}) \cdots ({}^{4l-1}V_{i(2l)}\,{}^{4l-1}W^{j(2l)})({}^{4l}V_{i(2l)}\,{}^{4l}W^{j(1)}) \\
&= \mathrm{ALT}({}^1W^{i(1)}\,{}^1W^{j(1)})({}^2W^{i(1)}\,{}^2W^{j(2)}) \cdots \\
&\qquad ({}^{4l-1}W^{i(2l)}\,{}^{4l-1)}W^{j(2l)}({}^{4l}W^{i(2l)}\,{}^{4l}W^{j(1)})
\end{aligned}$$

by Eq. (*), 3.3.2. But this last expression is zero since the function to be alternated is invariant under the permutation of the indices $1, \ldots, 4l$ given by

$$1 \to 2 \to 3 \to \cdots \to 4l-1 \to 4l \to 1,$$

which is an odd permutation.

4. Some History, Related Work, and Open Problems

4.1. *The Dilogarithm.*

4.1.1. The dilogarithm function

$$\mathrm{Li}(x) = \sum_{n=1}^{\infty} \frac{x^n}{n^2}$$

was apparently first considered by Leibnitz in 1696, then by Euler in 1779 [11]. It and elementary transformations of it were studied by a number of 19th-century mathematicians including Legendre, Lobachevsky, Lindelof, Jonquière (see [18] for an excellent survey). The form $\Phi(x) = \mathrm{Li}(x) - \mathrm{Li}(1-x)$ is from [22] and [14].

Recently there has been a renaissance of interest in the dilogarithm. It has been used in the study of algebraic k-theory by Bloch [6] Gross [13] and Beilinson [4]. It has been related to Chen's iterated integrals by Aomoto [1, 2].

4.1.2. In 0.3.1, we interpret $\Phi(x)$ as the volume of a certain region in G_2^2. This relation was conjectured in an attempt to reconcile the original proof of the combinatorial formula for the first Pontrjagin class of

Gabrielov, Gelfand, and Losik [14] with MacPherson's proof [19]. The original proof proceeded by successive integrations; Φ and its properties were used. MacPherson's proof used geometry in Grassmann manifolds instead.

Our original proof of the interpretation of $\Phi(x)$ as a volumn was obtained by direct integration. A later proof by Damiano [8] uses the theory of webs. A third proof would use the well-known fact that $\Phi(x)$ is essentially characterized by its functional equation. (We have found no published proof of this fact, however.)

4.1.3. The imaginary part of $\Phi(k)$ occurs in the formula for the volume of a tetrahedron in real hyperbolic three-space whose vertices lie in the complex projective line at infinity and have cross-ratio k [12, 20]. In this paper, we have interpreted the real part of $\Phi(k)$ as a volume in the Grassmann manifold G_2^2. Since both hyperbolic three-space and G_2^2 are symmetric spaces, there should be a conceptual explanation for this coincidence. We do not know of one.

4.1.4. From our present point of view, the key property of the dilogarithm is the functional equation of 0.3.2 for which we have given a geometric interpretation. This functional equation has many equivalent forms involving five dilogarithm terms in two variables plus various logarithmic terms (see [18], p. 7). It was discovered first by Spence in 1809 [23], then independently rediscovered by Abel, Hill, Kummer, and others.

4.1.5. One modern interpretation of this functional equation is as an Abelian equation for a particular web (of five foliations of the two-disk by lines) called the exceptional web. This is due to Bol [7]. Damiano has extended this theory to higher dimensions in a way related to the generalized dilogarithm forms discussed in this paper [8, 9].

4.1.6. A result of Wigner relates the function Φ to the first Pontrjagin class [24]. There is a continuous function Φ' from the real projective line $\mathbb{R}P'$ to $\mathbb{R}/\pi^2\mathbb{Z}$ such that $d\Phi' = d\Phi$. A continuous cohomology class in $H^3(SL_2(\mathbb{R}), \mathbb{R}/\pi^2\mathbb{Z})$ is represented by the cochain whose value on $X_0,..., X_3$ is $\Phi'(k)$, where k is the cross-ratio of the projections of $X_0,..., X_3$ to $\mathbb{R}P^1$. The functional equation for Φ gives the cocycle condition. The class c maps to the first Pontrjagin class by the connecting homomorphism of the coefficient sequence $0 \to \mathbb{Z} \to \mathbb{R} \to \mathbb{R}/\pi^2\mathbb{Z} \to 0$. It would be interesting to relate this to the connection between $d\Phi$ and the first Pontrjagin class of 0.3.2.

4.2. *The Tetrahedral Complex*

4.2.1. The study of particular algebraic families of lines in three-space was highly developed in the 19th century. A three dimensional family was called a complex. The degree of a complex C was defined to be the degree of the set of lines in C passing through a given point, considered as a curve in the projective plane of all lines through that point. (In modern terms, this amounts to the homology class of the complexification of C in the complex Grassmannian).

The tetrahedral complex of 0.2.3 (the Zariski closure of a Grassmannian simplex Γ_2^2) was first considered by Binet in 1911 [5] as the family of axes of inertia of a solid body. It was studied by Charles, Plücker, and Reye, among others. Various other characterizations of it were found: for example, the family of axes of the ellipses obtained as planar sections of a fixed ellipsoid, or the family of lines joining x and Tx for a fixed projective transformation T. A good survey is [26], where 16 such characterizations of the tetrahedral complex are listed.

4.2.2. The first paper of Lie on Lie groups was a note in 1870 with Klein on H acting on G_1^3, G_2^2, and G_3^1 (see [17]). In it, they prove that the tetrahedral complex is the closure of an H orbit in G_2^2. They used what amount to the exponential map and the fact that H is Abelian to prove geometric facts about the tetrahedral complex, and they saw the fact that their proofs were easier than the usual ones as justufication for the new tool of continuous groups.

4.2.3. The characterization of Klein and Lie of the tetrahedral complex leads to a generalization of it to the space of k planes in projective N-space. The generalized tetrahedral complex is the closure of an H orbit—in our language, the Zariski closure of a Grassmannian simplex. Since this is an N dimensional family, there is a natural inversion problem in integral geometry for it: Given a real valued function f on N space, can one reconstruct f from the knowledge of its integrals over all the k-planes of the generalized tetrahedral complex?

4.2.4. The Lie theoretic formulation allows a generalization of the whole set-up of this paper from $PGL(n+m)$ to an arbitrary semisimple Lie group. Let G be a split real form of a semisimple Lie group, let P be a parabolic subgroup, and let H be a split Cartan subgroup. Then tetrahedral complexes in G_n^m generalize to closures of H orbits in G/P.

4.3. *Grassmannian Simplices*

4.3.1. The hypersimplices Δ_n^m originated in the work of Gabrielov, Gelfand, and Losik [14]. The basis reason for their usefulness there, was that their face relations (2.1.3), are formally dual to the formulas for the exterior derivatives of certain differential forms $S^{n-1,m-1}$ one of which is essentially $d\Phi$. (see [14, Sect. 2; 25]).

The fact that Δ_n^m is homeomorphic to the closure Γ_n^m of an H^0 orbit in G_n^m in a way that takes the face structure of Δ_n^m to the orbit structure of Γ_n^m (2.3.4) was first proved by MacPherson [19, Sect. 6]. (In [19] a polyhedral structure on all the closures of H^0 orbits, including the non-generic ones, was determined.)

Although the present work originated in an attempt to unite these two points of view, it is still an open question to relate the forms $S^{n,m}$ (aside from $d\Phi$) to the Grassmannian geometry of this paper.

4.3.2. The homeomorphism of 2.3.4 between polyhedra and closures of H^0 orbits has been extended to the general semisimple Lie group context of 4.2.4 by Heckmann [15, Theorem 3, p. 25] and Atiyah [1, Theorem 3]. The polyhedra in question are convex hulls in the Lie algebra of H of Weyl group orbits, the choice of the orbit depending on the choice of the parabolic subgroup $P \subset G$. The closures of H^0 orbits are sections of the projection of G/P to H of Kostant's convexity theorem. (Compare [27].)

4.3.3. More generally still, the study of toric varieties of Demazure etc. (see [10]) shows that for any algebraic $(C^*)^n$ action on an complex algebraic variety, if the closure of the orbit through x contains only finitely many other orbits, then the closure of the $(\mathbb{R}^+)^n$ orbit through x is naturally modeled on a polyhedron with faces corresponding to orbits.

4.4. *The Vanishing Theorem*

4.4.1. The vanishing Theorem 1.2.4 ammounts to a set of differential equations satisfied by the generalized tetrahedral complexes in G_n^m.

It was first proved by MacPherson for P_2^3, where P is the first Pontrjagin class, then by Damiano [8] for P_2^{2i+1}, where P is the ith Pontrjagin class. Both of these proofs used the expression for P_n^m in terms of connections and curvature.

4.4.2. A beautiful conceptual proof of Theorem 1.2.4, together with a strong generalization has been found by Atiyah [1, Theorem 4]. This generalization applies to the situation of 4.3.3, where the $(\mathbb{C}^*)^n$ action is assumed to have a fixed point and to differential forms that are harmonic on the complex variety.

In particular, this generalization applies to the general semisimple Lie group context of 4.2.4 and 4.3.2. Therefore one has all the ingredients to carry out the program of this paper in that context—that is, one can find generalized dilogarithm forms which satisfy functional equations similar to those in 1.3.2.

4.4.3. In order to identify leading or coleading differential forms (as in 1.3.1), one wants to identify which forms $\hat{P}_n^m$ of positive degree vanish. So one wants stronger vanishing theorems than 1.2.4. One wants to identify for which triples (P, n, m) is the following true:

P_n^m is zero when evaluated on a set of tangent vectors of G_n^m, of which $n+m-1$ are tangent to an H orbit.

This vanishing has been shown to hold by Damiano for $n = 2$ and degree $P < 2m$ using curvature.

One approach would be to use the formula of 3.2.1 for P_n^m, and the expression of 3.3.2 for the tangent space to the H orbits. Using this method, we can duplicate Damiano's result above and, for example, we can prove vanishing for the triples $(Ph_2, 3, 5)$ or $(Ph_3, 3, 8)$. But the general case leads to a combinatorial problem that we cannot solve.

ACKNOWLEDGMENTS

We would like to thank A. Borel and D. Damiano for useful conversations.

REFERENCES

1. K. AOMOTO, Addition theorems of Abel type for hyper-logarithms, I, preprint, Nagoya University.
2. K. AOMOTO, A generalization of Poincaré normal function on a polarized manifold, *Proc. Japan Acad. Ser. A, Math. Sci.* **55**, No. 9 (1975), 353–358.
3. M. F. ATIYAH, Convexity and commuting Hamiltonians, preprint, Oxford University, 1981.
4. A. BEILINSON, Algebraic k-theory and values of L functions, preprint, Moscow, 1980.
5. J. BINET, Mémoire sur la théorie des axes conjugués et des moments d'inertie des corps, *J. École Polytechn.* **9**, No. 16 (1811), 41–67.
6. S. BLOCH, Applications of the dilogarithm function in algebraic k-theory and algebraic geometry, *in* "Proceedings, Int. Symp. on Algebraic Geometry, Kyoto," 1977, pp. 103–114.
7. G. BOL, Über ein bemerkensvertes Funfgewebe in der Ebene, *Abh. Math. Sem. Univ. Hamburg.* **11** (1935), 393.
8. D. B. DAMIANO, Webs, Abelian equations, and characteristic classes, Thesis, Brown University, May 1980.
9. D. B. DAMIANO, Webs and characteristic forms of Grassmann manifolds, preprint, Princeton University, 1981.
10. V. I. DANILOV, The geometry of toric varieties, *Usp. Mat. Nauk* **33**, No. 2 (1978), 85–134. [Trans., *Russian Math. Surveys* **33**, No. 2 (1978), 97–154].
11. L. EULER, De summatione serierum in hac formo contenlarum $a/1 + a^2/4 + a^3/9 + a^4/16 + \cdots$, *Mém. Acad. St. Petersberg Bd.* **3** (1779) 26–42.
12. T. GOODWILLIE, Volume and dilogarithm, notes (1979).
13. B. GROSS, On the values of Artin 1-functions, preprint.
14. A. M. GABRIELOV, I. M. GELFAND, AND M. V. LOSIK, "Combinatorial Calculation of Characteristic Classes, Functional Analysis and Its Applications," Vol. 9, No. 2, 1975; and Vol. No. 3, pp. 5–26, 1975. [In Russian].
15. G. J. HECKMAN, Projections of orbits and asymptotic behaviour of multiplicities for compact Lie groups, thesis, Leiden, 1980.
16. F. HIRZEBRUCH, "*Neue Topologischen Methoden in der Algebraischen Geometrie*," Springer-Verlag, Berlin, 1962.
17. F. KLEIN AND S. LIE, Sur une certaine famille de courbes et de surfaces, *C. R. Acad. Sci. Paris* (1870), 1222–1226, 1275–1279.

18. L. Lewin, "Dilogarithms and Associated Functions," MacDonald, London (1958).
19. R. D. MacPherson, The Combinatorial Formula of Gabrielov, Gelfand, and Losik for the First Pontrjagin Class, *Sem. Bourbaki*, No. 497, 1976/77.
20. J. Milnor, Computation of volume, in "The Geometry and Topology of Three-Manifolds" (W. P. Thurston, Ed.), Chap. 7, Princeton Mathematics Department 1980.
21. J. Milnor and J. Stasheff, "Characteristic Classes," Annals of Mathematics Studies No. 76, Princeton N.J., 1974.
22. L. J. Rogers, On function sum theorems connected with the series $\sum_{n=1}^{\infty} x^n/n^2$, *Proc. London Math. Soc.* **4** (1907), 169–189.
23. W. Spence, "An essay on logarithmic transcendents," London and Edinburgh, 1809.
24. D. Wigner, to appear.
25. B. V. Yusin, Sur les formes $S^{p,q}$ apparaissant dans le calcul combinatoire de la 2e classe de Pontrjagin par la méthode de Gabrielov, Gelfand et Losik, *C. R. Acad. Sci. Paris, Ser. A* 1981.
26. K. Zindler, Die Tetraedralen und die anderen Kollineationskomplexe, *in* "Encyklopädie der Mathematischen Wissenschaften" Band 3, Teil 2, pp. 1150–1161, 1928.
27. F. A. Berezin and I. M. Gelfand, Some remarks on the theory of spherical functions on symmetric Riemannian manifolds, *Trudy Moskov. Mat. Obšč.* **5** (1956), 311–351.

Part III

Functional integration; probability; information theory

1.
Generalized random processes

Dokl. Akad. Nauk SSSR **100** (1955) 853–856. Zbl. **64**:111

1. Usually, a random process is defined by probability distributions of random variables $(x(t_1),\dots,x(t_n))$ for n arbitrary moments of time. However, one can give many examples of random processes which are important in practice for which such probability distributions do not exist. An example of such a process is the white noise that can be obtained, roughly speaking, as a superposition of all frequencies with random amplitudes; these amplitudes are independent, identically distributed Gaussian variables. In this note we introduce a description of random processes that appears to cover all practically important examples. If $x(t)$ is a random process, then any "linear apparatus" gives us a probability distribution not for the variable $x(t)$, but for a variable $F(\varphi)=\int_{-\infty}^{\infty}\varphi(t)x(t)\,dt$ (where $\varphi(t)$ is a function describing the apparatus), because the apparatus cannot be switched on and off instantaneously and it operates for some length of time. Therefore, a natural definition of a random process can be given by defining probability distributions for a sufficiently complete set of functions $\varphi(t)$ (apparatuses).

2. *Definition of a generalized random process.* Consider the space K of infinitely differentiable functions, each of which vanishes outside some finite interval[1]. Denote by $F(\varphi)$ linear functionals on K. We will say that we are given a random process (random functional F) if for any function $\varphi(t)$ and for any a and b we are given a probability of the event $a<F(\varphi)<b$. These probabilities should satisfy smoothness and compatibility conditions that will be stated below.

Let $\varphi=\alpha_1\varphi_1+\dots+\alpha_n\varphi_n$. Then for any $\alpha_1,\dots,\alpha_n$ and $\varphi_1,\dots,\varphi_n$ we know the distribution of the random variable $\xi=\alpha_1\xi_1+\dots+\alpha_n\xi_n$ where $\xi=F(\varphi)$, $\xi_i=F(\varphi_i)$. Measures (probabilities) of strips $a<\alpha_1\xi_1+\dots+\alpha_n\xi_n<b$ in the n-dimensional space $(\xi_1,\dots,\xi_n)$ uniquely determine the measure of any set (see, for example, [3]). Therefore if the probabilities of inequalities $a<F(\varphi)<b$ can be extended to the probabilities of simultaneous inequalities $a_i<F(\varphi_i)<b_i$ $(i=1,\dots,n)$ for any φ_i and any a_i, b_i, then this extension is unique.

We will require that the probabilities $a<F(\varphi)<b$ satisfy the following conditions:

1. *Compatibility condition.* The probabilities $a<F(\varphi)<b$ for all $\varphi\in K$ and all a,b can be extended to the probability distributions $a_i<F(\varphi_i)<b_i$, $i=1,\dots,n$, for all n, all $\varphi_i\in K$, and all numbers a_i, b_i.

[1] The space K introduced by L. Schwartz [1] can be replaced by any other space of basic functions [2].

2. *Smoothness condition.* To any elements $\varphi_1, \ldots, \varphi_n$ corresponds a probability distribution for $F(\varphi_1), \ldots, F(\varphi_n)$, i.e. a measure on the n-dimensional space. We require that for any n the mapping of n-tuples of functions from K to the space W_n of measures on the n-dimensional space were continuous with respect to the topology in K and the weak topology in W_n. This continuity requirement is quite natural; it means that a small change of "apparatus" φ_i only slightly changes the corresponding "indications".

The derivative of a generalized random process is constructed similarly to the case of usual generalized functions [1], namely, if a random process $F_1(\varphi)$ is the derivative of the process $F(\varphi)$ then the probability of the event $a < F_1(\varphi) < b$ is, by definition, the derivative of the probability of the event $a > F(\varphi') > b$.

3. *Correlation functional. Gaussian processes.* The correlation functional $c(\varphi)$ for a given process is the mean value of the random variable $|F(\varphi)|^2$, i.e. $c(\varphi) = \mathfrak{M}[|F(\varphi)|^2]$.[2] The matching and smoothness conditions imply that $c(\varphi)$ is a positive-definite quadratic functional; it is continuous in the topology of K.

A random process is said to be Gaussian if for any φ the probability density of $F(\varphi)$ is Gaussian, i.e. if it is of the form $(\pi c)^{-\frac{1}{2}} \exp(-\xi^2 c^{-1}(\varphi))$ where $c(\varphi)$ is a continuous functional. As $c(\varphi)$ is the correlation functional of the above Gaussian process we see that $c(\varphi)$ should be a quadratic functional and that any continuous positive-definite quadratic functional is a correlation functional for some (Gaussian) process [4].

A random process is said to be stationary if for any τ the probabilities of inequalities $a < F(\varphi(t)) < b$ and $a < F(\varphi(t+\tau)) < b$ are equal. One can prove that the correlation functional of a stationary process is of the form $T(\varphi^* \times \varphi)$, where T is a linear functional on the space K (i.e. a generalized function) and $\varphi^* \times \varphi$ is the convolution of $\varphi(t)$ and $\varphi^*(t) = \bar{\varphi}(-t)$. In particular, an arbitrary stationary Gaussian process is given by the probability density for the variable ξ in the form $\exp(-\xi^2/T(\varphi^* \times \varphi))$. As $T(\varphi^* \times \varphi) \geqq 0$ for any $\varphi \in K$ the Bochner–Schwartz theorem [1] implies that a generalized function T defining the correlation functional of a stationary process is the Fourier transform (in the generalized sense, see [1], or, for other spaces, [2]) of some monotone function $\sigma(\lambda)$ with the property that for some n and for large $\lambda_2 - \lambda_1$ we have $|\sigma(\lambda_2) - \sigma(\lambda_1)| < c|\lambda_2 - \lambda_1|^n$.

The function $\sigma(\lambda)$ is called the spectral function of the given process. For example, in the case of "white noise" we have $c(\varphi) = \int_{-\infty}^{\infty} \varphi^2(t)\,dt$, $T = \delta(t)$, so that $\sigma(\lambda) = \lambda$.

4. *Processes with stationary increments of n-th order.* A process is said to be a process with stationary increments of n-th order if its n-th derivative is a stationary process. In other words, if $F_n(\varphi) = (-1)^n F(\varphi^{(n)})$, then $F_n(\varphi)$ should be invariant under the substitution $\varphi(t) \to \varphi(t+\tau)$. If $c(\varphi)$ is the correlation functional of the given process, then $c(\varphi^{(n)})$ is invariant under the substitution; going over to Fourier

[2] In general, a polylinear functional $c(\varphi_1, \ldots, \varphi_k) = \mathfrak{M}[F(\varphi_1) \cdots F(\varphi_k)]$ is an analog of a k-th order moment of a generalized random process.

transforms, we can obtain that the functional $\tilde{T}(\psi)$ (Fourier transform of T) is positive on positive functions $\psi(\lambda)$ with zero of the order $2n$ at the origin. Any such functional is of the form

$$\tilde{T}(\psi) = \int_{-1}^{1} \frac{\psi(\lambda) - \sum_{k=0}^{2n-1} \psi^{(k)}(0)\frac{\lambda^k}{k!}}{\lambda^{2n}}\, d\sigma(\lambda) + \int_{1}^{\infty} \psi(\lambda)\, d\sigma(\lambda)$$

$$+ \int_{-\infty}^{-1} \psi(\lambda)\, d\sigma(\lambda) + \sum a_k \psi^{(k)}(0),$$

where $\sigma(\lambda)$ is a monotone function satisfying, for some m and for sufficiently large $\lambda_2 - \lambda_1$, the inequality $|\sigma(\lambda_2) - \sigma(\lambda_1)| < c|\lambda_2 - \lambda_1|^m$. The correlation functional $T(\varphi^* \times \varphi)$ is determined by the linear functional T which is the Fourier transform of $\tilde{T}$. This result can be considered a generalization of results due to A.N. Kolmogorov [5] and A.M. Yaglom and M.S. Pinsker [6].

5. *Characteristic functional.* Following A.N. Kolmogorov [7] we define the characteristic functional by means of the formula $L(\varphi) = \mathfrak{M}[e^{iF(\varphi)}]$ where $\mathfrak{M}$ denotes the mean value. To compute L it suffices to know the probabilities of the inequalities $a < F(\varphi) < b$. For example, the characteristic functional of a Gaussian process is $\exp(-c(\varphi))$. One can prove the following theorem.

A functional $L(\varphi)$ is a characteristic functional if and only if it is positive definite and continuous in the topology of K.

One can easily show that a random process is uniquely determined by its characteristic functional.

6. *Processes with independent values in all points.* We will say that a process has "independent values in all points" if the random values $F(\varphi_1)$ and $F(\varphi_2)$ are independent for all pairs φ_1, φ_2 such that $\varphi_1(t)\varphi_2(t) = 0$, or, in other words, if results of measurement in nonintersecting time intervals are independent. Examples of such processes are derivatives of processes with independent increments (for example, the rate of a particle in Brownian motion). If $L(\varphi)$ is the characteristic functional of a process with independent values then $L(\varphi_1 + \varphi_2) = L(\varphi_1)L(\varphi_2)$ in the case when $\varphi_1(t)\cdot\varphi_2(t) = 0$. Assuming that $L(\varphi)$ nowhere vanishes we get that the functional $M(\varphi) = \log L(\varphi)$ satisfies $M(\varphi_1 + \varphi_2) = M(\varphi_1) + M(\varphi_2)$ in the case $\varphi_1(t)\varphi_2(t) = 0$. Such a functional will be called a functional of the local type. Examples of such functionals are

$$M(\varphi) = \int_{-\infty}^{\infty} f(\varphi(t), \varphi'(t), \ldots, \varphi^{(n)}(t), t)\, dt \tag{1}$$

where $f(0, 0, \ldots, 0, t) = 0$. In particular, for stationary processes f does not depend on t, i.e. f is of the form $f(\varphi(t), \varphi'(t), \ldots, \varphi^{(n)}(t))$. It would be interesting to prove that these examples exhaust all stationary processes with independent values. One can prove that the logarithm of the characteristic functional has the form (1) with a function f depending only on $\varphi(t)$ but not on $\varphi'(t),\ldots,\varphi^{(n)}(t)$ (i.e. $f=f(\varphi(t))$) if and only if the function $\exp(f(x))$ is the characteristic function of some infinitely divisible law. Therefore, such processes (i.e. those with $f=f(\varphi(t))$) coincide with derivatives of processes with independent increments. During the proof of this statement one obtains in a natural way the proof of the theorem on the canonical form of an infinitely divisible law from [8]–[11].

A new class of stationary processes with independent values which is wider than the class of derivatives of processes with independent increments can be obtained if we let a function f in (1) depend also on derivatives, i.e. be of the form $f(\varphi(t),\varphi'(t),\ldots,\varphi^{(k)}(t))$. We can indicate here some sufficient conditions for the positive definiteness of the corresponding characteristic functional:

The characteristic functional corresponding to a function f is positive definite provided for any $s>0$ the function $\exp(sf(x_0,x_1,\ldots,x_k))$ is positive definite in the variables $x_0,x_1,\ldots,x_k$. This gives the following canonical decomposition for the function f:

$$f(x_0,x_1,\ldots,x_n)=\int\limits_{|s|\leq 1}\frac{e^{i\sum x_p s_p}-1-\sum\limits_{p=0}^{k}x_p s_p}{v\left(\sum\limits_{p=0}^{k}s_p^2\right)}+\int\limits_{|s|>1}e^{i\sum x_p s_p}\,d\sigma+a_0+i\sum_{i=0}^{k}c_p x_p$$

where σ is a finite measure in the $(k+1)$-dimensional space, and $v(z)=z$ for $z<1$, $v(z)=1$ for $z>1$.

Let us also indicate a general form of the correlation functional $c(\varphi)$ for a process with independent increments. Just as for processes with independent increments, we have $c(\varphi_1+\varphi_2)=c(\varphi_1)+c(\varphi_2)$ for $\varphi_1\cdot\varphi_2=0$; one can show the following.

The general form of the correlation functional for a process with independent increments is given by the formula

$$c(\varphi)=\int\limits_{-\infty}^{\infty}\sum_{i,k=0}^{\infty}Q_{ik}(t)\varphi^{(i)}(t)\varphi^{(k)}(t)\,dt,$$

where the $Q_{ik}(t)$ are continuous functions such that on any finite interval only a finite number of functions $Q_{ik}(t)$ do not vanish, and $c(\varphi)\geqq 0$.

Moreover, one can show that any moment $c(\varphi_1,\ldots,\varphi_n)$ of a process with independent values has the following form; this enables one to get a general form for a process with independent values having an analytical characteristic functional.

Received 3.XII.1954

References

1. Schwartz, L.: Théorie des distributions. Paris, 1951
2. Gelfand, I.M., Shilov, G.E.: Usp. Mat. Nauk **8** (4) (1953) 3

3. Khachaturov, A.A.: Usp. Mat. Nauk **9** (3) (1954) 205
4. Khinchin, A.Ya.: Math. Ann. **109** (1934) 604
5. Kolmogorov, A.N.: Dokl. Akad. Nauk SSSR **26** (1) (1940) 6
6. Yaglom, A.M., Pinsker, M.S.: Dokl. Akad. Nauk SSSR **90** (5) (1953) 731
7. Kolmogorov, A.N.: C.R. **200** (1935) 1717
8. Kolmogorov, A.N.: Atti Acad. Naz. Lincei **805** (1931) 808, 866–869
9. Finetti, Bruno de: Rendiconti R. Accad. Lincei **6** (1930) 12, 278
10. Levy, P.: Annali R. Sc. Sup. Pisa (2), **3** (1934) 337
11. Khinchin, A.Ya.: Bull. Mosk. Univ. **1** (1) (1937)

2.

(with A. M. Yaglom)

Integration in functional spaces and its applications in quantum physics*†‡

Usp. Mat. Nauk 11 (1) (1956) 77–114 [J. Math. Phys. 1 (1960) 48–69]. Zbl. 92:451

This translation of the survey article by I. M. Gel'fand and A. M. Yaglom on the theory and applications of integration in functional spaces in problems of quantum physics was prepared because it was felt that such a review would be of interest and of use to mathematical physicists working in several different fields.
The article begins with a discussion of Wiener measure, after which the extension is made to the complex measure introduced by Feynman in his formulation of quantum mechanics, and examples are given of the use of these methods in quantum mechanics, quantum field theory, and quantum statistical physics. A comprehensive bibliography of works devoted to the theory and applications of functional integration methods is included.

At the present time, the methods of measure theory and integration in functional spaces are widely applied only in the theory of random processes. It is already apparent today, however, that these methods can have great importance for a number of other branches of mathematics and theoretical physics (primarily for the theory of the differential equations of quantum physics and hydrodynamics). In the present paper, we attempt briefly to throw light on the question of the utilization of these methods in quantum physics[1]; at the same time, certain mathematical questions connected with this application will also be examined.

The beginning of the penetration of the methods of integration in functional spaces into quantum physics, apparently must be considered as having occurred in 1942 when the dissertation of R. Feynman on the principle of least action in quantum physics, containing a new derivation of Schrödinger's equation, was defended in Princeton. This dissertation was never published in full, but the part of greatest interest to us is contained in Feynman's article [1a§], which appeared in 1948 and recently was translated into the Russian language. In 1949, an interesting mathematical work of Kac appeared [2a] (cf. also [2b]), written, as the author indicates, under the strong influence of Feynman's dissertation. It is devoted to the calculation of the mean values of certain functionals over the trajectories of a Brownian particle with the help of the reduction of this problem to the solution of differential equations related to Schrödinger's equation. A special measure in the space of continuous functions which gives the distribution of probabilities of distinct trajectories of a Brownian particle (with neglect of its inertia) plays an extremely essential role in Kac's work. This measure had already been thoroughly studied in the twenties by N. Wiener, and is generally called "Wiener measure." Following Kac's article a series of works by scholars in different countries appeared developing these same ideas further (cf., for example, [4]–[8]). We note further that independently of these works, beginning in 1944, a long series of works was published by Cameron, Martin, and their collaborators devoted to the investigation of separate questions connected with Wiener measure in functional space (cf., for example, [9]–[11], as well as a series of other works by these same authors); in recent years, these investigations also partially coincided with works based on the ideas of Feynman and Kac (cf., references [10b] and [2], [4]).

We enumerated above a series of mathematical works developing the ideas of Feynman's work, [1a]. Among theoretical physicists, however, this work was not sufficiently appreciated in the beginning, apparently due to the novelty and unusualness of the idea of the possibility of integration in functional space. In any event, in the first several years after the appearance of reference [1a] containing in it a new mathematical apparatus, its utilization in physical investigations was almost only by its author. However, in 1954, suddenly two works appeared almost simultaneously which in fact were devoted to the transference of the methods which were utilized in [1a] for the case of nonrelativistic quantum mechanics to the quantum theory of fields (these are the works of Edwards and Peierls [13] and I. M. Gel'fand and R. A. Minlos [14]). It is curious to note that there are no references to [1a] in both of these works; in both cases, the authors were led by a completely different approach[2] to the necessity of introducing the apparatus of integration in functional spaces. Following [13] and [14], there immediately appeared a series of analogous works devoted to the exposition of the fundamental facts of quantum field theory with the utilization of functional integrals (references [15]–[23a] and host of others); notwith-

---

* A review article, presented at the Third All-Union Conference on the Theory of Probability and its Applications, in Leningrad, May 30 to June 4, 1955, under the title "Methods of the Theory of Random Processes in Quantum Physics."

† Translated from Uspekhi Matematicheskikh Nauk, Vol. XI (Jan.–Feb., 1956), p. 77, by A. A. Maradudin, U. of Maryland.

‡ This translation was partially supported by the U. S. Air Force Office of Scientific Research under Contract AF18(600)1315.

[1] Certain ways of utilizing analogous methods in statistical hydrodynamics (in the theory of turbulence) were mentioned in reference [12].

§ Numbers in brackets refer to references in the Bibliography at the end of the paper.

[2] In essence, the apparatus of functional integration appeared in the these works in solving differential equations in the infinite dimensional space of quantum field theory (the so-called "Schwinger equations" in variational derivatives).

standing the fact that no particularly valuable results have been obtained as yet with the aid of these methods, at the present time it is becoming ever more apparent that the future development of quantum field theory (connected, apparently, with the overcoming of a number of serious difficulties of a physical character) will find its mathematical language in just this apparatus.

We will emphasize again that after the appearance of Feynman's work [1a] investigations related to it were in fact carried out independently by mathematicians and physicists. In the present article, we will examine, in parallel as far as possible, works which are relevant here.

1. INTEGRATION OVER WIENER MEASURE IN FUNCTIONAL SPACE

Wiener Measure

We will begin with a definition of Wiener measure in the space of continuous functions and with an examination of certain purely mathematical questions connected with integration over this measure; such an order of exposition will be most convenient for all that follows. We examine a particle undergoing Brownian motion along the x axis under the influence of random impulses (for example, molecular) in the absence of any kind of systematic forces and being found at time $t=0$ at the origin of coordinates. Then, if we neglect the particle's inertia the distribution of probabilities for its position at time $t\neq 0$ will be described by the fundamental solution of the simplest diffusion equation,

$$\partial\psi/\partial t = D\partial^2\psi/\partial x^2 \tag{1.1}$$

(D, the diffusion coefficient, is connected in the case of Brownian motion under the influence of molecular forces to the mass and dimensions of the particle, the temperature and viscosity of the medium by the well-known Einstein relation), i.e., it will have a density equal to

$$p_t(x) = \frac{1}{(4\pi Dt)^{\frac{1}{2}}}\exp\left(-\frac{x^2}{4Dt}\right). \tag{1.2}$$

For simplicity, we will everywhere in the following assume that $D=\frac{1}{4}$; clearly, this can always be achieved with the help of a choice of a suitable system of units. Then, for the probability that the coordinate $x(t_1)$ of the particle at time t_1 will be found within the limits $a_1<x(t_1)<b_1$, at time t_2 within the limits $a_2<x(t_2)<b_2$, $\cdots$, at time t_n within the limits $a_n<x(t_n)<b_n$, where $0<t_1<t_2\cdots<t_n$, we obtain the formula,

$$\frac{1}{[\pi^n t_1(t_2-t_1)\cdots(t_n-t_{n-1})]^{\frac{1}{2}}}\int_{a_1}^{b_1}\int_{a_2}^{b_2}\cdots\int_{a_n}^{b_n}\exp\left[-\frac{x_1^2}{t_1}-\frac{(x_2-x_1)^2}{t_2-t_1}-\cdots-\frac{(x_n-x_{n-1})^2}{t_n-t_{n-1}}\right]dx_1dx_2\cdots dx_n. \tag{1.3}$$

The measure (1.3) of the cylindrical set ("quasi-intervals" in the terminology of Wiener) in the space of all functions, dictated by the conditions

$$a_1<x(t_1)<b_1,\ \cdots,\ a_n<x(t_n)<b_n,$$

can be extended (by a well-known theorem of A. N. Kolmogorov [24]) to the class of all functions $x(\tau)$ of a continuous parameter τ; this is Wiener measure. It was demonstrated by N. Wiener [3] that the corresponding measure is completely concentrated on the class of continuous (but not differentiable) functions $x(\tau)$, equal to zero at $\tau=0$, and satisfying the Lipshitz condition with an exponent of $\frac{1}{2}-\epsilon$ for any $\epsilon>0$. It was also demonstrated by him that for a wide class of functionals $F[x(\tau)]$ (in particular for all bounded and continuous functionals) in a space C of continuous functions in the interval $[0,t]$, such that $x(0)=0$, there exists an integral over Wiener measure, which can be calculated in the following manner: the curve $x(\tau)$ is replaced by the broken one $x_n(\tau)$, which coincides with $x(\tau)$ at the points,

$$x(0)=0,\quad x(t_1)=x_1,\quad x(t_2)=x_2,\quad \cdots,\quad x(t_n)=x_n$$

(where $t_n=t$ and the points $t_1, t_2, \cdots, t_{n-1}$ divide the interval $[0,t]$ into n equal parts of length $\Delta t=t/n$), after which the integral of $F[x(\tau)]$ over Wiener measure is defined as

$$\int_C F[x(\tau)]d_wx = \lim_{n\to\infty}\frac{1}{(\pi\Delta t)^{n/2}}\int_{-\infty}^{\infty}\cdots\int_{-\infty}^{\infty}F(x_1,\cdots,x_n) \times\exp\left[-\frac{x_1^2}{\Delta t}-\sum_{j=1}^{n-1}\frac{(x_{j+1}-x_j)^2}{\Delta t}\right]dx_1\cdots dx_n, \tag{1.4}$$

where $\int_C\cdots d_wx$ denotes an integral over the Wiener measure d_wx, extended over the entire space C, while the function $F(x_1\cdots x_n)$ of n variables in the right hand side of the formula is the value of our functional on substituting in place of $x(\tau)$ the broken curve $x_n(\tau)$,

$$F(x_1,\cdots,x_n)=F[x_n(\tau)]. \tag{1.5}$$

Symbolically, the expression (1.4) of the integral over Wiener measure can be written in the form,

$$\int_C F[x(t)]d_wx = \frac{1}{N}\int_{-\infty}^{\infty}\cdots\int F[x(t)] \times\exp\left\{-\int_0^t\left[\frac{dx(\tau)}{d\tau}\right]^2 d\tau\right\}\prod_0^t dx(t), \tag{1.4'}$$

where N is a constant normalization factor which is determined by the condition that

$$\int_C d_w x = 1.$$

Indeed, if

$$\prod_0^t dx(\tau)$$

is replaced by the product of differentials of the "coordinates" $x(\tau)$ at the finite number of points $\tau = t_1, t_2, \cdots, t_n$, and the integral

$$\int_0^t [dx(\tau)/d\tau]^2 d\tau$$

in the exponent by the corresponding sum of difference ratios

$$\sum_{j=0}^n [x(t_{j+1}) - x(t_j)]^2/(t_{j+1}-t_j),$$

then (1.4′) goes over into (1.4).

Examples

1. In the simplest case, when $F[x(\tau)]$ depends only on the value of the function $x(\tau)$ at a finite number of points, the integral (1.4) by definition becomes an ordinary finite dimensional integral; its evaluation in this case does not involve any difficulties in principle. In the particular case when $F[x(\tau)] = x(t_1)x(t_2)\cdots x(t_k)$, we obtain the well known expressions for the moments of the Brownian motion process which we have described,

$$\int_C x(t_1) d_w x \equiv 0$$

and in general

$$\int_C x(t_1)x(t_2)\cdots x(t_{2k+1}) d_w x \equiv 0, \qquad (1.6)$$

$$\int_C x(t_1)x(t_2) d_w x \equiv b(t_1,t_2) = \tfrac{1}{2}\min(t_1,t_2), \qquad (1.7)$$

$$\int_C x(t_1)x(t_2)\cdots x(t_{2k}) d_w x \equiv b(t_1,t_2,\cdots,t_{2k})$$
$$= \sum b(t_{i_1},t_{i_2})\cdot b(t_{i_3},t_{i_4})\cdots b(t_{i_{2k-1}},t_{i_{2k}}), \qquad (1.8)$$

where the summation in the last formula extends over all possible partitions of the $2k$ indices $1, 2, \cdots, 2k$ into k pairs; $(i_1, i_2), (i_3, i_4), \cdots, (i_{2k-1}, i_{2k})$.

2. We examine now one instructive example in which $F[x(\tau)]$ depends on all the values of the function $x(\tau)$. Let

$$F[x(\tau)] = \exp\left\{\lambda \int_0^t p(\tau)x^2(\tau)d\tau\right\},$$

where λ is a real number and $p(\tau) \geqslant 0$. By substituting into (1.4),

$$F(x_1,x_2,\cdots,x_n) = \exp\{\lambda \sum_{j=1}^n p(j\Delta t)x^2(j\Delta t)\Delta t\}$$
$$\equiv \exp\{\lambda\Delta t \sum_{j=1}^n p_j x_j^2\}, \qquad (1.9)$$

where

$$\Delta t = t/n, \quad p_j = p(j\Delta t), \quad x_j = x(j\Delta t) \qquad (1.10)$$

and making use of the well-known formula,

$$\int_{-\infty}^{\infty}\cdots\int_{-\infty}^{\infty} \exp(-\sum_{j,k=1}^n a_{jk}x_jx_k)dx_1\cdots dx_n$$
$$= \pi^{n/2}[\det(a_{jk})]^{-\frac{1}{2}}, \qquad (1.11)$$

(which is valid on the assumption that (a_{jk}) is a positive definite matrix) we easily find that

$$\int_C \exp\left\{\lambda \int_0^t p^2(\tau)x^2(\tau)d\tau\right\} d_w x = \lim_{n\to\infty} [D_1^{(n)}]^{-\frac{1}{2}}, \qquad (1.12)$$

where $D_1^{(n)}$ is the nth order determinant,

$$D_1^{(n)} = \begin{vmatrix} 2-\lambda(\Delta t)^2 p_1 & -1 & 0 & \cdot & 0 & 0 \\ -1 & 2-\lambda(\Delta t)^2 p_2 & -1 & \cdot & 0 & 0 \\ 0 & -1 & 2-\lambda(\Delta t)^2 p_3 & \cdot & 0 & 0 \\ \cdot & \cdot & \cdot & \cdot & \cdot & \cdot \\ \cdot & \cdot & \cdot & \cdot & \cdot & \cdot \\ \cdot & \cdot & \cdot & \cdot & \cdot & \cdot \\ 0 & 0 & 0 & \cdot & 2-\lambda(\Delta t)^2 p_{n-1} & -1 \\ 0 & 0 & 0 & \cdot & -1 & 1-\lambda(\Delta t)^2 p_n \end{vmatrix}. \qquad (1.13)$$

To find the limit on the right hand side of (1.12), one can make use of the following argument. We denote by $D_k^{(n)}$ the principal minor of $(n-k+1)$st order of the determinant (1.13) situated in the lower right hand corner; then, expanding the determinant $D_k^{(n)}$ in terms of the elements of the first row, we readily obtain the recurrence relation,

$$D_k^{(n)} = [2-\lambda(\Delta t)^2 p_k]D_{k+1}^{(n)} - D_{k+2}^{(n)},$$
$$(1 \leqslant k \leqslant n-2). \qquad (1.14)$$

If instead of $D_k^{(n)}$ we write $D^{(n)}(k\Delta t)$, then Eq. (1.14) can be rewritten in the form,

$$\frac{D^{(n)}(k\Delta t)-2D^{(n)}((k+1)\Delta t)+D^{(n)}((k+2)\Delta t)}{(\Delta t)^2} = -\lambda p(k\Delta t)D^{(n)}((k+1)\Delta t). \quad (1.14')$$

We note further that $D_n^{(n)}=1-\lambda(\Delta t)^2p_n$ and $D_{n-1}^{(n)} =1-\lambda(\Delta t)^2(2p_n+p_{n-1})+\lambda^2(\Delta t)^4p_np_{n-1}$, so that

$$D^{(n)}(n\Delta t)=1-\lambda(\Delta t)^2p_n$$
$$\frac{D^{(n)}(n\Delta t)-D^{(n)}((n-1)\Delta t)}{\Delta t}=\lambda(\Delta t)[p_{n-1}+p_n -\lambda(\Delta t)^2p_np_{n-1}]. \quad (1.15)$$

In this way, $D^{(n)}(k\Delta t)$ will be the solution of the difference Eq. (1.14′) satisfying the conditions (1.15). From this, it may be derived that as $n\to\infty$ the value of $D_k^{(n)}$ approaches the value at the point $k\Delta t$ of the continuous function $D(\tau)$ given by the solution to the differential equation,

$$[d^2D(\tau)/d\tau^2]+\lambda p(\tau)D(\tau)=0 \quad (1.16)$$

subject to the conditions,

$$D(t)=1,\quad D'(t)=0. \quad (1.17)$$

Corresponding to this, in view of Eq. (1.12),

$$\int_C \exp\left\{\lambda\int_0^t p(\tau)x^2(\tau)d\tau\right\}d_wx=[D(0)]^{-\frac{1}{2}}. \quad (1.18)$$

It is natural that all of these arguments will be legitimate only when, in representing (1.4) in the form (1.11), we indeed have in the exponent a positive definite quadratic form, i.e., when $\lambda<\min(\lambda_1,\lambda_2,\cdots,\lambda_n)$ where $\lambda_1, \lambda_2, \cdots, \lambda_n$ are the zeros of the determinant (1.13); it can be shown that in the limit as $n\to\infty$, these zeros become the characteristic values of Eq. (1.16) subject to the boundary conditions $D'(t)=0$, $D(0)=0$, so that λ must be smaller than the smallest of these characteristic values. In the simplest particular case $p(\tau)\equiv 1$, we will have that $D(\tau)=\cos(\lambda)^{\frac{1}{2}}(t-\tau)$ from which we obtain

$$\int_C \exp\left\{\lambda\int_0^t x^2(\tau)d\tau\right\}d_wx=[\sec t(\lambda)^{\frac{1}{2}}]^{\frac{1}{2}} \quad \text{for } \lambda^{\frac{1}{2}}<\pi/2t. \quad (1.19)$$

The result (1.18) was first obtained by Cameron and Martin [9c] with the aid of a method of which we will say more below. The derivation presented here was pointed out in the work of Montroll [25], which also contains certain other examples of the direct evaluation of integrals over Wiener measure.

"Conditional" Wiener Measure

If in Eq. (1.3) we put $t_n=t$, fix the value of $x_n=x(t) =X$ and do not integrate over this coordinate, then the corresponding integral defines a certain measure of the cylindrical set,

$$a_1<x(t_1)<b_1, \cdots, a_{n-1}<x(t_{n-1})<b_{n-1}$$

of a functional space, which can then be extended to the class of all functions $x(\tau)$, $0<\tau<t$. It is not difficult to verify that the measure of all space here equals $(1/(\pi t)^{\frac{1}{2}})\exp(-X^2/t)$; the class of functions corresponding to the measure in this case will be the class of continuous functions $x(\tau)$, which are equal to zero for $\tau=0$, equal to X at $\tau=t$, and satisfy the very same conditions of smoothness as in the case of ordinary Wiener measure. We will denote the integral of a functional $F[x(\tau)]$ over this new measure extending over the class $C_{t,X}$ of all continuous functions satisfying the conditions $x(0)=0$, $x(t)=X$, by

$$\int_{C_{t,X}} F[x(\tau)]d_{w(t,X)}x; \quad (1.20)$$

this integral can be evaluated with the help of a formula differing from (1.4) only in the absence of an integration over dx_n on the right-hand side. It is clear that

$$\int F[x(\tau)]d_wx=\int_{-\infty}^{\infty}\left\{\int_{C_{t,X}} F[x(\tau)]d_{w(t,X)}x\right\}dX; \quad (1.21)$$

so that knowing the integral over $d_{w(t,X)}x$, we can determine the integral of the same functional over the measure d_wx with the help of a single quadrature. Together with the measure $d_{w(t,X)}x$, we will also investigate the normalized measure

$$d_{w(t,X)}{}^*x=(\pi t)^{\frac{1}{2}}\exp(X^2/t)d_{w(t,X)}x; \quad (1.22)$$

the integral of a functional over the measure $d_{w(t,X)}{}^*x$ will have the meaning of a conditional mathematical expectation of the corresponding functional over the trajectory of a Brownian particle under the condition that this trajectory passes through the given point X at the time t.

Together with the measures $d_{w(t,X)}x$ and $d_{w(t,X)}{}^*x$ assigned on the class $C_{t,X}$ of continuous functions satisfying the conditions $x(0)=0$ and $x(t)=X$, it sometimes is necessary to investigate likewise the analogous measures $d_{w(t_0,X_0;\,t,X)}x$ and $d_{w(t_0,X_0;\,t,X)}{}^*x$, assigned on the classes $C_{t_0,X_0;\,t,X}$ of continuous functions satisfying the conditions $x(t_0)=X_0$ and $x(t)=X$. Integration over these new measures, however, easily reduces to the integration studied by us over the set of curves starting at the point $x(0)=0$, namely[3]

$$\int_{C_{t_0,X_0;t,X}} F[x(\tau)]d_{w(t_0,X_0;\,t,X)}x$$

$$=\int_{C_{t-t_0;X-X_0}} F[X_0+x(\tau)]d_{w(t-t_0;\,X-X_0)}x \quad (1.23)$$

and analogously for the normalized measures "with an asterisk" (these equations can be taken, if necessary, as the definition of the measures $d_{w(t_0,X_0;\,t,X)}x$ and $d_{w(t_0,X_0;\,t,X)}x$).

Examples

1. Just as in the case of ordinary Wiener measure the integral over $d_{w(t,X)}x$ or $d_{w(t,X)}{}^*x$ of a functional $F[x(\tau)]$, which depends only on the value of the function $x(\tau)$ at a finite number of points, reduces to a simple finite dimensional integral. In the case when $F[x(\tau)]=x(t_1)x(t_2)\cdots x(t_k)$, the integrals over the measure $d_{w(t,X)}{}^*x$ are easily handled and give conditional moments of the trajectory of a Brownian particle (under the condition that at the time t this trajectory passes through the point X):

$$\int_{C_{t,X}} x(t_1)d_{w(t,X)}{}^*x=\frac{t_1}{t}X, \quad (1.24)$$

and in general,

$$\int_{C_{t,X}}\left[x(t_1)-\frac{t_1}{t}X\right]\left[x(t_2)-\frac{t_2}{t}X\right]\cdots$$

$$\times\left[x(t_{2k+1})-\frac{t_{2k+1}}{t}X\right]d_{w(t,X)}{}^*x\equiv 0, \quad (1.24')$$

$$\int_{C_{t,X}}\left[x(t_1)-\frac{t_1}{t}X\right]\left[x(t_2)-\frac{t_2}{t}X\right]d_{w(t,X)}{}^*x$$

$$\equiv b_t(t_1,t_2)=\frac{t_1(t_1-t_2)}{2t}, \quad t_1\leqslant t_2, \quad (1.25)$$

$$\int_{C_{t,X}}\left[x(t_1)-\frac{t_1}{t}X\right]\left[x(t_2)-\frac{t_2}{t}X\right]\cdots$$

$$\times\left[x(t_{2k})-\frac{t_{2k}}{t}X\right]d_{w(t,X)}{}^*x\equiv b_t(t_1,t_2,\cdots,t_{2k})$$

$$=\sum b_t(t_{i_1},t_{i_2})b_t(t_{i_3},t_{i_4})\cdots b_t(t_{i_{2k-1}},t_{i_{2k}}), \quad (1.26)$$

where the summation in the last formula extends over all possible partitions of the $2k$ indices $1, 2, \cdots, 2k$ into k pairs: (i_1,i_2), (i_3,i_4), $\cdots$, (i_{2k-1},i_{2k}).

[3] If the functional $F[x(\tau)]=F[x(\tau),\tau]$ depends explicitly on τ, then $F[X_0+x(\tau),t_0+\tau]$ must appear on the right side of this formula under the integral sign.

2. We also examine the integral

$$\int_{C_{t,X}}\exp\left[\lambda\int_0^t p(\tau)x^2(\tau)d\tau\right]d_{w(t,X)}x=I, \quad (1.27)$$

where

$$A=(a_{jk}),\quad c^2=\sum_{j,k} a_{jk}b_jb_k,\quad (a_{jk})^{-1}=A^{-1},$$

similar to (1.18). By making use again of formulas (1.9) and (1.4) (in formula (1.4) it must now be considered that $x_n=X$, and that it is not necessary to carry out an integration over x_n), we are led to a finite dimensional integral differing from (1.11) only in the presence of the term $2(\Delta t)^{-1}Xx_{n-1}$, linearly dependent on x_{n-1}, in the exponential under the integral sign. We now make use of the general formula,

$$\int_{-\infty}^{\infty}\cdots\int_{-\infty}^{\infty}F\left(\sum_{j=1}^{n-1}b_jx_j\right)\exp\left(-\sum_{j,k=1}^{n-1}a_{jk}x_jx_k\right)dx_1\cdots dx_{n-1}$$

$$=\frac{\pi^{(n-2)/2}}{[c^2\det A]^{\frac12}}\int_{-\infty}^{\infty}F(u)\exp(-u^2/c^2)du \quad (1.28)$$

(this formula is most easily obtained from probability theoretic considerations[4]). With its aid we introduce a finite dimensional integral of interest to us in the form,

$$I_n=\frac{\exp(-X^2/\Delta t)}{\pi[c^2D_1{}^{(n-1)}]}\int_{-\infty}^{\infty}\exp[u-(u^2/c^2)]du$$

$$=\frac{\exp[-(X^2/\Delta t)+(c^2/4)]}{[\pi D_1{}^{(n-1)}]^{\frac12}},\quad c^2=\frac{4X^2}{\Delta t}\frac{D_1{}^{(n-2)}}{D_1{}^{(n-1)}},$$

where $D_1{}^{(n-1)}$ is the minor of $(n-1)$st order lying in the upper left hand corner of the determinant (1.13), the lower row of which is moreover multiplied by Δt, while $D_1{}^{(n-2)}$ is the analogously transformed upper left hand principal minor of $(n-2)$nd order of the same determinant (1.13). The subsequent limiting process is carried out analogously to the limiting process in formula (1.12); by taking into account that the exponent in the formula for I_n after the substitution of the value of c^2 transforms into

$$-\frac{X^2}{D_1{}^{(n-1)}}\frac{D_1{}^{(n-1)}-D_1{}^{(n-2)}}{\Delta t},$$

[4] After the appropriate normalization the integral (1.28) expresses the mean value of the function $F(u)$ where $u=\sum b_jx_j$ is a linear combination of random variables distributed according to a multidimensional normal law with a zero vector of mean values and a matrix of second moments equal to $\frac12(a_{jk})^{-1}$; from this, it develops that u is also normally distributed with a mean value 0 and a dispersion equal to $c^2/2$.

we easily obtain the final result,

$$I=[\pi D_1(0)]^{-\frac{1}{2}}\exp\left[-\frac{D(0)}{D_1(0)}X^2\right], \qquad (1.29)$$

where $D(0)$ has the same meaning as in formula (1.18), while $D_1(0)$ is the value at zero of the solution to the differential equation (1.16) satisfying the conditions,

$$D_1(t)=0,\quad D_1'(t)=-1. \qquad (1.30)$$

By integrating the left side of formula (1.29) over X from $-\infty$ to $+\infty$, in view of (1.21), we obtain (1.18) again.

In the special case $p(\tau)=1$, we have

$$D_1(\tau)=\frac{1}{(\lambda)^{\frac{1}{2}}}\sin(\lambda)^{\frac{1}{2}}(t-\tau)$$

from which

$$\int_{C_{t,X}}\exp\left[\lambda\int_0^t x^2(\tau)d\tau\right]d_{w(t,X)}x$$

$$=\left[\frac{(\lambda)^{\frac{1}{2}}}{\pi}\operatorname{cosec}t(\lambda)^{\frac{1}{2}}\right]^{\frac{1}{2}}\exp\{-(\lambda)^{\frac{1}{2}}\operatorname{ctg}t(\lambda)^{\frac{1}{2}}X^2\} \qquad (1.31)$$

(it is obvious that this formula is correct only for $\lambda^{\frac{1}{2}}<\pi/t$).

Integral over Wiener Measure of the Functional $\exp\{-\int_0^t V[x(\tau)]d\tau\}$

The direct evaluation of the integral over Wiener measure of the functional $F[x(\tau)]$ with the aid of formula (1.4) can be accomplished only in extremely special isolated cases. As a rule, an explicit formula for the finite dimensional integral on the right hand side of (1.4) will be absent, and it will not be possible to carry out the limiting process in this formula in an effective way.[5] For this reason, those results are very interesting which relate the evaluation of such integrals to more traditional mathematical problems. The most important of such results is the result of Kac [2], which is concerned with the connection of integrals over Wiener measure of functionals of the type

$$\exp\left\{-\int_0^t V[x(\tau)]d\tau\right\},$$

where $V(x)$ is a sufficiently "good" function (for example, continuous and bounded from below[6]) with the solution of a certain special parabolic equation.

We will examine firstly the integral of our functional over the "conditional" Wiener measure $d_{w(t,X)}x$:

$$\psi(X,t)=\int_{C_{t,X}}\exp\left\{-\int_0^t V[x(\tau)]d\tau\right\}d_{w(t,X)}x. \qquad (1.32)$$

Then, according to reference [2] the function $\psi(X,t)$ for $t>0$ will be the solution which approaches zero as $X\to\pm\infty$ of the following parabolic differential equation,

$$\frac{\partial^2\psi(X,t)}{\partial t}=\frac{1}{4}\frac{\partial^2\psi(X,t)}{\partial X^2}-V(X)\psi(X,t), \qquad (1.33)$$

satisfying the condition $\psi(X,0)=\delta(X)$, where $\delta(X)$ is Dirac's δ function. In view of (1.21), we also obtain from this a simple formula for the integral of

$$\exp\left\{-\int_0^t V[x(\tau)]d\tau\right\}$$

over ordinary Wiener measure,

$$\int_C\exp\left\{-\int_0^t V[x(\tau)]d\tau\right\}d_wx$$

$$=\frac{1}{N}\int_{-\infty}^{\infty}\cdots\int_{-\infty}^{\infty}\exp\left(-\int_0^t\{V[x(\tau)]+\dot{x}^2(\tau)\}d\tau\right)$$

$$\times\prod_0^t dx(\tau)=\int_{-\infty}^{\infty}\psi(X,t)dX. \qquad (1.34)$$

We note further that the solution $\psi(X,t;X_0)$ of Eq. (1.33) satisfying the condition $\psi(X,0)=\delta(X-X_0)$ may also be introduced,

$$\psi(X,t;X_0)$$

$$=\int_{C_{t;X-X_0}}\exp\left\{-\int^t V[X_0+x(\tau)]d\tau\right\}d_{w(t;X-X_0)}x, \qquad (1.35)$$

and, in general, the fundamental solution $\psi(X,t;X_0,t_0)$ of the differential Eq. (1.33) satisfying the condition $\psi(X,t_0)=\delta(X-X_0)$ for $t>t_0$ will be equal to

$$\psi(X,t;X_0,t_0)$$

$$=\int_{C_{t_0,X_0;t,X}}\exp\left\{-\int_{t_0}^t V[x(\tau)]d\tau\right\}d_{w(t_0,X_0;t,X)}x \qquad (1.36)$$

[cf. (1.23)]. A general solution to Eq. (1.33) for arbitrary initial values $\psi(X,t_0)=\psi_0(X)$ is also easily constructed from the function (1.36); such a solution is

$$\psi(X,t)=\int_{-\infty}^{\infty}\psi_0(X_0)\psi(X,t;X_0,t_0)dX_0. \qquad (1.37)$$

[5] It is possible, of course, to raise the question of the numerical integration of integrals over Wiener measure; the first results in this direction are contained in work of Cameron [10a]. However, we will not concern ourselves with this question here.

[6] We note that it is entirely possible to consider that the function $V(x)$ also depends explicitly on τ; in what follows we will write $V[x(\tau)]$, and not $V[x(\tau),\tau]$ only for simplification of the notation.

The simplest method of deriving the results indicated here apparently is the method proposed by Feynman in applying it to Schrödinger's equation; a rigorous proof of this method (in application to the case examined by us here) was shown by Blanc-Lapierre and Fortet [6] and independently by E. B. Dynkin [7a]. Moreover, underlying this method is the circumstance that in view of the independence of the increments of the process $x(\tau)$, the Brownian motion we have described, for non-intersecting time intervals, the function (1.36), which gives the mean value of the functional,

$$\exp\left\{-\int_{t_0}^{t} V[x(\tau)]d\tau\right\}$$

along all trajectories of this process starting at time t_0 at point X_0 and ending at time t at point X, will for $t_0<t_1<t$ satisfy the relation,

$$\psi(X,t;X_0,t_0) = \int_{-\infty}^{\infty} \psi(X_1,t_1;X_0,t_0)\psi(X,t;X_1,t_1)dX_1, \quad (1.38)$$

analogous to the well-known Smoluchowski-Kolmogorov relation for the transition probabilities of a random Markov process. From this with certain simple conditions of regularity placed on $V(x)$, it is possible to obtain Eq. (1.33) in a completely analogous fashion to Kolmogorov's derivation of the basic differential equations for the transition probabilities of Markov random processes (cf., for example, [26], §52); the term $V(X)\psi(X,t)$, which distinguishes (1.33) from the ordinary diffusion equation (the Fokker-Planck-Kolmogorov equation for the process of Brownian motion) arises here from the fact that the function $\psi(X_1, t+\Delta t; X,t)$ does not satisfy the normalization condition,

$$\int_{-\infty}^{\infty} \psi(t+\Delta t, X_1; t, X)dX_1 \neq 1$$

(but on the other hand in view of (1.36),

$$\lim_{\Delta t\to 0}\left[1-\int_{-\infty}^{\infty} \psi(X_1, t+\Delta t; X, t)dX_1\right] = V(x)\Bigg).$$

Another very general way of obtaining this and a series of results related to it is found in the work of E. B. Dynkin [7b].

It may be thought that the possibility of representing the solution to the differential Eq. (1.33) in the form of a functional integral, following from formulas (1.35) and (1.37), has a very general character. If we first replace the differential equation by a difference equation and solve it step by step starting from the initial values, then the value of the solution at the point (X,t) appears in the form of a sum extending over all preceding points of the net; this sum may be considered as the sum of the contributions to the solution from the individual broken curves connecting the region of initial values with the point (X,t) through a sequence of neighboring points of the net.

The final expression of the solution in the form of a functional integral can be considered as a similar sum, obtained after a limiting process, of contributions to the solution from all possible continuous paths connecting the region of initial values with the final point. It is natural to suppose that such an expression for the solution will be possible for a wide class of problems and may present a definite interest for the theory of differential equations.[7] It would be interesting to examine the question of a similar expression for the solution of the Cauchy problem for a number of differential equations differing from (1.33), for which this problem is well-posed (for example, for the solution of the equation $\partial u/\partial t = -\partial^4 u/\partial x^4$). In the following, in the example of Schrödinger's equation, we will see that investigations of even simple differential equations by this means can meet with very serious difficulties.

The derivation of Eq. (1.33), which has been indicated here, will not in fact be changed if instead of the integral of

$$\exp\left\{-\int_{0}^{t} V[x(\tau)]d\tau\right\}$$

over Wiener measure, i.e., of the mean value of this functional over the trajectories of the process $x(\tau)$, of the previously described Brownian motion, we examine the mean value of the same functional over the trajectories of an arbitrary continuous Markov process $x(\tau)$. In this case, and in the general case, we arrive at a parabolic equation of the second order differing from the differential equation for the transition probabilities of the corresponding process only in the term $-V(X)\psi(X,t)$ on the right side; it is significant that the equation obtained (as is apparent from its derivation) depends only on the first and second moments of the increments of the process for small time intervals, but does not depend on all the other statistical characteristics of $x(\tau)$. This circumstance makes it possible to utilize results relating to integrals of functionals over Wiener measure for the proof of a wide class of meaningful limit theorems for the sums of independent terms (see the remarks on the "arcsine law" at the end of example 1 below, and also the interesting work of U. V. Prohorov [27], which is devoted to a general analysis of such an approach to the proof of limit theorems). On the other hand, from here it is possible to obtain one more method of proof of theorems about the connection of the function (1.32) with the solution to the differential

[7] It is possible, moreover, that in a number of cases the measure itself will not exist, and there will be only a certain method of integration for a wide class of "sufficiently good" functionals, so that in this case instead of speaking of measure we must speak of "distributions" in functional spaces in the sense of Schwartz, i.e., of "generalized measure".

equation (1.33): by replacing the process $x(\tau)$ by a sum of discrete independent random values, the increments of this process for nonintersecting small time intervals, and arguing that for the limit relations only the first two moments of the corresponding quantities are important but not their distribution functions, one can assume that all of these quantities assume only two values $+\Delta$ and $-\Delta$ (each with probability one-half), after which the equation for $\psi(X,t;X_0,t_0)$ is easily obtained from a probability theoretic argument in terms of finite differences, going over in the limit into the sought for differential equation. This method of deriving Eq. (1.33) was actually made use of in the work of Kac [2a].

In the case that the function $V(X)$ does not depend explicitly on τ, the calculation of the solutions $\psi(X,t;X_0)=\psi(X,t;X_0,0)$ and

$$\psi(X,t;X_0,t_0)=\psi(X,t-t_0;X_0,0)=\psi(X,t-t_0;X_0)$$

of the parabolic equation (1.33) can be related to the solution of the ordinary differential equation,

$$\frac{1}{4}\frac{d^2\varphi(X)}{dX^2}-[V(X)+E]\varphi(X)=0. \qquad (1.39)$$

Actually, solving Eq. (1.33) with the aid of Laplace transforms, we find that the function,

$$\varphi_E(X)=\int_0^\infty \psi(X,t;X_0)e^{-Et}dt \qquad (1.40)$$

will satisfy the differential equation (1.39); the conditions that $\psi(X,t;X_0)$ approach zero as $X\to\pm\infty$ and that it reduce to $\delta(X-X_0)$ as $t\to 0$ go over into the following conditions on $\varphi_E(X)$:

$$\varphi_E(X)\to 0 \text{ as } X\to\pm\infty;$$

$$\left.\frac{d\varphi_E}{dX}\right|_{X=X_0+0}-\left.\frac{d\varphi_E}{dX}\right|_{X=X_0-0}=-4. \qquad (1.41)$$

(The function $\varphi(X)$ itself will be continuous at the point X_0.[8]) The conditions (1.41) uniquely specify the solution of Eq. (1.39) we need; after it is found, the function $\psi(X,t;X_0)$ is determined from the formulas for Laplace integral transforms.

[8] Making use of formula (1.40) and the conditions which $\psi(X,t;X_0)$ satisfies, it is easy to prove that

$$|\varphi(X_0+\epsilon)-\varphi(X_0-\epsilon)|\to 0$$

as $\epsilon\to 0$; after that from the equality,

$$\int_{X_0-\epsilon}^{X_0+\epsilon}\int_0^\infty\left[\frac{\partial\psi(X,t;X_0)}{\partial t}-\frac{1}{4}\frac{\partial^2\psi}{\partial X^2}+V(X)\psi\right]e^{-Et}dtdX=0$$

and the condition $\psi(X,t;X_0)\to\delta(X-X_0)$ as $t\to 0$, the relation

$$\lim_{\epsilon\to 0}\left[\frac{\partial\varphi_E(X_0+\epsilon)}{\partial X}-\frac{\partial\varphi_E(X_0-\epsilon)}{\partial X}\right]=-4$$

is derived.

A different method of reducing the problem of finding the function $\psi(X,t;X_0)$ to the solution of the differential Eq. (1.39) is connected with the use of the expansion of this function in a series of characteristic functions. Let us suppose for simplicity that Eq. (1.39) has a purely discrete spectrum of characteristic values $E_1, E_2, \cdots, E_n, \cdots$, to which correspond normalized characteristic functions $\varphi_1(X), \varphi_2(X), \cdots, \varphi_n(X), \cdots$. Then, the system of functions $\{\varphi_n(X)\}$ constitutes an orthonormal basis in L^2, while the system of functions $\{\varphi_n(X)\cdot\varphi_m(X_0)\}$ is an orthonormal basis in the analogous space of functions of the two variables X and X_0. By expanding $\psi(X,t;X_0)$ in a series on this basis and taking into account that $L_X\psi=\partial\psi/\partial t$, $L_{X_0}\psi=-\partial\psi/\partial t$, where $L_X=\frac{1}{4}d^2/dX^2-V(X)$, so that $L_X\varphi_n(X)=-E_n\varphi_n(X)$, $L_{X_0}\varphi_m(X_0)=-E_m\varphi_m(X_0)$, we easily obtain that the coefficient $a_{nm}(t)$ of $\varphi_n(X)\varphi_m(X_0)$ in this expansion will have the form $a_{nm}(t)=\delta_{nm}a_ne^{-E_nt}$.[9] Moreover, inasmuch as it must be that

$$\psi(X,0,X_0)=\delta(X-X_0)=\textstyle\sum_n \varphi_n(X)\varphi_n(X_0),$$

then $a_n\equiv 1$, i.e.,

$$\psi(X,t;X_0)=\sum_{n=1}^\infty e^{-E_nt}\varphi_n(X)\varphi_n(X_0). \qquad (1.42)$$

This formula permits us to determine $\psi(X,t;X_0)$ in terms of the known characteristic functions and characteristic values of the differential equation (1.39).

Examples (Kac [2a])

1. Let

$$V(X)=\frac{1+\text{sign}X}{2}=\begin{cases}1 & \text{for } X>0,\\ 0 & \text{for } X<0.\end{cases} \qquad (1.43)$$

Equation (1.39) here takes the form,

$$\left.\begin{aligned}&\frac{1}{4}\frac{d^2\varphi}{dX^2}-(1+E)\varphi=0, \quad X>0\\ &\frac{1}{4}\frac{d^2\varphi}{dX^2}-E\varphi=0, \quad X<0.\end{aligned}\right\} \qquad (1.44)$$

[9] With the aid of integrations by parts we easily obtain

$$\begin{aligned}\frac{\partial a_{nm}(t)}{\partial t}&=\int_{-\infty}^{\infty}\int_{-\infty}^{\infty}\frac{\partial\psi(X,t;X_0)}{\partial t}\varphi_n(X)\varphi_m(X_0)dXdX_0\\ &=\int_{-\infty}^{\infty}\int_{-\infty}^{\infty}L_X\psi(X,t;X_0)\varphi_n(X)\varphi_m(X_0)dXdX_0\\ &=\int_{-\infty}^{\infty}\int_{-\infty}^{\infty}\psi(X,t;X_0)L_X\varphi_n(X)\varphi_m(X_0)dXdX_0\\ &=-E_na_{nm}(t),\end{aligned}$$

and analogously (using the equality $-\partial\psi/\partial t=L_{X_0}\psi$)

$$\frac{\partial a_{nm}(t)}{\partial t}=-E_ma_{nm}(t).$$

From this follows the relation written down.

The solution to this equation, which goes to zero as $X \to \pm\infty$, will be:

$$\varphi(X)=\begin{cases} A\exp(-2(1+E)^{1/2}X) & \text{for } X>0 \\ B\exp(2(E)^{1/2}X) & \text{for } X<0. \end{cases} \tag{1.45}$$

The condition of continuity of $\varphi(X)$ at $X=0$ and the condition that $\varphi'(+0)-\varphi'(-0)=-4$ leads to the following values for the constants A and B:

$$A=B=2/[(E)^{1/2}+(1+E)^{1/2}]. \tag{1.46}$$

In virtue of (1.40), (1.32), and (1.34) we will have

$$\int_0^\infty \left\{\int_C \exp\left\{-\int_{\substack{x(\tau)>0\\ \tau\leqslant t}} d\tau\right\} d_w x\right\} e^{-Et}dt$$
$$=\frac{2}{(E)^{1/2}+(1+E)^{1/2}}\left\{\int_0^\infty \exp[-2(1+E)^{1/2}X]dX + \int_{-\infty}^0 \exp[2(E)^{1/2}X]dX\right\}$$
$$=\frac{1}{[E(1+E)]^{1/2}}=\sum_{k=0}^\infty \binom{-\frac{1}{2}}{k}\frac{1}{E^{k+1}} \quad (E>1). \tag{1.47}$$

From this, we immediately obtain

$$\int \exp\left\{-\int_{\substack{x(\tau)>0\\ \tau\leqslant t}} d\tau\right\} d_w x=\sum_{k=0}^\infty \frac{1}{k!}\binom{-\frac{1}{2}}{k} t^k$$
$$=\frac{2}{\pi}\int_0^{\pi/2}\exp(-t\cos^2\theta)d\theta. \tag{1.48}$$

But, inasmuch as

$$\int_C \exp\left\{-\int_{\substack{x(\tau)>0\\ \tau\leqslant t}} d\tau\right\} d_w x=\int_0^t e^{-t_1}dF(t_1), \tag{1.49}$$

where $F(t_1)$ is the probability that the point $x(\tau)$, undergoing Brownian motion along the x axis and found at the origin at time $t=0$, after a time t will be found on the right (positive) semi-axis during a time not less than t_1, we see that $F(t_1)$ must coincide with the distribution function of the magnitude of $t\cos^2\theta$, where θ has a uniform distribution in the interval $0\leqslant\theta\leqslant\pi/2$, i.e.,

$$F(t_1)=(2/\pi)^{1/2}\arcsin(t_1/t)^{1/2}. \tag{1.50}$$

In accordance with that stated on p. 54, this will also be the limiting (as $t\to\infty$) distribution law for the value of positive partial sums of any sequence of t mutually independent uniformly distributed random values with a zero mean value and finite dispersion ("the arcsine law"); see, for example [28], ch XII, §5).

2. Let $V(X)=CX^2$. In this case, Eq. (1.39) becomes

$$\tfrac{1}{4}\varphi''(X)-CX^2\varphi(X)=E\varphi(X); \tag{1.51}$$

its characteristic values and characteristic functions, as is well known (see the solution of Schrödinger's equation for an oscillator in any course on quantum mechanics), are equal to

$$E_n=(n+\tfrac{1}{2})(C)^{1/2} \quad (n=0,1,2\cdots) \tag{1.52}$$

and correspondingly,

$$\varphi_n(X)=\left(\frac{4C}{\pi^2}\right)^{\frac{1}{8}}\frac{1}{(2^n n!)^{1/2}}H_n[(4C)^{1/4}X]\exp[-(C)^{1/2}X^2], \tag{1.53}$$

where $H_n(z)$ are Hermite polynomials. In virtue of (1.42), we obtain from this

$$\psi(X,t;X_0)=\left(\frac{4C}{\pi^2}\right)^{\frac{1}{4}}\sum_{n=0}^\infty\frac{\exp[-(n+\frac{1}{2})(C)^{1/2}t]}{2^n n!}H_n[(4C)^{1/4}X]$$
$$\times H_n[(4C)^{1/4}X_0]\exp[-(C)^{1/2}(X_0^2+X^2)]. \tag{1.54}$$

By making use now of a well-known formula from the theory of Hermite polynomials (see, for example, [29], p. 104),

$$\sum_{n=0}^\infty\frac{\exp\{\frac{1}{2}(z^2+z_0^2)\}}{2^n n!}a^n H_n(z)H_n(z_0)$$
$$=\frac{1}{(1-a^2)^{1/2}}\exp\left\{\frac{z^2-z_0^2}{2}-\frac{(z-az_0)^2}{1-a^2}\right\} \tag{1.55}$$

and assuming here $z=(4C)^{1/4}X$, $z_0=(4C)^{1/4}X_0$, $a=e^{-(C)^{1/2}t}$ we will have

$$\psi(X,t;X_0)=(C/\pi^2)^{1/4}[\operatorname{cosech}(C)^{1/2}t]^{1/2}\exp\{-(C)^{1/2}\operatorname{cth}(C)^{1/2}tX^2$$
$$+2(C)^{1/2}\operatorname{cosech}(C)^{1/2}tXX_0-(C)^{1/2}\operatorname{cth}(C)^{1/2}tX_0^2\}. \tag{1.56}$$

In the particular case $X_0=0$, the last formula gives

$$\psi(X,t)=(C/\pi^2)^{1/4}[\operatorname{cosech}(C)^{1/2}t]^{1/2}$$
$$\times\exp\{-(C)^{1/2}\operatorname{cth}(C)^{1/2}tX^2\}; \tag{1.57}$$

for $C=-\lambda$ it goes over into formula (1.31) as it should. From (1.31), by integration over X, it is also possible to obtain formula (1.18).

Many Dimensional Generalization

If we now understand $x(\tau)$ to be a many dimensional random process $\{x_1(\tau), x_2(\tau), \cdots, x_N(\tau)\}$ describing the Brownian motion of a particle without inertia in an N-dimensional space, then all of our arguments and all of the formulas (1.32)–(1.42) will remain valid with the only difference being that the integration over Wiener measure now must be understood everywhere as an integration over an N-dimensional measure

$d_w x = d_w x_1 d_w x_2 \cdots d_w x_N$, the quantities X, X_0, etc. as N-dimensional vectors, while Eqs. (1.33) and (1.39) and the last of their conditions (1.41) change correspondingly into

$$\frac{\partial\psi(X,t)}{\partial t} = \tfrac{1}{4}\Delta\psi(X,t) - V(X)\psi(X,t), \qquad (1.33')$$

$$\tfrac{1}{4}\Delta\varphi(X) - [V(X)+E]\varphi(X) = 0, \qquad (1.39')$$

and

$$\lim_{\epsilon\to 0}\int_{|X-X_0|=\epsilon}\frac{\partial\varphi(x)}{\partial n}ds = -4, \qquad (1.41')$$

where Δ is the N-dimensional Laplace operator, $\partial\varphi/\partial n$ the derivative of the function φ along the normal to the sphere $|X-X_0|=\epsilon$, while ds is a surface element of this sphere (see reference [4], which is specially devoted to the N-dimensional case). If we wish to obtain the solution to Eq. (1.33′) for a finite region Ω (for zero boundary conditions on the boundary of Ω), then we only have to stipulate that $V=\infty$ outside Ω, i.e., in (1.33) and (1.34)–(1.36) integration is carried out only over those trajectories $x(\tau)$, which do not intersect the boundary of Ω (see [5]); choosing the set of trajectories along which the integration is carried out in a different way it is possible to obtain other boundary conditions. Later on we will again return to the question of how it is possible to utilize these results for obtaining certain mathematical derivations concerning the behavior of the differential equation (1.33′); now, however, we will not delay ourselves on this point, but will proceed immediately to an examination of analogous methods in quantum mechanics.

2. INTEGRATION IN A FUNCTIONAL SPACE IN PROBLEMS OF QUANTUM MECHANICS

Case of Quantum Mechanics of a System with a Finite Number of Degrees of Freedom

In an investigation of problems of quantum mechanics, it is natural to begin with the simplest case of the nonrelativistic mechanics of a system with a finite number of degrees of freedom; we will proceed in just this way. As is well known, the state of the system is completely given by a wave function ψ, a complex function of the generalized coordinates of the system; the integral of the square of the modulus of this function over some region gives the probability of finding the system in this region. The change of the wave function with time is determined by Schrödinger's linear differential equation, which from the value of ψ at time $t=t_0$ uniquely allows the wave function to be determined at all subsequent times; for the case of the motion of a particle of mass m in a force field with potential $V(X)$ the equation has the form,

$$-\hbar i\partial\psi(X,t)/(\partial t) = \hbar^2/(2m)\Delta\psi - V(X)\psi, \qquad (2.1)$$

i.e., in fact it differs from Eq. (1.33′) only in the factor $-i$ before the time derivative (the constants $\hbar$ and $\hbar^2/2m$ clearly can be made equal to 1 and $\frac{1}{4}$ by choosing appropriate units for measuring mass, length, and time). The value of the wave function $\psi(X,t)$ at time $t>t_0$ can be determined by the value $\psi(X,t_0)$ with the aid of the formula,

$$\psi(X,t) = \int_{-\infty}^{\infty}\psi(X_0,t_0)\psi(X,t;X_0,t_0)dX_0, \qquad (2.2)$$

analogous to (1.37) (by $\psi(X,t;X_0,t_0)$ we denote here the fundamental solution of Schrödinger's equation (2.1), i.e., the value of the wave function under the condition that at time t_0 the system was found at the point X_0.[10] The basic problem of quantum mechanics consists of finding the function $\psi(X,t)$ or $\psi(X,t;X_0,t_0)$.

By following from the fact that $\psi(X,t)$ is a solution of the differential equation (2.1), it can be supposed that a representation of this function in the form of some "integral over a functional space," which gives the contribution of each separate trajectory in the final solution, is also possible here. This supposition was adopted by Feynman [1a] in the nature of a postulate. After this, it is already easy to understand how one must choose the numerical contribution from the separate trajectories in order to obtain a result equivalent to that following from Schrödinger's equation: since Eq. (1.33′) changes into (2.1) by a change of the time t into $it/\hbar$ and the coordinate X into $(m)^{\frac{1}{2}}X/(2)^{\frac{1}{2}}\hbar$, then we will obtain the correct formula for the wave function $\psi(X,t;X_0,t_0)$ by carrying out the same substitutions in formula (1.36),

$$\psi(X,t;X_0,t_0) = \frac{1}{N}\int\cdots\int\exp\left(\frac{i}{\hbar}\int_{t_0}^{t}\left\{\frac{m\dot{x}^2(\tau)}{2} - V[x(\tau)]\right\}d\tau\right)\times\prod_{t_0}^{t}dx(\tau). \qquad (2.3)$$

The integral over $dx(\tau)$ extends over all continuous paths connecting the points (X_0,t_0) and (X,t); N is a certain normalization constant, determined, for ex-

[10] We note that according to (2.2) the function $\psi(X,t)$ transforms analogously to the probability distribution density for a Markov process in which $\psi(X,t;X_0,t_0)$ plays the role of a "transition probability." By examining the state of the physical system at successive moments of time $t_0, t_1, t_2, \cdots$, we will have a chain analogous to a Markov chain with the only difference being that the role of the probability density in this case is played by a complex function $\psi(X)$ determined at time t_n from the known states of the system at preceding instants of time $t_0, t_1, \cdots, t_{n-1}$ by only one value of ψ, at the last of these instants t_{n-1}; the square of the modulus of this complex function will be equal to the probability density for locating the system. Discussions of the physical meaning of the differences indicated here between "quantum chains" and Markov chains basic for problems of classical statistical physics can be found in references [1a], [25].

ample, by the "unitarity condition"

$$\int_{-\infty}^{\infty} |\psi(X,t)|^2 dX = 1.$$

The meaning of the "symbolic expression" (2.3) is the same as of (1.4′): it is necessary to break up the interval $[t_0,t]$ into n equal parts by the points $t_0, t_1, \cdots, t_n=t$, to consider the n fold integral as analogous to that which figures in the right side of (1.4) (with the addition in appropriate places of the factors i, $\hbar$, and m), and to pass then to the limit as $n\to\infty$. We note immediately that in an exact accomplishment of this program, one will come up against definite difficulties: the integrals of rapidly oscillating functions which arise, generally speaking will not be convergent, and in order to assign to them a definite meaning it will be necessary to resort to certain devices, for example, to consider $\hbar$ as having a small purely imaginary addition, $-i\delta$, in the lower half-plane, and only after carrying out all integrations to pass to the limit as $\delta\to 0$ (compare the footnote on p. 184 of the Russian translation of reference [1a]).

The convention indicated here for the evaluation of oscillating integrals (2.3) can be illustrated in the following manner. On replacing $\hbar$ by a complex quantity $\hbar-i\delta$, we arrive at the following "complex measure" in functional space: to the quasi-interval given by the conditions $a_i<x(t_i)<b_i$ $(i=1, 2, \cdots, n)$ is related a quantity differing from (1.3) only by the presence in the exponents of the exponential under the integral sign of a constant complex multiplier $2m^{-1}(\delta+i\hbar)$ in the right half-plane (and an appropriate change in the normalization constant in front of the integral). It is natural that such a complex measure for arbitrary $m>0$, $\delta>0$, and R will be just as "good" as Wiener measure, i.e., it will have just as precise a meaning as measure in the space C of continuous functions, and it will allow integration over it of a wide class of functionals, including all continuous and bounded functionals.[11] Moreover, it will be possible to prove that a functional integral over this new measure, defined by the symbolic expression (2.3) (with complex $\hbar$), will represent a fundamental solution of an equation of the type of (1.33), namely Eq. (2.1) with a complex $\hbar$. If now we let the imaginary part of $\hbar$ tend to zero, then the corresponding fundamental solutions will tend to the fundamental solution of the original Eq. (2.1). In this way, although from a parabolic equation with a purely imaginary "diffusion coefficient" (as, for example, Schrödinger's equation) it is hardly possible to construct any arbitrary reasonable measure in a functional space, however, the solution of the corresponding equation can be obtained as the limit of a sequence of functional integrals corresponding to complex diffusion coefficients, as the imaginary part of this coefficient tends to zero.[12] This limit here will play the role of a representation of the solution in the form of a sum of "contributions" from separate trajectories, of which we have spoken earlier, and will "symbolically" be expressed by formula (2.3).

We turn now from the question of a precise mathematical meaning for the expression (2.3) and will argue further purely formally. We note first of all that the expression under the integral sign in the exponent in (2.3) completely coincides with Lagrange's function,

$$L(\dot{x},x)=(m\dot{x}^2/2)-V(x)$$
$$=\text{kinetic energy}-\text{potential energy} \quad (2.4)$$

for a particle of mass m in a force field with a potential $V(x)$; therefore, the integral from t_0 to t_1 is equal to the classical action $S(t_0,t_1)$ (to the integral of the Lagrangeian) along the trajectory $x(\tau)$. From this we see that formula (2.3) can be rewritten in a still more concise form,

$$\psi(X,t;X_0,t_0)\sim\int_{-\infty}^{\infty}\cdots\int \exp\left(\frac{i}{\hbar}S[x(\tau)]\right)\prod_{t_0}^{t} dx(\tau). \quad (2.5)$$

In other words, it should be considered that all trajectories give the same contribution to $\psi(X,t;X_0,t_0)$ in absolute magnitude and differ only in the argument ("phase") which is equal to the classical action along the corresponding trajectory taken with a factor $i/\hbar$. After this, just as in the case of integrals over Wiener measure, it may be proved that $\psi(X,t,X_0,t_0)$ is in fact a fundamental solution of Schrödinger's equation, so that the determination of this function indicated here is equivalent to the usual construction of quantum mechanics (see reference [1a]).

Until now, we have spoken only of Schrödinger's equation (2.1) for the simplest case of the motion of a particle in a potential field. However, the formulated method of writing the wave function in the form of an "integral over trajectories" requires only a knowledge of the classical action function for the problem, and can easily be extended to the case of motion of any system with a finite number of degrees of freedom, for which the Lagrangeian is known. Thus, for example, in the case of the motion of a charged particle (with a charge e) in a constant electromagnetic field in a three dimensional space, which is characterized by a scalar potential $\varphi(X)=1/(e)V(X)$, $X=(X_1,X_2,X_3)$, and a vector potential $\mathbf{A}(X)=(c/e)\mathbf{a}(X)$ (where c is the speed of light), Schrödinger's equation also contains terms with a first

[11] A strict proof of this fact does not differ from the corresponding proof for the case of Wiener measure.

[12] It is probable, in general, that for a wide class of functionals the integrals over a complex measure related to a complex diffusion coefficient $D=D_1+iD_2$, $D_1>0$ will tend to definite limits as $D_1\to 0$. In this way, even for $D_1=0$ we will have a certain method for the integration of functionals but the corresponding "integrals" will no longer be related to a completely additive measure.

derivative with respect to the coordinates,

$$-i\hbar\frac{\partial\psi}{\partial t}=\frac{\hbar^2}{2m}\Delta\psi-\frac{i\hbar}{m}\left(a_1\frac{\partial\psi}{\partial X_1}+a_2\frac{\partial\psi}{\partial X_2}+a_3\frac{\partial\psi}{\partial X_3}\right)$$
$$-\left(\frac{i\hbar}{2m}\,\mathrm{div}\,a+\frac{1}{2m}a^2+V\right)\psi. \quad (2.6)$$

The expression for the solution of this equation in the form of a functional integral obtained with the aid of the expression, known from electrodynamics, for the corresponding action function, has the form

$$\psi(X,t;\,X_0,t_0)$$
$$=\frac{1}{N}\int_{-\infty}^{\infty}\cdots\int \exp\left[\frac{i}{\hbar}\int_{t_0}^{t}\left\{\frac{m\dot{x}^2}{2}-V(x)+a\dot{x}\right\}d\tau\right]$$
$$\times\prod_{t_0}^{t}dx(\tau), \quad (2.7)$$

where $dx(\tau)=dx_1(\tau)dx_2(\tau)dx_3(\tau)$. It is curious to note that passing here anew from Schrödinger's equation to an ordinary parabolic equation with the aid of a change of variables, the reverse of the one which was used above for the passage from (1.33′) to (2.1), we find that the solution of the equation,

$$\frac{\partial\psi}{\partial t}=\tfrac{1}{4}\Delta\psi-a_1\frac{\partial\psi}{\partial X_1}-a_2\frac{\partial\psi}{\partial X_2}-a_3\frac{\partial\psi}{\partial X_3}-V\psi, \quad (2.8)$$

satisfying the condition $\psi(X,0)=\delta(X)$, is given by the functional integral

$$\psi(X,t)=\int_{C_{t,x}}\exp\left(-\int_0^t\{V([x(\tau)]-\tfrac{1}{2}\,\mathrm{div}\,a[x(\tau)]\right.$$
$$\left.+\tfrac{1}{2}a^2[x(\tau)])d\tau+a\,dx(\tau)\}\right)d_{w(t,x)}x, \quad (2.9)$$

where $d_{w(t,x)}x$ is an integration over "conditional" Wiener measure in the space of continuous vector functions $x(\tau)$ in three dimensional space. The last result apparently has not been encountered in the literature; it may be of interest in investigations of a series of mathematical questions concerning the solution of Eq. (2.8).[13]

[13] An analogous result for a one dimensional parabolic equation containing a term with a first derivative with respect to the coordinate can be obtained from (1.32) with the aid of the simple substitution $\psi(X,t)=e^{-a(X)}\varphi(X,t)$. This fact is closely related with that circumstance that in electrodynamics the vector potential having a single nonvanishing component A_1, depending only on the coordinate x_1, does not have a physical meaning and can always be eliminated with the aid of a "gradient transformation."

Case of the Quantum Field Theory[14]

We pass now to an investigation of the quantum mechanics of systems with an infinite number of degrees of freedom, quantum field theory. The most important example of such a theory is quantum electrodynamics, the theory of the interaction one with another of an electromagnetic and an electron field; in this case, however, the original classical equations (the systems of Maxwell and Dirac equations) are comparatively cumbersome and cannot be considered as common knowledge among mathematicians. Inasmuch as all investigations for arbitrary fields (in any event if the case of half-integer spin is excluded) are carried out completely analogously, then in the beginning we restrict ourselves to an investigation of the simplest model of quantum field theory, which retains all of the most characteristic features of such theories, in particular the presence of an infinite number of degrees of freedom and nonlinearity. A quantum theory of a field $u(x)$ in three-dimensional space $x=(x_1,x_2,x_3)$ whose potential energy is given by the formula,

$$V[u(x)]=\int\int\int_{-\infty}^{\infty}[(\mathrm{grad}\,u)^2+\kappa^2u^2+\lambda u^4]dx \quad (2.10)$$

is obtained with such a model. In the case of a one-dimensional space x the analogous field $u(x)$ with potential energy

$$V[u(x)]=\int_{-\infty}^{\infty}\left[\left(\frac{\partial u}{\partial x}\right)^2+\kappa^2u^2+\lambda u^4\right]dx \quad (2.10')$$

admits of a simple mechanical interpretation: $u(x)$ can be regarded as the displacement from equilibrium of an elastic string in tension, at every point of which a force is applied, which depends nonlinearly on displacement (with a density of force equal to $\kappa^2u+2\lambda u^3$). In the three dimensional case for descriptiveness, we may analogously speak of the oscillations of a three dimensional elastic volume with an applied nonlinear force. The corresponding mechanical "equation of vibration" has the form

$$\frac{\partial^2 u}{\partial t^2}=\Delta u+\kappa^2u+2\lambda u^3; \quad (2.11)$$

this will be the "classical" (not quantum) equation of the problem to be investigated. The model of a field having a potential energy given by (2.10) is very often

[14] The construction of a quantum field theory by Feynman's method [1a] presented here was described in a report of I. M. Gel'fand and R. A. Minlos at an All Union Congress on Quantum Electrodynamics and the Theory of Elementary Particles in Moscow, in March of 1955. In very recent times an approach to quantum field theory similar to this one is reported also in references [21] and [22]. Another approach to the introduction of functional integrals into the quantum theory of fields is contained in the works of N. N. Bogolyubov [15], E. C. Fradkin [16] and Matthews and Salem [17].

made use of in contemporary physical literature in the nature of a very simple model very well suited for an initial investigation of new methods and for the clarification of questions of principle in the general theory (it is commonly called "Thirring's model" after the name of the physicist who first studied this model).

We turn now to the formulation of the quantum theory of this field. Instead of a wave function $\psi(u_1,u_2,\cdots,u_n)=\psi(u)$ (for the case of a system with n degrees of freedom), we will have here a "wave functional" $\psi[u(x)]$, depending on the function $u(x)$.[15] Further relations are likewise completely analogous to the relations of quantum mechanics with a finite number of degrees of freedom, and we emphasize this by writing down the relations for ordinary quantum mechanics and for the quantum mechanics of our system in parallel (for simplicity we assume here a system of units in which $\hbar=1$ and we consider the "masses" corresponding to all degrees of freedom to be equal to 2):

Laplacian Operator

$$\Delta\psi=\sum_{k=1}^{n}\frac{\partial^2\psi(u)}{\partial u_k^2}, \qquad (2.12)$$

$$\Delta\psi=\int_{-\infty}^{\infty}\frac{\delta^2\psi[u(x)]}{\delta[u(x)]^2}dx, \qquad (2.12')$$

where in (2.12′) the expression under the integral sign represents the second variational derivative of the functional $\psi[u(x)]$.

Wave Equation (Schrödinger's Equation)

$$i\frac{\partial\psi}{\partial t}=-\tfrac{1}{4}\Delta\psi+V(u)\psi, \qquad (2.13)$$

$$i\frac{\partial\psi}{\partial t}=-\tfrac{1}{4}\Delta\psi+V[u(x)]\psi, \qquad (2.13')$$

where in (2.13′) $\Delta\psi$ is given by (2.12′) while $V[u(x)]$ is given by (2.10).

Representation of ψ in the Form of a Functional Integral

$$\psi(u)=\int\exp\left(i\int_{t_0}^{t}\left\{\sum_{k=1}^{n}\dot{u}_k^2(\tau)-V[u(\tau)]\right\}d\tau\right)$$
$$\times\prod_{(k,\tau)=(1,t_0)}^{(n,t)}du_k(\tau), \qquad (2.14)$$

$$\psi[u(x)]=\int\exp\left(i\int_{t_0}^{t}\int\int\int_{-\infty}^{\infty}\left\{\left(\frac{\partial u}{\partial t}\right)^2\right.\right.$$
$$\left.\left.-(\mathrm{grad}u)^2-\kappa^2u^2+\lambda u^4\right\}d\tau dx\right)$$
$$\times\prod_{(x,\tau)=(-\infty,t_0)}^{(+\infty,t)}du(x,\tau). \qquad (2.14')$$

We note that the expression in the exponent under the integral sign in (2.14′),

$$S[u]=\int\int\int\int(\partial u/\partial t)^2dxd\tau-\int V[u(x,\tau)]d\tau, \qquad (2.15)$$

is none other than the "classical action" for our field which can be written in the form

$$S[u]=\int\int\int\int Ldxd\tau, \qquad (2.16)$$

where

$$L=(\partial u/\partial t)^2-(\mathrm{grad}u)^2-\kappa^2u^2-\lambda u^4 \qquad (2.17)$$

is the so called Lagrangeian function for the field which is equal (making use of a mechanical interpretation) to the density of kinetic energy of the field minus the density of potential energy. The "classical" equations of the field (2.11) arise from the variational principle $\delta S[u]=0$ in precisely the same way as in the case of a system with a finite number of degrees of freedom.

The value of the wave functional $\psi[u(x)]$ completely determines all physical quantities connected with the quantum field $u(x)$. In particular, the so called matrix elements of the S matrix which are of the greatest interest to physicists and which give the transition probabilities from one state into another appear as variational derivatives of the functional $\psi[u(x)]$ (see, for example, [14]).

An analogous expression for the wave functional is obtained for an arbitrary field with integer spin. In particular, if we have a certain system of coupled scalar, vector, or tensor fields $u(x)$, $v(x)$, $\cdots$, $w(x)$ in three dimensional space with a given classical Lagrangeian $L[u,v,\cdots,w]$, then the corresponding wave functional will be given by the formula,

$$\psi[u,v,\cdots,w]=\int\exp\left\{i\int\int\int\int L[u,v,\cdots,w]dxd\tau\right\}$$
$$\times\prod_{x,\tau}du(x,\tau)dv(x,\tau)\cdots dw(x,\tau). \qquad (2.18)$$

In the case of the so called half-integer spin (i.e., in the presence of spinor fields), the matter is considerably more complicated. The additional difficulties encountered here were overcome in the works of N. N. Bogolyubov [15], E. C. Fradkin [16], Matthews and

[15] It would be more precise to write $\psi(u,t)$ and $\psi[u(x),t]$ inasmuch as both the wave function and the wave functional depend explicitly on t.

Salam [17], and in a work especially devoted to this question by I. M. Khalatnikov [23a].

3. CHANGE OF VARIABLES IN FUNCTIONAL INTEGRALS

Certain Mathematical Applications

Change of Variables in Functional Integrals

In the preceding part of this report, it was shown that just as the solution to the ordinary parabolic equation (1.33′) can be represented in the form of a functional integral, so also can the solution to Schrödinger's equation (2.1); it was also pointed out that this circumstance to all appearance has a rather general character. Nevertheless, the usefulness of this notation for the theory of equations was by no means asserted: because we have written the solution in the form of a special symbol, the meaning of which is not very simply presented, it appears that there is little more which can be extracted.[16] In what follows, we will show in several examples that in fact the situation is not quite like this; it appears to us, however, that all of the usefulness which is gained by the introduction of functional integrals can really be appraised only later on when the algorithmic rules of operating with functional integrals will be sufficiently worked out and a sufficient experience in utilizing these rules will be accumulated.[17] At the present time, the technique of operating with functional integrals is only beginning to be developed; nevertheless, a series of interesting results in this direction has already been obtained in certain cases by mathematicians (by Cameron, Martin, and their collaborators) and independently by physicists (by the latter usually without strong proofs). It is on these results that we now pause.

It is natural to begin with the question of the rules for transforming functional integrals by a change of the "variable of integration" $x(\tau)$; this question is the only one that we will examine here. For definiteness, we will examine only integrals over Wiener measure $d_w x$; at the same time, it will be convenient for us to utilize a "symbolic" notation for this measure, which is indicated in (1.4′). We begin with the simplest question of the transformation of integrals over Wiener measure by means of a *parallel translation* in the functional space C, i.e., under the following transformation of functions in C

$$x(\tau)\to y(\tau)=x(\tau)+x_0(\tau) \qquad (3.1)$$

[$x_0(\tau)$ is a fixed function in C]. Formally carrying out the substitution (3.1) into the right side of the formula (1.4′), we obtain

$$\frac{1}{N}\int\cdots\int F[x(\tau)]\exp\left\{-\int_0^t \dot{x}^2(\tau)d\tau\right\}\prod_0^t dx(\tau)$$

$$=\frac{1}{N}\int\cdots\int F[x+x_0]\exp\left\{-\int_0^t [\dot{x}^2(\tau)+2\dot{x}(\tau)\dot{x}_0(\tau)+\dot{x}_0^2(\tau)]\right\}\prod_0^t dx(\tau). \qquad (3.2)$$

On removing

$$\exp\left\{-\int_0^t \dot{x}_0^2(\tau)d\tau\right\}$$

outside the integral sign on the right-hand side of this formula, replacing $\dot{x}(\tau)d\tau$ by $dx(\tau)$, and transforming back into the usual notation for an integral over Wiener measure, we have

$$\int_S F[x(\tau)]d_w x =\exp\left[-\int_0^t \dot{x}_0^2(\tau)d\tau\right]\int_{TS} F[x+x_0] \times\exp\left[-2\int_0^t x_0(\tau)dx(\tau)\right]d_w x. \qquad (3.3)$$

[For generality, we assumed that the integral on the left-hand side is extended only over a subset S of C; then on the right-hand side we will have an integral over the set TS, which transforms into S under the transformation (3.1).] Formula (3.3) defines the transformation of Wiener integrals under a parallel translation; it is correct when the right-hand side of (3.3) has a meaning, i.e., when $x_0(\tau)$ is a differentiable function on C, having a derivative $\dot{x}_0(\tau)$ with a bounded variation. (If these conditions are not satisfied, then the transformation (3.1), generally speaking, will transform sets which are measurable over Wiener measure into nonmeasurable sets; see [10c].) The proof of formula (3.3) presented here at first glance appears weak; it is not difficult to realize, however, that a strong proof is obtained only by a detailed deciphering of the "symbolic transformations" contained in (3.2) and (3.3) (see reference [9a]).

For the next example, we examine the transformation of Wiener integrals under a general *linear transformation* in the space C,

$$x(\tau)\to y(\tau)=x(\tau)+\int_0^t K(\tau,\sigma)x(\sigma)d\sigma\equiv x+Ax. \qquad (3.4)$$

[16] It is rather clear that on the other hand the connection of functional integrals with differential equations can be useful for the study of functional integrals themselves (see p. 53 and the examples on p. 55).

[17] In certain respects, the passage from the task of solving partial differential equations with the aid of difference schemes to the task of obtaining them in the form of functional integrals can be compared to the passage from finite sums to integrals in finding areas, the solutions to an ordinary differential equation $dy/dx=f(x)$. It is natural, however, that in order that this passage be utilized properly it is necessary first of all to learn to handle the resulting expressions.

By reasoning just as in the derivation of (3.3), it is necessary in this case to keep in mind that under a linear transformation of the "coordinates" $x(\tau)$ according to formula (3.4), the "volume element" $\prod_0^t dx(\tau)$ is multiplied by the "Jacobian" for the transformation; the role of such a Jacobian, which is obtained by a limiting process from the Jacobian of a finite dimensional linear transformation of coordinates $x(t_1), x(t_2), \cdots, x(t_n)$, will be played, as is well known, by the Fredholm determinant D of the kernel $K(t,s)$,

$$D=1+\sum_{n=1}^{\infty}\frac{1}{n!}\int_0^t\cdots\times\int_0^t\begin{vmatrix}K(s_1,s_1), & \cdots, & K(s_1,s_n)\\ \cdot & \cdot & \cdot\\ K(s_n,s_1), & \cdots, & K(s_n,s_n)\end{vmatrix}ds_1\cdots ds_n. \quad (3.5)$$

(In fact, it is in this very connection that the Fredholm determinant appears in the theory of integral equations.) By taking this circumstance into account, we have

$$\int F[x(\tau)]\exp\left[-\int_0^t \dot{x}^2(\tau)d\tau\right]\prod_0^t dx(\tau)$$

$$=|D|\int F[x+Ax]\exp\left\{-\int_0^t [\dot{x}(\tau)+A\dot{x}(\tau)]^2 d\tau\right\}\prod_0^t dx(\tau)$$

$$=|D|\int F\left[x(\tau)+\int_0^t K(\tau,\sigma)x(\sigma)d\sigma\right]\exp\left\{-\int_0^t\left[\frac{d}{d\tau}\int_0^t K(\tau,\sigma)x(\sigma)d\sigma\right]^2 d\tau\right.$$

$$\left.-2\int_0^t\left[\frac{d}{d\tau}\int_0^t K(\tau,\sigma)x(\sigma)d\sigma\right]dx(\tau)-\int_0^t \dot{x}^2(\tau)d\tau\right\}\prod_0^t dx(\tau), \quad (3.6)$$

i.e.,

$$\int_S F[x(\tau)]d_w\tau=|D|\int_{TS} F[x+Ax]\exp\left\{-\int_0^t [A\dot{x}]^2 dx-2\int_0^t A\dot{x}(\tau)dx(\tau)\right\}d_w x. \quad (3.7)$$

[The symbol TS has a meaning here analogous to its meaning in (3.3).] A strong proof of formula (3.7) [under reasonable restrictions on the kernel $K(t,s)$] is contained in reference [9b]; it is rather long, but not complicated in principle [in fact it consists of proving the validity of the transformation (3.6) in formulas which occur prior to passing to the limit $n\to\infty$]. In applications, it is essential that the result which is obtained also remains valid in the case of a kernel $K(t,s)$, which undergoes a discontinuity on the diagonal $t=s$ (for example in the case of a Volterra kernel); for this, it is only necessary during the construction of Fredholm's determinant to equate the value of the kernel on the diagonal to half the sum of its right and left hand limiting values.

We point out further that formula (3.7) is found to be correct also for a wide class of *nonlinear transformations* of the form,

$$x(\tau)\to y(\tau)=x(\tau)+A[x(\tau);\tau], \quad (3.8)$$

where, together with certain other restrictions of the type of "conditions of regularity," we also need to require that the transformation $A[x;\tau]$ has a nondegenerate "linear part," i.e., that $A[x;\tau]$ has a "functional derivative" (variation),

$$\delta A=\frac{\partial}{\partial h}A[x+hy;\tau]_{h=0}, \quad (3.9)$$

which can be written in the form,

$$\delta A=\int_0^t K[x;\tau,\sigma]y(\sigma)d\sigma. \quad (3.10)$$

The role of the "Jacobian" in this case is played by the Jacobian of the "linear part" of the transformation, i.e., the Fredholm determinant of the kernel $K[x;\tau,\sigma]$, which is clearly a functional of $x=x(\tau)$. Therefore, in the corresponding formula (3.7) it is no longer possible to remove $D=D[x]$ outside the integral sign. In all other respects, the outward form of formula (3.7) even for nonlinear A, as a rule will remain unchanged; in connection with this see references [9d] and [11], which are devoted to the derivation of this formula for different restrictions on $A[x;\tau]$ (and also the applications at the end of reference [20a]).

In a number of problems, the utilization of only the simplest rules of transformation of functional integrals which have been presented here (which, besides, are the only known rules at the present time) allows a great simplification and formalization of all derivations.

Example. We examine the problem of evaluating the functional integral

$$J=\int_C \exp\left[\lambda\int_0^t p(\tau)x^2(\tau)d\tau\right]d_w x \quad (p(\tau)\geqslant 0). \quad (3.11)$$

We have already studied this problem previously and saw that a direct evaluation of this integral with the aid of formula (1.4) leads to cumbersome investigations, particularly if it is desired to make such an evaluation sufficiently rigorous. We will show now that making use of formula (3.7) enables us to obtain the result (1.18) very quickly with the aid of purely formal transformations. Actually, we examine the following linear transformation of continuous functions on the interval $0\leqslant\tau\leqslant t$,

$$x(\tau)\rightarrow y(\tau)-\int_0^\tau\frac{D'(\sigma)}{D(\sigma)}x(\sigma)d\sigma\quad(0\leqslant\tau\leqslant t),\quad(3.12)$$

where $D(\sigma)$ is the same as in formulas (1.16) and (1.17). [We assume that λ is smaller than the smallest of the characteristic values of (1.16) subject to the boundary conditions $D(0)=0$, $D'(t)=0$; in such a case, $D(\sigma)$ will not be identically zero on the interval $0\leqslant\tau\leqslant t$.] Fredholm's determinant here will be equal to

$$D=\exp\left[\tfrac{1}{2}\int_0^t-\frac{D'(\sigma)}{D(\sigma)}d\sigma\right]$$
$$=\exp\{-\tfrac{1}{2}[\log D(t)-\log D(0)]\}=[D(0)]^{\frac{1}{2}}\quad(3.13)$$

(because for Volterra's kernel, Fredholm's determinant is equal to

$$\exp\left\{\tfrac{1}{2}\int_0^t K(\sigma,\sigma)d\sigma\right\}\Bigg),$$

and

$$-\int_0^t\left[\frac{d}{d\tau}\int_0^\tau\frac{D'(\sigma)}{D(\sigma)}x(\sigma)d\sigma\right]^2d\tau$$
$$-2\int_0^t\left[\frac{d}{d\tau}\int_0^\tau\frac{D'(\sigma)}{D(\sigma)}x(\sigma)d\sigma\right]dx(\tau)$$
$$=-\int_0^t\left[\frac{D'(\tau)}{D(\tau)}x(\tau)\right]^2d\tau+\int_0^t\frac{D'(\tau)}{D(\tau)}d[x^2(\tau)]$$
$$=-\int^t\frac{D'^2(\tau)}{D^2(\tau)}x^2(\tau)d\tau-\int_0^t\left[\frac{D'(\tau)}{D(\tau)}\right]'x^2(\tau)d\tau$$
$$=-\int_0^t\frac{D''(\tau)}{D(\tau)}x^2(\tau)d\tau=\lambda\int_0^t p(\tau)x^2(\tau)d\tau\quad(3.14)$$

[in view of (1.16)]. Inasmuch as it is very easy to demonstrate that the transformation (3.12) reciprocally and single valuedly reflects the space C of continuous functions on itself (the reverse transformation here has the form

$$y(\tau)\rightarrow x(\tau)=y(\tau)+D(\tau)\int_0^t D'(\sigma)/[D^2(\sigma)]y(\sigma)d\sigma$$

and is also continuous), then (1.18) immediately follows from formula (3.7) by substituting into this formula $F[x(\tau)]=1$ and $S=C$,

$$1=\int_C d_w x=|D|\int_C\exp\left\{-\int_0^t\left[\frac{d}{d\tau}Ax(\tau)\right]^2d\tau\right.$$
$$\left.-2\int_0^t\left[\frac{d}{d\tau}Ax(\tau)\right]dx(\tau)\right\}d_w x.\quad(3.15)$$

It was just in this manner that formula (1.18) was first derived in reference [9c].

In an analogous way, formula (3.3) can also be utilized for the straightforward evaluation of certain special functional integrals (see for example reference [9a]).

We note that similar formal transformations can also be utilized for proving a theorem on the connection of functional integrals of

$$\exp\left\{-\int_0^t V[x(\tau)]d\tau\right\}$$

with the solution of the partial differential Eq. (1.33), of which we have spoken in the preceding section. (See in this connection the work of Cameron, reference [10b], where a representation for the solution to Cauchy's problem for Eq. (1.33) in the form of a functional integral is obtained in this very way by the use of a certain special nonlinear transformation $A[x;s]$.)

Use of Functional Integrals for Obtaining Asymptotics of Characteristic Functions and Characteristic Values of an Elliptic Differential Equation

To end the present section, we present one simple example which shows that in certain respects the use of functional integrals can be very helpful for the study of differential equations. In particular, following the work of Ray [5], we examine the question of obtaining with the aid of functional integrals a series of results which concern the asymptotic behavior of the characteristic functions and characteristic values of the elliptic differential equation (1.39′) for a nonnegative $V(x)$.

As we have already seen, the characteristic functions $\varphi_n(X)$ of Eq. (1.39′) are connected with the functional integral (1.35) over conditional Wiener measure $d_{w(t,X)}x$ by the relation (1.42); in this, it is supposed that Eq. (1.39′) has a discrete spectrum,[18] but it is assumed that (1.39′) is studied only in a certain region Ω of an N dimensional space (with zero boundary conditions); in the last case, it is only necessary to extend the integration over $d_{w(t,X)}x$ in (1.35) only over con-

[18] We note that the discreteness of the spectrum for the case when $V(X)\rightarrow\infty$ for $X\rightarrow\infty$ or when the equation is studied only in a finite region can also be established from probability theoretic considerations relevant to the introduction of functional integrals (see reference [5]).

tinuous curves $x(\tau)$ which do not leave the region Ω. We note now that

$$\int_{t,0} d_{w(t,0)}x = 1/[(\pi t)^{N/2}]$$

for arbitrary $t>0$[19]; from this, it is easily obtained that

$$\psi(X,t;X) = \int_{C_{t,0}} \exp\left\{-\int_0^t V[X+x(\tau)]d\tau\right\} d_{w(t,0)}x$$

$$\sim \frac{1}{(\pi t)^{N/2}} \quad \text{as } t\to 0. \qquad (3.16)$$

(In the case of integration over only the set of curves which are distributed in the interior and on the boundary of Ω, the last expression will be exact for any arbitrary internal point X of the region Ω.) In this way, we obtain

$$\sum_n \varphi_n{}^2(X)e^{-E_n t} = \psi(X,0;X) \sim \frac{1}{(\pi t)^N} \quad \text{as } t\to 0. \qquad (3.17)$$

By making use of the Tauberian theorem of Hardy-Littlewood-Karamata (see, for example, reference [30] page 208), we immediately obtain from this the following important asymptotic relation,

$$\sum_{E_n<E} \varphi_n{}^2(X) \sim \frac{E^{N/2}}{\pi^{N/2}\Gamma[(N/2)+1]} \quad \text{as } E\to\infty. \qquad (3.18)$$

This relation can also be obtained by other means; however, the derivation presented here seems to be the simplest and shortest.

For Eq. (3.18) in a finite region Ω, the sum on the right-hand side of (3.18) can be integrated term by term over dX (or we can integrate both sides of (3.17) over dX and apply the corresponding Tauberian theorem to the result); in this way we come to an asymptotic formula for the number $N(E)$ of characteristic values smaller than E,

$$N(E) = \sum_{E_n<E} 1 \sim \frac{E^{N/2}\,\text{mes}\Omega}{\pi^{N/2}\Gamma[(N/2)+1]} \quad \text{as } E\to\infty, \qquad (3.19)$$

where mesΩ is the Lebesgue measure of the region Ω. Thus, we have very rapidly obtained the most important asymptotic relation of the theory of elliptic differential equations and immediately in a very general form (see the derivation of particular cases of (3.19) in reference [31], chap. 6).

In the case of an unbounded region Ω (in particular for an equation in all space), the considerations presented here can also be made use of for obtaining [under certain conditions concerning the smoothness of the function $V(X)$] the asymptotic form of the function $N(E)$ for large E, but here it is already not as simple (see reference [5]).

[19] This relation was written above only for the case $N=1$; it is not difficult to see, however, that it will also be valid for arbitrary N.

4. THE UTILIZATION OF FUNCTIONAL INTEGRALS IN QUANTUM PHYSICS

In Sec. 2, it was shown that the wave function of a quantum system can be represented in the form of a certain functional integral, whereupon its determination is completely equivalent to the usual determination of a wave function as a solution of Schrödinger's equation. From this, it is clear that all results obtained with the aid of functional integrals will exactly coincide with the results obtained by older methods. It is natural to think, however, as is usual in similar situations, that in investigations of certain questions one of the indicated approaches will appear more convenient, while in investigations of other problems another, so that both of these approaches can be of definite practical interest. At the present time, the utilization of the functional formulation in the solution of problems of quantum physics is only just beginning, and moreover, it is strongly held back by a general lack of development of the corresponding mathematical apparatus; none the less, a series of questions in the study of which this formulation offers notable advantages has already been cleared up now. In what follows, we examine several simple examples of this type; other examples of obtaining physically interesting conclusions with the aid of functional integrals can be found in references [1a]–[1d], [13]–[23a].

Classical Limit and Quasi-Classical Approximation

It is well known that the square of the modulus of a wave function of a particle at any moment of time t gives the probability density for finding the particle at this moment at different points of the space (in the case of an arbitrary system it is necessary here only to replace the ordinary three dimensional physical space by a space of generalized coordinates). Formula (2.3) or, more generally, (2.5) presents this probability density in the form of a sum of contributions over different trajectories of the motion of the particle which end at the investigated point. If in the above mentioned formula the integrals are restricted only to the trajectories which lie in a certain portion Ω of the space, then the square of this integral gives us the probability density for the particle, at the point X_0 at time t_0, to appear at time t at the point X, and, moreover, for all times between t_0 and t not once to go outside the limits of Ω.[20] This circumstance enables us, proceeding from

[20] The neglect in (2.5) of the trajectories which go outside the limits of Ω, in view of the meaning of this functional integral indicated on p. 57, is equivalent to the assumption that $V(X)=\infty$ outside Ω, i.e., to the assumption of the existence along the boundary of Ω of an absolutely inpenetrable barrier (compare with the remarks on the representation of the fundamental solution of Eq. (1.33′) in a finite region in the form of a functional integral).

formula (2.5) to establish in the shortest way the connection between classical and quantum mechanics. Since for small $\hbar$ the exponential function $\exp\{(i/\hbar) \times S[x(\tau)]\}$ oscillates extremely rapidly even for small changes of S, for such $\hbar$ the contributions from neighboring trajectories, generally speaking, will mutually cancel out; the exception is that trajectory in the neighborhood of which the "phase" $S[x(\tau)]$ changes most slowly, i.e., for which $\delta S[x(\tau)]=0$ (compare with the method of stationary phase in mathematical physics). But the last condition determines precisely the trajectory of classical mechanics, and moreover constitutes Hamilton's principle.

In this way, the terms which correspond to the classical trajectory, or which are close to it play the basic role in the integral (2.5) for very small $\hbar$ (practically speaking the entire integral arises from the contributions of those trajectories which are enclosed in a narrow tube which surrounds the classical trajectory); in the limit $\hbar \to 0$, the particle definitely will move along some arbitrary classical trajectory. We note further that it is clear from this that in the "classical" approximation it is necessary to regard $\psi(X,t) \sim e^{S_{\text{cl}}}$, where S_{cl} is the classical action of the system.

This result, clearly, can also be obtained from Schrödinger's equation (see for example [32], p. 149), but there it is much less simple and obvious.

The foregoing considerations also enable us to obtain the further corrections to the classical result, i.e., the succeeding term in an expansion of the wave function in a series of powers of $\hbar$. Indeed, expanding the functional $S[x(\tau)]$ about the classical trajectory $x_{\text{cl}}(\tau)$ in a functional "Taylor series" we have

$$S[x(\tau)]=S_{\text{cl}}+\tfrac{1}{2}\delta^2 S[x_{\text{cl}}(\tau)]+\tfrac{1}{6}\delta^3 S[x_{\text{cl}}(\tau)]+\cdots, \tag{4.1}$$

where $\delta^2 S[x_{\text{cl}}(\tau)]$ is a quadratic functional with regard to the function

$$y(\tau)=x(\tau)-x_{\text{cl}}(\tau), \tag{4.2}$$

namely,

$$\delta^2 S[x_{\text{cl}}(\tau)]=\int_{t_0}^{t_1}\left[\frac{\delta^2 L}{\delta x^2}\bigg|_{x=x_{\text{cl}}} y^2(\tau)+2\frac{\delta^2 L}{\delta x \delta \dot{x}}\bigg|_{x=x_{\text{cl}}} y(\tau)\dot{y}(\tau) + \frac{\delta^2 L}{\delta \dot{x}^2}\bigg|_{x=x_{\text{cl}}} \dot{y}^2(\tau)\right]d\tau, \tag{4.3}$$

where L is Lagrange's function. Furthermore, taking $e^{S_{\text{cl}}}$ outside the functional integral sign and retaining only two terms of the "Taylor's series" (4.1), we obtain the functional integral of an exponential which contains in the exponent a quadratic functional (it is clearly necessary first of all to carry out the "parallel translation" (4.2), p. 61). The last integral can be evaluated with the aid of the formulas for finite dimensional integrals, (1.11) and (1.28); it gives the correction of the next order in $\hbar$ to the wave function equivalent to the so called "quasi-classical approximation" of quantum mechanics.[21]

In courses on quantum mechanics the quasi-classical approximation is usually studied primarily for problems of determining characteristic values; it is just here that it gives the most concrete and practically interesting results (it leads to the quantization conditions of Bohr). Mathematically, these results coincide with the asymptotic formulas for the large characteristic values of the differential equation (1.39′); in this way, the theory of the quasi-classical approximation ties in here with the question studied on p. 63. In particular, if in Eq. (1.39′) we multiply all terms by $2\hbar^2/m$, then the result (3.19) for a three-dimensional space transforms into the following: in the case of Schrödinger's equation in a finite region Ω for small $\hbar$, the number of characteristic values smaller than E will be equal to $(2)^{1/2}m^{3/2}/(3\pi^2\hbar^3)E^{3/2}\,\text{mes}\Omega$ where $\text{mes}\Omega$ is the volume of the region Ω. Since for a region of volume $\text{mes}\Omega$, a volume $8(2)^{1/2}\pi m^{3/2}E^{3/2}/(3)^{-1}\,\text{mes}\Omega$ corresponds to the values of energy less than E in the six dimensional phase space of a particle, the space of the pair (x,p), where x is the coordinate and p the momentum, so that $p=(2mE)^{1/2}$, we see from here that to each characteristic value there corresponds a cell of volume $\Delta v=1/(2\pi\hbar)^3$ in phase space, while the number of characteristic values related to the element of volume $\Delta p\cdot\Delta x$ in phase space is equal to $\Delta p\cdot\Delta x/(2\pi\hbar)^3$. This result is one of the important conclusions of the quasi-classical theory; it would be very interesting to consider the possibility of extending it simply, with the aid of "functional" methods, to the case of motion in an unbounded space (the first steps in this direction are contained in reference [5]).

The Evaluation of the Statistical Sum of Quantum Statistics [37]

Statistical integrals in the field of statistical physics can have applications similar to those investigated above. As is well known, in a statistical study of the thermodynamic properties of quantum systems a basic role is played by the "statistical sum,"

$$Z=\textstyle\sum_n e^{-\beta E_n}, \quad \beta=1/kT, \tag{4.4}$$

where k is Boltzmann's constant, T is the absolute temperature, and the sum is carried out over all charac-

[21] A proof of the fact that the approximation obtained with the aid of two terms of the series (4.1) gives us a wave function exact to terms of order $\hbar$ (i.e., equivalent to the "quasi-classical approximation") is contained in reference [33]; some examples are also given there. We note, however, that the strictness of the quoted proof is substantially lowered due to the fact that the question of the precise meaning of the functional integrals studied was not discussed.

Further examples of the "quasi-classical approximation" obtained in this manner for concrete quantum mechanical problems are contained in a recent large work [36].

teristic values E_n of the energy operator H of the system under investigation.[22] As an example, we investigate the simplest problem of the one dimensional motion of a particle of mass m in a force field with a potential $V(X)$; here clearly,

$$H=-\frac{\hbar^2}{2m}\frac{\partial^2}{\partial X^2}+V(X). \tag{4.5}$$

By making a change of variable $X\to[(2)^{1/2}\hbar/(m)^{1/2}]X$, we bring this operator into the form

$$H_1=-\frac{1}{4}\frac{\partial^2}{\partial X^2}+V\left(\frac{(2)^{1/2}\hbar}{(m)^{1/2}}X\right), \tag{4.6}$$

which coincides with the form of the operator in Eq. (1.39). According to Eqs. (1.35) and (1.42), we obtain from this the result that

$$\begin{aligned}\psi(X,\beta;X)&=\int_{C\beta;0}\exp\left\{-\int_0^\beta V\left[\frac{(2)^{1/2}\hbar}{(m)^{1/2}}(X+x(\tau))\right]d\tau\right\}d_{w(\beta,0)}x\\&=\sum_n \varphi_n{}^2(X)e^{-\beta E_n}\end{aligned} \tag{4.7}$$

and (since the functions $\varphi_n(X)$ are normalized),

$$Z=\sum_n e^{-\beta E_n}=\int_{-\infty}^{\infty}\left\{\int_{C\beta;0}\exp\left\{-\int_0^\beta V\left[\frac{(2)^{1/2}\hbar}{(m)^{1/2}}\times(X+x(\tau))\right]d\tau\right\}d_{w(\beta,0)}x\right\}dX. \tag{4.8}$$

The formula obtained for Z appears at first glance to be comparatively complicated, but nevertheless on solving some questions it turns out to be convenient and significantly simplifies derivations.

In a number of cases, the system can be regarded as classical in a first approximation, while quantum corrections can be regarded as small additions to the classical results; to this approach corresponds the evaluation of Z with the aid of expanding this quantity in a series of powers of $\hbar$ (analogous to the quasi-classical approximation). If we proceed from formula (4.8), then with this aim it is natural first of all to make a change of variable $(2)^{1/2}\hbar/(m)^{1/2}X=X_0$, and then, in the formula so obtained,

$$Z=\frac{(m)^{1/2}}{(2)^{1/2}\hbar}\int_{-\infty}^{\infty}\left\{\int_{C\beta;0}\exp\left[-\int_0^\beta V\left(X_0+\frac{(2)^{1/2}\hbar}{(m)^{1/2}}\times x(\tau)\right)d\tau\right]d_{w(\beta,0)}x\right\}dX_0 \tag{4.9}$$

to make use of the expansion of the function

$$\exp\left\{-\int_0^\beta V\left[X_0+\frac{(2)^{1/2}\hbar}{(m)^{1/2}}x(\tau)\right]d\tau\right\}$$

in a Taylor series in $\hbar$:

$$\begin{aligned}\exp&\left[-\int_0^\beta V\left(X_0+\frac{(2)^{1/2}\hbar}{(m)^{1/2}}x(\tau)\right)d\tau\right]\\&=e^{-\beta V(X_0)}-\frac{(2)^{1/2}\hbar}{(m)^{1/2}}\int_0^\beta x(\tau)d\tau\cdot V'(X_0)e^{-\beta V(X_0)}\\&\quad+\frac{\hbar^2}{2m}\left\{\int_0^\beta\int_0^\beta x(\tau)x(\sigma)d\tau d\sigma\cdot V'^2(X_0)\right.\\&\quad\left.-\int_0^\beta x^2(\tau)d\tau\cdot V''(X_0)\right\}e^{-\beta V(X_0)}+\cdots.\end{aligned} \tag{4.10}$$

The zero-order term of this expansion after substitution into (4.9) gives

$$Z\approx\frac{(m)^{1/2}}{(2\pi\beta)^{1/2}\hbar}\int_{-\infty}^{\infty}e^{-\beta V(X_0)}dX_0, \tag{4.11}$$

so that apart from a nonessential multiplicative factor[23] we arrive at the statistical integral of classical statistics,

$$Z_0=\int_{-\infty}^{\infty}e^{-\beta V(X_0)}dX_0. \tag{4.12}$$

All odd terms in (4.10) after substitution into (4.9) drop out in view of (1.24′). Therefore, the succeeding terms in the expansion of (4.10) give us a representation for Z in the form of the series

$$Z=\frac{(m)^{1/2}}{(2\pi\beta)^{1/2}\hbar}(Z_0+\hbar^2Z_2+\hbar^4Z_4+\cdots); \tag{4.13}$$

moreover, each term of this series contains only moments (1.26) of conditional Wiener measure $d_{w(\beta,0)}x$ and can be evaluated very simply. In particular, the

[22] The quantities $e^{-\beta E_n}$ in quantum statistics determine the probabilities of finding the system in a state with energy E_n, so that Z will enter into the mean value $\bar{A}$ of any physical quantity A:

$$\bar{A}=Z^{-1}\sum_n A_{nn}e^{-\beta E_n},$$

where A_{nn} are matrix elements of A. The connection of Z with thermodynamic quantities is determined by the formula, $-kT\times\log Z=F$, where F is the free energy of the system (see [34]).

[23] In the formula for $\bar{A}$, the constant factor in Z will be cancelled in the numerator and denominator; in the formula for F, it gives an unessential additive constant in the expression for the free energy.

second term of (4.13) in view of (1.25) gives

$$Z_2=\frac{1}{m}\left[\int_{-\infty}^{\infty} V'^2(X_0)e^{-\beta V(X_0)}dX_0\cdot\int_0^\beta\int_0^\beta\left\{\int_{C_{\beta;0}} x(\tau)x(\sigma)d^*_{w(\beta,0)}x\right\}d\tau d\sigma-\int_{-\infty}^{\infty} V''(X_0)e^{-\beta V(X_0)}dX_0 \times\int_0^\beta\left\{\int_{C_{\beta,0}} x^2(\tau)d^*_{w(\beta,0)}x\right\}d\tau\right]$$

$$=\frac{1}{m}\left\{\int_{-\infty}^{\infty} V'^2(X_0)e^{-\beta V(X_0)}dX_0\cdot 2\int_0^\beta\int_0^\tau \frac{\tau(\beta-\sigma)}{2\beta}d\tau d\sigma-\int_{-\infty}^{\infty} V''(X_0)e^{-\beta V(X_0)}dX_0\cdot\int_0^\beta\frac{\tau(\beta-\tau)}{2\beta}d\tau\right\}$$

$$=\frac{\beta^3}{24m}\int_{-\infty}^{\infty} V'^2(X_0)e^{-\beta V(X_0)}dX_0-\frac{\beta^2}{12m}\int_{-\infty}^{\infty} V''(X_0)e^{-\beta V(X_0)}dX_0=-\frac{\beta^3}{24m}\int_{-\infty}^{\infty} V'^2(X_0)e^{-\beta V(X_0)}dX_0; \qquad (4.14)$$

The further terms of the series (4.13) are evaluated in slighly more difficult fashion (see reference [37]).

The results obtained here coincide with the results already obtained at the beginning of the thirties by Wigner and Kirkwood proceeding from Schrödinger's equation (see [34], Sec. 33); yet another elegant derivation of these same results was recently put forth by I. M. Khalatnikov [23b]. The method indicated by us for obtaining these results with the aid of functional integrals appears, however, to be substantially simpler than all the others.

We note further that formula (4.8) for Z (in application to significantly more complicated systems) was applied by R. Feynman [1c]–[1d] to the problem of the behavior of liquid helium at low temperatures.

Determination of the Smallest Characteristic Value of Schrödinger's Equation

Functional integrals can be made use of also for the evaluation of the lowest characteristic value of Schrödinger's equation (we note that applying the "quasi-classical approximation" we on the other hand obtained estimates for large characteristic values). Let

$$i\partial\psi/\partial t=H\psi \qquad (4.15)$$

be such an equation in a system of units where $\hbar=1$ (for the time being we do not specify the form of the operator H); then, an arbitrary solution of this equation can be represented in the form

$$\psi=\sum_n C_n\varphi_n e^{-iE_nt}, \qquad (4.16)$$

where the E_n are the characteristic values of the operator H, and the φ_n are the corresponding characteristic functions. In formula (4.16), it is comparatively difficult to separate the lowest characteristic value from all the rest, but if we proceed from (4.15), to the corresponding equation of the "heat conduction type,"

$$\partial\psi/\partial t=-H\psi, \qquad (4.17)$$

and notice that the characteristic values and functions for (4.15) and (4.17) are identical and that the solution to (4.17) can be represented in the form

$$\psi=\sum_n C_n\varphi_n e^{-E_nt}, \qquad (4.18)$$

then, it is clear that to find the lowest E_n it is only necessary to study the asymptotic behavior of the solution (4.18) for large t. But (4.15) is transformed into (4.17) by a change of t into $-it$. By carrying out the same transformation in formula (2.5), we find that the solution to Eq. (4.17) will be given by a functional integral over Wiener measure

$$\psi=\int_{C_{t,X}} e^{S_0[t_0,t]}d_{w(t,X)}x, \qquad (4.19)$$

where $S_0[t_0,t]$ is that part of the classical action of the system remaining after the evaluation of the integral of the "kinetic energy" (the latter after the transformation $t\to -it$ in (2.5) immediately gives Wiener measure). For the investigation of the asymptotic behavior of this integral for large t Feynman in the work of reference [1e] applied the following method. We choose in the role of an approximation to the functional S_0 some simple real functional S_1 whose integral over Wiener measure can readily be evaluated (for example, with the aid of the reduction to the solution of some arbitrary well-studied differential equation). We rewrite (4.19) in the form

$$\psi=\int_{C_{t,X}} e^{(S_0-S_1)}e^{S_1}d_{w(t,X)}x. \qquad (4.19')$$

Then this expression can be considered as some mean of the functional $e^{S_0-S_1}$, evaluated with a positive "weight function" e^{S_1}. But in view of the general inequality between the geometric mean and the arithmetic mean ("Jensen's inequality"), the mean value of $e^{S_0-S_1}$ is never less than the exponential of the mean value of S_0-S_1; therefore, replacing S_0-S_1 in the exponent in (4.19') by the mean value $(S_0-S_1)_{Av}$ of this functional over the measure $e^{S_1[x]}d_wx$ and then removing $e^{(S_0-S_1)_{Av}}$ outside the functional integral sign, we obtain an upper bound on the value of ψ, i.e., (passing to the case of large t) the upper bound therefore for E_1. If the functional S_1 contains in addition several numerical parameters, then to obtain the best approximation to E_1 we must certainly solve a variational problem—to choose the parameters in such a way that the value obtained is the smallest.

In reference [1e] this method was applied to the problem of the motion of an electron in a polar crystal and led after several transformations to the investigation of the functional integral

$$\int_{C_{t,X}} \exp\left[\alpha \int\int \frac{e^{-|t-s|}}{|x(t)-x(s)|} dt ds\right] d_{w(t,X)}x. \quad (4.20)$$

In the role of "approximating functionals" S_1 in [1e] are studied functionals of the form $\exp\{-\int V[x(t)]dt\}$, where either $k/|x|$ or kx^2 is assumed for $V(x)$. The asymptotic form of integrals over Wiener measure of $\exp[-\int V(x)dt]$ for such V is naturally related to the determination of the characteristic values of Schrödinger's equat on for an electron in a Coulomb field and for a harmonic oscillator. The results obtained appear to be very accurate; comparison with the exact value for E_1 obtained by S. I. Pekar [35] for the case of very large α shows that choosing for $V(x)$ the functional kx^2 and determining the value of the parameter k from the solution of a variational problem we obtain E_1 with an error of less than 3%.

Perturbation Theory

We begin with an investigation of the simplest problem of the motion of a particle in a potential field. Let the potential $V[x(\tau)]$ be represented in the form,

$$V[x(\tau)] = V_0[x(\tau)] + \epsilon V_1[x(\tau)], \quad (4.21)$$

where $V_0[x(\tau)]$ is the "unperturbed" potential energy for which the solution of Schrödinger's equation is known [equal, say, to $\psi_0(X,t)$] while $\epsilon V_1(x)$ is a small perturbation (ϵ is a parameter of smallness). Then, according to (2.3),

$$\psi(X,t;X_0,t_0) = \frac{1}{N}\int_{t_0}^{t} \exp\left(\frac{i}{\hbar}\int_{t_0}^{t}\{L_0[x(\tau)] - \epsilon V_1[x(\tau)]\}d\tau\right)\prod_{\tau=t_0}^{t} dx(\tau), \quad (4.22)$$

where $L_0[x(\tau)]$ is the unperturbed Lagrangeian. From here, we easily obtain an expansion of the function ψ in a series of powers of ϵ:

$$\psi(X,t;X_0,t_0) = \frac{1}{N}\int \exp\left\{\frac{i}{\hbar}\int_{t_0}^{t} L_0[x(\tau)]d\tau\right\}\left\{1 - \frac{i\epsilon}{\hbar}\int_{t_0}^{t} V_1[x(\tau_1)]d\tau_1 + \frac{1}{2!}\left(\frac{i\epsilon}{\hbar}\right)^2 \int_{t_0}^{t}\int_{t_0}^{t} V_1[x(\tau_1)]V_1[x(\tau_2)]d\tau_1 d\tau_2 + \cdots\right\}\prod_{t_0}^{t} dx(\tau)$$

$$= \psi_0(X,t;X_0,t_0) - \frac{i\epsilon}{\hbar}\int_{-\infty}^{\infty}\int_{t_0}^{t} \psi_0(x_1,\tau_1;X_0,t_0)V_1(x_1)\psi_0(X,t;x_1,\tau_1)dx_1 d\tau_1$$

$$+ \left(\frac{i\epsilon}{\hbar}\right)^2 \int_{-\infty}^{\infty}\int_{-\infty}^{\infty}\int_{t_0}^{t}\int_{t_0}^{\tau_1} \psi_0(x_2,\tau_2;X_0,t_0)V_1(x_2)\psi_0(x_1,\tau_1;x_2,\tau_2)V_1(x_1) \times \psi_0(X,t;x_1,\tau_1)dx_1 dx_2 d\tau_1 d\tau_2 + \cdots \quad (4.23)$$

and in general

$$\psi(X,t;X_0,t_0) = \psi_0(X,t;X_0,t_0) + \sum_{n=1}^{\infty}\left(-\frac{i\epsilon}{\hbar}\right)^n \int_{-\infty}^{\infty}\int_{-\infty}^{\infty}\cdots\int_{-\infty}^{\infty}\int_{0}^{t}\int_{0}^{\tau_1}\cdots\int_{0}^{\tau_{n-1}} \psi_0(x_n,\tau_n;X_0,t_0)V_1(x_n) \times \psi_0(x_{n-1},\tau_{n-1};x_n,\tau_n)V_1(x_{n-1})\cdots V_1(x_1)\psi(X,t;x_1,\tau_1)dx_1 dx_2\cdots dx_n d\tau_1 d\tau_2\cdots d\tau_n. \quad (4.23')$$

This is the usual result of the nonstationary perturbation theory of quantum mechanics.

The perturbation theory developed here has received a particularly wide application in problems of quantum field theory. We take as an example the same model of Thirring of a quantum field theory which we studied on page 59. Here the wave functional $\psi[u(x)]$ is represented by formula (2.14′). This functional is given as an integral of an exponential function in the exponent of which is a biquadratic form in $u(x)$. The evaluation of such an integral presents a very great difficulty, and at the present time there exist no general approaches to the solution of this problem. However, if we consider the term of the fourth order in the exponential to be small and make use of the expansion of the wave functional in a series of powers of λ (we note that the parameter λ in our theory plays the role of an elementary charge), then all terms of the resulting series will represent functional integrals of the type of moments of a Gaussian distribution which can be evaluated directly.[24] It

[24] We do not concern ourselves here with complications connected with the fact that in concrete evaluations very often infinite expressions and divergent integrals appear which must be treated by a special regularization; these complications have a physical but not a mathematical origin.

is obvious that the last circumstance possesses a general character and is not connected with the choice of a special model of a quantum field/ theory. By their nature, almost all investigations of a quantum field theory published up to the present time are concerned with the investigation of perturbation series which are obtained in this way.[25] For evaluating the terms of these series a very unusual and interesting mathematical technique has been developed in recent times (the so called "Feynman diagrams"); however, we cannot here go more deeply into these important but not simple questions.

BIBLIOGRAPHY

1. R. P. Feynman, (a) "Space-time approach to nonrelativistic quantum mechanics," Revs. Modern Phys. **20**, No. 2 (1948), pp. 367–387; see also the Russian translation in the collection "Questions of Causality in Quantum Mechanics," IL (1955); (b) "Mathematical formulation of the quantum theory of electromagnetic interaction," Phys. Rev. **80**, 440–457 (1950); (c) "Atomic theory of the λ transition in helium," Phys. Rev. **91**, 1291–1301 (1953); (d) "Atomic theory of liquid helium near absolute zero," Phys. Rev. **91**, 1301–1308 (1953); (e) "Slow electrons in a polar crystal," Phys. Rev. **97**, 660–665 (1955).
2. M. Kac, (a) On distributions of certain Wiener functionals, Trans. Am. Math. Soc. **65**, No. 1, 1–13 (1949); (b) On some connections between probability theory and differential and integral equations, Proc. 2nd Berkeley Symposium Math. State. and Probab., Berkeley, 189–215 (1951).
3. N. Wiener, (a) Differential Space, J. Math. and Phys. **2**, 131–174 (1923); (b) "The average value of a functional," Proc. London Math. Soc. Ser. 2, **22**, No. 6, 454–467 (1924); (c) "Generalized harmonic analysis," Acta Math. **55**, 117–258 (1930).
4. M. Rosenblatt, "On a class of Markov processes," Trans. Am. Math. Soc. **77**, No. 1, 120–135 (1951).
5. D. Ray, On spectra of second-order differential operators, Trans. Am. Math. Soc. **77**, No. 2, 299–321 (1954).
6. A. Blanc-Lapierre and R. Fortet, "Theorie des fonctions aleatoires," Paris, 1953, Chap. VII, Sec. III.
7. E. B. Dynkin, (a) "On certain limit theorems for Markov chains," Ukr. Math. Zhur. **6**, No. 1, 21–27 (1954); (b) "Functionals of the trajectories of Markov random processes," DAN **104**, No. 5, 691–694 (1955).
8. R. Z. Hac'minskii, "Distribution of probabilities for functionals of trajectories of a random process of the diffusion type," DAN **104**, No. 1, 22–25 (1955).
9. R. H. Cameron and W. T. Martin, (a) "Transformations of Wiener integrals under translations," Ann. Math. **45**, No. 2 386–396 (1944); (b) "Transformations of Wiener integrals under a general class of linear transformations," Trans. Am. Math. Soc. **58**, No. 2, 184–219 (1945); (c) "Evaluation of various Wiener integrals by use of certain Sturm-Liouville differential equations," Bull. Am. Math. Soc. **51**, No. 2, 73–90 (1945); (d) "Transformations of Wiener integrals by nonlinear transformations," Trans. Am. Math. Soc. **66**, No. 2, 253–283 (1949).
10. R. H. Cameron, (a) "A Simpson rule for the numerical evaluation of Wiener integrals in function space," Duke Math. J. **18**, No. 1, 111–130 (1951); (b) "The general heat flow equation and a corresponding Poisson formula," Ann. Math. **59**, No. 3, 434–461 (1954); (c) "The translation pathology of Wiener space," Duke Math. J. **21**, No. 4, 623–627 (1954).
11. R. H. Cameron and R. E. Fagen, "Nonlinear transformations of Volterra type in Wiener space," Trans. Am. Math. Soc. **75**, No. 3, 552–575 (1953).
12. E. Hopf, "Statistical hydromechanics and functional calculus," J. ration. mech. and analysis **1**, No. 1, 87–123 (1952).
13. S. F. Edwards and R. E. Peierls, "Field equations in functional form," Proc. Roy. Soc. London, **A224**, No. 1156, 24–33 (1954) [see also the Russian translation in the collection "Problems of contemporary physics," IL, No. 3 (1955)].
14. I. M. Gel'fand and R. A. Minlos, "The solution of the equations of quantized fields," DAN U.S.S.R. **97**, No. 2, 209–212 (1954).
15. N. N. Bogolyubov, "On the representation of Green-Schwinger functions with the aid of functional integrals," DAN U.S.S.R. **99**, No. 1, 225–226 (1954).
16. E. C. Fradkin, "Green's function for the interaction of nucleons with mesons," DAN **98**, No. 1, 47–50 (1954).
17 P. T. Matthews and A. Salam, "The Green's functions of quantized fields," Nuovo cimento **12**, No. 4, 563–565 (1954).
18. N. P. Kelpikov, (a) "On the theory of a vacuum functional, DAN **98**, No. 6, 937–940 (1954); (b) "The solution of systems of equations for a vacuum functional," DAN **100**, No. 6, 1057–1059 (1955).
19. U. A. Gol'fand, "Construction of distribution functions by the method of quasi-fields," J. Exptl. Theoret. Phys. **28**, No. 2, 140–150 (1955).
20. S. F. Edwards, (a) "The nucleon Green function in charged meson theory, Proc. Roy. Soc. London **A228**, No. 1174, 411–424 (1955); (b) "The nucleon Green function in pseudoscalar meson theory," I–II, Proc. Roy. Soc. London **A232**, No. 1190, 371–389 (1955).
21. J. Hamilton, "Functional analysis and strong-coupling theory," Phys. Rev. **97**, No. 5, 1390–1391 (1955).
22. T. H. R. Skyrme, "Quantum field theory," Proc. Roy. Soc. (London) **A231**, 321–335 (1955).
23. I. M. Khalatnikov, (a) "The representation of Green's function in quantum electrodynamics in the form of continual integrals," J. Exptl. Theoret. Phys. **28**, No. 5, 633–636 (1955); (b) "On one method of evaluating the statistical sum," DAN **87**, No. 4, 539–542 (1952).
24. A. N. Kolmogorov, "Basic ideas of the theory of probability," M.-L., ONTI (1936).
25. E. W. Montroll, "Markoff chains, Wiener integrals, and quantum theory," Commun. pure appl. math. **5**, 415–453 (1952).
26. B. V. Gnedenko, "A course in the theory of probability," M.-L., Gostekhizdat (1954).
27. U. V. Prohorov, "The distribution of probabilities in functional spaces," UMN **8**, No. 3 (55) 165–167 (1953).
28. W. Feller, "Introduction to the theory of probability and its applications," M., IL, (1952).
29. E. Titchmarsh, "Introduction to the theory of Fourier integrals," M.-L. Gostekhizdat (1948).
30. G. Doetsch, "Theorie und Anwendung der Laplace-Transformationen," Berlin, (1937).
31. R. Courant and D. Hilbert, "Methods of mathematical Physics," M.-L., GTTI (1934), Vol. 1.
32. W. Pauli, "Basic principles of Wave Mechanics," M.-L., Gostekhizdat (1947).
33. C. Morette, "On the definition and approximation of Feynman's path integrals," Phys. Rev. **81**, 848–852 (1951).
34. L. D. Landau and E. M. Lifshitz, "Statistical physics," M.-L., Gostekhizdat (1951).
35. S. I. Pekar, "Theory of polarons," J. Exptl. Theoret. Phys. **19**, No. 9, 796–806 (1949).
36. P. Choquard, "Traitement semi-classique des forces generales dans la representation de Feynman," Helv. Phys. Acta **28**, No. 2–3, 89–157 (1955).
37. A. M. Yaglom, "Application of functional integrals to the evaluation of the statistical sum of quantum statistics," Theory of probability and its applications **1**, No. 1 (1956).

[25] Exceptions to this are found in certain works in which the result is sought in the form of a series in powers of the constant $1/\lambda$ (the "method of strong coupling"). See also reference [13], in which for one very schematic model problem the evaluation of the functional integrals which arise is carried out in a general form.

3.

(with A.N. Kolmogorov and A.M. Yaglom)

To the general definition of the amount of information

Dokl. Akad. Nauk SSSR **111** (1956) 745–748. Zbl. **71**:345

The definitions (2) and (4) and the properties I–IV of $I(\xi, \eta)$ given below were presented by A.N. Kolmogorov in his talk in the meeting on probability theory and mathematical statistics (Leningrad, June 1954). Theorem 4 and Theorem 5, on the semi-continuity of $I(\xi, \eta)$ under the weak convergence of distributions, were found by I.M. Gelfand and A.M. Yaglom. Subsequently A.N. Kolmogorov suggested the final version of this note; in this version it is noted that both the generalization from the finite case to the general case and the computation of the amount of information are absolutely trivial if they are considered from the point of view of normed Boolean algebras. It would be, of course, quite easy to consider more general limits, not for $n \to \infty$, but for some partially ordered sets.

1. In this section the system $\mathfrak{S}$ of all "random events" $A, B, C, \ldots$ is assumed to be a Boolean algebra with the complementary element A' for any A, with operations $A \cup B$ (union) and AB (product), with identity element ε and zero element N. The probability distributions $P(A)$ considered below are nonnegative functions on $\mathfrak{S}$ satisfying the additivity condition $P(A \cup B) = P(A) + P(B)$ for $AB = N$ and the normalization condition $P(\varepsilon) = 1$.

The classical notion of the "test" is naturally identified with the notion of a subalgebra of the algebra $\mathfrak{S}$ (roughly speaking, $\mathfrak{A}$ consists of all events whose outcome becomes known after the given test). If subalgebras $\mathfrak{A}$ and $\mathfrak{L}$ are finite then "the amount of information about the results of the test $\mathfrak{L}$ contained in the results of the test $\mathfrak{A}$" is defined by Shannon's formula,

$$I(\mathfrak{A}, \mathfrak{L}) = \sum_{i,j} P(A_i B_j) \log \frac{P(A_i B_j)}{P(A_i) P(B_j)}. \tag{1}$$

In the general case it is natural to set

$$I(\mathfrak{A}, \mathfrak{L}) = \sup_{\mathfrak{A}_1 \subseteq \mathfrak{A}, \mathfrak{L}_1 \subseteq \mathfrak{L}} I(\mathfrak{A}_1, \mathfrak{L}_1) \tag{2}$$

where the supremum is taken over all finite subalgebras $\mathfrak{A}_1 \subseteq \mathfrak{A}$ and $\mathfrak{L}_1 \subseteq \mathfrak{L}$. Symbolically, the definition (2) can be written in the form

$$I(\mathfrak{A}, \mathfrak{L}) = \int_{\mathfrak{A}} \int_{\mathfrak{L}} P(d\mathfrak{A}\, d\mathfrak{L}) \log \frac{P(d\mathfrak{A}\, d\mathfrak{L})}{P(d\mathfrak{A}) P(d\mathfrak{L})}. \tag{3}$$

For finite $\mathfrak{A}$ and $\mathfrak{L}$ the amount of information $I(\mathfrak{U}, \mathfrak{L})$ is real and nonnegative.

In the general case the only other value that can appear besides real nonnegative values of I is the value $I=+\infty$. The following (well known for the finite case) properties of I remain true (here $[\mathfrak{M}]$ denotes the smallest subalgebra of the algebra $\mathfrak{S}$ that contains a set $\mathfrak{M}$):

1. $I(\mathfrak{A}, \mathfrak{L}) = I(\mathfrak{L}, \mathfrak{A})$.
2. $I(\mathfrak{A}, \mathfrak{L}) = 0$ if and only if the systems of events $\mathfrak{A}$ and $\mathfrak{L}$ are independent.
3. If $[\mathfrak{A}_1 \cup \mathfrak{L}_1]$ and $[\mathfrak{A}_2 \cup \mathfrak{L}_2]$ are independent then
$$I([\mathfrak{A}_1 \cup \mathfrak{A}_2], [\mathfrak{L}_1 \cup \mathfrak{L}_2]) = I(\mathfrak{A}_1, \mathfrak{L}_1) + I(\mathfrak{A}_2, \mathfrak{L}_2).$$
4. If $\mathfrak{A}_1 \subseteq \mathfrak{A}$ then $I(\mathfrak{A}_1, \mathfrak{L}) \leqq I(\mathfrak{A}, \mathfrak{L})$.

The following theorems are almost obvious but, in some cases, they yield useful methods for computation and estimation of the amount of information by passage to the limit.

Theorem 1. *If an algebra $\mathfrak{A}_1 \subseteq \mathfrak{A}$ is everywhere dense in $\mathfrak{A}$ in the metric*
$$\rho(A, B) = P(AB' \cup A'B),$$
then
$$I(\mathfrak{A}_1, \mathfrak{L}) = I(\mathfrak{A}, \mathfrak{L}).$$

Theorem 2. *If*
$$\mathfrak{A}_1 \subseteq \mathfrak{A}_2 \subseteq \dots \mathfrak{A}_n \subseteq \dots,$$
then the algebra $\mathfrak{A} = \bigcup_n \mathfrak{A}_n$ satisfies
$$I(\mathfrak{A}, \mathfrak{L}) = \lim_{n\to\infty} I(\mathfrak{A}_n, \mathfrak{L}).$$

Theorem 3. *If the sequence of distributions $P^{(n)}$ converges on $[\mathfrak{A} \cup \mathfrak{L}]$ to a distribution P, i.e. if $\lim_{n\to\infty} P^{(n)}(C) = P(C)$ for $C \in [\mathfrak{A} \cup \mathfrak{L}]$ then the amounts of information $I^{(n)}(\mathfrak{A}, \mathfrak{L})$ and $I(\mathfrak{A}, \mathfrak{L})$ defined by (1) and (2) for distributions $P^{(n)}$ and P satisfy the following inequality:*
$$\liminf_{n\to\infty} I^{(n)}(\mathfrak{A}, \mathfrak{L}) \geqq I(\mathfrak{A}, \mathfrak{L}).$$

2. Now we will assume that the main Boolean algebra $\mathfrak{S}$ is a σ-algebra, and all the distributions we will consider are σ-additive distributions. We will denote by a single letter X a "measurable space" consisting of a set X with a σ-algebra S_X of its subsets containing the identity element X (this is a slight deviation from the terminology in [1] where S_X may be a σ-ring). By a "random element" ξ of the space X we will mean a homomorphism
$$\xi^*(A) = B$$
of the Boolean algebra S_X into the Boolean algebra $\mathfrak{S}$. Intuitively, $\xi^*(A)$ is "the event $\xi \in A$".

The homomorphism ξ^* maps the algebra S_X onto a subalgebra of the algebra $\mathfrak{S}$ which will be denoted

$$\mathfrak{S}_\xi = \xi^*(S_X).$$

Then it is natural to define

$$I(\xi, \eta) = I(\mathfrak{S}_\xi, \mathfrak{S}_\eta). \tag{4}$$

Following [1] we define in the usual way the measurable space $X \times Y$. The formula

$$(\xi, \eta)^*(A \times B) = \xi^*(A)\eta^*(B)$$

uniquely defines a homomorphism $(\xi, \eta)^*$ of the Boolean σ-algebra $S_{X \times Y}$ into $\mathfrak{S}$. This homomorphism is used to define the pair (ξ, η) of random objects ξ and η as a new random object, namely as a random element of the space $X \times Y$. The formula

$$\mathscr{P}_\xi(A) = P(\xi^*(A))$$

defines a measure in the space X that is called the distribution of the random object ξ. Finally define, according to [1], the following measure in the space $X \times Y$:

$$\prod = \mathscr{P}_\xi \times \mathscr{P}_\eta.$$

Applying to two measures $\prod$ and $\mathscr{P}(\xi, \eta)$ on $X \times Y$ the Radon–Nikodim theorem, we get

$$\mathscr{P}_{\xi,\eta}(C) = \iint\limits_C a(x, y)\, d\mathscr{P}_\xi\, d\mathscr{P}_\eta + S(C),$$

where the measure S is singular with respect to the measure $\prod$.

Theorem 4. *A necessary condition for the amount of information $I(\xi, \eta)$ to be finite is $S(C) = 0$.*

If this is the case then

$$I(\xi, \eta) = \int\limits_X \int\limits_Y a(x, y) \log a(x, y)\, d\mathscr{P}_\xi\, d\mathscr{P}_\eta = \int\limits_{X \times Y} \log a(x, y)\, d\mathscr{P}(\xi, \eta). \tag{5}$$

Let us note also that for any random element ξ of the space X and for any measurable Borel mapping $y = f(x)$ of the space X into the space Y a random element $\eta = f(\xi)$ of the space Y is defined by the mapping

$$\eta^*(A) = \xi^* f^{-1}(A)$$

of the algebra S_Y into the algebra $\mathfrak{S}$. With the usual definition of the expectation M the formula (5) can be written in the following form:

$$I(\xi, \eta) = M \log a(\xi, \eta).$$

Properties 1–4 of $I(\mathfrak{A}, \mathfrak{L})$ imply the following properties of $I(\xi, \eta)$:

I. $I(\xi, \eta) = I(\eta, \xi)$.
II. $I(\xi, \eta) = 0$ if and only if ξ and η are independent.

III. If pairs (ξ_1, η_1) and (ξ_2, η_2) are independent then

$$I((\xi_1, \xi_2), (\eta_1, \eta_2)) = I(\xi_1, \eta_1) + I(\xi_2, \eta_2).$$

IV. If ξ_1 is a function in ξ (in the above sense) then

$$I(\xi_1, \eta) \leqq I(\xi, \eta).$$

Similarly to the definition of the pair (ξ, η), a sequence $\xi = (\xi_1, \xi_2, \ldots, \xi_n, \ldots)$ can easily be defined as a random element of the space $X = \times_{n=1}^{\infty} X_n$. From Theorems 1 and 2 one can easily deduce that

$$I(\xi, \eta) = \lim_{n \to \infty} I((\xi_1, \ldots, \xi_n), \eta).$$

To conclude, let us show one more application of Theorem 3. Let X and Y be complete metric spaces. Taking for S_X and S_Y their algebras of Borel sets, we make X and Y measurable spaces. In this case we have the following theorem.

Theorem 5. *If for two random elements $\xi \in X$, $\eta \in Y$ distribution $\mathscr{P}^{(n)}(\xi, \eta)$ weakly converge to a distribution $\mathscr{P}(\xi, \eta)$ then the corresponding amounts of information satisfy*

$$\liminf_{n \to \infty} I^n(\xi, \eta) = I(\xi, \eta)$$

To prove this theorem one has to apply Theorem 3 to algebras $\mathfrak{A}$ and $\mathfrak{L}$ consisting, respectively, of the events $\xi \in A$ and $\eta \in B$ where A and B run over all continuity set for distributions $\mathscr{P}_\xi$ and, respectively, $\mathscr{P}_\eta$ (i.e. those sets whose boundaries satisfy the condition $\mathscr{P}_\xi = 0$ and, respectively, $\mathscr{P}_\eta = 0$), and to note that the algebra $[\mathfrak{A} \cup \mathfrak{L}]$ is everywhere dense in $\mathfrak{S}_{X \times Y}$ in the metric generated by the limiting distribution P.

Received 8.X.1956

References

1. Halmos, P.R.: Measure Theory. IL, 1953

4.
(with A. M. Yaglom)

Calculation of the amount of information about a random function contained in another such function

Usp. Mat. Nauk 12 (1) (1957) 3-52. [Transl., II. Ser., Am. Math. Soc. 12 (1959) 199-246]

Contents

Introduction 199
Chapter I. A general definition of information 200
§1. Information about a random vector contained in another such vector 200
§2. Information about a random function contained in another such function 210
Chapter II. Calculation of information for Gaussian random functions 215
§1. Information about a Gaussian random vector contained in another such vector 215
§2. Information about a Gaussian random process contained in another such process 220
§3. Examples 223
Literature 244

Introduction

The theory of information – which may be regarded as a new chapter in a general theory of probability – originated several years ago in the work of Shannon and other investigators. The development of this theory was prompted by practical problems, specifically, certain purely technological problems in the fields of electrical and radio communication. Because of its great practical as well as theoretical significance, information theory has recently been attracting considerable attention on the part of both engineers and mathematicians. Nevertheless, up to the present relatively little progress has been made in the treatment of the basic problem of calculating the amount of information for various types of random objects.

In Chapter I of the present work a general definition of information for a relatively broad class of random objects including random functions and generalized random functions (in the sense of [9], [10]) is considered. In §1 a definition of information for arbitrary random vectors based on Appendix 7 of Shannon's work (which was not included in the Russian translation) is given. The only new results here are Property II of the quantity $I(\xi, \eta)$ (see p. 206) and Theorem 1.1. The notion of information for arbitrary generalized random functions is considered in §2. Here we make use of the paper presented by A. N. Kolmogorov at the All-Union Conference on Probability Theory and its Applications (May, 1955) in which the notion of information was formulated in abstract terms and in a very general form applicable to objects of any nature (see also [32]). The basic idea, which is systematically employed in Chapter I to derive a general definition of information, consists in a successive reduction of the notion of information for increasingly general forms of random objects to the simplest case of information for discrete random variables. As will be seen in the sequel, this approach also simplifies the investigation of the general properties of information.

Part I of the present work is introductory in nature. Here the general definition and properties of information are discussed in elementary terms and in just sufficient detail needed for subsequent calculations in Part II. The main concern of this paper is with the calculation of the amount of information contained in Gaussian generalized random functions. In §1 and §2 of Part II general results pertaining to the information associated with such functions are obtained, while in Part III several examples in which the numerical computation of information is feasible are given.

The basic results of the present work were presented by the authors at the Conference on Functional Analysis and its Applications which was held in Moscow in January 1956. Some of these results were included in a joint paper with A. N. Kolmogorov which was presented at the Third All-Union Mathematics convention (Moscow, June – July, 1956).

Chapter I.
A general definition of information

§1. Information about a random vector contained in another such vector. We begin with a consideration of the simple case of discrete random variables ξ and η ranging over finite sets. Let ξ take on n values (say $x_1, x_2, \cdots, x_n$) with respective probabilities $P_\xi(1), P_\xi(2), \cdots, P_\xi(n)$, and let η take on m values (say $y_1, y_2, \cdots, y_m$) with respective probabilities $(P_\eta(1), P_\eta(2), \cdots, P_\eta(m)$. The variables ξ and η will, in general, be assumed to be interdependent, and the joint probability of ξ and η taking on the values x_i and y_k, respectively, will be denoted by $P_{\xi\eta}(i, k)$. According to Shannon [1], in this case the *amount of information about the variable* ξ *contained in the variable* η (or, more precisely, the average amount of information about the variable ξ conveyed by specifying the value of the variable η) *is measured by the following number*:

$$J(\xi, \eta) = \sum_{i=1}^{n} \sum_{k=1}^{m} P_{\xi\eta}(i, k) \log \frac{P_{\xi\eta}(i, k)}{P_\xi(i)\, P_\eta(k)}. \tag{1 1}$$

(Note: when $P_{\xi\eta}(i, k) = 0$, the corresponding term in the summation is taken to be zero; the base of logarithms is arbitrary – its selection being equivalent to the choice of a particular unit of information.) The function defined by (1.1) is symmetrical in ξ and η, so that we may speak, more simply, of the *amount of informa-*

tion contained in one of two random variables ξ and η about the other variable.

The function (1.1) of probabilities

$$P_\xi(i),\ P_\eta(k) \text{ and } P_{\xi\eta}(i, k)$$

possesses a number of general properties which make it a logical choice as a measure of information. Moreover, it is not difficult to introduce it axiomatically (see [2], [3]).

In the following we establish those basic properties of (1.1) which will be needed subsequently.

I) $$J(\xi, \eta) \geq 0, \tag{1.2}$$

with equality if and only if ξ and η are independent.

Proof of this assertion is well-known (see, for example, [1], [2]); we reproduce it here for completeness. Using the relations

$$\left.\begin{aligned} \sum_{i=1}^{n} P_{\xi\eta}(i, k) = P_\eta(k), \\ \sum_{k=1}^{m} P_{\xi\eta}(i, k) = P_\xi(i) \end{aligned}\right\} \tag{1.3}$$

equation (1.1) can be expressed in the form

$$J(\xi, \eta) = \sum_{i=1}^{n}\sum_{k=1}^{m}\left\{P_{\xi\eta}(i, k)\log\frac{P_{\xi\eta}(i, k)}{P_\xi(i)}\right\} - \sum_{k=1}^{m} P_\eta(k)\log P_\eta(k) =$$

$$= \sum_{k=1}^{m}\left\{\sum_{i=1}^{n} P_\xi(i)\,\Phi\left[\frac{P_{\xi\eta}(i, k)}{P_\xi(i)}\right] - \Phi[P_\eta(k)]\right\}, \tag{1.4}$$

where

$$\Phi(x) = x\log x. \tag{1.5}$$

Now for $x \geq 0$ the function $\Phi(x)$ is strictly concave (since $\Phi''(x) > 0$); therefore

$$\sum_{i=1}^{n}\lambda_i\Phi(x_i) \geqslant \Phi\left(\sum_{i=1}^{n}\lambda_i x_i\right) \quad \text{with} \quad \sum_{i=1}^{n}\lambda_i = 1, \tag{1.6}$$

with equality if and only if all the x_i are equal ([4], p. 96). Setting

$$\lambda_i = P_\xi(i), \qquad x_i = \frac{P_{\xi\eta}(i, k)}{P_\xi(i)},$$

we have

$$\sum_{i=1}^{n} P_\xi(i)\,\Phi\left[\frac{P_{\xi\eta}(i, k)}{P_\xi(i)}\right] \geqslant \Phi[P_\eta(k)], \tag{1.7}$$

by virtue of (1.3), with equality if and only if the conditional probabilities

$$\frac{P_{\xi\eta}(i, k)}{P_\xi(i)} = P\{\eta = y_k \mid \xi = x_i\} \qquad (j = 1, 2, \cdots, n)$$

are independent of k. ($P\{\eta = y_k \mid \xi = x_i\}$ denotes as usual the conditional probability that η is equal to y_k given that ξ is equal to x_i). It is clear that the latter condition is equivalent to the independence of ξ and η, from which it follows that

$J(\xi, \eta) = 0$ if and only if ξ and η are independent.

Ia) It is easy to show that we always have

$$J(\xi, \eta) \leq J(\eta, \eta) \tag{1.8}$$

(and hence $J(\xi, \eta) \leq J(\xi, \xi)$), with equality only when ξ is a random variable such that each conditional probability $P(\eta = k \mid \xi = i)$ is either zero or one.

Indeed, from (1.4) it follows that

$$J(\xi, \eta) - J(\eta, \eta) = \sum_{i=1}^{n} \sum_{k=1}^{m} P_{\xi}(i) \Phi\left[\frac{P_{\xi\eta}(i, k)}{P_{\xi}(i)}\right] \tag{1.9}$$

(here we make use of the equality $\Phi(1) = 0$) and since

$$\Phi(0) = 0 \text{ and } \Phi(\xi) < 0 \text{ for } 0 < \xi < 1,$$

property Ia follows immediately. The quantity $J(\eta, \eta) = -\sum_{k=1}^{m} \Phi[P_{\eta}(k)]$ is usually called the *entropy* of η and is denoted by $H(\eta)$; it plays a central role in applications of information theory to discrete random variables (see [1], [2], [5]).

II. *Let* ξ_1 *be a "refinement" of* ξ *in the sense that* ξ_1 *assumes* $n_1 > n$ *distinct values which can be divided into* n *(non-empty) groups*

$$(x_{i_1}, x_{i_2}, \cdots, x_{i_{k_i}}); \; i = 1; 2, \cdots, n; \; k_1 + k_2 + \cdots + k_n = n_1,$$

such that the values from ith group are assumed if and only if ξ *has its ith value (or, in other words, if* ξ *is a function of* ξ_1*). Then*

$$J(\xi_1, \eta) \geq J(\xi, \eta). \tag{1.10}$$

with equality if and only if the conditional probability of η *given any fixed value of* ξ_1 *is independent of this value, i.e.*

$$P(\eta = y_k \mid \xi_1 = x_{i_1}) = P(\eta = y_k \mid \xi_1 = x_{i_2}) = \cdots = P(\eta = y_k \mid \xi_1 = x_{i_{k_i}}) \tag{1.11}$$

for all values y_k *of* η *and all* $i = 1, 2, \cdots, n$.

The meaning of this property of $J(\xi, \eta)$ is quite apparent when $J(\xi, \eta)$ is interpreted as the amount of information concerning ξ which is contained in η. Evidently this interpretation of $J(\xi, \eta)$ is justified only if $J(\xi, \eta)$ has the property stated above.

It is clear that in proving inequality (1.10) ξ_1 can be regarded as obtainable from ξ by "breaking down" one of the values of ξ (say $\xi = x_1$) into two possible values x_1' and x_2'' (with probabilities $P_{\xi_1}(1')$ and $P_{\xi_1}(1'')$, where $P_{\xi_1}(1') + P_{\xi_1}(1'') = P_{\xi}(1)$); and setting $\xi_1 = \xi$ for $\xi \neq x_1$. (The general case can readily be reduced to a finite succession of such "breakdowns".) For the simple case considered here, the proof of inequality (1.10) is quite easy. The difference $J(\xi_1, \eta) - J(\xi, \eta)$ is by virtue of (1.4) equal to

$$\sum_{k=1}^{m} \Big\{ P_{\xi_1}(1') \Phi\left[\frac{P_{\xi_1\eta}(1', k)}{P_{\xi_1}(1')} \right] +$$

$$+ P_{\xi_1}(1'') \Phi\left[\frac{P_{\xi_1\eta}(1'', k)}{P_{\xi_1}(1'')} \right] - P_{\xi}(1) \Phi\left[\frac{P_{\xi\eta}(1, k)}{P_{\xi}(1)} \right] \Big\},$$

where $\Phi(x)$ is given by (1.5). From the inequality (1.6) with $n = 2$ and

$$\lambda_1 = \frac{P_{\xi_1}(1')}{P_{\xi}(1)}, \qquad \lambda_2 = \frac{P_{\xi_1}(1'')}{P_{\xi}(1)},$$

$$x_1 = \frac{P_{\xi_1\eta}(1', k)}{P_{\xi_1}(1') P_{\eta}(k)}, \quad x_2 = \frac{P_{\xi_1\eta}(1'', k)}{P_{\xi_1}(1'') P_{\eta}(k)}$$

(1.10) follows immediately. The condition (1.11) is then readily derived by noting that equality in (1.6) obtains if and only if all the x_i are equal.

We proceed now to the determination of the amount of information $J(\xi, \eta)$ for arbitrary real-valued random variables ξ and η. To make use of the definition of $J(\xi, \eta)$ given above for discrete random variables, it is expedient to employ the following approach. We first partition the ranges of values of ξ and η into a finite number of non-overlapping intervals $\Delta_1, \Delta_2, \cdots, \Delta_n$ and $\Delta_1', \Delta_2', \cdots, \Delta_m'$. (It is immaterial whether the intervals are finite or infinite, open or closed or semi-closed.) Consider now the discrete random variables $\xi(\Delta_1, \Delta_2, \cdots, \Delta_n)$ and $\eta(\Delta_1', \Delta_2', \cdots, \Delta_m')$ whose values are respectively equal to the subscripts of those intervals Δ_i $(i = 1, 2, \cdots, n)$ and Δ_k' $(k = 1, 2, \cdots, m)$ to which the values of ξ and η belong. For this case the amount of information $J[\xi(\Delta_1, \cdots, \Delta_n), \eta(\Delta_1', \cdots, \Delta_m')]$ can be calculated through the use of (1.1). Then we define the amount of information $J(\xi, \eta)$ about ξ contained in η as the supremum of $J[\xi(\Delta_1, \cdots, \Delta_n), \eta(\Delta_1', \cdots, \Delta_m')]$; that is, we adopt the following

Definition 1. *The amount of information $J(\xi, \eta)$ for arbitrary random variables ξ and η is given by*

$$J(\xi, \eta) = \sup J[\xi(\Delta_1, \cdots, \Delta_n), \eta(\Delta_1', \cdots, \Delta_m')], \tag{1.12}$$

where the supremum is taken over all possible positions of the ranges of values of ξ and η into a finite number of non-overlapping intervals.

It should be remarked that our definition is by no means the only one possible. We have restricted ourselves to partitions of the ranges of values of ξ and η into non-overlapping intervals. We could have considered partitions into non-intersecting sets of a more general nature, for example, Borel sets or sets which are measurable with respect to P_ξ and P_η, where P_ξ and P_η are measures defined by the probability distributions of ξ and η, respectively (i.e., more generally, into any sets for which the probability that the value of the random variable falls into a specified set is defined).

In any case, the quantity $J(\xi, \eta)$ defined by (1.12) (where $\Delta_1, \cdots, \Delta_n$; $\Delta_1', \cdots, \Delta_m'$ should be interpreted as elements of an admissible family of sets) may be called the amount of information about ξ contained in η. Clearly, the use of more inclusive classes of sets can only increase $J(\xi, \eta)$; in any case, it is not difficult to see that under some sufficiently broad restrictions on the distribution functions P_ξ, P_η and the joint distribution function $P_{\xi\eta}$, all the suprema on the right of (1.12) are equal to one another (and hence to the integral (1.19) below), so that the different definitions of information expressed by (1.12) are equivalent.

We shall not consider the interesting question of what are the most general conditions guaranteeing equality of the suprema for different admissible classes of sets Δ_i and Δ_k'. In the sequel, we shall generally employ the most obvious way of partitioning the ranges of values of random variables into intervals, and only in one place (see Theorem 1.1 on p. 207) shall we also use a definition of information based on partitions into all possible families of measurable sets.

Before proceeding further, we shall make one more remark concerning Definition 1. In this definition we employ the term "random variable" which usually signifies that the range of values of ξ and η is the real line. It is clear, however, that the definition preserves its meaning when the ranges of these variables are multi-dimensional Cartesian spaces, i.e., when $\xi \equiv \xi = \{\xi_1, \cdots, \xi_k\}$ and $\eta \equiv \eta = \{\eta_1, \cdots, \eta_l\}$ are random vectors, since this merely requires that intervals be interpreted as multi-dimensional intervals. In the following this will always be assumed to be the case, and the notation $J(\xi, \eta)$ (in place of $J(\xi, \eta)$ will be employed the throughout except where it is explicitly stated that ξ and η are scalars.

From Property I of the information $J(\xi, \eta)$ for discrete variables ξ and η, it follows at once that the following property of $J(\xi, \eta)$ holds for arbitrary ξ and η:

I) $$0 \leq J(\xi, \eta) \leq \infty, \tag{1.13}$$

with $J(\xi, \eta) = 0$ if and only if ξ and η are independent.

It is readily seen, however, that the information $J(\xi, \eta)$ in the general case of arbitrary random vectors (or random variables) can be equal to $+\infty$. For example, if ξ and η have components (say ξ_1 and η_1) for which the equality $\xi_1 = \eta_1$ occurs with positive probability, and the conditional distribution of ξ_1 (and hence also η_1) given $\xi_1 = \eta_1$ is continuous in some interval, then $J(\xi, \eta) = \infty$ (for, on partitioning the given interval into n equiprobable parts and selecting $\Delta_1, \cdots, \Delta_n$ and $\Delta_1', \cdots, \Delta_n'$ as sets corresponding to the falling of ξ_1 and η_1 in one of these parts while the values of other components of ξ and η are arbitrary, we see that

$J(\xi, \eta) > - p_0 \log \frac{p_0}{n}$, where $p_0 > 0$, and hence the right-hand member of the latter inequality can be arbitrarily large for sufficiently large n). Necessary and sufficient conditions for finiteness of $J(\xi, \eta)$ will be derived below. For the present, let us note that, although the inequality $J(\xi, \eta) \leq J(\xi, \xi)$ is also valid for arbitrary ξ and η, its significance in the general case is relatively small since, as the example considered above demonstrates, in most cases of interest (in particular in cases with continuous distribution functions) we have $J(\xi, \xi) = +\infty$.

We turn now to other properties of the information $J(\xi, \eta)$. In the case of discrete random variables, the information (1.1) is clearly a continuous function of the probabilities $P_\xi(i)$, $P_\eta(k)$ and $P_{\xi\eta}(i, k)$. The case where the variable ξ is discrete and η is arbitrary is considered in [6], where it is shown that in this case $J(\xi, \eta)$ is continuous in $P_\xi(i)$; the general case, however, does not seem to have been investigated from this viewpoint. It is not difficult to see, though, that in the general case the dependence of $J(\xi, \eta)$ on the distribution functions of the random vector (ξ, η) is not, in general, continuous. More precisely, if (ξ_n, η_n) $(n = 1, 2, \cdots)$ is a sequence of pairs of random vectors (one k- and the other l-dimensional) which converges in distribution to the pair (ξ, η) (where by convergence in distribution is meant, as usual, convergence of the sequence of probabilities $P_{\xi_n\eta_n}(A)$ $(n = 1, 2, \cdots)$ to $P_{\xi\eta}(A)$ as $n \to \infty$, with A being any $(k + l)$-dimensional continuity interval of $P_{\xi\eta}(A)$; see, for example, [7], [8]), then, in general, $J(\xi, \eta) \neq \lim_{n\to\infty} J(\xi_n, \eta_n)$; moreover, in general, $\lim_{n\to\infty} J(\xi_n, \eta_n)$ may not exist in this case. Indeed, consider a sequence of scalar random variables ξ_n and η_n such that for any two-dimensional interval A

$$P_{\xi_n\eta_n}(A) = \left(1 - \frac{1}{n}\right) P_{\xi\eta}(A) + \frac{1}{n} P_{\xi\eta}^{(1)}(A), \tag{1.14}$$

where $P_{\xi\eta}(A)$ is some fixed probability distribution with finite information $J(\xi, \eta)$, and $P_{\xi\eta}^{(1)}(A)$ is a distribution of (ξ, η) such that $\xi = \eta$ with probability 1 and the probability density of ξ (and of η) is constant over some interval. Clearly, $\lim_{n\to\infty} P_{\xi_n\eta_n}(A) = P_{\xi\eta}(A)$ for any interval (and also for any Borel set) A, while $J(\xi, \eta)$ is finite and $J(\xi_n, \eta_n) = +\infty$ for any n. By changing the values of $P_{\xi\eta}^{(1)}(A)$ for odd n it is a simple matter to construct also an example of a convergent sequence of distributions $P_{\xi_n\eta_n}(A)$ for which $\lim_{n\to\infty} J(\xi_n, \eta_n)$ does not exist.[1)]

1) It is easy to see that if we require only the convergence of (ξ_n, η_n) to (ξ, η) in distribution (in the usual sense of probability theory), then we can construct a simpler example illustrating that information is not continuously dependent on probability distributions. For this purpose, it is sufficient to choose the ξ_n $(n = 1, 2, \cdots)$ as elements of a sequence of continuously distributed random variables which converges in distribution to a discrete variable ξ; then $J(\xi_n, \eta_n) = \infty$ for all n while $J(\xi, \xi)$ is finite. The example given in the text, however, has the advantage of clearly showing that the crux of the matter does not lie in fine points of measure theory (such as the requirement that $P_{\xi_n\eta_n}$ converge to $P_{\xi\eta}$ only for continuity intervals of the distribution $P_{\xi\eta}$); in our example the convergence of $P_{\xi_n\eta_n}(A)$ to $P_{\xi\eta}(A)$ takes place for all Borel sets A.

From the general definition of information given above it follows that $J(\xi, \eta)$, regarded as a function of the distribution $P_{\xi\eta}(A)$, has the important property of semi-continuity – which is all that is necessary in most applications. This property can be formulated as follows.

II. *If the sequence* (ξ_n, η_n) $(n = 1, 2, \cdots)$ *of pairs of random vectors converges in distribution (in the sense indicated above) to a pair* (ξ, η), *then*

$$J(\xi, \eta) \leqslant \varliminf_{n\to\infty} J(\xi_n, \eta_n). \tag{1.15}$$

For demonstration of this property it is sufficient to remark that according to the definition of information $J(\xi, \eta)$, given any $\epsilon > 0$ we can always partition the range of ξ (k-dimensional space R_k) into a finite number of k-dimensional intervals $\Delta_1, \Delta_2, \cdots, \Delta_{n_1}$, and similarly partition the range of η (l-space R_e) into a finite number of l-dimensional intervals $\Delta_1', \cdots, \Delta_{n_2}'$, such that the inequality

$$J[\xi(\Delta_1, \ldots, \Delta_{n_1}), \eta(\Delta_1', \ldots, \Delta_{n_2}')] > J(\xi, \eta) - \frac{\varepsilon}{2} \tag{1.16}$$

is satisfied.

Now we can always shift slightly in one direction or another the boundaries of intervals Δ_i, Δ_k' in such a way that after the shift all $(k+l)$-dimensional intervals $\Delta_i \cdot \Delta_k'$ $(i = 1, \cdots, n_1;\ k = 1, \cdots, n_2)$ become continuity intervals of $P_{\xi\eta}$, and at the same time all probabilities $P_\xi(\Delta_i)$, $P_\eta(\Delta_k')$, $P_{\xi\eta}(\Delta_i \cdot \Delta_k')$ change arbitrarily slightly as a result of the shift (since for any $\delta > 0$ there exist at most a finite number of points on coordinate axes which on traversing the boundaries of the intervals give rise to changes in at least one of these probabilities by amounts exceeding δ). Inasmuch as the information $J[\xi(\Delta_1, \cdots, \Delta_{n_1}), \eta(\Delta_1', \cdots, \Delta_{n_2}')]$ is continuously dependent on the probabilities $P_\xi(\Delta_i)$, $P_\eta(\Delta_k')$ and $P_{\xi\eta}(\Delta_i \cdot \Delta_k')$, we can replace the intervals $\Delta_i \cdot \Delta_k$ $(i = 1, \cdots, n_1;\ k = 1, \cdots, n_2)$ with the intervals $\overline{\Delta}_i \cdot \overline{\Delta}_k'$ $(i = 1, \cdots, n_1;\ k = 1, \cdots, n_2)$ which constitute continuity intervals for $P_{\xi\eta}$, so that the inequality

$$J[\xi(\overline{\Delta}_1, \ldots, \overline{\Delta}_{n_1}), \eta(\overline{\Delta}_1', \ldots, \overline{\Delta}_{n_2}')] > J(\xi, \eta) - \varepsilon \tag{1.17}$$

is satisfied. However, from the convergence of $P_{\xi_n\eta_n}(A)$ to $P_{\xi\eta}(A)$ for continuity intervals A it follows that

$$\lim_{n\to\infty} J[\xi_n(\overline{\Delta}_1, \ldots, \overline{\Delta}_{n_1}), \eta_n(\overline{\Delta}_1', \ldots, \overline{\Delta}_{n_2}')] = J[\xi(\overline{\Delta}_1, \ldots, \overline{\Delta}_{n_1}); \eta(\overline{\Delta}_1', \ldots, \overline{\Delta}_{n_2}')],$$

and consequently that $\varliminf_{n\to\infty} J(\xi_n, \eta_n)$ cannot be less that $J(\xi, \eta)$, which is what we set out to prove.

A third property of information which we shall establish here is in some respects related to Property II of the information $J(\xi, \eta)$ for discrete variables ξ and η.

III. *The amount of information* $J(\xi, \eta)$ *can only increase with increase in the number of components of* ξ *or* η *or both. Thus*

$$J(\xi_1, \eta_1) \geq J(\xi, \eta) \tag{1.18}$$

where

$$\xi = (\xi_1, \ldots, \xi_k), \quad \eta = (\eta_1, \ldots, \eta_l),$$

$$\xi_1 = (\xi_1, \ldots, \xi_k, \xi_{k+1}, \ldots, \xi_{k+r}), \quad \eta_1 = (\eta_1, \ldots, \eta_l, \eta_{l+1}, \ldots, \eta_{l+s}).$$

Proof of this property follows at once from the fact that, for arbitrary intervals $\Delta \in R_k$ and $\Delta' \in R_l$ (where R_n denotes a Cartesian n-space), the following equalities hold: $P_\xi(\Delta) = P_{\xi_1}(\Delta \cdot R_r)$, $P_\eta(\Delta') = P_{\eta_1}(\Delta' \cdot R_s)$ and $P_{\xi\eta}(\Delta \cdot \Delta') = P_{\xi_1\eta_1}[(\Delta \cdot R_r) \cdot (\Delta' \cdot R_s)]$, where $\Delta \cdot R_r$ and $\Delta' \cdot R_s$ are cylinder sets (infinite intervals) in R_{k+r} and R_{l+s} respectively, comprising points whose projections on the subspace spanned by the first k (respectively l) coordinate axes belong to the interval Δ (respectively Δ'). (In this connection, see also Theorem 1.2 on p. 209).

We shall now formulate an important theorem which makes it possible to reduce the determination of information $J(\xi, \eta)$ to the evaluation of a certain integral.

Theorem 1.1. *In order that the amount of information $J(\xi, \eta)$ contained in one of two random vectors about the other vector be finite, it is necessary that the probability distribution $P_{\xi\eta}$ be absolutely continuous with respect to the distribution $P_\xi \cdot P_\eta$. Under this condition, the information $J(\xi, \eta)$ defined as the supremum* (1.12) *over all partitions of the ranges of values of ξ and η into a finite number of sets measurable with respect to P_ξ and P_η respectively, is equal to the following Lebesgue-Stieltjes integral:*

$$J(\xi, \eta) = \int \alpha(\mathbf{x}, \mathbf{y}) \log \alpha(\mathbf{x}, \mathbf{y})\, dP_\xi(\mathbf{x})\, dP_\eta(\mathbf{y}), \tag{1.19}$$

where

$$\alpha(\mathbf{x}, \mathbf{y}) = \frac{dP_{\xi\eta}(\mathbf{x}, \mathbf{y})}{dP_\xi(\mathbf{x})\, dP_\eta(\mathbf{y})} \tag{1.20}$$

(In particular, $J(\xi, \eta)$ is finite or infinite according as the integral on the right of (1.19) *is finite or infinite.)*

As regards the information $J(\xi, \eta)$ in the sense generally employed in this paper (i.e., defined through partitioning the ranges of values of ξ and η into intervals), all we can assert in the general case is that it does not exceed the value of the integral (1.19). However, under fairly broad conditions (e.g., when the integral (1.19) exists as a Riemann-Stieltjes integral in the broadest sense of the term, i.e., in the sense of a generalized Stiletjes integral [31]) the latter information coincides, as will be shown later, with the information defined through the partition into arbitrary measurable sets – that is, it coincides with the integral (1.19). Since in all cases of practical interest the condition for existence of the integral (1.19) in the Riemann-Stieltjes sense is always satisfied, the expression (1.19) can be used in all such cases without a detailed analysis of the basic definition of information.

Proof of Theorem 1.1. We first assume that the probability distribution $P_{\xi\eta}$ is not absolutely continuous with respect to $P_\xi P_\eta$. This implies that there exists some $\epsilon > 0$ and a sequence of positive numbers $\delta_1, \delta_2, \cdots, \delta_n, \cdots$ with $\delta_n \to 0$ as $n \to \infty$ such that for any n one can choose a finite number (say N_n) of $k+l$-dimensional intervals $S_1^{(n)}, S_2^{(n)}, \cdots, S_{N_n}^{(n)}$ satisfying the conditions

$$P_{\xi\eta}\left(\bigcup_{i=1}^{N_n} S_i^{(n)}\right) > \varepsilon, \quad P_\xi P_\eta\left(\bigcup_{i=1}^{N_n} S_i^{(n)}\right) < \delta_n \tag{1.21}$$

in consequence of which we have the limiting relation

$$\sum_{i=1}^{N_n} P_{\xi\eta}(S_i^{(n)}) \log \frac{P_{\xi\eta}(S_i^{(n)})}{P_\xi P_\eta(S_i^{(n)})} \to \infty \quad \text{for} \quad n \to \infty \tag{1.22}$$

(for, by virtue of the cancavity of the function $\Phi(x) = x \log x$ and its boundedness from below by the number $-\dfrac{\log e}{e}$, we can write the inequality

$$\sum_i P_{\xi\eta}(S_i^{(n)}) \log \frac{P_{\xi\eta}(S_i^{(n)})}{P_\xi P_\eta(S_i^{(n)})} = \sum_i P_\xi P_\eta(S_i^{(n)}) \cdot \Phi\left(\frac{P_{\xi\eta}(S_i^{(n)})}{P_\xi P_\eta(S_i^{(n)})}\right) \geqslant$$

$$\geqslant \sum_i P_\xi P_\eta(S_i^{(n)}) \cdot \Phi\left(\frac{\sum_i P_{\xi\eta}(S_i^{(n)})}{\sum_i P_\xi P_\eta(S_i^{(n)})}\right) =$$

$$-\sum_i P_{\xi\eta}(S_i^{(n)}) \cdot [\log(\sum_i P_{\xi\eta}(S_i^{(n)})) - \log(\sum_i P_\xi P_\eta(S_i^{(n)}))] \geqslant -\frac{\log e}{e} - \varepsilon \log \delta_n,$$

in which the term $-\epsilon \log \delta_n$ tends to infinity as $n \to \infty$). From this it follows immediately that in the case under consideration the information $J(\xi, \eta)$ is equal to to infinity.

Let us now assume that the distribution $P_{\xi\eta}$ is absolutely continuous with respect to $P_\xi \cdot P_\eta$, so that according to the Radon-Nikodym theorem there exists a density

$$\alpha(\mathbf{x}, \mathbf{y}) = \frac{dP_{\xi\eta}(\mathbf{x}, \mathbf{y})}{dP_\xi(\mathbf{x})\, dP_\eta(\mathbf{y})},$$

which is measurable with respect to $dP_\xi(\mathbf{x})\,dP_\eta(\mathbf{y})$, and hence the Lebesgue-Stieltjes integral (1.19) is meaningful. Partitioning the range of variation of the integrand $\alpha(\mathbf{x}, \mathbf{y}) \log \alpha(\mathbf{x}, \mathbf{y})$ (or truncated integrand, if $\alpha \log \alpha$ is not bounded) into equal intervals induces a partition of the ranges of random vectors ξ and η into sets Δ_i and Δ_k' which are measurable with respect to dP_ξ and dP_η, respectively; then $J[\xi(\Delta_1 \cdots, \Delta_{n_1}), \eta(\Delta_1', \cdots, \Delta_{n_2}')]$ will coincide with the integral sum sum of the Lebesgue integral (1.19), and it is clear that the supremum (1.12) taken over all partitions of the ranges of ξ and η into a finite number of measurable sets is identical with the Lebesgue-Stieltjes integral (1.19). This completes the proof of Theorem 1.1.

In the very important special case where the integral (1.19) exists in the Riemann-Stieltjes sense (i.e., as a "generalized Stieltjes integral" in the terminology of [31]), the sum $J[\xi(\Delta_1, \cdots, \Delta_{n_1}), \eta(\Delta_1', \cdots, \Delta_{n_2}')]$, for some partition of the ranges of vectors ξ and η into intervals is bounded by the upper and lower Darboux sums for the Riemann-Stieltjes integral. From this it follows that the supremum (1.12) over all partitions into intervals will be finite together with integral (1.19) and will have the same value. In the general case, where the Riemann-Stieltjes integral (1.19) may not exist, all we can assert is that the information $J(\xi, \eta)$ in the sense of Definition 1 is certainly not greater than the value of integral (1.19) (since the supremum (1.12) over all partitions into intervals is never greater than the supremum taken over partitions into arbitrary measurable sets).

The latter result can also be obtained quite simply without any reference to the fact that integral (1.19) is equal to the supremum of (1.12) taken over all partitions of the ranges of ξ and η into arbitrary measurable sets. For this we make use of the inequality

$$\int_S \Phi(\alpha(\mathbf{x}, \mathbf{y}))\, dP(\mathbf{x})\, dP(\mathbf{y}) \geqslant \Phi\left\{\int_S \alpha(\mathbf{x}, \mathbf{y})\, dP(\mathbf{x})\, dP(\mathbf{y})\right\} \quad \text{for} \quad \int_S dP(\mathbf{x})\, dP(\mathbf{y}) = 1, \tag{1.23}$$

which holds in the case of a concave function $\Phi(x)$ for any function $\alpha(\mathbf{x}, \mathbf{y})$ which is measurable with respect to $dP(\mathbf{x})dP(\mathbf{y})$ (compare with inequality (1.6)). Setting $S = \Delta_i \cdot \Delta_k' =$ some $(k + l)$-dimensional interval,

$$dP(\mathbf{x})\, dP(\mathbf{y}) = \frac{dP_\xi(\mathbf{x})\, dP_\eta(\mathbf{y})}{P_\xi(\Delta_i)\, P_\eta(\Delta_k')}$$

and

$$\Phi(x) = x \log x,$$

and summing over all intervals $\Delta_i \cdot \Delta_k'$ contained in some partition of the range of definition of (ξ, η), we have:

$$J[\xi(\Delta_1, \ldots, \Delta_{n_1}),\ \eta(\Delta_1', \ldots, \Delta_{n_2}')] \leqslant \int \alpha(\mathbf{x}, \mathbf{y}) \log \alpha(\mathbf{x}, \mathbf{y})\, dP_\xi(x)\, dP_\eta(\mathbf{y}), \tag{1.24}$$

so that

$$J(\xi, \eta) \leq \int \alpha(\mathbf{x}, \mathbf{y}) \log \alpha(\mathbf{x}, \mathbf{y})\, dP_\xi(\mathbf{x})\, dP_\eta(\mathbf{y}). \tag{1.25}$$

On writing information in the form of integral (1.19), it follows, in particular, that $J(\xi, \eta)$ is independent of the choice of bases in the spaces of vectors ξ and η. More specifically, we can state the following theorem, proof of which follows immediately from equation (1.19):

Theorem 1.2. *Let A be a linear transformation in k-dimensional vector space and let ξ be a k-dimensional random vector. Then*

$$J(\xi, \eta) \geq J(A\xi, \eta) \tag{1.26}$$

holds for any random vector η, with equality if the transformation A is non-singular.

A particular case of this theorem is the previously stated Property III of information $J(\xi, \eta)$. In the simplest case where the probability distributions P_ξ, P_η and $P_{\xi\eta}$ are defined in terms of densities $p_\xi(\mathbf{x})$, $p_\eta(\mathbf{y})$ and $p_{\xi\eta}(\mathbf{x}, \mathbf{y})$, the Stieltjes integral (1.19) reduces to the ordinary integral:

$$J(\xi, \eta) = \iint p_{\xi\eta}(x, y) \log \frac{p_{\xi\eta}(x, y)}{p_\xi(x)\, p_\eta(y)}\, dx\, dy \qquad (1.27)$$

(where $dx = dx_1 \cdots dx_k$, $dy = dy_1 \cdots dy_l$). This is the formula for the information about one random vector contained in another such vector which is employed in the work of Shannon [1]. From the foregoing discussion it follows that when integral (1.27) exists in the Riemann sense (as it does, for example, when all the densities are continuous or piecewise continuous) the formula (1.27) is valid regardless of which one of the definitions of information $J(\xi, \eta)$ given on p. 203 is adopted.

§2. Information about a random function contained in another such function. Before defining what is meant by information in this case, we should make one remark. In practical problems of communication theory – which were responsible for the development of information theory – an important role is played by so-called "white noise". which is an "improper" stationary random process with constant spectral density; in some practical problems other types of improper random processes are also encountered. For such processes the notion of "the value, $\xi(t)$, of the process ξ at the point t" loses its meaning. Thus in dealing with such processes one can speak only of "linear functionals $\xi(\phi)$ with respect to the process ξ" (or, in physical terminology, of "the result of measurement of the process ξ by the use of a linear device with dynamic characteristic ϕ"). A theory of generalized random processes which include the "improper" as well as ordinary processes was described in recent papers by Itô [9], and one of the authors of the present paper [10]. Here we shall formulate a definition of information $J(\xi, \eta)$ in such a way as to make it applicable to any generalized processes ξ and η,

Actually all of the following analyses relate to random functions ξ and η defined on arbitrary differentiable manifolds; however, in order to avoid complications in terminology we shall assume that ξ and η are random processes over the real line ("time axis") or a segment of it. To define a generalized random process we must first specify a space Φ of testing functions $\phi(t)$ on a set T of values of t, i.e., the space of all infinitely differentiable functions on T which vanish outside some closed interval [1]; as regards a topology in Φ, we introduce it in the same way as it is done in [11].

Following [10] we adopt the

Definition 2. *By a generalized random process is meant a random linear functional $\xi(\phi)$ defined on Φ, i.e., a family of random variables $\xi(\phi)$ generated by the functions $\phi \in \Phi$ and linearly dependent on ϕ. The finite-dimensional probability distributions of random variables $\xi(\phi_1), \xi(\phi_2), \cdots, \xi(\phi_n)$, which charac-*

1) Instead of the space Φ introduced by Schwartz [11], one can also use any other space of testing functions (see [12]).

terize the $\xi(\phi)$ process must satisfy the usual conditions of symmetry and consistency and, in addition, must be continuously dependent on the functions $\phi_k \in \Phi$ in the sense of weak topology in the space of finite-dimensional distributions and the topology defined in Φ. 1)

Clearly the notion of a generalized random process included as special cases the usual continuous (in stochastic sense) random processes which are integrable over any finite interval in the sense of convergence in probability of the integral sum to the limiting value (in the sequel this will be abbreviated to "integrable"). In this case $\xi(\phi)$ can be expressed in the form

$$\xi(\phi) = \int_{-\infty}^{\infty} \xi(t)\phi(t)\,dt, \tag{1.28}$$

which subsumes all the "improper" processes encountered in practice. For example, in the case of "white noise" $\xi(\phi)$ is a normally distributed random variable with zero mean and variance $\int_{-\infty}^{\infty} \phi^2(t)\,dt$.

Definition 3. *Let $\xi(\phi)$ and $\eta(\psi)$ be two generalized random processes defined on the spaces Φ and Ψ of testing functions. In this case the amount of information concerning one of these processes contained in the other is given by*

$$J(\xi, \eta) = \sup J[\xi(\phi_1, \cdots, \phi_n), \eta(\psi_1, \cdots, \psi_m)], \tag{1.29}$$

where $\xi(\phi_1, \cdots, \phi_n) = \{\xi(\phi_1), \cdots, \xi(\phi_n)\}$ and $\eta(\psi_1, \cdots, \psi_m) = \{\eta(\psi_1), \cdots, \eta(\psi_m)\}$ are ordinary random vectors (of n and m dimensions respectively) and the supremum is taken over all integers n and m and all families of functions $\phi_1, \cdots, \phi_n \in \Phi$ and $\psi_1, \cdots, \psi_m \in \psi$.

From Properties I and II of the information $J(\xi, \eta)$ for random vectors ξ and η there follow at once analogous properties of the information $J(\xi, \eta)$ for random processes:

I)
$$0 \le J(\xi, \eta) \le \infty, \tag{1.30}$$

with $J(\xi, \eta) = 0$ if and only if the processes ξ and η are mutually independent (i.e., when the variables $\xi(\phi)$ and $\eta(\psi)$ are independent for any $\phi \in \Phi$ and $\psi \in \Psi$).

II) *If a sequence of pairs of random processes $\{\xi_k, \eta_k\}$ converges to $\{\xi, \eta\}$ in the sense that, for any n, m, $\phi_1, \cdots, \phi_n \in \Phi$ and $\psi_1, \cdots, \psi_m \in \Psi$, the $(n+m)$-dimensional distribution function for the vector*

$$\{\xi_k(\phi_1), \cdots, \xi_k(\phi_n), \eta_k(\psi_1), \cdots, \eta_k(\psi_m)\}$$

converges weakly to the distribution function for the vector

1) **As shown in [10], it is actually sufficient to know only the one-dimensional distribution functions for the variables $\xi(\phi)$. As regards the condition of continuity, it was pointed out by A. N. Kolmogorov that it can be formulated in the following way: for each $\phi_2 \in \Phi$ and $\epsilon > 0$ there should exist a neighborhood Φ_{ϕ_0} of the function ϕ_0 such that $P\{|\xi(\phi) - \xi(\phi_0)| > \epsilon\} < \epsilon$ for $\phi \in \Phi_{\phi_0}$.**

$$\{\xi(\phi_1), \cdots, \xi(\phi_n), \eta(\psi_1), \cdots, \eta(\psi_m)\},$$

as $k \to \infty$, *then*

$$\varliminf_{k\to\infty} J(\xi_k, \eta_k) \geqslant J(\xi, \eta). \tag{1.31}$$

It is not difficult to construct simple examples showing that the left-hand member of the latter inequality can indeed be greater than the right-hand member; we shall not dwell on this point.

Theorem 1.3. *Let* $\xi(t)$ *and* $\eta(s)$ *be ordinary integrable random processes with parameters* $t \in T$ *and* $s \in S$, *where* T *and* S *are sets of real numbers. Then*

$$J(\xi, \eta) = \sup J[\xi(t_1, \cdots, t_n), \eta(s_1, \cdots, s_m)], \tag{1.32}$$

where $\xi(t_1), \cdots, t_n) = \{\xi(t_1), \cdots, \xi(t_n)\}$, $\eta(s_1, \cdots, s_m) = \{\eta(s_1), \cdots, \eta(s_m)\}$ *and the supremum is taken over all integers* n *and* m *and finite sets of points* $t_1, \cdots, t_n \in T$ *and* $s_1, \cdots, s_m \in S$.

Proof. Denote the right-hand side of (1.32) by $\widetilde{J}(\xi, \eta)$, reserving the symbol $J(\xi, \eta)$ for the quantity (1.29). We must prove that $\widetilde{J}(\xi, \eta) = J(\xi, \eta)$. Choose an $\epsilon > 0$ and let $t_1, \cdots, t_n$ and $s_1, \cdots, s_m$ be finite sets of points such that

$$J[\xi(t_1, \cdots, t_n), \eta(s_1, \cdots, s_m)] > \widetilde{J}(\xi, \eta) - \epsilon. \tag{1.33}$$

(According to the definition of $\widetilde{J}(\xi, \eta)$ such n, m, and $t_1, \cdots, t_n, s_1, \cdots, s_m$ can always be found for any $\epsilon > 0$.) Let us now select a sequence $\phi_1^{(k)}, \cdots, \phi_n^{(k)}$ $(k = 1, 2, \cdots)$ of functions in the space Φ which converges, as $k \to \infty$, to a family of Dirac delta-functions $\delta(t - t_1), \cdots, \delta(t - t_n)$ in the sense that

$$\int_T \phi_i^{(k)}(t)\,dt = 1 \text{ and } \phi_i^{(k)}(t) = 0 \text{ for } k > k_0(t) \text{ with } t \neq t_1.$$

Similarly, select a sequence $\psi_1^{(k)}(s), \cdots, \psi_m^{(k)}(s)$, converging to $\delta(s - s_1), \cdots, \delta(s - s_m)$. In this case, by virtue of the stochastic continuity of $\xi(t)$ and $\eta(s)$, the sequence of random variables $\xi(\phi_i^{(k)}), \eta(\psi_i^{(k)})$ will converge in probability to $\xi(t_i)$ and $\eta(s_i)$, respectively, as $k \to \infty$, i.e., the probability distribution of $\{\xi(\phi_1^{(k)}), \cdots, \xi(\phi_n^{(k)}), \eta(\psi_1^{(k)}), \cdots, \eta(\psi_m^{(k)})\}$ will converge weakly to the distribution of $\{\xi(t_1), \cdots, \xi(t_n), \eta(s_1), \cdots, \eta(s_m)\}$. In view of Property II of the information $J(\xi, \eta)$ for vectors ξ and η, it follows immediately that

$$J(\xi, \eta) \geq \widetilde{J}(\xi, \eta) - \epsilon \tag{1.34}$$

for any $\epsilon > 0$; hence

$$J(\xi, \eta) \geq \widetilde{J}(\xi, \eta). \tag{1.35}$$

To prove the reverse inequality

$$J(\xi, \eta) \leq \widetilde{J}(\xi, \eta) \tag{1.36}$$

we choose integers n, m and a family of functions $\phi_1, \cdots, \phi_n \in \Phi$ and $\psi_1, \cdots, \psi_m \in \Psi$ such that the inequality

$$J[\xi(\phi_1, \cdots, \phi_n), \eta(\psi_1, \cdots, \psi_m)] > J(\xi, \eta) - \epsilon; \tag{1.37}$$

is satisfied. (This can be done for any $\epsilon > 0$.) Now choose a sequence of integers

$N^{(k)}$, $M^{(k)}$ $(k = 1, 2, \cdots)$ and points $t_1^{(k)}, \cdots, t_{N(k)}^{(k)}$, $s_1^{(k)}, \cdots, s_{M(k)}^{(k)}$ such that the Riemann sums

$$\sum_{l=1}^{N(k)} \varphi_i(t_l^{(k)})\,\xi(t_l^{(k)})\,\Delta t_l^{(k)}, \qquad \sum_{l=1}^{M(k)} \psi_j(s_l^{(k)})\,\eta(s_l^{(k)})\,\Delta s_l^{(k)}$$

$(i = 1, \cdots, n;\ j = 1, \cdots, m)$ converge in probability to the integrals $\int_T \phi_i(t)\,\xi(t)\,dt$, $\int_S \psi_j(s)\eta(s)\,ds$ as $k \to \infty$. Then by virtue of Property II of the information $J(\xi, \eta)$, given any $\epsilon > 0$ it is always possible to find a finite number of points $t_1, \cdots, t_N$ and $s_1, \cdots, s_M$ such that

$$J[\xi(\hat{\varphi}_1, \ldots, \hat{\varphi}_n), \eta(\hat{\psi}_1, \ldots, \hat{\psi}_m)] > J[\xi(\varphi_1, \ldots, \varphi_n), \eta(\psi_1, \ldots, \psi_m)] - \varepsilon, \tag{1.38}$$

where $\xi(\hat{\phi}_1), \cdots, \xi(\hat{\phi}_n)$ are the Riemann sums for the integrals $\xi(\phi_1), \cdots, \xi(\phi_n)$, which in turn are linear combinations of the values of $\xi(t)$ at the points $t_1, \cdots, t_N$, and $\eta(\hat{\psi}_1), \cdots, \eta(\hat{\psi}_m)$ are the Riemann sums for the integrals $\eta(\psi_1), \cdots, \eta(\psi_m)$ which are linear combinations of the values of $\eta(s)$ at the points $s_1, \cdots, s_M$. Let us now select a new basis in the N-space of vectors $\{\xi(t_1), \cdots, \xi(t_N)\}$, using as base vectors those linear combinations of $\xi(\hat{\phi}_1), \cdots, \xi(\hat{\phi}_n)$ which are linearly independent, and, if necessary, extending this set to a basis by using some linear combinations of $\xi(t_1), \cdots, \xi(t_N)$. In a similar manner a basis $\{\eta(s_1), \cdots, \eta(s_M)\}$ is constructed in the space of vectors $\eta(\hat{\psi}_1), \cdots, \eta(\hat{\psi}_m), \cdots$. Then from Theorem 1.2 it follows that

$$J[\xi(t_1, \ldots, t_N), \eta(s_1, \ldots, s_M)] \geqslant J[\xi(\hat{\varphi}_1, \ldots, \hat{\varphi}_n), \eta(\hat{\psi}_1, \ldots, \hat{\psi}_m)] > J(\xi, \eta) - 2\varepsilon \tag{1.39}$$

(by (1.37), (1.38)), which leads at once to the inequality (1.36).

It is natural to expect that a theorem analogous to Theorem 1.1 holds in the case of (generalized) random processes ξ and η. Specifically, the information $J(\xi, \eta)$ is finite only if the probability distribution $P_{\xi\eta}$ (associated with the pair $\{\xi(\phi), \eta(\psi)\}$) on the space of pairs $\{x(\phi), y(\psi)\}$ of ordinary (non-random) generalized functions is absolutely continuous with respect to the distribution $P_\xi \cdot P_\eta$, and the function $\alpha(x, y)\log\alpha(x, y)$, where $\alpha(x, y)$ is defined by (1.20), is integrable with respect to the measure $dP_\xi dP_\eta$. In this case $J(\xi, \eta)$ is given by an integral over function space (i.e., integral with respect to a measure on function space) analogous to that appearing on the right side of (1.19). However, demonstration of this theorem requires the use of a result concerning the extension of finite-dimensional probability distributions for generalized random processes to a completely additive measure on the space of generalized functions, proof of which is not yet available.

Theorem 1.2 can readily be extended to the case of the information $J(\xi, \eta)$ for generalized processes $\xi(\phi)$ and $\eta(\psi)$.

Theorem 1.4. *Let A be a continuous linear transformation in the space Φ of testing functions. For a generalized random process $\xi(\phi)$ we define A by the relation*

$$(A\xi)(\phi) = \xi(A\phi). \tag{1.40}$$

Then for an arbitrary generalized process $\eta(\psi)$

$$J(\xi, \eta) \geq J(A\xi, \eta), \tag{1.41}$$

with equality if the transformation A *has an inverse* A^{-1}.

Proof of this theorem follows at once from the fact that any functional $A\xi(\phi)$ of the process $A\xi$ is simultaneously a functional of the process ξ (specifically, the functional $\xi(A\phi)$).

We shall now prove a theorem which makes it possible to represent the information $J(\xi, \eta)$ as a limit (rather than the supremum) of expressions having the form of the right-hand member of (1.29). This theorem is useful for purposes of calculating $J(\xi, \eta)$.

Theorem 1.5. *Let* $\Phi_1 \subset \Phi_2 \subset \cdots$ *be a monotone increasing sequence of finite-dimensional linear subspaces of* Φ *such that* $[\bigcup_{n=1}^{\infty} \Phi_n] = \Phi$ *(i.e., the closure of the union of all* Φ_n *coincides with* Φ*); similarly, let* $\Psi_1 \subset \Psi_2 \subset \cdots$ *and* $[\bigcup_{n=1}^{\infty} \Psi_n] = \Psi$. *Denote by* $\phi_1^{(k)}, \cdots, \phi_{n_k}^{(k)}$ *a basis in* Φ_k *and by* $\psi_1^{(k)}, \cdots, \psi_{m_k}^{(k)}$ *a basis in* Ψ_k. *Then*

$$\lim_{h\to\infty,\, l\to\infty} J[\xi(\varphi_1^{(k)}, \ldots, \varphi_{n_k}^{(h)}), \eta(\psi_1^{(l)}, \ldots, \psi_{m_l}^{(l)})] = J(\xi, \eta). \tag{1.42}$$

Note that by Theorem 1.2, the information

$$J[\xi(\phi_1^{(k)}, \cdots, \phi_{n_k}^{(k)}), \eta(\psi_1^{(l)}, \cdots, \psi_{m_l}^{(l)})]$$

is independent of the choice of bases in subspaces Φ_k and Ψ_l, and is uniquely determined by these subspaces; thus it can properly be denoted by $J[\xi(\Phi_n), \eta(\Psi_l)]$, and the equality (1.42) can be expressed in the form

$$\lim_{k\to\infty,\, l\to\infty} J[l(\Phi_k), \eta(\Psi_l)] = J[\xi(\Phi), \eta(\Psi)] \tag{1.42'}$$

where $J[\xi(\Phi), \eta(\Psi)]$ represents the information $J(\xi, \eta)$.

Proof. **Fixing $\epsilon > 0$, we choose integers n, m and finite families of functions $\phi_1, \cdots, \phi_n$ in Φ and $\psi_1, \cdots, \psi_m$ in Ψ such that**

$$J[\xi(\phi_1, \cdots, \phi_n), \eta(\psi_1, \cdots, \psi_m)] > J(\xi, \eta) - \epsilon. \tag{1.43}$$

Since $[\bigcup_{k=1}^{\infty} \Phi_k] = \Phi$ and $[\bigcup_{l=1}^{\infty} \Psi_l] = \Psi$, it is possible to find sequences $\hat{\phi}_1^{(k)}, \cdots, \hat{\phi}_n^{(k)}$ $(k = 1, 2, \cdots)$ and $\hat{\psi}_1^{(l)}, \cdots, \hat{\psi}_m^{(l)}$ $(l = 1, 2, \cdots)$ of linear combinations of functions $\phi_i^{(k)}$ and $\psi_j^{(l)}$ respectively, which as $k \to \infty$ and $l \to \infty$ converge to $\phi_1, \cdots, \phi_n$ and $\psi_1, \cdots, \psi_m$ respectively. Then the $(n+m)$-dimensional probability distribution of the random vector $\{\xi(\hat{\phi}_1^{(k)}), \cdots, \xi(\hat{\phi}_n^{(k)}), \eta(\hat{\phi}_1^{(l)}), \cdots, \eta(\hat{\psi}_m^{(l)})\}$ will converge weakly to the probability distribution of the vector $\{\xi(\phi_1), \cdots, \xi(\phi_n), \eta(\psi_1), \cdots, \eta(\phi_m)\}$ as $k \to \infty$, $l \to \infty$; (see the definition of a generalized random process). Recalling Properties II and III of the information $J(\xi, \eta)$ for vectors ξ and η, it follows that

$$\lim_{k\to\infty,\, l\to\infty} J[\xi(\Phi_k), \eta(\Psi_l)] \geqslant J[\xi(\varphi_1, \ldots, \varphi_n), \eta(\psi_1, \ldots, \psi_m)] > J(\xi, \eta) - \varepsilon \tag{1.44}$$

(we use here the symbol lim rather than $\underline{\lim}$ since the information $J[\xi(\Phi_k), \eta(\Psi_l)]$ does not decrease with increase in k and l, and, in any case, $J[\xi(\Phi_k), \eta(\Psi_l)]$ is not greater than $J(\xi, \eta)$; hence the limit exists). Thus, the inequality (1.44) holds for any $\epsilon > 0$; from this (1.42) follows immediately.

Chapter II.

Calculation of information for Gaussian random functions

§1. Information about a Gaussian random vector contained in another such vector. Up to this point our discussion was concerned with the definition and general properties of the information $J(\xi, \eta)$. Explicit calculation of $J(\xi, \eta)$ requires the knowledge of multi-dimensional probability distribution functions for the random processes ξ and η. Clearly such calculations cannot be carried out in general terms, and every concrete case requires its own treatment. Here we shall consider the most important (and also one of the simplest) of such cases – the case where the processes ξ and η are Gaussian, i.e., when all the multi-dimensional probability distributions for ξ, η and the pair (ξ, η) are multivariate normal. For this case the calculation of $J(\xi, \eta)$ can be carried considerably further than in the general case; some results in this direction are given in the present chapter.

As a preliminary to the consideration of Gaussian random processes, we shall examine the form assumed by $J(\xi, \eta)$ for Gaussian random vectors ξ and η. A method of calculating $J(\xi, \eta)$ for this case is indicated in the work of Shannon [1]; there, however, the calculation of $J(\xi, \eta)$ is left incomplete. Actually, for the case in question an explicit expression for $J(\xi, \eta)$ can be obtained quite simply; moreover, the result has a nice geometrical significance which materially facilitates further generalizations.

Thus let $\xi = \{\xi_1, \cdots, \xi_k\}$, $\eta = \{\eta_1, \cdots, \eta_l\}$ and $\zeta = \{\xi, \eta\} = \{\xi_1, \cdots, \xi_k, \eta_1, \cdots, \eta_l\}$ be normally distributed random vectors. For the present we shall assume that the associated multi-dimensional Gaussian distributions are non-singular. (The general case can readily be reduced to this case.) Furthermore, we assume that the expectations of all components of ξ and η are equal to zero (this assumption is not restrictive since the information $J(\xi, \eta)$ is independent of the values of these expectations). In this case, the probability distributions for ξ, η, and ζ will have densities of the form

$$\left.\begin{aligned} p_\xi(\mathbf{x}) &= \frac{1}{(2\pi)^{k/2}(\det A)^{1/2}} e^{-\frac{1}{2}(A^{-1}\mathbf{x}, \mathbf{x})}, \quad p_\eta(\mathbf{y}) = \frac{1}{(2\pi)^{l/2}(\det B)^{1/2}} e^{-\frac{1}{2}(B^{-1}\mathbf{y}, \mathbf{y})}, \\ p_\zeta(\mathbf{z}) &= p_{\xi,\eta}(\mathbf{z}) = \frac{1}{(2\pi)^{\frac{k+l}{2}}(\det C)^{1/2}} e^{-\frac{1}{2}(C^{-1}\mathbf{z}, \mathbf{z})}, \end{aligned}\right\} \qquad (2.1)$$

where $\mathbf{x}$ and $\mathbf{y}$ are points in k– and l– spaces respectively,

$\mathbf{z} = (\mathbf{x}, \mathbf{y}) = (x_1, \cdots, x_k, y_1, \cdots, y_l)$,

$$C = \begin{Vmatrix} A & D \\ D' & B \end{Vmatrix} \tag{2.2}$$

(prime denotes transposition), and A, B, D are moment matrices, i.e.,

$$A = \| \mathbf{M}\xi_i\xi_j \| = \| a_{ij} \|, \quad B = \| \mathbf{M}\eta_i\eta_j \| = \| b_{ij} \|, \quad D = \| \mathbf{M}\xi_i\eta_j \| = \| d_{ij} \| \tag{2.3}$$

(the existence of inverse matrices A^{-1}, B^{-1} and C^{-1} follows from the assumption of non-singularity of the probability distributions). On substituting these expressions in (1.27), the information $J(\xi, \eta)$ assumes the form

$$J(\xi, \eta) = \int \left\{ \frac{1}{2} \log \frac{\det A \cdot \det B}{\det C} - \frac{1}{2} [(C^{-1}\mathbf{z}, \mathbf{z}) - (A^{-1}\mathbf{x}, \mathbf{x}) - (B^{-1}\mathbf{y}, \mathbf{y})] \right\} p_\zeta(\mathbf{z})\, d\mathbf{z}. \tag{2.4}$$

For the calculation of this integral it is sufficient to note that by virtue of the definition of second moments we have

$$\left.\begin{aligned} \int x_i x_j p_\zeta(\mathbf{z})\, d\mathbf{z} &= \int x_i x_j p_\xi(\mathbf{x})\, d\mathbf{x} = \mathbf{M}\xi_i\xi_j = a_{ij}, \\ \int y_i y_j p_\zeta(\mathbf{z})\, d\mathbf{z} &= \int y_i y_j p_\eta(\mathbf{y})\, d\mathbf{y} = \mathbf{M}\eta_i\eta_j = b_{ij}, \\ \int x_i y_j p_\zeta(\mathbf{z})\, d\mathbf{z} &= \mathbf{M}\xi_i\eta_j = d_{ij}, \end{aligned}\right\} \tag{2.5}$$

from which the following relations ensue:

$$\left.\begin{aligned} \int (A^{-1}\mathbf{x}, \mathbf{x})\, p_\zeta(\mathbf{z})\, d\mathbf{z} &= \int (A^{-1}\mathbf{x}, \mathbf{x})\, p_\xi(\mathbf{x})\, d\mathbf{x} = \mathrm{Sp}\,(A^{-1}A) = k, \\ \int (B^{-1}\mathbf{y}, \mathbf{y})\, p_\zeta(\mathbf{z})\, d\mathbf{z} &= \int (B^{-1}\mathbf{y}, \mathbf{y})\, p_\eta(\mathbf{y})\, d\mathbf{y} = \mathrm{Sp}\,(B^{-1}B) = l, \\ \int (C^{-1}\mathbf{z}, \mathbf{z})\, p_\zeta(\mathbf{z})\, d\mathbf{z} &= \mathrm{Sp}\,(C^{-1}C) = k + l, \end{aligned}\right\} \tag{2.6}$$

where $\mathrm{Sp}\,K$ denotes the trace of matrix K, i.e., the sum of its diagonal elements. In view of the importance of the final result – which is obtained by substituting (2.6) in (2.4) – we state it as a theorem.

Theorem 2.1. *The information $J(\xi, \eta)$ contained in one of two Gaussian random vectors ξ, η about the other is given by the formula*

$$J(\xi, \eta) = \frac{1}{2} \log \frac{\det A \cdot \det B}{\det C}, \tag{2.7}$$

where A, B, and C are moment matrices for ξ, η and $\zeta = (\xi, \eta)$ respectively, which we assume to be non-singular.

In the simplest case where ξ and η are one-dimensional random variables, expression (2.7) becomes

$$J(\xi, \eta) = -\frac{1}{2} \log[1 - r^2(\xi, \eta)], \tag{2.8}$$

where $r(\xi, \eta)$ is the correlation coefficient of ξ and η. The unique functional relationship between the information and the correlation coefficient r is completely natural since both these quantities are measures of "mutual dependence" of ξ and

η. In particular, (2.8) shows that for one-dimensional Gaussian variables ξ and η the equality $J(\xi, \eta) = \infty$ can take place if and only if $r(\xi, \eta) = \pm 1$ (in this case η coincides with probability 1 with a linear function of ξ, and $J(\xi, \eta)$ coincides with $J(\xi, \xi) = \infty$).

Equation (2.8) can be extended directly to the general case of arbitrary multi-dimensional vectors (without the assumption of non-singularity of distributions, which is required for the application of Theorem 2.1). For this we make use of a known result in the theory of multivariate correlation (see [13] and particularly [14], [15]), according to which there always exist linear coordinate transformations in the spaces of vectors ξ, and η such that all the components of the vector

$$\zeta = \{\xi, \eta\} = \{\tilde{\xi}_1, \cdots, \tilde{\xi}_k, \tilde{\eta}_1, \cdots, \tilde{\eta}_l\}$$

(where $\tilde{\xi}_i$ and $\tilde{\eta}_j$ denote the components of ξ and η in new coordinate systems) with the exception of pairs $(\tilde{\xi}_j, \tilde{\eta}_j)$ $(j = 1, 2, \cdots, m)$ (where m is an integer not exceeding the smaller of the integers k, l) are mutually independent. (For a discussion of this result see pp. 219–220.) In this case, from the general properties of information $J(\xi, \eta)$ it readily follows that

$$J(\xi, \eta) = \sum_{j=1}^{m} J(\tilde{\xi}_j, \tilde{\eta}_j),$$

and by virtue of (2.8)

$$J(\xi, \eta) = -\frac{1}{2} \sum_{j=1}^{m} \log(1 - r_j^2), \tag{2.8'}$$

where $m \leq \min(k, l)$, and the r_j

$$r_j = r(\tilde{\xi}_j, \tilde{\eta}_j) = \frac{\mathrm{M}\tilde{\xi}_j \tilde{\eta}_j}{\sqrt{\mathrm{M}\tilde{\xi}_j^2 \cdot \mathrm{M}\tilde{\eta}_j^2}}$$

are correlation invariants for the pair of vectors ξ and η (see [14], [15]). Expression (2.8′) is applicable to any pair of Gaussian random vectors (ξ, η) and is the desired generalization of (2.8).

Let us remark further that the expression for $J(\xi, \eta)$ derived above has a simple geometrical interpretation. Following A. N. Kolmogorov [16], [17], we regard random variables (more precisely, classes of equivalent random variables) having finite variance as elements of a real unitary space $\mathfrak{H}$ with inner product $(\xi_1, \xi_2) = \mathrm{M}\xi_1, \xi_2$. In this case to a random vector $\xi = \{\xi_1, \xi_2, \cdots, \xi_k\}$ will correspond a finite-dimensional linear subspace H_1 (which is k-dimensional when the corresponding multi-dimensional probability distribution is non-singular) of $\mathfrak{H}$ which is spanned by the vectors $\xi_1, \cdots, \xi_k$; similarly, to the vector $\eta = \{\eta_1, \eta_2, \cdots, \eta_l\}$ will correspond another finite-dimensional subspace H_2 of the same space. According to Theorem 1.2, the information $J(\xi, \eta)$ depends only on these subspaces (and not on the choice of bases in them) so that in place of $J(\xi, \eta)$ we can use the notation $J(H_1, H_2)$. In the case of one-dimensional variables ξ and η, the correlation coefficient $r(\xi, \eta)$ can be interpreted as the cosine of the angle α between the vectors ξ and η, so that

$$J(\xi, \eta) = -\frac{1}{2}\log(1 - \cos^2\alpha) = -\log|\sin\alpha|. \tag{2.9}$$

In the general case where ξ and η are vectors, the k- and l-dimensional subspaces H_1 and H_2 form l (if $l \leq k$) stationary angles $\alpha_1, \alpha_2, \cdots, \alpha_l$ and the information (2.7) can be expressed in terms of these angles thus:

$$J(\xi, \eta) = -\frac{1}{2}\log(\sin^2\alpha_1 \sin^2\alpha_2, \cdots \sin^2\alpha_l) = -\log|\sin\alpha_1 \sin\alpha_2 \cdots \sin\alpha_l| \tag{2.10}$$

(see, for example, [18], p. 410). In case the distribution of ξ or of η is singular, the dimensions of subspaces H_1 and H_2 will be less than k and l, respectively. Then the number m of stationary angles $\alpha_1, \alpha_2, \cdots, \alpha_m$ between the subspaces H_1 and H_2 will be less than both k and l, and equation (2.10) will assume the form:

$$J(\xi, \eta) = -\frac{1}{2}\log(\sin^2\alpha_1 \sin^2\alpha_2 \ldots \sin^2\alpha_m) = \\ = -\log|\sin\alpha_1 \sin\alpha_2 \ldots \sin\alpha_m|. \tag{2.10'}$$

It should be noted that the quantities $\cos^2\alpha_1, \cos^2\alpha_2, \cdots, \cos^2\alpha_m$ are identical with the correlation invariants $r_1, r_2, \cdots, r_m$ of the pair of vectors (ξ, η), so that the expression (2.10') represents merely a different way of writing (2.8').

The result expressed by (2.10) (or (2.10')) can also be framed as follows. Denote by P_1 the operation of projection on subspace H_1 in the space $\mathfrak{H}$ (or, equivalently, in the finite-dimensional space spanned by all the ξ_i, η_j) and let P_2 be the projection on H_2. Consider the operators

$$B_1 = P_1P_2P_1, \quad B_2 = P_2P_1P_2, \tag{2.11}$$

which are self-adjoint operators with norms less than or equal to unity. Since B_1 and B_2 are null operators in the orthogonal complements of H_1 and H_2, they can be regarded as operators in H_1 and H_2 respectively (and correspondingly can be written as $B_1 = P_1P_2$ and $B_2 = P_2P_1$ [1])). In this case, the respective non-zero proper values of the operators B_1 and B_2 will coincide with each other (while the corresponding proper vectors are simply projections of one another). These proper values – which constitute the only geometrical invariants of the pair of subspaces (H_1, H_2) – determine the angles between these subspaces (more specifically, are equal to the squares of cosines of these angles; see [18]). Thus Theorem (2.1) can be formulated also in the following manner.

Theorem 2.2. *Let H_1 and H_2 be two finite-dimensional subspaces of the space $\mathfrak{H}$ of random variables, and let P_1 and P_2 be the projections of $\mathfrak{H}$ on these subspaces. Then, if all the multi-dimensional probability distributions of the variables in the subspaces H_1 and H_2 are multivariate normal, the information*

1) Every vector in H_1 is projected by the operator B_1 first on the subspace H_2 and then on H_1. The operator B_2 projects a vector in H_2 first on the subspace H_1 and then on H_2.

$J(H_1, H_2)$ contained in the points of one of these subspaces concerning the points of the other is given by the formula

$$J(H_1, H_2) = -\frac{1}{2}\log\det(E - B_1) = -\frac{1}{2}\log\det(E - B_2), \tag{2.12}$$

where E is the identity matrix, and B_1 and B_2 are matrices defined by (2.11).

Clearly, the result expressed by Theorem 2.2 is completely independent of how the variables $\xi_1, \cdots, \xi_k$ and $\eta_1, \cdots, \eta_l$ defining the subspaces H_1 and H_2 are chosen, and, in particular, of whether or not these variables are linearly independent (i.e., whether the corresponding Gaussian probability distributions are non-singular or singular).

Proof of formula (2.12) can be obtained quite simply without any reference to relation (2.7) or (2.8'). Indeed, choose as bases in the subspaces H_1 and H_2 the vectors $\tilde{\xi}_1, \cdots, \tilde{\xi}_k$ and $\tilde{\eta}_1, \cdots, \tilde{\eta}_l$ which comprise the proper vectors of B_1 and B_2 respectively, assigning identical subscripts to the proper vectors of B_1 and B_2 having the same proper value. In this way it is clearly seen that the vector $\tilde{\xi}_i$, for example, is orthogonal to all the vectors $\tilde{\xi}_j$ and $\tilde{\eta}_j$ with $i \neq j$. This implies that, in the Gaussian case, the variable $\tilde{\xi}_i$ is independent of all the other variables $\tilde{\xi}_j$ and $\tilde{\eta}_j$ with the exception of $\tilde{\eta}_i$ (which is the projection of $\tilde{\xi}_i$ on H_2: $\tilde{\eta}_i = P_2\tilde{\xi}_i$); as for the variables $\tilde{\xi}_{(l+1)}, \cdots, \tilde{\xi}_k$ (with $k \geq l$) which correspond to the null proper value of the operator B_1, they are generally independent of all the $\tilde{\eta}_j$. Thus the bases $(\tilde{\xi}_1, \cdots, \tilde{\xi}_k)$ and $(\tilde{\eta}_1, \cdots, \tilde{\eta}_l)$ are indentical with those which entered in the derivation of (2.8') (in which reference was made to [13]–[15]), and the existence of the latter is established thereby. It is now sufficient to remark that the mutual information $J(H_1, H_2)$ reduces simply to the sum of the amounts of information concerning $\tilde{\eta}_i$ contained in the corresponding variables $\tilde{\xi}_i$ $(i = 1, 2, \cdots, l)$. Since in the one-dimensional case the information is given by (2.9), we arrive at once at the desired result (2.10) which is equivalent to (2.12).

From expression (2.10) it follows, in particular, that in the case of multi-dimensional Gaussian random variables the equality $J(\xi, \eta) = \infty$ takes place if and only if at least one of the stationary angles between the finite-dimensional subspaces corresponding to the variables ξ and η is equal to zero, i.e., when there exists a line common to both subspaces. (Note that in this case the subspace spanned by the variables $\xi_1, \cdots, \xi_k, \eta_1, \cdots, \eta_l$ will obviously be less than $(k+l)$-dimensional, i.e., the corresponding $(k+l)$-dimensional probability distribution will certainly be singular.) The reason why $J(\xi, \eta) = \infty$ in this case is quite clear; there is a variable $\hat{\xi} = a_1\xi_1 + \cdots + a_k\xi_k$ which can also be represented as a linear combination $\hat{\eta} = b_1\eta_1 + \cdots + b_l\eta_l$; hence $J(\hat{\xi}, \hat{\eta}) = J(\hat{\xi}, \hat{\xi}) = \infty$, and a fortiori $J(\xi, \eta) = \infty$.

In conclusion we remark that the general expression (2.7) for the information $J(\xi, \eta)$ in terms of the corresponding moment matrices can be cast into a form analogous to (2.12) which is more convenient for purposes of numerical computation than (2.7). Indeed, let us introduce the square matrices

$$C_1 = \begin{pmatrix} A & 0 \\ 0 & B \end{pmatrix}, \quad C_1^{-1} = \begin{pmatrix} A^{-1} & 0 \\ 0 & B^{-1} \end{pmatrix} \tag{2.13}$$

(where 0 denotes a null matrix; the order of 0, as well as that of the identity matrix E, will always be inferred from the context). By virtue of (2.13), the argument of the logarithm in the right-hand member of (2.7) can be rewritten in the form

$$\frac{\det A \det B}{\det C} = \{\det (C \cdot C_1^{-1})\}^{-1} = \begin{vmatrix} E & DB^{-1} \\ D'A^{-1} & E \end{vmatrix}^{-1}. \tag{2.14}$$

On making use of standard formulae for the determinant of a matrix partitioned into four parts (see, for example, [19], p. 45, also [14], formulae (52) and (53)), we obtain:

$$J(\xi, \eta) = -\frac{1}{2} \log \det (E - DB^{-1}D'A^{-1}) = -\frac{1}{2} \log \det (E - D'A^{-1}DB^{-1}), \tag{2.15}$$

which is the desired expression for the information $J(\xi, \eta)$.

§2. Information about a Gaussian random process contained in another such process. A random process $\xi(\phi)$ [1)] is called *Gaussian* if all the finite-dimensional distribution functions for the variables $\xi(\phi_1), \cdots, \xi(\phi_n)$ are multivariate normal. In order that this condition be satisfied, it is sufficient that all the one-dimensional distribution functions for the variables $\xi(\phi)$, $\phi \in \Phi$, be univariate normal. A pair of processes $\{\xi(\phi), \eta(\psi)\}$ is called *Gaussian* if both these processes are Gaussian, and all the joint distribution functions for the variables $\xi(\phi)$ and $\eta(\psi)$ are normal. Clearly, a Gaussian process $\xi(\phi)$ with zero expectation $\mathrm{M}\xi(\phi) \equiv 0$ (the latter can always be assumed in the calculation of information) is completely characterized by the *correlation functional*

$$b^{\xi\xi}(\phi, \phi_1) = \mathrm{M}\xi(\phi)\xi(\phi_1). \tag{2.16}$$

(Note that it is sufficient to know $b_{\xi\xi}(\phi, \phi) \equiv b_\xi(\phi) = \mathrm{M}[\xi(\phi)]^2$ – the variance of the random variable $\xi(\phi)$; the functional $b_{\xi\xi}(\phi, \phi_1)$ can be obtained from $b_\xi(\phi)$ with the aid of the well-known formula expressing a bilinear form in terms of the corresponding quadratic form.) In the case of a pair of Gaussian processes $\{\xi(\phi), \eta(\psi)\}$ it is necessary to have, in addition to $b_{\xi\xi}(\phi, \phi_1)$, the correlation functional of the $\eta(\psi)$:

$$b_{\eta\eta}(\psi, \psi_1) = \mathrm{M}\eta(\psi)\eta(\psi_1), \tag{2.17}$$

as well as the joint correlation functional

$$b_{\xi\eta}(\phi, \psi_1) = \mathrm{M}\xi(\phi)\eta(\psi_1). \tag{2.18}$$

These functionals are bilinear in their arguments, positive definite (in the sense that the corresponding quadratic functionals are always non-negative) and continuous under the topologies in Φ and Ψ respectively, while the functional $b_{\xi\eta}$ is

1) As was stated previously, by a random process we always mean a generalized random process.

bilinear, continuous in both arguments and such that the sum

$$b_{\xi\xi}(\phi, \phi_1) + b_{\xi\eta}(\phi, \psi_1) + b_{\xi\eta}(\phi_1, \psi) + b_{\eta\eta}(\psi, \psi_1), \tag{2.19}$$

which is a bilinear functional in the pairs $\{\phi, \psi\}$ and $\{\phi_1, \psi_1\}$, is positive-definite. Conversely, any functionals $b_{\xi\xi}$, $b_{\eta\eta}$ and $b_{\xi\eta}$ satisfying these conditions are correlation functionals for some pair of Gaussian random processes (see [9], [10]). In this case, the multi-dimensional probability distributions for the variables $\xi(\phi_1), \cdots, \xi(\phi_k), \eta(\psi_1), \cdots, \eta(\psi_l)$ which characterize the pair $\{\xi(\phi), \eta(\psi)\}$ are multivariate normal with zero expectations and moment matrix of the form

$$\left\| \begin{matrix} b_{\xi\xi}(\varphi_i, \varphi_j) & b_{\xi\eta}(\varphi_i, \psi_n) \\ b_{\eta\xi}(\psi_m, \varphi_j) & b_{\eta\eta}(\psi_m, \psi_n) \end{matrix} \right\|, \quad b_{\eta\xi}(\psi_m, \varphi_j) \equiv b_{\xi\eta}(\varphi_j, \psi_m). \tag{2.20}$$

We proceed now to the determination of information $J(\xi, \eta)$ for a pair of Gaussian random processes. In view of the separability of the spaces Φ and Ψ of testing functions, we can represent the linear spaces $\xi(\Phi)$ and $\eta(\Psi)$ which are spanned by the variables $\xi(\phi)$, $\phi \in \Phi$ and $\eta(\psi)$, $\psi \in \Psi$ respectively, as limits of increasing sequences of their finite-dimensional subspaces $\xi(\Phi_n)$ $(n = 1, 2 \cdots)$ and $\eta(\Psi_m)$ $(m = 1, 2, \cdots)$. According to Theorem 1.5, the information $J(\xi, \eta)$ can in this case be obtained as a limit (as $n \to \infty$, $m \to \infty$) of finite-dimensional information functions $J[\xi(\Phi_n), \eta(\Psi_m)]$ expressed by (2.7), (2.8′) or (2.12). From this the following theorem can readily be deduced.

Theorem 2.3. *Let* $\{\xi(\phi), \eta(\psi)\}$, $\phi \in \Phi$, $\psi \in \Psi$ *be a pair of generalized Gaussian random processes, and let* $\xi(\Phi)$ *and* $\eta(\Psi)$ *be subspaces of the space* $\mathfrak{H}$ *of random variables having finite variance, which are spanned by the variables* $\xi(\phi)$ *and* $\eta(\psi)$, *respectively. Denote by* P_1 *and* P_2 *the projections of* $\mathfrak{H}$ *on the subspaces* $\xi(\Phi)$ *and* $\eta(\Psi)$, *and by* B_1 *and* B_2 *the operators*

$$B_1 = P_1 P_2 P_1, \quad B_2 = P_2 P_1 P_2. \tag{2.21}$$

Then, in order that the information $J(\xi, \eta)$ *be finite it is necessary and sufficient that at least one of the operators* B_1, B_2 *be a completely continuous operator with finite trace. Under this condition the other operator will possess the same property, and the information* $J(\xi, \eta)$ *will be given by the expression*

$$J\{\xi(\phi), \eta(\psi)\} = -\frac{1}{2}\log\det(E - B_1) = -\frac{1}{2}\log\det(E - B_2), \tag{2.22}$$

where E *is the identity operator.*

Here the *trace* of an operator represents, as usual, the sum of all its proper values, while its *determinant* is their product. As is well-known, finiteness of the determinant of $E - B$ depends on the convergence of the infinite product $\prod_1^\infty (1 - \lambda_k)$ and hence a necessary and sufficient condition for finiteness of the determinant is finiteness of the trace of B. Let us remark further that the operator $E - B_1$ can be regarded as an operator in the subspace $\xi(\Phi)$ (since it maps this subspace into it itself and is the identity operator on its orthogonal complement); similarly, $E - B_2$ can be regarded as an operator in $\eta(\psi)$. Thus, in place of (2.21) we may use the expressions: $B_1 = P_1P_2$, $B_2 = P_2P_1$.

We proceed to prove Theorem 2.3. First, assume that the operator B_1 is not a completely continuous operator with finite trace. Write $B_1^{(n)} = P_1^{(n)}P_2^{(n)}$, where $P_1^{(n)}$ and $P_2^{(n)}$ are projections on the subspaces $\xi(\Phi_n)$ and $\eta(\psi_n)$ respectively. It is easy to see that the sequence of operators $B_1^{(n)}$ in Φ converges weakly to the operator B_1 as $n \to \infty$ (i.e., that $\lim_{n\to\infty} (B_1^{(n)}\xi, \xi) = (B_1\xi, \xi)$ for any $\xi \in \xi(\Phi)$). Since the trace of the operator B_1 (respectively $B_1^{(n)}$) is equal to $\sum_k (B_1\xi_k, \xi_k)$ (respectively $\sum_k (B_1^{(n)}\xi_k, \xi_k)$), where the summation extends over all the elements ξ of an arbitrary orthonormal basis in the space $\xi(\Phi)$, it follows that under the above assumption on B_1 there exists for any (arbitrarily large) positive M such an $n_0 = n_0(M)$ that $\operatorname{Sp} B_1^{(n)} > M$ (where Sp denotes the trace) for $n > n_0$. But for $1 > \lambda_k > 0$ it is obvious that $\frac{1}{1-\lambda_k} > 1 + \lambda_k$ and $\prod_k (1+\lambda_k) > \sum_k \lambda_k$; consequently, for the finite-dimensional non-negative operator $B_1^{(n)}$ we have the bound

$$\det [E - B_1^{(n)}]^{-1} > \operatorname{Sp} B_1^{(n)}. \tag{2.23}$$

Thus in the case under consideration

$$J[\xi(\Phi_n), \eta(\Psi_n)] = \frac{1}{2}\log\{\det[E - B_1^{(n)}]^{-1}\}$$

increases without bound as $n \to \infty$ and hence $J[\xi(\phi), \eta(\psi)] = \infty$.

Next, assume that B_1 is a completely continuous operator with finite trace. As an orthonormal basis in $\xi(\Phi)$ take the proper vectors ξ_k of B_1, and for an orthonormal basis in $\eta(\Psi)$ choose a basis comprising the vectors $\eta_k = P_2\xi_k$ and some vectors ζ_k which are orthogonal to the subspace $\xi(\Phi)$ (if such vectors exist in $\eta(\Psi)$). Denote by $\xi(\Phi_n)$ the subspace of $\xi(\Phi)$ which is spanned by the vectors $\xi_1, \xi_2, \cdots, \xi_n$, and by $\eta(\Psi_n)$ the subspace of $\eta(\Phi)$ which is spanned by the vectors $\eta_1, \eta_2, \cdots, \eta_m$ and the first n of the vectors ζ_n. (If there are fewer than n ζ_k vectors, then include all these vectors in $\eta(\Psi_n)$. Now by virtue of Theorem 1.5 we have

$$J[\xi(\phi), \eta(\psi)] = \lim_{n\to\infty} J[\xi(\Phi_n), \eta(\Psi_n)], \tag{2.24}$$

but it is clear that

$$J[\xi(\Phi_n), \eta(\Psi_n)] = -\frac{1}{2}\log\det(E - B_1^{(n)}) = -\frac{1}{2}\log\prod_{k=1}^{n}(1-\lambda_k), \tag{2.25}$$

where $\lambda_1, \cdots, \lambda_n$ are the first n proper values of B_1. Hence from (2.24) and (2.25) it follows immediately that

$$J[\xi(\varphi), \eta(\psi)] = -\frac{1}{2}\log\left\{\prod_{k=1}^{\infty}(1-\lambda_k)\right\} = -\frac{1}{2}\log\det(E - B_1) \tag{2.26}$$

in which the convergence of the infinite product is insured by finiteness of the trace of B_1. This clearly shows that in this case $\operatorname{Sp} B_2$ is finite and the information $J[\xi(\phi), \eta(\psi)]$ is equal to $-\frac{1}{2}\log\det(E - B_2)$; conversely, if $\mathrm{Sp}\, B_2$ is finite, so is $\mathrm{Sp}\, B_1$, and the information $J\{\xi(\phi), \eta(\psi)\}$ is given by either of the expressions (2.22). This completes the proof of Theorem 2.3.

In the general case of arbitrary Gaussian random processes the result expressed by Theorem 2.3 provides, it seems, a definite solution of the problem of calculating the information $J(\xi, \eta)$. Needless to say, in concrete particular cases we can proceed even further and pose the problem of numerically evaluating the determinants in (2.22), that is, explicitly calculating the information $J(\xi, \eta)$. In the following we consider several examples in which such calculations are feasible. In all these examples it will be assumed – but not always stated explicitly – that the pair $\{\xi, \eta\}$ is Gaussian.

§3. Examples

A. *Information contained in a random process about a random variable (or in a random variable about a random process).* Let $\xi(\phi)$ be a random process (ordinary or generalized) over some (possibly infinite) interval T, and let η be a random variable. According to Theorem (2.3), in order to calculate the information $J\{\xi(\phi), \eta\}$ it is necessary to project the vector η in the space $\mathfrak{H}$ of random variables on the subspace $\xi(\Phi)$ spanned by the vectors $\{\xi(\phi), \phi \in \Phi)$, and then project the vector $\tilde{\eta} = P_1\eta$ back on η. On writing $\hat{\eta} = P_2\tilde{\eta} = P_2P_1\eta$, the desired information can be expressed as

$$J\{\xi(\varphi), \eta\} = -\frac{1}{2}\log\left(1 - \frac{\|\hat{\eta}\|}{\|\eta\|}\right) = -\log|\sin\alpha|, \qquad (2.27)$$

where $\|\eta\| = M\eta^2$ is the length of η and α is the angle between η and the generally infinite dimensional) subspace $\xi(\Phi)$. We note that the random variable $\tilde{\eta} = P_1\eta$ coincides with the best linear estimate of η based on the values of the process $\xi(\phi)$ in the interval T, i.e., is the solution of the problem of linear filtration of the process $\xi(\phi)$ (see, for example, [20] – [23]). If we denote by σ_η^2 the mean square filtration error, it is easy to see that

$$\sigma_\eta^2 = \mathrm{M}|\eta - \tilde{\eta}|^2; \quad \hat{\sigma}_\eta^2 \equiv \frac{\sigma_\eta^2}{\|\eta\|^2} = 1 - \frac{\mathrm{M}|\tilde{\eta}|^2}{\mathrm{M}|\eta|^2} = 1 - \cos^2\alpha = \sin^2\alpha; \qquad (2.28)$$

and thus

$$J\{\xi(\varphi), \eta\} = -\frac{1}{2}\log\frac{\sigma_\eta^2}{\|\eta\|^2} = -\frac{1}{2}\log\hat{\sigma}_\eta^2, \qquad (2.29)$$

where σ_η^2 is a normalized mean square filtration error.

We see that in the case here the calculation of information $J\{\xi(\phi), \eta\}$ reduces to the solution of a problem in filtration (more precisely, to the determination of the mean square filtration error) to which a rather extensive specialized literature is devoted. Particularly much attention has been given to the case where $\xi(\phi) \equiv \xi(t)$ is an ordinary stationary random process over the half-line $T = (-\infty, t_1]$ or over an interval $T = [t_0, t_1]$ (where $t_0 < t_1$), and $\eta = \xi(s)$ is the value of the same process at some point s outside T (extrapolation of stationary processes). For $T = (-\infty, t_1]$ a general expression for $\sigma^2_{\xi(s)}$ (and hence for the information $J\{\xi(t), t \in T;\ \xi(s)\}$) was found by A. N. Kolmogorov [17] and M. G. Kreĭn [24] (see also monograph [25]). For $T = [t_0, t_1]$ and $\eta = \xi(s)$, $s > t_1$ or $s < t_0$, certain conditions under which $\sigma^2_{\xi(s)} = 0$ (i.e., $J\{\xi(t);\ \xi(s)\} = \infty$) were indicated by M. G. Kreĭn (see references in [21]–[23]). An effective method of calculating $\sigma^2_{\xi(s)}$ (i.e., $J\{\xi(t);\ \xi(s)\}$) for the case of a process $\xi(t)$ with rational spectral density was given by L. A. Zadeh and J. R. Ragazzini [26] (see also [22], [23]). The case where $\xi(\phi) \equiv \xi(t)$ is a stationary random process over the half-line $T = (-\infty, t_1]$ and η is a correlated random variable was first considered by N. Wiener [20] (see also [22], [25]). For the case where the spectral density of $\xi(\phi)$ is rational and the correlation of $\xi(\phi)$ and η has a special form, effective methods of calculating σ^2_η are given in [26], [23], [22]. Analogous problems for a process $\xi(t)$ with stationary increments of a certain order are considered in [27] (see also [23]). The general case of a non-stationary random process $\xi(t)$ over a finite interval is considered in [28]; however, the results derived there are not fully effective.

A'. *Information contained in a random process about a random vector (or in a random vector about a random process).* We can similarly analyze the case where $\xi \equiv \xi(\phi)$ is a random process and $\eta \equiv \eta = \{\eta_1, \cdots, \eta_k\}$ is a random vector having a finite number of components. In this case, according to Theorem 2.3 it is first necessary to project all the random variables $\eta_1, \cdots, \eta_k$ on the subspace $\xi(\Phi)$; after that the projections $\tilde{\eta}_1 = P_1\eta_1, \cdots, \tilde{\eta}_k = P_1\eta_k$ are again projected on the finite subspace H_2 which is spanned by $\{\eta_1, \cdots, \eta_k\}$. (The determination of the projections $\hat{\eta}_i = P_2\tilde{\eta}_i$ for given $\tilde{\eta}_i$ reduces to the solution of a simple system of linear equations: $(\tilde{\eta}_i - \sum_{j=1}^{k} a_{ij}\eta_j, \eta_l) = 0$ $(l = 1, \cdots, k)$, where $\sum_j a_{ij}\eta_j \equiv \hat{\eta}_i$). Finally, the information $J\{\xi(\phi), \eta\}$ is found from the relation

$$J\{\xi(\phi), \eta\} = -\frac{1}{2}\log\det\|\delta_{ij} - a_{ij}\|. \tag{2.30}$$

We observe that the main difficulty here consists in the necessity of solving k filtration problems for the process $\xi(\phi)$ and finding the best estimates $\tilde{\eta}_1, \cdots, \tilde{\eta}_k$; once this is done, the calculation of $J\{\xi(\phi), \eta\}$ reduces to a problem in elementary algebra.

B. *Information about a stationary random process contained in another such process which is stationarily related to it.* [1]) We begin with the case of a process with discrete spectrum. Let $\xi(\phi)$ be a stationary Gaussian process over the real line $(-\infty, \infty)$ with zero expectation and discrete spectrum, i.e.,

1) This example can be omitted on first reading.

$$\xi(\varphi) = \sum_{h} \left\{ z_k^{(1)} \int_{-\infty}^{\infty} \varphi(t) \cos \lambda_k t \, dt + z_k^{(2)} \int_{-\infty}^{\infty} \varphi(t) \sin \lambda_k t \, dt \right\}, \tag{2.31}$$

where λ_k $(k = 1, 2, \cdots)$ is a finite or enumerable sequence of non-negative numbers, and $\{z_k^{(1)}, z_k^{(2)}\}$ $(k = 1, 2, \cdots)$ is respectively a finite or enumerable sequence of normally distributed two-dimensional vectors satisfying the conditions

$$\left.\begin{aligned} &\mathbf{M}z_k^{(1)} = \mathbf{M}z_k^{(2)} = 0; \quad \mathbf{M}z_k^{(1)} z_l^{(2)} = 0 \quad \text{for all} \quad k, l; \\ &\mathbf{M}z_k^{(1)} z_l^{(1)} = \mathbf{M}z_k^{(2)} z_l^{(2)} = 0 \quad \text{for} \quad k \neq l \end{aligned}\right\} \tag{2.32}$$

and

$$\mathbf{M}[z_k^{(1)}]^2 = \mathbf{M}[z_k^{(2)}]^2 = f_k, \tag{2.33}$$

where the f_k are non-negative numbers such that $\sum_{\lambda_k < \Lambda} f_k$ increases as $\Lambda \to \infty$ not faster than a finite power of Λ. (In the case where $\sum_k f_k < \infty$ and only in this case, the process $\xi(\phi)$ will have the form (1.28), where $\xi(t)$ is an ordinary stationary random process with discrete spectrum.) Assume that $\eta(\psi)$ is another process of the same type:

$$\eta(\psi) = \sum_k w_k^{(1)} \int_{-\infty}^{\infty} \psi(t) \cos \lambda_k t \, dt + w_k^{(2)} \int_{-\infty}^{\infty} \psi(t) \sin \lambda_k t \, dt , \tag{2.34}$$

where

$$\left.\begin{aligned} &\mathbf{M}w_k^{(1)} = \mathbf{M}w_k^{(2)} = 0; \quad \mathbf{M}w_k^{(1)} w_l^{(2)} = 0 \quad \text{for all} \quad k, l; \\ &\mathbf{M}w_k^{(1)} w_l^{(1)} = \mathbf{M}w_k^{(2)} w_l^{(2)} = 0 \quad \text{for} \quad k \neq l, \end{aligned}\right\} \tag{2.35}$$

and

$$\mathbf{M}[w_k^{(1)}]^2 = \mathbf{M}[w_k^{(2)}]^2 = g_k. \tag{2.36}$$

We assume that the process $\eta(\psi)$ is stationarily related to $\xi(\phi)$, i.e.,

$$\mathbf{M}\{\xi(\phi(t))\eta(\psi(t))\} = \mathbf{M}\{\xi(\phi(t+\tau))\eta(\psi(t+\tau))\}$$

for all ϕ, ψ and τ.

In this case, if the frequencies λ_k in (2.31) and (2.34) are indexed in such a way that the same indices represent identical frequencies (this, of course, can always be done if it is assumed that the variances f_k and g_k of z_k and w_k may be equal to zero), then the following equalities will hold

$$\left.\begin{aligned} &\mathbf{M}z_k^{(1)} w_l^{(1)} = \mathbf{M}z_k^{(1)} w_l^{(2)} = \mathbf{M}z_k^{(2)} w_l^{(1)} = \mathbf{M}z_k^{(2)} w_l^{(2)} = 0 \quad \text{for} \quad k \neq l, \\ &\mathbf{M}z_k^{(1)} w_k^{(1)} = \mathbf{M}z_k^{(2)} w_k^{(2)} = h_k^{(1)}, \quad \mathbf{M}z_k^{(1)} w_k^{(2)} = -\mathbf{M}z_k^{(2)} w_k^{(1)} = h_k^{(2)}, \end{aligned}\right\} \tag{2.37}$$

where $h_k^{(1)}$ and $h_k^{(2)}$ are real numbers satisfying the inequalities

$$[h_k^{(1)}]^2 + [h_k^{(2)}]^2 \leq f_k g_k \qquad (k = 1, 2, \cdots). \tag{2.38}$$

(The conditions (2.37) are necessary and sufficient in order that the stationary processes (2.31) and (2.34) be stationarily related; conditions (2.38) stem from (2.32)–(2.37).) In this case, the information $J\{\xi(\phi), \eta(\psi)\}$ can be determined quite simply. Indeed, as we have already seen the information $J\{\xi(\phi), \eta(\psi)\} \equiv J\{\xi(\Phi), \eta(\Psi)\}$ depends only on the linear subspace $\xi(\Phi)$ and $\eta(\Psi)$ of the space $\mathfrak{H}$ of random variables, and not on any particular choice of families $\{\xi(\phi), \phi \in \Phi$ and $\{\eta(\psi), \psi \in \Psi\}$ of vectors in these subspaces (by which, in fact, the subspaces

$\xi(\Phi)$ and $\eta(\Psi)$ are determined). In the present case, it is most convenient to choose as bases for the spaces $\xi(\Phi)$ and $\eta(\Psi)$ the families of mutually orthogonal complex vectors

$$z_k = \frac{1}{2}(z_k^{(1)} - iz_k^{(2)}),\quad z_{-k} = \frac{1}{2}(z_k^{(1)} + iz_k^{(2)})\quad (k = 1, 2, \cdots) \tag{2.39}$$

and

$$w_k = \frac{1}{2}(w_k^{(1)} - iw_k^{(2)}),\quad w_{-k} = \frac{1}{2}(w_k^{(1)} + iw_k^{(2)})\quad (k = 1, 2, \cdots) \tag{2.40}$$

respectively.

(In terms of these vectors, expressions (2.31) and (2.34) can be written as

$$\xi(\varphi) = \sum_k z_k \int_{-\infty}^{\infty} \varphi(t) e^{i\lambda_k t} dt;\quad \eta(\psi) = \sum_k w_k \int_{-\infty}^{\infty} \psi(t) e^{i\lambda_k t} dt, \tag{2.41}$$

where the summation extends over both positive and negative values of the index k and $\lambda_{-k} = -\lambda_k$.) Now define the inner product for complex random variables by the relation $(x, y) = \mathbf{M}x\bar{y}$; then, (2.32), (2.33), (2.35), (2.36), (2.37) and (2.38) become

$$(z_k, w_l) = (z_k, z_l) = (w_k, w_l) = 0 \text{ for } k \neq l, \tag{2.42}$$

$$(z_k, z_k) = (z_{-k}, z_{-k}) = \frac{f_k}{2},\quad (w_k, w_k) = (w_{-k}, w_{-k}) = \frac{g_k}{2},$$

$$(z_k, w_k) = (w_{-k}, z_{-k}) = \frac{h_k^{(1)} + ih_k^{(2)}}{2}. \tag{2.43}$$

With this choice of bases both the operator $B_1 = P_1P_2$ (in subspace $\xi(\Phi)$) and the operator $B_2 = P_2P_1$ (in subspace $\eta(\Psi)$) become diagonalized; moreover,

$$B_1 z_k = \frac{(h_k^{(1)})^2 + (h_k^{(2)})^2}{f_k g_k} z_k,\quad B_1 z_{-k} = \frac{(h_k^{(1)})^2 + (h_k^{(2)})^2}{f_k g_k} z_{-k}\quad (k = 1, 2. \ldots) \tag{2.44}$$

and correspondingly

$$B_2 w_k = \frac{(h_k^{(1)})^2 + (h_k^{(2)})^2}{f_k g_k} w_k,\quad B_2 w_{-k} = \frac{(h_k^{(1)})^2 + (h_k^{(2)})^2}{f_k g_k} w_{-k}\quad (k = 1, 2, \ldots). \tag{2.45}$$

By (2.22) these relations lead to the

Theorem 2.4. *The information $J\{\xi(\phi), \eta(\psi)\}$ contained in one of two stationary and stationarily related Gaussian processes $\xi(\phi)$ and $\eta(\psi)$ (defined by (2.31) and (2.34)) about the other is given by the expression*

$$J\{\xi(\varphi), \eta(\psi)\} = \sum_k \log \frac{f_k g_k}{f_k g_k - |h_k|^2}, \tag{2.46}$$

where $h_k = h_k^{(1)} + ih_k^{(2)}$, and f_k, g_k, $h_k^{(1)}$ and $h_k^{(2)}$ are defined by equations (2.33), (2.36) and (2.37).

In the special case where both processes $\xi(\phi)$ and $\eta(\psi)$ are "periodic with period T," i.e., when the frequencies λ_k are of the form $\lambda_k = k \cdot \frac{2\pi}{T}$ $(k=0, 1, 2, \cdots)$, it is natural to consider that these processes are defined not over the entire real line $(-\infty, +\infty)$, but over some segment of length T. (Note that under the assumption that $\xi(\phi)$ and $\eta(\psi)$ have the form (1.28), the values of $\xi(t)$ and $\eta(t)$ for all t are uniquely determined by virtue of their periodicity by the values of these processes over the segment of length T.) In this case, we can also introduce the notion of average information per unit time which is defined by the relation

$$i\{\xi(\phi),\ \eta(\psi)\} = \frac{1}{T} J\{\xi(\phi),\ \eta(\psi)\}, \tag{2.47}$$

where $J\{\xi(\phi),\ \eta(\psi)\}$ is given by (2.46).

Now let $\xi(\phi)$ and $\eta(\psi)$ be arbitrary stationary Gaussian processes with zero expectations. It is not difficult to see that for such processes the amount of information $J\{\xi(\phi),\ \eta(\psi)\}$ contained in one about the other is in general infinite. However, based on the previous remark, we can define for the pair $\xi(\phi)$, $\eta(\psi)$ an average (per unit time) amount of information $i\{\xi(\phi),\ \eta(\psi)\}$ which is finite in many cases of interest. Indeed, let us make use of the fact that for arbitrary processes $\xi(\phi)$, $\eta(\psi)$ of the type considered here we have:

$$\xi(\varphi) = \int_0^\infty \left\{ \int_{-\infty}^\infty \varphi(t) \cos \lambda t \right\} dZ^{(1)}(\lambda) + \int_0^\infty \left\{ \int_{-\infty}^\infty \varphi(t) \sin \lambda t \, dt \right\} dZ^{(2)}(\lambda), \tag{2.48}$$

$$\eta(\psi) = \int_0^\infty \left\{ \int_{-\infty}^\infty \psi(t) \cos \lambda t \, dt \right\} dW^{(1)}(\lambda) + \int_0^\infty \left\{ \int_{-\infty}^\infty \psi(t) \sin \lambda t \, dt \right\} dW^{(2)}(\lambda), \tag{2.49}$$

where $Z^{(1)}(\lambda)$, $Z^{(2)}(\lambda)$ and $W^{(1)}(\lambda)$, $W^{(2)}(\lambda)$ are Gaussian processes with uncorrelated (and hence independent) increments such that for any sets $\mathcal{E}_1$, $\mathcal{E}_2$, and $\mathcal{E}_3$ the following relations hold:

$$\left.\begin{aligned}
&MZ^{(1)}(\mathcal{E}) = MZ^{(2)}(\mathcal{E}) = MW^{(1)}(\mathcal{E}) = MW^{(2)}(\mathcal{E}) \equiv 0,\\
&MZ^{(1)}(\mathcal{E}_1) Z^{(2)}(\mathcal{E}_2) = MW^{(1)}(\mathcal{E}_1) W^{(2)}(\mathcal{E}_2) = 0,\\
&MZ^{(1)}(\mathcal{E}_1) Z^{(1)}(\mathcal{E}_2) = MZ^{(2)}(\mathcal{E}_1) Z^{(2)}(\mathcal{E}_2) = F(\mathcal{E}_1 \cap \mathcal{E}_2),\\
&MW^{(1)}(\mathcal{E}_1) W^{(1)}(\mathcal{E}_2) = MW^{(2)}(\mathcal{E}_1) W^{(2)}(\mathcal{E}_2) = G(\mathcal{E}_1 \cap \mathcal{E}_2),\\
&MZ^{(1)}(\mathcal{E}_1) W^{(1)}(\mathcal{E}_2) = MZ^{(2)}(\mathcal{E}_1) W^{(2)}(\mathcal{E}_2) = H^{(1)}(\mathcal{E}_1 \cap \mathcal{E}_2),\\
&MZ^{(1)}(\mathcal{E}_1) W^{(2)}(\mathcal{E}_2) = -MZ^{(2)}(\mathcal{E}_1) W^{(1)}(\mathcal{E}_2) = H^{(2)}(\mathcal{E}_1 \cap \mathcal{E}_2).
\end{aligned}\right\} \tag{2.50}$$

In equations (2.50) we have set

$$\left.\begin{aligned}
&Z^{(i)}(\mathcal{E}) = \int_{\mathcal{E}} dZ^{(i)}(\lambda), \qquad W^{(i)}(\mathcal{E}) = \int_{\mathcal{E}} dW^{(i)}(\lambda) \qquad (i = 1, 2);\\
&F(\mathcal{E}) = \int_{\mathcal{E}} dF(\lambda), \quad G(\mathcal{E}) = \int_{\mathcal{E}} dG(\lambda), \quad H^{(i)}(\mathcal{E}) = \int_{\mathcal{E}} dH^{(i)}(\lambda) \quad (i = 1, 2),
\end{aligned}\right\} \tag{2.51}$$

where $F(\lambda)$ and $G(\lambda)$ are monotone non-decreasing functions satisfying for some integers n, m the inequalities

$$\int_0^\infty \frac{dF(\lambda)}{1+\lambda^n} < \infty, \quad \int_0^\infty \frac{dG(\lambda)}{1+\lambda^m} < \infty, \tag{2.52}$$

in which $H^{(1)}(\lambda)$ and $H^{(2)}(\lambda)$ are real-valued functions of bounded variation such that

$$[H^{(1)}(\mathcal{E})]^2 + [H^{(2)}(\mathcal{E})]^2 \le F(\mathcal{E})G(\mathcal{E}) \tag{2.53}$$

(see [9], where the case of complex-valued generalized random processes is considered in detail). The functions $F(\lambda)$ and $G(\lambda)$ are called *spectral functions* of the processes $\xi(\phi)$ and $\eta(\psi)$ respectively, while the complex-valued function of bounded variation

$$H(\lambda) = H^{(1)}(\lambda) + iH^{(2)}(\lambda) \tag{2.54}$$

is called the *mutual spectral function* of these processes.

The $\xi(\phi)$ and $\eta(\psi)$ processes can be arbitrarily closely approximated (in the sense made more precise below) by "periodic stationary processes". For this it is sufficient to choose an arbitrary sequence $T_1, T_2, \cdots$ of positive numbers T_n such that $T_n \to \infty$ as $n \to \infty$, and set

$$\left.\begin{aligned} \xi^{(n)}(\varphi) &= \sum_{k=0}^{\infty} \Big\{ \int_{(2k-1)\pi/T_n}^{(2k+1)\pi/T_n} dZ^{(1)}(\lambda) \cdot \int_{-\infty}^{\infty} \varphi(t)\cos\frac{2k\pi t}{T_n}\,dt + \\ &\quad + \int_{(2k-1)\pi/T_n}^{(2k+1)\pi/T_n} dZ^{(2)}(\lambda) \int_{-\infty}^{\infty} \varphi(t)\sin\frac{2k\pi t}{T_n}\,dt \Big\}, \\ \eta^{(n)}(\psi) &= \sum_{k=0}^{\infty} \Big\{ \int_{(2k-1)\pi/T_n}^{(2k+1)\pi/T_n} dW^{(1)}(\lambda) \cdot \int_{-\infty}^{\infty} \psi(t)\cos\frac{2k\pi t}{T_n}\,dt + \\ &\quad + \int_{(2k-1)\pi/T_n}^{(2k+1)\pi/T_n} dW^{(2)}(\lambda) \cdot \int_{-\infty}^{\infty} \psi(t)\sin\frac{2k\pi t}{T_n}\,dt \Big\} \end{aligned}\right\} \tag{2.55}$$

in which $Z^{(1)}(\lambda) \equiv Z^{(1)}(0)$ and $W^{(1)}(\lambda) \equiv W^{(1)}(0)$ for $\lambda < 0$). Indeed, the random variables $\xi^{(n)}(\phi)$, $\eta^{(n)}(\psi)$ will converge in the mean (and hence with probability 1) to $\xi(\phi)$ and $\eta(\psi)$, respectively, for all ϕ, ψ as $n \to \infty$. The convergence is uniform for all ϕ and ψ which vanish outside a fixed finite interval. Thus, $\xi(\phi)$ and $\eta(\psi)$ will be approximated arbitrarily closely over any fixed interval of time by $\xi^{(n)}(\phi)$ and $\eta^{(n)}(\psi)$, respectively, with respect to the metric in $\mathfrak{H}$.

Based on the foregoing results, the average (per unit time) amount of information $i\{\xi(\phi), \eta(\psi)\}$ contained in one of two (generalized) stationary processes $\xi(\phi)$, $\eta(\psi)$ about the other may be defined as follows:

$$i\{\xi(\phi), \eta(\psi)\} = \lim_{n\to\infty} i\{\xi^{(n)}(\phi), \eta^{(n)}(\psi)\} = \lim_{n\to\infty} \frac{1}{T_n} J\{\xi^{(n)}(\phi), \eta^{(n)}(\psi)\}. \tag{2.56}$$

In consequence of (2.46), the information $J\{\xi^{(n)}(\phi), \eta^{(n)}(\psi)\}$ is given by

$$J\{\xi^{(n)}(\varphi), \eta^{(n)}(\psi)\} = \sum_{k=0}^{\infty} \log \frac{\int_{(2k-1)\pi/T_n}^{(2k+1)\pi/T_n} dF(\lambda) \cdot \int_{(2k-1)\pi/T_n}^{(2k+1)\pi/T_n} dG(\lambda)}{\int_{(2k-1)\pi/T_n}^{(2k+1)\pi/T_n} dF(\lambda) \cdot \int_{(2k-1)\pi/T_n}^{(2k+1)\pi/T_n} dG(\lambda) - \left| \int_{(2k-1)\pi/T_n}^{(2k+1)\pi/T_n} dH(\lambda) \right|^2}, \tag{2.57}$$

where the functions $F(\lambda)$, $H(\lambda)$ and $G(\lambda)$ are given by (2.51) and (2.54). By letting $T_n \to \infty$, we obtain for the particular case where the functions $F(\lambda)$, $G(\lambda)$ and $H(\lambda)$ are absolutely continuous and have continuous derivatives (or differ from such functions only by jump functions) the following expression for the *average (per unit time) amount of information about one of two Gaussian (generalized) stationary processes* $\xi(\phi)$, $\eta(\psi)$ *contained in the other:*

$$i\{\xi(\varphi),\ \eta(\psi)\} = \frac{1}{2\pi}\int_0^\infty \log \frac{f(\lambda)g(\lambda)}{f(\lambda)g(\lambda) - |h(\lambda)|^2}\, d\lambda, \tag{2.58}$$

where

$$f(\lambda) = F'(\lambda), \qquad g(\lambda) = G'(\lambda), \qquad h(\lambda) = H'(\lambda), \tag{2.59}$$

and $F(\lambda)$, $G(\lambda)$ *and* $H(\lambda)$ *are, respectively, the spectral functions of* $\xi(\phi)$ *and* $\eta(\psi)$ *and the mutual spectral function of* $\xi(\phi)$ *and* $\eta(\psi)$. It is believed that the result expressed by (2.58) is valid under considerably broader conditions than those indicated here; this question, however, was not investigated by us, since we feel that the definition (2.56) of the average amount of infomation is not definitive.

It is understood that in (2.58) the integrand should be taken as zero when $f(\lambda)g(\lambda) = 0$, For the important special case in which $\eta(\psi) = \xi(\psi) + \zeta(\psi)$, where $\zeta(\psi)$ is a stationary random process ("noise") with spectral function $F^{(1)}(\lambda)$, we have:

$$G(\lambda) = F(\lambda) + F^{(1)}(\lambda), \qquad H(\lambda) = F(\lambda), \tag{2.60}$$

so that (2.58) reduces to

$$i\{\xi(\varphi),\ \eta(\psi)\} = \frac{1}{2\pi}\int_0^\infty \log\left(1 + \frac{f(\lambda)}{f^{(1)}(\lambda)}\right) d\lambda, \quad f^{(1)}(\lambda) = \frac{d}{d\lambda}F^{(1)}(\lambda). \tag{2.61}$$

The result expressed by (2.58) was obtained previously by M. S. Pinsker [29] for ordinary (not generalized) stationary processes $\xi(t)$, $\eta(t)$ at least one of which is regular, on the basis of a different definition of the average (per unit time) amount of information $i\{\xi, \eta\}$. Specifically, in Pinsker's work the stationary processes $\xi(t)$, $\eta(t)$ are replaced by stationary sequences

$$\xi_n(t) = \xi(t/T_n) \qquad (t = 0, \pm 1, \pm 2, \cdots)$$

and

$$\eta_n(t) = \eta(t/T_n) \qquad (t = 0, \pm 1, \pm 2, \cdots),$$

where $T_n \to \infty$ as $n \to \infty$. Then the average information is taken as the limit when $n \to \infty$,

$$i\{\xi(t), \eta(t)\} = \lim_{n\to\infty} i\{\xi_n(t), \eta_n(t)\}, \tag{2.62}$$

of the unambiguously defined average (per unit time) amount of information $i\{\xi_n(t), \eta_n(t)\}$ for the sequences $\xi_n(t)$, $\eta_n(t)$. However, a more direct definition, different from (2.56) and (2.62), is the following

$$i\{\xi(\phi), \eta(\psi)\} = \lim_{T\to\infty} \frac{1}{T} J\{\xi_T(\phi), \eta_T(\psi)\}, \tag{2.63}$$

where $J(\xi_T(\phi), \eta_T(\psi)\}$ is the amount of infomation about the process $\xi(\phi)$ over an interval of length T, contained in a process $\eta(\psi)$ over the same interval. In what follows we consider the problem of calculating $J\{\xi_T(\phi), \eta_T(\psi)\}$ and show that, in certain simple cases in which explicit expressions for this function can be obtained, the limit (2.63) is given by (2.58), i.e., is equal to the limit (2.56) (as well as (2.62)). It appears that in the general case as well, equation (2.58) gives the limit (2.63) (and perhaps also (2.56) and (2.62)); this, however, has not yet been proved.

B. *Information contained in a stationary process over a finite interval about another stationary process over the same interval.* Let $\xi(\phi)$ and $\eta(\phi)$ be stationary Gaussian processes which are stationarily and normally related to one another. Denote by $\xi_T(\phi)$ and $\eta_T(\phi)$ the processes $\xi(\phi)$ and $\eta(\phi)$ over a finite interval $-\frac{T}{2} \leq t \leq \frac{T}{2}$ of length T, i.e., a family of random variables $\xi(\phi)$ and $\eta(\phi)$ generated by the functions $\phi \in \Phi$ which vanish identically outside this interval. We are interested in calculating the information $J\{\xi_T(\phi), \eta_T(\psi)\}$ contained in one of these processes about the other.

Our starting point is the expression (2.15) for the mutual information in two Gaussian random vectors. For the calculation of $J\{\xi_T(\phi), \eta_T(\phi)\}$ it is necessary to choose some complete system of functions $\phi_1(t), \phi_2(t), \cdots$, in the space Φ_T which are infinitely differentiable on the interval $\left[-\frac{T}{2}, \frac{T}{2}\right]$, and vanish at the end points together with all their derivatives. Then the information $J_n\{\xi, \eta\}$ is obtained by substituting for A, B, D, and D' in (2.15) the following nth order matrices: ,

$$A_n = \| b_{\xi\xi}(\varphi_i, \varphi_j)\|_1^n, \quad B_n = \| b_{\eta\eta}(\varphi_i, \varphi_j)\|_1^n,$$

$$D_n = \| b_{\xi\eta}(\varphi_i, \varphi_j)\|_1^n, \quad D' = \| b_{\eta\xi}(\varphi_i, \varphi_j)\|_1^n \tag{2.64}$$

(where $b_{\xi\xi}$, $b_{\eta\eta}$, $b_{\xi\eta}$ and $b_{\eta\xi}$ are the correlation functionals for $\xi(\phi)$ and $\eta(\psi)$: see p. 220) and passing to the limit as $n \to \infty$. Let us now determine the limiting forms as $n \to \infty$ of the nth order matrices (i.e., linear operators on n-space) A_n, B_n, D_n and D'_n. We begin with A_n, which defines the following linear transformation:

$$A_n \xi(\phi_i) = \sum_{j=1}^{n} b_{\xi\xi}(\phi_i, \phi_j) \cdot \xi(\phi_j). \tag{2.65}$$

In virtue of the linear dependence of the random variable $\xi(\phi)$ on the function ϕ, we can regard A_n as a linear transformation in the n-dimensional function space spanned by the functions $\phi_1(t), \cdots, \phi_n(t)$, which is defined by the relation

$$A_n \phi = \sum_{j=1}^{n} b_{\xi\xi}(\phi, \phi_j) \cdot \phi_j. \tag{2.66}$$

We now make use of the known relation ([9], [10]),

$$b_{\xi\xi}(\varphi, \psi) = b_{\xi\xi}(\varphi * \psi^*) \equiv b_{\xi\xi}\left(\int_{-\infty}^{\infty} \varphi(t-t')\psi(-t')\,dt'\right) = b_{\xi\xi}\left(\int_{-\infty}^{\infty} \varphi(t+t')\psi(t')\,dt'\right), \tag{2.67}$$

where $\phi * \psi^*$ is a convolution of ϕ with $\psi^*(s) = \psi(-s)$, and $b_{\xi\xi}(\phi)$ is an ordinary (non-random) generalized function, i.e., a linear functional on the space Φ of test-ing functions of Schwartz. The function $b_{\xi\xi}(\phi)$ is called a *generalized correlation function* for the process $\xi(\phi)$.

On account of linearity of the functional $b_{\xi\xi}(\phi)$, expression (2.67) can also be written in the form

$$b_{\xi\xi}(\phi, \psi) = \int b_{\xi\xi}(\tau_t, \phi)\psi(t')\,dt', \tag{2.68}$$

where τ_h is a translation operator defined by:

$$\tau_h\phi(s) = \phi(s+h). \tag{2.69}$$

From the foregoing it is easy to obtain the limiting form of (2.66) as $n \to \infty$. The set of functions $\{\phi_k(t)\}$ $(k = 1, 2, \cdots)$ which is complete for the space Φ_T will also be complete for the space L_T^2 of square integrable functions on the interval $\left[-\frac{T}{2}, \frac{T}{2}\right]$. Furthermore, $b_{\xi\xi}(\tau_t\phi)$ is a functional on L_T^2 which is continuous in the parameter t, and hence

$$\sum_{j=1}^{\infty} \int b_{\xi\xi}(\tau_t\phi)\phi_j(t)\,dt \cdot \phi_j(t) = b_{\xi\xi}(\tau_T\phi). \tag{2.70}$$

Thus, as $n \to \infty$ the operator A_n tends to the following operator in L_T^2 (which is defined on an everywhere dense set of functions $\phi \in \Phi_T$):

$$A\phi = b_{\xi\xi}(\tau_t\phi). \tag{2.71}$$

In a similar way, it can be shown that as $n \to \infty$ the matrices B_n, D_n and D'_n tend respectively to the following operators in L_T^2:

$$B\phi = b_{\eta\eta}(\tau_t\phi), \qquad D\phi = b_{\xi\eta}(\tau_t\phi), \qquad D'\phi = b_{\eta\xi}(\tau_t\phi) = b_{\xi\eta}(\tau_{-t}\phi), \tag{2.72}$$

where $b_{\eta\eta}(\phi)$, $b_{\xi\eta}(\phi)$ and $b_{\eta\xi}(\phi)$ are respectively the generalized correlation function for $\eta(\phi)$ and the generalized mutual correlation functions for $\xi(\phi)$ and $\eta(\phi)$, and $\eta(\phi)$ and $\xi(\phi)$; to put it differently, they are generalized functions in terms of which the correlation functionals $b_{\eta\eta}(\phi, \psi)$, $b_{\xi\eta}(\phi, \psi)$, and $b_{\eta\xi}(\phi, \psi)$ are given by the relations

$$\left.\begin{aligned} b_{\eta\eta}(\varphi, \psi) &= b_{\eta\eta}(\varphi * \psi^*) \equiv b_{\eta\eta}\left(\int_{-\infty}^{\infty} \varphi(t-t')\,\psi(-t')\,dt'\right), \\ b_{\xi\eta}(\varphi, \psi) &= b_{\xi\eta}(\varphi * \psi^*) \equiv b_{\xi\eta}\left(\int_{-\infty}^{\infty} \varphi(t-t')\,\psi(-t')\,dt'\right), \\ b_{\eta\xi}(\varphi, \psi) &\equiv b_{\xi\eta}(\psi, \varphi) = b_{\eta\xi}(\varphi * \psi^*) \equiv b_{\eta\xi}\left(\int_{-\infty}^{\infty} \varphi(t-t')\,\psi(-t')\,dt'\right). \end{aligned}\right\} \tag{2.73}$$

Thus we arrive at the following theorem.

Theorem 2.5. *In order that the information* $J\{\xi_T(\phi), \eta_T(\phi)\}$ *contained in a stationary Gaussian process* $\xi(\phi)$ *over an interval* $\left[-\frac{T}{2}, \frac{T}{2}\right]$ *about another such process* $\eta(\phi)$ *(over the same interval) which is stationarily and normally related* *to* $\xi(\phi)$ *be finite, it is necessary and sufficient that at least one of the operators in* L_T^2

$$B^{(1)} = DB^{-1}D'A^{-1}, \quad B^{(2)} = D'A^{-1}DB^{-1} \tag{2.74}$$

(where A, B, D and D' are operators defined by (2.71) and (2.72)) be a completely continuous operator with finite trace. If this condition is satisfied, the other operator will possess the same property and the information $J\{\xi_T(\phi), \eta_T(\phi)\}$ *will be given by the relation*

$$J\{\xi_T(\phi), \eta_T(\phi)\} = -\frac{1}{2}\log\det(E - B^{(1)}) = -\frac{1}{2}\log\det(E - B^{(2)}). \tag{2.75}$$

It should be remarked that, despite the fact that the operators B_1 and B_2 in (2.74) are not self-adjoint, the spectra of these operators lie on the real axis, so that the condition of finiteness of trace does not require further elaboration. Indeed, since A and B are positive-definite self-adjoint operators, the operators whose spectra coincide with the spectra of operators $B^{(1)}$ and $B^{(2)}$ will also be self-adjoint and hence will have real spectra.

Theorem 2.5 makes it possible to obtain an explicit expression for the information $J\{\xi_T(\phi), \eta_T(\phi)\}$ in the important special case where the spectral functions of $\xi(\phi)$ and $\eta(\phi)$ processes (including the mutual spectral function $H(\lambda)$) are all absolutely continuous, and the derivatives of these functions (which in the case of absolutely continuous spectral functions are called *spectral densities*) are rational functions of λ. However, we shall not consider here the general case of rational spectral densities and will restrict ourselves to the simplest case where η is the sum of a process ξ and a white noise uncorrelated with ξ.

C. *Information about a stationary process over a finite interval, contained in a sum of this process and white noise.* We shall now examine a special case of the problem considered in B, in which

$$\eta(\phi) = \xi(\phi) + \zeta(\phi), \tag{2.76}$$

where $\zeta(\phi)$ is a "white noise" uncorrelated with $\xi(\phi)$, i.e., a generalized random process whose correlation functional is of the form

$$b_{\zeta\zeta}(\phi, \psi) = 2\pi f \int_{-\infty}^{\infty} \phi(t)\psi(t)\,dt \tag{2.77}$$

(in which f is the spectral density of white noise). Clearly, the corresponding generalized correlation function is given by

$$b_{\zeta\zeta} = 2\pi f\cdot\delta, \tag{2.78}$$

where δ is the Dirac δ-function. Here we have:

$$D\phi = D'\phi = A\phi, \quad B\phi = (A + 2\pi f\cdot E)\phi \tag{2.79}$$

E being the identity operator.

Consequently

$$B^{(1)} = DB^{-1}D'A^{-1} = A\cdot(A + 2\pi f E)^{-1} = E - \left(E + \frac{1}{2\pi f}A\right)^{-1}, \tag{2.80}$$

and hence

$$J\{\xi_T(\varphi),\ \eta_T(\varphi)\} = -\frac{1}{2}\log\det\left(E+\frac{1}{2\pi f}A\right)^{-1} = \frac{1}{2}\log\det\left(E+\frac{1}{2\pi f}A\right). \quad (2.81)$$

In the very important special case where $\xi(\phi)$ is a process of the type (1.28), i.e., is generated by an ordinary random process $\xi(t)$, the operator A is of the form

$$A\varphi = \int_{-\infty}^{\infty} B(s)\varphi(t+s)\,ds = \int_{-T/2}^{T/2} B(t-t')\varphi(t')\,dt', \quad (2.82)$$

where $B(s) \equiv B(-s) = \mathrm{M}\xi(t+s)\xi(t)$ is the correlation function of $\xi(t)$ 1). In this case the calculation of information reduces to evaluating the determinant of an operator of the form $E + A_1$, where E is the identity operator and $A_1 = \frac{1}{2\pi f}A$ is a Fredholm operator.

Before considering some examples in which the determinant of the operator $E + \frac{1}{2\pi f}A$ can be evaluated, we shall briefly dwell on the important case where the operator $A_1 = \frac{1}{2\pi f}A$ is small compared with the identity operator 2).

Here it is convenient to use the well-known formula

$$\det K = e^{\mathrm{Sp}\log K} \quad (2.83)$$

which is an immediate consequence of the fact that the proper values of the operator $\log K$ are equal to the logarithms of the proper values of K. Application of this formula leads to the following expression for the information $J\{\xi_T(\phi), \eta_T(\phi)\}$:

$$J\{\xi_T(\phi), \eta_T(\phi)\} = \frac{1}{2}\,\mathrm{sp}\,\log(E + A_1),\quad A_1 = \frac{1}{2\pi f}A. \quad (2.84)$$

On expanding the logarithmic function in Taylor's series we obtain:

$$J\{\xi_T(\varphi), \eta_T(\varphi)\} = \frac{1}{2}\left[\mathrm{Sp}\,A_1 - \frac{1}{2}\mathrm{Sp}\,A_1^2 + \frac{1}{3}\mathrm{Sp}\,A_1^3 + \dots\right], \quad (2.85)$$

or, after substitution of explicit expressions for the traces of various powers of the

1) If $\xi(\phi)$ is not an ordinary stationary random process but rather a generalized process with unbounded spectral function, then the operator $A\phi$ will not be, in general, a completely continuous operator with finite trace, and therefore in this case the information $J\{\xi_T(\phi), \eta_T(\phi)\}$ will be infinite.

2) When the spectral density $f(\lambda)$ of $\xi(t)$ exists (here, for simplicity, we restrict ourselves to this case only), it is given by $f(\lambda) = \frac{1}{2\pi}\int_{-\infty}^{\infty} e^{-it\lambda}B(t)\,dt$. This entails, in general, the approximate relation $f_0 \approx B_0 T_0$, where B_0 is a characteristic value of the correlation function (say $B(0)$), f_0 is the maximum (or some other characteristic) value of the spectral density $f(\lambda)$, and T_0 is the time-constant of the correlation function $B(t)$. From this it follows that, in terms of the order of magnitude, we can write

$$\frac{1}{2\pi f}A \sim \frac{f_0}{f}\cdot\frac{T}{T_0}$$

(provided $T \ll T_0$ or $T \sim T_0$; for $T \gg T_0$ we will have $\frac{1}{2\pi f}A \sim \frac{f_0}{f}$). Thus the condition of "smallness" of the operator $\frac{1}{2\pi f}A$ can be taken to be $f_0/f\cdot T/T_0 \ll 1$ or (for $T \gg T_0$) $f_0/f \ll 1$; in either case, the right-hand member provides a measure of the order of smallness of the operator $\frac{1}{2\pi f}A$, and, in the theory of disturbances which is developed below, it will play the role of a basic dimensionless parameter.

Fredholm operator A_1,

$$J\{\xi_T(\varphi), \eta_T(\varphi)\} = \frac{TB(0)}{4\pi f} - \frac{1}{16\pi^2 f^2} \int_{-T/2}^{T/2} \int_{-T/2}^{T/2} B^2(t-s)\,dt\,ds +$$

$$+ \frac{1}{48\pi^3 f^3} \int_{-T/2}^{T/2} \int_{-T/2}^{T/2} \int_{-T/2}^{T/2} B(t-t_1)\,B(t_1-t_2)\,B(t_2-t)\,dt\,dt_1\,dt_2 - \ldots = \tag{2.86}$$

$$= \frac{B(0)}{4\pi f}T - \frac{T}{16\pi^2 f^2} \int_{-T}^{T} \left(1 - \frac{\tau}{T}\right) B^2(\tau)\,d\tau + \ldots.$$

When the operator A_1 is small compared with the identity operator, this series for J can serve as a basis for a theory of disturbances, since it is a power series in a parameter characterizing the smallness of the operator A_1 and thus affords a simple means of calculating J to any desired degree of accuracy.

We turn now to examples illustrating the calculation of the information (2.81) in explicit form.

§3. Examples

1) Let

$$B(\tau) = \int_{-\infty}^{\infty} e^{i\lambda\tau} f(\lambda)\,d\lambda, \tag{2.87}$$

where the spectral density $f(\lambda)$ has the form

$$f(\lambda) = \frac{1}{|a_0(i\lambda)^n + a_1(i\lambda)^{n-1} + \ldots + a_n|^2} = \frac{1}{|P(i\lambda)|^2}, \tag{2.88}$$

in which all the roots of the polynomial $P(z)$ can be assumed, without any loss of generality, to lie in the left half-plane.

Our problem consists in evaluating the expression (2.89), where the

$$\det\left(E + \frac{1}{2\pi f}A\right) = \prod_{k=1}^{\infty}\left(1 + \frac{1}{2\pi f\lambda_k}\right), \tag{2.89}$$

$\lambda_k (k = 1, 2, \cdots)$ are the proper values of the Fredholm equation

$$\varphi(t) = \lambda \int_{-T/2}^{T/2} B(t-s)\,\varphi(s)\,ds. \tag{2.90}$$

(Note that in accordance with tradition the parameter λ is placed on the right side of the integral equation; as a result, in (2.89) the proper values λ_k appear in the denominator rather than in the numerator.)

First, we shall show that when $f(\lambda)$ is of the form (2.88) the correlation function (2.87) is the Green's function of the ordinary differential equation:

$$P\left(\frac{d}{dt}\right)P\left(-\frac{d}{dt}\right)\varphi(t) \equiv \left[a_0\frac{d^n}{dt^n} + a_1\frac{d^{n-1}}{dt^{n-1}} + \ldots + a_n\right] \times$$

$$\times \left[(-1)^n a_0 \frac{d^n}{dt^n} + (-1)^{n-1} a_1 \frac{d^{n-1}}{dt^{n-1}} + \ldots + a_n\right] \cdot \varphi(t) = 0, \tag{2.91}$$

and, accordingly, the proper values of the integral equation (2.90) coincide with the proper values of a certain Sturm-Liouville problem. For the derivation of this result (which is discussed, for example, in papers by M. G. Kreĭn) it is simplest to use the following approach. Rewrite equation (2.90) in the form

$$\varphi(t)=\lambda \int_{-T/2}^{T/2} \int_{-\infty}^{\infty} \frac{e^{i(t-s)\lambda}}{|P(i\lambda)|^2} \varphi(s)\, d\lambda\, ds \tag{2.92}$$

and apply to both sides of this equation the operator $P\left(\frac{d}{dt}\right) P\left(-\frac{d}{dt}\right)$. Using the relation $\int_{-\infty}^{\infty} e^{i(t-s)\lambda} d\lambda = 2\pi\delta(t-s)$, (where $\delta(t)$ is the Dirac δ-function), we get:

$$P\left(\frac{d}{dt}\right)P\left(-\frac{d}{dt}\right)\varphi(t)=2\pi\lambda\varphi(t). \tag{2.93}$$

Thus, every solution of the integral equation (2.92) is also a solution of the linear differential equation (2.93) with constant coefficients. In order to find the boundary conditions associated with equation (2.93), we apply to both sides of (2.92) the operators $P\left(\frac{d}{dt}\right)$, $\frac{d}{dt}P\left(\frac{d}{dt}\right)$, $\ldots$, $\frac{d^{n-1}}{dt^{n-1}}P\left(\frac{d}{dt}\right)$, after which we set $t = T/2$. This gives:

$$\frac{d^k}{dt^k} P\left(\frac{d}{dt}\right)\varphi(t)\Big|_{t=T/2}=\lambda \int_{-T/2}^{T/2} \left\{ \int_{-\infty}^{\infty} \frac{e^{i\left(\frac{T}{2}-s\right)\lambda}(i\lambda)^k}{P(-i\lambda)} d\lambda \right\} \varphi(s)\, ds. \tag{2.94}$$

Taking into account that all roots of the polynomial $P(-i\lambda)$ lie in the left half-plane, we have by virtue of Cauchy's theorem

$$\int_{-\infty}^{\infty} \frac{e^{i(T/2-s)\lambda}(i\lambda)^k}{P(-i\lambda)} d\lambda=0 \qquad \text{for} \qquad s<\frac{T}{2} \quad (k=0,1,\ldots,n-1), \tag{2.95}$$

from where it follows

$$\frac{d^k}{dt^k} P\left(\frac{d}{dt}\right)\varphi(t)\Big|_{t=\frac{T}{2}}=0 \qquad (k=0,1,\ldots,n-1). \tag{2.96}$$

Similarly:

$$\frac{d^k}{dt^k} P\left(-\frac{d}{dt}\right)\varphi(t)\Big|_{t=-\frac{T}{2}}=0 \qquad (k=0,1,\ldots,n-1). \tag{2.97}$$

In this way, we obtain a Sturm-Liouville system (2.93), (2.96), (2.97) which is equivalent to the integral equation (2.90).

Explicit determination of all the proper values of such a Sturm-Liouville system can not be accomplished: as is well-known this problem reduces in general to the solution of an unsolvable transcendental equation. Remarkably enough, however, the corresponding Fredholm determinant (2.89) can be evaluated explicitly with the aid of certain theorems in the theory of entire functions. We illustrate below how this can be accomplished.

1) Consider the case where

$$B(t) = Ce^{-\alpha|t|}, \quad f(\lambda) = \frac{C\alpha}{\pi(\lambda^2+\alpha^2)}. \tag{2.98}$$

Here in virtue of (2.93), (2.96), and (2.97), we have

$$\det\left(E + \frac{1}{2\pi f}A\right) = \prod_{k=1}^{\infty}\left(1 + \frac{1}{2\pi f\lambda_k}\right),$$

where the λ_k $(k = 1, 2, \cdots)$ are the proper values of the equation

$$\phi''(t) - (\alpha^2 - 2C\alpha\lambda)\phi(t) = 0 \tag{2.99}$$

subject to the boundary conditions

$$\left.\begin{aligned} \varphi'\left(\frac{T}{2}\right) + \alpha\varphi\left(\frac{T}{2}\right) &= 0, \\ \varphi'\left(-\frac{T}{2}\right) - \alpha\varphi\left(-\frac{T}{2}\right) &= 0. \end{aligned}\right\} \tag{2.100}$$

The general solution of (2.99) has the form:

$$\left.\begin{aligned} \varphi(t) &= C_1e^{\Lambda t} + C_2e^{-\Lambda t}, \\ \varphi'(t) &= C_1\Lambda e^{\Lambda t} - C_2\Lambda e^{-\Lambda t}, \quad \Lambda = \sqrt{\alpha^2 - 2C\alpha\lambda}; \end{aligned}\right\} \tag{2.101}$$

and it will satisfy the boundary conditions (2.100) if

$$\begin{vmatrix} (\alpha+\Lambda)e^{\Lambda T/2} & (\alpha-\Lambda)e^{-\Lambda T/2} \\ (\alpha-\Lambda)e^{-\Lambda T/2} & (\alpha+\Lambda)e^{\Lambda T/2} \end{vmatrix} = (\alpha+\Lambda)^2e^{\Lambda T} - (\alpha-\Lambda)^2e^{-\Lambda T} = 0, \tag{2.102}$$

where $\Lambda = \sqrt{\alpha^2 - 2C\alpha\lambda}$. Thus, in our case the λ_k $(k = 1, 2, \cdots)$ are the roots of the transcendental equation (2.102), with one exception: if λ is such that $\alpha^2 - 2C\alpha\lambda = 0$ (i.e., if the characteristic equation $\Lambda^2 - (\alpha^2 - 2C\alpha\lambda) = 0$ has repeated roots), then equation (2.102) will obviously be satisfied, but the general solution of equation (2.99) will have a form different from (2.101), and the boundary conditions (2.100) will not be satisfied. Consequently, the root $\lambda = \frac{\alpha}{2C}$ of equation (2.102) is not a proper value.

It is expedient to divide the right-hand member of (2.102) by $4\alpha\Lambda = 4\alpha\sqrt{\alpha^2 - 2C\alpha\lambda}$, since this eliminates the extraneous root $\lambda = \frac{\alpha}{2C}$. The result is a function

$$\Phi(\lambda) = \operatorname{ch}\Lambda T + (\alpha^2 + \Lambda^2)\frac{\operatorname{sh}\Lambda T}{2\alpha\Lambda}, \quad \Lambda = \sqrt{\alpha^2 - 2C\alpha\lambda}, \tag{2.103}$$

which is an entire function of λ, of exponential type and order $\frac{1}{2}$. Now, according to a well-known theorem of Hadamard concerning the factorization of entire functions of finite order (see, for example, [30], p. 284) a function of order less than unity such as $\Phi(\lambda)$ can be expressed in the form

$$\Phi(\lambda) = \Phi(0)\prod_{k=1}^{\infty}\left(1 - \frac{\lambda}{\lambda_k}\right), \tag{2.104}$$

where the λ_k $(k = 1, 2, \cdots)$ are the zeros of $\Phi(\lambda)$ (which in our case coincide with the proper values of integral equation (2.90)). Thus, we have

$$\det\left(E + \frac{1}{2\pi f}A\right) = \frac{\Phi(-1/2\pi f)}{\Phi(0)} \tag{2.105}$$

and

$$J\{\xi_T(\varphi),\ \eta_T(\varphi)\} = \frac{1}{2}\log\frac{\Phi(-1/2\pi f)}{\Phi(0)}. \tag{2.106}$$

After substituting (2.103) in (2.106) the expression for $J\{\xi_T(\phi), \eta_T(\phi)\}$ can be put into a convenient form through the use of two dimensionless parameters:

$$\tau = \alpha T = \frac{T}{T_0}, \qquad k = \frac{C}{\pi\alpha f} = \frac{f_0}{f} \tag{2.107}$$

where $T_0 = \frac{1}{\alpha}$ is the time-constant of the correlation function and $f_0 = C/\pi\alpha$ is the value of the spectral density of $\xi(t)$ process for $\lambda = 0$. In this way we obtain the final result:

$$J\{\xi,\ \eta\} = \frac{1}{2}\log\left\{\frac{\sqrt{1+k}+\left(1+\frac{k}{2}\right)}{2\sqrt{1+k}}e^{\tau(\sqrt{1+k}-1)} + \frac{\sqrt{1+k}-\left(1+\frac{k}{2}\right)}{2\sqrt{1+k}}e^{-\tau(\sqrt{1+k}+1)}\right\}. \tag{2.108}$$

For $k \ll 1$ ("strong noise") or $\tau \ll 1$ (short observation time) the expression (2.108) becomes equivalent to the result yielded by the series expansion (2.86) of the theory of disturbances:

$$J\{\xi,\ \eta\} = \begin{cases} \frac{k\tau}{4} - \frac{k^2\tau}{16}\left\{1 - \frac{1-e^{-2\tau}}{2\tau}\right\} + \dots & \text{for } k \ll 1, \\ \frac{k\tau}{4} - \frac{k^2\tau^2}{16} + \dots & \text{for } \tau \ll 1. \end{cases} \tag{2.109}$$

In another limiting case, $\tau \gg 1$ (long observation time), we have

$$J\{\xi,\ \eta\} \approx \frac{\sqrt{1+k}-1}{2}\tau, \tag{2.110}$$

which verifies for this case the general result (2.61). Finally, for $k \gg 1$ (weak noise) and not too small τ (specifically, for $\tau \gg 1/\sqrt{k}$) we have:

$$J\{\xi,\ \eta\} \approx \frac{1}{2}\sqrt{k}\,\tau. \tag{2.111}$$

2) Consider now the case where

$$f(\lambda) = \frac{A}{|-\lambda^2 + 2i\alpha\lambda + \omega^2|^2} = \frac{A}{(\lambda^2-\omega^2)^2 + 4\alpha^2\lambda^2} \tag{2.112}$$

and consequently

$$B(t) = \frac{\pi A}{2\alpha\omega^2}e^{-\alpha|\tau|}\left(\cos\beta\tau + \frac{\alpha}{\beta}\sin\beta|\tau|\right), \quad \beta = \sqrt{\omega^2-\alpha^2} \tag{2.113}$$

(for concreteness we assume that in (2.112) $\omega^2 > \alpha^2$). Here $(E + \frac{1}{2\pi f}A)$ is given by the expression (2.89), where the λ_k are the proper values of the differential equation

$$\phi^{IV}(t) + 2(\omega^2 - 2\alpha^2)\phi''(t) + (\omega^4 - 2\pi A\lambda)\phi(t) = 0 \tag{2.114}$$

subject to the boundary conditions

$$\left.\begin{aligned}\varphi''\left(\frac{T}{2}\right)+2\alpha\varphi'\left(\frac{T}{2}\right)+\omega^2\varphi\left(\frac{T}{2}\right)=0,\\ \varphi'''\left(\frac{T}{2}\right)+2\alpha\varphi''\left(\frac{T}{2}\right)+\omega^2\varphi'\left(\frac{T}{2}\right)=0,\\ \varphi''\left(-\frac{T}{2}\right)-2\alpha\varphi'\left(-\frac{T}{2}\right)+\omega^2\varphi\left(-\frac{T}{2}\right)=0,\\ \varphi'''\left(-\frac{T}{2}\right)-2\alpha\varphi''\left(-\frac{T}{2}\right)+\omega^2\varphi'\left(-\frac{T}{2}\right)=0.\end{aligned}\right\} \tag{2.115}$$

It is readily seen that these proper values are given by the roots of the transcendental equation

$$\Phi_0(\lambda)=\begin{vmatrix} P(\theta_1)\,e^{\theta_1 T/2} & P(-\theta_1)\,e^{-\theta_1 T/2} & P(\theta_2)\,e^{\theta_2 T/2} & P(-\theta_2)\,e^{-\theta_2 T/2}\\ P(-\theta_1)\,e^{-\theta_1 T/2} & P(\theta_1)\,e^{\theta_1 T/2} & P(-\theta_2)\,e^{-\theta_1 T/2} & P(\theta_2)\,e^{\theta_2 T/2}\\ \theta_1 P(\theta_1)\,e^{\theta_1 T/2} & -\theta_1 P(-\theta_1)\,e^{-\theta_1 T/2} & \theta_2 P(\theta_2)\,e^{\theta_2 T/2} & -\theta_2 P(-\theta_2)\,e^{-\theta_2 T/2}\\ -\theta_1 P(-\theta_1)\,e^{-\theta_1 T/2} & \theta_1 P(\theta_1)\,e^{\theta_1 T/2} & -\theta_2 P(-\theta_2)\,e^{-\theta_2 T/2} & \theta_2 P(\theta_2)\,e^{\theta_2 T/2}\end{vmatrix}=0, \tag{2.116}$$

where

$$P(z) = z^2 + 2\alpha z + \omega^2, \tag{2.117}$$

and $\pm\theta_1$ and $\pm\theta_2$ are the four roots of the biquadratic equation

$$\theta^4 + 2(\omega^2 - 2\alpha^2)\theta^2 + (\omega^4 - 2\pi A\lambda) = 0, \tag{2.118}$$

that is,

$$\theta_{1,\,2}=\sqrt{2\alpha^2-\omega^2\pm\sqrt{2\pi A\lambda-4\alpha^2(\omega^2-\alpha^2)}}. \tag{2.119}$$

As before, the values of λ corresponding to repeated roots of (2.116) will be the "extraneous roots" of (2.116) which are not proper values of the Sturm-Liouville system (2.114), (2.115). With these λ the general solution of (2.114) will not be the sum $C_1e^{\theta_1 t} + C_2e^{-\theta_1 t} + C_3e^{\theta_2 t} + C_4e^{-\theta_2 t}$, which it was assumed to be in the derivation of (2.116). Consequently, the desired values of the $\lambda_k = (k = 1, 2, \cdots)$ will coincide with the zeros of the function

$$\Phi(\lambda)=\frac{\Phi_0(\lambda)}{\theta_1\theta_2\,[2\pi A\lambda-4\alpha^2(\omega^2-\alpha^2)]} \tag{2.120}$$

It now remains to show that $\Phi(\lambda)$ is an entire function of order less than unity. Once this is established, the expression (2.89) can readily be evaluated by the use of Hadamard's theorem.

With reference to (2.116), consider the function $\Phi_0(\lambda)/\theta_1\theta_2$. On expanding the determinant and substituting power series for the exponential functions, the function in question assumes the form of a power series in θ_1, θ_2. In this series the terms of like power in θ_1, θ_2 are symmetrical functions of the four roots of equation (2.118). Hence, such terms will constitute a polynomial in the parameter λ which enters linearly in the last coefficient of equation (2.118). From this it follows that the expression under consideration is an entire function of λ, and since it vanishes for $2\pi A\lambda = 4\alpha^2(\omega^2 - \alpha^2)$, the right-hand member is likewise an entire function of λ. In virtue of (2.119), it is clear that this entire function is a function of exponential type of order $\frac{1}{4}$. On employing Hadamard's theorem and performing some algebraic operations on the determinant $\Phi_0(\lambda)$, we are lead to the following final result:

$$J\{\xi_T, \eta_T\} = \frac{1}{2}\log\frac{\Phi(-{}^1/2\pi f)}{\Phi(0)}, \tag{2.121}$$

where

$$\Phi(\lambda) = \frac{(2\pi A\lambda)^2 - \sum_{k=1}^{4} P_k(\lambda)\varphi_k(\lambda)}{2\pi A\lambda - 4\alpha^2(\omega^2 - \alpha^2)}, \tag{2.122}$$

$\phi_k(\lambda)$ (λ = 1, 2, 3, 4) are entire transcendental functions of λ given by the equations

$$\left.\begin{aligned} &\varphi_1(\lambda) = \operatorname{ch}\theta_1 T\cdot\operatorname{ch}\theta_2 T, \quad \varphi_2(\lambda) = \frac{\operatorname{sh}\theta_1 T}{\theta_1}\cdot\frac{\operatorname{sh}\theta_2 T}{\theta_2},\\ &\varphi_3(\lambda) = \theta_1\operatorname{sh}\theta_1 T\cdot\operatorname{ch}\theta_2 T + \theta_2\operatorname{sh}\theta_2 T\cdot\operatorname{ch}\theta_1 T,\\ &\varphi_4(\lambda) = \frac{\operatorname{sh}\theta_1 T}{\theta_1}\cdot\operatorname{ch}\theta_2 T + \frac{\operatorname{sh}\theta_2 T}{\theta_2}\cdot\operatorname{ch}\theta_1 T, \end{aligned}\right\} \tag{2.123}$$

and the $P_k(\lambda)$ (λ = 1, 2, 3, 4) are the following polynomials in λ:

$$\left.\begin{aligned} &P_1(\lambda) = (2\pi A\lambda)^2 + 32\alpha^2\omega^2\cdot 2\pi A\lambda + 128\alpha^4\omega^2(\omega^2 - \alpha^2),\\ &P_2(\lambda) = (\omega^2 + 14\alpha^2)(2\pi A\lambda)^2 - 32\alpha^2(\omega^4 + 2\alpha^2\omega^2 - 2\alpha^4)\,2\pi A\lambda + 128\alpha^4\omega^4(\omega^2 - \alpha^2),\\ &P_3(\lambda) = 8\alpha(3\alpha^2 - \omega^2)\cdot 2\pi A\lambda,\\ &P_4(\lambda) = 8\pi A\lambda\alpha(2\pi\alpha\lambda - 8\alpha^4 + 6\alpha^2\omega^2 - 2\omega^4). \end{aligned}\right\} \tag{2.124}$$

At first glance this result appears to be quite complicated. Actually, with its aid it is possible to calculate the information $J\{\xi_T, \eta_T\}$ for any prescribed values of the parameters α, ω, A, and T without excessive labor.

From the expressions (2.121)–(2.124) one can deduce a number of fairly simple asymptotical formulae which are applicable to various limiting cases. Thus, for example, when $\alpha T \gg 1$ ("long observation time") the dominant term in (2.121) is the logarithm of $e^{[(\theta_1+\theta_2)-(\theta_1^0-\theta_2^0)T]}$, where

$$\left.\begin{aligned} &\theta_1 = \sqrt{2\alpha^2 - \omega^2 + i\sqrt{4\alpha^2(\omega^2 - \alpha^2) + \omega^4 k}},\\ &\theta_2 = \sqrt{2\alpha^2 - \omega^2 - i\sqrt{4\alpha^2(\omega^2 - \alpha^2) + \omega^4 k}}\\ &\theta_1^0 = \alpha + i\sqrt{\omega^2 - \alpha^2},\ \theta_2^0 = \alpha - i\sqrt{\omega^2 - \alpha^2},\ k = \frac{A}{\omega^4 f} = \frac{f_0}{f} \end{aligned}\right\} \tag{2.125}$$

(the square root being chosen in each case so as to render the real part positive). Consequently, for $\alpha T \gg 1$

$$J\{\xi_T, \eta_T\} \approx \frac{1}{2}[(\theta_1 + \theta_2) - (\theta_1^0 + \theta_2^0)]T = \frac{1}{2}\left\{\sqrt{2(2\alpha^2 - \omega^2) + 2\omega^2\sqrt{1+k}} - 2\alpha\right\}T, \tag{2.126}$$

from which it is evident that in this case the limit (2.63) coincides with the expression (2.61). In another limiting case – when $\omega T \ll 1$ – it is expedient to expand the hyperbolic functions in (2.123) into Taylor's series; this yields

$$J\{\xi_T, \eta_T\} \approx \frac{\omega^2}{4\alpha^2} k \cdot \alpha T, \tag{2.127}$$

which is in accord with (2.86). The same result is obtained for $k \ll 1$, the "theory of disturbances" also being applicable in this case. Finally, for $k \gg 1$ the dominant term in $\Phi(\lambda)$ is $e^{(\theta_1+\theta_2)T}$, and we have:

$$J\{\xi_T, \eta_T\} \approx \frac{\sqrt{2}}{2} \omega \sqrt{k} T. \tag{2.128}$$

The problem of calculating information for any other case where the spectral density is of the form (2.88) can be resolved in a similar fashion. In particular, it is easy to see that as $T \to \infty$ the information will grow asymptotically as a linear function of T. More specifically, in the case of any spectral density of the form (2.88), we have

$$\lim_{T\to\infty} \frac{1}{T} J\{\xi_T, \eta_T\} = \frac{1}{2} \sum_{k=1}^{n} (\theta_k - \theta_k^0), \tag{2.129}$$

where the θ_k ($k = 1, 2, \cdots, n$) are those roots of the characteristic equation

$$[a_0\theta^n + a_1\theta^{n-1} + \cdots + a_n]\,[a_0(-\theta)^n + a_1(-\theta)^{n-1} + \cdots + a_n] = -1/f, \tag{2.130}$$

which have positive real parts, while the θ_k^0 ($k = 1, 2, \cdots, n$) are the corresponding roots of the same equation for $f = \infty$. It is not difficult to verify that, in this case, the right-hand member of (2.129) coincides with the right-hand member of (2.61). Similarly, when $f \to 0$ the dominant term in $(E + \frac{1}{2\pi f} A)$ is $e^{(\theta_1 + \cdots + \theta_n)T}$, and hence for very small f

$$J\{\xi_T, \eta_T\} \approx \frac{c}{\sqrt[2n]{a_n^2 f}} \cdot T = c\left(\frac{a_n^2}{a_0^2}\right)^{\frac{1}{2n}} \cdot \sqrt[2n]{k}\, T, \tag{2.131}$$

where

$$k = \frac{1}{a_n^2 f} = \frac{f(0)}{f}, \qquad c = 1/\sin\frac{\pi}{2n} \tag{2.132}$$

k being the signal-to-noise ratio ($k \gg 1$ in the case under consideration), and c the sum of real parts of those roots of $x^{2n} = (-1)^{n-1}$ which are located in the right half-plane. Finally, as $T \to 0$ or $f \to \infty$ (i.e., $k \to 0$) the asymptotic behavior of $J\{\xi_T, \eta_T\}$ can be obtained approximately with the aid of the series (2.86) of the theory of disturbances.

II. We now consider briefly the case where the correlation function (2.87) corresponds to a rational spectral density of the general form

$$f(\lambda) = \frac{|b_0(i\lambda)^m + \ldots + b_m|^2}{|a_0(i\lambda)^n + \ldots + a_n|^2} = \frac{|Q(i\lambda)|^2}{|P(i\lambda)|^2}, \quad m < n. \tag{2.133}$$

Here the problem reduces to evaluating the infinite product (2.89) for the case where the λ_k are the proper values of the following Fredholm equation:

$$\varphi(t) = \lambda \int_{-T/2}^{T/2} \int_{-\infty}^{\infty} e^{i(t-s)\lambda} \frac{|Q(i\lambda)|^2}{|P(i\lambda)|^2} \varphi(s)\, d\lambda\, ds. \tag{2.134}$$

Applying to both sides of this equation the differential operator $P\left(\frac{d}{dt}\right)P\left(-\frac{d}{dt}\right)$ and making use of

$$\int_{-\infty}^{\infty}(i\lambda)^k e^{i(t-s)\lambda}d\lambda = 2\pi\delta^{(k)}(t-s),$$

we obtain

$$P\left(\frac{d}{dt}\right)P\left(-\frac{d}{dt}\right)\varphi(t) = 2\pi\lambda Q\left(\frac{d}{dt}\right)Q\left(-\frac{d}{dt}\right)\varphi(t). \tag{2.135}$$

Thus, in this case as well, every solution of the integral equation (2.90) is also a solution of a linear differential equation with constant coefficients.

The boundary conditions for the equation (2.135) corresponding to the integral equation (2.134) can be found in a manner analogous to that used in the case of the spectral density (2.88). We apply to both sides of equation (2.134) the operator

$$\frac{d^k}{dt^k}P\left(\frac{d}{dt}\right) \quad (k = 0,1,\ldots,n-1);$$

after which we extract a polynomial term $\sum_{j=0}^{2m+k-n} c_j^{(k)}(i\lambda)$ from the rational function $\frac{Q(i\lambda)Q(-i\lambda)(i\lambda)^k}{P(-i\lambda)}$ which is under the integral sign in (2.134), and set $t = \frac{T}{2}$. (Note that this can be done provided $2m + k \geq n - 1$; otherwise the polynomial is zero.) The polynomial term yields in the right side of the equation terms of the form $2\pi\lambda \sum_{j=0}^{2m+k-n} c_j^{(k)}\phi^{(j)}(t)$, while the integral term vanishes for $t = \frac{T}{2}$ by virtue of (2.95). Consequently, we have

$$\frac{d^k}{dt^k}P\left(\frac{d}{dt}\right)\varphi(t)\Big|_{t=T/2} = 2\pi\lambda \sum_{j=0}^{2m+k-n} c_j^{(k)}\frac{d^j}{dt^j}\varphi(t)\Big|_{t=T/2}; \tag{2.136}$$

and in an analogous manner it can be shown that

$$(-1)^k\frac{d^k}{dt^k}P\left(-\frac{d}{dt}\right)\varphi(t)\Big|_{t=-\frac{T}{2}} = 2\pi\lambda \sum_{j=0}^{2m+k-n}(-1)^j c_j^{(k)}\frac{d^j}{dt^j}\varphi(t)\Big|_{t=-\frac{T}{2}}, \tag{2.137}$$

where $\sum_{j=0}^{2m+k-n}(-1)^j c_j^{(k)}(i\lambda)^j$ is the polynomial term in $\frac{Q(i\lambda)Q(-i\lambda)(-i\lambda)^k}{P(i\lambda)}$. The proper values λ_k of (2.134) are also the proper values of the Sturm-Liouville system (2.135), (2.136), (2.137) which is analogous to the system (2.93), (2.96),(2.97).

Further steps in the calculation of $J\{\xi_T, \eta_T\}$ are in no way different from those followed previously in the case of the spectral density (2.88). Here, as there, the λ_k are the zeros of an entire function which can be expressed as a sum of exponential functions whose arguments are linear combinations – with coefficients that are proportional to T and linearly dependent on λ – of the roots of the characteristic equation

$$P(\theta)P(-\theta) - 2\pi\lambda Q(\theta)Q(-\theta) = 0. \tag{2.138}$$

Since $m < n$, it is evident that any such function will be of order less than unity, so that $\Pi(1 + \frac{1}{2\pi f\lambda_k})$ can be evaluated with the aid of Hadamard's theorem concerning the factorization of entire functions. Clearly, as $T \to \infty$,

$$\frac{J\{\xi_T, \eta_T\}}{T} \to \frac{1}{2}\sum_{k=1}^{n}(\theta_k - \theta_k^0), \tag{2.139}$$

where the θ_k $(k = 1, \cdots, n)$ are the roots of the equation (2.138) (with $\lambda = \frac{-1}{2\pi f}$) located in the right half-plane, and the θ_k^0 $(k = 1, \cdots, n)$ are the corresponding roots with $\lambda = 0$. This result coincides with the result yielded by equation (2.61).

An asymptotic form for $J\{\xi_T, \eta_T\}$ when $f \to 0$ is given by

$$J\{\xi_T, \eta_T\} \approx \frac{1}{2}\sum_{k=1}^{n}\theta_k(f)\cdot T, \tag{2.140}$$

where the $\theta_k(f)$ $(k = 1, \cdots, n)$ are the right half-plane roots of the equation

$$P(\theta)P(-\theta) + \frac{1}{f}Q(\theta)Q(-\theta) = 0. \tag{2.141}$$

Equivalently

$$J\{\xi_T, \eta_T\} \approx c\sqrt[2(n-m)]{\frac{b_0^2}{a_0^2 f}}\cdot T = cA\sqrt[2(n-m)]{k}\,T \quad \text{for} \quad k \gg 1, \tag{2.142}$$

where

$$A = \left(\frac{a_n^2 b_0^2}{a_0^2 b_m^2}\right)^{\frac{1}{2(n-m)}}, \quad k = \frac{b_m^2}{a_n^2 f} = \frac{f(0)}{f}, \quad c = \frac{1}{\sin \pi/2(n-m)} \tag{2.143}$$

(k being the signal-to-noise ratio and c the sum of the right half-plane roots of the equation $x^{2n} + (-1)^{n-m}x^{2m} = 0$). When $T \to 0$ or $f \to \infty$, the asymptotic behavior of $J\{\xi_T, \eta_T\}$ can readily be found from the series (2.86) of the theory of disturbances.

It should be remarked that the exponent $2(n - m)$ in (2.142) characterizes the "degree of smoothness" of the correlation function $B(\tau)$ (and hence also of the random process $\xi(t)$). (In the case of spectral density (2.133) the function (2.87) has $2(n - m) - 2$ continuous derivatives; its $(2n - 2m - 1)$st derivative has a jump at $\tau = 0$, and all its derivatives exist for $\tau \neq 0$.) Thus the analytic behavior of $J\{\xi, \eta\}$ for $f \to 0$ is determined by the smoothness of $B(\tau)$. It is not difficult to show that this result does not depend on the special form (2.133) of the spectral density, and is valid under quite general conditions. Indeed, assume that the correlation function $B(\tau)$ admits of the representation

$$B(\tau) = B_0(\tau) + B_1(\tau), \tag{2.144}$$

where the Fourier transform of $B_0(\tau)$ is of the form (2.88), and $B_1(\tau)$ is a smooth function possessing at least $2n$ continuous derivatives. [1] Then

[1] Functions of the form (2.144) comprise a relatively broad class of correlation functions which have $2n - 2$ continuous derivatives with $(2n - 1)$st derivative being discontinuous at $\tau = 0$. The representation (2.144) will always exist if the spectral density $f(\lambda)$ behaves as

$$f(\lambda) = \frac{a_1}{\lambda^{2n}} + O\left(\frac{1}{\lambda^{2n+1+\varepsilon}}\right)$$

(where $\epsilon > 0$) when $|\lambda| \to \infty$.

$$E+\frac{1}{2\pi f}A=\left(E+\frac{1}{2\pi f}A_0\right)\left[E+\left(E+\frac{1}{2\pi f}A_0\right)^{-1}\cdot\frac{1}{2\pi f}A_1\right], \tag{2.145}$$

where A_0 and A_1 are Fredholm operators over $\left[-\frac{T}{2}, \frac{T}{2}\right]$, with kernels $B_0(t-s)$ and $B_1(t-s)$, respectively. As we have seen previously, $\det(E+\frac{1}{2\pi f}A_0)$ increases as $f^{-1/2n}$ when $f\to 0$:

$$\det(E+\frac{1}{2\pi f}A_0)\sim c_0 f^{-1/2n\,T} \tag{2.146}$$

(see (2.131)). Hence it is sufficient to investigate the behavior of

$$\det\left[E+\left(E+\frac{A_0}{2\pi f}\right)^{-1}\cdot\frac{A_1}{2\pi f}\right]$$

as $f\to 0$. The operator $(E+\frac{A_0}{2\pi f})^{-1}$ takes the function $\psi(t)$ into a function $\phi(t)$ such that

$$(E+\frac{1}{2\pi f}A_0)\phi=\psi. \tag{2.147}$$

From the foregoing analysis it follows that equation (2.147) is equivalent to the differential equation

$$P\left(\frac{d}{dt}\right)P\left(-\frac{d}{dt}\right)\varphi+\frac{1}{f}\varphi=P\left(\frac{d}{dt}\right)P\left(-\frac{d}{dt}\right)\psi \tag{2.148}$$

with the boundary conditions

$$\left.\begin{aligned}&\frac{d^k}{dt^k}P\left(\frac{d}{dt}\right)\varphi\Big|_{t=\frac{T}{2}}=\frac{d^k}{dt^k}P\left(\frac{d}{dt}\right)\psi\Big|_{t=\frac{T}{2}} \quad (k=0,\ 1,\ \ldots,\ n-1),\\&\frac{d^k}{dt^k}P\left(-\frac{d}{dt}\right)\varphi\Big|_{t=-\frac{T}{2}}=\frac{d^k}{dt^k}P\left(-\frac{d}{dt}\right)\psi\Big|_{t=-\frac{T}{2}} \quad (k=0,\ 1,\ \ldots,\ n-1).\end{aligned}\right\} \tag{2.149}$$

In order to obtain the operator $(E+\frac{1}{2\pi f}A_0)^{-1}\cdot\frac{A_1}{2\pi f}$ it is sufficient to substitute for $\psi(t)$ in (2.148) and (2.149) the function

$$\frac{A_1}{2\pi f}\psi(t)=\frac{1}{2\pi f}\int_{-T/2}^{T/2}B_1(t-s)\psi(s)\,ds.$$

Consequently, the operator $(E+\frac{A_0}{2\pi f})^{-1}\cdot\frac{A_1}{2\pi f}$ is an integral operator of the Fredholm type with kernel $\phi_1(t-s)$ defined over the interval $\left[-\frac{T}{2}, \frac{T}{2}\right]$, where $\phi_1(t)$ is given by the solution of the differential equation

$$P\left(\frac{d}{dt}\right)P\left(-\frac{d}{dt}\right)\varphi_1+\frac{1}{f}\varphi_1=\frac{1}{2\pi f}P\left(\frac{d}{dt}\right)P\left(-\frac{d}{dt}\right)B_1 \tag{2.150}$$

subject to the boundary conditions which result from substituting $\psi=\frac{1}{2\pi f}B_1$ into (2.149). (Here obvious use is made of the assumption that B_1 has $2n$ continuous derivatives.) It is easy to see that the solution of (2.150) which is of interest to us is given by the sum of a function $\phi_2=\frac{1}{2\pi}P\left(\frac{d}{dt}\right)P\left(-\frac{d}{dt}\right)B_1$ independent of f and a function $\phi_3(t)$ (a solution of the homogeneous differential equation) which, as $f \to 0$, tends to zero at all interior points of the interval $[-T/2, T/2]$ (and uniformly on any interval $\left[-\frac{T}{2}+\epsilon, \frac{T}{2}-\epsilon\right]$). Consequently, when $f \to 0$ the operator $(E + \frac{A_0}{2\pi f})^{-1} \cdot \frac{A_0}{2\pi f}$ tends to a fixed (independent of f) Fredholm operator A_2 (with kernel $\phi_2(t-s)$, and $\det\left[E + (E + \frac{A_0}{2\pi f})^{-1} \cdot \frac{A_1}{2\pi f}\right]$ tends to a constant which is the determinant of the Fredholm operator $E + A_2$. Thus, for any correlation function which admits the representation (2.144) (that is, in effect, for the broad class of correlation functions having $2n-2$ continuous derivatives and discontinuous $(2n-1)$st derivative at $r = 0$), we have the result

$$J\{\xi_T, \eta_T\} \cdot \sqrt[2n]{f/T} \to c \quad \text{for } f \to 0 \tag{2.151}$$

where c is a constant independent of f and T.

LITERATURE

[1] C. E. Shannon and W. Weaver, *The mathematical theory of communication*, Univ. of Illinois Press, Urbana, Ill., 1949. An incomplete Russian translation of this work is included in the collection *The theory of transmission of electrical signals in the presence of noise*, Izdat. Inostr. Lit., Moscow, 1953.

[2] A. Ya. Hinčin, *The concept of entropy in the theory of probability*, Uspehi Matem. Nauk (N.S.) 8 (1953), no. 3(55), 3–20. (Russian)

[3] D. K. Fadeev, *On the concept of entropy of a finite probabilistic scheme*, Uspehi Mat. Nauk (N.S.) 11 (1956), no. 1(67), 227–231. (Russian)

[4] G. H. Hardy, J. E. Littlewood, and G. Pólya, *Inequalities*, Cambridge, at the University Press, 1952.

[5] S. Goldman, *Information theory*, Prentice-Hall, New York, 1953.

[6] A. Feinstein, *A new basic theorem of information theory*, Trans. IRE, PGIT-4 (1954), 2–22.

[7] H. Cramer, *Random variables and probability distributions*, Cambridge Tracts in Math. no. 36, Cambridge, 1937.

[8] H. Cramer, *Mathematical methods of statistice*, Princeton University Press, Princeton, N.J., 1946.

[9] K. Itô, *Stationary random distributions*, Mem. College Sci. Univ. Kyoto, Ser. A. Math. 28 (1954), 209–223.

[10] I. M. Gel'fand, *Generalized random processes*, Dokl. Akad. Nauk SSSR (N.S.) 100 (1955), 853–856. (Russian)

[11] L. Schwartz, *Théorie des distributions*, I-II, Hermann, Paris, 1950, 1951.

[12] I. M. Gel'fand and G. E. Šilov, *Forrier transformation of rapidly increasing*

functions and questions of uniqueness of the solution of Cauchy's problem, Uspehi Mat. Nauk 8, no. 6(58) (1953), 3–54. (Russian)

[13] H. Hotelling, *Relations between two sets of variates*, Biometrica 28 (1936), 321–377.

[14] A. M. Obukhov, *Normal correlation of vectors*, Izv. Akad. Nauk SSSR. Ser. Mat. Nat. Sci. 3 (1938), 339–370. (Russian)

[15] A. M. Obukhov, *Theory of correlation of vectors*, Moscov. Gos. Univ. Uč. Zap. Mat. 45 (1940), 73–92. (Russian)

[16] A. N. Kolmogorov, *Stationary sequences in Hilbert space*, Byull. Moscov. Gos. Univ. 2, no. 6 (1941), 3–40. (Russian)

[17] A. N. Kolmogorov, *Interpolation and extrapolation of stationary sequences*, Izv. Akad. Nauk SSSR. Ser. Mat. 5 (1941), 3–14. (Russian)

[18] P. A. Širokov, *Tensor calculus*, ONTI, Moscow-Leningrad, 1934.

[19] F. R. Gantmacher, *Theory of matrices*, Gosudarstv. Izdat. Tehn.-Teor. Lit. Moscow, 1953.

[20] N. Wiener, *Extrapolation, interpolation, and smoothing of stationary time series*, Wiley, New York, 1949.

[21] A. M. Yaglom, *Theory of extrapolation and filtering of random processes*, Ukrain. Mat. Ž. 6 (1954), 43–57. (Russian)

[22] A. M. Yaglom, *Extrapolation, interpolation and filtering of stationary random processes with rational spectral density*, Trudy Moscov. Mat. Obšč. 4 (1955), 333–374. (Russian)

[23] M. G. Kreĭn, *On a basic approximation problem of the theory of extrapolation and filtration of stationary random processes*, Dokl. Akad. Nauk SSSR (N.S.) 94 (1954), 13–16. (Russian)

[24] M. G. Kreĭn, *On a problem of extrapolation due to A. N. Kolmogorov*, Dokl. Akad. Nauk SSSR (N.S.) 46 (1945), 339–342. (Russian)

[25] J. L. Doob, *Stochastic processes*, Wiley, New York, 1953.

[26] L. A. Zadeh and J. R. Ragazzini, *An extension of Wiener's theory of prediction*, J. Appl. Phys. 21 (1950), 645–655.

[27] A. M. Yaglom, *Correlation theory of processes with random stationary n^{th} increments*, Mat. Sb. N.S. 37(79) (1955), 141–196. (Russian) = Amer. Math. Soc. Transl. (2) 8 (1958), 87–141.

[28] R. C. Davis, *On the theory of prediction of non-stationary stochastic processes*, J. Appl. Phys. 23 (1952), 1047–1053.

[29[M. S. Pinsker, *The quantity of information about a Gaussian random stationary process, contained in a second process connected with it in a stationary manner*, Dokl. Akad. Nauk SSSR (N.S.) 99 (1954), 213–216. (Russian)

[30] E. C. Titchmarsh, *The theory of functions*, Oxford Univ. Press, London. 1939.

[31] V. I. Smirnov, *A course in higher mathematics*, vol. 5, Gosudarstv. Izdat. Tehn.-Teor. Lit., Moscow-Leningrad, 1947.

[32] I. M. Gel'fand, A. N. Kolmogorov and A. M. Yaglom, *On the general definition of amount of information*, Dokl. Akad. Nauk SSSR (N.S.) 111 (1956), 745–748. (Russian)

Translated by:
L. A. Zadeh

Part IV

Mathematics of computation; cybernetics; biology

1.

(with N.N. Chentsov, S.M. Fejnberg and A.S. Frolov)

On the application of the method of random tests (the Monte Carlo method) for the solution of the kinetic equation*

Proceedings of the Second International Conference on the Peaceful Use of Atomic Energy. (Geneva, 1958), 2 (1960) 628–683

§1. This paper covers some of the problems connected with reactor calculations. In general reactors and other multiplication systems are calculated by means of the kinetic equation. Furthermore the greater part of the information obtained consists of only some numerical characteristics, such as: the effective multiplication factor, the fast fission factor, the effective neutron lifetime, etc. The calculation of the required characteristics in this paper will come to the calculation of mean values of different functionals for neutron paths.

If we neglect neutron–neutron interactions, the motion of a separate neutron in a six-dimensional phase space will be a Markov stochastic process. The corresponding probability distribution $P\{\cdot\}$ in neutron path space is completely described by the neutron position at the initial instant of time and by the transient probabilities. In their turn the latter are determined by the physical properties of the media: its composition, nuclear reaction cross sections and the fission and scattering laws. From such a standpoint every functional φ of $\xi(t)$ path becomes a random variable. Its mean value $\overline{\varphi[\xi(\cdot)]}$ being its mathematical expectation. The integral $\int[x(\cdot)]P\{dx(\cdot)\}$ corresponding to this mathematical expectation is the integral over measure P in the functional space of six-dimensional functions $x(t)$, i.e., a path integral.[1] The motion of a neutron avalanche may be considered in a similar way. In this case the probability distribution will be given in the space of "trees" of neutron paths.

§2. Examples. Let function $K(x, B)$ be the mathematical expectation of neutron numbers spontaneously generated in region B by a single neutron which appeared itself at point x of the phase space. One can write down

$$K(x, B) = \overline{\nu[\xi_x(\cdot)]\chi_B(\xi_x(\tau))}, \tag{1}$$

where $\nu[\xi_x(\cdot)] = \nu(\xi_x(\tau))$ is a random number of neutrons produced after the absorption of the neutron following the fission of the neutron absorber nucleous (in case of the emergence from the system $\nu[\xi(\cdot)] = 0$); $\tau = \tau[\xi(\cdot)]$ is a random moment of absorption (or emergence), $\chi_B(y) = 1$ at $y \in B$, $\chi_B(y) = 0$ at $y \notin B$.

---

* Translated through the courtesy of the USSR Government.

[1] Due to peculiarities of neutron motion such a path integral may be written in the form of an infinite sum of multiple integrals corresponding to the number of links in the neutron path.

The average is taken along all possible paths $\xi_x(t)$ of the neutron generated at the instant $t=0$ at point x. Then the distribution density of the functional $\xi_x(\tau)v(\xi)$ will be the production density of neutrons generated by the source placed at point x, the function $k(x)=v[\xi_x(\cdot)]$ being the neutron multiplication factor of this source. The averaging of the function $k(x)$ over all production positions with the weight proportional to the production density of neutrons gives an effective multiplication factor of the k_{eff} system.

It is often required to calculate the mean value of some functional as well as all its distribution. For example, one should be acquainted with the production density of neutrons in order to calculate the critical system. Let us choose some system of the orthonormal functions $h_k(z)$. The coefficients a_k of the expansion by the system $h_k(z)$ of an unknown distribution density $p(z)$ of the functional $\varphi[\xi(\cdot)]$ can be written down as follows:

$$a_k=\int h_k(z)p(z)dz=\int h_k(\varphi[x(\cdot)]dP=\overline{h_k(\varphi[\xi(\cdot)])}$$

By correctly choosing the functions $h_k(z)$ the approximating polinoms $p_n(z)=\sum_{k=1}^{n}a_k h_k(z)$ give a fairly good approximation to the unknown graph $p(z)$. Thus the approximate calculation of functional φ distribution should come to the definition of mean values of some other functionals $h_k(\varphi)$.

§3. To calculate the mean values of functionals the random process is simulated numerically. The results obtained from random experiments permit to plot the broken lines imitating particle paths. The only requirement that should be observed in the construction is that the probability distributions in the space of functions $x(t)$ induced by random paths should correspond to the probability distributions for a simulated process. The arithmetical mean of φ values of experimental paths is taken as an estimation. The error of the unknown mean $\varphi[\xi(\cdot)]$ resulting from this specific quadrature formula will be a random one though a confidence limits may be given. It is assumed that the error does not exceed $3\sqrt{D(\varphi)/N}$ where $D(\varphi)=\int\varphi^2dP-(\int\varphi dP)^2$, N is the number of constructed paths.

The functional $v[\xi(\cdot)]$ like many other important functionals defined on neutron paths takes integers only. For such integral-valued functionals a deviation of the experimental mean value from the true value is great, so that the calculation of the integral by direct simulation takes too much time. Therefore the corresponding path integral should be transformed to a mean value of some other functional over the paths of some fictitious process.

There exists quite a number of transformation methods but an optimal model should be found for every separate problem. It is well to base these researches not only on theoretical analysis but on practical tests too. The quality of the model must be estimated by the quantity of machine time necessary for obtaining the given accuracy.

This approach allows the computer to make the best choice of the optimal model.

§4. It is best to construct such paths by following consistently the history of the particle. Infinite sequences of independent random numbers distributed uniformly in the interval [0, 1] are used for such constructions. The element with the nucleous of which the particle has collided, the type of collision, the new velocity and length of the next range are determined according to these numbers and the laws of particle motion. The capture or emergence of a particle from the system brings the calculation of the path to an end. Then the calculation of a new path must be based on a new sequence. Thus every infinite sequence of internal points [0, 1] is confronted by a corresponding path. Thereby the mapping S of a single hypercube $H=[0,1]^{\infty}$ on the path space for which the Lebesque volume V in H turns into the probability distribution P determining the simulated process, each functional of path $\varphi[\xi]$ passing into the function $\tilde{\varphi}(y)=\varphi[Sy]$ of hypercube point.

An important property of the $\tilde{\varphi}$ type functions should be noted. Almost all the neutron paths consist of a finite number of elements. Therefore for almost all sequences of arguments the function $\tilde{\varphi}$ will only depend upon the finite number of the first arguments. Infinite paths can occur in fictitious processes as well. However even here the $\tilde{\varphi}$ type functions but slightly will depend on arguments, whose numbers are sufficiently great remote variables. In order to calculate the mean values we should only learn to construct quadrature formulae for functions of the type $\tilde{\varphi}$ on H as the integral for distribution $P=VS$ is reduced to volume V integral.

For a number of reasons machine computing hinders the use of any outer transmitters of random numbers. Thus so-called pseudorandom numbers generated in the machine itself are used instead of random numbers.

An interesting technique of obtaining a sequence of pseudorandom points was offered by N.M. Korobov. For smooth function $f(y)$ on H depending on the first r variables he gives an deterministic estimate

$$\left|\frac{1}{N}\sum_{k=1}^{N} f(y_k)-\int f(y)dV(y)\right|=O\left(\frac{r-1}{\sqrt{N}}\right)$$

Here the $y_k^{(s)}$ coordinates of y_k points are given by formula $y_k^{(s)}=\{g^{(2k+1)s}/p\}$, where $p=N$ is the prime number $p=2p_1+1$ type, g is a primitive root modulo p, $\{z\}$ is the fractional part of value z. Yet another method of obtaining a sequence of pseudorandom points was pointed out by I.I. Pyatetski-Shapiro. If one were to choose $y_k^{(s)}=\{k\theta_s\}$, then for every function $f(y)$ with a finite sum Q of absolute values of Fourier's coefficients and for almost all sets of $\theta_s, 0<\theta_s<1$ a more rapid convergence of the order $O(Q(\ln N/N))$ takes place. The mentioned sequences have the advantage of guaranteeing an exact convergence:

$$\lim\frac{1}{N}\sum_{k=1}^{N}\varphi[Sy_k]=\int\varphi[x(\cdot)]P\{dx(\cdot)\}$$

and not only a convergence in probability.

§5. Calculation of a small-size reactor. For a critical reactor (in a steady state) the following "conservation law" is put down

$$P(B)=\int K(x,B)P(dx) \tag{2}$$

where the number of neutrons produced in region B per unit of time is defined by

$P(B)$ and, the function $K(x, B)$ determined by the formula (1). Let the operator K acting in the space of measures of bounded variation and given by the kernel $K(x, B)$ be introduced by the formula

$$Kq(\cdot) = \int K(x, \cdot)q(dx) \tag{3}$$

Then formula (2) indicates that measure $P(\cdot)$ is the operator K eigenfunction corresponding to the characteristic value $k = 1$ (it is clear that $k = 1$ is the greatest characteristic value of operator K).

Let us assume now that kernel $K_c(x, B)$ and correspondingly the operator K_c are dependent upon the parameter "c", the reactor becoming critical at some value $c = c_0$. For the poorest assumption the operators K_c are positive therefore they have a positive characteristic value k_c which is the largest eigenvalue in modulus, corresponding at the same time to the positive eigenmeasure $P_c(\cdot)$. Otherwise this follows for K_c

$$K_c P_c(\cdot) = k_c P_c(\cdot) = \int K_c(x, \cdot) P_c(dx) \tag{4}$$

and if

$$c = c_0 \quad \text{then} \quad k_{co} = 1$$

Thus if we are able to establish the characteristic values of K_c under "c" fixed an approximate definition of c_0 will present no great difficulties. On the other hand the determination of the operator K_c eigenfunction is no easy matter therefore it is more convenient to examine the conjugated operator K_c^* of the same spectrum and acting in the space of functions from a point according to the formula

$$K_c^* f(\cdot) = \int K_c(\cdot, dy) f(y) \tag{5}$$

In the problems presenting some interest to us the kernel $K_c(x, B)$ appears to be a piece-wize smooth function of x. That is why the operator K_c^* converts continuous functions into smooth ones; therefore its eigenfunctions appear to be piece-wize smooth too.

For each fixed x the integral $\int K(x, dy) f(y)$ has a simple sense

$$g(x) = \int K(x, dy) f(y) = \overline{v \cdot f(\xi_x(\tau))} \tag{6}$$

where the mean value is taken over all possible neutron paths beginning with $t = 0$ at point x. This value is easily calculated by the Monte Carlo method. So the following procedure for determining k_c is proposed. m of points $x_1, \ldots, x_m$ and n of functions $f_1(x), \ldots\ldots, f_n(x)$ are taken, the latter being linearly independent on a set of points $\{x_j\}_{j=1}^m n \leqq m$. Let $g_i(x) = \int K(x, dy) f_i(y)$ be known under all $x = x_j$, $j = 1, \ldots, m$. Then, in some way, we can approximate $g_i(x)$ by the function $g_i(x) = \sum_{s=1}^n \alpha_s^{(i)} f_s(x)$ and, accordingly approximate operator K^* by the finite-dimensional operator K_n^* operating in the space of linear combinations of basic function $f_i(x)$ according to formula

$$K_n^* f_i(\cdot) = \sum_{s=1}^n \alpha_s^{(i)} f_s(\cdot),$$

n being sufficiently large the largest characteristic value of operator K_n^* will approximate to the characteristic value k of operator K^*. Calculating $g_i(x_j)$ by the

Monte-Carlo method it is not difficult to calculate at the same time their mean squares of deviations δ_{ij}. Therefore it is expedient to approximate the functions $g_i(x)$ by the polinoms $\tilde{g}_i(x)$ according to the least square method. The largest characteristic value of the matrix representing operator K_n^* in the base of $f_i(x)$ is calculated by some direct method. Moreover the eigenfunction of operator K_n^{**} is calculated which can be used for an approximate determination of mean values of some functions of $U(x)$ according to the eigen-measure $P_c(dx)$, in particular for the calculation of the production density expansion coefficients $P_c(dx)/dx$ by a set of orthogonal functions.

The calculation of only the mean values of $\overline{v \cdot f(\xi(\tau))}$ is convenient in that the calculation of these path integrals can be greatly speeded up first of all by the transformation of the integral itself.

A great effect is obtained from the simulation of the motion of particles with the fictitious mass. During this process particle absorption and emergence do not occur but are taken into account at each collision by the weight factor. The value $v \cdot f(\xi(\tau))$ is replaced by

$$\sum_k m_{k-1} \frac{v \cdot \Sigma_f(E_k, \xi(t_k))}{\Sigma_t(E_k; \xi(t_k))} f(\xi(t_k)) \tag{7}$$

where m_k is the mass left after collision number k; $\xi(t_k)$ is the particle position at the moment of collision number k; Σ_t and Σ_f are the total and fission cross sections. The path calculation is finished as soon as mass m_k becomes less than the constant given before.

The technique mentioned above gives good results in the case when the neutron migration length is of the same order as the size of the reactor. Then the eigenfunction of operator K^* is well approximated by the linear combination of a small number of basic functions f_i.

If we neglect the dependence of fission neutron velocity spectrum from the velocity of the neutron inducing fission, it will be possible to separate then the variables in the equation (5) thus greatly simplifying the solution of the problem.

§6. Calculation of the fast fission factor. Specifying the common definition let us consider the fast fission factor ε to be the ratio of the number of neutrons moderated to the energy E_0,—the fission threshold of U^{238} to the number of neutrons produced after thermal fission. The distribution $p(B)$ of the number of neutrons generated in region B per unit of time by slow neutrons is assumed to be known (from previous tests).

Let us introduce function $L(x, B)$, which is the mathematical expectation of number of neutrons spontaneously produced in region B by a neutron generated at point x of a phase space previously to its energy becoming smaller to that of E_0. Let us introduce function $l(x)$ as well. It is a probability for the neutron generated at point x of slow down to an energy smaller than E_0.

Let us consider the operator L given by the kernel $L(x, B)$. We shall assume the initial neutrons to be the neutrons of zero generation. Then the number of neutrons of the number k generation produced in the region per unit of time is given by the measure $q_k(\cdot) = L^k p(\cdot)$. Denoting $\int f(x) q(dx) = (f, q)$ we can write down the fast fission factor ε in the following way

$$\varepsilon = \sum_{k=0}^{\infty} \int l(x) q_k(dx) = \left(l, \sum_{k=0}^{\infty} L^k p \right) \tag{8}$$

Passing over to the conjugated operators we shall have

$$\varepsilon = ([\mathbb{1} - L^*]^{-1} l, p) \tag{9}$$

For rough calculations operator L^* is approximated by finite-dimensional operator L^* in a way similar to that used at the reactor calculation. Function $l(x)$ is calculated by the Monte-Carlo method for several arguments and after which it is replaced as well by an approximate linear combination $l_n(x)$ of the basic functions $f_i(x)$, $i = 1, \ldots, n$. As an evaluation of ε we take value

$$\varepsilon_n = ([\mathbb{1} - L_n^*]^{-1} l_n, p) \tag{10}$$

where integrating over p is carried out in the usual way.

When calculating values $l(x)$ and $g_i(x) = L^* f_i(\cdot)|_x$ a great effect is achieved by simulation of the motion of particles with the fictitious mass. Under these circumstances particle absorptions and the decrease of their energies lower than E_0 do not occur but are taken into the account by a weight factor. There are other ways of accelerating calculations, for instance the use of dependent tests.

It is to be noted that factor ε can be defined by direct simulation of the fast neutron cascade process. Errors arising in such a process of calculation of path "trees" are not difficult to eliminate but it takes much time to obtain the given accuracy.

The problem is greatly simplified if we assume that the velocity spectra of fission neutrons though different for U^{238} and U^{235} do not depend nevertheless upon the velocity of the neutron undergoing fission.

2.I

(with O. V. Lokytsievskij)

On difference schemes for the solution of the equation of thermal conductivity

Appendix I to 'Theory of difference schemes. An introduction.' by S. K. Godunov and V. S. Ryaben'kij

Moscow: Fizmatgiz 1962. Zbl. 106:319.
(Amsterdam: North-Holland 1964, pp. 232–262)

Two questions are discussed in this study of an example of a difference scheme for the approximate solution of the equation of thermal conductivity

$$\frac{\partial u}{\partial t} = f \frac{\partial^2 u}{\partial x^2} + g .$$

The first is connected with the formulation of difference schemes, the second with the solution of the system of algebraic equations which result from these schemes.

1. *Formulation of difference schemes*

A difference scheme must take into account the many aspects of the physical problem which is being solved (conservation laws, correct account of discontinuities, etc.). Therefore, the art of formulating difference schemes requires a certain resourcefulness which must not be limited by an a priori given framework for the formulation. Various methods which may be suggested help to account for only certain factors and are useful, as a rule, only for preliminary evaluations. One of these methods, essentially a modification of the method of undetermined coefficients, is discussed below.

The essence of this method is the expansion of the solution in a series in x and t whose coefficients are chosen such as to cancel the greatest number of terms. For a given distribution of points a formula of the best order of accuracy is thus obtained. Curiously, however, this optimum order of accuracy is sometimes not attained, in particular, in the example to be discussed below, if the formulation of the difference scheme is based (as is usually done) upon the substitution of finite difference relations for the derivatives in the differential equation.

We point out here that the question of order of accuracy is not of major importance, since an increase in accuracy can be obtained by a decrease in the size of the step, while the possibilities for doing so are very great with the use of modern electronic computers. Of course, these possibilities are to some extent limited, so that the question of accuracy has some interest all the same.

Derivation of the computational formula. We derive a computational formula for solution of the equation

$$\frac{\partial u}{\partial t} = \frac{\partial^2 u}{\partial x^2} \tag{1}$$

for the distribution of points shown in fig. 30.

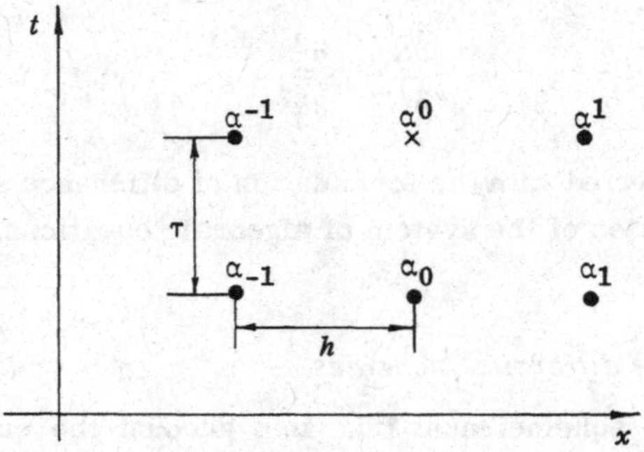

Fig. 30

Placing the origin of the coordinates at the point α_0, we express the solution $u(x, t)$ in an expansion in powers of x and t limited to terms of fourth order in x and second order in t, i.e.,

$$u(x, t) = a_0 + a_1 x + a_2 x^2 + a_3 x^3 + a_4 x^4 + b_0 t + b_1 tx + b_2 tx^2 + c_0 t^2 + \ldots . \tag{2}$$

Substituting this expansion into (1) gives the equalities

$$b_0 = 2a_2\,, \qquad b_1 = 6a_3\,, \qquad b_2 = c_0 = 12a_4\,. \tag{3}$$

We denote the values of the function $u(x, t)$ at the points $\alpha_{-1}, \alpha_0, \alpha_1, \alpha^{-1}, \alpha^1$ by $u_{-1}, u_0, u_1, u^{-1}, u^1$, respectively. Substituting successively the coordinates of the points $\alpha^0, \alpha^{-1}, \alpha_0, \alpha_1, \alpha^{-1}, \alpha^1$ into (2), we obtain

$$\begin{aligned}
u(0,\tau) &= a_0 + b_0\tau + c_0\tau^2 + \dots ,\\
u_{-1} &= a_0 - a_1h + a_2h^2 - a_3h^3 + a_4h^4 + \dots ,\\
u_0 &= a_0 ,\\
u_1 &= a_0 + a_1h + a_2h^2 + a_3h^3 + a_4h^4 + \dots ,\\
u^{-1} &= u_{-1} + b_0\tau - b_1\tau h + b_2\tau h^2 + c_0\tau^2 + \dots ,\\
u^1 &= u_1 + b_0\tau + b_1\tau h + b_2\tau h^2 + c_0\tau^2 + \dots .
\end{aligned} \tag{4}$$

Eliminating from here and from (3) all quantities a_i, b_i and c_0 we are led to the relation *

$$u(0,\tau) = u_0 + \tfrac{1}{2}(r+\tfrac{1}{6})(u_{-1} - 2u_0 + u_1) + \tfrac{1}{2}(r-\tfrac{1}{6})(u^{-1} - 2u(0,\tau) + u^1) + \dots ,$$

where $r = \tau/h^2$. Neglecting higher order terms, we obtain the equality

$$u^0 = u_0 + \tfrac{1}{2}(r+\tfrac{1}{6})(u_{-1} - 2u_0 + u_1) + \tfrac{1}{2}(r-\tfrac{1}{6})(u^{-1} - 2u^0 + u^1) . \tag{5}$$

(Here u^0 denotes the *calculated* value $u(0,\tau)$.)

Remark. The usual method of forming different schemes (i.e., replacing derivatives by finite differences) gives for the same distribution of points the equality

$$\tilde{u}^0 = u_0 + \tfrac{1}{2}r(u_{-1} - 2u_0 + u_1) + \tfrac{1}{2}r(u^{-1} - 2\tilde{u}^0 + u^1) . \tag{6}$$

We shall prove that formula (5) yields a higher order of accuracy than (6) (see the estimate of the errors).

The values of $u(x,t)$ at the $(n+1)$-st level (i.e., for $t = (n+1)\tau$) are usually calculated from its values at the n-th level. In this case (5) becomes the system of equations

$$u^k = u_k + \tfrac{1}{2}(r+\tfrac{1}{6})(u_{k-1} - 2u_k + u_{k+1}) + \tfrac{1}{2}(r-\tfrac{1}{6})(u^{k-1} - 2u^k + u^{k+1}) , \tag{7}$$

where u_l (correspondingly u^l) is the value of $u(x,t)$ at the point of the n-th (correspondingly of the $(n+1)$-st) level with abscissa $x = lh$. We now consider a method for the solution of this system (all u_l are known and u^l are to be determined).

Iteration method for the solution of system (7). Direct iteration of (7) gives convergence only for $r < \frac{2}{3}$. We shall show that the iteration process

* We maintain the right-hand side of $u(0,t)$ in a form convenient for comparison with (6).

converges for any r, if each of the equations (7) is first solved for u^k, i.e., if we solve by iteration the system

$$u^k = \frac{6r-1}{6r+5}\,\tfrac{1}{2}(u^{k-1} + u^{k+1}) + \frac{6r+1}{6r+5}\,\tfrac{1}{2}(u_{k-1} + u_{k+1}) + \frac{5-6r}{5+6r}\,u_k\,. \qquad (8)$$

Let $u^{(s)} = \{u^{k,s}\}$ be the s-th approximation. Then according to (8), the $(s+1)$-st approximation is obtained from the equations

$$u^{k,(s+1)} = \frac{6r-1}{6r+5}\,\tfrac{1}{2}(u^{(k-1),s} + u^{(k+1),s}) + \frac{6r+1}{6r+5}\,\tfrac{1}{2}(u_{k-1} + u_{k+1}) + \frac{5-6r}{5+6r}\,u_k\,.$$

As is known, it is sufficient for convergence of the iteration process that the spectrum of the operator defined by

$$\bar{u}^{k,(s+1)} = \frac{6r-1}{6r+5}\,\tfrac{1}{2}(\bar{u}^{(k-1),s} + \bar{u}^{(k+1),s})\,,$$

which transforms the vector $\bar{u}^{(s)} = \{\bar{u}^{k,s}\}$ into the vector $\bar{u}^{(s+1)} = \{\bar{u}^{k,(s+1)}\}$, lies within the circle $|\lambda| < 1$.

One can show that to each eigenvalue λ there corresponds an eigenvector of the form $\tilde{u}^{(s)} = \{e^{ik\varphi}\}$ *. Then in our search for λ we are led to the equation

$$\lambda\, e^{ik\varphi} = e^{ik\varphi}\,\frac{6r-1}{6r+5}\cos\varphi\,,$$

from which we find

$$|\lambda| \leqslant \left|\frac{6r-1}{6r+5}\right| < 1\,.$$

Hence convergence of the iteration process is shown.

Estimate of the error. The orders of the discarded terms in the expansion of (2) imply that formula (5) has an error

$$\Delta = u^0 - u(0, \tau)$$

of order h^5 (since $r = \tau/h^2$ is assumed finite). In fact, it can be shown that Δ is of order $O(h^6)$. Without carrying through the calculation, we point out that

* Here $\varphi = \varphi(\lambda)$.

$$u^0 - u(0,\tau) = \frac{1}{6r+5}\left[\frac{rh^6}{60}\,\frac{\partial^6 u}{\partial x^6} - \frac{\partial^3 u}{\partial\ 3}\,\tau^3\right.$$

$$\left. + (6r-1)\left(\frac{1}{24}\,\frac{\partial^5 u}{\partial t\,\partial x^4}\,h^4\tau + \frac{1}{4}\,\frac{\partial^4 u}{\partial t^2\,\partial x^2}\,h^2\tau^2\right)\right]_{x=t=0} + \ldots\ . \qquad (9)$$

At the same time we can show that the error of formula (6) is $O(h^4)$, i.e., that

$$\tilde{u}^0 - u(0,\tau) = \frac{1}{12(r+1)}\left.\frac{\partial^3 u}{\partial x^2\,\partial t}\right|_{x=t=0}\tau h^2 + \ldots\ . \qquad (10)$$

Proof of stability for difference scheme (7). We consider system (7) as an operator which transforms the vector $u = \{u_k\}$ into the vector $\bar{u} = \{u^k\}$. Stability is guaranteed if the spectrum of this operator lies within the circle $|\lambda| \leq 1$. Assuming (as before) that each point of the spectrum (λ) corresponds to an eigenvector $\bar{u} = \{e^{ik\varphi}\}$, we obtain for λ the equation

$$\lambda - 1 = (r+\tfrac{1}{6})\,(\cos\varphi - 1) + \lambda(r-\tfrac{1}{6})\,(\cos\varphi\ - 1)\ .$$

From this we find

$$|\lambda| = \left|\frac{5+\cos\varphi - 12r\sin^2\tfrac{1}{2}\varphi}{5+\cos\varphi + 12r\sin^2\tfrac{1}{2}\varphi}\right| \leq 1\ .$$

Thus, the calculation is stable for all r.

Summary. For the distribution of points we are studying, the best approximation of the function is given by the solution of the system of equations

$$u^k = \frac{5-6r}{5+6r}\,u_k + \frac{6r+1}{6r+5}\,\tfrac{1}{2}(u_{k-1}+u_{k+1}) + \frac{6r-1}{6r+5}\,\tfrac{1}{2}(u^{k-1}+u^{k+1})\ .$$

This system can be solved by an iteration procedure which converges for any r. The calculation is always stable. The error to leading order is given by

$$u^0 - u(0,\tau) = \frac{1}{6r+5}\left[\tfrac{1}{60}r\Delta^6_{x^6}u + \Delta^3_{t^3}u + (6r-1)\left(\tfrac{1}{24}\Delta^5_{tx^4}u + \tfrac{1}{4}\Delta^4_{t^2x^2}u\right)\right] + \ldots\ .$$

2. *Stability of difference schemes and the number of iterations*

We now introduce a necessary condition for the stability of the calculation according to a formula of the form *

$$u^k = l_0 u_k + \sum_{\nu=1}^{s} l_\nu (u_{k+\nu} + u_{k-\nu}) . \tag{11}$$

Namely, setting $H = sh$ and assuming that as $h \to 0$ the process approaches the solution of equation (1), we shall show that the following basic assertion holds. *No matter what the values of the coefficients* $l_0, l_1, \ldots, l_s$, *the calculation according to formula* (11) *is unstable for* $\tau/H^2 > \frac{1}{2}$.

Considering (11) as an operator which transforms the vector $u = \{u_k\}$ into the vector $\bar{u} = \{u^k\}$, we obtain for the eigenvalues the equation

$$\lambda(\varphi) = l_0 + \sum_{\nu=1}^{s} l_\nu (e^{i\nu\varphi} + e^{-i\nu\varphi}) .$$

The stability condition is $|\lambda(\varphi)| \leqslant 1$, from which, in view of the S. N. Bernshtein theorem **, there results the inequality

$$|\lambda''(\varphi)| \leqslant s^2 . \tag{12}$$

On the other hand, we will show that convergence of the process to the solution of (1) yields the equality

$$\lambda''(0) = -2r = -2\tau/h^2 . \tag{13}$$

Since our basic hypothesis easily follows from (12) and (13), it remains only to prove (13).

To do this we notice that for initial conditions

$$u(x, 0) = e^{ix}$$

the true solution of equation (1) is the function

$$u(x, t) = e^{ix-t} . \tag{14}$$

Under those same initial conditions we obtain as solution at the point $x = 0$, $t_0 = n\tau$ according to scheme (11) the result

$$\tilde{u}(0, t_0) = \lambda^n(h) . \tag{15}$$

* u_k and u^k are the calculated values of the function $u(x,t)$ at the points with coordinates $x = kh$, $t = n\tau$ and $x = kh$, $t = (n+1)\tau$, respectively.

** If for all φ, $|\lambda(\varphi)| \leqslant L$, then $|\lambda'(\varphi)| \leqslant sL$. The proof may be found on page 242 of the book by V. L. Goncharov [1].

Since $n = t_0/\tau = t_0/rh^2$, (14) and (15), together with the stability condition, give the limiting relation

$$\lim_{h\to 0} [\lambda(h)]^{t_0/rh^2} = e^{-t_0}, \tag{16}$$

from which it follows that $\lambda(0) = 1$. Since, in addition, $\lambda(\varphi)$ is an even function, we have

$$\lambda(\varphi) = 1 + \tfrac{1}{2}\lambda''(0)\varphi^2 + \dots .$$

Substituting this expression into (16), we obtain

$$\lim_{h\to 0} (1 + \tfrac{1}{2}\lambda''(0)h^2 + \dots)^{1/rh^2} = e^{-1} .$$

Letting the left-hand side of this last equation go to the limit, we obtain our proof of (13).

We now note that by solving system (8) by an iteration procedure, we have really performed the calculation according to a formula of the form (11), while s is the number of iterations. Therefore, it follows from our basic assertion that not less than $\sqrt{2r}$ iterations are needed to guarantee stability of the calculation according to formula (8).

The following question then arises. Which is more advantageous, to solve the equation of thermal conductivity according to the simplest explicit scheme using small time steps, or to use the implicit scheme just described and perform a large number of iterations? Apparently, it is more sensible when applying an implicit scheme to use some noniterative method, for example, an elimination method.

Bibliography

[1] V. L. Goncharov, Theory of interpolation and approximation of functions, 2nd ed. (Goztekhizdat, Moscow, 1954).

2. II
(with O. V. Lokytsievskij)

The double sweep method for solution of difference equations*

Appendix II to 'Theory of difference schemes. An introduction'.
by S. K. Godunov and V. S. Ryaben'kij

Moscow: Fizmatgiz 1962. Zbl. **106**:319.
(Amsterdam : North-Holland 1964, pp. 232-262)

§ 1. Introduction

Difference methods for the approximate integration of boundary-value problems for both ordinary and partial differential equations are based on the replacement of the differential equations by an approximating system of difference equations. The values of the unknowns are calculated at a discrete number of points called the computation grid, which together with the system of difference equations **, forms the difference scheme.

Difference schemes may be of two essentially different types. The first type are the so-called explicit or local schemes. They are constructed such that each value of the unknown function is expressed in terms of the values already found (the matrix of the system of difference equations is triangular). Implicit schemes, in distinction to explicit schemes, require the solution of a large system of algebraic equations. Obviously, explicit schemes are to be preferred on account of their simplicity, but they are often inconvenient because of limitations which they place on the ratio of the grid steps corresponding to the different variables. These limitations result from the requirement of stability. Therefore, calculations are very often performed by the use of implicit schemes.

Systems of equations corresponding to difference schemes usually have a special form (the matrices contain many zeros), which sometimes allows

* This paper was written in the years 1953-1954. A good deal of literature connected with sweep methods has appeared since then. References are given in the bibliography attached to the main part of the book (cf. [2], [24], [30], [31], [8]).

** The coefficients of the system depend, of course, on the distribution of the points of the computation grid. It is often convenient to consider difference schemes for variable grids (with steps that approach zero).

one to use relatively simple methods for their solution. One of these methods presently used in practical calculations, known as a "sweep" method, is described in this paper. We shall limit ourselves to relatively simple examples from which, however, the essence of the method will become clear. All considerations will be with respect to linear equations. The transition to quasi-linear equations present no complications.

§ 2. Ordinary differential equations

We consider the equation

$$\epsilon y'' = p(x)y + F(x) , \tag{1}$$

where $\epsilon > 0$, $p(x) > 0$, with boundary conditions *

$$y' = \alpha_0 y + \beta_0 \big|_{x=0} , \tag{2a}$$

$$y' = \psi_0 y + \omega_0 \big|_{x=l} . \tag{2b}$$

We shall see later in § 3 that this problem arises in particular during the solution by difference methods of a mixed boundary-value problem for the equation of thermal conductivity. The coefficient ϵ from eq. (1) is to be considered in this case as a small parameter **, proportional to the time step ‡, which multiplies the highest derivative. We describe a relatively simple method for the solution of this problem which does not lead to a loss of accuracy even in this most unfavorable ‡‡, but important case.

* We limit ourselves for simplicity to boundary conditions of this form, although they need not be solvable for y'; moreover, one of them (for example, the right-hand one) could be taken as non-linear, i.e., $\varphi(y, y')\big|_{x=l} = 0$.

** The "smallness" of the parameter ϵ can be characterized by the dimensionless constant

$$A = \frac{1}{\sqrt{\frac{l}{\epsilon}\int\limits_0^l p(x)\,dx}} .$$

‡ The small parameter occurs in the boundary conditions only if they contain a derivative of the unknown function with respect to time. We disregard this case for simplicity.

‡‡ This unfavorable character arises because the solution of the homogeneous equation $\epsilon y'' = p(x)y$ goes as $\exp\left[\int\limits_0^x \sqrt{p(x)/\epsilon}\,dx\right]$, i.e., it sharply increases toward the right-hand limit of the interval.

This method is based on the following considerations. The left-hand boundary condition

$$y' = \alpha_0 y + \beta_0 \big|_{x=0} \tag{2a}$$

extracts from the general solution of eq. (1) a family of one-parameter integral curves. Thus, we induce the linear relation

$$y'(x) = \alpha(x)y(x) + \beta(x) \tag{3}$$

between $y(x)$ and $y'(x)$ at each point of the segment $0 \leqslant x \leqslant l$, where $\alpha(x)$ and $\beta(x)$ are uniquely determined by eq. (1) and the left-hand boundary condition (2a). In this way the left-hand boundary condition is translated to all points of the initial interval. This translation of the boundary condition is called direct sweep.

We will show that the functions $\alpha(x)$ and $\beta(x)$ which produce the direct sweep are determined from the first-order differential equations

$$\epsilon(\alpha' + \alpha^2) = p(x)\,, \tag{4}$$

$$\epsilon(\beta' + \alpha\beta) = F(x) \tag{5}$$

(the initial conditions for this are obviously $\alpha(0) = \alpha_0$ and $\beta(0) = \beta_0$).

To obtain these equations, we must differentiate relation (3),

$$y' = \alpha y + \beta\,,$$

and eliminate from the new equation and (3) itself the derivative y'. We are thus led to the relation

$$y'' = (\alpha' + \alpha^2)y + (\beta' + \alpha\beta)\,. \tag{6}$$

Comparing its coefficients with those of the initial equation (1),

$$\epsilon y'' = p(x)y + F(x)\,,$$

we obtain eqs. (4) and (5).

The problem posed in this section is now solved in the following manner.

1) We find $\alpha(x)$ $(0 \leqslant x \leqslant l)$ from eq. (4),

$$\alpha' + \alpha^2 = p(x)/\epsilon$$

(the equation of Riccati), and the initial condition $\alpha(0) = \alpha_0$.

2) Knowing $\alpha(x)$, we find $\beta(x)$ $(0 \leqslant x \leqslant l)$ from eq. (5),

$$\beta' + \alpha\beta = F(x)/\epsilon$$

(which for known $\alpha(x)$ is a linear first-order equation), and the initial condition $\beta(0) = \beta_0$.

We accomplish the direct sweep in this way and at the point $x = l$ obtain another boundary condition

$$y'(l) = \alpha(l)y(l) + \beta(l),$$

"swept" from the point $x = 0$. This condition, together with condition (2b) given at the right-hand boundary, allows us to find $y(l)$.

3) Therefore, we find $y(l) = y_l$ from the system of equations *

$$y' = \psi_0 y + \omega_0 \big|_{x=l}, \qquad y'(l) = \alpha(l)y(l) + \beta(l). \tag{2b}$$

We could now find $y(x)$ by solving the problem given by eq. (1) with the initial conditions

$$y(l) = y_l, \qquad y' = \psi_0 y + \omega_0 \big|_{x=l}. \tag{7}$$

However, this would lead to a drastic loss of accuracy for the same reasons mentioned above. On the other hand, it is not necessary to solve the original differential equation (1) in order to obtain $y(x)$, since for this purpose we can use the simpler equation (3).

4) Thus, we find $y(x)$ $(0 \leqslant x \leqslant l)$ from eq. (3),

$$y'(x) = \alpha(x)y(x) + \beta(x)$$

(which for known $\alpha(x)$ and $\beta(x)$ is a linear first-order equation), and the initial condition $y(l) = y_l$. The determination of $y(x)$ from eq. (3) is called inverse sweep.

In this way the process of solution of our problem consists in the consecutive solution of the three equations

$$\alpha'(x) + \alpha^2(x) = p(x)/\epsilon, \tag{4'}$$

$$\beta'(x) + \alpha(x)\beta(x) = F(x)/\epsilon, \tag{5'}$$

$$y'(x) - \alpha(x)y(x) = \beta(x), \tag{3'}$$

* As a rule the determinant of this system is different from zero. If it is close to zero, it means that the original problem has been poorly posed.

while the initial conditions for the first two equations are given at the left-hand boundary and for the third at the right-hand boundary. The described sweep method is valid provided that α_0 is significantly larger * than $(-\chi/\sqrt{\epsilon})$, where χ is defined in the following.

We now consider the basis of its validity, while for simplicity we restrict ourselves to the case of $p(x)$ constant. We set $p(x) = \chi^2 > 0$. The solution of eq. (4'),

$$\alpha'(x) + \alpha^2(x) = \chi^2/\epsilon \ ,$$

with the initial condition $\alpha(0) = \alpha_0$, can then be written in the explicit form

$$\alpha(x) = \begin{cases} \dfrac{\chi}{\sqrt{\epsilon}} \operatorname{cth}\left\{\dfrac{\chi}{\sqrt{\epsilon}}x + \tfrac{1}{2}\ln\dfrac{\chi + \alpha_0\sqrt{\epsilon}}{-\chi + \alpha_0\sqrt{\epsilon}}\right\}, & |\alpha_0| > \chi/\sqrt{\epsilon}\ , \\[2ex] \dfrac{\chi}{\sqrt{\epsilon}} \operatorname{th}\left\{\dfrac{\chi}{\sqrt{\epsilon}}x - \tfrac{1}{2}\ln\dfrac{\chi + \alpha_0\sqrt{\epsilon}}{\chi - \alpha_0\sqrt{\epsilon}}\right\}, & |\alpha_0| < \chi/\sqrt{\epsilon}\ , \\[2ex] \pm\chi/\sqrt{\epsilon}\ , & |\alpha_0| = \chi/\sqrt{\epsilon}\ . \end{cases} \tag{8}$$

First let $\alpha_0 \geqslant 0$. Then from (8) it is clear that $\alpha(x) \geqslant 0$ over the entire interval $0 \leqslant x \leqslant l$ **. When solving eq. (5'), we can consider computational errors as errors in $F(x)$. The initial condition $\beta(0) = \beta_0$ for this equation is given at the left-hand boundary. Using a variational formula for the arbitrary constant and the fact that the solution of the homogeneous equation $\beta' + \alpha\beta = 0$ $(\alpha > 0)$ decreases with increase in x, we see that the influence of the initially admitted error disappears as x increases.

We turn now to eq. (3'),

$$y' - \alpha y = \beta \ ,$$

which is also a linear equation of first order, while the coefficient of the corresponding homogeneous equation fulfills the condition $(-\alpha) < 0$ and its solution, in contrast to that of (5'), decreases with decreasing x. However, the initial condition $y(l) = y_l$ is given at the right instead of at the left-hand boundary and therefore, neither is any accuracy lost in the solution of this equation.

* This limitation, naturally, is removed if ϵ is a small parameter.

** It can be shown that this is also true for variable $p(x)$. All further considerations then follow.

It remains to investigate whether we lose accuracy while solving eq. (4'),

$$\alpha' + \alpha^2 = p(x)/\epsilon \ .$$

We allow an error δ_0 in the boundary condition $\alpha(0) = \alpha_0$ (at the left-hand boundary). The magnitude of the error $\delta(x)$ which results as a consequence of δ_0 satisfies up to small deviations of higher order the equation

$$\delta' + 2\alpha\delta = 0 \tag{9}$$

obtained by variation of (4'). For the coefficient of this equation we have $2\alpha > 0$. Since the initial condition is given at the left-hand boundary, the situation is the same as in the case of eq. (5').

Let now α_0 be negative such that* $|\alpha_0|$ is considerably less than $\chi/\sqrt{\epsilon}$. Solution of eq. (4'),

$$\epsilon(\alpha' + \alpha^2) = \chi^2 \ ,$$

then has according to (8) the form

$$\alpha = \frac{\chi}{\sqrt{\epsilon}} \operatorname{th} \left\{ \frac{\chi}{\sqrt{\epsilon}} x - \tfrac{1}{2} \ln \frac{\chi + \alpha_0\sqrt{\epsilon}}{\chi - \alpha_0\sqrt{\epsilon}} \right\} .$$

It is clear from this that $\alpha(x)$ can change sign only at the point ** $x = x_0 > 0$. Everywhere to the right of this point, $\alpha(x) > 0$. Therefore, a loss of accuracy is possible only in the interval $0 \leqslant x \leqslant x_0$. We estimate this loss of accuracy for the function $\alpha(x)$. As mentioned previously, the error $\delta\alpha = \delta(x)$ satisfies, up to small deviations of higher order, eq. (9). We find from this equation that the value of the ratio $\delta(x)/\delta(0)$ nowhere in the interval $0 \leqslant x \leqslant x_0$ exceeds the quantity

$$\Delta = \exp\left[-2\int_0^{x_0} \alpha(x)\, dx\right] . \tag{10}$$

We now calculate Δ. As a consequence of (8), we have

$$\alpha(x) = \frac{\chi}{\sqrt{\epsilon}} \operatorname{th}\left(\frac{\chi}{\sqrt{\epsilon}} x - \tfrac{1}{2} \ln \frac{\chi + \sqrt{\epsilon}\alpha_0}{\chi - \sqrt{\epsilon}\alpha_0} \right) , \tag{11}$$

* This limitation, naturally, is removed if ϵ is a small parameter.
** This can also be proven for variable $p(x)$.

and consequently,

$$x_0 = \frac{\sqrt{\epsilon}}{2\chi} \ln \frac{\chi + \alpha_0\sqrt{\epsilon}}{\chi + \alpha_0\sqrt{\epsilon}} . \tag{12}$$

Thus we obtain

$$\int_0^{x_0} \alpha(x)\, dx = \int_0^{x_0} \frac{1}{\sqrt{\epsilon}}\, \chi \,\mathrm{th} \left(\frac{\chi}{\sqrt{\epsilon}}\, x - \tfrac{1}{2} \ln \frac{\chi + \alpha_0\sqrt{\epsilon}}{\chi - \alpha_0\sqrt{\epsilon}} \right) dx$$

$$= \ln \mathrm{ch} \left(\frac{\chi}{\sqrt{\epsilon}}\, x - \tfrac{1}{2} \ln \frac{\chi + \alpha_0\sqrt{\epsilon}}{\chi - \alpha_0\sqrt{\epsilon}} \right) \Bigg|_0^{x_0} = - \ln \frac{\chi}{\sqrt{\chi^2 - \alpha_0^2\epsilon}} .$$

Therefore,

$$\Delta = \exp \left[-2 \int_0^{x_0} \alpha(x)\, dx \right] = \exp \left[2 \ln \frac{\chi}{\sqrt{\chi^2 - \alpha_0^2}} \right] = \frac{\chi^2}{\chi^2 - \epsilon\alpha_0^2} ,$$

or equivalently,

$$\Delta = \frac{1}{1 - \epsilon(\alpha_0/\chi)^2} . \tag{13}$$

It is obvious from this relation that Δ cannot be very large, provided that ϵ is considerably less than $(\chi/\alpha_0)^2$ *. We note that when solving eq. (5'),

$$\beta' + \alpha\beta = F(x)/\epsilon ,$$

and eq. (3'),

$$y' - \alpha y = \beta ,$$

the increase in the error can only be less, since in this case the increase is determined by the value of the quantity

$$\Delta_1 = \exp \left[- \int_0^{x_0} \alpha(x)\, dx \right] .$$

Therefore, validity of the sweep is also guaranteed for negative values of α_0 which satisfy the condition $\alpha_0 \gg -\chi/\sqrt{\epsilon}$ formulated above.

* Actually, ϵ need only be somewhat less than $(\chi/\alpha_0)^2$. Thus, even if $\epsilon = 0.9(\chi/\alpha_0)^2$, $\Delta = 10$, and the calculation can be performed by carrying two extra decimal places, one of which will be lost during the direct sweep and the other during the inverse sweep.

One might think that the method just described is only applicable to an equation of a special form. In fact, this is not so. In order to illustrate the general applicability of this method, we shall consider its application to the system of second-order differential equations

$$\boldsymbol{u}'' - P(x)\boldsymbol{u} + \boldsymbol{F}(x) = 0 , \tag{14}$$

where $\boldsymbol{u}(x)$ and $\boldsymbol{F}(x)$ are n-dimensional vectors, and P is a square matrix which we assume to be symmetric and positive-definite. To solve this system we use the difference scheme

$$\frac{\boldsymbol{u}_{m+1} - 2\boldsymbol{u}_m + \boldsymbol{u}_{m-1}}{h^2} - P(x_m)\boldsymbol{u}_m + \boldsymbol{F}(x_m) = 0 , \tag{15}$$

which can be written in the form

$$\boldsymbol{u}_{m+1} - 2B_m\boldsymbol{u}_m + \boldsymbol{u}_{m-1} + \boldsymbol{d}_m = 0 , \tag{16}$$

where

$$B_m = E + \tfrac{1}{2}h^2P(x_m) \quad * , \tag{17}$$

$$\boldsymbol{d}_m = h^2\boldsymbol{F}(x_m) . \tag{18}$$

In order that system (16) define a unique solution, we must add to it the equations representing the boundary conditions of system (14), which we assume for $x = 0$ and $x = l$ have the form

$$\left.\frac{\mathrm{d}\boldsymbol{u}}{\mathrm{d}x}\right|_{x=0} - A\boldsymbol{u}\big|_{x=0} + \boldsymbol{\beta} = 0 , \tag{19a}$$

$$\left.\frac{\mathrm{d}\boldsymbol{u}}{\mathrm{d}x}\right|_{x=l} + \Psi\boldsymbol{u}\big|_{x=l} + \boldsymbol{\omega} = 0 . \tag{19b}$$

Matrices A and Ψ in these conditions will be square, although this need not be so in general. In fact, the number of boundary conditions at each boundary is not the same in all cases.

The number of boundary conditions for $x = 0$ must as a rule be equal to the number of linearly independent solutions, decreasing to the right, of the homogeneous equation $\boldsymbol{u}'' - P\boldsymbol{u} = 0$, and the number of boundary conditions for $x = l$ equal to the number of solutions decreasing to the left. Since in

* E is the unit matrix.

our case P is positive-definite, there will be n of each kind, i.e., the matrices A and Ψ are square.

Replacing the derivatives in (19) by difference relations, we are led to the difference boundary conditions which can be written in the form

$$\boldsymbol{u}_0 = X_1\boldsymbol{u}_1 + \boldsymbol{y}_1 , \tag{20a}$$

$$\boldsymbol{u}_{M-1} = G\boldsymbol{u}_M + \boldsymbol{g} \tag{20b}$$

(X_1 and G are matrices, $M = l/h$).

Difference equations (16), together with the boundary conditions (20), form already the complete system which we are to solve. The method of solution, as in the case of the single equation, will be to translate successively the left-hand boundary condition (20a) to the right until it reaches the right-hand boundary of the interval and allows us, together with (20b), to determine $\boldsymbol{u}_M$.

Suppose we have already obtained the relation

$$\boldsymbol{u}_{m-1} = X_m\boldsymbol{u}_m + \boldsymbol{y}_m \tag{21}$$

at the mth point. Substituting $\boldsymbol{u}_{m-1}$ from (21) into (16), we obtain

$$\boldsymbol{u}_{m+1} - 2B_m\boldsymbol{u}_m + X_m\boldsymbol{u}_m + \boldsymbol{y}_m + \boldsymbol{d}_m = 0 .$$

Then,

$$\boldsymbol{u}_m = (2B_m - X_m)^{-1}\,\boldsymbol{u}_{m+1} + (2B_m - X_m)^{-1}\,(\boldsymbol{y}_m + \boldsymbol{d}_m) .$$

Thus, we have for X_m and $\boldsymbol{y}_m$ the recursion relations

$$X_{m+1} = (2B_m - X_m)^{-1} , \tag{22}$$

$$\boldsymbol{y}_{m+1} = X_{m+1}(\boldsymbol{y}_m + \boldsymbol{d}_m) . \tag{23}$$

Using these relations and knowing X_1 and $\boldsymbol{y}_1$ (cf. (20a)), we can determine successively $X_1, X_2, \ldots, X_M$ and $\boldsymbol{y}_1, \boldsymbol{y}_2, \ldots, \boldsymbol{y}_M$. In particular, for $m = M$ we have

$$\boldsymbol{u}_{M-1} = X_M\boldsymbol{u}_M + \boldsymbol{y}_M .$$

On the other hand, from (20b) we have

$$\boldsymbol{u}_{M-1} = G\boldsymbol{u}_M + \boldsymbol{g} .$$

We can use these two relations to determine u_M, from which we can then obtain $u_{M-1}, u_{M-2}, \dots, u_0$ from (21). This method can be shown to be valid, although we will not linger with this proof, since it is completely analogous to that given below (§ 4), where we study the validity of the matrix sweep method for Poisson's equation *.

§ 3. The double sweep method for the equation of thermal conductivity

We consider for the equation of thermal conductivity

$$\frac{\partial u}{\partial t} = \mu^2 \frac{\partial^2 u}{\partial x^2} \qquad (\mu = \mu(x, t)), \tag{1}$$

the mixed boundary-value problem

$$u(x, 0) = \varphi(x),$$

$$\frac{\partial u}{\partial x} = \alpha_0(t) u + \beta_0(t)\Big|_{x=0}, \tag{2a}$$

$$\frac{\partial u}{\partial x} = \psi_0(t) u + \omega_0(t)\Big|_{x=l}. \tag{2b}$$

As is well known, implicit schemes are often used to obtain an approximate solution of this problem.

A convenient scheme of this type is given by the relation

$$u_m^{n+1} = u_m^n + \mu^2(x_m, t_n) \frac{\tau}{h^2} (u_{m+1}^{n+1} - 2u_m^{n+1} + u_{m-1}^{n+1}). \tag{3}$$

We turn to the seemingly most unfavorable case with respect to stability, that of an infinitely small space step. In the limit, eq. (3) then reduces to the ordinary differential equation

$$\mu^2(x, t_n)\tau \frac{d^2 u^{n+1}(x)}{dx^2} = u^{n+1}(x) - u^n(x) \tag{4}$$

for $u^{n+1}(x)$ with a small parameter τ multiplying the highest derivative.

* We mentioned in the first footnote on p. 239 that there exists a large literature connected with this method. In particular, there are the papers of Abramov and Godunov in which the changes necessary to apply the method to systems of first order are described (cf. in the bibliography [1] and [8]). The study of the validity of the matrix sweep method is discussed by Rusanov and by Babenko and Chentsov (cf. the third footnote on p. 259).

By virtue of (2) we get for this equation the boundary conditions

$$\frac{du^{n+1}}{dx} = \alpha_0^{n+1} u^{n+1} + \beta_0^{n+1} \big|_{x=0} , \tag{5a}$$

$$\frac{dx^{n+1}}{dx} = \psi_0^{n+1} u^{n+1} + \omega_0^{n+1} \big|_{x=l} . \tag{5b}$$

Thus, knowing $u^n(x)$, we can determine $u^{n+1}(x)$ from the boundary-value problem (4), (5) *.

The sweep method for the solution of this problem was described in § 2. Application of this method to our problem consists of the following. A linear relation between $u^{n+1}(x)$ and du^{n+1}/dx is induced from the left-hand boundary condition (5a) at all points in the interval $0 \leq x \leq l$, i.e.,

$$\frac{du^{n+1}(x)}{dx} = \alpha(x)u^{n+1}(x) + \beta(x) . \tag{6}$$

The coefficients $\alpha(x)$ and $\beta(x)$ of this relation are determined from the differential equations

$$\alpha' + \alpha^2 = \frac{1}{\mu^2(x, t_n)\tau} , \tag{7}$$

$$\beta' + \alpha\beta = - \frac{u^n(x)}{\mu^2(x, t_n)\tau} , \tag{8}$$

and the initial conditions $\alpha(0) = \alpha_0{}^{n+1}$ and $\beta(0) = \beta_0{}^{n+1}$. Having determined $\alpha(x)$ and $\beta(x)$ $(0 \leq x \leq l)$, we find $u^{n+1}(l)$ from the system

$$\frac{du^{n+1}(l)}{dx} = \alpha(l)u^{n+1}(l) + \beta(l) ,$$

$$\frac{du^{n+1}}{dx} = \psi_0^{n+1} u^{n+1} + \omega_0^{n+1} \big|_{x=l} . \tag{5b}$$

Then, knowing $u^{n+1}(l)$, we determine $u^{n+1}(x)$ $(0 \leq x \leq l)$ from eq. (6).

The basis for the validity of this method was given in § 2, to which we add the following remark. For the purpose of calculation, τ is assumed to be a finite quantity. We saw in § 2 that in this approach to the problem the

* The timewise stability of this process can be verified by the method of Fourier (for constant μ).

sweep method is not valid for negative values of α_0 of order $\sqrt{1/\mu^2\tau}$, while it is valid for all larger values of α_0^{n+1}, in particular, for $\alpha_0^{n+1} > 0$.

We look to the left-hand boundary condition

$$\frac{du^{n+1}}{dx} = \alpha_0^{n+1} u^{n+1} + \beta_0^{n+1} \Big|_{x=0} \tag{5a}$$

for the physical meaning of the loss of validity for $\alpha_0 \sim -\sqrt{1/\mu^2\tau}$. Multiplying this equality by $\mu^2\tau$, we obtain

$$\mu^2\tau \frac{du^{n+1}}{dx} = \alpha_0^{n+1} \mu^2 \tau u^{n+1} + \mu^2 \tau \beta_0^{n+1} \Big|_{x=0} .$$

The left-hand side of this expression represents the quantity of heat which has passed through the left-hand boundary after time τ ($\mu^2 du/dx$ is the heat flux). If $\alpha_0 \sim 1/\mu\sqrt{\tau}$, this quantity of heat is of order $\sqrt{\tau}$ instead of order τ. It is clear from physical considerations that such a step is inadmissible. Therefore, in order to guarantee the validity of the sweep, the time step should be chosen considerably smaller than $1/\alpha_0^2(t)\mu^2$.

This method should be somewhat altered for practical applications. Namely, the sweep relations should be given for the difference equation (3),

$$\mu^2(x, t_n) \frac{d^2u^{n+1}}{dx^2} = u^{n+1}(x) - u^n(x) ,$$

instead of for the ordinary differential equation (4),

$$u_m^{n+1} = u_m^n + \mu^2(x_m, t_n) \frac{\tau}{h^2} (u_{m+1}^{n+1} - 2u_m^{n+1} + u_{m-1}^{n+1}) .$$

It is convenient in this case to choose the computational grid as the set of points with the coordinates

$$x_m = \tfrac{1}{2}h + mh \qquad (h = l/M,\ M \text{ integer}) ,$$
$$t_n = n\tau . \tag{9}$$

Two fictitious points ($x = -\frac{1}{2}h$, $x = l + \frac{1}{2}h$) lying outside the interval $[0, l]$ are added to this grid for each instant of time t_n. The values of $u(x, t)$ at these points must be determined from the boundary conditions.

For fixed n relation (3) must be fulfilled for all interior computational points of the interval $0 < x < l$, i.e., for $m = 0, 1, \ldots, M-1$. Thus we have

M equations altogether containing $M+2$ unknowns, i.e., $u_{-1}^{n+1}, u_0^{n+1}, \ldots, u_M^{n+1}$. We must obtain the two missing equations from boundary conditions (2),

$$\frac{\partial u}{\partial x} = \alpha_0(t)u + \beta_0 \big|_{x=0} , \tag{2a}$$

$$\frac{\partial u}{\partial x} = \psi_0(t)u + \omega_0 \big|_{x=l} , \tag{2b}$$

which are sufficiently approximated by the relations

$$\frac{u_0^{n+1} - u_{-1}^{n+1}}{h} = \tfrac{1}{2}\alpha_0^{n+1}(u_0^{n+1} + u_{-1}^{n+1}) + \beta_0^{n+1} , \tag{10a}$$

$$\frac{u_M^{n+1} - u_{M-1}^{n+1}}{h} = \tfrac{1}{2}\psi_0^{n+1}(u_M^{n+1} + u_{M-1}^{n+1}) + \omega_0^{n+1} . \tag{10b}$$

Equalities (10), together with eq. (3) for the points $m = 0, 1, 2, \ldots, M-1$, give for a fixed n a system of $M+2$ equations for $M+2$ unknowns. The sweep method for this system of equations is carried out by means of quite simple formulae. We rewrite relation (3) in the form (omitting the index $n+1$ for the quantities u_m^{n+1})

$$a_m u_{m+1} - 2b_m u_m + c_m u_{m-1} + d_m = 0 \quad (m = 0, 1, 2, \ldots, M-1) , \tag{11}$$

where in our case *

$$a_m = c_m = 1 , \quad b_m = 1 + \frac{h^2}{2\tau\mu^2(x_m, t_n)} , \quad d_m = \frac{h^2}{\tau\mu^2(x_m, t_n)} u_m^n . \tag{12}$$

The boundary conditions (10) are written in the form

$$u_{-1} = X_0 u_0 + y_0 , \tag{13a}$$

$$u_{M-1} = p u_M + q , \tag{13b}$$

where

$$X_0 = \frac{2 - \alpha_0^{n+1}h}{2 + \alpha_0^{n+1}h} , \quad y_0 = \frac{2\beta_0^{n+1}h}{2 + \alpha_0^{n+1}h} , \tag{13*}$$

$$p = \frac{2 - \psi_0^{n+1}h}{2 + \psi_0^{n+1}h} , \quad q = \frac{2\omega_0^{n+1}h}{2 + \psi_0^{n+1}h} . \tag{13**}$$

* If the equation of thermal conductivity is written in the form $\partial u/\partial t = (\partial/\partial x)(\mu^2 \partial\mu/\partial x)$, then for variable μ, $a_m \neq c_m$.

To solve systems (11) and (13), we "sweep" relation (13a) from left to right, i.e., we obtain its value at point $m+1$ from the known value at point m.

We thus suppose that we have obtained a relation of the form

$$u_{m-1} = X_m u_m + y_m \tag{14}$$

between u_{m-1} and u_m. Substituting u_{m-1} from this relation into (11), we find (after solving for u_m)

$$u_m = \left(\frac{a_m}{2b_m - c_m X_m}\right) u_{m+1} + \left(\frac{c_m y_m + d_m}{2b_m - c_m X_m}\right).$$

Setting in this relation

$$X_{m+1} = \frac{a_m}{2b_m - c_m X_m}, \tag{15}$$

$$y_{m+1} = \frac{c_m y_m + d_m}{a_m} X_{m+1}, \tag{16}$$

we can rewrite it in the form

$$u_m = X_{m+1} u_{m+1} + y_{m+1}.$$

Our system can now be solved according to the following procedure.

1) Calculate according to formula (12) the "local" coefficients

$$a_m, b_m, c_m, d_m \qquad (m = 0, 1, 2, \ldots, M-1).$$

2) Obtain from formula (13*) the quantities

$$X_0, y_0.$$

3) From formulae (15) and (16) find successively the "sweep coefficients"

$$X_1, X_2, \ldots, X_M, \qquad y_1, y_2, \ldots, y_M.$$

4) Calculate p and q from formula (13**) and then find u_M by solving the system

$$u_{M-1} = p u_M + q, \qquad u_{M-1} = X_M u_M + y_M.$$

5) Find successively from formula (14)

$$u_{M-1}, u_{M-2}, \ldots, u_1, u_0, u_{-1}.$$

The validity of the method results from the following consideration. If $|X_m| \leqslant 1$, then since $b_m > 1$ and $a_m = c_m = 1$ (cf. (12), (15)), we have $|X_{m+1}| \leqslant 1$. Therefore, if $|X_0| \leqslant 1$, then $|X_m| \leqslant 1$ for any m. However, in this case the error admitted in y_m and u_m is multiplied in the calculation process by a coefficient whose absolute magnitude is less than unity (cf. (16), (14)). As for the quantity X_m itself, varying (15), we find (since $a_m = c_m = 1$)

$$\delta X_{m+1} = \frac{\delta X_m}{(2b_m - c_m X_m)^2} = X_{m+1}^2 \delta X_m .$$

Thus, if $|X_{m+1}| \leqslant 1$, the error δX_m will not increase as we proceed to the following point. Recalling that

$$X_0 = \frac{2 - \alpha_0^{n+1} h}{2 + \alpha_0^{n+1} h},$$

(cf. (13)), we have consequently $|X_0| \leqslant 1$ for $\alpha_0 > 0$. Thus the method is valid in any case if $\alpha_0{}^{n+1} > 0$. The validity of the method can be proven for negative $\alpha_0{}^{n+1}$ if τ is small compared with $\frac{1}{2}/\alpha_0{}^2(t)\mu_0{}^2(t)$, where $\mu_0(t) = \min\limits_{0 \leqslant x \leqslant l} \mu(x, t)$. However, we will omit this proof.

§ 4. Matrix sweep for Poisson's equation *

We consider in the rectangle D $(0 \leqslant x \leqslant a,\ 0 \leqslant y \leqslant b)$ the equation

$$\frac{\partial^2 u}{\partial x^2} + \frac{\partial^2 u}{\partial y^2} = f(x, y) \tag{1}$$

with the condition

$$\partial u/\partial n = S(x, y)u + q(x, y) \tag{2}$$

at the boundary. ($\partial/\partial n$ denotes differentiation with respect to the outward normal.) We shall describe the application of the sweep method to this problem.

* The possibility of applying the sweep method to boundary-value problems for elliptical equations was first discussed by M.V. Keldish.

We first approximate (1) and (2) by a system of difference equations. The steps in x and y (denoted by h and k) are chosen such that $a/h = M$ and $b/k = N$ are integers. As computation grid we take the aggregate of points x_m and y_m whose coordinates are

$$x_m = (m+\tfrac{1}{2})h\ , \quad y_n = (n+\tfrac{1}{2})k\ ,$$

where m and n are integers. We denote the value of any function $\varphi(x, y)$ at the point (x_m, y_n) by $\varphi_{m,n}$. In particular,

$$u(x_m, y_n) = u_{m,n}\ , \quad f(x_m, y_n) = f_{m,n}\ .$$

Eq. (1) is approximated by the difference relation

$$\frac{u_{m+1,n} - 2u_{m,n} + u_{m-1,n}}{h^2} + \frac{u_{m,n+1} - 2u_{m,n} + u_{m,n-1}}{k^2} = f_{m,n}\ . \tag{3}$$

These relations are to be fulfilled for all points of our computation grid lying inside the rectangle D, i.e., for

$$m = 0, 1, 2, \ldots, M\text{-}1\ , \quad n = 0, 1, 2, \ldots, N\text{-}1\ .$$

Thus we have MN equations. The number of unknowns $\{(M+2)(N+2)\}$ is larger than the number of equations, since $u_{-1,n}$ and $u_{M,n}$ $(n = 0, 1, \ldots, N\text{-}1)$ and $u_{m,-1}$ and $u_{m,N}$ $(m = 0, 1, \ldots, M\text{-}1)$ are also unknowns. We obtain the extra equations from boundary conditions (2):

for $m = 0$

$$\frac{u_{-1,n} - u_{0,n}}{h} = \tfrac{1}{2}S(0, y_n)(u_{0,n} + u_{-1,n}) + q(0, y_n)\ , \tag{4a}$$

for $m = M - 1$

$$\frac{u_{M,n} - u_{M-1,n}}{h} = \tfrac{1}{2}S(a, y_n)(u_{M,n} + u_{M-1,n}) + q(a, y_n) \tag{4b}$$

$$(n = 0, 1, 2, \ldots, N\text{-}1)\ ,$$

for $n = 0$

$$\frac{u_{m,-1} - u_{m,0}}{k} = \tfrac{1}{2}S(x_m, 0)(u_{m,0} + u_{m,-1}) + q(x_m, 0)\ , \tag{4c}$$

for $n = N - 1$

$$\frac{u_{m,N} - u_{m,N-1}}{k} = \tfrac{1}{2}S(x_m, b)(u_{m,N} + u_{m,N-1}) + q(x_m, b) \tag{4d}$$

$$(m = 0, 1, 2, \ldots, M-1) .$$

Thus, we add $2(M+N)$ equations to eqs. (3). The totality of (3) and (4) forms the closed system.

It is convenient in our description of the solution of this system by the sweep method to write eqs. (3) and (4) in vector form. Beginning with (4a) and (4b), we consider the N-dimensional vectors

$$\begin{gathered} \boldsymbol{u}_m = \{u_{m,0}, u_{m,1}, \ldots, u_{m,N-1}\} , \\ \boldsymbol{\beta}_0 = \{q(0, y_0), q(0, y_1), \ldots, q(0, y_{N-1})\} \end{gathered} \tag{5}$$

and the diagonal matrix of order N

$$A_0 = \begin{bmatrix} S(0, y_0) & 0 & \ldots & 0 \\ 0 & S(0, y_1) & \ldots & 0 \\ \cdots & \cdots & \cdots & \cdots \\ 0 & 0 & \ldots & S(0, y_{N-1}) \end{bmatrix} . \tag{6}$$

With this notation the system of relations (4a) is written in the form

$$\frac{\boldsymbol{u}_{-1} - \boldsymbol{u}_0}{h} = \tfrac{1}{2}A_0(\boldsymbol{u}_{-1} + \boldsymbol{u}_0) + \boldsymbol{\beta}_0 , \tag{7}$$

and system (4b) in the form

$$\frac{\boldsymbol{u}_M - \boldsymbol{u}_{M-1}}{h} = \tfrac{1}{2}\Psi(\boldsymbol{u}_M + \boldsymbol{u}_{M-1}) + \boldsymbol{\omega} , \tag{8}$$

where

$$\boldsymbol{\omega} = \{q(a, y_0), q(a, y_1), \ldots, q(a, y_{N-1})\}$$

and

$$\Psi = \begin{bmatrix} S(a, y_0) & 0 & \ldots & 0 \\ 0 & S(a, y_1) & \ldots & 0 \\ \cdots & \cdots & \cdots & \cdots \\ 0 & 0 & & S(a, y_{N-1}) \end{bmatrix} .$$

Turning to eqs. (4c), (3) and (4d), we have for fixed m

$$\frac{u_{m,-1} - u_{m,0}}{k} = \tfrac{1}{2}S(x_m,0)(u_{m,0}+u_{m,-1}) + q(x_m,0) \ , \tag{4c}$$

$$\frac{u_{m+1,n} - 2u_{m,n} + u_{m-1,n}}{h^2} + \frac{u_{m,n+1} - 2u_{m,n} + u_{m,n-1}}{k^2} = f_{m,n} \tag{3}$$

$$(m = 0,1,2,\dots,N-1) \ ,$$

$$\frac{u_{m,N} - u_{m,N-1}}{k} = \tfrac{1}{2}S(x_m,b)(u_{m,N}+u_{m,N-1}) + q(x_m,b) \ . \tag{4d}$$

The first of relations (3) contains $u_{m,-1}$, and the last $u_{m,N}$. We can eliminate these quantities by the use of (4c) and (4d), respectively. We find

$$u_{m,-1} = \frac{1+\frac{1}{2}kS(x_m,0)}{1-\frac{1}{2}kS(x_m,0)}\,u_{m,0} + \frac{q(x_m,0)\,k}{1-\frac{1}{2}kS(x_m,0)} \ ,$$

$$u_{m,N} = \frac{1+\frac{1}{2}kS(x_m,b)}{1-\frac{1}{2}kS(x_m,b)}\,u_{m,N-1} + \frac{q(x_m,b)\,k}{1-\frac{1}{2}kS(x_m,b)} \ .$$

With the use of these values we can rewrite (3) in the form

$$\frac{u_{m+1,n} - 2u_{m,n} + u_{m-1,n}}{h^2} + \frac{\lambda_n^m u_{m,n+1} - 2(1-\mu_n^m)u_{m,n} + \nu_n^m u_{m,n-1}}{k^2} = f_{m,n} - F_{m,n} \ ,$$

where

$$\lambda_n^m = \begin{cases} 1 & \text{for } 0 \le n \le N-2 \ , \\ 0 & \text{for } n = N-1 \ , \end{cases} \qquad \nu_n^m = \begin{cases} 0 & \text{for } n = 0 \ , \\ 1 & \text{for } 1 \le n \le N-1 \ , \end{cases}$$

$$\mu_n^m = \begin{cases} \dfrac{1}{2}\,\dfrac{1+\frac{1}{2}kS(x_m,0)}{1-\frac{1}{2}kS(x_m,0)} & \text{for } n = 0 \ , \\ 0 & \text{for } 1 \le n \le N-2 \ , \\ \dfrac{1}{2}\,\dfrac{1+\frac{1}{2}kS(x_m,b)}{1-\frac{1}{2}kS(x_m,b)} & \text{for } n = N-1 \ , \end{cases}$$

$$F_n^m = \begin{cases} \dfrac{1}{k^2}\,\dfrac{q(x_m,0)k}{1-\frac{1}{2}kS(x_m,0)} & \text{for } n=0\,, \\ 0 & \text{for } 1 \leqslant n \leqslant N-2\,, \\ \dfrac{1}{k^2}\,\dfrac{q(x_m,b)k}{1-\frac{1}{2}kS(x_m,b)} & \text{for } n=N-1\,. \end{cases}$$

These relations can be written in the vector form

$$\boldsymbol{u}_{m+1} - 2B_m\boldsymbol{u}_m + \boldsymbol{u}_{m-1} + \boldsymbol{d}_m = 0\,, \tag{9}$$

where

$2B_m =$

$$\begin{bmatrix} 2\left(1+\frac{h^2}{k^2}\right)-\frac{h^2}{k^2}\,\mu_0^m & -\frac{h^2}{k^2} & \cdots & 0 & 0 \\ -\frac{h^2}{k^2} & 2\left(1+\frac{h^2}{k^2}\right) & -\frac{h^2}{k^2}\ \cdots & 0 & 0 \\ 0 & -\frac{h^2}{k^2} & 2\left(1+\frac{h^2}{k^2}\right)\cdots & 0 & 0 \\ \cdots & \cdots & \cdots & \cdots & \cdots \\ 0 & 0 & 0\ \cdots & 2\left(1+\frac{h^2}{k^2}\right) & -\frac{h^2}{k^2} \\ 0 & 0 & 0\ \cdots & -\frac{h^2}{k^2} & 2\left(1+\frac{h^2}{k^2}\right)-\frac{h^2}{k^2}\,\mu_{N-1}^m \end{bmatrix}$$

and

$$\boldsymbol{d}_m = \{-h^2(f_{m,0}-F_{m,0}), -h^2f_{m,1}, \ldots, -h^2f_{m,N-2} - h^2(f_{m,N-1}-F_{m,N-1})\}\,.$$

Thus, according to (7), (8) and (9), we have the system of vector equations

$$\frac{\boldsymbol{u}_{-1}-\boldsymbol{u}_0}{h} = \tfrac{1}{2}A_0(\boldsymbol{u}_{-1}+\boldsymbol{u}_0) + \boldsymbol{\beta}_0\,, \tag{7}$$

$$\boldsymbol{u}_{m+1} - 2B_m\boldsymbol{u}_m + \boldsymbol{u}_{m-1} + \boldsymbol{d}_m = 0 \tag{9}$$

$$(m = 0,1,2,\ldots,M-1)\,,$$

$$\frac{\boldsymbol{u}_M-\boldsymbol{u}_{M-1}}{h} = \tfrac{1}{2}\Psi(\boldsymbol{u}_M+\boldsymbol{u}_{M-1}) + \boldsymbol{\omega}\,. \tag{8}$$

Solving eq. (7) for u_{-1}, we obtain

$$u_{-1} = X_0 u_0 + y_0 ,$$

where the matrix X_0 is defined by the equality

$$X_0 = \left(\frac{1}{h} E - \tfrac{1}{2} A_0\right)^{-1} \left(\frac{1}{h} E + \tfrac{1}{2} A_0\right) , \tag{10}$$

and the vector y_0 by the equality

$$y_0 = \left(\frac{1}{h} E - \tfrac{1}{2} A_0\right)^{-1} \beta_0 . \tag{11}$$

The direct sweep for this case is completely analogously to the one-dimensional case described in § 2. That is, suppose we have already obtained for the mth point the relation

$$u_{m-1} = X_m u_m + y_m . \tag{12}$$

Substituting u_{m-1} from (12) into (9), we obtain

$$u_{m+1} - 2B_m u_m + X_m u_m + y_m + d_m = 0 ,$$

and consequently,

$$u_m = (2B_m - X_m)^{-1} u_{m+1} + (2B_m - X_m)^{-1} (d_m + y_m) .$$

Thus, for X_m and y_m we have the recursion relations

$$X_{m+1} = (2B_m - X_m)^{-1} , \tag{13}$$

$$y_{m+1} = X_{m+1}(y_m + d_m) . \tag{14}$$

Woth these relations and the known quantities X_0 and y_0 (cf. (10), (11)), we can determine successively

$$X_1, X_2, \ldots, X_M , \quad y_1, y_2, \ldots, y_M .$$

In particular, for $m = M$ we have

$$u_{M-1} = X_M u_M + y_M .$$

On the other hand, by virtue of (8) we have

$$u_{M-1} = \left(\frac{1}{h} E + \tfrac{1}{2}\Psi\right)^{-1} \left(\frac{1}{h} E - \tfrac{1}{2}\Psi\right) u_M - \left(\frac{1}{h} E + \tfrac{1}{2}\Psi\right)^{-1} \omega .$$

We can determine u_M from these two relations, after which we can obtain $u_{M-1}, u_{M-2}, \ldots, u_0$ from (12).

The major part of the computing time for this method is consumed by the inversions of the matrices (formulae (12), (14)). Instead of solving a system of $MN+2(M+N)$ equations, $M+1$ inversions of a matrix of order N are required. Thus, a considerable amount of time is saved by this process. This gain in time is made particularly noticeable by a judicious choice of the sweep direction. The axes are chosen, of course, so that $M > N$.

We shall give the basis for the validity of this method *, while supposing for simplicity that

$$\mu_0^m = \mu_{N-1}^m = 0 .$$

It is easy to see that this corresponds to the boundary condition which fixes the value of $u(x, y)$ at the upper and lower edges of the rectangle **. We will assume that the function $S(x, y)$ is negative at the left-hand boundary of the rectangle. As in the case of the equation of thermal conductivity, validity of the method follows from the inequality

$$\|X_m\| \leqslant 1 ,$$

which we shall prove.

We show first that all eigenvalues of the matrix $2B_m = Q$ are larger than two *** (for $\mu_0{}^m = \mu_{N-1}{}^m = 0$ this matrix does not depend on m). We must calculate the determinant $|Q - \lambda E| = |R|$. Let $\delta = -h^2/k^2$ and $\gamma = 1 - \delta - \frac{1}{2}\lambda$. Then

$$Q - \lambda E = R_N = \begin{bmatrix} 2\gamma & \delta & 0 & 0 & \ldots & 0 & 0 \\ \delta & 2\gamma & \delta & 0 & \ldots & 0 & 0 \\ 0 & \delta & 2\gamma & \delta & \ldots & 0 & 0 \\ \cdots & \cdots & \cdots & \cdots & \cdots & \cdots & \cdots \\ 0 & 0 & 0 & 0 & \ldots & \delta & 2\gamma \end{bmatrix}$$

* An analogous study of the validity of the application of the matrix sweep method to the solution of some problems in hydrodynamics was given in 1954 by K. I. Babenko and N. N. Chentsov.

** We must somewhat change the computation grid, viz., we suppose that $u_{m,-1}$ and $u_{m,N}$ are the values of $u(x,y)$ at the lower and upper boundaries, respectively, of the rectangle D.

*** Since Q is symmetric, all of its eigenvalues are real.

(N is the order of the matrix R_N). By decomposing $|R_N|$ with respect to elements in the first row, we obtain

$$|R_N| = 2\gamma|R_{N-1}| - \delta^2|R_{N-2}| .$$

This difference equation has a general solution of the form

$$|R_N| = C_1(\gamma + \sqrt{\gamma^2 - \delta^2})^N + C_2(\gamma - \sqrt{\gamma^2 - \delta^2})^N .$$

The constants C_1 and C_2 are found from the conditions

$$|R_1| = 2\gamma , \qquad |R_2| = 4\gamma^2 - \delta^2 ,$$

from which we obtain

$$\begin{aligned} 2\gamma &= C_1(\gamma + \sqrt{\gamma^2 - \delta^2}) + C_2(\gamma - \sqrt{\gamma^2 - \delta^2}) , \\ 4\gamma^2 - \delta^2 &= C_1(\gamma + \sqrt{\gamma^2 - \delta^2})^2 + C_2(\gamma - \sqrt{\gamma^2 - \delta^2})^2 . \end{aligned} \tag{15}$$

Multiplying the first equality by 2γ and subtracting it from the second, we obtain $-\delta^2 = (C_1 + C_2)(-\delta^2)$, from which it follows that $C_1 + C_2 = 1$.

Using this equality and the first of eqs. (15), we find

$$C_1 - C_2 = \frac{\gamma}{\sqrt{\gamma^2 - \delta^2}} .$$

Thus,

$$C_1 = \frac{\gamma + \sqrt{\gamma^2 - \delta^2}}{2\sqrt{\gamma^2 - \delta^2}} , \qquad C_2 = -\frac{\gamma - \sqrt{\gamma^2 - \delta^2}}{2\sqrt{\gamma^2 - \delta^2}} ,$$

and consequently,

$$|R_N| = \frac{1}{2\sqrt{\gamma^2 - \delta^2}} \{(\gamma + \sqrt{\gamma^2 - \delta^2})^{N+1} - (\gamma - \sqrt{\gamma^2 - \delta^2})^{N+1}\} .$$

It is then obvious that $|R_N|$ can be equal to zero only if *

$$(\gamma + \sqrt{\gamma^2 - \delta^2})^{N+1} = (\gamma - \sqrt{\gamma^2 - \delta^2})^{N+1} ,$$

which is only possible if

* In fact, $|R_N|$ has only N roots; the $(N + 1)$-st node of the numerator is obtained for $\gamma = \delta$, when the denominator also goes to zero. In this case $|R_N| = (N + 1)\gamma^N$.

$$\frac{\gamma + \sqrt{\gamma^2 - \delta^2}}{\gamma - \sqrt{\gamma^2 - \delta^2}} = e^{2i\varphi_n} ,$$

where $\varphi_n = n\pi/(N+1)$ is real. However, then

$$\gamma = \pm\delta \cos \varphi_n ,$$

and since

$$\gamma = 1 - \delta - \tfrac{1}{2}\lambda ,$$

we have

$$\lambda = 2 - \delta(1 \mp \cos \varphi_n) .$$

Remembering that $\delta = -h^2/k^2 < 0$, it is immediately clear that

$$\lambda \geqslant 2 .$$

We now show by the use of this inequality that if

$$\|X_m\| \leqslant 1 , \quad \text{then} \quad \|X_{m+1}\| \leqslant 1 .$$

Actually, since the matrix $2B = Q$ is symmetric, we have for any N-dimensional vector ϱ_1, the inequality

$$\|2B\varrho_1\| \geqslant \min|\lambda|\,\|\varrho_1\| \geqslant 2\|\varrho_1\| \quad * . \tag{16}$$

On the other hand, from the assumption that $\|X_m\| \leqslant 1$ it follows that

$$\|X_m\varrho_1\| \leqslant \|\varrho_1\| . \tag{17}$$

Now let ϱ, another N-dimensional vector, be

$$\varrho = (2B - X_m)\varrho_1 .$$

Then, according to (13) we have

$$\varrho_1 = (2B - X_m)^{-1}\varrho = X_{m+1}\varrho .$$

However, in view of (16) and (17), we obtain

$$\|\varrho\| = \|(2B - X_m)\varrho_1\| = \|2B\varrho_1 - X_m\varrho_1\| \geqslant \|2B\varrho_1\| - \|X_m\varrho_1\|$$

$$\geqslant 2\|\varrho_1\| - \|\varrho_1\| = \|\varrho_1\| .$$

* The norm of the vector is Euclidean.

Thus, $\|\varrho_1\| \leqslant \|\varrho\|$, i.e.,

$$\|X_{m+1}\| \leqslant 1 ,$$

and our assertion is proven.

We recall that the matrix X_0, defined by formula (10), can be reduced with the help of (6) to the form

$$X_0 = \begin{bmatrix} \dfrac{1 + \frac{1}{2}hS(0, y_1)}{1 - \frac{1}{2}hS(0, y_1)} & \dots & 0 \\ \dots\dots & \dots & \dots\dots \\ 0 & \dots & \dfrac{1 + \frac{1}{2}hS(0, y_N)}{1 - \frac{1}{2}hS(0, y_N)} \end{bmatrix},$$

from which it is obvious that its norm does not exceed the value

$$\max_{0 \leqslant y \leqslant b} \left| \frac{1 + \frac{1}{2}hS(0, y)}{1 - \frac{1}{2}hS(0, y)} \right| .$$

Obviously, for negative $S(0, y)$, $\|X_0\| < 1$ *. Thus, the validity of the sweep method has been demonstrated. The present example has shown that the double sweep method can be applied successfully to the solution of multidimensional difference equations. M. V. Keldish is responsible for initiating such an application of this method.

Bibliography

[1] S. K. Godunov, On the numerical solution of boundary-value problems for systems of linear ordinary differential equations, Usp. Mat. Nauk 16 (1961).
[2] V.V. Rusanov, On the stability of the matrix sweep method, Vychisl. Mat. 6 (1960).
[3] V.V. Rusanov, On the solution of systems of difference equations, Dokl. Akad. Nauk 136 (1961).

* Since the derivative in the boundary conditions was taken with respect to the outward normal, the assumption $S(0, y) < 0$ corresponds to the case $\alpha_0 > 0$ of the one-dimensional problem.

3.

(with M. L. Tsetlin)

Some methods of control for complex systems

Usp. Mat. Nauk **17** (1) (1962) 3–25 [Russ. Math. Surv. **17** (1) (1962) 95–117].
Zbl. **107** :299

Introduction 95
§1. The problem of finding the minimum of a function of many variables 96
§2. An example of a minimum problem 103
§3. The tactics of movement 108
References 115

Introduction

The problems leading to the necessity of studying complex control systems are extremely varied and arise in the most diverse parts of modern science and technology.

The peculiarities of these systems force us to redefine the word 'studying' itself. The fact is that a complete isomorphic description, allowing us to take every particular of a phenomenon into account, proves unsatisfactory for complex systems just because of their complexity. There are many examples to show the inadequacy of such descriptions. Thus, describing a gas by means of the differential equations of motion of its particles and of their initial coordinates and velocities does not substantially increase our knowledge of the macroscopic properties of the gas.

For complex systems the typical situation is that in which the method of description is dictated by the problem to be solved with the help of the description. We remark that in complex systems in which there is a problem to be solved, for example in situations of importance to a living organism, it makes sense to speak of the quality of a solution, of its degree of satisfactoriness.

We shall examine such a situation in detail in a particular example. This will be the problem of finding the minimum of a function of many variables. Here such notions as complexity of a problem, organization, search, tactics, hypothesis, and several others arise naturally. In the same section we shall more fully describe one of the methods of non-local search, the so-called 'ravine' method. An application of this method to the problem of phase shift analysis of proton-proton scattering will be given in §2, where we shall describe a series of cases typical of those that occur in applications of the ravine method.

Perhaps the clearest examples of complex control systems are encountered when one investigates the interaction of an animal with its environment. For such problems the inadvisability and even the practical impossibility of isomorphic description is particularly evident. This possibly explains why statements in physiology often have the character of analogies.

Among the questions that arise when one studies the behaviour of 'a small animal in a great world' one can distinguish two main problems more or less clearly.

The first is connected with the animal's behaviour, with the single indivisible process of studying the environment and taking and implementing decisions. Of the various aspects of behaviour we shall here touch on only one, the question of how movements are constituted. This question, important even in itself in physiology, attracts us by its resemblance to physics, for many parameters of movement can be measured and described quantitatively. Here it becomes necessary to use organization and to apply tactics pertaining, in view of the indivisibility of the process, both to the analysis of the incoming impressions and to forming the strictly motoric acts. Some considerations bearing on this field will be expounded in §3 of this paper.

The second problem arising from the study of behaviour is the question of what kind of mechanisms can be used in carrying out the control tactics. These mechanisms must possess an extremely high degree of reliability, quick action, flexibility of control, economy and some other properties sharply distinguishing them from modern technical equipment. The very possibility of constructing such machines would seem to us utterly incredible, but for the fact of the existence and working of the central nervous system of man and the higher animals.

There have been numerous well known attempts to describe the working of the brain by means of algebraic logical models or by far-reaching analogies with electronic computing machines. Other models use the notions of continuous control systems. The first, perhaps still very naive, steps in studying the possibilities of continuous control systems were taken in the papers [1]-[4]. But in this paper we shall not touch on these questions.

We have discussed the main propositions of this paper with N.A. Bernshtein, V.S. Gurfinkel', L.N. Ivanova and I.I. Shapiro-Pyatetskii. Our conversations with them have been very useful.

§1. The problem of finding the minimum of a function of many variables.

We shall consider the functioning of a complex control system whose working is directed towards the achievement of some definite objective. It will be assumed that the system is able to determine its degree of proximity to the objective. The information necessary for the successful working of the system is obtained by it in the process of activity directed towards attaining the objective.[1] The complexity of a system is defined to be the number of parameters giving its state. The study of the control processes of many physiological mechanisms and of complex engineering systems leads us to such problems. Any movements of animals or men can serve as examples of systems of this kind. We shall return to the description of movement in §3 below.

Many problems in the study of complex systems have been, it would seem, sufficiently well investigated by means of classical mathematics and functional analysis. But often the algorithms offered there, while theoretically allowing us to solve the problem, turn out to be not feasible in practice. Thus, for example, for finding an extremum of a function of many variables, classical mathematics offers the following method. Differentiate the function in turn with respect to each of its arguments, and equate to zero the derivatives obtained. Thus, the problem leads to the solution of a system of equations, the number of equations being equal to the number of arguments. But, in practical problems of computation, it turns out that the solution of such a system is by no means a simpler problem than the direct search for an extremum. For any considerable number of arguments, both problems are in general very unwieldy and their solution exceeds the resources of modern computing techniques. Direct application of other algorithmic methods is also not feasible in practice.[2]

It would be possible to give other problems too where, even if it is possible to construct an algorithm suitable for all cases, it turns out that applying the algorithm is impracticable, in view of limitations of computing resources and of the time within which the problem must be solved. The limitation of the time that can be spent on the solution of the problem is especially important. The fact is that for practical problems, physiological say, a situation changing with time is typical, so that a delayed solution may turn out to be plain wrong. In this sense a comparatively inexact approximate solution obtained quickly may turn out to be preferable to an exact but delayed solution.[3] In such situations

---

[1] In this sense the systems studied here are dual control systems according to A.A. Fel'dbaum [5], [6].

Generally speaking, the more natural case is that in which some amount of exact knowledge about the objective is also obtained in the course of solving the problem. The question of working out the objective (and the corresponding evaluation function or system of such evaluations) must be even more important. But this problem is more difficult, and we limit ourselves here to the problems in which the objective and the evaluation function are known. An example of a problem where working out the evaluation function is very important for the solution is the game of chess.

[2] S. Ulam, [7], remarking that the search for the minimum of a function of more than four or five variables on a computing machine is in practice often impossible, proposes constructing a synergesis (coordinated collaboration) of man and machine. The machine shows on a screen some two-dimensional section or other of the function and the man, looking at this relief, decides how to proceed further, i.e. what sections to take, what to look at under magnification, and so on. In this very clever proposal we can see a clear understanding of the imperfection of exact isomorphic algorithms. But this proposed division of labour between man and machine is based on existing machines and methods of using them.

[3] 'It is possible to compute tomorrow's weather accurately, but the computation would require a month.' (L.F. Richardson).

an acceptable solution may be obtained only by making use of the organization present to some extent in the problems encountered in the practical activity of man and perhaps in physiology.

To give a complete definition of the notion of organization is extremely complicated. In essence, by organization one may understand the characteristic features of the problem or situation that may facilitate achieving a solution. These particulars are not exactly known in advance, but only more or less probable. Therefore, to make use of the organization, one proceeds by way of advancing some hypotheses or other and constructing tactics based on these hypotheses. The hypotheses advanced cannot, in general, be immediately verified. They are verified, so to speak, in practice. The nature of the solution serves as a criterion for the correctness of the hypothesis.

Let us take a simple example. A solution by differences of any complicated system of partial differential equations (in hydrodynamics, say) rests on an unverified hypothesis. Surely nobody proves that the solution of the difference system with a chosen computing step δ differs from the exact solution by less than a given number ε.

Using a hypothesis entails deliberately leaving out of consideration the totality of all possible cases, and indeed leaving out the chaotic cases which, in the formal mathematical sense, are the most probable ones.

In this section we shall try to construct a method for solving the problem of finding the minimum of a function of many variables, using a conjectural limitation of the problem. The kind of limitation on the function that will be assumed (its description will be given below) is quite typical of many cases and is probably one of the most widespread assumptions.

The extremal problems are interesting in themselves, as they constitute an important part of the modern theory of regulation, namely automatic optimization (cf. [8]-[10]).

Let $F(x_1, \ldots, x_n, y_1, \ldots, y_n)$ be the function whose minimal values are to be looked for. It is assumed that the system has the possibility of measuring the values of F. It is also assumed that the system can measure the values of the variables $x_1, \ldots, x_n$, which we shall call its working parameters. The arguments $y_1, \ldots, y_n$ are the hidden parameters of the system, depending on time and perhaps on the variables $x_1, \ldots, x_n$. The system has no way of measuring nor altering the values of the hidden parameters. Let us write $F(x_1, \ldots, x_n, y_1, \ldots, y_n)$ in the form $\Phi(x_1, \ldots, x_n, t)$; then Φ will be called the evaluation function of the system.

We remark that it is not assumed that the function Φ is given analytically nor in any other fashion, so the choice of the required values of the working parameters must be made experimentally. The dependence of the evaluation function on time (this dependence is by no means assumed to be known) leads to the necessity of continual search for the required values of the arguments.[1]

[1] Moreover, if Φ were not dependent on the time and if the time available for the search were infinite, it would be possible to confine oneself to a full survey of its values and, using only a single memory cell containing the least of the preceding values, to obtain the absolute minimum value.

An important property of such a search is its speed. A method of search can be acceptable only if, roughly speaking, satisfactory values of the evaluation function are obtained in an interval of time not long enough for the function to alter materially. Thus, the speed of search turns out to be bound up with the rate of change of the evaluation function.

We remark that the dependence of Φ on time makes the problem of finding the absolute minimum meaningless. We must therefore limit ourselves only to seeking a region of relatively small values of the evaluation function, a level we are to maintain.

An automatic search for values of the working arguments giving sufficiently small values of Φ can be carried out in various ways. It will be convenient to divide these into three groups.

It is natural to put in the first group the so-called methods of blind search. For them it is characteristic that all the points of the space of working parameters are either looked over in a definite order (scanning), or are chosen at random (homeostatic principle [11], [12]). On attaining sufficiently small values of Φ, the search is halted until such time as these values go outside the permitted limits.

In the methods of blind search the special properties of the evaluation function (its organization) are used to a very slight degree. The results of an individual experiment are not used for the next search, so that one loses information about Φ extracted from the measurement of its value. Therefore the values of Φ do not improve from experiment to experiment, and on the average turn out to be relatively high. Methods of blind search use only one value of Φ, the one it has at the moment in question. In this sense, systems using blind search have no memory.

The second group is formed of methods of local search. There are quite a number of such methods. To this group belong the gradient method, relaxation method, method of steepest descent and some others. Their common feature is that they are local. The working point moves continuously in the space of the working parameters. In preparing for the next experiment one uses the knowledge of the values of Φ in a small neighbourhood of the previous experiment. Methods of this kind allow one to obtain a systematic decrease in the values of Φ in the process of search, and this is the essential advantage of the local methods in comparison with the methods of blind search.

Local search systems, as applied to problems of automatic optimization, are described in detail in the works of A.A. Fel'dbaum [8]-[10], where methods are also given for making the diagrams of electronic automata of this kind.

In using any specific local method one has to select experimentally some constants on which the search depends, e.g. the size of step in the gradient method of steepest descent. The values of these constants giving rise to the most rapid search are important properties of the function to be minimized. But since the local search methods use only local properties of Φ, these constants are different in different places, and they describe Φ very incompletely.

Using only local properties of Φ limits the effectiveness of the local search methods, creating a constant danger of a cyclic search in some minor shallow depression. For small values of the gradient the search wanders and its effectiveness becomes vanishingly small.

The third group of methods of automatic search will be called methods of non-local search. It is characteristic of these methods that the trajectory of the working point moving in the parameter space is not continuous. Therefore the volume of the region looked over per unit time is materially increased and the search itself is considerably speeded up, giving rise to a significant advantage over the local methods.

The simplest of the non-local search methods is a combination of the homeostatic principle with some local method. Methods of this kind, often applied in practice in computing, reduce to the following. Choosing a point at random, they proceed to a search (descent) using some local method. When the search 'cycles', i.e. when with further movements the change in the value of the evaluation function becomes small, the next point is chosen at random and the process continues. Thus we get the simplest non-local gradient method.

Methods of this kind also use only local properties of the function to be minimized. The information obtained during the local descent is not used further and is lost. Therefore at the beginning of each descent, one inevitably goes out into the region of large values of Φ with a consequent long local descent.

We now describe a non-local search method that was called in [13] the ravine method. This method allows one to use deeper properties of the organization of the function than merely its local behaviour.

The ravine method proves effective in the cases where the working parameters $x_1, \ldots, x_n$ can be divided into two groups. The first group (it contains almost all these parameters) consists of the parameters whose change leads to significant change in the value of Φ. Therefore the selection of the values of these parameters (we call them inessential) is carried out comparatively simply and quickly.

The second group of parameters contains a small group of variables (e.g. one, two, or three). These variables can themselves be some of the working parameters $x_1, \ldots, x_n$ or, more often, are functions of them. A change in the variables of this group (essential variables) leads to a relatively small change in the values of the evaluation function. The number of essential variables will be called the dimension of the ravine.

Of course such a division of the parameters is not possible for every function a mathematician could give. But for functions that occur in one's practical activity (here we have in view reasonable problems of physics, technology and physiology), there obviously is such a division in a very considerable number, probably in an overwhelming majority, of the cases. While understanding all the difficulty of exactly defining these concepts, we nevertheless allow ourselves to call a function well organized if it permits such a division of parameters. And it is this hypothesis, that the evaluation function Φ is well organized, that is fundamental to the ravine method.

The division of the parameters into essential and inessential ones must of course be carried out automatically in the process of search itself. Here it is important to note that the division of the parameters into groups depends in general on the time and on the point $x = (x_1, \ldots, x_n)$ in the space of working parameters.

Let us turn now to the description of the search itself.

First an arbitrary point x_0 is selected. From this point a descent is carried out by the gradient method or by any other local method. This descent continues as long as the relative decrease $\frac{\Delta\Phi}{\Phi}$ exceeds some previously fixed value Δ called the gradient criterion. In using the ravine method the local descent should be carried out roughly, taking the value of Δ relatively large, e.g. $\Delta = 20\%$. The point is that as soon as the local search ceases to decrease the value of Φ significantly, we get into a region where the variables of the first and second groups become equally important, i.e. where the function ceases to be well organized. Therefore, if we continue the local search, we shall not make any appreciable headway with the significant variables, but will wander confusedly, changing the inessential variables. Indeed this is the reason the local search methods are of so little effect.

So, let the gradient descent take us to the point A_0 (fig. 1). After this a point X_1 is chosen near X_0 at a distance considerably greater than the interval of the gradient descent, for example in a direction perpendicular to the gradient. From X_1 we carry out a local descent to a point A_1. After the points A_0 and A_1 have been obtained, the point X_2 is found by means of the so called step along the ravine. The points A_0 and A_1 are joined by a straight line on which X_2 is selected at a distance L from A_1, where L is called the length of the ravine step.

For well organized functions this length is chosen considerably greater than the size of the gradient step. The ravine step is chosen experimentally. This quantity largely determines the quality of the automatic search. For a fixed size of L we cross over small ridges and skirt the high mountains. The scales of these are determined by the magnitude of the ravine step.

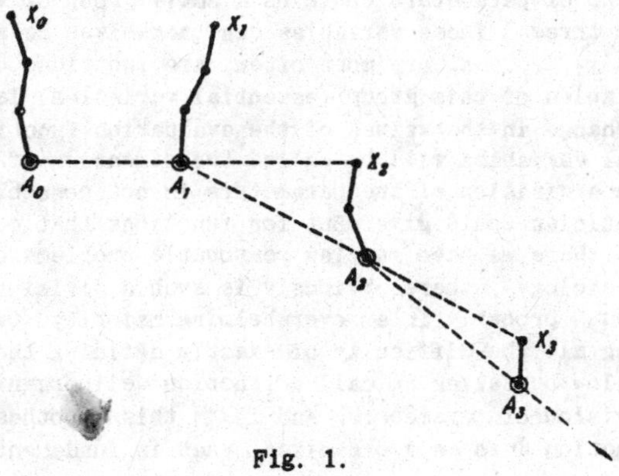

Fig. 1.

After X_2 is chosen, a gradient descent to A_2 is carried out. X_3 is chosen using A_1 and A_2, just as X_2 was chosen using A_0 and A_1, and then the process repeats.

Thus the points X_i are chosen at or near[1] places where small values of the evaluation function are to be expected, so that the whole search is carried out mainly in regions where the values of the function are small.

We note one more important circumstance. With the right choice of the length of the ravine step, as the motion proceeds along the ravine, it adapts itself to the ravine's direction, so that the lengths of the gradient descents become far smaller than the length of the ravine step. This adaptation is connected with the fact that, as the motion proceeds along the ravine, its direction becomes more precisely determined. Thus, as the essential variables become more precisely distinguished, the portion of time spent on the gradient descents decreases. This leads both to speeding up the process of search itself and to a considerable decrease in the computed values of Φ. Such an adaptation, such an accommodation of the tactics, has certain features that can be associated with such terms as learning and adaptive behaviour.

The quality of the search depends in a large measure on the choice of the length L of the ravine step and the size of the gradient criterion Δ. The values of these parameters assuring the most effective search are important characteristics of the function Φ. Thus the use of the ravine method enables one to obtain considerably more information about the structure of this function than the use of local methods.

We have described the ravine method in its simplest form, i.e. when the ravine is one-dimensional. In the case of a many-dimensional ravine this method, especially at the beginning of the search, proves insufficiently effective. In these cases it is useful at the beginning of the search to take several points, e.g. two for each of the working parameters. When one thus obtains descending points, one can pick out from them the correct direction of the ravine. It is often expedient to begin the search with a 'pencil of ravines' going out of one region.

We have described here the ravine tactics of the first rank, in the sense that the variables are divided into only two groups. It is not difficult to construct tactics of higher rank also. Thus, for example, in the tactics of the second rank, the variables are divided into inessential, essential of the first rank and essential of the second rank. Correspondingly one chooses lengths L_1 and L_2 of ravine steps with respect to these variables.

Gradient methods, and methods of steepest descent, require the computation of the gradient, i.e. the computation of values the of Φ, at $n + 1$ points, and this involves a considerable number of operations for large values of n.

Here it may turn out to be useful to apply finite automata having adaptive behaviour [14], [15]. In this connection one can use automata $L_{2kn,n}$ where k is the number of working parameters and n is the complexity of the automaton. 'Penalty' and 'non-penalty' are defined according to the given value of the gradient criterion and to each ray of the state diagram of the automaton there corresponds a motion in a definite direction following one of the working parameters.

It is important to note that for functions rapidly changing in time, in a motion along a ravine the correlation between the values of the function at points whose distance is equal to the ravine step will be small. Therefore the efficiency of the ravine method begins to approach the efficiency of the most simple non-local gradient method. Further increase in the dependence on time leads to a decrease even in the local correlation, i.e. the correlation between values of the function at points whose distance

---

1 The values of Φ at the points X_i (on the "slopes of the ravine") need not be small because of the influence of the inessential variables.

apart is equal to the gradient step. Thus the tactics begin to approach those of blind search. A similar situation arises also with a lesser degree of organization of the function.

Essentially the ravine method includes in itself both the non-local gradient method and the blind search, giving a considerable gain in those cases where the evaluation function is well organized and the rate of search exceeds the rate of change of Φ with respect to time. In the remaining cases the ravine method is not inferior to the other methods.

In using various methods of automatic search for an extremum, there arises the question of their relative merits. Here it is natural to use functionals of the form

$$\frac{1}{T}\int_0^T \Psi(\Phi(x_1, \ldots, x_n, t))\, dt \tag{1}$$

The function Ψ can be chosen in various ways depending on the choice of a criterion for the quality of the search. Thus, for example, with $\Psi(\Phi) = \Phi$, the value of this functional corresponds to the so-called 'search fee' or 'search payment' defined for the simplest systems of automatic minimization [16].

In problems where the aim is to attain values of Φ not exceeding some level C ('minimizing with respect to the level C'), a convenient functional is that obtained from (1) by setting

$$\Psi(\Phi) = \begin{cases} 0 & \text{if } \Phi \leqslant C, \\ 1 & \text{if } \Phi > C. \end{cases}$$

With the ravine method the values of these functionals turn out to be lower than with the use of other methods, since owing to the adaptation the whole search takes place in a region of small values of Φ.

§2. Example of a minimum problem

In this section we discuss a minimum problem, which is solved by the ravine method described in §1. This problem is similar to many others that arise when one works up the results of observations. It is characteristic of problems of this kind that a certain number of parameters in a theoretical formula are determined by requiring the closest possible agreement between the values of some quantity, computed by means of the theoretical formula, and the results of the corresponding experiments. Usually the results of the experiments turn out to be in excess, that is, the number of observations considerably exceeds the number of parameters being determined. In this sort of problem it is natural to determine the parameters themselves by the requirement that the quantity

$$\chi^2 = \sum_i (f_i^{exp_v} - f_i^{theor})^2 \alpha_i$$

should be a minimum, where f_i^{exp} and f_i^{theor} are the observed and computed results respectively and α_i is the weighting factor of the i-th observation.

We remark that exact computation of the minimum is here known to be unnecessary because of the scatter of the results which is unavoidable with experiments. One need only find and investigate those regions of the parameters where χ^2 takes sufficiently small values.

There are numerous examples of problems of this kind. Apart from the problem to whose description this paragraph is devoted, we shall mention only the problem of X-ray structural analysis which consists in determining from given X-ray scattering the arrangement of atoms in a crystal. Here one must find small values of a function χ^2 depending on a large number of arguments, namely the coordinates of the atoms.

Let us pass on to the problem of phase shift analysis of scattering. According to the general formulae of quantum mechanics, the scattering of one particle by another can be expressed [17], [18] by means of parameters $\delta_1, \delta_2, \ldots, \delta_k$, changing within the limits 0 to 2π, called the phase shifts of the scattering. In view of the absence of any precise theory of nuclear forces, these parameters can be determined only experimentally.

We shall investigate this problem in the case of scattering of a proton with energies of the order of a hundred Mev [19], [20]. We shall here consider altogether nine different phase shifts of the scattering. Thus, we consider the problem of finding regions in the space of the arguments $\delta_1, \ldots, \delta_k$ in which the value of a certain function $f(\delta_1, \ldots, \delta_k)$ should be small. For the problem of proton-proton scattering, typical values of this function are of the order of 10^6.

We shall not write out the formulae giving this function. They are too long and complicated to be of any help in understanding the problem.

Therefore the hypothesis that the function to be minimized is well organized must be based, not on the simplicity of the formulae from which its values are computed, but on the reasonableness of the statement of the physical problem itself.[1]

It was on this hypothesis that the search for regions of small values of the function was carried out by the ravine method. We remark once more that for success in applying this method, what is important is not a knowledge of the pattern of the organization of the function, but only the fact that there is such an organization. The structure indeed becomes revealed to some extent in the process of computation.

The search for a minimum by the ravine method is in a certain sense a game between the computing machine and the function to be minimized. Just for this reason, a description of such a search is necessarily an analysis of certain typical situations occurring in the course of the search, resembling, very likely, a commentary on chess problems and games.

It is convenient to depict the trajectory of the movement of the working point in the space of the variables $\delta_1, \ldots, \delta_k$ in the form of a curve in the plane, taking as axes some pair of variables δ_i, δ_j. The picture obtained is particularly clear if these variables are essential ones.

We introduce below several diagrams illustrating the course of a search. In these diagrams we mark with a circle or cross the points at the foot of a descent (A_i in fig. 1), the other points are on or at the top of the descent (X_i in fig. 1). Next to the points we write the corresponding values of the function being minimized.

[1] It is sufficient to change some of the coefficients in these formulae for the well organized function to turn into a collection of hillocks and ruts which it is difficult to traverse. It is interesting to note that such a case actually occurred in practical computation, because of a slip of the pen in writing the formulae defining the function.

Fig. 2 shows a typical case; the motion is along a ravine in the plane of the essential variables δ_1, δ_3, Here the value L of the ravine step is chosen to be 0.4, the gradient criterion $\Delta = 20\%$. In a search with such values of L and Δ, the values of the function at the bottom points have magnitudes of the order of a hundred, at the top points they are of the order of several hundred, while typical values of $f(\delta_1, \ldots, \delta_k)$ are of the order of 10^6.

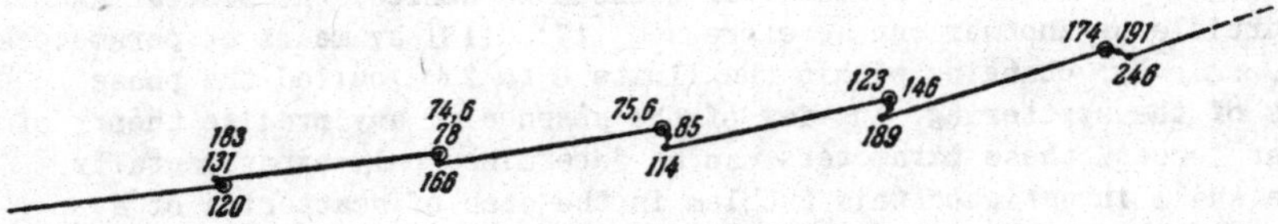

Fig. 2

In fig. 2 the adaptation is clearly visible. The lengths of the gradient descents decrease as one proceeds along the ravine and they become considerably shorter than the length of the ravine step. The difference between the values of the function at the top points and those at the bottom points also decreases. At first glance it may therefore appear as if the presence of the gradient descents generally plays no great role. But this is far from so. The gradient descents adjust the direction of the ravine motion to the structure of the function and continuously control its correctness. Leaving out some gradient descents is equivalent to increasing the length of the ravine step. It inevitably leads to increasing the values of the function obtained during the search. If the descents are not resumed at all, the search shortly turns into a banal motion across the plane and the values of $f(\delta_1, \ldots, \delta_k)$ thus computed will be of the order of 10^6.

Fig. 3 shows ravine motion at a bend skirting an elevation. This figure shows a ravine passing through two depressions with typical values of the function about 50 and going over a gentle rise with typical values of the order of 150-200. In this figure it is clearly seen that while

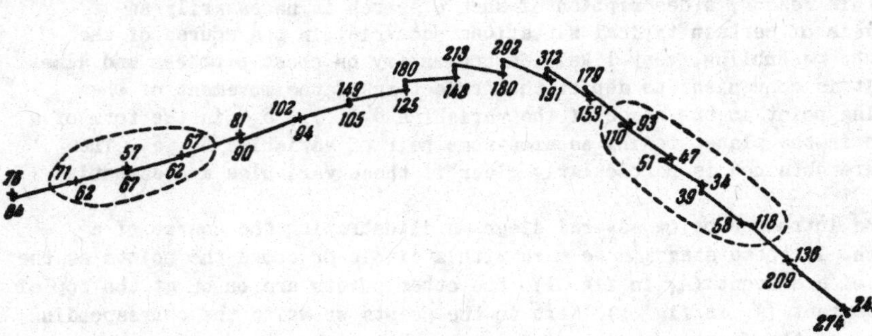

Fig. 3

skirting the elevation the lengths of the gradient descents increase. The difference between the values of the function at the top and bottom of a descent also increases. This property of the search, connected with a certain slowing down, illustrates the necessity of deciding on the direction of the further ravine motion. Essentially we have to do here with a change in the ravine behaviour under a change in the external conditions.

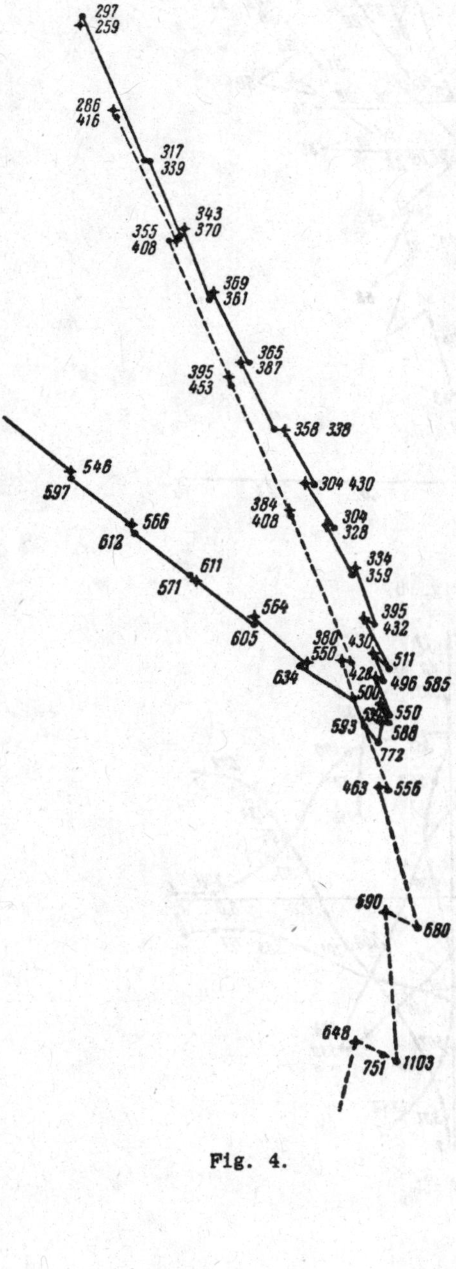

Fig. 4.

We remark again that the non-local method of search is camouflaged under stable conditions and becomes always less noticeable because of the increasing adaptation, and only a change in the external conditions reveals its existence and the mechanism of its action.

Fig. 4 shows the influence of the magnitude L of the ravine step on the course of the search. The continuous line in this diagram represents a ravine for which $L = 0.2$. It is clear that this ravine proceeds fairly straight at first with comparatively short gradient descents. Then the ravine meets a ridge and, making some long gradient descents, turns through a sharp angle and takes up a new direction.

In the same diagram the dotted line represents a ravine with $L = 0.4$. This value of the ravine step turns out to be too crude. The working point dashes over the ridge in a rush, not noticing the turn. By the way, one can see in the diagram that in the region where the ravine with the smaller step turns, there is an appreciable increase in the lengths of the gradient descents in the ravine with the large step. At the turn the ravine, as it were, adapts itself to the changing role of the inessential arguments.

Fig. 5 shows the behaviour of a pencil of ravines starting in a region of small values

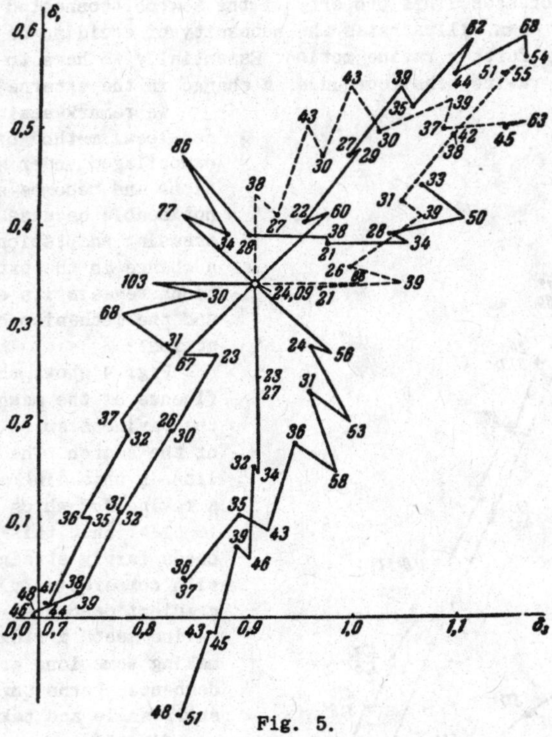

Fig. 5.

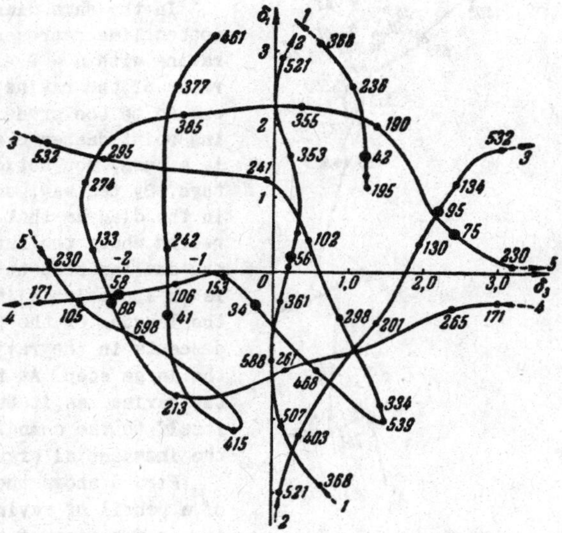

Fig. 6.

of the function $f(\delta_1, \ldots, \delta_k)$ and in eight different directions. Such a pencil of ravines allows one to obtain additional information about the structure of the function in the region being investigated. In particular one can see in this diagram that the eight ravines coming out have a tendency to be drawn out into two pencils in the direction $\delta_1 = \delta_3$.

Finally, in fig. 6, there is a schematic representation of a ravine motion, certain fragments of which have been shown in magnified form in fig. 3 and fig. 4 (continuous line). This ravine ($L = 0.2$, $\Delta = 20\%$) passes continuously through the space of the arguments $\delta_1, \ldots, \delta_k$ and crosses the low regions eight times. In these regions there are the minima of the function $f(\delta_1, \ldots, \delta_k)$ computed earlier, in the papers [21], [22], by the simplest non-local method.

§3. Tactics of movement

In this section we set forth certain preparatory considerations on the use of tactics in forming the movements of man and the higher animals. A summary of this section was published by us jointly with V.S. Gurfinkel' in [30]. Considering how unusual this material is, we thought it appropriate to give the mathematician reader a short account of certain facts about the physiology of movement.

First of all we remark that the control system for the movements of man and the higher animals is indeed complex. At any instant of time the position of some organ or other in this system is a function of an extremely large number of arguments.

Thus, for example, when a man walks, almost all the muscles of the body (some hundreds of them) are involved to some extent so that movements take place not only in the main joints of the lower extremities, but also in the numerous joints of the trunk and the upper extremities. The whole movement thus adds up to dozens of degrees of freedom and the work of muscles corresponding to nearly every joint.

Movement is effected under the direction of a control system which is specific to man and the higher animals, namely the central nervous system. This system proves capable of fairly rapidly processing great masses of information received by it and of forming the control signals, which are communicated along conducting paths to the effector endings (in this case the motor endings) and which in the end lead to contractions of some muscles or other. Here it is important to emphasize the possibility of simultaneous and independent control of a whole series of muscles.

Information is received by the central nervous system from numerous sensory nerve endings called receptors. The organs of movement, namely the muscles, tendons, joint capsules and ligaments, are especially rich in receptors. These receptors, known as proprioceptors, play an important role in the control system of movement, allowing information to be received concerning the angular position of the joints and the magnitudes of the muscular forces tending to alter these.

Also of great importance in the movement control system are those receptors perceiving influences generated in the outer environment. These exteroceptors, including the teloreceptors of sight, hearing and smell, enable one to take in the more general integrated characteristics of motor acts. The majority of these receptors are tuned organs. Their sensitivities are determined by the efferent (centrifugal) influences generated by the central nervous system.

Fig. 7 contains a sketch of a proprioceptor, the so-called muscle spindle. The intramuscular nerve trunk 1 contains motor fibres innervating the principal mass of the muscle fibres 9 and the fibres 2 going to the muscle poles of the proprioceptor. The sensory nerve fibres 10, 11 send the nervous system signals about the tension of the muscle. The frequency of these signals is determined by the degree of tension of the proprioceptor and also by its sensitivity. The sensitivity of the receptor is regulated by the central nervous system by means of the system of so called gamma efferents 2 which control the tension of the proper muscle apparatus of the receptor. Besides the muscle spindles sending signals about changes in the length of the muscle, there are still other types of proprioceptors, giving information about the muscle tension developed (the Golgi tendon organs), about the magnitude and direction of the mutual displacements of the parts of a joint and the speed of these displacements. The sensitivity of these receptors is fairly great; in the case of passive displacements in his joints, a man detects changes measured in fractions of a degree.

Thus the working parameters of the movement control system are the tensions of the muscles and perhaps the sensitivities of the receptors. But in order to estimate the results of the

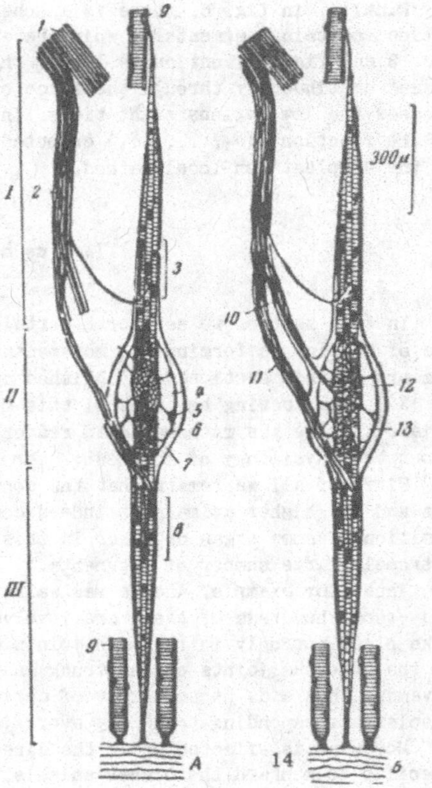

Fig. 7. Sketch of a proprioceptor (after D. Barker).

I – proximal pole; II – equatorial region; III – distal pole. 1 – intramuscular nerve trunk; 2 – spindle nerve trunk (sensory fibres cut short); 3 – motor endplates, innervated by a single large motor fibre; 4 – cluster of spherical nuclei (nuclear bag); 5 – cluster of oval nuclei; 6 – lymph space, bridged across by trabecular connective tissue; 7 – capsule; 8 – motor endplates innervated by three small motor fibres; 9 – extrafuzal muscle fibre; 10 – primary fibre in thick endoneurial sheath; 11 – secondary fibre; 12 – primary ending; 13 – secondary ending; 14 – aponeurosis.

movements in the course of realizing them, use is made of information from the numerous proprioceptors and exteroceptors.

The complexity of the movement system in man and the higher animals provides indeed astonishing reliability and precision in solving the various problems of movement.

Thus, for example, for a man living in the conditions of a modern city and having to make the journey twice a day to his place of work and back, the probability of his suffering an accident remains all the same insignifcantly small. By the way, it is quite impossible to imagine any technical device, whose operation would be so dependable that it could solve this problem reliably.

The precision of a man's movements is sometimes so great that attempting to estimate it leads to paradoxes.

It seems to us appropriate to introduce here an elementary example. The resolving power of the human eye is estimated to be about one minute of arc [23]. With the radius of the eye of the order of 1 cm. this gives a minimal distance of distinguishable images on the retina of the order of some thousands of a millimetre, which corresponds to the diameter of a photosensitive segment or element (cone). Consider now a simple problem of movement, typical for the game of billiards. With one ball of diameter about 5 cms. it is required to hit another such ball in such a way that the second ball will fall in a pocket whose diameter exceeds that of the ball by about 1 mm. Let us further assume that the distance between the balls and the distance from the second ball to the pocket are both, for example, 1 metre.

An elementary computation shows that the allowable error in the direction in which the cue strikes is an angle of not more than a few seconds, that is, at least an order of magnitude less than the least angle which can be detected by the human eye. During a discussion of this example with A.N. Kolmogorov, he drew our attention to the existence of just such a 'precision paradox' in shooting. It is certainly difficult to explain why it is all the same possible to play billiards and possible, in shooting, to hit the bull's eye.

Perhaps the paradoxical increase in precision occurs owing to the use of a suitable system of hypotheses. Thus, for example, throwing a circumference of large radius on to a millimetre mesh, one can determine its diameter with a precision considerably better than 1 mm. Here the use of the hypothesis that the figure in question is a circumference accounts for the 'precision paradox'. This hypothesis organizes the measurements and allows one to work up their results so that great precision is attained (in determining the centre of the corresponding distribution).

It is possible to suppose that the precision of a complex system of proprioceptors considerably exceeds the resolving power of the eye or the precision of a single proprioceptor. We have such a situation typically in the case of physiological measuring systems. Their precision considerably exceeds the precision of the separate elements on account of the use of a suitable system of organizing hypotheses.

The fundamental concepts and results concerning the physiology of movement are fully presented in a series of works [23]-[29] of which we wish to recommend especially the excellent monograph of N.A. Bernshtein.

Considering the structure of movement, N.A. Bernshtein introduced a series of useful concepts closely related to those used in the study of systems for automatic regulation. In particular, he studied in detail the question of the utilization of the information received from the various kinds of receptors for correcting the movement in the course of forming it (sensory correction). And he noted the important property of the movement control system that it has many levels. Of great importance also is the role of the so-called 'feedback', mentioned by P.K. Anokhin.

The movement control system functions effectively in complex, often rapidly changing circumstances.[1] One and the same problem of movement is often solved by many completely different methods. It is therefore hard to believe that movements are exactly programmed in advance. If indeed one takes into consideration the complexity of the dynamical system involved in executing a movement and the colossal diversity of the acts of movement themselves, then it turns out that a priori programming of movements entails the necessity of using a memory of gigantic size, and selecting from it would take up too much time.

From these considerations, one has to assume that the organism must form each new movement experimentally, selecting from the set of all possible combinations of muscle contractions those that enable it to attain its objective within an acceptable time and by means of a sufficiently economic method.

We hold that each movement is directed towards the accomplishment of some task and that at each moment of time there is a possibility of determining the nearness to its solution, owing to the use of information from the receptors. The totality of signals from the receptors is known in physiology as the afferentation.

Thus, the problem of constructing movements leads to a search for the regions of the working parameters that permit attaining the best approximation to the objective, and in this sense it recalls the problem described in §1 of finding an extremum of a function of many variables.

It seems to us likely that the large number of working parameters of the movement system are controlled by using the organization of this system through the application of hypotheses and of the corresponding tactics. Perhaps the tactics described above are suitable as models. But one must not forget that, even if essential features of these tactics are indeed present in physiological systems, the mathematical language, such expressions as, say 'gradient' or 'linear extrapolation' are of course completely foreign to physiology.

We must make one more important remark. It concerns the formulation of the movement problem itself. The fact is that, in forming movements, the specification of the very problem, of the objective of a particular act of movement, clearly arises in the process of the movement itself. (See also the first footnote to §1 above.)

In the course of carrying out a movement an enormous mass of diverse information is received from the receptors. All this information must be organized in short a time into a satisfactory evaluation system, approximation to which permits solving the movement problem. Working out such an

[1] Let us recall the complex set of problems of movement that appear in the course of athletic games.

evaluation system is impossible without using the organization, which the movement problem and the surrounding circumstances introduce into the mass of raw afferentation, thus determining which are the essential variables. It is the organization which thus arises that creates the possibility of applying the corresponding tactics.[1]

The similarity of the problem of forming movements and the problem of finding an extremum of a function of many variables allows us to apply certain of the concepts we introduced above in §1.

The method of blind search, as adapted to the problem of forming movements, consists in choosing combinations of muscle contractions at random until acceptable ones are found. It is clear that in a complicated situation the use of such tactics cannot ensure the setting up of a satisfactory movement within a reasonable time. Nevertheless the use of tactics of blind search is unavoidable in cases when there is no time for using more complex tactics.

When blind search is used, information about the results of a trial are not remembered. The characteristic times (duration of the choice, realization and evaluation of any single combination of muscle contractions) can be small. What is important is that during this interval of time it should be possible to form a combination of muscle contractions and to evaluate the results of the trial. The duration of such an elementary act of search is clearly about 0.1 sec. for man.

Local search methods, like those used for the problem of automatic optimization, can also be used in the process of forming movements. In the physiology of movement these methods are characterized by the fact that fixing on any combination of muscle contraction is accompanied by checking some nearby combinations. This checking permits determining a direction for further selection which will assure systematic improvement of the movement in the course of carrying it out. Taking decisions about the further choices on the basis of analyzing a series of nearby situations creates the necessity for remembering these situations. Since each of these decisions consists in a whole series of trials, the use of local tactics must be characterized by the presence of low frequency components along with high frequency components accompanying the elementary acts of search. Here it is again important to remark that the decisions are taken while the movement is going on, so that a relative slowness in taking decisions does not slow down the movement itself.

There are reasons for supposing that local tactics occur quite often in forming comparatively simple movements. The application of local tactics for solving more complex problems of movement is hampered by the slowness of the search, for the search goes from some combination of muscle contractions to another one nearby. Also of importance is the typical danger with local tactics, the danger of breaking off the search when some combination of muscle contractions is not sufficiently effective but there is no better one in its neighbourhood. We have already remarked on this above in §1.

[1] Such considerations arise not only in the investigation of the mechanism organizing the movement afferentation. We believe that the analysis of any somewhat complicated afferentation, e.g. of hearing or seeing, is connected with the use of tactics. We discussed these questions, especially in connection with the work of the acoustic analyzer, with G.V. Gershuni. It was in the course of these conversations that the point of view expressed here was formulated.

For more complex movements the most suitable model tactic is the ravine tactic. The hypotheses on which this tactic is based and its basic features have already been fully described above in §1. Therefore we shall here dwell only on some questions connected with the application of this tactic in forming movements.

The essential variables must here be understood to be a small number of muscle contraction combinations which are characteristic of the movement problem in question.

The method of non-local search consists in a random choice of some combination K_1 and a local improvement of it to a combination K_2. Just as above, it is important here that the local improvement should be carried on only so long as it is effective. that is, as long as the improvement in the results proceeds quickly encugh. The further search is carried on as follows. There is a random choice of a combination K_3, somehow related to K_1 but not lying in its immediate neighbourhood. Then a local improvement leads to K_4. The choice of K_5 depends on K_2 and K_4. It is not made at random, but by extrapolation in the direction of the better result. It is natural to call this extrapolation the ravine step, by analogy with §1. A local improvement of K_5 leads to K_6. Using K_4 and K_6, K_7 is chosen and is improved by a local method to K_8, and so on.

Thus, an alternation of locally improved values is characteristic of this non-local search tactic. Here each extrapolation step (ravine step) somehow disturbs the system in its inessential variables, and these are set right by means of local improvement. Therefore non-monotoneity of results is typical for non-local search. The combinations K_3, K_5, K_7, ... can be worse that the combinations K_2, K_4, K_6, ...

Just as in the problem of automatic optimization, in the course of the search the essential variables become more precisely distinguished. Moreover, the share of time spent on the local improvement decreases and the search becomes more rapid; adaptation takes place.

When the non-local search method is used, two kinds of decisions are taken. The decisions of the first kind concern the direction of the local improvement; they are typical of local search methods. The decisions of the second kind concern breaking off the local improvement and going over to the next ravine step; these decisions are characteristic of non-local search tactics.

Non-local tactics contain as elements local tactics and tactics of blind search. A frequency spectrum for non-local tactics must contain components typical of these simpler tactics as well as lower frequencies coming from taking decisions of the second kind.

Tactics typical for their characteristic times may also be connected with different levels in the formation of movements. Here the lower levels correspond to simple tactics and to smaller characteristic intervals of time. The application of more complex tactics is connected with the analysis of events spread over greater intervals of time. Such an analysis takes place on a relatively high level and may proceed relatively slowly. Its results (decisions) are used for the control of simpler tactics which are on lower levels. On the other hand, data about the carrying out of these simplest tactics will serve as information for the analysis.

Let us now consider certain experimental results which, it seems to us, help in judging whether various movement tactics are present.

These results were obtained by V.S. Gurfinkel' [31]-[33] when he investigated the mechanism by which a man maintains his body in an upright position.

Maintaining upright posture is a complex task related to the control of the kinematic system which has some tens of degrees of freedom.

This is a suitable subject for investigation, because the process of maintaining a vertical position can be satisfactorily described by a single function $r(t)$, the magnitude of the displacement from equilibrium position of the projection of the centre of gravity on a horizontal plane. Graphs of this quantity are called stabilograms. Analysis of stabilograms [31]-[33] shows that holding a vertical position is accompanied by continuous oscillation reflecting the man's activity directed towards maintaining equilibrium.

The frequencies of these oscillations in man are quite stable and scarcely change with changes in such biomechanical factors as weight (whether carrying a load or under conditions of partial weightlessness [34]), height and size of support. On the other hand, the frequencies of these oscillations change noticeably under such influences as insufficient oxygen (hypoxia), or taking chloral hydrate or alcohol, that is, changing the state of functioning of the nervous system. Proof that the oscillations of the projection of the centre of gravity are connected with muscular activity is given by direct synchronized measurements of $r(t)$ and of the electric potentials (electromyograms) of the muscles taking part in maintaining vertical position [31]-[33], [35].

The main background of the curves $r(t)$ consists of oscillations of small amplitude (of the order of 0.1 mm) and of frequency of the order of 8 to 10 oscillations per second. Against this background one distinguishes oscillations of greater amplitude (about 2 to 3 mm) and frequency about 1 per second. A third kind consists of slow oscillations of frequency about 1 to 3 per minute.

Thus the stabilogram frequency spectrum has typical maxima in the regions of 10, 1 and 0.05 oscillations per second. They give, it seems to us, an indication that complex search tactics are used in forming movements. These tactics are realized by a structure on many levels (three levels at least) containing parts of the nervous apparatus, distinguished by functional properties.

Other reasons indicating the use of non-local search methods arise out of analysis of the forms of the curves $r(t)$. A typical stabilogram is given in fig. 8. Here, first of all, it strikes one that the displacement from equilibrium has a general tendency to decrease, but in general it is not a monotone decreasing function of time. What is typical here is a relatively swift and smooth climb accompanied by a slow descent. On the descent one can clearly see the oscillations of relatively high frequency and small amplitude. Using the model expounded above to represent movement tactics, we venture to conjecture that the swift climb corresponds to the ravine step and the oscillations on the descent correspond to local improvements of equilibrium.

We hope that further data on the tactical structure of movements may be obtained from investigating disturbances to the working of the

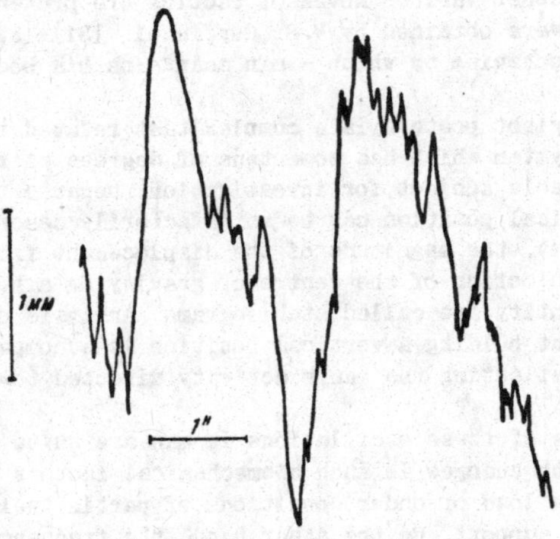

Fig. 8.

movements system with various diseases of the nervous system (Parkinson's disease, cerebellar tremor, etc.).

Received by the Editor, 20th July 1961.

References

[1] N. Wiener and A. Rosenblueth, Mathematical formulation of problem of conduction of impulses in cardiac muscle, Arch. Inst. Cardiol. Mexico 16 (1946), 205-265.

[2] I.M. Gel'fand and M.L. Tsetlin, Continuous models of control systems, Dokl. Akad. Nauk SSSR 131 (1960), 1242-1245; Soviet Math. Dokl. 1 (1960), 409-412.

[3] E.R. Caianiello, Outline of a theory of thought-processes and thinking machines, J. Theoret. Biol. 2 (1961), 204-235.

[4] I.S. Balakhovskii, The possibility of modelling the simplest acts of behaviour by a discrete homogeneous medium (in Russian), Probl. Kibernet, no. 5, (1961), 271-277.

[5] A.A. Fel'dbaum, Dual control theory I, Automat. i Telemekh. 21 (1960), 1240-1249; Automat. Remote Control 21 (1960), 874-880.

[6] A.A. Fel'dbaum, Dual control theory II, Automat, i Telemekh. 21 (1960), 1453-1464; Automat. Remote Control 21 (1960), 1033-1039.

[7] S. Ulam, A collection of mathematical problems, New York - London, 1960.

[8] A.A. Fel'dbaum, *Vychislitel'nye ustroistva v avtomaticheskikh sistemakh* (Computing equipment in automatic systems), Moscow, 1959.

[9] A.A. Fel'dbaum, Application of computing equipment in automatic systems (in Russian), Avtomat. i Telemekh. 17 (1956), 1046-1056.

[10] A.A. Fel'dbaum, Automatic optimalizer, Avtomat. i Telemekh 19 (1958), 731-743; Automat. Remote Control 19 (1958), 718-728.

[11] W.R. Ashby, *An introduction to cybernetics*, London, 1956; Russian translation, Moscow, 1960.

[12] G.V. Savinov, Electric modelling of homeostatic systems (in Russian), Probl. Kibernet., no. 4, (1960), 37-44.

[13] I.M. Gel'fand and M.L. Tsetlin, The principle of non-local search in automatic optimization systems, Dokl. Akad. Nauk SSSR 137 (1961), 295-298; Soviet Physics Dokl. 6 (1961), 192-194.

[14] M.L. Tsetlin, On the behaviour of finite automata in a random environment (in Russian), Avotmat. i Telemekh. 22 (1961), 1345-1354.

[15] M.L. Tsetlin, Some problems in the behaviour of finite automata, Dokl. Akad. Nauk SSSR 139 (1961), 830-833; Soviet Physics Dokl. 6 (1961), 670-673.

[16] H.S. Tsien, *Engineering cybernetics*, New York, 1954; Russian translation, Moscow, 1956.

[17] L.D. Landau and E.M. Lifshits, *Kvantovaya mekhanika*, vol. 1, Moscow-Leningrad, 1948; *Quantum mechanics, non-relativistic theory*, London-Paris, 1958.

[18] N.S. Mott and H.S.W. Massey, *The theory of atomic collisions*, Oxford, 1949; Russian translation, Moscow, 1951.

[19] V.A. Borovikov, I.M. Gel'fand, A.F. Grashin and I.Ya. Pomeranchuk, Phase shift analysis of *pp*-scattering at 95 Mev, Zhurnal Eksper. Teor. Fiz. 40 (1961), 1106-1111; Soviet Physics JETP 13 (1961), 780-784.

[20] I.M. Gel'fand, A.F. Grashin and L.N. Ivanova, Phase shift analysis of *pp*-scattering at an energy of 150 Mev. Zhurnal Eksper. Teor. Fiz. 40 (1961), 1338-1342; Soviet Physics JETP 13 (1961), 942-945.

[21] R.C. Stabler and E.L. Lomon, Proton-proton scattering phase shifts at 150 Mev, Nuovo Cimento 15 (1960), 150-152.

[22] H.P. Stapp, T.J. Ypsilantis and N. Metropolis, Phase-shift analysis of 310 Mev proton-proton scattering experiments, Phys. Rev. 105 (1957), 302-310.

[23] A.G. Ginetsinskii and A.V. Lebedinskii, *Kurs normal'noi fiziologii* (Course of normal physiology), Moscow, 1956.

[24] N.A. Bernshtein, Clinical methods in modern biomechanics (in Russian), Sb. trudov gos. inst. po usovershenstvovaniyu vrachei v Kazani (Collected works of the Kazan post-graduate medical institute), 1 (1929).

[25] N.A. Bernshstein, Problem der Wechselbeziehungen der Koordination und der Lokalisation (Russian with German summary), Archives Sci. Biol. 38 (1935), 1-34.

[26] N.A. Bernshstein, *O postroenii dvizhenii* (On the structure of movements), Moscow, 1947.

[27] N.A. Bernshstein, Some pressing problems on the regulation of motor acts (in Russian), Voprosy Psikhologii, no. 6, (1957), 70-90.

[28] P.K. Anokhin, Problems of the centre and the periphery (in Russian), Introductory article, Gorkii, 1935.

[29] P.K. Anokhin, *Problemy vyshei nervoi deyatel'nosti* (Problems of higher nerve activity), Moscow, 1949.

[30] I.M. Gel'fand, V.S. Gurfinkel' and M.L. Tsetlin, Tactical structure of movements, Dokl. Akad. Nauk SSSR 139 (1961), 1250-1253; Dokl. Biol. Sci. Sect. 139 (1961), 703-706.

[31] V.S. Gurfinkel', Standing of healthy people and of those provided with artificial limbs after amputation of the lower extremities (in Russian), Abstract of doctoral dissertation, Akad. Medits. Nauk, 1961.

[32] V.S. Gurfinkel', Materials on the significance of visual and vestibular apparatus in the steadiness of standing in man (in Russian), Sborn. III nauchn. Sessi Tsentral. N.-I- Inst. Protezirovaniya i Protezostroeniya (Proc. 3rd sci. session Centr. Artificial Limb Research Inst.), Moscow, 1953.

[33] V.S. Gurfinkel', The problem of steadiness of standing and its significance for prosthesis (in Russian), Sborn. II nauchn. sessi Tsentral. N.-I. Inst. Protezirovaniya i Protezostroeniya (Proc. 2nd sci. session Centr. Artifical Limb Research Inst.), Moscow, 1952.

[34] V.S. Gurfinkel', P.K. Isakov, V.B. Malkin and V.I. Popov, Coordination of human posture and motion in conditions of increased and reduced gravitation, Byull. Eksper. Biol. Med. 48, no. 11, (1959), 12-18; Bull. Exper. Biol. Med. 48 (1959), 1320-1325.

[35] F.A. Hellebrandt and L.E. Kelso, Synchronizing biplane stance photography with centre of gravity observations, Physiotherapy Rev. 22 (1942), 83-87.

Translated by C.H. Dowker.

4.

(with Yu. G. Fedorov and I. I. Piatetski-Shapiro)

Determination of crystal structure by the method of nonlocal search

Dokl. Akad. Nauk SSSR 152 (1963) 1045–1048 [Sov. Math. Dokl. 4 (1963) 1487–1490]

The basic problem in the x-ray analysis of the structure of crystals is to determine the coordinates of the atoms in the elementary cell of the crystal from the results of measurements on the intensities of scattering of the x-rays. Up to the present there exists no regular method of solving this problem. In the present note we shall show that it is possible to formulate the problem of determining the global minimum of a certain function of many variables which is such that knowledge of this minimum will also permit us to determine the coordinates of the atoms in the crystal cell.

The problem of finding the extremum of a function of many variables is in itself a very difficult problem of computation. The ordinary methods of local search (the gradient method, the method of steepest descent and the like) do not accomplish the aim. However, certain ideas about the "good organization" of the corresponding functions and a method, based on these ideas, of nonlocal search (the method of "ravines") [1] give us confidence in the effectiveness of the proposed computational method, which has already been successfully applied in problems of phase analysis [2, 3].

1°. The distribution of atoms in crystal substances is characterized by the electron density $\rho(x, y, z)$, whose maxima correspond to the positions of the atoms; the magnitudes of the maxima are proportional to the number of electrons in the corresponding atoms. The function $\rho(x, y, z)$ is periodic with respect to the arguments x, y, z, with periods a, b, c respectively. The domain $0 \le x \le a$, $0 \le y \le b$, $0 \le z \le c$ is called the elementary cell. For the sake of simplicity, we shall assume in what follows that $a = b = c = 1$.

Let us represent the function $\rho(x, y, z)$ by the Fourier series

$$\rho(x, y, z) = \sum_{h,k,l} F_{hkl} \exp[-2\pi i (hx + ky + lz)]. \tag{1}$$

From experiments we find only the quantities $F_{hkl}^{\text{exper}} = F_{hkl}$ (the absolute values of the structural amplitudes of the corresponding reflections), which obviously do not suffice to determine the function $\rho(x, y, z)$. Additional information is necessary regarding the structure of the function. It is customary to assume [4, 5] that the electron density of the substance contained in the elementary cell is representable in the form of the sum of the electron densities of the individual atoms or ions. Thus

$$\rho(x, y, z) = \sum_{i=1}^{N} \rho_i(x - x_i, y - y_i, z - z_i), \tag{2}$$

where the functions $\rho_i(\xi, \eta, \zeta)$ are known functions of their arguments (the electron densities of the individual atoms or ions) and N is the total number of atoms in the elementary cell. The coordinates x_i, y_i, z_i of the atoms are unknown and are subject to determination. In general, the number of atoms subject to determination in the elementary cell can be reduced if one takes into account the requirement of group invariance. However, this is not essential for our purposes, and we shall not make use of it in what follows.

It follows from (1) and (2) that

$$F_{hkl} = \sum_{j=1}^{N} f_j(h, k, l) \exp[2\pi i (hx_j + ky_j + lz_j)], \tag{3}$$

$$f_j(h, k, l) = \int \rho_j(\xi, \eta, \zeta) \exp[-2\pi i (h\xi + k\eta + l\zeta)]\, d\xi\, d\eta\, d\zeta. \tag{4}$$

The functions $f_j(h, k, l)$ are called atomic factors. There are detailed tables for them. Consider the function $\Phi_{\alpha\beta}$ of $3N$ variables, namely the coordinates of the atoms x_i, y_i, z_i; $i = 1, 2, \cdots, N$:

$$\Phi_{\alpha\beta}(x_1, y_1, z_1, x_2, \ldots, y_N, z_N) = \sum_{h,k,l} |F_{hkl}|^{\alpha} - (F^{\text{exper}}_{hkl})^{\alpha}|^{\beta}, \tag{5}$$

where F_{hkl} are the structural amplitudes defined in (3), F^{exper}_{hkl} is a set of experimental results for the absolute values of the structural amplitudes, and the summation is carried out with respect to the corresponding values of h, k, l. Thus the function $\Phi_{\alpha\beta}$ can be formed with respect to various sets of values of h, k, l. The parameters α and β can be variously chosen depending on the nature of the specific problem.

This function has a minimum at the point $(x_1, y_1, z_1, x_2, \cdots, y_N, z_N)$ corresponding to the desired position of the atoms in the cell. In general, the magnitude of this minimum differs from zero, and the minimum point itself corresponds to the desired position of the atoms only with a certain margin of error, which is determined by the accuracy of the experiment.

The function $\Phi_{\alpha\beta}(x_1, y_1, z_1, x_2, \cdots, y_N, z_N)$ has, as a rule, a complicated "poorly organized" structure with an immense number of minima, most of which do not correspond to the desired distribution of atoms. Therefore, solving the problem by means of the $\Phi_{\alpha\beta}(x_1, y_1, z_1, x_2, \cdots, y_N, z_N)$ is, generally speaking, extremely difficult.

2°. For a large class of crystal structures (for example, for crystals of many organic compounds), the problem of finding the desired minimum can be substantially simplified. In such crystals the totality of atoms in the elementary cell falls into a small number of parts, in each of which the atoms are connected with one another by rigid or semi-rigid bonds (the molecules). In other words, each of these parts can be considered as a rigid body, possibly with additional degrees of freedom. In this case, the position of the atoms in the elementary cell is given by a comparatively small number of parameters (the degrees of freedom of rigid bodies and the additional degrees of freedom). Let us denote the new variables by $\kappa_1, \kappa_2, \cdots, \kappa_p$ and express the coordinates of the atoms in terms of them. Set

$$\tilde{\Phi}_{\alpha\beta}(\varkappa_1, \varkappa_2, \ldots, \varkappa_p) = \Phi_{\alpha\beta}[x_1(\varkappa_1, \varkappa_2, \ldots, \varkappa_p), \ldots, z_N(\varkappa_1, \varkappa_2, \ldots, \varkappa_p)]. \tag{6}$$

As a result of its "natural organization", the function $\tilde{\Phi}_{\alpha\beta}$ is much simpler to construct than the function $\Phi_{\alpha\beta}$ originally chosen.

We propose to reduce the problem of the x-ray analysis of structure to the determination of the global minimum of the function $\tilde{\Phi}_{\alpha\beta}$ in the whole domain of variation of the variables $\kappa_1, \kappa_2, \cdots, \kappa_p$.

We observe that the function $\Phi_{\alpha\beta}$ (normalized in an appropriate manner) is frequently used in the x-ray analysis of structure [5]; for example, in the problem of determining the coordinates of the atoms more precisely and of testing the regularity of the structure. In this connection, it is usual to take the values $\alpha = \beta = 2$ or $\alpha = \beta = 1$. Note also that in these cases it is usual to employ for the construction of the function $\Phi_{\alpha\beta}$ all the measured reflections (100–1000 experimental data). Instead of this customary method of exploiting the function $\Phi_{\alpha\beta}$, we propose: first, to seek the minimum of the function $\tilde{\Phi}_{\alpha\beta}$, and not that of $\Phi_{\alpha\beta}$; second, to seek it in the large and not locally; third, in constructing the function $\widetilde{\Phi}_{\alpha\beta}$ for preliminary determination of the possible range of values of the minimum in the variables $\kappa_1, \kappa_2, \cdots, \kappa_p$, to use a relatively small number of reflections (considerably less than the total number ordinarily measured).

This last point is connected with the fact that the object (the molecule) whose position in the elementary cell we are determining, has considerably larger dimensions than the individual atoms.

It may turn out that the function $\widetilde{\Phi}_{\alpha\beta}$, formed with respect to a small collection of reflections, has several minima that "enjoy equal rights". The selection of the minimum that corresponds to the desired distribution of atoms involves a considerable increase in the number of reflections measured. But then we can also make use of supplementary (geometrical) conditions concerning the admissible arrangement of molecules with respect to one another. The more detailed use of these conditions we propose to describe in further publications.

3°. To determine the minima of the function $\widetilde{\Phi}_{\alpha\beta}$ we propose to use the method of nonlocal search (the method of "ravines"), which was first applied to problems of phase analysis [2, 3]. The basic concepts of the method of "ravines" ("essential" and "nonessential" variables, well-organized function, step along the ravine, gradient test etc.) are described in [1]. In this note, we give only a brief description of certain essential supplements to the method of search used earlier.

The total volume of calculations involved in solving the problem of finding a minimum for the function $f(x_1, x_2, \cdots, x_n)$ is, roughly speaking, proportional to the number of times the function f is calculated. To find the gradient of the function f it is necessary to compute its value $n+1$ times, which for large n constitutes a considerable number of operations. Finding a minimum of the function f on a straight line is much less "expensive". To do this, it is necessary to compute the values of f at 3 or 4 points of the straight line.

The methods suggested here allow us to reduce the total number of calculations of the function f and, what is essential, make it possible to judge the nature of the contour of the function f in a large domain of variation of the variables $x_1, x_2, \cdots, x_n$ by comparing the local structure of the function f at various points of that domain.

Let us recall that by points of a ravine or lowered points are meant those points at which a gradient descent is completed (the points A_i in Figure 1 of [1]). By departure points or nonlowered points for a movement along a ravine are meant those points at which a gradient descent begins (the points X_i in the same reference).

I. Let $A_i\,(\bar{x}_1^i, \bar{x}_2^i, \cdots, \bar{x}_n^i)$ be the last of the existing points of the ravine; $\mathbf{g}_1^i, \mathbf{g}_2^i \cdots, \mathbf{g}_k^i$ the sequence of directions of the descent by means of which that point was obtained from its nonlowered point $X_i\,(x_1^i, x_2^i, \cdots, x_n^i)$. Let $X_{i+1}\,(x_1^{i+1}, x_2^{i+1}, \cdots, x_n^{i+1})$ be a new departure point. For the descent from the point X_{i+1} let us use first the vectors $\mathbf{g}_1^i, \mathbf{g}_2^i, \cdots, \mathbf{g}_k^i$, that is, we decrease the value of the function $f(x_1, x_2, \cdots, x_n)$ successively along the straight lines given by the vectors $\mathbf{g}_1^i, \mathbf{g}_2^i, \cdots, \mathbf{g}_k^i$. From the point $(x_1^{i+1}, x_2^{i+1}, \cdots, x_n^{i+1})$ thus obtained we carry out in addition an ordinary gradient descent and obtain a new point of the ravine $A_{i+1}\,(\bar{x}_1^{i+1}, \bar{x}_2^{i+1}, \cdots, \bar{x}_n^{i+1})$. Let $\mathbf{h}_1, \mathbf{h}_2, \cdots, \mathbf{h}_m$ be the sequence of vectors of the additional gradient descent. From the aggregate of the vectors $\mathbf{g}_1^i, \mathbf{g}_2^i, \cdots, \mathbf{g}_k^i; \mathbf{h}_1, \mathbf{h}_2, \cdots, \mathbf{h}_m$ there are selected, according to some "reasonable" principle, the most "useful" vectors which we denote by $\mathbf{g}_1^{i+1}, \mathbf{g}_2^{i+1}, \cdots, \mathbf{g}_l^{i+1}$. These vectors we utilize for computation at the next point and so forth.

One can, for example, make use of such principles of selection as these:

1. From the $k+m$ directions of descent, the k "best" are chosen (according to the magnitude

of Δ, see [1]) and are utilized at the next point.

2. Of the $k+m$ directions we exclude the "nonresultant" ones; the remaining ones are utilized at the following point and are supplemented by at least one new direction.

In making the selection, it is also possible to take into account the usefulness of the given direction at several preceding points ("memory").

Thus, it is possible to let the number of vectors k vary at various points of the ravine. Important nonlocal information about the function f is given, for example, by the "nonremovability" of certain vectors from the aggregates $g_1^i, g_2^i, \cdots, g_k^i$.

II. With the vectors $g_1^i, g_2^i, \cdots, g_k^i$ let us span the subspace G_k^i (we shall assume that $g_1^i, g_2^i, \cdots, g_k^i$ constitutes an orthogonal basis in G_k^i). To compute the directions of descent from the new departure point $X_{i+1}(x_1^{i+1}, x_2^{i+1}, \cdots, x_n^{i+1})$, we use at first only vectors from G_k^i (that is, we find only the projections of the gradient in the complete space onto the subspace G_k^i). Here the gradient descent (we shall call it the descent in G_k^i) is made with its own gradient test. From the point $(\tilde{x}_1^{i+1}, \tilde{x}_2^{i+1}, \cdots, \tilde{x}_n^{i+1})$ thus obtained, we perform in addition the usual gradient descent and obtain the point of the ravine $A_{i+1}(\bar{x}_1^{i+1}, \bar{x}_2^{i+1}, \cdots, \bar{x}_n^{i+1})$. Let $\tilde{g}_1^i, \tilde{g}_2^i, \cdots, \tilde{g}_s^i$ be the vectors of descent in the subspace G_k^i; and let $h_1, h_2, \cdots, h_m$ be the vectors of the additional gradient descent. With the vectors $\tilde{g}_1^i, \tilde{g}_2^i, \cdots, \tilde{g}_s^i; h_1, h_2, \cdots, h_m$ we span the subspace G_l^{i+1} and in it take the orthogonal base $g_1^{i+1}, g_2^{i+1}, \cdots, g_l^{i+1}$ (using for this purpose as much as possible of the base $g_1^i, g_2^i, \cdots, g_k^i$). We use the new subspace G_l^{i+1} for the calculation at the following point, and so forth. In general, the dimensions of subspaces G_k^i and G_l^{i+1} almost always coincide and are not large in comparison with n, except for places of abrupt change in the contour of the function f. The character of the bases of the subspaces G_k^i contains essential information about the behavior of the function f in the large domain.

As a rule, for the additional gradient descents in these methods it is necessary to compute only 1 or 2 gradients, which essentially reduces the total volume of computation.

In conclusion, we point out that the proposed procedures for determining crystal structure were tested in a preliminary way on a simple known structure (naphthaline).

We express our thanks to Academician N. N. Semenov for having drawn our attention to the desirability of creating new direct methods in the x-ray analysis of structure. N. S. Andreeva, B. K. Vaĭnšteĭn, A. I. Kitaĭgorodskiĭ, M. A. Poraĭ-Košic, and A. A. Levin took part in the discussion of basic points in this article. An active part in carrying out the calculations was taken by L. N. Ivanova, S. L. Ginzburg, E. I. Dinaburg, and M. M. Voronovickiĭ. We express our profound gratitude to all of them.

Received 13/JULY/63

BIBLIOGRAPHY

[1] I. M. Gel'fand and M. A. Cetlin, Uspehi Mat. Nauk 17 (1962), no. 1 (103), 3. MR 25 #1081.

[2] I. M. Gel'fand, A. F. Grašin, I. Ja. Pomerančuk and V. A. Borovikov, Ž. Èksper. Teoret. Fiz. 40 (1961), no. 4.

[3] I. M. Gel'fand, A. F. Grašin and L. N. Ivanova, ibid. 40 (1961), no. 5.

[4] A. I. Kitaĭgorodskiĭ, *x-ray analysis of structures*, Moscow, 1950. (Russian)

[5] ———, *The theory of structural analysis*, Izdat. Akad. Nauk SSSR, Moscow 1957. (Russian)

Translated by:
Martin Chancey

5.

(with Yu.G. Fedorov, R.A. Kayushina, and B.K. Vainstein)

Determination of crystal structures by the method of the *R*-factor minimization

Dokl. Akad. Naūk SSSR **153** (1963) 93–96

X-ray studies of crystal atomic structures are based on the measurements of intensities I_{hkl} of *X*-ray reflection from crystals; from these intensities one immediately finds the squares of the moduli of the structure amplitudes $|F_{hkl}|^2$. One usually agrees that in the first stage of the structure determination one finds a preliminary model in which all atoms, or most of them, are placed in positions that differ from the proper ones by not more than 0.1–0.2 Å. In the second stage the structure is determined more precisely.

As a criterion for the correctness of a preliminary model one usually takes the value of the reliability factor (*R*-factor)

$$R = \frac{\sum_{k,h,l} ||F_{hkl}| - |F_{hkl}|_{\text{exper}}|}{\sum_{h,k,l} |F_{hkl}|_{\text{exper}}}, \tag{1}$$

which should be of order 15–20%. The *R*-factor *R* is a function of the 3*N* coordinates of the atoms in the cell in the expression

$$F_{hkl} = \sum_{j=1}^{N} f_j \exp 2\pi i(hx_j + ky_j + lz_j). \tag{2}$$

A long time ago several authors [1, 2] had already expressed the idea that the function *R* itself, or similar functions, can be used in the search for a preliminary model, because the proper structure corresponds to the minimum of *R*. However, the application of known optimization methods to be function *R* or to other similar functions is only useful in solving the problem of more precise structural determination if a preliminary model has already been found [3, 4]. On the other hand, it has not been possible up to now to apply either these methods, or any other method, to the determination of a preliminary model by means of the function *R*, the reason being that we still do not have a general method for finding a minimum of a function in many variables.

In [5, 6] the "ravine" method or the method of nonlocal search for the determination of the global minimum of a function in many variables was developed; it was also indicated how to apply this method to problems of structural analysis. The idea of the method is to quickly find regions where the function takes on small values and then to study these regions. It is essential here that the function be "well organized". Let us note that the ravine method contains two important numerical parameters: "the gradient try" and "the step size along the ravine"; the choice of their values depends on the specific minimized function. After this choice is made all the computations are completely left to the computer.

Instead of the function R itself, which depends on $3N$ variables and has a rather complicated structure, it is reasonable to use a "parametrized" function R depending on a smaller number p of variables [6]. The parametrization is performed most easily for those molecular structures where the shape of a molecule is known beforehand. In this case one can take for the parameters the usual parameters of a solid, namely the coordinates of the center of the molecule (taking any of its atoms as the center) and the Euler rotation angles. Expressing the coordinates of all atoms in these parameters and substituting into (1) and (2) we get the "parametrized" function $\tilde{R}$.

It is essential here not only that $\tilde{R}$ depend on a small number of variables but also that it "behaves better", because this function already accounts for our knowledge of the mutual positions of the atoms.

The number of reflections required for the preliminary search depends on the number of independent variables in the function $\tilde{R}$. It appears that if the reflections are selected reasonably then at the beginning of the search it is convenient to take approximately $7p$–$10p$ reflections. In selecting the reflections one has to take into account the following. Reflections with small values of h, k, l correspond to low harmonics in the Fourier decomposition, and so they are less sensitive to inaccuracies in determination of the coordinates of atoms in a model of the molecule. It is therefore reasonable to use them in the first stage of the search. The search should include strong reflections containing the largest amount of information, whereas weak reflections may be excluded from the search. However the intensity of some strong reflections might be decreased significally because of extinction, so we impose on them the following restriction: if $|F_{hkl}|_{\text{exper}} < |F_{hkl}|$ at some point then the term $||F_{hkl}| - F_{hkl}|_{\text{exper}}|$ in the function $\tilde{R}$ should be given the value 0 at this point.

In this choice process and in the estimation of minima, one has to increase the number of reflections in the function $\tilde{R}$ near these minima. Moreover, in the choice of minima one can use the criterion of admissible intermolecular distances [7]. We have constructed a function that helps to exclude those minima of $\tilde{R}$ that do not satisfy this criterion.

We have chosen naphthalene ($C_{10}H_8$), with the spatial group $P2_1/a$, $z = 2$ [8], as the first test structure. The position of a centrally symmetric naphthalene model is determined by three angular parameters. The use of 30 reflections clearly distinguished two mutually symmetric minima that correspond to the real positions of naphthalene molecules in a crystal.

The possibilities of the new method were also tested on the already known non-centrally-symmetric structure of hydroxyproline ($C_5H_9O_3N$) [9], with the spatial group $P2_12_12_1$, $z = 4$. The search from some arbitrary position of a molecule in a cell, first with 40 ($\sin\theta/\lambda < 0.35$) and then with 64 reflections ($\sin\theta/\lambda$ 0.4), and with the use of extinction restrictions, gave 4 symmetrically placed minima regions that correspond to the true structure. The function $\tilde{R}$ in this case was more complicated than for naphthalene.

It is important to note that in constructing the function R (for naphthalene as well) we used the true positions of atoms in a molecule that were already known with high precision. Of course, in the case of an unknown structure this cannot be done. However, small distortions of a molecule resulting in small shifts of atoms from their true positions in a molecule (up to 0.1–0.2 Å) did not shift the regions of minima of the function $\tilde{R}$, but only somewhat increased its value. At the same time greater distortions might lead to the distorian of the "true" form of the function.

Finally, the method of nonlocal search was used for the determination of the previously unknown structure of one of the natural amino acids, namely of l-proline ($C_5H_9O_2N$). An elementary cell has $a = 11.44$ Å, $b = 9.02$ Å, $c = 5.20$ Å, and the spatial group is $P2_12_12_1$, $z = 4$ (this agrees with data from [10]). The values of approximately 500 moduli of structure constants were obtained with a precision of about 10–15%. For the parametrization we used the model obtained earlier in the determination of the structure of $Cu(C_5O_9N)_2 2H_2O$ [11] (see Fig. 1). Besides 6 parameters of the solid (coordinates x, y, z of the atom C_2 and Euler angles $\varphi_1, \theta, \varphi_2$) we introduced one more angular parameter χ characterizing the rotation of the carboxy group $C_1O_1O_2$ around the axis C_1C_2. Other atoms of a molecule were fixed (the atom C_4 was in trans-conformation to $C_1O_1O_2$ with respect to the pyrolidine ring plane).

The initial search was performed with 30 reflections, $\sin\theta/\lambda < 0.2$. It gave many minima which could not be easily distinguished from one another. Increasing the number of reflections to 60 ($\sin\theta/\lambda < 0.35$) by including most of the strong reflections, we found a wide region with low values of the function $\tilde{R}$ (of order 30%). This region appeared to be unacceptable for crystallochemical reasons. Then we introduced into the search the intermolecular distance function mentioned above. However, the joint use of this distance function and of the $\tilde{R}$-function did not allow us to leave this wide region.

This phenomenon seemed to be related to wrongly fixing one or several atoms in the initial model (the oxyproline example shows that the inaccuracy should exceed 0.2–0.3 Å). So we introduced one more parameter, namely the displacement of the atom C_4 with respect to the ring plane (because for structures containing

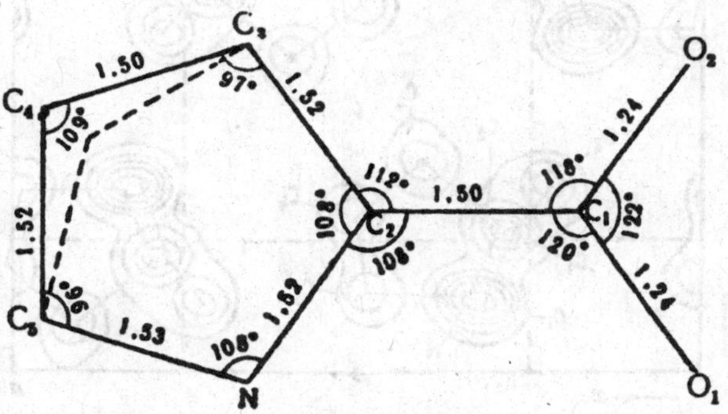

Fig. 1. The original model of l-proline. The C_4-atom is in trans-conformation to $C_1O_1O_2$ with respect to the pyrolidine ring plane.

proline different positions of the C_4-atom were indicated). We added 10 new strong reflections with the above extinction restriction. This led to a minimum with $\tilde{R}_{200} = 32\%$ (for 200 reflections). and this minimum was acceptable from the point of view of intermolecular distances. The C_4-atom appeared to be in the cis-conformation. Attempts to improve this structure using the least squares method did not lead to an increase of the R-factor. The construction of two-dimensional Fourier series and the computation of R separately for $hk0$, $h0l$ and $0kl$ reflections have shown that the orientation of the molecule and the yz-coordinates of its center seem to be determined correctly. The movement of the ravine along the x-axis gives a deeper minimum in which $\tilde{R}_{200} = 20\%$ for 200 reflections and $\tilde{R} = 29\%$ for all reflections. The width of $\tilde{R}_{200} \leqq 25\%$ of this point is about 0.3 Å. Contour lines $\tilde{R}_{200} = 25\%$ and $\tilde{R}_{200} = 30\%$ in the planes xy and $\varphi_1\varphi_2$ characterizing the shape of the function $\tilde{R}_{200}$ near the minimum are shown in Fig. 2a, b. In this region we

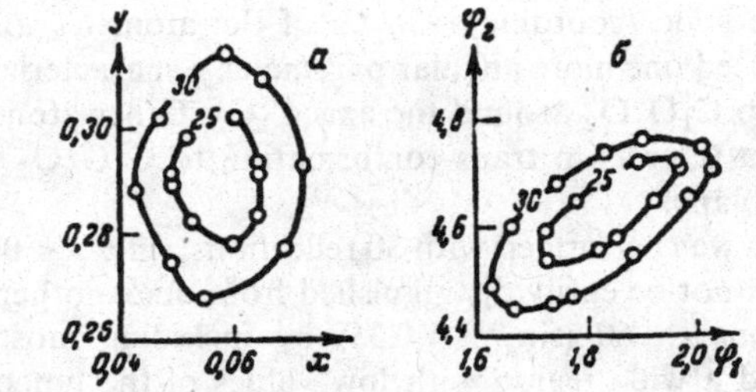

Fig. 2. (a) Contour lines of the function $\tilde{R}_{200}$ on the plane xy (x and y are measured in fractions of the cell parameter); (b) contour lines of the function $\tilde{R}_{200}$ on the plane $\varphi_1\varphi_2$ (φ_1 and φ_2 are measured in radians).

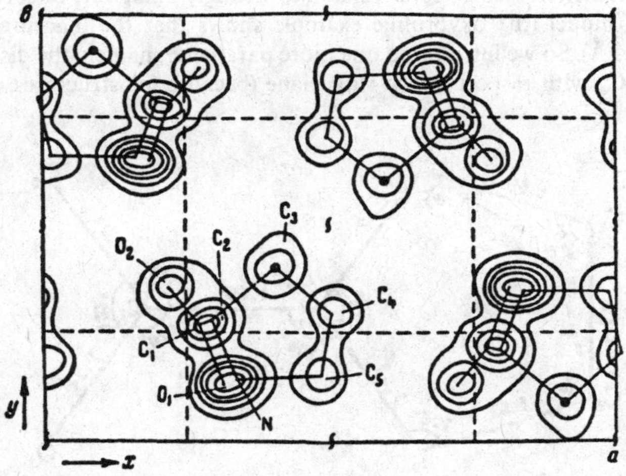

Fig. 3. The projection of electronic density on the plane xy (signs are computed by coordinates of the model with $R = 20.7\%$). Contour lines are drawn with spacing 4 el/Å^2.

have used the ravine method to make the positions of all atoms more precise by going from $\tilde{R}$ to R. This method gave $\tilde{R} = 21\%$ for 420 reflections and $R = 23.5\%$ for all reflections. This completes the search by the ravine method.

The additional criteria we used for the choice of minima and for the fixation of the final solution were the following: (a) the fulfilment of crystallochemical laws; (b) the convergence of the least squares method; (c) the analysis of projections of electronic density (the absence of false peaks, and so on) and separate computation of R in different zones.

Projections of the electronic density on the *xy*-plane for the model with $R = 20.7\%$ are shown in Fig. 3.

It is interesting that there exists, apparently, the possibility of finding the correct orientation of a molecule when its position in a cell is still wrong (or, more precisely, irrespective of this position). This can be shown as follows. The quantities $|F_{hkl}|^2$ are Fourier coefficients of a function of interatomic distances. If the orientation of a molecule is correct we get the equality of all intermolecular atomic vectors corresponding to peaks of the Patterson function. This means that in this case the computed quantity $|F_{hkl}|$ (2) has a correct component, whereas for the wrong orientation this component is incorrect. Therefore the correct orientation corresponds to the minimum of the function $\tilde{R}$ in angle variables.

Thus, the minimization of the *R*-factor by the ravine method appears to be very effective in the search of preliminary models of molecular structures. One might assume that further improvements of this method will enable it to be applied to more complicated structures.

The authors express their gratitude to A.I. Kitaigorodskii for the suggestion that we use the criterion of intermolecular constants, and to I.I. Pyatesky-Shapiro for valuable advice and discussions on related matters. The authors also thank L.N. Ivanova and S.L. Ginsburg for their active help in the development of the method and for the computations of structures of hydroxyproline and of *l*-proline.

Received 16.VII.1963

References

1. Booth, A.D.: Nature **160** (1947) 196
2. Vand, V., Niggley, A, Repinsky, R.: Acta Crystallogr. **13**, **12**, (1960) 1001, 1002
3. Porai-Koshitz, M.I.: Practical Course of *X*-ray Analysis, 1960
4. Vainstein, B.K.: Structural Electronography, 1956
5. Gelfand, I.M., Tsetlin, M.L.: Usp. Mat. Nauk **17**, no. 1 (1962) 103
6. Gelfand, I.M., Pyateski-Shapiro, I.I., Fedorov, Yu.G.: Dokl. Akad. Nauk **152**, no. 5 (1963)
7. Kitaigorodskii, A.I.: Organic Crystallochemistry, 1955
8. Abrahams, S.G., Robertson, J.M., White, J.G.: Acta Crystallogr. **5** (1959) 238
9. Donohue, J., Trueblood, K.N.: Acta Crystallogr. **5** (1952) 414
10. Wright, B.A., Cole, P.A.: Acta Crystallogr. **2** (1949) 129
11. Matthieson, D.M., Welsh, H.K.: Acta Crystallogr. **5** (1952) 599

6.

(with I. I. Piatetski-Shapiro and M. L. Tsetlin)

On certain classes of games and automata games

Dokl. Akad. Nauk SSSR **152** (1963) 845–848. [Transl., II. Ser., Am. Math. Soc. **87** (1970) 275–208]. MR **28**: 1068. Zbl. **137**:143

Problems of the collective behavior of automata naturally arise in the construction of mathematical models of the simplest forms of collective behavior. The collective behavior of automata is generated by their interrelationship. A convenient way of expressing this relationship is the language of the theory of games [1, 2]. The application of this language, which restricts the class of forms of behavior studied, leads, however, to the construction of a number of interesting models. In the construction of such models we have made use of construction of automata having appropriate behavior in stationary random means (in "games with nature"), and also of the definition of games between automata, given in the papers [6, 7]. A zero-sum game between two automata was studied in the papers [5, 7].

Attempts to construct games simulating comparatively uncomplicated forms of collective behavior lead naturally to the distinguishing of certain relatively simple classes of games. This note is devoted to the description of such classes and the concepts arising in connection with them.

1. Suppose that N players $A_1, \cdots, A_N$ participate in the game Γ. The player A_k, $k = 1, \cdots, N$, has available n_k strategies $f_1^k, \cdots, f_{n_k}^k$. A play f of the game Γ is a collection $f = (f_{i_1}^1, \cdots, f_{i_N}^N)$ of strategies chosen by the players. There are given N functions $m_k(f)$ on the set of plays. The function $m_k(f)$ is said to be the payoff function for the player A_k and has the meaning of an amount of winnings (mathematical expectation of winnings) of the player A_k in the play f.

In the study of the behavior of a collection of a large number of players it is natural to distinguish classes of games for which the payoff functions depend only on a small number of parameters. Consider in particular games in which the payoff function for each player is determined only by the choice of a small number of players—his "neighbors". We present examples of such games.

Example 1. *Games on the circumference.*

1a. The payoff function depends on the player's own strategy and that of his left neighbor.

1b. The payoff function depends on the player's own strategy and on the strategies of his left and right neighbors.

Example 2. *A game on the plane.* In the game there are players $A_{11}, A_{12}, \cdots, A_{NN}$. The payoff to the player A_{ik} is determined by his strategy and by the strategies of the players $A_{i+1,\,k}$, $A_{i-1,\,k}$, $A_{i,\,k+1}$ and $A_{i,\,k-1}$.

One can define an analogous game by using an arbitrary difference scheme.

Games with a bounded number of players may conveniently be given using special graphs of the game. For this the player A_k is assigned the kth vertex of the graph. If the payoff $m_k(f)$ depends on the strategy of player i, then one draws an arrow from the vertex i to the vertex k.

Graphs of games on the circumference (Example 1) are presented in Figure 1 (a and b).

2. Suppose that in the game Γ there is a play f^0 in which none of the players can profitably change his strategy given that the other players do not change theirs. We shall call such a game a Nash game, and the play f^0 a Nash play.* If $f^0 = (f^1_{i_1}, \cdots, f^N_{i_N})$, then the inequality

$$m_k(f^0) \geqslant m_k(f^1_{i_1}, \ldots, f^{k-1}_{i_{k-1}}, f^k_j, f^{k+1}_{i_{k+1}}, \ldots, f^N_{i_N}) \tag{1}$$

holds for all $k = 1, \cdots, N$ and $j = 1, \cdots, n_k$.

Example 3. There are two participants A_1 and A_2, each having two strategies 1 and 2. Denote by m_{ik} the payoff of a player using his ith strategy against the kth strategy of his opponent. If $m_{11} = m_{22} = 0.25$, $m_{12} = 0.9$, and $m_{21} = 0.1$, then the play $f = (1, 1)$ is a Nash play. Indeed, changing the strategy of either player changes his payoff from 0.25 to -0.1.

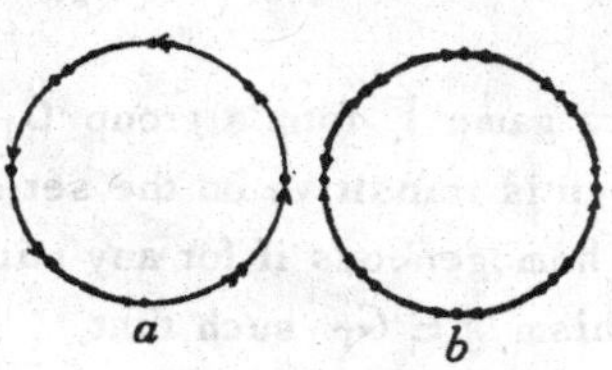

Figure 1

Now suppose that in a Nash game the players may choose their strategies in such a way that the payoff functions are simultaneously maximized for all players. Such a game is called a K-game. In K-games there exists a play f_0 such that $m_k(f_0) \geq m_k(f)$ for all $k = 1, \cdots, N$ and all plays f. The play f_0 will be called a K-play.

---

* *Translator's note.* Usually called a Nash equilibrium point, after John Nash's paper, *Non-cooperative games*, Ann. of Math. (2) 54 (1951), 286–295.

K-games admit a full solidarity of the interests of all players. We present the simplest example of such a game.

Example 4. Suppose that $m_k(f) = i_1 + \cdots + i_N$, where $f = (f^1_{i_1}, \cdots, f^N_{i_N})$. Obviously $f^0 = (f^1_{n_1}, \cdots, f^N_{n_N})$ is a K-play.

For any game it is possible to construct a K-game corresponding to it as follows. Suppose that $m_k(f)$ is the system of payoffs of some game Γ. Then the payoff functions $M_k(f)$ of the K-game Γ_K are defined by the relations

$$M_k(f) = \frac{1}{N} \sum m_j(f), \quad k = 1, \ldots, N. \tag{2}$$

The functions $M_k(f)$ coincide for all players and therefore simultaneously reach a maximus on some K-play. We shall return to this procedure again in the consideration of games between automata.

3. In this section we consider a class of games in which all the participants are equal—the so-called homogeneous games. A homogeneous game is given completely by one payoff function, which essentially simplifies the description. Particularly interesting are the homogeneous games with a small number of "neighbors".

We present some definitions. We will say that a mapping of the game g onto itself is given, if there are given: 1) a one-to-one mapping of the set of players onto itself, $gA_i = A_j$; 2) for each i a one-to-one mapping $f^i_k \longrightarrow f^{gi}_{gk}$ of the set of strategies of player A_i into the set of strategies of player A_{gi}. The mapping g defines in a natural way a mapping $f \longrightarrow gf$ of the set of plays into itself.

A mapping g which preserves the payoff functions, i.e. $m_k(f) = m_{gk}(gf)$, is called an *automorphism* of the game.

It is easy to see that the automorphisms of a game Γ form a group G_Γ. The game is said to be *homogeneous* if this group is transitive on the set of players. In other words, the game is said to be homogeneous if for any pair of players A_i and A_k there exists an automorphism $g \in G_\Gamma$ such that $g(A_i) = A_k$.

4. In this section we consider homogeneous games between identical automata, called in what follows homogeneous automata games. An automata game, understood in the sense of [5-7], consists of multiply repeated plays. Each of the participants in the automata game obtains only information on his own payoff or loss in a given play. This information is used for the choice of actions (strategies) in the succeeding plays. One supposes that the automata do not have a priori information on the game in which they are participating. The payoff functions of the game describe the interactions

of the automata. For ergodic automata games there exists a final probability distribution of plays. If $p(f)$ is the final probability of the play f in an ergodic homogeneous automata game Γ, then $p(gf) = p(f)$, so that in a homogeneous automata game the mathematical expectations of the payoffs of all the automata coincide.

In a homogeneous game, along with each play f it is natural to consider the set $\{gf\}$ of all plays of the form gf, $g \in G_\Gamma$, which is invariant relative to the group G_Γ. For each player his average payoff $U(f)$ on such an invariant set is equal to the arithmetical average payoff of all players on the play f. We shall call the quantity $U(f)$ the value of the play. The maximal value of a play will be called the maximal payoff, and the minimal value of a play the minimal payoff. Thus, for example, the game of Example 3 is homogeneous and has 3 invariant sets. The first of them consists of the play (1, 1), the second of the play (2, 2), and the third of the plays (1, 2) and (2, 1). The maximal payoff is equal to 0.4 and the minimal 0.25.

In a simulation of this example on a computing machine the mean payoff, for a sufficient memory capacity, was close to 0.25. Generally speaking, automata receiving information only on their payoffs and losses at separate plays of the game do not obtain the maximal payoff. The point is that if the play f on which the maximal payoff is achieved is not a Nash payoff, then at least one of the players changes his strategy, and the common payoff decreases.

Now we consider the invariant set of plays generated by the Nash play. If the maximal payoff is achieved on this set, then the set is said to be a Mora set, and the plays generated by it mora plays.

The game of Example 2 does not have mora sets. However, the maximal payoff 0.4 would be achieved if the players agreed among themselves to play only the plays (1, 2) and (2, 1) with equal probabilities.

For homogeneous automata games one may propose a procedure which in a certain sense replaces such an arrangement. This procedure, which we call "common account",* is equivalent to the construction relative to a game Γ of a homogeneous game Γ' with the same players and strategies as Γ, and in which the payoff functions are defined by formula (2). A game Γ' with common account may be understood as a game Γ in which the automata receive information not only on their own payoffs, but also on the payoffs of

* The definition of "common account" [обшая касса] arose in connection with models in the work of M. L. Cetlin, S. L. Ginzburg and B. Yu. Krylov [8]:

all the other playing automata, and use this information in order to determine their behavior. Such a large quantity of information on the game makes it possible for the automata to raise their payoffs up of the maximal. In the game Γ' at each play the payoffs of all the players are the same. Therefore in the game Γ' there necessarily exists a mora point [8].

For the game Γ described in Example 2, the game Γ' is given by the following payoff functions: $m'_{11} = m_{11} = 0.25$, $m'_{22} = m_{22} = 0.25$, $m'_{12} = m'_{21} = 0.4$. In simulation on a computing machine the maximal payoff 0.4 was achieved, given that the memory of the playing automata is sufficiently large. Other examples of achieving the maximal payoff by application of "common account" are presented in the paper [8].

In a number of important cases the maximal payoff is achieved if the play contains a mora point and the automata form an asymptotically optimal sequence in the sense of [3]. If on the other hand the game does not have a mora point, then the average payoff of the players coincides with the maximal value of a Nash play.

5. Now we shall describe homogeneous games on the circumferences whose graphs were depicted in Figures 1a and 1b. Each of the players has a choice of two actions 1 and 2.

For the game of Figure 1a the payoff of each of them is determined by his strategy and the strategy of the left neighbor: $m_k(f^1_{i_1}, \cdots, f^N_{i_N}) = m_k(f^k_{i_k}, f^{k+1}_{i_k+1})$. If the number N is even, this game always has a Nash play. Indeed, the numbers $M(\lambda_1, \lambda_2)$, $\lambda_1, \lambda_2 = 1, 2$, satisfy at least one of the following groups of inequalities:

$$\text{I. } m(1,1) > m(2,1). \quad \text{II. } m(2,2) > m(1,2).$$
$$\text{III. } m(1,1) < m(2,1) \text{ and } m(2,2) < m(1,2).$$

For case I the Nash play has the form $f^0 = (1, 1, \cdots, 1)$. For case 2, $f^0 = (2, 2, \cdots, 2)$. For case 3, $f^0 = (1, 2, 1, 2, \cdots, 1, 2)$.

For the game whose graph is given in Figure 1b the payoff of each of the players is given by the formulas

$$m_k(f^1_{i_1}, \ldots, f^N_{i_N}) = m(f^{k-1}_{i_{k-1}}, f^k_{i_k}, f^{k+1}_{i_{k+1}}),\ k = 1, \ldots, N, \tag{3}$$

and, accordingly, the game is determined by the weight numbers $m(\lambda_1, \lambda_2, \lambda_3)$, where $\lambda_i = 1, 2$, $i = 1, 2, 3$. This game may also fail to have a Nash play.

A sequence $f = (\lambda_1, \cdots, \lambda_N)$ which is a Nash play must have the following properties: if in the sequence there occurs a triple $\lambda_1, \lambda_2, \lambda_3$, then there is in it no triple $\lambda_1\mu\lambda_3$, where $\mu \neq \lambda_2$. From this remark it follows that a Nash play can only have the following form: either 1) all the λ are equal, or 2) the 1 appears only one at a time, and the 2 not more than twice running, or, finally, 3) the other way around.

BIBLIOGRAPHY

[1] R. Luce and H. Raiffa, *Games and decisions: Introduction and critical survey*, Wiley, New York, 1957; Russian transl., IL, Moscow, 1961. MR 19, 373.

[2] J. C. C. McKinsey, *Introduction to the theory of games*, McGraw-Hill, New York, 1952; Russian transl., IL, Moscow, 1960. MR 14, 300.

[3] M. L. Cetlin, *On the behavior of finite automata in random media*, Avtomat. i Telemeh. 22 (1961), 1345–1354 = Automat. Remote Control 22 (1962), 1210–1219. MR 25 #4973.

[4] ———, *Finite automata and the simulation of the simplest forms of behavior*, Uspehi Mat. Nauk 18 (1963), no. 4 (112), 3–28 = Russian Math. Surveys 18 (1963), no. 4, 1–27. MR 28 #2951.

[5] ———, *Remark on a finite-automation game with an opponent using a mixed strategy*, Dokl. Akad. Nauk SSSR 149 (1963), 52–53; English transl., Amer. Math. Soc. Transl. (2) 87 (1970), 271–274. MR 26 #3531.

[6] M. L. Cetlin and V. Ju. Krylov, *Examples of games with automata*, Dokl. Akad. Nauk SSSR 149 (1963), 284–287 = Soviet Physics Dokl. 8 (1963), 232–234. MR 29 #6945.

[7] V. Ju. Krylov and M. L. Cetlin, *On games for automata*, Avtomat. i Telemeh. 24 (1963), 975–987 = Automat. Remote Control 24 (1963), 889–899. MR 27 #3482.

[8] S. L. Ginzburg, V. Ju. Krylov and M. L. Cetlin, *One example of a game for many identical automata*, Avtomat. i Telemeh. 25 (1964), 668–671 = Automat. Remote Control 25 (1964), 608–611.

Translated by:
J. M. Danskin

7.

(with V. S. Gurfimkel, M. L. Shik, and M. L. Tsetlin)

Some problems in the analysis of movements

In the collection: Models of structure-functional organisation of certain biological systems, pp. 264–276, Moscow: Nauka 1966 (In: Models of the Structural-Functional Organization of Certain Biological Systems, pp. 329–345, Cambridge, Massachusetts, London: MIT Press 1971)

This article serves as a type of preface to data dealing with an experimental investigation of movements, but not in the slightest degree is it a systematic review of the literature on this question. We wanted to set forth in it some general concepts which seemed to us helpful in the pursuit of the physiology of motor activity, and which assist us in choosing the direction of further work. Many of these views are closely connected with studies on the structure of movements as developed in the remarkable works of N. A. Bernstein [2–4].

The control of movements is one of the most important functions of the nervous system. The structure and function of the nervous system undoubtedly is determined to a large extent by this problem. The physiology of movements is basically a study of the purposeful activity of the nervous system as a whole. Therefore, the control of movements seems to us one of the most natural objects for study of those integral functions of the nervous system which are connected, so to say, with *operative control*.

The final result of the work of the nervous system on the control of movements is the sending of impulses to the muscles and the basic question of the physiology of movements is the study of the mechanism for the development of expedient combinations and sequences of such signals.

Here the simplest viewpoint is the concept of the presence of some higher nerve center (situated, for example, in the cortex), where there stems the generation of commands of the sort which completely determine movement, so that the role of all the remaining nerve mechanisms is only the transmission of these commands. The inconclusiveness of this viewpoint is obvious, and in works on the physiology of movements, beginning with the discovery of Fritsch and Hitzig, and after that in the classical works of Sherrington, Magnus, Pavlov, Ukhtomskiy, and in the publications of contemporary research, the question concerning the interaction of different nerve mechanisms in the process of the realization of movement takes the center place.

The fact is that in natural movements dozens of different muscles are involved working coordinately. The system of commands necessary for movement realization cannot but be in this case very complicated. During its development calculation of a rich and varied afferentation is essential, including also ones originating in the course of the movement itself. The concrete realization of movements to a large extent depends on initial conditions—the original pose, etc.

Undoubtedly, a whole series of nerve centers are occupied in the development of commands for the muscles and processing for the purpose of afferentation. The study of their interaction led us to the attempt to describe the features of complicated systems of control from a single viewpoint which will be presented below. Some results of experimental verification of this viewpoint are written up in subsequent articles of the collection. However, many and occasionally important features of the ideas presented in this article still require experimental verification; they are examined by us as a working hypothesis for further research.

This point of view—we will call it the principle of the least interaction [see 11, and also Chapter 13] is a complicated, multilevel system of control examined as an aggregate of subsystems having relative autonomy. Each of the such subsystems has its own "individual" problem consisting of the decrease of interaction with the "outer medium"; the latter for a given subsystem is made up of the medium outside as regards the whole system and the remaining subsystems. Complex systems of control can consist of several levels each of which includes a series of such subsystems.

For subsystems of a certain level actions of the outer, as regards this level, medium include the afferentation coming from below, and the organization of their interaction is determined by the interaction of higher levels. For the lowest level afferentation is exclusively prescribed; the subsystems of this level have outputs for the effectors. In section 2 we will dwell on these ideas in greater detail.

In the organization of the control of movements, utilization of such features of the motor problems which can simplify control, reduce the number of independently controlled effector parameters, and simplify the processing of the incoming afferentation, plays an important part. The problems having such features are, so to say, organized.

In the control of movements organization becomes apparent first of all in that for each motor act it is possible to single out a relatively small number of leading effector parameters and to determine the basic afferentation necessary for the realization of this movement. The tendency for such simplification is also a manifestation of the principle of the least interaction. The decrease of the number of controlled parameters lowers the overall level essential for impulse control.

This article is made up of three sections. The first of them is devoted to synergies and certain other mechanisms simplifying control of movements. In the second we will set forth our general concepts about the arrangement at the spinal level of the structure of movements. In the last part, model representations connected with the function of a pool of motor neurons are described.

1. In order for the higher levels of the central nervous system to solve effectively the problems of the organization of motor acts in the time required, it is essential that the number of controlled parameters not be too large and the afferentation requiring analysis not be too high. The so-called synergies play an important role in the establishment of such conditions of work. It is customary to call *synergies* those classes of movements which have similar kinematic characteristics, coinciding active muscle groups and conducting types of afferentation.

For each synergy there are distinctive, specific connections imposed on certain muscle groups, a subdivision of all the muscles participating in a movement into a small number of connected groups. Because of this, for realization of a movement it is enough to control a small number of independent parameters, although the number of muscles participating in movement can be large.

Although there are but a few synergies, they make possible almost the whole variety of voluntary movements to be included. We can distinguish relatively simple synergies of postural preservation (the stabilization synergy), cyclic locomotive synergies (walking, running, swimming, etc.), synergies of throwing, blowing, jumping, and a certain (small) number of others.

The synergies listed here for an adult human appear already fully worked out; the biomechanical side of the majority of them have been studied. A more detailed research of one special synergy—the respiratory synergy of standing—is written up in Chapter 13.

We will mention this respiratory synergy here very briefly. The fact is that with respiration noticeable displacements of different parts of the body take place. However, on the position of the overall center of gravity these displacements hardly show up. The reason, as explained, is that cophasally with deflection of the torso backward (during inspiration) deflection of the pelvis forward takes place; during expiration the paired displacements occur in the opposite directions. This synergy is specific and does not take place, for example, during external disturbance (a light push in the back). It breaks down during certain neurological illnesses, and then the body's center of gravity oscillates in accordance with the phase of respiration. We can suppose that the described respiratory synergy of a vertical pose is not an exception but an example of a typical mechanism participating in the most diverse natural movements.

It is natural to assume that movement training is the development of corresponding synergies which reduce the number of parameters requiring individual control. A new synergy like that is made, of course, each time not on an empty place but on the framework of a small number of basic synergies and inherent neurophysiological mechanisms which lower the number of independent parameters of the controlled system. Some of the mechanisms of this sort, although not examined by research in this project, are already well known. We refer here to such well-studied examples of functional organization as the system of interaction of the muscle motor neurons—antagonists, acting on the same joint; the system of postural reactions, using a fixed system of the interaction of different kinds of receptions (labyrinth, otolith, proprioceptive of the neck and limbs); and also the important mechanism of development of temporary connections.

The basic synergies and the simplest neurophysiological mechanisms enumerated here form, so to say, "a dictionary of movements." Using this analogy, we can say that the efforts of the muscles are the letters of the language of movements, and the synergies combine these letters into words, the number of which are much less than simply the number of combinations of letters. Moreover, the wealth of the dictionary provides a variety of admissible movements. The majority of motor problems arising before an organism are contained in the scope of this dictionary and only in exceptional cases does there arise a need for its enrichment.

Until now the question has chiefly been the effector aspect of the synergies. Undoubtedly, however, with each synergy there are connected afferent currents in which are singled out the conduction signals and the addresses characteristic of it. The language of the synergies is in this sense not only the external language of movements but also the internal language of the nervous system during control of the movements. The synergies are able to simplify the processing of afferentation, having organized it according to the motor problem.

From this viewpoint in such a problem, as for example the identification of images, in the first place it is determined to which synergy this problem belongs, which in turn predetermines the subsequent course of identification.

We can assume that new movement training consists of the development of a simple process of movement control and leads to the search and correction of suitable synergies or groups of synergies, including here also isolation of conducting afferentation. In the work of V. I. Krinskiy and M. L. Shik [30] the role of this last factor was studied in conditions when the motor problem had to be carried out under the control of specially distorted visual afferentation. It turned out that to some transformations of the visual field the available system of utilization of visual afferentation in the problem of pose retention is resistant. We can assume that such distortions of the visual field are still able to use the available synergy. The more marked distortions made the realization of the problem impossible, and some training time was needed in order that the problem again become executable. It is interesting that during some transformations of a visual signal about deviations from an assigned pose the retraining, although allowing the problem to be executed, did not reestablish that exactness of its solution available to the examinee during utilization of undistorted visual afferentation.

Utilization of the mechanism of synergies is, of course, only one of the ways of simplifying the problem of the control of movements. Another possible approach is connected with the existing similarity of the problem of movement control with the mathematical problem concerning the search for the minimum function of many variables. The language natural for this mathematical problem [14] was found suitable also for a description of the structure of movement. The combination of local adaptations with extrapolations characteristic for nonlocal search methods of the extremum apparently is also typical for the process of admission of solutions in motor problems, and certain peculiarities of such a search were successfully detected in the experiments. Such an approach in connection with the problem of preservation of the erect posture was used in the work of Gelfand, Gurfinkel and Tsetlin [10], and stimulated further research devoted to the study of tremor.

In V. I. Krinskiy and M. L. Shik's work, an attempt was made to investigate the solution method of a simple motor problem which presented, however, high demands of exactness for the execution of the assignment. The examinee was asked to find such a position of two joints of the upper extremities so that the pointer on the galvanometer, the position of which was a certain function of the angles of the indicated joints, would come into a zero position. The experimenter could simultaneously record the trajectory of the cathode ray on the screen of the oscilloscope in coordinates of the limb angles. The change of one of the limb angles caused movement of the ray along the horizontal; the other, along the vertical line. The examinee did not see the screen. The experimenter could also change the coefficients in order that during successive executions of the problem (each took 10–60 sec) the desired position of the joints was different. The function employed had the shape of a "small boat" with a single lowest point

$$f(x;y) = |(x - a) - (y - b)| + \alpha|x - a| + d|y - b|.$$

In the initial attempts all the examinees were allowed only one method: they made successive changes of the joint angles bringing the position of the ray step by step along the horizontal and vertical to the assigned point (a change of the joint was made when the pointer, passing over the minimum, again indicated a withdrawal from the required pose). But later on, together with that sort of solution method, another was applied which utilized the organization of the function, although for the examinee it remained unknown (regardless of the level of theoretical preparedness). On the screen it was seen how the examinee going to the "bottom," descended along it to the lowest point, not climbing "on board" (this is feasible only during coordinated changes of the angles of both joints), and only in the immediate vicinity of the deepest point, made the desired pose more precise by successive changes of the angles of the joints. The tactics of behavior of the examinee in this case were similar to the so-called ravine tactics [15].

If we agree with the idea stated here that the higher levels of the nervous system, revising the conducting afferentation, control the work only of a small group of muscles determining a given movement (or synergy); if, moreover, we still consider that this control is effected not immediately but by means of reorganization of interaction at the lower levels, then it becomes clear that the basic "manual" work with respect to movement realization is carried out by precisely these lower levels, while the higher ones form only functional synergies and reorganize the interaction system of the elements of the lower levels.

In many papers on modeling of pattern recognition, the aim of the model is the formation of a generalized image in which are reflected only the most significant attributes of the real object. During control of movements the reverse problem, so to say, is solved—by abstract representation the structure of a real movement with all the essential details is reproduced.

The realization of an image as a real movement requires its translation from three-dimensional, kinematic language into the language of muscle dynamics: the motor composition, the number of motor units, the spatial and temporal order of their recruitment, etc.

2. For a further account it is convenient to single out the intermediate neural structures in order to dwell at greater length, later, on the function of the last effector link—the motorneuron pool and the interneuronal structures connected with it.

We will analyse the working principles of the intermediate neural structures in an example of the organization of movements at this spinal level.

Characteristic for the spinal level is an extremely large volume of transformed afferentation and a large number of efferent outputs. Moreover, afferentation is received first hand at the spinal level, and the efferent commands are immediately realized. On all the activity at the spinal level are acting supraspinal influences organizing its activity in order to realize the assigned movement. The presence of relatively autonomous subsystems, also divided spatially, is an essential feature of the structure at this level. The presence of a series of innate (so-called "soldered") interactions is typical for those subsystems.

The autonomy (although relative) of the individual subsystems at the spinal level allow effective control to be carried out in a short time interval—the solution should not assume a complex or long concordance. Obviously, the organization of any complex movement by no means leads to the working of only one such subsystem or several soldered subsystems. Therefore, in our ideas about the organization of movements at the spinal level, interaction among individual subsystems plays a central role.

When we talk about the possibility of the relatively autonomous working of the individual subsystems, then inevitably the question arises which particular, "individual" goal in a concrete problem one or another subsystem pursues. It seems likely to us that such a goal in the first approach is the decrease of the total of impulses received by the subsystem both from the periphery and from the sides of the other subsystems and higher levels of the nervous system (the *principle of least interaction*). This total of impulses, serving as an estimate of function of one or another subsystem, is developed apparently by special mechanisms the operation of which depends on the assigned interaction system. It is natural to assume that so-called "usual" or "expected" afferentation adds a relatively small contribution to this total afferentation in comparison with the "unusual" or "unexpected." Perhaps a "nonequilibrium," "unbalanced" afferentation plays a special role. In other words, we can assume that the goal which the subsystem pursues is the minimization of external interactions tending to disturb it from that position in which it finds itself at a given moment. The tendency to minimization of interaction leads to the coordinated working of the individual subsystems subordinating the autonomous activity of each of them in the interest of the solution of the overall problem assigned by the supraspinal afferentation.

Moreover, the basic role of the supraspinal influences is the appropriate rearrangement of the interaction organization of the individual subsystems at the spinal level [34].[1] Such a rearrangement may find expression in the motor effect; however, this result is not obligatory. Rearrangement can create a "readiness for movement" for realization of which, however, a supplementary action is required. Possibly some supraspinal influences (for example, connected with the operation of the labyrinthine apparatus) generally only in this way even affect spinal activity.

The activity of each relatively autonomous subsystem, as we assumed, is directed to the reduction of the total afferentation formed from proprioceptive afferentation and afferentation appearing external as regards the given subsystem—from the neighboring subsystems of the same level and supraspinal ones. A change of the mode of function of the subsystem leads first to a change of the characteristic afferentation and is directed to minimization of the total afferentation. If the contribution of its afferentation into the total one is relatively large then the role of the rest, and among them the supraspinal afferent actions, is reduced. We can see this in conditions of special experiments [17, 18, 37, 40] or in certain situations arising during athletic activity. Probably it is also natural to examine the phenomenon of a dominant, discovered by Ukhtomskiy, from this point of view. The transmission of the control of habitual movements to the lower levels (the automation of motor habits according to N. A. Bernstein) from this viewpoint is the result of the tendency of the higher levels to minimize their own interaction with the lower levels.

The individual details of movement are formed and completed in the process of the interaction of the subsystems forming the spinal level of

[1] Automata games [39] are a natural mathematical model for the study of control methods by the assignment of interaction. On the digital computer, behavior of a system consisting of a large number of automata was modeled, the interaction among which was corrected by some higher level. The problem of the higher level was to develop an interaction guaranteeing beforehand the assigned behavior of the automata. The modeling on the digital computer showed that a correction of interaction can be made without waiting for the complete correspondence of the behavior of the collection of automata with the interaction assigned from above.

regulation of movement. If we somewhat schematize the Wells classification [41], then in typical movements of a human being and higher animals we can single out the muscles which carry out the basic active part of the motor problem (relatively few of them) and the muscles which stabilize the position of the basic mass of kinematic links of the body (the majority of them).

The activity of the corresponding controlling neural mechanisms is basically autonomous, and the dominant afferentation is peripheral; the interaction of these mechanisms with the rest will be relatively small. Hence, there results the typicalness of such "stabilizing" modes of function in the movements. The relatively small role of interaction in such modes of this sort guarantees their "freedom," permitting the higher levels of the central nervous system not to be responsible for the control of the corresponding cerebrospinal subsystems. Apparently in some of the simplest cases such stabilization is already realized at the segmental level [23]; if afferentation is not minimized on this lowest level, the higher levels are included. The work of a large number of stabilizing mechanisms guarantees the smoothness, the evenness of the movement; without them the movement would become ataxic.

Of course the regulation of an active movement is by no means reduced to the operation of the stabilizing mechanisms. The system of interaction assigned to the higher levels should guarantee nonequilibrium functioning of the muscles responsible for the active part of the movement, and almost always an expedient change of the angles of the joints. Although voluntary movements have been studied relatively little, we can nevertheless state some considerations resulting from the concepts stated above. In particular, if the role of the higher levels of the nervous system is not in the sending of direct commands but in the reorganization of the interaction system (the tuning) of the neural mechanisms at the spinal level, then naturally such a tuning should take more time than for the usual transmission of commands.

We know that the latent time of a simple motor reaction is virtually constant, and that even systematic training does not lessen it more than 10 to 15 msec. Moreover, of the 120 to 180 msec of latent time the communication time is by no means more than 50 to 80 msec. Inasmuch as any of the active movements for the majority of the neural structures of the spinal level signify the inclusion of stabilization mechanisms required by the given problem, so such a tuning should for the majority of spinal mechanisms be in general outline the same. Of course, the tuning may change substantially even in the course of the movement itself. However, study of such tuning is simplest when movement has not yet begun. We will call *pretuning* the phenomena connected with the preparation of the spinal neural mechanisms for movement, and we will dwell on them in somewhat greater detail.

One of the first observations, which, as is now clear, indicates the presence of pretuning, belongs to Hufschmidt [26], who discovered that as much as 60 msec before the start of voluntary contraction of the muscle, inhibition of the muscle activity of the antagonist occurs.

Our experiment consisted of the following procedure: We instructed the examinee by a signal to make a specific movement, for example, to straighten the foot. Within a certain time after the signal (but before the start of the movement) a tendon reflex was produced and the amplitude of the electromyographic response was measured. It turned out that the size of the reflex essentially depends on what time remained before the start of the movement. This dependence has the same character for the tendon reflexes and the monosynaptic H-reflex [25] for the corresponding muscle. It is interesting to note that for relatively long intervals before the start of a movement (70–50 msec) the changes of amplitude of the spinal reflexes are approximately the same both for those muscles which are to participate in the movement and for those which in this movement are not activated. As the time to the start of movement decreases, these changes show up considerably sharper in regard to precisely those muscles which will participate in the movement. In this way, apparently, tuning of the spinal mechanisms bears at the beginning a so to say diffuse nature, and as the moment of initiation of the movement approaches, the change of the system of interaction at the spinal level is localized. True, it is not ruled out that the diffuse changes of the condition of the spinal-controlling mechanisms are connected with an orienting reaction arising in the experiments described. These experiments are described in greater detail in Chapter 11 in this collection.

3. In this section we want to talk about some ideas connected with the function of the last effector link of the control system of movements—the motorneuronal pool. At the same time the mode of function of the individual motor units will interest us, so that in the center of attention are found the working features of the pool connected with its atomicity—a common characteristic of all neural structures.

The nerve cells themselves do not have and cannot have any complex behavior: the information obtained by the individual neuron is immeasurably more meager than that which the whole organism receives and its reactions are very stereotyped. Therefore the central problem in neurophysiology is the study of how the expedient behavior of the organism in interaction with the changing environment is formed from the interaction of different nerve structures and, all things considered, from the behavior of individual neurons. In connection with this the search for such principles of interaction of the neurons which would guarantee execution of the integral physiological acts is very important.

The tendency to minimization of interaction, about which we have already spoken above, gives rise to the possibility of the nonindividualized control of the neurons of one or another nerve structure by means of the influence on their own system of interaction. Of course, action on the interaction system of the neurons does not exclude direct influences on these neurons. Nonindividualized action on the neurons of some center gives rise to the possibility of a simple description of their operation.

The determining role of the autonomous collective function indicates, in particular, that the key place in the solution of any physiological problem belongs to the so-called "horizontal" interaction of the neurons as yet studied experimentally only in the simplest examples.

These ideas are applicable, probably, to the working organization of both the motor and sensory systems.

Suppose that there is a homogeneous medium of neurons connected together so that each, being excited, exerts a facilitatory influence on its neighbors. Then such a condition of the system will be stable when all the active neurons are working synchronously. Moreover, interaction of them will be minimal. The applicability of this assertion to a real neural structure is difficult to check immediately, in the first place because of methodical difficulties, and in the second, because there is no basis to think that the interaction of neurons is only positive and symmetrical.[2]

Among the neurons of a homogeneous system—the motor neurons of one muscle—there cannot but be interaction, simply because of the proximity of their electrical fields, the overlapping of the branching zones of the dendrites, and also in view of the facts that each muscle receptor is projected not on one motor neuron but on several, and that each motor neuron receives an impulse from several muscle receptors. The different receptors exert their influence on the motor neurons either directly or by interneurons. The impulse of an individual muscle receptor depends on the activity of the many motor units of the muscle. But once interference of the motor neurons exists, then if it is facilitatory the motor neurons should become excited synchronously. If they work independently, then this indicates the existence of a special mechanism preventing synchronization.

An experimental study of the activity of individual motor units of a human muscle developing moderate tension in the postural mode showed that these motor units work virtually independently [9, 20]. This is all the more surprising as the mean frequency of impulse of all the active motor units is approximately the same 7–11 per second (activity with a different frequency is evidently not stable), and the impulse of the individual motor unit in the course of several tens of beats is very stable (the ratio of the standard deviation to the mean duration of the interval between two impulses is about 0.2–0.3).

The independent impulse of different motor units of one muscle creates the principal possibility of separate control of the units. In special conditions of artificial visual and auditory control of activity of the motor units, a human being can actually voluntarily "switch on" or "switch off" the assigned motor unit (speaking more strictly, the group of motor units containing the assigned one), without changing the impulse of the second randomly chosen motor unit of the same muscle [1, 9].

[2] A mathematical model of such a structure was examined by Gelfand and Tsetlin [13]. The pacemaker of the heart, all the elements of which work synchronously, can serve as a physiological model. Moreover, the element, the impulse frequency of which is the greatest, is the "leader" [12].

The fact of independent activity of the motor units makes it possible to understand the genesis of a physiological tremor, to predict the dependence of its amplitude on the effort built up, and the spectral composition [7, 8].

The actual arrangement of the mechanism guaranteeing asynchronous activity and contrast of the active motor units, is not known. However, available information about the characteristics of certain elements of the segmental apparatus of the spinal cord and the nature of the connections among them [6, 19, 24, 27, 28, 32, 33, 35, 36, 38, 42, 43] makes it possible to suggest a model of the arrangement of this mechanism. The arrangement of the model is examined in reference [9], and a more detailed account will be given in Kotov's paper as yet unpublished [29]. Here we will note only that important for a normal mode of function of the motor units of the muscle are, according to these model representations, besides the known characteristics of motor neurons themselves, reciprocal inhibition, in particular the dependence of the volley of the Renshaw cells on the frequency of its activation, and the hypothesis concerning the collaborative activation of the corresponding alpha and gamma motor neurons. For an explanation of certain pathological modes of function of the muscle (Parkinson's disease, post-poliomyelitis paralysis) during which the activity of the motor units differs substantially from the normal, it is enough to assume in the limits of this model wholly concrete and small changes of the characteristics of the elements of the model. The study of the behavior of a model makes it possible to distinguish certain significant parameters of the architecture of the segmental apparatus and its elements, from change of which the mode of function of the system varies substantially, and parameters variations of which little effect the working of the system. Such conclusions would undoubtedly be much more difficult to obtain in direct experiments.

REFERENCES

1.
J. V. Basmajian, *Science* 1963:141, 440.
2.
N. A. Bernstein, *Arkhiv Biol. Nauk* 1935:38, 1.
3.
———, *O postroenii dvizheniy* (*Concerning the Construction of Movements*), Moscow: Medgiz, 1947.
4.
———, *Problemy kibernetiki* 1961:6, 101.
5.
V. I. Bryzgalov, I. I. Pyatetskiy-Shapiro, and M. L. Shik, *Dokl. AN, SSSR* 1965:160, 1039.
6.
J. Eccles, *The Physiology of the Nerve Cells*, Baltimore: Johns Hopkins Press, 1957.
7.
A. G. Feldman, *Biofizika* 1964:9, 726.
8.
I. M. Gelfand, V. S. Gurfinkel, Ya. M. Kots, M. L. Krinskiy, M. L. Tsetlin, and M. L. Shik, *Biofizika* 1964:9, 710.
9.
———, V. S. Gurfinkel, Ya. M. Kots, M. L. Tsetlin, and M. L. Shik, *Biofizika* 1963:8, 475.
10.
———, V. S. Gurfinkel, and M. L. Tsetlin, *Dokl. AN, SSSR* 1961:139, 1250.
11.
———, V. S. Gurfinkel, M. L. Tsetlin, collection *Biol. aspekty kibernetiki* (*Biological Aspects of Cybernetics*), Moscow: Izdvo AN SSSR (Publishing House of the Academy of Medical Sciences of the USSR), 1962, p. 66.
12.
———, S. A. Kovalev, and L. M. Chaylakhyan, *Dokl. AN, SSSR* 1963:148, 973.
13.
——— and M. L. Tsetlin, *Dokl. AN, SSSR* 1960:131, 1242.
14.
——— and M. L. Tsetlin, *Dokl. AN, SSSR* 1961:137, 295.
15.
——— and M. L. Tsetlin, *Uspekhi Matem. Nauk* 1962:17, 3.

16.
——— and M. L. Tsetlin, Chapter 11.
17.
B. E. Gernandt and H. W. Ades, *Exp. Neurol.* 1964:10, 52.
18.
———, J. Katsuki, and R. B. Livingston, *J. Neurophysiol.* 1957:20, 453.
19.
R. Granit, *Receptors and Sensory Perception: A Discussion of Aims, Means, and Results of Electrophysiological Research and the Process of Reception*, New Haven: Yale University Press, 1955.
20.
V. S. Gurfinkel, A. N. Ivanova, Ya. M. Kots, I. I. Pyatetskiy-Shapiro, and M. L. Shik, *Biofizika* 1964:9, 636.
21.
———, Ya. M. Kots, V. I. Krinskiy, Ye. I. Paltsev, A. G. Feldman, M. L. Tsetlin, and M. L. Shik, Chapter 11.
22.
———, Ya. M. Kots, Ye. I. Paltsev, and A. G. Feldman, Chapter 13.
23.
———, Ya. M. Kots, and M. L. Shik, *Regulyatsiya pozy cheloveka*, (*The Regulation of Human Posture*), Moscow: Izd-vo Nauka, 1965.
24.
J. Haase, *Pflüg. Arch. Ges. Physiol.* 1963:276, 471.
25.
P. Hoffman, *Untersuchungen über die Eigenreflexe menschlichen Muskeln*, Berlin, 1922.
26.
H. Hufschmidt, *Pflüg. Arch. Ges. Physiol.* 1962:275, 463.
27.
C. C. Hunt, *J. Gen. Physiol.* 1955:38, 801.
28.
P. G. Kostyuk, *Fiziol. Zh. SSSR* 1961:47, 1241.
29.
Yu. B. Kotov, in press.
30.
V. I. Krinskiy, M. L. Shik, *Biofizika* 1963:8, 513.
31.
——— and M. L. Shik, *Biofizika* 1964:9, 607.
32.
D. P. C. Lloyd, *J. Neurophysiol.* 1941:4, 525.

33.
P. B. C. Matthews, *Physiol. Rev.* 1964:44, 219.
34.
I. I. Pyatetskiy-Shapiro and M. L. Shik, *Biofizika* 1964:9, 488.
35.
B. Renshaw, *J. Neurophysiol.* 1946:9, 191.
36.
A. I. Shapovalov, *Dokl. AN, SSSR* 1961:141, 1267.
37.
S. M. Sverdlov and Ye. V. Maksimova, *Biofizika* 1965:10, 161.
38.
J. Szentágothai, in *Basic Research in Paraplegia*, ed. J. D. French and R. W. Porter, Springfield, Ill.: Thomas, 1962, p. 51.
39.
M. L. Tsetlin, *Uspekhi Matem. Nauk* 1963:18, 3.
40.
N. V. Veber, L. M. Rodioriov, and M. L. Shik, *Biofizika* 1965:10, 334.
41.
J. Wells, *Kinesiologie* 1955.
42.
V. J. Wilson, *Basic Research in Paraplegia*, ed. J. D. French and R. W. Porter, 1962, Springfield, Ill.: Thomas, 1962, p. 74.
43.
G. P. Zhukova, *Arkhiv Arat.* 1958:35, 43.

8.

(with Yu. I. Arshavskij, M. B. Berkenblit, O. I. Fukson, and G. N. Orlovskij)

Recordings of neurones of the dorsal spinocerebellar tract during evoked locomotion

Brain Res. 43 (1972) 272–275

The functional organization of connexions to neurones of the dorsal spinocerebellar tract (DSCT) was studied in detail in decerebrated, immobilised and anaesthetised cats[2,5,6,8] but there are no data about their activity during natural movements. This study deals with the activity of DSCT neurones during evoked locomotion.

Fifteen mesencephalic cats were used in our experiments. Transection of the brain stem was performed under ether anaesthesia approximately at A5 level, and a

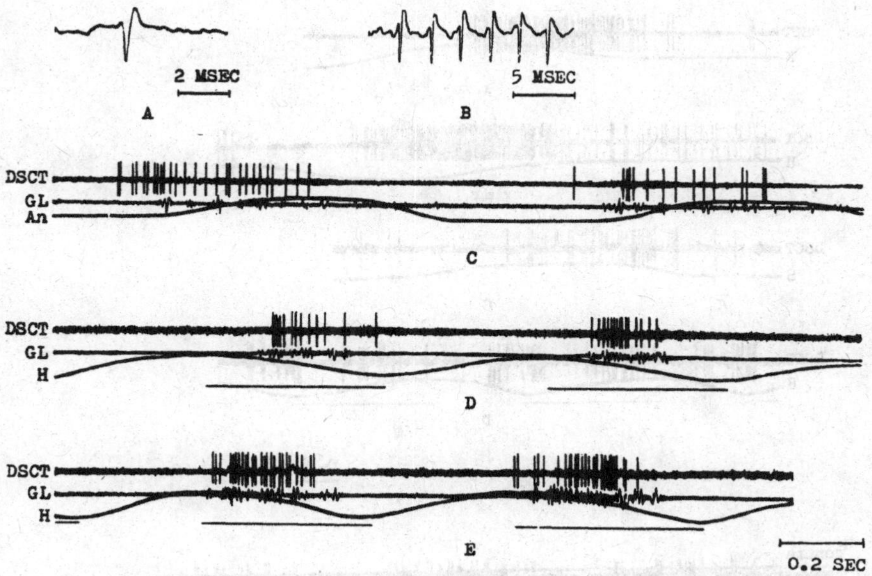

Fig. 1. Activity of a DSCT neurone. Antidromic invasion (from the ipsilateral intermediate part of the cerebellar cortex) of the DSCT neurone is shown in A (single stimulus) and in B (370/sec). In C, the activity of a DSCT neurone is shown during passive movements in the ipsilateral ankle joint (An, flexion — up) together with the EMG activity of the gastrocnemius lateralis (GL). In D and E is shown the activity of the neurone during locomotion evoked by weak (D) and increased (E) stimulation of the 'locomotor region'. Stance phases of the ipsilateral hindlimb are marked by horizontal lines, H-ipsilateral hip movement (flexion — up). To record joint movements, either the femur, or the ankle, or the foot were connected with the potentiometer transducer with a thread. This method could provide rather high sensitivity in the recording of hip movements, but was less sensitive with the recording of knee and ankle movements. This might be one reason for some delay of the ankle recording in C as compared with DSCT.

laminectomy performed at the 1st–4th lumbar vertebrae. In 3 experiments a wide laminectomy was carried out to expose the whole lumbosacral spinal cord, and all dorsal roots from L3 on both sides were cut. The head of the cat, the 1st and the 5th lumbar vertebrae and the pelvis were fixed in a stereotaxic device. The spinal cord was covered with mineral oil. The legs of the cat were lowered to the belt of a treadmill which could be set in motion at a speed of 2–5 km/h. Locomotion was evoked by stimulation (30 pulses/sec, 1 msec pulse duration, 10–20 V) of the 'locomotor region' of the mesencephalon[7,12].

Extracellular recording of the DSCT neurones was performed with platinum microelectrodes[3] (tip diameter 10–20 μm, 100–200 kΩ) in Clarke's column at the 1st–3rd lumbar segments. The neurones were identified by their antidromic activation from lobules II–III of the ipsilateral intermediate part of the cerebellar cortex where DSCT axons terminate[4] (Fig. 1A, B).

Seventeen DSCT neurones were examined in cats with intact dorsal roots. In most cases we could find the muscle (or a few synergists) from which the neurone was

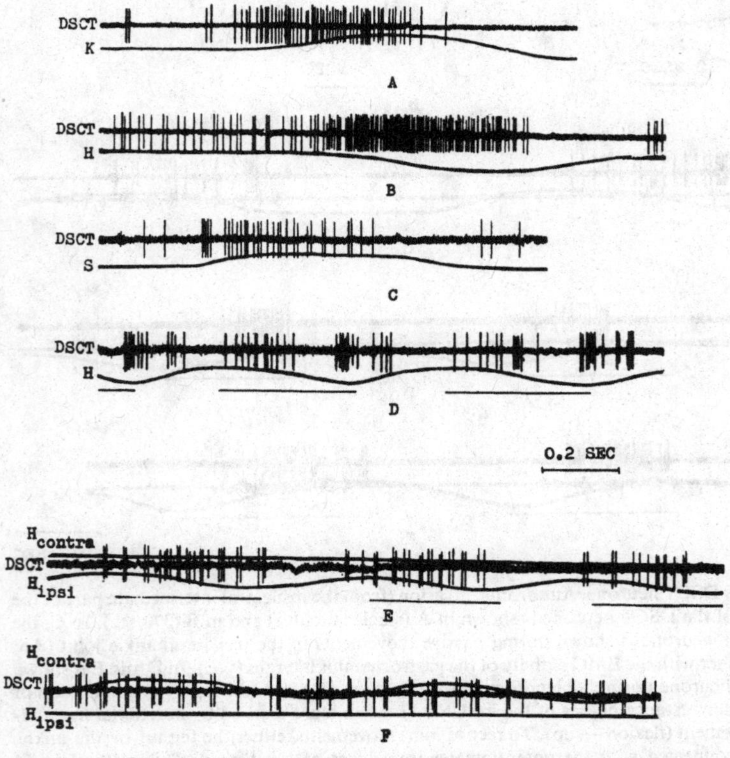

Fig. 2. Activity of two DSCT neurones. Activation of a DSCT neurone by passive knee (K) flexion is shown in A, by hip (H) extension in B, and by pressing (S) of muscles on the rostral femur surface in C. In D is shown the activity of the neurone during evoked locomotion. Stance phases of the ipsilateral hindlimb are marked by horizontal lines. Activity of another DSCT neurone during locomotion is shown in E and F (the contralateral hindlimb was stopped by force in E, and the ipsilateral one in F).

supplied with an afferent input. Fig. 1C shows a DSCT neurone, which was activated by passive flexion of the ankle joint. It was not influenced by passive movements in the knee and the hip joints. One may conclude that this neurone receives afferents from the ankle extensors, for example from the gastrocnemius lateralis, which was also activated with the ankle flexion (GL in Fig. 1C). During locomotion (Fig. 1D) the DSCT neurone showed a distinct modulation in relation to the locomotor cycle — it was active during the stance phase of the step when the gastrocnemius lateralis was active, and was not active during the swing phase. With stronger stimulation of the 'locomotor region' (Fig. 1E), the extensor activity increased, the movements of the limb became more forceful and the activity of the neurone also increased markedly. Corresponding results were obtained in all 8 neurones supplied by the afferents from one-joint muscles. These neurones received afferents from extensors and during locomotion they were active in the stance phase.

Some neurones were activated by passive movements in two joints (knee and hip), as a neurone in Fig. 2A–D. This DSCT neurone was activated by both knee flexion (A) and hip extension (B), as well as by pressing the muscles at the rostral surface of the femur (C). One may conclude that this neurone receives afferents from muscles which act as knee extensor and hip flexor. Correspondingly, during locomotion (D) this neurone was activated both in the swing and in the stance phases of the step. Corresponding results were obtained in 9 neurones supplied by the afferents from double-joint muscles.

The modulation described above is exclusively determined by afferent activity in the ipsilateral hindlimb. The arrest of this limb during locomotion (Fig. 2F) results in the disappearance of modulation, while during the arrest of the contralateral hindlimb (E) the neurone has a distinct modulation. Besides, 11 DSCT neurones were recorded during locomotion of cats with deafferented hindlimbs. In spite of the fact that all limbs in these preparations participated in locomotion[11], rhythmic modulation of DSCT neurones was not revealed. In this respect the DSCT greatly differs from the VSCT[1] in which all recorded neurones were modulated during locomotion of cats with deafferented hindlimbs.

DSCT neurones are known to receive monosynaptic connexions from group I muscle afferents[2,5,6,8]. During locomotion these afferents (both Ia and Ib) are found to be simultaneously active in phase with corresponding muscles[10], that is, for Ia afferents due to the gamma-modulation[9]. This may be why the DSCT neurones in this study were found to be active simultaneously with the corresponding muscles.

In conclusion, our findings confirm the view, based on the study of the afferent connexions of DSCT neurones[2,5,6,8], that the DSCT transmits information to the cerebellum about the activity (*i.e.* the phase and the strength of the contraction) of separate muscles or of a few synergists.

Institute of Information Transmission Problems,
Academy of Sciences, and
Interfaculty Laboratory
of Mathematical Methods in Biology,
M. V. Lomonosov Moscow State University,
Moscow (U.S.S.R.)

YU. I. ARSHAVSKY
M. B. BERKINBLIT
O. I. FUKSON
I. M. GELFAND
G. N. ORLOVSKY

1 ARSHAVSKY, YU. I., BERKINBLIT, M. B., FUKSON, O. I., GELFAND, I. M., AND ORLOVSKY, G. N., Origin of modulation in neurones of the ventral spinocerebellar tract during locomotion, *Brain Research*, 43 (1972) 276–279.

2 ECCLES, J. C., OSCARSSON, O., AND WILLIS, W. D., Synaptic action of group I and II afferent fibres of muscle on the cells of the dorsal spinocerebellar tract, *J. Physiol. (Lond.)*, 158 (1961) 517–547.

3 GESTELAND, R. C., HOWLAND, B., LETTWIN, J. Y., AND PITTS, W. H., Comments on microelectrodes, *Proc. Inst. Radio Engng.*, 47 (1959) 1856–1862.

4 GRANT, G., Spinal course and somatotopically localized termination of the spinocerebellar tracts. An experimental study in the cat, *Acta physiol. scand.*, 56, Suppl. 193 (1962) 1–45.

5 KOSTYUK, P. G., On the functions of dorsal spino-cerebellar tract in cat. In R. LLINÁS (Ed.), *Neurobiology of Cerebellar Evolution and Development*, AMA Press, Chicago, 1969, pp. 539–548.

6 LUNDBERG, A., AND OSCARSSON, O., Functional organization of the dorsal spino-cerebellar tract in the cat. VII. Identification of units by antidromic activation from the cerebellar cortex with recognition of five functional subdivisions, *Acta physiol. scand.*, 50 (1960) 356–374.

7 ORLOVSKY, G. N., SEVERIN, F. V., AND SHIK, M. L., Locomotion evoked by stimulation of the brain stem, *Dokl. Akad. Nauk SSSR*, 169 (1966) 1223–1226.

8 OSCARSSON, O., Functional organization of the spino- and cuneocerebellar tracts, *Physiol. Rev.*, 45 (1965) 495–522.

9 SEVERIN, F. V., On the role of gamma-motor system for extensor alpha-motoneurons activation during the controlled locomotion, *Biofizika*, 15 (1970) 1096–1102.

10 SEVERIN, F. V., ORLOVSKY, G. N., AND SHIK, M. L., Activity of the muscle receptors during controlled locomotion, *Biofizika*, 12 (1967) 502–511.

11 SHIK, M. L., ORLOVSKY, G. N., AND SEVERIN, F. V., Organization of the locomotor synergism, *Biofizika*, 11 (1966) 879–886.

12 SHIK, M. L., SEVERIN, F. V., AND ORLOVSKY, G. N., Structures of the brain stem responsible for evoked locomotion, *Sechenov Physiol. J. (U.S.S.R.)*, 53 (1967) 1125–1132.

(Accepted April 14th, 1972)

9.

(with Yu. I. Arshavskij, M. B. Berkenblit, O. I. Fukson, and G. N. Orlovskij)

Origin of modulation in neurones of the ventral spinocerebellar tract during locomotion

Brain Res. 43 (1972) 276–279

The principle that an ascending pathway may signal information regarding not only peripheral events but also events in reflex pathways was first proposed by Lundberg[5,6] for the various pathways (including ventral spinocerebellar tract, VSCT), influenced from FRA. More recently, Oscarsson[9,16,17] suggested that some of the ascending pathways might monitor the level of interneuronal activity and thus might serve as feedback channels for information concerning the effectiveness of descending

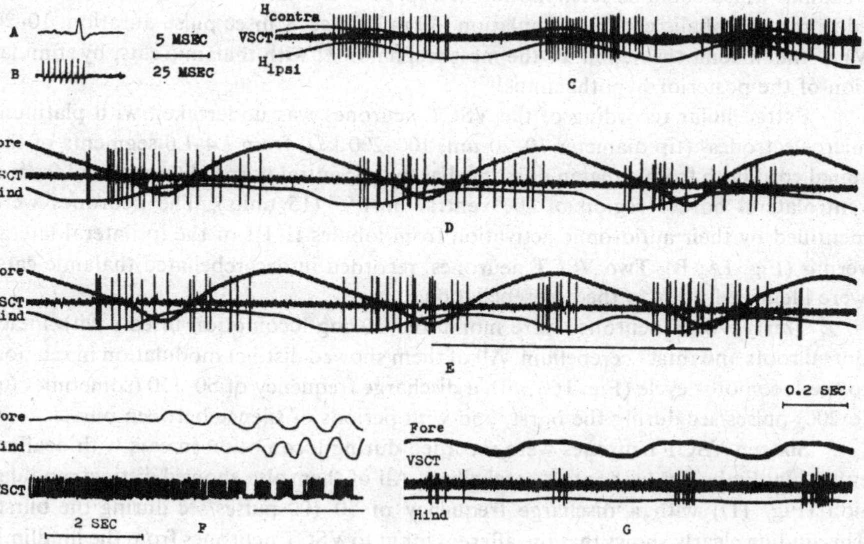

Fig. 1. Activity of VSCT neurones. The antidromic invasion (from the ipsilateral vermis) of the VSCT neurone is shown in A (single stimulus) and in B (400/sec). In C is shown the activity of a VSCT neurone together with the movements in both ipsilateral hip joint (H_{ipsi}, flexion—up) and contralateral one (H_{contra}, extension—up). Locomotion of this mesencephalic cat (dorsal roots intact) was evoked by stimulation of the 'locomotor region' of the brain stem (pulses 1 msec, 30/sec, 20 V). In D and E is shown the activity of a VSCT neurone during locomotion of a mesencephalic cat (both hindlimbs are deafferented), as well as movements in the ipsilateral hip joint (Hind; flexion—up) and in the contralateral shoulder joint (Fore; flexion—up). Two steps of the hindlimb during the period of one step of the forelimb are marked by a horizontal line in E. In F is shown another example of different periods of the fore- and the hindlimb movements, observed in a mesencephalic cat (both hindlimbs are deafferented) during the beginning of locomotion; horizontal line indicates 3 steps of the forelimb during 4 steps of the hindlimb (inkwriter recording). In G is shown the activity of a VSCT neurone during locomotion of thalamic cat evoked by stimulation of the posterior hypothalamus (cerebellum is removed, both hindlimbs are deafferented).

pathways. In a new hypothesis concerning the function of the VSCT Lundberg[7] supposed that the VSCT activity depends heavily upon interneuronal activity, especially upon the activity of inhibitory interneurones within the spinal cord.

We have examined the activity of VSCT neurones during locomotion of mesencephalic[14,19] and thalamic[10] cats. These preparations are capable of locomotion when their hindlimbs are deafferented[18], thus we have an opportunity to compare activities of VSCT neurones in preparations with intact and deafferented hindlimbs to estimate the role of the afferent input to VSCT neurones.

Two thalamic and 10 mesencephalic cats were used in our experiments. Transection of the brain stem was performed under ether anaesthesia at approximately A5 level (mesencephalic cats) and A12 level (thalamic cats), a laminectomy at the lumbosacral vertebrae (to expose the spinal cord from L3) was also performed. In 5 mesencephalic and 2 thalamic cats all dorsal roots from L3 on both sides were cut; the entire cerebellum was removed in both thalamic cats. The head of the cat, the 1st and the 5th lumbar vertebrae and the pelvis were fixed in a stereotaxic device. The spinal cord was covered with mineral oil. The legs of the cat were lowered to the belt of a treadmill which could be set in motion at a speed of 2–5 km/h. Locomotion was evoked in mesencephalic cats by stimulation (30 pulses/sec, 1 msec pulse duration, 10–20 V) of the 'locomotor region' of the mesencephalon[19]; with thalamic cats, by stimulation of the posterior hypothalamus[10].

Extracellular recording of the VSCT neurones was undertaken with platinum microelectrodes[3] (tip diameter 10–20 μm, 100–200 kΩ) from L4–L6 segments of the spinal cord both from a region dorsomedial to the ventral horn[4] (33 units) and from a ventrolateral border region of the ventral horn[1,2] (15 units). The neurones were identified by their antidromic activation from lobules II–III of the ipsilateral lateral vermis (Fig. 1A, B). Two VSCT neurones, recorded in decerebellated thalamic cats, were identified prior to the decerebellation.

Thirty VSCT neurones were monitored during locomotion in cats with intact dorsal roots and intact cerebellum. All of them showed distinct modulation in relation to the locomotor cycle (Fig. 1C) with a discharge frequency of 50–150 (sometimes up to 200) pulses/sec during the burst, and with periods of silence between bursts.

Sixteen VSCT neurones were recorded during locomotion in cats with deafferented hindlimbs but with intact cerebellum. All of them also showed distinct modulation (Fig. 1D) with a discharge frequency of 50–100 pulses/sec during the burst. This finding clearly shows that the afferent input to VSCT neurones from the hindlimb is not the main basis for their modulation.

As a rule, the locomotor rhythm of the deafferented hindlimbs is equal to that of the forelimbs (*cf.* ref. 18). Hence, the periodic influences from the forelimbs (which were not deafferented) might be a basis for the modulation of VSCT neurones. However, in two cases, we managed to observe different periods of the fore- and hindlimb movements (Fig. 1E, F). In both cases, the activity of VSCT neurones was related to the movement of the hind- but not to the forelimbs.

VSCT neurones are known to have direct supraspinal control[8,15]; therefore their modulation could also be explained by periodic changes of the descending influences. To examine this possibility, we tried to record VSCT neurones in decerebellated cats, because decerebellation results in the disappearance of rhythmic modulation (in relation to the locomotor cycle) in all descending pathways tested (*i.e.*, reticulospinal[11], vestibulospinal[12] and rubrospinal[13]). These experiments were found to be very difficult to perform because locomotor movements in deafferented and decerebellated cats were either weak or could not be evoked at all. In two cats we have recorded only two VSCT neurones during locomotion, and both neurones showed a distinct modulation (Fig. 1G).

Thus we have obtained some data indicating that the descending influences (from the forelimbs and from the supraspinal structures) are not responsible for modulation of VSCT neurones.

The most striking finding of this study, *i.e.* modulation of VSCT neurones after the complete deafferentation of the hindlimbs, strongly confirms Lundberg and Oscarsson's point of view that the VSCT is not an ordinary afferent pathway transmitting information as regards peripheral events. It is more likely that it informs the cerebellum about the active processes within the spinal cord; in our case, about the activity of the spinal mechanism generating stepping movements.

Institute of Information Transmission Problems,
Academy of Sciences, and
Interfaculty Laboratory
of Mathematical Methods in Biology,
M. V. Lomonosov Moscow State University,
Moscow (U.S.S.R.)

YU. I. ARSHAVSKY
M. B. BERKINBLIT
O. I. FUKSON
I. M. GELFAND
G. N. ORLOVSKY

1 Burke, R., Lundberg, A., and Weight, F., Spinal border cell origin of the ventral spinocerebellar tract, *Exp. Brain Res.*, 12 (1971) 283–294.
2 Cooper, S., and Sherrington, C. S., Gower's tract and spinal border cells, *Brain*, 63 (1940) 123–134.
3 Gesteland, R. C., Howland, B., Lettwin, J. Y., and Pitts, W. H., Comments on microelectrodes, *Proc. Inst. Radio Engng.*, 47 (1959) 1856–1862.
4 Hubbard, J. I., and Oscarsson, O., Localization of the cell bodies of the ventral spino-cerebellar tract in lumbar segments of the cat, *J. comp. Neurol.*, 118 (1962) 199–204.
5 Lundberg, A., Integrative significance of patterns of connections made by muscle afferents on the spinal cord, *Symp. XXI int. Physiol. Congr., Buenos Aires*, (1959) 1–5.
6 Lundberg, A., Integration in the reflex pathway. In R. Granit (Ed.), *Nobel Symposium I. Muscular Afferents and Motor Control*, Almqvist and Wiksell, Stockholm, 1966, pp. 275–305.
7 Lundberg, A., Function of the ventral spinocerebellar tract — a new hypothesis, *Exp. Brain Res.*, 12 (1971) 317–330.
8 Lundberg, A., and Weight, F., Functional organization of connexions to the ventral spinocerebellar tract, *Exp. Brain Res.*, 12 (1971) 295–316.
9 Miller, S., and Oscarsson, O., Termination and functional organization of spino-olivocerebellar paths. In W. S. Fields and W. D. Willis (Eds.), *The Cerebellum in Health and Disease*, Warren H. Green, St. Louis, 1970, pp. 172–200.
10 Orlovsky, G. N., Spontaneous and evoked locomotion of the thalamic cat, *Biofizika*, 14 (1969) 1095–1102.
11 Orlovsky, G. N., Cerebellar influences on the reticulo-spinal neurons during locomotion, *Biofizika*, 15 (1970) 894–901.
12 Orlovsky, G. N., Activity of vestibulospinal neurons during locomotion, *Brain Research*, 46 (1972) in press.
13 Orlovsky, G. N., Activity of rubrospinal neurons during locomotion, *Brain Research*, 46 (1972) in press.

14 Orlovsky, G. N., Severin, F. V., and Shik, M. L., Locomotion evoked by stimulation of the brain stem, *Dokl. Akad. Nauk SSSR*, 169 (1966) 1223–1226.
15 Oscarsson, O., Functional organization of the spino- and cuneocerebellar tracts, *Physiol. Rev.*, 45 (1965) 495–522.
16 Oscarsson, O., Functional significance of information channels from the spinal cord to the cerebellum. In M. D. Yahr and D. P. Purpura (Eds.), *Neurophysiological Basis of Normal and Abnormal Motor Activities, 3rd Symposium of the Parkinson's Disease Information and Research Center*, Raven Press, Hewlett, N.Y., 1967, pp. 93–117.
17 Oscarsson, O., The sagittal organization of the cerebellar anterior lobe as revealed by the projection patterns of the climbing fiber system. In R. Llinás (Ed.), *Neurobiology of Cerebellar Evolution and Development*, AMA Press, Chicago, 1969, pp. 525–537.
18 Shik, M. L., Orlovsky, G. N., and Severin, F. V., Organization of the locomotor synergism, *Biofizika*, 11, (1966) 879–886.
19 Shik, M. L., Severin, F. V., and Orlovsky, G. N., Structures of the brain stem responsible for evoked locomotion, *Sechenov Physiol. J. (U.S.S.R.)*, 53 (1967) 1125–1132.

(Accepted April 14th, 1972)

10.

(with M. B. Berkinblit, T. G. Delyagina, A. G. Fel'dman, and G. N. Orlovskij)

Generation of scratching
I. Activity of spinal interneurons during scratching

J. Neurophysiol. 41 (1978) 1040–1057

SUMMARY AND CONCLUSIONS

1. In decerebrate, curarized cats, stimulation of the cervical spinal cord evoked fictitious scratching (9), i.e., periodical activity of the hindlimb motoneurons with a discharge pattern typical of actual scratching (cycle duration about 250 ms, flexor phase about 200 ms, extensor phase about 50 ms). During fictitious scratching, extracellular records were obtained from 182 spinal neurons located in different regions of the gray matter cross section (except for the motor nuclei), from segments L_4 and L_5.

2. The firing rate of 73% of neurons was rhythmically modulated in relation with the scratch cycle. Most of the modulated neurons fired in bursts and were silent between bursts. They were located mainly in Rexed's (22) layer VII.

3. Burst onsets ("switchings on" of the neurons) were distributed rather evenly throughout the scratch cycle except for a small maximum at the very beginning of the cycle (the cycle was assumed to start with the termination of the extensor phase). Burst terminations ("switchings off") in the overwhelming majority of the neurons were distributed over the last-third part of the cycle. As a result, those neurons which began to fire earlier in the cycle usually had longer bursts, compared to the neurons which began to fire later. Besides, since there were very few switchings off in the first half of the cycle, the number of simultaneously active neurons increased during the first half of the cycle, reached the maximum somewhat later than the middle of the cycle, and considerably decreased by the end of the cycle.

4. With more intensive scratching, the firing rate in the bursts considerably increased in all neurons tested, while the duration of the scratch cycle changed only slightly.

5. A correlation between the burst position in the cycle and the behavior during the latent period of scratching (when stimulation of the cervical spinal cord had already been started but rhythmical oscillations had not yet appeared) was found in many neurons. Most of the neurons which began to fire at the beginning of the scratch cycle and had long bursts were tonically activated during the latent period. On the contrary, most of the neurons which fired in short bursts at the end of the cycle were either inhibited or not affected during this period.

6. A correlation between the burst position in the cycle and the frequency pattern was found in many neurons. In most of the neurons which began to fire in the first half of the cycle (except for the very beginning), the discharge rate increased in the course of the burst. In the remaining neurons, the discharge rate changed only slightly during the burst.

7. Hypotheses concerning organization of the spinal mechanism of scratching are discussed.

INTRODUCTION

Sherrington (25, 27) found that the scratch reflex was easy to evoke in spinal animals.

Received for publication August 25, 1977.

This finding means that the neuronal mechanism of the reflex is located in the spinal cord. Sherrington also found that scratching movements were preserved after deafferentation of the limb (27). Therefore, the neuronal mechanism of the spinal cord is capable of generating rhythmical oscillations in the absence of sensory feedback from the moving limb. But deafferentation leads to some disturbances in motor coordination; the animal often does not scratch the point which is irritated (15).

Detailed study of joint movements and muscular activity in the intact and deafferented limb (9) has revealed that most of the essential features of scratching movements are preserved after deafferentation. In both the intact and deafferented limb, almost all muscles can be referred to one of two reciprocal groups, i.e., to either flexors or extensors. Flexors are active during the greater part of the scratch cycle (L phase, about 200 ms) and extensors are active during the short interval (S phase, about 50 ms). Deafferentation usually does not affect either the duration of the cycle or the durations of the L and S phases. In both the intact and deafferented limb, the scratch reflex usually starts with the tonic activation of flexors which determines an initial limb posture; then the limb begins to oscillate rhythmically. In both the intact and deafferented limb, duration of the scratch cycle only slightly depends on the intensity of scratching. Thus, these essential features of scratching are determined by the central spinal mechanism.

In the present experiments, we investigated the central spinal mechanism of scratching. For this purpose, we recorded activity of spinal interneurons during fictitious scratching (9). This process can be evoked in curarized cats by stimulating either the pinna or upper cervical segments of the spinal cord. Besides, for the generation of rhythmical oscillations it is necessary to put the limb into the proper position ("scratch posture").

The study of activity of motoneurons (7) and muscle nerves (9) has shown that the efferent pattern of fictitious scratching is similar to that of normal scratching: tonic activity of flexor motoneurons appears before rhythmical oscillations begin, duration of the cycle and those of the flexor and extensor phases are also preserved after an animal has been immobilized.

Not only activity of motoneurons, but also that of some spinocerebellar pathways (2–5) and of vestibulospinal or reticulospinal tracts (1, 18) change only slightly after immobilization of an animal. This assured us that activity of many neuronal mechanisms during fictitious scratching is similar to that during normal scratching.

It is evident that the mechanism generating the rhythm and temporal efferent pattern of scratching is located in the lumbosacral segments of the spinal cord since Sherrington (25, 28) managed to evoke scratching in animals spinalized even at the middle thoracic level. So, in the present study, we tried to obtain a general picture of the activity of interneurons from lumbar segments during fictitious scratching. For this purpose, we tended to explore evenly a cross section of the gray matter (except for the motor nuclei). Activity of interneurons was recorded not only during rhythmical generation, but also during the latent period of scratching when the cervical spinal cord had already been stimulated but rhythmical oscillations had not yet appeared. We supposed that the behavior of interneurons during this period reflected changes in the spinal network which were essential for the rise of rhythmical oscillations.

METHODS

The experimental procedures were partly described earlier (9). The cats were decerebrated under ether anesthesia at the intercollicular level. Then the head, spine, and pelvis were rigidly fixed, and laminectomies at the levels of C_1–C_3 and L_1–S_2 were performed. The exposed spinal cord was covered with warm mineral oil. Animals were immobilized with Flaxedil (4–5 mg/kg, iv) and artificially respirated. Scratching (in the subsequent text we shall use this word to denote fictitious scratching) was evoked by electrical stimulation of the cervical spinal cord (C_2) or, sometimes, by tactile stimulation of the pinna. Before stimulating, the hindlimb was put into the scratch posture, i.e., it was partly flexed at the knee and ankle joints and deflected forward by flexing the hip joint. Electrical activity (ENG) of the nerve supplying m. gastrocnemius lateralis and, sometimes, of other muscle nerves was recorded bipolarly. Extra-

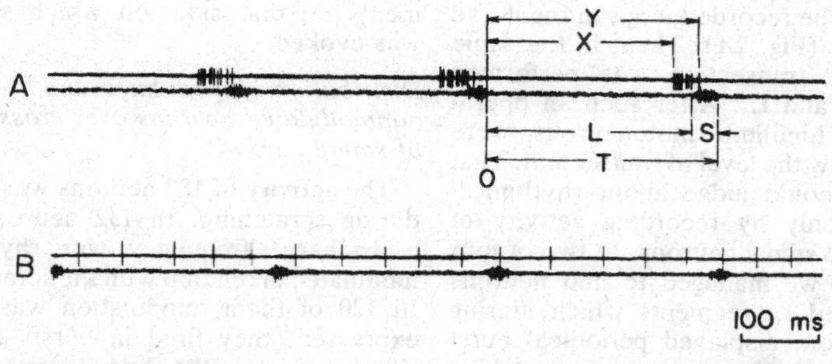

FIG. 1. Activity of a modulated (*A*) and nonmodulated (*B*) neurons during scratching. The lower trace is ENG of n. gastrocnemius lateralis. Marked are: termination of the extensor efferent activity (0) which is assumed to be the beginning of the cycle; cycle duration (T), flexor phase (L), extensor phase (S), onset of the burst (X), and termination of the burst (Y).

cellular recording of spinal neurons was undertaken with platinum microelectrodes (tip diameter, 10–20 μm; resistance, 100–500 kΩ). The electrode was inserted into the L_4–L_5 segments (this region was chosen after preliminary experiments, see RESULTS). Coordinates of each neuron (a distance from the middle line and a depth of the microelectrode insertion) were marked on the map (22) of the spinal cord cross section. We tended to explore evenly the cross section of the gray matter (except for the motor nuclei) on that side on which scratching was evoked. All neurons found while moving the microelectrode were tested during scratching. The search for neurons was often carried out initially during scratching to detect units without background activity; 182 neurons were recorded in 15 experiments. There was no identification of neurons; most of the recorded cells seem to be interneurons since they were located outside the motor nuclei (see RESULTS). However, one cannot exclude the possibility that some of the cells are motoneurons.

The following procedures were used for the data analyses. For each neuron, a part of the record containing regular cycles of intensive scratching was selected. For neurons having distinct modulation of the discharge during scratching (i.e., firing in bursts like a neuron in Fig. 1*A*), a burst position in the normalized scratch cycle was determined. The termination of the burst of gastrocnemius ENG was assumed to be the beginning of the cycle (0 in Fig. 1*A*), and the cycle duration (T), burst onset (X), and termination (Y) were measured. These measurements were performed for five successive cycles, and mean values of X, Y, and T were calculated. Then mean X and Y were divided by mean T to determine phases of switching on and off of a neuron. Such normalization allowed us to compare phases of activity for different neurons in spite of the fact that the cycle duration in various experiments ranged from 180 to 330 ms, with mean value of 243 ± 28 (SD) ms. Mean phase of the extensor burst onset (border between L and S phases) was equal to 0.84 ± 0.1 (6).

Besides the values noted above, mean frequency in the burst (the number of spikes divided by the burst duration) was calculated for each neuron. To estimate the frequency pattern in the burst, the number of spikes in the first and second halves of the burst were counted (m_1 and m_2) and their ratio $\alpha = m_2/m_1$ was obtained. These values were also averaged over five successive cycles.

In some of the neurons, the measurements and calculations were performed not only for intensive, but also for weak scratching in which the ENG amplitude was 2–3 times smaller.

RESULTS

Choice of region for neuron recordings

To choose a region of the spinal cord for recording the neuronal activity, preliminary experiments were carried out. The first series of experiments revealed that most neurons in the segments L_4–S_1 are rhythmically modulated during scratching, while in more rostral segments the percentage of modulated neurons decreases so sharply that in the L_1 segment such neurons are difficult to find.

In the second series (four experiments), transections of the spinal cord at different levels were performed. First, transection was performed between L_5 and L_6; in all four cases stimulation of the C_2 segment resulted in the rhythmical efferent activity, which could be recorded, e.g., in the nerve of m. vastus (Fig. 2A). Then, in the same experiments, transection was performed between L_4 and L_5. After such an operation, most hindlimb motoneurons were situated below the level of transection. That is why we could judge about rhythmical generation only by recording activity of nonidentified spinal neurons. In two of four experiments we managed to find neurons in the L_3 and L_4 segments which, during C_2 stimulation, displayed periodical burst firing with a rhythm typical of scratching (Fig. 2B). Finally, transection was performed between L_3 and L_4. After the transection, in one of four experiments, we managed to find a neuron in the L_3 segment which generated bursts during C_2 stimulation, but generation was not regular (Fig. 2C).

Thus, stable generation of rhythmical oscillations is observed provided the L_5 and more rostral segments are intact. On the other hand, the amount of modulated neurons sharply decreases in the upper lumbar segments. Thus, we have chosen the L_4 and L_5 segments for further analysis.

In two experiments, we destroyed the gray matter of the L_3–L_5 segments on the contralateral side. In both cases, scratching could be evoked after the destruction. Apparently, neurons from the contralateral side of the spinal cord are not necessary for generation of rhythmical oscillations. This is why, in the continuing study, we recorded neurons from the L_4 and L_5 segments on that side, on which scratching was evoked.

Distribution of modulated and nonmodulated neurons over cross section of spinal cord

The activity of 182 neurons was recorded during scratching. In 132 neurons (73%), a discharge frequency was rhythmically modulated in relation with the scratch cycle. In 120 of them, modulation was strongly expressed; they fired in bursts and were silent between the bursts, as a neuron in Fig. 1A. In 50 neurons, modulation was not observed, as in a neuron in Fig. 1B. Frequencies of the resting discharge in nonmodulated neurons ranged from 0 to 40 Hz, with a mean value of 15 Hz. During scratching, the discharge frequency in most of these neurons increased 2–3 times, mean value during scratching was 26 Hz. In four neurons, an inhibition was observed during scratching.

Figure 3 shows the distribution of modulated (A) and nonmodulated (B) neurons over the cross section of the spinal cord gray matter. Nonmodulated neurons are located mainly in the dorsal horn, but they can be found in the ventral horn as well. The neurons with strongly expressed modulation (black circles in Fig. 3A) are located mainly in the intermediate part of the gray matter, which approximately corresponds to the Rexed's layer VII (22), but they can also be found in the layers V, VI, and VIII. The neurons having weak

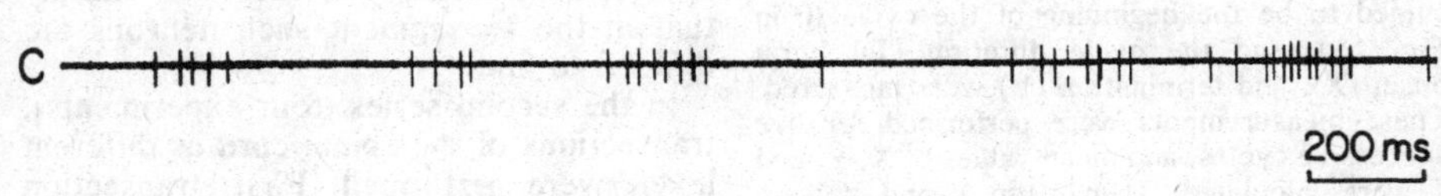

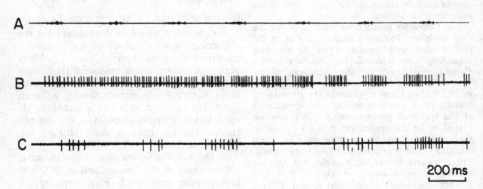

FIG. 2. Effects of spinal cord transections between L_5 and L_6 (A), L_4 and L_5 (B), L_3 and L_4 (C). Activity of n. vastus (A) and of a neuron from L_3 (B, C) are shown. In A and C, C_2 stimulation was started earlier than recording; in B, simultaneously with recording.

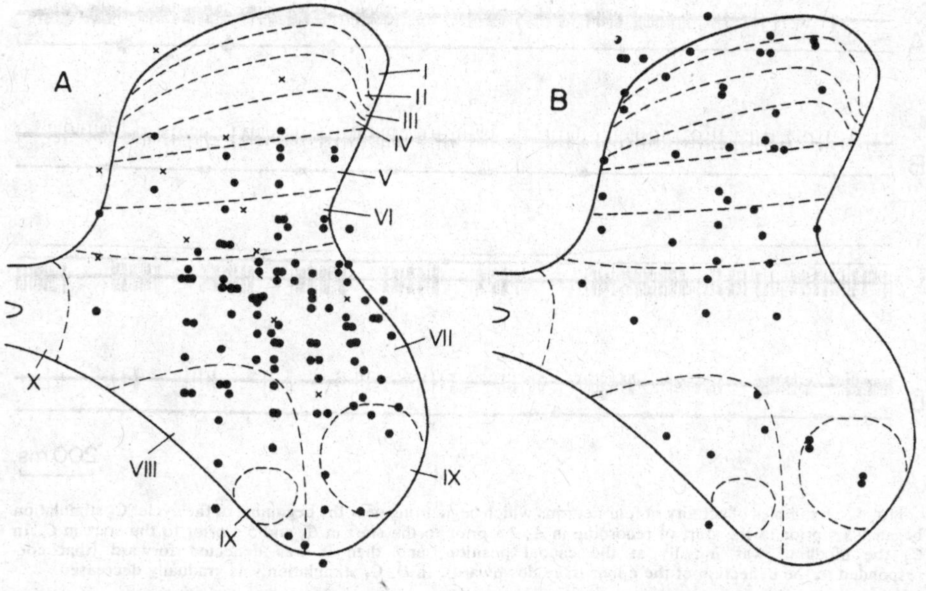

FIG. 3. Distribution of all recorded neurons over the cross section of the spinal cord gray matter. Modulated neurons are presented in *A* (neurons with weak modulation are marked by crosses), nonmodulated ones in *B*. Numbers of Rexed's layers (I–X) are indicated.

modulation (crosses in Fig. 3*A*) are located mainly in the dorsal horn. Figure 3 shows that most neurons were recorded outside the motor layer IX, and they may thus be considered as interneurons. For further analyses we used 120 cells with strongly expressed discharge modulation.

Peculiarities of neurons firing in different phases

Examples presented in Figs. 1*A*, 4, 5, and 6 show that burst positions in the cycle greatly differ in different neurons. But for each individual neuron, the burst position was rather stable both in successive cycles and during repeated tests. The burst position slightly changed in the minority of neurons only while changing the intensity of scratching (Fig. 5*B*, *C*). Therefore, the burst position can be considered as a stable parameter of each neuron. The burst position is determined by two values, i.e., by the phases of switching on and off, the former one being especially diverse in different neurons (Figs. 1*A*, 4, 5, and 6). Besides, further analysis has revealed that the phase of switching on is linked with some other parameters of a neuron. Therefore, we shall consider separately the neurons which begin firing at the beginning, in the middle, and at the end of the cycle.

The records of activity of four neurons firing from the beginning of the cycle are presented in Fig. 4. These neurons fired in long bursts (lasting more than a half of the cycle), the bursts terminating shortly before the onset of the extensor ENG. Thus, the burst positions coincided almost completely with the flexor phase of the cycle. During the latent period of scratching (when C_2 stimulation had already been started but rhythmical oscillations had not yet appeared), a tonic activation was observed in most neurons. Such activation is seen in *A*, where the discharge rate reaches 130 Hz by the beginning of oscillations. With the appearance of ENG bursts, a tonic firing of the neuron is periodically interrupted. A similar example is presented in *B* (the beginning of activation is not shown). The discharge rate of the neuron is about 50 Hz both by the end of the latent period

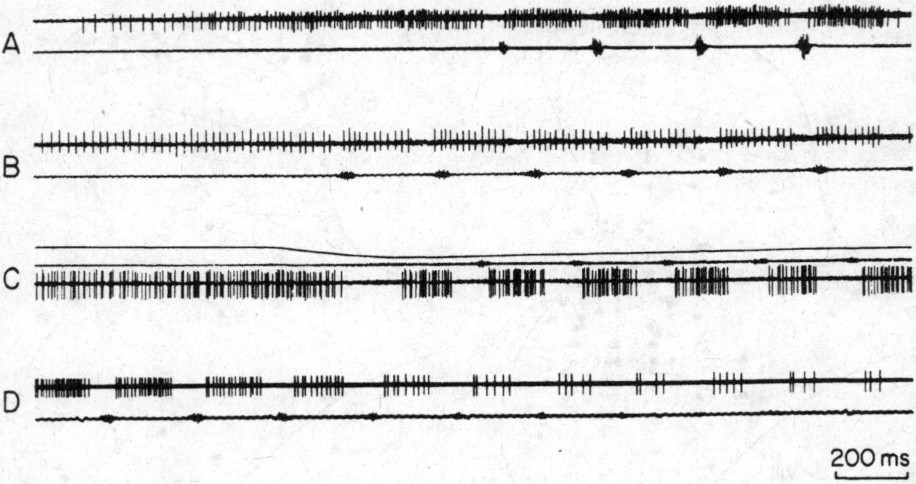

FIG. 4. Examples of activity of four neurons which begin firing near the beginning of the cycle. C_2 stimulation began 0.5 s prior to the start of recording in *A*, 2 s prior to the start in *B*, and 5 s prior to the start in *C*. In *C*, the hindlimb was initially at the caudal position, and then it was deflected forward (that corresponded to the deflection of the upper trace downward). In *D*, C_2 stimulation was gradually decreased.

and in the bursts during scratching. In the records *A* and *B*, the hindlimb was at all times at the scratch posture (deflected forward) that was necessary for generation of rhythmical oscillations (9). In *C*, the hindlimb was initially deflected backward, and rhythmical oscillations did not arise in spite of continuous C_2 stimulation; the only effect of stimulation was strong continuous firing of the neuron. Then the limb was rapidly deflected forward, and generation of rhythmical oscillations immediately began.

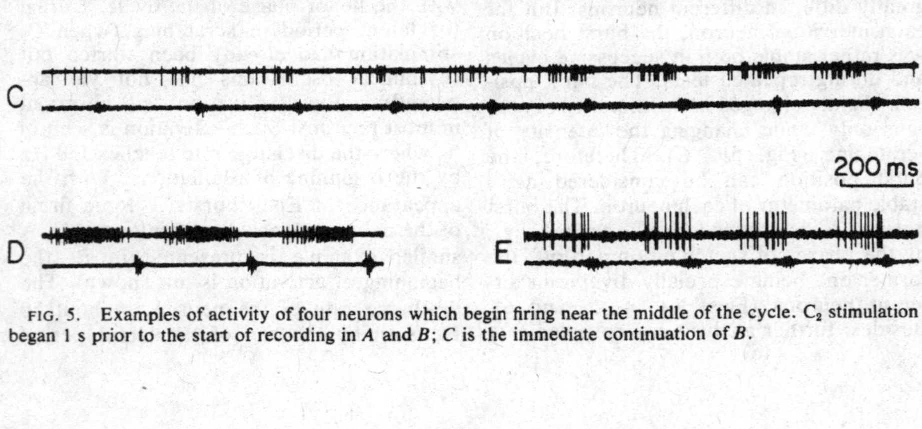

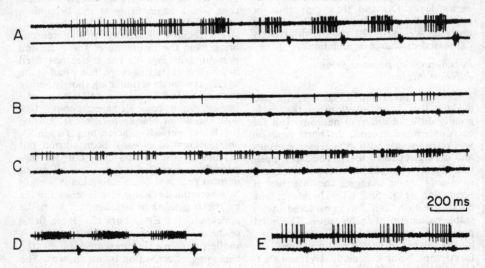

FIG. 5. Examples of activity of four neurons which begin firing near the middle of the cycle. C_2 stimulation began 1 s prior to the start of recording in *A* and *B*; *C* is the immediate continuation of *B*.

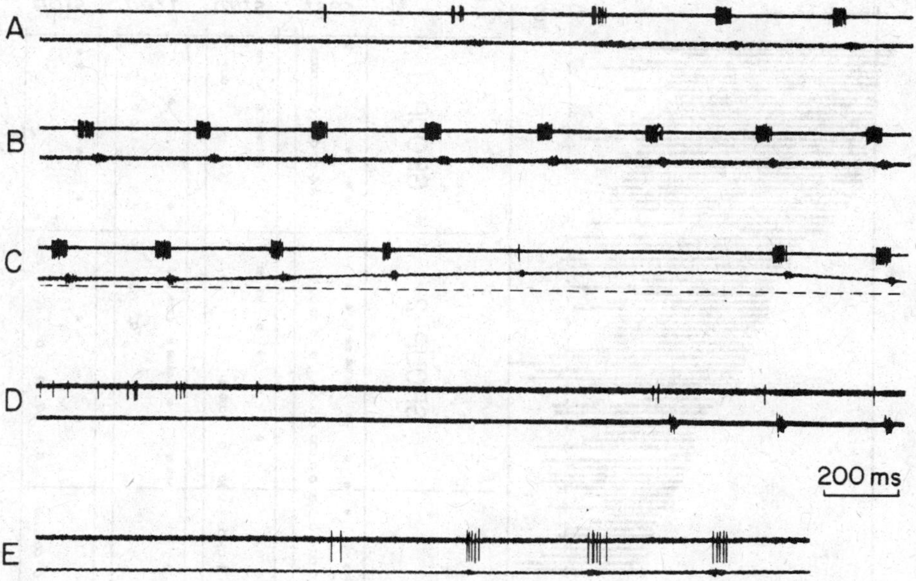

FIG. 6. Examples of activity of three neurons (*A–C*, *D*, and *E* correspondingly) firing at the end of the cycle. C_2 stimulation began 1 s prior to the start of recording in *A*, *D*, and *E*. *B* is the continuation of *A*, *C* is the continuation of *B*. In *C*, the hindlimb was deflected backward (that corresponded to the deflection of the lower trace upward) and then returned to the initial position.

In *D*, C_2 stimulation was gradually decreased, which resulted in reduction and subsequent disappearance of extensor bursts, as well as in decrease of rhythmical activity of the neuron. But rhythmical bursts in the neuron continued longer than efferent bursts. Corresponding effects are seen in *C*: the first pause in the neuron discharge was not accompanied by an efferent burst.

Now we shall consider the neurons which begin to fire not from the very beginning of the cycle, but closer to the middle of the cycle (Fig. 5). These neurons usually fired in bursts shorter than those of neurons firing from the beginning of the cycle. Bursts usually terminated shortly before or during the extensor phase. In a large number of neurons, the firing rate increased in the course of the burst, sometimes considerably (*D*). During the latent period of scratching, some neurons were activated (*A*) while others were not (*B*). With more intensive scratching, rhythmical activity of the neurons increased; the frequency became higher (*A*) and, in some neurons, the burst became longer (cf. *B* and *C*). In some cases, rhythmical bursting began earlier in the neuron than in the muscle nerve (*A*).

Finally, we shall consider those neurons which begin firing closer to the end of the cycle (Fig. 6). They fired in bursts even shorter than the neurons described above; sometimes the burst comprised only two to three spikes. During the latent period of scratching, these neurons were not activated (*A*) or, in many cases, their resting discharge was inhibited (*D*). With more intensive scratching, the rhythmical activity of the neurons increased; the frequency in the burst became higher and the bursts longer (*A*). In *C*, the effect of deflection of the hindlimb backward is shown; both the efferent activity and that of the neuron decreased and then disappeared. They reappeared immediately when the limb was deflected forward.

Phase distribution of neurons

Figure 7 shows burst positions of 120 neurons in the normalized scratch cycle;

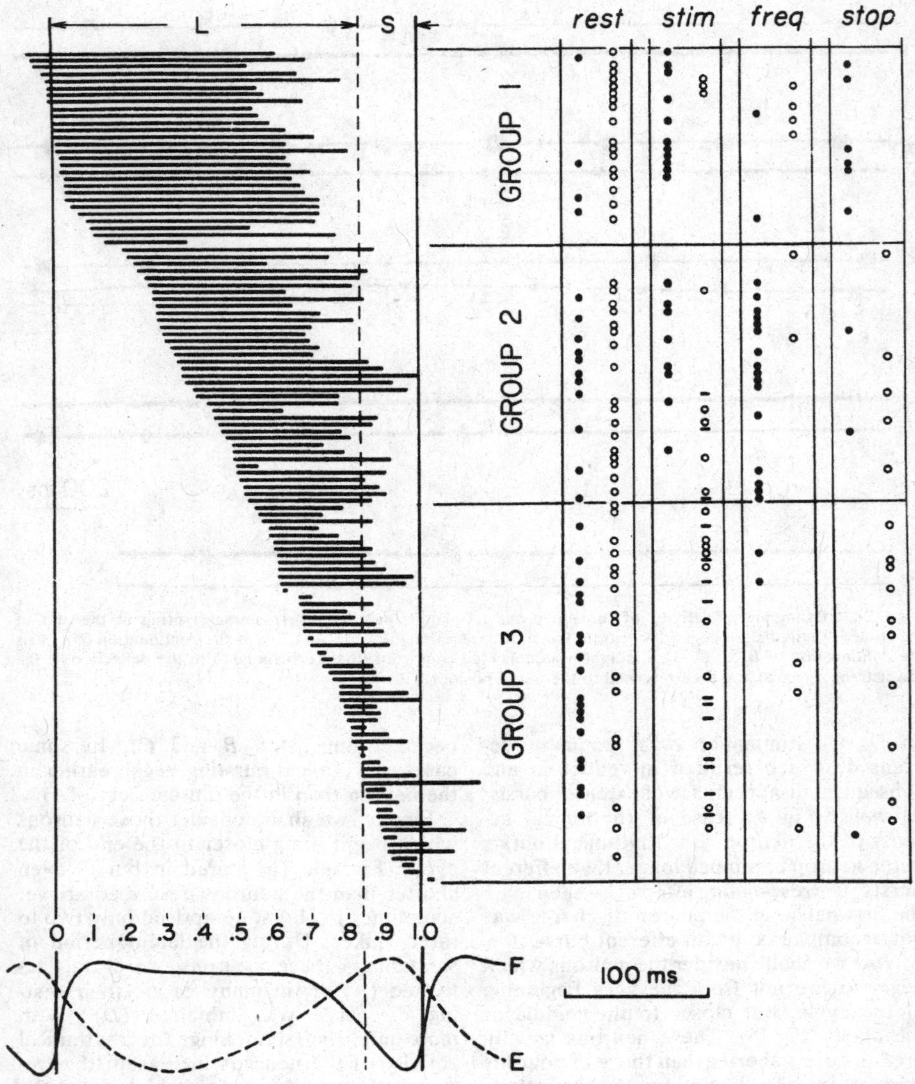

FIG. 7. Phase distribution of 120 neurons and table of their parameters. The burst position of each neuron in the normalized scratch cycle is shown by a line. Curves F and E show approximate dependence between the membrane potential of flexor and extensor motoneurons and the phase of the cycle (7). Time calibration is presented for "average" scratch cycle. In the rest column, the behavior of 83 neurons at resting conditions is shown (filled circles, neurons with resting discharge; open circles, without resting discharge). In the stim column, the behavior of 52 neurons during the latent period of scratching is shown (filled circles, neurons are activated; open circles, neurons have no resting discharge and are not activated; minuses, resting discharge of neurons is inhibited). In the freq column, the frequency pattern of all neurons is being characterized (filled circles, the frequency increases in the course of the burst; open circles, the frequency decreases; no sign, the frequency does not change significantly). In the stop column, the effect of the backward deflection of the limb is shown for 24 neurons (filled circles, tonic activation; open circles, inhibition).

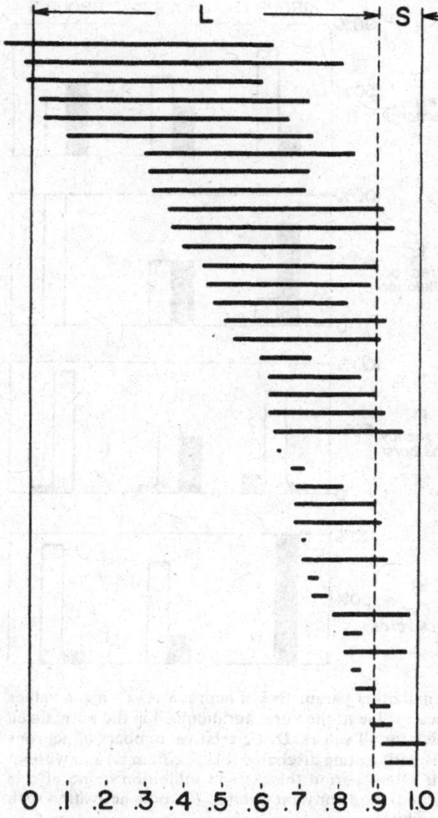

FIG. 8. Phase distribution of 40 neurons recorded in one experiment.

the flexor ("long") and extensor ("short") phases of the cycle are also indicated (L and S). In this graph, the neurons are put in order according to the phase of switching on; the later the neuron is switched on in the cycle, the lower it is presented on the graph. This order is broken for only a few neurons that begin firing at the very end of the cycle but are shown in the upper part of the graph.

The neurons presented on the graph were recorded in 15 experiments. Phase distributions obtained in various experiments were rather similar and resembled the total distribution. Figure 8 shows the phase distribution obtained in the most successful experiment in which 40 neurons were recorded. In this experiment, the rhythm was very stable; the mean value of cycle duration was 268 ± 28 (SD) ms, mean phase of the extensor burst onset was 0.89 ± 0.03. One can see that this distribution is very similar to the total one (Fig. 7).

Phases of switching on, as one can see in Fig. 7, are distributed rather evenly throughout the cycle, except for the very beginning where about 15 neurons begin firing almost simultaneously. This is also distinctly seen in a histogram of Fig. 9*A* showing the number of switchings on in different phases of the cycle. Phases of switching off are distributed mainly in the second half of the cycle. A histogram in Fig. 9*B* shows the number of switchings off in different phases. Starting from the middle of the cycle, the amount of switchings off increases, reaches the maximum at the beginning of the S phase, and sharply decreases by the end of the cycle.

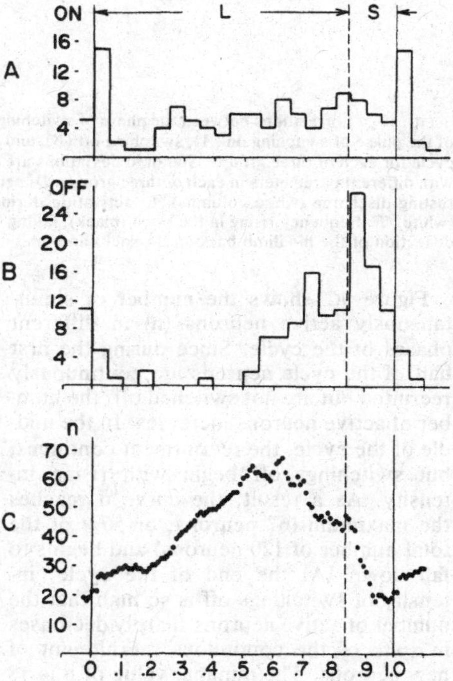

FIG. 9. Numbers of switchings on (*A*), switchings off (*B*), and number of active neurons (*C*) as functions of the phase of the cycle.

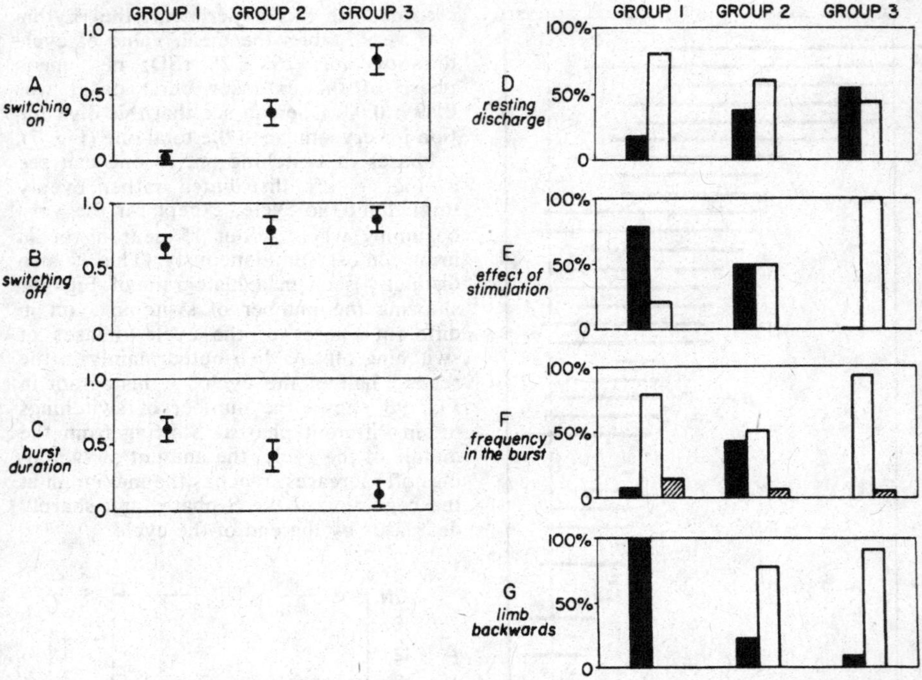

FIG. 10. Correlations between the phase of switching on and other parameters of neurons. *A–C*: mean values of the phase of switching on (*A*), switching off (*B*), and a mean value of the burst duration (*C*) in the normalized cycle for each of three groups. Standard deviations are shown for all values. *D–G*: relative numbers of neurons with different parameters in each of three groups. *D*: neurons with resting discharge (black columns) and without resting discharge (white columns). *E*: activation during the latent period (black) and inhibition or no effects (white). *F*: frequency rising in the burst (black), falling (hatched), and constant (white). *G*: tonic activation with deflection of the hindlimb backward (black) and inhibition (white).

Figure 9*C* shows the number of simultaneously active neurons (n) in different phases of the cycle. Since during the first half of the cycle neurons are continuously recruited but are not switched off, the number of active neurons increases. In the middle of the cycle, the recruitment continues, but switchings off begin with rising intensity. As a result, the curve n reaches the maximum (67 neurons, or 56% of the total number of 120 neurons) and begins to fall down. At the end of the cycle, intensity of switchings off is so high that the number of active neurons sharply decreases in spite of the continuous recruitment of new neurons. The minimal value of n is 18 (15% of the total number). Thus, the amount of active neurons changes by a factor of 3.7 during the cycle.

Parameters of neurons correlated with phase of switching on

While describing neurons firing in different phases of the cycle, we noted that phases of switching on were correlated with some other parameters (burst duration, frequency pattern, etc.). These correlations can be also seen in Fig. 7 where a table of parameters is presented. For the analyses of the correlations, we divided all neurons into three groups according to the phase of switching on (Fig. 7). To group *1* we referred the neurons firing in phase with flexor motoneurons. Curve F in the lower part of Fig. 7 schematically shows changes in the membrane potential of flexor motoneurons during the scratch cycle; motoneurons fire during the depolarization lasting for the greater part of the L phase (7).

Neurons of group *1* thus discharge in phase both with the depolarization of flexor motoneurons and with their bursts. Group *1* comprises 28 neurons (23% of the total number) with phases of switching on ranging from −0.07 (or 0.93 if measuring from the beginning of the preceding cycle) to 0.14.

We experienced some difficulties in attempting to pick out the neurons with the pattern of activity similar to that of extensor motoneurons. Curve E in the lower part of Fig. 7 schematically shows changes in the membrane potential of extensor motoneurons during the scratch cycle (7). It is seen that the depolarization continuously increases during the second half of the L phase, but these changes of the membrane potential are not reflected in the firing of motoneurons; they discharge only on the peak depolarization, in the S phase. Thus, in extensor motoneurons, there exists no simple correlation between the membrane potential and the discharge, unlike the case of flexor motoneurons. Thus, it is not clear which neurons should be considered to be similar to extensor motoneurons—those which fire in the S phase or those which begin firing in the second half of the L phase. This is why we decided to use another criterion for further grouping the neurons which had not fallen into the group *1*. We used the phase 0.55 as a border between groups *2* and *3*. This phase is remarkable for being the beginning of switchings off (Figs. 9*B* and 12). Besides, there were practically no neurons in which the discharge rate began to decrease earlier than at 0.55. Group *2* comprises 37 neurons (31%) with the phases of switching on ranging from 0.18 to 0.55. Group *3* comprises 52 neurons with the phases of switching on from 0.56 to 0.93, and three neurons whose phases of switching on overlapped with those in group *1* (these three neurons are shown in the lower part of the phase distribution in Fig. 7). Group *3* thus comprises 55 neurons (46% of the total number). Mean values of the phase of switching on for groups *1–3* are: 0.02 ± 0.05, 0.37 ± 0.11, and 0.77 ± 0.12 (Fig. 10*A*).

The following parameters were found to be correlated with the phase of switching on.

PHASE OF SWITCHING OFF AND BURST DURATION. One can see in Fig. 7 that, on the average, the neurons which begin firing later in the cycle, stop firing also somewhat later. Mean values of the phase of switching off for the groups *1–3* are: 0.65 ± 0.09, 0.8 ± 0.11, and 0.9 ± 0.1 (Fig. 10*B*). Since mean values of the phases of switching off differ considerably less than those of switching on (Fig. 10*A*), there is a large difference between the burst durations in groups *1–3*. Mean values of the burst duration are: 0.63 ± 0.1, 0.43 ± 0.13, and 0.13 ± 0.1 (Fig. 10*C*).

RESTING DISCHARGE. Recording of the resting discharge was carried out in 84 of 120 neurons; 50 neurons (60% of the total number of tested neurons) discharged at rest with frequences ranged from 1 to 30 (sometimes up to 50) Hz. In the "rest" column (Fig. 7), neurons having resting discharge are shown by filled circles, while those without resting discharge, by open circles. One can see that the relative number of filled circles increases in the lower part of the rest column. These numbers for the groups *1–3* are: 19, 39, and 54% (Fig. 10*D*).

BEHAVIOR DURING LATENT PERIOD. Activity during the latent period of scratching was recorded in 52 neurons. The results are presented in the "stim" column (Fig. 7). The neurons which were activated (as neurons in Figs. 4*A*, *B* and 5*A*) are shown by filled circles. The neurons in which an inhibition of the resting discharge was observed (as in a neuron in Fig. 6*D*) are shown by minuses. Finally, the neurons which had no resting discharge and were not activated by C_2 stimulation (as neurons in Figs. 5*B* and 6*A*, *E*) are shown by open circles. One can see that the neurons which were activated during the latent period prevail in the upper part of the graph, while those which were not activated or were inhibited, in the lower part. In group *1*, 79% of neurons were activated; in group *2*, 50%; in group *3*, no neurons were activated (Fig. 10*E*).

FREQUENCY PATTERN. The parameter α which characterizes frequency pattern (see METHODS) was determined for all neurons. The results are presented in "freq" column (Fig. 7). The neurons with $\alpha > 1.2$, i.e., those in which the frequency increased in the course of the burst (as in neurons in Fig. 5*A*, *C*, *D*), are shown by filled circles, while those with decreasing frequency ($\alpha < 0.8$) (as a neuron in Fig. 1*A*) are shown by open circles. The neurons with $0.8 \leq \alpha \leq 1.2$, i.e., discharging with rather constant frequency, are not marked. One can see that a great number of neurons in the middle part of the graph have rising frequencies. The overwhelming majority of neurons in groups *1* and *3* has rather constant frequencies, while 43% of neurons in group *2* have rising frequencies (Fig. 10*F*).

EFFECTS OF LIMB DEFLECTION. Twenty-four neurons were tested by a deflection of the hindlimb backward, which resulted in the cessation of rhythmical generation. The results are presented in the "stop" column (Fig. 7). All six neurons from group *1* displayed tonic activity when rhythmical generation was interrupted (as a neuron in Fig. 4*C*). In group *2*, a tonic firing was observed in two of seven tested neurons. In group *3*, 10 of 11 neurons became completely silent with the limb deflection (as a neuron in Fig. 6*C*). These results are also shown on the diagram in Fig. 10*G*.

Neuron parameters not correlated with phase of switching on

LOCATION OF NEURON IN SPINAL CORD. We failed to find any difference between distributions of groups *1–3* neurons over the cross section of the spinal cord. Besides, in a number of cases, pairs of neurons were recorded simultaneously by the same microelectrode. Various combinations could be found among the pairs: a neuron of group *1* and a neuron of group *1*; a neuron of group *1* and a neuron of group *2*, etc. Thus, neurons of different groups seem to be "intermingled" in the spinal cord.

DISCHARGE FREQUENCY. Mean frequencies in bursts during intensive scratching were found to be almost the same in groups *1–3*; 84 ± 41, 82 ± 42, and 84 ± 57 Hz. The activity of neurons depended on the intensity of scratching; 10 neurons of group *1*, 19 neurons of group *2*, and 12 neurons of group *3* were recorded both during intensive and during weak scratching (ENG amplitudes differed 2–3 times). During intensive scratching, all recorded neurons fired with frequencies 1.5–3 times higher than during weak scratching. Figure 11 shows activity of groups *1–3* in different phases of the cycle, during weak (open circles) and intensive (filled circles) scratching. To obtain the curves, the cycle was divided into 20 intervals, and numbers of spikes generated by each neuron in the given interval were summed up. This procedure was carried out separately for each group. One can see that during intensive scratching the overall activity in each group increases 1.7–2.5 times, compared to weak scratching. In *C*, mean value of the cycle duration and that of the S phase are shown for intensive and weak scratching. The cycle of intensive scratching is 10% shorter than that of weak scratching, while the S phase is somewhat longer. Thus, the twofold increase of the activity of neurons during intensive scratching is accompanied by only minor change of the cycle duration.

RHYTHMICAL BURSTING IN ABSENCE OF RHYTHMICAL EFFERENT ACTIVITY. In

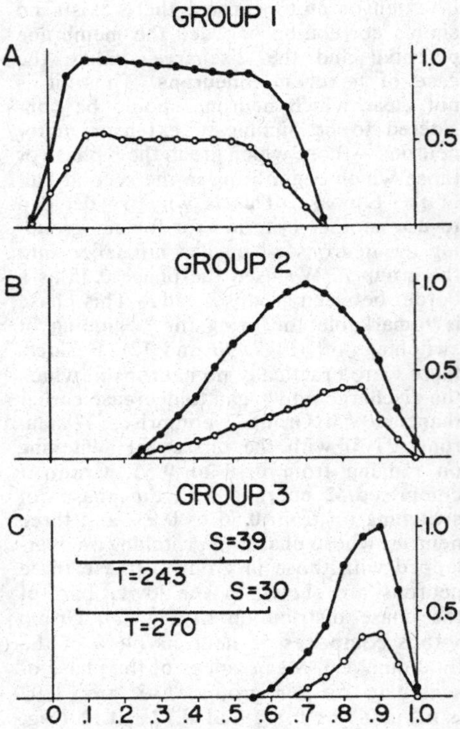

FIG. 11. Overall activity in each of three groups during intensive scratching (filled circles) and during weak scratching (open circles) as a function of the phase of the cycle (arbitrary units). Durations of the cycle and of the S phase are shown in *C* for intensive (upper line) and weak (lower line) scratching (see text).

some neurons, rhythmical burst firing began earlier than periodical efferent activity (as in a neuron in Fig. 5*A*). Such cases were observed in all three groups.

DISCUSSION

Localization of rhythm generator

Our experiments have shown that regular rhythmical oscillations can be generated by the spinal cord after removing the segments below L_5 or, sometimes, below L_4. On the other hand, the amount of rhythmically active neurons sharply decreases in L_1–L_3. Therefore, it seems likely that the segments L_4 and L_5 are responsible for generation of rhythmical oscillation when lower segments are removed. Are lower segments also capable of generating rhythmical oscillations? Our data fail to answer this question. However, it was demonstrated that the segments below L_5 were capable of generating the rhythm of stepping (S. Grillner and P. Zangger personal communication). These data suggest that both upper and lower parts of the lumbosacral enlargement of the spinal cord are capable of generating rhythmical processes.

We have found that neurons with the rhythmical burst firing are very sparse in the dorsal horn (Fig. 3*A*). Corresponding results were obtained in the study of locomotion (17); rhythmical modulation of neurons in the dorsal horn was observed when the hindlimb innervation was intact and was not observed when the hindlimbs were deafferented. These data are in accordance with the general point of view that the dorsal horn neurons belong to the afferent system.

Most modulated neurons were located in Rexed's (22) layer VII. Similar localization of modulated neurons was found in the study of locomotion of cats with deafferented hindlimbs (17). In the same area there were localized neurons which responded by long bursts (resembling those during locomotion) to peripheral stimulation after administrating dopa (16), i.e., of the drug enhancing locomotion (14). These data suggest that generation of both the rhythm of scratching and that of stepping is produced by neurons of this area.

Behavior of neurons of different groups

The present study has shown that the burst position in the scratch cycle is rather stable both in successive cycles and when scratching is repeatedly evoked. Also, the burst position in most neurons changes only slightly with considerable changes of the intensity of scratching. This is why the burst position can be considered as the stable parameter of a neuron. Our data have also shown that the burst position is not only the stable parameter but also the essential one, in the sense that it is connected with some other parameters (resting discharge, behavior during the latent period, frequency pattern, and effect of the limb deflection). These correlations are of a statistical nature since not all neurons firing in a given phase have identical parameters. The arbitrary division of all neurons into three groups used above is also convenient to describe typical features of the behavior of neurons firing in different phases.

GROUP *1*. The behavior of group *1* neurons resembles, in many respects, that of flexor motoneurons: *a*) Tonic firing of the neurons appears (or increases) during the latent period of scratching. Correspondingly, a tonic depolarization is observed in flexor motoneurons during this period (7). *b*) During scratching, tonic firing of the neurons is periodically interrupted. The pauses begin shortly before the end of the L phase and terminate at the end of the S phase or at the beginning of the subsequent L phase. Correspondingly, a hyperpolarization and cessation of the discharge is observed in flexor motoneurons at the end of the L phase and during the S phase (curve F in Fig. 7) (7). *c*) The firing rate in the burst is rather constant and almost equal to that reached by the end of the latent period. Correspondingly, the level of depolarization in flexor motoneurons changes only slightly during the greater part of the L phase; this level is equal to that reached by the end of the latent period (7). *d*) When the hindlimb is deflected backward and the rhythm generation ceases, group *1* neurons fire tonically. The same behavior is observed in flexor motoneurons (9). *e*) Most of the group *1* neurons begin to fire almost simultaneously, while terminations of their bursts are considerably less synchronized (Figs. 7 and 10*A*, *B*). Correspondingly, the speed of depolarization in flexor motoneurons at the end of the S phase is considerably higher than that of hyperpolarization at the end of the L phase (7).

One can see in the table of Fig. 7 that the group *1* is rather homogeneous: the parameters of most neurons are the same. The behavior of the "typical" group *1* neuron is, in many respects, similar to that of flexor motoneurons. But it should be emphasized that most of the recorded neurons seem not to be flexor motoneurons since *a*) they are located outside the motor nuclei and *b*) their discharge rate is considerably higher (on the average) than that of flexor motoneurons (84 against 40 Hz (7)).

GROUP *2*. One can see in the table of Fig. 7 and in Fig. 10*D–G* that this group is not homogeneous: 39% of neurons have a resting discharge, 50% are activated during the latent period, 43% discharge in bursts with rising frequencies. Activity of group *2* neurons correlates neither with that of flexor motoneurons nor with that of extensor ones: the neurons are continuously recruited in the L phase when the membrane potential of flexor motoneurons does not change; the discharge in most neurons terminates prior to the extensor burst (Fig. 7).

GROUP *3*. This group is rather homogeneous (see the table in Fig. 7 and Fig. 10*D–G*), and a description of the typical group *3* neuron can thus be presented: *a*) the neuron is not activated (or is inhibited) during the latent period of scratching; *b*) the neuron fires in short bursts at the end of the cycle. *c*) the discharge rate in the burst is rather constant; *d*) when the hindlimb is deflected backward and rhythm generation ceases, the neuron is silent. A part of group *3* neurons fire in phase with extensor motoneurons and out of phase with flexor ones.

It is worth noticing that though the line between groups *2* and *3* was drawn according to formal criteria, there are indirect data indicating that these groups might be of different functional meaning. Investigation of the activity of spinocerebellar pathways during fictitious scratching has shown that the parameters of neurons of the ventral spinocerebellar tract are similar to those of the groups *1* and *2* neurons (5), while parameters of neurons of the spinoreticulocerebellar pathway are similar to those of the group *3* neurons (4). These results raise the interesting possibility that different spinocerebellar pathways convey messages about different processes occuring in the spinal cord.

Functional considerations

Spinal interneurons recorded in this study can belong to different functional groups: they can generate rhythmical oscillations, control motoneurons, modulate transmission of signals in reflex and descending pathways, send messages to the cerebellum and other brain centers, etc. But our data have shown that the variety of neuron behavior during scratching is not very rich since different parameters of neurons are strongly linked. This is why an attempt to design a model of the system controlling scratching does not seem hopeless, so we shall consider possible roles of different neurons in the control of scratching.

CONTROL OF MOTONEURONS. The studies of muscular activity (9, 27) and of the activity of single motoneurons (7) have shown that flexor motoneurons are tonically activated during the latent period of scratching. When rhythmical oscillations begin, this tonic activity is periodically interrupted, the pauses approximately coinciding with the S phase of the cycle. One can suggest two ways of controlling flexor motoneurons: *a*) motoneurons receive a sustained excitatory inflow from one group of interneurons and a periodical inhibitory inflow from another group; and *b*) they receive only an excitatory inflow from a certain group of interneurons, this group being periodically inhibited. A combination of the two ways is also possible. But, in any case, the rhythm generator must provide an inhibitory drive either to motoneurons themselves or to interneurons which control motoneurons, the inhibition having to be in the S phase.

Let us consider interneurons which could participate in the control of flexor motoneurons. It was noted that the behavior of group *1* neurons was, in many respects, similar to that of flexor motoneurons. This is also seen in Fig. 12*A* which shows the number of simultaneously active group *1* neurons in different phases of the cycle. The overall activity of the group *1* gradually decreases by the end of the L phase, and sharply increases at the end of the cycle. There is striking similarity between this curve and the curve of the membrane potential of flexor motoneurons (curve F in

Fig. 7). It seems very likely that group *1* neurons either provide an excitatory drive to flexor motoneurons or are controlled in parallel with them. The process of switching off in group *1* begins from the phase of 0.55 (Fig. 12*A*). This was one reason why we referred to group *3* the neurons which were recruited starting from the phase of 0.55 (see RESULTS). A pause in the activity of group *1* neurons and flexor motoneurons at the end of the cycle can thus be the result of an inhibitory inflow from some neurons of group *3*.

In contrast to flexor motoneurons, the membrane potential of extensor motoneurons changes only slightly before rhythmical oscillations begin (7). Therefore, group *1* neurons, which are strongly activated during the latent period of scratching, can hardly participate in the control of extensor motoneurons. During scratching, extensor motoneurons are periodically depolarized (curve E in Fig. 7). The depolarization continuously increases during the second half of the L phase, reaches the maximum in the S phase, and falls down at the beginning of the cycle. The beginning of the depolarization can be the result of an excitatory inflow from neurons of group *2* which are continuously recruited during this very period (Fig. 7). But activity of group *2* falls down in the S phase; therefore, group *2* cannot be responsible for the peak depolarization of extensor motoneurons in the S phase. This depolarization can be explained by an excitatory inflow from some neurons of group *3* firing in the S phase.

GENERATION OF RHYTHMICAL OSCILLATIONS. The experimental data as well as some considerations presented above allow one to suggest that group *1* neurons should not comprise the network responsible for generation of rhythmical oscillations. It seems more likely that they are targets for periodical inhibitory influences from the generator. This is why we shall consider separately the behavior of neurons from groups *2* and *3*.

Figure 12*B* shows the number of simultaneously active neurons of groups *2* and *3* in different phases of the cycle (curve n). Starting from 0.18, the number of active neurons continuously increases until 0.63, then fluctuates near the maximal value (about 50 neurons, or 54% of the total number of neurons in groups *2* and *3*) until 0.85, and rapidly falls to almost zero during the S phase. In the interval from 0.18 to 0.63, not only does the number of neurons increase, but also the firing rate in many neurons rises (freq column in Fig. 7). Thus, groups *2* and *3* neurons (77% of all modulated neurons) exhibit striking behavior: during half of the cycle the overall activity increases and reaches a high level, then the activity completely disappears. Such great oscillations of the overall activity in groups *2* and *3* are determined by a specific distribution of burst onsets and terminations over the cycle. In Fig. 12*B* are presented the cumulative curves of switchings on (Σon) and of switchings off (Σoff). They show the amount of switchings on (off) which has occurred by a given phase of the cycle, starting from 0.18 when the first neuron of group *2* begins to fire. In the interval from 0.18 to 0.9, the curve Σon can be approximated by a straight line, i.e, in this interval new neurons are recruited rather evenly. In the interval from 0.18 to 0.55, the curve Σon coincides with the curve of the number of active neurons (n), since in this interval there are no switchings off. A process of switchings off begins from 0.55. In the interval from 0.7 to 0.85, a slope of the curve Σoff is almost equal to that of the curve Σon, i.e., in this interval, the rates of switchings on and off are the same. As a result, the number of active neurons does not change. In the S phase, the rate of switchings off becomes very high; that results in the rapid decrease of the number of active neurons in spite of the continuous recruitment of new neurons. It is worth noticing that the process of switchings off in groups *2* and *3* begins simultaneously with that in group *1* (marked by arrows in Fig. 12*A*, *B*). These processes could be thus determined by the inhibitory action of the same neurons from group *3*.

Undoubtedly, the main features of the rhythmical activity of groups *2* and *3* neurons considered above are important for understanding the process of rhythm generation. But we think that the activity of neurons during the latent period of scratching is not less important. Stimulation of the C_2 segment leads to the excitation of some

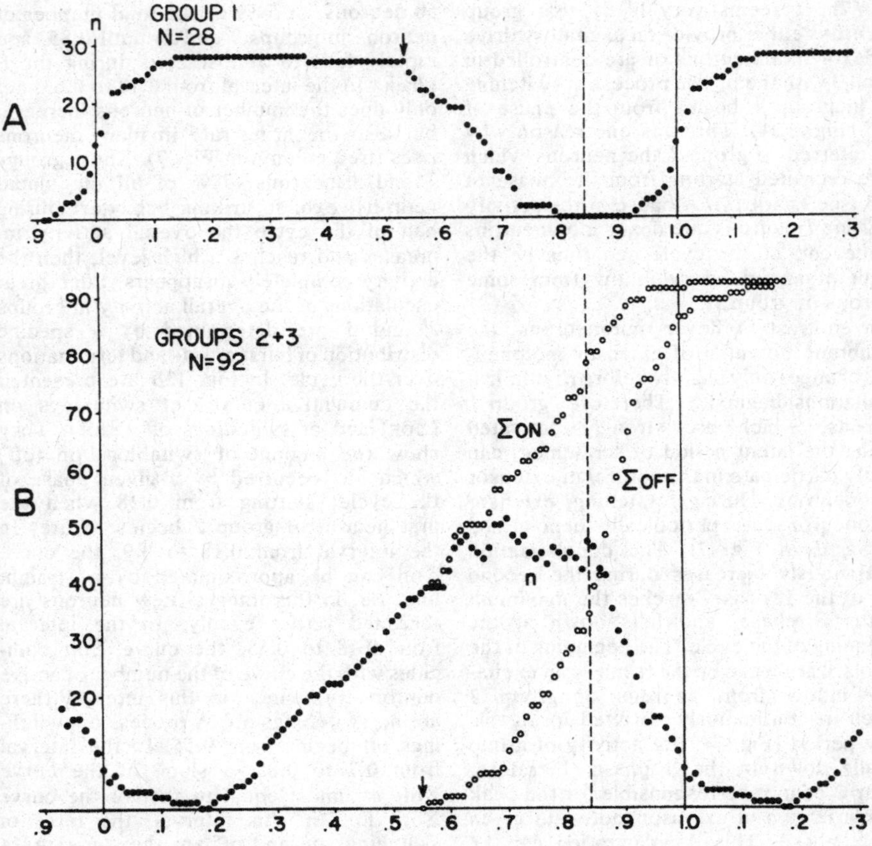

FIG. 12. *A*: number of active neurons of group *1* as a function of the phase of the cycle. *B*: number of active neurons of groups *2* and *3* as a function of the phase of the cycle (curve n), and cumulative curves of switchings on and off (Σon and Σoff) (see text). Arrows in *A* and *B* show the beginning of the process of switchings off.

group *2* neurons (as a neuron in Fig 5*A*). The remaining group *2* neurons as well as all group *3* neurons either are inhibited by stimulation (if they have resting discharge, as a neuron in Fig. 6*D*) or remain silent (if they have no resting discharge, as neurons in Figs. 5*D* and 6*A*, *E*). As a result, by the beginning of the first scratch cycle, the majority of groups *2* and *3* neurons are silent, and only a minority of them are active (the stim column in Fig. 7). It is worth noticing that the activation is observed in those neurons which begin discharging earlier in the cycle than other neurons of groups *2* and *3*. We are inclined to interpret the events occuring during the latent period of scratching in the following way: *a*) the inhibition of the neurons which discharge at the end of the cycle releases the network generating rhythmical oscillations; and *b*) generation of the first and subsequent cycles begins with the excitation of the neurons, which begin discharging earlier in the scratch cycle than other neurons of the groups *2* and *3*.

We think that the experimental data obtained in this study can be satisfactorily explained within the limits of the following hypotheses about organization of the network generating rhythmical oscillations.

There is a group of neurons with mutual excitatory connections (E-neurons). A part of these neurons also exerts an excitatory action on the group of inhibitory neurons (I-neurons). The latter ones, in turn, strongly inhibit E-neurons. The whole system is in the state of a stable equilibrium and has low activity since any increase of E-neurons activity is accompanied by a large increase of the inhibitory inflow to E-neurons from I-neurons. The system can be transferred into the state of rhythmical oscillations by inhibiting I-neurons. E-neurons are thus released from the inhibition, and their mutual excitation results in a regenerative process: new neurons are continuously recruited and discharge frequencies increased. When activity of E-neurons is high enough, the excitatory inflow to I-neurons reaches their thresholds, and I-neurons begin firing. While firing, I-neurons inhibit E-neurons and, therefore, eliminate the source of their own excitation. The firing of both E- and I-neurons stops, and the system returns to the same state as before the regenerative process. A new cycle begins.

The proposed system thus exhibits the following main features of the behavior of spinal neurons of groups *2* and *3*: *a*) The number of simultaneously active neurons continuously increases and discharge frequencies rise in one part of the cycle. The overall activity rapidly decreases and then disappears in another part of the cycle. *b*) During the latent period of scratching, some of the neurons are inhibited, namely, those which fire at the end of the cycle. On the contrary, tonic activity appears during the latent period in those neurons which fire earlier in the cycle. *c*) The cycle duration can be independent of the "amplitude of oscillations" (i.e., on the firing rate in the bursts). One of the possible ways for stabilizing the frequency of oscillations is the following. To increase the amplitude of oscillations, a tonic excitatory inflow is sent to E-neurons and, at the same time, an additional tonic inhibitory inflow is sent to I-neurons. Due to the increased excitability of E-neurons, the regenerative process becomes faster. But the thresholds of I-neurons become higher. It is evident that these two factors influence the cycle duration in opposite ways and can thus compensate each other.

The proposed system controls motoneurons in the following way: *a*) E-neurons (which correspond to the neurons of group *2* and, partly, of group *3*) excite extensor motoneurons, E-neurons from group *3* providing the greater part of the excitatory drive. This is why the peak depolarization of extensor motoneurons occurs in the S phase. *b*) I-neurons (which correspond to some of the group *3* neurons) produce pauses in the activity of flexor motoneurons either directly, by inhibiting their tonic firing, or indirectly, by inhibiting a tonic discharge of group *1* neurons which were supposed to excite flexor motoneurons. The tonic discharge of group *1* neurons and flexor motoneurons seems to be accounted for by a tonic excitatory inflow from a system activating the rhythm generator, apparently, from propriospinal neurons responsible for the scratch reflex (8, 28).

Essential features of this model of the rhythm generator (regenerative process in the group of mutually exciting neurons and termination of the cycle resulting from the excitation of inhibitory neurons) can be found in models proposed to explain some other rhythmical processes in the nervous system (10, 12, 20, 21, 23, 29). However, until now there were no direct experimental data in favor of these assumptions. Moreover, the data obtained in invertebrates present direct evidence that, at least in some cases, the cyclic acitivity is not a network property but an internal property of single "oscillatory" neurons (13, 19, 24). Such a possibility should not be neglected while considering rhythm generators in vertebrates. But until such oscillatory neurons are found in vertebrates, we prefer to explain experimental data on the basis of interaction between "normal" neurons.

In the following paper (6) we shall describe nonregular regimes of activity of the central spinal mechanism controlling scratching. Beside which, we shall compare the mechanism controlling scratching with those controlling some other limb movements.

Present address of T. G. Deliagina, I. M. Gelfand, and G. N. Orlovsky: Belozersky Interfaculty Laboratory, Corpus A, Moscow State University, Moscow 117234, USSR.

REFERENCES

1. ARSHAVSKY, Y. I., GELFAND, I. M., ORLOVSKY, G. N., AND PAVLOVA, G. A. Origin of modulation in vestibulospinal neurons during scratching. *Biofizika* 20: 946–947, 1975.
2. ARSCHAVSKY, Y. I., GELFAND, I. M., ORLOVSKY, G. N., AND PAVLOVA, G. A. Activity of neurones of the ventral spinocerebellar tract during "fictive scratching." *Biofizika* 20: 748–749, 1975.
3. ARSHAVSKY, Y. I., GELFAND, I. M., ORLOVSKY, G. N., AND PAVLOVA, G. A. Activity of neurones of the lateral reticular nucleus during scratching. *Biofizika* 22: 177–179, 1977.
4. ARSHAVSKY, Y. I., GELFAND, I. M., ORLOVSKY, G. N., AND PAVLOVA, G. A. Messages conveyed by spinocerebellar pathways during scratching in the cat. I. Activity of neurons of the lateral reticular nucleus. *Brain Res.* In press.
5. ARSHAVSKY, Y. I., GELFAND, I. M., ORLOVSKY, G. N., AND PAVLOVA, G. A. Messages conveyed by spinocerebellar pathways during scratching in the cat. II. Activity of neurons of the ventral spinocerebellar tract. *Brain Res.* In press.
6. BERKINBLIT, M. B., DELIAGINA, T. G., FELDMAN, A. G., GELFAND, I. M., AND ORLOVSKY, G. N. Generation of scratching. II. Nonregular regimes of generation. *J. Neurophysiol.* 41: 1058–1069, 1978.
7. BERKINBLIT, M. B., DELIAGINA, T. G., ORLOVSKY, G. N., AND FELDMAN, A. G. On oscillations of the membrane potential of motoneurones during generation of scratching. *Neirofiziologiya*. In press.
8. BERKINBLIT, M. B., DELIAGINA, T. G., ORLOVSKY, G. N., AND FELDMAN, A. G. Activity of propriospinal neurons during scratch reflex in the cat. *Neirofiziologiya* 9: 504–511, 1977.
9. DELIAGINA, T. G., FELDMAN, A. G., GELFAND, I. M., AND ORLOVSKY, G. N. On the role of central program and afferent inflow in the control of scratching movements in the cat. *Brain Res.* 100: 297–313, 1975.
10. DUNIN-BARKOVSKII, V. L. AND JAKOBSON, V. S. Analyses of transient processes and oscillations in the net of "counting" neurons. *Biofizika* 16: 1080–1084, 1971.
11. EDGERTON, V. R., GRILLNER, S., SJÖSTRÖM, A., AND ZANGGER, P. Central generation of locomotion in vertebrates. In: *Neural Control of Locomotion*, edited by R. M. Herman, S. Grillner, P. S. G. Stein, and D. G. Stuart. New York: Plenum, 1976, p. 439–464.
12. FELDMAN, J. L. AND COWAN, J. D. Large-scale activity in neural nets. II. A model for the brainstem respiratory oscillator. *Biocybernetics* 17: 39–51, 1975.
13. FOURTNER, C. R. Central nervous control of cockroach walking. In: *Neural Control of Locomotion*, edited by R. M. Herman, S. Grillner, P. S. G. Stein, and D. G. Stuart. New York: Plenum, 1976, p. 401–418.
14. GRILLNER, S. Locomotion in the spinal cat. In: *Control of Posture and Locomotion*, edited by R. B. Stein, K. G. Pearson, R. S. Smith, and J. B. Redford. New York: Plenum, 1973, p. 515–535.
15. JANKOWSKA, E. Instrumental scratch reflex of the deafferented limb in cats and rats. *Acta Biol. Exptl. Warszawa* 19: 233–247, 1959.
16. JANKOWSKA, E. JUKES, M. G. M., LUND, S., AND LUNDBERG, A. The effect of DOPA on the spinal cord. 6. Half-centre organization of interneurones transmitting effects from the flexor reflex afferents. *Acta Physiol. Scand.* 70: 389–402, 1967.
17. ORLOVSKY, G. N. AND FELDMAN, A. G. Classification of lumbosacral neurones according to their discharge patterns during evoked locomotion. *Neirofiziologiya* 4: 410–417, 1972.
18. PAVLOVA, G. A. Activity of reticulospinal neurons during scratching. *Biofizika* 22: 740–742, 1977.
19. PEARSON, K. G. AND FOURTNER, C. R. Nonspiking interneurons in the walking system of the cockroach. *J. Neurophysiol.* 38: 33–52, 1975.
20. POMPEIANO, O. AND VALENTINUZZI, M. A. Mathematical model for the mechanisms of rapid eye movements induced by an anticholinesterase in the decorticate cat. *Arch. Ital. Biol.* 114: 100–154, 1976.
21. RASHEVSKY, N. Neural circuit which exhibits some features of epileptic attacks. *Bull. Math. Biophys.* 34: 71–78, 1972.
22. REXED, B. A. Cytoarchitectonic atlas of the spinal cord of the cat. *J. Comp. Neurol.* 100: 297–397, 1954.
23. ROSE, R. M. Analysis of burst activity of the buccal ganglion of *Aplysia depilans*. *J. Exptl. Biol.* 64: 385–404, 1976.
24. SELVERSTON, A. I. Neural mechanisms for rhythmic motor pattern generation in a simple system. In: *Neural Control of Locomotion*, edited by R. M. Herman, S. Grillner, P. S. G. Stein, and D. G. Stuart. New York: Plenum, 1976, p. 377–399.
25. SHERRINGTON, C. S. Observations on the scratch-reflex in the spinal dog. *J. Physiol. London* 34: 1–50, 1906.
26. SHERRINGTON, C. S. *The Integrative Action of the Nervous System*. New Haven, Conn.: Yale Univ. Press, 1906.
27. SHERRINGTON, C. S. Notes on the scratch-reflex of the cat. *Quart. J. Exptl. Physiol.* 3: 213–220, 1910.
28. SHERRINGTON, C. S. AND LASLETT, E. E. Observations on some reflexes and interconnection of spinal segments. *J. Physiol. London* 29: 58–96, 1903.
29. WILSON, D. M. AND WALDRON, I. Models for the generation of the motor output in flying locusts. *Proc. IEEE* 56: 1058–1064, 1968.

11.

(with Yu. I. Arshavskij and G. N. Orlovskij)

The cerebellum and control of rhythmical movements

Trends Neurosci. **6** (10) (1983) 417–422

During rhythmical locomotory and scratching movements the cerebellum receives information both about the current state of the peripheral motor apparatus and about the activity of the spinal rhythmical generator. Comparison of cerebellar input and output signals suggests that the cerebellum 'selects' essential information concerning the activity of the motor mechanisms. On the basis of this information, the cerebellum regulates the transmission of signals from various motor brain centres and receptors to the spinal cord. This paper elaborates the hypothesis that the cerebellum co-ordinates different motor synergisms and adapts them to the environment.

The spino-cerebellar loop

Recently, a new approach to the study of cerebellar functions has been developed which involves recording cerebellar input and output signals accompanying movements. This approach is based on two different methods. In chronic experiments the activity of cerebellar neurons is recorded in animals which are awake and which have been trained to perform certain movements[11]. Alternatively, neurons in the cerebellum and structures related to it are recorded from acute decerebrate cats 'automatically' performing locomotor or scratching movements[2–7,10,17,18]. Though this second technique limits investigations to brain-stem–cerebellar and spinal mechanisms, it has the advantage of providing greater possibilities for analytical studies. Studies of the scratch reflex were especially fruitful. This reflex can be easily evoked in immobilized cats ('the fictitious scratch reflex')[12], and immobility of the animal facilitates microelectrode recordings. In addition, by comparing the activity of neurons of the cerebellum (and of other structures) during actual and fictitious scratching, one can estimate the relative roles of central and peripheral factors in generating cerebellar inputs and outputs.

Fig. 1 illustrates the main structures in decerebrate cats concerned with the control of hindlimb movements during locomotion and scratching. Spinal structures generating rhythmical movements are 'switched on' by signals arriving either from the supraspinal structures or from the upper spinal cord. When these spinal mechanisms are operating and the animal is performing rhythmical movements, spinocerebellar pathways convey signals about the activity at the spinal level of the motor control to the medial and intermediate parts of the cerebellum, and affect cerebellar neurons which in turn influence the neurons of descending brainstem–spinal tracts – the spino-cerebellar loop. In addition to receiving inputs from the 'preceding' elements of the loop, both the descending tracts and the cerebellum receive inputs from other parts of the brain ('external inputs').

Cerebellar input signals

Rhythmical signals related to hindlimb movements reach the cerebellum via the dorsal spinocerebellar tract (DSCT), the ventral spinocerebellar tract (VSCT), and the spinoreticulocerebellar pathway (SRCP). The DSCT transmits detailed information about the operation of the peripheral motor apparatus such as the phase and strength of contraction of single muscles, the joint angles, and the time at which the limb touches the ground[2,19]. Fig. 2B, C shows the activity of a DSCT neuron during locomotion. The analysis of the neuron's responses to passive limb movements showed that it received afferent signals from the ankle extensor(s). During locomotion the neuron fired periodically in the stance phase of the step (i.e. the phase when the foot touches the ground and all extensors are active) (Fig. 2B, C). When the extensor activity increased (cf. the electromyograms, EMG, in Fig. 2B, C) and limb movements became more forceful, the neuron activity also increased.

The rhythmical activity of DSCT neurons is determined by afferent signals from the peripheral motor apparatus rather than from central structures; and during locomotion of cats with deafferented hindlimbs, as well as during fictitious scratching, any rhythmical modulation in DSCT neurons was absent. In contrast, information conveyed by the VSCT and SRCP is of central origin. In 1971, Lundberg advanced a hypothesis that the VSCT conveys information about the activity of central spinal mechanisms but not about peripheral events[16]. This hypothesis has been confirmed experimentally by demonstration that the activity of VSCT neurons is rhythmically modulated during locomotion after deafferentation of the hindlimbs[3,5] (Fig. 3A). Also, comparison of the activity of VSCT neurons during actual and fictitious scratching has shown that cessation of the rhythmical afferent inflow from limb receptors (after an animal had been immobilized) does not affect the pattern of modulation of VSCT neurons (Fig. 3B, C). This persistent modulation after immobilization and

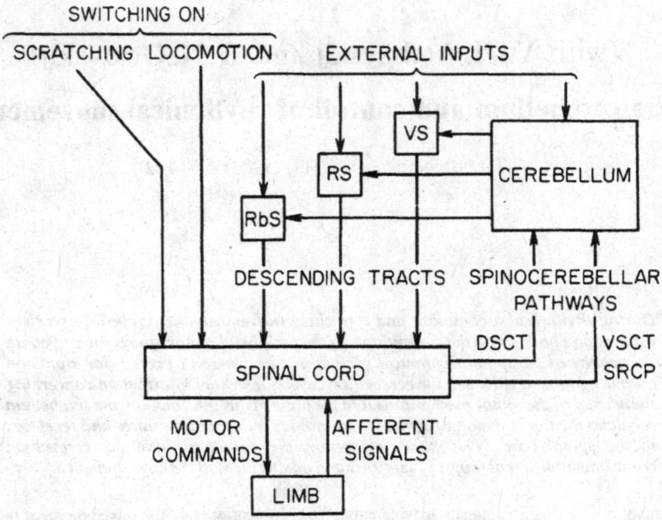

Fig. 1. *Spino-cerebellar loop. Abbreviations: VS = vestibulospinal tract; RS = reticulospinal tract; RbS = rubrospinal tract; DSCT = dorsal spinocerebellar tract; VSCT = ventral spinocerebellar tract; SRCP = spinoreticulocerebellar pathway. The spino-olivocerebellar pathway is not shown since it does not participate in transmitting the rhythmical signals during locomotion and scratching*[9, 10]. *Explanations in the text.*

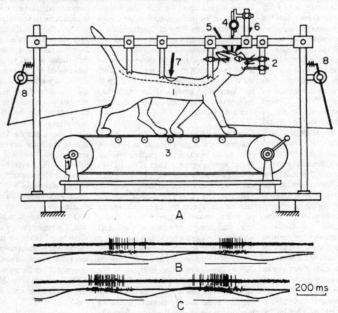

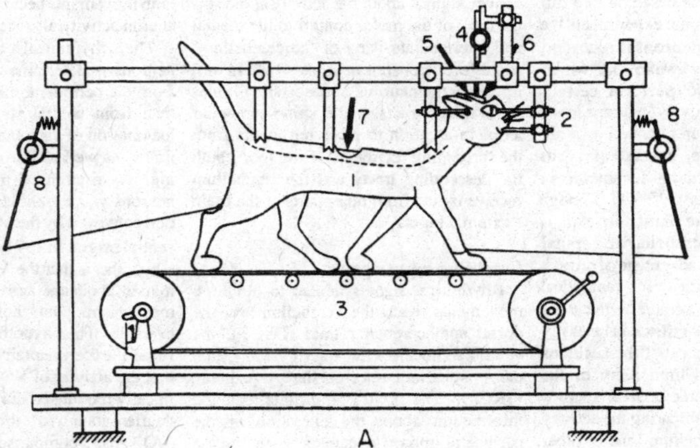

Fig. 2. *The DSCT conveys messages about the activity of the peripheral motor apparatus.* **(A)** *The experimental arrangement for investigation of locomotion. The thalamic or mesencephalic cat (1) is fixed in a stereotaxic device (2) with its limbs on the treadmill (3); electrodes (4–7) are inserted into the spinal cord and into the brain for stimulation and recordings. Limb movements are recorded by potentiometric transducers (8).* **(B, C)** *Activity of a DSCT neuron that is excited by afferents from ankle extensor(s) during weak* **(B)** *and intense* **(C)** *locomotion. The middle trace is the gastrocnemius EMG, the lower one the angle of the ankle (flexion – up). Horizontal lines indicate the stance phases.*

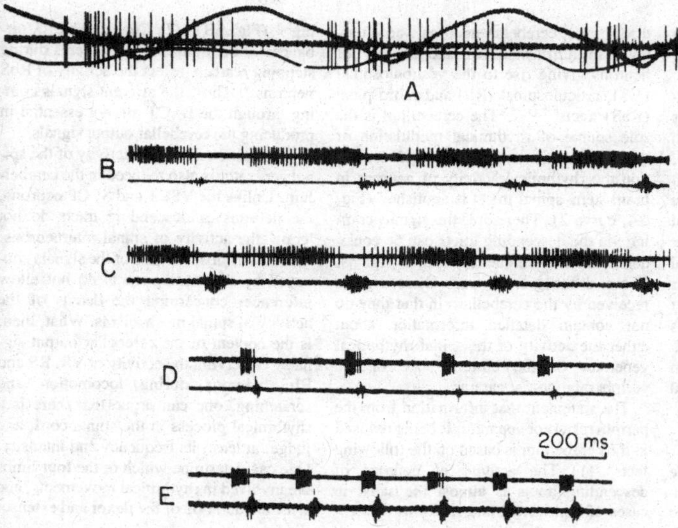

Fig. 3. *The VSCT and SRCP convey messages about intraspinal processes.* **(A)** *activity of a VSCT neuron during locomotion (the hindlimbs of the cat were deafferented); the upper and lower traces are movements of the contralateral shoulder joint and of the ipsilateral hip joint.* **(B–E)** *Activity of a VSCT neuron* **(B, C)** *and a reticulocerebellar neuron from the lateral reticular nucleus* **(D, E)** *recorded initially during actual scratching* **(B, D)** *and then during fictitious scratching* **(C, E)** *after the animals had been immobilized with Flaxedil (5 mg kg*$^{-1}$*, i.v.). The lower trace is the gastrocnemius EMG* **(B, D)** *or ENG* **(C, E)**.

deafferentation shows that intraspinal processes are the main source of rhythmical modulation of these neurons. The same result has been obtained for neurons of the SRCP[4] (Fig. 3D, E).

What are the spinal neuronal mechanisms whose activity is monitored by the VSCT and SRCP? To answer this question, part of the spinal hindlimb centre was 'switched off' by means of cooling (Fig. 4A, B). Fig. 4 shows that cooling of the L5 segment, resulting in cessation of the rhythmical activity in L5 and more caudal segments, affected neither the rhythmical generation in the L4 segment (see the sartorius electroneurogram, ENG, in Fig. 4C–F) nor the firing pattern of VSCT (Fig. 4C, D) and SRCP (Fig. 4E, F) neurons. Thus, 'switching off' of the greater part of the lumbosacral enlargement, where the majority of the motoneurons and interneurons participating in the control of hindlimb movements are located, does not influence the firing pattern of VSCT or SRCP neurons. It follows that the VSCT and SRCP convey information about activity of the rhythm generatory mechanisms

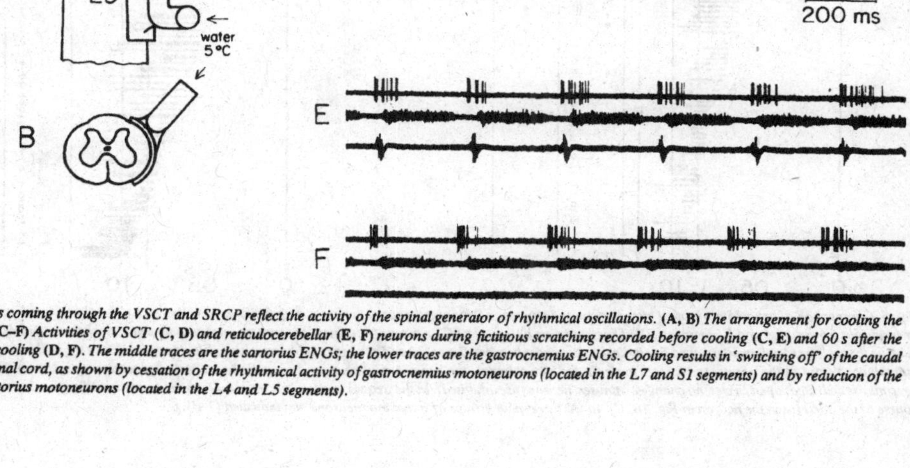

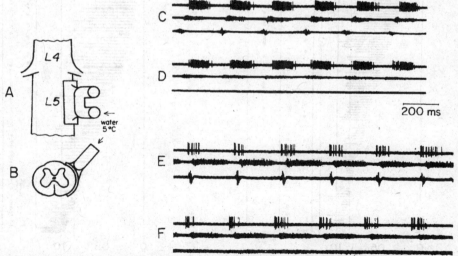

Fig. 4. *Signals coming through the VSCT and SRCP reflect the activity of the spinal generator of rhythmical oscillations.* **(A, B)** *The arrangement for cooling the spinal cord.* **(C–F)** *Activities of VSCT* **(C, D)** *and reticulocerebellar* **(E, F)** *neurons during fictitious scratching recorded before cooling* **(C, E)** *and 60 s after the beginning of cooling* **(D, F)**. *The middle traces are the sartorius ENGs; the lower traces are the gastrocnemius ENGs. Cooling results in 'switching off' of the caudal part of the spinal cord, as shown by cessation of the rhythmical activity of gastrocnemius motoneurons (located in the L7 and S1 segments) and by reduction of the activity of sartorius motoneurons (located in the L4 and L5 segments).*

rather than activity of 'output' neuron mechanisms (motoneurons, interneurons of various reflex chains, etc.).

Fig. 5, which depicts the phase distributions of the activity of spinal interneurons from the L4 and L5 segments (Fig. 5A) and of VSCT and SRCP neurons (Fig. 5B), recorded during fictitious scratching, shows the striking similarity between the behaviour of the spinal interneurons and that of the neurons of the spinocerebellar pathways. The three main groups of spinal interneurons[8] (Fig. 5A, 1–3) have their 'counterparts' among the neurons of spinocerebellar pathways. This supports the hypothesis[5] that the VSCT and SRCP convey information about processes in spinal mechanisms generating rhythmical oscillations.

Cerebellar output signals

The rhythmical signals reaching the cerebellum via the spinocerebellar pathways produce rhythmical modulation of the discharge of cerebellar neurons whose outputs, in turn, modulate the activity of neurons giving rise to the vestibulospinal (VS), reticulospinal (RS) and rubrospinal (RbS) tracts[6,7,17,18]. The cerebellum is the sole source of rhythmical modulation of these neurons, since after cerebellar ablation the rhythmical activity of neurons in brain-stem–spinal tracts is abolished (Fig. 6A, curve 2). Therefore, the signals coming via the descending tracts can be considered as the cerebellar output signals. These differ greatly from those signals received by the cerebellum in that they do not contain detailed information about either the activity of the spinal rhythmical generator or the current state of the peripheral motor apparatus.

The statement that information from the peripheral motor apparatus is being reduced in the cerebellum is based on the following facts. (1) The activity of neurons of descending tracts is almost the same in cases of both actual and fictitious scratching[6,7] (Fig. 6B). (2) Considerable disturbances of the hindlimb movements during stepping scarcely affect the activity of RbS neurons[17]. Thus, the afferent signals coming through the DSCT are not essential in producing the cerebellar output signals.

Information about the activity of the spinal generator is also reduced in the cerebellum. Unlike the VSCT and SRCP neurons, the neurons of descending tracts do not 'copy' the activity of spinal interneurons. This is why observations of the signals conveyed by descending tracts do not allow inferences concerning the details of the activity of spinal mechanisms. What, then, is the content of the cerebellar output signals? Observing the activity of VS, RS and RbS neurons during locomotion and scratching, one can argue that there is a rhythmical process in the spinal cord, and judge, at least, its frequency and intensity. One can determine which of the four limbs are involved in rhythmical movements, and what the duration of the flexor and extensor

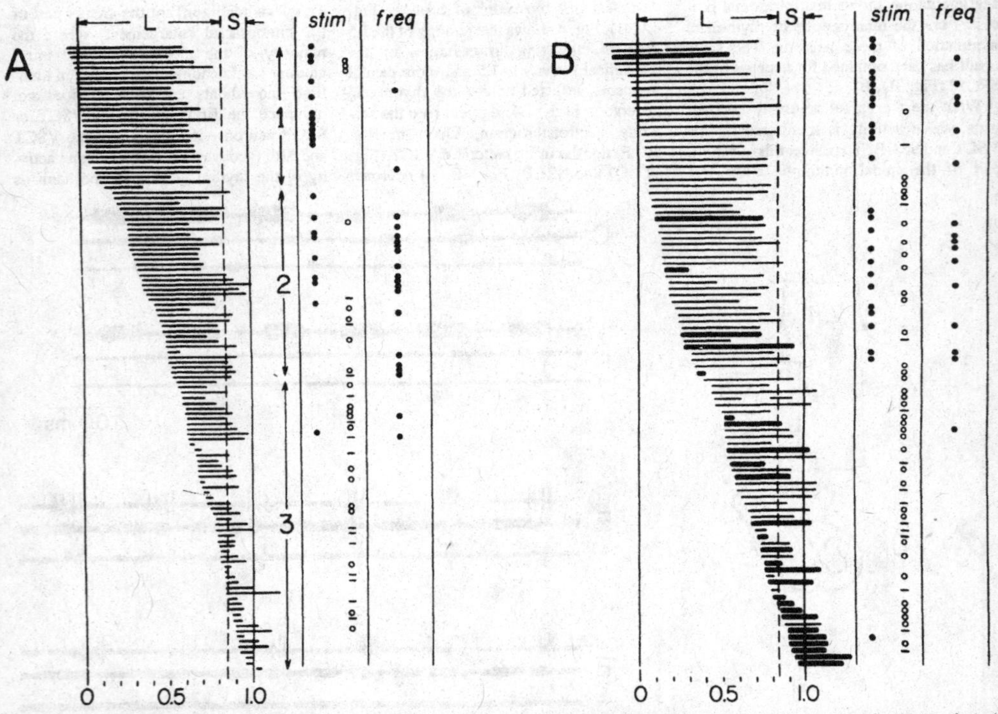

Fig. 5. Comparison of behaviour of spinal interneurons and neurons of spinocerebellar pathways during fictitious scratching. (A) The phase distribution of spinal interneurons from the L4 and L5 segments. (B) The phase distribution of VSCT neurons (thin lines) and reticulocerebellar neurons (thick lines). Solid vertical lines indicate the cycle borders; an interrupted line indicates the border between flexor (L) and extensor (S) phases of the cycle. Horizontal lines show phases of activity of individual neurons in the normalized scratch cycle. In the stim *columns, the behaviour of neurons during the latent period of scratching is shown (filled circles, neurons are activated; open circles, no changes; minuses, neurons are inhibited). In the* freq *columns, filled circles indicate neurons whose frequencies increase in the course of the burst (as in the neuron in Fig. 3B, C). In (A) three main groups of spinal interneurons[8] are indicated (1–3).*

phases of the cycle is. It is especially important that signals coming through the descending tracts depend on the phase of movement. Neurons of the VS tract, that exerts an excitatory action on extensor motoneurons[14,21], are maximally active in the extensor phase of the cycle (Figs 6, 7C). In contrast, the majority of neurons of the RS and RbS tracts that exert an excitatory action mainly on flexor motoneurons[14,15,21] are active in the flexor phase[7,17,18].

Thus, the cerebellar output signals reflect the most essential characteristics of rhythmical movements. A similar conclusion was reached by Robertson and Grimm[20] on the basis of the activity of neurons of the cerebellar dentate nucleus in monkeys trained to touch three buttons in a certain sequence. It was found that the discharge pattern of the cerebellar neurons was related to general aspects of the motor performance but not to the concrete trajectories of movements.

One can suppose that the cerebellum 'selects' the most relevant data concerning the activity of peripheral and central motor mechanisms out of an immense inflow of information. Apparently, the cerebellum also performs the analysis of signals coming from distant receptors and 'selects' essential data about the environment. In this connection the results obtained while studying reactions of cerebellar neurons to binaural acoustic stimuli[1] are of great interest. These reactions strongly depended on those parameters of stimuli which were important for the acoustic orientation (i.e. the spatial localization of the sound source and direction of its motion) but did not depend greatly on the intensity, duration or frequency of the sound.

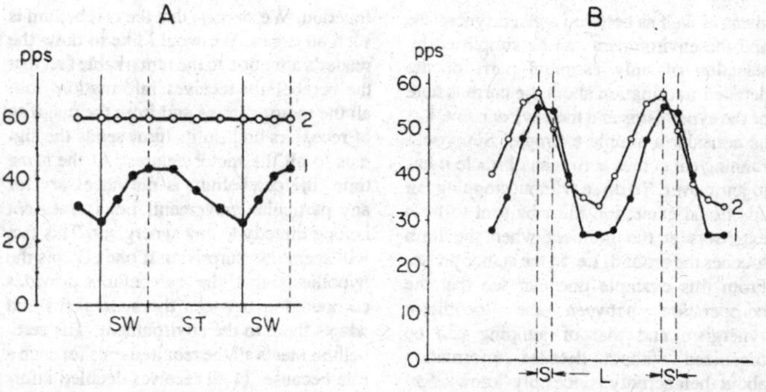

Fig. 6. *Activity in the vestibulospinal tract during locomotion and scratching.* **(A)** *Discharge frequency (pulses per second) of the 'average' VS neuron as a function of the phase of the step. The swing (SW) and stance (ST) phases of the step are indicated. Curve 1 was obtained from cats with the cerebellum intact; curve 2 was obtained from decerebellate cats.* **(B)** *The discharge frequency of the 'average' VS neuron as a function of the phase of the scratch cycle. Curve 1 was obtained for actual scratching; curve 2 was obtained for fictitious scratching. To obtain each of the curves, the discharge frequencies of many neurons were averaged.*

The cerebellum regulates the transmission of signals from various motor brain centres to the spinal cord

Various brain motor centres (cerebral cortex, basal ganglia, hypothalamus, reticular formation, etc.) can influence the spinal cord via descending tracts either directly or through the cerebellum. The cerebellum receives signals both from the motor centres and from various receptors and, in its turn, can affect the neurons of all the descending tracts[13]. During locomotion and scratching, the signals coming from the spinal cord evoke rhythmical activity in the cerebellar neurons whose outputs result in rhythmical modulation of the neurons of descending tracts. As a result, the sensitivity of both cerebellar and descending tracts' neurons to the signals arriving through 'external inputs' of the spino-cerebellar loop (Fig. 1) from other brain centres and various receptors is rhythmically changing in relation to the activity of spinal mechanisms. This was experimentally demonstrated for the 'external input' from the vestibular apparatus. The effects of this input upon the VS neurons were studied during fictitious scratching. To stimulate vestibular receptors, the cat was sinusoidally tilted in the frontal plane (Fig. 7A). Most of the VS neurons were excited during the ipsilateral tilt and inhibited during contralateral tilt (Fig. 7B). During fictitious scratching, these vestibular reactions were rhythmically modulated (Fig. 7D). Thus, transmission of signals from the vestibular apparatus to the spinal cord depends on the phase of the cycle.

A hypothesis: the cerebellum co-ordinates different motor synergisms and adapts them to the environment

In real life, animals commonly perform several motor acts simultaneously under changing external conditions. This means that the nervous mechanism controlling a given motor act (the motor synergism[9]) must operate concordantly with other synergisms and must modify its activity in accordance with the signals coming from distant receptors. It is evident that the problem of interaction between different syner-

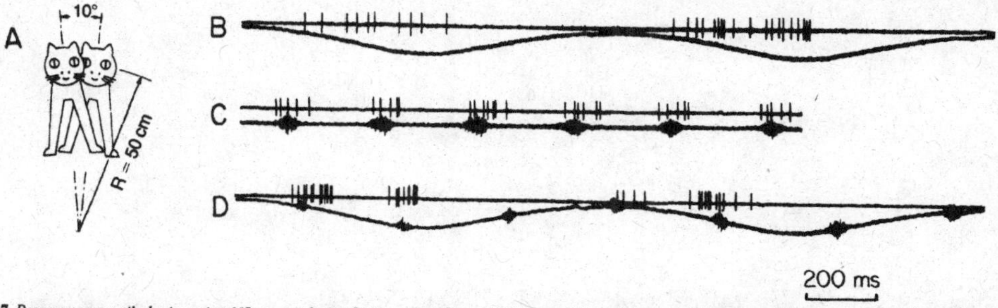

Fig. 7. *Responses to vestibular input in a VS neuron during fictitious scratching.* **(A)** *Scheme of the experiment.* **(B)** *Reactions of the neuron to tilts before elicitation of scratching.* **(C)** *Activity of the neuron during fictitious scratching.* **(D)** *Reactions of the neuron to tilts during fictitious scratching. The lower trace is the tilt angle (ipsilateral tilt – down) and the gastrocnemius ENG.*

gisms as well as between a given synergism and the environment can be simplified by selection of only essential parts of the detailed information about the current state of the synergisms and the environment. Let us consider a simple example. Suppose a running cat comes across an obstacle it has to jump over. To do so without stopping, an additional excitation must be sent to limb extensors at the moment when the limb touches the ground, i.e. in the stance phase. From this example one can see that the co-operation between the locomotor synergism and that of jumping can be organized without detailed information about their activity, since only 'knowledge' of the phase of the locomotor cycle is necessary for triggering the nervous mechanisms controlling jumping.

What are the possible mechanisms of the interaction between synergisms and between a synergism and the environment? Mechanisms controlling different motor acts and analysing the environment may be located in different parts of the nervous system (for instance, the locomotor synergism is located in the spinal cord, while the synergism of jumping over an obstacle seems, at least partly, to be located in the cerebral cortex). To organize the interaction between synergisms, each synergism could have its 'representation' ('receptors') in the domains of all other synergisms. But such a mode of interaction seems to be vulnerable from the evolutionary point of view, since development of synergisms and an increase in the number of synergisms and possible combinations would have made the system of the interaction overcomplicated. The problem of interaction could be solved much more easily if there was a special organ responsible for this function. We suggest that the cerebellum is such an organ. We would like to draw the reader's attention to the remarkable fact that the cerebellum receives information from all the motor centres and from the majority of receptors and, in its turn, sends the signals to all the motor centres. At the same time, the cerebellum is not necessary for any particular movement, i.e. it does not belong directly to any synergism. This fact will seem less surprising if one accepts the hypothesis that the cerebellum provides co-operation between the synergisms and adapts them to the environment. The cerebellum meets all the requirements for such a role because: (1) it receives detailed information about the state of motor synergisms and the environment; (2) out of this information it selects essential data concerning both the activity of motor synergisms and the state of the environment; and (3) it can regulate the transmission of signals from one part of the nervous system to another.

Acknowledgements

The authors thank Dr E. V. Evarts for his valuable comments on the manuscript.

Reading list

1 Altman, J. A., Bechterev, N. N., Radionova, E. A., Shmigidina, G. N. and Syka, J. (1976) *Exp. Brain Res.* 26, 285–298
2 Arshavsky, Yu. I., Berkinblit, M. B., Fukson, O. I., Gelfand,, I. M. and Orlovsky, G. N. (1972) *Brain Res.* 43, 272–275
3 Arshavsky, Yu. I., Berkinblit, M. B., Fukson, O. I., Gelfand, I. M. and Orlovsky, G. N. (1972) *Brain Res.* 43, 276–279
4 Arshavsky, Yu. I., Gelfand, I. M., Orlovsky, G. N. and Pavlova, G. A. (1978) *Brain Res.* 151, 479–491
5 Arshavsky, Yu. I., Gelfand, I. M., Orlovsky, G. N. and Pavlova, G. A. (1978) *Brain Res.* 151, 493–506
6 Arshavsky, Yu. I., Gelfand, I. M., Orlovsky, G. N. and Pavlova, G. A. (1978) *Brain Res.* 159, 99–110
7 Arshavsky, Yu. I., Orlovsky, G. N., Pavlova, G. A. and Perret, C. (1978) *Brain Res.* 159, 111–123
8 Berkinblit, M. B. Deliagina, T. G., Feldman, A. G., Gelfand, I. M. and Orlovsky G. N. (1978) *J. Neurophysiol.* 41, 1040–1057
9 Bernstein, N. (1967) *The Coordination and Regulation of Movements*, Pergamon, Oxford
10 Boylls, C. C. (1977) *Soc. Neurosci. Abstr.* 3, 55
11 Brooks, V. B. and Thach, W. T. (1981) in *Handbook of Physiology: The Nervous System*, Vol. II, pp. 877–946, Williams and Wilkins, Baltimore
12 Deliagina, T. G., Feldman, A. G., Gelfand, I. M. and Orlovsky, G. N. (1975) *Brain Res.* 100, 297–313
13 Dow, R. S. and Moruzzi, G. (1958) *The Physiology and Pathology of the Cerebellum*, University of Minnesota Press, Minneapolis
14 Grillner, S., Hongo, T. and Lund, S. (1971) *Exp. Brain Res.* 12, 457–479
15 Hongo, T., Jankowska, E. and Lundberg, A. (1969) *Exp. Brain Res.* 7, 344–364
16 Lundberg, A. (1971) *Exp. Brain Res.* 12, 317–330
17 Orlovsky, G. N. (1972) *Brain Res.* 46, 99–112
18 Orlovsky, G. N. and Shik, M. L. (1976) in *International Review of Physiology: Neurophysiology II* (Porter, R., ed.), Vol. 10, pp. 281–317, University Park Press, Baltimore
19 Oscarsson, O. (1973) in *Handbook of Sensory Physiology* (Iggo, A., ed.), Vol. 11, pp. 339–380, Springer-Verlag, Berlin
20 Robertson, L. T. and Grimm, R. J. (1975) *Exp. Brain Res.* 23, 447–462
21 Shapovalov, A. I. (1975) *Rev. Physiol., Biochem. Pharmacol.* 72, 1–54

***Yu. I. Arshavsky** is a Senior Scientific Worker at the Laboratory of Intercellular Interaction, Institute of Problems of Information Transmission, Academy of Sciences, Moscow 101447, USSR.*

I. M. Gelfand is Chief and G. N. Orlovsky is a Senior Scientific Worker at the Department of Mathematical Methods in Biology, Belozersky Interfaculty Laboratory, Moscow State University, Moscow 117234, USSR.

12.

(with V. I. Guelstein, A. G. Malenkov, and Yu. M. Vasil'ev)

Interrelationships of contacting cells in the cell complexes of mouse ascites hepatoma

Int. J. Cancer 1 (1966) 451–462

Ascitic fluid formed after intraperitoneal implantation of mouse ascites hepatoma 22a contains single cells (I-cells) and complexes consisting of two, three or more cells (II, III, IV-complexes). Relative numbers of I-cells and of the complexes of various types were the same in all samples of ascitic fluid taken from various generations of tumour transplantation. Non-separation of sister cells after mitosis seems to play a leading role in the formation of the complexes. There are two stages in the mitotic cycle at which the complex may be broken: a) prophase, beginning synchronously in several contacting cells of the complex and, b) beginning of G1 phase in cells formed after mitoses of a I-cell or after asynchronous mitosis in the complex.

There are two discrete types of the II-complexes in the ascitic fluid: complexes with synchronous mitotic cycles of contacting cells and those with considerable asynchrony of the cycle. Both cells of " synchronous " complexes start and finish S phase and G2 phase simultaneously.

The mean generation time of the cells forming II-complexes is shorter than that of I-cells; this shortening is due to differing mean duration of the G1 phase which is 10-11 hours for the cells in II-complexes as compared with 17-18 hours for I-cells.

Studies on interactions between contacting cells are of considerable interest in elucidating the mechanism of tissue structure formation and the processes regulating cell multiplication in these structures (Moscona, 1962; Steinberg, 1963; Stoker, 1964; Weiss, 1963). The main aim of the experiments described in this paper was to investigate cell interactions in the cell complexes of mouse ascites hepatoma 22a. The cells of this ascites hepatoma as well as those of some other hepatoma strains (Yoshida, 1964) preserved one of the typical properties of epithelial cells: they formed stable cell complexes which might be regarded as simplified analogues of normal epithelial structures. Small mean size of the complexes was characteristic for mouse ascites hepatoma 22a: most of these complexes contained from two to five cells. Therefore this tumour strain was very convenient for quantitative studies of cell interactions in small cell groups.

MATERIAL AND METHODS

Ascitic hepatoma 22a was obtained in this laboratory in 1962 by serial intraperitoneal transplantation of solid hepatoma 22a. The solid strain was obtained 10 years earlier by subcutaneous inoculation of primary hepatoma induced by o-aminoazotoluene (Guelstein, 1954). Both solid and ascitic variants of hepatoma 22a are carried in inbred mice of C3HA strain.

Date received: 27 December, 1965.
Approved for publication: 8 February, 1966.

Five days after intraperitoneal transplantation of 10^7 cells of ascites hepatoma, about 2 ml of ascitic fluid are found; before the death of tumour-bearing animals (at about 12-13 days) the volume of fluid is approximately 5-7 ml. Numerous small tumour nodules are seen at this time on the peritoneal wall and on the mesenterium. Ascitic fluid is haemorrhagic and contains about 10^8 cells per ml. Most of these cells (90-95%) are typical of tumour tissue elements with large basophilic nuclei. Small numbers of mesothelial and polynuclear cells are also present. Tumour cells in the ascitic fluid are single or form complexes consisting of two, three or more cells. In some of the two-cell complexes the surfaces of the contacting cells adhere to each other at only a few points and non-staining " lumen " is seen between the cells. In other two-cell complexes the area of contact is wider and the " lumen " is not seen. Electron microscopic study of the complexes (Chenzov and Olshevskaja, 1966) revealed the formation of special " adhesion zones " at the points of cell contacts.

Three kinds of measurement were performed in the experiments with ascitic hepatoma: *a)* determination of relative numbers of cell complexes of various types; *b)* determination of mitotic indices, *c)* autoradiographic studies with the use of tritium-labelled thymidine. Unless otherwise stated, samples of ascitic fluid withdrawn 5 days after tumour transplantation were used in all experiments.

Determination of relative numbers of cell types

Cell counts were made in a haemocytometer filled with fresh ascitic fluid diluted 1:20 with saline. First, the relative number of single cells per two-cell complex was counted, then the number of two-cell complexes per three-cell complex, and so on. Two hundred complexes were counted at each step and each such count was repeated five or more times. Average ratios and their standard errors were estimated for each sample from the results of these counts.

Mitotic indices were counted in smears of the ascitic fluid. Smears of fresh, non-diluted ascitic fluid were dried, fixed in methanol, hydrolyzed 2 minutes in 1 N HCl at 56° C and stained with 0.05% methylene blue. Special comparative counts had shown that relative numbers of various complex types in the fresh ascitic fluid and in the smears made from this fluid were similar; thus, no significant clumping or destruction of the complexes took place in the process of smearing. To determine mitotic indices separately for various types of complexes, not less than 1,000 cells of each group were usually counted.

Criteria used for the determination of the early mitotic phases deserve special comment. Various investigators have used different criteria while counting these stages in smears of ascitic tumours. Thus, according to Baserga *et al.* (1963) the prophase index in growing Ehrlich ascites cells was about 1-2%, while, according to Edwards *et al.* (1960), the index in the same tumour was about 18-20%.

In smears stained with methylene blue, it was easy to see all the stages of nuclear changes preceding mitosis. The earliest changes of this type (increased basophilia of the chromatin clumps) were seen in about 15-20% of all tumour cells: we counted as prophases only the cells with more pronounced nuclear changes (disappearance of the nuclear membrane and of the nucleoli) which usually were about 1-2% of the whole population.

Many prophases in the complexes were synchronous, being seen in two or more cells of the same complexes (Fig. 1). It was important to differentiate these synchronous prophases in two contacting cells from two daughter cells in late

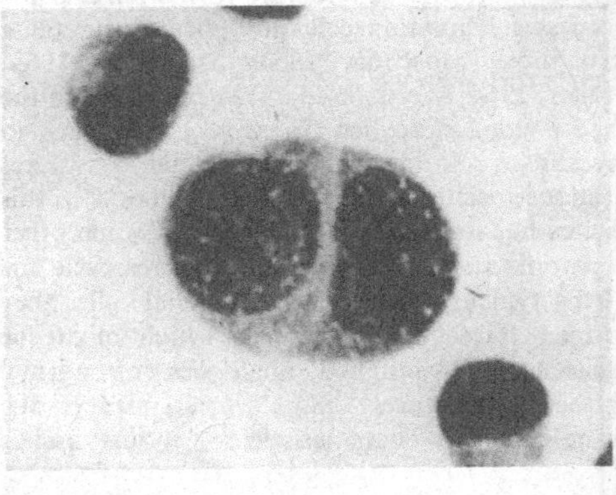

FIGURE 1

Synchronous early prophases in both cells of II-complexes. Smear stained with methylene blue. ×1,500.

telophase. Morphologically, the telophasic nuclei were smaller and chromatin clumps less dense when compared with prophases. Special counts were made in order to estimate the limits of possible error in counting synchronous prophases and late telophases.

These counts were made in the smears of the ascitic fluid withdrawn 4 hours after the single injection of thymidine-^{3}H. About half of all the prophases were labelled at this time. Obviously, among the cells that were counted as synchronous prophases but were in fact late telophases, the percentage of labelled cells could not be higher than in the group of single cells in anaphase and early telophase. On the basis of this assumption one could estimate the minimum percentage of true synchronous prophases. These minimum values were found to be close to the observed ones, the difference being less than 10% of all the prophases.

The mitotic indices for all tumour cells in the ascitic fluid varied from 1 to 3% at 5 days: at 9 days the same index was lower (0.5-1%). Distribution of mitotic phases was as follows: prophases, 25-40% of all mitotic cells; metaphases, 40-50%; anaphases, 7-8%; telophases 10-20%.

The percentage of synchronous prophases was estimated in many experiments. The number of prophases occurring in contacting cells was expressed as a percentage of all prophases in cells of the same type. The percentage of synchronous metaphases and of synchronously labelled cells (in autoradiography) were calculated in the same way.

Autoradiographic experiments

Tritium - labelled thymidine (Radiochemical Centre, Amersham, England; specific activity—8,600 mc/mM) was injected intraperitoneally in a single dose of 10 μc per mouse. Smears were hydrolyzed and stained as described above, coated with the emulsion (M-type, NIKFI, Moscow), and exposed for 7 days.

The following values were determined:

1. Labelling index (LI), that is, the percentage of labelled cells to all cells of a given type. A cell was counted as labelled if there were 6 or more grains over its nucleus. LI were counted in smears from the ascitic fluid withdrawn *(a)* 0.5-20 hours after a single injection of thymidine-^{3}H; *(b)* 0.5 hours after the second of two injections at an interval of 3 hours or 1 hour; *(c)* 0.5 hours after the last of nine injections at intervals of 4 hours. Differential counts of LI for various cell groups were always made in the same areas of the same smear; from 500 to 10,000 cells were counted in each group.

2. Numbers of silver grains per nucleus were counted in various cell groups. Here again comparisons were made between cells counted in the same smear. One hundred labelled cells were counted in each group.

3. The percentage of labelled metaphases was estimated in the smears made from the samples of the ascitic fluid withdrawn at 0.5-20 hours after single injection of thymidine-^{3}H.

The following abbreviations are used for various types of complexes in the ascitic fluid; "I-cells" for single cells; "II-complexes" for complexes consisting of two cells; "III-complexes" for complexes consisting of three cells, etc. P-prophases, M-metaphases, A-anaphases, T-telophases.

RESULTS AND DISCUSSION

Principal results of the cell counts and of the mitotic counts as well as those of autoradiographic experiments are presented in Tables I-V. Some of the conclusions based on these data will be discussed in the following paragraphs. We shall also describe results of experiments performed to clarify certain specific points.

Ratio of complexes of various types in the ascitic fluid

Table I shows that the proportions of various complexes were very similar in all samples of fluid withdrawn at 5 days after transplantation although each sample was taken from a different mouse at times corresponding to different passages of the transplanted tumor.

Relative frequency of the complexes of a given type was inversely proportional to the number of cells in this complex. The proportion of the number of complexes containing "N" cells remained constant or increased slightly with the increase of " N ". In samples taken at 9 days after transplantation, the relative number of larger complexes was somewhat higher than in the corresponding counts made at 5 days. This shift in the ratio of the complexes observed between 5 and 9 days after transplantation was obviously reversible, because in each subsequent passage the original distribution of the complexes was restored.

TABLE I

FREQUENCY OF COMPLEXES CONTAINING VARIOUS NUMBERS OF CELLS IN THE ASCITIC FLUID OF MICE WITH HEPATOMA 22A[1]

Mouse No.	Relative numbers of complexes		
	I/II	II/III	III/IV
5 days after transplantation			
3	3.6±0.4	2.8±0.4	3.3±1.0
4	4.2±0.3	4.2±0.4	4.0±1.1
5	6.7±0.2	3.3±0.4	1.9±1.2
6	4.4±0.3	3.7±0.3	1.9±1.2
7	4.6±0.4	4.5±0.4	2.7±0.3
8	4.0±0.5	5.1±0.4	3.3±0.3
9	3.4±0.4	3.7±0.4	2.8±0.3
Mean	4.4±0.13	3.9±0.12	2.8±0.1
9 days after transplantation			
12	2.7±0.2	2.8±0.4	2.6±1.0
13	2.6±0.2	2.3±0.3	2.9±1.0
14	3.8±0.1	3.8±0.3	2.0±1.0
15	3.8±0.1	3.8±0.2	2.0±0.2
16	3.4±0.3	4.0±0.3	2.5±0.3
17	4.2±0.3	3.0±0.2	2.2±0.3
18	3.3±0.2	2.9±0.3	2.3±0.2
Mean	3.4±0.1	3.2±0.1	2.3±0.1

[1] Number of complexes containing " N " cells per complex containing " N+1 " cells is given in each column (± standard errors). Thus, the first column shows the number of I-cells per II-complex; the second column shows the number of II-complexes per III-complex, etc.

Thus, relative numbers of complexes of various types remained constant from one generation to another in spite of the multiplication of tumour cells. To explain this constancy it is necessary to assume that some of the cells leave their " parent " complexes at certain stages of mitotic cycle, while some I-cells form new complexes.

Phases of mitotic cycles when complexes are broken and formed

One of the stages at which the complex may be broken is the prophase of synchronous mitosis. This conclusion is based on the analysis of the distribution of mitotic figures in the complexes of various types. Table II shows typical results of the mitotic counts in one sample of ascitic fluid (see also Table III and Figures 1-3). A considerable proportion (40-50% or more) of prophases found in II-complexes were synchronous, that is, they were found in both cells of the same complex. The percentage of syn-

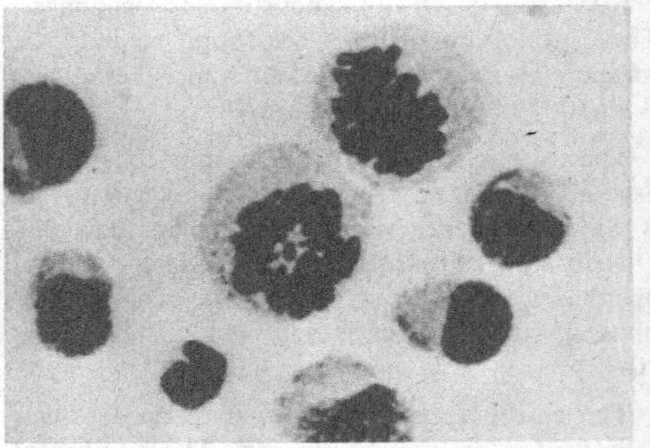

FIGURE 2

Synchronous metaphases in both cells of a II-complex. Smear stained with methylene blue. ×1,500.

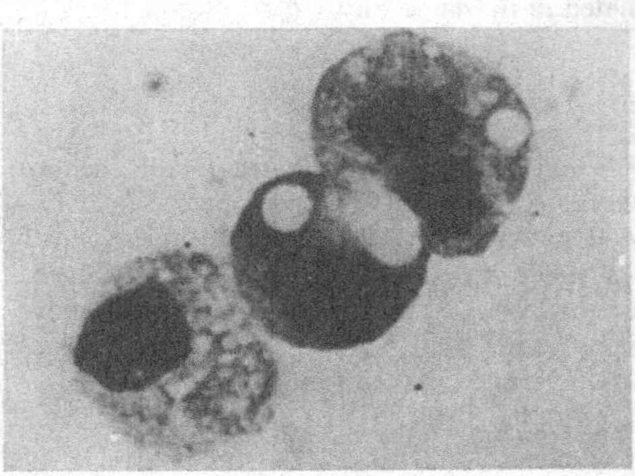

FIGURE 3

Asynchronous metaphase in one of the cells forming III-complex. Smear stained with methylene blue. ×1,500.

TABLE II

MITOTIC INDICES AND SYNCHRONIZATION OF MITOTIC PHASES IN THE COMPLEXES OF VARIOUS TYPES (SAMPLE No. 27)

Type of complex	Number of cells counted	Index of prophases	Index of M+A+T	Percentage of prophases					Percentage of M, A and T			Calculated index of M+A+T[1]
				Asynchronous	Synchronous in 2 cells	Synchronous in 3 cells	Synchronous in 4 cells	Synchronous with M, A or T	Asynchronous	Synchronous with M, A or T	Synchronous with P	
I	14665	1.07±0.08	2.17±0.11	—	—	—	—	—	—	—	—	2.7
II	8240	1.20±0.11	1.43±0.11	40	57	—	—	3	90	8	2	1.2
III	3975	1.26±0.15	2.14±0.22	20	64	12	—	4	88	9	3	1.8
IV	2884	0.94±0.11	1.11±0.17	14	66	—	16	4	85	12	3	0.6

[1] Calculations of the indices of M+A+T were based on the scheme shown in Figure 5; see text for explanations.

TABLE III

PERCENTAGE OF SYNCHRONOUS MITOSES AND OF SYNCHRONOUS LABEL IN II-COMPLEXES IN VARIOUS SAMPLES [1]

No. of sample	Labelling index for II-cells in II-complexes	Percentage of synchronous label	Index of prophases for cells in II-complexes	Percentage of synchronous prophases	Index of M+A+T for cells in II-complexes	Percentage of synchronous M, A and T
26	—	—	1.1±0.1	76±6	0.9±0.1	0
27	—	—	1.2±0.1	57±5	1.4±0.1	8±3
24	—	—	1.5±0.2	72±6	1.3±0.8	14±6
22	21.0±0.4	56±1	0.5±0.1	44±11	0.3±0.1	0
23	30.2±0.7	73±1	1.3±0.2	62±7	2.3±0.3	0
45	33.3±0.9	62±1.5	1.4±0.1	75±5	2.1±0.1	3±1
06	24.1±0.7	59±2	1.6±0.1	40±10	1.9±0.3	0

[1] Thymidine-^{3}H was not injected into mice Nos. 24, 27 and 26. Thymidine-^{3}H was injected into four other mice half an hour (sample 06) 1 hour (samples 23 and 45) or 2 hours (sample 22) before the withdrawal of ascitic fluid.

chronous metaphases, anaphases and telophases was low (9-13% or lower). Combinations of prophase with metaphase, anaphase or telophase, within the same complex were also rare. A considerable difference between the percentage of synchronous prophases and that of synchronous metaphases was also observed in larger (III and IV) complexes. Frequency of cells found in prophase (prophasic index) was not higher than that of the cells found in all the later mitotic stages. Therefore, decreased frequency of synchronous prophases could not be a result of different probabilities of random combinations of cells passing these stages in the same complex. The only possible explanation of these results was that most complexes with synchronous mitoses were broken before the beginning of metaphase. The cells of broken complexes passing metaphase or telophase were then counted in the groups of I-cells or as parts of smaller complexes (see Figure 4). Complexes with asynchronous mitoses were not broken during prophase. Therefore most metaphases in the complexes were asynchronous.

Separation of contacting cells during synchronous prophase also explained the peculiar pattern of mitotic indices observed in the cell complexes. In all the samples counted the index of metaphases and later mitotic stages (metaphases plus anaphases plus telophases per hundred cells of a given type) was higher in the complexes containing an uneven number of cells than in those with an even number of cells: this index was higher for I-cells than for cells in II-complexes, higher for III-complexes than for II-complexes, and lower for IV-complexes than for III-complexes (see Table II). These differences were not characteristic for prophasic indices. This pattern was easily explained with the aid of a diagram of the transformations of complexes during mitoses based on the assumption that complexes with synchronous mitoses were broken during prophase (Fig. 5). Using this diagram and knowing the frequencies and distribution of prophases in various types of complexes, one could easily calculate the frequency and distribution of later mitotic stages. These calculated indices followed the same pattern as experimentally observed ones: they were lower for the complexes with uneven cell numbers (I and III); see Table II.

If the prophase of synchronous mitosis were the only stage at which the cells could become separated, the relative numbers of I-cells in the population would be very small because the cells separated after prophase would have formed II-complexes after telophase. The relative number of I-cells in the population is high; therefore another period must exist when the cells may become separated. This period probably coincides with the beginning of G1 stage. If the cells of II-complexes formed after mitosis were separated at some other stage (say, at the end of G1 or during S phase), then these complexes would contain relatively more cells passing the early stages and, consequently, prophasic indices as well as LI for cells in II-complexes would be lower than for I-cells. This was not the case (see Tables II and IV).

Theoretically, there may be two processes leading to the formation of new II-complexes from I-cells: non-separation of two daughter cells after mitosis of I-cell, and aggregation of two I-cells. The following experiment was made with the aim of studying the role of cell aggregation. Ascitic hepatomas had been implanted intraperitoneally into two mice simultaneously. Five days later one of the mice received nine injections of thymidine-^{3}H at 4-hourly intervals. Ascitic fluid of this mouse was withdrawn half an

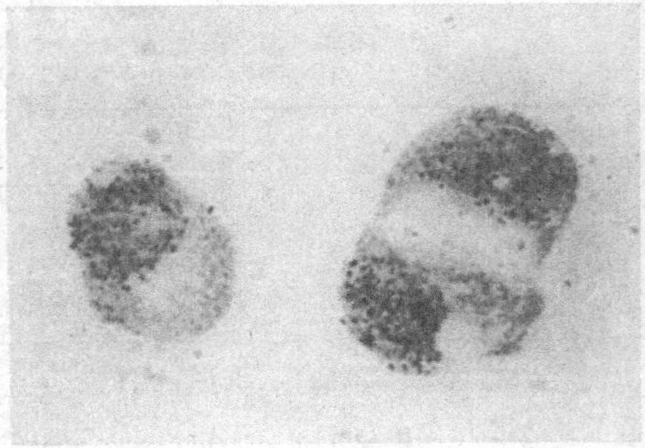

FIGURE 4

Labelled I-cell and II-complex with both cells labelled. " Lumen " between the cells of II-complex. Autoradiograph of the smear of ascitic fluid after the injection of thymidine-^{3}H. $\times$ 1,500.

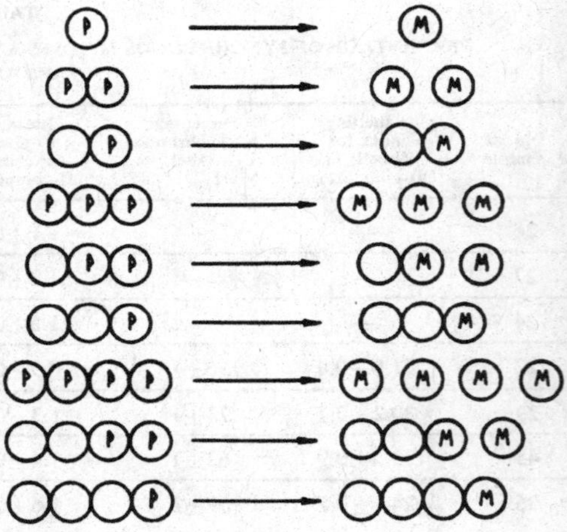

FIGURE 5

Diagram illustrating the dissociation of complexes in the course of mitosis. Left: complexes containing cells in prophase (P); right: I-cells and smaller complexes formed in metaphase (M).

TABLE IV

LABELLING INDICES IN THE COMPLEXES CONTAINING VARIOUS NUMBERS OF CELLS

No. of sample	Interval between the single injection of thymidine-^{3}H and withdrawal of ascitic fluid (hours)	LI for			
		I-cells	II-complexes	III-complexes	IV-complexes
12	2	13.8±0.3	21.0±0.4	20.0±0.5	24.3±0.9
23	1	27.0±0.4	30.2±0.7	30.2±1.2	31.7±1.4
34	1	18.0±0.4	28.5±0.6	18.0±1.2	23.5±1.3
45	1	24.0±0.4	33.3±0.9	27.6±1.6	30.3±1.9
06	0.5	16.4±0.7	24.1±0.7	21.0±1.0	28.7±2.7
07	1.5	17.9±0.8	17.2±0.8	Not counted	Not counted
37	2	14.8±0.6	19.6±1.0	19.8±1.7	Not counted
38	2	16.7±0.6	21.8±1.6	18.0±2.3	Not counted
39	0.5	17.9±0.5	25.6±1.0	19.6±1.7	16.0±2.0
Mean		18.5	24.5	21.8	25.7

TABLE V

DURATION OF S-PHASE DETERMINED IN THE EXPERIMENTS WITH TWO INJECTIONS OF THYMIDINE

Mouse No.	Interval between the first and second injection of thymidine-^{3}H	LI of I-cells after first injection		LI of I-cells after second injection		Duration of S-phase (in hours)
		Found	Corrected for exponential growth	Found	Corrected for exponential growth	
49	3	16	19	23	29	5.7
50	3	24	29	35	43	6.1
51	1	25	30	29	36	5.0

hour after the last injection. LI for I-cells in this fluid was found to be 75% and for II-complexes, 74%; while the amount of synchronous label in II-complexes was 66%. 0.5 ml of this fluid were injected intraperitoneally into the second mouse with ascitic hepatoma that did not receive thymidine. Five hours later ascitic fluid was withdrawn from this mouse. In this fluid LI for I-cells was 17%; that is, transfused cells formed only about 1/5 of the whole population. If after the transfusion an appreciable part of the II-complexes were formed from the aggregated I-cells, then the percentage of synchronous label in II-complexes in the second mouse would have become much lower as compared to that found in the first mouse: the probability of the aggregation of I-cells could be found in this experiment. Non-separation of daughter cells after mitosis of I-cells seems to play a leading role in the formation of II-complexes.

Duration of various phases of mitotic cycles in single cells and in the cells forming complexes

LI after a single injection of thymidine-^{3}H varied from 14% to 27% in different mice, the average being 18% (see Table IV and Figure 4). Experiments in which LI were compared in the same mouse after one and two injections of thymidine-^{3}H have shown that the duration of S-phase in single cells was about 6 hours (Table V). After nine injections (10 mc at each injection) of thymidine-^{3}H at 4-hourly intervals, LI varied from 68 to 75% in various mice. Obviously, at this stage of tumour growth, not less than 70% per cent of all cells in the ascitic fluid participated in the mitotic cycle. More exact determination of the size of the proliferative pool was difficult because this size could change in the course of the experiment with multiple injections of thymidine-^{3}H. Assuming that the size of the proliferative pool is at maximum (100%) and making the correction for exponential growth (Baserga and Lisco, 1963) the mean generation time for I-cells was about 28 hours. If one assumed that the pool size is at its minimum (70%) then generation time was 20 hours. The first of these figures will be used in the following text: there is no difficulty in calculating all the cycle parameters also for the minimum size of the pool.

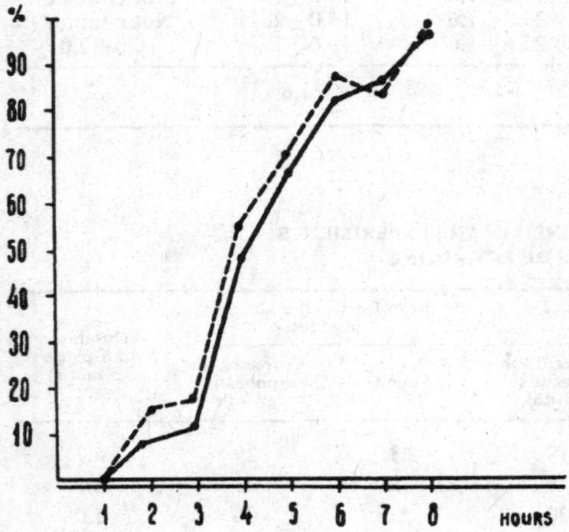

FIGURE 6

Labelled mitoses at various times after the injection of thymidine-^{3}H. Percentage of labelled mitoses (except prophases) in I-cells: (———) percentage of labelled mitoses in the cells forming II-complexes; (------). Each point of the curve is the mean of measurements in three to six samples.

Duration of G2 stage plus prophase in I-cells varied from 2 to 6 hours (see Figure 6). Thus, average duration of the phases of cell cycle in I-cells was as follows: G1-phase: 18 hours; S-phase: 6 hours; G2 plus prophase: 4 hours; mitosis without prophase: about half an hour.

After a single injection of thymidine-^{3}H in all but one mouse, LI for II-complexes was higher than that for I-cells, these differences being highly significant for most samples (see Table IV). There may be several explanations for these differences:

a) Duration of generation time (T_c) was the same for I-cells and for cells in II-complexes, but S-phase took more time in II-complexes. This explanation was not correct because the average number of silver grains over the labelled nuclei was found to be the same for both cell types (see Table VI).

TABLE VI

MEAN GRAIN COUNT PER NUCLEUS IN LABELLED CELLS OF VARIOUS TYPES[1]
(SAMPLE No. 49)

Cell type	Number of cells counted	Mean number of grains ± mean error
I-cells	200	45.3 ± 1.6
Cells of synchronous II-complexes	100	45.4 ± 2.3
Cells of asynchronous II-complexes	100	48.9 ± 1.9

[1] All counts were made in the same part of the same smear. Grain counts made in smears from two other mice gave similar results, i.e. no significant differences between cells of various types were observed.

b) T_c was the same for both cell types, but the size of the proliferative pool was different. This explanation was not correct because, after multiple injections of thymidine, LI of I-cells and of II-complexes were always similar (about 70%).

c) Generation time of I-cells was longer than that of cells in II-complexes. Other suggestions being incompatible with the experimental data, this seems to be the only possible explanation for the observed differences. If T_c for I-cells was 28 hours, then T_c for cells in II-complexes would be about 21 hours.

As we mentioned earlier, the duration of S-phase was the same in I-cells and in II-complexes. Analysis of the curve of labelled mitoses (see Figure 6) had shown that difference in the duration of G2 phase plus prophase for these two cell types was less than half an hour. Obviously, shortening of generation time in the cells forming II-complexes must be due to the shortening of G1 phase. If duration of G1 phase for I-cells was 17-18 hours, then the corresponding figure for cells in II-complexes was 10-11 hours.

Relative values of LI for III- and IV-complexes were found to be more variable than those for II-complexes (see Table IV). In most samples LI for IV-complexes were similar to those for II-complexes, while LI for III-complexes had intermediate values: they were lower than LI for II-complexes but higher than those for I-cells.

Comparison of the LI for the III- and IV-complexes permits the suggestion that here processes leading to the shortening of the generation time are acting preferentially within separate cell pairs. For instance, in III-complexes generation time may be shortened in two cells but not in the third cell of the same complex.

Cell complexes with synchronous and asynchronous mitotic cycles

Prophase took only about 1% of the whole mitotic cycle, therefore even small differences in the time when the cells of the complex start prophase would lead to an almost complete disappearance of synchronous prophases. From 40 to 70% of prophases were synchronous in II-complexes (see Tables II and IV). Obviously the cells of these complexes finished G2 periods synchronously.

In the samples taken 2-6 hours after the injection of thymidine some cells in prophase were already labelled. In these samples we counted the distribution of the label in the II-complexes with synchronous prophases. We found that in most of these complexes both cells were either labelled or unlabelled (see Table VII). Only about 10% of all the II-complexes with synchronous prophases had one cell labelled and another unlabelled. Obviously the cells that synchronously entered prophase had synchronously finished S-phases. Differences in the duration of G2 period between two cells of the same complex of this type did not exceed about 10% of the duration of prophase, that is a few minutes. Distribution of label in the II-complexes with synchronous prophases 10-14 hours after injection of thymidine-^{3}H (when percentage of labelled prophases decreased) was counted in a similar way and gave similar results. These counts had shown that the cells synchronously entering prophase had synchronously started S-periods.

The percentage of synchronous label in II-complexes in the samples taken 0.5-2.0 hours after thymidine-^{3}H was similar to the percentage of synchronous prophases in the same smears (see Table III). This fact was in good agreement with the suggested synchrony of the S-phases and of the G2 phases of both cells in certain II-complexes.

Another group of II-complexes were "asynchronous" cell pairs: in these complexes the cells did not pass the mitotic cycle synchronously. There must have been considerable mean differences between the time when S-phase began in the first and in the second cell of these complexes. If this difference were small, many asynchronous complexes would have the label in both cells

TABLE VII

SYNCHRONIZATION OF LABEL IN II-COMPLEXES WITH SYNCHRONOUS PROPHASES

No. of sample	Number of II-complexes with synchronous prophases having: both cells labelled	both cells unlabelled	one cell labelled and one unlabelled
2 hours after injection of thymidine-^{3}H			
6	9	14	3
35	12	17	2
36	3	11	1
Total	24	42	6
4 hours after injection of thymidine-^{3}H			
6a	12	2	4
37	7	9	2
34	8	1	2
38	9	6	0
42	24	11	4
Total	60	29	12
6 hours after injection of thymidine-^{3}H			
37a	2	2	1
31	50	5	1
Total	52	7	2

because of the overlapping of S-periods in these cells. Then the observed percentage of synchronous label would have been considerably higher than that of synchronous prophases, but this was not the case. Calculations showed that the mean interval between the times of onset of DNA synthesis in the first and in the second cells exceeded 6-8 hours.

Relation of the number of II-complexes with asynchronous prophases to that of all II-complexes containing cells in prophase was about 0.3-0.6 in most samples. Tentatively one might assume that the relation of the number of synchronous II-complexes to that of all II-complexes is near to these figures.

It was more difficult to find the relation between the numbers of larger complexes synchronized in various ways. In some III- and IV-complexes all three or four cells synchronously entered prophase. In most III-complexes two cells are synchronous (see Tables II and VIII). In these complexes either the two synchronous cells or the third " odd " cell may enter prophase or become labelled. Therefore, all III-complexes with two cells in prophase and a corresponding number of III-complexes with one cell in prophase belong to this group. Synchrony of two cells is often observed even in IV-complexes. Thus synchronous cells were most often seen in pairs; these pairs either formed II-complexes or were parts of III- and IV-complexes.

CONCLUSIONS

Both formation and dissociation of cell complexes are linked with mitosis. Separation of contacting cells entering prophase may be a result of some alterations of the cell surface accompanying mitosis (see Weiss, 1962).

Each mitosis in the ascitic hepatoma cells leads to mutual transformations of various cell types, e.g. I-cells are transformed into II-complexes and vice versa. Obviously, ascitic hepatoma is one of the variants of biological systems in which constant relation between the cells belonging to various classes is a result of constant probabilities of cell transitions between these classes. Processes of this type may regulate cell numbers not only in tumours but also in certain organs, e.g., in haematopoietic tissues (Till *et al.*, 1964). Effects of contacting cells upon the mitotic cycles of their neighbours have two aspects: *(a)* synchronization of cycles in the contacting cells and *(b)* effect of the contacting cells upon the length of certain phases of the cycle.

Contacting cells of certain complexes synchronously pass S and G2 phases of the cycle. This synchrony may be the result of simultaneous beginning of the cycle in both sister cells which after mitosis form the II-complex. However, this synchrony is so precise that it is natural to suggest that it is supported by some interactions of the contacting cells which counteract random

TABLE VIII

SYNCHRONIZATION OF LABEL IN COMPLEXES OF VARIOUS TYPES

Mouse No.	Synchronous label in [1]:					
	II-complexes	III-complexes		IV-complexes		
	in II cells	in II cells	in III cells	in II cells	in III cells	in IV cells
12	59	39	22	28	29	24
23	73	24	41	19	20	53
45	63	35	27	35	28	19
39	33	34	9	28	6	8
06	60	49	15	25	17	43
34	66	35	34	24	26	37
07	65	29	16	18	21	44

[1] Number of labelled cells seen simultaneously in the same complex as a percentage of all labelled cells in complexes of given type. All samples were taken ½-1 hour after injection of thymidine-^{3}H.

fluctuations of the cycles. Numerous examples of the natural synchrony of cell divisions in syncytia and in the contacting cells of plants and embryos have been described (Agrell, 1964; Erickson, 1964; Rusch and Sachsenmaier, 1964). It has been suggested that in some of these systems synchrony of division is maintained by the accumulation in the cells of a common pool of certain substances which are of critical importance for mitosis (Rusch and Sachsenmaier, 1964). Possibly, a common pool of these hypothetical substances is formed also in the contacting cells of synchronous complexes. Mechanisms responsible for synchronization of S and G2 phases in the ascitic complexes may be similar to the compensatory mechanisms regulating the length of various parts of the mitotic cycle in cultured human cells (Sisken and Morasca, 1965).

In asynchronous complexes, as contrasted with the synchronous ones, the difference between the times when S-phase begins (or ends) in two contacting cells is not less than 6-8 hours. Thus II-complexes are most often in one of two discrete states: they are either characterized by complete synchrony or by considerable asynchrony of the contacting cells. Intermediate states, characterized by relatively small differences between the cells, seem to be unstable. To explain the existence of these two states one may suggest that the " synchronizing mechanism " discussed above can counteract small fluctuations of the cycles but cannot eliminate more considerable differences between the cells.

Synchronous beginning of mitosis in contacting cells leads to disintegration of the complex. Therefore, factors causing synchronization or desynchronization of mitotic cycles may be of great importance for the stability of cell complexes in normal and neoplastic tissues.

It is interesting to note that in the early stages of normal embryogenesis in various animal species the onset of differentiation is correlated with the desynchronization of cell divisions (Agrell, 1964). It would be important to study the role of mitotic synchrony in the development of tissue anaplasia characteristic for malignant tumours.

Contact with another cell can change the duration of the mitotic cycle: cells forming II-complexes have a shorter generation time than single cells. This shortening is due to the different duration of G1 stage in I-cells and in II-complexes. Many facts indicate that the presence of other cells may have a " conditioning " effect upon the multiplication of normal and neoplastic cells *in vivo* and *in vitro*; " conditioning " cells may liberate into the environment certain metabolites which are of critical importance for cell proliferation (see Eagle, 1963; Stoker and Sussman, 1965; Vasiliev, 1962). Shortening of the mitotic cycle in II-cell complexes may be a manifestation of the same effect, that is of the stabilization of local environment by the neighbour cells. In the case of ascites hepatoma complexes " local cell environment " may be the space enclosed by the contacting cells. Cell interactions which result in synchronization and those which result in the shortening of generation times may be correlated in the same way. It is natural to suggest that the same factors acting in the cell complexes may induce a simultaneous and earlier start of DNA synthesis in paired cells. However, at present this possibility cannot be confirmed or rejected because we cannot determine the duration of the phases of the mitotic cycle separately for " synchronous " and " asynchronous " complexes; we can only obtain average figures for all the II-complexes.

RÉSUMÉ

Le liquide ascitique formé après implantation intrapéritonéale d'hépatome ascitique de souris 22a contient des cellules isolées (cellules-I) et des agrégats formés de deux ou trois cellules ou davantage (agrégats-II, III, IV). On a constaté que les nombres relatifs des cellules-I et des agrégats des différents types étaient les mêmes dans tous les spécimens de liquide ascitique prélevé sur diverses générations de tumeurs transplantées. La non-séparation des cellules sœurs après la mitose semble jouer un rôle déterminant dans la formation des agrégats. Il y a deux moments du cycle mitotique auxquels l'agrégat peut se défaire: a) à la prophase commençant de façon synchrone dans plusieurs cellules en contact au sein de l'agrégat, et b) au commencement de la phase G1 dans les cellules formées après des mitoses d'une cellule-I ou après mitose asynchrone de l'agrégat.

On trouve dans le liquide d'ascite deux types discrets d'agrégats II: ceux dont les cellules en contact ont des cycles mitotiques synchrones et ceux dont les cycles sont fortement asynchrones. Les deux cellules des agrégats " synchrones " commencent et finissent simultanément la phase S et la phase G2.

Le temps moyen de génération des cellules qui forment des agrégats-II est plus court que celui des cellules-I; cette différence provient de la durée moyenne de la phase G1 qui est de 10 à 11 heures pour les cellules des agrégats-II, alors qu'elle est de 17 à 18 heures pour les cellules-I.

REFERENCES

AGRELL, I., Natural division synchrony and mitotic gradients in metazoan tissues, *in* E. Zeuthen (ed.) *Synchrony and mitotic gradients in metazoan tissues*, p. 39-67, Interscience Publishers, New York and London (1964).

BASERGA, R., and LISCO, E., Duration of DNA synthesis in Ehrlich ascites-cells as estimated by double-labeling with C^{14} and H^{3} thymidine and autoradiography. *J. nat. Cancer Inst.*, **31**, 1559-1571 (1963).

BASERGA, R., TYLER, S. A., and KIESELSKI, W. E., The kinetics of growth of the Ehrlich tumor. *Arch. Path. (Chicago)*, **76**, 9-13 (1963).

CHENZOV, G. S., and OLSHEVSKAJA, L. V., Electron-microscopic study of the solid and ascitic cells of hepatoma 22. *Cytologia, (Russ.)*, to be published (1966).

EAGLE, H., Population density and the nutrition of cultured mammalian cells, *in* D. Mazia and A. Tyler (ed.), *General physiology of cell specialization*, p. 151-170, McGraw-Hill Book Co., New York (1963).

EDWARDS, J. L., KOCH, A. L., YOUCIS, P., FREESE, H.L., LAITE, M. B., and DONALSON, J. T., Some characteristics of DNA synthesis and the mitotic cycle in Ehrlich ascites tumor cells. *J. biophys. biochem. Cytol.*, **7**, 273-282 (1960).

ERICKSON, R. O., Synchronous cell and nuclear division in tissues of the higher plants, *in* E. Zeuthen (ed.) *Synchrony in cell division and growth*, p. 11-38, Interscience Publishers, New York and London (1964).

GUELSTEIN, V. I., The transplanted strain of experimental liver cancer (hepatoma 22). *Vop. Onkol., (Russ.)*, **7**, 172-180 (1954).

MOSCONA, A., Cellular interactions in experimental histogenesis. *Int. Rev. Path.*, **1**, 351-428 (1962).

RUSCH, N. P., and SACHSENMAIER, M., Time of mitosis in relation to synthesis of DNA and RNA in Physarum polycephalum. *Acta Un. int. Cancr.*, **20**, 1282-1284 (1964).

SISKEN, J. E., and MORASCA, L., Intrapopulation kinetics of the mitotic cycle. *J. cell. Biol.*, **25**, 179-189 (1965).

STEINBERG, M. S., Reconstruction of tissue by dissociated cells. *Science*, **141**, 401-408 (1963).

STOKER, M., Regulation of growth and orientation in hamster cells transformed by polyoma virus. *Virology*, **24**, 164-174 (1964).

STOKER, M. G. P., and SUSSMAN, M., Studies on the action of feeder layers in cell culture. *Exp. cell Res.*, **38**, 645-653 (1965).

TILL, J. E., MCCULLOCH, E. A., and SIMONOWITCH, L., A stochastic model of stem cell proliferation, based on the growth of spleen colony-forming cells. *Proc. nat. Acad. Sci. (Wash.)*, **51**, 17-24 (1964).

VASILIEV, JU. M., The local stimulatory effect of normal tissues upon the growth of tumor cells, *in* M. Brennan and W. Simpson (ed.), *Biological interactions in normal and neoplastic growth, Henry Ford Hospital Symposium*, p. 299-309, Little Brown and Co., Boston (1962).

WEISS, L., Cell movement and cell surfaces: a working hypothesis. *J. theoret. Biol.*, **2**, 236-250 (1962).

WEISS, P. A., Cell interactions, *in* R. W. Begg (ed.), *Proceedings of the Canadian Cancer Conference*, Vol. 5, p. 241-276, Academic Press, New York (1963).

YOSHIDA, T., (ed.), Ascites tumors. Yoshida sarcomas and ascites hepatomas. *Nat. Cancer Inst. Monograph* No. 16, p. 51-94 (1964).

13.

(with V. I. Guelstein and Yu. M. Vasil'ev)

Initiation of DNA synthesis in cell cultures by colcemid

Proc. Natl. Acad. Sci. USA **68** (1971) 977–979

Communicated February 16, 1971

ABSTRACT Mitotic inhibitors (colcemid, colchicine, and vinblastine) initiate DNA synthesis in dense stationary cultures of mouse embryo fibroblast-like cells. This initiation is not due to any changes in local cell population density. Relationships between the activation of proliferation and other changes of interphase fibroblast-like cells produced by the mitotic inhibitors (disappearance of cytoplasmic microtubules and activation of movements of cell surface) are discussed.

In dense cultures of embryo fibroblast-like cells proliferation is inhibited; this inhibition is local and seems to be associated with the changes of the cell surface (1). Experiments presented in this paper show that several substances interfering with the formation of microtubules in the cells (the so-called "metaphase inhibitors" colchicine, colcemid, vinblastine) are able to initiate DNA synthesis in these stationary cultures.

The starting point of these experiments was the finding that one of these inhibitors, colcemid, produces considerable changes in the locomotor activity of interphase fibroblast-like cells (2). In normal moving cells of this type it was found that only part of the edge (the so-called leading edge) actively changed its form by forming and withdrawing local protrusions; other parts of the edge, especially those that were in contact with neighbor cells, remained stable. In colcemid-treated cells, however, all of the edge eventually became active. The first stage of contact inhibition was unaffected by colcemid: the forward movement of the surface protrusion stopped when the protrusion touched the surface of another cell. However, subsequent formation of the stable areas of the surface in the vicinity of the contact did not take place; formation and withdrawal of protrusions continued to go on in these zones for a long time. This suppression of "stabilization" of cell surfaces was accompanied by the inhibition of the formation of microtubules under these surfaces. These data suggest that microtubule formation is essential for stabilization.

Activation of the whole cell edge made colcemid-treated cells unable to translocate directionally, e.g., to migrate into the wound in the culture. These colcemid-induced changes were completely reversible (2). Colcemid did not affect actions that were dependent on local movements of small parts of cell surface, e.g., phagocytosis (3) or migration from the grooves (4).

Activation of the movements of the cell surfaces in colcemid-treated cultures suggests that substances of this type can initiate proliferation in stationary monolayers. This suggestion is based on the assumption (see detailed discussion in ref. 5) that a set of reactions leading to growth and mitosis is switched on by surface changes which temporarily decrease the isolation of the cell interior from the external environment. Diverse agents that may cause surface changes of this type (e.g., by increasing the permeability of the cell membrane or by removing the outer cell coats) can be expected to initiate the proliferation. Activation of the surface movements is likely to be accompanied by these changes (5).

These considerations induced us to perform the experiments described below.

MATERIALS AND METHODS

Stationary cultures of the first passage of mouse embryo cells grown on coverslips in flasks containing penicillin were used. The following inhibitors were used: colcemid (Ciba, Switzerland) or chemically identical colchamine (Sojusreactiv, USSR); colchicine (Merck, DBR); vinblastine (Richter, Hungary). Culture medium consisted of 45% basal Eagle medium, 45% lactalbumin hydrolysate, and 10% bovine serum; the medium was changed every 48 hr.

Mitotic inhibitors, dissolved in 0.1 or 0.2 ml of saline, were added to the medium of the cultures 174 hr after seeding and 30 hr after the last medium change; the same volume of saline was added to control cultures. No change of the medium was made during the period of incubation. [^{3}H]thymidine (Radiochemical Centre, Amersham, England; 8.6 Ci/mmol) was added to the cultures to assay DNA synthesis; cultures were then fixed and examined autoradiographically. In the experiments with pulse labeling, [^{3}H]dT (1 μCi/ml) was added 30 min before fixation. In the experiments with continuous labeling, [^{3}H]dT (1 μCi/ml) was added together with the inhibitor and remained in the medium throughout the incubation (3–30 hr). The cultivation procedures and autoradiography were identical with those described earlier (6).

The percentage of labeled interphase cells (labeling index) was determined for each culture. The percentage of labeled mitoses and the mean number of grains per labeled nucleus were also determined in some cultures. Three or more cultures were used to test each concentration of an inhibitor in a single experiment; each experiment was repeated three or more times.

RESULTS

The labeling index (Table 1) of the cultures incubated for 24 hr with colcemid (effective concentrations (0.05–0.4 μg/ml), colchicine (0.05–0.1 μg/ml), or vinblastine (0.05 μg/ml) was in all the experiments considerably higher than in those of the control cultures. This increase was observed in the experiments with pulse-labeled and with continuously-labeled cultures. It was not accompanied by any significant changes in the mean number of grains per labeled nucleus. These results show that the three metaphase inhibitors are able to induce in-

TABLE 1. *Effects of mitotic inhibitors on DNA synthesis in stationary cultures of mouse embryo fibroblast-like cells*

Inhibitor	Concentration (μg/ml)	% Interphase cells labeled with [³H]dT after 24 hr of incubation: Pulse labeling (30 min) Expt.	Pulse labeling (30 min) Control	Continuous labeling (24 hr) Expt.	Continuous labeling (24 hr) Control
Colcemid (Ciba)	0.1	13.3 ± 1.52	5.0 ± 0.78	42.1 ± 3.41	32.1 ± 2.16
	0.2	24.6 ± 1.86	4.6 ± 0.35	47.1 ± 2.48	
	0.4	26.4 ± 2.68		46.9 ± 2.38	
(Sojuzreaciv)	0.01*	5.0 ± 0.68	4.5 ± 0.40	NT	NT
	0.05	14.7 ± 0.96	4.3 ± 0.71	39.0 ± 2.55	29.2 ± 1.46
	0.1	19.9 ± 1.08	4.6 ± 0.51	43.3 ± 1.70	27.6 ± 1.28
Colchicine	0.05	24.8 ± 2.13	4.6 ± 0.35	40.0 ± 2.37	32.1 ± 2.16
	0.1	29.8 ± 3.07		43.6 ± 3.00	
Vinblastine	0.01	8.3 ± 0.86	5.7 ± 0.30	24.8 ± 1.03	27.4 ± 1.52
	0.02	7.5 ± 0.56	3.4 ± 0.12	28.6 ± 2.12	27.7 ± 1.22
	0.05	17.1 ± 2.66	4.9 ± 0.51	42.5 ± 3.21	28.1 ± 2.17

Each horizontal row summarizes the results of 3–4 Expts.; means ± SE shown.
NT, not tested.
* Metaphase block was not observed.

creased entry of the cell into the S phase of the mitotic cycle. All concentrations of the inhibitors that induced significant increases of the labeling index also caused complete metaphase block: anaphases and telophases were not seen in these cultures. These concentrations also produce changes in locomotion and in cell form in the same system (2).

To test the effect of cell population density on the cell entry into S phase, "density-adjusted labeling indices" were determined in several paired control and colcemid-treated cultures. The total number of cells and the number of labeled cells were counted separately in each of 100 randomly chosen fields of view of the microscope, the area of each being 8.7×10^{-3} mm². Then, labeling indices for areas with various cell densities were calculated. The indices in colcemid-treated cultures were always considerably higher than those in areas having the same cell densities in paired control cultures (Fig. 1). Thus, increased cell entry into S phase in colcemid-treated cultures was not a result of any changes in cell population densities.

Experiments with pulse-labeled cultures had shown that the fraction of DNA-synthesizing cells in the cultures increased significantly about 12 hr after the addition of the inhibitor and continued to increase thereafter. Fig. 2 shows the time course of the effects of colcemid in detail. In continuously labeled control cultures (Fig. 2*B*), the labeling index increased with time not only because new cells entered S phase but also because previously labeled cells were doubled in number after mitosis. This doubling did not take place in colcemid-treated cultures and, therefore, the labeling index was lower than in control cultures until 16 hr. Later this effect was overruled by increased entry of cells into S phase.

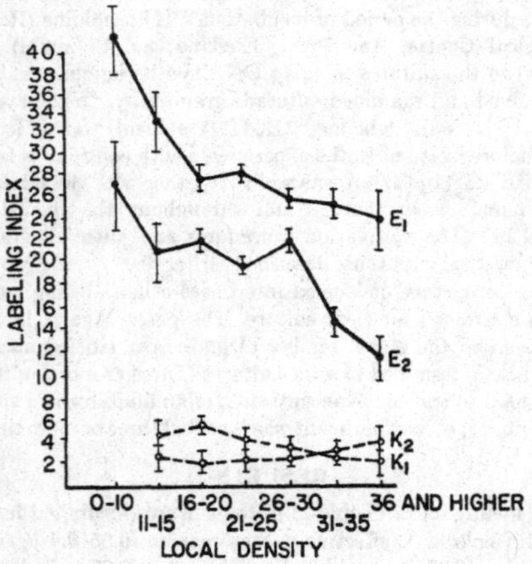

FIG. 1. Labeling indices in areas of cultures having various local cell densities after pulse labeling with [³H]dT. E_1, E_2, experimental cultures (incubated 24 hr with 0.1 μg/ml of Colcemid). K_1, K_2, controls. Abscissa, number of cells per field of view. Values are means ±SE.

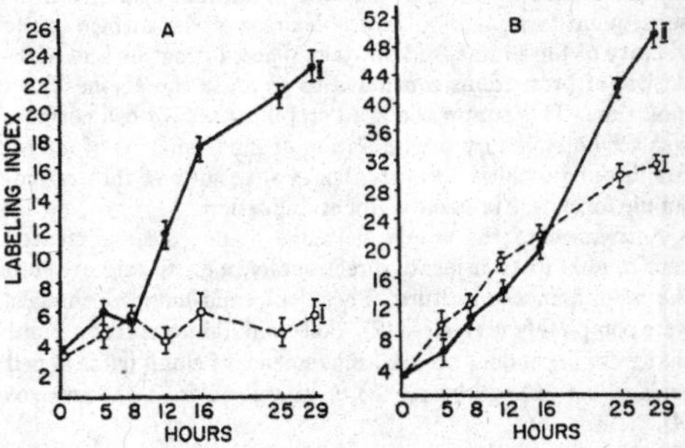

FIG. 2. Time course of the changes in [³H]dT incorporation induced by colcemid (0.1 μg/ml) in stationary cultures (means ±SE). Abscissa, time of incubation with colcemid. ○, Control cultures. ●, Colcemid-treated cultures. *A*, Pulse labeling. *B*, Continuous labeling.

In the experiments with continuous labeling, all the mitotic figures in control cultures became labeled at about 8 hr; in colcemid-treated cultures, 75–100% of the mitoses became labeled after 12–16 hr. This result seems to indicate that cells in colcemid-treated cultures enter mitosis some time after passage through S phase. More exact determination of the kinetics of this entry into mitosis was impossible because the destruction of some mitotic cells and (or) reversal of blocked mitoses in colcemid-treated cultures (7) could affect the counts of mitoses.

CONCLUSION

Initiation of DNA synthesis in stationary cultures of fibroblast-like cells can be induced by agents of several types: (*a*) by wounding of the culture (1); (*b*) by the factor (or factors) contained in fresh serum (1, 6, 8); (*c*) by various enzymes such as proteases (9), hyaluronidases (6), and ribonuclease (6); (*d*) by detergents such as lysolecithin (10) and digitonin (6). Proliferative reactions induced by the agents of the last three groups are not preceded by any changes in cell population density. Experiments described in this paper show that a new group has to be added to this list of activating agents, namely substances that affect microtubular structures (mitotic inhibitors) (11). The time course of the initiating effect of these substances is similar to a few other agents (6).

Many activating agents were shown to become toxic when the conditions of these experiments are changed (6). In these and previous experiments (2), we observed no significant toxic changes in the interphase cells treated with mitotic inhibitors. However, in a few other experimental systems, certain toxic effects of these inhibitors, e.g., inhibition of DNA synthesis, have been described (12).

Thus, mitotic inhibitors, besides producing a characteristic metaphase block, cause at least three types of alterations in the interphase fibroblast-like cells: disappearance of cytoplasmatic microtubules (2), disappearance of the stable parts of the cell surface (2), and initiation of the DNA synthesis in dense cultures. Similar concentrations of inhibitors produce each of these alterations.

It would be interesting to discover the relationships between these alterations. We have mentioned above the suggestion that the initiation of proliferation by all types of activating agents is a consequence of surface changes (5). This generalization seems to have some heuristic value: it has helped to reveal the ability of mitotic inhibitors and of a few other activating agents (6) to initiate DNA synthesis. Little is known, however, about the real mechanisms of action of these agents.

1. Todaro, G. J., G. K. Lazar, and H. Green, *J. Cell. Comp. Physiol.*, **66**, 325 (1965); Vasiliev, J. M., I. M. Gelfand, and L. V. Erofeeva, *Dokl. Akad. Nauk SSSR* (in Russian), **171**, 721 (1966); Vasiliev, J. M., I. M. Gelfand, L. V. Domnina, and R. I. Rappoport, *Exp. Cell Res.*, **54**, 83 (1969); Macieira-Coelho, A., *Int. J. Cancer*, **2**, 297 (1967); Eagle, H., and E. M. Levine, *Nature*, **213**, 1102 (1967); Dulbecco, R., *Nature*, **227**, 802 (1970); Clarke, G. D., M. G. P. Stoker, A. Ludlow, and M. Thornton, *Nature*, **227**, 798 (1970).
2. Vasiliev, J. M., I. M. Gelfand, L. V. Domnina, O. J. Ivanova, S. G. Komm, and L. V. Olshevskaja, *J. Embryol. Exp. Morphol.*, **24**, 625 (1970).
3. Mojzess, T. G., *Tsitologiya* (in Russian), **13**, 493 (1969).
4. Rovensky, J. A., I. L. Slavnaja, and J. M. Vasiliev, *Exp. Cell Res.* (1971) in press.
5. Vasiliev, J. M., and I. M. Gelfand, *Curr. Mod. Biol.*, **2**, 43 (1968).
6. Vasiliev, J. M., I. M. Gelfand, V. I. Guelstein, and E. K. Fetisova, *J. Cell. Physiol.*, **75**, 305 (1970).
7. Eigsti, O. J., and P. Dustén, *Colchicine in Agriculture, Medicine, Biology, and Chemistry* (Iowa State College Press, Ames, Iowa, 1955).
8. Todaro, G. Y., Y. Matsuja, S. Bloom, A. Robbins, and H. Green, in *Growth-Regulating Substances for Animal Cells in Culture, Symposium monograph no. 7*, ed. V. Defendi and M. Stoker (The Wistar Institute Press, Philadelphia, 1967), p. 87; Temin, H. M., *J. Nat. Cancer Inst.*, **35**, 167 (1966); Temin, H. M., *J. Cell. Physiol.*, **69**, 77 (1967); Holley, R. W., and J. A. Kiernan, *Proc. Nat. Acad. Sci. USA*, **60**, 300 (1968); Yeh, J., and H. W. Fisher, *J. Cell Biol.*, **40**, 382 (1969).
9. Burger, M. M., *Nature*, **227**, 170 (1970); Sefton, B. M., and H. Rubin, *Nature*, **227**, 843 (1970).
10. Nilausen, K., *Nature*, **217**, 268 (1968).
11. Stimulation of DNA synthesis in the root-tips of *Allium cepa* was observed by Chakraborty, A., and A. A. Biswas, *Exp. Cell Res.*, **38**, 57 (1965).
12. Hell, E., and D. C. Cox, *Nature*, **197**, 287 (1963); Ilan, J., and J. H. Quastler, *Biochem. J.*, **100**, 448 (1966); Gustafsson, M., *Exp. Cell Res.*, **50**, 1 (1968); Fitzgerald, P. H., and L. A. Brehaut, *Exp. Cell Res.*, **59**, 27 (1970).

14.

(with Yu. M. Vasil'ev)

Mechanisms of morphogenesis in cell cultures

Int. Rev. Cytol. 50 (1977) 159–274

I.	Introduction	159
II.	Basic Morphogenetic Reactions of Cultured Cells	162
	A. The Main Morphological States of Cells in Cultures	162
	B. Submembranous Cortical Layer	163
	C. Reactions of Active Attachment	169
	D. Contact Inhibition of the Formation of Pseudopods	185
	E. Stabilization Reactions	188
	F. Conclusion	193
III.	Shape and Behavior of Normal Cells in Culture	194
	A. Fibroblasts	194
	B. Epithelial Cells	231
	C. Comparison of Morphogenetic Reactions of Fibroblasts and of Epithelial Cells	235
IV.	Alterations in the Morphogenetic Reactions Accompanying Cell Transformation	237
	A. Introduction	237
	B. Basic Morphogenetic Reactions of Transformed Cells	240
	C. Shape and Behavior of Transformed Fibroblasts	244
	D. Shape and Behavior of Transformed Epithelial Cells	263
V.	Concluding Remarks	267
	References	268

I. Introduction

The aim of this article is to describe and discuss mechanisms of morphogenesis and locomotion in cell cultures of two main tissue types: fibroblasts and epithelium. Common features of both types of cells are their ability to attach themselves to solid substrates and to form organized multicellular structures on these substrates. However, there are considerable differences between the structures formed by epithelial and by fibroblastic cells.

Epithelial cells usually form coherent monolayered cell sheets, and the cells in these sheets are firmly attached to each other. Fibroblasts may form several variants of structures: monolayers of mutually oriented cells not linked firmly to each other, multilayered sheets, or spherical aggregates.

Multicellular structures formed by epithelial cells and fibroblasts in cultures are similar in many aspects to the tissue structures formed by cells of the same type *in vivo*. This similarity is further increased by the ability of multicellular structures in culture to regenerate their initial structure after damage; the well-known phenomenon of wound healing in culture is an example of such regeneration.

Formation of an organized multicellular histological structure is obviously the end result of a long series of alterations in individual cells, such as cell spreading on the substrate, cell polarization, and translocation. Each of these cellular alterations in turn is the end result of a series of more simple cellular reactions which are referred to as basic morphogenetic reactions. In this article we distinguish three groups of basic morphogenetic reactions:

1. Reactions of active attachment, which consist of several stages: the formation of pseudopods at the cell surface, the attachment of these pseudopods to other surfaces, and the development of tension within the attached pseudopod.
2. Contact paralysis, that is, cessation of the formation of pseudopods at the site of cell-cell contact.
3. Stabilization reactions controlling distribution of the sites of formation and of retraction of pseudopods in the cell.

These three groups of basic morphogenetic reactions are discussed in Section II. The behavior of fibroblasts and of epithelial cells under different conditions is described in Section III. We discuss here cell spreading, polarization, translocation, cell-cell interactions, and contact guidance. We also try to determine how each behavioral act is composed of combinations of basic morphogenetic reactions. Section IV is devoted to the alterations in morphogenetic reactions and in the locomotory behavior observed in transformed cultures.

The nature of the cells used in studies of morphogenesis in cultures needs special comment. Two groups of cell cultures are used in such experiments:

1. Primary and secondary cultures usually obtained from dissociated embryonic tissues or, more rarely, from those of adult animals. The old-fashioned technique involving the explantation of tissue fragments also remains useful for certain purposes, especially to obtain epithelial cultures. The advantage of primary cultures is their relative normalcy, that is, absence of the morphological alterations that develop in the course of long-term cultivation.
2. Permanent cell lines and strains. The advantages of these lines are obvious. However, when the aim of an investigator is to study

normal morphogenesis, one has to take into consideration the fact that the morphology of these lines is usually somewhat different from that of progenitor primary cultures. These alterations are often somewhat similar to those observed after transformation of cultures by oncogenic viruses and chemical carcinogens (see Section IV). The degree of these alterations may be different in different lines. One may say that all the cell lines are transformed but some lines are more transformed than the others. Of course, this does not mean that continuous cell lines should not be used in studies of normal morphogenesis. However, the degree of normalcy of the cells used should be taken into account in the interpretation of results.

It is also obvious that cells of the same tissue type obtained from different sources may share certain common properties but are not identical. Fibroblastlike cells in general, as well as epithelial cells in general, are nothing more than abstract archetypes of actual cell varieties. It is known that explants of different epithelia have a somewhat different morphology. Fibroblastlike cells that are morphologically similar in cultures may belong to different subclasses of mechanocytes (Willmer, 1965). Special experimental procedures may reveal different potentialities in morphologically similar cells from different sources. For instance, transplantation to isogenic animals of cultured fibroblastic clones originally obtained from bone marrow revealed the ability of these cells to undergo osteogenesis; fibroblastlike cells originally isolated from the spleen did not have this ability (Friedenstein and Lalykina, 1973). Glial cells in cultures may be morphologically very similar to fibroblasts (Pontén, 1975). The properties of each cell type may also vary, depending on the age and species of the animals from which they were derived.

The analysis of all these intraclass differences among many variants of epithelial and fibroblastic cells is mostly a task for future studies. In this article we deal mainly with general interclass differences between these two cell types.

In our experiments the main types of cultures used in the studies of normal morphogenesis were (1) secondary cultures of mouse embryo fibroblastlike cells, (2) explants of mouse kidney epithelium, and (3) a continuous MPTR line of mouse kidney cells. Although these cells have been transformed by SV40 virus and contain the genome of this virus, they retain the normal ability to form a coherent epithelial sheet. Only detailed examination of the properties of this line (see Section IV,D) reveals some alteration in their attachment to the substrate.

Standard tissue culture methods and other techniques (scanning electron microscopy, microcinematography, etc.) used for studies of morphogenesis are not discussed here; their detailed description can be found in original reports cited in the text.

II. Basic Morphogenetic Reactions of Cultured Cells

A. The Main Morphological States of Cells in Cultures

Cells in culture may exist in several different morphological states; transition from one state to another occurs when cell-substrate relationships change, for example, when cells are seeded on the substrate or detached from it (Fig. 1). These transitions are fully reversible and can be repeated many times. One may distinguish two main morphological states of an isolated fibroblast: spherical and polarized. The spherical state is characteristic of cells not attached to solid substrate. In particular, it is characteristic of cells suspended in a fluid medium. When spherical cells contact an appropriate substrate, they are gradually transformed into polarized cells. This transformation usually passes through an intermediate state: that of the radially spread cell. Accordingly, transition from the spherical to the polarized state may be subdivided into two consecutive stages: (1) radial spreading, that is, transition from the spherical to the radially spread state; and (2) polarization, that is, transition from the radially spread to the polarized state.

Cells in the radially spread state are discoid in shape and are firmly attached to the substrate. These cells have two structurally and functionally different zones: (1) a central zone (the endoplasm), containing the nucleus and all the main vesicular organelles, and (2) lamellar cytoplasm (the lamelloplasm), that is, a pheripheral zone which contains no particulate organelles. Sites of cell-substrate attachment are localized preferentially in the lamellar cytoplasm. The lamellar cytoplasm of radially spread cells has a circular shape; it forms a ring surrounding the endoplasm. The external edge of the lamellar cytoplasm of these cells is active, that is, pseudopods are continuously formed and retracted along this edge. The radially spread state is not stable, and radially spread cells undergo spontaneous polarization. Under certain conditions (see Section III,A,5) spreading and polarization may proceed simultaneously, that is, a spherical cell may be transformed into a polarized one without passing through the radially spread state.

A polarized cell, like a radially spread one, is attached to the substrate, but its external contour is not circular. These cells can have a variety of shapes; they may be fusiform, fanlike, stellate, and so on. The shape of an individual polarized cell is not stable but may change with time, especially during locomotion. An essential general characteristic of the polarized state, distinguishing it from the radially spread state, is a division of the external cellular edge into active and nonactive parts. Because of this division polarized cells can move directionally on the substrate.

Polarized cells, like radially spread ones, have endoplasm and lamellar cytoplasm. However, the lamellar cytoplasm of polarized cells does not form a single ring but is divided into several discrete areas. The active parts of the external edge usually delimit areas of the lamellar cytoplasm. Nonactive parts of the edge may delimit either the lamellar cytoplasm or the endoplasm. When a polarized cell moves directionally on the substrate, the largest active part of the edge (the leading edge) is usually localized at the anterior end. This leading edge delimits the largest area of the lamellar cytoplasm (the anterior lamella). When polarized or radially spread cells are detached from a substrate, they return immediately to the spherical state.

Fibroblasts spread in dense cultures may be regarded as special variants of polarized cells. These cells are elongated and have areas of lamellar cytoplasm. It is not clear, however, whether or not they have active areas along the edge (see Section III,A,4).

Morphological transformations of epithelial cells have not yet been studied in detail. The main morphological states of these cells are probably, the same as those of fibroblasts: spherical, radially spread, and polarized. To these one may add a special state characteristic of the central cells of epithelial sheets. All the lateral edges of these cells are not active and are firmly attached to other cells; attachment of the lower surface to the substrate may be absent.

Presumably, transitions of cells from one morphological state to another, for example, cell spreading and polarization, may be regarded as end results of a long series of few basic morphogenetic reactions. Before discussing these reactions we review briefly the structure of the cell part that plays a leading role in these reactions, namely, the submembranous cortical layer.

B. Submembranous Cortical Layer

The complex of structures designated the cell surface or, more rarely, the cell periphery (L. Weiss, 1967), consists of an extramembranous cell coat, a cell membrane, and a submembranous cortical

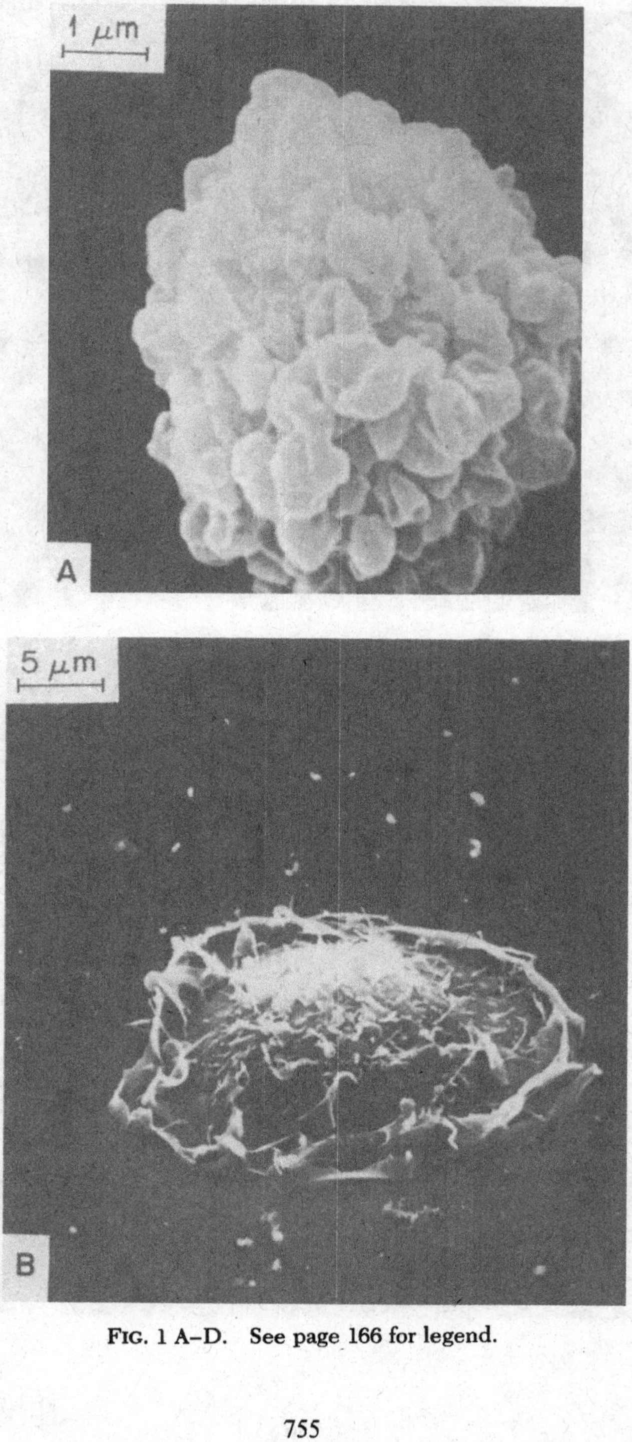

FIG. 1 A–D. See page 166 for legend.

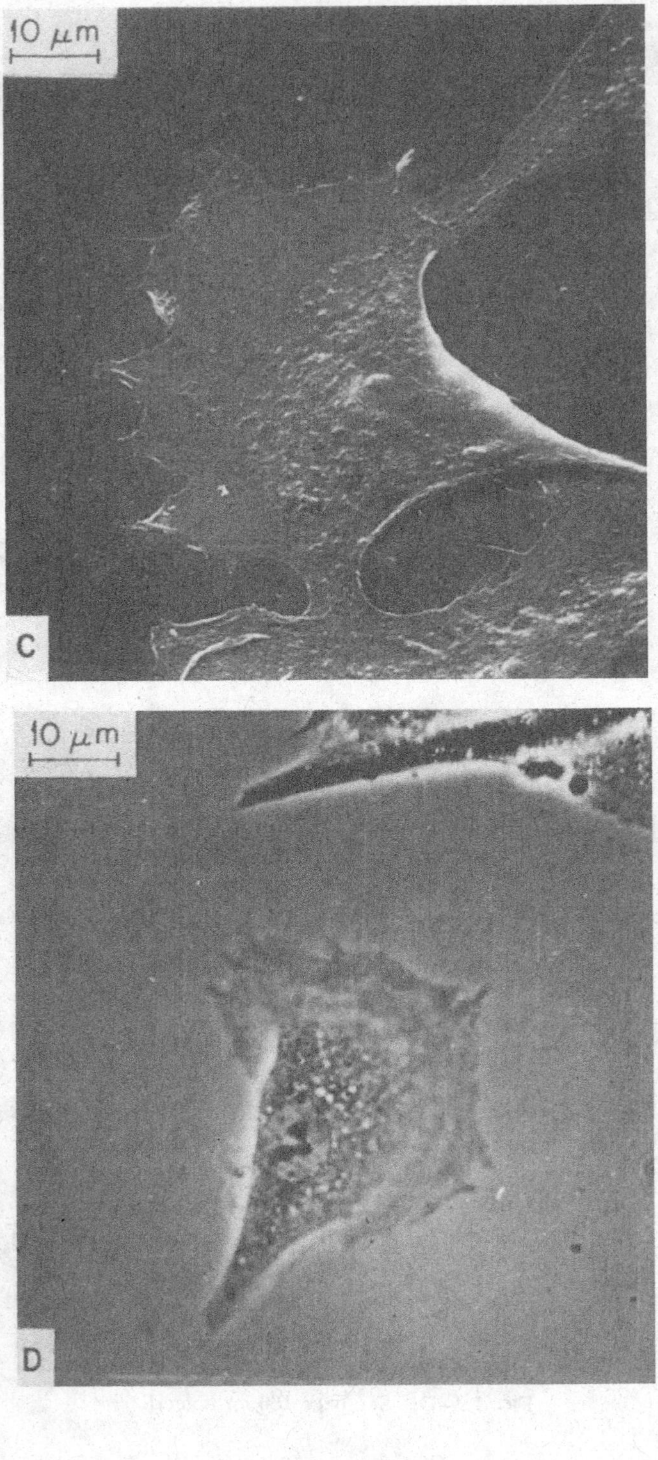
10 μm
C
10 μm
D

layer. The cortical layer is a zone under the cell membrane having a thickness of about 0.1–0.5 μm and can be seen in all cell parts. In particular, the cortical layer is almost the only structural component of the cytoplasm of the lamelloplasm and of the pseudopods extended by the cells. In sections the cortical layer appears as a zone of medium electron density from which ribosomes and other organelles are excluded. The structures most often seen in this zone are microfilaments 4–7 nm in diameter. These microfilaments contain polymerized actin, as shown by their ability to bind heavy meromyosin (Ishikawa *et al.*, 1969; Wessels *et al.*, 1973; Goldman and Knipe, 1973). The presence of actin in the bundles of microfilaments was confirmed by immunofluorescence studies using antiactin antibodies (Lazarides and Weber, 1974; Lazarides, 1975a,b; Pollack *et al.*, 1975). Large amounts of actin-like protein were found in several types of nonmuscle cells (Tilney and Mooseker, 1971; Bray, 1972; Allison, 1973; Pollard, 1975).

Microtubules and 10- to 11-nm filaments can also be seen in sections of some parts of the cortical layer; the chemical nature of these filaments is not clear.

Actin microfilaments may form two types of configurations in the cortical layer (Spooner *et al.*, 1971; Wessels *et al.*, 1973):

1. A three-dimensional matrix of microfilaments without visible regular pattern. (In sections certain areas of the cortical layer seem to consist of an amorphous substance of medium electron density. Probably, the cortical layer in these areas also consists of a matrix of microfilaments, but individual microfilaments are not discernable. In particular, one cannot exclude that in these areas the microfilament matrix is embedded in some other amorphous component.)
2. Bundles of parallel microfilaments (the microfilament sheath) corresponding to the stress fibers visible with light microscopy (Buckley and Porter, 1967; Goldman *et al.*, 1975).

FIG. 1. The main morphological states of normal mouse embryo fibroblasts in culture. (A) Spherical cell with the surface covered with blebs. (B) Radially spread cell 1 hour after seeding on glass. Note the ruffles near the cell edge. (C and D) Polarized cells. Note division of the cell body into peripheral lamelloplasm and central endoplasm; the endoplasm but not the lamelloplasm contains vesicular organelles. Scanning electron micrographs (A–C). Phase-contrast micrograph of a living cell (D). All scanning electron micrographs in this and following illustrations were made with a Cambridge Stereoscan-S4; glutaraldehyde-fixed cultures were critical-point dried; carbon dioxide was used as a transitional fluid. Photographs were taken at 10 kV; the tilt angle was 45°.

Electron microscope observations suggest that most of the polymerized actin in cultured cells is localized in the cortical layer. The only exceptions to this rule are microfilament bundles passing through the internal parts of the cell; however, even in this case the ends of the bundles tend to be localized in the cortical layer. At present we do not know the reason for the predominantly cortical localization of the actin microfilaments. Possibly, conditions favorable to the polymerization of actin are found in this zone.

Besides actin, the cortical layer also contains several other proteins which are able to interact with actin. They include:

1. Myosinlike protein. The protein isolated from cultured nonmuscle cells seems to be more similar to the myosin in smooth muscle and in platelets than to skeletal muscle myosin (Adelstein *et al.*, 1972; Groeschel-Stewart, 1971; Ostlund *et al.*, 1974; Stossel and Pollard, 1973; Pollard, 1975; Chi *et al.*, 1975). Immunomorphological studies indicate that myosin is localized in striated structures probably identical to the microfilament bundles; myosin may also be present outside the bundles (Weber and Groeschel-Stewart, 1974; Lazarides, 1975a,b; Pollack *et al.*, 1975).
2. Tropomyosin, visualized in microfilament bundles and also in a diffuse form (Lazarides, 1975a,b).
3. α-Actinin, found especially near the ends of microfilament bundles and also in striations along these bundles (Lazarides, 1975a,b).

Probably several other proteins interacting with actin are also present in the cortical layer.

Spherical cells have only a matrix cortical layer. Microfilament bundles appear during spreading and disappear again during cell rounding after detachment from the substrate. Thus, the pattern of organization of the cortical structures changes quickly and reversibly in the course of cellular transition from one state to another. Besides the development and disappearance of microfilament bundles there are probably many other alterations in the organization of the cortical layer, which we are unable to distinguish. Possibly, the matrix-type cortical layer has a definite pattern of microfilament arrangement which is not revealed in cell sections. The framework of cortical actin microfilaments provides sufficient mechanical stability, but at the same time the cell is able to move and to change its shape; therefore this framework is probably responsible for the maintenance of and alterations in cell shape and surface topography.

Mechanical interrelationships between the cell membrane and the cortical layer may be compared with those between a cloth and the framework on which it has been stretched. At the molecular level these interrelationships are much more complex and diversified. Several experimental results indicate that submembranous cortical components may interact with integral membrane proteins. As a result of these transmembrane interactions alterations in the distribution of the external surface structures may affect the distribution of cortical components, and vice versa. In particular, it is probable that reversible binding of membrane proteins to cortical structures may inhibit free diffusion of these proteins within the plane of the membrane. It may also cause directional translocation of these proteins into certain areas of the membrane (see review and discussion in Berlin *et al.,* 1974; Nicholson, 1974; de Petris, 1975; Yahara and Edelman, 1975; Poste *et al.*, 1975). There is a growing conviction, shared by us, that these transmembrance interactions are real and play important roles in cell physiology. However, one must stress that very little is known at present about the exact phenomenology of these interactions and about their molecular mechanisms. Besides being involved in direct structural interactions with the cortical layer, the cell membrane may of course affect the state of this layer through the products of its enzymes, such as adenylcyclase, as well as through alterations in permeability to ions and molecules.

One particular group of pharmacologically active substances which profoundly affect the state of the cortical layer deserves special brief comment. These are cytochalasins, widely used in studies of cellular morphogenesis. Cytochalasins produce reversible inhibition of most types of cell movement and movement within the cell (see review in Allison, 1973). They also produce striking and characteristic alterations in cell morphology. In particular, treatment of polarized fibroblasts with cytochalasins leads to the disappearance of the lamellar cytoplasm at the cell periphery; instead, a system of branched cytophalstic cords becomes visible; this morphological alteration has been termed arborization (Spooner *et al.*, 1971; Wessels *et al.*, 1973; Sanger, 1974; Croop and Holtzer, 1975). Electron microscopy of cytochalasin-treated cells reveals the disappearance of microfilaments, especially in the cortical matrix; certain microfilaments, especially those in the bundles, may be preserved in cytochalasin-treated cells (Wessels *et al.*, 1973; Goldman and Knipe, 1973). Cytochalasin-treated cells lose the ability to acquire a spherical shape after detachment from the substrate (Vasiliev *et al.*, 1975b).

Cytochalasin B inhibits sugar transport in mammalian cells; this effect is observed at lower concentrations than those affecting cell shape and movement (Kletzien *et al.*, 1972; Estensen and Plagemann, 1972; Plagemann and Estensen, 1972). However, cell incubation in glucose-free medium does not lead to morphological alterations similar to those induced by cytochalasin, so these alterations are not a result of glucose deprivation (Yamada and Wessels, 1973; Taylor and Wessels, 1973). Cytochalasin B is bound with a high affinity by certain membrane proteins and with a lower affinity by certain other unidentified cell components (Lin and Spudich, 1974; Lin *et al.*, 1974). Cytochalasin B also alters the structure of the microfilaments formed by purified actin *in vitro* (Spudich, 1972). It is not clear which of these molecular effects, if any, is responsible for characteristic reversible alterations produced by this drug at the cellular level.

The cell membrane and the cortical layer play essential roles in all three types of basic morphogenetic reactions. We now discuss the first group of these reactions, namely, reactions of active attachment.

C. Reactions of Active Attachment

1. *Experimental Data Suggesting That Special Active Reactions Are Needed for Formation of the Attachments between the Cell Surface and Other Surfaces*

It is common knowledge that fibroblasts and epithelial cells are able to attach themselves to surfaces by forming local contact structures with these surfaces: specialized cell-cell contacts or cell-substrate attachment sites. Another variant of cell attachment to a surface is adhesion of various particles; this adhesion is the first stage of phagocytosis.

Several facts indicate that various parts of the cell surface have a different ability to form new attachments with other surfaces. More specifically, this ability seems to be characteristic only of the surface of pseudopods actively extended by the cell. The special role of pseudopods in mediating cell adhesion to various surfaces had been suggested by theoretical considerations and experimental data obtained from various systems (Bangham and Pethica, 1960; Pethica, 1961; Lesseps, 1963; Garrod and Born, 1971). With regard to cultured epithelial and fibroblastic cells two groups of facts support this suggestion.

The first group is related to the formation of contact structures by various parts of the surface of epithelial sheets. Numerous microcinematographic observations indicate that the upper surfaces of these sheets, as well as the lateral surfaces of the central cells locked by firm

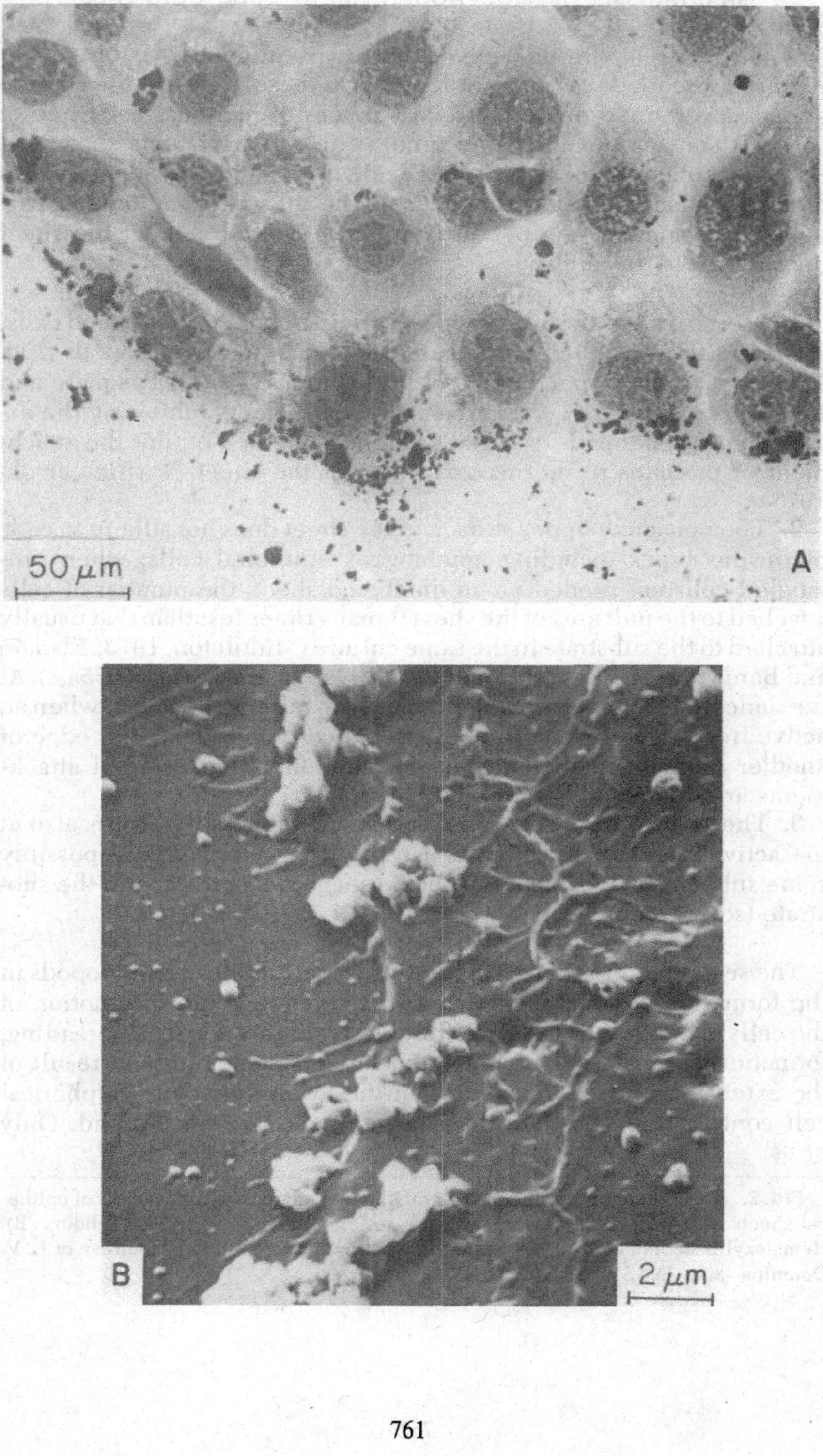
50 μm
A
B
2 μm

cell-cell contacts, are nonactive; that is, pseudopods, visible at the light microscope level, are not formed on these surfaces. Such pseudopods are formed only on lateral surfaces of marginal cells free of cell-cell contacts. At the same time, various experiments show that only the surface of these active free edges of marginal cells is able to form new attachments to other cells, to the substrate, and to inert particles. The nonactive upper surface of the central cells of the sheet does not have this ability:

1. Inert particles of various types (carmine particles, red blood cells, etc.) are easily attached to the surfaces of marginal pseudopods (Fig. 2) but not to the upper surface of the central cells (Di Pasquale and Bell, 1974; Vasiliev *et al.*, 1975a,c). Various agents inhibiting the extension of pseudopods (sorbitol, cytochalasin, etc.) inhibit the attachment of particles to the marginal area of the sheet (Vasiliev *et al.*, 1975a).

2. The nonactive upper surface of the sheet does not adhere to cells of various types including homologous epithelial cells; when prelabeled cells are seeded on an unlabeled sheet, the number of cells attached to the unit area of the sheet is many times less than that usually attached to the substrate in the same culture (Middleton, 1973; Elsdale and Bard, 1974; Di Pasquale and Bell, 1974; Vasiliev *et al.*, 1975a,c). At the same time, microcinematographic observations show that, when an active free edge of a marginal epithelial cell meets a similar edge of another homologous cell during locomotion stable cell-cell attachments are immediately formed.

3. The formation of cell-substrate contacts probably occurs also at the active edges of marginal cells; only the marginal (and possibly some submarginal) cells of the sheets seem to be attached to the substrate (see Section II,B).

The second group of facts suggesting a special role of pseudopods in the formation of attachments concerns spreading and locomotion of the cells of another tissue type, fibroblasts. In the course of spreading, formation of local cell-substrate attachments seems to be a result of the extension of pseudopods. When the lower surface of a spherical cell contacts the substrate, local attachments are not formed. Only

FIG. 2. Attachment of carmine particles to the free edges of marginal cells of epithelial sheets of strain MPTR. Incubation with carmine for 24 hours (A) and 2 hours (B). Hematoxylin-stained culture (A). Scanning electron micrograph (B). Courtesy of L. V. Domnina and O. S. Zacharova.

later, when pseudopods have been extended by the cell, are local attachment sites formed at their ends; microfilament bundles associated with the contact structures are also formed in the cytoplasm of these pseudopods (Bragina *et al.*, 1976). A polarized fibroblast translocating on the substrate also forms new attachment sites at the active cell edge, that is, at the edge where new pseudopods are formed (Abercrombie *et al.*, 1971). When the active edges of two moving fibroblasts contact each other, specialized cell-cell contacts are immediately formed (Heaysman and Pegrum, 1973).

These facts give reason to single out a group of special morphogenetic reactions, reactions of active attachment, which have three main stages: extension of pseudopods, attachment of these pseudopods to other surfaces, and development of tension within the attached pseudopods (Vasiliev *et al.*, 1975a; Vasiliev and Gelfand, 1976a). We now discuss each of these stages in more detail.

2. *Extension of Pseudopods*

a. *Morphology of Pseudopods.* Primary surface extensions (pseudopods) should be distinguished from composite cytoplasmic outgrowths. A primary extension is formed as the result of a rapid one-step extension. By "rapid" we mean that the extension lasts only a few minutes. By "one-step" we mean that the extension is not interrupted when observed microcinematographically on the usual time scale, that is, at intervals of the order of several seconds. In contrast, composite outgrowths, for example, stable cytoplasmic processes of polarized fibroblasts, are formed as the result of a long series of extensions and attachments of primary extensions. In discussing reactions of active attachment we consider only those surface extensions for which there is sufficient reason to assume that they are of primary character.

Pseudopods formed by cultured fibroblasts and epithelial cells may have different morphological shapes (Fig. 3). There is no unified terminology for the designation of these structures (see discussion in Vesely and Boyde, 1973). The following main groups of processes are usually distinguished.

1. Cylindrical or conical processes having a length much greater than their width. Larger processes of this type with a diameter of about 0.4–0.5 μm are usually called microspikes. Filopodium is the term used to describe a process of somewhat smaller diameter; these processes often are quite long (up to 10–20 μm). The processes of smallest diameter (0.1–0.2 μm) are usually designated microvilli. The usage of these three terms by different investigators varies considerably.

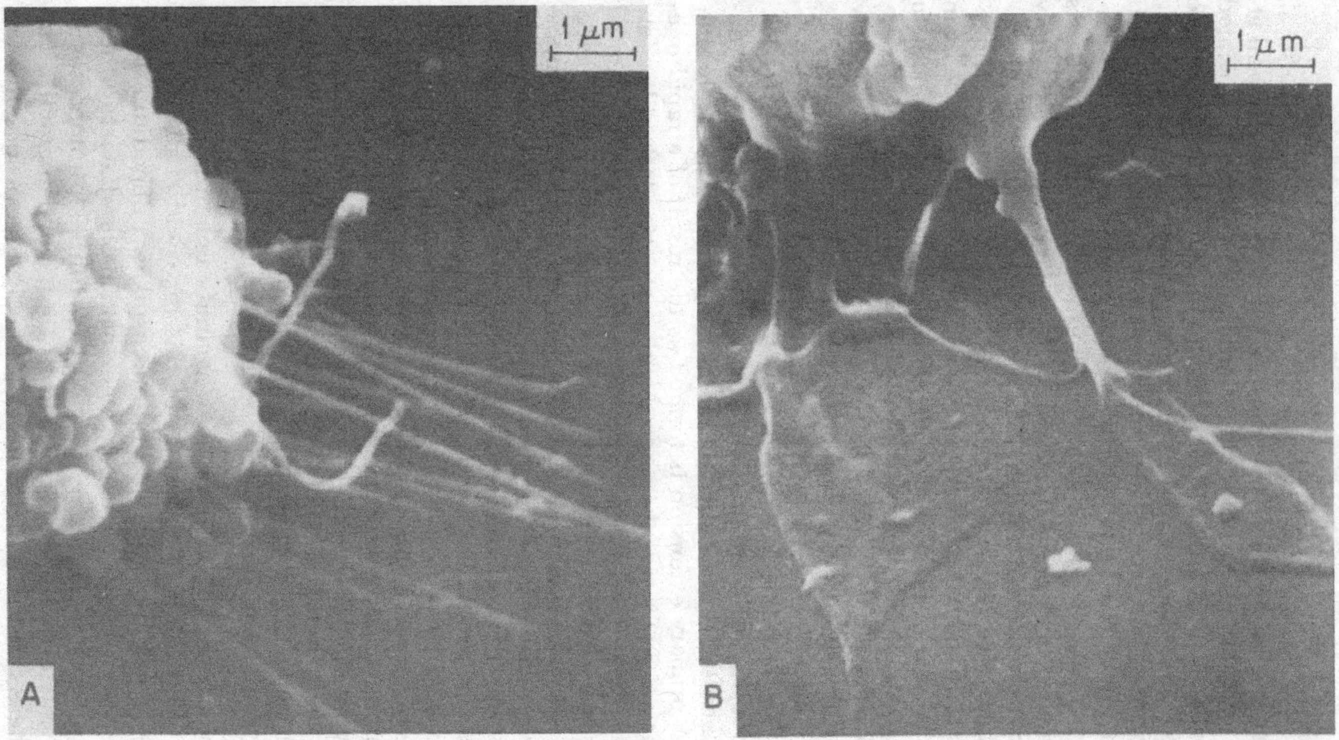

FIG. 3. Cytoplasmic processes formed by spherical normal mouse fibroblasts at an early stage of attachment (30 minutes after seeding). (A) Left: Spherical cell body covered with blebs. Right: Filopodial processes attached to the substrate. (B) Left: Lamellipodium. Right: Pseudopod of mixed morphology with cylindrical proximal part and lamellar distal part. Scanning electron micrographs.

2. Flattened extensions usually about 0.1–0.5 μm thick. These processes are designated lamellipodia when they are formed at the cell edge near the substrate. When they are found at the upper surface of the cell, they are called ruffles.
3. Spherical or hemispherical blebs usually about 1–2 μm in diameter.

Extensions may also have shapes intermediate between these three types. When micrographs of cells are examined, one should take into account that the morphology of a pseudopod at the moment of fixation may be different from that of the same pseudopod at the moment of formation. For example, as a result of the developed tension, a lamellipodium after attachment may be transformed into a cylindrical filopodium or microspike. Likewise, an attached filopodium may eventually become flattened and transformed into a lamellipodium. Some extensions, for example, long filopodia, may be formed in several stages, that is, they may not be primary pseudopods but composite outgrowths. These considerations show that any detailed morphological classification of extensions at present would inevitably be rather artificial. It is not clear whether or not microvilli and blebs are able to initiate the formation of contact structures. Unless otherwise mentioned, in discussing reactions of active attachment we are dealing with extensions that have the morphology of microspikes, filopodia, and lamellipodia.

An important study of the time course of expansion and contraction of lamellipodia formed at the anterior edge of fibroblasts was performed by Ingram (1969), who succeeded in obtaining side-view photographs of these cells. Expansion was rapid (about 4–5 μm per minute); its direction in normal cells was usually parallel to the substrate. Expansion was followed by contraction. Ingram (1969) suggests that contraction may follow several different courses. If contraction of a lamellipodium is stronger on the upper surface, it will lift upward and ruffling will result. If contraction is equally strong on both surfaces, simple retraction will be the result. If contraction of the lower surface is stronger, the pseudopod will curl toward the substrate and eventually make contact.

Discussion of the mechanisms of the formation of pseudopods may be reduced to several more specific questions: What signals induce the extension of a pseudopod? What cytoplasmic changes are responsible for the extension? What changes in the cell surface accompany the extension?

b. *What Signals Induce the Extension?* Pseudopods at the free cell surface are usually formed near the sites of this surface contacting the substrate. Thus, at early stages of spreading filopodia and lamellipodia sprout from the lower parts of cell surfaces whose distance from the substrate does not exceed a few micrometers. The same is true for radially spread and polarized cells; here pseudopods are formed at the external edges of the lamellar cytoplasm near the sites of cell contact with the substrate. These data suggest that cell contact with the substrate somehow induces extension of a pseudopod in a nearby surface area. One may speculate that some surface receptors activated by contact with the substrate produce signals causing alterations in the cortical layer. These reactions may be similar to the activation of blood platelets induced by various agents, including contact with thrombogenic surfaces. Formation of pseudopods is one of the main manifestations of this activation (Born, 1972; Walsh and Barnhart, 1973). However, at present one cannot be sure that large pseudopods (such as lamellipodia or microspikes) are formed only after cell contact with the substrate. Examination of cells in suspension sometimes reveals those with a ruffled surface. Certain humoral factors (such as ADP) may activate the formation of pseudopods by blood platelets in suspension (Born, 1972). It is important to find out whether or not similar activating agents can be found for suspended tissue cells.

c. *What Cytoplasmic Changes Lead to the Extension of Pseudopods?* Extension of a pseudopod is obviously the result of local movements of cytoplasmic components. It seems probable that this extension is accompanied by alterations in the state and position of microfilaments in the cortical layer. Several possible mechanisms for these alterations can be imagined:

1. Actin microfilaments may slide with regard to each other as a result of their interactions with myosin molecules. This interaction may be similar to that responsible for muscle contraction. A variation on such a hypothesis has been developed by Bray (1973).
2. Actin microfilaments may change their positions as a result of alterations in their packing. A mechanism of this type was reported to be responsible for the formation of acrosomal processes by the horseshoe crab (*Limulus*) sperm; the interaction of actin microfilaments with certain nonmyosin proteins seems to be essential for this control of packing (Tilney, 1975a,b).
3. Extension may be due to local polymerization of actin microfilaments from monomeric actin; this mechanism may be similar to that responsible for the formation of acrosomal processes in exchinoderm spermatozoa (Tilney, 1975a,b).

4. Extension may be associated with local depolymerization of preexisting microfilaments in the cortical layer. Considerable hydrostatic pressure existing in the internal parts of the cells is probably a result of tension exerted by the cortical layer. Local depolymerization of microfilaments may produce a hole in the cortical layer. As a result of internal pressure, cytoplasm would start to flow externally through this hole and produce local extension of the surface (Harris, 1973b). The formation of pseudopods may be inhibited reversibly by an increase in the osmotic pressure of the medium (Di Pasquale, 1975b). Possibly, in this case the external pressure would equilibrate with the internal one and stop the flow through the hypothetical holes in the cortical layer.

We do not know which, if any, of these hypotheses is correct.

d. *Surface Changes Accompanying the Formation of Pseudopods.* Movements of the entire cell surface or, at least, of certain components of this surface have special characteristics in areas where pseudopods are formed, that is, in the lamellar cytoplasm near the active cell edges. This is indicated by two groups of experimental data:

1. Insert solid particles attached to the upper or lower cell surface near the active edge or to filopodia move directionally from this edge to the surface of the central part of the cell (Abercrombie *et al.*, 1970c; Harris and Dunn, 1972; Harris, 1973a,b; Albrecht-Buehler and Goldman, 1976).

2. Surface components that have bound molecules of the plant lectin concanavalin A migrate directionally from the surface of areas located near active edges. These results were obtained in experiments with various cells: chick fibroblasts (Abercrombie *et al.*, 1972), mouse fibroblasts (Vasiliev *et al.*, 1976), transformed L fibroblasts (Weller, 1974), transformed 3T3 cells (Ukena *et al.*, 1974), and epithelial MPTR cells (Vasiliev *et al.*, 1976). Concanavalin-A-labeled surface components were visualized in these experiments by various immunomorphological methods. Let us describe in more detail typical experiments of this type as performed by Vasiliev *et al.* (1976). When prefixed fibroblasts or epithelial cells are incubated in medium containing concanavalin A, surface receptors binding this lectin are found to be distributed diffusely over the whole cell surface. If, however, living cells are incubated with concanavalin A for 30–60 minutes and then fixed, two types of alterations in the distribution of surface receptors can be observed. First, concanavalin-binding receptors are collected in small patches over the surface. The formation of these patches is probably due to cross-linking of receptors by multivalent

lectin molecules. Second, the patches are selectively removed from the surface near the active cell edges (Fig. 4). In contrast, surface areas near nonactive cell edges are not freed from these patches.

Thus both solid particles and patches of concanavalin-A receptors move directionally from the active edges into the central parts of the cell. Several different explanations of this phenomenon have been proposed (see discussions in Harris, 1973b; de Petris and Raff, 1973; de Petris, 1975):

1. The extension of pseudopods is accompanied by exteriorization of new cell surface. This surface is formed at the active edges from intracellular sources (Abercrombie *et al.*, 1970c). As the number of

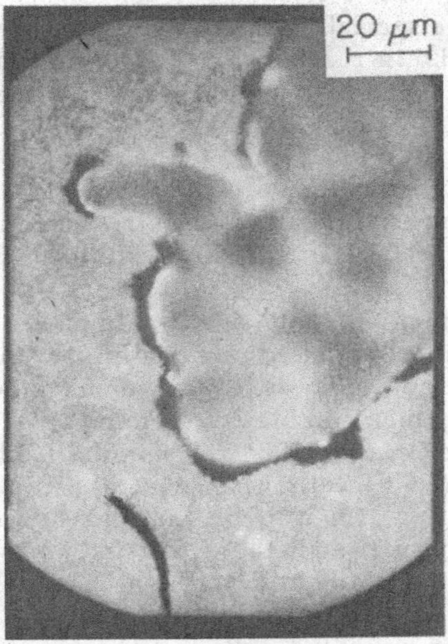

FIG. 4. Selective removal of concanavalin-A receptors from the surface of the active edges of an island of epithelial cells of strain MPTR. The cells were incubated 20 minutes with concanavalin and then fixed; the distribution of concanavalin was revealed by the indirect immunofluorescence method. Dark zones are areas near the active edges from which concanavalin receptors were removed during incubation. Other areas of the upper surface of the epithelial cells remain fluorescent. The surface of the glass around the island is also fluorescent because of the attachment of serum components reacting with concanavalin A. Courtesy of L. V. Domnina and N. A. Dorphman.

formed pseudopods always exceeds that of attached ones, a surplus of new membrane is present near the active edge. This surplus creates centripetal flow of the cell membrane, which is visualized in experiments with particles or with concanavalin A.

2. There is no bulk membrane flow from the active edge. However, in this area certain components of the membrane start to migrate directionally toward the central parts of the cell. Perhaps this migration is a result of reversible binding of these components to cortical microfilaments which then propel them centripetally along the surface. When these membrane components reach the surface above the endoplasm, they are detached from the cortical structures and start to move by passive diffusion back into the lamellar cytoplasm. Cross-linking of these membrane components by concanavalin A or their binding to particles does not prevent their centripetal migration but stops their passive backward diffusion.

3. Under normal conditions neither the whole membrane nor its components move centripetally. However, when several membrane molecules are "patched," that is, glued into a group by a ligand or by their attachment to the surface of a particle, directional migration of the patches located in the area near the active edge is induced by some mechanism. Possibly, cortical microfilaments are preferentially attached to patches located in this area but not to those located near the nonactive edges.

It is difficult to choose between hypotheses 1, 2, and 3. It is important to stress that all these hypotheses, especially, 2 and 3, imply that interrelationships between the membrane components and those of the cortical layer in areas of pseudopod formation are different from those in other areas of the cell surface.

3. *Attachment of Pseudopods*

When an extending pseudopod collides with a surface, it may become attached to that surface; this attachment is accompanied by the formation of local contact structures. Obviously, the probability of formation of these structures, as well as their character, depend on the specificity of the cell that extends the pseudopod and on the nature of the surface that it meets. Thus complexes consisting of various types of intercellular contact structures are formed between the epithelial cells in the sheets (Fig. 5); these structures include occluding junctions, gap junctions, and desmosomes (Middleton, 1973; Guillouzo *et al.*, 1972; Neupert, 1972; Orci *et al.*, 1973; Pickett *et al.*, 1975). These complexes are rather similar to the contact complexes formed

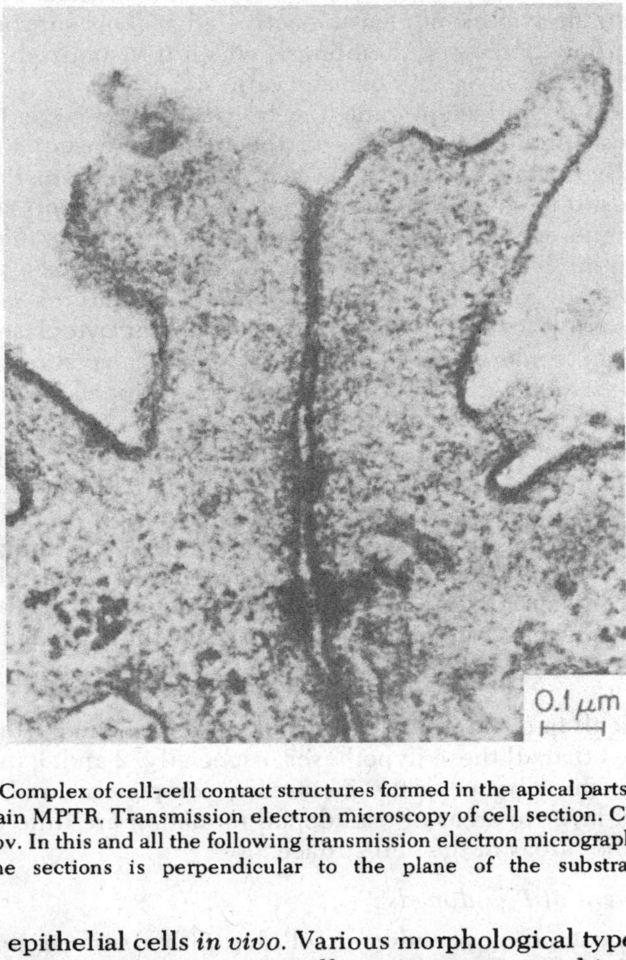

FIG. 5. Complex of cell-cell contact structures formed in the apical parts of epithelial cells of strain MPTR. Transmission electron microscopy of cell section. Courtesy of A. V. Ljubimov. In this and all the following transmission electron micrographs, the direction of the sections is perpendicular to the plane of the substrate.

between epithelial cells *in vivo*. Various morphological types of intercellular contact structures, especially, gap junctions and intermediate junctions, have been observed in fibroblast cultures (Martinez-Palomo *et al*., 1969; Pinto da Silva and Gilula, 1972; Cherny *et al*., 1975). We do not discuss here the possible functional significance and morphology of various contact structures (see review in Gilula, 1974; Revel, 1974). At least some of these structures seem to consist of specific protein molecules orderly packed in the membrane. The formation of stable cell-cell contacts is a highly selective process depending on the tissue specificity of the participating cells. This specificity in its simplest form is demonstrated by observations showing that the

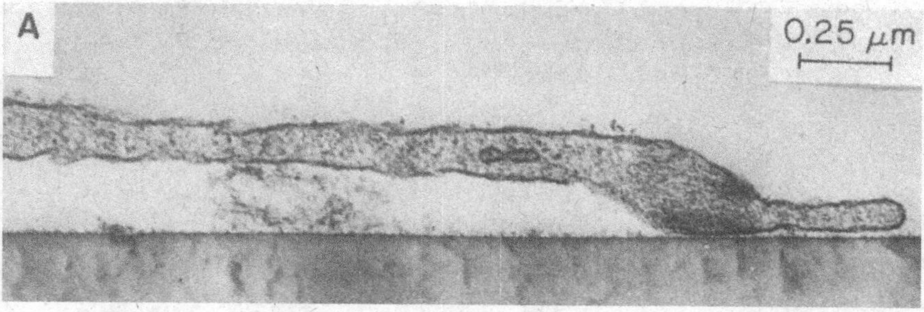

FIG. 6. Cell-substrate attachment sites of normal mouse fibroblasts. (A) Distal end of substrate-attached filopodium 30 minutes after seeding. (B) Leading edge of polarized fibroblast. Transmission electron micrograph. Courtesy of E. E. Bragina.

duration of lateral adhesions between chick fibroblasts is significantly greater than that between chick fibroblasts and chick epithelial cells (Di Pasquale and Bell, 1975). The duration of lateral contacts between two homotypic epithelial cells is many times greater than between fibroblasts; epithelial cells almost never separate spontaneously from their contacts. We do not know, however, at what stage of contact formation this selectivity operates, whether contact structures between heterotypic cells are not formed, less frequently formed, or more easily destroyed than those between homotypic cells. One can find in the literature various theoretical considerations of possible molecular mechanisms determining specificity of cell-cell contacts (see reviews in Moscona, 1974; Roseman, 1974), however, we do not discuss them here.

It is also important to stress that protein molecules of the contact structures embedded in the lipid layer of the membrane do not form a mechanically stable system able to resist stresses unless they are linked in some way to underlying cortical structures.

Cell contact with the substrate also leads to the formation of local structures, so-called plaques or sites of close attachment (Fig. 6). The morphology of these plaques has been described by several investigators (Cornell, 1969; Abercrombie *et al*., 1971; Brunk *et al*., 1971; Revel and Wolken, 1973; Revel, 1974), but is less well known than that of intercellular contacts. The characteristic features of these structures in sections are close apposition of the membrane to the substrate so that the distance between them does not exceed 10–15 nm, increased electron density of the cytoplasm near the membrane, and attachment of the bundle of filaments. Increased electron density may

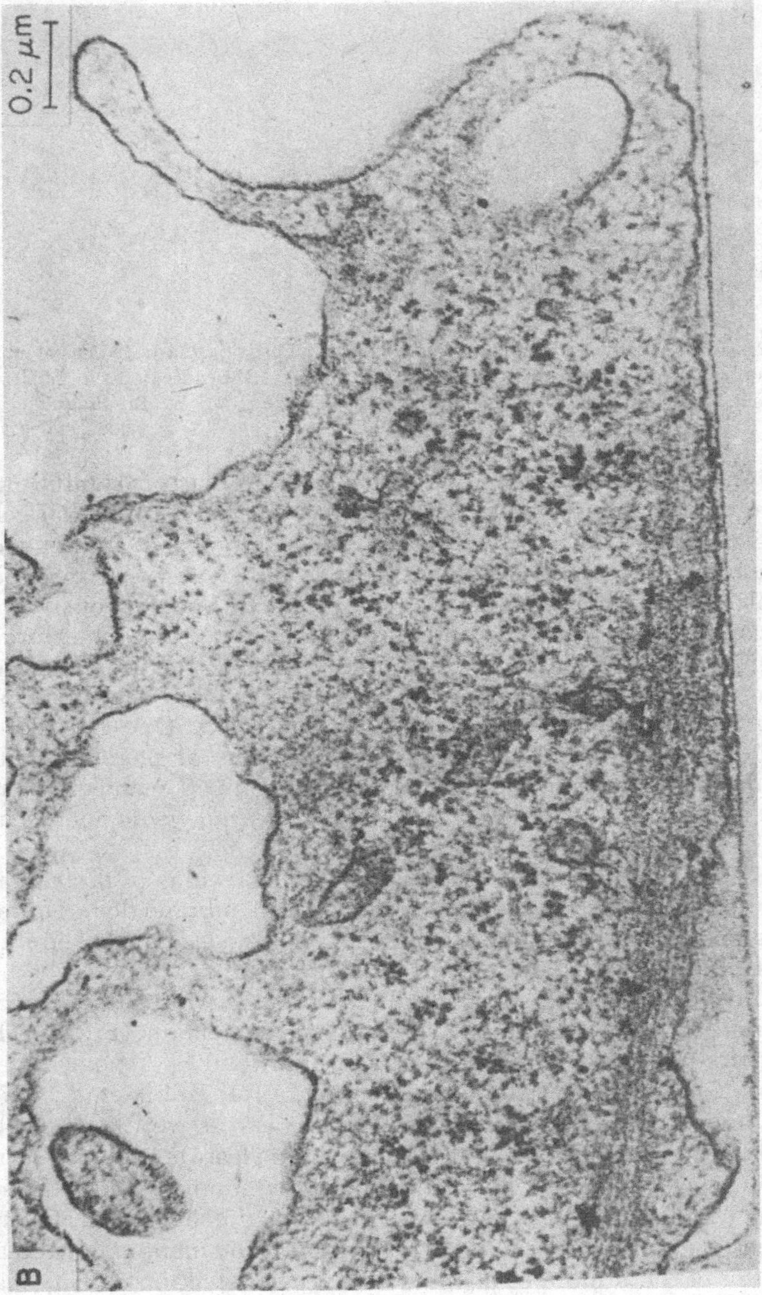

FIG. 6B. See facing page for legend.

be not obvious in certain cases; this may complicate the exact determination of the location of these structures. The specificity of cell-substrate attachments seems to be less strict than that of cell-cell contacts; fibroblasts and epithelial cells are able to spread a variety of substrates. However, the nature and limits of this specificity are far from clear (see Section III,A,2,c). What are the mechanisms of pseudopod attachment to various surfaces? More specifically, what are the distinctive properties of the surface of pseudopods responsible for their ability to form attachments with other surfaces? In order to bring together two cell surfaces before the formation of contact structures it is necessary to overcome an electrostatic repulsion between these surfaces. Theoretical considerations show that the surface of a pseudopod may overcome this barrier more easily than a flat nonmoving surface because of its small radius (Pethica, 1961) and/or its active forward movement (Weiss, 1962). It seems probable that these physical factors play a certain role in the mechanism of attachment of pseudopods. However, these considerations do not explain why, after overcoming the electrostatic barrier, the surface of a pseudopod is able to form a specialized local contact structure with a specific morphology.

At least two hypothetical explanations of these properties of pseudopods can be proposed:

1. The external side of the pseudopod membrane may contain specific components absent from the external side of the membrane of other cell parts. These components may be exteriorized during the extension of pseudopods. For instance, it is possible that this extension is accompanied by penetraton of the membrane by the ends of actin microfilaments. Another possibility is secretion of an adhesive material in the area of pseudopod formation. Association of pseudopod formation with the secretion of membrane-coated vesicles had been suggested by Bray (1973).

2. Membrane components participating in the formation of contact structures may be present not only on the external surface of pseudopods but also on that of other cell parts. However, interactions of these components with the underlying cortical structures may have a specific character in pseudopods. Such a specific character of membrane-cortex interactions in pseudopods is suggested by the data on surface movements discussed above. For instance, it is possible that membrane proteins in pseudopods may be anchored to actin microfilaments, while in other parts of the membrane these proteins move freely in the lipid layer. As mentioned before, only anchored membrane structures can form a stable contact; nonanchored mem-

brane proteins attached to other surfaces are easily displaced and even removed from the membranes by mechanical stress. Thus a difference in the anchorage of membrane receptors is sufficient to explain different adhesive properties of active and nonactive parts of the cell surface.

We favor the second suggestion but, obviously, all the possibilities should be tested in further experiments.

4. *Tension in the Attached Pseudopod*

a. *Development of Centripetal Tension within an Extended Pseudopod.* It is probable that forward extension of a pseudopod is always followed by the development of centripetal tension within the cytoplasm of this pseudopod. This tension may be responsible for the retraction of unattached pseudopods, upward bending of these pseudopods, and subsequent movement of the ruffle on the upper cell surface. The tension continues to exist after attachment of the pseudopod to another surface. We do not know whether or not the degree of tension is the same within attached and unattached pseudopods. Depending on the conditions of attachment, tension in an attached pseudopod may lead to various consequences:

1. To a break in the contact followed by retraction of the pseudopod.
2. To displacement of the cell body in the direction of the contact site; this happens, for instance, during cell movement on the substrate.
3. To the centripetal displacement of the contact site; this may happen when a small particle is attached to the surface (see above).

In radially spread and polarized cells centripetal tension continues to act on all the peripheral contact structures. The existence of this tension is shown by several facts:

1. When a cell-cell or cell-substrate contact is destroyed by any treatment, for example, mechanically, contraction of the cytoplasmic area near the contact immediately takes place.
2. In dense cultures each cell exerts tension on the contacting neighboring cells. If a wound is made in such a culture, nonequilibrated tension at the edge of the wound leads to the retraction of this edge (Vasiliev *et al.*, 1969). Tension in dense fibroblast cultures was measured by James and Taylor (1969) and found to be 3.4×10^4 dynes per cm^2 of culture cross-sectional area.

3. Cells spread on small threads are able to bend these threads (Harris, 1973b). This observation shows that tension does not arise after the destruction of contacts but acts permanently within the attached cells. We do not know, however, whether or not the degree of tension is the same before and after the destruction of contacts.

b. *Structural Alterations within the Attached Pseudopods.* The main structural alteration accompanying attachment is the formation of a bundle of parallel microfilaments (Fig. 6B). One end of this new bundle is usually located at the attachment structure; the location of the other end is not clear. In the early stages of spreading the central ends of newly formed bundles seem to be located somewhere in the matrix cortical layer of the cell body (Bragina *et al.*, 1976). The development of bundles near the sites of attachment of particles in the first stage of phagocytosis has been observed in experiments with macrophages (Reaven and Axiline, 1973). Formation of the attachment site probably provides a stimulus for development of the bundle, for example, for ordered polymerization of actin. However, we know neither the nature of this stimulus nor the exact nature of the alteration it induces.

The functional role of the bundles also is not quite clear. It is usually assumed that they are responsible for tension acting on contact structures. This assumption is probably correct, but one cannot exclude at present another possibility: Tension may be created by the microfilaments of the matrix, while bundles may play the role of skeletal elements counteracting this tension. It is also possible that various bundles or even the same bundle in various states may have different functions.

5. *Concluding Remarks on Active Attachment Reactions*

a. *Active and Nonactive Attachment.* We have discussed considerations suggesting that specific local attachment structures are formed only by active extended pseudopods. Of course these suggestions need further tests in various experimental systems. As we tried to show above, these considerations may also help to formulate questions for further studies. One of the questions deserving special comment is the following. Are cells able to attach themselves to certain surfaces without forming specialized contact structures?

Various data indicate that contact interactions of this type, that is, attachments not accompanied by conspicuous morphological changes, are possible. For example, an unspread spherical cell may become attached to the substrate by its lower surface before the formation of pseudopods and specialized contacts. This attachment is sufficiently strong to resist washing of the culture. Electron microscopy shows that flattening of the lower surface is the only morphological change accompanying this attachment (Bragina *et al.*, 1976). It is possibly identical to an initial attachment described by Taylor (1961); in his experiments even fixed cells demonstrated this attachment. This nonspecialized diffuse attachment probably does not require active reactions of the cell. In spread cells similar interactions may be involved in the substrate attachment of the lower surface of areas of the lamellar cytoplasm between plaques. "Parallel contacts" between cells may be of the same nature. Obviously, this group of nonactive diffuse attachments deserves further study.

b. *Ambiguity of the Term Adhesiveness.* An understanding of the complex and active nature of attachment reactions is important in interpreting the results of numerous published experiments in which cell adhesiveness was assessed under various conditions, for example, adhesion of cells to various substrates was compared, mutual adhesiveness between cells of different types was determined, and the effect of various agents on cell adhesion to the substrate was measured. To measure adhesiveness various parameters were used in different experiments: number of cells attached to the surface per unit area per given time, size of cell aggregates, minimal force required to detach cells from the surface, and so on. Obviously, none of these parameters measures the elementary interaction of the cell surface with another surface; at best, they reflect summarized results of a long series of active attachment reactions. In various situations similar alteration in some of these parameters may be due to quite different causes: alterations in the stability of individual contact structures, in the number of these structures, in the frequency of pseudopod extension, in the probability of attachment of these pseudopods, in the strength of the tension within the attached pseudopods, and so on. Therefore one should be very cautious when drawing conclusions about the molecular mechanisms of cell attachment from experiments in which alterations in adhesiveness have been found.

D. Contact Inhibition of the Formation of Pseudopods

Contact inhibition of pseudopod formation (contact paralysis) may be described as cessation of the formation of pseudopods in parts of the cell edge that have contacted the surface of another cell.

1. *Terminology*

Before describing the phenomenon of contact inhibition we shall discuss briefly rather complicated terminology problems with regard to this phenomenon (see also discussions in Martz and Steinberg, 1973; Armstrong and Lackie, 1975). As seen from the definition given above, contact paralysis (synonyms: contact inhibition of the formation of pseudopods, contact inhibition of pseudopodial activity) may be regarded as a basic morphogenetic reaction. It was first described by Abercrombie and Ambrose (1958). The term contact paralysis was introduced by Gustafson and Wolpert (1967) and by Wolpert and Gingell (1968). The terms contact inhibition of ruffling and contact inhibition of blebbing describe particular manifestations of the same reaction (Harris, 1974). The concept of contact paralysis is a further development of Abercrombie's notion of contact inhibition of movement (Abercrombie and Heaysman, 1954; Abercrombie and Ambrose, 1962; Abercrombie, 1961, 1965, 1970). Contact inhibition of movement may be defined as directional restriction of cell displacement on contact (Abercrombie, 1970). In contrast to contact paralysis, contact inhibition of movement is not a basic morphogenetic reaction; it may be a consequence of different reactions. Usually contact inhibition of movement is a consequence of contact paralysis. However, one can imagine contact inhibition not accompanied by contact paralysis. After cell-cell contact an active edge continues to form pseudopods but does not overlap the surface of other cells, presumably, because the pseudopods are unable to attach themselves to this surface. It was proposed that these special cases of contact inhibition be designated as contact inhibition of the second kind (Harris, 1974). In this article the term contact inhibition of movement is used only to describe the cessation of locomotion associated with contact paralysis, that is, according to the terminology mentioned above for the description of contact inhibition of the first kind. Besides contact inhibition of movement, contact inhibition of pinocytosis (Vesely and Weiss, 1973) and of phagocytosis (Vasiliev *et al.*, 1975a) can be regarded as special corollaries of contact paralysis.

As seen from the definitions given above, the presence or absence of contact paralysis and of contact inhibition of movement can be revealed only by direct microcinematographic observation of living cells. Sometimes conclusions about the presence or absence of contact inhibition are made on the basis of such criteria as the degree of mutual overlapping of cell nuclei or the degree of mutual cell orientation. However, these parameters reflect only the end state of large cell groups; this state is a statistical result of a long series of morphogenetic reactions. Besides the changes in contact inhibition, alterations in these parameters may be due to many other causes (see Sections III,A,4, III,A,5, and IV,B).

2. *Phenomenology*

Contact paralysis is most easily observed when two active edges of two fibroblasts or epithelial cells collide with each other; almost immediately, the formation of pseudopods and ruffling stops at the site of contact. This paralysis is often accompanied by the retraction of contacting cell edges (Abercrombie, 1970). If a cell-cell attachment formed at the site of the contact is firm enough, mutual retraction brings both cells closer to each other. In the opposite case it leads to separation of the contact. Contact paralysis may be also observed after the contact of an active edge of one cell with the stable edge of another; in this case only the paralyzed active edge retracts. All observations on contact paralysis were made in experiments with radially spread or polarized cells, that is, with cells contacting both the substrate and another cell. It is not clear whether or not a similar phenomenon can be observed when a cell contacts only another cell but not the substrate, for example, when a spherical cell contacts another spherical cell or the upper surface of a spread cell.

Contact paralysis is a local phenomenon; pseudopodial activity stops only in the part of an active edge that contacts another cell; nearby parts of the same edge remain active (Trinkaus *et al.*, 1971). Contact paralysis is reversible. It is effective as long as cell-cell contact is preserved but, when this contact is broken, pseudopodial activity resumes rapidly. Only when the effect of contact paralysis is stabilized by another morphogenetic reaction (see the following paragraph) can the immobility of the edge be preserved after the break of a cell-cell contact. Physical contact between two cells seems to be sufficient for the development of contact paralysis. The formation of specialized contact structures is not essential. Thus we observed contact paralysis in cultures of transformed fibroblasts of the L strain which do not form specialized cell-cell structures (Domnina *et al.*, 1972).

Contact paralysis is observed only when a living cell collides with another living cell. It is not observed when contact is made with the surface of a fixed dead cell or with other nonliving material (Harris, 1974). Of considerable interest are the recent results of Abercrombie and Dunn (1975) showing that contact paralysis is not accompanied by the disappearance of cell-substrate attachment sites in the nearby area of the leading lamella. These results indicate that retraction of the edge accompanying contact paralysis is not a result of the detachment of this edge from the substrate.

Nothing is known about the mechanisms of contact paralysis; various hypotheses on this subject have been critically discussed in previous reviews (Abercrombie, 1970; Harris, 1974) and are not considered here. It seems probable that contact paralysis is a surface-induced reaction leading to alterations in the cortical layer that prevent those unknown changes involved in the formation of pseudopods.

E. STABILIZATION REACTIONS

1. *Phenomenology*

Both morphogenetic reactins discussed above (reactions of active attachment and contact paralysis) are local, that is, each reaction involves only relatively small areas of the cell periphery. We now discuss another group of morphogenetic reactions, those involving the whole cell periphery or, at least, its major parts. These reactions control the distribution of sites at which pseudopods are extended, as well as the distribution of tension produced by the attached pseudopods. The main experimental results suggesting the existence of these reactions were obtained in investigations of the effects of antitubulins (Colcemid, colchicine, and vinblastine) on the transition of fibroblasts from one morphological state to another. Antitubulins are selectively bound by tubulin—the main structural molecule of microtubules. As a result of this binding antitubulins prevent the polymerization of microtubules. Probably, most cellular effects of Colcemid and other antitubulins are consequences of the disorganization of microtubular structures. However, at present one cannot be sure that microtubules are the only structural targets of antitubulins. Therefore we use the neutral term antitubulin-sensitive structures. Antitubulins prevent and reverse the polarization of fibroblasts (see details and references in Section III,A,3) and disorganize the course of the spreading (see Section III,A,2). Analysis of the effects produced by these agents had shown that they do not inhibit active attachment reactions; cells treated with antitubulins continue to form and to attach pseudopods, and these pseudopods are able to exert tension. Also, contact paralysis is not prevented by these agents. This analysis of the effects of antitubulins suggests that, besides these two groups of local morphogenetic reactions, one should distinguish another group of reactions which are selectively inhibited by antitubulins (Vasiliev and Gelfand, 1976a,b). These morphogenetic reactions are referred to as stabilization reactions. Two main variants of stabilization reactions can be distinguished:

1. Processes responsible for stable division of the cellular edge into active and nonactive parts. Inhibition of this process by antitubulins

reverses the polarization of spread cells; the entire external edge of these cells remains active (see details in Section III,A,3).

2. Processes responsible for the equal distribution of pseudopodial activity between the various parts of the active edge. Antitubulin-treated cells, in the course of spreading, form pseudopods of much more variable size and shape than normal cells. Retractions of the attached pseudopods of these cells often lead to their detachment from the substrate; these retractions seem to be a result of nonequilibrated tensions existing in antitubulin-treated cells (see details in Section III,A,2).

Possibly, these two variants can be regarded as different manifestations of the same process, of the stabilization of the distribution of the sites of active attachment reactions along the cell edge. Needless to say, the term stabilization means organization of local activities preventing their excessive fluctuations but not their inhibition.

Alterations in cell shape and position induced by external factors, such as spreading, polarization, and locomotion of the polarized cell, proceed in an organized way and lead to stable results only if the cell is able to undergo antitubulin-sensitive stabilization reactions. Let us discuss, as examples, interrelationships among external factors, local morphogenetic reactions, and global stabilization reactions involved in the orientation of polarized fibroblasts. It is well known that fibroblasts are able to orient themselves with regard to other cells and to substrate structures (see Section III,A,5). These external factors usually create conditions under which various parts of an active edge either form pseudopods with a different intensity or attach these pseudopods with a different efficiency. The surface of another cell is an example of an external factor of the first type; if some part of an active edge touches another cell, the formation of pseudopods is stopped in this part by contact paralysis; the formation of pseudopods in the nearby parts of the edge may be continued (Fig. 7B). The boundary between the adhesive and nonadhesive parts of the substrate is an example of an external factor of the second type. A cell moving on the adhesive substrate stops of course when it touches this boundary (Fig. 7A). The active edge that makes contact continues to extend pseudopods but is unable to attach them. Other parts of the active edge continue to extend and attach pseudopods. In both cases temporary differences among various parts of the edge created by external factors are made more permanent as a result of stabilization reactions. The main result of this reaction is that the cell continues to extend pseudopods only from those parts of the edge at which the attachment of pseudopods was the most efficient during the preceding time interval.

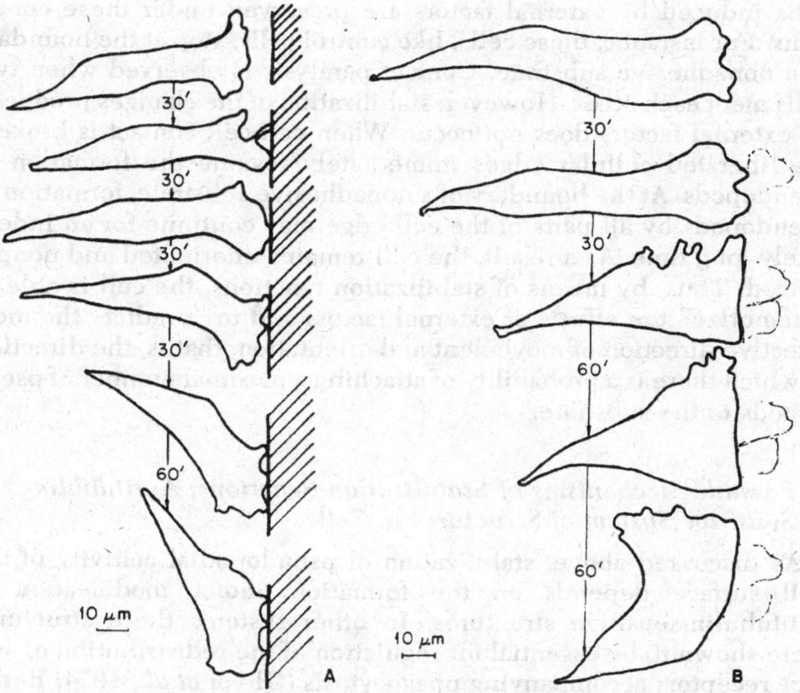

FIG. 7. Alterations in the direction of movement of polarized normal mouse fibroblasts after contact of their leading edges with the boundaries of a nonadhesive lipid film (A) or of the contact-inhibiting edge of an epithelial sheet (B). Each drawing is a series of tracings of the contours of one cell from a time-lapse film. Figures near the arrows show the time between two consecutive drawings. Courtesy of O. J. Ivanova and L. B. Margolis.

As a result of stabilization those parts of the edge where the extension of pseudopods was either inhibited or inefficient are made permanently nonactive. In the first example given above (contact with another cell), the result of stabilization is the continuation of contact paralysis; the paralyzed part of the cell edge remains inactive for a long time even after the break in cell-cell contact. In the second example (contact with the boundary of a nonadhesive substrate), the result of stabilization is the cessation of pseudopodial activity at the edge contacting this boundary. The role of stabilizaton reactions is demonstrated by the observation of moving fibroblasts incubated in medium containing antitubulins (Vasiliev *et al.*, 1970; Vasiliev and Gelfand, 1976b, and unpublished results). Primary alterations in the cells induced by external factors are preserved under these conditions. For instance, these cells, like control cells, stop at the boundary of a nonadhesive substrate. Contact paralysis is observed when two cells meet each other. However, stabilization of the changes produced by external factors does not occur. When cell-cell contact is broken, the liberated cellular edges immediately resume the formation of pseudopods. At the boundary of a nonadhesive substrate, formation of pseudopods by all parts of the cell edge may continue for an indefinitely long time. As a result, the cell remains unoriented and nonpolarized. Thus, by means of stabilization reactions, the cell is able to "memorize" the effects of external factors and to "predict" the most effective direction of movement and orientation, that is, the direction in which there is a probability of attaching a maximal number of pseudopods to the substrate.

2. *Possible Mechanisms of Stabilization Reactions; Antitubulin-Sensitive System of Structures in Cells*

As discussed above, stabilization of pseudopodial acitivity of the cell surface depends on the formation and/or modification of antitubulin-sensitive structures. In other systems these structures were shown to be essential for regulation of the redistribution of surface receptors accompanying phagocytosis (Oliver *et al.*, 1974; Berlin *et al.*, 1974) and capping (Edelman *et al.*, 1973; Yahara and Edelman, 1973, 1975; Ukena *et al.*, 1974).

What is the possible pattern of the organization of the antitubulin-sensitive structures responsible for stabilization reactions? Stabilization reactions have a global character, that is, they coordinate and integrate local reactions occurring in various parts of the cell periphery. Therefore it is probable that these structures form a centralized system uniting various parts of the cell. We suggested (Vasiliev and Gelfand, 1976b) that this system has a predominantly radial pattern of organization, that is, that it consists of elements connecting various parts of the cell periphery with the central part of the cell. Such a system of structures may determine the predominantly radial direction of the intracellular movements of various intracellular components (Freed and Lebowitz, 1970; Bhisey and Freed, 1971), as well as the radial distribution of tension in the cortical layer. An antitubulin-treated cell with a destroyed radial system retains the ability to transport and to develop tension. However, the direction of these movements and this tension is randomized; their predominantly radial pattern is lost. Distribution of the sites of extension of pseudopods is perhaps one of the functions controlled by the radial system. This control may be somehow connected with the control of intracellular movements; the extension of pseudopods obviously requires some movement of cytoplasmic components. The radial system may also control radial distribution of tension exerted by the attached pseudopods. As a result, tension from opposite parts of the periphery may be completely (in the radially spread cell) or partially (in the moving polarized cell) equilibrated in the central part of the cell.

This hypothesis may explain the above-mentioned effects of antitubulins on cell spreading and polarization, as well as on secretion (Allison, 1973; Stein *et al.*, 1974).

A radiating pattern of microtubules in cultured normal cells was observed in the experiments of Brinkley *et al.*, (1975), Osborn and Weber (1976), and Frankel (1976); in these experiments tubulin was visualized by immunofluorescent staining. The treatment of polarized fibroblasts with cytochalasin B reveals a system of branching cytoplasmic radial cords radiating from the central part of the cell; this system is destroyed by Colcemid (see Section III,A,3). One may suggest that the pattern of these cords somehow reflects the pattern of radial structure that existed in the cell before treatment with cytochalasin.

The data on the orientation of polarized fibroblasts, discussed above, indicate that the position of antitubulin-sensitive elements in the cell may be changed by external factors. By alterations in the efficiency of the formation and/or attachment of pseudopods these factors eventually may alter distribution of the contact structures and of the tension within the cortical layer; for example, if the cell is near the boundary of a nonadhesive substrate, the new attachment sites will form most often along a line approximately perpendicular to this boundary. As a result, the predominant orientation of tension will become approximately parallel to the boundary. This altered distribution of the contact sites and/or of tension may lead to redistribution of the antitubulin-sensitive structures. Thus interactions between antitubulin-sensitive systems and local reactions of active attachment seem to be bilateral in nature. The combined results of many local reactions gradually change the position of the elements of the radial system. Once these elements have acquired a certain orientation, they stabilize the existing distribution of tension and of the sites of pseudopodial extension.

It would be easy to construct a plausible morphological model of an antitubulin-sensitive system. The main elements of this system are most probably microtubules which may be connected by some central element (centriole?). Mechanically rigid microtubules interact with the tension-producing structures of the cortical layer (microfilament bundles?). Changes in the distribution of microfilaments and/or in the attachment sites may alter the distribution of microtubles, for example, by favoring polymerization of tubulin at preferred sites near the cell membrane (Albertini and Clark, 1975), or by changing the orientation of preexisting microtubules. Reciprocally, microtubules support and stabilize the existing pattern of the distribution of microfilaments. However, it would be premature to make detailed schemes of the organization and of the functions of the radial system until we know more about the interactions of microtubules with each other and with microfilaments, as well as about the conditions of polymerization and depolymerization of microtubules.

F. Conclusion

Two systems of interacting structures play leading roles in the preservation and alteration of the shape of tissue cells: (1) a peripheral system of surface structures, especially, the cortical layer, and (2) a centralized system of antitubulin-sensitive structures.

Both systems participate in each morphogenetic reaction. Alterations in the cortical layer and in the membrane probably play the leading role in local reactions of active attachment and of contact paralysis. Global stabilization reactions are performed mainly by the centralized antitubulin-sensitive system. As a result of the interactions of the peripheral and centralized system, organization of the cell is both stable and dynamic.

The stability of this organization is demonstrated by the ability of the cell to maintain its shape, distribution of pseudopodial activities and, probably, pattern of intracellular movements. The dynamic nature of this organization is demonstrated by the ability of cells to undergo rapid and organized changes in shape and position when subjected to appropriate external stimuli. Cell transitions from one state into another, as well as directional movements and contact paralysis, may be performed by enucleated cells (Goldman *et al*., 1973), as well as by cells with inhibited protein synthesis (Goldman and Knipe, 1973). These data suggest that alterations in the main reactions of the synthesis of macromolecules are not directly involved in the mechanisms of basic morphogenetic reactions. The mechanisms of these reactions probably include polymerization and depolymerization of fibrillar structures, as well as movement of these structures. At present we know a little about the main types of cell structures participating in these reactions; however, almost nothing is known about the exact nature of the molecular mechanisms of these reactions.

III. Shape and Behavior of Normal Cells in Culture

After discussing the basic morphogenetic reactions we now review cell movements and alterations in cell shape involving various combinations of these basic reactons. We first consider behavior of fibroblasts, and then the behavior of epithelial cells.

A. Fibroblasts

As mentioned before, there are three main morphological states of fibroblastic cells in culture: spherical, radially spread, and polarized. We begin by discussing cell transitions from one state to another: cell rounding, cell spreading, and cell polarization. In Section III,A,3 we also discuss the locomotion of a polarized cell on a flat surface in a sparse culture where cell-cell interactions are minimal. In Section III,A,3 we discuss cell-cell interactions leading to the formation of organized multicellular structures in dense cultures on flat substrates. Sections III,A,4–III,A,6 are devoted to a description of the behavior of fibroblasts on two special types of substrates: those controlling direction of cell orientation and those inducing cell aggregation.

1. *Cell Transition to a Spherical State; Cell Rounding*

The rounding of a radially spread or polarized cell can be induced by any treatment detaching it from the substrate: by proteases, calcium-chelating agents, and mechanical stress. A morphologically similar rounding process often accompanies cell entry into mitosis. Two morphological stages can be distinguished during rounding. At the first stage the cell body is rounded, but the cell remains attached to the substrate by long retraction fibers morphologically similar to filopodia. At the second stage these fibers are detached from the substrate and retract. Cell rounding can be stopped at the first stage and followed by rapid spreading. This usually happens during mitosis or after short-term treatment with chelating agents. Cell rounding is accompanied by alterations in surface topography. In spherical cells the surface rarely remains smooth but is usually covered with blebs, microvilli, or ruffles. Our unpublished observations show that in suspensions of mouse fibroblasts detached from glass with EDTA a large proportion of cells (60–80%) are covered with blebs; there are also cells covered with microvilli (20–30%) and with ruffles (5–10%), as well as some cells of mixed topography, for example, cells with blebs and microvilli. These differences in topography probably reflect differences in the organization of the cortical layer, but their nature is not clear. We do not know yet how the cell state, especially the phase in the mitotic cycle, as well as conditions of rounding, affect surface topography in this population.

Transition into the spherical state is also accompanied by conspicuous alterations in the structure of the cortical layer: bundles of microfilaments disappear; only the matrix cortical layer (Fig. 8) is revealed in the spherical cells (Goldman and Knipe, 1973; Bragina *et al.*, 1976). The disappearance of the bundles may result from two types of alterations; either detachment from the substrate is accompanied by depolymerization of the microfilaments forming the bundle, or in the course of contraction their mutual orientation is lost so that they become

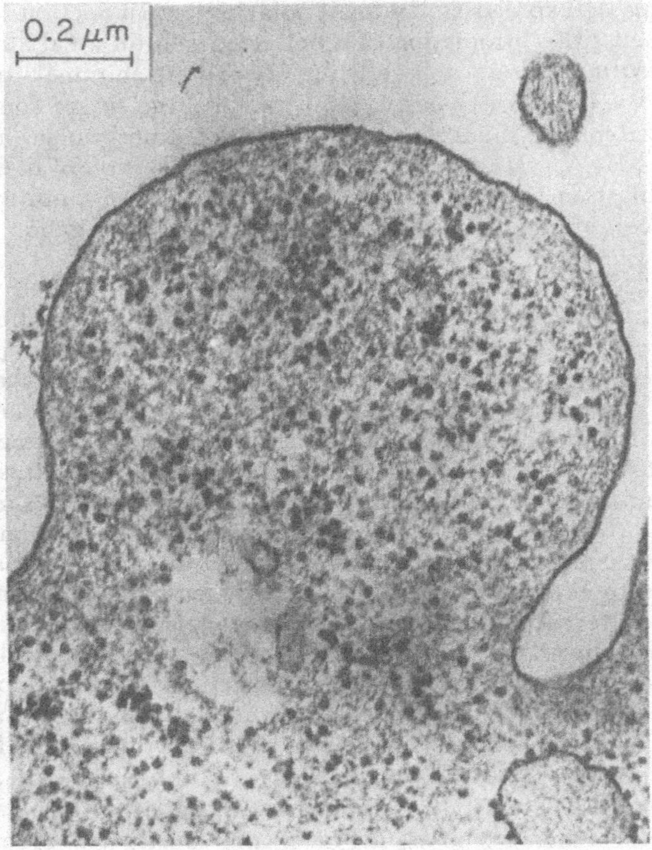

FIG. 8. Matrix-type cortical layer and ribosomes in the cytoplasm of a bleb of a spherical normal fibroblast fixed in suspension. Transmission electron micrograph. Courtesy of E. E. Bragina.

undistinguishable in sections from the microfilaments of the matrix-type cortex. Microtubules are found with difficulty in suspended cells (Goldman *et al.*, 1973). Possibly they are depolymerized during rounding.

It is also important to determine whether or not suspended fibroblasts retain polarity of surface structure and of internal organization. Cell rounding is an active ATP-requiring reaction. This is indicated by the results of experiments (Vasiliev *et al.*, 1975b) in which spread mouse fibroblasts were treated with glycerin solutions. These cells with a partially destroyed plasma membrane were not rounded after detachment from glass; they retained their flattened shape in suspension. The addition of ATP to the suspension immediately induced rounding of the cells. Detachment of the living cells from the substrate probably led to contraction of the cortical layer which had been in a state of tension when the cell was spread. In a spherical cell the cortical layer probably remains in a contracted state. As indicated by the experiments of Izzard and Izzard (1975), the level of intracellular calcium and magnesium may be essential for maintenance of the contracted state.

Almost nothing is known about the nature of the effects of proteases and chelators leading to cell detachment. These agents may destroy some component of the attachment sites. Another possibility is that, by destroying some external surface component, these agents induce a membrane reaction leading to an increase in tension in the cortical layer. Perhaps a trypsin-sensitive glycoprotein of high molecular weight plays a role in these changes. This glycoprotein was found on the external surface of fibroblasts (Weston and Hendricks, 1972; Hynes, 1973; Vaheri and Ruoslahti, 1974; Yamada and Weston, 1975). We know very little about the time course of the changes accompanying detachment. How are the retraction fibers formed? Are their ends the preserved sites of the cell attachment to the substrate? If so, during the first stage of rounding, what happens to the microfilament bundles associated with these sites? Are they preserved in the retraction fibers or do they contract and detach themselves from the cell-substrate contact? These simple questions remain unanswered.

2. *Radial Spreading*

Radial spreading begins when a spherical cell contacts an appropriate substrate. Spreading consists of a long series of active attachment reactions: extension and attachment of pseudopods, and the development of tension after attachment. This series of reactions leads to transition of the cell from a spherical to a discoid shape. As shown by the experiments discussed below, the normal state of the antitubulin-sensitive system of structures is essential for a coordinated course of spreading.

a. *Morphology of Spreading*. Morphological alterations accompanying spreading on glass and other adhesive smooth surfaces has been studied by various investigators (Taylor, 1961; Witkowski and Brighton, 1971; Domnina *et al.*, 1972; Vasiliev and Gelfand, 1973; Rajamaran *et al.*, 1974; Bragina *et al.*, 1976). On the basis of these studies one can distinguish three consecutive stages in these changes (Fig. 9):

I. Attachment of a spherical cell without the extension of pseudopods.
II. Attachment of discrete pseudopods to the substrate.
III. Formation and further expansion of the ring of lamellar cytoplasm.

At the first stage the only observed morphological change is flattening of the lower cell surface contacting the sute. This initial "diffuse contact" (see Section II,C,5) possibly induces the extension of pseudopods.

At the beginning of stage II scanning electron microscopy reveals mostly cells attached by cylindrical processes (filopodia and microspikes). Later one can see cells that, besides filopodia, also have lamellar processes at their periphery. Some of these lamellar processes are possibly formed from cylindrical ones; one often sees at this stage pseudopods of a mixed type, with distal lamellar and proximal cylindrical parts, or vice versa (Fig. 3B). However, even at the early stages of spreading some lamellipodia have no "birthmarks" suggesting filopodial origin. Possibly, these lamellipodia are formed without the preceding filopodial stage. At the end of stage II the cell body is gradually flattened and the upper cell surface becomes smooth except for a few microvilli and ruffles.

Transition from stage II to stage III is gradual; a single ring of lamellar cytoplasm is formed at the cell periphery from numerous attached lamellipodia. This ring surrounds the flattened central part. The extension of pseudopods continues along the external edge of the lamellar cytoplasm.

Cell spreading is accompanied by the formation of microfilament bundles in the cytoplasm (Fig. 10). These bundles first appear within the attached pseudopods at stage II. At stage III microfilament bundles form a sheet which fills almost the entire lamellar cytoplasm. The direction of the bundles in this sheet varies from radial to tangen-

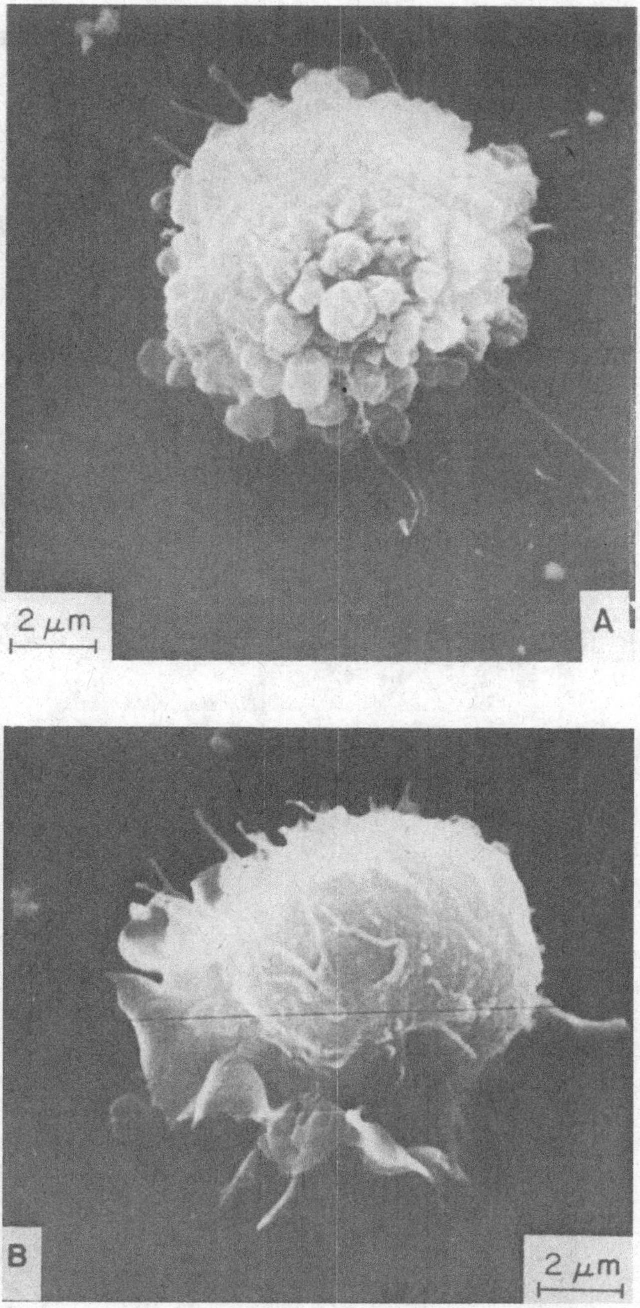

FIG. 9. Stages of radial spreading of a normal mouse fibroblast. (A) Unspread cell covered with blebs and having few attached filopodia. This cell is in the very beginning of stage II of spreading. (B) Cell with numerous lamellipodia and ruffles around the cell body. Stage II of spreading. (C and D) Cells with a rim of lamellar cytoplasm at the periphery (stage III of spreading). Partially (C) and completely (D) flattened central cell parts. Scanning electron micrographs.

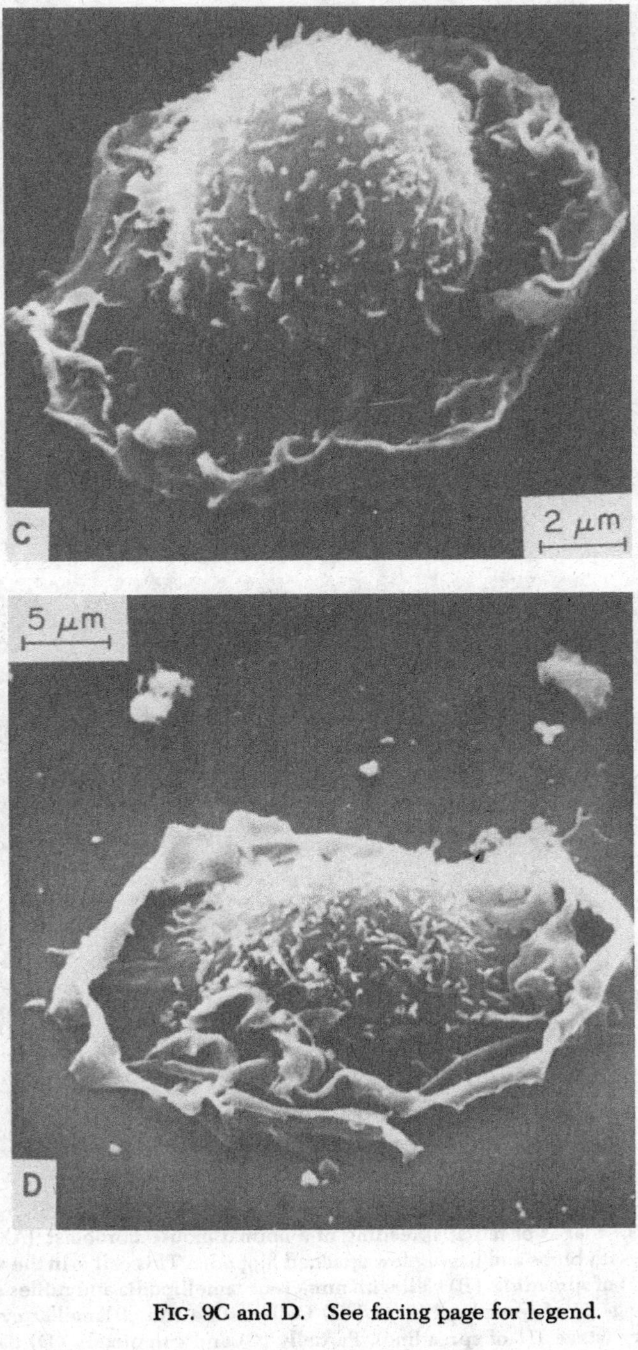

FIG. 9C and D. See facing page for legend.

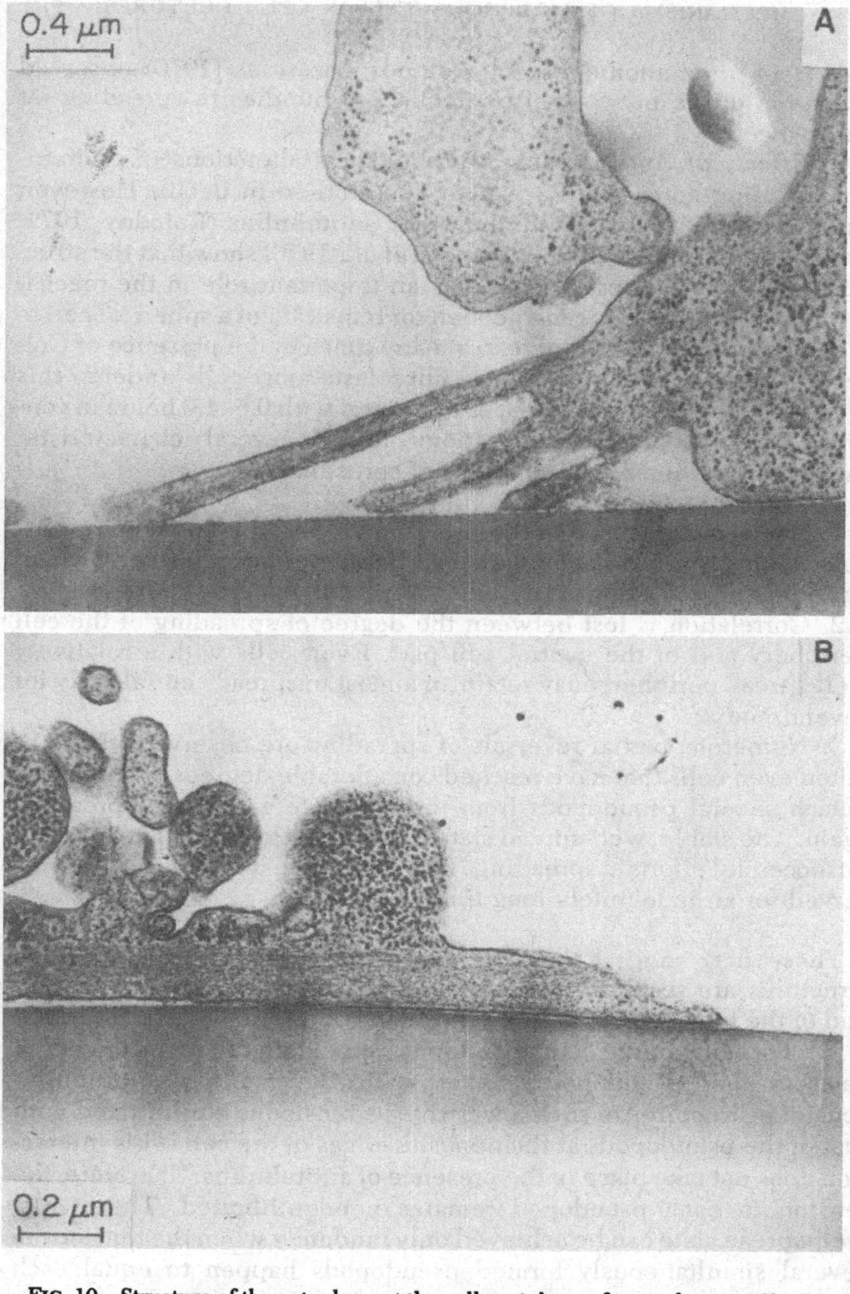

FIG. 10. Structure of the cytoplasm at the cell periphery of normal mouse fibroblasts during radial spreading. (A) Stage II of spreading. Three substrate-attached filopodia. Microfilament bundle penetrates from the cytoplasm of the upper filopodium into the cell body. (B) Stage III of spreading. Sheath of microfilament bundles in the lamellar cytoplasm. Transmission electron micrographs. Courtesy of E. E. Bragina.

tial. Using immunofluorescent methods Lazarides (1976) observed formation of the network of microfilament bundles in spreading rat embryo cells.

b. *Effects of Antitubulins.* Morphological alterations in microtubules during spreading have not been studied in detail. However, experiments with Colcemid and other antitubulins (Kolodny, 1972; Vasiliev and Gelfand, 1976b; Ivanova *et al.*, 1976) show that the structures sensitive to these agents play an important role in the mechanism of spreading. The time needed for transition of a spherical cell to the radially spread state increases many times in the presence of Colcemid; in a population of mouse fibroblasts most cells undergo this transition only after 6–8 hours, as compared with 0.5–1.0 hours in controls. Morphological investigations reveal several characteristic changes in the intermediate stages of spreading:

1. The size and shape of the pseudopods become much more variable; their distribution along the cell perimeter becomes very irregular.
2. Correlation is lost between the degree of spreading of the cell periphery and of the central cell part. Even cells with a relatively well-spread periphery may retain an almost unspread central body for several hours.
3. Numerous partial reversals of spreading are observed (Fig. 11). Often even cells that have reached considerable degrees of spreading detach several pseudopods from the substrate and begin spreading again. The stable, well-spread state is often reached only after several unsuccessful efforts at spreading. This state, once reached, may be preserved for an indefinitely long time.

These data suggest that, during spreading, antitubulin-sensitive structures are responsible for the equal distribution of pseudopods and of the tension produced by the attached pseudopods (see Section II,E). Possibly, in the course of normal spreading the cortical layer of each attached pseudopod interacts with the system of antitubulin-sensitive structures in such a way that its tension is equilibrated with that of the pseudopods at the opposite edges of the cell. This interaction does not take place in the presence of antitubulins. Therefore, the tension in each pseudopod remains nonequilibrated. The stable, well-spread state can be achieved only randomly when the tensions in several simultaneously formed pseudopods happen to equal each other.

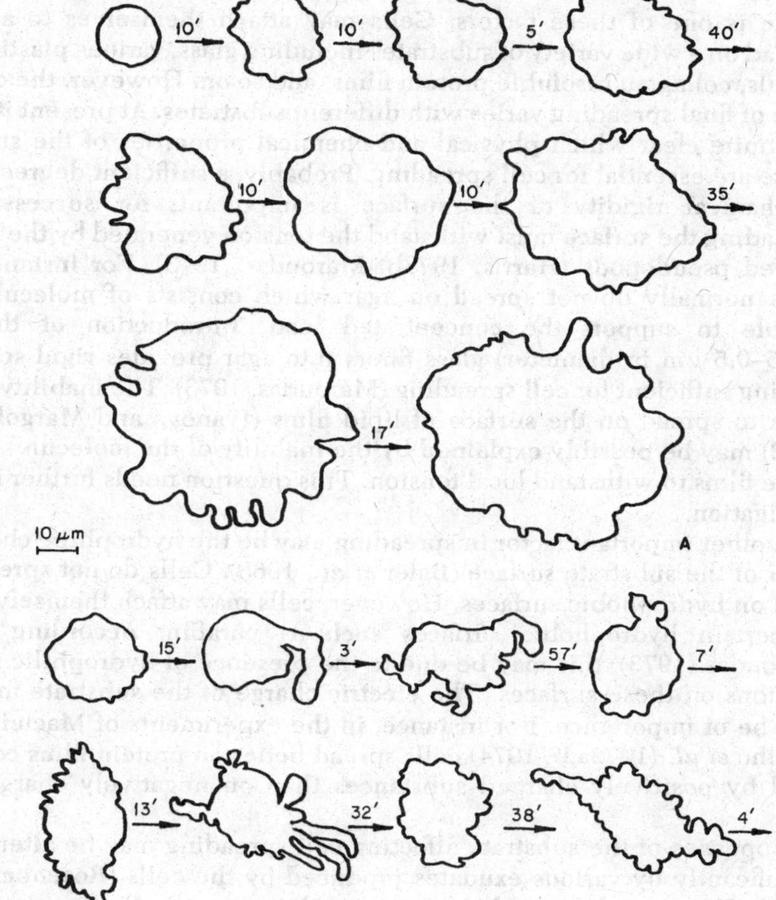

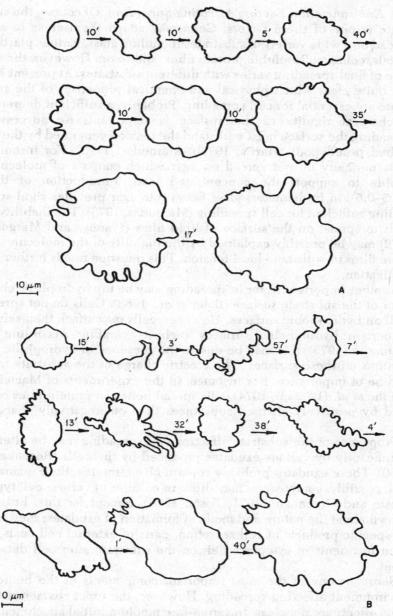

FIG. 11. Effect of Colcemid on radial spreading of normal mouse fibroblast. Each series of drawings shows tracings of the contours of one cell from a time-lapse film. (A) Cell in control medium. (B) Cell in medium containing Colcemid (0.1 mg/ml). Note repeated formation and detachment of long processes. The first drawings were made 15 minutes (A) and 30 minutes (B) after seeding. Courtesy of O. Ivanova.

c. *Environmental Factors Affecting Spreading.* Of course, the substrate is one of these factors. Cells may attach themselves to and spread on a wide variety of substrates including glass, various plastics, metals, collagen, insoluble protein films, and so on. However, the degree of final spreading varies with different substrates. At present it is not quite clear which physical and chemical properties of the substrate are essential for cell spreading. Probably, a sufficient degree of mechanical rigidity of the surface is important; for successful spreading the surface must withstand the tension generated by the attached pseudopods (Harris, 1973b; Maroudas, 1973). For instance, cells normally do not spread on agar which consists of molecules unable to support the concentrated load. Introduction of thin (0.05–0.5 μm in diameter) glass fibers into agar provides rigid scaffolding sufficient for cell spreading (Maroudas, 1973). The inability of cells to spread on the surface of lipid films (Ivanova and Margolis, 1972) may be possibly explained by the inability of the molecules of these films to withstand local tension. This question needs further investigation.

Another important factor in spreading may be the hydrophilic character of the substrate surface (Baier *et al.*, 1968). Cells do not spread well on hydrophobic surfaces. However, cells may attach themselves to certain hydrophobic surfaces such as paraffin. According to Maroudas (1973), this may be due to the presence of hydrophilic inclusions on these surfaces. The electric charge of the substrate may also be of importance. For instance, in the experiments of Macieira-Coelho *et al.* (1972a,b; 1974) cells spread better on protein films covered by positively charged substances than on negatively charged films.

Properties of the substrate affecting cell spreading may be altered significantly by various exudates produced by the cells (Rosenberg, 1960). These exudates probably contain glycoproteins; their quantity and, possibly, composition may differ in cultures of various cell types (Poste and Greenham, 1971; Poste, 1973). Except for this, little is known about the nature and mode of formation of exudates; they may be specific products of cell secretion, parts of external cell coats, or even fragments of cytoplasm left on the substrate after cell detachment.

Serum is one of the most important components of the humoral environment affecting spreading. However, the exact characteristics of its effects are not clear. In serum-free medium initial attachment of unspread cells (stage I) may be more firm (Taylor, 1961) and more resistant to EDTA and trypsin (Unhjem and Prydz, 1973) than that in serum-containing medium. The course of further spreading becomes abnormal in serum-free medium (Witkowski and Brighton, 1972). Investigation of the end results of cell spreading in serum-free medium is complicated by the poor survival of cells; it is not easy to distinguish early stages of cell degeneration from the more specific alterations in spreading. Certain serum components are easily absorbed on substrates (Revel and Wolken, 1973) and may alter considerably the properties of this substrate. For instance, the ability of hydrophobic surfaces to form attachments with cells is increased in serum-containing medium as compared with serum-free medium (Weiss and Blumenson, 1967). The presence of calcium, and probably of magnesium, is essential for spreading, but a detailed analysis of spreading in media containing various concentrations of these ions has not yet been made.

Naturally, besides alterations in the substrate and in the concentrations of normal medium components, many other experimental interventions are able to affect initial cell attachment and spreading. We mention here only a few examples of the results obtained with various kinds of cells. If the suspended cells are pressed to the substrate by centrifugal force, the rate of their attachment to this substrate increases considerably (Milam *et al.*, 1973). Possibly, centrifugation facilitates establishment of the initial cell-substrate contact (stage I). The strength of the cell-substrate attachment can also be increased by concanavalin A (Grinnel, 1973; Sato and Takasawa-Nishizawa, 1974). Perhaps this lectin cross-links the cell surface receptors with the serum components adsorbed on the substrate. It is interesting that the rate of cell attachment can also be increased by various metabolic inhibitors such as ouabain, actinomycin, puromycin, and cyclohexamide (Weiss, 1972, 1974; Weiss and Chang, 1973). Unfortunately, none of these interesting effects has been analyzed morphologically, and we do not know which components of spreading are affected in each case. In summary, the effects of environmental factors on spreading are poorly understood. The main difficulty is that the possible mechanisms of action of each of these factors are manifold. External factors may alter various components of the active attachment reaction, they may affect the antitubulin-sensitive structures, and they may also change the production of cellular exudates. Besides cellular effects, these factors may alter the properties of the original substrate as well as those of serum components and of the cellular exudates adsorbed on this substrate. Differential analysis of all these effects remains a task for the future.

3. *Polarization on a Flat Substrate*

a. *Main Characteristics.* A cell radially spread on a flat glass or plastic surface is spontaneously polarized several hours later. In cultures of mouse fibroblasts this process usually takes 3–4 hours. Irregularities in the substrate structure as well as the presence of other cells may determine the direction of polarization; they may also affect the rate of polarization (see Sections III,A,4 and 5). It is not clear which factors determine the rate and direction of polarization under conditions in which the effects of substrate structure and of other cells are minimal, that is, on a homogeneous surface in a sparse culture. There are several possible mechanisms that may determine the direction of polarization under these conditions:

1. The direction may be determined by the hidden polarization of intracellular structures (e.g., of the centrioles), which already exists in radially spread cells and possibly even in spherical cells.
2. The direction may be determined by hidden irregularities in the substrate. It would be interesting to compare the rate of polarization and its direction on the usual substrates and on substrates with an artificially increased degree of surface smoothness.
3. The direction of polarization may be established randomly. We do know which of these mechanisms really works in cultures.

The characteristic feature of polarized cells is the differentiation of their edge into active and nonactive zones; the number of discrete active zones rarely exceeds four. Reactions of active attachment occurring at the active edges are chainlike; when one pseudopod is attached, other pseudopods are usually formed at its edges. As a result, each active area has a tendency to expand and to move centrifugally on the substrate, that is, to increase its distance from the center.

The sizes of various active edges, as well as the efficiencies of attachment reactions at these edges, are usually unequal. These initial differences may be greatly increased because of the chainlike nature of the active attachment reactions. As a result, one active edge becomes the leading one; it becomes considerably larger than the other ones and moves more actively on the substrate. Other active areas cannot equilibrate the tension from the leading one. From time to time the tension from the leading edge detaches lamellar zones near nonleading edges from the substrate; the cell body contracts and is moved toward the leading edge. Thus cell translocation on the substrate seems to be a net result of several chain reactions of spreading occurring in different active areas and competing with each other. Stabilization reactions, maintaining the differentation of the edge into active and nonactive areas (see Section II,E), are obviously essential for this mechanism of locomotion. Fanlike cells having only two active zones at the opposite poles usually move most efficiently. It was shown that even on a flat substrate without predetermined directions moving fibroblasts have a tendency to maintain a definite direction of movement for time intervals up to 2.5 hours (Gail and Boone, 1970).

The moving cell may change the direction of its translocation in two ways:

1. The leading status of one active area may be lost and acquired by another area. This happens when the efficiency of spreading in the preexisting leading edge decreases, for example, because of contact with another cell. In this way the fanlike cell may reverse the direction of its translocation by expanding the size of its rear leading area and reducing the size of its previous leading edge.
2. The leading edge may begin to spread asymmetrically to one side.

One large active zone may be divided into two zones when its central part is inactivated. A new active zone is not usually formed in the middle of a nonactive one. Thus the following rule seems to be true for polarized cells. Each active area arises from a preexisting active area. Detailed quantitative studies of the movements of the leading active edge of fibroblasts during locomotion have been performed by Abercrombie *et al.* (1970a,b). They found that any point on the leading edge undergoes repetitive extension and withdrawal; the average duration of each fluctuation is about 4 minutes. The net forward displacement of the fluctuating edge results not from more rapid movement forward than backward, but from a greater time spent moving forward than backward.

b. *Surface of Polarized Fibroblasts*. The upper surface of most interphase polarized fibroblasts is relatively smooth except for a few microvilli and ruffles. The ruffles are seen mostly near the active edges; they move centripetally and normally to these edges (Abercrombie *et al.*, 1970b). The particles attached to the surface move in the same direction. There are no surface movements in the direction normal to the nonactive cell edges.

Cell-substrate attachments are localized mainly in the lamellar areas near active edges. Some cells also have local cell-substrate attachments in the anterior part of the endoplasm, which lies in front of the nucleus. There are no local contact structures in the posterior part of the endoplasm (Bragina *et al.*, 1973). In lamellar areas the lower cell surface between the attachment sites is usually smooth and often a relatively small distance (20–40 nm) from the substrate. The distance of the lower surface from the substrate in the central part of the cell may be much greater.

Obviously, cell locomotion is accompanied not only by the formation of new cell-substrate attachments but also by the destruction of old ones. The mechanism of this destruction is not clear. The tension from new attachments formed at the leading edge probably plays the major role in this destruction.

c. *Fibrillar Structures in Polarized Cells.* A well-organized system of microfilament bundles (Fig. 12) is a characteristic feature of polarized fibroblasts (Buckley and Porter, 1967; Spooner *et al.*, 1971; Wessels *et al.*, 1973; Goldman and Knipe, 1973). The predominant direction of these bundles is parallel to the stable cell edges and perpendicular to the active ones.

The bundles pass through the lamelloplasm and endoplasm; the rear part of the cell may be almost completely filled with bundles. The bundles are located mostly in cortical layers near the upper and lower surfaces; correspondingly, lower and upper bundles can be distinguished. Large bundles usually run along the nonactive lateral cell edges; they form the boundary between the upper and lower bundles. In the lamellar cytoplasm the upper bundles approach the lower surface; both upper and lower bundles end at cell-substrate attachment sites. Bundles are separated from each other by matrix cortical layers. Groups of two to four parallel microtubules are often seen in the sections near the bundles. They are never found within the bundles. It is not clear whether or not there are structural connections between the microtubules and the bundles. As seen from this description, the transition of fibroblasts from the radially spread into the polarized state is accompanied by reorganization of the system of microfilament bundles. In the same area of the lamellar cytoplasm of radially spread cells the direction of bundles with regard to the center may vary from radial to tangential. In polarized cells most bundles have a predominantly radial pattern of orientation. It is interesting to compare these alterations in the distribution of bundles accompanying polarization with the effects of cytochalasin B on radially spread and polarized cells (T. M. Svitkina, J. M. Vasiliev, and I. M. Gelfand, unpublished). After the incubation of

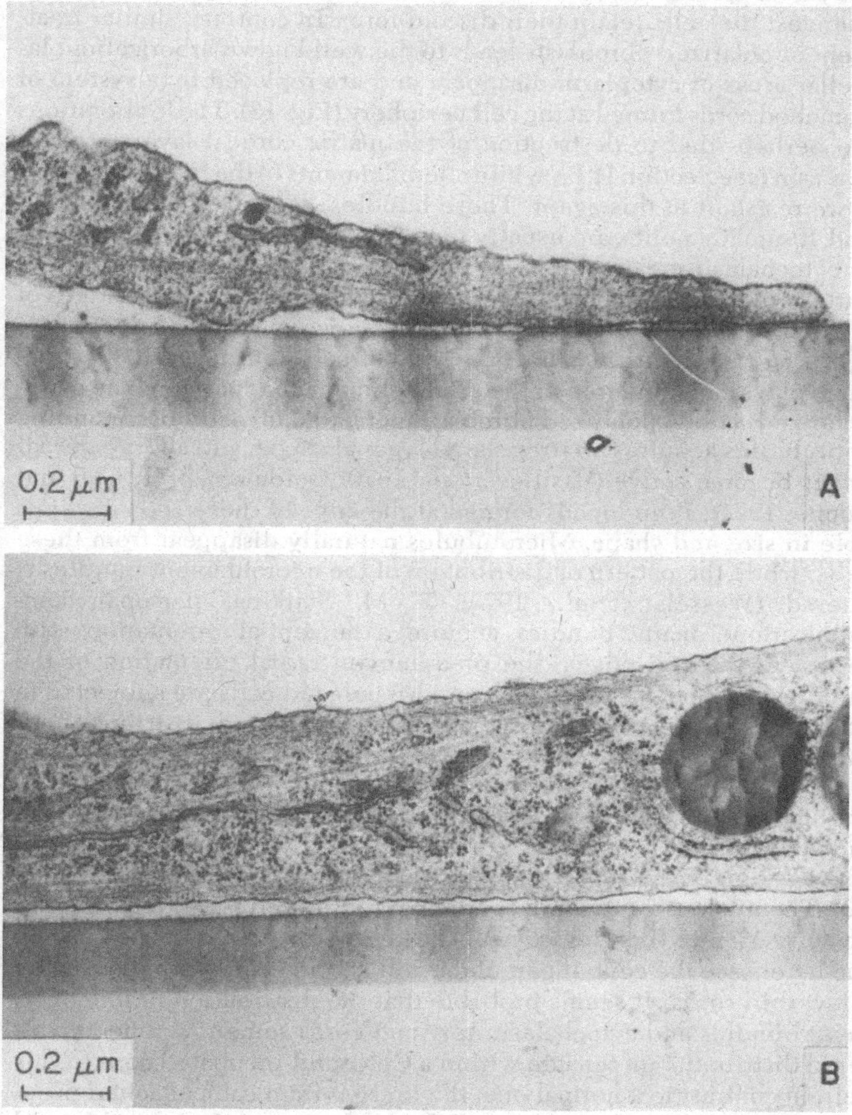

FIG. 12. Fibrillar structures in the cytoplasm of polarized mouse fibroblasts. (A) Microfilament bundle in the anterior lamella. (B) Microfilament bundles near the upper and lower cell surface of the endoplasm. Microtubules are also seen in the cytoplasm. Courtesy of E. E. Bragina.

radially spread mouse fibroblasts in medium containing cytochalasin B (10 μg/ml for 1 hour) the shape of these cells undergoes no striking changes; the cells retain their discoid form. In contrast, similar treatment of polarized fibroblasts leads to the well-known arborization: lamellar areas of cytoplasm disappear and are replaced by a system of branched cords formed at the cell periphery (Fig. 13). These alterations are perhaps due to destruction of the matrix cortical layer by cytochalasin (see Section II,B), while microfilaments in the bundles may be more resistant to this agent. These bundles, as well as microtubules and 10-nm filaments, are usually seen in sections of cytoplasmic cords of cytochalasin-treated cells. In any case, it seems possible that the pattern of these cords may reflect the predominantly radial pattern of fibrillar structures established during polarization.

d. *Effects of Antitubulins.* A system of antitubulin-sensitive structures plays a leading role in the establishment and maintenance of the polarized state. A polarized fibroblast incubated in medium containing antitubulins acquires an irregular polygonal shape, and all its external edges become active (Vasiliev *et al.*, 1970; Goldman, 1971; Gail and Boone, 1971). Pseudopods formed at the edge of these cells are variable in size and shape. Microtubules naturally disappear from these cells, while the pattern of distribution of the microfilament bundles is altered (Wessels *et al.*, 1973; T. M. Svitkina, personal communication); many bundles acquire a tangential orientation with regard to the cell edges; the predominant radial orientation of the bundles disappears. When Colcemid-incubated cells are subjected to additional treatment with cytochalasin, an altered distribution of cytochalasin-resistant cords is revealed (Vasiliev and Gelfand, 1976b; T. M. Svitkina, J. M. Vasiliev, and V. I. Gelfand, unpublished). In contrast to normal fibroblasts Colcemid-incubated cells have no system of radial cytochalasin-resistant cords. Instead, cytochalasin reveals in these cells a peripheral ring consisting of small cords and plates (Fig. 13). The nucleus-containing central part of the cell is located either in some section of this ring or near the geometric center of the ring; in the latter case the central part of the cell is connected with the ring by a few thin cords. It seems probable that the distribution of microfilament bundles and cytochalasin-resistant cords somehow reflects a altered distribution of tension within a Colcemid-incubated cell. In this cell, in contrast to a normal one, the tension from each attached pseudopod cannot be transmitted radially to the center and equilibrated there with the tension from other active parts of the edge (see Section II,E). A relatively stable well-spread state is maintained only in cases in which the tension of each attached pseudopod is equilibrated locally

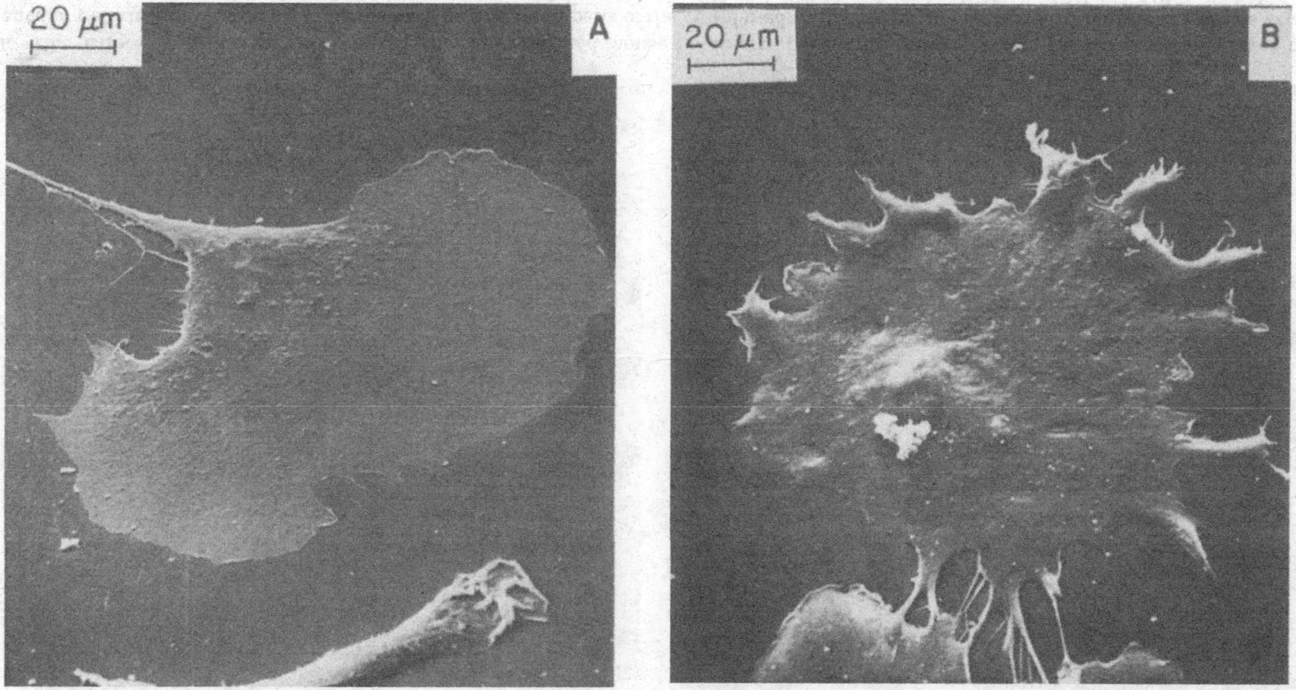

FIG. 13 A and B.

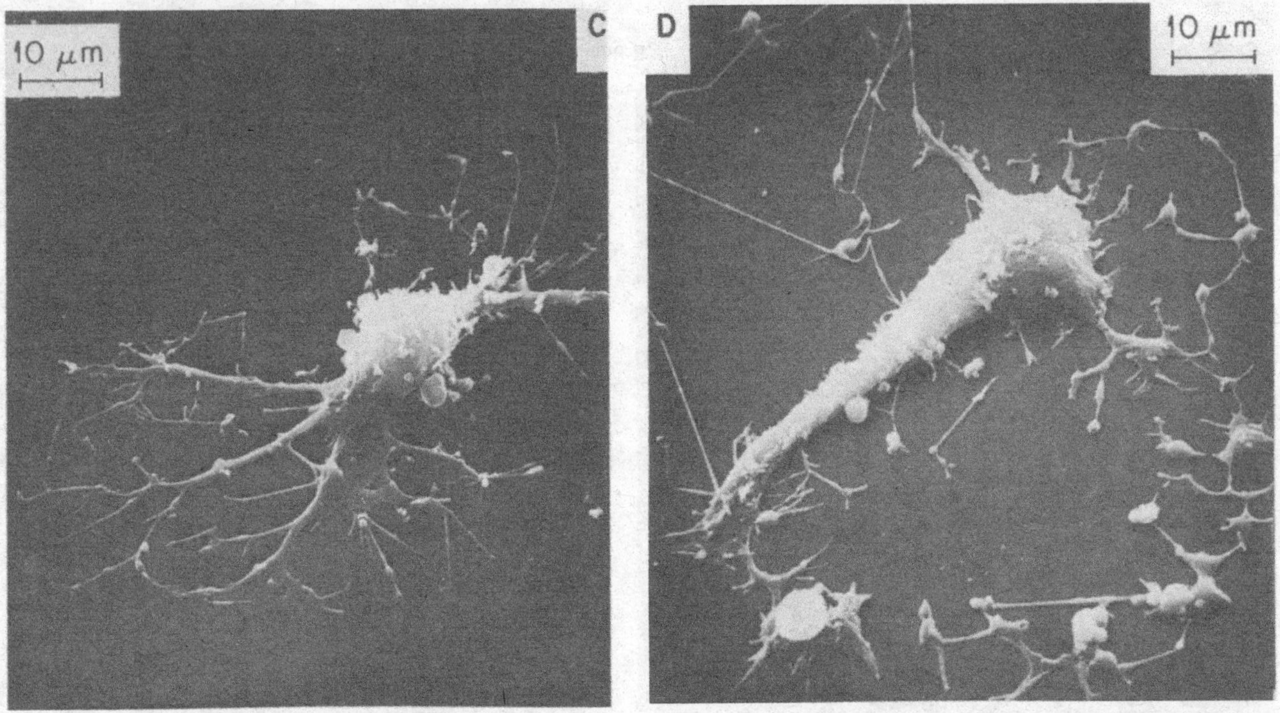

FIG. 13. Effects of Colcemid and cytochalasin B on polarized mouse fibroblasts. (A) Control polarized cell. (B) Colcemid-incubated cell. Absence of polarization. Large pseudopods have formed in all parts of the cellular edge. (C) Polarized cell treated 1 hour with cytochalasin B. Branching cytoplasmic cords radiate from the central part of the cell body. (D) Colcemid-incubated cells treated 1 hour with cytochalasin B. Radial branching system of cords is absent. The cell body and cytoplasmic cords form a circle. Courtesy of T. M. Svitkina.

with the tension of pseudopods attached in the nearby area. This may lead to the development of a circumferential peripheral system of equilibrated tension, and of the microfilament bundles visualized by cytochalasin.

e. *Polarization of Intracellular Organelles.* Besides alterations in external cell shape, pseudopodial activity and intracellular fibrillar structures, polarization is also accompanied by alterations in the shape and distribution of intracellular organelles. Some cytoplasmic organelles, for example, mitochondria, may acquire predominantly radial orientation. The same seems to be true with regard to the saltatory movements of intracellular organelles (Freed and Lebowitz, 1970). Elongation and orientation of the nucleus is one of the characteristic features of polarized cells. A projection of the nucleus of these cells onto the plane of the substrate usually has the shape of an ellipse. In fan-shaped or fusiform cells with two active edges, the long axis of this ellipse is usually located on a line connecting the centers of these two edges (Weiss and Garber, 1952). Elongation of the nucleus is probably caused by tension in the cortical layers; this tension flattens the cell and compresses its internal structures. In polarized cells the degree of compression is different in various directions; this anisotropy leads to elongation of the nucleus. It seems reasonable to assume that the direction of the long axis of the nucleus corresponds to the direction of maximal tension in the cortical layer acting on the central cell part. The degree of nuclear elongation can be measured as the relation of the length of the long axis of the same projection of the nucleus (A) to that of the short axis of the same projection (B). A/B probably reflects the degree of asymmetry of the tension acting in the cortical layer. A/B is easily measured in all cells attached to the substrate. The direction of the axis A may be easily determined in all cells with the projection of a nucleus different from the circle ($A/B > 1.0$). In contrast, determination of the position of the cellular axis is not always easy in cells with two active edges. For cells with three or more active edges it is difficult even to give an exact definition of the cellular axis. Therefore the only safe way to measure orientation of polarized cells is to determine the direction of the long nuclear axis (Weiss and Garber, 1952; Margolis *et al.*, 1975).

Very little is known about the intracellular position of certain organelles that may be essential for maintenance of the polarized state. There are, first of all, centrioles, which may connect microtubules with each other. The centrioles are probably located somewhere near the nucleus, but their position with regard to the nuclear axis and to the direction of locomotion has not been investigated systematically.

When the cell enters mitosis, it loses all the obvious manifestations of polarity; it usually acquires an almost spherical shape, and most preexisting microtubules and microfilament bundles disappear. Nevertheless, some manifestations of the preexisting orientation of intracellular organelles are preserved in mitotic cells. This is suggested by the results of microcinematographic analysis of mitotic fibroblasts (Ivanova *et al.*, 1974). In these experiments the direction of the division furrows was found to be strictly correlated with the direction of the long axis of the nucleus of the same cell before division; for most cells the angle between these two directions was found to be about 90°. We do not know which intracellular organelles retain during mitosis, a "memory" of the orientation of the cell in the previous interphase; possibly these organelles are centrioles.

f. *Effects of Environmental Factors on the Shape and Locomotion of Polarized Cells.* As discussed above, cell translocation can be regarded as a result of asymmetrical spreading occurring preferentially at the leading edge. Obviously, all the factors affecting radial spreading should also affect the translocation of fibroblasts as well as their shape. In fact, a lower concentration of calcium in the medium considerably decreases the rate of cell movement and changes the shape of the cells (Gail, 1973; Gail *et al.*, 1973). The rate of movement is also decreased by procaine which presumably binds membrane calcium (Gail and Boone, 1972b). Cell shape and locomotion may also be affected by alterations in serum concentration in the medium (Gail, 1973), as well as by alterations in the concentration of certain individual serum components such as insulin, fibroblast growth factor, and others (Lipton *et al.*, 1971; Gospodarowicz and Moran, 1974).

Analysis of the effects of all these factors on locomotion is even more difficult than of the effects of the same factors on cell spreading. For instance, a change leading to an increase in the strength and/or number of cell-cell substrate attachments can be expected to promote radial spreading. However, it is difficult to predict the effect of such a change on the locomotion of polarized cells, as it includes not only the formation of new contacts but also the destruction of old ones. In fact, in the experiments of Gail and Boone (1972a) the rate of cell movement decreased both on poorly adhesive and on very adhesive substrates; substrates of intermediate adhesiveness were optimal for cell motility. In these experiments the force needed to detach cells from the substrate was regarded as the measure of adhesiveness of the substrate.

4. *Cell-Cell Interactions on Flat Surfaces*

All known cell-cell interactions leading to morphological changes in fibroblastic cultures are mediated by local intercellular contacts, that is, are observed only in cases in which one cell touches another. The final distribution of cells in a culture is the statistical result of a great number of individual cell-cell interactions. This final distribution depends on cell density. At lower densities cells are usually distributed as monolayers; at higher densities they may form a multilayered sheet. In this section we first describe interactions of individual cells observed in low-density cultures and later discuss processes leading to the formation of multilayered dense cultures.

a. *Cell-Cell Collisions in Sparse Cultures.* Individual collisions of polarized fibroblasts can be divided into two groups: (1) collisions between an active leading edge of one cell and an active edge of another cell (head-head collisions), and (2) collisions between an active leading edge of one cell and a lateral nonactive edge of another cell (head-side collision; Fig. 14). Microcinematographic statistical analysis of the collisions of normal mouse fibroblasts (Guelstein *et al.*, 1973) has shown that most head-head collisions result in a halt, that is, a temporary cessation of forward movement of both active edges; this cessation is usually accompanied by contact paralysis. About one-half of all head-side collisions (Fig. 14) also result in a halt of the active edge, and the other half lead to underlapping, that is, to forward translocation of the leading edge of one cell under the lateral side of the other. The degree of underlapping in sparse cultures of mouse fibroblasts is not high; forward translocation of the underlapping active edge soon stops. Both head-head and head-side collisions rarely (in less than 10%) resulted in overlapping, that is, in forward translocation of an active edge of one cell over the upper surface of another; the degree of observed overlapping was always small. Thus contact inhibi-

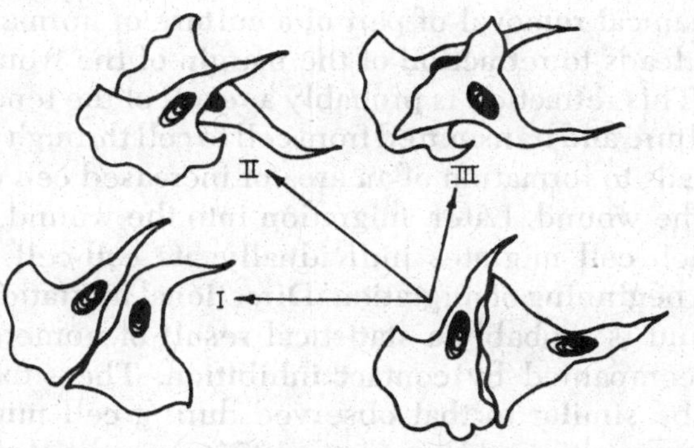

FIG. 14. Scheme of the possible results of head-side collisions of two fibroblasts. (I) Halt leading to alteration in the direction of movement. (II) Underlapping. (III) Overlapping.

tion of movements was very efficient in sparse cultures of mouse fibroblasts. The same also seems to be true of fibroblasts of other species (see Abercrombie, 1970; Harris, 1974).

The statistical result of many individual cell-cell collisions is a monolayered distribution of cells. When cell density in a culture increases, partial overlaps of the cytoplasm of different cells become frequent. A detailed analysis of the processes leading to the formation of these overlaps has not been made, but microcinematographic observations suggest that they arise as a result of underlapping.

b. *Morphology of Dense Cultures.* Fibroblasts in dense cultures may form multilayered sheets (Elsdale and Bard, 1972a; Cherny *et al.*, 1975). Dense cultures of mouse fibroblasts were shown to consist of 6 to 10 cell layers (Fig. 15). The cells forming these sheets are well spread; they have a well-developed lamellar cytoplasm (Fig. 16), and the mean area of the projection of one cell onto the plane of the substrate is almost the same as in sparse cultures (Cherny *et al.*, 1975). Various types of intercellular contact structures are formed between the cells in these cultures (see references in Section II,C,3). Extracellular substances revealed in these cultures include collagen fibers and mucopolysaccharides.

The vertical distribution of cells in these cultures has a certain regularity. An examination of sections shows that central parts of the cells in two adjacent layers are rarely located immediately over each other; usually the central part of one cell contacts the lamellar cytoplasm of the cell located over or under it. Due to the nonrandom distribution of nuclei, the multilayered character of the cultures may be easily overlooked with light microscopy.

c. *Cell Migration into a Wound from Dense Cultures.* Wounding, that is, mechanical removal of part of a culture of normal fibroblasts, immediately leads to retraction of the margin of the wound (Vasiliev *et al.*, 1969). This retraction is probably a result of the tension existing in a dense culture and transmitted from cell to cell through the contacts. Retraction leads to formation of an area of increased cell density near the edge of the wound. Later, migration into the wound begins from this area. Each cell migrates individually; its cell-cell contacts are broken at the beginning of migration. Directional migration of the cells into the wound is probably a statistical result of numerous cell-cell collisions accompanied by contact inhibition. The situation in the wound may be similar to that observed during cell migration from explants (Abercrombie and Heaysman, 1954). A gradient of cell density is soon established in the wound. It is interesting that for several days after wounding one can still see the original margin of the wound, that

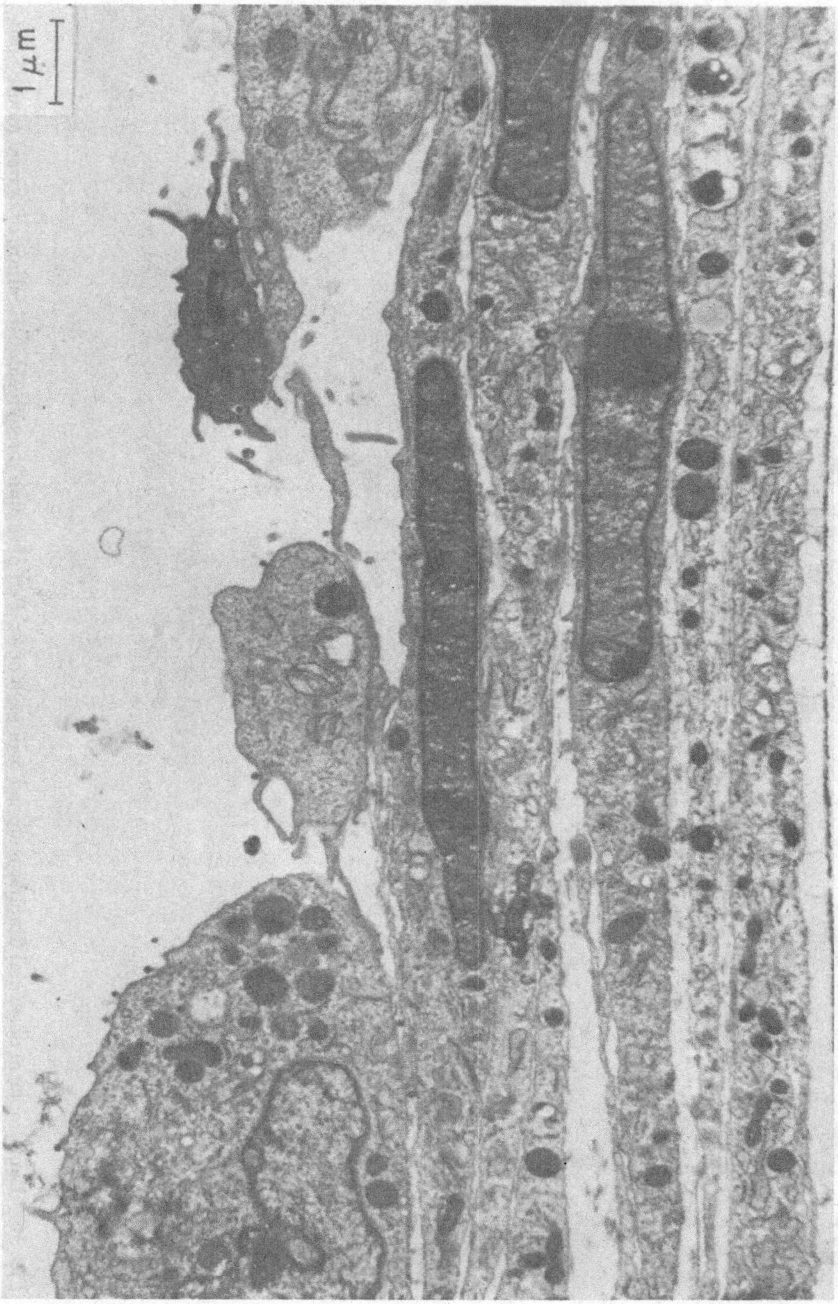

FIG. 15. Mixed culture of normal mouse fibroblasts and transformed L cells. Flattened normal fibroblasts form a multilayered structure. Poorly spread transformed cells occupy the upper surface of the multilayered sheet of normal cells. Courtesy of A. P. Cherny.

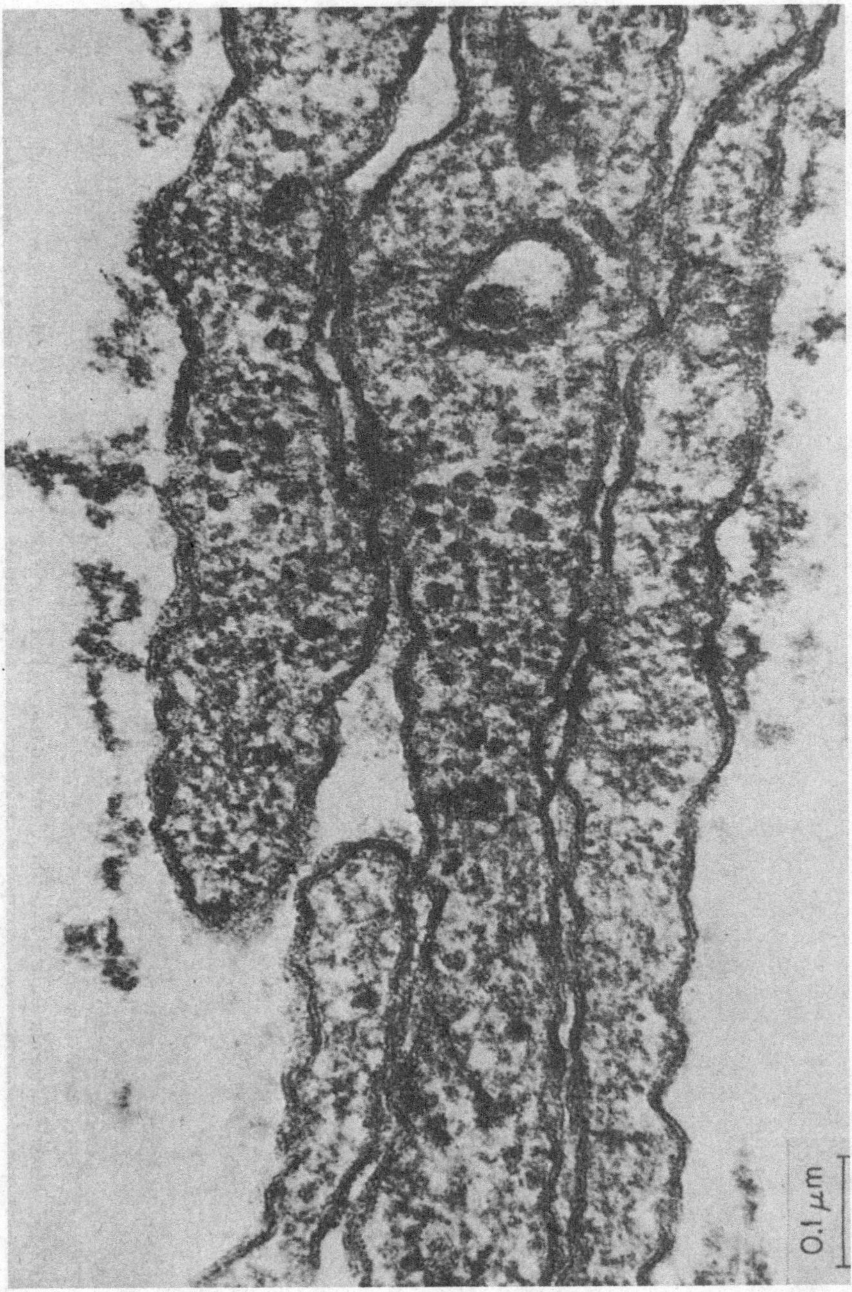

FIG. 16. Lamelloplasm of several fibroblasts spread over each other. Part of section through a dense culture of normal mouse fibroblasts. Courtesy of A. P. Cherny.

is, a stepwise transition from the high-density area of the old wound to a nearby area of lower density filled with migrating cells (Vasiliev *et al.*, 1969). These observations suggest that the maximal cell density achieved as a result of migration is much lower than saturation density in dense cultures. This interesting phenomenon has not been studied in detail.

d. *Mechanisms of Formation of Dense Cultures.* As discussed above, effective contact inhibition of movement is characteristic of sparse cultures of fibroblasts. Nevertheless, in dense cultures these cells spread over each other and form multilayered structures. How are these structures formed? There are several possible mechanisms of multilayering:

1. Contact inhibition of movement may become less effective as cell density increases. Because of the obvious technical difficulties a microcinematographic analysis of cell movements in dense cultures has not yet been made. Conditions existing in a narrow strip of adhesive substrate filled with fibroblasts may be regarded as an approximation to the conditions existing in these cultures. Our unpublished observations show that numerous underlappings may be observed in these strips; active overlapping of one cell by another are rare. These data do not support the suggestion that contact inhibition of movement decreases in dense cultures.

2. It is possible that the cells in dense cultures are spread over collagen fibers and over other extracellular substances. Elsdale and Bard (1972a) observed the disappearance of multilayering in dense cultures of human fibroblasts after treatment with collagenase. However, our results (Cherny *et al.*, 1975) indicate that cell spreading over extracellular matrices is not the only mechanism of multilayering in cultures of mouse fibroblasts. Direct numerous cell-cell contacts are seen in these cultures between cells located in adjacent layers. The multilayered structure in these cultures is preserved after incubation with collagenase and hyaluronidase.

3. The multilayered structure may be a result of cell-cell underlapping followed by detachment of the upper cell from the substrate and its subsequent spreading over the surface of the lower cell. As mentioned above, underlapping is common in cultures of medium density; it probably becomes even more common in high-density cultures. It is not clear whether or not detachment of upper cells underlapped by the other cells takes place in these cultures. Detachment of transformed cells underlapped by normal fibroblasts had been ob-

served in mixed cultures of these cells (Domnina *et al.*, 1972). Detached cells subsequently spread over the surface of lower fibroblasts (Fig. 15). Therefore this mechanism of multilayering seems plausible.

4. Poorly attached mitotic cells are perhaps detached from the substrate by more firmly attached interphase cells. After mitosis detached cells may be forced to spread over the surface of other cells.

It seems probable at present that the mechanisms listed above under points 2 through 4 may be involved in the formation of multilayered structures; the role of each of these mechanisms is not clear. As a result of their actions, a multilayered structure can be formed without a loss of contact inhibition.

e. *Adhesiveness of the Upper Surface of Dense Cultures.* The upper surface of dense cultures of mouse fibroblasts adheres to cells and inert particles; living cells (fibroblasts and epithelial cells), as well as carmine particles (Fig. 17), are readily attached to this surface (Domnina *et al.*, 1972; Vasiliev *et al.*, 1975a). How can this adhesiveness be explained if we assume that formation of cell-cell attachments is an active process involving the formation of pseudopods?

Cell-cell contacts in dense cultures are not firm; they are readily broken during cell migration into a wound. Possibly, because of tension in the cells, contacts in the upper cell layer of dense cultures may be broken from time to time. These breaks may lead to the activation of separated cell edges, that is, to the formation of adhesive pseudopods. Another possibility is that the entire upper surface of the lamellar cytoplasm of the fibroblasts remain adhesive, even at a time when the cells do not form new pseudopods at their edges. The second suggestion seems to be a natural one, since the lamellar cytoplasm is a structure formed as a result of the fusion of many attached pseudopods. Finally, the accumulation of extracellular substances may play a certain role in the adhesiveness of the upper surface of dense cultures.

None of these suggestions has been tested experimentally.

f. *Mutual Cell Orientation in Dense Cultures.* Mutual cell orientation is a characteristic feature of dense cultures. To measure this orientation it is convenient to use indexes based on determination of the direction of the long axis of the projection of nuclei (Margolis *et al.*, 1975). One should add that the term orientation is a statistical one, in the sense that it describes the behavior not of an individual cell but of a group of cells. Within a culture, the degree of mutual orientation in groups containing the same number of cells increase with time

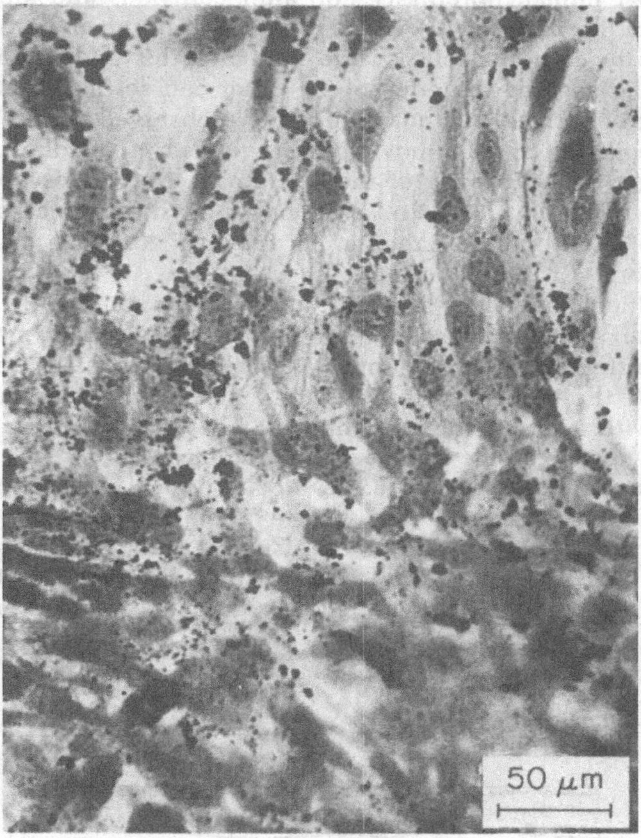

FIG. 17. Carmine particles attached to cells in a wounded dense culture of mouse fibroblasts. The particles attach both to the dense part of culture (lower part of the photograph) and to the cells migrating into the wound (upper part). Compare with Fig. 2A. Hematoxylin. Courtesy of O. S. Zacharova.

(Margolis *et al.*, 1975). The size of cell groups having the same direction of orientation also increases with time (Elsdale and Bard, 1972a,b).

There may be at least two types of processes that may lead to mutual cell orientation: contact inhibition of movement and contact guidance of underlying cells. In cultures of relatively low cell density orientation is probably a result of cell-cell collisions leading to alterations in the direction of locomotion. Observations of Elsdale and Bard (1972a) suggest that human fibroblasts stop after collisions and alter their direction of locomotion only in the case in which the angle between the axes of the colliding cells is large enough. If this angle is too small, one cell slides along the edge of the other cell so that direction of movement of both cells is not changed. We observed similar sliding after the collision of cells moving on narrow strips of adhesive substrate (Fig. 18). Narrowing of the colliding active edges obviously facilitates the sliding. These observations suggest that, once the cells in a group acquire mutual orientation, they can continue to move in the direction of their orientation.

An additional mechanism of mutual orientation possibly acts in dense multilayered cultures; here the surface of lower cells may serve as an orienting substrate for upper ones. In fact, labeled fibroblasts seeded on the upper surface of unlabeled homotypic cells are oriented with regard to their neighbors (Stoker, 1964). Living labeled fibroblasts seeded on the surface of glutaraldehyde-fixed cells are also oriented in parallel to the lower dead cells (V. I. Samoilov, personal communication). In this case the surface relief preserved after fixation obviously acts as an orienting factor. A special type of orientation was observed by Elsdale and Bard (1972a) in multilayered cultures of human fibroblasts; the cells in the adjacent layers were oriented not

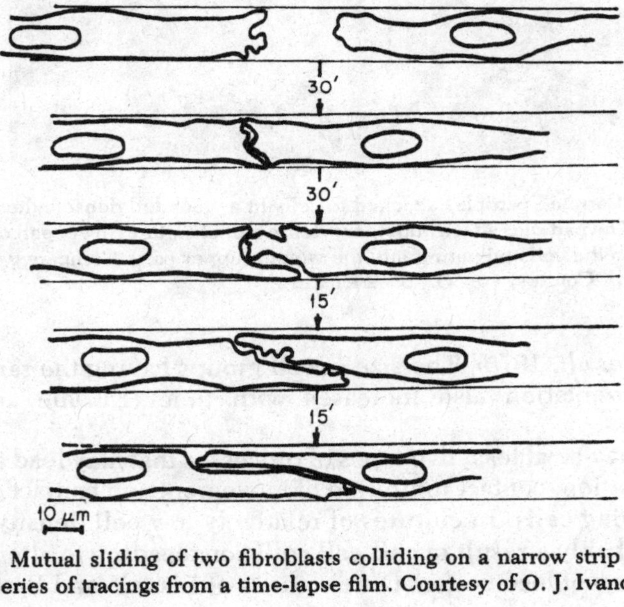

FIG. 18. Mutual sliding of two fibroblasts colliding on a narrow strip of adhesive substrate. Series of tracings from a time-lapse film. Courtesy of O. J. Ivanova.

parallel but perpendicular to each other. The mechanism of this "orthogonal orientation" is not clear.

5. *Orientation of Fibroblasts by the Substrate; Contact Guidance*

Cell orientation based on the heterogeneities of the substrate was first observed by Harrison (1914) and studied in detail by P. Weiss (1929, 1934, 1961) who introduced the term contact guidance to describe this phenomenon. Substrates able to induce definite cell orientation may be divided into two groups:

1. Substrates with a chemically heterogeneous surface. These surfaces have areas of two types: those that are more preferred and those that are less preferred by cells for spreading. The transition from one type of surface to another may be discrete or gradual [substrates with surface gradients (Carter, 1967)]. In the first case the boundary between the two areas acts as an orienting factor.
2. Chemically homogeneous substrates with an ordered anisotropic geometric surface relief.

To measure the orienting effect of a substrate it is convenient to determine the variation of the angles between the long nuclear axes and the direction of the orienting structure, for example, that of the groove or of the boundary of a nonadhesive area (Margolis *et al.*, 1975).

We now describe in more detail several examples of cell orientation on both varieties of substrates.

a. *Cell Orientation on the Boundary of a Lipid Film.* A lipid film covering a glass is an example of a nonadhesive substrate to which cell pseudopods are not attached. By removing some parts of this lipid mechanically it is easy to make margins of various shapes between the adhesive and nonadhesive parts of the substrate (Ivanova and Margolis, 1972). A specific variant of such a system is a narrow (about 20- to 30-μm-wide) strip of glass surface made by scratching the lipid film with a microneedle. On such strips the cells reach a maximal degree of orientation with regard to the direction of the strip (Fig. 19A). After seeding on these strips the cells reach the polarized state at a much faster rate than on flat homogeneous glass: 1–1.5 hours after seeding as compared with 3–4 hours (Ivanova *et al.*, 1976). Spreading and polarization on the strips proceed simultaneously. Thus heterogeneities of the substrate may affect not only the direction of cell orientation but also the rate of polarization.

Cell polarization on narrow strips is prevented by Colcemid; in medium containing this inhibitor some cells are able to acquire an

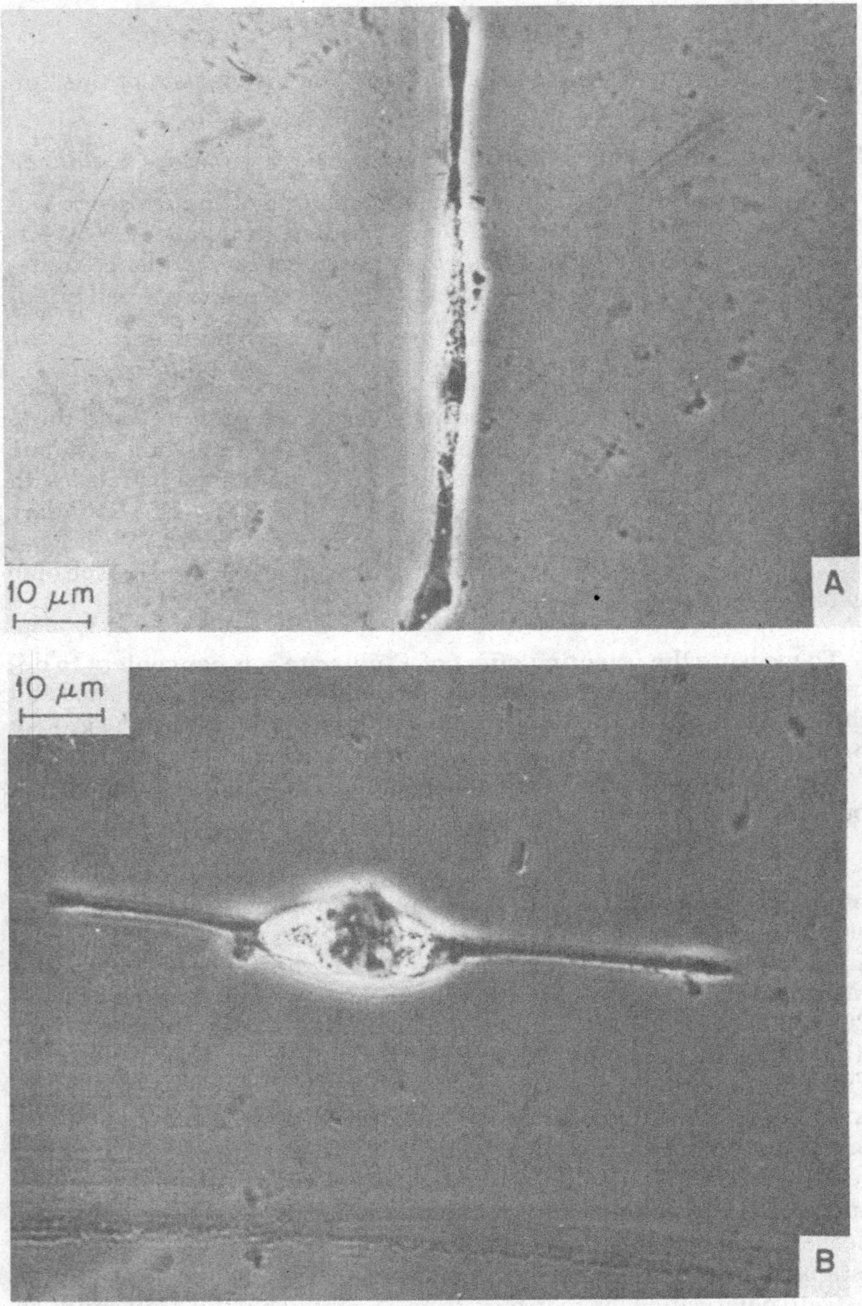

FIG. 19. Elongated normal mouse fibroblast (A) and transformed L fibroblast (B) on a narrow strip of glass between two lipid films. Phase-contrast micrograph. Courtesy of O. J. Ivanova and L. B. Margolis.

elongated shape; however, all the cells continue to form pseudopods in all directions; their edge is not divided into active and nonactive zones (Ivanova *et al.,* 1976). These observations suggest that an antitubulin-sensitive system is essential for polarization, even under conditions in which external factors maximally enhance polarization and strictly determine its direction. One boundary of lipid is also able to induce the orientation of large cell groups (Fig. 20). Orientation is acquired not only by cells contacting the margin but also by several rows of adjacent cells which do not contact the margin directly. Microcinematographic observations (L. B. Margolis, personal communication) show that cells contacting the boundary acquire orientation by gradually shifting the position of their active edge (see Section II,E). The orientation of other cells is a result of collisions with those contacting the boundary. Thus, because of cell-cell interactions, an orienting effect of some local structure on the substrate may gradually spread on large cell groups which do not contact this structure directly.

b. *The Orienting Effect of Grooves.* The effect of a groove is the

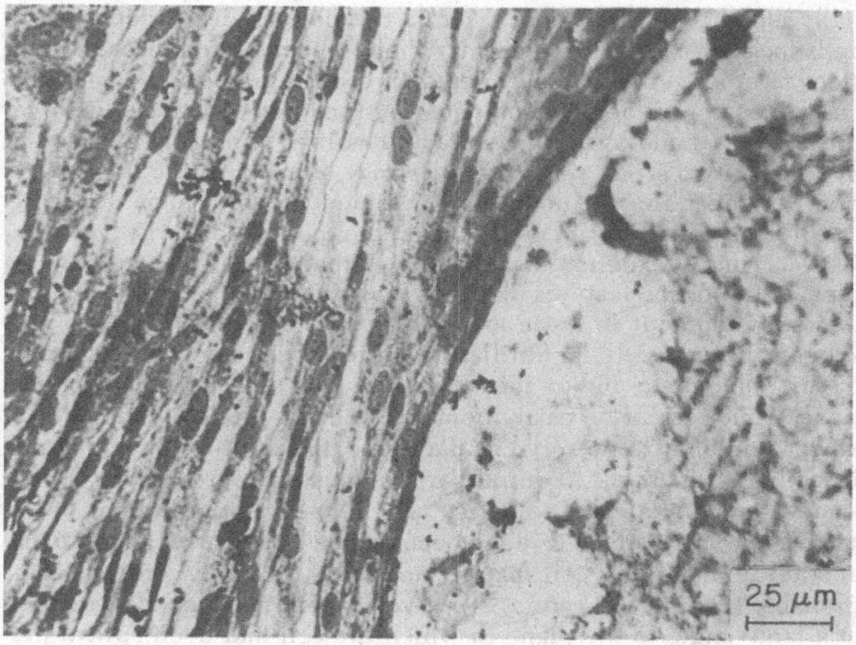

FIG. 20. Orientation of normal mouse fibroblasts near the border of a lipid film. Hematoxylin-stained culture. Courtesy of L. B. Margolis.

best studied example of orientation induced by a geometric surface structure. The orienting effect of grooves of various types has been observed by several investigators (P. Weiss, 1961; Curtis and Varde, 1964; Rovensky *et al.*, 1971). Orientation may be induced even by small scratches on glass with a width and depth of the order of a few micrometers. Another type of orienting structure is a large groove with a width and depth of the order of several dozens of micrometers. What is the mechanism of cell orientation by grooves? The bottom of a large groove was shown to be unfavorable for cell spreading (Rovensky *et al.*, 1971; Rovensky and Slavnaja, 1974). By 15–30 minutes after seeding most spherical cells are located at the bottom of the grooves (Fig. 21). Later, simultaneously with spreading, the cells begin to migrate on the sides, and 2–3 hours after seeding the bottoms of the grooves are almost completely cleared of cells. This migration was observed in experiments with fibroblasts of various species (Slavnaja and Rovensky, 1975). Normal hamster fibroblasts migrated poorly from the grooves; the reason for this atypical behavior is not clear. One may suggest that cell orientation in the areas between the grooves (Figs. 21 and 22) is a result of the avoidance of spreading at the bottom of these grooves; active cell edges are spread preferentially along the sides but not across the bottom. The orientation of cells located near the grooves may then be transmitted through collisions to other cells. In experiments with small grooves, in contrast to large ones, it is difficult to compare the spreading of active edges over these grooves and over a flat surface. Possibly, spreading is different over these two areas; this difference may be responsible for orientation. However, there is no direct experimental evidence on this topic.

Why do cells spread poorly over the bottom of large grooves? Various control experiments have shown that this effect cannot be explained by special features of the surface microrelief or of the humoral microenvironment in these areas (Rovensky *et al.*, 1971). One may suggest that intracellular fibrillar structures (the system of the bundles of microfilaments and/or microtubules) have a certain mechanical rigidity and cannot be easily bent. Therefore a well-spread cell may have a tendency to preserve its flat shape, that is, the shape in which its active edges and its central part are located approximately in the same plane. When the cell tries to spread over the bottom of the groove, various parts of its edge become positioned at an angle to one another (Fig. 24); this cell shape is not stable. In other words, it seems probable that the cell cannot effectively attach pseudopods extended at a large angle to the plane in which the cell had been spread previously.

One should note that besides cell orientation considerable elonga-

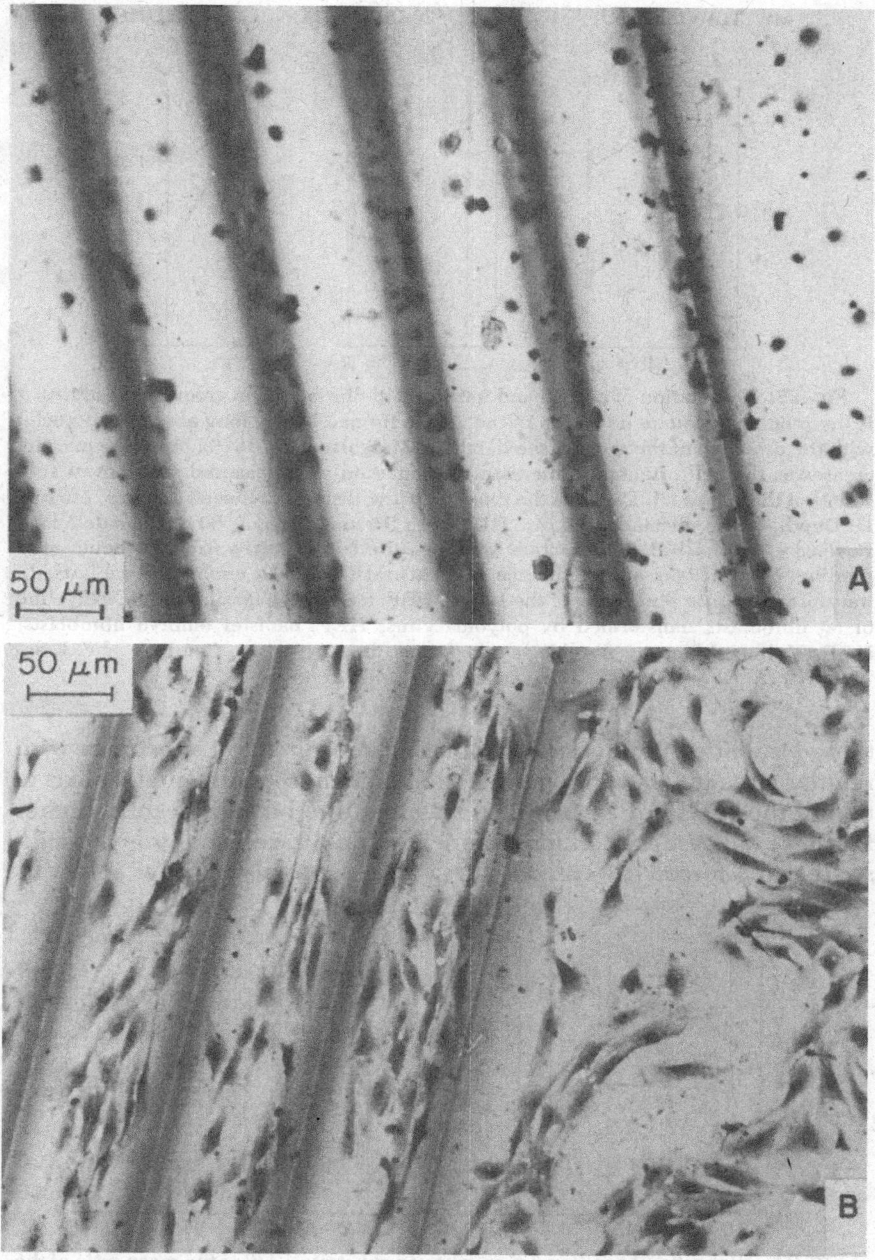

FIG. 21. Behavior of normal mouse fibroblasts on a plastic substrate with large grooves. (A) Thirty minutes after seeding. Most cells are attached near the bottom of the groove. (B) Twenty-four hours after seeding. Most cells have migrated from the grooves and are located on the cylindrical prominences between the grooves. The flat part of the substrate without grooves is in the right part of photograph. Hematoxylin-stained cultures. Courtesy of J. A. Rovensky and I. L. Slavnaja.

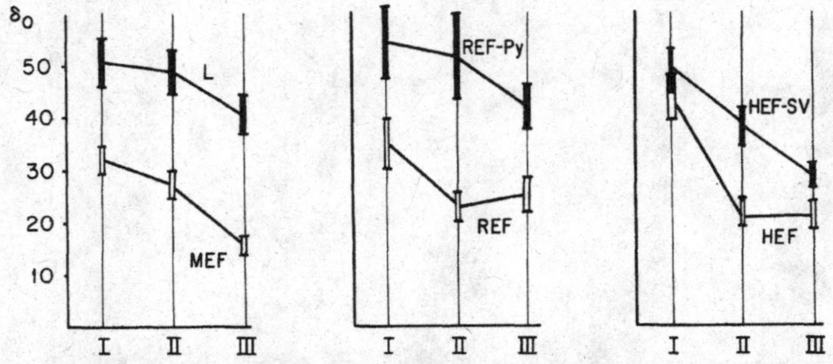

FIG. 22. Orientation of normal and transformed fibroblasts on grooved substrates. σ is the root mean square deviation of the angles formed by the long axes of cell nuclei with the direction of the groove (see details in Margolis *et al.*, 1975); the value of σ decreases as the cell orientation increases. Orientation was measured on grooved substrates of three types. I: Depth of the grooves, 5 μm; distance between grooves, 115 μm. II: Depth, 30 μm; distance, 120 μm. III: Depth, 15 μm; distance, 80 μm (see details in Rovensky *et al.*, 1971). Orientation was measured in cultures fixed 24 hours after seeding. Vertical bars are confidence intervals. MEF, Mouse embryo fibroblasts; L, transformed mouse fibroblasts of the L line; REF, rat embryo fibroblasts; REF-Py, line of rat fibroblasts transformed by polyoma virus; HEF, hamster embryo fibroblasts; HEF-SV, line of hamster fibroblasts transformed by SV40 virus. Courtesy of V. I. Samoilov, J. A. Rovensky, I. L. Slavnaja, and M. S. Slovachevsky.

tion of the nuclei of normal fibroblasts is observed on substrates with regular large grooves; the ratio of the length of the long nuclear axis to that of the short one (*A/B* index) was found to be significantly higher for these substrates than for flat ones (Fig. 23). Cells on grooved substrates are forced to spread along the relatively narrow intervals

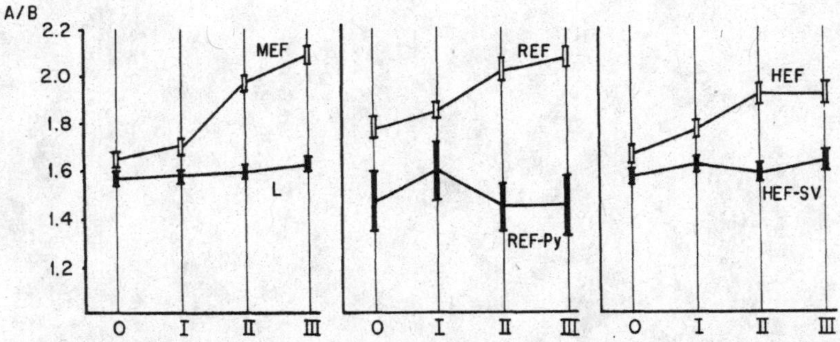

FIG. 23. Elongation of cell nuclei on grooved substrates. *A/B* is the mean ratio of the length of the long nuclear axis to that of the short nuclear axis. 0, Flat area of the substrate without grooves. All other designations are the same as in Fig. 22. Courtesy of V. I. Samoilov, J. A. Rovensky, and I. L. Slavnaja.

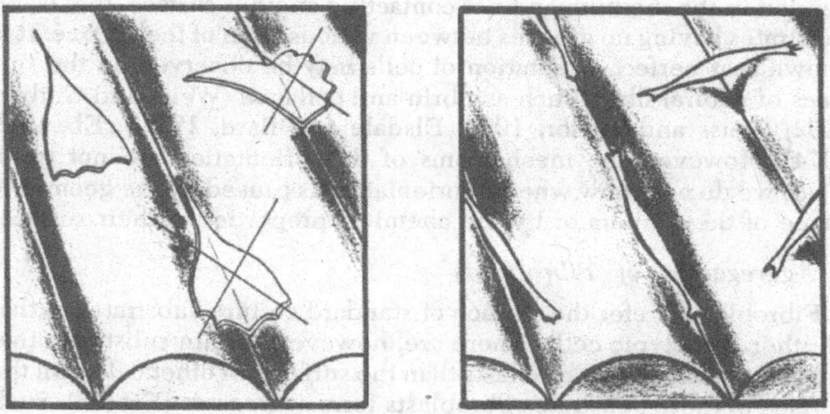

FIG. 24. Schemes illustrating behavior of normal (left) and transformed (right) fibroblasts on grooved substrates. Unstable positions of cells are crossed. See text for explanations. Drawing by A. D. Bershadsky.

between the grooves. Therefore the cells usually acquire a bipolar shape on the groove; to be accommodated on the groove their leading edges have to be more narrow than on the flat substrate. As a result, the ratio of the tension stretching the cell along the long axis of the nucleus to that stretching it in the perpendicular direction will be higher than on the plane.

c. *Unsolved Problems.* Many problems arising with regard to phenomenology and mechanisms of contact guidance remain unsolved. For instance, we do not know whether or not cell orientation can be induced by a flat surface consisting of anisotropic molecules oriented in the plane of the substrate. Very little is known about the effects of the geometrical shape of the substrate. Orienting effects of the grooves were described above. However, the dependence of orientation on the parameters of these grooves (depth, angle between the slopes) has not been systematically investigated. Exciting data were obtained by Rosenberg (1963) who observed an orienting effect of very small heterogeneities of the substrate: troughs with depths as small as 6 nm. However, these investigations were not continued. It is not clear whether or not convex and concave cylindrical surfaces are able to induce cell orientation. Contact guidance was observed in all the experiments in which fibroblasts were seeded on cylindrical surfaces. However, all these surfaces also contained grooves which were certainly able to induce orientation. This question will be answered only when cell orientation is tested on the surfaces of cylinders suspended in the medium and not contacting another surface, that is, on substrates having no grooves between various parts of the surface. It is known that perfect orientation of cells may be observed on the surfaces of natural fibers such as fibrin and collagen (Weiss and Garber, 1952; Weiss and Taylor, 1956; Elsdale and Bard, 1972b; Ebendal, 1974). However, the mechanisms of this orientation are not quite clear; we do not know whether orientation is caused by the geometric shape of these fibers or by the chemical properties of their surface.

6. *Aggregation of Fibroblasts*

Fibroblasts prefer the surface of standard culture substrates to that of other homotypic cells. There are, however, certain substrates that are less preferred by fibroblasts than the surfaces of other cells. On the surface of these substrates fibroblasts form aggregates (Fig. 25). Such aggregation has been observed on the surface of Millipore filters, as well as on the glutaraldehyde-fixed surface of dense cultures of fibroblasts (Friedenstein *et al.*, 1967; Ambrose and Ellison, 1968; Bershadsky and Guelstein, 1973). Quantitative methods for the assessment of aggregation have been developed (Bershadsky and Guelstein, 1973). Fibroblasts in suspension are also able to form aggregates (Waddell *et al.*, 1974). When normal mouse or hamster fibroblasts are seeded on glutaraldehyde-fixed monolayers, aggregation starts almost immediately and continues for several days until most cells are trapped in large (up to 1 mm in diameter) spherical or oval aggregates. Examination of sections of these aggregates has shown that they consist of several cellular layers, the cells of an external layer being spread on the surface of more internal fibroblasts. The formation of aggregates seems to be a result of active and organized cell movements. This is indicated by the results of experiments showing that cytochalasin B and Colcemid inhibit their formation (Waddell *et al.*, 1974; Bershadsky and Guelstein, 1976).

7. *Conclusion: The Choice of Substrate by Fibroblasts*

In previous sections we have described various types of alterations in shape and in locomotory behavior of normal fibroblasts in culture. As seen from these descriptions, fibroblasts are able to solve different and often rather complex problems by using various combinations of only a few basic morphogenetic reactions. In particular, one should stress the ability of fibroblasts to discriminate between different substrates. The surface of other fibroblasts may be regarded in this context as one of the substrates available for spreading. We have discussed above several examples of substrate choices made by fibro-

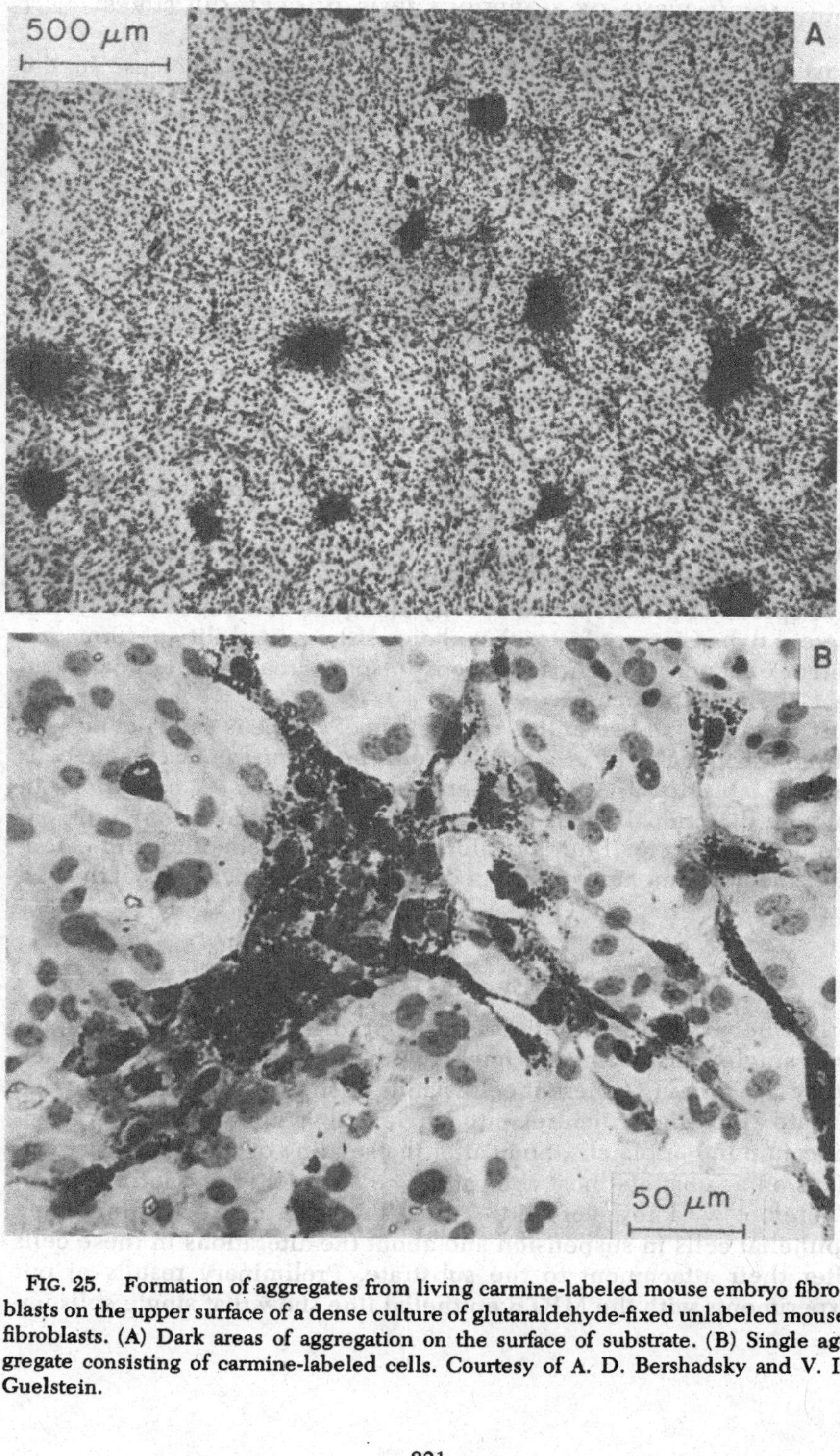

FIG. 25. Formation of aggregates from living carmine-labeled mouse embryo fibroblasts on the upper surface of a dense culture of glutaraldehyde-fixed unlabeled mouse fibroblasts. (A) Dark areas of aggregation on the surface of substrate. (B) Single aggregate consisting of carmine-labeled cells. Courtesy of A. D. Bershadsky and V. I. Guelstein.

blasts. Among these examples were two opposite choices: In the margin of a wound cells migrated from the surface of other fibroblasts on a glass surface, while the same cells spread over the surface of each other in dense cultures when the area of the glass became too small to accommodate all the fibroblasts. Comparison of these two situations shows that the results of a choice between the two substrates depend not only on the nature of the substrates but also on their relative area. Other examples of substrate choice discussed above include migration of cells from the sides of grooves, as well as cell aggregation, that is, movement from the surface of poorly adhesive substrates to that of other cells. Several other examples of substrate choices made by fibroblasts can be found in the report by Harris (1973a). The choice of substrate as well as other variants of directional locomotion are based on differential spreading of various parts of active cell edges. Small initial differences in the efficiency of active attachment reactions occurring in various parts of the edge may be increased as a result of their chainlike nature. Differences between various edges are stabilized by antitubulin-sensitive reactions which also prevent an excessive spread of pseudopodial activity along the cell edge. This system is very dynamic; any factor that changes the probability of formation and/or of attachment of pseudopods in some area of the edge may affect the substrate choice.

It seems probable that this mechanism of choice is used not only by fibroblasts but also by other cells *in vivo* and *in vitro*. As a result, cells may be able to discriminate between different local conditions in moving directionally and in selecting an optimal microenvironment on the basis of small initial differences in the probabilities of extension and attachment of pseudopods formed on various parts of the cell surface.

B. Epithelial Cells

The behavior of epithelium can be studied either in experiments with single cells or in experiments with cells firmly attached to each other and forming coherent cell islands or sheets. It is difficult to dissociate epithelial structures into single cells without damaging a significant number of cells. Soon after the seeding of single-cell suspensions on a substrate most cells attach to each other and form islands. Therefore we know very little about the surface structure of single epithelial cells in suspension and about the alterations in these cells after their attachment to the substrate. Preliminary results of our experiments with the MPTR epithelial line show that single cells un-

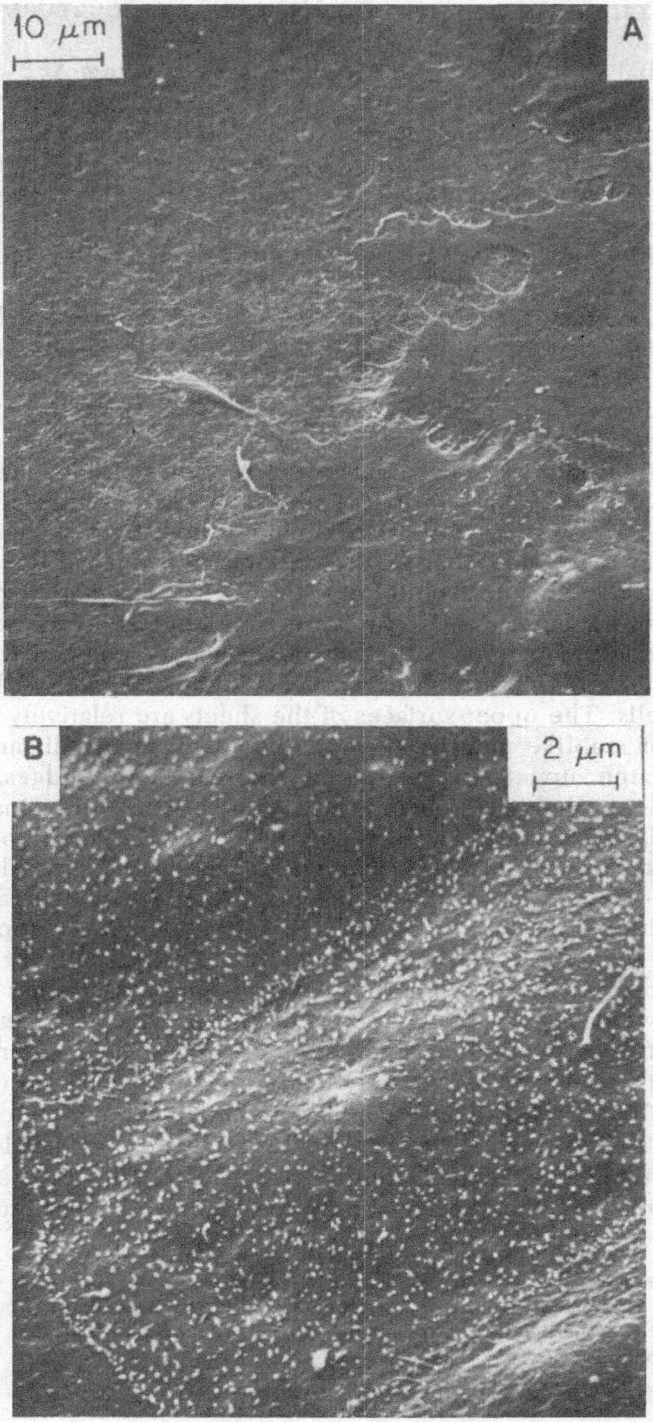
10 μm
A
B
2 μm

dergo radial spreading during 2–4 hours after seeding. At 5–6 hours they become polarized; however, considerable elongation of these cells is not observed. Polarization of single cells can be inhibited and reversed by Colcemid. One cannot exclude that isolated life outside a coherent cell group is abnormal for epithelial cells. This suggestion is in agreement with the results of Di Pasquale (1975a), who observed vigorous blebbing on the surface of single epithelial cells accidentally detached from a sheet.

The locomotory behavior of epithelial cells in groups is much better known than that of single cells (Middleton, 1973; Di Pasquale, 1975a,b; Vasiliev *et al.*, 1975a,c). Sheets of certain epithelia, for example, those of the epidermis and of corneal epithelium are bilayered; their upper layer consists of squamous cells and does not participate in locomotion. The sheets of many other epithelia are monolayered. The main characteristics of these sheets as observed in our experiments as well as in those of the other investigators mentioned above, may be described as follows. Contacting edges of the cells in the sheets are firmly attached to each other; intercellular gaps are rarely seen not only in the light microscope but also in the scanning electron microscope (Fig. 26). Belts of specialized contact structures (see Section III,C,3) are observed near the apical parts of the contacting cells. The upper surfaces of the sheets are relatively smooth except for stublike microvilli. Parallel rows of microvilli are often seen on the upper surface along the contacting edges.

Special morphological features are characteristic of marginal cells, that is, of cells that have a free edge. Pseudopods (lamellipodia and microspikes) are continuously formed at these free edges. Flattened and well-attached areas of lamellar cytoplasm are seen near these edges. The morphology of the surface of this lamellar cytoplasm is similar to that of fibroblasts. The structure of its cortical layer has not been studied in detail. Numerous cell-substrate contacts are located on the lower surface of lamellar cytoplasm near the free edges. A few local contacts may also be found on the lower surface of submarginal cells, that is, cells located directly behind the marginal ones (Di Pasquale, 1975a). Probably only the marginal and submarginal cells are attached to the substrate. However, this conclusion needs additional experimental verification. Quantitative characteristics of the movements of ruffles and of the extension and withdrawal of pseudopods at

---

FIG. 26. Upper surface of a sheet of normal kidney epithelial cells. (A) Free edge of marginal cell. Lamellipodia and microspikes. (B) Central cells of the sheet. Courtesy of O. S. Zacharova.

the active edges of marginal epithelial cells were found to be rather similar to those of fibroblasts (Di Pasquale, 1975a). The forward movement of free edges on the substrate is due to the greater duration of extensions as compared with withdrawals. Besides the free edges of marginal cells pseudopods may occasionally form on the lower surface of submarginal cells (Di Pasquale, 1975a). All the other central cells of the sheets do not form surface extensions larger than microvilli.

When the cell sheet moves directionally on the substrate, for instance, after wounding, many rows of central cells follow actively translocating marginal cells; as a result, considerable areas of the sheet are stretched. It is not clear how these central cells move. Microcinematographic observations show that this movement is accompanied neither by the formation of a significant number of intercellular gaps nor by the extension of pseudopods by central cells. One has to suppose that central cells passively follow the marginal cells. It is not clear which structures (matrix cortical layer? bundles of microfilaments?) create the pulling force exerted by marginal cells and transmit it from cell to cell in the sheet. When the marginal cells of two islands touch each other, further extension of pseudopods immediately stops in the contact zone. This contact paralysis is accompanied by retraction, pulling both cells toward each other and leading to the disappearance of areas of the lamellar cytoplasm.

As discussed above (Section II,C,1), the upper surface of the central cells of epithelial sheets adheres neither to other cells nor to particles (see discussion on the generality of this conclusion in Prop, 1975; Elsdale and Bard, 1975; Di Pasquale and Bell, 1975). In contrast, lamellar cytoplasm at the free edge of marginal cells is highly adhesive. In other words, pseudopodial contact of the edges of two marginal cells leads to stable differentiation of the cell surface into two areas: areas of specialized contact structures and nonspecialized areas of the upper surface. It is not clear whether the lower surface of the central cells of the sheet is adhesive or not.

Colcemid does not affect locomotion of epithelial cells (Vasiliev *et al.*, 1975a; Di Pasquale, 1975b). In particular, it does not affect directional translocation of marginal cells into a wound and does not activate movement of the stable edges of central cells. This absence of effects is not due to insensitivity of epithelial cells to Colcemid; this inhibitor effectively blocks mitotic epithelial cells in metaphase.

Reactions of epithelial cells to various substrate structures has not been adequately studied. When a moving epithelial sheet contacts the boundary of a lipid film, its further translocation is stopped. However, extension of pseudopods along the margin of the sheet contacting the film may continue for several days; stabilization of this margin, as well as orientation of the nuclei of marginal cells, are not observed. Each cell of the sheet is stretched by the tension in neighboring cells. Therefore alteration in the distribution of tension that can be induced by the substrate in each cell is minimal. In this connection it is important to study experimentally two problems: (1) To determine how the shape of an epithelial island is affected by the substrate structure. Possibly, although the shape of each individual cell may be minimally changed by these structures, the shape of the whole island may become elongated and acquire a definite orientation. (2) To test the ability of single isolated epithelial cells to become oriented on various substrates.

C. Comparison of Morphogenetic Reactions of Fibroblasts and of Epithelial Cells

The main behavioral features of epithelial cells and fibroblasts are summarized in Table I. Several conclusions about the differences in basic morphogenetic reactions of the cells of these two types can be drawn from these data. Probably the main difference is in reactions of active attachment. This difference becomes obvious when consequences of the contacts of active edges of two homologous cells are

TABLE I
Comparative Characteristics of Epithelial and Fibroblastic Cultures

Epithelium	Fibroblasts
Morphology of a dense culture	
Monolayer sheet (sometimes bilayered)	Multilayered structure
Cells are not elongated	Cells are elongated and mutually oriented
Migration into a wound	
Cell-cell contacts are not broken during migration	Contacts are easily broken
Migration is not sensitive to antitubulins	Migration is sensitive to antitubulins
Adhesiveness of the upper surface of a dense culture	
Labeled homologous and heterologous cells are not attached to the surface	Labeled cells readily attach to the surface
Inert particles are attached only to the surface of marginal cells of the sheet	Inert particles are attached to the surface of all the cells of the upper layer

compared. Under these conditions active attachment reactions of epithelial cells lead to formation of much more complex, widespread, firm contact structures than those of fibroblasts. This different stability of cell-cell contacts probably determines all the main differences in morphology between epithelial and fibroblastic cultures. In epithelial cultures the stability of the contact structures leads to permanent contact paralysis of movements of contacting cellular edges. This contact inhibition of pseudopodial activity is in turn responsible for the poor adhesiveness of the upper surface of the epithelial sheet.

In fibroblastic cultures cell-cell contacts are less stable. It is possible that the greater adhesiveness of the upper surface of dense cultures is a consequence of the poor stability of these contacts. In turn, this adhesiveness may be a prerequisite for the formation of multilayered structures. In fibroblastic cultures the weakness of cell-cell contacts leads to the separation of cells during their migration into a wound. An antitubulin-sensitive system is essential for maintenance of the direction of translocation by these individually moving cells. This system is also essential for maintenance of the elongated shape of fibroblastic cells. In contrast, in epithelial cultures moving cells are not detached from their neighbors during locomotion. Here the directional character of the cell translocation is simply a result of differences in the pseudopodial activity at the free edges of marginal cells and at their other edges locked by contacts. These considerations show why an antitubulin-sensitive system is not essential for the movement of epithelial sheets. Possibly, this system plays a greater role in the behavior of single epithelial cells, but this question has not been adequately studied. One should also note that all the experiments performed thus far in cultures examined the effects of antitubulins only on cellular movements that take place in the plane parallel to that of the substrate; the extension of pseudopods is an example of movement of this type. However, in highly organized epithelial tissues *in vivo* antitubulin-sensitive structures were shown to play a part in the organization of dorsoventral intracellular movements, that is, in the organization of movement between the apical and basal parts of epithelial cells. This role is shown, in particular, by numerous experiments revealing the inhibition of various types of secretion in epithelial tissues *in vivo* by antitubulins (see review in Allison, 1973). Dorsoventral transcellular transport of fluids has also been observed in certain types of epithelial cultures (Leighton *et al.*, 1969, 1970; McGrath, 1971; McGrath *et al.*, 1972; Pickett *et al.*, 1975). The effects of antitubulins on these transcellular movements in cultured epithelial cells have not been investigated.

IV. Alterations in the Morphogenetic Reactions Accompanying Cell Transformation

A. Introduction

1. *The Concept of Morphological Transformation*

The action of various oncogenic agents, such as viruses and chemical carinogens, on fibroblastic and epithelial cultures leads to the appearance of cells with relatively stable alterations in morphology. In some cases the spontaneous appearance of such cells in cultures may be observed during long-term cultivation. Morphological alterations are transmitted to the progeny of these cells. These genetically stable alterations in cell morphology are usually designated morphological transformations. Morphological alterations accompanying transformation are manifold. The transformation of cells of one type by various agents may lead to somewhat different alterations. Even after the action of a particular agent on a particular culture, the degree and character of the alterations may vary from one clone to another. The same clone may undergo several consecutive transformations. Obviously, the use of the term morphological transformation to designate all these manifold and multistep alterations is justified only if we assume that they all have something in common, that they all affect the same morphogenetic processes. The possible nature of these alterations has rarely been discussed at length in recent literature. Various investigators, ourselves included (Barker and Sanford, 1970; Sanford *et al.*, 1970; Domnina *et al.*, 1972; Vasiliev and Gelfand, 1973), have suggested that a deficiency of cell-cell and cell-substrate attachments is a general morphological feature of transformed cells. Each transformation is accompanied by alterations in these processes. The degree of deficiency increases with each consecutive transformation. Of course, besides defective attachment, each particular line may also have additional secondary alterations.

The term transformed cells is used in this article in a broad sense, this is, for designation of all cells with some degree of morphological difference from their normal progenitors. According to this usage, nontransformed cells are cells of primary normal cultures. Among continuous cell lines possibly only those of diploid human fibroblasts have a nontransformed morphology identical to that of primary cultures. However, this question needs reexamination.

It is useful to distinguish cultures with minimal and advanced manifestations of transformation. Primary transformation may be defined as a transition of nontransformed cells into cells with minimal or advanced transformation. Secondary transformation is a transition of minimally transformed cells into cells with advanced transformation. The reverse transition from more transformed to less transformed cells is usually designated reversion.

Certain special variants of transformed cells are particularly useful in many investigations. These are, first, temperature-sensitive transformants, that is, cells in which the expression of transformation can be considerably and reversibly altered by temperature shifts. Also of interest are variants of transformed cells adapted to growth in serumless medium. In experiments with one of these variant lines it was shown that the degree of expression of its transformed phenotype is dependent on the presence of serum; it is considerably greater in serum-containing medium than in serum-free medium (Gelfand, 1974).

Morphological transformation is often correlated with alterations in several other cell properties, in particular, with various manifestations of deficient growth regulation and with the development of an ability to form tumors after implantation into syngeneic hosts. These alterations, as well as genetic mechanisms of transformation, are beyond the scope of this article and are not discussed here. We only mention that it is natural to try to find correlations between morphological deficiencies of transformed cells *in vitro* and certain characters of neoplastic cells *in vivo,* such as the formation of atypical tissue structures or invasiveness. However, little specific is known about these interrelationships. In particular, the degree of correlation between the oncogenicity of cells *in vivo* and morphological alterations in the same cells *in vitro* needs further detailed analysis.

2. *Diagnosis of Morphological Transformation*

Various investigators use different criteria in the diagnosis of morphological transformation. Often a negative or positive diagnosis is based on the presence or absence of a particular symptom which is not specific for all transformed cells. This is true, in particular, for such widely used criteria as loss of mutual cell orientation and alteration in number of cell layers in dense cultures (see Section IV,C,4,b). Much more meaningful criteria of transformation are provided by examination of the morphology of cells spread on the substrate in sparse cultures, especially by visual or morphometric assessment of the mean projected area of whole cells and of their lamellar cytoplasm, of the size of their active edges, and of the scanning electron microscope morphology of the cell surface. Alterations in these characters reflect more or less directly a deficiency of cell-substrate interactions which, as noted above, is possibly a general feature of transformed cells. However, until we know with certainty which cellular changes are specific to transformation, it would be safer to base the diagnosis of transformation not on a single test but on the application of a system of different tests. We think that ideally this system of tests should include examination of the morphology of single cells in three main morphological states (suspended, radially spread, and polarized), examination of the morphology of dense cultures, examination of the cells' ability to orient and elongate on various substrate structures, as well as examination of their ability to aggregate. Of course, to conclude that a culture is transformed it is not always necessary to make all these tests. Positive diagnosis can be made on the basis of several symptoms. Often, examination of substrate-attached cells in the light microscope may be sufficient for positive diagnosis. However, in order to exclude transformation in some cultures one would require a more rigorous system of tests. This system is also essential when we wish to compare the degree of transformation of several cultures. Finally, it would be useful to apply a system of tests to many lines of cultured cells in order to determine the degree of correlation between different indications of morphological transformation. At present, there is no group of normal and transformed cells that has been subjected to all these tests.

Obscure tissue origin of the cell line may sometimes considerably complicate the diagnosis of transformation. During long-term cultivation the morphology of transformed cells may be altered so profoundly that they begin to resemble those of the other tissue type. For instance, because of a deficiency of cell-cell contacts epithelioid cultures may acquire a morphology somewhat resembling that of fibroblastic cultures. Transformed fibroblasts with deficient attachments to the substrate may acquire polygonal or hemispherical shapes somewhat resembling those of epithelial cells. These profoundly changed lines are sometimes described as fibroblastlike or epithelial; these terms may be misleading, as they may suggest wrong ideas about the tissue origin of these lines. The most difficult problems of tissue origin arise with regard to the widely used BALB/3T3 line. Often it is described as a fibroblastic line; however, morphologically these cells are similar to endothelial cells and possibly developed from endothelium (Porter *et al.*, 1973). We discuss briefly the morphology of these cells in Section IV,D.

In our experiments we used several cell lines obtained by the transformation of mouse and hamster fibroblasts by oncogenic viruses, as well as by long-term cultivation (see description in Vasiliev and Gelfand, 1973; Guelstein *et al.*, 1973; Cherny *et al.*, 1975). We also worked with several lines obtained from cultures of mouse sarcomas induced by implanted plastic films. Another type of transformed cells used in our experiments were primary cultures of mouse embryo fibroblasts transformed by mouse sarcoma virus (see Guelstein *et al.*, 1973). In experiments with transformed epithelial cells we used cultures of several strains of mouse hepatomas originally induced by chemical carcinogens, as well as the MPTR line of transformed mouse kidney epithelium (see description in Vasiliev *et al.*, 1975a).

B. Basic Morphogenetic Reactions of Transformed Cells

1. *Deficiency of Active Attachment Reactions*

Numerous data show that many lines of transformed cells attach poorly to the substrate and to each other:

1. In the polarized state these cells are poorly spread on the substrate. The area occupied by the cells on the substrate, especially the area of lamellar cytoplasm, is decreased. Their bodies are less flat than those of normal cells.
2. The size of their active edges is often decreased, and the morphology of the pseudopods formed at these edges may be altered.
3. The formation of bundles of microfilaments accompanies normal spreading and polarization of fibroblasts. The formation of these bundles becomes deficient in transformed cultures.
4. Transformed cells often have a decreased ability to form specialized cell-cell contacts in dense cultures; transformed fibroblasts in dense cultures spread poorly over the surface of each other.

These data, discussed in more detail in Section IV,C, suggest that morphological transformation is usually accompanied by some degree of deficiency of active attachment reactions. At present we are unable to compare separately the characteristics of each particular stage of active attachment reactions of normal and transformed cells. However, analysis of the available facts suggests that all three stages may be deficient in transformed cells. Deficient extension of pseudopods is observed during spreading and at the active edges. Abnormality in the attachment stage is suggested by the deficient formation of specialized cell-cell and cell-substrate contacts in transformed cultures. Abnormality in the tension developed within the extended pseudopod is suggested by the absence of the bundles of microfilaments.

These considerations suggest that all three stages of active attachment reactions have some common mechanism which becomes deficient during transformation. In particular, it seems probable that this deficiency may be due to alterations in the organization of microfilaments in the cortical layer. Disorganization of the system of microfilaments may affect not only the extension of pseudopods but also their attachment and tension. A decrease in membrane-associated actin was observed in fibroblasts transformed by Rous sarcoma virus (Wickus *et al.,* 1975).

Several other suggestions about the possible mechanism of deficient attachment of transformed cells have been described in the literature:

1. It was shown that certain types of transformed fibroblasts liberate into the medium an activator of serum plasminogen. It was suggested that activated serum plasmin is a proteolytic enzyme responsible for transformed phenotype traits including deficient attachment (Unkeless *et al.,* 1973, 1974; Ossowski *et al.,* 1973, 1974; Pollack *et al.,* 1974). Some investigators express doubt about the universality of this trait of transformed cells and about the exclusive role of serum plasmin in morphological alterations (Mott *et al.,* 1974; Chen and Buchanan, 1975).

It was suggested that, besides activating serum plasminogens, transformed cells may directly secrete proteases into the medium (Chen and Buchanan, 1975). It is not clear whether or not this property is common to all the lines and, if so, whether or not these proteases are involved in the deficient attachment.

Surface proteins of high molecular weight, possibly involved in cell-substrate and cell-cell attachment (see Section III,A,1), were found to be lacking on the surfaces of many types of transformed cells (Hynes, 1973; Gahmberg and Hakomori, 1973; Hogg, 1974; Vaheri and Ruoslahti, 1974). The disappearance of these proteins may be a result of the activity of the above-mentioned proteolytic enzymes. The possible role of these proteins and mechanism of their disappearance are currently the topic of active studies.

2. The addition of a high concentration of cAMP analogs to the medium of transformed cells in several experiments caused alterations in cell morphology and increased the strength of cell-substrate attachments (Hsie and Puck, 1971; Johnson *et al.,* 1971; Sheppard, 1971; Johnson and Pastan, 1972). Therefore it was suggested that morphological abnormalities in the transformed cells are somehow linked with the deficient formation of cAMP in these cells (see review in

Pastan and Johnson, 1974). One should note, however, that analogs of cAMP are not the only agents able to alter temporarily the spreading of transformed cells on the substrate. Other factors of a diverse nature have been reported to "normalize" the morphology of transformed cultures, for example, medium containing galactose instead of glucose (Gahmberg and Hakomori, 1973; Kalckar *et al.*, 1973), dimethyl sulfoxide (Kish *et al.*, 1973), and sodium butyrate (Wright, 1973). Recently a cell line has been described in which the addition of cAMP to the medium did not inhibit transformation but, on the contrary, was essential for its expression (Somers *et al.*, 1975). Detailed morphological analysis of the alterations produced by all these agents has not been made. Therefore the nature and specificity of the effects produced by external cAMP are not clear at present.

The suggestions listed above are not mutually exclusive. Possibly, each of the postulated changes (alterations in the cortical layer, deficient production of cAMP, activation or secretion of proteases) is one of the steps in a chain of events distorting the normal course of active attachment reactions. At present, however, an essential role for any of these changes has not been proved.

2. *Contact Paralysis*

Alterations in mutual cell distribution in transformed cultures are often regarded as a result of complete or partial loss of the contact inhibition of movement. However, direct microcinematographic analysis of cell-cell collisions in several types of transformed cultures (Domnina *et al.*, 1972; Bell, 1972; Vasiliev and Gelfand, 1973; Guelstein *et al.*, 1973) led to another conclusion. In these experiments it was observed that the leading active edge of a transformed cell usually stopped after collision with an active edge of a homologous cell or a normal fibroblast. Thus in head-head collisions these transformed fibroblasts demonstrated the same contact inhibition of movement as their normal progenitors. This contact inhibition was usually accompanied by contact paralysis and retraction. Retraction of the edges of transformed fibroblasts after collision often led to the complete detachment and disappearance of large elongated cytoplasmic processes adjacent to these edges. These frequent detachments were probably a corollary to the poor attachment of the peripheral cell areas to the substrate. Normally formation of cell-cell contacts tends to keep collided cells together; deficient formation of cell-cell contacts may also be a factor leading to more violent retractions. Thus at least some transformed cells retain the ability to undergo contact inhibition of movements. Abnormal morphology in the colonies of these cells is probably not a consequence of the loss of contact inhibition but a secondary result of deficient cell-substrate attachment (see Section IV,C,4). At the same time, microcinematography of certain lines of transformed cells revealed the absence of contact paralysis after collisions with homologous cells and, especially, after collisions with normal cells (Vesely and Weiss, 1973). These descriptions possibly indicate that certain transformed cells, besides defective attachment, may also have impaired ability to undergo contact paralysis after cell-cell contact. Comparative studies of contact inhibition in various transformed lines should be continued. In these studies one should take into consideration that analysis of the results of cell-cell collisions in cultures with deficient substrate attachments involves several problems. As stressed above, collisions of these cells often result in severe retractions of the active edges. Retractions may lead to the disappearance of the areas of contact and to reactivation of the paralyzed edges a short time after collision. Besides this, pseudopods of transformed cells are often more narrow than those of normal cells. At the site of contact of a narrow pseudopod with the surface of another cell, contact paralysis may easily be unnoticed. In sum, because of deficiencies of active attachment reactions, contact paralysis may become more short-lived and restricted to smaller areas of the cell edge. Under these conditions, to find out whether or not the cells have retained the ability to undergo contact paralysis after cell-cell collision, it is essential to analyze the movements of colliding cellular edges at high magnification and at high time resolution.

3. *Reactions of Stabilization*

Stabilization reactions are responsible for coordination of the local reactions of active attachment. A deficiency of those local reactions considerably impedes an evaluation of the state of a stabilizing system in transformed cells. This situation may be compared with that of an investigator who tries to evaluate the ability of the central nervous system to coordinate locomotion in an animal with peripheral muscular dystrophy. Nevertheless, certain data suggest that transformed fibroblasts are able to undergo antitubulin-sensitive reactions. Our unpublished experiments show that all the examined transformed fibroblasts, like normal cells, were polarized after radial spreading on the substrate. Polarized fibroblasts often acquire an elongated shape (Fig. 19) and are able to translocate directionally. Colcemid inhibits and reverses the polarization of transformed cells. Colcemid-treated cells lose their elongated shape, and all their edges become active. At the same time, the morphologies of transformed and normal fibroblasts do not become identical in a Colcemid-containing medium; transformed cells remain less spread on the substrate, and many areas of their periphery are not attached to the substrate. Thus these observations do not reveal the disappearance of the ability to stabilize a polarized state in transformed cells. We do not know, however, the state of another antitubulin-sensitive reaction in transformed cells, namely, the ability to normalize distribution of the sites of formation of pseudopods along the active edge. Nothing is known about the other functions of the antitubulin-sensitive system, for example, about the regulation of movements of intracellular organelles.

The morphology of Colcemid-sensitive structures in transformed cells also requires investigation. In the experiments of Brinkley *et al.* (1975) microtubules were not revealed in the cytoplasm of intermitotic transformed cells of several types by an immunomorphological method. In contrast, microtubules were found in all types of normal cells examined.

It would be hardly correct to conclude from these experiments that transformed fibroblasts do not contain cytoplasmic microtubules. This conclusion would be in contradiction to the above-described effects of Colcemid on transformed cells. It is possible that transformed cells retain the general pattern of an antitubulin-sensitive system but that the number of microtubules in this system is decreased so that they are not revealed by an immunomorphological method. The tension exerted by the attached peripheral parts of the transformed cells is possibly considerably lower than in normal cells. Therefore a much smaller number of microtubules may be sufficient for coordination of this tension.

In summary, there is reason to suggest that deficient reactions of active attachment are characteristic of transformed cells. At least some of the transformed cell lines retain the ability to undergo two other morphogenetic reactions: contact paralysis and stabilization. Possibly, advanced degrees of deficiency of active attachment reactions may lead to certain secondary alterations in the manifestation of contact paralysis and of the structure of an antitubulin-sensitive system. This possibility needs further investigation.

C. SHAPE AND BEHAVIOR OF TRANSFORMED FIBROBLASTS

We now discuss the shape and locomotion of transformed fibroblasts in the spherical, radially spread, and polarized states.

1. *Spherical Cells*

Little is known about the morphology of transformed fibroblasts in the spherical state. Willingham and Pastan (1975) regard numerous surface microvilli as a distinctive character of these cells (see also Ko-

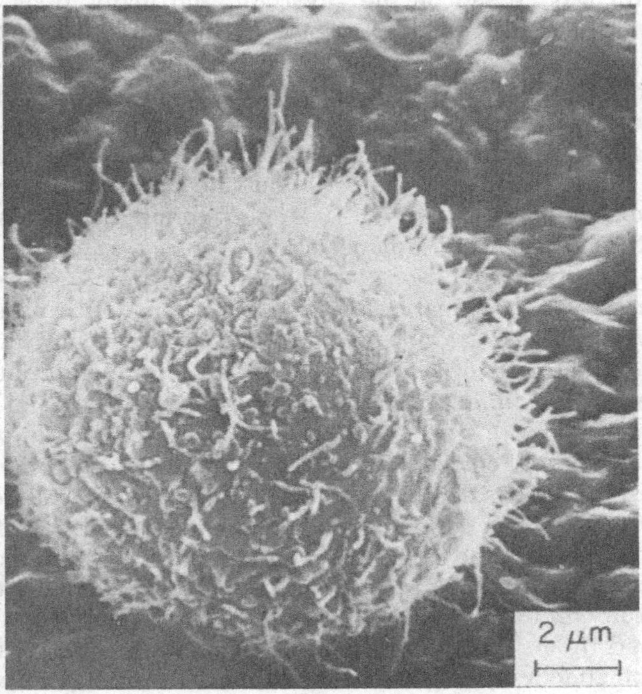

FIG. 27. Transformed cell of the LSF substrain of the L strain; the cell was grown in serum and then removed from the substrate and fixed in suspension. Cell surface is covered with microvilli.

lata, 1975). Our unpublished data indicate that, in contrast to normal mouse fibroblasts (see Section III,A,1), a suspension of transformed L cells consists mostly of cells covered by microvilli (Fig. 27); only a small minority of cells (5%) have surfaces covered with blebs. In our experiments with the LSF subline of L cells adapted to growth in serum-free medium it was found that the percent of cells with surfaces covered with microvilli decreased from about 90% in serum-containing medium to 20–30% in serum-free medium. As mentioned above (Section IV,A,1), this decrease was correlated with an inhibition of the expression of several other transformed traits (Bershadsky *et al.*, 1976).

These data are obviously not sufficient for general conclusions. However, they indicate that alterations in the surface topography of transformed cells in the spherical state deserve further study.

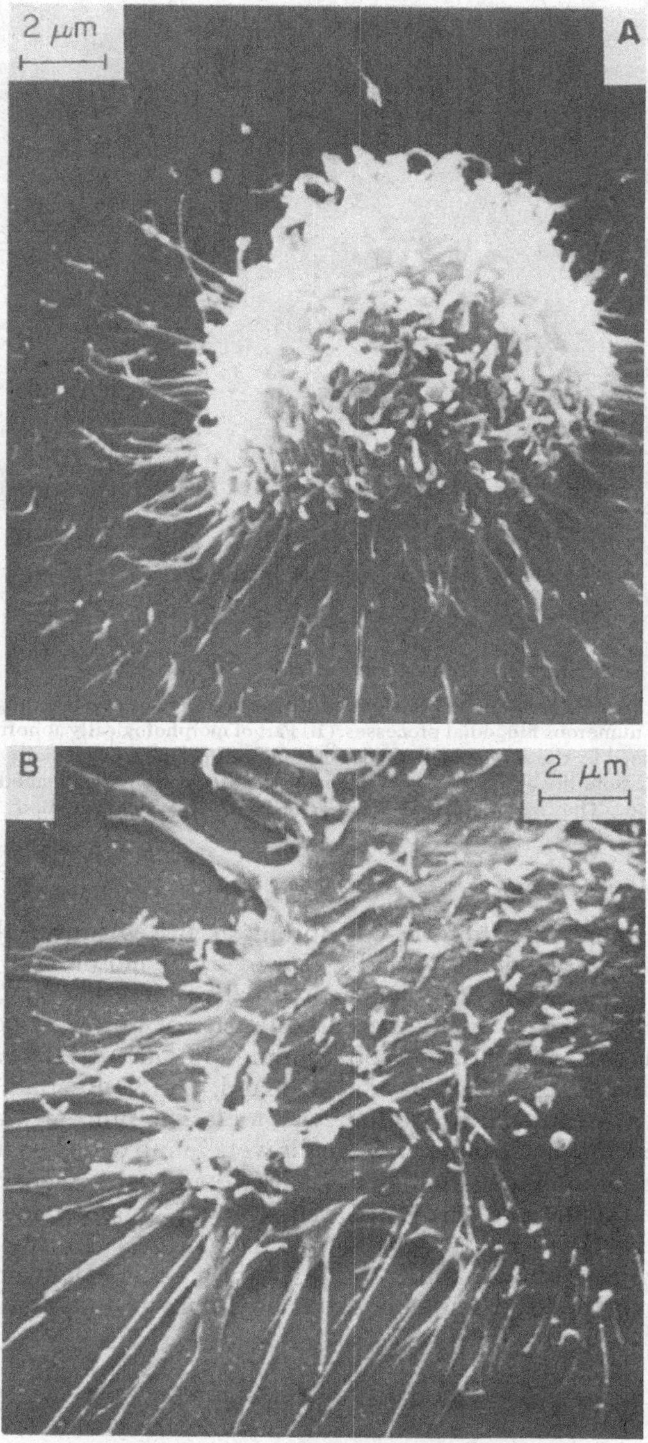

FIG. 28 A and B. See facing page for legend.

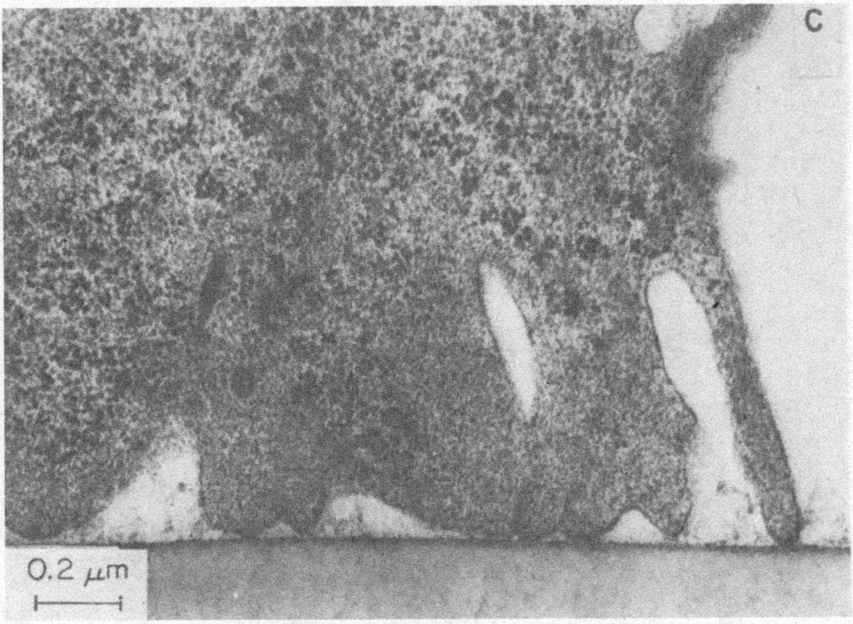

FIG. 28. Radial spreading of transformed L fibroblast 1 hour after seeding. (A) Cell attached by numerous filopodial processes. (B) Part of morphologically abnormal lamelloplasm formed by the spread cell. (C) Structure of cell periphery of radially spreading L fibroblast. Note the absence of microfilament bundles within the attached filopodia on the right. The lower surface of the cell body is not flattened.

2. *Radial Spreading*

Radial spreading of transformed fibroblasts passes through the same three stages as that of normal cells. However, the morphology of cells at each of these stages may have certain abnormalities. We describe briefly alterations in spreading characteristic of fibroblasts of the L strain (Fig. 28), which have been examined in more detail than other transformed lines (Vasiliev and Gelfand, 1973; Bragina, 1975).

At stage I of spreading, when new cytoplasmic processes have not yet formed, the lower surfaces of L cells were found to be less flattened than those of normal fibroblasts. Microvilli and folds were often retained on the lower surfaces of L cells; they were never seen on the lower surfaces of normal cells. These data suggest that transformed cells may have abnormalities in the "diffuse" attachment preceding formation of the first pseudopods. It is important to study these abnormalities in more detail.

At stage II of the spreading of L cells one can see around the cell body pseudopods of the same two types as those seen during normal spreading: cylindrical filopodial processes and lamellipodia. However, in contrast to normal spreading, filopodia around the L cells are not rapidly replaced by lamellipodia, but may persist up to the final stage of spreading. For 1–2 hours many cells do not form lamellar areas at the periphery; these cells are shaped like hemispheres surrounded by numerous filopodia (Fig. 28A). Other cells develop lamellar areas at their periphery but, in contrast to normal cells, have filopodia at the edges of these lamellae (Fig. 28B). Examination of sections of these cells shows that neither filopodia nor lamellipodia contain bundles of microfilaments (Fig. 28C). Development of a regular ring of lamellar cytoplasm is a characteristic feature of stage III of normal spreading. Many L cells do not reach this stage even after 2–3 hours. In other cells a ring of lamellar cytoplasm is formed, but the morphology of this cytoplasm often has various abnormalities; its thickness and diameter vary considerably in different parts of the same cell, and filopodia are often seen on the external edges and microvilli on the upper surface. Bundles of microfilaments are not formed in the lamellar cytoplasm. The spreading of the other transformed fibroblastic lines studied in our experiments was less abnormal, and most cells of these lines reached stage III although their lamellar cytoplasm often had various abnormalities similar to those listed above. Thus the main abnormalities of radial spreading typical of at least some lines of transformed cells include: predominance of filopodial processes over lamellipodia, decreased formation and abnormal morphology of lamellar cytoplasm, absence of bundles of microfilaments within the attached pseudopods and in the lamellar cytoplasm.

3. *Polarized Cells*

The transition of transformed cells from the radially spread to the polarized state has not been studied. More is known about the morphology of cells in the polarized state. We first discuss the general morphology of these cells, then the topography of their upper and lower surfaces and, last, their fibrillar structure.

a. *Cell Shape.* Deficient spreading of polarized cells is manifested by a smaller mean area on the substrate as well as by a smaller percent of this area being occupied by lamellar cytoplasm (Domnina *et al.*, 1972; Vasiliev and Gelfand, 1973). Depending on the degree of development of the lamellar cytoplasm one can distinguish three main classes of transformed cells:

1. Cells without lamellar cytoplasm. They have an elongated and/or hemispherical shape (Figs. 29A and B and 30).

2. Cells with significantly decreased areas of lamellar cytoplasm. These areas are often located at the ends of long cytoplasmic processes; the lateral sides of these processes are unattached to the substrate. Relatively more spread cells often have a jagged external contour; instead of one large anterior lamella these cells may have at their anterior end several discrete lamellar bands. The morphology of each lamellar area often shows abnormalities similar to those described above in Section III,A,2 (Fig. 29C and D).

3. Cells with almost normal areas of lamellar cytoplasm (quasi-normal cells). Even these cells often demonstrate abnormalities in the morphology of lamellar areas.

Microcinematographic observations show that the degree of spreading of one cell may vary considerably with time, so that this cell may pass from one morphological class into another. Interesting periodic transitions of transformed cells into an almost unspread spherical shape were recently described by Paranjpe and Boone (1975).

Of course, there are no clear-cut boundaries between various morphological classes. Cultures of various lines differ from one another in the relative number of cells of various classes, as well as in details of cellular morphology within each class. Typical transformed lines described in the literature and observed in our experiments contain mostly cells without lamellar cytoplasm and/or with decreased areas of lamellar cytoplasm. The proportion of quasi-normal cells in these cultures is small. However, certain mouse fibroblastic lines derived from plastic-induced sarcomas contain a high proportion of quasi-normal cells (Vasiliev and Gelfand, 1973).

b. *The Upper Cell Surface.* Polarized normal fibroblasts usually have a relatively smooth upper surface. In contrast, many types of transformed fibroblasts are characterized by the presence of various extensions such as microvilli, ruffles, blebs, and so on (Hodges, 1970; Hodges and Muir, 1972; Boyde *et al.*, 1972; Domnina *et al.*, 1972; Vesely and Boyde, 1973; Porter and Fonte, 1973; Perecko *et al.*, 1973; Hale *et al.*, 1975). The shape of these extensions may vary in various parts of the surface of one cell and from cell to cell within the same culture. Often, but not always, the number of extensions on the upper surface seems to be correlated with a deficiency of spreading; extensions are characteristic, especially of cells of the first and second morphological class described above. At the same time topography of the upper surface has distinctive features in each particular cell line.

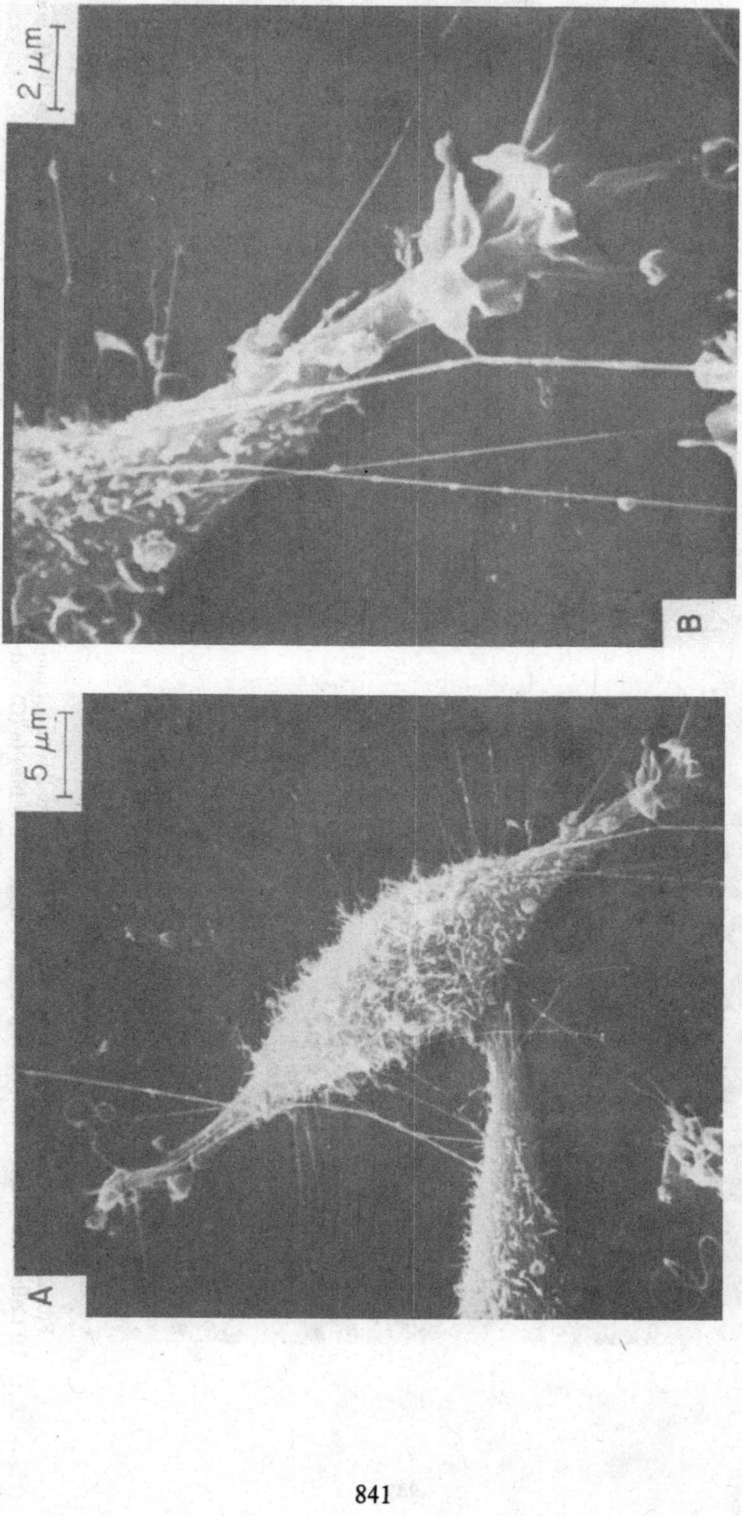

FIG. 29 A and B.

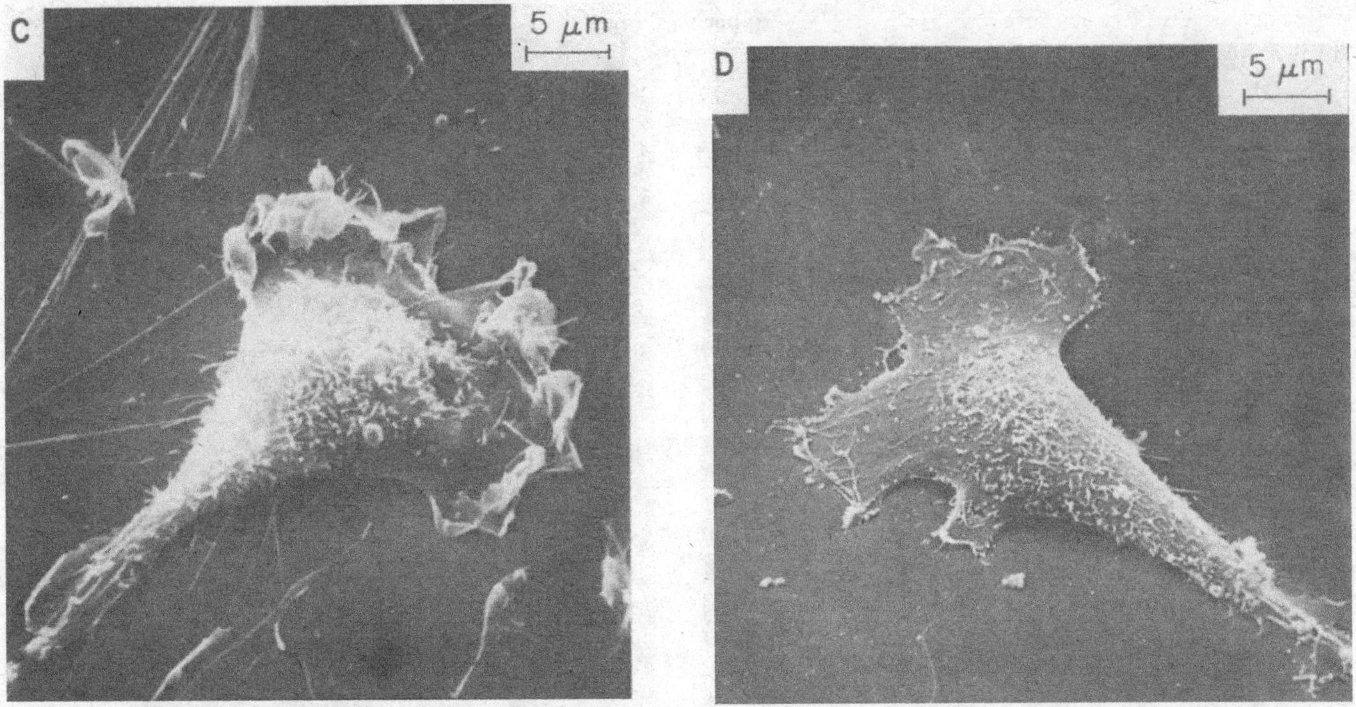

FIG. 29. Polarized L cells. (A and B) Spindle-shaped cell without lamellar areas. Note the ruffles near the attached cell pole in (B). (C and D) Cells with anterior lamella of decreased area. Note ruffles in (C) and jagged contour of the lamella in (D).

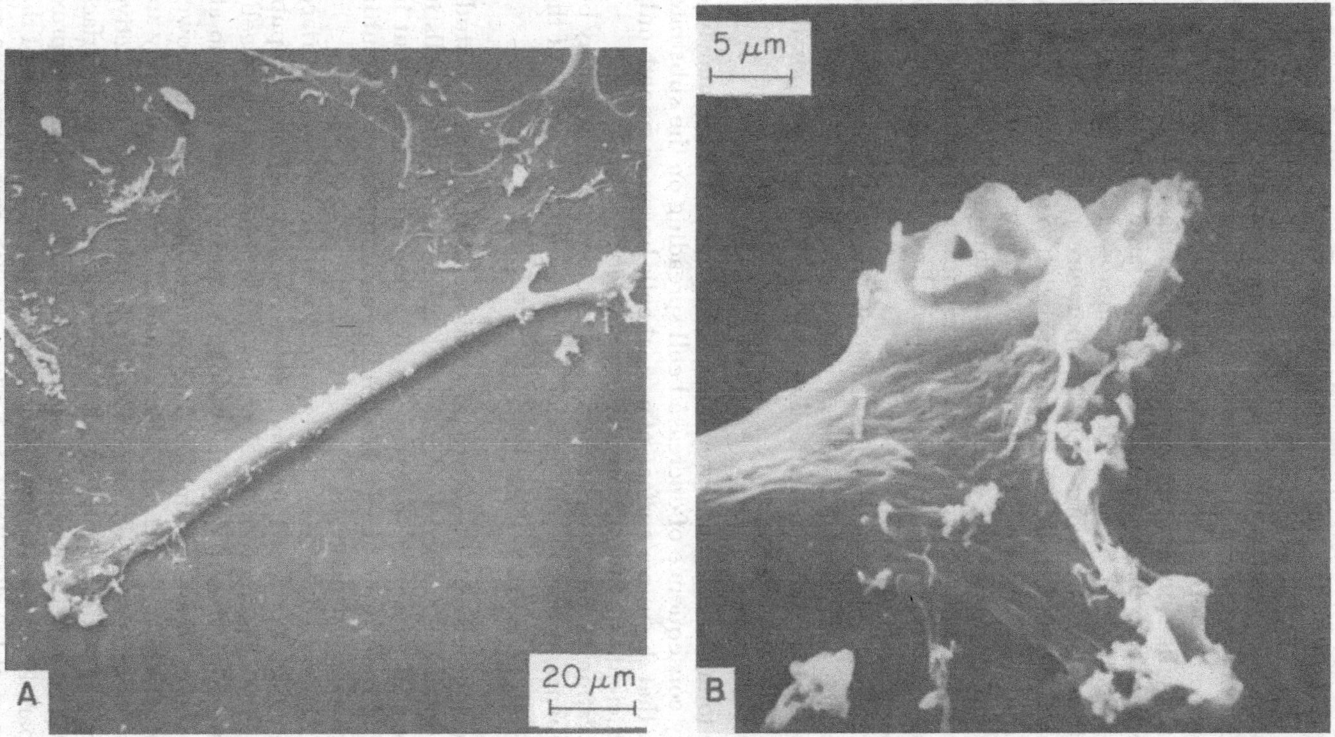

FIG. 30. Elongated mouse fibroblast in a culture transformed by mouse sarcoma virus. Lamellar areas are almost absent. Note ruffles near the attached cell pole in (B). Courtesy of V. I. Guelstein.

For instance, our observations show that the presence of numerous microvilli is characteristic of certain transformed lines, for example, L cells. However, poorly spread transformed cells of other types do not have a microvillous upper surface. The upper surface of certain types of transformed cells has been reported to have large ruffles (Ambros *et al.*, 1975; Pontén, 1975). According to Ambros *et al.* (1975), when chicken fibroblasts transformed by temperature-sensitive Rous sarcoma virus are transferred from a nonpermissive to a permissive temperature, the development of large ruffles on the upper surface is the earliest manifestation of the expression of transformed phenotype. Extensions of the upper cell surface of transformed cells perform active movements which can be observed by special microcinematographic techniques (Vesely, 1972).

It seems probable that the presence of microextensions on the upper surface and the deficiency of active attachment reaction are somehow interrelated. Theoretically there are at least two possible types of interrelationships:

1. The formation of microextensions on the upper surface may be a secondary consequence of decreased cell spreading on the substrate. We know that the upper surface of normal cells becomes smooth only after sufficient spreading on the substrate.
2. Activation of the upper surface and deficient attachment may be two independent manifestations of some general abnormality of the organization of the cortical layer.

In choosing between these two possibilities it is important to study in more detail the surface topography of spherical unspread cells in suspension. If it can be shown that this topography is different in normal and transformed cells, it would be an important argument in favor of the second suggestion.

c. *The Lower Cell Surface.* The topography of the lower cell surface has been studied in detail only in L cells (Bragina, 1975 and unpublished results). The morphology of cell-surface contacts in these cells is different from that of normal fibroblasts, these zones have no attached bundles of microfilaments. The number of zones on the lower surface of poorly spread L cells (class 1 and 2) is very small; they are seen only near active cell edges. Zones located behind the active edges are greatly diminished in number or absent. The lower surface of the anterior lamella of normal cells is relatively smooth and approximately parallel to the substrate. In contrast, the lower surface of the anterior part of L cells often forms a kind of arch with numerous microvilli (Fig. 31). Microvilli are absent on the lower surface of the posterior parts of the same cells. These data indicate that abnormal surface activity may be characteristic not only of the upper surface of L cells but also of the lower surface. It is significant that this abnormal activity at the lower surface is polarized. These cells, like normal cells, form surface extensions mainly in their direction of locomotion, although the shape and localization of these extensions are altered. Probably, other lines of transformed cells also have altered morphology of the lower cell surface; however, special studies of this morphology have not yet been made.

d. *Cytoplasmic Fibrillar Structures.* As mentioned above, deficient formation of microfilament bundles is observed in many types of transformed cells. The absence of these bundles in the cytoplasm of substrate-attached transformed fibroblasts was first revealed by electron microscope studies (Ambrose *et al.*, 1970; McNutt *et al.*, 1971, 1972; Domnina *et al.*, 1972) and then confirmed by immunomorphological investigations using antiactin antibodies (Pollack *et al.*, 1975). When better spread revertant lines were obtained from transformed fibroblasts, these revertants, in contrast to the original transformants, had well-developed microfilament bundles in their cytoplasm (McNutt *et al.*, 1972). Suppression of the transformed phenotype obtained by a temperature shift in cells transformed by temperature-sensitive virus mutants was accompanied by the development of microfilament bundles; these bundles disappeared again after the transfer of cells to a permissive temperature (Pollack *et al.*, 1975). These data indicate that the disappearance of microfilament bundles is a significant trait of advanced transformation. Alterations in the matrix-type cortex in transformed cells have not been described. The possible state of microtubules in these cells was discussed above.

4. *Cell-Cell Interactions on Flat Substrates*

a. *Cell-Cell Collisions.* In our experiments (Domnina *et al.*, 1972; Vasiliev and Gelfand, 1973; Guelstein *et al.*, 1973) cell-cell collisions were analyzed microcinematographically in several types of transformed fibroblastic cultures. The efficiency of contact paralysis after head-head collisions was discussed in Section IV,B,2. Head-side collisions in transformed cultures, like those in normal ones, lead either to a halt in an active edge or to the underlapping of one cell by another (Fig. 32). In certain transformed lines the degree of underlapping is significantly greater than that observed in normal cultures. In these transformed cultures the whole body of one cell often passes under an unattached part of another cell. This underlapping may lead to forma-

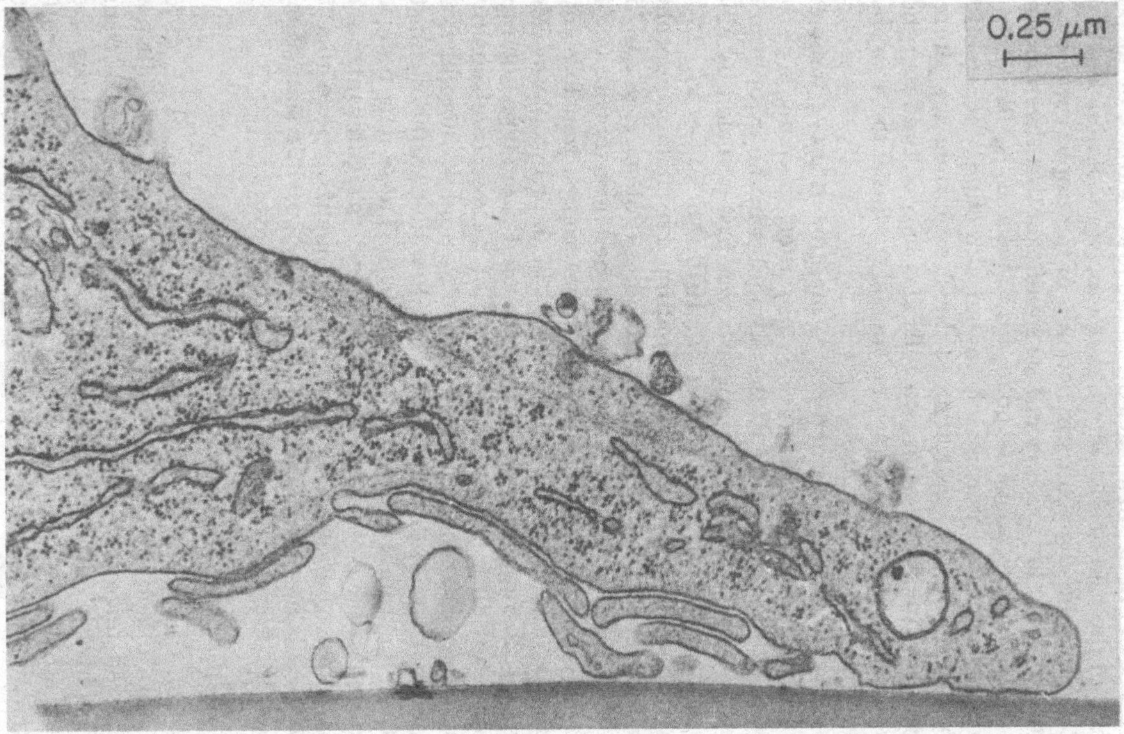

FIG. 31. Anterior lamella of a transformed L fibroblast. Note the absence of microfilament bundles approaching the cell-substrate attachment site and the presence of microvilli on the lower surface. Courtesy of E. E. Bragina.

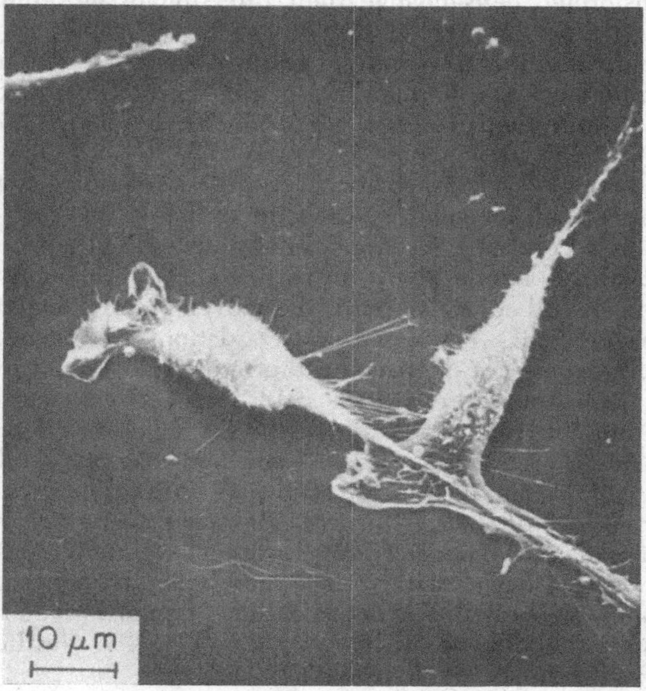

FIG. 32. Anterior lamella of L fibroblast underlapping the unattached part of the rear process of another L cell.

tion of the criss-cross pattern typical of many transformed cultures. Obviously, increased ease of underlapping is due to poor attachment of large parts of cell bodies to the substrate. Head-side and head-head collisions in transformed cultures often lead to retractions which involve larger parts of the cytoplasm than retractions observed in normal cultures. Retraction of upper or lower cells often occurs during underlapping. These retractions may lead to the detachment of large parts of the cytoplasm from the substrate. These intense retractions probably are a result of deficient attachment of the cells to the substrate. Retractions are especially prominent in certain cell lines, for example, in L cells. Microcinematographic observations of these cells suggest that even their active edges are so poorly attached to the substrate that they are unable to underlap other cells efficiently. Comparative studies of locomotion in relatively sparse cultures reveal several differences between transformed cells and their normal prototypes:

1. Transformed cells change their direction of movement more often (Andrianov *et al.,* 1971).

2. In normal cultures the rate of cell movement decreased as cell density increased. At a similar range of cell density the rate of movement of transformed cells did not decrease (Gail, 1973).

We do not know exactly what causes these alterations in statistical parameters of cell locomotion. It seems probable that they may be regarded as secondary consequences of deficient cell attachment to the substrate and to other cells. In particular, in normal cultures formation of cell-cell contacts retards the movement of contacting cells (Abercrombie, 1961). Perhaps these retardations are responsible for the density-dependent decrease in the rate of movement in normal cultures. Transformed cells form stable cell-cell contacts less frequently; therefore increased cell density may have less effect on their motility. Intense retractions of poorly attached cells may lead to more frequent alterations in the direction of movement.

b. *Morphology of Dense Cultures.* There are many variants in the distribution of cells in dense cultures of transformed fibroblasts. Each transformed line has a characteristic morphological pattern. Among the morphological variants of dense cultures of transformed mouse fibroblasts are the following (Cherny *et al.,* 1975, J. M. Vasiliev and I. M. Gelfand, unpublished results):

1. Multilayered or monolayered cultures consisting of poorly spread cylindrical fibroblasts forming streams in which neighboring cells are parallel to each other. In our material this morphology was found to be characteristic of mouse fibroblasts transformed by mouse sarcoma virus (Fig. 33).

2. Multilayered cultures consisting of poorly spread cylindrical, polygonal, and stellate cells crossing under each other at various angles without mutual orientation. In our material this morphology was found to be characteristic of several lines transformed by SV40 lines and derived from mouse sarcomas induced by plastic films.

3. Monolayered cultures consisting of almost spherical cells contacting each other by microvilli. In our material this morphology was found to be characteristic of the L line. A monolayered morphology has also been described by several investigators in certain other transformed lines (Defendi and Lehman, 1966; Diamandopoulos, 1968).

A general feature of different variants of dense cultures of transformed fibroblasts is their deficient cell spreading; a mean projected

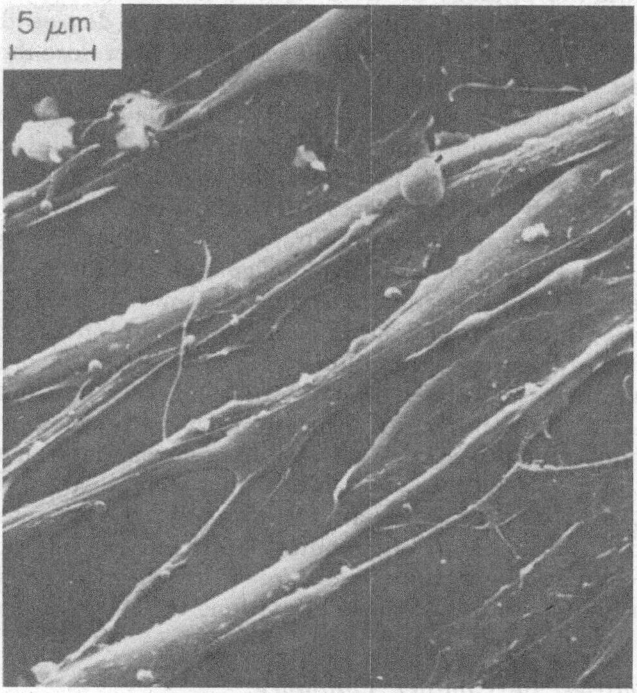

FIG. 33. Mutual orientation of poorly attached cells in a culture of mouse fibroblasts transformed by mouse sarcoma virus.

area of the cells onto the substrate is smaller than that of normal fibroblasts (Cherny *et al.*, 1975). Deficient spreading of cells in dense cultures, like that in sparse ones, is accompanied by deficient formation of lamellar cytoplasm. Another general feature of transformed fibroblasts in dense cultures seems to be deficient formation of specialized cell-cell contacts. These contacts are seen more rarely than in cultures of normal cells. Certain types of specialized contacts may disappear completely (Cornell, 1969; Martinez-Palomo *et al.*, 1969).

These data suggest that morphological abnormalities of transformed cells in dense and in sparse cultures are similar; they are due to the deficient attachment of these cells to the substrate and to each other.

Cells of certain transformed lines may retain the ability to underlap each other (see above). These cells in dense cultures form multilayered structures like normal cells. However, because of deficient spreading, the morphology of these structures is abnormal. In other lines the deficiency of attachments reaches such a degree that the cells are hardly able even to underlap each other. Therefore they form monolayered dense cultures.

The loss of mutual orientation, as well as multilayering, cannot be regarded as a general morphological feature of transformed cultures distinguishing them from normal ones. The absence of mutual orientation is observed in certain cultures; however, other variants of transformed cells, for example, cultures transformed by mouse sarcoma virus, may retain this character. At present we do not know what factors determine the presence or absence of mutual orientation in transformed cultures. Multilayering cannot be regarded as a distinctive character of transformed cultures, as it is observed in normal fibroblastic cultures. Moreover, in some cases, during transformation cell may lose the ability to form multilayered structures.

Before concluding this section we should mention that transformed fibroblasts, like normal ones, are able to migrate into wounds from dense cultures. However, in certain transformed lines the rate of migration and the orientation of migrating cells are considerably decreased as compared with normal cultures (Vasiliev *et al.*, 1969); detailed quantitative studies of these alterations in migration have not yet been made.

5. *Abnormalities in Contact Guidance*

The behavior of transformed fibroblasts on the boundary of a lipid film has been studied only in experiments with one type of cells, those of the L line (unpublished results of O. J. Ivanova and L. M. Margolis). L cells, like normal fibroblasts, oriented themselves and acquired an elongated shape on narrow strips of glass between two boundaries of lipid films (Fig. 19B). In contrast, contact with one boundary of lipid film, which was sufficient for orientation of normal fibroblasts, did not induce orientation in L cells.

The reaction of several lines of transformed cells on substrates with large grooves and cylindrical prominences was examined in experiments performed by our group (Rovensky *et al.*, 1971; Rovensky and Slavnaja, 1974; V. I. Samoilov, Y. A. Rovensky, and I. L. Slavnaja, unpublished results). Behavior of all the examined transformed lines on these substrates was found to differ in several respects from that of their normal prototypes:

1. Decreased migration. Transformed chicken, mouse, rat, and human fibroblasts migrated from the grooves much less efficiently

than their normal counterparts. As mentioned above, normal hamster fibroblasts migrated poorly from the grooves; therefore differences in migration between normal and transformed hamster cells were minimal.

2. Decreased orientation. Degree of orientation of transformed cells on grooved substrates was much less than that of homologous normal cells on the same substrates (Fig. 22). However, in other experiments, orienting structures of smaller size (bundles of polymer threads about 10–20 μm in diameter) were found to induce orientation of L cells (Slavnaja *et al.*, 1974).

3. Absence of elongation of nuclei. The greatest difference between normal and transformed cells observed in these experiments was in the reaction of the shapes of their nuclei on the grooved substrate (Fig. 23). Elongation of the nuclei of all types of normal fibroblasts was considerably greater on grooved substrates than on flat ones. In contrast, none of the examined transformed lines demonstrated any degree of elongation.

What are the mechanisms of these alterations in the reactions of transformed cells to contact with substrate structures? It seems reasonable to assume that these changes are consequences of deficient active attachment reactions. A cell changes the direction of its movement and its orientation after contact of its active leading edge with the orienting line of the substrate, for example, with the boundary of a lipid film or a groove. Because of deficient attachment, active edges of transformed cells are often more narrow than those of normal fibroblasts. This decreases the probability of these edges contacting the orienting lines. As a result, the orientation of cell groups may be decreased or absent on substrates with relatively large distances between orienting lines. However, even the narrow active edges of transformed cells may be sufficient to produce an orientation of these cells on substrates with small distances between orienting structures, for example, on narrow strips between two lipid films or on bundles of thin threads.

When we discussed the behavior of normal cells on grooved substrates (Section III,A,5), we suggested that increased elongation of their nuclei is a result of the narrowing of their active edges and of the increased tension in the direction of the long nuclear axis. Transformed cells have narrow active edges even on flat substrates. These edges may be accommodated on grooved substrates without further narrowing. This circumstance may be a cause of the absence of additional elongation of nuclei on grooved substrates. The tension developing at the attachment sites of transformed cells can be decreased as compared with normal cells. This can be another factor responsible for defective elongation of nuclei.

Poor migration of transformed cells from grooves may be due to the fact that these poorly spread cells are able to attach themselves along the grooves or across them without considerable bending or folding of their bodies (Fig. 24). In addition, decrease of the tension exerted by the attached parts of transformed cells may also impair their ability to migrate on one of the two sides of the groove.

6. *Inability to Form Aggregates*

Many lines of transformed fibroblasts have decreased ability to form aggregates on substrates that induce aggregation of normal fibroblasts, for example, the surface of Millipore filters or the glutaraldehyde-fixed surface of normal fibroblastic cultures (Friedenstein *et al.*, 1967; Ambrose and Ellison, 1968). A quantitative comparison of the aggregation of various normal fibroblasts and transformed fibroblasts on these substrates was made by Bershadsky and Lustig (1974). In these experiments all the examined normal fibroblasts demonstrated effective aggregation during the first 24 hours after seeding. Most examined transformed lines showed various degrees of deficiency of aggregation. Several lines did not show any aggregation at all; the cells remained randomly distributed on the substrate even after 3–6 days of cultivation. The cells of other lines formed aggregates but more slowly than normal cells, and these aggregates were smaller. There was, however, a particular cell line (HEK-40 hamster cells) that formed aggregates as efficiently as normal fibroblasts. We do not know what particular property of this line determined its quasi-normal behavior in aggregation experiments. The morphology of these cells in sparse and dense cultures was typical of transformed fibroblasts. This example shows once again that deficient aggregation, like any other single trait, is not absolutely specific for all transformed cells.

7. *Conclusion*

We have discussed above several morphological and behavioral abnormalities of transformed fibroblasts. The list of the most characteristic traits of these cells includes:

1. Alteration in the area and shape of polarized cells in sparse cultures and, in particular, a decreased area of lamelloplasm.
2. Alteration in cell shape and cell distribution in dense cultures.
3. Deficient elongation and orientation of cells on substrates with

large grooves, as well as poor migration of these cells from the grooves.

4. Deficient aggregation on poorly adhesive substrates.

As we have tried to show, these manifold alterations are possibly consequences of a deficiency of one morphogenetic reaction—active attachment. A general consequence of this deficiency may be described as decreased ability of a cell to choose a substrate. In cases in which normal fibroblasts quickly and efficiently migrate on one of two available substrates, transformed cells choose more slowly and less efficiently. For instance, these cells migrate poorly from the surface of Millipore filters on that of another cell, or, in other experiments, from the bottom of the groove on its side, and so on. The choice of substrate made by a normal cell is probably a result of differences in the efficiencies of active attachment reactions performed by various parts of the active cell edges. Initial differences may be increased as a result of the chainlike nature of active attachment reactions. Attachment reactions of transformed cells, like those of normal cells, may have different efficiencies in various parts of the active edges. In other words, pseudopods formed by these cells probably attach themselves better to one substrate than to another. However, because of the general deficiency in active attachment reactions, these initial differences may not be sufficiently increased by a chain of further attachment reactions. As a result, final differences in the sizes of lamellar zones attached to various substrates, in the total strength of attachment of these zones, and in the tension produced by them may be insufficient for directional cell migration on the preferred substrate. Deficient attachment reactions may also be essential for the behavior of transformed fibroblasts in mixed cultures containing normal cells. In these cultures better attached normal fibroblasts may completely detach transformed cells from the substrate surface; detached transformed cells then must spread on the upper surface of normal fibroblasts. These interactions of normal and transformed cells have been discussed in more detail elsewhere (Vasiliev and Gelfand, 1976a).

Of course, explanations of typical differences between normal and transformed cells, outlined in this and previous sections, should be regarded only as working hypotheses requiring further experimental tests. Further experiments are also needed to explain atypical behavioral traits of certain normal and transformed cells, for example, poor migration of normal hamster fibroblasts from grooves, and the ability of a few transformed lines to form aggregates.

D. Shape and Behavior of Transformed Epithelial Cells

Much less is known about the manifestations of transformation in epithelial cultures as compared with fibroblastic cultures. Light microscope studies often do not reveal significant differences between certain epithelial lines and their normal prototypes. Tumor-producing lines of liver epithelial cells described by Weinstein *et al.* (1975), as well as the kidney line MPTR used in our experiments, belong to this group. These lines form monolayered coherent sheets, indistinguishable from normal epithelial sheets at the level of the light microscope. However, a detailed study of the morphology of the MPTR line with the scanning electron microscope made by Zacharova (1976) revealed several differences between the surface morphology of these cells and that of normal kidney epithelial cells: (1) The cell surface of marginal and submarginal cells of transformed cultures is less flattened. More numerous microvilli are seen on the upper surface of these cells. (2) The lamellar cytoplasm at the free edges of marginal cells in transformed cultures is smaller and less flattened; the contours of the free edges of these cells are less smooth. The shape of cellular processes seen at the free edges is somewhat changed; filopodial processes are often seen here, in contrast to the situation in normal cells, where wide, flat lamellopodia predominate (Fig. 34). These data indicate that MPTR can be regarded as a minimally transformed line with certain alterations in cell-substrate attachment; the cell-cell attachment reactions of these cells remain relatively normal. It is important to study in detail scanning electron microscope morphology of other quasi-normal epithelial transformed lines and to find out whether or not attachment reactions are also slightly different from those of normal epithelium.

Another group of epithelial cultures includes those with more profoundly altered morphology; these are cells that do not form coherent sheets. In our material (Vasiliev *et al.*, 1975a) the cell line derived from anaplastic mouse hepatoma 22a belong to this group. The cultures of this line consisted of pleomorphic cells which only occasionally formed specialized cell-cell contacts. Most cells were poorly spread and had small areas of lamellar cytoplasm (Fig. 35). Deficiency of cell-cell contacts was correlated with the adhesiveness of the upper cell surface in dense cultures; inert particles (Fig. 36) and labeled cells easily attached to this surface. Multilayered groups were often seen in dense cultures. In contrast to normal epithelium, the cells of hepatoma 22a migrated poorly and individually into wounds. In coherent sheets only marginal and submarginal cells seem to be at-

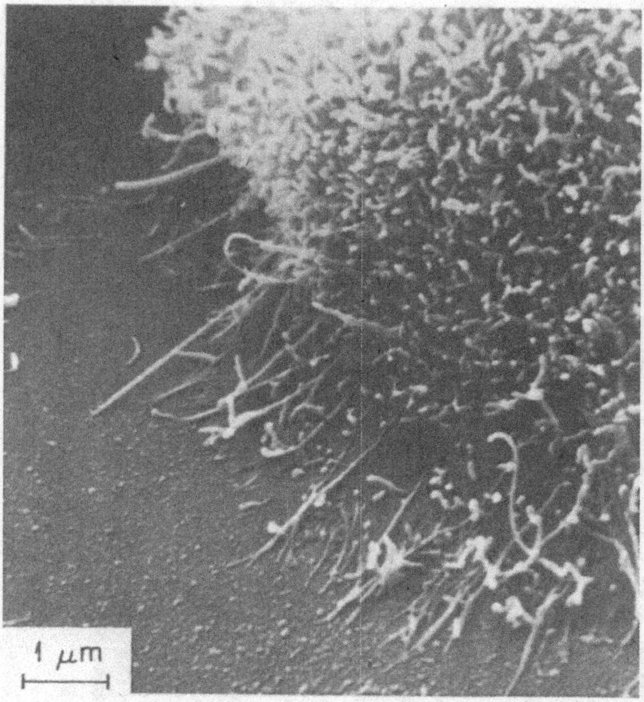

FIG. 34. Area of defective lamelloplasm at the free edge of an MPTR epithelial sheet. Compare with Fig. 25A.

tached to the substrate. In contrast, in sparse and dense cultures of anaplastic epithelium each cell, except those in the upper layers, is attached to the substrate; attachment is not limited to marginal cells.

The time course of attachment of transformed epithelial cells had been studied in detail only in experiments with one anaplastic line, namely, HeLa cells (Springer *et al.*, 1976). It was found that, while suspended HeLa cells are covered with microvilli 3 hours after seeding, attached cells had surfaces blebs and large processes but no microvilli. During the following 9 hours asymmetrical formation of lamelloplasm around the cell bodies was observed. The surface of this lamelloplasm contained numerous blebs. It is difficult to compare the morphology and size of the lamelloplasm of these cells with those of their normal prototypes. To make valid comparisons one has to obtain normal epithelial cultures consisting of single cells. As mentioned in

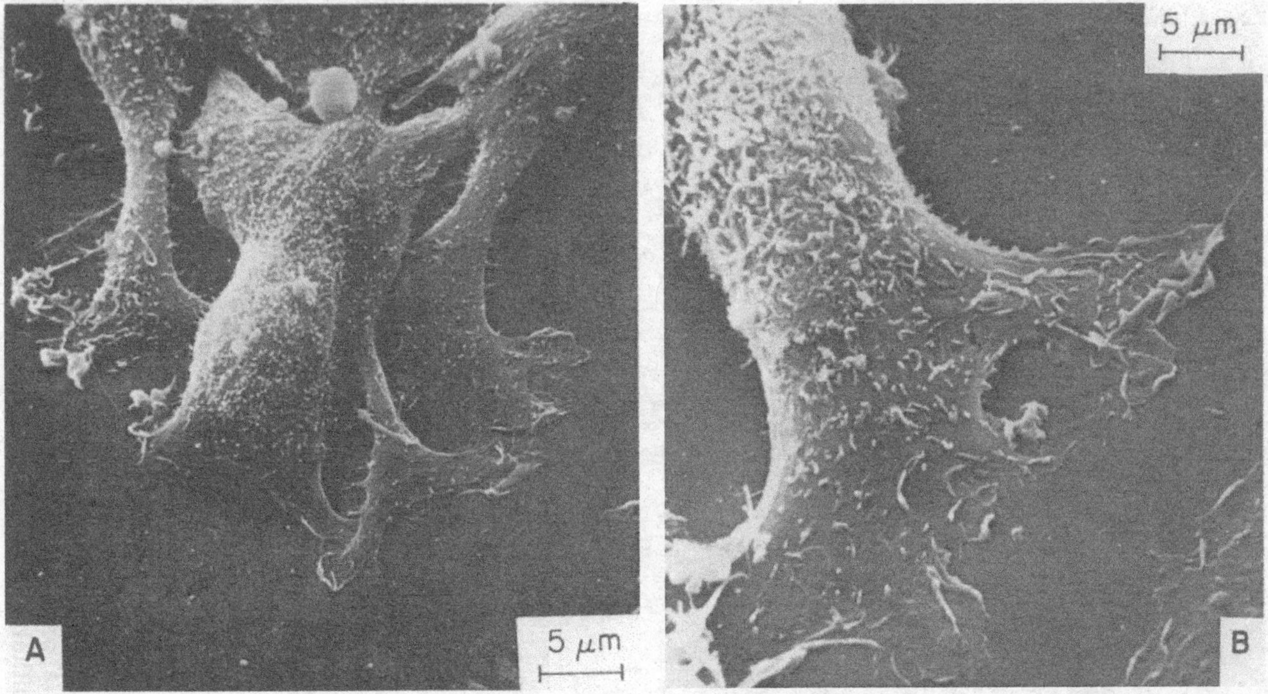

FIG. 35. Cells at the free edge of a dense culture of mouse hepatoma 22a. The cells are poorly attached to each other. Small anterior lamellae with jagged contours are seen in (B). Courtesy of O. S. Zacharova.

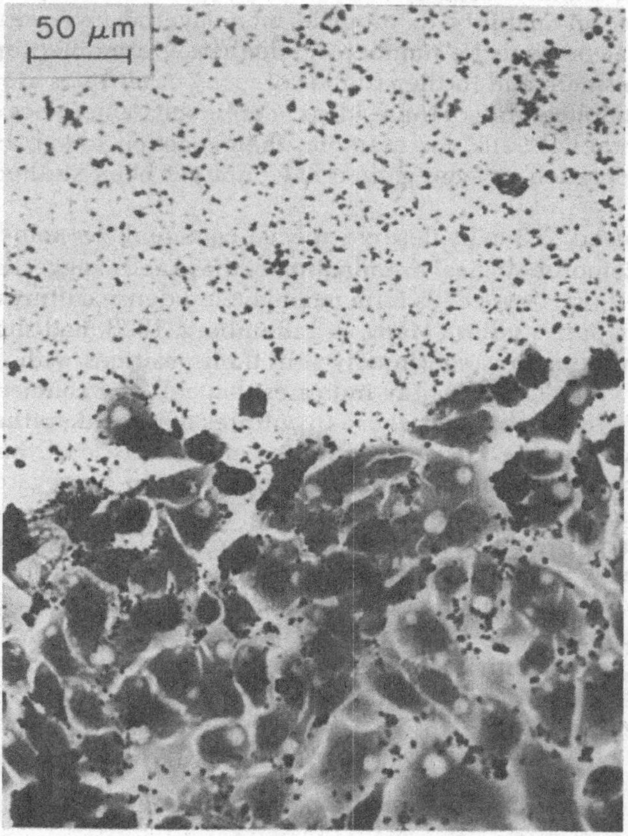

FIG. 36. Adhesion of carmine particles to the upper surface of a wounded culture of hepatoma 22a. Twenty-four hours of incubation with carmine. The particles are attached equally to marginal and central cells. Compare with Fig. 2A.

Section III,B, the preparation of these suspensions involves several difficulties.

In summary, the data available on the morphology of transformed epithelial cells are in agreement with the suggestion that a general feature of these cultures may be a deficiency of attachment reactions. This deficiency may be of varying degree. Minimally transformed lines may be characterized by some deficiency of cell-substrate attachment but retain the ability to form firm cell-cell contacts and to produce coherent sheets. More advanced degrees of transformation may also be characterized by deficient formation of cell-cell contacts, resulting in an inability to produce monolayered coherent sheets. However, at present the number of adequately examined epithelial lines is too small, and further studies of many other lines are needed to test the applicability of these ideas. Certain characteristics of transformed epithelial cells, for example, their reaction to substrates inducing orientation or aggregation, have not yet been studied in any line.

Widely used 3T3 cells deserve special comment. As noted in Section IV,A,2, morphological examination of these cells suggested an endothelial origin. These cells form monolayered dense cultures. Microcinematographic studies (Martz and Steinberg, 1973) had shown that in dense cultures they continually shift their positions with regard to each other. This fact probably indicates that cell-cell contacts in 3T3 cultures are less stable than those in cultures of normal epithelium. It is not clear whether this feature is also characteristic of the normal progenitors of endothelial cells or whether it can be regarded as a manifestation of minimal transformation developed during continuous cultivation. Additional transformation of 3T3 cells may occur spontaneously or may be induced by oncogenic viruses. These transformations lead to formation of cells weakly attached to the substrate and to each other, which form multilayered structures. In general, these cultures are rather similar to those of anaplastic epithelial cells. Cell shape and surface topography may vary considerably from one transformed 3T3 line to another (Porter *et al.*, 1973).

V. Concluding Remarks

In this article we attempted to describe the main facts related to morphogenesis and locomotory behavior of fibroblasts and epithelium in cell cultures. We also tried to present and to elaborate on a consistent interpretation of these facts, to find out how different combinations of a few basic morphogenetic reactions may lead to complex alterations in individual cells and to the formation of organized multicellular structures. Of course, this interpretation cannot be regarded as final. Each type of morphogenetic reaction distinguished in this report is, at best, an approximate designation of a group of kindred phenomena. Nevertheless, efforts to develop such a language may be important and even essential for the progress of our understanding of the mechanisms of morphogenesis. Alterations in large cell groups leading to the formation of tissue structures are complex and have many stages. One can hardly hope to understand the molecular mechanisms of these alterations without first reducing them to a few basic cellular reactions.

References

Abercrombie, M. (1961). *Exp. Cell Res., Suppl.* **8,** 188–198.

Abercrombie, M. (1965). *In* "Cells and Tissues in Culture" (E. Willmer, ed.), Vol. 1, pp. 177–202. Academic Press, New York.

Abercrombie, M. (1970). *In Vitro* **6,** 128–142.

Abercrombie, M., and Ambrose, E. J. (1958). *Exp. Cell Res.* **15,** 332–345.

Abercrombie, M., and Ambrose, E. J. (1962). *Cancer Res.,* **22,** 525–548.

Abercrombie, M., and Dunn, G. A. (1975). *Exp. Cell Res.* **92,** 57–62.

Abercrombie, M., and Heaysman, J. E. M. (1954). *Exp. Cell Res.* **6,** 293–306.

Abercrombie, M., Heaysman, J. E. M., and Pegrum, S. M. (1970a). *Exp. Cell Res.* **59,** 393–398.

Abercrombie, M., Heaysman, J. E. M., and Pegrum, S. M. (1970b). *Exp. Cell Res.* **60,** 437–444.

Abercrombie, M., Heaysman, J. E. M., and Pegrum, S. M. (1970c). *Exp. Cell Res.* **62,** 389–398.

Abercrombie, M., Heaysman, J. E. M., and Pegrum, S. M. (1971). *Exp. Cell Res.* **67,** 359–367.

Abercrombie, M., Heaysman, J. E. M., and Pegrum, S. M. (1972). *Exp. Cell Res.* **73,** 539–539.

Adelstein, A. C., Conti, M. A., Johnson, G. S., Pastan, J., and Pollard, T. D. (1972). *Proc. Natl. Acad. Sci. U.S.A.* **69,** 3693–3697.

Albertini, D. F., and Clark, J. I. (1975). *Proc. Natl. Acad. Sci. U.S.A.* **72,** 4976–4980.

Albrecht-Buehler, G., and Goldman, R. D. (1976). *Exp. Cell Res.* **97,** 329–339.

Allison, A. C. (1973). *Locomotion Tissue Cells, Ciba Found. Symp., 1972* Vol. 14, pp. 109–148.

Ambros, V. R., Chen, L. B., and Buchanan, J. M. (1975). *Proc. Natl. Acad. Sci. U.S.A.* **72,** 3144–3148.

Ambrose, E. J., and Ellison, M. (1968). *Eur. J. Cancer* **4,** 459–462.

Ambrose, E. J., Batzdorf, U., Osborne, J. S., and Stuart, P. R. (1970). *Nature (London)* **227,** 397–398.

Andrianov, L. A., Belitzky, G. A., Ivanova, O. J., Vasiliev, J. M., Domnina, L. V., Olshevskaja, L. V., Panov, M. A., Slavnaja, I. L., and Khesina, A. Y. (1971). "Effects of Carcinogenic Hydrocarbons on Cells." Medicina, Moscow (in Russian).

Armstrong, P. B., and Lackie, J. M. (1975). *J. Cell Biol.* **65,** 439–462.

Baier, R. E., Shafrin, E. A., and Zisman, W. A. (1968). *Science* **162,** 1360–1368.

Bangham, A. D., and Pethica, B. A. (1960). *Proc. R. Soc. Edinburgh* **28,** 43–52.

Barker, B. E., and Sanford, K. K. (1970). *J. Natl. Cancer Inst.* **44,** 39–64.

Bell, P. B., Jr. (1972). *J. Cell Biol.* **55,** 16A.

Berlin, R. D., Oliver, J. M., Ukena, T. E., and Yin, H. H. (1974). *Nature (London)* **247,** 45–46.

Bershadsky, A. D., Gelfand, V. I., Guelstein, V. J., Vasiliev, J. M., and Gelfand, I. M. (1976). *Int. J. Cancer,* **18,** 83–92.

Bershadsky, A. D., and Guelstein, V. I. (1973). *Ontogenez* **4,** 472–480.

Bershadsky, A. D., and Guelstein, V. I. (1976). *Byull. Eksp.* **81,** 49–51.

Bershadsky, A. D., and Lustig, T. M. (1974). *Tsitologiya* **17,** 639–646.

Bhisey, A. N., and Freed, J. J. (1971). *Exp. Cell Res.* **64,** 430–438.

Born, G. V. R. (1972). *Ann. N.Y. Acad. Sci.* **201,** 4–12.
Boyde, A., Weiss, R. A., and Vesely, P. (1972). *Exp. Cell Res.* **71,** 265–272.
Bragina, E. E. (1975). *Tsitologiya* **17,** 248–253.
Bragina, E. E., Vasiliev, J. M., and Chentsov, I. S. (1973). *Tsitologiya* **15,** 1370–1374.
Bragina, E. E., Vasiliev, J. M., and Gelfand, I. M. (1976). *Exp. Cell Res.* **97,** 241–248.
Bray, D. (1972). *Cold Spring Harbor Symp. Quant. Biol.* **37,** 567–571.
Bray, D. (1973). *Nature (London)* **244,** 93–96.
Brinkley, B. R., Fuller, G. M., and Highfield, D. P. (1975). *Proc. Natl. Acad. Sci. U.S.A.* **72,** 4981–4985.
Brunk, U., Ericsson, J. L. E., Pontén, J., and Westermark, B. (1971). *Exp. Cell Res.* **67,** 407–415.
Buckley, I. K., and Porter, K. R. (1967). *Protoplasma* **64,** 349–380.
Carter, S. B. (1967). *Nature (London)* **213,** 256–261.
Chen, L. B., and Buchanan, J. M. (1975). *Proc. Natl. Acad. Sci. U.S.A.* **72,** 131–135.
Cherny, A. P., Vasiliev, J. M., and Gelfand, I. M. (1975). *Exp. Cell Res.* **90,** 317–327.
Chi, J. C., Fellini, S. A., and Holtzer, H. P. (1975). *Proc. Natl. Acad. Sci. U.S.A.* **72,** 4999–5003.
Cornell, R. (1969). *Exp. Cell Res.* **58,** 289–295.
Croop, J., and Holtzer, H. (1975). *J. Cell Biol.* **65,** 271–285.
Curtis, A. S. G., and Varde, M. (1964). *J. Natl. Cancer Inst.* **33,** 15–26.
Defendi, V., and Lehman, J. M. (1966). *J. Cell. Comp. Phys.* **66,** 351–409.
de Petris, S. (1975). *J. Cell Biol.* **65,** 125–146.
de Petris, S., and Raff, M. C. (1973). *Locomotion Tissue Cells, Ciba Found. Symp., 1972* Vol. 14, pp. 27–52.
Diamandopoulos, G. T. (1968). *Am. J. Pathol.* **52,** 633–647.
Di Pasquale, A. (1975a). *Exp. Cell Res.* **94,** 191–215.
Di Pasquale, A. (1975b). *Exp. Cell Res.* **95,** 425–439.
Di Pasquale, A., and Bell, P. B., Jr. (1974). *J. Cell Biol.* **62,** 198–214.
Di Pasquale, A., and Bell, P. B., Jr. (1975). *J. Cell Biol.* **66,** 216–218.
Domnina, L. V., Ivanova, O. J., Margolis, L. B., Olshevskaja, L. V., Rovensky, J. A., Vasiliev, J. M., and Gelfand, I. M. (1972). *Proc. Natl. Acad. Sci. U.S.A.* **69,** 248–262.
Ebendal, T. (1974). *Zoon* **2,** 99–104.
Edelman, G. M., Yahara, I., and Wang, J. L. (1973). *Proc. Natl. Acad. Sci. U.S.A.* **70,** 1442–1446.
Elsdale, T., and Bard, J. (1972a). *Nature (London)* **236,** 152–155.
Elsdale, T., and Bard, J. (1972b). *J. Cell Biol.* **54,** 626–637.
Elsdale, T., and Bard, J. (1974). *J. Cell Biol.* **63,** 343–349.
Elsdale, T., and Bard, J. (1975). *J. Cell Biol.* **66,** 218–219.
Estensen, R. D., and Plagemann, P. G. W. (1972). *Proc. Natl. Acad. Sci. U.S.A.* **69,** 1430–1434.
Frankel, F. R. (1976). *Proc. Natl. Acad. Sci. U.S.A.* **73,** 2798–2802.
Freed, J. J., and Lebowitz, M. M. (1970). *J. Cell Biol.* **45,** 334–354.
Friedenstein, A. I., and Lalykina, K. S. (1973). "Bone Induction and Osteogenic Precursor Cells." Medicina, Moscow (in Russian).
Friedenstein, A. I., Rapoport, R. I., and Luria, E. A. (1967). *Dokl. Akad. Nauk SSSR* **176,** 452–455.
Gahmberg, C. G., and Hakomori, S. (1973). *Proc. Natl. Acad. Sci. U.S.A.* **70,** 3329–3333.
Gail, M. H. (1973). *Locomotion Tissue Cells, Ciba Found. Symp., 1972* Vol. 14, pp. 287–310.
Gail, M. H., and Boone, C. W. (1970). *Biophys. J.* **10,** 980–993.

Gail, M. H., and Boone, C. W. (1971). *Exp. Cell Res.* **65**, 221–227.
Gail, M. H., and Boone, C. W. (1972a). *Exp. Cell Res.* **70**, 33–40.
Gail, M. H., and Boone, C. W. (1972b). *Exp. Cell Res.* **73**, 252–255.
Gail, M. H., Boone, C. W., and Thompson, C. S. (1973). *Exp. Cell Res.* **79**, 386–390.
Garrod, D. R., and Born, G. V. R. (1971). *J. Cell Sci.* **8**, 751–765.
Gelfand, V. I. (1974). *Dokl. Akad. Nauk SSSR* **215**, 460–463.
Gilula, N. B. (1974). *In* "Cell Communication" (R. D. Cox, ed.), pp. 1–29. Wiley, New York.
Goldman, R. D. (1971). *J. Cell Biol.* **51**, 752–762.
Goldman, R. D., and Knipe, D. M. (1973). *Cold Spring Harbor Symp. Quant. Biol.* **37**, 523–534.
Goldman, R. D., Pollack, R., and Hopkins, N. H. (1973). *Proc. Natl. Acad. Sci. U.S.A.* **70**, 750–754.
Goldman, R. D., Lazarides, E., Pollack, R., and Weber, K. (1975). *Exp. Cell Res.* **90**, 333–344.
Gospodarowicz, D., and Moran, J. (1974). *Proc. Natl. Acad. Sci. U.S.A.* **71**, 4648–4652.
Grinell, F. (1973). *J. Cell Biol.* **58**, 602–607.
Groeschel-Stewart, U. (1971). *Biochim. Biophys. Acta* **229**, 322–334.
Guelstein, V. I., Ivanova, O. Y., Margolis, L. B., Vasiliev, J. M., and Gelfand, I. M. (1973). *Proc. Natl. Acad. Sci. U.S.A.* **70**, 2011–2044.
Guillouzo, A., Oudea, Y., Le Guilly, Y., Oudea, M. C., Lenoir, P., and Bourel, M. (1972). *Exp. Mol. Pathol.* **16**, 1–15.
Gustafson, T., and Wolpert, I. (1967). *Biol. Rev. Cambridge Philos. Soc*, **42**, 169–198.
Hale, A. H., Winkelhake, J. L., and Weber, M. J. (1975). *J. Cell Biol.* **64**, 398–407.
Harris, A. K. (1973a). *Exp. Cell Res.* **77**, 285–297.
Harris, A. K. (1973b). *Locomotion Tissue Cells, Ciba Found. Symp., 1972* Vol. 14, pp. 3–26.
Harris, A. K. (1974). *In* "Cell Communication" (R. P. Cox, ed.), pp. 147–186. Wiley, New York.
Harris, A. K., and Dunn, G. (1972). *Exp. Cell Res.* **73**, 519–523.
Harrison, R. G. (1914). *J. Exp. Zool.* **17**, 521–544.
Heaysman, J. E. M., and Pegrum, S. M. (1973). *Exp. Cell Res.* **78**, 71–78.
Hodges, G. M. (1970). *Eur. J. Cancer* **6**, 235–239.
Hodges, G. M., and Muir, M. D. (1972). *J. Cell Sci.* **11**, 233–247.
Hogg, N. M. (1974). *Proc. Natl. Acad. Sci. U.S.A.* **71**, 489–492.
Hsie, A. W., and Puck, T. T. (1971). *Proc. Natl. Acad. Sci. U.S.A.* **68**, 358–361.
Hynes, R. O. (1973). *Proc. Natl. Acad. Sci. U.S.A.* **70**, 3170–3174.
Ingram, V. M. (1969). *Nature (London)* **222**, 641–644.
Ishikawa, H., Bischoff, R., and Holtzer, H. (1969). *J. Cell Biol.* **43**, 312–328.
Ivanova, O. Y., and Margolis, L. B. (1972). *Nature (London)* **242**, 200–201.
Ivanova, O. Y., Margolis, L. B., Vasiliev, J. M., and Gelfand, I. M. (1974). *Proc. Natl. Acad. Sci. U.S.A.* **71**, 2032.
Ivanova, O. Y., Margolis, L. B., Vasiliev, J. M., and Gelfand, I. M. (1976). *Exp. Cell Res.* **101**, 207–219.
Izzard, C. S., and Izzard, S. L. (1975). *J. Cell Sci.* **18**, 241–256.
James, D. W., and Taylor, J. F. (1969). *Exp. Cell Res.* **54**, 107–110.
Johnson, G. S., and Pastan, I. (1972). *Nature (London), New Biol.* **236**, 247–249.
Johnson, G. S., Friedman, R. M., and Pastan, I. (1971). *Proc. Natl. Acad. Sci. U.S.A.* **68**, 425–429.
Kalckar, H. M., Ullrey, D., Kijomato, S., and Hakomori, S. (1973). *Proc. Natl. Acad. Sci. U.S.A.* **70**, 839–843.

Kish, A. L., Kelley, R. O., Crissman, H., and Paxton, L. (1973). *J. Cell Biol.* **57,** 38–53.
Kletzien, R. F., Perdue, J. F., and Springer, A. (1972). *J. Biol. Chem.* **247,** 2964–2966.
Kolata, G. B. (1975). *Science* **188,** 819–820.
Kolodny, G. M. (1972). *Exp. Cell Res.* **70,** 196–202.
Lazarides, E. (1975a). *J. Cell Biol.* **65,** 549–561.
Lazarides, E. (1975b). *J. Histochem. Cytochem.* **23,** 507–528.
Lazarides, E. (1976). *J. Cell Biol.* **68,** 202–219.
Lazarides, L., and Weber, K. (1974). *Proc. Natl. Acad. Sci. U.S.A.* **71,** 2268–2272.
Leighton, J., Brada, F., Estes, L. W., and Justh, G. (1969). *Science* **163,** 472–473.
Leighton, J., Estes, L. W., Mansukhani, S., and Brada, F. (1970). *Cancer* **26,** 1022–1028.
Lesseps, R. J. (1963). *J. Exp. Zool.* **153,** 171–182.
Lin, S., and Spudich, J. A. (1974). *Biochim. Biophys. Acta* **61,** 1471–1476.
Lin, S., Santi, D. V., and Spudich, J. A. (1974). *J. Biol. Chem.* **249,** 2268–2274.
Lipton, A., Linger, I., Paul, D., and Holley, R. W. (1971). *Proc. Natl. Acad. Sci. U.S.A.* **68,** 2799–2801.
McGrath, C. M. (1971). *J. Natl. Cancer Inst.* **47,** 455–467.
McGrath, C. M., Nandi, S., and Young, L. (1972). *J. Virol.* **9,** 367–376.
Macieira-Coelho, A., and Avrameas, S. (1972a). *Proc. Natl. Acad. Sci. U.S.A.* **69,** 2469–2473.
Macieira-Coelho, A., and Avrameas, S. (1972b). *Proc. Soc. Exp. Biol. Med.* **139,** 1374–1378.
Macieira-Coelho, A., Berumen, L., and Avrameas, S. (1974). *J. Cell. Physiol.* **83,** 379–388.
McNutt, N. S., Culp, L. A., and Black, P. H. (1971). *J. Cell Biol.* **50,** 691–708.
McNutt, N. S., Culp, L. A., and Black, P. H. (1972). *J. Cell Biol.* **56,** 412–428.
Margolis, L. B., Samoilov, V. I., Vasiliev, J. M., and Gelfand, I. M. (1975). *J. Cell Sci.* **17,** 1–10.
Maroudas, N. G. (1973). *Nature (London)* **244,** 353–354.
Martinez-Palomo, A., Braislovsky, C., and Bernhard, W. (1969). *Cancer Res.* **29,** 925–937.
Martz, E., and Steinberg, M. S. (1973). *J. Cell. Physiol.* **81,** 25–38.
Middleton, C. A. (1973). *Locomotion Tissue Cells, Ciba Found. Symp., 1972* Vol. 14, pp. 251–270.
Milam, M., Grinnell, F., and Srere, P. (1973). *Nature (London), New Biol.* **244,** 83–84.
Moscona, A. A. (1974). *In* "The Cell Surface in Development" (A. Moscona, ed.), pp. 67–100. Wiley, New York.
Mott, D. M., Fabisch, P. H., Sani, B. P., and Sorof, S. (1974). *Biochem. Biophys. Res. Commun.* **61,** 571–577.
Neupert, G. (1972). *Z. Mikrosk.-Anat. Forsch.* **86,** 443–464.
Nicholson, G. L. (1974). *Int. Rev. Cytol.* **39,** 89–190.
Oliver, J. M., Ukena, T. E., and Berlin, R. D. (1974). *Proc. Natl. Acad. Sci. U.S.A.* **71,** 394–398.
Orci, L., Like, A. A., Amherdt, M., Blondel, B., Kanazawa, Y., Marliss, E. B., Lambert, A. E., Wollheim, C. B., and Renold, A. E. (1973). *J. Ultrastruct. Res.* **43,** 270–297.
Osborn, M., and Weber, K. (1976). *Proc. Natl. Acad. Sci. U.S.A.* **73,** 867–871.
Ossowski, L., Quigley, J. P., Kellerman, G. M., and Reich, E. (1973). *J. Exp. Med.* **138,** 1056–1064.
Ossowski, L., Quigley, J. P., and Reich, E. (1974). *J. Biol. Chem.* **249,** 4306–4311.
Ostlund, R. E., Pastan, I., and Adelstein, R. S. (1974). *J. Biol. Chem.* **249,** 3903–3907.
Paranjpe, M. S., and Boone, C. W. (1975). *Exp. Cell Res.* **94,** 147–151.
Pastan, I., and Johnson, G. S. (1974). *Adv. Cancer Res.* **19,** 303–329.

Perecko, J. P., Berezesky, J. K., and Grimley, P. M. (1973). *In* "Scanning Electron Microscopy/1973" (O. Johari and I. Corwin, eds.), pp. 521–528. IIT Res. Inst., Chicago, Illinois.
Pethica, B. A. (1961). *Exp. Cell Res., Suppl.* **8,** 123–140.
Pickett, P. B., Pitelka, D. R., Hamamoto, S. T., and Misfeldt, D. S. (1975). *J. Cell Biol.* **66,** 316–322.
Pinto da Silva, P., and Gilula, N. B. (1972). *Exp. Cell Res.* **71,** 393–401.
Plagemann, P. G. W., and Estensen, R. D. (1972). *J. Cell Biol.* **55,** 179–185.
Pollack, R., Risser, R., Conlon, S., and Rifkin, D. (1974). *Proc. Natl. Acad. Sci. U.S.A.* **71,** 4792–4796.
Pollack, R., Osborn, M., and Weber, K. (1975). *Proc. Natl. Acad. Sci. U.S.A.* **72,** 994–998.
Pollard, T. D. (1975). *In* "Molecules and Cell Movement" (S. Inoué and R. D. Stephens, eds.), pp. 259–284. Raven, New York.
Pontén, J. (1975). *In* "Human Tumor Cells In Vitro" (J. Fogh, ed.), pp. 175–206. Plenum, New York.
Porter, K. R., and Fonte, V. G. (1973). *In* "Scanning Electron Microscopy/1973" (O. Johari and I. Corwin, eds.), pp. 683–688. IIT Res. Inst., Chicago, Illinois.
Porter, K. R., Todaro, G. J., and Fonte, V. G. (1973). *J. Cell Biol.* **59,** 633–642.
Poste, G. (1973). *Exp. Cell Res.* **77,** 264–270.
Poste, G., and Greenham, L. W. (1971). *Cytobios* **3,** 5–15.
Poste, G., Papahadjopoulos, D., and Nicholson, G. L. (1975). *Proc. Natl. Acad. Sci. U.S.A.* **72,** 4430–4434.
Prop, F. J. A. (1975). *J. Cell Biol.* **66,** 215–216.
Rajaraman, R., Rounds, D. E., Yen, S. P., and Rembaum, A. (1974). *Exp. Cell Res.* **88,** 327–329.
Reaven, E. P., and Axiline, S. G. (1973). *J. Cell Biol.* **59,** 12–27.
Revel, J. P. (1974). *Symp. Soc. Exp. Biol.* **28,** 447–461.
Revel, J. P., and Wolken, K. (1973). *Exp. Cell Res.* **78,** 1–14.
Roseman, S. (1974). *In* "The Cell Surface in Development" (A. Moscona, ed.), pp. 255–272. Wiley, New York.
Rosenberg, M. D. (1960). *Biophys. J.* **1,** 137–159.
Rosenberg, M. D. (1963). *Science* **139,** 411–412.
Rovensky, J. A., and Olshevskaja, L. V. (1973). *Tsitologiya* **15,** 1029–1036.
Rovensky, J. A., and Slavnaja, I. L. (1974). *Exp. Cell Res.* **84,** 199–206.
Rovensky, J. A., Slavnaja, I. L., and Vasiliev, J. M. (1971). *Exp. Cell Res.* **65,** 193–201.
Sanford, K. K., Barker, B. E., Parshad, R., Westfall, B. B., Woods, M. W., Jackson, J. L., King, D. L., and Peppers, E. V. (1970). *J. Natl. Cancer Inst.* **45,** 1071–1096.
Sanger, J. W. (1974). *Proc. Natl. Acad. Sci. U.S.A.* **71,** 3621–3625.
Sato, C., and Takasawa-Nishizawa, K. (1974). *Exp. Cell Res.* **89,** 121–126.
Sheppard, J. R. (1971). *Proc. Natl. Acad. Sci. U.S.A.* **68,** 1316–1320.
Slavnaja, I. L., and Rovensky, J. A. (1975). *Tsitologiya* **17,** 309–313.
Slavnaja, I. L., Rovensky, J. A., Smurova, E. V., and Novikova, S. P. (1974). *Tsitologiya* **16,** 1296–1300.
Somers, K. D., Rachmeler, M., and Christensen, M. (1975). *Nature (London)* **257,** 58–60.
Spooner, B. S., Yamada, K. M., and Wessels, N. K. (1971). *J. Cell Biol.* **49,** 595–613.
Springer, E. L., Hackett, A. J., and Nelson-Rees, W. A. (1976). *Int. J. Cancer* **17,** 407–415.
Spudich, J. A. (1972). *Cold Spring Harbor Symp. Quant. Biol.* **37,** 585–593.
Stein, O., Sanger, L., and Stein, I. (1974). *J. Cell Biol.* **62,** 90–103.
Stoker, M. (1964). *Virology* **24,** 165–174.

Stossel, T. P., and Pollard, T. D. (1973). *J. Biol. Chem.* **248,** 8288–8294.
Taylor, A. C. (1961). *Exp. Cell Res.* **74,** 21–26.
Taylor, E. L., and Wessels, N. K. (1973). *Dev. Biol.* **31,** 421–425.
Tilney, L. G. (1975a). *J. Cell Biol.* **64,** 289–310.
Tilney, L. G. (1975b). *In* "Molecules and Cell Movement" (S. Inoué and R. E. Stephens, eds.), pp. 339–387. Raven, New York.
Tilney, L. G., and Mooseker, M. (1971). *Proc. Natl. Acad. Sci. U.S.A.* **68,** 2611–2615.
Trinkaus, J. P., Betchaku, T., and Krulikowski, L. S. (1971). *Exp. Cell Res.* **64,** 437–444.
Ukena, T. E., Borisenko, J. Z., Karnovsky, M. J., and Berlin, R. D. (1974). *J. Cell Biol.* **61,** 70–82.
Unhjem, O., and Prydz, H. (1973). *Exp. Cell Res.* **83,** 418–420.
Unkeless, J. C., Tobia, A., Ossowski, L., Quigley, J. P., Rifkin, D. B., and Reich, E. (1973). *J. Exp. Med.* **137,** 85–111.
Unkeless, J. C., Dan, K., Kellerman, G. M., and Reich, E. (1974). *J. Biol. Chem.* **249,** 4295–4305.
Vaheri, A., and Ruoslahti, E. (1974). *Int. J. Cancer* **13,** 579–586.
Vasiliev, J. M., and Gelfand, I. M. (1973). *Locomotion Tissue Cells, Ciba Found. Symp., 1972* Vol. 14, pp. 311–332.
Vasiliev, J. M., and Gelfand, I. M. (1976a). *In* "Fundamental Aspects of Metastasis" (L. Weiss, ed.), pp. 71–98. North-Holland Publ., Amsterdam.
Vasiliev, J. M., and Gelfand, I. M. (1976b). *In* "Cell Motility," pp. 279–304. Cold Spring Harbor Laboratory, Cold Spring Harbor, New York.
Vasiliev, J. M., Gelfand, I. M., Domnina, L. V., and Rappoport, R. I. (1969). *Exp. Cell Res.* **54,** 83–93.
Vasiliev, J. M., J. M., Gelfand, I. M., Domnina, L. V., Ivanova, O. Y., Komm, S. G., and Olshevskaja, L. V. (1970). *J. Embryol. Exp. Morphol.* **24,** 625–640.
Vasiliev, J. M., Gelfand, I. M., Domnina, L. V., Zacharova, O. S., and Ljubimov, A. V. (1975a). *Proc. Natl. Acad. Sci. U.S.A.* **72,** 719–722.
Vasiliev, J. M., Gelfand, I. M., and Tint, I. S. (1975b). *Tsitologiya* **17,** 633–638.
Vasiliev, J. M., Gelfand, I. M., Zakharova, O. S., and Ljubimov, A. V. (1975c). *Tsitologiya* **17,** 1400–1406.
Vasiliev, J. M., Gelfand, I. M., Domnina, L. V., Dorfman, N. A., and Pletjushkina, O. J. (1976). *Proc. Natl. Acad. U.S.A.* **73,** 4085–4089.
Vesely, P. (1972). *Folia Biol. (Prague)* **18,** 395–401.
Vesely, P., and Boyde, A. (1973). *In* "Scanning Electron Microscopy/1973" (O. Johari and I. Corwin, eds.), pp. 689–696. IIT Res. Inst., Chicago, Illinois.
Vesely, P., Weiss, R. A. (1973). *Int. J. Cancer* **11,** 64–76.
Waddell, A. W., Robson, R. T., and Edwards, J. G. (1974). *Nature (London)* **248,** 239–241.
Walsh, R. W., and Barnhart, M. I. (1973). *In* "Scanning Electron Microscopy/1973" (O. Johari and I. Corvin, eds.), pp. 481–488. IIT Res. Institute, Chicago, Illinois.
Weber, K., and Groeschel-Stewart, U. (1974). *Proc. Natl. Acad. Sci. U.S.A.* **71,** 4561–4564.
Weinstein, I. B., Orenstein, J. B., Gebert, R., Kaighn, M. E., and Stadler, U. C. (1975). *Cancer Res.* **35,** 253–263.
Weiss, L. (1962). *J. Theor. Biol.* **2,** 236–250.
Weiss, L. (1967). "The Cell Periphery, Metastasis and Other Contact Phenomena." North-Holland Publ., Amsterdam.
Weiss, L. (1972). *Exp. Cell Res.* **71,** 281–288.
Weiss, L. (1974). *Exp. Cell Res.* **86,** 223–234.

Weiss, L., and Blumenson, L. (1967). *J. Cell. Physiol.* **70**, 23–31.
Weiss, L., and Chang, M. K. (1973). *J. Cell Sci.* **12**, 655–664.
Weiss, P. (1929). *Wilhelm Roux' Arch. Entwicklungsmech. Org.* **116**, 438–554.
Weiss, P. (1934). *J. Exp. Zool.* **68**, 393–448.
Weiss, P. (1961). *Exp. Cell Res., Suppl.* **8**, 260–281.
Weiss, P., and Garber, B. (1952). *Proc. Natl. Acad. Sci. U.S.A.* **38**, 264–280.
Weiss, P., and Taylor, A. C. (1956). *Anat. Rec.* **124**, 381–382.
Weller, N. K. (1974). *J. Cell Biol.* **63**, 699–706.
Wessels, N. K., Spooner, B. S., and Luduena, M. A. (1973). *Locomotion Tissue Cells, Ciba Found. Symp., 1972* Vol. 14, pp. 53–77.
Weston, J. A., and Hendricks, K. L. (1972). *Proc. Natl. Acad. Sci. U.S.A.* **69**, 3727–3731.
Wickus, G., Gruenstein, E., Robbins, P. W., and Rich, A. (1975). *Proc. Natl. Acad. Sci. U.S.A.* **72**, 746–749.
Willingham, M. C., and Pastan, I. (1975). *Proc. Natl. Acad. Sci. U.S.A.* **72**, 1263–1267.
Willmer, E. N. (1965). *In* "Cells and Tissues in Culture" (E. N. Willmer, ed.), Vol. 1, pp. 143–176. Academic Press, New York.
Witkowski, J. A., and Brighton, W. E. (1971). *Exp. Cell Res.* **68**, 372–380.
Witkowski, J. A., and Brighton, W. E. (1972). *Exp. Cell Res.* **70**, 41–48.
Wolpert, L., and Gingell, D. (1968). *Symp. Soc. Exp. Biol.* **22**, 169–198.
Wright, J. A. (1973). *Exp. Cell Res.* **78**, 456–460.
Yahara, I., and Edelman, G. M. (1973). *Nature (London)* **236**, 152–155.
Yahara, I., and Edelman, G. M. (1975). *Exp. Cell Res.* **91**, 125–142.
Yamada, K. M., and Wessels, N. K. (1973). *Devel. Biol.* **31**, 413–420.
Yamada, K. M., and Weston, J. A. (1975). *Cell* **5**, 75–81.
Zacharova, O. S. (1976). *Tsitologiya* **18**, 1311–1314.

15.
(with Yu. M. Vasil'ev)

Possible common mechanism of morphological and growth-related alterations accompanying neoplastic transformation

Proc. Natl. Acad. Sci. USA **79** (1982) 2594–2597

Contributed by I. M. Gelfand, January 11, 1982

ABSTRACT Two main groups of phenotypic alterations usually accompany neoplastic transformations of cultured fibroblastic and epithelial cells: alterations of growth control and decreased formation of cell–substrate and cell–cell contacts. We suggest that both types of alterations are due to change of cell response to clustered membrane receptors. Clustering of certain receptors by corresponding ligands possibly induces a dual set of cellular reactions: (*i*) activation of cell proliferation preceded by an ordered sequence of prereplicative changes and (*ii*) attachment–clearing reactions that are associated with anchoring of ligand–receptor complexes by cytoskeletal components and may eventually lead to internalization of these complexes. The first stages of these two subsets of dual reactions are probably identical, but later the two chains of reactions are separated; completion of clearing reaction may stop further progress of the activation of proliferation. Attachment of the cell membrane to another surface can be regarded as a variant of unfinished clearing reaction. Transformed cells may be characterized by alterations of the dual reactions; the primary change may affect either the attachment–clearing branch or the common early stages preceding the separation of branches. These alterations may lead to facilitated activation of proliferation and also to primary or secondary decrease of the ability to perform attachment–clearing reactions in response to external ligands.

Neoplastic transformations of cultured fibroblasts and epithelial cells are usually accompanied by characteristic changes of structure (morphological transformations) and of growth regulation (loss of anchorage dependence, loss of serum dependence, etc.). Both groups of changes can arise simultaneously in the same cell. As shown by the experiments with oncogenic viruses, all the manifestations of transformation can be induced by an increase of the intracellular concentration of a single oncogene-coded protein. These data indicate that morphological and growth-related changes are intrinsically linked; dissociations between the expression of various transformed characters observed in many experiments are, probably, due to secondary changes of transformed cells (see review in ref. 1). Structure and growth regulation are also closely interlinked in normal cells (1–3). The nature of interrelationships between growth and structure in normal and transformed cells remains, however, obscure.

The aim of this paper is to present the hypothesis that different transformed characters can be regarded as various manifestations of the alterations of one particular group of reactions—namely, of reactions to ligand-induced clustering of membrane receptors. In the first and second sections of the paper we discuss the main characteristics of these reactions; the third section is devoted to their alterations in neoplastic cells.

Surface-clearing and attachment reactions have many common features

Receptor-mediated endocytosis and receptor capping are two groups of closely related processes used by the cell to clear its surface of the groups of membrane components clustered after binding the corresponding ligands; collectively they can be designated as surface-clearing reactions. In the course of these reactions clustered receptors are internalized in membrane vesicles; further processing of these clusters (degradation and recycling) will not be discussed here. Internalization can be preceded by translocation of clusters in the plane of the membrane, that is, by the formation of a cap, by transfer of receptors into coated pits, etc.

Pseudopodial reactions (extension and attachment of pseudopods) are used by the cells to attach themselves to other surfaces. In particular, spreading of the cultured cells on the substrate can be regarded as a series of pseudopodial attachment reactions leading to formation of cytoplasmic lamellae. Extension, attachment, and contraction of pseudopods is also the mechanism by which the spread cell moves directionally on the substrate (1, 4). Aggregation and adhesion of platelets are also mediated by pseudopods (reviewed in refs. 5 and 6).

Phagocytosis is another closely related process in which the cell repeatedly extends and attaches its pseudopods. As stressed many times before, spreading can be regarded as an unsuccessful effort to phagocytosize an infinitely large foreign surface. At the same time phagocytosis is closely related to receptor-mediated endocytosis and to capping (7).

More detailed analysis of surface-clearing reactions and of pseudopodial attachment reactions confirms that they have many common features:

(*i*) Almost by definition, attachment of cell to another surface involves the formation of a group of membrane molecules linked to that surface directly or via some intermediate molecules. Such a group of immobilized membrane molecules formed in the course of attachment can be regarded as the counterpart of the cluster of receptors crosslinked by a soluble ligand.

(*ii*) Formation of pseudopods can be induced not only by cell surface contact with another surface but also by interaction of membrane receptors with soluble ligands. Thus, induction of pseudopods can be induced by protein growth factors and by other protein hormones (8–12). Suspended platelets treated by various activating agents also extend pseudopods (6).

(*iii*) Centripetal translocation of clustered receptors in the plane of the membrane and into the cell interior is characteristic of clearing reactions. Cell–substrate contacts in the spread cells are under tension from the cell interior; depending on conditions, this tension may detach the contact, deform the substrate, etc. (13). Thus, both in clearing reactions and in attachment reactions the clustered membrane components are subjected to the action of centripetally directed forces.

(*iv*) The centripetal force mentioned above is probably exerted by actin microfilaments and other cytoskeleton components anchoring the clustered receptors from the inside. Con-

Abbreviation: EGF, epidermal growth factor.

siderable indirect evidence supports the suggestion that clustering of receptors may lead to such anchoring (reviewed in refs. 1, 14, and 15). On the other hand, certain types of cell–substrate contacts (focal contacts) are always associated with the ends of microfilament bundles (16). Accumulations of α-actinin and, especially, of vinculin (17) are found at the inner side of these contacts and are thought to be essential for the linkage of microfilaments to the membrane components.

(*v*) Translocation of clustered receptors as well as formation of contact structures usually takes place only in certain areas of cell surface, especially on the surface of pseudopods and lamellae induced by the same cluster or by some other factor (reviewed in ref. 1). Possibly, the pseudopodial surface is especially adapted for anchoring of the clustered receptors. This adaptation may be a result of special composition of cytoskeleton (18, 19) and membrane (20) in these zones. Coated pits are other special areas of cell surface adapted for anchoring and the internalization of certain types of receptors (21, 22); these structures, like pseudopods, may be preformed or formed *de novo* (23) in the course of a given reaction. Interrelationships between various surface specializations, such as pseudopods and coated pits, are at present unknown.

In summary, pseudopodial attachment reactions and surface-clearing reactions belong to a large family of processes that may be collectively designated as attachment–clearing reactions. Possibly, anchoring of clustered receptors by cytoskeleton structures is the general basis of all these reactions. In the case of clearing this anchoring eventually leads to internalization of cluster. In the case of attachment to the substrate the anchored clusters remain on the cell surface; in other words, spreading on the substrate can be regarded as a series of unfinished clearing reactions. Needless to say, there are also many other differences between the various subclasses of reactions and between the various reactions belonging to the same subclass. For instance, there is considerable diversity in the ways of endocytosis of various receptors (24).

Activation of proliferation by receptor-bound ligands and attachment–clearing reactions

Studies of the action of protein growth factors on cultured cells suggest that activation of proliferation and attachment–clearing reactions may be closely interlinked. Typical growth factors—such as epidermal growth factor (EGF), platelet growth factor, insulin-like factors, and others—added to the medium in low concentrations stimulate the proliferation of "resting" cultures of sensitive cell types (reviewed in refs. 25–28). The entry of stimulated cells into S phase is preceded by a lag period; minimal duration of this period is usually within the range 12–20 hr.

Transition of cells through the whole lag period or at least through a considerable part of it requires the continuous presence of the inducing growth factor in the medium; possibly, in certain cases nonidentical growth factors can stimulate progress of cells through various parts of the latent period. An ordered set of "prereplicative changes" of many metabolic characteristics is observed within the cell during the latent period (28–31). Different growth factors are bound to different specific membrane receptors. Ligand binding induces clustering of corresponding receptors and their internalization, with subsequent degradation of ligands (32–34). At the same time various growth factors are able to induce pseudopodial reactions.

Induction of the protrusions of various types of pseudopodia by the cells treated by growth factors (insulin, EGF, and others) was observed by several authors (8–11). Addition of certain growth factors into the medium increases cell migration into a "wound" in a layer of cultured cells (35); platelet-derived growth factor was shown to act as a chemotactic attractant for smooth muscle cells (36). These effects are possibly due to stimulation of pseudopodial attachment reactions by growth factors. Thus, binding of growth factors to their receptors may induce not only activation of proliferation but also various manifestations of clearing–attachment reactions. Possibly, besides the receptors of growth factors, there are also other receptors that, when clustered by corresponding ligands, may induce both types of reactions. For instance, unknown receptors involved in cell attachment to usual culture substrates may belong to this group. Attachment of these receptors to the surface of the substrate may promote the proliferation or, in other words, it may be the basis of "anchorage dependence" of replication. Induction of fibroblast proliferation by particles of diverse nature attaching to the surface of dense cultures (37, 38) may be a similar phenomenon. Antibodies against surface immunoglobulins stimulate proliferation of B lymphocytes (39); this may be yet another example of growth and division induced by ligands clustering corresponding receptors.

At present we do not know how many types of receptor–ligand complexes can induce both attachment–clearing and growth activation; probably their number is considerable. Clustering of certain other receptors induces only clearing reactions and does not produce any significant alterations of metabolism (24). We do not know at present whether there are any ligand–receptor complexes that can induce growth activation but not clearing reactions.

Experimental analysis of the interactions of growth factor molecules with their receptors supports the suggestion that clustering of these receptors is essential for induction of prereplicative changes (34, 40).

Thus, induction of proliferation and attachment–clearing reactions may have a common initiating membrane change, that is, clustering of certain receptors. Possibly these reactions may also have other common early steps. In particular, induction of pseudopodial extension or some closely related change may be essential not only for the attachment–clearing but also for the development of prereplicative changes. In this connection, it may be significant that migration of fibroblasts from a densely populated area into a wound is accompanied both by the stimulation of pseudopodial activity at the leading cell edge and by the induction of proliferation (reviewed in ref. 1). In contrast to cell–substrate attachment, cell–cell contacts paralyze pseudopodial activity (41) and may inhibit proliferation. Microtubule-destroying drugs, such as colchicine or vinblastine, induce both nondirectional pseudopodial activity (1, 4) and DNA synthesis (42–44). In contrast, cytochalasin B may inhibit pseudopodial activity and also DNA synthesis (45). These correlations indicate that induction of pseudopodial activity may be closely related to stimulation of replication. The relationship of the final stages of clearing reactions to the induction of prereplicative changes is not clear. Inhibition of internalization of EGF by various alkylamines was reported to increase the stimulation of DNA synthesis (33), but later this result was disputed (46). Further experiments are needed to settle this controversy and to find out whether the alkylamines specifically affect only internalization. As mentioned above, the growth factors induce DNA synthesis only if they are present in the medium for a long time. Therefore it seems probable that this response is induced not by some final product of intracellular degradation of growth factors but by the ligand–receptor complexes present on the surface, the complexes undergoing the early stages of attachment–clearing reactions, or both. The growth-stimulating effect of cell–substrate attachment obviously cannot be due to internalization of corresponding receptors.

In summary, binding of certain ligands to their surface receptors induces a dual set of reactions: attachment–clearing reactions and activation of proliferation. The first stages of these two subsets may be identical, but later the two chains of reactions are separated; possibly completion of the clearing reaction may stop the progress of the activation of proliferation. Attachment of the receptors to another surface prevents the completion of the clearing reaction. These supposed interrelationships are summarized in Fig. 1. It should be stressed that attachment-clearing may be a local reaction of a part of cell membrane; in contrast, induction of proliferation is always a global reaction of the whole cell. Effects of many similar or different ligand–receptor complexes have to be integrated in order to produce prereplicative changes.

Dual reactions to clustered ligands possibly developed in evolution as two facets of processes restoring the normal state of the membrane after various local disturbances of its structure. The clearing reactions obviously removed these local disturbances, whereas activation of metabolic processes probably was directed toward restoration of cellular structures damaged as a result of membrane changes (see discussion in refs. 1 and 47). In the course of evolution these restorative mechanisms could have been adapted for many additional purposes such as adhesion to various surfaces, activation of growth, etc.

Reactions of transformed cells to clustered receptors

Analysis of diverse morphological alterations accompanying neoplastic transformations of fibroblasts and of epithelial cells had shown that they all can be regarded as various consequences of the decreased formation of cell–substrate and cell–cell contacts, that is, as manifestations of altered pseudopodial attachment reactions (1, 4). Various transformed characters related to growth can be regarded as manifestations of the increased ability of cells to undergo activation of proliferation; in the course of neoplastic evolution the range of environmental conditions in which cell proliferation is activated becomes broader (1).

Developing the considerations discussed in previous paragraphs, we suggest that alteration of the dual set of reactions to clustered receptors can be the common cause both of decreased attachment and of facilitated growth activation in transformed cells. Either the attachment–clearing branch of these reactions or their first stages common for both branches (see Fig. 1) can be primarily altered in transformed cells. Let us discuss in more detail several suggestions about the possible nature of these alterations:

(i) Decreased anchoring. Transformation may primarily change a certain step of the attachment–clearing—e.g., anchoring of clustered receptors to cytoskeletal components. Then the clearing of clustered receptors of growth-activating factors will become deficient and such clusters will remain at the surface for a longer time, more effectively inducing growth activation. Simultaneously, attachment reactions will become deficient. Decreased spreading may diminish the number of substrate-attached receptors at the cell surface. However, due to deficient clearing reactions the total number of growth-stimulating receptor clusters present at the surface will be increased. Vinculin, a protein present in cell–substrate contacts and possibly involved in the anchoring of surface receptors, was shown to be phosphorylated by the oncogene-coded proteins of Rous sarcoma virus and of Abelson virus (48). This finding indirectly supports the suggestion that primary alterations of anchoring may be characteristic of certain transformed cells.

(ii) Increased load of clusters. Transformation may increase the number of clustered receptors present on the surface and inducing dual reactions. This increase may be due to increased production of endogenous growth factors (49, 50) or to altered properties and numbers of receptors. Then the growth will be activated by these clusters. Probably the total intensity of attachment–clearing reactions per cell cannot exceed a certain maximum. Therefore, an increased number of clusters may overload cellular systems performing these reactions. As a result, the cell will have diminished capacity to perform additional attachment–clearing reactions—e.g., to spread on the substrate.

(iii) Spontaneous dual reactions. Transformation may change the mechanisms performing the first common steps of dual reactions to clustered receptors in such a way that the cell will overreact to the normal numbers of ligand–bound receptors or even will set in motion dual reactions in the absence of stimulating clusters. In this case proliferation may be perpetually activated. The cell may have the high rates of surface activities associated with attachment–clearing reactions—e.g., of pinocytosis or of pseudopodial extension. However, this excessive spontaneous activity may lead to exhaustion of attachment–clearing systems, and therefore the cell may have diminished ability to perform normal reactions to exogenous stimuli, such as spreading on the substrate, receptor-mediated endocytosis of certain ligands, etc. The mechanisms performing the first steps of dual reactions are at present unknown, but it seems probable that they may be associated with ionic changes and with activation of some membrane-bound enzymes—e.g., special protein kinases (32). Spontaneous stimulation of these changes may be characteristic of transformed cells.

Suggestions listed above are not mutually exclusive. They are given here as examples of possible specific variants of the gen-

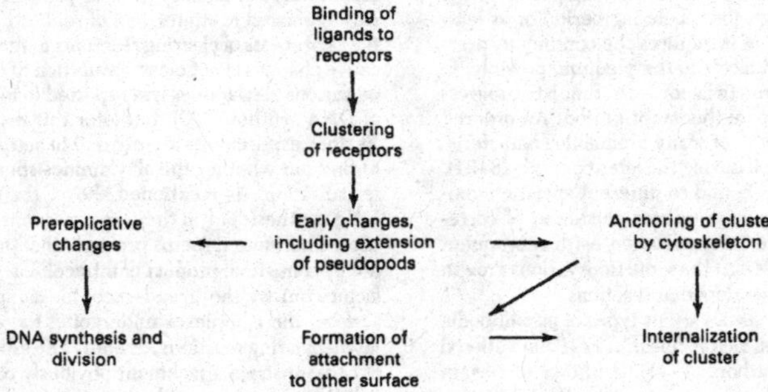

FIG. 1. Hypothetical interrelationships between activation of growth and attachment–clearing reactions induced by certain ligands binding to corresponding membrane receptors. See text for further explanation.

eral hypothesis that the main phenotypic traits of transformed cells are consequences of the altered dual reaction to clustered receptors. According to this hypothesis, primary or secondary decrease of the ability to perform attachment–clearing reactions in response to external ligands is a characteristic feature of transformed cells. Deficient attachment of neoplastic cells to the substrate (1, 4) has been already mentioned. Little is known about the relative ability of normal and transformed cells to perform other variants of attachment–clearing reactions—e.g., phagocytosis or receptor-mediated endocytosis. The fraction of cell surface cleared from crosslinked concanavalin A receptors is significantly smaller in substrate-attached transformed fibroblasts as compared with normal ones (51, 52).

Inefficient internalization of receptor-bound low density lipoprotein in human epithelioid carcinoma A-431 cells had been observed by Anderson and coauthors (53). Brown and others (54) had found that the total binding of EGF and the rate of its degradation by the cell are higher in BALB 3T3 cells than in the cells of a benzo[*a*]pyrene-transformed daughter line; these authors suggest that one cause of the diminished serum requirement of transformed cells is a reduced rate of utilization of serum growth factors. Systematic studies of various attachment–clearing processes in normal and transformed cells and of their interrelationships with growth activation remain a task for future experiments. The hypothesis discussed in this paper may be useful for planning and interpreting these experiments.

1. Vasiliev, J. M. & Gelfand, I. M. (1981) *Neoplastic & Normal Cells in Culture* (Cambridge Univ. Press, London).
2. Stoker, M., O'Neill, C., Berryman, Y. & Waxman, V. (1968) *Int. J. Cancer* **3**, 683–693.
3. Folkman, J. & Moscona, A. (1978) *Nature (London)* **273**, 345–349.
4. Vasiliev, J. M. & Gelfand, I. M. (1977) *Int. Rev. Cytol.* **50**, 159–274.
5. Allen, R. D., Zacharski, L. R., Widirstky, S. T., Rosenstein, R., Zaitlin, L. M. & Burgess, D. R. (1979) *J. Cell Biol.* **83**, 126–142.
6. Barnhart, M. I. (1978) *Mol. Cell. Biochem.* **22**, 113–136.
7. Berlin, R. D. & Oliver, J. M. (1978) *J. Cell Biol.* **77**, 789–804.
8. Evans, R. B., Morhenn, V., Jones, A. L. & Tomkins, G. M. (1974) *J. Cell Biol.* **61**, 95–106.
9. Brum, K. U., Schellens, J. & Westermark, B. (1976) *Exp. Cell Res.* **103**, 295–302.
10. Connolly, J. L., Greene, L. A., Viscarello, R. R. & Riley, W. D. (1979) *J. Cell Biol.* **82**, 820–827.
11. Chinkers, M., McKanna, J. A. & Cohen, S. (1979) *J. Cell Biol.* **83**, 260–265.
12. Ericson, L. E. (1981) *Mol. Cell. Endocrinol.* **22**, 1–24.
13. Harris, A. K., Stopak, D. & Wild, P. W. (1980) *Science* **208**, 177–179.
14. Condeelis, J. (1981) in *International Cell Biology 1980–1981*, ed. Schweiger, H. (Springer, Berlin), pp. 306–320.
15. Koch, G. L. E. (1981) in *International Cell Biology 1980–1981*, ed. Schweiger, H. (Springer, Berlin), pp. 321–330.
16. Heath, J. P. & Dunn, G. A. (1978) *J. Cell Sci.* **29**, 197–212.
17. Geiger, B. (1981) in *International Cell Biology 1980–1981*, ed. Schweiger, H. (Springer, Berlin), pp. 761–773.
18. Gottlieb, A. I., Heggeness, M. H., Ash, J. F. & Singer, S. J. (1979) *J. Cell. Physiol.* **100**, 563–578.
19. Bretscher, A. & Weber, K. (1980) *J. Cell Biol.* **86**, 335–340.
20. Robinson, J. M. & Karnovsky, M. J. (1980) *J. Cell Biol.* **87**, 562–568.
21. Pearse, B. (1980) *Trends Biochem. Sci.* **5**, 131–134.
22. Willingham, M. C., Haigler, H. T., Dickson, R. B. & Pastan, I. H. (1981) in *International Cell Biology 1980–1981*, ed. Schweiger, H. G. (Springer, Berlin), pp. 613–621.
23. Salisbury, J. L., Condeelis, J. S. & Satir, P. (1980) *J. Cell Biol.* 132–141.
24. Kaplan, J. (1981) *Science* **212**, 14–20.
25. Gospodarowicz, D. & Moran, J. S. (1976) *Annu. Rev. Biochem.* **45**, 531–558.
26. Carpenter, G. & Cohen, S. (1979) *Annu. Rev. Biochem.* **48**, 193–216.
27. Scher, C. D., Shepard, R. C., Antoniades, H. N. & Stiles, C. D. (1979) *Biochim. Biophys. Acta* **560**, 217–241.
28. Holley, R. W. (1980) *J. Supramol. Struct.* 13, 191–197.
29. Rozengurt, E. (1976) *J. Cell. Physiol.* **89**, 627–632.
30. Epifanova, O. I. (1977) *Int. Rev. Cytol. Suppl.* **5**, 303–335.
31. Baserga, R. (1978) *J. Cell. Physiol.* **95**, 377–386.
32. Heigler, H. T. & Cohen, S. (1979) *Trends Biochem. Sci.* **4**, 132–134.
33. Maxfield, F. R., Davies, P. J. A., Klempner, L., Willingham, M. C. & Pastan, I. (1979) *Proc. Natl. Acad. Sci. USA* **76**, 5731–5735.
34. Schlessinger, J. (1980) *Trends Biochem. Sci.* **5**, 210–214.
35. Bürk, R. R. (1976) *Exp. Cell Res.* **101**, 293–298.
36. Grotendorst, G. R., Seppä, H. E. J., Kleinman, H. K. & Martin, I. R. (1981) *Proc. Natl. Acad. Sci. USA* **78**, 3669–3672.
37. Barnes, D. W. & Colowick, S. P. (1977) *Proc. Natl. Acad. Sci. USA* **74**, 5593–5597.
38. Rubin, A. H. & Bowen-Pipe, D. F. (1979) *J. Cell. Physiol.* **98**, 81–94.
39. Potash, M. J. (1981) *Cell* **23**, 7–8.
40. Kahn, C. R., Baird, K. L., Jarrett, D. B. & Flier, J. S. (1978) *Proc. Natl. Acad. Sci. USA* **75**, 4209–4213.
41. Abercrombie, M. (1970) *In Vitro* **6**, 128–142.
42. Vasiliev, J. M., Gelfand, I. M. & Guelstein, V. I. (1971) *Proc. Natl. Acad. Sci. USA* **68**, 977–979.
43. McClain, D. A. & Edelman, G. M. (1980) *Proc. Natl. Acad. Sci. USA* **77**, 2748–2752.
44. Crossin, K. L. & Carney, D. H. (1981) *Cell* **23**, 61–71.
45. O'Neill, F. J. (1979) *J. Cell. Physiol.* **101**, 201–218.
46. King, A. C., Hernaez-Davis, L. & Cuatrecasas, P. (1981) *Proc. Natl. Acad. Sci. USA* **78**, 717–721.
47. Vasiliev, J. M. & Gelfand, I. M. (1968) *Curr. Mod. Biol.* **2**, 43–55.
48. Sefton, B. M., Hunter, T., Ball, E. H. & Singer, S. J. (1981) *Cell* **24**, 165–174.
49. De Larco, J. E. & Todaro, G. J. (1978) *Proc. Natl. Acad. Sci. USA* **75**, 4001–4005.
50. Ozanne, B., Fulton, R. J. & Kaplan, P. (1980) *J. Cell. Physiol.* **105**, 163–180.
51. Domnina, L. V., Pletyushkina, O. Y., Vasiliev, J. M. & Gelfand, I. M. (1977) *Proc. Natl. Acad. Sci. USA* **74**, 2865–2868.
52. Alexandrova, A. Y. & Domnina, L. V. (1980) *Byull. Eksp. Biol. Med.* **91**, 741–743.
53. Anderson, R. G. W., Brown, M. S. & Goldstein, J. L. (1981) *J. Cell Biol.* **88**, 441–452.
54. Brown, K. D., Yeh, Y.-C. & Holley, R. W. (1979) *J. Cell. Physiol.* **100**, 227–238.

16.

(with V. B. Dugina, T. M. Svitkina, and Yu. M. Vasil'ev)

Special type of morphological reorganization induced by phorbol ester: Reversible partition of cell into mobile and stable domains

Proc. Natl. Acad. Sci. USA **84** (1987) 4122–4125

Contributed by I. M. Gelfand, January 29, 1987

ABSTRACT **The phorbol ester phorbol 12-myristate 13-acetate (PMA) induced reversible alteration of the shape of fibroblastic cells of certain transformed lines—namely, partition of the cells into two types of domains: motile body actively extending large lamellas and stable narrow cytoplasmic processes. Dynamic observations have shown that stable processes are formed from partially retracted lamellas and from contracted tail parts of cell bodies. Immunofluorescence microscopy and electron microscopy of platinum replicas of cytoskeleton have shown that PMA-induced narrow processes are rich in microtubules and intermediate filaments but relatively poor in actin microfilaments; in contrast, lamellas and cell bodies contained numerous microfilaments. Colcemid-induced depolymerization of microtubules led to contraction of PMA-induced processes; cytochalasin B prevented this contraction. It is suggested that PMA-induced separation of cell into motile and stable parts is due to directional movement of actin structures along the microtubular framework. Similar movements may play an important role in various normal morphogenetic processes.**

Reversible morphological reorganizations of fibroblasts and other cultured cells are actively studied as prototypes of normal morphogenetic processes; alterations of cytoskeleton regulated by the membrane-generated signals seem to play central roles in these reorganizations (for reviews, see refs. 1–3). In this paper we describe a special type of reversible alteration of cell shape and cytoskeleton induced in certain fibroblastic lines by the tumor promoter phorbol 12-myristate 13-acetate (PMA). This agent caused reversible cell partition into a motile pseudopod-forming body and nonmotile elongated processes rich in microtubules and intermediate filaments. Analysis of this reorganization suggests that alterations of interactions between the actin cortex and microtubules are of key importance in its development. It is possible that reorganizations of a similar type are involved in normal morphogenetic processes—e.g., in the formation of neurites by neural cells, in forward translocation of polarized fibroblasts, etc.

MATERIALS AND METHODS

Several types of cultured fibroblastic cells were used: secondary cultures of mouse embryo fibroblasts, minimally transformed mouse BALB/3T3 line, rat NRK (normal rat kidney) line, transformed PS-103 and PS-104 lines obtained in this laboratory from mouse sarcomas induced by implanted plastic films, and cloned subline 152/15 of the spontaneously transformed CAK-7 line of mouse fibroblasts (4). The cells were grown at 37°C in Eagle's medium with 10% fetal calf serum. The cells were seeded on the coverslips placed into Petri dishes at a concentration of 10^4 cells per ml. The experiments were started at 48 hr after seeding.

Colcemid, PMA, and its analogues were from Serva (Heidelberg), and cytochalasin B was from Aldrich. Rhodamine-labeled phalloidin was a gift of T. Wieland (Max Planck Institute for Medical Research, Heidelberg). Polyclonal antibodies to tubulin and myosin were obtained from V. I. Gelfand, F. K. Gioeva, and A. D. Bershadsky (Moscow State University), and monoclonal antibodies to vimentin were from O. S. Rochlin (Cardiological Center, Moscow). The methods of indirect immunofluorescent examination of cytoskeleton (5) and electron microscopic examination of the platinum replicas (6) were performed as described in previous publications (5, 6). To perform dynamic-phase contrast microscopy of living cells, the cultures were grown in special glass chambers at 37°C and photographed at various time intervals, usually each 5–10 min.

RESULTS

Alterations of Cell Shape. Treatment of several types of mouse tumorigenic lines (PS-103, PS-104, 152/15) with PMA (10–20 ng/ml) leads to similar morphological alterations: contraction of cell body and lamellas, accompanied by the formation of narrow cytoplasmic processes. These changes were observed in 70–90% of cells after 2–4 hr of incubation. Similar changes were induced in the minimally transformed BALB/3T3 and NRK lines, but here the effective concentrations of PMA were higher (25–50 ng/ml) and percentages of the altered cells were lower (20–40%).

The cultures of mouse embryo fibroblasts were the least sensitive to PMA: a high concentration of this agent (100 ng/ml) induced only partial contraction of lamellas in 10–20% of the cells. The 152/15 line had the highest sensitivity to PMA. Most cells of this line underwent drastic morphological changes after the 1- to 2-hr incubation with low concentrations of the drug (2–5 ng/ml). Therefore, these cells treated with a standard concentration of PMA (5 ng/ml) were selected for further detailed study of the effects of the tumor promoter. Control cells of this line, spread on the substratum, had typical fibroblast-like shape with small leading lamellas and short tail processes; formation of small ruffles at the leading edges and contraction of the tail accompanied their translocation. Dynamic observations of individual cells (Fig. 1) had shown that the first morphological changes were observed at about 20–30 min after the addition of PMA. The large fan-like lamellas were formed at the cell edge; simultaneously, some contraction of the cell body took place. Later, at 30–60 min, the newly formed lamellas contracted into the narrow cylindrical processes; these processes occasionally had thick parts ("bulbs") and small lamellar zones. New lamellas were extended from other sites of the edge simultaneously with the retraction of the old ones. These extensions and retractions went on for many hours. Often they were not accompanied by any considerable displacements of the cell body; this body became surrounded by several narrow processes formed from the retracted lamellas.

Abbreviation: PMA, phorbol 12-myristate 13-acetate.

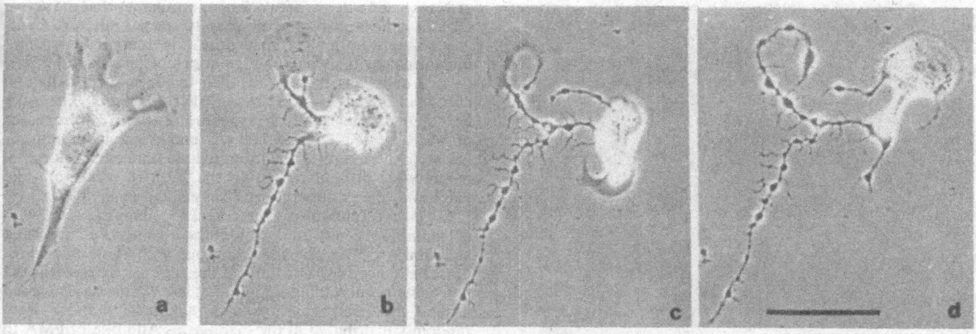

FIG. 1. Phase-contrast micrographs of the cell of the 152/15 line immediately before the addition of PMA to the medium (*a*) and 25 min (*b*), 70 min (*c*), and 90 min (*d*) after the addition. In *b* and *c*, two large lamellas were formed at the leading edge, and the cell body moved into the right lamella; the tail part of the body and a part of the left lamella contracted into narrow processes. In *b–d*, the left lamella was completely transformed into the narrow process. The cell body formed new lamellas and moved toward the right upper corner. The tail narrow process was elongated at its proximal end but otherwise retained the stable shape. (Bar = 50 μm.)

In other cases translocations of the cell body were observed; the body was displaced from time to time toward the newly extended lamellas, while the tail part of the body contracted. These contractions lead to elongation of the proximal parts of narrow processes. These processes, once formed by contraction of lamellas and cell bodies, usually did not change their shape during the next several hours, except for occasional formation and disappearance of bulbs.

Morphological alterations induced by PMA were not accompanied by any significant changes of the substratum area occupied by one cell. At 16–20 hr, formation of narrow processes was gradually reversed in spite of continued presence of PMA in the medium; these processes became wider and shorter. At 24–30 hr, the morphology of PMA-treated cultures became indistinguishable from that of control cultures. When, after 4–6 hr of incubation, the PMA-containing medium was substituted by the control medium, the original morphology was restored 2–3 hr later. Two analogues of PMA devoid of tumor-promoting activity, 4-oxymethyl-PMA (100 ng/ml) and phorbol (100 ng/ml), did not induce any morphological changes in the cultures of the 152/15 line. In contrast, the PMA-related tumor promotor, mezerein (1 μg/ml), induced formation of narrow processes in these cultures.

Alterations of Cytoskeleton. Immunofluorescence examination (Fig. 2) revealed thin actin bundles in the cytoplasm of control 152/15 cells. These cells had well-developed systems of microtubules and intermediate filaments. The cells incubated with PMA for 2 hr had newly formed lamellas with bands of intensely stained actin near their outer edges; proximal parts of lamellas and bodies also contained large intensely stained areas, whereas discrete bundles were usually absent. The intensity of actin staining of the narrow processes was similar to or weaker than that of the cell body; the bulbs often were stained more intensely than other parts of the processes. Staining for myosin was more intense in the central part of the body than in lamellas and narrow processes. Microtubules were numerous in all parts of the PMA-treated cell; they formed a dense network in lamellas and compact cables in the narrow processes. The distribution of intermediate filaments was similar to that of microtubules except that these filaments were absent from the distal parts of lamellas.

Examination of platinum replicas (Fig. 3) of the control

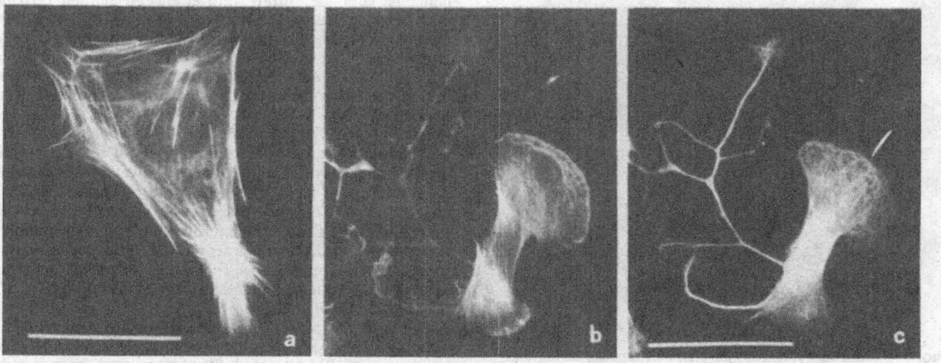

FIG. 2. Fluorescence microscopy after rhodamine-phalloidin treatment (*a* and *b*) or after tubulin antibody revealed by fluorescein isothiocyanate-labeled second antibody (*c*). (*a*) Actin bundles in the control 152/15 cell. (*b* and *c*) PMA-treated cell. This cell has an actin-rich body and lamellas; narrow processes contain less actin (*b*). Numerous microtubules are present in all the parts of the same cell (*c*). (Bars = 50 μm.)

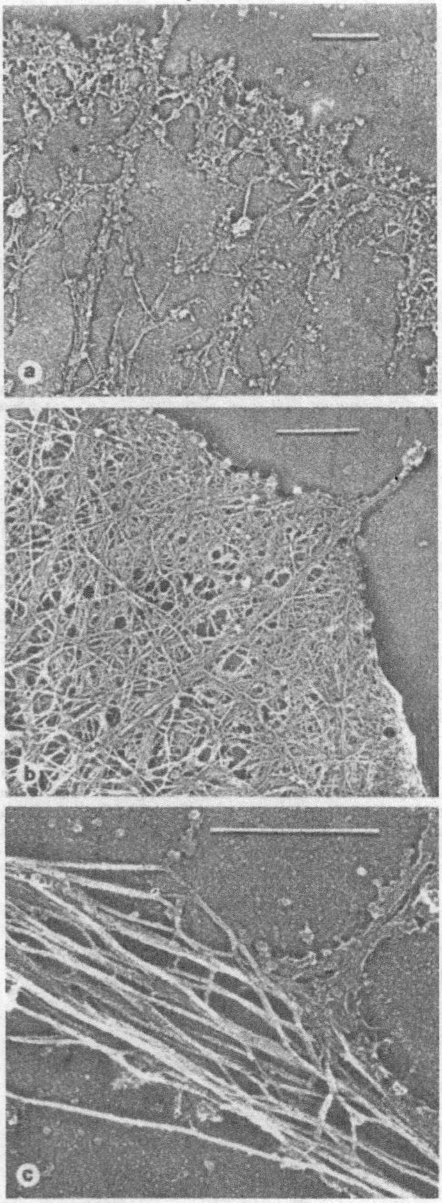

FIG. 3. Electron microscopy of platinum replicas of the cytoskeleton. (*a*) Microfilament meshwork near the leading edge of a control cell. (*b*) More dense and wide network in the lamella of a PMA-treated cell. (*c*) Microtubules and intermediate filaments in the narrow process of a PMA-treated cell. (Bars = 0.5 μm.)

cells revealed a dense network of microfilaments at the active edge of lamella; the zone of sparsely spaced microfilaments was adjacent to this marginal zone. The marginal microfilamental meshwork at the active edges of lamellas of PMA-treated cells was more wide than in the control cells and consisted of longer fragments of microfilaments. The adjacent sparse zone usually disappeared. The narrow processes of PMA-treated cells contained numerous longitudinally oriented microtubules and intermediate filaments. Microfilaments were present mostly in the form of meshwork-like patches at the periphery of processes and in the bulbs.

Effects of Inhibitors. The 152/15 cells incubated with colcemid (1 μg/ml) for 1 hr lost their elongated shape and acquired irregular polygonal contour with short thick processes extended from all parts of the edge (1). These cells did not contain microtubules; intermediate filaments collapsed around the nucleus. Short and thin actin bundles were randomly distributed in the cytoplasm. Addition of PMA to the colcemid-containing medium did not induce any additional morphological changes of these cells.

In other experiments the cells were first incubated with PMA for 2–3 hr and then colcemid was added to the same medium. In this case colcemid caused gradual disappearance of large lamellas and of narrow processes (Fig. 4), so that the cells incubated for 1–2 hr in the medium containing both PMA and colcemid became morphologically indistinguishable from those treated with colcemid alone. Cytochalasin B (10 μg/ml) caused characteristic arborization of the cells. Subsequent addition of PMA to the medium did not induce further changes of shape. When the cells were pretreated with PMA for 2 hr and then cytochalasin B was added to the medium for 1 hr, arborization of lamellas was observed, but the narrow processes were not considerably changed. When colcemid was added to the medium of these cells, it did not induce any additional morphological alterations.

DISCUSSION

Numerous effects of PMA on various cultured cells have been described (see reviews in refs. 7–9). The list of these effects includes some morphological alterations, such as disappearance of actin bundles (9), induction of ruffles (9), and contraction of the cell body (10). In our experiments these changes were associated with another previously unknown effect—namely, with the reversible division of cell into several parts: motile cell body and stable elongated processes. These processes were formed from the retracted lamellas; later they often were elongated at their proximal ends because of contraction of the body, which moved away from these ends. The immunomorphological and electron microscopic examination had shown that lamellas and the cell body of PMA-treated cells contain extensive actin networks. The relationship of the number of actin microfilaments to that of microtubules was much lower in the narrow processes than in the lamellas. We have to conclude that, when the processes are formed from lamellas and tail parts of the body, polymerized or depolymerized actin moves away from them along the microtubular framework. PMA did not cause any shape alterations in the cells pretreated either with colcemid or with cytochalasin B. Thus, both microtubules and microfilaments are essential for the development of PMA-induced alterations of shape. One possible suggestion is that PMA somehow detaches actin microfilaments from the microtubules and/or intermediate filaments.

The network of actin microfilaments in the lamellas of untreated fibroblasts develops centripetal tension (1–3). Actin cortex of PMA-treated fibroblasts may be reversibly detached from the other components of cytoskeleton; when this cortex contracts, microtubules and intermediate filaments are left behind. Stable narrow processes are, possibly,

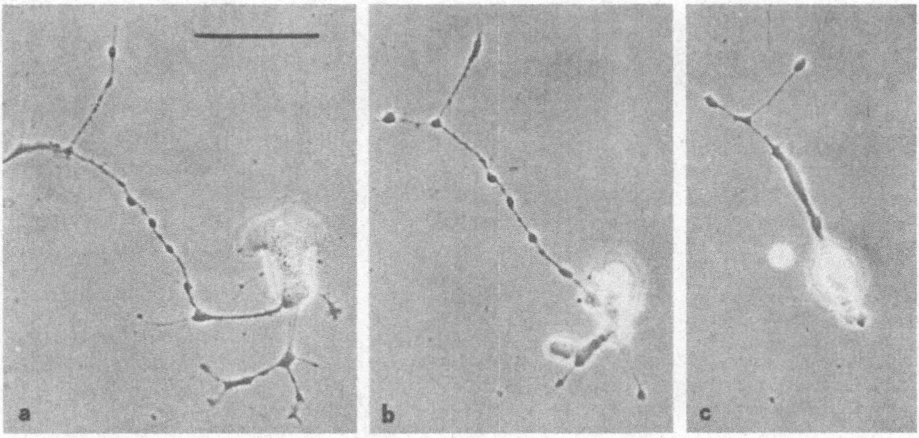

FIG. 4. Phase-contrast micrographs of the cell of 152/15 line preincubated for 1 hr with PMA before the addition of colcemid (1 μg/ml) (*a*) and 15 min and 45 min after the addition (*b* and *c*, respectively). Note the progressing shortening of the long process and translocation of the cell body toward the end of the process. (Bar = 50 μm.)

formed in this way. Actin-rich "bulbs" are, possibly, the fragments of contracted cortex remaining on these processes. Actin transported from the old lamellas may be used for the formation of microfilamental network within the newly extended lamellas.

Microtubules seem to be essential not only for the transport of actin from lamellas but also for the prevention of its return into the stable processes. Destruction of microtubules by colcemid removes the ban for the restoration of actin network and leads to contraction of the processes; the inhibitory effect of cytochalasin B confirms the essential role of microfilaments in this contraction (see also ref. 11).

The molecular mechanisms of the effects of PMA on the cytoskeleton remain completely obscure. PMA is a specific activator of protein kinase C participating in the transduction of membrane-generated signals (12). The long list of proteins phosphorylated by this enzyme contains, among other substrates, some cytoskeleton proteins such as vinculin, certain microtubule-associated proteins, etc. (13). Possibly, phosphorylation of one of these substrates modifies in some way the interactions of microtubules with actin cortex.

Factors determining the relative sensitivity of various types of fibroblasts to the induction of shape changes by PMA remain unknown. The alterations of cell shape, like many other effects of PMA, are spontaneously reversed after prolonged incubation. Possibly, this reversion is due to the down-regulation of the main cellular target of PMA, protein kinase C (14).

We suppose that movements of actin cortex along the microtubular framework take place not only in PMA-treated fibroblasts of certain lines but also in many cell types during their normal morphogenesis. For instance, the growing processes of the neuron consist of stable parts rich in microtubules and of motile lamellar growth cones rich in microfilaments. When the growth cone moves forward, its proximal part retracts and is transformed into the more narrow stable processes. The moving untreated fibroblast continually forms actin-rich lamellas and narrow tails; possibly, actin moves continually toward the leading edge. Dissociation of actin from the other components of cytoskeleton may be less complete in these untreated cells than in PMA-treated ones because their tails remain contractile and repeatedly disappear during movement. In other words, PMA possibly exaggerates normal reorganizations essential for locomotion. Thus, experiments described in this paper suggest that the cells have a membrane-activated mechanism that reversibly separates large or small motile parts from stable cell parts. This mechanism possibly operates by moving actin along microtubules.

1. Vasiliev, J. M. & Gelfand, I. M. (1981) *Neoplastic and Normal Cells in Culture* (Cambridge Univ. Press, Cambridge).
2. Vasiliev, J. M. (1985) *Biochim. Biophys. Acta* **780**, 21–65.
3. Bellairs, R., Curtis, A. & Dunn, G., eds. (1982) *Cell Behaviour: A Tribute to Michael Abercrombie* (Cambridge Univ. Press, Cambridge).
4. Bershadsky, A. D., Brodskaya, R. M., Mansurov, P. G., Stavrovskaya, A. A. & Stromskaya, T. B. (1984) *Eksp. Onkol.* **6**, 27–32.
5. Bershadsky, A. D., Gelfand, V. I., Svitkina, T. M. & Tint, I. S. (1980) *Exp. Cell Res.* **127**, 423–431.
6. Svitkina, T. M., Shevelev, A. A., Bershadsky, A. D. & Gelfand, V. I. (1984) *Eur. J. Cell Biol.* **34**, 64–74.
7. Borzsonyi, M., Lapis, K., Day, N. E. & Yamasaki, H., eds. (1984) *Models, Mechanisms and Etiology of Tumor Promotion* (Int. Agency Res. Cancer, Lyon, France).
8. Slaga, T. J., ed. (1984) *Mechanisms of Tumor Promotion: Cellular Responses to Tumor Promoters* (CRC, Boca Raton, FL), Vol. 4.
9. Schliwa, M., Nakamura, T., Porter, K. R. & Euteneuer, U. (1984) *J. Cell Biol.* **99**, 1045–1059.
10. Driedger, P. E. & Blumberg, P. M. (1977) *Cancer Res.* **37**, 3257–3265.
11. Solomon, F. & Magendantz, M. (1981) *J. Cell Biol.* **89**, 157–161.
12. Nishizuka, Y. (1984) *Nature (London)* **308**, 693–698.
13. Takai, Y., Kaibuchi, K., Tsuda, T. & Hoshijima, M. (1985) *J. Cell. Biochem.* **29**, 153–155.
14. Vara, F., Schneider, J. A. & Rosengurt, E. (1985) *Proc. Natl. Acad. Sci. USA* **82**, 2384–2388.

Part V

General theory of hypergeometric functions

1.
General theory of hypergeometric functions

Dokl. Akad. Nauk SSSR **288** (1) (1986) 14–18 [Sov. Math., Dokl. **33** (1986) 573–577]

The aim of this and subsequent publications is to present a general theory of hypergeometric functions of several variables.

1. Definition of the hypergeometric function. The natural domain of the general hypergeometric function is a certain line bundle over a Grassmann manifold. Let $G_{k,n}$ be a real Grassmann manifold, i.e., a manifold of k-dimensional linear subspaces of the real n-dimensional space $\mathbf{R}^n$. Let $\tilde{G}_{k,n}$ be a manifold whose points are the pairs (ς, ω), where $\varsigma \in G_{k,n}$ and ω is a nonsingular exterior k-form on the subspace ς, i.e., a nonzero element of the highest exterior power of the dual space ς', $\omega \in \wedge^k \varsigma'$. The natural mapping $\theta: \tilde{G}_{k,n} \to G_{k,n}$ is a fibering with the fiber $\mathbf{R}^*$.

The general hypergeometric function depends on two sets of parameters: a sequence of complex numbers $\alpha = (\alpha_1, \dots, \alpha_n)$ satisfying the condition

$$\sum \alpha_j = n - k, \tag{1}$$

and a sequence $\varepsilon = (\varepsilon_1, \dots, \varepsilon_n)$, $\varepsilon_j = \pm 1$.

We define for each j a homogeneous generalized function π_j on $\mathbf{R}$ of degree of homogeneity $\alpha_j - 1$ by setting

$$\pi_j(x) = \begin{cases} x_+^{\alpha_j - 1}, & \text{if } \varepsilon_j = +1, \\ x_-^{\alpha_j - 1}, & \text{if } \varepsilon_j = -1. \end{cases}$$

Further, for each $(\varsigma, \omega) \in \tilde{G}_{k,n}$, we define a differential form $\hat{\omega}$ of order $k-1$ on ς as follows. Let $I = \sum x_j \partial/\partial x_j$ be a radial vector field on $\mathbf{R}^n$ (an Euler operator). We set $\hat{\omega} = i_I \omega$ (i_I denoting inner derivation along I). In the system of coordinates $t_1, \dots, t_k$ on ς, in which $\omega = c\,dt_1 \wedge \cdots \wedge dt_k$, $c \neq 0$, we have

$$\hat{\omega} = c \sum_{i=1}^{k} (-1)^{i+1} t_i dt_1 \wedge dt_2 \wedge \cdots \wedge dt_{i-1} \wedge dt_{i+1} \wedge \cdots \wedge dt_k. \tag{2}$$

DEFINITION. The *general hypergeometric function* $\Phi(\alpha, \varepsilon; \tilde{\varsigma})$ on $\tilde{G}_{k,n}$ is defined by the integral

$$\Phi(\alpha, \varepsilon; \tilde{\varsigma}) = \int_S \prod_{j=1}^{n} \pi_j(x_j(v))\hat{\omega}(v), \tag{3}$$

where $\tilde{\varsigma} = (\varsigma, \omega) \in \tilde{G}_{k,n}$, $x_j(v)$ being the jth coordinate of the point $v \in \mathbf{R}^n$ and S being an arbitrary $(k-1)$-dimensional hypersurface in ς enclosing the point 0 (for example, a sphere). It follows from the homogeneity of π_j, condition (1), and the definition of $\hat{\omega}$ that the integral does not depend on the choice of S.(1)

1980 *Mathematics Subject Classification* (1985 *Revision*). Primary 33A30; Secondary 14M15, 51M35, 14L30.

(1) Actually, the integration is carried out over that part $S^{(\varepsilon)}$ of the surface S where $\operatorname{sgn} x_j(v) = \varepsilon_j$ for $j = 1, \dots, n$. For some ε and ς, this region may be empty so that $\Phi(\alpha, \varepsilon; \tilde{\varsigma}) = 0$ for all α. Questions concerning the structure of the regions $S^{(\varepsilon)}$ have been considered in another connection by A. M. Vershik.

2. Analytic continuation. Formula (3) gives $\Phi(\alpha,\varepsilon;\tilde{\varsigma})$ only for those α for which the integral converges. To define $\Phi(\alpha,\varepsilon;\tilde{\varsigma})$ for other α, it is necessary to use analytic continuation in α. We describe it more explicitly.

a) Suppose first that $\tilde{\varsigma}=(\varsigma,\omega)\in\tilde{G}_{k,n}$ is such that the subspace ς is in general position relative to the coordinate axes in $\mathbf{R}^n$ (i.e., the intersection of ς with any subspace $x_{j_1}=\dots=x_{j_p}=0$ in $\mathbf{R}^n$ has dimension not greater than $k-p$). In this case, integral (3) converges if $\operatorname{Re}\alpha_j>0$ for all $j=1,\dots,n$, and defines an analytic function of α. We define $\Phi(\alpha,\varepsilon;\tilde{\varsigma})$ as the analytic continuation of this function.

b) For some special subspaces ς, it may turn out that the integral (3) is divergent for all α satisfying (1). In this case, we use the following method of partitions. Let S_j be the region in S given by the condition $|x_j(v)|>a$ for some $a>0$. For sufficiently small a, the regions S_j form a covering of S. Let $1=\sum\varphi_j$ be a partition of unity corresponding to this covering. Clearly, if $\alpha_1,\dots,\alpha_n$ satisfying (1) are such that $\operatorname{Re}\alpha_j>0$ for $j\neq p$, then

$$I_p(\alpha,\varepsilon;\tilde{\varsigma})=\int_S\varphi_p(v)\prod_{j=1}^{n}\pi_j(x_j(v))\hat{\omega}(v)$$

converges and defines an analytic function of α. We continue $I_p(\alpha,\varepsilon;\tilde{\varsigma})$ analytically, and set $\Phi(\alpha,\varepsilon;\tilde{\varsigma})=\sum I_p(\alpha,\varepsilon;\tilde{\varsigma})$. Standard arguments show that the function $\Phi(\alpha,\varepsilon;\tilde{\varsigma})$ thus defined does not depend on a or the partition of unity. By using the explicit form of analytic continuation of functions of the type $x_{\pm}^{\alpha-1}$, one can prove the following proposition.

PROPOSITION 1. *For each $\tilde{\varsigma}\in\tilde{G}_{k,n}$ and each ε, the function $\Phi(\alpha,\varepsilon;\tilde{\varsigma})$ is a single-valued meromorphic function of α.*

3. Other methods of defining the hypergeometric function. These methods are connected with the fact that every k-dimensional subspace in $\mathbf{R}^n$ can be defined either as the image of an imbedding of $\mathbf{R}^k$ in $\mathbf{R}^n$ or as the kernel of a projection of $\mathbf{R}^n$ onto $\mathbf{R}^{n-k}$. We give the formulas relating to the first method of definition. Let $t_1,\dots,t_k$ be coordinates in $\mathbf{R}^k$. Any imbedding $\mathbf{R}^k\to\mathbf{R}^n$ is given by a $k\times n$ matrix $x=(x_{ij})$ of rank k. We define a $(k-1)$-form $\hat{\omega}$ on $\mathbf{R}^k$ by formula (2) with $c=1$, and define generalized functions π_j as in §1. Let

$$\Psi(\alpha,\varepsilon;x)=\int_S\prod_{j=1}^{n}\pi_j\left(\sum_{i=1}^{k}x_{ij}t_i\right)\hat{\omega}(t),$$

where S is a hypersurface in $\mathbf{R}^k$ enclosing the origin. It is easy to see that $\Psi(\alpha,\varepsilon;x)=\Phi(\alpha,\varepsilon;\tilde{\varsigma})$, where $\tilde{\varsigma}=(\varsigma,\omega)$, ς being the image of $\mathbf{R}^k$ under the imbedding given by the matrix x and ω being the image of the form $dt_1\wedge\dots\wedge dt_k$.

4. Symmetry properties. a) We shall denote by H the group of diagonal matrices of order n with positive elements. The group H acts naturally on $\mathbf{R}^n$, and this action can be extended to the action of H on $G_{k,n}$ and $\tilde{G}_{k,n}$. It follows easily from the definition of the hypergeometric function that

$$\Phi(\alpha,\varepsilon;h\tilde{\varsigma})=\prod_{j=1}^{n}\delta_j^{-\alpha_j+1}\Phi(\alpha,\varepsilon;\varsigma)\tag{4}$$

for $h=\operatorname{diag}(\delta_1,\dots,\delta_n)\in H$ and $\tilde{\varsigma}\in\tilde{G}_{k,n}$. In particular, taking $h=\operatorname{diag}(\delta,\dots,\delta)$ we find that on each fiber of the bundle $\tilde{G}_{k,n}\to G_{k,n}$ the function Φ is a homogeneous function of degree $+1$ relative to $\delta\in\mathbf{R}$, $\delta>0$.

b) The symmetric group S_n of degree n acts naturally on $\mathbf{R}^n$ (by permutations of basis vectors), and this action can be extended to the action of S_n on $G_{k,n}$ and $\tilde{G}_{k,n}$. In addition, S_n acts on the collections α and ε (by permutations of elements). We have

$$\Phi(w\alpha, w\varepsilon; w\tilde{\varsigma}) = \Phi(\alpha, \varepsilon; \tilde{\varsigma}), \quad w \in S_n. \tag{5}$$

Both the above symmetries express important properties of the hypergeometric function. Equality (4) shows the necessity of passing to the orbits of the action of H in $\tilde{G}_{k,n}$, which we shall do in §5 below. Equalities (5) give the discrete symmetries of general hypergeometric functions. We shall not take them up in this paper.

5. The action of H on a Grassmann manifold. Strata. For each subset $J \subset I_n = \{1, \ldots, n\}$, we shall denote by $\mathbf{R}^J$ the subspace in $\mathbf{R}^n$ spanned by the basis vectors e_j, $j \in J$. We shall say that $\varsigma \in G_{k,n}$ is not singular relative to a k-element subset $J \subset I_n$ if $\varsigma \cap \mathbf{R}^{I_n \setminus J} = \{0\}$. We shall call any nonempty collection of k-element subsets of I_n a list Ξ.

DEFINITION. A *stratum of type* Ξ in $G_{k,n}$ is a subset $G_{k,n}^{\Xi} \subset G_{k,n}$ consisting of all ς which are not singular for precisely all the $J \in \Xi$. A stratum of type Ξ in $\tilde{G}_{k,n}$ is a subset $\tilde{G}_{k,n}^{\Xi} \subset \tilde{G}_{k,n}$ consisting of all (ς, ω) with $\varsigma \in G_{k,n}^{\Xi}$.

PROPOSITION. *Every stratum $\tilde{G}_{k,n}^{\Xi}$ is invariant relative to the action of H in $\tilde{G}_{k,n}$, and the mapping $\tilde{G}_{k,n}^{\Xi} \to \tilde{G}_{k,n}^{\Xi}/H$ is a fibering.*

A similar assertion is valid for $G_{k,n}^{\Xi}$.

Another description of strata in $G_{k,n}$ can be given with the help of the moment mapping. For this, let us recall the definition of Plücker coordinates in $\tilde{G}_{k,n}$. Let $J = \{j_1, \ldots, j_k\}$ be a k-element subset of I_n and let $E_J: \varsigma \to \mathbf{R}^J$ be the operator of projection on $\mathbf{R}^J$ parallel to the complementary subspace $\mathbf{R}^{I_n \setminus J}$. The Plücker coordinate p_J of the point $\tilde{\varsigma} = (\varsigma, \omega)$ is defined as a quotient of volume forms $p_J = E_J^*(dx_{j_1} \wedge \cdots \wedge dx_{j_k})/\omega$ (it should be recalled that the form ω is not singular). The Plücker coordinates on $G_{k,n}$ are the corresponding homogeneous coordinates (sets p_J to within a factor). Plücker coordinates satisfy a set of well-known quadratic relations (see [1]).

The action of H on $\tilde{G}_{k,n}$ enables one to determine a moment mapping μ from $\tilde{G}_{k,n}$ into the Lie algebra of the group H (which we shall identify with $\mathbf{R}^n$) (the general definition can be seen in [2] and [3]). An explicit description of μ in our case is as follows. For each $J = \{j_1 < \cdots < j_k\} \subset I_n$, let us denote by e_J the vector in $\mathbf{R}^n$ in which the coordinates with the indices j_s are equal to 1 while the other coordinates are equal to 0. Then $\mu(\tilde{\varsigma}) = \sum |p_J|^2 e_J / \sum |p_J|^2 \in \mathbf{R}^n$, where the sum is taken over all k-element subsets of I_n, p_J being the Plücker coordinates of $\tilde{\varsigma}$. The image of any H-orbit under the mapping μ is a convex polyhedron in $\mathbf{R}^n$ spanned by some of the vectors e_J (see [4]; the general convexity theorem for the action of any Lie group is proved in [2] and [3]). It is easy to verify that the stratum is the union of all H-orbits in $\tilde{G}_{k,n}$ with the same image relative to μ (more precisely, the image of μ for orbits from the stratum $\tilde{G}_{k,n}^{\Xi}$ is the convex hull of e_J, $J \in \Xi$).

6. Configurations. The orbits of the action of H on $\tilde{G}_{k,n}$ and on $G_{k,n}$ are conveniently described with the help of configurations, i.e., sets of $p \leq n$ points in $(k-1)$-dimensional projective space defined to within a projective transformation. Let $\varsigma \in G_{k,n}$. We choose an imbedding $\psi: \mathbf{R}^k \to \mathbf{R}^n$ with image ς. Let $y_j = \psi'(x_j) \in (\mathbf{R}^k)'$ be the images of the coordinate functions $x_j \in (\mathbf{R}^n)'$ under the transformation $\psi': (\mathbf{R}^n)' \to (\mathbf{R}^k)'$ adjoint to ψ. Let us first assume that all the y_j are different from 0. Clearly, the images $[y_j]$ of the points y_j in the $(k-1)$-dimensional projective space $\mathbf{P}((\mathbf{R}_k)')$ do not change under the action of H on $G_{k,n}$, and, for a different choice of the imbedding ψ with the same image ς, this set undergoes a projective transformation in $\mathbf{P}^{k-1}$. This collection of n points $[y_1], \ldots, [y_n] \in \mathbf{P}^{k-1}$ will be called *the configuration corresponding to the orbit* *of ς under the action of H.* With those ς for which some of the y_j are equal to 0 we associate a configuration of $p < n$ points in $\mathbf{P}^{k-1}$ corresponding to only the nonzero $y_j \in (\mathbf{R}^k)'$. The fact that the H-orbits in $G_{k,n}$ and $\tilde{G}_{k,n}$ belong to strata is described by linear dependences between the pooints of the configuration.

7. Hypergeometric functions on strata.

DEFINITION. By a *hypergeometric function of type* Ξ we understand the restriction $\Phi^{\Xi}(\alpha, \varepsilon; \tilde{\varsigma})$ of the function $\Phi(\alpha, \varepsilon; \tilde{\varsigma})$ to $\tilde{G}^{\Xi}_{k,n}$.

PROPOSITION. 1) *For fixed α and ε, the hypergeometric function $\Phi^{\Xi}(\alpha, \varepsilon; \tilde{\varsigma})$ is a real analytic function of $\tilde{\varsigma} \in \tilde{G}^{\Xi}_{k,n}$.*

2) *If the isotropy subgroup of the point $\tilde{\varsigma} \in \tilde{G}^{\Xi}_{k,n}$ is nontrivial in H (the stratum $\tilde{G}^{\Xi}_{k,n}$ is called singular in this case), then $\Phi^{\Xi}(\alpha, \varepsilon; \tilde{\varsigma}) = 0$ for any α and ε.*

It follows from the homogeneity condition (4) that the function $\Phi^{\Xi}(\alpha, \varepsilon; \tilde{\varsigma})$ can be uniquely recovered from its values on any second Γ of the fibering $\tilde{G}^{\Xi}_{k,n} \to \tilde{G}^{\Xi}_{k,n}/H$. For a suitable choice of k, n, Ξ, and Γ, one can obtain many classical hypergeometric functions of one and several variables.

8. Examples. a) $k = 2, n = 4, \varepsilon_j = +1$ for all j, $\tilde{G}^{\Xi}_{2,4}$ being the stratum of planes in a general position in $\tilde{G}_{2,4}$, and p_{ij} being Plücker coordinates in $\tilde{G}_{2,4}$. The quotient space $\tilde{G}^{\Xi}_{2,4}/H$ is one-dimensional and is parametrized by the cross-ratio of four points on $\mathbf{P}^1$ forming the configuration corresponding to $\tilde{\varsigma}$. On taking as Γ the section defined by the equalities $p_{12} = p_{32} = p_{42} = p_{31} = 1$, we get the Gauss hypergeometric function

$$\Phi^{\Xi}(\alpha, \varepsilon; \tilde{\varsigma}) = (\Gamma(b)\Gamma(c-b)/\Gamma(c))F(a, b, c; p_{41}),$$

with $\alpha_1 = a - c + 1, \alpha_2 = b, \alpha_3 = c - b, \alpha_4 = -a + 1$, and $0 < p_{41} < 1$.

b) We indicate with which strata certain other classical hypergeometric functions are connected. For $k = 2$, Ξ being all two-subsets in $\{1, \ldots, n\}$ (planes in general position in $\mathbf{R}^n$), we get the Appell function F_1 for $n = 5$ and the Lauricella function F_D for $n > 5$ (the definitions of F_1 and F_D can be found in [5]). The Appell functions F_2 and F_3 correspond to the stratum $\tilde{G}^{\Xi}_{3,6}$, where Ξ consists of all 3-subsets in $\{1, \ldots, 6\}$ except $\{1, 2, 3\}$ and $\{1, 4, 5\}$ (F_2 and F_3 differ in the choice of the section Γ).

9. Concluding remarks. We shall briefly indicate some further problems which relate to general hypergeometric functions. Some of these problems will be considered in subsequent publications.

a) The continuation of hypergeometric functions into the complex domain (in the variable $\tilde{\varsigma}$) is very interesting. One method consists in defining an integral analogous to (3) over certain $(k-1)$-dimensional cycles in a complex k-dimensional space (instead of an integral over S). A second method consists in defining hypergeometric functions as the solutions of a holonomic system of differential equations.

b) As a special case of the hypergeometric function we have the so-called continuous analogue of the Kostant function, introduced in [6].

c) The general hypergeometric function Φ can be regarded as the Radon transformation of the function $\prod (x_j)_{\pm}^{\alpha_j - 1}$ on $\mathbf{R}^n$. This remark points to an important connection between the theory of hypergeometric functions and integral geometry.

d) An interesting class is constituted by the hypergeometric functions Φ^{Ξ}, connected with the strata $\tilde{G}^{\Xi}_{k,n}$, consisting of a single H-orbit. As the simplest example here we have the B-function of Euler (the corresponding configuration being three points on the projective line).

e) The general hypergeometric functions constructed above admit q-analogues which lead to general hypergeometric q-series.

In conclusion, I would like to thank S. I. Gel'fand, M. I. Graev, and V. V. Serganov, discussions with whom were extremely essential for writing this paper.

Keldysh Institute of Applied Mathematics
Academy of Sciences of the USSR
Moscow

Received 24/FEB/86

BIBLIOGRAPHY

1. Phillip Griffiths and Joseph Harris, *Principles of algebraic geometry*, Wiley, 1978.
2. V. Guillemin and S. Sternberg, Invent. Math. **67** (1982), 491–513.
3. M. F. Atiyah, Bull. London Math. Soc. **14** (1982), 1–15.
4. I. M. Gel'fand and R. D. MacPherson, Advances in Math. **44** (1982), 279–312.
5. A. Erdélyi et al., *Higher transcendental functions*. Vol. 1, McGraw-Hill, 1953.
6. F. A. Berezin and I. M. Gel'fand, Trudy Moskov. Mat. Obshch. **5** (1956), 311–351; English transl. in Amer. Math. Soc. Transl. (2) **21** (1962).

Translated by V. N. SINGH

2.
(with S. I. Gelfand)

Generalized hypergeometric equations

Dokl. Akad. Nauk SSSR **288** (2) (1986) 279–283
[Sov. Math., Dokl. **33** (1986) 643–646]

It is the purpose of this note to study a system of linear partial differential equations satisfied by the generalized hypergeometric functions introduced in [1].

Notation. $Z = Z_{k,n}$, $k < n$, is the space of $k \times n$ complex matrices

$$z = \begin{pmatrix} z_{11} \cdots z_{1n} \\ \cdots\cdots\cdots \\ z_{k1} \cdots z_{kn} \end{pmatrix}.$$

On $Z_{k,n}$ we define a left action of $\mathrm{GL}(k, \mathbf{C})$ and a right action of the diagonal subgroup $H_n \subset \mathrm{GL}(n, \mathbf{C})$ (by matrix multiplication).

$Z' \subset Z_{k,n}$ is the open dense subset consisting of matrices z of rank k (i.e., the rows of z are linearly independent); $Z'' \subset Z'$ is the open dense subset consisting of those z such that every minor of order k in z is different from 0.

$G_{k,n}$ is the Grassmann manifold of k-dimensional subspaces in $\mathbf{C}^n$, and $\psi\colon Z' \to G_{k,n}$ is the natural projection ($\psi(z) = \{$the subspace spanned by the rows of $z\}$); ψ is the principal bundle with fiber $\mathrm{GL}(k, \mathbf{C})$; in particular, ψ is an affine morphism.

1. Hypergeometric equations. We let $\alpha_1, \ldots, \alpha_n$ be n complex numbers. The hypergeometric equations in $\Phi(z)$, $z \in Z_{k,n}$, split into three groups:

a) H_n-homogeneity:

$$\sum_{i=1}^{k} z_{ip}\frac{\partial \Phi}{\partial z_{ip}} - (\alpha_p - 1)\Phi = 0, \qquad p = 1, \ldots, n; \tag{1}$$

b) $\mathrm{GL}(k, \mathbf{C})$-invariance:

$$\sum_{p=1}^{n} z_{ip}\frac{\partial \Phi}{\partial z_{jp}} - n^{-1}\sum(\alpha_p - 1)\delta_{ij}\Phi = 0, \qquad i, j = 1, \ldots, k, \tag{2}$$

where δ_{ij} is the Kronecker symbol;

c) second-order equations:

$$\frac{\partial^2 \Phi}{\partial z_{ip}\partial z_{jq}} - \frac{\partial^2 \Phi}{\partial z_{iq}\partial z_{jp}} = 0, \qquad i, j = 1, \ldots, k,\ p, q = 1, \ldots, n. \tag{3}$$

It is not difficult to verify that the generalized hypergeometric function defined (in a complex region) in [1] satisfies (1)–(3). In addition, the equations (2) guarantee that every solution Φ is invariant under the action of $\mathrm{SL}(k, \mathbf{C}) \subset \mathrm{GL}(k, \mathbf{C})$ on Z. Therefore the restriction of Φ to Z' gives a function $\tilde{\Phi}$ on $G_{k,n}$; we set $\tilde{\Phi}(\psi(z)) = \Phi(z)$.

2. $\mathcal{D}$-modules. We denote by M the left module over the ring of analytic differential operators on Z (more precisely, the sheaf of modules over the sheaf of rings of differential

---

1980 *Mathematics Subject Classification* (1985 *Revision*). Primary 58G07, 32C38.

operators on Z) which is the quotient of $\mathcal{D}_Z$ by the ideal spanned by the left sides of (1)–(3), i.e., by the operators

$$\sum_{i=1}^{k} z_{ip}\frac{\partial}{\partial z_{ip}} - (\alpha_p - 1), \qquad p = 1, \dots, n,$$

$$\sum_{p=1}^{n} z_{ip}\frac{\partial}{\partial z_{jp}} - n^{-1}\sum(\alpha_p - 1)\delta_{ij}, \qquad i, j = 1, \dots, k;$$

$$\frac{\partial^2}{\partial z_{ip}\partial z_{jq}} - \frac{\partial^2}{\partial z_{iq}\partial z_{jp}}, \qquad i, j = 1, \dots, k,\ p, q = 1, \dots, n.$$

We call M a *hypergeometric module.*

Further, let M' be the restriction of M to Z', and let $M_G = \psi_*(M')$ be a sheaf of modules over the ring $\mathcal{D}_G$ of analytic differential operators over $G_{k,n}$ (here ψ_* is the direct image functor in the category of $\mathcal{D}$-modules [2]; since ψ is an affine morphism, ψ_* is an exact functor). It is easy to verify that M_G is a module with a single generator; therefore it is given by a system of differential equations on $G_{k,n}$. This system is uniquely determined by the fact that it is satisfied by those and only those locally analytic functions f on $G_{k,n}$ whose liftings over Z' (i.e., the same functions f regarded as functions of z) satisfy system (1)–(3). We present these equations on $G_{k,n}$ below for $k = 2$.

3. The following theorem constitutes the main result of this note.

THEOREM. *M' and M_G are holonomic $\mathcal{D}$-modules over Z' and $G_{n,k}$, respectively.*

We get the following assertion from this theorem (more precisely, from the lemma of §6 below).

COROLLARY. *The solutions of system* (1)–(3) *in an arbitrary open region $U \subset Z''$ form a finite-dimensional space $E_{k,n}(U)$ of analytic functions; in particular, the analytic continuations of the hypergeometric integrals in* [1] *generate a finite-dimensional space.*

For $k = 2$ (for sufficiently small U), $\dim E_{2,n}(U) = n - 2$; for arbitrary k we have only the estimate $\dim E_{k,n}(U) \le C_{n-2}^{k-1}$.

4. The proof of the theorem occupies §§4–8. It suffices to show that M' is holonomic, since the functor ψ_* carries holonomic modules into holonomic modules (see [2]).

Let T^*Z' be a cotangent bundle in Z'; let the point $y \in T^*Z'$ be the pair $y = (z, \xi)$, where

$$z \in Z', \qquad \xi = \begin{pmatrix} \xi_{11} \cdots \xi_{1n} \\ \cdots\cdots\cdots \\ \xi_{k1} \cdots \xi_{kn} \end{pmatrix},$$

and the standard symplectic 2-form ω on T^*Z has the form $\omega = \sum_{i,p} dz_{ip} \wedge d\xi_{ip}$.

The fact that M is holonomic means that the dimension of the characteristic variety $V \subset T^*Z'$ is half that of T^*Z, i.e.,

(4) $$\dim V = nk.$$

By definition, V is the set of common zeros of the principal symbols of equations (1)–(3) and the equations that are (differential) consequences of (1)–(3). Instead of V, we study the variety $W \subset T^*Z'$ which is defined as the set of common zeros of the principal symbols of (1)–(3) themselves. Thus we define W as the set of pairs $(z, \xi) \in T^*X$ which satisfy the following equations:

(5) $$\sum_{i=1}^{k} z_{ip}\xi_{ip} = 0, \qquad p = 1, \dots, n,$$

$$\sum_{p=1}^{n} z_{ip}\xi_{jp} = 0, \qquad i,j = 1,\dots,k, \tag{6}$$

$$\xi_{ip}\xi_{jq} - \xi_{iq}\xi_{jp} = 0, \qquad i,j = 1,\dots,k,\ p,q = 1,\dots,n. \tag{7}$$

Since $W \subset V$, to prove the theorem it suffices to verify that $\dim W = nk$.

5. First of all, (7) implies that the rank of ξ is not greater than 1. Therefore there exist vectors $a = (a_1,\dots,a_k)$ and $b = (b_1,\dots,b_n)$ such that

$$\xi_{ip} = a_i b_p. \tag{8}$$

Moreover, if $\xi \neq 0$, then a and b are determined uniquely to within a transformation $(a,b) \to (\lambda a, \lambda^{-1} b)$, $\lambda \neq 0$.

If we substitute (8) into (5) and (6), we have

$$b_p \sum_{i=1}^{k} z_{ip} a_i = 0, \qquad p = 1,\dots,n; \tag{9}$$

$$a_j \sum_{p=1}^{n} z_{ip} b_p = 0, \qquad i,j = 1,\dots,k. \tag{10}$$

It is clear that $\xi = 0$ always satisfies (5)–(7). Therefore, in particular, $\dim W \geq \dim Z' = nk$. We show first of all that if $z \in Z''$, then there are no other solutions.

6. LEMMA. *Suppose that $z \in Z''$. Then $\xi = 0$ is the only solution of* (5)–(7).

PROOF. Suppose, on the contrary, that there exists a $\xi \neq 0$ which satisfies (5)–(7). Then $\xi_{ip} = a_i b_p$, $a \neq 0$, $b \neq 0$; taking j such that $a_i \neq 0$ in (10), we get

$$\sum_{p=1}^{n} z_{ip} b_p = 0, \qquad i = 1,\dots,k. \tag{11}$$

We let $P = P_b$ be the set of indices p for which $b_p \neq 0$. Then from the condition $z \in Z''$ and from (11) we have that $|P| > k$. On the other hand, by virtue of (9), $\sum z_{ip} a_i = 0$ for $p \in P$, so that again in view of the conditions $a \in Z''$ and $a \neq 0$, we get that $|P| < k$. This contradiction proves the lemma.

7. We turn now to the general case. For each subset $P \subset \{1,\dots,n\}$ and for $z \in Z$, we let z_P be the $k \times |P|$ submatrix of z consisting of those columns of z whose indices are in P. We also put $r_P(z) = \operatorname{rk} z_P$. For fixed P and a fixed number r we let $Z(P,r)$ be the subset of Z' consisting of those matrices z which satisfy the following two conditions:

a) $r_P(z) = r$;

b) $r_{P-\{p\}}(z) = r$ for every $p \in P$.

We note that $Z(P,r)$ is nonempty if and only if $r < |P|$, $r \leq k$, and $k - r \leq n - |P|$. Moreover, $Z(P,r)$ is a locally closed algebraic subset of Z' whose closure consists of all matrices which satisfy a).

8. LEMMA. a) *Suppose that for a given z, there exists a nonzero ξ which satisfies* (5)–(7). *Let a and b be the vectors which correspond to ξ under* (8), *and let $P = P_b$ be the set of indices p for which $b_p \neq 0$. Then $z \in Z(P,r)$ for some r.*

b) *Suppose that $z \in Z(P,r)$. Then the set of pairs (z,ξ), $\xi \neq 0$, which satisfy* (5)–(7) *and for which $P_b = P$ is empty for $r = k$ and has dimension $|P| + k - 2r - 1$ for $r < k$.*

c) $\dim Z(P,r) = k(n - |P|) + r(k - r) + r|P|$.

Assertions a) and b) can be proved by starting with (9)–(11) and making use of simple estimates of the dimensions of the solution spaces of the linear homogeneous equations; c) can be verified directly.

The inequality $\dim W \leq nk$ which proves the holonomy theorem follows from the lemma in §8 since for $r < k$ we have that $\dim Z(P,r) + |P| + k - 2r - 1 = nk - (k-r-1)(|P|-r-1) \leq nk$, and for $r = k$ we have that $\dim Z(P,r) \leq \dim Z' = nk$.

9. A first-order system. The lemma in §6 shows that, on the submanifold $Z'' \subset Z'$, the system (1)–(3) is equivalent to a system of first-order equations in a certain vector-valued function $f(z)$. In other words, on Z'' system (1)–(3) gives an integrable connection ∇ in some vector bundle L over Z''. We obtain certain properties of ∇ and L in the first nontrivial case $k = 2$. We will assume that the degrees $\alpha_1, \dots, \alpha_n$ of homogeneity satisfy the condition $\sum \alpha_p = n - 2$; this condition is satisfied for equations which are satisfied by the hypergeometric integrals in [1].

a) For $n \geq 3$, we have that $\dim E_{2,n} = n - 2$, so that, locally on Z'', system (1)–(3) has $n - 2$ linearly independent analytic solutions.

b) The singularities of the solutions are concentrated on

$$\overline{Z - Z''} = \left(\bigcup_{|P|=1} Z(P,0)\right) \cup \left(\bigcup_{|P|=2} Z(P,1)\right).$$

In addition, $Z(P,1)$ with $|P| = 2$ has codimension 1 in Z' (for $P = \{p_1, p_2\}$, $Z(P,1)$ consists of those matrices z whose p_1th and p_2th columns are linearly dependent). In a circuit around $Z(P,1)$ for $P = \{p_1, p_2\}$, the local monodromy operator of ∇ has $n-3$ eigenvalues 1, and one eigenvalue $\exp\{2\pi i(\alpha_{p_1} + \alpha_{p_2})\}$. The dimension (for general α_p) of the space of solutions which are analytically continuable to $Z(P,1)$, $|P| = 2$, is thereby equal to $n - 3$. For different P, these spaces of solutions are of course different. The restrictions of the extended solutions to $Z(P,1)$ are generalized hypergeometric functions corresponding to degenerate strata (see [1]).

10. Equations on a Grassmann manifold. Again we restrict ourselves to the case $k = 2$ and $\sum \alpha_p = n-2$. On the Grassmann manifold $G_{2,n}$ of two-dimensional subspaces of an n-dimensional space, we introduce the Plücker coordinates $p_{ij} = z_{1i}z_{2j} - z_{1j}z_{2i}$, $1 \leq i < j \leq n$. It is convenient to assume that $p_{ii} = 0$ and $p_{ij} = -p_{ji}$. One can verify that, in these coordinates, the $\mathcal{D}$-module M_G on $G_{2,n}$ is given by the operators

$$\sum_j p_{ij} \frac{\partial}{\partial p_{ij}} - (\alpha_i - 1) = 0, \qquad j = 1, \dots, n,$$

$$\sum_{i,j} p_{ij} \Delta_{ijlm} = 0, \qquad l, m = 1, \dots, n,$$

where

$$\Delta_{ijlm} = \frac{\partial^2}{\partial p_{ij} \partial p_{lm}} - \frac{\partial^2}{\partial p_{il} \partial p_{jm}} + \frac{\partial^2}{\partial p_{im} \partial p_{jl}}.$$

In conclusion, we wish to thank M. I. Graev and A. V. Zelevinskiĭ, whose discussions were very useful in writing this note.

Keldysh Institute of Applied Mathematics
Academy of Sciences of the USSR
Moscow
Institute for Problems of Information Transmission
Academy of Sciences of the USSR
Moscow

Received 24/FEB/86

BIBLIOGRAPHY

1. I. M. Gel'fand, Dokl. Akad. Nauk SSSR **288** (1986), 14–18; English transl. in Soviet Math. Dokl. **33** (1986).

2. J.-E. Björk, *Rings of differential operators*, North-Holland, 1979.

Translated by H. T. JONES

3.

(with M. I. Graev)

A duality theorem for general hypergeometric functions

Dokl. Akad. Nauk SSSR **289** (1) (1986) 19–23 [Sov. Math., Dokl. **34** (1987) 9–13
Zbl. **619**:33006

In this note, we study properties of the general hypergeometric function introduced in [1], and we establish its relation to certain classical hypergeometric functions.

1. We give a definition of a general hypergeometric function. Let $G_{k,n}$ be the manifold of k-dimensional linear subspaces of $\mathbf{R}^n$. We denote by $\tilde{G}_{k,n}$ the manifold pairs (τ, ω), where $\tau \in G_{k,n}$ and ω is a nondegenerate exterior k-form on τ. We define a sequence of complex numbers $\alpha = (\alpha_1, \dots, \alpha_n)$ such that $\sum \alpha_j = n - k$, and a sequence $\varepsilon = (\varepsilon_1, \dots, \varepsilon_n)$, $\varepsilon_j = \pm 1$. Let π_j, $j = 1, \dots, n$, be a generalized homogeneous function on $\mathbf{R}$ such that

$$\pi_j(t) = \begin{cases} t_+^{\alpha_j - 1} & \text{if } \varepsilon_j = 1, \\ t_-^{\alpha_j - 1} & \text{if } \varepsilon_j = -1. \end{cases}$$

For any pair $(\tau, \omega) \in \tilde{G}_{k,n}$ we define a differential $(k-1)$-form $\tilde{\omega}$ on τ by putting $\omega = i_I \omega$, where i_I is the inner derivation operator along the vector field $I = \sum x_j \partial/\partial x_j$ (for an explicit expression for $\tilde{\omega}$, see [1]. We define the following differential $(k-1)$-form on τ: $\prod_1^n \pi_j(x_j(v))\tilde{\omega}(v)$, where $x_j(v)$ is the jth coordinate of $v \in \tau$. This form has homogeneity degree 0 and descends onto the manifold of rays issuing from the point O in τ. The *general hypergeometric function* Φ is given by the following integral:

$$\Phi(\alpha, \varepsilon; \varsigma)k = \int\limits_S \prod_{j=1}^n \pi_j(x_j(v))\tilde{\omega}(v), \qquad \varsigma = (\tau, \omega), \tag{1}$$

where the integration extends over the manifold of rays issuing from the point O in τ, i.e., over an arbitrary surface $S \subset \tau$ which intersects each of these rays exactly once.

We now introduce the submanifolds $\tilde{G}_{k,n}^{\Xi} \subset \tilde{G}_{k,n}$ (see [1]). Let $I = \{1, \dots, n\}$. We let $\mathbf{R}^{\mathfrak{J}}$, $\mathfrak{J} \subset I$, denote the subspace in $\mathbf{R}^n$ spanned by the basis vectors e_j, $j \in \mathfrak{J}$. We call any nonempty collection of k-element subsets of I a *list* Ξ. A *type* Ξ *stratum* is a subset $\tilde{G}_{k,n}^{\Xi} \subset \tilde{G}_{k,n}$ consisting of pairs $(\tau, \omega) \in \tilde{G}_{k,n}$ such that for any k-element subset $\mathfrak{J} \subset I$ we have $\tau \cap R^{I \setminus \mathfrak{J}} = \{0\}$ if and only if $\mathfrak{J} \in \Xi$. The restriction of a general hypergeometric function Φ to the stratum $\tilde{G}_{k,n}^{\Xi}$ is called a *type* Ξ *hypergeometric function* and is denoted by Φ^{Ξ} [1].

2. Dual strata. Let $(\mathbf{R}^n)'$ be the dual space of $\mathbf{R}^n$. The *dual element of* $\varsigma = (\tau, \omega) \in \tilde{G}_{k,n}$ is the pair $\varsigma' = (\tau', \omega')$, where τ' is the $(n-k)$-dimensional subspace of $(\mathbf{R}^n)'$ which is orthogonal to τ, and ω' is the nondegenerate exterior $(n-k)$-form on τ' defined as follows. We choose a basis $\{f_i\}$ in $\mathbf{R}^n$ such that the vectors $f_1, \dots, f_k$ belong to τ and such that $f_1 \wedge \dots \wedge f_n = e_1 \wedge \dots \wedge e_n$, where $\{e_i\}$ is the standard basis in $\mathbf{R}^n$. If $\{f_i'\}$ is the dual basis to $\{f_i\}$ in $(\mathbf{R}^n)'$, then the vectors $f_{k+1}', \dots, f_n'$ belong to τ', and we put

$$\omega'(f_{k+1}', \dots, f_n') = (\omega(f_1, \dots, f_k))^{-1}.$$

1980 *Mathematics Subject Classification* (1985 *Revision*). Primary 33A35.

It is not difficult to verify that the definition of ω does not depend on the choice of $\{f_i\}$. We call the manifold of pairs $\varsigma' = (\tau', \omega')$ the *dual of* $\tilde{G}_{k,n}$, and we denote it by $\tilde{G}'_{l,n}$, where $l = n - k$. The duality relation gives a canonical isomorphism of $\tilde{G}'_{k,n}$ and $\tilde{G}'_{l,n}$.

We say that two lists Ξ and Ξ' of k-element and l-element subsets of $I = \{1, \ldots, n\}$ where $l = n - k$, are *dual* if they consist of mutually complementary subsets. We call the stratum $\tilde{G}^{\Xi'}_{l,n} \subset \tilde{G}'_{l,n}$ the *dual* of the stratum $\tilde{G}^{\Xi}_{k,n} \subset \tilde{G}_{k,n}$.

LEMMA. *Under the isomorphism $\tilde{G}_{k,n} \cong \tilde{G}'_{l,n}$, every stratum $\tilde{G}^{\Xi}_{k,n} \subset \tilde{G}_{k,n}$ goes into the dual stratum.*

3. A duality theorem. Let π_j, $j = 1, \ldots, n$, be arbitrary generalized homogeneous functions on $\mathbf{R}$ whose degrees of homogeneity are $\alpha_j - 1$, where $\sum \alpha_j = n - k$, i.e., functions of the form $\pi_j(t) = c_{j,1} t_+^{\alpha_j - 1} + c_{j,-1} t_-^{\alpha_j - 1}$ (see [2]). In analaogy with $\Phi(\alpha, \varepsilon; \varsigma)$ we define the following function on $\tilde{G}_{k,n}$:

$$\Phi_k(\pi; \varsigma) = \int_S \prod_{j=1}^{n} \pi_j(x_j(v)) \tilde{\omega}(v), \qquad \pi = (\pi_1, \pi_2, \ldots, \pi_n) \tag{2}$$

(the notation is the same as in (1)). We denote the restriction of Φ_k to a stratum of type Ξ by Φ_k^{Ξ}. From the equalities $\pi_j(t) = c_{j,1} t_+^{\alpha_j - 1} + c_{j,-1} t_-^{\alpha_j - 1}$ it follows that

$$\Phi_k^{\Xi}(\pi; \varsigma) = \sum_{\varepsilon} c_{\varepsilon} \Phi^{\Xi}(\alpha, \varepsilon; \varsigma), \qquad c_{\varepsilon} = \prod_j c_{j, \varepsilon_j}. \tag{3}$$

We let $\tilde{\pi}_j$ denote the Fourier transform of the generalized homogeneous function π_j:

$$\tilde{\pi}_j(s) = (2\pi)^{-1/2} \int_{-\infty}^{+\infty} \pi_j(t) e^{its} dt$$

(see [2] for explicit expressions for the $\tilde{\pi}_j$). We put $\tilde{\pi} = (\tilde{\pi}_1, \ldots, \tilde{\pi}_n)$.

THEOREM. *For any pair of dual elements $\varsigma \in \tilde{G}_{k,n}$, $\varsigma' \in \tilde{G}'_{l,n}$, $l = n - k$,*

$$\Phi_k(\pi; \varsigma) = \Phi_l(\tilde{\pi}, \varsigma'). \tag{4}$$

COROLLARY. *For any pair of dual lists Ξ, Ξ', and any pair of dual elements $\varsigma \in \tilde{G}^{\Xi}_{k,n}$, $\varsigma' \in \tilde{G}^{\Xi'}_{l,n}$, $k + l = n$,*

$$\Phi_k^{\Xi}(\pi; \varsigma) = \Phi_l^{\Xi'}(\tilde{\pi}; \varsigma'). \tag{5}$$

REMARK. In the case where $n = 2k$ and the lists Ξ and Ξ' are equivalent, i.e., Ξ' is obtained from Ξ by some permutation of the indices $1, \ldots, n$, (5) is a functional relation for Φ^{Ξ}.

From (2), (5), and the explicit expressions for the Fourier transforms of $t_+^{\alpha_j - 1}$ and $t_-^{\alpha_j - 1}$, it follows that

$$\Phi(\alpha; \varepsilon; \varsigma) = \sum_{\varepsilon'} c_{\varepsilon'}(\alpha, \varepsilon) \Phi(1 - \alpha, \varepsilon', \varsigma'),$$

where

$$c_{\varepsilon'}(\alpha, \varepsilon) = (2\pi)^{-n/2} \prod_{j=1}^{n} \Gamma(\alpha_j) \exp\left(i \frac{\pi \lambda_j}{2} \varepsilon_j \varepsilon'_j\right), \qquad 1 - \alpha = (1 - \alpha_j, \ldots, 1 - \alpha_n).$$

By virtue of the duality theorem, it is possible to represent the hypergeometric function on $\tilde{G}_{k,n}$ defined by the $(k-1)$-fold integral (2) as an $(l-1)$-fold integral, where $l = n - k$. Hence, in particular, we obtain well-known representations for Appell's function F_1 and Lauricella's function F_D (see [4]) as single Euler-type integrals.

4. The duality theorem for general hypergeometric functions is a consequence of a duality theorem for transformations in integral geometry related to $\tilde{G}_{k,n}$ and $\tilde{G}'_{l,n}$, $l = n-k$. We formulate this theorem. Let $F_k(\mathbf{R}^n)$ be the space of C^∞-functions on $\mathbf{R}^n\backslash\{0\}$ which satisfy the condition $f(\lambda x) = x^{-k}f(x)$ for every $\lambda > 0$, and let $\Phi_k(\tilde{G}_{k,n})$ be the space of C^∞-functions on $\tilde{G}_{k,n}$ which satisfy $\varphi(\tau, \lambda\omega) = \lambda\varphi(\tau, \omega)$ for every $\lambda > 0$. We define the integral transformation $J_k\colon F_k(\mathbf{R}^n) \to \Phi_k(\tilde{G}_{k,n})$ as follows:

$$(J_k f)(\varsigma) = \int_S f(x_1(v), x_2(v), \ldots, x_n(v))\tilde{\omega}(v), \qquad \varsigma = (\tau, \omega)$$

(where the notation is the same as in (1)). In an analogous way, we define the function spaces $F_l((\mathbf{R}^n)')$ and $\Phi_l(\tilde{G}'_{l,n})$ on $(\mathbf{R}^n)'$ and $\tilde{G}'_{l,n}$, respectively, and the integral transformation $J'_l\colon F_l((\mathbf{R}^n)') \to \Phi_l(\tilde{G}'_{l,n})$. We note that the isomorphism of manifolds $\tilde{G}_{k,n} \cong \tilde{G}'_{l,n}$, $l = n-k$, induces an isomorphism of the linear spaces $i\colon \Phi_k(\tilde{G}_{k,n}) \to \Phi_l(\tilde{G}'_{l,n})$. We define the isomorphism $\mathcal{F}\colon F_k(\mathbf{R}^n) \to F_l((\mathbf{R}^n)')$:

$$(\mathcal{F}f)(\xi) = (2\pi)^{-n/2}\int_{\mathbf{R}^n} f(x)e^{i\langle\xi,x\rangle}\,dx$$

(the Fourier transform of a generalized homogeneous function on $\mathbf{R}^n$; see [2]).

THEOREM. *The following diagram is commutative*:

$$\begin{array}{ccc} F_k(\mathbf{R}^n) & \xrightarrow{\ \mathcal{F}\ } & F_l((\mathbf{R}^n)') \\ {\scriptstyle J_k}\big\downarrow & & \big\downarrow{\scriptstyle J'_l} \\ \Phi_k(\tilde{G}_{k,n}) & \xrightarrow[\ i\ ]{} & \Phi_l(\tilde{G}'_{l,n}) \end{array}$$

The theorem is easy to prove if we rely on the fact that the operators J_k, J'_l, i, and $\mathcal{F}$ are intertwining for the representations of the group $\mathrm{GL}(n, \mathbf{R})$ in the respective spaces.

5. An expression for Φ and a duality relation in the coordinates on $G_{k,n}$. We fix a k-element subset $\mathfrak{J} \subset I$, for example, $\mathfrak{J} = \{1, \ldots, k\}$, and we let $G^0_{k,n}$ denote the submanifold of k-dimensional subspaces $\tau \in G_{k,n}$ such that $\tau \cap \mathbf{R}^{I\backslash\mathfrak{J}} = \{0\}$. It is clear that 1) $G^0_{k,n}$ is open in $G_{k,n}$ and is the union of those and only those strata $G^\Xi_{k,n}$ whose lists Ξ contain $\mathfrak{J}$, and that 2) $G^0_{k,n}$ consists of those and only those k-subspaces in $\mathbf{R}^n$ which are given by equations of the form $x_{k+i} = \sum_{j=1}^k z_{ij}x_j$, $i = 1, \ldots, n-k$, in the coordinates of $\mathbf{R}^n$. We take the elements of the matrix $z = \|z_{ij}\|$ as coordinates on $G^0_{k,n}$. We note that in the coordinates z_{ij} the action of the group H of diagonal matrices $h = \mathrm{diag}(h_1, \ldots, h_n)$, $h_j > 0$, is given by $z_{ij} \to h_{k+i}h_j^{-1}z_{ij}$.

We pass from the function $\Phi_k(\pi; \varsigma) = \Phi_k(\pi; \tau, \omega)$ to a function on $G^0_{k,n}$, putting

$$\Phi_k(\pi; \tau) = (\omega(e^\tau_1, e^\tau_2, \ldots, e^\tau_k))^{-1}\Phi_k(\pi; \tau, \omega), \qquad \tau \in G^0_{k,n},$$

where the e^τ_i are the projections of the vectors e_i of the standard basis in $\mathbf{R}^n$ onto the subspace τ parallel to $\mathbf{R}^{I\backslash\mathfrak{J}}$. In the coordinates z_{ij}, the function $\Phi_k(\pi; \tau) = \Phi_k(\pi; , z)$ is given by the integral

$$\Phi_k(\pi; z) = \int_{S_k}\prod_{j=1}^k \pi_j(x_j)\prod_{i=1}^{n-k}\pi_{k+i}(z_{i1}x_1 + z_{i2}x_2 + \cdots + z_{ik}x_k)\omega_k(x),$$

where

$$\omega_k(x) = \sum_j (-1)^{j-1}x_j\,dx \wedge dx_2 \wedge \cdots \wedge dx_{j-1} \wedge dx_{j+1} \wedge \cdots \wedge dx_k,$$

and the integral is taken over an arbitrary surface $S_k \subset \mathbf{R}^3$ which intersects each ray issuing from O exactly once.

PROPOSITION. *In the coordinates z_{ij} the duality relation* (4) *has the form*

$$\Phi_k(\pi_1, \pi_2, \dots, \pi_n; z) = \Phi_l(\tilde{\pi}_{k+1}, \dots, \tilde{\pi}_n, \tilde{\pi}_1, \dots, \tilde{\pi}_k; -z'),$$

where z' is the transposed matrix, i.e.,

$$\int_{S_k} \prod_{j=1}^{k} \pi_j(x_j) \prod_{i=1}^{l} \pi_{k+i}(z_{i1}x_1 + \cdots + z_{ik}x_k)\omega_k(z)$$

$$= \int_{S_l} \prod_{i=1}^{l} \tilde{\pi}_{k+i}(y_i) \prod_{j=1}^{k} \tilde{\pi}_j(-z_{1j}y_1 - z_{2j}ky_2 - \cdots - z_{lj}y_l)\omega_l(y), \qquad l = n - k.$$

6. Examples. 1°. $k = 2, n = 4$,

$$\pi_j(t) = \begin{cases} (t + i0)^{\alpha_j - 1} & \text{for } j = 1, 4, \\ t_+^{\alpha_j - 1} & \text{for } j = 2, 3, \end{cases} \qquad \sum \alpha_j = 2$$

and $\Phi_2(\pi; z)$ is a function of the 2×2 matrix z. For $z_{11} = z_{21} = -z_{12} = 1$, we get a function of $x = -z_{22}$, which for $x < 1$ is given by the integral

$$\mathcal{F}(\alpha; x) = \int_0^1 t^{\alpha_2 - 1}(1 - t)^{\alpha_3 - 1}(1 - xt)^{\alpha_4 - 1} dt.$$

To within a factor, $\mathcal{F}$ coincides with the Gauss hypergeometric function $F(a, b, c; x)$ [4], where $a = 1 - \alpha_4$, $b = \alpha_2$, and $c = \alpha_2 + \alpha_3$. The duality relation is equivalent to the well-known relation $F(a, b, c; x) = F(b, a, c; x)$.

2°. $k = 3, n = 5$,

$$\pi_j(t) = \begin{cases} (t + i0)^{\alpha_j - 1} & \text{for } j = 1, 5, \\ t_+^{\alpha_j - 1} & \text{for } j = 2, 3, 4, \end{cases} \qquad \sum \alpha_j = 2,$$

and $\Phi_3(\pi, z)$ is a function of the 2×3 matrix z. For $z_{11} = z_{21} = -z_{12} = 1$, we get a function of the two variables $x = -z_{22}$ and $y = -z_{23}$, which for $x < 1$ and $y < 1$ is given by the integral

$$\mathcal{F}(\alpha; x, y) = \int\int u_+^{\alpha_2 - 1} v_+^{\alpha_3 - 1}(1 - u - v)_+^{\alpha_4 - 1}(1 - xu - yv)^{\alpha_5 - 1} du\, dv.$$

To within a factor, $\mathcal{F}$ coincides with Appel's function $F_1(a, b, b', c; x, y)$ [4], where $a = 1, -\alpha_5$, $b = \alpha_2$, $b' = \alpha_3$, and $c = \alpha_2 + \alpha_3 + \alpha_4$. The duality relation implies the representation of F_1 as a single Euler-type integral (this representation was first obtained by Picard).

3°. $k = 3, n = 6$, $\pi_j(t) = t_+^{\alpha_j - 1}$, $j = 1, 2, \dots, 6$; the list Ξ contains all 3-element subsets of $I = \{1, 2, \dots, 6\}$ except $\{1, 2, 4\}$ and $\{1, 3, 5\}$. $\Phi_3^{\Xi}(\pi; z)$ is a function on the submanifold of 3×3 matrices z given by $z_{13} = z_{22} = 0$. If we put $z_{11} = z_{21} = z_{31} = -z_{12} = -z_{23} = 1$ and $z_{11} = z_{21} = z_{31} = -z_{32} = -z_{33} = 1$, we get the following functions of the two variables x and y, respectively:

$$\mathcal{F}_1(\alpha; x, y) = \int\int u_+^{\alpha_2 - 1} v_+^{\alpha_3 - 1}(1 - u)_+^{\alpha_4 - 1}(1 - v)_+^{\alpha_5 - 1}(1 - xu - yu)_+^{\alpha_6 - 1} du\, dv,$$

$$\mathcal{F}_2(\alpha; x, y) = \int\int u_+^{\alpha_2 - 1} v_+^{\alpha_3 - 1}(1 - xu)_+^{\alpha_4 - 1}(1 - yv)_+^{\alpha_5 - 1}(1 - u - v)^{\alpha_6 - 1} du\, dv.$$

To within a factor, for $x < 1$ and $y < 1$, $\mathcal{F}_1$ coincides with Appell's function $F_2(a, b, b', c, c'; x, y)$ (see [4]), where $a = 1 - \alpha_5$, $b = \alpha_2$, $b' = \alpha_3$, $c = \alpha_2 + \alpha_4$, and

$c' = \alpha_3 + \alpha_5$, and $\mathcal{F}_2$ with Appell's function $F_3(a, a', b, b', c; x, y)$, where $a = 1 - \alpha_4$, $a' = 1 - \alpha_5$, $b = \alpha_2$, $b' = \alpha_3$, and $c = -\alpha_2 - \alpha_3 - \alpha_5$. It is easy to see that the list Ξ dual to Ξ is obtained from Ξ by a permutation of the indices 1, 2, 3, 4, 5, 6; therefore the functional equation for Φ_3^{Ξ} follows from the duality relation.

4°. $n = 2k$, $k \geq 3$, $\pi_j(t) = t_+^{\alpha_j - 1}(\sum \alpha_j = k)$; Ξ is given by a configuration of n points $A_1, \dots, A_n$ in P^{k-1} (see [1]) such that $A_1, \dots, A_k$ are points in general position, A_{k+i} belongs to the line (A_i, A_{i+1}) for $i < k$, and A_{2k} belongs to the line (A_1, A_k). Restricting Φ^{Ξ} to a suitable cross-section of the bundle $\tilde{G}_{k,n} \to \tilde{G}_{k,n}/H$, we get a function of one variable which coincides, to within a factor, with the generalized hypergeometric function ${}_kF_{k-1}(a_1, \dots, a_k; b_1, \dots, b_{k-1}; x)$. The functional equation for ${}_kF_{k-1}$ follows from the duality theorem.

REMARK. There is a study of hypergeometric functions defined by integrals in the important work of K. Aomoto. In our terms, Aomoto has studied mainly functions on a generic stratum. Instead of the differential equations in [5], Aomoto has used the Gauss-Manin connection.

Keldysh Institute of Applied Mathematics
Academy of Sciences of the USSR
Moscow

Received 4/APR/86

BIBLIOGRAPHY

1. I. M. Gel'fand, Dokl. Akad. Nauk SSSR **288** (1986), 14–18; English transl. in Soviet Math. Dokl. **33** (1986).

2. I. M. Gel'fand and G. E. Shilov, *Generalized functions*. Vol. 1: *Operations on them*, 2nd ed., Fizmatgiz, Moscow, 1959; English transl. of 1st ed., Academic Press, 1964.

3. I. M. Gel'fand, S. G. Gindikin, and M. I. Graev, Itogi Nauki i Tekhn.: Sovremennye Problemy Mat., vol. 16, VINITI, Moscow, 1980, pp. 53–226; English transl. in J. Soviet Math. **18** (1982), no. 1.

4. A. Erdélyi et al., *Higher transcendental functions*. Vol. I, McGraw-Hill, 1953.

5. I. M. Gel'fand and S. I. Gel'fand, Dokl. Akad. Nauk SSSR **288** (1986), 279—283; English transl. in Soviet Math. Dokl **33** (1986).

Translated by H. T. JONES

4.

(with A. V. Zelevinskij)

Algebraic and combinatorial aspects of the general theory of hypergeometric functions

Funkts. Anal. Prilozh. **20** (3) (1986) 17-34 [Funct. Anal. Appl. **20** (1986) 183-197].
Zbl. **619**:33004

INTRODUCTION

1. This article is a part of a program of study of the general hypergeometric functions ntroduced in [6]. Basically, we set forth here the algebraic and combinatorial aspects of he theory; other results on general hypergeometric functions are given in [7, 9]. We also ite some applications, including some applications to the study of the continuous analogue f the partition functions of Kostant [21] introduced in [4].

General hypergeometric functions are essentially functions on a Grassmannian. A Grassannian is a manifold of k-dimensional subspaces in an n-dimensional vector space V over the ield **R** or **C**. At the present time, these manifolds play an important role in a large number f problems which are closely related to each other (see [8, 10], for example; we note also hat the well-known twistor program of Penrouz [12] is based on a study of Gassmannians of wo-dimensional planes in C^4). The Grassmannian of k-dimensional subspaces in V will be deoted by $G_k(V)$.

Although the Grassmannian $G_k(V)$ is a homogeneous space with respect to the group GL(V), e will first be interested in the action of a maximal torus H in the group GL(V) on it. The hoice of a maximal torus fixes a basis $e_1,\dots,e_n$ in V, i.e., it allows us to identify V with the coordinate space $\mathbf{R}^n$ or $\mathbf{C}^n$; moreover, in the complex case, this torus is the group of all diagonal matrices, and, in the real case, the group of diagonal matrices with positive elements. The orbits of H on $G_k(\mathbf{C}^n)$ are toroidal manifolds; in the real case these orbits reduce to interesting manifolds, so-called Grassman simplexes [19] (see also [13, 20]). In this article we will concern ourselves only with real Grassmannians, which makes it possible for us to give a well-rounded treatment of the combinatorial and geometric aspects. However, a complete understanding of the situation is impossible without entering the complex domain; the complex theory will be treated in a subsequent article.

2. Definition of a General Hypergeometric Function. The existence of the torus H in $GL(\mathbf{R}^n)$ allows us to distinguish the class of homogeneous functions $\pi(x)$ in $\mathbf{R}^n$. For each set $\alpha=(\alpha_1,\dots,\alpha_n)$ of complex numbers with sum n − k, we consider the homogeneous function $\pi_\alpha(x)=\prod_{1\le i\le n}(x_i)_+^{\alpha_i-1}$; here $x=(x_1,\dots,x_n)\in\mathbf{R}^n$, and $(x_i)_+^{\alpha_i-1}=x_i^{\alpha_i-1}$ for $x_i>0$ and $(x_i)_+^{\alpha_i-1}=0$ for $x_i\leqslant 0$.

Now let $\zeta\in G_k(\mathbf{R}^n)$ be a k-dimensional subspace in $\mathbf{R}^n$, and let $S(\zeta)$ be a "sphere" in ζ (by a "sphere" in ζ we understand a factor space $\zeta\setminus 0$ with respect to the action of the multiplicative group $\mathbf{R}_+\setminus 0$, or an arbitrary smooth surface in ζ which intersects every ray from 0 in one point). Let ω be a nonzero exterior k-form on ζ. The general hypergeometric function $\Phi(\alpha;\zeta,\omega)$ is defined by

$$\Phi(\alpha;\zeta,\omega)=\int_{S(\zeta)}\pi_\alpha(x)\tilde\omega(x). \tag{1}$$

Here $\tilde\omega(x)$ is a $(k-1)$-form on $\zeta\setminus 0$ induced by ω in a natural way: if $t_1,\dots,t_k$ are coordinates in ζ and $\omega=dt_1\wedge\dots\wedge dt_k$, then $\tilde\omega=\sum_i(-1)^{i-1}t_i dt_1\wedge\dots dt_{i-1}\wedge dt_{i+1}\wedge\dots dt_k$. Since $\sum_i\alpha_i=n-k$, the form $\pi_\alpha(x)\tilde\omega(x)$ has degree of homogeneity 0, which means it decreases on the "sphere" $S(\zeta)$; in addition, $S(\zeta)$ is furnished with a natural orientation which is induced in a natural way by ω. An expresssion for $\Phi(\alpha;\zeta,\omega)$ in the local coordinates on the Grassmannian is given in Sec. 1.

Moscow State University. Cybernetic Science Society, Academy of Sciences of the USSR.
Translated from Funktsional'nyi Analiz i Ego Prilozheniya, Vol. 20, No. 3, pp. 17-34, July-September, 1986. Original article submitted March 24, 1986.

Since the integral in (1) diverges for some α and ζ, the definition of $\Phi(\alpha; \zeta, \omega)$ requires some revision for these values; the precise definition, using analytic continuation with respect to α and the method of partitionings, is given in [6]. We note that, along with $\Phi(\alpha; \zeta, \omega)$, the function $\Phi(\alpha, \varepsilon; \zeta, \omega)$ is also defined for all sets $\varepsilon = (\varepsilon_1,\ldots,\varepsilon_n)$, where $\varepsilon_i = \pm 1$; we have the relation $\Phi(\alpha, \varepsilon; \zeta, \omega) = \Phi(\alpha; \varepsilon\zeta, \varepsilon\omega)$, where the pair $(\varepsilon\zeta, \varepsilon\omega)$ is obtained by acting on (ζ, ω) with the diagonal matrix $\varepsilon = \mathrm{diag}(\varepsilon_1,\ldots,\varepsilon_n)$. A holonomic system of differential equations which all of the functions $\Phi(\alpha, \varepsilon; \zeta, \omega)$ satisfy is constructed in [7].

Every function $\Phi(\alpha, \varepsilon; \zeta, \omega)$ satisfies the intrinsic homogeneity condition with respect to the action of the group H: for $\lambda = \mathrm{diag}(\lambda_1, \ldots, \lambda_n) \in H$ we have that $\Phi(\alpha, \varepsilon; \lambda\zeta, \lambda\omega) = \pi_\alpha(\lambda)\Phi(\alpha, \varepsilon; \zeta, \omega)$. Thus if the value of this function is known at any point ζ whatever of the Grassmannian, then it is known on the entire orbit of ζ under the action of H, i.e., it is actually a function on the space of orbits $G_k(\mathbf{R}^n)/H$.

3. Decomposition of a Grassmannian into Strata. In view of the fact that we have identified a maximal torus of H, the coordinate subspaces $\mathbf{R}^{i_1\ldots i_m} = \mathbf{R}^J$ play an important role in $\mathbf{R}^n$. We say that two points ζ and ζ' of the Grassmannian $G_k(\mathbf{R}^n)$ are equivalent if $\dim \times (\zeta \cap \mathbf{R}^J) = \dim(\zeta' \cap \mathbf{R}^J)$ for all $\mathbf{R}^J$. The equivalence classes are called strata in $G_k(\mathbf{R}^n)$. There exists a unique open stratum called the general stratum; it consists of subspaces $\zeta \in G_k(\mathbf{R}^n)$, which are in general position with respect to all coordinate subspaces.

It is possible to define strata in many ways: as the intersection of Schubert cells related to various orderings of the basis vectors in $\mathbf{R}^n$; by means of Plücker coordinates; in terms of mappings of moments [19, 13, 20]; and from the point of view of combinatorial geometry (see below). The relationships among these definitions are studied in an article of Segranov and Gel'fand, and they are carried over to "Grassmannians" and flag manifolds for all semisimple groups.

As shown in [6], the restriction of every function $\Phi(\alpha, \varepsilon; \zeta, \omega)$ to any stratum in $G_k(\mathbf{R}^n)$ is real analytic. In addition, on a general stratum the number of linearly independent ones among them is $\leqslant \binom{n-2}{k-1}$ (see [7]); the homological interpretation of this number is given in the present article. By considering the restrictions of general hypergeometric functions to various strata, it is possible to obtain many classical hypergeometric functions of one and several variables (see [6]; this question is treated in greater detail in [9]). Thus the Gauss hypergeometric function $F(a, b; c; z)$ turns out to be related to the general stratum in $G_2(\mathbf{R}^4)$ (see [6] and Sec. 2 of the present article).

4. In this and subsequent articles we study the analytic behavior of the restrictions of the general hypergeometric function to various strata in $G_k(\mathbf{R}^n)$ [henceforth, we will speak only of the function $\Phi(\alpha; \zeta, \omega)$, but everything we say carries over with obvious modifications to all functions $\Phi(\alpha, \varepsilon; \zeta, \omega)$]. Since we are interested here in the algebraic and combinatorial aspects of the theory, we consider only special integer sets of the indices α called polynomial sets (see Sec. 4). In this case $\Phi(\alpha; \delta, \omega)$ is actually piecewise rational. The case of general indices will be treated in a subsequent article.

The restriction of $\Phi(\alpha; \zeta, \omega)$ to a stratum Γ in $G_k(\mathbf{R}^n)$ has singularities on the boundaries of Γ; therefore it behaves differently on different connected components of Γ. The connected components of Γ are called cells. The description of cells in the general case is very difficult and can even be an unsolvable problem. It turns out, however, that in the study of $\Phi(\alpha; \zeta, \omega)$, not individual cells, but certain unions of them, which we call large cells, are essential. Large cells on the stratum Γ are parametrized by the $(n - k + 1)$-dimensional coordinate subspaces $\mathbf{R}^J$ in $\mathbf{R}^n$ in general position with respect to Γ, i.e., such that $\dim(\zeta \cap \mathbf{R}^J) = 1$ for $\zeta \in \Gamma$; the corresponding large cell $\Gamma(J)$ consists of subspaces $\zeta \in \Gamma$ such that the direction vector of the line $\zeta \cap \mathbf{R}^J$ has positive coordinates.

As we shall see in Sec. 4, for polynomial α the function $\Phi(\alpha; \zeta, \omega)$ admits an expansion of the form

$$\Phi(\alpha; \zeta, \omega) = \sum_J \Phi_J(\alpha; \zeta, \omega), \tag{2}$$

on Γ, where every function $\Phi_J(\alpha; \zeta, \omega)$ is rational in the large cell $\Gamma(J)$ and equal to 0 outside of $\Gamma(J)$. However this expansion is not unique. A central result of this article consists in the choice of a "basis in the space of large cells," i.e., a system B of large cells ch that the expansion of the form (2), where the sum extends over all $J \in B$, exists and is ique; we call such a system B fundamental.

For the case of a general stratum Γ as B, it is possible to take the system of all large lls Γ(J) such that $\mathbf{R}^J$ contains a certain fixed two-dimensional coordinate subspace in $\mathbf{R}^n$. us the number of functions $\Phi_J(\alpha;\ \zeta,\ \omega)$ in the expansion (2) is $\binom{n-2}{k-1}$ in this case.

An expansion of the form (2) also has meaning for general indices α. Every function $_J(\alpha;\ \zeta,\ \omega)$ is equal to 0 outside of Γ(J) as before, and "has unique analytic behavior" in (J); this means that its analytic continuations into the complex domain from different cells Γ(J) are consolidated into a single branch of an analytic function. If the stratum Γ is eneral, we therefore get $\binom{n-2}{k-1}$ linearly independent solutions of the holonomic system of quations in [7]; taking into account an estimate in [7], we get that these functions form basis in the space of solutions. Explicit expressions for the Φ_J can be given in terms of ntegral representations in the complex domain; this will be done in a subsequent article.

A precise statement of the result concerning the expansion (2) is given in Sec. 4 (Theorem 4.2). Roughly speaking, this result is as follows: we construct a group of homological rigin with respect to the stratum Γ, and we define a class of special bases in it; to each asis in this class there corresponds an expansion of $\Phi(\alpha;\ \zeta,\ \omega)$ on Γ in a sum (2) whose terms re parametrized by elements of this basis.

The above homology group can be defined in several different ways topologically (see 2, 15]), algebraically, and geometrically (see Sec. 3). It is remarkable that, making use f results of Orlik and Solomon [22], it is possible to give a purely combinatorial definition n terms of so-called combinatorial geometry (or the theory of matroids). One of the conclusions of the present article is the fact that this theory, which has been developed in the ast 30 years by Whitney and by Birkhoff and MacLane and which has received a new impetus hanks to the work of Rota and his school (see [1, 17, 23]), is a natural combinatorial basis or the theory of general hypergeometric functions.

5. Theorem 4.2 is closely related to a result concerning the expansion of rational functions which is of interest in itself (Theorem 5.2); this result gives a multidimensional generalization of the expansion of a rational function in simple fractions. Let ξ be an l-dimensional vector space (over **R** or **C**) in which there is given a finite family $\mathcal{F} = (f_i)_{i\in I}$ of nonzero linear forms. For each set $\alpha = (\alpha_i)_{i\in I}$ of nonnegative integers, we let F_α denote a rational function $(\prod_{i\in I} f_i^{\alpha_i})^{-1}$ on ξ. The problem solved in Theorem 5.2 consists in the construction of (linear) basis in the family of all functions F_α. The rank of a subset $J \subset I$ is defined to be the rank of the family of linear forms $(f_j)_{j\in J}$. We will assume that I has rank l, and we will consider functions F_α only for sets $\alpha = (\alpha_i)_{i\in I}$, such that $\operatorname{supp}\alpha = \{i \in I\colon \alpha_i \neq 0\}$ has rank l (the general case can easily be reduced to this one).

We denote the vector space $H = H(\xi, \mathcal{F})$ generated by the functions $F_J = \prod_{j\in J} f_j^{-1}$, where J runs through all l-subsets of rank I in l.

We call a system B consisting of l-subsets in I of rank l *fundamental* if the functions F_J for $J \in B$ form a basis in the space H. Theorem 5.2 asserts that, for every fundamental system B, the functions F_{α}' with $\operatorname{supp}\alpha' \in B$ form a basis in the space spanned by all F_α.

There is also an explicit method for constructing fundamental systems in arbitrary linearly ordered sets I (Theorem 3.1), which makes Theorem 5.2 more effective.

We note that (for a suitable choice of ξ and $\mathcal{F}$) the space $H(\xi, \mathcal{F})$ is one of the realizations of the homology group with which we were concerned in the preceding section; indeed, the special bases mentioned there are bases of functions F_J, where J runs through any fundamental system. It is interesting that the concept of a fundamental system also admits a purely combinatorial definition; the method of constructing them given in Theorem 3.1 in a combinatorial situation is due essentially to Bjorner [14] (more precisely, it is obtained by adapting a result of Bjorner to a construction of Orlik and Solomon [22]).

6. Another application of Theorem 4.2 relates to the following beautiful geometry problem. In $\mathbf{R}^n$ we consider the family of k-dimensional affine planes ζ + f which are parallel to a given subspace $\zeta \in G_k(\mathbf{R}^n)$. We consider the polyhedra obtained by taking the intersection of these planes with the positive octant $\mathbf{R}^n_+$. The problem consists in studying the volume of such

a polyhedron as a function of f; we denote this function by $\Psi(f)$. This function has a complicated piecewise polynomial behavior. In Sec. 5 we show that it can be obtained as a restriction of the general hypergeometric function $\Phi(\alpha; \zeta, \omega)$ (for special values of α) to some submanifold. Applying Theorem 4.2, we get that the behavior of $\Psi(f)$ is governed by the same homology group as above, and that there is a class of explicit expressions for it corresponding to the various choices of a basis in this homology group.

It turns out that the continuous analogue of Kostant's partition function [21] which is introduced in [4] and which plays an important role in representation theory can also be defined as a function of the form $\Psi(f)$ for some special choice of $\mathbf{R}^n$ and of ζ; this means that all of what has been described above is also applicable to it. For systems of roots of type A_l the continuous analogue of Kostant's function is calculated in [11]; this result is included in our general scheme.

A very interesting question is whether it is possible to apply the various methods described here to a study of Kostant's function itself. In geometric terms, it revolves around the study of the "discrete analogue" of $\Psi(f)$ obtained by replacing the volume of the polyhedron by the number of integer points in it. This question will be taken up in another article, where explicit expressions will be obtained for various systems of roots.

7. This article is organized in the following way. The necessary definitions and notation are brought together in Sec. 1. In Sec. 2, we establish useful functional relations for general hypergeometric functions, which we call Gauss relations (for the Gauss hypergeometric function they reduce to the classical Gauss relations [3]). In Sec. 3 we study a homology group which plays a central role in the article, and we obtain a number of realizations of it. In Sec. 4, we obtain the main theorem, Theorem 4.2 and in Sec. 5 we bring together some of its applications.

The material from combinatorial geometry which we need is developed in the Appendix. We describe there the construction of Orlik and Solomon [22] and Bjorner's theorem [14] in forms which are convenient for our purposes. So that this article can be read independently, we give a new proof of this theorem.

1. DEFINITIONS AND NOTATION

Instead of $\mathbf{R}^n$, it will be convenient to consider the vector space $\mathbf{R}^I$ with a preferred basis $(e_i)_{i\in I}$, indexed by the finite set I. For each $J \subset I$ we let $\mathbf{R}^J$ denote the subspace of $\mathbf{R}^I$ spanned by the vectors $(e_j)_{j\in J}$, and we let $\mathbf{R}^J_+$ be the (open) positive octant in $\mathbf{R}^J$, i.e., $\mathbf{R}^J_+ = \{\sum_{j\in J} x_j e_j : x_j > 0\}$. We denote the number of elements in the finite set J by $|J|$; if $|J| = m$, then we say that J is an m-set.

We denote by $G_k(\mathbf{R}^I)$ the Grassmannian of k-dimensional vector subspaces in $\mathbf{R}^I$. The codimension of the subspaces, i.e., the number $|I| - k$, will always be denoted by l.

Let $\zeta \in G_k(\mathbf{R}^I)$. We denote by $\zeta^\perp$ the subspace of linear forms on $\mathbf{R}^I$ whose restrictions to ζ are equal to 0. We put $L = L(\zeta) = \mathbf{R}^I/\zeta$. We denote the projection $\mathbf{R}^I \to L$ by q, and we put $f_i = q(e_i) \in L$ for $i \in I$. We identify L with the dual space of $\zeta^\perp$; in particular, the f_i will be regarded as linear forms on $\zeta^\perp$. The family $(f_i)_{i\in I}$ of vectors in L will be denoted by $\mathcal{F} = \mathcal{F}(\zeta)$.

We consider the pregeometry of rank l on I corresponding to $\mathcal{F}(\zeta)$ (see the Appendix); the rank function of this pregeometry is given by $r(J) = \operatorname{rk}(f_j)_{j\in J} = \dim(\mathbf{R}^J/\zeta \cap \mathbf{R}^J)$. It is clear that two points ζ and ζ' lie on a single stratum Γ in $G_k(\mathbf{R}^I)$ if and only if the pregeometries on I corresponding to them coincide. We will use the terminology of the Appendix relative to this pregeometry, adding, if necessary, a designation to ζ (or Γ). Thus a subset J of I is independent for ζ if $\zeta \cap \mathbf{R}^J = 0$, and is a basis (of the pregeometry) for ζ if $\zeta \oplus \mathbf{R}^J = \mathbf{R}^I$. The set of all bases of the pregeometry for ζ is denoted by $B(\zeta)$. We note that the list of Γ defined in [6] consists of the k-subsets of I which are complements of subsets of $B(\zeta)$.

For every l-subset $J \subset I$ we put $\Gamma^J = \{\zeta \in G_k(\mathbf{R}^I) : \zeta \oplus \mathbf{R}^J = \mathbf{R}^I\}$. The set Γ^J is a coordinate neighborhood in $G_k(\mathbf{R}^I)$: the elements of Γ^J are parametrized by the real matrices $Z = (z_{ij})_{i\in I\setminus J,\, j\in J}$, and the subspace $\zeta(Z) \in \Gamma^J$ with basis $(e_i + \sum_{j\in J} z_{ij} e_j)_{i\in I\setminus J}$ corresponds to the matrix Z.

We denote by $\tilde{G}_k(\mathbf{R}^I)$ the set of pairs $\tilde{\zeta} = (\zeta, \omega)$, where $\zeta \in G_k(\mathbf{R}^I)$, and ω is a nonzero skew symmetric k-linear form on ζ. The projection $\tilde{\zeta} \to \zeta$ converts $\tilde{G}_k(\mathbf{R}^I)$ into a fiber bundle ver $G_k(\mathbf{R}^I)$ with fiber $\mathbf{R}\setminus 0$; for each subset $\Gamma\subset G_k(\mathbf{R}^I)$ we will denote the preimage of Γ under his projection by $\tilde{\Gamma}\subset\tilde{G}_k(\mathbf{R}^I)$. In particular, the sets $\tilde{\Gamma}$, where Γ is a stratum in $G_k(\mathbf{R}^I)$, are alled strata in $\tilde{G}_k(\mathbf{R}^I)$.

Suppose that $\tilde{\zeta}=(\zeta,\omega)\in\tilde{G}_k(\mathbf{R}^I)$, and let $\alpha=(\alpha_i)_{i\in I}$ be a set of complex numbers whose um is l. The general hypergeometric function $\Phi(\alpha;\tilde{\zeta}) = \Phi(\alpha;\zeta,\omega)$ is defined in the Introuction [see (1)]. For fixed $\tilde{\zeta}$, the function $\Phi(\alpha;\tilde{\zeta})$ is univalent and meromorphic in α, and or fixed α it is a real analytic function of $\tilde{\zeta}$ if we restrict it to any stratum in $\tilde{G}_k(\mathbf{R}^I)$ see [6]).

From our agreement concerning the choice of an orientation on the "sphere" $S(\zeta)$ it folows that $\Phi(\alpha;\zeta,\lambda\omega)=|\lambda|\Phi(\alpha;\zeta,\omega)$ for $0\neq\lambda\in\mathbf{R}$. In particular, the function $\Phi(\alpha;\zeta,\omega)$ emains unchanged if we replace ω by $-\omega$. Making use of this, we will often give ω only to ithin its sign.

Example 1.1. Let $k = 1$. We choose a generator $\sum_{i\in I} b_ie_i$ of the line $\zeta\in G_1(\mathbf{R}^I)$ so that at east one of the coordinates b_i is positive. Then

$$\Phi(\alpha;\zeta,\omega)=\prod_{i\in I}(b_i)_+^{\alpha_i-1}\Big|\omega\Big(\sum_i b_ie_i\Big)\Big|.$$

We write $\Phi(\alpha;\tilde{\zeta})$ in the local coordinates introduced above. Let J be an l-subset in I, nd let $Z=(z_{ij})_{i\in I\setminus J,\,j\in J}$ be a real matrix. Let $\omega(Z)$ be a k-form on $\zeta(Z)$ taking on the values 1 on the set of vectors $(e_i+\sum_{j\in J} z_{ij}e_j)_{i\in I\setminus J}$. Let $I\setminus J = \{i_1,\ldots,i_k\}$. In this notation, (1) can e written

$$\Phi(\alpha;\zeta(Z),\omega(Z))=\int_{S(\mathbf{R}^{I\setminus J})}\prod_{i\in I\setminus J}(x_i)_+^{\alpha_i-1}\prod_{j\in J}\Big(\sum_{i\in I\setminus J} z_{ij}x_i\Big)_+^{\alpha_j-1}\varpi(x), \tag{3}$$

where

$$\varpi(x)=\sum_{1\leq r\leq k}(-1)^{r-1}x_{i_r}dx_{i_1}\wedge\ldots\wedge dx_{i_{r-1}}\wedge dx_{i_{r+1}}\wedge\ldots\wedge dx_{i_k}. \tag{4}$$

For brevity, we will write $\Phi(\alpha;Z)$ instead of $\Phi(\alpha;\zeta(Z),\omega(Z))$.

If $\alpha=(\alpha_i)_{i\in I}$ and $\beta=(\beta_i)_{i\in I}$ are two sets of indices, then the set $(\alpha_i+\beta_i)_{i\in I}$ will be denoted by $\alpha+\beta$. For each $J\subset I$, we use 1_J to denote the set $(\iota_i)_{i\in I}$, where $\iota_j=1$ for $j\in J$ and $\iota_i=0$ for $i\notin J$. We will write 1_i instead of $1_{\{i\}}$. For example, $\alpha-1_j$ denotes the set $(\alpha_i')_{i\in I}$, where $\alpha_i'=\alpha_i-\delta_{ij}$. Finally, we put $|\alpha|=\sum_i\alpha_i$ and $\operatorname{Supp}\alpha=\{i\in I:\alpha_i\neq 0\}$.

2. GAUSS RELATIONS

THEOREM 2.1. Let $\tilde{\zeta}=(\zeta,\omega)\in\tilde{G}_k(\mathbf{R}^I)$.

a) Let $\beta=(\beta_i)_{i\in I}$ be a set of complex numbers with sum $l-1$, and let $v=\sum_{i\in I}a_ix_i\in\zeta^\perp$ be a linear form which is identically 0 on ζ. Then

$$\sum_{i\in I}a_i\Phi(\beta+1_i;\tilde{\zeta})=0. \tag{5}$$

b) Let $\gamma=(\gamma_i)_{i\in I}$ be a set of complex numbers with sum $l+1$, and let $y=\sum_{i\in I}b_ie_i\in\zeta$. Then

$$\sum_{i\in I}b_i(\gamma_i-1)\Phi(\gamma-1_i;\tilde{\zeta})=0. \tag{6}$$

The relations (5) and (6) are called the Gauss relations for the general hypergeometric function on an arbitrary stratum (see example 2.1 below); altogether, they give $k + l$ $I|$* independent relations which relate the values of $\Phi(\alpha;\tilde{\zeta})$ for fixed $\tilde{\zeta}$ and "contiguous" indices α. The relations (5) follow directly from the definitions, and the relations (6) follow from (5) and a duality theorem in [9]. For a general stratum, it is convenient to construct the proof in local coordinates. We choose an l-subset $J\subset I$, and we consider the function $\Phi(\alpha;Z)$ defined in (3).

Proposition 2.1. a) For fixed $j\in J$, the function $\Phi(\alpha;Z)$, as a function of the column $(z_{ij})_{i\in I\setminus J}$, has degree of homogeneity (α_j-1) (i.e., for $\lambda>0$, if we make the replacement $z_{ij}\to\lambda z_{ij}$ for all $i\in I\setminus J$ in $\Phi(\alpha;Z)$, this multiplies the functions by λ^{α_j-1}).

*Something missing in the Russian original — Publisher.

b) For fixed $i \in I \setminus J$, the function $\Phi(\alpha; Z)$, as a function of the row $(z_{ij})_{j\in J}$, has degree of homogeneity $(-\alpha_i)$.

c) For all $i \in I \setminus J$ and all $j \in J$,

$$\partial\Phi(\alpha; Z)/\partial z_{ij} = (\alpha_j - 1)\,\Phi(\alpha + 1_i - 1_j; Z). \tag{7}$$

All of these assertions follow immediately from (3).

COROLLARY. a) For every $j \in J$,

$$\Phi(\alpha; Z) = \sum_{i\in I\setminus J} z_{ij}\Phi(\alpha + 1_i - 1_j; Z). \tag{8}$$

b) For every $i \in I \setminus J$,

$$\alpha_i\Phi(\alpha; Z) + \sum_{j\in J}(\alpha_j - 1)\, z_{ij}\Phi(\alpha + 1_i - 1_j; Z) = 0. \tag{9}$$

To prove (8), we write the homogeneity condition of $\Phi(\alpha; Z)$ with respect to the column $(z_{ij})_{i\in I\setminus J}$ in the Euler form

$$\sum_{i\in I\setminus J} z_{ij}\partial\Phi(\alpha; Z)/\partial z_{ij} = (\alpha_j - 1)\,\Phi(\alpha; Z).$$

Substituting this into (7), we get (8). The result in (9) is proved in exactly the same way.

Equations (8) and (9) are the Gauss relations for the general hypergeometric function on a general stratum. We note that (7) and (8) (in our terms, for a general stratum) have been proved by Aomoto [24, 25].

Example 2.1. Let $J = \{1, 2\}$, $I \setminus J = \{0, 3\}$, and let $k = 2$, so that we are dealing with the Grassmannian of two-dimensional subspaces of $\mathbf{R}^4$. By virtue of Proposition 2.1 (a), (b), to calculate $\Phi(\alpha; Z)$ we may assume that all of the matrix elements of Z, except the first, are fixed. We put $z_{31} = z_{32} = -1$, $z_{01} = 1$, $z_{02} = z$; by virtue of (3) we have for such a matrix Z

$$\Phi(\alpha; Z) = \int_0^1 x^{\alpha_3 - 1}(1 - x)^{\alpha_1 - 1}(z - x)_+^{\alpha_2 - 1}\, dx.$$

Hence it follows that

$$\left.\begin{aligned} &\Phi(\alpha; Z) = 0 \quad \text{for} \quad z \leqslant 0,\\ &\Phi(\alpha; Z) = \frac{\Gamma(\alpha_2)\,\Gamma(\alpha_3)}{\Gamma(\alpha_2 + \alpha_3)} z^{\alpha_2+\alpha_3-1}F(1 - \alpha_1, \alpha_3; \alpha_2 + \alpha_3; z) \quad \text{for} \quad 0 \leqslant z \leqslant 1,\\ &\Phi(\alpha; Z) = \frac{\Gamma(\alpha_1)\,\Gamma(\alpha_3)}{\Gamma(\alpha_1 + \alpha_3)} z^{\alpha_2-1}F(1 - \alpha_2, \alpha_3; \alpha_1 + \alpha_3; z^{-1}) \quad \text{for} \quad z > 1, \end{aligned}\right\} \tag{10}$$

where $F(a, b; c; z)$ is the classical Gauss hypergeometric function.

If we make a suitable choice of a basis of four relations in the space of relations of the form (5) and (6), and if we transform them by making use of (10), we get the classical Gauss relations for F ([3], 2.8. (38), (42), (35), (43)); it is easy to verify that all 15 of the Gauss relations ([3], 2.8. (31)-(45)) are linear combinations of these four relations (applied to various sets of indices α_i).

3. THE SPACE H(ζ) AND ITS REALIZATIONS

We fix a stratum Γ in $G_k(\mathbf{R}^I)$ and a point $\zeta \in \Gamma$. We construct a finite-dimensional space $H(\zeta)$ and we describe a special class of bases in it; these concepts play a central role in what follows. We obtain several different realizations of $H(\zeta)$. If ζ contains any basis vector e_i, then we put $H(\zeta) = 0$; we note that the strata Γ with this property are said to be degenerate in the terminology of [6]. Thus in what follows we assume that this is not the case; in other words, all of the vectors f_i in the space $L = L(\zeta)$ are different from 0.

Algebraic Definition. For each $J \in B(\zeta)$ we let F_J denote the rational function $\prod_{j\in J} f_j^{-1}$ on $\zeta^\perp$ (see Sec. 1). By definition $H(\zeta)$ is the vector space generated by all such functions F_J.

Topological Definition. Let $\zeta_{\mathbf{C}}^\perp$ be the complexification $\zeta^\perp$, and suppose that $X = \zeta_{\mathbf{C}}^\perp \setminus \bigcup_{i\in I} f_i^\perp$ is obtained from $\zeta_{\mathbf{C}}^\perp$ by deleting the hyperplanes orthogonal to all of the f_i. The space

(ζ) is isomorphic to the highest cohomology group $H^l(X)$ of X. More precisely, we have the following result.

Proposition 3.1 [2, 15]. The mapping $H(\zeta)\otimes\Lambda^l(L)\to H^l(X)$, which sends each element $\otimes\,\omega$ into the class of cohomologies of the differential l-form $F\omega$ on X is an isomorphism.

Combinatorial Definition. We consider the pregeometry of rank l on I constructed with respect to ζ, i.e., corresponding to the family of vectors $\mathcal{F}=(f_i)_{i\in I}$ in L. Let $\mathcal{A}_l=\mathcal{A}_l(\zeta)$ be the space corresponding to this pregeometry under the construction of Orlik and Solomon (see the Appendix). Then $H(\zeta)$ and $\mathcal{A}_l(\zeta)$ are isomorphic. More precisely, we can construct a natural isomorphism between $\mathcal{A}_l(\zeta)$ and $H(\zeta)\otimes\wedge^l(L)$.

The requirement that $J\in B(\zeta)$ means that the restriction of the projection $q\colon \mathbf{R}^I\to L$ to $\mathbf{R}^J$ is an isomorphism of $\mathbf{R}^J$ with L; for each $J\in B(\zeta)$, we again denote by $q=q_{J,\zeta}$ the isomorphism $\Lambda^l(\mathbf{R}^J)\simeq\Lambda^l(L)$, introduced by this isomorphism. We recall (see Appendix) that there is a natural epimorphism $\varepsilon\colon\mathcal{C}_l\to\mathcal{A}_l$ with kernel $\partial\mathcal{C}_{l+1}$, where the space $\mathcal{C}_l$ is identified with $\bigoplus_{J\in B(\zeta)}\Lambda^l(\mathbf{R}^J)$.

Proposition 3.2. The mapping $\mathcal{C}_l\to H(\zeta)\otimes\Lambda^l(L)$, which carries $\omega\in\Lambda^l(\mathbf{R}^J)\subset\mathcal{C}_l$ into $F_J\otimes q_{J,\zeta}(\omega)$, is an epimorphism, and its kernel coincides with $\partial\mathcal{C}_{l+1}$. Thus this mapping induces a natural isomorphism of $\mathcal{A}_l$ with $H(\zeta)\otimes\Lambda^l(L)$.

This result is essentially due to Orlik and Solomon [22]. Combining Propositions 3.1 and 3.2, we get an isomorphism between $\mathcal{A}_l$ and $H^l(X)$; in [22] it is shown that the algebra $\mathcal{A}=\bigoplus_{m\leq l}\mathcal{A}_m$ constructed in the Appendix is isomorphic in a natural way to the cohomology ring $H^*(X)$.

It is possible to restate Proposition 3.2 in a purely algebraic fashion in terms of $H(\zeta)$. Indeed, let $\hat{J}$ be an $(l+1)$-subset in I having rank l (for ζ); this means that $\dim(\zeta\cap\mathbf{R}^{\hat{J}})=1$. Let $y=\sum_{j\in\hat{J}}b_je_j$ be a nonzero vector in ζ; in other words, this means that $\sum_{j\in\hat{J}}b_jf_j=0$. Dividing this last equality by $\prod_{j\in\hat{J}}f_j$, we get the following linear relation in $H(\zeta)$:

$$\sum_{j\in\hat{J}}b_jF_{\hat{J}\setminus j}=0 \tag{11}$$

(it is easy to see that $b_j\neq 0$ if and only if $\hat{J}\setminus j\in B(\zeta)$).

Proposition 3.3. Every linear relation among the elements $F_J\ (J\in B(\zeta))$ in $H(\zeta)$ is a consequence of relations of the form (11).

This follows directly from Proposition 3.2.

We note a similarity between (11) and the Gauss relations (6). This similarity is not accidental; we will make the connection with the hypergeometric function later.

Geometric Definition. We assume that ζ satisfies the additional restriction that $\zeta\cap\mathbf{R}^I_+=\varnothing$ (it is easy to see that, for any stratum Γ, the set of such ζ is nonempty and open in Γ). In other words, this means that all vectors f_i lie in some semispace in $L(\zeta)$; if the f_i are regarded as linear forms on $\zeta^\perp$, then this means that there exists a point $x\in\zeta^\perp$, for which $f_i(x)>0$ for all $i\in I$. For each subset $J\subset I$ of rank l (for ζ), we let C_J denote the open cone in L generated by the f_j for $j\in J$ (in other words, C_J is the image of $\mathbf{R}^J_+$ under the projection $q\colon\mathbf{R}^I\to L$). Let L^0 be the subset in L consisting of those vectors f which are in general position with respect to the system $\mathcal{F}=(f_i)_{i\in I}$ (i.e., f does not lie in any characteristic subspace in L spanned by a subsystem of $\mathcal{F}$). We put $C^0_J=C_J\cap L^0$ and we let $\chi_J=\chi_{J,\zeta}$ denote the characteristic function of the set C^0_J. Let $H'=H'(\zeta)$ be the vector space of functions on L generated by the functions χ_J for $J\in B(\zeta)$. We choose a nonzero skew-symmetric l-linear form ω on L, and, for each $J=\{j_1,\ldots,j_l\}\in B(\zeta)$ we put $c_J(\omega)=|\omega(f_{j_1},\ldots,f_{j_l})|$.

Proposition 3.4. The mapping which carries χ_J into $c_J(\omega)F_J$ for all $J\in B(\zeta)$, can be extended to an isomorphism of $H'(\zeta)$ and $H(\zeta)$.

Proof. We put $C^*_I=\{x\in\zeta^\perp: f(x)>0 \text{ for } f\in C_I\}$; from the condition that $\zeta\cap\mathbf{R}^I_+=\varnothing$ is a nonempty open convex cone in $\zeta^\perp$. We consider the Laplace transformation which carries a function φ on C_I into the function $P\varphi$ on C^*_I given by

$$P\varphi(x)=\int_L\varphi(f)e^{-f(x)}\omega(f). \tag{12}$$

We can see immediately that $P\chi_J = c_J(\omega)F_J$ for all $J \subseteq B(\zeta)$. In addition, it is easy to see that the restriction of P to $H'(\zeta)$ is injective; therefore it gives the required isomorphism.

We pass to the special basis in $H(\zeta)$. We call a subset $B \subseteq B(\zeta)$ a fundamental system for ζ if the functions F_J for $J \in B$ form a basis in $H(\zeta)$. By virtue of Proposition 3.2, thi definition agrees with the combinatorial definition in the Appendix; according to Proposition 3.4, a system B is fundamental if and only if the function χ_J for $J \in B$ form a basis in $H'(\zeta$ We describe a general method for constructing fundamental systems.

We say that a subset $I' \subseteq I$ is a circuit (for ζ) if the vectors f_i for $i \in I'$ are linearl dependent, but for any characteristic subset of I' this is not true. Now suppose that a linear ordering has been introduced on I; we call a subset of I an open circuit if it is obtaine from some circuit in I by deleting the maximal element.

THEOREM 3.1. The system of all subsets in $B(\zeta)$ which do not contain open circuits (with respect to an arbitrary given linear ordering of I) is fundamental.

In the combinatorial situation, this result (and an even more general one) has been established by Bjorner [14] (Theorem II.1 in the Appendix). Theorem 3.1 follows from Theorem II.1 if we make use of Proposition 3.2.

In particular, if ζ is a point of a general stratum in $G_k(\mathbf{R}^I)$, then $\dim H(\zeta) = \dim H'(\zeta) = \binom{|I|-1}{l-1}$, and we may take as a basis in $H(\zeta)$ $(H'(\zeta))$ the family of functions F_J (χ_J), where J runs through all l-subsets of I which contain a certain fixed element $i \in I$ (Appendix, Example 2).

4. FUNDAMENTAL SYSTEMS AND GENERAL HYPERGEOMETRIC FUNCTIONS

In this section we apply the concepts developed in Sec. 3 to a study of the general hypergeometric function. Again, let Γ be some fixed stratum in $G_k(\mathbf{R}^I)$. We consider the set $\hat{I} = I \cup \{0\}$, obtained by adjoining a distinguished point, denoted here by 0, to I. A technicality that arises here is that the space $H(\zeta)$ and the fundamental system constructed with reference to the stratum Γ are to be applied to a study of the general hypergeometric function, not on Γ, but on some stratum $\hat{\Gamma}^0$ in $G_{k+1}(\mathbf{R}^{\hat{I}})$. More precisely, we put $\hat{\Gamma} = \{\hat{\zeta} \in G_{k-1}(\mathbf{R}^{\hat{I}}): (\hat{\zeta} \cap \mathbf{R}^I) \in \Gamma\}$, and we let $p: \hat{\Gamma} \to \Gamma$ denote the projection given by $p(\hat{\zeta}) = \hat{\zeta} \cap \mathbf{R}^I$. It is not difficult to see that $\hat{\Gamma}$ is a fiber bundle over Γ whose fiber at each point $\zeta \in \Gamma$ is isomorphic in a natural way to the space $L(\zeta) = \mathbf{R}^I/\zeta$. Indeed, every subspace $\hat{\zeta} \in G_{k+1}(\mathbf{R}^{\hat{I}})$ such that $\hat{\zeta} \cap \mathbf{R}^I = \zeta$, is obtained by adjoining to ζ some vector of the form $e_0 + f$, where f is in $\mathbf{R}^I$ and is defined uniquely modulo ζ; thus it is possible to assume that $f \in L(\zeta)$, and the mapping $\hat{\zeta} \to f$ is the required isomorphism between $p^{-1}(\zeta)$ and $L(\zeta)$. Taking this isomorphism into account, we will write the elements of $\hat{\Gamma}$ as pairs (ζ, f), where $\zeta \in \Gamma$, and $f \in L = L(\zeta)$. It is easy to see that there is precisely one open stratum $\hat{\Gamma}^0 = \{(\zeta, f) \in \hat{\Gamma}: f \in L^0\}$ in $\hat{\Gamma}$ (we recall that an open subset $L^0 \subset L(\zeta)$ consists of vectors in general position with respect to the family of vectors $(f_i)_{i \in I}$; see Sec. 3).

We put $\hat{\Gamma}_+ = \{(\zeta, f) \in \Gamma: \zeta \cap \mathbf{R}_+^I = \varnothing\}$, and we let $\hat{\alpha} = (\alpha_i)_{i \in \hat{I}}$ be a set of indices with $|\hat{\alpha}| = l = |I| - k$. We will be concerned with the restriction of the function $\Phi(\hat{\alpha}; \hat{\zeta}, \hat{\omega})$ to the open subset $\hat{\Gamma}_+^0 = \hat{\Gamma}^0 \cap \hat{\Gamma}_+$ of $\hat{\Gamma}^0$.

We recall that, for each subset $J \subset I$ of rank l (for ζ), we use C_J to denote the open convex cone in $L = L(\zeta)$ generated by the vectors f_j for $j \in J$. We put $\hat{\Gamma}_+(J) = \{(\zeta, f) \in \hat{\Gamma}_+: f \in C_J\}$ and $\hat{\Gamma}_+^0(J) = \hat{\Gamma}^0 \cap \hat{\Gamma}_+(J)$. It is clear that $\hat{\Gamma}_+(I)$ consists of those subspaces $\hat{\zeta} \in \hat{\Gamma}_+$, for which $\hat{\zeta} \cap \mathbf{R}_+^{\hat{I}} \neq \varnothing$; hence it follows that the restriction of $\Phi(\hat{\alpha}; \hat{\zeta}, \hat{\omega})$ to $\hat{\Gamma}_+^0$ is concentrated on the subset $\hat{\Gamma}_+^0(I)$. We note that if $|J| = l$, then $\hat{\Gamma}_+^0(J)$ is the intersection of $\hat{\Gamma}_+^0$ with the large cell in $\hat{\Gamma}^0$ corresponding to the coordinate subspace $\mathbf{R}^{J \cup \{0\}}$ (see the Introduction).

As we have already remarked, we will be concerned with special integer indices $\hat{\alpha}$. But first we discuss the situation for general $\hat{\alpha}$ in an informal way. Let B be an arbitrary fundamental-system of l-subsets in I for the stratum Γ (see Sec. 3). We assert that the function $\Phi(\hat{\alpha}; \hat{\zeta}, \hat{\omega})$ on $\hat{\Gamma}_+^0$ can be expanded in a sum

$$\Phi(\hat{\alpha}; \hat{\zeta}, \hat{\omega}) = \sum_{J \in B} \Phi_J^{(B)}(\hat{\alpha}; \hat{\zeta}, \hat{\omega}), \qquad (*)$$

here $\Phi_J^{(B)}$ is equal to 0 for $\hat{\zeta} \notin \hat{\Gamma}_+^0(J)$ and "has unique analytic behavior" in $\hat{\Gamma}_+^0(J)$; this means hat if we continue the restrictions of $\Phi_J^{(B)}$ to the various cells of $\hat{\Gamma}_+^0(J)$ analytically into he complex domain, we get a unique branch of an analytic function in some region U in the omplexification $\hat{\Gamma}_{\mathbf{C}}^0$ of the stratum $\hat{\Gamma}^0$. If U is a sufficiently small region and E(U) is the finite-dimensional) space of restrictions to U of solutions of the holonomic system of equaions in [7], then it is possible to suggest that $\dim E(U) = \dim H(\zeta)$ (where ζ lies in the tratum Γ, covered by $\hat{\Gamma}^0$) and that it is possible to obtain a basis in E(U) by continuing $\Phi_J^{(B)}$ analytically into the complex domain. These questions will be treated in another article.

We pass to the precise statements. First of all, we introduce the normalization of the unction $\Phi(\hat{\alpha}; \hat{\zeta}, \hat{\omega})$ which is suitable for continuing it analytically with respect to $\hat{\alpha}$. Let $= (\zeta, f) \in \hat{\Gamma}_+^0$, and let ω be a nonzero k-form on ζ such that $\tilde{\zeta} = (\zeta, \omega) \in \tilde{\Gamma}$ (see Sec. 1). From we construct the (k + 1)-form $\hat{\omega}$ on $\hat{\zeta}$ given by $\hat{\omega}(y_1, \ldots, y_k, e_0 + \bar{f}) = \omega(y_1, \ldots, y_k)$, where $y_1, \ldots, y_k\}$ is a basis in ζ, and $\bar{f}$ is an arbitrary representative of the vector $f \in L = \mathbf{R}^I/\zeta$ n $\mathbf{R}^I$. We will write the point $(\hat{\zeta}, \hat{\omega}) \in \hat{\Gamma}_+^0$ in the form $(\tilde{\zeta}, f) = (\zeta, \omega, f)$, and the set of inices $\hat{\alpha}$ as a pair (α, α_0), where $\alpha = (\alpha_i)_{i \in I}$ is an arbitrary set of complex numbers, and $\alpha_0 = -\sum_{i \in I} \alpha_i$. We put

$$\Psi(\alpha;\ \zeta,\ \omega,\ f) = \Phi(\hat{\alpha}; \zeta, \omega, f) / \prod_{i \in I} \Gamma(\alpha_i). \tag{13}$$

It will be important to clarify the behavior of $\Psi(\alpha; \zeta, \omega, f)$ as a function of α in a eighborhood of a point where some of the α_i can vanish. More precisely, let $\alpha^{(0)} = (\alpha_i^{(0)})_{i \in I}$ e a set such that $\operatorname{Supp} \alpha^{(0)} = J$, where J is some m-subset in I which has rank l for ζ; we enote the subset $(\alpha_j^{(0)})_{j \in J}$ by $\alpha_J^{(0)}$. We put $\zeta_J = \zeta \cap \mathbf{R}^J$; thus $\zeta_J \in G_{m-l}(\mathbf{R}^J)$. From the k-form ω n ζ we construct the (m − 1)-form ω_J on ζ_J whose value at the vectors $y_1, \ldots, y_{m-l} \in \zeta_J$ is $(y_1, \ldots, y_{m-l}, \bar{e}_{i_1}, \ldots, \bar{e}_{i_r})$, where $\{i_1, \ldots, i_r\} = I \setminus J$, and the $\bar{e}_i$ are vectors in ζ congruent to e_i nodulo $\mathbf{R}^J$ (it is easy to see that such a form ω_J is uniquely determined up to its sign; see Sec. 1). We note that there is a natural isomorphism between $L(\zeta_J) = \mathbf{R}^J/\zeta_J$ and $L(\zeta) = \mathbf{R}^I/ = (\mathbf{R}^J + \zeta)/\zeta$.

<u>Proposition 4.1.</u> Under the above assumptions, $\Psi(\alpha; \zeta, \omega, f)$ can be continued analytically with respect to α to the point $\alpha^{(0)}$, and its value at this point is $\Psi(\alpha_J^{(0)}; \zeta_J, \omega_J, f)$.

The proof of this proposition will be given in another article.

<u>COROLLARY.</u> We suppose that the set $J = \operatorname{Supp} \alpha$ consists of l elements. Let $I \setminus J = \{i_1, \ldots, i_k\}$, and suppose that the vectors $\bar{e}_i \in \zeta$ for $i \in I \setminus J$ are as above. Then

$$\Psi(\alpha; \zeta, \omega, f) = |\omega(\bar{e}_{i_1}, \ldots, \bar{e}_{i_k})| \cdot \prod_{j \in J} (x_j)_+^{\alpha_j - 1} / \Gamma(\alpha_j), \tag{14}$$

where the x_j are the coordinates in the expansion of f with respect to the basis $(f_j)_{j \in J}$ of L. In particular, $\Psi(\alpha; \zeta, \omega, f)$ is concentrated on $\hat{\Gamma}_+^0(J)$, and can be continued analytically to $\hat{\Gamma}_+(J)$.

This follows directly from Proposition 4.1 and Example 1 of Sec. 1.

We call a set $\alpha = (\alpha_i)_{i \in I}$ *polynomial* for ζ (or for the stratum Γ) if all of the α_i are nonnegative integers and $\operatorname{Supp} \alpha$ has rank l for ζ. The following theorem gives a sharpening of the expansion (*) for polynomial α.

<u>THEOREM 4.2.</u> Let the set $\alpha = (\alpha_i)_{i \in I}$ be polynomial for Γ, and let $B \subset B(\zeta)$ be some fundamental system of l-subsets in I. Then the restriction of the function $\Psi(\alpha; \zeta, \omega, f)$ to Γ_+^0 can be written as a linear combination of the form

$$\Psi(\alpha; \zeta, \omega, f) = \sum_{\alpha'} c_{\alpha\alpha'}(\zeta, \omega)\, \Psi(\alpha'; \zeta, \omega, f),$$

where α' runs through the polynomial sets for Γ such that $|\alpha'| = |\alpha|$ and $\operatorname{Supp} \alpha' \in B$, and the coefficients $c_{\alpha\alpha'}(\zeta, \omega)$ do not depend on f but are analytic functions of $(\zeta, \omega) \in \tilde{\Gamma}$. The expansion with these properties is uniquely determined.

We note that every function $\Psi(\alpha'; \zeta, \omega, f)$ in the theorem is given by (14); hence it follows, in particular, that $\Psi(\alpha; \zeta, \omega, f)$ is piecewise polynomial with respect to f of degree $|\alpha| - l$. If we collect the terms corresponding to those sets α' with the same support, $\operatorname{Supp} \alpha'$, we get an expansion of the form (*).

Main Steps in the Proof of Theorem 4.2. 1. Let Γ' be a general stratum in $G_k(\mathbf{R}^I)$. We adjoin to B certain l-subsets which are dependent for Γ in order to obtain a system B' which is fundamental for Γ' (see Proposition II.1). We show that Theorem 4.2 holds for Γ' and the fundamental system B', and that, in addition, we have the following refinement: all of the coefficients $c_{\alpha\alpha'}(\zeta, \omega)$ can be continued analytically from $\tilde{\Gamma}'$ to $\tilde{\Gamma}$, and this continuation is equal to 0 if $\operatorname{Supp}\alpha'$ is dependent for Γ. It is clear that the existence of the expansion in Theorem 4.2 for arbitrary Γ follows from this refinement.

2. We fix the system B'. We call a polynomial set of indices α good if the refinement in Step 1 holds for $\Psi(\alpha; \zeta, \omega, f)$. We must show that every set α is good.

LEMMA 1. If the set α is good and $i \in \operatorname{Supp}\alpha$, then $\alpha + \mathbf{1}_i$ is also good.

LEMMA 2. Suppose that $\hat{J} \subset I$ and that $|\hat{J}| > l$. We assume that all of the sets α' such that $\operatorname{Supp}\alpha'$ is obtained from $\hat{J}$ by deleting some elements are good. Then $\mathbf{1}_{\hat{J}}$ is good.

Lemma 1 can be deduced from (7) in Sec. 2, and Lemma 2 from the Gauss relations (6) (or (9)).

3. By virtue of Lemmas 1 and 2, it suffices to verify the assertion that all of the α are good for $\alpha = \mathbf{1}_J$, where $J \in B(\zeta)$. But by virtue of (14) the function $\Psi(\mathbf{1}_J; \zeta, \omega, f)$, regarded as a function of f for fixed ζ and ω, is proportional to the function χ_J introduced in Sec. 3. Therefore the fact that $\mathbf{1}_J$ is good follows from the results in Sec. 3. This proves the existence of the expansion in Theorem 4.2. The uniqueness requires a separate argument, which we do not give here.

We postpone a detailed proof of Theorem 4.2 to a later article, where we will also consider the case of general indices.

5. AN APPLICATION OF THEOREM 4.2

We regard the function $\Psi(\alpha; \zeta, \omega, f)$ defined by (13) as a function of $f \in L = L(\zeta)$ for fixed ζ and ω. For brevity, we will write $\Psi(\alpha; f)$ instead of $\Psi(\alpha; \zeta, \omega, f)$. We note that, with respect to the family $\mathcal{F} = (f_i)_{i\in I}$ of rank l in L, the subspace $\zeta \in G_k(\mathbf{R}^I)$ is restored as the kernel of the projection $q: \mathbf{R}^I \to L$ which carries e_i into f_i. As before, we will assume that all of the f_i are different from 0 and lie in some open halfspace in L. We will be interested only in the case where all of the indices α_i are positive integers. In this case, $\Psi(\alpha; f)$ admits a completely "elementary" definition.

For each vector $f \in L$ we put $\Delta(f) = q^{-1}(f) \cap \mathbf{R}^I_+$, which is a bounded (open) polyhedron in the k-dimensional affine plane $q^{-1}(f)$ in $\mathbf{R}^I$; it is obvious that $\Delta(f) \neq \varnothing$ if and only if $f \in C_I$, where C_I is the open convex cone in L generated by all of the f_i. We denote by ω the k-form on $q^{-1}(f)$ obtained from the form ω on $\zeta = q^{-1}(0)$ by parallel translation, as well as its restriction to $\Delta(f)$.

Proposition 5.1. Using the above notation, we can write $\Psi(\alpha; f)$ in the form

$$\Psi(\alpha; f) = \Big(\prod_{i\in I} \Gamma(\alpha_i)\Big)^{-1} \int_{\Delta(f)} \Big(\prod_{i\in I} x_i^{\alpha_i - 1}\Big)\, \omega(x). \tag{15}$$

This follows directly from the definition.

In particular, the function $\Psi(\mathbf{1}_I; f)$ is simply the volume of the polyhedron $\Delta(f)$ with respect to the form ω.

THEOREM 5.1. Let $\alpha = (\alpha_i)_{i\in I}$ be a set of positive integral indices. Let B be some fundamental system of l-subsets in I for ζ. Then for each $J \in B$ there exists a unique polynomial $\Psi_J^{(B)}(\alpha; f)$ on the space L of degree $|\alpha| - l$ such that for $f \in L^0$ we have the expansion

$$\Psi(\alpha; f) = \sum_{J\in B} \Psi_J^{(B)}(\alpha; f)\, \chi_J(f)$$

(for the definitions of L^0 and χ_J, see Sec. 3).

This theorem follows immediately from Theorem 4.2.

A theorem on the expansion of rational functions is another interesting application of Theorem 4.2.

From the form ω on ζ, we construct the l-form ω_L on L in the following way: for $v_1, \ldots, v_l \in L$ we put $\omega_L(v_1, \ldots, v_l) = \omega_I(y_1, \ldots, y_k, \bar{v}_1, \ldots, \bar{v}_l)/\omega(y_1, \ldots, y_k)$, where $y_1, \ldots, y_k$ is some basis n ζ, where $\bar{v}_i \in q^{-1}(v_i)$ for i = 1,...,l, and where ω_I is a (k + l)-form on $\mathbf{R}^I$ which takes on ne values ±1 on the set $(e_i)_{i\in I}$ (it is obvious that the form ω_L is defined uniquely up to its ign). As in Sec. 3, we identify vectors in L with linear forms on $\zeta^\perp$. Let P be the Laplace ransformation relative to ω_L [see (12) in Sec. 3].

Proposition 5.2. We have that $P\Psi(\alpha; f) = \prod_{i\in I} f_i^{-\alpha_i}$.

The proof follows directly from the definitions.

For each polynomial set α, we define the rational function F_α on $\zeta^\perp$ by putting $F_\alpha = \prod_{i\in I} f_i^{-\alpha_i}$.

THEOREM 5.2. Let B be a fundamental system of l-subsets for ζ. Every rational function α can be represented in the form of a linear combination $\sum_{\alpha'} c_{\alpha\alpha'} F_{\alpha'}$ with constant coefficients αα', where α' runs through a set such that $|\alpha'| = |\alpha|$ and Supp $\alpha' \in B$. This representation s unique.

This follows immediately from Theorem 4.2 and Proposition 5.2.

The expansions in Theorem 5.1 and 5.2 are closely related to each other.

Proposition 5.3. Let $F_\alpha = \sum_{\alpha'} c_{\alpha\alpha'} F_{\alpha'}$ be the expansion in Theorem 5.2. Then the polynomial $\Psi_J^{(B)}(\alpha; f)$ in Theorem 5.1 is given by

$$\Psi_J^{(B)}(\alpha; f) = c_J^{-1} \sum_{\alpha'} c_{\alpha\alpha'} \prod_{j\in J} x_j^{\alpha'_j - 1}/\Gamma(\alpha'_j),$$

where $c_J = |\omega_L((f_j)_{j\in J})|$, α' runs through the sets such that $|\alpha'| = |\alpha|$ and Supp $\alpha' = J$, and the x_j are the coordinates in the expansion of f with respect to the basis $(f_j)_{j\in J}$.

This follows immediately from Proposition 5.2 and (14) in Sec. 4.

Remarks. a) It is possible to derive an algorithm for calculating the $\Psi_J^{(B)}(\alpha; f)$ from the proof of Theorem 4.2; another method is to apply Proposition 5.3. Since we have a general method for constructing fundamental systems (Theorem 3.1), it is possible in principle to obtain an explicit formula for Ψ(α; f). It is clear that, in concrete examples, obtaining such a formula may not be a simple matter.

b) Proposition 5.3 is closely related to the duality theorem for general hypergeometric functions obtained in [9].

We turn to Kostant's partition function. Let $\mathcal{F} = (f_i)_{i\in I}$ be the positive roots of some system R of roots in L; we normalize the form ω so that the l-form ω_L on L corresponding to it takes on the values ±1 on the set of simple roots in R. We recall that, by definition, Kostant's function $K_R(f)$ is the number of representations of f in the form $\sum_{i\in I} m_i f_i$, where all of the m_i are nonnegative integers [21]. Using the terminology introduced above, we can reformulate this definition thus: $K_R(f)$ is the number of integer points in the closure of the polyhedron Δ(f). We define the *continuous analog* of $K_R(f)$ as the volume of the polyhedron Δ(f); it is easy to verify that this definition is equivalent to that given in [4]. Thus, in the notation introduced above, the continuous analog of Kostant's function is $\Psi(\mathbf{1}_I; f)$. All of the results obtained above are applicable to this function. In particular, Theorem 5.1 shows that to each choice of a fundamental system B for the pregeometry on I given by $\mathcal{F}$, there is related an expansion $\Psi(\mathbf{1}_I; f) = \sum_{J\in B} \Psi_J^{(B)}(\mathbf{1}_I; f)\chi_J(f)$, where each function $\Psi_J^{(B)}(\mathbf{1}_I; f)\chi_J(f)$ is concentrated in the cone C_J and coincides with the restriction of the polynomial $\Psi_J^{(B)}(\mathbf{1}_I; f)$ in it; the degree of these polynomials in the case at hand is equal to the number of positive roots in the system R minus its rank.

We give a number of examples of choices of fundamental systems in this situation. The number of elements of B is given by the following proposition.

Proposition 5.4. The dimension of the space $\mathcal{A}_I$, corresponding to the pregeometry with respect to a family of positive roots of some system of roots R of rank l is equal to product $m_1 \ldots m_l$, where the numbers m_i are the indices of R (see [5]).

In view of Propositions 3.1 and 3.2, this result is proved in [2, 15] (see also [16]).

Examples. 1. Let R be of type A_l. In this case, the set I consists of pairs (i, j), where $1 \leqslant i < j \leqslant l + 1$, and the family $\mathcal{F}$ consists of vectors $\varepsilon_i - \varepsilon_j$, where $\{\varepsilon_1, \ldots, \varepsilon_{l+1}\}$ is the standard basis in $\mathbf{R}^{l+1}$. We choose the lexicographic order on I: (i, j) < (i', j') if i < i' or if i = i' and j < j'. It is not difficult to verify that a subset J of I does not contain open circuits (with respect to this order) if and only if, for each i = 1,...,l, there is not more than one vector of the form $\varepsilon_i - \varepsilon_j$. According to Theorem 3.1, the system B of all such l-subsets is fundamental. Hence $|B| = l!$, which agrees with Proposition 5.4.

2. Let R be of type A_l as before. Let w be a permutation of the set {1,...,l + 1} such that w(l + 1) = l + 1. For each r = 1,...,l, we put i(w, r) = min (w(r), w(r + 1)) and j(w, r) = max (w(r), w(r + 1)); we define the l-subset $J_w \subset I$, by putting $J_w = \{(i(w, r), j(w, r)): r = 1, \ldots, l\}$. It is not difficult to verify that the system of all subsets J_w is fundamental; we note that it is not obtained by means of the construction in Theorem 3.1. It is possible to interpret a result in Lidskii [11] as an explicit calculation of the polynomials $\Psi_{J}^{(B)}(\mathbf{1}_I; f)$ for this fundamental system.

3. Let R be of type B_l. The set I consists of the symbols $\{i, (i, j)^+, (i, j)^- : 1 \leqslant i < j \leqslant l\}$, and the family $\mathcal{F}$ consists of the vectors $f_i = \varepsilon_i$, $f_{(i,j)^\pm} = \varepsilon_i \pm \varepsilon_j$. We denote the subset of I consisting of the elements i, $(i, j)^+$, and $(i, j)^-$ for fixed i and all possible j > i by I_i. We introduce a certain linear ordering on I such that if i < i', the elements of I_i precede those of $I_{i'}$. It is not difficult to verify that the l-subset J of I does not contain open circuits relative to such an order if and only if J intersects every subset I_i in precisely one element. By virtue of Theorem 3.1, the system B of such l-subsets is fundamental. It is clear that $|B| = \prod_{1 \leqslant i \leqslant l} |I_i| = 1 \cdot 3 \cdot 5 \cdot \ldots \cdot (2l-1)$, which again agrees with Proposition 5.4.

4. For type C_l, the set I and the pregeometry on it induced by the family of positive roots is the same as for B_l. In particular, the system B constructed in Example 3 is also fundamental for C_l.

5. Suppose that R is of type G_2. In this case, $\mathcal{F}$ consists of 6 vectors in $\mathbf{R}^2$ in general position. By virtue of Example 2 of the Appendix, the system B of all 2-subsets in I containing an arbitrary fixed element is fundamental. In particular, |B| = 5.

Other examples, as well as explicit expressions for $\Psi(\mathbf{1}_I; f)$ related to various choices of fundamental systems, will be considered in a separate article.

APPENDIX. COMBINATORIAL GEOMETRY

The concept of a (combinatorial) pregeometry (or matroid) admits many equivalent definitions ([1, 17, 23]). For our purposes, a definition in terms of rank functions is convenient.

Definition. Let I be a finite set. We say that a pregeometry is given on I if there is defined on the set of all subsets of I an integer-valued function r which satisfies the following conditions:

(i) $0 \leqslant r(J) \leqslant |J|$ for all $J \subset I$;

(ii) $r(J_1) \leqslant r(J_2)$ for $J_1 \subset J_2$;

(iii) $r(J_1 \cap J_2) + r(J_1 \cup J_2) \leqslant r(J_1) + r(J_2)$ for all $J_1, J_2 \subset I$.

The number r(J) is called the rank of J, and r(I) the rank of the pregeometry.

Example. Let L be a vector space over some field. Then to every family of vectors $(f_i)_{i \in I}$ in L there corresponds a pregeometry on I: the rank r(J) is defined to be the dimension of the vector space spanned by the vectors f_j for $j \in J$. Pregeometries of this type are called *linear*. This example makes the terminology introduced below seem natural.

Let I be a pregeometry defined on I with rank function r and rank r(I) = l. A subset $J \subset I$ is said to be *independent* if r(J) = |J| and *dependent* if r(J) < |J|. Maximal independent subsets in I are called *bases of the pregeometry*; it is well known that every basis has rank l. Minimal dependent subsets of I are called *circuits* (this terminology originates in graph theory).

With every pregeometry on I of rank l is associated a graded commutative superalgebra $\mathcal{A} = \bigoplus_{0 \leqslant m \leqslant l} \mathcal{A}_m$. Indeed, let $\mathcal{E} = \bigoplus_{0 \leqslant m \leqslant |I|} \mathcal{E}_m$ be the Grassman algebra generated by the elements

$_{i})_{i\in I}$ (over the field $\mathbf{R}$, for definiteness). For each $J\subset I$ we put $[J]=\Lambda^{|J|}(\mathbf{R}^J)$ for brevity, so that $\mathscr{E}_m=\bigoplus_{|J|=m}[J]$. Let $\partial:\mathscr{E}\to\mathscr{E}$ be the (super)differentiation of $\mathscr{E}$ which carries all e_i into i. We put $\mathscr{I}^0=\bigoplus[J]$, where the sum is taken over all *dependent* $J\subset I$, and we let $\mathscr{I}=\mathscr{I}^0+\partial\mathscr{I}^0$; it is easy to see that $\mathscr{I}^0$ and $\mathscr{I}$ are graded ideals in $\mathscr{E}$, and $\mathscr{I}^0_m=\mathscr{I}_m=\mathscr{E}_m$ for $m>l$. We define the algebra $\mathscr{A}=\bigoplus_{0\leqslant m\leqslant l}\mathscr{A}_m$ as the factor algebra $\mathscr{E}/\mathscr{I}$.

We describe a construction of special bases in $\mathscr{A}$. Let m be an integer between 0 and l, and let B be some system of independent m-subsets in I. We call a system *fundamental* if the restriction of the natural projection $\mathscr{E}_m\to\mathscr{A}_m$ to the subspace $\bigoplus_{J\in B}[J]$ gives its isomorphism with $\mathscr{A}_m$ (in other words, if e_J is a generator of the space [J], then the images in $\mathscr{A}_m$ of the elements e_J for $J\in B$ constitute a basis in $\mathscr{A}_m$). In particular, $|B|=\dim\mathscr{A}_m$ is an invariant of the pregeometry.

We choose an arbitrary linear ordering "<" on I. We say that a subset J' of I is an *open circuit* if J' is obtained from some circuit in I by deleting a maximal element; we call $J\subset I$ *proper* if J does not contain any open circuits. It is easy to see that all proper subsets are independent.

<u>THEOREM II.1</u>. For every linear ordering on I and for all $0\leqslant m\leqslant l$, the system of all proper m-subsets in I is fundamental.

<u>Remarks.</u> The definition of a pregeometry (or matroid) is due to Whitney, MacLane, and Birkhoff (see the historical notes in [17]). The construction of the algebra $\mathscr{A}$ is due to Orlik and Solomon [22], and Theorem II.1 is essentially due to Bjorner [14], but it is described there in different terms. As given in [14], the concept of an open circuit and the combinatorial ideas at the foundation of Theorem II.1 originated with Whitney and Rota. A more general construction of fundamental sets is given in [14], but Theorem II.1 is sufficient for our purposes.

We give a new proof of Theorem II.1 which is independent of [14].

First of all, it is not difficult to show that the proof of the theorem for arbitrary m reduces to the case m = l.

Now let m = l, i.e., B is the family of all proper bases of our pregeometry.

1. Let e_J be a generator of the space [J]. We show that the images of the vectors e_J $(J\in B)$ under the projection $\mathscr{E}_l\to\mathscr{A}_l$ generate all of $\mathscr{A}_l$. In other words, we must show that the subspace $\bigoplus_{J\in B}[J]+\mathscr{I}^0_l+\partial\mathscr{E}_{l+1}$ coincides with the entire space $\mathscr{E}_l$, i.e., contains all of the e_{J^0}, where J^0 is an arbitrary l-subset in I. If J^0 is dependent or $J^0\in B$, then there is nothing to prove, so we suppose that J^0 is a basis which is not proper. By definition, this means that there are a subset $J'\subset J^0$ and an element $i\in I$ larger than all of the elements of J' such that $J'\cup\{i\}$ is a circuit in I. It is clear that $i\notin J^0$; we put $\hat{J}=J^0\cup\{i\}$. The element $\partial e_{\hat{J}}$ is a linear combination with nonzero coefficients of elements $e_{\hat{J}\setminus j}$ for $j\in\hat{J}$; in addition, for $j\in J^0\setminus J'$, the set $\hat{J}\setminus j$ is dependent. Therefore the element e_{J^0} is congruent modulo $\mathscr{I}^0_l+\partial\mathscr{E}_{l+1}$ to a linear combination of elements of the form $e_{\hat{J}\setminus j}$ for $j\in J'$. But it is easy to see that there exists a linear order on the set of l-subsets of I with respect to which all sets $\hat{J}\setminus j$ for $j\in J'$ are less than $J^0=\hat{J}\setminus i$. Applying induction on J with respect to this order, we get that $e_{\hat{J}\setminus j}$ for $j\in J'$ lies in $\bigoplus_{J\in B}[J]+\mathscr{I}^0_l+\partial\mathscr{E}_{l+1}$; this means that this is also true for e_{J^0}, which is what is required.

2. For each m with $0\leqslant m\leqslant|I|$, we let $\mathscr{B}_m\subset\mathscr{E}_m$ denote the sum of subspaces of [J] with respect to all m-subsets J of rank < l. It is obvious that $\mathscr{B}_l=\mathscr{I}^0_l$, $\mathscr{B}_m=\mathscr{E}_m$ for m < l and $\partial(\mathscr{B}_m)\subset\mathscr{B}_{m-1}$ for all m. We put $\mathscr{C}_m=\mathscr{E}_m/\mathscr{B}_m$ and we denote the mapping $\mathscr{C}_m\to\mathscr{C}_{m-1}$ induced by ∂ also by ∂. Since $\mathscr{C}_l=\mathscr{E}_l/\mathscr{I}^0_l$, and $\mathscr{A}_l=\mathscr{E}_l/(\mathscr{I}^0_l+\partial\mathscr{E}_{l+1})$, there is a natural projection $\mathscr{C}_l\to\mathscr{A}_l$; we denote it by ε. Then the sequence

$$0\to\mathscr{C}_{|I|}\xrightarrow{\partial}\mathscr{C}_{|I|-1}\xrightarrow{\partial}\dots\xrightarrow{\partial}\mathscr{C}_l\xrightarrow{\varepsilon}\mathscr{A}_l\to 0. \qquad (*)$$

arises. We assert that this sequence is exact.

The exactness of the terms $\mathscr{A}_l$ and $\mathscr{C}_l$ is obvious. From Folkman's theorem on homologies of geometric lattices [18] it follows easily that the complex $((\mathscr{B}_m),\partial)$ is acyclic everywhere except for $\mathcal{B}_{l-1}$ (see [16], no. 17); in addition, it is known that the complex $((\mathcal{E}_m), \partial)$ is acyclic. The exactness of (*) follows immediately from this with the help of the exact sequence of the pair.

3. It is clear that $\mathcal{C}_m = \mathcal{E}_m/\mathcal{B}_m$ can be identified, as a vector space, with $\oplus [J]$, where the sum is taken over all m-subsets $J \subset I$ of rank l. In particular, $\dim \mathcal{C}_m$ is the number of such subsets; we denote it by r_m. Taking the Euler–Poincaré characteristic of the exact sequence (*), we get that $\dim \mathcal{A}_l = \sum_{m \geqslant l} (-1)^m r_m$.

4. It remains to show that $|B| = \sum_{m \geqslant l} (-1)^m r_m$. Let E be the set of all subsets of rank l in I which do not belong to B. For $J \in E$, we put $p(J) = (-1)^{|J|}$. We must show that

$$|\{J \in E:\ p(J) = 1\}| = |\{J \in E : p(J) = -1\}|. \qquad (**)$$

For the proof of (**), it suffices to construct an involution $\sigma: E \to E$ such that $p(\sigma(J)) = -p(J)$ for $J \in E$.

Suppose that $J \in E$. By definition, J contains an open circuit, i.e., there exists a circuit J^0 with maximal element $i \in I$ such that $J^0 \setminus i \subset J$. We denote the maximal element i having this property by i(J). We define σ(J) thus: if $i(J) \notin J$, then $\sigma(J) = J \cup \{i(J)\}$; but if $i(J) \in J$ then $\sigma(J) = J \setminus \{i(J)\}$. Obviously $\sigma(J) \in E$ and $p(\sigma(J)) = -p(J)$. In addition, it follows easily from the construction that $i(\sigma(J)) = i(J)$, from which it is clear that $\sigma(\sigma(J)) = J$. Thus σ is the required involution, which proves (**) and completes the proof of the theorem.

Examples. 1. We suppose that there is an element $i \in I$ such that $r(\{i\}) = 0$. In this case $\mathcal{A} = 0$. In fact, the empty set $\emptyset$ is an open circuit, so there are no proper subsets.

2. Suppose that the rank function is defined by $r(J) = \min(l, |J|)$, so all m-subsets are independent for $m \leqslant l$. In this case, $\mathcal{A}_m = \mathcal{E}_m$ for $m < l$. We can take all l-subsets containing some fixed element $i \in I$ as a fundamental system. In particular, $\dim \mathcal{A}_l = \binom{|I|-1}{l-1}$.

Other examples are given in Sec. 5 of the main text.

Proposition II.1. Suppose that some pregeometry of rank l is given, and suppose that B is a fundamental system of l-subsets in I relative to this pregeometry. Then B can be extended by means of certain independent l-subsets to a fundamental system of l-subsets relative to the pregeometry in Example 2.

This follows directly from the definition.

LITERATURE CITED

1. M. Aigner, Combinatorial Theory, Springer-Verlag, New York (1982).
2. V. I. Arnol'd, "The cohomology ring of the group of dyed braids," Mat. Zametki, 5, No. 1, 227-231 (1969).
3. H. Bateman and A. Erdelyi, Higher Transcendental Functions, Vol. 1, McGraw-Hill, New York (1953).
4. F. A. Berezin and I. M. Gel'fand, "Some remarks on the theory of spherical functions on symmetric Riemann manifolds," Trudy Moskv. Mat. Obshsch., 5, 311-352 (1956).
5. N. Bourbaki, Lie Groups and Lie Algebras [Russian translation], Mir, Moscow (1972), Chaps. IV-VI.
6. I. M. Gel'fand, "General theory of hypergeometric functions," Dokl. Akad. Nauk SSSR, 288, No. 1, 14-18 (1986).
7. I. M. Gel'fand and S. I. Gel'fand, "Generalized hypergeometric equations," Dokl. Akad. Nauk SSSR, 288, No. 2, 279-283 (1986).
8. I. M. Gel'fand, S. G. Gindikin, and M. I. Graev, "Integral geometry in affine and projective spaces," Sov. Prob. Mat. (VINITI), 16, 53-226 (1980).
9. I. M. Gel'fand and M. I. Graev, "A duality theorem for general hypergeometric functions," Dokl. Akad. Nauk SSSR, 289, No. 1 (1986).
10. I. M. Gel'fand and A. V. Zelevinskii, "Models for representations of classical groups and their latent symmetries," Funkts. Anal. Prilozhen., 18, No. 3, 14-31 (1984).
11. B. V. Lidskii, "On the Kostant function of a system of roots A_n," Funkts. Anal. Prilozhen., 18, No. 1, 76-77 (1984).
12. Twistors and Gauge Fields [Russian translation], Mir, Moscow (1983).
13. M. F. Atiyah, "Convexity and commuting Hamiltonians," Bull. London Math. Soc., 14, 1-15 (1982).

4. A. Bjorner, "On the homology of geometric lattices," Algebra Univ., 14, No. 1, 107-128 (1982).
5. E. Briskorn, "Sur les groupes de tresses (d'apres V. I. Arnold)," Sém. Bourbaki, 1971/72, Lecture Notes Math., No. 317, Springer-Verlag, New York (1973), pp. 21-44.
6. P. Cartier, "Les arrangements d'hyperplans; un chapitre de géométrie combinatoire," Sém. Bourbaki, 1980/81, Lecture Notes Math., No. 901, Springer-Verlag, New York (1981), pp. 1-22.
7. H. Crapo and G. C. Rota, On the Foundations of Combinatorial Theory: Combinatorial Geometries, MIT Press, Cambridge (1970).
18. J. Folkman, "The homology groups of a lattice," J. Math. Mech., 15, 631-636 (1966).
19. I. M. Gel'fand and R. MacPherson, "Geometry in Grassmannians and in generalization of the dilogarithm," Adv. Math., 44, No. 3, 279-312 (1982).
20. V. Guillemin and S. Sternberg, "Convexity properties of the moment mapping," Invent. Math., 67, 491-513 (1982).
21. B. Kostant, "A formula for the multiplicity of a weight," Trans. Am. Math. Soc., 93, 53-73 (1959).
22. P. Orlik and L. Solomon, "Combinatorics and topology of complements of hyperplanes," Invent. Math., 56, 167-189 (1980).
23. D. J. A. Welsh, Matroid Theory, Academic Press, New York (1976).
24. K. Aomoto, "Les équations aux différences linéaires et les intégrales des fonctions multiformes," J. Fac. Sci. Univ. Tokyo, Math., 22, 271-297 (1975).
25. K. Aomoto, "Configurations and invariant Gauss–Manin connections of integrals. I," Tokyo J. Math., 5, 249-287 (1982).

5.

(with M. Goresky, R. D. MacPherson, V. V. Serganova)

Combinatorial geometries, convex polyhedra, and Schubert cells

Adv. Math. **63** (3) (1987) 301–316. Zbl. **622**:57014

Introduction

This paper is a continuation of [GM] which was published in the same journal. We will explore a remarkable connection between the geometry of the Schubert cells in the Grassmann manifold, the theory of convex polyhedra, and the theory of combinatorial geometries in the sense of Crapo and Rota [CR]. The results in this paper were obtained simultaneously and independently by Gelfand and Serganova (in Moscow) and by Goresky and MacPherson (at the I.H.E.S. in Paris) as part of larger programs with different purposes (see below). The geometry of this simple example is so beautiful that we decided to publish it independently of the applications. We believe that combinatorial methods will play an increasing role in the future of geometry and topology.

We consider the Grassmann manifold G^k_{n-k} of all $(n-k)$-dimensional subspaces of $\mathbb{C}^n$. By fixing the standard basis in $\mathbb{C}^n$ we obtain an action of the torus $H=(\mathbb{C}^*)^n$ on G^k_{n-k} which is induced from stretching the coordinate axes in $\mathbb{C}^n$ (see also Sect. 1). We will describe not only the trajectories, but also the "strata" of a new and interesting decomposition of the

Grassmanian (which is finer than the usual stratification by isotropy subgroup of H). Understanding the geometry of the strata and the quotient space of this action is useful in many situations, and this paper may be considered as an introduction to these other situations: (1) for understanding the generalized hypergeometric functions and the Kostant partition function [G, GG, GZ], (2) for understanding the dilogarithm and the polylogarithms and their functional equations [GM, HM], (3) for the study of combinatorial geometries which are associated to other Lie groups and parabolic subgroups [GS], (4) for construction of combinatorial Chern and Pontrjagin classes [GGL, M], (5) for the study of he representability of matroids [GoM], and (6) for the study of algebraic K-theory [BMS].

According to [GM] the trajectories of the action of $(\mathbb{C}^*)^n$ on the Grassmannian G^k_{n-k} correspond to projective configurations of n points in $\mathbb{P}^{k-1}(\mathbb{C})$. This torus action also gives rise to a moment map $\mu\colon G^k_{n-k} \to \mathbb{R}^n$ (see [GM] for the case of the Grasmannian, and [A] or [GuS] for an important generalization) with the property that the image of each trajectory is a convex polyhedron. Our main result is that the following three different decompositions of the Grassmannian into strata all coincide:

(1) The set of points in G^k_{n-k} such that the corresponding projective configuration represents a fixed combinatorial geometry (see Sect. 1).

(2) The union of the orbits of $(\mathbb{C}^*)^n$ whose projection under μ is a fixed convex polyhedron (see Sect. 2).

(3) A multi-intersection of translates of Schubert cells which are obtained by permuting the coordinate axes (see Sect. 3).

The equivalence of (1) and (2) establishes a one to one correspondence between representable (over $\mathbb{C}$) combinatorial geometries (or matroids) and certain convex polyhedra. In Section 4 we extend this to a correspondence between all matroids and certain polyhedra which are characterized by a restriction on their vertices and edges (1-dimensional faces). This characterization is equivalent to the Steiner exchange axiom. The marriage of matroid theory and convex set theory should have interesting consequences. The polyhedron corresponding to the Fano plane is particularly beautiful.

We would like to thank S. I. Gelfand for his valuable suggestions concerning this manuscript.

1. The Grassmann Strata and Combinatorial Geometries

1.1. Definitions. Throughout this paper we fix the standard unit vectors $e_1, e_2, \ldots, e_n$ of $\mathbb{C}^n$ and let G^k_{n-k} denote the Grassmann manifold of

$(n-k)$-dimensional subspaces of $\mathbb{C}^n$. For each plane $P \in G_{n-k}^k$ the projection

$$\pi_p: \quad \mathbb{C}^n \to \mathbb{C}^n/P$$

determines n vectors (some of which may be 0), $\pi_p(e_1)$, $\pi_p(e_2), \ldots, \pi_p(e_n)$ in the quotient $\mathbb{C}^n/P \cong \mathbb{C}^k$. We obtain in this way a (representable over $\mathbb{C}$) *matroid* (or combinatorial geometry) of rank k on the set $\{1, 2, 3, \ldots, n\}$, i.e., a "rank function" defined on subsets $J \subset \{1, 2, \ldots, n\}$, which is given by

$$\operatorname{rank}(J) = \dim_{\mathbb{C}}(\operatorname{span}\{\pi_p(e_j) \mid j \in J\})$$

and which satisfies the following matroid axions: [Wh, VW, CR, W]:

(R1) $\operatorname{rank}(\phi) = 0$,

(R2) $I \subseteq J \Rightarrow \operatorname{rank}(I) \leqslant \operatorname{rank}(J)$,

(R3) $\operatorname{rank}(I \cup J) + \operatorname{rank}(I \cap J) \leqslant \operatorname{rank}(I) + \operatorname{rank}(J)$.

Remark. Given any k-dimensional complex vectorspace V and any n vectors $v_1, v_2, \ldots, v_n$ which span V, there is a plane $P \in G_{n-k}^k$ and an isomorphism $F: \mathbb{C}^n/P \cong V$ such that $F(\pi_p(e_i)) = v_i$ (for $i = 1, 2, \ldots, n$). In fact, F is induced by the surjective homomorphism $\tilde{F}: \mathbb{C}^n \to V$ which is defined by $\tilde{F}(e_i) = v_i$.

1.2. *Grassmann Strata*

DEFINITION. Two points $P_1, P_2 \in G_{n-k}^k$ are said to lie in the same *Grassmann stratum* Γ of G_{n-k}^k if they give rise to the same matroid, i.e., if for each subset $J \subset \{1, 2, \ldots, n\}$ we have,

$$\dim_{\mathbb{C}} \operatorname{span}\{\pi_{P_1}(e_j) \mid j \in J\} = \dim_{\mathbb{C}} \operatorname{span}\{\pi_{P_2}(e_j) \mid j \in J\}.$$

1.3. *Torus Action*

The algebraic torus $H = (\mathbb{C}^*)^n$ acts on $\mathbb{C}^n$ by stretching the coordinate axes, i.e., if $\lambda = (\lambda_1, \lambda_2, \ldots, \lambda_n) \in H$ and if $x \in \mathbb{C}^n$ then

$$\lambda \cdot x = (\lambda_1 x_1, \lambda_2 x_2, \ldots, \lambda_n x_n).$$

The action of each $\lambda \in H$ is linear so it takes subspaces to subspaces and therefore induces an action on G_{n-k}^k. The fixed points of this action are easily described: for each k-element subset $J \subset \{1, 2, \ldots, n\}$ there are coordinate k and $n-k$ planes,

$$R_J = \operatorname{span}\{e_j \mid j \in J\},$$
$$R_J^\perp = \operatorname{span}\{e_j \mid j \notin J\}.$$

It is easy to see that the fixed points of the action of H on G^k_{n-k} are precisely the coordinate $n-k$ planes $R_J^\perp$ (for arbitrary k-element subsets J).

Remark. The closure (in G^k_{n-k}) of an orbit of H is a normal algebraic subvariety of G^k_{n-k} which is H-stable and consists of finitely many H-orbits, i.e. it is a toric variety [D].

1.4. LEMMA. *Fix $P \in G^k_{n-k}$ and let Φ denote the corresponding matroid. Let $\overline{H \cdot P}$ denote the closure (in G^k_{n-k}) of the orbit of H which contains P. Then the fixed points of H which lie in $\overline{H \cdot P}$ are precisely those coordinate $n-k$ planes $R_J^\perp$ such that J is a basis (i.e., a maximal independent subset) of Φ.*

Proof. First, suppose that J is a basis of Φ. This means that $\{\pi_P(e_j) \mid j \in J\}$ are linearly independent in $\mathbb{C}^n/P$, i.e., that $P \cap R_J = \{0\}$, where

$$R_J = \operatorname{span}\{e_j \mid j \in J\}.$$

Thus the plane P can be realized as the *graph* of a linear transformation

$$f\colon \quad R_J^\perp \to R_J$$

in the product space $\mathbb{C}^n = R_J^\perp \oplus R_J$. Now consider the action of $\mathbb{C}^* \subset H$ on the Grassmannian G^k_{n-k}, which is induced by the following action on $\mathbb{C}^n$:

$$\lambda \cdot e_j = \begin{cases} \lambda e_j & \text{if } j \in J, \\ e_j & \text{if } j \notin J. \end{cases}$$

It follows that for any plane $P \in G^k_{n-k}$, the induced action satisfies

$$\lambda \cdot P = \operatorname{graph}(\lambda f),$$

so

$$\lim_{\lambda \to 0} (\lambda \cdot P) = \operatorname{graph}(0) = R_J^\perp,$$

i.e., the coordinate plane $R_J^\perp$ is in the closure of $H \cdot P$.

On the other hand, suppose that J is not a basis of Φ, but suppose there exists a sequence $\lambda_i \in H$ such that $\lambda_i \cdot P \to R_J^\perp$. Then for sufficiently large i we have

$$\lambda_i \cdot P \cap R_J = 0$$

since any such $n-k$ plane which is sufficiently close to $R_J^\perp$ will necessarily be transverse to R_J. However, this implies that J must be an independent set of Φ: if it were dependent then $\{\pi_p(e_j) \mid j \in J\}$ would be linearly dependent which would mean that $P \cap R_J \neq 0$, and so the same would be true for $\lambda \cdot P \cap R_J$.

1.5. *Remarks on Projective Configurations*

For any $r \geqslant k$, we let $C_r^n(\mathbb{P}^{k-1})$ denote the set of maps $c\colon S \to \mathbb{P}^{k-1}$ from an r-element subset $S \subset \{1, 2,..., n\}$ to $\mathbb{P}^{k-1}$, whose image $c(S)$ spans $\mathbb{P}^{k-1}$. (Thus an element of $C_r^n(\mathbb{P}^{k-1})$ is r points, not necessarily distinct, and labelled by certain integers between 1 and n). The group $PGl_k(\mathbb{C})$ acts on the space $C_r^n(\mathbb{P}^{k-1})$. A *projective configuration* is an element of the quotient space $C_r^n(\mathbb{P}^{k-1})/PGl_k(\mathbb{C})$.

Fix a plane $P \in G_{n-k}^k$ and let r denote the number of nonzero vectors in the collection $\{\pi_p(e_i) \mid 1 \leqslant i \leqslant n\} \subset \mathbb{C}^n/P$. We thus obtain a configuration $\Lambda(P)$ of r ordered points (which are labelled by r of the integers between 1 and n) in the projective space $\mathbb{P}(\mathbb{C}^n/P) \cong \mathbb{P}^{k-1}$. The following proposition indicates that we may transform questions involving the action of $PGL_k(\mathbb{C})$ on the space of ordered r-tuples of points in $\mathbb{P}^{k-1}$ into questions involving the action of the torus $H = (\mathbb{C}^*)^n$ on G_{n-k}^k:

PROPOSITION [GM]. *The association Λ induces a one-to-one correspondence between the factor spaces*

$$G_{n-k}^k/H \quad \text{and} \quad \left(\coprod_{r=k}^{n} C_r^n(\mathbb{P}^{k-1})\right)\Big/ PGL_k(\mathbb{C}).$$

Remark. There is a natural (non-Hausdorff) topology on each of these spaces.

Proof of Proposition. We repeat the essential idea behind the proof in [GM]. Choose an r-element subset $J \subset \{1, 2,..., n\}$. Let

$$\tilde{G}_J = \{P \in G_{n-k}^k \mid \pi_p(e_j) = 0 \Leftrightarrow j \in J\}.$$

It suffices to show that Λ induces a bijection

$$\tilde{\Lambda}_J\colon \quad \tilde{G}_J/H \to C_J(\mathbb{P}^{k-1})/PGL_k(\mathbb{C}),$$

where $C_J(\mathbb{P}^{k-1}) \subset C_r^n(\mathbb{P}^{k-1})$ denotes the set of r-tuples of points in $\mathbb{P}^{k-1}$ which span $\mathbb{P}^{k-1}$ and are labelled by the integers in the set J.

Any element $P \in \tilde{G}_J$ is the kernel of a surjective linear map $\pi\colon \mathbb{C}^n \to \mathbb{C}^k$ which is uniquely determined up to composition with elements of $GL_k(\mathbb{C})$ because the induced map $\mathbb{C}^n/P \to \mathbb{C}^k$ is an isomorphism. Thus P determines a unique GL_k-equivalence class of r nonzero vectors in $\mathbb{C}^k$. The action of H stretches these vectors but does not change their directions, so the corresponding points in $\mathbb{P}^{k-1}$ are well defined (modulo PGL_k equivalence). Thus $\tilde{\Lambda}_J$ is well defined and we have already remarked (Sect. 1.1) that it is surjective. To see that $\tilde{\Lambda}_J$ is injective, suppose that $P_1, P_2 \in G^k_{n-k}$ are kernels of surjective homomorphisms $\pi_1, \pi_2 \colon \mathbb{C}^n \to \mathbb{C}^k$ and that $\tilde{\Lambda}_J(P_1) = \tilde{\Lambda}_J(P_2)$, i.e., there exists an invertible linear transformation $F\colon \mathbb{C}^k \to \mathbb{C}^k$ such that, for each $j \in J$ there exists $\lambda_j \in \mathbb{C}^*$ with $F\pi_1(e_j) = \lambda_j \pi_2(e_j)$. If $\Delta\colon \mathbb{C}^n \to \mathbb{C}^n$ is given by the diagonal matrix

$$\Delta_{ii} = \begin{cases} \lambda_j & \text{if } j \in J, \\ 1 & \text{if } j \notin J. \end{cases}$$

then the following diagram commutes:

$$\begin{array}{ccc} \mathbb{C}^n & \xrightarrow{\pi_1} & \mathbb{C}^k \\ {\scriptstyle \Delta}\downarrow & & \downarrow{\scriptstyle F} \\ \mathbb{C}^n & \xrightarrow{\pi_2} & \mathbb{C}^k \end{array}$$

and therefore $\Delta(P_2) = P_1$.

2. Moment Map

2.1. Definition of the Moment Map. Associated to the torus action (Sect. 1.3) of H on the Grassmannian G^k_{n-k}, there is a moment map

$$\mu\colon \quad G^k_{n-k} \to \mathbb{R}^n$$

which was defined first (in this case of the Grassmannian) in [M] and [GM], and later, for arbitrary group actions on symplectic manifolds in [A] and [GuS]. In this section we will give an explicit expression for the moment map.

A plane $P \in G^k_{n-k}$ can be realized as the kernel of a surjective homomorphism $F\colon \mathbb{C}^n \to \mathbb{C}^k$ which corresponds to a matrix M with n columns and k rows. For any subset $J \subset \{1, 2, \ldots, n\}$ of cardinality k, we obtain a $k \times k$ matrix $M(J)$ consisting of the columns of M which are indexed by J. There are $\binom{n}{k}$ such subsets.

Proposition. *The coordinates $\mu_i\colon G^k_{n-k} \to \mathbb{R}$ of the moment map are given by*

$$\mu_i(P) = \frac{\sum_{i \in J} |\det M(J)|^2}{\sum_J |\det M(J)|^2},$$

where the summation in the numerator is over all k-element subsets J which contain the index i, and where the summation in the denominator is over all k-element subsets J.

Proof. The association $P \to \{|\det M(J)|\}$ (where J varies over the k element subsets of $\{1, 2, \ldots, n\}$) gives rise to the Plücker embedding

$$G_{n-k}^{k} \to \mathbb{P}^{\binom{n}{k}-1}$$

on which the moment map is computed as in [*K*].

2.2. *The Hypersimplex*

For any $P \in G_{n-k}^{k}$ we have

$$0 \leqslant \mu_i(P) \leqslant 1 \qquad \text{and} \qquad \sum_{i=1}^{n} \mu_i(P) = k.$$

Thus the image of the moment map μ is the hypersimplex Δ_{n-k}^{k} of [GGL] and [GM], i.e., the set of all points $x \in \mathbb{R}^n$ such that $0 \leqslant x_i \leqslant 1$ and $\sum_{i=1}^{n} x_i = k$. The hypersimplex Δ_{n-k}^{k} is the convex hull of the $\binom{n}{k}$ vectors $e(J) \in \mathbb{R}^n$ which are indexed by k-element subsets $J \subset \{1, 2, \ldots, n\}$ and are given by

$$e(J)_j = \begin{cases} 1 & \text{if } j \in J \\ 0 & \text{if } j \notin J \end{cases}$$

2.3. Convexity Theorem. *We recall the convexity theorem of* [GuS], *and* [A]: *Let $\overline{H \cdot P}$ denote the closure in G_{n-k}^{k} of the orbit of the point P under the action of $H = (\mathbb{C}^*)^n$. Then the image $\mu(\overline{H \cdot P})$ is the convex hull of the points $\mu(Q)$ where Q varies over the fixed points in the closure $\overline{H \cdot P}$. (In other words, $\mu(\overline{H \cdot P})$ is the convex hull of a certain subset of the vertices of the hypersimplex.)*

Lemma. *The preimage of each vertex of the hypersimplex is the H-fixed point $\mu^{-1}(e(J)) = R_J^{\perp}$.*

Proof. By [A] the preimage $\mu^{-1}(e(J))$ of any vertex of Δ_{n-k}^{k} consists of a single fixed point. However the coordinate $n-k$ plane $R_J^{\perp}$ may be represented as the kernel of a matrix $M: \mathbb{C}^n \to \mathbb{C}^k$ such that the minor $M(J)$ is the identity and the remaining columns of M are all zero. Therefore, for any k-element subset $K \subset \{1, 2, \ldots, n\}$ we have

$$\det M(K) = \begin{cases} 1 & \text{if } K = J \\ 0 & \text{if } K \neq J, \end{cases}$$

so

$$\mu_i(R_J^{\perp}) = \begin{cases} 1 & \text{if } i \in J \\ 0 & \text{if } i \notin J \end{cases}$$

which shows that $\mu(R_J^{\perp}) = e(J)$.

2.4. SECOND DEFINITION OF THE STRATIFICATION. We shall say that two points $P, Q \in G_{n-k}^k$ are in the same stratum of the second stratification of G_{n-k}^k if the image under the moment map of the closure of the H-trajectory of P coincides with the image under the moment map of the closure of the H-trajectory of Q, i.e. if

$$\mu(\overline{H \cdot P}) = \mu(\overline{H \cdot Q})$$

THEOREM. *The second stratification of* G_{n-k}^k *coincides with the first stratification of* G_{n-k}^k *which was defined in Sect.* 1.2.

COROLLARY. *We have therefore assigned, to each representable combinatorial geometry* Φ, *a unique convex polyhedron*

$$\Delta(\Phi) = \text{closure}(\mu(\Gamma)),$$

where Γ *is the stratum in* G_{n-k}^k *which corresponds to* Φ. *Moreover,* (*by Lemma* 1.4 *and Lemma* 2.3), *the polyhedron* $\Delta(\Phi)$ *has the simple description as the convex hull of the vectors*

$$\{e(J) \mid J \text{ is a basis of } \Phi\}.$$

Proof of Theorem. If two points $P, Q \in G_{n-k}^k$ lie in the same stratum Γ (as defined in Sect. 1.2) then they determine the same matroid so (by Lemma 1.4 and the convexity theorem) they have the same bases, so $\mu(\overline{H \cdot P})$ and $\mu(\overline{H \cdot Q})$ are the convex hulls of the same collection of vectors, so they coincide. On the other hand, suppose that P and Q have the property that $\mu(\overline{H \cdot P}) = \mu(\overline{H \cdot Q})$. Then the matroids corresponding to P and Q have the same bases. However the bases of a matroid determine the matroid [W] so P and Q are in the same stratum Γ of G_{n-k}^k.

3. SCHUBERT CELLS AND STRATA IN THE GRASSMANNIAN

3.1. *Schubert Cells*

The standard ordering $\{e_1, e_2, \ldots, e_n\}$ of the standard basis of $\mathbb{C}^n$ gives rise to the standard flag

$$F^1 \subset F^2 \subset \cdots \subset R^n = \mathbb{C}^n,$$

where $F^i = \text{span}\{e_1, e_2, \ldots, e_i\}$.

A *Schubert symbol* is a sequence of k numbers,

$$1 \leqslant i_1 < i_2 < \cdots < i_k \leqslant n$$

and determines the Schubert cell

$$\Omega[i_1 i_2 \cdots i_k] = \left\{ P \in G^k_{n-k} \,\middle|\, \begin{array}{l} \dim(P \cap F^{i_j - 1}) < j \\ \dim(P \cap F^{i_j}) = j \end{array} \right\},$$

i.e., the numbers i_j label the subspaces for which the dimension of the intersection with P jumps up.

The Schubert cells form a decomposition of the Grassmannian into even-dimensional cells. [MS].

Now let $\sigma \in \Sigma_n$ be a permutation on $\{1, 2, ..., n\}$, and consider the new ordering $\{e_{\sigma(1)}, e_{\sigma(2)}, ..., e_{\sigma(n)}\}$ of the basis vectors of $\mathbb{C}^n$. This gives rise to a new flag

$$F^1_\sigma \subset F^2_\sigma \subset \cdots \subset F^n_\sigma = \mathbb{C}^n,$$

where $F^i_\sigma = \operatorname{span}\{e_{\sigma(1)}, e_{\sigma(2)}, ..., e_{\sigma(i)}\}$. We obtain a new decomposition of the Grassmannian into Schubert cells,

$$\Omega^\sigma[i_1 i_2 \cdots i_k]$$

by replacing F^i with F^i_σ in the above definition.

3.2. *The Third Stratification of the Grassmannian*

We define the third stratification of the Grassmannian to be the common refinement of the $n!$ decompositions into Schubert cells $\Omega^\sigma[i_1 i_2 \cdots i_k]$, where σ is allowed to vary over all permutations and $i_1 i_2 \cdots i_k$ is allowed to vary over all Schubert symbols.

THEOREM. The third decomposition of the Grassmannian coincides with the decomposition of G^k_{n-k} into the strata of Section 1.2.

Proof. If $P \in \Omega^\sigma[i_1 i_2 \cdots i_k]$ then the rank function r of the corresponding matroid satisfies

$$r(\sigma(1), \sigma(2), ..., \sigma(m)) = m - j,$$

where j is uniquely determined by

$$i_j \leqslant m < i_{j+1}$$

because $\dim(F^m_\sigma / F^m_\sigma \cap P) = m - \dim(F^m_\sigma \cap P) = m - j$. In other words, the rank function is not completely determined, however, its value on the particular subsets

$$\{\sigma(1)\}, \{\sigma(1), \sigma(2)\}, ..., \{\sigma(1), \sigma(2), ..., \sigma(n)\}$$

is determined. Now a stratum in the third decomposition of G^k_{n-k} has the form

$$S = \bigcap_{\sigma \in \Sigma_n} \Omega^\sigma[L_\sigma],$$

where each L_σ is some Schubert symbol. (Most such intersections will be empty, of course, and a given stratum may have many such representations.) Thus a point $P \in S$ corresponds to a matroid whose rank function is completely determined: if $J \subset \{1, 2,..., n\}$ is any subset then we can find a permutation σ such that $J = \{\sigma(1), \sigma(2),..., \sigma(|J|)\}$. If $L_\sigma = [i_1 i_2 \cdots i_k]$ then the value of the rank function is $r(J) = |J| - j$, where

$$i_j \leqslant |J| < i_{j+1}.$$

(The permutation σ is not unique. However, if another permutation τ is found such that $J = \{\tau(1), \tau(2),..., \tau(|J|)\}$ and if the resulting computation for $r(J)$ differs from the above, then this will imply $S = \phi$.) This shows that the intersection S is contained in at most one stratum of the stratification from Section 1.2.

On the other hand, suppose that Γ is a stratum of the stratification from Section 1.2. Fix a permutation $\sigma \in \Sigma_n$. For each $P \in \Gamma$ the ranks of the sets

$$\{\sigma(1)\}, \{\sigma(1), \sigma(2)\},..., \{\sigma(1), \sigma(2),..., \sigma(n)\}$$

are determined by the rank function r of the matroid associated to Γ. However,

$$r\{\sigma(1), \sigma(2),..., \sigma(m)\} = \dim(F^m_\sigma / F^m_\sigma \cap P) = m - \dim(F^m_\sigma \cap P)$$

so the dimensions $\dim(F^m_\sigma \cap P)$ are also determined by Γ. This means that P is in a certain Schubert cell of type $\Omega^\sigma[L_\sigma]$ and the Schubert symbol L_σ is determined by the matroid associated to Γ. Thus $\Gamma \subset \Omega^\sigma[L_\sigma]$. If we allow the permutation σ to vary, we conclude that each stratum Γ is contained in a unique intersection,

$$\Gamma \subset \bigcap_{\sigma \in \Sigma_n} \Omega^\sigma[L_\sigma]$$

which completes the proof.

4. Matroids and Convex Polyhedra

4.1. *Introduction.*

We can extend the correspondence (Corollary 2.4) between representable matroids and certain convex polyhedra, to all matroids. Thus, to any matroid Φ (of rank k, defined on the set $\{1, 2, \ldots, n\}$), we associate the convex polyhedron $\Delta(\Phi)$,

$$\Delta(\Phi) = \text{convex hull}\{e(I) \mid I \text{ is a basis of } \Phi\}$$

In this section we will investigate which polyhedra can occur.

DEFINITION. We will say that a convex polyhedron Δ which is contained in the hypersimplex Δ^k_{n-k} is a *matroid polyhedron* if the vertices of Δ are a subset of the vertices of the hypersimplex Δ^k_{n-k} and if each edge (i.e., 1-dimensional face) of Δ is a translation of one of the vectors $e_i - e_j$ (for $i \neq j$).

4.1. THEOREM. *Suppose Δ is a convex polyhedron which is contained in the hypersimplex Δ^k_{n-k}. Then there exists a matroid Φ such that $\Delta = \Delta(\Phi)$ iff Δ is a matroid polyhedron, and in this case the matroid Φ is uniquely determined.*

Remarks. (1) The vectors $e_i - e_j$ are the "roots" of the group $GL_n(\mathbb{C})$.

(2) Isomorphic matroids determine congruent polyhedra.

(3) This theorem implies, for example, that if Φ_1 and Φ_2 are matroids such that $\Delta(\Phi_1) \subset \Delta(\Phi_2)$ then the edges (and the vertices) of $\Delta(\Phi_1)$ are a subset of the edges (and the vertices) in $\Delta(\Phi_2)$.

(4) The essential observation in the proof is that an edge which is a translate of $e_i - e_j$ joins two bases which are related by a Steiner exchange.

4.3. *Proof of* ($\Rightarrow$)

Fix a matroid Φ. For each basis $B \subset \{1, 2, \ldots, n\}$ of Φ we denote the corresponding vertex of Δ^k_{n-k} by $e(B)$, i.e.,

$$e(B)_i = \begin{cases} 1 & \text{if } i \in B \\ 0 & \text{if } i \notin B. \end{cases}$$

Now suppose that I and J are bases of the matroid Φ, and that the vertices $e(I)$ and $e(J)$ are joined by an edge in the convex set $\Delta(\Phi)$. By reordering the elements of the matroid, we may suppose that the vectors $e(I)$ and $e(J)$ differ only in the first $2p$ coordinates and that

$$e(I) = (1, 1, \ldots, 1, 0, 0, \ldots, 0, \text{other}),$$

$$e(J) = (0, 0, \ldots, 0, 1, 1, \ldots, 1, \text{other})$$

(there are p ones and p zeroes in each case). We will show that unless $p = 1$, the midpoint

$$m = (\tfrac{1}{2}, \tfrac{1}{2}, \ldots, \tfrac{1}{2}, \text{other})$$

of the segment joining $e(I)$ and $e(J)$ is a nontrivial convex combination of other vertices of $\Delta(\Phi)$ and therefore this segment is not an edge of $\Delta(\Phi)$. For this discussion we can ignore the "other" coordinates, i.e., we may take $I=\{1, 2, 3,..., p\}$ and $J=\{p+1, p+1,..., 2p\}$. We will repeatedly apply the Steiner exchange axiom to these two bases.

Step 1 = *Step* 1b. Exchange the element $1 \in I$ with the basis J, obtaining a new basis B_1 of Φ which, by reordering the elements in J, can be assumed to be

$$B_1 = B_{1b} = I - \{1\} + \{p+1\}.$$

Step 2a. Exchange the element $p+1 \in J$ with the basis I, obtaining one of two possibilities (up to a reordering of the elements $\{2, 3,..., p\}$): $B_{2a} = J - \{p+1\} + \{1\}$ or else $J - \{p+1\} + \{2\}$. In the first case we have

$$m = \tfrac{1}{2}[e(B_{2a}) + e(B_{1b})]$$

so we are finished. Thus, we can assume

$$B_{2a} = J - \{p+1\} + \{2\}$$

is a basis of Φ.

Step 2b. Exchange $2 \in I$ with the basis J, obtaining one of two possible bases (up to a reordering of the elements $\{p+2, p+3,..., 2p\}$): $B_{2b} = I - \{2\} + \{p+1\}$ or $I - \{2\} + \{p+2\}$. In the first case,

$$m = \tfrac{1}{2}[e(B_{2b}) + e(B_{2a})]$$

so we are finished. Thus, we can assume

$$B_{2b} = I - \{2\} + \{p+2\}$$

is a basis of Φ.

Continuing in this way, we either prove that m does not lie on an edge, or else we construct a sequence of bases $B_{1b}, B_{2a}, B_{2b}, B_{3a},...,$ of Φ. At the kth step (part a) we exchange $p+k-1 \in J$ with the basis I, obtaining one of k possibilities (up to a reordering of the elements $\{e_k, e_{k+1},..., e_p\}$),

$$B_{ka} = J - \{p+k-1\} + \{i\},$$

where $1 \leqslant i \leqslant k$. However, one checks (by a straightforward but messy computation) that if $i \neq k$ then

$$m = \frac{1}{2i}[e(B_{ka}) + e(B_{(k-1)b}) + e(B_{(k-1)a}) + \cdots + e(B_{(k-i)b})]$$

and so m does not lie on an edge. This leaves only the possibility that

$$B_{ka} = J - \{p+k+1\} - \{k\}$$

is a basis of Φ.

Similarly the kth step (part b) gives a basis

$$B_{kb} = I - \{k\} + \{p+k\}.$$

This process terminates after p steps when we exchange (in step $(p+1)$ part (a)) the element $2p \in J$ with the basis I. There are one of p possible results,

$$B_{(p+1)a} = J - \{p\} + \text{one of } \{1, 2, 3, \dots, p\}$$

and in each case the point m can be written as a nontrivial convex combination of previous vertices, as above. This completes the proof that the edges of any $\Delta(\Phi)$ must be translates of vectors $e_i - e_j$.

4.4. *Proof of* ($\Leftarrow$)

Suppose that Δ is a convex hull of some vertices in the hypersimplex Δ^k_{n-k}, and that each edge of Δ is a translation of some vector $e_i - e_j$. We must show that the vertices of Δ constitute the bases of a matroid. By [W] this is equivalent to verifying the Steiner exchange axiom: if I and J are k-element subsets of $\{1, 2, \dots, n\}$ such that $e(I)$ and $e(J)$ are vertices of Δ, and if $m \in I - J$, then there exists $l \in J - I$ such that the vector $e(I - \{m\} + \{l\})$ is a vertex of Δ. By relabelling the coordinate axes in $\mathbb{R}^n$, we may assume that $e(I)$ and $e(J)$ differ only in the first $2p$ positions, and that

$$e(I) = (1, 1, \dots, 1, 0, 0, \dots, 0, \text{other}),$$
$$e(J) = (0, 0, \dots, 0, 1, 1, \dots, 1, \text{other}).$$

We may further assume that the "other" coordinates are arranged so that all the 1's appear before the 0's. In this way we have divided the set $\{1, 2, \dots, n\}$ into four intervals:

$$A = \{1, 2, \dots, p\}, \qquad B = \{p+1, p+2, \dots, 2p\},$$
$$C = \{2p+1, 2p+2, \dots, p+k\}, \quad D = \{p+k+1, p+k+2, \dots, n\}$$

such that $I = A \cup C$, $J = B \cup C$, and $m \in A$.

Since Δ is convex, the line segment joining $e(I)$ to $e(J)$ is completely contained in Δ, which is in turn contained in the convex cone which is spanned by the edges $E_1, E_2, \dots, E_r$ which emanate from the vertex $e(I)$. Thus there are nonnegative real numbers $a_1, a_2, \dots, a_r$ such that

$$e(J) - e(I) = (-1, -1, \dots, -1; 1, 1, \dots, 1; 0, 0, \dots, 0; 0, 0, \dots, 0) = \sum_{i=1}^{r} a_i E_i \qquad (*)$$

(where the semicolons are used to separate the coordinates which are in A, B, C, and D). By assumption, each such edge vector E which emanates from the vertex $e(I)$ is of the form $e_l - e_k$, for some l and k in the set $\{1, 2, ..., n\}$. Since the vertex $e(I) + E$ lies in the hypersimplex, which is contained in the region $0 \leqslant |x_i| \leqslant 1$ (for $1 \leqslant i \leqslant n$), we must have

$$l \notin A \cup C \quad \text{and} \quad k \notin B \cup D \qquad (**)$$

Furthermore, if such an edge vector $E = e_l - e_k$ appears with nonzero coefficient in the above sum (*), then $l \notin D$: otherwise this would give a positive value to the coordinate x_l, which could not be cancelled by any other terms in the sum, because of the condition (**). Similarly, we must have $k \notin C$. In conclusion, each of the vectors $E = e_l - e_k$ which appear with nonzero coefficient in the sum (*), must satisfy $l \in B$ and $k \in A$.

Now consider the particular coordinate $m \in A$. Since $(e(J) - e(I))_m = -1$, at least one of the vectors (say, E_s) in the sum (*) has -1 in the mth coordinate. For this particular vector, we have

$$E_s = e_l - e_m \quad \text{and} \quad l \in B \subset J - I.$$

Thus, the vertex of Δ which is given by

$$e(I) + E_s = e(I - \{m\} + \{l\})$$

verifies the desired Steiner exchange.

4.5. *The Fano Polyhedron*

Associated to the Fano configuration (which is not representable over $\mathbb{C}$),

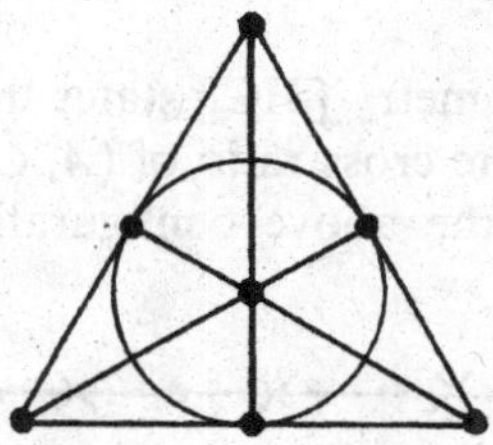

we obtain a beautiful, highly symmetric 6-dimensional convex polyhedron with 28 vertices, 126 edges, 245 2-dimensional faces, 238 3-dimensional faces, 112 four-dimensional faces and 21 5-dimensional faces. The full symmetry group of this polyhedron is the finite simple group $PGL_3(\mathbb{F}_2)$. This example has obvious generalizations to other finite projective spaces.

5. Remarks

5.1. *Topology of the Strata*

We do not know whether each stratum $\Gamma \subset G^k_{n-k}$ is nonsingular. We do not know whether each stratum Γ is a $K(\pi, 1)$ space.

5.2. *Degeneration of Matroids*

If $\Gamma \subset G^k_{n-k}$ is a stratum and if $P \in \bar{\Gamma} - \Gamma$, we shall say that the matroid corresponding to Γ degenerates to the matroid corresponding to P. In this case, we have for any subset $J \subset \{1, 2, ..., n\}$ the following relation on their corresponding rank functions:

$$r_\Gamma(J) \leqslant r_P(J).$$

However, the closure of the stratum Γ is not necessarily a union of strata Γ', and may for example contain a proper subset of a stratum Γ', as the following example shows

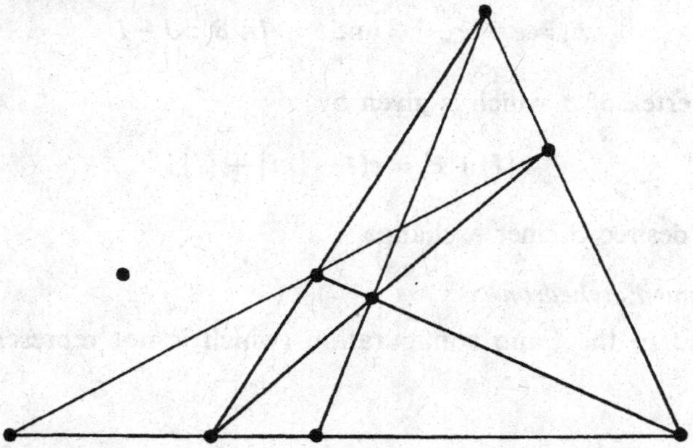

A theorem in projective geometry [HC] states that the four points A, B, C, and D are harmonic, i.e., the cross ratio of $(A, C; B, D)$ is -1. However, it is possible to degenerate the above configuration to the following configuration:

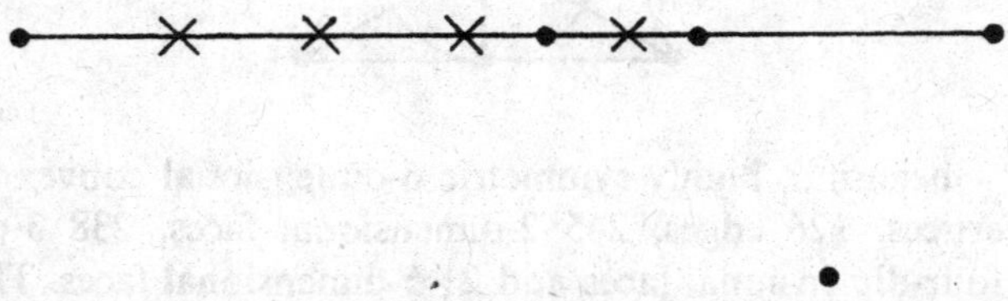

but in doing so we will only obtain 8-tuples of points such that A, B, C, and D are harmonic.

5.3. *Other Lie Groups and Parabolics*

For any complex algebraic Lie group G and parabolic subgroup P, the moment map associated to the torus action, $\mu: G/P \rightarrow \mathfrak{g}^*$ gives rise to new combinatorial geometries and interesting convex polyhedra. These will be explored in [GS].

REFERENCES

[A] M. F. ATIYAH, Convexity and commuting Hamiltonians, *Bull. London Math. Soc.* **14** (982), 1–15.

[BMS] A. A. BEILINSON, R. D. MACPHERSON, AND V. SCHECHTMAN, Notes on motivic cohomology, *Duke Math. J.*, in press.

[CR] H. H. CRAPO AND G. C. ROTA, "On the Foundations of Combinatorial Theory: Combinatorial Geometries," MIT Press, Cambridge, Mass., 1970.

[D] V. I. DANILOV, The geometry of toric varieties, *Uspekhi Mat. Nauk.* **33** (1978), 85–134; translated in *Russian Math Surveys*.

[GGL] A. M. GABRIELOV, I. M. GELFAND, AND M. V. LOSIK, Combinatorial calculation of characteristic classes, *Funct. Anal. Appl.* **9**, No. 2 (1975), 12–28; **9**, No. 3 (1975), 5–26.

[G] I. M. GELFAND, General theory of hypergeometric functions, *Dokl.* (1986), in press.

[GG] I. M. GELFAND AND S. I. GELFAND, Generalized hypergeometric equations, *Dokl.* (1986), in press.

[GM] I. M. GELFAND AND R. MACPHERSON, Geometry in Grassmannians and a generalization of the dilogarithm. *Advan. in Math.* **44** (1982), 279–312.

[GS] I. M. GELFAND AND V. SERGANOVA, to appear.

[GZ] I. M. GELFAND AND A. ZELEVINSKY, Hypergeometric functions, *Funct. Anal. Appl.*, in press.

[GoM] M. GORESKY AND R. MACPHERSON, On representations of matroids, to appear.

[GuS] V. GUILLEMIN AND S. STERNBERG, Convexity properties of the moment map, *Invent. Math.* **67** (1982), 491–513.

[HC] D. HILBERT AND S. COHN–VOSSEN, "Geometry and the Imagination," Chelsea, New York, 1965.

[HM] R. HAIN AND R. MACPHERSON, to appear.

[K] F. KIRWAN, "Cohomology of Quotients in Symplectic and Algebraic Geometry," Princeton Univ. Press, Princeton, N.J., 1984.

[M] R. D. MACPHERSON, The combinatorial formula of Gabrielov, Gelfand, and Losik for the first Pontrjagin class, *in* "Sem. Bourbaki No. 497," 1976/77; Lecture Notes in Mathematics, No. 677, Springer-Verlag, New York, 1978.

[MS] J. MILNOR AND J. STASHEFF, "Characteristic Classes," Ann. Math. Stud. No. 56, Princeton Univ. Press, Princeton, N.J., 1974.

[VW] B. L. VAN DER WAERDEN, "Moderne Algebra," Springer-Verlag, Berlin, 1937.

[W] D. J. A. WELSH, "Matroid Theory," Academic Press, New York, 1976.

[Wh] H. WHITNEY, On the abstract properties of linear dependence, *Amer. J. Math.* **57** (1935), 509–533.

6.

(with V. V. Serganova)

Strata of a maximal torus in a compact homogeneous space

Dokl. Akad. Nauk SSSR **292** (3) (1987) 524–528 [Sov. Math., Dokl. **35** (1987) 63–66]

Let G be a complex semisimple Lie group, and M a compact homogeneous space connected with G. Then M has the form $G/\mathcal{P}$, where $\mathcal{P}$ is some parabolic subgroup of G. Fix a maximal torus $H \subset \mathcal{P} \subset G$. In this note we consider a partition of the manifold M into strata consisting of orbits of H whose closures are isomorphic as torus manifolds (see [1]–[3]).

In the case when $G = \mathrm{SL}(n)$ and M is the Grassmannian $G_k(\mathbf{C}^n)$ of k-dimensional subspaces of $\mathbf{C}^n$, a description of the starta and their closures is very important for the general theory of hypergeometric functions (see [4]–[6] and [8]). Connected with each nonsingular stratum on the Grassmannian is a hypergeometric function whose singularities lie in the closure of the stratum.

In this note we give some equivalent definitions of a stratum, show that every stratum in $G/\mathcal{P}$ is a (W, P)-matroid [7], and describe how the strata are connected with convex polytopes.

1. Generalized Plücker coordinates and first definition of a stratum. Let $\mathfrak{g}$ and $\mathfrak{h}$ be the Lie algebras of the respective groups G and H, and let $\mathfrak{h}_{\mathbf{R}}$ be a real form of $\mathfrak{h}$. Choose a Borel subgroup B such that $H \subset B \subseteq \mathcal{P}$. Let $\Sigma \subset \mathfrak{h}^*_{\mathbf{R}}$ be the root system of $\mathfrak{g}$. Then the subgroup B determines a system of simple roots $\{\sigma_1, \dots, \sigma_n\} \subset \Sigma$. Let W be the Weyl groups of $\mathfrak{g}$. Then $W \cong N(H)/H$, where $N(H)$ is the normalizer of H in G. The subgroup $W_P \cong N_{\mathcal{P}}(H)/H$, where $N_{\mathcal{P}}(H) = N(H) \cap \mathcal{P}$, is a parabolic subgroup of W, i.e., W_P is generated by reflections with respect to some subset P of the simple roots $\{\sigma_1, \dots, \sigma_n\}$.

Consider the $\omega_P \in \mathfrak{h}^*_{\mathbf{R}}$ given by the conditions $(\omega_P, \sigma_i)/(\sigma_i, \sigma_i) = 1$ for $\sigma_i \notin P$ and $= 0$ for $\sigma_i \in P$. The weight ω_P is the highest weight of some finite-dimensional irreducible representation ρ_P of G. The highest vector of ρ_P is an eigenvector with respect to the subgroup $\mathcal{P}$; therefore, an imbedding p of the homogeneous manifold $M = G/\mathcal{P}$ into the projectivization of the space V of the representation ρ_P is defined. The image $p(M)$ coincides with the orbit of the highest vector of ρ_P under the action of G.

Let $\mathcal{A}$ be the set of weights of ρ_P, with multiplicities taken into account. Choose some weight basis $\{e_\alpha,\ \alpha \in \mathcal{A}\}$ in V. An arbitrary point $X \in M$ is determined uniquely to within a factor by the collection of numbers $\{p^\alpha(X), \alpha \in \mathcal{A}\}$, where $p(X) = \sum_{\alpha \in \mathcal{A}} p^\alpha(X) e_\alpha$. If $W\omega_P$ is the orbit of the highest weight ω_P in $\mathfrak{h}^*_{\mathbf{R}}$ under the action of the Weyl group, then $W\omega_P \subseteq \mathcal{A}$. The collection of numbers $p^\alpha(X)$, where $\alpha \in W\omega_P$, is said to comprise *generalized Plücker coordinates of the point* $X \in M$. In the case when $G = \mathrm{SL}(n)$ and $M = G_k(\mathbf{C}^n)$, the generalized Plücker coordinates coincide with the usual Plücker coordinates in $G_k(\mathbf{C}^n)$.

Let $L_X = \{\alpha \in W \cdot \omega_P | p^\alpha(X) \neq 0\}$ for $X \in M$. Since the weights in $W \cdot \omega_P$ have multiplicity 1, the subset L_X does not depend on the choice of a weight basis in V.

DEFINITION 1. Two points X and Y in M are said to be *equivalent* if $L_X = L_Y$. The equivalence classes in M are called *strata*.

1980 *Mathematics Subject Classification* (1985 *Revision*). Primary 22E46, 43A85; Secondary 14M15, 33A35.

REMARK. The partition of M into strata is not a stratification in the usual sense, since the closure of a stratum need not be a union of strata.

2. The moment mapping and the second definition of a stratum. Let T and K be compact real forms of the groups H and G, with $T \subset K$. The group K acts transitively on M, and hence there is a K-invariant Kähler metric on M. This metric makes it possible to define the moment mapping $\mu: M \to LT^*$ (see [1] and [2]), where LT is the Lie algebra of the group T. We identify LT with $\mathfrak{h}_{\mathbf{R}}$. Then $\mu(M)$ is a convex polytope with vertices at the points in $W \cdot \omega_P$ (see [1] and [2]).

THEOREM 1. *Let $X \in M$, let $H \cdot X$ be the orbit of X in M under the action of H, and let $\overline{H \cdot X}$ be its closure in M. The image $\mu(\overline{H \cdot X})$ under the moment mapping is a convex polytope in $\mathfrak{h}^*_{\mathbf{R}}$ with vertices at the points of L_X. The moment mapping implements a one-to-one correspondence between the k-dimensional orbits of the group H in $\overline{H \cdot X}$ and the k-dimensional open faces of the polytope $\mu(\overline{H \cdot X})$.*

Theorem 1 follows from results in [1] and [2].

DEFINITION 2. A *stratum* in M is defined to be a union of all the orbits of the torus H whose images under the moment mapping coincide.

It follows from Theorem 1 that Definitions 1 and 2 are equivalent.

3. Connection of the strata with Schubert cells. Connected with each Borel subgroup $C \subset G$ is a partition of M into Schubert cells: the orbits of C in M. Denote by $\hat{C}$ the set of all Borel subgroups of G containing H.

DEFINITION 3. The *strata* in M are defined to be the nonempty intersections $\bigcap_{C \in \hat{C}} U(C)$, where $U(C)$ is some Schubert cell connected with the subgroup $C \subset G$.

PROPOSITION 1. *Definition* 3 *is equivalent to Definitions* 1 *and* 2.

REMARK. Definition 3 carries over without difficulty to the case of an arbitrary field, and all the subsequent assertions are also true for an arbitrary field.

4. Strata and (W, P)-matroids. Let W^P be the set of left cosets of W with respect to the subgroup W_P. We identify W^P with the orbit $W \cdot \omega_P$ in $\mathfrak{h}^*_{\mathbf{R}}$ in the obvious way.

THEOREM 2. *Let Γ be an arbitrary stratum in M, and X an arbitrary point in Γ. Then (W, P, L_X) is a (W, P)-matroid.*

As is clear from simple examples, not every (W, P)-matroid can be obtained from some stratum in G/P.

5. Strata and hypersimplexes. It follows from Theorem 1 that an arbitrary stratum Γ is mapped under the moment mapping onto the interior of a convex polytope $\Delta_\Gamma \subset \mathfrak{h}^*_{\mathbf{R}}$. The set of vertices of Δ_Γ is contained in $W \cdot \omega_P$. It turns out that Δ_Γ has another interesting property.

DEFINITION. A polytope $\Delta \subset \mathfrak{h}^*_{\mathbf{R}}$ with vertices at certain points of the orbit $W \cdot \omega_P$ is called a *hypersimplex* if all its edges are parallel to roots of the Lie algebra $\mathfrak{g}$.

PROPOSITION 2. *Let Γ be a stratum in M. Then the polytope Δ_Γ is a hypersimplex.*

6. Nondegeneracy. A stratum $\Gamma \subset M$ is said to be *nonsingular* if and only if it consists of orbits of dimension equal to $\dim H$. By Theorem 1, a stratum Γ is nonsingular if $\dim \Delta_\Gamma = \dim H$.

PROPOSITION 3. *A stratum $\Gamma \subset M$ is nonsingular if and only if the (W, P)-matroid corresponding to it is nonsingular* (*see* [7]). *Every singular stratum $\Gamma \subset G/P$ is isomorphic to some stratum $\Gamma' \subset G'/P'$, where G' is a regular semisimple subgroup of G, and $P' = P \cap G'$.*

7. Configurations. A *configuration of n vectors in a k-dimensional vector space F* is defined to be a collection of vectors $x_1, \dots, x_n \in F$ generating the whole of F which is determined to within the action of the group $\mathrm{GL}(k)$. The manifold of all configurations of n vectors in a k-dimensional space is isomorphic to $G_k(\mathbf{C}^n)$ (see [3]), since each k-dimensional space $X \subset \mathbf{C}^n$ can be given as the image under the nonsingular linear mapping $x: F^* \to \mathbf{C}^n$ defined by $(x(v))_i = x_i(v)$, where $v \in F^*$ and $x_1, \dots, x_n \in F$, and two mappings x and y give the same subspace of $\mathbf{C}^n$ if $y = x \cdot g$ for some $g \in \mathrm{GL}(k)$.

Two configurations $x = (x_1, \dots, x_n)$ and $y = (y_1, \dots, y_n)$ are said to be *equivalent* if $\mathrm{rk}(x_{i_1}, \dots, x_{i_p}) = \mathrm{rk}(y_{i_1}, \dots, y_{i_p})$ for any $i_1, \dots, i_p \leq n$.

Let $x^1 = (x_1^1, \dots, x_n^1), \dots, x^m = (x_1^m, \dots, x_n^m)$ be a given collection of configurations of n vectors in spaces $F_1, \dots, F_m$ of respective dimensions $k_1, \dots, k_m$, $k_1 < \dots < k_m$. A collection of configurations is said to be *compatible* if $\sum_{i=1}^n \lambda_i x_i^p = 0$ for all $p < j$ whenever $\sum_{i=1}^n \lambda_i x_i^j = 0$. It is easy to see that the manifold of all compatible collections of configurations is isomorphic to the manifold of flags of subspaces of dimensions $k_1, \dots, k_m$ in $\mathbf{C}^n$.

Two compatible collections of configurations are said to be *equivalent* if the configurations x^i aned y^i are equivalent for all $i = 1, \dots, m$.

PROPOSITION 4. *An equivalence class of compatible collections of configurations is isomorphic to some stratum in the flag manifold.*

For an orthogonal group and for a symplectic group G the homogeneous space $G/\mathcal{P}$, $\mathcal{P}$ some maximal parabolic subgroup, can be realized as the Grassmannian of isotropic and Lagrangian subspaces, respectively. In these cases we give constructions analogous to configurations.

a) An *orthogonal configuration* of N vectors in a k-dimensional vector space F is defined to be a configuration $x_1, \dots, x_n, y_1, \dots, y_n \in F$ if $N = 2n$, or a configuration $x_1, \dots, x_n, y_1, \dots, y_n, z \in F$ if $N = 2n + 1$, such that $\sum_{i=1}^n x_i \cdot y_i + z \cdot z = 0$, where $v \cdot w = v \otimes w + w \otimes v \in S^2 F$. In the case $N = 2n$ we set $z = 0$.

The manifold of orthogonal configurations of N vectors in a k-dimensional space is isomorphic to the Grassmannian of k-dimensional isotropic subspaces of $\mathbf{C}^n$.

b) A *symplectic configuration* of $2n$ vectors in a k-dimensional space F is defined to be a configuration $x_1, \dots, x_n, y_1, \dots, y_n \in F$ such that

$$\sum_{i=1}^n x_i \wedge y_i = 0.$$

The manifold of symplectic configurations of $2n$ vectors in a k-dimensional space is isomorphic to the Grassmannian of k-dimensional Lagrangian subspaces of $\mathbf{C}^n$.

A subset A of vectors in an arbitrary orthogonal or symplectic configuration is said to be *admissible* if the configuration obtained from the given one by zeroing out all the vectors not in A is orthogonal or symplectic, respectively. Admissible subsets of linearly independent vectors of a configuration are said to be *independent*. Obviously, independent subsets do not contain the vectors x_i and y_i simultaneously, nor the vector z in the case of an orthogonal configuration. Two configurations are said to be *equivalent* if they have the same collection of independent sets.

It is possible to define a compatible collection of orthogonal and symplectic configurations for an arbitrary parabolic subgroup $\mathcal{P} \subset G$ just as for $\mathrm{SL}(n)$.

PROPOSITION 5. *For an orthogonal or symplectic group G an equivalence class of compatible collections of configurations is isomorphic to a stratum in $G/\mathcal{P}$.*

We analyze the case of the exceptional Lie group G_2. This group contains two nonconjugate parabolic subgroups $\mathcal{P}_1$ and $\mathcal{P}_2$ corresponding to the short and long simple roots.

The case $M = G_2/P_1$ is not interesting, since G_2/P_1 is isomorphic to a quadric in $\mathbf{C}P^6$, i.e., to a homogeneous space with the group SO(7). Let $M = G_2/P_2$. Then any point $X \in M$ is given by a configuration of seven vectors $x_1, x_2, x_3, y_1, y_2, y_3, z$ in $\mathbf{C}^2$ such that

$$\sum_{i=1}^{3} x_i \cdot y_i + z \cdot z = 0, \quad \sum_{i=1}^{3} x_i \wedge y_i = 0, \quad 2x_i \wedge x_j = y_k \wedge z, \quad 2y_i \wedge y_j = x_k \wedge z,$$

where (i, j, k) is a cyclic permutation of $(1, 2, 3)$.

The manifold G_2/B, B a Borel subgroup, is isomorphic to a pair of compatible configurations of seven vectors on the line and on the plane which satisfy the equations given above. Equivalence of configurations is defined just as for the classical groups, and Proposition 5 is true for the group G_2.

Moscow State University
Scientific Council on the Complex Problem "Cybernetics"
Academy of Sciences of the USSR
Moscow

Received 26/SEPT/86

BIBLIOGRAPHY

1. M. F. Atiyah, Bull. London Math. Soc. **14** (1982), 1–15.
2. V. Guillemin and S. Sternberg, Invent. Math. **67** (1982), 491–513.
3. I. M. Gel'fand and R. D. MacPherson, Advances in Math. **44** (1982), 279–312.
4. I. M. Gel'fand, Dokl. Akad. Nauk SSSR **288** (1986), 14–18; English transl. in Soviet Math. Dokl. **33** (1986).
5. I. M. Gel'fand and A. V. Zelevinskiĭ, Funktsional. Anal. i Prilozhen. **20** (1986), no. 3, 17–34; English transl. in Functional Anal. Appl. **20** (1986).
6. I. M. Gel'fand and S. I. Gel'fand, Dokl. Akad. Nauk SSSR **288** (1986), 279–283; Enlgish transl. in Soviet Math. Dokl. **33** (1986).
7. I. M. Gel'fand and V. V. Serganova, Dokl. Akad. Nauk SSSR **292** (1987), 15–20; English transl. in Soviet Math. Dokl. **35** (1987).
8. V. A. Vasil'ev, I. M. Gel'fand, and A. V. Zelevinskiĭ, Dokl. Akad. Nauk SSSR **290** (1986), 277–281; English transl. in Soviet Math. Dokl. **34** (1987).

Translated by H. H. MCFADEN

7.

(with V.V. Serganova)

Combinatorial geometries and torus strata on homogeneous compact manifolds

Usp. Mat. Nauk **42** (2) (1987) 107–133

Contents

Introduction 926
1. Torus orbits and Grassmannian strata 928
2. Matroids and Grassmannian strata 935
3. Moment mapping and toric varieties 941
4. Geometry of compact homogeneous spaces 943
5. Definition of stratum through the moment mapping 945
6. Strata and Schubert cells 946
7. Polyhedra corresponding to strata 946
8. General (W, Q)-matroid 947
9. Examples of strata 952
References 957

Introduction

This paper considers decomposition of the compact homogeneous space of a complex semi-simple group G into orbits with respect to the Cartan subgroup action. The geometry of these orbits is very interesting. Even more interesting is the decomposition of manifolds into strata of orbits having the same geometry.

The first part of the paper (Sects. 1 and 2) considers the Grassmannian manifold $G_k(\mathbb{C}^n)$ of k-dimensional subspaces in the n-dimensional space. The results presented there are mainly those obtained simultaneously and independently by Goresky, MacPherson and the authors, in [17]. Grassmannian strata can be defined in three different ways: (a) as a set of the Cartan subgroup orbits which under the momentum mapping yield one and the same polyhedron (see [4]); (b) as a set of projective configurations giving the same combinatorial geometry; (c) as an intersection of Schubert cells corresponding to those Borel subgroups that contain a given Cartan subgroup.

The study of strata and their closures on Grassmannians is also important for the general theory of hypergeometric functions ([15], [11], [18], [19]). For each non-degenerate stratum on a Grassmannian there is a corresponding hypergeometric function which has singularities on strata lying in the closure of a given stratum.

Strata on Grassmannians are studied for a different reason in a paper by Vershik and Mnev.

No previous knowledge is required in the first part of the paper. We believe that the elementary exposition has its advantages here because it provides a more clear description of the geometry of the situation.

The decompositions of Grassmannians constructed here turn out to be directly linked to the theory of matroids, or combinatorial geometries. In particular, the problem of describing all strata on a Grassmannian is reduced to the problem of representability of matroids (see [6]–[8]). We strongly recommend the book by Crapo and Rota [8] as a very good introduction to this fascinating theory.

Section 2 explores an important relation between matroids and convex polyhedra of a special type which apparently has not yet been discussed in the matroid literature.

In the second part of this paper (Sects. 3–9) the results of the first part are transferred to the case of compact homogeneous space corresponding to an arbitrary complex semi-simple Lie group. It also contains a review of results concerning toric varieties, moment mappings and hypersimplexes.

The Grassmannian $G_k(\mathbb{C}^n)$ is a compact homogeneous space with respect to the group $SL(n)$. Evidently, $G_k(\mathbb{C}^n) = SL(n)/P$ where P is a parabolic subgroup in $SL(n)$ that leaves invariant a k-dimensional subspace in $\mathbb{C}^n$.

Now let G be an arbitrary complex semisimple group acting transitively on a compact manifold M. Then M can be identified with a homogeneous subspace G/P where P is a parabolic subgroup in G, i.e. a subgroup containing a maximal solvable subgroup in G.

For $G = SL(n)$, all compact homogeneous manifolds can be described as flag manifolds (in general, incomplete) in $\mathbb{C}^n$. Grassmannians correspond under that correspondence to maximal parabolic subgroups.

The Kähler metric χ exists on the manifold $G = G/P$ and is invariant under the action of the compact form K of the group G. The imaginary part of the metric χ defines a simplectic form on M considered as a real manifold. The resulting simplectic structure makes it possible to construct the moment mapping $\mu: M \to \mathfrak{k}^*$ where $\mathfrak{k}$ is the Lie algebra of the group K (see [2], [3], [10], [16]). With the use of the mapping μ one can obtain orbits of the Cartan subgroup H in M, whose closures are isomorphic toric varieties, defining in that way the decomposition of M into strata.

The classical matroid notion corresponds to $SL(n)$-action on a Grassmannian. Replacing the group $SL(n)$ with another semi-simple group, and the Grassmannian with another homogeneous space we obtain new combinatorial geometries which we call general matroids. For example, replacing the Grassmannian with an arbitrary flag manifold, one gets supermatroids (see [7]) satisfying special restrictions described in Sect. 9.

Section 3 contains the necessary information on toric varieties and the moment mapping (see [1], [5], [10]) and Sect. 4 is devoted to the geometry of compact homogeneous spaces corresponding to complex semi-simple groups.

Section 5 gives a general definition of strata, and Sect. 6 studies their relation to the Schubert cells decomposition.

Exactly as for Grassmannians, strata on an arbitrary homogeneous space G/P define convex polyhedra which we call hypersimplexes. For Grassmannians, this property is equivalent to the exchange axiom for matroids. In the general case, this makes it possible to define matroids corresponding to the pair (G, P) where G is a semi-simple group and P is its parabolic subgroup.

In Sect. 8 we define a (W, Q)-matroid for an arbitrary Coxeter group W and an arbitrary subset of generators Q which coincides with the usual definition of matroid if $W = S_n$ and Q defines a maximal parabolic subgroup. For an arbitrary parabolic subgroup of the group S_n the notion of (W, Q)-matroid turns out to be closely connected with the recently introduced notion of a greedoid (see [20], [21]). This relation will be described in more detail in future publications. In our opinion the notion of a (W, Q)-matroid is itself an interesting one in combinatorics.

In Sect. 9 we consider examples of strata on homogeneous spaces corresponding to semi-simple groups $SO(n)$, $Sp(2n)$, G_2.

We are grateful to A.V. Zelevinsky for his help and attention to the work and to A.B. Goncharov and B.L. Feigin for useful comments.

1. Torus orbits and Grassmannian strata

1.1. Denote by $G_k(E)$ the manifold of all k-dimensional subspaces in the complex n-dimensional space E.

The group $GL(E)$ of all linear transformations of the space E naturally acts on $G_k(E)$. What is essential for our purposes is the action of a Cartan subgroup $H \subset GL(E)$ on $G_k(E)$. Choose a basis $\{e_1, \ldots, e_n\}$ in E, thus identifying E with the coordinate space $\mathbb{C}^n$. Then one can take for H the subgroup of all diagonal matrices in that basis. Denote by $H \cdot X$ the orbit of a point $X \in G_k(\mathbb{C}^n)$ under the action of H and by $\overline{H \cdot X}$ its closure in $G_k(\mathbb{C}^n)$. One can show that $H \cdot X$ is a compact algebraic variety consisting of a finite number of orbits of the group H in which $H \cdot X$ is the only orbit in $\overline{H \cdot X}$ which is open and dense everywhere. Moreover, $\overline{H \cdot X}$ is a toric variety (see definition in Sect. 3).

Each orbit $H \cdot X$ in $G_k(\mathbb{C}^n)$ is isomorphic to $(\mathbb{C}^*)^r$ for some $r \leqq n-1$. Orbits of dimension $n-1$ are called non-degenerate. Stable points with respect to H are k-dimensional coordinate subspaces in $\mathbb{C}^n$.

In this section we study subvarieties in $G_k(\mathbb{C}^n)$, consisting of orbits of the group H, which are such that their closures are isomorphic as toric varieties.

1.2. Definition of stratum. Let $I_n = \{1, \ldots, n\}$. Denote by $B(I_n)$ the set of all subsets in I_n. Let $B_p(I_n) = \{J \in B(I_n) \,|\, |J| = p\}$. For any $J = \{i_1, \ldots, i_p\}$ $B_p(I_n)$ denote by $\mathbb{C}^J$ the coordinate subspace in $\mathbb{C}^n$ spanned by $\{e_{i_1}, \ldots, e_{i_p}\}$.

Two subspaces X and Y are called equivalent if $\dim(X \cap \mathbb{C}^J) = \dim(Y \cap \mathbb{C}^J)$ for all $J \in B(I_n)$.

Definition 1. Classes of equivalent subspaces in $G_k(C^n)$ are called strata.

Evidently, each stratum Γ can be defined by a function $s: B(I_n) \to \mathbb{Z}$ in the following way: $\Gamma = \{X \in G_k(\mathbb{C}^n) \,|\, \dim(X \cap \mathbb{C}^J) = s(J)\}$. We will define each stratum by means of another function r such that

$$\Gamma_r = \{X \in G_k(\mathbb{C}^n) \,|\, \dim(X / X \cap \mathbb{C}^{I_n \setminus J}) = r(J)\}.$$

Occasionally it is more convenient to define a stratum by means of the function r^*:

$$\Gamma^{r^*}=\{X\in G_k(G^n)\,|\,\dim(C^J/C^J\cap X)=r^*(J)\}.$$

Both ways of defining a stratum are equally possible, and, as we shall see below (Sect. 2), lead to dual combinatorial geometries.

The functions s, r and r^* defining the same stratum are related in the following way:

$$r(J)=k-s(I_n\backslash J)=|J|+r^*(I_n\backslash J)+k-n.$$

Example. Let $r_0(J)=\min(k,|J|)$. Then the stratum Γ_{r_0} in $G_k(\mathbb{C}^n)$ consists of those subspaces that are in general position with all coordinate subspaces. Γ_{r_0} is the only stratum in $G_k(\mathbb{C}^n)$ which is dense everywhere. Let us call it the general stratum.

We call a function $r: B(I_n)\to\mathbb{Z}$ admissible for the Grassmannian $G_k(\mathbb{C}^n)$ if the set Γ_r is non-empty.

Proposition 1. *Suppose that a function $r: B(I_n)\to\mathbb{Z}$ is admissible for $G_k(\mathbb{C}^n)$. Then:*

(1) $r(I_n)=k$,

(2) $0\leqq r(J)\leqq|J|$,

(3) *if* $J\subseteq I$ *then* $r(J)\leqq r(I)$,

(4) $r(I)+r(J)\geqq r(I\cap J)+r(I\cup J)$.

The proof is evident.

1.3. Strata and Schubert cells. A traditional decomposition of the Grassmannian $G_k(\mathbb{C}^n)$ is into Schubert cells. It is defined by specifying a coordinate flag $0\subset\mathbb{C}^{J_1}\subset\cdots\subset C^{J_n}=C^n$ where $|J_i|=i$. Schubert cells corresponding to a given flag consist of those subspaces in $G_k(\mathbb{C}^n)$ that have fixed dimensions of intersection with all $\mathbb{C}^{J_i}$.

For each of $n!$ coordinate flags in $\mathbb{C}^n$ let us choose one of the corresponding Schubert cells. The intersection of all chosen Schubert cells will be called a thin cell.

Proposition 2. *Decomposition of a Grassmannian into thin cells coincides with decomposition into strata.*

The proof is almost evident (see [17]).

1.4. Configurations. Let F be a k-dimensional complex vector space. Any k-dimensional subspace in $\mathbb{C}^n$ can be defined as an image of a non-degenerate linear mapping $x: F\to\mathbb{C}^n$. Two mappings x_1 and x_2 have the same image if and only if $g\in\mathrm{GL}(F)$ exists such that $x_1=x_2\circ g$.

The mapping x is of the form $x(v)=\sum_{1\leqq j\leqq n}x^j(v)e_j$ with respect to the basis $\{e_1,\dots,e_n\}$, where $x^1,\dots,x^n\in F^*$ and $x^1,\dots,x^n$ span the whole of F^*.

Denote the collection of all sets of n vectors spanning a given k-dimensional space by $C_{n,k}$. Points of $C_{n,k}$ considered modulo $\mathrm{GL}(k)$-action will be called configurations. Thus one has a principal $\mathrm{GL}(k)$-bundle $q: C_{n,k}\to C_k(\mathbb{C}^n)$.

Let X be defined by a configuration $x^1, \ldots, x^n$ in F. Then $\dim X/(X \cap C^{I_n \setminus J}) = \operatorname{rk}(x^{j_1}, \ldots, x^{j_p})$ for all $J = (j_1, \ldots, j_p) \subset I_n$.

Choosing a basis in the space F^*, one can identify $C_{n,k}$ with the space of $(n \times k)$-matrices of rank k. Therefore, points in the Grassmannian $G_k(\mathbb{C}^n)$ may be defined by specifying an $(n \times k)$-matrix of rank k up to multiplication by a non-singular square matrix of order k on the right. With the help of configurations it is possible to specify points in the space $G_k(\mathbb{C}^n)/H$. The action of the group H on $C_{n,k}$ is introduced in the following way: $h(x^1, \ldots, x^n) = (h_1 x^1, \ldots, h_n x^n)$ where $(x^1, \ldots, x^n) \in C_{n,k}$, $h = \operatorname{diag}(h_1, \ldots, h_n) \in H$. The action of the group H in $C_{n,k}$ agrees with the action in $G_k(\mathbb{C}^n)$, i.e. $q \circ h = h \circ q$ for any $h \in H$. Thus, the projection $q: C_{n,k} \to G_k(\mathbb{C}^n)$ induces the mapping $\tilde{q}: C_{n,k}/H \to G_k(\mathbb{C}^n)/H$.

Consider an open subset $C^0_{n,k}$ in $C_{n,k}$ consisting of sets of non-vanishing vectors. Then $C^0_{n,k}/H$ can be identified with the variety of sets of n points in the projective space $\mathbb{C}P^{k-1}$. A set of n points in $\mathbb{C}P^{k-1}$ modulo $PGL(k)$-action will be called a projective configuration. Thus, each projective configuration defines an orbit of the group H in $G_k(\mathbb{C}^n)$.

1.5. The Plücker coordinates. Let $P(\Lambda^k(E))$ be a projectivisation of the k-th exterior power of the space E. Recall the construction of the Plücker map $p: G_k(E) \to P(\Lambda^k(E))$. Let $X \subset E$ be an arbitrary k-dimensional subspace. Then $\Lambda^k(X)$ is a one-dimensional subspace in $\Lambda^k(E)$, i.e. a point in $P(\Lambda^k(E))$. Then $p(X) = \Lambda^k(X)$.

Choose a basis in $\Lambda^k(E)$ of the form

$$\{e_J = e_{j_1} \wedge \cdots \wedge e_{j_k}, J = \{j_1, \ldots, j_k\} \in B_k(I_n), j_1 < \cdots < j_k\}.$$

Then $p(X)$ can be written in the form $p(X) = \sum_{J \in B_k(I_n)} p^J(X) e_J$ where the numbers $p^J(X)$ are defined up to a scalar factor. The numbers $p^J(X)$ are called the Plücker coordinates of the subspace X.

If a subspace $X \in G_k(C^n)$ is given by an $(n \times k)$-matrix, then $p^J(X) = p^{j_1, \ldots, j_k}(X)$ is its minor composed of columns with numbers $j_1, \ldots, j_k$.

In the Plücker coordinates the image of $C_k(\mathbb{C}^n)$ is given by the following equations (see [14]): $\sum_{l=0}^{k} (-1)^l p^{i_1, \ldots, i_{k-1}, j_l} \cdot p^{j_0, \ldots, \hat{j}_l, \ldots, j_k} = 0$ for all $\{i_1, \ldots, i_{k-1}\} \in B_{k-1}(I_n)$, $\{j_0, \ldots, j_k\} \in B_{k+1}(I_n)$.

1.6. Affine coordinates. Let $J = \{j_1, \ldots, j_k\}$ be a k-element subset in I_n. Consider the subset C_J of $(n \times k)$-matrices such that the minor consisting of columns $\{j_1, \ldots, j_k\}$ is the unit matrix. Then the map q gives an isomorphism of C_J on an open dense subset in $G_k(\mathbb{C}^n)$. The open set $q(C_J)$ is called an affine chart in $G_k(C^n)$ and the elements of the matrix $q^{-1}(X) \in C_J$ are called affine coordinates of the point $X \in G_k(\mathbb{C}^n)$. Evidently, $G_k(\mathbb{C}^n)$ is covered by affine charts $q(C_J)$ where J runs over all k-element subsets in I_n.

1.7. The moment mapping in $G_k(\mathbb{C}^n)$. For each subset $J \subset I_n$ consider the point δ_J in $\mathbb{R}^n$ whose coordinates are

$$(\delta_J)_i = \begin{cases} 1, & i \in J, \\ 0, & i \notin J. \end{cases}$$

Consider the mapping $\mu(X)\colon G_k(\mathbb{C}^n)\to\mathbb{R}^n$ which takes a point $X\in G_k(\mathbb{C}^n)$ having the Plücker coordinates $p^J(X)$ into a convex sum

$$\mu(X)=\sum_{J\in B_k(I_n)}|p^J(X)|^2\cdot\delta_J\Big/\sum_{J\in B(I_n)}|p^J(X)|^2. \tag{1}$$

The formula (1) shows that $\mu(G_k(C^n))$ is a convex polyhedron in $\mathbb{R}^n$ defined by the following relations:

$$\sum_{1\leq i\leq n}\xi_i=k,\quad 0\leq\xi_i\leq 1.$$

Denote this polyhedron by $\Delta_{n,k}$. It was considered for the first time in [4], where it was called a hypersimplex. The mapping $\dot{\mu}$ has been constructed in [4] for the real Grassmannian $G_k(\mathbb{R}^n)$ and it is proved there that the closure of the general orbit $(\mathbb{R}^+)^n$ is taken homeomorphically onto $\Delta_{n,k}$.

A generalisation of the mapping to a general compact Kähler manifold with the toric action $(\mathbb{C}^*)^n$ has been constructed in [2] and [3]. It is called the moment mapping (see Sect. 3). A very important convexity theorem which strongly generalises the results of [4] is also proved there.

Theorem 1. *Let $H\cdot X$ be an orbit of a point $X\in G_k(\mathbb{C}^n)$ under the action of the group H. Then $\mu(\overline{H\cdot X})$ is a convex polyhedron in $\mathbb{R}^n$ having the set of vertices* Vert $\mu(\overline{H\cdot X})=\{\delta_J|p^J(X)\neq 0\}$. *The mapping μ defines a one-to-one correspondence between p-dimensional orbits of the group H in $\overline{H\cdot X}$ and p-dimensional open faces of the polyhedron $\mu(\overline{H\cdot X})$.*

All these statements, except the description of Vert $\mu(\overline{H\cdot X})$ in Theorem 1, are of a general nature, and the proofs are omitted for this reason (see [2], [3], Sects. 3.2, 4.3).

Definition 2. Two orbits $H\cdot X$ and $H\cdot Y$ are called equivalent if $\mu(H\cdot X)=\mu(H\cdot Y)$. Equivalence classes of orbits of the group H in $G_k(\mathbb{C}^n)$ are called strata.

Note that for any stratum Γ its image $\mu(\Gamma)$ is the interior of a convex polyhedron. Denote it by Δ_Γ.

Proposition 3. *Definitions 1 and 2 are equivalent.*

The proof is based on an important combinatorial notion of a list.

For any subspace $X\in G_k(\mathbb{C}^n)$ its list is the subset $L_X=\{J\in B_k(I_n)|X\cap C^{I_n\setminus J}=\{0\}\}$.

Now let Γ be a stratum in $G_k(\mathbb{C}^n)$ in the sense of Definition 1 and $X\in\Gamma$. Evidently, the list L_X is uniquely defined by the function r which describes the stratum (see Sect. 1.2). On the other hand, the definition of list clearly implies that $L_X=\{J\in B_k(I_n)|p^J(X)\neq 0\}$. Therefore Theorem 1 implies that Γ is contained in some stratum in the sense of Definition 2.

Conversely, one has to prove that the function $r_X(J)=\dim(X/X\cap C^{I_n\setminus J})$ can be uniquely reconstructed from the list L_X. This follows from the following lemma:

Lemma 1. $r_X(J) = \max_{B \in L_X} |B \cap J|$.

Let X be defined by a configuration of vectors $x^1, \ldots, x^n \in F^*$. Then $r_X(\{i_1, \ldots, i_p\}) = \mathrm{rk}(x^{i_1}, \ldots, x^{i_p})$. But $\mathrm{rk}(x^{i_1}, \ldots, x^{i_p})$ equals the maximal number of linearly independent vectors among $x^{i_1}, \ldots, x^{i_p}$. Let $x^{j_1}, \ldots, x^{j_q}$ be any such maximal independent subsystem of vectors. Let us add to it some other vectors from the configuration so that the resulting system forms a basis. This can be done since $\mathrm{rk}(x^1, \ldots, x^n) = k$. Let B be the set consisting of numbers of vectors included in that basis. Then, obviously, $\mathrm{rk}(x^{i_1}, \ldots, x^{i_p}) = |B \cap \{i_1, \ldots, i_p\}|$. On the other hand, it is clear that $r_X(J) \geqq |B \cap J|$ for any B in L_X.

Lemma 1 and Proposition 3 are proved.

As we shall see in Sect. 2, Lemma 1 has a purely combinatorial generalisation which concerns arbitrary matroids. Note that each stratum Γ in $G_k(\mathbb{C}^n)$ can be defined by a list L_Γ, namely

$$\Gamma = \{X \in G_k(\mathbb{C}^n) | L_X = L_\Gamma\}.$$

1.8. Non-degenerate strata. A stratum Γ in $G_k(\mathbb{C}^n)$ consisting of non-degenerate orbits of the group H (i.e. of orbits of dimension $n-1$) will be called non-degenerate. Non-degeneracy of Γ means that the quotient of the group H with respect to the subgroup of scalar matrices acts freely on Γ.

Proposition 4. *The following statements are equivalent*:

(a) *Stratum $\Gamma \in G_k(\mathbb{C}^n)$ is degenerate.*

(b) *There exists $J \in B(I_n)$ such that any $X \in \Gamma$ can be represented in the form $X \cap C^J \oplus X \cap C^{I_n \setminus J}$.*

This proposition will be proved in Sect. 2.

It is clear that non-degenerate orbits are always defined by projective configurations.

Example. Let an orbit $H \cdot X \subset G_3(\mathbb{C}^n)$ be given by a projective configuration in $\mathbb{C}P^2$. Then $H \cdot X$ is non-degenerate if the configuration has at least four points that are in general position.

1.9. Admissible polyhedra and hypersimplexes. A polyhedron Δ in $\mathbb{R}^n$ is called admissible for $G_k(\mathbb{C}^n)$ if $\Delta = \Delta_\Gamma$ for some stratum $\Gamma \subset G_k(\mathbb{C}^n)$. Theorem 1 implies that any face of an admissible polyhedron is again an admissible polyhedron.

A polyhedron Δ in $\mathbb{R}^n$ is called an (n, k)-hypersimplex if all its edges and vertices are edges and vertices of the polyhedron $\Delta_{n,k}$. Note that all edges of the polyhedron $\Delta_{n,k}$ are parallel to vectors $\delta_i - \delta_j$ for some $i, j \in I_n$.

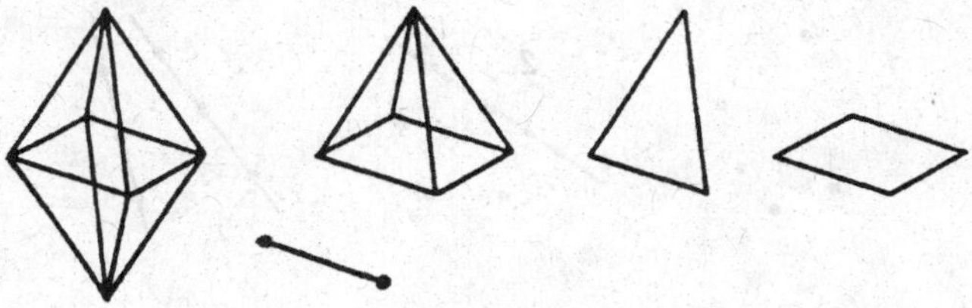

Fig. 1.

Figure 1 shows all (4, 2)-hypersimplexes up to the action of the symmetry group: octahedron $\Delta_{4,2}$, its upper half and all its faces.

Proposition 5. *Any admissible polyhedron is an (n, k)-hypersimplex.*

Proof. By Theorem 1, any face, in particular any edge l, of an admissible polyhedron Δ is also an admissible polyhedron.

Let δ_I and δ_J be the vertices of the edge l. Then $L_Y = \{I, J\}$ for some $Y \in G_k(\mathbb{C}^n)$. Suppose that l is not an edge of $\Delta_{n,k}$. Then $|I \cap J| \leqq k-2$. Therefore, there exist $i, j \in I \setminus J$ and $p \in J \setminus I$. Then Lemma 1 implies that $r_Y(\{i,p\}) = r_Y(\{j,p\}) = 1 = r_Y(\{p\})$, $r_Y(\{i,j,p\}) = 2$. But then Condition (4) of Proposition 1 is not satisfied for the sets $\{i,p\}$ and $\{j,p\}$ which contradicts the assumption that the edge l is admissible.

Remark. Not every (n, k)-hypersimplex is an admissible polyhedron (see example in Sect. 2.2).

1.10. Example. There is no difficulty in describing strata in $G_2(\mathbb{C}^n)$ since this can be reduced to describing projective configurations in $\mathbb{C}P^1$. Strata are numbered by partitions of the set I_n into non-intersecting subsets $A_0, A_1, \ldots, A_p$ where $p \geqq 3$. The function r_Γ is defined by the partition in the following way:

$$r_\Gamma(J) = \begin{cases} 0 & \text{for } J \subset A_0, \\ 1 & \text{for } J \subset A_0 \cup A_i \quad \text{where} \quad i \neq 0, \\ 2 & \text{in other cases.} \end{cases}$$

We now give a list of all non-degenerate strata in $G_3(\mathbb{C}^6)$ (degenerate strata correspond to Grassmannians of lesser dimension). To each non-degenerate stratum there corresponds a special type of configuration of six points in $\mathbb{C}P^2$. All of these are shown on Fig. 2. (A straight line is shown if it contains more than two points of the configuration.)

1.11. Closure of strata. Let Γ be a stratum in $G_k(\mathbb{C}^n)$. It turns out that its closure is not necessarily a union of strata although that is the case for $G_2(\mathbb{C}^n)$ and $G_3(\mathbb{C}^n)$.

Example 1. Consider a one-orbit stratum in $G_3(\mathbb{C}^7)$ given by the projective configuration shown on Fig. 3. Evidently, the orbit given by the configuration with the additional condition $x^7 = 0$ lies in the closure of Γ. However, this orbit belongs

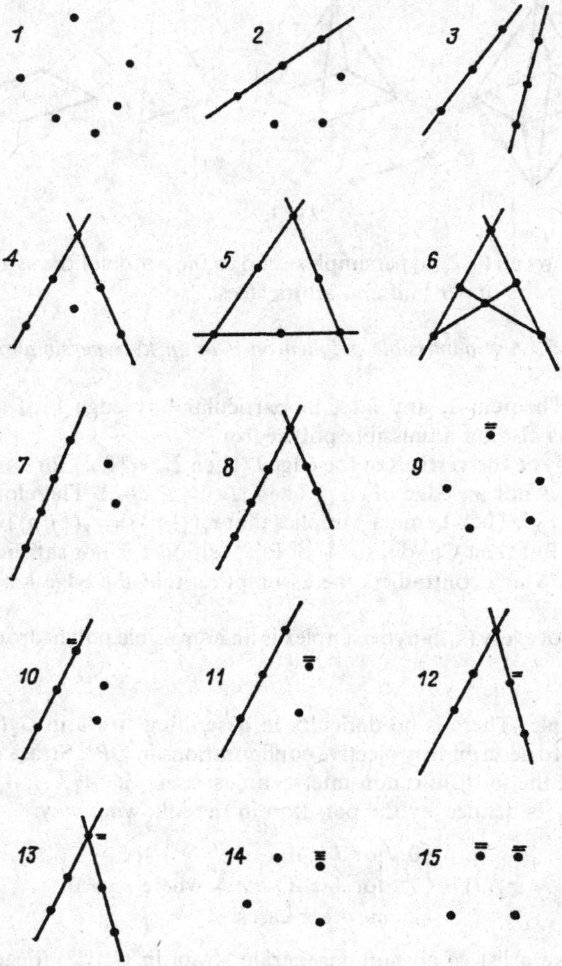

Fig. 2. Non-degenerate strata in $G_3(\mathbf{C}^6)$. The = sign denotes the coincidence of points.

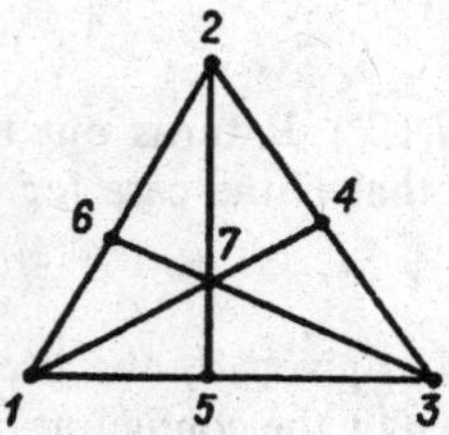

Fig. 3.

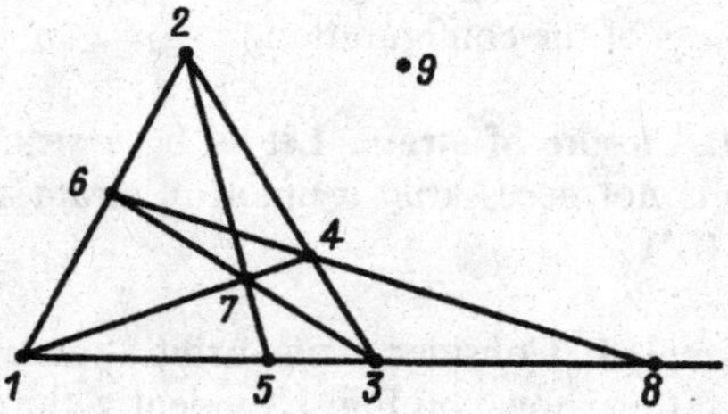

Fig. 4.

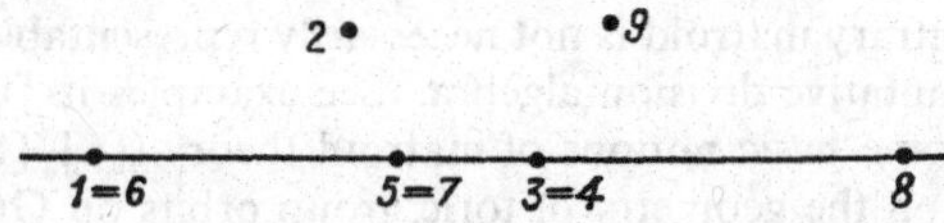

Fig. 5.

to a one-parameter family of orbits, none of which (with the exception of the first one) lies in the closure of Γ.

Example 2. Consider the stratum Γ in $G_3(\mathbb{C}^8)$ obtained from the above example by adding the eighth point, namely, that lying on the intersection of the straight lines (64) and (135) (Fig. 4). When the points 1 and 6, 5 and 7, 4 and 3 coincide, the configuration which has orbits lying in the closure of Γ defines the stratum of 6-dimensional orbits (Fig. 5). However, an orbit belongs to $\bar{\Gamma}$ only if the double ratio satisfies the condition

$$(|x^1x^5|\cdot|x^3x^8|)/(|x^5x^3|\cdot|x^1x^8|)=1.$$

Adding to both configurations one more point in the general position (see Figs. 4, 5) one obtains an example of the situation when both strata are non-degenerate.

2. Matroids and Grassmannian strata

2.1. Matroids. Basic definitions. Definition 1. A matroid of rank k is a pair (S,r) where S is a set and r is an integer-valued function on $B(S)$ such that:

(1) $r(S)=k$,
(2) $0\leqq r(A)\leqq|A|$,
(3) if $A\subseteq B$ then $r(A)\leqq r(B)$,
(4) $r(A)+r(B)\geqq r(A\cup B)+r(A\cap B)$.

Sometimes, instead of "matroid" one says "combinatorial pregeometry" (see, e.g. [8]). The function r is called the rank function of the matroid (S,r). Below we consider only finite sets S.

Proposition 1 of Sect. 1 implies that each stratum Γ in $G_k(\mathbb{C}^n)$ corresponds to a matroid (I_n,r_Γ) where r_Γ is an admissible function defining stratum Γ.

Two matroids (S_1,r_1) and (S_2,r_2) are called isomorphic ($(S_1,r_1)\cong(S_2,r_2)$) if there exists a one-to-one mapping of S_1 onto S_2 taking r_1 into r_2.

Example. A configuration of n vectors $x^1,\ldots,x^n$ in a vector space F^k over an arbitrary field defines a matroid with the rank function $r(\{i_1,\ldots,i_p\})=\mathrm{rk}(x^{i_1},\ldots,x^{i_p})$.

A matroid is said to be representable over the field F if it is isomorphic to a matroid defined by a configuration of vectors in a vector space over F. In particular, the problem of describing all strata in Grassmannians is reduced to describing matroids representable over $\mathbb{C}$. This is a very difficult combinatorial problem which is still awaiting solution.

Note that an arbitrary matroid is not necessarily representable over some field or even a non-commutative division algebra. (See examples in [6], [7].)

We now recall some basic notions of matroid theory ([6]–[8]) and state the corresponding facts on the geometry of toric group orbits on Grassmannians.

A set $A \subseteq S$ is called a flat of matroid (S, r) if $r(A \cup a) > r(A)$ for all $a \in S \backslash A$. Denote the set of all flats of a matroid (S, r) by $\mathrm{Fl}(S, r)$. The properties of the rank function imply that the set $\mathrm{Fl}(S, r)$ is closed with respect to intersections. Introduce a new operation $\vee$ in $\mathrm{Fl}(S, r)$:

$$A \vee B = \bigcap_{J \in \mathrm{Fl}(S,r), A, B \subseteq J} J.$$

Then $\mathrm{Fl}(S, r)$ turns into a geometric lattice, i.e.

$$r(A) + r(B) \geqq r(A \cap B) + r(A \vee B).$$

Let $X \in G_k(\mathbb{C}^n)$. Denote by $R(X)$ the geometric lattice of all subspaces in X obtained by intersecting X with coordinate subspaces $\mathbb{C}^J$ and denote by $R^*(X)$ the lattices of subspaces in X dual to the subspaces of $R(X)$. One can easily check that the following statement is true:

Proposition 1. $R^*(X) \cong \mathrm{Fl}(I_n, r_\Gamma)$ *for all* $X \in \Gamma$.

A set $A \subseteq S$ is called independent if $r(A) = |A|$. Maximal independent sets are called bases of a matroid (S, r). For each matroid (S, r) the family of its bases will be denoted by $\mathscr{B}(S, r)$.

Consider a stratum $\Gamma \subset G_k(\mathbb{C}^n)$. It is evident that its list L_Γ coincides with $\mathscr{B}(I_n, r_\Gamma)$. It has been proved in Sect. 1.7 that a stratum is uniquely defined by its list. This statement can be generalised to arbitrary matroids: any matroid (S, r) is uniquely defined by the family of its bases. Moreover, the properties of the family of bases can be used as a starting point for a definition of a matroid.

Proposition 2. *Let* (S, r) *be a matroid of rank* k. *Then*

(1) $|B| = k$ *for any* $B \in \mathscr{B}(S, r)$.

(2) *Let* $B, B' \in \mathscr{B}(S, r)$. *For each* $b \in B \backslash B'$ *there exists* $b' \in B'$ *such that* $(B \backslash b) \cup b' \in \mathscr{B}(S, r)$. *(The exchange axiom.)*

If a family $\mathscr{B} \subset B(S)$ *satisfies conditions* (1) *and* (2), *then it is a family of bases of a uniquely defined matroid* (S, r).

For a proof see [8].

2.2. Hypersimplexes and matroids. As with Grassmannian strata one can associate to each matroid (S, r) a polyhedron Δ_r. Let $\mathscr{L}(S)$ be the vector space of all real-valued functions on S. Define a polyhedron Δ_r as a convex sum of characteristic functions δ_B for all bases B of matroid (S, r). Define a scalar product in $\mathscr{L}(S)$ by the formula $(f, g) = \sum_{a \in S} f(a) \cdot g(a)$. Then the function r on $B(S)$ is given by the formula

$$r(J) = \max_{\xi \in \Delta_r} (\delta_J, \xi).$$

Note that two matroids are isomorphic if and only if the corresponding polyhedra are congruent.

A question arises which polyhedra in $\mathscr{L}(S)$ correspond to matroids.

Let us call any polyhedron in $\mathscr{L}(S)$ that has vertices of the form δ_B, where $B \in B_k(S)$, and edges parallel to $\delta_a - \delta_b$ for some $a, b \in S$ (S, k)-hypersimplex.

Theorem 1. *A polyhedron Δ in $\mathscr{L}(S)$ corresponds to some matroid (S, r) of rank k if and only if Δ is an (S, k)-hypersimplex.*

Theorem 1 follows immediately from Lemma 1.

Lemma 1. *Let $\mathscr{B}$ be a family of k-element subsets of the set S and $\Delta_{\mathscr{B}}$ a polyhedron in $\mathscr{L}(S)$ that is a convex sum of the characteristic functions δ_B for all $B \in \mathscr{B}$. Then condition (2) of Proposition 2 is equivalent to the condition that Δ is an (S, k)-hypersimplex.*

Proof. Let l be an edge of the polyhedron $\Delta_{\mathscr{B}}$ and δ_{B_1} and δ_{B_2} its vertices. There exists a linear function χ on $\mathscr{L}(S)$ which is constant on the edge l and has values that are less than those on l in all other points of $\Delta_{\mathscr{B}}$. Let $\chi = \sum_{b \in S} v_b \delta_b^*$. Take $b \in (B_1 \cup B_2) \backslash (B_1 \cap B_2)$ so that the value of v_b is minimal. One can assume without any loss of generality that $b \in B_1$. Condition (2) implies that there exists $a \in B_2$ such that $B = (B_1 \backslash b) \cup a \in \mathscr{B}$. Then $\chi(\delta_{B_1}) \leqq \chi(\delta_B)$. Hence $B = B_2$ and the edge l is parallel to the vector $\delta_a - \delta_b$.

Conversely, let $\Delta_{\mathscr{B}}$ be a hypersimplex. Consider a pair of its vertices δ_{B_1} and δ_{B_2} and an arbitrary $a \in B_1 \backslash B_2$.

Let $\Delta'_{\mathscr{B}}$ be a face of the hypersimplex $\Delta_{\mathscr{B}}$ on which the linear function $\chi = \sum_{b \in B_1 \cup B_2} \delta_b^*$ takes its maximum value. Evidently, $\Delta'_{\mathscr{B}}$ is a hypersimplex, and

$$\operatorname{Vert} \Delta'_{\mathscr{B}} = \{\delta_B | B \subset B_1 \cup B_2\}.$$

Consider the linear function δ_a for $a \in B_1 \backslash B_2$. Since $\delta_a^*(\delta_{B_1}) > \delta_a^*(\delta_{B_2})$ there exists an edge $l \subseteq \Delta'_{\mathscr{B}}$ with the vertex δ_{B_1} such that δ_a^* is decreasing on l. Let δ_B be another vertex of the edge l. Since $\delta_a^*(\delta_B) < \delta_a^*(\delta_{B_1})$ one has $\delta_a^*(\delta_B) = 0$, i.e. $a \notin B$. But the edge l is parallel to the vector $\delta_b - \delta_a$, and therefore $B = (B_1 \backslash a) \cup b$. Now $b \in B_1 \cup B_2$ whence $b \in B_2$. The statement is proved.

Example. Consider the matroid corresponding to the configuration (Fig. 6) consisting of seven points in the projective plane over the two-element field F_2. The configuration cannot be realised over the fields $\mathbb{C}$ or $\mathbb{R}$. The corresponding hypersimplex Δ_F has the full symmetry group $\mathrm{PGL}(3, F_2)$ and for each dimension the number of faces of that dimension is:

Face dimension	0	1	2	3	4	5	6
Number of faces	28	126	245	238	112	21	1

Note that to each face there corresponds a configuration of points in $\mathbb{R}P^2$, i.e. all faces are admissible polyhedra. All the resulting configurations are shown in Fig. 7.

Taking configurations of points in the projective space over the finite field F_q

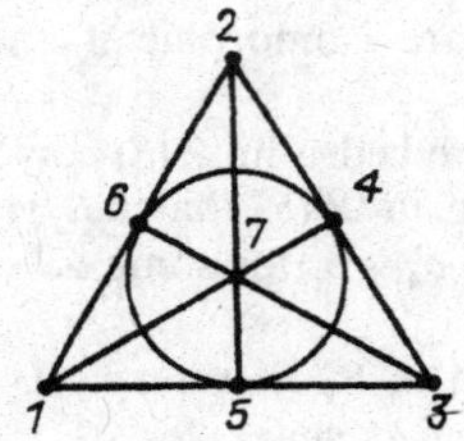

Fig. 6.

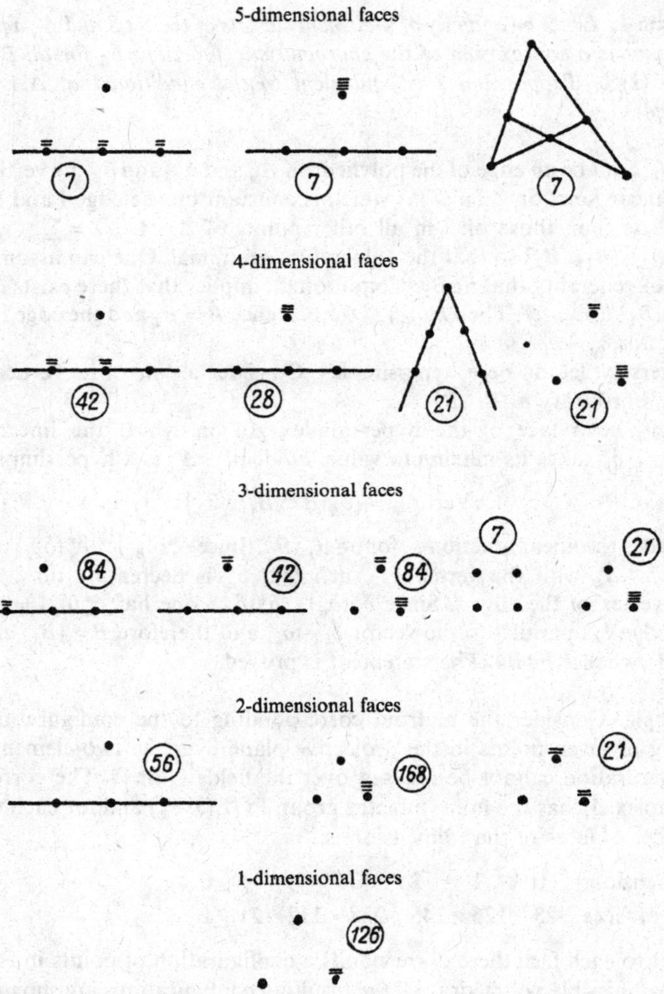

Fig. 7. The numbers in circles give the general number of faces of each type.

one obtains the series of hypersimplexes $\Delta(m, F_q)$ with the full symmetry group $\mathrm{PGL}(m, F_q)$.

2.3. Operations on matroids. The restriction of a matroid (S,r) on $A \subseteq S$ is the matroid (A,r) obtained by restricting the rank function r of (S,r) to $B(A)$. The matroid (A,r) is said to be a submatroid of (S,r).

The retraction $(S,r)/A$ of a matroid (S,r) by $A \subseteq S$ is the matroid $(S\backslash A, r)$ with the rank function r defined on $B(S\backslash A)$ by the formula: $r(J) = r(J \cup A) - r(A)$. The matroid $(S,r)/A$ is called a quotient matroid.

Let Γ be a stratum in $G_k(\mathbb{C}^n)$ and (I_n, r_Γ) the corresponding matroid, $J \subseteq I_n$, $r(J) = p$. Consider the strata $\Gamma_1 \subset G_p(\mathbb{C}^J)$ and $\Gamma_2 \subset G_{k-p}(\mathbb{C}^{I_n\backslash J})$ obtained by taking the projections of subspaces of the stratum Γ on $\mathbb{C}^J$ along $C^{I_n\backslash J}$ and by taking intersections with $\mathbb{C}^{I_n\backslash J}$, respectively. Then the matroids (J, r_{Γ_1}) and $(I_n\backslash J, r_{\Gamma_2})$ are isomorphic to the restriction of the matroid (I_n, r_Γ) on J and to its retraction by J, respectively.

Let (S,r) be a matroid of rank k, $|S| = n$. The matroid (S, r^*) of rank $(n-k)$ with the rank function r^* such that $r^*(A) = |A| - k + r(S\backslash A)$ is called the dual matroid of (S,r). Dual strata $\Gamma \subset G_k(\mathbb{C}^n)$ and $\Gamma^* = \{X^\perp \in G_{n-k}((\mathbb{C}^n)^*)$ where $X \in \Gamma\}$, evidently correspond to dual matroids.

Let (S_1, r_1) and (S_2, r_2) be two matroids. Then the product $(S_1, r_1) \times (S_2, r_2)$ is the matroid $(S_1 \cup S_2, r)$ with rank function $r(J) = r_1(J \cap S_1) + r_2(J \cap S_2)$.

If Γ_1 and Γ_2 are arbitrary strata in $G_p(\mathbb{C}^n)$ and $G_q(\mathbb{C}^m)$, then $\Gamma_1 \times \Gamma_2 = \{X \in G_{p+q}(\mathbb{C}^n \oplus \mathbb{C}^m) | X = X_1 \oplus X_2$ for some $X_1 \in \Gamma_1, X_2 \in \Gamma_2\}$ is a stratum. Evidently, $(I_n \cup I_m, r_{\Gamma_1 \times \Gamma_2}) = (I_n, r_{\Gamma_1}) \times (I_m, r_{\Gamma_2})$.

2.4. Hypersimplex dimension. Let Δ_r be a hypersimplex corresponding to a matroid (S,r). Let us find its dimension. For this purpose we need some other notions of matroid theory.

A subset $U \subseteq S$ is called a separator of the matroid (S,r) if $(S,r) = (U,r) \times (S\backslash U, r)$, i.e. $r(A) = r(A \cap U) + r(A\backslash U)$ for all $A \subseteq S$.

Proposition 3. (*See* [6].) *A union or intersection of separators and the complement of each separator in S is again a separator. The family of non-empty minimal separators forms a partition of the set S.*

Denote by $K(S,r)$ the number of minimal non-empty separators of (S,r). A matroid (S,r) is called connected if $K(S,r) = 1$.

Proposition 4. *Let (S,r) be a matroid with the family of bases B and the rank function r; let Δ_r be the corresponding hypersimplex. Then* $\dim \Delta_r = -K(S,r) + |S|$.

Proof. Consider the subspace W in $\mathscr{L}(S)$ spanned by the vectors $\alpha - \beta$ where α and β are vertices connected by an edge. Theorem 1 implies that all edges of Δ_r are of the form $\delta_a - \delta_b$. Introduce an equivalence relation on the set S: $a \sim b$ if and only if $\delta_a - \delta_b$ is an edge of the hypersimplex Δ_r. Consider a partition of S into a non-connected union of equivalence classes $J_1, \ldots, J_q$. Evidently, $\dim W = \dim \Delta_r =$

$n-q$ and the subspace W is defined by q equations

$$\sum_{b\in J_i} \xi_b = k_i \quad (i=1,\dots,q).$$

Let $(J_i, r_i) = (J_i, r)$, let Δ_{r_i} be hypersimplexes corresponding to matroids (J_i, r_i) and let $\mathscr{B}_i$ be the family of bases for (J_i, r_i). Evidently, (J_i, r_i) is a matroid of rank k_i and for each $B\in\mathscr{B}$ one has $B\cap J_i\in\mathscr{B}_i$. Therefore, $\Delta_r \subseteq \Delta_{r_1}\times\cdots\times\Delta_{r_q}$. If a vertex $\alpha_1\times\cdots\times\alpha_q\in\Delta_r$ and if $\beta_i\in\Delta_{r_i}$ is a vertex connected to α_i by an edge, then $\alpha_1\times\cdots\times\beta_i\times\cdots\times\alpha_q\in\Delta_r$ because Δ_r is a hypersimplex. This easily implies that $\Delta_r = \Delta_{r_1}\times\cdots\times\Delta_{r_q}$, i.e. $\mathscr{B} = \{B_1\cup\cdots\cup B_q \mid B_i\in\mathscr{B}_i\}$.

Thus $(S,r) = (J_1, r_1)\times\cdots\times(J_q, r_q)$. Hence, $J_1,\dots,J_q$ are separators of the matroid (S,r).

Matroids (J_i, r_i) correspond to hypersimplexes of maximum dimension. Let us prove that each matroid (J_i, r_i) is connected. Indeed, let U be a separator of (J_i, r_i), $J_i \neq U$ and $\varnothing \neq U$. Then the function $\chi(\xi) = (\delta_U, \xi)$ is constant on Δ_{r_i}, which contradicts the assumption that Δ_{r_i} has maximum dimension.

Consequently, $J_1,\dots,J_q$ are minimal separators of the matroid (S,r), $q = K(S,r)$, $\dim\Delta_r = n-q$.

Note that Proposition 4 implies Proposition 4 of Sect. 1.

2.5. Faces of hypersimplexes. Let (S,r) be a connected matroid. The non-empty subset $A\subset S$ will be called non-degenerate if the matroids $(S,r)/A$ and (A,r) are connected.

Theorem 2. *Let (S,r) be a connected matroid and Δ_r the corresponding hypersimplex. The family of faces of co-dimension 1 of the hypersimplex Δ_r is in one-to-one correspondence with the family of non-degenerate subsets A in S. A non-degenerate subset A is associated under this correspondence to the face Δ_{r_A}, where $(S, r_A) = (A,r)\times(S,r)/A$.*

Proof. As we have already seen in the proof of the preceding theorem, each face of the hypersimplex Δ_r is defined by an equation of the form $\sum_{b\in A}\xi_b = k_1$ where k_1 is the maximum value of the function (δ_A, ξ) on Δ_r. The face is evidently a hypersimplex corresponding to some matroid (S, r_A). The set $\mathscr{B}_A$ of all bases of the matroid (S, r_A) consists of the elements of the form $B_1\cup B_2$, where B_1 and B_2 are some bases of matroids (A,r) and $(S,r)/A$ respectively. Thus $(S,r_A) = (A,r)\times(S,r)/A$. Now, since $\dim\Delta_{r_A} = |S|-2$, one has $K(S,r)=2$. Thus, the set A is non-degenerate.

2.6. The exchange axiom and an ordering on $B(I_n)$. Introduce a partial ordering into $B_k(I_n)$ in the following way. Let $A, B\in B_k(I_n)$ and $A = \{i_1,\dots,i_k\}$, $i_1<\cdots<i_k$, $B=\{j_1,\dots,j_k\}$, $j_1<\cdots<j_k$. Then $A\leqq B$ if and only if $i_p\leqq j_p$ for all $p=1,\dots,k$.

Denote by S_n the permutation group for the set I_n. Then to each $s\in S_n$ one can associate an ordering on $B_k(I_n)$, setting $A\overset{s}{\leqq} B$ if and only if $s^{-1}(A)\leqq s^{-1}(B)$. Evidently, under that definition $\overset{e}{\leqq}$ coincides with $\leqq$.

Let $\mathscr{B} \subset B_k(I_n)$. An element $B \in \mathscr{B}$ is called s-minimal in B if $B \overset{s}{\leqq} J$ for any $J \in \mathscr{B}$. The family $\mathscr{B}$ is said to satisfy the minimality condition if for each $s \in S_n$ there exists an s-minimal element in $\mathscr{B}$.

Proposition 5. *The exchange condition of Proposition 2 is equivalent to the minimality condition.*

This reformulation of the exchange axiom will be required below. The minimality condition is equivalent for the convergence condition of the greedy algorithm (see [21]).

3. Moment mapping and toric varieties

3.1. Let M be a real simplectic manifold, i.e. let there be a non-degenerate closed 2-form ω defined on M. Let a compact Lie group K with the Lie algebra $\mathfrak{k}$ be acting on M leaving the form ω invariant. Denote by TM the Lie algebra of vector fields on M and by $\Omega^1 M$ the set of 1-forms on M. Then a unique Lie algebra homomorphism $a: \mathfrak{k} \to TM$ exists and the dual mapping $a^*: \Omega^1 M \to \mathfrak{k}^*$.

For each point $X \in M$ the 2-form ω defines the isomorphism ω_X of the tangent space $T_X M$ and the cotangent space $\Omega_X^1 M$ in the following way: $\omega_X(v)(v') = \omega_X(v, v')$ for any $v' \in T_X M$.

For any smooth mapping $\theta: M_1 \to M_2$ denote by $d\theta_X$ the differential of θ at X.

Definition 1. A mapping $\mu: M \to \mathfrak{k}^*$ is called a moment mapping if
(1) μ is equivariant under the action of the group K;
(2) $d\mu_X = a^* \circ \omega_X$ for each $X \in M$.

Remark. Below, we always consider the case when M is a complex manifold with a K-invariant Kähler metric χ. Then, by definition, the imaginary part ω of the metric χ is a simplectic form on M considered as a real manifold. Clearly, ω is invariant under the action of K. Therefore, a moment mapping can be defined in this case.

Example. Let $K = U(n)$, $M = \mathbb{C}P^n$. Evidently, there exists a K-invariant Kähler form χ which in homogeneous coordinates is of the form: $\chi = \sum dz_i\, d\bar{z}_i / \sum z_i \bar{z}_i$. Let $z_i = x_i + \sqrt{-1}\, y_i$. Then the 2-form $\omega = \operatorname{Im} \chi$ is of the form $\omega = \sum dx_i \wedge dy_i / \sum |z_i|^2$. Let us identify $\mathfrak{k}^*$ with the set of Hermitian matrices of order $n+1$. Then the moment mapping takes the point $X \in \mathbb{C}P^n$ with coordinates $(z_0, z_1, \ldots, z_n)$ into a Hermitian matrix (z_{ij}) where $z_{ij} = z_i \bar{z}_j / \sum |z_i|^2$.

Let K be an arbitrary compact Lie group acting transitively on a simplectic manifold M. Then the moment mapping $\mu: M \to \mathfrak{k}^*$ makes it possible to identify M with an orbit of the group K in the co-adjoint representation of $\mathfrak{k}^*$.

3.2. Convexity theorem. Let M be a compact Kähler manifold. Suppose that a complex group H isomorphic to $(\mathbb{C}^*)^n$ acts on M in such a way that the Kähler metric is invariant with respect to the compact form T of the group H. Consider a moment mapping $\mu: M \to \mathbb{R}^n$ (in this case the Lie algebra of the group T is isomorphic to $\mathbb{R}^n$).

Theorem 1. *Under the moment mapping the image of the manifold M is a convex polyhedron. The closure $\overline{H \cdot X}$ of a general orbit of the group H in M is mapped onto $\mu(M)$. The mapping μ provides a one-to-one correspondence between the set of orbits of the group H in $\overline{H \cdot X}$ and the set of faces of the polyhedron $\mu(M)$. Under this correspondence, a q-dimensional (over $\mathbb{C}$) orbit is mapped onto a q-dimensional (over $\mathbb{R}$) open face of the polyhedron $\mu(M)$.*

For the proof see [2], [3].

Example. Let $M = \mathbb{C}P^n$ and let H be the subgroup of diagonal matrices in $\mathrm{SL}(n+1)$. Then $\mu(M)$ is an n-dimensional simplex in R^n.

Remark. Theorem 1 is also valid for singular compact Kähler manifolds.

3.3. Toric varieties. A complex algebraic variety M is called toric if the group $H = (\mathbb{C}^*)^n$ acts on it in such a way that there is an open dense orbit in M isomorphic to H.

Suppose that on a compact toric variety M there is a Kähler metric χ invariant with respect to the compact form T of the group H. Let a moment mapping $\mu: M \to \mathbb{R}^n$ be given on M. Then all the statements of Theorem 1 can, evidently, be applied to M (see [5]).

A fan Σ in $\mathbb{R}^n$ is a set of convex polyhedral cones in $\mathbb{R}^n$ such that, together with each cone in Σ, all its faces also belong to Σ and any two cones may intersect only by a common face. A fan is said to be integral if all its one-dimensional cones are spanned by vectors with integer coordinates.

It is known (see [1]) that any toric variety is defined by some integer fan Σ_M. One can show that the fan Σ_M is dual to the polyhedron $\mu(M)$ in the sense that each cone of the fan Σ_M in $\mathbb{R}^n$ consists of linear functions taking their maximum values on faces of $\mu(M)$.

The moment mapping makes it possible to reformulate many properties of compact toric varieties in terms of convex polyhedra (see [5]).

A convex polyhedron in $\mathbb{R}^n$ is said to be simple if each of its vertices is an intersection point for exactly n edges.

Proposition 1. *A toric variety is non-singular if and only if its image under the moment mapping is a simple polyhedron.*

For the proof see [5].

By the F-polynomial for a simple n-dimensional polyhedron we mean the generating polynomial $F = \sum_{0 \leq i \leq n} F_i t^i$ where F_i is the number of i-dimensional faces of the polyhedron. Let $H(t)=F(t-1)$. The polynomial H defined for the polyhedron $\mu(M)$ makes it possible to compute Betti numbers of the manifold M.

Proposition 2. *The polynomial H for any simple polyhedron is reflexive, i.e. i $H_i=H_{n-i}$. A smooth toric variety M has only even non-zero Betti numbers b_{2i}, so that $b_{2i}=H_i$ for the H-polynomial of the polyhedron $\mu(M)$.*

The proof of Proposition 2 is given in [5]. It makes use of the following elegant theorem. Let φ be a general linear function for the polyhedron Δ, i.e. φ is not constant on any edge of Δ. For each vertex α of a simple polyhedron Δ its index is the number $i_\varphi(\alpha)$ of edges going out of α on which the function φ decreases. Evidently, $0\leqq i_\varphi(\alpha)\leqq n$. Denote by h_i the number of vertices having the same index i.

Proposition 3. *The H-polynomial of a simple polyhedron is of the form*

$$H(t)=\sum_{0\leqq i\leqq n} h_i t^i.$$

Example. Let G be a complex semi-simple Lie group and B its Borel subgroup. Let $M=G/B$ and H be a Cartan subgroup in G. Then there is a Kähler metric in M invariant with respect to the compact form T of the group H. One can define the moment mapping $\mu: M\to\mathfrak{h}^*_{\mathbb{R}}$ where $\mathfrak{h}^*_{\mathbb{R}}$ is the real form of the Lie algebra $\mathfrak{h}$ of the group H. Let $\Delta=\mu(M)$. We show in Sect. 5.1 that Δ is the convex hull for the set of points $W\cdot\lambda$ where $\lambda\in\mathfrak{h}^*_{\mathbb{R}}$ is a point in general position and $W\cdot\lambda$ is the orbit of λ under the action of the Weyl group W.

Consider a general orbit $H\cdot X$ of the group H in M. Then $\mu(\overline{H\cdot X})=\Delta$. Each vertex α of the polyhedron Δ lies in its own Weyl chamber and hence uniquely defines a system of simple roots $\sigma_1(\alpha),\dots,\sigma_n(\alpha)$ in $\mathfrak{h}^*_{\mathbb{R}}$ given by the condition $(\sigma_i(\alpha),\alpha)>0$ for all $i=1,\dots,n$. One can easily see that all the edges of the polyhedron Δ, containing α, have a second vertex $r_{\sigma_i(\alpha)}(\alpha)$ where r_σ denotes the reflection with respect to the root σ. Therefore the polyhedron Δ is simple, and hence $\overline{H\cdot X}$ is a smooth compact manifold. Let us compute its Betti numbers using Proposition 3. (This has been done in [12].)

Let α be an arbitrary vertex of Δ, $\sigma_1,\dots,\sigma_n$ the corresponding system of simple roots and $R^{(\pm)}$ the set of positive (negative) roots of that system. Any other vertex of Δ is of the form $w(\alpha)$ for some $w\in W$. Let $\varphi(\xi)=(\xi,\alpha)$ for all $\xi\in\mathfrak{h}^*_{\mathbb{R}}$. We now compute the index of the vertex $w(\alpha)$. The edges going out of $w(\alpha)$ are scalar multiples of $w(\sigma_j)$. Thus $i_\varphi(w(\alpha))=|\{\sigma_j|(w(\sigma_j),\alpha)<0\}|=|\{\sigma_j|w(\sigma_j)\in R^-\}|$. Therefore, the coefficient h_i of the polynomial H equals the number of Weyl group elements that take exactly i simple roots into negative ones.

4. Geometry of compact homogeneous spaces

4.1. Parabolic subgroups. Let G be a complex semi-simple Lie group with Lie algebra $\mathfrak{G}$, and let H be a fixed Cartan subgroup in G. Denote by $\mathfrak{h}$ the Lie algebra of the group H, and by $\mathfrak{h}_{\mathbb{R}}$ its real part. Fix a Borel subgroup B containing H. The choice of B defines a system of simple roots $\sigma_1, \ldots, \sigma_n$ of the Lie algebra $\mathfrak{G}$ in $\mathfrak{h}_{\mathbb{R}}$. Closed subgroups in G containing B will be called standard parabolic subgroups.

Any compact homogeneous space M of the group G is of the form G/P where P is a standard parabolic subgroup in G.

Let W be the Weyl group of the group G, i.e. the group generated by the reflections $r_1, \ldots, r_n$ of $\mathfrak{h}_{\mathbb{R}}^*$ with respect to the roots $\sigma_1, \ldots, \sigma_n$. Let S be a subset of simple roots. The subgroup $W_S \subset W$ generated by the elements r_i for all $\sigma_i \in S$ is called the parabolic subgroup of the group W.

Let $N(H)$ be the normalizer of the group H in G. Then the group W is canonically isomorphic to $N(H)/H$. Let P be a standard parabolic subgroup in G. One can show that the subgroup $(N(H) \cap P)/H$ is parabolic in W. Any standard parabolic subgroup P can be uniquely reconstructed from $(N(H) \cap P)/H$. Thus there is a one-to-one correspondence between standard parabolic subgroups in G, parabolic subgroups in W and subsets of simple roots. The subset of simple roots corresponding to a standard parabolic subgroup P will be denoted by S_P and the corresponding parabolic subgroup W_{S_P} by W_P.

4.2. Generalised Plücker coordinates. Let $P \subset G$ be a standard parabolic subgroup. Consider a finite-dimensional irreducible representation ρ of the group G defined in the representation space V and having the highest weight $\lambda = \sum_{\sigma_i \in S_P} \omega_i$, where ω_i is the fundamental weight corresponding to a simple root σ_i. Then the homogeneous space G/P can be identified with the highest weight orbit in the projectivisation $P(V)$ of the space V.

Let $\mathscr{A}$ be the set of the weights of the representation ρ taken according to their multiplicities. Choose a weight basis $\{e_\alpha | \alpha \in A\}$ in V. Then any point $X \in G/P$ is uniquely (up to a scalar multiple d) defined by a set of numbers $p^\alpha(X)$ where $X = d \cdot \sum_{\alpha \in A} p^\alpha(X) e_\alpha$.

Let $W \cdot \lambda$ be the orbit of the highest weight λ in $\mathfrak{h}_{\mathbb{R}}^*$ under the action of W. Then the set $W \cdot \lambda$ can be identified with the set W^P of left cosets W/W_P. The points $W \cdot \lambda$ lie in the vertices of some convex polyhedron $\Delta_P \in \mathfrak{h}_{\mathbb{R}}^*$. The remaining points of A lie inside the polyhedron Δ_P (see [2]).

The set of numbers $\{p^2(X) | \alpha \in W \cdot \lambda\}$ defined up to a scalar multiple is called the generalized Plücker coordinates of the point $X \in M$.

Example 1. Let $G = \mathrm{SL}(n)$, $M = G_k(\mathbb{C}^n)$. Then $W \cong S_n$, $W_P \cong S_k \times S_{n-k}$ and $V = \Lambda^k E$ where E is the standard representation of $\mathrm{SL}(n)$. The set $\mathscr{A} = W^P$ can be identified with $B_k(I_n)$. The basis elements e_α are $e_{i_1} \wedge \cdots \wedge e_{i_k}$ where $\alpha = \{i_1, \ldots, i_k\}$ and $p^\alpha(X)$ are the usual Plücker coordinates of the space $X \in G_k(\mathbb{C}^n)$ (see Sect. 1.5).

Example 2. Let $G = \mathrm{SL}(n)$, $P = B$. Then M is the full flag manifold in $\mathbb{C}^n$. In this case $W_P = \{1\}$ and $W^P \cong S_n$. The elements of $W \cdot \lambda$ can, therefore, be identified with permutations. Thus, the generalized Plücker coordinates of a flag are numbered by permutations. Let X be a flag of subspaces $X_1 \subset \cdots \subset X_n$. Then $p^{(j_1, \ldots, j_n)} = p^{j_1}(X_1), \ldots, p^{j_1 \cdots j_n}(X_n)$ for any permutation $(j_1, \ldots, j_n) \in S_n$.

4.3. Kähler structure on G/P. Choose a compact form K of the group G such that it contains the compact form T of the group H. Then there exists a K-invariant Hermitian metric $\bar{\chi}$ in V, which induces a Kählerian metric in the projective space $P(V)$. Denote its restriction on M by χ. The metric χ defines a Kählerian structure on the manifold M.

One can easily see that it is possible to choose a weight basis $\{e_\alpha\}$ in V such that $\bar{\chi}(\sum p_1^\alpha e_\alpha, \sum p_2^\alpha e_\alpha) = \sum p_2^\alpha \bar{p}_2^\alpha$ where $\alpha \in \mathscr{A}$.

Below we always work in a weight basis chosen in this way.

4.4. Definition of stratum. Let $X \in G/P$. The list of X is a subset $L_X \subset W \cdot \lambda$ given by the condition $L_X = \{\alpha \in W \cdot \lambda \mid p^\alpha(X) \neq 0\}$. Two points X and Y in M are called equivalent if $L_X = L_Y$.

Definition 1. The equivalence classes Γ in M are called strata.
The set $L_\Gamma = L_X$ for any $X \in \Gamma$ is called the list of the stratum Γ.

Note that the definition of the list L_X does not depend on the choice of a weight basis in V since the weights $\alpha \in W \cdot \lambda$ are of multiplicity one.

5. Definition of stratum through the moment mapping

5.1. The moment mapping for $M = G/P$. In Sect. 4.2 we constructed a Kähler form χ on the manifold $M = G/P$. The imaginary part of the form χ is a simplectic form on the manifold M which is invariant under the action of K and, in particular, under the action of the compact part T of the group H. Let us identify the Lie algebra of the group T with $\mathfrak{h}_{\mathbb{R}}$. One can easily see that the mapping $\mu: M \to \mathfrak{h}_{\mathbb{R}}^*$ defined by the formula $\mu(X) = \sum_{\alpha \in A} |p^\alpha(X)|^2 \cdot \alpha / \sum_{\alpha \in A} |p^\alpha(X)|^2$, where $X = d \cdot \sum_{\alpha \in A} p^\alpha(X) e_\alpha$, is the moment mapping.

5.2. Images of orbits of the group H. Let $X \in M$. Denote by $H \cdot X$ the orbit of the point X under the action of H and by $\overline{H \cdot X}$ its closure.

Proposition 1. *The image $\mu(\overline{H \cdot X})$ is a convex polyhedron having vertices in points α for all $\alpha \in L_X$.*

Proof. Since $\overline{H \cdot X}$ is a compact Kähler manifold (which may be singular), Theorem 1 of Sect. 3 can be applied to it. Therefore $\mu(\overline{H \cdot X})$ is a convex polyhedron whose vertices are images of stable points in $\overline{H \cdot X}$ under the action of H. For a weight vector e_α denote by v_α the straight line defined by e_α. Then all the stable points of the group H in M are given by those v_α for which $\alpha \in W \cdot \lambda$. Since $\mu(v_\alpha) = \alpha$, it remains for us to find out when $v_\alpha \in \overline{H \cdot X}$. It is clear that if $p^\alpha(X) = 0$ then $p^\alpha(Y) = 0$ for any $Y \in \overline{H \cdot X}$. Let $p^\alpha(X) \neq 0$. Note that since $\mathscr{A} \subseteq \Delta_P$, $h \in \mathfrak{h}_{\mathbb{R}}$ exists such that $\alpha(h) > \beta(h)$ for any $\beta \in \mathscr{A}$, $\beta \neq \alpha$. Consider a one-parameter subgroup $h_t = \exp ht$ in the group H. A simple calculation shows that $\lim_{t \to \infty} h_t(X) = v_\alpha$. Therefore $v_\alpha \in \overline{H \cdot X}$.

Let us call two orbits $H\cdot X$ and $H\cdot Y$ equivalent if $\mu(H\cdot X)=\mu(H\cdot Y)$. Then Proposition 1 immediately implies the following.

Proposition 2. *An equivalence class of orbits of the group H in M is a stratum.*

6. Strata and Schubert cells

6.1. Schubert cells in $M=G/P$. Let C be a Borel subgroup of the group G containing H. A Schubert cell on the manifold $M=G/P$ with respect to a Borel subgroup C is, by definition, an orbit of the group C in M. In other words, a Schubert cell is a two-sided coset in $C\backslash G/P$. It is a known fact (see, e.g. [9]) that the number of Schubert cells is the same as the number of cosets W/W_P and all of them are of the form $C\cdot\alpha\cdot P$ where $\alpha\in W_P=W/W_P$. A Schubert cell $C\cdot\alpha\cdot P$ will be denoted simply by $C\cdot\alpha$.

6.2. Strata and Schubert cells. For each Borel subgroup $C\supset H$ let us choose exactly one Schubert cell in M corresponding to C. The intersection of all the chosen cells will be called a thin cell, provided it is not empty.

Theorem 1. *Decomposition of M into small cells coincides with decomposition into strata.*

Proof. Suppose that two points X and Y belong to the same thin cell. Let us prove that $L_X=L_Y$. Assuming that this is not true, an $\alpha\in W\cdot\gamma$ exists such that $p^\alpha(X)=0$, $p^\alpha(Y)\neq 0$. Choose a Borel subgroup $C\supset H$ with respect to which α is a lowest weight. Then $p^\alpha(c(X))=0$ for any $c\in C$. Then Y and X cannot belong to the same orbit of the group C, and, consequently, to the same thin cell. Conversely, let $X, Y\in M$ and $L_X=L_Y$. Suppose that X and Y belong to different thin cells. Then a Borel subgroup C exists containing H, such that $X\in C\cdot\alpha$, $Y\in C\cdot\beta$, $\alpha\neq\beta$. Let $\sigma_1,\dots,\sigma_n$ be a system of simple roots of the Lie algebra $\mathfrak{G}$ corresponding to the given Borel subgroup C. There exists $h\in\mathfrak{h}_{\mathbb{R}}$ such that $\sigma_i(h)<0$ for all $i=1,\dots,n$ and $\alpha(h)\neq\beta(h)$. Let $h_t=\exp h\cdot t$. Then $\lim_{t\to\infty}h_tch_t^{-1}$ exists and belongs to H for any $c\in C$. Therefore, $\lim_{t\to\infty}h_t(X)=v_\alpha$ and $\lim_{t\to\infty}h_t(Y)=v_\beta$. However, the proof of Proposition 1 (Sect. 5) implies that $\lim_{t\to\infty}h_t(X)=v_\gamma$ where $\gamma\in L_X$ such that $\gamma(h)>\delta(h)$ for $\delta\in L_X$, $\delta\neq\gamma$. Since $L_X=L_Y$, one has $\gamma=\alpha=\beta$ contradicting the assumption.

7. Polyhedra corresponding to strata

7.1. Consider in more detail combinatorial properties of polyhedra corresponding to toric varieties in G/P.

A polyhedron Δ in $\mathfrak{h}_{\mathbb{R}}^*$ is called admissible for G/P if it is an image of the closure of some orbit $H\cdot X$ in $M=G/P$ under the moment mapping $\mu: M\to\mathfrak{h}_{\mathbb{R}}^*$. Proposition 1 of Sect. 5 implies that for any admissible polyhedron Δ one has $\operatorname{Vert}\Delta\subset W\cdot\lambda$ where $W\cdot\lambda$ is the orbit of the highest weight λ of representation ρ under the action of the Weyl group and $\operatorname{Vert}\Delta$ is the set of vertices of the polyhedron Δ.

Definition 1. A polyhedron Δ with vertices in $W \cdot \lambda$ will be called a (G, P)-hypersimplex if all its edges are parallel to roots of the Lie algebra $\mathfrak{G}$ of the group G.

Definition 1 is a generalisation of the definition of a hypersimplex for the Grassmannian $G_k(\mathbb{C}^n)$ given in Sect. 2.4.

Theorem 1. *Any polyhedron admissible for G/P is a (G, P)-hypersimplex.*

Proof. Let Δ be an admissible polyhedron. Then any of its edges l is also admissible. Consider an orbit $H \cdot X \subset M$ for which $\mu(\overline{H \cdot X}) = l$. Let H_X be a subgroup in H leaving X fixed and let $\mathfrak{h}_X$ be the corresponding Lie subalgebra. Evidently, $\dim \mathfrak{h}_X = \dim \mathfrak{h} - 1$ and the subspace $\mathfrak{h}_X$ in $\mathfrak{h}$ is given by the equation $\gamma(h) = \beta(h)$ for all $h \in \mathfrak{h}_X$, where β and γ are vertices of l.

On the other hand, consider the coset gP corresponding to the point $X \in G/P$. Then the condition that $h \in \mathfrak{h}_X$ can be written in the form $(\exp ht) \cdot g \in g \cdot P$. Since $\exp ht \in P$ one has

$$(\exp ht) \cdot g (\exp ht)^{-1} \in gP \tag{1}$$

for all $t \in C$.

Let $\mathfrak{P}$ be the Lie algebra of the group P, R the system of roots of $\mathfrak{G}$, g_α the root vector in G of the weight α, and $R^P = \{\alpha \in R \mid g_\alpha \in \mathfrak{P}\}$. Then any $g \in G$ can be represented in the form $g = \prod_{\alpha \in R \setminus R^P} (\exp g_\alpha) \cdot p_0$ where $p_0 \in P$ and some of the g_α may vanish. Let $g_0 = \prod_{\alpha \in R \setminus R^P} \exp g_\alpha$. Then condition (1) is equivalent to the following condition on g_0:

$$(\exp ht) \cdot g_0 \cdot (\exp ht)^{-1} = g_0 .$$

This condition can also be written in the following way:

$$\prod_{\alpha \in R \setminus R^P} (\exp ht) \cdot \exp g_\alpha \cdot (\exp ht)^{-1} = \prod_{\alpha \in R \setminus R^P} \exp\,[ht, g_\alpha]$$

$$= \prod_{\alpha \in R \setminus R^P} \exp \alpha(h) t \cdot g_\alpha = \prod_{\alpha \in R \setminus R^P} \exp g_\alpha .$$

This equality has to be satisfied for all $t \in C$, and, therefore, $\alpha(h) = 0$ for $\alpha \in R \setminus R^P$, $g_\alpha \neq 0$. Since R has no vectors of multiplicity greater that one and since $\dim \mathfrak{h}_X = \dim \mathfrak{h} - 1$, there is only one non-vanishing vector g_α. It follows, therefore, that the weight $\beta - \gamma$ is a scalar multiple of α which proves Theorem 1.

Theorem 1 makes it possible to formulate a necessary condition for a polyhedron with vertices in $W \cdot \lambda$ to be admissible. As we saw in Sect. 2.2, for $M = G_k(\mathbb{C}^n)$, is equivalent to the exchange axiom for matroid bases.

8. General (W, Q)-matroid

8.1. Definition of (W, Q)-matroid. Let W be a Coxeter group. We remind the reader that a Coxeter group is a group with a finite set of generators R defined by the relations $r^2 = 1$ for all $r \in R$, $(r_i r_j)^{m(r_i r_j)} = 1$ for all $r_i, r_j \in R$ where $m(r_i, r_j) \in \mathbb{N} \cup \infty$.

Let $w \in W$. The minimal number of factors in the decomposition $w = r_1 \cdots r_l$, where $r_i \in R$, is called the length of the element w and is denoted by $l(w)$. Partial Bruhat ordering on the group W is defined in the following way: $w_1 \leqq w_2$ if s_1 and $s_2 \in W$ exist such that $w_2 = s_1 w_1 s_2$ and $l(w_2) = l(s_1) + l(w_1) + l(s_2)$.

We associate a new ordering on W to each $w \in W$ in the following way: $w_1 \overset{w}{\leqq} w_2$ if $w^{-1}w_1 \leqq w^{-1}w_2$. Evidently, the Bruhat ordering coincides with $\overset{1}{\leqq}$.

Let L be an arbitrary subset of the group W. An element $s \in L$ is called w-minimal in L if for any $u \in L$ the inequality $s \overset{w}{\leqq} u$ holds. We say that L satisfies the minimality condition if for any $w \in W$ there exists a w-minimal element in L.

Definition 1. A pair (W, L), where W is a Coxeter group and L is a subset in W satisfying the minimality condition, is called a flag W-matroid.

The set L will be called the set of bases for a flag W-matroid (W, L).

Let Q be an arbitrary subset in R. Denote by W_Q the subgroup in W generated by the elements of Q. The subgroups W_Q are called parabolic subgroups of the group W. Denote by W^Q the set of left cosets W/W_Q. Let $\alpha \in W^Q$; then the left coset α considered as a subset in W satisfies the minimality condition. For an arbitrary class $\alpha \in W^Q$, denote by α_w the w-minimal element in α. Introduce a partial ordering $\overset{w}{\leqq}$ into the set W^Q by letting $\alpha \overset{w}{\leqq} \beta$ if $\alpha_w \overset{w}{\leqq} \beta_w$. Then the minimality condition makes sense for an arbitrary subset $L \subseteq W^Q$.

Definition 2. A triple (W, Q, L), where W is a Coxeter group, Q a set of generators and L is a subset in W^Q satisfying the minimality condition, is called a (W, Q)-matroid.

The set L is called the set of bases of that (W, Q)-matroid.

A flag W-matroid (W, L) will be called W_Q-invariant if $W_Q \cdot L = L$.

Proposition 1. *Let W be a Coxeter group, $Q \subset R$. The natural projection of W on W^Q gives a one-to-one correspondence between the set of W_Q-invariant flag W-matroids and the set of (W, Q)-matroids.*

8.2. Examples. Example 1. Let $W = S_n$ be the group of permutations of the set $I_n = \{1, \ldots, n\}$. The set R consists of transpositions $(i, i+1)$, $i = 1, \ldots, n-1$ and $Q = R \backslash (k, k+1)$. Then $W_Q = \{w \in W \mid w(I_k) = I_k\}$. We now show that in this case the definition of a (W, Q)-matroid coincides with that of a usual matroid.

The set W^Q of left cosets W/W_Q is naturally identified with the set $B_k(I_n)$ of all k-element subsets in I_n. The Bruhat ordering in W^Q is then of the form $A \leqq B$ for all $A, B \in B_k(I_n)$, $A = \{a_1, \ldots, a_k\}$, $B = \{b_1, \ldots, b_k\}$, where $a_1 < \cdots < a_k$, $b_1 < \cdots < b_k$, if $a_i \leqq b_i$ for all $i = 1, \ldots, k$ (see Sect. 2.6). An arbitrary element $w \in W$ defines another linear ordering in I_n, $w(1) < \cdots < w(n)$, which defines the ordering $\overset{w}{\leqq}$ in $B_k(I_n)$.

Proposition 2. *Let $L \subseteq B_k(I_n) = W^Q$. A triple (W, Q, L) is a (W, Q)-matroid if and only if L is a set of bases for some matroid (I_n, r) of rank k.*

The proof follows immediately from Proposition 5 of Sect. 2.

Example 2. Let W again be S_n and $Q = Q_l = \{(l+1, l+2), \ldots, (n-1, n)\}$. In this case, the notion of (W, Q_l)-matroid is closely connected with the notion of greedoid (see [20], [21]).

Consider a set of letters S (an alphabet). We consider only the case $|S| < \infty$ and identify S with I_n. An arbitrary sequence of letters from S is called a word. The length of a word α we denote by $|\alpha|$. An arbitrary set of words $\mathscr{L}$ is called a language. A language $\mathscr{L}$ is called a greedoid when the following conditions are satisfied:

(1) $\varnothing \in \mathscr{L}$;
(2) no word $\alpha \in \mathscr{L}$ contains any repeating letters;
(3) for any $\alpha \in \mathscr{L}$, if $\alpha = \beta\gamma$ then $\beta \in \mathscr{L}$;
(4) if $\alpha, \beta \in \mathscr{L}$ and $|\alpha| < |\beta|$ then $x \in \beta$ exists such that $\alpha x \in \mathscr{L}$.

Axiom (4) implies that all maximal words of a greedoid are of the same length, which is called the rank of a greedoid $\mathscr{L}$. Maximal words of a greedoid $\mathscr{L}$ are called bases.

Let $\mathscr{L}$ be a greedoid of rank l with the alphabet I_n. Then the set L of its bases is naturally identified with a subset of the set of left cosets W^{Q_l}. On the other hand, to each subset $L \subseteq W^{Q_l}$ there corresponds a language $\hat{L}$ consisting of the words obtained from the words of L by dropping the end elements.

Proposition 3. (a) *Let L be a set of bases of a (W, Q)-matroid. Then $\hat{L}$ is a greedoid of rank l.* (b) *Let $\mathscr{L}$ be a greedoid of rank l with the alphabet I_n such that $i \in \mathscr{L}$ for any $i \in I_n$. Then the set L of its bases is the set of bases for some (W, Q)-matroid.*

Example 3. Let $J_n = \{1, \ldots, n, 1^*, \ldots, n^*\}$, and let an involution $*$ be given in J_n such that $(i)^* = i^*$, $(i^*)^* = i$. A subset $A \subseteq J_n$ will be called admissible if $A \cap A^* = \varnothing$. The set of all admissible subsets in J_n will be denoted by $R(J_n)$, and the set of k-element admissible subsets by $R_k(J_n)$.

Let $L \subseteq R_k(J_n)$. A pair (J_n, L) is called a symplectic matroid of rank k provided the following conditions are satisfied:

(1) for any $A, B \in L$ and $a \in A \backslash B$ there exists $b \notin A$ such that either $(A \backslash \{a\}) \cup \{b\} \in L$ or $(A \backslash \{a, b^*\}) \cup \{a^*, b\} \in L$;
(2) for any $A, B \in L$ and $b \in B \backslash (A \cup A^*)$ there exists $a \in A$ such that $(A \backslash \{a\}) \cup \{b\} \in L$.

The set L is called the set of bases of the symplectic matroid (J_n, L).

Let W be the group of permutations of the set J_n commuting with the involution $*$, i.e. the Weyl group of the Lie algebras $sp(2n)$ and $o(2n+1)$. Then R consists of permutations of the form $r_i = (i, i+1) \cdot (i^*, (i+1)^*)$ $(i = 1, \ldots, n-1)$ and of the transposition $r_n = (n, n^*)$. Let $Q = R \backslash \{r_k\}$. Then $W_Q = \{w \in W \mid w(I_k) = I_k\}$ and the set W^Q is naturally identified with $R_k(J_n)$. A linear ordering on the set J_n, $1 < \cdots < n < n^* < \cdots < 1$, induces a partial ordering on $R_k(J_n)$ in a similar fashion to Example 1. The identification of $R_k(I_n)$ with W^Q turns it into a Bruhat ordering. Exactly as in Example 1, any $w \in W$ gives another ordering in $R_k(J_n)$.

Proposition 4. *Let $L \subseteq R_k(J_n) = W^Q$. A triple (W, Q, L) is a (W, Q)-matroid if and only if L is a set of bases for some symplectic matroid of rank k.*

It turns out that symplectic matroids can be defined through the rank function in a similar fashion to usual matroids (see Sect. 2.1). The role of the Boolean algebra $B(I_n)$ is then played by the set of admissible sets $R(J_n)$ which is made a lattice after the maximal element 1 is added to it. We associate a rank function $r: R(J_n) \to \mathbb{Z}$ to each symplectic matroid (J_n, L) by setting $r(A) = \max_{B \in L} |A \cap B|$.

A symplectic matroid (J_n, L) is said to be without loops if, for any $a \in J_n$, $A \in L$ exists such that $a \in A$.

Proposition 5. (a) *Let (J_n, L) be a symplectic matroid without loops. Then its rank function satisfies the following conditions:*

(1) $0 < r(A) \leqq |A|$ *for any* $A \in R(J_n) \setminus \varnothing$;

(2) $r(A) \leqq r(B)$ *for* $A \subseteq B$;

(3) $r(A) + r(B) \geqq r(A \cap B) + r(A \cup B)$ *for any* A, $B \in R(J_n)$ *such that* $A \cup B \in R(J_n)$.

(b) *Conversely, let $r: R(J_n) \to \mathbb{Z}$ satisfy conditions* (1)–(3). *Then r is a rank function for some symplectic matroid (J_n, L) without loops.*

Remark. The lattice $R(J_n) \cup 1$ is dual to the lattice formed by the faces of the n-dimensional cube. This lattice can be defined axiomatically [22]. Other interesting non-distributive lattices are apparently associated to matroids corresponding to other Coxeter groups.

Example 4. A symplectic matroid of rank k with the set of bases L is called orthogonal if for any $A \in L$ and $a \in A$ such that $(A \setminus \{a\}) \cup \{a^*\} \in L$ there exists $b \in A \cup A^*$ for which $(A \setminus \{a\}) \cup \{b\} \in L$, $(A \setminus \{a\}) \cup \{b^*\} \in L$.

Let W be the group of those permutations of J_n that commute with the involution $*$ such that $|w(I_n) \setminus I_n|$ is even for any $w \in W$ (the Weyl group of the Lie algebra $O(2n)$). In this case $R = \{r_1, \ldots, r_n\}$, where $r_1, \ldots, r_{n-1}$ are the same as in Example 3, and $r_n = (n^*, n-1) \cdot (n, (n-1)^*)$. Let $Q = R \setminus r_k$. One has $W_Q = \{w \in W \mid w(I_k) = I_k\}$ for $k \neq n-1$ and $W_Q = \{w \in W \mid w(I') = I'\}$ for $k = n-1$ where $I' = \{1, \ldots, n-1, n^*\}$. The set of left cosets W^Q will be identified with $R_k(J_n)$ for $k < n-1$ and with $R_n^+(J_n)$ (or with $R_n^-(J_n)$) for $k = n$ (or for $k = n-1$, respectively) where $R_n^{+(-)}(J_n) = \{X \in R_n(J_n) \mid |X \setminus I_n| = 0(1) \bmod 2\}$. The set $R_{n-1}(J_n)$ will be identified with W^Q for $Q = R \setminus \{r_{n-1}, r_n\}$. With this identification, the Bruhat ordering on the set $R_k(J_n)$ is induced (in the same way as in Examples 1 and 3) by the following partial ordering on J_n:

$$1 < \cdots < n, \quad n^* < \cdots < 1^*, \quad 1 < \cdots < n-1 < n^*,$$

where there is no relation between n and n^*.

Proposition 6. *Let $L \subseteq R_k(J_n) = W^Q$. A triple (W, Q, L) is a (W, Q)-matroid if and only if L is a set of bases for some orthogonal matroid of rank k.*

8.3. Let W be the Weyl group of a Lie group G. Let $Q \subseteq \sigma_1, \ldots, \sigma_n$ where $\sigma_1, \ldots, \sigma_n \in \mathfrak{h}_{\mathbb{R}}^*$ is a system of simple roots. Consider a point $\omega_Q \in \mathfrak{h}_{\mathbb{R}}^*$ defined by the condition

$$\frac{(\omega_Q,\sigma_i)}{(\sigma_i,\sigma_i)}=\begin{cases}1 & \text{for } \sigma_i\notin Q,\\ 0 & \text{for } \sigma_i\in Q.\end{cases}$$

Since the group W_Q is a stationary subgroup of the point ω_Q there is a mapping $\bar{\mu}\colon W^Q\to\mathfrak{h}_{\mathbb{R}}^*$ taking a left coset $w\cdot W_Q$ into the point $w(\omega_Q)$. We associate to each subset $L\subseteq W^Q$ a polyhedron Δ_L which is the convex hull of the points $\bar{\mu}(L)$. Let P be a standard parabolic subgroup of the group G for which $S_P=Q$ (see Sect. 4.2). Then $\omega_Q=\lambda$, and the moment mapping μ maps $M=G/P$ onto Δ_{W^Q}.

The mapping $\bar{\mu}$ identifies the elements of the set W^Q with the points of the orbit $W\cdot\omega_Q$. This identification makes it possible to define the partial ordering $\overset{w}{\leqq}$ in a geometrical way. Let C_w be a convex cone in $\mathfrak{h}_{\mathbb{R}}^*$ consisting of vectors $y=\sum_{i=1}^n m_i w(\sigma_i)$ such that $m_i\geqq 0$ for all $i=1,\dots,n$. The ordering $\overset{w}{\leqq}$ is then defined by setting $x\overset{w}{\leqq}y$ if $y-x\in C_w$. The restriction of this ordering to the set $W\cdot\omega_Q$ coincides with the ordering $\underset{w}{\leqq}$ on W^Q.

Theorem 1. *Let W be the Weyl group of a semi-simple group G, and let P be its standard parabolic subgroup. The following conditions are equivalent*:

(1) *(W,S_P,L) is a (W,S_P)-matroid,*
(2) *Δ_L is a (G,P)-hypersimplex.*

Proof. (1)⇒(2). Let (W,S_P,L) be a (W,S_P)-matroid. Suppose that Δ_L is not a hypersimplex. Then an edge l exists with vertices α and β such that there is no root to which l is parallel. Consider a linear function χ in $\mathfrak{h}_{\mathbb{R}}^*$ constant on l and taking greater values in all other points of the polyhedron Δ_L. A unique system of simple roots $\tilde{\sigma}_1,\dots,\tilde{\sigma}_n$ exists for which $\chi(\tilde{\sigma}_i)>0$. Since the Weyl group acts transitively on simple root systems, $w\in W$ exists, taking $\{\sigma_1,\dots,\sigma_n\}$ into $\{\tilde{\sigma}_1,\dots,\tilde{\sigma}_n\}$. Then for any $\gamma\neq\alpha,\gamma\in L$, the vector $\alpha-\gamma$ has at least one negative coefficient in its decomposition with respect to the basis $\tilde{\sigma}_1,\dots,\tilde{\sigma}_n$. The same is valid for β. Thus, there is no w-minimal element in L, contradicting the assumption.

(2)⇒(1). Let Δ_L be a hypersimplex and w an arbitrary element in W. Let $\tilde{\sigma}_1=w(\sigma_1),\dots,\tilde{\sigma}_n=w(\sigma_n)$. Choose a linear function χ on $\mathfrak{h}_{\mathbb{R}}^*$ in such a way that $\chi(\tilde{\sigma}_i)>0$ for all $i=1,\dots,n$. Let α be a vertex of the polyhedron Δ_L in which χ has the minimal value. Let $\beta_1,\dots,\beta_n$ be vertices on Δ_L having a common edge with α. Since each edge of Δ_L is parallel to some root, one has $\beta_i-\alpha=k_i\gamma_i$ where γ_i are roots and $k_i\geqq 0$. Then for any $\beta\in L$ we have $\beta-\alpha=\sum_{i=1}a_ik_i\gamma_i$ where $a_i\geqq 0$.

Since $\chi(\alpha)$ is the minimum of χ one has $\chi(\gamma_i)\geqq 0$. Therefore all the roots are γ_i-positive with respect to the simple root system $\tilde{\sigma}_1,\dots,\tilde{\sigma}_n$. This means that $\beta-\alpha=\sum b_i\tilde{\sigma}_i$ where $b_i\geqq 0$. Therefore, $\beta\overset{w}{\geqq}\alpha$. Thus α is a w-minimal element.

Theorem 1 and Theorem 1 of Sect. 7 imply the following:

Theorem 2. *Let Γ be an arbitrary stratum in G/P with the list L_Γ. Then (W, S_P, L) is a (W, S_P)-matroid.*

Example 5. Consider a polyhedron Δ_W corresponding to an open dense stratum on the manifold G/B. The polyhedron Δ_W has been considered in the papers [23], [12]. One can show that there is a one-to-one correspondence between the set of its faces and the set of all left cosets with respect to all parabolic subgroups in $W: \bigcup_{Q \subseteq R} W^Q$. The dimension of a face corresponding to a coset $w \cdot W_Q$ equals $|Q|$. In combinatorics one calls the polyhedron Δ_W a Coxeter complex (see [24]).

For $W = S_3 \Delta_W$ is a regular hexahedron; for $W = S_4$ it is a semi-regular polyhedron in $\mathbb{R}^3$ having 24 vertices, 8 hexagonal and 6 square faces.

Example 6. Let (J_n, L) be a simplectic matroid of rank k. Then the hypersimplex Δ_L has its vertices in some of the vertices of the cube $E_n = \{\xi \in R^n \mid |\xi_i| \leqq 1\}$, and its edges coincide either with edges or with diagonals of two-dimensional faces of the cube. A simplectic matroid corresponding to Δ_L is orthogonal if and only if all the edges of the polyhedron Δ_L are diagonals of the two-dimensional faces of the cube.

8.4. Non-degenerate (W, Q)-matroids and strata. Let (W, Q, L) be a (W, Q)-matroid. A subgroup $\bar{W} \subseteq W$ conjugate to a parabolic subgroup of the group W is called a separator of the (W, Q)-matroid if $L \subseteq \bar{W} \cdot \alpha$ for some $\alpha \in L$.

Proposition 7. *Let $\bar{W}$ be a separator of a (W, Q)-matroid (W, Q, L), $L \subseteq W \cdot w \cdot W_Q$, $W_{\bar{Q}} = \bar{W} \cap w W_Q w^{-1}$, $\bar{L} = \{\alpha w^{-1} \cap \bar{W} \mid \alpha \in L\}$. Then $(\bar{W}, \bar{Q}, \bar{L})$ is a $(\bar{W}, \bar{Q})$-matroid.*

A (W, Q)-matroid is said to be non-degenerate if it has no separators other than W. A stratum $\Gamma \subset G/P$ is called non-degenerate if Γ consists of orbits of the torus H of maximum dimension.

Proposition 8. *A stratum $\Gamma \subset G/P$ is non-degenerate if and only if the corresponding (W, S_P)-matroid is also non-degenerate.*

Any non-degenerate stratum $\Gamma \subset G/P$ is isomorphic to a stratum of the form $\Gamma' \subset G'/P'$, where G' is a regular semi-simple subgroup of the group G and $P' = P \cap G'$.

9. Examples of strata

9.1. Strata on the flag manifolds. Denote by $F_{k_1, \ldots, k_m}(E)$ $(k_1 < \cdots < k_m)$ the manifold of flags of subspaces $(X_1 \subset \cdots \subset X_m)$ of dimensions $k_1, \ldots, k_m$, respectively, in the complex n-dimensional space E. In particular, $G_k(E) = F_k(E)$. Choose a basis $\{e_1, \ldots, e_n\}$ in the space E, identifying in this way E with $\mathbb{C}^n$ and the subgroup H in $GL(n)$ diagonal in this basis. Consider different ways of defining a stratum in this case.

Theorem 1 of Sect. 6 implies that any stratum Γ in $F_{k_1, \ldots, k_m}(\mathbb{C}^n)$ is an intersection of Schubert cells for all coordinate flags. We remained the reader that the Schubert cell corresponding to a coordinate flag $\mathbb{C}^{I_1} \subset \cdots \subset \mathbb{C}^{I_n}$, $I_j \in B_j(I_n)$ consists of all the flags $(X_1, \ldots, X_m)$ such that $\dim X_i \cap \mathbb{C}^{I_j}$ has a fixed value for all $i = 1, \ldots, m$, $j = 1, \ldots, n$.

Thus, any stratum Γ on a flag manifold is defined by dimensions of intersections of the flag subspaces with the coordinate subspaces.

As for Grassmannians, a stratum Γ will be defined by the functions $r_1, \ldots, r_m$: $B(I_n) \to \mathbb{Z}$ given by the formula $r_i(J) = \dim(X_i / X_i \cap \mathbb{C}^{I_n \setminus J})$ for any $(X_1, \ldots, X_m) \in \Gamma$. Consider an embedding

$$F_{k_1, \ldots, k_m}(\mathbb{C}^n) \subsetneq G_{k_1}(\mathbb{C}^n) \times \cdots \times G_{k_m}(\mathbb{C}^n).$$

In Sect. 1.4 we described a principal bundle $q \colon G_{n,k} \to G_k(\mathbb{C}^n)$. Denote by q the natural bundle

$$C_{n,k_1} \times \cdots \times C_{n,k_m} \to G_{k_1}(\mathbb{C}^n) \times \cdots \times G_{k_m}(\mathbb{C}^n).$$

Let us now describe $q^{-1}(F_{k_1, \ldots, k_m}(\mathbb{C}^n))$.

Definition 1. A sequence of n-tuples of vectors $\{x_j^i, i = 1, \ldots, n. j = 1, \ldots, m\}$ lying in the spaces $F_1, \ldots, F_m$ of dimensions $k_1, \ldots, k_m$, respectively, is said to be compatible if the condition $\sum_{1 \leqq j \leqq n} \lambda_j x_p^j = 0$ implies $\sum_{1 \leqq j \leqq n} \lambda_j x_i^j = 0$ for all $i \leqq p$.

The submanifold of all compatible sequences in $C_{n,k_1} \times \cdots C_{n,k_m}$ will be denoted by $CF_{n,k_1, \ldots, k_m}$. One can easily see that $q^{-1}(F_{k_1, \ldots, k_m}(\mathbb{C}^n)) = CF_{n,k_1, \ldots, k_m}$.

The bundle $q \colon CF_{n,k_1, \ldots, k_m} \to F_{k_1, \ldots, k_m}(\mathbb{C}^n)$ is reduced (in the same way as for Grassmannians) to the bundle

$$\tilde{q} \colon CF_{n,k_1, \ldots, k_m}/H \to F_{k_1, \ldots, k_m}(\mathbb{C}^n)/H.$$

While for Grassmannians strata are matroids, the above considerations show that on flag manifolds a stratum can be defined by a sequence of matroids.

Definition 2. A sequence of matroids $(S, r_1), \ldots, (S, r_m)$ together with a sequence of families of flats $\mathrm{Fl}_1, \ldots, \mathrm{Fl}_m$ is called compatible if $\mathrm{Fl}_1 \subseteq \cdots \subseteq \mathrm{Fl}_m$.

Proposition 1. *Let Γ be a stratum in $F_{k_1, \ldots, k_m}(\mathbb{C}^n)$ given by a sequence of functions $r_1, \ldots, r_m$. Then the sequence of matroids $(I_n, r_1), \ldots, (I_n, r_m)$ is compatible.*

The proof is obvious.

For any $A \subseteq S$ denote by $\bar{A}^i$ the closure of A in the matroid (S, r_i), i.e. $A^i = \bigcap_{C \in \mathrm{Fl}_i, A \subseteq C} C$.

Lemma 1. *Let $(S, r_1), \ldots, (S, r_m)$ be a compatible sequence of matroids of rank $k_1, \ldots, k_m$, respectively. Then*

(a) *$A^i \subseteq A^j$ for any $A \subseteq S$ and $j \leqq i$,*

(b) *$r_1 \leqq \cdots \leqq r_m$.*

Proof. (a) follows from the definition.

(b) Suppose that i and j, $i < j$, such that $r_i(J) > r_j(J)$ for some $J \subseteq S$. Take a set

J containing the minimal number of elements and denote it by J_0. Note that $J_0 \neq \emptyset$. Then $r_i(J_0 \backslash b) = r_j(J_0 \backslash b)$ for any $b \in J_0$. Therefore $b \in \overline{(J_0 \backslash b)}^j$ and $b \notin \overline{(J_0 \backslash b)}^i$ which contradicts (a).

Let $(X_1, \ldots, X_m) = X$ be a flag in $\mathbb{C}^n$ and p_i^J the Plücker coordinates of X_i. We call the numbers

$$p^{J_1, \ldots, J_m}(X) = \prod_{i=1}^{m} p_i^{J_i}$$

where $J_i \in B_{k_i}(I_n)$, $J_1 \subset \cdots \subset J_m$ the generalised Plücker coordinates of the flag X. One can uniquely reconstruct the Plücker coordinates for each X_i from a set of the generalised Plücker coordinates.

For each flag $X \in F_{k_1, \ldots, k_m}(\mathbb{C}^n)$ the set

$$L_X = \{(J_1, \ldots, J_m) \mid p^{J_1, \ldots, J_m}(X) \neq 0\}$$

will be called its list. It follows from Sect. 4 that any stratum Γ in $F_{k_1, \ldots, k_m}(\mathbb{C}^n)$ can be defined by a list L^Γ:

$$\Gamma = \{X \in F_{k_1, \ldots, k_m}(\mathbb{C}^n) \mid L_X = L^\Gamma\}.$$

Let us define a list of bases $\mathscr{B}$ for a compatible sequence of matroids $(S, r_1), \ldots, (S, r_m)$ in the following way:

$$\mathscr{B} = \{(B_1, \ldots, B_m) \mid B_1 \subset \cdots \subset B_m,\ r_i(S) = r_i(B_i) = |B_i|\}.$$

Hence, similarly to Sect. 2.2, where we associated to each matroid a hypersimplex, we can associate to each compatible sequence of matroids a polyhedron $\Delta_{\mathscr{B}}$ in $\mathscr{L}(S)$, namely, Vert $\Delta_{\mathscr{B}} = \{\delta_{B_1} + \cdots + \delta_{B_m} \mid (B_1, \ldots, B_m) \in \mathscr{B}\}$.

Let $(S, r_1), \ldots, (S, r_m)$ be a compatible sequence of matroids of ranks $k_1, \ldots, k_m$, respectively. Define the function r on S in the following way: $r = r_1 + \cdots r_m$. The function r is monotone, $r(\emptyset) = 0$, and $r(A \cup B) + r(A \cap B) \leqq r(A) + r(B)$. Thus, a pair (S, r) is, by definition, a supermatroid (see [8]). Evidently, two different compatible sequences of matroids define different supermatroids.

Consider a supermatroid (S, r) and a set of integers $k_1 < \cdots < k_m$ such that $r(S) = \sum_{i=1}^m k_i$. When can a supermatroid (S, r) be obtained from a compatible sequence of matroids of ranks $k_1, \ldots, k_m$? Let $\mathscr{B}_{k_1, \ldots, k_m}(I_n, r) = \{(B_1, \ldots, B_m) \mid B_i \in B_{k_i}(I_n), B_1 \subset \cdots \subset B_m, r(B_i) = \sum_{1 \leqq j \leqq i} k_j + k_i(m - i)\}$. Choose a linear ordering in S, identifying S in this way with I_n.

Proposition 2. *Let (I_n, r) be a supermatroid and $k_1 < \cdots k_m$ a set of integers such that $r(I_n) = k_1 + \cdots + k_m$. Then the following conditions are equivalent*:

(a) *A compatible sequence of matroids $(I_n, r_1), \ldots, (I_n, r_m)$ of ranks $k_1, \ldots, k_m$ exists such that $r = r_1 + \cdots + r_m$.*

(b) *$B_{k_1, \ldots, k_m}(I_n, r)$ is a general matroid (see Example 2 in Sect. 8.3).*

Proof. (a)$\Rightarrow$(b). Let $\mathscr{B}_i$ be the set of bases of the matroid (I_n, r_i). Then, evidently, $\mathscr{B}_{k_1, \ldots, k_m}(I_n, r) = \{(B_1, \ldots, B_m) \mid B_i \in \mathscr{B}_i,\ B_1 \subset \cdots \subset B_m\}$. Let s be an arbitrary linear ordering on the set I_n. Then there is an s-minimal element C_i in each family of bases $\mathscr{B}_i$ (see Sect. 2.6). We now show that $C_1 \subset \cdots \subset C_m$. Suppose that this is not the case. Then $C_i \not\subset C_{i+1}$ for some $i < m$. Let $A = C_i \cap C_{i+1}$, and let c be an s-minimal element in $C_i \backslash C_{i+1}$. Then $C_{i+1} \backslash C_i = D_0 \cup D_1$, where all the elements of D_0 are less than c (in the sense of s-ordering) and all the elements of D_1 are greater than c. Then s-minimality of C_i implies that $(C_i \backslash d) \cup b_0 \notin \mathscr{B}_i$ for any $b_0 \in D_0$, $d \in C_i \backslash C_{i+1}$. Hence $D_0 \in \bar{A}^i$. Similarly, $(C_{i+1} \backslash b_1) \cup c \notin \mathscr{B}_{i+1}$ for any $b_1 \in D_1$. Hence $c \in \overline{(C_{i+1} \backslash D_0)}^{i+1}$. Since $\overline{(C_{i+1} \backslash D_0)}^{i+1} \subseteq \overline{(C_{i+1} \backslash D_0)}^i$ one has $c \in \overline{(C_{i+1} \backslash D_0)}^i = \overline{(A \cup D_0)}^i$. But $D_0 \subset \bar{A}^i$ and, consequently, $c \in \bar{A}^i$, which is an impossibility because $A \cup c$ is an independent subset of the matroid (S, r_i). Therefore, $C_1 \subset \cdots \subset C_m$. But then $(C_1, \ldots, C_m)$ is an s-minimal element in $\mathscr{B}_{k_1,\ldots,k_m}(I_n, r)$.

(b)$\Rightarrow$(a). Let $\mathscr{B}_i = \{B \in B_{k_i}(I_n) |$ and let $(B_1, \ldots, B_{i-1}, B, B_{i+1}, \ldots, B_m) \in \mathscr{B}_{k_1,\ldots,k_m}(I_n, r)\}$. exist. Since for any $s \in S_n$ there is an s-minimal element in $\mathscr{B}_i$, then B_i is a set of bases for some matroid (I_n, r_i) (see Sect. 2.6). Let Fl_i be the set of all flats for the matroid (I_n, r_i). Then for any $A \in \mathrm{Fl}_i$ and $b \in I_n \backslash A$, $B \in \mathscr{B}_i$ exists such that $b \in B$ and $r_i(A) = |A \cap B|$. Let $|A| = p$. Let $s \in S_n$ be such that $s(I_n) = A$, $s(p+1) = b$. Let $(C_1, \ldots, C_m)$ be an s-minimal element in $\mathscr{B}_{k_1,\ldots,k_m}(I_n)$. Then $C_i \overset{s}{\leqq} B$, which implies that $r_i(A) = |A \cap C_i|$ and $b \in C_i$. It is now clear that $r_j(A) = |A \cap C_j|$ and $b \in C_j$ for any $j \geqq i$. Since $b \in I_n \backslash A$ is an arbitrary element, the definition of a flat implies that $A \in \mathrm{Fl}_j$ for any $j \geqq i$. Thus we have proved that $\mathrm{Fl}_1 \subset \cdots \subset \mathrm{Fl}_m$, i.e. (b)$\Rightarrow$(a).

9.2. Strata for Grassmannians corresponding to the orthogonal and symplectic groups. (a) Let $G = SO(2n)$. G acts in the $2n$-dimensional space E leaving invariant a non-degenerate scalar product $(,)$. Choose a basis $\{e_1, \ldots, e_n, f_1, \ldots, f_n\}$ such that $(e_i, e_j) = (f_i, f_j) = 0$, $(e_i, f_j) = \delta_{ij}$, identifying in this way E with $\mathbb{C}^{2n}$. We fix a Cartan subgroup that is H-diagonal in this basis. Consider the Grassmannian $I_k(\mathbb{C}^{2n})$ of k-dimensional isotropic subspaces in $\mathbb{C}^{2n}$ $(k \leqq n)$. For $k = n$ we take for $I_k(\mathbb{C}^{2n})$ a connected component of the Grassmannian of isotropic subspaces.

Recall that J_n is the set consisting of $2n$ elements $1, \ldots, n, 1^*, \ldots, n^*$ with involution $*: J_n \to J_n$. A subset $A \subseteq J_n$ is called admissible if $A \cap A^* = \varnothing$ (see Example 3 of Sect. 8.3). Denote by $R(J_n)$ the set of all admissible subsets J_n and by $\bar{R}(J_n)$ the set of subsets complementary to the admissible subsets. For any $A \subseteq J_n$ denote by $\mathbb{C}^A$ the coordinate subspace in $\mathbb{C}^{2n}$ spanned by the basis vectors e_i, f_j for all $i \in A$, $j^* \in A$. If $A \in R(J_n)$ then $\mathbb{C}^A$ is isotropic.

Proposition 3. *An arbitrary stratum Γ in $I_k(\mathbb{C}^{2n})$ is defined by the function $r_\Gamma: R(J_n) \to \mathbb{Z}$ in the following way:*

$$\Gamma = \{X \in I_k(\mathbb{C}^{2n}) | \dim (X / X \cap \mathbb{C}^{J_n \backslash A}) = r_\Gamma(A)\}.$$

Proof. By Theorem 1 of Sect. 6 each stratum is an intersection of Schubert cells $\cap C \cdot \alpha(C)$ corresponding to all Borel subgroups C containing a given Cartan subgroup H. The subgroups C are numbered by the coordinate flags $\mathbb{C}^{D_1} \subset \cdots \subset \mathbb{C}^{D_n} \subset \mathbb{C}^{D_n \cup D_1^*} \subset \cdots \subset \mathbb{C}^{D_n \cup D_n^*}$ where $D_i \in R(J_n)$, $|D_i| = i$, and the corresponding Schubert cells $C \cdot \alpha$ are defined by the dimensions of the intersections of $X \in C \cdot \alpha$ with the subspaces $\mathbb{C}^{D_i}$ and $\mathbb{C}^{D_n \cup D_i^*}$. This easily implies Proposition 3.

One can show that the dimension $\dim(X \cap \mathbb{C}^A)$ for $A \notin R(J_n) \cup \bar{R}(J_n)$ can change within the stratum Γ.

The Grassmannian $I_k(\mathbb{C}^{2n})$ can be realised as an orbit of the highest weight vector in $P(\Lambda^k(\mathbb{C}^{2n}))$. Thus, any $X \in I_k(\mathbb{C}^{2n})$ can be represented in the form

$$X = c \cdot \sum_{J \in B_k(J_n)} p^J(X) e_J \quad \text{where} \quad e_J = e_{j_1} \wedge \cdots \wedge e_{j_k}$$

and $J = \{j_1, \ldots, j_k\}$.

The generalised Plücker coordinates (see Sect. 4.2) are given in this case by the set of numbers $\{p^J(X) | J \in R_k(J_n)\}$ defined up to a scalar multiple. Thus, the list of a subspace $X \in I_k(\mathbb{C}^{2n})$ consists of admissible k-element subsets in J_n.

As we have shown in Sect. 1.3, any k-element subspace in $\mathbb{C}^{2n}$ is defined by a configuration of $2n$ vectors $x^1, \ldots, x^n, \tilde{x}^1, \ldots, \tilde{x}^n$ in the k-dimensional space F^*. A configuration $(x^1, \ldots, x^n, \tilde{x}^1, \ldots, \tilde{x}^n)$ defines an isotropic subspace in $\mathbb{C}^{2n}$ if and only if

$$\sum_{1 \leq i \leq n} x^i \cdot \tilde{x}^i = 0 \qquad (2)$$

where $x^i \cdot \tilde{x}^i = x^i \otimes \tilde{x}^i + \tilde{x}^i \otimes x^i$ is a vector in $S^2(F^*)$.

Denote the set of $2n$-tuples of vectors in the k-dimensional space satisfying equation (2) by $CI_{2n,k}$. Then, similarly to the case of Grassmannians, one has a $GL(k)$-bundle $q: CI_{2n,k} \to I_k(\mathbb{C}^{2n})$.

(b) The case $G = \mathrm{SP}(2n)$ is similar to that of $\mathrm{SO}(2n)$. The only difference is that a symmetric form $(,)$ in E is replaced by a skew-symmetric one $\langle , \rangle$. Any isotropic k-dimensional subspace in E with respect to the form $\langle , \rangle$ can be defined by a configuration of $2n$ vectors $x^1, \ldots, x^n, \tilde{x}^1, \ldots, \tilde{x}^n$ in a k-dimensional space satisfying the equation

$$\sum_{1 \leq i \leq n} x^i \wedge \tilde{x}^i = 0.$$

(c) Let $G = \mathrm{SO}(2n+1)$. Then G is the group of linear transformations of the space E leaving invariant a non-degenerate symmetric bilinear form $(,)$. Take for G/P, where P is a maximal parabolic subgroup, the Grassmannian $I_k(E)$ of k-dimensional isotropic subspaces in E. Choose a basis $\{e_1, \ldots, e_n, f_1, \ldots, f_n, g_0\}$ in E such that $(e_i, e_j) = (f_i, f_j) = (g_0, e_i) = (g_0, f_j) = 0$, $(g_0, g_0) = 1$, $(e_i, f_j) = \delta_{ij}$. For H let us take the subgroup of matrices diagonal in this basis.

It can be shown that each stratum Γ in $I_k(E)$ is defined by the function $r_\Gamma: R(J_n) \to \mathbb{Z}$ as in Proposition 3. The proof makes use of the fact that $\dim(X \cap \mathbb{C}^A) = \dim(X \cap (\mathbb{C}^A \oplus \mathbb{C} g_0))$ for any $X \in I_k(E)$ and $A \in R_n(J_n)$. The generalised Plücker coordinates and the list L_X of an isotropic space X are also defined on the same lines.

We also note that an arbitrary k-dimensional isotropic subspace X in G^{2n+1} can be defined by a configuration of $2n+1$ vectors $z, x^1, \ldots, x^n, \tilde{x}^1, \ldots, \tilde{x}^n$ in $\mathbb{C}^k$ satisfying the condition

$$\sum_{1 \leq i \leq n} x^i \tilde{x}^i + 2z \cdot z = 0.$$

Note that the vector $\pm Z$ is uniquely defined by $x^1, \ldots, x^n, \tilde{x}^1, \ldots, \tilde{x}^n$.

Fig. 8.

9.3. Grassmannians and Strata for G_2. There are three standard parabolic subgroups in G_2. We now describe the corresponding homogeneous spaces.

Let $\mathbb{Q}$ be a Cayley algebra over $\mathbb{C}$. Choose a basis $\{1, e_0, e_1, e_2, e_3, f_1, f_2, f_3\}$ in $\mathbb{Q}$ satisfying the following relations:

$$e_0 \cdot f_q = -f_q \cdot e_0 = -f_q; \quad e_0 \cdot e_q = -e_q \cdot e_0 = e_q; \quad e_0^2 = 1; \quad e_q^2 = f_q^2 = 0;$$
$$e_p \cdot f_q = (1 - e_0)/2\delta_{pq}; \quad e_i \cdot e_j = -e_j \cdot e_i = f_k; \quad f_i \cdot f_j = -f_j \cdot f_i = e_k,$$

where (i, j, k) is a cyclic permutation of $(1, 2, 3)$, $q \neq 0$, $p \neq 0$.

Let E be a subspace in $\mathbb{Q}$ spanned by $e_0, \ldots, e_3, f_1, \ldots, f_3$. G_2 is the group of automorphisms of Q, and E is its faithful irreducible representation [13].

Let P_1, P_2 be maximal parabolic subgroups in G_2 corresponding to simple roots of the numbers 1 and 2, respectively (see Fig. 8). Then the homogeneous spaces G_2/P_1 and G_2/P_2 can be realised as sets of one-dimensional (two-dimensional respectively) subalgebras with trivial multiplication in E. The homogeneous space G_2/B where $B = P_1 \cap P_2$ is the Borel subgroup in G_2 is realised as the space of flags (X_0, X_1) such that $X_0 \subset X_1 \subset E$ and $X_1 \cdot X_1 = 0$. Note that $G_2/P_1 \cong I_1(\mathbb{C}^7)$.

One can easily verify that points in G_2/P_i are defined by configurations of seven vectors $z, x_1, x_2, x_3, y_1, y_2, y_3$ in $\mathbb{C}^i$ satisfying conditions

$$\sum_{1 \leq j \leq 3} x_i \cdot y_i + 2z \cdot z = 0; \quad x_i \wedge x_j = y_k \wedge z;$$
$$\sum_{1 \leq j \leq 3} x_i \wedge y_i = 0; \quad y_i \wedge y_j = x_k \wedge z,$$

where (i, j, k) is a cyclic permutation of $(1, 2, 3)$.

Let $\mathscr{K}_3 = \{1, 2, 3, \tilde{1}, \tilde{2}, \tilde{3}\}$. One-element subsets in $\mathscr{K}_3$ and two-element subsets of the form $\{p, \tilde{j}\}$ where $p \neq j$. Admissible subsets and they only correspond to coordinate subalgebras with trivial multiplication $\mathbb{C}^J \subset \mathbb{C}^7 = E$, i.e. to subalgebras spanned by the basis vectors. It can be shown that any stratum in G_2/P_i is given by a function r_Γ, defined on admissible subsets in $\mathscr{K}_3$, namely,

$$\Gamma = \{X \in G_2/P_i \mid \dim(X/X \cap \mathbb{C}^{\mathscr{K}_3 \setminus I}) = r_\Gamma(I).$$

References

1. Danilov, I.V.: Geometry of toric varieties. Usp. Mat. Naūk 33 (1978) 2 (Russian)
2. Atiyah, M.F.: Convexity and commuting hamiltonians. Bull. London Math. Soc. **14** (1982) 1–15
3. Guillemin, V., Sternberg, S.: Convexity properties of the moment mapping. Invent. Math. **67** (1982) 491–513
4. Gelfand, I.M. MacPherson R.: Geometry in Grassmannians and a generalisation of the dilogarithm. Adv. Math. **44** (3) (1982) 279–312
5. Khovansky, A.G.: Sections of polyhedra, toric varieties and discrete groups of the Lobachevsky space. Funct. Anal. Appl. **20** (1) (1986) 50–61 (Russian)
6. Aigner, M.: Combinatorial theory. Mir, Moscow, 1982 (Russian translation)

7. Welsh, D.J.A.: Matroid theory. Academic Press, 1976
8. Crapo, H.H., Rota G.C.: On the foundation of combinatorial theory. MIT Press, Cambridge, Mass., 1970
9. Bourbaki, N.: Groupes et algebres de Lie. Hermann, Paris, 1960–1975.
10. Brugnières A.: Propriétés de convexité de l'application moment. Seminaire N. Bourbaki 1985 Vol. 1, p. 654
11. Gelfand, I.M., Zelevinskij, A.V.: Algebraic and combinatorial aspects of the general theory of hypergeometric functions. Funct. Analy. Appl. **20** (3) (1986) 17–35 (Russian)
12. Klyachko, A.A.: Orbits of the maximal torus on the flag space. Funct. Anal. Appl. **19** (1) (1985) 77–78 (Russian)
13. Jacobson, N.: Lie Algebras. Wiley, New York, 1962
14. Griffits P. Harris J. Principles of Algebraic Geometry. Wiley, New York, 1978
15. Gelfand, I.M.: General theory of hypergeometric functions. Dokl. Akad Naūk SSSR **228** (1) (1986) 14–18 (Russian)
16. Kirwan, F.: Cohomology of quotients in simplectic and algebraic geometry. Princeton University Press, Princeton, N.J. 1984
17. Gelfand, I.M., Goresky, R.M., MacPherson, R.D., Serganova, V.V.: Combinatorial geometries, convex polyhedra, and Schubert cells Adv. Math. (1986)
18. Gelfand, I.M., Gelfand, S.I.: Generalised hypergeometric equations. Dokl. Akad. Naūk SSSR **288** (1) (1986) 279–283 (Russian)
19. Gelfand, I.M., Graev, M.I.: Duality theorem for general hypergeometric functions. Dokl. Akad. Naūk SSSR **289** (1) (1986) 19–23 (Russian)
20. Korte, B., Lovasz, L.: Non-interval greedoids and the transposition property. Discrete Math. **59** (1986) 297–314
21. Crapo, H.H: Selectors: a theory of formal languages, semimodular lattices, and branching and shelling processes. Adv. Math. **54** (1984) 233–277
22. Metropolis, N., Rota, G.C.: Combinatorial structure of the faces of the *n*-cube. SIAM J. Appl. Math. **54** (3) (1978) 689–694
23. Gabrielov, A.M., Gelfand, I.M., Losik, M.V.: Combinatorial computation of characteristic classes. Funct. Anal. Appl. **9** (2) (1975) 12–28 (Russian)
24. Bjorner, A.: Some combinatorial and algebraic properties of Coxeter complexes and Tits building. Adv. Math. **52** (1984) 173–212

8.

(with V.A. Vasiliev and A.W. Zelevinskij)

General hypergeometric functions on complex Grassmannians

Funkts. Anal. Prilozh. **21** (1) (1987) 23–38. Zbl. **614**:33008

Introduction

This is one in the series of papers starting with [1] (the other works in the series are [2–5]), devoted to the study of general hypergeometric functions. In [1] a decomposition of a real Grassmannian manifold into strata has been constructed, associating to each stratum its own hypoergeometric functions that are analytical on that stratum. As is always the case for analytical functions, a fully-fledged investigation requires a study of the situation in the complex area. This is the subject of the present paper.

Grassmannian strata can be defined in many equivalent ways; they have been considered in detail in [5]. For our purposes the following version of definition is convenient. Let $G_k(CP^n)$ be the Grassmannian of k-dimensional subspaces in an n-dimensional complex space. Let $(x_0:x_1:\cdots:x_n)$ be homogeneous coordinates in CP^n. For each set of indices $J=\{j_1,\dots,j_r\}\subset[0,n]$ denote by σ_J the coordinate plane in CP^n defined by the condition $x_{j_1}=\cdots=x_{j_r}=0$. Two planes ξ and ζ in $G_k(CP^n)$ are called equivalent if $\dim(\xi\cap\sigma_J)=\dim(\zeta\cap\sigma_J)$ for all coordinate planes σ_J. The equivalence classes are called strata in $G_k(\mathrm{CP}^n)$. There is a unique open stratum, called generic, consisting of those planes that are in general position with all the planes σ_J.

To each stratum Γ in $G_k(\mathrm{CP}^n)$ we associate a set of general hypergeometric functions that are analytical on that stratum. They are defined by integrals of the form

$$\Phi_h(\alpha;\zeta,\omega)=\int_h \pi_\alpha\omega.$$

Precise definitions will be given in §2; here we note only that $\zeta\in\Gamma$, $\alpha=(\alpha_0,\alpha_1,\dots,\alpha_n)$ is a set of complex numbers such that $\alpha_0+\alpha_1+\cdots+\alpha_n=n-k$, and π_α denotes the multivalued analytical function $\prod_{0\leq j\leq n}x_j^{\alpha_j-1}$, while the integrals are evaluated along k-dimensional cycles. For the case of a general stratum in $G_1(CP^3)$ the integrals are of the form $\int\pi_\alpha(t)\,dt$ where $\pi_\alpha(t)=t^{\alpha_1-1}(1-t)^{\alpha_2-1}(x-t)^{\alpha_3-1}$ and one has the Gaussian hypergeometric function.

The main results of this paper concern hypergeometric functions defined on special strata in $G_k(CP^n)$, which we call basic, in the vicinity of their real points. A stratum Γ is called basic if an index j_0 exists such that for any $\zeta\in\Gamma$ the set of planes $\zeta\cap\sigma_j\,(j\neq j_0)$ has only normal intersections in the affine chart $\zeta-\sigma_{j_0}$. For example, the generic stratum is basic. One can easily see that the mapping associating to each

$\zeta \in \Gamma$ the intersection $\zeta \cap \sigma_{j_0}$ gives the projection mapping of a basic stratum Γ on some stratum Γ_0 in $G_{k-1}(\sigma_{j_0})$, and that almost all strata in $G_{k-1}(\sigma_{j_0})$ can be obtained in such a way (in particular, all non-degenerate strata in the sense of [1]). Thus, the supply of general strata is quite extensive.

A point $\zeta \in \Gamma$ is called a real point if $\dim(\zeta \cap RP^n) = k$. The set of real points of a stratum Γ is denoted by Γ_R. One can assume without loss of generality that j_0 in the definition of a basic stratum Γ is equal to 0. For $\zeta \in \Gamma_R$ let $K_\zeta = \zeta - \sigma_0$ and $S_j = (\zeta - \sigma_0) \cap \sigma_j$ for all $j \in [1, n]$. Thus K_ζ is a k-dimensional affine space and $S_1, \dots, S_n$ are affine hyperplanes in K_ζ having only normal intersections. There is a real k-dimensional affine space $K_{\zeta,R} = (\zeta - \sigma_0) \cap RP^n$ in K_ζ such that $\dim(S_j \cap K_{\zeta,R}) = k-1$ for $j \in [1, n]$. The study of hypergeometric functions on a stratum Γ in a neighbourhood of a point ζ is based on the geometrical and topological analysis of the arrangement of hyperplanes S_j in K_ζ. This is the subject of §1 of the paper; a point ζ is assumed there to be fixed, and one writes K instead of K_ζ.

Three types of integrals play an important role in the theory of the Gaussian hypergeometric function: $\int_0^{i\infty} \pi_\alpha(t)dt$, $\int_0^1 \pi_\alpha(t)dt$ and $\int_\Delta \pi_\alpha(t)dt$ where Δ is a double loop (see [6], Section 12.43, Fig. 2). In §1 we construct k-dimensional cycles in the space $K - S$, where $S = \cup S_j$, which provide multidimensional analogues for those three integration paths. We call them "imaginary cones", "visible cycles", and "double loops", respectively. There is a fundamental difference between the first two types of cycles and the third type: the former end in singularities while the latter encircle them. That is why for their formal description one has to use two different homology theories, using locally finite and finite singular chains, respectively. To take into account that the integrals also depend on the exponents $\alpha = (\alpha_0, \dots, \alpha_n)$ one has to consider the homology of the space $K - S$ with coefficients in a local system L_τ of rank 1 on $K - S$ having non-trivial monodromy exponents $\tau = (\tau_1, \dots, \tau_n)$ where $\tau_j = \exp(-2\pi i \alpha_j)$. Accordingly, in §1 we study finite homology groups $H_r(K - S, L_\tau)$ and locally finite homology groups $H_r^{lf}(K - S, L_\tau)$.

The main tools for studying the groups $H_r^{lf}(K - S, L_\tau)$ are "imaginary cones". Since all the intersections of the planes S_j are normal, for each k-element subset $J \subset [1, n]$ the intersection $S_J = \bigcap_{j \in J} S_j$ is either an empty set or a point in K_R. Points of the form S_J are called vertices. In Section 1.5 we associate to each vertex S_J the element $\tilde{\nabla}_J \in H_k^{lf}(K - S, L_\tau)$. It is defined by choosing a section of the local system L_τ over some k-dimensional polyhedronal cone in K which intersects the real space K_R only in the vertex S_J. We call the element $\tilde{\nabla}_J$ "an imaginary cone with vertex S_J" for this reason. The elements $\tilde{\nabla}_J$ generate the group $H_k^{lf}(K - S, L_\tau)$ but they are not linearly independent. Using them, one can explicitly construct a (finite-dimensional) complex which makes it possible to compute the groups $H_r^{lf}(K - S, L_\tau)$. Remarkably, if all $\tau_j \neq 1$, this complex turns out to be isomorphic to another complex constructed according to totally different considerations in [4]. This makes it possible to state the first major result of §1: if all $\tau_j \neq 1$, then $H_r(K - S, L_\tau) = H_r^{lf}(K - S, L_\tau) = 0$ for $r \neq k$ (Theorem 1.7a). In addition, a set of vertices (called regular) is singled out such that the corresponding "imaginary cones" form a basis in $H_k^{lf}(K - S, L_\tau)$ (Proposition 1.8.1).

Using an elegant geometrical idea from [7] we show that if $\tau_j \neq 1$ then the dimension of each of the groups $H_k(K - S, L_\tau)$ and $H_k^{lf}(K - S, L_\tau)$ equals the number of bounded connected components of the space $K_R - S$ (Theorem 1.7b). To each such component we associate two important homology classes: "visible cycles" $\tilde{\Delta} \in H_k^{lf}(K - S, L_\tau)$ and "double loops" $[\Delta] \in H_k(K - S, L_\tau)$. A geometrical construction of "double loops" will be given in a separate paper; in Section 1.6 we provide an "axiomatic" characterisation of cycles $[\Delta]$ by specifying their intersection indices with all "imaginary cones" (Theorem 1.6.1). Theorem 1.7c states that if all $\tau_j \neq 1$ then the set of "double loops" $[\Delta]$ forms a basis in $H_k(K - S, L_\tau)$.

The last result of §1 is Theorem 1.10 in which we find out when "visible cycles" $\tilde{\Delta}$ form a basis in $H_k^{lf}(K - S, L_\tau)$. The condition that all $\tau_j \neq 1$ is not sufficient; there is a finite number of conditions of the form $\prod_{1 \leqq j \leqq n} \tau_j^{a_j} \neq 1$ which have to be satisfied for some integer $a_1, \ldots, a_n$. We call them non-resonance conditions for τ.

We also note that a number of results from §1 for the case of the generic stratum are contained in the interesting papers by Aomoto [8, 9].

Our main results on hypergeometric functions are stated in §§2 and 3. Let Γ be a basic stratum in $G_k(CP^n)$ and $\zeta \in \Gamma_R$ its real point. In §2 we show that for fixed $\alpha = (\alpha_0, \ldots, \alpha_n)$, hypergeometric functions in a neighbourhood of the point ζ are parameterised by the elements $h \in H_k^{lf}(K_\zeta - S, L_\tau)$. Theorem 2.6.1 states that if all α_i are non-integer, then linearly independent elements h correspond to linearly independent hypergeometric functions. Thus, the results of §1 make it possible to find the number of linearly independent hypergeometric functions in the vicinity of a real point of each basic stratum. The proof of Theorem 2.6.1 is based on the computation of monodromy for hypergeometric functions on generic strata. It will be given in a separate paper.

All hypergeometric functions constructed here for the generic stratum satisfy a holonomic system of differential equations given in [2]; using the estimate of the number of solutions from [2] we show that they form a full system of solutions of this system of equations. It follows therefrom that the number of solutions of the system of equations from [2] on the generic stratum in $G_k(CP^n)$ equals $\binom{n-1}{k}$.

Another major result of §2 (Theorem 2.7) concerns the behaviour of a general hypergeometric function on a basic stratum as a function of α_j for a fixed $\zeta \in \Gamma_R$. It states that if the set of numbers $\{\tau_j = \exp(-2\pi_i \alpha_j)\}$ satisfies the non-resonance conditions mentioned above, then hypergeometric functions have no singularities at the point α.

In §3 the developed technique is applied to the study of a continuous analogue of the (vector) partition function. Let L be an l-dimensional vector space over R and $\mathscr{F}$ a system of vectors $\{f_1, \ldots, f_n\}$ in L generating the whole of L and lying in an open half-space in L. Let $L^0 \subset L$ be an open set of vectors that are in general position with vectors from $\mathscr{F}$. To each $f \in L^0$ we associate an $(n - l)$-dimensional polyhedron in R^n defined by the inequalities $x_1 \geqq 0, \ldots, x_n \geqq 0$ and an equation $\sum x_i f_i = f$. A continuous analogue of the partition function is defined as a function $P_{\mathscr{F}}$ on L^0 associating to each vector f the value of the corresponding polyhedron. The function $P_{\mathscr{F}}$ has been introduced in [11] for the case when $\mathscr{F}$ is a system of positive roots. This function is a continuous analogue of the Kostant partition function and it plays an important role in representation theory. The function $P_{\mathscr{F}}$ for an arbitrary system $\mathscr{F}$ has been considered in [13].

As shown in [4], function $P_{\mathscr{F}}$ is a special case of a general hypergeometric function. In the terms introduced above this result can be formulated in the following way: for each pair $(L, \mathscr{F})$ one constructs a basic stratum Γ in $G_{n-l}(CP^n)$ and a submanifold Γ^0 in Γ_R which is naturally identified with L^0; then the function $P_{\mathscr{F}}$ is identified with the restriction of some hypergeometric function on that stratum to Γ^0.

The function $P_{\mathscr{F}}$ on L^0 is of a complicated piecewise polynomial character. A class of decompositions of that function into "elementary terms" of a more simple nature has been found in [4]. To be more precise, for each subset $I = \{i_1, \ldots, i\} \subset [1, n]$ such that $\operatorname{rk}(f_{i_1}, \ldots, f_{i_l}) = l$, denote by C_I^0 the intersection of L^0 with the open convex cone spanned by the vectors $f_{i_1}, \ldots, f_{i_l}$. Let χ_I be a characteristic function of the set C_I^0 and $H(F)$ the vector space of piecewise constant functions on L^0 spanned by all the functions χ_I[1]. A system B of subsets in I is called fundamental if the functions χ_I for $I \in B$ form a basis in $H(\mathscr{F})$. In [4] it has been proved that for each fundamental system B and for all $I \in B$ there exist polynomials $P_I^{(B)}$ uniquely defined on the whole of L such that the function $P_{\mathscr{F}}$ on L^0 is of the form $\sum_{I \in B} \chi_I P_I^{(B)}$. The paper [4] also provides an algebraic procedure for finding polynomials $P_I^{(B)}$.

In §3 we give a geometrical interpretation of the polynomials $P_I^{(B)}$ expressing them in terms of volumes of certain polyhedra; this geometrical interpretation is based on the results from [7]. Our result is naturally formulated in the language of hypergeometric functions: we specify those hypergeometric functions on Γ whose restrictions to the submanifold Γ^0 are identified with the functions $\chi_I P_I^{(B)}$.

Some results of this paper have been announced in [14].

§1. Homology of local systems in an affine complex space with deleted hyperplanes

1.1. Fix n different affine hyperplanes $S_1, \ldots, S_n$ in a k-dimensional complex affine space K. Let $S = \cup S_j$ and for any $J \subset \{1, \ldots, n\}$ let $S_J = \bigcap_{j \in J} S_j$.

Let $\tau = (\tau_1, \ldots, \tau_n)$ be a set of non-zero complex numbers.

Definition (see [15]). A local system L_τ of rank 1 on $K - S$ with monodromy exponents $\tau = (\tau_1, \ldots, \tau_n)$ is a vector bundle on $K - S$ with fibre C^1 endowed with a flat connection such that a loop in $K - S$ around S_j in the positive direction yields multiplication by τ_j in fibres.

Horizontal sections of a system L_τ will be called, for brevity, sections. We denote by L_τ^V a local system dual to L_τ. Its monodromy coefficients are equal to τ_j^{-1}.

Denote by $C_r(K - S, L_\tau)$ the space of singular r-chains with coefficients in L. By definition, $C_r(K - S, L_\tau)$ consists of finite linear combinations of the form $\sum c_\Delta \cdot \Delta$ where Δ is an oriented simplex in $K - S$ and c_Δ is a section of the system L_τ over it.

---

[1] Several different realisations of the space $H(\mathscr{F})$ are given in [4], including a purely combinatorial one, as a cohomology group of some matroid.

There is a standard boundary operator $d: C_r(K-S, L_\tau) \to C_{r-1}(K-S, L_\tau)$. Homology of the resulting complex $C_*(K-S, L_\tau)$ is called (finite) homology of a local system L_τ and denoted $H_r(K-S, L_\tau)$.

Consider also the complex $C_*^{lf}(K-S, L_\tau)$ of locally finite (in a neighbourhood of any point in $K-S$) chains with coefficients in L_τ. Its homology groups are denoted by $H_r^{lf}(K-S, L_\tau)$.

We say that a pair (K, S) is real if there is a real subspace K_R in K such that $K = K_R \otimes C$ and $\dim(S_j \cap K_R) = k-1$ for $j = 1, \ldots, n$.

1.2. Proposition. (a) *Intersection indices of cycles define an identification between $H_r(K-S, L_\tau)$ and the space dual to $H_{2k-r}^{lf}(K-S, L_\tau^\vee)$.*

(b) *If a pair (K, S) is real then $H_r(K-S, L_\tau) \cong H_r(K-S, L_\tau^\vee)$.*

(c) *The group $H_r(K-S, L_\tau)$ is 0 for $r > k$, and the group $H_r^{lf}(K-S, L_\tau)$ is 0 for $r < k$.*

The first statement is the Poincaré duality, the second follows from the fact that complex conjugation on $K-S$ takes the local system L_τ into $L_\tau^\vee$, and the third is a consequence of the fact that $K-S$ is a Stein manifold and is, therefore, homotopically equivalent to a cellular complex of dimension $\leqq k$.

1.3. Proposition. *For any $r = 0, 1, \ldots, 2k$ and for any positive integer p the set of such n-tuples $\tau = (\tau_1, \ldots, \tau_n)$ that $\dim H_r(K-S, L_\tau) \geqq p$ forms an algebraic subset in $(C^*)^n$. In particular, for almost all τ, the dimension of the group $H_r(K-S, L_\tau)$ takes one and the same (the lowest possible) value. The same is true for the group $H_r^f(K-S, L_\tau)$.*

This can be proved in a standard way.

1.4. We shall assume up to the end of this section (§1) that the pair (K, S) is real and that the planes S_j have only normal intersections. In other words, for $|J| > k$ the intersection S_J is empty while for $|J| \leqq k$ it is either empty or its codimension in K is equal to $|J|$. For $|J| = k$ the set S_J is either empty or is a point in K_R. Points of the form S_J are called vertices.

We now construct a (finite-dimensional) chain complex $C_*(\tau)$ which makes it possible to compute the groups $H_r^{lf}(K-S, L_\tau)$. For any integer r we define $C_r(\tau)$ as a linear space with a fixed basis $\{\tilde{\nabla}_J\}$ whose elements correspond to all non-empty subsets S_J in K such that $|J| = 2k-r$. The differential $d: C_r(\tau) \to C_{r-1}(\tau)$ is defined by the formula

$$d(\tilde{\nabla}_J) = \sum \varepsilon_j (1 - \tau_j^{-1}) \tilde{\nabla}_{J+j} \tag{1}$$

where the sum is taken over all $j \notin J$ such that S_{J+j} is non-empty, and ε_j is either 1 or -1 according to whether the number of elements of the set J that are less than j is even or odd.

1.5. Proposition. *For any τ and r the group $H_r^{lf}(K-S, L_\tau)$ is isomorphic to $H_r(C_*(\tau))$, i.e. the homology group of the complex $C_*(\tau)$.*

To prove this, we decompose the space $K-S$ into cells in the following way. Take an arbitrary connected component of the set $K_R - S$ and denote it by Δ_+. Choose a set of linear functions $q_1, \ldots, q_n$ on K such that each plane S_j is defined by an equation of a the form $q_j = 0$, where all functions q_j are real on K_R and positive on Δ_+. To each non-empty set S_J we associate the submanifold $\bar{\nabla}_J$ in K defined by the conditions $\operatorname{Re} q_j = 0$, $\operatorname{Im} q_j > 0$ for $j \in J$. Deleting from this set its intersection with S and all similar subsets $\bar{\nabla}_{J'}$ for $J' \supset J$, one obtains the set ∇_J which is a cell of dimension $2k - |J|$ (in particular, if $J = \varnothing$ then ∇_J is an open cell in $K-S$). Indeed, the natural projection of ∇_J on K_R coincides with the real plane $S_{J,R} = S_J \cap K_R$ and a fibre of that projection over any point is convex. Evidently, $K - S = \bigcup_J \nabla_J$, and we obtain a finite semi-algebraic decomposition of the space $K-S$ into cells. Let $j_1 < \cdots < j_{|J|}$ be the indices constituting the set J. Define an orientation of the cell ∇_J, choosing at each of its points a $(2k-|J|)$-form ω such that the form $d \operatorname{Re} q_{j_1} \wedge \cdots \wedge d \operatorname{Re} q_{j_{|J|}} \wedge \omega$ defines the canonical (complex) orientation of the space K.

Let $C_r(\tau)$ be the group consisting of linear combinations $\sum c_J \nabla_J$ of r-dimensional cells ∇_J where each coefficient c_J is a section of the system L_τ over the cell ∇_J. The boundary operator $C_r(\tau) \to C_{r-1}(\tau)$ is defined in a standard way, and standard considerations show that homology of the resulting finite-dimensional complex coincide with the homology groups $H_r^{lf}(K-S, L_\tau)$.

Now we choose a special section of the system L_τ over each cell ∇_J. First we choose an arbitrary non-zero section over Δ_+. Consider some interval I in general position in K_R beginning in the set S_J and ending in Δ_+. Consider a path in $K-S$ connecting some point of the set ∇_J with the endpoint of I such that (a) the path lies on a complex line containing the interval I, (b) the projection along the purely imaginary direction maps it onto I, (c) it coincides with I outside some neighbourhood of the set S, and (d) it makes a half-loop of the form $\{q_j = \varepsilon \exp(it), \varepsilon > 0, t \in [\pi, 2\pi]\}$ around each intersection point of the interval I with the planes S_j, $j \notin J$ (see the figure). Let us continue the section chosen over Δ_+ along that path to its starting point, and, consequently, on the whole set ∇_J. Call the resulting section c_J and let $\tilde{\nabla}_J = c_J \cdot \nabla_J$. An easy computation shows that the boundary operator acts on $\tilde{\nabla}_J$ according to formula (1), and Proposition 1.5 is proved.

If the intersection S_J is a vertex, then the corresponding element $\tilde{\nabla}_J \in C_k(\tau)$ is a cycle. We denote the corresponding homology class in $H_k^{lf}(K-S, L_\tau)$ by the same letter $\tilde{\nabla}_J$. The classes $\tilde{\nabla}_J$ are called "imaginary cones" (see Introduction). They are defined up to a common constant depending on a choice of the section of L_τ over Δ_+.

By Proposition 1.5 "imaginary cones" generate the group $H_k^{lf}(K-S, L_\tau)$; however, there are relations between them.

Now we define "visible cycles" in $H_k^{lf}(K-S, L_\tau)$. Fix an orientation in the space K_R, and, thereby, in all connected components of the set $K_R - S$. To each connected component Δ endowed with a section of the system L_τ chosen over it there corresponds a homology class $\tilde{\Delta} \in H_k^{lf}(K-S, L_\tau)$ (defined by Δ uniquely up to a scalar multiple). Cycles of this form are called "visible cycles" (cf. [10]).

1.6. To each bounded connected component Δ of the space $K_R - S$ we associate a "double loop" $[\Delta] \in H_k(K-S, L_\tau)$. A geometric construction of cycles $[\Delta]$ will be given in the next paper; it is similar to the construction given in [16]. Here we define cycles $[\Delta]$ "axiomatically". Fix a section c^V of the local system L_τ^V over the chamber Δ_+ as in Section 1.5, and let $\tilde{\nabla}_J$ be the corresponding "imaginary cones" in $H_k^{lf}(K-S, L_\tau^V)$. For each vertex S_J and each bounded component in $K_R - S$ we define the coefficient $\varepsilon(\Delta, J)$ in the following way: it is equal to 0 if S_J is not a vertex of Δ, and if S_J is a vertex of Δ then $\varepsilon(\Delta, J)$ is equal to either 1 or -1 depending on whether the number of functions q_j, $j \in J$ negative in Δ is even or odd. Let $c(\Delta, J) = \operatorname{sgn} J \cdot \varepsilon(\Delta, J) \cdot \prod(1-\tau_j)$ where the product is taken over all $j \notin J$ such that S_j intersects the boundary of Δ, and the number sign J is either 1 or -1 depending on whether the orientation fixed on K_R coincides with that defined by the form $dq_{j_1} \wedge \cdots \wedge dq_{i_k}$ (where $J = \{j_1 < \cdots < j_k\}$) or not.

1.6.1. Theorem. *For each bounded connected component Δ in $K_R - S$ there exists a unique element $[\Delta] \in H_k(K-S, L_\tau)$ such that its intersection index with any class $\tilde{\nabla}_J \in H_k^{lf}(K-S, L_\tau^V)$ equals $c(\Delta, J)$.*

Consider now how the cycles $[\Delta]$ are related to "visible cycles" $\tilde{\Delta} \in H_k^{lf}(K-S, L_\tau)$. Let us normalise $\tilde{\Delta}$ in the following way Choose a section c of the system L_τ over Δ_+ on which the section c^V taken above takes the value 1. Continuing section c into the component Δ in the same way as was done for "imaginary cones", one obtains a section of the system L_τ over Δ. We now take for $\tilde{\Delta}$ the "visible cycle" which corresponds to this section.

1.6.2. Theorem. *The image of the cycle $[\Delta]$ under the natural homomorphism $H_k(K-S, L_\tau) \to H_k^{lf}(K-S, L_\tau)$ is $\prod(1-\tau_j)\tilde{\Delta}$ where the product is taken over all j such that S_j intersects the boundary of Δ.*

1.7. The main theorem. *Let (K, S) be a real pair and suppose that all intersections of the planes S_j are normal. Let all τ_j be different from 1. Then*

(a) $H_r^{lf}(K-S, L_\tau) = 0$ *for* $r > k$ *and* $H_r(K-S, L_\tau) = 0$ *for* $r < k$;

(b) *the dimension of the group $H_k^{lf}(K-S, L_\tau)$ equals the number of bounded components of the set $K_R - S$;*

(c) *the cycles $[\Delta]$ corresponding to bounded components in $K_R - S$ form a basis in $H_k(K-S, L_\tau)$.*

Warning. In general, the conditions of Theorem 1.7 do not imply that the cycles $\tilde{\Delta}$ form a basis in the group $H_k^{lf}(K-S, L_\tau)$. This is true only if some additional restrictions on τ are satisfied (see Section 1.10).

1.8. *Proof of Theorem* 1.7. Replace the basis in the complex $C_*(\tau)$ constructed in 1.4 with another basis defined by the formula $\nabla'_J = \tilde{\nabla}_J \prod_{j \in J}(1-\tau_j^{-1})$. Evidently, the action of the differential in the basis $\{\nabla'_J\}$ does not depend on τ and is obtained by letting all τ_j in formula (1) in 1.5 equal ∞. Denote the resulting "universal" complex by $\mathscr{S}'$. It turns out that the complex $\mathscr{S}'$ coincides up to a grading with the complex $\mathscr{S} = (0 \to \mathscr{S}_n \to \cdots \to \mathscr{S}_{n-k} \to 0)$ constructed in [4, Appendix, formula (*)]. Recall that the space $\mathscr{S}_m$ is generated by symbols $[I]$ for some m-tuples $I \subset [1, n]$.

Proposition 3.1.2(c) in §3 implies that the relevant subsets I are exactly those for which $S_{[1,n]-I} \neq \varnothing$. Therefore, the mapping $\nabla'_J \to [[1,n]-J]$ yields an isomorphism of spaces $\mathscr{S}'_{2k-1}$ and $\mathscr{S}_{n-r}$; it follows easily from the definitions that this mapping is compatible with differentials. It is proved in [4] that the complex $\mathscr{S}$ is acyclic in every term except $\mathscr{S}_{n-k}$. This proves statement (a) of Theorem 1.7.

In order to prove the statements (b) and (c) we use the Bjorner theorem which provides a description of some special basis in $H_{n-k}(\mathscr{S}_*)$ (see [4, Theorem II.1]). Identifying $H_{n-k}(\mathscr{S}_*)$ with $H_k^{lf}(K-S, L_\tau)$ one can formulate this theorem in the following convenient form. A vertex S_J will be called regular if for any $j \in J$ there exists $j' < j$ such that $S_{J-j+j'}$ is a vertex (note that the notion of regular vertex depends on an ordering of the set of planes $S_1, \ldots, S_n$).

1.8.1. Proposition. *If $\tau_j \neq 1$ for all j then the elements $\tilde{\nabla}_J$ corresponding to regular vertices S_J form a basis in $H_k^{lf}(K-S, L_\tau)$.*

Following [7] we define a linear ordering on the set of all vertices S_J by setting $S_J < S_{J'}$ if $j \in [1,n]$ exists such that $|q_i(S_J)| = |q_i(S_{J'})|$ for $i < j$ and $|q_j(S_J)| < |q_j(S_{J'})|$.

1.8.2. Proposition [7]. *The mapping associating to each bounded component in $K_{\mathbf{R}} - S$ its maximal vertex yields a one-to-one correspondence between the set of bounded components and the set of regular vertices.*

The statement (b) of Theorem 1.7 follows immediately from Propositions 1.8.1 and 1.8.2. To prove the statement (c), denote by Δ_J the bounded component with the maximal vertex S_J and consider the matrix of intersection indices $\langle [\Delta_J], \nabla_{J'} \rangle$ whose rows and columns are numbered by regular vertices. In view of Proposition 1.8.2 and Theorem 1.6.1 this matrix is triangular, and if all τ_j are different from 1 then all its diagonal elements are different from 0. Therefore the statement (c) is a consequence of Propositions 1.2 (a) and 1.8.1.

Corollary. *The number of bounded connected components of the set $K_{\mathbf{R}} - S$ depends only on the fact as to which of the sets S_J are non-empty. In the case of a general position it is equal to $\binom{n-1}{k}$.*

A more detailed study of the complex $C_*(\tau)$ makes it possible to formulate the following stronger version of parts (a) and (b) of Theorem 1.7.

1.8.3. Theorem. *Suppose that a pair (K, S) is real and that at least one of the following two conditions is satisfied:*

(1) *the planes $S_1, \ldots, S_n$ are in general position and at least one of the numbers τ_j is different from* 1;

(2) *the planes $S_1, \ldots, S_n$ have only normal intersections and there is a vertex S_J such that $\tau_j \neq 1$ for $j \in J$.*

Then the statements (a) and (b) of Theorem 1.7 hold.

1.9. Following [7] consider relative homology spaces $H_k(K,S;C)$ and cohomology spaces with trivial coefficients $H^k(K,S;C)$. Let again a pair (K,S) be real and the planes S_j have only normal intersections. Each bounded component Δ in $K_R - S$ with orientation inherited from the space K_R defines an element in $H_k(K,S,C)$ which will be denoted simply by Δ; those elements form a basis in $H_k(K,S;C)$. The space $H^k(K,S;C)$ can be identified with the dual space of $H_k(K,S;C)$. Following [7] we associate to each vertex S_J an element Δ^J of the group $H^k(K,S;C)$ whose value on each element Δ is $\varepsilon(\Delta, J)$ (see Section 1.6).

1.9.1. Theorem. *Suppose that all the conditions of Theorem 1.7 are satisfied. Then there is an isomorphism* $H_k(K,S;C) \xrightarrow{\sim} H_k(K-S,L_\tau)$ *taking each generator* Δ *into the class* $[\Delta]/\prod(1-\tau_j)$ *where the product is taken over all* j *such that* S_j *intersects the boundary of* Δ. *The corresponding isomorphism of dual spaces* $H^k(K,D;C) \xrightarrow{\sim} H_k^{lf}(K-S,L_\tau^\vee)$ *takes each element* Δ^J *into* $\nabla_j^* = \operatorname{sgn} J \cdot \prod_{j\in J}(1-\tau_j)\cdot\tilde{\nabla}_J$.

This theorem is an immediate consequence of Theorems 1.7 and 1.6.1. Note that by Theorem 1.6.2 the image of the element Δ under the homomorphism $H_k(K,S;C) \to H_k(K-S,L) \to H_k^{lf}(K-S,L_\tau)$ is equal to $\tilde{\Delta}$.

1.10. Definition. Let A be a set of vectors in the integer lattice Z^n. A vector $\tau = (\tau_1,\dots,\tau_n)$ is called A-non-resonant if $\tau_1^{a_1}\cdots\tau_n^{a_n} \neq 1$ for all $(a_1,\dots,a_n)\in A$.

Theorem. *Suppose that a pair* (K,S) *is real and that all the intersections of the planes* S_j *are normal. Then there exists a finite set* $A(S)\subset Z^n$ *such that if a set* τ *is* $A(S)$*-non-resonant, then the visible cycles corresponding to all bounded components of the set* $K_R - S$ *form a basis of the group* $H_k^{lf}(K-S,L_\tau)$ *and the natural homomorphism* $H_k(K-S,L_\tau)\to H_k^{lf}(K-S,L_\tau)$ *is an isomorphism. The set* $A(S)$ *is fully defined by the fact as to which of the sets* S_J *are empty. It always includes* n *unit vectors and the vector* $(1,\dots,1)$, *and in the case when all* S_j *are in general position it coincides with this set of* $(n+1)$ *vectors.*

A proof of this theorem and an explicit construction of the set $A(S)$ will be given in the next paper.

1.10.1. Example. Let $k=2$, $n=4$, $S_1\parallel S_2$, $S_3\parallel S_4$, $S_1\parallel S_3$ and let the pair (K,S) be real. Then there exists a unique bounded component Δ in $K_R - S$. Let all τ_j be different from 1 so that $H_k^{lf}(K-S,L_\tau)\cong C^1$. One can easily check that under these assumptions the element $\tilde{\Delta}\in H_k^{lf}(K-S,L_\tau)$ is different from 0 if and only if $\tau_1\tau_2\neq 1$, $\tau_3\tau_4\neq 1$ so that the set $A(S)$ includes, besides the five standard ones, two more vectors: $(1,1,0,0)$ and $(0,0,1,1)$.

Remark. For the case when the S_j are in general position the statements of Theorems 1.7(a), (b) and 1.10 are given in [8, 9], but the conditions imposed there on τ_j are more restrictive than those formulated above.

§2. General hypergeometric functions on complex Grassmannians

2.1. Denote by $G_k(CP^n)$ the Grassmannian manifold of k-dimensional planes in the complex projective space CP^n. For each $\zeta \in G_k(CP^n)$ denote by $\bar{\zeta}$ a $(k+1)$-dimensional subspace in C^{n+1} such that its projectivisation is ζ. (The correspondence $\zeta \to \bar{\zeta}$ yields an isomorphism between $G_k(CP^n)$ and the Grassmannian of vector subspaces $G_{k+1}(C^{n+1})$.) Let $x_0, \ldots, x_n$ be coordinates in C^{n+1}. For each $j = 0, \ldots, n$ denote by $\bar{\sigma}_j$ the coordinate hyperplane $x_j = 0$ in C^{n+1} and by $\sigma_j \subset CP^n$ its projectivisation. Let $\bar{\sigma} = \cup \bar{\sigma}_j$, $\sigma = \cup \sigma_j$. If $J = \{j_1, \ldots, j_r)$ is a subset of $[0, n]$ let $\sigma_J = \sigma_{j_1} \cap \cdots \cap \sigma_{j_r}$. Two points ζ, $\xi \in G_k(CP^n)$ are called equivalent if $\dim(\zeta \cap \sigma_J) = \dim(\xi \cap \sigma_J)$ for all J. (We denote points of a Grassmannian and the corresponding subspaces in CP^n by the same letters ζ, ξ, ...)

The equivalence classes with respect to this relation are called strata in $G_k(CP^n)$. In the sequel we assume that ζ is not contained in σ_j for any j, and that all intersections $\zeta \cap \sigma_j$ are different. We shall see that these assumptions do not involve any loss of generality in our subsequent study of hypergeometric functions.

2.2. Let $\zeta \in G_k(CP^n)$. Denote by $D(\zeta)$ the set of k-forms ω on $\bar{\zeta}$ of the form

$$\omega = \sum_{0 \leqq j \leqq k} (-1)^j t_j \, dt_0 \wedge \cdots \wedge dt_{j-1} \wedge dt_{j+1} \wedge \cdots \wedge dt_k$$

where $t_0, \ldots, t_k$ is an arbitrary system of linear coordinates in $\bar{\zeta}$. (Evidently, all forms in $D(\zeta)$ are proportional.)

For each set of complex numbers $\alpha = (\alpha_0, \ldots, \alpha_n)$ such that its sum equals $n-k$ denote by π_α the multivalued analytic function $\prod_{0 \leqq j \leqq n} x_j^{\alpha_j - 1}$ on C^{n+1} having branch points on $\bar{\sigma}$. We now construct a covering P_ζ over $\zeta - \sigma$ and define a single-valued analytical form $\pi_\alpha \cdot \omega$ on it which is analytically dependent on parameter α.

Let $\bar{P}$ be a covering over $C^{n+1} - \bar{\sigma}$ whose fibre over a point $x = (x_0, \ldots, x_n)$ consists of all n-tuples $(a_0, \ldots, a_n)$ such that $\exp a_i = x_i$ where two n-tuples define the same point of the fibre if they differ by a scalar multiple of $2\pi i(1, \ldots, 1)$. For any α the function π_α lifts to a single-valued analytical function π_α on the covering $\bar{P}$ by the formula $\pi_\alpha(x, a_0, \ldots, a_n) = \exp(\sum a_j(\alpha_j - 1))$. Denote the lift of the form ω to the covering space $\bar{P}|_{\bar{\zeta}-\bar{\sigma}}$ by the same letter ω; thus one has a k-form $\pi_\alpha \cdot \omega$ on $\bar{P}|_{\bar{\zeta}-\bar{\sigma}}$.

The group C^* acts by scalar multiplication on the space $\bar{\zeta} - \bar{\sigma}$. Lifting that action to the covering $\bar{P}|_{\bar{\zeta}-\bar{\sigma}}$ define the space P_ζ as a quotient of the space $\bar{P}|_{\bar{\zeta}-\bar{\sigma}}$ with respect to that action. The natural projection $P_\zeta \to \zeta - \sigma$ makes P_ζ into a covering over $\zeta - \sigma$. The condition $\sum \alpha_j = n - k$ implies that the form $\pi_\alpha \cdot \omega$ goes into a (single-valued) k-form on P_ζ which we also denote by $\pi_\alpha \cdot \omega$.

2.2.1. Proposition. (a) *A loop in $\zeta - \sigma$ lifts to a loop in the covering P_ζ if and only if its class in the group $H_1(\zeta - \sigma)$ is 0.*

(b) *The group $H_1(\zeta - \sigma)$ is isomorphic to Z^n: it is generated by positive loop circuits around the planes σ_j. The only relation between them is that their sum equals* 0.

Proof is trivial.

2.3. Denote by $C_*^{rf}(P_\zeta)$ the complex of such locally finite chains on P_ζ whose projections on $\zeta-\sigma$ are also locally finite. The homology of that complex will be denoted by $H_*^{rf}(P_\zeta)$.

Again let $\alpha=(\alpha_1,\dots,\alpha_n)$ be a set of complex numbers such that their sum equals $n-k$. Let $\tau_j=\exp(-2\pi i\alpha_j)$ so that $\prod\tau_j=1$. Denote by $B_*(\zeta,\alpha)$ a subcomplex in $C_*^{rf}(P_\zeta)$ generated by the chains of the form $s\Delta-(\prod_{0\leq j\leq n}\tau_j^{s_j})\Delta$, where $\Delta\in C_*^{rf}(P_\zeta)$, $s\in H_1(\zeta-\sigma)$ and the s_j are linkage indices of s with the planes σ_j, and $s\Delta$ is the image of the chain Δ under the natural action of the element s. Let $C_*(\zeta,\alpha)$ be a corresponding quotient chain complex $C_*^{rf}(P_\zeta)/B_*(\zeta,\alpha)$. Denote the homology of the complex $C_*(\zeta,\alpha)$ by $H_*(\zeta,\alpha)$.

Denote by K_ζ the affine chart $\zeta-\sigma_0$ in ζ and let $S_j=K_\zeta\cap\sigma_j$ for all $j=1,\dots,n$. Thus, K_ζ is a k-dimensional affine space, $S_1,\dots,S_n$ hyperplanes in it, and we are in the situation considered in §1. Therefore, the set of exponents $\tau_1,\dots,\tau_n$ introduced above defines a local system L_τ on $K_\zeta-S$. Clearly, $K_\zeta-S=\zeta-\sigma$, i.e. P_ζ covers the whole of $K_\zeta-S$. There is a natural surjection $p\colon C_*^{rf}(P_\zeta)\to C_*^{lf}(K_\zeta-S,L_\tau)$ (see Section 1.1) defined up to a scalar multiple. In order to define it uniquely one has to specify a point $x\in P_\zeta$, its projection y in $K_\zeta-S$ and a trivialisation of the fibre of the system L_τ at the point y.

2.3.1. Proposition. (a) $\operatorname{Ker} p=B_*(\zeta,\alpha)$ *and* p *induces an isomorphism between* $H_*(\zeta,\alpha)$ *and* $H_*^{lf}(K_\zeta-S,L_\tau)$.

(b) *The natural projection* $H_k^{rf}(P_\zeta)\to H_k(\zeta,\alpha)$ *is a surjection.*

Statement (a) is an immediate consequence of the definitions. In order to prove (b) note that one can compute homology groups $H_*^{rf}(P_\zeta)$ and $H_*(B_*(\zeta,\alpha))$ by decomposing $K_\zeta-S$ into cells (similarly to the groups $H_*^{lf}(K-S,L_\tau)$ in §1). Since $K_\zeta-S$ is a Stein manifold, there is a decomposition that has no cells of dimension less than k. Hence $K_{k-1}(B_*(\zeta,\alpha))=0$. The statement is then implied from the exact homology sequence of the pair.

Now let Γ be a stratum in $G_k(CP^n)$. There is a natural fibre bundle $\mathscr{H}$ over Γ with a fibre over ζ equal to $H_k^{rf}(P_\zeta)$. There is a canonical flat connection in this bundle called the Gauss–Manin connection. For each set α consider the quotient bundle with the fibre over a point ζ equal to $H_k(\zeta,\alpha)$. The Gauss–Manin connection on $\mathscr{H}$ induces a flat connection in the quotient bundle which we also call the Gauss–Manin connection.

2.4. Fix a point $\zeta\in G_k(CP^n)$, a form $\omega\in D(\zeta)$ and a homology class $h\in H_k^{rf}(P_\zeta)$. Recall that for each set $\alpha=(\alpha_0,\dots,\alpha_n)$ with the sum $n-k$ there is a k-form $\pi_\alpha\omega$ on P_ζ. Let $\int_h\pi_\alpha\omega$ be the integral of the form $\pi_\alpha\omega$ over a cycle representing the class h. The cycle h is, in general, non-compact, and the integral may diverge. In order to give a correct definition we use, as in [1], an approach which considers decompositions and analytic continuation along the parameter α.

Fix a semi-algebraic cell decomposition of the space ζ such that $\zeta-\sigma$ is a union of cells and for each cell its boundary does not intersect all σ_j simultaneously. According to Proposition 2.3.1 the element h can be represented as a linear combination of cells of this decomposition. For any cell there exists a domain in the space of exponents α such that for each α in it the integral of the form $\pi_\alpha\omega$ over that cell converges absolutely and is an analytic function of α. Analytically prolonging these functions along α and adding them up, one obtains a meromorphic function depending on α; we take this function for $\int_h \pi_\alpha \omega$. It is easy to see that the integral defined in this way does not depend on the choice of cell decomposition and cycle representing h.

2.4.1. Proposition. *The integral $\int_h \pi_\alpha \omega$ depends only on the image of the class h in the group $H_k(\zeta, \alpha)$.*

Proof is a direct consequence of definitions.

2.5. Finally, we can define hypergeometric functions on a stratum Γ in $G_k(CP^n)$. Fix a point $\zeta \in \Gamma$, a class $h \in H_k^{rf}(P_\zeta)$ and a simply connected neighbourhood U of ζ in Γ. For each $\xi \in U$, $\omega \in D(\xi)$ and a set of exponents α with the sum $n-k$ we define a general hypergeometric function $\phi_h(\alpha; \xi, \omega)$ by means of the formula

$$\phi_h(\alpha; \xi, \omega) = \int_{h(\xi)} \pi_\alpha \omega,$$

where $h(\xi) \in H_k^{rf}(P_\xi)$ is the class obtained by transporting h in a parallel fashion from the point ζ into the point ξ according to the Gauss–Manin connection.

For fixed ζ and $\omega \in D(\zeta)$ the function $\phi_h(\alpha; \xi, \omega)$ is a meromorphic function of α. For fixed α the function $\phi_h(\alpha; \xi, \omega)$ is an analytic function of the pair (ξ, ω). In view of Proposition 2.4.1 this function depends only on the image h_0 of the class h in the group $H_k(\zeta, \alpha)$. Hence for a fixed α one can assume that a general hypergeometric function is defined by the element $h_0 \in H_k(\zeta, \alpha)$, and for this reason we will also denote it by $\Phi_{h_0}(\alpha; \xi, \omega)$. Identifying $H_k(\zeta, \alpha)$ with $H_k^{lf}(K_\zeta - S, L_\tau)$ as in Proposition 2.3.1(a), one can regard general hypergeometric functions for a fixed α as being parameterized by the group $H_k^{lf}(K_\zeta - S, L_\tau)$.

Let us now show that general hypergeometric functions on a real Grassmannian defined in [1] are a special case of functions we have just introduced. Let Γ be a stratum in $G_k(CP^n)$; denote by Γ_R the set of its real points. Let $\zeta \in \Gamma_R$ and let Δ_+ be the component in $\zeta_R - \sigma$ defined by the condition $x_i/x_0 > 0$ for $i = 1, \ldots, n$. Take the sheet of P_ζ over Δ_+ into which the points $(x, a_0, \ldots, a_n) \in \bar{P}|_{\bar{\zeta} - \bar{\sigma}}$ with real a_j are projected (see Section 2.2). Lifting the chamber Δ_+ onto this sheet, denote the corresponding element of the group $H_k^{rf}(P_\zeta)$ by the same letter Δ_+. The general hypergeometric function constructed in [1] coincides, by definition, with the function $\phi_{\Delta_+}(\alpha; \xi, \omega)$ in a neighbourhood of the point ζ in the set Γ_R. Note that under the projection $H_k^{rf}(P_\zeta) \to H_k^{lf}(K_\zeta - S, L_\tau)$ the class Δ_+ is taken into the visible cycle $\tilde{\Delta}_+$ (see Section 1.5).

2.6. Let Γ be a stratum in $G_k(CP^n)$, $\zeta \in \Gamma$ and K_ζ and S be defined as in Section 2.3. Recall that a stratum Γ is called basic if the set of planes $S_1, \ldots, S_n$ have only normal intersections in the affine space K_ζ (see Introduction).

2.6.1. Theorem. *Let Γ be a basic stratum in $G_k(CP^n)$, $\zeta \in \Gamma_R$, $\alpha = (\alpha_0, \ldots, \alpha_n)$ a fixed set of complex numbers with the sum $n-k$ such that none of α_j is an integer, and*

$\tau_j = \exp(-2\pi i\alpha_j)$ *for* $j \in [1, n]$. *Then if the classes* $h_1, \ldots, h_N$ *in* $H_k^{lf}(K_\zeta - S, L_\tau)$ *are linearly independent, the corresponding hypergeometric functions* $\phi_{h_i}(\alpha; \xi, \omega)$ *are also linearly independent.*

Proof will be given in a following paper.

2.6.2. Corollary. *Under the conditions of Theorem 2.6.1 the number of linearly independent hypergeometric functions on a stratum* Γ *equals the number of bounded connected components in* $K_{R,\zeta} - S$.

This corollary is a consequence of Theorem 1.7.

Now let Γ be the generic stratum. One can easily verify that functions $\phi_h(\alpha; \xi, \omega)$ satisfy the holonomic system of equations from [2]. In [2] it is proved that the number of linearly independent solutions of that system does not exceed $\binom{n-1}{k}$. On the other hand, according to the Corollary to Proposition 1.8.2, the number of bounded components in $K_{R,\zeta} - S$ in that case equals $\binom{n-1}{k}$. We obtain the following corollary.

2.6.3. Corollary. *If none of* α_j *is an integer, then the system of hypergeometric equations on the generic stratum in* $G_k(CP^n)$ *has exactly* $\binom{n-1}{k}$ *linearly independent solutions.*

2.7. Let Γ again be a basic stratum in $G_k(CP^n)$, $\zeta \in \Gamma_R$ and $S = (S_1, \ldots, S_n)$ the corresponding set of planes in the affine space $K_\zeta = \zeta - \sigma_0$. By Theorem 1.10 the finite set $A(S) \subset Z^n$ introduced there depends only on the stratum Γ.

Theorem. *Let* $h \in H_k^{lf}(P_\zeta)$, $\omega \in D(\zeta)$. *Then the function* $\phi_h(\alpha; \zeta, \omega)$, *considered as a function of* α *for fixed* ζ, ω, *has no singularities at those points* α *for which the set of numbers* $\tau_j = \exp(-2\pi i\alpha_j)$ *is* $A(S)$*-nonresonant.*

The statement of the theorem follows from the fact that in the non-resonant case Theorem 1.10 ensures that a class h can be realised by a compact cycle.

§3. Dual description of strata and a continuous analogue of the partition function

3.1. In this section we use the notation introduced above. As in §2 let $\zeta \in G_k(CP^n)$ and let $\bar{\zeta} \in G_{k+1}(C^{n+1})$ be a vector space whose projectivisation is ζ. Let $l = n - k$. Denote by $\bar{\zeta}^\perp \in G_l((C^{n+1})^*)$ the annulator of $\bar{\zeta}$ in the dual space $(C^{n+1})^*$. If $\bar{\zeta}$ runs over some stratum Γ in $G_{k+1}(C^{n+1})$ then $\bar{\zeta}^\perp$ runs over a stratum $\Gamma^\perp$ in $G_l((C^{n+1})^*)$ which is called dual to Γ (see [3]). Vectors $e_0, e_1, \ldots, e_n$ of the standard basis in C^{n+1} define linear forms on the space $\overline{\zeta}^{\perp}$ which we denote by $f_0(\zeta), \ldots, f_n(\zeta)$. For all $I \subset [0, n]$ denote the family of forms $(f_i(\zeta))_{i \in I}$ by $\mathscr{F}_I(\zeta)$. The following proposition is an immediate consequence of the definitions.

3.1.1. Proposition. *For all subsets J in $[0, n]$ one has $|J| - \operatorname{codim}_\zeta(\zeta \cap \sigma_J) = l - rk \mathscr{F}_{[0,n]-J}(\zeta)$.*

This proposition shows that strata on Grassmannians can be described in terms of the rank function $I \to \operatorname{rk} \mathscr{F}_I(\zeta)$, an approach which has been taken in [4]. Now we consider how the notions and results of [4] are related to the notions and results formulated above.

As in §2 we assume that the space ζ does not lie in any coordinate hyperplane in CP^n. By Proposition 3.1.1 this condition is equivalent to the relation $\operatorname{rk} F_I(\zeta) = l$ for all n-element subsets I in $[0, n]$.

3.1.2. Proposition. (a) *The hyperplanes $\zeta \cap \sigma_j$ are different if and only if $\operatorname{rk} F_I(\zeta) = l$ for all $(n-1)$-element subsets I in $[0, n]$.*

(b) *The hyperplanes $\zeta \cap \sigma_1, \ldots, \zeta \cap \sigma_n$ have only normal intersections in the affine space $K_\zeta = \zeta - \sigma_0$ if and only if the vector $f_0(\zeta)$ is in general position with respect to the family of vectors $F_{[1,n]}(\zeta)$, i.e. if it does not belong to any hyperplane spanned by those vectors.*

(c) *Let conditions* (a) *and* (b) *be satisfied and let $J \subset [1, n]$. Then $S_J \neq \varnothing$ if and only if $\operatorname{rk} F_{[1,n]-J}(\zeta) = l$. In particular, S_J is a vertex if and only if the system of vectors $F_{[1,n]-J}(\zeta)$ forms a basis in the space $(\overline{\zeta}^{\perp})^*$.*

All these statements follow immediately from Proposition 3.1.1.

Proposition 3.1.2 shows that basic strata on Grassmannians are exactly those studied in Section 4 in [4].

3.2. Now let ζ be a real point of a basic stratum Γ, i.e. the space $\overline{\zeta}$ is a complexification of the real space $\overline{\zeta}_R \in G_{k+1}(R^{n+1})$. Denote the real vector space spanned by linear forms $f_0(\zeta), \ldots, f_n(\zeta)$ in $\overline{\zeta}^{\perp}$ by $L(\zeta)$. The space $L(\zeta)$ is l-dimensional and is naturally identified with the quotient space $R^{n+1}/\overline{\zeta}_R$ (under this identification $f_i(\zeta)$ is the image of the basis vector $e_i \in R^{n+1}$ in the quotient space $L(\zeta)$). Again, let $K_\zeta = \zeta - \sigma_0$ and $q_i = x_i/x_0$ $(1 \leqq i \leqq n)$ be affine linear functions defining hyperplanes $S_1, \ldots, S_n$ in K_ζ. Let $K_{\zeta,R}$ be the real part of K_ζ. It consists of the points in which all q_i are real. For each set $\varepsilon = (\varepsilon_1, \ldots, \varepsilon_n)$ which each ε_i is either 1 or -1 denote by $\Delta(\varepsilon, \zeta)$ the part of $K_{\zeta,R}$ defined by the set of inequalities $\varepsilon_i q_i > 0$ $(1 \leqq i \leqq n)$. Evidently, non-empty sets $\Delta(\varepsilon, \zeta)$ are exactly connected components of the space $K_{\zeta,R} - S$.

3.2.1. Proposition. *Let ζ be a real point of a basic stratum in $G_k(CP^n)$ and $\varepsilon = (\varepsilon_1, \ldots, \varepsilon_n)$ a set of signs. Then:* (a) *Chamber $\Delta(\varepsilon, \zeta)$ is non-empty if and only if the vector $(-f_0(\zeta))$ lies in the open convex cone generated by the vectors $\varepsilon_i f_i(\zeta)$ $(i = 1, \ldots, n)$.*

(b) *A vertex S_J is an extreme point of the chamber $\Delta(\varepsilon, \zeta)$ if and only if $(-f_0(\zeta))$ lies in the open convex cone generated by the vectors $\varepsilon_i f_i(\zeta)$ $(i \in [1, n] - I)$.*

(c) *Chamber $\Delta(\varepsilon, \zeta)$ is bounded if and only if all the vectors $\varepsilon_1 f_1(\zeta), \ldots, \varepsilon_n f_n(\zeta)$ lie in an open half-space in $L(\zeta)$.*

All these statements follow immediately from the definitions.

3.3. Fix a basic stratum Γ in $G_k(CP^n)$, and let Γ_R be a set of its real points. Define the mapping $p: \Gamma_R \to G_{k-1}(CP^{n-1}) = G_{k-1}(\sigma_0)$ by the formula $p(\zeta) = \zeta \cap \sigma_0$. Evidently, p is the projection of Γ_R on the set of real points of some stratum in $G_{k-1}(CP^{n-1})$. Fix $\zeta_0 \in p(\Gamma_R)$ and let $\Gamma^0 = p^{-1}(\zeta_0) = \{\zeta \in \Gamma_R : \zeta \cap \sigma_0 = \zeta_0\}$. Denote by L the space $L(\zeta_0) = R^n / \bar{\zeta}_{0,R}$ and by $\mathscr{F} = (f_1, \ldots, f_n)$ the family of vectors in L where f_i is the image in L of the basis vector e_i. Let L^0 be an open set in L consisting of those vectors that are in general position with the vectors $f_1, \ldots, f_n$. We now show that the set L^0 can be naturally identified with the submanifold Γ^0 in Γ_R.

For each $\xi \in \Gamma^0$ the embeddings $R^n \to R^{n+1}$ and $\bar{\zeta}_0 \to \bar{\xi}$ induce the isomorphism $L \tilde{\to} L(\xi)$. Let us identify the spaces $L(\xi)$ with L by those isomorphisms. For each $i \in [0, n]$ let $F_i: \Gamma^0 \to L$ be the mapping which takes a point ξ into the image of the vector $f_i(\xi) \in L(\xi)$ under the above identification of $L(\xi)$ with L. The definitions immediately imply:

3.3.1. Proposition. *The mappings $F_1, \ldots, F_n$ are constant on Γ^0, and $F_i(\xi) = f_i$ for all $i \in [1, n]$. The mapping F_0 defines a one-to-one correspondence between Γ^0 and L^0.*

3.4. In addition to the assumptions formulated in Section 3.3 suppose that all the vectors $f_1, \ldots, f_n$ lie in an open half-space in L. By Propositions 3.2.1(c) and 3.3.1 this is equivalent to the condition that the chamber $\Delta_+(\xi) = \Delta((1, \ldots, 1), \xi)$ be bounded in the space K_ξ for all $\xi \in \Gamma^0$. Let $H(\mathscr{F})$ be the vector space of piecewise constant functions on L^0, spanned by the characteristic functions χ_I for all subsets C_I^0 in L^0 (see Introduction).

3.4.1. Theorem. *For all $\xi \in \Gamma^0$ there is a natural isomorphism $\chi: H^k(K_\xi, S; C) \to H(\mathscr{F})$ taking each generator Δ^J corresponding to the vertex S_J into $\chi_{[1,n]-J}$ (see Section 1.9).*

This theorem follows from Theorem 1.9.1, the argument given in Section 1.8 and the results of Section 3 in [4]. Now we give a direct geometric construction of the isomorphism χ.

First of all we note that there are natural bundles $\mathscr{H}_k$ and H^k on Γ^0 with fibres at $\xi \in \Gamma^0$ equal, respectively, to $H_k(K_\xi, S; C)$ and $H^k(K_\xi, S; C)$. In the fibre bundles $\mathscr{H}_k$ and $\mathscr{H}^k$ there is the natural Gauss–Manin connection.

3.4.2. Proposition [7]. (a) *Monodromy in $\mathscr{H}^k$ and $\mathscr{H}_k$ is trivial, and, consequently, the spaces $H_k(K_\xi, S; C)$ can be naturally identified for all $\xi \in \Gamma^0$, the same being true for the spaces $H^k(K_\xi, S; C)$.*

(b) *The elements $\Delta^J \in H^k(K_\xi, S; C)$ are covariantly constant.*

By virtue of Proposition 3.4.2(a) we can denote each of the spaces $H^k(K_\xi, S; C)$ simply by $H^k(K, S, C)$, thus omitting any mention of the point ξ (and doing likewise for the spaces $H_k(K_\xi, S, C)$). To each $h \in H^k(K, S, C)$ we associate a function X_h on Γ^0, associating to each point $\xi \in \Gamma^0$ the value $h(\Delta_+(\xi))$ of the class h on the element $\Delta_+(\xi) \in H_k(K_\xi, S, C)$ (see Section 1.9). Identifying Γ^0 with L^0 by the mapping $(-F_0)$ (Proposition 3.3.1), we see that the correspondence $h \to X_h$ defines a linear mapping X from $H^k(K, S, C)$ into the space of functions on L^0. It is the isomorphism X of Theorem 3.4.1. The equality $X(\Delta^J) = X_{[1,n]-J}$ follows immediately from the defition of Δ^J in Section 1.9 and Proposition 3.2.1(b).

3.5. Under the assumptions of Sections 3.3 and 3.4, consider a continuous analogue of the partition function, i.e. the function $P_{\mathscr{F}}$ on L^0 defined in the Introduction. Identifying L^0 with Γ^0 by the mapping $(-F_0)$ one has a function on Γ^0 which will also be denoted by $P_{\mathscr{F}}$. We now show that the function $P_{\mathscr{F}}(\xi)$ is a restriction to Γ^0 of a hypergeometric function on the stratum Γ.

First, we fix some volume forms on L and R^n and note that for all $\xi \in \Gamma^0$ they define a volume form on K_ξ, and, consequently, an element ω_ξ of the space $D(\xi)$ (see Section 2.2). Define the set of exponents $\alpha^0 = (\alpha_0, \ldots, \alpha_n)$ by setting $\alpha^0 = (-k, 1, \ldots, 1)$. All the corresponding exponents τ_j are equal to 1 so that the local system L_τ coincides with the trivial system C on each space $K_\xi - S$. Let $\widetilde{\Delta_+(\xi)} \in H_k^{lf}(K_\xi - S, C)$ be a "visible cycle" corresponding to the bounded chamber $\Delta_+(\xi)$ in the space $K_{\xi,R} - S$ (see Section 1.5).

3.5.1. Proposition. *The function $P_{\mathscr{F}}(\xi)$ coincides with the hypergeometric function $\Phi_{\widetilde{\Delta_+(\zeta)}}(\alpha^0; \xi, \omega)$ in a neighbourhood of each point $\zeta \in \Gamma^0$ (see Section 2.5).*

This follows directly from the definitions.

3.6. Recall that by a fundamental system B we mean a set of l-subsets I in $[1, n]$ which are such that the functions X_I for $I \in B$ form a basis in the space $H(\mathscr{F})$. Fix a fundamental system B and denote by B^c the system of k-subsets $J = [1, n] - I$ for all $I \in B$. By virtue of Theorem 3.4.1, the elements Δ^J for $J \in B^c$ form a basis in $H^k(K, S; C)$. Denote by $\{\Delta_{J,B}\}_{J \in B^c}$ the dual basis in the space $H_k(K, S; C)$. Proposition 3.4.2(b) ensures that all elements of $\Delta_{J,B}$ are covariantly constant.

For each $\xi \in \Gamma^0$ consider the homomorphism $v_\xi: H_k(K_\xi, S; C) \to H_k^{lf}(K_\xi - S, C)$ taking each bounded component into the "visible cycle" $\tilde{\Delta}$ (see Sections 1.9, 1.5). Let $h_{J,B}(\xi) = v_\xi(\Delta_{J,B})$. For each $I \in B$ define the function $P_I^{(B)}$ on Γ^0 by the formula $P_I^{(B)}(\xi) = \Phi_{h_{J,B}(\xi)}(\alpha^0, \xi, \omega_\xi)$ where $J = [1, n] - I$ and α^0 and ω_ξ are defined in Section 3.5. As usual, identifying Γ^0 with L^0 by the mapping $(-F_0)$ one gets functions on L^0 which will also be denoted by $P_I^{(B)}$.

3.6.1. Theorem. (a) *There is a decomposition*

$$P_{\mathscr{F}} = \sum_{I \in B} \chi_I \cdot P_I^{(B)}.$$

(b) *Each function $P_I^{(B)}$ is a restriction on L^0 of some polynomial in the space L.*

Proof. It follows immediately from the definitions that $\widetilde{\Delta_+(\xi)} = \sum_{J \in B^c} \Delta^J(\Delta_+(\xi)) \cdot h_{J,B}(\xi)$. The construction given in Section 3.4 shows that $\Delta^J(\Delta_+(\xi)) = X_{[1,n]-J}(-F_0(\xi))$, and, therefore, part (a) follows from Proposition 3.5.1. Part (b) is a consequence of Proposition 3.4.2 (see [7]).

Remarks. 1. It follows from the definitions that the cycle $h_{J,B}(\xi)$ is a linear combination of "visible cycles" in $H_k^{lf}(K_\xi - S, C)$. Hence each function $P_I^{(B)}$ can be interpreted in a geometrical way as a linear combination of volumes of some k-dimensional polyhedra. In order to find these polyhedra one has to decompose the elements $\Delta_{J,B}$ with respect to the basis consisting of bounded components in the group $H_k(K_\xi, S; C)$.

2. The decomposition of Theorem 3.6.1(a) coincides with the decomposition of Theorem 5.1 in [4]. Thus, we have obtained a geometrical interpretation and a proof of that theorem.

3. Proposition 3.5.1 and Theorem 3.6.1 are trivially extended to the case when the set of exponents α^0 is replaced by an arbitrary set $\alpha = (\alpha_0, \alpha_1, \ldots, \alpha_n)$ with the sum $n - k$ such that all the numbers $\alpha_1, \ldots, \alpha_n$ are positive integers. This provides a full geometrical interpretation of Theorem 5.1 in [4].

4. The decomposition of Theorem 3.6.1(a) can be extended to the case of arbitrary exponents α (including non-integer exponents).

References

1. Gelfand, I.M.: General theory of hypergeometric functions. Dokl. Akad. Nauk SSSR **288** (1) (1986) 14–18 (Russian).
2. Gelfand, I.M., Gelfand, S.I.: General hypergeometric equations. Dokl. Akad. Naūk SSSR **288** (2) (1986) 279–283 (Russian).
3. Gelfand, I.M., Graev, M.I.: Duality theorem for general hypergeometric functions. Dokl. Akad. Naūk SSSR **289** (1) (1986) 19–23 (Russian).
4. Gelfand, I.M., Zelevinskij, A.V.: Algebraic and combinatorial aspects of the general theory of hypergeometric functions. Funct. Anal. Appl. **20** (3) (1986) 17–34 (Russian).
5. Gelfand, I.M., Serganova, V.V.: Combinatorial geometries and torus strata on homogeneous compact manifolds. Usp. Mat. Naūk **42** (2) (1987) (Russian).
6. Whittaker, E., Watson, G.: A course of modern analysis. Cambridge University Press, 1927.
7. Varchenko, A.N.: Combinatorics and topology of positioning affine hyperplanes in a real space. Funct. Anal. Appl. **21** (1) (1987) 11–22 (Russian).
8. Aomoto, K.: On vanishing of cohomology attached to certain many valued functions meromorphic. J. Math. Soc. Japan **27** (1975) 248–255.
9. Aomoto, K.: On the structure of integrals of power product of linear functions. Sci. papers, Coll. Gen. Ed., Univ. Tokyo, 1977, Vol. 27, pp. 49–61.
10. Aomoto, K.: Les équations aux differences linéaires et les intégrales des functions multiforms. J. of the Fac. of Sci., Univ. of Tokyo, 1975, Vol. 22, pp. 271–297.
11. Berezin, F.A., Gelfand, I.M.: Some remarks on the theory of spherical functions on symmetric Riemannian manifolds. Trudy MMO **5** (1956) 311–352 (Russian).
12. Kostant, B.: A formula for the multiplicity of a weight. Trans. Amer. Math. Soc. **93** (1959) 53–73.
13. Heckman, G.J.: Projections of orbits and asymptotic behavior of multiplicities for compact connected Lie groups. Invent. Math. **67** (2) (1982) 333–356.
14. Vasiliev, V.A., Gelfand, I.M., Zelerinskij, A.V.: Behaviour of general hypergeometric functions in complex domains. Dokl. Akad. Naūk SSSR **290** (2) (1986) 277–281 (Russian).

15. Deligne, P., Mostow, G.D.: Monodromy of hypergeometric functions and non-lattice integral monodromy. Publ. Math. IHES **63** (1986) 5–90.
16. Pham, F.: Formules de Picard-Lefschetz generalisées et ramification des integrales. Bull. Soc. Math. France **93** (1965) 333–367.

Moscow State University

Received 10 October 1986

9.

(with M.I. Graev and A.V. Zelevinskij)

Holonomic systems of equations and series of hypergeometric type

Dokl. Akad. Naūk SSSR **295** (1) (1987) 14–19

This note joins a series of papers on the theory of general hypergeometric functions (see [1–10]); however, it can be read independently. According to [1] general hypergeometric functions can be defined as functions on the Grassmannian $G_k(\mathbf{C}^n)$ satisfying some holonomic system of linear partial differential equations. Some of these equations form the homogeneity conditions with respect to the natural action of the commutative group $(\mathbf{C}^*)^n$ on $G_k(\mathbf{C}^n)$. The group $(\mathbf{C}^*)^n$ acts in the standard local coordinates on $G_k(\mathbf{C}^n)$ by linear transformations.

We will show that the system from [1] can be significantly generalized, namely, we associate to any complex torus $H \simeq (\mathbf{C}^*)^n$ in the group $GL(V)$ of linear transformations of the complex N-dimensional vector space V such that H contains all scalar transformations a holonomic system of linear differential equations on V. This system will be called the system of hypergeometric type associated to the subgroup $H \subset GL(V)$.

Solutions of a system of hypergeometric type are functions on V that are homogeneous under H. Therefore they actually depend on $N-n$ variables. We study a class of solutions that are a special power series on $r = N - n$ variables: their coefficients are products of N factors, each defined as a value of the Γ-function at some point determined as a linear function on indices (see formula (4) below). A lot of classical hypergeometric functions can be expressed in such a form. For example, for the Gauss function we have $r=1$, $N=4$, for the generalized hypergeometric function ${}_pF_{p-1}$ we have $r=1$, $N=2p$, for the Appel functions F_2 and F_3 we have $r=2$, $N=7$. Our approach to series consists, therefore, in representing the r-dimensional space of variables as the quotient of an N-dimensional space by the action of an N-dimensional complex torus.

For the system of equations from [1] on the Grassmannian $G_k(\mathbf{C}^n)$, the methods developed here enable $\binom{n-2}{k-1}$ linearly independent solutions to be explicitly constructed; this gives a simple proof of the exactness of the bound from [1] for the number of solutions. Another proof of this fact based on the integral representations of general hypergeometric functions is given in [3].

Let us note that the class of systems of hypergeometric type includes, in particular, a lot of systems related to various strata on Grassmannians (see [2, 6, 7]).

1. Equations of hypergeometric type

Let H be an n-dimensional complex torus included in the group $GL(V)$ of linear transformations of an N-dimensional complex space V and containing the group of scalar transformations.

Fix a basis $e_1,\ldots,e_N$ in V in which all transformations from H are diagonal. Let $\chi_1,\ldots,\chi_N$ be characters of the group H such that $he_i=\chi_i(h)e_i$ for $h\in H$; the characters $\chi_1,\ldots,\chi_N$ are called roots of the group H. We will write vectors $v\in V$ in the form $v=\sum v_ie_i$ and elements of the Lie algebra $\mathfrak{h}$ of the group H as vector fields on V. Let $\chi'_1,\ldots,\chi'_N$ be the linear functionals on $\mathfrak{h}$ corresponding to the roots; then any field $X\in\mathfrak{h}$ can be written in the form $\sum_{1\leq i\leq N}\chi'_i(X)v_i(\partial/\partial v_i)$.

Choosing the coordinates $z_1,\ldots,z_n$ in the group $H\simeq(\mathbf{C}^*)^n$ we can write each root χ_i in the form $\chi_i=\prod_{1\leq k\leq n}Z_k^{\chi_{ki}}$ where (χ_{ki}) is some $(n\times N)$ matrix of rank n with integral entries; the condition that H contains multiplications by scalars is equivalent to the following condition: $(*)$ there exist integers $c_1,\ldots,c_n$ such that $\sum_{1\leqslant k\leqslant n}c_k\chi_{ki}=1$ for $i=1,\ldots,N$. It is easy to see that vector fields $X_1,\ldots,X_n$ such that $\chi'_i(X_k)=\delta_{ki}$ form a basis in $\mathfrak{h}$.

For any linear functional $\beta\in\mathfrak{h}^*$, a system of hypergeometric type with the parameter β is the following system of linear differential equations of the function $\Phi(v)$, $v\in V$.

(1) H-homogeneity: $X\Phi=\beta(X)\Phi$ for all $X\in\mathfrak{h}$.

(2) Higher order equations: for any relation among roots of the form $\chi_1^{a_1},\ldots,\chi_N^{a_N}=1$, $a_i\in\mathbf{Z}$ we have

$$\left[\prod_{i:a_i>0}\left(\frac{\partial}{\partial v_i}\right)^{a_i}\right]\Phi=\left[\prod_{i:a_i<0}\left(\frac{\partial}{\partial v_i}\right)^{-a_i}\right]\Phi.$$

Let, as above, take coordinates in the group H. Then the equations (1) are equivalent to the following system of n equations:

$$\sum_{1\leq i\leq N}\chi_{ki}v_i\frac{\partial\Phi}{\partial v_i}=\beta_k\Phi(v),\quad 1\leq k\leq n,$$

depending on n complex parameters $\beta_k=\beta(X_k)$. Let $L=\{a=(a_1,\ldots,a_N)\in\mathbf{Z}^N$: $\chi_1^{a_1},\ldots\chi_N^{a_N}=1\}$ be the integral lattice of relations among the roots; in other words, points $a\in L$ are integral solutions of the system $\sum_{1\leq i\leq N}a_i\chi_{ki}=0, 1\leq k\leq n$. The condition $(*)$ easily gives $\sum_{1\leq i\leq N}a_i=0$ for all $a\in L$; therefore all equations (2) are homogeneous. One can easily check that all equations (2) follow from the finite number of them.

Example. Let V be the space of complex $(k\times l)$ matrices $v=(v_{ij})$, $1\leq i\leq k$, $1\leq j\leq l$, and let H be the group of linear transformations of V generated by all dilatations of rows and columns, so that in this case $N=kl$, $n=k+l-1$. The equations (1) mean that the function $\Phi(v)$ is homogeneous of given homogeneity degree in all rows and columns of a matrix v. The lattice L consists of integral $(k\times l)$ matrices with zero sums in all rows and columns. The equations (2) are equivalent to the following system:

$$\frac{\partial^2 \Phi}{\partial v_{ij}\partial v_{i'j'}}=\frac{\partial^2 \Phi}{\partial v_{i'j}\partial v_{ij'}}$$

for all i,j,i',j'. So in this case the system (1), (2) coincides with the system from [1] for hypergeometric functions on the generic strata in the Grassmannian $G_k(\mathbf{C}^{k+1})$ written in local coordinates (see [4]).

2. Holonomicity

Theorem 1. *The system* (1), (2) *determines a holonomic D-module on the space V.*

Proof. Let $W\subset T^*V$ be the set of zeros of the principal symbols of equations (1), (2). To prove the theorem it suffices to check that $\dim W\leqq N$ (cf. [1]). For all $v,\xi\in\mathbf{C}^N$ let $\xi\cdot v=(\xi_1 v_1,\xi_2 v_2,\ldots,\xi_N v_N)\in\mathbf{C}^N$. Also let $L_{\mathbf{C}}\subset\mathbf{C}^N$ be the complexification of the lattice L. Equations (1), (2) imply that W can be considered as the set of all pairs $(v,\xi)\in\mathbf{C}^N\times\mathbf{C}^N$ satisfying the following conditions:

(1′) $\xi\cdot v\in L_{\mathbf{C}}$;

(2′) $\prod_{i:a_i>0}\xi_i^{a_i}=\prod_{i:a_i<0}\xi_i^{-a_i}$ for all $a\in L$.

For any subset $I\subset[1,n]$ let $\mathbf{C}^I=\{\xi\in\mathbf{C}^N:\xi_i=0 \text{ for } i\notin I\}$, $(\mathbf{C}^*)^I=\{\xi\in\mathbf{C}^I:\xi_i\neq 0$ for $i\in I\}$, and $W_I=\{(v,\xi)\in W:\xi\in(\mathbf{C}^*)^I\}$. As W is the union of all W_I's it is enough to prove that $\dim W_I\leqq N$ for all I.

Let $L^I=L\cap\mathbf{C}^I$, $L^I_{\mathbf{C}}=L_{\mathbf{C}}\cap\mathbf{C}^I$ be the complexification of the lattice L^I and $H_I\subset(\mathbf{C}^*)^I$ be the subgroup defined by equations $\prod_{i\in I}\xi_i^{a_i}=0$ for all $a\in L^I$ (in particular, $H_{[1,N]}=H$). If $(v,\xi)\in W_I$ then, by (1′) and (2′), we have $\xi\cdot v\in L^I_{\mathbf{C}}$ and $\xi\in H_I$. As $\dim H_I=|I|-\operatorname{rk}L^I$ and $\dim\{v\in\mathbf{C}^N:\xi\cdot v\in L^I_{\mathbf{C}}\}=\dim L^I_{\mathbf{C}}+(N-|I|)$ for all $\xi\in H_I$, we have $\dim W_I\leqq(|I|-\operatorname{rk}L^I)+\dim L^I_{\mathbf{C}}+(N-|I|)=N$, as required.

Using the above notation, let $D_I=(H_I\cdot L^I_{\mathbf{C}})\times\mathbf{C}^{[1,N]-I}\subset\mathbf{C}^N=V$ for any $I\subset[1,N]$. Let $D\subset V$ be the union of all D_I and all coordinate hyperplanes in V.

Theorem 2. (a) *The dimension of D is equal to* $N-1$.

(b) *Solutions of the system* (1), (2) *in any open region in* $V\backslash D$ *form a finite-dimensional space of analytic functions.*

3. Power series

For $v=\sum v_i e_i\in V$, $\gamma=(\gamma_1,\ldots,\gamma_N)\in\mathbf{C}^N$ with all v_i's nonzero, let $v^\gamma=\prod_i v_i^{\gamma_i}$.

Proposition 1. *Let* $\gamma\in\mathbf{C}^N$ *satisfy the condition* $\prod_{1\leq i\leq N}\gamma_i\chi_i'=\beta\in\mathfrak{f}^*$. *Then the formal series*

(3) $\Phi(\gamma;v)=v^\gamma\sum_{a\in L}(\prod_{1\leq i\leq N}\Gamma(\gamma_i+a_i+1))^{-1}v^a$ *satisfies equations* (1) *and* (2).

Proof. Obvious.

Vectors $\gamma \in \mathbf{C}^N$ satisfying the conditions of Proposition 1 will be called compatible with β. To pass from formal solutions to analytic ones we choose a basis $A = \{a^1, \ldots, a^r\}$ in the lattice L and define a neighbourhood $U(A, \varepsilon) \subset V$ for any $\varepsilon > 0$ by means of conditions $0 < |v^{a^j}| < \varepsilon$, $1 \leqq j \leqq r$. For all $m = (m_1, \ldots, m_r) \in \mathbf{Z}^r$ we define

$$c_m(\gamma, A) = \left(\prod_{1 \leqq i \leqq N} \Gamma\left(\gamma_i + \sum_j a_i^j m_j + 1 \right) \right)^{-1}.$$

We will say that a vector γ satisfies the break-off condition with respect to A if $c_0(\gamma, A) \neq 0$ and $c_m(\gamma, A) = 0$ whenever at least one of m_j is negative. Let $x_j = v^{a^j}$; it is clear that if γ satisfies the break-off condition with respect to A the sum over the lattice L in (3) can be written as a power series

$$\sum_{m_1, \ldots, m_r \geqq 0} c_m(\gamma, A) x_1^{m_1} \cdots x_r^{m_r}.$$

Proposition 2. If a vector $\gamma \in \mathbf{C}^N$ is compatible with β and satisfies the break-off condition with respect to A then the series $\Phi(\gamma; v)$ converges in the region $U(A, \varepsilon)$ for sufficiently small $\varepsilon > 0$ and gives there a (multi-valued) analytic function satisfying (1) and (2). For different γ's these functions are linearly independent.

Let us note that the series (4) can be naturally considered as the restriction of the function $\Phi(\gamma; v)$ to some section of the fibration $U(A, \varepsilon) \to U(A, \varepsilon)/H$.

By Theorem 2, for fixed β and A only a finite number of vectors γ exist which are compatible with β and satisfy the break-off condition with respect to A. Their determination is a purely combinatorial problem. Let us introduce a class of lattices L for which the description of these vectors is particularly simple. Let a, b be nonzero vectors of L; b is said to be subordinate to a if $|b_i| \leqq |a_i|$ and $a_i b_i \geqq 0$ for all $i = 1, \ldots, N$. A lattice L is said to be simple if for any nonzero $a \in L$ there exists a subordinate to a vector $b \in L$ with all coordinates equal to 0 or ± 1.

An n-element subset $I \subset [1, N]$ is called a base if the functionals χ_i' with $i \in I$ form a basis in $\mathfrak{h}^*$ (so bases are bases of the matroid associated to the family of vectors $\chi_1', \ldots, \chi_N'$, see [5, 6]). By definition, for any $\beta \in \mathfrak{h}^*$ and for any base I there exists exactly one vector $\gamma \in \mathbf{C}^N$ compatible with β and satisfying the condition $\gamma_i = 0$ for $i \in I$; denote this vector by $\gamma(\beta; I)$. A vector $\beta \in \mathfrak{h}^*$ is said to be nondegenerate if for all bases I and for all $i \in I$ the i-th coordinate of the vector $\gamma(\beta; I)$ is different from $0, -1, -2, \ldots$; this implies, in particular that all vectors $\gamma(\beta; I)$ are distinct.

Let a lattice L be simple. Denote by $B = B(L)$ the set of all nonzero vectors $b \in L$ such that b has no subordinates; by the simplicity of L, all coordinates of any vector $b \in B$ are 0 or ± 1. For $b \in B$, write $S_-(b) = \{i \in [1, N]: b_i = -1\}$. A base I is said to be A-admissible if I does not contain any of the sets $S_-(b)$ for those vectors $b \in B$ that have at least one negative coefficient in the decomposition with respect to the basis A.

Theorem 3. *Let a lattice L be simple, let A be a basis of L and let $\beta \in \mathfrak{h}^*$ be nondegenerate. Then the vectors γ compatible with β and satisfying the break-off condition with respect to A are exactly the vectors $\gamma(\beta; I)$ for all A-admissible bases I.*

Corollary 1. *Under the assumptions of Theorem 3, the number of linearly independent analytic solutions of the system* (1), (2) *in the region* $U(A, \varepsilon)$ *for all sufficiently small* $\varepsilon > 0$ *is not less than the number of A-admissible bases.*

Example. Let $V = \mathbf{C}^{2p}$ and let H be the group of all dilatations $v \mapsto c.v = (c_1 v_1, \ldots, c_{2p} v_{2p})$ such that $c_1 c_2 \cdots c_p = c_{p+1} \cdots c_{2p}$. Then L is a simple lattice of rank 1 generated by the vector $a = (1, 1, \ldots, 1, -1, \ldots, -1)$. Let us write $\beta \in \mathfrak{h}^*$ in the form $\beta = \sum_{1 \leq i \leq p} (\beta_i \chi'_i - \beta'_i \chi_{i+p})$; the nondegeneracy of β means that no two of the numbers $\beta_1, \ldots, \beta_p, \beta'_1, \ldots, \beta'_p$ differ by an integer. The equations (2) are reduced to a single equation of order p:

$$\frac{\partial^p \Phi}{\partial v_1 \cdots \partial v_p} = \frac{\partial^p \Phi}{\partial v_{p+1} \cdots \partial v_{2p}}.$$

Let $A = \{a\}$, and $x = (-1)^p v^a = (-1)^p v_1 \cdots v_p / v_{p+1} \cdots v_{2p}$. Then A-admissible bases are sets $I_i = [1, 2p] \setminus \{i\}$ where $i = 1, \ldots, p$; corresponding vectors $\gamma^{(i)} = \gamma(\beta; I_i)$ are equal to $\gamma^{(i)} = \beta - \beta_i a$. One can easily see that the function $\Phi(\gamma^{(i)}, v)$ is proportional to

$$v^{\gamma^{(i)}} \cdot {}_pF_{p-1} \begin{bmatrix} \beta'_1 - \beta_i, \beta'_2 - \beta_i, \ldots, \beta'_p - \beta_i; x \\ \beta_1 - \beta_i + 1, \ldots, \beta_{i-1} - \beta_i + 1, \beta_{i+1} - \beta_i + 1, \ldots, \beta_p - \beta_i + 1 \end{bmatrix}$$

where ${}_pF_{p-1}$ is the classical generalized hypergeometric series (see [11]). These functions give p linearly independent solutions of the system (1), (2) in the region $\{0 < |x| < \varepsilon\}$.

4. Application to General Hypergeometric Functions

Proposition 3. *Let V and H be as in the example from Sect 1. Then*

(a) *the lattice L is simple;*

(b) *bases are* $(k + l - 1)$*-element subsets* $I \subset [1, k] \times [1, l]$ *intersecting all rows and columns and such that any two points from I can be joined in I by a chain of a rook's moves;*

(c) *the set* $A = \{a^{ij} = e_{ij} - e_{i+1,j} - e_{i,j+1} + e_{i+1,j+1}\colon 1 \leqq i \leqq k-1,\ 1 \leqq j \leqq l-1\}$ *is a basis of L; corresponding admissible bases are all shortest rook paths from the point* $(k, 1)$ *to the point* $(1, l)$. *The number of such bases is* $\binom{k+l-2}{k-1}$.

Using the bound for the number of solutions of the system (1), (2) from [1], we get

Corollary 2. *For any nondegenerate* β *the dimension of the space of analytic solutions of the system* (1), (2) *in the region* $U(A, \varepsilon)$ *for sufficiently small* $\varepsilon > 0$ *equals* $\binom{k+l-2}{k-1}$ *and functions* $\Phi(\gamma(\beta, I); v)$ *for all A-admissible bases I form a basis in this space.*

Remarks. (1) Many other solutions of our system have been constructed in [4] for the case when $k=l=3$, corresponding to the Grassmannian $G_3(\mathbf{C}^6)$. All these solutions are of the form $\Phi(\gamma(\beta; I); v)$ for some basis A and some A-admissible base I.

(2) Many nondegenerate strata of the Grassmannian $G_k(\mathbf{C}^{k+l})$ are represented by open subsets in $\mathbf{C}^J$ for some $J \subset [1, k] \times [1, l]$ (for example, in the case of $G_3(\mathbf{C}^6)$ only one stratum, up to the action of the Weyl group, is not of such a form). Hypergeometric functions on such strata satisfy the system of the form (1), (2) for some subgroup H_J and some lattice L^J (see the proof of Theorem 1). It is clear that all lattices L^J are simple, so that hypergeometric functions on these strata can be constructed with the use of Proposition 2 and Theorem 3. For example, if $k=l=p$ and J consists of $2p$ points $(1,1), (2,2), \ldots, (p,p), (1,2), (2,3), \ldots, (p-1,p)$, $(p,1)$ then the action of H_J on $\mathbf{C}^J$ is isomorphic to the one from the examaple in Sect. 3, so that the hypergeometric functions constructed in the above example correspond to a special stratum in $G_p(\mathbf{C}^{2p})$.

Received 26.III.1987

References

1. Gelfand, I.M., Gelfand, S.I.: Dokl. Akad. Naūk **288** (2) (1986) 279–283
2. Gelfand, I.M.: Dokl. Akad. Naūk SSSR **288** (1) (1986) 14–18
3. Vasiliev, V.A., Gelfand, I.M., Zelevinskij, A.V.: Funkz. Anal. Prill. **21** (1) (1987) 23–28.
4. Gelfand, I.M., Graev, M.I.: Dokl. Akad. Naūk SSSR **293** (2) (1987) 288–293
5. Gelfand, I.M., Zelevinskij, A.V.: Funkz. Anal. Pril. **20** (3) (1986) 17–24
6. Gelfand, I.M., Serganova, V.V.: Usp. Mat. Nauk **42** (2) (1987) 107–133
7. Gelfand, I.M., Serganova, V.V.: Dokl. Akad. Naūk SSSR **292** (1) (1987) 15–20
8. Gelfand, I.M., Serganova, V.V.: Dokl. Akad. Naūk SSSR **292** (3) (1987) 524–528
9. Vasiliev, V.A., Gelfand, I.M., Zelevinskij, A.V.: Dokl. Akad. Naūk SSSR **290** (2) (1986) 277–281
10. Gelfand, I.M., Graev, M.I.: Dokl. Akad. Naūk SSSR **289** (1) (1986) 19–23
11. Bateman, H., Erdelyi, A.: Higher Transcendental Functions, Vol. 1. New York Toronto London: McGraw-Hill, 1953

10.

(with A.N. Varchenko)

On Heaviside functions of configuration of hyperplanes

Funkts. Anal. Prilozh. 21 (4) (1987) 1–18

§1. Introduction

The ring of integer-valued functions defined on M, the complement of a finite union of hyperplanes in a real affine space, whose values on the components are constants, has been considered. The ring is denoted by P; it contains distinguished multiplicative generators, namely, Heaviside functions of hyperplanes defined in the following manner: for a given hyperplane, fix a function which is equal to 1 on one side of the hyperplane and 0 on the other. Each element of the ring is a polynomial in Heaviside functions. A filtration of the ring by degrees of polynomials, denoted by $\{P^k\}$, $k \geqq 0$, introduces some properties into P which are close to those of the cohomology ring of the complement $M_{\mathbf{C}}$ to a union of complexified hyperplanes in a complexified affine space.

The ring $H^*(M_{\mathbf{C}})$ has been described by V.I. Arnold [1], E. Brieskorn [2], and P. Orlik and L. Solomon [3]. Orlik and Solomon have attracted attention to the fact that the dimension of the space $H^*(M_{\mathbf{C}})$ is equal to the number of components of the set M. The present paper proposes an explanation of this fact based on comparison of the rings P and H^*. P is a commutative ring endowed with an increasing filtration $\{P^k\}$, H^* is an anticommutative ring endowed with a graduation $\{H^k\}$. We formulate the properties of the ring P by referring to similar known properties of the ring H^*.

The rings P and H^* can possibly be included in a one-parameter family of rings which have independent meanings in themselves.

This paper is related to investigations of general hypergeometric functions [4–10] and deals with the geometrical aspects of the theory.

The authors express their gratitude to V.I. Arnold for useful discussions.

1. Definition. Consider a finite set of linear functions $\{f_i\}$, $i \in I$, on an n-dimensional affine space V over the field $\mathbf{R}$. Denote by S the union of hyperplanes $A_i = \{v \in V \mid f_i(v) = 0\}$, $i = I$. We call the couple S and $\{f_i\}$ a configuration of hyperplanes. Consider the ring $P(S, \mathbf{Z})$ of integer-valued functions on $M = V \setminus S$, which are constant on every connected component. We consider, in the ring P, the multiplicative generators, i.e., the Heaviside functions x_i, $i \in I$, determined by the conditions: $x_i(v) = 1$ if $f_i(v) > 0$ and $x_i(v) = 0$ if $f_i(v) < 0$. Every function $x \in P(S, \mathbf{Z})$ is written as a polynomial, with integer coefficients, in $\{x_i\}, i \in I$. The minimum degree of polynomials in $\{x_i\}$ representing x is called the degree of the function $x \in P(S, \mathbf{Z})$.

Define an increasing filtration

$$O \subset P^0 \subset P^1 \subset \cdots \subset P,$$

where P^k is a subspace of functions which can be represented by polynomials of degree not exceeding k. In particular, the P^0 are the constant functions. Obviously, $P^k P^l \subset P^{k+l}$. We call $\{P^k\}$ the degree filtration.

Example. Consider a configuration of lines on a plane, $\{A_i\}$, $i \in I$. The degree filtration in the ring P consists of three terms: the constant functions P^0, the linear combinations of Heaviside functions P^1 and $P^2 = P$. Suppose that no three lines intersect at one point. A basis over $\mathbf{Z}$ in the ring consists of the constant function which is equal to 1, the Heaviside functions and all the monomials $x_i x_j$ with intersecting lines A_i and A_j. The dimension of P^0 is equal to 1. The dimension of P^1/P^0 is equal to the number of lines. The dimension of P^2/P^1 is equal to the number of points of intersection of lines.

We say that if the alternating sum of four of the values of the function $x \in P$ on four components of the complement approaching the point of intersection of two lines is equal to zero, then the function has a zero index at the point. A function $x \in P$ has degree $\leqq 1$ if and only if it has zero index at every point of intersection.

2. Properties of the ring P

Theorem 1. $P^n = P$, *that is, any piecewise constant function on the complement to the union of hyperplanes in an n-dimensional affine space is a polynomial of degree not exceeding n in Heaviside functions of hyperplanes.*

Let $V_{\mathbf{C}}$ be the complexification of the space V, $A_{i,\mathbf{C}}$ the complexification of a hyperplane A_i, $i \in I$, $S_{\mathbf{C}}$ the union of hyperplanes $\{A_{i,\mathbf{C}}\}$, $i \in I$, and $M_{\mathbf{C}} = V_{\mathbf{C}} \backslash S_{\mathbf{C}}$. Theorem 1 is an analogue of the statement: $H^k(M_{\mathbf{C}}) = 0$ for $k > n$.

The configuration S on V naturally induces on every affine subspace $U \subset V$ a new configuration denoted by S_U. S_U consists of hyperplanes $\{A \cap U \mid A \in S, U \not\subset A\}$ determined by linear functions $\{f_i|_U\}, i \in I$.

If an affine subspace U is not contained in S then a natural homomorphism $j_U: P(S) \to P(S_U)$ is defined so as to restrict functions of $P(S)$ on $U \backslash S_U$.

Any non-empty intersection F of hyperplanes of a configuration is called an edge. The Codimension of an edge is denoted by $r(F)$. In particular, hyperplanes are the edges of codimension 1. The set of all edges is denoted by $\mathscr{L}$.

A d-dimensional affine subspace $U \subset V$ is called a generally positioned space if U is transversre with respect to all the edges and crosses all the edges whose codimensions do not exceed d.

Theorem 2. *If $U \subset V$ is a generally positioned subspace, then the homomorphism j_U restricted to $P^k(S)$ defines an isomorphism between $P^k(S)$ and $P^k(S_U)$, for $k \leqq d$.*

This theorem is an analogue of Brieskorn's theorem [2]: if $U_{\mathbf{C}} \subset V_{\mathbf{C}}$ is a sufficiently general d-dimensional subspace, then $H^k(M_{\mathbf{C}}) \to H^k(M_{\mathbf{C}} \cap U_{\mathbf{C}})$ is an isomorphism for $k \leqq d$.

Let F be an edge of a configuration and $I^F \subset I$ the set of all the indices i such that $F \subset A_i$. We denote by S^F the configuration composed of hyperplanes $\{A_i\}$, $i \in I^F$, that is the hyperplanes containing F. We say that S^F is a localization of the configuration S at the edge F. Consider the ring $P(S^F, \mathbf{Z})$ of the configuration S^F. A natural inclusion $P(S^F, \mathbf{Z}) \to P(S, \mathbf{Z})$ exists, defined by restricting the functions of $P(S^F, \mathbf{Z})$ to M. Its image is the subring generated by functions $\{x_i\}$, $i \in I^F$. The inclusion pressrves the degree filtration.

Theorem 3. *The natural mapping*

$$\bigoplus_{\substack{F \in \mathscr{L} \\ r(F) = k}} P^k(S^F)/P^{k-1}(S^F) \to P^k(S)/P^{k-1}(S)$$

is an isomorphism for every $k > 0$.

Corollary 1. *If the configuration hyperplanes have only normal intersections, then* $\dim_{\mathbf{Z}} P^k(S)/P^{k-1}(S)$ *is equal to the number of k-codimensional edges.*

Let $M_{\mathbf{C}}^F = V_{\mathbf{C}} \setminus \bigcup_{i \in I^F} A_{i,\mathbf{C}}$. According to Brieskorn [2], the natural mapping $\bigoplus_{\substack{F \in \mathscr{L} \\ r(F)=k}} H^k(M_{\mathbf{C}}^F, \mathbf{Z}) \to H^k(M_{\mathbf{C}}, \mathbf{Z})$ is an isomorphism for every $k > 0$. Theorem 3 is an analogue to Brieskorn's theorem.

Corollary 2. $\dim_{\mathbf{Z}} P^k(S, \mathbf{Z})/P^{k-1}(S, \mathbf{Z}) = \dim_{\mathbf{Z}} H^k(M_{\mathbf{C}}, \mathbf{Z})$ *for* $k \geqq 0$.

This corollary can easily be deduced by induction with respect to the dimension of the containing space, from an observation by Orlik and Solomon: $\dim_{\mathbf{Z}} H^*(M_{\mathbf{C}}) = \dim_{\mathbf{Z}} C_n(S)$, from Theorem 3, and from Brieskorn's theorem.

Remarks: 1. In particular, the corollary implies that $\sum_{k \geqq 0} (-1)^k \dim_{\mathbf{Z}} P^k(S)/P^{k-1}(S)$ is equal to the number of bounded components of M (cf. the combinatorial formulae for $\dim H^k(M_{\mathbf{C}})$ and the number of bounded components [3, 13]). 2. If k, i are positive numbers, then the unique decomposition of k exists:

$$k = \binom{n_i}{i} + \binom{n_{i-1}}{i-1} + \cdots + \binom{n_j}{j},$$

where $n_i > n_{i-1} > \cdots > n_j \geqq j \geqq 1$. Following [14], we define

$$k^{\langle i \rangle} = \binom{n_i + 1}{i+1} + \binom{n_{i-1}+1}{i} + \cdots + \binom{n_j + 1}{j+1}, \quad 0^{\langle i \rangle} = 0.$$

An integer vector $(k_0, k_1, \ldots, k_d)$ is called an M-vector if $k_0 = 1$ and $0 \leqq k_{i+1} \leqq k_i^{\langle i \rangle}$ for $1 \leqq i \leqq d-1$. It follows from [15] that the sequence of numbers $\dim_{\mathbf{Z}} P^k(S)/P^{k-1}(S)$, $k \geqq 0$, is the M-vector.

We call a monomial $x_{i_1} \cdots x_{i_k} \in P$ an admissible monomial if $df_{i_1} \wedge \cdots \wedge df_{i_k} \neq 0$.

A k-dimensional simplicial cone multiplied by an $(n-k)$-dimensional affine subspace is a support of an admissible monomial.

Corollary 3. *The set of admissible monomials generates P as a module over* **Z**.

3. Dual degree filtration. Consider the ring $P(S, \mathbf{Z})$ of a configuration of hyperplanes in an n-dimensional affine space as a linear space over **Z**. We define on the dual space P^* a decreasing filtration

$$O \subset P_n^* \subset P_{n-1}^* \subset \cdots \subset P_0^* = P^*$$

by making use of the condition $P_k^* = \operatorname{Ann} P^{k-1}$. We call $\{P_k^*\}$ a degree filtration.

We present here another construction of the filtration. We point out a finite set of vectors of P^* which are called flag cochains. Every flag cochain has a degree. Then P_k^* coincides with the linear hull of all the flag cochains of degree not less than k. Now we proceed to the construction itself.

The connected components of M are called the regions. The regions are the n-dimensional polyhedra (not necessarily bounded). Open facets of any dimensions of these polyhedrons are called the facets of the configuration. In particular, n-dimensional facets are regions. Zero-dimensional facets are called vertices.

Let $F_{n-k} \subset F_{n-k+1} \subset \cdots \subset F_n = V$ be a sequence of edges of a configuration S, the dimension of F_j being equal to j, and let the coorientation of the edge F_j in the edge F_{j+1} be given. We call this the flag of the edges of degree k and we denote it by F. Let Δ be a $(n-k)$-dimensional facet of the configuration lying in F_{n-k}. The flag F, together with the facet Δ, is called a distinguished flag.

2^k regions are related to a distinguished flag, with Δ being included in the closures of every region. Indeed, there are exactly 2^k regions that can be reached, first by a small move in any direction along F_{n-k+1}, then by a still smaller move in any direction along F_{n-k+2}, etc. until a move is made from F_{n-1} in any direction. To any such region corresponds an ordered sequence α of length k, which consists of pluses and minuses, $+$ or $-$, occupying the position j, depending on whether the motion into the region was along or against the coorientation of $F_{n-k-j-1}$ in F_{n-k-j}. The region with the index α is denoted Δ_α (see Fig. 1).

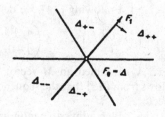

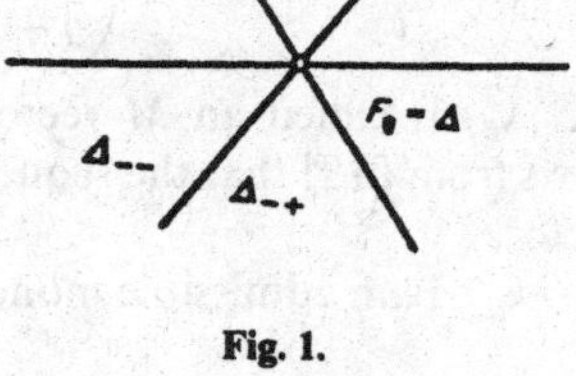

Fig. 1.

A vector $\psi_{F,\Delta} \in P^*$ is defined as follows: for any $x \in P$,

$$\psi_{F,\Delta}(x) = \sum_{\alpha} (-1)^{\varepsilon(\alpha)} x(\Delta_\alpha),$$

where $\varepsilon(\alpha)$ is the number of minuses in the sequence α, is called a flag cochain of the distinguished flag. k is called the degree of the flag cochain.

Theorem 4. *The linear hull of flag cochains of degree not less than k coincides with* Ann P^{k-1}.

Theorem 4 can be used to determine the degree of a given function (cf. the example in Sect. 1.1).

4. Relations between heaviside functions. Here we suppose that V is a linear n-dimensional space, the $\{f_i\}$, $i \in I$, are linear functions on V and that all the hyperplanes $\{A_i\}$ pass through the origin.

Let $\alpha_1 f_{j_1} + \cdots + \alpha_s f_{j_s} = 0$ be a linear relation; the J_+ are the numbers of all the linear functions with positive factors in the relation, and the J_- are the numbers of those with negative factors.

Theorem 5. *The Heaviside functions of the configuration* $\{f_i\}$, $i \in I$, *satisfy the relations:*

(1) $x_i^2 - x_i = 0$, $i \in I$,

and

(2) $\prod_{j \in J_+} x_j \prod_{k \in J_-} (x_k - 1) - \prod_{j \in J_+} (x_j - 1) \prod_{k \in J_-} x_k = 0$

for any linear dependence $\alpha_1 f_{j_1} + \cdots + \alpha_s f_{j_s} = 0$.

If all the coefficients in the linear relation differ from zero, then there is a polynomial in (2) of degree $s-1$ having precisely s monomials of degree $s-1$. Equation (2) is an even analogue of the Orlik–Solomon relation [3] for differential forms, namely, consider differential forms $w_i = df_i/2\pi\sqrt{-1} f_i$. If $f_{j_1}, \ldots, f_{j_s}$ are linearly dependent, then

$$\sum_{l=1}^{s} (-1)^{l-1} w_{j_1} \wedge \cdots \hat{w}_{j_l} \cdots \wedge w_{j_s} = 0.$$

Theorem 6. *Equations (1) and (2) determine P. More precisely, if* ϑ *is an ideal in the ring of polynomials* $\mathbf{Z}[X_i; i \in I]$ *generated by the left-hand sides of the relations listed in Theorem 5, then the natural homomorphism* $\mathbf{Z}[X_i; i \in I]/\vartheta \to P$ *is an isomorphism.*

Theorem 6 is an even analogue of an Orlik-Solomon theorem [3], which describes the ring $H^*(M_{\mathbf{C}}, \mathbf{Z})$, namely, consider the external algebra of the vector space whose basis is e_i, $i \in I$. Consider its ideal generated by the elements $\sum_{l=1}^{s} (-1)^{l-1} e_{j_1} \cdots \hat{e}_{j_l} \cdots e_{j_s}$ where $(f_{j_1}, \ldots, f_{j_s}) \subset \{f_i\}$, $i \in I$, is an arbitrary subset of linearly dependent elements. Then the factor algebra of the external algebra with respect to the ideal is naturally isomorphic to $H^*(M_{\mathbf{C}}, \mathbf{Z})$. Under the isomorphism, elements e_i are transformed into the cohomology classes of forms w_i. Consider in a Euclidean space with the coordinates $\{x_i\}$, $i \in I$, a unit cube whose boundaries are the hyperplanes $x_i = 0, x_i = 1, i \in I$. A subset of vertices of the cube is related to the configuration $\{f_i\}$: for every component M_0 of the set M, we mark the vertex of the cube at which $x_i = 1$ if $f_i(M_0) > 0$, and we mark the vertex at which $x_i = 0$ if $f_i(M_0) < 0$. Let X be the set of all marked vertices of the cube. Obviously, the ring of integer-valued functions on X is isomorphic to P. Theorem 6 yields the system of equations defining X as a subset of the Euclidean space: if, in Euclidean space, the system of equations cited in Theorem 5 is considered, the set of its solutions coincides with X.

We point out a basis over $\mathbf{Z}$ in the ring P. A subset $J = (j_1, \dots, j_k) \subset I$ is called a circuit if covectors $f_{j_1}, \dots, f_{j_k}$ are linearly dependent, but this does not hold for any proper subset of J.

Fix a linear ordering in I. A subset $J \subset I$ is called a broken circuit if there exists an index $j_0 \in I$ such that $(j_0, j_1, \dots, j_k)$ is a circuit, j_0 being less than any element of J.

We make a monomial $x_{j_1} \cdots x_{j_k} \in P$ correspond to any subset $I \subset J$. $1 \in P$ will correspond to the empty set.

Theorem 7. *The system of all the monomials corresponding to subsets of I which do not contain broken circuits forms a basis over $\mathbf{Z}$ in P. Moreover, the system of all distinguished monomials of degree not exceeding k is a basis in $P^k, k \geqq 0$.*

Theorem 7 is an analogue of Theorem II.1 of [7], which in turn goes back to [11]. By Theorem II.1, the system of differential forms $w_{j_1} \wedge \cdots \wedge w_{j_k}$ for all the subsets $J \subset I$ not containing broken circuits is a basis in $H^*(M_{\mathbf{C}}, \mathbf{Z})$. See also the theorem in [12].

6. A comparison with cohomologies. Here we define a non-canonical linear mapping $\pi_k: P^k \to H^k(M_{\mathbf{C}}, \mathbf{Z})$ whose kernel coincides with P^{k-1}. π_k is determined by a choise of coorientations of all the edges of codimension k.

Fix coorientations of all the k-codimensional edges. Determine the image of a monomial $x = x_{i_1} \cdots x_{i_l}$, $l \leqq k$. Set $\pi_k(x) = 0$ when $l < k$ or when $l = k$ and $df_{i_1} \wedge \cdots \wedge df_{i_k} = 0$. If $l = k$ and $df_{i_1} \wedge \cdots \wedge df_{i_k} \neq 0$ set $\pi_k(x) = \pm [w_{i_1} \wedge \cdots \wedge w_{i_k}]$, where the plus sign is chosen when the fixed coorientation of an edge $f_{i_1} = \cdots = f_{i_k} = 0$ coincides with the coorientation induced by the form $df_{i_1} \wedge \cdots \wedge df_{i_k}$, and the minus sign is chosen otherwise.

Theorem 8. *π_k can be continued so as to become a well-defined linear mapping $P^k \to H^k(M_{\mathbf{C}}, \mathbf{Z})$ whose kernel coincides with P^{k-1}.*

Define π_k in a geometrical way. Set $\pi_k(x) = 0$ if $l < k$, or if $l = k$ and $df_{i_1} \wedge \cdots \wedge df_{i_k} = 0$. If $l = k$ and $df_{i_1} \wedge \cdots \wedge df_{i_k} \neq 0$ then set $\pi_k(x)$ equal to the following linear function on $H_k(M_{\mathbf{C}}, \mathbf{Z})$: the index of intersection of classes of $H_k(M_{\mathbf{C}}, \mathbf{Z})$ with a non-compact $(2n-k)$-dimensional cycle $\{v \in M_{\mathbf{C}} | f_{i_1}(v) > 0, \dots,$

$f_{i_k}(v)>0\}$ whose orientation we define with the help of the complex orientation of $\{v\in M_{\mathbf{C}}|f_{i_1}(v)=0,\ldots,f_{i_k}(v)=0\}$ and of the previously fixed coorientation of the edge $\{v\in V|f_{i_1}(v)=0,\ldots,f_{i_k}(v)=0\}$ in V.

7. The ring of functions which are constant on facets. Let a configuration S of hyperplanes in an n-dimensional real affine space V be given. Consider the ring $Q(S,\mathbf{Z})$ of integer-valued functions on V which are constant on every facet of the configuration. Consider, in the ring Q, the multiplicative generators which are the Heaviside functions x_i, X_i, $i\in I$, given by the relations: $x_i(v)=1$ if $f_i(v)>0$, $x_i(v)=0$ if $f_i(v)\leqq 0$, and $X_i(v)=1$ if $f_i(v)=0$, $X_i(v)=0$ if $f_i(v)\neq 0$. In other words, $\{x_i\}$ are the functions defined by the conditions of Sect. 1.1 and $\{X_i\}$ are the characteristic functions of hyperplanes of the configuration. Every function $x\in Q(S,\mathbf{Z})$ is written in the form of a polynomial in $\{x_i,X_i\}$, $i\in I$, with integer coefficients. We call the minimum degree of polynomials in $\{x_i,X_i\}$ representing x the degree of the function x. Define a degree filtration

$$O\subset Q^0\subset Q^1\subset\cdots\subset Q$$

where Q^k is the subspace of functions which can be presented as polynomials of degree not exceeding k.

The properties of the ring Q and its filtration are analogous to the properties of the ring P. We have discussed in detail in this paper the analogues of Theorems 1–4 for Q. It is not difficult to produce the analogues of Theorems 5–8.

8. Chains. An integer linear combination of facets is called the integer chain of the configuration. An integer-valued linear function on a linear space of chains is called an integer cochain. The functions of $Q(S,\mathbf{Z})$ are in 1–1 correspondence to the chains of the configuration $x\in Q\mapsto\sum x(\Delta)\Delta$ with the summation running over all the facets Δ of the configuration. The functions of $P(S,\mathbf{Z})$ are in 1–1 correspondence to n-dimensional chains of the configuration. This is a linear correspondence. The present paper uses the geometrical language of chains and cochains.

Denote by $C_k(S)$ the space linear over $\mathbf{Z}$ of integer k-dimensional chains, by $C^k(S)$ the space linear over $\mathbf{Z}$ of integer k-dimensional cochains, and by $C_k^{\mathrm{comp}}(S)\subset C_k(S)$ the subspace of integer linear combinations of bounded k-dimensional facets.

Key issues in the present paper are dimensional and degree filtrations in the space of chains $C_*(S)=\bigoplus_{k=0}^n C_k(S)$ defined below.

$$\text{Set } D_k(S)=C_0(S)\oplus C_1(S)\oplus\cdots\oplus C_k(S),\ k\geqq 0.$$

Then

$$O\subset D_0(S)\subset D_1(S)\subset\cdots\subset D_n(S)=C_*(S).$$

This filtration is called the dimensional filtration.

We say that a configuration S_1 is included in a configuration S_2 if the union of hyperplanes of the first configuration is contained in the union of hyperplanes of the second configuration. Any facet of the configuration S_1, regarded as a set, is represented as a sum of the facets of the configuration S_2. Thus, a natural inclusion $C_*(S_1) \subsetneq C_*(S_2)$ of the chains is defined which preserves the degree filtration.

Example. Let V be one-dimensional, S_1 a point a, and S_2 two points $a<b$. Then the facet $\{v \in V | a<v\}$ of the configuration S_1 is the sum of the three facets of configuration S_2: $\{v \in V | a<v<b\}+\{b\}+\{v \in V | b<v\}$.

We call a configuration which has all its hyperplanes passing through one point and intersecting normally an elementary configuration.

Let S_1 be an elementary configuration consisting of k planes. Clearly, $k \leqq n$. Suppose that S_1 is included in S_2. Images of the facets of the configuration S_1 in $C_*(S_2)$ will be called the elementary chains of degree k of the configuration S_2. The space V itself will be called the elementary chain of degree 0.

Examples. A hyperplane and the open subspace bounded by it are elementary chains of degree 1. A vertex of a configuration is an elementary chain of degree n.

Define a subspace $W_k \subset C_*(S)$ as the linear hull of elementary chains of degree $\leqq k$, $0 \leqq k \leqq n$. Then

$$O \subset W_0 \subset W_1 \subset \cdots \subset W_n \subset C_*(S).$$

This filtration will be called the degree filtration.

Set $gr_l D = D_l/D_{l-1}$, $W_k gr_l D = (W_k \cap D_l + D_{l-1})/D_l$, $gr_k W gr_l D = W_k gr_l D / W_{k-1} gr_l D$. $gr_l D$ is canonically isomorphic to $C_l(S)$, $\{W_k gr_l D\}$, $k \geqq 0$, is the degree filtration induced on $gr_l D$, and $\{gr_k W gr_l D\}$, $k \geqq 0$, its factor-spaces.

It is rather easy to see that Theorems 1–8 are statements on degree filtration induced on $gr_n D$. It is not difficult to produce generalizations of Theorems 5–8, while the generalizations of Theorems 1–4 are in Sect. 4.

Remark. Theorems 1–4 are proved in Sect. 4, Theorems 5–8 in Sect. 5.

The appendix to this paper (Sect. 6) contains a multidimensional generalization of the theorem on decomposition of a rational function into simple fractions, the idea behind which is linked to geometric constructions of the present paper.

§2. Chains of a configuration

This section contains some combinatorial information, preparatory to proving the theorems formulated above.

1. Cones and angles. A configuration having at least one vertex is called a regular configuration. If a configuration has a non-empty intersection of all its hyperplanes, it is called a central configuration. Any chains of a regular central configuration will be called a linear cone. Any chain of a non-regular central configuration will be called an angle. The form of the angle is the direct product

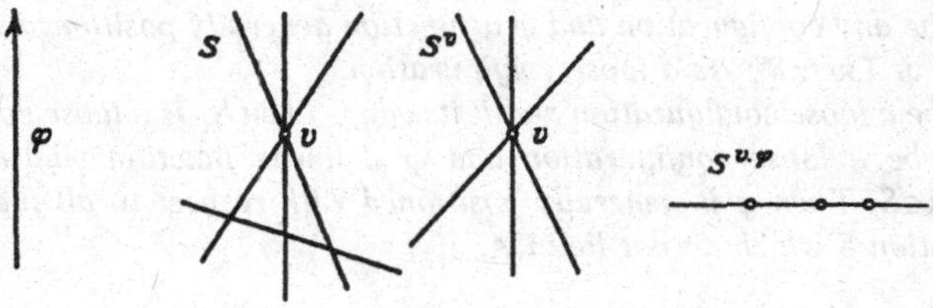

Fig. 2.

of a line and a chain of the configuration induced on a generally situated hyperplane by the given configuration.

Any chain of the configuration S^v considered as a chain in S is called a cone with vertex v of the configuration S. Any chain of the configuration S^F considered as a chain in S is called an angle with edge F of the configuration S. We recall that S^v and S^F are localizations of the configuration in the edges v and F.

2. Linear functions and configurations. A facet of a configuration is said to be bounded from above in relation to a linear function φ defined on V if the facet lies in an appropriate half-space $\varphi \leqq \text{const}$. The set of all facets bounded from above is called a skeleton of the configuration S with respect to φ, and is written S_φ. The space of integer linear combinations of the facets of S_φ is denoted by $C_*(S_\varphi)$.

We call an affine localization $S^{v,\varphi}$ of the configuration S with respect to a point $v \in V$ and a linear function φ, a configuration cut out from the configuration S^v on a level hyperplane $\{x \in V \mid \varphi(x) = \varphi(v) - 1\}$ (see Fig. 2). Let Γ be a bounded facet of the configuration $S^{v,\varphi}$. Consider a cone with vertex v and guide Γ, the vertex v being deleted from it. This set is denoted by $K(\Gamma, v)$. The mapping $p: \Gamma \mapsto K(\Gamma, v)$ gives a monomorphism of the space $C_*^{\text{comp}}(S^{v,\varphi})$ into the subspace of chains of the configuration S^v bounded from above.

A linear function on V is called a function in general position with respect to the vertex v of the configuration S if it is not constant on edges of positive dimension passing through v, and it is called a function in general position with respect to the configuration S if it is not constant on all the edges with positive dimension and has pairwise different values on vertices.

Lemma 1. *Let φ be a generally positioned function with respect to the vertex v. Then p sets an isomorphism $C_k^{\text{comp}}(S^{v,\varphi}) \simeq C_{k+1}(S^v_\varphi)$.*

The proof is obvious.

3. Loose configurations. We call a configuration a loose configuration if every one of its hyperplanes intersects with every one of its edges of positive dimension. We list its obvious properties.

Lemma 2. 1. *Let S be a loose configuration and U a hyperplane which transversally crosses all its edges. Then S_U is a loose configuration.*

2. *Let S be any configuration and φ a function generally positioned with respect to the vertex v. Then $S^{v,\varphi}$ is a loose configuration.*

3. *Let S be a loose configuration and F its edge. Then S_F is a loose configuration.*

4. *Let S be a loose configuration and φ a linear function vanishing on the hyperplane $A \in S$. Then φ is generally positioned with respect to all the vertices of the configuration S which do not lie in A.*

4. Distinguished substar (Cf. [9]). Let $f_1, \ldots, f_N$ be an ordered set of linear functions on the n-dimensional affine space V and S the configuration determined by the functions. Define a linear order of vertices of the configuration $S: v < w$ if for a certain k, $(f_j(v))^2 = (f_j(w))^2$ if $j < k$, $(f_k(v))^2 < (f_k(w))^2$. The maximal vertex of a bounded facet will be said to be distinguished. The set of all bounded facets with a common distinguished vertex is called a distinguished substar of the vertex. The dimension of a distinguished substar is called the multiplicity of the corresponding vertex. We describe a distinguished substar of a vertex of a loose configuration.

Let v be a vertex of the configuration S. For any facet Γ^v of the configuration S^v there exists a unique facet Γ of the configuration S whose germ in v coincides with the germ of the facet Γ^v in v. The facets Γ^v, Γ will be said to be mutually induced in v.

Lemma 3. *Let v be a vertex of the loose ordered configuration $f_1(v) > 0$. Then an edge Γ belongs to a distinguished substar of a vertex v if and only if Γ is induced from the facet of the configuration S^v bounded from above with respect to f_1.*

The proof is obvious.

Similarly, let $v \in A_1 \cap A_2 \cap \cdots \cap A_{k-1}$, $f_k(v) > 0$. Consider a configuration $\tilde{S}$ cut out from S by $A_1 \cap \cdots \cap A_{k-1}$.

Lemma 4. *The facet Γ belongs to a distinguished substar of the vertex v if and only if Γ is induced from a facet of the configuration $\tilde{S}^v$ bounded from above with respect to f_k.*

Corollary. *A loose regular configuration has exactly one vertex of zero multiplicity.*

5. Euler characteristic of some chains. Let Γ be a facet of the configuration S, Γ^* a cochain which is equal to 1 on Γ and 0 on other facets, and $d(\Gamma)$ the dimension of a facet. The cochain $\chi = \sum(-1)^{d(\Gamma)}\Gamma^*$, summation being carried out over all the facets, will be called the Euler characteristic cochain.

Let Δ be a chain which is equal to the sum of all bounded facets of the configuration S. Let F be an edge of the configuration S, Γ a non-closed facet of the configuration S^F, and Δ_Γ the chain which is equal to the sum of all bounded facets of the configuration S which lie in Γ.

Theorem 9. *Let S be a regular loose configuration. Then $\chi(\Delta) = 1$, $\chi(\Delta_\Gamma) = 0$.*

The proof is obtained by induction on the dimension of the configuration. If $\dim V=1$ Theorem 9 is obvious. Suppose that the theorem is proved for loose configurations in a space of dimension not exceeding $n-1$. Prove it for dimension n, the case $d(\Gamma)=n$ being sufficient.

Enumerate the hyperplanes of S so that the first hyperplanes are those containing $(n-1)$-dimensional facets of the polyhedron Γ.

Prove that $\chi(\Delta)=1$. Let $\Delta(v)$ be a distinguished substar of the vertex v. We have $\Delta=\sum_v \Delta(v)$. For only one vertex v with multiplicity 0, we have $v=\Delta(v)$. Prove that $\chi(\Delta(v))=0$ if the multiplicity of the vertex is positive. Indeed, in this case, by virtue of Lemmas 1–4, k-dimensional facets of the distinguished substar correspond 1–1 to $(k-1)$-dimensional bounded facets of a suitable loose configuration in a space whose dimension is less than n. By assumption in the induction, it follows that the Euler characteristic of the distinguished substar is 0.

Prove that $\chi(\Delta_\Gamma)=0$. Denote by $\Delta_\Gamma(v)$ the sum of the facets of the distinguished substar which fall within Γ. We have $\Delta_\Gamma=\sum_v \Delta_\Gamma(v)$, summation being carried out over all vertices which are in the closure of the set Γ. If a vertex v belongs to the interior of the set Γ, its multiplicity is positive, its distinguished substar coincides with $\Delta_\Gamma(v)$ and, as proved above, $\chi(\Delta_\Gamma(v))=0$. Prove that $\chi(\Delta_\Gamma(v))=0$ if v belongs to the boundary of the set Γ and $\Delta_\Gamma(v)$ is non-empty. In this case v does not belong to the first of the hyperplanes and $\Delta_\Gamma(v)$ consists of positive-dimensional facets. Let, to be certain, $f_1(v)>0$. By Lemmas 3 and 1, the facets of $\Delta_\Gamma(v)$ correspond 1–1 to bounded facets of the affine localization S^{v,f_1} on the hyperplane $H=\{x\in V \mid f_1(x)=f_1(v)-1\}$ which are in a set $\Gamma(v)$ determined by v and Γ. Describe $\Gamma(v)$ and prove that Theorem 1 is applicable to it. Thus Theorem 1 for $\dim V=n$ will be proved.

Let a facet Γ be defined by the inequalities $f_{i_1}>0, f_{i_2}>0,\ldots,f_{i_k}>0$ in a small neighbourhood of the point v. The hyperplanes $A_{i_1},\ldots,A_{i_k}\in S^F$ and v belong to $A=A_{i_1}\cap\cdots\cap A_{i_k}$. The conditions $f_{i_1}>0,\ldots,f_{i_k}>0$ define a subset $\Gamma(v)$ of H. The set A contains F lying in A_1. Thus $\Gamma(v)$ is bounded by hyperplanes whose intersection is non-empty, and, therefore, $\Gamma(v)$ is a union of non-closed facets of the configuration $(S^{v,\varphi})^{A\cap H}$. For $\Gamma(v)$ in H, Theorem 1 implies that $\chi(\Delta_\Gamma(v))=0$.

§3. Decomposition into cones

1. Theorem 10. *Let φ be a linear generally positioned function with respect to the configuration S and Δ the chain composed of facets bounded from above. Then Δ can be represented as the sum of cones of the configuration S bounded from above: (3) $\Delta=\sum_v K_v$ where the summation is over the vertices of the configuration. The decomposition (3) is unique. Every cone in (3) has dimension not greater than the dimension of the chain Δ.*

Proof. The values of the function at the vertices determine the order of the vertices. In a neighbourhood of the greatest of the vertices, v, of the chain Δ, the chain has the form of a cone K_v of the configuration S^v bounded from above. Subtract the cone from Δ. We do the same with the rest of the chain. The uniqueness of the representation is obvious.

For a description of the cone K_v of the decomposition (3) in terms of the behaviour of the chain Δ in a neighbourhood of the vertex v, see Sect. 3.5.

Corollary. *The natural mapping*

$$\bigoplus_{F \in \mathscr{L}} C_*(S^F) \to C_*(S)$$

is an epimorphism.

Proof. It is easy to show, by making use of Theorem 10, that every facet of the configuration S is a sum of the cones (provided it has a vertex) or a sum of angles.

Denote by $I(S)$ the subspace of chains generated by the angles of the configuration. In other words, $I(S)$ is the image of the natural mapping

$$\bigoplus_{\substack{F \in \mathscr{L} \\ r(F) < n}} C_*(S^F) \to C_*(S).$$

The dimension filtration induces a filtration on $I: I_l(S) = I(S) \cap D_l(S)$, $l \geqq 0$. Denote by $CI(S)$ the factor-space $C_*(S)/I(S)$. The dimension filtration $CI_l(S) = (D_l + I)/I$ is defined on $CI(S)$.

Theorem 11. *Let φ be a linear function, and Δ the chain whose dimension does not exceed l, $l \geqq 0$. Then there exists a linear combination of angles $\sum \Gamma_\alpha$, with dimensions not greater than l such that the chain $\Delta - \sum \Gamma_\alpha$ is bounded from above with respect to φ. In other words, the natural mapping $C_l(S_\varphi) \to CI_l(S)$ is an epimorphism.*

Proof. Let a number t_0 be given, such that for any $t > t_0$, the level hyperplane of the level t of φ crosses transversally all the edges of the configuration S. Consider a configuration S_U cut out on the hyperplane $U = \{x \in V \mid \varphi(x) = t_0\}$.

Let Γ be a cone of the configuration S_U with vertex at v. Then there exists an angle $\tilde{\Gamma}$ of the configuration S such that $\Gamma = \tilde{\Gamma} \cap U$. The edge of $\tilde{\Gamma}$ is one-dimensional and passes through v. Similarly, let Γ be an angle of the configuration S_U. Then there exists an angle $\tilde{\Gamma}$ of the configuration S whose intersection with U is Γ. The intersection of an edge of the angle $\tilde{\Gamma}$ is an edge of the angle Γ.

Let Δ be the original chain and $\Delta \cap U$ its intersection with U. By the corollary of Theorem 10, $\Delta \cap U$ can be represented as a linear combination of angles and cones of the configuration $S_U: \Delta \cup U = \sum \Gamma_\alpha$. Let $\tilde{\Gamma}_\alpha$ be the angle of the configuration for which $\Gamma_\alpha = \tilde{\Gamma}_\alpha \cap U$. It is easy to see that the chain $\Delta - \sum \tilde{\Gamma}_\alpha$ is bounded from above.

2. Skew cochains. A cochain of the configuration is said to be localized at a given vertex if it is equal to zero on any facet which is not included in the star of the vertex. A cochain is said to be a skew cochain if it is equal to zero on any angle of the configuration.

We will produce a great store of skew local chains.

Let φ be a linear function on V. The cochain $\chi_\varphi = \sum_{\Gamma^*:\varphi(\Gamma)\leqq 0}(-1)^{d(\Gamma)}\Gamma^*$ is called the cochain associated to φ. The value of the cochain on an arbitrary chain is equal to the Euler characteristics of the part of the chain which falls into the semispace $\varphi \leqq 0$. Any cochain associated to a linear function in general position is called an Euler cochain of configuration. The set of all Euler cochains is finite.

Theorem 12. *An Euler cochain is equal to zero on any angle.*

Proof. Let χ_φ be an Euler cochain and Δ an angle with an edge F. It suffices to investigate the case when Δ is a facet of dimension n of the configuration S^F.

Let Γ be a facet of the configuration S lying in $\Delta \cap \{\varphi \leqq 0\}$. Let the maximum of the function φ, which is bounded on the closure of the facet Γ be reached at the vertex v. Denote by $\Delta(v)$ the sum of all such facets. Then $\chi(\Delta) = \sum_v \chi(\Delta(v))$. Prove that $\chi(\Delta(v)) = 0$. If v belongs to the interior of the set Δ, then $\Delta(v)$ consists of all the facets of the star of the vertex v on which the maximum of the function φ is reached at v. These positive-dimensional facets correspond 1–1 to bounded facets of the affine localization $S^{v,\varphi}$. Now the equality $\chi(\Delta(v)) = 0$ follows from the first part of Theorem 9.

If v belongs to the boundary of the set Δ, then the equality similarly follows from the second part of Theorem 9.

3. Linear combinations of Euler cochains. Let φ be a generally positioned function and $t_1 < t_2 < \dots < t_N$ its critical values, i.e. its values at the vertices of the configuration S. Consider an Euler cochain $\chi_{\varphi - t}$. It does not change when $t \in [t_j, t_{j+1})$. $\chi_{\varphi - t} \equiv 0$ for $t < t_1$. If v is a vertex such that $\varphi(v) = t$, set $\chi_{\varphi,v} = \chi_{\varphi - t + \varepsilon} - \chi_{\varphi - t - \varepsilon}$ where ε is a small positive number. $\chi_{\varphi,v}$ is the cochain localized at v. More precisely, $\chi_{\varphi,v} = \sum(-1)^{d(\Gamma)}\Gamma^*$ with the summation taken over all the facets of the star of the vertex v such that the maximum of φ is reached at v. That is, all the facets going from v in the direction of decreasing φ are taken into the summation. The cochain $\chi_{\varphi,v}$ will be called the local Euler cochain centred at v.

Theorem 13. *Let u, v be two different vertices of the configuration S, K_u the cone of the configuration S with vertex at u, and $\chi_{\varphi,v}$ the local Euler cochain centred at v. Then $\chi_{\varphi,v}(K_u) = 0$.*

Proof. The cone K_u in a neighbourhood of v looks like an angle. Thus, by Theorem 12, $\chi_{\varphi,v}(K_u) = 0$.

4. There is a sufficient number of Euler cochains.

Theorem 14. *Let Δ be a non-zero cochain of the configuration S, bounded from above with respect to a function φ which is in general position. Then there exists a linear combination of Euler cochains whose value on Δ is not equal to zero.*

Corollaries. 1. *A non-zero chain which is bounded from above is not a linear combination of angles, that is the natural mapping $C_*(S_\varphi) \to CI(S)$ is an isomorphism.*

2. *Any skew cochain is a linear combination of Euler cochains. In other words, Euler cochains generate a space dual to* $CI(S)$.

Proof. This is proved by induction in the configuration's dimension. If $\dim V = 1$ the theorem is obvious. We prove an inductive step.

Decompose Δ into a sum of cones bounded from above with respect to $\varphi: \Delta = \sum_v K_v$. Choose a vertex v at which the cone does not equal zero. By Theorem 13, it suffices to prove the existence of a linear combination of local Euler cochains centred at v whose value on K_v differs from zero.

If the vertex v enters K_v with the coefficient λ, then $\chi_{-\varphi,v}(K_v) = \lambda$. If $\lambda \neq 0$, then the theorem is proved. Further, we assume that v enters K_v with zero coefficient.

Let $\chi_{\alpha,v}$ be a local Euler cochain. The isomorphism of Lemma 1 transforms it into a cochain ψ on bounded chains of the localization $S^{v,\varphi}$.

Lemma 5. ψ *is an Euler cochain of the configuration* $S^{v,\varphi}$ *associated to a linear function* $\alpha - \alpha(v)$ *restricted to the space of affine localization.*

Proof. Lemma is obvious.

Turn back to the cone K_v. It induces a non-zero bounded chain K in $S^{v,\varphi}$. By inductive assumption, for the configuration $S^{v,\varphi}$ there exists a linear combination of its Euler cochains having non-zero value at K. This fact and Lemma 5 imply the theorem.

5. Characterization of the cones of decomposition (3). Let φ be a linear function in general position on S, Δ a chain bounded from above, and $\Delta = \sum_v K_v$ a decomposition into a sum of cones of the configuration S bounded from above with respect to φ.

Theorem 15. 1. *For any local Euler cochain* $\chi_{\alpha,v}$ *centred at* v

$$\chi_{\alpha,v}(\Delta) = \chi_{\alpha,v}(K_v).$$

2. *Let* K *be a cone of the configuration* S *bounded from above with respect to* φ. *Suppose that, for any local Euler cochain* $\chi_{\alpha,v}$ *centred at* v, $\chi_{\alpha,v}(\Delta) = \chi_{\alpha,v}(K)$. *Then* $K = K_v$.

The first part of the theorem is a corollary of Theorem 13, the second is a corollary of Theorem 14.

6. Combinatorial connection (Corollaries of Theorem 14). Let φ_1, φ_2 be linear functions in general position. The following two statements result from Corollary 1.

Corollary 3. $C_*(S_{\varphi_1})$ *and* $C_*(S_{\varphi_2})$ *are canonically isomorphic to each other.*

Let S be a central regular configuration with vertex v and let φ_1, φ_2 be linear functions in general position.

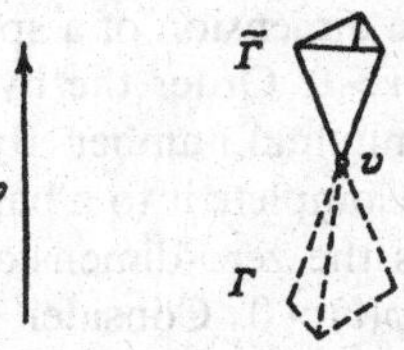

Fig. 3.

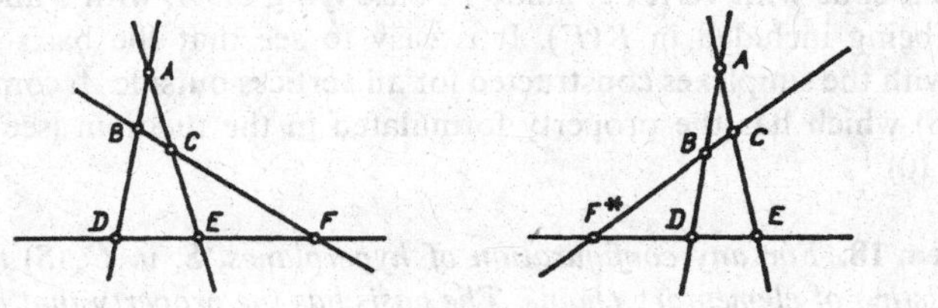

Fig. 4.

Corollary 4. *$C_*^{\text{comp}}(S^{v,\varphi_1})$, $C_*^{\text{comp}}(S^{v,\varphi_2})$ are canonically isomorphic to each other.*

Example 1. Let S be a regular central configuration and φ a function in general position. We point out an isomorphism $\pi: C_*(S_\varphi) \to C_*(S_{-\varphi})$ consisting in the fact that the cone going down with respect to φ is transformed into a cone going upwards, by adding angles.

Theorem 16 (*cf. Theorem 7 of* [9]). *Let $\Gamma \in C_*(S_\varphi)$ be a k-dimensional facet. Then $\pi(\Gamma) = (-1)^k \tilde{\Gamma}$, where $\tilde{\Gamma}$ is the closure of the reflection in the vertex v of the facet Γ (see Fig. 3).*

Example 2. Consider a configuration in $\mathbf{R}^3$ consisting of four planes passing through zero, and being in a general position. If φ is a function in general position, then $S^{v,\varphi}$ is a configuration on a plane consisting of four generally positioned lines. For suitable φ_1, φ_2, the configurations are given by Fig. 4. We point out the isomorphism of bounded chains: $F \mapsto -F^*$, $CF \mapsto -CF^* - C$, $EF \mapsto -EF^* - E$, $CEF \mapsto -CEF^* - CE$, where CF, CF^*, EF, EF^* are open intervals and CEF, CEF^* are the open triangles. The rest of the facets are transformed into the facets with the same names: $A \mapsto A$, etc.

4. Dimensional and degree filtrations

1. Decomposition into simplexes.

Theorem 17. *If S is a loose configuration then there exists a basis of $C_*^{\text{comp}}(S)$ consisting of open simplexes of different dimensions. The basis has the following property: if there is a chain whose dimension does not exceed l, the dimensions of simplexes of its decomposition in terms of the basis elements are not greater than l.*

Proof is by induction on the dimension of a space. When $\dim V = 0$, the point is the only facet. Let $\dim V = n > 0$. Order the hyperplanes of the configuration. Let A be the hyperplane with minimal number. By inductive assumption, there is a simplicial basis of $C_*^{\text{comp}}(S_A)$. Complete it to a basis of $C_*^{\text{comp}}(S)$. Let v be a vertex outside A. Include in the basis the zero-dimensional chain v. Let φ be a linear function such that $\varphi(A) = 0$, $\varphi(v) > 0$. Consider an affine localization $S^{v,\varphi}$. Its hyperplanes form an ordered set. Using induction, we choose a simplicial basis of $C_*^{\text{comp}}(S^{v,\varphi})$. We make a simplex $K(\Gamma)$ correspond to each of its elements; this simplex is a cone with vertex v, guide Γ, base lying on A, with v and the bottom base not being included in $K(\Gamma)$. It is easy to see that the basis of $C_*^{\text{comp}}(S_A)$, together with the simplexes constructed for all vertices outside A, composes a basis of $C_*^{\text{comp}}(S)$ which has the property formulated in the theorem (see the proof of Theorem 10).

Theorem 18. *For any configuration of hyperplanes, S, in $C_*(S)$ there exists a basis consisting of elementary chains. The basis has the property that a chain whose dimension does not exceed l can be decomposed into a sum of elementary chains with dimensions not greater than l.*

The proof is by induction on the dimension of the configuration and follows easily from Theorem 17 (cf. Theorems 10, 11, and Lemma 1).

2. Properties of dimensional and weight filtrations in chains.

1. $W_n = C_*(S)$.
2. D_l is a linear hull of elementary chains whose dimensions do not exceed l, for $l \geqq 0$.
3. If $S_1 \subset S_2$ and $i: C_*(S_1) \to C_*(S_2)$ is the natural inclusion, then $i(W_k(S_1)) \subset W_k(S_2)$, $i(D_l(S_1)) \subset D_l(S_2)$ for $k, l \geqq 0$.
4. $W_k(S)$ coincides with the image under natural mapping:

$$\bigoplus_{\substack{F \in \mathcal{L} \\ r(F) \leqq k}} C_*(S^F) \to C_*(S), \tag{4}$$

where S^F is the localization of the configuration at the edge F and $r(F)$ is the codimension of the edge.

5. If $U \subset V$ is a subspace and $j_U: C_*(S) \to C_*(S_U)$ is the natural epimorphism, then $j_U(W_k(S)) = W_k(S_U)$.

The third property is a corollary of the definition of filtrations. The first and second properties are corollaries of Theorem 18.

We prove Property 4. Obviously, $W_k(S)$ is contained in the image of mapping (4). By Property 1, $C_*(S^F) = W_{r(F)}(S^F)$. Thus, $W_k(S)$ coincides with the image of the mapping (4).

Property 4 can serve as a definition of a degree filtration.

We prove Property 5. Each elementary chain of degree k of $C_*(S_U)$ is an image of an elementary chain of degree k of $C_*(S)$. Thus $j_U(W_k(S)) \supset W_k(S_U)$. An image of an elementary chain of degree k is a facet of a configuration on U consisting of no more than k hyperplanes. Property 4 implies that the image belongs to $W_k(S_U)$.

6. If $\Delta \in W_k \cap D_l$ then Δ can be represented as a linear combination of elementary chains whose degrees do not exceed k and dimensions do not exceed l.

The proof is by induction with respect to the dimension of the configuration. For $n=0$, the theorem is valid for any k, l. Let $n>0$. Let φ be a linear function in general position with respect to the configuration S. Let t_0 be a number such that, for any $t>t_0$, the hyperplane of constant level t of φ crosses all the edges transversally. Consider the configuration S_U cut out on the hyperplane $U=\{x \in V \mid \varphi(x)=t_0\}$. Then the dimension of the chain $\Delta \cap U$ is $\leqq l-1$.

Let $k<n$. Then $\Delta \cap U \in W_k(S_U) \cap D_{l-1}(S_U)$. By the inductive assumptions, $\Delta \cap U=\sum a_m \Delta_m$, where $a_m \in \mathbf{Z}$ and $\{\Delta_m\} \subset W_k(S_U) \cap D_{l-1}(S_U)$ are elementary chains. For every elementary chain Δ_m there exists an elementary chain $\tilde{\Delta}_m \in W_k(S) \cap D_l(S)$ such that $\tilde{\Delta}_m \cap U=\Delta_m$. The chain $\tilde{\Delta}=\Delta-\sum a_m \tilde{\Delta}_m$ has no intersection with U. By the choice of U, the chain $\tilde{\Delta}$ is bounded from above. By Theorems 12 and 14, $\tilde{\Delta}=0$. Property 6 is proved. For $k=n$, the property follows from Theorem 18.

7. $W_k \cap D_l=0$ for $k+l<n$.

8. Let $U \subset V$ be a d-dimensional subspace in general position with respect to the configuration S. Then the natural homomorphism $j_U: C_*(S) \to C_*(S_U)$ reduces the dimension of any chain by $n-d$ and, for any $0 \leqq k, l \leqq d$, sets an isomorphism

$$W_k(S) \cap D_{l+n-d}(S) \to W_k(S_U) \cap D_l(S_U).$$

Proof. It suffices to consider the case of U being a hyperplane. The fact of the generality of the hyperplane U obviously implies $D_{l+1}(S) \to D_l(S_U)$, $W_k(S) \to W_k(S_U)$. We prove the absence of the kernel of $j_U|_{W_{n-1}}$. Indeed, if $\Delta \in W_{n-1}(S)$ belongs to the kernel of the homomorphism j_U, then $\Delta=\Delta_+ + \Delta_-$, with $\Delta_\pm$ lying at different sides of the hyperplane U. It follows from Theorems 10–13 and 14 that $\Delta_+=\Delta_-=0$. Thus, j_U sets an inclusion $W_k(S) \cap D_{l+1}(S) \subsetneq W_k(S_U) \cap D_l(S_U)$, for $k \leqq n-1$. If $\Delta \in W_k(S_U) \cap D_l(S_U)$ then, by Property 6, there exists $\tilde{\Delta} \in W_k(S) \cap D_{l+1}(S)$ such that $j_U(\tilde{\Delta})=\Delta$. Property 8 is proved.

9. Under the conditions of Property 8, j_U sets an isomorphism $W_k gr_{l+n-d} D(S) \to W_k gr_l D(S_U)$ for any $0 \leqq k, l \leqq d$.

Property 9 is a corollary of Property 8, since $W_k gr_m D=W_k \cap D_m / D_{m-1} \cap W_k$.

10. For any $k \geqq 0$, the natural mapping

$$\bigoplus_{\substack{F \in \mathscr{L} \\ r(F)=k}} W_k(S^F)/W_{k-1}(S^F) \to W_k(S)/W_{k-1}(S)$$

is an isomorphism.

Proof. Let φ be a linear function in general position. Then, by Theorems 10–14, $C_*(S_\varphi) = \bigoplus_{\substack{F\in\mathcal{L}\\ r(F)=n}} C_*(S_\varphi^F)$. According to Corollary 1 of Theorem 14, this equality implies Property 10, for $k=n$. The case of arbitrary k can be reduced to the case just considered, with the help of Property 8. The same argument proves Property 11.

11. For any $k, l \geqq 0$, set $D_l gr_k W = D_l \cap W_k / D_l \cap W_{k-1}$. Then the natural mapping

$$\bigoplus_{\substack{F\in\mathcal{L}\\ r(F)=k}} D_l gr_k W(S^F) \to D_l gr_k W(S)$$

is an isomorphism.

12. A chain from $W_k(S)$ is uniquely determined by its general k-dimensional cross-section. More precisely, if $U \subset V$ is a subspace in general position with respect to S, $\dim U = k$, and $c \in W_k(S)$ is the chain with $c \cap U = 0$, then $c = 0$.

Property 12 follows from Property 8.

13. Let $U_1, U_2 \subset V$ be subspaces in general positions with respect to the configuration S, $\dim U_1 = \dim U_2$. Then $C_*(S_{U_1})$ is canonically isomorphic to $C_*(S_{U_2})$. The isomorphism is defined by Property 8.

The isomorphism will be called the combinatorial connection. In [9] the combinatorial connection is defined for a similar situation.

3. The ring $P(S, \mathbf{Z})$ defined in the introduction. Properties 1, 9 and 10 yield Theorems 1–3.

4. The filtration dual to degree filtration. Define a degree filtration in $C^*(S)$ by making use of the condition $W^k = \operatorname{Ann} W_{k-1}$. We have

$$O \subset W^n \subset W^{n-1} \subset \cdots \subset W^0 = C^*(S).$$

We give another construction of the filtration. Let $\chi(S) \subset C^*(S)$ be a linear hull of Euler cochains. Let $U_1, \ldots, U_{n-1}, U_n = V$ be affine subspaces whose dimensions are, respectively, $1, \ldots, n-1, n$. Suppose that the subspaces are in general positions with respect to the configuration S. Let $j_k: C^*(S_{U_k}) \to C^*(S)$ be the natural monomorphism.

Theorem 19. *For any* $k \geqq 0$,

$$W^k(S) = \chi(S) + j_{n-1}(\chi(S_{U_{n-1}})) + \cdots + j_k(\chi(S_{U_k})).$$

Theorem 19 follows from Corollary 2 of Theorem 14 and from Property 8 of Sect. 4.2.

5. Flag cochains.

Lemma 7. *An arbitrary flag cochain of degree n is a linear combination of Euler cochains.*

The proof is by induction with respect to the dimension of the configuration. In carrying out this proof, one has to descend to an affine localization of a zero-dimensional edge of a flag and to use Lemma 5.

Lemma 8. *Flag cochains of degree n generate the space which is dual to* $C_*(S)/(C_{n-1}(S)+W_{n-1}(C_*(S)))$.

The proof is by induction with respect to the dimension of the space. Let φ be a function in general position. By Theorems 10–14, it suffices to prove that the flag cochains of degree n generate $C^n(S_\varphi)$. Moreover, it suffices to investigate the case of S being a regular central configuration. We denote its vertex by v. Pass to the affine localization $S^{v,\varphi}$. Then $C_n(S_\varphi)\simeq C_{n-1}^{\mathrm{comp}}(S^{v,\varphi})$. Flag cochains of degree n on S transform into flag cochains of degree $n-1$ on $S^{v,\varphi}$. By the inductive assumption, for a non-zero chain from $C_{n-1}^{\mathrm{comp}}(S^{v,\varphi})$, there exists a flag cochain of degree $n-1$ which does not vanish on the former. The lemma is proved.

Proof of Theorem 4. Lemma 8 coincides with the statement of Theorem 4 for $k=n$. The case of arbitrary k follows from Lemma 8 and from Property 9 of Sect. 4.2.

§5. Comparison to cohomologies and relations

1. Comparison to cohomologies. Let S be a configuration of hyperplanes in a real n-dimensional affine space V.

Every oriented n-dimensional facet $\Delta\in C_n(S)$ determines a cohomology class $[\Delta]\in H^n(M_{\mathbf{C}},\mathbf{Z})$ which is equal to the index of intersection with a non-compact cycle Δ (we suppose $M_{\mathbf{C}}$ to be complex-oriented).

Fix an orientation on V and thus on every n-dimensional facet. Define a linear mapping $\pi: C_n(S)\to H^n(M_{\mathbf{C}},\mathbf{Z})$ by establishing correspondence between linear combinations of facets, $\sum a_\alpha\Delta_\alpha$, and classes $\sum a_\alpha[\Delta_\alpha]$.

In order to describe the kernel of the mapping π, we consider the natural isomorphism $i: C_n(S)\to gr_nD$.

Theorem 20. $\ker\pi=i^{-1}(W_{n-1}gr_nD)$.

Proof. Relate a class of homologies in $H_n(M_{\mathbf{C}},\mathbf{Z})$ of the torus defined below to each flag $F=\{F_0\subset F_1\subset\cdots\subset F_n)$, namely, let $\varepsilon_1,\ldots,\varepsilon_n>0$. Consider a torus $T(\varepsilon)=\{z\in\mathbf{C}^n \mid |z_j|=\varepsilon_j\}$ in $\mathbf{C}^n$. Fix its orientation. Affinely map $\mathbf{R}^n$ with coordinates $z_1,\ldots,z_n$ onto V so that the standard flag $\{z_1=\cdots=z_n=0\}\subset\{z_2=\cdots=z_n=0\}\subset\cdots\subset\{Z_n=0\}\subset\mathbf{R}^n$ be mapped onto the flag F. Consider a complexification $\mathbf{C}^n\to V_{\mathbf{C}}$ of the mapping. Torus $T(\varepsilon)$ for $0<\varepsilon_1\ll\varepsilon_2\ll\cdots\ll\varepsilon_n\ll 1$ is mapped into $M_{\mathbf{C}}$ and defines a homology class not depending on ε. Denote it by T_F.

Lemma 9. *Consider a cochain on $C_n(S)$ which is equal to the index of intersection with T_F. Then the cochain is equal to the flag cochain ψ_F up to a multiplication by ± 1.*

The proof is obvious.

The linear hull of the classes T_F obtained for different flags F of degree n will be denoted by L. By Lemmas 8 and 9, $\dim_{\mathbf{Z}} L = \dim_{\mathbf{Z}} C_n(S)/i^{-1}(W_{n-1} gr_n D)$. By Corollary 2 of Theorem 3, $\dim_{\mathbf{Z}} H_n(M_{\mathbf{C}}, \mathbf{Z}) = \dim_{\mathbf{Z}} C_n(S)/i^{-1}(W_{n-1} gr_n D)$. As a corollary, we obtain Theorem 20 as well as the statement $L = H_n(M_{\mathbf{C}}, \mathbf{Z})$.

Example. Consider a configuration of coordinate hyperplanes $A_j = \{z_j = 0\}$, $j = 1, \ldots, n$, in $\mathbf{R}^n$. By making use of the form $dz_1 \wedge \cdots \wedge dz_n$, set an orientation in $\mathbf{R}^n$. M consists of 2^n octants. The space $C_n(S)/i^{-1}(W_{n-1} gr_n D) = gr_n W gr_n D$ is one-dimensional and is generated by a positive octant $\{z_1 > 0, \ldots, z_n > 0\}$. Under the mapping π, the positive octant is transformed into a cohomology class of the form $(-1)^{n(n-1)/2} w_1 \wedge \cdots \wedge w_n$ where $w_j = dz_j / 2\pi\sqrt{-1}\, z_j$.

Notice a useful corollary of Theorem 20. Let φ be a linear function in general position. Then the mapping π restricted on $C_n(S_\varphi)$ gives an isomorphism of n-dimensional chains of the configuration and the space $H^n(M_{\mathbf{C}}, \mathbf{Z})$ which are bounded from above.

Proof of Theorem 8. For $k = n$, Theorem 8 is a corollary of Theorem 20, of the previous example and of Properties 10 and 11 of Sect. 4.2. The case $k < n$ follows from the case $k = n$, by making use of Property 9 of Sect. 4.2.

2. Relations.

Proof of Theorem 5. The statement is reduced to the case of the circuit $f_1, \ldots, f_{k-1}, f_k = -(f_1 + \cdots + f_{k-1})$ for which the relations $x_1 \cdots x_k = 0$, $(x_1 - 1) \cdots (x_k - 1) = 0$ hold.

Proof of Theorem 6. By Theorem 5, it suffices to show that $\dim \mathbf{Z}[x]/\vartheta = \dim H^*(M_{\mathbf{C}})$. The relations $x_i^2 = x_i$ annihilate the monomials which include at least one of variables raised to a power greater than 1. One has to prove that the work of the relation concerning the rest of the monomials can be deduced from (2). Under the isomorphism $P^s/P^{s-1} \to H^s(M_{\mathbf{C}})$, relation (2) is transformed into a homogeneous relation of degree $s-1$:

$$w_{i_2} \wedge \cdots \wedge w_{i_s} - w_{i_1} \wedge w_{i_3} \wedge \cdots \wedge w_{i_s} + \cdots + (-1)^{s-1} w_{i_1} \wedge \cdots \wedge w_{i_{s-1}}. \qquad (5)$$

According to [3], the external algebra spanned on $\{w_{i \in I}\}$ and factorized by making use of relations (5) for all the circuits is isomorphic to $H^*(M_{\mathbf{C}})$. Thus $\mathbf{Z}[x]/\vartheta$ has the needed dimension.

Theorem 7 is a corollary of Theorems 6 and 8 and of the Theorem based on [7, 11, 12]

§6. Appendix. Decomposition into elementary fractions

There exists an elementary analytical analogue of Theorem 18 on the decomposition into simplicial cones that is a generalization of the theorem on the decomposition of rational functions into elementary fractions.

1. Decomposition. Consider a rational function $R = P/Q$ on an n-dimensional linear space V, where P and Q are polynomials. Suppose that Q can be decomposed into a product of polynomials of degree 1: $Q = \prod_{i \in I} f_i^{k_i}$. Let $z_1, \ldots, z_n$ be the linear coordinates in V.

Theorem 21. *The function R can be represented in the form:*

$$R = \sum_{t,\alpha} (t_1)^{\alpha_1} \cdots (t_n)^{\alpha_n} A_{t,\alpha}, \tag{6}$$

where either $\alpha_j \geqq 0$ and $t_j = z_j$, or $\alpha_j < 0$ and t_j is a polynomial of degree 1 in z_j, $z_{j+1}, \ldots, z_n$, with 1 as the coefficient of z_j; the $A_{t,\alpha}$ are numbers. The representation is unique.

Proof. Restrict R to each of the lines parallel to the z_1-axis and decompose it into elementary fractions in z_1. We thus obtain a representation

$$R = \sum_{i \in I} \sum_{-k_i \leqq p < 0} f_i^p B_{p,i} + \sum_l z_1^l C_l,$$

where $B_{p,i}$ and C_l are rational functions in $z_2, \ldots, z_n$, whose denominators are products of polynomials whose degrees are equal to 1. This representation is unique. The functions $B_{p,i}$ and C_l can be treated similarly.

Example. $1/(z_1 - z_3)\ (z_1 - z_2)\ (z_1 - z_2 - z_3) = 1/(z_1 - z_3)\ (z_3 - z_2)\ (-z_2) + 1/(z_1 - z_2)\ (z_2 - z_3)\ (-z_3) + 1/(z_1 - z_2 - z_3) z_2 z_3 = 1/(z_1 - z_3)\ (z_3 - z_2)\ (-z_3) + 1/(z_1 - z_3)\,(-z_2) z_3 + 1/(z_1 - z_2)\,(z_2 - z_3)\,(-z_3) + 1/(z_1 - z_2 - z_3) z_2 z_3$.

Theorem 21 is close to Theorem 5.2 of [7].

Let all the linear functions $\{f_{i \in I}\}$ be homogeneous and the set $t_1, \ldots, t_n$ enter (6) with negative coefficients $\alpha_1, \ldots, \alpha_n$. We point out the relation between $t_1, \ldots, t_n$ and the covectors $\{f_{i \in I}\}$. t_1 is proportional to one of the $\{f_{i \in I}\}$. We project the rest of the covectors of $\{f_{i \in I}\}$ along t_1, onto a hyperplane in V^* which is orthogonal to the vector $(1, 0, \ldots, 0)$. Thus we obtain a set of covectors $\{g\}$. Then t_2 is proportional to one of those covectors. Project along t_2 the rest of the covectors of $\{g\}$, onto an $(n-2)$-dimensional plane orthogonal to the plane in V, spanned on $(1, 0, \ldots, 0)$, $(0, 1, 0, \ldots, 0)$. We get a set of covectors $\{h\}$. Then t_3 is proportional to one of those, etc.

If the set of covectors $\{f_{i \in I}\}$ is closed with respect to the projection operations just described, then, for $\alpha_j < 0$, the polynomial t_j of (6) is proportional to one of $\{f_{i \in I}\}$. The following are examples of such families:

1. Type A. Q is the product of degrees of polynomials $z_j = z_k$, $j < k$.
2. Type B. Q is the product of degrees of polynomials $z_j, z_j \pm z_k, j < k$.

2. Application. Consider a linear mapping $h: \mathbf{R}^N \to \mathbf{R}^n$, under which the j-th basis vector is transformed into a vector h_j. Suppose that all the h_j lie in the half-space $x_1 > 0$, where $x_1, \ldots, x_n$ are the coordinates in $\mathbf{R}^n$. Define the function U on $\mathbf{R}^n$. $U(x)$ is the $(N-n)$-dimensional volume of the intersection of the fibre lying above x and of the positive octant in $\mathbf{R}^N$, where, as a form of volume, the ratio of, respectively, volume forms on $\mathbf{R}^N$ and $\mathbf{R}^n$ is taken. For details of the function see [7].

Theorem 22. *For the set* $\{h_{j \leqq N}\}$ *in general position,*

$$U(x) = \sum_{\substack{1 \leqq s_1, \ldots, s_{n-1} \leqq N \\ s_i \neq s_j}} \frac{(h_{s_1}, \ldots, h_{s_{n-1}}, x)^{N-n} \chi_{s_1, \ldots, s_{n-1}}(x)}{(N-n)! \prod\limits_{\substack{s_n \notin \{s_1, \ldots, s_{n-1}\} \\ 1 \leqq s_n \leqq N}} (h_{s_1}, \ldots, h_{s_n})}, \tag{7}$$

where $(v_1, \ldots, v_n)$ *is the determinant whose columns are* $v_1, \ldots, v_n$ *and* $\chi_{s_1, \ldots, s_{n-1}}$ *is a characteristic function of the simplicial cone generated by the vectors* $h_{s_1}, h_{s_1, s_2}, \ldots, h_{s_1, \ldots, s_n}$, *where* $h_{s_1, \ldots, s_p}$, *for* $p > 1$, *is the projection of the vector* h_p *along* $\{h_1, \ldots, h_{p-1}\}$ *on the coordinate plane* $\{(0, \ldots, 0, 1_p, 0, \ldots, 0), \ldots, (0, \ldots, 0, 1)\}$, *its sign being taken so that its p-th coordinate is positive.*

In particular, for $n = 2$, χ_j *is the characteristic function of the cone generated by the vectors* h_j, $(0, 1)$.

For arbitrary n and $N = n$, (7) assumes the form:

$$U(x) = \sum_{s \in S_n} \chi_{s_1, \ldots, s_{n-1}}(x) / (h_{s_1}, \ldots, h_{s_n}).$$

In this case, $U(x)$ is piecewise constant and is proportional to the characteristic function of the cone generated by the vectors $h_1, \ldots, h_n$.

Remark. If we drop χ from (7), the sum will be identically equal to zero. For example, for $n = 2$, $f_j = (1, a_j)$, we have

$$\sum_{i=1}^{N} \frac{(x_2 - a_i x_1)^{N-2}}{(a_1 - a_i) \cdots (a_{i-1} - a_i)(a_{i+1} - a_i) \cdots (a_N - a_i)} \equiv 0.$$

To prove the theorem, it suffices to consider the Laplace transformation of the function $U(x)$ (see [7]), to decompose the rational function thus obtained into elementary fractions and to perform the inverse Laplace transformation of the terms of the sum.

References

1. Arnold, V.I.: The cohomology ring of the group of colored braids. Mat. Zametki **5** (1) (1969) 227–231 (Russian)
2. Brieskorn, E.: On groups of braids (according to V.I. Arnold). Séminaire Bourbaki, 24ᵉ année 1971/1972. Lecture Notes in Mathematics, vol. 317. Springer, Berlin Heidelberg New York
3. Orlik, P., Solomon, L.: Combinatorics and topology of complements of hyperplanes. Inv. Math. **56** (1980) 167–169

4. Gelfand, I.M.: General theory of hypergeometrical functions. Sov. Mat. Dokl. **288** (1) (1986) 14–18
5. Gelfand, I.M., Gelfand, S.I.: Generalized hypergeometrical equations. Sov. Mat. Dokl. **288** (2) (1986) 279–283
6. Gelfand, I.M., Graev, M.I.: Duality theorem for general hypergeometrical functions. Sov. Mat. Dokl. **289** (1) (1986) 19–23
7. Gelfand, I.M., Zelevinskij, A.V.: Algebraic and combinatorial aspects of general theory of hypergeometrical functions. Funct. Anal. Appl. **20** (3) (1986) 17–34 (Russian)
8. Vasiliev, V.A., Gelfand, I.M., Zelevinskij, A.V.: General hypergeometrical functions on complex Grassmannians. Funct. Anal. Appl. **21** (1) (1987) 23–38 (Russian)
9. Varchenko, A.N.: Combinatorics and topology of configuration of affine hyperplanes in a real space. Funct. Anal. Appl. **21** (1) (1987) 11–22 (Russian)
10. Kohno, T.: Homology of a local system on the complement of hyperplanes. Proc. Japan. Acad., Ser. A **62** (1) (1986) 144–147
11. Bjorner, A.: On the homology of geometric lattices. Algebra Universalis **14** (1) (1982) 107–128
12. Iambu, M., Leborgne, D.: Möbius's function and arrangement of hyperplanes. C. R. Acad. Sci. **303** (7) (1986) 311–319
13. Barnabei, M., Brini, A., Rota, G.-C.: Theory of Möbius functions. Usp. Mat. Nauk **41** (3) (1986) 113–157
14. McMullen, P.: The numbers of faces of simplicial polytopes. Israel J. Math. **9** (1971) 559–570
15. Stanley, R.: Hilbert functions of graded algebras. Adv. Math. **28** (1978) 57–83

11.

(with T.V. Alekseevskaya and A.V. Zelevinskij)

Arrangements of real hyperplanes and the associated partition function

Dokl. Akad. Naūk SSSR **297** (6) (1987) 1289–1293

A lot of important problems of the geometry of Lie groups, as well as dual problems of representation theory, depend on the structure of a simple function, the so-called partition function, in its continuous [1] and discrete [2] versions. In studying this function some geometrical problems arise that are undoubtedly interesting in themselves. These problems are closely related to the calculus of Heaviside functions constructed in [7]. This paper is mostly devoted to these geometrical problems; their applications to the study of the partition function is given at the end of the paper.

I. Statements of geometrical problems: chambers and simplicial cones. We are given a finite set of one-dimensional subspaces $L_1, \dots, L_n$ and a finite set of subspaces $W_1, \dots, W_m$ of codimension 1 in an l-dimensional real vector space V. We assume that the following condition holds:

(C1) Any subspace in V spanned by some of $L_1, \dots, L_n$ is the intersection of some of $W_1, \dots, W_m$.

Condition (C1) includes, in particular, the requirement that $W_1 \cap \dots \cap W_m = 0$. The most important case in applications is when $L_1 + \dots + L_n = V$. In this case, condition (C1) is equivalent to the requirement that any subspace of codimension 1 spanned by some of $L_1, \dots, L_n$ be one of the subspaces $W_1, \dots, W_m$.

Let us choose for any $i = 1, \dots, n$ an open half-line $L_i^+ \subset L_i$ beginning at the origin, in such a way that

(C2) All half-lines L_i^+ lie at one side of some hyperplane in V passing through the origin.

In an example that is important for applications the role of $L_1, \dots, L_n$ is played by half-lines spanned by positive roots of some root system in V.

We will be interested in geometrical objects of two types: simplicial cones C_I generated by half-lines L_i^+ and chambers Γ determined by subspaces $W_1, \dots, W_m$.

A subset $I \subset [1, n]$ is said to be independent if subspaces L_i, $i \in I$, are in general position, i.e. generate a $|I|$-dimensional subspace in V. Denote by C_I the closed convex cone spanned by half-lines L_i, $i \in I$. For example, $C_\varnothing$ is the cone consisting of the single point 0. It is clear that all cones C_I corresponding to independent subsets I are simplicial.

We will say that two points $x, y \in W$ lie in the same chamber Γ if for any $j = 1, \ldots, m$ the closed segment $[x, y]$ either does not intersect W_j or lies in W_j. For example, l-dimensional chambers in V are connected components of $V \setminus (\bigcup_{1 \leqq j \leqq m} W_j)$. It is clear that any chamber is a polyhedral convex cone in V. They are nonclosed, expect for the chamber consisting of the single point 0.

Condition (C1) implies that each cone C_I is a union of some chambers. Let us introduce the incidence matrix M. Its rows are indexed by chambers Γ and its columns are indexed by independent subsets $I \subset [1, n]$; the entry (Γ, C_I) at the intersection of the row Γ and of the column I equals 1 if $\Gamma \subset C_I$ and 0 otherwise.

We consider the following problems:

1. Find a complete system of linear relations between the columns of the matrix M.

2. Construct a basis in the vector space generated by the columns of M.

1′ and 2′. The same problems for the rows of M.

Let $M_r (r = 0, \ldots, 1)$ be the submatrix of M consisting of rows Γ and columns I with $|I| = \dim \Gamma = r$. We will give solutions for all the above problems, not only for M but for all submatrices M_r as well. The most important problem for applications is the study of the submatrix M_1 in M corresponding to chambers and cones of the maximum dimension.

Denote by φ_I the column of the matrix M corresponding to an independent subset I and by ψ_Γ the row of M corresponding to a chamber Γ. If $|I| = \dim \Gamma = r$ we denote by φ_I^r and ψ_Γ^r the corresponding column and row of the matrix M_r. Let Φ be the vector space generated by all columns φ_I and Ψ be the vector space generated by all rows ψ_Γ. Similarly, let Φ_r and Ψ_r be the vector spaces generated by columns and rows of M_r.

It is clear from the definitions that Φ can be naturally identified with the space of functions on V generated by the characteristic functions of all cones C_I. Elements of the space Ψ can be naturally represented as linear functionals on Φ, i.e. as "distributions" on V corresponding to the function space Φ.

2. Linear relations between columns φ_I, φ_I^r. A subset $J \subset [1, n]$ is said to be weakly dependent if $\dim \sum_{i \in J} L_i = |J| - 1$. To any weakly dependent subset J we associate a linear relation between the columns φ_I corresponding to independent subsets $I \subset J$.

Let us choose a vector $v_i \in L_i^+$ for each $i \in J$. As J is weakly dependent, there exists exactly one (up to a scalar factor) linear relation $\sum_{i \in J} a_i v_i = 0$ between the vectors v_i. Let $J_+ = \{i \in J : a_i < 0\}$, $J_- = \{i \in J : a_i > 0\}$ and $J_0 = \{i \in J : a_i = 0\}$. It is clear that the decomposition $J = J_+ \cup J_- \cup J_0$ does not depend on the choice of vectors v_i and is determined by J uniquely up to exchange of J_+ and J_-. Let us note that J_+ and J_- are nonempty due to (C2). It is clear that for any $i \in J_+ \cup J_-$ the subset $J \setminus i$ is independent.

Theorem 1. (a) *For any weakly dependent subset $J = J_+ \cup J_- \cup J_0$ the following relation holds:*

$$\sum_{\emptyset \neq \Omega \subset J_+} (-1)^{|\Omega| - 1} \varphi_{J \setminus \Omega} = \sum_{\emptyset \neq \Omega \in J_-} (-1)^{|\Omega| - 1} \varphi_{J \setminus \Omega}. \tag{1}$$

(b) *Any linear relation between the elements* $\varphi_I \in \Phi$ *is a linear combination of relations* (1).

Theorem 2. (a) *For any weakly dependent subset* $J = J_+ \cup J_- \cup J_0$ *with* $|J| = r+1$ *the following relation holds*:

$$\sum_{i \in J_+} \varphi^r_{J \setminus i} = \sum_{i \in J_-} \varphi^r_{J \setminus i}. \tag{2}$$

(b) *Any linear relation between the elements* $\varphi^r_I \in \Phi_r$ *is a linear combination of relations* (2).

3. Linear relations between the rows of ψ_Γ and ψ^r_Γ. First, it is clear that

$$\psi_\Gamma = 0 \text{ if a chamber } \Gamma \text{ does not lie in at least one cone } C_I. \tag{3}$$

To describe the remaining relation we give several definitions. A subspace $U \subset V$ is said to be a wall if it is the intersection of some of the subspaces $W_1, \ldots, W_m$. Walls spanned by some of the subspaces $L_1, \ldots, L_n$ are said to be essential, otherwise they are called unessential. For example, essential one-dimensional walls are exactly subspaces $L_1, \ldots, L_n$.

Let Γ be a chamber and $\bar{\Gamma}$ be the closure of Γ in V. We will say that a chamber Γ abuts on a wall U if $\bar{\Gamma} \cap U$ contains a chamber that is open in U.

Theorem 3. (a) *Let* L *be an unessential one-dimensional wall,* W *an arbitrary subspace of codimension 1 in* V *containing* L *and not containing any other wall, and* V^+_W *one of two closed half-spaces bounded by* W. *Then*

$$\sum_\Gamma (-1)^{\dim \Gamma - 1} \psi_\Gamma = 0 \tag{4}$$

where Γ *runs over all chambers abutting on* L *and contained in* V_W.

(b) *Any linear relation between* $\psi_\Gamma \in \Psi$ *is a linear combination of relations* (3) *and* (4).

When we talk about an r-flag we mean an increasing chain of walls $F = (0 = U_0 \subset U_1 \subset \cdots \subset U_r)$ with $\dim U_s = s$ for all s. An r-flag is said to be oriented if for all $s = 1, \ldots, r$ one of two connected components of $U_s \setminus U_{s-1}$ is chosen; we call it U^+_s. An oriented flag F will be denoted $\vec{F}$. A chamber Γ is said to be abutting on a flag F if it abuts on all its walls. For a chamber Γ abutting on an oriented r-flag $\vec{F}$ we set $\varepsilon(\Gamma, \vec{F}) = +1$ or -1 depending on the parity of the number of those $s \in [1, r]$ for which $\Gamma \cap U^+_s = \varnothing$.

Following [7], define for any oriented r-flag $\vec{F}$ an element $\psi^r(\vec{F}) \in \Psi_r$ by means of the formula $\psi^r(\vec{F}) = \sum_\Gamma \varepsilon(\Gamma, \vec{F}) \psi^r_\Gamma$ where Γ runs over all r-dimensional chambers abutting on F (one can easily see that there are exactly 2^r such chambers). It is clear that the elements $\psi^r(\vec{F})$ associated to different orientations of one flag may differ from each other in sign only.

Theorem 4. (a) *For any oriented* r-*flag* $\vec{F}$ *whose one-dimensional wall is unessential, we have*

$$\psi^r(\vec{F})=0. \tag{5}$$

(b) *Any linear relation amongst elements* $\psi^r{}_\Gamma \in \Psi_r$ *is a linear combination of relations* (5) *and relations* $\psi^r{}_\Gamma = 0$ *for all r-dimensional chambers* Γ *not contained in any cone* C_I.

Remark. Relations (4) and (5) can be included in a unified system of relations. We do not present it here because of lack of space.

4. Bases in spaces Φ_r and Φ. Let us fix a mapping τ of the set of nonzero essential walls into the set $[1,n]$ satisfying the following condition:

(C3) $L_i \subset U$ for $\tau(U)=i$.

Define a class $\mathscr{I}_\tau$ of independent subsets by means of the following requirements: (a) $\varnothing \in \mathscr{I}_\tau$. (b) Let I be a nonempty independent subset in $[1,n]$ and $\tau(\sum_{i\in I} L_i)=i_0$. Then $I \in \mathscr{I}_\tau$ if and only if $i_0 \in I$ and $I \backslash i_0 \in \mathscr{I}_\tau$.

Theorem 5. *Elements* φ^r_I *for all* $I \in \mathscr{I}_\tau$ *with* $|I|=r$ *form a basis in* Φ.

Theorem 6. *Elements* φ_I *for all* $I \in \mathscr{I}_\tau$ *form a basis in* Φ.

Remark. The definition of $\mathscr{I}_\tau$ was given essentially in [3]. For a special choice of τ, $\mathscr{I}_\tau$ becomes a class of "sets without open cycles" considered in [4, 7].

5. Bases in Ψ_r and Ψ. Let τ be the same as in Sect. 4. Denote by $\mathscr{F}^r_\tau$ the set of all oriented r-flags $\vec{F}=(U_0 \subset \cdots \subset U_r)$ such that : (a) all walls U_s are essential; (b) if $\tau(U_s)=i_s$ then $L^+_{i_s} \subset U^+_s$ $(s=1,\ldots,r)$.

Theorem 7. *Elements* $\psi^r(\vec{F})$ *for all* $\vec{F} \in \mathscr{F}^r_\tau$ *form a basis in* Φ.

For any oriented r-flag $\vec{F}$ define an element $\psi(\vec{F}) \in \Psi$ by means of the formula $\psi(\vec{F})=\sum_\Gamma \varepsilon(\Gamma,\vec{F})\psi_\Gamma$, where Γ runs over all r-dimensional chambers abutting on $\vec{F}$. Let also $\mathscr{F}_\tau = \bigcup_{0\le r\le l} \mathscr{F}^r_\tau$.

Theorem 8. *Elements* $\psi(\vec{F})$ *for all* $\vec{F} \in \mathscr{F}_\tau$ *form a basis in* Ψ.

The basis in Theorem 7 is formed not by elements $\psi^r{}_\Gamma$ themselves, but by linear combinations of them; in this sense, it is more "complicated" than the basis in Theorem 5. On the other hand, it has the following pleasant property:

Theorem 9. *For any r-dimensional chamber* Γ *all coefficients in the decomposition of* $\psi^r{}_\Gamma$ *with respect to the basis from Theorem 7 are equal to* 0 *or to* 1.

6. Applications to the structure of the partition function. Let us consider the "positive octant" $\mathbf{R}^n_+$ in $\mathbf{R}^n$. Let $V_0 \subset \mathbf{R}^n$ be an $(n-1)$-dimensional vector subspace

satisfying the condition $V_0 \cap \mathbf{R}^n_+ = 0$. Then for all $x \in \mathbf{R}^n$ the parallel plane $V_0 + x$ intersects $\mathbf{R}^n_+$ in a compact (possibly empty) convex polyhedron $\Delta_x = (V_0 + x) \cap \mathbf{R}^n_+$. The partition function is defined as the "volume" of the polyhedron Δ_x; it depends, clearly, only on the image of the point x in the quotient space $V = \mathbf{R}^n / V_0$. By "volume" we mean the following. Let us choose a differential $(n-1)$-form ω in $\mathbf{R}^n$ with polynomial coefficients, and fix compatible orientations in all planes $V_0 + x$. Then the partition function is defined by the formula $P_\omega(v) = \int_{\Delta_x} \omega$ where $v \in V$ is the image of $x \in \mathbf{R}^n$ under the natural projection $p: \mathbf{R}^n \to V$.

The function $P_\omega(v)$ is a piecewise polynomial function. To study it we apply the above methods in the following situation. Let $e_1, \ldots, e_n$ be the canonical basis in $\mathbf{R}^n$. Define one-dimensional subspaces $L_1, \ldots, L_n$ in V by $L_i = p(\mathbf{R}e_i)$. As $W_1, \ldots, W_m$ we take all subspaces of codimension 1 in V of the form $p(\mathbf{R}^I)$ for some coordinate subspace $\mathbf{R}^I$ in $\mathbf{R}^n$. Finally, let $L_i^+ \subset L_i$ $(i = 1, \ldots, n)$ be the half-line passing through the point $p(e_i)$. Conditions (C1) and (C2) are easily checked.

Let $V' = V \setminus (\bigcup_j W_j)$ be the union of all l-dimensional chambers in V. For any independent l-element subset $I \subset [1, n]$ denote by χ_I the characteristic function of the open convex cone $C_I \cap V'$.

In [4] it is proved that the function P_ω on V' admits the decomposition

$$P_\omega = \sum_I P^I_\omega \chi_I \tag{6}$$

where P^I are some polynomial functions on V (see also [5, 6]). The decomposition (6) is, in general, nonunique, because the functions χ_I can be linearly dependent. The complete system of linear relations between the functions χ_I is given by Theorem 2 (it is clear from the definitions that these relations are identical to the ones between the elements $\varphi^l_I \in \Phi_l$). By Theorem 5, decomposition 6 becomes unique if we leave only terms with $I \in \mathscr{I}_\tau$.

From decomposition (6) it follows, in particular, that the restriction of the function P_ω to any chamber Γ is equal to some polynomial P^Γ_ω. The list of all polynomials P^Γ_ω describes the function P_ω "dually" with respect to the decomposition (6). Theorem 4 gives a complete system of universal (i.e. not depending on ω) linear relations between polynomials P^Γ_ω (it is clear from the definitions that these relations are identical to the ones between the elements $\psi^l{}_\Gamma \in \Psi_l$). Theorem 7 shows that in order to compute all polynomials P^Γ_ω it suffices to compute "flag" linear combinations $\sum_\Gamma \varepsilon(\Gamma, \vec{F}) P^\Gamma_\omega$ for all oriented l-flags $\vec{F} \in \mathscr{F}^l_\tau$; by Theorem 9, each P^Γ_ω is a sum of some of these "flag" combinations.

References

1. Berezin, F.A., Gelfand, I.M.: Trudy Mosk. Mat. Ob-va **5** (1956) 311–352.
2. Kostant, B.: Trans. Amer. Math. Soc. **93** (1959) 55–73.
3. Björner, A.: Algebra Universalis **14** (1) (1982) 107–128.
4. Gelfand, I.M., Zelevinskij, A.V.: Funkts. Anal. Priloz. **20** (3) (1986) 17–34.
5. Varchenko, A.N.: Funkts. Anal. Priloz. **21** (1) (1987) 11–22.
6. Vasiliev, V.A., Gelfand, I.M., Zelevinskij, A.V.: Funkts. Anal. Priloz. **21** (1) (1987) 23–38.
7. Varchenko, A.N., Gelfand, I.M.: Funkts. Anal. Priloz. **21** (4) (1987) 1–18.

12.

(with V.S. Retakh and V.V. Serganova)

Generalized Airy functions, Schubert cells and Jordan groups

Dokl. Akad. Naŭk SSSR **298** (1) (1988) 17–21

The aim of this note is to generalize the theory of Airy functions to several variables. Our approach is similar to the one adopted in the theory of hypergeometric functions, see [1–3].

Generalized Airy functions are defined as solution of some holonomic systems of differential equations on a vector bundle on the Grassmann manifold.

Denote by G_k^n the Grassmannian of k-dimensional subspaces of the n-dimensional spaces. An important role in [1–3] was played by the decomposition of G_k^n into so-called strata. In our theory, a similar role is played by the decomposition of G_k^n into Schubert cells.

Similarly to the general theory of hypergeometric functions which, in the case of G_2^4, is reduced to the classical hypergeometric equation and to the Gauss function [1, 2], the theory of generalized Airy functions in the case of G_2^4 is reduced to the classical Airy equation. The solution of this equation is given by the Airy function which describes the asymptotical behavior of the wave field near a smooth caustic. For the open Schubert cell in G_2^5 we obtain the Pearcey function [8] describing this asymptotic behavior near a cusp of a caustic, and for G_2^6 and G_2^7 we obtain the asymptotics near a "swallow tail" and near a "butterfly" respectively. Asymptotics near focal points of a wave field [5] are related to Schubert cells of nonzero codimension in G_3^9 and G_3^{12}.

Let us note that the role played in [1–3] by the subgroup of diagonal matrices in played in our theory by another maximum Abelian subgroup of the group of nondegenerate matrices, namely, by the Jordan subgroup (see Sect. 1).

The Jordan subgroup is the centralizer of a maximum Jordan matrix. Centralizers of other matrices that are built from Jordan matrices of smaller dimensions lead to generalizations of various confluent hypergeometric functions (Kummer's, Whitteker's, Bessel's, Weber's, etc.)

1. Let Z_k^n be the manifold of all rank k real or complex matrices $z = (z_{ij})$, $i = 1, \ldots, k; j = 1, \ldots, n$. Let $\alpha = (\alpha_1, \ldots, \alpha_{n-1})$ be an $(n-1)$-tuple of complex numbers. Denote by δ_{ij} the Kronecker delta.

Definition 1. By the system of generalized Airy equations on Z_k^n for $k < n$ with homogeneity indices $\alpha_1, \ldots, \alpha_{n-1}$ we mean the following system of differential equations:

$$\sum_{j=1}^{n} z_{ij} \frac{\partial F}{\partial z_{l_j}} = -\delta_{il} F, \quad i, l = 1, \ldots, k, \qquad (I_1)$$

$$(I)\qquad \sum_{j=1}^{n-p}\sum_{i=1}^{k} z_{ij}\frac{\partial F}{\partial z_{i,j+p}}=\alpha_p F, \qquad (I_2)$$

$$\frac{\partial^2 F}{\partial z_{ip}\partial z_{jq}}=\frac{\partial^2 F}{\partial z_{iq}\partial z_{jp}},\quad i,j=1,\dots,k,\quad p,q=1,\dots,n. \qquad (I_3)$$

Let us clarify the meaning of the differential equations (I_1).

Denote by GL(n) the group of all invertible $(n\times n)$ matrices. The formula $z\mapsto gzh$, $g\in \mathrm{GL}(k)$, $h\in\mathrm{GL}(n)$ determines the left action of GL(k) and the right action of GL(n) on Z_k^n.

Equations (I_1) imply the following.

Proposition 1. *For any solution F of the system (I) we have* $F(gz)=(\det g)^{-1}F(z)$.

Proposition 1 shows that a solution of (I) depends essentially on the k-dimensional subspace spanned by the rows of a matrix $z\in Z_k^n$ in the n-dimensional space.

Consider now the equations (I_2). We need several definitions. Let J be a commutative Lie group. A function χ on J with values in $\mathbf{C}\backslash\{0\}$ is said to be a character of the group J if $\chi(g_1g_2)=\chi(g_1)\chi(g_2)$ for $g_1,g_2\in J$. We refer to the subgroup of GL(n) consisting of all matrices of the form $[a_0,\dots,a_{n-1}]=\sum_{i=0}^{n-1}a_i\tau^i$, where $a_0,\dots,a_{n-1}$ belong to the base field and the matrix $\tau=(\tau_{ij})$ is given by $\tau_{i,i-1}=1$, $\tau_{ij}=0$ for $j-i\neq 1$, as the Jordan group $J(n)$. We denote by $J_0(n)$ the subgroup of $J(n)$ consisting of matrices of the form $[a_0,\dots,a_{n-1}]$ with $a_0=1$.

Let also $u=1+b_1T+b_2T^2+\cdots$ be a formal series in the indeterminate T. Define polynomials $\theta_i(b_1,\dots,b_i)$ from the decomposition

$$\log(1-u)=\sum_{i=1}^{\infty}\theta_i(b_1,\dots,b_i)T^i. \qquad (1)$$

It is clear that $\theta_1=b_1$, $\theta_2=b_2-b_1^2/2$, $\theta_3=b_3-b_1b_2+b_1^3/3$.

Let us determine now all characters of the group $J_0(n)$.

Proposition 2. *For any $(n-1)$-tuple $\alpha=(\alpha_1,\dots,\alpha_{n-1})$ the formula*

$$\chi_\alpha([1,a_1,\dots,a_{n-1}])=\exp\left(\sum_{i=1}^{n-1}\alpha_i\theta_i(a_1,\dots,a_i)\right) \qquad (2)$$

gives a character of the group $J_0(n)$. Conversely, for any character χ there exists an α such that χ is given by the formula (2).

Proposition 3. *For any solution F of the system (I) and for any $c=[c_0,c_1,\dots,c_{n-1}]\in J(n)$ we have*

$$F(zc)=F(z)\chi_\alpha([1,c_1c_0^{-1},\dots,c_nc_0^{-1}])$$

The proof immediately follows from the equations (I_2).

2. **Theorem 1.** *The system (I) is holonomic; in particular, in a neighborhood of any point of Z_k^n it has a finite number of linearly independent solutions.*

Therefore the system (I) is somewhat similar to a system of ordinary differential equations.

To prove the theorem let us consider the characteristic variety $M_F \subset T^*Z_k^n$ (see [2]) of the system (I). For any $z \in Z_k^n$ denote by z_t the matrix composed by the first t columns of the matrix z. Let $W = \{z \in Z_k^n \mid \det z_k \neq 0\}$. It is clear that W is open and dense in Z_k^n.

Theorem 1 results from the following proposition.

Proposition 4. *Let $z \in W$. A point (z, ξ) lies in M_F if an only if $\xi = 0$. Therefore* $\dim W \cap M_F = kn$. *The dimension of M_F near any point $z \in Z_k^n$ does not exceed kn.*

Let $M_{F,z} = \{(a, \xi) \in M_F \mid a = z\}$. Let us formulate the statement about the dimension of $M_{F,z}$. Let us associate to any point $z \in Z_k^n$ a number sequence $\pi(z) = (s_0, \ldots, s_k)$ by means of the formula $s_i = \max\{t \mid \operatorname{rk} z_t = i\}$, $i = 0, \ldots, k$. Let π be a nondecreasing sequence of $k+1$ positive integers. The set $W_\pi = \{z \in Z_k^n \mid \pi(z) = \pi\}$ is called the Schubert cell of index π. The set of subspaces spanned by rows of all matrices from W_π is the Schubert cell of index π in the manifold of all k-dimensional subspaces of the n-dimensional space.

Proposition 5. *The dimension of $M_{F,z}$ depends only on the Schubert cell containing z, and is given by the formula*

$$\max_{i=0,\ldots,k-1} [(s_i + k - 2i - 1) \operatorname{sgn}(s_i - i)]$$

Let us determine now the dimension of the space of solutions on the open dense subset $W \subset Z_k^n$. In our notations W is the Schubert cell of index $(0, 1, \ldots, k)$. We call W the highest cell because W is the only open Schubert cell.

Theorem 2. *The system (I) with $\alpha_{n-1} \neq 0$ has $\binom{n-2}{k-1}$ linearly independent regular solutions in a neighborhood of any point $z \in W$.*

3. To prove Theorem 2 we rewrite the system (I) in coordinates (v_{ij}) that are transversal to the action of $\mathrm{GL}(k)$. The resulting system (II) is of interest for its own sake.

Let V_k^n be the space of all matrices $v = (v_{ij})$, $i = 1, \ldots, k$, $j = k+1, \ldots, n$. Under the action of $\mathrm{GL}(k)$ each matrix $z \in W$ is equivalent to a unique matrix of the form (E_k, v) where E_k is the identity matrix of order k and $v \in V_k^n$. Thus we get a one-to-one correspondence between k-dimensional subspaces spanned by rows of matrices from W and matrices from V_k^n.

Proposition 6. *The system (I) determines the following system of differential equations on V_k^n (we agree that $v_{ij} = \delta_{ij}$ for $i \leqslant k$, $v_{ij} = 0$ for $i > k$):*

$$\sum_{j=p+1}^{n} \sum_{i=1}^{k} v_{i,j-p} \frac{\partial F}{\partial v_{ij}} = \alpha_p F, \quad p = k, \ldots, n-1,$$

$$\sum_{j=p+1}^{n}\sum_{i=1}^{k}(v_{i,j-p}-(v_{i+p,j}-v_{i+p,j-p}))\frac{\partial F}{\partial v_{ij}}=\alpha_p F,\quad p=1,\dots,k-1, \tag{II}$$

$$\frac{\partial^2 F}{\partial v_{ip}\partial v_{jq}}=\frac{\partial^2 F}{\partial v_{iq}\partial v_{jp}},\quad i,j=1,..,k;\quad p,q=k+1,\dots,n.$$

Theorem 2 is equivalent to the following theorem.

Theorem 2′. *The number d of linearly independent solutions of the system (II) with $\alpha_{n-1}\neq 0$ in a neighborhood of any point of V_k^n is equal to* $\binom{n-2}{k-1}$.

The estimate $d\leqq\binom{n-2}{k-1}$ follows from the general theory of holonomic systems (cf. [2]); the estimate for d from below follows from Theorem 3 in Sect. 4.

4. We begin with constructions from [6, 7]. Let $P_v(t)$ be an arbitrary polynomial in variables $t\in\mathbf{C}^p$, $v\in\mathbf{C}^N$. Denote by μ_v the sum of Milnor numbers of all singular points of P_v [4] for fixed v and variable t.

Let $M_{a,v}=\{t\,|\,\operatorname{Re}P_v(t)\geqq a\}$, $N_{a,v}=\{t\,|\,\operatorname{Re}P_v(t)=a\}$. Denote by $H_p(a,v)$ the relative homology group $H_p(M_{a,v},N_{a,v})$. For any v there exists an a_0 such that $H_p(a,v)\cong\mathbf{Z}\oplus\cdots\oplus\mathbf{Z}$ (μ_v times) for $a<a_0$.

For $b<a$ we have the canonical mapping $H_p(b,v)\to H_p(a,v)$. Let $H_p(v)$ be the inductive limit of $H_p(a,v)$ as $a\to-\infty$. It is clear that $H_p(v)$ is isomorphic to the direct sum of μ_v copies of $\mathbf{Z}$.

For any point v in a sufficiently small neighborhood of a fixed point v_0 there exists a canonical isomorphism $\kappa_v\colon H_p(v_0)\rightrightarrows H_p(v)$ inducing the Gauss–Manin connection [7].

Let $v=(v_{ij})$, $i=1,\dots,k$; $j=k+1,\dots,n$. Let $t=(t_1,\dots,t_{k-1})$. Let $w_1=t_1$ for $l=1,\dots,k-1$, $w_{l-1}=v_{1l}+v_{2l}t_1+v_{3l}t_2+\cdots+v_{kl}t_{k-1}$ for $l=k+1,\dots,n$. Set

$$P_v(t)=\sum_{l=1}^{n-1}\alpha_l\theta_l(w_1,\dots,w_l),$$

where the polynomials θ_1 are defined by the decomposition (1).

Theorem 3. *Let $\gamma_1,\dots,\gamma_\mu$ be a basis in $H_{k-1}(v_0)$, $v_0\in V_k^n$ For $1\leqq m\leqq\mu$ the functions*

$$\Phi_m(v)=\int_{\kappa_v(\gamma_m)}\exp(P_v(t))dt_1\cdots dt_{k-1}$$

are defined in some neighborhood of the point v_0 and form in this neighborhood a set of linearly independent solutions of the system (II).

Proposition 7. *If the last homogeneity index α_{n-1} is nonzero then* $\mu_v=\binom{n-2}{k-1}$.

The proof follows from the Koushnirenko theorem [4] or from an explicit construction of generating cycles of the group $H_{k-1}(v)$. This construction is interesting in itself, but we omit it because of the lack of space.

The most simple is the structure of the basis of $H_{k-1}(v)$ for $k=2$. Let $\Gamma'(p,q)$ be the ray in **C** passing through $(\varepsilon_q)^p$ and through 0, where $\varepsilon_q=\sqrt[q]{-1}$ is a primitive root. Denote $\Gamma(p,q)=\Gamma'(p,q)\cup[0,+\infty]$. The paths $\Gamma(p,n-1)$ for $p=1,\ldots,n-2$ form a basis of the group $H_1(v)$.

The generalization of this statement is the following proposition.

Proposition 8. *Let all numbers $(n-1)/(m-1)$, $m=2,\ldots,k$ be integers. Then the paths $\prod_{m=2}^{k}\Gamma(p_m,(n-1)/(m-1))$ for $p_m\leqq(n-1)/(m-1)$ form a basis in $H_{k-1}(v)$.*

5. Let us study the system (*II*) for $k=2$. To do this we rewrite it in coordinates that are transversal to the action of $J(n)$. Let us assume that $\alpha=\alpha_{n-1}\neq 0$.

Let $P=\sum_{i=1}^{n-1}\alpha_i\theta_i(t,v_{13}+tv_{23},\ldots,v_{1,i+1}+tv_{2,i+1})$ where the θ_i are polynomials determined by the formula (1). Let $t_1=t+(\alpha_{n-2}/\alpha)v_{1,n-1}$. Then P can be written in the form

$$P=u_0+u_1t_1+\cdots+u_{n-3}t_1^{n-3}+\frac{(-1)^n}{n-1}\alpha t_1^{n-1}.$$

the coefficients u_i for $i\geqq 1$ are constant on orbits of the group $J(n)$ in V_2^n.

Let $f=e^{-u_0}F$.

Theorem 4. *For $k=2$ the system (II) can be transformed to the form*

$$\begin{aligned}&\frac{\partial^2 f}{\partial u_1\partial u_i}=\frac{\partial f}{\partial u_{i+1}},\quad 1\leqq i\leqq n-4,\\ &\alpha\frac{\partial^2 f}{\partial u_1\partial u_{n-3}}=u_1f+\sum_{j=2}^{n-3}ju_j\frac{\partial f}{\partial u_{j-1}}.\end{aligned}\tag{III}$$

For $n=4$ this system is reduced to the Airy equation $f''=u_1f$. For $n=5$ this system is reduced to two equations, one of which is the heat equation.

Let us show that the system (*III*) can be reduced to an ordinary differential equation.

Let $i=(i_2,\ldots,i_{m-1})$ be a multi-index, $i_j\geqq 0$ for $2\leqq j\leqq m-1$. Write $i!=i_2!i_3!\cdots i_{m-1}!$, $l(i)=2i_2+3i_3+\cdots(m-1)i_{m-1}$, $u=u_2^{i_2}u_3^{i_3}\cdots u_{m-1}^{i_{m-1}}$. As usual, we denote by $\varphi^{(p)}(t)$ the derivative of order p of a function $\varphi(t)$.

Proposition 9. The series $\sum_i\varphi^{(l(i))}(u_1)(u_i/i!)$ *is a solution of the system (III) if and only if $\varphi(u_1)$ is a solution of the equation $\varphi^{(n-2)}-u_1\varphi=0$.*

References

1. Gelfand, I.M.: Dokl. Akad. Naūk SSSR **288** (1) (1986) 14–18
2. Gelfand, I.M., Gelfand, S.I.: Dokl. Akad. Naūk SSSR **288** (2) (1986) 279–283

3. Vasiliev, V.A., Gelfand, I.M., Zelevinskij, A.V.: Funkts. Anal. Priloz. **21** (1) (1987) 23–38
4. Koushnirenko, A.G.: Funkts. Anal. Priloz. **9** (1) (1975) 74–75
5. Lukin, D.S., Palkin, E.A.: Experimental and numerical study of diffraction structures of a wave field in focal regions. In: Propagation of decimeter waves. Moscow, IZMIRAN, 1980, pp. 37–46
6. Fedorjuk, M.V.: Saddle-point method. Moscow, Nauka, 1977
7. Malgrange, B.: Intégrals asymptotiques et monodromie. Ann. Sci. Ec. Norm. Sup. 4ᵉ Serie **7** (1974) 405–430
8. Pearcey, T.: The structure of an electromagnetic field in the neighborhood of a cusp of a caustic. Phil. Mag. **37** (1946) 311–317

Appendix

V.I. Arnold

Cardiac arrhythmias and circle mappings

In 1958 I.M. Gelfand and M.L. Tsetlin introduced a mathematical model explaining the origin of some cardiac arrhythmias by the phase locking of a periodically excited relaxator. Here I reproduce the description of the Gelfand–Tsetlin model excerpted from my diploma dissertation on the mappings of a circle to itself (Moscow University, 1959); the applications to cardiac arrhythmias were omitted in the published version of this work (Izv. Akad. Nauk SSSR 1961; Am. Math. Soc. Transl. (2) **46** (1965), 213–284).

The numbering of the sections and figures below is that of the 1959 dissertation.

...§2. Synchronization of the relaxators

We call a device transforming an *input* $x(t)$ into an *output* $y(t)=\{x(t)\}$, where $\{x\}=x-[x]$ is the fractional part of x, a *relaxator*. If $x=\int g\,dt$, the relaxator stores the incoming quantity g until y reaches the value 1; then a jump to $y=0$ occurs and y starts growing again until the next jump.

We consider the synchronization by an *impulse* f of period T. This means that $x=t+f(t)$, where f is a T-periodic function with one sharp maximum on a period (and frequently equal to 0 over an important part of the period).

Let us consider the graph of $F(t)=1-f(t)>0$. The straight line with inclination 45° starting at $(t,0)$ intersects the graph at the point $(t',F(t'))$. We define a line mapping P by $P(t)=t'$; it defines a mapping of the circle $t \bmod T$ to itself. This mapping may be discontinuous. The moments at which jumps occur are $t_n=P_n t_0$. The rotation number of P is the limit of t_n/nT for $n\to\infty$. The rotation number

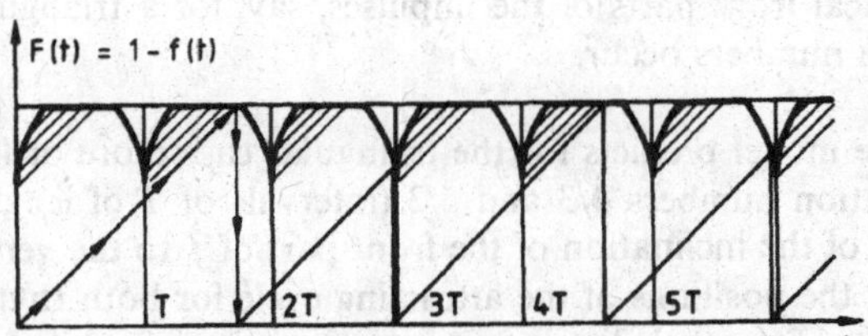

Fig. 4.

being equal to m/n signifies that to m impulses f there correspond n jumps (on average).

[Subject to some mild restrictions, omitted here], if the front part of the impulse f is nonvertical, the mapping P^{-1} is continuous and (nonstrictly) monotone. [In the previous part of the work it was proved that] any continuous and (nonstrictly) monotone orientation-preserving circle mapping has a rotation number independent of the initial point and continuously dependent on the mapping, and [that] any mapping whose inverse has a rotation number has itself an (opposite) rotation number.

Hence P has a rotation number, which depends continuously of f (provided that its front part is nonvertical).

In some of the examples considered below the front part is vertical, but the rotation number still exists and is a rational number [according to other theorems omitted here] and hence discontinuously depends on f.

... §4. On a model of Wenckebach cycles

The heart contains auricles and ventricles. Normally, every auricular beat is followed by a ventricular beat. Other rhythms are called *arrhythmias.* Recently I.M. Gelfand and M.L. Tsetlin introduced the following model, explaining the results of Rosenbluth's experiments [9], described below.

The Gelfand–Tsetlin model. The ventricular beat follows the jump of some relaxator—let us call it AV, because it is located at the atrioventricular knot. AV receives an impulse from the sinus knot.

The Rosenbluth experiments. One substitutes for the sinus knot an artificial generator of rectangular impulses of period T. For $T > T_1 = 233$ ms one observes the normal synchronization 1:1. For $T < T_2 = 210$ ms the ventricular beats are two times less frequent than those of the generator. For $T_2 < T < T_1$ one observes the Wenckebach cycles. For some T, every T, every 20th, ..., 4th, 3rd of the synchronizing impulses are lost. For a fixed T, this rhythm is conserved over minutes, but a small variation in T may change it.

The model prediction (for a rectangular impulse) is (2:1) 210 ms (3:2) 222 ms (4:3) 225 ms (5:4) 227 ms (6:5) 228 ms (7:6) 229 ms $\cdots$ 233 ms (1:1). (For 222 ms $< T <$ 225 ms every fourth beat is lost.)

For nonvertical front parts of the impulses, say, for a triangular impulse, all rational rotation numbers occur.

Remark. The model predicts for the triangular case more or less comparable tongues for rotation numbers 4/3 and 5/3 (intervals of T of length $\sim Q^2$, where Q is the tangent of the inclination of the front part of f to the vertical). But there is a difference in the positions of the attracting cycle for both rhythms. For a 4/3 rhythm the points of the cycle lie closer to the summit of f, while for a 5/3 rhythm they are concentrated at the start of the front part.

Now if we approximate a more realistic f by a triangular impulse, the effective

Q of the triangular approximation for the 5/3 rhythm will be smaller than for the 4/3 rhythm (the slope of the impulse at the points of the cycle is larger for the 5/3 case). Hence the interval of T in which the 5/3 rhythm will be observed is much smaller than that for the 4/3 rhythm.

The resonance zones for some particular impulses f. **Notation:** T-period, A-amplitude, $a = A/T$.

1. *A momentary impulse.* The resonance zones are given in Figs. 7 and 8.

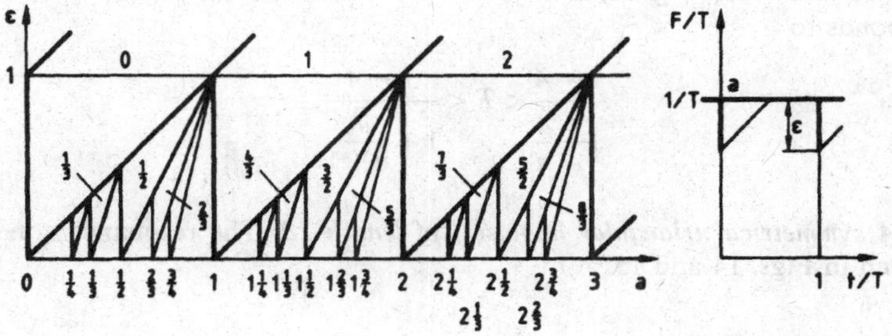

Fig. 7.

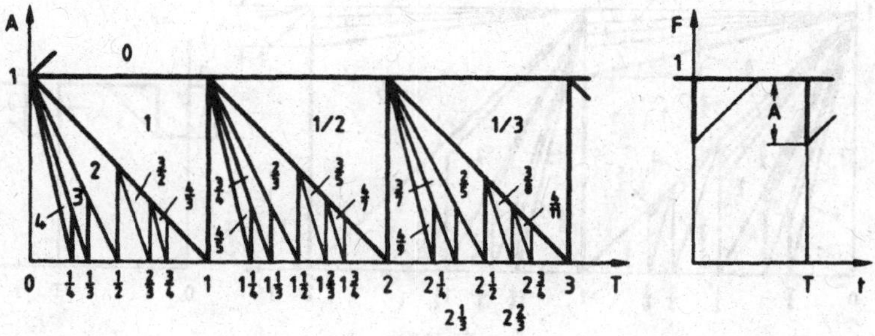

Fig. 8.

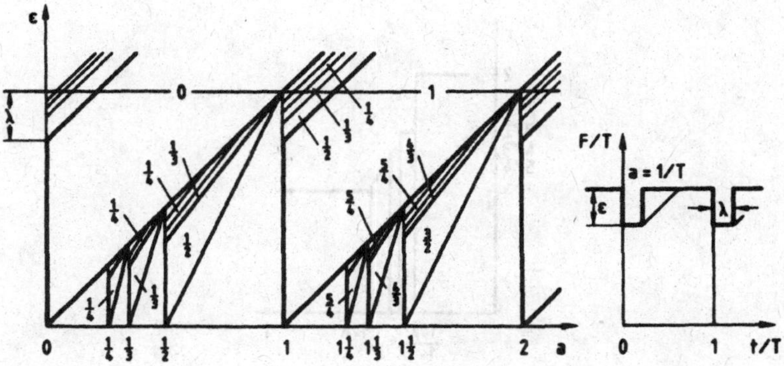

Fig. 10.

The periodicities follow from the following general lemmata:

(a) If $P(t) = t + p(t)$ has rotation number r (and p has period T), then $P_n(t) = t + nT + p(t)$ has rotation number $r + n$.

(b) Let the T-periodic impulse f correspond to the rotation number m/n. Then the $(T + k)$-periodic impulse, obtained from f by the insertion of voids ($f = 0$) of integer length k, corresponds to the rotation number $m/(n + km)$.

These lemmata also explain the following figures.

2. *A rectangular impulse f of length $\Delta = \lambda T$.* The resonance zones are presented in Figs. 10 and 11. In Fig. 12, $A = 1/2$ and $\lambda = 1/2$. The rotation number $(n+1)/n$ corresponds to

$$\frac{1-A}{1+\dfrac{\lambda}{n}} < T < \frac{1-A}{1+\dfrac{\lambda}{n+1}}.$$

3. *A symmetrical triangular impulse f of length 2Δ.* The resonance zones are presented in Figs. 14 and 15.

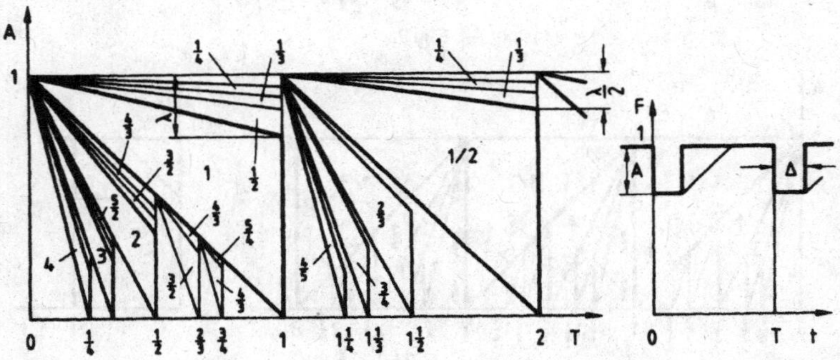

Fig. 11.

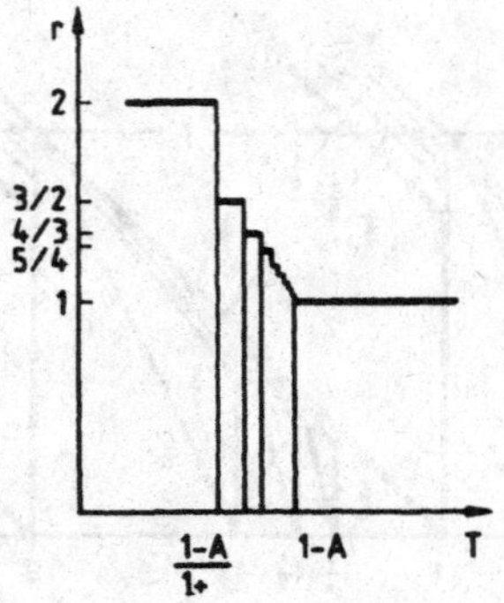

Fig. 12.

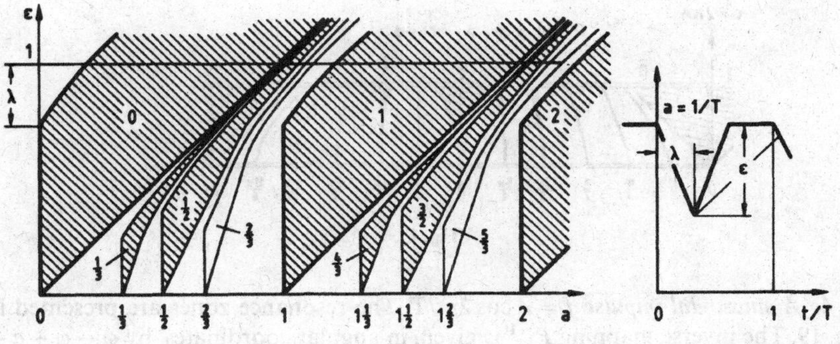

Fig. 14.

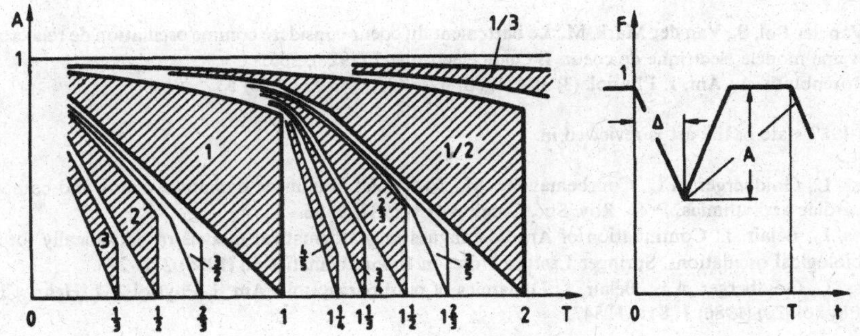

Fig. 15.

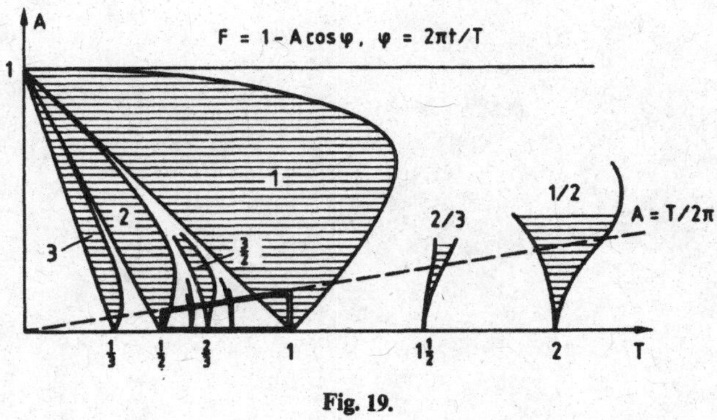

Fig. 19.

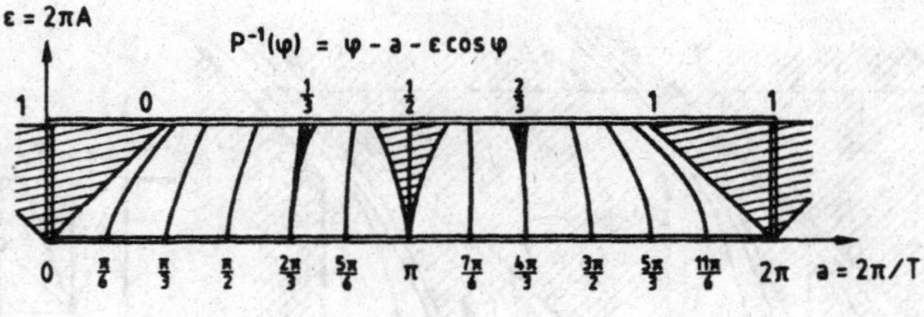

Fig. 17.

4. *A sinusoidal impulse* $f = A\cos 2\pi t/T$. The resonance zones are presented in Fig. 19. The inverse mapping P^{-1} is given, in angular coordinates, by $\varphi \mapsto \varphi + a + \varepsilon\cos\varphi$, $a = -2\pi/T$, $\varepsilon = -A\cdot 2\pi/T$. The resonance zones in coordinates a, ε are presented in Fig. 17.

References

8. Van der Pol, B., Van der Mark, M.: Le battement du coeur consideré comme oscillation de relaxation et une modèle electrique de coeur. L'Onde electrique **7** (1928), 365.
9. Rosenbluth, A.: Am. J. Physiol. (3) **194** (1958) 491, 495; (1) **195** (1958) 53.

The 1987 state of the art is reviewed in:

Glass, L., Goldberger, A.L., Courbemanche, M., Shrier, A.: Nonlinear dynamics, chaos and complex cardiac arrhythmias. Proc. Roy. Soc. Lond. **A413** (1987) 9–26.

Glass, L., Bélair, J.: Continuation of Arnold tongues in mathematical models of periodically forced biological oscillations. Springer Lecture Notes in Biomathematics **66** (1986) 232–243.

Glass, L., Goldberger, A.L., Bélair, J.: Dynamics of pure parasystole. Am. J. Physiol **251** (Heart Circ. Physiol. 20) (1986) H 841–H 847.

An Editorial Perspective

It seems to me to be a correct assessment to say that I. M. Gelfand has produced seminal papers in more areas of mathematics than any other mathematician – certainly in the latter half of this century. In the first part of the twentieth century Hilbert and Weyl are comparable in this regard. In discussing his work with other mathematicians I am struck by the number of times that I have learned that a key idea in their field goes back to Gelfand.

In viewing this huge collection of papers there are, among others, two aspects to his work which I would like to mention and which, in my opinion, make him an extraordinarily remarkable mathematician. The first of these is that his results by and large convey concepts rather than just technical information. More often than not, one comes away with an illuminating picture in one's head and a strikingly new way of looking at things. Rather than finishing off matters his mathematics generally points the way to new lines of development. The second aspect is his almost uncanny ability to see connections between previously unrelated matters. In my life I have never met or even read about anybody who has this talent to the extent that Gelfand has it. His instinct, it appears to me, is to always unify and to look for a single idea which would lead to an understanding of a multitude of diverse matters. In this regard – in a lighter vein – I have been personally much encouraged by a statement he has been quoted as having made, "all of mathematics is some kind of representation theory".

As an example of a work which unifies much of what has gone before one might study his papers (with other authors) in the present volume on integral geometry. Here in one fell swoop, using the lovely device of a double fibration, one comes away with a clarifying picture of what is the nature of a "transform" and why one can expect to have an "inverse transform".

The papers in the three volumes speak for themselves. To get an idea where his mind has been at over the years it is really necessary to read them. It would create a distortion to isolate any one such paper for a discussion here. Nonetheless it is interesting to track the evolution of some of his ideas from the time when they first appeared to the present. As an example consider the representation theory of reductive Lie groups. The still fundamental ideas of the "Bruhat decomposition" and the principal series go back to Gelfand. Making Cartan's theory of symmetric spaces into a vehicle for harmonic analysis goes back to Gelfand's theory of spherical functions. The symplectic-orbit picture received a considerable boost with the Gelfand-Kirillov conjecture and general algebra was the beneficiary of the notion of the Gelfand-Kirillov dimension. The introduction of functors and categories in representation theory is traceable to "BGG". Among other remarkable developments this revolutionary innovation set the stage for the Kazhdan-Lusztig theory.

March 1989 Bertram Kostant

Table of contents for volumes I and II

Volume I

Part I. Survey lectures and articles of general content

1. Some aspects of functional analysis and algebra
2. On some problems of functional analysis
3. Some questions of analysis and differential equations
4. Integral geometry and its relation to the theory of group representations
5. On elliptic equations
6. Automorphic functions and the theory of representations
7. The cohomology of infinite dimensional Lie algebras; some questions of integral geometry

Part II. Banach spaces and normed rings

1. Sur un lemme de la théorie des espaces linéaires
2. Abstrakte Funktionen und lineare Operatoren
3. On one-parametrical groups of operators in a normed space
4. On normed rings
5. To the theory of normed rings. II. On absolutely convergent trigonometrical series and integrals
6. To the theory of normed rings. III. On the ring of almost periodic functions
7. On the theory of characters of commutative topological groups (with D. A. Rajkov)
8. Normierte Ringe
9. Über verschiedene Methoden der Einführung der Topologie in die Menge der maximalen Ideale eines normierten Ringes (mit G. E. Shilov)
10. Ideale und primäre Ideale in normierten Ringen
11. Zur Theorie der Charaktere der Abelschen topologischen Gruppen
12. Über absolut konvergente trigonometrische Reihen und Integrale
13. On the imbedding of normed rings into the ring of operators in Hilbert space (with M. A. Najmark)
14. Commutative normed rings (with D. A. Rajkov and G. E. Shilov)
15. Normed rings with an involution and their representations (with M. A. Najmark)

Part III. Differential equations and mathematical physics

1. Eigenfunction expansions for equations with periodic coefficients
2. On the determination of a differential equation from its spectral function (with B. M. Levitan)
3. On a simple identity for eigenvalues of second order differential operator (with B. M. Levitan)
4. Solution of quantum field equations (with R. A. Minlos)
5. On the structure of the regions of stability of linear canonical systems of differential equations with periodic coefficients (with V. B. Lidskij)
6. Eigenfunction expansions for differential and other operators (with A. G. Kostyuchenko)
7. On identities for eigenvalues of a second order differential operator
8. Some problems in the theory of quasilinear equations
9. On a theorem of Poincaré (with I. I. Piatetski-Shapiro)
10. Fractional powers of operators and Hamiltonian systems (with L. A. Dikij)
11. A family of Hamiltonian structures related to nonlinear integrable differential equations (with L. A. Dikij)
12. Asymptotic behaviour of the resolvent of Sturm-Liouville equations and the algebra of the Korteweg-de Vries equations (with L. A. Dikij)
13. The resolvent and Hamiltonian systems (with L. A. Dikij)
14. Integrable nonlinear equations and the Liouville theorem (with L. A. Dikij)
15. Hamiltonian operators and algebraic structures related to them (with I. Ya. Dorfman)
16. The Schouten bracket and Hamiltonian operators (with I. Ya. Dorfman)
17. Hamiltonian operators and the classical Yang-Baxter equation (with I. Ya. Dorfman)

Essays in honour of Izrail M. Gelfand

On his fiftieth birthday (by M. I. Vishik, A. N. Kolmogorov, S. V. Fomin, and G. E. Shilov)

On his sixtieth birthday (by S. G. Gindikin, A. A. Kirillov, and D. B. Fuks; O. V. Lokytsievskij and N. N. Chentsov; M. B. Berkinblit, Yu. M. Vasil'ev, and M. L. Shik)

On his seventieth birthday (by N. N. Bogolyubov, S. G. Gindikin, A. A. Kirillov, A. N. Kolmogorov, S. P. Novikov, and L. D. Faddeev)

Some remarks on I. M. Gelfand's works
(by V. W. Guillemin and S. Sternberg)

Tentative table of contents for volumes II and III

Bibliography

Acknowledgements

Volume II

Part I. General problems of representation theory

1. Irreducible unitary representations of locally bicompact groups (with D. A. Rajkov)
2. Unitary representations of the group of linear transformations of the straight line (with M. A. Najmark)
3. Center of the infinitesimal group ring
4. Spherical functions on symmetric Riemannian spaces

Part II. Infinite-dimensional representations of semisimple Lie groups

1. Unitary representations of the Lorentz group (with M. A. Najmark)
2. Unitary representations of the Lorentz group (with M. A. Najmark)
3. On unitary representations of the complex unimodular group (with M. A. Najmark)
4. The principal series of irreducible representations of the complex unimodular group (with M. A. Najmark)
5. Complementary and degenerate series of representations of the complex unimodular group (with M. A. Najmark)
6. The trace in principal and complementary series representations of the complex unimodular group (with M. A. Najmark)
7. On the connection between representations of a complex semisimple Lie group and its maximal compact subgroup (with M. A. Najmark)
8. An analogue of the Plancherel formula for the complex unimodular group (with M. A. Najmark)
9. General relativistically invariant equations and infinite-dimensional representations of the Lorentz group (with A. M. Yaglom)
10. On the structure of the ring of rapidly decreasing functions on a Lie group (with M. I. Graev)
11. Unitary representations of classical groups (with M. A. Najmark)
12. Unitary representations of the real simple Lie groups (with M. I. Graev)
13. Unitary representations of real unimodular groups (Principal non-degenerate series) (with M. I. Graev)
14. On a general method for decomposition of the regular representation of a Lie group into irreducible representations (with M. I. Graev)

Part III. Geometry of homogeneous spaces; spherical functions; automorphic functions

1. Some remarks on the theory of spherical functions on symmetric Riemannian manifolds (with F. A. Berezin)
2. Geodesic flows on manifolds of constant negative curvature (with S. V. Fomin)
3. Geometry of homogeneous spaces, representations of groups in homogeneous spaces and related questions of integral geometry (with M. I. Graev)
4. Unitary representations in homogeneous spaces with discrete stationary groups (with I. I. Piatetski-Shapiro)
5. Unitary representations in a space G/Γ, where G is a group of n-by-n real matrices and Γ is a subgroup of integer matrices (with I. I. Piatetski-Shapiro)

Part IV. Models of representations; representations of groups over various fields

1. Categories of group representations and the problem of classifying irreducible representations (with M. I. Graev)
2. Construction of irreducible representations of simple algebraic groups over a finite field (with M. I. Graev)
3. Representations of quaternion groups over locally compact and functional fields (with M. I. Graev)
4. A new model for representations of finite semisimple algebraic groups (with I. N. Bernstein and S. I. Gelfand)
5. Irreducible unitary representations of the group of unimodular second-order matrices with elements from a locally compact field (with M. I. Graev)
6. Plancherel's formula for the groups of the unimodular second-order matrices with elements in a locally compact field (with M. I. Graev)
7. On the representation of the group *GL (n, K)*, where *K* is a local field (with D. A. Kazhdan)
8. Models of representations of Lie groups (with I. N. Bernstein and S. I. Gelfand)
9. Complex manifolds whose skeletons are semisimple real Lie groups, and analytic discrete series of representations (with S. G. Gindikin)
10. Models of representations of classical groups and their hidden symmetries (with A. V. Zelevinskij)

Part V. Verma modules; resolutions of finite-dimensional representations

1. Differential operators on the base affine space and a study of $\mathfrak{g}$-modules (with I. N. Bernstein and S. I. Gelfand)

2. Structure of representations generated by vectors of heighest weight (with I. N. Bernstein and S. I. Gelfand)
3. Differential operators on a cubic cone (with I. N. Bernstein and S. I. Gelfand)
4. Schubert cells and cohomology of the spaces *G/P* (with I. N. Bernstein and S. I. Gelfand)
5. Category of g-modules (with I. N. Bernstein and S. I. Gelfand)
6. Structure locale de la catégorie des modules de Harish-Chandra I (avec I. N. Bernstein et S. I. Gelfand)
7. Structure locale de la catégorie des modules de Harish-Chandra II (avec I. N. Bernstein et S. I. Gelfand)
8. Algebraic bundles over P^n and problems of linear algebra (with I. N. Bernstein and S. I. Gelfand)

Part VI. Enveloping algebras and their quotient skew-fields

1. Fields associated with enveloping algebras of Lie algebras (with A. A. Kirillov)
2. Sur les corps liés aux algèbres enveloppantes des algèbres de Lie (avec A. A. Kirillov)
3. On the structure of the field of quotients of the enveloping algebra of a semisimple Lie algebra (with A. A. Kirillov)
4. The structure of the Lie field connected with a split semisimple Lie algebra (with A. A. Kirillov)

Part VII. Finite-dimensional representations

1. Finite-dimensional representations of the group of unimodular matrices (with M. L. Tsetlin)
2. Finite-dimensional representations of the group of orthogonal matrices (with M. L. Tsetlin)
3. Finite-dimensional irreducible representations of the unitary and the full linear groups, and related special functions (with M. I. Graev)

Part VIII. Indecomposable representations of semisimple Lie groups and of finite-dimensional algebras; problems of linear algebra

1. Indecomposable representations of the Lorentz group (with V. A. Ponomarev)
2. Remarks on the classification of a pair of commuting linear transformations in a finite-dimensional space (with V. A. Ponomarev)

3. The classification of the linear representations of the group $SL(2, \mathbf{C})$ (with M. I. Graev and V. A. Ponomarev)
4. Quadruples of subspaces of a finite-dimensional vector space (with V. A. Ponomarev)
5. Coxeter functors and Gabriel's theorem (with I. N. Bernstein and V. A. Ponomarev)
6. Model algebras and representations of graphs (with V. A. Ponomarev)

Part IX. Representations of infinite-dimensional groups

1. Representations of the group $SL(2, \mathbf{R})$, where $\mathbf{R}$ is a ring of functions (with M. I. Graev and A. M. Vershik)
2. Representations of the group of smooth mappings of a manifold X into a compact Lie group (with M. I. Graev and A. M. Vershik)
3. Irreducible representations of the group G^X and cohomologies (with M. I. Graev and A. M. Vershik)
4. Representations of the group of diffeomorphisms (with M. I. Graev and A. M. Vershik)
5. Representations of the group of functions taking values in a compact Lie group (with M. I. Graev and A. M. Vershik)
6. A commutative model of representation of the group of flows $SL(2, \mathbf{R})^X$ that is connected with a unipotent subgroup (with M. I. Graev and A. M. Vershik)

Appendix

1. Two papers on representation theory (by G. Segal)
2. Four papers on problems in linear algebra (by C. M. Ringel)

Table of contents for volumes I and III

Bibliography

Acknowledgements

Bibliography

The bibliography in these *Collected Papers* is a revised and updated version of the "List of I.M. Gelfand's Publications" published in the birthday addresses which can be found at the end of volume I. Hence the numbering of this Bibliography does not match that of these original 'Lists'.

The articles and monographs are listed in chronological order. The numbers in the right-hand column indicate where an article can be found in these Collected Papers, for example I.II.1 means volume I, part II, article 1.

From 1933–1947 the *Doklady* were published both in Russian and in a foreign language edition entitled *Comptes Rendus (Doklady) de l'Académie de l'URSS*.

Articles marked with an asterisk (*) were originally published in Russian and translated especially for this publication.

References to the reviews published in Mathematical Reviews (MR) and Zentralblatt für Mathematik (Zbl.) have been given as far as could be ascertained.

1936

1. Sur un lemme de la théorie des espaces linéaires. Izv. Nauchno-Issled. Inst. Mat. Khar'kov Univ., Ser. 4, **13** (1936) 35–40. Zbl. **14**:162 I.II.1

1937

2. Zur Theorie abstrakter Funktionen. Dokl. Akad. Nauk SSSR **17** (1937) 243–245. Zbl. **18**:71
3. Operatoren und abstrakte Funktionen. Dokl. Akad. Nauk SSSR **17** (1937) 245–248. Zbl. **18**:72

1938

4. Abstrakte Funktionen und lineare Operatoren. Mat. Sb., Nov. Ser. **4** (46) (1938) 235–284. Zbl. **20**:367 I.II.2

1939

5. (with A.N. Kolmogorov) On rings of continuous functions. Dokl. Akad. Nauk SSSR **22** (1939) 11–15. Zbl. **21**:411
6. On normed rings. Dokl. Akad. Nauk SSSR **23** (1939) 430–432. Zbl. **21**:294 I.II.4

7. To the theory of normed rings. II. On absolutely convergent trigonometrical series and integrals. Dokl. Akad. Nauk SSSR **25** (1939) 570–572. Zbl. **22**:357 I.II.5

8. To the theory of normed rings. III. On the ring of almost periodic functions. Dokl. Akad. Nauk SSSR **25** (1939) 573–574. Zbl. **22**:357 I.II.6

9. On one-parametrical groups of operators in a normed space. Dokl. Akad. Nauk SSSR **25** (1939) 713–718. Zbl. **22**:358 I.II.3

1940

10. (with D.A. Rajkov) On the theory of characters of commutative topological groups. Dokl. Akad. Nauk SSSR **28** (1940) 195–198. Zbl. **24**:120 I.II.7

1941

11. Normierte Ringe. Mat. Sb., Nov. Ser. **9** (51) (1941) 3–23. Zbl. **24**:320 I.II.8

12. (with G.E. Shilov) Über verschiedene Methoden der Einführung der Topologie in die Menge der maximalen Ideale eines normierten Ringes. Mat. Sb., Nov. Ser. **9** (51) (1941) 25–38. Zbl. **24**:321 I.II.9

13. Ideale und primäre Ideale in normierten Ringen. Mat. Sb., Nov. Ser. **9** (51) (1941) 41–48. Zbl. **24**:322 I.II.10

14. Zur Theorie der Charaktere der Abelschen topologischen Gruppen. Mat. Sb., Nov. Ser. **9** (51) (1941) 49–50. Zbl. **24**:323 I.II.11

15. Über absolut konvergente trigonometrische Reihen und Integrale. Mat. Sb., Nov. Ser. **9** (51) (1941) 51–66. Zbl. **24**:323 I.II.12

1942

16. (with M.A. Najmark) On the embedding of normed rings into the ring of operators in Hilbert space. Mat. Sb., Nov. Ser. **12** (54) (1942) 197–213. Zbl. **60**:270 I.II.13

17. (with D.A. Rajkov) Irreducible unitary representations of locally bicompact groups. Mat. Sb. 13 (55), 301–316 (1942) [Transl., II. Ser., Am. Math. Soc. **36** (1964) 1–15]. Zbl. **166**:401 II.I.1

1943

18. (with D.A. Rajkov) Irreducible unitary representations of locally compact groups. Dokl. Akad. Nauk SSSR **42** (1943) 199–201. Zbl. **61**:253

1946

19. (with M.A. Najmark) Unitary representations of the Lorentz group. J. Phys., Acad. Sci. USSR **10** (1946) 93–94. Zbl. **61**:253 II.II.1

20. (with M.A. Najmark) On unitary representations of the complex unimodular group. Dokl. Akad. Nauk SSSR **54** (1946) 195–198. Zbl. **29**:5 II.II.3

21. (with D.A. Rajkov and G.E. Shilov) Commutative normed rings. Usp. Mat. Nauk **1**, 2 (1946) 48–146 [Transl., II. Ser., Am. Math. Soc. 5 (1957) 115–220]. Zbl. **201**:457 I.II.14

22. (with M.A. Najmark) Unitary representations of semisimple Lie groups I. Mat. Sb., Nov. Ser. **21** (63) (1946) 405–434 [Transl., II. Ser., Am. Math. Soc. (reprinted) **9** (1962) 1–14]. Zbl. **38**:17

*23. (with M.A. Najmark) Complementary and degenerate series of representations of the complex unimodular group. Dokl. Akad. Nauk SSSR **58** (1946) 1577–1580. Zbl. **37**:304 II.II.5

1947

24. (with M.A. Najmark) Unitary representations of the group of linear transformations of the straight line. Dokl. Akad. Nauk SSSR **55** (1947) 567–570. Zbl. **29**:5 II.I.2

25. (with M.A. Najmark) The principal series of irreducible representations of the complex unimodular group. Dokl. Akad. Nauk SSSR **56** (1947) 3–4. Zbl. **29**:5 II.II.4

*26. (with M.A. Najmark) Unitary representations of the Lorentz group. Izv. Akad. Nauk SSSR, Ser. Mat. **11** (1947) 411–504. Zbl. **37**:153 II.II.2

1948

27. Lectures on linear algebra. Moscow-Leningrad: Gostekhizdat 1948 (in Russian). Zbl. **38**:156

28. (with M.A. Najmark) Normed rings with an involution and their representations. Izv. Akad. Nauk SSSR, Ser. Mat. **12**, 445–480 (1948) (English translation in: Commutative normed rings, I.M. Gelfand, D.A. Rajkov and G.E. Shilov, pp. 240–274. Chelsea 1964). Zbl. **31**:34 I.II.15

29. (with A.M. Yaglom) General relativistically invariant equations and infinite-dimensional representations of the Lorentz group. Dokl. Akad. Nauk SSSR **59** (1948) 655–659 (in Russian). Zbl. **37**:127

30. Integral equations. Article in the Great Soviet Encyclopedia (1948) (in Russian)

*31. (with A.M. Yaglom) General relativistically invariant equations and infinite-dimensional representations of the Lorentz group. Zh. Ehksp. Teor. Fiz. **18** (1948) 703–733 II.II.9

*32. (with M.A. Najmark) The trace in principal and complementary series representations of the complex unimodular group. Dokl. Akad. Nauk SSSR **61** (1948) 9–11. Zbl. **35**:299 II.II.6

33. (with A.M. Yaglom) Relativistically invariant equations corresponding to a definite charge and a definite energy. Dokl. Akad. Nauk SSSR **63** (1948) 371–374 (in Russian). Zbl. **31**:95

34. (with A.M. Yaglom) Pauli's theorem for general relativistically invariant equations. Zh. Ehksp. Teor. Fiz. **18** (1948) 1096–1104 (in Russian)

35. (with A.M. Yaglom) Charge conjugacy for general relativistically invariant equations. Zh. Ehksp. Teor. Fiz. **18** (1948) 1105–1111 (in Russian)

*36. (with M.A. Najmark) On the connection between representations of complex semisimple Lie groups and its maximal compact subgroup. Dokl. Akad. Nauk SSSR **63** (1948) 225–228. Zbl. **35**:15 II.II.7

*37. (with M.A. Najmark) An analogue of the Plancherel formula for the complex unimodular group. Dokl. Akad. Nauk SSSR **63** (1948) 609–612. Zbl. **38**:18 II.II.8

1950

*38. Center of the infinitesimal group ring. Mat. Sb., Nov. Ser. **26** (68) (1950) 103–112. Zbl. **35**:300 II.I.3

39. (with M.A. Najmark) The connexion between the unitary representations of the complex unimodular group and those of its unitary subgroup. Izv. Akad. Nauk SSSR, Ser. Mat. **14** (1950) 239–260 (in Russian). Zbl. **37**:15

40. Spherical functions on symmetric Riemannian spaces. Dokl. Akad. Nauk SSSR 70, (1950) 5–8 [Transl., II. Ser., Am. Math. Soc. **37** (1964) 39–43]. Zbl. **38**:274 II.I.4

*41. (with M.L. Tsetlin) Finite-dimensional representations of the group of unimodular matrices. Dokl. Akad. Nauk SSSR **71** (1950) 825–828. Zbl. **37**:153 II.VII.1

*42. (with M.L. Tsetlin) Finite-dimensional representations of the group of orthogonal matrices. Dokl. Akad. Nauk SSSR **71** (1950) 1017–1020. Zbl. **37**:153 II.VII.2

*43. Eigenfunction expansions for equations with periodic coefficients. Dokl. Akad. Nauk SSSR **73** (1950) 1117–1120. Zbl. **37**:345 I.III.1

44. (with M.A. Najmark) Unitary representations of classical groups. Tr. Mat. Inst. Steklova **36** (1950) 1–288 (in Russian). Zbl. **41**:362

(English transl. of the Introduction, Chap. 9 'Spherical functions' and Chap 18 'Transitivity classes for the set of pairs. Another way of describing representations of the complementary series') II.II.11

1951

45. (with S.V. Fomin) Unitary representations of Lie groups and geodesic flows on surfaces of constant negative curvature. Dokl. Akad. Nauk SSSR **76** (1951) 771–774 (in Russian). Zbl. **45**:388

46. Lectures on linear algebra, 2nd ed. Moscow-Leningrad: Gostekhizdat 1951 (English transl.: New York: Interscience 1961). Zbl. **98**:11

47. (with B.M. Levitan) On the determination of a differential equation from its spectral function. Izv. Akad. Nauk SSSR, Ser. Mat. **15** (1951) 309–361 [Transl., II. Ser., Am. Math. Soc. **1** (1955) 253–304]. Zbl. **44**:93 I.III.2

48. Remark on N.K. Bari's paper 'Biorthogonal system and bases in a Hilbert space'. Uch. Zap. Mosk. Gos. Univ., Ser. Mat. **140** (4) (1951) 224–225 (in Russian)

1952

49. (with Z.Ya. Shapiro) Representations of the group of rotations of 3-dimensional space and their applications. Usp. Mat. Nauk **7** (1) (1952) 3–117 (in Russian). Zbl. **49**:157

50. (with S.V. Fomin) Geodesic flows on manifolds of constant negative curvature. Usp. Mat. Nauk **7** (1) (1952) 118–137 [Transl., II. Ser., Am. Math. Soc. **1** (1955) 49–65]. Zbl. **66**:361 II.III.2

51. (with M.A. Najmark) Unitary representations of the unimodular group that contain an identity representation of the unitary subgroup. Tr. Mosk. Mat. O.-va **1** (1952) 423–473 (in Russian). Zbl. **49**:358

*52. (with M.I. Graev) Unitary representations of the real simple Lie groups. Dokl. Akad. Nauk SSSR **86** (1952) 461–463. Zbl. **49**:358 II.II.12

53. On the spectrum of non-selfadjoint differential operators. Usp. Mat. Nauk **7** (6) (1952) 183–184 (in Russian). Zbl. **48**:96

1953

*54. (with B.M. Levitan) On a simple identity for eigenvalues of second order differential operator. Dokl. Akad. Nauk SSSR **88** (1953) 593–596. Zbl. **53**:60 I.III.3

*55. (with M.I. Graev) Unitary representations of real unimodular groups (Principal non-degenerate series). Izv. Akad. Nauk SSSR, Ser. Mat. **17** (1952) 189–249. Zbl. **52**:341 II.II.13

*56. (with M.I. Graev) On a general method for decomposition of the regular representation of a Lie group into irreducible representations. Dokl. Akad. Nauk SSSR **92** (1952) 221–224. Zbl. **53**:15 II.II.14

57. (with M.I. Graev) The analogue of Plancherel's theorem for real unimodular groups. Dokl. Akad. Nauk SSSR **92** (1952) 461–464 (in Russian). Zbl. **53**:15

58. (with G.E. Shilov) Fourier transforms of rapidly increasing functions and questions of the uniqueness of the solution of Cauchy's problem. Usp. Mat. Nauk **8** (6) (1952) 3–54 [Transl., II. Ser., Am. Math. Soc. **5** (1957) 221–274]. Zbl. **52**:116

1954

*59. (with R.A. Minlos) Solution of quantum field equations. Dokl. Akad. Nauk SSSR **97** (1954) 209–212. Zbl. **58**:232 I.III.4

1955

60. (with V.B. Lidskij) On the structure of the regions of stability of linear canonical systems of differential equations with periodic coefficients. Usp. Mat. Nauk **10** (1) (1955) 3–40 [Transl., II. Ser., Am. Math. Soc. **8** (1958) 143–181]. Zbl. **64**:89 I.III.5

*61. Generalized random processes. Dokl. Akad. Nauk SSSR**100** (1955) 853–856. Zbl. **68**:112 III.III.1

62. (with M.I. Graev) The traces of unitary representations of the real unimodular group. Dokl. Akad. Nauk SSSR **100** (1955) 1037–1040 (in Russian). Zbl. **64**:111

63. (with M.I. Graev) The analogue of Plancherel's formula for the classical groups. Tr. Mosk. Mat. O.-va **4** (1955) 375–404 [Transl., II. Ser., Am. Math. Soc. **9** (1958) 123–154. Zbl. **66**:20

64. (with Z.Ya. Shapiro) Homogeneous functions and their applications. Usp. Mat. Nauk **10** (3) (1955) 3–70 [Transl., II. Ser., Am. Math. Soc. **8** (1958) 21–85]. Zbl. **65**:101

65. (with G.E. Shilov) On a new method in theorems concerning the uniqueness of the solution of Cauchy's problem for systems of linear partial differential equations. Dokl. Akad. Nauk SSSR **102** (1955) 1065–1068 (in Russian). Zbl. **67**:72

*66. (with A.G. Kostyuchenko) Eigenfunction expansions for differential and other operators. Dokl. Akad. Nauk SSSR **103** (1955) 349–352. Zbl. **65**:104 I.III.6

1956

67. (with A.M. Yaglom) Integration in functional spaces and its applications in quantum physics. Usp. Mat. Nauk **11** (1) (1956) 77–114 [J. Math. Phys. **1** (1960) 48–69]. Zbl. **92**:451 III.III.2

*68. On identities for eigenvalues of a second order differential operator. Usp. Mat. Nauk **11** (1) (1956) 191–198. Zbl. **70**:83 I.III.7

69. (with F.A. Berezin) Some remarks on the theory of spherical functions on symmetric Riemannian manifolds. Tr. Mosk. Mat. O.-va **5** (1956) 311–351 [Transl., II. Ser., Am. Math. Soc. **21** (1962) 193–238]. Zbl. **72**:17 II.III.1

70. (with G.E. Shilov) Quelques applications de la théorie des fonctions généralisées. J. Math. Pures et Appl., IX. Ser. **35** (1956) 383–413. Zbl. **75**:285

*71. (with A.N. Kolmogorov and A.M. Yaglom) To the general definition of the amount of information. Dokl. Akad. Nauk SSSR **111** (1956) 745–748 (German transl. in: Arbeiten zur Informationstheorie II, Mathematische Forschungsberichte.

Berlin: VEB Deutscher Verlag der Wissenschaften 1958). Zbl. **71**:345 III.III.3

72. On some problems of functional analysis. Usp. Mat. Nauk **11** (6) (1956) 3–12 [Transl., II. Ser., Am. Math. Soc. **16** (1960) 315–324]. Zbl. **100**:321 I.I.2
73. (with F.A. Berezin, M.I. Graev, and M.A. Najmark) Group representations. Usp. Mat. Nauk **11** (6) (1956) 13–40 [Transl., II. Ser., Am.Math. Soc. **16** (1960) 325–353]. Zbl. **74**:103
74. (with N.N. Chentsov) On the numerical evaluation of continuous integrals. Zh. Ehksp. Teor. Fiz. **31** (1956) 1106–1107 (in Russian)
75. (with M.L. Tsetlin) On quantities with anomalous parity. Zh. Ehksp. Teor. Fiz. **31** (1956) 1107–1109 (in Russian)

1957

76. (with A.M. Yaglom) Calculation of the amount of information about a random function contained in another such function. Usp. Mat. Nauk **12** (1) (1957) 3–52. [Transl., II. Ser., Am. Math. Soc. **12** (1959) 199–246] III.III.4
77. On the subrings of the ring of continuous functions. Usp. Mat. Nauk **12** (1) (1957) 247–251 [Transl., II. Ser., Am. Math. Soc. **16** (1957) 477–479]. Zbl. **100**:322
78. Some aspects of functional analysis and algebra. Proc. Int. Congr. Math. 1954, Amsterdam **1** (1957) 253–276. Zbl. **79**:326 I.I.1

1958

79. (with G.E. Shilov) Generalized functions 1. Properties and operations. Moscow: Fizmatgiz 1958 (English transl.: New York: Academic Press 1964; German transl.: Berlin: VEB Deutscher Verlag der Wissenschaften 1960; French transl.: Paris: Dunod 1962). Zbl. **91**:111
80. (with G.E. Shilov) Generalized functions 2. Spaces of fundamental and generalized functions. Moscow: Fizmatgiz 1958 (English transl.: New York: Academic Press 1968; German transl.: Berlin: VEB Deutscher Verlag der Wissenschaften 1960; French transl.: Paris: Dunod 1964). Zbl. **91**:111
81. (with G.E. Shilov) Generalized functions 3. Theory of differential equations. Moscow: Fizmatgiz 1958 (English transl.: New York: Academic Press 1967; German transl.: Berlin: VEB Deutscher Verlag der Wissenschaften 1964; French transl.: Paris: Dunod 1964). Zbl. **91**:111
82. (with K.I. Babenko) Some observations on hyperbolic systems. Nauchn. Dokl. Vyssh. Shk. **1** (1958) 12–18 (in Russian). Zbl. **144**:139

83. (with R.A. Minlos and Z.Ya. Shapiro) Representations of the rotation and Lorentz groups. Moscow: Fizmatgiz 1958 (English transl.: London: Pergamon Press 1963). Zbl. **108**:220

84. (with R.A. Minlos and A.M. Yaglom) Path integrals. Proceedings of the Third All-Union Mathematics Congress **3** (1958) 521–531 (in Russian)

85. (with F.A. Berezin, M.I. Graev and M.A. Najmark) Representations of Lie groups. Proceedings of the Third All-Union Mathematics Congress **3** (1958) 246–254 (in Russian). Zbl. **97**:109

86. (with A.N. Kolmogorov and A.M. Yaglom) The amount of information and entropy for continuous distributions. Proceedings of the Third All-Union Mathematics Congress **3** (1958) 300–320 (in Russian). Zbl. **92**:340

87. (with I.G. Petrovskij and G.E. Shilov) The theory of systems of partial differential equations. Proceedings of the Third All-Union Mathematics Congress **3** (1958) 65–72 (in Russian). Zbl. **107**:74

88. (with S.I. Braginskij and R.P. Fedorenko) The theory of the compression and pulsation of a plasma column under a powerful pulse discharge. Fiz. Plazmy Probl. Upr. Termoyad. Reakts. **4** (1958) 201–222 (in Russian)

89. (with N.N. Chentsov and A.S. Frolov) The computation of continuous integrals by the Monte Carlo method. Izv. Vyssh. Uchebn. Zaved., Mat. **5** (6) (1958) 32–45 (in Russian). Zbl. **139**:323

1959

90. Some problems in the theory of quasilinear equations. Usp. Mat. Nauk **14** (2) (1959) 87–158 [Transl., II. Ser., Am. Math. Soc. **29** (1963) 295–381]. Zbl. **96**:66 I.III.8

91. (with I.I. Piatetski-Shapiro) Theory of representations and theory of automorphic functions. Usp. Mat. Nauk **14** (2) (1959) 171–194 [Transl., II. Ser., Am. Math. Soc. **26** (1963) 173–200]. Zbl. **121**:306

92. Some questions of analysis and differential equations. Usp. Mat. Nauk **14** (3) (1959) 3–19 [Transl., II. Ser., Am. Math. Soc. **26** (1963) 201–219]. Zbl. **91**:88 I.I.3

93. (with M.I. Graev) Geometry of homogeneous spaces, representations of groups in homogeneous spaces and related questions of integral geometry. Tr. Mosk. Mat. O.-va **88** (1959) 321–390 [Transl., II. Ser., Am. Math. Soc. **37** (1964) 351–429]. Zbl. **136**:434 II.III.3

94. (with M.I. Graev) The decomposition into irreducible components of representations of the Lorentz group in the spaces

of functions defined on symmetric spaces. Dokl. Akad. Nauk SSSR **127** (1959) 250–253 (in Russian). Zbl. **99**:321

*95. (with I.I. Piatetski-Shapiro) On a theorem of Poincaré. Dokl. Akad. Nauk SSSR **127** (3) (1959) 490–493. Zbl. **107**:171 I.III.9

*96. (with M.I. Graev) On the structure of the ring of rapidly decreasing functions on a Lie group. Dokl. Akad. Nauk SSSR **124** (1959) 19–21. Zbl. **103**:336 II.II.10

1960

97. (with N.N. Chentsov, S.M. Fejnberg, and A.S. Frolov) On the application of the method of random tests (the Monte Carlo method) for the solution of the kinetic equation. Proceedings of the Second International Conference on the Peaceful Use of Atomic Energy. (Geneva, 1958), **2** (1960) 628–683 III.IV.1

98. (with Sya Do-Shin) On positive definite distributions. Usp. Mat. Nauk **15** (1) (1960) 185–190 (in Russian). Zbl. **97**:314

99. Integral geometry and its relation to the theory of group representations. Usp. Mat. Nauk **15** (2) (1960) 155–164 [Russ. Math. Surv. **15** (2) (1960) 143–151]. Zbl. **119**:177 I.I.4

100. (with M.I. Graev) Fourier transforms of rapidly decreasing functions on complex semisimple groups. Dokl. Akad. Nauk SSSR **131** (1960) 496–499 (in Russian). Zbl. **103**:337

101. (with M.L. Tsetlin) On continuous models of control systems. Dokl. Akad. Nauk SSSR **131** (1960) 1242–1245 (in Russian)

102. On elliptic equations. Usp. Mat. Nauk **15** (3) (1960) 121–132 [Russ. Math. Surv. **15** (1960) 113–123]. Zbl. **95**:78 I.I.5

103. On a paper by K. Hoffmann and I.M. Singer. Usp. Mat. Nauk **15** (3) (1960) 239–240 (in Russian). Zbl. **154**:386

104. (with D.A. Rajkov and G.E. Shilov) Commutative normed rings. Moscow: Fizmatgiz 1960 (English transl.: New York: Chelsea 1964; German transl.: Berlin: VEB Deutscher Verlag der Wissenschaften 1964; French transl.: Paris: Gauthier-Villars 1964). Zbl. **134**:321

105. (with M.I. Graev) Integrals over hyperplanes of fundamental and generalized functions. Dokl. Akad. Nauk SSSR **135** (1960) 1307–1310 [Sov. Math., Dokl. **1** (1960) 1369–1372]. Zbl. **108**:296

1961

106. (with N.Ya. Vilenkin) Generalized functions 4. Applications of harmonic analysis. Moscow: Fizmatgiz 1961 (English transl.: New York: Academic Press 1964; German transl.: Berlin: VEB Deutscher Verlag der Wissenschaften 1960; French transl.: Paris: Dunod 1967). Zbl. **136**:112

107. (with S.V. Fomin) Calculus of Variations. Moscow: Fizmatgiz (1961) (English transl.: Englewood Cliffs, N.J.: Prentice Hall· 1963). Zbl. **127**:54

108. (with V.A. Borovikov, A.F. Grashin, and I.Ya. Pomeranchuk) Phase-shift analysis of *pp*-scattering at 95 *Mev*. Zh. Ehksp. Teor. Fiz. **40** (1961) 1106–111 [Soviet Physics **13** (4) (1961) 780–784]

109. (with A.F. Grashin and L.N. Ivanova) Phase-shift analysis of *pp*-scattering at an energy of 150 Mev. Zh. Ehksp. Teor. Fiz. **40** (5) (1961) 1338–1342 (in Russian)

110. (with V.S. Gurfinkel and M.L. Tsetlin) Some considerations on the tactics of making movements. Dokl. Akad. Nauk SSSR **139** (1961) 1250–1253 (in Russian)

111. (with M.I. Graev et al.) Magnetic surfaces of the 3-path helical magnetic field excited by a crimped field. Zh. Tekh. Fiz. **31** (1961) 1164–1169 (in Russian)

112. (with M.I. Graev) Integral transformations connected with straight line complexes in a complex affine space. Dokl. Akad. Nauk SSSR **138** (1961) 1266–1269 [Sov. Math., Dokl. **2** (1961) 809–812]. Zbl. **109**:151 III.I.1

113. (with M.L. Tsetlin) The principle of non-local search in automatic optimization systems. Dokl. Akad. Nauk SSSR **137** (1961) 295–298 [Sov. Phys., Dokl. **6** (1961) 192–194]

1962

114. (with M.L. Tsetlin) Some methods of control for complex systems. Usp. Mat. Nauk **17** (1) (1962) 3–25 [Russ. Math. Surv. **17** (1) (1962) 95–117]. Zbl. **107**:299 III.IV.3

115. (with A.I. Morozov, N.M. Zueva et al.) An example of the theoretical determination of a magnetic field that does not have magnetic surfaces. Dokl. Akad. Nauk SSSR **143** (1962) 81–83 (in Russian)

116. (with M.I. Graev) An application of the horysphere method to the spectral analysis of functions in real and imaginary Lobachevskii space. Tr. Mosk. Mat. O.-va **11** (1962) 243–308 (in Russian). Zbl. **176**:443

117. (with M.I. Graev and N.Ya. Vilenkin) Generalized functions 5. Integral geometry and representation theory. Moscow: Fizmatgiz 1962 (English transl.: New York: Academic Press 1966; French transl.: Paris: Dunod 1970). Zbl. **115**:167

118. (with M.I. Graev) Categories of group representations and the problem of classifying irreducible representations. Dokl. Akad. Nauk SSSR **146** (1962) 757–760 [Sov. Math., Dokl. **3** (1962) 1378–1381] II.IV.1

119. (with I.I. Piatetski-Shapiro) Unitary representations in homogeneous spaces with discrete stationary groups. Dokl. Akad. Nauk SSSR **147** (1962) 17–20 [Sov. Math., Dokl. **3** (1962) 1528–1531]. Zbl. **119**:270. II.III.4

120. (with I.I. Patetski-Shapiro) Unitary representations in a space G/Γ, where G is a group of n-by-n real matrices and Γ is a subgroup of integer matrices. Dokl. Akad. Nauk SSSR **147** (1962) 275–278 [Sov. Math., Dokl. **3** (1962) 1574–1577]. Zbl. **119**:271 II.III.5

121. (with M.I. Graev) Construction of irreducible representations of simple algebraic groups over a finite field. Dokl. Akad. Nauk SSSR **147** (1962) 529–532 [Sov. Math., Dokl. **3** (1962) 1646–1649]. Zbl. **119**:269 II.IV.2

122. (with V.S. Gurfinkel and M.L. Tsetlin) On the techniques of control of complex systems and their relation to physiology. Symposium 'Biological aspects of cybernetics', pp. 66–73. Moscow: Publ. Akad. Nauk SSSR (1962) [Transl., II. Ser., Am. Math. Soc. **111** (1978) 213–219]

123. (with O.V. Lokytsievskij) On difference schemes for the solution of the equation of thermal conductivity. The 'double sweep' method for the solution of difference equations. Appendices I and II to 'Theory of difference schemes. An introduction.' by S.K. Godunov and V.S. Ryaben'kij. Moscow: Fizmatgiz 1962. Zbl. **106**:319. (Amsterdam: North-Holland 1964, pp. 232–262) III.IV.2

1963

124. (with L.M. Chailakhyan and S.A. Kovalev) Intracellular irritation of the various compartments of a frog's heart. Dokl. Akad. Nauk SSSR **148** (1963) 973–976 (in Russian)

125. (with M.I. Graev, A.I. Morozov et al.) On the structure of a toroidal magnetic field that does not have magnetic surfaces. Dokl. Akad. Nauk SSSR **148** (1963) 1286–1289 (in Russian)

126. (with M.I. Graev) Irreducible unitary representations of the group of unimodular second-order matrices with elements from a locally compact field. Dokl. Akad. Nauk SSSR **149** (1963) 499–502 [Sov. Math., Dokl. **4** (1963) 397–400]. Zbl. **119**:270 II.IV.5

127. (with I.I. Piatetski-Shapiro) Automorphic functions and the theory of representations. Tr. Mosk. Mat. O.-va **12** (1963) 389–412 [Trans. Mosc. Math. Soc. **12** (1965)]. Zbl. **136**:73

128. Automorphic functions and the theory of representations. Proc. Int. Congr. Math. Stockholm (1962) 74–85. Zbl. **138**:71 I.I.6

129. (with M.I. Graev) Plancherel's formula for the groups of the unimodular second-order matrices with elements in a locally compact field. Dokl. Akad. Nauk SSSR **151** (1963) 262–264 [Sov. Math., Dokl. **4** (1963) 397–400]. Zbl. **204**:141 II.IV.6

130. (with M.I. Graev) Representations of a group of second order matrices with elements from a locally compact field and special functions on locally compact fields. Usp. Mat. Nauk **18** (4) (1963) 29–99 [Russ. Math. Surv. **18** (4) (1963) 29–100]. Zbl. **166**:402

131. (with A.Ya. Fridenstein) On the possible mechanism of change of immunological tolerance. Usp. Sov. Biol. **55** (1963) 428–429 (in Russian)

132. (with V.S. Gurfinkel, Ya.M. Kots, M.L. Shik, and M.L. Tsetlin) On the synchronization of motor units and some related ideas. Biofizika **8** (1963) 475–488 (in Russian)

133. (with Yu.G. Fedorov and I.I. Piatetski-Shapiro) Determination of crystal structure by the method of nonlocal search. Dokl. Akad. Nauk SSSR **152** (1963) 1045–1048 [Sov. Math. Dokl. **4** (1963) 1487–1490] III.IV.4

*134. (with Yu.G. Fedorov, R.A. Kayushina, and B.K. Vainstein) Determination of crystal structures by the method of the *R*-factor minimization. Dokl. Akad. Nauk SSSR **153** (1963) 93–96 III.IV.5

135. (with I.I. Piatetski-Shapiro and M.L. Tsetlin) On certain classes of games and automata games. Dokl. Akad. Nauk SSSR **152** (1963) 845–848. [Transl., II. Ser., Am. Math. Soc. **87** (1970) 275–280]. MR **28**:1068. Zbl. **137**:143 III.IV.6

136. (with M.I. Graev) The structure of the ring of finite functions on the group of second-order unimodular matrices with elements from a disconnected locally compact field. Dokl. Akad. Nauk SSSR **153** (1963) 512–515 [Sov. Math. Dokl. **4** (1963) 1679–1700]. MR **33**:4183. Zbl. **199**:200

137. (with G.E. Shilov) Categories of finite-dimensional spaces. Vestn. Mosk. Univ., Ser. I. **4** (1963) 27–48 (in Russian). MR **28**:1223. Zbl. **161**:27

1964

138. (with V.I. Guelstein, A.G. Malenkov, and Yu.M. Vasil'ev) Characteristics of cell complexes of ascitic mouse hepatoma 22. Dokl. Akad. Nauk SSSR **156** (1964) 168–170 (in Russian)

139. (with M.I. Graev and I.I. Piatetski-Shapiro) Representations of adèle groups. Dokl. Akad. Nauk SSSR **156** (1964) 487–490. [Sov. Math., Dokl. **5** (1964) 657–661]. MR **29**:2237. Zbl. **133**:294

140. (with V.S. Gurfinkel) Investigation of recognition activity. Biofizika **9** (1964) 710–717 (in Russian)

141. (with V.I. Bryzgalov, V.S. Gurfinkel, and M.L. Tsetlin) Homogeneous automata games and their simulation on digital

computers. Avtom. Telemekh. **25** (1964) 1572–1580 (in Russian). MR **30**:1897. Zbl. **141**:339

142. (with E.G. Glagoleva and A.A. Kirillov) The coordinate method. Moscow: Nauka 1964 (English transl.: New York: Gordon & Breach 1967; German transl.: Leipzig: Teubner 1968; Czech. Transl.: Bratislava: ALFA 1976)

1965

143. (with M.I. Graev) Finite-dimensional irreducible representations of the unitary and the full linear groups, and related special functions. Izv. Akad. Nauk SSSR, Ser. Mat. **29** (1965) 1329–1356 [Transl., II. Ser., Am. Math. Soc. **64** (1965) 116–146]. MR **34**:1450. Zbl. **139**:307 II.VII.3
144. (with V.I. Guelstein, A.G. Malenkov, and Yu.M. Vasil'ev) Cell complexes in ascitic hepatomata of mice and rats, in the collection "Cell differentiation and induction mechanisms". Moscow: Nauka (1965) 220–232 (in Russian)
145. (with E.G. Glagoleva and E.E. Schnol) Functions and their graphs. Moscow: Nauka 1965 (English transl.: New York: Gordon & Breach 1967; German transl.: Leipzig: Teubner 1971). Zbl. **129**:267

1966

146. (with M.I. Graev and I.I. Piatetski-Shapiro) Theory of representations and automorphic functions. Moskau: Nauka 1966. (English transl.: Philadelphia London Toronto: Saunders 1969). MR **36**:3725. Zbl. **138**:72
147. (with A.A. Kirillov) Fields associated with enveloping algebras of Lie algebras. Dokl. Akad. Nauk SSSR **167** (1966) 503–505 [Sov. Math., Dokl. **7** (1966) 407–409]. Zbl. **149**:29 II.VI.1
148. (avec A.A. Kirillov) Sur les corps liés aux algèbres enveloppantes des algèbres de Lie., Publ. Math., Inst. Hautes Etud. Sci. **31** (1966) 509–523. MR **33**:7731. Zbl. **144**:21 II.VI.2
149. (with V.I. Guelstein, A.G. Malenkov, and Yu.M. Vasil'ev) Local interactions of cells in cell complexes of ascitic hepatoma 22. Dokl. Akad. Nauk SSSR **167** (1966) 437–439 (in Russian)
150. (with V.I. Guelstein, A.G. Malenkov, and Yu.M. Vasil'ev) Interrelationships of contacting cells in the cell complexes of mouse ascites hepatoma. Int. J. Cancer **1** (1966) 451–462 III.IV.12
151. (with M.I. Graev and E.Ya. Shapiro) Integral geometry on a manifold of k-dimensional planes. Dokl. Akad. Nauk SSSR **168** (1966) 1236–1238 [Sov. Math., Dokl. **7** (1966) 801–804]. Zbl. **168**:201
152. (with L.V. Erofeeva, and Yu.M. Vasil'ev) The behaviour of fibroblasts of a cell culture on removal of part of the mono-

layer. Dokl. Akad. Nauk SSSR **171** (1966) 721–724 (in Russian)

153. (with M.L. Tsetlin) Mathematical simulation of mechanisms of the central nervous system. In the collection: Models of structure-functional organisation of certain biological systems, pp. 9–26. Moscow: Nauka 1966 (in Russian)

154. (with V.S. Gurfinkel, M.L. Tsetlin, and M.L. Shik) Some problems in the analysis of movements. In the collection: Models of structure-functional organisation of certain biological systems, pp. 264–276, Moscow: Nauka 1966 (In: Models of the Structural-Functional Organization of Certain Biological Systems, pp. 329–345, Cambridge, Massachusetts, London: MIT Press 1971) III.IV.7

155. (with Yu.G. Fedorov, S.L. Ginzburg, and E.B. Vul) The ravine method in problems of X-ray structure analysis, pp. 1–77, Moscow: Nauka 1966 (in Russian)

156. Lectures on linear algebra, pp. 1–280. Moscow: Nauka 1966 (in Russian). MR **34**:4274. Zbl. **158**:297

1967

157. (with M.I. Graev and Z.Ya. Shapiro) Integral geometry on *K*-dimensional planes. Funkts. Anal. Prilozh. **1** (1) (1967) 15–31 [Funct. Anal. Appl. **1** (1967) 14–27]. MR **35**:3620. Zbl. **164**:231 III.I.2

158. (with V.A. Ponomarev) Categories of Harish-Chandra models over the Lie algebra of the Lorentz group. Dokl. Akad. Nauk SSSR **176** (1967) 243–246 [Sov. Math., Dokl. **8** (1967) 1065–1068]. MR **36**:6552. Zbl. **241**:22025

159. (with V.A. Ponomarev) Classification of indecomposable infinitesimal representations of the Lorentz group. Dokl. Akad. Nauk SSSR **176** (1967) 502–505 [Sov. Math., Dokl. **8** (1967) 114–1117]. MR **36**:2739. Zbl. **246**:22013

160. (with D.B. Fuks) Cohomology of Lie groups with real coefficients. Dokl. Akad. Nauk SSSR **176** (1967) 24–27 [Sov. Math., Dokl. **8** (1967) 1031–1034]. MR **37**:2252a. Zbl. **169**:547

161. (with D.B. Fuks) Topology of noncompact Lie groups. Funkts. Anal. Prilozh. **1** (4) (1967) 33–45 [Funct. Anal. Appl. **1** (1967) 285–295]. MR **37**:2253. Zbl. **169**:547 III.II.2

162. (with M.I. Graev) Representations of the quaternion group over a disconnected locally compact continuous field. Dokl. Akad. Nauk SSSR **177** (1967) 17–20 [Sov. Math., Dokl. **8** (1967) 1346–1349]. MR **36**:2742. Zbl. **225**:22009

163. (with Yu.I. Arshavskij, M.B. Berkinblit, and V.S. Yakobson) Functional organization of afferent connections of Purkinje cells of the paramedian lobe of the cerebellum. Dokl. Akad. Nauk SSSR **177** (1967) 732–753 (in Russian)

164. (with D.B. Fuks) Topological invariants of non-compact Lie

groups connected with infinite-dimensional representations. Dokl. Akad. Nauk SSSR **177** (1967) 763–766 [Sov. Math., Dokl. **8** (1967) 1483–1486]. MR **37**:2252b. Zbl. **169**:548

165. (with V.S. Imshennik, L.G. Khazin, O.V. Lokytsievskij, V.S. Ryaben'kij, and N.M. Zueva) The theory of non-linear oscillation of electron plasma. Zh. Vychisl. Mat. Mat. Fiz. **7** (1967) 322–347 (in Russian). Zbl. **181**:575
166. (with M.I. Graev) Irreducible representations of the Lie algebras of the group $U(p, q)$. In the collection: High energy physics and the theory of elementary particles, pp. 216–226. Kiev: Naukova Dumka 1967 (in Russian). MR **37**:3814

1968

167. (with M.I. Graev) Complexes of k-dimensional planes in the space $\mathbf{C}^n$ and Plancherel's formula for the group $GL(n, \mathbf{C})$. Dokl. Akad. Nauk SSSR **179** (1968) 522–525 [Sov. Math., Dokl. **9** (1968) 394–398]. MR **37**:4764. Zbl. **198**:271 III.I.3
168. (with M.I. Graev) Complexes of straight lines in the space $\mathbf{C}^n$. Funkts. Anal. Prilozh. **2**(3) (1968) 39–52 [Funct. Anal. Appl. **2** (1968) 219–229]. MR **38**:6522. Zbl. **179**:509 III.I.4
169. (with V.A. Ponomarev) Indecomposable representations of the Lorentz group. Usp. Mat. Nauk **23** (2) 3–60 (1968) [Russ. Math. Surv. **23** (2) (1968) 1–58]. MR **38**:5325. Zbl. **236**:22012 II.VIII.1
170. (with Yu.M. Vasil'ev) Surface changes disturbing intracellular homeostasis as a factor inducing cell growth and division. Curr. Mod. Biol. **2** (1968) 43–55
171. (with A.A. Kirillov) On the structure of the field of quotients of the enveloping algebra of a semisimple Lie algebra. Dokl. Akad. Nauk SSSR **180** (1968) 775–777 [Sov. Math., Dokl. **9** (1968) 669–671]. MR **37**:5260. Zbl. **244**:17006 II.VI.3
172. (with D.B. Fuks) On classifying spaces for principal fiberings with Hausdorff bases. Dokl. Akad. Nauk SSSR **181** (1968) 515–518 [Sov. Math., Dokl. **9** (1968) 851–854]. MR **38**:716. Zbl. **181**:266 III.II.1
173. (with D.B. Fuks) The cohomologies of the Lie algebra of the vector fields in a circle. Funkts. Anal. Prilozh. **2** (4) (1968) 92–93 [Funct. Anal. Appl. **2** (1968) 342–343]. MR **39**:6348a. Zbl. **176**:115 III.II.3
174. (with Yu.M. Vasil'ev) Change of cellular surface – the basis of biological singularities of a tumor cell. Vestn. Akad. Med. Nauk SSSR **3** (1968) 45–49 (in Russian)
175. (with M.I. Graev) Representations of quaternion groups over locally compact and functional fields. Funkts. Anal. Prilozh. **2** (1) (1968) 20–35 [Funct. Anal. Appl. **2** (1968) 19–33]. MR **38**:4611. Zbl. **233**:20016 II.IV.3

176. (with L.V. Domnina, R.I. Rapoport, and Yu.E.M. Vasil'ev) Wound healing in cell cultures. Exp. Cell Res. **54** (1968) 83–93

177. (with A.B. Fel'dman, V.S. Gurfinkel, G.N. Orlovskij, F.V. Severin, and M.L. Shik) The control of certain types of movement. In the collection: Material of the international symposium IFAK on technical and biological problems of control (1968) (in Russian)

178. (with L.V. Erofeeva, O.Yu. Ivanova, I.L. Slavnaya, Yu.M. Vasil'ev, and A.A. Yaskovets) Factors controlling the proliferation of normal and tumour cells. In: Connective tissue in normal and pathological conditions, pp. 212–215. Novosibirsk: Nauka 1969 (in Russian)

1969

179. (with A.A. Kirillov) The structure of the Lie field connected with a split semisimple Lie algebra. Funkts. Anal. Prilozh. **3** (1) (1969) 7–26 [Funct. Anal. Appl. **3** (1969) 6–21]. MR **39**:2827. Zbl. **244**:17007 II.VI.4

180. (with M.I. Graev and Z.Ya. Shapiro) Differential forms and integral geometry. Funkts. Anal. Prilozh. **3** (2) (1969) 24–40 [Funct. Anal. Appl. **3** (1969) 101–114]. MR **39**:6232. Zbl. **191**:528 III.I.5

181. (with D.B. Fuks) Cohomology of the Lie algebra of vector fields on a manifold. Funkts. Anal. Prilozh. **3** (2) (1969) 87 [Funct. Anal. Appl. **3** (1969) 155]. MR **39**:6348 b. Zbl. **194**:246

182. (with D.B. Fuks) Cohomologies of the Lie algebra of tangential vector fields of a smooth manifold. Funkts. Anal. Prilozh. **3** (3) (1969) 32–52 [Funct. Anal. Appl. **3** (1969) 194–210]. MR **41**:1067. Zbl. **216**:203 III.II.4

183. (with A.S. Mishchenko) Quadratic forms over commutative group rings and the K-theory. Funkts. Anal. Prilozh. **3** (4) (1969) 28–33 [Funct. Anal. Appl. **3** (1969) 277–281]. MR **41**:9243. Zbl. **239**:55004 III.II.8

184. (with V.A. Ponomarev) Remarks on the classification of a pair of commuting linear transformations in a finite-dimensional space. Funkts. Anal. Prilozh. **3** (4) (1969) 81–82 [Funct. Anal. Appl. **3** (1969) 325–329]. MR **40**:7279. Zbl. **204**:453 II.VIII.2

185. (with Yu.I. Arshavskij, M.B. Berkinblit, and V.S. Yakobson) Two types of granular cell in the cortex of the cerebellum. Nejrofiziologiya **1** (1969) 167–176 (in Russian)

186. (with E.K. Fetisova, V.I. Guelstein, and Yu.M. Vasil'ev) Stimulation of synthesis of DNA in mouse embryo fibroblastlike cells in vitro by factors of different character. Dokl. Akad. Nauk SSSR **187** (1969) 913–915 (in Russian)

187. (with Sh.A. Guberman, M.L. Shik, and Yu.M. Vasil'ev) Interaction in biological systems. Priroda **6**, 13–21; **7**, 24–33 (1969) (in Russian)

188. (with N.M. Chebotareva, Sh.A. Guberman, M.L. Izvekova, E.I. Kandel, T.V. Lebedeva, D.K. Luhev, and I.F. Nikolaeva) Prognostic matematic alevelutiei ictusorilor hemorogice in scepul preciazavii indicatiiler tratementului chirurgical. K. Accidouteler vasculare cerebrale, pp. 44–45, Bucuresti (1969)

189. (with N.M. Chebotareva, Sh.A. Guberman, M.L. Izvekova, E.I. Kandel, T.V. Lebedeva, D.K. Lunev, and I.F. Nikolaeva) Computer prognosis of spontaneous intracerebral haemorrhage for the purpose of its surgical treatment. IV. Int. Congr. Neurosurg., Excerpta Med. **32** (1969)

1970

190. (with M.I. Graev and Z.Ya. Shapiro) Integral geometry in projective space. Funkts. Anal. Prilozh. **4** (1) (1970) 14–32 [Funct. Anal. Appl. **4** (1970) 12–28]. MR **43**:6856. Zbl. **199**:255 III.I.6

191. (with M.I. Graev and Z.Ya. Shapiro) A problem of integral geometry connected with a pair of Grassmann manifolds. Dokl. Akad. Nauk SSSR **193** (1970) 259–262 [Sov. Math., Dokl. **11** (1970) 892–896]. MR **42**:3728. Zbl. **209**:267 III.I.7

192. (with M.I. Graev and V.A. Ponomarev) The classification of the linear representations of the group SL (2, **C**). Dokl. Akad. Nauk SSSR **194** (1970) 1002–1005 [Sov. Math., Dokl. **11** (1970) 1319–1323]. MR **43**:2162. Zbl. **229**:22024 II.VIII.3

193. (with D.B. Fuks) Cohomology of the Lie algebra of formal vector fields. Dokl. Akad. Nauk SSSR **190** (1970) 1267–1270 [Sov. Math., Dokl. **11** (1970) 268–271]. MR **44**:2247. Zbl. **264**:17005

194. (with D.B. Fuks) Cohomology of the Lie algebra of formal vector fields. Izv. Akad. Nauk SSSR, Ser. Mat. **34** (1970) 322–337 [Math. USSR, Izv. **34** (1970) 327–342]. MR **44**:1103. Zbl. **216**:203 III.II.5

195. (with D.B. Fuks) Cohomologies of Lie algebra of tangential vector fields II. Funkts. Anal. Prilozh. **4** (4) (1970) 23–31 [Funct. Anal. Appl. **4** (1970) 110–116]. MR **44**:2248. Zbl. **208**:514 III.II.6

196. (with D.B. Fuks) Cohomologies of Lie algebra of vector fields with nontrivial coefficients. Funkts. Anal. Prilozh. **4** (3) (1970) 10–25 [Funct. Anal. Appl. **4** (1970) 181–192]. MR **44**:7752. Zbl. **222**:58001 III.II.7

197. The cohomology of infinite dimensional Lie algebras; some questions of integral geometry. Int. Congr. Math., Nice 1970, **1** (1970) 95–111. Zbl. **239**:58004 I.I.7

198. (with I.N. Bernstein and S.I. Gelfand) Differential operators on the base affine space. Dokl. Akad. Nauk SSSR **195** (1970) 1255–1258 [Sov. Math., Dokl. **11** (1970) 1646–1649]. MR **43**:3402. Zbl. **217**:369
199. (with E.K. Fetisova, V.I. Guelstein, and J.M. Vasil'ev) Stimulation of DNA synthesis in cultures of mouse embryo fibroblastlike cells. J. Cell Physiol. **75** (1970) 305–313
200. (with Yu.I. Arshavskij, B.M. Berkinblit, and V.S. Yakobson) Organization of afferent connections of intercalary neurons in the paramedian lobe of the cerebellum of a cat. Dokl. Akad. Nauk SSSR **193** (1970) 250–253 (in Russian)
201. (with D.B. Fuks) Cycles representing cohomology classes of the Lie algebra of formal vector fields. Usp. Mat. Nauk **25** (5) (1970) 239–240. MR **45**:2737 (in Russian). Zbl. **216**:204
202. (with D.B. Fuks) Upper bounds for cohomology of infinite-dimensional Lie algebras. Funkts. Anal. Prilozh. **4** (4) (1970) 70–71 [Funct. Anal. Appl. **4** (1970) 323–324]. MR **44**:4792. Zbl. **224**:18013 III.II.9
203. (with N.M. Chebotareva, Sh.A. Guberman, M.L. Izvekova, E.I. Kandel', N.V. Lebedeva, D.K. Lunev, and I.F. Nikolaeva) Mathematical prognosis of the outcome of haemorrhages with the aim of determining evidence for surgical treatment. Zh. Nevropatol. Psikhiatr. **2** (1970) 177–181 (in Russian)
204. (with Yu.I. Arshavskij, M.B. Berkenblit, O.I. Fukson, and V.S. Yakobson) Features of the influence of the lateral reticular nucleus of medulla oblongata on the cortex of the cerebellum. Nejrofiziologiya **2** (1970) 581–586 (in Russian)
205. (with L.V. Domnina, O.Yu. Ivanova, S.G. Komm, L.V. Ol'shevskaya, and Yu.M. Vasil'ev) Effect of colcemid on the locomotory behaviour of fibroblasts. J. Embryol. Exp. Morphol. **24** (1970) 625–640
206. (with V.A. Ponomarev) Problems of linear algebra and classification of quadruples of subspaces in a finite-dimensional vector space. Colloq. Math. Soc. Janos Bolyai **5**. Hilbert space operators. Tihany, Hungary 1970 (1972). Zbl. **294**:15002
207. (with S.L.Ginzburg, G.V. Gurskaya, G.M. Lobanova, M.G. Nejgauz, and L.A. Novakovskaya) Crystal structure of paroxyacetophenon. Dokl. Akad. Nauk SSSR **195** (1970) 341–344 [Sov. Phys., Dokl. **15** (1970) 999–1002]

1971

208. (with V.A. Ponomarev) Quadruples of subspaces of a finite-dimensional vector space. Dokl. Akad. Nauk SSSR **197** (1971)

762–765 [Sov. Math., Dokl. **12** (1971) 535–539]. MR **44**:2762. Zbl. **294**:15001 II.VIII.4

209. (with I.N. Bernstein and S.I. Gelfand) Structure of representations generated by vectors of heighest weight. Funkts. Anal. Prilozh. **5** (1) (1971) 1–9 [Funct. Anal. Appl. **5** (1971) 1–8]. MR **45**:298. Zbl. **246**:17008 II.V.2

210. (with V.I. Guelstein and Yu.M. Vasil'ev) Initiation of DNA synthesis in cultures of mouse fibroblastlike cells under the action of substances that disturb the formation of microtubes. Dokl. Akad. Nauk SSSR **197** (1971) 1425–1428 (in Russian)

211. (with Yu.I. Arshavskij, M.B. Berkinblit, and O.I. Fukson) Organization of projections of somatic nerves in different regions of the cortex of the cerebellum of a cat. Nejrofiziologiya **3** (2) (1971) (in Russian)

212. (with Yu.I. Arshavskij, M.B. Berkinblit, I.A. Keder-Stepanova, E.M. Smelyanskaya, and V.S. Yakobson) Background activity of Purkinje cells in intact and deafferentized frontal lobes of the cortex of the cerebellum of a cat. Biofizika **16** (1971) 684–691 (in Russian)

213. (with Yu.I. Arshavskij, M.B. Berkinblit, O.I. Fukson, and V.S. Yakobson) Afferent connections and interaction of neurons of the cortex of the cerebellum. In the collection: Structural and functional organization of the cerebellum, pp. 40–47. Moscow: Nauka (1971) (in Russian)

214. (with Yu.I. Arshavskij, M.B. Berkinblit, O.I. Fukson, and V.S. Yakobson) The reticular afferent system of the cerebellum and its functional significance. Izv. Akad. Nauk SSSR, Ser. Biol. **3** (1971) 375–383 (in Russian)

215. (with L.B. Margolis, V.I. Samojlov, and Yu.M. Vasil'ev) A quantitative estimate of the form and orientation of cell nuclei in a culture. Ontogenez **2** (1971) 138–144 (in Russian)

216. (with T.A. Fajn, Sh.A. Guberman, G.G. Guelstein, and I.M. Rotvajn) An estimate of the pressure in the pulmonary artery from electro- and phonocardiographical data under a defect of the intraventricular partition. Kardiologiya **5** (1971) 84–87 (in Russian)

217. (with Yu.I. Arshavskij, M.B. Berkinblit, O.I. Fukson, and V.S. Yakobson) Functional role of the reticular afferent system of the cerebellum. Prepr. IPM Akad. Nauk SSSR (1971) (in Russian)

218. (with V.I. Guelstein and Yu.M. Vasil'ev) Initiation of DNA synthesis in cell cultures by colcemid. Proc. Natl. Acad. Sci. USA **68** (1971) 977–979 III.IV.13

219. (with V.Ya. Brodskij, L.V. Domnina, V.I. Guelstein, L.B. Klempner, T.L. Marshak, and Yu.M. Vasil'ev) The kinetics

of proliferation in cultures of mouse embryo fibroblastlike cells. Tsitologiya **13** (1971) 1362–1377 (in Russian)

220. (with D.A. Kazhdan) Representations of the group *GL*(*n*, *K*), where *K* is a local field. Prepr. **71**, IPM Akad. Nauk SSSR (1971) (English transl. in: Lie groups and their representations. Proc. Summer School in Group Representations. Bolyai Janos Math. Soc., Budapest 1971, pp. 95–118. New York: Halsted 1975). Zbl. **348**:22011

221. (with D.A. Kazhdan) Some questions of differential geometry and the computation of the cohomology of Lie algebras of vector fields. Dokl. Akad. Nauk SSSR **200** (1971) 269–272 [Sov. Math., Dokl. **12** (1971) 1367–1370]. MR **44**:4770. Zbl. **238**:58001

222. (with T.L. Fajn, Sh.A. Guberman, G.G. Guelstein, I.M. Rotvajn, V.A. Silin, and V.K. Sukhov) Recognition of the degree of pulmonary hypertonia under a defect of the intraventricular partition with the aid of the EVM. Krovoobrashchenie **6** (1971) (in Russian)

1972

223. (with L.V. Domnina, O.Yu. Ivanova, and Yu.M. Vasil'ev) The action of metaphase inhibitors on the form and movement of interphase fibroblasts in a culture. Tsitologiya **14** (1972) 80–88 (in Russian)

224. (with Yu.I. Arshavskij, M.B. Berkinblit, O.I. Fukson, and V.S. Yakobson) Suppression of reactions of Purkinje cells under preceding activation of the reticulo-cerebellar path. Fiziol. Zh. SSSR Im. I.M. Sechenova **58** (1972) 208–214 (in Russian)

225. (with I.N. Bernstein and S.I. Gelfand) Differential operators on a cubic cone. Usp. Mat. Nauk **27** (1) (1972) 185–190 [Russ. Math. Surv. **27** (1) (1972) 169–174]. Zbl. **257**:58010 II.V.3

226. (with D.B. Fuks and D.A. Kazhdan) The actions of infinite-dimensional Lie algebras. Funkts. Anal. Prilozh. **6** (1) (1972) 10–15 [Funct. Anal. Appl. **6** (1972) 9–13]. MR **46**:922. Zbl. **267**:18023 III.II.10

227. (with I.N. Bernstein and S.I. Gelfand) Differential operators on the base affine space and a study of g-modules. Prepr. **77**, IPM Akad. Nauk SSSR (1972) (English transl. in: Lie groups and their representations. Proc. Summer School in Group Representations. Bolyai Janos Math. Soc., Budapest 1971, pp. 21–64. New York: Halsted 1975). Zbl. **338**:58019 II.V.1

228. (with L.V. Domnina, O.Yu. Ivanova, L.B. Margolis, L.V. Ol'shevskaya, Yu.A. Rovenskij, and Yu.M. Vasil'ev) Defective formation of the lamellar cytoplasm in neoplastic fibroblasts. Proc. Natl. Acad. Sci. USA **69** (1972) 248–252

229. (with Sh.A. Guberman, M.L. Izvekova, V.J. Kejlis-Borok, and E.Ya. Rantsman) Criteria of high seismicity, determined by pattern recognition. Proc. Final Symp. Upper Mantle Project **13** (1972)

230. (with Yu.I. Arshavskij, M.B. Berkinblit, O.I. Fukson, and G.N. Orlovskij) Activity of the neurons of the dorsal spinocerebellar tract under locomotion. Biofizika **17** (1972) 487–494 (in Russian)

231. (with Yu.A. Arshavskij, M.B. Berkinblit, O.I. Fukson, and G.N. Orlovskij) Activity of the neurons of the ventral spinocerebellar tract under locomotion. Biofizika **17** (1972) 883–896 (in Russian)

232. (with Yu.I. Arshavskij, M.B. Berkenblit, O.I. Fukson, and G.N. Orlovskij) Activity of the neurons of the ventral spinocerebellar tract under locomotion of cats with deafferentized hind legs. Biofizika **17** (1972) 1113–1119 (in Russian)

233. (with Sh.A. Guberman, M.L. Izvekova, V.I. Kejlis-Borok, E.Ya. Rantsman) Criteria of high seismicity. Dokl. Akad. Nauk SSSR **202** (1972) 1317–1320 (in Russian)

234. (with Yu.I. Arshavskij, M.B. Berkenblit, O.I. Fukson, and G.N. Orlovskij) Recordings of neurones of the dorsal spinocerebellar tract during evoked locomotion. Brain Res. **43** (1972) 272–275 III.IV.8

235. (with Yu.I. Arshavskij, M.B. Berkenblit, O.I. Fukson, and G.N. Orlovskij) Origin of modulation in neurones of the ventral spinocerebellar tract during locomotion. Brain Res. **43** (1972) 276–279 III.IV.9

236. (with D.B. Fuks and D.I. Kalinin) Cohomology of the Lie algebra of Hamiltonian formal vector fields. Funkts. Anal. Prilozh. **6** (63) (1972) 25–29 [Funct. Anal. Appl. **6** (1972) 193–196]. MR **47**:1088. Zbl. **259**:57023 III.II.11

237. (with D.A. Kazhdan) On the representation of the group *GL*(*n*, *K*), where *K* is a local field. Funkts. Anal. Prilozh. **6** (4) (1972) 73–74 [Funct. Anal. Appl. **6** (1972) 315–317]. Zbl. **288**:22024 II.IV.7

238. (with Sh.A. Guberman, M.S. Kaletskaya, V.I. Kejlis-Borok, E.Ya. Rantsman, and M.P. Zhidkov) An attempt to carry over criteria of heigh seismicity from Central asia to Anatolia and adjoining regions. Dokl. Akad. Nauk SSSR **210** (1972) 327–330 (in Russian)

239. (with L.V. Domnina, V.I. Guelstein, and Yu.M. Vasil'ev) Regulation of the behaviour of connective tissue cells in multicell systems. In: Histophysiology of connective tissue, vol. **1**, pp. 31–36. Novosibirsk 1972

240. (with L.V. Domnina, O.Yu. Ivanova, S.G. Komm, L.V. Ol'shevskaya, and Yu.M. Vasil'ev) The action of metaphase inhibitors on the form and movement of fibroblasts in a culture. Tsitologiya **14** (1972) 80–88 (in Russian)

1973

241. (with Sh.A. Guberman, M.L. Izvekova, V.I. Kejlis-Borok, and E.Ya. Rantsman) Recognition of places of possible origin of powerful earthquakes (in Eastern Central Asia). Vychisl. Seismol. **6** (1973) (in Russian)

242. (with Yu.I. Arshavskij, M.V. Berkenblit, O.I. Fukson, and G.N. Orlovskij) Activity of neurons of the cuneo-cerebellar tract under locomotion. Biofizika **18** (1973) 126–131 (in Russian)

243. (with Yu.M. Vasil'ev) Interactions of normal and neoplastic fibroblasts with the substratum. Ciba Foundation Symposium on Cell Locomotion (1973) 312–331

244. (with I.N. Bernstein and V.A. Ponomarev) Coxeter functors and Gabriel's theorem. Usp. Mat. Nauk **28** (2) (1973) 19–33 [Russ. Math. Surv. **28** (2) (1973) 17–32]. Zbl. **269**:08001 II.VIII.5

245. (with I.N. Bernstein and S.I. Gelfand) Schubert cells and cohomology of flag spaces. Funkts. Anal. Prilozh. **7** (1) (1973) 64–65 [Funct. Anal. Appl. **7** (1973) 53–55]. MR **47**:6713. Zbl. **282**:20035

246. (with L.V. Domnina, O.Yu. Ivanova, L.B. Margolis, and Yu.M. Vasil'ev) Intracellular interaction in cultures of transformed fibroblasts of strain L and normal mouse fibroblasts.Tsitologiya **15** (1973) 1024–1028 (in Russian)

247. (with I.N. Bernstein and S.I. Gelfand) Schubert cells and cohomology of the spaces G/P. Usp. Mat. Nauk **28** (3) (1973) 3–26 [Russ. Math. Surv. **28** (3) (1973) 1–26]. Zbl. **289**:57024 II.V.4

248. (with M.I. Graev and A.M. Vershik) Representations of the group $SL(2, \mathbf{R})$, where $\mathbf{R}$ is a ring of functions. Usp. Mat. Nauk. **28** (5) (1973) 82–128 [Russ. Math. Surv. **28** (5) (1973) 87–132]. Zbl. **297**:22003 II.IX.1

249. (with Yu.M. Vasil'ev) Disturbance of morphogenetic reactions of cells under tumorous transformation. Vestn. Akad. Med. Nauk SSSR **4** (1973) 61–69 (in Russian)

250. (with V.I. Guelstein, O.Yu. Ivanova, L.B. Margolis, and Yu.M. Vasil'ev) Contact inhibition of movement in the cultures of transformed cells. Proc. Natl. Acad. Sci. USA **70** (1973) 2011–2014

251. (with L.V. Domnina, E.E. Krivitska, L.V. Ol'shevskaya, Yu.A. Rovenskij, and Yu.M. Vasil'ev) The structure of the lamellar cytoplasm of normal and tumorous fibroblasts.

Papers from a Soviet-French symposium. In: Ultrastructure of cancerous cells, pp. 49–71. Moscow: Nauka 1973 (in Russian)

252. (with L.V. Domnina, E.K. Fetisova, O.Yu. Pletyushkina, and Yu.M. Vasil'ev) Comparative study of density dependent inhibition of growth in the cultures of normal and neoplastic fibroblast-like cells. Abstracts 6th meeting of the European study group for cell proliferation, p. 15. Moscow: Nauka 1973

253. (with Yu.M. Vasil'ev) Factors inducing DNA synthesis and mitosis in normal and neoplastic cell culture. Abstracts 6th meeting of the European study group for cell proliferation, p. 61. Moscow: Nauka 1973

254. (with Sh.A. Guberman, V.I. Kejlis-Borok, E.Ya. Rantsman, I.M. Rotvajn, and M.I. Zhidkov) Determination of criteria of high seismism by means of recognition algorithms. Vestn. Mosk. Gos. Univ. **5** (1973) 78–83 (in Russian)

255. (with A.D. Bershadskij, L.V. Domnina, V.I. Guelstein, O.Yu. Ivanova, S.G. Komm, L.B. Margolis, and Yu.M. Vasil'ev) Interactions of normal and neoplastic cells with various surfaces. Neoplasma **20** (1973) 583–585

256. (with D.B. Fuks) PL Foliations. Funkts. Anal. Prilozh. **7** (4) (1973) 29–37 [Funct. Anal. Appl. **7** (1973) 278–284]. MR **49**:3958. Zbl. **294**:57016. III.II.12

257. (with Yu.I. Arshavskij, M.B. Berkinblit, O.I. Fukson, G.N. Orlovskij, and B.S. Yakobson) Some peculiarities of the organization of afferent links of the cerebellum. In: 4th International Biophysical Congress, Pushchino Symp. **3** (1973) 327–346 (in Russian)

258. (with Yu.I. Arshavskij, O.I. Fukson, and G.N. Orlovskij) Activity of the neurons of the cuneo-cerebellar tract for locomotion. Biofizika **18** (1973) 126–131 (in Russian)

1974

259. (with M.I. Graev and A.M. Vershik) Irreducible representations of the group G^X and cohomologies. Funkts. Anal. Prilozh. **8** (2) (1974) 67–69 [Funct. Anal. Appl. **8** (1974) 151–153]. MR **50**:530. Zbl. **299**:22004. II.IX.3

260. (with B.L. Feigin and D.B. Fuks) Cohomologies of the Lie algebra of formal vector fields with coefficients in its adjoint space and variations of characteristic classes of foliations. Funkts. Anal. Prilozh. **8** (2) (1974) 13–29 [Funct. Anal. Appl. **8** (1974) 99–112]. MR **50**:8553. Zbl. **298**:57011. III.II.13

*261. (with I.N. Bernstein and S.I. Gelfand) A new model for representations of finite semisimple algebraic groups. Usp. Mat. Nauk **29** (3) (1974) 185–186. MR 53:5760. Zbl. **354**:20031 II.IV.4

262. (with Yu.N. Arshavskij, M.B. Berkinblit, A.M. Smelyanskij, and V.S. Yakobson) Background activity of Pourkynje cells of the paramedian part of the cortex of the cerebellum of a cat. Biofizika **19** (1974) 903–907 (in Russian)

263. (with Yu.I. Arshavskij, M.B. Berkinblit, O.I. Fukson, and G.N. Orlovskij) Differences in the working of spino-cerebral tracts in artificial irritation and locomotion. In: Mechanisms of the union of neurons in the nerve centre, pp. 99–105. Leningrad: Nauka 1974 (in Russian)

264. (with Sh.A. Guberman, M.S. Kaletska, V.I. Kejlis-Borok, E.Ya. Rantsman, I.M. Rotvajn, and M.P. Zhidkov) Recognition of places where strong earthquakes are possible. II. Four regions of Asia Minor and South-East Europe. Vychisl. Seismol. **7** (1974) 3–39 (in Russian)

265. (with Sh.A. Guberman, V.I. Kejlis-Borok, E.Ya. Rantsman, I.M. Rotvajn, and M.P. Zhidkov) Recognition of places where strong earthquakes are possible. III. The case when the boundaries of disjunctive nodes are not known. Vychisl. Seismol. **7** (1974) 41–62 (in Russian)

266. (with V.I. Guelstein, O.Yu. Ivanova, S.G. Komm, and L.B. Margolis, and Yu.M. Vasil'ev) The results of intercellular impacts in cultures of normal and transformed fibroblasts. Tsitologiya **16** (1974) 752–756 (in Russian)

267. (with D.B. Fuks) PL Foliations. II. Funkts. Anal. Prilozh. **8** (3) (1974) 7–11 [Funct. Anal. Appl. **8** (1974) 197–200]. MR **54**:6159. Zbl. **316**:57010 III.II.15

268. (with V.A. Ponomarev) Free modular lattices and their representations. Usp. Mat. Nauk **29** (6) (1974) 3–58 [Russ. Math. Surv. **29** (6) (1974) 1–56]. MR **53**:5393. Zbl. **314**:15003

269. (with O.Yu. Ivanova, L.B. Margolis, and Yu.M. Vasil'ev) Orientation of mitosis of fibroblasts is determined in the interphase. Proc. Natl. Acad. Sci. USA **71** (1974) 2032

270. (with Yu.I. Arshavskij, M.B. Berkinblit, O.I. Fukson, and G.N. Orlovskij) Peculiarities of information entering the cortex of the cerebellum via different afferent paths, structural and functional organization of the cerebellum, pp. 34–41. Kiev: Naukova Dumka 1974 (in Russian)

1975

271. (with L.V. Domnina, A.V. Lyubimov, Yu.M. Vasil'ev, and O.S. Zakharova) Contact inhibition of phagocytosis in epithelial sheets: alterations of cell surface properties induced by cell-cell contacts. Proc. Natl. Acad. Sci. USA **72** (1975) 719–722

272. (with A.P. Chern and Yu.M. Vasil'ev) Spreading of normal and transformed fibroblasts in dense cultures. Exp. Cell Res. **90** (1975) 317–327

273. (with A.M. Gabrielov and M.V. Losik) The combinatorial computation of characteristic classes. Funkts. Anal. Prilozh. **9** (1975) 54–55 [Funct. Anal. Appl. **9** (1975) 48–49]. MR **51**:1839. Zbl. **312**:57016

274. Quantitative evaluation of cell orientation in culture. J. Cell. Sci. **17** (1975) 1–10

275. (with M.I. Graev and A.M. Vershik) Representations of the group of diffeomorphisms connected with infinite configurations. Prepr. Inst. Appl. Math. **46** (1975) 1–62 (in Russian)

276. (with M.I. Graev and A.M. Vershik) The square roots of quasiregular representations of the group $SL(2, k)$. Funkts. Anal. Prilozh. **9** (2) (1975) 64–66 [Funct. Anal. Appl. **9** (1975) 146–148]. MR **51**:8338. Zbl. **398**:22010

277. (with I.N. Bernstein and S.I. Gelfand) Models of representations of compact Lie groups. Funkts. Anal. Prilozh. **9** (4) (1975) 61–62 [Funct. Anal. Appl. **9** (1975) 322–324]. MR **54**:2884. Zbl. **339**:22009

278. (with I.N. Bernstein and S.I. Gelfand) Models of representations of Lie groups. Proc. Petrovskij Semin. **2** (1976) 3–21. [Sel. Math. Sov. **1** (2) (1981) 121–142] Zbl. **499**:22004 II.IV.8

279. (with A.M. Gabriélov and M.V. Losik) Combinatorial computation of characteristic classes. Funkts. Anal. Prilozh. **9** (2) (1975) 12–28 [Funct. Anal. Appl. **9** (1975) 103–115]. MR **53**:14504a. Zbl. **312**:57016 III.II.16

280. (with A.M. Gabrielov and M.V. Losik) Combinatorial computation of characteristic classes. Funkts. Anal. Prilozh. **9** (3) (1975) 5–26 [Funct. Anal. Appl. **9** (1975) 186–202]. MR **53**:14504a. Zbl. **341**:57017 III.II.17

281. (with L.A. Dikij) Asymptotic behaviour of the resolvent of Sturm-Liouville equations and the algebra of the Korteweg-de Vries equations. Usp. Mat. Nauk **30** (5) (1975) 67–100 [Russ. Math. Surv. **30** (5) (1975) 77–113]. MR **58**:22746. Zbl. **461**:35072. I.III.12

282. (with I.S. Tint and Yu.M. Vasil'ev) Processes determining the changes of shape of a cell after its separation from the epigastrium. Tsitologiya **5** (1975) 633–638 (in Russian)

283. (with O.Yu. Pletyushkina and Yu.M. Vasil'ev) Neoplastic fibroblasts sensitive to growth inhibition by parent normal cells. Br. J. Cancer **31** (1975) 535–543

284. (with L.B. Margolis, V.I. Samojlov, and Yu.M. Vasil'ev) Methods of measuring the orientation of cells. Ontogenez **6** (1) (1975) 105–110 (in Russian)

285. (with E.K. Fetisoba, O.Yu. Pletyushkina, and Yu.M. Vasil'ev) Insensibility of dense cultures of transformed mice fibroblasts to the action of agents, stimulating the synthesis of DNA in cultures of normal cells. Tsitologiya **17** (1975) 442–446 (in Russian)
286. (with Yu.I. Arshavskij, G.N. Orlovskij, and G.A. Pavlova) The activity of neurons of the ventral spino-cerebral tract in "fictitious scratching". Biofizika **20** (1975) 748–749 (in Russian)
287. (with Yu.I. Arshavskij, G.N. Orlovskij, and G.A. Pavlova) Origin of the modulation of the activity of vestibular-spinal neurons in scratching. Biofizika **20** (1975) 946–947 (in Russian)
288. (with Yu.I. Arshavskij, M.B. Berkinblit, T.G. Delyagina, A.G. Fel'dman, O.I. Fukson, G.N. Orlovskij, and G.A. Pavlova) On the role of the cerebellum in regulating some rhythmic movements (locomotion, scratching). Summaries 12th meeting of the All-Union Physiological Society, pp. 15–16. Tbilisi 1975 (in Russian)
289. (with T.G. Delyagina, A.G. Fel'dman, and G.N. Orlovskij) On the role of the central program and afferent inflow in generation of scratching movements in the cat. Brain Res. **100** (1975) 297–313
290. (with M.I. Graev and A.M. Vershik) Representations of the group of diffeomorphisms. Usp. Mat. Nauk **30** (6) (1975) 3–50 [Russ. Math. Surv. **30** (6) (1975) 1–50]. MR **53**:3188. Zbl. **317**:58009 II.IX.4
291. (with Sh.A. Guberman, M.S. Kaletska, V.I. Kejlis-Borok, E.Ya. Rantsman, I.M. Rotvajn, and L.P. Zhidkov) Prognosis of a place where strong earthquakes occur, as a problem of recognition. In: Modelling of training and behaviour, pp. 18–25. Moscow: Nauka 1975 (in Russian)

1976

292. (with D.B. Fuks and A.M. Gabrielov) The Gauss-Bonnet theorem and the Atiyah-Patodi-Singer functionals for the characteristic classes of foliations. Topology **15** (1976) 165–188. MR **55**:4201. Zbl. **347**:57009 III.II.14
293. (with A.M. Gabrielov and M.V. Losik) A local combinatorial formula for the first class of Pontryagin. Funkts. Anal. Prilozh. **10** (1) (1976) 14–17 [Funct. Anal. Appl. **10** (1976) 12–15]. MR **53**:14504b. Zbl. **328**:57006 III.II.18
294. (with L.A. Dikij) A Lie algebra structure in a formal variational calculation. Funkts. Anal. Prilozh. **10** (1) (1976) 1–8

[Funct. Anal. Appl. **10** (1976) 16–22]. MR **57**:7670. Zbl. **347**:49023

295. (with I.N. Bernstein and S.I. Gelfand) Category of g-modules. Funkts. Anal. Prilozh. **10** (2) (1976) 1–8 [Funct. Anal. Appl. **10** (1976) 87–92]. MR **53**:10880. Zbl. **353**:18013 II.V.5

296. (with A.M. Gabrielov and M.V. Losik) Atiyah-Patodi-Singer functionals for characteristic functionals for tangent bundles. Funkts. Anal. Prilozh. **10** (2) (1976) 13–28 [Funct. Anal. Appl. **10** (1976) 95–107]. MR **54**:1245. Zbl. **344**:57008 III.II.19

297. (with L.A. Dikij) Fractional powers of operators and Hamiltonian systems. Funkts. Anal. Prilozh. **10** (4) (1976) 13–29 [Funct. Anal. Appl. **10** (1976) 259–273]. MR **55**:6484. Zbl. **346**:35085 I.III.10

298. (with Yu.I. Manin and M.A. Shubin) Poisson brackets and the kernel of the variational derivative in the formal calculus of variations. Funkts. Anal. Prilozh. **10** (4) (1976) 30–34 [Funct. Anal. Appl. **10** (1976) 274–278]. MR **55**:13486. Zbl. **395**:58005

299. (with V.A. Ponomarev) Lattices, representations, and algebras connected with them. I. Usp. Mat. Nauk **31** (5) (1976) 71–88 [Russ. Math. Surv. **31** (5) (1976) 67–85]. MR **58**:16779a. Zbl. **358**:06020

*300. (with M.V. Losik) Computing characteristic classes of combinatorial vector bundles. Prepr. Inst. Appl. Mat. **99** (1976) III.II.20

301. (with Sh.A. Guberman, V.I. Kejlis-Borok, L. Knopov, E. Press, E.Ya. Rantsman, I.M. Rotvajn, and A.M. Sadovskij) Conditions for the occurence of strong earthquakes (California and some other areas). Vychisl. Seismol. **9** (1976) 3–92 (in Russian)

302. (with Sh.A. Guberman, V.I. Kejlis-Borok, L. Knopov, F. Press, E.Ya. Rantsman, I.M. Rotvajn, and A.M. Sadovskij) Pattern recognition applied to earthquake epicenters in California. Phys. Earth Planet. Inter. **11** (1976) 277–283

303. (with M.A. Alekseevskaya, I.V. Martynov, and V.M. Sablin) First results of the prognostication of the effect of transmural (large focal) myocardial infarcts. Aktual'nye voprosy kardiologii, Otdelennye rezul'taty lecheniya elokachestvennykh opukholej, 19–24. Moscow: Nauka 1976 (in Russian)

304. (with E.E. Bragina and Yu.M. Vasil'ev) Formation of bundles of microfilaments during spreading of fibroblasts on the substratum. Exp. Cell Res. **97** (1976) 241–248

305. (with A.D. Bershadskij, V.I. Gelfand, V.I. Guelstein, and Yu.M. Vasil'ev) Serum dependence of expression of the transformed phenotype experiments with subline of mouse *L* fibro-

blasts adapted to growth in serum-free medium. Int. J. Cancer (1976) 84–92

306. (with O.Yu. Ivanova, L.B. Margolis, and Yu.M. Vasil'ev) Effect of colcemid on the spreading of fibroblasts in culture. Exp. Cell Res. **101** (1976) 207–219
307. (with L.V. Domnina, N.A. Dorfman, O.Yu. Pletyushkina, and Yu.M. Vasil'ev) Active cell edge and movements of concanavalin *A* receptors on the surface of epithelial and fibroblastic cells. Proc. Natl. Acad. Sci. USA **73** (1976) 4085–4089
308. (with N.M. Chebotareva, Sh.A. Guberman, M.L. Izvekova, E.I. Kandel, N.V. Lebedeva, D.K. Lunev, I.F. Nikolaeva, and E.V. Shmidt) A computer study of prognosis of cerebral haemorrhage for choosing optimal treatment. Eur. Congr. Neurosurg., pp. 71–72. Edinburgh (1976)
309. (with N.M. Chebotareva, Sh.A. Guberman, M.L. Izvekova, E.I. Kandel, and E.V. Shmidt) Prognostication of the results of surgical treatment of haemorrhaging lesions by means of a computer. Vopr. Nejrokhir. **3** (1976) 20–23 (in Russian)
310. (with O.Yu. Ivanova, L.B. Margolis, and Yu.M. Vasil'ev) Effect of colcemid on spreading of fibroblast in culture. Exp. Cell Res. **101** (1976) 207–219
311. (with Yu.M. Vasil'ev) Effects of colcemid on morphogenetic processes and locomotion of fibroblasts. Cell Motility **3** (1976) 279–304

1977

312. (with M.A. Alekseevskaya, Sh.A. Guberman, I.V. Martynov, I.M. Rotvajn, and V.M. Sablin) Prognostication of the result of a large focal myocardial infarct by means of learning program. Kardiologiya **17** (1977) 26–31 (in Russian)
313. (with M.A. Alekseevskaya, L.D. Golovnya, Sh.A. Gubermann, M.L. Izvekova, and A.L. Syrkin) Prognostication of the result of a myocardial infarct by means of the program "Cortex-3". Kardiologiya **17** (6) (1977) 13–23 (in Russian)
314. (with M.Yu. Melikova, S.G. Gindikin, and M.L. Izvekova) Prognostication of the healing of duodenal ulcers. Aktual. Vopr. Gastroenterol. **10** (1977) 42–51 (in Russian)
315. (with M.A. Alekseevskaya, L.D. Golovnya, M.L. Izvekova, I.V. Martynov, V.M. Sablin, and A.L. Syrkin) A general guide-line or a general method for creating one (On ways of applying mathematical methods in medicine). Summaries of lectures at the All-Union Conf. on the theory and practice of automatic electrocardiological and clinical investigations, pp. 3–5. Kaunas (1977) (in Russian)

316. (with M.A. Alekseevskaya, E.S. Klyushin, A.V. Nedostup and A.L. Syrkin) On the methodology of creating a formalized description of the patient (using the example of prognostication of remote results of electro-impulsive treatment of a constant form of flickering arrhythmy). Summaries of lectures at the All-Union Conf. on the theory and practice of automatic electrocardiological and clinical investigations, pp. 5–8. Kaunas (1977) (in Russian)
317. (with L.V. Domnina, O.Yu. Pletyushkina, and Yu.M. Vasil'ev) Effects of antitubilins on redistribution of cross-linked receptors on the surface of fibroblasts and epithelial cells. Proc. Natl. Acad. Sci. USA **74** (1977) 2865–2868
318. (with O.Yu. Ivanova, S.G. Komm, L.B. Margolis, and Yu.M. Vasil'ev) The influence of colcemid on the polarization of cells on narrow strips of the adhesive substratum. Tsitologiya **19** (1977) 357–360 (in Russian)
319. (with A.D. Bershadskij, V.I. Gelfand, and Yu.M. Vasil'ev) The influence of serum on the development of cell transformation. Vestn. Akad. Mech. Nauk **3** (1977) 55–59 (in Russian)
320. (with A.D. Bershadskij, A.D. Lyubimov, Yu.A. Rovenskij, and Yu.M. Vasil'ev) Contact interaction of cell surfaces. Lectures at the Soviet-Italian symposium "Tissue proteinases in normal and pathological state", Moscow 22–27 September 1977 (in Russian)
321. (with Yu.M. Vasil'ev) Mechanisms of morphogenesis in cell cultures. Int. Rev. Cytol. **50** (1977) 159–274 III.IV.14
322. (with Yu.I. Arshavskij, M.B. Berkinblit, and V.S. Yakobson) A formula for the analysis of histograms of intercellular intervals of Pourkine cells. Biofizika **22** (1977) (in Russian)
323. (with M.A. Alekseevskaya, A.M.Gabrielov, A.D. Gvishiani, and E.Ya. Rantsman) Morphological division of mountainous countries by formalized criteria. Vychisl. Seismol. **10** (1977) 33–79 (in Russian)
324. (with M.A. Alekseevskaya, A.M. Gabrielov, A.D. Gvishiani, and E.Ya. Rantsman) Formal morphostructural zoning at mountain territories. J. Geophys. **43** (1977) 227–235
325. (with L.A. Dikij) The Resolvent and Hamiltonian systems. Funkts. Anal. Prilozh. **11** (2) (1977) 11–27 [Funct. Anal. Appl. **11** (1977) 93–105]. MR **56**:1359. Zbl. **357**:58005. I.III.13
326. (with S.G. Gindikin) Nonlocal inversion formulas in real integral geometry. Funkts. Anal. Prilozh. **11** (3) (1977) 12–19 [Funct. Anal. Appl. **11** (1977) 173–179]. MR **56**:16265. Zbl. **385**:53056 III.I.8
327. (with V.A. Ponomarev) Representation lattices and the algebras connected with them. Usp. Mat. Nauk **32** (1) (1977)

85–107 [Russ. Math. Surv. **32** (1) (1977) 91–114]. Zbl. **358**:06021

328. (with S.G. Gindikin) Complex manifolds whose skeletons are semisimple real Lie groups, and analytic discrete series of representations. Funkts. Anal. Prilozh. **11** (4) (1977) 20–28 [Funct. Anal. Appl. **11** (1977) 258–265]. MR **58**:11230. Zbl. **444**:22006. II.IV.9
329. (with M.I. Graev and A.M. Vershik) Representations of the group of smooth mappings from a manifold *X* into a compact Lie group. Dokl. Akad. Nauk SSSR **323** (1977) 745–748 [Sov. Math., Dokl. **18** (1977) 118–121]. MR **55**:10602. Zbl. **393**:22012
330. (with M.I. Graev and A.M. Vershik) Representations of the group of smooth mappings of a manifold *X* into a compact Lie group. Compos. Math. **35** (1977) 299–334. MR **58**:28257. Zbl. **368**:53034 II.IX.2
331. (with Yu.I. Arshavskij, G.N. Orlovskij, and G.A. Pavlova) Activity of neurons of the lateral reticular nucleus in scratching. Biofizika **22** (1) (1977) (in Russian)

1978

332. (with L.V. Domnina, O.Yu. Pletyushkina, and Yu.M. Vasil'ev) Influence of agents, destroying microtubules, on the distribution of receptors of the surface of cultured cells. Tsitologiya **20** (1978) 796–801 (in Russian)
333. (with Yu.I. Arshavskij, G.N. Orlovskij, and G.A. Pavlova) Messages conveyed by spino-cerebellar pathways during scratching in the cat. 1. Activity of neurons of lateral reticular nucleus. Brain Res. **151** (1978) 479–491
334. (with Yu.I. Arshavskij, G.N. Orlovskij, and G.A. Pavlova) Messages conveyed by spino-cerebellar pathways during scratching in the cat. 2. Activity of neurons of the ventral spino-cerebellar tract. Brain Res. **151** (1978) 493–506
335. (with M.B. Berkinblit, T.G. Delyagina, A.G. Fel'dman, and G.N. Orlovskij) Generation of scratching. I. Activity of spinal interneurons during scratching. J. Neurophysiol. **41** (1978) 1040–1057 III.IV.10
336. (with M.B. Berkinblit, T.G. Delyagina, A.G. Fel'dman, and G.N. Orlovskij) Generation of scratching. 2. Non-regular regimes of generation. J. Neurophysiol. **41** (1978) 1058–1069
337. (avec I.N. Bernstein et S.I. Gelfand) Structure locale de la catégorie des modules de Harish-Chandra I. C.R. Acad. Sci., Paris, Ser. A **286** (1978) 435–437. MR **58**:16966, Zbl. **416**:22018 II.V.6

338. (avec I.N. Bernstein et S.I. Gelfand) Structure locale de la catégorie des modules de Harish-Chandra II. C.R. Acad. Sci., Paris, Ser. A **286** (1978) 495–497. MR **81e**:22026. Zbl. **431**:22013 II.V.7

339. (with Yu.M. Vasil'ev) Mechanisms of non-adhesiveness of endothelial and epithelial surfaces. Nature **275** (1978) 710–711

340. (with Yu.M. Vasil'ev, A.D. Bershadskij, V.A. Rozenblat, and I.S. Tint) Microtubular system in cultured mouse epithelial cells. Cell Biol. Int. Rep. **2** (1978) 345–351

341. (with L.A. Dikij) Variational calculus and the Korteweg-de Vries equations. Partial differential equations. Proc. All-Union Conf., Moscow 1976, dedic. I.G. Petrovskij pp. 81–83 (1978) (in Russian). Zbl. **498**:35074

342. (with L.A. Dikij) Calculus of jets and non-linear Hamiltonian systems. Funkts. Anal. Prilozh. **12** (2) (1978) 8–23 [Funct. Anal. Appl. **12** (1978) 81–94]. MR **58**:18561, Zbl. **388**:58009.

*343. (with L.A. Dikij) A family of Hamiltonian structures related to nonlinear integrable differential equations. Prepr. Inst. Appl. Mat. **136** (1978). MR **81**:58027 I.III.11

344. (with I.N. Bernstein and S.I. Gelfand) Algebraic bundles over P^n and problems of linear algebra. Funkts. Anal. Prilozh. **12** (3) (1978) 66–68 [Funct. Anal. Appl. **12** (1978) 212–214]. MR **80c**:14010a, Zbl. **402**:14005. II.V.8

345. (with B.L. Feigin and D.B. Fuks) Cohomology of infinite-dimensional Lie algebras and Laplace operators. Funkts. Anal. Prilozh. **12** (4) (1978) 1–5 [Funct. Anal. Appl. **12** (1978) 243–247]. MR **80i**:58050, Zbl. **396**:17008. III.II.21

346. (with S.G. Gindikin, M.L. Izvekova, and M.Yu. Melikova) On one approach to formalization of the diagnostic attitude of a doctor (using the prognosis of the healing of duodenal ulcers). Summaries of lectures at the All-Union Conf. biological and medical cybernetics, vol. 2, pp. 27–31 (1978) (in Russian)

347. (with S.G. Gindikin, M.L. Izvekova, and M.Yu. Melikova) On some questions of mathematical diagnostics: examples of problems from gastroenterology. Summaries of lectures at the Second All-Union Cong. on Gastroenterology, vol. 2, pp. 57–58, Moscow-Leningrad: Nauka (1978) (in Russian)

348. (with Yu.I. Manin) Dualità, Enciclopedia Einaudi, Vol. 5, pp. 126–178. Einaudi, Torino (1978)

349. (with Yu.I. Arshavskij, G.N. Orlovskij, and G.A. Pavlova) Messages conveyed by descending tract during scratching in the cat. I. Activity of vestibulospinal neurons. Brain Res. **159** (1978) 88–110

350. (with M.A. Alekseevskaya, E.S. Klyushin, A.V. Nedostup, and A.L. Syrkin) A new approach to the problem of the choice of information and formalization of the description of the patient for the solution of medical problems on a computer. Prepr. Inst. Appl. Math. **144** (1978) (in Russian)

1979

351. (with M.A. Alekseevskaya, E.S. Klyushin, and A.V. Nedostup) The gathering of medical information for processing on a computer (manual). Inst. Appl. Math. (1979) (in Russian)
352. (with S.G. Gindikin, M.L. Izvekova, and M.Yu. Melikova) One method of formalizing the diagnostic attitude of a doctor (examples of prognosis of the healing of a duodenal ulcer). Prepr. Acad. Sci. USSR Sci. Committee on the complex problem "Cybernetics" (1979) (in Russian)
353. (with L.B. Margolis, E.J. Vasil'eva, and Yu.M. Vasil'ev) Upper surfaces of epithelial sheets and of fluid lipid films are non-adhesive for platelets. Proc. Natl. Acad. Sci. USA **76** (1979) 2303–2305
354. (with A.D. Bershadskij, V.I. Gelfand, V.A. Rozenblat, I.S. Tint, and Yu.M. Vasil'ev) Morphology of microtubular systems in epithelial cells in the kidney of a mouse. Ontogenez **10** (1979) 231–235 (in Russian)
355. (with L.A. Dikij) Integrable nonlinear equations and the Liouville theorem. Funkts. Anal. Prilozh. **13** (1) (1979) 8–20 [Funct. Anal. Appl. **13** (1979) 6–15]. MR **80i**:58027. Zbl. **423**:34003. I.III.14
356. (with S.G. Gindikin and Z.Ya. Shapiro) A local problem of integral geometry in a space of curves. Funkts. Anal. Prilozh. **13** (2) (1979) 11–31 [Funct. Anal. Appl. **13** (1980) 87–102]. MR **80k**:53100. Zbl. **415**:53046. III.I.9
357. (with V.A. Ponomarev) Model algebras and representations of graphs. Funkts. Anal. Prilozh. **13** (3) (1979) 1–12. [Funct. Anal. Appl. **13** (1980) 157–166]. Zbl. **437**:16020 II.VIII.6
358. (with I.Ya. Dorfman) Hamiltonian operators and algebraic structures related to them. Funkts. Anal. Prilozh. **13** (4) (1979) 13–31 [Funct. Anal. Appl. **13** (1980) 248–262]. MR **81c**:58035. Zbl. **428**:58009 I.III.15
359. (with S.G. Gindikin and M.I. Graev) A problem of integral geometry in RP^n, connected with the integration of differential forms. Funkts. Anal. Prilozh. **13** (4) (1979) 64–67 [Funct. Anal. Appl. **13** (1980) 288–290]. MR **83a**:43006. Zbl. **423**:58001 III.I.10
360. (with Yu.I. Arshavskij) The role of the brain stem and cerebellum in the regulation of rhythmic movements. Proc. 13 Congr. Pavlov Physiol. Soc., vol. 1, pp. 474–475. Leningrad: Nauka 1979 (in Russian)

361. (with Yu.I. Arshavskij, M.B. Berkinblit, and G.N. Orlovskij) The significance of signals passing along the various spinocerebral pathways for the work of locomotive centres of the brain stem in scratching. In: Nejronnye mekhanizmy integrativnoj deyatel'nosti mozzhechka, pp. 88–91. Erevan 1979 (in Russian)

362. (with Yu.I. Arshavskij, M.B. Berkinblit, G.N. Orlovskij) Signalling mechanisms of the scratching reflex and their interaction with the cerebellum. In: Nejronnye mekhanizmy integrativnoj deyatel'nosti mozzhechka, pp. 92–96. Erevan 1979 (in Russian)

1980

363. (with M.Ya. Ratner, B.I. Rozenfeld, and V.V. Serov) The problem of classifying glomerule kidneys. Prepr. Acad. Sci. USSR Sci. Committee on the complex problem "Cybernetics" (1980) (in Russian)

364. (with V.A. Ponomarev) Representations of graphs. Perfect sub-representations. Funkts. Anal. Prilozh. **14** (3) (1980) 14–31 [Funct. Anal. Appl. **14** (1980) 177–190]. MR **83c**:05113. Zbl. **453**:05027

365. (with I.Ya. Dorfman) The Schouten bracket and Hamiltonian operators. Funkts. Anal. Prilozh. **14** (3) (1980) 71–74 [Funct. Anal. Appl. **14** (1980) 223–226]. MR **82e**:58039. Zbl. **444**:58010. I.III.16

366. (with M.I. Graev) Admissible *n*-dimensional complexes of curves in $\mathbf{R}^n$. Funkts. Anal. Prilozh. **14** (4) (1980) 36–44 [Funct. Anal. Appl. **14** (1980) 274–281]. MR **82**:53013. Zbl. **454**:53042.

367. (with S.G. Gindikin and M.I. Graev) Integral geometry for one-dimensional fibrations of general form over $\mathbf{RP}^n$. Prepr. Inst. Appl. Math. **60** (1980) 1–24 (in Russian). MR **82g**:53081

368. (with S.G. Gindikin and M.I. Graev) Integral geometry in affine and projective spaces. Itogi Nauki Tekh., Ser. Sovrem. Probl. Mat. 16, 53–226, Moscow: VINITI (1980) [J. Sov. Math. **18** (1980) 39–167]. MR **82m**:43017. Zbl. **465**:52005. III.I.11

369. (with I.V. Cherednik and S.A. Chernyakevich) A formalized differentiated description of the motor of the stomach and the duodenal intestine. Prepr. Acad. Sci. USSR Sci. Committee on the complex problem "Cybernetics", pp. 1–34 (1980) (in Russian)

370. (with G.G. Guelstein, I.P. Lukashevich, and M.A. Shifrin) Study of the correlation between electrocardiograph and coronary data. Prepr. Acad. Sci. USSR Sci. Committee on the complex problem "Cybernetics", pp. 1–28 (1980) (in Russian)

371. (with Zh.L. Bliokh, L.V. Domnina, O.Yu. Ivanova, O.Yu. Pletyushkina, T.S. Svitkina, V.V. Smolyaninov, and Yu.M. Vasil'ev) Spreading of fibroblasts in a medium containing cytochalasin *B*: Formation of lamellar cytoplasm as a combination of several functionally different processes. Proc. Natl. Acad. Sci. USA **77** (1980) 5919–5922

1981

372. (with E.V. Pomerantsev, V.M. Sablin, M.N.Starkova, V.A. Sulimova, A.L. Syrkin, and V.L. Vakhlyaev) Prognostication of complications and classification of patients with severe myocardial infarcation. Summaries of lectures at the second All-Union Conf. "Theory and practice of automation of electrocardiological and clinical studies", pp. 274–276. Kaunas 1981 (in Russian)

373. (with G.G. Guelstein, I.P. Lukashhevich, M.A. Shifrin, and L.S. Zingerman) Expressibility of electrocardiograph changes in severe disease of the coronary artery in patients with chronic ischemic heart disease. Summaries of lectures at the second All-Union Conf. "Theory and practice of automation of electron-cardiological and clinical studies", pp. 304–307. Kaunas 1981 (in Russian)

374. (with S.M. Khoroshkin, E.V. Pomerantsev, B.I. Rozenfel'd, V.A. Sulimov, A.L. Syrkin, and V.L. Vaklyaev) Choice of information for the classification of patients with myocardial infarcation and choice of medical tactics. Summaries of lectures at the second All-Union Conf. "Theory and practice of automation of electrocardiological and clinical studies", pp. 267–278. Kaunas 1981 (in Russian)

375. (with Yu.M. Vasil'ev) Neoplastic and normal cells in culture. London-Sydney: Cambridge University Press 1981

376. (with S.G. Gindikin, M.L. Izvekova, and M.Yu. Melikova) The immediate prognosis for healing of duodenal ulcers (control of classification). Trans. Second Moscow Med. Inst. Ser. "Surgery" **32** (1981) 73–80 (in Russian)

377. (with Yu.M. Vasil'ev) Interaction of normal and tumorous cells with the medium. Moscow: Nauka 1981 (in Russian)

378. (with V.M. Alekseev, M.A. Alekseevskaya, E.E. Gogin, L.D. Golovnya, M.L. Izvekova, E.S. Klyushin, I.V. Martynov, V.A. Ponomarev, I.V. Sablin, A.L. Syrkin, and R.M. Zaslavska) Multi-purpose chart of a patient with myocardial infarction (for setting up a data bank in a computer). Prepr. Acad. Sci. USSR Sci. Committee on the complex problem "Cybernetics" (1981) (in Russian)

379. (with V.A. Ponomarev) Gabriel's theorem is also true for representations of graphs endowed with relations. Funkts. Anal.

Prilozh. **15** (2) (1981) 71–22 [Funct. Anal. Appl. **15** (1981) 132–133]. Zbl. **479**:18003

380. (with I.Ya. Dorfman) Hamiltonian operators and infinite-dimensional Lie algebras. Funkts. Anal. Prilozh. **15** (3) (1981) 23–40 [Funct. Anal. Appl. **15** (1982) 173–187]. MR **82j**:58045. Zbl. **478**:58013
381. (with Yu.I. Manin) Simmetria, Enciclopedia Einaudi, Vol. 12, pp. 916–943. Einaudi, Torino 1981
382. (with A.D. Bershadskij, Zh.L. Bliokh, L.V. Domnina, O.Yu. Ivanova, V.V. Smolyahinov, T.M. Svitkina, I.S. Tint, and Yu.M. Vasil'ev) Mechanisms of morphological reactions determining the shape and movement of normal and transformed cells in culture. In: Nemyshechnie sistemy, pp. 65–75. Moscow: Nauka 1981 (in Russian)
383. (with O.Yu. Ivanova, S.G., Komm, and Yu.M. Vasil'ev) Stabilization independent of micropipelets of the cell surface of normal and transformed connective tissue cells. Tsitologiya **23** (1981) 62–65 (in Russian)
384. (with M.I. Graev and A.M. Vershik) Representations of the group of functions taking values in a compact Lie group. Compos. Math. **42**, 217–243 (1981). MR **83g**:22002. Zbl. **449**:22019. II.IX.5

1982

385. (with M.I. Graev) Integral transformations connected with two remarkable complexes in projective space. Prepr. Inst. Appl. Math. **93** (1982) (in Russian)
386. (with M.I. Graev and R. Rosu) Non-local inversion formulae in a problem of integral geometry connected with *p*-dimensional planes in real projective space. Funkts. Anal. Prilozh. **16** (3) (1982) 49–51 [Funct. Anal. Appl. **16** (1982) 196–198]. Zbl. **511**:53072
387. (with M.N. Starkova and A.L. Syrkin) Classification of patients and prognosis of healing in myocardial infarction. Prepr. Acad. Sci. USSR Sci. Committee on the complex problem "Cybernetics" (1982) (in Russian)
388. (with M.L. Izvekova, M.N. Starkova, and A.L. Syrkin) The methodology of comparing material from two hospitals and the construction of a single guide-line for the prognosis of the effect of a strong focal myocardial infarction. Prepr. Acad. Sci. USSR Sci. Committee on the complex problem "Cybernetics" (1982) (in Russian)
389. (with M.A. Brodskij, M.Ya. Ratner, B.I. Rozenfel'd, V.V. Serov, I.I. Stenina, and V.A. Varshavskij) Determination of a morphological picture of glomerule kidney from clinical-

functional data (by means of a formal scheme modelling the diagnosis of kidney consultants). Prepr. Acad. Sci. USSR Sci. Committee on the complex problem "Cybernetics" (1982) (in Russian)

390. (with S.M. Khorshkin, B.I. Rozenfeld, V.A. Sulimov, A.L. Syrkin, and V.D. Vakhlyaev) Selection of information for the classification of patients with myocardial infarcation and choice of medical tactics. Prepr. Acad. Sci. USSR Sci. Committee on the complex problem "Cybernetics" (1982) (in Russian)

391. (with I.Ya. Dorfman) Hamiltonian operators and the classical Yang-Baxter equation. Funkts. Anal. Prilozh. **16**(4) (1982) 1–9 [Funct. Anal. Appl. **16** (1982) 241–248]. Zbl. **527**:58018 I.III.17

392. (with Yu.I. Arshavskij and G.N. Orlovskij) The cerebellum and control of rhythmical movements. Trends Neurosci. **6** (10) (1983) 417–422 III.IV.11

393. (with Yu.I. Arshavskij, I.N. Beloozerova, G.N. Orlovskij, and Yu.V. Panchin) Neural mechanisms in the generation of nutritional rhythmics in molluscs. Lecture at the First All-Union Biophysical Congress. Moscow 1982 (in Russian)

394. (with Yu.M. Vasil'ev) Possible common mechanism of morphological and growth-related alterations accompanying neoplastic transformation. Proc. Natl. Acad. Sci. USA **79** (1982) 2594–2597 III.IV.15

395. (with M.I. Graev and A.M. Vershik) A commutative model of the basic representation of the group $SL(2, R)X$ connected with a unipotent subgroup. Prepr. Inst. Appl. Math. **169** (1982) (in Russian)

396. (with B.I. Rozenfeld and M.A. Shifrin) Structural organisation of data in problems of medical diagnosis and prognosis. Prepr. Acad. Sci. USSR Sci. Committee on the complex problem "Cybernetics" (1982) (in Russian)

397. (with R.G. Ajrapetyan, M.I. Graev, and G.R. Oganesyan) The Plancherel formula for the integral transformation connected with a complex of lines intersecting an algebraic straight line in $\mathbf{C}^3$ and $\mathbf{CF}^3$. Dokl., Akad. Nauk Arm. SSR **75** (1) (1982) 9–15 (in Russian). Zbl. **504**:43009

398. (with L.V. Domnina, V.I. Gelfand, O.Yu. Ivanova, O.Yu. Pletyushkina, and Yu.M. Vasil'ev) Effects of small doses of cytochalasins on fibroblasts: preferential changes of active edges and focal contacts. Proc. Natl. Acad. Sci. USA **79** (1982) 7754–7757

399. (with R.D. MacPherson) Geometry in Grassmannians and a generalization of the dilogarithm. Adv. Math. **44** (1982) 279–312. Zbl. **504**:57021 III.II.22

1983

400. (with I.V. Cherednik) An abstract Hamiltonian formalism for the classical Yang-Baxter bundles. Usp. Mat. Nauk **38** (3) (1983) 3–21 [Russ. Math. Surv. **38** (3) (1983) 1–22]. Zbl. **536**:58006
401. (with R.G. Ajrapetyan, M.I. Graev, and G.R. Oganesyan) Plancherel theorem for the integral transformation connected with a complex of *p*-dimensional planes in $\mathbf{CF}^n$. Dokl. Akad. Nauk SSSR **268**, 265–268 (1983) [Sov. Math., Dokl. **27** (1983) 47–50]. Zbl. **527**:53045
402. (with G.S. Shmelev) Geometric structures of double bundles and their relation to certain problems in integral geometry. Funkts. Anal. Prilozh. **17** (2) (1983) 7–22 [Funct. Anal. Appl. **17** (1983) 84–96]. Zbl. **519**:53058 III.I.12
403. (with M.I. Graev and A.M. Vershik) A commutative model of representation of the group of flows $SL(2, \mathbf{R})^X$ that is connected with a unipotent subgroup. Funkts. Anal. Prilozh. **17** (2) (1983) 70–72 [Funct. Anal. Appl. **17** (1983) 137–139]. Zbl. **536**:22008 II.IX.6
404. (with A.Yu. Lyuiko, M.N. Starkova, and A.L. Syrkin) Retrospective estimate of non-stable cardiac angina in various forms of myocardial infarction. Klin. Med. **3** (1983) 28–31 (in Russian)
405. (with A.A. Grinberg, M.L. Izvekova, and V.P. Lakhtina) Prognosis of recidive haemorrhaging in patients with ulcerous disease of the stomach and duodenal intestine. Vestn. Khir., Grekov **130** (4) (1983) 21–24 (in Russian)

1984

406. (with A.V. Zelevinskij) Models of representations of classical groups and their hidden symmetries. Funkts. Anal. Prilozh. **18** (3) (1984) 14–31 [Funct. Anal. Appl. **18** (1984) 183–198]. Zbl. **556**:22003 II.IV.10
407. (Yu.L. Daletskij) Some formal differential structures related to Lie superalgebras. Prepr. Inst. Math. **85** (1984) (in Russian)
408. (with R.G. Ajrapetyan, M.I. Graev, and G.R. Oganesyan) The Plancherel theorem for the integral transformation connected to a pair of Grassmannians. Izv. Akad. Nauk Arm. SSR, Mat. **19** (6) (1984) 467–483 [Sov. J. Contemp. Math. Anal., Arm. Acad. Sci. **18** (4) (1983) 21–32]. MR **86c**:53046. Zbl. **577**:44002
409. (with M.I. Graev and R. Rosu). The problem of integral geometry and intertwining operators for a pair of real Grassmannian manifolds. J. Oper. Theory **12** (2) (1984) 359–383. MR **86c**:22016. Zbl. **551**:53034 III.I.13

410. (with M.A. Brodskij, M.Ya. Ratner, B.I. Rozenfeld, I.I. Stenina, and V.A. Varshavskij) Morphological-clinical variants of chronic glomerulonephritis and their role in evaluation of serenity of disease. Arkh. Patol. **11** (1984) 46–52 (in Russian)
411. Functions of the cerebellum for the control of rhythmic movements. Today's views on the function of the cerebellum, pp. 181–188. Erevan 1984 (in Russian)
412. (with Yu.I. Arshavskij and G.N. Orlovskij) The cerebellum and the control of rhythmic movements. Moscow: Nauka 1984 (in Russian)
413. (with Yu.I. Arshavskij, G.N. Orlovskij, G.A. Pavlova, and L.B. Popova) Origin of signals convate by the ventral spinocerebellar tract and spino-reticulo-cerebellar pathway. Exp. Brain Res. **54** (3) (1984) 426–431
414. (with L.V. Domnina, O.Yu. Ivanova, O.Yu. Pletyushkina, T.M. Svitkina, and Yu.M. Vasil'ev) Formation of processes in the spreading of fibroblasts in a medium with Cytochalasin B in vitro. Ontogenez **15** (3) (1984) 275–282 (in Russian)
415. (with Yu.M. Vasil'ev) Membrane-cytoskeleton interactions during cell spreading on non-cellular surfaces. 16th meeting of the Federation of European Biochemical Societies, Abstracts, p. 34 (1984) (in Russian)
416. (with A.D. Bershadskij, V.I. Gelfand, L.A. Lyass, A.S. Serpinskaya, and Yu.M. Vasil'ev) Multinucleation induced improvements of the spreading of the transformed cells on the substratum. Proc. Natl. Acad. Sci. USA **81** (1984) 3098–3102

1985

417. (with M.I. Graev) On some families of irreducible unitary representations of the group $U(\infty)$. Prepr. Inst. Appl. Math. **51** (1985) (in Russian)
418. (with A.V. Zelevinskij) Polyhedra in the scheme space and the canonical basis for irreducible representations of gl_3. Funkts. Anal. Prilozh. **19**(2) (1985) 72–75 [Funct. Anal. Appl. **19** (1985) 141–144]. Zbl. **606**:17006
419. (with A.V. Zelevinskij) The canonical basis in irreducible representations of gl_3 and applications. In: Proc. III. Int. Semin. on Group-Theoretical Methods in Physics, Yurmala, 1985 (in Russian)
420. (with A.V. Zelevinskij) Multiplicities and good bases for gl_n. In: Proc. III. Int. Semin. on Group-Theoretical Methods in Physics, Yurmala, 1985 (in Russian)
421. (with B.I. Rozenfel'd and M.A. Shifrin) Structural organization of data in medical diagnostics and prognosis. In: Problems of medical diagnostics and prognosis from the point of

view of a mathematician, I.M. Gelfand (ed.). Vopr. Kibern., Mosk. 112 (1985) 5–64 (in Russian)

422. (with N.M. Chebotareva, S.G. Gindikin, Sh.A. Guberman, M.L. Izvekova, E.I. Kandel, M.Yu. Melikova, B.I. Rozenfel'd, M.N. Starkova, and A.L. Syrkin) Some problems of classification and prognosis from various area of the medicin. In: Problems of medical diagnostics and prognosis from the point of view of a mathematician, I.M. Gelfand (ed.). Vopr. Kibern., Mosk. **112** (1985) 65–127 (in Russian)
423. (with G.I. Dzuba, Sh.A. Guberman, and L.V. Kuznetsov) Applications of the global approach to the discrimination of objects in the automatized analysis of chest x-rays. In: Problems of medical diagnostic and prognosis from the point of view of mathematician. I.M. Gelfand (ed.). Vopr. Kibern., Mosk. **112** (1985) 148–171 (in Russian)
424. (with B.I. Rozenfeld, and M.A. Shifrin) "Diagnostic games" in medical diagnostics and prognosis. Psikhol. Zh. **5** (1985) (in Russian)
425. (with Yu.I. Arshavskij, G.N. Orlovskij, Yu.V. Panchin, G.A. Pavlova, and L.B. Popova) Regeneration of neurons in pedal ganglia of pteropodial mollusc Clione limacina. Nejrofiziologiya **17** (4) (1985) 449–455 (in Russian)
426. (with A.V. Zelevinskij) Representation models for classical groups and their higher symmetries. In: Elie Cartan et les mathématiques d'aujourdhui. The mathematical heritage of Elie Cartan, Sémin. Lyon 1984, Astérisque, No. Hors Sér., 117–128 (1985). Zbl. **594**:22007

1986

427. (with G.N. Orlovskij and M.L. Shik) Locomotion and stratching in tetrapods. In: Neural control of rhythmic movements, J. Wiley, N.Y., 1986
428. General theory of hypergeometric functions. Dokl. Akad. Nauk SSSR **288** (1) (1986) 14–18 [Sov. Math., Dokl. **33** (1986) 573–577] III.V.1
429. (with S.I. Gelfand) Generalized hypergeometric equations. Dokl. Akad. Nauk SSSR **288** (2) (1986) 279–283 [Sov. Math., Dokl. **33** (1986) 643–646] III.V.2
430. (with M.I. Graev) A duality theorem for general hypergeometric functions. Dokl. Akad. Nauk SSSR **289** (1) (1986) 19–23 [Sov. Math., Dokl. **34** (1987) 9–13]. Zbl. **619**:33006 III.V.3
431. (with A.B. Goncharov) On a characterization of Grassmann manifolds. Dokl. Akad. Nauk SSSR **289** (5) (1986) 1047–1052 [Sov. Math., Dokl. **34** (1987) 189–193] III.I. 14

432. (with A.V. Zelevinskij) Algebraic and combinatorial aspects of the general theory of hypergeometric functions. Funkts. Anal. Prilozh. **20** (3) (1986) 17–34 [Funct. Anal. Appl. **20** (1986) 183–197]. Zbl. **619**:33004 III.V.4
433. (with V.A. Vasil'ev and A.V. Zelevinskij) Behaviour of general hypergeometric functions in complex domain. Dokl. Akad. Nauk SSSR **290** (2) (1986) 277–281 [Sov. Math., Dokl. **34** (1987) 268–272]. Zbl. **619**:33005
434. (with A.B. Goncharov) Reconstruction of a function with compact support by its integrals over lines intersecting the given set of points in the space. Dokl. Akad. Nauk SSSR **290** (5) (1986) 1037–1040 [Sov. Math., Dokl. **34** (1987) 373–376]. Zbl. **621**:53052
435. (with V.V. Minakhin and V.N. Shander) Integration on supermanifolds and Radon supertransformations. Funkts. Anal. Prilozh. **20** (4) (1986) 67–69 [Funct. Anal. Appl. **20** (1986) 310–312]
436. (with A.V. Zelevinskij) Canonical basis in irreducible representations of gl_3 and its applications. In: Group theoretic methods in physics, vol. 2, pp. 31–45. Moscow: Nauka 1986 (in Russian)
437. (with A.V. Zelevinskij) Multiplicities and regular bases for gl_n. In: Group theoretic methods in physics. Moscow: Nauka, vol. 2 (1986) 22–31 (in Russian)
438. (with G.I. Dzuba, Sh.A. Guberman, and L.V. Kuznetsov) The experience in the development of the system for the automatized chest radiograph processing. Proc. 1st All-Union Seminar "Algorithms and Software for the Data Analysis in Medical and Biological Studies", Pushchino (1986) (in Russian)
439. (with B.I. Rozenfeld, A.L. Syrkin, and M.A. Shifrin) Clinical course types of the myocardium infarction and their prognostic value. Kardiologiya **9** (1986) 9–12 (in Russian)
440. (with A.V. Alekseevskij, M.A. Shifrin, and M.A. Rainer) Algorithms for differential diagnosis of purulent meningitis of various etiologies in children of the first year of life. Preprint of the Scientific Council on Cybernetics, Moscow (1986) 49 pp. (in Russian)
441. (with V.B. Dugina and Yu. M. Vasil'ev) Reversible reorganization of cultured cell cytoskeleton induce byphorbol ether. Dokl. Akad. Nauk SSSR **291** (1) (1986) 985–988 (in Russian)
442. (with M.B. Berkenblit and A.G. Fel'dman) A model for the control of polyartrial extremity movements. Biofizika **31** (1) (1986) 128–138 (in Russian)
443. (with M.B. Berkenblit and A.G. Fel'dman) A model for the aiming phase of the wiping reflex. In: Neurobiology of verte-

brate locomotion. Ed. S. Grillner. Macmillan Press (1986) 217–230

444. (with Yu.I. Arshavskij, T.G. Delyagina, G.N. Orlovskij, G.A. Pavlova, Yu.V. Panchin, and L.B. Popova) The influence of the locomotion system of pedal ganglia of the pteropodial mollusc upon isolated neurons. Nejrofiziologiya **18** (6) (1986) 756–763 (in Russian)

1987

445. (with M. Goresky, R.D. MacPherson, V.V. Serganova) Combinatorial geometries, convex polyhedra, and Schubert cells. Adv. Math. **63** (3) (1987) 301–316. Zbl. **622**:57014 III.V.5

446. (with V.V. Serganova) On the general definition of a matroid and a greedoid. Dokl. Akad. Nauk SSSR **292** (1) (1987) 15–19 [Sov. Math., Dokl. **35** (1987) 6–10]

447. (with V.V. Serganova) Strata of a maximal torus in a compact homogeneous space. Dokl. Akad. Nauk SSSR **292** (3) (1987) 524–528 [Sov. Math., Dokl. **35** (1987) 63–66] III.V.6

*448. (with V.V. Serganova) Combinatorial geometries and torus strata on homogeneous compact manifolds. Usp. Mat. Nauk **42** (2) (1987) 107–133 III.V.7

*449[1]. (with V.A. Vasil'ev, A.V. Zelevinskij) General hypergeometric funcions on complex Grassmannians. Funkts. Anal. Prilozh. **21** (1) (1987) 23–38. Zbl. **614**:33008 [Funct. Anal. Appl. **21** (1988) 19–31] III.V.8

450. (with M.I. Graev) General hypergeometric functions on the Grassmannian $G_{3,6}$. Preprint IFM **123** (1987) 27pp. (in Russian)

451. (with M.I. Graev) Strata in $G_{3,6}$ and corresponding hypergeometric functions. Preprint IFM **127** (1987) 25pp. (in Russian)

*452[1]. (with M.I. Graev and A.V. Zelevinskij) Holonomic systems of equations and series of hypergeometric type. Dokl. Akad. Nauk SSSR **295** (1) (1987) 14–19 [Sov. Math., Dokl. 36 (1988) 5–10] III.V.9

*453[1]. (with A.N. Varchenko) On Heaviside functions of configuration of hyperplanes. Funkts. Anal. Prilozh. **21** (4) (1987) 1–18 [Funct. Anal. Appl. **21** (1988) 255–270] III.V.10

*454. (with T.V. Alekseevskaya and A.V. Zelevinskij) Arrangements of real hyperplanes and the associated partition function. Dokl. Akad. Nauk SSSR **297** (6) (1987) 1289–1293 III.V.11

455. (with B.I. Rozenfeld and M.A. Shifrin) Data structuring in medical problems. Preprint of the Scientific Council on Cybernetics, Moscow (1987) 45 pp. (in Russian)

[1] Shortly before publication of this volume translations of these articles also appeared in the sources named.

456[1]. (with V.B. Dugina, T.M. Svitkina, and Yu.M. Vasil'ev) Special type of morphological reorganization induced by phorbol ester: Reversible partition of cell into mobile and stable domains. Proc. Natl. Acad. Sci. USA **84** (1987) 4122–4125 III.IV.16

457. (with M.I. Graev) Hypergeometric functions related to the grassmannian $G_{3,6}$. Dokl. Akad. Nauk SSSR **293** (2) (1987) 288–292 [Sov. Math., Dokl. **35** (1987) 298–303]

458. (with B.V. Lidskij and V.A. Ponomarev) Preprojective reduction of the free modular lattice D^r. Dokl. Akad. Nauk SSSR **293** (3) (1987) 524–528 [Sov. Math., Dokl. **35** (1987) 334–338]

459. (with Yu.M. Vasil'ev) Membrane and cytoskeleton interrelations for cells binding to non-cellular surfaces. In: Proc. 16th FEBS Conf., Moscow, vol. 1 (1987) 164–166 (in Russian)

460. (with Yu.I. Arshavskij, T.G. Delyagina, E.S. Meiserov, G.N. Orlovskij, G.A. Pavlova, Yu.V. Panchin, and L.B. Popova) Growth of neuritis and development of connections in the neuron culture of pteropodial mollusc. Nejrofiziologiya **19** (1) (1987) 81–86 (in Russian)

461. (with Yu.I. Arshavskij, T.G. Delyagina, E.S. Meiserov, G.N. Orlovskij, G.A. Pavlova, Yu.V. Panchin, and L.B. Popova) Neuronal mechanisms controlling locomotion in the pteropodial mollusc Clione Limacina. Zh. Ob. Biologii **48** (3) (1987) 325–339 (in Russian)

1988

462. (with V.S. Retakh and V.V. Serganova) Generalized Airy functions, Schubert cells and Jordan groups. Dokl. Akad. Nauk SSSR **298** (1) (1988) 17–21 III.V.12

Acknowledgements

We would like to thank the original publishers of I. M. Gelfand's papers for granting permission to reprint them here.

The numbers following each source correspond to the numbering of the article in this volume.

Reprinted from *Advances in Mathematics.* © Academic Press: II.22, V.5
Reprinted from *Brain Research.* © Elsevier Science Publishers B.V.: IV.8, IV.9
Reprinted from *Funct. Anal. Appl.* © Consultants Bureau: I.2, I.4, I.5, I.6, I.8, I.9, I.10, I.12, II.2, II.3, II.4, II.6, II.7, II.8, II.9, II.10, II.11, II.12, II.13, II.15, II.16, II.17, II.18, II.19, II.21, IV.10, V.4
Reprinted from *Int. J. of Cancer.* © Alan R. Liss Inc.: IV.12
Reprinted from *Int. Rev. of Cytology.* © Academic Press: IV.14
Reprinted from *J. Math. Physics.* © American Institute of Physics: III.2
Reprinted from *J. Neurophysiology.* © American Physiological Society: IV.10
Reprinted from *J. Operator Theory.* © INCREST: I.13
Reprinted from *J. Sov. Math.* © Plenum Publishing Corporation: I.11
Reprinted from *Math. USSR, Izv.* © American Mathematical Society: II.5
Reprinted from *Models of Some Biological Systems.* © MIT Press: IV.7
Reprinted from *Russ. Math. Surv.* © British Library: IV.3
Reprinted from *Sov. Math., Dokl.* © American Mathematical Society: I.1, I.3, I.7, I.14, II.1, IV.4, V.1, V.2, V.3, V.6
Reprinted from *Theory of Difference Schemes.* © North-Holland: IV.2
Reprinted from *Topology.* © Pergamon Press: II.14
Reprinted from *Transl., II. Ser., Am. Math. Soc.* © American Mathematical Society: III.4, IV.6
Reprinted from *Trends in Neurosciences.* © Elsevier Science Publishers B.V.: IV.11

Printed in the United States
By Bookmasters